New Pattern

JEE PROBLEMS

Mathematics

for JEE Main & Advanced

More than 8000+ JEE Level Problems Including All Types of Questions

Only One Correct Option, One & More than One Correct Option, Comprehension Type, Numerical Value Type, Single Integer Type & Matrix Match Type

New Pattern

JEE PROBLEMS

Mathematics

for JEE Main & Advanced

More than 8000+ JEE Level Problems Including All Types of Questions

Only One Correct Option, One & More than One Correct Option, Comprehension Type, Numerical Value Type, Single Integer Type & Matrix Match Type

Dr. SK Goyal

ARIHANT PRAKASHAN (Series), MEERUT

arihant

Arihant Prakashan (Series), Meerut

Administrative & Production Offices

Regd. Office

'Ramchhaya' 4577/15, Agarwal Road, Darya Ganj, New Delhi -110002
Tele: 011- 47630600, 43518550

Head Office

Kalindi, TP Nagar, Meerut (UP) - 250002, Tel: 0121-7156203, 7156204

Sales & Support Offices

Agra, Ahmedabad, Bengaluru, Bareilly, Chennai, Delhi, Guwahati, Hyderabad, Jaipur, Jhansi, Kolkata, Lucknow, Nagpur & Pune.

ISBN 978-93-25792-25-8

PRICE 830.00

PO No : TXT-XX-XXXXXXX-X-XX

Published by Arihant Publications (India) Ltd.

For further information about the books published by Arihant, log on to www.arihantbooks.com or e-mail at info@arihantbooks.com

Follow us on    

PREFACE

The JEE (Joint Entrance Examination) one of the toughest entrance exam of the world is conducted for admission into 23 IITs spread across the country. Making it to the cut-off list of the JEE is dream come true. The level of competition can be gauged from the fact that more than two lakh students appear at the JEE every year for a handful of seats available in these institutions.

The questions put forward in the JEE are designed to test comprehension, reasoning and analytical ability of the candidates in an objective way. Generally, it is seen that students are more fearful of the Mathematics section of the test. This book, New Pattern Mathematics of is an effort to keep all the misgivings about JEE Mathematics which a student may bear down in his/her mind. It aims at completely banishing the fear of JEE Mathematics from the minds of students.

The book in your hand, gives you a collection of all the types of Objective Questions which are being asked in JEE these days as well as many more types which can be 'surprise' elements in the coming JEEs, thus making the book futuristic too. Solutions in this book are presented to make you (If you go minutely with pen, paper and pencil) learn how to 'strategize' for a particular type of problems along with the ways to move 'tactically' to get the correct answer to a question which you are solving. A revolutionary approach towards problem solving in Mathematics has been adopted in this book.

It is my firm belief that this book will step-by-step infuse self-confidence in the students as required to tackle the JEE Mathematics and the preparation for JEE Main & Advanced will no longer remain a nightmare for the students. Suggestions for further improvement of the book are welcome.

Dr. SK GOYAL

CONTENTS

01

Complex Numbers

Objective Questions Type I [Only one correct answer]

In each of the questions below, four choices are given of which only one is correct. You have to select the correct answer which is the most appropriate.

1. The value of $\left(\dfrac{1+\sin\frac{2\pi}{9}+i\cos\frac{2\pi}{9}}{1+\sin\frac{2\pi}{9}-i\cos\frac{2\pi}{9}}\right)^3$ is

(a) $-\frac{1}{2}(\sqrt{3}-i)$ (b) $-\frac{1}{2}(1-i\sqrt{3})$

(c) $\frac{1}{2}(\sqrt{3}-i)$ (d) $\frac{1}{2}(1-i\sqrt{3})$

2. If $\text{Im}\left(\dfrac{az+z_1}{bz+z_2}\right)=1$ (where z, z_1 and z_2 are complex numbers and a, b are real numbers). Then z lies on

(a) a straight line (b) a circle

(c) a parabola (d) an ellipse

3. If m and x are two real numbers, then $e^{2mi\cot^{-1}x}\left(\dfrac{xi+1}{xi-1}\right)^m$, (where $i=\sqrt{-1}$) is equal to

(a) $\cos x+i\sin x$ (b) $m/2$

(c) 1 (d) $(m+1)/2$

4. Let S be the set of all complex numbers z satisfying $|z-2+i|\geq\sqrt{5}$. If the complex number z_0 is such that $\dfrac{1}{|z_0-1|}$ is the maximum of the set $\left\{\dfrac{1}{|z-1|}: z\in S\right\}$, then the principal argument of $\dfrac{4-z_0-\bar{z}_0}{z_0-\bar{z}_0+2i}$ is

[JEE Advanced 2019]

(a) $\frac{\pi}{4}$ (b) $\frac{3\pi}{4}$

(c) $-\frac{\pi}{2}$ (d) $\frac{\pi}{2}$

5. If $|z-3+2i|\leq 4$, (where $i=\sqrt{-1}$) then the difference of greatest and least values of $|z|$ is

(a) $2\sqrt{11}$ (b) $3\sqrt{11}$

(c) $2\sqrt{13}$ (d) $3\sqrt{13}$

6. The value of $(\sqrt{z}+\sqrt{\bar{z}})^4$ and $(\sqrt{z}-\sqrt{\bar{z}})^4$ are respectively (where $z=4+3\sqrt{20}\,i$, $i=\sqrt{-1}$)

(a) 1296,400 (b) 72,16

(c) 36,9 (d) 216,25

7. If z lies on the circle centred at origin. If area of the triangle whose vertices are z, ωz and $z+\omega z$ where ω is the cube root of unity, is $4\sqrt{3}$ sq unit. Then radius of the circle is

(a) 1 unit (b) 2 unit

(c) 3 unit (d) 4 unit

8. If ω is a complex cube root of unity, then $\cos\left[\{(1-\omega)(1-\omega^2)+\ldots+(10-\omega)(10-\omega^2)\}\dfrac{\pi}{900}\right]$

(a) -1 (b) 0 (c) 1 (d) $\sqrt{3}/2$

9. If $|z-i\,\text{Re}(z)|=|z-\text{Im}(z)|$, (where $i=\sqrt{-1}$) , then z lies on

(a) $\text{Re}(z)=2$ (b) $\text{Im}(z)=2$

(c) $\text{Re}(z)+\text{Im}(z)=2$ (d) None of these

10. For a complex number z, the minimum value of $|z|+|z-\cos\alpha-i\sin\alpha|$ (where $i=\sqrt{-1}$) is

(a) 0 (b) 1

(c) 2 (d) None of these

11. If a, b, c, p, q, r are six complex numbers, such that $\dfrac{p}{a}+\dfrac{q}{b}+\dfrac{r}{c}=1+i$ and $\dfrac{a}{p}+\dfrac{b}{q}+\dfrac{c}{r}=0$, where$i=\sqrt{-1}$ then value of $\dfrac{p^2}{a^2}+\dfrac{q^2}{b^2}+\dfrac{c^2}{r^2}$ is

(a) 0 (b) -1 (c) $2i$ (d) $-2i$

12. If z be complex number such that equation $|z-a^2|+|z-2a|=3$ always represents an ellipse, then range of $a\ (\in R^+)$ is

(a) $(1,\sqrt{2})$ (b) $[1,\sqrt{3}]$

(c) $(-1, 3)$ (d) $(0, 3)$

13. If $A(z_1), B(z_2)$ and $C(z_3)$ are three points in the argand plane where $|z_1 + z_2| = ||z_1| - |z_2||$ and $|(1-i)z_1 + iz_3| = |z_1| + |z_3 - z_1|$, where $i = \sqrt{-1}$ then

(a) A, B and C lie on a fixed circle with centre $\frac{(z_2 + z_3)}{2}$

(b) A, B and C are collinear points

(c) ABC form an equilateral triangle

(d) ABC form an obtuse angle triangle

14. The least positive integer n for which $\left(\frac{1+i}{1-i}\right)^n = \frac{2}{\pi}\sin^{-1}\frac{1+x^2}{2x}$, where $x > 0$ and $i = \sqrt{-1}$ is

(a) 2 (b) 4

(c) 8 (d) 12

15. Let $u = \frac{2z+i}{z-ki}$, $z = x + iy$ and $k > 0$. If the curve represented by $\text{Re}(u) + \text{Im}(u) = 1$ intersects the Y-axis at the points P and Q where $PQ = 5$, then value of k is **[JEE Main 2020, 4 September]**

(a) 1/2 (b) 4

(c) 2 (d) 3/2

16. Common roots of the equations : $z^3 + 2z^2 + 2z + 1 = 0$, $z^{1985} + z^{100} + 1 = 0$

(a) $1, \omega$ (b) ω, ω^2

(c) ω^4 (d) $3, \omega$

17. If $x_r = \cos(\pi/3^r) - i\sin(\pi/3^r)$, (where $i = \sqrt{-1}$), then value of $x_1 \cdot x_2 \cdot \ldots \infty$, is

(a) 1 (b) -1

(c) $-i$ (d) i

18. $(1+i)(1+2i)(1+3i)\ldots(1+ni) = \alpha + i\beta$, then value of $2 \cdot 5 \cdot 10 \ldots (1+n^2)$ is

(a) $\alpha^2 + \beta^2$ (b) $\alpha + \beta$

(c) $\alpha - \beta^2$ (d) $\alpha^2 - \beta$

19. Let $z_1 = 6 + i$ and $z_2 = 4 - 3i$ (where $i = \sqrt{-1}$). Let z be a complex number such that $\arg\left(\frac{z - z_1}{z_2 - z}\right) = \frac{\pi}{2}$, then z satisfies

(a) $|z - (5-i)| = 5$ (b) $|z - (5-i)| = \sqrt{5}$

(c) $|z - (5+i)| = 5$ (d) $|z - (5+i)| = \sqrt{5}$

20. The number of solutions of the equation $z^2 + |z|^2 = 0$, where $z \in C$ is

(a) one (b) two

(c) three (d) infinitely many

21. The continued product of all the four values of the complex number $(1+i)^{3/4}$ is

(a) $2^3(1+i)$ (b) $2(1-i)$

(c) $2(1+i)$ (d) $2^3(1-i)$

22. The locus of z which satisfies the inequality $\log_{0.3}|z-1| > \log_{0.3}|z-i|$ is given by

(a) $x + y < 0$ (b) $x + y > 0$

(c) $x - y > 0$ (d) $x - y < 0$

23. If z_1 and z_2 are any two complex numbers, then $|z_1 + \sqrt{(z_1^2 - z_2^2)}| + |z_1 - \sqrt{(z_1^2 - z_2^2)}|$ is equal to

(a) $|z_1|$ (b) $|z_2|$

(c) $|z_1 + z_2|$ (d) None of these

24. If z_1, z_2 are complex numbers such that $\text{Re}(z_1) = |z_1 - 1|$, $\text{Re}(z_2) = |z_2 - 1|$ and $\arg(z_1 - z_2) = \frac{\pi}{6}$, then $\text{Im}(z_1 + z_2)$ is equal to **[JEE Main 2020, 3 September]**

(a) $\frac{\sqrt{3}}{2}$ (b) $\frac{1}{\sqrt{3}}$

(c) $\frac{2}{\sqrt{3}}$ (d) $2\sqrt{3}$

25. The centre of square $ABCD$ is at $z = 0$. A is z_1. Then, the centroid of triangle ABC is

(a) $z_1(\cos\pi \pm i\sin\pi)$

(b) $\frac{z_1}{3}(\cos\pi \pm i\sin\pi)$

(c) $z_1(\cos\pi/2 \pm i\sin\pi/2)$

(d) $\frac{z_1}{3}(\cos\pi/2 \pm i\sin\pi/2)$

26. Let the lines $(2-i)z = (2+i)\bar{z}$ and $(2+i)z + (i-2)\bar{z} - 4i = 0$, (here $i^2 = -1$) be normal to a circle C. If the line $iz + \bar{z} + 1 + i = 0$ is tangent to this circle C, then its radius is **[JEE Main 2021, 16 March]**

(a) $\frac{3}{\sqrt{2}}$ (b) $3\sqrt{2}$

(c) $\frac{3}{2\sqrt{2}}$ (d) $\frac{1}{2\sqrt{2}}$

27. If z and w are two complex numbers such that $|zw| = 1$ and $\arg(z) - \arg(w) = \frac{\pi}{2}$, then **[JEE Main 2019, 10 April]**

(a) $\bar{z}w = -i$ (b) $z\bar{w} = \frac{1-i}{\sqrt{2}}$

(c) $\bar{z}w = i$ (d) $z\bar{w} = \frac{-1+i}{\sqrt{2}}$

28. Let complex numbers α and $1/\bar{\alpha}$ lies on circles $(x - x_0)^2 + (y - y_0)^2 = r^2$ and $(x - x_0)^2 + (y - y_0)^2 = 4r^2$, respectively. If $z_0 = x_0 + iy_0$ satisfies the equation $2|z_0|^2 = r^2 + 2$, then $|\alpha|$ is equal to

(a) $\frac{1}{\sqrt{2}}$ (b) $\frac{1}{2}$

(c) $\frac{1}{\sqrt{7}}$ (d) $\frac{1}{3}$

29. If $\text{Re}\left(\frac{z-1}{2z+i}\right)=1$, where $z = x + iy$, then the point (x, y) lies on a **[JEE Main 2020, 7 January]**

(a) straight line whose slope is $-\frac{2}{3}$.

(b) circle whose centre is at $\left(-\frac{1}{2}, -\frac{3}{2}\right)$.

(c) straight line whose slope is $\frac{3}{2}$.

(d) circle whose diameter is $\frac{\sqrt{5}}{2}$.

30. If z_1 and $\overline{z_1}$ represent adjacent vertices of a regular polygon of n sides whose centre is origin and if $\frac{\text{Im}(z_1)}{\text{Re}(z_1)} = \sqrt{2}-1$, then n is equal to

(a) 8 (b) 16
(c) 24 (d) 32

31. Let z_1, z_2 be two complex numbers such that $\bar{z}_1 - i\bar{z}_2 = 0$ and $\arg(z_1 z_2) = \frac{3\pi}{4}$, then $\arg(z_1) =$

(a) $\frac{\pi}{4}$ (b) $\frac{-\pi}{8}$
(c) $\frac{\pi}{8}$ (d) $\frac{\pi}{3}$

32. For $x_1, x_2, y_1, y_2 \in R$. If $0 < x_1 < x_2$, $y_1 = y_2$ and $z_1 = x_1 + iy_1$, $z_2 = x_2 + i\,y_2$, (where $i = \sqrt{-1}$) and $z_3 = \frac{1}{2}(z_1 + z_2)$, then z_1, z_2 and z_3 satisfy

(a) $|z_1| = |z_2| = |z_3|$
(b) $|z_1| < |z_2| < |z_3|$
(c) $|z_1| > |z_2| > |z_3|$
(d) $|z_1| < |z_3| < |z_2|$

33. Let a complex number $z, |z| \neq 1$, satisfy $\log_{\frac{1}{\sqrt{2}}}\left(\frac{|z|+11}{(|z|-1)^2}\right) \leq 2$. Then the largest value of $|z|$ is equal to **[JEE Mains 2021, 16 March]**

(a) 8 (b) 7
(c) 6 (d) None of these

34. If z is a non-real complex number, then the minimum value of $\frac{\text{Im}\, z^5}{\text{Im}^5\, z}$ is

(a) -1 (b) -2
(c) -4 (d) -5

35. Let z_1 and z_2 be any two non-zero complex numbers such that $3|z_1| = 4|z_2|$. If $z = \frac{3z_1}{2z_2} + \frac{2z_2}{3z_1}$, then **[JEE Main 2019, 10 January]**

(a) $|z| = \sqrt{\frac{17}{2}}$ (b) $\text{Im}(z) = 0$
(c) $\text{Re}(z) = 0$ (d) $|z| = \sqrt{\frac{5}{2}}$

36. The argument of $(z-a)(\bar{z}-b)$ is equal to that of $\frac{(\sqrt{3}+i)(1+\sqrt{3}i)}{1+i}$, where a, b are real numbers. If locus of z in the arg and diagram is a circle with its centre at $(3+i)/2$, then the ordered pair (a, b) is

(a) (3, 2) (b) (2, 1)
(c) (2, 3) (d) (2, 4)

37. If ω is a complex cube root of unity and $(1+\omega)^7 = A + B\omega$, then A and B are respectively equal to

(a) 0, 1 (b) 1, 1
(c) 1, 0 (d) -1, 1

38. If 1, ω and ω^2 are the three cube roots of unity, then the roots of the equation $(x-1)^3 - 8 = 0$ are

(a) $-1, -1-2\omega, -1+2\omega^2$ (b) $3, 2\omega, 2\omega^2$
(c) $3, 1+2\omega, 1+2\omega^2$ (d) None of these

39. If $1, \omega, \omega^2, \ldots, \omega^{n-1}$ are n, nth roots of unity, the value of $(9-\omega)(9-\omega^2)\ldots(9-\omega^{n-1})$ will be

(a) n (b) 0
(c) $\frac{9^n-1}{8}$ (d) $\frac{9^n+1}{8}$

40. If $8iz^3 + 12z^2 - 18z + 27i = 0$, (where $i = \sqrt{-1}$) then

(a) $|z| = 3/2$ (b) $|z| = 2/3$
(c) $|z| = 1$ (d) $|z| = 3/4$

41. If $z = re^{i\theta}$, then $|e^{iz}|$ is equal to

(a) $e^{-r\sin\theta}$ (b) $re^{-r\sin\theta}$
(c) $e^{-r\cos\theta}$ (d) $re^{-r\cos\theta}$

42. If z_1, z_2, z_3 are three distinct complex numbers and a, b, c are three positive real numbers such that $\frac{a}{|z_2-z_3|} = \frac{b}{|z_3-z_1|} = \frac{c}{|z_1-z_2|}$, then $\frac{a^2}{(z_2-z_3)} + \frac{b^2}{(z_3-z_1)} + \frac{c^2}{(z_1-z_2)}$ is equal to

(a) 0 (b) abc
(c) $3abc$ (d) $a+b+c$

43. For all complex numbers z_1, z_2 satisfying $|z_1| = 12$ and $|z_2 - 3 - 4i| = 5$, the minimum value of $|z_1 - z_2|$ is

(a) 0 (b) 2 (c) 7 (d) 17

44. The equation of any .... in the complex plane is of the form $z\bar{z} + b\bar{z} + \bar{b}z + c = 0$, where $(b \in \mathbf{C}, c \in \mathbf{R})$

(a) Circle (b) Straight line
(c) Parabola (d) Hyperbola

45. Let $A = \left\{\theta \in \left(-\frac{\pi}{2}, \pi\right) : \frac{3+2i\sin\theta}{1-2i\sin\theta}\right.$ is purely imaginary$\left.\right\}$. Then, the sum of the elements in A is **[JEE Main 2019, 9 January]**

(a) $\frac{3\pi}{4}$ (b) $\frac{5\pi}{6}$
(c) π (d) $\frac{2\pi}{3}$

46. If z_1, z_2, z_3 are the vertices of an equilateral triangle with centroid z_0, then $z_1^2 + z_2^2 + z_3^2$ equals

(a) z_0^2 (b) $2z_0^2$
(c) $3z_0^2$ (d) $9z_0^2$

47. If a complex number z lies on a circle of radius 1/2, then the complex number $(-1+4z)$ lies on a circle of radius

(a) 1/2 (b) 1
(c) 2 (d) 4

48. The least value of $|z|$ where z is complex number which satisfies the inequality

$$\exp\left(\frac{(|z|+3)(|z|-1)}{||z|+1|}\log_e 2\right) \geq \log_{\sqrt{2}}|5\sqrt{7}+9i|$$

$i = \sqrt{-1}$, is equal to **[JEE Main 2021, 16 March]**

(a) 3 (b) $\sqrt{5}$
(c) 2 (d) 8

49. If the vertices of a triangle are $8+5i, -3+i, -2-3i$, the modulus and the argument of the complex number representing the centroid of this triangle respectively are

(a) $2, \frac{\pi}{4}$ (b) $\sqrt{2}, \frac{\pi}{4}$
(c) $2\sqrt{2}, \frac{\pi}{4}$ (d) $2\sqrt{2}, \frac{\pi}{2}$

50. The value of $\sum_{k=1}^{10}\left(\sin\frac{2\pi k}{11} - i\cos\frac{2\pi k}{11}\right)$ is

(a) -1 (b) 0
(c) $-i$ (d) i

51. All complex numbers 'z' which satisfy the relation $|z-|z+1|| = |z+|z-1||$ on the complex plane line on the

(a) $y = x$
(b) $y = -x$
(c) circle $x^2 + y^2 = 1$
(d) line $x = 0$ or on a line segment joining $(-1, 0)$ to $(1, 0)$

52. If the points represented by complex numbers $z_1 = a + ib$, $z_2 = c + id$ (where $i = \sqrt{-1}$) and $z_1 - z_2$ are collinear, then

(a) $ad + bc = 0$ (b) $ad - bc = 0$
(c) $ab + cd = 0$ (d) $ab - cd = 0$

53. Let A, B and C represent the complex numbers z_1, z_2, z_3 respectively on the complex plane. If the circumcentre of the triangle ABC lies at the origin, then the nine point centre is represented by the complex number

(a) $\frac{z_1+z_2}{2} - z_3$ (b) $\frac{z_1+z_2-z_3}{2}$
(c) $\frac{z_1+z_2+z_3}{2}$ (d) $\frac{z_1-z_2-z_3}{2}$

54. Let α and β be two distinct complex numbers such that $|\alpha| = |\beta|$. If real part of α is positive and imaginary part of β is negative, then the complex number $(\alpha+\beta)/(\alpha-\beta)$ may be

(a) zero
(b) real and negative
(c) real and positive
(d) purely imaginary

55. The complex number z satisfies the condition $\left|z - \frac{25}{z}\right| = 24$. The maximum distance from the origin of coordinates to the point z is

(a) 25 (b) 30
(c) 32 (d) None of these

56. The points A, B and C represent the complex numbers $z_1, z_2, (1-i)z_1 + iz_2$ (where $i = \sqrt{-1}$) respectively on the complex plane. The triangle ABC is

(a) isosceles but not right angled
(b) right angled but not isosceles
(c) isosceles and right angled
(d) None of the above

57. The system of equations $\left.\begin{matrix}|z+1+i| = \sqrt{2}\\ |z| = 3\end{matrix}\right\}$, (where $i = \sqrt{-1}$) has

(a) no solution (b) one solution
(c) two solutions (d) None of these

58. The centre of the circle represented by $|z+1| = 2|z-1|$ on the complex plane is

(a) 0 (b) 5/3
(c) 1/3 (d) None of these

59. The value of the expression $2(1+\omega)(1+\omega^2) + 3(2\omega+1)(2\omega^2+1) + 4(3\omega+1)(3\omega^2+1) + \ldots + (n+1)(n\omega+1)(n\omega^2+1)$ is (ω is the cube root of unity)

(a) $\frac{n^2(n+1)^2}{4}$ (b) $\left(\frac{n(n+1)}{2}\right)^2 + n$
(c) $\left(\frac{n(n+1)}{2}\right)^2 - n$ (d) None of these

60. If $\sum_{k=0}^{200} i^k + \prod_{p=1}^{50} i^p = x + iy$, (where $i = \sqrt{-1}$) then (x, y) is

(a) (0, 1) (b) (1, −1)
(c) (2, 3) (d) (4, 8)

61. If $|z_1 - 1| < 1, |z_2 - 2| < 2, |z_3 - 3| < 3$, then $|z_1 + z_2 + z_3|$

(a) is less than 6 (b) is more than 3
(c) is less than 12 (d) lies between 6 and 12

62. If $|z| = \max\{|z-1|, |z+1|\}$, then

(a) $|z + \bar{z}| = \frac{1}{2}$ (b) $z + \bar{z} = 1$
(c) $|z + \bar{z}| = 1$ (d) $z - \bar{z} = 5$

63. If z is point on the Argand plane such that $|z-1|=1$, then $\frac{z-2}{z}$ is equal to

(a) $\tan(\arg z)$ (b) $\cot(\arg z)$
(c) $i\tan(\arg z)$ (d) None of these

64. The area of the triangle with vertices $A(z)$, $B(iz)$ and $C(z+iz)$ is **[JEE Main 2021, 17 March]**

(a) 1 (b) $\frac{1}{2}|z|^2$
(c) $\frac{1}{2}$ (d) $\frac{1}{2}|z+iz|^2$

65. Let S_1, S_2 and S_3 be three sets defined as

$S_1=\{z\in C:|z-1|\le\sqrt{2}\}$
$S_2=\{z\in C:\text{Re}((1-i)z)\ge 1\}$
$S_3=\{z\in C:\text{Im}(z)\le 1\}$

Then, the set $S_1\cap S_2\cap S_3$ **[JEE Main 2021, 17 March]**

(a) is a singleton
(b) has exactly two elements
(c) has infinitely many elements
(d) has exactly three elements

66. If X be the set of all complex numbers z such that $|z|=1$ and define relation R on X by $z_1 R z_2$ is $|\arg z_1-\arg z_2|=\frac{2\pi}{3}$, then R is

(a) reflexive (b) symmetric
(c) transitive (d) anti-symmetric

67. The number of points in the complex plane that satisfying the conditions $|z-2|=2$, $z(1-i)+\bar{z}(1+i)=4$, (where $i=\sqrt{-1}$) is

(a) 0 (b) 1
(c) 2 (d) more than 2

68. If $\frac{3+i\sin\theta}{4-i\cos\theta}$, $\theta\in[0,2\pi]$, is a real number, then an argument of $\sin\theta+i\cos\theta$ is **[JEE Main 2020, 7 January]**

(a) $\pi-\tan^{-1}\left(\frac{4}{3}\right)$ (b) $-\tan^{-1}\left(\frac{3}{4}\right)$
(c) $\tan^{-1}\left(\frac{4}{3}\right)$ (d) $\pi-\tan^{-1}\left(\frac{3}{4}\right)$

69. If $\alpha_0,\alpha_1,\alpha_2,\ldots,\alpha_{n-1}$ be the n, nth roots of the unity, then the value of $\sum_{i=0}^{n-1}\frac{\alpha_i}{(3-\alpha_i)}$ is equal to

(a) $\frac{n}{3^n-1}$ (b) $\frac{n-1}{3^n-1}$
(c) $\frac{n+1}{3^n-1}$ (d) $\frac{n+2}{3^n-1}$

70. If $|z-1|+|z+3|\le 8$, then the range of values of $|z-4|$, (where $i=\sqrt{-1}$) is

(a) $(0,7)$ (b) $(1,8)$
(c) $[1,9]$ (d) $[2,5]$

71. $\sin^{-1}\left\{\frac{1}{i}(z-1)\right\}$, where z is non real and $i=\sqrt{-1}$, can be the angle of a triangle if

(a) $\text{Re}(z)=1$, $\text{Im}(z)=2$
(b) $\text{Re}(z)=1$, $-1\le\text{Im}(z)\le 1$
(c) $\text{Re}(z)+\text{Im}(z)=0$
(d) None of the above

72. If for the complex numbers z_1 and z_2, $|z_1+z_2|=|z_1-z_2|$, then $\arg z_1-\arg z_2=$

(a) π (b) $\pi/2$
(c) $\pi/4$ (d) 4

73. Let $A=\frac{2}{\sqrt{3}}e^{i\pi/2}$, $B=\frac{2}{\sqrt{3}}e^{-i\pi/6}$, $C=\frac{2}{\sqrt{3}}e^{-i5\pi/6}$ (where $i=\sqrt{-1}$) be three points forming a triangle ABC in the argand plane. Then ΔABC is

(a) equilateral (b) isosceles
(c) scalene (d) None of these

74. If $|z|=1$, then $\left(\frac{1+z}{1+\bar{z}}\right)^n+\left(\frac{1+\bar{z}}{1+z}\right)^n$ is equal to

(a) $2\cos n(\arg(z))$ (b) $2\sin(\arg(z))$
(c) $2\cos n(\arg(z/2))$ (d) $2\sin n(\arg(z/2))$

75. If all the roots of $z^3+az^2+bz+c=0$ are of unit modulus, then

(a) $|a|\le 3$ (b) $|b|>3$
(c) $|c|\le 3$ (d) None of these

76. The trigonometric form of $z=(1-i\cot 8)^3$ (where $i=\sqrt{-1}$) is

(a) $\text{cosec}^3 8\cdot e^{i(24-3\pi/2)}$ (b) $\text{cosec}^3 8\cdot e^{-i(24-3\pi/2)}$
(c) $\text{cosec}^3 8\cdot e^{i(36-\pi/2)}$ (d) $\text{cosec}^2 8\cdot e^{-24i+\pi/2}$

77. Let z be a complex number such that $\left|\frac{z-i}{z+2i}\right|=1$ and $|z|=\frac{5}{2}$. Then, the value of $|z+3i|$ is **[JEE Main 2020, 9 January]**

(a) $\sqrt{10}$ (b) $\frac{7}{2}$
(c) $\frac{15}{4}$ (d) $2\sqrt{3}$

78. Perimeter of the locus represented by $\arg\left(\frac{z+i}{z-i}\right)=\frac{\pi}{4}$, (where $i=\sqrt{-1}$) is equal to

(a) $\frac{3\pi}{2}$ (b) $\frac{3\pi}{\sqrt{2}}$
(c) $\frac{\pi}{\sqrt{2}}$ (d) None of these

79. If $z=\frac{\sqrt{3}}{2}+\frac{i}{2}(i=\sqrt{-1})$, then $(1+iz+z^5+iz^8)^9$ is equal to **[JEE Main 2019, 8 April]**

(a) 1 (b) $(-1+2i)^9$
(c) -1 (d) 0

80. If $z_1 \neq -z_2$ and $|z_1 + z_2| = \left|\frac{1}{z_1} + \frac{1}{z_2}\right|$, then
(a) at least one of z_1, z_2 is unimodular
(b) both z_1, z_2 are unimodular
(c) $z_1 \cdot z_2$ is unimodular
(d) None of the above

81. If $z = x + iy$ and $\omega = \frac{1 - iz}{z - i}$, (where $i = \sqrt{-1}$) then $|\omega| = 1$ implies that in the complex plane
(a) z lies on imaginary axis
(b) z lies on real axis
(c) z lies on unit circle
(d) None of the above

82. All the points in the set $S = \left\{\frac{\alpha + i}{\alpha - i} : \alpha \in \mathbf{R}\right\}$ $(i = \sqrt{-1})$ lie on a
(a) circle whose radius is $\sqrt{2}$.
(b) straight line whose slope is –1.
(c) circle whose radius is 1.
(d) straight line whose slope is 1.

83. For complex numbers $z_1 = x_1 + iy_1$ and $z_2 = x_2 + iy_2$, (where $i = \sqrt{-1}$) we write $z_1 \cap z_2$ of $x_1 \leq x_2$ and $y_1 \leq y_2$, then for all complex numbers z with $1 \cap z$, we have $\frac{1 - z}{1 + z} \cap \ldots$ is
(a) 0 (b) 1
(c) 2 (d) 3

84. Let $z \in C$ be such that $|z| < 1$. If $\omega = \frac{5 + 3z}{5(1 - z)}$, then
(a) $4\,\text{Im}(\omega) > 5$ (b) $5\,\text{Re}(\omega) > 1$
(c) $5\,\text{Im}(\omega) < 1$ (d) $5\,\text{Re}(\omega) > 4$

85. The distances of the roots of the equation $|\sin\theta_1| z^3 + |\sin\theta_2| z^2 + |\sin\theta_3| z + |\sin\theta_4| = 3$, from $z = 0$, are
(a) greater than 2/3
(b) less than 2/3
(c) greater than $|\sin\theta_1| + |\sin\theta_2| + |\sin\theta_3| + |\sin\theta_4|$
(d) less than $|\sin\theta_1| + |\sin\theta_2| + |\sin\theta_3| + |\sin\theta_4|$

86. Let S be the set of complex number z which satisfy $\log_{1/3}\{\log_{1/2}(|z|^2 + 4|z| + 3)\} < 0$, then S is (where $i = \sqrt{-1}$)
(a) $1 \pm i$ (b) $3 - i$
(c) $\frac{5}{2} + 4i$ (d) empty set

87. Let $f(z) = \sin z$ and $g(z) = \cos z$. If * denotes a composition of functions, then the value of $(f + ig) * (f - ig)$ (where $i = \sqrt{-1}$) is
(a) $ie^{e^{iz}}$ (b) $ie^{-e^{iz}}$
(c) $ie^{e^{-iz}}$ (d) $ie^{-e^{-iz}}$

88. Let $f_p(\alpha) = e^{i\alpha/p^2} \cdot e^{2i\alpha/p^2} \cdot e^{3i\alpha/p^2} \ldots e^{i\alpha/p}$, (where $i = \sqrt{-1}$ and $p \in N$), then $\lim_{n \to \infty} f_n(\pi)$ is
(a) 1 (b) – 1
(c) i (d) $-i$

Objective Questions Type II [One or more than one correct answer(s)]

In each of the questions below four choices of which one or more than one are correct. You have to select the correct answer(s) accordingly.

1. If z satisfies $|z - 1| < |z + 3|$ then $\omega = 2z + 3 - i$, (where $i = \sqrt{-1}$) satisfies
(a) $|\omega - 5 - i| < |\omega + 3 + i|$ (b) $|\omega - 5| < |\omega + 3|$
(c) $\text{Im}(i\omega) > 1$ (d) $|\arg(\omega - 1)| < \pi/2$

2. Let a, b, x and y be real numbers such that $a - b = 1$ and $y \neq 0$. If the complex number $z = x + iy$ satisfies $\text{Im}\left(\frac{az + b}{z + 1}\right) = y$, then which of the following is(are) possible value(s) of x? **[JEE Advanced 2017]**
(a) $1 - \sqrt{1 + y^2}$ (b) $-1 - \sqrt{1 - y^2}$
(c) $1 + \sqrt{1 + y^2}$ (d) $-1 + \sqrt{1 - y^2}$

3. If $\frac{z + 1}{z + i}$ is a purely imaginary number, (where $i = \sqrt{-1}$) then z lies on a
(a) straight line
(b) circle
(c) circle with radius $= 1/\sqrt{2}$
(d) circle passing through the origin

4. The equation whose roots are nth power of the roots of the equation, $x^2 - 2x\cos\theta + 1 = 0$, is given by
(a) $(x + \cos n\,\theta)^2 + \sin^2 n\theta = 0$
(b) $(x - \cos n\theta)^2 + \sin^2 n\theta = 0$
(c) $x^2 + 2x\cos n\theta + 1 = 0$
(d) $x^2 - 2x\cos n\theta + 1 = 0$

5. If the imaginary part of $\frac{z - 3}{e^{i\theta}} + \frac{e^{i\theta}}{z - 3}$ is zero, then z can lie on
(a) a circle with unit radius
(b) a circle with radius 3 units
(c) a straight line through the point (3, 0)
(d) a parabola with the vertex (3, 0)

6. If $|z_1 + z_2|^2 = |z_1|^2 + |z_2|^2$, then
(a) $\frac{z_1}{z_2}$ is purely real (b) $\frac{z_1}{z_2}$ is purely imaginary
(c) $z_1\bar{z}_2 + z_2\bar{z}_1 = 0$ (d) $\text{amp}\left(\frac{z_1}{z_2}\right) = \frac{\pi}{2}$

7. If z_1, z_2, z_3, z_4 are the four complex numbers represented by the vertices of a quadrilateral taken in order such that $z_1 - z_4 = z_2 - z_3$ and amp $\left(\frac{z_4 - z_1}{z_2 - z_1}\right) = \frac{\pi}{2}$, then the quadrilateral is a

(a) rhombus (b) square
(c) rectangle (d) cyclic quadrilateral

8. For a non-zero complex number z, let $\arg(z)$ denote the principal argument with $-\pi < \arg(z) \le \pi$. Then, which of the following statement(s) is (are) FALSE?

(a) $\arg(-1-i) = \frac{\pi}{4}$, where $i = \sqrt{-1}$

(b) The function $f : R \to (-\pi, \pi]$, defined by $f(t) = \arg(-1+it)$ for all $t \in R$, is continuous at all points of R, where $i = \sqrt{-1}$.

(c) For any two non-zero complex numbers z_1 and z_2, $\arg\left(\frac{z_1}{z_2}\right) - \arg(z_1) + \arg(z_2)$ is an integer multiple of 2π.

(d) For any three given distinct complex numbers z_1, z_2 and z_3, the locus of the point z satisfying the condition $\arg\left(\frac{(z - z_1)(z_2 - z_3)}{(z - z_3)(z_2 - z_1)}\right) = \pi$, lies on a straight line.

9. If α is a complex constant such that $\alpha z^2 + z + \bar{\alpha} = 0$ has a real root, then

(a) $\alpha + \bar{\alpha} = 1$
(b) $\alpha + \bar{\alpha} = 0$
(c) $\alpha + \bar{\alpha} = -1$
(d) the absolute value of the real root is 1

10. If z_1, z_2, z_3, z_4 are roots of the equation $a_0 z^4 + a_1 z^3 + a_2 z^2 + a_3 z + a_4 = 0$ where a_0, a_1, a_2, a_3 and a_4 are real, then

(a) $\bar{z}_1, \bar{z}_2, \bar{z}_3, \bar{z}_4$ are also roots of the equation
(b) z_1 is equal to at least one of $\bar{z}_1, \bar{z}_2, \bar{z}_3, \bar{z}_4$
(c) $-\bar{z}_1, -\bar{z}_2, -\bar{z}_3, -\bar{z}_4$ are also roots of the equation
(d) None of the above

11. The reflection of the complex number $\frac{2-i}{3+i}$, (where $i = \sqrt{-1}$) in the straight line $z(1+i) = \bar{z}(i-1)$ is

(a) $\frac{-1-i}{2}$ (b) $\frac{-1+i}{2}$
(c) $\frac{i(i+1)}{2}$ (d) $\frac{-1}{1+i}$

12. The common roots of the equations $z^3 + (1+i)z^2 + (1+i)z + i = 0$, (where $i = \sqrt{-1}$) and $z^{1993} + z^{1994} + 1 = 0$ are

(a) 1 (b) ω
(c) ω^2 (d) ω^{981}

13. Let $a, b \in R$ and $a^2 + b^2 \neq 0$.

Suppose $S = \left\{z \in C : z = \frac{1}{a + i\,bt}, t \in R, t \neq 0\right\}$, where $i = \sqrt{-1}$. If $z = x + iy$ and $z \in S$, then (x, y) lies on

(a) the circle with radius $\frac{1}{2a}$ and centre $\left(\frac{1}{2a}, 0\right)$ for $a > 0, b \neq 0$

(b) the circle with radius $-\frac{1}{2a}$ and centre $\left(-\frac{1}{2a}, 0\right)$ for $a < 0, b \neq 0$

(c) the X-axis for $a \neq 0, b = 0$

(d) the Y-axis for $a = 0, b \neq 0$

14. If z_1, z_2 are two complex numbers satisfying the equation $\left|\frac{z_1 + z_2}{z_1 - z_2}\right| = 1$, then z_1/z_2 is a number which is

(a) positive real (b) negative real
(c) zero (d) purely imaginary

15. If $z_1 = a + ib$ and $z_2 = c + id$ (where $i = \sqrt{-1}$) are two complex numbers such that $|z_1| = |z_2| = 1$ and $\text{Re}(z_1 \bar{z}_2) = 0$, then the pair of complex numbers, $\omega_1 = a + ic$ and $\omega_2 = b + id$ satisfy

(a) $|\omega_1| = 1$ (b) $|\omega_2| = 1$
(c) $|\omega_1 \bar{\omega}_2| = 1$ (d) $\text{Re}(\bar{\omega}_1 \omega_2) = 0$

16. The real value of θ for which the expression $\frac{1 + i\cos\theta}{1 - 2i\cos\theta}$ (where $i = \sqrt{-1}$) is a real number is

(a) $2n\pi + \frac{\pi}{2}, n \in I$ (b) $2n\pi - \frac{\pi}{2}, n \in I$
(c) $2n\pi \pm \frac{\pi}{2}, n \in I$ (d) $2n\pi \pm \frac{\pi}{4}, n \in I$

17. If $\cos\alpha + \cos\beta + \cos\gamma = \sin\alpha + \sin\beta + \sin\gamma = 0$, then

(a) $\cos(2\alpha) + \cos(2\beta) + \cos(2\gamma) = 0$
(b) $\sin(2\alpha) + \sin(2\beta) + \sin(2\gamma) = 0$
(c) $\cos(\beta + \gamma) + \cos(\gamma + \alpha) + \cos(\alpha + \beta) = 0$
(d) $\sin(\beta + \gamma) + \sin(\gamma + \alpha) + \sin(\alpha + \beta) = 0$

18. If $1, \omega, \omega^2, \ldots, \omega^{n-1}$ are the n, nth roots of unity, then $(2 - \omega)(2 - \omega^2) \ldots (2 - \omega^{n-1})$ equals

(a) $2^n - 1$
(b) ${}^nC_1 + {}^nC_2 + \ldots + {}^nC_n$
(c) $[{}^{2n}C_0 + {}^{2n+1}C_1 + {}^{2n+1}C_2 + \ldots + {}^{2n+1}C_n]^{1/2} - 1$
(d) $2^n + 1$

19. If z_1, z_2, z_3, z_4 are the vertices of a square in that order, then

(a) $z_1 + z_3 = z_2 + z_4$
(b) $|z_1 - z_2| = |z_2 - z_3| = |z_3 - z_4| = |z_4 - z_1|$
(c) $|z_1 - z_3| = |z_2 - z_4|$
(d) $(z_1 - z_3)/(z_2 - z_4)$ is purely imaginary

20. If $2\cos\theta = x + \dfrac{1}{x}$ and $2\cos\phi = y + \dfrac{1}{y}$, then

(a) $\dfrac{x}{y} + \dfrac{y}{x} = 2\cos(\theta - \phi)$

(b) $x^m y^n + \dfrac{1}{x^m y^n} = 2\cos(m\theta + n\phi)$

(c) $\dfrac{x^m}{y^n} + \dfrac{y^n}{x^m} = 2\cos(m\theta - n\phi)$

(d) $xy + \dfrac{1}{xy} = 2\cos(\theta + \phi)$

21. If $(1+x)^n = C_0 + C_1 x + \ldots + C_n x^n$, where n is a positive integer, then

(a) $C_0 - C_2 + C_4 - \ldots = 2^{n/2}\cos\left(\dfrac{n\pi}{4}\right)$

(b) $C_1 - C_3 + C_5 - \ldots = 2^{n/2}\sin\left(\dfrac{n\pi}{4}\right)$

(c) $C_0 + C_4 + C_8 + \ldots = 2^{n-2} + 2^{(n-2)/2}\cos\left(\dfrac{n\pi}{4}\right)$

(d) $C_0 + C_3 + C_6 + \ldots = \dfrac{1}{3}\left(2^n + 2\cos\left(\dfrac{n\pi}{3}\right)\right)$

Linked-Comprehension Type

In these questions, a passage (paragraph) has been given followed by questions based on each of the passages. You have to answer the questions based on the passage given.

PASSAGE 1

Complex numbers z satisfy the equation $|z - (4/z)| = 2$

On the basis of above information, answer the following questions

1. The difference between the least and the greatest moduli of complex numbers is

(a) 2 (b) 4
(c) 1 (d) 3

2. The value of $\arg(z_1/z_2)$, where z_1 and z_2 are complex numbers with the greatest and the least moduli, can be

(a) 2π (b) π
(c) $\pi/2$ (d) None of these

3. Locus of z, if $|z - z_1| = |z - z_2|$, where z_1 and z_2 are complex numbers with the greatest and the least moduli, is

(a) line parallel to the real axis
(b) line parallel to the imaginary axis
(c) line having a positive slope
(d) line having a negative slope

PASSAGE 2

Let A, B, C be three sets of complex number as defined below

$$A = \{z : \text{Im}(z) \geq 1\}$$
$$B = \{z : |z - 2 - i| = 3\}$$
$$C = \{z : \text{Re}((1-i)z) = \sqrt{2}\}$$

On the basis of above information, answer the following questions

1. The number of elements in the set $A \cap B \cap C$ is

(a) 0 (b) 1
(c) 2 (d) ∞

2. Area of B is equal to

(a) $\dfrac{10\pi}{3}$ (b) 9π
(c) $\dfrac{16\pi}{3}$ (d) $\dfrac{32\pi}{3}$

3. Let z be any point in $A \cap B \cap C$ and let w be any point satisfying $|w - 2 - i| < 3$. Then, $|z| - |w| + 3$ lies between

(a) -6 and 3 (b) -3 and 6
(c) -6 and 6 (d) -3 and 9

PASSAGE 3

If $x^2 + 1 = 0 \Rightarrow x^2 = -1$ or $x = \pm\sqrt{-1} = \pm i$ (*iota*) is called the imaginary unit.

Also, $i^2 = -1, i^3 = i^2 \cdot i = (-1)i = -i$ and $i^4 = (i^2)^2 = (-1)^2 = 1$.

i.e. $i^n + i^{n+1} + i^{n+2} + i^{n+3} = 0 \;\forall\; n \in I$ (Integer) and

$x^3 - 1 = 0 \Rightarrow (x-1)(x^2 + x + 1) = 0$

$\Rightarrow \quad (x-1)(x-\omega)(x-\omega^2) = 0$

$\therefore \quad x = 1, \omega, \omega^2$ are the cube roots of unity.

i.e. $\quad \omega^n + \omega^{n+1} + \omega^{n+2} = 0 \;\forall\; n \in I$ (Integer)

Now let $\quad z = a + ib$ if $|a : b| = \sqrt{3} : 1$ or $1 : \sqrt{3}$

Then, convert z in terms of ω or ω^2.

Also, $|1 - \omega| = |1 - \omega^2| = \sqrt{3}$

On the basis of above information, answer the following questions

1. The value of the sum $\sum_{n=1}^{16} (i^n + i^{n+1})$, where $i = \sqrt{-1}$, equals

(a) i (b) $i - 1$
(c) $-i$ (d) 0

2. For positive integers n_1, n_2; the value of the expression, $(1+i)^{n_1} + (1+i^3)^{n_1} + (1+i^5)^{n_2} + (1+i^7)^{n_2}$, where $i = \sqrt{-1}$, is real if

(a) $n_1 = n_2 + 1$
(b) $n_1 = n_2 - 1$
(c) $n_1 = n_2$
(d) $n_1 > 0, n_2 > 0$

3. If ($\omega \neq 1$) is a cube root of unity and $i = \sqrt{-1}$, then

$$\begin{vmatrix} 1 & 1+i+\omega^2 & \omega^2 \\ 1-i & -1 & \omega^2 - 1 \\ -i & -i+\omega-1 & -1 \end{vmatrix}$$ is equal to

(a) 0 (b) 1
(c) i (d) ω

4. Let $\omega = -\frac{1}{2} + i\frac{\sqrt{3}}{2}$, where $i = \sqrt{-1}$, then the value of the determinant $\begin{vmatrix} 1 & 1 & 1 \\ 1 & -1-\omega^2 & \omega^2 \\ 1 & \omega^2 & \omega^4 \end{vmatrix}$ is

(a) 3ω (b) $3\omega(\omega - 1)$
(c) $3\omega^2$ (d) $3\omega(1-\omega)$

5. If $i = \sqrt{-1}$, then

$$4 + 5\left(-\frac{1}{2} + i\frac{\sqrt{3}}{2}\right)^{334} + 3\left(-\frac{1}{2} + \frac{i\sqrt{3}}{2}\right)^{365}$$ is equal to

(a) $1 - i\sqrt{3}$ (b) $-1 + i\sqrt{3}$
(c) $i\sqrt{3}$ (d) $-i\sqrt{3}$

6. The complex numbers z_1, z_2 and z_3 satisfying $\frac{z_1 - z_3}{z_2 - z_3} = \frac{1 - i\sqrt{3}}{2}$ (where $i = \sqrt{-1}$) are the vertices of a triangle which is
(a) of area zero
(b) right angled isosceles
(c) equilateral
(d) obtuse angled isosceles

7. The value of $\sum_{r=2}^{n} (r-1)(r-\omega)(r-\omega^2)$, where ω is an imaginary cube root of unity, is
(a) $(\Sigma n)^2 - 2n$ (b) $(\Sigma n)^2 - n$
(c) $(\Sigma n)^2 + n$ (d) $(\Sigma n)^2 + 2n$

8. The smallest positive integer n for which $\left(\frac{1+i}{1-i}\right)^n = 1$, where $i = \sqrt{-1}$, is
(a) 1 (b) 2
(c) 3 (d) 4

9. If α, β and γ are the roots of $x^3 - 3x^2 + 3x + 7 = 0$, then $\Sigma\,((\alpha - 1)/(\beta - 1))$ is
(a) 0 (b) 3ω
(c) $3/\omega$ (d) $2\omega^2$
(where ω is cube root of unity)

10. If $\left(\frac{3}{2} + \frac{i\sqrt{3}}{2}\right)^{50} = 3^{25}(x - iy)$ where x, y are real and $i = \sqrt{-1}$, then the ordered pair (x, y) is given by

(a) $(0, 3)$ (b) $\left(\frac{1}{2}, -\frac{\sqrt{3}}{2}\right)$

(c) $(-3, 0)$ (d) $\left(-\frac{1}{2}, \frac{\sqrt{3}}{2}\right)$

11. The roots of the cubic equation $(z + \alpha\beta)^3 = \alpha^3\ (\alpha \neq 0)$, represent the vertices of a triangle of sides of length
(a) $\frac{1}{\sqrt{3}}|\alpha\beta|$ (b) $\sqrt{3}|\alpha|$
(c) $\sqrt{3}|\beta|$ (d) $\frac{1}{\sqrt{3}}|\alpha|$

12. If $z = \frac{\sqrt{3} + i}{2}$ where $i = \sqrt{-1}$, then $(z^{101} + i^{103})^{105}$ is equal to
(a) z (b) z^2
(c) z^3 (d) z^4

PASSAGE 4

Let $z = a + ib = re^{i\theta}$ where $a, b, \theta \in R$ and $i = \sqrt{-1}$

Then, $r = \sqrt{(a^2 + b^2)} = |z|$

and $\theta = \tan^{-1}\left(\frac{b}{a}\right) = \arg(z)$

Now, $|z|^2 = a^2 + b^2 = (a + ib)(a - ib) = z\bar{z}$

$\Rightarrow \frac{1}{z} = \frac{\bar{z}}{|z|^2}$

and $|z_1 z_2 z_3 \ldots z_n| = |z_1||z_2||z_3|\ldots|z_n|$

If $|f(z)| = 1$, then $f(z)$ is called unimodular. In this case $f(z)$ can always be expressed as $f(z) = e^{i\alpha}, \alpha \in R$

Also, $e^{i\alpha} + e^{i\beta} = e^{i\left(\frac{\alpha+\beta}{2}\right)} 2\cos\left(\frac{\alpha - \beta}{2}\right)$ and

$e^{i\alpha} - e^{i\beta} = e^{i\left(\frac{\alpha+\beta}{2}\right)} 2i\sin\left(\frac{\alpha - \beta}{2}\right)$ where $\alpha, \beta \in R$.

On the basis of above information, answer the following questions

1. If z_1, z_2, z_3 are complex numbers such that $|z_1| = |z_2| = |z_3| = |z_1 + z_2 + z_3| = 1$, then $\left|\frac{1}{z_1} + \frac{1}{z_2} + \frac{1}{z_3}\right|$ is
(a) equal to 1 (b) less than 1
(c) greater than 3 (d) equal to 3

2. If $|z_1| = 1, |z_2| = 2, |z_3| = 3$ and $|z_1 + z_2 + z_3| = 1$, then $|9z_1z_2 + 4z_3z_1 + z_2z_3|$ is equal to
(a) 6 (b) 36
(c) 216 (d) 1296

3. If $|z_1|=|z_2|=\dots=|z_n|=1$, then the value of $|z_1+z_2+z_3+\dots+z_n|$ is equal to
(a) 1
(b) $|z_1|+|z_2|+|z_3|+\dots+|z_n|$
(c) $\left|\frac{1}{z_1}+\frac{1}{z_2}+\frac{1}{z_3}+\dots+\frac{1}{z_n}\right|$
(d) n

4. If $z=x+iy$ is a complex number with rationals x and y and $i=\sqrt{-1}$ and $|z|=1$, then $|z^{2n}-1|$ is $(n\in N)$
(a) an irrational number
(b) a rational number
(c) non terminating non recurring
(d) a positive real number

5. The value of $\tan\left(i\ln\left(\frac{a-ib}{a+ib}\right)\right)$, (where $i=\sqrt{-1}$)is equal to
(a) $\frac{2ab}{b^2-a^2}$ (b) $\frac{2ab}{a^2-b^2}$
(c) $\frac{a^2+b^2}{a^2-b^2}$ (d) $\frac{a^2-b^2}{a^2+b^2}$

6. If $|z-3i|=3$, (where $i=\sqrt{-1}$) and $\arg z\in(0,\pi/2)$, then $\cot(\arg z)-\frac{6}{z}$ is equal to
(a) 0 (b) $-i$ (c) i (d) π

7. If z_1 and z_2 are complex numbers satisfying $\left|\frac{z_1+z_2}{z_1-z_2}\right|=1$ and $\arg\left(\frac{z_1-z_2}{z_1+z_2}\right)\neq m\pi\ (m\in I)$, then $\frac{z_1}{z_2}$ is
(a) zero
(b) a rational number
(c) a positive real number
(d) a purely imaginary

PASSAGE 5

Let a quadratic equation $az^2+bz+c=0$ where $a,b,c\in R$ and $a\neq 0$. If one root of this equation is $p+iq$, then other must be the conjugate $p-iq$ and *vice-versa*. ($p,q\in R$ and $i=\sqrt{-1}$). But if a,b,c are not real, then roots of $az^2+bz+c=0$ are not conjugate to each other.

ie, if one root is real, then other may be non real, Now, combining both cases we can say that $az^2+bz+c=0$ where $a,b,c\in C$ and $a\neq 0$.

On the basis of above information, answer the following questions

1. If one root of the quadratic equation $(1+i)x^2-(7+3i)x+(6+8i)=0$, where $i=\sqrt{-1}$ is $4-3i$, then the other root must be
(a) $4+3i$ (b) $1-i$
(c) $1+i$ (d) $i(i-1)$

2. The condition that the equation $z^2+az+b=0$ has a purely imaginary root where a and b are complex constants, is
(a) $(a+\bar{a})(a\bar{b}-\bar{a}b)+(b+\bar{b})^2=0$
(b) $(a+\bar{a})(a\bar{b}+\bar{a}b)+(b-\bar{b})^2=0$
(c) $(a-\bar{a})(a\bar{b}+\bar{a}b)+(b-\bar{b})^2=0$
(d) $(a+\bar{a})(a\bar{b}+\bar{a}b)+(b+\bar{b})^2=0$

3. The condition that the equation $z^2+az+b=0$ has a purely real root where a and b are complex constants, is
(a) $\frac{a\bar{b}+b\bar{a}}{b-\bar{b}}=\frac{b+\bar{b}}{\bar{a}-a}$ (b) $\frac{a\bar{b}+b\bar{a}}{b+\bar{b}}=\frac{b+\bar{b}}{\bar{a}+a}$
(c) $\frac{a\bar{b}-b\bar{a}}{b+\bar{b}}=\frac{b-\bar{b}}{\bar{a}+a}$ (d) $\frac{a\bar{b}-b\bar{a}}{b-\bar{b}}=\frac{b-\bar{b}}{\bar{a}-a}$

4. The condition that the equation $az^2+bz+c=0$ has both real roots where a, b, c are complex constants is
(a) $\frac{a}{\bar{a}}=\frac{b}{\bar{b}}=\frac{c}{\bar{c}}$ (b) $\frac{a}{\bar{a}}=\frac{b}{\bar{b}}=-\frac{c}{\bar{c}}$
(c) $-\frac{a}{\bar{a}}=\frac{b}{\bar{b}}=\frac{c}{\bar{c}}$ (d) $\frac{a}{\bar{a}}=-\frac{b}{\bar{b}}=\frac{c}{\bar{c}}$

5. The condition that the equation $az^2+bz+c=0$ has both purely imaginary roots where a, b, c are complex constants is
(a) $\frac{a}{\bar{a}}=\frac{b}{\bar{b}}=\frac{c}{\bar{c}}$ (b) $\frac{a}{\bar{a}}=-\frac{b}{\bar{b}}=-\frac{c}{\bar{c}}$
(c) $\frac{a}{\bar{a}}=\frac{b}{\bar{b}}=-\frac{c}{\bar{c}}$ (d) $\frac{a}{\bar{a}}=-\frac{b}{\bar{b}}=\frac{c}{\bar{c}}$

6. If equations $az^2+bz+c=0$ and $z^2+2z+3=0$ have a common root where $a,b,c\in R$, then $a:b:c$ is
(a) $2:3:1$ (b) $1:2:3$
(c) $3:1:2$ (d) $3:2:1$

7. If α is a non real complex number and $x^2+\alpha x+\bar{\alpha}=0$ has a real root γ, then
(a) $\gamma=\alpha+\bar{\alpha}$ (b) $\gamma=\alpha-\bar{\alpha}$
(c) $\gamma=1$ (d) $\gamma=|\alpha-\bar{\alpha}|$

8. If the quadratic equation $z^2+(a+ib)z+c+id=0$, where a, b, c, d are non-zero real and $i=\sqrt{-1}$, has a real root, then
(a) $abd=b^2c+d^2$ (b) $abc=bc^2+d^2$
(c) $abd=bc^2+ad^2$ (d) $abc=bc^2+ad^2$

PASSAGE 6

Let $z=a+ib=(a,b)$ be any complex number $\forall\ a,b\in R$ and $i=\sqrt{-1}$. If $(a,b)\neq(0,0)$, then $\arg(z)=\tan^{-1}\left(\frac{b}{a}\right)$

where $-\pi<\arg(z)\leq\pi$ and

$$\arg(\bar{z})+\arg(-z)=\begin{cases}\pi, & \text{if } \arg(z)<0\\ -\pi, & \text{if } \arg(z)>0\end{cases}$$

On the basis of above information, answer the following questions

1. If $\arg(z) < 0$, then $\arg(-z) - \arg(z)$ is equal to
(a) π (b) $-\pi$
(c) $-\frac{\pi}{2}$ (d) $\frac{\pi}{2}$

2. Let z and w be two non zero complex numbers such that $|z| = |w|$ and $\arg(z) + \arg(w) = \pi$, then z equals
(a) w (b) $-w$
(c) $\overline{w}$ (d) $-\overline{w}$

3. If $\arg(2z_1) - \arg(3z_2) = \pi$, then the value of $\left|\frac{z_1}{z_2}\right|$ is equal to
(a) 0.5 (b) 1.0
(c) 1.5 (d) 2.0

4. If $\arg(z) > 0$, then $\arg(-z) - \arg(z) = \lambda_1$ and if $\arg(z) < 0$ then $\arg(z) - \arg(-z) = \lambda_2$, then
(a) $\lambda_1 + \lambda_2 = 0$ (b) $\lambda_1 - \lambda_2 = 0$
(c) $3\lambda_1 - 2\lambda_2 = 0$ (d) $2\lambda_1 - 3\lambda_2 = 0$

5. The value of
$\sqrt{\{\arg(z) + \arg(-\overline{z}) - 2\pi\}\{\arg(-z) + \arg(\overline{z})\}}$
$\forall\ z = x + iy$, (where $i = \sqrt{-1}$) $x, y > 0$ is
(a) π (b) $-\pi$
(c) 0 (d) not defined

PASSAGE 7

Let $z_1 = a_1 + ib_1 \equiv (a_1, b_1)$ and $z_2 = a_2 + ib_2 \equiv (a_2, b_2)$; where $i = \sqrt{-1}$, be two complex numbers

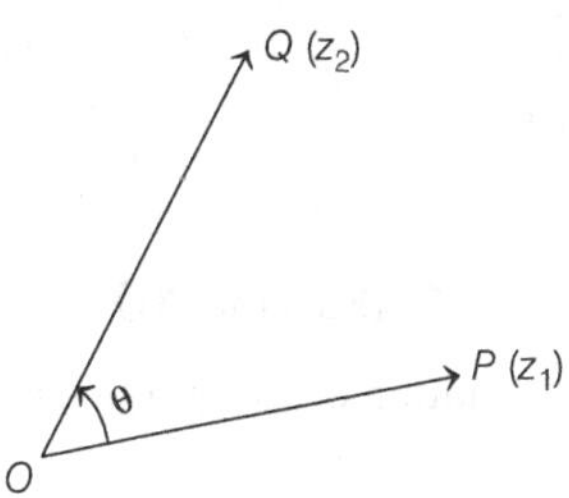

If $\angle POQ = \theta$, From Rotation theorem

$$\frac{z_2 - 0}{z_1 - 0} = \frac{|z_2|}{|z_1|} e^{i\theta} \Rightarrow \frac{z_2\,\overline{z}_1}{z_1\,\overline{z}_1} = \frac{|z_2|}{|z_1|} e^{i\theta}$$

$$\Rightarrow \frac{z_2\,\overline{z}_1}{|z_1|^2} = \frac{|z_2|}{|z_1|} e^{i\theta} \Rightarrow z_2\,\overline{z}_1 = |z_1||z_2| e^{i\theta}$$

$$\Rightarrow z_2\,\overline{z}_1 = |z_1||z_2|(\cos\theta + i\sin\theta)$$

$\therefore$ $\text{Re}(z_2\,\overline{z}_1) = |z_1||z_2|\cos\theta$ ...(i)

and $\text{Im}(z_2\,\overline{z}_1) = |z_1||z_2|\sin\theta$ ...(ii)

The dot product of z_1 and z_2 is defined by $z_1 o z_2 = |z_1||z_2|\cos\theta = \text{Re}(z_2\,\overline{z}_1)$ [from (i)] and cross product of z_1 and z_2 is defined by

$z_1 \times z_2 = |z_1||z_2|\sin\theta = \text{Im}(z_2\,\overline{z}_1)$ [from Eq. (ii)]

On the basis of above information, answer the following questions

1. If $z_1 = 2 + 5i$, $z_2 = 3 - i$, then the value of $\sqrt{(z_1 \cdot z_2 + z_2 \times z_1)}$ is equal to
(a) 2 (b) 3
(c) $2\sqrt{3}$ (d) $3\sqrt{2}$

2. If $z_1 = 3 + 4i$ and $z_2 = 4 + 3i$, then the value of $\sin\theta\left(\pi < \theta < \frac{3\pi}{2}\right)$ is equal to
(a) $-\frac{1}{7}$ (b) $-\frac{7}{25}$
(c) $-\frac{24}{25}$ (d) $-\frac{1}{25}$

3. If $z_1 = 5 + 12i$ and $z_2 = 3 + 4i$, then (the projection of z_1 on z_2 + projection of z_2 on z_1) is equal to
(a) $\frac{4131}{65}$ (b) $\frac{3411}{65}$
(c) $\frac{1134}{65}$ (d) $\frac{1341}{65}$

4. If a and b are real numbers between 0 and 1 such that the points $z_1 = a + i$, $z_2 = 1 + bi$ and $z_3 = 0$ form an equilateral triangle, then (a, b) is
(a) $(2 - \sqrt{3}, \sqrt{3} - 1)$ (b) $(\sqrt{3} - 1, \sqrt{3} - 1)$
(c) $(2 - \sqrt{3}, 2 - \sqrt{3})$ (d) $(\sqrt{3} - 1, 2 - \sqrt{3})$

5. Let z_1 and z_2 be roots of the equation $z^2 + z + 1 = 0$. If $\angle POQ = \alpha \neq 0\ (0 < \alpha < \pi)$ and $OP = OQ$ where O is the origin, then α is equal to
(a) $\pi/4$ (b) $\pi/2$
(c) $\pi/3$ (d) $2\pi/3$

PASSAGE 8

The equation $z^n - 1 = 0$ has n roots which are called the nth roots of unity. The n, nth roots of unity are $1, \alpha, \alpha^2, \ldots, \alpha^{n-1}$ which are in GP, where $\alpha = \cos\left(\frac{2\pi}{n}\right) + i\sin\left(\frac{2\pi}{n}\right)$; $i = \sqrt{-1}$

then we have following results :

(i) $\sum_{r=0}^{n-1} \alpha^r = 0$ or $\sum_{r=0}^{n-1} \cos\left(\frac{2\pi r}{n}\right) = 0$

and $\sum_{r=0}^{n-1} \sin\left(\frac{2\pi r}{n}\right) = 0$

(ii) $z^n - 1 = \prod_{r=0}^{n-1} (z - \alpha^r)$

(iii) $\prod_{r=0}^{n-1} \alpha^r = (-1)^{n-1}$

(iv) $\sum_{r=0}^{n-1} \alpha^{kr} = \begin{cases} n, & \text{if } k \text{ is multiple of } n \\ 0, & \text{if } k \text{ is not multiple of } n \end{cases}$

On the basis of above information, answer the following questions

1. The value of $\sum_{r=1}^{n-1} \frac{1}{(2-\alpha^r)}$ is equal to
   (a) $(n-2)\,2^n$ (b) $\frac{(n-2)\,2^{n-1}+1}{2^n-1}$
   (c) $\frac{(n-2)\,2^{n-1}}{2^n-1}$ (d) $\frac{(n-1)\,2^{n-1}}{2^n-1}$

2. If ω be non real complex cube root of unity, then the value of $\frac{\prod_{p=1}^{4}(\omega+\alpha^p)}{\prod_{q=1}^{7}(\omega^2-\alpha^q)}$ is equal to
   (a) $\left(\frac{-1+i\sqrt{3}}{2}\right)$ (b) $\left(\frac{1-i\sqrt{3}}{2}\right)$
   (c) $\left(\frac{-1-i\sqrt{3}}{2}\right)$ (d) $\left(\frac{1+i\sqrt{3}}{2}\right)$

3. The algebraic sum of perpendicular distances from the points $1, \alpha, \alpha^2, \alpha^3, \dots \alpha^{n-1}$ to the line $\bar{a}z + a\bar{z} + b = 0$, (where a is complex number and b is real) is equal to
   (a) $\frac{n}{2|a|}$ (b) $\frac{n|b|}{2a}$
   (c) $\frac{nb}{|a|}$ (d) $\frac{nb}{2|a|}$

4. If $\alpha = \cos\left(\frac{2\pi}{7}\right) + i\sin\left(\frac{2\pi}{7}\right)$, then equation whose roots are $\alpha+\alpha^2+\alpha^4$ and $\alpha^3+\alpha^5+\alpha^6$ is
   (a) $z^2 - z - 2 = 0$ (b) $z^2 - z + 2 = 0$
   (c) $z^2 + z - 2 = 0$ (d) $z^2 + z + 2 = 0$

5. If $n \in N$, then the value of $\prod_{r=1}^{n-1} \sin\left(\frac{r\pi}{n}\right)$ is equal to
   (a) $\frac{n}{2^{n-1}}$ (b) $n \cdot 2^{n-1}$
   (c) $\frac{n}{2^{n+1}}$ (d) $n \cdot 2^{n+1}$

6. If $n \in I, n \ge 2$, then the value of $\sum_{r=1}^{n-1}(n-r)\cos\left(\frac{2r\pi}{n}\right)$ is equal to
   (a) $-n$ (b) n
   (c) $-\frac{n}{2}$ (d) $\frac{n}{2}$

7. The expression
   $$\sum_{m=1}^{4n+1}\left\{\sum_{k=1}^{m-1}\left\{\sin\left(\frac{2\pi k}{m}\right) - i\cos\left(\frac{2\pi k}{m}\right)\right\}\right\}^m$$
   (a) has real part equal to 1
   (b) purely real
   (c) purely imaginary
   (d) has imaginary part equal to 2

PASSAGE 9

Let $A(z_1), B(z_2), C(z_3)$ be the vertices of an equilateral triangle ABC such that $|z_1| = |z_2| = |z_3| = 2$. A circle is inscribed in the triangle ABC which touches the sides AB, BC and CA at $D(z_4), E(z_5)$ and $F(z_6)$ respectively. $P(z)$ be any point on its incircle other than D, E, F.

On the basis of above information, answer the following questions

1. The value of $(AB)^2 + (BC)^2 + (CA)^2$ is equal to
   (a) 9 (b) 18
   (c) 27 (d) 36

2. The value of $(PA)^2 + (PB)^2 + (PC)^2$ is equal to
   (a) 9 (b) 12
   (c) 15 (d) 18

3. The value of $(DE)^2 + (EF)^2 + (FD)^2$ is equal to
   (a) 12 (b) 3
   (c) $3\sqrt{3}$ (d) 9

4. The value of Re $(z_1\bar{z}_2 + z_2\bar{z}_3 + z_3\bar{z}_1)$ is equal to
   (a) 0 (b) 6
   (c) -6 (d) -3

5. If $z_1 = \sqrt{3} + i, i = \sqrt{-1}$, then the value of $\sqrt{|z_2 - z_3|^2 + |z_2 + z_3|^2}$ is equal to
   (a) 0 (b) 2
   (c) 4 (d) 6

6. $\frac{z_1}{z_3}$ is equal to
   (a) $1 - i\sqrt{3}$ (b) $1 + i\sqrt{3}$
   (c) $\frac{-1+i\sqrt{3}}{2}$ (d) $\frac{1+i\sqrt{3}}{2}$

PASSAGE 10

The general equation of straight line is $\bar{a}z + a\bar{z} + b = 0$ ...(i)

where a is complex number and b is real number. The real and complex slopes of the line are $-i\left(\frac{a+\bar{a}}{a-\bar{a}}\right)$ and $-\frac{a}{\bar{a}}$, (where $i = \sqrt{-1}$) respectively. If adding $z\bar{z}$ in LHS (i), then (i) convert in general equation of circle

ie, $z\bar{z} + a\bar{z} + \bar{a}z + b = 0$

with centre $-a$ and radius $\sqrt{|a|^2 - b}$ if $a = 0$, then circle $|z|^2 + b = 0$

which is defined only when $b < 0$

or $|z|^2 = -b = r^2$ (say)

$\therefore$ $|z| = r$ (radius of circle)

On the basis of above information, answer the following questions

1. If lines $a\bar{z} + \bar{a}z + b = 0$ and $a_1\bar{z} + \bar{a}_1 z + b_1 = 0$ where a, a_1 are complex numbers and $b, b_1 \in R$; are perpendicular then

(a) $a\bar{a}_1 - \bar{a}a_1 = 0$ (b) $a\bar{a}_1 + a_1\bar{a} = 0$
(c) $a\bar{a} - a_1\bar{a}_1 = 0$ (d) $aa_1 + \bar{a}\bar{a}_1 = 0$

2. Reflection of the line $(2+3i)z + (2-3i)\bar{z} = 0, i = \sqrt{-1}$, on the real axis is

(a) $(2-3i)z - (2+3i)\bar{z} = 0$
(b) $(2+3i)z - (2-3i)\bar{z} = 0$
(c) $(2+3i)z + (2-3i)\bar{z} = 0$
(d) $(2-3i)z + (2+3i)\bar{z} = 0$

3. Two different non parallel lines meet the circle $|z| = r$ in the points a, b and c, d; where a, b, c, d are complex numbers respectively, then these lines meet in the point z is given by

(a) $\dfrac{a^{-1} + c^{-1} - b^{-1} - d^{-1}}{a^{-1}c^{-1} - b^{-1}d^{-1}}$

(b) $\dfrac{a^{-1} + b^{-1} - c^{-1} - d^{-1}}{a^{-1}b^{-1} - c^{-1}d^{-1}}$

(c) $\dfrac{a^{-1} + d^{-1} - b^{-1} - c^{-1}}{a^{-1}d^{-1} - b^{-1}c^{-1}}$

(d) $\dfrac{b^{-1} + c^{-1} - a^{-1} - d^{-1}}{b^{-1}c^{-1} - a^{-1}d^{-1}}$

4. If $|z - i| \le 2$ and $z_0 = 5 + 3i$, the maximum value of $|iz + z_0|$, where $i = \sqrt{-1}$, is

(a) $2 + \sqrt{31}$ (b) $\sqrt{31} - 2$
(c) 7 (d) -7

5. The number of values of z which satisfies both the equations $|z - 1 - i| = \sqrt{2}$ and $|z + 1 + i| = 2$, where $i = \sqrt{-1}$, is

(a) 0 (b) 1
(c) 2 (d) infinitely many

6. If z_1, z_2 and z_3 be three points on $|z| = 1$. If θ_1, θ_2, and θ_3 be the arguments of z_1, z_2 and z_3 respectively, then $\Sigma \cos(\theta_1 - \theta_2)$ is

(a) $\ge -\dfrac{3}{2}$ (b) $\le -\dfrac{3}{2}$
(c) $\ge \dfrac{3}{2}$ (d) ≥ 1

7. If the circles $z\bar{z} + \bar{a}_1 z + a_1\bar{z} + b_1 = 0$ and $z\bar{z} + \bar{a}_2 z + a_2\bar{z} + b_2 = 0$ (where $b_1, b_2, \in R$) intersect orthogonally, then

(a) $a_1a_2 + \bar{a}_1\bar{a}_2 = b_1 + b_2$ (b) $a_1a_2 - \bar{a}_1\bar{a}_2 = b_1 - b_2$
(c) $a_1\bar{a}_2 + \bar{a}_1a_2 = b_1 + b_2$ (d) $a_1\bar{a}_2 - \bar{a}_1a_2 = b_1 - b_2$

PASSAGE 11

If $a\cos\alpha + b\cos\beta + c\cos\gamma = 0$
$= a\sin\alpha + b\sin\beta + c\sin\gamma$;
where $a, b, c \in R$ and $-\pi < \alpha, \beta, \gamma \le \pi$
Then, let $A = e^{i\alpha}, B = e^{i\beta}$ and $C = e^{i\gamma}$ where $i = \sqrt{-1}$

$\therefore\ aA + bB + cC = 0$ and $\dfrac{a}{A} + \dfrac{b}{B} + \dfrac{c}{C} = 0$

we get $(aA)^3 + (bB)^3 + (cC)^3 = 3\,abc\,ABC$,

$$\left(\frac{a}{A}\right)^3 + \left(\frac{b}{B}\right)^3 + \left(\frac{c}{C}\right)^3 = \frac{3\,abc}{ABC}$$

and $aBC + bCA + cAB = 0$

On the basis of above information, answer the following questions

1. If $\dfrac{A}{B} + \dfrac{B}{C} + \dfrac{C}{A} = 1$, then $\Sigma\cos(\alpha - \beta)$ is equal to

(a) $-\dfrac{3}{2}$ (b) 0
(c) 1 (d) $\dfrac{3}{2}$

2. If $\cos 3\alpha + \lambda\cos 3\beta + \mu\cos 3\gamma = \nu\cos(\alpha + \beta + \gamma)$, then the value of $|\mu - 3\lambda| + |\nu - \mu|$ is equal to

(a) 0 (b) 6
(c) 12 (d) 16

3. If $\sin^n\alpha + \sin^n\beta + \sin^n\gamma = \dfrac{3}{2}$

and $\cos(\alpha + \beta) + \cos(\beta + \gamma) + \cos(\gamma + \alpha) = \lambda$, then the value of (n, λ) is

(a) (3, 2) (b) (2, 3)
(c) (2, 0) (d) (3, 0)

4. If $\sin\alpha + 2\sin\beta + 3\sin\gamma = 0$, then the value of $\alpha + \beta - \gamma$ is equal to

(a) -2α (b) -2β
(c) -2γ (d) 0

5. If $\sin(2\alpha - \beta - \gamma) + 8\sin(2\beta - \gamma - \alpha)$
$+ 27\sin(2\gamma - \alpha - \beta) = N_1$
and $\sin(\alpha + \beta) + \sin(\beta + \gamma) + \sin(\gamma + \alpha) = N_2$, then the value of $N_1 + N_2$ is equal to

(a) 36 (b) 18
(c) 9 (d) 0

PASSAGE 12

Let $A(z_1), B(z_2)$ and $C(z_3)$ be the vertices of a triangle ABC on the complex plane which is circumscribed by a circle $|z| = 1$. If the altitude of the triangle through the vertex $A(z_1)$ meets BC at D and circle $|z| = 1$ at P.

On the basis of above information, answer the following questions

1. The complex number associated with the point D is equal to
(a) $\frac{1}{2}\left\{z_1+z_2+z_3-\frac{z_1z_2}{z_3}\right\}$
(b) $\frac{1}{2}\left\{z_1+z_2+z_3-\frac{z_2z_3}{z_1}\right\}$
(c) $\frac{1}{2}\left\{z_1+z_2+z_3-\frac{z_3z_1}{z_2}\right\}$
(d) $\frac{1}{2}\left\{z_1-z_2+z_3-\frac{z_1z_2}{z_3}\right\}$

2. The complex number associated with the point P is equal to
(a) $-\frac{z_1z_2}{z_3}$ (b) $\frac{z_1z_2}{z_3}$
(c) $-\frac{z_2z_3}{z_1}$ (d) $\frac{z_2z_3}{z_1}$

3. If Q be the image of P about the line BC, then complex number associated with the point Q is equal to
(a) $z_1+z_2+z_3$ (b) $\frac{z_1+z_2+z_3}{2}$
(c) $\frac{z_1+z_2+z_3}{3}$ (d) $\frac{2}{3}(z_1+z_2+z_3)$

4. If R be the image of P about the origin (O), then the distance between the points B and R is equal to
(a) $|z_1-z_2|$ (b) $|z_2-z_3|$
(c) $|z_3-z_1|$ (d) $|z_1+z_3|$

5. The orthocentre of the triangle ABC associated with the point O' is equal to
(a) $\frac{2z_1-z_2+z_3}{3}$ (b) $\frac{z_1+2z_2-z_3}{3}$
(c) $z_1-z_2+z_3$ (d) $z_1+z_2+z_3$

Numerical Value Type

Solve the following problems and write your answer.

1. If the least and the largest real values of α, for which the equation $z+\alpha|z-1|+2i=0$ ($z\in C$ and $i=\sqrt{-1}$) has a solution, are p and q respectively; then $4(p^2+q^2)$ is equal to **[JEE Main 2021, 24 February]**

2. If a, b, c are distinct integers and $\omega\neq 1$ is a cube root of unity and if minimum value of $|a+b\omega+c\omega^2|+|a+b\omega^2+c\omega|=n^{1/4}$, then the value of n must be equal to

3. If $\omega\neq 1$ is a cube root of unity, and $a+b=21$, $a^3+b^3=8001$, then the value of $(a\omega^2+b\omega)(a\omega+b\omega^2)$ must be equal to

4. If the equation of all the circles which are orthogonal to $|z|=1$ and $|z-1|=4$ is $|z+7-ib|=\sqrt{(\lambda+b^2)}$, $i=\sqrt{-1}$ and $b\in R$, then the value of λ must be equal to

5. If $\sum_{p=1}^{32}(3p+2)\left\{\sum_{q=1}^{10}\left\{\sin\left(\frac{2q\pi}{11}\right)-i\cos\left(\frac{2q\pi}{11}\right)\right\}\right\}^{4p}=\lambda$ (where $i=\sqrt{-1}$), then the value of $\frac{\lambda}{515}$ must be equal to

6. Let $i=\sqrt{-1}$. If $\frac{(-1+i\sqrt{3})^{21}}{(1-i)^{24}}+\frac{(1+i\sqrt{3})^{21}}{(1+i)^{24}}=k$ and $n=[|k|]$ be the greatest integral part of $|k|$. Then, $\sum_{j=0}^{n+5}(j+5)^2-\sum_{j=0}^{n+5}(j+5)$ is equal to **[JEE Main 2021, 25 February]**

7. The sum of maximum and minimum modulus of a complex number z satisfying $|z-25i|\leq 15$, $i=\sqrt{-1}$ must be equal to

8. If z_1 and z_2 are complex numbers, such that $|15z_1-13z_2|^2+|13z_1+15z_2|^2=\lambda(|z_1|^2+|z_2|^2)$, then the value of $\sqrt{\lambda\sqrt{\lambda\sqrt{\lambda\sqrt{\lambda\ldots\ldots\infty}}}}$ must be equal to

9. If z is a complex number and the minimum value of $|z|+|z-1|+|2z-3|$ is λ and if $y=2[x]+3=3[x-\lambda]$, then the value of $[x+y]$ must be equal to
(where [.] denotes the greatest integer function)

10. If $\omega(\neq 1)$ is a cube root of unity, then the value of $(1+\omega)(1+\omega^2)\quad(1+\omega^3)(1+\omega^4)$ $(1+\omega^5)\ldots(1+\omega^{30})$ must be equal to

11. Let z be those complex number which satisfy $|z+5|\leq 4$ and $z(1+i)+\bar{z}(1-i)\geq -10$, $i=\sqrt{-1}$. If the maximum value of $|z+1|^2$ is $\alpha+\beta\sqrt{2}$, then the value of $\left(\frac{\alpha+\beta}{20}\right)$ is **[JEE Main 2021, 25 February]**

12. Let $A=\{a\in R\}$ and n be the number of real roots of the equation $(1+2i)x^3-2(3+i)x^2+5(4-i)x+2a^2=0$, which has at least one real root, then the value of $\frac{\Sigma a^2}{10n}$ is

Single Integer Type

Solve the following problems and write your answer.

1. If $\left(\frac{1+i}{1-i}\right)^{m/2} = \left(\frac{1+i}{i-1}\right)^{n/3} = 1$, $(m, n \in N)$ then the greatest common divisor of the least values of m and n is ....

[JEE Main 2020]

2. For any integer k, let $\alpha_k = \cos\left(\frac{k\pi}{7}\right) + i\sin\left(\frac{k\pi}{7}\right)$, where $i = \sqrt{-1}$. The value of the expression $\frac{\sum_{k=1}^{12}|\alpha_{k+1} - \alpha_k|}{\sum_{k=1}^{3}|\alpha_{4k-1} - \alpha_{4k-2}|}$ is

[JEE Advanced 2016]

3. For a complex number z, let Re(z) denote the real part of z. Let S be the set of all complex numbers z satisfying $z^4 - |z|^4 = 4iz^2$, where $i = \sqrt{-1}$. Then, the minimum possible value of $|z_1 - z_2|^2$, where $z_1, z_2 \in S$ with Re$(z_1) > 0$ and Re$(z_2) < 0$ is ....

4. Let $z_1 = 10 + 6i$ and $z_2 = 4 + 6i$. If z is any complex number such that the argument of $(z - z_1)/(z - z_2)$ is $\pi/4$, then $|z - 7 - 9i|$ is equal to $k\sqrt{2}$, then k is ......

5. If centre of regular hexagon is at origin and one of the vertices on Argand diagram is $1 + 2i$ and its perimeter is $x\sqrt{5}$, then x is

6. If $|z_1| = 2$ and $(1 - i)z_2 + (1 + i)\bar{z}_2 = 8\sqrt{2}$, then the minimum value of $|z_1 - z_2|$ is

Matrix-Match Type

Given below are Matching Type Questions, with two columns (each having some items) each . Each item of Column I has to be matched with the items of Column II, by encircling the correct match(es).

NOTE *An item of Column I can be matched with more than one items of Column II.*

1. Match the following lists:

	Column I		Column II
(A)	For complex number z, the minimum value of $\lvert z\rvert + \lvert z - \cos\alpha - i(\sin\alpha)\rvert$ is	(P)	0
(B)	Number of complex numbers z satisfying $\lvert z+2\rvert + \lvert z-2\rvert = 8$ and $\lvert z-1\rvert + \lvert z+1\rvert = 2$ is	(Q)	1
(C)	If $\left\lvert\frac{z_1 - 3z_2}{3 - z_1\bar{z}_2}\right\rvert = 1$ and $\lvert z_1\rvert \neq 1$, then $\lvert z_1\rvert$ is equal to	(R)	2
(D)	Let $z = \frac{(2\sqrt{3}+2i)^8}{(1-i)^6} + \frac{(1+i)^6}{(2\sqrt{3}-2i)^8}$. If θ is argument of z such that $\theta \in (-\pi, \pi]$, then $\lvert 4\sin\theta\rvert$ is equal to	(S)	3
		(T)	5

2. If $z = \cos\alpha + i\sin\alpha$, $0 < \alpha < \pi/6$, then the argument of

	Column I		Column II
(A)	$1 + z^3$	(P)	$2\alpha - \frac{\pi}{2}$
(B)	$1 - z^4$	(Q)	$\frac{\pi}{2} + \frac{\alpha}{2}$
(C)	$\frac{1+z^3}{1-z^4}$	(R)	$\frac{3\alpha}{2}$
(D)	$\frac{z^4-1}{z^3+1}$	(S)	$3\alpha + \frac{\pi}{2}$
		(T)	$\frac{\pi}{2} - \frac{\alpha}{2}$

3. Observe the following columns :

	Column I		Column II
(A)	If z_1, z_2, z_3 are the vertices of an equilateral triangle with z_0 as its nine point centre, then $z_1^2 + z_2^2 + z_3^2$ is	(P)	$(z_1 - z_2)^2 = 2(z_1 - z_3)(z_3 - z_2)$
(B)	If z_1, z_2, z_3 are the vertices A, B, C respectively of an isosceles right angled triangle with right angle at C, then	(Q)	$3z_0^2$
(C)	If z_1, z_2, z_3 are the vertices A, B, C respectively of an isosceles triangle and the angles at B and C are each equal to $\pi/6$, then	(R)	$(z_2 - z_3)^2 = 3(z_3 - z_1)(z_1 - z_2)$
		(S)	$z_1z_2 + z_2z_3 + z_3z_1$
		(T)	$\frac{z_2 - z_3}{z_1 - z_3} = e^{i\pi/2}$

4. Observe the following columns :

	Column I		Column II
(A)	If $\lvert a_i\rvert < 1; \lambda_i \geq 0$ for $i = 1, 2, 3, \ldots, n$ and $\lambda_1 + \lambda_2 + \ldots + \lambda_n = 1$ and ω is a complex cube root of unity, then $\lvert \lambda_1 a_1\omega + \lambda_2 a_2\omega^2 + \ldots + \lambda_n a_n\omega^n\rvert$ cannot exceed	(P)	$\lvert z\rvert^n + \frac{1}{\lvert z\rvert}$

	Column I		Column II
(B)	If Re $(z) < 0$, then the value of $(1 + z + z^2 + \ldots + z^n)$ can not exceed	(Q)	2
(C)	If $\omega (\neq 1)$ is a cube root of unity, then $\frac{1}{\sqrt{3}}$ $\lvert 1 + 2\omega + 3\omega^2 + \ldots + 3n\omega^{3n-1} \rvert$ $(n \in N)$ cannot exceed	(R)	n
		(S)	1
		(T)	$\lvert a_1 \rvert + \lvert a_2 \rvert + \ldots \lvert a_n \rvert$

5. Observe the following columns :

	Column I		Column II
(A)	If G be the greatest and L be the least values of $\lvert z - 1 \rvert$, if $\lvert z + 2 + i \rvert \leq 1$ where $i = \sqrt{-1}$, then	(P)	$LG = 9$
(B)	If G be the greatest and L be the least values of $\lvert z + 2i \rvert$, where $i = \sqrt{-1}$, if $1 \leq \lvert z - 1 \rvert \leq 3$, then	(Q)	$L + G = 6$
(C)	If G be the greatest and L be the least values of $\lvert z - 2 \rvert$, if $\lvert z + i \rvert \leq 1$ where $i = \sqrt{-1}$, then	(R)	$(\sqrt{2G} - \sqrt{2L})^2 = 4$
		(S)	$LG = 4$
		(T)	$G - L = 2$

Answers

Objective Questions Type I [Only one correct answer]

1. (a)	**2.** (b)	**3.** (c)	**4.** (c)	**5.** (c)	**6.** (a)	**7.** (d)	**8.** (b)	**9.** (d)	**10.** (b)
11. (c)	**12.** (d)	**13.** (a)	**14.** (b)	**15.** (c)	**16.** (b)	**17.** (c)	**18.** (a)	**19.** (b)	**20.** (d)
21. (b)	**22.** (c)	**23.** (d)	**24.** (d)	**25.** (d)	**26.** (c)	**27.** (a)	**28.** (c)	**29.** (d)	**30.** (a)
31. (c)	**32.** (d)	**33.** (b)	**34.** (c)	**35.** (b)	**36.** (b)	**37.** (b)	**38.** (c)	**39.** (c)	**40.** (a)
41. (a)	**42.** (a)	**43.** (b)	**44.** (a)	**45.** (d)	**46.** (c)	**47.** (c)	**48.** (a)	**49.** (b)	**50.** (d)
51. (b)	**52.** (b)	**53.** (c)	**54.** (d)	**55.** (a)	**56.** (c)	**57.** (a)	**58.** (b)	**59.** (b)	**60.** (b)
61. (c)	**62.** (c)	**63.** (c)	**64.** (b)	**65.** (c)	**66.** (b)	**67.** (c)	**68.** (a)	**69.** (a)	**70.** (c)
71. (b)	**72.** (b)	**73.** (a)	**74.** (a)	**75.** (a)	**76.** (a)	**77.** (b)	**78.** (b)	**79.** (c)	**80.** (c)
81. (b)	**82.** (c)	**83.** (a)	**84.** (b)	**85.** (a)	**86.** (d)	**87.** (b)	**88.** (c)		

Objective Questions Type II [One or more than one correct answer(s)]

1. (b, c, d)	**2.** (b, d)	**3.** (b, c, d)	**4.** (b, d)	**5.** (a, c)
6. (b, c, d)	**7.** (c, d)	**8.** (a, b, d)	**9.** (a, c, d)	**10.** (a, b)
11. (b, c, d)	**12.** (b, c)	**13.** (a, c, d)	**14.** (c, d)	**15.** (a, b, c, d)
16. (a, b, c)	**17.** (a, b, c, d)	**18.** (a, b, c)	**19.** (a, b, c, d)	**20.** (a, b, c, d)
21. (a, b, c, d)				

Linked-Comprehension Type

Passage 1 **1.** (a) **2.** (b) **3.** (b)

Passage 2 **1.** (b) **2.** (b) **3.** (d)

Passage 3 **1.** (d) **2.** (d) **3.** (a) **4.** (b) **5.** (c) **6.** (c) **7.** (b) **8.** (d) **9.** (c) **10.** (b) **11.** (b) **12.** (c)

Passage 4 **1.** (a) **2.** (a) **3.** (c) **4.** (b) **5.** (b) **6.** (c) **7.** (d)

Passage 5 **1.** (c) **2.** (b) **3.** (d) **4.** (a) **5.** (d) **6.** (b) **7.** (c) **8.** (a)

Passage 6 **1.** (a) **2.** (d) **3.** (c) **4.** (b) **5.** (a)

Passage 7 **1.** (d) **2.** (b) **3.** (c) **4.** (c) **5.** (d)

Passage 8 **1.** (b) **2.** (b) **3.** (d) **4.** (d) **5.** (a) **6.** (c) **7.** (c)

Passage 9 **1.** (d) **2.** (c) **3.** (d) **4.** (b) **5.** (c) **6.** (c)

Passage 10 **1.** (b) **2.** (d) **3.** (b) **4.** (c) **5.** (c) **6.** (a) **7.** (c)

Passage 11 **1.** (c) **2.** (c) **3.** (c) **4.** (c) **5.** (d)

Passage 12 **1.** (b) **2.** (c) **3.** (a) **4.** (c) **5.** (d)

Numerical Value Type

1. (10) **2.** (144) **3.** (381) **4.** (48) **5.** (3.2)
6. (310) **7.** (50) **8.** (394) **9.** (30) **10.** (1024)
11. (2.4) **12.** (0.6)

Single Integer Type

1. (4) **2.** (4) **3.** (8) **4.** (3) **5.** (6) **6.** (2)

Matrix-Match Type

1. A→(Q), B→(P), C→(S), D→(R)
2. A→(R), B→(P), C→(T), D→(Q)
3. A→(Q,S), B→(P,T), C→(R)
4. A→(Q,S,T), B→(P), C→(R)
5. A→(P,T), B→(Q,R,S), C→(S,T)

Solutions

Objective Questions Type [Only one correct answer]

1. Given expression, $\left(\dfrac{1+\sin\dfrac{2\pi}{9}+i\cos\dfrac{2\pi}{9}}{1+\sin\dfrac{2\pi}{9}-i\cos\dfrac{2\pi}{9}}\right)^3$

$$=\left[\frac{\left(\sin^2\frac{2\pi}{9}-i^2\cos^2\frac{2\pi}{9}\right)+\left(\sin\frac{2\pi}{9}+i\cos\frac{2\pi}{9}\right)}{1+\sin\frac{2\pi}{9}-i\cos\frac{2\pi}{9}}\right]^3$$

$$=\left[\frac{\left(\sin\frac{2\pi}{9}+i\cos\frac{2\pi}{9}\right)\cdot\left\{\sin\frac{2\pi}{9}-i\cos\frac{2\pi}{9}+1\right\}}{1+\sin\frac{2\pi}{9}-i\cos\frac{2\pi}{9}}\right]^3$$

$$=\left(\sin\frac{2\pi}{9}+i\cos\frac{2\pi}{9}\right)^3=\left(-i^2\sin\frac{2\pi}{9}+i\cos\frac{2\pi}{9}\right)^3$$

$$=i^3\left[\cos\frac{2\pi}{9}-i\sin\frac{2\pi}{9}\right]^3=-i\left[\cos\left(-\frac{2\pi}{9}\right)+i\sin\left(-\frac{2\pi}{9}\right)\right]^3$$

$$=-i\left[\cos\frac{2\pi}{3}-i\sin\frac{2\pi}{3}\right]$$

[according the De-Movier's theorem]

$$=-i\left[\left(-\frac{1}{2}\right)-i\frac{\sqrt{3}}{2}\right]$$

$$=-\frac{\sqrt{3}}{2}+\frac{i}{2}=-\frac{1}{2}(\sqrt{3}-i)$$

2. $\text{Im}\left(\dfrac{az+z_1}{bz+z_2}\right)=1$

$$\Rightarrow \frac{\left(\dfrac{az+z_1}{bz+z_2}\right)-\left(\dfrac{a\bar{z}+\bar{z}_1}{b\bar{z}+\bar{z}_2}\right)}{2i}=1$$

$$\Rightarrow (az+z_1)(b\bar{z}+\bar{z}_2)-(a\bar{z}+\bar{z}_1)(bz+z_2)=2i\,(bz+z_2)(b\bar{z}+\bar{z}_2)$$

$$\Rightarrow abz\bar{z}+az\bar{z}_2+bz_1\bar{z}+z_1\bar{z}_2-abz\bar{z}-a\bar{z}\,z_2-b\bar{z}_1z-\bar{z}_1z_2=2i\,(b^2z\bar{z}+bz\bar{z}_2+bz_2\bar{z}+z_2\bar{z}_2)$$

$$\Rightarrow 2ib^2z\,\bar{z}+z\,(2ib\bar{z}_2-a\bar{z}_2+b\bar{z}_1)+\bar{z}\,(2ibz_2+az_2-bz_1)+2i\,|z_2|^2-z_1\bar{z}_2+\bar{z}_1z_2=0$$

It is a circle.

3. Let $\cot^{-1}x=\theta\Rightarrow\cot\theta=x$

$$\Rightarrow \left(\frac{xi+1}{xi-1}\right)=\frac{i\cot\theta+1}{i\cot\theta-1}=\frac{\cot\theta-i}{\cot\theta+i}$$

$$=\frac{\cos\theta-i\sin\theta}{\cos\theta+i\sin\theta}=\frac{e^{-i\theta}}{e^{i\theta}}=\frac{1}{e^{2i\theta}}.$$

$$\Rightarrow e^{2i\theta}\left(\frac{xi+1}{xi-1}\right)=1\Rightarrow e^{2mi\theta}\left(\frac{xi+1}{xi-1}\right)^m=1$$

$$\therefore \quad e^{2mi\cot^{-1}x}\left(\frac{xi+1}{xi-1}\right)^m=1.$$

4. The complex number z satisfying $|z-2+i|\geq\sqrt{5}$, which represents the region outside the circle (including the circumference) having centre $(2,-1)$ and radius $\sqrt{5}$ units.

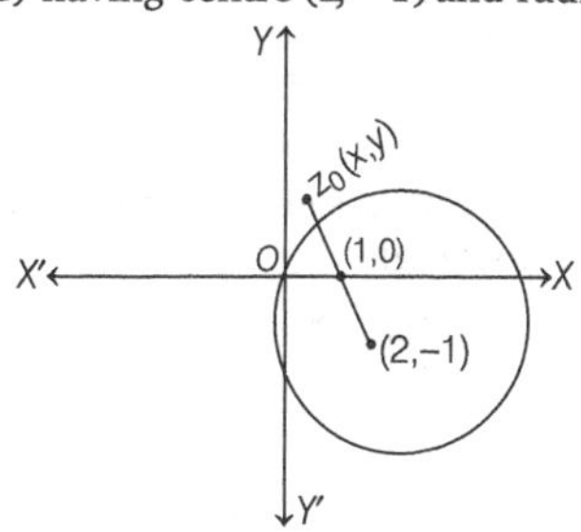

Now, for $z_0\in S$ $\dfrac{1}{|z_0-1|}$ is maximum.

$|z_0-1|$ is minimum. And for this it is required that $z_0\in S$, such that z_0 is collinear with the points $(2,-1)$ and $(1,0)$ and lies on the circumference of the circle $|z-2+i|=\sqrt{5}$.

So, let $z_0=x+iy$, and from the figure $0<x<1$ and $y>0$.

So, $\dfrac{4-z_0-\bar{z}_0}{z_0-\bar{z}_0+2i}=\dfrac{4-x-iy-x+iy}{x+iy-x+iy+2i}$

$$=\frac{2(2-x)}{2i(y+1)}=-i\left(\frac{2-x}{y+1}\right)$$

$\because \dfrac{2-x}{y+1}$ is a positive real number, so $\dfrac{4-z_0-\bar{z}_0}{z_0-\bar{z}_0+2i}$ is purely negative imaginary number.

$\Rightarrow \arg\left(\dfrac{4-z_0-\bar{z}_0}{z_0-\bar{z}_0+2i}\right)=-\dfrac{\pi}{2}$

5. $|z-3+2i|\le 4 \Rightarrow |z-(3-2i)|\le 4$

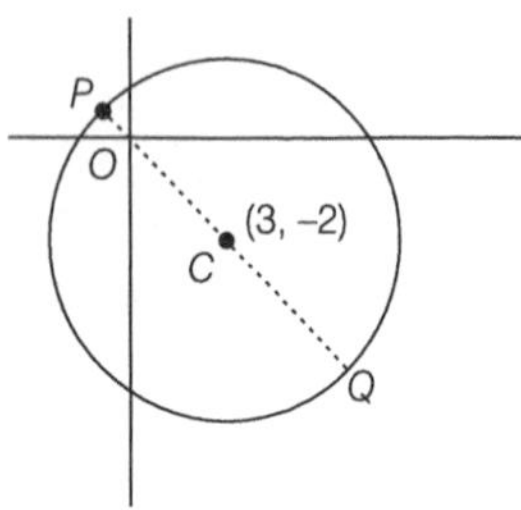

Least value of $|z|=|z-0|$

$$=OP=CP-CO$$
$$=4-\sqrt{13}$$

and greatest value of $|z|=|z-0|$

$$=OQ=OC+CQ$$
$$=\sqrt{13}+4$$

$\therefore$ Difference $=(4+\sqrt{13})-(4-\sqrt{13})=2\sqrt{13}.$

6. $\sqrt{z}=\sqrt{(4+3\sqrt{20}\,i)}=\sqrt{4+2\sqrt{(-45)}}$

$$=\sqrt{9-5+2\sqrt{-45}}=\pm(3+i\sqrt{5})$$

$\therefore \quad \sqrt{\bar{z}}=\pm(3-i\sqrt{5})$

$\therefore \quad (\sqrt{z}+\sqrt{\bar{z}})=\pm 6$ or $\pm 2i\sqrt{5}$

$\Rightarrow \quad (\sqrt{z}+\sqrt{\bar{z}})^4=1296$ or 400

and $\quad \sqrt{z}-\sqrt{\bar{z}}=\pm 2i\sqrt{5}$ or ± 6

$\Rightarrow \quad (\sqrt{z}-\sqrt{\bar{z}})^4=400$ or $1296.$

7. Circle be $|z|=r$

$\therefore \quad |\omega z|=|\omega||z|=1\cdot r=r$

and $|z+\omega z|=|(1+\omega)z|=|-\omega^2 z|=|-\omega^2||z|=1\cdot r=r$

Hence z, ωz and $z+\omega z$ are the vertices of an equilateral triangle.

$\therefore \quad \text{Area}=\dfrac{\sqrt{3}}{4}r^2=4\sqrt{3}$ (given)

$\therefore \quad r^2=16$

$\therefore \quad r=4$ units

8. $\cos\left[\{(1-\omega)(1-\omega^2)+\ldots.+(10-\omega)(10-\omega^2)\}\dfrac{\pi}{900}\right]$

$$=\cos\left[\left\{\sum_{r=1}^{10}(r-\omega)(r-\omega^2)\right\}\frac{\pi}{900}\right]$$

$$=\cos\left[\left\{\sum_{r=1}^{10}(r^2+r+1)\right\}\frac{\pi}{900}\right]$$

$$=\cos\left[\left(\frac{10\times 11\times 21}{6}+\frac{10\times 11}{2}+10\right)\frac{\pi}{900}\right]=\cos\left(\frac{\pi}{2}\right)=0$$

9. $|z-i\,\text{Re}(z)|=|z-\text{Im}(z)|$

If $z=x+iy$,

then $\quad |x+iy-ix|=|x+iy-y|$

$$\sqrt{x^2+(y-x)^2}=\sqrt{(x-y)^2+y^2}$$

or $\quad x^2=y^2$

$\therefore \quad x=\pm y$

$\Rightarrow \quad \text{Re}(z)=\pm\,\text{Im}(z)$

$\Rightarrow \quad \text{Re}(z)+\text{Im}(z)=0$ and $\text{Re}(z)-\text{Im}(z)=0.$

10. $\because |z|+|z-\cos\alpha-i\sin\alpha|=|z|+|z-e^{i\alpha}|$

$\therefore \quad |z-(z-e^{i\alpha})|\le|z|+|z-e^{i\alpha}|$

$\Rightarrow \quad 1\le|z|+|z-e^{i\alpha}|$

$\Rightarrow \quad |z|+|z-\cos\alpha-i\sin\alpha|\ge 1$

$\therefore$ Minimum value $=1$.

11. Let $\dfrac{p}{a}=A, \dfrac{q}{b}=B, \dfrac{r}{c}=C$

$\therefore \quad A+B+C=1+i, \dfrac{1}{A}+\dfrac{1}{B}+\dfrac{1}{C}=0$

or $\quad AB+BC+CA=0$

$\Rightarrow \quad (A+B+c)^2=(1+i)^2$

$\Rightarrow \quad A^2+B^2+C^2+2(AB+BC+CA)$

$$=1+i^2+2i=1-1+2i$$

$\Rightarrow \quad A^2+B^2+C^2+0=2i$

$\Rightarrow \quad \dfrac{p^2}{a^2}+\dfrac{q^2}{b^2}+\dfrac{r^2}{c^2}=2i$

12. $|a^2-2a|<3$

$\Rightarrow \quad -3<a^2-2a<3$

$\Rightarrow \quad -3+1<a^2-2a+1<3+1$

$\Rightarrow \quad -2<(a-1)^2<4$

$\therefore \quad 0\le(a-1)^2<4$

$\Rightarrow \quad -2<a-1<2$

or $\quad -1<a<3$

But $a\in R^+$

$\therefore \quad 0<a<3$

$\Rightarrow \quad a\in(0, 3).$

13. $\because |z_1+z_2|=||z_1|-|z_2|| \Rightarrow \arg\left(\dfrac{z_1}{z_2}\right)=\pm\pi$

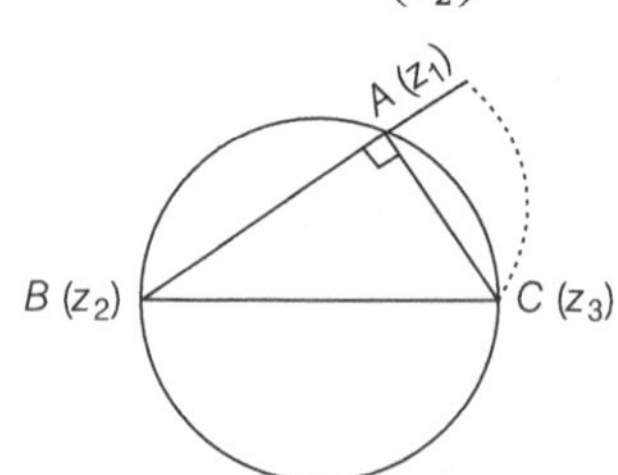

and $|z_1+i(z_3-z_1)|=|z_1|+|z_3-z_1| \Rightarrow \arg\left(\dfrac{z_1}{z_3-z_1}\right)=\dfrac{\pi}{2}$

$\therefore$ Centre of circle is $\left(\dfrac{z_2+z_3}{2}\right)$

and ΔABC is right angled triangle $\angle A=90°$.

14. $\because \quad -1 \le \dfrac{1+x^2}{2x} \le 1$

$\Rightarrow \quad \left|\dfrac{1+x^2}{2x}\right| \le 1$

$\Rightarrow \quad \dfrac{1+x^2}{2|x|} \le 1$

$\Rightarrow \quad \dfrac{1+|x|^2}{2|x|} - 1 \le 0$

$\Rightarrow \quad \dfrac{(|x|-1)^2}{|x|} \le 0$

$\therefore \quad |x| > 0,$

$\therefore \quad (|x|-1)^2 \le 0$

$\Rightarrow \quad (|x|-1)^2 = 0$

$\therefore \quad |x| = 1 \Rightarrow x = \pm 1$

$\therefore \quad x = 1 \qquad [\because x > 0]$

$\left(\dfrac{1+i}{1-i}\right)^n = \dfrac{2}{\pi} \cdot \sin^{-1}(1) = \dfrac{2}{\pi} \cdot \dfrac{\pi}{2} = 1$

$\Rightarrow \quad \left(\dfrac{(1+i)^2}{2}\right)^n = 1$

$(i)^n = 1$

$\Rightarrow \quad (-1)^{n/2} = (-1)^2, (-1)^4, (-1)^6, \ldots$

$\Rightarrow \quad \dfrac{n}{2} = 2$

$\therefore \quad n = 4$ (least +ve value)

15. It is given $u = \dfrac{2z+i}{z-ki}$, $z = x+iy$, $k > 0$

$\Rightarrow u = \dfrac{2x + i(2y+1)}{x + i(y-k)} \times \dfrac{x - i(y-k)}{x - i(y-k)}$

$= \dfrac{[2x^2 + (2y+1)(y-k)] + i\,[x(2y+1) - 2x(y-k)]}{x^2 + (y-k)^2}$

$\therefore \text{Re}(u) + \text{Im}(u)$

$= \dfrac{2x^2 + (2y+1)(y-k) + x(2y+1) - 2x(y-k)}{x^2 + (y-k)^2}$

$= 1 \qquad$ [given]

$\Rightarrow 2x^2 + 2y^2 + (1-2k)y - k + 2xy + x - 2xy + 2xk$

$= x^2 + y^2 - 2ky + k^2$

$\because$ The above curve intersect the Y-axis.

So put $x = 0$; then we get

$2y^2 + (1-2k)y - k = y^2 - 2ky + k^2$

$\Rightarrow \quad y^2 + y - (k^2 + k) = 0 \qquad \ldots(i)$

Now, the y-coordinates of points P and Q are roots of the Eq. (i).

So, PQ = Difference of roots

$= \dfrac{\sqrt{D}}{|\text{coefficient of } x^2|} = \sqrt{1 + 4(k^2+k)} = 5 \qquad$ [given]

$\Rightarrow \quad 4k^2 + 4k + 1 = 25$

$\Rightarrow \quad k^2 + k - 6 = 0$

$\Rightarrow \quad (k+3)(k-2) = 0 \Rightarrow k = -3, 2$

$\Rightarrow \quad k = 2 \qquad [\because k > 0]$

16. The first equation can be written as $(z+1)(z^2+z+1) = 0$.

Its roots are -1, ω and ω^2.

Now, let $f(z) = z^{1985} + z^{100} + 1$

$f(-1) = -1 + 1 + 1 = 1 \Rightarrow f(-1) \ne 0$

$\therefore\ -1$ is not a root of the equation $f(z) = 0$

Again, $f(\omega) = \omega^{1985} + \omega^{100} + 1$

$= (\omega^3)^{661} . \omega^2 + (\omega^3)^{33} . \omega + 1$

$= \omega^2 + \omega + 1 = 0$

$\therefore\ \omega$ is a root of the equation $f(z) = 0$.

Similarly, we can show that $f(\omega^2) = \theta$ so that ω^2 is also a root of the equation $f(z) = 0$.

Thus, out of the three roots of the first equation, ω and ω^2 are also the roots of the second equation.

Hence, ω and ω^2 are the common roots.

17. Since $x_r = \cos\left(\dfrac{\pi}{3^r}\right) - i \sin\left(\dfrac{\pi}{3^r}\right)$

$\therefore x_1 \cdot x_2 \cdot x_3 \ldots \infty = \cos\left(\dfrac{\pi}{3^1} + \dfrac{\pi}{3^2} + \dfrac{\pi}{3^3} + \ldots \infty\right)$

$- i \sin\left(\dfrac{\pi}{3^1} + \dfrac{\pi}{3^2} + \dfrac{\pi}{3^3} + \ldots \infty\right)$

$= \cos\left(\dfrac{\pi/3}{1 - 1/3}\right) - i \sin\left(\dfrac{\pi/3}{1-1/3}\right)$

$= \cos(\pi/2) - i \sin(\pi/2) = -i$

18. We have, $(1+i)(1+2i)(1+3i) \ldots. (1+ni) = \alpha + i\beta$

$\Rightarrow \quad |(1+i)(1+2i)(1+3i) \ldots. (1+ni)| = |\alpha + i\beta|$

$\Rightarrow |(1+i)| \cdot |(1+2i)| \cdot |(1+3i)| \ldots. |(1+ni)| = |\alpha + i\beta|$

$\Rightarrow \quad |1+i|^2 \cdot |1+2i|^2 \cdot |1+3i|^2 \ldots. |1+ni|^2 = |\alpha + i\beta|^2$

$\Rightarrow \quad (1^2+1^2)(1^2+2^2)(1^2+3^2) \ldots. (1^2+n^2) = \alpha^2 + \beta^2$

$\Rightarrow \quad 2 \cdot 5 \cdot 10 \ldots.. (1+n^2) = \alpha^2 + \beta^2$

19. $\arg\left(\dfrac{z - z_1}{z_2 - z}\right) = \dfrac{\pi}{2}$.

$\therefore\ z_1, z_2, z_3$ lie on a circle.

$\Rightarrow z_1$ and z_2 are the end points of diameter

$\therefore$ centre $(z_0) = \dfrac{z_1 + z_2}{2}$

$\therefore \quad z_0 = 5 - i$

and radius $= |z_1 - z_0| = |1 + 2i|$

$r = \sqrt{5}$

$\therefore$ Equation circle is $|z - z_0| = r$ or $|z - (5-i)| = \sqrt{5}$.

20. $\quad z^2 + |z|^2 = 0$

$\Rightarrow \quad z^2 + z\bar{z} = 0$

$\Rightarrow \quad z(z + \bar{z}) = 0$

or $\quad z \cdot 2\,\text{Re}(z) = 0$

$\therefore \quad z = 0$ and $\text{Re}(z) = 0$

If $z = a + ib$

$\therefore$ Solutions are $z = 0, ib\ (b \in R)$

21. Let $z = 1 + i = \sqrt{2}\left(\cos\dfrac{\pi}{4} + i \sin\dfrac{\pi}{4}\right)$

Therefore,

$2^{3/4} = 2^{3/8}\left[\cos\left(2k\pi + \dfrac{\pi}{4}\right)\dfrac{3}{4} + i \sin\left(2k\pi + \dfrac{\pi}{4}\right)\dfrac{3}{4}\right]$

for $k = 0, 1, 2, 3$. The product of the values of this is equal to

$$2^{3/2}\left[\cos\left\{\left(\frac{\pi}{4}+\frac{9\pi}{4}+\frac{17\pi}{4}+\frac{25\pi}{4}\right)\frac{3}{4}\right\} + i\sin\left\{\left(\frac{\pi}{4}+\frac{9\pi}{4}+\frac{17\pi}{4}+\frac{25\pi}{6}\right)\frac{3}{4}\right\}\right]$$

$$= 2^{3/2}\left[\cos\left(\frac{52\pi}{4}\times\frac{3}{4}\right)+i\sin\left(\frac{52\pi}{4}\times\frac{3}{4}\right)\right]$$

$$= 2^{3/2}\left[\cos\frac{39\pi}{4}+i\sin\frac{39\pi}{4}\right]$$

$$= 2\sqrt{2}\left[\cos\left(10\pi-\frac{\pi}{4}\right)+i\sin\left(10\pi-\frac{\pi}{4}\right)\right]$$

$$= 2\sqrt{2}\left[\cos\frac{\pi}{4}-i\sin\frac{\pi}{4}\right]$$

$$= 2\sqrt{2}\left[\frac{1}{\sqrt{2}}-i\times\frac{1}{\sqrt{2}}\right] = 2(1-i)$$

22. $\log_{0.3}|z-1| > \log_{0.3}|z-i|$

$\Rightarrow |z-1| < |z-i|$

$\Rightarrow |z-1|^2 < |z-i|^2$

$\Rightarrow (z-1)(\bar{z}-1) < (z-i)(\bar{z}+i)$

$\Rightarrow z\bar{z}-z-\bar{z}+1 < z\bar{z}+iz-i\bar{z}+1$

$\Rightarrow (1+i)z+(1-i)\bar{z} > 0$

$\Rightarrow (z+\bar{z})+i(z-\bar{z}) > 0$

$\Rightarrow \left(\frac{z+\bar{z}}{2}\right)-\left(\frac{z-\bar{z}}{2i}\right) > 0$ [if $z = x+iy$]

$\Rightarrow x-y > 0$

23. Let $A = z_1$, $B = \sqrt{z_1^2 - z_2^2}$

$\Rightarrow |z_1+\sqrt{z_1^2-z_2^2}|+|z_1+\sqrt{z_1^2-z_2^2}| = |A+B|+|A-B|$

$= \sqrt{|A+B|^2+|A-B|^2+2|A+B||A-B|}$

$= \sqrt{2\{|A|^2+|B|^2\}+2|A+B||A-B|}$

$= \sqrt{2\{|z_1|^2+|z_1^2-z_2^2|+0|z_2|^2\}}$

$= \sqrt{2\{|z_1|^2+|z_2|^2+|z_1+z_2||z_1-z_2|\}}$

$= \sqrt{|z_1+z_2|^2+|z_1-z_2|^2+2|z_1+z_2||z_1-z_2|}$

$= |z_1+z_2|+|z_1-z_2|$.

24. Let the complex numbers

$z_1 = x_1+iy_1$ and $z_2 = x_2+iy_2$

Now, it is given that

$\text{Re}(z_1) = |z_1-1| \Rightarrow x_1^2 = (x_1-1)^2+y_1^2$

$\Rightarrow y_1^2+1 = 2x_1$ ...(i)

and $\text{Re}(z_2) = |z_2-1| \Rightarrow x_2^2 = (x_2-1)^2+y_2^2$

$\Rightarrow y_2^2+1 = 2x_2$ ...(ii)

and $\arg(z_1-z_2) = \frac{\pi}{6} \Rightarrow \frac{y_1-y_2}{x_1-x_2} = \frac{1}{\sqrt{3}}$ ...(iii)

From Eqs. (i) and (ii), we get

$y_1^2-y_2^2 = 2(x_1-x_2)$

$\Rightarrow \frac{y_1-y_2}{x_1-x_2}(y_1+y_2) = 2 \Rightarrow \frac{1}{\sqrt{3}}(y_1+y_2) = 2$

$\Rightarrow y_1+y_2 = 2\sqrt{3} \Rightarrow \text{Im}(z_1+z_2) = 2\sqrt{3}$

25. $OA = OB = OC$

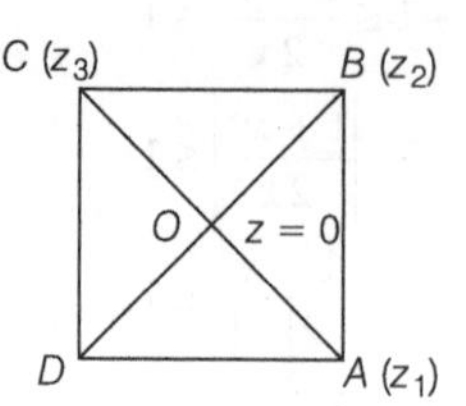

Let $B(z_2)$ and $C(z_3)$

$\therefore \frac{z_2-0}{z_1-0} = \frac{OB}{OA}e^{i\pi/2}$

$\therefore z_2 = iz_1$

and $\frac{z_3-0}{z_1-0} = \frac{OC}{OA}e^{i\pi} \Rightarrow z_3 = -z_1$

$\therefore$ Centroid $= \frac{z_1+z_2+z_3}{3} = \frac{z_1(1+i-i)}{3} = \frac{z_1}{3}$

If B at D, then $z_2 = -z_1/3$

$\therefore$ Centroid $= \pm\frac{z_1}{3} = \frac{z_1}{3}[\cos(\pi/2)\pm i\sin(\pi/2)]$.

26. $(2-i)z = (2+i)\bar{z}$

$\Rightarrow (2-i)(x+iy) = (2+i)(x-iy)$

$\Rightarrow 2x-ix+2iy+y = 2x+ix-2-iy+y$

$\Rightarrow 2ix-4iy = 0$

$L_1 : x-2y = 0$

$\Rightarrow (2+i)z+(i-2)\bar{z}-4i = 0$

$\Rightarrow (2+i)(x+iy)+(i-2)(x-iy)-4i = 0$

$\Rightarrow 2x+ix+2iy-y+ix-2x+y+2iy-4i = 0$

$\Rightarrow 2ix+4iy-4i = 0$

$L_2 : x+2y-2 = 0$

Solve L_1 and L_2 $4y = 2$, $y = \frac{1}{2}$

$\therefore x = 1$

Centre $\left(1, \frac{1}{2}\right)$

$L_3 : iz+\bar{z}+1+i = 0$

$\Rightarrow i(x+iy)+x-iy+1+i = 0$

$\Rightarrow ix-y+x-iy+1+i = 0$

$\Rightarrow (x-y+1)+i(x-y+1) = 0$

Radius = distance from $\left(1, \frac{1}{2}\right)$ to $x-y+1 = 0$

$\therefore r = \frac{1-\frac{1}{2}+1}{\sqrt{2}} = \frac{3}{2\sqrt{2}}$

27. Let $z = r_1e^{i\theta_1}$ and $w = r_2e^{i\theta_2}$, $\bar{z} = r_1e^{-i\theta_1}$

Given, $|zw| = 1 \Rightarrow |r_1e^{i\theta_1}\cdot r_2e^{i\theta_2}| = 1$

$\Rightarrow r_1r_2 = 1$ ...(i)

and $\arg(z)-\arg(w) = \frac{\pi}{2}$

$\theta_1-\theta_2 = \frac{\pi}{2}$ ...(ii)

$\bar{z}w = r_1e^{-i\theta_1}\cdot r_2e^{i\theta_2} = r_1r_2\,e^{-i(\theta_1-\theta_2)} = 1\times e^{-i\frac{\pi}{2}}$

[from Eqs. (i) and (ii)]

$= \cos\frac{\pi}{2}-i\sin\frac{\pi}{2}$

$\therefore \bar{z}w = -i$

28. Here, $(x - x_0)^2 + (y - y_0)^2 = r^2$

and $(x - x_0)^2 + (y - y_0)^2 = 4r^2$ can be written as,

$$|z - z_0|^2 = r^2 \text{ and } |z - z_0|^2 = 4r^2$$

Since, α and $\frac{1}{\bar{\alpha}}$ lies on first and second respectively.

$\therefore$ $|\alpha - z_0|^2 = r^2$ and $\left|\frac{1}{\bar{\alpha}} - z_0\right|^2 = 4r^2$

$\Rightarrow$ $(\alpha - z_0)(\bar{\alpha} - \bar{z}_0) = r^2$

$\Rightarrow$ $|\alpha|^2 - z_0\bar{\alpha} - \bar{z}_0\alpha + |z_0|^2 = r^2$ ...(i)

and $\left|\frac{1}{\bar{\alpha}} - z_0\right|^2 = 4r^2$

$\Rightarrow$ $\left(\frac{1}{\bar{\alpha}} - z_0\right)\left(\frac{1}{\alpha} - \bar{z}_0\right) = 4r^2$

$\Rightarrow$ $\frac{1}{|\alpha|^2} - \frac{z_0}{\alpha} - \frac{\bar{z}_0}{\bar{\alpha}} + |z_0|^2 = 4r^2$

Since, $|\alpha|^2 = \alpha \cdot \alpha$

$\Rightarrow$ $\frac{1}{|\alpha|^2} - \frac{z_0 \cdot \bar{\alpha}}{|\alpha|^2} - \frac{\bar{z}_0}{|\alpha|^2} \cdot \alpha + |z_0|^2 = 4r^2$

$\Rightarrow$ $1 - z_0\bar{\alpha} - \bar{z}_0\alpha + |\alpha|^2|z_0|^2 = 4r^2|\alpha|^2$ ...(ii)

On subtracting Eqs. (i) and (ii), we get

$$(|\alpha|^2 - 1) + |z_0|^2(1 - |\alpha|^2) = r^2(1 - 4|\alpha|^2)$$

$\Rightarrow$ $(|\alpha|^2 - 1)(1 - |z_0|^2) = r^2(1 - 4|\alpha|^2)$

$\Rightarrow$ $(|\alpha|^2 - 1)\left(1 - \frac{r^2 + 2}{2}\right) = r^2(1 - 4|\alpha|^2)$

$\left(|z_0|^2 = \frac{r^2 + 2}{2}\right)$

$\Rightarrow$ $(|\alpha|^2 - 1) \cdot \left(\frac{-r^2}{2}\right) = r^2(1 - 4|\alpha|^2)$

$\Rightarrow$ $|\alpha|^2 - 1 = -2 + 8|\alpha|^2$

$\Rightarrow$ $7|\alpha|^2 = 1$

$\therefore$ $|\alpha| = 1/\sqrt{7}$

29. For $z = x + iy$, $\frac{z - 1}{2z + i} = \frac{(x - 1) + iy}{2x + (2y + 1)i}$

$$= \frac{[(x-1)+iy][2x - i(2y+1)]}{(2x)^2 + (2y+1)^2}$$

[on rationalization]

$\therefore$ $\text{Re}\left(\frac{z - 1}{2z + i}\right) = \frac{2x(x - 1) + y(2y + 1)}{4x^2 + (2y + 1)^2}$

Now, it is given that $\text{Re}\left(\frac{z - 1}{2z + i}\right) = 1$

$\Rightarrow$ $\frac{2x(x - 1) + y(2y + 1)}{4x^2 + (2y + 1)^2} = 1$

$\Rightarrow$ $2x^2 - 2x + 2y^2 + y = 4x^2 + 4y^2 + 4y + 1$

$\Rightarrow$ $2x^2 + 2y^2 + 2x + 3y + 1 = 0$, is a circle

whose centre is $\left(-\frac{1}{2}, -\frac{3}{4}\right)$ and radius is $\sqrt{\frac{1}{4} + \frac{9}{16} - \frac{1}{2}}$

$= \frac{\sqrt{5}}{4}$, so diameter is $\frac{\sqrt{5}}{2}$.

30.

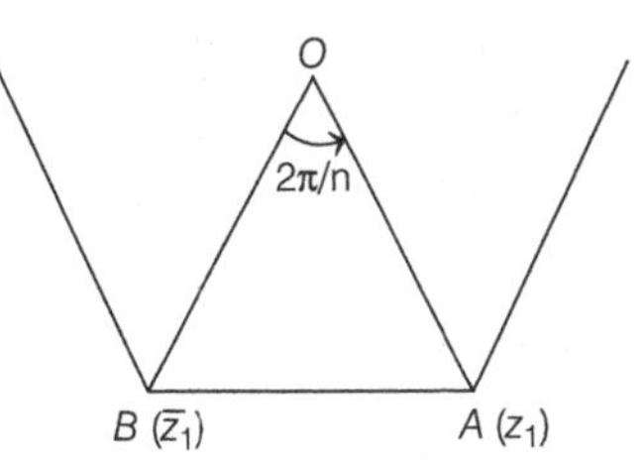

From rotation theorem (Coni method)

$$\frac{\bar{z}_1 - 0}{z_1 - 0} = \frac{|\bar{z}_1 - 0|}{|z_1 - 0|} e^{2i\pi/n} = e^{2\pi i/n} \Rightarrow \frac{z_1}{\bar{z}_1} = e^{2\pi i/n}$$

Let $z_1 = re^{i\theta}$

$\therefore$ $\frac{z_1}{\bar{z}_1} = \frac{re^{i\theta}}{re^{-i\theta}} = e^{2i\theta} = e^{2\pi i/n}$

$\therefore$ $\theta = \pi/n$

$\Rightarrow$ $\frac{\text{Im}(z_1)}{\text{Re}(z_1)} = \sqrt{2} - 1$

$\Rightarrow$ $\frac{r\sin(\pi/n)}{r\cos(\pi/n)} = \sqrt{2} - 1$

$\Rightarrow$ $\tan \pi/n = \tan \pi/8$

$\therefore$ $n = 8$

31. Given, $\bar{z}_1 - i\bar{z}_2 = 0 \Rightarrow \bar{z}_1 = i\bar{z}_2 \Rightarrow \bar{z}_1 = \overline{iz_2} \Rightarrow z_1 = -iz_2$

Clearly argument of z_1 = argument of $z_2 - \frac{\pi}{2}$

or argument z_1 = argument $z_2 - \frac{\pi}{2}$

$\Rightarrow$ argument z_2 = argument $z_1 + \frac{\pi}{2}$

Let argument $z_1 = \alpha$

Then, given argument $(z_1z_2) = \frac{3\pi}{4}$

$\Rightarrow$ argument z_1 + argument $z_2 = \frac{3\pi}{4}$

$$\alpha + \alpha + \frac{\pi}{2} = \frac{3\pi}{4}$$

$\Rightarrow$ $2\alpha = \frac{3\pi}{4} - \frac{\pi}{2} \Rightarrow \alpha = \frac{\pi}{8}$

32. Since $x_1 < x_2 \Rightarrow \sqrt{x_1^2 + y_1^2} < \sqrt{x_2^2 + y_2^2}$ $[\because y_1 = y_2]$

$\therefore$ $|z_1| < |z_2|$ ...(i)

$$|z_3| = \frac{1}{2}|z_1 + z_2| < \frac{1}{2}|z_1| + \frac{1}{2}|z_2| < \frac{1}{2}|z_2| + \frac{1}{2}|z_2|$$

$\Rightarrow$ $|z_3| < |z_2|$ ...(ii)

From Eqs. (i) and (ii),

$$|z_1| < |z_3| < |z_2|.$$

33. $\log_{\frac{1}{\sqrt{2}}}\left(\frac{|z|+11}{(|z|-1)^2}\right) \le 2$

$$\frac{|z|+11}{(|z|-1)^2} \ge \frac{1}{2}$$

$$2|z|+22 \ge (|z|-1)^2$$

$$2|z|+22 \ge |z|^2+1-2|z|$$

$$|z|^2-4|z|-21 \le 0$$

$\Rightarrow$ $|z| \le 7$

$\therefore$ Largest value of $|z|$ is 7.

34. Let $z = a + ib, b \neq 0$ where $\text{Im}(z) = b$

$$z^5 = (a+ib)^5$$
$$= a^5 + {}^5C_1 a^4 bi + {}^5C_2 a^3 b^2 i^2 + {}^5C_3 a^2 b^3 i^3 + {}^5C_4 ab^4 i^4 + i^5 b^5$$
$$\text{Im}\, z^5 = 5a^4 b - 10a^2 b^3 + b^5$$
$$y = \frac{\text{Im}\, z^5}{\text{Im}^5 z} = 5\left(\frac{a}{b}\right)^4 - 10\left(\frac{a}{b}\right)^2 + 1$$

Let $\left(\frac{a}{b}\right)^2 = x$ (say), $x \in R^+$

$$y = 5x^2 - 10x + 1 = 5[x^2 - 2x] + 1 = 5[(x-1)^2] - 4$$

Hence, $y_{\min} = -4$

35. Given, $3|z_1| = 4|z_2| \Rightarrow \frac{|z_1|}{|z_2|} = \frac{4}{3}$ $[\because z_2 \neq 0 \Rightarrow |z_2| \neq 0]$

$$\therefore \quad \frac{z_1}{z_2} = \left|\frac{z_1}{z_2}\right| e^{i\theta} \text{ and } \frac{z_2}{z_1} = \left|\frac{z_2}{z_1}\right| e^{-i\theta}$$

$[\because z = |z|(\cos\theta + i\sin\theta) = |z|\, e^{i\theta}]$

$$\Rightarrow \quad \frac{z_1}{z_2} = \frac{4}{3} e^{i\theta} \text{ and } \frac{z_2}{z_1} = \frac{3}{4} e^{-i\theta}$$
$$\Rightarrow \quad \frac{3}{2}\frac{z_1}{z_2} = 2e^{i\theta} \text{ and } \frac{2}{3}\frac{z_2}{z_1} = \frac{1}{2} e^{-i\theta}$$

On adding these two, we get

$$z = \frac{3}{2}\frac{z_1}{z_2} + \frac{2}{3}\frac{z_2}{z_1} = 2e^{i\theta} + \frac{1}{2}e^{-i\theta}$$
$$= 2\cos\theta + 2i\sin\theta + \frac{1}{2}\cos\theta - \frac{1}{2}i\sin\theta$$

$[\because e^{\pm i\theta} = (\cos\theta \pm i\sin\theta)]$

$$= \frac{5}{2}\cos\theta + \frac{3}{2}i\sin\theta$$
$$\Rightarrow \quad |z| = \sqrt{\left(\frac{5}{2}\right)^2 + \left(\frac{3}{2}\right)^2} = \sqrt{\frac{34}{4}} = \sqrt{\frac{17}{2}}$$

Note that z is neither purely imaginary and nor purely real.

'*' None of the options is correct.

36. $(z - a)(\bar{z} - b) = z\bar{z} - a\bar{z} - bz + ab \; (\because z = x + iy)$

$x^2 + y^2 - ax - by + ab + iy(a - b)$

Argument of $(z - a)(\bar{z} - b)$

$$= \tan^{-1}\left(\frac{y(a-b)}{x^2 + y^2 - ax - bx + ab}\right) \quad \ldots\text{(i)}$$

$$\text{Also, } \frac{(\sqrt{3}+i)(1+\sqrt{3}i)}{1+i} = \frac{\sqrt{3} - \sqrt{3} + i(1+3)}{1+i} = 2 + 2i$$

Argument of $2 + 2i = \tan^{-1}(2/2) = \pi/4$ ...(ii)

Comparing Eqs. (i) and (ii), we get

$\Rightarrow x^2 + y^2 - (a+b)x - (a-b)y + ab = 0$

Thus, the locus of z is a circle

Given centre is $\frac{3+i}{2}$.

$$\Rightarrow \quad \frac{a+b}{2} = \frac{3}{2} \text{ and } \frac{a-b}{2} = \frac{1}{2}$$

$\Rightarrow a = 2, b = 1$

37. $\because \quad 1 + \omega + \omega^2 = 0$

$\therefore \quad 1 + \omega = -\omega^2$

$\therefore \quad (1+\omega)^7 = -\omega^{14}$

$= -\omega^2 = A + B\omega$

$\Rightarrow \quad 1 + \omega = A + B\omega$

$\therefore \quad A = 1, B = 1$

38. $(x-1)^3 - 8 = 0$

$$\Rightarrow \quad \left(\frac{x-1}{2}\right)^3 = 1 \Rightarrow \frac{x-1}{2} = (1)^{1/3} = 1, \omega, \omega^2$$

$\Rightarrow \quad x - 1 = 2, 2\omega, 2\omega^2$

$\Rightarrow \quad x = 3, 1 + 2\omega, 1 + 2\omega^2.$

39. Let $\quad x = (1)^{1/n}$

$x^n - 1 = 0$

or $\quad x^n - 1 = (x-1)(x-\omega)(x-\omega^2)\ldots.(x-\omega^{n-1})$

$$\Rightarrow \quad \frac{x^n - 1}{x - 1} = (x-\omega)(x-\omega^2)\ldots.(x-\omega^{n-1})$$

Putting $x = 9$ in both sides, we have

$$(9-\omega)(9-\omega^2)(9-\omega^3)\ldots(9-\omega^{n-1}) = \frac{9^n - 1}{8}.$$

40. $8iz^3 + 12z^2 - 18z + 27i = 0$

$(2iz + 3)(4z^2 + 9i) = 0$

$2iz + 3 = 0$ and $4z^2 + 9i = 0$

$$\Rightarrow \quad z = \frac{-3}{2i} \text{ and } z^2 = -\frac{9i}{4}$$

$\Rightarrow \therefore \quad |z| = 3/2, |z|^2 = 9/4$

41. $\quad z = re^{i\theta} = r(\cos\theta + i\sin\theta)$

$= (r\cos\theta) + i(r\sin\theta)$

$\therefore \quad iz = -(r\sin\theta) + i(r\cos\theta)$

$e^{iz} = e^{-r\sin\theta} e^{i(r\cos\theta)}$

$\therefore \quad |e^{iz}| = e^{-r\sin\theta} |e^{i(r\cos\theta)}|$

$= e^{-r\sin\theta} \cdot 1 = e^{-r\sin\theta}.$

42. Let $\frac{a}{|z_2 - z_3|} = \frac{b}{|z_3 - z_1|} = \frac{c}{|z_1 - z_2|} = \lambda$ (say)

$\Rightarrow \quad a = \lambda|z_2 - z_3|, b = \lambda|z_3 - z_1|, c = \lambda|z_1 - z_2|$

$\Rightarrow \quad a^2 = \lambda^2|z_2 - z_3|^2 = \lambda^2(z_2 - z_3)(\bar{z}_2 - \bar{z}_3)$

$$\therefore \quad \frac{a^2}{(z_2 - z_3)} = \lambda^2(\bar{z}_2 - \bar{z}_3)$$

$$\text{Similarly} \quad \frac{b^2}{z_3 - z_1} = \lambda^2(\bar{z}_3 - \bar{z}_1)$$

$$\text{and} \quad \frac{c^2}{(z_1 - z_2)} = \lambda^2(\bar{z}_1 - \bar{z}_2)$$

$$\therefore \quad \frac{a^2}{(z_2 - z_3)} + \frac{b^2}{(z_3 - z_1)} + \frac{c^2}{(z_1 - z_2)} = 0.$$

43. $\because \quad |z_1 - z_2| = |z_1 - (z_2 - 3 - 4i) - (3 + 4i)|$

$\geq |z_1| - |z_2 - 3 - 4i| - |3 + 4i|$

$= 12 - 5 - 5 = 2$

$\therefore \quad |z_1 - z_2| \geq 2$

44. Let $z = x + iy$ and $b = g + if$

$$z\bar{z} + b\bar{z} + \bar{b}z + c = 0$$
$$(x+iy)(x-iy)+b(x-iy)+\bar{b}(x+iy)+C=0$$
$$x^2+y^2+x(b+\bar{b})+iy(\bar{b}-b)+C=0$$
$$x^2+y^2+x(2g)+iy(-2if)+C=0$$
$$x^2+y^2+2gx+2fy+C=0$$

Hence, the equation of any circle in a complex plane is of the form $z\bar{z}+b\bar{z}+\bar{b}z+C=0$.

45. Let $z=\left(\dfrac{3+2i\sin\theta}{1-2i\sin\theta}\right)\times\left(\dfrac{1+2i\sin\theta}{1+2i\sin\theta}\right)$

[rationalising the denominator]

$$=\frac{3-4\sin^2\theta+8i\sin\theta}{1+4\sin^2\theta}$$

$[\because a^2-b^2=(a+b)(a-b) \text{ and } i^2=-1]$

$$=\left(\frac{3-4\sin^2\theta}{1+4\sin^2\theta}\right)+\left(\frac{8\sin\theta}{1+4\sin^2\theta}\right)i$$

Since, real part of $z=0$

$$\therefore \quad \frac{3-4\sin^2\theta}{1+4\sin^2\theta}=0$$
$$\Rightarrow \quad 3-4\sin^2\theta=0$$
$$\Rightarrow \quad \sin^2\theta=\frac{3}{4}$$
$$\Rightarrow \quad \sin\theta=\pm\frac{\sqrt{3}}{2}$$

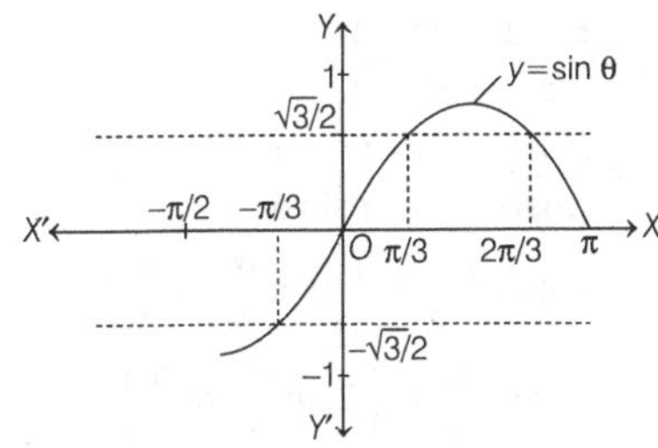

$$\Rightarrow \quad \theta\in\left\{-\frac{\pi}{3},\frac{\pi}{3},\frac{2\pi}{3}\right\}$$

Sum of values of $\theta=\dfrac{2\pi}{3}$.

46. $z_0=\dfrac{z_1+z_2+z_3}{3}$

$$\Rightarrow \quad 9{z_0}^2={z_1}^2+z_2^2+{z_3}^2+2(z_1z_2+z_2z_3+z_3z_1)$$

(For equilateral $z_1^2+{z_2}^2+z_3^2=z_1z_2+z_2z_3+z_3z_1$)

$$\therefore \quad {z_1}^2+{z_2}^2+{z_3}^2=3{z_0}^2.$$

47. Let $|z|=1/2$ ...(i)

Let $z_1=-1+4z$

$$\therefore \quad z=\frac{z_1+1}{4}$$

From (i), $\left|\dfrac{z_1+1}{4}\right|=1/2$

$$\Rightarrow \quad |z_1+1|=2$$

z_1 lies on a circle with radius 2.

48. $\exp\left(\dfrac{(|z|+3)(|z|-1)}{||z|+1|}\ln 2\right)\geq \log_{\sqrt{2}}|5\sqrt{7}+9i|$

$$\Rightarrow 2^{\frac{(|z|+3)(|z|-1)}{(|z|+1)}}\geq\log_{\sqrt{2}}(16)$$
$$\Rightarrow 2^{\frac{(|z|+3)(|z|-1)}{(|z|+1)}}\geq 2^3 \Rightarrow \frac{(|z|+3)(|z|-1)}{(|z|+1)}\geq 3$$
$$\Rightarrow (|z|+3)(|z|-1)\geq 3(|z|+1)$$
$$\Rightarrow (|z|^2+2|z|-3\geq 3|z|+3)$$
$$\Rightarrow |z|^2+|z|-6\geq 0|z|+3$$
$$\Rightarrow (|z|-3)(|z|+2)\geq 0\Rightarrow |z|-3\geq 0$$
$$\Rightarrow |z|\geq 3\Rightarrow |z|_{\min}=3$$

49. Centroid $=\dfrac{8+5i-3+i-2-3i}{3}$

$p(\text{say})=1+i$

$|p|=\sqrt{1+1}=\sqrt{2}$

$\arg(p)=\tan^{-1}(1/1)=\pi/4.$

50. $\displaystyle\sum_{k=1}^{10}\left(\sin\frac{2\pi k}{11}-i\cos\frac{2\pi k}{11}\right)$

$$=-i\sum_{k=1}^{10}\left(\cos\frac{2\pi k}{11}+i\sin\frac{2\pi k}{11}\right)$$
$$=-i\sum_{k=1}^{10}e^{i(2\pi k/11)}$$
$$=-i\left\{-1+\sum_{k=0}^{10}e^{i(2\pi k/11)}\right\}$$

$=-i\,(-1+$ sum of 11, 11th roots of unity).

$=-i\,(-1+0)=i$

51. $|z-|z+1||=|z+|z-1||$

Let $z+1=\alpha$ and $z-1=\beta$

$|z-\alpha|=|z+\beta|$

α and β are real numbers, locus of z is a straight line and it is a perpendicular bisector of line joining points (–1, 0) and (1, 0).

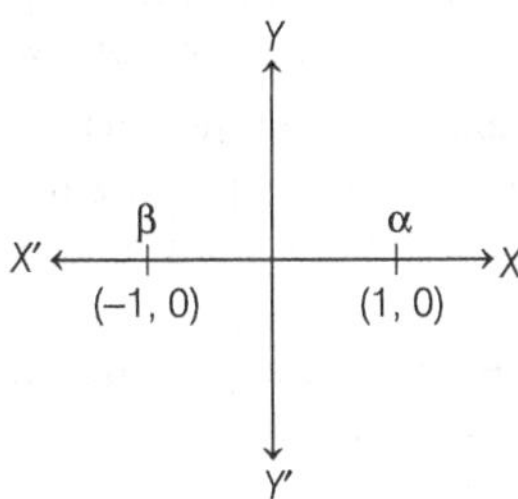

Perpendicular bisector of line is Y-axis (imaginary axis) i.e. $x=0$

Hence, option (d) is correct answer.

52. $z_1=a+ib,\ z_2=c+id,$

$\therefore \quad z_1-z_2=(a-c)+i\,(b-d)$

Since z_1, z_2, z_1-z_2 are collinear

$\therefore \quad \arg\{z_1-(z_1-z_2)\}=\arg\{z_2-(z_1-z_2)\}$

$\arg(z_2)=\arg(2z_2-z_1)$

$$\tan^{-1}\left(\frac{b}{a}\right)=\tan^{-1}\left(\frac{2d-b}{2c-a}\right)$$
$$\Rightarrow \quad ad-bc=0$$

53. Circumcentre = 0, centroid = $\frac{z_1 + z_2 + z_3}{3}$

Let orthocentre be γ.

$\therefore$ Centroid divides orthocentre and circumcentre as in the ratio 2 : 1 (internally)

$\therefore \quad \frac{z_1 + z_2 + z_3}{3} = \frac{1 \cdot \gamma + 2 \cdot 0}{2+1}$

$\therefore \quad \gamma = z_1 + z_2 + z_3$

But nine point centre is the mid point of orthocentre and circumcentre.

$\therefore$ Nine point centre $= \frac{\gamma + 0}{2} = \frac{z_1 + z_2 + z_3}{2}$.

54. Let $\alpha = re^{i\theta}$ and $\beta = re^{i\phi}$ $\forall \theta \in \left(-\frac{\pi}{2}, \frac{\pi}{2}\right)$ and $\phi \in (\pi, 2\pi)$

$$\therefore \quad \frac{\alpha + \beta}{\alpha - \beta} = \frac{e^{i\theta} + e^{i\phi}}{e^{i\theta} - e^{i\phi}} = \frac{2\cos\left(\frac{\theta - \phi}{2}\right)}{2i \sin\left(\frac{\theta - \phi}{2}\right)}$$

$$= -i \cot\left(\frac{\theta - \phi}{2}\right) = \text{Purely imaginary}$$

55. $\left|z - \frac{25}{z}\right| \geq \left||z| - \frac{25}{|z|}\right| \Rightarrow 24 \geq \left||z| - \frac{25}{|z|}\right|$

$\Rightarrow \quad -24 \leq |z| - \frac{25}{|z|} \leq 24$ or $-24|z| \leq |z|^2 - 25 \leq 24|z|$

$\therefore |z|^2 + 24|z| - 25 \geq 0$ and $|z|^2 - 24|z| - 25 \leq 0$

$\Rightarrow (|z| + 25)(|z| - 1) \geq 0$ and $(|z| - 25)(|z| + 1) \leq 0$

$\therefore |z| - 1 \geq 0$ and $|z| - 25 \leq 0$

Hence, $\quad 1 \leq |z| \leq 25$ or $1 \leq |z - 0| \leq 25$

56. Since $A \equiv z_1, B \equiv z_2, C \equiv (1 - i) z_1 + iz_2$

$\therefore \quad AB = |z_1 - z_2|, BC = |(1 - i) z_1 + iz_2 - z_2|$

$= |1 - i||z_1 - z_2| = \sqrt{2}|z_1 - z_2|$

and $\quad CA = |(1 - i) z_1 + iz_2 - z_1|$

$= |i||-z_1 + z_2| = |z_1 - z_2|$

$\therefore \quad AB = CA$, and $(AB)^2 + (CA)^2 = (BC)^2$.

57. $\because$ Centres and radius of the circles

$|z + 1 + i| = \sqrt{2}$ and $|z| = 3$ are respectively

$C_1 \equiv (1, 1),\ r_1 = \sqrt{2}$ and $C_2 \equiv (0, 0),\ r_2 = 3$

$\therefore \quad |C_1 C_2| = \sqrt{2} < r_2 - r_1$

Hence, one circle totally inside the other circle.

Hence, no solution.

58. $\quad |z + 1| = 2|z - 1|$

$\Rightarrow \quad |z + 1|^2 = 4|z - 1|^2$

$\Rightarrow \quad (z + 1)(\bar{z} + 1) = 4(z - 1)(\bar{z} - 1)$

$\Rightarrow \quad z\bar{z} + z + \bar{z} + 1 = 4z\bar{z} - 4z - 4\bar{z} + 4$

$\Rightarrow \quad 3z\bar{z} - 5z - 5\bar{z} + 3 = 0$

$\Rightarrow \quad z\bar{z} - \frac{5}{3}z - \frac{5}{3}\bar{z} + 1 = 0$

$\therefore$ Centre $= \frac{5}{3}$.

59. $2(1 + \omega)(1 + \omega^2) + 3(2\omega + 1)(2\omega^2 + 1)$

$+ 4(3\omega + 1)(3\omega^2 + 1) + \ldots + \ldots.$

$+ (n + 1)(n\omega + 1)(n\omega^2 + 1)$

nth term $\quad T_n = (n + 1)(n\omega + 1)(n\omega^2 + 1)$

$= (n + 1)(n^2 - n + 1)$

$= n^3 + 1$

$$\therefore \quad S_n = \Sigma T_n = \Sigma n^3 + \Sigma 1 = \left\{\frac{n(n + 1)}{2}\right\}^2 + n.$$

60. $\because \quad \sum_{k=0}^{200} i^k + \prod_{p=1}^{50} i^p = x + iy$

$\Rightarrow \quad i^0 + 0 + i^{1+2+\ldots+50} = x + iy$

$\Rightarrow \quad 1 + i^{1275} = x + iy$

$\Rightarrow \quad 1 - i = x + iy$

On comparing $x = 1$ and $y = -1$

$\therefore \quad (x, y) = (1, -1)$.

61. $\because \quad |z_1 - 1| < 1, |z_2 - 2| < 2, |z_3 - 3| < 3$

$\therefore \quad |z_1 + z_2 + z_3| = |(z_1 - 1) + (z_2 - 2) + (z_3 - 3) + 6|$

$\leq |z_1 - 1| + |z_2 - 2| + |z_3 - 3| + 6$

$< 1 + 2 + 3 + 6$

$\therefore \quad |z_1 + z_2 + z_3| < 12$

62. If $|z| = |z - 1| \Rightarrow |z|^2 = |z - 1|^2$

$\Rightarrow z\bar{z} = (z - 1)(\bar{z} - 1) \Rightarrow z + \bar{z} = 1$

and if $|z| = |z + 1| \Rightarrow |z|^2 = |z + 1|^2$

$\Rightarrow \quad z\bar{z} = (z + 1)(\bar{z} + 1) \Rightarrow z + \bar{z} = -1$

$\therefore \quad z + \bar{z} = \pm 1$

$\therefore \quad |z + \bar{z}| = 1$

63. Given, $|z - 1| = 1$

$\Rightarrow \quad z - 1 = e^{i\alpha} \Rightarrow z = 1 + e^{i\alpha}$

Let $z_1 = \frac{z - 2}{z} = \frac{e^{i\alpha} - 1}{e^{i\alpha} + 1}$

$$= \frac{(\cos\alpha - 1) + i\sin\alpha}{(\cos\alpha + 1) + i\sin\alpha}$$

$$= \frac{-2\sin^2\frac{\alpha}{2} + i\left(2\sin\frac{\alpha}{2}\cos\frac{\alpha}{2}\right)}{2\cos^2\frac{\alpha}{2} + i\left(2\sin\frac{\alpha}{2}\cos\frac{\alpha}{2}\right)}$$

$$= \tan\frac{\alpha}{2}\left[\frac{-\sin\frac{\alpha}{2} + i\cos\frac{\alpha}{2}}{\cos\frac{\alpha}{2} + i\sin\frac{\alpha}{2}}\right]$$

$$= \tan\frac{\alpha}{2}\left[\frac{i^2\sin\frac{\alpha}{2} + i\cos\frac{\alpha}{2}}{\cos\frac{\alpha}{2} + i\sin\frac{\alpha}{2}}\right]$$

$$i\left(\tan\frac{\alpha}{2}\right)\left[\frac{\cos\frac{\alpha}{2} + i\sin\frac{\alpha}{2}}{\cos\frac{\alpha}{2} + i\sin\frac{\alpha}{2}}\right] = i\tan\frac{\alpha}{2}$$

$z = 1 + e^{i\alpha} = (1 + \cos\alpha) + i\sin\alpha$

$= 2\cos^2\frac{\alpha}{2} + i\,2\sin\frac{\alpha}{2}\cos\frac{\alpha}{2} = 2\cos\frac{\alpha}{2}e^{i\alpha/2}$

$\arg(z) = \alpha/2 \quad \ldots\text{(i)}$

Using Eq. (i), we get

$z_1 = i\tan\frac{\alpha}{2} = i\tan(\arg(z))$

64. 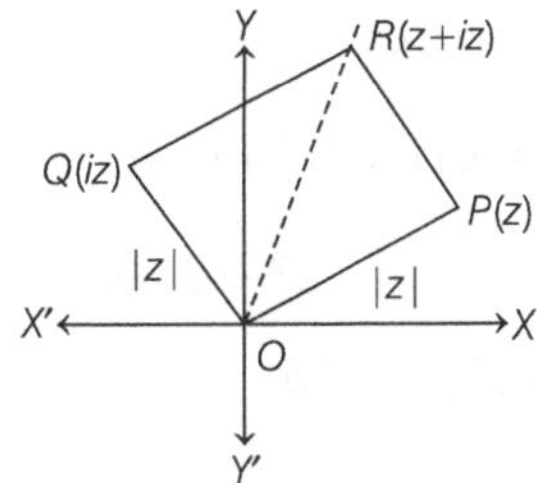

$$A = \frac{1}{2}|z||iz| = \frac{|z|^2}{2}$$

65. For $|z-1| \le \sqrt{2}$, z lies on and inside the circle of radius $\sqrt{2}$ units and centre (1, 0).

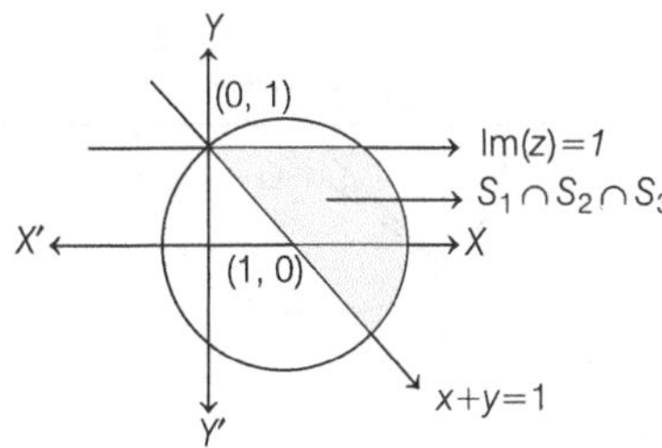

For S_2, let $z = x + iy$

Now, $(1-i)(z) = (1-i)(x+iy)$

$\text{Re}((1-i)z) = x + y \Rightarrow x + y \ge 1$

$\Rightarrow S_1 \cap S_2 \cap S_3$ has infinity many elements

66. $\because |z| = 1$, $\therefore$ $z = \cos\theta + i\sin\theta$

$\therefore$ $\arg(z) = \theta$

Then $\arg(z_1) = \theta_1$ and $\arg(z_2) = \theta_2$

$\therefore$ $z_1 R z_2 \Leftrightarrow |\arg z_1 - \arg z_2| = \dfrac{2\pi}{3}$

$z_1 R z_2 \Leftrightarrow |\theta_1 - \theta_2| = \dfrac{2\pi}{3}$

$\Leftrightarrow$ $|\theta_2 - \theta_1| = 2\pi/3$

$\Leftrightarrow$ $z_2 R z_1$ but $z_1 \ne z_2$

[if $z_1 = z_2$, then $0 = 2\pi/3$ (impossible)]

$\therefore$ Symmetric.

67. $\because$ Centre of the circle $|z-2| = 2$ i.e. 2.

lie on $z(1-i) + \bar{z}(1+i) = 4$,

Hence given line $z(1-i) + \bar{z}(1+i) = 4$ pass through the centre of circle i.e. intersect at two points.

$\therefore$ Number of solutions = 2

68. Given complex expression

$$\frac{3 + i\sin\theta}{4 - i\cos\theta}, \theta \in [0, 2\pi] = \frac{3 + i\sin\theta}{4 - i\cos\theta} \times \frac{4 + i\cos\theta}{4 + i\cos\theta}$$

[on rationalization]

$$= \frac{(12 - \sin\theta\cos\theta) + i(4\sin\theta + 3\cos\theta)}{16 + \cos^2\theta}$$

$\because \dfrac{3 + i\sin\theta}{4 - i\cos\theta}$ is a real number,

so $4\sin\theta + 3\cos\theta = 0$ ...(i)

Now, argument of $\sin\theta + i\cos\theta$ lies in either quadrant second or fourth because according to Eq. (i) $\sin\theta$ and $\cos\theta$ are opposite in sign,

so either, $\arg(\sin\theta + i\cos\theta) = \pi - \tan^{-1}\left|\dfrac{\cos\theta}{\sin\theta}\right|$

[for second quadrant]

$= \pi - \tan^{-1}\left(\dfrac{4}{3}\right)$

$\left[\text{from Eq. (i)} \dfrac{\cos\theta}{\sin\theta} = -\dfrac{4}{3}\right]$

or, $\arg(\sin\theta + i\cos\theta) = -\tan^{-1}\left|\dfrac{\cos\theta}{\sin\theta}\right| = -\tan^{-1}\left(\dfrac{4}{3}\right)$

69. Let $P = \sum_{i=0}^{n-1} \dfrac{\alpha_i}{(3-\alpha_i)} = -\sum_{i=0}^{n-1} \dfrac{(3-\alpha_i) - 3}{(3-\alpha_i)}$

$= 3\sum_{i=0}^{n-1} \dfrac{1}{(3-\alpha_i)} - \sum_{i=0}^{n-1} 1$ ...(i)

$\because$ $z^n - 1 = \sum_{i=0}^{n-1} (z - \alpha_i)$

$\therefore$ $\ln(z^n - 1) = \sum_{i=0}^{n-1} \ln(z - \alpha_i)$

Differentiating both sides w.r.t. z, then

$$\frac{nz^{n-1}}{(z^n - 1)} = \sum_{i=0}^{n-1} \frac{1}{(z - \alpha_i)}$$

and put $z = 3$, then

$$\frac{n \cdot 3^{n-1}}{(3^n - 1)} = \sum_{i=0}^{n-1} \frac{1}{(3 - \alpha_i)}$$

From Eq. (i), $3\left(\dfrac{n \cdot 3^{n-1}}{3^n - 1}\right) - n = \dfrac{n\,3^n}{3^n - 1} - n = \dfrac{n}{(3^n - 1)}$.

70. $\because$ $|z-1| + |z+3| \le 8$

$\therefore$ z lies inside or on the ellipse whose foci are (1, 0) and (– 3, 0) and vertices are (– 5, 0) and (3, 0)

Now minimum and maximum value of $|z-4|$ are 1 and 9 respectively

$\therefore$ $|z-4| \in [1, 9]$

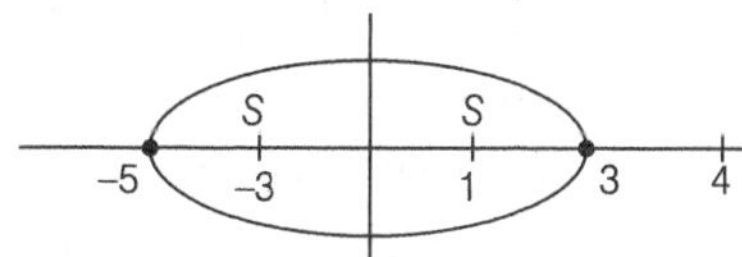

71. $\dfrac{z-1}{i}$ = real, $\dfrac{x + iy - 1}{i}$ = real.

$\dfrac{x-1}{i} + y$ = real

$\therefore$ $x - 1 = 0$

$\therefore$ $x = 1$

Then, $\sin^{-1}\left(\dfrac{z-1}{i}\right) = \sin^{-1}(y)$

So, $-1 \le y \le 1$

$\therefore$ $\text{Re}|z| = x = 1, -1 \le \text{Im}(z) \le 1$.

72. $|z_1 + z_2| = |z_1 - z_2| \Rightarrow |z_1 + z_2|^2 = |z_1 - z_2|^2$

$\Rightarrow |z_1|^2 + |z_2|^2 + (z_1\bar{z}_2 + \bar{z}_1 z_2) = |z_1|^2 + |z_2|^2 - (z_1\bar{z}_2 + \bar{z}_1 z_2)$

$\Rightarrow z_1\bar{z}_2 + \bar{z}_1 z_2 = 0 \Rightarrow z_1\bar{z} = -\bar{z}_1 z_2$

$\Rightarrow \dfrac{z_1}{z_2} = \dfrac{\bar{z}_1}{z_2} \Rightarrow \dfrac{z_1}{z_2} + \overline{\left(\dfrac{z_1}{z_2}\right)} = 0$

$\Rightarrow \text{Re}\left(\frac{z_1}{z_2}\right) = 0 \Rightarrow \frac{z_1}{z_2}$ is purely imaginary number.

$\Rightarrow \frac{z_1}{z_2} = ki$

$\Rightarrow \arg\left[\frac{z_1}{z_2}\right] = \arg(z_1) - \arg(z_2) = \pm\frac{\pi}{2}$

73. $A = \frac{2}{\sqrt{3}} e^{i\pi/2}, B = \frac{2}{\sqrt{3}} e^{-i\pi/6}, C = \frac{2}{\sqrt{3}} e^{-i5\pi/6}$

$A = \frac{2}{\sqrt{3}}(i), B = \frac{2}{\sqrt{3}}\left(\frac{\sqrt{3}-i}{2}\right), C = \frac{2}{\sqrt{3}}\left(\frac{-\sqrt{3}-i}{2}\right)$

$|A - B| = 2, |B - C| = 2, |C - A| = 2$

$\therefore$ ΔABC is equilateral.

74. If $|z| = 1$, $\left(\frac{1+z}{1+\bar{z}}\right)^n + \left(\frac{1+\bar{z}}{1+z}\right)^n = \left(\frac{z(1+z)}{z+z\bar{z}}\right)^n + \left(\frac{z+z\bar{z}}{z(1+z)}\right)^n$

$= \left(\frac{z(1+z)}{z+|z|^2}\right)^n + \left(\frac{z+|z|^2}{z(1+z)}\right)^n$

$= \left(\frac{z(1+z)}{z+1}\right)^n + \left(\frac{(z+1)}{z(1+z)}\right)^n$

$= z^n + \frac{1}{z^n} = 2\cos n(\arg z).$

75. $z^3 + az^2 + bz + c = 0$

if roots are α, β, γ,

then $\alpha + \beta + \gamma = -a, \alpha\beta + \beta\gamma + \gamma\alpha = b, \alpha\beta\gamma = c$

$|-a| = |\alpha + \beta + \gamma|$

$|a| \le |\alpha| + |\beta| + |\gamma|$

$|a| \le 1 + 1 + 1$

$|a| \le 3$

$$\left\{\begin{aligned} |b| &= |\alpha\beta + \beta\gamma + \gamma\alpha| \\ |b| &\le |\alpha||\beta| + |\beta||\gamma| + |\gamma||\alpha| \\ |b| &\le 1\cdot1 + 1\cdot1 + 1\cdot1 \\ |b| &\le 3 \end{aligned}\right.$$

$|c| = |\alpha\beta\gamma| = |\alpha||\beta||\gamma| = 1\cdot1\cdot1 = 1$

76. $z = (1 - i\cot 8)^3 = \text{cosec}^3 8(\sin 8 - i\cos 8)^3$

$= \text{cosec}^3 8\left(\cos\left(\frac{\pi}{2} - 8\right) - i\sin\left(\frac{\pi}{2} - 8\right)\right)^3$

$= \text{cosec}^3 8\left(\cos\left(\frac{3\pi}{2} - 24\right) - i\sin\left(\frac{3\pi}{2} - 24\right)\right)$

$= \text{cosec}^3 8 \cdot e^{-i(3\pi/2 - 24)} = \text{cosec}^3 8 \cdot e^{i(24 - 3\pi/2)}$

77. It is given for complex number z,

$\left|\frac{z-i}{z+2i}\right| = 1$ ...(i)

and $|z| = \frac{5}{2}$ ...(ii)

From Eq. (i), $|z - i| = |z + 2i|$

$\Rightarrow$ Locus of z is a straight line and it is perpendicular bisector of line joining points (0, – 2) and (0, 1), so locus of z is $y = -\frac{1}{2}$, where $z = x + iy$.

From Eq. (ii), $\sqrt{x^2 + y^2} = \frac{5}{2} \Rightarrow x^2 + \left(\frac{-1}{2}\right)^2 = \left(\frac{5}{2}\right)^2$

$\Rightarrow \quad x^2 + \frac{1}{4} = \frac{25}{4} \Rightarrow x^2 = 6$

Now, $|z + 3i| = |x + (y+3)i| = \sqrt{x^2 + (y+3)^2}$

$= \sqrt{6 + \left(-\frac{1}{2} + 3\right)^2}$ $[\because x^2 = 6 \text{ and } y = -1/2]$

$= \sqrt{6 + \frac{25}{4}} = \sqrt{\frac{24 + 25}{4}} = \sqrt{\frac{49}{4}} = \frac{7}{2}$

78.

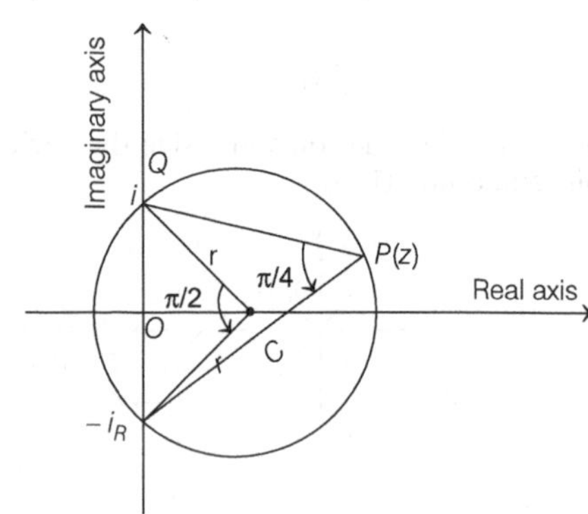

$\because \angle QPR = \pi/4$

$\therefore \angle QCR = \pi/2$, where C is centre of circle.

Now in ΔQCR,

$r^2 + r^2 = (QR)^2 = |2i|^2 = 4$

$\therefore \quad r = \sqrt{2}$

$\therefore$ Required parimeter $= 2\pi r - \frac{1}{4}(2\pi r)$

$= \frac{3}{2}\pi r = \frac{3}{2}\pi(\sqrt{2}) = \frac{3\pi}{\sqrt{2}}$

79. Given, $z = \frac{\sqrt{3}}{2} + \left(\frac{1}{2}\right)i = \cos\frac{\pi}{6} + i\sin\frac{\pi}{6} = e^{i\frac{\pi}{6}}$

so, $(1 + iz + z^5 + iz^8)^9 = \left(1 + ie^{i\frac{\pi}{6}} + e^{i\frac{5\pi}{6}} + ie^{i\frac{8\pi}{6}}\right)^9$

$= \left(1 + e^{i\frac{\pi}{2}} \cdot e^{i\frac{\pi}{6}} + e^{i\frac{5\pi}{6}} + e^{i\frac{\pi}{2}} \cdot e^{i\frac{4\pi}{3}}\right)^9$ $\left[\because i = e^{i\frac{\pi}{2}}\right]$

$= \left(1 + e^{i\frac{2\pi}{3}} + e^{i\frac{5\pi}{6}} + e^{i\frac{11\pi}{6}}\right)^9$

$= \left[1 + \left(\cos\frac{2\pi}{3} + i\sin\frac{2\pi}{3}\right) + \left(\cos\frac{5\pi}{6} + i\sin\frac{5\pi}{6}\right) + \left(\cos\frac{11\pi}{6} + i\sin\frac{11\pi}{6}\right)\right]^9$

$= \left(1 - \frac{1}{2} + \frac{i\sqrt{3}}{2} - \frac{\sqrt{3}}{2} + \frac{1}{2}i + \frac{\sqrt{3}}{2} - \frac{i}{2}\right)^9$

$= \left(\frac{1}{2} + \frac{\sqrt{3}i}{2}\right)^9 = \left(\cos\frac{\pi}{3} + i\sin\frac{\pi}{3}\right)^9$

$= \cos 3\pi + i\sin 3\pi$ [$\because$ for any natural number 'n' $(\cos\theta + i\sin\theta)^n = \cos(n\theta) + i\sin(n\theta)$]

$= -1$

80. $|z_1 + z_2| = \left|\frac{1}{z_1} + \frac{1}{z_2}\right|$

$$= \left|\frac{z_1 + z_2}{z_1 z_2}\right|$$

$$\Rightarrow \quad |z_1 + z_2|\left[1 - \frac{1}{|z_1 z_2|}\right] = 0$$

$$\therefore \quad |z_1 z_2| = 1 \qquad [\because z_1 \neq -z_2]$$

81. $\omega = \frac{1 - iz}{z - i}$

$$\Rightarrow \quad |\omega| = \left|\frac{1 - iz}{z - i}\right| \Rightarrow 1 = \left|\frac{1 - iz}{z - i}\right|$$

$$\Rightarrow |z - i|^2 = |1 - iz|^2$$

$$(z - i)(\bar{z} + i) = (1 - iz)(1 + i\bar{z})$$

$$\Rightarrow \quad z\bar{z} + iz - i\bar{z} + 1 = 1 + i\bar{z} - iz + z\bar{z}$$

$$2iz - 2i\bar{z} = 0$$

$$\therefore \quad z = \bar{z}$$

$\Rightarrow$ z lies on real axis.

82. Let $x + iy = \frac{\alpha + i}{\alpha - i} \Rightarrow x + iy = \frac{(\alpha + i)^2}{\alpha^2 + 1} = \frac{(\alpha^2 - 1) + (2\alpha)i}{\alpha^2 + 1}$

$$= \frac{\alpha^2 - 1}{\alpha^2 + 1} + \left(\frac{2\alpha}{\alpha^2 + 1}\right) i$$

On comparing real and imaginary parts, we get

$$x = \frac{\alpha^2 - 1}{\alpha^2 + 1} \text{ and } y = \frac{2\alpha}{\alpha^2 + 1}$$

Now, $x^2 + y^2 = \left(\frac{\alpha^2 - 1}{\alpha^2 + 1}\right)^2 + \left(\frac{2\alpha}{\alpha^2 + 1}\right)^2$

$$= \frac{\alpha^4 + 1 - 2\alpha^2 + 4\alpha^2}{(\alpha^2 + 1)^2} = \frac{(\alpha^2 + 1)^2}{(\alpha^2 + 1)^2} = 1$$

$$\Rightarrow \quad x^2 + y^2 = 1$$

So, $S = \left\{\frac{\alpha + i}{\alpha - i}; \alpha \in \mathbf{R}\right\}$ lies on a circle with radius 1.

83. As given for $z_1 = x_1 + iy_1$ and $z_2 = x_2 + iy_2$, the symbol $z_1 \cap z_2$ is used if $x_1 \leq x_2$ and $y_1 \leq y_2$.

Therefore for $1 \cap z$, we have $1 \leq x$ and $0 \leq y$ as $1 = 1 + 0 \cdot i$ and $z = x + iy$

Now, $\frac{1 - z}{1 + z} = \frac{1 - (x + iy)}{1 + (x + iy)} = \frac{(1 - x - iy)(1 + x - iy)}{(1 + x)^2 + y^2}$

$$= \frac{(1 - x^2 - y^2)}{(1 + x)^2 + y^2} - \frac{2iy}{(1 + x)^2 + y^2}$$

But $x \geq 1$ and $y \geq 0$

$$\therefore \quad x^2 + y^2 \geq 1 \text{ and } -2y \leq 0$$

$$\Rightarrow \frac{1 - x^2 - y^2}{(1 + x)^2 + y^2} \leq 0 \text{ and } \frac{-2y}{(1 + x)^2 + y^2} \leq 0$$

$$\therefore \quad \frac{1 - z}{1 + z} \cap 0$$

84. Given complex number

$$\omega = \frac{5 + 3z}{5(1 - z)} \Rightarrow 5\omega - 5\omega z = 5 + 3z$$

$$\Rightarrow \quad (3 + 5\omega)z = 5\omega - 5 \Rightarrow |3 + 5\omega||z| = |5\omega - 5| \quad \ldots(i)$$

[applying modulus both sides and $|z_1 z_2| = |z_1||z_2|$]

$$\because \quad |z| < 1$$

$$\therefore \quad |3 + 5\omega| > |5\omega - 5| \qquad \text{[from Eq. (i)]}$$

$$\Rightarrow \quad \left|\omega + \frac{3}{5}\right| > |\omega - 1|$$

Let $\omega = x + iy$, then $\left(x + \frac{3}{5}\right)^2 + y^2 > (x - 1)^2 + y^2$

$$\Rightarrow \quad x^2 + \frac{9}{25} + \frac{6}{5}x > x^2 + 1 - 2x$$

$$\Rightarrow \quad \frac{16x}{5} > \frac{16}{25} \Rightarrow x > \frac{1}{5} \Rightarrow 5x > 1$$

$$\Rightarrow \quad 5\,\text{Re}(\omega) > 1$$

85. $\because |\sin\theta_1|z^3 + |\sin\theta_2|z^2 + |\sin\theta_3|z + |\sin\theta_4| = 3$

$$\therefore \quad |3| = ||\sin\theta_1|z^3 + |\sin\theta_2|z^2 + |\sin\theta_3|z + |\sin\theta_4||$$

$$\leq |\sin\theta_1||z|^3 + |\sin\theta_2||z|^2 + |\sin\theta_3||z| + |\sin\theta_4|$$

$$\leq |z|^3 + |z|^2 + |z| + 1 \qquad [\because |\sin\theta_i| \leq 1]$$

$$< 1 + |z| + |z|^2 + \ldots \infty$$

$$= \frac{1}{1 - |z|}$$

$$\therefore \quad 3 < \frac{1}{1 - |z|} \Rightarrow \frac{1}{3} > 1 - |z| \Rightarrow |z| > \frac{2}{3}$$

$$\therefore \quad |z - 0| > \frac{2}{3}$$

86. $\because \log_{1/3}\{\log_{1/2}(|z|^2 + 4|z| + 3)\} < 0$

$$\Rightarrow \quad \log_{1/2}(|z|^2 + 4|z| + 3) > \left(\frac{1}{3}\right)^0$$

$$\Rightarrow \quad \log_{1/2}(|z|^2 + 4(z) + 3) > 1$$

$$\Rightarrow \quad |z|^2 + 4|z| + 3 < \left(\frac{1}{2}\right)^1$$

$$\Rightarrow \quad |z|^2 + 4|z| + \frac{5}{2} < 0$$

Positive quantity < 0 (Impossible)

Hence, S be an empty set or null set.

87. $\because \quad (f - ig)z = f(z) - ig(z) = \sin z - i\cos z$

$$= -i(\cos z + i\sin z)$$

$$\therefore \ (f - ig)z = -ie^{iz} = z_1 \text{ (say)} \qquad \ldots(i)$$

Then $(f + ig) * (f - ig)z = (f + ig) * (f(z) - ig(z))$

$$= (f + ig) * z_1$$

$$= f(z_1) + ig(z_1)$$

$$= \sin z_1 + i\cos z_1$$

$$= i(\cos z_1 - i\sin z_1)$$

$$= ie^{-iz_1}$$

$$= ie^{-i(-ie^{iz})} \qquad \text{[from Eq. (i)]}$$

$$= ie^{-e^{iz}}$$

88. $f_p(\alpha) = e^{\frac{i\alpha}{p^2}(1 + 2 + 3 + 4 + \ldots + p)} = e^{\frac{i\alpha}{p^2} \cdot \frac{p(p+1)}{2}} = e^{\frac{i\alpha}{2}\left(1 + \frac{1}{p}\right)}$

$$\therefore \quad \lim_{n \to \infty} f_n(\pi) = \lim_{n \to \infty} e^{\frac{i\pi}{2}\left(1 + \frac{1}{n}\right)}$$

$$= e^{\frac{i\pi}{2}(1 + 0)} = e^{\frac{i\pi}{2}} = \cos\frac{\pi}{2} + i\sin\frac{\pi}{2} = i$$

Objective Questions Type II [One or more than one correct answer(s)]

1. Given $|z-1|<|z+3|$

$\Rightarrow \quad |z-1|^2<|z+3|^2$

$\Rightarrow \quad |z|^2+1-2\operatorname{Re}(z)<|z|^2+9+2\operatorname{Re}(3z)$

$\Rightarrow \quad 2\operatorname{Re}(4z)>-8$

$\therefore \quad 4z+4\bar{z}>-8$

or $\quad z+\bar{z}>-2$

and $\quad \omega=2z+3-i \quad$ ...(i)

$\therefore \quad \omega+\bar{\omega}=2z+3-i+2\bar{z}+3+i$

$=2(z+\bar{z})+6$

or $\quad \omega+\bar{\omega}>2 \quad$ ...(ii)

Alternate (a) :

$|\omega-5-i|<|\omega+3+i|$

$\Rightarrow |2z+3-i-5-i|<|2z+3-i+3+i|$

$\Rightarrow |2z-2-2i|<|2z+6|$

$\Rightarrow |z-1-i|<|z+3|$

which is false.

Alternate (b) :

$|\omega-5|<|\omega+3|$

$\Rightarrow |2z+3-i-5|<|2z+3-i+3|$

$\Rightarrow |Z-1-i/2|<|z+3-i/2|$

or $\quad |z-1|<|z+3|$

which is true.

Alternate (c) :

$\operatorname{Im}(i\,\omega)>1$

$\Rightarrow \dfrac{i\omega-i\bar{\omega}}{2i}>1 \Rightarrow \dfrac{i\,\omega+i\,\bar{\omega}}{2i}>1$

$\Rightarrow \omega+\bar{\omega}>2$ [which is true from Eq. (ii)]

Alternate (d) : $\quad |\arg(\omega-1)|<\pi/2$

$|\arg(2z+2-i|<\pi/2$

$$\left|\tan^{-1}\left(\frac{\operatorname{Im}(2z+2-i)}{\operatorname{Re}(2z+2-i)}\right)\right|<\pi/2$$

$\therefore \quad \operatorname{Re}(2z+2-i)>0$

$\Rightarrow \quad \dfrac{(2z+2-i)+(2\bar{z}+2+i)}{2}>0$

$\Rightarrow z+\bar{z}>-2$ [which is true from Eq. (i)].

2. $\dfrac{az+b}{z+1}=\dfrac{ax+b+aiy}{(x+1)+iy}=\dfrac{(ax+b+aiy)((x+1)-iy)}{(x+1)^2+y^2}$

$\therefore \quad \operatorname{Im}\left(\dfrac{az+b}{z+1}\right)=\dfrac{-(ax+b)y+ay(x+1)}{(x+1)^2+y^2}$

$\Rightarrow \quad \dfrac{(a-b)y}{(x+1)^2+y^2}=y$

$\because \quad a-b=1$

$\therefore \quad (x+1)^2+y^2=1$

$\therefore \quad x=-1\pm\sqrt{1-y^2}$

3. $\operatorname{Re}\left(\dfrac{z+1}{z+i}\right)=0 \Rightarrow \dfrac{\dfrac{z+1}{z+i}+\dfrac{\bar{z}+1}{\bar{z}+\bar{i}}}{2}=0$

$\Rightarrow \quad \dfrac{z+1}{z+i}+\dfrac{\bar{z}+1}{\bar{z}-i}=0$

$(z+1)(\bar{z}-i)+(\bar{z}+1)(z+i)=0$

$\Rightarrow \quad 2z\bar{z}+(1-i)z+(1+i)\bar{z}=0$

$z\bar{z}+\left(\dfrac{1-i}{2}\right)z+\left(\dfrac{1+i}{2}\right)\bar{z}=0$ which is a circle

whose radius $=\sqrt{\left(\dfrac{1+i}{2}\right)^2}=\sqrt{\dfrac{1}{2}}=\dfrac{1}{\sqrt{2}}$

and passing through $z=0$, [$\because$ no constant term]

4. $\quad x^2-2x\cos\theta+1=0$

$\therefore \quad x=\cos\theta\pm i\sin\theta$

If $\alpha=\cos\theta+i\sin\theta$, then $\beta=\cos\theta-i\sin\theta$

Equation $\quad x^2-(\alpha^n+\beta^n)x+\alpha^n\beta^n=0$

$\Rightarrow \quad x^2-2x\cos n\theta+1=0$

$\Rightarrow \quad (x-\cos n\theta)^2+\sin^2 n\theta=0$

5. Let $z-3=re^{i\phi}$

$\therefore \quad \dfrac{z-3}{e^{i\theta}}+\dfrac{e^{i\theta}}{z-3}=re^{i(\phi-\theta)}+\dfrac{1}{r}e^{i(\theta-\phi)} \quad$ ...(i)

Imaginary part of Eq. (i) $=r\sin(\phi-\theta)-\dfrac{1}{r}\sin(\phi-\theta)$

$r\sin(\phi-\theta)-\dfrac{1}{r}\sin(\phi-\theta)=0$ (given)

$\Rightarrow \quad r-\dfrac{1}{r}=0$

or $\quad \sin(\phi-\theta)=0$

$\Rightarrow \quad r-\dfrac{1}{r}=0$

$\Rightarrow \quad r=1$

$\Rightarrow \quad |z-3|=1$

$\Rightarrow z$ lies on a circle with unit radius

$\sin(\phi-\theta)=0$

$\Rightarrow \quad \phi=\theta$

$\therefore \quad z-3=re^{i\theta}$

$z-3=r\cos\theta,\ y=r\sin\theta$

$\Rightarrow \quad x-3=r\cos\theta,\ y=r\sin\theta$

$\dfrac{x-3}{\cos\theta}=\dfrac{y}{\sin\theta}=r$

Which is a straight line passing through (3, 0).

6. $|z_1+z_2|^2=|z_1|^2+|z_2|^2$

$\Rightarrow \quad |z_1|^2+|z_2|^2+2\operatorname{Re}(z_1\bar{z}_2)=|z_1|^2+|z_2|^2$

$\Rightarrow \quad 2\operatorname{Re}(z_1\bar{z}_2)=0$

$\Rightarrow \quad z_1\bar{z}_2+\bar{z}_1 z_2=0$

$\Rightarrow \quad \dfrac{z_1}{z_2}=-\dfrac{\bar{z}_1}{\bar{z}_2};\ \dfrac{z_1}{z_2}=-\left(\overline{\dfrac{z_1}{z_2}}\right)$

$\therefore \dfrac{z_1}{z_2}$ is purely imaginary.

Then, $\quad \arg\left(\dfrac{z_1}{z_2}\right)=\pm\dfrac{\pi}{2}.$

7. $\text{amp}\left(\frac{z_4 - z_1}{z_2 - z_1}\right) = \frac{\pi}{2}$

but $\quad z_1 - z_4 = z_2 - z_3$

and $\quad z_3 - z_4 = z_2 - z_1$

$$\text{amp}\left(\frac{-(z_2 - z_3)}{z_3 - z_4}\right) = \frac{\pi}{2}$$

$$\Rightarrow \quad \text{amp}\left(\frac{z_2 - z_3}{z_4 - z_3}\right) = \frac{\pi}{2}$$

$\therefore ABCD$ is rectangle and cyclic quadrilateral ($\because A + C = 180°$)

8. (a) Let $\quad z = -1 - i$ and $\arg(z) = \theta$

Now, $\quad \tan\theta = \left|\frac{\text{im}(z)}{\text{Re}(z)}\right| = \left|\frac{-1}{-1}\right| = 1$

$$\Rightarrow \quad \theta = \frac{\pi}{4}$$

Since, $x < 0, y < 0$

$$\therefore \quad \arg(z) = -\left(\pi - \frac{\pi}{4}\right) = -\frac{3\pi}{4}$$

(b) We have, $f(t) = \arg(-1 + it)$

$$\arg(-1 + it) = \begin{cases} \pi - \tan^{-1} t, & t \geq 0 \\ -(\pi + \tan^{-1} t), & t < 0 \end{cases}$$

This function is discontinuous at $t = 0$.

(c) We have,

$$\arg\left(\frac{z_1}{z_2}\right) - \arg(z_1) + \arg(z_2)$$

Now, $\arg\left(\frac{z_1}{z_2}\right) = \arg(z_1) - \arg(z_2) + 2n\pi$

$$\therefore \quad \arg\left(\frac{z_1}{z_2}\right) - \arg(z_1) + \arg(z_2)$$
$$= \arg(z_1) - \arg(z_2) + 2n\pi - \arg(z_1) + \arg(z_2)$$
$$= 2n\pi$$

So, given expression is multiple of 2π.

(d) We have, $\arg\left(\frac{(z - z_1)(z_2 - z_3)}{(z - z_3)(z_2 - z_1)}\right) = \pi$

$$\Rightarrow \left(\frac{z - z_1}{z - z_3}\right)\left(\frac{z_2 - z_3}{z_2 - z_1}\right) \text{ is purely real}$$

Thus, the points $A(z_1)$, $B(z_2)$, $C(z_3)$ and $D(z)$ taken in order would be concyclic if purely real.

Hence, it is a circle.

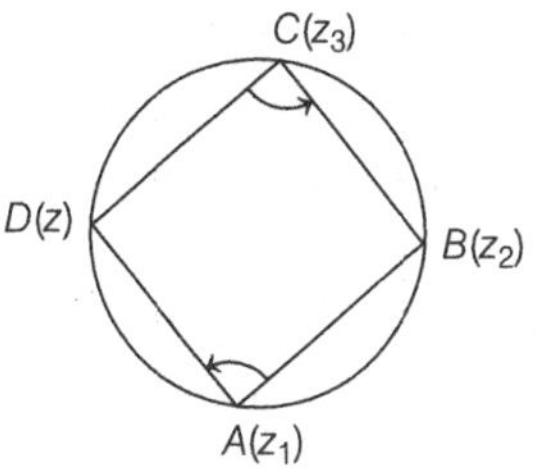

$\therefore$ (a), (b), (d) are false statement.

9. $\alpha z^2 + z + \bar{\alpha} = 0$

Let $z = a$ be a real root,

then $\quad \alpha a^2 + a + \bar{\alpha} = 0 \quad$ ...(i)

and let $\quad \alpha = p + iq$

$\therefore \quad (p + iq)a^2 + a + p - iq = 0$

$\Rightarrow \quad pa^2 + a + p = 0$ and $a^2 q - q = 0$

$\therefore \quad a = \pm 1 \quad [\because q \neq 0]$

$\therefore$ From (i), $\alpha \pm 1 + \bar{\alpha} = 0$

Also, $\quad |a| = 1$

10. $a_0 z^4 + a_1 z^3 + a_2 z^2 + a_3 z + a_4 = 0$

Taking conjugate on both sides.

$$a_0(\bar{z})^4 + a_1(\bar{z})^3 + a_2(\bar{z})^2 + a_3(\bar{z}) + a_4 = 0$$

$\therefore \bar{z}_1, \bar{z}_2, \bar{z}_3, \bar{z}_4$ are the roots of the equation if z_1 is real, then $z_1 = \bar{z}_1$ and is z_1 if non real, then $\bar{z}_1$ is also root because imaginary roots occur in conjugate pair.

11. $\frac{2-i}{3+i} = \frac{(2-i)(3-i)}{10} = \frac{5-5i}{10} = \frac{1}{2} - \frac{i}{2}$

i.e. $\left(\frac{1}{2}, -\frac{1}{2}\right)$

and $\quad z(1 + i) = \bar{z}(i - 1)$

$\Rightarrow \quad (z + \bar{z}) + i(z - \bar{z}) = 0$

$\Rightarrow \quad \left(\frac{z + \bar{z}}{2}\right) + i\left(\frac{z - \bar{z}}{2}\right) = 0$

$\Rightarrow \quad x + i(iy) = 0 \quad (z = x + iy)$

$\Rightarrow \quad x - y = 0$

$\therefore$ Reflection of $\left(\frac{1}{2}, -\frac{1}{2}\right)$ w.r.t. $y = x$ is $\left(-\frac{1}{2}, \frac{1}{2}\right)$

ie, $\quad -\frac{1}{2} + \frac{i}{2} = \frac{-1 + i}{2} \quad$ (Alternate (b))

$$= \frac{i^2 + i}{2}$$

$$= \frac{i(i + 1)}{2} \quad \text{(Alternate (c))}$$

$$= \frac{i(1 + i)^2}{2(1 + i)}$$

$$= \frac{i(1 + i^2 + 2i)}{2(1 + i)} = \frac{i^2}{1 + i}$$

$$= \frac{-1}{1 + i} \quad \text{(Alternate (d))}$$

12. $z^3 + (1+i) z^2 + (1+i) z + i = 0$

$\Rightarrow \quad z^2 (z+i) + z(z+i) + (z+i) = 0$

$\Rightarrow \quad (z+i)(z^2 + z + 1) = 0$

$\Rightarrow \quad (z+i)(z-\omega)(z-\omega^2) = 0$

$\therefore \quad z = -i, \omega, \omega^2$

Now in $z^{1993} + z^{1994} + 1$ put $z = -i, \omega, \omega^2$

then $(-i)^{1993} + (-i)^{1994} + 1 = -i - 1 + 1 = -i \neq 0$

and $(\omega)^{1993} + \omega^{1994} + 1 = \omega + \omega^2 + 1 = 0$

and $(\omega^2)^{1993} + (\omega^2)^{1994} + 1 = \omega^2 + \omega + 1 = 0$

Hence ω and ω^2 are common roots.

13. Here, $x + iy = \dfrac{1}{a + ibt} \times \dfrac{a - ibt}{a - ibt}$

$\therefore \quad x + iy = \dfrac{a - ibt}{a^2 + b^2t^2}$

Let $a \neq 0, b \neq 0$

$\therefore \quad x = \dfrac{a}{a^2 + b^2t^2}$ and $y = \dfrac{-bt}{a^2 + b^2t^2}$

$\Rightarrow \quad \dfrac{y}{x} = \dfrac{-bt}{a} \Rightarrow t = \dfrac{-ay}{bx}$

On putting $t = \dfrac{-ay}{bx}$ in $x = \dfrac{a}{a^2 + b^2t^2}$, we get

$$x\left(a^2 + b^2 \cdot \frac{a^2y^2}{b^2x^2}\right) = a \Rightarrow a^2(x^2 + y^2) = ax$$

or $\quad x^2 + y^2 - \dfrac{x}{a} = 0 \quad \ldots\text{(i)}$

or $\quad \left(x - \dfrac{1}{2a}\right)^2 + y^2 = \dfrac{1}{4a^2}$

$\therefore$ Option (a) is correct.

For $a \neq 0$ and $b = 0$,

$$x + iy = \frac{1}{a} \Rightarrow x = \frac{1}{a}, y = 0$$

$\Rightarrow z$ lies on X-axis.

$\therefore$ Option (c) is correct.

For $a = 0$ and $b \neq 0$, $x + iy = \dfrac{1}{ibt}$

$\Rightarrow \quad x = 0, y = -\dfrac{1}{bt}$

$\Rightarrow z$ lies on Y-axis.

$\therefore$ Option (d) is correct.

14. $\left|\dfrac{z_1 + z_2}{z_1 - z_2}\right| = 1 \Rightarrow \dfrac{z_1 + z_2}{z_1 - z_2} = \cos\alpha + i\sin\alpha$

where α is the argument of $(z_1 + z_2)/(z_1 - z_2)$.
Applying componendo and dividendo, we get

$$\frac{z_1}{z_2} = \frac{1 + \cos\alpha + i\sin\alpha}{-1 + \cos\alpha + \sin\alpha} = \frac{1 + e^{i\alpha}}{-1 + e^{i\alpha}} = \frac{e^{i\alpha/2} \cdot 2\cos\left(\dfrac{\alpha}{2}\right)}{e^{i\alpha/2} \cdot 2i\sin\left(\dfrac{\alpha}{2}\right)}$$

$$= -i\cot(\alpha/2)$$

which is zero, if $\alpha = k\pi (k \in I)$ and otherwise purely imaginary in nature.

15. Since $|z_1| = |z_2| = 1$, we have

$z_1 = \cos\theta_1 + i\sin\theta_1$ and $z_2 = \cos\theta_2 + i\sin\theta_2$

where $\theta_1 = \arg(z_1)$ and $\theta_2 = \arg(z_2)$.

Also, $z_1 = a + ib$ and $z_2 = c + id$.

$\therefore \quad a = \cos\theta_1, b = \sin\theta_1, c = \cos\theta_2$ and $d = \sin\theta_2$

Also, $\text{Re}(z_1 \overline{z_2}) = 0$

$\Rightarrow \text{Re}[(\cos\theta_1 + i\sin\theta_1)(\cos\theta_2 - i\sin\theta_2)] = 0$

$\Rightarrow \quad \text{Re}[\cos(\theta_1 - \theta_2) + i\sin(\theta_1 - \theta_2)] = 0$

$\Rightarrow \quad \cos(\theta_1 - \theta_2) = 0$

$\Rightarrow \quad \theta_1 - \theta_2 = \pi/2$

or $\quad \theta_1 = \theta_2 + \pi/2$

Now $\omega_1 = a + ic = \cos\theta_1 + i\cos\theta_2$

$= \cos\theta_1 + i\sin\theta_1 = z_1$

$\Rightarrow |\omega_1| = 1$

Similarly $\omega_2 = b + id = \sin\theta_1 + i\sin\theta_2$

$= \cos\theta_2 + i\sin\theta_2 = z_2$

$\Rightarrow \quad |\omega_2| = 1$

Also $|\omega_1 \overline{\omega_2}| = |\omega_1||\overline{\omega_2}| = |\omega_1||\omega_2| = 1.1 = 1$

$\text{Re}(\overline{\omega_1}\omega_2) = \text{Re}(\overline{z_2} z_2) = \text{Re}(z_1\overline{z_2}) = 0$

16. We have

$\dfrac{1 + i\cos\theta}{1 - 2i\cos\theta} = \dfrac{(1 + i\cos\theta)(1 + 2i\cos\theta)}{(1 - 2i\cos\theta)(1 + 2i\cos\theta)}$

$= \dfrac{(1 - 2\cos^2\theta) + i\,3\cos\theta}{1 + 4\cos^2\theta}$

Thus $(1 + i\cos\theta)/(1 - 2i\cos\theta)$ is a real number if $\cos\theta = 0$

$\Rightarrow \quad \theta = 2n\pi \pm \dfrac{\pi}{2}$,

where n is an integer.

17. Let $a = \cos\alpha + i\sin\alpha$, $b = \cos\beta + i\sin\beta$
and $c = \cos\gamma + i\sin\gamma$
We have,

$a + b + c = (\cos\alpha + \cos\beta + \cos\gamma)$
$+ i(\sin\alpha + \sin\beta + \sin\gamma) = 0 + i0 = 0$

Also $\dfrac{1}{a} + \dfrac{1}{b} + \dfrac{1}{c} = a^{-1} + b^{-1} + c^{-1}$

$= (\cos\alpha + \cos\beta + \cos\gamma) - i(\sin\alpha + \sin\beta + \sin\gamma)$

[using De Moivre's theorem]

$= 0 - i \cdot 0 = 0$

$\Rightarrow \quad bc + ca + ab = 0$

$\Rightarrow [\cos(\beta + \gamma) + i\sin(\beta + \gamma)] + [\cos(\gamma + \alpha)] + i\sin(\gamma + \alpha)$
$+ [\cos(\alpha + \beta) + i\sin(\alpha + \beta)] = 0$

$\Rightarrow \quad \cos(\beta + \gamma) + \cos(\gamma + \alpha) + \cos(\alpha + \beta) = 0$

and $\sin(\beta + \gamma) + \sin(\gamma + \alpha) + \sin(\alpha + \beta) = 0$

Also $\quad a + b + c = 0$

$\Rightarrow \quad (a + b + c)^2 = 0$

$a^2 + b^2 + c^2 + 2(ab + bc + ca) = 0$

$\Rightarrow \quad a^2 + b^2 + c^2 = 0 \quad [\because ab + bc + ca = 0]$

$\Rightarrow [\cos(2\alpha) + i\sin(2\alpha)] + [\cos(2\beta) + i\sin(2\beta)]$
$+ [\cos(2\gamma) + i\sin(2\gamma)] = 0$ [using De Moivre's theorem]

$\Rightarrow \quad \cos(2\alpha) + \cos(2\beta) + \cos(2\gamma) = 0$

and $\quad \sin(2\alpha) + \sin(2\beta) + \sin(2\gamma) = 0$

18. We have

$$z^n - 1 = (z-1)(z-\omega)(z-\omega^2)\dots(z-\omega^{n-1})$$

$$\Rightarrow \quad (z-\omega)(z-\omega^2)\dots(z-\omega^{n-1}) = \frac{z^n - 1}{z-1}$$

Putting $z = 2$, we get

$$(2-\omega)(2-\omega^2)\dots(2-\omega^{n-1}) = \frac{2^n - 1}{2-1} = 2^n - 1$$

Also, $\quad {}^nC_1 + {}^nC_2 + \dots + {}^nC_n = 2^n - 1$

Let $\quad E = {}^{2n+1}C_0 + {}^{2n+1}C_1 + \dots + {}^{2n+1}C_n$

$$\Rightarrow \quad 2E = 2({}^{2n+1}C_0) + 2({}^{2n+1}C_1) + \dots + 2({}^{2n+1}C_n)$$

$$= ({}^{2n+1}C_0 + {}^{2n+1}C_{2n+1}) + ({}^{2n+1}C_1 + {}^{2n+1}C_{2n}) + \dots$$

$$\dots + ({}^{2n+1}C_n + {}^{2n+1}C_{n+1})$$

$$= 2^{2n+1} \Rightarrow E = 2^{2n}$$

$$\therefore \; {}^{2n}C_0 + {}^{2n+1}C_1 + {}^{2n+1}C_2 \dots + {}^{2n+1}C_n$$

$$= {}^{2n+1}C_0 + {}^{2n+1}C_1 + {}^{2n+1}C_2 + \dots + {}^{2n+1}C_n$$

$$= 2^{2n}$$

$$\therefore \; ({}^{2n}C_0 + {}^{2n+1}C_1 + {}^{2n+1}C_2 + \dots + {}^{2n+1}C_n)^{1/2} - 1 = 2^n - 1$$

19. Let the four points represented by z_1, z_2, z_3 and z_4 be A, B, C and D respectively.

Since $ABCD$ is a square.

Mid-point of AC = mid point of BD

$$\Rightarrow \quad \frac{1}{2}(z_1 + z_3) = \frac{1}{2}(z_2 + z_4) \text{ or } z_1 + z_3 = z_2 + z_4$$

Also, $\quad AB = BC = CD = DA$

$$\Rightarrow \quad |z_1 - z_2| = |z_2 - z_3| = |z_3 - z_4| = |z_4 - z_1|$$

Since diagonals of the square $ABCD$ are equal $AC = BD$

or $\quad |z_1 - z_3| = |z_2 - z_4|$

Also, since $AC \perp BD$,

$(z_1 - z_3)/(z_2 - z_4)$ is purely imaginary.

20. $\because 2\cos\theta = x + \dfrac{1}{x}$

$$x^2 - 2x\cos\theta + 1 = 0$$

$$\therefore \; x = \frac{2\cos\theta \pm \sqrt{(4\cos^2\theta - 4)}}{2} = \cos\theta \pm i\sin\theta$$

Let $\quad x = e^{i\theta}$

Similarly $\quad y = e^{i\phi}$

Alternate (a) : $\dfrac{x}{y} + \dfrac{y}{x} = e^{i(\theta-\phi)} + e^{-i(\theta-\phi)}$

$$= 2\cos(\theta - \phi)$$

Alternate (b) : $x^m y^n + \dfrac{1}{x^m y^n} = e^{i(m\theta + n\phi)} + e^{-i(m\theta + n\phi)}$

$$= 2\cos(m\theta + n\phi)$$

Alternate (c) : $\dfrac{x^m}{y^n} + \dfrac{y^n}{x^m} = e^{i(m\theta - n\phi)} + e^{-i(m\theta - n\phi)}$

$$= 2\cos(m\theta - n\phi)$$

Alternate (d) : $xy + \dfrac{1}{xy} = e^{i(\theta+\phi)} + e^{-i(\theta+\phi)}$

$$= 2\cos(\theta + \phi)$$

21. $\because$ Put $x = i$ in the given expression, then

$$C_0 + C_1 i + C_2 i^2 + C_3 i^3 + C_4 i^4 + C_5 i^5 + \dots = (1+i)^n$$

$$= (\sqrt{2})^n \left(\frac{1}{\sqrt{2}} + \frac{i}{\sqrt{2}}\right)^n$$

$$\Rightarrow (C_0 - C_2 + C_4 \dots) + i(C_1 - C_3 + C_5 \dots)$$

$$= 2^{n/2}(\cos\pi/4 + i\sin\pi/4)^n$$

$$= 2^{n/2}\left(\cos\left(\frac{n\pi}{4}\right) + i\sin\left(\frac{n\pi}{4}\right)\right)$$

And also put $x = 1, \omega, \omega^2$ in (i) and then addiing, we get

$$3(C_0 + C_3 + C_6 + \dots) = 2^n + (1+\omega)^n + (1+\omega^2)^n$$

$$= 2^n + (-\omega^2)^n + (-\omega)^n$$

$$= 2^n + (-1)^n\{e^{4\pi ni/3} + e^{2\pi ni/3}\}$$

$$= 2^n + (-1)^n \cdot e^{n\pi i} \cdot 2\cos\left(\frac{n\pi}{3}\right)$$

$$= 2^n + 2\cos\left(\frac{n\pi}{3}\right)$$

$$\therefore \quad C_0 + C_3 + C_6 + \dots = \frac{1}{3}\left(2^n + 2\cos\left(\frac{n\pi}{3}\right)\right)$$

On comparing real and imaginary parts, then we get

$$C_0 - C_2 + C_4 \dots\dots = 2^{n/2}\cos\left(\frac{n\pi}{4}\right) \qquad \dots\text{(i)}$$

$$\text{and } C_1 - C_3 + C_5 \dots\dots = 2^{n/2}\sin\left(\frac{n\pi}{4}\right) \qquad \dots\text{(ii)}$$

$$\text{Also } C_0 + C_2 + C_4 + C_6 + C_8 + \dots\dots \; 2^{n-1} \qquad \dots\text{(iii)}$$

Adding (i) and (iii), then we get

$$2(C_0 + C_4 + C_8 \dots\dots) = 2^{n-1} + 2^{n/2}\cos\left(\frac{n\pi}{4}\right)$$

$$C_0 + C_4 + C_8 \dots\dots = 2^{n-2} + 2^{(n-2)/2}\cos\left(\frac{n\pi}{4}\right)$$

Linked Comprehension Type

PASSAGE 1

Sol. (Q.No. 1-3)

We have, $|z-(4/z)|=2$

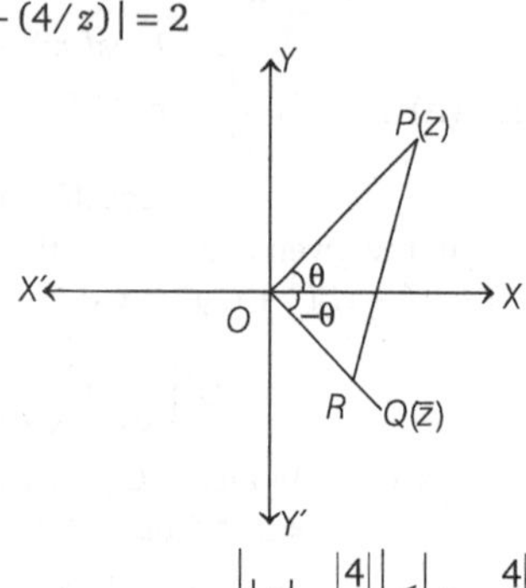

$$\Rightarrow \quad \left||z|-\left|\frac{4}{z}\right|\right| \le \left|z-\frac{4}{z}\right| = 2$$

$$\Rightarrow \quad -2 \le |z| - \frac{4}{|z|} \le 2$$

$$\Rightarrow \quad |z|^2 + 2|z| - 4 \ge 0$$

and $\quad |z|^2 - 2|z| - 4 \le 0$

$$\Rightarrow \quad (|z|+1)^2 - 5 \ge 0$$

and $\quad (|z|-1)^2 \ge 5$

$$\Rightarrow \quad (|z|+1+\sqrt{5})(|z|+1-\sqrt{5}) \le 0$$

$$\Rightarrow \quad |z| \ge \sqrt{5}-1$$

and $(|z|-1+\sqrt{5})\times(|z|-1-\sqrt{5}) \le 0$

$$\Rightarrow \quad \sqrt{5}-1 \le |z| \le \sqrt{5}+1$$

$$\Rightarrow \quad \sqrt{5}-1 \le |z| \le \sqrt{5}+1$$

$\therefore$ The least modulus is $\sqrt{5}-1$ and the greatest modulus is $\sqrt{5}+1$.

and also $|z| + \sqrt{5}+1$

$$\Rightarrow \quad \frac{4}{|z|} = \sqrt{5}-1$$

Now, $\frac{4}{z} = \frac{4\bar{z}}{|z|^2}$

$\Rightarrow 4/z$ lies in the direction of $\bar{z}$.

$$\left|z-\frac{4}{z}\right| = PR = 2 \qquad \text{[given]}$$

Now, we have

$$OP = \sqrt{5}+1 \text{ and } OR = \sqrt{5}-1$$

$$\Rightarrow \cos 2\theta = \frac{OP^2 + OR^2 - PR^2}{2\,OP \cdot OR}$$

$$= \frac{(\sqrt{5}+1)^2 + (\sqrt{5}-1)^2 - 4}{2(5-1)} = 1$$

$$\Rightarrow \quad 2\theta = 0, 2\pi$$

$$\Rightarrow \quad \theta = 0, \pi$$

$\Rightarrow z$ is purely real

$$\Rightarrow \quad z = \pm(\sqrt{5}+1)$$

Similarly for $|z| = \sqrt{5}-1$, we have

$$z = \pm(\sqrt{5}-1)$$

1. (a) 2. (b) 3. (b)

PASSAGE 2

1. Let $z = x+iy$

A = set of points on and above the line $y=1$ in the agrand plane

B = set of points on the circle $(x-2)^2+(y-1)^2=3^2$

C = set of the point lies on the line $x+y=\sqrt{2}$

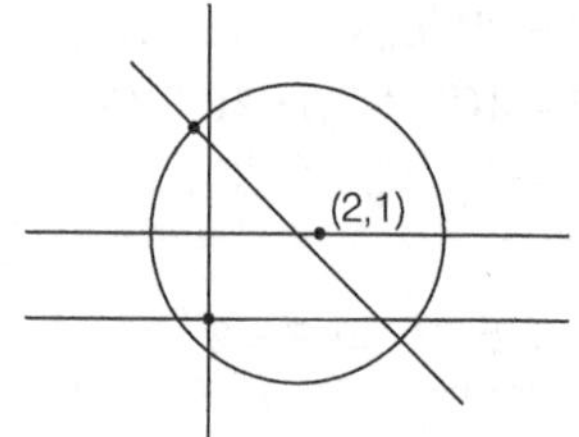

Hence, there is only one element in the set $A \cap B \cap C$

2. B denotes the circle with radius 3.

Area of B is $\pi r^2 = \pi(3)^2 = 9\pi$

3. Given, $|w-2-i|<3$

$|w-(2+i)|<3$

$$-3 < |w| - |2+i| < 3$$

$$-3 < |w| - \sqrt{5} < 3$$

$$-3+\sqrt{5} < |w| < 3+\sqrt{5}$$

$$-(3+\sqrt{5}) < -|w| < -(-3+\sqrt{5})$$

$$-3-\sqrt{5} < -|w| < 3-\sqrt{5}$$

$|z-2-i|=3$

$|z-(2+i)|=3$

$|z|-|2+i|<3$

$$-3 < |z| - \sqrt{5} < 3$$

$$-3+\sqrt{5} < |z| < 3+\sqrt{5}$$

$$-6 < |z| - |w| < 6$$

$$-6+3 < |z| - |w|^{+3} < 6+3$$

$$-3 < |z| - |w| + 3 < 9$$

PASSAGE 3

1. $\sum_{n=1}^{16}(i^n + i^{n+1}) = \sum_{n=1}^{16} i^n + \sum_{n=1}^{16} i^{n+1} = 0+0 = 0$

2. $(1+i)^{n_1} + (1+i^3)^{n_1} + (1+i^5)^{n_2} + (1+i^7)^{n_2}$

$$= (1+i)^{n_1} + (1-i)^{n_1} + (1+i)^{n_2} + (1-i)^{n_2}$$

$$= (\sqrt{2})^{n_1}\left\{\left(\frac{1}{\sqrt{2}}+\frac{i}{\sqrt{2}}\right)^{n_1} + \left(\frac{1}{\sqrt{2}}-\frac{i}{\sqrt{2}}\right)^{n_1}\right\}$$

$$+ (\sqrt{2})^{n_2}\left\{\left(\frac{1}{\sqrt{2}}+\frac{i}{\sqrt{2}}\right)^{n_2} + \left(\frac{1}{\sqrt{2}}-\frac{i}{\sqrt{2}}\right)^{n_2}\right\}$$

$$= (\sqrt{2})^{n_1}\left\{\left(\cos\frac{\pi}{4}+i\sin\frac{\pi}{4}\right)^{n_1} + \left(\cos\frac{\pi}{4}-i\sin\frac{\pi}{4}\right)^{n_1}\right\}$$

$$+ (\sqrt{2})^{n_2}\left\{\left(\cos\frac{\pi}{4}+i\sin\frac{\pi}{4}\right)^{n_2} + \left(\cos\frac{\pi}{4}-i\sin\frac{\pi}{4}\right)^{n_2}\right\}$$

$$= (\sqrt{2})^{n_1}\left\{e^{\frac{n_1\pi}{4}i} + e^{\frac{-n_1\pi i}{4}}\right\} + (\sqrt{2})^{n_2}\left\{e^{\frac{n_2\pi i}{4}} + e^{\frac{-n_2\pi i}{4}}\right\}$$

$$= (\sqrt{2})^{n_1}\cdot 2\cos\left(\frac{n_1\pi}{4}\right) + (\sqrt{2})^{n_2}\cdot 2\cos\left(\frac{n_2\pi}{4}\right)$$

$$= (2)^{(n_1/2)+1}\cdot\cos\left(\frac{n_1\pi}{4}\right) + (2)^{(n_2/2)+1}\cdot\cos\left(\frac{n_2\pi}{4}\right)$$

= Real (given)

$\therefore \quad n_1 > 0, n_2 > 0$

3. Applying $R_2 \to R_2 - (R_1 + R_3)$, then

$$\begin{vmatrix} 1 & 1+i+\omega^2 & \omega^2 \\ 0 & 0 & 0 \\ -i & -i+\omega-1 & -1 \end{vmatrix} = 0$$

4. Applying $C_1 \to C_1 + C_2 + C_3$, then $\begin{vmatrix} 3 & 1 & 1 \\ 0 & \omega & \omega^2 \\ 0 & \omega^2 & \omega \end{vmatrix}$

$$= 3(\omega^2 - \omega^4) = 3\omega(\omega - 1)$$

5. $\because 4 + 5\left(-\frac{1}{2} + \frac{i\sqrt{3}}{2}\right)^{334} + 3\left(-\frac{1}{2} + \frac{i\sqrt{3}}{2}\right)^{365}$

$$= 4 + 5\omega^{334} + 3\omega^{365}$$
$$= 4 + 5\omega + 3\omega^2$$
$$= 1 + 2\omega + 3(1 + \omega + \omega^2)$$
$$= 1 + 2\left(\frac{-1 + i\sqrt{3}}{2}\right) + 0$$
$$= 1 - 1 + i\sqrt{3} = i\sqrt{3}$$

6. $\because \frac{z_1 - z_3}{z_2 - z_3} = -\left(\frac{-1 + i\sqrt{3}}{2}\right) = -\omega$ ...(i)

$\therefore \quad \left|\frac{z_1 - z_3}{z_2 - z_3}\right| = |-\omega| = 1$

$\Rightarrow \quad \frac{|z_1 - z_3|}{|z_2 - z_3|} - 1$

$\therefore \quad |z_1 - z_3| = |z_2 - z_3|$ ...(ii)

Also, from Eq. (i),

$$\frac{z_1 - z_3}{z_2 - z_3} = 1 + \omega^2$$

$\Rightarrow \quad \frac{z_1 - z_3}{z_2 - z_3} - 1 = \omega^2$

$\Rightarrow \quad \frac{z_1 - z_2}{z_2 - z_3} = \omega^2$

$\therefore \quad \left|\frac{z_1 - z_2}{z_2 - z_3}\right| = |\omega^2| = 1$

$\Rightarrow \quad \frac{|z_1 - z_2|}{|z_2 - z_3|} = 1$

$\therefore \quad |z_1 - z_2| = |z_2 - z_3|$ ...(iii)

From Eqs. (ii) and (iii), we get

$$|z_1 - z_2| = |z_2 - z_3| = |z_1 - z_3|$$

Hence, triangle is equilateral.

Alternate Method

$\because$ Given,

$$\frac{z_1 - z_3}{z_2 - z_3} = \frac{1 - i\sqrt{3}}{2} = e^{-i\pi/3}$$

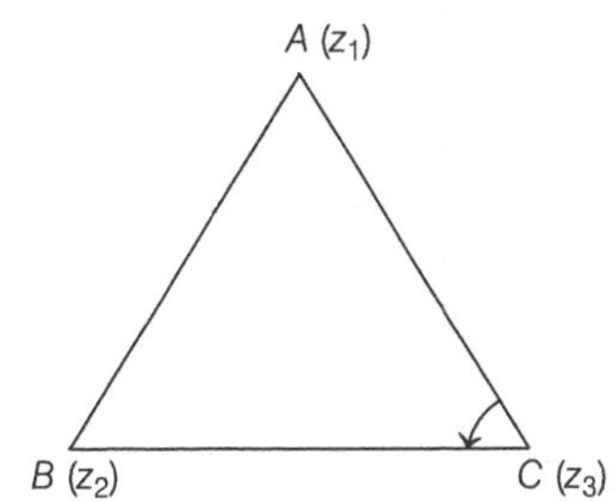

$\therefore \quad \frac{z_2 - z_3}{z_1 - z_3} = e^{i\pi/3} = \frac{BC}{AC}\, e^{i\pi/3}$

$\therefore \quad \angle C = \pi/3$ and $AC = BC$,

Then, $\quad \angle A = \angle B = \pi/3$

Hence, ΔABC is equilateral

7. $\sum_{r=2}^{n}(r-1)(r-\omega)(r-\omega^2) = \sum_{r=2}^{n}(r-1)(r^2+r+1)$

$$= \sum_{r=2}^{n}(r^3 - 1^3)$$
$$= \sum_{r=2}^{n}(r^3 - 1)$$
$$= \sum_{r=2}^{n} r^3 - \sum_{r=2}^{n} 1$$
$$= \Sigma n^3 - n = (\Sigma n)^2 - n$$

8. $\because \left(\frac{1+i}{1-i}\right)^n = 1 \Rightarrow \left(\frac{(1+i)^2}{(1-i^2)}\right)^n = 1 \Rightarrow \left(\frac{1+i^2+2i}{2}\right)^n = 1$

$\Rightarrow \quad (i)^n = 1$

$\Rightarrow \quad (\sqrt{-1})^n = 1$

$\Rightarrow \quad (-1)^{n/2} = (-1)^2$

$\therefore \quad \frac{n}{2} = 2 \Rightarrow n = 4$

9. $x^3 - 3x^2 + 3x + 7 = 0$

$\Rightarrow \quad (x-1)^3 + 8 = 0$

$\Rightarrow \quad (x-1)^3 = (-2)^3$

$\Rightarrow \quad \left(\frac{x-1}{-2}\right)^3 = 1 \Rightarrow \frac{x-1}{-2} = (1)^{1/3} = 1, \omega, \omega^2$

$\Rightarrow \quad x - 1 = -2, -2\omega, -2\omega^2$

$\therefore \quad x = -1, 1 - 2\omega, 1 - 2\omega^2$

Then, $\alpha = -1, \beta = 1 - 2\omega, \gamma = 1 - 2\omega^2$

$$\Sigma\left(\frac{\alpha-1}{\beta-1}\right) = \left(\frac{\alpha-1}{\beta-1}\right) + \left(\frac{\beta-1}{\gamma-1}\right) + \left(\frac{\gamma-1}{\alpha-1}\right)$$
$$= \frac{-2}{-2\omega} + \frac{-2\omega}{-2\omega^2} + \frac{-2\omega^2}{-2}$$
$$= \frac{2}{\omega} + \omega^2 = \frac{2}{\omega} + \frac{1}{\omega} = \frac{3}{\omega}$$

10. $\left(\frac{3}{2}+\frac{i\sqrt{3}}{2}\right)^{50}=\left(\sqrt{3}\left(\frac{\sqrt{3}+i}{2}\right)\right)^{50}=3^{25}\left(\frac{-1+i\sqrt{3}}{2i}\right)^{50}$

$$=3^{25}\left(\frac{\omega}{i}\right)^{50}=3^{25}\frac{\omega^2}{i^2}=-3^{25}\left(\frac{-1-i\sqrt{3}}{2}\right)$$

$$=3^{25}\left(\frac{1}{2}+\frac{i\sqrt{3}}{2}\right)=3^{25}(x-iy) \text{ (given)}$$

$\therefore \quad x=\frac{1}{2}, y=-\frac{\sqrt{3}}{2}$

Then, $(x, y)\equiv\left(\frac{1}{2}, \frac{-\sqrt{3}}{2}\right)$

11. $\because (z+\alpha\beta)^3=\alpha^3$

$\Rightarrow \quad \left(\frac{z+\alpha\beta}{\alpha}\right)^3=1$

$\Rightarrow \quad \left(\frac{z+\alpha\beta}{\alpha}\right)=(1)^{1/3}=1, \omega, \omega^2$

$\Rightarrow \quad z+\alpha\beta=\alpha, \alpha\omega, \alpha\omega^2$

$\therefore \quad z=\alpha-\alpha\beta, \alpha\omega-\alpha\beta, \alpha\omega^2-\alpha\beta$

$\therefore \quad AB=|(\alpha-\alpha\beta)-(\alpha\omega-\alpha\beta)|$

A $(\alpha-\alpha\beta)$

B $(\alpha\omega-\alpha\beta)$ C $(\alpha\omega^2-\alpha\beta)$

$$=|\alpha-\alpha\omega|$$
$$=|\alpha||1-\omega|$$
$$=\sqrt{3}|\alpha|$$

$BC=|(\alpha\omega-\alpha\beta)-(\alpha\omega^2-\alpha\beta)|=|\alpha\omega-\alpha\omega^2|=|\alpha||\omega||1-\omega|$

$$=|\alpha|\cdot 1\sqrt{3}=\sqrt{3}|\alpha|$$

and $CA=|(\alpha-\alpha\beta)-(\alpha\omega^2-\alpha\beta)|=|\alpha-\alpha\omega^2|=|\alpha||1-\omega^2|$

$$=\sqrt{3}|\alpha|$$

Hence, $AB=BC=CA=\sqrt{3}|\alpha|$

12. $\because z=\frac{\sqrt{3}+i}{2}=\frac{i^2+i\sqrt{3}}{2i}=\frac{1}{i}\left(\frac{-1+i\sqrt{3}}{2}\right)=\frac{\omega}{i}$

$\therefore \quad z^{101}=\frac{\omega^{101}}{i^{101}}=\frac{\omega^2}{i}=-i\omega^2$

and $\quad i^{103}=i^3=-i$

$\therefore \quad z^{101}+i^{103}=-i\omega^2-i$

$$=-i(\omega^2+1)$$
$$=-i(-\omega)=i\omega$$

$\Rightarrow \quad (z^{101}+i^{103})^{105}=(i\omega)^{105}=i^{105}\omega^{105}=i=z^3$

$$\left[\because z^3=\frac{\omega^3}{i^3}=-\frac{1}{i}=i\right]$$

PASSAGE 4

1. $\because \quad |z_1|=|z_2|=|z_3|=|z_1+z_2+z_3|=1$

$\therefore \quad |z_1|^2=1, |z_2|^2=1, |z_3|^2=1$

$\Rightarrow \quad z_1\bar{z}_1=1, z_2\bar{z}_2=1, z_3\bar{z}_3=1$

$$z_1=\frac{1}{\bar{z}_1}, z_2=\frac{1}{\bar{z}_2}, z_3=\frac{1}{\bar{z}_3}$$

Now, $|z_1+z_2+z_3|=1$

$\Rightarrow \quad \left|\frac{1}{\bar{z}_1}+\frac{1}{\bar{z}_2}+\frac{1}{\bar{z}_3}\right|=1$

$\Rightarrow \quad \left|\overline{\frac{1}{z_1}+\frac{1}{z_2}+\frac{1}{z_3}}\right|=1$

or $\quad \left|\frac{1}{z_1}+\frac{1}{z_2}+\frac{1}{z_3}\right|=1$

2. $\because \quad |z_1|=1, |z_2|=2, |z_3|=3$

$\therefore \quad |z_1|^2=1, |z_2|^2=4, |z_3|^2=9$

$\Rightarrow \quad z_1\bar{z}_1=1, z_2\bar{z}_2=4, z_3\bar{z}_3=9$

$\therefore \quad \frac{1}{z_1}=\bar{z}_1, \frac{4}{z_2}=\bar{z}_2, \frac{9}{z_3}=\bar{z}_3$

Now, $|9z_1z_2+4z_3z_1+z_2z_3|$

$$=\left|z_1z_2z_3\left(\frac{9}{z_3}+\frac{4}{z_2}+\frac{1}{z_1}\right)\right|=|z_1||z_2||z_3|\left|\frac{9}{z_3}+\frac{4}{z_2}+\frac{1}{z_1}\right|$$

$$=|z_1||z_2||z_3||\bar{z}_1+\bar{z}_2+\bar{z}_3|$$
$$=|z_1||z_2||z_3||\overline{z_1+z_2+z_3}|$$
$$=|z_1||z_2||z_3||z_1+z_2+z_3|=1\cdot 2\cdot 3\cdot 1=6$$

$$[\because |z_1+z_2+z_3|=1]$$

3. $\because \quad |z_1|=|z_2|=|z_3|=\ldots=|z_n|=1$

$\Rightarrow \quad |z_1|^2=|z_2|^2=|z_3|^2=\ldots=|z_n|^2=1$

$\therefore \quad z_1\bar{z}_1=z_2\bar{z}_2=z_3\bar{z}_3=\ldots=z_n\bar{z}_n=1$

$\Rightarrow \quad z_1=\frac{1}{\bar{z}_1}, z_2=\frac{1}{\bar{z}_2}, z_3=\frac{1}{\bar{z}_3}, \ldots, z_n=\frac{1}{\bar{z}_n}$

$\therefore |z_1+z_2+z_3+\ldots+z_n|=\left|\frac{1}{\bar{z}_1}+\frac{1}{\bar{z}_2}+\frac{1}{\bar{z}_3}+\ldots+\frac{1}{\bar{z}_n}\right|$

$$=\left|\overline{\frac{1}{z_1}+\frac{1}{z_2}+\frac{1}{z_3}+\ldots+\frac{1}{z_n}}\right|$$

$$=\left|\frac{1}{z_1}+\frac{1}{z_2}+\frac{1}{z_3}+\ldots+\frac{1}{z_n}\right|$$

4. $\because |z|=1$

$\therefore \quad z=e^{i\alpha}=\cos\alpha+i\sin\alpha=x+iy$ (given)

$\Rightarrow \quad x=\cos\alpha, y=\sin\alpha$

$\because$ x and y are rational

$\therefore \cos\alpha, \sin\alpha$ are also rational

$\therefore |z^{2n}-1|=|e^{2ni\alpha}-1|=|e^{ni\alpha}(e^{ni\alpha}-e^{-ni\alpha})|$

$$=|e^{ni\alpha}||(e^{ni\alpha}-e^{-ni\alpha}|$$
$$=1.|2i\sin n\alpha|$$
$$=2|\sin n\alpha|$$

[$\because \sin\alpha, \cos\alpha$ are rational, then $\sin n\alpha$ is also rational]

= Rational

5. $\left|\dfrac{a-ib}{a+ib}\right| = \dfrac{|a-i\,b|}{|a+i\,b|} = \dfrac{\sqrt{(a^2+b^2)}}{\sqrt{(a^2+b^2)}} = 1$

$\therefore$ Let $\dfrac{a-ib}{a+ib} = e^{i\alpha} \Rightarrow \dfrac{(a-ib)^2}{a^2+b^2} = e^{i\alpha}$

$\Rightarrow \quad \left(\dfrac{a^2-b^2}{a^2+b^2}\right) - i\left(\dfrac{2ab}{a^2+b^2}\right) = e^{i\alpha}$

$\Rightarrow \quad \ln\left(\dfrac{a-ib}{a+ib}\right) = i\alpha$

$\therefore \tan\left(i\ln\left(\dfrac{a-ib}{a+ib}\right)\right) = \tan(i\cdot i\alpha) = \tan(-\alpha)$

$= -\tan\alpha = -\dfrac{\sin\alpha}{\cos\alpha}$

$= -\left\{\dfrac{\dfrac{-2ab}{a^2+b^2}}{\dfrac{a^2-b^2}{a^2+b^2}}\right\} = \dfrac{2ab}{(a^2-b^2)}$

6. $\because \; |z-3i| = 3$

$\therefore \quad z - 3i = 3e^{i\alpha}$

$\Rightarrow \quad z = 3i + 3e^{i\alpha} = 3(e^{i\pi/2} + e^{i\alpha})$

$= 3e^{i\left(\frac{\pi}{4}+\frac{\alpha}{2}\right)} \cdot 2\cos\left(\dfrac{\pi}{4} - \dfrac{\alpha}{2}\right)$

$\therefore \; \arg(z) = \left(\dfrac{\pi}{4} + \dfrac{\alpha}{2}\right)$

Now, $\cot(\arg z) - \dfrac{6}{z} = \cot\left(\dfrac{\pi}{4}+\dfrac{\alpha}{2}\right) - \dfrac{6}{6e^{i\left(\frac{\pi}{4}+\frac{\alpha}{2}\right)}\cos\left(\dfrac{\pi}{4}-\dfrac{\alpha}{2}\right)}$

$= \cot\left(\dfrac{\pi}{4}+\dfrac{\alpha}{2}\right) - \dfrac{e^{-i\left(\frac{\pi}{4}+\frac{\alpha}{2}\right)}}{\cos\left(\dfrac{\pi}{2} - \left(\dfrac{\pi}{4}+\dfrac{\alpha}{2}\right)\right)}$

$= \cot\left(\dfrac{\pi}{4}+\dfrac{\alpha}{2}\right) - \dfrac{\left\{\cos\left(\dfrac{\pi}{4}+\dfrac{\alpha}{2}\right) - i\sin\left(\dfrac{\pi}{4}+\dfrac{\alpha}{2}\right)\right\}}{\left\{\sin\left(\dfrac{\pi}{4}+\dfrac{\alpha}{2}\right)\right\}}$

$= \cot\left(\dfrac{\pi}{4}+\dfrac{\alpha}{2}\right) - \cot\left(\dfrac{\pi}{4}+\dfrac{\alpha}{2}\right) + i$

$= i$

7. $\because \quad \left|\dfrac{z_1+z_2}{z_1-z_2}\right| = 1 \Rightarrow \dfrac{z_1+z_2}{z_1-z_2} = \dfrac{e^{i\alpha}}{1}$

$\Rightarrow \quad \dfrac{2z_1}{2z_2} = \dfrac{e^{i\alpha}+1}{e^{i\alpha}-1}$

$= \dfrac{e^{i\alpha/2} + e^{-i\alpha/2}}{e^{i\alpha/2} - e^{-i\alpha/2}} = \dfrac{2\cos(\alpha/2)}{2i\sin(\alpha/2)}$

$\dfrac{z_1}{z_2} = -i\cot(\alpha/2)$

= Purely imaginary [$\because \cot(\alpha/2) \neq 0$ (given)]

PASSAGE 5

1. $4 - 3i + \alpha = \dfrac{7+3i}{1+i}$

$= \dfrac{(7+3i)(1-i)}{(1+i)(1-i)}$

$= \dfrac{10-4i}{2} = 5 - 2i$

$\therefore \quad \alpha = 1 + i$

2. $\because z^2 + az + b = 0$ ...(i)

$\overline{z^2} + \bar{a}\bar{z} + \bar{b} = 0$

$\because$ For purely imaginary roots $\bar{z} = -z$,

then, $\quad z^2 - \bar{a}z + \bar{b} = 0$ ...(ii)

From Eqs. (i) and (ii), we get

z^2	z	1	z^2	z
1	a	b	1	a
1	$-\bar{a}$	$\bar{b}$	1	$\bar{a}$

$\therefore \quad \dfrac{z^2}{(a\bar{b}+\bar{a}b)} = \dfrac{z}{(b-\bar{b})} = \dfrac{1}{(-\bar{a}-a)}$

from first two members, $z = \dfrac{(a\bar{b}+\bar{a}b)}{(a-\bar{b}}$

and from last two members, $z = -\dfrac{(b-\bar{b})}{(a+\bar{a})}$

$\therefore \quad \dfrac{(a\bar{b}+\bar{a}b)}{(b-\bar{b})} = -\dfrac{(b-\bar{b})}{(a+\bar{a})}$

$\Rightarrow \quad (a\bar{b}+\bar{a}b)(a+\bar{a}) + (b-\bar{b})^2 = 0$

$\Rightarrow \quad (a+\bar{a})(a\bar{b}+\bar{a}b) + (b-\bar{b})^2 = 0$

3. $\because z^2 + az + b = 0$ ...(i)

$\overline{z^2} + \overline{az} + \bar{b} = 0$

$\therefore \quad (\bar{z})^2 + \bar{a}\bar{z} + \bar{b} = 0$

For purely real roots $\bar{z} = z$,

Then, $\quad z^2 + \bar{a}z + \bar{b} = 0$ ...(ii)

From Eqs. (i) and (ii), we get

z^2	z	1	z^2	z
1	a	b	1	a
1	$\bar{a}$	$\bar{b}$	1	$\bar{a}$

$\therefore \quad \dfrac{z^2}{(a\bar{b}-\bar{a}b)} = \dfrac{z}{(b-\bar{b})} = \dfrac{1}{(\bar{a}-a)}$

From first two members

$z = \left(\dfrac{a\bar{b}-\bar{a}b}{b-\bar{b}}\right)$

and from last two members

$z = \dfrac{(b-\bar{b})}{(\bar{a}-a)}$

Hence, $\quad \dfrac{(a\bar{b}-\bar{a}b)}{(b-\bar{b})} = \dfrac{(b-\bar{b})}{(\bar{a}-a)}$

4. $\because az^2 + bz + c = 0$ ...(i)

has both real roots, then $\bar{z} = z$

From Eq. (i), $\overline{az^2 + bz + c} = \bar{0}$

$\Rightarrow \bar{a}(\bar{z})^2 + \bar{b}\bar{z} + \bar{c} = 0$

$\Rightarrow \bar{a}z^2 + \bar{b}z + \bar{c} = 0$ ...(ii)

Eqs. (i) and (ii) are identical

$\therefore \frac{a}{\bar{a}} = \frac{b}{\bar{b}} = \frac{c}{\bar{c}}$

5. $\therefore az^2 + bz + c = 0$ ...(i)

has both purely imaginary roots, then $\bar{z} = -z$

From Eqs. (i),

$\overline{az^2 + bz + c} = \bar{0}$

$\bar{a}(\bar{z})^2 + \bar{b}\bar{z} + \bar{c} = 0$

$\bar{a}(-z)^2 + \bar{b}(-z) + \bar{c} = 0$

$\Rightarrow \bar{a}z^2 - \bar{b}z + \bar{c} = 0$ ...(ii)

Eqs. (i) and (ii) are identical

$\therefore \frac{a}{\bar{a}} = -\frac{b}{\bar{b}} = \frac{c}{\bar{c}}$

6. $\because z^2 + 2z + 3 = 0$

$\Rightarrow (z+1)^2 + 2 = 0$

roots are complex and given $a, b, c \in R$

$\therefore$ Roots are conjugate

Hence, both roots are common

$\therefore \frac{a}{1} = \frac{b}{2} = \frac{c}{3}$

$\Rightarrow a : b : c = 1 : 2 : 3$

7. Let other root is β

$\therefore \beta + \gamma = -\alpha \Rightarrow \beta = -\gamma - \alpha$

and $\beta\gamma = \bar{\alpha} \Rightarrow \beta = \frac{\bar{\alpha}}{\gamma}$

or $-\gamma - \alpha = \frac{\bar{\alpha}}{\gamma}$

$\Rightarrow -\gamma = \alpha + \frac{\bar{\alpha}}{\gamma}$

for $\gamma = 1$, $-1 = \alpha + \bar{\alpha}$ which is true $\left[\because \frac{\alpha + \bar{\alpha}}{2} = \text{Re}(\alpha)\right]$

8. $z^2 + (a + ib)z + (c + id) = 0$ ...(i)

Let $z = x \in R$

$\because$ Eq. (i) has real roots.

$x^2 + (a + ib)x + (c + id) = 0 + 0 \cdot i$

On comparing, $x^2 + xa + c = 0$ ...(ii)

and $bx + d = 0$

or $x = -\frac{d}{b}$ ...(iii)

From (ii) and (iii),

$\frac{d^2}{b^2} - \frac{ad}{b} + c = 0$

$\Rightarrow d^2 - abd + b^2c = 0$

$\Rightarrow abd = b^2c + d^2$

PASSAGE 6

1. $\arg(-z) - \arg(z) = -\{\arg(z) - \arg(-z)\}$

$= -(-\pi)$ $(\because \arg(z) < 0)$

$= \pi$

2. $\because \arg(z) + \arg(w) = \pi$

$\Rightarrow \arg(z) - \arg(\bar{w}) = \pi$

$\because |z| = |\bar{w}| = |w|$ $[\because \text{given } |z| = |w|]$

$\therefore z + \bar{w} = 0$

or $z = -\bar{w}$

3. $\because \arg(2z_1) - \arg(3z_2) = \pi$

then $|2z_1| = |3z_2|$ and $2z_1 + 3z_2 = 0$

$\Rightarrow \left|\frac{z_1}{z_2}\right| = 1.5$ and $\frac{z_1}{z_2} = -1.5$

which is true.

Hence, $\left|\frac{z_1}{z_2}\right| = 1.5$

4. $\because \arg(z) > 0$ and $\arg(-z) - \arg(z) = \lambda_1$,

then $-(\arg(z) - \arg(-z)) = \lambda_1$

$\Rightarrow -\pi = \lambda_1$

$\therefore \lambda_1 = -\pi$

and $\because \arg(z) < 0$, $\arg(z) - \arg(-z) = \lambda_2$

$\Rightarrow -\pi = \lambda_2$

$\therefore \lambda_2 = -\pi$

$\Rightarrow \lambda_1 = \lambda_2 \Rightarrow \lambda_1 - \lambda_2 = 0$

5. $\sqrt{\{\arg(z) + \arg(-\bar{z}) - 2\pi\}\{\arg(-z) + \arg(\bar{z})\}}$

$= \sqrt{\{\arg(z) + \pi - \arg(z) - 2\pi\}\{\arg(-z) - \arg(z)\}}$

$= \sqrt{(-\pi)\{-(\arg z - \arg(-z)\}}$ $(\because \arg(z) > 0)$

$= \sqrt{(-\pi)(-\pi)}$

$= \pi$

PASSAGE 7

1. $\because z_1 = 2 + 5i$, $z_2 = 3 - i$

$\therefore z_1 \cdot z_2 = \text{Re}(z_2\bar{z}_1) = \text{Re}(3 - i)(2 - 5i) = 1$

and $z_2 \times z_1 = (3)(5) - (2)(-1) = 17$

$\therefore \sqrt{(z_1 \cdot z_2) + (z_2 \times z_1)} = \sqrt{(1 + 17)} = 3\sqrt{2}$

2. $\because z_1 = 3 + 4i$, $z_2 = 4 + 3i$

$\sin\theta = \frac{z_1 \times z_2}{|z_1||z_2|} = \frac{(3)(3) - (4)(4)}{5 \cdot 5} = -\frac{7}{25}$

3. $\therefore z_1 = 5 + 12i$, $z_2 = 3 + 4i$

$\therefore z_1 \cdot z_2 = (5)(3) + (12)(4) = 63$

and $z_2 \cdot z_1 = (5)(3) + (12)(4) = 63$

and $|z_1| = 13, |z_2| = 5$

(Projection of z_1 on z_2 + Projection of z_2 on z_1)

$= \frac{z_1 \cdot z_2}{|z_2|} + \frac{z_2 \cdot z_1}{|z_1|}$

$= \frac{63}{5} + \frac{63}{13}$

$= \frac{63 \times 18}{65} = \frac{1134}{65}$

4. $\because AB = BC = CA$ and $0 < a, b < 1$

By Rotation theorem,

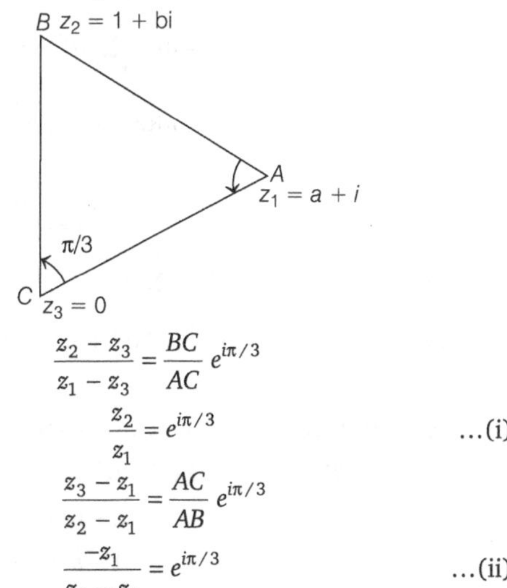

$$\frac{z_2 - z_3}{z_1 - z_3} = \frac{BC}{AC} e^{i\pi/3}$$

$$\frac{z_2}{z_1} = e^{i\pi/3} \quad \ldots(i)$$

and $$\frac{z_3 - z_1}{z_2 - z_1} = \frac{AC}{AB} e^{i\pi/3}$$

$$\Rightarrow \frac{-z_1}{z_2 - z_1} = e^{i\pi/3} \quad \ldots(ii)$$

From Eqs. (i) and (ii), we get

$$\frac{z_2}{z_1} = -\frac{z_1}{z_2 - z_1}$$

$$\Rightarrow z_2^2 - z_1 z_2 = -z_1^2 \Rightarrow z_1^2 + z_2^2 = z_1 z_2$$

$$\Rightarrow (a + i)^2 + (1 + bi)^2 = (a + i)(1 + bi)$$

$$\Rightarrow (a^2 - 1 + 1 - b^2) + i(2a + 2b) = (a - b) + i(ab + 1)$$

comparing real and imaginary parts

$\therefore \quad a^2 - b^2 = a - b$ and $2(a + b) = ab + 1$

$\Rightarrow (a - b)(a + b - 1) = 0$ and $2(a + b) = ab + 1$

If $a = b$, then $2(2a) = a^2 + 1$

$$\Rightarrow a^2 - 4a + 1 = 0$$

$$\therefore \quad a = \frac{4 \pm \sqrt{16 - 4}}{2} = 2 \pm \sqrt{3}$$

$$\therefore \quad a = b = 2 - \sqrt{3} \quad [\because 0 < a, b < 1]$$

Also, if $a + b = 1$, then $2 = a(1 - a) + 1$

$\Rightarrow a^2 - a + 1 = 0$

$$a = \frac{1 \pm \sqrt{-3}}{2} \text{ (impossible)}$$

Hence, $a = b = 2 - \sqrt{3}$

$\therefore \quad (a, b) \equiv (2 - \sqrt{3}, 2 - \sqrt{3})$

5. $\because z_1$ and z_2 be roots of the equation

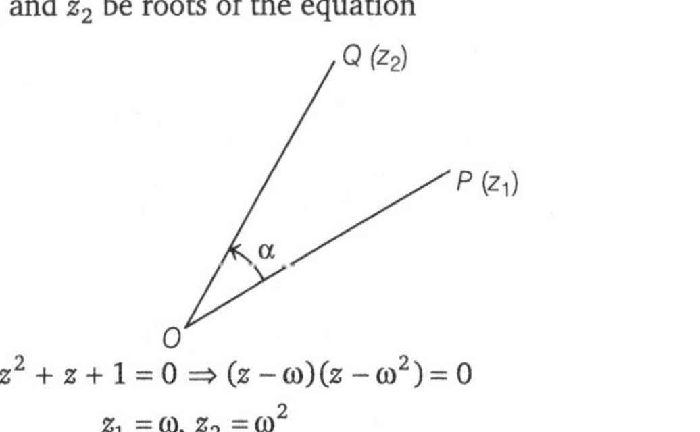

$$z^2 + z + 1 = 0 \Rightarrow (z - \omega)(z - \omega^2) = 0$$

$\therefore \quad z_1 = \omega, z_2 = \omega^2$

$\therefore \quad OP = OQ$

$$\therefore \quad \frac{z_2 - 0}{z_1 - 0} = \frac{OQ}{OP} e^{i\alpha}$$

$$\frac{z_2}{z_1} = e^{i\alpha} \Rightarrow \omega = e^{i\alpha} \Rightarrow e^{2\pi i/3} = e^{i\alpha}$$

$$\therefore \quad \alpha = \frac{2\pi}{3}$$

PASSAGE 8

1. $\because z^n - 1 = \prod_{r=0}^{n-1} (z - \alpha^r)$

$$\Rightarrow \ln(z^n - 1) = \sum_{r=0}^{n-1} \ln(z - \alpha^r)$$

Differentiate both sides w.r.t. z, then

$$\Rightarrow \frac{nz^{n-1}}{z^n - 1} = \sum_{r=0}^{n-1} \frac{1}{(z - \alpha^r)}$$

Put $z = 2$, then

$$\frac{n(2)^{n-1}}{2^n - 1} = \sum_{r=0}^{n-1} \frac{1}{(2 - \alpha^r)} = \frac{1}{1} + \sum_{r=1}^{n-1} \frac{1}{(2 - \alpha^r)}$$

$$\therefore \quad \sum_{r=1}^{n-1} \frac{1}{(2 - \alpha^r)} = \frac{n(2)^{n-1}}{2^n - 1} - 1$$

$$= \frac{n(2)^{n-1} - 2^n + 1}{(2^n - 1)} = \frac{(n - 2)2^{n-1} + 1}{(2^n - 1)}$$

2. $\because z^n - 1 = \prod_{r=0}^{n-1} (z - \alpha^r)$

$$(z^n - 1) = (z - 1) \prod_{r=1}^{n-1} (z - \alpha^r)$$

or $$\frac{(z^n - 1)}{(z - 1)} = \prod_{r=1}^{n-1} (z - \alpha^r) \quad \ldots(i)$$

Put $n = 5$ and $z = -\omega$, then

$$\frac{-\omega^5 - 1}{-\omega - 1} = \prod_{r=1}^{4} (-\omega - \alpha^r)$$

$$\frac{\omega^2 + 1}{\omega + 1} = \prod_{r=1}^{4} (\omega + \alpha^r) \text{ or } \frac{-\omega}{-\omega^2} = \prod_{p=1}^{4} (\omega + \alpha^p)$$

i.e. $$\prod_{p=1}^{4} (\omega + \alpha^p) = \frac{1}{\omega} \quad \ldots(ii)$$

Also in Eq. (i) put $z = \omega^2$, $n = 8$, $r = q$ then

$$\frac{\omega^{16} - 1}{\omega^2 - 1} = \prod_{q=1}^{7} (\omega^2 - \alpha^q)$$

$$\Rightarrow \frac{1}{\omega + 1} = \prod_{q=1}^{7} (\omega^2 - \alpha^q)$$

$$\Rightarrow -\frac{1}{\omega^2} = \prod_{q=1}^{7} (\omega^2 - \alpha^q) \quad \ldots(iii)$$

From Eqs. (ii) and (iii), we get

$$\frac{\prod_{p=1}^{4} (\omega + \alpha^p)}{\prod_{q=1}^{7} (\omega^2 - \alpha^q)} = \frac{\frac{1}{\omega}}{-\frac{1}{\omega^2}} = -\omega = -\left(\frac{-1 + i\sqrt{3}}{2}\right)$$

$$= \frac{1 - i\sqrt{3}}{2}$$

3. According to question

$$\sum_{r=0}^{n-1} \frac{\bar{a}(\alpha)^r + a(\bar{\alpha})^r + b}{|\bar{a}| + |a|}$$

$$\Rightarrow \quad \sum_{r=0}^{n-1} \frac{\bar{a}(\alpha)^r + a(\bar{\alpha})^r + b}{2|a|}$$

$$\Rightarrow \quad \frac{1}{2|a|}\left\{\bar{a}\sum_{r=0}^{n-1}(\alpha)^r + a\sum_{r=0}^{n-1}(\bar{\alpha})^r + b\sum_{r=0}^{n-1}1\right\}$$

$$= \frac{1}{2|a|}\{0 + 0 + nb\} = \frac{nb}{2|a|}$$

[$\because$ sum of n, nth roots of unity is zero]

4. $\because \alpha^7 = 1$

and $\quad 1 + \alpha + \alpha^2 + \alpha^3 + \alpha^4 + \alpha^5 + \alpha^6 = 0$

$\Rightarrow \quad 1 + (\alpha + \alpha^2 + \alpha^4) + (\alpha^3 + \alpha^5 + \alpha^6) = 0$

or $\quad (\alpha + \alpha^2 + \alpha^4) + (\alpha^3 + \alpha^5 + \alpha^6) = -1$

and $(\alpha + \alpha^2 + \alpha^4)(\alpha^3 + \alpha^5 + \alpha^6)$

$$= \alpha^4 + \alpha^6 + \alpha^7 + \alpha^5 + \alpha^7 + \alpha^8 + \alpha^7 + \alpha^9 + \alpha^{10}$$

$$= \alpha^4 + \alpha^6 + 1 + \alpha^5 + 1 + \alpha + 1 + \alpha^2 + \alpha^3$$

$$= 2 + (1 + \alpha + \alpha^2 + \alpha^3 + \alpha^4 + \alpha^5 + \alpha^6)$$

$$= 2 + 0 = 2$$

Then required equation is

$$z^2 - (-1)z + 2 = 0 \text{ i.e. } z^2 + z + 2 = 0$$

5. $\because z^n - 1 = (z - 1)\prod_{r=1}^{n-1}(z - \alpha^r)$

$$\Rightarrow \quad \frac{z^n - 1}{z - 1} = \prod_{r=1}^{n-1}(z - \alpha^r)$$

$$\lim_{z \to 1}\left(\frac{z^n - 1}{z - 1}\right) = \lim_{z \to 1}\prod_{r=1}^{n-1}(z - \alpha^r)$$

$$\Rightarrow \quad n = \prod_{r=1}^{n=1}(1 - \alpha^r)$$

$$= \prod_{r=1}^{n-1}\left(1 - e^{\frac{2\pi ri}{n}}\right)$$

$$= \prod_{r=1}^{n-1}\left\{-e^{\frac{\pi ri}{n}}\left(e^{\frac{\pi ri}{n}} - e^{\frac{-\pi ri}{n}}\right)\right\}$$

$$= \prod_{r=1}^{n-1} -e^{\frac{\pi ri}{n}} \cdot 2i\sin\left(\frac{\pi r}{n}\right)$$

$$|n| = \prod_{r=1}^{n-1} 2\sin\left(\frac{\pi r}{n}\right)$$

$$\therefore \quad |n| = 2^{n-1}\prod_{r=1}^{n-1}\sin\left(\frac{\pi r}{n}\right)$$

$$\therefore \quad \prod_{r=1}^{n-1}\sin\left(\frac{\pi r}{n}\right) = \frac{n}{2^{n-1}} \quad (\because n > 1)$$

6. $\displaystyle\sum_{r=1}^{n-1}(n - r)\cos\left(\frac{2r\pi}{n}\right) = \text{Re}\sum_{r=1}^{n-1}(n - r)e^{\frac{2\pi ri}{n}}$

$$= \text{Re}\sum_{r=1}^{n-1} ne^{\left(\frac{2\pi ri}{n}\right)} - \text{Re}\sum_{r=1}^{n-1} re^{\left(\frac{2\pi ri}{n}\right)}$$

$$= n\text{Re}\sum_{r=1}^{n-1} e^{\left(\frac{2\pi ri}{n}\right)} - \text{Re}\sum_{r=1}^{n-1} re^{\left(\frac{2\pi ri}{n}\right)}$$

$$= n\text{Re}\sum_{r=1}^{n-1}\alpha^r - \text{Re}\sum_{r=1}^{n-1} r\alpha^r$$

$$= n\text{Re}(-1) - \text{Re}\sum_{r=1}^{n-1} r\alpha^r$$

$$= -n - \text{Re}\sum_{r=1}^{n-1} r\alpha^r \quad \ldots\text{(i)}$$

Let $\quad P = \displaystyle\sum_{r=1}^{n-1} r\alpha^r$

Let $P = \alpha + 2\alpha^2 + 3\alpha^3 + 4\alpha^4 + \ldots + (n - 2)\alpha^{n-2} + (n - 1)\alpha^{n-1}$

$\alpha P = \alpha^2 + 2\alpha^3 + 3\alpha^4 + \ldots + (n - 2)\alpha^{n-1} + (n - 1)\alpha^n$

$\therefore \quad (1 - \alpha)P = (\alpha + \alpha^2 + \alpha^3 + \alpha^4 + \ldots + \alpha^{n-1}) - (n - 1)\alpha^n$

$$= -1 - (n - 1)\alpha^n$$

$$= -1 - n + 1 \; (\because \alpha^n = 1) = -n$$

$$\Rightarrow \quad P = -\frac{n}{(1 - \alpha)} = -\frac{n}{(1 - e^{2\pi i/n})}$$

$$= \frac{n}{e^{\pi i/n}(e^{\pi i/n} - e^{-\pi i/n})}$$

$$= \frac{ne^{-\pi i/n}}{2i\sin\left(\frac{\pi}{n}\right)}$$

$$= \frac{n\left(\cos\left(\frac{\pi}{n}\right) - i\sin\left(\frac{\pi}{n}\right)\right)}{2i\sin\left(\frac{\pi}{n}\right)}$$

$$= \frac{n}{2}\left(-i\cot\left(\frac{\pi}{n}\right) - 1\right)$$

$$\therefore \quad \text{Re}(P) = -\frac{n}{2}$$

From Eq. (i),

$$\text{Required sum} = -n - \left(-\frac{n}{2}\right)$$

$$= -\frac{n}{2}$$

7. $\because \displaystyle\sum_{k=1}^{m-1}\left\{\sin\left(\frac{2\pi k}{m}\right) - i\cos\left(\frac{2\pi k}{m}\right)\right\}$

$$= \sum_{k=1}^{m-1}(-i)\left\{\cos\left(\frac{2\pi k}{m}\right) + i\sin\left(\frac{2\pi k}{m}\right)\right\}$$

$$= -i\sum_{k=1}^{m-1} e^{\frac{2\pi ki}{m}} = -i\left(\sum_{k=0}^{m-1} e^{\frac{2\pi ki}{m}} - 1\right)$$

$$= -i\,(0 - 1) = i$$

Now, $\displaystyle\sum_{m=1}^{4n+1}\left\{\sum_{k=1}^{m-1}\left\{\sin\left(\frac{2\pi k}{m}\right) - i\cos\left(\frac{2\pi k}{m}\right)\right\}\right\}^m$

$$= \sum_{m=1}^{4n+1}(i)^m = i^1 + 0 = i = \text{Purely imaginary}$$

PASSAGE 9

1. $\because \angle AOB = \angle BOC = \angle COA = \frac{2\pi}{3}$

$\therefore \ (AB)^2 = (OA)^2 + (OB)^2 - 2 \cdot OA \cdot OB \cos\left(\frac{2\pi}{3}\right)$

$= 4 + 4 - 2 \cdot 2 \cdot 2 \cdot \left(\frac{-1}{2}\right) = 12$

$\because \Delta ABC$ is equilateral

$\therefore \quad (AB)^2 = (BC)^2 = (CA)^2 = 12$

$\Rightarrow \ (AB)^2 + (BC)^2 + (CA)^2 = 12 + 12 + 12 = 36$

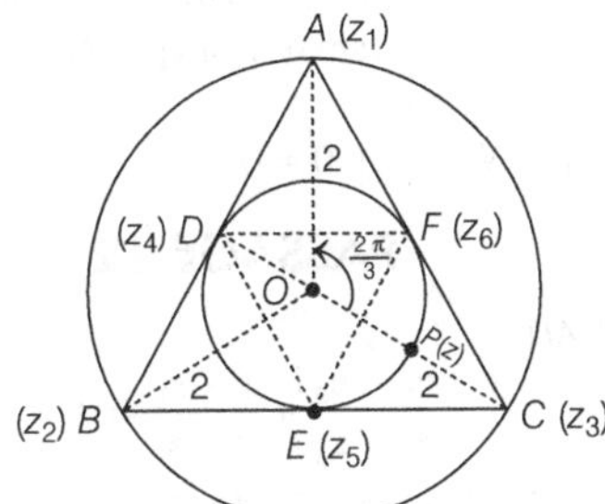

2. $(PA)^2 = (OA)^2 + (OP)^2 - 2OA.OP\cos 120°$

$= 4 + 1 - 2 \cdot 2 \cdot 1 \cdot \left(\frac{-1}{2}\right) = 7 = (PB)^2$

and $PC = OC - OP = 2 - 1 = 1$

Hence, $(PA)^2 + (PB)^2 + (PC)^2 = 7 + 7 + 1 = 15$

$\because \ AD = AF = DF = OA\cos 30 = 2 \times \frac{\sqrt{3}}{2} = \sqrt{3}$

$\Rightarrow DE = EF = FD = \sqrt{3}$

Then, $(DE)^2 + (EF)^2 + (FD)^2 = 3 + 3 + 3 = 9$

4. $\because |z_1 - z_2|^2 = |z_1|^2 + |z_2|^2 - 2\,\text{Re}(|z_1\bar{z}_2|)$

$\Rightarrow \quad (AB)^2 = (OA)^2 + (OB)^2 - 2\text{Re}(z_1\bar{z}_2)$

$\therefore 2\text{Re}(z_1\bar{z}_2) = 12 - 4 - 4 = 4$

$\therefore \ \text{Re}(z_1\bar{z}_2) = 2$

Similarly, $\text{Re}(z_2\bar{z}_3) = 2$ and $\text{Re}(z_3\bar{z}_1) = 2$

Hence,

$\text{Re}(z_1\bar{z}_2 + z_2\bar{z}_3 + z_3\bar{z}_1) = \text{Re}(z_1\bar{z}_2) + \text{Re}(z_2\bar{z}_3) + \text{Re}(z_3\bar{z}_1)$

$= 2 + 2 + 2 = 6$

5. $\because z_1 = \sqrt{3} + i$

$\sqrt{|z_2 - z_3|^2 + |z_2 + z_3|^2} = \sqrt{2(|z_2|^2 + |z_3|^2)}$

$= \sqrt{2(2^2 + 2^2)} = \sqrt{16} = 4$

6. $\because \angle AOC = \frac{2\pi}{3}$

$\therefore \quad \frac{z_1 - 0}{z_3 - 0} = \frac{2}{2} e^{-\frac{2\pi i}{3}}$

$\therefore \quad \frac{z_1}{z_3} = e^{\frac{2\pi i}{3}} = \cos\left(\frac{2\pi}{3}\right) + i\sin\left(\frac{2\pi}{3}\right)$

$= \left(\frac{-1 + i\sqrt{3}}{2}\right)$

PASSAGE 10

1. Given lines an $a\bar{z} + \bar{a}z + b = 0$ ...(i)

and $\quad a_1\bar{z} + \bar{a}_1 z + b_1 = 0$ ...(ii)

Complex slopes of (i) and (ii) are $-\frac{a}{\bar{a}}$ and $-\frac{a_1}{\bar{a}_1}$.

$\because$ Lines (i) and (ii) are perpendicular

$\therefore \quad \left(-\frac{a}{\bar{a}}\right) + \left(-\frac{a_1}{\bar{a}_1}\right) = 0$

$\Rightarrow \quad a\bar{a}_1 + a_1\bar{a} = 0$

2. For replection on the real axis replacing z by $\bar{z}$, then

$(2 + 3i)\bar{z} + (2 - 3i)z = 0$

or $\quad (2 - 3i)z + (2 + 3i)\bar{z} = 0$

3. $\because a, b, c, d$ are lie on $|z| = r$

Then, $|a| = r, |b| = r, |c| = r$ and $|d| = r$

$\therefore \ |a|^2 = r^2, |b|^2 = r^2, |c|^2 = r^2$

and $\quad |d|^2 = r^2$

$\therefore \quad \bar{a} = \frac{r^2}{a}, \bar{b} = \frac{r^2}{b}, \bar{c} = \frac{r^2}{c}$

and $\quad \bar{d} = \frac{r^2}{d}$

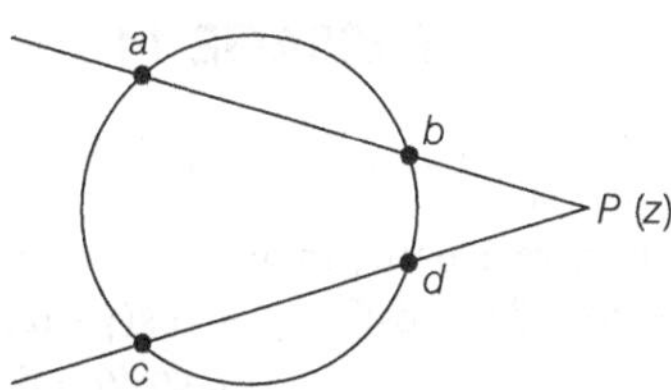

$\because \ a, b, z$ are collinear

$\therefore \quad \begin{vmatrix} z & \bar{z} & 1 \\ a & \bar{a} & 1 \\ b & \bar{b} & 1 \end{vmatrix} = 0$

$\Rightarrow \quad z(\bar{a} - \bar{b}) - \bar{z}(a - b) + (a\bar{b} - \bar{a}b) = 0$

$\Rightarrow z\left(\frac{r^2}{a} - \frac{r^2}{b}\right) - \bar{z}(a - b) + \left(\frac{ar^2}{b} - \frac{r^2 b}{a}\right) = 0$

$\Rightarrow \quad \frac{z}{ab} + \frac{\bar{z}}{r^2} = \frac{1}{a} + \frac{1}{b}$ ...(i)

similarly, c, d, z are collinear, then

$\frac{z}{cd} + \frac{\bar{z}}{r^2} = \frac{1}{c} + \frac{1}{d}$ ...(ii)

subtracting Eq. (ii) from Eq. (i), then

$z(a^{-1}b^{-1} - c^{-1}d^{-1}) = a^{-1} + b^{-1} - c^{-1} - d^{-1}$

$\therefore \quad z = \frac{(a^{-1} + b^{-1} - c^{-1} - d^{-1})}{(a^{-1}b^{-1} - c^{-1}d^{-1})}$

4. $\because |z - i| \le 2$ and $z_0 = 5 + 3i$

$\therefore \quad |iz + z_0| = |i(z - i) + i^2 + z_0| = |i(z - i) + z_0 - 1|$

$= |i(z - i) + 4 + 3i| < |i(z - i)| + |4 + 3i|$

$= |i||z - i| + 5$

$= |z - i| + 5 \le 2 + 5 = 7$

$\therefore \quad |iz + z_0| \le 7$

$\therefore$ Maximum value of $|iz + z_0|$ is 7.

5. $|z-1-i|=\sqrt{2} \Rightarrow$ centre $C_1(1+i)$, radius $r_1=\sqrt{2}$
and $|z+1+i|=2 \Rightarrow$ centre $C_2(-1-i)$, radius $r_2=2$
$\therefore |C_1C_2|=|1+i-(-1-i)|=|2+2i|=\sqrt{4+4}=2\sqrt{2}$
$\therefore \quad r_2-r_1<|C_1C_2|<r_1+r_2$
$2-\sqrt{2}<2\sqrt{2}<2+\sqrt{2}$
Hence, point of intersection is 2.

6. $\because |z_1+z_2+z_3|^2 \geq 0$
$\Rightarrow |z_1|^2+|z_2|^2+|z_3|^2+2\text{Re}(z_1\bar{z}_2+z_2\bar{z}_3+z_3\bar{z}_1)\geq 0$
$\Rightarrow 1+1+1+2\cos(\theta_1-\theta_2)+2\cos(\theta_2-\theta_3)$
$+\cos(\theta_3-\theta_1)\geq 0$
$\therefore \Sigma\cos(\theta_1-\theta_2)\geq -\frac{3}{2}$

7. Circles, $z\bar{z}+\bar{a}_1z+a_1\bar{z}+b_1=0$ ...(i)
and $z\bar{z}+\bar{a}_2z+a_2\bar{z}+b_2=0$ ...(ii)
Centre and radius of Eq. (i) are $-a_1$ and $\sqrt{|a_1|^2-b_1}$
and of Eq. (ii) are $-a_2$ and $\sqrt{|a_2|^2-b_2}$
For orthogonally,
$|a_1-a_2|^2=|a_1|^2-b_1+|a_2|^2-b_2$
$\Rightarrow \quad |a_1|^2+|a_2|^2-(a_1\bar{a}_2+\bar{a}_1a_2)=|a_1^2|+|a_2|^2-b_1-b_2$
$\Rightarrow \quad a_1\bar{a}_2+\bar{a}_1a_2=b_1+b_2$

PASSAGE 11

1. $\because \frac{A}{B}+\frac{B}{C}+\frac{C}{A}=1 \Rightarrow e^{i(\alpha-\beta)}+e^{i(\beta-\gamma)}+e^{i(\gamma-\alpha)}=1$
comparing real parts, then
$\cos(\alpha-\beta)+\cos(\beta-\gamma)+\cos(\gamma-\alpha)=1$
$\therefore \quad \Sigma\cos(\alpha-\beta)=1$

2. $\because \quad (aA)^3+(bB)^3+(cC)^3=3abcABC$
$\Rightarrow a^3e^{3i\alpha}+b^3e^{3i\beta}+c^3e^{3i\gamma}=3abce^{i(\alpha+\beta+\gamma)}$
Comparing real parts, then
$a^3\cos 3\alpha+b^3\cos 3\beta+c^3\cos 3\gamma=3abc\cos(\alpha+\beta+\gamma)$
$\therefore \quad a^3=1,\ b^3=\lambda,\ c^3=\mu,\ 3abc=\nu$
Now, $|\mu-3\lambda|+(\nu-\mu|=|c^3-3b^3|+|3abc-c^3|$
$=|c^3-3b^3|+|3bc-c^3|$
For $b=2, c=3$
$=|27-24|+|18-27|=3+9=12$

3. $\because \sin^n\alpha+\sin^n\beta+\sin^n\gamma=3/2$
and $\cos(\alpha+\beta)+\cos(\beta+\gamma)+\cos(\gamma+\alpha)=\lambda$
For $a=b=c=1$,
$\cos\alpha+\cos\beta+\cos\gamma=0=\sin\alpha+\sin\beta+\sin\gamma$
$\therefore \quad A+B+C=0$ and $\frac{1}{A}+\frac{1}{B}+\frac{1}{C}=0$
$\Rightarrow \quad A+B+C=0$ and $AB+BC+CA=0$
$\therefore \quad A^2+B^2+C^2=0$ and $AB+BC+CA=0$
$e^{2i\alpha}+e^{2i\beta}+e^{2i\gamma}=0$ and $e^{i(\alpha+\beta)}+e^{i(\beta+\gamma)}+e^{i(\gamma+\alpha)}=0$
Comparing real parts on both equations, then
$\cos 2\alpha+\cos 2\beta+\cos 2\gamma=0$
and $\cos(\alpha+\beta)+\cos(\beta+\gamma)+\cos(\gamma+\alpha)=0$
$\Rightarrow 1-2\sin^2\alpha+1-2\sin^2\beta+1-2\sin^2\gamma=0$
$\therefore \quad \sin^2\alpha+\sin^2\beta+\sin^2\gamma=3/2$
Hence, $n=2, \lambda=0 \quad \therefore (n,\lambda)\equiv(2,0)$

4. Let $\alpha+\beta+\gamma=0$
$\therefore \quad \sin(-\beta-\gamma)+2\sin(-\gamma-\alpha)+3\sin(-\alpha-\beta)=0$
$\therefore \quad \sin(\beta+\gamma)+2\sin(\gamma+\alpha)+3\sin(\alpha+\beta)=0$
which is true.
$\therefore \quad \alpha+\beta-\gamma=-\gamma-\gamma=-2\gamma$

5. Let $\alpha+\beta+\gamma=0$
$\therefore \sin(2\alpha-\beta-\gamma)+8\sin(2\beta-\gamma-\alpha)+27\sin(2\gamma-\alpha-\beta)$
$=\sin 3\alpha+8\sin 3\beta+27\sin 3\alpha$
$=3\cdot1\cdot2\cdot3\sin(\alpha+\beta+\gamma)=18\sin(0)=0$
$\therefore \quad N_1=0$
and $\sin(\alpha+\beta)+\sin(\beta+\gamma)+\sin(\gamma+\alpha)$
$=\sin(-\gamma)+\sin(-\alpha)+\sin(-\beta)$
$=-(\sin\alpha+\sin\beta+\sin\gamma)=-(0)=0$
$\therefore \quad N_2=0,$
then $N_1+N_2=0+0=0$

PASSAGE 12

1. $\because BC\perp AD$
$\therefore \quad \frac{z_2-z_3}{\bar{z}_2-\bar{z}_3}+\frac{z-z_1}{\bar{z}-\bar{z}_1}=0$
$\Rightarrow \quad \frac{(z_2-z_3)}{\left(\frac{1}{z_2}-\frac{1}{z_3}\right)}+\frac{z-z_1}{\bar{z}-\bar{z}_1}=0$

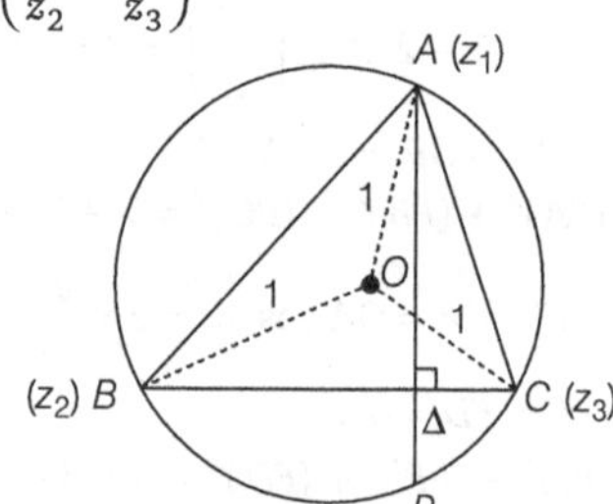

$\Rightarrow \quad -z_2z_3+\frac{z-z_1}{\bar{z}-\bar{z}_1}=0$ ...(i)
$\Rightarrow \quad \bar{z}-\bar{z}_1=\frac{z}{z_2z_3}-\frac{z_1}{z_2z_3}$ ...(ii)
and also $BD\perp AD$
$\therefore \quad \frac{z-z_2}{\bar{z}-\bar{z}_2}+\frac{z-z_1}{\bar{z}-\bar{z}_1}=0$ ...(iii)
From Eqs. (ii) and (iii), we get
$z_2z_3=-\frac{z-z_2}{\bar{z}-\bar{z}_2}$
$\Rightarrow \quad \bar{z}-\bar{z}_2=-\frac{z}{z_2z_3}+\frac{1}{z_3}$ ...(iv)
Subtracting Eq. (ii) from Eq. (iv), then
$\bar{z}_1-\bar{z}_2=-\frac{2z}{z_2z_3}+\frac{1}{z_3}+\frac{z_1}{z_2z_3}$
$\Rightarrow \quad \frac{2z}{z_2z_3}=-\frac{1}{z_1}+\frac{1}{z_2}+\frac{1}{z_3}+\frac{z_1}{z_2z_3}$
$\Rightarrow \quad 2z=\left(-\frac{z_2z_3}{z_1}+z_3+z_2+z_1\right)$
$\therefore \quad z=\frac{1}{2}\left(z_1+z_2+z_3-\frac{z_2z_3}{z_1}\right)$

2. $\because AP \perp BC$

$\therefore \quad \dfrac{z_2 - z_3}{\bar{z}_2 - \bar{z}_3} + \dfrac{z - z_1}{\bar{z} - \bar{z}_1} = 0$

$\Rightarrow \quad \dfrac{(z_2 - z_3)}{\left(\dfrac{1}{z_2} - \dfrac{1}{z_3}\right)} + \dfrac{(z - z_1)}{\left(\dfrac{1}{z} - \dfrac{1}{z_1}\right)} = 0$

[$\because z, z_1, z_2, z_3$ lie on circle]

$\Rightarrow \quad -z_2z_3 - zz_1 = 0$

$\therefore \quad z = -\dfrac{z_2z_3}{z_1}$

3. $\because Q$ be the image of P w.r.t. BC, then D is the mid point of P and Q.

$z_1 + z_2 + z_3 - \dfrac{z_2z_3}{z_1} - \left(-\dfrac{z_2z_3}{z_1}\right) = z$

$\therefore \quad z = z_1 + z_2 + z_3$

4. $\because R$ is the image of P about the origin

$\therefore \quad R \equiv \dfrac{z_2z_3}{z_1}$

then distance between B and R is $\left|z_2 - \dfrac{z_2z_3}{z_1}\right|$

$= \dfrac{|z_2|}{|z_1|}|z_1 - z_3| = \dfrac{1}{1}|z_3 - z_1| = |z_3 - z_1|$

5. $\because$ Centroid $G\left(\dfrac{z_1 + z_2 + z_3}{3}\right)$, Circumcentre (0)

Let orthocentre $O'(z)$, then

$G = \dfrac{1.\,O' + 2\,(\text{Circumcentre})}{1 + 2}$

$\dfrac{z_1 + z_2 + z_3}{3} = \dfrac{z + 0}{3}$

$\therefore \quad z = z_1 + z_2 + z_3$

Numerical Value Type

1. $x + iy + \alpha\sqrt{(x-1)^2 + y^2} + 2i = 0$

$\therefore \quad y + 2 = 0$ and $x + \alpha\sqrt{(x-1)^2 + y^2} = 0$

$y = -2$ and $x^2 = \alpha^2(x^2 - 2x + 1 + 4)$

$\alpha^2 = \dfrac{x^2}{x^2 - 2x + 5}$

$\Rightarrow \quad x^2(\alpha^2 - 1) - 2x\alpha^2 + 5\alpha^2 = 0$

$\Rightarrow \quad x \in R,\ D \geq 0$

$\alpha\alpha^4 - 4(\alpha^2 - 1)5\alpha^2 \geq 0$

$\alpha^2[4\alpha^2 - 2\alpha^2 + 20] \geq 0$

$\alpha^2[-16\alpha^2 + 20] \geq 0$

$\alpha^2\left[\alpha^2 - \dfrac{5}{4}\right] \leq 0$

$0 \leq \alpha^2 \leq \dfrac{5}{4}$

$\therefore \quad \alpha^2 \in \left[0, \dfrac{5}{4}\right]$

$\therefore \quad \alpha \in \left[-\dfrac{\sqrt{5}}{2}, \dfrac{\sqrt{5}}{2}\right]$

Then, $4[(q)^2 + (p)^2] = 4\left[\dfrac{5}{4} + \dfrac{5}{4}\right] = 10$

2. Let $z = (a + b\omega + c\omega^2)$

$\therefore \quad \bar{z} = \overline{(a + b\omega + c\omega^2)} = (a + b\bar{\omega} + \overline{\omega^2}) = (a + b\omega^2 + c\omega)$

and $z\bar{z} = (a + b\omega + c\omega^2)(a + b\omega^2 + c\omega)$

$= (a^2 + b^2 + c^2 - ab - bc - ca)$

$= \dfrac{1}{2}\{(a-b)^2 + (b-c)^2 + (c - a^2)\}$

or $\quad |z|^2 = \dfrac{1}{2}\{(a-b)^2 + (b-c)^2 + (c-a)^2\}$

$\therefore \quad |a + b\omega + c\omega^2| + |a + b\omega^2 + c\omega| = |z| + |\bar{z}|$

$= |z| + |z| = 2|z|$

$= 2\cdot\dfrac{1}{\sqrt{2}}\sqrt{(a-b)^2 + (b-c)^2 + (c-a)^2}$

$= \sqrt{2}\sqrt{(a-b)^2 + (b-c)^2 + (c-a)^2}$

$\geq \sqrt{2}\sqrt{(1^2 + 1^2 + 2^2)}$

$= \sqrt{12}$

$= (144)^{1/4}$

$\left[\begin{array}{l}\because a, b, c \text{ are distinct integers, minimum value of} \\ (a-b)^2 + (b-c)^2 + (c-a)^2 = 1^2 + 1^2 + 2^2 = 6\end{array}\right]$

$\therefore |a + b\omega + c\omega^2| + |a + b\omega^2 + c\omega| \geq (144)^{1/4}$

$\therefore$ Minimum value of

$|a + b\omega + c\omega^2| + |a + b\omega^2 + c\omega| = (144)^{1/4}$

$\therefore \quad n = 144$

3. $\because (a\omega^2 + b\omega)(a\omega + b\omega^2) = a^2 - ab + b^2$

$= \dfrac{(a+b)(a^2 - ab + b^2)}{(a+b)}$

$= \dfrac{a^3 + b^3}{a + b}$

$= \dfrac{8001}{21} = 381$

4. Let $|z-\alpha| = k$ ...(i)

(where $\alpha = a + ib$ and $a, b, k \in R$) be a circle which cuts the circle $|z| = 1$ ...(ii)

and $|z-1| = 4$ ...(iii)

orthogonally, then

$$k^2 + 1 = |\alpha - 0|^2 = \alpha\bar{\alpha}$$

and $$k^2 + 16 = |\alpha - 1|^2 = (\alpha - 1)(\bar{\alpha} - 1)$$

$$= \alpha\bar{\alpha} - (\alpha + \bar{\alpha}) + 1$$

or $1 - (\alpha + \bar{\alpha}) - 15 = 0$

or $\alpha + \bar{\alpha} = -14$

or $2a = -14$

$\Rightarrow$ $a = -7$

$\Rightarrow$ $\alpha = a + ib = -7 + ib$

Also, $k^2 = |\alpha|^2 - 1$

$= 49 + b^2 - 1$

$= 48 + b^2$

$\Rightarrow$ $k = \sqrt{(48 + b^2)}$

Therefore, required family of circles is given by

$$|z + 7 - ib| = \sqrt{(48 + b^2)}$$

$\therefore$ $\lambda = 48$

5. $\sum_{q=1}^{10}\left(\sin\frac{2q\pi}{11} - i\cos\frac{2q\pi}{11}\right)$

$$= -i\sum_{q=1}^{10}\left(\cos\left(\frac{2q\pi}{11}\right) + i\sin\left(\frac{2q\pi}{11}\right)\right)$$

$$= -i\sum_{q=1}^{10} e^{2q\pi i/11}$$

$$= -i\left(\sum_{q=0}^{10} e^{2q\pi\, i/11} - 1\right)$$

$= -i$ (sum of 11, 11th roots of unity -1)

$= -i(0-1)$

$= i$

$$\therefore \sum_{p=1}^{32}(3p+2)\left(\sum_{q=1}^{10}\left(\sin\left(\frac{2q\pi}{11}\right) - i\cos\left(\frac{2q\pi}{11}\right)\right)\right)^{4p}$$

$$= \sum_{p=1}^{32}(3p+2)(i)^{4p}$$

$$= \sum_{p=1}^{32}(3p+2)$$

$$= 3\sum_{p=1}^{32} p + 2\sum_{p=1}^{32} 1$$

$$= \frac{3\cdot 32\cdot 33}{2} + 2\cdot 32$$

$= 1648$

The value of $\lambda = 1648$

$$\frac{\lambda}{515} = \frac{1648}{515}$$

$$= \frac{16}{5} = 3.2$$

6. $$\frac{\left(2e^{i\frac{2\pi}{3}}\right)^{21}}{\left(\sqrt{2}e^{-i\frac{\pi}{4}}\right)^{24}} + \frac{\left(2e^{i\frac{\pi}{3}}\right)^{21}}{\left(\sqrt{2}e^{i\frac{\pi}{4}}\right)^{24}}$$

$$= \frac{2^{21}\cdot e^{i14\pi}}{2^{12}\cdot e^{-i6\pi}} + \frac{2^{21}(e^{i7\pi})}{2^{12}(e^{i6\pi})}$$

$$= 2^9 e^{i(20\pi)} + 2^9 e^{i\pi}$$

$$= 2^9 + 2^9(-1) = 0$$

$\Rightarrow$ $n = 0$

$$\sum_{j=0}^{5}(j+5)^2 - \sum_{j=0}^{5}(j+5)$$

$\Rightarrow [5^2 + 6^2 + 7^2 + 8^2 + 9^2 + 10^2] - [5+6+7+8+9+10]$

$\Rightarrow [(1^2 + 2^2 + \ldots + 10^2) - (1^2 + 2^2 + 3^2 + 4^2)]$

$- [(1+2+3+\ldots+10) - (1+2+3+4)]$

$\Rightarrow (385-30) - [55-10]$

$\Rightarrow 355 - 45 = 310$

7. $\because$ $|z - 25i| \geq ||z| - |25i|| = ||z| - 25|$

$\Rightarrow$ $|z - 25i| \geq ||z| - 25|$ ...(i)

But, given

$||z| - 25i| \leq 15$ ...(ii)

From Eq. (i) and Eq. (ii)

$$||z| - 25| \leq |z - 25i| \leq 15$$

$\Rightarrow$ $||z| - 25| \leq 15$

$\Rightarrow$ $-15 \leq |z| - 25 \leq 15$

or $10 \leq |z| \leq 40$

$\therefore$ Minimum value of $|z|$ is 10 and maximum value is 40.

$\therefore$ Required sum $= 10 + 40 = 50$

8. $\because |az_1 - bz_2|^2 + |bz_1 + az_2|^2 = (a^2 + b^2)(|z_1|^2 + |z_2|^2)$

then $\lambda = a^2 + b^2 = (15)^2 + (13)^2 = 394$

$$\therefore \sqrt{\lambda\sqrt{\lambda\sqrt{\lambda\sqrt{\lambda\ldots\ldots\infty}}}} = \lambda^{\frac{1}{2}+\frac{1}{4}+\frac{1}{8}+\ldots\infty}$$

$$= \lambda^{\frac{1/2}{1-1/2}}$$

$= \lambda$

$= 394$

9. $|z| + |z-1| + |2z-3| = |z| + |z-1| + |3-2z|$

$\geq |z + z - 1 + 3 - 2z| = 2$

$\therefore$ $|z| + |z-1| + |2z-3| \geq 2$

$\therefore$ $\lambda = 2$,

then $2[x] + 3 = 3[x - \lambda] = 3[x-2]$

$\Rightarrow$ $2[x] + 3 = 3([x] - 2)$

or $[x] = 9$, then $y = 2\cdot 9 + 3 = 21$

$\therefore$ $[x+y] = [x+21] = [x] + 21$

$= 9 + 21 = 30$

10. We have, $\omega = \omega^4 = \omega^7 = \omega^{10} = \ldots = \omega^{28}$

$$\omega^2 = \omega^5 = \omega^8 = \omega^{11} = \ldots = \omega^{29}$$

and $\omega^3 = \omega^6 = \omega^9 = \omega^{12} = \ldots = \omega^{30} = 1$

$$\therefore \quad (1+\omega)(1+\omega^2)(1+\omega^3)(1+\omega^4)(1+\omega^5)\ldots(1+\omega)^{30}$$
$$= \{(1+\omega)(1+\omega^2)(1+\omega^3)\}^{10}$$
$$= \{(-\omega^2)(-\omega)(2)\}^{10}$$
$$= (\omega^3 \cdot 2)^{10}$$
$$= (1 \cdot 2)^{10}$$
$$= 2^{10}$$
$$= 1024$$

11. Given, $|z+5| \le 4$

$\Rightarrow \quad (x+5)^2 + y^2 \le 16 \quad \ldots$(i)

Also, $\quad z(1+i) + \bar{z}(1-i) \ge -10$

$\Rightarrow \quad x - y \ge -5 \quad \ldots$(ii)

From Eqs. (i) and (ii) Locus of z is the shaded region in the diagram.

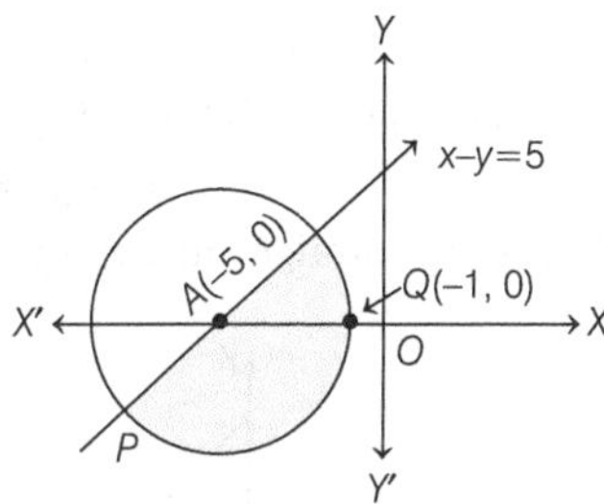

$|z+1|$ represents distance of 'z' from $Q(-1, 0)$. Clearly 'p' is the required position of 'z' when $|z+1|$ is maximum.

$\therefore P \equiv (-5-2\sqrt{2}, -2\sqrt{2})$

$\therefore (PQ)^2|_{max} = 32 + 16\sqrt{2}$

$\Rightarrow \quad \alpha = 32$

$\Rightarrow \quad \beta = 16$

Thus, $\dfrac{\alpha+\beta}{20} = \dfrac{48}{20} \Rightarrow 2.4$

12. $A \equiv (1+2i)x^3 - 2(3+i)x^2 + (5-4i)x + 2a^2$

Let the real root of the equation be α, then

$(1+2i)\alpha^3 - 2(3+i)\alpha^2 + (5-4i)\alpha + 2a^2 = 0$

Equating imaginary part equal to zero, we get

$$2\alpha^3 - 2\alpha^2 - 4\alpha = 0$$
$$2\alpha(\alpha^2 - \alpha - 2) = 0$$
$$\alpha = 0, -1, 2$$

Now equating real part equal to zero, we get

$$\alpha^3 - 6\alpha^2 + 5\alpha + 2a^2 = 0$$

If $\alpha = 0$, then $a = 0$

$\alpha = -1$, then $a = \pm\sqrt{6}$

$\alpha = 2$, then $a = \pm\sqrt{3}$

$n = 3$

$\Sigma a^2 = 0^2 + (\sqrt{6})^2 + (-\sqrt{6})^2 + (\sqrt{3})^2 + (-\sqrt{3})^2$

$= 6 + 6 + 3 + 3 = 18$

$\dfrac{\Sigma a^2}{10n} = \dfrac{18}{3\times 10} = \dfrac{6}{10} = 0.6$

Single Integer Type

1. It is given that,

$$\left(\frac{1+i}{1-i}\right)^{m/2} = \left(\frac{1+i}{i-1}\right)^{n/3} = 1, (m, n \in N)$$

$\Rightarrow \quad i^{m/2} = 1 = (-i)^{n/3}$

So, the least values of 'm' and 'n' are 8 and 12 respectively.

$\therefore \quad$ GCD {8, 12} = 4

Hence, answer is 4.00.

2. Given, $\alpha_k = \cos\left(\dfrac{k\pi}{7} + i\sin\left(\dfrac{k\pi}{7}\right)\right)$

$$= \cos\left(\frac{2k\pi}{14}\right) + i\sin\left(\frac{2k\pi}{14}\right)$$

$\therefore a_k$ are vertices of regular polygon having 14 sides.

Let the side length of regular polygon be a.

$\therefore |a_{k+1} - a_k|$ = length of a side of the regular polygen

$= a \quad \ldots$(i)

and $|a_{4k-1} - a_{4k-2}|$ = length of a side of the regular polygon

$= a \quad \ldots$(ii)

$$\therefore \quad \frac{\sum_{k=1}^{12} |a_{k+1} - a_k|}{\sum_{k=1}^{3} |a_{4k-1} - a_{4k-2}|} = \frac{12(a)}{3(a)} = 4$$

3. For a complex number z, it is given that

$$z^4 - |z|^4 = 4iz^2$$

$\Rightarrow \quad z^4 - z^2\bar{z}^2 = 4iz^2$

$\Rightarrow \quad z^2(z-\bar{z})(z+\bar{z}) = 4iz^2$

So, either $z^2 = 0$ or $(z-\bar{z})(z+\bar{z}) = 4i$

Now, **Case I**, if $z^2 = 0$ and $z = x + iy$

So, $\quad x^2 - y^2 + 2ixy = 0$

$\Rightarrow \quad x^2 - y^2 = 0$

and $\quad xy = 0$

$\Rightarrow \quad x = y = 0$

$\Rightarrow x = 0$ which is not possible according to given conditions.

Case-II, if $(z-\bar{z})(z+\bar{z}) = 4i$ and $z = x + iy$

So, $(2iy)(2x) = 4i$

$\Rightarrow xy = 1$ is an equation of rectangular hyperbola and for minimum value of $|z - z_2|^2$, the z_1 and z_2 must be vertices of the rectangular hyperbola.

Therefore, $z_1 = 1 + i$

and $\quad z_2 = -1 - i$

$\therefore$ Minimum value of $|z_1 - z_2|^2 = (1+1)^2 + (1+1)^2$

$= 4 + 4 = 8$

4. Let $z = x + iy$, $x, y \in R$.
Then, $z - z_1 = (x-10) + i(y-6)$
and $z - z_2 = (x-4) + i(y-6)$.
Therefore, $\frac{z-z_1}{z-z_2} = \frac{(x-10)+i(y-6)}{(x-4)+i(y-6)}$
$$= \frac{[(x-10)+i(y-6)][(x-4)-i(y-6)]}{(x-4)^2+(y-6)^2}$$
Therefore, real part of
$$\frac{z-z_1}{z-z_2} = \frac{(x-10)(x-4)+(y-6)^2}{(x-4)^2+(y-6)^2}$$
Imaginary part of
$$\frac{z-z_1}{z-z_2} = \frac{(x-4)(y-6)-(x-10)(y-6)}{(x-4)^2+(y-6)^2}$$
$$= \frac{6(y-6)}{(x-4)^2+(y-6)^2}$$
Now,
$$\frac{\pi}{4} = \arg\left(\frac{z-z_1}{z-z_2}\right) = \tan^{-1}\left[\frac{6(y-6)}{(x-10)(x-4)+(y-6)^2}\right]$$
$\therefore\ (x-10)(x-4)+(y-6)^2 = 6(y-6)$
$x^2+y^2-14x-18y+112=0$
Now, $|z-7-9i|^2 = (x-7)^2+(y-9)^2$
$= x^2-14x+y^2-18y+130$
$= -112+130 = 18$
Therefore, $|z-7-9i| = \sqrt{18} = 3\sqrt{2}$
$\therefore\ k = 3$

5.

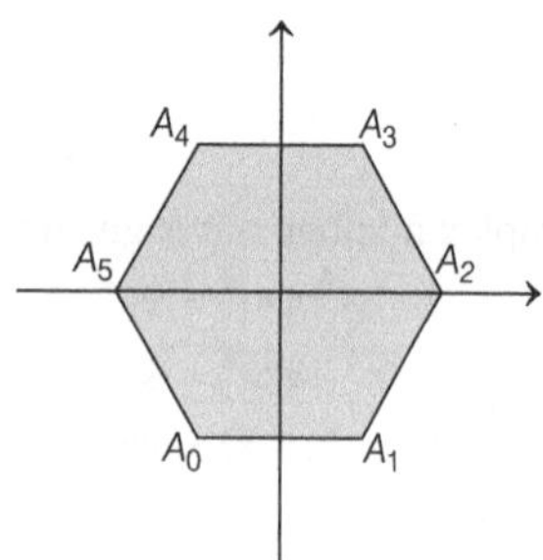

Let the vertices of regular hexagon are be $z_0, z_1 \ldots\ldots z_5$ w.r.t. centre O as origin and $|z_0| = \sqrt{5}$
$\therefore A_0A_1 = |z_1 - z_0| = |z_0e^{i\theta} - z_0$
$= |z_0||\cos\theta + i\sin\theta - 1|$
$= \sqrt{5}\sqrt{(\cos\theta-1)^2+\sin^2\theta}$
$= \sqrt{5}\sqrt{2(1-\cos\theta)}$
$= \sqrt{5}(2)\sin\theta/2$
$\Rightarrow A_0A_1 = 2\sqrt{5}\sin\frac{\pi}{6}$
$= \sqrt{5}(2)\left(\frac{1}{2}\right) = \sqrt{5}$
Hence perimeter $= 6\sqrt{5}$
$\therefore\ x = 6$

6. We have, $|z_1| = 2$
$\Rightarrow z_1$ lies on the circle having center at origin and radius 2.
and $(1-i)z_2 + (1+i)\bar{z}_2 = 8\sqrt{2}$
$\therefore\ (1-i)(x_2+iy_2)+(1+i)(x_2-iy_2) = 8\sqrt{2}$
$\Rightarrow\ x_2 + y_2 = 4\sqrt{2}$
So, z_2 lies on the straight line $x + y = 4\sqrt{2}$

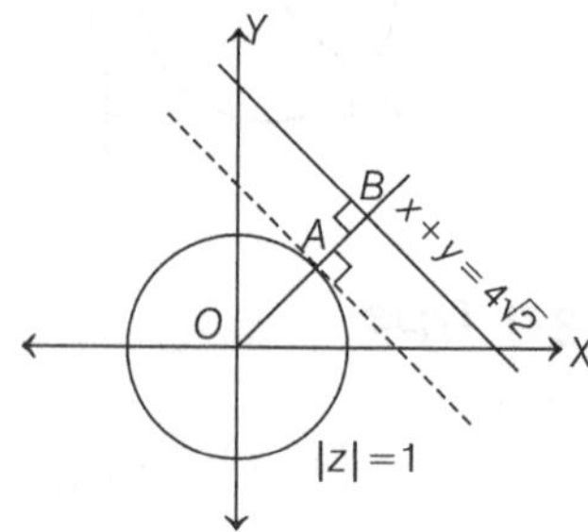

$|z_1 - z_2|_{min}$ = shortest distance between circle and straight line $= AB = OB - OA = 4 - 2 = 2$

Matrix-Match Type

1. A. Minimum value of equal to distance between (0, 0) and $(\cos\alpha, \sin\alpha)$ which is equal to 1.

B. $|z-1|+|z+1| = 2$ represents the line segment joining points '1' and '–1'. $|z+2|+|z-2| = 8$ represents the ellipse with foci '–2' and '2'.
Clearly, line segment does not intersect the ellipse.

C. $(z_1 - 3z_2)(\bar{z}_1 - 3\bar{z}_2) = (3 - z_1\bar{z}_2)(3 - \bar{z}_1z_2)$
$\Rightarrow |z_1|^2 - |z_1|^2|z_2|^2 + 9|z_2|^2 = 9 - 0$
$\Rightarrow\ (1-|z_2|^2)(|z_1|^2 - 9) = 0 \Rightarrow |z_1| = 3$

D. $z = \frac{(2\sqrt{3}+2i)^8}{(1-i)^6} + \frac{(1+i)^6}{(2\sqrt{3}-2i)^8}$
$= \frac{z_1^8}{\bar{z}_2^6} + \frac{z_2^6}{\bar{z}_1^8} = \frac{|z_1|^{16}+|z_2|^{12}}{(\bar{z}_2)^6(\bar{z}_1)^8}$
$\Rightarrow \arg(z) = -8\arg(\bar{z}_1) - 6\arg(\bar{z}_2)$
$= 8\arg(z_1) + 6\arg(z_2)$
$= 8.\frac{\pi}{6} + 6.\frac{\pi}{4} = \frac{4\pi}{3} + \frac{3\pi}{2} = \frac{17\pi}{6}$

$\therefore\ |4\sin\theta| = 4\left|\sin\dfrac{17\pi}{6}\right| = 2$

2. **(A)** $1 + z^3 = 1 + \cos 3\alpha + i\sin 3\alpha$

$$= 2\cos^2\frac{3\alpha}{2} + 2i\sin\frac{3\alpha}{2}\cos\frac{3\alpha}{2}$$

$$= 2\cos\left(\frac{3\alpha}{2}\right)\left[\cos\left(\frac{3\alpha}{2}\right) + i\sin\left(\frac{3\alpha}{2}\right)\right]$$

$\Rightarrow \arg(1+z^3) = 3\alpha/2$

(B) $1 - z^4 = 1 - \cos 4\alpha - i\sin 4\alpha$

$$= 2\sin^2 2\alpha - 2i\cos 2\alpha \sin 2\alpha$$

$$= 2\sin 2\alpha[\cos(2\alpha - \pi/2) + i\sin(2\alpha - \pi/2)]$$

$\Rightarrow \arg(1 - z^4) = 2\alpha = \pi/2$

(C) $\dfrac{1+z^3}{1-z^4} = \dfrac{\cos(3\alpha/2)}{\sin(2\alpha)}\left[\cos\left(\dfrac{\pi}{2} - \dfrac{\alpha}{2}\right) + i\sin\left(\dfrac{\pi}{2} - \dfrac{\alpha}{2}\right)\right]$

$\Rightarrow \arg\left(\dfrac{1+z^3}{1-z^4}\right) = \dfrac{\pi}{2} - \dfrac{\alpha}{2}$

(D) $\dfrac{z^4-1}{z^3+1} = \dfrac{\sin(2\alpha)}{\cos(3\alpha/2)}\left[\cos\left(\dfrac{\alpha}{2} - \dfrac{\pi}{2}\right) + i\sin\left(\dfrac{\alpha}{2} - \dfrac{\pi}{2}\right)\right]$

$$= \frac{\sin(2\alpha)}{\cos(3\alpha/2)}\left[\cos\left(\frac{\pi}{2} + \frac{\alpha}{2}\right) + i\sin\left(\frac{\pi}{2} + \frac{\alpha}{2}\right)\right]$$

Thus, $\arg\left(\dfrac{z^4-1}{z^3+1}\right) = \dfrac{\pi}{2} + \dfrac{\alpha}{2}$

3. **(A)** Since triangle is equilateral then

$$z_0 = \text{Nine point centre} = \text{centroid}$$

$$= \frac{z_1 + z_2 + z_3}{3}$$

$\therefore\quad z_1 + z_2 + z_3 = 3z_0$

$$AB = BC = CA = a \text{ (say)}$$

$\angle B = \pi/3$, then by rotation theorem

$$\therefore\quad \frac{z_1 - z_2}{z_3 - z_2} = \frac{a}{a}e^{e^{i\pi/3}} \quad \text{...(i)}$$

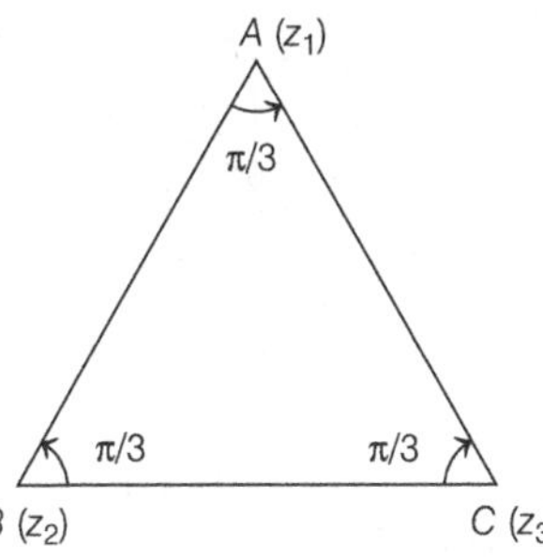

$\angle A = \pi/3$, then by rotation theorem

$$\therefore\quad \frac{z_3 - z_1}{z_2 - z_1} = \frac{a}{a}e^{\pi/3} \quad \text{...(ii)}$$

From Eq. (i) and Eq. (ii)

$$\frac{z_1 - z_2}{z_3 - z_2} = \frac{z_3 - z_1}{z_2 - z_1}$$

$$\Rightarrow\quad z_1^2 + z_2^2 + z_3^2 = z_1z_2 + z_2z_3 + z_3z_1 \text{ (S)} \quad \text{...(iii)}$$

and $\quad (z_1 + z_2 + z_3)^2 = (3z_0)^2$

$\Rightarrow z_1^2 + z_2^2 + z_3^2 + 2(z_1z_2 + z_2z_3 + z_3z_1) = 9z_0^2$

$\Rightarrow z_1^2 + z_2^2 + z_3^2 + 2(z_1^2 + z_2^2 + z_3^2) = 9z_0^2$ [from Eq. (iii)]

or $\quad 3(z_1^2 + z_2^2 + z_3^2) = 9z_0^2$

or $\quad z_1^2 + z_2^2 + z_3^2 = 3z_0^2$ (Q)

(B) $\because \angle C = \pi/2$, then by rotation theorem

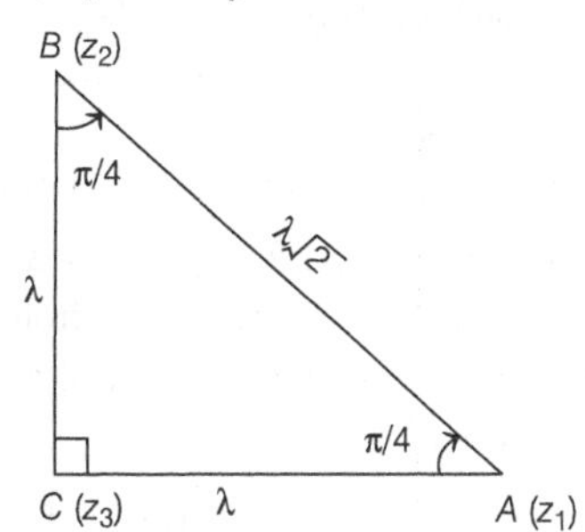

$$\therefore\quad \frac{z_2 - z_3}{z_1 - z_3} = \frac{\lambda}{\lambda}e^{i\pi/2}$$

or $\quad \dfrac{z_2 - z_3}{z_1 - z_3} = e^{i\pi/2}$ (T)

Also, $\because \angle B = \pi/4$ then by rotation theorem

$$\frac{z_1 - z_2}{z_3 - z_2} = \frac{\lambda\sqrt{2}}{\lambda}e^{i\pi/4}$$

or $\quad \dfrac{z_1 - z_2}{z_3 - z_2} = \sqrt{2}\,e^{i\pi/4} \quad$ ...(i)

and $\angle A = \pi/4$, then by rotation theorem

$$\frac{z_3 - z_1}{z_2 - z_1} = \frac{\lambda}{\lambda\sqrt{2}}e^{i\pi/4}$$

or $\quad \dfrac{z_3 - z_1}{z_2 - z_1} = \dfrac{1}{\sqrt{2}}\cdot e^{i\pi/4} \quad$ ...(ii)

Dividing Eq. (i) by Eq. (ii), then we get

$$(z_1 - z_2)^2 = 2(z_1 - z_3)(z_3 - z_2) \text{ (P)}$$

(C) Applying sine rule in ΔABC, then

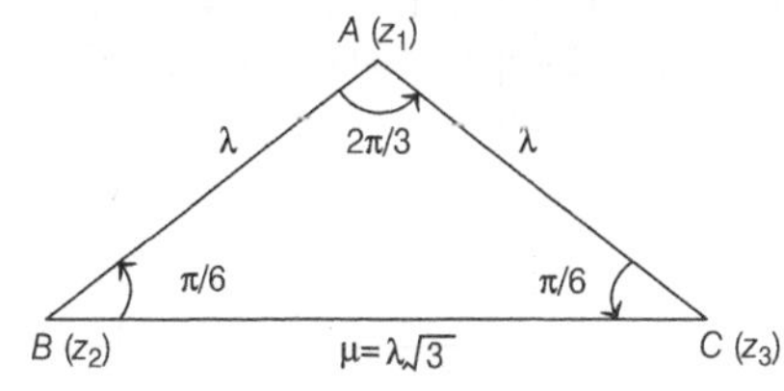

$$\frac{\mu}{\lambda} = \frac{\sin(2\pi/3)}{\sin(\pi/6)} = \sqrt{3}$$

$\therefore\quad \mu = \lambda\sqrt{3}$

$\because\quad \angle B = \pi/6$, applying rotation theorem

$$\frac{z_1 - z_2}{z_3 - z_2} = \frac{\lambda}{\lambda\sqrt{3}}e^{i\pi/6} \quad \text{...(i)}$$

and $\angle C = \pi/6$, applying rotation theorem

$$\frac{z_2 - z_3}{z_1 \quad z_3} = \frac{\lambda\sqrt{3}}{\lambda}e^{i\pi/6} \quad \text{...(ii)}$$

Dividing Eq. (ii) by Eq. (i), then

$$\frac{(z_2 - z_3)^2}{(z_1 - z_2)(z_3 - z_1)} = 3$$

or $\quad (z_2 - z_3)^2 = 3(z_1 - z_2)(z_3 - z_1)$ (R)

4. **(A)** $|\lambda_1 a_1\omega + \lambda_2 a_2\omega^2 + \dots + \lambda_n a_n\omega^n|$

$$\le |\lambda_1 a_1\omega| + |\lambda_2 a_2\omega^2| + \dots + |\lambda_n a_n\omega^n|$$

$$= |\lambda_1||a_1||\omega| + |\lambda_2||a_2||\omega|^2 + \dots + |\lambda_n||a_n||\omega|^n \quad (\because |\omega| = 1, \lambda_i \ge 0)$$

$$= \lambda_1|a_1| + \lambda_2|a_2| + \dots + \lambda_n|a_n| \quad \dots(\text{i})$$

$$< (\lambda_1 + \lambda_2 + \dots + \lambda_n) \; [\because |a_i| < 1]$$

$$= 1 < 2 \text{ (Q)}$$

$\therefore$ $|\lambda_1 a_1\omega + \lambda_2 a_2\omega^2 + \dots + \lambda_n a_n\omega^n| < 1$ (S)

Also, as $\lambda_i \ge 0$ and $\lambda_1 + \lambda_2 + \lambda_3 + \dots + \lambda_n = 1$, none of λ_i can exceed 1, thus $0 \le \lambda_i \le 1$ therefore, from Eq. (i), we get $|\lambda_1 a_1\omega + \lambda_2 a_2\omega^2 + \dots + \lambda_n a_n\omega^n| < |a_1| + |a_2| + - + |a_n|$ (T)

(B) $\because |1 + z + z^2 + \dots + z^n|$

$$= \left|\frac{z^{n+1} - 1}{z - 1}\right| < \left|\frac{z^{n+1} - 1}{z}\right| \le |z|^n + \frac{1}{|z|} \text{ (P)}$$

$$[\because \text{Re}(z) < 0, |z - 1| > |z|]$$

(C) $\because 1 + 2x + 3x^2 + \dots + 3nx^{3n-1}$

$$= \frac{d}{dx}(x + x^2 + x^3 + \dots + x^{3n})$$

$$= \frac{d}{dx}\left\{\frac{x(1 - x^{3n})}{1 - x}\right\}$$

$$= \frac{(1 - x)(1 - (3n + 1)x^{3n}) + x(1 - x^{3n})}{(1 - x)^2}$$

Put $x = \omega$, then

$$1 + 2\omega + 3\omega^2 + \dots + 3n\omega^{3n-1}$$

$$= \frac{(1 - \omega)(1 - (3n + 1)) + \omega(1 - 1)}{(1 - \omega)^2} = \frac{-3n}{(1 - \omega)}$$

$\therefore$ $|1 + 2\omega + 3\omega^2 + \dots + 3n\omega^{3n-1}| = \left|\frac{-3n}{1 - \omega}\right| = n\sqrt{3}$

$\therefore$ $\frac{1}{\sqrt{3}}|1 + 2\omega + 3\omega^2 + \dots + 3n\omega^{3n-1}| = n$ (R)

5. **(A)** $|z + 2 + i| \le 1 \Rightarrow |(z - 1) + 3 + i| \le 1$

or $|w + 3 + i| \le 1$ (where $w = z - 1$)

Clearly least value of

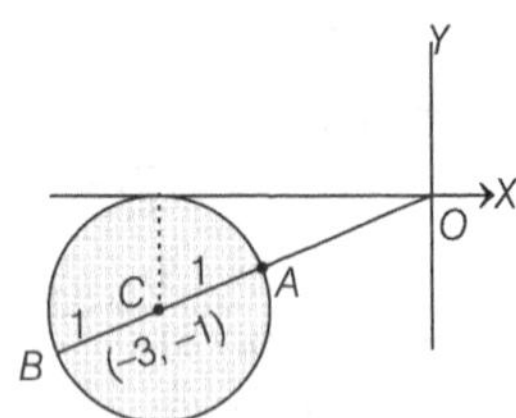

$$|w| = |z - 1| = OA = OC - CA$$
$$= \sqrt{10} - 1$$

$\therefore$ $L = \sqrt{10} - 1$

and greatest value of

$$|w| = |z - 1| = OB = OC + CB = \sqrt{10} + 1$$

$\therefore$ $G = \sqrt{10} + 1$

$\Rightarrow$ $LG = 9, G - L = 2$ (P,T)

(B) $1 \le |z - 1| \le 3$

$\Rightarrow$ $1 \le |(z + 2i) - 2i - 1| \le 3$

$\Rightarrow$ $1 \le |w - 2i - 1| \le 3$ (where $w = z + 2i$)

$PA = PB = 3$

Greatest value of $|w| = OA$

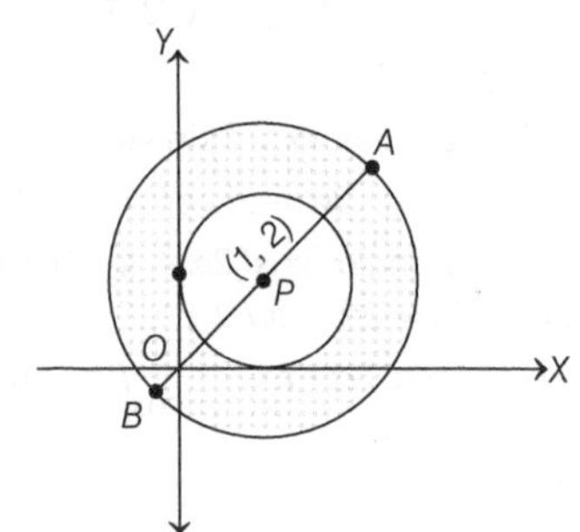

$$= OP + PA = \sqrt{5} + 3$$

$\therefore$ $G = \sqrt{5} + 3$

and least value of $|w| = OB = BP - OP$

$$= 3 - \sqrt{5}$$

$\therefore$ $L = 3 - \sqrt{5}$

$\Rightarrow$ $L + G = 6, LG = 4$ (Q, S)

and $\sqrt{2G} = \sqrt{5} + 1$

and $\sqrt{2L} = \sqrt{5} - 1$

$\therefore$ $\sqrt{2G} - \sqrt{2L} = 2$

or $(\sqrt{2G} - \sqrt{2L})^2 = 4$ (R)

(C) $|z + i| \le 1 \Rightarrow |(z - 2) + (2 + i)| \le 1$

or $|w + 2 + i| \le 1$ (where $w = z - 2$)

Greatest value of $|w| = OB$

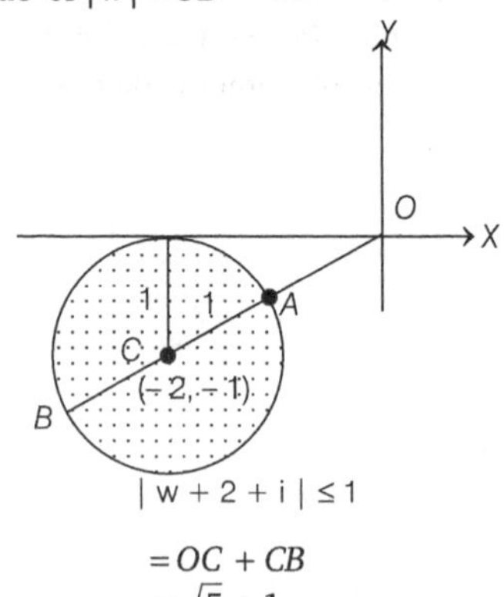

$$= OC + CB$$
$$= \sqrt{5} + 1$$

$\therefore$ $G = \sqrt{5} + 1$

and least value of $|w| = OA$

$$= OC - CA$$
$$= \sqrt{5} - 1$$

$\therefore$ $L = \sqrt{5} - 1$

then $LG = 4, G - L = 2$ (S,T)

02

Theory of Equations

Objective Questions Type I [Only one correct answer]

In each of the questions below, four choices are given of which only one is correct. You have to select the correct answer which is the most appropriate.

1. If α, β are the roots of the equation $ax^2 + bx + c = 0$ such that $\beta < \alpha < 0$, then the quadratic equation whose roots are $|\alpha|, |\beta|$ is given by

(a) $|a|x^2 + |b|x + |c| = 0$ (b) $ax^2 - |b|x + c = 0$
(c) $|a|x^2 - |b|x + |c| = 0$ (d) $a|x|^2 + b|x| + c = 0$

2. If one root of the equation $x^2 - \lambda x + 12 = 0$ is even prime while $x^2 + \lambda x + \mu = 0$ has equal roots, then μ is

(a) 8 (b) 16
(c) 24 (d) 32

3. Let p and q be two positive number such that $p + q = 2$ and $p^4 + q^4 = 272$. Then, p and q are roots of the equation **[JEE Main 2021, 24 February]**

(a) $x^2 - 2x + 2 = 0$
(b) $x^2 - 2x + 8 = 0$
(c) $x^2 - 2x + 136 = 0$
(d) $x^2 - 2x + 16 = 0$

4. If t_n denotes the nth term of an AP and $t_p = \frac{1}{q}$ and $t_q = \frac{1}{p}$, then which of the following is necessarily a root of the equation
$(p + 2q - 3r)x^2 + (q + 2r - 3p)x + (r + 2p - 3q) = 0$ is

(a) t_p (b) t_q
(c) t_{pq} (d) t_{p+q}

5. The value of a for which the equations $x^3 + ax + 1 = 0$ and $x^4 + ax^2 + 1 = 0$ have a common root is

(a) -2 (b) -1
(c) 1 (d) 2

6. The curve $y = (\lambda + 1)x^2 + 2$ intersects the curve $y = \lambda x + 3$ in exactly one point, if λ equals

(a) $\{-2\}$ (b) $\{2\}$
(c) $\{-2, 2\}$ (d) $\{1\}$

7. Let a, b, c be in arithmetic progression. Let the centroid of the triangle with vertices (a, c), $(2, b)$ and (a, b) be $\left(\frac{10}{3}, \frac{7}{3}\right)$. If α, β are the roots of the equation $ax^2 + bx + 1 = 0$, then the value of $\alpha^2 + \beta^2 - \alpha\beta$ is **[JEE Main 2021, 24 February]**

(a) $\frac{71}{256}$ (b) $-\frac{69}{256}$
(c) $\frac{69}{256}$ (d) $-\frac{71}{256}$

8. If α, β are the roots of $ax^2 + c = bx$, then the equation $(a + cy)^2 = b^2y$ in y has the roots

(a) α^{-1}, β^{-1} (b) α^2, β^2
(c) $\alpha\beta^{-1}, \alpha^{-1}\beta$ (d) α^{-2}, β^{-2}

9. If the equation $ax^2 + bx + c = 0$ and $x^3 + 3x^2 + 3x + 2 = 0$ have two common roots, then

(a) $a = b \neq c$ (b) $a \neq b = c$
(c) $a = b = c$ (d) $a = -b = c$

10. Number of ordered pair (x, y) satisfying $x^2 + 1 = y$ and $y^2 + 1 = x$, is

(a) 0 (b) 1
(c) 2 (d) 3

11. In the quadratic equation $p(x) = 0$ with real coefficients has purely imaginary roots. Then, the equation $p[p(x)] = 0$ has **[JEE Advanced 2014]**

(a) only purely imaginary roots
(b) all real roots
(c) two real and two purely imaginary roots
(d) neither real nor purely imaginary roots

12. The number of positive integral solutions of $x^4 - y^4 = 3789108$ is

(a) 0 (b) 1
(c) 2 (d) 4

13. If λ be an integer and α, β be the roots of $x^2 - 16x + \lambda = 0$ such that $1 < \alpha < 2$ and $2 < \beta < 3$, then possible values of λ are

(a) $\{60, 64, 68\}$ (b) $\{61, 62, 63\}$
(c) $\{49, 50, \ldots, 62, 63\}$ (d) $\{62, 65, 68, 71, 75\}$

14. The value of k for which the expression $x^2 + 2xy + ky^2 + 2x + k = 0$ can be resolved into linear factors is given by

(a) $\{0, 2\}$ (b) $\{0\}$
(c) $\{-2, 0\}$ (d) $\{2\}$

15. If $S = \{a \in N, 1 \le a \le 100\}$ and $[\tan^2 x] - \tan x - a = 0$ has real roots, where [.] denotes the greatest integer function, then number of elements in set A equals

(a) 2 (b) 5 (c) 6 (d) 9

16. Let α, β and γ be the roots of $f(x) = 0$, where $f(x) = x^3 + x^2 - 5x - 1 = 0$. Then $[\alpha] + [\beta] + [\gamma]$, where [.] denotes the greatest integer function, is equal to

(a) 1 (b) 4
(c) -1 (d) -3

17. Let $f(x) = x^3 + x^2 + 100x + 7\sin x$, then equation $\frac{1}{y - f(1)} + \frac{2}{y - f(2)} + \frac{3}{y - f(3)} = 0$ has

(a) no real root
(b) one real root
(c) two real roots
(d) more than two real roots

18. Let $-\frac{\pi}{6} < \theta < -\frac{\pi}{12}$. Suppose α_1 and β_1 are the roots of the equation $x^2 - 2x\sec\theta + 1 = 0$, and α_2 and β_2 are the roots of the equation $x^2 + 2x\tan\theta - 1 = 0$. If $\alpha_1 > \beta_1$ and $\alpha_2 > \beta_2$, then $\alpha_1 + \beta_2$ equals **[JEE Advanced 2016]**

(a) $2(\sec\theta - \tan\theta)$ (b) $2\sec\theta$
(c) $-2\tan\theta$ (d) 0

19. If α and β are the roots of the equation $x^2 - p(x+1) - q = 0$, then the value of $\frac{\alpha^2 + 2\alpha + 1}{\alpha^2 + 2\alpha + q} + \frac{\beta^2 + 2\beta + 1}{\beta^2 + 2\beta + q}$ is

(a) 2 (b) 1
(c) 0 (d) None of these

20. The value of $4 + \cfrac{1}{5 + \cfrac{1}{4 + \cfrac{1}{5 + \cfrac{1}{4 + \ldots\ldots \infty}}}}$ **[JEE Main 2021, 17 March]**

(a) $2 + \frac{2}{5}\sqrt{30}$ (b) $2 + \frac{4}{\sqrt{5}}\sqrt{30}$
(c) $4 + \frac{4}{\sqrt{5}}\sqrt{30}$ (d) $5 + \frac{2}{5}\sqrt{30}$

21. The real roots of the equation $5^{\log_5 (x^2 - 4x + 5)} = x - 1$ are

(a) 1 and 2 (b) 2 and 3
(c) 3 and 4 (d) 4 and 5

22. The number of real solutions of the equation $2|x|^2 - 5|x| + 2 = 0$ is

(a) 0 (b) 2
(c) 4 (d) infinite

23. If $x^2 + px + 1$ is a factor of $2\cos^2\theta\, x^3 + 2x + \sin 2\theta$, then

(a) $\theta = n\pi, n \in I$ (b) $\theta = n\pi + \frac{\pi}{2}, n \in I$
(c) $\theta = 2n\pi, n \in I$ (d) $\theta = \frac{n\pi}{2}, n \in I$

24. The number of real roots of the equation $x^3 + x^2 + 2x + \sin x = 0$ in $[-2\pi, 2\pi]$ is (are)

(a) zero (b) one
(c) two (d) three

25. The values of a for which the equation $2(\log_3 x)^2 - |\log_3 x| + a = 0$ possess four real solutions

(a) $-2 < a < 0$ (b) $0 < a < \frac{1}{8}$
(c) $0 < a < 5$ (d) None of these

26. If α and β are the roots of the equation $x^2 + px + 2 = 0$ and $\frac{1}{\alpha}$ and $\frac{1}{\beta}$ are the roots of the equation $2x^2 + 2qx + 1 = 0$, then $\left(\alpha - \frac{1}{\alpha}\right)\left(\beta - \frac{1}{\beta}\right)\left(\alpha + \frac{1}{\beta}\right)\left(\beta + \frac{1}{\alpha}\right)$ is equal to **[JEE Main 2020, 3 September]**

(a) $\frac{9}{4}(9 + q^2)$ (b) $\frac{9}{4}(9 - q^2)$
(c) $\frac{9}{4}(9 + p^2)$ (d) $\frac{9}{4}(9 - p^2)$

27. Let $f(x)$ be a quadratic polynomial such that $f(-1) + f(2) = 0$. If one of the roots of $f(x) = 0$ is 3, then its other root lies in **[JEE Main 2020, 2 September]**

(a) $(-1, 0)$ (b) $(1, 3)$
(c) $(-3, -1)$ (d) $(0, 1)$

28. The solution set of the equation $(x+1)^2 + [x-1]^2 = (x-1)^2 + [x+1]^2$, where $[x]$ and (x) are the greatest integer and nearest integer to x, is **[JEE Main 2015]**

(a) $x \in R$ (b) $x \in N$
(c) $x \in I$ (d) $x \in Q$

29. The number of values of a for which $(a^2 - 3a + 2)x^2 + (a^2 - 5a + 6)x + a^2 - 4 = 0$ is an identity in x is

(a) 0 (b) 1
(c) 2 (d) 3

30. The solution of $x-1=(x-[x])(x-\{x\})$ (where $[x]$ and $\{x\}$ are the integral and fractional part of x) is
(a) $x \in R$ (b) $x \in R \sim [1, 2)$
(c) $x \in [1, 2)$ (d) $x \in R \sim [1, 2]$

31. The number of solutions of $2^{\sin(|x|)} = 4^{|\cos x|}$ in $[-\pi, \pi]$ is equal to
(a) 0 (b) 2
(c) 4 (d) 6

32. Let $P(x) = x^2 + bx + c$ be a quadratic polynomial with real coefficient such that $\int_0^1 P(x)\,dx = 1$ and $P(x)$ leaves remainder 5 when it is divided by $(x-2)$. Then the value of $9(b+c)$ is equal to **[JEE Main 2021, 16 March]**
(a) 9 (b) 15
(c) 7 (d) 11

33. The value of α for which the equation $(\alpha+5)x^2 - (2\alpha+1)x + (\alpha-1) = 0$ has roots equal in magnitude but opposite in sign, is
(a) 7/4 (b) 1
(c) $-1/2$ (d) -5

34. The number of elements in the set
$\{x \in R : (|x|-3)|x+4| = 6\}$ is equal to
[JEE Main 2021, 16 March]
(a) 3 (b) 2
(c) 4 (d) 1

35. If α, β are the roots of the equation $x^2 + x\sqrt{\alpha} + \beta = 0$, then the values of α and β are
(a) $\alpha = 1$ and $\beta = -1$ (b) $\alpha = 1$ and $\beta = -2$
(c) $\alpha = 2$ and $\beta = 1$ (d) $\alpha = 2$ and $\beta = -2$

36. The coefficients a, b and c of the quadratic equation, $ax^2 + bx + c = 0$ are obtained by throwing a dice three times. The probability that this equation has equal roots is **[JEE Main 2021, 25 February]**
(a) $\frac{1}{54}$ (b) $\frac{1}{72}$
(c) $\frac{1}{36}$ (d) $\frac{5}{216}$

37. For $a \neq b$, if the equations $x^2 + ax + b = 0$ and $x^2 + bx + a = 0$ have a common root, then the value of $(a+b)$ is
(a) -1 (b) 0
(c) 1 (d) 2

38. Let α, β be the roots of the equation $(x-a)(x-b) = c$, $c \neq 0$. Then, the roots of the equation $(x-\alpha)(x-\beta) + c = 0$ are
(a) a, c (b) b, c
(c) a, b (d) $a+c, b+c$

39. If α, β are the roots of the equation $x^2 + x + 1 = 0$, then the equation whose roots are α^{19} and β^7 is
(a) $x^2 + x + 1 = 0$ (b) $x^2 - x + 1 = 0$
(c) $x^2 + x + 2 = 0$ (d) $x^2 + 19x + 7 = 0$

40. The number of real solutions of
$1 + |e^x - 1| = e^x(e^x - 2)$ is
(a) 0 (b) 1
(c) 2 (d) 4

41. The number of solutions of the equation $|x| = \cos x$ is
(a) one (b) two
(c) three (d) zero

42. The value of $3 + \cfrac{1}{4 + \cfrac{1}{3 + \cfrac{1}{4 + \cfrac{1}{3 + \dots \infty}}}}$ is equal to
[JEE Main 2021, 18 March]
(a) $1.5 + \sqrt{3}$ (b) $2 + \sqrt{3}$
(c) $3 + 2\sqrt{3}$ (d) $4 + \sqrt{3}$

43. The value of p for which both the roots of the equation $4x^2 - 20px + (25p^2 + 15p - 66) = 0$, are less than 2, lies in
(a) $(4/5, 2)$ (b) $(2, \infty)$
(c) $(-1, -4/5)$ (d) $(-\infty, -1)$

44. If the equation $ax^2 + 2bx - 3c = 0$ has non real roots and $(3c/4) < (a+b)$; then c is always
(a) < 0 (b) > 0
(c) ≥ 0 (d) zero

45. The root of the equation
$2(1+i)x^2 - 4(2-i)x - 5 - 3i = 0$, where $i = \sqrt{-1}$, which has greater modulus is
(a) $(3-5i)/2$ (b) $(5-3i)/2$
(c) $(3+i)/2$ (d) $(1+3i)/2$

46. Let $a, b, c \in R$ and $a \neq 0$. If α is a root of $a^2x^2 + bx + c = 0$, β is a root of $a^2x^2 - bx - c = 0$ and $0 < \alpha < \beta$, then the equation, $a^2x^2 + 2bx + 2c = 0$ has a root γ that always satisfies
(a) $\gamma = \alpha$ (b) $\gamma = \beta$
(c) $\gamma = (\alpha+\beta)/2$ (d) $\alpha < \gamma < \beta$

47. The roots of the equation, $2^{x+2} \cdot 3^{3x/(x-1)} = 9$ are given by
(a) $1 - \log_2 3, 2$ (b) $\log_2(2/3), 1$
(c) $2, -2$ (d) $-2, 1 - (\log 3)/(\log 2)$

48. Let α, β, γ be the roots of the equation $(x-a)(x-b)(x-c) = d$, $d \neq 0$, then the roots of the equation $(x-\alpha)(x-\beta)(x-\gamma) + d = 0$ are
(a) a, b, d (b) b, c, d
(c) a, b, c (d) $a+d, b+d, c+d$

49. Let α and β be the roots of $x^2 - 3x + p = 0$ and γ and δ be the roots of $x^2 - 6x + q = 0$. If $\alpha, \beta, \gamma, \delta$ form a geometric progression. Then, ratio $(2q+p):(2q-p)$ is
[JEE Main 2020, 4 September]
(a) 9 : 7 (b) 3 : 1
(c) 5 : 3 (d) 33 : 31

50. Let $\lambda \neq 0$ be in R. If α and β are the roots of the equation, $x^2 - x + 2\lambda = 0$ and α and γ are the roots of the equation, $3x^2 - 10x + 27\lambda = 0$, then $\frac{\beta\gamma}{\lambda}$ is equal to **[JEE Main 2020, 4 September]**

(a) 36 (b) 9
(c) 27 (d) 18

51. Let $[t]$ denote the greatest integer $\leq t$. Then, the equation in x, $[x]^2 + 2[x+2] - 7 = 0$ has **[JEE Main 2020, 4 September]**

(a) infinitely many solutions
(b) exactly four integral solutions
(c) no integral solution
(d) exactly two solutions

52. The solution set of $(x)^2 + (x+1)^2 = 25$, where (x) is the nearest integer greater than or equal to x, is

(a) $(2, 4)$ (b) $[-5, -4) \cup [2, 3)$
(c) $[-4, -3) \cup [3, 4)$ (d) None of these

53. If α, β are roots of $375x^2 - 25x - 2 = 0$ and $s_n = \alpha^n + \beta^n$, then $\lim_{n \to \infty} \sum_{r=1}^{n} s_r$ is **[JEE Main 2019, 10 January]**

(a) $\frac{7}{116}$ (b) $\frac{1}{12}$
(c) $\frac{29}{358}$ (d) None of these

54. Let α and β be the roots of the equation $x^2 - x - 1 = 0$. If $p_k = (\alpha)^k + (\beta)^k, k \geq 1$, then which one of the following statements is not true? **[JEE Main 2020, 7 January]**

(a) $p_3 = p_5 - p_4$
(b) $(p_1 + p_2 + p_3 + p_4 + p_5) = 26$
(c) $p_5 = p_2 \cdot p_3$
(d) $p_5 = 11$

55. How many roots does the following equation possess ? $3^{|x|} |2 - |x|| = 1$

(a) 1 (b) 2
(c) 3 (d) 4

56. If the roots of the equation, $x^2 + 2ax + b = 0$, are real and distinct and they differ by at most $2m$, then b lies in the interval

(a) $(a^2 - m^2, a^2)$ (b) $[a^2 - m^2, a^2)$
(c) $(a^2, a^2 + m^2)$ (d) None of these

57. If $x^2 + px + 1$ is a factor of the expression $ax^3 + bx^2 + c$, then

(a) $a^2 + c^2 = -ab$ (b) $a^2 - c^2 = -ab$
(c) $a^2 - c^2 = -bc$ (d) None of these

58. Suppose a, b denote the distinct real roots of the quadratic polynomial $x^2 + 20x - 2020$ and suppose c, d denote the distinct complex roots of the quadratic polynomial $x^2 - 20x + 2020$. Then the value of $ac(a-c) + ad(a-d) + bc(b-c) + bd(b-d)$ is **[JEE Advanced 2020]**

(a) 0 (b) 8000
(c) 8080 (d) 16000

59. Let α and β be the roots of $x^2 - 6x - 2 = 0$. If $a^n = \beta^n$ for $n \geq 1$, then the value of $\frac{a_{10} - 2a_8}{3a_9}$ is **[JEE Main 2021, 25 February]**

(a) 4 (b) 1
(c) 2 (d) 3

60. The integer 'k', for which the inequality $x^2 - 2(3k-1)x + 8k^2 - 7 > 0$ is valid for every x in R is **[JEE Main 2021, 25 February]**

(a) 3 (b) 2
(c) 4 (d) 0

61. Given that, for all $x \in R$, the expression $\frac{x^2 - 2x + 4}{x^2 + 2x + 4}$ lies between $\frac{1}{3}$ and 3, the value between which the expression $\frac{9 \cdot 3^{2x} + 6 \cdot 3^x + 4}{9 \cdot 3^{2x} - 6 \cdot 3^x + 4}$ lies are

(a) 3^{-1} and 3 (b) -2 and 0
(c) -1 and 1 (d) 0 and 2

62. The solution set of the equation $\log_x 2 \log_{2x} 2 = \log_{4x} 2$ is

(a) $\{2^{-\sqrt{2}}, 2^{\sqrt{2}}\}$ (b) $\left\{\frac{1}{2}, 2\right\}$
(c) $\left\{\frac{1}{4}, 2^2\right\}$ (d) None of these

63. The solution of $\left|\frac{x}{x-1}\right| + |x| = \frac{x^2}{|x-1|}$ is

(a) $x \geq 0$ (b) $x > 0$
(c) $x \in (1, \infty)$ (d) None of these

64. The number of negative integral solutions of $x^2 \cdot 2^{x+1} + 2^{|x-3|+2} = x^2 \cdot 2^{|x-3|+4} + 2^{x-1}$ is

(a) none (b) only one
(c) two (d) four

65. For the equation $|x^2 - 2x - 3| = b$ which statement or statements are true

(a) for $b < 0$ there are no solutions
(b) for $b = 0$ there are three solutions
(c) for $0 < b < 1$ there are four solutions
(d) for $b = 1$ there are two solutions

66. The roots of the equation $(3-x)^4 + (2-x)^4 = (5-2x)^4$ are

(a) all real
(b) all imaginary
(c) two real and two imaginary
(d) None of the above

67. The largest interval in which $x^{12} - x^9 + x^4 - x + 1 > 0$ is
(a) $[0, \infty)$ (b) $(-\infty, 0]$
(c) $(-\infty, \infty)$ (d) None of these

68. If $\alpha, \beta \in R$ are such that $1 - 2i$ (here $i^2 = -1$) is a root of $z^2 + \alpha z + \beta = 0$, then $(\alpha - \beta)$ is equal to
[JEE Main 2021, 5 February]
(a) 7 (b) -3
(c) 3 (d) -7

69. If α, β be the roots of $x^2 - x - 1 = 0$ and $A_n = \alpha^n + \beta^n$, then AM of A_{n-1} and A_n is
(a) $2A_{n+1}$ (b) $(1/2)A_{n+1}$
(c) $2A_{n-2}$ (d) None of these

70. Let S be the set of values of 'a' for which 2 lie between the roots of the quadratic equation $x^2 + (a + 2)x - (a + 3) = 0$, then S is given by
(a) $(-\infty, -5)$ (b) $(5, \infty)$
(c) $(-\infty, -5]$ (d) $[5, \infty)$

Objective Questions Type II [One or more than one correct answer(s)]

In each of the questions below four choices of which one or more than one are correct. You have to select the correct answer(s) accordingly.

1. The equation $x^{[(\log_3 x)^2 - (9/2)\log_3 x + 5]} = 3\sqrt{3}$ has
(a) at least one real solution
(b) exactly three real solutions
(c) exactly one irrational solution
(d) complex roots

2. For $a > 0$, the roots of the equation $\log_{ax} a + \log_x a^2 + \log_{a^2 x} a^3 = 0$, are given by
(a) $a^{-4/3}$ (b) $a^{-3/4}$
(c) $a^{-1/2}$ (d) a^{-1}

3. The real values of λ for which the equation, $3x^3 + x^2 - 7x + \lambda = 0$, has two distinct real roots in $[0, 1]$ lie in the interval(s)
(a) $(-2, 0)$ (b) $[0, 1]$
(c) $[1, 2]$ (d) $(-\infty, \infty)$

4. If α is one root of the equation $4x^2 + 2x - 1 = 0$, then its other root is given by
(a) $4\alpha^3 - 3\alpha$ (b) $4\alpha^3 + 3\alpha$
(c) $\alpha - (1/2)$ (d) $-\alpha - (1/2)$

5. The roots of the equation, $(x^2 + 1)^2 = x(3x^2 + 4x + 3)$, are given by
(a) $2 - \sqrt{3}$
(b) $(-1 + i\sqrt{3})/2, i = \sqrt{-1}$
(c) $2 + \sqrt{3}$
(d) $(-1 - i\sqrt{3})/2, i = \sqrt{-1}$

6. If A, G and H are the Arithmetic mean, Geometric mean and Harmonic mean between two unequal positive integers. Then the equation $Ax^2 - |G|x - H = 0$ has
(a) both roots are fractions
(b) at least one root which is negative fraction
(c) exactly one positive root
(d) at least one root which is an integer

7. If $0 < a < b < c$, and the roots α, β of the equation $ax^2 + bx + c = 0$ are non real complex roots, then
(a) $|\alpha| = |\beta|$ (b) $|\alpha| > 1$
(c) $|\beta| < 1$ (d) None of these

8. If $a, b, c \in R$ and the equality $ax^2 - bx + c = 0$ has complex roots which are reciprocal of each other, then one has
(a) $|b| \le |a|$ (b) $|b| \le |c|$
(c) $a = c$ (d) $b \le a$

9. If a, b, c are positive rational numbers such that $a > b > c$ and the quadratic equation $(a + b - 2c)x^2 + (b + c - 2a)x + (c + a - 2b) = 0$ has a root in the interval $(-1, 0)$, then
(a) $b + c > a$
(b) $c + a < 2b$
(c) both roots of the given equation are rational
(d) the equation $ax^2 + 2bx + c = 0$ has both negative real roots

10. The equation $|x + 1||x - 1| = a^2 - 2a - 3$ can have real solutions for x, if a belongs x to
(a) $(-\infty, -1] \cup [3, \infty)$
(b) $[1 - \sqrt{5}, 1 + \sqrt{5}]$
(c) $[1 - \sqrt{5}, -1] \cup [3, 1 + \sqrt{5}]$
(d) None of the above

11. If a, b, c are rational and no two of them are equal, then the equations $(b - c)x^2 + (c - a)x + (a - b) = 0$ and $a(b - c)x^2 + b(c - a)x + c(a - b) = 0$
(a) have rational roots
(b) will be such that at least one has rational roots
(c) have exactly one root common
(d) have at least one root common

12. If α, β are the roots of $ax^2 + bx + c = 0$ and $\alpha + h, \beta + h$ are the roots of $px^2 + qx + r = 0$, then
(a) $h = \frac{1}{2}\left(\frac{b}{a} - \frac{q}{p}\right)$ (b) $\frac{b^2 - 4ac}{a^2} = \frac{q^2 - 4pr}{p^2}$
(c) $\frac{a}{p} = \frac{b}{q} = \frac{c}{r}$ (d) None of these

13. The equation $x^2 + a^2 x + b^2 = 0$ has two roots each of which exceeds a number c, then
(a) $a^4 > 4b^2$ (b) $c^2 + a^2 c + b^2 > 0$
(c) $-a^2/2 > c$ (d) None of these

14. If $a + 3b + 9c = 0$, $ac < 0$ and one root of the equation $ax^2 + bx + c = 0$ is square of the other, then
(a) a and b have same sign
(b) b and c have opposite sign
(c) both roots are rational
(d) a, b, c are irrational

15. A quadratic equation whose difference of roots is 3 and the sum of the squares of the roots is 29 is given by
(a) $x^2 + 9x + 14 = 0$ (b) $x^2 + 7x + 10 = 0$
(c) $x^2 - 7x - 10 = 0$ (d) $x^2 - 7x + 10 = 0$

16. If the equation $ax^2 - 2bx + c = 0$ has real roots which are reciprocal of each other, then
(a) $a = c$ (b) $b \le a$
(c) $|b| \ge |a|$ (d) $|b| \ge |c|$

17. Let $a, b, c \in R$. If $ax^2 + bx + c = 0$ has two real roots A and B where $A < -1$ and $B > 1$, then
(a) $1 + \left|\frac{b}{a}\right| + \frac{c}{a} < 0$ (b) $1 - \left|\frac{b}{a}\right| + \frac{c}{a} < 0$
(c) $|c| < |a|$ (d) $|c| < |a| - |b|$

18. If a, b, c are distinct numbers in arithmetic progression, then both the roots of the quadratic equation
$(a + 2b - 3c)x^2 + (b + 2c - 3a)x + (c + 2a - 3b) = 0$
are
(a) real (b) positive
(c) negative (d) rational

19. If roots of $ax^2 + 2bx + c = 0$ $(a \ne 0)$ are non real complex and $a + c < 2b$, then
(a) $c > 0$ (b) $c < 0$
(c) $4a + c < 4b$ (d) $4a + c > 4b$

20. If $b^2 \ge 4ac$ for the equation $ax^4 + bx^2 + c = 0$, then all roots of the equation will be real, if
(a) $b > 0, a > 0, c > 0$ (b) $b < 0, a > 0, c > 0$
(c) $b > 0, a > 0, c < 0$ (d) $b > 0, a < 0, c < 0$

21. Let S be the set of all non-zero real numbers α such that the quadratic equation $\alpha x^2 - x + \alpha = 0$ has two distinct real roots x_1 and x_2 satisfying the inequality $|x_1 - x_2| < 1$. Which of the following interval(s) is/are a subset of S? **[JEE Advanced 2015]**
(a) $\left(-\frac{1}{2}, -\frac{1}{\sqrt{5}}\right)$ (b) $\left(-\frac{1}{\sqrt{5}}, 0\right)$
(c) $\left(0, \frac{1}{\sqrt{5}}\right)$ (d) $\left(\frac{1}{\sqrt{5}}, \frac{1}{2}\right)$

22. If $a, b, c \in R$ and $a + b + c = 0$, then the quadratic equation $3ax^2 + 2bx + c = 0$ has
(a) at least one root in $[0, 1]$
(b) at least one root in $[-1, 1]$
(c) at least one root in $[0, 2]$
(d) none of the above

23. If $ax + by = 1$, $cx^2 + dy^2 = 1$ have only one solution, then
(a) $\frac{a^2}{c} + \frac{b^2}{d} = 1$ (b) $x = \frac{a}{c}$
(c) $y = \frac{b}{d}$ (d) None of these

24. If the roots of the equation $x^3 + bx^2 + cx - 1 = 0$ form an increasing GP, then
(a) $b + c = 0$
(b) $b \in (-\infty, -3)$
(c) one of the roots is 1
(d) one root is smaller than one and one root is more than 1

25. If the equation $ax^2 + bx + c = 0$ $(a > 0)$ has two roots α and β such that $\alpha < -2$ and $\beta > 2$, then
(a) $b^2 - 4ac > 0$ (b) $c < 0$
(c) $a + |b| + c < 0$ (d) $4a + 2|b| + c < 0$

26. If $c \ne 0$ and the equation $\frac{p}{2x} = \frac{a}{x + c} + \frac{b}{x - c}$ has two equal roots, then p can be
(a) $(\sqrt{a} - \sqrt{b})^2$ (b) $(\sqrt{a} + \sqrt{b})^2$
(c) $a + b$ (d) $a - b$

27. Suppose $a, b \in R$ and $a \ne 0, b \ne 0$. Let α, β be the roots of $x^2 + ax + b = 0$. Then
(a) $\frac{1}{\alpha}, \frac{1}{\beta}$ are roots of $bx^2 + ax + 1 = 0$
(b) $-\alpha, -\beta$ are roots of $x^2 - ax + b = 0$
(c) α^2, β^2 are roots of $x^2 + (2b - a^2)x + b^2 = 0$
(d) $\frac{\alpha}{\beta}, \frac{\beta}{\alpha}$ are roots of $bx^2 + (2b - a^2)x + b = 0$

28. If α and β are the roots of the equation $x^2 + px + q = 0$ and α^4, β^4 are the roots of $x^2 - rx + s = 0$, then the equation $x^2 - 4qx + 2q^2 - r = 0$ has always
(a) two real roots
(b) two positive roots
(c) two negative roots
(d) one positive and one negative root

Linked-Comprehension Type

In these questions, a passage (paragraph) has been given followed by questions based on each of the passages. You have to answer the questions based on the passage given.

PASSAGE 1

Let p, q be integers and let α, β be the roots of the equation, $x^2 - x - 1 = 0$ where $a \neq \beta$. For $n = 0, 1, 2, \ldots$, let $a_n = p\alpha^n + q\beta^n$.

FACT If a and b are rational numbers and $a + b\sqrt{5} = 0$, then $a = 0 = b$.

On the basis of above information, answer the following questions **[JEE Advanced 2017]**

1. $a_{12} =$
   (a) $a_{11} + 2a_{10}$ (b) $2a_{11} + a_{10}$
   (c) $a_{11} - a_{10}$ (d) $a_{11} + a_{10}$
2. If $a_4 = 28$, then $p + 2q =$
   (a) 14 (b) 7
   (c) 21 (d) 12

PASSAGE-2

Consider the equation $2 + |x^2 + 4x + 3| = m, m \in R$

On the basis of above information, answer the following questions

1. Set of all real values of m so that the given equation has three solutions is
   (a) {3} (b) {2}
   (c) {1} (d) {0}
2. Set of all real values of m so that given equation have four distinct solutions, is
   (a) (0, 1) (b) (1, 2)
   (c) (1, 3) (d) [2, 3)
3. Set of all real values of m so that the given equation have two solutions is
   (a) $(3, \infty)$ (b) $(2, \infty)$
   (c) $\{2\} \cup (3, \infty)$ (d) None of these

PASSAGE-3

Consider the polynomial $f(x) = 1 + 2x + 3x^2 + 4x^3$. Let S be the sum of all distinct real root of $f(x)$ and let $t = |s|$.

On the basis of above information, answer the following questions **[JEE Advanced 2010]**

1. The real numbers s lies in the interval
   (a) $\left(-\frac{1}{4}, 0\right)$ (b) $\left(-11, -\frac{3}{4}\right)$
   (c) $\left(-\frac{3}{4}, -\frac{1}{2}\right)$ (d) $\left(0, \frac{1}{4}\right)$
2. The area bounded by the curve $y = f(x)$ and the lines $x = 0, y = 0$ and $x = t$, lies in the interval
   (a) $\left(\frac{3}{4}, 3\right)$ (b) $\left(\frac{21}{64}, \frac{11}{16}\right)$
   (c) (9, 10) (d) $\left(0, \frac{21}{64}\right)$
3. The function $f'(x)$ is
   (a) increasing in $\left(-t, -\frac{1}{4}\right)$ and decreasing in $\left(-\frac{1}{4}, t\right)$
   (b) decreasing in $\left(-t, -\frac{1}{4}\right)$ and increasing in $\left(-\frac{1}{4}, t\right)$
   (c) increasing in $(-t, t)$
   (d) decreasing in $(-t, t)$

PASSAGE 4

Let consider quadratic equation

$$ax^2 + bx + c = 0 \quad \ldots(i)$$

where $a, b, c \in R$ and $a \neq 0$. If Eq. (i) has roots, α, β

$\therefore \quad \alpha + \beta = -\frac{b}{a}, \alpha\beta = \frac{c}{a}$ and Eq. (i) can be written as $ax^2 + bx + c = a(x - \alpha)(x - \beta)$.

Also, if $a_1, a_2, a_3, a_4, \ldots$ are in AP, then $a_2 - a_1 = a_3 - a_2 = a_4 - a_3 = \ldots \neq 0$ and if $b_1, b_2, b_3, b_4, \ldots$ are in GP, then $\frac{b_2}{b_1} = \frac{b_3}{b_2} = \frac{b_4}{b_3} = \ldots \neq 1$ Now, if $c_1, c_2, c_3, c_4, \ldots$ are in HP, then $\frac{1}{c_2} - \frac{1}{c_1} = \frac{1}{c_3} - \frac{1}{c_2} = \frac{1}{c_4} - \frac{1}{c_3} = \ldots \neq 0$

On the basis of above information, answer the following questions

1. Let p and q be roots of the equation $x^2 - 2x + A = 0$ and let r and s be the roots of the equation $x^2 - 18x + B = 0$. If $p < q < r < s$ are in arithmetic progression. Then the values of A and B respectively are
   (a) – 5, 67 (b) – 3, 77
   (c) 67, – 5 (d) 77, – 3
2. Let α_1, α_2 be the roots of $x^2 - x + p = 0$ and α_3, α_4 be the roots of $x^2 - 4x + q = 0$. If $\alpha_1, \alpha_2, \alpha_3, \alpha_4$ are in GP, then the integral values, of p and q respectively are
   (a) – 2, – 32 (b) – 2, 3
   (c) – 6, 3 (d) – 6, – 32
3. Given that β_1, β_3 be roots of the equation $Ax^2 - 4x + 1 = 0$ and β_2, β_4 the roots of the equation $Bx^2 - 6x + 1 = 0$.

If $\beta_1, \beta_2, \beta_3, \beta_4$ are in HP, then the integral values of A and B respectively are

(a) $-3, 8$ (b) $-3, 16$
(c) $3, 8$ (d) $3, 16$

4. If a, b, c, d and x are distinct real numbers such that $(a^2+b^2+c^2)x^2-2(ab+bc+cd)x+(b^2+c^2+d^2)\le 0$, then a, b, c, d

(a) are in AP (b) are in GP
(c) are in HP (d) satisfy $ab = cd$

5. The harmonic mean of the roots of the equation $(5+\sqrt{2})x^2-(4+\sqrt{5})x+8+2\sqrt{5}=0$ is

(a) 2 (b) 4
(c) 6 (d) 8

6. If the roots of equation
$$a(b-c)x^2+b(c-a)x+c(a-b)=0$$
be equal, then a, b, c are in

(a) AP (b) GP
(c) HP (d) not in AP/GP/HP

7. Let α, β be the roots of the equation $(x-a)(x-b)+c=0, c\ne 0$, then the roots of the equation $(x-\alpha)(x-\beta)=c$ are

(a) a, c (b) b, c
(c) a, b (d) $a+c, b+c$

PASSAGE 5

Let $(a+\sqrt{b})^{Q(x)}+(a-\sqrt{b})^{Q(x)-2\lambda}=A$, where $\lambda\in N, A\in R$ and $a^2-b=1$.

$\therefore\ (a+\sqrt{b})(a-\sqrt{b})=1$

$\Rightarrow\ (a+\sqrt{b})=(a-\sqrt{b})^{-1}$ and $(a-\sqrt{b})=(a+\sqrt{b})^{-1}$

ie, $(a\pm\sqrt{b})=(a+\sqrt{b})^{\pm 1}$ or $(a-\sqrt{b})^{\pm 1}$

On the basis of above information, answer the following questions

1. If $(4+\sqrt{15})^{[x]}+(4-\sqrt{15})^{[x]}=62$, where [.] denotes the greatest integer function, then

(a) $x\in[-3,-2)\cup[1,2)$ (b) $x\in[-3,2)\cup[-2,1)$
(c) $x\in[-2,-1)\cup[2,3)$ (d) $x\in[-2,3)\cup[-1,2)$

2. Solutions of
$$(2+\sqrt{3})^{x^2-2x+1}+(2-\sqrt{3})^{x^2-2x-1}=\frac{4}{2-\sqrt{3}}$$ are

(a) $1\pm\sqrt{3}, 1$ (b) $1\pm\sqrt{2}, 1$
(c) $1\pm\sqrt{3}, 2$ (d) $1\pm\sqrt{2}, 2$

3. The number of real solutions of the equation $(15+4\sqrt{14})^t+(15-4\sqrt{14})^t=30$ are where $t=x^2-2|x|$

(a) 0 (b) 2 (c) 4 (d) 6

4. If α, β are the roots of the equation $1!+2!+3!+\ldots(x-1)!+x!=k^2$ and $k\in I$, where $\alpha<\beta$ and if $\alpha_1, \alpha_2, \alpha_3, \alpha_4$ are the roots of the equation
$$(a+\sqrt{b})^{x^2-[1+2\alpha+3\alpha^2+4\alpha^3+5\alpha^4]}+(a-\sqrt{b})^{x^2+[-5\beta]}$$
$=2a$, where $a^2-b=1$ and [.] denotes the greatest integer function, then the value of $|\alpha_1+\alpha_2+\alpha_3+\alpha_4-\alpha_1\alpha_2\alpha_3\alpha_4|$ is

(a) 216 (b) 221
(c) 224 (d) 209

5. If $\left(\sqrt{(49+20\sqrt{6})}\right)^{\sqrt{a\sqrt{a\sqrt{a\ldots\infty}}}}$
$$+(5-2\sqrt{6})^{x^2+x-3-\sqrt{x\sqrt{x\sqrt{x\ldots\infty}}}}=10$$
where $a=x^2-3$, then x is

(a) $-\sqrt{2}$ (b) $\sqrt{2}$
(c) -2 (d) 2

6. If α, β are the roots of the equation $x^2-4x+1=0$, where $\alpha>\beta$, then the number of real solutions of the equation $\alpha^{y^2-2y+1}+\beta^{y^2-2y-1}=\dfrac{101}{10\beta}$ are

(a) 0 (b) 2
(c) 4 (d) 6

PASSAGE 6

Let consider the quadratic equation $(1+m)x^2-2(1+3m)x+(1+8m)=0$, where $m\in R\sim\{-1\}$.

On the basis of above information, answer the following questions

1. The number of integral values of m such that given quadratic equation has imaginary roots are

(a) 0 (b) 1
(c) 2 (d) 3

2. The set of values of m such that the given quadratic equation has at least one root is negative is

(a) $m\in(-\infty,-1)$ (b) $m\in\left(-\dfrac{1}{8},\infty\right)$
(c) $m\in\left(-1,-\dfrac{1}{8}\right)$ (d) $m\in R$

3. The set of values of m such that the given quadratic equation has both roots are positive is

(a) $m\in R$ (b) $m\in(-1,3)$
(c) $m\in[3,\infty)$ (d) $(-\infty,-1)\cup[3,\infty)$

4. The set of values of m such that the given quadratic equation has at least one root is positive is

(a) $m\in(-\infty,-1)\cup\left(-1,-\dfrac{1}{8}\right)\cup\left(-\dfrac{1}{3},\infty\right)$

(b) $m\in(-\infty,-1)\cup\left(-\dfrac{1}{3},-\dfrac{1}{8}\right)\cup[3,\infty)$

(c) $m\in\left(-\dfrac{1}{3},-\dfrac{1}{8}\right)$

(d) $m\in(-\infty,-1)\cup\left(-1,-\dfrac{1}{8}\right)\cup[3,\infty)$

5. The number of real values of m such that the given quadratic equation has roots are in the ratio 2 : 3 are
(a) 0 (b) 2
(c) 4 (d) infinite many

PASSAGE 7

If α, β, γ be the roots of the equation $ax^3 + bx^2 + cx + d = 0$. To obtain the equation whose roots are $f(\alpha), f(\beta), f(\gamma)$, where f is a function, we put $y = f(\alpha)$ and obtain $\alpha = f^{-1}(y)$.

Now, α is a root of the equation $ax^3 + bx^2 + cx + d = 0$, then we obtain the desired equation which is $a\{f^{-1}(y)\}^3 + b\{f^{-1}(y)\}^2 + c\{f^{-1}(y)\} + d = 0$

For example, if α, β, γ are the roots of $ax^3 + bx^2 + cx + d = 0$. To find equation whose roots are $\alpha^2, \beta^2, \gamma^2$, we put $y = \alpha^2 \Rightarrow \alpha = \sqrt{y}$.

As α is a root of $ax^3 + bx^2 + cx + d = 0$,

we get $ay^{3/2} + by + c\sqrt{y} + d = 0$

or $\sqrt{y}(ay + c) = -(by + d)$

On squaring both sides, then

$y(a^2y^2 + 2acy + c^2) = b^2y^2 + 2bdy + d^2$ or

$a^2y^3 + (2ac - b^2)y^2 + (c^2 - 2bd)y - d^2 = 0$

This is desired equation.

On the basis of above information, answer the following questions

1. If α, β are the roots of the equation $ax^2 + bx + c = 0$, then the roots of the equation $a(x+1)^2 + b(x+1)(x-2) + c(x-2)^2 = 0$ are
(a) $\frac{2\alpha+1}{\alpha-1}, \frac{2\beta+1}{\beta-1}$ (b) $\frac{2\alpha-1}{\alpha+1}, \frac{2\beta-1}{\beta+1}$
(c) $\frac{\alpha+1}{\alpha-2}, \frac{\beta+1}{\beta-2}$ (d) $\frac{2\alpha+3}{\alpha-1}, \frac{2\beta+3}{\beta-1}$

2. If α, β are the roots of the equation $2x^2 + 4x - 5 = 0$, the equation whose roots are the reciprocals of $2\alpha - 3$ and $2\beta - 3$ is
(a) $x^2 + 10x - 11 = 0$ (b) $11x^2 + 10x + 1 = 0$
(c) $x^2 + 10x + 11 = 0$ (d) $11x^2 - 10x + 1 = 0$

3. If α, β are the roots of the equation $px^2 - qx + r = 0$, then the equation whose roots are $\alpha^2 + \frac{r}{p}$ and $\beta^2 + \frac{r}{p}$ is
(a) $p^3x^2 + pq^2x + r = 0$ (b) $px^2 - qx + r = 0$
(c) $p^3x^2 - pq^2x + q^2r = 0$ (d) $px^2 + qx - r = 0$

4. If α, β, γ are the roots of the equation $x^3 - x - 1 = 0$, then the value of $\Pi\left(\frac{1+\alpha}{1-\alpha}\right)$ is equal to
(a) -7 (b) -5
(c) -3 (d) -1

5. If α, β, γ are the roots of the equation $x^3 + qx - r = 0$ then the equation whose roots are $\beta\gamma + \frac{1}{\alpha}$, $\gamma\alpha + \frac{1}{\beta}$ and $\alpha\beta + \frac{1}{\gamma}$ is
(a) $rx^3 + q(r+1)x^2 - (r+1)^3 = 0$
(b) $rx^3 - q(r+1)x^2 - (r+1)^3 = 0$
(c) $rx^3 + q(r+1)x^2 + (r+1)^3 = 0$
(d) $rx^3 + q(r+1)x^2 + (r+1)^2 = 0$

6. If α, β, γ are the roots of the equation $x^3 + 3bx + c = 0$, then the value of $\Sigma(\alpha - \beta)^2(\beta - \gamma)^2$ is equal to
(a) $9b^2$ (b) $27b^2$
(c) $81b^2$ (d) $243b^2$

7. If α, β, γ are the roots of the equation $x^3 + x + 2 = 0$, then the equation whose roots are $(\alpha - \beta)(\alpha - \gamma)$, $(\beta - \gamma)(\beta - \alpha)$, $(\gamma - \alpha)(\gamma - \beta)$ is
(a) $x^3 - 6x^2 + 216 = 0$ (b) $x^3 - 3x^2 + 112 = 0$
(c) $x^3 + 6x^2 - 216 = 0$ (d) $x^3 + 3x^2 - 112 = 0$

Numerical Value Type

Solve the following problems and write your answer.

1. Let α and β be two real numbers such that $\alpha + \beta = 1$ and $\alpha\beta = -1$. Let $P_n = (\alpha)^n + (\beta)^n$, $P_{n-1} = 11$ and $P_{n+1} = 29$ for some integer $n \geq 1$. Then, the value of P_n^2 is
[JEE Main 2021, 26 February]

2. If α, β are the roots of $x^2 - 8x + A = 0$ and γ, δ are the roots of $x^2 - 72x + B = 0$. If $\alpha < \beta < \gamma < \delta$ are in GP, then the value of $A + B$ must be equal to

3. If $a, b, c > 0$, $a^2 = bc$ and $a + b + c = abc$, then the least value of $a^4 + a^2 + 7$ must be equal to

4. The sum of squares of all integral values of a for which the quadratic expression $(x - a)(x - 10) + 1$ can be factored as a product $(x + \alpha)(x + \beta)$ of two factors and $\alpha, \beta \in I$, must be equal to

5. The maximum value of $\frac{6x^2 - 22x + 21}{5x^2 - 18x + 17}$ for real x

6. If $ax^2 - bx + c = 0$ have two distinct roots lying in the interval $(0, 1)$, $a, b, c \in N$, then the least value of abc must be equal to

7. Let $f(x) = x^2 - px + q$, p is an odd positive integer and the roots of the equation $f(x) = 0$ are two distinct prime numbers. If $p + q = 35$, and $f(10)\sum_{r=1}^{10} f(r) = 88n$, then the value of $\frac{p+q}{n}$ is

8. If α and β are the roots of the equation $x^2 + px + q = 0$ and also $x^{3900} + p^{1950}x^{1950} + q^{1950} = 0$ and if $\frac{\alpha}{\beta}, \frac{\beta}{\alpha}$ are the roots of $x^n + 1 + (x+1)^n = 0$, then the value of n must be equal to

Single Integer Type

Solve the following problems and write your answer.

1. The least positive value of 'a' for which the equation, $2x^2 + (a+10)x + \frac{33}{2} = 2a$ has real roots is **[JEE Main 2020, 8 January]**
2. If α and β are two roots of the equation $x^2 - 64x + 256 = 0$. Then, value of $\left(\frac{\alpha^3}{\beta^5}\right)^{1/8} + \left(\frac{\beta^3}{\alpha^5}\right)^{1/8}$ is
3. The number of real roots of the equation, $e^{4x} + e^{3x} - 4e^{2x} + e^x + 1 = 0$ is
4. The number of integral values of m for which the quadratic expression, $(1+2m)x^2 - 2(1+3m)x + 4(1+m), x \in R$, is always positive, is
5. The value of a for which the sum of the squares of the roots of the equation $x^2 - (a-2)x - a - 1 = 0$ assume the least value is
6. The number of all possible positive integral values of α for which the roots of the quadratic equation, $6x^2 - 11x + \alpha = 0$ are rational numbers is **[JEE Main 2019, 9 January]**
7. The number of solutions of the equation $\log_4(x-1) = \log_2(x-3)$ is **[JEE Main 2021, 26 February]**
8. If $f(x)$ and $g(x)$ are two polynomials such that the polynomial $P(x) = f(x^3) + xg(x^3)$ is divisible by $x^2 + x + 1$, then $P(1)$ is equal to .... **[JEE Main 2021, 18 March]**
9. The number of the real roots of the equation $(x+1)^2 + |x-5| = \frac{27}{4}$ is **[JEE Main 2021, 24 February]**

Matrix-Match Type

Given below are Matching Type Questions, with two columns (each having some items) each . Each item of Column I has to be matched with the items of Column II, by encircling the correct match(es).

NOTE *An item of Column I can be matched with more than one items of Column II.*

1. Observe the following columns :

	Column I		Column II
(A)	If a, b, c, d are four non zero real numbers such that $(d+a-b)^2 + (d+b-c)^2 = 0$ and roots of the equation $a(b-c)x^2 + b(c-a)x + c(a-b) = 0$ are real and equal, then	(P)	$a + b + c \neq 0$
(B)	If a, b, c are three non zero real numbers such that the roots of the equation $(b-c)x^2 + (c-a)x + (a-b) = 0$ are real and equal, then	(Q)	a, b, c are in AP
(C)	If the three equations $x^2 + px + 12 = 0$, $x^2 + qx + 15 = 0$ and $x^2 + (p+q)x + 36 = 0$ have a common positive root and a, b, c be their other roots, then	(R)	a, b, c are in GP
		(S)	a, b, c are in HP
		(T)	$a = b = c$

2. Observe the following columns :

	Column I		Column II
(A)	The equation $x^3 - 6x^2 + 9x + \lambda = 0$ have exactly one root in (1, 3), then $[\lambda + 1]$ is (where [.] denotes the greatest integer function)	(P)	–3
(B)	If $-3 < \frac{x^2 - \lambda x - 2}{x^2 + x + 1} < 2$ for all $x \in R$, then $[\lambda]$ is where [.] denotes the greatest integer function	(Q)	–2
(C)	If $x^2 + \lambda x + 1 = 0$ and $(b-c)x^2 + (c-a)x + (a-b) = 0$ have both the roots common, then $[\lambda - 1]$ is (where [.] denotes the greatest integer function)	(R)	–1
		(S)	0
		(T)	1

3. **Observe the following columns :**

	Column I		Column II
(A)	If N be the number of solutions and S be the sum of all roots of the equation $\|x-\|4-x\|\|-2x=4$, then	(P)	$S = 0$
(B)	If N be the number of solutions and S be the sum of all roots of the equation $\|x^2 - x - 6\| = x + 2$, then	(Q)	$N = 1$
(C)	If N be the number of solutions and S be the sum of all roots of the equation $x^2 - 3\|x\| - 4 = 0$, then	(R)	$N = 2$
		(S)	$N = 3$
		(T)	$S = 4$

4. **Match the Column I with Column II :**

	Column I		Column II
(A)	The real values of a for which the quadratic equation $2x^2 - (a^3 + 8a - 1)x + a^2 - 4a = 0$. Possesses roots of opposite signs are given by	(P)	$a \geq 6$
(B)	If the equation $x^2 + 2(a+1)x + 9a - 5 = 0$ has only negative roots, then	(Q)	$0 \leq a \leq 1$
(C)	The value of a for which the inequality $x^2 - 2(4a-1)x + 15a^2 - 2a - 7 > 0$ is valid for all $x \in R$, is	(R)	$0 < a < 4$
(D)	If $x \in R$, then $\dfrac{x^2 + 2x + a}{x^2 + 4x + 3a}$ can take all real values, if	(S)	$2 < a < 4$

5. **Match the following lists**

	Column I		Column II
(A)	If $x^2 + ax + b = 0$ has roots α, β and $x^2 + px + q = 0$ has roots $-\alpha, \gamma$, then	(P)	$(1 - bq)^2 = (a - pb)(p - aq)$
(B)	If $x^2 + ax + b = 0$ has roots α, β and $x^2 + px + q = 0$ has roots $1/\alpha, \gamma$, then	(Q)	$(4 - bq)^2 = (4a + 2pb)(-2p - aq)$
(C)	If $x^2 + ax + b = 0$ has roots α, β and $x^2 + px + q = 0$ has roots $-2/\alpha, \gamma$, then	(R)	$(1 - 4bq)^2 = (a + 2bp)(-2p - 4aq)$
(D)	If $x^2 + ax + b = 0$ has roots α, β and $x^2 + px + q = 0$ has roots $-1/(2\alpha), \gamma$, then	(S)	$(q - b)^2 = (aq + bp)(-p - a)$

Answers

Objective Questions Type [Only one correct answer]

1. (c)	**2.** (b)	**3.** (d)	**4.** (c)	**5.** (a)	**6.** (a)	**7.** (d)	**8.** (d)	**9.** (c)	**10.** (a)
11. (d)	**12.** (a)	**13.** (c)	**14.** (a)	**15.** (d)	**16.** (d)	**17.** (b)	**18.** (c)	**19.** (b)	**20.** (a)
21. (b)	**22.** (c)	**23.** (a)	**24.** (b)	**25.** (b)	**26.** (d)	**27.** (a)	**28.** (c)	**29.** (c)	**30.** (c)
31. (c)	**32.** (c)	**33.** (d)	**34.** (b)	**35.** (b)	**36.** (d)	**37.** (a)	**38.** (c)	**39.** (a)	**40.** (b)
41. (b)	**42.** (a)	**43.** (d)	**44.** (a)	**45.** (a)	**46.** (d)	**47.** (d)	**48.** (c)	**49.** (a)	**50.** (d)
51. (a)	**52.** (d)	**53.** (b)	**54.** (c)	**55.** (d)	**56.** (b)	**57.** (c)	**58.** (d)	**59.** (c)	**60.** (a)
61. (a)	**62.** (a)	**63.** (d)	**64.** (a)	**65.** (c)	**66.** (c)	**67.** (c)	**68.** (d)	**69.** (b)	**70.** (a)

Objective Questions Type II [One or more than one correct answer(s)]

1. (a, b, c)	**2.** (a, c)	**3.** (a, b, c, d)	**4.** (a, d)	**5.** (a, b, c, d)
6. (b, c)	**7.** (a, b)	**8.** (a, b, c)	**9.** (b, c, d)	**10.** (a, c)
11. (a, c)	**12.** (a, b)	**13.** (a, b, c)	**14.** (a, b)	**15.** (b, d)
16. (a, c, d)	**17.** (a, b)	**18.** (a, d)	**19.** (b, c)	**20.** (b, d)
21. (a, d)	**22.** (a, b, c)	**23.** (a, b, c)	**24.** (a, b, c, d)	**25.** (a, b, c, d)
26. (a, b)	**27.** (a, b, c, d)	**28.** (a, d)		

Linked Comprehension Type

Passage 1 **1.** (d) **2.** (d)
Passage 2 **1.** (a) **2.** (d) **3.** (c)
Passage 3 **1.** (c) **2.** (a) **3.** (b)
Passage 4 **1.** (b) **2.** (a) **3.** (c) **4.** (b) **5.** (b) **6.** (c) **7.** (c)
Passage 5 **1.** (c) **2.** (b) **3.** (c) **4.** (c) **5.** (d) **6.** (b)
Passage 6 **1.** (c) **2.** (c) **3.** (d) **4.** (d) **5.** (b)
Passage 7 **1.** (a) **2.** (b) **3.** (c) **4.** (d) **5.** (b) **6.** (c) **7.** (d)

Numerical Value Type

1. 324 **2.** 984 **3.** 19 **4.** 208 **5.** 1.25 **6.** 25 **7.** 3.5 **8.** 1950

Single Integer Type

1. (8) **2.** (2) **3.** (1) **4.** (7) **5.** (1) **6.** (5) **7.** (1) **8.** (0) **9.** (2)

Matrix-Match Type

1. A → (P, Q, R, S, T); B → (P, Q); C → (P)
2. A → (P, Q, R, S); B → (R, S, T); C → P
3. A → (P, Q); B → (S, T); C → (P, R)
4. A → (R); B → (P); C → (S); D → (Q)
5. A → (S); B → (P); C → (Q); D → (R)

Solutions

Objective Questions Type [Only one correct answer]

1. Sum of roots $= |\alpha| + |\beta| = -\alpha - \beta$ $[\because \alpha < 0 \text{ and } \beta < 0]$

$$= -(\alpha + \beta) = -\left(-\frac{b}{a}\right) = \frac{b}{a} = \left|\frac{b}{a}\right|$$

$[\because |\alpha| + |\beta| > 0]$

and product of roots $= |\alpha||\beta| = |\alpha\beta| = \left|\frac{c}{a}\right|$

Hence, required equation is

$$x^2 - \left(\left|\frac{b}{a}\right|\right)x + \left|\frac{c}{a}\right| = 0$$

or $|a|x^2 - |b|x + |c| = 0$

2. We know that only even prime is 2, then $(2)^2 - \lambda(2) + 12 = 0$

$\Rightarrow \quad \lambda = 8 \quad \ldots(\text{i})$

and $x^2 + \lambda x + \mu = 0$

has equal roots

$\therefore \quad \lambda^2 - 4\mu = 0$

or $(8)^2 - 4\mu = 0$ [from Eq. (i)]

$\therefore \quad \mu = 16$

3. $(p^2 + q^2)^2 - 2p^2q^2 = 272$

$\Rightarrow ((p + q)^2 - 2pq)^2 - 2p^2q^2 = 272$

$\Rightarrow \quad 16 - 16pq + 2p^2q^2 = 272$

$\Rightarrow \quad (pq)^2 - 8pq - 128 = 0$

$\Rightarrow \quad pq = \dfrac{8 \pm 24}{2} = 16, -8$

$\Rightarrow \quad pq = 16$

Now, $x^2 - (p + q)x + pq = 0$

$x^2 - 2x + 16 = 0$

4. If A = first term, D = common difference

$\therefore \quad t_p = A + (p - 1)D = \dfrac{1}{q} \quad \ldots(\text{i})$

and $\quad t_q = A + (q - 1)D = \dfrac{1}{p} \quad \ldots(\text{ii})$

Solving Eqs. (i) and (ii), we get

$$A = D = \frac{1}{pq}$$

$\therefore \quad t_{pq} = A + (pq - 1)D = 1$

Also, sum of the coefficients of the given equation

$(p + 2q - 3r) + (q + 2r - 3p) + (r + 2p - 3q) = 0$

Hence, $x = 1$ is a root of the equation. Hence, $x = 1 = t_{pq}$

5. Multiply first equation by x

then $\quad x^4 + ax^2 + x = 0 \quad \ldots(\text{i})$

and given $\quad x^4 + ax^2 + 1 = 0 \quad \ldots(\text{ii})$

Subtract Eq. (ii) from (i), then

$x - 1 = 0 \Rightarrow x = 1$ is a common root

then $\quad 1 + a + 1 = 0$

$\therefore \quad a = -2$

6. If $(\lambda + 1)x^2 + 2 = \lambda x + 3$ has only one solution.

If discriminant = 0

$\Rightarrow \quad \lambda^2 - 4(\lambda + 1)(-1) = 0$

$\Rightarrow \quad \lambda^2 + 4\lambda + 4 = 0 \Rightarrow (\lambda + 2)^2 = 0$

$\therefore \quad \lambda = -2$

7. $2b = a + c$

$\frac{2a+2}{3} = \frac{10}{3}$ and $\frac{2b+c}{3} = \frac{7}{3}$

$a = 4, \left.\begin{matrix} 2b + c = 7 \\ 2b - c = 4 \end{matrix}\right\}$, solving

$b = \frac{11}{4}, c = \frac{3}{2}$

$\therefore$ Quadratic equation is $4x^2 + \frac{11}{4}x + 1 = 0$

$\therefore$ The value of $(\alpha + \beta)^2 - 3\alpha\beta = \frac{121}{256} - \frac{3}{4} = -\frac{71}{256}$

8. $ax^2 - bx + c = 0$

$$\alpha + \beta = \frac{b}{a}, \alpha\beta = \frac{c}{a}$$

Also, $(a + cy)^2 = b^2y$

$\Rightarrow$ $c^2y^2 - (b^2 - 2ac)y + a^2 = 0$

or $\left(\frac{c}{a}\right)^2 y^2 - \left(\left(\frac{b}{a}\right)^2 - 2\left(\frac{c}{a}\right)\right) y + 1 = 0$

or $(\alpha\beta)^2 y^2 - (\alpha^2 + \beta^2)y + 1 = 0$

or $y^2 - (\alpha^{-2} + \beta^{-2})y + \alpha^{-2}\beta^{-2} = 0$

or $(y - \alpha^{-2})(y - \beta^{-2}) = 0$

$\therefore$ Roats are α^{-2}, β^{-2}.

9. $\because$ $x^3 + 3x^2 + 3x + 2 = 0 \Rightarrow (x+1)^3 + 1 = 0$

$\Rightarrow$ $(x + 1) = (-1)^{1/3} = -1, -\omega, -\omega^2$

$\Rightarrow$ $x = -2, -1 - \omega, -1 - \omega^2$

$\therefore$ $x = -2, \omega^2, \omega$

Since given equations have two common roots therefore ω and ω^2 are the common roots of the two equations.

$\therefore$ Equation whose roots are ω and ω^2 is $x^2 + x + 1 = 0$

Comparing with $ax^2 + bx + c = 0$, we get

$$\frac{a}{1} = \frac{b}{1} = \frac{c}{1}$$

i.e. $a = b = c$

10. The given equations are symmetrical about the line $y = x$, then $x^2 + 1 = x$

$\Rightarrow$ $x^2 - x + 1 = 0 \Rightarrow (x + \omega)(x + \omega^2) = 0$

$\therefore$ $x = -\omega, -\omega^2$

i.e. no real roots.

$\therefore$ Number of solutions = 0

11. If quadratic equation has purely imaginary roots, then coefficient of x must be equal to zero.

Let $p(x) = ax^2 + b$ with a, b of same sign and $a, b \in R$.

Then, $p[p(x)] - a(ax^2 + b)^2 + b$

$p(x)$ has imaginary roots say ix.

Then, also $ax^2 + b \in R$ and $(ax^2 + b)^2 > 0$

$\therefore$ $a(ax^2 + b)^2 + b \neq 0, \forall x$

Thus, $p[p(x)] \neq 0, \forall x$

12. $\because$ RHS is even integer

$\therefore$ LHS is an even integer

So, either both x and y are even integers, or both of them are odd integers

$\because$ $x^4 - y^4 = (x - y)(x + y)(x^2 + y^2)$

$\Rightarrow x - y, x + y, x^2 + y^2$ must be an even integer.

Therefore, $(x - y)(x + y)(x^2 + y^2)$ must be divisible by 8. But RHS is not divisible by 8.

Hence, the given equation has no solution.

13. $\because$ $x^2 - 16x + \lambda = 0$

$\Rightarrow$ $x = \frac{16 \pm \sqrt{(256 - 4\lambda)}}{8} = \frac{8 \pm \sqrt{(64 - \lambda)}}{4}$

$$\alpha, \beta = 2 \pm \frac{\sqrt{(64-\lambda)}}{4}$$

Here, $64 - \lambda > 0$

$\therefore$ $\lambda < 64$

Also, $1 < \alpha < 2$ and $2 < \beta < 3$

$\Rightarrow$ $1 < 2 - \frac{\sqrt{64-\lambda)}}{4} < 2$ and $2 < 2 + \frac{\sqrt{(64-\lambda)}}{4} < 3$

$\Rightarrow$ $-1 < -\frac{\sqrt{(64-\lambda)}}{4} < 0$ and $0 < \frac{\sqrt{(64-\lambda)}}{4} < 1$

$\Rightarrow$ $1 > \frac{\sqrt{(64-\lambda)}}{4} > 0$ and $0 < \frac{\sqrt{(64-\lambda)}}{4} < 1$

i.e. $0 < \frac{\sqrt{(64-\lambda)}}{4} < 1$

$\Rightarrow$ $0 < \sqrt{(64 - \lambda)} < 4$

$\Rightarrow$ $0 < (64 - \lambda) < 16$

$\Rightarrow$ $\lambda > 48$ or $48 < \lambda < 64$

$\therefore$ $\lambda = \{49, 50, 51, 52, \ldots, 63\}$.

14. If $ax^2 + 2hxy + by^2 + 2gx + 2fy + c = 0$ can be resolved into linear factors, then

$$\Delta = 0$$

i.e. $abc + 2fgh - af^2 - bg^2 - ch^2 = 0$

$$1 \times k \times k + 0 - 1 \times 0 - k \times 1^2 - k \times 1 = 0$$

$$k^2 - 2k = 0$$

$\therefore$ $k = 0, 2$

15. $\because$ $[\tan^2 x]$ = Integer, a = Integer

$\therefore$ $\tan x$ is also an integer.

Then, $\tan^2 x - \tan x - a = 0$

$\therefore$ $a = \tan x(\tan x - 1) = I(I - 1)$

= product of two consecutive integers

$\therefore a = 2, 6, 12, 20, 30, 42, 56, 72, 90.$

Hence, set S has 9 elements.

16. $\because$ $f(x) = x^3 + x^2 - 5x - 1$

$\therefore$ $f'(x) = 3x^2 + 2x - 5$

$= (x - 1)(3x + 5)$

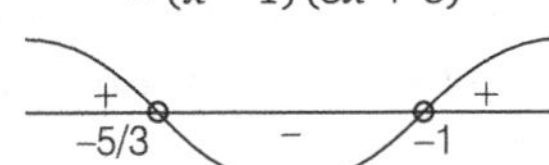

$\therefore$ Graph of $y = f(x)$

$\because \quad f(1) < 0$

and $\quad f(-5/3) > 0$

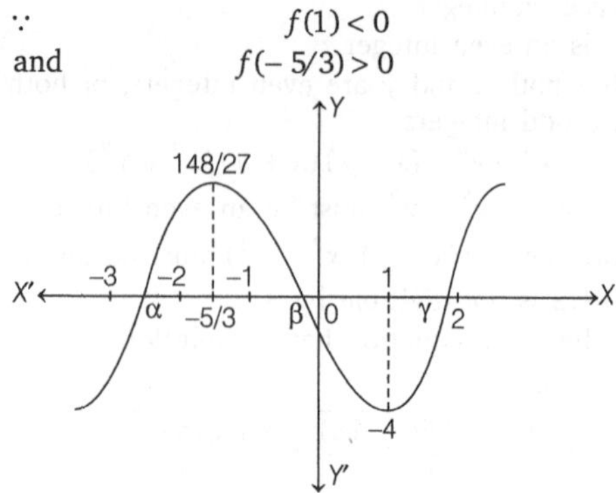

It is clear from figure

$-3 < \alpha < -2, -1 < \beta < 0$ and $1 < \gamma < 2$

$\therefore [\alpha] + [\beta] + [\gamma] = -3 - 1 + 1 = -3$

17. $\because \quad f(x) = x^3 + x^2 + 100x + 7 \sin x$

$\therefore \quad f'(x) = 3x^2 + 2x + 100 + 7 \cos x$

$= 3x^2 + 2x + 93 + 7(1 + \cos x)$

$= 3\left\{\left(x + \frac{1}{3}\right)^2\right\} + \frac{278}{3} + 7(1 + \cos x) > 0$

$\therefore f(x)$ is increasing function

$\Rightarrow \quad f(1) < f(2) < f(3)$

Let $\quad f(1) = a$, $f(2) = b$ and $f(3) = c$

i.e. $\quad a < b < c$

Now, given equation is $\frac{1}{y-a} + \frac{2}{y-b} + \frac{3}{y-c} = 0$

or $(y-b)(y-c) + 2(y-c)(y-a) + 3(y-a)(y-b) = 0$

Let $g(y) = (y-b)(y-c) + 2(y-c)(y-a) + 3(y-a)(y-b)$

$\therefore \quad g(a) > 0$, $g(b) < 0$ and $g(c) > 0$

Hence, one real root between a and b and other between b and c.

18. Here, $x^2 - 2x \sec\theta + 1 = 0$ has roots α_1 and β_1.

$\therefore \alpha_1, \beta_1 = \frac{2\sec\theta \pm \sqrt{4\sec^2\theta - 4}}{2\times 1} = \frac{2\sec\theta \pm 2|\tan\theta|}{2}$

Since, $\quad \theta \in \left(-\frac{\pi}{6}, -\frac{\pi}{12}\right)$,

i.e. $\quad \theta \in$ IV quadrant $= \frac{2\sec\theta \mp 2\tan\theta}{2}$

$\therefore \quad \alpha_1 = \sec\theta - \tan\theta$ and $\beta_1 = \sec\theta + \tan\theta \quad$ [as $\alpha_1 > \beta_1$]

and $x^2 + 2x\tan\theta - 1 = 0$ has roots α_2 and β_2.

i.e. $\quad \alpha_2, \beta_2 = \frac{-2\tan\theta \pm \sqrt{4\tan^2\theta + 4}}{2}$

$\therefore \quad \alpha_2 = -\tan\theta + \sec\theta$

and $\quad \beta_2 = -\tan\theta - \sec\theta \quad$ [as $\alpha_2 > \beta_2$]

Thus, $\quad \alpha_1 + \beta_2 = -2\tan\theta$

19. By hypothesis, $\alpha + \beta = p$; $\alpha\beta = -(p+q)$

Now, $(\alpha+1)(\beta+1) = (\alpha+\beta) + \alpha\beta + 1 = 1 - q \quad$ ...(i)

The given expression

$= \frac{(\alpha+1)^2}{(\alpha+1)^2 + (q-1)} + \frac{(\beta+1)^2}{(\beta+1)^2 + (q-1)}$

$= \frac{(\alpha+1)^2}{(\alpha+1)^2 - (\alpha+1)(\beta+1)} + \frac{(\beta+1)^2}{(\beta+1)^2 - (\alpha+1)(\beta+1)}$

$= \frac{(\alpha+1)}{(\alpha-\beta)} + \frac{(\beta+1)}{(\beta-\alpha)} = \frac{(\alpha-\beta)}{(\alpha-\beta)} = 1$

20. $\quad y = 4 + \frac{1}{\left(5 + \frac{1}{y}\right)}$

$\Rightarrow \quad y - 4 = \frac{y}{(5y+1)}$

$\Rightarrow \quad 5y^2 - 20y - 4 = 0$

$\Rightarrow \quad y = \frac{20 + \sqrt{480}}{10}$

$\Rightarrow \quad y = \frac{20 - \sqrt{480}}{10} \rightarrow$ rejected

$\Rightarrow \quad y = 2 + \sqrt{\frac{480}{100}} = 2 + \sqrt{\frac{24}{5}} = 2 + \frac{2\sqrt{6}\times\sqrt{5}}{\sqrt{5}\times\sqrt{5}}$

$= 2 + \frac{2\sqrt{30}}{5}$

21. $\quad 5^{\log_5 (x^2 - 4x + 5)} = x - 1$

$\Rightarrow \quad 5^{\log_5 \{(x-2)^2 + 1\}} = x - 1$

$\Rightarrow \quad (x-2)^2 + 1 = (x-1)$

$\Rightarrow \quad (x-2)^2 - (x-2) = 0 \Rightarrow (x-2)(x-3) = 0$

$\therefore \quad x = 2, 3$

22. $\quad 2|x|^2 - 5|x| + 2 = 0$

$\Rightarrow \quad (2|x| - 1)(|x| - 2) = 0$

$\therefore \quad |x| = 1/2, 2$

$\therefore \quad x = \pm 1/2, \pm 2$

23. $\because \quad 2\cos^2\theta\, x^3 + 2x + \sin 2\theta$

$= (x^2 + px + 1)(2\cos^2\theta\, x + \sin 2\theta)$

On comparing coefficients of x and x^2, then

$2 = p\sin 2\theta + 2\cos^2\theta$

$2\sin^2\theta = p\sin 2\theta$

or $\quad p = \tan\theta \quad$ ...(i)

and $\quad 0 = \sin 2\theta + 2p\cos^2\theta$

$\therefore \quad p = -\tan\theta \quad$ ...(ii)

From Eqs. (i) and (ii), $\tan\theta = -\tan\theta$

$2\tan\theta = 0$

$\tan\theta = 0$

$\therefore \quad \theta = n\pi, n \in I$

24. Let $f(x) = x^3 + x^2 + 2x + \sin x = 0$

$f'(x) = 3x^2 + 2x + 2 + \cos x$

Here, $3x^2 + 2x + 2$ is a quadratic function.

$f \min = -\frac{D}{4a}$,

[D is Discriminant and a is primary coefficient]

$= \frac{-(b^2 - 4ac)}{4a}$

$= -\frac{(4 - 4\times 3\times 2)}{4\times 3} = -\frac{(4-24)}{12} = \frac{20}{12} = \frac{5}{3}$

and $\cos x \in [-1, 1]$ for $x \in [-2\pi, 2\pi]$

Here, $f'(x)$ is always positive.
So, $f(x)$ is strictly increasing.
Also, $f(x) = x^3 + x^2 + 2x + \sin x \Rightarrow f(0) = 0$
Therefore, $f(x)$ is strictly increasing and no other root possible.
Number of real root = 1

25. **Case I** If $\log_3 x \geq 0 \Rightarrow x \geq 1$

$\therefore$ From given equation

$$2(\log_3 x)^2 - \log_3 x + a = 0$$

For real solution

$$(-1)^2 - 4 \cdot 2 \cdot a > 0$$

$$\therefore \quad a < \frac{1}{8} \quad \ldots\text{(i)}$$

Case II If $\log_3 x < 0$

$\therefore \quad x < 1$

$\therefore$ From given equation $2(\log_3 x)^2 + \log_3 x + a = 0$

For real solution $(1)^2 - 4 \cdot 2 \cdot a > 0$

$$\therefore \quad a < \frac{1}{8} \quad \ldots\text{(ii)}$$

From equation

$$\log_3 x = \frac{-1 \pm \sqrt{(1-8a)}}{4} \quad (\text{Here, } \log_3 x < 0)$$

$$\Rightarrow \frac{-1 \pm \sqrt{(1-8a)}}{4} < 0 \Rightarrow -1 \pm \sqrt{(1-8a)} < 0$$

$$\Rightarrow \sqrt{(1-8a)} < 1 \quad (\text{Taking + ve sign})$$

$$\therefore \quad a > 0 \quad \ldots\text{(iii)}$$

Hence, from Eqs. (i), (ii) and (iii)

$$0 < a < \frac{1}{8}$$

26. Given quadratic equations having roots α and β is $x^2 + px + 2 = 0$, so $\alpha + \beta = -p$ and $\alpha\beta = 2$

and having roots $\frac{1}{\alpha}$ and $\frac{1}{\beta}$ is

$2x^2 + 2qx + 1 = 0$, so $\frac{1}{\alpha} + \frac{1}{\beta} = -q$

$$\text{So, } \left(\alpha - \frac{1}{\alpha}\right)\left(\beta - \frac{1}{\beta}\right)\left(\alpha + \frac{1}{\beta}\right)\left(\beta + \frac{1}{\alpha}\right)$$

$$= \left(\alpha\beta - \frac{\alpha}{\beta} - \frac{\beta}{\alpha} + \frac{1}{\alpha\beta}\right)\left(\alpha\beta + 1 + 1 + \frac{1}{\alpha\beta}\right)$$

$$= \left(2 - \frac{\alpha^2 + \beta^2}{\alpha\beta} + \frac{1}{2}\right)\left(2 + 1 + 1 + \frac{1}{2}\right)$$

$$= \left[\frac{5}{2} - \frac{(\alpha+\beta)^2 - 2\alpha\beta}{2}\right]\left[\frac{9}{2}\right]$$

$$= \frac{9}{4}(5 - (-p)^2 + 4) = \frac{9}{4}(9 - p^2)$$

27. Let the other root of the quadratic equation $f(x) = 0$ be α.
So, according to the question;

$$f(x) = a(x-3)(x-\alpha),\ a \neq 0$$

$$\because \quad f(-1) + f(2) = 0$$

$$\Rightarrow a(-1-3)(-1-\alpha) + a(2-3)(2-\alpha) = 0$$

$$\Rightarrow \quad a[4(1+\alpha) + (\alpha - 2)] = 0$$

$$\Rightarrow \quad 5\alpha + 2 = 0 \quad [\because a \neq 0]$$

$$\Rightarrow \quad \alpha = -\frac{2}{5} \in (-1, 0)$$

28. $\because \quad (x+1)^2 + [x-1]^2 = (x-1)^2 + [x+1]^2$

$$\Rightarrow \{(x)+1\}^2 + \{[x]-1\}^2 = \{(x)-1\}^2 + ([x]+1)^2$$

$$\Rightarrow \quad 4(x) = 4[x]$$

$$\Rightarrow \quad (x) = [x]$$

Hence, $x \in I$.

29. $(a^2 - 3a + 2)x^2 + (a^2 - 5a + 6)x + a^2 - 4 = 0$

for identity $a^2 - 3a + 2 = 0$

$$\Rightarrow \quad a = 1, 2$$

$$a^2 - 5a + 6 = 0$$

$$\Rightarrow \quad a = 2, 3$$

$$\Rightarrow \quad a^2 - 4 = 0$$

$$\Rightarrow \quad a = +2, -2$$

and $\quad a^2 - 3a + 2 = a^2 - 5a + 6 = a^2 - 4$

$$\therefore \quad a = 2$$

30. $\because \quad (x-1) = (x - [x])(x - \{x\})$

$$\Rightarrow \quad x = 1 + \{x\}[x]$$

$$\Rightarrow \quad [x] + \{x\} = 1 + \{x\}[x]$$

$$\Rightarrow \quad (\{x\} - 1)([x] - 1) = 0$$

$$\{x\} - 1 \neq 0,\ \therefore [x] - 1 = 0$$

$$[x] = 1$$

$$\Rightarrow \quad x \in [1, 2)$$

31. It is clear from the graph
Number of solutions = 4

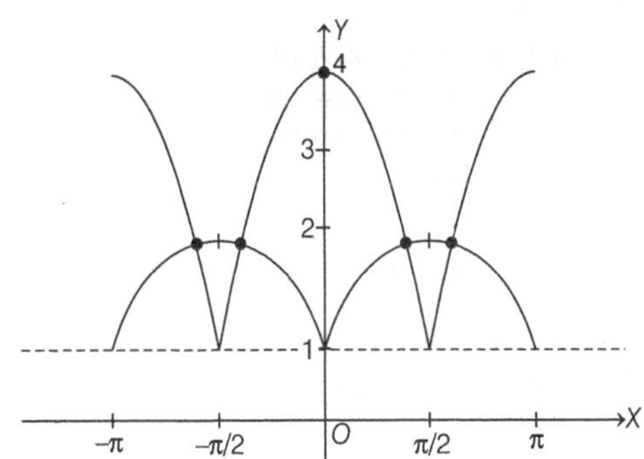

32. $\int_0^1 (x^2 + bx + c)\,dx = 1$

$$\frac{1}{3} + \frac{b}{2} + c = 1 \Rightarrow \frac{b}{2} + c = \frac{2}{3}$$

$$3b + 6c = 4 \quad \ldots\text{(i)}$$

$$\Rightarrow \quad P(2) = 5$$

$$\Rightarrow \quad 4 + 2b + c = 5$$

$$\Rightarrow \quad 2b + c = 1 \quad \ldots\text{(ii)}$$

From Eqs. (i) and (ii), we get

$$b = \frac{2}{9} \text{ and } c = \frac{5}{9}$$

$$\Rightarrow \quad 9(b + c) = 7$$

33. $(\alpha + 5)x^2 - (2\alpha + 1)x + (\alpha - 1) = 0$

Let roots of this equation are λ and $-\lambda$

$\therefore$ Sum of roots $= \lambda - \lambda = \dfrac{2\alpha + 1}{\alpha + 5}$

$\Rightarrow \quad \dfrac{2\alpha + 1}{\alpha + 5} = 0$

$\therefore \quad \alpha = -1/2, \alpha \neq -5$

34. $x \neq -4$

$(|x| - 3)(|x + 4|) = 6 \Rightarrow |x| - 3 = \dfrac{6}{|x + 4|}$

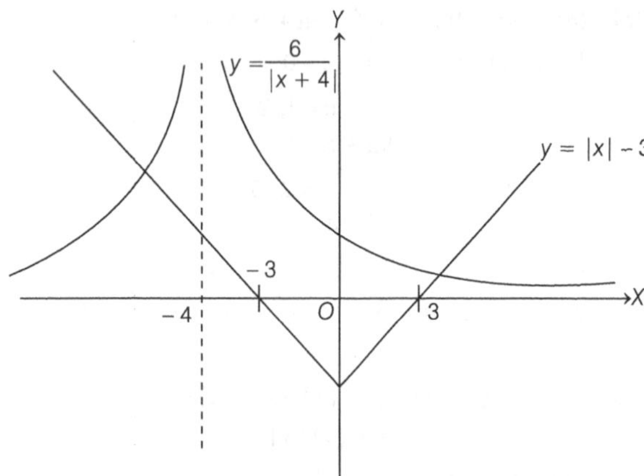

Number of solutions = 2

35. $\alpha + \beta = -\sqrt{\alpha}$ and $\alpha\beta = \beta$

From second relation $\beta \neq 0$

$\therefore \quad \alpha = 1$

From first relation, $\beta = -2$

Hence, $\alpha = 1$ and $\beta = -2$

36. $ax^2 + bx + c = 0$

$a, b, c \in \{1, 2, 3, 4, 5, 6\}$

$n(s) = 6 \times 6 \times 6 = 216$

$D = 0 \Rightarrow b^2 = 4ac$

$ac = \dfrac{b^2}{4}$

If $b = 2$, $ac = 1 \Rightarrow a = 1, c = 1$

If $b = 4$, $ac = 4 \Rightarrow a = 1, c = 4$

$a = 4, c = 1$

If $b = 6$, $ac = 9 \Rightarrow 2, c = 2$

$\therefore$ Probability $= \dfrac{5}{216}$

37. The equations $x^2 + ax + b = 0$ and $x^2 + bx + a = 0$ have a common root, then

$(x^2 + ax + b) - (x^2 + bx + a) = 0$

$\Rightarrow \quad (a - b)x - (a - b) = 0$

$\therefore \quad x = 1 \qquad [\because a \neq b]$

is a common which satisfies both equations, then

$1 + a + b = 0$

$\therefore \quad a + b = -1$

38. $\because \quad (x - a)(x - b) - c = (x - \alpha)(x - \beta)$

$\therefore \ (x - \alpha)(x - \beta) + c = (x - a)(x - b)$

Then, the roots of $(x - \alpha)(x - \beta) = c$ are

$x = a, b$

39. $\because \quad x^2 + x + 1 = 0$

$(x - \omega)(x - \omega^2) = 0$

$\therefore \quad \alpha = \omega, \beta = \omega^2$

$\therefore \quad \alpha^{19} = \omega^{19} = \omega$ and $\beta^7 = \omega^{14} = \omega^2$

Required equation is, $x^2 - (\alpha^{19} + \beta^7)x + \alpha^{19}\beta^7 = 0$

$\Rightarrow \quad x^2 + x + 1 = 0$

40. $1 + 1 + |e^x - 1| = e^{2x} - 2e^x + 1$

$\Rightarrow \quad 2 + |e^x - 1| = |e^x - 1|^2$

$\Rightarrow \ |e^x - 1|^2 - |e^x - 1| - 2 = 0$

$\therefore \quad |e^x - 1| = 2, -1 \qquad [\because |e^x - 1| \neq -1]$

$\Rightarrow \quad |e^x - 1| = 2$

$\Rightarrow \quad e^x - 1 = \pm 2$

$\therefore \quad e^x = 1 \pm 2 = 3, -1 \qquad [\because e^x \neq -1]$

$\Rightarrow \quad e^x = 3 \Rightarrow x = \log_e 3$

$\therefore$ Number of solutions is one.

41. Let $y = |x| = \cos x$

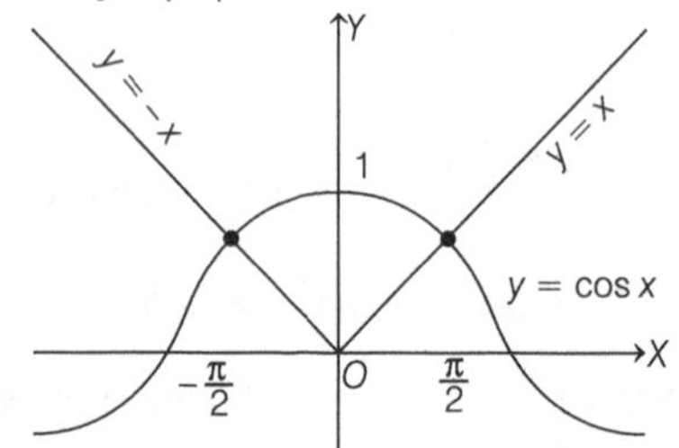

$\because$ Curve $y = |x|$ and $y = \cos x$ cuts at two points.

$\therefore$ Number of solutions is two.

42. Let $x = 3 + \cfrac{1}{4 + \cfrac{1}{3 + \cfrac{1}{4 + \cfrac{1}{3 + \dots \infty}}}}$

So, $x = 3 + \dfrac{1}{4 + \dfrac{1}{x}} = 3 + \dfrac{1}{\dfrac{4x + 1}{x}}$

$\Rightarrow \quad x - 3 = \dfrac{x}{4x + 1}$

$\Rightarrow \quad (4x + 1)(x - 3) = x$

$\Rightarrow \quad 4x^2 - 12x + x - 3 = 0$

$\Rightarrow \quad x = \dfrac{12 \pm \sqrt{(12)^2 + 12 \times 4}}{2 \times 4}$

$\Rightarrow \quad x = \dfrac{12 \pm \sqrt{12(16)}}{8} = \dfrac{12 \pm 4 \times 2\sqrt{3}}{8}$

$\Rightarrow \quad x = \dfrac{3 \pm 2\sqrt{3}}{2}$

$\Rightarrow \quad x = \dfrac{3}{2} \pm \sqrt{3}$

$\Rightarrow \quad x = 1.5 \pm \sqrt{3}$

But only positive value is accepted.

So, $x = 1.5 + \sqrt{3}$.

43. Discriminant ≥ 0

$\therefore$ $p \leq 22/5$

Roots less than 2

$\therefore$ $f(2) > 0$

$\therefore$ $p^2 - p - 2 > 0 \Rightarrow p > 2$ or $p < -1$

Combine both casses, we get $p \in (-\infty, -1) \in \left(2, \frac{22}{5}\right]$.

44. Let $f(x) = ax^2 + 2bx - 3c$

Since, $f(x) = 0$ has non real roots, $f(x)$ will have the same sign for values of x

Given $3c/4 < a + b$

$\Rightarrow 4a + 4b - 3c > 0$

$\Rightarrow f(2) > 0$

$\Rightarrow f(0) > 0 \Rightarrow c < 0$

45. Roots $= \dfrac{4(2-i) \pm \sqrt{16(2-i)^2 + 8(1+i)(5+3i)}}{4(1+i)}$

$= -\dfrac{i}{1+i}$ or $\dfrac{4-i}{1+i}$

$= \dfrac{-1-i}{2}$ or $\dfrac{3-5i}{2}$

$\therefore$ $\left|\dfrac{3-5i}{2}\right| = \sqrt{\left(\dfrac{9}{4} + \dfrac{25}{4}\right)} = \sqrt{\dfrac{17}{2}}$

and $\left|\dfrac{-1-i}{2}\right| = \sqrt{\left(\dfrac{1}{4} + \dfrac{1}{4}\right)} = \sqrt{\dfrac{1}{2}}$

Here, $\sqrt{\dfrac{17}{2}} > \sqrt{\dfrac{1}{2}}$

Hence, root is $(3 - 5i)/2$.

46. Let $f(x) = a^2x^2 + 2bx + 2c$

From the question, $a^2\alpha^2 + b\alpha + c = 0$ and $a^2\beta^2 - b\beta - c = 0$

Now, $f(\alpha) = a^2\alpha^2 + 2b\alpha + 2c = b\alpha + c = -a^2\alpha^2$

$f(\beta) = a^2\beta^2 + 2b\beta + 2c = 3(b\beta + c) = 3a^2\beta^2$

but $0 < \alpha < \beta \Rightarrow \alpha, \beta$ are real

$\therefore$ $f(\alpha) < 0,\ f(\beta) > 0$

Hence, $\alpha < \gamma < \beta$.

47. $2^{x+2} \cdot 3^{3x/(x-1)} = 9$

Taking logarithm

Then, $(x + 2)\log 2 + \dfrac{3x}{(x-1)}\log 3 = 2\log 3$

$\Rightarrow$ $(x^2 + x - 2)\log 2 + 3x\log 3 = 2(x - 1)\log 3$

$\Rightarrow$ $x^2\log 2 + (\log 2 + \log 3)x - 2\log 2 + 2\log 3 = 0$

$\Rightarrow$ $x = \dfrac{-(\log 2 + \log 3) \pm \sqrt{\{(\log 2 + \log 3)^2 - 4\log 2\,(-2\log 2 + 2\log 3)\}}}{2\log 2}$

$x = \dfrac{-(\log 2 + \log 3) \pm \sqrt{\{(3\log 2)^2 - 6\log 2\log 3 + (\log 3)^2\}}}{2\log 2}$

$= \dfrac{-(\log 2 + \log 3) \pm (3\log 2 - \log 3)}{2\log 2}$

$\therefore$ $x = -2, 1 - \dfrac{(\log 3)}{(\log 2)}$

48. $\because$ $(x - a)(x - b)(x - c) - d = (x - \alpha)(x - \beta)(x - \gamma)$

$\Rightarrow$ $(x - \alpha)(x - \beta)(x - \gamma) + d = (x - a)(x - b)(x - c)$

$\therefore$ Roots of the equation $(x - \alpha)(x - \beta)(x - \gamma) + d = 0$ are $x = a, b, c$.

49. It is given that α and β be the roots of $x^2 - 3x + p = 0$, so

$\alpha + \beta = 3$ and $\alpha\beta = p$ ...(i)

and γ and δ are be the roots of $x^2 - 6x + q = 0$.

So, $\gamma + \delta = 6$

and $\gamma\delta = q$ ...(ii)

$\because$ $\alpha, \beta, \gamma, \delta$ form a geometric progression

So, let $\alpha = a$, $\beta = ar$, $\gamma = ar^2$, $\delta = ar^3$

$\therefore$ $\dfrac{2q + p}{2q - p} = \dfrac{2\gamma\delta + \alpha\beta}{2\gamma\delta - \alpha\beta} = \dfrac{2a^2r^5 + a^2r}{2a^2r^5 - a^2r} = \dfrac{2r^4 + 1}{2r^4 - 1}$

Now, $\alpha + \beta = 3 \Rightarrow a(1 + r) = 3$

and $\gamma + \delta = 6 \Rightarrow ar^2(1 + r) = 6$

So, $r^2 = 2$

$\therefore$ $\dfrac{2q + p}{2q - p} = \dfrac{2(r^2)^2 + 1}{2(r^2)^2 - 1}$

$= \dfrac{2(2)^2 + 1}{2(2)^2 - 1} = \dfrac{9}{7}$

50. It is given that α is a common roots of given quadratic equations

$x^2 - x + 2\lambda = 0$ and $3x^2 - 10x + 27\lambda = 0$

$\therefore$ $3\alpha^2 - 10\alpha + 27\lambda = 0$

$3\alpha^2 - 3\alpha + 6\lambda = 0$

$\quad - \quad + \quad -$

$0 - 7\alpha + 21\lambda = 0 \Rightarrow \alpha = 3\lambda$

So, $9\lambda^2 - 3\lambda + 2\lambda = 0 \Rightarrow \lambda = \dfrac{1}{9}$ $[\because \lambda \neq 0]$

$\Rightarrow$ $\alpha = \dfrac{1}{3}$

As $\alpha\beta = 2\lambda \Rightarrow \beta = \dfrac{2\lambda}{\alpha} = \dfrac{2 \times \frac{1}{9}}{1/3} = \dfrac{2}{3}$

and $\alpha\gamma = 9\lambda \Rightarrow \gamma = \dfrac{9\lambda}{\alpha} = \dfrac{9 \times \frac{1}{9}}{1/3} = 3$

$\therefore$ $\dfrac{\beta\gamma}{\lambda} = \dfrac{\frac{2}{3} \times 3}{1/9} = 18$

51. Given equation in x is $[x]^2 + 2[x + 2] - 7 = 0$

$\because$ For any integer n, $[x + n] = [x] + n,\ \forall\ x \in R$.

So, $[x]^2 + 2[x] + 4 - 7 = 0$

$\Rightarrow$ $[x]^2 + 2[x] - 3 = 0$

$\Rightarrow$ $[x]^2 + 3[x] - [x] - 3 = 0$

$\Rightarrow$ $[x] = -3$ or 1

$\therefore$ $x \in [-3, -2) \cup [1, 2)$

$\therefore$ There are infinitely many solutions.

52. $(x)^2 + (x+1)^2 = 25$

$\Rightarrow (x)^2 + \{1 + (x)\}^2 = 25$

$\Rightarrow 2(x)^2 + 2(x) - 24 = 0$

$\Rightarrow (x)^2 + (x) - 12 = 0 \Rightarrow (x) = -4, 3$

If $(x) = -4$, then $-5 < x \le -4$

and if $(x) = 3$, then $2 < x \le 3$

$\therefore x \in (-5, -4] \cup (2, 3]$

53. Since α, β are the roots of $375x^2 - 25x - 2 = 0$

$\therefore \alpha + \beta = \frac{25}{375} = \frac{1}{15}$ and $\alpha\beta = -\frac{2}{375}$

$\therefore \lim_{n\to\infty} \sum_{r=1}^{n} S_r = \lim_{n\to\infty} \sum_{r=1}^{n} (\alpha^r + \beta^r)$

$= (\alpha + \alpha^2 + \alpha^3 + \dots \infty) + (\beta + \beta^2 + \beta^3 + \dots \infty)$

$= \frac{\alpha}{1-\alpha} + \frac{\beta}{1-\beta} = \frac{\alpha - \alpha\beta + \beta - \alpha\beta}{(1-\alpha)(1-\beta)}$

$= \frac{\alpha + \beta - 2\alpha\beta}{1 - (\alpha+\beta) + \alpha\beta} = \frac{\frac{1}{15} + \frac{4}{375}}{1 - \frac{1}{15} - \frac{2}{375}}$

$= \frac{25 + 4}{375 - 25 - 2} = \frac{29}{348} = \frac{1}{12}$

54. It is given that α and β are roots of quadratic equation $x^2 - x - 1 = 0$, so sum of roots $= \alpha + \beta = 1$, and product of roots $= \alpha\beta = -1$ and $p_k = \alpha^k + \beta^k, k \ge 1$

So, $p_1 = \alpha + \beta = 1$

$p_2 = \alpha^2 + \beta^2 = (\alpha+\beta)^2 - 2\alpha\beta = 1 + 2 = 3$

$p_3 = \alpha^3 + \beta^3 = (\alpha+\beta)^3 - 3\alpha\beta(\alpha+\beta) = 1 + 3 = 4$

$p_4 = \alpha^4 + \beta^4 = (\alpha+\beta)^4 - 4\alpha\beta(\alpha^2+\beta^2) - 6\alpha^2\beta^2$

$= 1 + 12 - 6 = 7$

and $p_5 = \alpha^5 + \beta^5$

$= (\alpha+\beta)^5 - 5\alpha\beta(\alpha^3+\beta^3) - 10\alpha^2\beta^2(\alpha+\beta)$

$= 1 + 20 - 10 = 11$

$\because p_3 = p_5 - p_4 = 4$

$p_1 + p_2 + p_3 + p_4 + p_5 = 1 + 3 + 4 + 7 + 11 = 26$

but $p_5 \ne p_2 \cdot p_3$

55. $\because 3^{|x|}|2 - |x|| = 1$ or $|2 - |x|| = 3^{-|x|} = y$ (say)

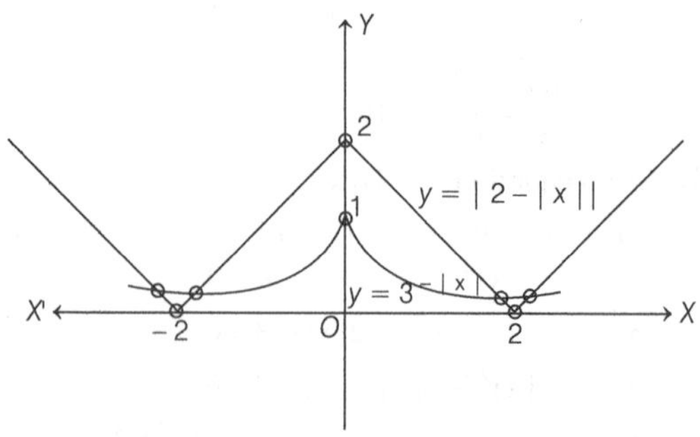

It is clear that graphs $y = |2 - |x||$ and $y = 3^{|x|}$ cuts at four points.

$\therefore$ Number of solutions is four.

56. Let roots be α, β

$\therefore \alpha + \beta = -2a$ and $\alpha\beta = b$

By hypothesis, $|\alpha - \beta| \le 2m$

$\Rightarrow |\alpha - \beta|^2 \le (2m)^2$

$(\alpha+\beta)^2 - 4\alpha\beta \le 4m^2$

$\Rightarrow 4a^2 - 4b \le 4m^2$

$\Rightarrow a^2 - m^2 \le b$ and discriminant > 0

$\Rightarrow 4a^2 - 4b > 0$

$\therefore b < a^2$

Hence, $b \in [a^2 - m^2, a^2)$

57. $\because ax^3 + bx^2 + c = (x^2 + px + 1)(ax + c)$ ...(i)

Comparing coefficient of x^2 and coefficient of x in Eq. (i), then

$b = c + ap \Rightarrow p = \frac{b-c}{a}$ ...(ii)

and $0 = pc + a \Rightarrow p = -\frac{a}{c}$ ...(iii)

From Eqs. (ii) and (iii),

$\frac{b-c}{a} = -\frac{a}{c} \Rightarrow a^2 - c^2 = -bc$

58. Given quadratic polynomials, $x^2 + 20x - 2020$ and $x^2 - 20x + 2020$ having a, b distinct real and c, d distinct complex roots respectively.

So, $a + b = -20$, $ab = -2020$

and $c + d = 20$, $cd = 2020$

Now, $ac(a-c) + ad(a-d) + bc(b-c) + bd(b-d)$

$= a^2(c+d) - a(c^2+d^2) + b^2(c+d) - b(c^2+d^2)$

$= (c+d)(a^2+b^2) - (c^2+d^2)(a+b)$

$= (c+d)[(a+b)^2 - 2ab] - (a+b)[(c+d)^2 - 2cd]$

$= 20[(20)^2 + 4040] + 20[(20)^2 - 4040]$

$= 2 \times 20 \times (20)^2 = 40 \times 400 = 16000$

59. $x^2 - 6x - 2 = 0$ with roots α, β: $\alpha + \beta = 6$, $\alpha\beta = -2$

and $\alpha^2 - 6\alpha - 2 = 0 \Rightarrow \alpha^2 - 2 = 6\alpha$

Now, $\beta^2 - 6\beta - 2 = 0 \Rightarrow \beta^2 - 2 = 6\beta$

$\Rightarrow \frac{a_{10} - 2a_8}{3a_9} = \frac{(\alpha^{10} - \beta^{10}) - 2(\alpha^8 - \beta^8)}{3(\alpha^9 - \beta^9)}$

$= \frac{(\alpha^{10} - 2\alpha^8) - (\beta^{10} - 2\beta^8)}{3(\alpha^9 - \beta^9)}$

$= \frac{\alpha^8(\alpha^2 - 2) - \beta^8(\beta^2 - 2)}{3(\alpha^9 - \beta^9)}$

$= \frac{\alpha^8(6\alpha) - \beta^8(6\beta)}{3(\alpha^9 - \beta^9)} = \frac{6(\alpha^9 - \beta^9)}{3(\alpha^9 - \beta^9)} = \frac{6}{3} = 2$

60. $D < 0$

$(2(3k-1))^2 - 4(8k^2 - 7) < 0$

$\Rightarrow 4(9k^2 - 6k + 1) - 4(8k^2 - 7) < 0$

$\Rightarrow k^2 - 6k + 8 < 0$

$\Rightarrow (k-4)(k-2) < 0$

$\Rightarrow 2 < k < 4$, then $k = 3$

61. Given, $\frac{1}{3} < \frac{x^2 - 2x + 4}{x^2 + 2x + 4} < 3$

Replacing x by 3^{x+1}

Then $\frac{1}{3} < \frac{3^{2x} \cdot 9 - 6 \cdot 3^x + 4}{3^{2x} \cdot 9 + 6 \cdot 3^x + 4} < 3$

$\Rightarrow \quad 3 > \frac{9 \cdot 3^{2x} + 6 \cdot 3^x + 4}{9 \cdot 3^{2x} - 6 \cdot 3^x + 4} > \frac{1}{3}$

or $\quad 3^{-1} < \frac{9 \cdot 3^{2x} + 6 \cdot 3^x + 4}{9 \cdot 3^{2x} - 6 \cdot 3^x + 4} < 3$

62. $\log_x 2 \log_{2x} 2 = \log_{4x} 2$

Here, $x > 0$ and $x \neq 1, \frac{1}{2}, \frac{1}{4}$

$\Rightarrow \quad \frac{1}{\log_2 x} \cdot \frac{1}{\log_2 2x} = \frac{1}{\log_2 4x}$

$\Rightarrow \quad \log_2 x \cdot (1 + \log_2 x) = (2 + \log_2 x)$

Put $\quad \log_2 x = t$

$\therefore \quad t^2 = 2$

$\therefore \quad t = \pm\sqrt{2}$

Then $\quad \log_2 x = \pm\sqrt{2}$

$x = 2^{\pm\sqrt{2}}$

$\therefore \quad x = 2^{-\sqrt{2}}, 2^{\sqrt{2}}$

63. $\left|\frac{x}{x-1}\right| + |x| = \frac{x^2}{|x-1|}$

$\Rightarrow \quad \left|\frac{x}{x-1}\right| + |x| = \left|x + \frac{x}{x-1}\right|$

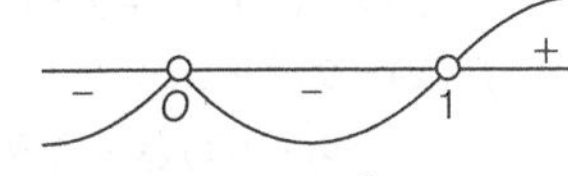

$\therefore \quad x \cdot \frac{x}{x-1} \geq 0 \Rightarrow \frac{x^2}{x-1} \geq 0$

$\therefore \quad x \in (1, \infty) \cup \{0\}$

64. We have,

$x^2 \cdot 2^{x+1} + 2^{|x-3|+2} = x^2 \cdot 2^{|x-3|+4} + 2^{x-1}$

If $x - 3 \geq 0$ *i.e.* $x \geq 3$ (impossible since x is negative)

Now, $\quad x - 3 < 0$

$\Rightarrow \quad x < 3$

Then, $\quad x^2 \cdot 2^{x+1} + 2^{-x+5} = x^2 \cdot 2^{-x+7} + 2^{x-1}$

$x^2 \cdot 2^{-x+7}(2^{2x-6} - 1) = 2^{-x+5}(2^{2x-6} - 1)$

$\Rightarrow \quad (2^{2x-6} - 1)(x^2 \cdot 4 - 1) = 0$

$\therefore \quad x = \pm\frac{1}{2}$ and $x = 3$

$\because x < 3$ and x is negative integer, then

$x \neq \pm\frac{1}{2}$ and $x \neq 3$.

Hence, number of negative integral solutions is zero.

65. $\because |x^2 - 2x - 3| = b$

Alternate (a) : For $b < 0$

$|x^2 - 2x - 3| < 0$

Impossible, there are no solutions.

Alternate (b) : For $\quad b = 0$

$x^2 - 2x - 3 = 0$

$\therefore x = 3, -1$, there are two solutions

Alternate (c) : For $0 < b < 1$

$0 < |x^2 - 2x - 3| < 1$

$\Rightarrow \quad 0 < x^2 - 2x - 3 < 1$

and $\quad -1 < x^2 - 2x - 3 < 0$

$\Rightarrow \quad x \in (1 - \sqrt{5}, -1) \cup (3, 1 + \sqrt{5})$

and $\quad x \in (-1, 1 - \sqrt{3}) \cup (1 + \sqrt{3}, 3)$

Alternate (d) : For $b = 1$

$|x^2 - 2x - 3| = 1 \Rightarrow x^2 - 2x - 3 = \pm 1$

$\therefore \quad x = 1 \pm \sqrt{5}$ and $x = 1 \pm \sqrt{3}$

There are four solutions.

66. Given equation is

$(3 - x)^4 + (2 - x)^4 = (5 - 2x)^4 \quad$ ...(i)

Let $\quad a = 3 - x$ and $b = 2 - x$

Eq. (i) becomes

$a^4 + b^4 = (a + b)^4$

$\Rightarrow \quad 2ab(2a^2 + 3ab + 2b^2) = 0$

$\therefore \quad a = 0 \Rightarrow 3 - x = 0 \Rightarrow x = 3$

$b = 0 \Rightarrow 2 - x = 0 \Rightarrow x = 2$

and $\quad 2a^2 + 3ab + 2b^2 = 0$

$2\left(\frac{a}{b}\right)^2 + 3\left(\frac{a}{b}\right) + 2 = 0$

$\frac{a}{b} = \frac{-3 \pm \sqrt{9 - 16}}{4}$

$\frac{3 - x}{2 - x} = \frac{-3 \pm i\sqrt{7}}{4}$

$\Rightarrow \quad \frac{3 - x}{2 - x} - 1 = \frac{-7 \pm i\sqrt{7}}{4}$

$\Rightarrow \quad \frac{1}{2 - x} = \frac{-7 \pm i\sqrt{7}}{4}$

$2 - x = \frac{4}{-7 \pm i\sqrt{7}}$

$x = 2 - \frac{4}{-7 \pm i\sqrt{7}}$

$x = \frac{-18 \pm 2i\sqrt{7}}{-7 \pm i\sqrt{7}}$

Taking '+' sign

$x = \frac{(-18 + 2i\sqrt{7})(-7 - i\sqrt{7})}{(-7 + i\sqrt{7})(-7 - i\sqrt{7})}$

$= \frac{140 + 4i\sqrt{7}}{56}$

$\therefore \quad x = \frac{35 + i\sqrt{7}}{14}$

Taking '–' sign, $x = \frac{35 - i\sqrt{7}}{14}$

Hence, roots of Eq. (i) are 2, 3, $\frac{35 \pm i\sqrt{7}}{14}$

i.e. Two real and two imaginary.

67. Let $f(x) = x^{12} - x^9 + x^4 - x + 1$

$$= (x^9 + x)(x^3 - 1) + 1 > 0, \ \forall \ x \geq 1 \quad \ldots\text{(i)}$$

Again, $f(x) = x^4(x^8 + 1) - x(x^8 + 1) + 1$

$$= x(x^8 + 1)(x^3 - 1) + 1 > 0, \ \forall \ x \leq 0 \quad \ldots\text{(ii)}$$

Next $f(x) = x^{12} + x^4(1 - x^5) + (1 - x) > 0, \ \forall \ 0 < x < 1 \ldots\text{(iii)}$

Combining Eqs. (i), (ii), (iii), we get

$$x \in (-\infty, \infty).$$

68. $(1 - 2i)^2 + \alpha(1 - 2i) + \beta = 0$

$\Rightarrow \quad 1 - 4 - 4i + \alpha - 2i\alpha + \beta = 0$

$\Rightarrow \quad (\alpha + \beta - 3) - i(4 + 2\alpha) = 0$

$\Rightarrow \alpha + \beta - 3 = 0$ and $4 + 2\alpha = 0$

$\Rightarrow \quad \alpha = -2, \beta = 5 \ \Rightarrow \alpha - \beta = -7$

69. $\alpha + \beta = 1$ and $\alpha\beta = -1$

AM of A_{n-1} and $A_n = \dfrac{A_{n-1} + A_n}{2}$

$$= \frac{\alpha^{n-1} + \beta^{n-1} + \alpha^n + \beta^n}{2}$$

$$= \frac{\alpha^{n-1}(1 + \alpha) + \beta^{n-1}(1 + \beta)}{2}$$

$$= \frac{\alpha^{n-1}(\alpha^2) + \beta^{n-1}(\beta^2)}{2}$$

$$[\because \alpha^2 = \alpha + 1 \text{ and } \beta^2 = \beta + 1]$$

$$= \frac{1}{2}(\alpha^{n+1} + \beta^{n+1}) = \frac{1}{2}A_{n+1}$$

70. Let $\alpha < \beta$

Then $\alpha < 2 < \beta$

$\therefore \quad D > 0, (a + 2)^2 - 4 \cdot 1 \cdot (-a - 3) > 0$

$\Rightarrow \quad a^2 + 4 + 4a + 4a + 12 > 0$

$\Rightarrow \quad a^2 + 8a + 16 > 0$

$\Rightarrow (a + 4)^2 > 0$ which is true.

Again, $f(2) < 0$

$\Rightarrow \quad 4 + (a + 2) \cdot 2 - (a + 3) < 0$

$\Rightarrow \quad 5 + a < 0$

$\Rightarrow \quad a < -5$

or $\quad a \in (-\infty, -5)$

Objective Questions Type II [One or more than one correct answer(s)]

1. Given equation is

$$x^{\{(\log_3 x)^2 - (9/2)\log_3 x + 5\}} = 3\sqrt{3}$$

or $\quad x^{\{(\log_3 x)^2 - (9/2)\log_3 x + 5\}} = 3^{3/2}$

or $\{(\log_3 x)^2 - (9/2)\log_3 x + 5\} \log_3 x = \dfrac{3}{2} \quad \ldots\text{(i)}$

Put $\log_3 x = t$

Then from Eq. (i),

$$t^3 - \frac{9}{2}t^2 + 5t = \frac{3}{2}$$

or $\quad 2t^3 - 9t^2 + 10t - 3 = 0$

or $\quad (t - 1)(2t^2 - 7t + 3) = 0$

or $\quad (t - 1)\left(t - \dfrac{1}{2}\right)(t - 3) = 0$

$\therefore \quad t = \dfrac{1}{2}, 1, 3$

or $\quad \log_3 x = \dfrac{1}{2}, 1, 3$

$\Rightarrow \quad x = \sqrt{3}, 3, 27.$

2. Given equation is

$$\log_{ax} a + \log_x a^2 + \log_{a^2x} a^3 = 0$$

$\Rightarrow \quad \dfrac{\log_e a}{\log_e a + \log_e x} + \dfrac{2\log_e a}{\log_e x} + \dfrac{3\log_e a}{2\log_e a + \log_e x} = 0$

$\Rightarrow \quad \dfrac{1}{1 + \log_a x} + \dfrac{2}{\log_a x} + \dfrac{3}{2 + \log_a x} = 0 \ \ldots\text{(i)}$

Put $\log_a x = t$

Then from Eq. (i),

$$\frac{1}{1 + t} + \frac{2}{t} + \frac{3}{2 + t} = 0$$

or $\quad 6t^2 + 11t + 4 = 0$

$\Rightarrow \quad (2t + 1)(3t + 4) = 0$

$\therefore \quad t = -\dfrac{1}{2}, -\dfrac{4}{3}$

$\Rightarrow \quad \log_a x = -\dfrac{1}{2}, -\dfrac{4}{3}$

$$x = a^{-1/2}, a^{-4/3}.$$

3. Given equation is $3x^3 + x^2 - 7x + \lambda = 0 \quad \ldots\text{(i)}$

Let $\quad f(x) = 3x^3 + x^2 - 7x + \lambda$

$\therefore \quad f'(x) = 9x^2 + 2x - 7$

$$= 9(x + 1)(x - 7/9)$$

For max or min

$$f'(x) = 0$$

$\therefore \quad x = -1, x = 7/9$

$$\frac{7}{9} \in [0, 1]$$

Hence Eq. (i) has two distinct real roots in [0, 1] for all values of λ.

$\therefore \quad \lambda \in (-\infty, \infty)$

4. If other root is β,

then $\quad \alpha + \beta = -2/4$

$\therefore \quad \beta = -\dfrac{1}{2} - \alpha$

and $\quad 4\alpha^2 + 2\alpha - 1 = 0$

$\Rightarrow \quad 4\alpha^2 = 1 - 2\alpha$

$\therefore \quad 4\alpha^3 = \alpha(4\alpha^2) = \alpha(1 - 2\alpha) = \alpha - 2\alpha^2$

$$= \alpha - 2\left[\frac{1 - 2\alpha}{4}\right]$$

$$= \alpha - \frac{1}{2} + \alpha = 2\alpha - \frac{1}{2}$$

$4\alpha^3 - 3\alpha = -\alpha - \dfrac{1}{2} = \beta.$

5. Given equation is

$$(x^2+1)^2 = x(3x^2+4x+3)$$

$$\Rightarrow \quad x^4-3x^3-2x^2-3x+1=0$$

$$\Rightarrow \quad x^2\left(x^2-3x-2-\frac{3}{x}+\frac{1}{x^2}\right)=0$$

$$\because \quad x \neq 0$$

$$\therefore \quad x^2+\frac{1}{x^2}-3\left(x+\frac{1}{x}\right)-2=0$$

$$\Rightarrow \quad \left(x+\frac{1}{x}\right)^2-3\left(x+\frac{1}{x}\right)-4=0$$

$$\Rightarrow \quad \left(x+\frac{1}{x}-4\right)\left(x+\frac{1}{x}+1\right)=0$$

or $$(x^2-4x+1)(x^2+x+1)=0$$

or $$\{(x-2)^2-3\}\left(\left(x+\frac{1}{2}\right)^2+\frac{3}{4}\right)=0$$

$$\therefore \quad x = 2\pm\sqrt{3}, \frac{-1\pm i\sqrt{3}}{2}.$$

6. Given equation is

$$Ax^2-|G|\,x-H=0 \quad \text{...(i)}$$

Let α, β are the roots

then $\alpha+\beta=\frac{|G|}{A}$ and $\alpha\beta=-\frac{H}{A}$

Since, $A>|G|>H$ or $1>\frac{|G|}{A}>\frac{H}{A}$

Hence, A is positive

$\therefore \alpha+\beta$ and $\alpha\beta$ has positive and negative fraction respectively.

Also, $|G|^2 = AH$ ...(ii)

Discriminant of Eq. (i) $=(-|G|)^2-4\cdot A\cdot(-H)$

$$=|G|^2+4AH$$

$$=5|G|^2>0 \quad \text{[from Eq. (ii)]}$$

Hence, roots of Eq. (i) are real and distinct.

$\because \quad \alpha+\beta>0$ and $\alpha\beta<0$

One root is positive and other is negative and at least one root is a fraction.

So, the equation has a negative fraction root.

7. $0<a<b<c, \alpha+\beta=-\frac{b}{a}$ and $\alpha\beta=\frac{c}{a}$

For non real complex roots

$$b^2-4ac<0$$

$$\Rightarrow \quad \frac{b^2}{a^2}-\frac{4c}{a}<0$$

$$\Rightarrow \quad (\alpha+\beta)^2-4\alpha\beta<0$$

$$\Rightarrow \quad (\alpha-\beta)^2<0$$

$$\because \quad 0<a<b<c$$

$\therefore$ Roots are conjugate, then $|\alpha|=|\beta|$

but $\alpha\beta=\frac{c}{a}$

$$|\alpha\beta|=\left|\frac{c}{a}\right|>1 \quad \left[\because a<c, \therefore \frac{c}{a}>1\right]$$

$$\Rightarrow \quad |\alpha||\beta|>1$$

$$|\alpha|^2>1 \text{ or } |\alpha|>1$$

8. If roots is $\alpha, \frac{1}{\alpha}$

$$\therefore \quad \alpha\cdot\frac{1}{\alpha}=\frac{c}{a}$$

$$\Rightarrow \quad c=a$$

Since, $(|a|-|b|)^2\geq 0$ or $|a|\geq|b|$

but $a=c$,

$\therefore |c|\geq|b|$.

9. $\because \quad a>b>c$ ...(i)

and given equation is

$$(a+b-2c)x^2+(b+c-2a)x+(c+a-2b)=0 \quad \text{...(ii)}$$

$\because$ Eq. (ii) has a root in the interval $(-1, 0)$

$$\therefore \quad f(-1)\,f(0)<0$$

$$\Rightarrow \quad (2a-b-c)(c+a-2b)<0 \quad \text{...(iii)}$$

From Eq. (i)

$$a>b \Rightarrow a-b>0$$

and $$a>c \Rightarrow a-c>0$$

$$\therefore \quad 2a-b-c>0 \quad \text{...(iv)}$$

From Eqs. (iii) and (iv),

$$c+a-2b<0 \text{ or } c+a<2b$$

Alternate (b) is correct.

and Discriminant of Eq. (ii)

$=(b+c-2a)^2-4(a+b-2c)(c+a-2b)$

$=\{(b-a)+(c-a)\}^2-4\{(a-c)+(b-c)\}\{c-b+a-b\}$

$=\{(c-a)-(a-b)\}-4\{(b-c)-(c-a)\}$

$\times\{(a-b)-(b-c)\}$

If $b-c=X, c-a=Y$ and $a-b=Z$

then, discriminant $=(Y-Z)^2-4(X-Y)(Z-X)$

$$=(2X-Y-Z)^2 \quad (\because X+Y+Z=0)$$

$$=(0+3X)^2=9X^2=9(b-c)^2$$

$\therefore$ Both roots of the given equation are rational.

Alternate (c) is correct.

If roots of $ax^2+2bx+c=0$ are α and β.

Then $\alpha+\beta=-\frac{2b}{a}<0$

$$\therefore \quad \alpha+\beta<0 \text{ and } \alpha\beta=\frac{c}{a}>0 \quad [\because a>b>c]$$

and discriminant of

$$ax^2+2bx+c=0$$

is $$4b^2-4ac>0 \quad [\because b>c \text{ and } b<a]$$

Hence, roots of $ax^2+2bx+c=0$ are negative and real.

Alternate (d) is correct.

10. $|x+1||x-1|=a^2-2a-3$

$$\Rightarrow \quad |x^2-1|=a^2-2a-3$$

$$\therefore \quad a^2-2a-3\geq 0 \Rightarrow (a+1)(a-3)\geq 0$$

$$\therefore \quad a\in(-\infty,-1)\cup[3,\infty)$$

11. Given equations are

$$(b-c)x^2+(c-a)x+(a-b)=0 \quad \text{...(i)}$$

and $$a(b-c)x^2+b(c-a)x+c(a-b)=0 \quad \text{...(ii)}$$

Since $x=1$ satisfies Eqs. (i) and (ii)

Hence, $x=1$ is exactly one common root.

Let $b - c = A, c - a = B, a - b = C$
and $a(b - c) = A', b(c - a) = B', c(a - b) = C'$
$\therefore$ $A + B + C = 0$ and $A' + B' + C' = 0$
Discriminant of Eq. (i) $= (c - a)^2 - 4(b - c)(a - b)$
$= B^2 - 4AC$
$= (-A - C)^2 - 4AC$
$[\because B = -A - C]$
$= (A - C)^2 = (2b - a - c)^2$
and discriminant of Eq. (ii)
$= \{b(c - a)\} - 4a(b - c) \cdot c(a - b)$
$= (B')^2 - 4A'C'$
$= (-A' - C')^2 - 4A'C'$ $[\because B' = -A' - C']$
$= (A' - C')^2$
$= (ab - ac - ac + bc)^2$
$= (ab + bc - 2ac)^2$
Hence, roots of Eqs. (i) and (ii) are rational.

12. $\alpha + \beta = -b/a, \alpha\beta = c/a, \alpha + h + \beta + h = -q/p$
and $(\alpha + h)(\beta + h) = r/p$.
Therefore, $-\frac{q}{p} = \alpha + \beta + 2h = -\frac{b}{a} + 2h$
$\Rightarrow$ $h = \frac{1}{2}\left(\frac{b}{a} - \frac{q}{p}\right)$
Now, $(\alpha - \beta)^2 = (\alpha + \beta)^2 - 4\alpha\beta = \left(-\frac{b}{a}\right)^2 - 4\frac{c}{a} = \frac{b^2 - 4ac}{a^2}$
and $(\alpha - \beta)^2 = [(\alpha + h) - (\beta + h)]^2 = \frac{q^2 - 4pr}{p^2}$
$\Rightarrow$ $\frac{b^2 - 4ac}{a^2} = \frac{q^2 - 4pr}{p^2}$

13. $\because$ Roots are real
$\therefore$ $B^2 - 4AC > 0$
$\Rightarrow a^4 > 4b^2$, (a) is correct.
If $f(x) = x^2 + a^2x + b^2$ ($\because c$ lie outside the roots)
$\therefore$ $f(c) > 0$, then $c^2 + a^2c + b^2 > 0$
(b) is correct.
Also (x – coordinate of vertex) $> c$ $\Rightarrow -\frac{a^2}{2} > c$
$\therefore$ Option (c) is correct.

14. Let $f(x) = ax^2 + bx + c$
$f\left(\frac{1}{3}\right) = \frac{a}{9} + \frac{b}{3} + c = \frac{1}{9}(a + 3b + 9c) = 0$
$\therefore \frac{1}{3}$ is the root of the given equation.
Also, product of the roots $= \frac{ac}{a^2} < 0$
Therefore, another root must be negative, hence it will be $-1/\sqrt{3}$ and required equation is
$$\left(x - \frac{1}{3}\right)\left(x + \frac{1}{\sqrt{3}}\right) = 0$$
or $3\sqrt{3}x^2 + (3 - \sqrt{3})x - 1 = 0$

15. $|\alpha - \beta| = 3$ and $\alpha^2 + \beta^2 = 29$.
$|\alpha - \beta|^2 = 9$
$\Rightarrow \alpha^2 + \beta^2 - 2\alpha\beta = 9$
$\therefore$ $\alpha\beta = 10$ $[\because \alpha^2 + \beta^2 = 29]$
then $(\alpha + \beta)^2 = \alpha^2 + \beta^2 + 2\alpha\beta$
$= 29 + 20 = 49$
$\therefore$ $\alpha + \beta = \pm 7$
$\therefore$ Required equation is
$$x^2 \pm 7x + 10 = 0$$

16. Roots are reciprocal, $\alpha\beta = \frac{c}{a} = 1$
$\Rightarrow$ $c = a$
Also, $D = 4b^2 - 4ac \geq 0$
$\Rightarrow$ $b^2 - a^2 \geq 0$ $(\because a = c)$
$\Rightarrow$ $|b| \geq |a| = |c|$.

17. $a > 0$, $f(-1) < 0$ and $f(1) < 0$
$\Rightarrow$ $a - b + c < 0$ and $a + b + c < 0$
$\Rightarrow$ $1 - \frac{b}{a} + \frac{c}{a} < 0$ and $1 + \frac{b}{a} + \frac{c}{a} < 0$
$\Rightarrow$ $1 \pm \left|\frac{b}{a}\right| + \frac{c}{a} < 0$

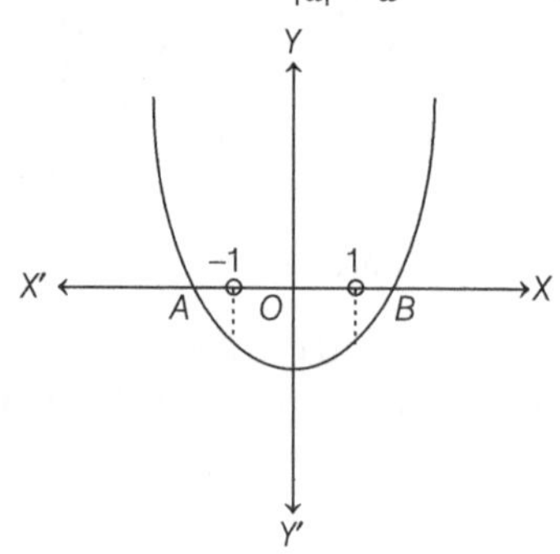

$a < 0$, $f(-1) > 0$ and $f(1) > 0$
$\Rightarrow$ $a \pm b + c > 0$

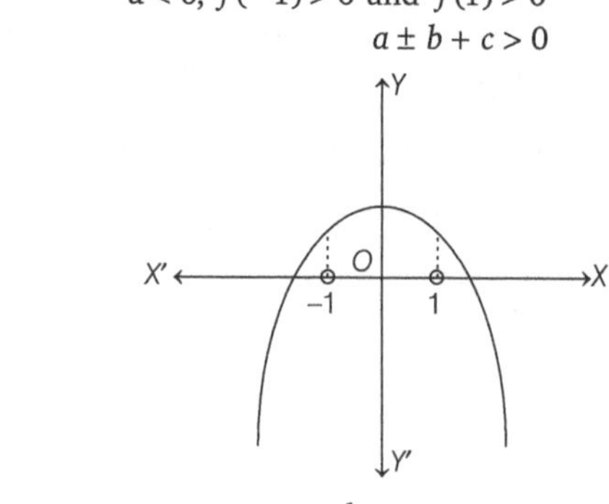

$\Rightarrow$ $1 \pm \frac{b}{a} + \frac{c}{a} < 0$ $(\because a < 0)$
$\Rightarrow$ $1 \pm \left|\frac{b}{a}\right| + \frac{c}{a} < 0$

18. Let $A = a + 2b - 3c, B = b + 2c - 3a, C = c + 2a - 3b$
$\therefore$ $A + B + C = 0$
Hence, roots are 1 and $\frac{C}{A}$.

19. Roots of $ax^2 + 2bx + c = 0$ are non real complex.

$\therefore f(x) = ax^2 + 2bx + c > 0$ or < 0 for all x

But $\quad f(-1) = a - 2b + c < 0$

$\therefore f(0)$ and $f(-2)$ must be less than zero

$$f(0) < 0 \Rightarrow c < 0$$

$\Rightarrow$ and $\quad f(-2) < 0 \Rightarrow 4a + c < 4b.$

20. All roots of equation $ax^4 + bx^2 + c = 0$ will be real if both roots of $ay^2 + by + c = 0$ will be + ve (let $x^2 = y$)

ie, sum of roots $= -\frac{b}{a} > 0$ and product of roots $= \frac{c}{a} > 0.$

Hence, a and b are of opposite sign, while a and c are of same sign.

21. Given, x_1 and x_2 are roots of $\alpha x^2 - x + \alpha = 0.$

$\therefore \quad x_1 + x_2 = \frac{1}{\alpha}$ and $x_1 x_2 = 1$

Also, $\quad |x_1 - x_2| < 1$

$\Rightarrow \quad |x_1 - x_2|^2 < 1 \Rightarrow (x_1 - x_2)^2 < 1$

or $\quad (x_1 + x_2)^2 - 4x_1 x_2 < 1$

$\Rightarrow \quad \frac{1}{\alpha^2} - 4 < 1$ or $\frac{1}{\alpha^2} < 5$

$\Rightarrow \quad 5\alpha^2 - 1 > 0$ or $(\sqrt{5}\alpha - 1)(\sqrt{5}\alpha + 1) > 0$

+   –   +

$-1/\sqrt{5}$ $1/\sqrt{5}$

$$\therefore \alpha \in \left(-\infty, -\frac{1}{\sqrt{5}}\right) \cup \left(\frac{1}{\sqrt{5}}, \infty\right) \quad \text{...(i)}$$

Also, $\quad D > 0 \Rightarrow 1 - 4\alpha^2 > 0$

or $\quad \alpha \in \left(-\frac{1}{2}, \frac{1}{2}\right) \quad$ ...(ii)

From Eqs. (i) and (ii), we get

$$\alpha \in \left(-\frac{1}{2}, \frac{-1}{\sqrt{5}}\right) \cup \left(\frac{1}{\sqrt{5}}, \frac{1}{2}\right)$$

22. Let $f(x) = ax^3 + bx^2 + cx + d$

f is continuous and derivable on R. Also, $f(0) = d$ and $f(1) = a + b + c + d = 0 + d = d.$

By the Rolle's theorem, there exists at least one $\alpha \in (0, 1)$ such that

$$f'(\alpha) = 0 \Rightarrow 3a\alpha^2 + 2b\alpha + c = 0$$

Thus, $3ax^2 + 2bx + c = 0$ has at least one root in $[0, 1]$.

Also, $[0, 1] \subseteq [-1, 1]$ and $[0, 1] \subseteq [0, 2]$.

23. $ax + by = 1 \Rightarrow y = \frac{1 - ax}{b}$

Putting this value in second equation, we get

$$cx^2 + \frac{d}{b^2}(1 - ax)^2 = 1$$

$\Rightarrow (b^2c + a^2d)x^2 - 2adx + d - b^2 = 0$

Equation will have equal roots, if

$$D = 4a^2d^2 - 4(b^2c + a^2d)(d - b^2) = 0$$

$\Rightarrow \quad b^2[b^2c + a^2d - cd] = 0$

$\Rightarrow \quad \frac{b^2}{d} + \frac{a^2}{c} = 1$

Also, in this case

$$x = \frac{2ad}{2(b^2c + a^2d)} = \frac{a}{c}$$

$$y = \frac{1 - ax}{b} = \frac{1}{b}\left(1 - \frac{a^2}{c}\right) = \frac{1}{b}\left(\frac{b^2}{d}\right) = \frac{b}{d}$$

24. Let the roots of the given equation be $\alpha/r, \alpha, \alpha r$, where $\alpha > 0$ and $r > 1$. Then

$$\frac{\alpha}{r} + \alpha + \alpha r = -b \quad \text{...(i)}$$

$$\frac{\alpha}{r} \cdot \alpha + \alpha \cdot \alpha r + \frac{\alpha}{r} \cdot \alpha r = c \quad \text{...(ii)}$$

and $\quad \left(\frac{\alpha}{r}\right)(\alpha)(\alpha r) = 1 \quad$ ...(iii)

From Eq. (iii), we get

$$\alpha^3 = 1 \text{ or } \alpha = 1$$

From Eq. (i), we get

$$\frac{1}{r} + 1 + r = -b \quad \text{...(iv)}$$

$\Rightarrow \quad \left(\frac{1}{\sqrt{r}} - \sqrt{r}\right)^2 + 3 = -b$

$\Rightarrow \quad -b - 3 > 0$ or $b < -3$

$\Rightarrow \quad b \in (-\infty, -3)$

Also, from Eq. (ii) $1/r + r + 1 = c \quad$ ...(v)

From Eqs. (iv) and (v) $-b = c$ or $b + c = 0$

As $r > 1$, $\frac{\alpha}{r} = \frac{1}{r} < 1$ and $\alpha r = r > 1$.

25. Since, the equation has two distinct roots α and β, the discriminant $b^2 - 4ac > 0$, we must have

$$f(x) = ax^2 + bx + c < 0 \text{ for } \alpha < x < \beta$$

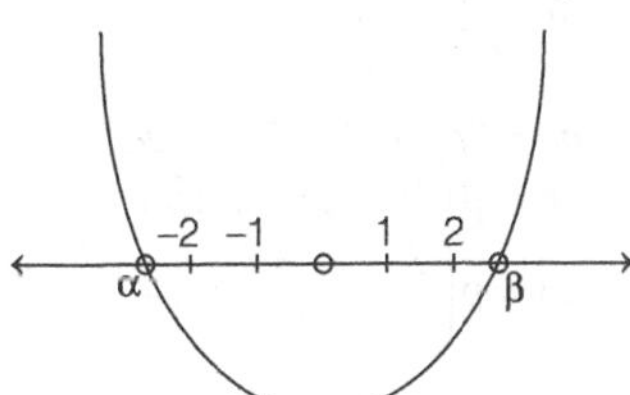

Since, $\alpha < 0 < \beta$ we must have $f(0) = c < 0$

Also, as $\alpha < -1, 1 < \beta$ we get

$$f(-1) = a - b + c < 0$$

and $f(1) = a + b + c < 0$, i.e. $a + |b| + c < 0$

Since, $\alpha < -2, 2 < \beta$; $f(-2) = 4a - 2b + c < 0$

and $f(2) = 4a + 2b + c < 0$ i.e. $4a + 2|b| + c < 0.$

26. $\frac{p}{2x} = \frac{(a + b)x + c(b - a)}{x^2 - c^2}$

or $\quad p(x^2 - c^2) = 2(a + b)x^2 - 2c(a - b)x$

or $\quad (2a + 2b - p)x^2 - 2c(a - b)x + pc^2 = 0$

Now, $\quad c^2(a - b)^2 - pc^2(2a + 2b - p) = 0$

[$\because$ equal roots]

$\Rightarrow \quad (a - b)^2 - 2p(a + b) + p^2 = 0 \quad [\because c^2 \neq 0]$

$\Rightarrow \quad [p - (a + b)]^2 = (a + b)^2 - (a - b)^2$

$\Rightarrow \quad p = a + b \pm 2\sqrt{ab} = (\sqrt{a} \pm \sqrt{b})^2.$

27. We have $\alpha + \beta = -a$ and $\alpha\beta = b$

To find the equation whose roots are $1/\alpha$ and $1/\beta$

$\frac{1}{\alpha} + \frac{1}{\beta} = -\frac{a}{b}$ and $\frac{1}{\alpha} \cdot \frac{1}{\beta} = \frac{1}{b}$

$\therefore$ Equation is $x^2 - \left(-\frac{a}{b}\right)x + \frac{1}{b} = 0$ i.e. $bx^2 + ax + 1 = 0$

To find the equation whose roots are $-\alpha$ and $-\beta$

So, $-\alpha - \beta = a$ and $(-\alpha)(-\beta) = b$

$\therefore$ Equation is $\quad x^2 - ax + b = 0$

To find the equation whose roots are α^2 and β^2

So, $\alpha^2 + \beta^2 = (\alpha + \beta)^2 - 2\alpha\beta = a^2 - 2b$ and $\alpha^2\beta^2 = b^2$

$\therefore$ Equation is $x^2 - (a^2 - 2b)x + b^2 = 0$

i.e. $\quad x^2 + (2b - a^2)x + b^2 = 0$

To find the equation whose roots are α/β and β/α.

So, $\quad \frac{\alpha}{\beta} + \frac{\beta}{\alpha} = \frac{(\alpha+\beta)^2 - 2\alpha\beta}{\alpha\beta} = \frac{a^2 - 2b}{b}$

and $\quad \frac{\alpha}{\beta} \cdot \frac{\beta}{\alpha} = 1$

$\therefore$ Equation is $x^2 - \left(\frac{a^2 - 2b}{b}\right)x + 1 = 0$

i.e. $\quad bx^2 + (2b - a^2)x + b = 0$

28. We have, $\alpha + \beta = -p$, $\alpha\beta = q$, $\alpha^4 + \beta^4 = r$ and $\alpha^4\beta^4 = s$

Therefore, $\alpha^2 + \beta^2 = (\alpha + \beta)^2 - 2\alpha\beta = p^2 - 2q$, so that

$r = \alpha^4 + \beta^4 = (\alpha^2 + \beta^2)^2 - 2\alpha^2\beta^2 = (p^2 - 2q)^2 - 2q^2$

i.e. $\quad (p^2)^2 - 4qp^2 + 2q^2 - r = 0$

This shows that p^2 is one root of

$x^2 - 4qx + 2q^2 - r = 0$. If its other root is γ, we have

$\gamma + p^2 = 4q$

i.e. $\gamma = 4q - p^2$.

$\therefore \quad D = (-4q)^2 - 4(2q^2 - r) = 8q^2 + 4[(p^2 - 2q)^2 - 2q^2]$

$= 4(p^2 - 2q)^2 \geq 0$

So, that both roots p^2 and $-p^2 + 4q$ are real. Since α and β are real, $p^2 - 4q \geq 0$ *ie*, $-p^2 + 4q \leq 0$.

Thus, the roots of $x^2 - 4qx + 2q^2 - r = 0$ are positive and negative.

Linked Comprehension Type

PASSAGE 1

1. $a^2 = \alpha + 1$

$\beta^2 = \beta + 1$

$a_n = p\alpha^n + q\beta^n$

$= p(\alpha^{n-1} + \alpha^{n-2}) + q(\beta^{n-1} + \beta^{n-2}) = a_{n-1} + a_{n-2}$

$\therefore a_{12} = a_{11} + a_{10}$

2. $\alpha = \frac{1+\sqrt{5}}{2}, \beta = \frac{1-\sqrt{5}}{2}$

$a_4 = a_3 + a_2 = 2a_2 + a_1 = 3a_1 + 2a_0$

$\Rightarrow \quad 28 = p(3\alpha + 2) + q(3\beta + 2)$

$\Rightarrow \quad 28 = p\left(\frac{3}{2} + \frac{3\sqrt{5}}{2} + 2\right) + q\left(\frac{3}{2} - \frac{3\sqrt{5}}{2} + 2\right)$

$\Rightarrow \quad 28 = (p+q)\left(\frac{3}{2} + 2\right) + (p-q)\left(\frac{3\sqrt{5}}{2}\right)$

$\therefore \quad p - q = 0$ and $(p+q) \times \frac{7}{2} = 28$

$\Rightarrow \quad p + q = 8 \Rightarrow p = q = 4$

$\therefore \quad p + 2q = 12$

PASSAGE 2

Given equation is $2 + |x^2 + 4x + 3| = m$

$\therefore \quad |x^2 + 4x + 3| = m - 2$

Given equation is meaningful, if $m - 2 \geq 0$ $\quad \ldots$(i)

$\therefore \quad x^2 + 4x + 3 = \pm(m - 2)$

$\therefore \quad x^2 + 4x + 5 - m = 0$

or $\quad x^2 + 4x + 1 + m = 0$

$\therefore \quad x = \frac{-4 \pm \sqrt{16 - 4(5 - m)}}{2}$

or $\quad x = \frac{-4 \pm \sqrt{16 - 4(1 + m)}}{2}$

$\therefore \quad x = -2 \pm \sqrt{m - 1}$

or $\quad x = -2 \pm \sqrt{3 - m}$

1. **Case I** $m = 1$ and $3 - m > 0$

$m = 1$ and $m < 3$

No values of m (as $m \geq 2$)

Case II $m = 3$ and $m - 1 > 0 \Rightarrow m > 1$

Hence, $m = 3$.

2. For four distinct solution.

$m - 1 > 0$ and $3 - m > 0$

$m > 1$ and $m < 3$

As $m \geq 2$

Hence, $m \in [2, 3)$.

3. For two solutions $m - 1 \geq 0$ and $3 - m \leq 0 \Rightarrow m \geq 3$

$m - 1 \leq 0$ and $3 - m \geq 0$, not possible

$m = 2$ $\quad$ [using (1)]

$m \in \{2\} \cup (3, \infty)$

PASSAGE 3

1. Given, $f(x) = 4x^3 + 3x^2 + 2x + 1$

$f'(x) = 2(6x^2 + 3x + 1) \Rightarrow D = 9 - 24 < 0$

Hence, $f(x) = 0$ has only one real root.

$f\left(-\frac{1}{2}\right) = 1 - 1 + \frac{3}{4} - \frac{4}{8} > 0$

$f\left(-\frac{3}{4}\right) = 1 - \frac{6}{4} + \frac{27}{16} - \frac{108}{64}$

$= \frac{64 - 96 + 108 - 108}{64} < 0$

$f(x)$ changes its sign in $\left(-\frac{3}{4}, -\frac{1}{2}\right)$

Hence, $f(x) = 0$ has a root in $\left(-\frac{3}{4}, -\frac{1}{2}\right)$.

2. $\int_0^{1/2} f(x)\,dx < \int_0^t f(x)\,dx < \int_0^{3/4} f(x)\,dx$

Now, $\int f(x)\,dx = \int (1 + 2x + 3x^2 + 4x^3)\,dx$

$= x + x^2 + x^3 + x^4$

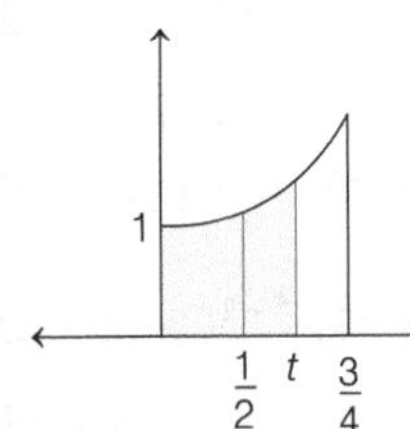

$\Rightarrow \int_0^{1/2} f(x)\,dx = \frac{15}{16} > \frac{3}{4}, \int_0^{3/4} f(x)\,dx = \frac{530}{256} < 3$

3. As, $f''(x) = 2(12x + 3)$

$f'(x) > 0$, when $x > -\frac{1}{4}$ and $f'(x) < 0$, when $x < -\frac{1}{4}$.

$\therefore$ It could be shown as

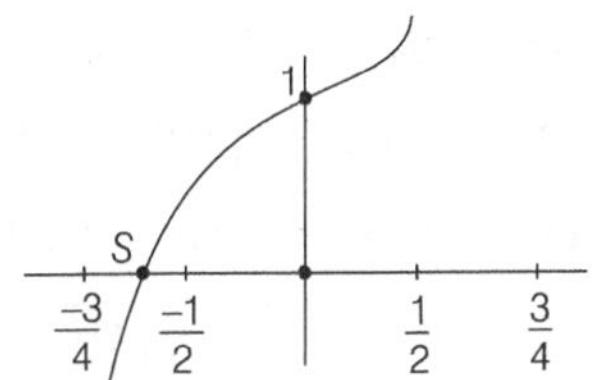

PASSAGE 4

1. $p + q = 2, pq = A$ ...(i)

and $r + s = 18, rs = B$ ...(ii)

$\because p, q, r, s$ are in AP.

Then $q = p + D, r = p + 2D$ and $s = p + 3D$

Then from Eq. (i), $2p + D = 2$

and from Eq. (ii), $2p + 5D = 18$

After solving, we get $D = 4$ and $p = -1$

$\therefore q = 3, r = 7$ and $s = 11$

From Eq. (i), $A = pq = -3$

From Eq. (ii), $B = rs = 77$

2. $\alpha_1 + \alpha_2 = 1, \alpha_1 \alpha_2 = p$ ....(i)

and $\alpha_3 + \alpha_4 = 4, \alpha_3 \alpha_4 = q$ ...(ii)

$\because \alpha_1, \alpha_2, \alpha_3, \alpha_4$ are in GP

$\therefore \alpha_2 = \alpha_1 R, \alpha_3 = \alpha_1 R^2, \alpha_4 = \alpha_1 R^3$

$\therefore$ From Eq. (i), $\alpha_1 + \alpha_1 R = 1$

$\Rightarrow \alpha_1(1 + R) = 1$ ...(iii)

and from Eq. (ii), $\alpha_1 R^2 + \alpha_1 R^3 = 4$ ...(iii)

$\Rightarrow \alpha_1 R^2(1 + R) = 4$ ...(iv)

Dividing Eq. (iv) by (iii), then

$R^2 = 4$ or $R = \pm 2$

If $R = 2$, then $\alpha_1 = \frac{1}{3}$ [from Eq. (iii)]

and if $R = -2$, then $\alpha_1 = -1$ [from Eq. (iii)]

$\because p$ and q are integer

$\therefore R = -2$ and $\alpha_1 = -1$

$\Rightarrow \alpha_2 = 2, \alpha_3 = -4, \alpha_4 = 8$

$p = \alpha_1 \alpha_2 = -2$ [from Eq. (i)]

and $q = \alpha_3 \alpha_4 = -32$ [from Eq. (i)]

$\therefore p = -2, q = -32$

3. $\beta_1 + \beta_3 = \frac{4}{A}, \beta_1 \beta_3 = \frac{1}{A}$ ...(i)

and $\beta_2 + \beta_4 = \frac{6}{B}, \beta_2\beta_4 = \frac{1}{B}$ ...(ii)

From Eq. (i), $\frac{\beta_1\beta_3}{(\beta_1 + \beta_3)} = \frac{1}{4}$ ....(iii)

and from Eq. (ii), $\frac{\beta_2\beta_4}{(\beta_2 + \beta_4)} = \frac{1}{6}$ ...(iv)

Also, $\beta_1, \beta_2, \beta_3, \beta_4$ are in HP.

Then $\beta_2 = \frac{2\beta_1\beta_3}{\beta_1 + \beta_3}$ and $\beta_3 = \frac{2\beta_2\beta_4}{\beta_2 + \beta_4}$

$\Rightarrow \beta_2 = \frac{1}{2}$ and $\beta_3 = \frac{1}{3}$ [from Eqs. (iii) and (iv)]

$\because \beta_3$ is a root of $Ax^2 - 4x + 1 = 0$

Then $A\left(\frac{1}{3}\right)^2 - 4\left(\frac{1}{3}\right) + 1 = 0 \Rightarrow \frac{A}{9} - \frac{1}{3} = 0$

$\therefore A = 3$

and β_2 is a root of $Bx^2 - 6x + 1 = 0$

then $B\left(\frac{1}{2}\right)^2 - 6\left(\frac{1}{2}\right) + 1 = 0 \Rightarrow B = 8$

Hence, $A = 3, B = 8$.

4. $(ax - b)^2 + (bx - c)^2 + (cx - d)^2 \le 0$

which is possible, $(ax - b)^2 + (bx - c)^2 + (cx - d)^2 = 0$

or $ax - b = 0, bx - c = 0$, and $cx - d = 0$

$\Rightarrow \frac{b}{a} = \frac{c}{b} = \frac{d}{c} = x$

Hence, a, b, c, d are in GP.

5. $\because \alpha + \beta = \frac{(4 + \sqrt{5})}{(5 + \sqrt{2})}$ and $\alpha\beta = \left(\frac{8 + 2\sqrt{5}}{5 + \sqrt{2}}\right) = 2(\alpha + \beta)$

$\therefore$ HM $= \frac{2\alpha\beta}{(\alpha + \beta)} = \frac{2 \times 2(\alpha + \beta)}{(\alpha + \beta)} = 4.$

6. $\because a(b - c) + b(c - a) + c(a - b) = 0$

$\therefore x = 1$ is a root of

$a(b - c)x^2 + b(c - a)x + c(a - b) = 0$

Since, roots are equal.

$\therefore$ Other root = 1

$\Rightarrow 1 \times 1 = \frac{c(a - b)}{a(b - c)} \Rightarrow b = \frac{2ac}{a + c}$

$\therefore a, b, c$ are in HP.

7. $\because (x - a)(x - b) + c = (x - \alpha)(x - \beta)$

$\Rightarrow (x - \alpha)(x - \beta) - c = (x - a)(x - b)$

$\therefore$ Roots of $(x - \alpha)(x - \beta) - c = 0$

or $(x - a)(x - b) = 0$

are $x = a, b.$

PASSAGE 5

1. $\because (4+\sqrt{15})^2 + (4-\sqrt{15})^2 = 62$

Then, $\frac{[x]}{2} = \pm 1$

$\therefore$ $[x] = \pm 2 \Rightarrow [x] = -2$ and $[x] = 2$

$\therefore$ $-2 \le x < -1$ and $2 \le x < 3$

$\Rightarrow$ $x \in [-2, -1) \cup [2, 3)$.

2. $\because (2+\sqrt{3})^{x^2-2x+1} + (2-\sqrt{3})^{x^2-2x-1} = \frac{4}{(2-\sqrt{3})}$

$(2+\sqrt{3})^{x^2-2x} + (2-\sqrt{3})^{x^2-2x} = 4$

$\because$ $(2+\sqrt{3}) + (2-\sqrt{3}) = 4$

$\therefore$ $x^2 - 2x = \pm 1$

or $(x-1)^2 = 1 \pm 1$

or $(x-1)^2 = 2, 0$

$\therefore$ $x - 1 = \pm\sqrt{2}, 0$

$x = 1 \pm \sqrt{2}, 1.$

3. $\because$ $(15+4\sqrt{14}) + (15-4\sqrt{14}) = 30$

$\therefore$ $t = \pm 1$

or $x^2 - 2|x| = \pm 1$

or $(|x|-1)^2 = 1 \pm 1 = 2, 0$

$\therefore$ $|x| - 1 = \pm\sqrt{2}, 0$

or $|x| = 1 \pm \sqrt{2}, 1$

$= 1 + \sqrt{2}, 1$ $[\because |x| \ge 0]$

$\therefore$ $x = \pm(1+\sqrt{2}), \pm 1.$

4. For $x \ge 4$ the last digit of the sum $1! + 2! + \ldots + x!$ is equal to 3 and therefore this sum cannot be equal to a square of a whole number k (because a square of a whole number cannot end with 3).

$\therefore$ For $x < 4$, the given equation has the only solutions

$x = 1, k = \pm 1$ and $x = 3, k = \pm 3$

$\therefore$ $\alpha = 1, \beta = 3$

$\therefore$ $1 + 2\alpha + 3\alpha^2 + 4\alpha^3 + 5\alpha^4 = 15$ and $-5\beta = -15$

Then $(a+\sqrt{b})^{x^2-15} + (a-\sqrt{b})^{x^2-15} = 2a$

$\because$ $(a+\sqrt{b}) + (a-\sqrt{b}) = 2a$

$\therefore$ $x^2 - 15 = \pm 1$

$\Rightarrow$ $x^2 = 16, 14 \Rightarrow x = \pm 4, \pm\sqrt{14}$

$\alpha_1 = -4, \alpha_2 = 4, \alpha_3 = -\sqrt{14}. \alpha_4 = \sqrt{14}$

$\therefore |\alpha_1 + \alpha_2 + \alpha_3 + \alpha_4 - \alpha_1\alpha_2\alpha_3\alpha_4| = |0 - 16 \times 14| = 224$

5. $\because \sqrt{a\sqrt{a\sqrt{a\ldots\infty}}} = a^{\frac{1}{2}+\frac{1}{4}+\frac{1}{8}+\ldots} = a$

$\therefore$ $\sqrt{x\sqrt{x\sqrt{x\ldots\infty}}} = x$ and $\sqrt{49+20\sqrt{6}} = (5+2\sqrt{6})$

$\because$ $a > 0$ and $x > 0$

$\therefore$ $x^2 - 3 > 0$ and $x > 0$

Then $x > \sqrt{3}$

Now, $\{\sqrt{(49+20\sqrt{6})}\}^{\sqrt{a\sqrt{a\sqrt{a\ldots\infty}}}}$

$+ (5-2\sqrt{6})^{x^2+x-3-\sqrt{x\sqrt{x\sqrt{x\ldots\infty}}}} = 10$

$\Rightarrow$ $(5+2\sqrt{6})^{x^2-3} + (5-2\sqrt{6})^{x^2-3} = 10$

Here, $5 + 2\sqrt{6} + 5 - 2\sqrt{6} = 10$

Then, $x^2 - 3 = \pm 1$

$x^2 = 4, 2 \Rightarrow x = \pm 2, \pm\sqrt{2}$

$\therefore$ $x = -2, 2, -\sqrt{2}, \sqrt{2}$

Hence, $x = 2$ $[\because x > \sqrt{3})]$

6. $\because x^2 - 4x + 1 = 0$

$\Rightarrow$ $x = \frac{4 \pm \sqrt{(16-4)}}{2} = 2 \pm \sqrt{3}$

$\therefore$ $\alpha = 2 + \sqrt{3}, \beta = 2 - \sqrt{3}$

$\Rightarrow$ $(2+\sqrt{3})^{y^2-2y+1} + (2-\sqrt{3})^{y^2-2y-1} = \frac{101}{10(2-\sqrt{3})}$

$\Rightarrow$ $(2+\sqrt{3})^{y^2-2y} + (2-\sqrt{3})^{y^2-2y} = \frac{101}{10}$

Let $(2+\sqrt{3})^{y^2-2y} = \lambda$

Then $\lambda + \frac{1}{\lambda} = \frac{101}{10}$ or $10\lambda^2 - 101\lambda + 10 = 0$

or $(\lambda - 10)(10\lambda - 1) = 0$

or $\lambda = 10, 10^{-1}$

$\Rightarrow$ $(2+\sqrt{3})^{y^2-2y} = 10, 10^{-1}$

$\Rightarrow$ $y^2 - 2y = \log_{2+\sqrt{3}}(10), -\log_{2+\sqrt{3}}(10)$

$\Rightarrow$ $(y-1)^2 = 1 + \log_{2+\sqrt{3}}(10), 1 - \log_{2+\sqrt{3}}(10)$

or $(y-1)^2 = 1 + \log_{2+\sqrt{3}}(10)$ $[\because 1 - \log_{2+\sqrt{3}}(10) < 0]$

or $y = 1 \pm \sqrt{1 + \log_{2+\sqrt{3}}(10)}$

PASSAGE 6

If α, β are the roots and D be the discriminant of the given quadratic equation, then

$$\alpha + \beta = \frac{2(1+3m)}{(1+m)}, \alpha\beta = \frac{(1+8m)}{(1+m)} \quad \ldots(i)$$

and $D = 4(1+3m)^2 - 4(1+m)(1+8m)$

$= 4(m^2 - 3m) = 4m(m-3)$

If roots are real, then $D \ge 0$

$\therefore$ $m \in (-\infty, 0] \cup [3, \infty)$ ....(ii)

1. $D < 0$

$\Rightarrow$ $4m(m-3) < 0 \Rightarrow 0 < m < 3$

$\therefore$ $m = 1, 2$

2. At least one root is negative *ie*, one root is negative or both roots are negative, then $\{(\alpha\beta < 0) \cup (\alpha + \beta < 0)\} \cap (D \ge 0)$

$\Rightarrow \left\{\left(\frac{(1+8m)}{(1+m)} < 0\right) \cup \left(\frac{2(1+3m)}{(1+m)} < 0\right)\right\}$

$\cap\, m \in (-\infty, 0] \cup [3, \infty)$

$\Rightarrow \left\{m \in \left(-1, -\frac{1}{8}\right) \cup m \in \left(-1, -\frac{1}{3}\right)\right\}$

$\cap\, m \in (-\infty, 0] \cup [3, \infty)$

$\Rightarrow \left\{m \in \left(-1, -\frac{1}{8}\right)\right\} \cap \{m \in (-\infty, 0] \cup [3, \infty)\}$

i.e. $m \in \left(-1, -\frac{1}{8}\right)$

3. $\alpha+\beta>0$ and $\alpha\beta>0$

$$\Rightarrow \quad (\alpha+\beta>0)\cap(\alpha\beta>0)\cap(D\geq 0)$$

$$\Rightarrow \left(\frac{2(1+3m)}{(1+m)}>0\right)\cap\left(\frac{1+8m}{1+m}>0\right)\cap\{4m(m-3)\geq 0\}$$

$$\therefore \quad m\in\left\{(-\infty,-1)\cup\left(-\frac{1}{3},\infty\right)\right\}\cap\left\{(-\infty,-1)\cup\left(-\frac{1}{8},\infty\right)\right\}$$

$$\cap\{m\in(-\infty,0]\cup[3,\infty)\}$$

$$\Rightarrow \quad m\in(-\infty,-1)\cup[3,\infty).$$

4. At least one root is positive i.e. one root is positive or both roots are positive.

Then, $\{(\alpha\beta<0)\cup(\alpha+\beta>0)\}\cap(D\geq 0)$

$$\Rightarrow\left\{\left(\frac{(1+8m)}{(1+m)}<0\right)\cup\left(\frac{2(1+3m)}{(1+m)}>0\right)\right\}\cap\{4m(m-3)\geq 0\}$$

$$\Rightarrow m\in\left[\left\{\left(-1,-\frac{1}{8}\right)\right\}\cup\left\{(-\infty,-1)\cup\left(-\frac{1}{3},\infty\right)\right\}\right]$$

$$\cap\, m\in(-\infty,0]\cup[3,\infty)$$

$$\therefore \quad m\in(-\infty,-1)\cup\left(-1,-\frac{1}{8}\right)\cup[3,\infty).$$

5. Roots are 2α and 3α

$$\therefore \quad 2\alpha+3\alpha=\frac{2(1+3m)}{(1+m)}\Rightarrow\alpha=\frac{2(1+3m)}{5(1+m)} \quad \text{...(i)}$$

$$\Rightarrow \quad 2\alpha\cdot 3\alpha=\frac{(1+8m)}{(1+m)}\Rightarrow\alpha^2=\frac{(1+8m)}{6(1+m)} \quad \text{...(ii)}$$

From Eqs. (i) and (ii), we get

$$\frac{4(1+3m)^2}{25(1+m)^2}=\frac{(1+8m)}{6(1+m)}$$

$$\Rightarrow \quad 24(1+3m)^2=25(1+m)(1+8m)$$

$$\Rightarrow \quad 16m^2-81m-1=0$$

$$\therefore \quad m=\frac{81\pm\sqrt{(81)^2+16\times 4}}{2\times 16}=\frac{81\pm\sqrt{(6625)}}{32}$$

PASSAGE 7

1. $a\left(\dfrac{x+1}{x-2}\right)^2+b\left(\dfrac{x+1}{x-2}\right)+c=0$

$$\Rightarrow \quad \frac{x+1}{x-2}=\alpha,\beta\Rightarrow x=\frac{2\alpha+1}{\alpha-1},\frac{2\beta+1}{\beta-1}$$

2. Put $y=\dfrac{1}{2\alpha-3}\Rightarrow\alpha=\dfrac{1}{2}\left(3+\dfrac{1}{y}\right)$

$$\therefore \quad 2\left\{\frac{1}{2}\left(3+\frac{1}{y}\right)\right\}^2+4\left\{\frac{1}{2}\left(3+\frac{1}{y}\right)\right\}-5=0$$

$$\Rightarrow \quad 11y^2+10y+1=0$$

$$\text{or} \quad 11x^2+10x+1=0$$

3. Put $y=\alpha^2+\dfrac{r}{p}\Rightarrow\alpha=\sqrt{\left(y-\dfrac{r}{p}\right)}$

$$\therefore \quad p\left(y-\frac{r}{p}\right)-q\sqrt{\left(y-\frac{r}{p}\right)}+r=0$$

$$\Rightarrow \quad py=q\sqrt{\left(y-\frac{r}{p}\right)}\Rightarrow p^2y^2=q^2y-\frac{q^2r}{p}$$

or $\quad p^3y^2-pq^2y+q^2r=0$ or $p^3x^2-pq^2x+q^2r=0$

4. Put $y=\dfrac{1+\alpha}{1-\alpha}\Rightarrow\alpha=\left(\dfrac{y-1}{y+1}\right)$

$$\therefore \quad \left(\frac{y-1}{y+1}\right)^3-\left(\frac{y-1}{y+1}\right)-1=0\Rightarrow y^3+7y^2-y+1=0$$

$$\therefore \quad \Pi\left(\frac{1+\alpha}{1-\alpha}\right)=-1$$

5. Put $y=\beta\gamma+\dfrac{1}{\alpha}=\dfrac{\alpha\beta\gamma+1}{\alpha}=\dfrac{r+1}{\alpha}$ or $\alpha=\left(\dfrac{r+1}{y}\right)$

Thus, $\left(\dfrac{r+1}{y}\right)^3+q\left(\dfrac{r+1}{y}\right)-r=0$

$$\Rightarrow \quad ry^3-q(r+1)y^2-(r+1)^3=0$$

$$\text{or} \quad rx^3-q(r+1)x^2-(r+1)^3=0$$

6. Put $y=(\alpha-\beta)^2=(\alpha+\beta)^2-4\alpha\beta$

$$=(\alpha+\beta+\gamma-\gamma)^2-\frac{4\alpha\beta\gamma}{\gamma}$$

$$=(0-\gamma)^2-\frac{4(-c)}{\gamma} \quad \left[\because \alpha+\beta+\gamma=0,\ \alpha\beta\gamma=-c\right]$$

$$=\frac{\gamma^3+4c}{\gamma}$$

$$=\frac{(\gamma^3+c)+3c}{\gamma} \quad [\because \gamma^3+3b\gamma+c=0]$$

$$=\frac{-3b\gamma+3c}{\gamma}$$

$$\text{or} \quad \gamma=\left(\frac{3c}{y+3b}\right)$$

Thus, $\left(\dfrac{3c}{y+3b}\right)^3+3b\left(\dfrac{3c}{y+3b}\right)+c=0$

$$\Rightarrow \quad 27c^3+9bc(y+3b)^2+c(y+3b)^3=0$$

$$\text{or} \quad (y+3b)^3+9b(y+3b)^2+27c^2=0$$

$$\text{or} \quad y^3+18by^2+81b^2y+(108b^2+27c^2)=0$$

$$\therefore \quad \Sigma(\alpha-\beta)^2(\beta-\gamma)^2=81b^2$$

7. Put $y=(\alpha-\beta)(\alpha-\gamma)=\alpha^2-(\beta+\gamma)\alpha+\beta\gamma$

$$=\alpha^2-(-\alpha)\alpha+\left(\frac{-2}{\alpha}\right) \quad \left[\because \alpha+\beta+\gamma=0,\ \alpha\beta\gamma=-2\right]$$

$$=\frac{2\alpha^3-2}{\alpha}$$

$$=\frac{2(-\alpha-2)-2}{\alpha} \quad [\because \alpha^3+\alpha+2=0]$$

$$=-2-\frac{6}{\alpha}$$

$$\text{or} \quad \alpha=-\frac{6}{(y+2)}$$

Thus, $\left(-\dfrac{6}{y+2}\right)^3+\left(-\dfrac{6}{y+2}\right)+2=0$

$$\Rightarrow \quad y^3+3y^2-112=0 \text{ or } x^3+3x^2-112=0$$

Numerical Value Type

1. Given, $\alpha + \beta = 1, \alpha\beta = -1$

$\therefore$ Quadratic equation with roots α, β is

$$x^2 - x - 1 = 0 \Rightarrow \alpha^2 = \alpha + 1$$

Multiplying both sides by α^{n-1} $\alpha^{n+1} = \alpha^n + \alpha^{n-1}$ ...(i)

Similarly, $\beta^{n+1} = \beta^n + \beta^{n-1}$ ...(ii)

Adding Eqs. (i) and (ii),

$$\alpha^{n+1} + \beta^{n+1} = (\alpha^n + \beta^n) + (\alpha^{n-1} + \beta^{n-1})$$

$\Rightarrow \quad P_{n+1} = P_n + P_{n-1}$

$\Rightarrow \quad 29 = P_n + 1 \quad$ [Given, $P_{n+1} = 29, P_{n-1} = 11$]

$\Rightarrow \quad P_n = 18$

$\therefore \quad P_n^2 = 18^2 = 324$

2. Let the common ratio be r, then

$\beta = \alpha r, \gamma = \alpha r^2, \delta = \alpha r^3$

$\therefore \quad \alpha + \beta = \alpha(1 + r) = 8$ ...(i)

and $\quad \gamma + \delta = \alpha r^2(1 + r) = 72$ ...(ii)

Dividing Eq. (ii) by (i), $r^2 = 9$

$\therefore \quad r = 3$

From Eq. (i), $\alpha = 2$

$\therefore \quad \beta = 6, \gamma = 18, \delta = 54$

$\Rightarrow \quad A = \alpha\beta = 2 \cdot 6 = 12$

and $\quad B = \gamma\delta = 18 \cdot 54$

$\therefore A + B = 12 + 18 \times 54 = 984$

3. $\because \quad bc = a^2$

and $\quad b + c = abc - a = a(a^2) - a = a^3 - a$

$\therefore b, c$ are the roots of the equation

$$x^2 - (b + c)x + bc = 0$$

i.e. $x^2 - (a^3 - a)x + a^2 = 0$

$\because$ roots (b, c) are real

$\therefore \quad (a^3 - a)^2 - 4 \cdot 1 \cdot a^2 \geq 0$

$\Rightarrow \quad (a^3 - a + 2a)(a^3 - a - 2a) \geq 0$

or $\quad a^2(a^2 + 1)(a^2 - 3) \geq 0$

$\Rightarrow \quad a^2 \geq 3$

or $\quad a^4 + a^2 + 7 \geq 3^2 + 3 + 7 = 19$

or $\quad a^4 + a^2 + 7 \geq 19$

4. We have,

$$(x - a)(x - 10) + 1 = (x + \alpha)(x + \beta)$$

putting $x = -\alpha$ in both sides, we obtain

$$(-\alpha - a)(-\alpha - 10) + 1 = 0$$

or $\quad (\alpha + a)(\alpha + 10) = -1$

$\alpha + a$ and $\alpha + 10$ are integers $\quad [\because a, \alpha \in I]$

$\therefore \quad \alpha + a = -1$ and $\alpha + 10 = 1$

or $\quad \alpha + a = 1$ and $\alpha + 10 = -1$

(I) If $\alpha + 10 = 1$

$\therefore \quad \alpha = -9$, then $a = 8$

(II) If $\alpha + 10 = -1$

$\Rightarrow \quad \alpha = -11$, then $a = 12$

Hence, $a = 8, 12$

$\therefore \quad 8^2 + 12^2 = 64 + 144 = 208$

5. Let $y = \dfrac{6x^2 - 22x + 21}{5x^2 - 18x + 17}$

$(6 - 5y)x^2 - 2x(11 - 9y) + 21 - 17y = 0$

Since, x is real

$$D \geq 0$$

$$4(11 - 9y)^2 - 4(6 - 5y)(21 - 17y) \geq 0$$

$\Rightarrow 121 + 81y^2 - 198y - 126 + 105y + 102y - 85y^2 \geq 0$

$\Rightarrow \quad -4y^2 + 9y - 5 \geq 0$

$\Rightarrow \quad 4y^2 - 9y + 5 \leq 0$

$\Rightarrow \quad 4(y - 1)\left(y - \dfrac{5}{4}\right) \leq 0$

$\Rightarrow \quad 1 \leq y \leq \dfrac{5}{4}$

The maximum value of y is $\dfrac{5}{4} = 1.25$

6. Let roots are α and β.

$\therefore \quad \alpha + \beta = \dfrac{b}{a}, \alpha\beta = \dfrac{c}{a}$

and $\quad 0 < \alpha < 1, 0 < \beta < 1$

$\therefore \quad 0 < 1 - \alpha < 1$ and $0 < 1 - \beta < 1$

$\dfrac{\alpha + (1 - \alpha)}{2} \geq \sqrt{\alpha(1 - \alpha)} \quad [\because AM \geq GM]$

$\Rightarrow \quad \dfrac{1}{4} \geq \alpha(1 - \alpha)$

Similarly $\dfrac{1}{4} \geq \beta(1 - \beta)$

$\therefore \quad \dfrac{1}{16} \geq \alpha\beta(1 - \alpha)(1 - \beta) > 0$

But α and β are distinct.

$\therefore \quad 0 < \alpha\beta(1 - \alpha)(1 - \beta) < \dfrac{1}{16}$

$\Rightarrow \quad 0 < \alpha\beta\{1 - (\alpha + \beta) + \alpha\beta\} < \dfrac{1}{16}$

$\Rightarrow \quad 0 < \dfrac{c}{a}\left(1 - \dfrac{b}{a} + \dfrac{c}{a}\right) < \dfrac{1}{16}$

$\Rightarrow \quad 0 < c(a - b + c) < \dfrac{a^2}{16}$

Now, $c(a - b + c) =$ Natural number $\quad [\because a, b, c \in N]$

$\therefore$ Minimum value of $c(a - b + c) = 1$

$\therefore \quad \dfrac{a^2}{16} > 1 \Rightarrow a \geq 5 \quad [\because a \in N]$

and condition for real roots,

$$b^2 - 4ac > 0$$

$\Rightarrow \quad b^2 > 4ac \geq 20c \quad [\because a \geq 5]$

$\Rightarrow \quad b^2 \geq 20c$

$\Rightarrow \quad b \geq 5 \quad [\because b \in N]]$

and minimum value of $c = 1$

Hence, $abc \geq 25$

$\therefore$ Minimum value of abc is 25.

7. $\because$ Roots of $x^2 - px + q = 0$ are two distinct prime numbers.
Sum of roots $= p =$ odd positive integer (given)
$\Rightarrow$ Both roots of the equation cannot be odd.
$\therefore$ One root must be 2 (prime number)
Then $(2)^2 - p(2) + q = 0$ or $2p - q = 4$ ...(i)
and given $p + q = 35$ ...(ii)
Solving Eqs. (i) and (ii), we get

$$p = 13, q = 22$$

$$\therefore \quad f(x) = x^2 - 13x + 22$$

$$\therefore \quad \sum_{r=1}^{10} f(r) = \sum_{r=1}^{10} (r^2 - 13r + 22)$$

$$= \sum_{r=1}^{10} r^2 - 13\sum_{r=1}^{10} r + 22\sum_{r=1}^{10} 1$$

$$= \frac{10 \cdot 11 \cdot 21}{6} - \frac{13 \cdot 10 \cdot 11}{2} + 22 \cdot 10$$

$$= -110 \text{ and } f(10) = -8$$

$$\therefore \quad f(10)\sum_{r=1}^{10} f(r) = (-8)(-110) = 880$$

$$\Rightarrow \quad 88n = 880$$

$$\Rightarrow \quad n = 10$$

$$\Rightarrow \quad \frac{p+q}{n} = \frac{13+22}{10} = \frac{35}{10} = 3.5$$

8. $\because$ α, β are the roots of $x^2 + px + q = 0$
Then, $\alpha + \beta = -p, \alpha\beta = q$ ...(i)
Also α, β are roots of $x^{3900} + p^{1950}x^{1950} + q^{1950} = 0$
$\therefore$ $\alpha^{1950} + \beta^{1950} = -p^{1950}$ and $\alpha^{1950}\beta^{1950} = q^{1950}$ ...(ii)
Now, α/β is a root of $x^n + 1 + (x+1)^n = 0$

Then, $$\left(\frac{\alpha}{\beta}\right)^n + 1 + \left(\frac{\alpha}{\beta} + 1\right)^n = 0$$

$$\Rightarrow \quad \alpha^n + \beta^n + (\alpha + \beta)^n = 0$$

$$\Rightarrow \quad \alpha^n + \beta^n + (-p)^n = 0 \quad [\because \alpha + \beta = -p]$$

or $$\alpha^n + \beta^n = -(-p)^n \quad \text{...(iii)}$$

From Eqs. (ii) and (iii), we get

$$n = 1950$$

Single Integer Type

1. Since the quadratic equation $2x^2 + (a-10)x + \frac{33}{2} = 2a$ has real roots, so $D \geq 0$

$$\Rightarrow \quad (a-10)^2 - 4(2)\left(\frac{33}{2} - 2a\right) \geq 0$$

$$\Rightarrow a^2 - 20a + 100 - 4(33 - 4a) \geq 0$$

$$\Rightarrow \quad a^2 - 4a - 32 \geq 0$$

$$\Rightarrow \quad a^2 - 8a + 4a - 32 \geq 0$$

$$\Rightarrow \quad a(a-8) + 4(a-8) \geq 0$$

$$\Rightarrow \quad (a+4)(a-8) \geq 0$$

$$\Rightarrow \quad a \in (-\infty, -4] \cup [8, \infty)$$

$\therefore$ The least positive value of 'a' for which the given quadratic equation has real roots is 8.

2. Given quadratic equation $x^2 - 64x + 256 = 0$ having roots α and β, so
$\alpha + \beta = 64$ and $\alpha\beta = 256$

Now, $$\left(\frac{\alpha^3}{\beta^5}\right)^{1/8} + \left(\frac{\beta^3}{\alpha^5}\right)^{1/8} = \frac{\alpha + \beta}{(\alpha\beta)^{5/8}}$$

$$= \frac{64}{(256)^{5/8}} = \frac{64}{2^5} = \frac{64}{32} = 2$$

3. Given equation is $e^{4x} + e^{3x} - 4e^{2x} + e^x + 1 = 0$
Let $e^x = y > 0$, so $y^4 + y^3 - 4y^2 + y + 1 = 0$

$$\Rightarrow \left(y^2 + \frac{1}{y^2}\right) + \left(y + \frac{1}{y}\right) - 4 = 0 \quad \text{[on dividing by } y^2\text{]}$$

$$\Rightarrow \left(y + \frac{1}{y}\right)^2 + \left(y + \frac{1}{y}\right) - 6 = 0$$

Again, let $y + \frac{1}{y} = t$, then $t^2 + t - 6 = 0$

$$\Rightarrow \quad t^2 + 3t - 2t - 6 = 0$$

$$\Rightarrow t(t+3) - 2(t+3) = 0$$

$$\Rightarrow \quad (t-2)(t+3) = 0 \Rightarrow t = -3, 2$$

$$\therefore \quad y + \frac{1}{y} = -3 \text{ or } 2$$

$$\Rightarrow \quad e^x + \frac{1}{e^x} = -3 \text{ or } 2$$

$$\Rightarrow \quad e^x + \frac{1}{e^x} = 2$$

$$\left[\because r^x > 0 \Rightarrow e^x + \frac{1}{e^x} \neq -3 \;\forall\; x \in R\right]$$

$$\Rightarrow \quad (e^x - 1)^2 = 0$$

$$\Rightarrow \quad e^x = 1 \Rightarrow x = 0$$

$\therefore$ Number of real roots of given equation is 1.

4. The quadratic expression $ax^2 + bx + c$, $x \in R$ is always positive, if $a > 0$ and $D < 0$.
So, the quadratic expression
$(1 + 2m)x^2 - 2(1 + 3m)x + 4(1 + m)$,
$x \in R$ will be always positive, if $1 + 2m > 0$ ...(i)
and $D - 4(1 + 3m)^2 - 4(2m + 1)4(1 + m) < 0$ ...(ii)
From inequality Eq. (i), we get

$$m > -\frac{1}{2} \quad \text{...(iii)}$$

From inequality Eq. (ii), we get
$1 + 9m^2 + 6m - 4(2m^2 + 3m + 1) < 0$

$$\Rightarrow \quad m^2 - 6m - 3 < 0$$

$$\Rightarrow [m - (3 + \sqrt{12})][m - (3 - \sqrt{12})] < 0 \; [\because m^2 - 6m - 3 = 0]$$

$$\Rightarrow \quad m = \frac{6 \pm \sqrt{36 + 12}}{2} = 3 \pm \sqrt{12}$$

$$\Rightarrow \quad 3 - \sqrt{12} < m < 3 + \sqrt{12} \quad \text{...(iv)}$$

From inequalities Eqs. (iii) and (iv), the integral values of m are 0, 1, 2, 3, 4, 5, 6.
Hence, the number of integral values of m is 7.

5. Let α and β be the roots of equation

$$x^2 - (a-2)x - a - 1 = 0$$

Then, $\alpha + \beta = a - 2$ and $\alpha\beta = -a - 1$

Now, $\alpha^2 + \beta^2 = (\alpha + \beta)^2 - 2\alpha\beta$

$\Rightarrow \quad \alpha^2 + \beta^2 = (a-2)^2 + 2(a+1)$

$\Rightarrow \quad \alpha^2 + \beta^2 = a^2 - 2a + 6$

$\Rightarrow \quad \alpha^2 + \beta^2 = (a-1)^2 + 5$

The value of $\alpha^2 + \beta^2$ will be least, if $a - 1 = 0$

$\Rightarrow a = 1$

6. For the roots of quadratic equation $ax^2 + bx + c = 0$ to be rational $D = (b^2 - 4ac)$ should be perfect square.

In the equation $6x^2 - 11x + \alpha = 0$

$$a = 6,\ b = -11 \text{ and } c = \alpha$$

$\therefore$ For roots to be rational

$D = (-11)^2 - 4(6)(\alpha)$ should be a perfect square.

$\Rightarrow D(\alpha) = 121 - 24\alpha$ should be a perfect square

Now, $D(1) = 121 - 24 = 97$ is not a perfect square.

$D(2) = 121 - 24 \times 2 = 73$ is not a perfect square.

$D(3) = 121 - 24 \times 3 = 49$ is a perfect square.

$D(4) = 121 - 24 \times 4 = 25$ is a perfect square.

$D(5) = 121 - 24 \times 5 = 1$ is a perfect square.

and for $\alpha \geq 6$, $D(\alpha) < 0$, hence imaginary roots.

$\therefore$ For 3 values of α ($\alpha = 3, 4, 5$), the roots are rational.

7. $\frac{1}{2}\log_2(x-1) = \log_2(x-3)$

$\Rightarrow \quad x - 1 = (x-3)^2$

$\Rightarrow x^2 - 6x + 9 = x - 1$

$\Rightarrow x^2 - 7x + 10 = 0$

$$x = 2, 5$$
$$X = 2$$

Not possible as $\log_2(x-3)$ is not defined.

$\Rightarrow$ Number of solution $= 1$

8. $P(x) = f(x^3) + xg(x^3)$

$P(1) = f(1) + g(1)$ ...(i)

Now, $P(x)$ is divisible by $x^2 + x + 1$

$\Rightarrow P(x) = Q(x)(x^2 + x + 1)$

$P(w) = 0 = P(w^2)$,

where w, w^2 are non-real cube roots of units.

$P(x) = f(x^3) + xg(x^3)$

$P(w) = f(w^3) + wg(w^3) = 0$

$f(1) + wg(1) = 2$ ...(ii)

$P(w^2) = f(w^6) + w^2g(w^6) = 0$

$f(1) + w^2g(1) = 0$ ...(iii)

On adding Eqs. (iii) and (iv),

$\Rightarrow \quad 2f(1) + (w + w^2)g(1) = 0$

$2f(1) = g(1)$ ...(iv)

On subtracting Eq. (iii) from Eq. (ii),

$\Rightarrow \quad (w - w^2)g(1) = 0$

$g(1) = 0 = f(1)$ [from Eq. (iv)]

From Eq. (i) $P(1) = f(1) + g(1) = 0$

9. $x \geq 5$

$(x+1)^2 + (x-5) = \frac{27}{4}$

$\Rightarrow \quad x^2 + 3x - 4 = \frac{27}{4}$

$\Rightarrow \quad x^2 + 3x - \frac{43}{4} = 0$

$\Rightarrow 4x^2 + 12x - 43 = 0$

$$x = \frac{-12 \pm \sqrt{144 + 688}}{8}$$

$\Rightarrow \quad x = \frac{-12 \pm \sqrt{832}}{8}$

$$= \frac{-12 \pm 28.8}{8}$$
$$= \frac{-3 \pm 7.2}{2}$$
$$= \frac{-3 + 7.2}{2}, \frac{-3 - 7.2}{2}$$ [Therefore no solution]

For $x \leq 5$

$$(x+1)^2 - (x-5) = \frac{27}{4}$$

$\Rightarrow \quad x^2 + x + 6 - \frac{27}{4} = 0$

$\Rightarrow \quad 4x^2 + 4x - 3 = 0$

$\Rightarrow \quad x = \frac{-4 \pm \sqrt{16 + 48}}{8}$

$\Rightarrow \quad x = \frac{-4 \pm 8}{8}$

$\Rightarrow \quad x = -\frac{12}{8}, \frac{4}{8}$

$\therefore$ 2 real root's.

Matrix-Match Type

1. **(A)** $\because a(b-c)+b(c-a)+c(a-b)=0$

$\therefore$ $x=1$ is a root of $a(b-c)x^2+b(c-a)x+c(a-b)=0$

$\because$ Roots are equal then other root is also 1.

$\therefore$ Product of roots $=\dfrac{c(a-b)}{a(b-c)}=1$

$\therefore \quad b=\dfrac{2ac}{a+c} \Rightarrow a, b, c$ an in HP. (S)

Also, $(d+a-b)^2+(d+b-c)^2=0$

which is possible only when

$d+a-b=0,\ d+b-c=0$

$\therefore \quad b-a=c-b=d$

i.e. $2b=a+c \Rightarrow a, b, c$ are in AP (Q)

$\because$ a, b, c are in AP and HP then $a=b=c$ (T)

then a, b, c are also in GP (R)

$\because \quad a=b=c$

$\therefore \quad a+b+c\neq 0$ (P)

(B) $\because (b-c)+(c-a)+(a-b)=0$

$\therefore x=1$ is a root of $(b-c)x^2+(c-a)x+(a-b)=0$

$\because$ Roots are equal

$\therefore$ Other root is also 1.

Then product of roots $=\dfrac{a-b}{b-c}=1$

$b=\dfrac{a+c}{2} \Rightarrow a, b, c$ are in AP. (Q)

and $\quad a+b+c\neq 0$ (P)

(C) Let α be a common positive root, then

$\alpha^2+p\alpha+12=0 \quad \ldots$(i)

$\alpha^2+q\alpha+15=0 \quad \ldots$(ii)

and $\quad \alpha^2+(p+q)\alpha+36=0 \quad \ldots$(iii)

Applying Eqs. (i) + (ii) − (iii), we get

$\alpha^2+27-36=0$

or $\quad \alpha^2=9 \Rightarrow \alpha=3 \quad [\because \alpha$ is positive]

Let other root of Eq. (i) is a (given)

then $\quad a\times 3=12 \Rightarrow a=4$

Let other root of Eq. (ii) is b (given)

then $\quad b\times 3=15 \Rightarrow b=5$

and let other root of Eq. (iii) is c (given)

then $\quad c\times 3=36 \Rightarrow c=12$

$\therefore \quad a+b+c=21\neq 0$ (P)

2. **(A)** Let $f(x)=x^3-6x^2+9x+\lambda$

$\therefore \quad f'(x)=3x^2-12x+9=3(x-1)(x-3)$

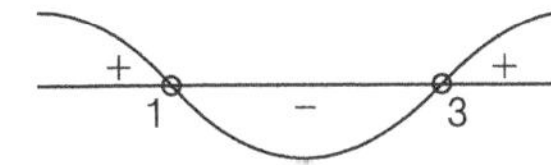

$\therefore \quad f'(x)<0$ in $(1, 3)$

But $f(1)=4+\lambda$ and $f(3)=\lambda$

for $f(x)=0$ to have exactly one root in $(1, 3)$

$\Rightarrow f(1)$ and $f(3)$ should have opposite signs

$\therefore \quad f(1)f(3)<0$

$\Rightarrow \quad \lambda(\lambda+4)<0$

$\Rightarrow \quad -4<\lambda<0$

$\therefore \quad -3<\lambda+1<1$

$\Rightarrow \quad [\lambda+1]=-3, -2, -1, 0$ (P, Q, R, S)

(B) $\because -3<\dfrac{x^2-\lambda x-2}{x^2+x+1}<2$

$\Rightarrow \underset{\text{(i)}}{-3x^2-3x-3<x^2-\lambda x-2}\underset{\text{(ii)}}{<2x^2+2x+2}$

$[\because x^2+x+1>0$ for all $x\in R]$

From inequality (i),

$4x^2-(\lambda-3)x+1>0$

$\Rightarrow \quad (\lambda-3)^2-4\cdot 4\cdot 1<0$

$\Rightarrow \quad -4<\lambda-3<4$

$\Rightarrow \quad -1<\lambda<7 \quad \ldots$(i)

From inequality (ii)

$x^2+(\lambda+2)x+4>0$

$\therefore \quad (\lambda+2)^2-4\cdot 1\cdot 4<0$

$\Rightarrow \quad -4<\lambda+2<4$

$\Rightarrow \quad -6<\lambda<2 \quad \ldots$(ii)

From Eqs. (i) and (ii), we get $-1<\lambda<2$

$\therefore \quad [\lambda]=-1, 0, 1$ (R, S, T)

(C) $\because$ $x=1$ satisfies

$(b-c)x^2+(c-a)x+(a-b)=0$

$\therefore x=1$ satisfies $\quad x^2+\lambda x+1=0$

then $\quad 1+\lambda+1=0$

$\Rightarrow \quad \lambda=-2$

$\therefore \quad [\lambda-1]=[-2-1]=-3$ (P)

3. **(A) Case I :** $x\geq 4$

Then $\quad |x+4-x|-2x=4$

$\Rightarrow \quad 4-2x=4$

$\therefore \quad x=0 \quad$ [Impossible]

Case II : $x<4$

Then $|x-(4-x)|-2x=4$

$\Rightarrow \quad |2x-4|=2x+4$

$\Rightarrow \quad |x-2|=x+2$

$$\begin{cases} x-2=x+2, & x\geq 2 \\ 2-x=x+2, & x<2 \end{cases}$$

$\Rightarrow \quad x=0,\ x<2$

is only solution,

$\therefore \quad S=0, N=1$ (P, Q)

(B) $\because |x^2-x-6|=x+2$

$\Rightarrow \quad |(x-3)(x+2)|=x+2$

$\Rightarrow \quad |x-3||x+2|=x+2$

$$\Rightarrow \begin{cases} (x-3)(x+2)=x+2, & x<-2 \\ -(x-3)(x+2)=x+2, & -2\leq x<3 \\ (x-3)(x+2)=x+2, & x>3 \end{cases}$$

$$\Rightarrow \begin{cases} x = 4, & x < -2 \\ x = -2, 2, & -2 \le x < 3 \\ x = 4, & x > 3 \end{cases}$$

Hence, $x = -2, 2, 4$

$\therefore \quad N = 3, S = 4$ (S, T)

(C) $x^2 - 3|x| - 4 = 0$

$\Rightarrow |x|^2 - 3|x| - 4 = 0$

$\Rightarrow \quad |x| = 4, -1$

$\therefore \quad |x| = 4, \quad |x| \ne -1$

$\therefore \quad x = \pm 4$

$\therefore \quad N = 2, S = 0$ (P, R)

4. (A) Roots of the given equation will be opposite sign, if $D \ge 0$ and product of roots < 0.

$\Rightarrow (a^3 + 8a - 1)^2 - 8(a^2 - 4a) \ge 0$ and $\dfrac{a^2 - 4a}{2} < 0$

$\Rightarrow \quad a^2 - 4a < 0$

$[\because a^2 - 4a < 0 \Rightarrow (a^3 + 8a - 1)^2 - 8(a^2 - 4a) \ge 0]$

$\Rightarrow \quad 0 < a < 4$ (R)

(B) Equation has negative roots of discriminant ≥ 0, $\alpha < 0, \beta < 0$

i.e. $\quad \alpha + \beta < 0$ and $f(0) > 0$

$\therefore \quad 4(a+1)^2 - 36a + 20 > 0$

$\Rightarrow \quad a^2 - 7a + 6 \ge 0$

$\Rightarrow \quad a \le 1$ or $a \ge 6 \quad$ ...(i)

$\alpha + \beta < 0$

$\Rightarrow \quad a + 1 > 0 \Rightarrow a > -1 \quad$ ...(ii)

and $\quad f(0) > 0$

$9a - 5 > 0 \Rightarrow a > \dfrac{5}{9} \quad$ ...(iii)

From Eqs. (i), (ii) and (iii), we get

$a \ge 6$ (P)

(C) Given,

$x^2 - 2(4a - 1)x + 15a^2 - 2a - 7 > 0$

$\therefore$ discriminant $< 0 \quad [\because$ coefficient of $x^2 > 0]$

$\Rightarrow 4(4a - 1)^2 - 4(15a^2 - 2a - 7) < 0$

$\Rightarrow \quad a^2 - 6a + 8 < 0$

$\Rightarrow \quad 2 < a < 4$ (S)

(D) Let $y = \dfrac{x^2 + 2x + a}{x^2 + 4x + 3a}$

$\Rightarrow x^2(y - 1) + 2(2y - 1)x + a(3y - 1) = 0$

As, $\quad x \in R, D \ge 0$

$\therefore \quad 4(2y - 1)^2 - 4(y - 1)a(3y - 1) \ge 0$

$\Rightarrow \quad (4 - 3a)y^2 - 4(1 - a)y + 1 - a \ge 0$

$\Rightarrow \quad a < \dfrac{4}{3}$ and $a(a - 1) \le 0$

$0 \le a \le 1$

$\therefore \quad 0 \le a \le 1$ (Q)

5. (A) $x^2 + ax + b = 0$ has root α.

Hence, $\alpha^2 + a\alpha + b = 0 \quad$ ...(i)

$x^2 + px + q = 0$ has root $-\alpha$,

Hence, $\alpha^2 + p\alpha + q = 0 \quad$ ...(ii)

Eliminating α from (i) and (ii), we get

$(q - b)^2 = (aq + bp)(-p - a)$

or $(q - b)^2 = -(aq + bp)(p + a)$ (S)

(B) $x^2 + ax + b = 0$ has root α.

Hence, $\alpha^2 + a\alpha + b = 0 \quad$ (i)

$x^2 + px + q = 0$ has root $1/\alpha$.

Hence, $q\alpha^2 + p\alpha + 1 = 0 \quad$ (ii)

Eliminating α from (i) and (ii), we get

$(1 - bq)^2 = (a - pb)(p - aq)$ (P)

(C) $x^2 + ax + b = 0$ has roots α, β.

Hence, $\alpha^2 + a\alpha + b = 0 \quad$ (i)

$x^2 + px + q = 0$ has roots $-2/\alpha, \gamma$.

Hence, $q\alpha^2 - 2p\alpha + 4 = 0$

Eliminating α from (i) and (ii), we get

$(4 - bq)^2 = (4a + 2pb)(-2p - aq)$ (Q)

(D) $x^2 + ax + b = 0$ has roots α, β.

Hence, $\alpha^2 + a\alpha + b = 0 \quad$ (i)

$x^2 + px + q = 0$ has roots $-1/2\alpha, \gamma$.

Hence, $4q\alpha^2 - 2p\alpha + 1 = 0$ (R) $\quad$ (ii)

03

Sequences and Series

Objective Questions Type I [Only one correct answer]

In each of the questions below, four choices are given of which only one is correct. You have to select the correct answer which is the most appropriate.

1. Number of identical terms in the sequence 2, 5, 8, 11, ... upto 100 terms and 3, 5, 7, 9, 11, ... upto 100 terms are

(a) 17 (b) 33
(c) 50 (d) 147

2. If $ab^2c^3, a^2b^3c^4, a^3b^4c^5$ are in AP $(a, b, c > 0)$, then the minimum value of $a + b + c$ is

(a) 1 (b) 3
(c) 5 (d) 9

3. Let S be the sum of the first 9 terms of the series $\{x + ka\} + \{x^2 + (k + 2)a\} + \{x^3 + (k + 4)a\} + \{x^4 + (k + 6)a\} + \dots$ where $a \neq 0$ and $x \neq 1$. If $S = \dfrac{x^{10} - x + 45a(x - 1)}{x - 1}$, then k is equal to **[JEE Main 2020, 2 September]**

(a) -5 (b) 1 (c) -3 (d) 3

4. The sum of the integers lying between 1 and 100 (both inclusive) and divisible by 3 or 5 or 7 is

(a) 818 (b) 1828 (c) 2838 (d) 3848

5. The sum of first n terms of the series $\dfrac{3}{5} + \dfrac{21}{25} + \dfrac{117}{125} + \dots$ is

(a) $n + \dfrac{2^{n+1}}{3 \times 5^n} - \dfrac{2}{3}$ (b) $n - \dfrac{2^{n+1}}{3 \times 5^n} - \dfrac{2}{3}$

(c) $n + \dfrac{2^{n+1}}{3 \times 5^n} + \dfrac{2}{3}$ (d) $n - \dfrac{2^{n+1}}{3 \times 5^n} + \dfrac{2}{3}$

6. If $a_i > 0, i = 1, 2, 3, \dots, 50$ and $a_1 + a_2 + a_3 + \dots + a_{50} = 50$, then the minimum value of $\dfrac{1}{a_1} + \dfrac{1}{a_2} + \dfrac{1}{a_3} + \dots + \dfrac{1}{a_{50}}$ is equal to

(a) 150 (b) 100 (c) 50 (d) $(50)^2$

7. If positive numbers a^{-1}, b^{-1}, c^{-1} are in AP, then the product of roots of the equation $x^2 - kx + 2b^{101} - a^{101} - c^{101} = 0, (k \in R)$ has

(a) > 0 (b) < 0
(c) $= 0$ (d) undefined

8. If $x^a = y^b = z^c$, where a, b, c are unequal positive numbers and x, y, z are in GP, then $a^3 + c^3$ is

(a) $\geq 2b^3$ (b) $> 2b^3$
(c) $\leq 2b^3$ (d) $< 2b^3$

9. The 1025th term in the sequence 1, 22, 4444, 88888888, ... is

(a) 2^9 (b) 2^{10}
(c) 2^{11} (d) 2^{12}

10. If a, b, c are non-zero real numbers such that $3(\Sigma a^2 + 1) = 2(\Sigma a + \Sigma ab)$, then a, b, c are in

(a) AP (b) GP
(c) HP (d) AP and GP

11. If a, b, c, d are positive real numbers such that $a + b + c + d = 2$, then $m = (a + b)(c + d)$ satisfies the relation

(a) $0 < m \leq 1$ (b) $1 \leq m \leq 2$
(c) $2 \leq m \leq 3$ (d) $3 < m \leq 4$

12. If $1 + \lambda + \lambda^2 + \dots + \lambda^n = (1 + \lambda)(1 + \lambda^2)(1 + \lambda^4)(1 + \lambda^8)(1 + \lambda^{16})$, then the value n is (where $n \in N$)

(a) 32 (b) 16
(c) 31 (d) 15

13. If $a_1, a_2, a_3, \dots, a_{20}$ are AMs between 13 and 67, then the maximum value of $a_1 a_2 a_3 \dots a_{20}$ is

(a) $(20)^{20}$ (b) $(40)^{20}$
(c) $(60)^{20}$ (d) $(80)^{20}$

14. If $a_1, a_2, \dots, a_n$ are positive real numbers whose product is a fixed number c, then the minimum value of $a_1 + a_2 + \dots + a_{n-1} + 2a_n$ is

(a) $n(2c)^{1/n}$ (b) $(n + 1)c^{1/n}$
(c) $2nc^{1/n}$ (d) $(n + 1)(2c)^{1/n}$

15. Suppose a, b, c are in AP and a^2, b^2, c^2 are in GP. If $a > b > c$ and $a + b + c = 3/2$, then the value of a is

(a) $\frac{1}{\sqrt{2}}$ (b) $\frac{1}{\sqrt{3}}$

(c) $\frac{1}{2}+\frac{1}{\sqrt{3}}$ (d) $\frac{1}{2}+\frac{1}{\sqrt{2}}$

16. If the sum of the first 100 terms of an AP is -1 and the sum of even terms lying in first 100 terms is 1, then which of the following is not true?

(a) Common difference of the sequence is $\frac{3}{50}$

(b) First term of the sequence is $-\frac{149}{50}$

(c) 100th term $= \frac{74}{25}$

(d) None of the above

17. If a, b, c, d are distinct integers in AP such that $d = a^2 + b^2 + c^2$, then $a + b + c + d$ is

(a) 0 (b) 1 (c) 2 (d) 3

18. An infinite GP has first term x and sum 5, then x belongs to

(a) $x < -10$ (b) $-10 < x < 0$

(c) $0 < x < 10$ (d) $x > 10$

19. If $\sum_{i=1}^{21} a_i = 693$, where $a_1, a_2, \ldots, a_{21}$ are in AP, then the value of $\sum_{r=0}^{10} a_{2r+1}$ is

(a) 361 (b) 363

(c) 365 (d) 398

20. Let α and β be the roots of the quadratic equation $ax^2 + bx + c = 0$ and $\Delta = b^2 - 4ac$. If $\alpha + \beta, \alpha^2 + \beta^2$ and $\alpha^3 + \beta^3$ and in GP, then

(a) $\Delta \neq 0$ (b) $b\Delta = 0$

(c) $c\Delta = 0$ (d) $\Delta = 0$

21. If a_1, a_2, a_3, a_4 are in HP, then $\frac{1}{a_1 a_4} \sum_{r=1}^{3} a_r a_{r+1}$ is a root of

(a) $x^2 + 2x + 15 = 0$ (b) $x^2 + 2x - 15 = 0$

(c) $x^2 - 2x + 15 = 0$ (d) $x^2 - 2x - 15 = 0$

22. If a, b, c are in AP, then $\frac{1}{\sqrt{b}+\sqrt{c}}, \frac{1}{\sqrt{c}+\sqrt{a}}, \frac{1}{\sqrt{a}+\sqrt{b}}$ are in

(a) AP (b) GP

(c) HP (d) no definite sequence

23. Let $a_1, a_2, \ldots, a_n$ be a given AP. Whose common difference is an integer and $S_n = a_1 + a_2 + \ldots + a_n$. If $a_1 = 1$, $a_n = 300$ and $15 \le n \le 50$, then the ordered pair (S_{n-4}, a_{n-4}) is equal to **[JEE Main 2020, 4 September]**

(a) (2490, 249) (b) (2480, 249)

(c) (2480, 248) (d) (2490, 248)

24. If $\log_3 2, \log_3 (2^x - 5)$ and $\log_3 (2^x - 7/2)$ are in AP, then the value of x is

(a) 2 (b) 3 (c) 4 (d) 5

25. Let $S = \frac{8}{5} + \frac{16}{65} + \frac{24}{325} + \ldots + \frac{128}{2^{18}+1}$, then

(a) $S = \frac{1088}{545}$ (b) $S = \frac{545}{1088}$

(c) $S = \frac{1056}{545}$ (d) $S = \frac{545}{1056}$

26. If $a, b, c, d \in R^+$ and a, b, c, d are in HP, then

(a) $a + d \ge b + c$ (b) $a + b \ge c + d$

(c) $a + c \ge b + d$ (d) None of these

27. An AP consist of even number of terms $2n$ having middle terms equal to 1 and 7 respectively. If n is the maximum value which satisfy $t_1 t_{2n} + 713 \ge 0$, then the value of the first term of the series is

(a) 17 (b) -15

(c) 21 (d) -23

28. If a, b, c are in AP, then the equation $(a - b)x^2 + (c - a)x + (b - c) = 0$ has two roots which are

(a) rational and equal (b) rational and distinct

(c) irrational conjugates (d) complex conjugates

29. In an increasing, geometric series, the sum of the second and the sixth term is $\frac{25}{2}$ and the product of the third and fifth term is 25. Then, the sum of 4th, 6th and 8th terms is equal to **[JEE Main 2021, 26 February]**

(a) 35 (b) 30

(c) 26 (d) 32

30. Five numbers are in AP, whose sum is 25 and product is 2520. If one of these five numbers is $-\frac{1}{2}$, then the greatest number amongst them is **[JEE Main 2020, 7 January]**

(a) 7 (b) 16 (c) 27 (d) $\frac{21}{2}$

31. If $x, (2x + 2), (3x + 3), \ldots$ are in GP, then the next term of this sequence is

(a) 27 (b) -27

(c) 13.5 (d) -13.5

32. If each term of a GP is positive and each term is the sum of its two succeeding terms, then the common ratio of the GP is

(a) $\left(\frac{\sqrt{5}-1}{2}\right)$ (b) $\left(\frac{\sqrt{5}+1}{2}\right)$

(c) $-\left(\frac{\sqrt{5}+1}{2}\right)$ (d) $\left(\frac{1-\sqrt{5}}{2}\right)$

33. Suppose a, b, c are in AP and a^2, b^2, c^2 are in GP. If $a < b < c$ and $a + b + c = 3/2$, then the value of a is

(a) $\frac{1}{2\sqrt{2}}$ (b) $\frac{1}{2\sqrt{3}}$

(c) $\frac{1}{2}-\frac{1}{\sqrt{3}}$ (d) $\frac{1}{2}-\frac{1}{\sqrt{2}}$

34. Three numbers, the third of which being 12, form decreasing GP. If the last term were 9 instead of 12, the three numbers would have formed an AP. The common ratio of the GP is

(a) 1/3 (b) 2/3
(c) 3/4 (d) 4/5

35. If $\sum_{r=1}^{n} T_r = n(2n^2 + 9n + 13)$, then the value of $\sum_{r=1}^{n} \sqrt{T_r}$

(a) $\sqrt{\frac{3}{2}}(n^2 + 3n)$ (b) $\sqrt{\frac{5}{2}}(n^2 + n)$
(c) $\sqrt{\frac{5}{2}}(n^2 + 3n)$ (d) $\frac{1}{2}(n^2 + 3n + 2)$

36. If a, b, c are in HP, b, c, d are in GP and c, d, e are in AP, then the value of e is

(a) $\frac{ab^2}{(2a-b)^2}$ (b) $\frac{a^2b}{(2a-b)^2}$
(c) $\frac{a^2b^2}{(2a-b)^2}$ (d) None of these

37. If a, b, c are digits, then the rational number represented by $0.cababab\ldots$ is

(a) $cab/990$ (b) $(99c + ba)/990$
(c) $(99c + 10a + b)/99$ (d) $(99c + 10a + b)/990$

38. If $\log_2(a+b) + \log_2(c+d) \geq 4$. Then the minimum value of the expression $a + b + c + d$ is

(a) 2 (b) 4
(c) 8 (d) None of these

39. The HM of two numbers is 4 and their AM and GM satisfy the relation $2A + G^2 = 27$, then the numbers are

(a) -3 and 1 (b) 5 and -25
(c) 5 and 4 (d) 3 and 6

40. Let $A_n = \frac{1\cdot2\cdot3 + 2\cdot3\cdot4 + 3\cdot4\cdot5 + \ldots + \text{upto } n \text{ terms}}{n(1\cdot2 + 2\cdot3 + 3\cdot4 + \ldots + \text{upto } n \text{ term})}$, then $\lim_{n\to\infty} A_n$ is

(a) $\frac{3}{4}$ (b) $\frac{1}{4}$
(c) $\frac{1}{2}$ (d) $\frac{5}{4}$

41. The natural numbers are grouped as follows {1}, {2, 3, 4}, {5, 6, 7, 8, 9}, ..., then the first element of the nth group is

(a) $n^2 - 1$ (b) $n^2 + 1$
(c) $(n-1)^2 - 1$ (d) $(n-1)^2 + 1$

42. Sum of 20 terms of the series

$\frac{3}{1^2\cdot2^2} + \frac{7}{1^3\cdot2^3} + \frac{5}{2^2\cdot3^2} + \frac{19}{2^3\cdot3^3} + \frac{7}{3^2\cdot4^2} + \frac{37}{3^3\cdot4^3} + \ldots$ is

(a) $\frac{2684}{1331}$ (b) $\frac{2674}{1331}$
(c) $\frac{2640}{1331}$ (d) $\frac{2650}{1331}$

43. The sum of the series

$1\cdot n + 2\cdot(n-1) + 3\cdot(n-2) + \ldots + n\cdot1$ is

(a) $\frac{n(n+1)(n+2)}{6}$ (b) $\frac{n(n+1)(n+2)}{3}$
(c) $\frac{n(n+1)(2n+1)}{6}$ (d) $\frac{n(n+1)(2n+1)}{3}$

44. If p, q, r are three positive real numbers are in AP, then the roots of the quadratic equation $px^2 + qx + r = 0$ are all real for

(a) $\left|\frac{r}{p} - 7\right| \geq 4\sqrt{3}$ (b) $\left|\frac{p}{r} - 7\right| < 4\sqrt{3}$
(c) all p and r (d) no p and r

45. If $\frac{a+bx}{a-bx} = \frac{b+cx}{b-cx} = \frac{c+dx}{c-dx}$ $(x \neq 0)$, then a, b, c, d are in

(a) AP (b) GP
(c) HP (d) none of these

46. Let S_1 be the sum of first $2n$ terms of an arithmetic progression. Let S_2 be the sum of first $4n$ terms of the same arithmetic progression. If $(S_2 - S_1)$ is 1000, then the sum of the first $6n$ terms of the arithmetic progression is equal to **[JEE Main 2021, 18 March]**

(a) 1000 (b) 7000
(c) 5000 (d) 3000

47. The sum of the products of ten numbers $\pm1, \pm2, \pm3, \pm4, \pm5$ taking two at a time is

(a) -55 (b) 55
(c) 165 (d) -165

48. Given that n arithmetic means are inserted between two sets of numbers $a, 2b$ and $2a, b$, where $a, b \in R$. Suppose further that mth mean between these two sets of numbers is same, then the ratio, $a : b$ equals

(a) $n - m + 1 : m$ (b) $n - m + 1 : n$
(c) $m : n - m + 1$ (d) $n : n - m + 1$

49. The sum of n-terms of the series $1 + \frac{4}{5} + \frac{7}{5^2} + \frac{10}{5^3} \ldots\ldots$

(a) $\frac{5}{4} + \frac{15}{16}\left(1 - \frac{1}{5^{n-1}}\right) - \frac{(3n-2)}{4(5^{n-1})}$

(b) $\frac{5}{4} + \frac{15}{16}\left(1 - \frac{1}{5^n}\right) - \frac{(3n+1)}{4.5^n}$

(c) $\frac{5}{4} + \frac{15}{16}\left(1 - \frac{1}{5^{n-2}}\right) - \frac{(3n-5)}{4.5^n}$

(d) $\frac{5}{4} + \frac{5}{16}\left(1 - \frac{1}{5^{n-2}}\right) - \frac{(3n-2)}{5^{n-2}}$

50. If $\frac{a+b}{1-ab}, b, \frac{b+c}{1-bc}$ are in AP, then $a, \frac{1}{b}, c$ are in

(a) AP (b) GP
(c) HP (d) None of these

51. If $a_1, a_2, a_3, \ldots, a_n$ are in HP and $f(k) = \sum_{r=1}^{n} a_r - a_k$, then $\frac{a_1}{f(1)}, \frac{a_2}{f(2)}, \frac{a_3}{f(3)}, \ldots, \frac{a_n}{f(n)}$ are in

(a) AP (b) GP
(c) HP (d) None of these

52. The sixth term of an AP is equal to 2. The value of the common difference of the AP which makes the product $a_1a_4a_5$ least is given by

(a) $\frac{8}{5}$ (b) $\frac{5}{4}$
(c) $\frac{2}{3}$ (d) None of these

53. The solution of

$\log_{\sqrt{3}} x + \log_{\sqrt[4]{3}} x + \log_{\sqrt[6]{3}} x + \ldots + \log_{\sqrt[16]{3}} x = 36$ is

(a) $x = 3$ (b) $x = 4\sqrt{3}$
(c) $x = 9$ (d) $x = \sqrt{3}$

54. If $a_1, a_2, a_3, \ldots, a_n$ are in HP, then

$\frac{a_1}{a_2 + a_3 + \ldots + a_n}, \frac{a_2}{a_1 + a_3 + \ldots + a_n}, \ldots,$

$\frac{a_n}{a_1 + a_2 + \ldots + a_{n-1}}$ are in

(a) AP (b) GP
(c) HP (d) AGP

55. If the arithmetic progression whose common difference is none zero, the sum of first $3n$ terms is equal to the sum of the next n terms. Then the ratio of the sum of the first $2n$ terms to the next $2n$ terms is

(a) 1/5 (b) 2/3
(c) 3/4 (d) None of these

56. Let a, b, c be three positive prime numbers. The progression in which $\sqrt{a}, \sqrt{b}, \sqrt{c}$ can be three terms (not necessarily consecutive) is

(a) AP (b) GP
(c) HP (d) None of these

57. If n is an odd integer greater than or equal to 1, then the value of $n^3 - (n-1)^3 + (n-2)^3 - \ldots + (-1)^{n-1} 1^3$ is

(a) $\frac{(n+1)^2 (2n-1)}{4}$ (b) $\frac{(n-1)^2 (2n-1)}{4}$
(c) $\frac{(n+1)^2 (2n+1)}{4}$ (d) None of these

58. If the sides of a right angled triangle form an AP, then the sines of the acute angles are

(a) $\frac{3}{5}, \frac{4}{5}$ (b) $\sqrt{3}, \frac{1}{3}$
(c) $\sqrt{\left(\frac{\sqrt{5}-1}{2}\right)}, \sqrt{\left(\frac{\sqrt{5}+1}{2}\right)}$ (d) $\frac{\sqrt{3}}{2}, \frac{1}{2}$

59. The coefficient of x^{n-2} in the polynomial $(x-1)(x-2)(x-3)\ldots(x-n)$ is

(a) $\frac{n(n^2+2)(3n+1)}{24}$ (b) $\frac{n(n^2-1)(3n+2)}{24}$
(c) $\frac{n(n^2+1)(3n+4)}{24}$ (d) None of these

60. Let $\{a_n\}$ be a GP such that $\frac{a_4}{a_6} = \frac{1}{4}$ and $a_2 + a_5 = 216$. Then a_1 is equal to

(a) 12 or $\frac{108}{7}$ (b) 10
(c) 7 or $\frac{54}{7}$ (d) None of these

61. Let a_n be the nth term of a GP of positive terms. If $\sum_{n=1}^{100} a_{2n+1} = 200$ and $\sum_{n=1}^{100} a_{2n} = 100$, then $\sum_{n=1}^{200} a_n$ is equal to **[JEE Main 2020, 9 January]**

(a) 300 (b) 175
(c) 225 (d) 150

62. The sum to infinity of the series,

$1 + 2\left(1 - \frac{1}{n}\right) + 3\left(1 - \frac{1}{n}\right)^2 + \ldots$ is

(a) n^2 (b) $n(n+1)$
(c) $n\left(1 + \frac{1}{n}\right)^2$ (d) None of these

63. If $a_1, a_2, \ldots, a_n$ are n non zero real numbers such that

$(a_1^2 + a_2^2 + \ldots + a_{n-1}^2)(a_2^2 + a_3^2 + \ldots + a_n^2)$
$\leq (a_1a_2 + a_2a_3 + \ldots + \ldots + a_{n-1}a_n)^2$, then

$a_1, a_2 \ldots, a_n$ are in

(a) AP (b) GP
(c) HP (d) None of these

64. The cubes of the natural numbers are grouped as $1^3, (2^3, 3^3), (4^3, 5^3, 6^3), \ldots$, then sum of the numbers in the nth group is

(a) $\frac{1}{8} n^3 (n^2+1)(n^2+3)$
(b) $\frac{1}{16} n^3 (n^2+16)(n^2+12)$
(c) $\frac{n^3}{12} (n^2+2)(n^2+4)$
(d) None of the above

65. If $x = \sum_{n=0}^{\infty} a^n, y = \sum_{n=0}^{\infty} b^n, z = \sum_{n=0}^{\infty} c^n$

where a, b, c are in AP such that $|a| < 1, |b| < 1$ and $|c| < 1$, then x, y, z are in

(a) AP (b) GP
(c) HP (d) None of these

66. If $\langle a_n \rangle$ and $\langle b_n \rangle$ be two sequences given by $a_n = (x)^{1/2^n} + (y)^{1/2^n}$ and $b_n = (x)^{1/2^n} - (y)^{1/2^n}$ for all $n \in N$. Then, $a_1a_2a_3 \ldots a_n$ is equal to

(a) $x - y$ (b) $\frac{x+y}{b_n}$
(c) $\frac{x-y}{b_n}$ (d) $\frac{xy}{b_n}$

67. If $a_1 = 0$ and $a_1, a_2, a_3, \ldots, a_n$ are real numbers such that $|a_i| = |a_{i-1} + 1|$ for all i, then the arithmetic mean of the numbers $a_1, a_2, \ldots, a_n$ has value x where

(a) $x < -1$ (b) $x < -\frac{1}{2}$

(c) $x \geq -\frac{1}{2}$ (d) $x = -\frac{1}{2}$

68. If a_1, a_2, a_3 $(a_1 > 0)$ are in GP with common ratio r, then the value of r for which the inequality $9a_1 + 5a_3 > 14a_2$ holds can not lie in the interval

(a) $[1, \infty)$ (b) $\left[1, \frac{9}{5}\right]$

(c) $\left[\frac{4}{5}, 1\right]$ (d) $\left[\frac{5}{9}, 1\right]$

69. The sum of the series $\sum_{n=1}^{\infty} \frac{n^2 + 6n + 10}{(2n+1)!}$ is equal to **[JEE Main 2021, 26 February]**

(a) $\frac{41}{8}e + \frac{19}{8}e^{-1} - 10$ (b) $-\frac{41}{8}e + \frac{19}{8}e^{-1} - 10$

(c) $\frac{41}{8}e - \frac{19}{8}e^{-1} - 10$ (d) $\frac{41}{8}e + \frac{19}{8}e^{-1} + 10$

70. If the sum of n terms of the series $\frac{1}{1^3} + \frac{1+2}{1^3+2^3} + \frac{1+2+3}{1^3+2^3+3^3} + \ldots$ is S_n, then S_n exceeds 199 for all n greater than

(a) 99 (b) 50

(c) 199 (d) 100

71. The numbers $3^{2\sin 2x - 1}$, 14, $3^{4 - 2\sin 2x}$ form first three terms of an AP, its fifth term is equal to

(a) -25 (b) -12

(c) 40 (d) 53

72. In a sequence of $(4n + 1)$ terms the first $(2n + 1)$ terms are in AP whose common difference is 2, and the last $(2n + 1)$ terms are in GP whose common ratio is 0.5. If the middle terms of the AP and GP are equal, then the middle term of the sequence is

(a) $\frac{n \cdot 2^{2n+1}}{2^{2n} - 1}$ (b) $\frac{n \cdot 2^{n+1}}{2^{2n-1}}$

(c) $n2^n$ (d) None of these

73. If $a, a_1, a_2, a_3, \ldots, a_{2n}, b$ are in AP and $a, g_1, g_2, g_3, \ldots, g_{2n}, b$ are in GP and h is the HM of a and b, then $\frac{a_1 + a_{2n}}{g_1 g_{2n}} + \frac{a_2 + a_{2n-1}}{g_2 g_{2n-1}} + \ldots + \frac{a_n + a_{n+1}}{g_n g_{n+1}}$ is equal to

(a) $\frac{2n}{h}$ (b) $2nh$

(c) nh (d) $\frac{n}{h}$

74. Let $b_i > 1$ for $i = 1, 2, \ldots, 101$. Suppose $\log_e b_1, \log_e b_2, \ldots, \log_e b_{101}$ are in AP with the common difference $\log_e 2$. Suppose $a_1, a_2, \ldots, a_{101}$ are in AP, such that $a_1 = b_1$ and $a_{51} = b_{51}$. If $t = b_1 + b_2 + \ldots + b_{51}$ and $s = a_1 + a_2 + \ldots + a_{51}$, then **[JEE Advanced 2016]**

(a) $s > t$ and $a_{101} > b_{101}$

(b) $s > t$ and $a_{101} < b_{101}$

(c) $s < t$ and $a_{101} > b_{101}$

(d) $s < t$ and $a_{101} < b_{101}$

75. If $a + b + c = 3$ and $a > 0, b > 0, c > 0$, then the greatest value of $a^2b^3c^2$ is

(a) $\frac{3^{10} \cdot 2^4}{7^7}$ (b) $\frac{3^8 \cdot 2^4}{7^7}$

(c) $\frac{3^9 \cdot 2^4}{7^7}$ (d) None of these

76. If $\ln(a + c)$, $\ln(c - a)$, $\ln(a - 2b + c)$ are in AP, then

(a) a, b, c are in AP (b) a^2, b^2, c^2 are in AP

(c) a, b, c are in GP (d) a, b, c are in HP

77. Let $a_1, a_2, a_3, \ldots$ be in an AP with common difference not a multiple of 3. Then, maximum number of consecutive terms so that all are primes is

(a) 2 (b) 3

(c) 5 (d) infinite

78. The sum of the product taken two at a time of the numbers $1, 2, 2^2, 2^3, \ldots 2^{n-2}, 2^{n-1}$ is

(a) $\frac{1}{3} \cdot 2^{2n} + \frac{2}{3}$ (b) $\frac{1}{3} \cdot 2^{2n} - 2^n + \frac{1}{3}$

(c) $\frac{1}{3} \cdot 2^{2n} - \frac{1}{3}$ (d) $\frac{1}{3} \cdot 2^{2n} - 2^n + \frac{2}{3}$

79. The sum of the infinite terms of the series $\frac{5}{3^2 \cdot 7^2} + \frac{9}{7^2 \cdot 11^2} + \frac{13}{11^2 \cdot 15^2} + \ldots$ is

(a) $\frac{1}{18}$ (b) $\frac{1}{36}$

(c) $\frac{1}{54}$ (d) $\frac{1}{72}$

80. Let a, b, c be positive real numbers, such that $bx^2 + (\sqrt{(a+c)^2 + 4b^2})\,x + (a + c) \geq 0, \forall\, x \in R$, then a, b, c are in

(a) GP (b) AP

(c) HP (d) None of these

81. $\left(1\frac{2}{3}\right)^2 + \left(2\frac{1}{3}\right)^2 + 3^2 + \left(3\frac{2}{3}\right)^2 + \ldots$ to 10 terms, the sum is

(a) $\frac{1390}{9}$ (b) $\frac{1790}{9}$

(c) $\frac{1990}{9}$ (d) None of these

82. The consecutive odd integers whose sum is $45^2 - 21^2$ are

(a) 43, 45, ..., 75 (b) 43, 45, ..., 79

(c) 43, 45, ..., 85 (d) 43, 45, ..., 89

83. If $0 < \theta, \phi < \frac{\pi}{2}$, $x = \sum_{n=0}^{\infty} \cos^{2n}\theta$, $y = \sum_{n=0}^{\infty} \sin^{2n}\phi$, and $z = \sum_{n=0}^{\infty} \cos^{2n}\theta \cdot \sin^{2n}\phi$, then **[JEE Main 2021, 25 February]**

(a) $xyz = 4$
(b) $xy - z = (x + y)z$
(c) $xy + yz + zx = z$
(d) $xy + z = (x + y)z$

84. If $x, |x+1|, |x-1|$ are the three terms of an AP. Its sum upto 20 terms is

(a) 90 or 175 (b) 180 or 350
(c) 360 or 700 (d) 720 or 1400

85. If $\Sigma n, \frac{\sqrt{10}}{3}\Sigma n^2, \Sigma n^3$ are in GP, then the value of n is

(a) 3 (b) 4
(c) 2 (d) non-existent

86. If $\sum_{r=1}^{n} t_r = \frac{n(n+1)(n+2)(n+3)}{8}$, where t_r denotes the r th term of a series, then $\lim_{n\to\infty} \sum_{r=1}^{n} \frac{1}{t_r}$ is

(a) $\frac{1}{8}$ (b) $\frac{1}{4}$
(c) $\frac{1}{2}$ (d) 1

87. If an AP, $a_7 = 9$ if $a_1\, a_2\, a_7$ is least, the common difference is

(a) $\frac{13}{20}$ (b) $\frac{23}{20}$
(c) $\frac{33}{20}$ (d) $\frac{43}{20}$

88. If the ratio of AM between two positive real numbers a and b to their HM is $m : n$; then $a : b$ is equal to

(a) $\frac{\sqrt{(m-n)}+\sqrt{n}}{\sqrt{(m-n)}-\sqrt{n}}$ (b) $\frac{\sqrt{n}+\sqrt{(m-n)}}{\sqrt{n}-\sqrt{(m-n)}}$
(c) $\frac{\sqrt{m}+\sqrt{(m-n)}}{\sqrt{m}-\sqrt{(m-n)}}$ (d) $\frac{\sqrt{(m-n)}+\sqrt{m}}{\sqrt{(m-n)}-\sqrt{m}}$

89. If $\sum_{n=1}^{k}\left[\frac{1}{3}+\frac{n}{90}\right] = 21$, where $[x]$ denotes the integral part of x, then k is equal to

(a) 84 (b) 80
(c) 85 (d) None of these

90. If $x \in \{1, 2, 3, \ldots, 9\}$ and $f_n(x) = xxx \ldots x$ (n digits), then $f_n^2(3) + f_n(2)$ is equal to

(a) $2f_{2n}(1)$ (b) $f_n^2(1)$
(c) $f_{2n}(1)$ (d) $-f_{2n}(4)$

91. If three positive real numbers a, b, c are in AP with $abc = 4$, then minimum value of b is

(a) 4 (b) 3
(c) 2 (d) 1/2

92. The sum $\sum_{k=1}^{20} k\frac{1}{2^k}$ is equal to **[JEE Main 2019, 8 April]**

(a) $2 - \frac{11}{2^{19}}$ (b) $1 - \frac{11}{2^{20}}$
(c) $2 - \frac{3}{2^{17}}$ (d) $2 - \frac{21}{2^{20}}$

93. The coefficient of x^{15} in the product $(1-x)(1-2x)(1-2^2x)(1-2^3x)\ldots(1-2^{15}x)$ is

(a) $2^{105} - 2^{121}$ (b) $2^{121} - 2^{105}$
(c) $2^{120} - 2^{104}$ (d) $2^{105} - 2^{104}$

94. The roots of equation $x^2 + 2(a-3)x + 9 = 0$ lie between -6 and 1 and $2, h_1, h_2, \ldots, h_{20}, [a]$ are in HP, where $[a]$ denotes the integral part of a and $2, a_1, a_2, \ldots, a_{20}, [a]$ are in AP, then a_3h_{18} is equal to

(a) 6 (b) 12
(c) 3 (d) None of these

95. Value of $L = \lim_{n\to\infty} \frac{1}{n^4}\left[1\cdot\left(\sum_{k=1}^{n} k\right) + 2\cdot\left(\sum_{k=1}^{n-1} k\right) + 3\cdot\left(\sum_{k=1}^{n-2} k\right) + \ldots + n\cdot 1\right]$ is

(a) 1/24 (b) 1/12
(c) 1/6 (d) 1/3

Objective Questions Type II [One or more than one correct answer(s)]

In each of the questions below four choices of which one or more than one are correct. You have to select the correct answer(s) accordingly.

1. If $S_n = \sum_{2}^{4n} (-1)^{\frac{k(k+1)}{2}} k^2$. Then, S_n can take vaue(s) **[JEE Advanced 2013]**

(a) 1056 (b) 1088 (c) 1120 (d) 1332

2. Let α and β be the roots of $x^2 - x - 1 = 0$, with $\alpha > \beta$. For all positive integers n, define

$$a_n = \frac{\alpha^n - \beta^n}{\alpha - \beta}, n \geq 1, b_1 = 1 \text{ and } b_n = a_{n-1} + a_{n+1}, n \geq 2$$

Then which of the following options is/are correct? **[JEE Advanced 2019]**

(a) $\sum_{n=1}^{\infty} \frac{b_n}{10^n} = \frac{8}{89}$

(b) $b_n = \alpha^n + \beta^n$ for all $n \geq 1$

(c) $a_1 + a_2 + a_3 + \ldots + a_n = a_{n+2} - 1$ for all $n \geq 1$

(d) $\sum_{n=1}^{\infty} \frac{a_n}{10^n} = \frac{10}{89}$

3. If three unequal numbers p, q, r are in HP and their squares are in AP, then the ratio $p:q:r$ is
(a) $1-\sqrt{3}:-2:1+\sqrt{3}$ (b) $1:\sqrt{2}:-\sqrt{3}$
(c) $1:-\sqrt{2}:\sqrt{3}$ (d) $1+\sqrt{3}:-2:1-\sqrt{3}$

4. For a positive integer n, let $\alpha(n) = 1 + 1/2 + 1/3 + 1/4 + \ldots + 1/(2^n - 1)$, then
(a) $\alpha(100) \le 100$ (b) $\alpha(100) > 100$
(c) $\alpha(200) \le 100$ (d) $\alpha(200) > 100$

5. The pth term T_p of HP is $q(p+q)$ and qth term T_q is $p(p+q)$ when $p>1$, $q>1$, then
(a) $T_{p+q} = pq$ (b) $T_{pq} = p+q$
(c) $T_{p+q} > T_{pq}$ (d) $T_{pq} > T_{p+q}$

6. If the first and $(2n-1)$th term of an AP, a GP and a HP are equal and their nth terms are a, b and c respectively, then :
(a) $a=b=c$ (b) $a+c=b$
(c) $a \ge b \ge c$ (d) $ac = b^2$

7. If a, b, c be three unequal positive quantities in HP, then
(a) $a^{100} + c^{100} > 2b^{100}$ (b) $a^3 + c^3 > 2b^3$
(c) $a^5 + c^5 > 2b^5$ (d) $a^2 + c^2 > 2b^2$

8. The sum of n terms of the series
$$\frac{1}{1\cdot2\cdot3\cdot4} + \frac{1}{2\cdot3\cdot4\cdot5} + \frac{1}{3\cdot4\cdot5\cdot6} + \ldots \text{ is}$$
(a) $\dfrac{n(n^2+6n+11)}{18(n+1)(n+2)(n+3)}$
(b) $\dfrac{n^3+6}{18(n+1)(n+2)(n+3)}$
(c) $\dfrac{1}{18} - \dfrac{1}{3(n+1)(n+2)(n+3)}$
(d) $\dfrac{1}{6} - \dfrac{1}{2(n+1)(n+2)(n+3)}$

9. Given that $0 < x < \pi/4$ and $\pi/4 < y < \pi/2$ and $\sum_{k=0}^{\infty} (-1)^k \tan^{2k} x = a$, $\sum_{k=0}^{\infty} (-1)^k \cot^{2k} y = b$, then $\sum_{k=0}^{\infty} \tan^{2k} x \cot^{2k} y$ is
(a) $\dfrac{1}{a} + \dfrac{1}{b} - \dfrac{1}{ab}$ (b) $a + b - ab$
(c) $\dfrac{1}{\frac{1}{a} + \frac{1}{b} - \frac{1}{ab}}$ (d) $\dfrac{ab}{a+b-1}$

10. If a, b, c are in HP, then the value of $\left(\dfrac{1}{b} + \dfrac{1}{c} - \dfrac{1}{a}\right)\left(\dfrac{1}{c} + \dfrac{1}{a} - \dfrac{1}{b}\right)$ is
(a) $\dfrac{2}{bc} - \dfrac{1}{b^2}$ (b) $\dfrac{1}{4}\left(\dfrac{3}{c^2} + \dfrac{2}{ca} - \dfrac{1}{a^2}\right)$
(c) $\dfrac{3}{b^2} - \dfrac{2}{ab}$ (d) None of these

11. $\sum_{i=1}^{n} \sum_{j=1}^{i} \sum_{k=1}^{j} 1$ is equal to
(a) $\dfrac{n(n+1)(n+2)}{6}$ (b) Σn^2
(c) ${}^nC_3$ (d) ${}^{n+2}C_3$

12. If $1, \log_y x, \log_z y, -15\log_x z$ are in AP, then
(a) $z^3 = x$ (b) $x = y^{-1}$
(c) $z^{-3} = y$ (d) $x = y^{-1} = z^3$

13. If b_1, b_2, b_3 $(b_1 > 0)$ are three successive terms of a GP with common ratio r, the value of r for which the inequality $b_3 > 4b_2 - 3b_1$ holds is given by
(a) $r > 3$ (b) $r < 1$
(c) $r = 3.5$ (d) $r = 5.2$

14. If $\log_x a$, $a^{x/2}$ and $\log_b x$ are in GP, then x is equal to
(a) $\log_a(\log_b a)$
(b) $\log_a(\log_e a) - \log_a(\log_e b)$
(c) $-\log_a(\log_a b)$
(d) $\log_a(\log_e b) - \log_a(\log_e a)$

15. If a, b, c are in HP, then
(a) $\dfrac{a}{b+c-a}, \dfrac{b}{c+a-b}, \dfrac{c}{a+b-c}$ are in HP
(b) $\dfrac{2}{b} = \dfrac{1}{b-a} + \dfrac{1}{b-c}$
(c) $a - \dfrac{b}{2}, \dfrac{b}{2}, c - \dfrac{b}{2}$ are in GP
(d) $\dfrac{a}{b+c}, \dfrac{b}{c+a}, \dfrac{c}{a+b}$ are in HP

16. If $\sum_{r=1}^{n} r(r+1)(2r+3) = an^4 + bn^3 + cn^2 + dn + e$, then
(a) $a = 1/2$ (b) $b = 8/3$
(c) $c = 9/2$ (d) $e = 0$

17. If $1, \log_9(3^{1-x} + 2)$ and $\log_3(4\cdot3^x - 1)$ are in AP, then x is equal to
(a) $\log_4 3$ (b) $\log_3 4$
(c) $1 - \log_3 4$ (d) $\log_3(0.75)$

18. The series of natural numbers is divided into groups $1; 2, 3, 4; 5, 6, 7, 8, 9; \ldots$ and so on. Then the sum of the numbers in the nth group is
(a) $(2n-1)(n^2-n+1)$ (b) $n^3 - 3n^2 + 3n - 1$
(c) $n^3 + (n-1)^3$ (d) $n^3 + (n+1)^3$

19. Given that α, γ are roots of the equation $Ax^2 - 4x + 1 = 0$ and β, δ are roots of the equation $Bx^2 - 6x + 1 = 0$. If α, β, γ and δ are in HP, then
(a) $A = 5$ (b) $A = 3$ (c) $B = 8$ (d) $B = -8$

20. If $a_i > 0$ for all $i \in N$, then
(a) $(a_1 + a_2 + a_3)\left(\dfrac{1}{a_1} + \dfrac{1}{a_2} + \dfrac{1}{a_3}\right) \ge 9$
(b) $\left(\dfrac{a_1}{a_2} + \dfrac{a_3}{a_4} + \dfrac{a_5}{a_6}\right)\left(\dfrac{a_2}{a_1} + \dfrac{a_4}{a_3} + \dfrac{a_6}{a_5}\right) \ge 9$
(c) $(a_1 a_2 + a_3 a_4)(a_1 a_3 + a_2 a_4) \ge 4a_1 a_2 a_3 a_4$
(d) None of the above

21. The pth, 2pth and 4pth terms of an AP are in GP the common ratio of GP is

(a) 2 (b) 1
(c) 4 (d) 1/2

22. If a, b, c are real and for some real x,
$(a^2 + b^2)x^2 + 2(ab + bc)x + (b^2 + c^2) \leq 0$, then

(a) a, b, c are in GP
(b) a, b, c are in AP
(c) $ax^2 + 2bx + c \geq 0$ for all x
(d) $ax^2 + 2bx + c = 0$ for all x

23. Which of the following are true

(a) $x + \frac{1}{x} \geq 2$, if $x > 0$ (b) $x + \frac{1}{x} \leq -2$, if $x < 0$
(c) $2^x + 2^{-x} \geq 2$ for all x (d) $n! > n^n$

24. If a, b, c, d are distinct positive numbers in AP, then

(a) $ad < bc$
(b) $a + c < b + d$
(c) $a + d = b + c$
(d) $(a + 1)(d + 1) < (b + 1)(c + 1)$

25. The value of x for which $\frac{1}{1+\sqrt{x}}, \frac{1}{1-x}, \frac{1}{1-\sqrt{x}}$ are in AP lie in

(a) $(0, 1)$ (b) $(1, \infty)$
(c) $(0, \infty)$ (d) None of these

26. If a, b, c are in AP and a^2, b^2, c^2 are in HP, then

(a) $a = b = c$ (b) $a^2 = b^2 = c^2/2$
(c) a, b, c are in GP (d) $-a/2, b, c$ are in GP

27. If $\log_2(5 \cdot 2^x + 1)$, $\log_4(2^{1-x} + 1)$ and 1 are in AP, then x is equal to

(a) $\frac{\log 5}{\log 2}$ (b) $\log_2(0.4)$
(c) $1 - \frac{\log 5}{\log 2}$ (d) $\frac{\log 2}{\log 5}$

28. If the arithmetic mean of two positive numbers a and b $(a > b)$ is twice their geometric mean, then $a : b$ is

(a) $2 + \sqrt{3} : 2 - \sqrt{3}$ (b) $7 + 4\sqrt{3} : 1$
(c) $1 : 7 - 4\sqrt{3}$ (d) $2 : \sqrt{3}$

29. Let a and b be two positive real numbers. Suppose A_1, A_2 are two arithmetic means; G_1, G_2 are two geometric means and H_1, H_2 are two harmonic means between a and b; then

(a) $\frac{G_1G_2}{H_1H_2} = \frac{A_1 + A_2}{H_1 + H_2}$
(b) $\frac{G_1G_2}{H_1H_2} = \frac{H_1 + H_2}{A_1 + A_2}$
(c) $\frac{G_1G_2}{H_1H_2} - \frac{5}{9} = \frac{2}{9}\left(\frac{a}{b} + \frac{b}{a}\right)$
(d) $\frac{H_1 + H_2}{A_1 + A_2} = \frac{9ab}{(2a + b)(a + 2b)}$

Linked-Comprehension Type

In these questions, a passage (paragraph) has been given followed by questions based on each of the passages. You have to answer the questions based on the passage given.

PASSAGE 1

The numbers a, b and c are between 2 and 18, such that

(i) their sum is 25
(ii) the number 2, a and b are consecutive terms of an AP.
(iii) the numbers b, c, 18 are consecutive terms of a GP.

On the basis of above information, answer the following questions

1. The value of abc is

(a) 500 (b) 460
(c) 720 (d) 480

2. Roots of the equation $ax^2 + bx + c = 0$ are

(a) real and positive
(b) real and negative
(c) imaginary
(d) real and of opposite sign

3. If a, b and c are roots of the equation $x^3 + qx^2 + rx + s = 0$, then the value of r is

(a) 184 (b) 196
(c) 224 (d) None of these

PASSAGE 2

Suppose p is the first of n $(n > 1)$ AM's between two positive numbers a and b; q the first of n HM's between the same two numbers.

On the basis of above information, answer the following questions

1. The value of p is

(a) $\frac{na + b}{n + 1}$ (b) $\frac{na - b}{n + 1}$
(c) $\frac{nb + a}{n + 1}$ (d) $\frac{nb - a}{n + 1}$

2. The value of q is

(a) $\frac{ab(n + 1)}{b + an}$ (b) $\frac{ab(n + 1)}{(a + bn)}$
(c) $\frac{ab(n - 1)}{b + an}$ (d) $\frac{ab(n - 1)}{a + bn}$

3. The value of $\left(\frac{p}{q}-1\right)$ is

(a) $\frac{n}{(n-1)^2}\left(\sqrt{\frac{a}{b}}-\sqrt{\frac{b}{a}}\right)^2$ (b) $\frac{n}{(n-1)^2}\left(\sqrt{\frac{a}{b}}+\sqrt{\frac{b}{a}}\right)^2$

(c) $\frac{n}{(n+1)^2}\left(\sqrt{\frac{a}{b}}-\sqrt{\frac{b}{a}}\right)^2$ (d) $\frac{n}{(n+1)^2}\left(\sqrt{\frac{a}{b}}+\sqrt{\frac{b}{a}}\right)^2$

4. If $p = 8$ and $n = 3$, then
(a) q lies between 8 and 32
(b) q lies between 8 and 16
(c) q does not lies between 8 and 32
(d) q does not lies between 8 and 16

5. Final conclusion is
(a) q lies between p and $\left(\frac{n+1}{n-1}\right)p$
(b) q lies between p and $\left(\frac{n+1}{n-1}\right)^2 p$
(c) q does not lies between p and $\left(\frac{n+1}{n-1}\right)p$
(d) q does not lies between p and $\left(\frac{n+1}{n-1}\right)^2$.

PASSAGE 3

If A, G and H are respectively arithmetic, geometric and harmonic means between a and b both being unequal and positive, then

$$A = \frac{a+b}{2} \Rightarrow a+b = 2A, G = \sqrt{ab} \Rightarrow ab = G^2 \text{ and}$$

$$H = \frac{2ab}{a+b} \Rightarrow G^2 = AH$$

From above discussion we can say that a, b are the roots of the equation $x^2 - 2A\,x + G^2 = 0$

Now, quadratic equation $x^2 - Px + Q = 0$ and quadratic equation $a(b-c)x^2 + b(c-a)x + c(a-b) = 0$ have a root common and satisfy the relation $b = \frac{2ac}{(a+c)}$, where a, b, c are real numbers.

On the basis of above information, answer the following questions

1. The value of $\frac{a(b-c)}{b(c-a)}$ is
(a) −2 (b) 2
(c) −1 / 2 (d) 1/2

2. The value of $[P]$ is (where [.] donotes the greatest integer function)
(a) −2 (b) −1
(c) 2 (d) 1

3. The value of $[2P - Q]$ is (where [.] denotes the greatest integer function)
(a) 2 (b) 3 (c) 5 (d) 6

4. If the geometric and harmonic means of two numbers are 16 and $12\frac{4}{5}$, then the ratio of one number to the other is
(a) 1 : 4 (b) 2 : 3
(c) 1 : 2 (d) 2 : 1

5. The sum of the AM and GM of two positive numbers is equal to the difference between the numbers. The numbers are in the ratio
(a) 1 : 3 (b) 1 : 6
(c) 9 : 1 (d) 1 : 12

6. The ratio of the AM, GM and HM of the roots of the given quadratic equation is
(a) 1 : 2 : 3 (b) 1 : 1 : 2
(c) 2 : 2 : 3 (d) 1 : 1 : 1

PASSAGE 4

If a sequence or series is not a direct form of an AP, GP, etc. Then its nth term can not be determined. In such cases, we use the following steps to find the nth term (T_n) of the given sequence.

Step-I : Find the differences between the successive terms of the given sequence. If these differences are in AP, then take $T_n = an^2 + bn + c$, where a, b, c are constants.

Step-II : If the successive differences finding in step I are in GP with common ratio r, then take

$T_n = a + bn + cr^{n-1}$, where a, b, c are constants.

Step-III : If the second successive differences (Differences of the differences) in step I are in AP, then take

$T_n = an^3 + bn^2 + cn + d$, where a, b, c, d are constants.

Step-IV : If the second successive differences (Differences of the differences) in step I are in GP, then take

$T_n = an^2 + bn + c + dr^{n-1}$, where a, b, c, d are constants.

Now let sequences :

A : 1, 6, 18, 40, 75, 126, ...
B : 1, 1, 6, 26, 91, 291, ...
C : ln 2 ln 4, ln 32, ln 1024, ...

On the basis of above information, answer the following questions

1. Second successive differences of the sequences A and B are :
(a) Both AP
(b) Both GP
(c) AP and GP respectively
(d) GP and AP respectively

2. If the nth term (T_n) of the sequence A is $an^3 + bn^2 + cn + d$, then $6a + 2b - d$ is
(a) ln 2 (b) 2
(c) ln 8 (d) 4

3. The format of nth term (T_n) of the sequence C is
(a) $an^2 + bn + c$ (b) $an^3 + bn^2 + cn + d$
(c) $an + b + cr^{n-1}$ (d) $an^2 + bn + c + dr^{n-1}$

4. The correct statement for sequence B is
   (a) second successive differences makes AP with common difference 3
   (b) second successive differences makes AP with common differences ln 4
   (c) second successive differences makes a GP with common ratio 3
   (d) second successive differences makes a GP with common ratio 4
5. The correct statement for sequence C is
   (a) First successive differences form an AP with common difference ln 4
   (b) First successive differences form a GP with common ratio 4
   (c) Second successive differences form an AP with common difference ln 2
   (d) Second successive differences form a GP with common ratio 2

PASSAGE 5

The sum of the squares of three distinct real numbers which are in strictly increasing GP is S^2. If their sum is αS.

On the basis of above information, answer the following questions

1. α^2 lies in
   (a) $\left(\frac{1}{3}, 1\right)$ (b) $(1, 2)$
   (c) $\left(\frac{1}{3}, 3\right)$ (d) $\left(\frac{1}{3}, 1\right) \cup (1, 3)$
2. If $\alpha^2 = 2$, then the value of $[r]$ is (where [.] denotes the greatest integer function and r is common ratio of GP)
   (a) 0 (b) 1
   (c) 2 (d) 3
3. If $r = 2$, then the value of (α^2) is (where (.) denotes the least integer function and r is common ratio of GP)
   (a) 0 (b) 1
   (c) 2 (d) 3
4. If $S = 10\sqrt{3}$, then the greatest value of the middle term is
   (a) 5 (b) $5\sqrt{3}$
   (c) 10 (d) $10\sqrt{3}$
5. If we drop the condition that the GP is strictly increasing and take $r^2 = 1$, (where r is common ratio of GP) then the value of α is
   (a) 0 (b) ± 1 (c) ± 2 (d) $\pm\sqrt{3}$

PASSAGE 6

We are giving the concept of arithmetic mean of mth power. Let $a, b > 0$ and $a \neq b$ and let m be a real number.

Then, $\frac{a^m + b^m}{2} > \left(\frac{a+b}{2}\right)^m$, if $m \in R \sim [0, 1]$

However if $m \in (0, 1)$, then $\frac{a^m + b^m}{2} < \left(\frac{a+b}{2}\right)^m$

Obviously if $m \in \{0, 1\}$, then $\frac{a^m + b^m}{2} = \left(\frac{a+b}{2}\right)^m$

On the basis of above information, answer the following questions

1. If a, b be positive and $a + b = 1$ $(a \neq b)$ and if $A = \sqrt[3]{a} + \sqrt[3]{b}$ then the correct statement is
   (a) $A > 2^{2/3}$ (b) $A = \frac{2^{2/3}}{3}$
   (c) $A < 2^{2/3}$ (d) $A = 2^{2/3}$
2. If x, y be positive real numbers such that $x^2 + y^2 = 8$, then the maximum value of $x + y$ is
   (a) 2 (b) 4
   (c) 6 (d) 8
3. If a, b, c are positive real numbers but not all equal such that $a + b + c = 1$, then best option of values $\frac{b^2 + c^2}{b + c} + \frac{c^2 + a^2}{c + a} + \frac{a^2 + b^2}{a + b}$ lie between
   (a) $\left(\frac{3}{2}, \infty\right)$ (b) $(1, \infty)$
   (c) $(0, \infty)$ (d) None of these
4. If a and b are positive $(a \neq b)$ and $a + b = 1$ and if $A = \left(a + \frac{1}{a}\right)^2 + \left(b + \frac{1}{b}\right)^2$, then the correct statement is
   (a) $A > 8$ (b) $A < 8$
   (c) $A > \frac{25}{2}$ (d) $A < \frac{25}{2}$
5. If a, b, c be positive and in Harmonic progression and if $\lambda = \frac{a^n + c^n}{b^n}, \forall\, n \notin (0, 1)$, then the correct statement is
   (a) $\lambda < 2$ (b) $\lambda > 2$
   (c) $\lambda = 2$ (d) None of these

PASSAGE 7

We are giving the concept of arithmetic mean of m th power. Let $a_1, a_2, a_3, \ldots, a_n$ be n positive real numbers (not all equal) and let m be real number.

Then,

$$\frac{a_1^m + a_2^m + a_3^m + \ldots + a_n^m}{n} > \left(\frac{a_1 + a_2 + a_3 + \ldots + a_n}{n}\right)^m;$$

if $m \in R \sim [0,1]$

However if $m \in (0,1)$, then

$$\frac{a_1^m + a_2^m + a_3^m + \ldots + a_n^m}{n} < \left(\frac{a_1 + a_2 + a_3 + \ldots + a_n}{n}\right)^m$$

Obviously if $m \in \{0, 1\}$, then

$$\frac{a_1^m + a_2^m + a_3^m + \ldots + a_n^m}{n} = \left(\frac{a_1 + a_2 + a_3 + \ldots + a_n}{n}\right)^m$$

On the basis of above information, answer the following questions

1. If sum of the mth powers of first n odd numbers is λ, $\forall\, m > 1$ then
   (a) $\lambda < n^m$ (b) $\lambda > n^m$
   (c) $\lambda < n^{m+1}$ (d) $\lambda > n^{m+1}$
2. If a, b, c be positive real numbers, then the possible best option of values $\frac{a}{b+c} + \frac{b}{c+a} + \frac{c}{a+b}$ lie between
   (a) $(1, \infty)$ (b) $(0, \infty)$
   (c) $\left[\frac{3}{2}, \infty\right)$ (d) $[2, \infty)$
3. If $x > 0, y > 0, z > 0$ and $x + y + z = 1$, then the minimum value of $\frac{x}{2-x} + \frac{y}{2-y} + \frac{z}{2-z}$ is
   (a) $\frac{9}{5}$ (b) $\frac{8}{5}$
   (c) $\frac{3}{5}$ (d) $\frac{2}{5}$
4. If $a_1, a_2, a_3, \ldots, a_n$ are all positive such that $a_1^2 + a_2^2 + a_3^2 + \ldots. + a_n^2 = A$ and if greatest and least values of $(a_1 + a_2 + a_3 + \ldots + a_n)^2$ are G and L respectively, then
   (a) $G - L = 2A$ (b) $G - L = nA$
   (c) $G - L = (n-1)A$ (d) $G - L = (n-2)A$
5. If a, b, c, d be positive and not all equal to one another such that $a + b + c + d = 4/3$, then the minimum value of
   $$\sqrt{\left(\frac{1}{b+c+d}\right)} + \sqrt{\left(\frac{1}{c+d+a}\right)} + \sqrt{\left(\frac{1}{d+a+b}\right)} + \sqrt{\left(\frac{1}{a+b+c}\right)}$$ is
   (a) 2 (b) 4
   (c) 6 (d) 8

PASSAGE 8

If $a_i > 0, i = 1, 2, 3, \ldots, n$ and $m_1, m_2, m_3, \ldots, m_n$ be positive rational numbers, then

$$\left(\frac{m_1a_1 + m_2a_2 + \ldots + m_na_n}{m_1 + m_2 + \ldots + m_n}\right)$$

$$\geq (a_1^{m_1}\, a_2^{m_2} \ldots a_n^{m_n})^{1/(m_1 + m_2 + \ldots + m_n)}$$

$$\geq \frac{(m_1 + m_2 + \ldots + m_n)}{\frac{m_1}{a_1} + \frac{m_2}{a_2} + \ldots + \frac{m_n}{a_n}}$$ is called weighted mean theorem

where $A^* = \frac{m_1a_1 + m_2a_2 + \ldots + m_na_n}{m_1 + m_2 + \ldots + m_n}$

= Weighted arithmetic mean

$G^* = (a_1^{m_1} a_2^{m_2} \ldots a_n^{m_n})^{1/(m_1 + m_2 + \ldots + m_n)}$

= Weighted geometric mean

and $H^* = \frac{m_1 + m_2 + \ldots + m_n}{\frac{m_1}{a_1} + \frac{m_2}{a_2} + \ldots \frac{m^n}{a_n}}$

= Weighted harmonic mean

ie., $A^* \geq G^* \geq H^*$

Now, let $a + b + c = 5$ $(a, b, c > 0)$

and $x^2y^3 = 6$ $(x > 0, y > 0)$

On the basis of above information, answer the following questions

1. The greatest value of ab^3c is
   (a) 3 (b) 9
   (c) 27 (d) 81
2. Which statement is correct
   (a) $\frac{1}{5} \geq \frac{1}{\frac{1}{a} + \frac{3}{b} + \frac{1}{c}}$ (b) $\frac{1}{25} \geq \frac{1}{\frac{1}{a} + \frac{9}{b} + \frac{1}{c}}$
   (c) $\frac{1}{5} \geq \frac{1}{\frac{1}{a} + \frac{9}{b} + \frac{1}{c}}$ (d) $\frac{1}{25} \geq \frac{1}{\frac{1}{a} + \frac{6}{b} + \frac{1}{c}}$
3. The least value of $3x + 4y$ is
   (a) 5 (b) 7
   (c) 10 (d) 17
4. Which statement is correct
   (a) $\frac{2x + 3y}{5} \geq (6)^{1/5} \geq \frac{5}{\frac{3}{x} + \frac{2}{y}}$
   (b) $\frac{2x + 3y}{5} \geq (6)^{1/5} \geq \frac{5xy}{3x + 2y}$
   (c) $\frac{2x + 3y}{5} \geq (6)^{1/5} \geq \frac{5xy}{3x + 4y}$
   (d) $\frac{2x + 3y}{5} \geq (6)^{1/5} \geq \frac{5xy}{2x + 3y}$
5. The maximum value of $\lambda\,\mu\,\upsilon$ when $\frac{\lambda^2}{d_1^2} + \frac{\mu^2}{d_2^2} + \frac{\upsilon^2}{d_3^2} = 1$ is
   (a) $\frac{d_1d_2d_3}{\sqrt[3]{3}}$ (b) $\frac{d_1d_2d_3}{3\sqrt{3}}$
   (c) $\frac{d_1d_2d_3}{3}$ (d) $\frac{d_1d_2d_3}{\sqrt{3}}$

Numerical Value Type

Solve the following problems and write your answer.

1. If $x_1, x_2, x_3, \ldots, x_{2008}$ are in HP and $\sum_{i=1}^{2007} x_i x_{i+1} = \lambda x_1 x_{2008}$, then λ must be equal to :
2. A three-digit number whose consecutive digits form a GP. If we subtract 792 from this number, we get a number consisting of the same digits written in the reverse order. Now, if we increase the second digit of the required number by 2, the resulting digits will form an AP. The three-digit number must be
3. If $\sum_{r=1}^{n} T_r = \frac{n(n+1)(n+2)(n+3)}{8}$, then the value of $\lim_{n\to\infty} \sum_{r=1}^{n} \frac{4016}{T_r}$ must be
4. If m arithmetic means (AMs) and three geometric means (GMs) are inserted between 3 and 243 such that 4th AM is equal to 2nd GM, then m is equal to **[JEE Main 2020, 3 September]**
5. If a_n denotes the coefficient of x^n in $P(x) = (1 + x + 2x^2 + 3x^3 + \ldots + n x^n)^2$, then the value of a_{24} must be
6. If $S = \sum_{r=1}^{80} \frac{r}{(r^4 + r^2 + 1)}$, then the value of $\frac{64815}{1000}$ is
7. The sum of two digit even numbers which do not end with zero is
8. The number of zeroes in the end in the product of $5^6 \times 6^7 \times 7^8 \times 8^9 \times \ldots\ldots \times 50^{51}$ must be
9. If $\log_{10} x + \frac{1}{2}\log_{10} x + \frac{1}{4}\log_{10} x + \ldots = y$ and $\frac{1+3+5+\ldots+(2y-1)}{4+7+10+\ldots+(3y+1)} = \frac{20}{7\log_{10} x}$ $\forall\, x, y \in N$, then the value of $(\log_y x)^5$ must be
10. Let a, b be positive real numbers. If a, A_1, A_2, b are in AP. a, G_1, G_2, b are in GP and a, H_1, H_2, b are in HP, then $\frac{G_1G_2}{H_1H_2} = \frac{A_1 + A_2}{H_1 + H_2} = \frac{(\lambda a + b)(a + \lambda b)}{\mu\, ab}$, then the value of $\lambda^{\mu} + \mu^{\lambda}$ must be
11. Let $AP(a; d)$ denote the set of all the terms of an infinite arithmetic progression with first term a and common difference $d > 0$.
If $AP(1; 3) \cap AP(2; 5) \cap AP(3; 7) = AP(a; d)$, then $a + d$ equals .... **[JEE Main 2019]**
12. If $1 \times 2020 + 2 \times 2019 + 3 \times 2018 + \ldots 2020 \times 1 = 2020 \times 2021 \times k$, then the value of $\frac{k}{100}$ is equal to

Single Integer Type

Solve the following problems and write your answer.

1. If the 20th term of an HP is 1 and the 30th term is $\frac{-1}{17}$, then its largest term.
2. The sides of a right angled triangle are in arithmetic progression. If the triangle has area 24, then what is the length of its smallest side? **[JEE Advanced 2017]**
3. If $(m + 1)$th term, $(n + 1)$th term and $(r + 1)$th terms of an AP are in GP and m, n and r are in HP, then the ratio of the common difference to the first term of AP is k / n, then value of $|k|$ is ....
4. Let $a_1, a_2, a_3, \ldots, a_{101}$ are in GP with $a_{101} = 25$ and $\sum_{i=1}^{201} a_i = 625$. Then, the value of $\sum_{i=1}^{201} \frac{1}{a_i}$ equals ....
5. Let $S_k;\ k = 1, 2, \ldots, 100$, denote the sum of the infinite geometric series whose first term is $\frac{k-1}{k!}$ and the common ratio is $\frac{1}{k}$. Then, the value of $\frac{100^2}{100!} + \sum_{k=1}^{100} |(k^2 - 3k + 1)S_k|$ is
6. Let $S_1, S_2, \ldots.$ be squares such that for each $n \ge 1$, the length of a side of S_n equals the length of a diagonal of S_{n+1}. If the length of a side of S_1 is 10 cm, then the smallest value of n for which Area $(S_n) < 1$ is
7. Let a, b, c be positive integers such that b / a is an integer. If a, b, c are in geometric progression and the arithmetic mean of a, b, c is $b + 2$, then the value of $\frac{a^2 + a - 14}{a + 1}$ is **[JEE Advanced 2014]**
8. Let $a_1, a_2, a_3, \ldots.$ be a sequence of positive integers in arithmetic progression with common difference 2. Also, let $b_1, b_2, b_3, \ldots.$ be a sequence of positive integers in geometric progression with common ratio 2. If $a_1 = b_1 = c$, then the number of all possible values of c, for which the equality
$$2(a_1 + a_2 + \ldots + a_n) = b_1 + b_2 + \ldots + b_n$$
holds for some positive integer n, is ......
9. If $S_n = 1 + \frac{1}{2} + \frac{1}{2^2} + \ldots + \frac{1}{2^{n-1}}$ and $2 - S_n < \frac{1}{100}$, then the least value of n must be :
10. Suppose that all the terms of an arithmetic progression are natural numbers. If the ratio of the sum of the first seven terms to the sum of the first eleven terms is 6 : 11 and the seventh term lies in between 130 and 140, then the common difference of this AP is **[JEE Advanced 2015]**
11. If a, b, c are in HP and if $\left(\frac{a+b}{2a-b}\right) + \left(\frac{c+b}{2c-b}\right) > \sqrt{\lambda\sqrt{\lambda\sqrt{\lambda\ldots\infty}}}$, then the value of λ must be
12. A pack contains n cards numbered from 1 to n. Two consecutive numbered cards are removed from the pack and the sum of the numbers on the remaining cards is 1224. If the smaller of the numbers on the removed cards is k, then $k - 20$ is equal to **[JEE Advanced 2013]**

Matrix-Match Type

Given below are Matching Type Questions, with two columns (each having some items) each. Each item of Column I has to be matched with the items of Column II, by encircling the correct match(es).

NOTE *An item of Column I can be matched with more than one items of Column II.*

1. Observe the following columns :

	Column I		Column II
(A)	If a, b, c are non-zero real numbers such that $3(a^2+b^2+c^2+1) = 2\times(a+b+c+ab+bc+ca)$, then a, b, c are in	(P)	AP
(B)	If the square of differences of three numbers be in AP, then their differences are in	(Q)	GP
(C)	If $a-b, ax-by, ax^2-by^2$ $(a, b \neq 0)$ are in GP, then $x, y, \dfrac{ax-by}{a-b}$ are in	(R)	HP
		(S)	Equal

2. Observe the following columns :

	Column I		Column II
(A)	If a, b, c, d are in AP, then	(P)	$a+d>b+c$
(B)	If a, b, c, d are in GP, then	(Q)	$ad>bc$
(C)	If a, b, c, d are in HP, then	(R)	$\frac{1}{a}+\frac{1}{d}>\frac{1}{b}+\frac{1}{c}$
		(S)	$bc>ad$
		(T)	$c^{-1}d^{-1}+a^{-1}b^{-1}>2(b^{-1}d^{-1}+a^{-1}c^{-1}-a^{-1}d^{-1})$

3. Observe the following columns :

	Column I		Column II
(A)	Three numbers a, b, c between 2 and 18 such that (i) $a+b+c=25$ (ii) 2, a, b are consecutive terms of an AP. (iii) b, C, 18 are consecutive terms of a GP. If $G=\text{Max}\{a, b, c\}$ and $L=\text{Min}\{a,b,c\}$ then	(P)	$G-L=4$
(B)	Three numbers a, b, c are in GP such that (i) $a+b+c=70$ (ii) $4a, 5b, 4c$ are in AP. If $G=\text{Max}\{a, b, c\}$ and $L=\text{Min}(a, b, c)$, then	(Q)	$\left[\frac{G}{L}\right]=4$ where [.] denotes the greatest integer function
(C)	If $(1+3+5+\ldots+a)+(1+3+5+\ldots+b)=(1+3+5+\ldots+c)$, where each set of parentheses contains the sum of consecutive odd integers as shown such that (i) $a+b+c=21$ (ii) $a>6$ If $G=\text{Max}\{a, b, c\}$ and $L=\text{Min}\{a, b, c\}$, then	(R)	$G-L=7$
		(S)	$\left(\frac{G}{L}\right)=3$ where (.) denotes the least integer function
		(T)	$\left[\frac{G}{L}\right]+\left(\frac{G}{L}\right)=3$ where [.] and (.) denotes the greatest and least integer functions respectively.

4. Observe the following columns :

	Column I		Column II
(A)	The AM of two positive numbers a and b exceed their GM by 3/2 and the GM exceed their HM by 6/5, such that $a+b=\alpha, \lvert a-b\rvert=\beta$, then	(P)	$\alpha+\beta^2=96$
(B)	The AM of two positive numbers a and b exceed their GM by 2 and HM is one fifth of the greater of a and b, such that $a+b=\alpha, \lvert a-b\rvert=\beta$, then	(Q)	$\alpha+\beta^2=74$
(C)	The HM of two positive numbers a and b is 4, their AM is A and Geometric mean G satisfy the relation $2A+G^2=27$. If $a+b=\alpha, \lvert a-b\rvert=\beta$, then	(R)	$\alpha^2+\beta=234$
		(S)	$\alpha^2+\beta=84$
		(T)	$1+\left(\frac{\alpha}{\beta}\right)+\left(\frac{\alpha}{\beta}\right)^2+\ldots\ldots+\left(\frac{\alpha}{\beta}\right)^5=364$

5. Observe the following columns :

	Column I		Column II
(A)	If sum of n terms of an AP is given by $S_n = a + bn + cn^2$ where a, b, c are independent of n, then	(P)	$T_n = a + b + (2n - 1)c$
(B)	If sum of n terms of an AP is given by $S_n = (a-1)n^3 + (b-2)n^2 + (c-3)n$, where a, b, c are independent of n, then	(Q)	$T_n = b + (2n - 1)c$
(C)	The sum of n terms of an AP is given by $S_n = bn + cn^2$ where b and c are independent of n, then	(R)	$T_n = c - a - 4n + b(2n - a)$
		(S)	Common difference $(D) = 2c$
		(T)	Common difference $(D) = (2b - 4)$

Answers

Objective Questions Type I [Only one correct answer]

1. (b)	**2.** (b)	**3.** (c)	**4.** (c)	**5.** (a)	**6.** (c)	**7.** (b)	**8.** (b)	**9.** (b)	**10.** (d)
11. (a)	**12.** (c)	**13.** (b)	**14.** (a)	**15.** (d)	**16.** (d)	**17.** (c)	**18.** (c)	**19.** (b)	**20.** (c)
21. (b)	**22.** (a)	**23.** (d)	**24.** (b)	**25.** (a)	**26.** (a)	**27.** (d)	**28.** (a)	**29.** (a)	**30.** (b)
31. (d)	**32.** (a)	**33.** (d)	**34.** (b)	**35.** (a)	**36.** (a)	**37.** (d)	**38.** (c)	**39.** (d)	**40.** (a)
41. (d)	**42.** (d)	**43.** (a)	**44.** (a)	**45.** (b)	**46.** (d)	**47.** (a)	**48.** (c)	**49.** (a)	**50.** (c)
51. (c)	**52.** (c)	**53.** (d)	**54.** (c)	**55.** (a)	**56.** (d)	**57.** (a)	**58.** (a)	**59.** (b)	**60.** (a)
61. (d)	**62.** (a)	**63.** (b)	**64.** (a)	**65.** (c)	**66.** (c)	**67.** (c)	**68.** (b)	**69.** (c)	**70.** (c)
71. (d)	**72.** (a)	**73.** (a)	**74.** (b)	**75.** (a)	**76.** (d)	**77.** (d)	**78.** (d)	**79.** (d)	**80.** (b)
81. (d)	**82.** (d)	**83.** (d)	**84.** (b)	**85.** (b)	**86.** (c)	**87.** (c)	**88.** (c)	**89.** (b)	**90.** (c)
91. (a)	**92.** (a)	**93.** (a)	**94.** (b)	**95.** (a)					

Objective Questions Type II [One or more than one correct answer(s)]

1. (a, d)	**2.** (b, c, d)	**3.** (a, d)	**4.** (a, d)	**5.** (a, b, c)
6. (c, d)	**7.** (a, b, c, d)	**8.** (a, c)	**9.** (c, d)	**10.** (a, b, c)
11. (a, d)	**12.** (a, b, c, d)	**13.** (a, b, c, d)	**14.** (a, b)	**15.** (a, b, c, d)
16. (a, b, c, d)	**17.** (c, d)	**18.** (a,c)	**19.** (b, c)	**20.** (a, b, c)
21. (a, b)	**22.** (a,c)	**23.** (a, b, c)	**24.** (a, c, d)	**25.** (a, b)
26. (a, c, d)	**27.** (b, c)	**28.** (a, b, c)	**29.** (a, c, d)	

Linked-Comprehension Type

Passage 1 **1.** (d) **2.** (c) **3.** (b)

Passage 2 **1.** (a) **2.** (b) **3.** (c) **4.** (c) **5.** (d)

Passage 3 **1.** (c) **2.** (c) **3.** (b) **4.** (a) **5.** (c) **6.** (d)

Passage 4 **1.** (c) **2.** (d) **3.** (a) **4.** (c) **5.** (a)

Passage 5 **1.** (d) **2.** (c) **3.** (d) **4.** (c) **5.** (d)

Passage 6 **1.** (c) **2.** (b) **3.** (b) **4.** (c) **5.** (b)

Passage 7 **1.** (d) **2.** (c) **3.** (c) **4.** (c) **5.** (b)

Passage 8 **1.** (c) **2.** (c) **3.** (c) **4.** (b) **5.** (b)

Numerical Value Type

1. 2007	**2.** 931	**3.** 2008	**4.** 39	**5.** 2600	**6.** 3.24
7. 1980	**8.** 285	**9.** 3125	**10.** 593	**11.** 157	**12.** 3.37

Single Integer Type Questions

1. (1) **2.** (6) **3.** (2) **4.** (1) **5.** (3) **6.** (8) **7.** (4) **8.** (3) **9.** (8) **10.** (9) **11.** (4) **12.** (5)

Matrix-Match Type

1. A → (P, Q, S); B → (R); C → (P, Q, S) **2.** A → (R, S, T); B → (P, R, T); C → (P, Q, R, T)

3. A → (R, S); B → (Q); C → (P, T) **4.** A → (P, R); B → (Q); C → (S, T) **5.** A → (P, Q, S); B → (R, T); C → (Q, S)

Solutions

Objective Questions Type [Only one correct answer]

1. Common terms are 5, 11, 17, ...

$\therefore$ nth term of common sequence

$$t_n = 5 + (n-1)6 = (6n-1)$$

Also 100th term of the first sequence

$$= 2 + (100-1)3 = 299$$

and 100th term of the second sequence

$$= 3 + (100-1)2 = 201$$

$\Rightarrow$ $t_n \le 201 \Rightarrow 6n - 1 \le 201$

$\Rightarrow$ $n \le 33\frac{2}{3}$

$\therefore$ $n = 33$ $[\because n \in N]$

2. $\because$ AM $\ge$ GM

$\Rightarrow$ $\frac{a+b+c}{3} \ge (abc)^{1/3}$

or $a + b + c \ge 3(abc)^{1/3}$ ...(i)

But given $ab^2c^3, a^2b^3c^4, a^3b^4c^5$ are in AP

$\Rightarrow 1, abc, a^2b^2c^2$ are also in AP $[\because abc \ne 0]$

$\Rightarrow$ $2abc = 1 + a^2b^2c^2$

$\Rightarrow$ $(abc-1)^2 = 0$

$\therefore$ $abc = 1$

Now from Eq. (i),

$$a + b + c \ge 3(1)^{1/3}$$

or $(a+b+c) \ge 3$

Hence, minimum value of $a + b + c$ is 3.

3. It is given that $S = \{x + ka\} + \{x^2 + (k+2)a\}$

$+\{x^3 + (k+4)a\} + \{x^4 + (k+6)a\} + \ldots\ldots +$ upto 9 terms

$\Rightarrow$ $S = \{x + x^2 + x^3 + x^4 + \ldots + x^9\}$

$+ a\{k + (k+2) + (k+4)\} + \ldots\ldots + (k+16)\}$

$$= \frac{x(x^9-1)}{x-1} + a\left(\frac{9}{2}\right)\{2k + (9-1)2\}$$

$$= \frac{x^{10}-x}{x-1} + 9a(k+8)$$

$\Rightarrow$ $\frac{x^{10} - x + 9a(k+8)(x-1)}{x-1} = \frac{x^{10} - x + 45a(x-1)}{x-1}$

So, on comparing, we get

$$k + 8 = 5 \Rightarrow k = -3$$

Hence, option (c) is correct.

4. Required sum $= S_3 + S_5 + S_7 - S_{15} - S_{35} - S_{21}$

$$= \frac{33}{2}(3+99) + \frac{20}{2}(5+100) + \frac{14}{2}(7+98)$$

$$- \frac{6}{2}(15+90) - (35+70) - \frac{4}{2}(21+84)$$

$$= 1683 + 1050 + 735 - 315 - 105 - 210$$

$$= 2838$$

where, $S_3 = 3 + 6 + 9 + \ldots + 99$

$S_5 = 5 + 10 + 15 + \ldots + 100$

$S_7 = 7 + 14 + 21 + \ldots + 98$

$S_{15} = 15 + 30 + 45 + \ldots + 90$

$S_{35} = 35 + 70$

$S_{21} = 21 + 42 + 63 + 84$

5. Given, series $\frac{3}{5} + \frac{21}{25} + \frac{117}{125} + \ldots +$ upto n terms

$$= \left(1 - \frac{2}{5}\right) + \left(1 - \frac{4}{25}\right) + \left(1 - \frac{8}{125}\right) + \ldots + \text{ upto } n \text{ terms}$$

$$= n - \left[\frac{2}{5} + \left(\frac{2}{5}\right)^2 + \left(\frac{2}{5}\right)^3 + \ldots + \text{ upto } n \text{ terms}\right]$$

$$= n - \frac{\frac{2}{5}\left(1 - \left(\frac{2}{5}\right)^n\right)}{1 - \frac{2}{5}}$$

$$\left[\because a + ar + ar^2 + \ldots + ar^{n-1} = \frac{a(1-r^n)}{1-r}\right]$$

$$= n - \frac{\frac{2}{5}\left[1 - \left(\frac{2}{5}\right)^n\right]}{3/5} = n + \frac{2^{n+1}}{3 \times 5^n} - \frac{2}{3}$$

Hence, option (a) is correct.

6. $\because$ AM $\ge$ HM

$\therefore$ $\frac{a_1 + a_2 + a_3 + \ldots + a_{50}}{50} \ge \frac{50}{\left(\frac{1}{a_1} + \frac{1}{a_2} + \frac{1}{a_3} + \ldots + \frac{1}{a_{50}}\right)}$

or $\frac{50}{50} \ge \frac{50}{\frac{1}{a_1} + \frac{1}{a_2} + \frac{1}{a_3} + \ldots + \frac{1}{a_{50}}}$

$\Rightarrow$ $\frac{1}{a_1} + \frac{1}{a_2} + \frac{1}{a_3} + \ldots + \frac{1}{a_{50}} \ge 50$

Hence, minimum value of $\frac{1}{a_1} + \frac{1}{a_2} + \frac{1}{a_3} + \ldots + \frac{1}{a_{50}}$ is 50.

7. $\because$ a^{-1}, b^{-1}, c^{-1} are in AP.

$\therefore$ a, b, c are in HP.

Now for any three numbers $a^{101}, b^{101}, c^{101}$

$\because$ AM $>$ GM

$\Rightarrow$ $\frac{a^{101} + c^{101}}{2} > (\sqrt{ac})^{101} > b^{101}$ $(\because \sqrt{ac} > b)$

$\Rightarrow$ $a^{101} + c^{101} - 2b^{101} > 0$

$\Rightarrow$ $2b^{101} - a^{101} - c^{101} < 0$ ...(i)

$\therefore$ Product of roots $= \frac{2b^{101} - a^{101} - c^{101}}{1} < 0$ [from Eq. (i)]

$\Rightarrow$ Product of roots < 0

8. $\because$ $x^a = y^b = z^c = \lambda$ (say)

$\therefore$ $x = \lambda^{1/a}, y = \lambda^{1/b}, z = \lambda^{1/c}$

Now, $\because$ x, y, z are in GP

$\therefore$ $y^2 = zx$

$\Rightarrow$ $\lambda^{2/b} = \lambda^{1/c} \cdot \lambda^{1/a}$

$\Rightarrow$ $\lambda^{2/b} = \lambda^{1/c + 1/a}$

$\Rightarrow$ $\frac{2}{b} = \frac{1}{a} + \frac{1}{c}$

$\therefore$ a, b, c are in HP .

Now, GM > HM

$\Rightarrow \sqrt{ac} > b$ ...(i)

Now, for three numbers a^3, b^3, c^3

AM > GM

$\Rightarrow \dfrac{a^3 + c^3}{2} > (\sqrt{ac})^3 > b^3$ [from Eq. (i)]

$\therefore a^3 + c^3 > 2b^3$

9. The number of digits in each term of the sequence are 1, 2, 4, 8, ... which are in GP.

Let 1025th term is 2^n.

Then, $1 + 2 + 4 + 8 + \ldots + 2^{n-1} < 1025 \le 1 + 2 + 4 + 8 + \ldots + 2^n$

$\Rightarrow 1.\dfrac{(2^n - 1)}{(2-1)} < 1025 \le 1.\dfrac{(2^{n+1} - 1)}{(2-1)}$

$\Rightarrow 2^n - 1 < 1025 \le 2^{n+1} - 1$

$\Rightarrow 2^n < 1026 \le 2^{n+1}$ ...(i)

or $2^{n+1} \ge 1026 > 1024$

$\Rightarrow 2^{n+1} > 2^{10}$

or $n + 1 > 10$

$\therefore n > 9$

$\therefore n = 10$ (which is always satisfy Eq. (i)),

$\therefore$ 1025th term is 2^{10}.

10. $\because 3(\Sigma a^2 + 1) = 2(\Sigma a + \Sigma ab)$

$\Rightarrow 3(a^2 + b^2 + c^2 + 1) = 2(a + b + c + ab + bc + ca)$

$\Rightarrow 2(a^2 + b^2 + c^2 - ab - bc - ca) + (a^2 + b^2 + c^2 - 2a - 2b - 2c + 1 + 1 + 1) = 0$

$\Rightarrow (a-b)^2 + (b-c)^2 + (c-a)^2 + (a-1)^2 + (b-1)^2 + (c-1)^2 = 0$

which is possible only when

$a - b = 0, b - c = 0, c - a = 0,$

$a - 1 = 0, b - 1 = 0, c - 1 = 0$

$\Rightarrow a = b = c = 1$

$\Rightarrow \therefore a, b, c$ are in AP and GP both.

11. $\because a, b, c, d$ are positive real numbers.

$\therefore m > 0$ ...(i)

Now, AM ≥ GM

$\Rightarrow \dfrac{(a+b)+(c+d)}{2} \ge \sqrt{(a+b)(c+d)}$

$\Rightarrow \dfrac{2}{2} \ge \sqrt{m}$

or $m \le 1$ ...(ii)

From Eqs. (i) and (ii), we get, $0 < m \le 1$

12. $\because$ LHS $= \dfrac{1(1-\lambda^{n+1})}{(1-\lambda)} = \left(\dfrac{1-\lambda^{n+1}}{1-\lambda}\right)$

and RHS $= (1+\lambda)(1+\lambda^2)(1+\lambda^4)(1+\lambda^8)(1+\lambda^{16})$

$(1-\lambda)(1+\lambda)(1+\lambda^2)(1+\lambda^4)$

$= \dfrac{(1+\lambda^8)(1+\lambda^{16})}{(1-\lambda)} = \dfrac{(1-\lambda^{32})}{(1-\lambda)}$

$\Rightarrow \dfrac{1-\lambda^{n+1}}{1-\lambda} = \dfrac{1-\lambda^{32}}{1-\lambda}$

$\Rightarrow 1 - \lambda^{n+1} = 1 - \lambda^{32}$

$\therefore n + 1 = 32 \Rightarrow n = 31$

13. $\because 13, a_1, a_2, a_3, \ldots, a_{20}, 67$ are in AP.

$\therefore a_1 + a_2 + a_3 + \ldots + a_{20} = 20\left(\dfrac{13+67}{2}\right) = 800$

Now, AM ≥ GM

$\because \dfrac{a_1 + a_2 + a_3 + \ldots + a_{20}}{20} \ge (a_1 a_2 a_3 \ldots a_{20})^{1/20}$

$\Rightarrow \left(\dfrac{800}{20}\right) \ge (a_1\, a_2\, a_3 \ldots a_{20})^{1/20}$

or $a_1\, a_2\, a_3 \ldots a_{20} \le (40)^{20}$

$\therefore$ Maximum value of $a_1\, a_2\, a_3 \ldots a_{20}$ is $(40)^{20}$.

14. Given, $a_1\, a_2\, a_3 \ldots a_{n-1}\, a_n = c$ ...(i)

$\because$ AM ≥ GM

$\Rightarrow \dfrac{a_1 + a_2 + \ldots + a_{n-1} + 2a_n}{n} \ge (a_1\, a_2 \ldots a_{n-1}\, 2a_n)^{1/n}$

$= (2c)^{1/n}$ [from Eq. (i)]

$\therefore a_1 + a_2 + \ldots + a_{n-1} + 2a_n \ge n\,(2c)^{1/n}$

Hence, minimum value of $a_1 + a_2 + \ldots + a_{n-1} + 2a_n$ is $n\,(2c)^{1/n}$.

15. Let $b = a + d, c = a + 2d$ ...(i)

$\because a^2, b^2, c^2$ are in GP.

$\therefore (b^2)^2 = a^2c^2$

or $\pm b^2 = ac$ ...(ii)

$\because a, b, c$ are in AP .

$\therefore 2b = a + c$

Given, $a + b + c = 3/2 \Rightarrow 3b = 3/2 \Rightarrow b = 1/2$

From Eq. (i),

$a = \dfrac{1}{2} - d, c = \dfrac{1}{2} + d$

$\therefore$ From Eq. (ii),

$\pm\dfrac{1}{4} = \left(\dfrac{1}{2} - d\right)\left(\dfrac{1}{2} + d\right)$

$\Rightarrow \pm\dfrac{1}{4} = \dfrac{1}{4} - d^2$

Taking (– ve) sign, $\therefore d = \pm\dfrac{1}{\sqrt{2}}$

$\therefore a = \dfrac{1}{2} - d = \dfrac{1}{2} \pm \dfrac{1}{\sqrt{2}}$

$\Rightarrow a = \dfrac{1}{2} + \dfrac{1}{\sqrt{2}}$ $(\because a > b)$

16. $x_1 + x_2 + x_3 + \ldots + x_{100} = \dfrac{100}{2}(x_1 + x_{100}) = -1$

$\Rightarrow x_1 + x_{100} = -\dfrac{1}{50}$

$\Rightarrow x_2 + x_4 + \ldots + x_{100} = \dfrac{50}{2}(x_1 + d + x_{100}) = 1$

$\Rightarrow x_1 + x_{100} + d = \dfrac{1}{25}$

$\Rightarrow d = \dfrac{3}{50}$

$\Rightarrow x_1 + x_1 + 99d = \dfrac{-1}{50} \Rightarrow x_1 = \dfrac{-149}{50}$

$\Rightarrow x_{100} = x_1 + 99d$

$= \dfrac{-149}{50} + 99 \times \dfrac{3}{50} = \dfrac{74}{25}$

17. $\because a, b, c, d \in I$

and let $b = a + \lambda$, $c = a + 2\lambda$, $d = a + 3\lambda$

Then, $\lambda \in I$

$\therefore a + 3\lambda = a^2 + (a+\lambda)^2 + (a + 2\lambda)^2$

$\Rightarrow \quad 5\lambda^2 + 3(2a-1)\lambda + 3a^2 - a = 0 \quad \ldots(i)$

$\because \lambda$ is real.

$\therefore \quad D \ge 0$

$\Rightarrow \quad 9(2a-1)^2 - 4\cdot 5\cdot (3a^2 - a) \ge 0$

$\Rightarrow \quad 24a^2 + 16a - 9 \le 0$

$\Rightarrow \quad -\frac{1}{3} - \frac{\sqrt{70}}{12} \le a \le -\frac{1}{3} + \frac{\sqrt{70}}{12}$

$\because \quad a \in I$

$\therefore \quad a = -1, 0$

If $a = -1$, then from Eq. (i), $\lambda = 1$ $\quad [\because \lambda \in I]$

and for $a = 0$, then from Eq. (i), $\lambda \notin I$

$\therefore \quad a + b + c + d = \frac{4}{2}(a + d) = 2(2a + 3\lambda)$

$= 2(-2 + 3) = 2$

18. $\because a = x, S = 5$

If r be the common ratio, then $\frac{a}{1-r} = 5$

$\Rightarrow \quad \frac{x}{1-r} = 5$

$\Rightarrow \quad 1 - r = \frac{x}{5}$

$\Rightarrow \quad r = \left(1 - \frac{x}{5}\right)$

$\because -1 < r < 1$ (for infinite GP)

$\Rightarrow \quad -1 < 1 - \frac{x}{5} < 1$

$\Rightarrow \quad -2 < -\frac{x}{5} < 0$

$\Rightarrow \quad 10 > x > 0$

$\therefore \quad 0 < x < 10$

19. $\because a_1, a_2, \ldots a_{21}$, are in AP

$\therefore \quad a_1 + a_2 + \ldots + a_{21} = \frac{21}{2}(a_1 + a_{21})$

$\Rightarrow \quad 693 = \frac{21}{2}(a_1 + a_{21})$ $\quad$ [given]

$\therefore \quad a_1 + a_{21} = 66 \quad \ldots(i)$

$\therefore \sum_{r=0}^{10} a_{2r+1} = a_1 + a_3 + a_5 + a_7 + a_9 + \ldots + a_{21}$

$= (a_1 + a_{21}) + (a_3 + a_{19}) + (a_5 + a_{17}) + (a_7 + a_{15}) + (a_9 + a_{13}) + a_{11}$

$= 5 \times (a_1 + a_{21}) + a_{11}$ $\quad [\therefore T_n + T_n{}' = a + l]$

$= 5 \times 66 + a_{11}$

$= 330 + a_{11}$

$= 330 + \left(\frac{a_1 + a_{21}}{2}\right)$ $\quad [\because a_{11}$ is middle term$]$

$= 330 + 33 = 363$

20. Since α and β be the roots of the given quadratic equation, we have

$$\alpha + \beta = \frac{-b}{a} \text{ and } \alpha\beta = \frac{c}{a}$$

Now, $\alpha^2 + \beta^2 = (\alpha+\beta)^2 - 2\alpha\beta = \frac{b^2}{a^2} - \frac{2c}{a}$

$\alpha^3 + \beta^3 = (\alpha+\beta)^3 - 3\alpha\beta(\alpha+\beta) = \frac{-b^3}{a^3} + \frac{3bc}{a^2}$

Suppose that $\alpha + \beta$, $\alpha^3 + \beta^3$ are in GP.

Then, $\quad (\alpha^2 + \beta^2)^2 = (\alpha + \beta)(\alpha^3 + \beta^3)$

$\left(\frac{b^2}{a^2} - \frac{2c}{a}\right)^2 = \frac{-b}{a}\left(-\frac{b^3}{a^3} + \frac{3bc}{a^2}\right)$

$(b^2 - 2ac)^2 = -b(-b^3 + 3abc)$

$4c^2a^2 + b^4 - 4b^2ac = b^4 - 3ab^2c$

$b^2ca - 4c^2a^2 = 0$

$ca(b^2 - 4ac) = 0$

Since, $\quad a \ne 0$, $c\Delta = 0$

21. $\because a_1, a_2, a_3, a_4$ are in HP.

$\therefore \frac{1}{a_1}, \frac{1}{a_2}, \frac{1}{a_3}, \frac{1}{a_4}$ are in AP.

$\Rightarrow \quad \frac{1}{a_2} - \frac{1}{a_1} = \frac{1}{a_3} - \frac{1}{a_2} = \frac{1}{a_4} - \frac{1}{a_3} = \lambda$ (say)

$\Rightarrow \quad \frac{a_1 - a_2}{a_1a_2} = \frac{a_2 - a_3}{a_2a_3} = \frac{a_3 - a_4}{a_3a_4} = \frac{a_1 - a_4}{a_1a_2 + a_2a_3 + a_3a_4}$

or $\quad \frac{a_1 - a_2}{a_1 a_2} = \frac{(a_1 - a_4)}{\sum_{r=1}^{3} a_r\, a_{r+1}}$

$\therefore \quad \frac{1}{a_1 a_4}\sum_{r=1}^{3} a_r\, a_{r+1} = \frac{1}{a_1 a_4} \times \frac{a_1 a_2 (a_1 - a_4)}{(a_1 - a_2)}$

$= \frac{\frac{1}{a_4} - \frac{1}{a_1}}{\frac{1}{a_2} - \frac{1}{a_1}} = \frac{3\lambda}{\lambda} = 3$

Hence, 3 is a root of $x^2 + 2x - 15 = 0$.

22. $b - a = c - b$

$\Rightarrow \quad (\sqrt{b} + \sqrt{a})(\sqrt{b} - \sqrt{a}) - (\sqrt{c} + \sqrt{b})(\sqrt{c} - \sqrt{b})$

$\Rightarrow \quad \frac{\sqrt{b} - \sqrt{a}}{(\sqrt{c} + \sqrt{b})} = \frac{\sqrt{c} - \sqrt{b}}{(\sqrt{b} + \sqrt{a})}$

$\Rightarrow \quad \frac{(\sqrt{b} + \sqrt{c}) - (\sqrt{c} + \sqrt{a})}{(\sqrt{b} + \sqrt{c})(\sqrt{c} + \sqrt{a})} = \frac{(\sqrt{c} + \sqrt{a}) - (\sqrt{a} + \sqrt{b})}{(\sqrt{a} + \sqrt{b})(\sqrt{c} + \sqrt{a})}$

$\Rightarrow \quad \frac{1}{\sqrt{c} + \sqrt{a}} - \frac{1}{\sqrt{b} + \sqrt{c}} = \frac{1}{\sqrt{a} + \sqrt{b}} - \frac{1}{\sqrt{c} + \sqrt{a}}$

$\therefore \frac{1}{\sqrt{b} + \sqrt{c}}, \frac{1}{\sqrt{c} + \sqrt{a}}, \frac{1}{\sqrt{a} + \sqrt{b}}$ are in AP.

23. For the given AP.

$S_n = \frac{n}{2}[a_1 + a_n] = \frac{n}{2}[1 + 300] \quad \ldots(i)$

As, $a_1 = 1$ and $a_n = 300$

and $300 = 1 + (n-1)d$, where d is the common difference of AP.

$\Rightarrow \quad d = \frac{299}{n-1} = \frac{23 \times 13}{n-1}$

$\because d$ is an integer so $n - 1 = 23$, or 13

$\Rightarrow \quad n = 24$ or 14

As, it is given that, $15 \le n \le 50$

So, $n = 24 \Rightarrow d = 13$

$\therefore \quad S_{n-4} = S_{20} = \frac{20}{2}[2 + 19 \times 13] = 10 \times 249 = 2490$

and $\quad a_{n-4} = a_{20} = 1 + (19 \times 13) = 248$

$\therefore$ The ordered pair $(S_{n-4}, a_{n-4}) = (S_{20}, a_{20})$

$= (2490, 248)$

24. $2\log_3(2^x - 5) = \log_3 2 + \log_3(2^x - 7/2)$

$\Rightarrow (2^x - 5)^2 = 2(2^x - 7/2)$

Put $\quad 2^x = t$

$\therefore \quad (t - 5)^2 = 2t - 7$

$\Rightarrow \quad t^2 - 12t + 32 = 0$

$\Rightarrow \quad (t - 8)(t - 4) = 0$

$\therefore \quad t = 8, 4 \quad [\because 2^x > 5]$

$\therefore \quad t > 5$

$\therefore \quad t = 8$

$2^x = 8 = 2^3$

$\Rightarrow \quad x = 3$

25. $S = \sum_{r=1}^{16} \frac{8r}{(4r^4 + 1)}$

$= \sum_{r=1}^{16} \frac{8r}{(2r^2 - 2r + 1)(2r^2 + 2r + 1)}$

$= 2\sum_{r=1}^{16}\left(\frac{1}{2r^2 - 2r + 1} - \frac{1}{2r^2 + 2r + 1}\right)$

$= 2\left\{\frac{1}{1} - \frac{1}{5} + \frac{1}{5} - \frac{1}{13} + \frac{1}{13} + \dots + \frac{1}{481} - \frac{1}{545}\right\}$

$= 2\left(1 - \frac{1}{545}\right) = \frac{1088}{545}$

26. We know, that AM $\ge$ HM

b is the harmonic mean of a and c $\frac{a + c}{2} \ge b$

c is the harmonic mean of b and d $\frac{b + d}{2} \ge c$

Now, $a + c \ge 2b$

$b + d \ge 2c$

Add the two

$a + c + b + d \ge 2(b + c)$

$a + d \ge b + c$

27. Given mid terms $t_n = 1$ and $t_{n+1} = 7$

$\therefore \quad t_n + t_{n+1} = 8 = t_1 + t_{2n}$

and $t_{n+1} - t_n = 6 = d \quad$ [common difference of AP]

$t_n + t_{n+1} = 8$

$\therefore a + (n - 1)d + a + nd = 8$

$\therefore \quad a + 6n = 7$

Now, $4t_1 t_{2n} = [(t_1 + t_{2n})^2 - (t_{2n} - t_1)^2]$

$= 64 - 36(2n - 1)^2 \quad$ [as $t_{2n} - t_1 = (2n - 1) \times 6$]

$\therefore \quad t_1 t_{2n} = 16 - 9(2n - 1)^2$

$\therefore \quad 16 - 9(2n - 1)^2 + 713 \ge 0$

$\therefore \quad -4 \le n \le 5$

$\therefore \quad n = 5$

Hence, from $a + 6n = 7$, $a = -23$

28. $\because a, b, c$ are in AP.

$\therefore \quad b - a = c - b$ or $a - b = b - c \quad$ ...(i)

Given equation is

$(a - b)x^2 + (c - a)x + (b - c) = 0 \quad$ ...(ii)

at $x = 1$,

$(a - b) + (c - a) + (b - c) = 0$

Hence, $x = 1$ is one root of Eq. (ii), if other root is β

Then, $\quad 1 \times \beta = \frac{b - c}{a - b} = 1 \quad$ [from Eq. (i)]

$\therefore \quad \beta = 1$

Hence, roots rational and equal.

29. $ar + ar^5 = \frac{25}{2}$

$ar^2 \times ar^4 = 25$

$a^2 r^6 = 25$

$ar^3 = 5 \Rightarrow a = \frac{5}{r^3}$

$\frac{5r}{r^3} + \frac{5r^5}{r^3} = \frac{25}{2}$

$\frac{1}{r^2} + r^2 = \frac{5}{2}$

Put $\quad r^2 = t$

$\frac{t^2 + 1}{t} = \frac{5}{2}$

$2t^2 - 5t + 2 = 0$

$2t^2 - 4t - t + 2 = 0$

$(2t - 1)(t - 2) = 0$

$t = \frac{1}{2}, 2 \Rightarrow r^2 = \frac{1}{2}, 2$

$r = \sqrt{2} = ar^3 + ar^5 + ar^7$

$= ar^3(1 + r^2 + r^4)$

$= 5[1 + 2 + 4] = 35$

30. Let five numbers, which are in AP is $a - 2d, a - d, a, a + d, a + 2d$.

According to given information,

$5a = 25 \Rightarrow a = 5$

and $\quad a(a^2 - d^2)(a^2 - 4d^2) = 2520$

$\Rightarrow \quad 5(25 - d^2)(25 - 4d^2) = 2520$

$\Rightarrow \quad (d^2 - 25)(4d^2 - 25) = 504$

$\Rightarrow \quad 4d^4 - 125d^2 + 625 = 504$

$\Rightarrow \quad 4d^4 - 125d^2 + 121 = 0$

$\Rightarrow \quad 4d^4 - 4d^2 - 121d^2 + 121 = 0$

$\Rightarrow \quad 4d^2(d^2 - 1) - 121(d^2 - 1) = 0$

$\Rightarrow \quad d^2 = 1, \frac{121}{4} \Rightarrow d = \pm 1, \pm\frac{11}{2}$

If $d = 1$, then terms are 3, 4, 5, 6, 7, and

if $d = -1$, then terms are 7, 6, 5, 4, 3, and

if $d = \frac{11}{2}$, then terms are $-6, -\frac{1}{2}, 5, \frac{21}{2}, 16$.

When $a = 5$ and $d = \pm\frac{11}{2}$, then one of these five numbers of AP is $-\frac{1}{2}$ and the greatest number amongst them is 16.

31. Since, $x, 2x+2, 3x+3, \dots$ are in GP.

$\therefore \quad \frac{2x+2}{x} = \frac{3x+3}{2x+2} = r \ (x \neq -1)$

$\therefore \quad r = 3/2$

and $\quad \frac{2x+2}{x} = \frac{3}{2}$

$\Rightarrow \quad 4x+4 = 3x$

$\therefore \quad x = -4$

Then, next term $= r(3x+3)$

$= \frac{3}{2}(-12+3) = -\frac{27}{2} = -13.5$

32. If $a, ar, ar^2, \dots$ are in GP.

$a = ar + ar^2 \Rightarrow r^2 + r - 1 = 0$

$\therefore \quad r = \frac{-1+\sqrt{5}}{2}$ or $r = \left(\frac{\sqrt{5}-1}{2}\right)$

33. We have, $a + c = 2b \Rightarrow 3b = a+b+c = 3/2 \Rightarrow b = 1/2$

Thus, $a + c = 1$ and $a^2c^2 = b^4 = 1/16$

$\Rightarrow \quad a^2(1-a)^2 = \frac{1}{16} \Rightarrow a(1-a) = \pm 1/4$

$\Rightarrow \quad a^2 - a \pm 1/4 = 0$

$\Rightarrow \quad 4a^2 - 4a \pm 1 = 0$

$\Rightarrow \quad (2a-1)^2 = 0$ or $(2a-1)^2 = 2$

$\Rightarrow \quad a = \frac{1}{2}$ or $a = \frac{1}{2} \pm \frac{1}{\sqrt{2}}$

Since, $a < b < c$ and $b = \frac{1}{2}$

$\therefore \quad a = \frac{1}{2} - \frac{1}{\sqrt{2}}$

34. $\because$ Numbers $a, b, 12$ are in GP.

$\therefore \quad b^2 = 12a \quad \dots$(i)

and $a, b, 9$ are in AP.

$\therefore \quad 2b = a + 9$

or $\quad a = 2b - 9 \quad \dots$(ii)

From Eqs. (i) and (ii),

$b^2 = 12(2b-9)$

$\Rightarrow \quad b^2 - 24b + 108 = 0$

$(b-18)(b-6) = 0$

$\therefore \quad b = 6, 18$

From Eq. (ii), $a = 3, 27$ respectively.

$\therefore$ Common ratio $= \frac{b}{a} = \frac{6}{3}$ and $\frac{18}{27} = 2$ and $\frac{2}{3}$

$\therefore$ Common ratio $= \frac{2}{3}$ (for decreasing GP common ratio $\neq 2$)

35. Given, $\quad S_n = \sum_{r=1}^{n} T_r = n(2n^2 + 9n + 13)$

Now, $\quad S_r = r(2r^2 + 9r + 13)$

and $\quad S_{r-1} = (r-1)[2(r-1)^2 + 9(r-1) + 13]$

$T_r = S_r - S_{r-1}$

$T_r = 6r^2 + 12r + 6 = 6(r+1)^2$

$\sqrt{T_r} = \sqrt{6}(r+1) \Rightarrow \sum_{r=1}^{n} \sqrt{T_r} = \sqrt{6} \sum_{r=1}^{n} (r+1)$

$= \sqrt{6} \sum_{r=1}^{n} r + \sum_{r=1}^{n} 1$

$\Rightarrow \quad \sqrt{6}\left[\frac{n(n+1)}{2} + n\right]$

$\Rightarrow \quad \frac{\sqrt{6}}{2}[n^2 + 3n] \Rightarrow \sqrt{\frac{3}{2}}(n^2 + 3n)$

36. Given, a, b, c are in HP $\Rightarrow b = \frac{2ac}{a+c} \quad \dots$(i)

b, c, d are in GP $\Rightarrow c^2 = bd \quad \dots$(ii)

and c, d, e are in AP $\Rightarrow d = \frac{c+e}{2} \quad \dots$(iii)

From Eq. (i), $ab + bc = 2ac$

$\Rightarrow \quad c = \frac{ab}{2a-b} \quad \dots$(iv)

From Eq. (iii) and (iv),

$d = \frac{1}{2}\left[\frac{ab}{2a-b} + e\right] \quad \dots$(v)

From Eqs. (ii), (iv) and (v),

$\frac{a^2b^2}{(2a-b)^2} = \frac{b}{2}\left[\frac{ab}{2a-b} + e\right]$

$\Rightarrow \quad e = \frac{ab^2}{(2a-b)^2}$

37. $\because P = 0 \cdot cab\,ab\,ab \dots\dots$

$\therefore \quad 10P = c \cdot ababab \quad \dots$(i)

and $\quad 10^2 \times 10P = cab \cdot ababab \quad \dots$(ii)

Subtracting Eq. (i) from Eq. (ii), then

$990P = cab - c$

or $\quad P = \frac{100c + 10a + b - c}{990}$

$= \frac{99c + 10a + b}{990}$

38. $\quad \log_2(a+b) + \log_2(c+d) \geq 4$

$\Rightarrow \quad \log_2\{(a+b)(c+d)\} \geq 4$

$\Rightarrow \quad (a+b)(c+d) \geq 2^4$

But AM $\geq$ GM

$\therefore \quad \frac{(a+b)+(c+d)}{2} \geq \sqrt{(a+b)(c+d)} = 2^2$

$\therefore \quad a + b + c + d \geq 8$

39. $\quad H = 4, G^2 = AH$

$\Rightarrow \quad 2A + G^2 = 27 \Rightarrow 2A + AH = 27$

$\Rightarrow \quad 6A = 27 \Rightarrow A = 9/2, G^2 = 18$

$a + b = 9, ab = 18$

$\therefore a = 3, b = 6$ or $a = 6, b = 3$

40. Let $N = 1\cdot2\cdot3 + 2\cdot3\cdot4 + 3\cdot4\cdot5 + \dots$ upto n terms

and $D = 1\cdot2 + 2\cdot3 + 3\cdot4 + \dots$ upto n terms

$N \Rightarrow \sum_{r=1}^{n} r(r+1)(r+2)$

$= \sum_{r=1}^{n} (r^3 + 3r^2 + 2r)$

$= \sum_{r=1}^{n} r^3 + 3\sum_{r=1}^{n} r^2 + 2\sum_{r=1}^{n} r$

$= \frac{n^2(n+1)^2}{4} + 3 \times \frac{n(n+1)(2n+1)}{6} + 2 \times \frac{n(n+1)}{2}$

$$\Rightarrow n(n+1)\left[\frac{n(n+1)}{4}+\frac{2n+1}{2}+1\right]$$

$$=n(n+1)\left[\frac{n^2+n+4n+2+4}{4}\right]$$

$$=\frac{n(n+1)(n^2+5n+6)}{4}=\frac{n(n+1)(n+2)(n+3)}{4}$$

$$D \Rightarrow \sum_{r=1}^{n} r(r+1)$$

$$=\sum_{r=1}^{n} r^2+\sum_{r=1}^{n} r$$

$$=\frac{n(n+1)(2n+1)}{6}+\frac{n(n+1)}{2}$$

$$=\frac{n(n+1)}{2}\left[\frac{2n+1}{3}+1\right]=\frac{n(n+1)}{2}\left[\frac{2n+4}{3}\right]$$

$$=\frac{n(n+1)(n+2)}{3}$$

$$A_n=\frac{N}{n\times D}$$

$$=\frac{\dfrac{n(n+1)(n+2)(n+3)}{4}}{n\times\dfrac{n(n+1)(n+2)}{3}}$$

$$\Rightarrow \frac{3(n+3)}{4n}\Rightarrow\frac{3}{4}\left[1+\frac{3}{n}\right]$$

$$\lim_{n\to\infty}A_n=\lim_{n\to\infty}\frac{3}{4}\left[1+\frac{3}{n}\right]=\frac{3}{4}$$

41. $\because$ Last term of Ist group is 1 *ie*, 1^2

Last term of 2nd group is 4 *ie*, 2^2

Last term of 3rd group is 9 *ie*, 3^2

....

....

$\therefore$ Last term of $(n-1)$ th group is $(n-1)^2$

$\therefore$ First term of nth group is $(n-1)^2+1$

42. $$\frac{3}{1^2\cdot2^2}+\frac{7}{1^3\cdot2^3}+\frac{5}{2^2\cdot3^2}+\frac{19}{2^3\cdot3^3}+\frac{7}{3^2\cdot4^2}+\frac{37}{3^3\cdot4^3}+\dots\text{ 20 terms}$$

$$=\left(\frac{3}{1^2\cdot2^2}+\frac{5}{2^2\cdot3^2}+\frac{7}{3^2\cdot4^2}\dots\text{10 terms}\right)+\left(\frac{7}{1^3\cdot2^3}+\frac{19}{2^3\cdot3^3}+\frac{37}{3^3\cdot4^3}\dots\text{10 terms}\right)$$

$$=\left(\frac{2^2-1^2}{1^2\cdot2^2}+\frac{3^2-2^2}{2^2\cdot3^2}+\frac{4^2-3^2}{3^2\cdot4^2}\dots\text{10 terms}\right)+\left(\frac{2^3-1^3}{1^3\cdot2^3}+\frac{3^3-2^3}{2^3\cdot3^3}+\frac{4^3-3^3}{3^2\cdot4^3}\dots\text{10 terms}\right)$$

$$=\left(\frac{1}{1^2}-\frac{1}{2^2}+\frac{1}{2^2}-\frac{1}{3^2}\dots+\frac{1}{10^2}+\frac{1}{11^2}\right)+\left(\frac{1}{1^3}-\frac{1}{2^3}+\frac{1}{2^2}-\frac{1}{3^3}\dots+\frac{1}{10^3}-\frac{1}{11^3}\right)$$

$$=1-\frac{1}{11^2}+1-\frac{1}{11^3}=2-\frac{1}{121}-\frac{1}{1331}=\frac{2650}{1331}$$

43. $$\sum_{r=1}^{n} r\cdot(n-r+1)=\sum_{r=1}^{n}(n+1)r-\sum_{r=1}^{n}r^2$$

$$=(n+1)\Sigma n-\Sigma n^2$$

$$=\frac{n(n+1)^2}{2}-\frac{n(n+1)(2n+1)}{6}$$

$$=\frac{n(n+1)}{6}\{3n+3-2n-1\}$$

$$=\frac{n(n+1)(n+2)}{6}$$

44. $\because$ p, q, r are in AP.

$\therefore$ $2q=p+r$ ...(i)

$\because$ Roots of $px^2+qx+r=0$ are all real, then

$$q^2-4pr\geq0$$

$$\Rightarrow \left(\frac{p+r}{2}\right)^2-4pr\geq0 \quad \text{[from Eq. (i)]}$$

$$\Rightarrow (p+r)^2-16pr\geq0$$

$$\Rightarrow p^2+r^2-14pr\geq0$$

$$\Rightarrow \left(\frac{r}{p}\right)^2-14\left(\frac{r}{p}\right)+1\geq0$$

$$\left(\frac{r}{p}-7\right)^2\geq48\Rightarrow\left|\frac{r}{p}-7\right|\geq4\sqrt{3}$$

45. $$\frac{a+bx}{a-bx}=\frac{b+cx}{b-cx}=\frac{c+dx}{c-dx}$$

$$\Rightarrow \frac{\frac{a}{b}+x}{\frac{a}{b}-x}=\frac{\frac{b}{c}+x}{\frac{b}{c}-x}=\frac{\frac{c}{d}+x}{\frac{c}{d}-x}$$

From first two relations

$$\frac{2\frac{a}{b}}{2x}=\frac{2\frac{b}{c}}{2x}$$

[by componendo and dividendo method]

$\therefore$ $\frac{a}{b}=\frac{b}{c}$ ...(i)

and from last two relations

$$\frac{2\frac{b}{c}}{2x}=\frac{2\frac{c}{d}}{2x}$$

$\therefore$ $\frac{b}{c}=\frac{c}{d}$ ...(ii)

From Eqs. (i) and (ii), $\frac{a}{b}=\frac{b}{c}=\frac{c}{d}$

Hence, a, b, c, d are in GP.

46. $S_{2n}=\frac{2n}{2}[2a+(2n-1)d]$, $S_{4n}=\frac{4n}{2}[2a+(4n-1)d]$

$$\Rightarrow S_2-S_1=\frac{4n}{2}[2a+(4n-1)d]-\frac{2n}{2}[2a+(2n-1)d]$$

$$=4an+(4n-1)2nd-2na-(2n-1)dn$$

$$=2na+nd[8n-2-2n+1]$$

$$\Rightarrow 2na+nd[6n-1]=1000$$

$$2a+(6n-1)d=\frac{1000}{n}$$

Now, $S_{6n}=\frac{6n}{2}[2a+(6n-1)d]=3n\cdot\frac{1000}{n}=3000$

47. $\because$ $(1-1+2-2+3-3+4-4+5-5)^2$

$$=1^2+1^2+2^2+2^2+3^2+3^2+\dots+5^2+2P$$

Where P is sum of the products taking two at a time

$\therefore$ $2P=-2(1^2+2^2+3^2+4^2+5^2)$

$\therefore$ $P=-55$

48. $a, A_1, A_2, \dots, A_n, 2b$

$$2b=(n+2)\text{ th term}=a+(n+2-1)d$$

$$d=\frac{2b-a}{n+1}$$

$$A_m=a+m\left(\frac{2b-a}{n+1}\right)$$

Again $2a, B_1, B_2, \dots, B_n, b$

$b=(n+2)$th term $=2a+(n+2-1)D$

$$\therefore \quad D=\frac{b-2a}{n+1}$$

$$\therefore\ B_m=2a+mD=2a+m\left(\frac{b-2a}{n+1}\right) \Rightarrow A_m=B_m$$

$$\Rightarrow a+m\left(\frac{2b-a}{n+1}\right)=2a+m\left(\frac{b-2a}{n+1}\right)$$

$\Rightarrow$ $m(a+b)=a(n+1)$ or $a(n-m+1)=bm$

$\therefore$ $a:b=m:n-m+1$

49. Let $S=1+\frac{4}{5}+\frac{7}{5^2}+\frac{10}{5^3}+\dots+\frac{(3n-2)}{5^{n-1}}$

$$\frac{1}{5}S=\frac{1}{5}+\frac{4}{5^2}+\frac{7}{5^3}+\dots+\frac{(3n-5)}{5^{n-1}}=\frac{(3n-2)}{5^n}$$

Now, subtract the two

$$\frac{4}{5}S=1+\left[\frac{3}{5}+\frac{3}{5^2}+\dots\dots(n-1)\text{ terms}\right]-\frac{(3n-2)}{5^n}$$

$$\frac{4}{5}S=1+\frac{\left(\frac{3}{5}\right)\left[1-\left(\frac{1}{5}\right)^{n-1}\right]}{\left[1-\frac{1}{5}\right]}-\frac{(3n-2)}{5^n}$$

$$=1+\frac{3}{4}\left[1-\frac{1}{5^{n-1}}\right]-\frac{(3n-2)}{5^n}$$

$$S=\frac{5}{4}+\frac{15}{16}\left[1-\frac{1}{5^{n-1}}\right]-\frac{(3n-2)}{4\cdot 5^{n-1}}$$

50. $\because$ $\frac{a+b}{1-ab}, b, \frac{b+c}{1-bc}$ are in AP.

$$\therefore \quad b-\frac{a+b}{1-ab}=\frac{b+c}{1-bc}-b$$

$$\Rightarrow \quad \frac{-a-ab^2}{1-ab}=\frac{c+b^2c}{1-bc}$$

$\Rightarrow$ $(-a-ab^2)(1-bc)=(1-ab)(c+b^2c)$

$\Rightarrow$ $(1+b^2)(a+c-2abc)=0$

$\because$ $1+b^2\neq 0$

$\therefore$ $a+c-2abc=0$

or $\frac{1}{b}=\frac{2ac}{a+c}$

$\therefore a, \frac{1}{b}, c$ are in HP.

51. $\because$ $f(k)=\sum_{r=1}^{n} a_r-a_k$

$$\sum_{r=1}^{n} a_r=a_k+f(k)$$

$\Rightarrow a_1+f(1)=a_2+f(2)=a_3+f(3)=\dots=a_n+f(n)$

Given, $a_1, a_2, a_3, \dots, a_n$ are in HP.

$\therefore \frac{1}{a_1}, \frac{1}{a_2}, \frac{1}{a_3}, \dots, \frac{1}{a_n}$ are in AP.

$\Rightarrow \frac{a_1+f(1)}{a_1}, \frac{a_2+f(2)}{a_2}, \frac{a_3+f(3)}{a_3}, \dots, \frac{a_n+f(n)}{a_n}$ are also in AP.

Subtracting 1 in each term, then

$\frac{f(1)}{a_1}, \frac{f(2)}{a_2}, \frac{f(3)}{a_3}, \dots, \frac{f(n)}{a_n}$ are also in AP.

$\therefore \frac{a_1}{f(1)}, \frac{a_2}{f(2)}, \frac{a_3}{f(3)}, \dots, \frac{a_n}{f(n)}$ are in HP.

52. $\because$ $a_1+5d=2$

$\therefore$ $P=a_1a_4a_5$

$$=a_1(a_1+3d)(a_1+4d)$$

$$=(2-5d)(2-2d)(2-d)$$

$$=2\{4-16d+17d^2-5d^3\}$$

$\therefore$ $P'=2\{-16+34d-15d^2]$

For max or min $P'=0$

$\therefore$ $d=2/3$ and $8/5$

For least $P''=+\text{ve}$

$\therefore$ $d=2/3$

53. $\log_3 x\{2+4+6+\dots+16\}=36$

$$\log_3 x\left\{\frac{8}{2}(2+16)\right\}=36 \Rightarrow \log_3 x=\frac{1}{2}$$

or $x=3^{1/2}=\sqrt{3}$

54. $\because a_1, a_2, a_3, \dots, a_n$ are in HP

$\therefore \frac{1}{a_1}, \frac{1}{a_2}, \frac{1}{a_3}, \dots, \frac{1}{a_n}$ are in AP.

$\therefore \frac{a_1+a_2+a_3+\dots+a_n}{a_1}, \frac{a_1+a_2+a_3+\dots+a_n}{a_2}, \dots, \frac{a_1+a_2+\dots+a_n}{a_n}$ are also in AP

$\Rightarrow 1+\frac{a_2+a_3+\dots+a_n}{a_1}, 1+\frac{a_1+a_3+\dots+a_n}{a_2}, \dots, 1+\frac{a_1+a_2+\dots+a_{n-1}}{a_n}$ are in AP.

Subtracting 1 from each term, then

$\frac{a_2+a_3+\dots+a_n}{a_1}, \frac{a_1+a_3+\dots+a_n}{a_2}, \dots, \frac{a_1+a_2+\dots+a_{n-1}}{a_n}$ are in AP.

Hence, $\frac{a_1}{a_2+a_3+\dots+a_n}, \frac{a_2}{a_1+a_3+\dots+a_n}, \dots \frac{a_n}{a_1+a_2+\dots+a_{n-1}}$

55. Let $S_n = Pn^2 + Qn$ = Sum of first n terms

According to question,

Sum of first $3n$ terms = sum of the next n terms

$$\Rightarrow \quad S_{3n} = S_{4n} - S_{3n}$$

or $$2S_{3n} = S_{4n}$$

or $$2[P(3n)^2 + Q(3n)] = P(4n)^2 + Q(4n)$$

$$\Rightarrow \quad 2Pn^2 + 2Qn = 0$$

or $$Q = -nP \qquad \text{...(i)}$$

Then, $$\frac{\text{Sum of first } 2n \text{ terms}}{\text{Sum of next } 2n \text{ terms}} = \frac{S_{2n}}{S_{4n} - S_{2n}}$$

$$= \frac{P(2n)^2 + Q(2n)}{[P(4n)^2 + Q(4n)] - [P(2n)^2 + Q(2n)]}$$

$$= \frac{2nP + Q}{6Pn + Q} = \frac{nP}{5nP} = \frac{1}{5} \qquad \text{[from Eq. (i)]}$$

56. If in AP,

$$\sqrt{b} = \sqrt{a} + (m-1)d$$

and $$\sqrt{c} = \sqrt{a} + (n-1)d$$

$\therefore \dfrac{\sqrt{b} - \sqrt{a}}{\sqrt{c} - \sqrt{a}} = \dfrac{m-1}{n-1}$ = a rational number but a, b, c are prime, then $\dfrac{\sqrt{b} - \sqrt{a}}{\sqrt{c} - \sqrt{a}}$ is irrational.

$\therefore$ irrational $\neq$ rational

So, $\sqrt{a}, \sqrt{b}, \sqrt{c}$ are not in AP similarly they are not in GP or HP.

57. As n is odd, the value of the given expression

$$= 1^3 - 2^3 + 3^3 - \ldots + n^3$$

$$= (1^3 + 2^3 + \ldots + n^3) - 2\{2^3 + 4^3 + \ldots + (n-1)^3\}$$

$$= \left[\frac{n(n+1)}{2}\right]^2 - 16\left\{1^3 + 2^3 + \ldots + \left(\frac{n-1}{2}\right)^3\right\}$$

$$= \left[\frac{n(n+1)}{2}\right]^2 - 16\left\{\frac{\left(\frac{n-1}{2}\right)\left(\frac{n+1}{2}\right)}{2}\right\}^2$$

$$\frac{n^2(n+1)^2}{4} - \frac{(n-1)^2(n+1)^2}{4} = \frac{(n+1)^2(2n-1)}{4}$$

58. Let sides be $a-d, a, a+d$

$$\therefore \quad (a+d)^2 = (a-d)^2 + a^2$$

$$\Rightarrow \quad 4ad = a^2$$

$$\therefore \quad a = 4d$$

Then sides are $3d$, $4d$ and $5d$

$$\sin A = \frac{a-d}{a+d} = \frac{3d}{5d} = \frac{3}{5}$$

and $$\sin C = \frac{a}{a+d} = \frac{4d}{5d} = \frac{4}{5}$$

59. $\because (x-1)(x-2)(x-3)\ldots(x-n)$

$$= x^n - x^{n-1}\Sigma n + x^{n-2}\Sigma 1 \cdot 2 - \ldots$$

$\therefore$ coefficient of $x^{n-2} = \Sigma 1 \cdot 2$

$$= 1 \cdot 2 + 1 \cdot 3 + \ldots + 1 \cdot n + 2 \cdot 3 + 2 \cdot 4$$

$$= \ldots + 2 \cdot n + \ldots + (n-1)n = S \text{ (say)} \qquad \text{...(i)}$$

$$\because (1 + 2 + 3 + 4 + \ldots + n)^2 = 1^2 + 2^2 + 3^2 + \ldots + n^2 + 2S$$

$$\therefore \quad S = \frac{(\Sigma n)^2 - \Sigma n^2}{2}$$

$$= \frac{\left\{\frac{n(n+1)}{2}\right\}^2 - \frac{n(n+1)(2n+1)}{6}}{2}$$

$$= \frac{n(n+1)}{4}\left[\frac{n(n+1)}{2} - \frac{2n+1}{3}\right]$$

$$= \frac{n(n+1)}{24}(3n^2 - n - 2) = \frac{n(n^2-1)(3n+2)}{24}$$

60. $\because \{a_n\}$ be a GP.

$\therefore \quad a_1, a_2, a_3, a_4, \ldots, a_n$ are in GP

$$\because \quad \frac{a_4}{a_6} = \frac{1}{4} \Rightarrow \frac{a_1 r^3}{a_1 r^5} = \frac{1}{4} \Rightarrow r^2 = 4$$

$$\therefore \quad r = \pm 2$$

and $$a_2 + a_5 = 216$$

$$a_1 r + a_1 r^4 = 216 \Rightarrow a_1(r + r^4) = 216$$

For $$r = 2, \; a_1(2 + 16) = 216$$

$$\therefore \quad a_1 = 12$$

and for $$r = -2, \; a_1(-2 + 16) = 216$$

$$\therefore \quad a_1 = \frac{216}{14} = \frac{108}{7}$$

61. It is given that a_n is the nth term of a GP of positive terms, such that

$$\sum_{n=1}^{100} a_{2n+1} = 200$$

$$\Rightarrow \quad a_3 + a_5 + a_7 + \ldots + a_{201} = 200$$

$$\Rightarrow \quad ar^2 + ar^4 + ar^6 + \ldots + ar^{200} = 200$$

[where a is the first term of GP and r is the common ratio]

$$\Rightarrow \quad ar^2(1 + r^2 + r^4 + \ldots + r^{198}) = 200$$

$$\Rightarrow \quad ar^2\left(\frac{1 - r^{200}}{1 - r^2}\right) = 200 \qquad \text{...(i)}$$

and $$\sum_{n=1}^{100} a_{2n} = 100 \qquad \text{(given)}$$

$$\Rightarrow \quad a_2 + a_4 + a_6 + \ldots + a_{200} = 100$$

$$\Rightarrow \quad ar + ar^3 + ar^5 + \ldots + ar^{199} = 100$$

$$\Rightarrow \quad ar\left(\frac{1 - r^{200}}{1 - r^2}\right) = 100 \qquad \text{...(ii)}$$

From Eqs. (i) and (ii), we get

$$r = 2 \text{ and } a = \frac{150}{2^{200} - 1}$$

$$\therefore \quad \sum_{n=1}^{200} a_n = \frac{a(r^{200} - 1)}{r - 1} = \frac{150}{2^{200} - 1} \times \frac{(2^{200} - 1)}{2 - 1} = 150$$

Hence, option (d) is correct.

62. $S=1+2\left(1-\frac{1}{n}\right)+3\left(1-\frac{1}{n}\right)^2+\ldots\ldots\infty$

$$\left(1-\frac{1}{n}\right)S=1\left(1-\frac{1}{n}\right)+2\left(1-\frac{1}{n}\right)^2+\ldots\ldots\infty$$

then, $S\left(1-1+\frac{1}{n}\right)=1+\left(1-\frac{1}{n}\right)+\left(1-\frac{1}{n}\right)^2+\ldots\infty$

$$=\frac{1}{1-\left(1-\frac{1}{n}\right)}$$

$$\frac{S}{n}=n$$

$\therefore$ $S=n^2$

63. $(a_1a_3-a_2a_2)^2+(a_2a_4-a_3a_3)^2+\ldots+(a_{n-2}a_n-a_{n-1}a_{n-1})^2\leq 0$ [by lagrange's identity]

It is possible only, if

$a_1a_3-a_2a_2=a_2a_4-a_3a_3=\ldots=a_{n-2}a_n-a_{n-1}\,a_{n-1}=0$

$\Rightarrow$ $\frac{a_1}{a_2}=\frac{a_2}{a_3}=\frac{a_3}{a_4}=\ldots=\frac{a_n}{a_{n-1}}$

Hence, $a_1, a_2, a_3, \ldots, a_n$ are in GP.

64. Last term of 1st group is 1^3

Last term of 2nd group is 3^3 i.e. $(1+2)^3$.

Last term of 3rd group is 6^3 i.e. $(1+2+3)^3$.

....

Last term of $(n-1)$ th group is

$$\{1+2+3+\ldots+(n-1)\}^3 \text{ i.e. } \left\{\frac{(n-1)n}{2}\right\}^3$$

$\therefore$ First term of the nth group is $\left\{1+\frac{(n-1)^n}{2}\right\}^3$

or $\left(\frac{n^2-n+2}{2}\right)^3$

and Last term of nth group is $(1+2+3+\ldots+n)^3$

i.e. $\left\{\frac{n\,(n+1)}{2}\right\}^3$

$\therefore$ Sum of the numbers in the nth group.

$$=\Sigma\left\{\frac{n\,(n+1)}{2}\right\}^3-\Sigma\left\{\frac{(n-1)n}{2}\right\}^3$$

$$=\left[\frac{\frac{n\,(n+1)}{2}\cdot\left\{\frac{n\,(n+1)}{2}+1\right\}}{2}\right]^2-\left[\frac{\frac{(n-1)\,n}{2}\left\{\frac{(n-1)n}{2}+1\right\}}{2}\right]^2$$

$$=\frac{1}{8}n^3\,(n^2+1)\,(n^2+3)$$

65. $x=\frac{1}{1-a}, y=\frac{1}{1-b}, z=\frac{1}{1-c}$

or $a=1-\frac{1}{x}, b=1-\frac{1}{y}, c=1-\frac{1}{z}$

$\because$ a, b, c are in AP

$1-\frac{1}{x}, 1-\frac{1}{y}, 1-\frac{1}{z}$ are in AP

$\Rightarrow$ $-\frac{1}{x}, -\frac{1}{y}, -\frac{1}{z}$ are in AP

or $\frac{1}{x}, \frac{1}{y}, \frac{1}{z}$ are in AP

$\therefore x, y, z$ are in HP.

66. $a_n=x^{1/2^n}+y^{1/2^n}$ and $b_n=x^{1/2^n}-y^{1/2^n}$

Now, $a_nb_n=(x^{1/2^n}+x^{1/2^n})(x^{1/2^n}-x^{1/2^n})$

$\Rightarrow$ $a_nb_n=(x^{1/2^n})^2-(x^{1/2^n})^2$

$\Rightarrow$ $a_nb_n=x^{1/2^{n-1}}-x^{1/2^{n-1}}=b_{n-1}$ ... (i)

Now, $a_1a_2a_3\ldots a_n=\left(\frac{a_1a_2a_3\ldots a_n}{b_n}\right)b_n$

$$=\frac{(a_1a_2a_3\ldots a_{n-1})(a_nb_n)}{b_n}$$

$$=\frac{a_1a_2a_3\ldots a_{n-1}b_{n-1}}{b_n} \quad \text{[using Eq. (i)]}$$

$$=\frac{(a_1a_2a_3\ldots a_{n-2})(a_{n-1}b_{n-1})}{b_n}$$

$$=\frac{a_1a_2a_3\ldots a_{n-2}b_{n-2}}{b_n}$$

... ...

... ...

$$=\frac{a_1b_1}{b_n}=\frac{b_0}{b_n}=\frac{x-y}{b_n}$$

67. Given, $|a_i|=|a_{i-1}+1|$

Squaring $a_i^2=a_{i-1}^2+2a_{i-1}+1$

We have, $a_1^2=0$

$$a_2^2=a_1^2+2a_1+1$$
$$a_3^2=a_2^2+2a_2+1$$

– – – – –

– – – – –

– – – – –

$$a_n^2=a_{n-1}^2+2a_{n-1}+1$$
$$a_{n+1}^2=a_n^2+2a_n+1$$

Adding, we get

$$\sum_{i=1}^{n+1}a_i^2=\sum_{i=1}^{n}a_i^2+2\sum_{i=1}^{n}a_i+n$$

$$2\sum_{i=1}^{n}a_i=-n+a_{n+1}^2\geq -n \Rightarrow \sum_{i=1}^{n}a_i\geq -\frac{n}{2}$$

$\Rightarrow$ $\frac{a_1+a_2+\ldots+a_n}{n}\geq -\frac{1}{2}$

$\Rightarrow$ $x\geq -1/2$

68. $\because a_1, a_2, a_3$ are in GP.

$\therefore \quad a_2 = a_1 r,\ a_3 = a_1 r^2$

$\because \quad 9a_1 + 5a_3 > 14a_2$

$\Rightarrow \quad 9a_1 + 5a_1 r^2 > 14a_1 r$

$\therefore \quad 9 + 5r^2 > 14r \qquad [\because a_1 \neq 0]$

$\Rightarrow \quad 5r^2 - 14r + 9 > 0$

$\Rightarrow \quad 5r^2 - 9r - 5r + 9 > 0$

$(5r - 9)(r - 1) > 0$

$\therefore \quad r \in (-\infty, 1) \cup \left(\frac{9}{5}, \infty\right)$

$\therefore \quad r \notin \left[1, \frac{9}{5}\right]$

69. $\sum_{n=1}^{\infty} \frac{n^2 + 6n + 10}{(2n+1)!}$

Put $2n + 1 = r$, where $r = 3, 5, 7, \ldots$

$\Rightarrow \quad n = \frac{r-1}{2}$

$$\frac{n^2 + 6n + 10}{(2n+1)!} = \frac{\left(\frac{r-1}{2}\right)^2 + 3r - 3 + 10}{r!} = \frac{r^2 + 10r + 29}{4r!}$$

Now, $\sum_{r=3,5,7}^{\infty} \frac{r(r-1) + 11r + 29}{4r!}$

$$= \frac{1}{4}\sum_{r=3,5,7,\ldots}\left(\frac{1}{(r-2)!} + \frac{11}{(r-1)!} + \frac{29}{r!}\right)$$

$$= \frac{1}{4}\left\{\left(\frac{1}{1!} + \frac{1}{3!} + \frac{1}{5!} + \ldots\right) + 11\left(\frac{1}{2!} + \frac{1}{4!} + \frac{1}{6!} + \ldots\right) + 29\left(\frac{1}{3!} + \frac{1}{5!} + \frac{1}{7!} + \ldots\right)\right\}$$

$$= \frac{1}{4}\left\{\frac{e - \frac{1}{e}}{2} + 11\left(\frac{e + \frac{1}{e} - 2}{2}\right) + 29\left(\frac{e - \frac{1}{e} - 2}{2}\right)\right\}$$

$$= \frac{1}{8}\left\{e - \frac{1}{e} + 11e + \frac{11}{e} - 22 + 29e - \frac{29}{e} - 58\right\}$$

$$= \frac{1}{8}\left\{41e - \frac{19}{e} - 80\right\}$$

70. $\therefore$ nth term $T_n = \frac{1 + 2 + 3 + \ldots + n}{1^3 + 2^3 + 3^3 + \ldots + n^3}$

$$= \frac{\frac{n(n+1)}{2}}{\left[\frac{n(n+1)}{2}\right]^2} = \frac{2}{n(n+1)} = 2\left(\frac{1}{n} - \frac{1}{n+1}\right)$$

Putting $n = 1, 2, 3, 4, \ldots, n$

$\therefore T_1 + T_2 + T_3 + \ldots + T_n$

$$= 2\left(1 - \frac{1}{2} + \frac{1}{2} - \frac{1}{3} + \frac{1}{3} - \frac{1}{4} + \ldots + \frac{1}{n} - \frac{1}{n+1}\right)$$

$$= 2\left(1 - \frac{1}{n+1}\right)$$

$$S_n = 2\left(1 - \frac{1}{n+1}\right)$$

$\therefore S_n + 199 = \frac{2n}{n+1} + 199 > 199$ for all n.

71. Since, $3^{2\sin 2x - 1}, 14, 3^{4 - 2\sin 2x}$ are in AP.

$\therefore \quad 28 = 3^{2\sin 2x} \cdot \frac{1}{3} + 3^4 \cdot \frac{1}{3^{2\sin 2x}}$

Put $3^{2\sin 2x} = t$

$\therefore \quad 28 = \frac{t}{3} + \frac{81}{t}$

$\Rightarrow \quad 84t = t^2 + 243$

$\Rightarrow \quad t^2 - 84t + 243 = 0$

$(t - 81)(t - 3) = 0$

$\therefore \quad t = 3, 81$

$3^{2\sin 2x} = 3^1, 3^4$

$\therefore \quad 2\sin 2x = 1, 4$

$\sin 2x = \frac{1}{2}, 2 \qquad [\because \sin 2x \neq 2]$

$\therefore \quad \sin 2x = \frac{1}{2}$

$\therefore$ First term $= 3^{1-1} = 1$

Second term $= 14$

Third term $= 27$

Here, common difference $= 13$

$\therefore$ Fifth term $= 1 + 4 \times 13 = 53$.

72. $(4n + 1)$ terms are

$a, a + 2, a + 4, a + 6, \ldots, a + 4n, \frac{1}{2}(a + 4n), \frac{1}{4}(a + 4n), \ldots,$

Middle term of AP $= (n + 1)$ th term

$= a + n(2) = a + 2n$

and middle term GP $= (a + 4n)\left(\frac{1}{2}\right)^{2n+1-1}$

$$= \frac{(a + 4n)}{2^{2n}}$$

According to the question,

$$a + 2n = \frac{(a + 4n)}{2^{2n}}$$

$a(2^{2n} - 1) + 4n \cdot 2^{2n-1} - 4n = 0$

$a(2^{2n} - 1) + 4n(2^{2n-1} - 1) = 0$

$\therefore$ Middle term of the sequence

$= a + 4n$

$$= -\frac{4n(2^{2n-1} - 1)}{(2^{2n} - 1)} + 4n$$

$$= 4n\left(1 - \frac{2^{2n-1} - 1}{2^{2n} - 1}\right)$$

$$= 4n\left(\frac{2^{2n} - 2^{2n-1}}{2^{2n} - 1}\right)$$

$$= \frac{4n \cdot 2^{2n-1}(2 - 1)}{(2^{2n} - 1)}$$

$$= \frac{2n \cdot 2^{2n}}{2^{2n} - 1} = \frac{n \cdot 2^{2n+1}}{2^{2n} - 1}.$$

73. $a+b=a_1+a_{2n}=a_2+a_{2n-1}=\ldots=a_n+a_{n+1}$

and $ab=g_1\cdot g_{2n}=g_2\cdot g_{2n-1}=\ldots=g_n\cdot g_{n+1}$ and $h=\dfrac{2ab}{a+b}$.

$$\therefore \frac{a_1+a_{2n}}{g_1g_{2n}}+\frac{a_2+a_{2n-1}}{g_2g_{2n-1}}+\ldots+\frac{a_n+a_{n+1}}{g_ng_{n+1}}$$
$$=n\left(\frac{a+b}{ab}\right)=2n\left(\frac{a+b}{2ab}\right)=\frac{2n}{h}$$

74. If $\log b_1, \log b_2, \ldots, \log b_{101}$ are in AP, with common difference $\log_e 2$, then $b_1, b_2, \ldots, b_{101}$ are in GP, with common ratio 2.

$\therefore\ b_1=2^0b_1,\ b_2=2^1b_1,\ b_3=2^2b_1,\ldots,\ b_{101}=2^{100}b_1$ ...(i)

Also, $a_1, a_2, \ldots, a_{101}$ are in AP.

Given, $a_1=b_1$ and $a_{51}=b_{51}$

$\Rightarrow\ a_1+50D=2^{50}b_1$

$\Rightarrow\ a_1+50D=2^{50}a_1$

$[\because a_1=b_1]$ ...(ii)

Now, $t=b_1+b_2+\ldots+b_{51}$

$\Rightarrow\ t=b_1\dfrac{(2^{51}-1)}{2-1}$ ...(iii)

and $s=a_1+a_2+\ldots+a_{51}$

$=\dfrac{51}{2}(2a_1+50D)$ ...(iv)

$\therefore\ t=a_1(2^{51}-1)$ $[\because a_1=b_1]$

or $t=2^{51}a_1-a_1<2^{51}a_1$ ...(v)

and $s=\dfrac{51}{2}[a_1+(a_1+50D)]$ [from Eq. (ii)]

$=\dfrac{51}{2}[a_1+2^{50}a_1]$

$=\dfrac{51}{2}a_1+\dfrac{51}{2}2^{50}a_1$

$\therefore\ s>2^{51}a_1$ ...(vi)

From Eqs. (v) and (vi), we get

$s>t$

Also, $a_{101}=a_1+100D$ and $b_{101}=2^{100}b_1$

$\therefore\ a_{101}=a_1+100\left(\dfrac{2^{50}a_1-a_1}{50}\right)$ and $b_{101}=2^{100}a_1$

$\Rightarrow\ a_{101}=a_1+2^{51}a_1-2a_1=2^{51}a_1-a_1$

$\Rightarrow\ a_{101}<2^{51}a_1$ and $b_{101}>2^{51}a_1$

$\Rightarrow\ b_{101}>a_{101}$

75. AM and GM of 7 numbers $\dfrac{a}{2},\dfrac{a}{2},\dfrac{b}{3},\dfrac{b}{3},\dfrac{b}{3},\dfrac{c}{2},\dfrac{c}{2}$

is $$\frac{2\cdot\frac{a}{2}+3\cdot\frac{b}{3}+2\cdot\frac{c}{2}}{7}\geq\left\{\left(\frac{a}{2}\right)^2\left(\frac{b}{3}\right)^3\left(\frac{c}{2}\right)^2\right\}^{\frac{1}{7}}$$

$$\Rightarrow\ \frac{3}{7}\geq\left(\frac{a^2b^3c^2}{2^23^32^2}\right)^{\frac{1}{7}}$$

$$\Rightarrow\ \frac{3^7}{7^7}\geq\frac{a^2b^3c^2}{2^2\cdot3^3\cdot2^2}$$

$$\Rightarrow\ a^2b^3c^2\leq\frac{3^{10}\cdot2^4}{7^7}$$

$\therefore$ Greatest value of $a^2b^3c^2=\dfrac{3^{10}\cdot2^4}{7^7}$

76. $\because \ln(a+c), \ln(c-a), \ln(a-2b+c)$ (are in AP).

$\therefore\ a+c,\ c-a,\ a-2b+c$ (are in GP)

$(c-a)^2=(a+c)(a-2b+c)$

$\Rightarrow\ 2b(a+c)=(a+c)^2-(c-a)^2=(2c)(2a)$

$\therefore\ b=\dfrac{2ac}{a+c}$

$\Rightarrow a, b, c$ are in HP.

77. $\because a_2-a_1=a_3-a_2=\ldots=$ common difference $\neq$ multiple of $3=d$ say, then consecutive terms

$a_1,\ a_1+d,\ a_1+2d,\ a_1+3d,\ldots$

$\because d$ is not multiple of 3.

Let $d=3\lambda+1$ or $3\lambda+2$

$\therefore$ Consecutive term

$a_1,\ a_1+1+3\lambda,\ a_1+2+6\lambda,\ a_1+3+9\lambda,\ldots$

or $a_1,\ a_1+2+3\lambda,\ a_1+4+6\lambda,\ a_1+6+9\lambda,\ldots$

are all prime

Hence, number of terms is infinite.

78. $(1+2+2^2+2^3+\ldots+2^{n-1})^2$

$=(1^2+2^2+2^4+2^6+\ldots+2^{2n-2})+2P$

[where P is sum of products taken two at a time]

$$\therefore\ P=\frac{(2^n-1)^2-\dfrac{1^2\cdot(2^{2n}-1)}{(2^2-1)}}{2}=\frac{1}{3}\cdot2^{2n}-2^n+\frac{2}{3}$$

79. $$T_n=\frac{5+(n-1)4}{[3+(n-1)4]^2[7+(n-1)4]^2}$$

$$=\frac{1}{8}\left\{\frac{1}{(4n-1)^2}-\frac{1}{(4n+3)^2}\right\}$$

$\therefore\ S_n=T_1+T_2+T_3+\ldots+T_n$

$$=\frac{1}{8}\left\{\frac{1}{3^2}-\frac{1}{7^2}+\frac{1}{7^2}-\frac{1}{11^2}+\frac{1}{11^2}-\frac{1}{15^2}+\ldots\ldots+\ldots+\frac{1}{(4n-1)^2}-\frac{1}{(4n+3)^2}\right\}=\frac{1}{8}\left[\frac{1}{3^2}-\frac{1}{(4n+3)^2}\right]$$

$$\therefore\ S_\infty=\frac{1}{8}\left\{\frac{1}{9}-0\right\}=\frac{1}{72}.$$

80. $bx^2+(\sqrt{(a+c)^2+4b^2})x+(a+c)\geq0$

$\therefore\ B^2-4AC\leq0$

$\Rightarrow\ (a+c-2b)^2\leq0$

Its possible only when

$a+c-2b=0$

$\therefore\ b=\dfrac{a+c}{2}$

Hence a, b, c are in AP.

81. $\left(1\dfrac{2}{3}\right)^2+\left(2\dfrac{1}{3}\right)^2+3^2+\left(3\dfrac{2}{3}\right)^2+\ldots$ to 10 terms

$$=\left(\frac{5}{3}\right)^2+\left(\frac{7}{3}\right)^2+\left(\frac{9}{3}\right)^2+\left(\frac{11}{3}\right)^2+\ldots+\left(\frac{23}{3}\right)^2$$

$$=\frac{5^2+7^2+9^2+\ldots+(23)^2}{9}=\frac{\sum_{r=2}^{11}(2r+1)^2}{9}$$

$$\frac{\sum_{r=1}^{10}(2r+3)^2}{9} = \frac{1}{9}\left\{4\sum_{r=1}^{10} r^2 + 12\sum_{r=1}^{10} r + 9\sum_{r=1}^{10} 1\right\}$$

$$= \frac{1}{9}\left\{4 \times \frac{10\times 11\times 21}{6} + \frac{12\times 10\times 11}{2} + 9\times 10\right\} = \frac{2290}{9}$$

82. Let n consecutive odd integers are

$$2m+1,\ 2m+3,\ 2m+5, \ldots, 2m+2n-1.$$

Given that, $(2m+1)+(2m+3)+(2m+5)$
$+ \ldots + (2m+2n-1) = 45^2 - 21^2$

$\Rightarrow \quad 2mn + (1+3+5+\ldots+2n-1) = 45^2 - 21^2$

$\Rightarrow \quad 2mn + \frac{n}{2}(1+2n-1) = 45^2 - 21^2$

$\Rightarrow \quad 2mn + n^2 = 45^2 - 21^2$

$\Rightarrow \quad (n+m)^2 - m^2 = 45^2 - 21^2$

On comparing $n+m=45$, $m=21$

$\therefore \quad n = 24$

$\therefore$ Consecutive odd integers are 43, 45, 47, ..., 89.

83. $x = 1 + \cos^2\theta + \ldots\ldots \infty$

$x = \frac{1}{1-\cos^2\theta} \Rightarrow \cos^2\theta = 1 - \frac{1}{x}$

$y = 1 + \sin^2\phi + \ldots\ldots \infty$

$y = \frac{1}{1-\sin^2\phi} \Rightarrow \sin^2\phi = 1 - \frac{1}{y}$

$z = 1 + \cos^2\theta\sin^2\phi + \ldots \infty$

$$z = \frac{1}{1-\cos^2\theta\cdot\sin^2\phi} = \frac{1}{1-\left(1-\frac{1}{x}\right)\left(1-\frac{1}{y}\right)} = \frac{xy}{xy-(x-1)(y-1)}$$

$\Rightarrow xz + yz - z = xy \quad \therefore\ xy + z = (x+y)z$

84. Since $x, |x+1|, |x-1|$ are in AP.

$\therefore \quad 2|x+1| = x + |x-1|$

Case I : If $x < -1$

Then, $\quad -2(x+1) = x - (x-1)$

$\Rightarrow \quad -2x - 2 = 1 \Rightarrow x = -3/2$

Then series $-\frac{3}{2}, \frac{1}{2}, \frac{5}{2}, \ldots$

$$S_{20} = \frac{20}{2}[-3 + (20-1)2] = 10(35) = 350$$

Case II : If $-1 \le x < 1$

Then, $\quad 2(x+1) = x - (x-1) \Rightarrow 2x + 2 = 1$

$\therefore \quad x = -\frac{1}{2}$

Then series $-\frac{1}{2}, \frac{1}{2}, \frac{3}{2}, \ldots$

$$S_{20} = \frac{20}{2}[-1 + (20-1)\cdot 1] = 10 \times 18 = 180$$

Case III : If $x \ge 1$

Then, $2(x+1) = x + (x-1) \Rightarrow 2 = -1$ (impossible)

Hence, sum is 180 or 350.

85. $\left(\frac{\sqrt{10}}{3}\Sigma n^2\right)^2 = \Sigma n \cdot \Sigma n^3$

$$\frac{10}{9}\cdot\left\{\frac{n(n+1)(2n+1)}{6}\right\}^2 = \frac{n(n+1)}{2}\times\left[\frac{n(n+1)}{2}\right]^2$$

$n \neq 0,\ n+1 \neq 0, \qquad [\because n \in N]$

$\therefore \quad \frac{10}{9}\cdot\frac{(2n+1)^2}{9} = \frac{n(n+1)}{2} \Rightarrow n^2 + n - 20 = 0$

or $\quad n = 4, -5$

$\therefore \quad n = 4, \qquad [\because n \neq -5]$

86. $\therefore \sum_{r=1}^{n} t_r = \frac{n(n+1)(n+2)(n+3)}{8} = S_n$ (say)

$\therefore \sum_{r=1}^{n-1} t_r = \frac{(n-1)n(n+1)(n+2)}{8} = S_{n-1}$

$\therefore \quad t_n = S_n - S_{n-1} = \frac{n(n+1)(n+2)}{2}$

$$\therefore \lim_{n\to\infty}\sum_{r=1}^{n}\frac{1}{t_r} = \lim_{n\to\infty}\sum_{r=1}^{n}\frac{2}{n(n+1)(n+2)}$$

$$= \lim_{n\to\infty}\sum_{r=1}^{n}\left(\frac{1}{n(n+1)} - \frac{1}{(n+1)(n+2)}\right)$$

$$= -\lim_{n\to\infty}\sum_{r=1}^{n}\left(\frac{1}{(n+1)(n+2)} - \frac{1}{n(n+1)}\right)$$

$$= -\lim_{n\to\infty}\left(\frac{1}{(n+1)(n+2)} - \frac{1}{2}\right)$$

$\therefore \quad -\left(0 - \frac{1}{2}\right) = \frac{1}{2}$

87. Let d be the common difference

$\because \quad a_7 = 9 \qquad \therefore\ a_1 + 6d = 9$

Let $\quad D = a_1 a_2 a_7 = (9-6d)(9-5d)9$

$$= 270\left\{\left(d - \frac{33}{20}\right)^2 - \frac{9}{400}\right\}$$

For least value of D, $d - \frac{33}{20} = 0$

$\therefore \quad d = 33/20$

88. Let $A = m\lambda$, $H = n\lambda$

$\therefore \quad G^2 = AH = mn\lambda^2$

$\because$ a, b are the roots of $t^2 - (a+b)t + ab = 0$

or $\quad t^2 - 2m\lambda t + mn\lambda^2 = 0$

$$\therefore \quad t = \frac{2m\lambda \pm \sqrt{(4m^2\lambda^2 - 4mn\lambda^2)}}{2}$$

$$= \lambda\sqrt{m}\{\sqrt{m} \pm \sqrt{(m-n)}\}$$

$\Rightarrow a : b = \sqrt{m} + \sqrt{(m-n)} : \sqrt{m} - \sqrt{(m-n)}$

or $\sqrt{m} - \sqrt{(m-n)} : \sqrt{m} + \sqrt{(m-n)}$

89. If $\left[\frac{1}{3} + \frac{n}{90}\right] = 0 \Rightarrow 0 \le \frac{1}{3} + \frac{n}{90} < 1 \Rightarrow 1 \le n < 60$

and if $\left[\frac{1}{3} + \frac{n}{90}\right] = 1 \Rightarrow 1 \le \frac{1}{3} + \frac{n}{90} < 2 \Rightarrow 60 \le n < 150$

$\because \quad \sum_{n=1}^{k}\left[\frac{1}{3}+\frac{n}{90}\right]=21$

$\Rightarrow \quad \sum_{n=1}^{59}\left[\frac{1}{3}+\frac{n}{90}\right]+\sum_{n=60}^{k}\left[\frac{1}{3}+\frac{n}{90}\right]=21$

$\Rightarrow \quad 0+\sum_{n=60}^{k}\left[\frac{1}{3}+\frac{n}{90}\right]=21$, which is possible

only when $k=80$ $\left(\begin{array}{l}\because 60\le n<150 \\ \therefore \left[\frac{1}{3}+\frac{n}{90}\right]=1\end{array}\right)$.

90. $f_n(x)=x\,x\,x\ldots x$ (n digits) $=x\left\{\frac{(10^n-1)}{9}\right\}$

$\therefore \quad f_n^2(3)=3^2\left\{\frac{10^n-1}{9}\right\}^2=\frac{(10^n-1)^2}{9}$

and $\quad f_n(2)=\frac{2(10^n-1)}{9}$

$\therefore \quad f_n^2(3)+f_n(2)=\left(\frac{10^n-1}{9}\right)(10^n-1+2)$

$=\frac{(10^n-1)(10^n+1)}{9}=\frac{10^{2n}-1}{9}=f_{2n}(1)$

91. $\because \quad \frac{a+b+c}{3}\ge(abc)^{1/3}\Rightarrow\frac{2b+b}{3}\ge(64)^{1/3}$

$\therefore \quad b\ge 4$

Minimum value of b is 4.

92. Let $S=\sum_{k=1}^{20}k\left(\frac{1}{2^k}\right)$

$S=\frac{1}{2}+\frac{2}{2^2}+\frac{3}{2^3}+\frac{4}{2^4}+\ldots+\frac{20}{2^{20}}$ ...(i)

On multiplying by $\left(\frac{1}{2}\right)$ both sides, we get

$\frac{S}{2}=\frac{1}{2^2}+\frac{2}{2^3}+\frac{3}{2^4}+\ldots+\frac{19}{2^{20}}+\frac{20}{2^{21}}$ ...(ii)

On subtracting Eq. (ii) from Eq. (i), we get

$S-\frac{S}{2}=\frac{1}{2}+\frac{1}{2^2}+\frac{1}{2^3}+\ldots+\frac{1}{2^{20}}-\frac{20}{2^{21}}$

$\Rightarrow \quad \frac{S}{2}=\frac{\frac{1}{2}\left(1-\frac{1}{2^{20}}\right)}{1-\frac{1}{2}}-\frac{20}{2^{21}}$ $\left[\because \text{ sum of GP}=\frac{a(1-r^n)}{1-r}, r<1\right]$

$=1-\frac{1}{2^{20}}-\frac{20}{2^{21}}=1-\frac{1}{2^{20}}-\frac{10}{2^{20}}=1-\frac{11}{2^{20}}$

$\Rightarrow \quad S=2-\frac{11}{2^{19}}$

93. $(1-x)(1-2x)(1-2^2x)(1-2^3x)\ldots(1-2^{15}x)$

$=(-(x-1))\left(-2\left(x-\frac{1}{2}\right)\right)\left(-2^2\left(x-\frac{1}{2^2}\right)\right)$

$\times\left(-2^3\left(x-\frac{1}{2^3}\right)\right)\ldots\left(-2^{15}\left(x-\frac{1}{2^{15}}\right)\right)$

$=2^{1+2+3+\ldots+15}(x-1)\left(x-\frac{1}{2}\right)\left(x-\frac{1}{2^2}\right)$

$\times\left(x-\frac{1}{2^3}\right)\ldots\left(x-\frac{1}{2^{15}}\right)$

$=2^{120}\left\{x^{16}-\left(1+\frac{1}{2}+\frac{1}{2^2}+\frac{1}{2^3}+\ldots+\frac{1}{2^{15}}\right)x^{15}+\ldots\right\}$

$\therefore$ Coefficient of x^{15} in RHS

$=-2^{120}\left(1+\frac{1}{2}+\frac{1}{2^2}+\frac{1}{2^3}+\ldots+\frac{1}{2^{15}}\right)$

$=-2^{120}\left\{\frac{1\left(1-\left(\frac{1}{2}\right)^{16}\right)}{1-\frac{1}{2}}\right\}=-2^{121}\left(1-\frac{1}{2^{16}}\right)=2^{105}-2^{121}$

94. $\because f(x)=x^2+2(a-3)x+9$

(i) $f(-6)>0$ (ii) $f(1)>0$ (iii) $D\ge 0$ (iv) $-6<\frac{-b}{2a}<1$

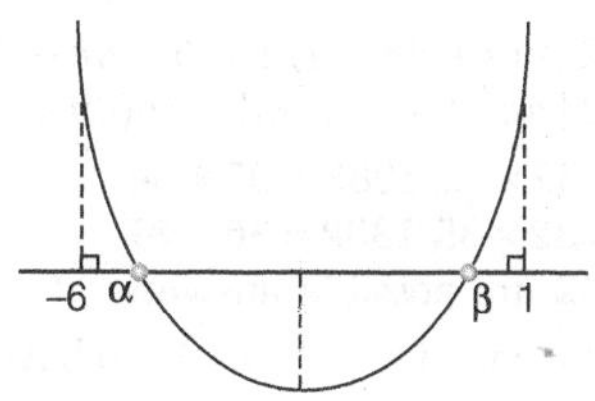

Then we get $6\le a<6.25$

$\therefore \quad [a]=6$

Now, 2, $h_1, h_2, \ldots, h_{20}$, $[a]$ are in HP

i.e. 2, $h_1, h_2, ,\ldots h_{20}$, 6 are in HP.

$\therefore \quad \frac{1}{h_{18}}=\frac{1}{2}+18\left(\frac{\frac{1}{6}-\frac{1}{2}}{21}\right)\Rightarrow h_{18}=\frac{14}{3}$

and 2, $a_1, a_2, \ldots, a_{20}$, 6 are in AP

$\therefore \quad a_3=2+3\left(\frac{6-2}{21}\right)=\frac{18}{7}$

$\therefore \quad a_3h_{18}=\frac{18}{7}\cdot\frac{14}{3}=12.$

95. $L=\frac{1}{2}\lim_{n\to\infty}\frac{1}{n^4}\sum_{r=1}^{n}r(n-r+1)(n-r+2)$

$=\frac{1}{2}\lim_{n\to\infty}\frac{1}{n^4}\sum_{r=1}^{n}\{r^3-(2n+3)r^2+(n+1)(n+2)r\}$

$=\frac{1}{2}\lim_{n\to\infty}\frac{1}{n^4}\{\Sigma n^3-(2n+3)\Sigma n^2+(n+1)(n+2)\Sigma n\}$

$=\frac{1}{2}\lim_{n\to\infty}\frac{1}{n^4}\left[\left(\frac{n(n+1)}{2}\right)^2-\frac{(2n+3)n(n+1)(2n+1)}{6}+\frac{(n+1)(n+2)n(n+1)}{2}\right]$

$=\frac{1}{2}\lim_{n\to\infty}\left[\left(\frac{1\cdot\left(1+\frac{1}{n}\right)}{2}\right)^2-\frac{\left(2+\frac{3}{n}\right)1\left(1+\frac{1}{n}\right)\left(2+\frac{1}{n}\right)}{6}+\frac{\left(1+\frac{1}{n}\right)^2 1\left(1+\frac{2}{n}\right)}{2}\right]$

$=\frac{1}{2}\left\{\frac{1}{4}-\frac{4}{6}+\frac{1}{2}\right\}=\frac{1}{8}-\frac{1}{3}+\frac{1}{4}=\frac{3-8+6}{24}=\frac{1}{24}$

Objective Questions Type II [One or more than one correct answer(s)]

1. Convert it into differences and use sum of n terms of an AP,

i.e. $S_n = \frac{n}{2}[2a + (n-1)d]$

Now, $S_n = \sum_{k=1}^{4n} (-1)^{\frac{k(k+1)}{2}} \cdot k^2$

$= -(1)^2 - 2^2 + 3^2 + 4^2 - 5^2 - 6^2 + 7^2 + 8^2 + \ldots$

$= (3^2 - 1^2) + (4^2 - 2^2) + (7^2 - 5^2) + (8^2 - 6^2) + \ldots$

$= 2\{\underbrace{(4 + 6 + 12 + \ldots)}_{n \text{ terms}} + \underbrace{(6 + 14 + 22 + \ldots)}_{n \text{ terms}}\}$

$= 2\left[\frac{n}{2}\{2 \times 4 + (n-1)8\} + \frac{n}{2}\{2 \times 6 + (n-1)8\}\right]$

$= 2[n(4 + 4n - 4) + n(6 + 4n - 4)]$

$= 2[4n^2 + 4n^2 + 2n] = 4n(4n+1)$

Here, $1056 = 32 \times 33$, $1088 = 32 \times 34$,
$1120 = 32 \times 35$, $1332 = 36 \times 37$

1056 and 1332 are possible answers.

2. Given quadratic equation $x^2 - x - 1 = 0$ having roots α and β, $(\alpha > \beta)$.

So, $\alpha = \frac{1+\sqrt{5}}{2}$ and $\beta = \frac{1-\sqrt{5}}{2}$

and $\alpha + \beta = 1$, $\alpha\beta = -1$

$\because \quad a_n = \frac{\alpha^n - \beta^n}{\alpha - \beta}, n \geq 1$

So, $a_{n+1} = \frac{\alpha^{n+1} - \beta^{n+1}}{\alpha - \beta} = \alpha^n + \alpha^{n-1}\beta + \alpha^{n-2}\beta^2 + \ldots + \alpha\beta^{n-1} + \beta^n$

$= \alpha^n - \alpha^{n-2} - \alpha^{n-3}\beta - \ldots - \beta^{n-2} + \beta^n$ [as $\alpha\beta = -1$]

$= \alpha^n + \beta^n - (\alpha^{n-2} + \alpha^{n-3}\beta + \ldots + \beta^{n-2})$

$= \alpha^n + \beta^n - a_{n-1}$

$\left[\text{as } a_{n-1} = \frac{\alpha^{n-1} - \beta^{n-1}}{\alpha - \beta} = \alpha^{n-2} + \alpha^{n-3}\beta + \ldots + \beta^{n-2}\right]$

$\Rightarrow a_{n+1} + a_{n-1} = \alpha^n + \beta^n = b_n, \forall n \geq 1$

So, option (b) is correct.

Now, $\sum_{n=1}^{\infty} \frac{b_n}{10^n} = \sum_{n=1}^{\infty} \frac{\alpha^n + \beta^n}{10^n}$ [as, $b_n = \alpha^n + \beta^n$]

$= \sum_{n=1}^{\infty}\left(\frac{\alpha}{10}\right)^n + \sum_{n=1}^{\infty}\left(\frac{\beta}{10}\right)^n \quad \left[\because \left|\frac{\alpha}{10}\right| < 1 \text{ and } \left|\frac{\beta}{10}\right| < 1\right]$

$= \frac{\frac{\alpha}{10}}{1 - \frac{\alpha}{10}} + \frac{\frac{\beta}{10}}{1 - \frac{\beta}{10}} = \frac{\alpha}{10-\alpha} + \frac{\beta}{10-\beta}$

$= \frac{10\alpha - \alpha\beta + 10\beta - \alpha\beta}{(10-\alpha)(10-\beta)} = \frac{10(\alpha+\beta) - 2\alpha\beta}{100 - 10(\alpha+\beta) + \alpha\beta}$

$= \frac{10(1) - 2(-1)}{100 - 10(1) - 1} = \frac{12}{89}$ [as $\alpha + \beta = 1$ and $\alpha\beta = -1$]

So, option (a) is not correct.

$\because \quad \alpha^2 = \alpha + 1$ and $\beta^2 = \beta + 1$

$\Rightarrow \quad \alpha^{n+2} = \alpha^{n+1} + \alpha^n$ and $\beta^{n+2} = \beta^{n+1} + \beta^n$

$\Rightarrow \quad (\alpha^{n+2} + \beta^{n+2}) = (\alpha^{n+1} + \beta^{n+1}) + (\alpha^n + \beta^n)$

$\Rightarrow \quad a_{n+2} = a_{n+1} + a_n$

Similarly, $a_{n+1} = a_n + a_{n-1}$

$a_n = a_{n-1} + a_{n-2}$

....

....

$a_3 = a_2 + a_1$

On adding, we get

$a_{n+2} = (a_n + a_{n-1} + a_{n-2} + \ldots + a_2 + a_1) + a_2$

$\left[\because a_2 = \frac{\alpha^2 - \beta^2}{\alpha - \beta} = \alpha + \beta = 1\right]$

So, $a_{n+2} - 1 = a_1 + a_2 + a_3 + \ldots + a_n$

So, option (c) is also correct.

And, now $\sum_{n=1}^{\infty} \frac{a_n}{10^n} = \sum_{n=1}^{\infty} \frac{\alpha^n - \beta^n}{(\alpha - \beta)10^n}$

$= \frac{1}{\alpha - \beta}\left[\sum_{n=1}^{\infty}\left(\frac{\alpha}{10}\right)^n - \sum_{n=1}^{\infty}\left(\frac{\beta}{10}\right)^n\right]$

$= \frac{1}{\alpha - \beta}\left[\frac{\frac{\alpha}{10}}{1 - \frac{\alpha}{10}} - \frac{\frac{\beta}{10}}{1 - \frac{\beta}{10}}\right], \quad \left[\text{as } \left|\frac{\alpha}{10}\right| < 1 \text{ and } \left|\frac{\beta}{10}\right| < 1\right]$

$= \frac{1}{\alpha - \beta}\left(\frac{\alpha}{10-\alpha} - \frac{\beta}{10-\beta}\right) = \frac{1}{\alpha - \beta}\left[\frac{10\alpha - \alpha\beta - 10\beta + \alpha\beta}{100 - 10(\alpha+\beta) + \alpha\beta}\right]$

$= \frac{10(\alpha - \beta)}{(\alpha - \beta)[100 - 10(\alpha+\beta) + \alpha\beta]} = \frac{10}{100 - 10 - 1} = \frac{10}{89}$

Hence, options (b), (c) and (d) are correct.

3. By hypothesis, $q = \frac{2pr}{p+r} \Rightarrow \frac{q}{2} = \frac{pr}{p+r} = k$ (say)

$\Rightarrow \quad q = 2k, pr = k(p+r)$

Also, p^2, q^2, r^2 are in AP .

$2q^2 = p^2 + r^2 = (p+r)^2 - 2pr$

$8k^2 = (p+r)^2 - 2k(p+r)$

$(p+r)^2 - 2(p+r)k - 8k^2 = 0$

$\Rightarrow \quad p + r = 4k, -2k$

$\Rightarrow \quad p + r = 4k, pr = 4k^2$

and $\quad (p-r)^2 = (p+r)^2 - 4pr = 0$

$\therefore \quad p = r$

This is against the hypothesis

$\therefore \quad p + r = -2k, pr = -2k^2$

Now, $(p-r)^2 = 12k^2 \Rightarrow p - r = \pm 2\sqrt{3}k$

Combine it with $p + r = -2k$ to get

$p = (-1 \pm \sqrt{3})k$ and $r = (-1 \mp \sqrt{3})k$

$\therefore \quad p : q : r :: -1 \pm \sqrt{3} : 2 : -1 \mp \sqrt{3}$

or $\quad p : q : r :: 1 \mp \sqrt{3} : -2 : 1 \pm \sqrt{3}$

4. $\because \alpha(n) = 1 + 1/2 + (1/3 + 1/4) + (1/5 + 1/6 + 1/7 + 1/8) + \ldots \left(\frac{1}{2^{n-1}+1} + \ldots + \frac{1}{2^n}\right) - \frac{1}{2^n}$

$\therefore \quad \alpha(n) > 1 + \frac{1}{2} + \left(\frac{1}{4} + \frac{1}{4}\right) + \left(\frac{1}{8} + \frac{1}{8} + \frac{1}{8} + \frac{1}{8}\right) + \ldots - \frac{1}{2^n}$

or $\alpha(n) > 1 + \left(\frac{1}{2} + \frac{1}{2} + \frac{1}{2} + \ldots n \text{ terms}\right) - \frac{1}{2^n}$

$\therefore \quad \alpha(n) > 1 + \frac{n}{2} - \frac{1}{2^n}$

$\Rightarrow \quad \alpha(200) > 1 + 100 - \frac{1}{2^{200}} > 100$

$\therefore \quad \alpha(200) > 100$

Also, $\alpha(n) = 1 + \left(\frac{1}{2} + \frac{1}{3}\right) + \left(\frac{1}{4} + \frac{1}{5} + \frac{1}{6} + \frac{1}{7}\right)$

$+ .. + \left(\frac{1}{2^{n-1}}\right) + .. + \frac{1}{2^n - 1} < 1 + 2\left(\frac{1}{2}\right) + 4\left(\frac{1}{4}\right) + ..$

$= 1 + 1 + 1 + \ldots + 1 = n$

$\therefore \quad \alpha(100) \le 100.$

5. T_p of AP $= \frac{1}{q(p+q)} = A + (p-1)D \quad \ldots(i)$

T_q of AP $= \frac{1}{p(p+q)} = A + (q-1)D \quad \ldots(ii)$

$\frac{1}{T_{p+q}} = A + (p+q-1)D$

and $\quad \frac{1}{T_{pq}} = A + (pq-1)D$

Now, solving Eqs. (i) and (ii), we get

$A = D = \frac{1}{pq(p+q)}$

$\therefore \quad \frac{1}{T_{p+q}} = A + (p+q-1)D = (p+q)D = \frac{1}{pq}$

and $\frac{1}{T_{pq}} = A + f\,(p+q-1)D = pqD = \frac{1}{p+q}$

$\Rightarrow T_{p+q} = pq$ and $T_{pq} = p + q$

Also, $\because p > 1, q > 1$

$\therefore \quad pq > p + q$ i.e. $T_{p+q} > T_{pq}$

6. Consider the AP since a is equidistant from the first term α and last term β of the AP therefore,

α, a, β are in AP $\Rightarrow a$ is the AM of α and β

$\therefore \quad a = \frac{\alpha + \beta}{2}$

Similarly, b and c are the GM and HM of α and β respectively, then

$b = \sqrt{\alpha\beta}$ and $c = \frac{2\alpha\beta}{a + \beta}$

$\therefore \quad (\text{GM})^2 = (\text{AM})(\text{HM})$

$\therefore \quad b^2 = ac$

and $\quad \text{AM} \ge \text{GM} \ge \text{HM}$

$\therefore \quad a \ge b \ge c.$

7. $\because \quad (a^{n/2} - c^{n/2})^2 > 0$

$\Rightarrow \quad a^n + c^n > 2a^{n/2}c^{n/2} \quad \ldots(i)$

$\text{GM} > \text{HM}$

$\sqrt{ac} > b$

$\Rightarrow \quad 2(ac)^{n/2} > 2b^n \quad \ldots(ii)$

From Eqs. (i) and (ii),

$a^n + c^n > 2b^n$

Putting $n = 2, 3, 5, 100$

8. $T_n = \frac{1}{n(n+1)(n+2)(n+3)}$

Let $\quad V_n = \frac{1}{(n+1)(n+2)(n+3)}$

$V_{n-1} = \frac{1}{n(n+1)(n+2)}$

$V_n - V_{n-1} = -3T_n$

$T_n = -\frac{1}{3}(V_n - V_{n-1})$

$S = T_1 + T_2 + \ldots + T_n$

$\therefore \quad S_n = -\frac{1}{3}(V_n - V_0)$

$= -\frac{1}{3}\left\{\frac{1}{(n+1)(n+2)(n+3)} - \frac{1}{6}\right\}$

$= \frac{1}{18} - \frac{1}{3(n+1)(n+2)(n+3)}$

$= \frac{n(n^2 + 6n + 11)}{18(n+1)(n+2)(n+3)}$

9. $a = \frac{1}{1 + \tan^2 x}$

$\Rightarrow \quad \tan^2 x = \frac{1}{a} - 1$

and $\quad b = \frac{1}{1 + \cot^2 y} = \cot^2 y = \frac{1}{b} - 1$

Then, $\sum_{k=0}^{\infty} \tan^{2k} x \cot^{2k} y$

$= \frac{1}{1 - \tan^2 x \cot^2 y}$

$= \frac{1}{1 - \left(\frac{1}{a} - 1\right)\left(\frac{1}{b} - 1\right)}$

$= \frac{ab}{ab - (1-a)(1-b)} = \frac{ab}{a + b - 1}$

or $\quad \frac{1}{\frac{1}{a} + \frac{1}{b} - \frac{1}{ab}}.$

10. $\because a, b, c$ are in HP.

$\therefore \quad \frac{1}{b} - \frac{1}{a} = \frac{1}{c} - \frac{1}{b} \quad \ldots(i)$

$\therefore \quad \left(\frac{1}{b} + \frac{1}{c} - \frac{1}{a}\right)\left(\frac{1}{c} + \frac{1}{a} - \frac{1}{b}\right) = \left(\frac{2}{c} - \frac{1}{b}\right)\left(\frac{1}{b}\right) = \frac{2}{bc} - \frac{1}{b^2} \quad \ldots(ii)$

Alternate. (a) is correct.

$= \frac{1}{c}\left(\frac{1}{a} + \frac{1}{c}\right) - \frac{1}{4}\left(\frac{1}{a} + \frac{1}{c}\right)^2$

[Putting value of b from Eq. (i)]

$= \frac{1}{4}\left\{\frac{4}{ca} + \frac{4}{c^2} - \frac{1}{a^2} - \frac{1}{c^2} - \frac{2}{ac}\right\}$

$= \frac{1}{4}\left\{\frac{3}{c^2} + \frac{2}{ac} - \frac{1}{a^2}\right\}$

Alternate. (b) is correct.

Substituting the value of $\frac{1}{c}$ from Eqs. (i) and (ii) since third alternate in terms of a and b

then, $\quad \frac{2}{b}\left(\frac{2}{b} - \frac{1}{a}\right) - \frac{1}{b^2}$

$= \frac{3}{b^2} - \frac{2}{ab}$

Alternate. (c) is correct.

11. $$\sum_{i=1}^{n}\sum_{j=1}^{i}\sum_{k=1}^{j} 1 = \sum_{i=1}^{n}\sum_{j=1}^{i} j = \sum_{i=1}^{n} \frac{i(i+1)}{2} = \frac{1}{2}\left[\sum_{i=1}^{n} i^2 + \sum_{i=1}^{n} i\right] = \frac{1}{2}\left\{\frac{n(n+1)(2n+1)}{6} + \frac{n(n+1)}{2}\right\} = \frac{n(n+1)(n+2)}{6}$$

12. Let $\log_y x = 1 + D$, $\log_z y = 1 + 2D$ and $-15\log_x z = 1 + 3D$

$\therefore$ $(1 + D)(1 + 2D)(1 + 3D) = -15$

$\Rightarrow$ $6D^3 + 11D^2 + 6D + 16 = 0$

$\Rightarrow$ $(D + 2)(6D^2 - D + 8) = 0$

$\therefore$ $D = -2,\ 6D^2 - D + 8 \neq 0$

Then $\log_y x = -1$, $\log_z y = -3$, $-15\log_x z = -5$

$\therefore$ $x = y^{-1}$, $y = z^{-3}$ and $x = z^3$

Also, $x = y^{-1} = z^3$

13. $b_2 = b_1 r$, $b_3 = b_1 r^2$

$\Rightarrow$ $b_3 > 4b_2 - 3b_1$

$\Rightarrow$ $b_1 r^2 > 4b_1 r - 3b_1$

$b_1 \neq 0$

$\therefore$ $r^2 - 4r + 3 > 0$

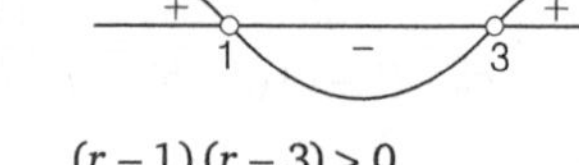

$\Rightarrow$ $(r - 1)(r - 3) > 0$

$\therefore$ $r \in (-\infty, 1) \cup (3, \infty)$

14. Since $\log_x a$, $a^{x/2}$, $\log_b x$ are in GP.

$\therefore$ $a^x = \log_x a \cdot \log_b x = \log_b a$

$\therefore$ $x = \log_a(\log_b a) = \log_a\left(\frac{\log_e a}{\log_e b}\right) = \log_a(\log_e a) - \log_a(\log_e b)$

15. Alternate (a) If $\frac{a}{b+c-a}, \frac{b}{c+a-b}, \frac{c}{a+b-c}$ are in HP.

Then, $\frac{b+c-a}{a}, \frac{a+c-b}{b}, \frac{a+b-c}{c}$ are in AP adding in each term is 2, then

$\frac{b+c+a}{a}, \frac{c+a+b}{b}, \frac{a+b+c}{c}$ are in AP .

Dividing in each term by $a + b + c$, then $\frac{1}{a}, \frac{1}{b}, \frac{1}{c}$ are in AP.

Hence, a, b, c are in HP.

Alternate (b) $\frac{2}{b} = \frac{1}{b-a} + \frac{1}{b-c}$

$2(b-a)(b-c) = b(2b - a - c)$

$\Rightarrow$ $2(b^2 - bc - ab + ac) = 2b^2 - b(a + c)$

$\Rightarrow 2b^2 - 2bc - 2ab + 2ac = 2b^2 - b(a + c)$

$\Rightarrow$ $-2bc - 2ab + 2ac = -b(a + c)$

$\Rightarrow$ $-2b(a + c) + 2ac = -b(a + c)$

$b(a + c) = 2ac$

$\therefore$ $b = \frac{2ac}{a+c}$

Hence a, b, c are in HP.

Alternate (c) If $a - \frac{b}{2}, \frac{b}{2}, c - \frac{b}{2}$ are in GP

$\therefore$ $\frac{b^2}{4} = \left(a - \frac{b}{2}\right)\left(c - \frac{b}{2}\right)$

$\frac{b^2}{4} = ac - \frac{ab}{2} - \frac{bc}{2} + \frac{b^2}{4}$

$\therefore$ $2ac = b(a + c)$

$\Rightarrow$ $b = \frac{2ac}{a+c}$

Hence, a, b, c are in HP.

Alternate (d) If $\frac{a}{b+c}, \frac{b}{c+a}, \frac{c}{a+b}$ are in HP

$\therefore \frac{b+c}{a}, \frac{c+a}{b}, \frac{a+b}{c}$ are in AP. Adding in each term is 1. Then $\frac{a+b+c}{a}, \frac{a+b+c}{b}, \frac{a+b+c}{c}$ are in AP.

Dividing in each term by $a + b + c$, then $\frac{1}{a}, \frac{1}{b}, \frac{1}{c}$ are in AP .

Hence, a, b, c are in HP.

16. $\sum_{r=1}^{n} (2r^3 + 5r^2 + 3r) = an^4 + bn^3 + cn^2 + dn + e$

$$\Rightarrow 2\left\{\left\{\frac{n(n+1)}{2}\right\}^2 + 5\left\{\frac{n(n+1)(2n+1)}{6}\right\}\right\} + 3\left[\frac{n(n+1)}{2}\right] = an^4 + bn^3 + cn^2 + dn + e.$$

Comparing similar terms, then we get

$a = \frac{1}{2}, b = \frac{16}{6} = \frac{8}{3}, c = \frac{9}{2}, d = \frac{7}{3}$ and $e = 0$

17. $1, \log_9(3^{1-x} + 2), \log_3(4 \cdot 3^x - 1)$ are in AP

$\therefore$ $2\log_9(3^{1-x} + 2) = 1 + \log_3(4 \cdot 3^x - 1)$

$\Rightarrow$ $2\log_{3^2}(3^{1-x} + 2) = \log_3 3 + \log_3(4 \cdot 3^x - 1)$

$\Rightarrow$ $\frac{2}{2} \cdot \log_3(3^{1-x} + 2) = \log_3(3\{4 \cdot 3^x - 1\})$

$\Rightarrow$ $3^{1-x} + 2 = 3 \cdot (4 \cdot 3^x - 1)$

$\Rightarrow$ $3 + 2 \cdot 3^x = 12 \cdot 3^{2x} - 3 \cdot 3^x$

$\Rightarrow$ $12(3^x)^2 - 5(3^x) - 3 = 0$

$\therefore$ $3^x = 3/4$, $3^x \neq -1/3 \Rightarrow x = 1 - \log_3 4 = \log_3\left(\frac{3}{4}\right)$.

18. Last term of $(n-1)$th group is $(n-1)^2$

$\therefore$ First term is nth group is $(n-1)^2 + 1$

and last term of nth group is n^2

Number of terms in nth group $= (2n - 1)$

$\therefore$ Sum of the numbers in the nth group

$= \frac{(2n-1)}{2}\{(n-1)^2 + 1 + n^2\}$

$= (2n-1)(n^2 - n + 1) = n^3 + (n-1)^3$.

19. Given, α, β, γ, δ are in HP.

$\Rightarrow \quad \frac{1}{\alpha}, \frac{1}{\beta}, \frac{1}{\gamma}, \frac{1}{\delta}$ are in AP.

Let d be the common difference of the AP.

Since α, γ are roots of $Ax^2 - 4x + 1 = 0$

$$\therefore \quad \frac{\alpha+\gamma}{\alpha\gamma} = \frac{4/A}{1/A} = 4 \Rightarrow \frac{1}{\alpha} + \frac{1}{\gamma} = 4$$

$$\Rightarrow \quad \frac{1}{\alpha} + \frac{1}{\alpha} + 2d = 4 \text{ or } \frac{1}{\alpha} + d = 2 \quad \ldots(i)$$

Also, β, δ are roots of $Bx^2 - 6x + 1 = 0$

$$\therefore \quad \frac{\beta+\delta}{\beta\delta} = \frac{1}{\beta} + \frac{1}{\delta} = \frac{6/B}{1/B} = 6 \text{ or } \frac{1}{\alpha} + d + 1 + 3d = 6$$

$$\Rightarrow \quad \frac{1}{\alpha} + 2d = 3 \quad \ldots(ii)$$

On solving Eqs. (i) and (ii), we get $\frac{1}{\alpha} = 1$ and $d = 1$

$$\therefore \quad \frac{1}{\alpha} = 1, \frac{1}{\beta} = 2, \frac{1}{\gamma} = 3 \text{ and } d = 1$$

$$\therefore \quad \frac{1}{\alpha} = 1, \frac{1}{\beta} = 2, \frac{1}{\gamma} = 3 \text{ and } \frac{1}{\delta} = 4$$

Since, $\frac{1}{\alpha\gamma} = A \Rightarrow A = 3$

Also, $\frac{1}{\beta\gamma} = B \Rightarrow B = 8$

20. (a) Using AM ≥ HM

$$\frac{a_1 + a_2 + a_3}{3} \geq \frac{3}{\left(\frac{1}{a_1} + \frac{1}{a_2} + \frac{1}{a_3}\right)}$$

$$\Rightarrow \quad (a_1 + a_2 + a_3)\left(\frac{1}{a_1} + \frac{1}{a_2} + \frac{1}{a_3}\right) \geq 9$$

(b) Similar to (a)

(c) $\frac{a_1 a_2 + a_3 a_4}{2} \geq \sqrt{a_1 a_2 a_3 a_4}$ and $\frac{a_1 a_3 + a_2 a_4}{2} \geq \sqrt{a_1 a_3 a_2 a_4}$

Multiplying, we get (c) is true.

21. $\because \quad \frac{t_{2p}}{t_p} = \frac{t_{4p}}{t_{2p}} = r$ (say)

If we start from t_p, then t_{2p} is the $(p+1)$ th term.

And if we start from t_{2p}, then t_{4p} is the $(2p+1)$ th term

$$\therefore \quad t_{2p} = t_p + pd \quad \ldots(i)$$

$$\text{and} \quad t_{4p} = t_{2p} + 2pd \quad (d = c \cdot d)$$

$$\Rightarrow \quad t_{4p} = t_{2p} + 2(t_{2p} - t_p) \quad \text{[from Eq. (i)]}$$

$$\Rightarrow \quad t_{4p} = 3t_{2p} - 2t_p \Rightarrow \frac{t_{4p}}{t_{2p}} = 3 - \frac{2t_p}{t_{2p}} \Rightarrow r = 3 - \frac{2}{r}$$

$$\Rightarrow \quad (r-1)(r-2) = 0 \Rightarrow r = 1, 2.$$

22. Since, for some real x (not for all real x)

$$(a^2 + b^2)x^2 + 2(ab + bc)x + (b^2 + c^2) \leq 0$$

$$\therefore \quad D \geq 0 \Rightarrow 4(ab + bc)^2 - 4(a^2 + b^2)(b^2 + c^2) \geq 0$$

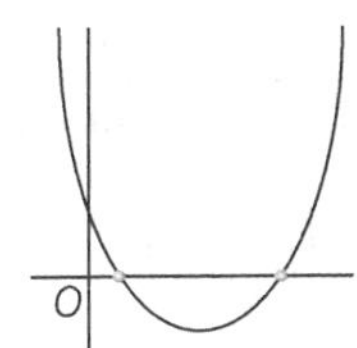

$$\Rightarrow \quad -4(ac - b^2)^2 \geq 0$$

$$\Rightarrow \quad ac - b^2 = 0 \quad [\because -4(ac - b^2)^2 < 0]$$

Thus a, b, c are in GP.

Now, let $f(x) = ax^2 + 2bx + c \ (a > 0)$.

Its discriminant $= 4(b^2 - ac) = 0$

$\therefore f(x) \geq 0$ for all x, i.e. (c) is true.

23. Note the fact that sum of a positive number and its reciprocal is greater than or equal to 2 (AM ≥ GM). Also, sum of a negative number and its reciprocal is less than or equal to -2. Thus (a), (b), (c) are true.

Also, $1 \cdot 2 \cdot 3 \cdot \ldots\ldots n < n \cdot n \cdot n \cdot n \cdot \ldots \cdot n \Rightarrow n! < n^n$

(d) is false.

24. Let common difference be y

(a) $\quad a(a+3y) < (a+y)(a+2y)$

$\Rightarrow \quad a^2 + 3ay < a^2 + 3ay + 2y^2$

$\Rightarrow \quad 0 < 2y^2$ true

(b) $\quad a + a + 2y < a + y + a + 3y$

$\quad 0 < 2y$, false if $a > b > c > d > 0$

(c) $a + a + 3y = a + y + a + 2y$, true

(d) $\because a, b, c, d$ are in AP

$\therefore \quad (a+1), (b+1), (c+1), (d+1)$ are in AP

$\therefore \quad (a+1)(d+1) < (b+1)(c+1)$.

25. For $\sqrt{x}$ to be defined we must have $x \geq 0$. Also, for $\frac{1}{1-x}$ and $\frac{1}{1-\sqrt{x}}$ to be defined we must have $x \neq 1$. For $x \in [0, 1) \cup (1, \infty)$, we have

$$\frac{2}{1-x} = \frac{1}{1-\sqrt{x}} + \frac{1}{1+\sqrt{x}}.$$

26. From the given conditions, we have $2b = a + c$ and

$$\frac{2}{b^2} = \frac{1}{a^2} + \frac{1}{c^2} \Rightarrow 2a^2c^2 = b^2(a^2 + c^2)$$

Eliminating b, we get

$$2a^2c^2 = \frac{1}{4}(a+c)^2(a^2 + c^2)$$

$$\Rightarrow \quad 8a^2c^2 = (a^2 + c^2 + 2ac)(a^2 + c^2)$$

$$\Rightarrow \quad (a^2 + c^2)^2 + 2ac(a^2 + c^2) - 8a^2c^2 = 0$$

$$\Rightarrow \quad (a^2 + c^2 - 2ac)(a^2 + c^2 + 4ac) = 0$$

$$\Rightarrow \quad a^2 + c^2 - 2ac = 0 \text{ or } a^2 + c^2 + 4ac = 0$$

$$\Rightarrow \quad (a-c)^2 = 0 \text{ or } (a+c)^2 + 2ac = 0$$

Hence, $a = c$ or $4b^2 + 2ac = 0$ ie, $b^2 = -\frac{1}{2}ac$. If $a = c$, then from $2b = a + c$, we get $b = a$, so that $a = b = c$, in which case a, b and c are evidently in GP. If $b^2 = -\frac{1}{2}ac$, then $-a/2, b$ and c are in GP.

27. From the given condition, we must have

$$2\log_4(2^{1-x} + 1) = \log_2(5 \cdot 2^x + 1) + 1$$

$$\Rightarrow \quad \frac{2\log(2^{1-x} + 1)}{\log 4} = \frac{\log(5 \cdot 2^x + 1)}{\log 2} + 1$$

$$\Rightarrow \quad \log(2^{1-x} + 1) = \log[(5 \cdot 2^x + 1)2]$$

$$\Rightarrow \quad 2^{1-x} + 1 = 5 \cdot 2^x \cdot 2 + 2$$

Now, put $2^x = y$, so that

$$\frac{2}{y} + 1 = 10y + 2 \Rightarrow 10y^2 + y - 2 = 0$$

$\Rightarrow \quad (5y - 2)(2y + 1) = 0$

Which means $y = 2/5$ or $y = -1/2$. Since, $y = 2^x$ cannot be negative, we have $y = 2^x = 2/5 = 0.4$ i.e., $x = \log_2 (0.4)$.

But $2^x = 2/5$ can also be expressed in the form

$$x = \frac{\log (2/5)}{\log 2} = 1 - \frac{\log 5}{\log 2}.$$

28. From the given condition

$$\frac{1}{2}(a+b) = 2\sqrt{ab} \Rightarrow \sqrt{\frac{a}{b}} + \sqrt{\frac{b}{a}} = 4$$

Putting $y = \sqrt{\frac{a}{b}}$

$$\Rightarrow \quad y + \frac{1}{y} = 4 \Rightarrow y = \frac{4 \pm 2\sqrt{3}}{2} \qquad \ldots(\text{i})$$

$$\Rightarrow \quad \sqrt{\frac{a}{b}} = 2 \pm \sqrt{3}$$

Since, the roots of Eq. (i) are $\sqrt{\frac{a}{b}}$ and $\sqrt{\frac{b}{a}}$, we have

$$\frac{\sqrt{a/b}}{\sqrt{b/a}} = \frac{2+\sqrt{3}}{2-\sqrt{3}} \Rightarrow \frac{a}{b} = \frac{2+\sqrt{3}}{2-\sqrt{3}} \Rightarrow a : b = 2+\sqrt{3} : 2-\sqrt{3}$$

From $\sqrt{a/b} = 2 + \sqrt{3}$, we get

$$\frac{a}{b} = (2+\sqrt{3})^2 = 7 + 4\sqrt{3} \Rightarrow a : b = 7 + 4\sqrt{3} : 1$$

From $\sqrt{\frac{b}{a}} = 2 - \sqrt{3}$, we get

$$\frac{b}{a} = (2-\sqrt{3})^2 = 7 - 4\sqrt{3}$$

$$\Rightarrow \quad a : b = 1 : 7 - 4\sqrt{3}$$

29. We have, $A_1 = a + \frac{1}{3}(b - a)$,

$$A_2 = b + \frac{2}{3}(b - a)$$

$\Rightarrow \quad A_1 + A_2 = a + b$

Similarly, $G_1 = a\left(\frac{b}{a}\right)^{1/3}, G_2 = a\left(\frac{b}{a}\right)^{2/3} \Rightarrow G_1G_2 = ab$

and $\quad \frac{1}{H_1} = \frac{1}{a} + \frac{1}{3}\left(\frac{1}{b} - \frac{1}{a}\right),\ H_2 = \frac{1}{a} + \frac{2}{3}\left(\frac{1}{b} - \frac{1}{a}\right)$

Now, $\quad \frac{1}{H_1} + \frac{1}{H_2} = \frac{1}{a} + \frac{1}{b}$

$$\Rightarrow \quad \frac{H_1 + H_2}{H_1H_2} = \frac{a+b}{ab} = \frac{A_1 + A_2}{G_1G_2}$$

$$\Rightarrow \quad \frac{G_1G_2}{H_1H_2} = \frac{A_1 + A_2}{H_1 + H_2}$$

Now, $\quad H_1 + H_2 = \frac{3ab}{a + 2b} + \frac{3ab}{2a + b} = \frac{9ab(a+b)}{(a+2b)(2a+b)}$

$$\Rightarrow \quad \frac{A_1 + A_2}{H_1 + H_2} = \frac{2(a^2 + b^2) + 5ab}{9ab}$$

Thus, $\quad \frac{G_1G_2}{A_1A_2} - \frac{5}{9} = \frac{2}{9}\left(\frac{a}{b} + \frac{b}{a}\right)$

Linked Comprehension Type

PASSAGE 1

Sol. 1. (d) 2. (c) 3. (b)

We have,

$$a + b + c = 25 \qquad \ldots(\text{i})$$

$$2a = b + 2 \qquad \ldots(\text{ii})$$

$$c^2 = 18b \qquad \ldots(\text{iii})$$

From Eqs. (i) and (ii), we get

$$b = 16 - \frac{2c}{3}$$

Then, from Eq. (iii)

$$c^2 = 18\left(16 - \frac{2c}{3}\right)$$

or $\quad c^2 + 12c - 18 \times 16 = 0$

or $\quad (c - 12)(c + 24) = 0$

Now, $c = -24$ is not possible since it does not lie between 2 and 18.

Hence, $c = 12$.

Then, from Eq. (iii), $b = 8$ and finally from Eq. (ii), $a = 5$.

Thus, $a = 5$, $b = 8$ and $c = 12$.

Hence, $abc = 5 \times 8 \times 12 = 480$.

Also, equation $ax^2 + bx + c = 0$ is $5x^2 + 8x + 12 = 0$, which has imaginary roots.

If a, b, c are roots of the equation $x^3 + qx^2 + rx + s = 0$, then the sum of product of roots taken two at a time is $r = 5 \times 8 + 5 \times 12 + 8 \times 12 = 196$.

PASSAGE 2

Since, $n + 1 > n - 1$

$$\Rightarrow \quad \left(\frac{n+1}{n-1}\right)^2 > 1 \Rightarrow p\left(\frac{n+1}{n-1}\right)^2 > p \qquad \ldots(\text{i})$$

Now, $p = a + d = a + \frac{(b-a)}{(n+1)} = \left(\frac{an+b}{n+1}\right)$

$\therefore$ **Ans.** $1 \to$ a

and $\quad \frac{1}{q} = \frac{1}{a} + D = \frac{1}{a} + \frac{\frac{1}{b} - \frac{1}{a}}{n+1} = \frac{(a+bn)}{ab(n+1)}$

$$\therefore \quad q = \frac{ab(n+1)}{(a+bn)}$$

$\therefore$ **Ans.** $2 \to$ b

Also, $\quad \frac{p}{q} = p \times \frac{1}{q} = \left(\frac{an+b}{n+1}\right)\left(\frac{a+bn}{ab(n+1)}\right)$

$$= \frac{(a^2n + abn^2 + ab + b^2n)}{ab(n+1)^2} = \frac{n\left(\frac{a}{b} + \frac{b}{a}\right) + n^2 + 1}{(n+1)^2}$$

$$\therefore \quad \frac{p}{q} - 1 = \frac{n\left(\frac{a}{b} + \frac{b}{a} - 2\right)}{(n+1)^2} = n\frac{\left(\sqrt{\frac{a}{b}} - \sqrt{\frac{b}{a}}\right)^2}{(n+1)^2}$$

$$= \frac{n}{(n+1)^2}\left(\sqrt{\frac{a}{b}} - \sqrt{\frac{b}{a}}\right)^2$$

$\therefore$ **Ans.** $3 \to$ c

$\because \quad \frac{p}{q} - 1 > 0 \Rightarrow \frac{p}{q} > 1$

$\Rightarrow \quad p > q \quad \ldots \text{(ii)}$

$\therefore$ From Eqs. (i) and (ii),

$$q < p < \left(\frac{n+1}{n-1}\right)^2 p$$

Ans. $5 \to$ (d)

Also, for $p = 8$, $n = 3$, then $q < 8 < 32$

$\therefore$ q does not lies between 8 and 32.

Ans. $4 \to$ (c)

PASSAGE 3

Given $a(b-c)x^2 + b(c-a)x + c(a-b) = 0 \quad \ldots \text{(i)}$

$\because \quad a(b-c) + b(c-a) + c(a-b) = 0$

$\therefore \quad x = 1$ is a root of Eq. (i)

Let other root is α, then

$$1 \times \alpha = \frac{c(a-b)}{a(b-c)} = \frac{c\left(a - \frac{2ac}{a+c}\right)}{a\left(\frac{2ac}{a+c} - c\right)} = 1$$

$\therefore \quad \alpha = 1$

Hence, both roots of Eq. (i) are 1, 1

Then, roots of $x^2 - Px + Q = 0$ are also 1, 1

Then, $\quad 1 + 1 = P,\ 1 \cdot 1 = Q$

$\therefore \quad P = 2, Q = 1$

$\therefore \quad [P] = [2] = 2$

$\therefore$ **Ans.** $2 \to$ (c)

Now, $[2P - Q] = [4 - 1] = 3$

$\therefore$ **Ans.** $3 \to$ (b)

$\because$ Roots of Eq. (i) are 1, 1, then

$$1 + 1 = -\frac{b(c-a)}{a(b-c)}$$

or $\quad \frac{a(b-c)}{b(c-a)} = -\frac{1}{2}$

$\therefore$ **Ans.** $1 \to$ (c)

$\because$ Roots of $x^2 - Px + Q = 0$ are 1, 1

$\therefore \quad \text{AM} = \text{GM} = \text{HM}$

$\Rightarrow \quad \text{AM : GM : HM} = 1 : 1 : 1$

4. $\quad \sqrt{ab} = 16$

$\Rightarrow \quad ab = 16^2 \Rightarrow ab = 8 \cdot 32$

and $\quad \frac{2ab}{a+b} = 12\frac{4}{5}$

$\Rightarrow \quad \frac{2 \times 256}{a+b} = \frac{64}{5}$

$\Rightarrow \quad a + b = 40 = 8 + 32$

$\therefore \quad \frac{a}{b} = \frac{8}{32} = \frac{1}{4}$

$\Rightarrow \quad a : b = 1 : 4$

5. If $a > b$

Then, $\quad \frac{a+b}{2} + \sqrt{ab} = a - b$

$\Rightarrow \quad 2\sqrt{ab} = a - 3b$

On squaring, then

$$4ab = a^2 - 6ab + 9b^2$$

$\Rightarrow \quad a^2 - 10ab + 9b^2 = 0$

$\Rightarrow \quad (a - 9b)(a - b) = 0$

$\therefore \quad a = 9b \quad [\because a \neq b]$

$a : b = 9 : 1$

6. AM : GM : HM = 1 : 1 : 1

PASSAGE 4

For sequence A :

1, 6, 18, 40, 75, 126, ...

5, 12, 22, 35, 51, ... (first successive difference)

7, 10, 13, 16, ... (second successive difference)

For sequence B :

1, 1, 6, 26, 91, 291, ...

0, 5, 20, 65, 200, ... (first successive difference)

5, 15, 45, 135,... (second successive difference)

For sequence C :

ln 2, ln 4, ln 32, ln 1024, ...

or ln 2, 2 ln 2, 5 ln 2, 10 ln 2, ...

ln 2, 3 ln 2, 5 ln 2, ... (first successive difference)

1. The second successive differences of A and B are in AP (common difference 3) and in GP (common ratio 3) respectively.

2. $\because T_n = an^3 + bn^2 + cn + d$

$\therefore \quad T_1 = a + b + c + d = 1 \quad \ldots \text{(i)}$

and $\quad T_2 = 8a + 4b + 2c + d = 6 \quad \ldots \text{(ii)}$

Subtraction $2\times$ Eq. (i) in Eq. (ii), then we get

$$6a + 2b - d = 4$$

3. $\because$ First successive difference of sequence C in AP, then

$$T_n = an^2 + bn + c$$

4. The second successive difference of sequence B in GP with common ratio 3.

5. The first successive difference of sequence C are in AP with common difference

$$= 3\ln 2 - \ln 2$$
$$= 2\ln 2 = \ln 4$$

PASSAGE 5

Let the three numbers in strictly increasing GP are $\frac{a}{r}$, a, ar $(r > 1)$

According to passage

$$\frac{a^2}{r^2} + a^2 + a^2r^2 = S^2$$

or $\quad a^2\left(\frac{1}{r^2} + 1 + r^2\right) = S^2$

or $\quad a^2\left(\frac{1}{r} + 1 + r\right)\left(\frac{1}{r} - 1 + r\right) = S^2 \quad \ldots \text{(i)}$

and $\quad \frac{a}{r} + a + ar = \alpha S$

or $\quad a\left(\frac{1}{r} + 1 + r\right) = \alpha S$

$\therefore \quad a^2\left(\frac{1}{r} + 1 + r\right)^2 = \alpha^2 S^2 \quad \ldots \text{(ii)}$

Dividing Eq. (ii) by (i), then

$$\left(\frac{\frac{1}{r}+1+r}{\frac{1}{r}-1+r}\right)=\alpha^2$$

$\Rightarrow \quad (1+r+r^2)=\alpha^2(1-r+r^2)$

$\Rightarrow \quad (\alpha^2-1)r^2-(\alpha^2+1)r+\alpha^2-1=0 \quad \ldots\text{(iii)}$

or $\quad r^2-\left(\frac{\alpha^2+1}{\alpha^2-1}\right)r+1=0 \quad [\because \alpha^2 \neq 1]$

$\because \quad r$ is real, $\left(\frac{\alpha^2+1}{\alpha^2-1}\right)^2-4\cdot1\cdot1>0$

[$\because$ the numbers in GP are distinct]

$\Rightarrow \quad \left(\frac{\alpha^2+1}{\alpha^2-1}+2\right)\left(\frac{\alpha^2+1}{\alpha^2-1}-2\right)>0$

$\Rightarrow \quad (3\alpha^2-1)(-\alpha^2+3)>0$

or $\quad \left(\alpha^2-\frac{1}{3}\right)(\alpha^2-3)<0$

$\therefore \quad \frac{1}{3}<\alpha^2<3$

But $\quad \alpha^2 \neq 1$

$\therefore \quad \alpha^2 \in \left(\frac{1}{3}, 1\right) \cup (1, 3) \quad \ldots\text{(iv)}$

1. From Eq. (iv), $\alpha^2 \in \left(\frac{1}{3}, 1\right) \cup (1, 3)$

2. Put $\alpha^2 = 2$ in Eq. (iii), then

$r^2-3r+1=0$

$\therefore \quad r=\frac{3\pm\sqrt{5}}{2}=\frac{3+\sqrt{5}}{2} \quad [\because r>1]$

$\therefore \quad [r]=2$

3. Put $r=2$ in Eq. (iii), then

$4(\alpha^2-1)-2(\alpha^2+1)+\alpha^2-1=0$

or $\quad 3\alpha^2-7=0$

$\therefore \quad \alpha^2=\frac{7}{3}$

$\therefore \quad (\alpha^2)=3$

4. Here, $S=10\sqrt{3}$

$\because \quad a^2\left(\frac{1}{r^2}+1+r^2\right)=S^2$

or $\quad \frac{1}{r^2}+1+r^2=\frac{S^2}{a^2}=\frac{300}{a^2}$

or $\quad \frac{1}{r^2}+r^2=\left(\frac{300}{a^2}-1\right)$

$\frac{1}{r^2}+r^2>2 \quad (\because r>1)$

$\Rightarrow \quad \frac{300}{a^2}-1>2$

or $\quad 300-3a^2>0$

or $\quad a^2<100$

$\therefore \quad -10<a<10$

$\Rightarrow$ Greatest value of a (middle term) is 10.

5. Put $r^2=1$ or $r=\pm 1$ in Eq. (iii), then

$(\alpha^2-1)\pm(\alpha^2+1)+\alpha^2-1=0$

Taking '+' sign, then $3\alpha^2-1=0 \Rightarrow \alpha=\pm\frac{1}{\sqrt{3}}$

and taking '–' sign, then $\alpha^2=3$

$\alpha=\pm\sqrt{3}$

PASSAGE 6

1. $\frac{a^{1/3}+b^{1/3}}{2}<\left(\frac{a+b}{2}\right)^{1/3}=\left(\frac{1}{2}\right)^{1/3} \quad [\because a+b=1]$

$\therefore \quad \frac{a^{1/3}+b^{1/3}}{2}<\frac{1}{2^{1/3}}$

$\Rightarrow \quad A<\frac{2}{2^{1/3}}$ or $A<2^{2/3}$

2. $\because \quad \frac{(x^2)^{1/2}+(y^2)^{1/2}}{2}\leq\left(\frac{x^2+y^2}{2}\right)^{1/2}=\left(\frac{8}{2}\right)^{1/2}=2$

$\Rightarrow \quad \frac{|x|+|y|}{2}\leq 2$ or $|x|+|y|\leq 4$

$\Rightarrow \quad x+y\leq 4 \quad [\because x$ and y are positive]

3. $\because \quad \frac{b^2+c^2}{2}>\left(\frac{b+c}{2}\right)^2 \Rightarrow \frac{b^2+c^2}{b+c}>\left(\frac{b+c}{2}\right) \quad \ldots\text{(i)}$

Similarly, $\quad \frac{c^2+a^2}{c+a}>\left(\frac{c+a}{2}\right) \quad \ldots\text{(ii)}$

and $\quad \frac{a^2+b^2}{a+b}>\left(\frac{a+b}{2}\right) \quad \ldots\text{(iii)}$

Adding Eqs. (i), (ii) and (iii), then

$\frac{b^2+c^2}{b+c}+\frac{c^2+a^2}{c+a}+\frac{a^2+b^2}{a+b}>a+b+c$

$=1 \quad [\because a+b+c=1]$

Hence, $\frac{b^2+c^2}{b+c}+\frac{c^2+a^2}{c+a}+\frac{a^2+b^2}{a+b}>1$

i.e. $\frac{b^2+c^2}{b+c}+\frac{c^2+a^2}{c+a}+\frac{a^2+b^2}{a+b}\in(1,\infty)$

4. $\because \quad \frac{\left(a+\frac{1}{a}\right)^2+\left(b+\frac{1}{b}\right)^2}{2}>\left(\frac{a+\frac{1}{a}+b+\frac{1}{b}}{2}\right)^2$

$\therefore \quad \left(a+\frac{1}{a}\right)^2+\left(b+\frac{1}{b}\right)^2>\frac{1}{2}(1+a^{-1}+b^{-1})^2 \quad \ldots\text{(i)}$

$[\because a+b=1]$

Again, $\frac{a^{-1}+b^{-1}}{2}>\left(\frac{a+b}{2}\right)^{-1}=\left(\frac{1}{2}\right)^{-1} \quad [\because a+b=1]$

$=2$

$\therefore \quad a^{-1}+b^{-1}>4$

or $\quad (1+a^{-1}+b^{-1})>5$

or $\quad \frac{1}{2}(1+a^{-1}+b^{-1})^2>\frac{25}{2} \quad \ldots\text{(ii)}$

From Eqs. (i) and (ii), we get

$\left(a+\frac{1}{a}\right)^2+\left(b+\frac{1}{b}\right)^2>\frac{25}{2}$ or $A>\frac{25}{2}$

5. $\because \dfrac{a^n + c^n}{2} > \left(\dfrac{a+c}{2}\right)^n$ or $a^n + c^n > 2\left(\dfrac{a+c}{2}\right)^n$ ...(i)

But we know that, AM > GM > HM

$\therefore$ AM > HM

$\Rightarrow \dfrac{a+c}{2} > b$

or $\left(\dfrac{a+c}{2}\right)^n > b^n \Rightarrow 2\left(\dfrac{a+c}{2}\right)^n > 2b^n$ ...(ii)

From Eqs. (i) and (ii), we get

$a^n + c^n > 2b^n$ or $\dfrac{a^n + c^n}{b^n} > 2$ or $\lambda > 2$

PASSAGE 7

1. $\because \dfrac{1^m + 3^m + 5^m + \ldots + (2n-1)^m}{n}$

$$> \left(\frac{1+3+5+\ldots+2n-1}{n}\right)^m = \left(\frac{\frac{n}{2}(1+2n-1)}{n}\right)^m = n^m$$

$\therefore 1^m + 3^m + 5^m + \ldots + (2n-1)^m > n^{m+1}$

or $\lambda > n^{m+1}$

2. $\because \dfrac{\left(\dfrac{b+c}{a+b+c}\right)^{-1}\left(\dfrac{c+a}{a+b+c}\right)^{-1} + \left(\dfrac{a+b}{a+b+c}\right)^{-1}}{3}$

$$\geq \left(\frac{\dfrac{b+c}{a+b+c} + \dfrac{c+a}{a+b+c} + \dfrac{a+b}{a+b+c}}{3}\right)^{-1}$$

$$= \left(\frac{2(a+b+c)}{3(a+b+c)}\right)^{-1} = \frac{3}{2}$$

or $\dfrac{\left(\dfrac{b+c}{a+b+c}\right)^{-1} + \left(\dfrac{c+a}{a+b+c}\right)^{-1} + \left(\dfrac{a+b}{a+b+c}\right)^{-1}}{3} \geq \dfrac{3}{2}$

or $\dfrac{a+b+c}{b+c} + \dfrac{a+b+c}{c+a} + \dfrac{a+b+c}{a+b} \geq \dfrac{9}{2}$

or $\dfrac{a}{b+c} + 1 + \dfrac{b}{c+a} + 1 + \dfrac{c}{a+b} + 1 \geq \dfrac{9}{2}$

or $\dfrac{a}{b+c} + \dfrac{b}{c+a} + \dfrac{c}{a+b} \geq \dfrac{3}{2}$

$\therefore \dfrac{a}{b+c} + \dfrac{b}{c+a} + \dfrac{c}{a+b} \in \left[\dfrac{3}{2}, \infty\right)$

3. $\because \dfrac{(2-x)^{-1} + (2-y)^{-1} + (2-z)^{-1}}{3}$

$$\geq \left(\frac{2-x+2-y+2-z}{3}\right)^{-1}$$

$$= \left(\frac{6-1}{3}\right)^{-1} \quad [\because x+y+z=1]$$

$$= \frac{3}{5}$$

$\therefore \dfrac{1}{2-x} + \dfrac{1}{2-y} + \dfrac{1}{2-z} \geq \dfrac{9}{5}$

or $\dfrac{2}{2-x} + \dfrac{2}{2-y} + \dfrac{2}{2-z} \geq \dfrac{18}{5}$

$\Rightarrow 1 + \dfrac{x}{2-x} + 1 + \dfrac{y}{2-y} + 1 + \dfrac{z}{2-z} \geq \dfrac{18}{5}$

$\Rightarrow \dfrac{x}{2-x} + \dfrac{y}{2-y} + \dfrac{z}{2-z} \geq \dfrac{3}{5}$

4. $\because \dfrac{a_1^2 + a_2^2 + a_3^2 + \ldots + a_n^2}{n} > \left(\dfrac{a_1 + a_2 + a_3 + \ldots + a_n}{n}\right)^2$

or $\dfrac{A}{n} > \dfrac{(a_1 + a_2 + a_3 + \ldots + a_n)^2}{n^2}$

Also, $(a_1 + a_2 + a_3 + \ldots + a_n)^2$

$= a_1^2 + a_2^2 + a_3^2 + \ldots + a_n^2 + 2\Sigma a_1 a_2$

or $(a_1 + a_2 + a_3 + \ldots + a_n)^2$

$= A + 2\Sigma a_1 a_2 > A$

$\therefore (a_1 + a_2 + a_3 + \ldots + a_n)^2 > A$ ...(ii)

From Eqs. (i) and (ii), we get

$A < (a_1 + a_2 + a_3 + \ldots + a_n)^2 < nA$

$\therefore G = nA$ and $L = A$

Hence, $G - L = (n-1)A$

5. $\because \dfrac{\left(\dfrac{b+c+d}{3}\right)^{-1/2} + \left(\dfrac{c+d+a}{3}\right)^{-1/2} + \left(\dfrac{d+a+b}{3}\right)^{-1/2} + \left(\dfrac{a+b+c}{3}\right)^{-1/2}}{4}$

$$> \left(\frac{\dfrac{b+c+d}{3} + \dfrac{c+d+a}{3} + \dfrac{d+a+b}{3} + \dfrac{a+b+c}{3}}{4}\right)^{-1/2}$$

$$= \left(\frac{a+b+c+d}{4}\right)^{-1/2}$$

$$= \left(\frac{1}{3}\right)^{1/2} \quad [\because a+b+c+d = 4/3]$$

or $\sqrt{\left(\dfrac{3}{b+c+d}\right)} + \sqrt{\left(\dfrac{3}{c+d+a}\right)} + \sqrt{\left(\dfrac{3}{d+a+b}\right)} + \sqrt{\left(\dfrac{3}{a+b+c}\right)} > 4\sqrt{3}$

$\therefore \sqrt{\left(\dfrac{1}{b+c+d}\right)} + \sqrt{\left(\dfrac{1}{c+d+a}\right)} + \sqrt{\left(\dfrac{1}{d+a+b}\right)} + \sqrt{\left(\dfrac{1}{a+b+c}\right)} > 4$

PASSAGE 8

1. $\because A^* \geq G^*$

$\Rightarrow \dfrac{a + \dfrac{b}{3} + \dfrac{b}{3} + \dfrac{b}{3} + c}{5} \geq \left(a \cdot \dfrac{b}{3} \cdot \dfrac{b}{3} \cdot \dfrac{b}{3} \cdot c\right)^{1/5}$

$$\Rightarrow \quad \frac{a+b+c}{5} \geq \left(\frac{ab^3c}{27}\right)^{1/5}$$

$$\Rightarrow \quad 1 \geq \left(\frac{ab^3c}{27}\right)^{1/5} \quad [\because a+b+c=5]$$

$$\therefore \quad ab^3c \leq 27$$

2. $\because A^* \geq H^*$

$$\Rightarrow \quad \frac{a+\frac{b}{3}+\frac{b}{3}+\frac{b}{3}+c}{5} \geq \frac{5}{\frac{1}{a}+\frac{3}{b}+\frac{3}{b}+\frac{3}{b}+\frac{1}{c}}$$

$$\Rightarrow \quad \frac{a+b+c}{5} \geq \frac{5}{\frac{1}{a}+\frac{9}{b}+\frac{1}{c}}$$

$$\Rightarrow \quad 1 \geq \frac{5}{\frac{1}{a}+\frac{9}{b}+\frac{1}{c}} \quad [\because a+b+c=5]$$

$$\text{or} \quad \frac{1}{5} \geq \frac{1}{\left(\frac{1}{a}+\frac{9}{b}+\frac{1}{c}\right)}$$

3. $\because A^* \geq G^*$

$$\therefore \quad \frac{\left(\frac{3x}{2}\right)+\left(\frac{3x}{2}\right)+\left(\frac{4y}{3}\right)+\left(\frac{4y}{3}\right)+\left(\frac{4y}{3}\right)}{5} \geq \left(\frac{3x}{2}\cdot\frac{3x}{2}\cdot\frac{4y}{3}\cdot\frac{4y}{3}\cdot\frac{4y}{3}\right)^{1/5}$$

$$\Rightarrow \quad \frac{3x+4y}{5} \geq \left(\frac{3^2\cdot 4^3 x^2 y^3}{2^2\cdot 3^3}\right)^{1/5}$$

$$= \left(\frac{6\cdot 4^3}{2^2\cdot 3}\right)^{1/5} \quad [\because x^2y^3 = 6]$$

$$= (32)^{1/5} = 2$$

$$\therefore \quad 3x+4y \geq 10$$

4. $\because A^* \geq G^* \geq H^*$

$$\frac{x+x+y+y+y}{5} \geq (x\cdot x\cdot y\cdot y\cdot y)^{1/5} \geq \frac{5}{\frac{1}{x}+\frac{1}{x}+\frac{1}{y}+\frac{1}{y}+\frac{1}{y}}$$

$$\Rightarrow \quad \frac{2x+3y}{5} \geq (x^2y^3)^{1/5} \geq \frac{5}{\frac{2}{x}+\frac{3}{y}}$$

$$\Rightarrow \quad \frac{2x+3y}{5} \geq (6)^{1/5} \geq \frac{5xy}{3x+2y}$$

5. $\because A^* \geq G^*$

$$\therefore \quad \frac{\frac{\lambda^2}{d_1^2}+\frac{\mu^2}{d_2^2}+\frac{v^2}{d_3^2}}{3} \geq \left(\frac{\lambda^2}{d_1^2}\cdot\frac{\mu^2}{d_2^2}\cdot\frac{v^2}{d_3^2}\right)^{1/3}$$

$$\Rightarrow \quad \left(\frac{1}{3}\right)^3 \geq \frac{\lambda^2\mu^2v^2}{d_1^2d_2^2d_3^2}$$

$$\text{or} \quad \lambda^2\mu^2v^2 \leq \frac{d_1^2d_2^2d_3^2}{27} \quad \text{or} \quad \lambda_1\lambda_2\lambda_3 \leq \frac{d_1d_2d_3}{3\sqrt{3}}$$

Numerical Value Type

1. Given $x_1, x_2, x_3, \ldots\, x_{2008}$ are in HP.

$\therefore \quad \frac{1}{x_1}, \frac{1}{x_2}, \frac{1}{x_3}, \ldots, \frac{1}{x_{2007}}, \frac{1}{x_{2008}}$ are in AP.

Then, $\frac{1}{x_2}-\frac{1}{x_1} = \frac{1}{x_3}-\frac{1}{x_2}$

$$= \ldots = \frac{1}{x_{2008}} - \frac{1}{x_{2007}} = D$$

$$\text{or} \quad \frac{x_1-x_2}{x_1x_2} = \frac{x_2-x_3}{x_2x_3} = \ldots = \frac{x_{2007}-x_{2008}}{x_{2007}\,x_{2008}} = D$$

$$= \frac{(x_1-x_2)+(x_2-x_3)+\ldots+(x_{2007}-x_{2008})}{x_1x_2+x_2x_3+\ldots+x_{2007}x_{2008}} = D$$

[by law of proportion]

$$\text{or} \quad x_1x_2+x_2x_3+\ldots+x_{2007}x_{2008} = \frac{x_1-x_{2008}}{D}$$

$$\text{or} \quad \sum_{i=1}^{2007} x_ix_{i+1} = \frac{x_1-x_{2008}}{D} \quad \ldots\text{(i)}$$

$$\text{Now,} \quad \frac{1}{x_{2008}} = \frac{1}{x_1} + (2008-1)D$$

$$\text{or} \quad \frac{x_1-x_{2008}}{D} = 2007\,x_1x_{2008} \quad \ldots\text{(ii)}$$

$\therefore$ From Eqs. (i) and (ii), we get

$$\sum_{i=1}^{2007} x_ix_{i+1} = 2007\,x_1x_{2008}$$

$$\therefore \quad \lambda = 2007$$

2. Let the three digits a, ar, ar^2, then according to hypothesis

$$100a+10\,ar+ar^2-792 = 100ar^2+10ar+a$$

$$\Rightarrow \quad a(1-r^2) = 8 \quad \ldots\text{(i)}$$

and a, $ar+2$, ar^2 are in AP.

Then, $\quad 2(ar+2) = a+ar^2 \Rightarrow a(1-r)^2 = 4 \quad \ldots$(ii)

Dividing Eq. (i) by Eq. (ii), then

$$\frac{1+r}{1-r} = 2 \Rightarrow r = \frac{1}{3}$$

From Eq. (i), $a = 9$

Thus, digits are 9, 3, 1 and so the required number is 931.

3. We have, for $n \geq 1$

$$T_n = \sum_{r=1}^{n} T_r - \sum_{r=1}^{n-1} T_r$$

$$= \frac{n(n+1)(n+2)(n+3)}{8} - \frac{(n-1)(n)(n+1)(n+2)}{8}$$

$$= \frac{1}{2}n(n+1)(n+2)$$

Thus, $\sum_{r=1}^{n} \frac{1}{T_r} = \sum_{r=1}^{n} \frac{2}{r(r+1)(r+2)} = \sum_{r=1}^{n} \frac{(r+2)-r}{r(r+1)(r+2)}$

$= \sum_{r=1}^{n} \left(\frac{1}{r(r+1)} - \frac{1}{(r+1)(r+2)}\right)$

$= -\sum_{r=1}^{n} \left(\frac{1}{(r+1)(r+2)} - \frac{1}{r(r+1)}\right)$

$= -\left(\frac{1}{(n+1)(n+2)} - \frac{1}{2}\right)$

$= \frac{1}{2}\left(1 - \frac{2}{(n+1)(n+2)}\right)$

$\therefore \quad \lim_{n\to\infty} \sum_{r=1}^{n} \frac{1}{T_r} = \frac{1}{2}(1-0) = \frac{1}{2}$

Hence, $\lim_{n\to\infty} \sum_{r=1}^{n} \frac{4016}{T_r} = \frac{4016}{2} = 2008$

4. The 4th AM out of m AMs inserted between 3 and 243 is

$$A_4 = 3 + 4\frac{243-3}{m+1} \quad \ldots(\text{i})$$

and the 2nd GM out of three GMs inserted between 3 and 243 is

$$G_2 = 3\left(\left(\frac{243}{3}\right)^{\frac{1}{3+1}}\right)^2 = 3(81)^{1/2} = 27$$

$\because \quad A_4 = G_2 \Rightarrow 3 + 4\left(\frac{240}{m+1}\right) = 27$

$\Rightarrow \quad 4\left(\frac{240}{m+1}\right) = 24 \Rightarrow m+1 = 40 \Rightarrow m = 39$

Hence, answer is 39.00.

5. $\because P(x) = (1 + x + 2x^2 + 3x^3 + \ldots + (n-1)x^{n-1} + nx^n)$
$\times (1 + x + 2x^2 + 3x^3 + \ldots + (n-1)x^{n-1} + nx^n)$

$\therefore \quad a_n = 1\cdot n + 2\cdot(n-1) + \ldots + (n-1)\cdot 1 + n\cdot 1$

$= \sum_{r=1}^{n} r(n-r+1)$

$\therefore \quad a_{24} = \sum_{r=1}^{24} r(25-r) = 25\sum_{r=1}^{24} r - \sum_{r=1}^{24} r^2$

$= \frac{25\cdot 24\cdot 25}{2} - \frac{24\cdot 25\cdot 49}{6}$

$= 7500 - 4900 = 2600$

6. $T_r = \frac{r}{r^4 + r^2 + 1}$

$= \frac{1}{2}\left\{\frac{(r^2+r+1)-(r^2-r+1)}{(r^2+r+1)(r^2-r+1)}\right\}$

$= \frac{1}{2}\left\{\frac{1}{r^2-r+1} - \frac{1}{r^2+r+1}\right\}$

$= \frac{1}{2}\left\{\frac{1}{r^2-r+1} - \frac{1}{(r+1)^2-(r+1)+1}\right\}$

$= \frac{1}{2}\{f(r) - f(r+1)\}$

$S = \Sigma T_r$

$= \sum_{r=1}^{80} \frac{1}{2}\{f(r) - f(r+1)\}$

$= \frac{1}{2}\{f(1) - f(2) + f(2) - f(3) + f(3) - f(4) + \ldots + f(80) - f(81)\}$

$= \frac{1}{2}\{f(1) - f(81)\}$

$= \frac{1}{2}\left(1 - \frac{1}{81^2 - 81 + 1}\right) = \frac{1}{2}\left(1 - \frac{1}{6481}\right)$

$= \frac{1}{2} \times \frac{6480}{6481} \Rightarrow \frac{3240}{6481}$

$\therefore \quad \frac{6481}{1000} \times \frac{3240}{6481} = 3.24$

7. Required sum

$= (12 + 14 + 16 + \ldots + 98) - (20 + 30 + 40 + \ldots + 90)$

$= \frac{44}{2}(12 + 98) - \frac{8}{2}(20 + 90)$

$= 22(110) - 4(110) = (110)(18) = 1980$

8. Since, one pair of 2 and 5 makes 10 (*i.e.*, $2 \times 5 = 10$) which gives one zero at the end.

Here, $5^6, 10^{11}, 15^{16}, 20^{21}, 25^{26}, 30^{31}, 35^{36}, 40^{41}, 45^{46}, 50^{51}$

$\therefore$ Total numbers of fives

$= 6 + 11 + 16 + 21 + 26 + 31 + 36 + 41 + 46 + 51$

$= \frac{10}{2}(6 + 51) = 5 \times 57 = 285$

Clearly number of 2's is greater than that of 5. Thus, number of zeroes at the end = 285.

9. From the first equation

$$\log_{10} x\left\{1 + \frac{1}{2} + \frac{1}{4} + \ldots \infty\right\} = y$$

$\Rightarrow \quad \log_{10} x\left\{\frac{1}{1-\frac{1}{2}}\right\} = y$

or $\quad 2\log_{10} x = y \quad \ldots(\text{i})$

From the second equation

$$\frac{1 + 3 + 5 + \ldots + (2y-1)}{4 + 7 + 10 + \ldots + (3y+1)} = \frac{20}{7\log_{10} x}$$

$\Rightarrow \quad \frac{\frac{y}{2}\{1 + 2y - 1\}}{\frac{y}{2}\{4 + 3y + 1\}} = \frac{20}{7\left(\frac{y}{2}\right)} \quad$ [from Eq. (i)]

$\Rightarrow \quad \frac{2y}{3y+5} = \frac{40}{7y} \Rightarrow 7y^2 = 60y + 100$

$\Rightarrow \quad 7y^2 - 60y - 100 = 0$

$\Rightarrow \quad (y-10)(7y+10) = 0$

$\therefore \quad y = 10, y \neq -\frac{10}{7} \quad [\because y \in N]$

Now, from Eq. (i), $\quad 2\log_{10} x = 10$

$\Rightarrow \quad \log_{10} x = 5$ or $x = 10^5$

$\therefore \quad \log_y x = \log_{10} 10^5 = 5$

$\therefore \quad (\log_y x)^5 = 5^5 = 3125$

10. $\because$ a, A_1, A_2, b are in AP.

$\therefore \quad A_1 + A_2 = a + b \quad \ldots(i)$

$\Rightarrow \quad a, G_1, G_2, b$ are in GP

$\therefore \quad G_1G_2 = ab \quad \ldots(ii)$

and a, H_1, H_2, b are in HP

$$\therefore \frac{1}{H_1} + \frac{1}{H_2} = \frac{1}{a} + \frac{1}{b} = \frac{a+b}{ab} = \frac{A_1 + A_2}{G_1G_2}$$

{from Eqs. (i) and (ii)}

$$\therefore \frac{G_1G_2}{H_1H_2} = \frac{A_1 + A_2}{H_1 + H_2} \quad \ldots(iii)$$

Also, $$\frac{1}{H_1} = \frac{1}{a} + \frac{\left(\frac{1}{b} - \frac{1}{a}\right)}{3} = \frac{(a+2b)}{3ab}$$

and $$\frac{1}{H_2} = \frac{1}{a} + \frac{2\left(\frac{1}{b} - \frac{1}{a}\right)}{3} = \frac{(2a+b)}{3ab}$$

$$\therefore \frac{1}{H_1H_2} = \frac{(a+2b)(2a+b)}{9a^2b^2}$$

or $$\frac{G_1G_2}{H_1H_2} = \frac{(a+2b)(2a+b)}{9ab} \quad \ldots(iv)$$

From Eqs. (iii) and (iv), we get

$$\frac{G_1G_2}{H_1H_2} = \frac{A_1 + A_2}{H_1 + H_2} = \frac{(a+2b)(2a+b)}{9ab}$$

Hence, $\lambda = 2$ and $\mu = 9$

Then, $\lambda^\mu + \mu^\lambda = 2^9 + 9^2 = 512 + 81 = 593$

11. Given that, $AP(a; d)$ denote the set of all the terms of an infinite arithmetic progression with first term 'a' and common difference $d > 0$.

Now, let mth term of first progression

$AP(1\,;3) = 1 + (m-1)3 = 3m - 2 \quad \ldots(i)$

and nth term of progression

$AP(2; 5) = 2 + (n-1)5 = 5n - 3 \quad \ldots(ii)$

and rth term of third progression $AP(3; 7)$

$= 3 + (r-1)7 = 7r - 4 \quad \ldots(iii)$

are equal.

Then, $3m - 2 = 5n - 3 = 7r - 4$

Now, for $AP(1\,;3) \cap AP(2;5) \cap AP(3;7)$, the common terms of first and second progressions, $m = \frac{5n-1}{3}$

$\Rightarrow n = 2, 5, 11, \ldots$ and the common terms of second and the third progressions, $r = \frac{5n+1}{7} \Rightarrow n = 4, 11, \ldots$

Now, the first common term of first, second and third progressions (when $n = 11$), so $a = 2 + (11-1)5 = 52$

and $d = \text{LCM}(3, 5, 7) = 105$

So, $AP(1\,;3) \cap AP(2;5) \cap AP(3\,;7) = AP(52\,;105)$

So, $a = 52$ and $d = 105 \Rightarrow a + d = 157.00$

12. $T_r = r(2021 - r)$

$$S = \sum_{r=1}^{2020} T_r = 2021 \sum_{r=1}^{2020} r - \sum_{r=1}^{2020} r^2$$

$$= 2021 \times \frac{2020 \times 2021}{2} - \frac{2020 \times 2021 \times 4041}{6}$$

$$= \frac{2020 \times 2021}{2}\left[2021 - \frac{4041}{3}\right]$$

$$= \frac{2020 \times 2021}{2}\left[\frac{6063 - 4041}{3}\right] = \frac{2020 \times 2021 \times 2022}{6}$$

$= 2020 \times 2021 \times 337$

It is given that $S = 2020 \times 2021 \times k$

Therefore, $k = 337 \Rightarrow \frac{k}{100} = \frac{337}{100} = 3.37$

Single Integer Type

1. Let the first term of AP be a and common difference be d

T_{20th} of HP $= 1$ T_{20th} of AP $= 1$

T_{30th} of HP $= \frac{-1}{17}$ T_{30th} of AP $= -17$

$a + 19d = 1 \quad \ldots(i)$

$a + 29d = -17 \quad \ldots(ii)$

Substract Eq. (i) from Eq. (ii), we get

$10d = -18$

$\Rightarrow \quad a = \frac{176}{5}, d = \frac{-9}{5}$

$$T_{nth} \text{ term of HP} = \frac{1}{a + (n-1)d}$$

$$= \frac{1}{\frac{176}{5} + (n-1)\left(\frac{-9}{5}\right)} = \frac{5}{185 - 9n}$$

For the largest term $185 - 9n$ will be least $n = 20$ the 20th term of the given HP is the largest term

$$T_{20} = \frac{5}{185 - 180} = 1.$$

2. Let the sides are $a - d$, a and $a + d$.

Then, $a(a-d) = 48 \quad \ldots(i)$

and $a^2 - 2ad + d^2 + a^2 = a^2 + 2ad + d^2$

$\Rightarrow \quad a^2 = 4ad \Rightarrow a = 4d \quad \ldots(ii)$

From Eqs. (i) and (ii), we get

$a = 8, d = 2$

Hence, $a - d = 6$

3. $(a + md)(a + rd) = (a + nd)^2$ and $n = \frac{2mr}{m+r}$

Therefore, $a^2 + ad(m+r) + mrd^2 = a^2 + 2$ and $+ n^2d^2$

$a(m+r) + mrd = 2an + n^2d$

$d(n^2 - mr) = a(m + r - 2n)$

$$\frac{d}{a} = \frac{(m+r) - 2n}{n^2 - mr} = \frac{(m+r) - 2n}{n^2 - [n(m+r)/2]} \quad \left[\because mr = \frac{n(m+r)}{2}\right]$$

$$= \frac{2[m + r - 2n]}{n(2n - m - r)} = \frac{-2}{n}$$

$\therefore |k| = 2$

4. Let a be the first term and r be the common ratio of GP.

Then, $\dfrac{a(1-r^{201})}{1-r}=625$

Now, $\sum_{i=1}^{201}\dfrac{1}{a_i}=\left(\dfrac{1}{a_1}+\dfrac{1}{a_2}+\ldots+\dfrac{1}{a_{201}}\right)$

$$=\frac{1}{a}+\frac{1}{ar}+\ldots+\frac{1}{ar^{200}}=\frac{\frac{1}{a}\left(\left(\frac{1}{r}\right)^{201}-1\right)}{\left(\frac{1}{r}-1\right)}$$

$$=\frac{1}{a}\left(\frac{1-r^{201}}{1-r}\right)\frac{1}{r^{200}}$$

$$=\frac{1}{a}\times\frac{625}{a}\times\frac{1}{r^{200}} \quad \text{[from Eq. (i)]}$$

$$=\frac{625}{(ar^{100})^2}=\frac{625}{(a_{101})^2}=\frac{625}{625}=1$$

5. $\sum_{k=2}^{100}|(k^2-3k+1)S_k|$

For $k=2$, $|(k^2-3k+1)S_k|=1$

$$\therefore \sum_{k=3}^{100}\left|\frac{k-1}{(k-2)!}-\frac{k-1+1}{(k-1)!}\right|$$

$$=\sum_{k=3}^{100}\frac{1}{(k-3)!}+\frac{1}{(k-2)!}-\frac{1}{(k-2)!}-\frac{1}{(k-1)!}$$

$$=\sum_{k=3}^{100}\left(\frac{1}{(k-3)!}-\frac{1}{(k-1)!}\right)$$

$$S=1+\left(1-\frac{1}{2!}\right)+\left(\frac{1}{1!}-\frac{1}{3!}\right)+\left(\frac{1}{2!}-\frac{1}{4!}\right)$$

$$+\left(\frac{1}{3!}-\frac{1}{5!}\right)+\left(\frac{1}{4!}-\frac{1}{6!}\right)+\ldots+\left(\frac{1}{94!}-\frac{1}{96!}\right)$$

$$+\left(\frac{1}{95!}-\frac{1}{97!}\right)+\left(\frac{1}{96!}-\frac{1}{98!}\right)+\left(\frac{1}{97!}-\frac{1}{99!}\right)$$

$$=2-\frac{1}{98!}-\frac{1}{99!}$$

$$\therefore \quad E=\frac{100^2}{100!}+3-\frac{1}{98!}-\frac{1}{99.98!}$$

$$=\frac{100^2}{100!}+3-\frac{100}{99!}=\frac{100^2}{100.99!}+3-\frac{100}{99!}=3$$

6. Let a_n denote the length of a side of S_n

Length of a side of S_n = Length of a diagonal of S_{n+1}

$\Rightarrow \quad a_n=\sqrt{2}\,a_{n+1}\Rightarrow\dfrac{a_{n+1}}{a_n}=\dfrac{1}{\sqrt{2}}$

Thus, $a_1, a_2, a_3, \ldots$ is a GP with first term 10 and common ratio $1/\sqrt{2}$.

Therefore, $a_n=10(1/\sqrt{2})^{n-1}$

Also, area $(S_n)=a^{2n}<1$

$\Rightarrow [10(1/\sqrt{2})^{n-1}]^2<1\Rightarrow 100<2^{n-1}$

$\Rightarrow$ Smallest possible value of n is 8.

7. Let a, b, c be a, ar, ar^2, where $r\in N$

Also, $\dfrac{a+b+c}{3}=b+2$

$\Rightarrow a+ar+ar^2=3(ar)+6$

$\Rightarrow ar^2-2ar+a=6$

$\Rightarrow \quad (r-1)^2=\dfrac{6}{a}$

Since, $6/a$ must be perfect square and $a\in N$. So, a can be 6 only.

$\Rightarrow \quad r-1=\pm1\Rightarrow r=2$

and $\dfrac{a^2+a-14}{a+1}=\dfrac{36+6-14}{7}=4$

8. Given arithmetic progression of positive integers terms $a_1, a_2, a_3, \ldots\ldots$ having common difference '2' and geometric progression of positive integers terms $b_1, b_2, b_3, \ldots\ldots$ having common ratio '2' with $a_1=b_1=c$, such that $2(a_1+a_2+a_3+\ldots+a_n)=b_1+b_2+b_3+\ldots+b_n$

$$\Rightarrow \quad 2\times\frac{n}{2}[2C+(n-1)2]=C\left(\frac{2^n-1}{2-1}\right)$$

$$\Rightarrow \quad 2nC+2n^2-2n=2^n\cdot C-C$$

$$\Rightarrow \quad C[2^n-2n-1]=2n^2-2n$$

$$\because \quad C\in N\Rightarrow 2n^2-2n\ge 2^n-2n-1$$

$$\Rightarrow \quad 2n^2+1\ge 2^n\Rightarrow n\le 6$$

and, also $\quad C>0\Rightarrow n>2$

$\therefore$ The possible values of n are 3, 4, 5, 6

So, at $n=3$, $C=\dfrac{(2\times9)-6}{8-6-1}=12$

at, $n=4$, $C=\dfrac{32-8}{16-8-1}=\dfrac{24}{9}=\dfrac{8}{3}\notin N$

at, $n=5$, $C=\dfrac{50-10}{32-10-1}=\dfrac{40}{21}\notin N$

and at, $n=6$, $C=\dfrac{72-12}{64-12-1}=\dfrac{60}{51}\notin N$

$\therefore$ The required value of $C=12$ for $n=3$.

9. $\therefore S_n=\dfrac{1\left(1-\left(\frac{1}{2}\right)^n\right)}{\left(1-\frac{1}{2}\right)}=2-\dfrac{1}{2^{n-1}}$

or $\quad 2-S_n=\dfrac{1}{2^{n-1}}<\dfrac{1}{100} \quad \left[\because 2-S_n<\dfrac{1}{100}\right]$

or $\quad 2^{n-1}>100>2^6\Rightarrow 2^{n-1}>2^6\Rightarrow n-1>6$

$\therefore \quad n>7$

Hence, least value of n is 8.

10. Given, $\dfrac{S_7}{S_{11}}=\dfrac{6}{11}$ and $130<t_7<140$

$$\Rightarrow \quad \frac{\frac{7}{2}[2a+6d]}{\frac{11}{2}[2a+10d]}=\frac{6}{11}\Rightarrow\frac{7(2a+6d)}{(2a+10d)}=6$$

$\Rightarrow \quad a=9d \quad \ldots(i)$

Also, $\quad 130<t_7<140$

$\Rightarrow \quad 130<a+6d<140$

$\Rightarrow \quad 130<9d+6d<140 \quad$ [from Eq. (i)]

$\Rightarrow \quad 130<15d<140$

$\Rightarrow \quad \dfrac{26}{3}<d<\dfrac{28}{3} \quad$ [since, d is a natural number]

$\therefore \quad d=9$

11. Let $P = \dfrac{a+b}{2a-b} + \dfrac{c+b}{2c-b}$

$$= \frac{a + \frac{2ac}{a+c}}{2a - \frac{2ac}{a+c}} + \frac{c + \frac{2ac}{a+c}}{2c - \frac{2ac}{a+c}} \qquad \left[\because b = \frac{2ac}{a+c}\right]$$

$$= \frac{a+3c}{2a} + \frac{3a+c}{2c}$$

$$= 1 + \frac{3}{2}\left(\frac{c}{a} + \frac{a}{c}\right) > 4 \qquad \left[\begin{array}{l}\because \frac{a}{c} + \frac{c}{a} > 2 \\ \therefore \frac{3}{2}\left(\frac{a}{c} + \frac{c}{a}\right) > 3\end{array}\right]$$

Also, $\sqrt{\lambda\sqrt{\lambda\sqrt{\lambda \ldots \infty}}} = \lambda^{\frac{1}{2}+\frac{1}{4}+\frac{1}{8}+\ldots\infty} = \lambda^{\frac{1/2}{1-1/2}} = \lambda$

$\therefore \qquad \lambda = 4$

12. Let number of removed cards be k and $(k+1)$.

$\therefore \qquad \dfrac{n(n+1)}{2} - k - (k+1) = 1224$

$\Rightarrow \qquad n^2 + n - 4k = 2450$

$\Rightarrow \qquad n^2 + n - 2450 = 4k \Rightarrow (n+50)(n-49) = 4k$

$\therefore \qquad n > 49$

Let $\qquad n = 50$

$\therefore \qquad 100 = 4k \Rightarrow k = 25$

Now, $\qquad k - 20 = 5$

Matrix-Match Type

1. (A) Given, $3(a^2 + b^2 + c^2 + 1) = 2(a + b + c + ab + bc + ca)$

$\Rightarrow \quad 2(a^2 + b^2 + c^2 - ab - bc - ca) + (a^2 + b^2 + c^2 - 2a - 2b - 2c + 3) = 0$

$\Rightarrow \quad \{(a-b)^2 + (b-c)^2 + (c-a)^2\} + \{(a-1)^2 + (b-1)^2 + (c-1)^2\} = 0$

$\Rightarrow a - b = b - c = c - a = a - 1 = b - 1 = c - 1 = 0$

$\Rightarrow \quad a = b = c = 1$

$\Rightarrow a, b, c$ are in AP and GP both. (P, Q)

Here, $a = b = c$, therefore they can not be in HP. (S)

(B) Let the three numbers be a, b, c. Then according to question,

$(a-b)^2, (b-c)^2, (c-a)^2$ are in AP.

$\Rightarrow \quad (b-c)^2 - (a-b)^2 = (c-a)^2 - (b-c)^2$

$\Rightarrow \quad (b - c + a - b)(b - c - a + b) = (c - a + b - c)(c - a - b + c)$

$\Rightarrow \quad (a-c)(b-c) + (a-c)(b-a) = (b-a)(c-a) + (b-a)(c-b)$

Dividing by $(a-b)(b-c)(c-a)$, then

$$-\frac{1}{(a-b)} + \frac{1}{(b-c)} = -\frac{1}{(b-c)} + \frac{1}{(c-a)}$$

$\Rightarrow \quad \dfrac{2}{(b-c)} = \dfrac{1}{(a-b)} + \dfrac{1}{(c-a)}$

$\therefore \quad \dfrac{1}{a-b}, \dfrac{1}{b-c}, \dfrac{1}{c-a}$ are in AP.

$\therefore \quad a - b, b - c, c - a$ are in HP. (R)

(C) Given, $(ax - by)^2 = (a-b)(ax^2 - by^2)$

$\Rightarrow \quad a^2x^2 + b^2y^2 - 2abxy = a^2x^2 + b^2y^2 - abx^2 - aby^2$

$\Rightarrow \quad ab(x^2 + y^2 - 2xy) = 0$

$\Rightarrow \quad ab(x-y)^2 = 0$

$\Rightarrow \quad x = y$

$\therefore \quad x = y = \dfrac{ax - by}{a - b}$. (S)

Hence, $x, y, \dfrac{ax-by}{a-b}$ are in AP and GP but not in HP.

(P, Q)

2. (A) $\because a, b, c, d$ are in AP.

Applying AM > GM

For first three numbers

$\Rightarrow \quad b > \sqrt{ac}$

$\Rightarrow \quad b^2 > ac \qquad \ldots(i)$

and for last three numbers

$c > \sqrt{bd}$

$\Rightarrow \quad c^2 > bd \qquad \ldots(ii)$

From Eqs. (i) and (ii),

$(b^2)(c^2) > (ac)(bd)$

$\Rightarrow \quad bc > ad$ (S)

Again applying, AM > HM

For first three numbers, $b > \dfrac{2}{\frac{1}{a} + \frac{1}{c}}$

$\Rightarrow \quad \dfrac{1}{a} + \dfrac{1}{c} > \dfrac{2}{b} \qquad \ldots(iii)$

And for last three numbers,

$c > \dfrac{2}{\frac{1}{b} + \frac{1}{d}}$

$\Rightarrow \quad \dfrac{1}{b} + \dfrac{1}{d} > \dfrac{2}{c} \qquad \ldots(iv)$

From Eqs. (iii) and (iv),

$\dfrac{1}{a} + \dfrac{1}{c} + \dfrac{1}{b} + \dfrac{1}{d} > \dfrac{2}{b} + \dfrac{2}{c}$

$\Rightarrow \quad \dfrac{1}{a} + \dfrac{1}{d} > \dfrac{1}{b} + \dfrac{1}{c}$ (R)

Also from Eq. (iii),

$bc + ab > 2ac \qquad \ldots(v)$

and from Eq. (iv),

$cd + bc > 2bd \qquad \ldots(vi)$

From Eqs. (v) and (vi),

$\Rightarrow \quad ab + cd + 2bc > 2(ac + bd)$

$\Rightarrow \quad ab + cd > 2(ac + bd - bc)$

Dividing by $abcd$, then we get

$c^{-1}d^{-1} + a^{-1}b^{-1} > 2(b^{-1}d^{-1} + a^{-1}c^{-1} - a^{-1}d^{-1})$ (T)

(B) $\because a, b, c, d$ are in GP.

Applying AM > GM

For first three numbers

$$\frac{a+c}{2} > b \Rightarrow a + c > 2b \qquad \ldots(i)$$

Similarly for last three terms

$$b + d > 2c \qquad \ldots(ii)$$

From Eqs. (i) and (ii),

$$a + c + b + d > 2b + 2c$$

$$\Rightarrow \quad a + d > b + c \text{ (P)}$$

Again, applying GM > HM

For first three numbers $b > \dfrac{2ac}{a+c}$

$$\Rightarrow \quad ab + bc > 2ac \qquad \ldots(iii)$$

Dividing by abc, we get

$$\frac{1}{c} + \frac{1}{a} > \frac{2}{b} \qquad \ldots(iv)$$

Similarly, for last three terms

$$bc + cd > 2bd \qquad \ldots(v)$$

Dividing by bcd, we get

$$\frac{1}{d} + \frac{1}{b} > \frac{2}{c} \qquad \ldots(vi)$$

Now, from Eqs. (iv) and (vi),

$$\frac{1}{a} + \frac{1}{d} > \frac{1}{b} + \frac{1}{c} \text{ (R)}$$

and from Eqs. (iii) and (v),

$$ab + cd + 2bc > 2ac + 2bd$$

$$\Rightarrow \quad ab + cd > 2(ac + bd - bc)$$

Dividing in each term by $abcd$, we get

$$c^{-1}d^{-1} + a^{-1}b^{-1} > 2(b^{-1}d^{-1} + a^{-1}c^{-1} - a^{-1}d^{-1}) \quad \text{(T)}$$

(C) $\because a, b, c, d$ are in HP.

Applying AM > HM

For first three numbers

$$\Rightarrow \quad \frac{a+c}{2} > b$$

$$\Rightarrow \quad a + c > 2b \qquad \ldots(i)$$

and for last three numbers

$$b + d > 2c \qquad \ldots(ii)$$

From Eqs. (i) and (ii),

$$a + c + b + d > 2b + 2c$$

$$\Rightarrow \quad a + d > b + c \text{ (p)}$$

Again, applying GM > HM

For first three numbers $\sqrt{ac} > b$

$$\Rightarrow \quad ac > b^2 \qquad \ldots(iii)$$

Similarly, for last three numbers

$$bd > c^2 \qquad \ldots(iv)$$

From Eqs. (iii) and (iv),

$$b^2 c^2 < (ac)(bd)$$

$$\Rightarrow \quad bc < ad \text{ (Q)}$$

Again, applying AM > HM for first three numbers,

$$b > \frac{2ac}{a+c}$$

$$\Rightarrow \quad bc + ab > 2ac \qquad \ldots(v)$$

Dividing by abc, we get

$$\frac{1}{a} + \frac{1}{c} > \frac{2}{b} \qquad \ldots(vi)$$

Similarly, for last three terms

$$bc + cd > 2bd \qquad \ldots(vii)$$

Dividing by bcd

$$\text{or} \quad \frac{1}{d} + \frac{1}{b} > \frac{2}{c} \qquad \ldots(viii)$$

Now, from Eqs. (vi) and (viii)

$$\frac{1}{a} + \frac{1}{d} > \frac{1}{b} + \frac{1}{c} \text{ (R)}$$

and from Eqs. (v) and (vii),

$$ab + cd + 2bc > 2ac + 2bd$$

$$\Rightarrow \quad ab + cd > 2(ac + bd - bc)$$

Dividing in each term by $abcd$, we get

$$c^{-1}d^{-1} + a^{-1}b^{-1} > 2(b^{-1}d^{-1} + a^{-1}c^{-1} - a^{-1}d^{-1}) \text{(T)}$$

3. **(A)** Given, $a + b + c = 25 \qquad \ldots(i)$

$$2a = 2 + b \qquad \ldots(ii)$$

$$\text{and} \quad c^2 = 18b \qquad \ldots(iii)$$

Substituting the values of a and b in terms of c from Eqs. (ii) and (iii) in Eq. (i), then

$$\frac{1}{2}\left(2 + \frac{c^2}{18}\right) + \frac{c^2}{18} + c = 25$$

$$\Rightarrow \quad c^2 + 12c - 288 = 0$$

$$\therefore \quad c = 12, -24$$

$\because a, b, c$ are between 2 and 18 $\qquad [\because c \neq -24]$

Hence, $c = 12$

From Eq. (iii), we get $b = \dfrac{(12)^2}{18} = 8$

From Eq. (ii), we get $a = \dfrac{2+b}{2} = \dfrac{10}{2} = 5$

$$\therefore \quad a = 5, b = 8, c = 12$$

Then, $\quad G = 12, L = 5$

$\therefore G - L = 7$ (R) and $\left(\dfrac{G}{L}\right) = (2.4) = 3$ (S)

(B) Let $b = ar, c = ar^2$

$$\therefore \quad a(1 + r + r^2) = 70 \qquad \ldots(i)$$

and $4a, 5ar, 4ar^2$ are in AP

$$\therefore \quad 10ar = 4a + 4ar^2$$

$$\Rightarrow \quad 2r^2 - 5r + 2 = 0$$

Then, $\quad r = 2, \dfrac{1}{2}$

From Eq. (i), $a = 10, 40$

then $a = 10, b = 20, c = 40$

or $\quad a = 40, b = 20, c = 10$

$\therefore \quad G = 40, L = 10$

Then, $\quad \left[\dfrac{G}{L}\right] = [4] = 4$ (Q)

(C) $\because 1 + 3 + 5 + \ldots + (2k - 1) = k^2$

Then the given equation can be written as

$$\left(\frac{a+1}{2}\right)^2 + \left(\frac{b+1}{2}\right)^2 = \left(\frac{c+1}{2}\right)^2$$

$$\Rightarrow \quad (a+1)^2 + (b+1)^2 = (c+1)^2$$

Therefore, $(a + 1, b + 1, c + 1)$ forms a pythagorean triplet.

as $a > 6 \Rightarrow a = 7$ (Minimum value of a)

The first pythagorean triplet is (6, 8, 10).

$\therefore$ We may take $a+1=8, b+1=6, c+1=10$

$\therefore \quad a=7, b=5, c=9$

$\Rightarrow \quad G=9, L=5$

$\therefore \quad G-L=4$ (P) and $\frac{G}{L}=1.8$

$\therefore \quad \left[\frac{G}{L}\right]+\left(\frac{G}{L}\right)=[1.8]+(1.8)=1+2=3$(T)

4. (A) $\because \quad A=G+\frac{3}{2}$ ...(i)

and $\quad G=H+\frac{6}{5}$ ...(ii)

$\because \quad G^2=AH$

$\Rightarrow \quad G^2=\left(G+\frac{3}{2}\right)\left(G-\frac{6}{5}\right)$ [from Eqs. (i) and (ii)]

$\therefore \quad G=6, A=\frac{15}{2}$ [from Eq. (i)]

a, b are the roots of $x^2-15x+36=0$

we get $a=12, b=3$ or $a=3, b=12$

$\therefore \quad \alpha=15, \beta=9$

$\Rightarrow \quad \alpha+\beta^2=15+81=96$ (P)

and $\quad \alpha^2+\beta=225+9=234$ (R)

(B) $\because \quad A=G+2$ ...(i)

and $\quad H=\frac{1}{5}b$ (if $b>a$)

$\because \quad G^2=AH=\frac{bA}{5}$

$\Rightarrow \quad 5ab=bA \Rightarrow A=5a$

$\Rightarrow \quad \frac{a+b}{2}=5a \Rightarrow b=9a$

From Eq. (i), $5a=\sqrt{ab}+2$

$\Rightarrow \quad 5a=3a+2$

$\therefore \quad a=1$, then $b=9$

$\Rightarrow \quad \alpha=a+b=10$

and $\quad b=|a-b|=8$

$\therefore \quad \alpha+\beta^2=10+64=74$ (Q)

(C) $\because H=4$

$\Rightarrow \quad 2A+G^2=27$

$\Rightarrow \quad 2A+AH=27 \; (\because G^2=AH)$

$\Rightarrow \quad 6A=27$

$\Rightarrow \quad A=\frac{9}{2}$ and $G^2=\frac{9}{2}\times 4=18$

$\therefore$ a,b are the roots of $x^2-9x+18=0$

$\therefore \quad a=6, b=3$ or $a=3, b=6$

$\therefore \quad \alpha=a+b=9, \beta=|a-b|=3$

$\therefore \quad \alpha^2+\beta=81+3=84$ (S)

and $1+\left(\frac{\alpha}{\beta}\right)+\left(\frac{\alpha}{\beta}\right)^2+\left(\frac{\alpha}{\beta}\right)^3+\left(\frac{\alpha}{\beta}\right)^4+\left(\frac{\alpha}{\beta}\right)^5$

$=1+3+3^2+3^3+3^4+3^5=364$ (T)

5. (A) For an AP, sum of n terms is a purely quadratic expression in n.

$\therefore \quad a=0$, then $S_n=bn+cn^2$

$\therefore \quad T_n=S_n-S_{n-1}$

$=(bn+cn^2)-\{b(n-1)+c(n-1)^2\}$

$=b+c(2n-1)$ (Q)

or $\quad T_n=a+b+c(2n-1)$ (P) $(\because a=0)$

and $D=T_n-T_{n-1}=\{b+c(2n-1)\}-\{b+c(2n-3)\}$

$=2c$ (S)

(B) According to reason in (A),

$a-1=0 \Rightarrow a=1$

$\therefore \quad S_n=(b-2)n^2+(c-3)n$

$\Rightarrow \quad T_n=S_n-S_{n-1}$

$=(b-2)\{n^2-(n-1)^2\}+(c-3)\{n-(n-1)\}$

$=(b-2)(2n-1)+(c-3)$ ...(i)

$=b(2n-1)-4n+2+c-3$

$=c-1-4n+b(2n-1)$

$=c-a-4n+b(2n-a)$ (R) $(\because a=1)$

and $D=T_n-T_{n-1}$ {from Eq. (i)}

$=(b-2)(2)+0$

$=2b-4$ (T)

(C) $\because S_n=bn+cn^2$

$\therefore \quad T_n=S_n-S_{n-1}$

$=(bn+cn^2)-\{b(n-1)+c(n-1)^2\}$

$=b+c(2n-1)$ (Q)

and $\quad D=T_n-T_{n-1}$

$=2c$ (S)

04

Permutations and Combinations

Objective Questions Type I [Only one correct answer]

In each of the questions below, four choices are given of which only one is correct. You have to select the correct answer which is the most appropriate.

1. 4 points out of 8 points in a plane are collinear. Number of different quadrilateral that can be formed by joining them is

(a) 56 (b) 53
(c) 76 (d) 60

2. Number of words of 4 letters that can be formed with the letters of the word IITJEE is

(a) 42 (b) 82
(c) 102 (d) 142

3. Let T_n denote the number of triangles which can be formed using the vertices of a regular polygon of n sides. If$T_{n+1} - T_n = 21$, then n equals

(a) 5 (b) 7
(c) 6 (d) 4

4. The number of arrangements of the letters of the word BANANA in which the two N' s do not appear adjacently is

(a) 40 (b) 60
(c) 80 (d) 100

5. The total number of integral solution for x, y, z such that $xyz = 24$, is

(a) 3 (b) 60
(c) 90 (d) 120

6. The number of permutations of the letters a, b, c, d such that b does not follow a, and c does not follow b and d does not follow c, is

(a) 9 (b) 11 (c) 13 (d) 14

7. A class has n students, we have to form a team of the students including at least two students and also excluding at least two students. The number of ways of forming the team is

(a) $2^n - 2n$ (b) $2^n - 2n - 2$
(c) $2^n - 2n - 4$ (d) $2^n - 2n - 6$

8. Two numbers are chosen from 1, 3, 5, 7, ..., 147, 149 and 151 and multiplied together in all possible ways. The number of ways which will give us the product a multiple of 5, is

(a) 75 (b) 1020
(c) 95 (d) 1030

9. If the number of ways of selecting n cards out of unlimited number of cards bearing the number 0, 9, 3, so that they cannot be used to write the number 903 is 93, then n is equal to

(a) 3 (b) 4
(c) 5 (d) 6

10. If a, b, c are odd positive integers, then number of integral solutions of $a + b + c = 13$, is

(a) 14 (b) 21
(c) 28 (d) 56

11. A rectangle with sides $2m - 1$ and $2n - 1$ divided into squares of unit length. The number of rectangle which can be formed with sides of odd length is

(a) m^2n^2 (b) $mn\,(m+1)\,(n+1)$
(c) 4^{m+n-1} (d) None of these

12. The lock of a safe consists of five discs each of which features the digits 0, 1, 2, ..., 9. The safe can be opened by dialing a special combination of the digits. The number of days sufficient enough to open the safe, if the work day lasts 13 h and 5 s are needed to dial one combination of digits is

(a) 9 (b) 10
(c) 11 (d) 12

13. The interior angles of a regular polygon measure 160° each. The number of diagonals of the polygon are
(a) 97 (b) 105
(c) 135 (d) 146

14. Let $S = \{1, 2, 3, ..., 9\}$. For $k = 1, 2, ... 5$, let N_k be the number of subsets of S, each containing five elements out of which exactly k are odd. Then $N_1 + N_2 + N_3 + N_4 + N_5 =$ **[JEE Advanced 2017]**
(a) 210 (b) 252
(c) 126 (d) 125

15. The number of 5 digit number of the form *abcba* in which $a < b$ is
(a) 320 (b) 340
(c) 360 (d) 380

16. Sum of all the odd divisors of 720 is
(a) 76 (b) 78
(c) 80 (d) 84

17. Let $A = \{x_1, x_2, x_3, x_4, x_5, x_6\}$, $B = \{y_1, y_2, y_3, y_4, y_5, y_6\}$. Then the number of one-one mappings from A to B such that $f(x_i) \neq y_i, i = 1, 2, 3, 4, 5, 6$ is
(a) 720 (b) 265
(c) 360 (d) 145

18. The remainder obtained, when $1! + 2! + 3! + \ldots + 175!$ is divided by 15 is
(a) 5 (b) 0
(c) 3 (d) 8

19. If the total number of m element subsets of the set $A = \{a_1, a_2, a_3, \ldots, a_n\}$ is λ times the number of 3 elements subsets containing a_4, then n is
(a) $(m-1)\lambda$ (b) $m\lambda$
(c) $(m+1)\lambda$ (d) 0

20. When simplified, the expression ${}^{47}C_4 + \sum_{n=1}^{5} {}^{52-n}C_3$ equals
(a) ${}^{47}C_5$ (b) ${}^{49}C_4$
(c) ${}^{52}C_5$ (d) ${}^{52}C_4$

21. How many 3×3 matrices M with entries from $\{0, 1, 2\}$ are there, for which the sum of the diagonal entries of M^TM is 5 ? **[JEE Advanced 2017]**
(a) 198 (b) 162
(c) 126 (d) 135

22. The least positive integral value of x which satisfies the inequality ${}^{10}C_{x-1} > 2 \cdot {}^{10}C_x$ is
(a) 7 (b) 8
(c) 9 (d) 10

23. The total number of three-letter words that can be formed from the letter of the word SAHARANPUR is equal to
(a) 210 (b) 237
(c) 247 (d) 227

24. The number of 6-digits numbers that can be formed using the digits 0, 1, 2,5, 7 and 9 which are divisible by 11 and no digit is repeated, is **[JEE Main 2019, 10 April]**
(a) 60 (b) 72
(c) 48 (d) 36

25. If n is an integer between 0 and 21, then the minimum value of $n!(21-n)!$ is
(a) $9!2!$ (b) $10!11!$
(c) $20!$ (d) $21!$

26. A debate club consists of 6 girls and 4 boys. A team of 4 members is to be selected from this club including the selection of a captain (from among these 4 members) for the team. If the team has to include atmost one boy, the number of ways of selecting the team is **[JEE Advanced 2016]**
(a) 380 (b) 320
(c) 260 (d) 95

27. The total number of three digit and five digit integers which can be formed by using the digits 0, 1, 2, 3, 4, 5 but using each digit not more than once in each number is
(a) 100 (b) 600
(c) 700 (d) 800

28. The maximum number of points into which 4 circles and 4 straight lines intersect, is
(a) 26 (b) 50
(c) 56 (d) 72

29. If 7 points out of 12 lie on the same straight line, then the number of triangles, thus formed, is
(a) 19 (b) 185
(c) 201 (d) 205

30. There are 3 sections in a question paper and each section contains 5 questions. A candidate has to answer a total of 5 questions, choosing at least one question from each section. Then, the number of ways, in which the candidate can choose the questions, is **[JEE Main 2020, 5 September]**
(a) 3000 (b) 1500
(c) 2255 (d) 2250

31. Every one of the 10 available lamps can be switched on to illuminate certain Hall. The total number of ways in which the hall can be illuminated, is
(a) 55 (b) 1023
(c) 2^{10} (d) $10!$

32. The number of ways in which 7 persons can be seated at a round table, if two particular persons are not to sit together, is
(a) 120 (b) 480
(c) 600 (d) 720

33. The number of three-digit numbers of the form *xyz* such that $x < y$ and $z \leq y$ is
(a) 276 (b) 285
(c) 240 (d) 244

34. The number of ways in which three students of a class may be assigned a grade of A, B, C or D, so that no two students receive the same grade, is
(a) 3^4 (b) 4^3 (c) ${}^4P_3$ (d) ${}^4C_3$

35. At an election a voter may vote for any number of candidates not exceeding the number to be elected. If 4 candidates are to be elected out of the 12 contested in the election and voter votes for at least one candidate, then the number of ways in which a voter can vote is

(a) 793 (b) 298
(c) 781 (d) 1585

36. If 5 parallel straight lines are intersected by 4 parallel straight lines, then the number of parallelograms, thus formed, is

(a) 20 (b) 60
(c) 101 (d) 126

37. The total number of numbers that can be formed by using all the digits 1, 2, 3, 4, 3, 2, 1, so that the odd digits always occupy the odd places, is

(a) 3 (b) 6
(c) 9 (d) 18

38. The sides AB, BC and CA of a triangle ABC have 3, 4 and 5 interior points respectively on them. The number of triangles that can be constructed using these interior points as vertices, is

(a) 205 (b) 208
(c) 220 (d) 380

39. Two families with three members each and one family with four members are to be seated in a row. In how many ways can they be seated so that the same family members are not separated? **[JEE Main 2020, 6 September]**

(a) $2!3!4!$ (b) $(3!)^3 \cdot (4!)$
(c) $(3!)^2 \cdot (4!)$ (d) $3!(4!)^3$

40. Ten different letters of an alphabet are given. Words with five letters (not necessarily meaningful or pronounceable) are formed from these letters. The total number of words which have atleast one letter repeated, is

(a) 21672 (b) 30240
(c) 69760 (d) 99748

41. All possible numbers are formed using the digits 1, 1, 2, 2, 2, 2, 3, 4, 4 taken all at a time. The number of such numbers in which the odd digits occupy even places is **[JEE Main 2019, 8 April]**

(a) 180 (b) 175
(c) 160 (d) 162

42. The number of order triplets of positive integers which are solutions of the equation $x + y + z = 100$ is

(a) 6005 (b) 4851
(c) 5081 (d) 6081

43. The total number of 3 digit even numbers that can be composed from the digits 1, 2, 3, ... , 9 when the repetition of digits is not allowed, is

(a) 224 (b) 280
(c) 324 (d) 405

44. The number of integers greater than 6000 that can be formed, using the digits 3, 5, 6, 7 and 8 without repetition, is **[JEE Main 2015]**

(a) 216 (b) 192
(c) 120 (d) 72

45. A natural number has prime factorisation given by $n = 2^x 3^y 5^z$, where y and z are such that $y + z = 5$ and $y^{-1} + z^{-1} = \frac{5}{6}$, $y > z$. Then the number of odd divisors of n, including 1, is **[JEE Mains 2021, 26 February]**

(a) 11 (b) 60
(c) 12 (d) 6

46. In an examination there are three multiple choice questions and each question has 4 choices of answers in which only one is correct. The total number of ways in which an examinee can fail to get all answers correct is

(a) 11 (b) 12
(c) 27 (d) 63

47. The sum of the digits in the unit's place of all the numbers formed with the digits 5, 6, 7, 8 when taken all at a time, is

(a) 104 (b) 126
(c) 127 (d) 156

48. Two straight lines intersect at a point O. Points $A_1, A_2, \ldots, A_n$ are taken on one line and points $B_1, B_2, \ldots, B_n$ on the other. If the point O is not to be used, the number of triangles that can be drawn using these points as vertices, is

(a) $n(n-1)$ (b) $n(n-1)^2$
(c) $n^2(n-1)$ (d) $n^2(n-1)^2$

49. How many different nine digit numbers can be formed from the number 22 33 55 888 by rearranging its digits, so that the odd digits occupy even positions ?

(a) 16 (b) 36
(c) 60 (d) 180

50. For $2 \le r \le n$, $\binom{n}{r} + 2\binom{n}{r-1} + \binom{n}{r-2}$ is equal to

(a) $\binom{n+1}{r-1}$ (b) $2\binom{n+1}{r+1}$
(c) $2\binom{n+r}{r}$ (d) $\binom{n+2}{r}$

51. The number of positive integers satisfying the inequality ${}^{n+1}C_{n-2} - {}^{n+1}C_{n-1} \le 100$ is

(a) nine (b) eight
(c) five (d) None of these

52. A class has 21 students. The class teacher has been asked to make n groups of r students each and go to zoo taking one group at a time. The size of group (*ie*, the value of r) for which the teacher goes to the maximum number of times is (no group can go to the zoo twice)

(a) 9 or 10 (b) 10 or 11
(c) 11 or 12 (d) 12 or 13

53. The number of ways in which a score of 11 can be made from a through by three persons, each throwing a single die once, is
(a) 45 (b) 18
(c) 27 (d) 68

54. The number of positive integers with the property that they can be expressed as the sum of the cubes of 2 positive integers in two different ways is
(a) 1 (b) 100
(c) infinite (d) 0

55. The number of triangles whose vertices are the vertices of an octagon but None of whose sides happen to come from the octagon is
(a) 16 (b) 28
(c) 56 (d) 70

56. There are n different books and p copies of each in a library. The number of ways in which one or more than one book can be selected is
(a) $p^n + 1$ (b) $(p+1)^n - 1$
(c) $(p+1)^n - p$ (d) p^n

57. In a plane there are 37 straight lines, of which 13 pass through the point A and 11 pass through the point B. Besides, no three lines pass through one point, no lines passes through both points A and B, and no two are parallel, then the number of intersection points the lines have is equal to
(a) 535 (b) 601
(c) 728 (d) 963

58. We are required to form different words with the help of the letters of the word INTEGER. Let m_1 be the number of words in which I and N are never together and m_2 be the number of words which begnning with I and end with R, then m_1/m_2 is given by
(a) 42 (b) 30
(c) 6 (d) 1/30

59. If a denotes the number of permutations of $x + 2$ things taken all at a time, b the number of permutations of x things taken 11 at a time and c the number of permutations of $x - 11$ things taken all at a time such that $a = 182\ bc$, then the value of x is
(a) 15 (b) 12
(c) 10 (d) 18

60. Some identical balls are arranged in rows to form an equilateral triangle. The first row consists of one ball, the second row consists of two balls and so on. If 99 more identical balls are added to the total number of balls used in forming the equilateral triangle, then all these balls can be arranged in a square whose each side contains exactly 2 balls less than the number of balls each side of the triangle contains. Then, the number of balls used to form the equilateral triangle is **[JEE Main 2019, 9 April]**
(a) 262 (b) 190
(c) 225 (d) 157

61. The number of ordered triples of positive integers which are solutions of the equation $x + y + z = 100$ is
(a) 5081 (b) 6005
(c) 4851 (d) 4987

62. The number of numbers less than 1000 that can be formed out of the digits, 0, 1, 2, 4 and 5, no digit being repeated is
(a) 69 (b) 68
(c) 130 (d) None of these

63. A is a set containing n elements. A subset P_1 is chosen and A is reconstructed by replacing the elements of P_1. The same process is repeated for subsets $P_1, P_2, \ldots, P_m$ with $m > 1$. The number of ways of choosing $P_1, P_2, \ldots, P_m$, so that $P_1 \cup P_2 \cup \ldots \cup P_m = A$ is
(a) $(2^m - 1)^{mn}$ (b) $(2^n - 1)^m$
(c) ${}^{m+n}C_m$ (d) None of these

64. If the sides AB, BC and CA of a triangle ABC have 3, 5 and 6 interior points respectively, then the total number of triangles that can be constructed using these points as vertices, is equal to **[JEE Mains 2021, 17 March]**
(a) 364 (b) 240 (c) 333 (d) 360

65. A is a set containing n elements. A subset P of A is chosen. The set A is reconstructed by replacing the element of P. A subset Q of A is again chosen. The number of ways of choosing P and Q, so that $P \cap Q = \phi$ is
(a) $2^{2n} - {}^{2n}C_n$ (b) 2^n
(c) $2^n - 1$ (d) 3^n

66. A father with 8 children takes 3 at a time to the zoological gardens, as often as he can without taking the same 3 children together more than once. The number of times he will go to the garden is
(a) 336 (b) 112
(c) 56 (d) None of these

67. If the $(n + 1)$ numbers $a, b, c, d, \ldots$, be all different and each of them a prime number, then the number of different factors (other than 1) of $a^m \cdot b \cdot c \cdot d \ldots$ is
(a) $m - 2^n$ (b) $(m + 1)\ 2^n$
(c) $(m + 1)\ 2^n - 1$ (d) None of these

68. The number of selections of four letters from the letters of the word ASSASSINATION is
(a) 72 (b) 71
(c) 66 (d) 52

69. The number of divisors a number 38808 can have, excluding 1 and the number itself is
(a) 70 (b) 72
(c) 71 (d) None of these

70. The letters of the word SURITI are written in all possible orders and these words are written out as in a dictionary. Then the rank of the word SURITI is
(a) 236 (b) 245
(c) 307 (d) 315

71. The total number of seven digit numbers the sum of whose digits is even, is
(a) 9×10^6 (b) 45×10^5
(c) 81×10^5 (d) 9×10^5

72. In a steamer there are stalls for 12 animals and there are cows, horses and calves (not less than 12 of each) ready to be shipped; the total number of ways in which the ship load can be made is
(a) 3^{12} (b) 12^3
(c) $^{12}P_3$ (d) $^{12}C_3$

73. The number of non-negative integral solutions of $x_1 + x_2 + x_3 + 4x_4 = 20$ is
(a) 530 (b) 532
(c) 534 (d) 536

74. The number of six digit numbers that can be formed from the digits 1, 2, 3, 4, 5, 6 and 7, so that digits do not repeat and the terminal digits are even is
(a) 144 (b) 72
(c) 288 (d) 720

75. Given that n is the odd, the number of ways in which three numbers in AP can be selected from 1, 2, 3, 4, ..., n is
(a) $\frac{(n-1)^2}{2}$ (b) $\frac{(n+1)^2}{4}$
(c) $\frac{(n+1)^2}{2}$ (d) $\frac{(n-1)^2}{4}$

76. A is a set containing n elements. A subset P of A is chosen. The set A is reconstructed by replacing the elements of P. A subset Q of A is again chosen. The number of ways of chosing P and Q, so that $P \cap Q$ contains exactly two elements is
(a) $9\,{}^nC_2$ (b) $3^n - {}^nC_2$
(c) $2\,{}^nC_n$ (d) None of these

77. Eight straight lines are drawn in the plane such that no two lines are parallel and no three lines are concurrent. The number of parts into which these lines divide the plane, is
(a) 29 (b) 32
(c) 36 (d) 37

78. Number of divisors of the form $4n + 2\ (n \ge 0)$ of the integer 240 is
(a) 4 (b) 8
(c) 10 (d) 3

79. An n digit number is a positive number with exactly n digits. Nine hundred distinct n digit numbers are to be formed using only the three digits 2, 5 and 7. The smallest value of n for which this is possible is
(a) 6 (b) 7
(c) 8 (d) 9

80. If a, b, c, d are odd natural numbers such that $a + b + c + d = 20$, then the number of values of the ordered quadruplet (a, b, c, d) is
(a) 165 (b) 310
(c) 295 (d) 398

81. The number of rectangles excluding squares from a rectangle of size 15×10 is
(a) 3940 (b) 4940
(c) 5940 (d) 6940

82. In a certain test, there are n questions. In this test 2^{n-i} students gave wrong answers to at least i questions, where $i = 1, 2, 3, 4, \ldots, n$. If the total number of wrong answers given is 2047, then n is equal to
(a) 10 (b) 11
(c) 12 (d) 13

83. If all the words (with or without meaning) having five letters, formed using the letters of the word SMALL and arranged as in a dictionary, then the position of the word SMALL is **[JEE Main 2016]**
(a) 46th (b) 59th
(c) 52nd (d) 58th

84. Let A and B be two sets containing four and two elements respectively. Then, the number of subsets of the set $A \times B$, each having atleast three elements are **[JEE Main 2015]**
(a) 219 (b) 256
(c) 275 (d) 510

85. The number of different ways the letters of the word VECTOR can be placed in the 8 boxes of the given below such that no row empty is equal to

(a) 26 (b) $26 \times 6!$
(c) $6!$ (d) $2! \times 6!$

86. In the next world cup of cricket there will be 12 teams, divided equally in two groups. Teams of each group will play a match against each other. From each group 3 top teams will qualify for the next round. In this round each team will play against others once. Four top teams of this round will qualify for the semifinal round, when each team will play against the others once. Two top teams of this round will go to the final round, where they will play the best of three matches. The minimum number of matches in the next world cup will be
(a) 54 (b) 53
(c) 52 (d) None of these

87. Two lines intersect at O. Points $A_1, A_2, \ldots, A_n$ are taken on one of them and $B_1, B_2, \ldots, B_n$ on the other the number of triangles that can be drawn with the help of these $(2n + 1)$ points is
(a) n (b) n^2 (c) n^3 (d) n^4

88. Seven different lecturers are to deliver lectures in seven periods of a class on a particular day. A, B and C are three of the lectures. The number of ways in which a routine for the day can be made such that A delivers his lecture before B, and B before C, is
(a) 210 (b) 420
(c) 840 (d) None of these

89. From 6 different novels and 3 different dictionaries, 4 novels and 1 dictionary are to be selected and arranged in a row on the shelf so that the dictionary is always in the middle. Then, the number of such arrangements is
(a) atleast 500 but less than 750
(b) atleast 750 but less than 1000
(c) atleast 1000
(d) less than 500

90. Consider a rectangle *ABCD* having 5, 7, 6, 9 points in the interior of the line segments *AB, CD, BC, DA* respectively. Let α be the number of triangles having these points from different sides as vertices and β be the number of quadrilaterals having these points from different sides as vertices. Then $(\beta - \alpha)$ is equal to **[JEE Main 2021, 16 March]**
(a) 795 (b) 1173
(c) 1890 (d) 717

91. The maximum number of different permutations of 4 letters of the word EARTHQUAKE is
(a) 1045 (b) 2190
(c) 4380 (d) 2348

92. In a city no persons have identical set of teeth and there is no person without a tooth. Also no person has more than 32 teeth. If we disregard the shape and size of tooth and consider only the positioning of the teeth, then the maximum population of the city is
(a) 2^{32} (b) $2^{32} - 1$
(c) $2^{32} - 2$ (d) $2^{32} - 3$

93. The number of ways in which a mixed double game can be arranged from amongst 9 married couples, if no husband and wife play in the same game is
(a) 756 (b) 1512
(c) 3024 (d) None of these

94. In a college examination, a candidate is required to answer 6 out of 10 questions which are divided into two sections each containing 5 questions further the candidate is not permitted to attempt more than 4 questions from either of the section. The number of ways in which he can make up a choice of 6 questions is
(a) 200 (b) 150
(c) 100 (d) 50

95. The number of ways in which 9 identical balls can be placed in three identical boxes is
(a) 55 (b) $\frac{9!}{(3!)^4}$
(c) $\frac{9!}{(3!)^3}$ (d) 12

96. If the number of arrangements of $(n-1)$ things taken from n different things is k times the number of arrangements of $(n-1)$ things taken from n things in which two things are identical, then the value of k is
(a) 1/2 (b) 2
(c) 4 (d) None of these

97. The number of different seven digit numbers that can be written using only the three digits 1, 2 and 3 with the condition that the digit 2 occurs twice in each number is
(a) ${}^7P_2 \cdot 2^5$ (b) ${}^7C_2 \cdot 2^5$
(c) ${}^7C_2 \cdot 5^2$ (d) None of these

98. The number of points (x, y, z) in space, whose each coordinate is a negative integer such that $x + y + z + 12 = 0$ is
(a) 385 (b) 55
(c) 110 (d) None of these

99. The number of divisors of $2^2 \cdot 3^3 \cdot 5^3 \cdot 7^5$ of the form $4n+1, n \in N$ is
(a) 46 (b) 47
(c) 96 (d) 94

100. The number of ways in which 30 coins of one rupee each be given to six persons, so that None of them receives less than 4 rupees is
(a) 231 (b) 462
(c) 693 (d) 924

101. The number of integral solutions of the equation $2x + 2y + z = 20$, where $x \geq 0$, $y \geq 0$ and $z \geq 0$ is
(a) 132 (b) 11
(c) 33 (d) 66

102. The number of ways in which we can choose 2 distinct integers from 1 to 100 such that difference between them is at most 10 is
(a) ${}^{10}C_2$ (b) 72
(c) ${}^{100}C_2 - {}^{90}C_2$ (d) None of these

103. Team 'A' consists of 7 boys and n girls and Team 'B' has 4 boys and 6 girls. If a total of 52 single matches can be arranged between these two teams when a boy plays against a boy and girl plays against a girl, then n is equal to **[JEE Mains 2021, 17 March]**
(a) 5 (b) 2
(c) 4 (d) 6

104. Number of positive integral solutions of $xyz = 30$ is
(a) 9 (b) 27
(c) 81 (d) 243

Objective Questions Type II [One or more than one correct answer(s)]

In each of the questions below four choices of which one or more than one are correct. You have to select the correct answer(s) accordingly.

1. Consider the word ALLAHABAD. Which of the following statements are true?
   (a) The total number of words that can be formed using all the letters of the word is 7560.
   (b) The number of words which begin with A and end with A is 1260.
   (c) The number of words in which vowels occupy the even places is 60.
   (d) The number of words in which all the four vowels occupy adjacent places is 360.

2. Suppose 5 red balls and 5 blue balls are arranged in a row. If the number of possible arrangements when no two red balls are together is $6(k!)^2$, then the number of 5-digit numbers that have the product of their digits as $(4k)$ is
   (a) $\dfrac{5\times 6!}{3\times 4!}$ (b) 75
   (c) 50 (d) $\dfrac{6!}{3\times 5!}$

3. Sanjay has 10 friends among whom two are married to each other. She wishes to invite 5 of the them for a party. If the married couple refuse to attend separately, then the number of different ways in which she can invite five friends is
   (a) ${}^8C_5$ (b) $2\times {}^8C_3$
   (c) ${}^{10}C_5 - 2\times {}^8C_4$ (d) None of these

4. There are n seats round a table marked $1, 2, 3, \ldots, n$. The number of ways in which $m\ (\le n)$ persons can take seats is
   (a) ${}^nP_m$ (b) ${}^nC_m \times (m-1)!$
   (c) ${}^nC_m \times m!$ (d) ${}^{n-1}P_{m-1}$

5. The number of ways in which 10 candidates $A_1, A_2, \ldots, A_{10}$ can be ranked, so that A_1 is always above A_2 is
   (a) $\dfrac{10!}{2}$ (b) $8!\times {}^{10}C_2$
   (c) ${}^{10}P_2$ (d) ${}^{10}C_2$

6. If a seven-digit number made up of all distinct digits 8, 7, 6, 4, 2, x, and y is divisible by 3, then
   (a) maximum value of $x-y$ is 9
   (b) maximum value of $x+y$ is 12
   (c) minimum value of xy is 0
   (d) minimum value of $x+y$ is 3

7. The number of ways of distributing 10 different books among 4 students $(S_1 - S_4)$ such that S_1 and S_2 get 2 books each and S_3 and S_4 get 3 books each is
   (a) 12600 (b) 25200
   (c) ${}^{10}C_4$ (d) $\dfrac{10!}{2!2!3!3!}$

8. Ten persons, amongst whom are A, B and C are speak at a function. The number of ways in which it can be done if A wants to speak before B, and B wants to speak before C is
   (a) $\dfrac{10!}{6}$ (b) 21870
   (c) $\dfrac{10!}{3!}$ (d) ${}^{10}P_7$

9. The number of ways to select 2 numbers from $\{0, 1, 2, 3, 4\}$ such that the sum of the squares of the selected numbers is divisible by 5 are (repitition of digits is allowed)
   (a) ${}^9C_1$ (b) ${}^9C_8$
   (c) 9 (d) 7

10. If n objects are arranged in a row, then the number of ways of selecting three of these objects so that no two of them are next to each other is
   (a) $\dfrac{(n-2)(n-3)(n-4)}{6}$ (b) ${}^{n-2}C_3$
   (c) ${}^{n-3}C_3 + {}^{n-3}C_2$ (d) None of these

11. The number of ways of arranging seven persons (having A, B, C and D among them) in a row so that A, B, C and D are always in order A-B-C-D (not necessarily together) is
   (a) 210 (b) 5040
   (c) $6\times {}^7C_4$ (d) ${}^7P_3$

12. Suppose $A_1, A_2, A_3, A_4, \ldots, A_{30}$ are thirty sets each with five elements and $B_1, B_2, \ldots, B_n$ are n sets each with three elements such that $\bigcup_{i=1}^{30} A_i = \bigcup_{i=1}^{n} B_i = S$.
   If each elements of S belongs to exactly ten of the A_i's and exactly 9 of B_i's, then the value of n is
   (a) 15 (b) 135
   (c) 45 (d) ${}^{10}C_2$

13. Triplet (x, y, z) is chosen from the set $\{1, 2, 3, , \ldots, n\}$, such that $x \le y < z$. The number of such triplets is
   (a) n^3 (b) ${}^{n+1}C_3$
   (c) ${}^nC_2$ (d) ${}^nC_2 + {}^nC_3$

14. Let $f(n)$ be the number of regions in which n coplanar circles can divided the plane. If it is known that each pair of circles intersect in two different points and no three of them have common point of intersection, then
   (a) $f(20) = 382$
   (b) $f(n)$ is always an even number
   (c) $f^{-1}(92) = 10$
   (d) $f(n)$ can be odd

15. Consider seven digit number $x_1, x_2, \ldots, x_7$, where $x_1, x_2, \ldots, x_7 \neq 0$ having the property that x_4 is the greatest digit and digits towards the left and right of x_4 are in decreasing order. Then total number of such numbers in which all digits are distinct is
(a) ${}^9C_7 \cdot {}^6C_3$ (b) ${}^9C_6 \cdot {}^5C_3$
(c) ${}^{10}C_7 \cdot {}^6C_3$ (d) ${}^9C_2 \cdot {}^6C_3$

16. A class has 30 students. The following prizes are to be awarded to the students of this class, first and second in Mathematics, first and second in Physics; first in Chemistry and first in Biology. If N denote the number of ways in which this can be done, then
(a) $400|N$
(b) $600|N$
(c) $8100|N$
(d) N is divisible by four distinct prime numbers

17. If ${}^nC_4$, ${}^nC_5$ and ${}^nC_6$ are in AP, the value of n can be
(a) 14 (b) 11
(c) 7 (d) 8

18. The letters of the word ARTICLE are arranged in all possible ways. Then, which of the following is (are) true?
(a) Number of words formed by using all the letters is 5040.
(b) The number of words with vowels in even places is 144.
(c) The number of words with vowels in odd places is 576.
(d) The number of words with I in the middle is 720

19. Let $S = \{1, 2, 3, \ldots, n\}$. If X denote the set of all subsets of S containing exactly two elements, then the value of $\sum_{A \in X} (\min A)$ is given by
(a) ${}^{n+1}C_3$ (b) ${}^nC_3$
(c) $\frac{n(n^2-1)}{6}$ (d) $\frac{(n-1)^n}{6}$

20. Let N denote the number of ways in which $3n$ letters can be selected from $2n$ A's, $2n$ B's and $2n$ C's. Then
(a) $3|(N-1)$ (b) $n|(N-1)$
(c) $(n+1)|(N-1)$ (d) $3n(n+1)|(N-1)$

21. Let N denote the greatest number of points in which m straight lines and n circles intersect, then
(a) $m|(N - {}^mC_2 - {}^nP_2)$
(b) $n|(N - {}^mC_2 - {}^nP_2)$
(c) $N - {}^mC_2 - {}^nP_2$ is an even integer
(d) $N - {}^mC_2 - {}^nP_2$ is an odd integer

22. m points on one straight line are joined to n points on another straight line. The number of points of intersection of the line segments thus formed is
(a) ${}^mC_2 \cdot {}^nC_2$ (b) $\frac{mn(m-1)(n-1)}{4}$
(c) $\frac{{}^mC_2 \cdot {}^nC_2}{2}$ (d) ${}^mC_2 + {}^nC_2$

23. If $\alpha = x_1 x_2 x_3$ and $\beta = y_1 y_2 y_3$ be two three digits numbers, the number of pairs of α and β can be formed so that α can be subtracted from β without borrowing is
(a) $2!10!10!$ (b) $(45)(55)^2$
(c) $3^2 \cdot 5^3 \cdot 11^2$ (d) 136125

24. A is a set containing n elements. A subset P_1 of A is chosen. The set A is reconstructed by replacing the elements of P_1. Next, a subset P_2 of A is chosen and again the set is reconstructed by replacing the elements of P_2. In this way, $m(>1)$ subset $P_1, P_2, \ldots, P_m$ of A are chosen. The number of ways of choosing $P_1, P_2, \ldots, P_m$ is
(a) $(2^m - 1)^n$ if $P_1 \cap P_2 \cap \ldots \cap P_m = \phi$
(b) 2^{mn} if $P_1 \cup P_2 \cup \ldots \cup P_m = A$
(c) 2^{mn} if $P_1 \cap P_2 \cap \ldots \cap P_m = \phi$
(d) $(2^m - 1)^n$ if $P_1 \cup P_2 \cup \ldots \cup P_m = A$

Linked-Comprehension Type

In these questions, a passage (paragraph) has been given followed by questions based on each of the passages. You have to answer the questions based on the passage given.

PASSAGE 1

Twelve persons are to be arranged around two round tables such that one table can accommodate seven persons and another five persons only.

On the basis of above information, answer the following questions

1. The number of ways in which these 12 persons can be arranged is
(a) ${}^{12}P_5 6!4!$ (b) $6!4!$
(c) ${}^{12}C_5 6!4!$ (d) None of these

2. The number of ways of arrangements, if two particular persons A and B do not want to be on the same table is
(a) ${}^{10}C_4 6!4!$ (b) $2\,{}^{10}C_6 6!4!$
(c) ${}^{11}C_6 6!4!$ (d) None of these

3. The number of ways of arrangement, if two particular persons A and B want to be together and consecutive is
(a) ${}^{10}C_7 6!3!2! + {}^{10}C_5 4!5!2!$ (b) ${}^{10}C_5 6!\,3! + {}^{10}C_7 4!5!$
(c) ${}^{10}C_7 6!2! + {}^{10}C_5 5!2!$ (d) None of these

PASSAGE 2

Suppose a lot of n objects contains n_1 objects of one kind, n_2 objects of second kind, n_3 objects of third kind, ..., n_k objects of kth kind. Such that $n_1 + n_2 + n_3 + \dots + n_k = n$, then the number of possible arrangements/permutations of r objects out of this lot is the coefficient of x^r in the expansion of

$$r!\prod\left(\sum_{\lambda=0}^{n_1}\frac{x^\lambda}{\lambda!}\right)$$

On the basis of above information, answer the following questions

1. The number of permutations of the letters of the word INDIA taken three at a time must be
(a) 27 (b) 30
(c) 33 (d) 57

2. If $n_1 = n_2 = n_3 = \dots = n_k = 1$, then the number of permutations of r objects must be
(a) ${}^nP_r$ (b) ${}^nC_r$
(c) ${}^kP_r$ (d) ${}^kC_r$

3. The number of permutations of the letters of the word INEFFECTIVE taken four at a time must be
(a) 2214 (b) 1422
(c) 5424 (d) 2454

4. If $n_1 + n_2 + n_3 + \dots + n_k = r$, then number of permutations must be
(a) ${}^nC_r$ (b) ${}^nP_r$
(c) $(k+r)!$ (d) $\dfrac{r!}{n_1!n_2!\dots n_k!}$

5. Five letter words are to be formed out of the letters of the word 'INFINITESIMAL', then number of permutations must be
(a) 16995 (b) 5665
(c) 11330 (d) 22660

PASSAGE 3

Different words are being formed by arranging the letters of the word "SUCCESS". All the words obtained by written in the form of a dictionary.

On the basis of above information, answer the following questions

1. The number of words in which the two C are together but no two S are together is
(a) 120 (b) 96
(c) 24 (d) 420

2. The number of words in which no two C and no two S are together is
(a) 120 (b) 96
(c) 24 (d) 420

3. The number of words in which the consonants appear in alphabatic order is
(a) 42 (b) 40
(c) 420 (d) 280

4. The rank of the word 'SUCCESS' in the dictionary is
(a) 328 (b) 329
(c) 330 (d) 331

5. The number of words in which the relative positions of vowels and consonants unaltered is
(a) 20 (b) 60
(c) 180 (d) 540

PASSAGE 4

Prizes are to be distributed among 6 students.

On the basis of above information, answer the following questions

1. The number of ways of distributing the prizes, if a student can receive any number of prizes, is
(a) 1296 (b) 16^3
(c) 15 (d) 30

2. The number of ways of distributing the prizes, if a student cannot receive all the prizes, is
(a) $16^3 - 16$ (b) 1290
(c) 11 (d) 26

3. If a particular student is to receive exactly 2 prizes, then the number of ways of distributing the prizes is
(a) 25 (b) 32
(c) 150 (d) 36

PASSAGE 5

Let $f(n)$ denotes the number of different ways the positive integer n can be expressed as the sum of 1's and 2's. For example

$$f(4) = 5$$

i.e., $$4 = 1+1+1+1$$
$$= 1+1+2$$
$$= 1+2+1$$
$$= 2+1+1 = 2+2$$

On the basis of above information, answer the following questions

1. The value of $f(6)$ is
(a) 10 (b) 13
(c) 16 (d) 19

2. The value of $f(f(6))$ is
(a) 356 (b) 377
(c) 389 (d) 427

3. The number of solutions of the equation $f(n) = n$, where $n \in N$ is
(a) 1 (b) 2
(c) 3 (d) 4

4. The number of solutions of the equation $f(n) = n + 1$, where $n \in N$ is
(a) 1 (b) 2
(c) 3 (d) 4

PASSAGE 6

Different words are being formed by arranging the letters of the word 'ARRANGE'. All the words obtained are written in the form of a dictionary.

On the basis of above information, answer the following questions

1. The number of words in which the two 'R' are not together
(a) 1260 (b) 660
(c) 900 (d) 240
2. The number of words in which neither two 'R' nor two 'A' come together is
(a) 1260 (b) 660
(c) 900 (d) 240
3. The number of words in which the consonants appear in alphabetic order is
(a) 100 (b) 105
(c) 360 (d) 240
4. The rank of the word 'ARRANGE' in the dictionary
(a) 340 (b) 341
(c) 342 (d) 343
5. The number of words in which at least one vowel is in between two consonant is
(a) 18 (b) 36
(c) 624 (d) 836

PASSAGE 7

Let $\theta = (a_1, a_2, a_3, \ldots, a_n)$ be a given arrangement of n distinct objects $a_1, a_2, a_3, \ldots, a_n$. A derangement of θ is an arrangement of these n objects in which one of the objects occupies its original position. Let D_n be the number of derangements of the permutation θ.

On the basis of above information, answer the following questions

1. D_n is equal to
(a) $(n-1)D_{n-1} + D_{n-2}$ (b) $D_{n-1} + (n-1)D_{n-2}$
(c) $n(D_{n-1} + D_{n-2})$ (d) $(n-1)(D_{n-1} + D_{n-2})$
2. The relation between D_n and D_{n-1} is given by
(a) $D_n - nD_{n-1} = (-1)^n$
(b) $D_n - (n-1)D_{n-1} = (-1)^{n-1}$
(c) $D_n - nD_{n-1} = (-1)^{n-1}$
(d) $D_n - D_{n-1} = (-1)^{n-1}$
3. There are 5 different colour balls and 5 boxes of colours same as those of the balls. The number of ways in which one can place the balls into boxes, one each in a box, so that no ball goes to a box of its own colour is
(a) 40 (b) 44
(c) 45 (d) 60

Numerical Value Type

Solve the following problems and write your answer.

1. A five digit number divisible by 3 is to be formed using the numericals 0, 1, 2, 3, 4 and 5 without repetition. If the total number of ways in which this can be done is n^3, then $\frac{n!}{100}$ must be
2. If the number of ways of selecting n coupons out of an unlimited number of coupons bearing the letters A, T, C so that they can not be used to spell to the used CAT is 189, then Σn^2 must be
3. A bouquet from 11 different flowers is to be made so that it contains not less than three flowers. The number of different ways of forming the bouquet must be
4. If ${}^{12}P_r = 11880$, then $\sum_{i=1}^{\lambda} {}^{\lambda}C_i$ must be (where $\lambda = r + 3$)
5. Number of words that can be formed with the letters of the word ALGEBRA so that all the vowels do not come together is
6. There are m men and two women participating in a chess tournament. Each participant plays two games with every other participant. If the number of games played between the men and women by 84, then the value $\frac{m}{5}$ is
7. The number of wards can be formed with the letters of the word PATALIPUTRA without changing the relative positions of vowels and consonants is
8. The sum of the digits in the tens place of all numbers formed with the help of 4, 5, 6, 7 taken all at a time is
9. The sum of all values of r in $\binom{18}{r-2} + 2\binom{18}{r-1} + \binom{18}{r} \geq \binom{20}{13}$ must be
10. The number of different numbers which are smaller than 2×10^8 and are divisible by 3, can be written by means of the digits 0, 1 and 2, must be
11. John has x children by his first wife. Mary has $(x+1)$ children by his first husband. They marry and have children of their own. The whole family has 24 children. Assuming that two children of the same parents do not fight. Then the maximum number of fights must be

12. If the letters of the word SCHOOL are arranged as in a dictionary, then the rank of the word SCHOOL will be

13. The number of integral solutions of $a + b + c = 0, a \geq -5, b \geq -5, c \geq -5$ must be

14. The number of ways in which 20 rupees can be distributed among 5 people such that each person gets at least Rs 3 is

15. Number of ways in which 5 boys and 4 -girls can be arranged on a circular table such that no two girls sit together and two particular boys are always together is

Single Integer Type

Solve the following problems and write your answer.

1. Words of length 10 are formed using the letters A, B, C, D, E, F, G, H, I, J. Let x be the number of such words where no letter is repeated; and let y be the number of such words where exactly one letter is repeated twice and no other letter is repeated. Then, $\frac{y}{9x} =$ **[JEE Advanced 2017]**

2. Let n be the number of ways in which 5 boys and 5 girls can stand in a queue in such a way that all the girls stand consecutively in the queue. Let m be the number of ways in which 5 boys and 5 girls can stand in a queue in such a way that exactly four girls stands consecutively in the queue. Then, the value of $\frac{m}{n}$ is **[JEE Advanced 2015]**

3. The number of ordered pairs (r, k) for which $6 \cdot {}^{35}C_r = (k^2 - 3) \cdot {}^{36}C_{r+1}$, where k is an integer, is **[JEE Main 2020, 7 January]**

4. If the number of five-digit numbers with distinct digits and 2 at the 10th place is 336 k, then k is equal to

5. There are 720 permutations of the digits 1, 2, 3, 4, 5, 6. Suppose these permutations are arranged for smallest to largest numerical values, beginning from 1 2 3 4 5 6 and ending with 6 5 4 3 2 1. Then, the digit in unit place of number at 267th position is ....

6. An n-digit number means a positive integer having n digits. A total of 900 distinct n-digit numbers are to be formed using only the numerals 2, 5 and 7. The smallest value of n for which this is possible is

7. Number of permutations of 1, 2, 3, 4, 5, 6, 7, 8, and 9 taken all at a time are such that the digit

1 appearing somewhere to the left of 2

2 appearing to the left of 4 and

5 somewhere to the left of 6, is $k \times 7!$ Then, the value of k is ....

Matrix-Match Type

Given below are Matching Type Questions, with two columns (each having some items) each . Each item of Column I has to be matched with the items of Column II, by encircling the correct match(es).

NOTE *An item of Column I can be matched with more than one items of Column II.*

1. Observe the following columns :

	Column I		Column II
(A)	The number of positive integral solutions of the equation $x_1x_2x_3x_4x_5 = 1050$ is λ , then λ is divisible by	(P)	3
(B)	Let y be the element of the set $A = \{1, 2, 3, 5, 6, 10, 15, 30\}$ and x_1, x_2, x_3 be integers such that $x_1x_2x_3 = y$. If λ be the number of integral solutions of $x_1x_2x_3 = y$, then λ is divisible by	(Q)	4
(C)	Let a be a factor of 120. If λ be the number of positive integral solutions of $x_1\,x_2x_3 = a$, then λ is divisible by	(R)	5
		(S)	8
		(T)	16

2. In a high school, a committee has to be formed from a group of 6 boys $M_1, M_2, M_3, M_4, M_5, M_6$ and 5 girls G_1, G_2, G_3, G_4, G_5. **[JEE Advanced 2018]**

(i) Let α_1 be the total number of ways in which the committee can be formed such that the committee has 5 members, having exactly 3 boys and 2 girls.

(ii) Let α_2 be the total number of ways in which the committee can be formed such that the committee has at least 2 members, and having an equal number of boys and girls.

(iii) Let α_3 be the total number of ways in which the committee can be formed such that the committee has 5 members, at least 2 of them being girls.

(iv) Let α_4 be the total number of ways in which the committee can be formed such that the committee has 4 members, having at least 2 girls such that both M_1 and G_1 are NOT in the committee together.

	Column I		Column II
(A)	The value of α_1 is	(P)	136
(B)	The value of α_2 is	(Q)	189
(C)	The value of α_3 is	(R)	192
(D)	The value of α_4 is	(S)	200
		(T)	381
		(U)	461

3. Observe the following columns :

	Column I		Column II
(A)	If n be the number of ways in which 12 different books can be distributed equally among 3 persons, then	(P)	4
	$\frac{(4!)^4 n}{12!}$ is divisible by	(Q)	6
(B)	If n be the number of ways in which 12 different things can be distributed in three groups. Then $\frac{(4!)^3\ (3!)^2}{12!}n$ is divisible by	(R)	12
(C)	If n be the number of ways in which 12 different things can be distributed in 5 sets of 2, 2, 2, 3, 3 things. Then	(S)	24
	$\frac{(3!)^3 (2!)^4\ 5!}{12!}n$ is divisible by	(T)	30

4. In Column I the types of distributions of playing cards and, in Column II, their corresponding number of distributions is given. Match the items in Column I with those in Column II.

	Column I		Column II
A.	52 playing cards are to be equally distributed among four players. The number of possible distributions is	P.	$\frac{(52)!}{[(13)!]^4}$
B.	52 cards are to be divided into four equal groups	Q.	$\frac{(52)!}{4![(13)!]^4}$
C.	52 cards are to be divided into 4 sets, three of them having 17 cards each and the 4th has just one card	R.	$\frac{(52)!}{3![(17)!]}$
D.	52 cards are to be divided equally into two sets	S.	$\frac{(52)!}{2!((26)!)^2}$

5. Observe the following columns :

	Column I		Column II
(A)	If λ_1 be the number of positive integral solutions of $x + y + z + w = 10$,	(P)	3
	then λ_1 is divisible by	(Q)	4
(B)	If λ_2 be the non-negative integral solutions of $x + y + z = 6$, then λ_2 is divisible by	(R)	5
(C)	If λ_3 be the number of solutions of $x + y + z = 15$ such that	(S)	7
	$x \geq 1, y \geq 2$ and $z \geq 3$, then λ_3 is divisible by	(T)	11

Answers

Objective Questions Type I [Only one correct answer]

1. (b)	**2.** (c)	**3.** (b)	**4.** (a)	**5.** (d)	**6.** (b)	**7.** (b)	**8.** (b)	**9.** (c)	**10.** (b)
11. (a)	**12.** (c)	**13.** (c)	**14.** (c)	**15.** (c)	**16.** (b)	**17.** (b)	**18.** (c)	**19.** (b)	**20.** (d)
21. (a)	**22.** (b)	**23.** (c)	**24.** (a)	**25.** (b)	**26.** (a)	**27.** (c)	**28.** (b)	**29.** (b)	**30.** (d)
31. (b)	**32.** (b)	**33.** (a)	**34.** (c)	**35.** (a)	**36.** (b)	**37.** (d)	**38.** (a)	**39.** (b)	**40.** (c)
41. (a)	**42.** (b)	**43.** (a)	**44.** (b)	**45.** (c)	**46.** (d)	**47.** (d)	**48.** (c)	**49.** (c)	**50.** (d)
51. (b)	**52.** (b)	**53.** (c)	**54.** (c)	**55.** (a)	**56.** (b)	**57.** (a)	**58.** (b)	**59.** (b)	**60.** (b)
61. (c)	**62.** (b)	**63.** (d)	**64.** (c)	**65.** (d)	**66.** (c)	**67.** (c)	**68.** (a)	**69.** (a)	**70.** (a)
71. (b)	**72.** (a)	**73.** (d)	**74.** (d)	**75.** (d)	**76.** (d)	**77.** (d)	**78.** (a)	**79.** (b)	**80.** (a)
81. (c)	**82.** (b)	**83.** (d)	**84.** (a)	**85.** (b)	**86.** (b)	**87.** (c)	**88.** (c)	**89.** (c)	**90.** (d)
91. (b)	**92.** (b)	**93.** (b)	**94.** (a)	**95.** (d)	**96.** (b)	**97.** (b)	**98.** (b)	**99.** (b)	**100.** (b)
101. (d)	**102.** (c)	**103.** (c)	**104.** (b)						

Objective Questions Type II [One or more than one correct answer(s)]

1. (a, b, c d)	**2.** (a, c)	**3.** (b, c)	**4.** (a, c)	**5.** (a, b)
6. (a, b, c, d)	**7.** (b, d)	**8.** (a, d)	**9.** (a, b, c)	**10.** (a, b, c)
11. (a, c, d)	**12.** (c, d)	**13.** (b, d)	**14.** (a, b, c)	**15.** (a, d)
16. (a, b, c, d)	**17.** (a, c)	**18.** (a, b, c, d)	**19.** (a, c)	**20.** (a, b, c, d)
21. (a, b, c)	**22.** (a, b)	**23.** (b, c, d)	**24.** (a, d)	

Linked-Comprehension Type

Passage 1 **1.** (c) **2.** (b) **3.** (a)
Passage 2 **1.** (c) **2.** (a) **3.** (b) **4.** (d) **5.** (d)
Passage 3 **1.** (c) **2.** (b) **3.** (a) **4.** (d) **5.** (a)
Passage 4 **1.** (a) **2.** (b) **3.** (c)
Passage 5 **1.** (b) **2.** (b) **3.** (b) **4.** (a)
Passage 6 **1.** (c) **2.** (b) **3.** (b) **4.** (c) **5.** (b)
Passage 7 **1.** (d) **2.** (a) **3.** (b)

Numerical Value Problems

1. (1.2)	**2.** (91)	**3.** (1981)	**4.** (127)	**5.** (2160)	**6.** (2.4)
7. (3600)	**8.** (132)	**9.** (70)	**10.** (4373)	**11.** (191)	**12.** (303)
13. (136)	**14.** (126)	**15.** (288)			

Single Integer Type

1. (5) **2.** (5) **3.** (4) **4.** (8) **5.** (6) **6.** (7) **7.** (9)

Matrix-Match Type

1. A → (P, R); B → (Q, S, T); C → (Q, R, S, T)
2. A → (S); B → (U); C → (T); D → (Q)
3. A → (P, Q, S); B → (P, Q, R, S, T); C → (P, Q, S)
4. A → (P); B → (Q); C → (R); D → (S)
5. A → (P, Q, S); B → (Q, S); C → (R, T)

Solutions

Objective Questions Type [Only one correct answer]

1. Number of ways = (collinear (0) and non collinear (4)) OR (collinear (1) and non collinear (3)) OR (collinear (2) and non collinear (2))

$$= {}^4C_0 \times {}^4C_4 + {}^4C_1 \times {}^4C_3 + {}^4C_2 \times {}^4C_2$$
$$= 1 + 16 + 36$$
$$= 53$$

2. Number of words = coefficient of x^4 in

$$4!\left(1 + \frac{x}{1!} + \frac{x^2}{2!}\right)^2 (1+x)^2$$
$$= \text{coefficient of } x^4 \text{ in } 6[(1+x)^2 + 1]^2 (1+x)^2$$
$$= \text{coefficient of } x^4 \text{ in}$$
$$6\{(1+x)^4 + 2(1+x)^2 + 1\} \times (1+x)^2$$
$$= \text{coefficient of } x^4 \text{ in}$$
$$6\{(1+x)^6 + 2(1+x)^4 + (1+x)^2\}$$
$$= 6\{{}^6C_4 + 2\cdot {}^4C_4 + 0\}$$
$$= 6\{15 + 2\} = 102$$

Alternate Method

	Selection	Arrangements
All different	${}^4C_4 = 1$	${}^4P_4 = 24$
2 same and 2 different	${}^2C_1 \times {}^3C_1 = 6$	$6 \times \frac{4!}{2!} = 72$
2 same of one kind and 2 same of other kind	${}^2C_2 = 1$	$1 \times \frac{4!}{2!\,2!} = 6$
		Total = 102

3. $${}^{n+1}C_3 - {}^nC_3 = 21$$
$\Rightarrow$ $${}^nC_2 + {}^nC_3 - {}^nC_3 = 21$$
$\Rightarrow$ $${}^nC_2 = 21 = \frac{21 \times 2}{2} = \frac{7 \times 6}{2} = {}^7C_2$$
$\therefore$ $$n = 7$$

4. Required number of arrangements = total ways – number of ways in which $2N'$ s come together.

$$= \frac{6!}{3!\,2!} - \frac{5!}{3!}$$
$$= 60 - 20 = 40$$

Alternate Method (Gap method)

$$\times B \times A \times A \times A \times$$
$$= {}^5C_2 \times \frac{4!}{3!} = 10 \times 4 = 40$$

5. $\because xyz = 2^3 \times 3^1$

Let $\alpha + \beta + \gamma = 3$, $\alpha + \beta + \gamma = 1$

Number of integral positive solutions

$$= {}^{3+3-1}C_{3-1} \times {}^{1+3-1}C_{3-1}$$
$$= {}^5C_2 \times {}^3C_2 = 30$$

Since, negative values of x, y, z is also allowed but since product is positive and hence any two of them may be negative.

$\therefore$ Number of negative integral solutions $= {}^3C_2 \times 30 = 90$

Hence, total number of integral solutions of $xyz = 24$ is

$$30 + 90 = 120$$

Alternate Method (for positive integral solutions)

$$24 = 2 \times 3 \times 4 = 2 \times 2 \times 6 = 1 \times 3 \times 8 = 1 \times 6 \times 4$$
$$= 1 \times 2 \times 12 = 1 \times 1 \times 24$$

$\therefore$ Number of positive integral solutions

$$= 3! + \frac{3!}{2!} + 3! + 3! + 3! + \frac{3!}{2!} = 30$$

6. **Case I**

a	x	

Here $x = c$ or d

Number of ways = 2

Case II

b	y	z	

Here $y = a$ or d

If $y = a$, then $z = c$ or d

$\therefore$ Number of ways = 3

Case III

c	u	p	

Here $u = a$ or b

If $u = a$, then $p = d$

and if $u = b$, then $p = a$ or d

$\therefore$ Number of ways = 3

Case IV

d	q	r	s

Here, $q = a$ or b or c

If $q = a$, then $r = c$ and $s = b$

If $q = b$, then $r = a$ and $s = c$

If $q = c$, then $r = a$ or b

If $r = a$, then $s = b$ (impossible)

and if $r = b$, then $s = a$ (possible)

$\therefore$ Number of ways = 3

$\therefore$ Total ways = 2 + 3 + 3 + 3 = 11 ways

7. Required number of ways

$$= {}^nC_2 + {}^nC_3 + {}^nC_4 + \ldots + {}^nC_{n-3} + {}^nC_{n-2}$$
$$= 2^n - ({}^nC_0 + {}^nC_1 + {}^nC_{n-1} + {}^nC_n)$$
$$= 2^n - 2(1+n) = 2^n - 2n - 2$$

8. In the given numbers 1, 3, 5, 7, ..., 147, 148, 151 the numbers which are multiple of 5 are 5, 15, 25, 35, ..., 145. which is an arithmetic sequence

$$T_n = a + (n-1)d$$
$\Rightarrow$ $$145 = 5 + (n-1)10$$
$\Rightarrow$ $$n = 15$$

and if total number of terms in the given sequence is m, then

$$151 = 1 + (m-1) \times 2 \Rightarrow m = 76$$

So, the number of ways in which product is a multiple of 5 = (both two numbers from 5, 15, 25, ..., 145) or (one number from 5, 15, 25, ..., 145 and one from remaining numbers)

$$= {}^{15}C_2 + {}^{15}C_1 \times {}^{76-15}C_1$$
$$= {}^{15}C_2 + {}^{15}C_1 \times {}^{61}C_1$$
$$= \frac{15 \times 14}{2} + 15 \times 61$$
$$= 105 + 915$$
$$= 1020$$

9. We can not write 903

If in the selection of n cards, we get either (9 or 3), (9 or 0), (0 or 3), (only 0), (only 3) or (only 9)

Can be selected $= 2 \times 2 \times 2 \times \ldots n$ times $= 2^n$

Similarly, (9 or 0) or (3 or 0) can be selected 2^n

In the above selection (only 0) or (only 3) or (only 9) is repeated twice

$\therefore$ Total ways $= 2^n + 2^n + 2^n - 3 = 93$

$\Rightarrow \quad 3 \cdot 2^n = 96$

$\Rightarrow \quad 2^n = 32 = 2^5$

$\therefore \quad n = 5$

10. Let $a = 2x + 1$, $b = 2y + 1$, $c = 2z + 1$

where $x, y, z \in$ whole number

$\therefore \quad a + b + c = 13$

$\Rightarrow \quad 2x + 1 + 2y + 1 + 2z + 1 = 13$

or $\quad x + y + z = 5$

The number of integrals solutions

$$= {}^{5+3-1}C_{3-1} = {}^7C_2$$
$$= \frac{7 \cdot 6}{1 \cdot 2} = 21$$

11. For length, number of choices is

$$(2m - 1) + (2m - 3) + \ldots. + 3 + 1 = m^2$$

Similarly for breadth number of choices is

$$(2n - 1) + (2n - 3) + \ldots. + 3 + 1 = n^2$$

Hence, required number of choices is m^2n^2.

12. Total time required

= (total number of dials required to sure open the lock) × 5 s

$$= 10^5 \times 5 \text{ s}$$
$$= \frac{500000}{60 \times 60 \times 13} \text{ days}$$
$$= 10.7 \text{ days}$$

Hence, 11 days are enough to open the safe.

13. Let n be the number of sides of the polygon

$n \cdot 160° = (n \quad 2)\ 180°$

or $\quad 20° \cdot n = 360°$

$\therefore \quad n = 18$

Then number of diagonal $= {}^{18}C_2 - 18$

$$= 153 - 18$$
$$= 135$$

14. $N_i = {}^5C_k \times {}^4C_{5-k}$

$N_1 = 5 \times 1$

$N_2 = 10 \times 4$

$N_3 = 10 \times 6$

$N_4 = 5 \times 4$

$N_5 = 1$

$N_1 + N_2 + N_3 + N_4 + N_5 = 126$

15. The first digit a can take any one of 1 to 8

The third digit c can take any one of 0 to 9

When $a = 1$, b can take any one of 2 to 9 = 8 values

When $a = 2$, b can take any one of 3 to 9 = 7 values

When $a = 3$, b can take any one of 4 to 9 = 6 values

....

When $a = 8$, b can take any one of (b = 1) = 1 value

Thus the total numbers $= (8 + 7 + 6 + \ldots. + 2 + 1) \times 10$

($\because z$ can take only one of 0 to 9 = 10 values)

$$= \frac{8 \times 9}{2} \times 10 = 360$$

16. $\because 720 = 2^4 \times 3^2 \times 5^1$

$\therefore$ Sum of all odd divisors $= (1 + 3 + 3^2)(1 + 5^1)$

$$= 13 \times 6 = 78$$

17. $\because f(x_i) \neq y_i$

i e, no object goes to its scheduled place. Then number of one-one mappings

$$= \lfloor 6 \left(1 - \frac{1}{\lfloor 1} + \frac{1}{\lfloor 2} - \frac{1}{\lfloor 3} + \frac{1}{\lfloor 4} - \frac{1}{\lfloor 5} + \frac{1}{\lfloor 6}\right)$$
$$= \lfloor 6 \left(\frac{1}{\lfloor 2} - \frac{1}{\lfloor 3} + \frac{1}{\lfloor 4} - \frac{1}{\lfloor 5} + \frac{1}{\lfloor 6}\right)$$
$$= 360 - 120 + 30 - 6 + 1$$
$$= 265$$

18. $\because \lfloor 5, \lfloor 6, \lfloor 7, \lfloor 8, \ldots., \lfloor 175$ each multiple of 15

and $\quad \lfloor 1 + \lfloor 2 + \lfloor 3 + \lfloor 4 = 33$

$\therefore$

```
15)33(2
   30
   --
    3
```

Hence, required remainder = 3.

19. Total number of m-elements subsets of $A = {}^nC_m$ ...(i)

and number of m-elements subsets of A each containing the element $a_4 = {}^{n-1}C_{m-1}$

According to question, ${}^nC_m = \lambda \cdot {}^{n-1}C_{m-1}$

$$\Rightarrow \quad \frac{n}{m} \cdot {}^{n-1}C_{m-1} = \lambda \cdot {}^{n-1}C_{m-1}$$

$$\Rightarrow \quad \lambda = \frac{n}{m}$$

or $\quad n - m\lambda$

20. ${}^{47}C_4 + \sum_{n=1}^{5} {}^{52-n}C_3$

$$= {}^{47}C_4 + {}^{51}C_3 + {}^{50}C_3 + {}^{49}C_3 + {}^{48}C_3 + {}^{47}C_3$$
$$= {}^{51}C_3 + {}^{50}C_3 + {}^{49}C_3 + {}^{48}C_3 + ({}^{47}C_3 + {}^{47}C_4)$$
$$= {}^{52}C_4$$

[using Pascal rule successively]

21. Sum of diagonal entries of M^TM is Σa_i^2.

$$\sum_{i=1}^{9} a_i^2 = 5$$

Possibilities

I. 2, 1, 0, 0, 0, 0, 0, 0, 0, which gives $\frac{9!}{7!}$ matrices

II. 1, 1, 1, 1, 1, 0, 0, 0, 0, which gives $\frac{9!}{4!\times 5!}$ matrices

Total matrices $= 9\times 8 + 9\times 7\times 2 = 198$

22. $10 \geq x-1 \Rightarrow x \leq 11$ and $10 \geq x$

$\therefore \quad x \leq 10$

$\because \quad {}^{10}C_{x-1} > 2\cdot{}^{10}C_x \Rightarrow 1 > 2\cdot\frac{{}^{10}C_x}{{}^{10}C_{x-1}}$

$\Rightarrow \quad 1 > 2\cdot\frac{10-x+1}{x}$

$\Rightarrow \quad x > 22-2x\,;\ \Rightarrow x > \frac{22}{3}$

$\Rightarrow \quad x > 7\frac{1}{3}$

$\therefore \quad x = 8$

23. 1 S, 3A, 1 H, 2 R, 1 N, 1 P, 1 U

When all letters are different, corresponding ways is ${}^7C_3\times 3! = {}^7P_3 = 210$. When two letters are of one kind and other is different, corresponding number of ways is ${}^2C_1\times{}^6C_1\times(3!/2!) = 36$. When all letters are alike, corresponding number of ways is 1. Thus, total number of words that can be formed is $210+36+1=247$.

24. Since, the sum of given digits

$$0+1+2+5+7+9=24$$

Let the six-digit number be *abcdef* and to be divisible by 11, so the difference of sum of odd placed digits and sum of even placed digits should be either 0 or a multiple of 11 means $|(a+c+e)-(b+d+f)|$ should be either 0 or a multiple of 11.

Hence, possible case is $a+c+e=12=b+d+f$ (only)

Now, ***Case I***

set $\{a, c, e\} = \{0, 5, 7\}$ and set $\{b, d, f\} = \{1, 2, 9\}$

So, number of 6-digits numbers $= (2\times 2!)\times(3!) = 24$

[$\because$ a can be selected in ways only either 5 or 7].

Case II

Set $\{a, c, e\} = \{1, 2, 9\}$ and set $\{b, d, f\} = \{0, 5, 7\}$

So, number of 6-digits numbers $= 3!\times 3! = 36$

So, total number of 6-digits numbers $= 24+36 = 60$

25. $n!(21-n)! = 21!\cdot\frac{n!(21-n)!}{21!} = 21!\cdot\frac{1}{{}^{21}C_n} = \frac{21!}{{}^{21}C_n}$

For minimum value, ${}^{21}C_n$ is maximum

Maximum value of ${}^{21}C_n = {}^{21}C_{\frac{21-1}{2}} = {}^{21}C_{10}$

$\therefore$ Minimum value $= \frac{21!}{{}^{21}C_{10}} = \frac{21!}{21!}\times 10!11! = 10!11!$

26. We have, 6 girls and 4 boys. To select 4 members (atmost one boy)

i.e. (1 boy and 3 girls) or (4 girls)

$= {}^6C_3\cdot{}^4C_1 + {}^6C_4 \quad \ldots$(i)

Now, selection of captain from 4 members $= {}^4C_1 \quad \ldots$(ii)

$\therefore$ Number of ways to select 4 members (including the selection of a captain, from these 4 members)

$= ({}^6C_3\cdot{}^4C_1 + {}^6C_4)\,{}^4C_1$

$= (20\times 4+15)\times 4 = 380$

27. Total number of three digit numbers $= 5\times 5\times 4 = 100$

Total number of five-digit numbers $= 5\times 5\times 4\times 3\times 2$

$= 600$

$\therefore$ Total number of required numbers $= 100+600 = 700$

28. 4 lines intersect each other in ${}^4C_2 = 6$ points. 4 circles intersect each other in ${}^4P_2 = 12$ points. Every lines cuts 4 circles into 8 points. Therefore, 4 lines cut four circles into 32 points.

Therefore, the maximum number of points

$= 6+12+32 = 50$

29. ${}^{12}C_3 - {}^7C_3 = 185$

30. As each section has 5 questions, so number of ways to select 5 questions are

	S_1	S_2	S_3
	1	1	3
	1	3	1
	3	1	1
	1	2	2
and	2	1	2
	2	2	1

$\therefore$ Total number of selection of 5 questions

$= 3\times({}^5C_1\times{}^5C_1\times{}^5C_3) + 3\times({}^5C_1\times{}^5C_2\times{}^5C_2)$

$= 3(5\times 5\times 10) + 3(5\times 10\times 10)$

$= 750+1500 = 2250$

31. $2^{10}-1 = 1023$

32. $6! - 2!\,5! = 480$

33. If zero is included it will be at $z \Rightarrow {}^9C_2$

If zero is excluded $\begin{cases} x, y, z \text{ all diff.} & \Rightarrow {}^9C_3\times 2! \\ x = z < y & \Rightarrow {}^9C_2 \\ x < y = z & \Rightarrow {}^9C_2 \end{cases}$

The total number of ways is 276.

34. Total grade $= 4\ (A, B, C, D)$

students $= 3$

$\therefore$ The number of ways $= {}^4P_3$

35. Number of ways of selections

$= {}^{12}C_1 + {}^{12}C_2 + {}^{12}C_3 + {}^{12}C_4$

$= 12 + \frac{12\times 11}{2\times 1} + \frac{12\times 11\times 10}{3\times 2\times 1} + \frac{12\times 11\times 10\times 9}{4\times 3\times 2\times 1}$

$= 12+66+220+495 = 793$

36. Number of parallelograms $= {}^5C_2 \times {}^4C_2 = 60$

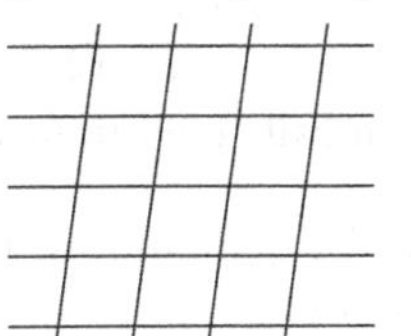

37. $\frac{4!}{2!\,2!} \times \frac{3!}{2!} = 18$

38. ${}^{12}C_3 - {}^3C_3 - {}^4C_3 - {}^5C_3 = 205$

39. Since two families has 3 members each and one family with four members. So that can be seated among themself, so same family members are not separated in 3!, 3! and 4! respectively.
Now the groups (means families) can arrange in 3! ways. So, required number of ways is

$$3! \times 3! \times 4! \times 3! = (3!)^3 \cdot 4!$$

40. $10 \times 10 \times 10 \times 10 \times 10 - {}^{10}P_5 = 69760$

41. Given digits are 1, 1, 2, 2, 2, 2, 3, 4, 4.
According to the question, odd numbers 1, 1, 3 should occur at even places only.

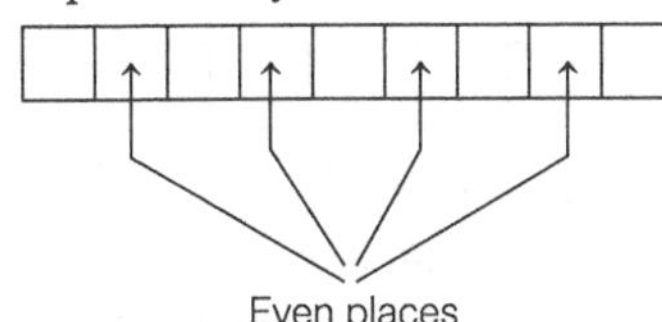

$\therefore$ The number of ways to arrange odd numbers at even places are ${}^4C_3 \times \frac{3!}{2!}$

and the number of ways to arrange remaining even numbers are $\frac{6!}{4!\,2!}$.

So, total number of 9-digit numbers, that can be formed using the given digits are

$${}^4C_3 \times \frac{3!}{2!} \times \frac{6!}{4!\,2!} = 4 \times 3 \times 15 = 180$$

42. $x + y + z = 100$, where $x \ge 1,\ y \ge 1,\ z \ge 1$
Let $x_1 = x - 1$, $y_1 = 4 - 1$ and $z_1 = z - 1$
Now, the equation becomes $x_1 + y_1 + z_1 = 97$.
The number of solution is ${}^{97+3-1}C_{3-1} = {}^{99}C_2 = 4851$

43.

8 type	7 type	2 or 4 or 6 or 8

Total number $= {}^8P_2 \times 4 = 224$

44. The integer greater than 6000 may be of 4 digits or 5 digits. So, here two cases arise.
***Case* I** When number is of 4 digits. Four-digit number can starts from 6, 7 or 8.
Thus, total number of 4-digit numbers, which are greater than $6000 = 3 \times 4 \times 3 \times 2 = 72$
***Case* II** When number is of 5 digits.
Total number of five-digit numbers which are greater than $6000 = 5! = 120$
$\therefore$ Total number of integers $= 72 + 120 = 192$

45. $y + z = 5$ ...(i)

$$\frac{1}{y} + \frac{1}{z} = \frac{5}{6}$$

$$\Rightarrow \quad \frac{y+z}{yz} = \frac{5}{6}$$

$$\Rightarrow \quad \frac{5}{yz} = \frac{5}{6}$$

$$\Rightarrow \quad yz = 6$$

Also, $(y - z^2) = (y + z)^2 - 4yz$

$$\Rightarrow \quad (y - z)^2 = 25 - 4(6) = 1$$

$$\Rightarrow \quad y - z = 1 \qquad \text{...(ii)}$$

From Eqs. (i) and (ii),

$$y = 3 \text{ and } z = 2$$

for calculating odd divisor of $p = 2^x \cdot 3^y \cdot 5^z$ x must be zero

$$P = 2^0 \cdot 3^3 \cdot 5^2$$

$\therefore$ Total odd divisors must be $(3+1)(2+1) = 12$

46. 4 choices for each question
total question $= 3$
All questions are correct $= 1$ way
$\therefore$ Required number of ways $= 4^3 - 1 = 63$

47. $(4-1)!\,(5 + 6 + 7 + 8) = 6 \times 26 = 156$

48. Number of triangles $= {}^{2n}C_3 - {}^nC_3 - {}^nC_3$

$$= \frac{2n\,(2n-1)\,(2n-2)}{6} - \frac{2n\,(n-1)\,(n-2)}{6}$$

$$= \frac{1}{3} n\,(n-1)\,(3n) = n^2\,(n-1)$$

49.

1	2	3	4	5	6	7	8	9

Number of even places $= 4$
Odd digits $= 3, 3, 5, 5$
Remaining digits on odd places $= 2\,2\,8\,8\,8$

$\therefore$ Required number of ways $= \frac{4!}{2!\,2!} \times \frac{5!}{2!\,3!}$

$$= 60$$

50. $\binom{n}{r} + 2\binom{n}{r-1} + \binom{n}{r-2}$

$$= {}^nC_r + 2 \cdot {}^nC_{r-1} + {}^nC_{r-2}$$

$$= ({}^nC_r + {}^nC_{r-1}) + ({}^nC_{r-1} + {}^nC_{r-2})$$

$$= {}^{n+1}C_r + {}^{n+1}C_{r-1} = {}^{n+2}C_r = \binom{n+2}{r}$$

51. ${}^{n+1}C_{n-2} - {}^{n+1}C_{n-1} \le 100$

$$\Rightarrow \quad {}^{n+1}C_3 - {}^{n+1}C_2 \le 100$$

$$\Rightarrow \quad \frac{1}{6}(n+1)\,n\,(n-1) - \frac{n\,(n+1)}{2} \le 100$$

$$\Rightarrow \quad \frac{1}{6} n\,(n+1)\,\{n - 1 - 3\} \le 100$$

$$\Rightarrow \quad n\,(n+1)\,(n-4) \le 600$$

It is true for $n = 2, 3, 4, 5, 6, 7, 8, 9$

52. Number of groups of r students in each when the number of times the teacher can go to the zoo = number of groups $= {}^{21}C_r$

For maximum value of ${}^{21}C_r$

$$r = \frac{21-1}{2} \text{ or } \frac{21+1}{2} = 10 \text{ or } 11$$

53. Coefficient of x^{11} in

$$(x^1 + x^2 + x^3 + x^4 + x^5 + x^6)^3$$

Coefficient of x^{11} in

$$x^3(1 + x + x^2 + x^3 + x^4 + x^5)^3$$

or Coefficient of x^{11} in

$$x^3(1 - x^6)^3(1-x)^{-3}$$

or Coefficient of x^8 in

$$(1 - 3x^6)(1 + {}^3C_1 x + {}^4C_2 x^2 + \ldots)$$
$$= {}^{10}C_8 - 3 \cdot {}^4C_2$$
$$= {}^{10}C_2 - 3 \cdot {}^4C_2 = 45 - 18 = 27$$

54. ${}^nC_2$, $n \in N$, infinite number.

55. Number of triangles = total number of triangles
– number of triangles having one side common
– number of triangles having two sides common

$$= {}^8C_3 - 8 \times 4 - 8$$
$$= 56 - 32 - 8 = 16$$

56. Total cases $= p + 1$ (if selected or not)

Required number of ways

$$= \{(p+1)(p+1)\ldots.\ n \text{ terms}\} - 1$$
$$= (p+1)^n - 1$$

57.

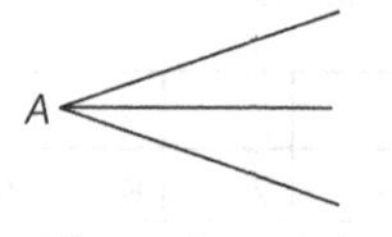

13 pass through A 11 Pass through B

∴ Number of intersection points

$$= {}^{37}C_2 - {}^{13}C_2 - {}^{11}C_2 + 2$$

[∵ two points A and B]

$$= 535$$

58. [IN] TEGER

$$\text{number of ways} = \frac{6!}{2!} \times 2! = 6! = 720$$

$$m_1 = \frac{7!}{2!} - 720 = 1800$$

and [I] NTEGE [R]

$$m_2 = \frac{5!}{2!} = 60$$

∴ $m_1/m_2 = 30$

59. ${}^{x+2}P_{x+2} = a \Rightarrow a = (x+2)!$

$${}^xP_{11} = b \Rightarrow b = \frac{x!}{(x-11)!}$$

and ${}^{x-11}P_{x-11} = c \Rightarrow c = (x-11)!$

∴ $a = 182\,bc$

$$(x+2)! = 182\frac{x!}{(x-11)!}(x-11)!$$

$\Rightarrow$ $(x+2)(x+1) = 182 = 14 \times 13$

∴ $x + 1 = 13$

∴ $x = 12$

60. Let there are n balls used to form the sides of equilateral triangle.

According to the question, we have

$$\frac{n(n+1)}{2} + 99 = (n-2)^2$$

$\Rightarrow$ $n^2 + n + 198 = 2[n^2 - 4n + 4]$

$\Rightarrow$ $n^2 - 9n - 190 = 0$

$\Rightarrow$ $n^2 - 19n + 10n - 190 = 0$

$\Rightarrow$ $(n-19)(n+10) = 0$

$\Rightarrow$ $n = 19, -10$

$\Rightarrow$ $n = 19$ [∵ number of balls $n > 0$]

Now, number of balls used to form an equilateral triangle is $\frac{n(n+1)}{2} = \frac{19 \times 20}{2} = 190$

61. Here $x \geq 1$, $y \geq 1$, $z \geq 1$

No. of solutions $= {}^{100-1}C_{3-1} = {}^{99}C_2$

$$= \frac{99 \cdot 98}{1 \cdot 2} = 4851$$

62. ☐ ☐☐ ☐☐

$${}^4C_1 + {}^4C_1 \times {}^4C_1 + {}^4C_1 \times {}^4C_1 \times {}^3C_1$$
$$= 4 + 16 + 48 = 68$$

63. Let $A = \{a_1, a_2, \ldots., a_n\}$ for each a_i $(1 \leq i \leq n)$

Either $a_i \in P_j$ or $a_i \notin P_j$ $(1 \leq j \leq m)$

∴ There are 2^m choices in which $a_i m a_j$ belongs to P_j,

Also there is exactly one choice, *ie*, $a_i \notin P_j$ for $j = 1, 2, \ldots., m$ for which $a_i \notin P_1 \cup P_2 \cup \ldots. \cup P_m$.

∴ $a_i \in P_1 \cup P_2 \cup \ldots \cup P_m$ in $(2^m - 1)$ ways.

Since, there are n elements in the set A, the number of ways of constructing subsets.

$P_1, P_2, \ldots., P_m$ is $(2^m - 1)^n$

64.

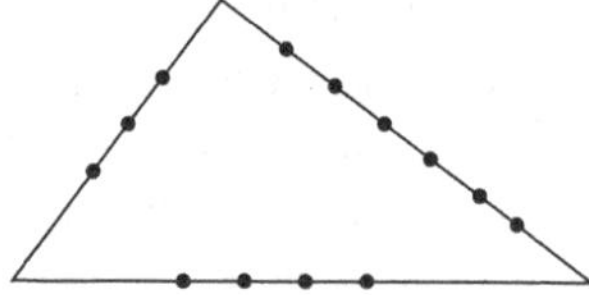

Total number of triangles formed

$$= {}^{14}C_3 - {}^3C_3 - {}^5C_3 - {}^6C_3 = 333$$

65. Let $A = \{a_1, a_2, a_3, \ldots., a_n\}$

$$1 \leq i \leq n$$

(i) $a_i \in P$ and $a_i \in Q$

(ii) $a_i \in P$ and $a_i \notin Q$

(iii) $a_i \notin P$ and $a_i \in Q$

(iv) $a_i \notin P$ and $a_i \notin Q$

Since, $P \cap Q = \phi$

[cases in favour 3 i.e., (ii), (iii), (iv)]

Total element in $A = n$

∴ Required number of ways $= 3^n$.

66. The number of times he will go to the garden
= the number of selecting 3 children from 8
$= {}^8C_3 = 56$

67. Number of different factors of $a^m \cdot b \cdot c \cdot d \ldots$
$= (m+1)\{(1+1)(1+1)(1+1)\ldots n \text{ factors}\} - 1$
$= (m+1)\,2^n - 1$

68. There are 13 letters
$A, A, A ; S, S, S, S ; I, I ; N, N ; T ; O$
Coefficient of x^4 in
$(1+x+x^2+x^3)(1+x+x^2+x^3+x^4) \times (1+x+x^2)^2 (1+x)^2$
$[\because 3A's, 4S's, 2I's, 2N's, 1T \text{ and } 1O]$
$\Rightarrow$ Coefficient of x^4 in
$$\frac{(1-x^4)}{(1-x)}\frac{(1-x^5)}{(1-x)}\frac{(1-x^3)^2}{(1-x)^2}(1+x)^2$$
$\Rightarrow$ Coefficient of x^4 in
$(1+2x+x^2-2x^3-5x^4)(1-x)^{-4}$
$\therefore$ Coefficient of $x^4 = 35+40+10-8-5 = 72$

69. Factorizing the given number, we have
$38808 = 2^3 \cdot 3^2 \cdot 7^2 \cdot 11$
Therefore the total number of divisors
$= (3+1)(2+1)(2+1)(1+1) - 1 = 71$
But this includes the division by the number itself.
Hence, the required number of divisors $= 71 - 1 = 70$.

70. The letters of SURITI can be arranged in $\frac{6!}{2!} = 360$ ways.
The alphabetic order is I, R, S, T and U. The number of words beginning I is $5! = 120$ and those beginning R is $\frac{5!}{2!} = 60$. Then come words beginning SI is $4! = 24$, SR is $\frac{4!}{2!} = 12$ and ST is $\frac{4!}{2!} = 12$. We get ${}^3P_3 = 3! = 6$
Words beginning SUI, after which come the words SURIIT and SURITI. Thus the rank of SURITI is
$120 + 60 + 24 + 12 + 12 + 6 + 2 = 236$

71. Suppose $a_1\, a_2\, a_3\, a_4\, a_5\, a_6\, a_7$ represents a seven digit number. Then a_1 takes the value 1, 2, 3, ..., 9 and $a_2, a_3, \ldots, a_7$ all take values 0, 1, 2, 3, ..., 9. If we keep $a_1a_2a_3\ldots a_6$ fixed, then the sum $a_1 + a_2 + a_3 + \ldots + a_6$ is either even or odd. Since a_7 takes 10 values 0, 1, 2, ..., 9 five of the numbers so formed will be even and five odd.
Hence, the required number of numbers
$= 9 \cdot 10 \cdot 10 \cdot 10 \cdot 10 \cdot 10 \cdot 5 = 45 \times 10^5$

72. Each stall can be occupied by a cow, a horse or a calf (i.e., in 3 ways)
Hence, required number of ways $= 3^{12}$

73. Coefficient of x^{20} in $(1-x)^{-3}(1-x^4)^{-1}$
Coefficient of x^2 in
$(1 + {}^3C_1 x + {}^4C_2 x^2 + \ldots)(1 + x^4 + x^8 + x^{12} + x^{16} + x^{20})$
$= 1 + {}^6C_4 + {}^{10}C_8 + {}^{14}C_{12} + {}^{18}C_{16} + {}^{22}C_{20}$
$= 1 + {}^6C_2 + {}^{10}C_2 + {}^{14}C_2 + {}^{18}C_2 + {}^{21}C_2$
$= 1 + 15 + 45 + 91 + 153 + 231 = 536$

74. Terminal digits are the first and last digits.
$\because$ Terminal digits are even
$\therefore$ Ist place can be filled in 3 ways and last place can be filled in 2 ways and remaining places can be filled in ${}^5P_4 = 120$ ways.
Hence, the number of six digit numbers, the terminal digits are even, is $= 3 \times 120 \times 2 = 720$

75. Let $n = 2m + 1$
If a, b, c are in AP,
then $2b = a + c$
ie, sum of two numbers is even.
Here, $m + 1$ is odd and m is even number.
$\because$ Sum of two numbers is even,
then both numbers are even or odd.
$\therefore$ Required number of ways $= {}^mC_2 + {}^{m+1}C_2$
$$= \frac{m(m-1)}{2} + \frac{(m+1)m}{2}$$
$$= m^2$$
$$= \left(\frac{n-1}{2}\right)^2 = \frac{(n-1)^2}{4}$$

76. $A = \{a_1, a_2, a_3, \ldots, a_n\}$
(i) $a_i \in P, a_i \in Q$
(ii) $a_i \notin P, a_i \notin Q$
(iii) $a_i \notin P, a_i \in Q$
(iv) $a_i \notin P, a_i \notin Q$
$P \cap Q$ contains exactly two elements, taking 2 elements in (i) and $(n-2)$ elements in (ii) or (iii) or (iv)
$\therefore$ Number of ways $= {}^nC_2 \times 3^{n-2}$

77. $1 + \sum_{n=1}^{8} n = 1 + \frac{8 \times 9}{2} = 37$

78. $\frac{240}{4n+2} = \lambda \in I$
$\therefore \quad \lambda = \frac{120}{2n+1} = \frac{2^3 \times 3 \times 5}{2n+1}$
Divisible by 1, 3, 5, 15 (odd numbers)

79. Each place can be filled in 3 ways
$\therefore n$ places can be filled in 3^n
$\therefore \quad 3^n \geq 900$
$\therefore \quad 3^6 = 729$ and $3^7 = 2187$
Hence $\quad n = 7$.

80. Let $a = 2x + 1, b = 2y + 1, c = 2z + 1, d = 2u + 1$
where $x, y, z, u \in W$
Then, $a + b + c + d = 20$
$\Rightarrow \quad 2x + 2y + 2z + 2u = 16$
$\Rightarrow \quad x + y + z + u = 8$
$\therefore$ Number of ordered quadruplet
$$= {}^{8+4-1}C_{4-1} = {}^{11}C_3 = \frac{11 \cdot 10 \cdot 9}{1 \cdot 2 \cdot 3}$$
$$= 165$$

81. $$= \frac{15 \cdot 10}{4}(15+1)(10+1) - \sum_{r=1}^{10}(11-r)(16-r)$$
$$= 6600 - 660 = 5940$$

82. The number of students answering exactly $i\ (1 \le i \le n-1)$ questions wrongly is $2^{n-i} - 2^{n-i-1}$. The number of students answering all n questions wrongly is 2°.

Hence, the total number of wrong answer is

$$\sum_{i=1}^{n-1} i\,(2^{n-i} - 2^{n-i-1}) + n\,(2)^{\circ} = 2047$$

$$\Rightarrow \quad 2^{n-1} + 2^{n-2} + 2^{n-3} + \dots + 2^{\circ} = 2047$$

$$\Rightarrow \quad 2^n - 1 = 2047$$

$$\Rightarrow \quad 2^n = 2048 \Rightarrow 2^n = 2^{11}$$

$$\therefore \quad n = 11$$

83. Clearly, number of words start with A $= \dfrac{4!}{2!} = 12$

Number of words start with L $= 4! = 24$

Number of words start with M $= \dfrac{4!}{2!} = 12$

Number of words start with SA $= \dfrac{3!}{2!} = 3$

Number of words start with SL $= 3! = 6$

Note that, next word will be "SMALL".

Hence, the position of word "SMALL" is 58th.

84. Given, $n(A) = 4$, $n(B) = 2 \Rightarrow n(A \times B) = 8$

Total number of subsets of set $(A \times B) = 2^8$

Number of subsets of set $A \times B$ having no element (i.e. ϕ) $= 1$

Number of subsets of set $A \times B$ having one element $= {}^8C_1$

Number of subsets of set $A \times B$ having two elements $= {}^8C_2$

$\therefore$ Number of subsets having atleast three elements

$$= 2^8 - (1 + {}^8C_1 + {}^8C_2)$$

$$= 2^8 - 1 - 8 - 28 = 2^8 - 37 = 256 - 37 = 219$$

85. Total ways $= ({}^8C_2 - 2) \times 6!$

$$= (28 - 2) \times 6! = 26 \times 6!$$

86. The number of matches in first round $= {}^6C_2 + {}^6C_2 = 30$

The number of matches in next round $= {}^6C_2 = 15$ and the number of matches in the semifinal round $= {}^4C_2 = 6$

Number of matches in the final round $= 2$

Hence, required number of matches

$$= 30 + 15 + 6 + 2 = 53$$

87. Number of triangles = (taking any two points in $B_1, B_2, \dots B_n$ and one point in $A_1, A_2, \dots A_n$) or (taking one point in $B_1, B_2, \dots, B_n$ and any two points in $A_1, A_2, \dots, A_n$) or (one point O one point in $A_1, A_2, \dots,$ A_n and one point in $B_1, B_2, \dots, B_n$)

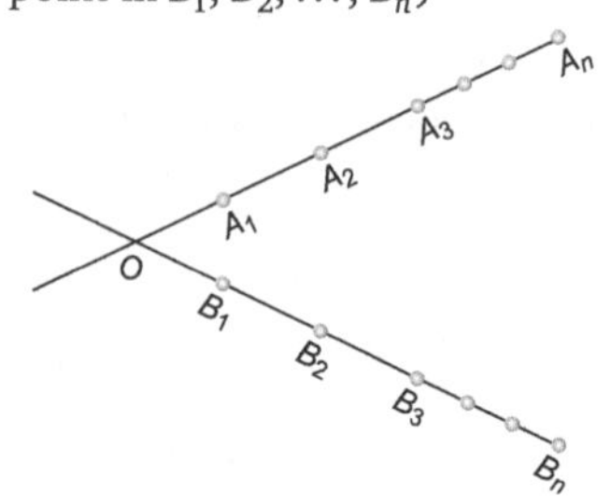

$$= {}^nC_2 \times {}^nC_1 + {}^nC_1 \times {}^nC_2 + 1 \times {}^nC_1 \times {}^nC_1$$

$$= \frac{n^2\,(n-1)}{2} + \frac{n^2\,(n-1)}{2} + n^2$$

$$= n^2\,(n-1) + n^2 = n^3$$

88. If order of A, B, C is not to change, then they are to be treated identical in arrangements.

So, the required number of ways $= \dfrac{7!}{3!}$

$$= 7 \cdot 6 \cdot 5 \cdot 4$$

$$= 840$$

89. Given, 6 different novels and 3 different dictionaries.

Number of ways of selecting 4 novels from 6 novels is

$${}^6C_4 = \frac{6!}{2!4!} = 15$$

Number of ways of selecting 1 dictionary is from 3 dictionaries is ${}^3C_1 = \dfrac{3!}{1!2!} = 3$

$\therefore$ Total number of arrangement of 4 novels and 1 dictionary where dictionary is always in the middle, is

$$15 \times 3 \times 4! = 45 \times 24 = 1080$$

90.

5 Points
A B
9 Points 6 Points
D C
7 Points

α = Number of triangles

$\alpha = 5 \cdot 6 \cdot 7 + 5 \cdot 7 \cdot 9 + 5 \cdot 6 \cdot 9 + 6 \cdot 7 \cdot 9$

$= 210 + 315 + 270 + 378 = 1173$

β = Number of Quadrilateral

$\beta = 5 \cdot 6 \cdot 7 \cdot 9 = 1890$

$\beta - \alpha = 1890 - 1173 = 717$

91. Here $2E, 2A, 1R, 1T, 1H, 1Q, 1U, 1K$

Maximum number of permutations

$= $ coefficient of x^4 in $4!\,(1+x)^6 \left(1 + \dfrac{x}{\underline{|1}} + \dfrac{x^2}{\underline{|2}}\right)^2$

$= $ coefficient of x^4 in $6\,(1+x)^6\,[(1+x)^2 + 1]^2$

$= $ coefficient of x^4 in $6\,\{(1+x)^{10} + 2\,(1+x)^8 + (1+x)^6\}$

$= 6 \cdot \{{}^{10}C_4 + 2 \cdot {}^8C_4 + {}^6C_4\}$

$= 6 \cdot \left\{\dfrac{10 \cdot 9 \cdot 8 \cdot 7}{1 \cdot 2 \cdot 3 \cdot 4} + 2 \cdot \dfrac{8 \cdot 7 \cdot 6 \cdot 5}{1 \cdot 2 \cdot 3 \cdot 4} + 15\right\}$

$= 2190$

92. We have 32 places for teeth. For each place we have two choices either there is a tooth or there is no tooth. Therefore the number of ways to fill up these places is 2^{32}. As there is no person without a tooth, then the maximum population is $2^{32} - 1$.

93. ${}^9C_2 \times {}^7C_2 \times \underline{|2} = \dfrac{9 \cdot 8}{1 \cdot 2} \times \dfrac{7 \cdot 6}{1 \cdot 2} \times 2 = 72 \times 21 = 1512$

94. The required number of ways

$= {}^5C_4 \cdot {}^5C_2 + {}^5C_3 \cdot {}^5C_3 + {}^5C_2 \cdot {}^5C_4$

$= 50 + 100 + 50 = 200$

95. Required ways $= ({}^3C_1 \times {}^2C_1 \times {}^1C_1) \times 2$

$= 12$

96. $${}^nP_{n-1} = k\left\{{}^1C_1 \times {}^{n-2}C_{n-3} \times \frac{(n-1)!}{2!} + {}^{n-1}P_{n-1}\right\}$$

$\Rightarrow \quad n = k\left\{\frac{n-2}{2} + 1\right\}$

$\Rightarrow \quad k = 2$

97. Other than 2, remaining five places by 1 and 3 for each place two conditions

number of ways for five places $= 2 \times 2 \times 2 \times 2 \times 2 = 2^5$

For 2, selecting 2 places out of $7 = {}^7C_2$

$\therefore$ Required number of ways ${}^7C_2 \cdot 2^5$

98. $\because \quad x + y + z = -12$

$\because \quad x < 0, y < 0, z < 0$

$\Rightarrow \quad -x > 0, -y > 0, -z > 0$

or $\quad -x \geq 1, -y \geq 1, -z \geq 1$

Let $\quad -x = \alpha, -y = \beta, -z = \gamma$

Then, $\alpha + \beta + \gamma = 12,\ \alpha \geq 1, \beta \geq 1, \gamma \geq 1$

Then number of points (x, y, z) in space

$= {}^{12-1}C_{3-1}$

$= {}^{11}C_2 = \frac{11 \cdot 10}{1 \cdot 2} = 55$

99. Number of divisors of $N = 2^2 \cdot 3^3 \cdot 5^3 \cdot 7^5$, which are of the form $4n + 1$ (odd) excluding 1

$= \{$number of terms in the product

$(1 + 3^2)(1 + 5 + 5^2 + 5^3)(1 + 7^2 + 7^4)\}$

$+\{$number of terms in the product

$(3 + 3^3)(7 + 7^3 + 7^5)(1 + 5 + 5^2 + 5^3)\} - 1$

$= 2 \times 4 \times 3 + 2 \times 3 \times 4 - 1$

$= 24 + 24 - 1 = 47$

100. $\because x_1 + x_2 + x_3 + x_4 + x_5 + x_6 = 30 \quad$ ...(i)

and $\quad x_i \geq 4, i = 1, 2, 3, 4, 5, 6$

Let $\quad x_i - 3 \geq 1$

(say) $\quad \alpha_i \geq 1, i = 1, 2, 3, 4, 5, 6$

Then Eq. (i) becomes, $\alpha_1 + \alpha_2 + \alpha_3 + \alpha_4 + \alpha_5 + \alpha_6 = 12$

and $\alpha_i \geq 1, i = 1, 2, 3, 4, 5, 6$

Hence, required number of ways

$= {}^{12-1}C_{6-1} = {}^{11}C_5 = 462$

101. $\because \ 2x + 2y = 20 - z$ or $x + y = \left(10 - \frac{z}{2}\right)$

$\therefore \quad$ Coefficient of $\alpha^{10 - z/2}$ in $(1 + \alpha + \alpha^2 + \ldots)^2$

or $\quad$ coefficient of $\alpha^{10 - z/2}$ in $(1 - \alpha)^{-2}$

$= {}^{11 - z/2}C_{10 - z/2}$

$= {}^{11 - z/2}C_1 = \left(11 - \frac{z}{2}\right)$

Putting $z = 0, 2, 4, 6, 8, 10, 12, 14, 16, 18, 20$

then, number of integral solutions

$= 11 + 10 + 9 + 8 + 7 + 6 + 5 + 4 + 3 + 2 + 1$

$= \frac{11 \times 12}{2} = 66$

102. $\because |x - y| \leq 10$

Total ways $= {}^{100}C_2$

and is $|x - y| \geq 10$

then, $\quad x - y \leq -10$ and $x - y \geq 10$

ie, $\quad x \leq y - 10$ and $x \geq y + 10$

ie, take two values out of 90 or ${}^{90}C_2$

$\therefore$ Required number of ways $= {}^{100}C_2 - {}^{90}C_2$

103. Total matches between boys of both team

$= {}^7C_1 \times {}^4C_1 = 28$

Total matches between girls of both team

$= {}^nC_1\ {}^6C_1 = 6n$

Now, $\quad 28 + 6n = 52$

$\Rightarrow \quad n = 4$

104. $\because \ xyz = 6 = 2 \times 3 \times 5$

2 will be a factor of x or y or z in 3 ways

3 will be a factor of x or y or z in 3 ways

and 5 will be a factor of x or y or z in 3 ways

$\therefore$ Total number of ways $= 3 \times 3 \times 3 = 27$ ways

Objective Questions Type II [One or more than one correct answer(s)]

1. The word ALLAHABAD consists of 4 As, 2 Ls, 1 B, 1 D and 1 H.

(a) out of the total 9 objects, 4 are alike of one kind and 2 are alike of another kind. Therefore, the number of words is

$$\frac{9!}{4!\ 2!} = \frac{9 \cdot 8 \cdot 7 \cdot 6 \cdot 5}{2} = 7560$$

(b) Put one A in the first place and another in the last place. In the remaining there are 2 As and 2 Ls. The number of such words is

$$\frac{7!}{2!\,2!} = 1260$$

(c) There are four even places and 4 vowels A, A, A, A. These can be put in eve places in only one way. The remaining 5 letters can be arranged in

$$\frac{5!}{2!} = 60 \text{ ways}$$

(d) Consider all the four as a single letter, so that among six objects, 2 are alike. Therefore, the number of such arrangements is

$$\frac{6!}{2!} = 360$$

2. Since, there is no condition on the arrangements of blue balls, the 5 balls can be arranged in 5! ways.

Suppose the arrangement of balls is done as

$\times B \times B \times B \times B \times B \times$

Then, there are six places for the red balls to arrange in a row such that no two red balls are together.
This can be done in ${}^6P_5$ ways
The total number of arrangement is

$$5!\times {}^6P_5 = 5!\times\frac{6!}{1!} = 6\times(5!)^2$$

Therefore $k = 5$ and $4k = 20$
There are two cases for 5-digit numbers that have the product of their digits as 20.
They are 5,4,1,1,1 and 5,2,2,1,1
So, the total number of ways of arranging these digits for both the cases is given as

$$\frac{5!}{3!}+\frac{5!}{2!\,2!} = 20+30 = 50$$

which can also be written as

$$\frac{5\times 6!}{3\times 4!}$$

3. The number of ways of inviting, with the couple not included $= {}^8C_5$

and the number of ways of inviting, with the couple included $= {}^8C_3$

$\therefore$ Required number of ways $= {}^8C_5 + {}^8C_3$

$$= {}^8C_3 + {}^8C_3 \qquad [\because {}^8C_5 = {}^8C_3]$$

$$= 2\times {}^8C_3$$

Also, $${}^{10}C_5 - 2\times {}^8C_4 = \frac{10}{5}\cdot {}^9C_4 - 2\times {}^8C_4$$

$$= 2({}^9C_4 - {}^8C_4)$$

$$= 2({}^8C_3 + {}^8C_4 - {}^8C_4) = 2\times {}^8C_3$$

4. The number of ways $= {}^nC_m \times m!$

{Here numbers 1, 2, 3, ..., n; $\therefore (m-1)!$ fail}

5. The number of ways of placing A_1 and A_2 in ten places so that A_1 is always above A_2 is ${}^{10}C_2$. There are 8 ! ways of arranging the eight other candidates.
Hence, total number of arrangements

$$= {}^{10}C_2\times 8! = \frac{10!}{2!\,8!}\,8! = \frac{10!}{2}$$

6. 8, 7, 6, 4, 2, x and y

Any number is divisible by 3, if sum of digits is divisible by 3, i.e., $x + y + 27$ is divisible by 3. x and y can take values from 0, 1, 3, 5, 9. Possible pairs are (5, 1) (3, 0) (9, 0) (9, 3) and (1, 5) (0, 3) (0, 9) (3, 9).
The maximum value of $x-y$ is at the pair (9, 0) which is 9

The maximum value of $x+y$ is at the pairs (9, 3) and (3, 9), which is 12

The minimum value of xy is at the pairs (3, 0), (9, 0), (0, 3) and (0, 9) which is 0.

The minimum value of $x+y$ is the pair (3, 0) and (0, 3) which is equal to 3.

7. $\dfrac{10!}{2!\,2!\,3!\,3!}$

[$\because S_1$ and S_2 get 2 books each and S_3 and S_4 get 3 books each]

8. A, B and C can be chosen in ${}^{10}C_3$ ways $= \dfrac{10!}{7!\,3!}$.

The remaining persons can speak in 7! ways. Hence, the number of ways in which they can speak is

$$\frac{10!}{7!\,3!}(7!) = \frac{10!}{6}$$

9. $$0^2+0^2=0,\ 1^2+2^2=5,\ 2^2+4^2=20,\ 3^2+4^2=25,$$
$$1^2+3^2=10,\ 2^2+1^2=5,\ 4^2+2^2=20,\ 4^2+3^2=25,$$
$$3^2+1^2=10$$

$\therefore$ Required number of ways $= {}^9C_1 = 9$

Also, ${}^9C_8 = {}^9C_{9-8} = {}^9C_1 = 9$

10. Let a_0 be the number of objects to the left of the first object chosen, a_1 be the number of object between first and second, the numbers of object between the second and the third is a_2 and number of object right to the third object is a_3.

$\therefore \qquad a_0, a_3 \ge 0$ and $a_1, a_2 \ge 1$

Also, $\qquad a_0 + a_1 + a_2 + a_3 = n - 3 \qquad$ ...(i)

$\because \qquad 1 + a_0, 1 + a_3 \ge 1$ and $a_1, a_2 \ge 1$

$\Rightarrow \qquad (1 + a_0) + a_1 + a_2 + (1 + a_3) = n - 1$

$\therefore$ Required number of ways $= {}^{n-1-1}C_{4-1} = {}^{n-2}C_3$

$$= \frac{(n-2)(n-3)(n-4)}{6}$$

Also, $\qquad {}^{n-2}C_3 = {}^{n-3}C_3 + {}^{n-3}C_2$

11. Total number of arrangements $= 7!$

Number of arrangements of A, B, C, D among themselves $= 4!$

$\therefore$ Number of arrangements when A, B, C, D occur in a particular order $= \dfrac{7!}{4!} = 210 = {}^7P_3 = 3!\times {}^7C_3$.

12. Since, A_i has 5 elements, we have

$$\sum_{i=1}^{30} n(A_i) = 5\times 30 = 150 \qquad \text{...(i)}$$

Suppose S has m distinct elements. Since, each elements of S belongs to exactly 10 of A_i' s we also have

$$\sum_{i=1}^{30} n(A_i) = 10\,m \qquad \text{... (ii)}$$

From Eqs. (i) and (ii), we get $m = 15$
Since B_i has 3 elements and each elements of S belongs to exactly 9 of B_i' s we have

$$\sum_{i=1}^{n} n(B_i) = 3n$$

and $$\sum_{i=1}^{n} n(B_i) = 9m$$

$\Rightarrow \qquad n = 3m = 45 = {}^{10}C_2$

13. Number of selections when $x < y < z = {}^nC_3$

Number of selections when $x = y < z = {}^nC_2$

(we have to select only two numbers out of n numbers)
$\therefore$ Required number $= {}^nC_3 + {}^nC_2 = {}^{n+1}C_3$

14. Let the number of regions for n circles be $f(n)$.

Clearly, $f(1) = 2$

Now, $f(n) = f(n-1) + 2(n-1), \forall n \geq 2$

$\Rightarrow f(n) - f(n-1) = 2(n-1)$

Putting $n = 2, 3, \ldots, n$, we get

$f(n) - f(1) = 2(1 + 2 + 3 + \ldots (n-1)) = (n-1)n$

$\Rightarrow f(n) = n(n-1) + 2 = (n^2 - n + 2)$

[which is always even]

$\Rightarrow f(20) = 20^2 - 20 + 2 = 382$

Also, $n^2 - n + 2 = 92$ or $n^2 - n - 90 = 0$ or $n = 10$

15. Number of selections of 7 digits out of the digit 1, 2, 3, ..., $9 = {}^9C_7$

Number of digits out of these 7 selected digits excluding the greatest digit = 6

These 6 digits can be divided in two groups each having 3 digits $= \dfrac{6!}{3!\,3!\,2!} = {}^6C_3 \times \dfrac{1}{2!}$

But the 3 digits on one side can go on the other side

∴ Required number of ways

$$= {}^9C_7 \cdot {}^6C_3 \cdot \frac{1}{2!} \cdot 2! = {}^9C_7 \cdot {}^6C_3 = {}^9C_2 \cdot {}^6C_3$$

16. First and second prizes in Mathematics (Physics) can be awarded in $({}^{30}P_2)({}^{30}P_2) = ({}^{30}P_2)^2 = (30)^2(29)^2$ ways.

First prize in Chemistry (Biology) can be awarded in $({}^{30}P_1)({}^{30}P_1) = (30)^2$ ways.

Hence

$N = (30)^2(29)^2(30)^2 = (30)^4(29)^2 = 2^4 \cdot 3^4 \cdot 5^4 \cdot 29^2$

Since, $400 = 2^4 \cdot 5^2$, $600 = 2^3 \cdot 3 \cdot 5^2$, $8100 = 2^2 \cdot 3^4 \cdot 5^2$

∴ N is divisible by 400, 600, and 8100.

Also, N is divisible by four distinct primes numbers 2, 3, 5, 29

17. $\because {}^nC_4, {}^nC_5, {}^nC_6$ are in AP

$\Rightarrow \dfrac{n(n-1)(n-2)(n-3)}{1\cdot 2\cdot 3\cdot 4}, \dfrac{n(n-1)(n-2)(n-3)(n-4)}{1\cdot 2\cdot 3\cdot 4\cdot 5},$

$\dfrac{n(n-1)(n-2)(n-3)(n-4)(n-5)}{1\cdot 2\cdot 3\cdot 4\cdot 5\cdot 6}$ are in AP.

Dividing each by $\dfrac{n(n-1)(n-2)(n-3)}{1\cdot 2\cdot 3\cdot 4}$ $(\because n \neq 0, 1, 2, 3)$

$\therefore \quad 1, \dfrac{(n-4)}{5}, \dfrac{(n-4)(n-5)}{5.6}$ are in AP

$\Rightarrow \dfrac{2(n-4)}{5} = 1 + \dfrac{(n-4)(n-5)}{5\cdot 6}$

$\Rightarrow 12n - 48 = 30 + n^2 - 9n + 20$

or $n^2 - 21n + 98 = 0$

or $(n-14)(n-7) = 0$

$\therefore \quad n = 7, 14$

18. (a) Seven different objects can be arranged in $7! = 5040$ ways.

(b) There are 3 even places and 3 vowels. Therefore, the vowels can be arranged in even places in 3! ways and after that the remaining 4 letters can be arranged in 4! ways. Therefore, the required numbers is $3! \times 4! = 144$.

(c) As in (B), the required number is ${}^4P_3 \times 4! = 4! \times 4! = 576$

(d) With I in the middle, the remaining 6 letters can be arranged in $6! = 720$ ways.

19. We have $(n-1)$ elements subsets of S having 1 as least element, $(n-2)$ elements subsets of S having 2 as least element, $(n-3)$ elements subsets of S having 3 as least element, ..., 1 elements subset of S having $(n-1)$ as least element.

Thus $\displaystyle\sum_{A \in X} (\min A) = \sum_{r=1}^{n-1} r(n-r) = n\sum_{r=1}^{n-1} r - \sum_{r=1}^{n-1} r^2$

$$= \frac{n(n-1)n}{2} - \frac{(n-1)n(2n-1)}{6}$$

$$= \frac{(n-1)n}{6}(3n - 2n + 1)$$

$$= \frac{(n+1)n(n-1)}{1\cdot 2\cdot 3}$$

$$= {}^{n+1}C_3 = \frac{n(n^2-1)}{6}$$

20. $\because x + y + z = 3n$

$\Rightarrow N =$ coefficient of α^{3n} in $(1 + \alpha + \alpha^2 + \ldots + \alpha^{2n})^3$

$=$ coefficient of α^{3n} in $(1 - \alpha^{2n+1})^3 (1-\alpha)^{-3}$

$=$ coefficient of α^{3n} in $(1 - 3\alpha^{2n+1} + 3\alpha^{4n+2} - \alpha^{6n+3})$
$\times (1 + {}^3C_1\alpha + {}^4C_2\alpha^2 + \ldots)$

$=$ coefficient of α^{3n} in
$(1 - 3\alpha^{2n+1})(1 + {}^3C_1\alpha + {}^4C_2\alpha^2 + \ldots)$

$= {}^{3n+2}C_{3n} - 3\cdot {}^{n+1}C_{n-1}$

$= {}^{3n+2}C_2 - 3\cdot {}^{n+1}C_2$

$= \dfrac{(3n+2)(3n+1)}{2} - \dfrac{3(n+1)n}{2}$

$= \dfrac{1}{2}\{(9n^2 + 9n + 2) - (3n^2 + 3n)\}$

$= \dfrac{1}{2}(6n^2 + 6n + 2)$

$= 3n^2 + 3n + 1$

$\therefore \quad N - 1 = 3n(n+1)$

21. m straight lines can intersect is at most ${}^mC_2$ points and n circles can intersect in at most $2 \times {}^nC_2 = n(n-1) = {}^nP_2$ points.

n straight lines and m circles can intersect is at most $2 \times m \times n = 2mn$ points

Hence, $N = {}^mC_2 + {}^nP_2 + 2mn$

or $(N - {}^mC_2 - {}^nP_2) = 2mn =$ is an even integer

Clearly $m \mid (N - {}^mC_2 - {}^nP_2)$ and $n \mid (N - {}^mC_2 - {}^nP_2)$

22. Each selection of 4 points, two on one line and two on the other will give one point of intersection.

∴ Required number of intersection $= {}^mC_2 \cdot {}^nC_2$

$$= \frac{m(m-1)}{2} \cdot \frac{n(n-1)}{2}$$

$$= \frac{mn(m-1)(n-1)}{4}$$

23. $\because \alpha$ can be subtracted from β without borrowing if $y_i \geq x_i$; for $i = 1, 2, 3$

Let $x_i = \lambda$

If $i = 1$, then $\lambda = 1, 2, 3, \ldots, 9$

and for $i = 2$ and 3, $\lambda = 0, 1, 2, 3, \ldots, 9$

Hence, total number of ways of choosing the pair α, β is

$$\left(\sum_{\lambda=1}^{9} (10-\lambda)\right)\left(\sum_{\lambda=0}^{9} (10-\lambda)\right)^2$$

$$= (45)(55)^2$$

24. Let $A = \{a_1, a_2, \ldots, a_n\}$. For each $a_i (1 \leq i \leq n)$, we have either $a_i \in P_j$ or $a_i \notin P_j (1 \leq j \leq m)$. That is, there are 2^m choices in which a_i $(1 \leq i \leq n)$ may belong to the P_j's. One of these, there is only one choice, in which $a_i \in P_j$ for all $j = 1, 2, \ldots, m$ which is not favourable for $P_1 \cap P_2 \cap \ldots \cap P_m$ to be ϕ. Thus, $a_i \notin P_1 \cap P_2 \cap \ldots \cap P_m$ in $2^m - 1$ ways.

Since there are n elements in set A, the total number of choices is $(2^m - 1)^n$.

Also, there is exactly one choice, in which, $a_i \notin P_j$ for all $j = 1, 2, \ldots, m$ which is not favourable for $P_1 \cup P_2 \cup \ldots \cup P_m$ to be equal to A.

Thus, a_i can belong to $P_1 \cup P_2 \cup \ldots \cup P_m$ in $(2^m - 1)$ ways.

Since there are n elements in set A, the number of ways in which $P_1 \cup P_2 \cup \ldots \cup P_m$ can be equal to A is $(2^m - 1)^n$.

Linked Comprehension Type

PASSAGE 1

1. Seven persons can be selected for the first table in ${}^{12}C_7$ ways. Now, these seven persons can be arranged in 6! ways. The remaining five persons can be arranged on the second table in 4! ways. Hence, total number of ways is ${}^{12}C_5 6!4!$

2. Here, A can sit on first table and B on the second or A on the second table and B on the first table.

   If A is on the first table, then remaining six for first table can be selected in ${}^{10}C_6$ ways. Now, these seven persons can be arranged in 6! ways. Remaining five can be arranged on the other table in 4! ways.

   Hence, total number of ways is $2\,{}^{10}C_6\, 6!\,4!$.

3. If A, B are on the first table, then remaining five can be selected in ${}^{10}C_5$ ways.

   Now, seven persons including A and B can be arranged on the first table in which A and B are together in 2!5! ways. Remaining five can be arranged on the second table in 4! ways. Total number of ways is ${}^{10}C_5 4!5!2!$.

   If A, B are on the second table, then remaining three can be selected in ${}^{10}C_3$ ways.

   Now, five persons including A and B can be arranged on the second table in which A and B are together in 2!3! ways. Remaining seven can be arranged on the first table in 6! ways. Hence, total number of ways ${}^{10}C_7 6!3!2!$.

PASSAGE 2

1. There are 5 letters

   A, D, I, I, N

   Number of permutations = coefficient of x^3 in

   $$3!\left(1+\frac{x}{1!}+\frac{x^2}{2!}\right)\left(1+\frac{x}{1!}\right)^3 \quad (\because 1A, 1D, 2I\text{'s}, 1N)$$

   = coefficient of x^3 in

   $$\frac{6(2+2x+x^2)(1+x)^3}{2}$$

   = coefficient of x^3 in $3[\{1+(1+x)^2\}(1+x)^3]$

   = coefficient of x^3 in $3\{(1+x)^3+(1+x)^5\}$

   $= 3\{{}^3C_3 + {}^5C_3\} = 3(1+10) = 33$

2. $\because n_1 = n_2 = n_3 = \ldots = n_k = 1$

   i.e. n distinct objects in a line taken r at a time is ${}^nP_r$.

3. There are 11 letters in the word INEFFECTIVE viz, 3E's, 2F's, 2I's, 1 C, 1 T, 1 N, 1 V

   Number of permutations = coefficient of x^4 in

   $$4!\left(1+\frac{x}{1!}+\frac{x^2}{2!}+\frac{x^3}{3!}\right)\times\left(1+\frac{x}{1!}+\frac{x^2}{2!}\right)^2\times\left(1+\frac{x}{1!}\right)^4$$

   $$= \text{coefficient of } x^4 \text{ in } 24\,\frac{(6+6x+3x^2+x^3)}{6}\times\frac{(2+2x+x^2)^2\times(1+x)^4}{4}$$

   $$= \text{coefficient of } x^4 \text{ in } (5+3x+(1+x)^3)\times(1+(1+x)^2)^2\times(1+x)^4$$

   = coefficient of x^4 in

   $$(5+3x+(1+x)^3)\times((1+x)^4+2(1+x)^6+(1+x)^8)$$

   $$= 5({}^4C_4+2\cdot{}^6C_4+{}^8C_4)+3({}^4C_3+2\cdot{}^6C_3+{}^8C_3) + ({}^7C_4+2\cdot{}^9C_4+{}^{11}C_4)$$

   $$= 5(101)+3(100)+617 = 1422$$

4. $\because\ n_1+n_2+n_3+\ldots+n_k = r \Rightarrow n = r$

   $\therefore$ Number of permutations = number of permutations of r objects in a line of which n_1 are of one kind, n_2 of second kind, n_3 of third kind, ..., n_k of kth kind

   $$= \frac{r!}{n_1!\,n_2!\ldots n_k!}$$

5. There are 13 letters in the word INFINITESIMAL viz, 4I's, 2N's, 1F, 1T, 1E, 1S, 1M, 1A, 1L

   Number of permutation = coefficient of x^5 in

   $$5!\left(1+\frac{x}{1!}+\frac{x^2}{2!}+\frac{x^3}{3!}+\frac{x^4}{4!}\right)\times\left(1+\frac{x}{1!}+\frac{x^2}{2!}\right)(1+x)^7$$

$= \text{coefficient of } x^5 \text{in } \frac{5}{2}(24 + 24x + 12x^2 + 4x^3 + x^4)$

$\times (2 + 2x + x^2)(1 + x)^7$

$= \text{coefficient of } x^5 \text{in } \frac{5}{2}(24 + 24x + 12x^2 + 4x^3 + x^4)$

$\times ((1 + x)^7 + (1 + x)^9)$

$= \frac{5}{2}\{24({}^7C_5 + {}^9C_5) + 24({}^7C_4 + {}^9C_4) + 12({}^7C_3 + {}^9C_3)$

$+ 4({}^7C_2 + {}^9C_2) + ({}^7C_1 + {}^9C_1)\}$

$= \frac{5}{2}\{3528 + 3864 + 1428 + 228 + 16\} = 22660$

PASSAGE 3

1. Considering CC as single object, U, CC, E can be arranged in 3! ways × U × CC× E ×
   Now the three S are to be placed in four available places.
   Hence, required number of ways = ${}^4C_3 \times 3! = 24$
2. Let us first find the words in no two S are together.
   (i) Arrange the remaining letters $= \frac{4!}{2!} = 12$ ways
   (ii) × U × C × C × E ×
   There are five available places for three SSS.
   Hence, total number of ways no two S together
   $= 12 \times {}^5C_3 = 120$
   ∴ Hence, number of words having CC separated and SSS separated = 120 − 24 = 96.
3. Total number of ways $= \frac{7!}{2!\,3!} = 420$
   consonants in SUCCESS are S, C, C, S, S
   Number of ways arranging consonants $= \frac{5!}{2!\,3!} = 10$
   Hence, number of words in which the consonants appear in alphabatic order $= \frac{420}{10} = 42$
4. The alphabatic order is C, E, S, U. The number of words beginning with C is $\frac{6!}{3!} = 120$ ways and those beginning with E is $\frac{6!}{2!\,3!} = 60$ ways.
   Then come wards beginning SC, numbering $\frac{5!}{2!} = 60$, SE, numbering $\frac{5!}{2!\,2!} = 30$ and SS, numbering $\frac{5!}{2!} = 60$.
   After which come the word SUCCESS.
   Thus the rank of SUCCESS is
   120 + 60 + 60 + 30 + 60 + 1 = 331
5. Vowels are EU
   These vowels can be arranged themselves in 2! = 2 ways.
   The consonants are CCSSS these consonants can be arranged themselves in $\frac{5!}{2!\,3!} = 10$ ways.
   ∴ Required number of words = 2×10 = 20 ways.

PASSAGE 4

1. Let the prizes be P_1, P_2, P_3 and P_4. P_1 can be given to any one of the 6 students and so are P_2, P_3 and P_4. Therefore, the number of distributions is $6^4 = 1296$.
2. The number of ways in which all the four prizes can be given to any one of the 6 students = 6. Therefore, the required number of ways is $6^4 - 6 = 1290$.
3. Give a set of two prizes to the particular student. Then, the remaining 2 can be distributed among 5 student in 5^2 ways. There are ${}^4C_2$ sets, each containing 2 prizes. Therefore, the required number of ways of distributing the prizes is
   $5^2 \times {}^4C_2 = 25 \times 6 = 150$

PASSAGE 5

1. $\because 6 = 1 + 1 + 1 + 1 + 1 + 1 = \frac{6!}{6!} = 1$
   $= 1 + 1 + 1 + 1 + 2 = \frac{5!}{4!} = 5$
   $= 1 + 1 + 2 + 2 = \frac{4!}{2!2!} = 6 = 2 + 2 + 2 = 1$
   $\therefore \quad f(6) = 13$
2. $\because f(f(6)) = f(13) = f(1 + 1 + 1 + \ldots + 1)$
   $= \frac{13!}{13!} = 1$
   $= f(1 + 1 + 1 + 1 + \ldots 11 \text{ times} + 2)$
   $= \frac{12!}{11!1!} = 12$
   $= f(1 + 1 + \ldots + 9 \text{ times} + 2 + 2)$
   $= \frac{11!}{9!2!} = 55$
   $= f(1 + 1 + \ldots + 7 \text{ times} + 2 + 2 + 2) = \frac{10!}{7!3!} = 120$
   $= f(1 + 1 + 1 + 1 + 1 + 2 + 2 + 2 + 2)$
   $= \frac{9!}{5!4!} = 126$
   $= f(1 + 1 + 1 + 2 + 2 + 2 + 2 + 2)$
   $= \frac{8!}{3!5!} = 56$
   $= f(1 + 2 + 2 + 2 + 2 + 2 + 2) = \frac{7!}{1!6!} = 7$
   Total = 377
3. For $n = 1$
   $1 = 1$
   $\therefore f(1) = 1$
   for $n = 2$
   i.e. $2 = 1 + 1 \to \frac{2!}{2!} = 1$
   $\therefore \quad f(2) = 1$ fail
   for $n = 3$
   i.e. $3 = 1 + 1 + 1 \to \frac{3!}{3!} = 1$
   $= 1 + 2 \to 2! = 2$
   $\therefore \quad f(3) = 3$

for $n = 4$

i.e. $4 = 1 + 1 + 1 + 1 \rightarrow \frac{4!}{4!} = 1$

$= 1 + 1 + 2 \rightarrow \frac{3!}{2!} = 3$

$= 2 + 2 \rightarrow \frac{2!}{2!} = 1$

$\therefore$ $f(4) = 5$ fail

for $n = 5$

i.e. $5 = 1 + 1 + 1 + 1 + 1 = \frac{5!}{5!} = 1$

$5 = 1 + 1 + 1 + 2 = \frac{4!}{3!} = 4$

$5 = 1 + 2 + 2 = \frac{3!}{2!} = 3$

$\therefore$ $f(5) = 8$

Thus we can say that $f(n) > n$ for $n = 4, 5, 6, \dots$

Hence number of solutions for $f(n) = n$ is 2.

4. Only $f(4) = 5$ (of above discussion)

i.e. $f(n) = n + 1$

PASSAGE 6

1. The number of ways in which two 'R' are not together $\times A \times A \times N \times G \times E \times$ is ${}^6C_2 \times \frac{5!}{2!} = 15 \times 60 = 900$

2. The number of ways in which two 'A' are together *ie*, $\times AA \times N \times G \times E \times$ is ${}^5C_2 \times 4! = 240$

Hence, number of ways in which neither two 'R' nor two 'A' come together is $900 - 240 = 660$

3. Total number of ways $= \frac{7!}{2!\,2!} = 1260$

consonants in ARRANGE are R, R, N, G.

$\therefore$ Number of ways arranging consonants $= \frac{4!}{2!} = 12$

Hence number of words in which the consonants appear in alphabatic order $= \frac{1260}{12} = 105$

4. The alphabetic order is A, E, G, N, R. The number of words beginning with AA is $\frac{5!}{2!} = 60$, beginning with AE is $\frac{5!}{2!} = 60$, beginning with AG is $\frac{5!}{2!} = 60$, beginning with AN is $\frac{5!}{2!} = 60$,

Then come words beginning ARA is $4! = 24$, beginning with ARE is $4! = 24$, beginning with ARG is $4! = 24$ and beginning with ARN is $4! = 24$, then commining words beginning ARRAE is $2! = 2$, beginning with ARRAG is $2! = 2$, then next words are ARRANEG and ARRANGE

Thus the rank of ARRANGE is

$60 + 60 + 60 + 60 + 24 + 24 + 24 + 24 + 2 + 2 + 2 = 342$

5. The consonants are N, G, R, R and vowels are A, A, E

$\times A \times A \times E \times$

Hence number of words at least one vowel is in between two consonants is

$$\frac{4!}{2!} \times \frac{3!}{2!} = 12 \times 3 = 36$$

PASSAGE 7

1. For every choice of $r = 1, 2, 3, \dots (n-1)$ when the nth object a_n goes to the rth place, there are $D_{n-1} + D_{n-2}$ ways of the other $(n-1)$ objects $a_1, a_2, \dots, a_{n-1}$ to be deranged.

Hence, $D_n = (n-1)(D_{n-1} + D_{n-2})$

2. $D_n - nD_{n-1} = (-1)(D_{n-1} - (n-1)D_{n-2})$

By implied induction on n, we obtain

$D_n - nD_{n-1} = (-1)^{n-2}(D_2 - 2D_1)$, when $D_1 = 0$

and $D_2 = 1 = (-1)^n$

3. $\frac{D_n}{n!} - \frac{D_{n-1}}{(n-1)!} = \frac{(-1)^n}{n!}$ gives

$$\frac{D_n}{n!} = \sum_{r=2}^{n}\left(\frac{D_r}{r!} - \frac{D_{r-1}}{(r-1)!}\right) = \sum_{r=2}^{n}\frac{(-1)^r}{r!}$$

$$\Rightarrow D_n = n!\sum_{r=2}^{n}\frac{(-1)^r}{r!}$$

$$\therefore D_5 = 5!\left(\frac{1}{2!} - \frac{1}{3!} + \frac{1}{4!} - \frac{1}{5!}\right) = 44$$

Numerical Value Type

1. Number of numbers when 1, 2, 3, 4, 5 are used $5! = 120$ and number of numbers when 0, 1, 2, 4, 5 are used

$= 5! - 4! = 96$

$\therefore$ Required number $= 120 + 96 = 216 = n^3$ (given)

$\Rightarrow$ $n = 6$

then $\frac{n!}{100} = \frac{6!}{100} = \frac{720}{100} = 7.2$

2. Number of ways of selecting n coupons consisting of C or A $= 2^n$ $\left(\because \text{the word CAT can not be written if at least one letters is not selected}\right)$

Now number of ways of selecting n coupons bearing only A $= 1^n$.

$\therefore$ Total number of ways $= 2^n + 2^n + 2^n - 1^n - 1^n - 1^n$

$= 3(2^n - 1)$

Given $3(2^n - 1) = 189$

$\Rightarrow$ $2^n - 1 = 63$

$\Rightarrow$ $2^n = 64 = 2^6$

$\therefore$ $n = 6$

then $\sum n^2 = \frac{n(n+1)(2n+1)}{6} = \frac{6 \cdot 7 \cdot 13}{6} = 91$

3. Required number of ways

$= 2^{11} - ({}^{11}C_0 + {}^{11}C_1 + {}^{11}C_2)$

$= 2048 - (1 + 11 + 55) = 2048 - 67 = 1981$

4. $\because {}^{12}P_r = 12 \times 11 \times 10 \times 9 = {}^{12}P_4$

$\therefore$ $r = 4$

then, $\lambda = r + 3 = 7$

$\therefore \sum_{i=1}^{7} {}^7C_i = 2^7 - 1 = 127$

5. Vowels are A, E, A

Consonants are L, G, B, R

According to question

A E A L G B R

1 2 3 4 5

$\therefore$ Required number of ways $= \frac{7!}{2!} - \frac{5!3!}{2!} = 2160$

6. The number of games played by men is ${}^mC_2 \times 2$

The number of games played between men and the women is ${}^mC_1 \times {}^2C_1 \times 2 = 84$

According to the question,

$$2 \times {}^mC_2 - {}^mC_1 \; {}^2C_1 \times 2 = 84$$

$$\Rightarrow \quad 2 \times \frac{m(m-1)}{2} - m \times 2 \times 2 = 84$$

$$\Rightarrow \quad m^2 - m - 4m - 84 = 0$$

$$\Rightarrow \quad m^2 - 5m - 84 = 0$$

$$\Rightarrow \quad (m-12)(m+7) = 0$$

$$\Rightarrow \quad m = 12, m = -7 \text{ (NP)}$$

$$\Rightarrow \quad \frac{m}{5} = \frac{12}{5} = 2.4$$

7. The word 'PATALIPUTRA' has eleven letters, in which 2P's, 3A's, 2T's 1L, 1U, 1R, 1I. Vowels are AAIUA.

These vowels can be arranged themselves in $\frac{5!}{2!} = 20$ ways.

The consonants are PTLPTR these consonants can be arranged themselves in $\frac{6!}{2!2!} = 180$ ways.

$\therefore$ Required number of words $= 20 \times 180 = 3600$ ways.

8. Sum of the digits in the tens place

$$= (4-1)!(4+5+6+7) = 6 \times 22 = 132$$

9. We have $\binom{18}{r-2} + 2\binom{18}{r-1} + \binom{18}{r} \geq \binom{20}{13}$ it means that

$${}^{18}C_{r-2} + 2 \cdot {}^{18}C_{r-1} + {}^{18}C_r \geq {}^{20}C_{13}$$

$$\Rightarrow \quad ({}^{18}C_{r-2} + {}^{18}C_{r-1}) + ({}^{18}C_{r-1} + {}^{18}C_r) \geq {}^{20}C_{13}$$

$$\Rightarrow \quad {}^{19}C_{r-1} + {}^{19}C_r \geq {}^{20}C_{13}$$

$$\Rightarrow \quad {}^{20}C_r \geq {}^{20}C_{13}$$

$$\text{or} \quad {}^{20}C_r \geq {}^{20}C_7$$

Hence $7 \leq r \leq 13$

$$\therefore \quad r = 7, 8, 9, 10, 11, 12, 13$$

$\therefore$ Required sum $= 7 + 8 + 9 + 10 + 11 + 12 + 13 = 70$

10. 12, 21 ... 122222222 are the required numbers we can assume all of them to be nine-digit in the form $a_1, a_2, a_3, a_4, a_5, a_6, a_7, a_8, a_9$ and can use 0 for a_1 ; a_2 and a_0 and a_0, a_1, a_2 and a_3... and so on to get 8 digit, 7 digit, 6 digit numbers etc. a_1 can assume one of the 2 values of 0 or 1. $a_2, a_3, a_4, a_5, a_6, a_7, a_8$ can assume any of 3 values 0, 1, 2. The number for which $a_1 = a_2 = a_3 = a_4 = a_5 = a_6 = a_7 = a_8 = a_9 = 0$ must be eleminated. The sum of first 8 digits *ie*, $a_1 + a_2 + \ldots + a_8$ can be in the form of $3n - 2$ or $3n - 1$ or $3n$. In each case a_9 can be chosen from 0, 1, 2 in only 1 way so that the sum of all 9 digits is equal to $3n$.

$\therefore$ Total numbers $= 2 \times 3^7 \times 1 - 1$

$$= 4374 - 1 = 4373$$

11. Let number of children of John and Mary is y and number of children of John and his first wife is x. Hence, number of children of Mary and his first husband is $(x + 1)$.

$$\Rightarrow \quad x + x + 1 + y = 24 \Rightarrow y = 23 - 2x$$

Total number of fights between two children $= {}^{24}C_2 = 276$

Now total number of fights between two children of same parents $= {}^{x+1}C_2 + {}^xC_2 + {}^yC_2$

$$= ({}^xC_1 + {}^xC_2) + {}^xC_2 + {}^{23-2x}C_2$$

$$= 2 \cdot {}^xC_2 + {}^xC_1 + {}^{23-2x}C_2$$

$$= 3x^2 - 45x + 253$$

Hence, the total number of fights subject to the condition that any two children of same parents do not fight.

$$N = 276 - (3x^2 - 45x + 253) = 23 - 3x^2 + 45x$$

$$\therefore \quad \frac{dN}{dx} = -6x + 45$$

For maximum or minimum of N, $\frac{dN}{dx} = 0 \Rightarrow x = 7.5$

$$\Rightarrow \quad x = 7 \qquad [\because x \in I]$$

$$\frac{d^2N}{dx^2} < 0 \text{ at } x = 7$$

$\therefore$ N will be maximum when $x = 7$

$\therefore$ $N = 23 - 3 \cdot 7^2 + 45 \cdot 7$

$$= 23 - 147 + 315 = 191$$

12. The alphabetic order is C, H, L,O and S. The number of words beginning with C is $\frac{5!}{2!} = 60$ ways and those beginning with H, L, O are $\frac{5!}{2!} = 60$ ways, 60 ways and 120 ways.

Total words $= 60 + 60 + 60 + 120 = 300$

Next 301th word is SCHLOO, 302th word is SCHOLO and 303th word is SCHOOL.

13. $\because a \geq -5 \Rightarrow a + 6 \geq 1$

similarly $b + 6 \geq 1$ and $c + 6 \geq 1$

let $a + 6 = \alpha$, $b + 6 = \beta$ and $c + 6 = \gamma$

then $\alpha + \beta + \gamma = 18$, $\alpha \geq 1$, $\beta \geq 1$, $\gamma \geq 1$

$\therefore$ Required number of solutions

$$= {}^{18-1}C_{3-1} = {}^{17}C_2 = \frac{17 \cdot 16}{2} = 136$$

14. Let the 5 people be given x, y, z, u, w rupees, then

$x + y + z + u + w = 20$

where $x, y, z, u, w \geq 3$

or $x - 2, y - 2, z - 2, u - 2, w - 2, \geq 1$

let $\alpha = x - 2, \beta = y - 2, \gamma = z - 2, \delta = u - 2, \theta = w - 2$

Then $\alpha + \beta + \gamma + \delta + \theta = 10$,

$$\alpha \geq 1, \beta \geq 1, \gamma \geq 1, \delta \geq 1, \theta \geq 1$$

$\therefore$ Required number $= {}^{10-1}C_{5-1} = {}^9C_4 = \frac{9 \cdot 8 \cdot 7 \cdot 6}{1 \cdot 2 \cdot 3 \cdot 4} = 126$

15. Let two particular boys as one boy, we have only four boys.

$\therefore$ 5 boys can be seated at a round table when two particular boys are always together $= 3!2!$

$\Rightarrow$ 4 girls have are 4 places

$\therefore$ 4 girls can be arranged in 4! ways

$\therefore$ Required number $= 3!2!4! = 6 \cdot 2 \cdot 24 = 288$

Single Integer Type

1. $x = 10!$

$y = {}^{10}C_1 \times {}^9C_8 \times \frac{10!}{2!} = 10 \times 9 \times \frac{10!}{2}$

$\Rightarrow \frac{y}{9x} = \frac{10}{2} = 5$

2. Here, $\_B_1\_B_2\_B_3\_B_4\_B_5\_$

Out of 5 girls, 4 girls are together and 1 girl is separate. Now to select 2 positions out of 6

Positions between boys $= {}^6C_2$ ...(i)

4 girls are to be selected out of $5 = {}^5C_4$ ...(ii)

Now, 2 groups of girls can be arranged in 2! ways. ...(iii)

Also, the group of 4 girls and 5 boys is arranged in $4! \times 5!$ ways. ...(iv)

Now, total number of ways $= {}^6C_2 \times {}^5C_4 \times 2! \times 4! \times 5!$

[from Eqs. (i), (ii), (iii) and (iv)]

$\therefore \quad m = {}^6C_2 \times {}^5C_4 \times 2! \times 4! \times 5!$

and $n = 5! \times 6!$

$\Rightarrow \frac{m}{n} = \frac{{}^6C_2 \times {}^5C_4 \times 2! \times 4! \times 5!}{6! \times 5!} = \frac{15 \times 5 \times 2 \times 4!}{6 \times 5 \times 4!} = 5$

3. Given relation, $6 \cdot {}^{35}C_r = (k^2 - 3) \cdot {}^{36}C_{r+1}$

$\because {}^nC_r \in N \ \forall \ n, r \in W$ such that $n \geq r$

$\therefore k^2 - 3 > 0$

$\Rightarrow k \in (-\infty, -\sqrt{3}) \cup (\sqrt{3}, \infty)$ ...(i)

and $k^2 - 3 = 6 \cdot \frac{{}^{35}C_r}{{}^{36}C_{r+1}} = 6 \cdot \frac{\frac{35!}{r!(35-r)!}}{\frac{36!}{(r+1)!(35-r)}}$, for $(35 \geq r)$

$\Rightarrow \quad k^2 - 3 = 6 \cdot \frac{r+1}{36} = \frac{r+1}{6}$

$\Rightarrow \quad 6k^2 - 18 = r + 1$

$\Rightarrow \quad r = 6k^2 - 19 \leq 35$ $[\because r \leq 35]$

$\Rightarrow \quad 6k^2 \leq 54 \Rightarrow k^2 \leq 9$

$\Rightarrow \quad k \in [-3, 3]$ ...(ii)

From intervals (i) and (ii), we get

$k \in [-3, -\sqrt{3}) \cup (\sqrt{3}, 3]$

$\therefore$ Possible integral values of $k = -3, -2, 2, 3$.

$\therefore$ Number of ordered pairs $(r, k) = (35, -3), (5, -2), (5, 2)$ and $(35, 3)$.

4. It is given that the 10th place of 5-digit number with distinct digits is '2', then the ten thousand place, thousand place, hundred place and unit place we can fill in 8, 8, 7 and 6 ways only.

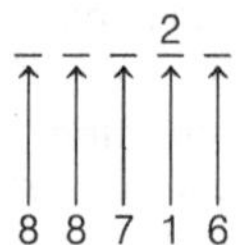

So, required number of five-digit numbers is

$8 \times 8 \times 7 \times 6 = 336k$ (given)

$\Rightarrow k = \frac{8 \times 8 \times 7 \times 6}{336} = \frac{8 \times 8 \times 7}{56} = 8$

5. Number of numbers beginning with 1 = 120 [1| | | | |]

Number of numbers beginning with 2 = 120 [2| | | | |]

Starting with 31 .... = 24 [3|1| | | |]

Starting with 3214 .... = 2 [3|2|1|4| |]

Finally = 1 [3|2|1|5|4|6]

Hence, unit place digit of 267th number is 6.

6. Each of the places can be filled in 3 ways (with 2, 5 or 7). The total number of ways

$3 \times 3 \times \ldots \times 3$ (n times) $= 3^n$

Now, $3^n \geq 900 \Leftrightarrow 3^{n-2} \geq 100 \Leftrightarrow n - 2 \geq 5 \Leftrightarrow n \geq 7$

Therefore, the smallest value of such $n = 7$

7. Number of digits are 9

Select 2 places for the digit 1 and 2 in ${}^9C_2$ ways.

From the remaining 7 places, select any two places for 3 and 4 in ${}^7C_2$ ways and from the remaining 5 places, select any two for 5 and 6 in ${}^5C_2$ ways.

Now, the remaining 3 digits can be filled in 3! ways

$\therefore$ Total ways $= {}^9C_2 \cdot {}^7C_2 \cdot {}^5C_2 \cdot 3!$

$= \frac{9!}{2! \cdot 7!} \cdot \frac{7!}{2! \cdot 5!} \cdot \frac{5!}{2! \cdot 3!} \cdot 3!$

$= \frac{9!}{8} = \frac{9 \cdot 8 \cdot 7!}{8} = 9 \cdot 7!$

Matrix-Match Type

1. (A) $\because x_1 x_2 x_3 x_4 x_5 = 2^1 \times 3^1 \times 3^2 \times 7^1$

$\alpha + \beta + \gamma + \delta + \theta = 1, 1, 2, 1$

$\therefore$ Number of solutions

$\lambda = {}^{1+5-1}C_{5-1} \times {}^{1+5-1}C_{5-1} \times {}^{2+5-1}C_{5-1} \times {}^{1+5-1}C_{5-1}$

$= {}^5C_4 \times {}^5C_4 \times {}^6C_4 \times {}^5C_4$

$= {}^5C_1 \times {}^5C_1 \times {}^6C_2 \times {}^5C_1$

$= 5 \times 5 \times 15 \times 5 = 1875$

$\therefore \lambda$ is divisible by 3 and 5. (P, R)

(B) Number of solutions of the given equation is the same as the number of solutions of the equation

$x_1 x_2 x_3 x_4 = 30 = 2 \times 3 \times 5$

$\because$ If $x_1 x_2 x_3 = 15$,

then $x_4 = 2$

and if $x_1 x_2 x_3 = 5$,

then $x_4 = 6$ etc.

x_4 is in fact a during variable.

Thus $x_1 x_2 x_3 x_4 = 2^1 \times 3^1 \times 5^1$

$\therefore \quad \alpha + \beta + \gamma + \delta = 1, 1, 1$

$\therefore$ Number of solutions

$$\lambda = {}^{1+4-1}C_{4-1} \times {}^{1+4-1}C_{4-1} \times {}^{1+4-1}C_{4-1}$$
$$= {}^4C_3 \times {}^4C_3 \times {}^4C_3$$
$$= {}^4C_1 \times {}^4C_1 \times {}^4C_1 = 64$$

$\therefore$ λ is divisible by 4, 8 and 16. (Q, S, T)

(C) Let $a = \dfrac{120}{x_4}$

or $x_1x_2x_3x_4 = 120 = 2^3 \times 3^1 \times 5^1$

$\therefore \alpha + \beta + \gamma + \delta = 3, 1, 1$

$\therefore$ Number of solutions

$$\lambda = {}^{3+4-1}C_{4-1} \times {}^{1+4-1}C_{4-1} \times {}^{1+4-1}C_{4-1}$$
$$= {}^6C_3 \times {}^4C_3 \times {}^4C_3$$
$$= {}^6C_3 \times 4 \times 4 = 20 \times 16 = 320$$

$\therefore \lambda$ is divisible by 4, 5, 8, and 16 (Q, R, S, T)

2. Given 6 boys $M_1, M_2, M_3, M_4, M_5, M_6$ and 5 girls G_1, G_2, G_3, G_4, G_5

(i) $\alpha_1 \to$ Total number of ways of selecting 3 boys and 2 girls from 6 boys and 5 girls.

i.e., ${}^6C_3 \times {}^5C_2 = 20 \times 10 = 200$

$\therefore \quad \alpha_1 = 200$

(ii) $\alpha_2 \to$ Total number of ways selecting at least 2 member and having equal number of boys and girls

i.e.,

$${}^6C_1{}^5C_1 + {}^6C_2{}^5C_2 + {}^6C_3{}^5C_3 + {}^6C_4{}^5C_4 + {}^6C_5{}^5C_5$$
$$= 30 + 150 + 200 + 75 + 6 = 461$$

$\Rightarrow \alpha_2 = 461$

(iii) $\alpha_3 \to$ Total number of ways of selecting 5 members in which at least 2 of them girls

i.e. ${}^5C_2{}^6C_3 + {}^6C_3{}^6C_2 + {}^5C_4{}^6C_1 + {}^5C_5{}^6C_0$

$= 200 + 150 + 30 + 1 = 381$

$\alpha_3 = 381$

(iv) $\alpha_4 \to$ Total number of ways of selecting 4 members in which at least two girls such that M_1 and G_1 are not included together.

G_1 is included $\to {}^4C_1 \cdot {}^5C_2 + {}^4C_2 \cdot {}^5C_1 + {}^4C_3$

$= 40 + 30 + 4 = 74$

M_1 is included $\to {}^4C_2 \cdot {}^5C_1 + {}^4C_3 = 30 + 4 = 34$

G_1 and M_1 both are not included

${}^4C_4 + {}^4C_3 \cdot {}^5C_1 + {}^4C_2 \cdot {}^5C_2$

$1 + 20 + 60 = 81$

$\therefore$ Total number $= 74 + 34 + 81 = 189$

$\alpha_4 = 189$

3. (A) $B\ G\ B\ G\ B\ G\ B\ G\ B\ G\ B$

Required number of ways = $6!5!$ (P, Q, S)

(B) Since here restriction on girls

$\times B \times B \times B \times B \times B \times B \times$ Let us seat the boys first, which can be done in $6!$ ways.

For five girls there are 7 places shown by '×' which can be done in ${}^7P_5$

Required number of ways $= {}^7P_5 \times 6!$

$$= \frac{7!}{2!} \times 6!$$
$$= \frac{7 \times 6 \times 5! \times 6!}{2}$$
$$= 21 \times 5!6! \text{ (P, Q, S)}$$

or $3 \times 5!7!$ (R, T)

$\therefore$ (P, Q, R, S, T)

(C) Total ways without restriction $= (11-1)! = 10!$

Number of ways in which all the girls can be seated together

$$= (7-1)! \times 5!$$
$$= 6!5!$$

$\therefore$ Required number of ways $= 10! - 6!5!$

$$= 41 \times 5!6! \text{ (P, Q, S)}$$

4. (A) The required number of distributions is

${}^{52}C_{13} \times {}^{39}C_{13} \times {}^{26}C_{13} \times {}^{13}C_{13}$

$$= \frac{(52)!}{(13)!(52-13)!} \times \frac{(39)!}{(13)!(39-13)!} \times \frac{(26)!}{(13)!(26-13)!} \times 1$$
$$= \frac{(52)!}{[(13)!]^4} \text{ (P)}$$

(B) To be divided into 4 equal groups. Elements of one group can be exchanged with another. This is possible in 4! ways. Therefore, the number of divisions is

$$\frac{(52)!}{4![(13)!]^4} \text{ (Q)}$$

(C) Three of them get 17 each. Again, cards can be exchanged among these 3 in 3! ways. Therefore, the number of divisions is

$$\frac{(52)!}{3![(17)!]^3} \text{ (R)}$$

(D) The number of distributions is

$$\frac{{}^{52}C_{26} \cdot {}^{26}C_{26}}{2!} = \frac{(52)!}{2![(26)!]^2} \text{ (S)}$$

5. (A) Given, $x \geq 1, y \geq 1, z \geq 1, w \geq 1$, then required number of solutions

$$= {}^{10-1}C_{4-1} = {}^9C_3 = \frac{9 \cdot 8 \cdot 7}{1 \cdot 2 \cdot 3} = 84$$

$\therefore \quad \lambda_1 = 84 = 3 \times 4 \times 7$ (P, Q, S)

(B) Given $x \geq 0, y \geq 0, z \geq 0$, then required number of solutions $= {}^{6+3-1}C_{3-1}$

$$= {}^8C_2 = 28$$

$\therefore \quad \lambda_2 = 28 = 4 \times 7$ (Q, S)

(C) Given $x \geq 1, y \geq 2, z \geq 3$ and $x + y + z = 15$

$\Rightarrow x \geq 1, y - 1 \geq 1, z - 2 \geq 1$

let $x = \alpha$, $y - 1 = \beta$ and $z - 2 = \gamma$

then $\alpha + \beta + \gamma = x + y + z - 3 = 15 - 3 = 12$

$\therefore \quad \alpha + \beta + \gamma = 12, \alpha \geq 1, \beta \geq 1, \gamma \geq 1$

then required number of solutions

$$= {}^{12-1}C_{3-1} = {}^{11}C_2 = 55$$

$\therefore \quad \lambda_3 = 55 = 5 \times 11$ (R, T)

05

Binomial Theorem

Objective Questions Type I [Only one correct answer]

In each of the questions below, four choices are given of which only one is correct. You have to select the correct answer which is the most appropriate.

1. If the number of terms in $\left(x+1+\frac{1}{x}\right)^n$ $(n \in N)$ is 301, then n is greater than

(a) 152 (b) 151
(c) 150 (d) 149

2. If $x+\frac{1}{x}=1$ and $\lambda = x^{4000}+\frac{1}{x^{4000}}$ and μ be the digit at unit place in the number $2^{2^n}+1, n \in N$ and $n > 1$, then $\lambda + \mu$ is equal to

(a) 8 (b) 7
(c) 6 (d) 5

3. If $\sum_{r=0}^{n}\left(\frac{r+2}{r+1}\right){}^nC_r = \frac{2^8-1}{6}$, then n is equal to

(a) 6 (b) 3
(c) 8 (d) 5

4. If $a^3+b^6=2$, then the maximum value of the term independent of x in the expansion of $(ax^{1/3}+bx^{-1/6})^9$ is $(a>0, b>0)$

(a) 42 (b) 68
(c) 84 (d) 148

5. Number of terms in the expansion of $\left(\frac{x^3+1+x^6}{x^3}\right)^{\Sigma n}$ (where $n \in N$) is

(a) $\Sigma n+1$ (b) ${}^{\Sigma n+2}C_2$
(c) $2n+1$ (d) n^2+n+1

6. $(115)^{96}-(96)^{115}$ is divisible by

(a) 15 (b) 17
(c) 19 (d) 21

7. If 5^{40} is divided by 11, then remainder is α and when 2^{2003} is divided by 17, then remainder is β, then the value of $\beta - \alpha$ is

(a) 3 (b) 5
(c) 7 (d) 8

8. If the sum of the coefficients in the expansion of $(1+2x)^m$ and $(2+x)^n$ are respectively 6561 and 243, then the position of the point (m, n) with respect to the circle $x^2+y^2-4x-6y-32=0$

(a) is inside the circle (b) is outside the circle
(c) is on the circle (d) cannot be fixed

9. The range of the values of the term independent of x in the expansion of $\left(x\sin^{-1}\alpha+\frac{\cos^{-1}\alpha}{x}\right)^{10}$, $\alpha \in [-1, 1]$ is

(a) $\left[\frac{{}^{10}C_5\cdot\pi^{10}}{2^5}, -\frac{{}^{10}C_5\pi^{10}}{2^{20}}\right]$ (b) $\left[-\frac{{}^{10}C_5\cdot\pi^{10}}{2^5}, \frac{{}^{10}C_5\cdot\pi^{10}}{2^{20}}\right]$

(c) $\left[\frac{{}^{10}C_5\cdot\pi^5}{2^5}, \frac{{}^{10}C_5\cdot\pi^5}{2^{20}}\right]$ (d) $\left[-\frac{{}^{10}C_5\cdot\pi^5}{2^5}, \frac{{}^{10}C_5\cdot\pi^5}{2^{20}}\right]$

10. Let $\alpha>0, \beta>0$ be such that $\alpha^3+\beta^2=4$. If the maximum value of the term independent of x in the binomial expansion of $\left(\alpha x^{\frac{1}{9}}+\beta x^{-\frac{1}{6}}\right)^{10}$ is $10\,k$, then k is equal to **[JEE Main 2020, 2 September]**

(a) 84 (b) 176 (c) 352 (d) 336

11. If ${}^{n-1}C_r = (k^2-3)\cdot{}^nC_{r+1}$, then k belongs to

(a) $(-\infty, -2]$ (b) $[2, \infty)$
(c) $[-\sqrt{3}, \sqrt{3}]$ (d) $(\sqrt{3}, 2]$

12. The value of $\binom{30}{0}\binom{30}{10}-\binom{30}{1}\binom{30}{11}+\binom{30}{2}\binom{30}{12} - \ldots + \binom{30}{20}\binom{30}{30}$, where $\binom{n}{r}={}^nC_r$, is

(a) $\binom{30}{10}$ (b) $\binom{30}{15}$

(c) $\binom{60}{30}$ (d) $\binom{31}{10}$

13. The coefficient of x^{20} in the expansion of $(1+x^2)^{40}.\left(x^2+2+\frac{1}{x^2}\right)^{-5}$ is

(a) ${}^{30}C_{10}$ (b) ${}^{30}C_{25}$
(c) 1 (d) 0

14. If S be the sum of coefficients in the expansion of $(\alpha x+\beta y-\gamma z)^n$, where $(\alpha,\beta,\gamma>0)$, then the value of $\lim_{n\to\infty}\frac{S}{\left\{S^{\frac{1}{n}}+1\right\}^n}$ is

(a) $e^{\left(\frac{\alpha\beta}{\gamma}\right)}$ (b) $e^{\left(\frac{\alpha+\beta-\gamma}{\alpha+\beta-\gamma+1}\right)}$
(c) $\frac{\alpha\beta}{\gamma}$ (d) 0

15. If $(5+2\sqrt{6})^n=I+f$; $n, I\in N$; and $0\le f<1$, then I equals

(a) $\frac{1}{f}-f$ (b) $\frac{1}{1+f}-f$
(c) $\frac{1}{1+f}+f$ (d) $\frac{1}{1-f}-f$

16. If $(1+x)^n=\sum_{r=0}^{n}a_rx^r$ and $b_r=1+\frac{a_r}{a_{r-1}}$ and $\prod_{r=1}^{n}b_r=\frac{(101)^{100}}{100!}$, then n is

(a) 99 (b) 100
(c) 101 (d) 102

17. If a, b, c, d be four consecutive coefficients in the binomial expansion of $(1+x)^n$, then the value of the expression $\left\{\left(\frac{b}{b+c}\right)^2-\frac{ac}{(a+b)(c+d)}\right\}$ (where $x>0$) is

(a) <0 (b) >0
(c) $=0$ (d) 2

18. Let $x\in\mathbf{R}$ be so small that the powers of x beyond two are insignificant and negligibly small. For such x, if $(1-x)^3(2+x)^6$ is approximated by $a+bx+cx^2$, then $a+b+c=$

(a) –80 (b) 144
(c) 80 (d) 127

19. The coefficients of x^2y^2, yzt^2 and $xyzt$ in the expansion of $(x+y+z+t)^4$ are in the ratio

(a) 4 : 2 : 1 (b) 1 : 2 : 4
(c) 2 : 4 : 1 (d) 1 : 4 : 2

20. Coefficient of x^6 in $((1+x)(1+x^2)^2(1+x^3)^3\ldots(1+x^n)^n)$ is

(a) 26 (b) 28
(c) 30 (d) 35

21. In the expansion of $\left(2^x+\frac{1}{4^x}\right)^n$, $n\in N$ if sum of the coefficients of 2nd and 3rd term is 36, then which of these are correct?

(a) $n=8$
(b) $n=9$
(c) $\frac{T_3}{T_2}=\frac{7}{4}$ when $x=-\frac{1}{3}$
(d) $\frac{T_3}{T_2}=7$ when $x=\frac{1}{3}$

22. Coefficient of x^{11} in the expansion of $(1+x^2)^4(1+x^3)^7(1+x^4)^{12}$ is **[JEE Advanced 2014]**

(a) 1051 (b) 1106
(c) 1113 (d) 1120

23. If the three successive coefficients in the binomial expansion of $(1+x)^n$ are 28, 56 and 70 respectively, then n equals

(a) 4 (b) 6
(c) 8 (d) 10

24. If the term independent of x in the expansion of $\left(\frac{3}{2}x^2-\frac{1}{3x}\right)^9$ is k, then 18 k is equal to **[JEE Main 2020, 3 September]**

(a) 5 (b) 7
(c) 9 (d) 11

25. If $p^4+q^3=2(p>0, q>0)$, then the maximum value of term independent of x in the expansion of $\left(px^{\frac{1}{12}}+qx^{-\frac{1}{9}}\right)^{14}$ is

(a) ${}^{14}C_4$ (b) ${}^{14}C_6$
(c) ${}^{14}C_7$ (d) ${}^{14}C_{12}$

26. The constant term in the expansion of $\left(x^2+\frac{1}{x^2}+y+\frac{1}{y}\right)^8$ is

(a) 4900 (b) 4950
(c) 5050 (d) 5151

27. If ${}^nC_0, {}^nC_1, {}^nC_2, \ldots, {}^nC_n$, respectively are the binomial coefficients in the expansion of $(1+x)^n$, then when $n=10$, $\sum_{r=1}^{10}{}^nC_r\cdot r(r-4)$ equals

(a) 5120 (b) 7680
(c) 20480 (d) 28160

28. The term independent of x in the expansion of $[(t^{-1}-1)x+(t^{-1}+1)^{-1}x^{-1}]^8$ is

(a) $56\left(\frac{1-t}{1+t}\right)^3$ (b) $56\left(\frac{1+t}{1-t}\right)^3$
(c) $70\left(\frac{1-t}{1+t}\right)^4$ (d) $70\left(\frac{1+t}{1-t}\right)^4$

29. The value of $\sum_{k=1}^{n}(C_k)(C_{k-1})$ is

(a) ${}^{2n}C_n$ (b) $\frac{1}{2}({}^{2n+2}C_{n+1}) - {}^{2n}C_n$

(c) ${}^{2n}C_{n+2}$ (d) None of these

30. If the sum of the binomial coefficients in the expansion of $\left(x+\frac{1}{x}\right)^n$ is 64, then the term independent of x is equal to

(a) 10 (b) 20

(c) 40 (d) 60

31. The coefficient of x^{11} in the expression $(1+x)^5(3+x)^4(7+x)^3$ equals

(a) 28 (b) 34

(c) 38 (d) 42

32. If C_r stands for ${}^nC_r$, then the sum of first $(n+1)$ terms of the series $aC_0 + (a+d)C_1 + (a+2d)C_2 + \ldots$, is

(a) 0 (b) $[a+nd]\,2^n$

(c) $[2a+(n-1)d]\,2^{n-1}$ (d) $[2a+nd]\,2^{n-1}$

33. The number of rational terms in $(\sqrt{2}+\sqrt[3]{3}+\sqrt[6]{5})^{10}$ is

(a) 6 (b) 4

(c) 3 (d) 1

34. If for some positive integer n, the coefficients of three consecutive terms in the binomial expansion of $(1+x)^{n+5}$ are in the ratio 5 : 10 : 14, then the largest coefficient in this expansion is

[JEE Main 2020, 4 September]

(a) 330 (b) 462

(c) 792 (d) 252

35. Let p and q be positive integers. Let

$${}^pC_q = \begin{cases} \dfrac{p!}{p!(p-q)!} & \text{when } p \geq q \\ 0 & \text{when } p < q \end{cases}$$

Then, the sum $\Sigma_{r=0}^{m}({}^{10}C_r \times {}^{20}C_{m-r})$ is maximum when m is

(a) 5 (b) 10

(c) 15 (d) 20

36. If $(1+x)^n = C_0 + C_1 + \ldots + C_n x^n$, then the value of $\sum\sum_{0 \leq r < s \leq n}(C_r + C_s)$ is equal to

(a) $n2^n$ (b) 2^{n+1}

(c) $(n-1)2^n$ (d) None of these

37. The value of ${}^{4n}C_0 + {}^{4n}C_4 + {}^{4n}C_8 + \ldots + {}^{4n}C_{4n}$ is

(a) $2^{4n-2}+(-1)^n 2^{2n-1}$ (b) $2^{4n-2}+2^{2n-1}$

(c) $2^{2n-1}+(-1)^n 2^{4n-2}$ (d) None of these

38. The sum of the series

$$\frac{1}{0!(n-1)!}+\frac{1}{3!(n-3)!}+\frac{1}{5!(n-5)!}+\ldots+\frac{1}{(n-1)!1!}$$ is

(a) $\frac{2^{n-1}}{(n-1)!}$ (b) $\frac{2^n}{(n-1)!}$

(c) $\frac{2^{n-1}}{n!}$ (d) $\frac{2^n}{n!}$

39. Given $(1-x^3)^n = \sum_{k=0}^{n} a_k x^k (1-x)^{3n-2k}$, then the value of $3 \cdot a_{k-1} + a_k$ is

(a) ${}^nC_k \cdot 3^k$ (b) ${}^{(n+1)}C_k \cdot 3^k$

(c) ${}^{(n+1)}C_k \cdot 3^{k-1}$ (d) ${}^nC_{k-1} \cdot 3^k$

40. For integer $n > 1$, the digit at unit place in the number $\sum_{r=0}^{100} r! + 2^{2^n}$ is

(a) 4 (b) 3

(c) 1 (d) 0

41. If $C_0, C_1, C_2, \ldots, C_n$ are the binomial coefficients in the expansion of $(1+x)^n$, n being even, then $C_0 + (C_0+C_1) + (C_0+C_1+C_2) + \ldots + (C_0+C_1+C_2+\ldots+C_{n-1})$ is equal to

(a) $n \cdot 2^n$ (b) $n \cdot 2^{n-1}$

(c) $n \cdot 2^{n-2}$ (d) $n \cdot 2^{n-3}$

42. If the fourth term in the expansion of $(x + x^{\log_2 x})^7$ is 4480, then the value of x where $x \in N$ is equal to

[JEE Main 2021, 17 March]

(a) 2 (b) 4

(c) 3 (d) 1

43. If α and β are the coefficients of x^4 and x^2 respectively in the expansion of $(x+\sqrt{x^2-1})^6 + (x-\sqrt{x^2-1})^6$, then

[JEE Main 2020, 8 January]

(a) $\alpha+\beta = -30$ (b) $\alpha-\beta = -132$

(c) $\alpha+\beta = 60$ (d) $\alpha-\beta = 60$

44. The expression $({}^{10}C_0)^2 - ({}^{10}C_1)^2 + \ldots - ({}^{10}C_9)^2 + ({}^{10}C_{10})^2$ equals

(a) ${}^{10}C_5$ (b) $-{}^{10}C_5$

(c) $({}^{10}C_5)^2$ (d) $(10!)^2$

45. The maximum value of the term independent of 't' in the expansion of $\left(tx^{\frac{1}{5}} + \frac{(1-x)^{\frac{1}{10}}}{t}\right)^{10}$ where $x \in (0, 1)$ is

[JEE Main 2021, 26 Febuary]

(a) $\frac{10!}{\sqrt{3}(5!)^2}$ (b) $\frac{2 \cdot 10!}{3(5!)^2}$

(c) $\frac{10!}{3(5!)^2}$ (d) $\frac{2 \cdot 10!}{3\sqrt{3}(5!)^2}$

46. $2(1+x^3)^{100} = \sum_{k=0}^{100}\left(a_k x^k - \cos\frac{\pi}{2}(x+k)\right)$, then the value of $a_0 + a_2 + a_4 + \ldots + a_{100}$

(a) 2^{99} (b) 2^{100}

(c) 2^{101} (d) None of these

47. $\sum_{m=1}^{n}\left(\sum_{k=1}^{m}\left(\sum_{p=k}^{m} {}^nC_m \cdot {}^mC_p \cdot {}^pC_k\right)\right) =$

(a) $3^n - 2^n$ (b) $4^n - 3^n$
(c) $3^n + 2^n$ (d) $4^n - 1$

48. The number of rational terms in the expansion of $(1 + \sqrt{2} + \sqrt[3]{3})^6$ is

(a) 6 (b) 7
(c) 5 (d) 8

49. The coefficient of a^3b^4c in the expansion of $(1 + a + b - c)^9$ is

(a) $2 \cdot {}^9C_7 \cdot {}^7C_4$ (b) $-2 \cdot {}^9C_2 \cdot {}^7C_3$
(c) ${}^9C_7 \cdot {}^7C_4$ (d) None of these

50. The greatest value of the term independent of x in the expansion of $(x \sin\alpha + x^{-1}\cos\alpha)^{10}, \alpha \in R$ is

(a) 2^5 (b) $\frac{10!}{(5!)^2}$
(c) $\frac{1}{2^5} \cdot \frac{10!}{(5!)^2}$ (d) None of these

51. If $(1 + x + 2x^2)^{20} = a_0 + a_1x + a_2x^2 + \ldots + a_{40}x^{40}$, then $a_0 + a_2 + a_4 + \ldots + a_{38}$ equals

(a) $2^{19}(2^{20} + 1)$ (b) $2^{19}(2^{20} - 1)$
(c) $2^{20}(2^{19} - 1)$ (d) None of these

52. If 7 divides $32^{32^{32}}$, the remainder is

(a) 1 (b) 0
(c) 4 (d) 6

53. $\sum_{r=0}^{n-1} \frac{{}^nC_r}{{}^nC_r + {}^nC_{r+1}}$ is equal to

(a) $\frac{n}{2}$ (b) $\frac{n+1}{2}$
(c) $\frac{n(n+1)}{2}$ (d) $\frac{n(n-1)}{2(n+1)}$

54. The largest term in the expansion of $\left(\frac{b}{2} + \frac{b}{2}\right)^{100}$ is

(a) b^{100} (b) $\left(\frac{b}{2}\right)^{100}$
(c) ${}^{100}C_{50} \cdot \left(\frac{b}{2}\right)^{100}$ (d) None of these

55. The coefficient of x^n in the polynomial $(x + {}^{2n+1}C_0)(x + {}^{2n+1}C_1)(x + {}^{2n+1}C_2)\ldots(x + {}^{2n+1}C_n)$ is

(a) 2^{n+1} (b) $2^{2n+1} - 1$
(c) 2^{2n} (d) None of these

56. If the fourth term of $\left(\sqrt{x^{\left(\frac{1}{1+\log_{10}x}\right)}} + \sqrt[12]{x}\right)^6$ is equal to 200 and $x > 1$, then x is equal to

(a) $10\sqrt{2}$ (b) 10
(c) 10^4 (d) None of these

57. If $(1 + x)^n = C_0 + C_1x + C_2x^2 + \ldots + C_nx^n$, then the value of $\sum_{k=0}^{n}(k+1)^2 \cdot C_k$ is

(a) $2^{n-3}(n^2 + 5n + 4)$ (b) $2^{n-2}(n^2 + 5n + 4)$
(c) $2^{n-2}(5n + 4)$ (d) None of these

58. If $\{x\}$ denotes the fractional part of x, then $\left\{\frac{3^{2n}}{8}\right\}, n \in N$ is

(a) 3/8 (b) 7/8
(c) 1/8 (d) None of these

59. The sum of the last ten coefficients in the expansion of $(1 + x)^{19}$, when expanded in ascending powers of x is

(a) 2^{18} (b) 2^{19}
(c) $2^{18} - {}^{19}C_{10}$ (d) None of these

60. If $a_n = \sum_{r=0}^{n}\frac{1}{{}^nC_r}$, then $\sum_{r=0}^{n}\frac{r}{{}^nC_r}$ equals

(a) $(n-1)a_n$ (b) $n\,a_n$
(c) $\frac{1}{2}n\,a_n$ (d) None of these

61. If $n \in N, n > 1$, then value of $E = a - {}^nC_1(a-1) + {}^nC_2(a-2) + \ldots (-1)^n(a-n)({}^nC_n)$ is

(a) a (b) 0
(c) a^2 (d) 2^n

62. The last two digits of the number 3^{400} are

(a) 39 (b) 29
(c) 01 (d) 43

63. The unit digit of $17^{1983} + 11^{1983} - 7^{1983}$ is

(a) 1 (b) 2
(c) 3 (d) 0

64. If $(1 + 2x + 3x^2)^{10} = a_0 + a_1x + a_2x^2 + \ldots + a_{20}x^{20}$, then a_1 equals

(a) 10 (b) 20
(c) 210 (d) 420

65. In the expansion of $(1 + x)^n(1 + y)^n(1 + z)^n$, the sum of the coefficients of the terms of degree r is

(a) $({}^nC_r)^3$ (b) $3 \cdot {}^nC_r$
(c) ${}^{3n}C_r$ (d) ${}^nC_{3r}$

66. If the second term in the expansion $\left(\sqrt[13]{a} + \frac{a}{\sqrt{a^{-1}}}\right)^n$ is $14a^{5/2}$, then the value of ${}^nC_3/{}^nC_2$ is

(a) 4 (b) 3
(c) 12 (d) 6

67. Let $(1 + x + 2x^2)^{20} = a_0 + a_1x + a_2x^2 + \ldots + a_{40}x^{40}$, then $a_1 + a_3 + a_5 + \ldots + a_{37}$ is equal to **[JEE Main 2021, 18 March]**

(a) $2^{20}(2^{20} - 21)$ (b) $2^{19}(2^{20} - 21)$
(c) $2^{19}(2^{20} + 21)$ (d) $2^{20}(2^{20} + 21)$

68. Coefficient of $1/x$ in the expansion of $(1+x)^n(1+1/x)^n$ is
(a) $\frac{n!}{(n-1)!(n+1)!}$ (b) $\frac{2n!}{(n-1)!(n+1)!}$
(c) $\frac{n!}{(2n-1)!(2n+1)!}$ (d) $\frac{2n!}{(2n-1)!(2n+1)!}$

69. For $n \in N$, in the expansion of $(4\sqrt{x^{-3}} + a^4\sqrt{x^5})^n$, the sum of all binomial coefficients lies between 200 and 400, and the term independent of x is 448. Then, the value of a is
(a) 1 (b) 2
(c) $\frac{1}{2}$ (d) 0

70. The value of x, for which the 6th term in the expansion of $\left\{2^{\log_2\sqrt{(9^{x-1}+7)}} + \frac{1}{2^{(1/5)\log_2(3^{x-1}+1)}}\right\}^7$ is 84, is equal to
(a) 4 (b) 3
(c) 2 (d) 5

71. $\sum_{r=1}^{n}\left(\sum_{p=0}^{r-1} {}^nC_r\,{}^rC_p\,2^p\right)$ is equal to
(a) 4^n-3^n+1 (b) 4^n-3^n-1
(c) 4^n-3^n+2 (d) 4^n-3^n

72. If $(1+x)^n = C_0 + C_1x + C_2x^2 + \dots + C_nx^n$, then $\sum_{0\le i<j\le n}\sum C_iC_j$ is
(a) $2^{2n-1} - \frac{2n!}{n!n!}$ (b) $2^{2n-1} - \frac{2\cdot 2n!}{n!n!}$
(c) $2^{2n+1} - \frac{2n!}{2\cdot n!}$ (d) $2^{2n-1} - \frac{{}^{2n}C_n}{2}$

73. The value of $99^{50} - 99\cdot 98^{50} + \frac{99\cdot 98}{1\cdot 2}(97)^{50} + \dots + 99$ is
(a) -1 (b) -2
(c) -3 (d) 0

74. If $x = (\sqrt{3}+1)^n$, then $[x]$ is (where $[x]$ denotes the greatest integer less than or equal to x)
(a) $2k$, where $k \in I$ (b) $2k+1$, where $k \in I$
(c) 4^n (d) 8^n

75. The number of irrational terms in the expansion of $(2^{1/5} + 3^{1/10})^{55}$ is
(a) 47 (b) 56
(c) 50 (d) 48

76. The coefficient of x^7 in the expression $(1+x)^{10} + x(1+x)^9 + x^2(1+x)^8 + \dots + x^{10}$ is
[JEE Main 2020, 7 January]
(a) 420 (b) 330
(c) 210 (d) 120

77. The value of the expression ${}^{n+1}C_2 + 2[{}^2C_2 + {}^3C_2 + {}^4C_2 + \dots + {}^nC_2]$ is
(a) Σn (b) Σn^2
(c) Σn^3 (d) $\frac{(n+1)}{2}$

78. If $n > 3$ and C_r stands for ${}^nC_r$, then $\sum_{r=0}^{n}(-1)^r(n-r)(n-r+1)(n-r+2)C_r$ is equal to
(a) 4 (b) 3
(c) 0 (d) 1

79. The coefficient of x^n in the polynomial $(x+{}^nC_0)(x+3\cdot{}^nC_1)(x+5\cdot{}^nC_2)\dots(x+(2n-1)\cdot{}^nC_n)$ is
(a) $n\cdot 2^n$ (b) $n\cdot 2^{n+1}$
(c) $(n+1)\cdot 2^n$ (d) $n\cdot 2^n+1$

80. If C_r stands for ${}^nC_r$, then the coefficient of $\lambda^n\mu^n$ in the expansion of $[(1+\lambda)(1+\mu)(\lambda+\mu)]^n$ is
(a) $\sum_{r=0}^{n} C_r^2$ (b) $\sum_{r=0}^{n} C_{r+2}^2$
(c) $\sum_{r=0}^{n} C_{r+3}^2$ (d) $\sum_{r=0}^{n} C_r^3$

81. If n is even positive integer, then the condition that the greatest term in the expansion of $(1+x)^n$ may have the greatest coefficient also is
(a) $\frac{n}{n+2} < x < \frac{n+2}{n}$ (b) $\frac{n+1}{n} < x < \frac{n}{n+1}$
(c) $\frac{n}{n+4} < x < \frac{n+4}{n}$ (d) None of these

82. For $1 \le r \le n$, the value of ${}^nC_r + {}^{n-1}C_r + {}^{n-2}C_r + \dots + {}^rC_r$ is
(a) ${}^nC_{r+1}$ (b) ${}^{n+1}C_r$
(c) ${}^{n+1}C_{r+1}$ (d) None of these

83. The remainder of 7^{103}, when divided by 25 is
(a) 7 (b) 25
(c) 18 (d) 9

84. If a_1, a_2, a_3, a_4 are the coefficients of any four consecutive terms in the expansion of $(1+x)^n$, then $\frac{a_1}{a_1+a_2} + \frac{a_3}{a_3+a_4}$ is equal to
(a) $\frac{a_2}{a_2+a_3}$ (b) $\frac{1}{2}\cdot\frac{a_2}{a_2+a_3}$
(c) $\frac{2a_2}{a_2+a_3}$ (d) $\frac{2a_3}{a_2+a_3}$

85. The coefficient of x^r $\{0 \le r \le (n-1)\}$ in the expansion of $(x+3)^{n-1} + (x+3)^{n-2}(x+2) + (x+3)^{n-3}(x+2)^2 + \dots + (x+2)^{n-1}$ is
(a) ${}^nC_r(3^r - 2^n)$ (b) ${}^nC_r(3^{n-r} - 2^{n-r})$
(c) ${}^nC_r(3^r + 2^{n-r})$ (d) None of these

86. If there is a term containing x^{2r} in $\left(x+\frac{1}{x^2}\right)^{n-3}$, then

(a) $n-2r$ is a positive integral multiple of 3
(b) $n-2r$ is even
(c) $n-2r$ is odd
(d) None of the abvoe

87. The value of the sum of the series
$3\cdot {}^nC_0 - 8\cdot {}^nC_1 + 13\cdot {}^nC_2 - 18\cdot {}^nC_3 + \ldots$ upto $(n+1)$ terms is
(a) 0 (b) 3^n
(c) 5^n (d) None of these

88. If $a+b=1$, then $\sum_{r=0}^{n} r\,{}^nC_r\,a^r b^{n-r}$ equals
(a) 1 (b) n
(c) na (d) nb

89. If n is a positive integer and $C_k = {}^nC_k$, then $\sum_{k=1}^{n} k^3\left(\frac{C_k}{C_{k-1}}\right)^2$ equals
(a) $\frac{n(n+1)(n+2)}{12}$ (b) $\frac{n(n+1)^2(n+2)}{12}$
(c) $\frac{n(n+1)(n+2)^2}{12}$ (d) $\frac{n^2(n+1)^2(n+2)^2}{144}$

90. The coefficient of x^{50} in the expansion of
$(1+x)^{1000} + 2x(1+x)^{999} + 3x^2(1+x)^{998} + \ldots + \ldots + 1001\,x^{1000}$ is
(a) ${}^{1000}C_{50}$ (b) ${}^{1001}C_{50}$
(c) ${}^{1002}C_{50}$ (d) ${}^{1000}C_{51}$

91. If $(6\sqrt{6}+14)^{n+1} = m$ and if f is the fractional part of m, then mf is equal to
(a) 15^{n+1} (b) 20^{n+1}
(c) 25^n (d) None of these

92. The number of terms in the expansion of $[(a+3b)^2(3a-b)^2]^3$ is
(a) 14 (b) 28
(c) 32 (d) 56

93. The coefficient of x^{n-2} in the polynomial $(x-1)(x-2)(x-3)\ldots(x-n)$ is
(a) $\frac{n(n^2+2)(3n+1)}{24}$ (b) $\frac{n(n^2-1)(3n+2)}{24}$
(c) $\frac{n(n^2+1)(3n+4)}{24}$ (d) None of these

94. If $n>3$ and $a, b \in R$, then the value of
$ab - n(a-1)(b-1) + \frac{n(n-1)}{1\cdot 2} \times (a-2)(b-2) - \ldots + (-1)^n(a-n)(b-n)$ is
(a) $a^n + b^n$ (b) $\frac{a^n - b^n}{a-b}$
(c) $(ab)^n$ (d) 0

95. The value of $\sum_{i=0}^{n}\sum_{j=1}^{n} {}^nC_j\,{}^jC_i, i \le j$ is
(a) $3^n - 1$
(b) 0
(c) 2^n
(d) None of the above

96. The coefficient of $a^{10}b^7c^3$ in the expansion of $(bc+ca+ab)^{10}$ is
(a) 30 (b) 60
(c) 120 (d) 240

97. If $(1+x+2x^2)^{20} = a_0 + a_1x + a_2x^2 + \ldots + a_{40}x^{40}$, then $a_1 + a_3 + a_5 + \ldots + a_{37}$ equals
(a) $2^{19}(2^{20}-21)$
(b) $2^{20}(2^{19}-19)$
(c) $2^{19}(2^{20}+21)$
(d) None of the above

98. The coefficient of x^{50} in $(1+x)^{41}(1-x+x^2)^{40}$ is
(a) 1 (b) 2
(c) 3 (d) 0

99. The number of terms in the expansion of $(\sqrt[4]{9}+\sqrt[6]{8})^{500}$, which are integers is given by
(a) 501 (b) 251
(c) 601 (d) 451

100. If n is an even integer and a, b, c are distinct, the number of distinct terms in the expansion of $(a+b+c)^n + (a+b-c)^n$ is
(a) $\left(\frac{n}{2}\right)^2$ (b) $\left(\frac{n+1}{2}\right)^2$
(c) $\left(\frac{n+2}{2}\right)^2$ (d) $\left(\frac{n+3}{2}\right)^2$

101. Let $a_0, a_1, a_2, \ldots$ be the coefficients in the expansion of $(1+x+x^2)^n$ arranged order of x. The value of $a_r - {}^nC_1\,a_{r-1} + {}^nC_2\,a_{r-2} - \ldots + (-1)^r\,{}^nC_r\,a_0$, where r is not divisible by 3, is
(a) 5 (b) 3
(c) 1 (d) 0

102. In the expansion of $(3^{-x/4} + 3^{5x/4})^n$ the sum of the binomial coefficient is 64 and the term with the greatest binomial coefficients exceeds the third by $(n-1)$ the value of x must be
(a) 0 (b) 1
(c) 2 (d) 3

103. $({}^mC_0 + {}^mC_1 - {}^mC_2 - {}^mC_3) + ({}^mC_4 + {}^mC_5 - {}^mC_6 - {}^mC_7) + \ldots = 0$, if and only if for some positive integer k, $m =$
(a) $4k$ (b) $4k+1$
(c) $4k-1$ (d) $4k+2$

Objective Questions Type II [One or more than one correct answer(s)]

In each of the questions below four choices of which one or more than one are correct. You have to select the correct answer(s) accordingly.

1. If $(1+x+x^2)^n = a_0 + a_1x + a_2x^2 + \dots + a_{2n}x^{2n}$, then
   (a) $a_0 - a_2 + a_4 - a_6 + \dots = 0$, if n is odd
   (b) $a_1 - a_3 + a_5 - a_7 + \dots = 0$, if n is even
   (c) $a_0 - a_2 + a_4 - a_6 + \dots = 0$, if $n = 4p$, $p \in I^+$
   (d) $a_1 - a_3 + a_5 - a_7 + \dots = 0$, if $n = 4p+1$, $p \in I^+$

2. The value of $C_0^2 + 3C_1^2 + 5C_2^2 + \dots$ to $(n+1)$ terms, (given that $C_r \equiv {}^nC_r$) is
   (a) ${}^{2n-1}C_{n-1}$
   (b) $(2n+1)\cdot{}^{2n-1}C_n$
   (c) $2(n+1)\cdot{}^{2n-1}C_n$
   (d) ${}^{2n-1}C_n + (2n+1)\cdot{}^{2n-1}C_{n-1}$

3. The value of the expression $C_0^2 - C_1^2 + C_2^2 - \dots + (-1)^n C_n^2$ is
   (a) 0 if n is odd
   (b) $(-1)^n$ if n is odd
   (c) $(-1)^{n/2}\,{}^nC_{n/2}$ if n is even
   (d) $(-1)^{n-1}\,{}^nC_{n-1}$ if n is even

4. If n is a positive integer and $(3\sqrt{3}+5)^{2n+1} = \alpha + \beta$, where α is an integer and $0 < \beta < 1$, then
   (a) α is an even integer
   (b) $(\alpha+\beta)^2$ is divisible by 2^{2n+1}
   (c) the integer just below $(3\sqrt{3}+5)^{2n+1}$ divisible by 3
   (d) α is divisible by 10

5. If $(8+3\sqrt{7})^n = P + F$, where P is an integer and F is a proper fraction, then
   (a) P is an odd integer (b) P is an even integer
   (c) $F\cdot(P+F) = 1$ (d) $(1-F)(P+F) = 1$

6. In the expansion of $\left(\sqrt[3]{4} + \dfrac{1}{\sqrt[4]{6}}\right)^{20}$
   (a) the number of rational terms = 4
   (b) the number of irrational terms = 19
   (c) the middle term is irrational
   (d) the number of irrational terms = 17

7. Let $a_n = \dfrac{(1000)^n}{n!}$ for $n \in N$. Then a_n is greatest, when
   (a) $n = 998$ (b) $n = 999$
   (c) $n = 1000$ (d) $n = 1001$

8. If $C_0, C_1, C_2, \dots, C_n$ are coefficients in the binomial expansion of $(1+x)^n$, then $C_0C_2 + C_1C_3 + C_2C_4 + \dots + C_{n-2}C_n$ is equal to
   (a) $\dfrac{2n!}{(n-2)!(n+2)!}$ (b) $\dfrac{2n!}{\{(n-2)!\}^2}$
   (c) $\dfrac{2n!}{\{(n+2)!\}^2}$ (d) ${}^{2n}C_{n-2}$

9. The expression $\{x + \sqrt{(x^3-1)}\}^5 + \{x - \sqrt{(x^3-1)}\}^5$ is a polynomial of degree
   (a) ${}^9C_2$ (b) ${}^7C_6$
   (c) 7 (d) ${}^8C_1$

10. In the expansion of $(x+y+z)^{25}$
    (a) every term is of the form ${}^{25}C_r \cdot {}^rC_k \cdot x^{25-r} \cdot y^{r-k} \cdot z^k$
    (b) the coefficient of $x^8y^9z^9$ is 0
    (c) the number of terms is 351
    (d) None of the above

11. If $(1+2x+3x^2)^{10} = a_0 + a_1x + a_2x^2 + \dots + a_{20}x^{20}$, then
    (a) $a_1 = 20$ (b) $a_2 = 210$
    (c) $a_4 = 8085$ (d) $a_{20} = 2^2 \cdot 3^7 \cdot 7$

12. The value of $\sum_{k=0}^{7}\left[\dfrac{\binom{7}{k}}{\binom{14}{5}}\sum_{r=k}^{14}\binom{r}{k}\binom{14}{r}\right]$, where $\binom{n}{r}$ denotes ${}^nC_r$ is
    (a) 6^7 (b) greater than 7^6
    (c) 8^7 (d) greater than 7^8

13. If $(6\sqrt{6}+14)^{2n+1} = P$, then
    (a) $[P]$ is an even integer (b) $PF = 2^{2n+1}$
    (c) $PF = 20^{2n+1}$ (d) $[P]$ is an odd integer
    Note [.] denotes integral part and F the fractional part of P.

14. The last digit of $3^{3^{4n}} + 1$, $n \in N$, is
    (a) ${}^4C_3$ (b) ${}^8C_7$
    (c) 8 (d) 4

15. Let $a_n = \left(1 + \dfrac{1}{n}\right)^n$. Then, for each $n \in N$
    (a) $a_n \ge 2$ (b) $a_n < 3$
    (c) $a_n < 4$ (d) $a_n < 2$

16. Which of the following is/are correct
    (a) $101^{50} - 99^{50} > 100^{50}$
    (b) $101^{50} - 100^{50} > 99^{50}$
    (c) $(1000)^{1000} > (1001)^{999}$
    (d) $(1001)^{999} > (1000)^{1000}$

17. In the expansion of $(2-2x+x^2)^9$
    (a) number of distinct terms is 10
    (b) coefficient of x^4 is 97
    (c) sum of coefficients is 1
    (d) number of distinct terms is 55

18. If n is a positive integer and if $(1 + x + x^2)^n = \sum_{r=0}^{2n} a_r x^r$, then
(a) $a_r = a_{2n-r}$ for $0 \le r \le 2n$
(b) $a_0 + a_1 + \dots + a_{n-1} = \frac{1}{2}(3^n - a_n)$
(c) $a_0^2 - a_1^2 + a_2^2 - a_3^2 + \dots + a_{2n}^2 = a_n$
(d) $a_0 + a_2 + \dots + a_{2n} = \frac{1}{2}(3^n + 1)$

19. If the term independent of x in the expansion of $(\sqrt{x} - \lambda/x^2)^{10}$ is 405, then value of λ is
(a) -3 (b) 9
(c) -9 (d) 3

20. In the expansion of $(x^2 + 2x + 2)^n$, (where n is a positive integer), then coefficient of
(a) x is $2^n \cdot n$ (b) x^2 is $n^2 \cdot 2^{n-1}$
(c) x^3 is $2^n \cdot {}^{n+1}C_3$ (d) None of these

Linked-Comprehension Type

In these questions, a passage (paragraph) has been given followed by questions based on each of the passages. You have to answer the questions based on the passage given.

PASSAGE 1

Let $(1 + x)^n = C_0 + C_1 x + C_2 x^2 + \dots + C_n x^n$, where $C_r = {}^nC_r$. For a given value of x, let $p = (n+1)|x|/(|x| + 1)$. If p is an integer, then the numerical values of pth and $(p + 1)$th terms are equal and they are the numerically greatest terms in the expansion of $(1 + x)^n$. If p is not an integer and [P] denotes the integral part of p, then $([p] + 1)$th term is numerically greatest term.

On the basis of above information, answer the following questions

1. The value of numerically in the expansion of $\sqrt{2}[1 + (1/\sqrt{2})]^{16}$ is
(a) $\frac{1}{16} \cdot {}^{16}C_7$ (b) $\frac{1}{16} \cdot {}^{16}C_8$
(c) $\frac{1}{18} \cdot {}^{16}C_8$ (d) $\frac{1}{8} \cdot {}^{16}C_7$

2. Numerically greatest term in the expansion of $(3 - 5x)^{15}$ when $x = 1/5$ is (are)
(a) fourth and fifth term (b) sixth term
(c) seventh term (d) eighth term

3. The greatest value of ${}^{20}C_r (0 \le r \le 20)$ is
(a) ${}^{20}C_8$ (b) ${}^{20}C_9$
(c) ${}^{20}C_{10}$ (d) ${}^{20}C_7$

PASSAGE 2

If n is positive integer and if $(1 + 4x + 4x^2)^n = \sum_{r=0}^{2n} a_r x^r$, where a_i's/ are $(i = 0, 1, 2, 3, \dots, 2n)$ real numbers.

On the basis of above information, answer the following questions

1. The value of $2\sum_{r=0}^{n} a_{2r}$ is
(a) $9^n - 1$ (b) $9^n + 1$
(c) $9^n - 2$ (d) $9^n + 2$

2. The value of $2\sum_{r=1}^{n} a_{2r-1}$ is
(a) $9^n - 1$ (b) $9^n + 1$
(c) $9^n - 2$ (d) $9^n + 2$

3. The value of a_{2n-1} is
(a) 2^{2n} (b) $(n-1) \cdot 2^{2n}$
(c) $n \cdot 2^{2n}$ (d) $(n+1) \cdot 2^{2n}$

4. The value of a_2 is
(a) $8n$ (b) $8n^2 - 4$
(c) $8n^2 - 4n$ (d) $8n - 4$

5. The correct statement is
(a) $a_r = a_{n-r}, 0 \le r \le n$ (b) $a_{n-r} = a_{n+r}, 0 \le r \le n$
(c) $a_r = a_{2n-r}, 0 \le r \le 2n$ (d) None of these

PASSAGE 3

If $S_1 = \sum_{i=1}^{n} a_i = a_1 + a_2 + a_3 + \dots + a_n$

$S_2 = \sum\sum_{1 \le i < j \le n} a_i a_j = a_1 a_2 + a_1 a_3 + \dots + a_{n-1} a_n$

$S_3 = \sum\sum\sum_{1 \le i < j < k \le n} a_i a_j a_k = a_1 a_2 a_3 + a_1 a_2 a_4 + \dots + a_{n-2} a_{n-1} a_n$

....

$S_n = a_1 a_2 a_3 \dots a_n$

Then, $(x + a_1)(x + a_2)(x + a_3) \dots (x + a_n)$ can be written as

$x^n + S_1 x^{n-1} + S_2 x^{n-2} + \dots + S_n$

On the basis of above information, answer the following questions

1. If $(1 + x)^n = C_0 + C_1 x + C_2 x^2 + C_3 x^3 + \dots + C_n x^n$, then the coefficient of x^{n-1} in the expression $(x + C_0)(x + C_1)(x + C_2)(x + C_3) \dots (x + C_n)$ is
(a) $2^{2n} - \frac{2n!}{2(n!)^2}$ (b) $2^{2n-1} - \frac{2n!}{2(n!)^2}$
(c) $2^{2n+1} - \frac{2n!}{2(n!)^2}$ (d) $2^{2n-2} - \frac{2n!}{2(n!)^2}$

2. If $(1+x)^n = C_0 + C_1x + C_2x^2 + C_3x^3 + \dots\dots + C_nx^n$, then the value of $\sum\sum_{0\le i<j\le n}(i+j)C_iC_j$ is

(a) $n\sum\sum_{0\le i<j\le n} C_iC_j$ (b) $(n+1)\sum\sum_{0\le i<j\le n} C_iC_j$
(c) $(n-1)\sum\sum_{0\le i<j\le n} C_iC_j$ (d) $2\sum\sum_{0\le i<j\le n} C_iC_j$

3. If $(1+x)^n = C_0 + C_1x + C_2x^2 + C_3x^3 + \dots\dots + C_nx^n$, then the value of $\sum\sum_{0\le i<j\le n}(i+j)(C_i+C_j+C_iC_j)$ is (where $\lambda = \sum\sum_{0\le i<j\le n} C_iC_j$)

(a) $n(2^n+\lambda)$ (b) $n^2\cdot 2^n+\lambda$
(c) $n(n\cdot 2^n+\lambda)$ (d) None of these

4. If $(1+x)^n = C_0 + C_1x + C_2x^2 + C_3x^3 + \dots\dots + C_nx^n$ and $\sum_{r=0}^{n}\frac{1}{{}^nC_r} = a$, then the value of $\sum\sum_{0\le i<j\le n}\left(\frac{i}{{}^nC_i}+\frac{j}{{}^nC_j}\right)$ in terms of a and n is

(a) $\frac{na}{2}$ (b) $\frac{n^2a}{2}$
(c) $\frac{na^2}{2}$ (d) $\frac{n^2a^2}{2}$

5. The coefficient of x^6 in the expression $(2+x)^3(3+x)^2(5+x)^3$ is

(a) 78 (b) 156 (c) 312 (d) 624

PASSAGE 4

Consider $(1+x+x^2)^{2n} = \sum_{r=0}^{4n} a_rx^r$, where $a_0, a_1, a_2, \dots,$ a_{4n} are real numbers and n is a positive integer.

On the basis of above information, answer the following questions

1. The value of $\sum_{r=0}^{n-1} a_{2r}$ is

(a) $\frac{9^n-2a_{2n}-1}{4}$ (b) $\frac{9^n+2a_{2n}+1}{4}$
(c) $\frac{9^n-2a_{2n}+1}{4}$ (d) $\frac{9^n+2a_{2n}-1}{4}$

2. The value of $\sum_{r=1}^{n} a_{2r-1}$ is

(a) $\left(\frac{9^n-1}{2}\right)$ (b) $\left(\frac{3^{2n}-1}{4}\right)$
(c) $\left(\frac{3^{2n}+1}{4}\right)$ (d) $\left(\frac{9^n+1}{2}\right)$

3. The value of a_2 is

(a) ${}^{4n+1}C_2$ (b) ${}^{3n+1}C_2$
(c) ${}^{2n+1}C_2$ (d) ${}^{n+1}C_2$

4. The value of a_{4n-1} is

(a) $2n$ (b) $2n^2+4n$
(c) $2n+3$ (d) $2n^2+3n$

5. The correct statement is

(a) $a_r = a_{n-r}, 0\le r\le n$
(b) $a_{n-r} = a_{n+r}, 0\le r\le n$
(c) $a_r = a_{2n-r}, 0\le r\le 2n$
(d) $a_r = a_{4n-r}, 0\le r\le 4n$

PASSAGE 5

If m, n, r are positive integers and if $r<m$, $r<n$, then ${}^mC_r + {}^mC_{r-1}\cdot{}^nC_1 + {}^mC_{r-2}\cdot{}^nC_2 + \dots + {}^nC_r =$ Coefficient of x^r in $(1+x)^m(1+x)^n$ = Coefficient of x^r in $(1+x)^{m+n}$

$$= {}^{m+n}C_r$$

On the basis of above information, answer the following questions

1. The value of ${}^nC_0\cdot{}^nC_n + {}^nC_1\cdot{}^nC_{n-1} + \dots + {}^nC_n\cdot{}^nC_0$ is

(a) ${}^{2n}C_{n-1}$ (b) ${}^{2n}C_n$
(c) ${}^{2n}C_{n+1}$ (d) ${}^{2n}C_2$

2. The value of r for which ${}^{30}C_r\cdot{}^{20}C_0 + {}^{30}C_{r-1}\cdot{}^{20}C_1 + \dots + {}^{30}C_0\cdot{}^{20}C_r$ is maximum, is

(a) 10 (b) 15
(c) 20 (d) 25

3. The value of r $(0\le r\le 30)$ for which ${}^{20}C_r\cdot{}^{10}C_0 + {}^{20}C_{r-1}\cdot{}^{10}C_1 + \dots + {}^{20}C_0\cdot{}^{10}C_r$ is minimum, is

(a) 0 (b) 1
(c) 5 (d) 15

4. If $S_n = {}^nC_0\cdot{}^nC_1 + {}^nC_1\cdot{}^nC_2 + \dots + {}^nC_{n-1}\cdot{}^nC_n$ and if $\frac{S_{n+1}}{S_n} = \frac{15}{4}$, then n equals

(a) 2, 4 (b) 4, 6
(c) 6, 8 (d) 8, 10

5. If $(1+x)^n = C_0 + C_1x + C_2x^2 + \dots + C_nx^n$ and n is odd, then the value of $C_0^2 - C_1^2 + C_2^2 - C_3^2 + \dots + (-1)^nC_n^2$ is

(a) 0 (b) ${}^{2n}C_n$
(c) $(-1)^n\,{}^{2n}C_{n-1}$ (d) ${}^{2n}C_{n-2}$

PASSAGE 6

If $a, b \in$ prime numbers and $n \in N$, then free from radical terms or rational terms in the expansion of $(a^{1/p} + b^{1/q})^n$ are the terms in which indices of a and b are integers.

On the basis of above information, answer the following questions

1. In the expansion of $(7^{1/3} + 11^{1/9})^{6561}$, the number of terms free from radicals is
   (a) 715 (b) 725
   (c) 730 (d) 745
2. In the expansion of $(\sqrt{2} + 3^{1/11})^{110}$, the number of irrational terms is
   (a) 85 (b) 95
   (c) 105 (d) 115
3. In the expansion of $(2^{1/5} + \sqrt{3})^{20}$, the sum of rational terms is
   (a) 21 (b) 84
   (c) 97 (d) None of these
4. The number of integral terms in the expansion of $(\sqrt{3} + \sqrt[8]{5})^{256}$ is
   (a) 32 (b) 33
   (c) 34 (d) 35
5. The number of terms with integral coefficient in the expansion of $(\sqrt[4]{9} + \sqrt[6]{8x})^{500}$ is
   (a) 501 (b) 250
   (c) 253 (d) 251

PASSAGE 7

If n is a positive integer and $a_1, a_2, a_3, \ldots, a_m \in C$, then

$$(a_1 + a_2 + a_3 + \ldots + a_m)^n = \sum \frac{n!}{n_1!\, n_2!\, n_3! \ldots n_m!} \cdot a_1^{n_1} \cdot a_2^{n_2} \cdot a_3^{n_3} \ldots a_m^{n_m}$$

where $n_1, n_2, n_3, \ldots, n_m$ are all non negative integers subject to the condition $n_1 + n_2 + n_3 + \ldots + n_m = n$

On the basis of above information, answer the following questions

1. The number of distinct terms in the expansion of $(x_1 + x_2 + x_3 + \ldots + x_n)^4$ is
   (a) ${}^{n+1}C_4$ (b) ${}^{n+2}C_4$
   (c) ${}^{n+3}C_4$ (d) ${}^{n+4}C_4$
2. The coefficient of x^3y^4z in the expansion of $(1 + x - y + z)^9$ is
   (a) 2320 (b) 2420
   (c) 2520 (d) 2620
3. The coefficient of $a^3b^4c^5$ in the expansion of $(bc + ca + ab)^6$ is
   (a) 40 (b) 60
   (c) 80 (d) 100
4. The coefficient of x^{39} in the expansion of $(1 + x + 2x^2)^{20}$ is
   (a) $5 \cdot 2^{19}$ (b) $5 \cdot 2^{20}$
   (c) $5 \cdot 2^{21}$ (d) $5 \cdot 2^{23}$
5. If coefficient of x^{20} in $(1 - x + x^2)^{20}$ and in $(1 + x - x^2)^{20}$ are respectively a and b, then
   (a) $a = b$ (b) $a > b$
   (c) $a < b$ (d) $a + b = 0$

PASSAGE 8

Suppose m divided by n, then quotient q and remainder r.

i.e. 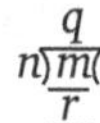

or we can say that, $m = nq + r \; \forall \; m, n, q, r \in I$ and $n \neq 0$

On the basis of above information, answer the following questions

1. 5^{97} is divided by 52, then the remainder is
   (a) 3 (b) 4
   (c) 5 (d) 0
2. $13^{99} - 19^{93}$ is divided by 162, then the remainder is
   (a) 3 (b) 4
   (c) 5 (d) 0
3. $27^{10} + 7^{51}$ is divided by 10, then the remainder is
   (a) 0 (b) 2
   (c) 4 (d) 5
4. $(106)^{85} - (85)^{106}$ is divided by 7, then the remainder is
   (a) 0 (b) 1
   (c) 2 (d) 3
5. $53^{53} - 33^3$ is divided by 10, then the remainder is
   (a) 0 (b) 2
   (c) 4 (d) 6

Numerical Value Type

Solve the following problems and write your answer.

1. Numerically the greatest term in the expansion of $(3-5x)^{11}$, where $x=\frac{1}{5}$ is $\lambda \times 729$, then λ must be

2. The coefficient of x^{50} in the polynomials after parenthesis have been removed and like terms have been collected in the expansion $(1+x)^{1000}+x(1+x)^{999}+x^2(1+x)^{998}+\ldots+x^{1000}$ is $\frac{\lambda!}{\mu!\nu!}$, then the value of $\lambda+2\mu+3\nu$ must be $(\nu>\mu)$

3. For a positive integer n, $\left(1+\frac{1}{x}\right)^n$ is expanded in increasing powers of x. If three consecutive coefficients in this expansion are in the ratio, 2 : 5 : 12, then n is equal to .... **[JEE Main 2020, 2 September]**

4. If $(1+x)^n = C_0+C_1x+C_2x^2+\ldots+C_nx^n$ and $\sum_{r=0}^{n}(r+1)^2C_r=2^{n-2}f(n)$, if roots of the equation $f(x)=0$ are α and β, then $\alpha^4+\beta^4$ must be

5. The expression $(\sqrt{(x^5-1)}+x)^9-(\sqrt{(x^5-1)}-x)^9$ is a polynomial of degree

6. The coefficient of x^{12} in the expansion of $(1+2x^2-x^3)^8$ must be

7. The number of permutations of the letters *a, b, c, d, e, f, g* so that neither the pattern beg nor cad appers must be

8. Coefficient of x^4 in $(1+x+x^2+x^3)^{11}$ is ......

9. If the sum of the coefficients in the expansion of $(1+3x)^n$ lies between 4000 and 10000, then the value of the greatest coefficient must be

10. Let $\lambda=(1992)^{1998}-(1955)^{1998}-(1938)^{1998}+(1901)^{1998}$, then λ is divisible by (In four digits)

11. Let coefficient of 3rd, 4th and 5th terms in the expansion of $\left(x+\frac{a}{x^2}\right)^n$, $x\neq 0$ be in the ratio 12 : 8 : 3, then the coefficient of the term independent of x in the expansion, is equal to

12. If $\sum_{r=0}^{n}{}^nC_r\sin(rx)\cos\{(n-r)x\}=f(n)\sin(nx)$, then the value of $f(13)$ must be

13. If $5^{81}+\lambda$ is divisible by 26, then minimum positive value of λ must be

14. If λ be the remainder, if 2^{2005} is divided by 17, then λ must be

15. Let $a_r=r^2\cdot\frac{{}^{100}C_r}{{}^{100}C_{r-1}}$ and λ be the coefficient of x^{99} in $\prod_{i=1}^{100}(x-a_i)$, then the value of $-\frac{\lambda}{50}$ must be

16. The sum of all two digit numbers when divided by 4, yield unity as remainder is

Single Integer Type Questions

Solve the following problems and write your answer.

1. Degree of the polynomial $(x^2-\sqrt{1-x^3})^4+(x^2+\sqrt{1-x^3})^4$, is

2. Let $(2x^2+3x+4)^{10}=\sum_{r=0}^{20}a_rx^r$. Then, $\frac{a_7}{a_{13}}$ is equal to ..... . **[JEE Main 2020, 4 September]**

3. Let $a=3^{\frac{1}{223}}+1$ and for all $n\geq 3$, let $f(n)={}^nC_0\cdot a^{n-1}-{}^nC_1\cdot a^{n-2}+{}^nC_2\cdot a^{n-3}-\ldots+(-1)^{n-1}\cdot{}^nC_{n-1}\cdot a^0$. If the value of $f(2007)+f(2008)=3^k$, where $k\in N$, then the value of k is ....

4. If $x=(7+4\sqrt{3})^{2n}=[x]+f$, then $x(1-f)$ is equal to

5. Let m be the smallest positive integer such that the coefficient of x^2 in the expansion of $(1+x)^2+(1+x)^3+\ldots+(1+x)^{49}+(1+mx)^{50}$ is $(3n+1)\,{}^{51}C_3$ for some positive integer n. Then, the value of n is **[JEE Advanced 2016]**

6. Let $(1+x)^{10}=\sum_{r=0}^{10}c_rx^r$ and $(1+x)^7=\sum_{r=0}^{7}d_rx^r$. If $P=\sum_{r=0}^{5}c_{2r}$ and $Q=\sum_{r=0}^{3}d_{2r+1}$, then $\frac{P}{Q}$ is equal to

7. If ninth term in the expansion of $\left(3^{\frac{1}{3}\log_3(9^{x-1}+7)}+\frac{1}{3^{\frac{1}{8}\log_3(3^{x-1}+1)}}\right)^{11}$ is 660, then the number of values of x is/are

Matrix-Match Type

Given below are Matching Type Questions, with two columns (each having some items) each . Each item of Column I has to be matched with the items of Column II, by encircling the correct match(es).

NOTE *An item of Column I can be matched with more than one items of Column II.*

1. Column I contains the different sum of the series and Column II contains the maximum value of these sums. Match the lists.

	Column I		Column II
A.	$\sum_{r=0}^{14} \frac{(15-r)}{r+1} {}^{15}C_r \, {}^{20}C_{m-r}$	P.	${}^{34}C_{17}$
B.	$\lvert {}^{35}C_0 - {}^{35}C_1 + {}^{35}C_2 - {}^{35}C_3 + \dots + (-1)^r \, {}^{35}C_r + \dots \rvert$	Q.	${}^{35}C_{18}$
C.	${}^{31}C_{r-5} + 5 \times {}^{35}C_{r-4} + 10 \times {}^{31}C_{r-3} + 10 \times {}^{31}C_{r-2} + 5 \times {}^{31}C_{r-1} = {}^{31}C_r$	R.	${}^{37}C_{18}$
D.	$\sum_{r=0}^{k} 2^{k-r} . (-1)^r . {}^{37}C_r \, {}^{37-r}C_{37-k}$	S.	${}^{36}C_{18}$

2. **Observe the following columns :**

	Column I		Column II
(A)	If λ be the number of terms in the expansion of $(1 + 5x + 10x^2 + 10x^3 + 5x^4 + x^5)^{20}$ and if unit's place and ten's place digits in 3^λ are O and T, then	(P)	$O + T = 3$
(B)	If λ be the number of terms in the expansion of $\left(x^2 + 1 + \frac{1}{x^2}\right)^{100}$ and if unit's place and ten's place digits in 7^λ are O and T, then	(Q)	$O + T = 7$
(C)	If λ be the number of terms in the expansion of $(1 + x)^{101} (1 + x^2 - x)^{100}$ and if unit's place and ten's place digits in 9^λ are O and T, then	(R)	$O + T = 9$
		(S)	$T - O = 7$
		(T)	$O - T = 7$

3. **Observe the following columns :**

	Column I		Column II
(A)	If last digit of the number 9^{9^9} is λ and last digit of $2^{\lambda^{100}}$ is μ, then	(P)	$\lambda + \mu = 9$
(B)	If last digit of the number 2^{999} is λ and last digit of $3^{\lambda\lambda\lambda}$ is μ, then	(Q)	$\lambda + \mu = 11$
(C)	Let $a = \frac{72!}{(36!)^2} - 1$ is divisible by $10\lambda + \mu$, then	(R)	$\lambda - \mu = 7$
		(S)	$\lambda - \mu = 4$
		(T)	$\lambda^\mu + \mu^\lambda = 9$

4. **Observe the following columns :**

	Column I		Column II
(A)	If n be the degree of the polynomial $\sqrt{(3x^2+1)}\{(x + \sqrt{(3x^2+1)})^7 - (x - \sqrt{(3x^2+1)})^7\}$, then n divisible by	(P)	2
(B)	In the expansion of $(x + a)^n$ there is only one middle term for $x = 3$, $a = 2$ and seventh term is numerically greatest term, then n divisible by	(Q)	4
		(R)	8
(C)	The sum of the binomial coefficients in the expansion of $(x^{-3/4} + n\,x^{5/4})^m$, where m is positive integer lies between 200 and 400 and the term independent of x equals 448. Then n^5 divisible by	(S)	16
		(T)	32

5. **Observe the following columns :**

	Column I		Column II
(A)	If sum of the coefficients of the first, second and third terms in the expansion of $\left(x^2 + \frac{1}{x}\right)^n$	(P)	9
	is 46, then coefficient of the term that does not contain x is	(Q)	15
(B)	For a positive integer n, if the expansion of $\left(\frac{5}{x^2} + x^4\right)^n$ has a term independent of x, then n can be	(R)	18
(C)	If in the expansion of $\left(x^3 - \frac{1}{x^2}\right)^n, n \in N$, sum of the coefficients of x^5 and x^{10} is zero, then the value of n is	(S)	27
		(T)	84

Answers

Objective Questions Type I [Only one correct answer]

1. (d)	**2.** (c)	**3.** (d)	**4.** (c)	**5.** (d)	**6.** (c)	**7.** (c)	**8.** (a)	**9.** (b)	**10.** (d)
11. (d)	**12.** (a)	**13.** (b)	**14.** (d)	**15.** (d)	**16.** (b)	**17.** (b)	**18.** (a)	**19.** (b)	**20.** (b)
21. (a)	**22.** (c)	**23.** (c)	**24.** (b)	**25.** (b)	**26.** (a)	**27.** (b)	**28.** (c)	**29.** (b)	**30.** (b)
31. (c)	**32.** (d)	**33.** (c)	**34.** (b)	**35.** (c)	**36.** (c)	**37.** (a)	**38.** (c)	**39.** (b)	**40.** (d)
41. (b)	**42.** (a)	**43.** (b)	**44.** (b)	**45.** (d)	**46.** (b)	**47.** (b)	**48.** (b)	**49.** (b)	**50.** (c)
51. (b)	**52.** (c)	**53.** (a)	**54.** (c)	**55.** (c)	**56.** (b)	**57.** (b)	**58.** (c)	**59.** (a)	**60.** (c)
61. (b)	**62.** (c)	**63.** (a)	**64.** (b)	**65.** (c)	**66.** (a)	**67.** (b)	**68.** (b)	**69.** (b)	**70.** (c)
71. (d)	**72.** (d)	**73.** (d)	**74.** (a)	**75.** (c)	**76.** (b)	**77.** (b)	**78.** (c)	**79.** (c)	**80.** (d)
81. (a)	**82.** (c)	**83.** (c)	**84.** (c)	**85.** (b)	**86.** (a)	**87.** (a)	**88.** (c)	**89.** (b)	**90.** (c)
91. (b)	**92.** (b)	**93.** (b)	**94.** (d)	**95.** (a)	**96.** (c)	**97.** (a)	**98.** (d)	**99.** (b)	**100.** (c)
101. (d)	**102.** (a)	**103.** (c)							

Objective Questions Type II [One or more than one correct answer(s)]

1. (a, b)	**2.** (c, d)	**3.** (a, c)	**4.** (a, d)	**5.** (a, d)
6. (b, c)	**7.** (b, c)	**8.** (a, d)	**9.** (b, c)	**10.** (a, b, c)
11. (a, b, c)	**12.** (a, b)	**13.** (a, c)	**14.** (a, d)	**15.** (a, b, c)
16. (a, b, c)	**17.** (c, d)	**18.** (a, b, c, d)	**19.** (a, d)	**20.** (a, b, c)

Linked-Comprehension Type

Passage 1 **1.** (d) **2.** (a) **3.** (c)

Passage 2 **1.** (b) **2.** (a) **3.** (c) **4.** (c) **5.** (d)

Passage 3 **1.** (b) **2.** (a) **3.** (c) **4.** (b) **5.** (c)

Passage 4 **1.** (c) **2.** (b) **3.** (c) **4.** (a) **5.** (d)

Passage 5 **1.** (b) **2.** (d) **3.** (a) **4.** (a) **5.** (a)

Passage 6 **1.** (c) **2.** (c) **3.** (d) **4.** (b) **5.** (d)

Passage 7 **1.** (c) **2.** (c) **3.** (b) **4.** (c) **5.** (b)

Passage 8 **1.** (c) **2.** (d) **3.** (b) **4.** (a) **5.** (d)

Numerical Value Type Questions

1. (1485)	**2.** (3954)	**3.** (118)	**4.** (257)
5. (21)	**6.** (6342)	**7.** (4806)	**8.** (990)
9. (1458)	**10.** (1998)	**11.** (3.75)	**12.** (4096)
13. (21)	**14.** (15)	**15.** (66.66)	**16.** (1210)

Single Integer Type Questions

1. (8)	**2.** (8)	**3.** (9)	**4.** (1)
5. (5)	**6.** (8)	**7.** (1 or 2)	

Matrix-Match Type

1. A $\rightarrow$ (Q); B $\rightarrow$ (P); C $\rightarrow$ (S); D $\rightarrow$ (T)

2. A $\rightarrow$ (P); B $\rightarrow$ (Q, T); C $\rightarrow$ (R, S)

3. A $\rightarrow$ (Q, R); B $\rightarrow$ (P, R, T); C $\rightarrow$ (S)

4. A $\rightarrow$ (P, Q, R); B $\rightarrow$ (P, Q, R, S); C $\rightarrow$ (P, Q, R, S, T)

5. A $\rightarrow$ (T); B $\rightarrow$ (P, Q, R, S, T); C $\rightarrow$ (Q)

Solutions

Objective Questions Type [Only one correct answer]

1. $\because \left(x+1+\frac{1}{x}\right)^n = \frac{(1+x+x^2)^n}{x^n}$

$\because (1+x+x^2)^n$ is of the form

$$a_0 + a_1 x + a_2 x^2 + \dots + a_{2n}x^{2n},$$

which contains $(2n+1)$ terms

Hence, $\left(x+1+\frac{1}{x}\right)^n$ contains $(2n+1)$ terms

$\therefore \quad 2n+1 = 301$

$\Rightarrow \quad n = 150$

Which is greater than 149.

2. $\because \quad x + \frac{1}{x} = 1 \Rightarrow x^2 - x + 1 = 0$

$\Rightarrow \quad (x+\omega)(x+\omega^2) = 0$

$\therefore \quad x = -\omega, -\omega^2$

$\therefore \quad \lambda = (-\omega)^{4000} + \frac{1}{(-\omega)^{4000}} = \omega^{4000} + \frac{1}{\omega^{4000}}$

$= \omega + \frac{1}{\omega} = \frac{\omega^2+1}{\omega} = \frac{-\omega}{\omega} = -1$

and also for $x = -\omega^2, \lambda = -1$

Now, for $\quad n > 1, 2^n = 4k, k \in N$

$\therefore \; 2^{2^n} = 2^{4k} = (16)^k =$ a number with last digit 6

$\therefore \quad \mu =$ the digit at unit place in the number

$$(2^{2^n}+1) = 6 + 1 = 7$$

Hence, $\quad \lambda + \mu = -1 + 7 = 6$

3. $\because \sum_{r=0}^{n}\left(\frac{r+2}{r+1}\right) {}^nC_r = \sum_{r=0}^{n}\left(1+\frac{1}{r+1}\right) {}^nC_r$

$= \sum_{r=0}^{n} {}^nC_r + \frac{1}{(n+1)}\sum_{r=0}^{n}\left(\frac{n+1}{r+1}\right) {}^nC_r$

$= \sum_{r=0}^{n} {}^nC_r + \frac{1}{(n+1)}\sum_{r=0}^{n} {}^{n+1}C_{r+1}$

$= 2^n + \frac{1}{(n+1)}(2^{n+1}-1)$

$= \frac{2^n(n+1) + 2^{n+1} - 1}{(n+1)}$

$= \frac{2^n(n+3)-1}{n+1} = \frac{2^8-1}{6}$ (given)

$\Rightarrow \quad \frac{2^n(n+3)-1}{n+1} = \frac{2^5(5+3)-1}{5+1}$

Hence, $\quad n = 5$

4. Let $(r+1)$th term be independent of x, then

$$T_{r+1} = {}^9C_r \left(ax^{\frac{1}{3}}\right)^{9-r} (bx^{-1/6})^r$$

$$= {}^9C_r a^{9-r} \cdot b^r \cdot x^{3-\frac{r}{3}-\frac{r}{6}}$$

As T_{r+1} is independent of x

$\therefore \quad 3 - \frac{r}{3} - \frac{r}{6} = 0$

$\Rightarrow \quad r = 6$

$\therefore \quad T_{6+1} = {}^9C_6 a^3 b^6 = {}^9C_3 a^3 b^6 = 84\, a^3 b^6$

$\because \quad \text{AM} \ge \text{GM}$

$\therefore \quad \frac{a^3+b^6}{2} \ge \sqrt{a^3b^6}$

$\Rightarrow \quad \frac{2}{2} \ge \sqrt{a^3b^6}$

$\Rightarrow \quad a^3b^6 \le 1$

or $\quad 84a^3b^6 \le 84$

or $\quad T_7 \le 84$

$\therefore$ Maximum value = 84

5. Let $\quad \Sigma n = \lambda \quad$ ...(i)

$\therefore \left(\frac{x^3+1+x^6}{x^3}\right)^{\Sigma n} = \left[1+\left(x^3+\frac{1}{x^3}\right)\right]^{\lambda}$

$= 1 + {}^{\lambda}C_1\left(x^3+\frac{1}{x^3}\right) + {}^{\lambda}C_2\left(x^3+\frac{1}{x^3}\right)^2 + \dots$

$\dots + {}^{\lambda}C_{\lambda}\left(x^3+\frac{1}{x^3}\right)^{\lambda}$

On expanding each term, two dissimilar terms are added in the expansion

$\because \quad \left(x^3+\frac{1}{x^3}\right)^3 = x^9 + \frac{1}{x^9} + 3\left(x^3+\frac{1}{x^3}\right)$

Only x^9 and $\frac{1}{x^9}$ are new terms. Coefficient of $x^3$and $\frac{1}{x^3}$ have occured earlier in ${}^{\lambda}C_1\left(x^3+\frac{1}{x^3}\right)$

Hence, number of terms

$= 1 + 2 + 2 + 2 + \dots$ upto λ

$= 1 + 2\lambda$

$= 1 + 2\Sigma n$

$= 1 + 2 \cdot \frac{n(n+1)}{2}$

$= 1 + n + n^2$

6. $\because \; (115)^{96} - (96)^{115} = (1+114)^{96} - (1+95)^{115}$

$= \{1 + {}^{96}C_1(114) + {}^{96}C_2(114)^2 + \dots\}$

$- \{1 + {}^{115}C_1(95) + {}^{115}C_2(95)^2 + \dots\}$

$= 114\{{}^{96}C_1 + {}^{96}C_2(114) + \dots\}$

$- 95\{{}^{115}C_1 + {}^{115}C_2(95) + \dots\}$

$= 19\{6({}^{96}C_1 + {}^{96}C_2(114) + \dots)$

$- 5({}^{115}C_1 + {}^{115}C_2(95) + \dots)\}$

Hence, given expression is divisible by 19.

7. $\because \quad 5^{40} = (5^2)^{20} = (22+3)^{20} = 22\lambda + 3^{20}, \lambda \in N$

Also, $3^{20} = (3^2)^{10} = (11-2)^{10} = 11\mu + 2^{10}, \mu \in N$

Now, $2^{10} = 1024 = 11 \times 93 + 1$

$\therefore$ Remainder $= 1$

i.e. $\alpha = 1$

Also, $2^{2003} = 2^3 \cdot 2^{2000} = 8(2^4)^{500} = 8\,(16)^{500}$

$= 8\,(17-1)^{500} = 8(17\nu + 1), \nu \in N$

$= 8 \times 17\nu + 8$

$\therefore$ Remainder $= 8$

i.e. $\beta = 8$

$\Rightarrow \quad \beta - \alpha = 8 - 1 = 7$

8. $\because \quad (1+2)^m = 6561 \Rightarrow 3^m = 3^8 \Rightarrow m = 8$

and $(2+1)^n = 243 \Rightarrow 3^n = 3^5 \Rightarrow n = 5$

The point (m, n) is $(8, 5)$.

Let $S = x^2 + y^2 - 4x - 6y - 32$

The value of S at $(8, 5)$, then

$$S_1 = 8^2 + 5^2 - 4 \times 8 - 6 \times 5 - 32$$

$$= -5 < 0$$

$\therefore$ Point (m, n) is inside the circle.

9. $\because \quad T_{r+1} = {}^{10}C_r(x \sin^{-1}\alpha)^{10-r}\left(\dfrac{\cos^{-1}\alpha}{x}\right)^r$

$$= {}^{10}C_r(\sin^{-1}\alpha)^{10-r}(\cos^{-1}\alpha)^r x^{10-2r}$$

Put $10 - 2r = 0 \Rightarrow r = 5$

$\therefore$ Term independent of x is

$$T_{5+1} = {}^{10}C_5(\sin^{-1}\alpha)^5(\cos^{-1}\alpha)^5$$

$$= {}^{10}C_5(\sin^{-1}\alpha \cos^{-1}\alpha)^5$$

Let $f(\alpha) = \sin^{-1}\alpha \cdot \cos^{-1}\alpha$

$$= \sin^{-1}\alpha\left(\frac{\pi}{2} - \sin^{-1}\alpha\right)$$

If $\sin^{-1}\alpha = t$

$$= t\left(\frac{\pi}{2} - t\right)$$

$$= -\left\{t^2 - \frac{\pi}{2}t\right\}$$

$$= -\left\{\left(t - \frac{\pi}{4}\right)^2 - \frac{\pi^2}{16}\right\} = \frac{\pi^2}{16} - \left(t - \frac{\pi}{4}\right)^2$$

$$\therefore \quad f(\alpha) = \frac{\pi^2}{16} - \left(\sin^{-1}\alpha - \frac{\pi}{4}\right)^2$$

Maximum value of $f(\alpha)$ is $\dfrac{\pi^2}{16}$, when $\sin^{-1}\alpha = \dfrac{\pi}{4}$

Also, $-1 \le \alpha \le 1$

$$\therefore \quad -\frac{\pi}{2} \le \sin^{-1}\alpha \le \frac{\pi}{2}$$

Minimum value $f(\alpha) = \dfrac{\pi^2}{16} - \left(-\dfrac{\pi}{2} - \dfrac{\pi}{4}\right)^2 = -\dfrac{\pi^2}{2}$

$\therefore$ Range $\left[{}^{10}C_5\left(-\dfrac{\pi^2}{2}\right)^5, {}^{10}C_5\left(\dfrac{\pi^2}{16}\right)^5\right]$

i.e. $\left[-\dfrac{{}^{10}C_5\pi^{10}}{2^5}, \dfrac{{}^{10}C_5\pi^{10}}{2^{20}}\right]$

10. Given binomial is $(\alpha x^{1/9} + \beta x^{-1/6})^{10}$

$\because$ The general term in the expansion of binomial is

$$T_{r+1} = {}^{10}C_r\,(\alpha\, x^{1/9})^{10-r}\,(\beta x^{-1/6})^r$$

$$= {}^{10}C_r\,\alpha^{10-r}\beta^r x^{\left(\frac{10-r}{9} - \frac{r}{6}\right)}$$

$$= {}^{10}C_r\,\alpha^{10-r}\beta^r x^{\left(\frac{20-2r-3r}{18}\right)}$$

For the term independent of x,

$$20 - 2r - 3r = 0 \quad \Rightarrow \quad r = 4$$

So, the value of the term independent of x is

$${}^{10}C_4\,\alpha^6\beta^4 = 210\alpha^6\beta^4$$

Now, it is given that $\alpha > 0, \beta > 0$, such that $\alpha^3 + \beta^2 = 4$, so according to AM-GM in equality, we have

$$\frac{\alpha^3 + \beta^2}{2} \ge \sqrt{\alpha^3\beta^2} \Rightarrow \frac{4}{2} \ge \sqrt{\alpha^3\beta^2} \Rightarrow \alpha^6\beta^4 \le 16$$

It is given that, the maximum value of

$$210\alpha^6\beta^4 = 10k$$

$$\Rightarrow \quad 210 \times 16 = 10k \Rightarrow k = 336$$

11. $\because \quad (k^2 - 3) = \dfrac{{}^{n-1}C_r}{{}^nC_{r+1}} = \dfrac{{}^{n-1}C_r}{\left(\dfrac{n}{r+1}\right){}^{n-1}C_r} = \left(\dfrac{r+1}{n}\right)$

$\because \quad 0 \le r \le n-1$

$\Rightarrow \quad 1 \le r+1 \le n$

$\Rightarrow \quad \dfrac{1}{n} \le \dfrac{r+1}{n} \le 1$

$\Rightarrow \quad \dfrac{1}{n} \le k^2 - 3 \le 1$

$\Rightarrow \quad 3 + \dfrac{1}{n} \le k^2 \le 4$

When $n \to \infty$, $3 \le k^2 \le 4$

or $\quad k \in [-2, -\sqrt{3}] \cup [\sqrt{3}, 2]$

12. To find $({}^{30}C_0)({}^{30}C_{10}) - ({}^{30}C_1)({}^{30}C_{11}) + ({}^{30}C_2)({}^{30}C_{12}) + \ldots + ({}^{30}C_{20})({}^{30}C_{30})$

[$\because$ difference of lower suffices $= 10$]

$\because (1+x)^{30} = {}^{30}C_0 + {}^{30}C_1 x + {}^{30}C_2 x^2 + \ldots + {}^{30}C_{20}x^{20} + \ldots + {}^{30}C_{30}x^{30}$ ...(i)

$\Rightarrow (x-1)^{30} = {}^{30}C_0 x^{30} - {}^{30}C_1 x^{29} + \ldots + {}^{30}C_{10}x^{20} - {}^{30}C_{11}x^{19} + {}^{30}C_{12}x^{18} + \ldots + {}^{30}C_{30}$ ...(ii)

Multiplying Eq. (i) and Eq. (ii) and comparing coefficient of x^{20}, then we get

$$({}^{30}C_0)({}^{30}C_{10}) - ({}^{30}C_1)({}^{30}C_{11}) + ({}^{30}C_2)({}^{30}C_{12}) + \ldots + ({}^{30}C_{20})({}^{30}C_{30})$$

$$= {}^{30}C_{10}$$

[$\because (1+x)^{30}(x-1)^{30} = (x^2-1)^{30} = (1-x^2)^{30}$]

$$= \binom{30}{10}$$

13. $\because (1+x^2)^{40}\cdot\left(x^2+2+\frac{1}{x^2}\right)^{-5} = (1+x^2)^{40}\,\frac{(1+x^2)^{-10}}{x^{-10}}$

$$= (1+x^2)^{30}\cdot x^{10}$$

$\therefore$ Coefficient of x^{20} in $(1+x^2)^{30}\cdot x^{10}$

$\Rightarrow$ Coefficient of x^{10} in $(1+x^2)^{30}$ is ${}^{30}C_5$ or ${}^{30}C_{25}$

14. $S = (\alpha\cdot 1+\beta\cdot 1-\gamma\cdot 1)^n = (\alpha+\beta-\gamma)^n$

$$\Rightarrow \quad S^{\frac{1}{n}} = \alpha+\beta-\gamma$$

$$\therefore \quad \lim_{n\to\infty}\frac{S}{\left\{S^{\frac{1}{n}}+1\right\}^n} = \lim_{n\to\infty}\left(\frac{\alpha+\beta-\gamma}{\alpha+\beta-\gamma+1}\right)^n$$

$$=0$$

$[\because \alpha+\beta-\gamma+1>\alpha+\beta-\gamma]$

15. $\because \; I+f+f' = (5+2\sqrt{6})^n + (5-2\sqrt{6})^n$

$= 2k$ [even integer]

$\therefore \quad f+f'=1$

Now, $(I+f)f' = (5+2\sqrt{6})^n(5-2\sqrt{6})^n$

$= (1)^n = 1$

$\Rightarrow \quad (I+f)(1-f)=1$

or $\quad I = \frac{1}{(1-f)} - f$

16. $\because (1+x)^n = \sum_{r=0}^{n} {}^nC_r x^r = \sum_{r=0}^{n} a_r x^r$ [given]

$\therefore \quad a_r = {}^nC_r$

Also, $b_r = 1+\frac{a_r}{a_{r-1}} = 1+\frac{{}^nC_r}{{}^nC_{r-1}} = \frac{{}^{n+1}C_r}{{}^nC_{r-1}}$

$$b_r = \left(\frac{n+1}{r}\right)$$

$$\because \quad \prod_{r=1}^{n} b_r = \prod_{r=1}^{n}\left(\frac{n+1}{r}\right) = \frac{(n+1)^n}{n!}$$

$= \frac{(101)^{100}}{100!}$ [given]

$\therefore \quad n = 100$

17. Let $a = {}^nC_{r-1},\ b = {}^nC_r,\ c = {}^nC_{r+1},$

and $d = {}^nC_{r+2}$

$\therefore \quad a+b = {}^{n+1}C_r,\ b+c = {}^{n+1}C_{r+1},\ c+d = {}^{n+1}C_{r+2}$

$$\Rightarrow \quad \frac{a+b}{a} = \frac{{}^{n+1}C_r}{{}^nC_{r-1}} = \frac{n+1}{r} \Rightarrow \frac{a}{a+b} = \frac{r}{n+1}$$

and $$\frac{b+c}{b} = \frac{{}^{n+1}C_{r+1}}{{}^nC_r} = \frac{n+1}{r+1} \Rightarrow \frac{b}{b+c} = \frac{r+1}{n+1}$$

and $$\frac{c+d}{c} = \frac{{}^{n+1}C_{r+2}}{{}^nC_{r+1}} = \frac{n+1}{r+2} \Rightarrow \frac{c}{c+d} = \frac{r+2}{n+1}$$

$\therefore \frac{a}{a+b}, \frac{b}{b+c}, \frac{c}{c+d}$ are in AP.

$\because \quad$ AM > GM

$$\Rightarrow \quad \frac{b}{b+c} > \sqrt{\frac{ac}{(a+b)(c+d)}}$$

or $$\left\{\left(\frac{b}{b+c}\right)^2 - \frac{ac}{(a+b)(c+d)}\right\} > 0$$

18. We have,

$(1-x)^3 = 1-3x+3x^2$ [neglecting higher term]

$(2+x)^6 = {}^6C_0 2^6 x^0 + {}^6C_1 2^5 x^1 + {}^6C_2 2^4 x^2$

[neglecting higher term]

$= 64+192x+240x^2$

Now,

$(1-x)^3(2+x)^6 = (1-3x+3x^2)(64+192x+240x^2)$

$= 64+192x+240x^2-192x-576x^2+192x^2$

(neglecting higher term)

$= -144x^2+64$

$\therefore \quad a = 64,\ b = 0$ and $c = -144$

$\therefore \; a+b+c = 64+0-144 = -80$

19. Coefficient of x^2y^2 in $(x+y+z+t)^4 = \frac{4!}{2!\,2!} = 6$

Coefficient of yzt^2 in $(x+y+z+t)^4 = \frac{4!}{1!\,1!\,1!} = 12$

and coefficient of $xyzt$ in

$(x+y+z+t)^4 = \frac{4!}{1!\,1!\,1!\,1!} = 24$

$\therefore$ Required ratio is 6 : 12 : 24

or $\quad$ 1 : 2 : 4

20. The coefficient of x^6 in the given expression

= coefficient of x^6 in

$(1+{}^6C_1x^6)(1+{}^5C_1x^5)(1+{}^4C_1x^4)$

$(1+{}^3C_1x^3+{}^3C_2x^6)(1+{}^2C_1x^2+{}^2C_2x^4)(1+x)$

= coefficient of x^6 in

$(1+6x^6+5x^5+4x^4)(1+2x^2+3x^3+$

$x^4+6x^5+3x^6)(1+x)$

= Coefficient of x^6 in $(11x^5+17x^6)(1+x) = 28$

21. ${}^nC_1 + {}^nC_2 = 36$

$$n+\frac{n(n-1)}{2} = 36$$

$$n^2+n-72 = 0$$

$n = -9,\ 8$

$\therefore \quad n = 8$

$$\frac{T_3}{T_2} = \frac{{}^8C_2(2^x)^6\left(\frac{1}{4^x}\right)^2}{{}^8C_1(2^x)^7\left(\frac{1}{4^x}\right)^1} = \frac{{}^8C_2}{{}^8C_1\cdot 2^x\cdot 4^x}$$

$$= \frac{8\times 7}{2}\times\frac{1}{8}\times\frac{1}{8^x} = \frac{7}{2}\times\frac{1}{8^x}$$

When $x = -\frac{1}{3} \Rightarrow \frac{T_3}{T_2} = \frac{7}{2}\times 8^{1/3} = 7$

When $x = \frac{1}{3}$

$$\frac{T_3}{T_2} = \frac{7}{2}\times 8^{-1/3} = \frac{7}{4}$$

22. Coefficient of x^r in $(1+x)^n$ is ${}^nC_r$.

In this type of questions, we find different composition of terms where product will give us x^{11}.

Now, consider the following cases for x^{11} in

$$(1+x^2)^4 (1+x^3)^7 (1+x^4)^{12}.$$

Coefficient of $x^0\ x^3\ x^8$; Coefficient of $x^2\ x^9\ x^0$

Coefficient of $x^4\ x^3\ x^4$; Coefficient of $x^8\ x^3\ x^0$

$$= {}^4C_0 \times {}^7C_1 \times {}^{12}C_2 + {}^4C_1 \times {}^7C_3 \times {}^{12}C_0 + {}^4C_2 \times {}^7C_1 \times {}^{12}C_1 + {}^4C_4 \times {}^7C_1 \times {}^{12}C_0$$

$$= 462 + 140 + 504 + 7 = 1113$$

23. ${}^nC_{r-1} = 28,\ {}^nC_r = 56,\ {}^nC_{r+1} = 70$

$\because$ $\dfrac{{}^nC_r}{{}^nC_{r-1}} = 2$

$\Rightarrow$ $\dfrac{n-r+1}{r} = 2$

$\Rightarrow$ $n - 3r + 1 = 0$ ...(i)

and $\dfrac{{}^nC_{r+1}}{{}^nC_r} = \dfrac{70}{56}$

$\Rightarrow$ $\dfrac{n-(r+1)+1}{r+1} = \dfrac{5}{4}$

$\Rightarrow$ $4n - 4r = 5r + 5$

$\Rightarrow$ $4n - 9r - 5 = 0$ ...(ii)

Solving Eqs. (i) and (ii), we get

$$n = 8,\ r = 3$$

24. Given binomial is $\left(\dfrac{3}{2}x^2 - \dfrac{1}{3x}\right)^9$, So the general term,

$$T_{r+1} = {}^9C_r\left(\frac{3}{2}x^2\right)^{9-r}\left(-\frac{1}{3x}\right)^r$$

$$= {}^9C_r\left(\frac{3}{2}\right)^{9-r}\left(-\frac{1}{3}\right)^r x^{18-3r}$$

If the term is independent of x, then

$$18 - 3r = 0 \Rightarrow r = 6$$

$\therefore (r+1)$th term $= 7$th term is independent of x.

Now, as $T_{6+1} = k$ [given]

$\Rightarrow$ ${}^9C_6\left(\dfrac{3}{2}\right)^3\left(-\dfrac{1}{3}\right)^6 = k$

$\Rightarrow$ $\dfrac{9\times 8\times 7}{3\times 2} \times \dfrac{1}{27\times 8} = k$

$\Rightarrow$ $k = \dfrac{7}{18}$

$\Rightarrow$ $18k = 7$

Hence, option (b) is correct.

25. Given expression, $\left(px^{\frac{1}{12}} + qx^{-\frac{1}{9}}\right)^{14}$

General term $T_{r+1} = 14C_r\left(px^{\frac{1}{12}}\right)^{14-r}\left(qx^{-\frac{1}{9}}\right)^r$

$$= {}^{14}C_r p^{14-r} q^r x^{\frac{14-r}{12} - \frac{r}{9}}$$

Term is independent of r, then $\dfrac{14-r}{12} - \dfrac{r}{9} = 0$

$\therefore$ $r = 6$

$\therefore$ Term independent of x is ${}^{14}C_5 p^8 q^6 = {}^{14}C_6 (p^4 q^3)^2$

Now, p^4, q^3 are positive.

Using AM $\geq$ GM

$$\frac{p^4+q^3}{2} \geq (p^4q^3)^{1/2} \Rightarrow (p^4q^3)^2 \leq 1$$

$\Rightarrow$ Maximum value of term independent of x is ${}^{14}C_6$.

26. Any element, is of the form of

$$\frac{8!}{t_1!t_2!t_3!t_4!}(x^2)^{t_1}\left(\frac{1}{x^2}\right)^{t_2}(y)^{t_3}\left(\frac{1}{y}\right)^{t_4}$$

where $t_1 + t_2 + t_3 + t_4 = 8,\ t_1 \geq 0$

The constant term occur when $t_1 = t_2$ and $t_3 = t_4$

So, $t_1 + t_3 = 4 \Rightarrow (0, 4), (4, 0); (1, 3); (3, 1); (2, 2)$

$\Rightarrow$ Constant term

$$\frac{8!}{0!0!4!4!} + \frac{8!}{4!4!0!0!} + \frac{8!}{1!1!3!3!} + \frac{8!}{3!3!1!1!} + \frac{8!}{2!2!2!2!}$$

$$= (2\times 70) + (2\times 1120) + 2520 = 4900$$

27. Given $n = 10$, then $\sum\limits_{r=1}^{10} {}^nC_r\, r(r-4)$

$$= \sum_{r=1}^{10} {}^nC_r\, r((r-1)-3)$$

$$= \sum_{r=1}^{10} (r(r-1)\,{}^nC_r - 3r\,{}^nC_r)$$

$$= \sum_{r=1}^{10} \left(r(r-1)\frac{n(n-1)}{r(r-1)}\,{}^{n-2}C_{r-2} - 3\cdot r\frac{n}{r}\cdot {}^{n-1}C_{r-1}\right)$$

$$= \sum_{r=1}^{10} (n(n-1)\,{}^{n-2}C_{r-2} - 3n\cdot {}^{n-1}C_{r-1})$$

$$= n(n-1)\sum_{r=1}^{10} {}^{n-2}C_{r-2} - 3n\sum_{r=1}^{10} {}^{n-1}C_{r-1}$$

Given, $n = 10$ and $\sum\limits_{r=1}^{10} {}^nC_r = 2^n$

$$\therefore 10(10-1)\cdot 2^{10-2} - 3\times 10\cdot 2^{10-1}$$

$$= 10\times 9\times 2^8 - 30\times 2^9$$

$$= (90-60)\cdot 2^8$$

$$= 30\times 256 = 7680$$

28. $T_{r+1} = {}^8C_r[(t^{-1}-1)x]^{8-r}[(t^{-1}+1)^{-1}x^{-1}]^r$

$$= {}^8C_r(t^{-1}-1)^{8-r}(t^{-1}+1)^{-r}x^{8-2r}$$

For independent of x, $8 - 2r = 0$

$\therefore$ $r = 4$

$\therefore$ $T_{4+1} = {}^8C_4(t^{-1}-1)^4(t^{-1}+1)^{-4}$

$$= 70\left(\frac{1-t}{1+t}\right)^4$$

29. We have,

$$C_0x + C_1x^2 + \ldots + C_nx^{n+1} = x(1+x)^n \quad \text{... (i)}$$

and $$C_0 + C_1\frac{1}{x} + \ldots + C_n\frac{1}{x^n} = \left(1+\frac{1}{x}\right)^n \quad \text{... (ii)}$$

Now, $\sum\limits_{k=1}^{n}(C_k)(C_{k-1}) =$ constant term in

$$\left[C_0 + C_1\frac{1}{x} + \ldots + C_n\frac{1}{x^n}\right]\times[C_0x + C_1x^2 + \ldots + C_nx^{n+1}]$$

$= \text{constant term in } \left(1+\frac{1}{x}\right)^n (1+x)^n x$

$= \text{coefficient of } x^{n-1} \text{ in } (1+x)^{2n}$

$= {}^{2n}C_{n-1}$

$\Rightarrow \quad {}^{2n}C_{n-1} + {}^{2n}C_n - {}^{2n}C_n$

$= \frac{1}{2}[2({}^{2n+1}C_n)] - {}^{2n}C_n$

$= \frac{1}{2}[{}^{2n+1}C_n + {}^{2n+1}C_n] - {}^{2n}C_n$

$= \frac{1}{2}[{}^{2n+1}C_n + {}^{2n+1}C_{n+1}] - {}^{2n}C_n$

$= \frac{1}{2}[{}^{2n+2}C_{n+1}] - {}^{2n}C_n$

30. $\left(1+\frac{1}{1}\right)^n = 64$

$\Rightarrow \quad 2^n = 2^6$

$\therefore \quad n = 6$

General term $T_{r+1} = {}^nC_r(x)^{n-r}\left(\frac{1}{x}\right)^r$

$= {}^6C_r x^{6-2r}$

For independent of x,

Let $6 - 2r = 0$

$\therefore \quad r = 3$

$\therefore \quad T_{3+1} = {}^6C_3 x^0$

$= {}^6C_3 = 20$

31. Coefficient of x^{11} in the expansion of

$(1+x)^5(3+x)^4(7+x)^3$

$= ({}^5C_0 + {}^5C_1 x + \ldots + {}^5C_4 \cdot x^4 + {}^5C_5 \cdot x^5)$

$({}^4C_0 \cdot 3^4 + {}^4C_1 \cdot 3^3 x + \ldots + {}^4C_3 \cdot 3x^3 + {}^4C_4 \cdot x^4)$

$({}^3C_0 \cdot 7^3 + {}^3C_1 \cdot 7^2 x + {}^3C_2 \cdot 7x^2 + {}^3C_3 + x^3)$

$= {}^5C_4 \cdot {}^4C_4 \cdot {}^3C_3 + ({}^4C_3 . 3){}^5C_5 \cdot {}^3C_3 + ({}^3C_2 \cdot 7){}^5C_5 \cdot {}^4C_4$

$= 5 \cdot 1 \cdot 1 + 4 \cdot 3 \cdot 1 \cdot 1 + 3 \cdot 7 \cdot 1 \cdot 1 = 38$

32. $aC_0 + (a+d)C_1 + (a+2d)C_2 + \ldots$

$= a(C_0 + C_1 + C_2 + \ldots) + d(C_1 + 2C_2 + \ldots) \quad \ldots(i)$

$\because \quad (1+x)^n = C_0 + C_1 x + C_2 x^2 + C_3 x^3 + \ldots + C_n x^n$

$\therefore n(1+x)^{n-1} = 0 + C_1 + 2C_2 x + 3C_3 x^2 + \ldots + nC_n x^{n-1}$

Putting $x = 1$

$\therefore \quad C_1 + 2C_2 + 3C_3 + \ldots = n \cdot 2^{n-1}$

From Eq. (i) $aC_0 + (a+d)C_1 + (a+2d)C_2 + \ldots$

$= a \cdot 2^n + d \cdot n\, 2^{n-1} = 2^{n-1}(2a + nd)$

33. General term in $(\sqrt{2} + \sqrt[3]{3} + \sqrt[6]{5})^{10}$ is

$\frac{10!}{\alpha!\beta!\gamma!}(\sqrt{2})^\alpha(\sqrt[3]{3})^\beta(\sqrt[6]{5})^\gamma$

or $\frac{10!}{\alpha!\beta!\gamma!}(2)^{\alpha/2}(3)^{\beta/3}(5)^{\gamma/6}$

where $\alpha + \beta + \gamma = 10$

For rational terms $\alpha = 0, 2, 4, 6, 8, 10$

$\beta = 0, 3, 6, 9$ and $\gamma = 0, 6$

$\because \alpha$ and γ are even numbers, then β should also be an even number $[\because \alpha + \beta + \gamma = 10]$

$\therefore \quad \beta = 0, 6$

For $\beta = 0, \alpha + \gamma = 10$, then $\alpha = 4, \gamma = 6$ and $\alpha = 10, \gamma = 0$

For $\beta = 6, \alpha + \gamma = 4$, then $\alpha = 4, \gamma = 0$

$\therefore$ Number of rational terms = 3.

34. Let the three consecutive terms in the binomial expansion of $(1+x)^{n+5}$ are

${}^{n+5}C_{r-1}, {}^{n+5}C_r$ and ${}^{n+5}C_{r+1}$

Now, according to the given information

${}^{n+5}C_{r-1} : {}^{n+5}C_r : {}^{n+5}C_{r+1} = 5 : 10 : 14$

$\Rightarrow \frac{(n+5)!}{(r-1)!(n-r+6)!} : \frac{(n+5)!}{r!(n-r+5)!} : \frac{(n+5)!}{(r+1)!(n-r+4)!} = 5 : 10 : 14$

$\Rightarrow \frac{1}{(n-r+6)(n-r+5)} : \frac{1}{r(n-r+5)} : \frac{1}{(r+1)r} = 5 : 10 : 14$

So, $\quad \frac{r}{n-r+6} = \frac{5}{10}$

$\Rightarrow \quad 2r = n - r + 6$

$\Rightarrow \quad n + 6 = 3r \quad \ldots(i)$

and $\quad \frac{r+1}{n-r+5} = \frac{5}{7}$

$\Rightarrow \quad 7r + 7 = 5n - 5r + 25$

$\Rightarrow \quad 5n + 18 = 12r \quad \ldots(ii)$

From Eqs. (i) and (ii), we have $n = 6$.

So, the largest coefficient in the expansion is same as the greatest binomial coefficient

$= {}^{11}C_5$ or ${}^{11}C_6$

$= \frac{11!}{5!6!} = \frac{11 \times 10 \times 9 \times 8 \times 7}{5 \times 4 \times 3 \times 2} = 462$

35. We have,

$(1+x)^{10} = C_0 + C_1 x + C_2 x^2 + \ldots + C_{10} x^{10}$

(where $C_r = {}^{10}C_r$)

$(1+x)^{20} = C_0 + C_1 x + C_2 x^2 + \ldots + C_{20} x^{20}$

(where, $C_r = {}^{20}C_r$)

Therefore, $\sum_{r=0}^{m}({}^{10}C_r \times {}^{20}C_{m-r})$ = coefficient of x^m in the expansion of $(1+x)^{10}(1+x)^{20}$ = coefficient of x^m in $(1+x)^{30} = {}^{30}C_m$

We know that ${}^nC_r$ is maximum, if $r = n/2$ when n is even. Therefore, the given sum is maximum, if

$m = \frac{30}{2} = 15$

36. $\sum_{r=0}^{n}\sum_{s=0}^{n}(C_r + C_s) = \sum_{r=0}^{n}(C_r + C_r) + 2\sum_{0 \le r < s \le n}\sum(C_r + C_s)$

$\Rightarrow (n+1)2^{n+1} = 2\left(\sum_{r=0}^{n} C_r\right) + 2\sum_{0 \le r < s \le n}\sum(C_r + C_s)$

$\Rightarrow (n+1)2^{n+1} = 2 \cdot 2^n + 2\sum_{0 \le r < s \le n}\sum(C_r + C_s)$

$\Rightarrow \quad \sum_{0 \le r < s \le n}\sum(C_r + C_s) = (n-1)2^n$

37. We have,

$$^{4n}C_0 + {^{4n}C_2}x^2 + {^{4n}C_4}x^4 + \ldots + {^{4n}C_{4n}}x^{4n}$$
$$= \frac{1}{2}[(1+x)^{4n} + (1-x)^{4n}]$$

On putting $x = 1$ and $x = i$, we get

$$^{4n}C_0 + {^{4n}C_2} + {^{4n}C_4} + \ldots + {^{4n}C_{4n}} = \frac{1}{2}[2^{4n}]$$

and $^{4n}C_0 - {^{4n}C_2} + {^{4n}C_4} - \ldots + {^{4n}C_{4n}}$

$$= \frac{1}{2}[(1+i)^{4n} + (1-i)^{4n}]$$

Thus, $2[^{4n}C_0 + {^{4n}C_4} + \ldots + {^{4n}C_{4n}}]$

$$= 2^{4n-1} + \frac{1}{2}[(1+i)^{4n} + (1-i)]^{4n}$$

Now, $(1+i)^{4n} + (1+i)^{4n}$

$$= \left[\sqrt{2}\left(\cos\frac{\pi}{4} + i\sin\frac{\pi}{4}\right)\right]^{4n} + \left[\sqrt{2}\left(\cos\frac{\pi}{4} - i\sin\frac{\pi}{4}\right)\right]^{4n}$$
$$= 2^{2n}(\cos n\pi + i\sin n\pi) + 2^{2n}(\cos n\pi - i\sin n\pi)$$
$$= 2^{2n+1}\cos n\pi = 2^{2n+1}(-1)^n$$
$$\therefore 2[^{4n}C_0 + {^{4n}C_4} + \ldots + {^{4n}C_{4n}}] = 2^{4n-1} + \frac{1}{2}2^{2n+1}(-1)^n$$
$$\Rightarrow \quad ^{4n}C_0 + {^{4n}C_4} + \ldots + {^{4n}C_{4n}} = 2^{4n-2} + (-1)^n 2^{2n-1}$$

38. $$\frac{1}{n!}\left\{\frac{n!}{1!(n-1)!} + \frac{n!}{3!(n-3)!} + \frac{n!}{5!(n-5)!} + \ldots + \frac{n!}{(n-1)!1!}\right\}$$
$$= \frac{1}{n!}\{^nC_1 + {^nC_3} + {^nC_5} + \ldots + {^nC_{n-1}}\} = \frac{2^{n-1}}{n!}$$

39. Given, $\{(1-x)(1+x+x^2)\}^n$

$$= \sum_{k=0}^{n} a_k x^k \frac{(1-x)^{3n}}{(1-x)^{2k}}$$
$$\left\{\frac{1+x+x^2}{(1-x)^2}\right\}^n = \sum_{k=0}^{n} a_k \frac{x^k}{(1-x)^{2k}}$$
$$\left\{\frac{(1-x)^2 + 3x}{(1-x)^2}\right\}^n = \sum_{k=0}^{n} a_k \left\{\frac{x}{(1-x)^2}\right\}^k$$
$$\left\{1 + \frac{3x}{(1-x)^2}\right\}^n = \sum_{k=0}^{n} a_k \left\{\frac{x}{(1-x)^2}\right\}^k$$

Put $\dfrac{x}{(1-x)^2} = t$

$$(1+3t)^n = \sum_{k=0}^{n} a_k t^k$$

$3.\, a_{k-1} + a_k = 3$ (coefficient of t^{k-1}) + coefficient of t^k

$$= 3 \cdot 3^{k-1}\, {^nC_{k-1}} + 3^k\, {^nC_k}$$
$$= 3^k(^nC_{k-1} + {^nC_k})$$
$$= 3^k(^{n+1}C_k)$$

40. $\because$ The digit at unit place in each of $5!, 6!, \ldots, 100!$ is 0

and $\quad 0! + 1! + 2! + 3! + 4! = 34$

$\therefore$ Digit at unit place in $\sum_{r=0}^{100} r!$ is 4

Now, $\quad 2^{2^n} = 2^{4k}$

($k \in N$, $= 2^n$ is multiple of 4, $n > 1$)

$\therefore$ The digit at unit place in $2^{2^n} = 2^{4k}$ is 6

$\therefore$ The digit at unit place in $\sum_{r=0}^{100} r! + 2^{2^n} = 0$

41. $\because C_0 + (C_0 + C_1) + (C_0 + C_1 + C_2) + \ldots + (C_0 + C_1 + C_2 + \ldots + C_{n-1})$

$$= nC_0 + (n-1)C_1 + (n-2)C_2 + \ldots \qquad \ldots(i)$$

$\because (x+1)^n = C_0 x^n + C_1 x^{n-2} + C_2 x^{n-2} + \ldots + C_{n-1}x + C_n$

Differentiating both sides w.r.t. x and then put $x = 1$, we get

$nC_0 + (n-1)C_1 + (n-2)C_2 + \ldots = n \cdot 2^{n-1}$

42.
$$^7C_3 x^4 x^{(3\log_2 x)} = 4480$$
$$\Rightarrow \quad x^{(4+3\log_2 x)} = 2^7$$
$$\Rightarrow \quad (4+3t)t = 7 \text{ and } t = \log_2 x$$
$$\Rightarrow \quad t = 1, \frac{-7}{3}$$
$$\therefore \quad x = 2$$

43. Given expression is $(x + \sqrt{x^2-1})^6 + (x - \sqrt{x^2-1})^6$

$$= 2[^6C_0 x^6 + {^6C_2}\, x^4(x^2-1) + {^6C_4}x^2(x^2-1)^2 + {^6C_6}(x^2-1)^3]$$

The coefficients of x^4 in above expansion

$$\alpha = 2[^6C_2(-1) + {^6C_4}\,{^2C_1}(-1) + {^6C_6}\,{^3C_1}(-1)]$$
$$= 2[-15 - (15 \times 2) - (1 \times 3)]$$
$$= -96$$

and the coefficient of x^2 in the expansion

$$\beta = 2[^6C_4\,{^4C_0} + {^6C_6}\,{^3C_2}]$$
$$= 2[15 + 3] = 36$$
$$\therefore \alpha - \beta = -96 - 36 = -132$$

Hence, option (b) is correct.

44. $\because (1+x)^{10} = {^{10}C_0} + {^{10}C_1}x + {^{10}C_2}x^2 + \ldots + {^{10}C_9}x^9 + {^{10}C_{10}}x^{10} \quad \ldots(i)$

and $(x-1)^{10} = {^{10}C_0}x^{10} - {^{10}C_1}x^9 + {^{10}C_2}x^8 - \ldots - {^{10}C_9}x + {^{10}C_{10}} \quad \ldots(ii)$

Multiplying Eq. (i) and Eq. (ii), then

$$(x^2-1)^{10} = (^{10}C_0 + {^{10}C_1}x + \ldots + {^{10}C_9}x^9 + {^{10}C_{10}}x^{10}) \times (^{10}C_0 x^{10} - {^{10}C_1}x^9 + \ldots - {^{10}C_9}x + {^{10}C_{10}})$$

Comparing the coefficient of x^{10}, then

$$^{10}C_5(-1)^5 = (^{10}C_0)^2 - (^{10}C_1)^2 + \ldots - (^{10}C_9)^2 + (^{10}C_{10})^2$$
$$\therefore \quad (^{10}C_0)^2 - (^{10}C_1)^2 + \ldots - (^{10}C_9)^2 + (^{10}C_{10})^2 = -{^{10}C_5}$$

45. $T_{r+1} = {}^{10}C_r(tx^{1/5})^{10-r}\left[\frac{(1-x)^{1/10}}{t}\right]^r$

$= {}^{10}C_r t^{(10-2r)} \times x^{\frac{10-r}{5}} \times (1-x)^{\frac{r}{10}}$

$\Rightarrow 10 - 2r = 0 \Rightarrow r = 5$

$\therefore \quad T_6 = {}^{10}C_5 x\sqrt{1-x}$

$\frac{dT_6}{dx} = {}^{10}C_5\left[\sqrt{1-x} - \frac{x}{2\sqrt{1-x}}\right] = 0 \Rightarrow 1 - x = \frac{x}{2}$

$\Rightarrow \quad 3x = 2 \Rightarrow x = \frac{2}{3}$

$T_6|_{\max} = \frac{10!}{5!5!} \times \frac{2}{3\sqrt{3}}$

46. $2(1+x^3)^{100} = \sum_{k=0}^{100}\left(a_k x^k - \cos\frac{\pi}{2}(x+k)\right)$

$= \sum_{k=0}^{100} a_k \cdot x^k - \sum_{k=0}^{100} \cos\frac{\pi}{2}(x+k)$

$= \sum_{k=0}^{100} a_k \cdot x^k - \left[\cos\frac{\pi}{2}x + \cos\left(\frac{\pi}{2}x + \frac{\pi}{2}\right) + \cos\left(\frac{\pi}{2}x + \pi\right) + \cos\left(\frac{\pi}{2}x + \frac{3\pi}{2}\right) + \ldots\right]$

$2(1+x^3)^{100} = \sum_{k=0}^{100} a_k \cdot x^k - (0)$

put $x = -1 \quad 0 = a_0 - a_1 + a_2 - a_3 + \ldots.$

put $x = 1 \quad 2^{101} = a_0 + a_1 + a_2 + a_3$

add $2^{101} = 2[a_0 + a_2 + a_4 + \ldots.]$

$a_0 + a_2 + a_4 + \ldots + a_{100} = 2^{100}$

47. $\sum_{m=1}^{n} {}^nC_m\left(\sum_{k=1}^{m}\left(\sum_{p=k}^{m}\frac{m!}{p!(m-p)!}\cdot\frac{p!}{k!(p-k)!}\right)\right)$

$= \sum_{m=1}^{n} {}^nC_m\left(\sum_{k=1}^{m}\left(\sum_{p=k}^{m} {}^{m-k}C_{p-k}\right)\frac{m!}{k!(m-k)!}\right)$

$= \sum_{m=1}^{n} {}^nC_m\left(\sum_{k=1}^{m} 2^{m-k} \cdot {}^mC_k\right)$

$= \sum_{m=1}^{n} {}^nC_m((1+2)^m - 2^m)$

$= \sum_{m=1}^{n} ({}^nC_m 3^m - {}^nC_m 2^m)$

$= (1+3)^n - 1 - (1+2)^n + 1 = 4^n - 3^n$

48. General term in $(1 + \sqrt{2} + {}^3\sqrt{3})^6$ is

$\frac{6!}{\alpha!\beta!\gamma!}(1)^\alpha(\sqrt{2})^\beta(\sqrt[3]{3})^\gamma$

or $\quad \frac{6!}{\alpha!\beta!\gamma!}(2)^{\beta/2}(3)^{\gamma/3}$

where $\alpha + \beta + \gamma = 6$

For rational terms $\beta = 0, 2, 4, 6$

and $\gamma = 0, 3, 6$

For $\gamma = 6, \alpha = 0, \beta = 0$

For $\gamma = 3, \alpha = 3, \beta = 0$

For $\gamma = 3, \alpha = 1, \beta = 2$

For $\gamma = 0, \alpha = 0, \beta = 6$

For $\gamma = 0, \alpha = 2, \beta = 4$

For $\gamma = 0, \alpha = 4, \beta = 2$

For $\gamma = 0, \alpha = 6, \beta = 0$

$\therefore$ Number of rational terms = 7

49. Coefficient of a^3b^4c in $(1 + a + b - c)^9$

$= \frac{9!}{3!4!1!}(-1)^1$

$= -\frac{9!}{2!7!} \times \frac{7!2!}{3!4!}$

$= -{}^9C_2 \times {}^7C_3 \times 2$

$= -2 \cdot {}^9C_2 \cdot {}^7C_3$

50. $\because T_{r+1} = {}^{10}C_r(x\sin\alpha)^{10-r}(x^{-1}\cos\alpha)^r$

$= {}^{10}C_r x^{10-2r}(\sin\alpha)^{10-r}(\cos\alpha)^r$

For independent of x

$10 - 2r = 0$

$\therefore \quad r = 5$

$\therefore \quad T_{5+1} = {}^{10}C_5 x^0(\sin\alpha)^5(\cos\alpha)^5$

$= {}^{10}C_5 \frac{1}{2^5}(\sin 2\alpha)^5$

For greatest value of independent term

$\sin 2\alpha = 1$

$\therefore$ Greatest value $= \frac{1}{2^5} \cdot {}^{10}C_5$

51. Putting $x = 1$ and -1,

Then, $\quad 4^{20} = a_0 + a_1 + a_2 + \ldots + a_{40}$

and $\quad 2^{20} = a_0 - a_1 + a_2 - \ldots + a_{40}$

$\therefore \quad 4^{20} + 2^{20} = 2(a_0 + a_2 + a_4 + \ldots. + a_{40})$

$\therefore \quad a_0 + a_2 + a_4 + \ldots. + a_{38} + a_{40} = \frac{4^{20} + 2^{20}}{2}$

$\Rightarrow \quad a_0 + a_2 + a_4 + \ldots. + a_{38} = \frac{4^{20} + 2^{20}}{2} - a_{40}$

$T_{R+1} =$ in $(1 + x + 2x^2)^{20}$

$= {}^{20}C_R(x + 2x^2)^{20}$

$a_{40} =$ coefficient of x^{40}

$= 2^{20}$

Then, $a_0 + a_2 + a_4 + \ldots. + a_{38} = \frac{4^{20} + 2^{20}}{2} - 2^{20}$

$= \frac{2^{40} - 2^{20}}{2} = 2^{19}(2^{20} - 1)$

52. $\quad 32 = 2^5$

$(32)^{32} = (2^5)^{32} = 2^{160}$

$= (3-1)^{160}$

$= 3m + 1, m \in I_+$

Now, $\quad (32)^{32^{32}} = 32^{3m+1} = 2^{5(3m+1)} = 2^{15m+5}$

$\therefore \quad 32^{32^{32}} = 2^{3(5m+1)} \cdot 2^2$

$= 4 \cdot (8)^{5m+1}$

$= 4 \cdot (7+1)^{5m+1}$

$= 4 \cdot [7n + 1], n \in I_+$

$= 28n + 4$

$\therefore \quad$ Remainder = 4

53. $\sum_{r=0}^{n-1} \frac{{}^nC_r}{{}^nC_r + {}^nC_{r+1}} = \sum_{r=0}^{n-1} \frac{{}^nC_r}{{}^{n+1}C_{r+1}}$

$= \sum_{r=0}^{n-1} \frac{{}^nC_r}{\left(\frac{n+1}{r+1}\right){}^nC_r}$

$= \sum_{r=0}^{n-1} \left(\frac{r+1}{n+1}\right)$

$= \frac{1+2+3+\ldots+n}{(n+1)}$

$= \frac{n(n+1)}{2(n+1)} = \frac{n}{2}$

54. If $T_{r+1} \geq T_r$

$\Rightarrow \frac{T_{r+1}}{T_r} \geq 1$

$\Rightarrow \frac{100-r+1}{r} \cdot \frac{b/2}{b/2} \geq 1$

$\Rightarrow 101 - r \geq r$

$\Rightarrow 2r \leq 101$

$\therefore r \leq 50.5$

$\therefore r = 50$

$\therefore$ Greatest term $T_{r+1} = T_{50+1}$

$= {}^{100}C_{50}\left(\frac{b}{2}\right)^{50}\left(\frac{b}{2}\right)^{50} = {}^{100}C_{50}\left(\frac{b}{2}\right)^{100}$

55. $(x + {}^{2n+1}C_0)(x + {}^{2n+1}C_1)(x + {}^{2n+1}C_2)\ldots(x + {}^{2n+1}C_n)$

$= x^{n+1} + x^n({}^{2n+1}C_0 + {}^{2n+1}C_1 + {}^{2n+1}C_2 + \ldots + {}^{2n+1}C_n) + \ldots$

$\therefore$ Coefficient of x^n

(say) $P = {}^{2n+1}C_0 + {}^{2n+1}C_1 + {}^{2n+1}C_2 + \ldots + {}^{2n+1}C_n$

$\Rightarrow P = {}^{2n+1}C_{2n+1} + {}^{2n+1}C_{2n} + {}^{2n+1}C_{2n-1} + \ldots + {}^{2n+1}C_{n+1}$

$\therefore 2P = 2^{2n+1}$ (on adding) $[\because {}^nC_r = {}^nC_{n-r}]$

$\therefore P = 2^{2n}$

56. Since, fourth term of

$$\left(\sqrt{x^{\left(\frac{1}{1+\log_{10}x}\right)}} + \sqrt[12]{x}\right)^6 = 200$$

$\therefore {}^6C_3\left(\sqrt{x^{\left(\frac{1}{1+\log_{10}x}\right)}}\right)^3 (\sqrt[12]{x})^3 = 200$

$\Rightarrow \frac{6\cdot5\cdot4}{1\cdot2\cdot3} \cdot x^{\frac{3}{2}\left(\frac{1}{1+\log_{10}x}\right)} \cdot x^{1/4} = 200$

$\Rightarrow x^{\frac{3}{2}\left(\frac{1}{1+\log_{10}x}\right)+\frac{1}{4}} = 10$

$\Rightarrow \left[\frac{3}{2}\left(\frac{1}{1+\log_{10}x}\right) + \frac{1}{4}\right]\log_{10}x = 1$

Let $\log_{10}x = t$

$\therefore \left\{\frac{3}{2(1+t)} + \frac{1}{4}\right\}t = 1$

$\Rightarrow \left\{\frac{6+1+t}{4(1+t)}\right\}t = 1 \Rightarrow t^2 + 7t = 4t + 4$

$t^2 + 3t - 4 = 0$

$(t+4)(t-1) = 0$

$t = -4, 1$

$\log_{10}x = -4, 1$

$\therefore x = 10^{-4}, 10^1$

57. $(1+x)^n = C_0 + C_1x + C_2x^2 + \ldots + C_nx^n$

or $x(1+x)^n = C_0x + C_1x^2 + C_2x^3 + \ldots + C_nx^{n+1}$

Differentiating both sides w.r.t. x, then

$x \cdot n(1+x)^{n-1} + (1+x)^n \cdot 1 = C_0 + 2C_1x + 3C_2x^2 + \ldots + (n+1)C_nx^n$

$\Rightarrow (1+x)^{n-1}(nx + 1 + x) = C_0 + 2C_1x + 3C_2x^2 + \ldots + (n+1)C_nx^n$

Again, multiplying by x, then

$(1+x)^{n-1}(nx^2 + x + x^2) = C_0x + 2C_1x^2 + 3C_2x^3 + \ldots + (n+1)C_nx^{n+1}$

Differentiating both sides w.r.t. x, then

$\therefore (1+x)^{n-1} \cdot (2nx + 1 + 2x) + (nx^2 + x + x^2) \times (n-1)(1+x)^{n-2}$

$= C_0 + 2^2C_1x + 3^2C_2x^2 + \ldots + (n+1)^2C_nx^n$

Putting $x = 1$ in both sides, then

$2^{n-1}(2n+3) + (n+2)(n-1)2^{n-2}$

$= 1^2 \cdot C_0 + 2^2C_1 + 3^2C_2 + \ldots + (n+1)^2C_n$

$\therefore \sum_{k=0}^{n} (k+1)^2C_k = 2^{n-2}(n^2 + 5n + 4)$

58. $3^{2n} = 9^n = (1+8)^n$

$= {}^nC_0 + {}^nC_1 \cdot 8 + {}^nC_2 \cdot 8^2 + \ldots + {}^nC_n 8^n$

$\therefore \frac{3^{2n}}{8} = \frac{1}{8} + ({}^nC_1 + {}^nC_2 \cdot 8 + \ldots + {}^nC_n \cdot 8^{n-1})$

$= \frac{1}{8} +$ integer

$\Rightarrow \left\{\frac{3^{2n}}{8}\right\} = \frac{1}{8}.$

59. Sum of the last ten coefficients in the expansion of $(1+x)^{19}$

= (sum of all coefficients) − (sum of first ten coefficients)

$= 2^{19} - ({}^{19}C_0 + {}^{19}C_1 + {}^{19}C_2 + \ldots + {}^{19}C_9)$

$= 2^{19} - \frac{1}{2}({}^{19}C_0 + {}^{19}C_1 + {}^{19}C_2 + \ldots + {}^{19}C_9 + {}^{19}C_{10} + {}^{19}C_{11} + \ldots + {}^{19}C_{19})$

$= 2^{19} - \frac{1}{2} \cdot 2^{19}$

$= 2^{19}\left(1 - \frac{1}{2}\right) = 2^{18}$

60. Let $P = \sum_{r=0}^{n} \frac{r}{^nC_r}$ ...(i)

Replacing r by $n - r$ in Eq. (i), then

$$P = \sum_{r=0}^{n} \frac{(n-r)}{^nC_{n-r}}$$

$$= \sum_{r=0}^{n} \frac{(n-r)}{^nC_r} \qquad [\because {}^nC_r = {}^nC_{n-r}] \ldots \text{(ii)}$$

Adding Eqs. (i) and (ii), then

$$2P = n\sum_{r=0}^{n} \frac{1}{^nC_r} = na_n \qquad \left[\because a_n = \sum_{r=0}^{n} \frac{1}{^nC_r}\right]$$

$$\therefore \; P = \frac{1}{2}na_n$$

61. We can write E as

$$a[{}^nC_0 - {}^nC_1 + {}^nC_2 - \ldots + (-1)^n({}^nC_n)$$
$$+ [{}^nC_1 - (2)({}^nC_2) + (3)({}^nC_3) - \ldots - (-1)^n(n)({}^nC_n)]$$
$$= 0 + F \text{ where}$$
$$F = {}^nC_1 - (2)({}^nC_2) + (3)({}^nC_3) - \ldots - (-1)^n(n)({}^nC_n)$$

We know that,

$$(1+x)^n = {}^nC_0 + {}^nC_1x + {}^nC_2x^2 + {}^nC_3\,x^3 + \ldots + {}^nC_nx^n \ldots \text{(i)}$$

On differentiating Eq. (i), we get

$$n(1+x)^{n-1} = {}^nC_1 + 2({}^nC_2)x + 3({}^nC_3)x^2 + \ldots + n({}^nC_n)x^{n-1}$$

Putting $x = -1$, we get

$$0 = {}^nC_1 - 2({}^nC_2) + 3({}^nC_3) - \ldots - (-1)^{n-1}\,(n)({}^nC_n)$$

$$\Rightarrow \quad 0 = F$$

Thus, $E = 0 + 0 = 0$

62. $3^{400} = (3^4)^{100}$

$$= (81)^{100} = (1+80)^{100}$$
$$= 1 + {}^{100}C_1(80) + {}^{100}C_2(80)^2 + \ldots + {}^{100}C_{100}(80)^{100}$$
$$= 1 + 8000 + \text{(last two digits in each term is 00)}$$

$\therefore$ Last two digits $= 01$

63. $17^{1983} + 11^{1983} - 7^{1983}$

$$= (17^{1983} - 7^{1983}) + 11^{1983}$$
$$= (17-7)(17^{1982} + 17^{1981} \cdot 7^1 + 17^{1980} \cdot 7^2 + \ldots + 7^{1982}) + 11^{1983}$$

$\therefore$ Last digit in $17^{1983} - 7^{1983}$ is 0

and last digit in 11^{1983} is 1

$\therefore$ Last digit of $17^{1983} + 11^{1983} - 7^{1983}$ is $0 + 1 = 1$

64. $\because (1 + 2x + 3x^2)^{10} = a_0 + a_1x + a_2x^2 + \ldots + a_{20}x^{20}$

$$\Rightarrow [1 + x(2+3x)]^{10} = 1 + {}^{10}C_1x(2+3x)^1 + {}^{10}C_2x^2(2+3x)^2 + \ldots$$

Comparing the coefficient of x in both sides, then

$$a_1 = 20$$

65. $\because (1+x)^n(1+y)^n(1+z)^n$

$$\underbrace{[(1+x)(1+x)\ldots(1+x)]}_{n \text{ factors}}$$
$$\underbrace{[(1+y)(1+y)\ldots(1+y)]}_{n \text{ factors}}$$
$$\underbrace{[(1+z)(1+z)\ldots.(1+z)]}_{n \text{ factors}}$$

There are $3n$ factors in $(1+x)^n(1+y)^n(1+z)^n$ for term of degree r. We choose r brackets out of $3n$ brackets and, then multiply second terms in each bracket. There are ${}^{3n}C_r$ such terms each having coefficient 1. Hence, the sum of the coefficients $= {}^{3n}C_r$.

66. $T_2 = 14a^{5/2}$

$$\Rightarrow \quad {}^nC_1(a^{1/13})^{n-1}\left(\frac{a}{a^{-1/2}}\right)^1 = 14a^{5/2}$$
$$\Rightarrow \quad na^{(n-1)/13} \cdot a^{3/2} = 14a^{5/2}$$
$$\Rightarrow \quad na^{(n-1)/13} = 14a^1$$
$$\Rightarrow \quad na^{(n-14)/13} = 14$$

Here, $n = 14$ (satisfied)

$$\therefore \quad \frac{{}^nC_3}{{}^nC_2} = \frac{{}^{14}C_3}{{}^{14}C_2} = \frac{\frac{14 \cdot 13 \cdot 12}{1 \cdot 2 \cdot 3}}{\frac{14 \cdot 13}{1 \cdot 2}} = 4$$

67. $(1 + x + 2x^2)^{20} = a_0 + a_1x + \ldots. + a_{40}x^{40}$

Put $x = 1$ and -1

Then, $a_0 + a_1 + a_2 + \ldots. + a_{40} = 4^{20}$

$$\Rightarrow \quad a_0 - a_1 + a_2 + \ldots. + a_{40} = 2^{20}$$
$$\Rightarrow \quad a_1 + a_3 + \ldots.\, a_{39} = \frac{4^{20} - 2^{20}}{2}$$
$$\Rightarrow \quad a_1 + a_3 + \ldots. + a_{37} = 2^{39} - 2^{19} - a_{39}$$

Here, $a_{39} = \dfrac{20!(2)^{19} \times 1}{19!} = 20 \times 2^{19}$

$$\Rightarrow \quad a_1 + a_3 + \ldots. + q_{37} = 2^{19}(2^{20} - 1 - 20)$$
$$= 2^{19}(2^{20} - 21)$$

68. Coefficient of $\frac{1}{x}$ in $(1+x)^n\left(1 + \frac{1}{x}\right)^n$

$\Rightarrow$ coefficient of x^{n-1} in $(1+x)^{2n} = {}^{2n}C_{n-1}$

$$= \frac{2n!}{(n-1)!(n+1)!}$$

69. $\because$ Sum of all binomial coefficients is 2^n.

and $200 < 2^n < 400 \Rightarrow n = 8$

$$\because \text{ General term } T_{r+1} = {}^nC_r\left(\frac{i}{x^{\frac{3}{4}}}\right)^{n-r} a^r\left(x^{\frac{5}{4}}\right)^r$$

$$= {}^nC_ra^r\,x^{-\frac{3}{4}n + \frac{3}{4}r + \frac{5}{4}r}$$

For independent of 'x'

$$-\frac{3}{4}n + 2r = 0 \qquad [\because n = 8]$$
$$\Rightarrow \quad 2r = 6$$
$$\Rightarrow \quad r = 3$$

According to the question,

$$^nC_ra^r = 448$$
$$\Rightarrow \quad {}^8C_3a^3 = 448$$
$$\Rightarrow \quad 56a^3 = 448$$
$$\Rightarrow \quad a^3 = 8$$
$$\Rightarrow \quad a = 2$$

70. Given, $T_6 = 84$

$\therefore \quad T_{5+1} = 84$

$${}^7C_5\left(2^{\log_2\sqrt{(9^{x-1}+7)}}\right)^2\left(\frac{1}{2^{1/5\log_2(3^{x-1}+1)}}\right)^5 = 84$$

$$\Rightarrow \quad 21\left\{\sqrt{(9^{x-1}+7)}\right\}^2\left(\frac{1}{2^{\log_2(3^{x-1}+1)}}\right) = 84$$

$$\Rightarrow \quad (9^{x-1}+7)\left(\frac{1}{3^{x-1}+1}\right) = 4$$

Put $\quad 3^{x-1} = t$

$$\therefore \quad \frac{t^2+7}{t+1} = 4$$

$$\Rightarrow \quad t^2 - 4t + 3 = 0$$

$$(t-1)(t-3) = 0$$

$$\therefore \quad t = 1, 3$$

$$\Rightarrow \quad 3^{x-1} = 1, 3^{x-1} = 3$$

$\therefore \quad x - 1 = 0$ and $x - 1 = 1$, $\therefore x = 1, 2$

71. $$\sum_{r=1}^{n}\left(\sum_{p=0}^{r-1}{}^nC_r\cdot{}^rC_p\cdot 2^p\right) = \sum_{r=1}^{n}{}^nC_r\left(\sum_{P=0}^{r-1}{}^rC_p\cdot 2^p\right)$$

$$= \sum_{r=1}^{n}{}^nC_r\{(1+2)^r - 2^r\}$$

$$= \sum_{r=1}^{n}{}^nC_r3^r - \sum_{r=1}^{n}{}^nC_r2^r$$

$$= (4^n - 1) - (3^n - 1)$$

$$= 4^n - 3^n$$

72. $C_0 + C_1 + C_2 + \ldots. + C_n = 2^n$

$$(C_0 + C_1 + C_2 + \ldots + C_n)^2 = (C_0^2 + C_1^2 + C_2^2 + \ldots + C_n^2) + 2\sum_{0\le i<j\le n}\sum C_iC_j$$

$$(2^n)^2 = {}^{2n}C_n + 2P \text{ (say)}$$

$$\therefore \quad P = 2^{2n-1} - \frac{{}^{2n}C_n}{2}$$

73. $99^{50} - 99\cdot 98^{50} + \frac{99\cdot 98}{1\cdot 2}(97)^{50} - \ldots. + 99$

$= 99^{50} - {}^{99}C_1(98)^{50} + {}^{99}C_2(97)^{50} - \ldots. + {}^{99}C_{98}\cdot 1$

$= {}^{99}C_0(99)^{50} - {}^{99}C_1(99-1)^{50} + {}^{99}C_2(99-2)^{50} - \ldots + {}^{99}C_{98}(99-98)^{50} - {}^{99}C_{99}(99-99)^{50}$

$= (99)^{50}\{{}^{99}C_0 - {}^{99}C_1 + {}^{99}C_2 - \ldots + {}^{99}C_{98} - {}^{99}C_{99}\} + {}^{50}C_1(99)^{49}\{{}^{99}C_1 - 2\cdot{}^{99}C_2 + 3\cdot{}^{99}C_3 - \ldots\} + \ldots\ldots$

$= 0 + 0 + \ldots = 0$

74. $x = (\sqrt{3}+1)^n = [x] + f$

$\therefore \quad 0 \le f < 1$

Let $\quad f' = (\sqrt{3}-1)^n$

$\therefore \quad 0 < f' < 1$

$[x] + f - f' = (\sqrt{3}+1)^n - (\sqrt{3}-1)^n$

$= 2k, k \in I$

$\because \quad -1 < f - f' < 1$

$\therefore \quad f - f' = 0$ (Integral value)

Then, $\quad [x] + 0 = 2k, k \in I$

$\therefore \quad [x] = 2k, k \in I$

75. $(2^{1/5} + 3^{1/10})^{55}$

Total terms $= 55 + 1 = 56$

$$T_{r+1} = {}^{55}C_r 2^{(55-r)/5}3^{r/10}$$

Here, $\quad r = 0, 10, 20, 30, 40, 50$

Number of rational terms = 6

$\therefore$ Number of irrational terms $= 56 - 6 = 50$

76. Given expression

$(1+x)^{10} + x(1+x)^9 + x^2(1+x)^8 + \ldots + x^{10}$

The above expression is geometric series having first term, $a = (1+x)^{10}$

common ratio, $r = \frac{x}{1+x}$ and number of terms, $n = 11$

$\because$ Sum of geometric series $= \frac{a(1-r^n)}{1-r}$

$$= \frac{(1+x)^{10}\left(1-\left(\frac{x}{1+x}\right)^{11}\right)}{1-\frac{x}{1+x}}$$

$$= \frac{(1+x)^{10}\left(\frac{(1+x)^{11}-x^{11}}{(1+x)^{11}}\right)}{\frac{1+x-x}{1+x}} = (1+x)^{11} - x^{11}$$

The coefficient of x^7 in the given expression

= coefficient of x^7 in the expansion of $[(1+x)^{11} - x^{11}]$

= coefficient of x^7 in the expansion of $(1+x)^{11}$

$$= {}^{11}C_7 = \frac{11\times 10\times 9\times 8}{4\times 3\times 2} = 11\times 10\times 3 = 330$$

77. ${}^{n+1}C_2 + 2\{{}^2C_2 + {}^3C_2 + {}^4C_2 + \ldots + {}^nC_2\}$

$= {}^{n+1}C_2 + 2\{{}^3C_3 + {}^3C_2 + {}^4C_2 + \ldots + {}^nC_2\}$

$= {}^{n+1}C_2 + 2\{{}^4C_3 + {}^4C_2 + \ldots + {}^nC_2\}$ (by Pascal rule)

$= {}^{n+1}C_2 + 2\{{}^5C_3 + \ldots + {}^nC_2\}$ (by Pascal rule)

$= {}^{n+1}C_2 + 2\cdot{}^{n+1}C_3$

$= {}^{n+1}C_2 + {}^{n+1}C_3 + {}^{n+1}C_3 = {}^{n+2}C_3 + {}^{n+1}C_3$

$$= \frac{(n+2)(n+1)n}{6} + \frac{(n+1)n(n-1)}{6}$$

$$= \frac{(n+1)n}{6}(n+2+n-1) = \frac{n(n+1)(2n+1)}{6} = \Sigma n^2$$

78. We have, $\quad (x-1)^n = \sum_{r=0}^{n}(-1)^rC_rx^{n-r}$

$$\Rightarrow \quad x^2(x-1)^n = \sum_{r=0}^{n}(-1)^rC_rx^{n-r+2}$$

Differentiating both sides w.r.t. x three times, then we get

$6n(x-1)^{n-1} + 6n(n-1)x(x-1)^{n-2} + n(n-1)(n-2)x^2(x-1)^{n-3}$

$$= \sum_{r=0}^{n}(-1)^r(n-r)(n-r+1)(n-r+2)C_r\cdot x^{n-r}$$

Put $x = 1$, then we get

$$\sum_{r=0}^{n}(-1)^r(n-r)(n-r+1)(n-r+2)C_r = 0$$

79. $(x + {}^nC_0)(x + 3\,{}^nC_1)(x + 5\,{}^nC_2)\ldots\{x + (2n+1)\cdot {}^nC_n\}$

$\Rightarrow$

$x^{n+1} + x^n\{{}^nC_0 + 3\cdot {}^nC_1 + 5\cdot {}^nC_2 + \ldots + (2n+1)\cdot {}^nC_n\} + \ldots$

$\therefore$ Coefficient of x^n

$$= {}^nC_0 + 3\,{}^nC_1 + 5\,{}^nC_2 + \ldots + (2n+1)\,{}^nC_n$$

$$= ({}^nC_0 + {}^nC_1 + {}^nC_2 + \ldots + {}^nC_n) + 2({}^nC_1 + 2\,{}^nC_2 + \ldots + n\,{}^nC_n)$$

$$= 2^n + 2\left\{n + \frac{n(n-1)}{1} + \ldots + n\right\}$$

$$= 2^n + 2n\left\{1 + (n-1) + \frac{(n-1)(n-2)}{1\cdot 2} + \ldots + 1\right\}$$

$$= 2^n + 2n({}^{n-1}C_0 + {}^{n-1}C_1 + {}^{n-1}C_2 + \ldots + {}^{n-1}C_{n-1})$$

$$= 2^n + 2n(1+1)^{n-1}$$

$$= 2^n + n\cdot 2^n = (n+1)2^n$$

80. $(1+\lambda)^n(\mu+1)^n(\lambda+\mu)^n = \sum_{r=0}^{n} C_r\lambda^r$

$$\sum_{s=0}^{n} C_s\mu^{n-s} \sum_{t=0}^{n} C_t\lambda^{n-t}\mu^t$$

Since, C_0^3 is the coefficient of $\lambda^0\mu^{n-0}\lambda^{n-0}\mu^0$

i.e. $\lambda^n\mu^n$ $(r = s = t = 0)$

Now, C_1^3 is the coefficient of $\lambda^1\mu^{n-1}\lambda^{n-1}\mu$

i.e. $\lambda^n\mu^n (r = s = t = 1)$ etc .

and C_k^3 is the coefficient of $\lambda^k\mu^{n-k}\lambda^{n-k}\mu^k$

i.e. $\lambda^n\mu^n (r = s = t = k)$

Hence, the coefficient of $\lambda^n\mu^n$ in

$$(1+\lambda)^n(\mu+1)^n(\lambda+\mu)^n$$

$$= C_0^3 + C_1^3 + C_2^3 + \ldots + C_n^3 = \sum_{r=0}^{n} C_r^3$$

81. If n is even, the greatest coefficient is ${}^nC_{n/2}$

$\therefore$ The greatest term $= {}^nC_{n/2}x^{n/2}$

$\therefore$ $\quad {}^nC_{n/2}x^{n/2} > {}^nC_{(n/2)-1}\,x^{(n/2)-1}$

and $\quad {}^nC_{n/2}x^{n/2} > {}^nC_{(n/2)+1}x^{(n/2)+1}$

$$\Rightarrow \quad \frac{n - \frac{n}{2} + 1}{\frac{n}{2}}x > 1$$

$$\text{and} \quad \frac{n - \left(\frac{n}{2}+1\right) + 1}{\frac{n}{2}+1}x < 1$$

$$\Rightarrow \quad x > \frac{n/2}{(n/2)+1} \text{ and } x < \frac{\frac{n}{2}+1}{\frac{n}{2}}$$

$$\Rightarrow \quad x > \frac{n}{n+2} \text{ and } x < \frac{n+2}{n}$$

$$\text{Hence,} \quad \frac{n}{n+2} < x < \frac{n+2}{n}$$

82. ${}^nC_r + {}^{n-1}C_r + {}^{n-2}C_r + \ldots + {}^rC_r$

= Coefficient of x^r in

$\{(1+x)^n + (1+x)^{n-1} + (1+x)^{n-2} + \ldots + (1+x)^r\}$

$$= \text{Coefficient of } x^r \text{ in } \left\{\frac{(1+x)^r\{(1+x)^{n-r+1} - 1\}}{(1+x-1)}\right\}$$

$\Rightarrow$ Coefficient of x^{r+1} in $\{(1+x)^{n+1} - (1+x)^r\}$

$$= {}^{n+1}C_{r+1}$$

83. $\because 7^{103} = 7(49)^{51} = 7(50-1)^{51}$

$$= 7\{50^{51} - {}^{51}C_1 50^{50} + {}^{51}C_2 50^{49} - \ldots {}^{51}C_{50}50 - 1\}$$

$$= 7(50^{51} - {}^{51}C_1 50^{50} + {}^{51}C_2 50^{49} - \ldots {}^{51}C_{50}50) + (-7 - 18 + 18)$$

$$= 7(50^{51} - {}^{51}C_1 50^{50} + {}^{51}C_2 50^{49} - \ldots + {}^{51}C_{50}50) - 25 + 18$$

$\therefore$ Remainder is 18.

84. Let $\quad a_1 = {}^nC_{r-2}$

$a_2 = {}^nC_{r-1}$

$a_3 = {}^nC_r$

$a_4 = {}^nC_{r+1}$

$$\because \quad \frac{a_1}{a_1+a_2} = \frac{{}^nC_{r-2}}{{}^{n+1}C_{r-1}} = \left(\frac{r-1}{n+1}\right)$$

$$\frac{a_2}{a_2+a_3} = \frac{{}^nC_{r-1}}{{}^{n+1}C_r} = \left(\frac{r}{n+1}\right)$$

$$\text{and} \quad \frac{a_3}{a_3+a_4} = \frac{{}^nC_r}{{}^{n+1}C_{r+1}} = \left(\frac{r+1}{n+1}\right)$$

It is clear that

$\frac{a_1}{a_1+a_2}, \frac{a_2}{a_2+a_3}$ and $\frac{a_3}{a_3+a_4}$ are in AP.

$$\therefore \frac{a_1}{a_1+a_2} + \frac{a_3}{a_3+a_4} = \frac{2a_2}{a_2+a_3}$$

85. $[(x+3)^{n-1} + (x+3)^{n-2}(x+2) + (x+3)^{n-3}(x+2)^2 + \ldots + (x+2)^{n-1}]$

$$\Rightarrow (x+3)^{n-1}\left\{1 + \left(\frac{x+2}{x+3}\right) + \left(\frac{x+2}{x+3}\right)^2 + \ldots + \left(\frac{x+2}{x+3}\right)^{n-1}\right\}$$

$$= (x+3)^{n-1}\left[\frac{1 - \left(\frac{x+2}{x+3}\right)^n}{1 - \left(\frac{x+2}{x+3}\right)}\right]$$

$$= (x+3)^n\left\{1 - \frac{(x+2)^n}{(x+3)^n}\right\} = (x+3)^n - (x+2)^n$$

$$= (3+x)^n - (2+x)^n$$

Coefficient of $x^r = {}^nC_r 3^{n-r} - {}^nC_r 2^{n-r} = {}^nC_r(3^{n-r} - 2^{n-r})$

86. $T_{R+1} = {}^{n-3}C_R x^{n-3-R}\left(\frac{1}{x^2}\right)^R = {}^{n-3}C_R x^{n-3-3R}$

Put $\quad n - 3 - 3R = 2r$

$\therefore \quad 3R = n - 3 - 2r$

$\therefore \quad T_{(n-2r)/3} = {}^{n-3}C_{(n-3-2r)/3}x^2$

$\frac{n-3-2r}{3}$ = whole number.

$n - 3 - 2r = 3$ (whole number)

$\therefore \quad n - 2r = 3$ (whole number + 1)

$= 3$ (natural number)

Hence, $n - 2r$ is positive integral multiple of 3.

87. $3\cdot{}^nC_0 - 8\cdot{}^nC_1 + 13\cdot{}^nC_2 - 18\cdot{}^nC_3 + \dots$

$$= 3({}^nC_0 - {}^nC_1 + {}^nC_2 - {}^nC_3 + \dots) + 5(-{}^nC_1 + 2\cdot{}^nC_2 - 3\cdot{}^nC_3 + \dots)$$

$\because \quad (1-x)^n = {}^nC_0 - {}^nC_1x + {}^nC_2x^2 - {}^nC_3x^3 + \dots \quad \dots$(i)

and $-n(1-x)^{n-1} = -{}^nC_1 + 2\,{}^nC_2x - 3\,{}^nC_3x^2 + \dots \quad \dots$(ii)

Putting $x = 1$ in Eqs. (i) and (ii), we get

$${}^nC_0 - {}^nC_1 + {}^nC_2 - {}^nC_3 + \dots = 0$$

and $$-{}^nC_1 + 2\cdot{}^nC_2 - 3\cdot{}^nC_3 + \dots = 0$$

Hence, $3\cdot{}^nC_0 - 8\cdot{}^nC_1 + 13\cdot{}^nC_2 - 18\cdot{}^nC_3 + \dots = 0$

88. $\because (1+x)^n = {}^nC_0 + {}^nC_1x + {}^nC_2x^2 + {}^nC_3x^3 + \dots + {}^nC_nx^n$

Differentiating both sides w.r.t. x

$$\therefore \quad n(1+x)^{n-1} = 0 + {}^nC_1 + 2\,{}^nC_2x + 3\,{}^nC_3x^2 + \dots + n\,{}^nC_nx^{n-1}$$

or $$nx(1+x)^{n-1} = {}^nC_1x + 2\,{}^nC_2x^2 + 3\,{}^nC_3x^3 + \dots + n\,{}^nC_nx^n$$

Putting $x = \dfrac{a}{b}$

Then, $$\frac{na}{b}\left(1+\frac{a}{b}\right)^{n-1} = {}^nC_1\left(\frac{a}{b}\right) + 2\,{}^nC_2\left(\frac{a^2}{b^2}\right) + 3\,{}^nC_3\left(\frac{a^3}{b^3}\right) + \dots + n\,{}^nC_n\left(\frac{a^n}{b^n}\right)$$

$$\Rightarrow \quad na(a+b)^{n-1} = {}^nC_1ab^{n-1} + 2\,{}^nC_2a^2b^{n-2} + 3\,{}^nC_3a^3b^{n-3} + \dots + n\,{}^nC_na^n$$

$\because \quad a + b = 1$

$$\therefore \quad na = \sum_{r=0}^{n} r\,{}^nC_ra^rb^{n-r}$$

89. $$\sum_{k=1}^{n} k^3\left(\frac{C_k}{C_{k-1}}\right)^2 = \sum_{k=1}^{n} k^3\left(\frac{n-k+1}{k}\right)^2$$

$$= \sum_{k=1}^{n} k(n-k+1)^2$$

$$= \sum_{k=1}^{n} k\{(n+1)^2 + k^2 - 2(n+1)k\}$$

$$= (n+1)^2\sum_{k=1}^{n} k + \sum_{k=1}^{n} k^3 - 2(n+1)\sum_{k=1}^{n} k^2$$

$$= (n+1)^2\cdot\frac{n(n+1)}{2} + \left[\frac{n(n+1)}{2}\right]^2 - 2(n+1)\cdot\frac{n(n+1)(2n+1)}{6}$$

$$= \frac{n(n+1)^2(n+2)}{12}$$

90. We have

$$(1+x)^{1000} + 2x(1+x)^{999} + 3x^2(1+x)^{998} + \dots + 1001x^{1000}$$

$$= (1+x)^{1000}\left\{1 + 2\left(\frac{x}{1+x}\right) + 3\left(\frac{x}{1+x}\right)^2 + \dots + 1001\left(\frac{x}{1+x}\right)^{1000}\right\} \quad \dots\text{(i)}$$

If $$P = 1 + 2\left(\frac{x}{1+x}\right) + 3\left(\frac{x}{1+x}\right)^2 + \dots + 1001\left(\frac{x}{1+x}\right)^{1000}$$

$$\therefore \quad \left(\frac{x}{1+x}\right)P = \left(\frac{x}{1+x}\right) + 2\left(\frac{x}{1+x}\right)^2 + \dots + 1000\left(\frac{x}{1+x}\right)^{1000} + 1001\left(\frac{x}{1+x}\right)^{1001}$$

$$\therefore \quad P\left(1 - \frac{x}{1+x}\right) = 1 + \left(\frac{x}{1+x}\right) + \left(\frac{x}{1+x}\right)^2 + \dots + \left(\frac{x}{1+x}\right)^{1000} - 1001\left(\frac{x}{1+x}\right)^{1001}$$

$$\Rightarrow \quad \frac{P}{1+x} = \frac{1\left\{1 - \left(\dfrac{x}{1+x}\right)^{1001}\right\}}{\left(1 - \dfrac{x}{1+x}\right)} - 1001\left(\frac{x}{1+x}\right)^{1001}$$

$$\therefore \quad P = (1+x)^2\left\{1 - \left(\frac{x}{1+x}\right)^{1001}\right\}$$

From Eq. (i),

$$= (1+x)^{1002}\left\{1 - \left(\frac{x}{1+x}\right)^{1001}\right\} - 1001\,x^{1001}$$

$\therefore$ Coefficient of $x^{50} = {}^{1002}C_{50}$

91. $(6\sqrt{6} + 14)^{n+1} = m = [m] + f \quad \dots$(i)

$$0 \le f < 1$$

and let $f' = (6\sqrt{6} - 14)^{n+1} \quad \dots$(ii)

$$0 < f' < 1$$

$\therefore \quad -1 < f - f' < 1 \quad \dots$(iii)

From Eqs. (i) and (ii)

$$[m] + f - f' = (6\sqrt{6} + 14)^{n+1} - (6\sqrt{6} - 14)^{n+1}$$

$$= 2k \text{ (even integer)}$$

$\therefore \quad f - f'$ is also an integer.

From Eq. (iii), $f - f' = 0$

$\therefore \quad f = f'$

Then, $mf = mf' = (6\sqrt{6} - 14)^{n+1}(6\sqrt{6} + 14)^{n+1}$

$$= (216 - 196)^{n+1} = (20)^{n+1}$$

92. Let $P = \{(a+3b)^2(3a-b)^2\}^3$

$$= [(a+3b)(3a-b)]^6$$

$$= (3a^2 - ab + 9ab - 3b^2)^6 = (3a^2 - 3b^2 + 8ab)^6$$

$\therefore$ Number of terms $= {}^{6+2}C_2 = {}^8C_2 = \dfrac{8\cdot 7}{1\cdot 2} = 28$

93. $E = (x-\alpha_1)(x-\alpha_2)(x-\alpha_3)\dots(x-\alpha_n)$,

where $\alpha_1 = 1, \alpha_2 = 2$ etc

$$= x^n - (\Sigma\alpha_1)x^{n-1} + (\Sigma\alpha_1\alpha_2)x^{n-2} + \dots$$

Hence, coefficient of x^{n-2} = sum of all the products of the first 'n' natural numbers taken two at a time

$$= \frac{(1+2+3+\dots+n)^2 - (1^2+2^2+\dots+n^2)}{2}$$

$$= \frac{n(n^2-1)(3n+2)}{24}$$

94. $ab - n(a-1)(b-1) + \frac{n(n-1)}{1 \cdot 2}(a-2)(b-2)$
$- \ldots + (-1)^n (a-n)(b-n)$ ...(i)

The $(k+1)$th term of Eq. (i) is given by
$$T_{k+1} = (-1)^k \cdot {}^nC_k (a-k)(b-k) = (-1)^k \; {}^nC_k[ab - k(a+b) + k^2]^2$$

Thus, the sum of series (i)
$$= \sum_{k=0}^{n} (-1)^k \cdot {}^nC_k [ab - k(a+b) + k^2]$$
$$= ab \sum_{k=0}^{n} (-1)^k \cdot {}^nC_k - (a+b) \sum_{k=0}^{n} (-1)^k k \cdot {}^nC_k + \sum_{k=0}^{n} (-1)^k k^2 \cdot {}^nC_k$$

We know that,
$$(1+x)^n = {}^nC_0 + {}^nC_1 x + {}^nC_2 x^2 + \ldots + {}^nC_n x^n \quad \ldots(ii)$$

Differentiating both sides w.r.t. x, we get
$$n(1+x)^{n-1} = 1 \cdot {}^nC_1 + 2 \cdot {}^nC_2 x + 3 \cdot {}^nC_3 x^2 + \ldots + n \cdot {}^nC_n x^{n-1} \quad \ldots(iii)$$

Multiplying both sides of Eq. (iii) by x, we get
$$nx(1+x)^{n-1} = 1 \cdot {}^nC_1 x + 2 \cdot {}^nC_2 x^2 + 3 \cdot {}^nC_3 x^3 + \ldots + n \cdot {}^nC_n x^n$$

Differentiating w.r.t. x, we get
$$n(1+x)^{n-1} + n(n-1)x(1+x)^{n-2} = 1^2 \cdot {}^nC_1 + 2^2 \cdot {}^nC_2 x + \ldots + n^2 \cdot {}^nC_n x^{n-1} \quad \ldots(iv)$$

Putting $x = -1$ in Eqs. (ii), (iii) and (iv), we get
$$\sum_{k=0}^{n} (-1)^k \cdot {}^nC_k = 0, \quad \sum_{k=0}^{n} (-1)^k k \cdot {}^nC_k = 0$$
and $\sum_{k=0}^{n} (-1)^k k^2 \cdot {}^nC_k = 0$

Thus, the sum of the series in Eq. (i) is 0.

95. $\sum_{i=0}^{n} \sum_{j=1}^{n} {}^nC_j \, {}^jC_i = {}^nC_1({}^1C_0 + {}^1C_1) + {}^nC_2({}^2C_0 + {}^2C_1 + {}^2C_2)$
$+ {}^nC_3({}^3C_0 + {}^3C_1 + {}^3C_2 + {}^3C_3)$
$+ {}^nC_4({}^4C_0 + {}^4C_1 + {}^4C_2 + {}^4C_3 + {}^4C_4)$
$+ \ldots + {}^nC_n({}^nC_0 + {}^nC_1 + {}^nC_2 + \ldots + {}^nC_n)$
$= {}^nC_1(2) + {}^nC_2(2)^2 + {}^nC_3(2)^3 + {}^nC_4(2)^4 + \ldots + {}^nC_n(2)^n$
$= (1+2)^n - 1$
$= 3^n - 1$

96. $(bc + ca + ab)^{10}$

General term $= \frac{10!}{p!q!r!}(bc)^p (ca)^q (ab)^r$
$= \frac{10!}{p!q!r!} a^{q+r} \cdot b^{r+p} \cdot c^{p+q}$

Let $\quad q + r = 10, \; r + p = 7, \; p + q = 3$
$\therefore \quad p + q + r = 10$
$\therefore \quad p = 0, q = 3, r = 7$
$\therefore$ Coefficient of $a^{10} b^7 c^3$ is
$$\frac{10!}{0!3!7!} = \frac{10 \cdot 9 \cdot 8 \cdot 7!}{1 \cdot 1 \cdot 2 \cdot 3 \cdot 7!} = 120$$

97. $\because (1 + x + 2x^2)^{20} = a_0 + a_1 x + a_2 x^2 + a_3 x^3 + a_4 x^4 + a_5 x^5 + \ldots + a_{37} x^{37} + a_{38} x^{38} + a_{39} x^{39} + a_{40} x^{40}$

Putting $x = 1$ and $x = -1$ and subtracting, we get
$$4^{20} - 2^{20} = 2(a_1 + a_3 + a_5 + \ldots + a_{37} + a_{39})$$
or $\quad a_1 + a_3 + a_5 + \ldots + a_{37} + a_{39} = 2^{19}(2^{20} - 1)$
$$a_1 + a_3 + a_5 + \ldots + a_{37} = 2^{19}(2^{20} - 1) - a_{39} \quad \ldots(i)$$

Now, $\quad (1 + x + 2x^2)^{20} = \frac{20!}{\alpha!\beta!\gamma!}(1)^\alpha (x)^\beta (2x^2)^\gamma$

For a_{39}, $\beta + 2\gamma = 39$ and $\alpha + \beta + \gamma = 20$
$$\gamma = 19, \beta = 1, \alpha = 0$$
$\therefore \quad a_{39} = \frac{20!}{0!1!19!} \cdot 2^{19}$
$= 20 \cdot 2^{19}$

$\therefore$ From Eq. (i) $a_1 + a_3 + a_5 + \ldots + a_{37} = 2^{19}(2^{20} - 21)$

98. Coefficient of x^{50} in $(1+x)^{41}(1 - x + x^2)^{40}$
or coefficient of x^{50} in $(1+x)(1+x^3)^{40}$
or coefficient of x^{50} in
$$(1+x)(1 + {}^{40}C_1 x^3 + {}^{40}C_2 x^6 + \ldots) = 0$$

99. $\because \quad (\sqrt[4]{9} + \sqrt[6]{8})^{500}$
$\Rightarrow \quad \left(9^{\frac{1}{4}} + 8^{\frac{1}{6}}\right)^{500}$
$\Rightarrow \quad \left(3^{\frac{1}{2}} + 2^{\frac{1}{2}}\right)^{500}$
$\therefore \quad T_{R+1} = {}^{500}C_R \left(3^{\frac{1}{2}}\right)^{500-R} \left(2^{\frac{1}{2}}\right)^R$
$= {}^{500}C_R \cdot 3^{250 - \frac{R}{2}} \cdot 2^{\frac{R}{2}}$
$\because \quad 0 \le R \le 500$
$\Rightarrow \quad 0 \le \frac{R}{2} \le 250$
$\therefore$ Number of terms which are integers are 251.

100. Let $n = 2m$
$\therefore \quad (a + b + c)^{2m} + (a + b - c)^{2m}$
$\Rightarrow \quad 2\{(a+b)^{2m} + {}^{2m}C_2(a+b)^{2m-2} \cdot c^2 + {}^{2m}C_4 \times (a+b)^{2m-4} \cdot c^4 + \ldots + {}^{2m}C_{2m} \cdot c^{2m}\}$

$\therefore$ Number of terms
$= (2m+1) + (2m-1) + (2m-3) + \ldots + 3 + 1$
$= \left(\frac{m+1}{2}\right)(2m + 1 + 1) = (m+1)^2 = \left(\frac{n}{2} + 1\right)^2$
$= \left(\frac{n+2}{2}\right)^2$

101. $\because (1 + x + x^2)^n = a_0 + \ldots + a_{r-2} x^{r-2} + a_{r-1} x^{r-1} + a_r x^r + \ldots$(i)

and $(1-x)^n = {}^nC_0 - {}^nC_1 x + {}^nC_2 x^2 - \ldots + (-1)^r \, {}^nC_r x^r + \ldots$(ii)

Multiplying Eqs. (i) and (ii) and comparing coefficient of x^r, then

$$(-1)^{r/3}\ {}^nC_{r/3} = a_r - {}^nC_1 \cdot a_{r-1} + {}^nC_2 \cdot a_{r-2} - \dots + (-1)^r\ {}^nC_r \cdot a_0$$

But r is not divisible by 3

$\therefore \quad {}^nC_{r/3} = 0$

Hence, $a_r - {}^nC_1 \cdot a_{r-1} + {}^nC_2 \cdot a_{r-2} - \dots + (-1)^r \cdot {}^nC_r \cdot a_r = 0$

102. $3^{-x/4} = 3^{5x/4} = 1$

$\therefore \quad 2^n = 64 = 2^6$

$\Rightarrow \quad n = 6$

$\therefore \quad \left(3^{\frac{-x}{4}} + 3^{\frac{5x}{4}}\right)^6$

$${}^6C_3\left(3^{-\frac{x}{4}}\right)^3\left(3^{\frac{5x}{4}}\right)^3 = {}^6C_2\left(3^{\frac{-x}{4}}\right)^4\left(3^{\frac{5x}{4}}\right)^2 + (6-1)$$

$\Rightarrow \quad 20 \cdot 3^{3x} = 15 \cdot 3^{\frac{3x}{2}} + 5$

Put $\quad 3^{\frac{3x}{2}} = t$

$\Rightarrow \quad 4t^2 - 3t - 1 = 0$

$\Rightarrow \quad (4t+1)(t-1) = 0$

$\quad t \neq \frac{-1}{4}$

$\therefore \quad t = 1$

$\Rightarrow \quad 3^{\frac{3x}{2}} = 1 = 3^0$

$\therefore \quad x = 0$

103. Consider,

$$(\cos\theta - i\sin\theta)^m = {}^mC_0 \cos^m\theta - {}^mC_1 \cos^{m-1}\theta i \sin\theta + \dots + {}^mC_m(-i\sin\theta)^m \quad \dots\text{(i)}$$

$$(\cos\theta + i\sin\theta)^m = {}^mC_0\cos^m\theta + {}^mC_1\cos^{m-1}\theta i\sin\theta + \dots + {}^mC_m(i\sin\theta)^m \quad \dots\text{(ii)}$$

On adding Eqs. (i) and (ii), we get

$$2\cos m\theta = 2[{}^mC_0\cos^m\theta - {}^mC_2 \cos^{m-2}\theta\sin^2\theta \dots] \quad \dots\text{(iii)}$$

On subtracting Eq. (i) from Eq. (ii), we get

$$2i\sin m\theta = 2i[{}^mC_1\cos^{m-1}\theta\sin\theta - {}^mC_3\cos^{m-3}\theta\sin^3\theta\dots] \quad \dots\text{(iv)}$$

On adding Eqs. (iii) and (iv), we get

$$\cos m\theta + \sin m\theta = [{}^mC_0\cos^m\theta + {}^mC_1\cos^{m-1}\theta\sin\theta - {}^mC_2\cos^{m-2}\theta\sin^2\theta - {}^mC_3\cos^{m-3}\theta\sin^3\theta\dots]$$

$$\Rightarrow \sqrt{2}\sin\left(m\theta + \frac{\pi}{4}\right) = [{}^mC_0\cos^m\theta + {}^mC_1\cos^{m-1}\theta\sin\theta - {}^mC_2\cos^{m-2}\theta\sin^2\theta - {}^mC_3\cos^{m-3}\theta\sin^3\theta\dots]$$

Putting $\theta = \frac{\pi}{4}$, we get

$$\sqrt{2}\sin\left(\frac{(m+1)\pi}{4}\right) = \frac{1}{2^{m/2}}[({}^mC_0 + {}^mC_1 - {}^mC_2 - {}^mC_3) + ({}^mC_4 + {}^mC_5 - {}^mC_6 - {}^mC_7) + \dots + ({}^mC_{m-3} + {}^mC_{m-2} - {}^mC_{m-1} - {}^mC_m)]$$

Hence, $m + 1 = 4k$, for given quantity to be 0.

$\Rightarrow m = 4k - 1$, where $k \in N$

Objective Questions Type II [One or more than one correct answer(s)]

1. $\because (1 + x + x^2)^n = a_0 + a_1x + a_2x^2 + \dots + a_{2n}x^{2n}$

Putting $x = i \ (i = \sqrt{-1})$

Then, we get

$$(1 + i + i^2)^n = (a_0 - a_2 + a_4 - a_6 + \dots) + i(a_1 - a_3 + a_5 - a_7 + \dots)$$

$$\Rightarrow \quad i^n = (a_0 - a_2 + a_4 - a_6 + \dots) + i(a_1 - a_3 + a_5 - a_7 + \dots)$$

If n is odd, then $\text{Re}(i^n) = 0$

$\Rightarrow \quad a_0 - a_2 + a_4 - a_6 + \dots = 0$

If n is even, then $\text{Im}(i^n) = 0$

$\Rightarrow \quad a_1 - a_3 + a_5 - a_7 + \dots = 0$

2. $C_0^2 + 3C_1^2 + 5C_2^2 + \dots + (2n+1)C_n^2$

$$\Rightarrow (C_0^2 + C_1^2 + C_2^2 + \dots + C_n^2) + 2(C_1^2 + 2C_2^2 + \dots + nC_n^2)$$

$$= {}^{2n}C_n + 2(C_1^2 + 2C_2^2 + \dots + nC_n^2) \quad \dots\text{(i)}$$

$$\because \quad (1+x)^n = C_0 + C_1x + C_2x^2 + C_3x^3 + \dots + C_nx^n$$

Differentiating both sides w.r.t. x, we get

$$n(1+x)^{n-1} = C_1 + 2C_2x + 3C_3x^2 + \dots + nC_nx^{n-1} \quad \dots\text{(ii)}$$

and $(x+1)^n = C_0x^n + C_1x^{n-1} + C_2x^{n-2} + \dots + C_n \quad \dots\text{(iii)}$

Multiplying Eqs. (ii) and (iii), we get

$$n(1+x)^{2n-1} = (C_1 + 2C_2x + 3C_3x^2 + \dots + nC_nx^{n-1}) \times (C_0x^n + C_1x^{n-1} + C_2x^{n-2} + \dots + C_n)$$

Comparing the coefficients of x^{n-1} in both sides, then

$$C_1^2 + 2C_2^2 + 3C_3^2 + \dots + nC_n^2 = n\ {}^{2n-1}C_{n-1} \quad \dots\text{(iv)}$$

From Eqs. (i) and (iv), we get

$$C_0^2 + 3C_1^2 + 5C_2^2 + \dots + (2n+1)C_n^2$$

$$= {}^{2n}C_n + 2n \cdot {}^{2n-1}C_{n-1}$$

$$= \frac{2n}{n} \cdot {}^{2n-1}C_{n-1} + 2n \cdot {}^{2n-1}C_{n-1}$$

$$= 2(n+1) \cdot {}^{2n-1}C_{n-1} = 2(n+1)\ {}^{2n-1}C_n$$

$$= {}^{2n-1}C_{n-1} + (2n+1)\ {}^{2n-1}C_{n-1}$$

$$= {}^{2n-1}C_n + (2n+1) \cdot {}^{2n-1}C_{n-1}$$

3. When n is odd, taken $n = 2m + 1$,

$\therefore S = C_0^2 - C_1^2 + C_2^2 - \dots + (-1)^{2m} C_{2m}^2 + (-1)^{2m+1} C_{2m+1}^2$

$= (C_0^2 - C_{2m+1}^2) - (C_1^2 - C_{2m}^2) + \dots + (-1)^m (C_m^2 - C_{m+1}^2)$

Since, ${}^{2m+1}C_0 = {}^{2m+1}C_{2m+1}, \; {}^{2m+1}C_1 = {}^{2m+1}C_{2m}$ etc.

Therefore, $S = 0$

When n is even, we take $n = 2m$. In this case

$C_0^2 - C_1^2 + C_2^2 - C_3^2 + \dots + (-1)^{2m} C_{2m}^2$

$= $ Coefficient of constant term in

$$[C_0 - C_1 x + C_2 x^2 - \dots + (-1)^{2m} C_{2m} x^{2m}]\left[C_0 + C_1 \frac{1}{x} + C_2 \frac{1}{x^2} + \dots + C_{2m} \frac{1}{x^{2m}}\right]$$

$= $ Coefficient of constant term in

$$(1 - x)^{2m}\left(1 + \frac{1}{x}\right)^{2m}$$

$= $ Coefficient of x^{2m} in $(1 - x)^{2m}(1 + x)^{2m}$

$= $ Coefficient of x^{2m} in $(1 - x^2)^{2m}$

$= (-1)^m ({}^{2m}C_m) = (-1)^{n/2} ({}^nC_{n/2})$

4. Given, $(3\sqrt{3} + 5)^{2n+1} = \alpha + \beta$ ...(i)

$\because \quad 0 < \beta < 1$

Let $\quad \beta' = (3\sqrt{3} - 5)^{2n+1}$ ...(ii)

$\therefore \quad 0 < \beta' < 1$

From Eqs. (i) and (ii)

$\alpha + \beta - \beta' = (3\sqrt{3} + 5)^{2n+1} - (3\sqrt{3} - 5)^{2n+1}$

$\Rightarrow \quad = 2[{}^{2n+1}C_1 (3\sqrt{3})^{2n} 5 + {}^{2n+1}C_3 (3\sqrt{3})^{2n-2} (5)^3 + \dots + {}^{2n+1}C_{2n+1} 5^{2n+1}]$

$\alpha + \beta - \beta' = 10I$

But $-1 < \beta - \beta' < 1$

$\therefore \beta - \beta'$ is an integer

$\therefore \quad \beta - \beta' = 0$

$\therefore \quad \alpha = 10\, I$

$\therefore \alpha$ divisible by 10

$\Rightarrow \alpha$ is an even integer.

$\Rightarrow \quad (\alpha + \beta)^2 = [(3\sqrt{3} + 5)^2]^{2n+1}$

$= (52 + 30\sqrt{3})^{2n+1}$

$= 2^{2n+1} (26 + 15\sqrt{3})^{2n+1}$

$\therefore (\alpha + \beta)^2$ is not divisible by 2^{2n+1}.

5. $\quad (8 + 3\sqrt{7})^n = P + F$ ...(i)

Here, $\quad 0 < F < 1$

Let $\quad F' = (8 - 3\sqrt{7})^n$ ...(ii)

$0 < F' < 1$

Adding Eqs. (i) and (ii), we get

$P + F + F' = (8 + 3\sqrt{7})^n + (8 - 3\sqrt{7})^n$

$= 2k$ (even integer) ...(iii)

and $\quad 0 < F + F' < 2$

$\because F + F'$ is an integer

$\therefore \quad F + F' = 1$

From Eq. (iii),

$P = 2k - 1$ (odd integer)

$\therefore (1 - F)(P + F) = F'(P + F)$

$= (8 - 3\sqrt{7})^n (8 + 3\sqrt{7})^n = (64 - 63)^n$

$= 1^n = 1$

6. $\left(\sqrt[3]{4} + \frac{1}{\sqrt[4]{6}}\right)^{20} = (4^{1/3} + 6^{-1/4})^{20}$

$= (2^{2/3} + 6^{-1/4})^{20}$

$T_{r+1} = {}^{20}C_r (2^{2/3})^{20-r} (6^{-1/4})^r$

$= {}^{20}C_r 2^{(160-11r)/12} \cdot 3^{-r/4}$

$\therefore$ For $r = 8, 20$; T_{r+1} is rational.

$\therefore$ Only two terms are rational.

So, $21 - 2 = 19$ terms are irrational.

7. $\frac{a_{n+1}}{a_n} = \frac{(1000)^{n+1}}{(n+1)!} \cdot \frac{n!}{(1000)^n} = \frac{1000}{(n+1)} \geq 1$

For $\quad n = 1, 2, 3, \dots, 999$

$\Rightarrow \quad a_{n+1} \geq a_n$

$a_{1000} = \frac{(1000)^{1000}}{1000!} = \frac{(1000)^{999}}{999!} = a_{999}$

Hence, a_{999} and a_{1000} are equal and are the greatest.

8. $(1 + x)^n \times (x + 1)^n = (C_0 + C_1 x + C_2 x^2 + \dots + C_{n-2} x^{n-2} + C_{n-1} x^{n-1} + C_n x^n)$

$\times (C_0 x^n + C_1 x^{n-1} + C_2 x^{n-2} + C_3 x^{n-3} + C_4 x^{n-4} + \dots + C_n)$

Coefficient of x^{n-2} in RHS

$= C_0 C_2 + C_1 C_3 + C_2 C_4 + \dots + C_{n-2} C_n$ ...(i)

Coefficient of x^{n-2} in LHS

$= {}^{2n}C_{n-2} = \frac{2n!}{(n-2)!(n+2)!}$ ...(ii)

From Eqs. (i) and (ii), we get

$C_0 C_2 + C_1 C_3 + C_2 C_4 + \dots + C_{n-2} C_n$

$= {}^{2n}C_{n-2} = \frac{2n!}{(n-2)!(n+2)!}$

9. Let $P = \{x + \sqrt{(x^3 - 1)}\}^5 + \{x - \sqrt{(x^3 - 1)}\}^5$ ...(i)

Let $\quad \sqrt{(x^3 - 1)} = \lambda$

$\therefore \quad \lambda^2 = x^3 - 1$ ...(ii)

From Eq. (i),

$P = (x + \lambda)^5 + (x - \lambda)^5$

$= 2\{x^5 + {}^5C_2 x^3 \lambda^2 + {}^5C_4 x \lambda^4\}$

$= 2\{x^5 + 10x^3(x^3 - 1) + 5x(x^3 - 1)^2\}$

$= 2\{5x^7 + 10x^6 + x^5 - 10x^4 - 10x^3 + 5x\}$

Hence, P is a polynomial of degree 7.

10. $(x + y + z)^{25} = [x + y + z]^{25}$

$= {}^{25}C_0 x^{25} + {}^{25}C_1 x^{24} (y + z)^1 + \dots + {}^{25}C_r x^{25-r} (y + z)^r + \dots$

$= \dots + {}^{25}C_r x^{25-r} (\dots + {}^rC_k y^{r-k} z^k + \dots) + \dots$

Hence, every term is of the form

${}^{25}C_r x^{25-r} \, {}^rC_k y^{r-k} z^k$

$\therefore \quad 25 - r + r - k + k = 25 \neq 8 + 9 + 9$

Hence, coefficient of $x^8y^9z^9$ is 0 and number of terms

$$= {}^{25+2}C_2$$
$$= \frac{27 \times 26}{2}$$
$$= 27 \times 13 = 351$$

11. Given, $(1 + 2x + 3x^2)^{10} = a_0 + a_1x + a_2x^2 + \dots a_{20}x^{20}$

$\therefore \quad [1 + x(2 + 3x)]^{10} = 1 + {}^{10}C_1x(2 + 3x) + {}^{10}C_2x^2 \times (2 + 3x)^2 + \dots + {}^{10}C_{10}x^{10}(2 + 3x)^{10}$

Alternate. (a) : a_1 = coefficient of x
$$= {}^{10}C_1 \cdot 2 = 20$$

Alternate. (b) : a_2 = coefficient of x^2
$$= {}^{10}C_1 \cdot 3 + {}^{10}C_2 \cdot 2^2$$
$$= 30 + 180 = 210$$

Alternate. (c) : a_4 = coefficient of x^4
$$= {}^{10}C_2 \cdot 3^2 + {}^{10}C_3 \cdot 36 + {}^{10}C_4 \cdot 2^4$$
$$= 8085$$

Alternate. (d) : a_{20} = coefficient of x^{20}
$$= {}^{10}C_{10} \cdot 3^{10}$$
$$= 3^{10} \neq 2^2 \cdot 3^7 \cdot 7$$

12. We have,

$$\sum_{k=0}^{7}\left(\frac{{}^7C_k}{{}^{14}C_k}\sum_{r=k}^{14} {}^rC_k \cdot {}^{14}C_r\right)$$
$$= \sum_{k=0}^{7}\left(\frac{{}^7C_k}{14!} \times k!(14-k)! \sum_{r=k}^{14} \frac{4!}{k!(r-k)!} \cdot \frac{14!}{r!(14-r)!}\right)$$
$$= \sum_{k=0}^{7}\left({}^7C_k \sum_{r=k}^{14} {}^{14-k}C_{r-k}\right)$$
$$= \sum_{k=0}^{7} {}^7C_k \cdot 2^{14-k}$$
$$= 2^{14} \sum_{k=0}^{7} {}^7C_k\left(\frac{1}{2}\right)^k$$
$$= 2^{14} \cdot \left(1 + \frac{1}{2}\right)^7 = 6^7 > 7^6$$

13. Let $(6\sqrt{6} + 14)^{2n+1} = P = I + F$

where $I = [P]$ and $F = P - [P]$.

Let $(6\sqrt{6} - 14)^{2n+1} = G$

so that $0 < G < 1$.

Now, $P - G = 2[(2n+1)C_1(6\sqrt{6})^{2n}(14) + (2n+1)C_3(6\sqrt{6})^{2n-2} \times (14)^2 + \dots]$

= 2 (some positive integer)

$\therefore I + F - G$ is an integer and $0 < F < 1$, $0 < G < 1$ implies that $F = G$.

Also, $[P] = I$ is an even integer and

$$PF = (I + F)F = (I + F)G$$
$$= (6\sqrt{6} + 14)^{2n+1}(6\sqrt{6} - 14)^{2n+1} = (20)^{2n+1}$$

14. $\because \quad 3^{4n} = 81^n = (1 + 80)^n = 1 + 80\lambda, \lambda \in N$

$\therefore \quad 3^{3^{4n}} = 3^{1+80\lambda} = 3 \cdot 3^{80\lambda} = 3 \cdot (9)^{40\lambda} = 3(10 - 1)^{40\lambda}$
$$= 3(1 + 10\mu) = 3 + 30\mu$$

$\therefore$ Last digit of $3^{3^{4n}} + 1$ is 4.

15. We have, $a_1 = 2$ and for $n \geq 2$,

$$a_n = \left(1 + \frac{1}{n}\right)^n = {}^nC_0 + {}^nC_1\left(\frac{1}{n}\right) + {}^nC_2\left(\frac{1}{n}\right)^2 + \dots + {}^nC_r\left(\frac{1}{n}\right)^r + \dots + {}^nC_n\left(\frac{1}{n}\right)^n$$
$$= 1 + 1 + \frac{n(n-1)}{2!}\frac{1}{n^2} + \dots + \frac{n(n-1)\dots(n-r+1)}{r!}\frac{1}{n^r} + \dots + \frac{n(n-1)\dots 2.1}{n!}\frac{1}{n^n} \quad \dots(i)$$
$$= 2 + \frac{1}{2!}\left(1 - \frac{1}{n}\right) + \frac{1}{3!}\left(1 - \frac{1}{n}\right)\left(1 - \frac{2}{n}\right) + \dots + \frac{1}{r!}\left(1 - \frac{1}{n}\right)\left(1 - \frac{2}{n}\right)\dots\left(1 - \frac{r-1}{n}\right) + \dots + \frac{1}{n!}\left(1 - \frac{1}{n}\right)\left(1 - \frac{2}{n}\right)\dots\left(1 - \frac{n-1}{n}\right)$$

Hence, from Eq. (i), $a_n \geq 2$ for all $n \in N$. Also,

$$a_n \leq 1 + 1 + \frac{1}{2!} + \frac{1}{3!} + \dots + \frac{1}{r!} + \dots + \frac{1}{n!}$$

For $2 \leq r \leq n$, we have $r! = 1 \cdot 2 \cdot 3 \dots r \geq 2^{r-1}$.

Thus, $a_n \leq 1 + 1 + \frac{1}{2} + \frac{1}{2^2} + \dots + \frac{1}{2^{r-1}} + \dots + \frac{1}{2^{n-1}}$

$$= 1 + \frac{1 - (1/2^n)}{1 - (1/2)}$$
$$= 1 + 2\left(1 - \frac{1}{2^n}\right) = 3 - \frac{1}{2^{n-1}}$$
$$a_n \leq 3 - \frac{1}{2^{n-1}} < 3 \ \forall \ n \geq 1$$
$$\Rightarrow \quad a_n < 4 \ \forall \ n \geq 1$$

16. $\because (101)^{50} - (99)^{50} = (100 + 1)^{50} - (100 - 1)^{50}$
$$= 2\{{}^{50}C_1(100)^{49} + {}^{50}C_3(100)^{47} + {}^{50}C_5(100)^{45} + \dots\}$$
$$= (100)^{50} + 2\{{}^{50}C_3(100)^{47} + {}^{50}C_5(100)^{45} + \dots\}$$
$$> (100)^{50}$$

$\Rightarrow \quad (101)^{50} - (99)^{50} > (100)^{50}$

or $\quad (101)^{50} - (100)^{50} > (99)^{50}$

Also, $\left(\frac{1001}{1000}\right)^{999} = \left(1 + \frac{1}{1000}\right)^{999}$

$$= 1 + {}^{999}C_1\left(\frac{1}{1000}\right) + {}^{999}C_2\left(\frac{1}{1000}\right)^2 + \dots\dots$$
$$< 1 + 1 + 1 + 1 + \dots + 1$$
$$= 1000$$

$\therefore \quad \left(\frac{1001}{1000}\right)^{999} < (1000)$

$\Rightarrow \quad (1001)^{999} < (1000)^{1000}$

17. $\because$ Number of distinct terms $= {}^{9+3-1}C_{3-1} = {}^{11}C_2 = 55$

Sum of coefficients $= (2 - 2 + 1)^9 = 1^9 = 1$

And $(2 - 2x + x^2)^9 = \Sigma \frac{9!}{\alpha!\beta!\gamma!}(2)^\alpha(-2x)^\beta(x^2)^\gamma$

Here, $\beta + 2\gamma = 4, \alpha + \beta + \gamma = 9$

$$\therefore \quad \begin{matrix} \alpha & \beta & \gamma \\ 5 & 4 & 0 \\ 6 & 2 & 1 \\ 7 & 0 & 2 \end{matrix}$$

$\therefore$ Coefficient of x^4 is

$$= \frac{9!}{5!4!0!} \cdot 2^5 \cdot (-2)^4 + \frac{9!}{6!2!1!} (2)^6 (-2)^2 + \frac{9!}{7!0!2!} (2)^7 (-2)^0$$

$$\Rightarrow \quad 2^9 (126 + 126 + 9) = 133632$$

18. $\sum_{r=0}^{2n} a_r \left(\frac{1}{x}\right)^r = \left(1 + \frac{1}{x} + \frac{1}{x^2}\right)^n$

$$= \frac{1}{x^{2n}} (x^2 + x + 1)^n$$

$$\Rightarrow \sum_{r=0}^{2n} a_r x^{2n-r} = (x^2 + x + 1)^n = \sum_{r=0}^{2n} a_r x^r$$

$$= \sum_{r=0}^{2n} a_{2n-r} x^{2n-r} \quad \text{...(i)}$$

Equating the coefficients of x^{2n-r} on both sides, we get $a_r = a_{2n-r}$ for $0 \le r \le 2n$

Now, putting $x = 1$ in Eq. (i), we get

$$a_0 + a_1 + a_2 + \ldots + a_{2n} = (1 + 1 + 1)^n = 3^n \quad \text{..(ii)}$$

But $a_r = a_{2n-r}$ for $0 \le r \le n - 1$

$$2(a_0 + a_1 + \ldots + a_{n-1}) + a_n = 3^n$$

$$a_0 + a_1 + \ldots + a_{n-1} = \frac{1}{2} (3^n - a_n)$$

$$(1 + x + x^2)^n = a_0 + a_1 x + a_2 x^2 + a_3 x^3 + \ldots + a_{2n} x^{2n} \text{...(iii)}$$

Replacing x by $-\frac{1}{x}$, then

$$\left(1 - \frac{1}{x} + \frac{1}{x^2}\right)^n = a_0 - \frac{a_1}{x} + \frac{a_2}{x^2} - \frac{a_3}{x^3} + \ldots + \frac{a_{2n}}{x^{2n}} \quad \text{...(iv)}$$

Multiplying Eqs. (iii) and (iv) and comparing constant term, then

RHS $= a_0^2 - a_1^2 + a_2^2 - a_3^2 + \ldots + a_{2n}^2$

Constant term in $(1 + x + x^2)^n \left(1 - \frac{1}{x} + \frac{1}{x^2}\right)^n$

$\therefore$ Coefficient of x^{2n} in $(1 + x^2 + x^4)^n$ is a_n.

Again putting $x = -1$ in Eq. (i), we get

$$a_0 - a_1 + a_2 - a_3 + \ldots + a_{2n} = 1 \quad \text{...(v)}$$

Adding Eqs. (ii) and (v) and dividing by 2, we get

$$a_0 + a_2 + a_4 + \ldots + a_{2n} = \frac{1}{2} (3^n + 1).$$

19. $T_{r+1} = {}^{10}C_r (\sqrt{x})^{10-r} \cdot \left(-\frac{\lambda}{x^2}\right)^r$

$$= {}^{10}C_r \cdot x^{5 - \frac{r}{2} - 2r} \cdot (-\lambda)^r$$

$\therefore$ Put $5 - \frac{r}{2} - 2r = 0$

$\Rightarrow \quad r = 2$

Then, $\quad T_{2+1} = {}^{10}C_2 \cdot (-\lambda)^2 = 45\lambda^2 = 405$ (given)

$\therefore \quad \lambda^2 = 9$

$\Rightarrow \quad \lambda = \pm 3$

20. $(x^2 + 2x + 2)^n = [2 + x (2 + x)]^n$

$$= 2^n + {}^nC_1 2^{n-1} x (2 + x) + {}^nC_2\, 2^{n-2} x^2 (2 + x)^2 + {}^nC_3 2^{n-3} x^3 (2 + x)^3 + {}^nC_4 2^{n-4} \cdot x^4 (2 + x)^4 + \ldots$$

$\therefore$ Coefficient of x is $n \cdot 2^{n-1} \cdot 2 = 2^n \cdot n$

Coefficient of x^2 is

$$n \cdot 2^{n-1} + \frac{1}{2} n (n - 1) 2^{n-2} \cdot 4 = n^2 \cdot 2^{n-1}$$

Coefficient of x^3 is ${}^nC_2 \cdot 2^{n-2} \cdot 4 + {}^nC_3 2^{n-3} \cdot 2^3$

$$= 2^n [{}^nC_2 + {}^nC_3] = 2^n \cdot {}^{n+1}C_3$$

Linked Comprehension Type

PASSAGE 1

1. We have,

$$p = \frac{(16 + 1)(1/\sqrt{2})}{(1/\sqrt{2}) + 1} = 17(\sqrt{2} - 1)$$

Therefore, $[p] = 7$. The eighth term is numerically greatest and its value is

$$\sqrt{2} \cdot {}^{16}C_7 \cdot \left(\frac{1}{\sqrt{2}}\right)^7 = \frac{1}{8} \cdot {}^{16}C_7$$

2. We have, $(3 - 5x)^{15} = 3^{15}\left(1 - \frac{5x}{3}\right)^{15}$

$$= 3^{15}\left(1 - \frac{1}{3}\right)^{15} \quad \left[\because x = \frac{1}{5}\right]$$

$$\therefore \quad p = \frac{(15 + 1)(1/3)}{1 + (1/3)} = \frac{16}{4} = 4$$

is an integer. Therefore, numerically greatest terms are fourth and fifth terms.

3. Consider, $(1 + 1)^{20}$.

Then, $p = \frac{(20 + 1)1}{1 + 1}$

Therefore, $[p] = 10$ and hence 11th term is greatest and its value is ${}^{20}C_{10}$.

PASSAGE 2

$$\because \quad 1 + 4x + 4x^2 = (1 + 2x)^2$$

Then, $\quad (1 + 2x)^{2n} = \sum_{r=0}^{2n} a_r x^r$

Put $x = 1$, then

$$3^{2n} = a_0 + a_1 + a_2 + a_3 + \ldots\ldots\ldots + a_{2n-1} + a_{2n} \quad \text{...(i)}$$

and put $x = -1$, then

$$1 = a_0 - a_1 + a_2 - a_3 + \ldots\ldots\ldots - a_{2n-1} + a_{2n} \text{...(ii)}$$

1. Adding Eqs. (i) and (ii), then $\sum_{r=0}^{n} a_{2r} = \dfrac{3^{2n}+1}{2}$

$\Rightarrow \qquad 2\sum_{r=0}^{n} a_{2r} = 9^n + 1$

2. Subtracting Eq. (ii) from (i), then $\sum_{r=1}^{n} a_{2r-1} = \dfrac{3^{2n}-1}{2}$

$\Rightarrow \qquad 2\sum_{r=0}^{n} a_{2r-1} = 9^n - 1$

3. a_{2n-1} = coefficient of x^{2n-1} in $(1 + 4x + 4x^2)^n$

ie, $\quad (1 + 2x)^{2n} = {}^{2n}C_{2n-1} \cdot 2^{2n-1}$

$$= {}^{2n}C_1 \cdot 2^{2n-1}$$
$$= 2n \cdot 2^{2n-1}$$
$$= n \cdot 2^{2n}$$

4. a_2 = coefficient of x^2 is $(1 + 2x)^{2n}$

$$= {}^{2n}C_2 \cdot 2^2$$
$$= \frac{2n(2n-1)}{2} \times 4$$
$$= 4n(2n-1)$$
$$= 8n^2 - 4n$$

5.
$$a_{n-r} = {}^{2n}C_{n-r} \cdot 2^{n-r}$$
$$a_r = {}^{2n}C_r \cdot 2^r$$
$$a_{n+r} = {}^{2n}C_{n+r} \cdot 2^{n+r}$$
$$a_{2n-r} = {}^{2n}C_{2n-r} \cdot 2^{2n-r}$$

PASSAGE 3

1. $\because (x + C_0)(x + C_1)(x + C_2) \ldots \ldots (x + C_n)$

$$= x^{n+1} + \left(\sum_{r=0}^{n} C_r\right) x^n + \left(\sum_{0 \le i<j \le n}\sum C_i C_j\right) x^{n-1} + \ldots\ldots$$

$\therefore$ Coefficient of x^{n-1} is $\sum_{0 \le i<j \le n}\sum C_i C_j$

$$= \frac{1}{2}\left\{\left(\sum_{r=0}^{n} C_r\right)^2 - \left(\sum_{r=0}^{n} C_r^{\,2}\right)\right\}$$

$$= \frac{1}{2}\{2^{2n} - {}^{2n}C_n\} = 2^{2n-1} - \frac{2n!}{2(n!)^2}$$

2. Let $P = \sum_{0 \le i<j \le n}\sum (i + j) C_i C_j$ ...(i)

Replacing i by $n - i$ and j by $n - j$, then

$$P = \sum_{0 \le i<j \le n}\sum (n - i + n - j) C_{n-i} C_{n-j}$$

$$P = \sum_{0 \le i<j \le n}\sum \{2n - (i + j)\} C_i C_j \ (\because C_r = C_{n-r}) \quad \ldots\text{(ii)}$$

Adding Eqs. (i) and (ii), then we get

$$2P = 2n \sum_{0 \le i<j \le n}\sum C_i C_j$$

$$P = n \sum_{0 \le i<j \le n}\sum C_i C_j$$

3. Let $P = \sum_{0 \le i<j \le n}\sum (i + j)(C_i + C_j + C_i C_j)$ ...(i)

Replacing i by $n - i$ and j by $n - j$, then

$$P = \sum_{0 \le i<j \le n}\sum (2n - i - j)(C_{n-i} + C_{n-j} + C_{n-i} C_{n-j})$$

$$P = \sum_{0 \le i<j \le n}\sum (2n - i - j)(C_i + C_j + C_i C_j) \quad \ldots\text{(ii)}$$

Adding Eqs. (i) and (ii), then

$$2P = 2n \sum_{0 \le i<j \le n}\sum (C_i + C_j + C_i C_j)$$
$$= 2n \sum_{0 \le i<j \le n}\sum (C_i + C_j) + 2n\lambda$$
$$= 2n \cdot (n \cdot 2^n) + 2n\lambda$$

$\therefore \qquad P = n^2 \cdot 2^n + n\lambda$

4. Let $P = \sum_{0 \le i<j \le n}\sum \left(\dfrac{i}{{}^nC_i} + \dfrac{j}{{}^nC_j}\right)$ ...(i)

Replacing i by $n - i$ and j by $n - j$, then

$$P = \sum_{0 \le i<j \le n}\sum \left(\frac{n-i}{{}^nC_{n-i}} + \frac{n-j}{{}^nC_{n-j}}\right)$$

$$= \sum_{0 \le i<j \le n}\sum \left(\frac{n-i}{{}^nC_i} + \frac{n-j}{{}^nC_j}\right) (\because C_r = C_{n-r}) \quad \ldots\text{(ii)}$$

Adding Eqs. (i) and (ii), then

$$2P = n \sum_{0 \le i<j \le n}\sum \left(\frac{1}{{}^nC_i} + \frac{1}{{}^nC_j}\right)$$

$$= n \cdot n \sum_{i=0}^{n} \frac{1}{{}^nC_i} = n^2 a$$

$\therefore \qquad P = \dfrac{n^2 a}{2}$

5. $(x + 2)^3 (x + 3)^2 (x + 5)^3 = (x + 2)(x + 2)(x + 2)(x + 3)$
$(x + 3) \times (x + 5)(x + 5)(x + 5)$

$$= x^8 + (2 + 2 + 2 + 3 + 3 + 5 + 5 + 5)x^7 + x^6(2 \cdot 2 + 2 \cdot 2 + 2 \cdot 3 + \ldots + 5 \cdot 5) + \ldots$$

$\therefore$ Coefficient of x^6

$$= \frac{(2 + 2 + 2 + 3 + 3 + 5 + 5 + 5)^2 - (2^2 + 2^2 + 2^2 + 3^2 + 3^2 + 5^2 + 5^2 + 5^2)}{2}$$

$$= \frac{729 - 105}{2} = 312$$

PASSAGE 4

$\because \qquad (1 + x + x^2)^{2n} = \sum_{r=0}^{4n} a_r x^r$ ...(i)

Replacing x by $\dfrac{1}{x}$ in Eq. (i), then

$$\left(1 + \frac{1}{x} + \frac{1}{x^2}\right)^{2n} = \sum_{r=0}^{4n} a_r \left(\frac{1}{x}\right)^r$$

or $\qquad (1 + x + x^2)^{2n} = \sum_{r=0}^{4n} a_r x^{4n-r}$ ...(ii)

From Eqs. (i) and (ii), we get

$$\sum_{r=0}^{4n} a_r x^r = \sum_{r=0}^{4n} a_r x^{4n-r}$$

Comparing coefficient of x^{4n-r} on both sides, then we get

$$a_r = a_{4n-r} \quad \ldots(\text{iii})$$

1. Put $x = 1$ and $x = -1$ in Eq. (i), then

$9^n = a_0 + a_1 + a_2 + a_3 + a_4 + \ldots + a_{2n} + \ldots + a_{4n}$ and

$1 = a_0 - a_1 + a_2 - a_3 + a_4 - \ldots + a_{2n} + \ldots + a_{4n}$

Adding and subtracting, then we get

$$\frac{9^n + 1}{2} = a_0 + a_2 + a_4 + \ldots + a_{2n} + \ldots + a_{4n-2} + a_{4n} \quad \ldots(\text{iv})$$

and $$\frac{9^n - 1}{2} = a_1 + a_3 + a_5 + \ldots + a_{2n-1} + \ldots \; a_{4n-1} \quad \ldots(\text{v})$$

Now, $\because$ $a_r = a_{4n-r}$

Put $r = 0, 2, 4, 6, \ldots, a_{2n-2}, a_{2n}$

$\therefore$ $a_0 = a_{4n}$

$a_2 = a_{4n-2}$

$a_4 = a_{4n-4}$

....

$a_{2n-2} = a_{2n+2}$

$\therefore$ $a_0 + a_2 + a_4 + \ldots + a_{2n-2}$

$= a_{2n+2} + \ldots + a_{4n-4} + a_{4n-2} + a_{4n}$

Now, from Eq. (iv),

$$\frac{9^n+1}{2} = 2(a_0 + a_2 + a_4 + \ldots. + a_{2n-2}) + a_{2n}$$

$$\Rightarrow \frac{9^n + 1 - 2a_{2n}}{4} = a_0 + a_2 + a_4 + \ldots + a_{2n-2}$$

$$\therefore \sum_{r=0}^{n-1} a_{2r} = \left(\frac{9^n + 1 - 2a_{2n}}{4}\right)$$

2. $\because$ $a_r = a_{4n-r}$

Put $r = 1, 3, 5, 7, \ldots, 2n-3, 2n-1$

$a_1 = a_{4n-1}$

$a_3 = a_{4n-3}$

$a_5 = a_{4n-5}$

....

$a_{2n-3} = a_{2n+3}$

$a_{2n-1} = a_{2n+1}$

$\therefore$ $a_1 + a_3 + a_5 + \ldots + a_{2n-1}$

$= a_{2n+1} + a_{2n+3} + \ldots + a_{4n-3} + a_{4n-1}$

Now, from Eq. (v)

$$\frac{9^n-1}{2} = 2(a_1 + a_3 + a_5 + \ldots + a_{2n-1})$$

$$\therefore \sum_{r=1}^{n} a_{2r-1} = \left(\frac{9^n-1}{4}\right)$$

3. a_2 = coefficient of x^2 in $(1 + x + x^2)^{2n}$

= coefficient of x^2 in

$\{1 + {}^{2n}C_1 (x + x^2) + {}^{2n}C_2 (x + x^2)^2 + \ldots\}$

$= {}^{2n}C_1 + {}^{2n}C_2$

$= {}^{2n+1}C_2$

4. $a_{4n-1} = a_1$ $\quad [\because a_r = a_{4n-r}]$

= coefficient of x in $(1 + x + x^2)^{2n}$

= coefficient of x in $\{1 + x(1 + x)\}^{2n}$

$= {}^{2n}C_1 = 2n$

5. From Eq. (iii), $a_r = a_{4n-r}$, $0 \le r \le 4n$

PASSAGE 5

1. Required sum = coefficient of x^n in $(1 + x)^n (1 + x)^n$

= coefficient of x^n in $(1 + x)^{2n} = {}^{2n}C_n$

2. $\because$ ${}^{30}C_r \cdot {}^{20}C_0 + {}^{30}C_{r-1} \cdot {}^{20}C_1 \cdot + \ldots + {}^{30}C_0 \cdot {}^{20}C_r$

= coefficient of x^r in $(1 + x)^{30} (1 + x)^{20}$

= coefficient of x^r in $(1 + x)^{50} = {}^{50}C_r$

$\because$ ${}^{50}C_r$ is maximum

$\therefore$ $r = \dfrac{50}{2} = 25$

3. $\because$ ${}^{20}C_r \cdot {}^{10}C_0 + {}^{20}C_{r-1} \cdot {}^{10}C_1 + \ldots + {}^{20}C_0 \cdot {}^{10}C_r$

= coefficient of x^r in $(1 + x)^{20} (1 + x)^{10}$

= coefficient of x^r in $(1 + x)^{30} = {}^{30}C_r$

$\because$ ${}^{30}C_r$ is minimum

$\therefore$ $r = 0$

4. $\because$ $S_n = {}^nC_0 \cdot {}^nC_1 + {}^nC_1 \cdot {}^nC_2 + \ldots + {}^nC_{n-1} \cdot {}^nC_n$

$= {}^nC_0 \cdot {}^nC_{n-1} + {}^nC_1 \cdot {}^nC_{n-2} + \ldots + {}^nC_{n-1} \cdot {}^nC_0$ $\quad [\because {}^nC_r = {}^nC_{n-r}]$

= coefficient of x^{n-1} in $(1 + x)^n (1 + x)^n$

= coefficient of x^{n-1} in $(1 + x)^{2n} = {}^{2n}C_{n-1}$

$$\because \frac{S_{n+1}}{S_n} = \frac{15}{4} \Rightarrow \frac{{}^{2n+2}C_n}{{}^{2n}C_{n-1}} = \frac{15}{4}$$

$$\Rightarrow \left(\frac{2n+2}{n}\right) \cdot \frac{{}^{2n+1}C_{n-1}}{{}^{2n}C_{n-1}} = \frac{15}{4}$$

$$\Rightarrow \left(\frac{2n+2}{n}\right) \cdot \frac{\dfrac{(2n+1)!}{(n-1)!(n+2)!}}{\dfrac{2n!}{(n-1)!(n+1)!}} = \frac{15}{4}$$

$$\Rightarrow \left(\frac{2n+2}{n}\right) \cdot \frac{(2n+1)}{(n+2)} = \frac{15}{4}$$

$$\Rightarrow 4(4n^2 + 6n + 2) = 15n^2 + 30n$$

$$\Rightarrow n^2 - 6n + 8 = 0 \Rightarrow (n-2)(n-4) = 0$$

$$\therefore n = 2, 4$$

5. $\because$ $C_0^2 - C_1^2 + C_2^2 - C_3^2 + \ldots + (-1)^n C_n^2$

$= C_0 C_0 - C_1 C_1 + C_2 C_2 - C_3 C_3 + \ldots + (-1)^n C_n C_n$

$= C_0 C_n - C_1 C_{n-1} + C_2 C_{n-2} - C_3 C_{n-3} + \ldots + (-1)^n C_n C_0$

$[\because C_r = C_{n-r}]$

= Coefficient of x^n in $(1 + x)^n (1 - x)^n$

= Coefficient of x^n in $(1 - x^2)^n$

$= (-1)^{n/2} \cdot {}^nC_{n/2} = 0$ $\quad [\because n \text{ is odd}]$

PASSAGE 6

1. $\because T_{r+1} = {}^{6561}C_r (7^{1/3})^{6561-r} \cdot (11^{1/9})^r$

$= {}^{6561}C_r \cdot 7^{2187-\frac{r}{3}} \cdot 11^{\frac{r}{9}}$

$\therefore T_{r+1}$ is rational.

If $\frac{r}{9}$ and $\frac{r}{3}$ are integers

$\therefore$ r is a multiple of 9

$\because \quad 0 \le r \le 6561$

$\Rightarrow \quad 0 \le \frac{r}{9} \le 729$

$\therefore \quad \frac{r}{9} = 0, 1, 2, 3, \dots, 729$

$\therefore$ Total terms $= 730$

2. $\because T_{r+1} = {}^{110}C_r (2^{1/2})^{110-r} \cdot (3^{1/11})^r$

$= {}^{110}C_r \cdot 2^{55-\frac{r}{2}} \cdot 3^{\frac{r}{11}}$

If T_{r+1} is rational, then $\frac{r}{2}$ and $\frac{r}{11}$ are integers

$\therefore$ r is a multiple of 22.

$\because \quad 0 \le r \le 110$

$\therefore \quad r = 0, 22, 44, 66, 88, 110$

The number of rational terms = 6.

Hence, number of irrational terms $= 111 - 6 = 105$

3. $\because \; T_{r+1} = {}^{20}C_r (2^{1/5})^{20-r} \cdot (\sqrt{3})^r$

$= {}^{20}C_r \cdot 2^{4-\frac{r}{5}} \cdot 3^{\frac{r}{2}}$

As 2 and 3 are relatively primes. T_{r+1} is rational, if $\frac{r}{5}$ and $\frac{r}{2}$ are integers.

$\therefore$ r is multiple of 10.

$\because \quad 0 \le r \le 20$

$\therefore \quad r = 0, 10, 20$

Thus, the sum of rational terms

$= T_1 + T_{11} + T_{21}$

$= {}^{20}C_0 \cdot 2^4 \cdot 3^0 + {}^{20}C_{10} \cdot 2^2 \cdot 3^5 + {}^{20}C_{20} \cdot 2^0 \cdot 3^{10}$

This is more than 21, 84, 97.

4. In $(3^{1/2} + 5^{1/8})^{256}$

$T_{r+1} = {}^{256}C_r (3^{1/2})^{256-r} \cdot (5^{1/8})^r$

$= {}^{256}C_r \cdot 2^{128-\frac{r}{2}} \cdot 5^{\frac{r}{8}}$

For T_{r+1} is integer, then $128 - \frac{r}{2} \ge 0, \frac{r}{8} \ge 0$ are integers, then

$r = 0, 8, 16, 24, 32, \dots, 256$

$\therefore$ Number of integral terms = 33

5. $\because (9^{1/4} + 8^{1/6} \cdot x)^{500} = (3^{1/2} + 2^{1/2} \cdot x)^{500}$

$\therefore \; T_{r+1} = {}^{500}C_r (3^{1/2})^{500-r} \cdot (2^{1/2} \cdot x)^r$

$= {}^{500}C_r \cdot 3^{250-\frac{r}{2}} \cdot 2^{\frac{r}{2}} \cdot x^r$

For integral coefficient,

$250 - \frac{r}{2} \ge 0, \frac{r}{2} \ge 0$ are integers.

Then, r is divisible by 2

$\because \quad 0 \le r \le 500$

$\Rightarrow \quad 0 \le \frac{r}{2} \le 250$

Number of integral coefficients = 251

PASSAGE 7

1. $(x_1 + x_2 + x_3 + \dots + x_n)^4$

$$= \sum \frac{4!}{\lambda_1!\lambda_2!\lambda_3!\dots\dots\lambda_n!}(x_1)^{\lambda_1}(x_2)^{\lambda_2}(x_3)^{\lambda_3}(x_4)^{\lambda_4}\dots\dots(x_n)^{\lambda_n}$$

where $\lambda_1 + \lambda_2 + \lambda_3 + \dots + \lambda_n = 4$ and $\lambda_1, \lambda_2, \dots, \lambda_n$ are whole numbers *ie*, $\lambda_i \ge 0$

Thus, number of distinct terms in the expansion of $(x_1 + x_2 + x_3 + \dots + x_n)^4$

= Number of non negative integral solutions of

$\lambda_1 + \lambda_2 + \lambda_3 + \dots + \lambda_n = 4 = {}^{4+n-1}C_{n-1}$

$= {}^{n+3}C_{n-1} = {}^{n+3}C_4$

2. $\because (1 + x - y + z)^9$

$$= \sum \frac{9!}{\lambda_1!\lambda_2!\lambda_3!\lambda_4!}(1)^{\lambda_1}(x)^{\lambda_2}(-y)^{\lambda_3}(z)^{\lambda_4}$$

$$= \sum \frac{9!}{\lambda_1!\lambda_2!\lambda_3!\lambda_4!} \cdot x^{\lambda_2} \cdot y^{\lambda_3} \cdot z^{\lambda_4}(-1)^{\lambda_3}$$

Here, $\lambda_2 = 3, \lambda_3 = 4, \lambda_4 = 1$ and $\lambda_1 + \lambda_2 + \lambda_3 + \lambda_4 = 9$

$\therefore \quad \lambda_1 = 1$

$\Rightarrow$ Required coefficient $= \frac{9!}{1!3!4!1!}(-1)^4$

$= \frac{9 \cdot 8 \cdot 7 \cdot 6 \cdot 5}{1 \cdot 6 \cdot 1} \cdot 1$

$= 9 \cdot 8 \cdot 7 \cdot 5 = 2520$

3. The coefficient of $a^3b^4c^5$ in $(bc + ca + ab)^6$

or the coefficient of $a^3b^4c^5$ in $a^6b^6c^6\left(\frac{1}{a}+\frac{1}{b}+\frac{1}{c}\right)^6$

or the coefficient of $\left(\frac{1}{a}\right)^3\left(\frac{1}{b}\right)^2\left(\frac{1}{c}\right)^1$ in $\left(\frac{1}{a}+\frac{1}{b}+\frac{1}{c}\right)^6$

$= \frac{6!}{3!\,2!\,1!} = \frac{6 \cdot 5 \cdot 4}{2 \cdot 1} = 60$

4. $\because \; (1 + x + 2x^2)^{20} = \sum \frac{20!}{\lambda_1!\lambda_2!\lambda_3!}(1)^{\lambda_1}(x)^{\lambda_2}(2x^2)^{\lambda_3}$

$= \sum \frac{20!}{\lambda_1!\lambda_2!\lambda_3!} \cdot x^{\lambda_2 + 2\lambda_3} \cdot 2^{\lambda_3}$

$\therefore \quad \lambda_1 + \lambda_2 + \lambda_3 = 20$ and $\lambda_2 + 2\lambda_3 = 39$

λ_1	λ_2	λ_3
0	1	19

(only)

$\therefore$ Coefficient of x^{37} is $\frac{20!}{1!\,1!\,19!} \cdot 2^{19} = (20) \cdot 2^{19} = 5 \cdot 2^{21}$

5. Let $(1-x+x^2)^{20} = a_0 + a_1x + a_2x^2 + \dots + a_{40}x^{40}$

Replacing x by $-x$, then

$(1+x+x^2)^{20} = a_0 - a_1x + a_2x^2 - \dots + a_{40}x^{40}$

$\therefore$ coefficient of x^{20} in $(1-x+x^2)^{20}$

= coefficient of x^{20} in $(1+x+x^2)^{20}$

Now, let

$(1+x+\lambda x^2)^{20} = b_0 + b_1x + b_2x^2 + b_3x^3 + \dots + b_{40}x^{40}$

Dividing both sides by x^{20}, then

$$\left(1+\lambda x+\frac{1}{x}\right)^{20} = \frac{b_0}{x^{20}} + \frac{b_1}{x^{19}} + \dots + b_{20} + b_{21}x + \dots + b_{40}x^{20}$$

$$\therefore\ b_{20} = \text{constant term in } \left(1+\lambda x+\frac{1}{x}\right)^{20}$$

$$= \text{constant term in } \left\{1 + {}^{20}C_1\left(\lambda x+\frac{1}{x}\right) + {}^{20}C_2\left(\lambda x+\frac{1}{x}\right)^2 + {}^{20}C_3\left(\lambda x+\frac{1}{x}\right)^3 + \dots + {}^{20}C_{20}\left(\lambda x+\frac{1}{x}\right)^{20}\right\}$$

$$= 1 + ({}^{20}C_2)({}^2C_1\lambda) + ({}^{20}C_4)({}^4C_2)\lambda^2 + \dots + {}^{20}C_{20}({}^{20}C_{10}\lambda^{10})$$

$\therefore\ a = b_{20}$ (at $\lambda = 1$) and $b = b_{20}$ (at $\lambda = -1$)

Then, we can say that

$\therefore\ a > b$

PASSAGE 8

1. $\because 5^{97} = 5\cdot 5^{96} = 5(25)^{48} = 5(26-1)^{48}$

$$= 5\{(26)^{48} - {}^{48}C_1(26)^{47} + {}^{48}C_2(26)^{46} - \dots - {}^{48}C_{47}(26) + 1\}$$

$= 5\{52\lambda + 1\}$ (λ is +ve integer)

$= 5(52)\lambda + 5$

$\therefore$ Remainder = 5

2. $13^{99} - 19^{93}$ = (odd number) − (odd number) = even number

$\therefore\ 13^{99} - 19^{93}$ is divisible by 2.

Also, $13^{99} - 19^{93} = (1+12)^{99} - (1+18)^{93}$

$$= (1 + {}^{99}C_1\cdot 12 + {}^{99}C_2\cdot 12^2 + {}^{99}C_3\cdot 12^3 + \dots) - (1 + {}^{93}C_1\cdot 18 + {}^{93}C_2\cdot 18^2 + {}^{93}C_3\cdot 18^3 + \dots)$$

$$= (99\times 12 - 93\times 18) + (12)^2\{{}^{99}C_2 + 12.{}^{99}C_3\} + 81\times \text{Integer}$$

$$= -486 + 144\{4851 + 12\times 156849\} + 81\times\text{Integer}$$

$= 81\times$ Integer

But 2 and 81 are coprime.

$\therefore\ 13^{99} - 19^{93}$ is divisible by 162.

$\therefore$ Remainder is 0.

3. $27^{10} + 7^{51} = \underbrace{27^{10} - 7^{10}}_{\substack{\text{divisible by}\\ 27-7=10}} + 7^{10} + 7^{51}$

$= 7^{10} + 7^{51} = (7^2)^5 + 7(7^2)^{25}$

$= (50-1)^5 + 7(50-1)^{25}$

$$= \{(50)^5 - {}^5C_1(50)^4 + {}^5C_2(50)^3 \dots + {}^5C_4(50) - {}^5C_5(1)\} + 7\{(50)^{25} - {}^{25}C_1(50)^{24} + \dots + {}^{25}C_{24}(50) - {}^{25}C_{25}(1)\}$$

$= (10\lambda - 1) + 7(10\mu - 1)$ (λ and μ are integers).

$= 10(\lambda + 7\mu) - 8 = 10(\lambda + 7\mu - 1) + 2$

$= 10\nu + 2$ (ν is integer)

$\therefore$ Remainder is 2.

4. $(106)^{85} - \{(85)^{106} = (106)^{85}\ 0 - \underbrace{\{(85)^{106} - 1\}}_{\substack{\text{is divisible by}\\ 85-1=84\\ \therefore\text{ also divisible by 7}}} = (106)^{85} - 1$

divisible by $106 - 1 = 105 = 15\times 7$

It is divisible by 7.

$\therefore$ Remainder is 0.

5. $53^{53} - 33^3 = \underbrace{53^{53} - 3^{53}}_{\substack{\text{divisible by 50,}\\ \text{then it is also}\\ \text{divisible by 10}}} + 3^{53} - \underbrace{(33^3 - 3^3}_{\substack{\text{divisible by 30,}\\ \text{then it is also}\\ \text{divisible by 10}}} + 3^3) = 3^{53} - 3^3$

$= 3^3\{3^{50} - 1\} = 3^3\{(10-1)^{25} - 1\}$

$= 3^3\{10\lambda - 2\} = 270\lambda - 54$

$= 270\lambda - 60 + 6 = 10\mu + 6$

where λ and μ are integers.

$\therefore$ Remainder is 6.

Numerical Value Type

1. Since, $(3-5x)^{11} = 3^{11}\left(1 - \frac{5x}{3}\right)^{11}$ $\left(\because x = \frac{1}{3}\right)$

$$\text{Now, } m = \frac{\left|-\frac{5x}{3}\right|(11+1)}{\left|\frac{-5x}{3}\right|+1} = \frac{\left|-\frac{1}{3}\right|\times 12}{\left|\frac{-1}{3}\right|+1} = \frac{\frac{1}{3}\times 12}{\frac{1}{3}+1} = 3$$

$\therefore$ The greatest terms in the expansion are T_3 and T_4.

$\therefore$ Greatest term (when $r = 2$)

$$= 3^{11}|T_{2+1}| = 3^{11}\cdot\left|{}^{11}C_2\left(-\frac{1}{3}\right)^2\right|$$

$= 55\times 3^9 = (55\times 27)\times 3^6$

$= (1485)\times 729$

$\therefore\ \lambda = 1485$

2. Using the formula for the sum of a geometric progression, we find

$(1+x)^{1000} + x(1+x)^{999} + x^2(1+x)^{998} + \dots + x^{1000}$

$$= \frac{(1+x)^{1000}\left[1 - \left(\frac{x}{1+x}\right)^{1001}\right]}{\left[1 - \left(\frac{x}{1+x}\right)\right]}$$

$$= \frac{(1+x)^{1000} - \frac{x^{1001}}{(1+x)}}{\frac{x+1-x}{(1+x)}} = (1+x)^{1001} - x^{1001}$$

Hence, the coefficient of $x^{50} = {}^{1001}C_{50}$

$$= \frac{1001!}{50!951!}$$

$\therefore \quad \lambda = 1001, \mu = 50$ and $\nu = 951$

$\therefore \quad \lambda + 2\mu + 3\nu = 1001 + 100 + 2853$

$$= 3954$$

3. Let the three consecutive coefficients ${}^nC_{r-1}$, ${}^nC_r$ and ${}^nC_{r+1}$ are in ratio 2 : 5 : 12 in the expansion of $\left(1 + \frac{1}{x}\right)^n$.

$$\therefore \quad {}^nC_{r-1} : {}^nC_r : {}^nC_{r+1} = 2 : 5 : 12$$

So, $$\frac{{}^nC_{r-1}}{{}^nC_r} = \frac{2}{5}$$

$$\Rightarrow \quad \frac{\dfrac{n!}{(r-1)!(n-r+1)!}}{\dfrac{n!}{r!(n-r)}} = \frac{2}{5}$$

$$\Rightarrow \quad \frac{r}{n-r+1} = \frac{2}{5}$$

$$\Rightarrow \quad 2n - 7r + 2 = 0 \quad \ldots\text{(i)}$$

and $$\frac{{}^nC_r}{{}^nC_{r+1}} = \frac{5}{12}$$

$$\Rightarrow \quad \frac{\dfrac{n!}{r!(n-r)!}}{\dfrac{n!}{(r+1)!(n-r-1)!}} = \frac{5}{12}$$

$$\Rightarrow \quad \frac{r+1}{n-r} = \frac{5}{12}$$

$$\Rightarrow \quad 5n - 17r - 12 = 0 \quad \ldots\text{(ii)}$$

On the elimination of 'r' from Eqs. (i) and (ii), we get

$$35n - 119r - 84 = 0$$
$$34n - 119r + 34 = 0$$
$$- \quad + \quad -$$
$$n - 118 = 0 \Rightarrow n = 118$$

Hence, answer is 118.00.

4. $\because \quad (1 + x)^n = \sum_{r=0}^{n} C_r x^r$

$\therefore \; x(1 + x)^n = \sum_{r=0}^{n} C_r x^{r+1}$

Differentiating both sides w.r.t. x, then

$$(1 + x)^{n-1}(1 + x + nx) = \sum_{r=0}^{n} (r + 1) C_r x^r$$

Again, multiplying both sides by x, then

$$(1 + x)^{n-1}(x + x^2 + nx^2) = \sum_{r=0}^{n} (r + 1) \cdot C_r \cdot x^{r+1}$$

Again, differentiating both sides w.r.t. x, then

$$(1 + x)^{n-2}\{(1 + x)(1 + 2x + 2nx) + (n - 1)(x + x^2 + nx^2)\}$$
$$= \sum_{r=0}^{n} (r + 1)^2 C_r x^r$$

Putting $x = 1$ on both sides, then we get

$$\sum_{r=0}^{n} (r + 1)^2 C_r = 2^{n-2}(n^2 + 5n + 4)$$

$\therefore \quad f(n) = n^2 + 5n + 4$

Then, $f(x) = x^2 + 5x + 4 = (x + 1)(x + 4)$

$\therefore \quad \alpha = -1, \beta = -4$

Then, $\alpha^4 + \beta^4 = (-1)^4 + (-4)^4$

$$= 1 + 256 = 257$$

5. Let $\sqrt{(x^5 - 1)} = \lambda$

$\therefore \; (\lambda + x)^9 - (\lambda - x)^9$

$$= 2\{{}^9C_1\lambda^8 x + {}^9C_3\lambda^6 x^3 + {}^9C_5\lambda^4 x^5 + {}^9C_7\lambda^2 x^7 + {}^9C_9 x^9\}$$
$$= 2\{{}^9C_1 x(x^5 - 1)^4 + {}^9C_3 x^3 (x^5 - 1)^3 + {}^9C_5 x^5 (x^5 - 1)^2 + {}^9C_7 (x^5 - 1) x^7 + {}^9C_9 x^9\}$$

$\therefore$ Degree of polynomial is 21.

6. $\because \quad (1 + 2x^2 - x^3)^8 = \sum \frac{8!}{\alpha!\beta!\gamma!}(1)^\alpha (2x^2)^\beta (-x^3)^\gamma$

$$= \sum \frac{8!}{\alpha!\beta!\gamma!} \cdot (2)^\beta \cdot (-1)^\gamma \cdot x^{2\beta + 3\gamma} \quad \ldots\text{(i)}$$

$\Rightarrow \quad 2\beta + 3\gamma = 12$

and $\quad \alpha + \beta + \gamma = 8$

α	β	γ
2	6	0
3	3	2
4	0	4

$\therefore$ Coefficient of x^{12} is

$$\frac{8!}{2!6!0!}(2)^6 \cdot (-1)^0 + \frac{8!}{3!3!2!}(2)^3 \cdot (-1)^2 + \frac{8!}{4!0!4!}(2)^0(-1)^4$$
$$= 1792 + 4480 + 70 = 6342$$

7. Let A be the property that 'beg' occurs

$\therefore \quad n(A) = 5! = 120$

and let B be the property that 'cad' occurs

$\therefore \quad n(B) = 5! = 120$

Now, $\quad n(A \cap B) = {}^3P_3 = 3! = 6$

$\Rightarrow \quad n(A \cup B) = n(A) + n(B) - n(A \cap B)$

$$= 120 + 120 - 6$$
$$= 234$$

Total number of permutations $= {}^7P_7 = 7! = 7040$

$\therefore$ Required number of permutations $= 5040 - 234$

$$= 4806$$

8. We have,

$$1 + x + x^2 + x^3 = 1 + x + x^2(1 + x) = (1 + x)(1 + x^2)$$

$$\Rightarrow \; (1 + x + x^2 + x^3)^{11} = (1 + x)^{11}(1 + x^2)^{11}$$
$$= (1 + {}^{11}C_1 x + {}^{11}C_2 x^2 + {}^{11}C_3 x^3 + {}^{11}C_4 x^4 + \ldots)(1 + {}^{11}C_1 x^2 + {}^{11}C_2 x^4 + \ldots)$$

Thus, coefficient of x^4 in $(1 + x + x^2 + x^3)^{11}$

$$= (1)({}^{11}C_2) + ({}^{11}C_2)({}^{11}C_1) + ({}^{11}C_4)(1)$$
$$= 55 + (55)(11) + 330 = 990$$

9. Put $x = 1$

Then, sum of coefficients $= 4^n$

Given $4000 < 4^n < 10000$

$\therefore \quad 4^n = 4^6 = 4096$

$\therefore \quad n = 6$

Then, $(1 + 3x)^6$

Now, $m = \dfrac{|3x|(6+1)}{(|3x|+1)} = \dfrac{3 \times 7}{4} = 5.25$

$\therefore$ Greatest term $T_{[m]+1} = T_{5+1} = {}^6C_5 (3x)^5$

$= 6 (3)^5 x^5$

$\therefore$ Greatest coefficient $= 6 \times 243 = 1458$

10. $\because \lambda = (1992^{1998} - 1955^{1998}) - (1938^{1998} - 1901^{1998})$

$\therefore \lambda$ is divisible by $1992 - 1955 = 37$

Also, $\lambda = (1992^{1998} - 1938^{1998}) - (1955^{1998} - 1901^{1998})$

$\therefore \quad \lambda$ is divisible by $= 1992 - 1938 = 54$

Hence, λ is divisible by $= 37 \times 54 = 1998$

11. $T_{r+1} = {}^nC_r (x)^{n-r} \left(\dfrac{a}{x^2}\right)^2 = {}^nC_r a^r x^{n-3r}$

According to the question,

${}^nC_2 a^2 : {}^nC_3 a^3 : {}^nC_4 a^4 = 12 : 8 : 3$

$\dfrac{{}^nC_2 a^2}{{}^nC_3 a^3} = \dfrac{12}{8} \Rightarrow a(n-2) = 2$

$\Rightarrow \quad an - 2a = 2 \quad \ldots(i)$

$\Rightarrow \dfrac{{}^nC_3 a^3}{{}^nC_4 a^4} = \dfrac{8}{3} \Rightarrow 2an - 6a = 3 \quad \ldots(ii)$

Solving Eqs. (i) and (ii), we get

$n = 6$ and $a = \dfrac{1}{2}$

For term, independent of x

$\Rightarrow \quad n = 3r \Rightarrow r = 2$

Coefficient of term independent of x is

$= {}^6C_2 \left(\dfrac{1}{2}\right)^2$

$= \dfrac{15}{4} = 3.75$

12. Let $P = \sum_{r=0}^{n} {}^nC_r \sin(rx) \cos\{(n-r)x\} \quad \ldots(i)$

Replacing r by $n - r$, then

$P = \sum_{r=0}^{n} {}^nC_{n-r} \sin\{(n-r)x\} \cos[\{n-(n-r)\} x]$

$= \sum_{r=0}^{n} {}^nC_r \sin\{(n-r)x\} \cos rx \; (\because {}^nC_r = {}^nC_{n-r}) \ldots(ii)$

Adding Eqs. (i) and (ii), then

$2P = \sum_{r=0}^{n} {}^nC_r \sin(nx) = \sin(nx) \cdot 2^n$

or $\quad P = 2^{n-1} \cdot \sin(nx)$

$\therefore \quad f(n) = 2^{n-1}$

Then, $\quad f(13) = 2^{12} = 4096$

13. $\because 5^{81} + \lambda = 5 \cdot 5^{80} + \lambda = 5 \cdot (5^2)^{40} + \lambda$

$= 5(26-1)^{40} + \lambda$

$= 5\{{}^{40}C_0(26)^{40} - {}^{40}C_1(26)^{39} + {}^{40}C_2(26)^{38} - \ldots - {}^{40}C_{39}(26) + 1\} + \lambda$

$= 5\{26\mu + 1\} + \lambda$, where $\mu \in N$

$= 5 \times 26\mu + 5 + \lambda$

$\because \; 5^{81} + \lambda$ is divisible by 26

$\therefore \quad 5 + \lambda = 26$

$\Rightarrow \quad \lambda = 21$

14. $\because \quad 2^{2005} = 2 \cdot 2^{2004} = 2 \cdot (2^4)^{501} = 2(17-1)^{501}$

$= 2\{{}^{501}C_0(17)^{501} - {}^{501}C_1(17)^{500} + {}^{501}C_2(17)^{499} - \ldots + {}^{501}C_{500}(17)^1 - {}^{501}C_{501} \cdot 1\}$

$= 2\{17\mu - 1\} = 2\{17\mu - 17 + 16\} \; (\mu \in N)$

$= 34(\mu - 1) + 32$

$= 34(\mu - 1) + 17 + 15$

$\therefore$ Remainder $(\lambda) = 15$

15. $\because \quad a_r = r^2 \cdot \dfrac{(100 - r + 1)}{r} = r \cdot (101 - r)$

$\because \prod_{i=1}^{100} (x - a_i) = (x - a_1)(x - a_2)(x - a_3) \ldots (x - a_{100})$

$= x^{100} - (a_1 + a_2 + a_3 + \ldots + a_{100}) x^{99} + \ldots$

$\therefore \quad \lambda = -(a_1 + a_2 + a_3 + \ldots + a_{100})$

$= -\sum_{r=1}^{100} a_r = -\sum_{r=1}^{100} r \cdot (100 - r)$

$= -100 \sum_{r=1}^{100} r + \sum_{r=1}^{100} r^2$

$= -505000 + 338350 = -166650$

$\therefore \quad -\dfrac{\lambda}{50} = \dfrac{166650}{2500} = 66.66$

16. Required sum $= 13 + 17 + 21 + 25 + 29 + \ldots + 97$

$= \dfrac{22}{2} \times (13 + 97) = 11 \times 110$

$= 1210$

Single Integer Type

1. We have, $(x-y)^4+(x+y)^4 = 2[{}^4C_0x^4 + {}^4C_2x^2y^2 + {}^4C_4y^4]$

Thus, $(x^2-\sqrt{1-x^3})^4 + (x^2+\sqrt{1-x^3})^4$

$= 2[{}^4C_0(x^2)^4 + {}^4C_2(x^2)^2(1-x^3) + {}^4C_4(1-x^3)^2]$

which is a polynomial of degree 8.

2. It is given that, $(2x^2+3x+4)^{10} = \sum_{r=0}^{20} a_r x^r$

Now, the general term in the expansion of trinomial $(2x^2+3x+4)^{10}$ is $\frac{10!}{n_1!n_2!n_3!}2^{n_1}3^{n_2}4^{n_3}x^{2n_1+n_2}$

such that $n_1+n_2+n_3 = 10$ and $n_1, n_2, n_3 \geq 0$

For a_7, $2n_1+n_2 = 7$

$$\therefore a_7 = \frac{10!}{0!7!3!}2^0 3^7 4^3 + \frac{10!}{1!5!4!}2^1 3^5 4^4 + \frac{10!}{2!3!5!}2^2 3^3 4^5 + \frac{10!}{3!1!6!}2^3 3^1 4^6$$

n_1	n_2	n_3
0	7	3
1	5	4
2	3	5
3	1	6

and, for a_{13}, $2n_1+n_2 = 13$

$$\therefore a_{13} = \frac{10!}{3!7!0!}2^3 3^7 + \frac{10!}{4!5!1!}2^4 3^5 4^1 + \frac{10!}{5!3!2!}2^5 3^3 4^2 + \frac{10!}{6!1!3!}2^6 3^1 4^3$$

n_1	n_2	n_3
0	13	Not possible
1	11	Not possible
2	9	Not possible
3	7	0
4	5	1
5	3	2
6	1	3

$$\therefore \frac{a_7}{a_{13}} = \frac{\frac{10!}{7!3!}3^7 4^3 + \frac{10!}{5!4!}2^1 3^5 4^4 + \frac{10!}{2!3!5!}2^2 3^3 4^5 + \frac{10!}{3!6!}2^3 3^1 4^6}{\frac{10!}{3!7!}2^3 3^7 + \frac{10!}{4!5!}2^4 3^5 4^1 + \frac{10!}{5!3!2!}2^5 3^3 4^2 + \frac{10!}{6!3!}2^6 3^1 4^3}$$

$$= 2^3\left[\frac{\frac{10!}{7!3!}2^3 3^7 + \frac{10!}{4!5!}2^6 3^5 + \frac{10!}{2!3!5!}2^9 3^3 + \frac{10!}{6!3!}2^{12} 3^1}{\frac{10!}{3!7!}2^3 3^7 + \frac{10!}{4!5!}2^6 3^5 + \frac{10!}{2!3!5!}2^9 3^3 + \frac{10!}{6!3!}2^{12} 3^1}\right]$$

$= 2^3 = 8$

Hence, answer is 8.00.

3. Given, $f(n) = {}^nC_0 a^{n-1} - {}^nC_1 a^{n-2} + {}^nC_2 a^{n-3} + \ldots + (-1)^{n-1}\,{}^nC_{n-1}a^0$

$$= \frac{1}{a}({}^nC_0 a^n - {}^nC_1 a^{n-1} + {}^nC_2 a^{n-2} + \ldots + (-1)^{-1}C_{n-1}a)$$

$$= \frac{1}{a}((a-1)^n - (-1)^n\,{}^nC_n)$$

$$= \frac{1}{a}\left(\left(3^{\frac{n}{223}} - (-1)^n\right)\right)$$

$$f(x) = \frac{3^{\frac{n}{223}} - (-1)^n}{\left(3^{\frac{1}{223}}+1\right)} \Rightarrow f(2007) = \frac{3^{\frac{2007}{223}}+1}{3^{\frac{1}{223}}+1}$$

$$\Rightarrow f(2008) = \frac{3^{\frac{2008}{223}}-1}{3^{\frac{1}{223}}+1}$$

$$\Rightarrow f(2007)+f(2008) = \frac{3^{\frac{2007}{223}} + 3^{\frac{2008}{223}}}{3^{\frac{1}{223}}+1}$$

$$= \frac{3^9 + 3^{9+\frac{1}{223}}}{3^{\frac{1}{223}}+1} = 3^9\frac{\left(1+3^{\frac{1}{223}}\right)}{\left(1+3^{\frac{1}{223}}\right)} = 3^9$$

$\Rightarrow \quad 3^9 = 3^k$ then $k = 9$

4. We have, $7-4\sqrt{3} = \frac{1}{7+4\sqrt{3}}$

$\therefore \quad 0 < 7-4\sqrt{3} < 1 \Rightarrow 0 < (7-4\sqrt{3})^{2n} < 1$

Let $F = (7-4\sqrt{3})^{2n}$. Then

$x + F = (7+4\sqrt{3})^{2n} + (7-4\sqrt{3})^{2n}$

$= 2[{}^{2n}C_0 7^{2n} + {}^{2n}C_2 7^{2n-2}(4\sqrt{3})^2 + {}^{2n}C_4 7^{2n-4}(4\sqrt{3})^4 + \ldots + {}^{2n}C_{2n}(4\sqrt{3})^{2n}]$

$= 2m$, where m is some positive integer.

$\Rightarrow [x] + f + F = 2m \Rightarrow f + F = 2m - [x]$

Since, $0 \leq f < 1$ and $0 < F < 1$, we get $0 < f + F < 2$.

Also, since $f + F$ is an integer, we must have $f + F = 1$

Thus, $x(1-f) = xF = (7+4\sqrt{3})^{2n}(7-4\sqrt{3})^{2n}$

$= (49-48)^{2n} = 1^{2n} = 1$

5. Coefficient of x^2 in the expansion of

$\{(1+x)^2 + (1+x)^3 + \ldots + (1+x)^{49} + (1+mx)^{50}\}$

$\Rightarrow {}^2C_2 + {}^3C_2 + {}^4C_2 + \ldots + {}^{49}C_2 + {}^{50}C_2 \cdot m^2 = (3n+1).{}^{51}C_3$

$\Rightarrow \quad {}^{50}C_3 + {}^{50}C_2 m^2 = (3n+1).{}^{51}C_3$

$[\because {}^rC_r + {}^{r+1}C_r + \ldots + {}^nC_r = {}^{n+1}C_{r+1}]$

$$\Rightarrow \frac{50\times49\times48}{3\times2\times1} + \frac{50\times49}{2}\times m^2 = (3n+1)\frac{51\times50\times49}{3\times2\times1}$$

$\Rightarrow \quad m^2 = 51n + 1$

$\therefore$ Minimum value of m^2 for which $(51n+1)$ is integer (perfect square) for $n=5$.

$\therefore\ m^2 = 51\times 5+1 \Rightarrow m^2 = 256 \Rightarrow m=16$ and $n=5$

Hence, the value of n is 5.

6. $P = \sum_{r=0}^{5} C_{2r} = {}^{10}C_0 + {}^{10}C_2 + {}^{10}C_4 + \ldots + {}^{10}C_{10} = \dfrac{2^{10}}{2} = 2^9$

$Q = \sum_{r=0}^{3} d_{2r+1} = d_1 + d_3 + d_5 + d_7$

$= {}^7C_1 + {}^7C_3 + {}^7C_5 + {}^7C_7 = \dfrac{2^7}{2} = 2^6$

$\therefore\ \dfrac{P}{Q} = 2^3 = 8$

7. We have, $\left((9^{x-1}+7)^{\frac{1}{3}} + \dfrac{1}{(3^{x-1}+1)^{1/8}}\right)$

$T_4 = {}^{11}C_3 \cdot (9^{x-1}+7)\cdot \dfrac{1}{(3^{x-1}+1)} = 660 \Rightarrow \dfrac{9^{x-1}+7}{3^{x-1}+1} = 4$

$\Rightarrow\ t^2 - 4t + 3 = 0 \Rightarrow t = 1, 3$

$\Rightarrow\ x = 1$ or 2

Matrix-Match Type

1. **(A)** We have, $\sum_{r=0}^{14} \dfrac{(15-r)}{r+1}\,{}^{15}C_r\,{}^{20}C_{m-r}$

$= \sum_{r=0}^{14} \dfrac{(15-r)}{r+1}\dfrac{15!}{r!(15-r)!}\,{}^{20}C_{m-r}$

$= \sum_{r=0}^{14} \dfrac{15!}{(r+1)!(14-r)!}\,{}^{20}C_{m-r}$

$= \sum_{r=0}^{14} {}^{15}C_{r+1}\,{}^{20}C_{m-r} = {}^{35}C_{m+1}$

Which has maximum value ${}^{35}C_{18}$.

(B) We have,

${}^{35}C_0 - {}^{35}C_1 + {}^{35}C_2 - {}^{35}C_3 + \ldots + (-1)^r\ {}^{35}C_r + \ldots$

= Coefficient of x^r in

$({}^{35}C_0 - {}^{35}C_1x + {}^{35}C_2x^2 - {}^{35}C_3x^3 + \ldots + (-1)^r\,{}^{35}C_r + \ldots)$
$\times (1 + x + x^2 + x^3 + \ldots + x^r + \ldots)$

= Coefficient of x^r in $(1-x)^{35}(1-x)^{-1}$

= Coefficient of x^r in $(1-x)^{34} = (-1)^r\,{}^{35}C_r$

Required maximum value is ${}^{34}C_{17}$.

(C) We have,

${}^{31}C_{r-5} + 5\times {}^{31}C_{r-4} + 10\times {}^{31}C_{r-3} + 10\times {}^{31}C_{r-2} + 5\times {}^{31}C_{r-1} + {}^{31}C_r$

$= {}^5C_5 \times {}^{31}C_{r-5} + {}^5C_4 \times {}^{31}C_{r-4} + {}^5C_3\times {}^{31}C_{r-3} + {}^5C_2 \times {}^{31}C_{r-2} + {}^5C_1\times {}^{31}C_{r-1} + {}^5C_0 \times {}^{31}C_r$

= Coefficient of x^r in $(1+x)^5(1+x)^{31}$

= Coefficient of x^r in $(1+x)^{36} = {}^{36}C_r$

Which has maximum value of ${}^{36}C_{18}$.

(D) We have,

$\sum_{r=0}^{k} 2^{k-r}\cdot(-1)^r\cdot {}^{37}C_r \cdot {}^{37-r}C_{37-k}$

$= \sum_{r=0}^{k} 2^{k-r}\cdot(-1)^r \cdot \dfrac{37!}{(37-r)!r!}\dfrac{(37-r)!}{(37-k)!(k-r)!}$

$= \dfrac{37!}{k!(37-k)!}\sum_{r=0}^{k} 2^{k-r}\cdot(-1)^r\cdot\dfrac{k!}{r!(k-r)!}$

$= {}^{37}C_k \sum_{r=0}^{k} {}^kC_r 2^{k-r}\cdot(-1)^r = {}^{37}C_k(2-1)^k = {}^{37}C_k$

Which has maximum value ${}^{37}C_{18}$.

2. **(A)** $\because (1+5x+10x^2+10x^3+5x^4+x^5)^{20} = \{(1+x)^5\}^{20} = (1+x)^{100}$

$\therefore\ \lambda = 100+1 = 101$

Then $3^\lambda = 3^{101} = 3\cdot 3^{100} = 3(9)^{50} = 3(10-1)^{50}$

$= 3\{(10)^{50} - {}^{50}C_1(10)^{49} + {}^{50}C_2(10)^{48} - \ldots - {}^{50}C_{49}(10) + 1\}$

$= 3\{100\mu + 1\} = 300\mu + 3$ (μ is +ve integer)

$\therefore$ Last two digits 03

$\therefore\ O = 3, T = 0 \Rightarrow O + T = 3$ (P)

(B) $\left(x^2 + 1 + \dfrac{1}{x^2}\right)^{100}$

$\therefore\ \lambda = 2\times 100 + 1 = 201$

Then $7^\lambda = 7^{201} = 7\cdot 7^{200} = 7\cdot(7^2)^{100} = 7\cdot(49)^{100}$

$= 7(50-1)^{100}$

$= 7\{(50)^{100} - {}^{100}C_1(50)^{99} + {}^{100}C_2(50)^{98} - \ldots - {}^{100}C_{99}(50) + 1\}$

$= 7\{100\mu + 1\} = 700\mu + 7$ ($\because \mu$ is +ve integer)

$\therefore$ Last two digits 07

$\therefore\ O = 7, T = 0$

$\Rightarrow O + T = 7$ and $O - T = 7$ (Q, T)

(C) $\because (1+x)^{101}(1+x^2-x)^{100}$

$= (1+x)\{(1+x)(1+x^2-x)\}^{100}$

$= (1+x)(1+x^3)^{100}$

$= (1+x)\{1 + {}^{100}C_1x^3 + {}^{100}C_2x^6 + {}^{100}C_3x^9 + \ldots + {}^{100}C_{100}x^{303}\}$

$= 1 + {}^{100}C_1x^3 + {}^{100}C_2x^6 + {}^{100}C_3x^9 + \ldots + {}^{100}C_{100}x^{303} + x + {}^{100}C_1x^4 + {}^{100}C_2x^7 + {}^{100}C_3x^{10} + \ldots + {}^{100}C_{100}x^{304}$

$\therefore\ \lambda = 1 + 100 + 101 = 202$

$\Rightarrow\ 9^\lambda = 9^{202} = (10-1)^{202}$

$= (10)^{202} - {}^{202}C_1(10)^{201} + {}^{202}C_2(10)^{200} - \ldots - {}^{202}C_{201}(10) + {}^{202}C_{202}$

$= 100\mu - 2020 + 1$ (μ is +ve integer)

$= 100(\mu - 21) + 81$

$= 100\nu + 81$ (ν is +ve integer)

$\therefore$ Last two digits 81

$\therefore\ O = 1$ and $T = 8$

$\Rightarrow O + T = 9$ and $T - O = 7$ (R, S)

3. (A) $\because$ Every even power of 9 can be represented in the form $9^{2r} = 81^r = \underbrace{81 \cdot 81 \cdot 81 \cdot \ldots \cdot 81}_{r \text{ times}}$

It's last digit is 1, every odd power of 9 can be written as $9^{2r+1} = 9 \cdot 81^r$, therefore its last digit is 9. In particular $9^{(9^9)}$ is an odd power of 9 and consequently the last digit of 9^{9^9} is 9.

$\therefore \quad \lambda = 9$

Now, $2^{\lambda^{100}} = 2^{9^{100}}$

$\because \quad 9^{100} = (2 \cdot 4 + 1)^{100} = 4n + 1$ (say)

$\therefore \quad 2^{9^{100}} = 2^{4n+1} = 2 \cdot (2^4)^n = 2 \cdot (16)^n$

The digit at units place in $(16)^n = 6$

$\therefore$ The digit at units place in $(16)^n \cdot 2 = 2$

$\therefore \quad \mu = 2 \Rightarrow \lambda + \mu = 11,\ \lambda - \mu = 7$ (Q, R)

(B) $2^1 = 2,\ 2^2 = 4,\ 2^3 = 8,\ 2^4 = 16$

we say that units digit in

$2^1\ 2^2, 2^3, 2^4, 2^5, 2^6, 2^7, 2^8, \ldots\ldots$ are

2, 4, 8, 6, 2, 4, 8, 6, ... (period4)

(1) (2) (3) (0) (1) (2) (3) (0)

$$4\overline{)999}(249$$
$$996$$
$$3$$

$\therefore$ Last digit of the number 2^{999} is 8

$\therefore \quad \lambda = 8$

Also, $3^{\lambda\lambda\lambda} = 3^{888} = (3^2)^{444} = (10-1)^{444} = 10n + 1$

$\therefore \quad \mu = 1$

$\Rightarrow \lambda + \mu = 9,\ \lambda - \mu = 7,\ \lambda^{\mu} + \mu^{\lambda} = 8^1 + 1^8 = 9$ (P, R, T)

(C) $a = \dfrac{72!}{(36!)^2} - 1 = \dfrac{(72 \cdot 71 \cdot 70 \cdot \ldots \cdot 37)\,36!}{(36!)^2} - 1$

$= \dfrac{72 \cdot 71 \cdot 70 \cdot \ldots \cdot 37}{36!} - 1$

$= \dfrac{(1-73)(2-73)(3-73)\ldots(36-73)}{36!} - 1$

$= \dfrac{(1 \cdot 2 \cdot 3 \cdot \ldots \cdot 36) - 73(m)}{36!} - 1$ [m is an integer]

$= 1 - \dfrac{73m}{36!} - 1 = -\dfrac{73m}{36!}$

which is divisible by 73

$= 70 + 3$

$\therefore \quad \lambda = 7, \mu = 3 \left(\begin{array}{l} \because a = \dfrac{72!}{(36!)^2} - 1 = {}^{72}C_{36} - 1 \\ = \text{even} - 1 = \text{odd} \end{array} \right)$

Then, $\lambda - \mu = 4$ (S)

4. (A) Given expression $= \sqrt{(3x^2+1)} \cdot 2\{{}^7C_1 x^6 (\sqrt{3x^2+1})$

$+ {}^7C_3 x^4 (\sqrt{3x^2+1})^3 + {}^7C_5 x^2 (\sqrt{(3x^2+1)})^5$

$+ {}^7C_7 (\sqrt{(3x^2+1)})^7\}$

$= (3x^2+1) \cdot 2\{7x^6 + 35x^4(3x^2+1)$

$+ 21x^2(3x^2+1)^2 + (3x^2+1)^3\}$

$\therefore \quad n = 8$ (P, Q, R)

(B) $\because \dfrac{T_7}{T_6} \ge 1 \Rightarrow \dfrac{{}^nC_6 x^{n-6} a^6}{{}^nC_5 x^{n-5} a^5} \ge 1 \Rightarrow \left(\dfrac{n-6+1}{6}\right) \cdot \left(\dfrac{2}{3}\right) \ge 1$

$\Rightarrow \quad 2n - 10 \ge 18 \Rightarrow n \ge 14$

Similarly, $\dfrac{T_7}{T_8} \ge 1 \Rightarrow n \le \dfrac{33}{2} \Rightarrow n = 16$

$\because$ There is only one middle term, n must be even. $n = 16$ gives the greatest term. (P, Q, R, S)

(C) $\because 200 < 2^m < 400 \Rightarrow m = 8\ (m \in I)$

$\therefore \quad T_{r+1} = {}^8C_r (x^{-3/4})^{8-r} \cdot (nx^{5/4})^r = 448\, x^0$

$\therefore \quad x^{-6 + \frac{3r}{4} + \frac{5r}{4}} = x^0 \Rightarrow -6 + 2r = 0 \Rightarrow r = 3$

$\Rightarrow \quad {}^8C_3 \cdot n^3 = 448 \Rightarrow n^3 = 8 \Rightarrow n = 2$

$\therefore \quad n^5 = 32$ (P, Q, R, S, T)

5. (A) $\because {}^nC_0 + {}^nC_1 + {}^nC_2 = 46 \Rightarrow 1 + n + \dfrac{n(n-1)}{2} = 46$

$\Rightarrow \quad 2 + 2n + n^2 - n = 92 \Rightarrow n^2 + n - 90 = 0$

$\Rightarrow \quad n = 9, n \ne -10$

$\therefore \quad \left(x^2 + \dfrac{1}{x}\right)^n = \left(x^2 + \dfrac{1}{x}\right)^9$

$\therefore \quad T_{r+1} = {}^9C_r (x^2)^{9-r} \cdot \left(\dfrac{1}{x}\right)^r = {}^9C_r \cdot x^{18-3r}$

$\because \quad 18 - 3r = 0 \Rightarrow r = 6$

$T_{6+1} = {}^9C_6 = {}^9C_3 = \dfrac{9 \cdot 8 \cdot 7}{1 \cdot 2 \cdot 3} = 84$ (T)

(B) $T_{r+1} = {}^nC_r \left(\dfrac{5}{x^2}\right)^{n-r} \cdot (x^4)^r = {}^nC_r \cdot (5)^{n-r} \cdot x^{-2n+2r+4r}$

$= {}^nC_r \cdot (5)^{n-r} \cdot x^{6r-2n}$

$\because 6r - 2n = 0$

$\therefore \quad r = \dfrac{n}{3}$

$\therefore \quad n = 9, 15, 18, 27, 84$ (P, Q, R, S, T)

(C) $\because T_{r+1} = {}^nC_r (x^3)^{n-r} \cdot \left(-\dfrac{1}{x^2}\right)^r = {}^nC_r \cdot (-1)^r \cdot x^{3n-5r}$

For coefficient of x^5, we get $3n - 5r = 5$

$\Rightarrow \quad r = \dfrac{3n-5}{5} = p$ (say)

For coefficient of x^{10}, we get $3n - 5r = 10$

$\Rightarrow \quad r = \dfrac{3n-10}{5} = q$ (say)

$\because \quad p - q = 1 \quad \ldots$(i)

Also, ${}^nC_p(-1)^p + {}^nC_q(-1)^q = 0$

${}^nC_p(-1)^p + {}^nC_{p-1}(-1)^{p-1} = 0$ [from Eq. (i)]

$\Rightarrow \quad (-1)^p\{{}^nC_p - {}^nC_{p-1}\} = 0 \Rightarrow {}^nC_p - {}^nC_{p-1} = 0$

$\Rightarrow \quad {}^nC_p = {}^nC_{p-1}$

or $\quad n = p + p - 1 \Rightarrow p = \dfrac{n+1}{2}$

$\Rightarrow \quad \dfrac{3n-5}{5} = \dfrac{n+1}{2}$

$\Rightarrow \quad 6n - 10 = 5n + 5$

$\therefore \quad n = 15$ (Q)

06

Determinant

Objective Questions Type I [Only one correct answer]

In each of the questions below, four choices are given of which only one is correct. You have to select the correct answer which is the most appropriate.

1. The number of distinct values of a 2×2 determinant whose entries are from the set $\{-1, 0, 1\}$ is

(a) 3 (b) 4

(c) 5 (d) 6

2. If λ is a non real cube root of -2, then the value of

$$\begin{vmatrix} 1 & 2\lambda & 1 \\ \lambda^2 & 1 & 3\lambda^2 \\ 2 & 2\lambda & 1 \end{vmatrix}$$ is equal to

(a) -11 (b) -12

(c) -13 (d) 0

3. Let $\begin{vmatrix} x & 2 & x \\ x^2 & x & 6 \\ x & x & 6 \end{vmatrix} = Ax^4 + Bx^3 + Cx^2 + Dx + E$, then the value of $5A + 4B + 3C + 2D + E$ is equal to

(a) -17 (b) -14

(c) -11 (d) -8

4. If $\Delta = \begin{vmatrix} my + nz & mq - nr & mb + nc \\ kz - mx & kr - mp & kb - ma \\ nx + ky & np + kq & na + kb \end{vmatrix}$, then

(a) $\Delta = 0$ (b) $\Delta = xyz + abc + pqr$

(c) $\Delta = f(x, y, z)$ (d) $\Delta = f(p, q, r)$

5. If $0 \le [x] < 2; -1 \le [y] < 1$ and $1 \le [z] < 3$. Where $[\cdot]$ denotes the greatest integer function, then the maximum value of the determinant

$$\begin{vmatrix} [x]+1 & [y] & [z] \\ [x] & [y]+1 & [z] \\ [x] & [y] & [z]+1 \end{vmatrix}$$ is

(a) 2 (b) 4 (c) 6 (d) 8

6. The value of θ in $(-2\pi, \pi)$ for which the system $(\sin\theta)x - y = 0, x + 2y = 0$ has infinity of solutions are given by

(a) $\left\{-\frac{\pi}{6}, -\frac{5\pi}{6}\right\}$ (b) $\left\{\frac{\pi}{6}, \frac{5\pi}{6}\right\}$

(c) $\left\{-\frac{7\pi}{6}, \frac{11\pi}{6}\right\}$ (d) Both (b) and (c)

7. The equations $\lambda x - y = 2, 2x - 3y = -\lambda$, $3x - 2y = -1$ are consistent for

(a) $\lambda = -4$ (b) $\lambda = -1, 4$

(c) $\lambda = -1$ (d) $\lambda = 1, -4$

8. If the minimum and the maximum values of the function $f: \left[\frac{\pi}{4}, \frac{\pi}{2}\right] \to R$, defined by

$$f(\theta) = \begin{vmatrix} -\sin^2\theta & -1 - \sin^2\theta & 1 \\ -\cos^2\theta & -1 - \cos^2\theta & 1 \\ 12 & 10 & -2 \end{vmatrix}$$

are m and M respectively, then the ordered pair (m, M) is equal to **[JEE Main 2020, 5 September]**

(a) $(0, 2\sqrt{2})$ (b) $(-4, 4)$

(c) $(0, 4)$ (d) $(-4, 0)$

9. If $f(x) = \begin{vmatrix} (1+x)^a & (1+2x)^b & 1 \\ 1 & (1+x)^a & (1+2x)^b \\ (1+2x)^b & 1 & (1+x)^a \end{vmatrix}$; a, b being positive integers, then

(a) constant term of $f(x)$ is 4

(b) coefficient of x in $f(x)$ is 0

(c) constant term in $f(x)$ is $(a - b)$

(d) constant term in $f(x)$ is $(a + b)$

10. If $\Delta_r = \begin{vmatrix} a^r & b^r & c^r \\ a & b & c \\ 1-a & 1-b & 1-c \end{vmatrix}$. If $\sum_{r=0}^{\infty} \Delta_r = 0$, then which statement is true.

(a) $a = b = c, a \in R$ (b) $a = b = c, |a| < 1$
(c) $a = b = c \neq 1$ (d) $a = b \neq c, |c| < 1$

11. If the value of the determinant $\begin{vmatrix} a & 1 & 1 \\ 1 & b & 1 \\ 1 & 1 & c \end{vmatrix}$ is positive, then

(a) $abc > 1$ (b) $abc > -8$
(c) $abc < -8$ (d) $abc > -2$

12. Given that $q^2 - pr < 0$, $p > 0$ the value of

$\begin{vmatrix} p & q & px+qy \\ q & r & qx+ry \\ px+qy & qx+ry & 0 \end{vmatrix}$ is

(a) zero (b) positive
(c) negative (d) $q^2 + pr$

13. If $f(n) = \alpha^n + \beta^n$ and $\begin{vmatrix} 3 & 1+f(1) & 1+f(2) \\ 1+f(1) & 1+f(2) & 1+f(3) \\ 1+f(2) & 1+f(3) & 1+f(4) \end{vmatrix}$

$= k(1-\alpha)^2(1-\beta)^2(\alpha-\beta)^2$, then k is equal to

(a) 1 (b) -1
(c) $\alpha\beta$ (d) $\alpha\beta\gamma$

14. Let α and β be the roots of the equation $x^2 + x + 1 = 0$. Then, for $y \neq 0$ in R,

$\begin{vmatrix} y+1 & \alpha & \beta \\ \alpha & y+\beta & 1 \\ \beta & 1 & y+\alpha \end{vmatrix}$ is equal to **[JEE Main 2019, 9 April]**

(a) $y(y^2-1)$ (b) $y(y^2-3)$
(c) $y^3 - 1$ (d) y^3

15. If $\Delta_1 = \begin{vmatrix} 1 & 1 & 1 \\ a & b & c \\ a^2 & b^2 & c^2 \end{vmatrix}$, $\Delta_2 = \begin{vmatrix} 1 & bc & a \\ 1 & ca & b \\ 1 & ab & c \end{vmatrix}$, then

(a) $\Delta_1 + \Delta_2 = 0$ (b) $\Delta_1 + 2\Delta_2 = 0$
(c) $\Delta_1 = \Delta_2$ (d) $\Delta_1 = 2\Delta_2$

16. If α, β are non real numbers satisfying $x^3 - 1 = 0$, then the value of $\begin{vmatrix} \lambda+1 & \alpha & \beta \\ \alpha & \lambda+\beta & 1 \\ \beta & 1 & \lambda+\alpha \end{vmatrix}$ is equal to

(a) 0 (b) λ^3 (c) $\lambda^3 + 1$ (d) $\lambda^3 - 1$

17. Let $\theta = \frac{\pi}{5}$ and $A = \begin{bmatrix} \cos\theta & \sin\theta \\ -\sin\theta & \cos\theta \end{bmatrix}$.

If $B = A + A^4$, then det (B) **[JEE Main 2020, 6 September]**

(a) is one (b) line in (2, 3)
(c) is zero (d) lies in (1, 2)

18. In a third order determinant a_{ij} denotes the element in the ith row and the jth column, ifz a_{ij} $= \begin{cases} 0, & i = j \\ 1, & i > j \\ -1, & i < j \end{cases}$, then the value of the determinant is

(a) 0 (b) 1
(c) -1 (d) None of these

19. If $\Delta(x) = \begin{vmatrix} x & 1+x^2 & x^3 \\ \log(1+x^2) & e^x & \sin x \\ \cos x & \tan x & \sin^2 x \end{vmatrix}$, then

(a) $\Delta(x)$ is divisible by x (b) $\Delta(x) = 0$
(c) $\Delta'(x) = 0$ (d) None of these

20. The value of the determinant

$\begin{vmatrix} \log_a(x/y) & \log_a(y/z) & \log_a(z/x) \\ \log_b(y/z) & \log_b(z/x) & \log_b(x/y) \\ \log_c(z/x) & \log_c(x/y) & \log_c(y/z) \end{vmatrix}$ is

(a) 1 (b) -1
(c) $\log_a xyz$ (d) None of these

21. If $\sqrt{-1} = i$ and ω is a non real cube root of unity, then the value of $\begin{vmatrix} 1 & \omega^2 & 1+i+\omega^2 \\ -i & -1 & -1-i+\omega \\ 1-i & \omega^2-1 & -1 \end{vmatrix}$ is equal to

(a) 1 (b) i
(c) ω (d) 0

22. If x, y, z are in arithmetic progression with common difference d, $x \neq 3d$, and the determinant of the matrix $\begin{bmatrix} 3 & 4\sqrt{2} & x \\ 4 & 5\sqrt{2} & y \\ 5 & k & z \end{bmatrix}$ is zero, then the value of k^2 is **[JEE Main 2021, 17 March]**

(a) 72 (b) 12
(c) 36 (d) 6

23. Using the factor theorem it is found that $b+c$, $c+a$ and $a+b$ are three factors of the determinant $\begin{vmatrix} -2a & a+b & a+c \\ b+a & -2b & b+c \\ c+a & c+b & -2c \end{vmatrix}$. The other factor in the value of the determinant is

(a) 4 (b) 2
(c) $a+b+c$ (d) None of these

24. The value of the determinant

$\begin{vmatrix} 1 & e^{i\pi/3} & e^{i\pi/4} \\ e^{-i\pi/3} & 1 & e^{2i\pi/3} \\ e^{-i\pi/4} & e^{-2i\pi/3} & 1 \end{vmatrix}$,

where $i = \sqrt{-1}$, is

(a) $2+\sqrt{2}$ (b) $-(2+\sqrt{2})$
(c) $-2+\sqrt{3}$ (d) $2-\sqrt{3}$

25. If $f(x)=\begin{vmatrix} 1 & x & x+1 \\ 2x & x(x-1) & x(x+1) \\ 3x(x-1) & x(x-1)(x-2) & x(x^2-1) \end{vmatrix}$, then $f(200)$ is equal to

(a) 1 (b) 0
(c) 200 (d) –200

26. For the system of linear equations

$x-2y=1,\ x-y+kz=-2,\ ky+4z=6,\ k\in R$

Consider the following statements: **[JEE Main 2021, 24 February]**

(A) The system has unique solution, if $k\neq 2,\ k\neq -2$.
(B) The system has unique solution, if $k=-2$.
(C) The system has unique solution, if $k=2$.
(D) The system has no solution, if $k=2$.
(E) The system has infinite number of solutions, if $k\neq -2$.

Which of the following statements are correct?

(a) (B) and (E) only
(b) (C) and (D) only
(c) (A) and (D) only
(d) (A) and (E) only

27. If a, b, c are sides of a triangle and $\begin{vmatrix} a^2 & b^2 & c^2 \\ (a+1)^2 & (b+1)^2 & (c+1)^2 \\ (a-1)^2 & (b-1)^2 & (c-1)^2 \end{vmatrix}=0$, then

(a) ΔABC is an equilateral triangle
(b) ΔABC is a right angled isosceles triangle
(c) ΔABC is an isosceles triangle
(d) None of the above

28. If the system of equations $2x-y+z=0$, $x-2y+z=0$, $tx-y+2z=0$ has infinitely many solutions and $f(x)$ be a continuous function such that $f(5+x)+f(x)=2$, then $\int_0^{-2t} f(x)\,dx$ is equal to

(a) 0 (b) $-2t$
(c) 5 (d) t

29. If $a+x=b+y=c+z+1$, where a, b, c, x, y, z are non-zero distinct real numbers, then $\begin{vmatrix} x & a+y & x+a \\ y & b+y & y+b \\ z & c+y & z+c \end{vmatrix}$ is equal to

[JEE Main 2020, 15 September]

(a) $y(b-a)$ (b) $y(a-b)$
(c) 0 (d) $y(a-c)$

30. For positive numbers x, y, z the numerical value of the determinant $\begin{vmatrix} 1 & \log_x y & \log_x z \\ \log_y x & 3 & \log_y z \\ \log_z x & \log_z y & 5 \end{vmatrix}$ is

(a) 0 (b) $\log x \log y \log z$
(c) 1 (d) 8

31. If $f(x)=ax^2+bx+c,\ a, b, c\in R$ and equation $f(x)-x=0$ has imaginary roots α and β and γ, δ be the roots of $f\{f(x)\}-x=0$, then $\begin{vmatrix} 2 & \alpha & \delta \\ \beta & 0 & \alpha \\ \gamma & \beta & 1 \end{vmatrix}$ is

(a) 0 (b) purely real
(c) purely imaginary (d) None of these

32. If n is a positive integer, then

$\begin{vmatrix} {}^{n+2}C_n & {}^{n+3}C_{n+1} & {}^{n+4}C_{n+2} \\ {}^{n+3}C_{n+1} & {}^{n+4}C_{n+2} & {}^{n+5}C_{n+3} \\ {}^{n+4}C_{n+2} & {}^{n+5}C_{n+3} & {}^{n+6}C_{n+4} \end{vmatrix}$ is equal to

(a) 3 (b) –1
(c) –5 (d) –9

33. Let $\Delta(x)=\begin{vmatrix} (x-2) & (x-1)^2 & x^3 \\ (x-1) & x^2 & (x+1)^3 \\ x & (x+1)^2 & (x+2)^3 \end{vmatrix}$, then the coefficient of x in $\Delta(x)$ is

(a) –3 (b) –2
(c) –1 (d) 0

34. If $f(x)=\begin{vmatrix} x & \cos x & e^{x^2} \\ \sin x & x^2 & \sec x \\ \tan x & 1 & 2 \end{vmatrix}$, then the value of $\int_{-\pi/2}^{\pi/2} f(x)\,dx$ is equal to

(a) 5 (b) 3
(c) 1 (d) 0

35. Let ω be a complex number such that $2\omega+1=z$, where $z=\sqrt{-3}$.

If $\begin{vmatrix} 1 & 1 & 1 \\ 1 & -\omega^2-1 & \omega^2 \\ 1 & \omega^2 & \omega^7 \end{vmatrix}=3k$, then k is equal to

[JEE Main 2017 (Offline)]

(a) $-z$ (b) z
(c) -1 (d) 1

36. If α, β, γ are the roots of the equation $x^3+px+q=0$, then the value of the determinant $\begin{vmatrix} \alpha & \beta & \gamma \\ \beta & \gamma & \alpha \\ \gamma & \alpha & \beta \end{vmatrix}$ is

(a) 4 (b) 2
(c) 0 (d) –2

37. If $a, b, c>0$ and $x, y, z\in R$, then the value of the determinant

$\begin{vmatrix} (a^x+a^{-x})^2 & (a^x-a^{-x})^2 & 1 \\ (b^y+b^{-y})^2 & (b^y-b^{-y})^2 & 1 \\ (c^z+c^{-z})^2 & (c^z-c^{-z})^2 & 1 \end{vmatrix}$ is

(a) 6 (b) 4
(c) 2 (d) 0

38. If $A+B+C=\pi$, then $\begin{vmatrix} \sin(A+B+C) & \sin B & \cos C \\ -\sin B & 0 & \tan A \\ \cos(A+B) & -\tan A & 0 \end{vmatrix}$ is equal to

(a) 1 (b) 0
(c) −1 (d) 2

39. Let $f(x)=\begin{vmatrix} \cos x & \sin x & \cos x \\ \cos 2x & \sin 2x & 2\cos 2x \\ \cos 3x & \sin 3x & 3\cos 3x \end{vmatrix}$, then $f'(\pi/2)$ is equal to

(a) 8 (b) 6
(c) 4 (d) 2

40. The value of $\begin{vmatrix} (-1)^n a & (-1)^{n+1} b & (-1)^{n+1} c \\ a+1 & 1-b & 1+c \\ a-1 & b+1 & 1-c \end{vmatrix} + \begin{vmatrix} (-1)^{n+1} a & a+1 & a-1 \\ (-1)^n b & 1-b & b+1 \\ (-1)^{n+2} c & 1+c & 1-c \end{vmatrix}$ is equal to

(a) 3 (b) 1
(c) −1 (d) None of these

41. If 3^n is a factor of the determinant $\begin{vmatrix} 1 & 1 & 1 \\ {}^nC_1 & {}^{n+3}C_1 & {}^{n+6}C_1 \\ {}^nC_2 & {}^{n+3}C_2 & {}^{n+6}C_2 \end{vmatrix}$, then the maximum value of n is

(a) 7 (b) 5 (c) 3 (d) 1

42. If a, b, c are non-zero real numbers and if the equations $(a-1)x=y+z$, $(b-1)y=z+x$, $(c-1)z=x+y$ has a non trivial solution, then $ab+bc+ca$ is equal to

(a) $a+b+c$ (b) abc
(c) 1 (d) None of these

43. The roots of the equation

$$\begin{vmatrix} 3x^2 & x^2+x\cos\theta+\cos^2\theta & x^2+x\sin\theta+\sin^2\theta \\ x^2+x\cos\theta+\cos^2\theta & 3\cos^2\theta & 1+\dfrac{\sin 2\theta}{2} \\ x^2+x\sin\theta+\sin^2\theta & 1+\dfrac{\sin 2\theta}{2} & 3\sin^2\theta \end{vmatrix}=0$$ are

(a) $\sin\theta, \cos\theta$ (b) $\sin^2\theta, \cos^2\theta$
(c) $\sin\theta, \cos^2\theta$ (d) $\sin^2\theta, \cos\theta$

44. In a triangle ABC, the value of the determinant

$$\begin{vmatrix} \sin\dfrac{A}{2} & \sin\dfrac{B}{2} & \sin\dfrac{C}{2} \\ \sin(A+B+C) & \sin\dfrac{B}{2} & \cos\dfrac{A}{2} \\ \cos\left(\dfrac{A+B+C}{2}\right) & \tan(A+B+C) & \sin\dfrac{C}{2} \end{vmatrix}$$

is less than or equal to

(a) 1/2 (b) 1/4
(c) 1/8 (d) None of these

45. Let $\lambda \in R$. The system of linear equations

$$2x_1-4x_2+\lambda x_3=1,$$
$$x_1-6x_2+x_3=2$$
$$\lambda x_1-10x_2+4x_3=3$$

is inconsistent for **[JEE Main 2020, 5 September]**

(a) exactly two values of λ
(b) exactly one positive value of λ
(c) every value of λ
(d) exactly one negative value of λ

46. If the system of linear equation

$$x+ky+3z=0$$
$$3x+ky-2z=0$$
$$2x+4y-3z=0$$

has a non-zero solution (x, y, z), then $\dfrac{xz}{y^2}$ is equal to **[JEE Main 2018]**

(a) −10 (b) 10 (c) −30 (d) 30

47. If all elements of a third order determinant are equal to 1 or −1, then the determinant itself is

(a) an odd number (b) an even number
(c) an imaginary number (d) a real number

48. The coefficient of x in the determinant $\begin{vmatrix} (1+x)^{a_1b_1} & (1+x)^{a_1b_2} & (1+x)^{a_1b_3} \\ (1+x)^{a_2b_1} & (1+x)^{a_2b_2} & (1+x)^{a_2b_3} \\ (1+x)^{a_3b_1} & (1+x)^{a_3b_2} & (1+x)^{a_3b_3} \end{vmatrix}$ is

(a) 4 (b) 2
(c) 0 (d) −2

49. The set of all values of λ for which the system of linear equations $2x_1-2x_2+x_3=\lambda x_1$, $2x_1-3x_2+2x_3=\lambda x_2$ and $-x_1+2x_2=\lambda x_3$ has a non-trivial solution **[JEE Main 2015]**

(a) is an empty set
(b) is a singleton set
(c) contains two elements
(d) contains more than two elements

50. If $\begin{vmatrix} x^k & x^{k+2} & x^{k+3} \\ y^k & y^{k+2} & y^{k+3} \\ z^k & z^{k+2} & z^{k+3} \end{vmatrix} = (x-y)(y-z)(z-x)\left(\dfrac{1}{x}+\dfrac{1}{y}+\dfrac{1}{z}\right)$, then

(a) $k=-2$ (b) $k=-1$
(c) $k=0$ (d) $k=1$

51. If $c<1$ and the system of equations $x+y-1=0$, $2x-y-c=0$, and $-bx+3by-c=0$ is consistent, then the possible real value of b are

(a) $b \in \left(-3, \dfrac{3}{4}\right)$ (b) $b \in \left(-\dfrac{3}{2}, 4\right)$

(c) $b \in \left(-\dfrac{3}{4}, 3\right)$ (d) None of these

52. If $[x]$ denotes the greatest integer $\leq x$, then the system of liner equations $[\sin\theta]x + [-\cos\theta]y = 0$, $[\cot\theta]x + y = 0$ **[JEE Main 2019, 12 April]**

(a) have infinitely many solutions if $\theta \in \left(\frac{\pi}{2}, \frac{2\pi}{3}\right)$ and has a unique solution if $\theta \in \left(\pi, \frac{7\pi}{6}\right)$.

(b) has a unique solution if $\theta \in \left(\frac{\pi}{2}, \frac{2\pi}{3}\right) \cup \left(\pi, \frac{7\pi}{6}\right)$

(c) has a unique solution if $\theta \in \left(\frac{\pi}{2}, \frac{2\pi}{3}\right)$ and have infinitely many solutions if $\theta \in \left(\pi, \frac{7\pi}{6}\right)$

(d) have infinitely many solutions if $\theta \in \left(\frac{\pi}{2}, \frac{2\pi}{3}\right) \cup \left(\pi, \frac{7\pi}{6}\right)$

53. The system of equations

$$(p\alpha + q)x + py + qz = 0$$
$$(q\alpha + r)x + qy + rz = 0$$
$$(p\alpha + q)y + (q\alpha + r)z = 0$$

has a non-trivial solution, if

(a) $2p = q + r$

(b) $\frac{p}{r} = \left(\frac{q}{p}\right)^2$

(c) $\frac{2}{q} = \frac{1}{p} + \frac{1}{r}$

(d) $p\alpha^2 + 2q\alpha + r = 0$

54. Let $\Delta = \begin{vmatrix} 1 & \sin\theta & 1 \\ -\sin\theta & 1 & \sin\theta \\ -1 & -\sin\theta & 1 \end{vmatrix}$; $0 \leq \theta < 2\pi$, then

(a) $\Delta = 0$ (b) $\Delta \in (0, \infty)$
(c) $\Delta \in [-1, 2]$ (d) $\Delta \in [2, 4]$

Objective Questions Type II [One or more than one correct answer(s)]

In each of the questions below four choices of which one or more than one are correct. You have to select the correct answer(s) accordingly.

1. The determinant

$$\Delta = \begin{vmatrix} a^2 + x^2 & ab & ac \\ ab & b^2 + x^2 & bc \\ ac & bc & c^2 + x^2 \end{vmatrix}$$ is divisible by

(a) x (b) x^2 (c) x^3 (d) x^4

2. If $D_k = \begin{vmatrix} 2^{k-1} & \frac{1}{k(k+1)} & \sin k\theta \\ x & y & z \\ 2^n - 1 & \frac{n}{n+1} & \frac{\sin\left(\frac{n+1}{2}\right)\theta \sin\left(\frac{n\theta}{2}\right)}{\sin\left(\frac{\theta}{2}\right)} \end{vmatrix}$,

then $\sum_{k=1}^{n} D_k$ is equal to

(a) 0
(b) independent of n
(c) independent of θ
(d) independent of x, y and z

3. If x is real and

$$\Delta(x) = \begin{vmatrix} x^2 + x & 2x - 1 & x + 3 \\ 3x + 1 & 2 + x^2 & x^3 - 3 \\ x - 3 & x^2 + 4 & 2x \end{vmatrix}$$
$$= a_0x^7 + a_1x^6 + a_2x^5 + \ldots + a_6x + a_7, \text{ then}$$

(a) $a_7 = 21$ (b) $\sum_{k=0}^{6} a_k = 111$
(c) $\Delta(-1) = -32$ (d) $\Delta(1) = 121$

4. The determinant $\begin{vmatrix} a & b & a\alpha + b \\ b & c & b\alpha + c \\ a\alpha + b & b\alpha + c & 0 \end{vmatrix}$ is equal to zero, if

(a) a, b, c are in AP
(b) a, b, c are in GP
(c) a, b, c are in HP
(d) α is a root of $ax^2 + bx + c = 0$

5. If $a > b > c$ and the system of equations $ax + by + cz = 0$, $bx + cy + az = 0$, $cx + ay + bz = 0$ has a non trivial solution, then both the roots of the quadratic equation $at^2 + bt + c = 0$ are

(a) real
(b) of opposite sign
(c) positive
(d) complex

6. Let $f(x) = \begin{vmatrix} n & n+1 & n+2 \\ {}^nP_n & {}^{n+1}P_{n+1} & {}^{n+2}P_{n+2} \\ {}^nC_n & {}^{n+1}C_{n+1} & {}^{n+2}C_{n+2} \end{vmatrix}$, where the symbols have their usual meanings. The $f(x)$ is divisible by

(a) $n^2 + n + 1$ (b) $(n + 1)!$
(c) $n!$ (d) None of these

7. Eliminating a, b, c from

$x = \frac{a}{b-c}, y = \frac{b}{c-a}, z = \frac{c}{a-b}$, we get

(a) $\begin{vmatrix} 1 & -x & x \\ 1 & -y & y \\ 1 & -z & z \end{vmatrix} = 0$ (b) $\begin{vmatrix} 1 & -x & x \\ 1 & 1 & -y \\ 1 & z & 1 \end{vmatrix} = 0$

(c) $\begin{vmatrix} 1 & -x & x \\ y & 1 & -y \\ -z & z & 1 \end{vmatrix} = 0$ (d) None of these

8. Which of the following values of α satisfy the equation

$$\begin{vmatrix} (1+\alpha)^2 & (1+2\alpha)^2 & (1+3\alpha)^2 \\ (2+\alpha)^2 & (2+2\alpha)^2 & (2+3\alpha)^2 \\ (3+\alpha)^2 & (3+2\alpha)^2 & (3+3\alpha)^2 \end{vmatrix} = -648\alpha?$$

[JEE Advanced 2015]

(a) –4 (b) 9 (c) –9 (d) 4

9. If $f(x) = \begin{vmatrix} 1+\sin^2 x & \cos^2 x & 4\sin 2x \\ \sin^2 x & 1+\cos^2 x & 4\sin 2x \\ \sin^2 x & \cos^2 x & 1+4\sin 2x \end{vmatrix}$, $f(a)$ and $f(b)$ be the least and greatest values of $f(x)$, then

(a) $f(a) = 2, f(b) = 6$ (b) $f(a) = -2, f(b) = 6$

(c) $f(a) = 2, f(b) = -6$ (d) period of $f(x)$ is π

10. If $f(\theta) = \begin{vmatrix} \sin\theta & \cos\theta & \sin\theta \\ \cos\theta & \sin\theta & \cos\theta \\ \cos\theta & \sin\theta & \sin\theta \end{vmatrix}$, then

(a) $f(\theta) = 0$ has exactly 2 real solutions in $[0, \pi]$

(b) $f(\theta) = 0$ has exactly 3 real solutions in $[0, \pi]$

(c) range of function $\frac{f(\theta)}{1-\sin 2\theta}$ is $[-\sqrt{2}, \sqrt{2}]$

(d) range of function $\frac{f(\theta)}{1-\sin 2\theta}$ is $[-3, 3]$

11. Let a, b, c be even natural numbers, then

$\Delta = \begin{vmatrix} a-x & a & a+x \\ b-x & b & b+x \\ c-x & c & c+x \end{vmatrix}$ is a multiple of

(a) 2 (b) 3

(c) 9 (d) 12

12. $\begin{vmatrix} 1 & a & a^2 \\ \cos(xz-yz) & \cos xz & \cos(xz-yz) \\ \sin(xz-yz) & \sin xz & \sin(xz-yz) \end{vmatrix}$ depends on

(a) x (b) y (c) z (d) a

13. If $x \in N$ and ${}^xC_i$, ${}^{x^2}C_i$ and ${}^{x^3}C_i$, $(i = 1, 2, 3)$ are binomial coefficients, then $12\begin{vmatrix} {}^xC_1 & {}^xC_2 & {}^xC_3 \\ {}^{x^2}C_1 & {}^{x^2}C_2 & {}^{x^2}C_3 \\ {}^{x^3}C_1 & {}^{x^3}C_2 & {}^{x^3}C_3 \end{vmatrix}$ is divisible by

(a) x^3 (b) x^6 (c) x^9 (d) x^{12}

14. Let a, b, c be real numbers and

$\Delta_i = \begin{vmatrix} 1 & 1 & 1 \\ a & b & c \\ a^2 & b^2 & c^2 \end{vmatrix}, \Delta_2 = \begin{vmatrix} 1 & 1 & 1 \\ b+c & c+a & a+b \\ b^2+c^2 & c^2+a^2 & a^2+b^2 \end{vmatrix}$

Then,

(a) $\Delta_1 + \Delta_2 = 0$

(b) $\Delta_1 = \Delta_2$

(c) $\Delta_2 = (ab + bc + ca)\Delta_1$

(d) $\Delta_1 = \Delta_2 = (b-c)(c-a)(a-b)$

15. Let $f(a, b) = \begin{vmatrix} a & a^2 & 0 \\ 1 & (2a+b) & (a+b)^2 \\ 0 & 1 & (2a+3b) \end{vmatrix}$, then

(a) $(a + b)$ is a factor of $f(a, b)$

(b) $(a + 2b)$ is a factor of $f(a, b)$

(c) $(2a + b)$ is a factor of $f(a, b)$

(d) a is a factor of $f(a, b)$

16. Let $f_1(x) = x + a$, $f_2(x) = x^2 + bx + c$ and

$\Delta = \begin{vmatrix} 1 & 1 & 1 \\ f_1(x_1) & f_1(x_2) & f_1(x_3) \\ f_2(x_1) & f_2(x_2) & f_2(x_3) \end{vmatrix}$, then

(a) Δ is independent of a

(b) Δ is independent of b and c

(c) Δ is independent of x_1, x_2, x_3

(d) none of the above

17. The determinant

$\Delta = \begin{vmatrix} b & c & b\lambda + c \\ c & d & c\lambda + d \\ b\lambda + c & c\lambda + d & a\lambda^3 + 3c\lambda \end{vmatrix}$ is equal to zero, if

(a) b, c, d are in AP

(b) b, c, d are in GP

(c) b, c, d are in HP

(d) λ is a root of $ax^3 - bx^2 + cx - d = 0$

18. If $a^2 + b^2 + c^2 = 1$, then

$$\begin{vmatrix} a^2 + (b^2+c^2)\cos\phi & ab(1-\cos\phi) & ac(1-\cos\phi) \\ ba(1-\cos\phi) & b^2 + (c^2+a^2)\cos\phi & bc(1-\cos\phi) \\ ca(1-\cos\phi) & cb(1-\cos\phi) & c^2 + (a^2+b^2)\cos\phi \end{vmatrix}$$

is independent of

(a) a (b) b (c) c (d) ϕ

19. If $f(x) = \begin{vmatrix} \sec^2 x & 1 & 1 \\ \cos^2 x & \cos^2 x & \operatorname{cosec}^2 x \\ 1 & \cos^2 x & \cot^2 x \end{vmatrix}$, then

(a) $\int_{-\pi/4}^{\pi/4} f(x)\,dx = \frac{1}{32}(3\pi + 8)$

(b) $f'\left(\frac{\pi}{2}\right) = 0$

(c) Maximum value of $f(x)$ is 1

(d) Minimum value of $f(x)$ is 0

20. If
$$(x_1 - x_2)^2 + (y_1 - y_2)^2 = a^2;$$
$$(x_2 - x_3)^2 + (y_2 - y_3)^2 = b^2;$$
$$(x_3 - x_1)^2 + (y_3 - y_1)^2 = c^2$$

and
$$4\begin{vmatrix} x_1 & y_1 & 1 \\ x_2 & y_2 & 1 \\ x_3 & y_3 & 1 \end{vmatrix}^2$$
$$= \lambda\{\lambda^3 - (\lambda_1 + \lambda_2 + \lambda_3)\lambda^2 + (\lambda_1\lambda_2 + \lambda_2\lambda_3 + \lambda_3\lambda_1)\lambda - \lambda_1\lambda_2\lambda_3\},$$ then

(a) $\lambda \ge \frac{3}{2}(\lambda_1\lambda_2\lambda_3)^{\frac{1}{3}}$
(b) $\lambda_1\lambda_2\lambda_3 = 8abc$
(c) $\Sigma\lambda_1\lambda_2 = 4\,\Sigma ab$
(d) $2\lambda = \lambda_1 + \lambda_2 + \lambda_3$

21. If $\begin{vmatrix} x^2 + x & x + 1 & x - 2 \\ 2x^2 + 3x - 1 & 3x & 3x - 3 \\ x^2 + 2x + 3 & 2x - 1 & 2x - 1 \end{vmatrix} = Ax + B$, where A and B are constants, then

(a) $A + B = 12$ (b) $A - B = 36$
(c) $A^2 + B^2 = 720$ (d) $A + 2B = 0$

22. If

$$A = \begin{vmatrix} 2 & \alpha + \beta + \gamma + \delta & \alpha\beta + \gamma\delta \\ \alpha + \beta + \gamma + \delta & 2(\alpha + \beta)(\gamma + \delta) & \alpha\beta(\gamma + \delta) + \gamma\delta(\alpha + \beta) \\ \alpha\beta + \gamma\delta & \alpha\beta(\gamma + \delta) + \gamma\delta(\alpha + \beta) & 2\alpha\beta\gamma\delta \end{vmatrix},$$

$$B = \begin{vmatrix} 2 & \alpha + \beta + \gamma + \delta & \alpha\gamma + \beta\delta \\ \alpha + \beta + \gamma + \delta & 2(\alpha + \gamma)(\beta + \delta) & \alpha\gamma(\beta + \delta) + \beta\delta(\alpha + \gamma) \\ \alpha\gamma + \beta\delta & \alpha\gamma(\beta + \delta) + \beta\delta(\alpha + \gamma) & 2\alpha\beta\gamma\delta \end{vmatrix}$$

and

$$C = \begin{vmatrix} 2 & \alpha + \beta + \gamma + \delta & \alpha\delta + \beta\gamma \\ \alpha + \beta + \gamma + \delta & 2(\alpha + \delta)(\beta + \gamma) & \alpha\delta(\beta + \gamma) + \beta\gamma(\alpha + \delta) \\ \alpha\delta + \beta\gamma & \alpha\delta(\beta + \gamma) + \beta\gamma(\alpha + \delta) & 2\alpha\beta\gamma\delta \end{vmatrix},$$

then

(a) $A^3 + B^3 + C^3 = 3ABC$
(b) $A^{91} + B^{38} - C^{92} = (ABC)^{54}$
(c) $\cos^A\theta + \cos^B\phi + \cos^C\psi = 3(\cos\theta)^A(\cos\phi)^B(\cos\psi)^C$
(d) None of the above

23. If $a + b + c = 0$, then
$$\Delta = \begin{vmatrix} a - x & c & b \\ c & b - x & a \\ b & a & c - x \end{vmatrix} = 0 \text{ is}$$

(a) $x = 0$ (b) $x = \sqrt{\frac{3}{2}(a^2 + b^2 + c^2)}$
(c) $x = -\sqrt{\frac{3}{2}(a^2 + b^2 + c^2)}$ (d) None of these

24. If $a_i, b_i, c_i \in R$ $(i = 1, 2, 3)$ and $x \in R$ and
$$\begin{vmatrix} a_1 + b_1x & a_1x + b_1 & c_1 \\ a_2 + b_2x & a_2x + b_2 & c_2 \\ a_3 + b_3x & a_3x + b_3 & c_3 \end{vmatrix} = 0,$$ then

(a) $x = 1$ (b) $x = -1$
(c) $\begin{vmatrix} a_1 & b_1 & c_1 \\ a_2 & b_2 & c_2 \\ a_3 & b_3 & c_3 \end{vmatrix} = 0$ (d) None of these

Linked-Comprehension Type

In these questions, a passage (paragraph) has been given followed by questions based on each of the passages. You have to answer the questions based on the passage given.

PASSAGE 1

If $S_k = a^k + b^k + c^k$, then

$$\begin{vmatrix} S_0 & S_1 & S_2 \\ S_1 & S_2 & S_3 \\ S_2 & S_3 & S_4 \end{vmatrix} = \begin{vmatrix} 3 & a + b + c & a^2 + b^2 + c^2 \\ a + b + c & a^2 + b^2 + c^2 & a^3 + b^3 + c^3 \\ a^2 + b^2 + c^2 & a^3 + b^3 + c^3 & a^4 + b^4 + c^4 \end{vmatrix}$$

$$= \begin{vmatrix} 1 & 1 & 1 \\ a & b & c \\ a^2 & b^2 & c^2 \end{vmatrix}^2 = (a - b)^2(b - c)^2(c - a)^2$$

If we can express a determinant as a product of two determinants, then it is easier for us.

On the basis of above information, answer the following questions

1. a, b, c and d are complex numbers, then the value of
$$\begin{vmatrix} 2 & a + b + c + d & ac + bd \\ a + b + c + d & 2(a + c)(b + d) & ac(b + d) + bd(a + c) \\ ac + bd & ac(b + d) + bd(a + c) & 2abcd \end{vmatrix} \text{ is}$$

(a) 0 (b) $a + b + c + d$
(c) $abcd$ (d) $ac + bd$

2. If $A, B, C, P, Q, R \in R$, a then the value of
$$\begin{vmatrix} \cos(A - P) & \cos(A - Q) & \cos(A - R) \\ \cos(B - P) & \cos(B - Q) & \cos(B - R) \\ \cos(C - P) & \cos(C - Q) & \cos(C - R) \end{vmatrix} \text{ is}$$

(a) $\cos A\cos P + \cos B\cos Q + \cos C\cos R$
(b) $\cos(A - P)\cos(B - Q)\cos(C - R)$
(c) $\cos(A + B + C) - \cos(P - Q - R)$
(d) None of the above

3. If $a_1, a_2, a_3, b_1, b_2, b_3 \in R$, then the value of

$$\begin{vmatrix} (a_1-b_1)^2 & (a_1-b_2)^2 & (a_1-b_3)^2 \\ (a_2-b_1)^2 & (a_2-b_2)^2 & (a_2-b_3)^2 \\ (a_3-b_1)^2 & (a_3-b_2)^2 & (a_3-b_3)^2 \end{vmatrix} \text{ is}$$

(a) $2(a_1-a_2)(a_2-a_3)(a_3-a_1)$
(b) $2(b_1-b_2)(b_2-b_3)(b_3-b_1)$
(c) $2(a_1-a_2)(a_2-a_3)(a_3-a_1) \times (b_1-b_2)(b_2-b_3)(b_3-b_1)$
(d) 0

4. Let α, β be the roots of the equation $ax^2+bx+c=0$. Let $S_n = \alpha^n + \beta^n$ for $n \geq 1$, then the value of

$$\begin{vmatrix} 3 & 1+S_1 & 1+S_2 \\ 1+S_1 & 1+S_2 & 1+S_3 \\ 1+S_2 & 1+S_3 & 1+S_4 \end{vmatrix} \text{ is}$$

(a) $\dfrac{(a+b+c)(b^2-4ac)^2}{a^4}$

(b) $\dfrac{(a+b+c)^2(b^2-4ac)}{a^4}$

(c) $\dfrac{(a+b+c)^2(b^2-4ac)^2}{a^4}$

(d) 0

5. If $a_1, a_2, a_3, b_1, b_2, b_3 \in R$ and are such that $a_ib_j \neq 1$ for $1 \leq i, j \leq 3$ and if $a_1 < a_2 < a_3$ and $b_1 < b_2 < b_3$, then the determinant

$$\begin{vmatrix} \frac{1-a_1^3b_1^3}{1-a_1b_1} & \frac{1-a_1^3b_2^3}{1-a_1b_2} & \frac{1-a_1^3b_3^3}{1-a_1b_3} \\ \frac{1-a_2^3b_1^3}{1-a_2b_1} & \frac{1-a_2^3b_2^3}{1-a_2b_2} & \frac{1-a_2^3b_3^3}{1-a_2b_3} \\ \frac{1-a_3^3b_1^3}{1-a_3b_1} & \frac{1-a_3^3b_2^3}{1-a_3b_2} & \frac{1-a_3^3b_3^3}{1-a_3b_3} \end{vmatrix} \text{ is}$$

(a) positive
(b) negative
(c) non-positive
(d) non-negative

PASSAGE 2

Let A be a 3×3 matrix,

$$X = \begin{bmatrix} x \\ y \\ z \end{bmatrix} \text{ and } b = \begin{bmatrix} b_1 \\ b_2 \\ b_3 \end{bmatrix}$$

Using elementary row operations on the matrix equation $AX = B$, we obtain an equation of the form $A'X = B$, which is equivalent to the system $AX = B$. That is either both systems are inconsistent or both have the same set of solutions in x, y and z.

Consider the following system of equations

$$x + y + z = 6,\ x + 2y + 2z = 10$$

and $x + 2y + \lambda z = \mu$

On the basis of above information, answer the following questions

1. The number of values of λ for which the system has unique solution is
(a) only one value
(b) all real values except two values
(c) only two real values
(d) all real values except one value

2. The system has no solution if
(a) $\lambda = 2, \mu \neq 10$ (b) $\lambda = 3, \mu = 10$
(c) $\lambda = -3, \mu = 10$ (d) $\lambda = -3, \mu = -10$

3. The system has infinitely many solutions, if
(a) $\lambda \neq 3, \mu \neq 10$ (b) $\lambda = 2, \mu = 10$
(c) $\lambda \neq 3, \mu = 10$ (d) $\lambda = 0, \mu = 10$

PASSAGE 3

Given that the system of equations $x = cy + bz$, $y = az + cx$, $z = bx + ay$ has non-zero solutions and at least one of the a, b, c is a proper fraction.

On the basis of above information, answer the following questions

1. $a^2 + b^2 + c^2$ is
(a) > 2 (b) > 3
(c) < 3 (d) < 2

2. abc is
(a) > -1 (b) > 1
(c) < 2 (d) < 3

3. System has solution such that
(a) $x : y : z \equiv (1-2a^2) : (1-2b^2) : (1-2c^2)$
(b) $x : y : z \equiv \dfrac{1}{1-2a^2} : \dfrac{1}{1-2b^2} : \dfrac{1}{1-2c^2}$
(c) $x : y : z \equiv \dfrac{a}{1-a^2} : \dfrac{b}{1-b^2} : \dfrac{c}{1-c^2}$
(d) $x : y : z \equiv \sqrt{1-a^2} : \sqrt{1-b^2} : \sqrt{1-c^2}$

PASSAGE 4

Let $\Delta \neq 0$ and Δ^c denotes the determinant of cofactors, then $\Delta^c = \Delta^{n-1}$, where n (> 0) is the order of Δ.

On the basis of above information, answer the following questions

1. If a, b, c are the roots of the equation $x^3 - px^2 + r = 0$, then the value of

$$\begin{vmatrix} bc-a^2 & ca-b^2 & ab-c^2 \\ ca-b^2 & ab-c^2 & bc-a^2 \\ ab-c^2 & bc-a^2 & ca-b^2 \end{vmatrix} \text{ is}$$

(a) p^2 (b) p^4
(c) p^6 (d) p^9

2. If $l_1, m_1, n_1; l_2, m_2, n_2; l_3, m_3, n_3$ are real quantities satisfying the six relations :
$l_1^2 + m_1^2 + n_1^2 = l_2^2 + m_2^2 + n_2^2 = l_3^2 + m_3^2 + n_3^2 = 1$
$l_2 l_3 + m_2 m_3 + n_2 n_3 = l_3 l_1 + m_3 m_1 + n_3 n_1$
$= l_1 l_2 + m_1 m_2 + n_1 n_2 = 0$, then the value of
$$\begin{vmatrix} l_1 & m_1 & n_1 \\ l_2 & m_2 & n_2 \\ l_3 & m_3 & n_3 \end{vmatrix}$$ is

(a) 0 (b) ± 1 (c) ± 2 (d) ± 3

3. If a, b, c are the roots of the equation $x^3 - 3x^2 + 3x + 7 = 0$, then the value of
$$\begin{vmatrix} 2bc - a^2 & c^2 & b^2 \\ c^2 & 2ac - b^2 & a^2 \\ b^2 & a^2 & 2ab - c^2 \end{vmatrix}$$ is

(a) 9 (b) 27
(c) 81 (d) 0

4. If $a^2 + b^2 + c^2 = \lambda^2$, then the value of
$$\begin{vmatrix} a^2 + \lambda^2 & ab + c\lambda & ca - b\lambda \\ ab - c\lambda & b^2 + \lambda^2 & bc + a\lambda \\ ac + b\lambda & bc - a\lambda & c^2 + \lambda^2 \end{vmatrix} \times \begin{vmatrix} \lambda & c & -b \\ -c & \lambda & a \\ b & -a & \lambda \end{vmatrix}$$ is

(a) $8\lambda^6$ (b) $27\lambda^9$
(c) $8\lambda^9$ (d) $27\lambda^6$

5. Suppose $a, b, c \in R$, $a + b + c > 0$, $A = bc - a^2$, $B = ca - b^2$ and $C = ab - c^2$ and $\begin{vmatrix} A & B & C \\ B & C & A \\ C & A & B \end{vmatrix} = 49$, then $\begin{vmatrix} a & b & c \\ b & c & a \\ c & a & b \end{vmatrix}$ equals

(a) -7 (b) 7
(c) -2401 (d) 2401

Numerical Value Type

Solve the following problems and write your answer.

1. Suppose $\det \begin{bmatrix} \sum_{k=0}^{n} k & \sum_{k=0}^{n} {}^nC_k k^2 \\ \sum_{k=0}^{n} {}^nC_k k & \sum_{k=0}^{n} {}^nC_k 3^k \end{bmatrix} = 0$ holds for some positive integer n. Then $\sum_{k=0}^{n} \frac{{}^nC_k}{k+1}$ equals ....
[JEE Advanced 2019]

2. If $D_p = \begin{vmatrix} p & 15 & 8 \\ p^2 & 35 & 9 \\ p^3 & 25 & 10 \end{vmatrix}$, then the value of $\left(-\frac{1}{100}\right) \sum_{p=1}^{5} D_p$ must be

3. If $\begin{vmatrix} (b+c)^2 & a^2 & a^2 \\ b^2 & (c+a)^2 & b^2 \\ c^2 & c^2 & (a+b)^2 \end{vmatrix}$ $= \lambda abc\,(a + b + c)^3$, then the value of $\sum_{r=1}^{10} \lambda^r$ must be

4. The value of $\begin{vmatrix} \log_3 512 & \log_4 3 \\ \log_3 8 & \log_4 9 \end{vmatrix} \times \begin{vmatrix} \log_2 3 & \log_8 3 \\ \log_3 4 & \log_3 4 \end{vmatrix}$ is

5. If the system of equations
$$3x - 2y + z = 0$$
$$\lambda x - 14y + 15z = 0$$
$$x + 2y + 3z = 0$$
have a non trivial solution, then the value of λ^2 must be

6. If $\Delta = \begin{vmatrix} 1 & 3\cos\theta & 1 \\ \sin\theta & 1 & 3\cos\theta \\ 1 & \sin\theta & 1 \end{vmatrix}$, then the $|\text{maximum value of } \Delta - \text{Minimum value of } \Delta|^3$ is equal to

7. If $A_r = \begin{vmatrix} r & r-1 \\ r-1 & r \end{vmatrix}$, where r is natural number, then the value of $\sqrt{\left(\sum_{r=1}^{2008} A_r\right)}$ must be

Single Integer Type

Solve the following problems and write your answer.

1. If p and q are two distinct real values of λ for which the system of equations
$$(\lambda - 1)x + (3\lambda + 1)y + 2\lambda z = 0$$
$$(\lambda - 1)x + (4\lambda - 2)y + (\lambda + 3)z = 0$$
$$2x + (3\lambda + 1)y + 3(\lambda - 1)z = 0$$
has non-zero solution, then the value of $p^2 + q^2 - pq$ is

2. If $\begin{vmatrix} 2bc - a^2 & c^2 & b^2 \\ c^2 & 2ac - b^2 & a^2 \\ b^2 & a^2 & 2ab - c^2 \end{vmatrix} = 0$ and a, b, c are different real number, then the value of $3(a + b + c) + 2$ is ....

3. a, b, c are distinct real numbers, not equal to one. If $ax + y + z = 0$, $x + by + z = 0$ and $x + y + cz = 0$ have a non-trivial solution, then the value of $\frac{1}{1-a} + \frac{1}{1-b} + \frac{1}{1-c}$ is equal to

4. If the system of equations

$$ax + hy + g = 0 \quad \ldots\text{(i)}$$

$$hx + by + f = 0 \quad \ldots\text{(ii)}$$

and $ax^2 + 2hxy + by^2 + 2gx + 2fy + c + t = 0 \quad \ldots\text{(iii)}$

has a unique solution, and

$\frac{abc + 2fgh - af^2 - bg^2 - ch^2}{h^2 - ab} = 8$, then the value of t is

5. Let P be a matrix of order 3×3 such that all the entries P are from the set $\{-1, 0, 1\}$. Then, the maximum possible value of the determinant of P is

6. For a real number α, if the system

$$\begin{bmatrix} 1 & \alpha & \alpha^2 \\ \alpha & 1 & \alpha \\ \alpha^2 & \alpha & 1 \end{bmatrix} \begin{bmatrix} x \\ y \\ z \end{bmatrix} = \begin{bmatrix} 1 \\ -1 \\ 1 \end{bmatrix}$$

of linear equations, has infinitely many solutions, then the value of $1 + \alpha + \alpha^2$ is **[JEE Advanced 2017]**

Matrix-Match Type

Given below are Matching Type Questions, with two columns (each having some items) each . Each item of Column I has to be matched with the items of Column II, by encircling the correct match(es).

NOTE *An item of Column I can be matched with more than one items of Column II.*

1. Observe the following columns :

	Column I		Column II
(A)	If $a^2 + b^2 + c^2 = 1$ and $\Delta = \begin{vmatrix} a^2 + (b^2 + c^2)\cos\phi & ab(1-\cos\phi) & ac(1-\cos\phi) \\ ba(1-\cos\phi) & b^2 + (c^2 + a^2)\cos\phi & bc(1-\cos\phi) \\ ca(1-\cos\phi) & cb(1-\cos\phi) & c^2 + (a^2 + b^2)\cos\phi \end{vmatrix}$, then Δ is independent of	(P)	a
(B)	If $\Delta = \begin{vmatrix} \sin a & \cos a & \sin(a+\phi) \\ \sin b & \cos b & \sin(b+\phi) \\ \sin c & \cos c & \sin(c+\phi) \end{vmatrix}$, then Δ is independent of	(Q)	b
(C)	If $\Delta = \begin{vmatrix} 1/c & 1/c & -(a+b)^2/c^2 \\ -(b+c)/a^2 & 1/a & 1/a \\ -\frac{b(b+c)}{a^2c} & \frac{(a+2b+c)}{ac} & \frac{-(a+b)b}{ac^2} \end{vmatrix}$, then Δ is independent of	(R)	c
		(S)	ϕ
		(T)	$\Delta = 0$

2. If A, B are two non-singular matrices of order $|B| = k$, a positive integer, then match the items of list-I with the items of list-II.

	List I		List II
(A)	$\lvert k^{-1} A^{-1}\rvert$	(P)	$BA^k + A^kB$
(B)	$\lvert \text{Adj}\,(A^{-1})\rvert$	(Q)	$\frac{B\text{Adj}(B)}{\lvert B\rvert}$
(C)	$BAB^{-1} = I, \Rightarrow BA^k B^{-1} =$	(R)	$\frac{1}{\lvert B\rvert^3 \lvert A\rvert}$
(D)	$\text{Adj}\,(\text{Adj}\,(A^{-1})) =$	(S)	$\frac{1}{\lvert A\rvert}(A^{-1})$
		(T)	$\frac{1}{\lvert A\rvert^2}$

3. Match the items of Column I with those of Column II.

	Column I		Column II
A.	The system of equations $\lambda x + y + z = 0$, $-x + \lambda y + z = 0$, $-x - y + \lambda z = 0$ will have non-zero solution, if real value of λ is	P.	1
B.	In the system of equations, given in (A), if $\lambda = 1$, then the number of solutions is	Q.	-1
C.	If the system of equations $x = cy + bz$, $y = az + cx$, $z = bx + ay$ has non-zero solution in x, y and z, then $a^2 + b^2 + c^2 + 2abc$ is equal to	R.	± 1
D.	If P is a matrix of order 3×3 such that $p^T P = I$ (unit matrix of order 3×3) and det $P = 1$, then det $(P - 1)$ equals	S.	0

4. Observe the following columns :

	Column I		Column II
(A)	If $\Delta(x) = \begin{vmatrix} x^2-5x+3 & 2x-5 & 3 \\ 3x^2+x+4 & 6x+1 & 9 \\ 7x^2-6x+9 & 14x-6 & 21 \end{vmatrix}$	(P)	$3a+4b+5c+d=141$
	$= ax^3+bx^2+cx+d$, then	(Q)	$a+2b+3c+5d=156$
(B)	If $\Delta(x) = \begin{vmatrix} x-1 & 5x & 7 \\ x^2-1 & x-1 & 8 \\ 2x & 3x & 0 \end{vmatrix} = ax^3+bx^2+cx+d$, then	(R)	$c-d=119$
(C)	If $\Delta(x) = \begin{vmatrix} 2x^3-3x^2 & 5x+7 & 2 \\ 4x^3-7x & 3x+2 & 1 \\ 7x^3-8x^2 & x-1 & 3 \end{vmatrix}$	(S)	$b-c=25$
	$= a+bx+cx^2+dx^3+ex^4$, then	(T)	$3a+2b+5c+5d=187$

Answers

Objective Questions Type I [Only one correct answer]

1. (c)	**2.** (c)	**3.** (c)	**4.** (a)	**5.** (b)	**6.** (a)	**7.** (b)	**8.** (d)	**9.** (b)	**10.** (b)
11. (b)	**12.** (c)	**13.** (a)	**14.** (d)	**15.** (a)	**16.** (b)	**17.** (d)	**18.** (a)	**19.** (a)	**20.** (d)
21. (d)	**22.** (a)	**23.** (a)	**24.** (b)	**25.** (b)	**26.** (c)	**27.** (c)	**28.** (b)	**29.** (b)	**30.** (d)
31. (b)	**32.** (b)	**33.** (b)	**34.** (d)	**35.** (a)	**36.** (c)	**37.** (d)	**38.** (b)	**39.** (c)	**40.** (d)
41. (c)	**42.** (b)	**43.** (a)	**44.** (c)	**45.** (d)	**46.** (b)	**47.** (b)	**48.** (c)	**49.** (c)	**50.** (b)
51. (c)	**52.** (a)	**53.** (d)	**54.** (d)						

Objective Questions Type II [One or more than one correct answer (a)]

1. (a, b, c, d)	**2.** (a, b, c, d)	**3.** (a, b, c)	**4.** (b, d)	**5.** (a, b)
6. (a, c)	**7.** (b, c)	**8.** (b, c)	**9.** (b, d)	**10.** (a, c)
11. (a, b, c, d)	**12.** (b, c, d)	**13.** (a, b, c)	**14.** (b, d)	**15.** (a, b, d)
16. (a, b)	**17.** (b, d)	**18.** (a, b, c)	**19.** (a, b, c, d)	**20.** (a, b, c, d)
21. (a, b, c, d)	**22.** (a, b, c)	**23.** (a, b, c)	**24.** (a, b, c)	

Linked-Comprehension Type

Passage 1 **1.** (a) **2.** (d) **3.** (c) **4.** (b) **5.** (a)
Passage 2 **1.** (d) **2.** (a) **3.** (b)
Passage 3 **1.** (c) **2.** (a) **3.** (d)
Passage 4 **1.** (c) **2.** (b) **3.** (d) **4.** (c) **5.** (b)

Numerical Value Based Problems

1. 6.20 **2.** 330 **3.** 2046 **4.** 10 **5.** 841 **6.** 1000 **7.** 2008

Single Integer Type Questions

1. 9 **2.** 2 **3.** 2 **4.** 8 **5.** 4 **6.** 1

Matrix-Match Type

1. A → (P, Q, R); B → (P, Q, R, S, T); C → (P, Q, R, S, T)
2. A → (R); B → (T); C → (Q): D → (S)
3. A → (S); B → (P); C → (P); D → (S)
4. A → (P); B → (Q, S); C → (R, T)

Solutions

Objective Questions Type [Only one correct answer]

1. Possible values are $-2, -1, 0, 1, 2$

i.e. $\begin{vmatrix} 1 & 0 \\ -1 & 1 \end{vmatrix} = 1, \begin{vmatrix} 1 & -1 \\ 0 & 0 \end{vmatrix} = 0, \begin{vmatrix} 0 & 1 \\ 1 & -1 \end{vmatrix} = -1$

$\begin{vmatrix} 1 & -1 \\ 1 & 1 \end{vmatrix} = 2, \begin{vmatrix} -1 & 1 \\ 1 & 1 \end{vmatrix} = -2$

2. $\begin{vmatrix} 1 & 2\lambda & 1 \\ \lambda^2 & 1 & 3\lambda^2 \\ 2 & 2\lambda & 1 \end{vmatrix}$ applying $R_3 \to R_3 - R_1$

Then $\begin{vmatrix} 1 & 2\lambda & 1 \\ \lambda^2 & 1 & 3\lambda^2 \\ 1 & 0 & 0 \end{vmatrix} = 6\lambda^3 - 1 = 6(-2) - 1 = -13$

$[\because \lambda = (-2)^{1/3}, \therefore \lambda^3 = -2]$

3. Let $\Delta(x) = \begin{vmatrix} x & 2 & x \\ x^2 & x & 6 \\ x & x & 6 \end{vmatrix} \Rightarrow \Delta(1) = 0$

$\therefore \quad \Delta'(x) = \begin{vmatrix} 1 & 0 & 1 \\ x^2 & x & 6 \\ x & x & 6 \end{vmatrix} + \begin{vmatrix} x^2 & 2 & x \\ 2x & 1 & 0 \\ x & x & 6 \end{vmatrix} + \begin{vmatrix} x & 2 & x \\ x^2 & x & 6 \\ 1 & 1 & 0 \end{vmatrix}$

$\Delta'(1) = 0 + \begin{vmatrix} 1 & 2 & 1 \\ 2 & 1 & 0 \\ 1 & 1 & 6 \end{vmatrix} + \begin{vmatrix} 1 & 2 & 1 \\ 1 & 1 & 6 \\ 1 & 1 & 0 \end{vmatrix}$

$= -17 + 6 = -11$

Now, $\Delta(1) + \Delta'(1) = 5A + 4B + 3C + 4D + E = -11$

Alternative method

Multiplying both sides by x, then

$\begin{vmatrix} x^2 & 2x & x^2 \\ x^2 & x & 6 \\ x & x & 6 \end{vmatrix} = Ax^5 + Bx^4 + Cx^3 + Dx^2 + Ex$

Differentiating both sides w.r.t.x, then

$\begin{vmatrix} 2x & 2 & 2x \\ x^2 & x & 6 \\ x & x & 6 \end{vmatrix} + \begin{vmatrix} x^2 & 2x & x^2 \\ 2x & 1 & 0 \\ x & x & 6 \end{vmatrix} + \begin{vmatrix} x^2 & 2x & x^2 \\ x^2 & x & 6 \\ 1 & 1 & 0 \end{vmatrix}$

$= 5Ax^4 + 4Bx^3 + 3Cx^2 + 2Dx + E$

Putting $x = 1$ in both sides, then

$\begin{vmatrix} 2 & 2 & 2 \\ 1 & 1 & 6 \\ 1 & 1 & 6 \end{vmatrix} + \begin{vmatrix} 1 & 2 & 1 \\ 2 & 1 & 0 \\ 1 & 1 & 6 \end{vmatrix} + \begin{vmatrix} 1 & 2 & 1 \\ 1 & 1 & 6 \\ 1 & 1 & 0 \end{vmatrix}$

$= 5A + 4B + 3C + 2D + E$

$\Rightarrow 0 + 1(6) - 2(12) + 1(2-1) + 1(-6) - 2(0-6) + 1(1-1)$

$= 5A + 4B + 3C + 2D + E$

Hence, $5A + 4B + 3C + 2D + E = -11$

4. Δ can be written as

$\Delta = \begin{vmatrix} my + nz & kz - mx & nx + ky \\ mq - nr & kr - mp & np + kq \\ mb + nc & kb - ma & na + kb \end{vmatrix}$

$= -\begin{vmatrix} x & y & z \\ p & q & r \\ a & b & c \end{vmatrix} \times \begin{vmatrix} 0 & m & n \\ -m & 0 & k \\ -n & -k & 0 \end{vmatrix} = 0$

$\left[\because \begin{vmatrix} 0 & m & n \\ -m & 0 & k \\ -n & -k & 0 \end{vmatrix} \text{ is skew symmetric}\right]$

5. $\because \quad 0 \le [x] < 2 \Rightarrow [x] = 0, 1$

$-1 \le [y] < 1 \Rightarrow [y] = -1, 0$

and $\quad 1 \le [z] < 3 \Rightarrow [z] = 1, 2$

Now, applying in the given determinant

$R_2 \to R_2 - R_1, R_3 \to R_3 - R_1$, then

$\begin{vmatrix} [x]+1 & [y] & [z] \\ -1 & 1 & 0 \\ -1 & 0 & 1 \end{vmatrix}$

$= ([x]+1)(1-0) - [y](-1-0) + [z](0+1)$

$= [x] + [y] + [z] + 1 = 1 + 0 + 2 + 1 = 4$

[$\because$ for maximum value, $[x] = 1, [y] = 0, [z] = 2$]

6. $\begin{vmatrix} \sin\theta & -1 \\ 1 & 2 \end{vmatrix} = 0 \Rightarrow 2\sin\theta + 1 = 0$

$\Rightarrow \quad \sin\theta = -\frac{1}{2} = -\sin\frac{\pi}{6}$

$\therefore \quad \theta = -\frac{\pi}{6}, -\frac{5\pi}{6}$

7. $\begin{vmatrix} \lambda & -1 & -2 \\ 2 & -3 & \lambda \\ 3 & -2 & 1 \end{vmatrix} = 0$

$\Rightarrow \quad \lambda(-3 + 2\lambda) + 1(2 - 3\lambda) - 2(-4 + 9) = 0$

$\Rightarrow \quad 2\lambda^2 - 6\lambda - 8 = 0 \Rightarrow \lambda^2 - 3\lambda - 4 = 0$

$\Rightarrow \quad (\lambda - 4)(\lambda + 1) = 0$

$\therefore \quad \lambda = -1, 4$

8. Given function

$f(\theta) = \begin{vmatrix} -\sin^2\theta & -1-\sin^2\theta & 1 \\ -\cos^2\theta & -1-\cos^2\theta & 1 \\ 12 & 10 & -2 \end{vmatrix}$

$= 2\begin{vmatrix} -\sin^2\theta & -1-\sin^2\theta & 1 \\ -\cos^2\theta & -1-\cos^2\theta & 1 \\ 6 & 5 & -1 \end{vmatrix}$

On applying $R_1 \to R_1 + R_3$ and $R_2 \to R_2 + R_3$, we get

$f(\theta) = 2\begin{vmatrix} 6-\sin^2\theta & 4-\sin^2\theta & 0 \\ 6-\cos^2\theta & 4-\cos^2\theta & 0 \\ 6 & 5 & -1 \end{vmatrix}$

$$= 2(-1)[(6-\sin^2\theta)(4-\cos^2\theta)-(4-\sin^2\theta)(6-\cos^2\theta)]$$

$$= -2[24-6\cos^2\theta-4\sin^2\theta+\sin^2\theta\cos^2\theta-24+4\cos^2\theta+6\sin^2\theta-\sin^2\theta\cos^2\theta]$$

$$= -2[-2\cos^2\theta+2\sin^2\theta] = 4\cos 2\theta$$

As, $\theta \in \left[\frac{\pi}{4}, \frac{\pi}{2}\right] \Rightarrow 2\theta \in \left[\frac{\pi}{2}, \pi\right]$

$\therefore \quad f(\theta)\big|_{\text{minimum}} = f\left(\frac{\pi}{2}\right) = -4 = m$

and $\quad f(\theta)\big|_{\text{maximum}} = f\left(\frac{\pi}{4}\right) = 0 = M$

$\therefore$ The ordered pair (m, M) is $(-4, 0)$.

9. Let $\begin{vmatrix} (1+x)^a & (1+2x)^b & 1 \\ 1 & (1+x)^a & (1+2x)^b \\ (1+2x)^b & 1 & (1+x)^a \end{vmatrix} = A + Bx + Cx^2 + \ldots$

Put $x = 0$, then $A = 0$

Differentiating both sides w.r.t.x and then put $x = 0$

$$\begin{vmatrix} a & 2b & 0 \\ 1 & 1 & 1 \\ 1 & 1 & 1 \end{vmatrix} + \begin{vmatrix} 1 & 1 & 1 \\ 0 & a & 2b \\ 1 & 1 & 1 \end{vmatrix} + \begin{vmatrix} 1 & 1 & 1 \\ 1 & 1 & 1 \\ 2b & 0 & a \end{vmatrix} = B$$

$\therefore \quad 0 + 0 + 0 = B$

$\Rightarrow \quad B = 0$

Hence, constant term of $f(x) = 0$ and coefficient of x in $f(x) = 0$.

10. $$\sum_{r=0}^{\infty} \Delta_r = \begin{vmatrix} \sum_{r=0}^{\infty} a^r & \sum_{r=0}^{\infty} b^r & \sum_{r=0}^{\infty} c^r \\ a & b & c \\ 1-a & 1-b & 1-c \end{vmatrix} = \begin{vmatrix} \frac{1}{1-a} & \frac{1}{1-b} & \frac{1}{1-c} \\ a & b & c \\ 1-a & 1-b & 1-c \end{vmatrix}$$

Its possible, where $|a|, |b| |c| < 1$

$C_2 \to C_2 - C_1,\ C_3 \to C_3 - C_1$

$$= \begin{vmatrix} \frac{1}{1-a} & \frac{b-a}{(1-a)(1-b)} & \frac{c-a}{(1-c)(1-a)} \\ a & b-a & c-a \\ 1-a & a-b & a-c \end{vmatrix}$$

$$= (b-a)(c-a)\begin{vmatrix} \frac{1}{1-a} & \frac{b-a}{(1-a)(1-b)} & \frac{1}{(1-c)(1-a)} \\ a & 1 & 1 \\ 1-a & -1 & -1 \end{vmatrix}$$

$C_3 \to C_3 - C_2$

$$= (b-a)(c-a)\begin{vmatrix} \frac{1}{1-a} & \frac{1}{(1-a)(1-b)} & \frac{c-b}{(1-c)(1-a)(1-b)} \\ a & 1 & 0 \\ 1-a & -1 & 0 \end{vmatrix}$$

$$= \frac{(b-a)(c-a)(b-c)}{(1-a)(1-b)(1-c)}$$

Now $\sum_{r=0}^{\infty} \Delta_r = 0$

$$\Rightarrow \quad \frac{(b-a)(c-a)(b-c)}{(1-a)(1-b)(1-c)} = 0$$

$a = b = c$ and $|a| < 1$

11. Let $\Delta' = \begin{vmatrix} a & 1 & 1 \\ 1 & b & 1 \\ 1 & 1 & c \end{vmatrix} = abc + 2 - a - b - c > 0$

or $\quad abc + 2 > a + b + c \quad$ ...(i)

$\because \quad$ AM > GM

$$\Rightarrow \quad \frac{a+b+c}{3} > (abc)^{1/3}$$

$\therefore \quad a + b + c > 3(abc)^{1/3} \quad$ ...(ii)

From Eqs. (i) and (ii),

then $\quad abc + 2 > 3(abc)^{1/3}$

let $\quad (abc)^{1/3} = x,$

then $\quad x^3 + 2 > 3x$

$\Rightarrow \quad (x-1)^2(x+2) > 0$

$\therefore \quad x + 2 > 0$

$x > -2$

$\Rightarrow \quad x^3 > -8$

$abc > -8$

12. Applying $C_3 \to C_3 - xC_1 - yC_2$, Then, we get

$$\begin{vmatrix} p & q & 0 \\ q & r & 0 \\ px+qy & qx+ry & -\{px^2+2qxy+ry^2\} \end{vmatrix}$$

$$= -(px^2+2qxy+ry^2)(pr-q^2)$$

$$= (px^2+2qxy+ry^2)(q^2-pr)$$

$$= \frac{1}{p}\{(px+qy)^2 + y^2(pr-q^2)\}(q^2-pr) < 0$$

13. Let $\quad \Delta = \begin{vmatrix} 3 & 1+\alpha+\beta & 1+\alpha^2+\beta^2 \\ 1+\alpha+\beta & 1+\alpha^2+\beta^2 & 1+\alpha^3+\beta^3 \\ 1+\alpha^2+\beta^2 & 1+\alpha^3+\beta^3 & 1+\alpha^4+\beta^4 \end{vmatrix}$

$$= \begin{vmatrix} 1 & 1 & 1 \\ 1 & \alpha & \beta \\ 1 & \alpha^2 & \beta^2 \end{vmatrix} \times \begin{vmatrix} 1 & 1 & 1 \\ 1 & \alpha & \beta \\ 1 & \alpha^2 & \beta^2 \end{vmatrix}$$

Applying $C_2 \to C_2 - C_1$ and $C_3 \to C_3 - C_1$, then

$$\begin{vmatrix} 1 & 0 & 0 \\ 1 & \alpha-1 & \beta-1 \\ 1 & \alpha^2-1 & \beta^2-1 \end{vmatrix}^2$$

$$= (\alpha-1)^2(\beta-1)^2(\beta-\alpha)^2$$

$$= (1-\alpha)^2(1-\beta)^2(\alpha-\beta)^2$$

Hence, $\quad k = 1$

14. Given, quadratic equation is $x^2 + x + 1 = 0$ having roots α, β.

Then, $\alpha + \beta = -1$ and $\alpha\beta = 1$

Now, given determinant

$$\Delta = \begin{vmatrix} y+1 & \alpha & \beta \\ \alpha & y+\beta & 1 \\ \beta & 1 & y+\alpha \end{vmatrix}$$

On applying $R_1 \to R_1 + R_2 + R_3$, we get

$$\Delta = \begin{vmatrix} y+1+\alpha+\beta & y+1+\alpha+\beta & y+1+\alpha+\beta \\ \alpha & y+\beta & 1 \\ \beta & 1 & y+\alpha \end{vmatrix}$$

$$= \begin{vmatrix} y & y & y \\ \alpha & y+\beta & 1 \\ \beta & 1 & y+\alpha \end{vmatrix} \quad [\because \alpha+\beta=-1]$$

On applying $C_2 \to C_2 - C_1$ and $C_3 \to C_3 - C_1$, we get

$$\Delta = \begin{vmatrix} y & 0 & 0 \\ \alpha & y+\beta-\alpha & 1-\alpha \\ \beta & 1-\beta & y+\alpha-\beta \end{vmatrix}$$

$$= y[(y+(\beta-\alpha))(y-(\beta-\alpha))-(1-\alpha)(1-\beta)]$$

[expanding along R_1]

$$= y[y^2-(\beta-\alpha)^2-(1-\alpha-\beta+\alpha\beta)]$$

$$= y[y^2-\beta^2-\alpha^2+2\alpha\beta-1+(\alpha+\beta)-\alpha\beta]$$

$$= y[y^2-(\alpha+\beta)^2+2\alpha\beta+2\alpha\beta-1+(\alpha+\beta)-\alpha\beta]$$

$$= y[y^2-1+3-1-1] = y^3 \quad [\because \alpha+\beta=-1 \text{ and } \alpha\beta=1]$$

15. $\because \quad \Delta_1 = \begin{vmatrix} 1 & 1 & 1 \\ a & b & c \\ a^2 & b^2 & c^2 \end{vmatrix} = (a-b)(b-c)(c-a)$

and $\quad \Delta_2 = \begin{vmatrix} 1 & bc & a \\ 1 & ca & b \\ 1 & ab & c \end{vmatrix}$

$$= \frac{1}{abc}\begin{vmatrix} a & abc & a^2 \\ b & abc & b^2 \\ c & abc & c^2 \end{vmatrix}$$

$$= \begin{vmatrix} a & 1 & a^2 \\ b & 1 & b^2 \\ c & 1 & c^2 \end{vmatrix} = -\begin{vmatrix} 1 & a & a^2 \\ 1 & b & b^2 \\ 1 & c & c^2 \end{vmatrix}$$

$$= -\begin{vmatrix} 1 & 1 & 1 \\ a & b & c \\ a^2 & b^2 & c^2 \end{vmatrix} = -\Delta_1$$

$\therefore \quad \Delta_1 + \Delta_2 = 0$

16. $x^3 - 1 = 0 \therefore x = 1, \omega, \omega^2$

Here, $\quad \alpha = \omega, \beta = \omega^2$

$\begin{vmatrix} \lambda+1 & \omega & \omega^2 \\ \omega & \lambda+\omega^2 & 1 \\ \omega^2 & 1 & \lambda+\omega \end{vmatrix}$ Applying $C_1 \to C_1 + C_2 + C_3$,

then $\begin{vmatrix} \lambda & \omega & \omega^2 \\ \lambda & \lambda+\omega^2 & 1 \\ \lambda & 1 & \lambda+\omega \end{vmatrix}$.

Applying $R_2 \to R_2 - R_1$ and $R_3 \to R_3 - R_1$, then we get

$$\begin{vmatrix} \lambda & \omega & \omega^2 \\ 0 & \lambda+\omega^2-\omega & 1-\omega^2 \\ 0 & 1-\omega & \lambda+\omega-\omega^2 \end{vmatrix}$$

$$= \lambda\{(\lambda+\omega^2-\omega)(\lambda+\omega-\omega^2)-(1-\omega)(1-\omega^2)\}$$

$$= \lambda(\lambda^2) = \lambda^3$$

17. Given matrix, $A = \begin{bmatrix} \cos\theta & \sin\theta \\ -\sin\theta & \cos\theta \end{bmatrix}$

So, $\quad A^2 = \begin{bmatrix} \cos\theta & \sin\theta \\ -\sin\theta & \cos\theta \end{bmatrix}\begin{bmatrix} \cos\theta & \sin\theta \\ -\sin\theta & \cos\theta \end{bmatrix}$

$$= \begin{bmatrix} \cos 2\theta & \sin 2\theta \\ -\sin 2\theta & \cos 2\theta \end{bmatrix}$$

$\therefore \quad A^4 = \begin{bmatrix} \cos 4\theta & \sin 4\theta \\ -\sin 4\theta & \cos 4\theta \end{bmatrix}$

Since, it is given matrix, $B = A + A^4$

$\Rightarrow \quad B = \begin{bmatrix} \cos\theta + \cos 4\theta & \sin\theta + \sin 4\theta \\ -(\sin\theta + \sin 4\theta) & \cos\theta + \cos 4\theta \end{bmatrix}$

$\therefore \quad \det(B) = (\cos\theta + \cos 4\theta)^2 + (\sin\theta + \sin 4\theta)^2$

$$= 2 + 2\cos(3\theta)$$

So, at $\theta = \dfrac{\pi}{5}$

$$\det(B) = 2 + 2\cos\frac{3\pi}{5} = 4\cos^2\left(\frac{3\pi}{10}\right)$$

$$= 4\left(\frac{\sqrt{10-2\sqrt{5}}}{4}\right)^2$$

$$= \frac{1}{4}(10-2\sqrt{5})$$

$$= \frac{10-2(2.23)}{4} \quad [\because \sqrt{5} \approx 2.23]$$

$$= \frac{10-4.46}{4}$$

$$= \frac{5.54}{4} \in (1, 2)$$

18. Let $\Delta = \begin{vmatrix} a_{11} & a_{12} & a_{13} \\ a_{21} & a_{22} & a_{23} \\ a_{31} & a_{32} & a_{33} \end{vmatrix} = \begin{vmatrix} 0 & -1 & -1 \\ 1 & 0 & -1 \\ 1 & 1 & 0 \end{vmatrix}$

$$= 0 + 1(0+1) - 1(1) = 0$$

19. Let $\quad \Delta(x) = A + Bx + Cx^2 + Dx^3 + \dots$

$\therefore \quad \Delta(0) = \begin{vmatrix} 0 & 1 & 0 \\ 0 & 1 & 0 \\ 1 & 0 & 0 \end{vmatrix} = 0 \Rightarrow A = 0$

$$\Delta'(0) = \begin{vmatrix} 1 & 0 & 0 \\ 0 & 1 & 0 \\ 1 & 0 & 0 \end{vmatrix} + \begin{vmatrix} 0 & 1 & 0 \\ 0 & 1 & 1 \\ 1 & 0 & 0 \end{vmatrix} + \begin{vmatrix} 0 & 1 & 0 \\ 0 & 1 & 0 \\ 0 & 1 & 0 \end{vmatrix} = 1$$

$\therefore \quad B = 1$

$\Rightarrow$ Then $\Delta(x) = x + Cx^2 + Dx^3 + \dots$.

$\therefore \Delta x$ is divisible by x.

20. Applying $C_1 \to C_1 + C_2 + C_3$

Then, $\begin{vmatrix} 0 & \log_a(y/z) & \log_a(z/x) \\ 0 & \log_b(z/x) & \log_b(x/y) \\ 0 & \log_c(x/y) & \log_c(y/z) \end{vmatrix} = 0$

21. Let $\Delta = \begin{vmatrix} 1 & \omega^2 & 1+i+\omega^2 \\ -i & -1 & -1-i+\omega \\ 1-i & \omega^2-1 & -1 \end{vmatrix}$

Applying $R_3 \to R_3 - (R_1 + R_2)$

Then, $\begin{vmatrix} 1 & \omega^2 & 1+i+\omega^2 \\ -i & -1 & -1-i+\omega \\ 0 & 0 & 0 \end{vmatrix} = 0$

22. $\begin{vmatrix} 3 & 4\sqrt{2} & x \\ 4 & 5\sqrt{2} & y \\ 5 & k & z \end{vmatrix} = 0$

$R_2 \to R_1 + R_3 - 2R_2$

$\Rightarrow \begin{vmatrix} 3 & 4\sqrt{2} & x \\ 0 & k-6\sqrt{2} & 0 \\ 5 & k & z \end{vmatrix} = 0 \qquad [\because 2y = x + z]$

$\Rightarrow (k - 6\sqrt{2})(3z - 5x) = 0$

If $3z - 5x = 0 \Rightarrow 3(x + 2d) - 5x = 0$

$\Rightarrow x = 3d$ (Not possible)

$\Rightarrow k = 6\sqrt{2} \Rightarrow k^2 = 72$

23. Let $\begin{vmatrix} -2a & a+b & a+c \\ b+a & -2b & b+c \\ c+a & c+b & -2c \end{vmatrix}$

$= k(b+c)(c+a)(a+b)$

Putting $a = b = c = \lambda$,

Then, $\begin{vmatrix} -2\lambda & 2\lambda & 2\lambda \\ 2\lambda & -2\lambda & 2\lambda \\ 2\lambda & 2\lambda & -2\lambda \end{vmatrix} = k(2\lambda)(2\lambda)(2\lambda)$

$\therefore \quad k = \begin{vmatrix} -1 & 1 & 1 \\ 1 & -1 & 1 \\ 1 & 1 & -1 \end{vmatrix}$

$= -1(0) - 1(-2) + 1(2) = 4$

24. Let $\Delta = \begin{vmatrix} 1 & e^{i\pi/3} & e^{i\pi/4} \\ e^{-i\pi/3} & 1 & e^{2i\pi/3} \\ e^{-i\pi/4} & e^{-2i\pi/3} & 1 \end{vmatrix}$

Taking $e^{-i\pi/3}$ common from R_2

$\therefore \quad \Delta = e^{-i\pi/3}\begin{vmatrix} 1 & e^{i\pi/3} & e^{i\pi/4} \\ 1 & e^{i\pi/3} & e^{i\pi} \\ e^{-i\pi/4} & e^{-2i\pi/3} & 1 \end{vmatrix}$

Multiplying $e^{-i\pi/3}$ in C_2, then

$\Delta = \begin{vmatrix} 1 & 1 & e^{i\pi/4} \\ 1 & 1 & e^{i\pi} \\ e^{-i\pi/4} & e^{-i\pi} & 1 \end{vmatrix}$

$= \begin{vmatrix} 1 & 1 & \frac{1+i}{\sqrt{2}} \\ 1 & 1 & -1 \\ \frac{1-i}{\sqrt{2}} & -1 & 1 \end{vmatrix}$

$= 1(1-1) - 1\left(1 + \frac{1-i}{\sqrt{2}}\right) + \left(\frac{1+i}{\sqrt{2}}\right)\left(-1 - \frac{1-i}{\sqrt{2}}\right)$

$= 0 - 1 - \left(\frac{1-i}{\sqrt{2}}\right) - \left(\frac{1+i}{\sqrt{2}}\right) - \frac{2}{2}$

$= -2 - \sqrt{2} = -(2 + \sqrt{2})$

25. $f(x) = x(x+1)\begin{vmatrix} 1 & 1 & 1 \\ 2x & x-1 & x \\ 3x(x-1) & (x-1)(x-2) & x(x-1) \end{vmatrix}$

$= x(x+1)(x-1)\begin{vmatrix} 1 & 1 & 1 \\ 2x & x-1 & x \\ 3x & x-2 & x \end{vmatrix}$

Applying $C_2 \to C_2 - C_1$ and $C_3 \to C_3 - C_1$, then

$f(x) = x(x+1)(x-1)\begin{vmatrix} 1 & 0 & 0 \\ 2x & -x-1 & -x \\ 3x & -2x-2 & -2x \end{vmatrix}$

$= x^2(x+1)^2(x-1)\begin{vmatrix} 1 & 0 & 0 \\ 2x & -1 & -1 \\ 3x & -2 & -2 \end{vmatrix} = 0$

$\therefore \quad f(200) = 0$

26. Given equations are

$x - 2y + 0\cdot z = 1$

$x - y + kz = -2$

$0\cdot x + ky + 4z = 6$

$\Delta = \begin{vmatrix} 1 & -2 & 0 \\ 1 & -1 & k \\ 0 & k & 4 \end{vmatrix} = 1(-4-k^2) - 1(-8)$

$= -4 - k^2 + 8 = 4 - k^2$

For unique solution,

$4 - k^2 \neq 0 \Rightarrow k \neq \pm 2$

For $k = 2$, $x - 2y + 0\cdot z = 1$

$x - y + 2z = -2$

$0\cdot x + 2y + 4z = 6$

$\Delta x = \begin{vmatrix} 1 & -2 & 0 \\ -2 & -1 & 2 \\ 6 & 2 & 4 \end{vmatrix} = (-8) + 2[-20]$

$\Delta x = -48 \neq 0$

For $k = 2$; $\Delta x \neq 0$

For $k = 2$

The system has no solution.

27. $\Delta = \begin{vmatrix} a^2 & b^2 & c^2 \\ (a+1)^2 & (b+1)^2 & (c+1)^2 \\ (a-1)^2 & (b-1)^2 & (c-1)^2 \end{vmatrix}$

Applying $R_2 \to R_2 - R_3$

$= 4\begin{vmatrix} a^2 & b^2 & c^2 \\ a & b & c \\ (a-1)^2 & (b-1)^2 & (c-1)^2 \end{vmatrix}$

Applying $R_3 \to R_3 - R_1 + 2R_2$

$$\therefore \quad \Delta = 4\begin{vmatrix} a^2 & b^2 & c^2 \\ a & b & c \\ 1 & 1 & 1 \end{vmatrix} = -4(a-b)(b-c)(c-a) = 0$$

If $\quad a-b=0$ or $b-c=0$ or $c-a=0$

$\therefore \quad a=b$ or $b=c$ or $c=a$

$\therefore \Delta ABC$ is an isosceles triangle.

28. For infinitely many solutions

$$\Delta = \Delta_1 = \Delta_2 = \Delta_3 = 0$$

$$\Delta = 0 \Rightarrow \begin{vmatrix} 2 & -1 & 1 \\ 1 & -2 & 1 \\ t & -1 & 2 \end{vmatrix} = 0 \Rightarrow t = 5$$

For $\quad t = 5, \Delta_1 = \Delta_2 = \Delta_3 = 0$

Now, $\int_0^{-2t} f(x)\,dx = \int_0^{-10} f(x)\,dx = \int_0^{-5} f(x)\,dx + \int_{-5}^{-10} f(x)\,dx$

$$= \int_{-5}^{-10} f(x+5)\,dx + \int_{-5}^{-10} f(x)\,dx$$

$$= \int_{-5}^{-10} \{f(x+5) + f(x)\}\,dx$$

$$= \int_{-5}^{-10} 2\,dx = -10 = -2t$$

29. Given determinant $D = \begin{vmatrix} x & a+y & x+a \\ y & b+y & y+b \\ z & c+y & z+c \end{vmatrix}$

$$= \begin{vmatrix} x & a+y & x \\ y & b+y & y \\ z & c+y & z \end{vmatrix} + \begin{vmatrix} x & a+y & a \\ y & b+y & b \\ z & c+y & c \end{vmatrix}$$

$$= 0 + \begin{vmatrix} x & a & a \\ y & b & b \\ z & c & c \end{vmatrix} + \begin{vmatrix} x & y & a \\ y & y & b \\ z & y & c \end{vmatrix} = y\begin{vmatrix} x & 1 & a \\ y & 1 & b \\ z & 1 & c \end{vmatrix}$$

On applying $C_1 \to C_1 + C_3$, we get

$$D = y\begin{vmatrix} x+a & 1 & a \\ y+b & 1 & b \\ z+c & 1 & c \end{vmatrix}$$

It is given that $a + x = b + y = c + z + 1 = \lambda$ (let), so

$$D = y\begin{vmatrix} \lambda & 1 & a \\ \lambda & 1 & b \\ \lambda - 1 & 1 & c \end{vmatrix}$$

On applying $R_1 \to R_1 - R_3$ and $R_2 \to R_2 - R_3$, we get

$$D = y\begin{vmatrix} 1 & 0 & a-c \\ 1 & 0 & b-c \\ \lambda-1 & 1 & c \end{vmatrix} = y(-1)[(b-c)-(a-c)]$$

$$= y(a-b)$$

30. Let $\quad \Delta = \begin{vmatrix} 1 & \log_x y & \log_x z \\ \log_y x & 3 & \log_y z \\ \log_z x & \log_z y & 5 \end{vmatrix}$

$$= \frac{1}{\log_e x \log_e y \log_e z}\begin{vmatrix} \log_e x & \log_e y & \log_e z \\ \log_e x & 3\log_e y & \log_e z \\ \log_e x & \log_e y & 5\log_e z \end{vmatrix}$$

$$= \begin{vmatrix} 1 & 1 & 1 \\ 1 & 3 & 1 \\ 1 & 1 & 5 \end{vmatrix} = 1(15-1) - 1(5-1) + 1(1-3)$$

$$= 14 - 4 - 2 = 8$$

31. $\because f(x) - x = 0$ has imaginary roots,

then $\quad f(x) - x > 0$ or $f(x) - x < 0, \forall x \in R$

for $\quad f(x) - x > 0, \forall x \in R$

then $\quad f\{f(x)\} - f(x) > 0, \forall x \in R$

Adding we get, $f\{f(x)\} - x > 0, \forall x \in R$

Similarly, $f\{f(x)\} - x < 0, \forall x \in R$

Thus, roots of equation $f\{f(x)\} - x = 0$ are imaginary.

Let $\quad z = \begin{vmatrix} 2 & \alpha & \delta \\ \beta & 0 & \alpha \\ \gamma & \beta & 1 \end{vmatrix}$

Then $\quad \bar{z} = \begin{vmatrix} 2 & \bar{\alpha} & \bar{\delta} \\ \bar{\beta} & 0 & \bar{\alpha} \\ \bar{\gamma} & \bar{\beta} & 1 \end{vmatrix} = \begin{vmatrix} 2 & \beta & \gamma \\ \alpha & 0 & \beta \\ \delta & \alpha & 1 \end{vmatrix} = \begin{vmatrix} 2 & \alpha & \delta \\ \beta & 0 & \alpha \\ \gamma & \beta & 1 \end{vmatrix} = z$

Hence, z is purely real.

32. $\because \quad {}^nC_r = {}^nC_{n-r}$

$$\therefore \begin{vmatrix} {}^{n+2}C_2 & {}^{n+3}C_2 & {}^{n+4}C_2 \\ {}^{n+3}C_2 & {}^{n+4}C_2 & {}^{n+5}C_2 \\ {}^{n+4}C_2 & {}^{n+5}C_2 & {}^{n+6}C_2 \end{vmatrix}$$

$$= \begin{vmatrix} \frac{(n+2)(n+1)}{2} & \frac{(n+3)(n+2)}{2} & \frac{(n+4)(n+3)}{2} \\ \frac{(n+3)(n+2)}{2} & \frac{(n+4)(n+3)}{2} & \frac{(n+5)(n+4)}{2} \\ \frac{(n+4)(n+3)}{2} & \frac{(n+5)(n+4)}{2} & \frac{(n+6)(n+5)}{2} \end{vmatrix}$$

$$\frac{1}{8}\begin{vmatrix} n^2+3n+2 & n^2+5n+6 & n^2+7n+12 \\ n^2+5n+6 & n^2+7n+12 & n^2+9n+20 \\ n^2+7n+12 & n^2+9n+20 & n^2+11n+30 \end{vmatrix}$$

Applying $C_2 \to C_2 - C_1$ and $C_3 \to C_3 - C_1$, then

$$\frac{1}{8}\begin{vmatrix} n^2+3n+2 & 2n+4 & 4n+10 \\ n^2+5n+6 & 2n+6 & 4n+14 \\ n^2+7n+12 & 2n+8 & 4n+18 \end{vmatrix}$$

Applying $R_2 \to R_2 - R_1$ and $R_3 \to R_3 - R_1$

$$\frac{1}{8}\begin{vmatrix} n^2+3n+2 & 2n+4 & 4n+10 \\ 2n+4 & 2 & 4 \\ 4n+10 & 4 & 8 \end{vmatrix}$$

Applying $R_3 \to R_3 - 2R_2$

$$\frac{1}{8}\begin{vmatrix} n^2+3n+2 & 2n+4 & 4n+10 \\ \vdots & & \\ 2n+4 & 2 & 4 \\ \vdots & & \\ 2 & \cdots\ 0\ \cdots & 0 \end{vmatrix}$$

$$= \frac{2}{8}\{(8n+16) - (8n+20)\} = -1$$

33. $\Delta(x)=\begin{vmatrix}(x-2) & (x-1)^2 & x^3\\ (x-1) & x^2 & (x+1)^3\\ x & (x+1)^2 & (x+2)^3\end{vmatrix}$

Applying $R_2 \to R_2 - R_1$ and $R_3 \to R_3 - R_1$, then

$$\Delta(x)=\begin{vmatrix}(x-2) & (x-1)^2 & x^3\\ 1 & (2x-1) & 3x^2+3x+1\\ 2 & 4x & 6x^2+12x+8\end{vmatrix}$$

Applying $R_3 \to R_3 - 2R_2$, then

$$\Delta(x)=\begin{vmatrix}(x-2) & (x-1)^2 & x^3\\ 1 & (2x-1) & 3x^2+3x+1\\ 0 & 2 & 6x+6\end{vmatrix}$$

Expand w.r.t. C_1, then

$$\Delta(x)=(x-2)\{(2x-1)(6x+6)-2(3x^2+3x+1)\} -1\{(x-1)^2(6x+6)-2x^3\}$$

$\therefore$ Coefficient of x in $\Delta(x)=-8+6=-2$

34. $\therefore f(-x)=\begin{vmatrix}-x & \cos x & e^{x^2}\\ -\sin x & x^2 & \sec x\\ -\tan x & 1 & 2\end{vmatrix}=-f(x)$

$\therefore \quad \int_{-\pi/2}^{\pi/2} f(x)\,dx=0$ [$\because f(x)$ is an odd function]

35. Given, $\quad 2\omega+1=z$

$\Rightarrow \quad 2\omega+1=\sqrt{-3} \quad [\because z=\sqrt{-3}]$

$\Rightarrow \quad \omega=\dfrac{-1+\sqrt{3}i}{2}$

Since, ω is cube root of unity.

$\therefore \quad \omega^2=\dfrac{-1-\sqrt{3}i}{2}$ and $\omega^{3n}=1$

Now, $\begin{vmatrix}1 & 1 & 1\\ 1 & -\omega^2-1 & \omega^2\\ 1 & \omega^2 & \omega^7\end{vmatrix}=3k \Rightarrow \begin{vmatrix}1 & 1 & 1\\ 1 & \omega & \omega^2\\ 1 & \omega^2 & \omega\end{vmatrix}=3k$

[$\because 1+\omega+\omega^2=0$ and $\omega^7=(\omega^3)^2\cdot\omega=\omega$]

On applying $R_1 \to R_1+R_2+R_3$, we get

$$\begin{vmatrix}3 & 1+\omega+\omega^2 & 1+\omega+\omega^2\\ 1 & \omega & \omega^2\\ 1 & \omega^2 & \omega\end{vmatrix}=3k$$

$\Rightarrow \quad \begin{vmatrix}3 & 0 & 0\\ 1 & \omega & \omega^2\\ 1 & \omega^2 & \omega\end{vmatrix}=3k$

$\Rightarrow \quad 3(\omega^2-\omega^4)=3k$

$\Rightarrow \quad (\omega^2-\omega)=k$

$\therefore \quad k=\left(\dfrac{-1-\sqrt{3}i}{2}\right)-\left(\dfrac{-1+\sqrt{3}i}{2}\right)=-\sqrt{3}i=-z$

36. Since α, β, γ are the roots of $x^3+px+q=0$

$\therefore \quad \alpha+\beta+\gamma=0$

Applying $C_1 \to C_1+C_2+C_3$, then

$$\begin{vmatrix}\alpha+\beta+\gamma & \beta & \gamma\\ \alpha+\beta+\gamma & \gamma & \alpha\\ \alpha+\beta+\gamma & \alpha & \beta\end{vmatrix}=\begin{vmatrix}0 & \beta & \gamma\\ 0 & \gamma & \alpha\\ 0 & \alpha & \beta\end{vmatrix}=0$$

37. Applying $C_1 \to C_1 - C_2$

Then, $\begin{vmatrix}4 & (a^x-a^{-x})^2 & 1\\ 4 & (b^y-b^{-y})^2 & 1\\ 4 & (c^y-c^{-y})^2 & 1\end{vmatrix}=0$

[since C_1 and C_3 are proportional]

38. Let $\quad \Delta=\begin{vmatrix}\sin\pi & \cos B & \cos C\\ -\sin B & 0 & \tan A\\ \cos(\pi-C) & -\tan A & 0\end{vmatrix}$

$$=\begin{vmatrix}0 & \sin B & \cos C\\ -\sin B & 0 & \tan A\\ -\cos C & -\tan A & 0\end{vmatrix}$$

$=0 \quad$ [$\because \Delta$ is skew-symmetric]

39. $f'(x)=\begin{vmatrix}-\sin x & \sin x & \cos x\\ -2\sin 2x & \sin 2x & 2\cos 2x\\ -3\sin 3x & \sin 3x & 3\cos 3x\end{vmatrix}$

$$+\begin{vmatrix}\cos x & \cos x & \cos x\\ \cos 2x & 2\cos 2x & 2\cos 2x\\ \cos 3x & 3\cos 3x & 3\cos 3x\end{vmatrix}+\begin{vmatrix}\cos x & \sin x & -\sin x\\ \cos 2x & \sin 2x & -4\sin 2x\\ \cos 3x & \sin 3x & -9\sin 3x\end{vmatrix}$$

$$f'(\pi/2)=\begin{vmatrix}-1 & 1 & 0\\ 0 & 0 & -2\\ 3 & -1 & 0\end{vmatrix}+0+\begin{vmatrix}0 & 1 & -1\\ -1 & 0 & 0\\ 0 & -1 & 9\end{vmatrix}$$

$=2(1-3)+0+1(9-1)=-4+8=4$

40. $\begin{vmatrix}(-1)^n a & a+1 & a-1\\ (-1)^{n+1}b & 1-b & b+1\\ (-1)^{n+1}c & 1+c & 1-c\end{vmatrix}+\begin{vmatrix}(-1)^{n+1}a & a+1 & a-1\\ (-1)^n b & 1-b & b+1\\ (-1)^{n+2}c & 1+c & 1-c\end{vmatrix}$

[interchanging rows in columns is I determinant]

$$=\begin{vmatrix}0 & a+1 & a-1\\ 0 & 1-b & b+1\\ 0 & 1+c & 1-c\end{vmatrix}=0$$

41. $\Delta=\begin{vmatrix}1 & 1 & 1\\ {}^nC_1 & {}^{n-3}C_1 & {}^{n+6}C_1\\ {}^nC_2 & {}^{n+3}C_2 & {}^{n+6}C_2\end{vmatrix}$

$$=\frac{1}{2}\begin{vmatrix}1 & 1 & 1\\ n & (n+3) & (n+6)\\ n(n-1) & (n+3)(n+2) & (n+6)(n+5)\end{vmatrix}$$

Applying $C_2 \to C_2 - C_1$ and $C_3 \to C_3 - C_1$, then

$$=\frac{1}{2}\begin{vmatrix}1 & 0 & 1\\ n & 3 & 6\\ n^2-n & 6n+6 & 12n+30\end{vmatrix}$$

$=27 \quad =3^3$

$\therefore n=3$

42. For non-trivial solution

$$\begin{vmatrix} a-1 & -1 & -1 \\ 1 & 1-b & 1 \\ 1 & 1 & 1-c \end{vmatrix} = 0$$

Applying $C_1 \to C_1 - C_3$ and $C_2 \to C_2 - C_3$, then

$$\begin{vmatrix} a & 0 & -1 \\ 0 & -b & 1 \\ c & c & 1-c \end{vmatrix} = 0$$

$\Rightarrow$ $a(-b + bc - c) - 0 + c(0 - b) = 0$

$\Rightarrow$ $ab + bc + ca = abc$

43. $\Rightarrow \begin{vmatrix} x^2 & x & 1 \\ \cos^2\theta & \cos\theta & 1 \\ \sin^2\theta & \sin\theta & 1 \end{vmatrix} \times \begin{vmatrix} 1 & x & x^2 \\ 1 & \cos\theta & \cos^2\theta \\ 1 & \sin\theta & \sin^2\theta \end{vmatrix} = 0$ [row by row]

or $\begin{vmatrix} 1 & x & x^2 \\ 1 & \cos\theta & \cos^2\theta \\ 1 & \sin\theta & \sin^2\theta \end{vmatrix} = 0$

Applying $R_2 \to R_2 - R_1$ and $R_3 \to R_3 - R_1$, then

$$\Rightarrow \begin{vmatrix} 1 & x & x^2 \\ 0 & \cos\theta - x & \cos^2\theta - x^2 \\ 0 & \sin\theta - x & \sin^2\theta - x^2 \end{vmatrix}$$

$\Rightarrow$ $(\cos\theta - x)(\sin\theta - x)\{\sin\theta + x - (\cos\theta + x)\} = 0$

$\therefore$ $x = \sin\theta, \cos\theta$

44. Let $\Delta = \begin{vmatrix} \sin\frac{A}{2} & \sin\frac{B}{2} & \sin\frac{C}{2} \\ \sin(A+B+C) & \sin\frac{B}{2} & \cos\frac{A}{2} \\ \cos\left(\frac{A+B+C}{2}\right) & \tan(A+B+C) & \sin\frac{C}{2} \end{vmatrix}$

$$= \begin{vmatrix} \sin\frac{A}{2} & \sin\frac{B}{2} & \sin\frac{C}{2} \\ 0 & \sin\frac{B}{2} & \cos\frac{A}{2} \\ 0 & 0 & \sin\frac{C}{2} \end{vmatrix} = \sin\frac{A}{2}\sin\frac{B}{2}\sin\frac{C}{2}$$

but $\sin\frac{A}{2}\sin\frac{B}{2}\sin\frac{C}{2} \le \frac{1}{8}$

$\therefore$ $\Delta \le \frac{1}{8}$

45. Given system of linear equation

$2x_1 - 4x_2 + \lambda x_3 = 1,\ x_1 - 6x_2 + x_3 = 2$

and $\lambda x_1 - 10x_2 + 4x_3 = 3$

is inconsistent, so $\Delta = 0$

$$\Rightarrow \begin{vmatrix} 2 & -4 & \lambda \\ 1 & -6 & 1 \\ \lambda & -10 & 4 \end{vmatrix} = 0$$

$\Rightarrow 2(-24 + 10) + 4(4 - \lambda) + \lambda(-10 + 6\lambda) = 0$

$\Rightarrow$ $6\lambda^2 - 14\lambda - 12 = 0$

$\Rightarrow$ $3\lambda^2 - 7\lambda - 6 = 0$

$\Rightarrow$ $3\lambda^2 - 9\lambda + 2\lambda - 6 = 0$

$\Rightarrow$ $3\lambda(\lambda - 3) + 2(\lambda - 3) = 0$

$\Rightarrow$ $\lambda = -\frac{2}{3}, 3$

On adding first and second given linear equations we have

$$3x_1 - 10x_2 + (\lambda + 1)x_3 = 3$$

And at $\lambda = 3$, it is same as the third linear equation, so for $\lambda = 3$ there are infinity many solution, but for $\lambda = \frac{-2}{3}$ both are different, so for $\lambda = \frac{-2}{3}$, the system of linear equations in inconsistent.

46. We have, $8x = 0$; $3x + ky - 2z = 0$; $2x + 4y - 2z = 0$

of equation has non-zero solution, if $\begin{vmatrix} 1 & k & 2 \\ 3 & k & -2 \\ 2 & 4 & -3 \end{vmatrix} = 0$

$(4h + 8) - k(9 + 4) + 3(12 - 2k) = 0$

$= 8k + 8 + 9k - 4k + 36 - 6k = 0$

$= 4k + 44 = 0 \Rightarrow k = 11$

$z = \lambda$, then we get

$x + 11y + 3\lambda = 0$ ...(i)

$3x + 11y - 2\lambda = 0$ ...(ii)

$2x + 4y - 3\lambda = 0$ ...(iii)

Solving Eq. (i) and (ii), we get

$$x = \frac{5\lambda}{2},\ y = \frac{-\lambda}{2},\ z = \lambda \Rightarrow \frac{xz}{y^2} = \frac{5\lambda^2}{2\times\left(-\frac{\lambda}{2}\right)^2} = 10$$

47. Let the third order determinant be

$$\Delta = \begin{vmatrix} a_{11} & a_{12} & a_{13} \\ a_{21} & a_{22} & a_{23} \\ a_{31} & a_{32} & a_{33} \end{vmatrix}$$

when $a_{ij} = 1$ or -1, then both a_{21}/a_{11} and a_{31}/a_{11} equal to ± 1

Applying $R_2 \to R_2 - \left(\frac{a_{21}}{a_{11}}\right)R_1$ and $R_3 \to R_3 - \left(\frac{a_{31}}{a_{11}}\right)R_1$

$$\Delta = \begin{vmatrix} a_{11} & a_{12} & a_{13} \\ 0 & 2\alpha & 2\beta \\ 0 & 2\gamma & 2\delta \end{vmatrix}$$

where α, β, γ and δ all equal to -1, 0 or 1. Now expand w.r.t. C_1

$\therefore$ $\Delta = 4a_{11}(\alpha\delta - \beta\gamma)$

which is an even integer.

$\because$ $\alpha\delta - \beta\gamma$ = integer.

48. Let $\begin{vmatrix} (1+x)^{a_1b_1} & (1+x)^{a_1b_2} & (1+x)^{a_1b_3} \\ (1+x)^{a_2b_1} & (1+x)^{a_2b_2} & (1+x)^{a_2b_3} \\ (1+x)^{a_3b_1} & (1+x)^{a_3b_2} & (1+x)^{a_3b_3} \end{vmatrix}$

$$= A + Bx + Cx^2 + \ldots$$

Differentiating both sides w.r.t. x and then put $x = 0$, we get

$$\begin{vmatrix} a_1b_1 & a_1b_2 & a_1b_3 \\ 1 & 1 & 1 \\ 1 & 1 & 1 \end{vmatrix} + \begin{vmatrix} 1 & 1 & 1 \\ a_2b_1 & a_2b_2 & a_2b_3 \\ 1 & 1 & 1 \end{vmatrix} + \begin{vmatrix} 1 & 1 & 1 \\ 1 & 1 & 1 \\ a_3b_1 & a_3b_2 & a_3b_3 \end{vmatrix} = B$$

$\Rightarrow$ $0 + 0 + 0 = B$

$\therefore$ $B = 0$

Hence, coefficient of $x = 0$.

49. Given system of linear equations

$$2x_1 - 2x_2 + x_3 = \lambda x_1 \quad \ldots(i)$$

$$\Rightarrow (2-\lambda)x_1 - 2x_2 + x_3 = 0 \quad \ldots(ii)$$

$$2x_1 - 3x_2 + 2x_3 = \lambda x_2 \quad \ldots(iii)$$

$$\Rightarrow 2x_1 - (3+\lambda)x_2 + 2x_3 = 0$$

$$-x_1 + 2x_2 = \lambda x_3$$

$$\Rightarrow -x_1 + 2x_2 - \lambda_3 = 0$$

Since, the system has non-trivial solution.

$$\therefore \begin{vmatrix} 2-\lambda & -2 & 1 \\ 2 & -(3+\lambda) & 2 \\ -1 & 2 & \lambda \end{vmatrix} = 0$$

$$\Rightarrow (2-\lambda)(3\lambda + \lambda^2 - 4) + 2(-2\lambda + 2) + 1(4 - 3 - \lambda) = 0$$

$$\Rightarrow (2-\lambda)(\lambda^2 + 3\lambda - 4) + 4(1-\lambda) + (1-\lambda) = 0$$

$$\Rightarrow (2-\lambda)(\lambda+4)(\lambda-1) + 5(1-\lambda) = 0$$

$$\Rightarrow (\lambda-1)[(2-\lambda)(\lambda+4) - 5] = 0$$

$$\Rightarrow (\lambda-1)(\lambda^2 + 2\lambda - 3) = 0$$

$$\Rightarrow (\lambda-1)[(\lambda-1)(\lambda+3)] = 0$$

$$\Rightarrow (\lambda-1)^2(\lambda+3) = 0$$

$$\Rightarrow \lambda = 1, 1, -3$$

Hence, λ contains two elements.

50. $\begin{vmatrix} x^k & x^{k+2} & x^{k+3} \\ y^k & y^{k+2} & y^{k+3} \\ z^k & z^{k+2} & z^{k+3} \end{vmatrix}$

$$= x^k \cdot y^k \cdot z^k (x-y)(y-z)(z-x)(xy + yz + zx)$$

$$= (xyz)^{k+1}(x-y)(y-z)(z-x)\left(\frac{1}{x} + \frac{1}{y} + \frac{1}{z}\right)$$

$\therefore \; k + 1 = 0 \Rightarrow k = -1$

51. The given system is consistent.

Therefore, $\Delta = \begin{vmatrix} 1 & 1 & -1 \\ 2 & -1 & -c \\ -b & 3b & -c \end{vmatrix} = 0$

or $\quad c + bc - 6b + b + 2c + 3bc = 0$

or $\quad 3c + 4bc - 5b = 0$

or $\quad c = \dfrac{5b}{4b+3}$

Now, $\quad c < 1$

$\Rightarrow \quad \dfrac{5b}{4b+3} < 1$

or $\quad \dfrac{5b}{4b+3} - 1 < 0$

or $\quad \dfrac{b-3}{4b+3} < 0$

$\Rightarrow \quad b \in \left(-\dfrac{3}{4}, 3\right)$

52. Given system of linear equations is

$$[\sin\theta]\, x + [-\cos\theta]\, y = 0 \quad \ldots(i)$$

and $\quad [\cot\theta]\, x + y = 0 \quad \ldots(ii)$

where, $[x]$ denotes the greatest integer $\le x$.

Here, $\quad \Delta = \begin{vmatrix} [\sin\theta] & [-\cos\theta] \\ [\cot\theta] & 1 \end{vmatrix}$

$\Rightarrow \quad \Delta = [\sin\theta] - [-\cos\theta][\cot\theta]$

When $\quad \theta \in \left(\dfrac{\pi}{2}, \dfrac{2\pi}{3}\right)$

$$\sin\theta \in \left(\frac{\sqrt{3}}{2}, 1\right)$$

$\Rightarrow \quad [\sin\theta] = 0 \quad \ldots(iii)$

$$-\cos\theta \in \left(0, \frac{1}{2}\right)$$

$\Rightarrow \quad [-\cos\theta] = 0 \quad \ldots(iv)$

and $\quad \cot\theta \in \left(-\dfrac{1}{\sqrt{3}}, 0\right)$

$\Rightarrow \quad [\cot\theta] = -1 \quad \ldots(v)$

So, $\Delta = [\sin\theta] - [-\cos\theta][\cot\theta] - (0 \times (-1)) = 0$

[from Eqs. (iii), (iv) and (v)]

Thus, for $\theta \in \left(\dfrac{\pi}{2}, \dfrac{2\pi}{3}\right)$, the given system have infinitely many solutions.

When $\theta \in \left(\pi, \dfrac{7\pi}{6}\right)$, $\sin\theta \in \left(-\dfrac{1}{2}, 0\right) \Rightarrow [\sin\theta] = -1$

$$-\cos\theta \in \left(\frac{\sqrt{3}}{2}, 1\right) \Rightarrow [\cos\theta] = 0$$

and $\cot\theta \in (\sqrt{3}, \infty) \Rightarrow [\cot\theta] = n,\; n \in N$.

So, $\quad \Delta = -1 - (0 \times n) = -1$

Thus, for $\theta \in \left(\pi, \dfrac{7\pi}{6}\right)$, the given system has a unique solution.

53. The given system of equation will have a non-trivial solution, if

$$\Delta = \begin{vmatrix} p\alpha + q & p & q \\ q\alpha + r & q & r \\ 0 & p\alpha + q & q\alpha + r \end{vmatrix} = 0$$

$R_3 \to R_3 - \alpha R_1 - R_2$

$$\begin{vmatrix} p\alpha + q & p & q \\ q\alpha + r & q & r \\ -(p\alpha^2 + 2q\alpha + r) & 0 & 0 \end{vmatrix} = 0$$

$$-(p\alpha^2 + 2q\alpha + r)(pr - q^2) = 0$$

So p, q, r in GP

or $\quad p\alpha^2 + 2q\alpha + r = 0$

54. $\because \Delta = \begin{vmatrix} 1 & \sin\theta & 1 \\ -\sin\theta & 1 & \sin\theta \\ -1 & -\sin\theta & 1 \end{vmatrix} = 1(1 + \sin^2\theta) - \sin\theta(0)$

$$+ 1(\sin^2\theta + 1)$$

$$= 2(1 + \sin^2\theta)$$

$\because \quad 0 \le \sin^2\theta \le 1$

$\Rightarrow \quad 1 \le 1 + \sin^2\theta \le 2$

$\Rightarrow \quad 2 \le 2(1 + \sin^2\theta) \le 4$

$\Rightarrow \quad 2 \le \Delta \le 4$

$\therefore \quad \Delta \in [2, 4]$

Objective Questions Type II [One or more than one correct answer(s)]

1. Taking a, b, c commons from C_1, C_2, C_3 respectively and then multiplying from a, b, c in R_1, R_2, R_3 respectively.

$$\therefore \quad \Delta = \begin{vmatrix} a^2+x^2 & a^2 & a^2 \\ b^2 & b^2+x^2 & b^2 \\ c^2 & c^2 & c^2+x^2 \end{vmatrix}$$

Applying $R_1 \to R_1 + R_2 + R_3$, then

$$\begin{vmatrix} x^2+a^2+b^2+c^2 & x^2+a^2+b^2+c^2 & x^2+a^2+b^2+c^2 \\ b^2 & b^2+x^2 & b^2 \\ c^2 & c^2 & c^2+x^2 \end{vmatrix}$$

Applying $C_2 \to C_2 - C_1$ and $C_3 \to C_3 - C_1$, then

$$\Delta = \begin{vmatrix} x^2+a^2+b^2+c^0 & 0 & 0 \\ b^2 & x^2 & 0 \\ c^2 & 0 & x^2 \end{vmatrix}$$

$$= x^4(x^2 + a^2b^2 + c^2)$$

2. $\sum\limits_{k=1}^{n} D_k =$

$$\begin{vmatrix} \sum\limits_{k=1}^{n} 2^{k-1} & \sum\limits_{k=1}^{n} \dfrac{1}{k(k+1)} & \sum\limits_{k=1}^{n} \sin k\theta \\ x & y & z \\ 2^n - 1 & \dfrac{n}{n+1} & \dfrac{\sin\left(\dfrac{n+1}{2}\right)\theta \sin\left(\dfrac{n\theta}{2}\right)}{\sin\left(\dfrac{\theta}{2}\right)} \end{vmatrix}$$

$$= \begin{vmatrix} 2^n - 1 & \dfrac{n}{n+1} & \dfrac{\sin\left(\dfrac{n+1}{2}\right)\theta \sin\left(\dfrac{n\theta}{2}\right)}{\sin\left(\dfrac{\theta}{2}\right)} \\ x & y & z \\ 2^n - 1 & \dfrac{n}{n+1} & \dfrac{\sin\left(\dfrac{n+1}{2}\right)\theta \sin\left(\dfrac{n\theta}{2}\right)}{\sin\left(\dfrac{\theta}{2}\right)} \end{vmatrix} = 0$$

$[\because R_1 = R_3]$

3. We have,

$$a_7 = \Delta(0) = \begin{vmatrix} 0 & -1 & 3 \\ 1 & 2 & -3 \\ -3 & 4 & 0 \end{vmatrix}$$

$$= 1(0-9) + 3(4+6)$$
$$= -9 + 30 = 21$$

Therefore, option (a) is true.

Again, $$\sum_{k=0}^{7} a_k = \Delta(1) = \begin{vmatrix} 2 & 1 & 4 \\ 4 & 3 & -2 \\ -2 & 5 & 2 \end{vmatrix}$$

$$= 2(6+10) - 1(8-4) + 4(20+6)$$
$$= 32 - 4 + 104 = 132$$

Therefore, $\sum\limits_{k=0}^{6} a_k = 132 - a_7$

$$= 132 - 21 = 111$$

Therefore, option (b) is true, Now

$$\Delta(-1) = \begin{vmatrix} 0 & -3 & 2 \\ -2 & 3 & -4 \\ -4 & 5 & -2 \end{vmatrix} = 3(4-16) + 2(-10+12)$$

$$= -36 + 4 = -32$$

Therefore, option (c) is true.

4. Applying $C_3 \to C_3 - \alpha C_1 - C_2$

$$\text{Then,} \quad \begin{vmatrix} a & b & 0 \\ b & c & 0 \\ a\alpha + b & b\alpha + c & (a\alpha^2 + 2b\alpha + c) \end{vmatrix} = 0$$

$$\Rightarrow \quad (a\alpha^2 + 2b\alpha + c)(b^2 - ac) = 0$$

a, b, c are in GP and $a\alpha^2 + 2b\alpha + c = 0$

5. $\because a > b > c$ and given equations are

$$ax + by + cz = 0$$
$$bx + cy + az = 0$$
$$cx + ay + bz = 0$$

For non trivial solution

$$\Delta = \begin{vmatrix} a & b & c \\ b & c & a \\ c & a & b \end{vmatrix} = 0$$

$$3abc - (a^3 + b^3 + c^3) = 0$$

$$\therefore \quad a + b + c = 0$$

If α, β be the roots of $at^2 + bt + c = 0$

$$\therefore \quad \alpha + \beta = -\frac{b}{a} \text{ and } \beta = \frac{c}{a}$$

$$b^2 - 4ac = (-a-c)^2 - 4ac$$
$$= (a-c)^2 > 0$$

For opposite sign $|\alpha - \beta| > 0$

$$\therefore \quad (\alpha+\beta)^2 - 4\alpha\beta > 0$$

$$\frac{b^2}{a^2} - \frac{4c}{a} > 0$$

$$b^2 - 4ac > 0$$

$$(-a-c)^2 - 4ac > 0$$

$$(a-c)^2 > 0 \text{ true.}$$

Hence, roots are real and have opposite sign.

6. $\because \quad f(x)=\begin{vmatrix} n & n+1 & n+2 \\ {}^nP_n & {}^{n+1}P_{n+1} & {}^{n+2}P_{n+2} \\ {}^nC_n & {}^{n+1}C_{n+1} & {}^{n+2}C_{n+2} \end{vmatrix}$

$=\begin{vmatrix} n & n+1 & n+2 \\ n! & (n+1)! & (n+2)! \\ 1 & 1 & 1 \end{vmatrix} \quad [\because {}^nP_n = n!, {}^nC_n = 1]$

Applying $C_2 \to C_2 - C_1$ and $C_3 \to C_3 - C_1$

Then $f(x)=\begin{vmatrix} n & 1 & 2 \\ n! & n\cdot n! & (n^2+3n+1)n! \\ 1 & 0 & 0 \end{vmatrix}$

$=\begin{vmatrix} 1 & 2 \\ n\cdot n! & (n^2+3n+1)n! \end{vmatrix}$

$= n!(n^2+n+1)$

7. $\because x=\dfrac{a}{b-c}, y=\dfrac{b}{c-a}, z=\dfrac{c}{a-b}$

or $-a+bx-cx=0,\ -ay-b+cy=0,\ az-bz-c=0$

Now, eliminate a, b, c, then

$$\begin{vmatrix} -1 & x & -x \\ -y & -1 & y \\ z & -z & -1 \end{vmatrix}=0$$

$\Rightarrow \quad (-1)^3\begin{vmatrix} 1 & -x & x \\ y & 1 & -y \\ -z & z & 1 \end{vmatrix}=0$ or $\begin{vmatrix} 1 & -x & x \\ y & 1 & -y \\ -z & z & 1 \end{vmatrix}=0$

Also, applying $C_1 \to C_1 + C_2 + C_3$, then we get

$$\begin{vmatrix} 1 & -x & x \\ 1 & 1 & -y \\ 1 & z & 1 \end{vmatrix}$$

8. Given determinant could be expressed as product of two determinants.

i.e. $\begin{vmatrix} (1+\alpha)^2 & (1+2\alpha)^2 & (1+3\alpha)^2 \\ (2+\alpha)^2 & (2+2\alpha)^2 & (2+3\alpha)^2 \\ (3+\alpha)^2 & (3+2\alpha)^2 & (3+3\alpha)^2 \end{vmatrix} = -648\alpha$

$\Rightarrow \begin{vmatrix} 1+2\alpha+\alpha^2 & 1+4\alpha+4\alpha^2 & 1+6\alpha+9\alpha^2 \\ 4+4\alpha+\alpha^2 & 4+8\alpha+4\alpha^2 & 4+12\alpha+9\alpha^2 \\ 9+6\alpha+\alpha^2 & 9+12\alpha+4\alpha^2 & 9+18\alpha+9\alpha^2 \end{vmatrix} = -648\alpha$

$\Rightarrow \begin{vmatrix} 1 & \alpha & \alpha^2 \\ 4 & 2\alpha & \alpha^2 \\ 9 & 3\alpha & \alpha^2 \end{vmatrix}\cdot\begin{vmatrix} 1 & 1 & 1 \\ 2 & 4 & 6 \\ 1 & 4 & 9 \end{vmatrix} = -648\alpha$

$\Rightarrow \alpha^3\begin{vmatrix} 1 & 1 & 1 \\ 4 & 2 & 1 \\ 9 & 3 & 1 \end{vmatrix}\cdot\begin{vmatrix} 1 & 1 & 1 \\ 2 & 4 & 6 \\ 1 & 4 & 9 \end{vmatrix} = -648\alpha$

$\Rightarrow \quad -8\alpha^3 = -648\alpha \Rightarrow \alpha^3 - 81\alpha = 0$

$\Rightarrow \quad \alpha(\alpha^2-81)=0$

$\therefore \quad \alpha = 0, \pm 9$

9. Applying $C_1 \to C_1 + C_2$

Then, $\quad f(x)=\begin{vmatrix} 2 & \cos^2 x & 4\sin 2x \\ 2 & 1+\cos^2 x & 4\sin 2x \\ 1 & \cos^2 x & 1+4\sin 2x \end{vmatrix}$

Now, applying $R_2 \to R_2 - R_1$ and $R_3 \to R_3 - R_1$, then

$f(x)=\begin{vmatrix} 2 & \cos^2 x & 4\sin 2x \\ 0 & 1 & 0 \\ -1 & 0 & 1 \end{vmatrix} = 2+4\sin 2x$

$\because \quad -1 \le \sin 2x \le 1$

$\Rightarrow \quad -2 \le 2+4\sin 2x \le 6$

$\Rightarrow \quad -2 \le f(x) \le 6$

$\therefore f(a)=-2$, $f(b)=6$ and $f(x)$ is periodic with period $=\dfrac{2\pi}{2}=\pi$

10. $f(\theta)=\sin^3\theta+\cos^3\theta-\cos\theta\sin\theta(\sin\theta+\cos\theta)$

$=(\sin\theta+\cos\theta)^3-4\sin\theta\cos\theta(\sin\theta+\cos\theta)$

$=(\sin\theta+\cos\theta)[1-\sin 2\theta]$

Now, $f(\theta)=0$

$\Rightarrow \quad \tan\theta=-1$ or $\sin 2\theta = 1$

$\Rightarrow \quad f(\theta)=0$ has 2 real solution in $[0, \pi]$

Also, $\dfrac{f(\theta)}{1-\sin 2\theta}=\sin\theta+\cos\theta \in [-\sqrt{2}, \sqrt{2}]$

11. Applying $C_3 \to C_3 + C_1$, then

$$\Delta=\begin{vmatrix} a-x & a & 2a \\ b-x & b & 2b \\ c-x & c & 2c \end{vmatrix}=0$$

12. Applying $C_3 \to C_3 - C_1$

Then $\begin{vmatrix} 1 & a & a^2-1 \\ \cos(xz-yz) & \cos xz & 0 \\ \sin(xz-yz) & \sin xz & 0 \end{vmatrix}$

$=(a^2-1)\sin yz$ which depends upon a, y, z

13. $12\begin{vmatrix} x & \dfrac{x(x-1)}{2} & \dfrac{x(x-1)(x-2)}{6} \\ x^2 & \dfrac{x^2(x^2-1)}{2} & \dfrac{x^2(x^2-1)(x^2-2)}{6} \\ x^3 & \dfrac{x^3(x^3-1)}{2} & \dfrac{x^3(x^3-1)(x^3-2)}{6} \end{vmatrix}$

$=x\cdot x^2\cdot x^3\begin{vmatrix} 1 & (x-1) & (x-1)(x-2) \\ 1 & (x^2-1) & (x^2-1)(x^2-2) \\ 1 & (x^3-1) & (x^3-1)(x^3-2) \end{vmatrix}$

$=x^6(x-1)^2\begin{vmatrix} 1 & 1 & x-2 \\ 1 & x+1 & (x+1)(x^2-2) \\ 1 & x^2+x+1 & (x^2+x+1)(x^3-2) \end{vmatrix}$

Now, applying $R_2 \to R_2 - R_1$ and $R_3 \to R_3 - R_1$

$$= x^6(x-1)^2\begin{vmatrix} 1 & 1 & x-2 \\ 0 & x & x(x^2+x-3) \\ 1 & x^2+x & x(x^4+x^3+x^2-2x-3) \end{vmatrix}$$

$$= x^6(x-1)\begin{vmatrix} x & x(x^2+x-3) \\ x(x+1) & x(x^4+x^3+x^2-2x-3) \end{vmatrix}$$

$$= x^8(x-1)\{x^4+x^3+x^2-2x-3-x^3-x^2+3x-x^2-x+3\}$$

$$= x^8(x-1)(x^4-x^2)$$

$$= x^{10}(x-1)(x^2-1) = x^{10}(x-1)^2(x+1)$$

14. We have, $\Delta_1 = \begin{vmatrix} 1 & 0 & 0 \\ a & b-a & c-a \\ a^2 & b^2-a^2 & c^2-a^2 \end{vmatrix}$

$$= (b-a)(c^2-a^2) - (c-a)(b^2-a^2)$$

$$= (a-b)(c-a)[-(c+a)+(b+a)]$$

$$= (a-b)(b-c)(c-a)$$

Also, $\Delta_2 = \begin{vmatrix} 1 & 1 & 1 \\ b+c & b+a & a+b \\ b^2+c^2 & c^2+a^2 & a^2+b^2 \end{vmatrix}$

$$= \begin{vmatrix} 1 & 0 & 0 \\ b+c & a-b & a-c \\ b^2+c^2 & a^2-b^2 & a^2-c^2 \end{vmatrix}$$

$$= (a-b)(a^2-c^2) - (a-c)(a^2-b^2)$$

$$= (a-b)(c-a)[-(c+a)+(a+b)]$$

$$= (a-b)(b-c)(c-a)$$

Therefore, $\Delta_1 = \Delta_2 = (a-b)(b-c)(c-a)$.

So, option (b) and option (d) are true.

15. Applying $C_2 \to C_2 - aC_1$

Then, $f(a, b) = \begin{vmatrix} a & 0 & 0 \\ 1 & (a+b) & (a+b)^2 \\ 0 & 1 & (2a+3b) \end{vmatrix}$

$$= a\begin{vmatrix} (a+b) & (a+b)^2 \\ 1 & (2a+3b) \end{vmatrix} = a(a+b)\begin{vmatrix} 1 & a+b \\ 1 & 2a+3b \end{vmatrix}$$

$$= a(a+b)(a+2b)$$

16. Applying $C_2 \to C_2 - C_1$ and $C_3 \to C_3 - C_1$, then

$$\Delta = \begin{vmatrix} 1 & 0 & 0 \\ f_1(x_1) & f_1(x_2) - f_2(x_1) & f_2(x_3) - f_2(x_1) \\ & x_2 - x_1 & x_3 - x_1 \end{vmatrix}$$

$$= \begin{vmatrix} x_2 - x_1 & x_3 - x_1 \\ (x_2-x_1)(x_1+x_2+b) & (x_3-x_1)(x_1+x_3+b) \end{vmatrix}$$

$$= (x_2-x_1)(x_3-x_1)\begin{vmatrix} 1 & 1 \\ x_1+x_2+b & x_1+x_3+b \end{vmatrix}$$

$$= (x_2-x_1)(x_3-x_1)(x_3-x_2)$$

17. Applying $C_3 \to C_3 - \lambda C_1 - C_2$

Then $\Delta = \begin{vmatrix} b & c & 0 \\ c & d & 0 \\ b\lambda + c & c\lambda + d & a\lambda^3 - b\lambda^2 + c\lambda - d \end{vmatrix}$

$$= (a\lambda^3 - b\lambda^2 + c\lambda - d)(bd - c^2) = 0 \text{ (given)}$$

$\therefore bd - c^2 = 0 \Rightarrow b, c, d$ are in GP,

and if $a\lambda^3 - b\lambda^2 + c\lambda - d = 0$,

then λ is a root of $ax^3 - bx^2 + cx - d = 0$

18. Taking a, b, c common from R_1, R_2 and R_3 respectively and then multiplying a, b, c in C_1, C_2, C_3 respectively, then

$$\begin{vmatrix} a^2+(b^2+c^2)\cos\phi & b^2(1-\cos\phi) & c^2(1-\cos\phi) \\ a^2(1-\cos\phi) & b^2+(c^2+a^2)\cos\phi & c^2(1-\cos\phi) \\ a^2(1-\cos\phi) & b^2(1-\cos\phi) & c^2+(a^2+b^2)\cos\phi \end{vmatrix}$$

Applying $C_1 \to C_1 + C_2 + C_3$, then

$$\begin{vmatrix} 1 & b^2(1-\cos\phi) & c^2(1-\cos\phi) \\ 1 & b^2+(c^2+a^2)\cos\phi & c^2(1-\cos\phi) \\ 1 & b^2(1-\cos\phi) & c^2+(a^2+b^2)\cos\phi \end{vmatrix}$$

$[\therefore a^2 + b^2 + c^2 = 1]$

Applying $R_2 \to R_2 - R_1$ and $R_3 \to R_3 - R_1$, then

$$\begin{vmatrix} 1 & b^2(1-\cos\phi) & c^2(1-\cos\phi) \\ 0 & \cos\phi & 0 \\ 0 & 0 & \cos\phi \end{vmatrix}$$

$[\because a^2 + b^2 + c^2 = 1]$

$= \cos^2\phi$, which is independent of a, b, and c.

19. Applying $C_2 \to C_2 - \cos^2 x\, C_1$, we get

$$f(x) = \begin{vmatrix} \sec^2 x & 0 & 1 \\ \cos^2 x & \cos^2 x - \cos^4 x & \operatorname{cosec}^2 x \\ 1 & 0 & \cot^2 x \end{vmatrix}$$

$$= (\cos^2 x - \cos^4 x)(\sec^2 x \cot^2 x - 1)$$

[expanding along C_2]

$$= \cos^2 x \sin^2 x (\operatorname{cosec}^2 x - 1)$$

$$= \cos^2 x \sin^2 x \cot^2 x = \cos^4 x$$

$$= \frac{1}{8}(3 + 4\cos 2x + \cos 4x)$$

Thus, $\int_{-\pi/4}^{\pi/4} f(x)\, dx$

$$= \frac{1}{8}\left(3x + 2\sin 2x + \frac{1}{4}\sin 4x\right)_0^{\frac{\pi}{4}} = \frac{1}{32}(3\pi + 8)$$

$$f'(x) = -4\cos^3 x \times \sin x$$

$$f'\left(\frac{\pi}{2}\right) = 0$$

It can be easily checked that each of (b), (c) and (d) also holds.

20. $\therefore$ Area of $\Delta PQR = \frac{1}{2}\begin{vmatrix} x_1 & y_1 & 1 \\ x_2 & y_2 & 1 \\ x_3 & y_3 & 1 \end{vmatrix}$

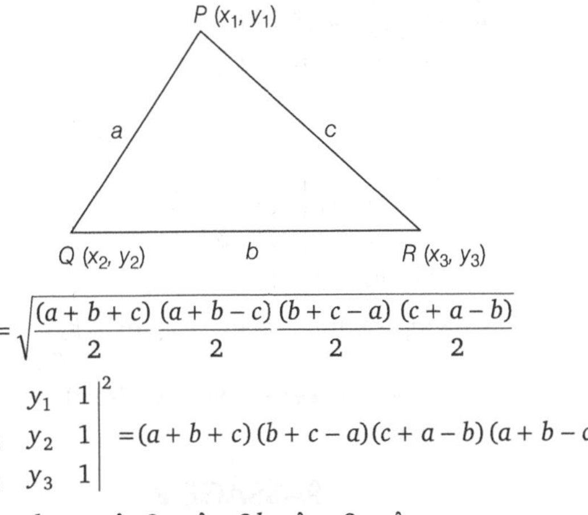

$$= \sqrt{\frac{(a+b+c)}{2}\frac{(a+b-c)}{2}\frac{(b+c-a)}{2}\frac{(c+a-b)}{2}}$$

$$4\begin{vmatrix} x_1 & y_1 & 1 \\ x_2 & y_2 & 1 \\ x_3 & y_3 & 1 \end{vmatrix}^2 = (a+b+c)(b+c-a)(c+a-b)(a+b-c)$$

i.e. $a+b+c=\lambda$, $2a=\lambda_1$, $2b=\lambda_2$, $2c=\lambda_3$

$\therefore \quad 2\lambda = \lambda_1+\lambda_2+\lambda_3$

$\Sigma\lambda_1\lambda_2 = 4(ab+bc+ca) = 4\,\Sigma ab$

$\lambda_1\lambda_2\lambda_3 = 8abc$

and $\quad \frac{a+b+c}{3} \geq (abc)^{1/3}$

$\Rightarrow \quad \frac{\lambda}{3} \geq \left(\frac{\lambda_1}{2}\cdot\frac{\lambda_2}{2}\cdot\frac{\lambda_3}{2}\right)^{1/3}$

$\Rightarrow \quad \lambda \geq \frac{3}{2}(\lambda_1\lambda_2\lambda_3)^{1/3}$

21. Put $x=0$ in both sides, then

$$\begin{vmatrix} 0 & 1 & -2 \\ -1 & 0 & -3 \\ 3 & -1 & -1 \end{vmatrix} = B$$

$\Rightarrow \quad 0-1(1+9)-2(1)=B$

$\Rightarrow \quad B=-12$

and put $x=1$ in both sides, then

$$\begin{vmatrix} 2 & 2 & -1 \\ 4 & 3 & 0 \\ 6 & 1 & 1 \end{vmatrix} = A+B$$

$\Rightarrow \quad 2(3)-2(4)-1(4-18)=A-12$

$\Rightarrow \quad -2+14=A-12$

$\therefore \quad A=24$

i.c. $\quad A-24, B--12$

22. $A = \begin{vmatrix} 1 & 1 & 0 \\ \gamma+\delta & \alpha+\beta & 0 \\ \gamma\delta & \alpha\beta & 0 \end{vmatrix} \times \begin{vmatrix} 1 & 1 & 0 \\ \alpha+\beta & \gamma+\delta & 0 \\ \alpha\beta & \gamma\delta & 0 \end{vmatrix} = 0\times 0 = 0$

Similarly, $B=0, C=0$

$\therefore \quad A+B+C=0 \Rightarrow A^3+B^3+C^3=3ABC$

23. Applying $R_1 \to R_1+R_2+R_3$ and taking $a+b+c-x$ common from the first row, then

$$(a+b+c-x)\begin{vmatrix} 1 & 1 & 1 \\ c & b-x & a \\ b & a & c-x \end{vmatrix} = 0$$

Applying $C_2 \to C_2-C_1$ and $C_3 \to C_3-C_1$, then

$$(a+b+c-x)\begin{vmatrix} 1 & 0 & 0 \\ c & b-x-c & a-c \\ b & a-b & c-x-b \end{vmatrix} = 0$$

$\Rightarrow \quad (a+b+c-x)\{(b\cdot x-c)(c-x-b)$
$\quad -(a-c)(a-b)\}=0$

$\Rightarrow \quad (0-x)\{x^2-(b-c)^2+ab+ca-bc-a^2\}=0$
$[\because a+b+c=0]$

$\Rightarrow \quad x\{x^2-a^2-b^2-c^2+ab+bc+ca\}=0$

$\Rightarrow \quad x\left\{\begin{array}{l} x^2-a^2-b^2+c^2 \\ \quad +\frac{1}{2}\{(a+b+c)^2-a^2-b^2-c^2\}\end{array}\right\}=0$

$\Rightarrow \quad x\left\{x^2-\frac{3}{2}(a^2+b^2+c^2)\right\}=0$

$[\because a+b+c=0]$

$\therefore \quad x=0, \pm\sqrt{\frac{3}{2}(a^2+b^2+c^2)}$

24. Applying $C_1 \to C_1 - xC_2$, then

$$\begin{vmatrix} a_1(1-x^2) & a_1x+b_1 & c_1 \\ a_2(1-x^2) & a_2x+b_2 & c_2 \\ a_3(1-x^2) & a_3x+b_3 & c_3 \end{vmatrix} = 0$$

$\Rightarrow \quad (1-x^2)\begin{vmatrix} a_1 & a_1x+b_1 & c_1 \\ a_2 & a_2x+b_2 & c_2 \\ a_3 & a_3x+b_3 & c_3 \end{vmatrix} = 0$

Now, applying $C_2 \to C_2 - xC_1$, then

$$(1-x^2)\begin{vmatrix} a_1 & b_1 & c_1 \\ a_2 & b_2 & c_2 \\ a_3 & b_3 & c_3 \end{vmatrix} = 0$$

$\therefore \quad x=\pm 1$ and $\begin{vmatrix} a_1 & b_1 & c_1 \\ a_2 & b_2 & c_2 \\ a_3 & b_3 & c_3 \end{vmatrix} = 0$

Linked Comprehension Type

PASSAGE 1

1. $\because \begin{vmatrix} 2 & a+b+c+d & ac+bd \\ a+b+c+d & 2(a+c)(b+d) & ac(b+d)+bd(a+c) \\ ac+bd & ac(b+d)+bd(a+c) & 2abcd \end{vmatrix}$

$$= \begin{vmatrix} 1 & 1 & 0 \\ b+d & a+c & 0 \\ bd & ac & 0 \end{vmatrix} \times \begin{vmatrix} 1 & 1 & 0 \\ a+c & b+d & 0 \\ ac & bd & 0 \end{vmatrix}$$

$= 0 \times 0 = 0$

2. $\begin{vmatrix} \cos(A-P) & \cos(A-Q) & \cos(A-R) \\ \cos(B-P) & \cos(B-Q) & \cos(B-R) \\ \cos(C-P) & \cos(C-Q) & \cos(C-R) \end{vmatrix}$

$$= \begin{vmatrix} \cos A & \sin A & 0 \\ \cos B & \sin B & 0 \\ \cos C & \sin C & 0 \end{vmatrix} \times \begin{vmatrix} \cos P & \sin P & 0 \\ \cos Q & \sin Q & 0 \\ \cos R & \sin R & 0 \end{vmatrix} = 0 \times 0 = 0$$

3. $\begin{vmatrix} (a_1-b_1)^2 & (a_1-b_2)^2 & (a_1-b_3)^2 \\ (a_2-b_1)^2 & (a_2-b_2)^2 & (a_2-b_3)^2 \\ (a_3-b_1)^2 & (a_3-b_2)^2 & (a_3-b_3)^2 \end{vmatrix}$

$$= \begin{vmatrix} a_1^2 & -2a_1 & 1 \\ a_2^2 & -2a_2 & 1 \\ a_3^2 & -2a_3 & 1 \end{vmatrix} \times \begin{vmatrix} 1 & b_1 & b_1^2 \\ 1 & b_2 & b_2^2 \\ 1 & b_3 & b_3^2 \end{vmatrix}$$

$$= 2\begin{vmatrix} 1 & a_1 & a_1^2 \\ 1 & a_2 & a_2^2 \\ 1 & a_3 & a_3^2 \end{vmatrix} \times \begin{vmatrix} 1 & b_1 & b_1^2 \\ 1 & b_2 & b_2^2 \\ 1 & b_3 & b_3^2 \end{vmatrix}$$

$= 2(a_1-a_2)(a_2-a_3)(a_3-a_1)(b_1-b_2)(b_2-b_3)(b_3-b_1)$

4. $\because \begin{vmatrix} 3 & 1+S_1 & 1+S_2 \\ 1+S_1 & 1+S_2 & 1+S_3 \\ 1+S_2 & 1+S_3 & 1+S_4 \end{vmatrix}$

$$= \begin{vmatrix} 3 & 1+\alpha+\beta & 1+\alpha^2+\beta^2 \\ 1+\alpha+\beta & 1+\alpha^2+\beta^2 & 1+\alpha^3+\beta^3 \\ 1+\alpha^2+\beta^2 & 1+\alpha^3+\beta^3 & 1+\alpha^4+\beta^4 \end{vmatrix}$$

$$= \begin{vmatrix} 1 & 1 & 1 \\ 1 & \alpha & \beta \\ 1 & \alpha^2 & \beta^2 \end{vmatrix} \times \begin{vmatrix} 1 & 1 & 1 \\ 1 & \alpha & \beta \\ 1 & \alpha^2 & \beta^2 \end{vmatrix}$$

$$= \begin{vmatrix} 1 & 1 & 1 \\ 1 & \alpha & \beta \\ 1 & \alpha^2 & \beta^2 \end{vmatrix}^2$$

$= (\alpha-1)^2(\beta-1)^2(\alpha-\beta)^2$

$= \{\alpha\beta-(\alpha+\beta)+1\}^2\{(\alpha+\beta)^2-4\alpha\beta\}$

$$= \left(\frac{c}{a}+\frac{b}{a}+1\right)^2\left(\frac{b^2}{a^2}-\frac{4c}{a}\right)$$

$$= \frac{(a+b+c)^2(b^2-4ac)}{a^4}$$

5. Let $P = \begin{vmatrix} 1+a_1b_2+a_1^2b_1^2 & 1+a_1b_2+a_1^2b_2^2 & 1+a_1b_3+a_1^2b_3^2 \\ 1+a_2b_1+a_2^2b_1^2 & 1+a_2b_2+a_2^2b_2^2 & 1+a_2b_3+a_2^2b_3^2 \\ 1+a_3b_1+a_3^2b_1^2 & 1+a_3b_2+a_3^2b_2^2 & 1+a_3b_3+a_3^2b_3^2 \end{vmatrix}$

$$\left[\because \frac{a^3-b^3}{a-b} = a^2+ab+b^2\right]$$

$$= \begin{vmatrix} 1 & a_1 & a_1^2 \\ 1 & a_2 & a_2^2 \\ 1 & a_3 & a_3^2 \end{vmatrix} \times \begin{vmatrix} 1 & b_1 & b_1^2 \\ 1 & b_2 & b_2^2 \\ 1 & b_3 & b_3^2 \end{vmatrix}$$

$= (a_1-a_2)(a_2-a_3)(a_3-a_1)(b_1-b_2)(b_2-b_3)(b_3-b_1)$

$= (-\text{ve})(-\text{ve})(+\text{ve})(-\text{ve})(-\text{ve})(+\text{ve})$

$= +\text{ ve}$

PASSAGE 2

1. The given system is

$$\begin{bmatrix} 1 & 1 & 1 \\ 1 & 2 & 2 \\ 1 & 2 & \lambda \end{bmatrix} X = \begin{bmatrix} 6 \\ 10 \\ \mu \end{bmatrix}$$

$$\begin{bmatrix} 1 & 1 & 1 \\ 0 & 1 & 1 \\ 0 & 1 & \lambda-1 \end{bmatrix} X = \begin{bmatrix} 6 \\ 4 \\ \mu-6 \end{bmatrix}$$

[by $R_2 \to R_2 - R_1$, $R_3 \to R_3 - R_1$]

$$\begin{bmatrix} 1 & 0 & 0 \\ 0 & 1 & 1 \\ 0 & 0 & \lambda-2 \end{bmatrix} X = \begin{bmatrix} 2 \\ 4 \\ \mu-10 \end{bmatrix}$$

[by $R_1 \to R_1 - R_2$, $R_3 \to R_3 - R_2$]

Let $A = \begin{bmatrix} 1 & 1 & 1 \\ 1 & 2 & 2 \\ 1 & 2 & \lambda \end{bmatrix}$,

then det $A = \lambda - 2$. Therefore, $\lambda \neq 2 \Rightarrow A$ is non-singular matrix and hence

$$AX = B \Rightarrow X = A^{-1}B$$

is the unique solution.

2. If $\lambda = 2$ and $\mu \neq 10$, then the system reduces to $x = 6$, $y + z = 4$, $0x + 0y = 0z = \mu - 10 \neq 0$

Hence, no solution, if $\lambda = 2$ and $\mu \neq 10$

3. For $\lambda = 2$, $\mu = 10$; then system has infinite number of solutions.

PASSAGE 3

The system of equations

$$-x + cy + bz = 0 \quad \ldots\text{(i)}$$

$$cx - y + az = 0 \quad \ldots\text{(ii)}$$

$$bx + ay - z = 0 \quad \ldots\text{(iii)}$$

has a non-zero solution,

if $$\Delta = \begin{vmatrix} -1 & c & b \\ c & -1 & a \\ b & a & -1 \end{vmatrix} = 0$$

$\Rightarrow \quad a^2 + b^2 + c^2 + 2abc - 1 = 0$

or $\quad a^2 + b^2 + c^2 + 2abc = 1 \quad \ldots(\text{iv})$

Then, clearly the system has infinitely many solution. From Eqs. (i) and (ii), we have

$$\frac{x}{ac+b} = \frac{y}{bc+a} = \frac{z}{1-c^2}$$

$$\therefore \quad \frac{x^2}{(ac+b)^2} = \frac{y^2}{(bc+a)^2} = \frac{z^2}{(1-c^2)^2}$$

or $$\frac{x^2}{(1-a^2)(1-c^2)} = \frac{y^2}{(1-b^2)(1-c^2)} = \frac{z^2}{(1-c^2)^2}$$

[from Eq. (iv)]

or $$\frac{x^2}{1-a^2} = \frac{y^2}{1-b^2} = \frac{z^2}{1-c^2} \quad \ldots(\text{v})$$

From Eq. (v), we see that $1-a^2, 1-b^2, 1-c^2$ are all positive or all negative.

Given, that one of a, b, c is proper fraction, so $1-a^2 > 0, 1-b^2 > 0, 1-c^2 > 0$, which gives

$$a^2 + b^2 + c^2 < 3 \quad \ldots(\text{vi})$$

Using Eqs. (iv) and (v), we get

$$1 < 3 + 2abc$$

or $\quad abc > -1 \quad \ldots(\text{vii})$

1. (c) **2.** (a) **3.** (d)

PASSAGE 4

1. $a + b + c = p, ab + bc + ca = 0$

$\therefore \quad a^2 + b^2 + c^2 = (a+b+c)^2 - 2(ab+bc+ca)$

$= p^2 - 0 = p^2$

If $\Delta = \begin{vmatrix} a & b & c \\ b & c & a \\ c & a & b \end{vmatrix}$

$$\therefore \Delta^c = \begin{vmatrix} bc-a^2 & ca-b^2 & ab-c^2 \\ ca-b^2 & ab-c^2 & bc-a^2 \\ ab-c^2 & bc-a^2 & ca-b^2 \end{vmatrix} = \Delta^{3-1}$$

$$= \Delta^2 = \begin{vmatrix} a & b & c \\ b & c & a \\ c & a & b \end{vmatrix}^2$$

$$= \begin{vmatrix} a & b & c \\ b & c & a \\ c & a & b \end{vmatrix} \times \begin{vmatrix} a & b & c \\ b & c & a \\ c & a & b \end{vmatrix}$$

$$= \begin{vmatrix} a^2+b^2+c^2 & ab+bc+ca & ab+bc+ca \\ ab+bc+ca & a^2+b^2+c^2 & ab+bc+ca \\ ab+bc+ca & ab+bc+ca & a^2+b^2+c^2 \end{vmatrix}$$

$$= \begin{vmatrix} p^2 & 0 & 0 \\ 0 & p^2 & 0 \\ 0 & 0 & p^2 \end{vmatrix} = p^6$$

2. $$\because \begin{vmatrix} l_1 & m_1 & n_1 \\ l_2 & m_2 & n_2 \\ l_3 & m_3 & n_3 \end{vmatrix} \times \begin{vmatrix} l_1 & m_1 & n_1 \\ l_2 & m_2 & n_2 \\ l_3 & m_3 & n_3 \end{vmatrix}$$

$$= \begin{vmatrix} l_1^2+m_1^2+n_1^2 & l_1l_2+m_1m_2+n_1n_2 & l_1l_3+m_1m_3+n_1n_3 \\ l_1l_2+m_1m_2+n_1n_2 & l_2^2+m_2^2+n_2^2 & l_2l_3+m_2m_3+n_2n_3 \\ l_3l_1+m_3m_1+n_3n_1 & l_2l_3+m_2m_3+n_2n_3 & l_3^2+m_3^2+n_3^2 \end{vmatrix}$$

$$= \begin{vmatrix} 1 & 0 & 0 \\ 0 & 1 & 0 \\ 0 & 0 & 1 \end{vmatrix} = 1$$

$$\therefore \quad \begin{vmatrix} l_1 & m_1 & n_1 \\ l_2 & m_2 & n_2 \\ l_3 & m_3 & n_3 \end{vmatrix}^2 = 1$$

$$\therefore \quad \begin{vmatrix} l_1 & m_1 & n_1 \\ l_2 & m_2 & n_2 \\ l_3 & m_3 & n_3 \end{vmatrix} = \pm 1$$

3. $\because x^3 - 3x^2 + 3x + 7 = 0$

$\Rightarrow \quad (x-1)^3 + 8 = 0$

$\Rightarrow \quad (x-1)^3 = (-2)^3$

$\Rightarrow \quad \left(\frac{x-1}{-2}\right)^3 = 1$

$\Rightarrow \quad \frac{x-1}{-2} = (1)^{1/3} = 1, \omega, \omega^2$

$\Rightarrow \quad x - 1 = -2, -2\omega, -2\omega^2$

or $\quad x = -1, \; 1-2\omega, 1-2\omega^2$

$\therefore \quad a = -1, b = 1-2\omega, c = 1-2\omega^2$

$$\because \begin{vmatrix} a & b & c \\ b & c & a \\ c & a & b \end{vmatrix}^2 = \begin{vmatrix} a & b & c \\ b & c & a \\ c & a & b \end{vmatrix} \times \begin{vmatrix} a & b & c \\ b & c & a \\ c & a & b \end{vmatrix}$$

$$= \begin{vmatrix} a & b & c \\ b & c & a \\ c & a & b \end{vmatrix} \times \begin{vmatrix} -a & c & b \\ -b & a & c \\ -c & b & a \end{vmatrix} \quad \text{[row by row]}$$

$$= \begin{vmatrix} 2bc-a^2 & c^2 & b^2 \\ c^2 & 2ac-b^2 & a^2 \\ b^2 & a^2 & c^2 \end{vmatrix}$$

$$= \begin{vmatrix} a & b & c \\ b & c & a \\ c & a & b \end{vmatrix}^2$$

$= (a^3 + b^3 + c^3 - 3abc)^2$

$= \{(a+b+c)(a^2+b^2+c^2-ab-bc-ca)\}^2$

$= \frac{1}{4}(a+b+c)^2\{(a-b)^2+(b-c)^2+(c-a)^2\}^2$

$= \frac{9}{4}\{-12(1+\omega+\omega^2)\} = 0$

4. Let $\Delta = \begin{vmatrix} \lambda & c & -b \\ -c & \lambda & a \\ b & -a & \lambda \end{vmatrix}$

$$\therefore\ \Delta^c = \begin{vmatrix} a^2+\lambda^2 & ab+c\lambda & ca-b\lambda \\ ab-c\lambda & b^2+\lambda^2 & bc+a\lambda \\ ac+b\lambda & bc-a\lambda & c^2+\lambda^2 \end{vmatrix} = \Delta^2$$

$$\Rightarrow \begin{vmatrix} a^2+\lambda^2 & ab+c\lambda & ca-b\lambda \\ ab-c\lambda & b^2+\lambda^2 & bc+a\lambda \\ ac+b\lambda & bc-a\lambda & c^2+\lambda^2 \end{vmatrix} \times \begin{vmatrix} \lambda & c & -b \\ -c & \lambda & a \\ b & -a & \lambda \end{vmatrix}$$

$$= \Delta^2 \times \Delta = \Delta^3$$

$$= \Delta^3 = \lambda^3 (\lambda^2 + a^2 + b^2 + c^2)^3$$

$$= \lambda^3 (2\lambda^2)^3 = 8\lambda^9$$

5. $\begin{vmatrix} A & B & C \\ B & C & A \\ C & A & B \end{vmatrix}$

$$= \begin{vmatrix} bc-a^2 & ca-b^2 & ab-c^2 \\ ca-b^2 & ab-c^2 & bc-a^2 \\ ab-c^2 & bc-a^2 & ca-b^2 \end{vmatrix}$$

$$= \begin{vmatrix} a & b & c \\ b & c & a \\ c & a & b \end{vmatrix}^2 = 49$$

$$\therefore \begin{vmatrix} a & b & c \\ b & c & a \\ c & a & b \end{vmatrix} = 7 \qquad [\because a+b+c>0]$$

Numerical Value Type

1. It is given that

$$\begin{vmatrix} \sum\limits_{k=0}^{n} k & \sum\limits_{k=0}^{n} {}^nC_k k^2 \\ \sum\limits_{k=0}^{n} {}^nC_k . k & \sum\limits_{k=0}^{n} {}^nC_k 3^k \end{vmatrix} = 0$$

$$\Rightarrow \begin{vmatrix} \dfrac{n(n+1)}{2} & n(n+1)2^{n-2} \\ n\cdot 2^{n-1} & 4^n \end{vmatrix} = 0$$

$$\left[\because \sum_{k=0}^{n} k = \frac{n(n+1)}{2}, \quad \sum_{k=0}^{n} {}^nC_k k = n\cdot 2^{n-1}, \quad \sum_{k=0}^{n} {}^nC_k k^2 = n(n+1)2^{n-2} \text{ and } \sum_{k=0}^{n} {}^nC_k 3^k = 4^n\right]$$

$$\Rightarrow \frac{n(n+1)}{2} 4^n - n^2(n+1)\, 2^{2n-3} = 0$$

$$\Rightarrow \frac{4^n}{2} - n\frac{4^{n-1}}{2} = 0 \Rightarrow n = 4$$

$$\therefore \sum_{k=0}^{n} \frac{{}^nC_k}{k+1} = \sum_{k=0}^{4} \frac{{}^4C_k}{k+1} = \frac{1}{5}\sum_{k=0}^{4} {}^5C_{k+1} = \frac{1}{5}(2^5-1)$$

$$= \frac{1}{5}(32-1) = \frac{31}{5} = 6.20$$

2. $\left(-\dfrac{1}{100}\right)\sum\limits_{p=1}^{5} D_p$

$$= \left(-\frac{1}{100}\right)\begin{vmatrix} 1+2+3+4+5 & 15 & 8 \\ 1^2+2^2+3^2+4^2+5^2 & 35 & 9 \\ 1^3+2^3+3^3+4^3+5^3 & 25 & 10 \end{vmatrix}$$

$$= \left(-\frac{1}{100}\right)\begin{vmatrix} 15 & 15 & 8 \\ 55 & 35 & 9 \\ 225 & 25 & 10 \end{vmatrix}$$

(Taking common 5 from each c_1 and c_2)

$$= \left(-\frac{1}{100}\right) \times 5 \times 5 \begin{vmatrix} 3 & 3 & 8 \\ 11 & 7 & 9 \\ 45 & 5 & 10 \end{vmatrix}$$

$$= \left(-\frac{1}{4}\right)(5)\begin{vmatrix} 3 & 3 & 8 \\ 11 & 7 & 9 \\ 9 & 1 & 2 \end{vmatrix} \qquad [\text{Taking 5 from } R_3]$$

$$= \left(-\frac{5}{4}\right)\{3(14-9) - 3(22-81) + 8(11-63)\}$$

$$= \left(-\frac{5}{4}\right)\{15 + 177 - 416\} = 330$$

3. Applying $C_2 \to C_2 - C_1$ and $C_3 \to C_3 - C_1$ and take out $a+b+c$ from C_2 as well as from C_3 to get

$$\text{Let } \Delta = (a+b+c)^2 \begin{vmatrix} (b+c)^2 & a-b-c & a-b-c \\ b^2 & c+a-b & 0 \\ c^2 & 0 & a+b-c \end{vmatrix}$$

Applying $R_1 \to R_1 - (R_2 + R_3)$

$$\Delta = (a+b+c)^2 \begin{vmatrix} 2bc & -2c & -2b \\ b^2 & c+a-b & 0 \\ c^2 & 0 & a+b-c \end{vmatrix}$$

Now, applying $C_2 \to C_2 + \frac{1}{b}C_1$ and $C_3 \to C_3 + \frac{1}{c}C_1$

$$\text{Then, } \Delta = (a+b+c)^2 \begin{vmatrix} 2bc & 0 & 0 \\ b^2 & c+a & \dfrac{b^2}{c} \\ c^2 & \dfrac{c^2}{b} & a+b \end{vmatrix}$$

$$= 2bc\,(a+b+c)^2 \{(c+a)(a+b) - bc\}$$

$$= 2abc\,(a+b+c)^3 \ \therefore\ \lambda = 2, \text{ then}$$

$$\sum_{r=1}^{10} \lambda^r = \sum_{r=1}^{10} 2^r = 2^1 + 2^2 + 2^3 + \ldots + 2^{10}$$

$$= \frac{2(2^{10}-1)}{2-1} = 2(1023) = 2046$$

4. $\Delta = \begin{vmatrix} \log_3 512 & \log_4 3 \\ \log_3 8 & \log_4 9 \end{vmatrix} \times \begin{vmatrix} \log_2 3 & \log_8 3 \\ \log_3 4 & \log_3 4 \end{vmatrix}$

$= \begin{vmatrix} \log_3 2^9 & \log_{2^2} 3 \\ \log_3 2^3 & \log_{2^2} 3^2 \end{vmatrix} \times \begin{vmatrix} \log_2 3 & \log_{2^3} 3 \\ \log_3 2^2 & \log_3 2^2 \end{vmatrix}$

$= \begin{vmatrix} 9\log_3 2 & \frac{1}{2}\log_2 3 \\ 3\log_3 2 & \frac{2}{2}\log_2 3 \end{vmatrix} \times \begin{vmatrix} \log_2 3 & \frac{1}{3}\log_2 3 \\ 2\log_3 2 & 2\log_3 2 \end{vmatrix}$

$= \left(9 \times 1 - \frac{3}{2}\right) \times \left(2 - \frac{2}{3}\right) = \frac{15}{2} \times \frac{4}{3} = 10$

5. $\begin{vmatrix} 3 & -2 & 1 \\ \lambda & -14 & 15 \\ 1 & 2 & 3 \end{vmatrix} = 0$

$\Rightarrow 3(-42 - 30) + 2(3\lambda - 15) + 1(2\lambda + 14) = 0$

$\Rightarrow \quad 8\lambda = 232$

$\therefore \quad \lambda = 29$

$\Rightarrow \quad \lambda^2 = 29^2 = 841$

6. $\because \Delta = 1(1 - 3\sin\theta\cos\theta) - 3\cos\theta(\sin\theta - 3\cos\theta) + 1(\sin^2\theta - 1)$

$= 1 - 6\sin\theta\cos\theta + 9\cos^2\theta + \sin^2\theta - 1$

$= (\sin\theta - 3\cos\theta)^2$

$\because -\sqrt{10} \le \sin\theta - 3\cos\theta \le \sqrt{10}$

$\therefore \quad 0 \le (\sin\theta - 3\cos\theta)^2 \le 10$

$\Rightarrow \quad 0 \le \Delta \le 10$

$|\text{Maximum value of } \Delta - \text{Minimum value of } \Delta|^3$

$= |10 - 0|^3 = 1000$

7. $\because \quad A_r = r^2 - (r-1)^2$

$\therefore \quad \sum_{r=1}^{2008} A_r = (2008)^2 - (1-1)^2 = (2008)^2$

$\therefore \quad \sqrt{\sum_{r=1}^{2008} A_r} = 2008$

Single Integer Type

1. We have,

$$(\lambda - 1)x + (3\lambda + 1)y + 2\lambda z = 0$$
$$(\lambda - 1)x + (4\lambda - 2)y + (\lambda + 3)z = 0$$
$$2x + (3\lambda + 1)y + 3(\lambda - 1)z = 0$$

Now, it can be express as $\quad AX = 0$

Where, $A = \begin{bmatrix} \lambda - 1 & 3\lambda + 1 & 2\lambda \\ \lambda - 1 & 4\lambda - 2 & \lambda + 3 \\ 2 & 3\lambda + 1 & 3(\lambda - 1) \end{bmatrix}$, $X = \begin{bmatrix} x \\ y \\ z \end{bmatrix}$

For non-zero solution $|A| = 0$

$$\begin{bmatrix} \lambda - 1 & 3\lambda + 1 & 2\lambda \\ \lambda - 1 & 4\lambda - 2 & \lambda + 3 \\ 2 & 3\lambda + 1 & 3(\lambda - 1) \end{bmatrix} = 0$$

On applying $R_2 \to R_2 - R_1$, $R_3 \to R_3 - R_1$

$$\begin{bmatrix} \lambda - 1 & 3\lambda + 1 & 2\lambda \\ 0 & \lambda - 3 & -\lambda + 3 \\ -\lambda + 3 & 0 & \lambda - 3 \end{bmatrix} = 0$$

$\Rightarrow \quad (\lambda - 1)[(\lambda - 3)^2] + (-\lambda + 3)((3\lambda + 1)(-\lambda + 3) - 2\lambda(\lambda - 3)) = 0$

$\Rightarrow \quad (\lambda - 1)(\lambda - 3)^2 - (\lambda - 3)[(\lambda - 3)(-3\lambda - 1 - 2\lambda)) = 0$

$\Rightarrow \quad (\lambda - 3)^2[\lambda - 1 + 5\lambda + 1] = 0$

$\Rightarrow \quad (\lambda - 3)^2(6\lambda) = 0$

$\Rightarrow \quad \lambda = 3, 0$

$p = 3, q = 0$

$\therefore \quad p^2 + q^2 - pq = 9 + 0 - 0 = 9$

2. $\begin{vmatrix} 2bc - a^2 & c^2 & b^2 \\ c^2 & 2ac - b^2 & a^2 \\ b^2 & a^2 & 2ab - c^2 \end{vmatrix}$

$= \begin{vmatrix} b & c & a \\ c & a & b \\ a & b & c \end{vmatrix} \times \begin{vmatrix} c & a & b \\ b & c & a \\ -a & -b & -c \end{vmatrix} = 0$

$\Rightarrow \quad \begin{vmatrix} a & b & c \\ b & c & a \\ c & a & b \end{vmatrix}^2 = (3abc - a^2 - b^3 - c^3)^2 = 0$

$\Rightarrow a^3 + b^3 + c^3 = 3abc \Rightarrow a + b + c = 0$

$\therefore 3(a + b + c) + 2 = 2$

3. Since, the system has non-trivial solution, we have

$$\begin{vmatrix} a & 1 & 1 \\ 1 & b & 1 \\ 1 & 1 & c \end{vmatrix} = 0$$

Applying $R_1 \to R_1 - R_2$, $R_2 \to R_2 - R_3$,
we get

$$\Delta = \begin{vmatrix} a - 1 & 1 - b & 0 \\ 0 & b - 1 & 1 - c \\ 1 & 1 & c \end{vmatrix} = 0$$

or $c(1 - a)(1 - b) + (1 - b)(1 - c) - (1 - c)(a - 1) = 0$

Dividing throughout by $(1 - a)(1 - b)(1 - c)$, we get

$$\frac{c}{1 - c} + \frac{1}{1 - a} + \frac{1}{1 - b} = 0$$

or $\quad -1 + \frac{1}{1 - c} + \frac{1}{1 - a} + \frac{1}{1 - b} = 0$

or $\quad \frac{1}{1 - c} + \frac{1}{1 - a} + \frac{1}{1 - b} = 1$

4. Since, we should be able to solve Eq. (i) and Eq. (ii) x and y, we must have $ab - h^2 \ne 0$. We can rewrite Eq. (iii) as

$x(ax + hy + g) + y(hx + by + f) + gx + fy + c + t = 0$

Since, any solution of Eqs. (i)and (ii) must be a solution of Eq. (iii), the given system of equations, can be rewritten as

$$ax + hy + g = 0$$
$$hx + by + f = 0$$

and $\quad gx + fy + c + t = 0$

Since, the above system of equations is consistent, we must have

$$\begin{vmatrix} a & h & g \\ h & b & f \\ g & f & c+t \end{vmatrix} = 0$$

$$\begin{vmatrix} a & h & g \\ h & b & f \\ g & f & c \end{vmatrix} + \begin{vmatrix} a & h & 0 \\ h & b & 0 \\ g & f & t \end{vmatrix} = 0$$

$\Rightarrow \quad [abc + 2fgh - af^2 - bg^2 - ch^2] + t(ab - h^2) = 0$

$\Rightarrow \quad t = \dfrac{abc + 2fgh - af^2 - bg^2 - ch^2}{h^2 - ab} = 8$

5. Let $\text{Det}(P) = \begin{vmatrix} a_1 & b_1 & c_1 \\ a_2 & b_2 & c_2 \\ a_3 & b_3 & c_3 \end{vmatrix}$

$$= a_1(b_2c_3 - b_3c_2) - a_2(b_1c_3 - b_3c_1) + a_3(b_1c_2 - b_2c_1)$$

Now, maximum value of Det $(P) = 6$

If $a_1 = 1$, $a_2 = -1$, $a_3 = 1$, $b_2c_3 = b_1c_3 = b_1c_2 = 1$

and $b_3c_2 = b_3c_1 = b_2c_1 = -1$

But it is not possible as

$$(b_2c_3)(b_3c_1)(b_1c_2) = -1$$

and $\quad (b_1c_3)(b_3c_2)(b_2c_1) = 1$

i.e., $\quad b_1b_2b_3c_1c_2c_3 = 1$ and -1

Similar contradiction occurs when

$$a_1 = 1,\ a_2 = 1,\ a_3 = 1,$$
$$b_2c_1 = b_3c_1 = b_1c_2 = 1$$

and $\quad b_3c_2 = b_1c_3 = b_1c_2 = -1$

Now, for value to be 5 one of the terms must be zero but that will make 2 terms zero which means answer cannot be 5.

Now, $\begin{vmatrix} 1 & 1 & 1 \\ -1 & 1 & 1 \\ 1 & -1 & 1 \end{vmatrix} = 4$

Hence, maximum value is 4.

6. If $\begin{vmatrix} 1 & \alpha & \alpha^2 \\ \alpha & 1 & \alpha \\ \alpha^2 & \alpha & 1 \end{vmatrix} = 0$

$\Rightarrow \quad \alpha^4 - 2\alpha^2 + 1 = 0$

$\Rightarrow \quad \alpha^2 = 1$

$\Rightarrow \quad \alpha = \pm 1$

But $a = 1$ not possible [not satisfying equation]

$\therefore \quad \alpha = -1$

Hence, $\quad 1 + \alpha + \alpha^2 = 1$

Matrix-Match Type

1. (A) Taking common a, b, c from R_1, R_2 and R_3 respectively and then multiplying in C_1, C_2 and C_3 by a, b, c respectively, then

$$\Delta = \begin{vmatrix} a^2 + (b^2 + c^2)\cos\phi & b^2(1 - \cos\phi) & c^2(1 - \cos\phi) \\ a^2(1 - \cos\phi) & b^2 + (c^2 + a^2)\cos\phi & c^2(1 - \cos\phi) \\ a^2(1 - \cos\phi) & b^2(1 - \cos\phi) & c^2 + (a^2 + b^2)\cos\phi \end{vmatrix}$$

Applying $C_1 \to C_1 + C_2 + C_3$, then

$$\Delta = \begin{vmatrix} 1 & b^2(1 - \cos\phi) & c^2(1 - \cos\phi) \\ 1 & b^2 + (c^2 + a^2)\cos\phi & c^2(1 - \cos\phi) \\ 1 & b^2(1 - \cos\phi) & c^2 + (a^2 + b^2)\cos\phi \end{vmatrix}$$

$[\because a^2 + b^2 + c^2 = 1]$

Applying $R_2 \to R_2 - R_1$ and $R_3 \to R_3 - R_1$, then

$$\Delta = \begin{vmatrix} 1 & b^2(1 - \cos\phi) & c^2(1 - \cos\phi) \\ 0 & \cos\phi & 0 \\ 0 & 0 & \cos\phi \end{vmatrix}$$

$[\because a^2 + b^2 + c^2 = 1]$

$= \cos^2\phi$ (P, Q, R)

(B) Applying $C_3 \to C_3 - \cos\phi\, C_2 - \sin\phi\, C_1$, then

$$\Delta = \begin{vmatrix} \sin a & \cos a & 0 \\ \sin b & \cos b & 0 \\ \sin c & \cos c & 0 \end{vmatrix} = 0 \text{ (P, Q, R, S, T)}$$

(C) Multiplying C_1 by a, C_2 by b and C_3 by c, then

$$\Delta = \frac{1}{abc}\begin{vmatrix} \frac{a}{c} & \frac{b}{c} & -\frac{(a+b)}{c} \\ -\frac{(b+c)}{a} & \frac{b}{a} & \frac{c}{a} \\ -\frac{b(b+c)}{ac} & \frac{b(a+2b+c)}{ac} & -\frac{(a+b)b}{ac} \end{vmatrix}$$

Now, applying $C_1 \to C_1 + C_2 + C_3$, then

$$\Delta = \frac{1}{abc}\begin{vmatrix} 0 & \frac{b}{c} & -\frac{(a+b)}{c} \\ 0 & \frac{b}{a} & \frac{c}{a} \\ 0 & \frac{b(a+2b+c)}{ac} & -\frac{(a+b)b}{ac} \end{vmatrix} = 0 \text{ (P, Q, R, S, T)}$$

2. It is given that the matrices A and B are non-singular of order 3 and $|B| = k$, a positive integer, so

$$|k^{-1}A^{-1}| = (k^{-1})^3|A^{-1}| = \frac{1}{k^3|A|} \qquad \left[\because |A^{-1}| = \frac{1}{|A|}\right]$$

$$= \frac{1}{|B|^3|A|}$$

$\because \quad |\text{adj}(A^{-1})| = |A^{-1}|^{3-1} = |A^{-1}|^2 = \dfrac{1}{|A|^2}$

$[\because |\text{adj}(A)| = |A|^{n-1}$, where n is the order of matrix A.]

Now, since it is given that $BAB^{-1} = I$

$\Rightarrow \quad AB^{-1} = B^{-1}$

$\therefore \quad BA^kB^{-1} = BA^{k-1}(AB^{-1}) = BA^{k-1}B^{-1}$

$$= BA^{k-2}(AB^{-1}) = BA^{k-2}B^{-1}$$

$\therefore \quad BA^kB^{-1} = BAB^{-1} = I$

and , $B\dfrac{\text{adj}(B)}{|B|} = BB^{-1} = I$

Therefore, $BA^kB^{-1} = B\dfrac{\text{adj}(B)}{|B|}$, if $BAB^{-1} = I$

Now, the adj (adj $(A^{-1})) = |A^{-1}|^{3-2}(A^{-1})$

[$\because$ adj (adj $A) = |A|^{n-2}A$, where n is the order of matrix A]

$$= \frac{1}{|A|}(A^{-1})$$

$A \to R, B \to T, C \to Q, D \to S.$

3. (A) The system has non-zero solution, if

$$\begin{vmatrix} \lambda & 1 & 1 \\ -1 & \lambda & 1 \\ -1 & -1 & \lambda \end{vmatrix} = 0$$

Solving this we get

$$\lambda(\lambda^2 + 3) = 0$$

$$\Rightarrow \quad \lambda = 0$$

(B) In the above system, if $\lambda = 1$, then the matrix

$$\begin{bmatrix} \lambda & 1 & 1 \\ -1 & \lambda & 1 \\ -1 & -1 & \lambda \end{bmatrix}$$

is non-singular and hence $x = 0$, $y = 0$, $z = 0$ is the only solution. This is called trivial solution.

(C) The system has non-zero solution. This implies

$$\begin{vmatrix} -1 & c & b \\ c & -1 & a \\ b & a & -1 \end{vmatrix} = c$$

$$-1(1 - a^2) - c(-c - ab) + b(ca + b) = 0$$

$$a^2 + b^2 + c^2 + 2abc = 1$$

(D) $|P - I1| = |P^T(P - I)| = |P^TP - P^T| = |I - P^T|$

$= |(I - P)^T| + |I - P|$

$= -|P - I|$

Therefore, $|P - I| = 0$

4. (A) $\Delta'(x) = \begin{vmatrix} 2x-5 & 2x-5 & 3 \\ 6x+1 & 6x+1 & 9 \\ 14x-6 & 14x-6 & 21 \end{vmatrix} + \begin{vmatrix} x^2-5x+3 & 2 & 3 \\ 3x^2+x+4 & 6 & 9 \\ 7x^2-6x+9 & 14 & 21 \end{vmatrix}$

$= 0 \quad (\because C_1 = C_2) + 0$ (C_2 and C_3 are proportional)

$= 0$

$\therefore \quad \Delta(0) = \text{constant}$

$\therefore \quad a = 0, b = 0, c = 0$

$$\therefore \Delta(0) = \begin{vmatrix} 3 & -5 & 3 \\ 4 & 1 & 9 \\ 9 & -6 & 21 \end{vmatrix} = d$$

$\therefore \quad d = 141$

(B) At $x = 0$

$$\Delta(0) = \begin{vmatrix} -1 & 0 & 7 \\ -1 & -1 & 8 \\ 0 & 0 & 0 \end{vmatrix} = d$$

$\therefore \quad d = 0$

and $\Delta'(x) = \begin{vmatrix} 1 & 5x & 7 \\ 2x & x-1 & 8 \\ 2 & 3x & 0 \end{vmatrix} + \begin{vmatrix} x-1 & 5 & 7 \\ x^2-1 & 1 & 8 \\ 2x & 3 & 0 \end{vmatrix}$

$= 3ax^2 + 2bx + c$

Put $x = 0$, then

$$\begin{vmatrix} 1 & 0 & 7 \\ 0 & -1 & 8 \\ 2 & 0 & 0 \end{vmatrix} + \begin{vmatrix} -1 & 5 & 7 \\ -1 & 1 & 8 \\ 0 & 3 & 0 \end{vmatrix} = c$$

$\therefore \quad 2(0 + 7) - 3(-8 + 7) = c$

$\Rightarrow \quad c = 17$

and $\Delta''(x) = \begin{vmatrix} 0 & 5x & 7 \\ 2 & x-1 & 8 \\ 0 & 3x & 0 \end{vmatrix} + \begin{vmatrix} 1 & 5 & 7 \\ 2x & 1 & 8 \\ 2 & 3 & 0 \end{vmatrix} + \begin{vmatrix} 1 & 5 & 7 \\ 2x & 1 & 8 \\ 2 & 3 & 0 \end{vmatrix}$

$= 6ax + 2b$

$\Rightarrow \; -2(0 - 21x) + 2(33) - 3(8 - 14x) + 2(33) - 3(8 - 14x) = 6ax + 2b$

$\Rightarrow \; 42x + 132 - 48 + 84x = 6ax + 2b$

$\therefore \quad a = 21, b = 42$

(C) Applying $R_1 \to R_1 - 2R_2$ and $R_3 \to R_3 - 3R_2$

Then,

$$\Delta(x) = \begin{vmatrix} -6x^3 - 3x^2 + 14x & -x + 3 & 0 \\ 4x^3 - 7x & 3x + 2 & 1 \\ -5x^3 - 8x^2 + 21x & -8x - 7 & 0 \end{vmatrix}$$

$= -\{(-6x^3 - 3x^2 + 14x)(-8x - 7) - (-5x^3 - 8x^2 + 21x)(-x + 3)\}$

$= -\{(6x^3 + 3x^2 - 14x)(8x + 7) - (5x^3 + 8x^2 - 21x)(x - 3)\}$

$= -\{43x^4 + 73x^3 - 46x^2 - 161x\}$

$= -43x^4 - 73x^3 + 46x^2 + 161x$

$\therefore a = 0, b = 161, c = 46$

$d = -73, e = -43$

07

Matrices

Objective Questions Type I [Only one correct answer]

In each of the questions below, four choices are given of which only one is correct. You have to select the correct answer which is the most appropriate.

1. If $P = \begin{bmatrix} \frac{\sqrt{3}}{2} & \frac{1}{2} \\ -\frac{1}{2} & \frac{\sqrt{3}}{2} \end{bmatrix}$, $A = \begin{bmatrix} 1 & 1 \\ 0 & 1 \end{bmatrix}$ and $Q = PAP^T$, then $P^T (Q^{2005}) P$ is equal to

(a) $\begin{bmatrix} 1 & 2005 \\ 0 & 1 \end{bmatrix}$ (b) $\begin{bmatrix} \frac{\sqrt{3}}{2} & 2005 \\ 1 & 0 \end{bmatrix}$

(c) $\begin{bmatrix} 1 & 2005 \\ \frac{\sqrt{3}}{2} & 1 \end{bmatrix}$ (d) $\begin{bmatrix} 1 & \frac{\sqrt{3}}{2} \\ 0 & 2005 \end{bmatrix}$

2. What is wrong in the following computation ?

$$\begin{bmatrix} 1 & 0.01 \\ 1 & 1 \end{bmatrix}^n = \left\{ \begin{bmatrix} 1 & 0 \\ 1 & 1 \end{bmatrix} + 10^{-2} \begin{bmatrix} 0 & 1 \\ 0 & 0 \end{bmatrix} \right\}^n$$

$$= \begin{bmatrix} 1 & 0 \\ 1 & 1 \end{bmatrix}^n + n \times 10^{-2} \begin{bmatrix} 1 & 0 \\ 1 & 1 \end{bmatrix}^{n-1} \begin{bmatrix} 0 & 1 \\ 0 & 0 \end{bmatrix}$$

Since, $\begin{bmatrix} 0 & 1 \\ 0 & 0 \end{bmatrix}^k = \begin{bmatrix} 0 & 0 \\ 0 & 0 \end{bmatrix}$, for $k \geq 2$.

(a) $\begin{bmatrix} 0 & 1 \\ 0 & 0 \end{bmatrix}^k = \begin{bmatrix} 0 & 0 \\ 0 & 0 \end{bmatrix}$, for $k \geq 2$ is not true

(b) Computation of second term on RHS is not valid

(c) First term should be calculated completely

(d) None of the above

3. If $A^5 = O$ such that $A^n \neq I$ for $1 \leq n \leq 4$, then $(I - A)^{-1}$ is equal to

(a) A^4 (b) A^3

(c) $I + A$ (d) None of these

4. If $A = \begin{bmatrix} \cos\theta & \sin\theta \\ -\sin\theta & \cos\theta \end{bmatrix}$, then $\lim_{n \to \infty} \frac{A^n}{n}$ is (where $\theta \in R$)

(a) a zero matrix (b) an identity matrix

(c) $\begin{bmatrix} 0 & 1 \\ 0 & -1 \end{bmatrix}$ (d) $\begin{bmatrix} 0 & 1 \\ -1 & 0 \end{bmatrix}$

5. If $A = \text{diag}\,(d_1, d_2, d_3, \ldots, d_n)$, then A^n is equal to

(a) $\text{diag}\,(d_1^{n-1}, d_2^{n-1}, d_3^{n-1}, \ldots, d_n^{n-1})$

(b) $\text{diag}\,(d_1^n, d_2^n, d_3^n, \ldots, d_n^n)$

(c) $\text{diag}\,(d_1, d_2, d_3, \ldots, d_n)$

(d) $\text{diag}\,(d_1, d_2^2, d_3^3, \ldots, d_n^n)$

6. The rank of the matrix $\begin{pmatrix} -1 & 2 & 5 \\ 2 & -4 & a-4 \\ 1 & -2 & a+1 \end{pmatrix}$ is (where $a = -6$)

(a) 1 (b) 2

(c) 3 (d) 4

7. If A is a square matrix of order 2×2 such that $A^2 = O$, then

(a) $A = \begin{pmatrix} \alpha & \beta \\ \gamma & -\alpha \end{pmatrix}$, where α, β, γ are numbers such that $\alpha^2 + \beta\gamma = 0$

(b) $A = \begin{pmatrix} \alpha & \beta \\ \beta & -\alpha \end{pmatrix}$ with $\alpha = \pm \beta$

(c) $A = \begin{pmatrix} \alpha & -\alpha \\ -\beta & \beta \end{pmatrix}$ with $\alpha^2 + \beta^2 = 1$

(d) None of the above

8. If $2x - y = \begin{bmatrix} 3 & -3 & 0 \\ 3 & 3 & 2 \end{bmatrix}$ and $2y + x = \begin{bmatrix} 4 & 1 & 5 \\ -1 & 4 & -4 \end{bmatrix}$, then

(a) $x + y = \begin{bmatrix} 3 & 0 & -1 \\ 0 & 3 & -2 \end{bmatrix}$ (b) $x = \begin{bmatrix} 2 & -1 & 1 \\ 1 & 2 & 0 \end{bmatrix}$

(c) $x - y = \begin{bmatrix} 1 & -2 & 3 \\ 2 & 1 & 2 \end{bmatrix}$ (d) $y = \begin{bmatrix} 1 & 1 & 2 \\ -1 & 1 & -2 \end{bmatrix}$

9. If $A = \begin{pmatrix} \frac{-1+i\sqrt{3}}{2i} & \frac{-1-i\sqrt{3}}{2i} \\ \frac{1+i\sqrt{3}}{2i} & \frac{1-i\sqrt{3}}{2i} \end{pmatrix}$, $i = \sqrt{-1}$ and $f(x) = x^2 + 2$, then $f(A)$ equals

(a) $\begin{bmatrix} 1 & 0 \\ 0 & 1 \end{bmatrix}$
(b) $\left(\frac{3-i\sqrt{3}}{2}\right)\begin{bmatrix} 1 & 0 \\ 0 & 1 \end{bmatrix}$
(c) $\left(\frac{5-i\sqrt{3}}{2}\right)\begin{bmatrix} 1 & 0 \\ 0 & 1 \end{bmatrix}$
(d) $(2+i\sqrt{3})\begin{bmatrix} 1 & 0 \\ 0 & 1 \end{bmatrix}$

10. If $A = \begin{bmatrix} 1 & 0 \\ -1 & 7 \end{bmatrix}$ and $A^2 = 8A + kI_2$, then k is equal to

(a) -1
(b) 1
(c) -7
(d) 7

11. If $\begin{bmatrix} \cos^2\theta & \cos\theta\sin\theta \\ \cos\theta\sin\theta & \sin^2\theta \end{bmatrix}\begin{bmatrix} \cos^2\phi & \cos\phi\sin\phi \\ \cos\phi\sin\phi & \sin^2\phi \end{bmatrix}$ is zero matrix, then θ and ϕ differ by

(a) even multiple of $\frac{\pi}{2}$
(b) odd multiple of $\frac{\pi}{2}$
(c) even multiple of π
(d) odd multiple of π

12. Let B, C be $n \times n$ matrices such that $A = B + C$, $BC = CB$ and C^2 is a null matrix. Then, $B^{2020}[B + (2021)C] =$

(a) A^{2020}
(b) null zero matrix of order $n \times n$
(c) A^{2021}
(d) B^{2021}

13. If a matrix A satisfy $A^2 - 5A + 7I = 0$ and $A^8 = pA + qI$, then the value of p is

(a) 9621
(b) 1265
(c) 5299
(d) undefined

14. Let A and B be 3×3 real matrices such that A is symmetric matrix and B is skew-symmetric matrix. Then, the system of linear equations $(A^2B^2 - B^2A^2)X = O$, where X is a 3×1 column matrix of unknown variables and O is a 3×1 null matrix, has

[JEE Main 2021, 24 February]

(a) a unique solution
(b) exactly two solutions
(c) infinitely many solutions
(d) no solution

15. Let $a, b, c \in R$ be all non-zero and satisfy $a^3 + b^3 + c^3 = 2$. If the matrix $A = \begin{pmatrix} a & b & c \\ b & c & a \\ c & a & b \end{pmatrix}$ satisfies $A^TA = I$, then a value of abc can be

[JEE Main 2020, 2 September]

(a) $-\frac{1}{3}$
(b) $\frac{1}{3}$
(c) 3
(d) $\frac{2}{3}$

16. If the value of $\prod_{k=1}^{50}\begin{bmatrix} 1 & 2k-1 \\ 0 & 1 \end{bmatrix}$ is equal to $\begin{bmatrix} 1 & r \\ 0 & 1 \end{bmatrix}$, then r is equal to

(a) 62500
(b) 2500
(c) 1250
(d) 12500

17. If $M = \begin{pmatrix} 1 & 1 & 1 \\ 1 & 1 & 1 \\ 1 & 1 & 1 \end{pmatrix}$, then $M^{50} =$

(a) 3^{49} M
(b) 0
(c) -1
(d) 1

18. Let $A = \begin{bmatrix} 0 & -\tan\theta/2 \\ \tan\theta/2 & 0 \end{bmatrix}$ $(\theta \neq n\pi)$

and $B = \begin{bmatrix} \cos\theta & -\sin\theta \\ \sin\theta & \cos\theta \end{bmatrix}$, $I = \begin{bmatrix} 1 & 0 \\ 0 & 1 \end{bmatrix}$

Then, the matrix $I + A$ is equal to

(a) $(I - A)B$
(b) $(I - A)^2B$
(c) $(I + A)^2B$
(d) $(I - A)^2$

19. The number of solutions of the matrix equation $X^2 = \begin{bmatrix} 1 & 1 \\ 2 & 3 \end{bmatrix}$ is

(a) more than 2
(b) 2
(c) 0
(d) 1

20. If $A = \begin{bmatrix} \cos\theta & i\sin\theta \\ i\sin\theta & \cos\theta \end{bmatrix}$, $\left(\theta = \frac{\pi}{24}\right)$ and $A^5 = \begin{bmatrix} a & b \\ c & d \end{bmatrix}$, where $i = \sqrt{-1}$, then which one of the following is not true? **[JEE Main 2020, 4 September]**

(a) $a^2 - d^2 = 0$
(b) $a^2 - c^2 = 1$
(c) $a^2 - b^2 = 1$
(d) $0 \leq a^2 + b^2 \leq 1$

21. If $\begin{bmatrix} 2 & 1 & 1 \\ 0 & 3 & -1 \\ 1 & -1 & 1 \end{bmatrix}\begin{bmatrix} x \\ y \\ z \end{bmatrix} = \begin{bmatrix} 1 \\ 1 \\ 0 \end{bmatrix}$ then $\begin{bmatrix} x \\ y \\ z \end{bmatrix} =$

(a) $\begin{bmatrix} 1 \\ 1 \\ 0 \end{bmatrix} + k\begin{bmatrix} 3 \\ 1 \\ -2 \end{bmatrix}, k \in R$
(b) $\begin{bmatrix} 1 \\ 1 \\ 0 \end{bmatrix} + k\begin{bmatrix} 1 \\ -2 \\ 3 \end{bmatrix}, k \in R$
(c) $\begin{bmatrix} 1 \\ 1 \\ 0 \end{bmatrix} + k\begin{bmatrix} -2 \\ 1 \\ 3 \end{bmatrix}, k \in R$
(d) $\begin{bmatrix} 1 \\ 0 \\ -1 \end{bmatrix} + k\begin{bmatrix} -2 \\ 1 \\ 3 \end{bmatrix}, k \in R$

22. Let $E(\alpha) = \begin{bmatrix} \cos^2\alpha & \cos\alpha\sin\alpha \\ \cos\alpha\sin\alpha & \sin^2\alpha \end{bmatrix}$. If α and β differs by an odd multiple of $\pi/2$, then $E(\alpha)\ E(\beta)$ is a

(a) null matrix
(b) unit matrix
(c) diagonal matrix
(d) orthogonal matrix

23. $A = \frac{1}{3}\begin{bmatrix} 1 & 2 & 2 \\ 2 & 1 & -2 \\ x & 2 & y \end{bmatrix}$

If $AA^T = I_3$, then

(a) $x = 2, y = -1$ (b) $x = 2, y = 1$
(c) $x = -2, y = -1$ (d) $x = -2, y = 1$

24. For each real x, $-1 < x < 1$. Let $A(x)$ be the matrix $(1-x)^{-1}\begin{bmatrix} 1 & -x \\ -x & 1 \end{bmatrix}$ and $z = \frac{x+y}{1+xy}$. Then,

(a) $A(z) = A(x)A(y)$ (b) $A(z) = A(x) - A(y)$
(c) $A(z) = A(x) + A(y)$ (d) $A(z) = A(x)[A(y)]^{-1}$

25. If $A = \begin{bmatrix} i & -i \\ -i & i \end{bmatrix}$ and $B = \begin{bmatrix} 1 & -1 \\ -1 & 1 \end{bmatrix}$, then A^8 equals

(a) $64B$ (b) $-64B$
(c) $-128B$ (d) $128B$

26. The matrix $\begin{bmatrix} 1 & 2 & 3 \\ 1 & 2 & 3 \\ -1 & -2 & -3 \end{bmatrix}$ is

(a) idempotent (b) nilpotent
(c) involutory (d) orthogonal

27. Let α be a root of the equation $x^2 + x + 1 = 0$ and the matrix $A = \frac{1}{\sqrt{3}}\begin{bmatrix} 1 & 1 & 1 \\ 1 & \alpha & \alpha^2 \\ 1 & \alpha^2 & \alpha^4 \end{bmatrix}$, then the matrix A^{31} is equal to **[JEE Main 2020, 7 January]**

(a) A^3 (b) I_3
(c) A^2 (d) A

28. Let $l, m, n \in R$ and $A = \begin{bmatrix} 1 & r & r^2 & l \\ r & r^2 & 1 & m \\ r^2 & 1 & r & n \end{bmatrix}$. Then, the set of all real values of r for which the rank of A is 3, is

(a) $(0, \infty)$ (b) R
(c) $R - \{1\}$ (d) $R - \{0\}$

29. If Z is an idempotent matrix, then $(I + Z)^n$

(a) $I + 2^n Z$ (b) $I + (2^n - 1)Z$
(c) $I - (2^n - 1)Z$ (d) None of these

30. The matrix $\begin{bmatrix} 0 & 1 \\ 1 & 0 \end{bmatrix}$ is the matrix reflection in the line

(a) $x = 1$ (b) $x + y = 1$
(c) $y = 1$ (d) $x = y$

31. Let $P = \begin{bmatrix} 1 & 0 & 0 \\ 4 & 1 & 0 \\ 16 & 4 & 1 \end{bmatrix}$ and I be the identity matrix of order 3. If $Q = [q_{ij}]$ is a matrix, such that $P^{50} - Q = I$, then $\frac{q_{31} + q_{32}}{q_{21}}$ equals **[JEE Advanced 2016]**

(a) 52 (b) 103
(c) 201 (d) 205

32. If $A = \begin{bmatrix} 2 & -2 & -4 \\ -1 & 3 & 4 \\ 1 & -2 & x \end{bmatrix}$ is an idempotent matrix, then x is equal to

(a) -5 (b) -1
(c) -3 (d) -4

33. Let $M = \begin{bmatrix} \sin^4\theta & -1 - \sin^2\theta \\ 1 + \cos^2\theta & \cos^4\theta \end{bmatrix} = \alpha I + \beta M^{-1}$, where $\alpha = \alpha(\theta)$ and $\beta = \beta(\theta)$ are real numbers, and I is the 2×2 identity matrix. If α^* is the minimum of the set $\{\alpha(\theta) : \theta \in [0, 2\pi)\}$ and β^* is the minimum of the set $\{\beta(\theta) : \theta \in [0, 2\pi)\}$, then the value of $\alpha^* + \beta^*$ is **[JEE Advanced 2019]**

(a) $-\frac{17}{16}$ (b) $-\frac{31}{16}$
(c) $-\frac{37}{16}$ (d) $-\frac{29}{16}$

34. The total number of matrices $A = \begin{bmatrix} 0 & 2y & 1 \\ 2x & y & -1 \\ 2x & -y & 1 \end{bmatrix}$, $(x, y \in R, x \neq y)$ for which $A^T A = 3I_3$ is **[JEE Main 2019, 9 April]**

(a) 2 (b) 4
(c) 3 (d) 6

35. If the matrix $A = \frac{1}{\sqrt{2}}\begin{bmatrix} 1 & i \\ -i & a \end{bmatrix}$, where $i = \sqrt{-1}$, is unitary, then a is equal to

(a) -2 (b) -1
(c) 0 (d) 1

36. If $3A = \begin{bmatrix} 1 & 2 & 2 \\ 2 & 1 & -2 \\ x & 2 & y \end{bmatrix}$ and $AA' = I$, then $x + y$ is equal to

(a) -5 (b) -4
(c) -3 (d) -2

37. The sum of two idempotent matrices A and B is dempotent, if $AB = BA =$

(a) 4 (b) 3
(c) 2 (d) O

38. The rank of $\begin{bmatrix} 1 & 2 & 3 \\ 2 & 4 & 6 \\ 3 & 6 & 9 \end{bmatrix}$ is equal to

(a) 1 (b) 2
(c) 3 (d) none of these

39. If $A = \begin{bmatrix} 1 & 0 \\ 0 & -1 \end{bmatrix}$, $P = \begin{bmatrix} 1 & 1 \\ 0 & 1 \end{bmatrix}$ and $X = APA^T$, then $A^T X^{50} A =$

(a) $\begin{bmatrix} 0 & 1 \\ 1 & 0 \end{bmatrix}$ (b) $\begin{bmatrix} 2 & 1 \\ 0 & -1 \end{bmatrix}$
(c) $\begin{bmatrix} 25 & 1 \\ 1 & -25 \end{bmatrix}$ (d) $\begin{bmatrix} 1 & 50 \\ 0 & 1 \end{bmatrix}$

40. Let $A + 2B = \begin{bmatrix} 1 & 2 & 0 \\ 6 & -3 & 3 \\ -5 & 3 & 1 \end{bmatrix}$ and $2A - B = \begin{bmatrix} 2 & -1 & 5 \\ 2 & -1 & 6 \\ 0 & 1 & 2 \end{bmatrix}$.

If $Tr(A)$ denotes the sum of all diagonal elements of the matrix A, then $Tr(A) - Tr(B)$ has value equal to **[JEE Main 2021, 18 March]**

(a) 1 (b) 2
(c) 0 (d) 3

41. Let $\alpha = \pi/5$ and $A = \begin{bmatrix} \cos\alpha & \sin\alpha \\ -\sin\alpha & \cos\alpha \end{bmatrix}$, then $B = A + A^2 + A^3 + A^4$ is

(a) singular (b) non-singular
(c) skew-symmetric (d) None of these

42. If $A = \begin{bmatrix} 0 & 5 \\ 0 & 0 \end{bmatrix}$ and $f(x) = x + x^2 + x \dots + x^{2018}$, then $f(A) + I =$

(a) $\begin{bmatrix} 0 & 0 \\ 0 & 0 \end{bmatrix}$ (b) $\begin{bmatrix} 1 & 5 \\ 0 & 0 \end{bmatrix}$
(c) $\begin{bmatrix} 0 & 5 \\ 1 & 1 \end{bmatrix}$ (d) $\begin{bmatrix} 1 & 5 \\ 0 & 1 \end{bmatrix}$

43. If $A = [a \ b]$, $B = [-b \ -a]$ and $C = \begin{bmatrix} a \\ -a \end{bmatrix}$, then the correct statement is

(a) $A = -B$ (b) $A + B = A - B$
(c) $AC = BC$ (d) $CA = CB$

44. If $X = \begin{bmatrix} 3 & -4 \\ 1 & -1 \end{bmatrix}$, then the value of X^n is

(a) $\begin{bmatrix} 3n & -4n \\ n & -n \end{bmatrix}$ (b) $\begin{bmatrix} 2+n & 5-n \\ n & -n \end{bmatrix}$
(c) $\begin{bmatrix} 3^n & (-4)^n \\ 1^n & (-1)^n \end{bmatrix}$ (d) None of these

45. Matrix A such that $A^2 = 2A - I$, where I is the identity matrix. Then for $n \geq 2$, A^n is equal to

(a) $nA - (n-1)I$ (b) $nA - I$
(c) $2^{n-1}A - (n-1)I$ (d) $2^{n-1}A - I$

Objective Questions Type II [One or more than one correct answer(s)]

In each of the questions below four choices of which one or more than one are correct. You have to select the correct answer(s) accordingly.

1. If $AB = A$ and $BA = B$, then which of the following is/are true?

(a) A is idempotent (b) B is idempotent
(c) A^T is idempotent (d) None of these

2. For all values of λ, the rank of the matrix

$$A = \begin{bmatrix} 1 & 4 & 5 \\ \lambda & 8 & 8\lambda - 6 \\ 1+\lambda^2 & 8\lambda + 4 & 2\lambda + 21 \end{bmatrix}$$

(a) for $\lambda = 2$, $\rho(A) = 1$ (b) for $\lambda = -1$, $\rho(A) = 2$
(c) for $\lambda \neq 2, -1$, $\rho(A) = 3$ (d) None of these

3. Let A and B be two matrices different from I such that $AB = BA$ and $A^n - B^n = A^{n+1} - B^{n+1} = A^{n+2} - B^{n+2}$, then

(a) $I - A$ is singular
(b) $I - B$ is singular
(c) $A + B = AB + I$
(d) $(I - A)(I - B)$ is non-singular

4. Let $P_1 = I = \begin{bmatrix} 1 & 0 & 0 \\ 0 & 1 & 0 \\ 0 & 0 & 1 \end{bmatrix}$, $P_2 = \begin{bmatrix} 1 & 0 & 0 \\ 0 & 0 & 1 \\ 0 & 1 & 0 \end{bmatrix}$,

$P_3 = \begin{bmatrix} 0 & 1 & 0 \\ 1 & 0 & 0 \\ 0 & 0 & 1 \end{bmatrix}$, $P_4 = \begin{bmatrix} 0 & 1 & 0 \\ 0 & 0 & 1 \\ 1 & 0 & 0 \end{bmatrix}$

$P_5 = \begin{bmatrix} 0 & 0 & 1 \\ 1 & 0 & 0 \\ 0 & 1 & 0 \end{bmatrix}$, $P_6 = \begin{bmatrix} 0 & 0 & 1 \\ 0 & 1 & 0 \\ 1 & 0 & 0 \end{bmatrix}$

where, P_k^T denotes the transpose of the matrix P_k. Then, which of the following option is/are correct? **[JEE Advanced 2019]**

(a) X is a symmetric matrix
(b) The sum of diagonal entries of X is 18
(c) $x - 30I$ is an invertible matrix
(d) If $X \begin{bmatrix} 1 \\ 1 \\ 1 \end{bmatrix} = \alpha \begin{bmatrix} 1 \\ 1 \\ 1 \end{bmatrix}$, then $\alpha = 30$

5. Let $A = \begin{bmatrix} 1 & 0 \\ 1 & 1 \end{bmatrix}$, then

(a) $A^{-n} = \begin{bmatrix} 1 & 0 \\ -n & 1 \end{bmatrix} \forall n \in N$ (b) $\lim_{n \to \infty} \dfrac{A^{-n}}{n} = \begin{bmatrix} 0 & 0 \\ -1 & 0 \end{bmatrix}$
(c) $\lim_{n \to \infty} \dfrac{A^{-n}}{n^2} = \begin{bmatrix} 0 & 0 \\ 0 & 0 \end{bmatrix}$ (d) None of these

6. Let A, B, C be square matrices of same order and I the unit matrix of the same order such that $A + B + C = AB + BC + CA$. Consider the following three statements

(i) $ABC = AC - CA$
(ii) $BCA = BA - AB$
(iii) $CBA = CB - BC$

Then,

(a) (i) and (ii) equivalent
(b) (ii) and (iii) are equivalent
(c) (iii) and (i) are equivalent
(d) all the three statements are equivalent

7. Let $P_1 = I = \begin{bmatrix} 1 & 0 & 0 \\ 0 & 1 & 0 \\ 0 & 0 & 1 \end{bmatrix}$, $P_2 = \begin{bmatrix} 1 & 0 & 0 \\ 0 & 0 & 1 \\ 0 & 1 & 0 \end{bmatrix}$,

$P_3 = \begin{bmatrix} 0 & 1 & 0 \\ 1 & 0 & 0 \\ 0 & 0 & 1 \end{bmatrix}$, $P_4 = \begin{bmatrix} 0 & 1 & 0 \\ 0 & 0 & 1 \\ 1 & 0 & 0 \end{bmatrix}$,

$P_5 = \begin{bmatrix} 0 & 0 & 1 \\ 1 & 0 & 0 \\ 0 & 1 & 0 \end{bmatrix}$, $P_6 = \begin{bmatrix} 0 & 0 & 1 \\ 0 & 1 & 0 \\ 1 & 0 & 0 \end{bmatrix}$

and $X = \sum_{k=1}^{6} P_k \begin{bmatrix} 2 & 1 & 3 \\ 1 & 0 & 2 \\ 3 & 2 & 1 \end{bmatrix} P_k^T$

where, P_k^T denotes the transpose of the matrix P_k. Then which of the following option is/are correct? **[JEE Advanced 2019]**

(a) X is a symmetric matrix
(b) The sum of diagonal entries of X is 18
(c) $X - 30I$ is an invertible matrix
(d) If $X\begin{bmatrix} 1 \\ 1 \\ 1 \end{bmatrix} = \alpha \begin{bmatrix} 1 \\ 1 \\ 1 \end{bmatrix}$, then $\alpha = 30$

8. If $ad - bc \neq 0$ and $A = \begin{bmatrix} a & b \\ c & d \end{bmatrix}$ and $A^2 + xA + yI_2 = O$, then

(a) $x = -(a+b)$
(b) $x = -(a+d)$
(c) $y = ad - bc$
(d) $y = bc - ad$

9. Let $x \in R$ and let $P = \begin{bmatrix} 1 & 1 & 1 \\ 0 & 2 & 2 \\ 0 & 0 & 3 \end{bmatrix}, Q = \begin{bmatrix} 2 & x & x \\ 0 & 4 & 0 \\ x & x & 6 \end{bmatrix}$ and $R = PQP^{-1}$, then which of the following options is/are correct? **[JEE Advanced 2019]**

(a) There exists a real, number x such that $PQ = QP$
(b) For $x = 0$, if $R\begin{bmatrix} 1 \\ a \\ b \end{bmatrix} = 6\begin{bmatrix} 1 \\ a \\ b \end{bmatrix}$, then $a + b = 5$
(c) For $x = 1$, there exists a unit vector $\alpha\hat{\mathbf{i}} + \beta\hat{\mathbf{j}} + \gamma\hat{\mathbf{k}}$ for which $R\begin{bmatrix} \alpha \\ \beta \\ \gamma \end{bmatrix} = \begin{bmatrix} 0 \\ 0 \\ 0 \end{bmatrix}$
(d) $\det R = \det \begin{bmatrix} 2 & x & x \\ 0 & 4 & 0 \\ x & x & 5 \end{bmatrix} + 8$, for all $x \in R$

10. Which of the following is correct ?

(a) If A is a square matrix, $(A + A')$ is a symmetric matrix
(b) If A is a square matrix, $(A - A')$ is a skew symmetric matrix
(c) Every square matrix can be expressed as the sum of a symmetric and skew symmetric matrix
(d) Some elements of the skew symmetric matrix must be zero

11. If D_1 and D_2 are two 3×3 diagonal matrices, then

(a) D_1D_2 is a diagonal matrix
(b) $D_1^2 + D_2^2$ is a diagonal matrix
(c) $D_1D_2 = D_2D_1$
(d) D_1^n is a diagonal matrix $\forall\, n \in N$

12. Let A, B and C be 2×2 matrices with entries from the set of real numbers. Define o as follows

$AoB = \frac{1}{2}(AB + BA)$, then

(a) $AoB = BoA$
(b) $AoI = A$, I is an identity matrix of order 2
(c) $AoA = A^2$
(d) $Ao(B + C) = AoB + AoC$

Linked-Comprehension Type

In these questions, a passage (paragraph) has been given followed by questions based on each of the passages. You have to answer the questions based on the passage given.

PASSAGE 1

Let A and B are two matrices of same order 3×3, where

$A = \begin{pmatrix} 1 & 3 & \lambda + 2 \\ 2 & 4 & 8 \\ 3 & 5 & 10 \end{pmatrix}, B = \begin{pmatrix} 3 & 2 & 4 \\ 3 & 2 & 5 \\ 2 & 1 & 4 \end{pmatrix}$

On the basis of above information, answer the following questions

1. If A is singular matrix , then $tr(A + B)$ is equal to

(a) 6
(b) 12
(c) 24
(d) 17

2. If matrix $2A + 3B$ is singular, then the value of 2λ is

(a) 11
(b) 13
(c) 15
(d) 17

3. If $\lambda = 3$, then $\frac{1}{7}(tr(AB) + tr(BA))$ is equal to

(a) 34
(b) 42
(c) 84
(d) 63

4. If $A = \begin{pmatrix} \lambda + 2 & 3 & 1 \\ 8 & 4 & 2 \\ 10 & 5 & 3 \end{pmatrix}$ and $B = \begin{pmatrix} 4 & 2 & 3 \\ 5 & 2 & 3 \\ 4 & 5 & 2 \end{pmatrix}$, if $tr(A - 2B) = 0$, then the value of λ is

(a) 3
(b) 5
(c) 7
(d) 9

5. The correct statement is
(a) $(A+B)(A-B) = A^2 + B^2 - 2AB$
(b) $(A+B)^2 = A^2 + B^2 + AB + BA$
(c) $(A+B)^2 = A^2 + B^2 + 2AB$
(d) None of the above

PASSAGE 2

Let A be a matrix of order 2×2 such that $A^2 = O$.

On the basis of above information, answer the following questions

1. $A^2 - (a+d)A + (ad - bc)I$ is equal to
(a) I (b) O
(c) $-I$ (d) None of these

2. tr(A) is equal to
(a) 1 (b) 0
(c) -1 (d) None of these

3. $(I + A)^{100} =$
(a) $100A$ (b) $100(I + A)$
(c) $100I + A$ (d) $I + 100A$

Numerical Value Type

Solve the following problems and write your answer.

1. If $M = \begin{pmatrix} 1 & 2 \\ 2 & 3 \end{pmatrix}$ and $M^2 - \lambda M - I_2 = O$, then 6^λ must be

2. Let $a_r = r({}^7C_r)$, $b_r = (7-r)({}^7C_r)$ and $A_r = \begin{pmatrix} a_r & 0 \\ 0 & b_r \end{pmatrix}$. If $A = \sum_{r=0}^{7} A_r = \begin{pmatrix} a & 0 \\ 0 & b \end{pmatrix}$, then the value of $\left(\frac{a+b}{35}\right)$ must be

3. The number of all 3×3 matrices A, with enteries from the set $\{-1, 0, 1\}$ such that the sum of the diagonal elements of AA^T is 3, is .... **[JEE Main 2020, 8 January]**

4. Suppose a matrix A satisfies $A^2 - 5A + 7I = O$. If $A^5 = aA + bI$, then the value of $2a - 3b$ must be

Single Integer Type Questions

Solve the following problems and write your answer.

1. If M is the matrix $\begin{bmatrix} 1 & -3 \\ -1 & 1 \end{bmatrix}$, then

$$M - \frac{1}{3}M^2 + \frac{1}{9}M^3 \ldots + \left(\frac{-1}{3}\right)^n M^{n+1} + \ldots \infty = \frac{3}{13}\begin{bmatrix} 1 & \alpha \\ \beta & 1 \end{bmatrix},$$

where $\left|\frac{\alpha}{\beta}\right|$ is equal to

2. If $\begin{bmatrix} 2 & -1 \\ 1 & 0 \\ -3 & 4 \end{bmatrix} A = \begin{bmatrix} -1 & -8 & -10 \\ 1 & -2 & -5 \\ 9 & 22 & 15 \end{bmatrix}$, then sum of all the elements of matrix A is ....

3. If $A = \begin{bmatrix} 1 & -1 \\ 2 & 1 \end{bmatrix}$, $B = \begin{bmatrix} a & 1 \\ b & -1 \end{bmatrix}$ and $(A+B)^2 = A^2 + B^2 + 2AB$, then $|a + b|$ equal to

4. If x, y, z are real and $A = \begin{bmatrix} 0 & 2y & z \\ x & y & -z \\ x & -y & z \end{bmatrix}$ is such that $AA^T = I$, then $x^2 + y^2 + z^2$ is equal to ...

5. If A is an idempotent matrix satisfying, $(I - 0.4A)^{-1} = I - \alpha A$, where I is the unit matrix of the same order as that of A, then the value of $|9\alpha|$ is equal to ....

Matrix-Match Type

Given below are Matching Type Questions, with two columns (each having some items) each . Each item of Column I has to be matched with the items of Column II, by encircling the correct match(es).

NOTE *An item of Column I can be matched with more than one items of Column II.*

1. Consider a matrix $A = [a_{ij}]$ of order 3×3 such that $a_{ij} = (k)^{i+j}$ where $k \in I$.

Match List I with List II and select the correct answer using the code given below the lists.

	Column I		Column II
(A)	A is singular if	(P)	$k \in \{0\}$
(B)	A is null matrix if	(Q)	$k \in \phi$
(C)	A is skew-symmetric which is not null matrix if	(R)	$k \in I$
(D)	$A^2 = 3A$ if	(S)	$k \in \{-1, 0, 1\}$
		(T)	$k \in \{-\infty, \infty\}$

2. Observe the following columns :

	Column I		Column II
(A)	If A, B and C be 2×2 matrices with entries from the set of real numbers. Define * as follows : $A * B = \frac{1}{2}(AB + BA)$, then	(P)	$A * B = B * A$
		(Q)	$A * (B + C) = A * B + A * C$
(B)	If A, B and C be 2×2 matrices with entries from the set of real numbers.	(R)	$A * A = A^2$
	Define * as follows : $A * B = \frac{1}{2}(AB' + A'B)$, then	(S)	$A * I = A$
(C)	If A, B and C be 2×2 matrices with entries from the set of real numbers. Define * as follows : $A * B = \frac{1}{2}(AB - BA)$, then	(T)	$A * I = O$

Answers

Objective Questions Type I [Only one correct answer]

1. (a)	**2.** (b)	**3.** (d)	**4.** (a)	**5.** (b)	**6.** (a)	**7.** (a)	**8.** (d)	**9.** (d)	**10.** (c)
11. (b)	**12.** (c)	**13.** (b)	**14.** (c)	**15.** (b)	**16.** (b)	**17.** (a)	**18.** (a)	**19.** (b)	**20.** (c)
21. (d)	**22.** (a)	**23.** (c)	**24.** (a)	**25.** (d)	**26.** (b)	**27.** (a)	**28.** (c)	**29.** (b)	**30.** (d)
31. (b)	**32.** (c)	**33.** (d)	**34.** (b)	**35.** (b)	**36.** (c)	**37.** (d)	**38.** (a)	**39.** (d)	**40.** (b)
41. (c)	**42.** (d)	**43.** (c)	**44.** (d)	**45.** (a)					

Objective Questions Type II [One or more than one correct answer(s)]

1. (a, b, c)	**2.** (a, b, c)	**3.** (a, b, c)	**4.** (a, b, d)	**5.** (a, b, c)
6. (a, b, c, d)	**7.** (a, b, d)	**8.** (b, c)	**9.** (b, d)	**10.** (a, b, c, d
11. (a, b, c, d)	**12.** (a, b, c, d)			

Linked-Comprehension Type

Passage 1 **1.** (c) **2.** (d) **3.** (a) **4.** (c) **5.** (b)

Passage 2 **1.** (b) **2.** (b) **3.** (d)

Numerical Value Problems

1. 1296 **2.** 25.6 **3.** 672 **4.** 1453

Single Integer Type Quesitons

1. (3) **2.** (1) **3.** (3) **4.** (1) **5.** (6)

Matrix-Match Type

1. A → (R); B → (P); C → (Q); D → (S)

2. A → (P, Q, R, S); B → (P, Q); C → (Q, T)

Solutions

Objective Questions Type [Only one correct answer]

1. $\because \quad Q = PAP^T$

$\Rightarrow \quad P^T Q = AP^T \quad (\because PP^T = I)$

$\therefore \quad P^T Q^{2005} P = AP^T Q^{2004} P$

$= A^2 P^T Q^{2003} P = A^3 P^T Q^{2002} P = \dots.$

$= A^{2004} P^T (QP) = A^{2004} P^T (PA)$ $[\because Q = PAP^T \Rightarrow QP = PA]$

$= A^{2005} = \begin{bmatrix} 1 & 2005 \\ 0 & 1 \end{bmatrix}$

2. If $A = \begin{bmatrix} 1 & 0 \\ 1 & 1 \end{bmatrix}$ and $B = \begin{bmatrix} 0 & 1 \\ 0 & 0 \end{bmatrix}$, then $AB = \begin{bmatrix} 0 & 1 \\ 0 & 1 \end{bmatrix}$ and $BA = \begin{bmatrix} 1 & 1 \\ 0 & 0 \end{bmatrix}$

$\therefore \quad AB \neq BA$

i.e. A and B are not commute.

$(A + B)^n$ can be expanded using binomial theorem, if A and B commute. Hence, we cannot use binomial theorem.

3. $\because \quad A^4 (I - A) = A^4 I - A^5 = A^4 - O = A^4 \neq I$

$A^3 (I - A) = A^3 I - A^4 = A^3 - A^4 \neq I$

and $(I + A)(I - A) = I^2 - A^2 = I - A^2 \neq I.$

4. $\because \quad A = \begin{bmatrix} \cos\theta & \sin\theta \\ -\sin\theta & \cos\theta \end{bmatrix}$

$\therefore \quad A^n = \begin{bmatrix} \cos n\theta & \sin n\theta \\ -\sin n\theta & \cos n\theta \end{bmatrix}$

$\Rightarrow \quad \dfrac{A^n}{n} = \begin{bmatrix} \dfrac{\cos n\theta}{n} & \dfrac{\sin n\theta}{n} \\ -\dfrac{\sin n\theta}{n} & \dfrac{\cos n\theta}{n} \end{bmatrix}$

$\therefore \quad \lim_{n\to\infty} \dfrac{A^n}{n} = \begin{bmatrix} \lim_{n\to\infty} \dfrac{\cos n\theta}{n} & \lim_{n\to\infty} \dfrac{\sin n\theta}{n} \\ -\lim_{n\to\infty} \dfrac{\sin n\theta}{n} & \lim_{n\to\infty} \dfrac{\cos n\theta}{n} \end{bmatrix}$

$= \begin{bmatrix} \dfrac{-1 \text{ to } 1}{\infty} & \dfrac{-1 \text{ to } 1}{\infty} \\ -\dfrac{(-1 \text{ to } 1)}{\infty} & \dfrac{-1 \text{ to } 1}{\infty} \end{bmatrix}$

$[\because -1 < \sin\infty < 1 \text{ and } -1 < \cos\infty < 1]$

$= \begin{bmatrix} 0 & 0 \\ 0 & 0 \end{bmatrix} = a$ zero matrix.

5. $\because \quad A = \begin{pmatrix} d_1 & 0 & 0 & \cdots & 0 \\ 0 & d_2 & 0 & \cdots & 0 \\ 0 & 0 & d_3 & \cdots & 0 \\ \vdots & \vdots & \vdots & \vdots & \vdots \\ 0 & 0 & 0 & \cdots & d_n \end{pmatrix}$

$\therefore \quad A^2 = \begin{pmatrix} d_1 & 0 & 0 & \cdots & 0 \\ 0 & d_2 & 0 & \cdots & 0 \\ 0 & 0 & d_3 & \cdots & 0 \\ \vdots & \vdots & \vdots & \vdots & \vdots \\ 0 & 0 & 0 & \cdots & d_n \end{pmatrix} \begin{pmatrix} d_1 & 0 & 0 & \cdots & 0 \\ 0 & d_2 & 0 & \cdots & 0 \\ 0 & 0 & d_3 & \cdots & 0 \\ \vdots & \vdots & \vdots & \vdots & \vdots \\ 0 & 0 & 0 & \cdots & d_n \end{pmatrix}$

$= \begin{pmatrix} d_1^2 & 0 & 0 & \cdots & 0 \\ 0 & d_2^2 & 0 & \cdots & 0 \\ 0 & 0 & d_3^2 & \cdots & 0 \\ \vdots & \vdots & \vdots & \vdots & \vdots \\ 0 & 0 & 0 & \cdots & d_n^2 \end{pmatrix}$

$= \text{diag }(d_1^2, d_2{}^2, d_3^2, \dots, d_n{}^2)$

$A^3 = A^2 \cdot A = \begin{pmatrix} d_1^2 & 0 & 0 & \cdots & 0 \\ 0 & d_2^2 & 0 & \cdots & 0 \\ 0 & 0 & d_3^2 & \cdots & 0 \\ \vdots & \vdots & \vdots & \vdots & \vdots \\ 0 & 0 & 0 & \cdots & d_n^2 \end{pmatrix} \times \begin{pmatrix} d_1 & 0 & 0 & \cdots & 0 \\ 0 & d_2 & 0 & \cdots & 0 \\ 0 & 0 & d_3 & \cdots & 0 \\ \vdots & \vdots & \vdots & \vdots & \vdots \\ 0 & 0 & 0 & \cdots & d_n \end{pmatrix}$

$= \begin{pmatrix} d_1^3 & 0 & 0 & \cdots & 0 \\ 0 & d_2^3 & 0 & \cdots & 0 \\ 0 & 0 & d_3^3 & \cdots & 0 \\ \vdots & \vdots & \vdots & \vdots & \vdots \\ 0 & 0 & 0 & \cdots & d_n^3 \end{pmatrix} = \text{diag }(d_1^3, d_2{}^3, d_3^3, \dots, d_n^3)$

....

....

$\therefore \quad A^n = \text{diag }(d_1^n, d_2{}^n, d_3^n, \dots, d_n^n)$

6. Let $A = \begin{pmatrix} -1 & 2 & 5 \\ 2 & -4 & a-4 \\ 1 & -2 & a+1 \end{pmatrix} \underset{\sim}{R_1 \to (-1) R_1} \begin{pmatrix} 1 & -2 & -5 \\ 2 & -4 & a-4 \\ 1 & -2 & a+1 \end{pmatrix}$

$\underset{R_3 \to R_3 - R_1}{\overset{R_2 \to R_2 - 2R_1}{\sim}} \begin{pmatrix} 1 & -2 & -5 \\ 0 & 0 & a+6 \\ 0 & 0 & a+6 \end{pmatrix}$

For $\quad a = -6$,

$A = \begin{pmatrix} 1 & -2 & -5 \\ 0 & 0 & 0 \\ 0 & 0 & 0 \end{pmatrix}$

$\therefore \quad \rho(A) = 1.$

7. If $\quad A = \begin{pmatrix} \alpha & \beta \\ \gamma & -\alpha \end{pmatrix}$

$\therefore \quad A^2 = \begin{pmatrix} \alpha & \beta \\ \gamma & -\alpha \end{pmatrix} \begin{pmatrix} \alpha & \beta \\ \gamma & -\alpha \end{pmatrix} = \begin{pmatrix} \alpha^2 + \beta\gamma & 0 \\ 0 & \alpha^2 + \beta\gamma \end{pmatrix}$

$= \begin{pmatrix} 0 & 0 \\ 0 & 0 \end{pmatrix} = O \qquad [\because \alpha^2 + \beta\gamma = 0]$

8. $\because \quad 2x - y = \begin{bmatrix} 3 & -3 & 0 \\ 3 & 3 & 2 \end{bmatrix}$ ...(i)

$\Rightarrow \quad 4x - 2y = \begin{bmatrix} 6 & -6 & 0 \\ 6 & 6 & 4 \end{bmatrix}$ ...(ii)

and $\quad x + 2y = \begin{bmatrix} 4 & 1 & 5 \\ -1 & 4 & -4 \end{bmatrix}$ ....(iii)

Adding Eqs. (ii) and (iii), then

$$5x = \begin{bmatrix} 10 & -5 & 5 \\ 5 & 10 & 0 \end{bmatrix}$$

$$\therefore \quad x = \begin{bmatrix} 2 & -1 & 1 \\ 1 & 2 & 0 \end{bmatrix}$$

From Eq. (iii),

$2x + 4y = \begin{bmatrix} 8 & 2 & 10 \\ -2 & 8 & -8 \end{bmatrix}$ ...(iv)

Subtracting Eq. (i) from (iv), then

$$5y = \begin{bmatrix} 5 & 5 & 10 \\ -5 & 5 & -10 \end{bmatrix}$$

$$\therefore \quad y = \begin{bmatrix} 1 & 1 & 2 \\ -1 & 1 & -2 \end{bmatrix}$$

9. $\because \omega = \dfrac{-1 + i\sqrt{3}}{2}$ and $\omega^2 = \dfrac{-1 - i\sqrt{3}}{2}$

Also, $\quad \omega^3 = 1$ and $\omega + \omega^2 = -1$

Then, $\quad A = \begin{pmatrix} \dfrac{\omega}{i} & \dfrac{\omega^2}{i} \\ -\dfrac{\omega^2}{i} & -\dfrac{\omega}{i} \end{pmatrix}$

$$= \frac{\omega}{i}\begin{pmatrix} 1 & \omega \\ -\omega & -1 \end{pmatrix}$$

$$\therefore \quad A^2 = \frac{\omega^2}{i^2}\begin{pmatrix} 1 & \omega \\ -\omega & -1 \end{pmatrix}\begin{pmatrix} 1 & \omega \\ -\omega & -1 \end{pmatrix}$$

$$= -\omega^2\begin{pmatrix} 1-\omega^2 & 0 \\ 0 & 1-\omega^2 \end{pmatrix}$$

$$= \begin{pmatrix} -\omega^2 + \omega^4 & 0 \\ 0 & -\omega^2 + \omega^4 \end{pmatrix}$$

$$= \begin{pmatrix} -\omega^2 + \omega & 0 \\ 0 & -\omega^2 + \omega \end{pmatrix}$$

$\because \quad f(x) = x^2 + 2$

$$\therefore \quad f(A) = A^2 + 2I = \begin{pmatrix} -\omega^2 + \omega & 0 \\ 0 & -\omega^2 + \omega \end{pmatrix} + \begin{pmatrix} 2 & 0 \\ 0 & 2 \end{pmatrix}$$

$$= \begin{pmatrix} -\omega^2 + \omega + 2 & 0 \\ 0 & -\omega^2 + \omega + 2 \end{pmatrix}$$

$$= (-\omega^2 + \omega + 2)\begin{pmatrix} 1 & 0 \\ 0 & 1 \end{pmatrix} = (3 + 2\omega)\begin{pmatrix} 1 & 0 \\ 0 & 1 \end{pmatrix}$$

$$= \left(3 + 2\left(\frac{-1 + i\sqrt{3}}{2}\right)\right)\begin{pmatrix} 1 & 0 \\ 0 & 1 \end{pmatrix}$$

$$= (2 + i\sqrt{3})\begin{pmatrix} 1 & 0 \\ 0 & 1 \end{pmatrix}$$

10. $\because \quad A^2 = \begin{bmatrix} 1 & 0 \\ -1 & 7 \end{bmatrix}\begin{bmatrix} 1 & 0 \\ -1 & 7 \end{bmatrix} = \begin{bmatrix} 1 & 0 \\ -8 & 49 \end{bmatrix}$

and $\quad 8A + kI_2 = 8\begin{bmatrix} 1 & 0 \\ -1 & 7 \end{bmatrix} + k\begin{bmatrix} 1 & 0 \\ 0 & 1 \end{bmatrix} = \begin{bmatrix} 8+k & 0 \\ -8 & 56+k \end{bmatrix}$

$$\therefore \quad A^2 = 8A + kI_2$$

$$\therefore \quad \begin{bmatrix} 1 & 0 \\ -8 & 49 \end{bmatrix} = \begin{bmatrix} 8+k & 0 \\ -8 & 56+k \end{bmatrix}$$

On comparing, we get $8 + k = 1 \Rightarrow k = -7$.

11. $\because \begin{bmatrix} \cos^2\theta & \cos\theta\sin\theta \\ \cos\theta\sin\theta & \sin^2\theta \end{bmatrix}\begin{bmatrix} \cos^2\phi & \cos\phi\sin\phi \\ \cos\phi\sin\phi & \sin^2\phi \end{bmatrix}$

$$= \begin{bmatrix} \cos\theta\cos\phi\cos(\theta \sim \phi) & \cos\theta\sin\phi\cos(\theta \sim \phi) \\ \cos\phi\sin\theta\cos(\theta \sim \phi) & \sin\theta\sin\phi\cos(\theta \sim \phi) \end{bmatrix}$$

$$= \begin{bmatrix} 0 & 0 \\ 0 & 0 \end{bmatrix} = O \ \left(\text{if } \theta \sim \phi = (2n+1)\frac{\pi}{2}, n \in I\right)$$

12. Given that, $\quad A = B + C$ ...(i)

$$BC = CB \text{ and } C^2 = 0$$

From Eq. (i) we get

$$A^{n+1} = (B + C)^{n+1}$$

$$\Rightarrow \quad A^{n+1} = {}^{n+1}C_0 \cdot B^{n+1} + {}^{n+1}C_1\, C \cdot B^n + \dots$$

But, $\quad C^2 = 0, C^3 = C^4 \dots = C^r = 0$

$$\therefore \quad A^{n+1} = {}^{n+1}C_0\, B^{n+1} + {}^{n+1}C_1 B^n \cdot C$$

$$= B^{n+1} + (n+1)B^n C$$

$$A^{n+1} = B^n[B + (n+1)C]$$

Put $\quad n = 2020$

$$\therefore \quad B^{2020}[B + (2020+1)C] = A^{2020+1}$$

$$\Rightarrow \quad B^{2020}[B + (2021)C] = A^{2021}$$

13. $A^4 = (5A - 7I)^2 = 25A^2 - 70A + 49I$

$$= 25(5A - 7I) - 70A + 49I$$

$$= 55A - 126I$$

Square both side

$$A^8 = 3025A^2 - 13860A + 15876I$$

$$= 3025(5A - 7I) - 13860A + 15876I$$

$$= 1265A - 5299I$$

So, $\quad p = 1265$

14. $A^T = A, B^T = -B$

Let $A^2B^2 - B^2A^2 = P$

$P^T = (A^2B^2 - B^2A^2)^T = (A^2B^2)^T - (B^2A^2)^T$

$= (B^2)^T(A^2)^T - (A^2)^T(B^2)^T = B^2A^2 - A^2B^2$

$\Rightarrow P$ is skew-symmetric matrix

$$\begin{bmatrix} 0 & a & b \\ -a & 0 & c \\ -b & -c & 0 \end{bmatrix}\begin{bmatrix} x \\ y \\ z \end{bmatrix} = \begin{bmatrix} 0 \\ 0 \\ 0 \end{bmatrix}$$

$$\therefore \quad ay + bz = 0$$

$$-ax + cz = 0$$

$$-bx - cy = 0$$

From Eqs. (i), (ii) and (iii), we get

$$\Delta = 0 \text{ and } \Delta_1 = \Delta_2 = \Delta_3 = 0$$

$\therefore$ Equation have infinite number of solution.

15. Given matrix, $A = \begin{bmatrix} a & b & c \\ b & c & a \\ c & a & b \end{bmatrix}$

So, $A^T A = \begin{bmatrix} a & b & c \\ b & c & a \\ c & a & b \end{bmatrix} \begin{bmatrix} a & b & c \\ b & c & a \\ c & a & b \end{bmatrix}$

$$= \begin{bmatrix} a^2+b^2+c^2 & ab+bc+ca & ac+ba+cb \\ ba+cb+ac & b^2+c^2+a^2 & bc+ca+ab \\ ca+ab+bc & cb+ac+ba & c^2+a^2+b^2 \end{bmatrix}$$

$\because \quad A^T A = I$ [given]

$\therefore \; a^2+b^2+c^2 = 1$ and $ab+bc+ca = 0$ ...(i)

Now, as $(a+b+c)^2 = a^2+b^2+c^2+2(ab+bc+ca)$

$\Rightarrow \quad (a+b+c)^2 = 1$

$\Rightarrow \quad a+b+c = \pm 1$ ...(ii)

As we know,

$$a^3+b^3+c^3-3abc = (a+b+c)(a^2+b^2+c^2-ab-bc-ca)$$

$\Rightarrow \quad 2-3abc = (a+b+c)(1-0)$

[from Eq. (i) and $a^3+b^3+c^3 = 2$ [given]]

$\Rightarrow \quad -3abc = (a+b+c)-2$

$\Rightarrow \quad abc = \dfrac{2-(a+b+c)}{3}$

If $\quad a+b+c = 1 \Rightarrow abc = \dfrac{1}{3}$, or

If $\quad a+b+c = -1 \Rightarrow abc = 1$

Now, from the options $abc = \dfrac{1}{3}$

16. We have, $\begin{bmatrix} 1 & 1 \\ 0 & 1 \end{bmatrix}\begin{bmatrix} 1 & 3 \\ 0 & 1 \end{bmatrix} = \begin{bmatrix} 1 & 4 \\ 0 & 1 \end{bmatrix}$

$\begin{bmatrix} 1 & 4 \\ 0 & 1 \end{bmatrix}\begin{bmatrix} 1 & 5 \\ 0 & 1 \end{bmatrix} = \begin{bmatrix} 1 & 9 \\ 0 & 1 \end{bmatrix}$

$\begin{bmatrix} 1 & 9 \\ 0 & 1 \end{bmatrix}\begin{bmatrix} 1 & 7 \\ 0 & 1 \end{bmatrix} = \begin{bmatrix} 1 & 16 \\ 0 & 1 \end{bmatrix}$

If n is numbers of matrices that are multiplied, then product is $\begin{bmatrix} 1 & n^2 \\ 0 & 1 \end{bmatrix}$.

$\therefore \quad r = 2500$

17. $M = \begin{bmatrix} 1 & 1 & 1 \\ 1 & 1 & 1 \\ 1 & 1 & 1 \end{bmatrix}$

$$M \times M = M^2 = \begin{bmatrix} 3 & 3 & 3 \\ 3 & 3 & 3 \\ 3 & 3 & 3 \end{bmatrix} = 3\begin{bmatrix} 1 & 1 & 1 \\ 1 & 1 & 1 \\ 1 & 1 & 1 \end{bmatrix} = 3M$$

$$M \times M \times M = M^3 = 3 \times 3\begin{bmatrix} 1 & 1 & 1 \\ 1 & 1 & 1 \\ 1 & 1 & 1 \end{bmatrix} = 3^2 M$$

$\therefore \quad M^{50} = 3^{49} \times M$

18. Put $\tan(\theta/2) = a$ so that

$$B = \begin{bmatrix} \dfrac{1-a^2}{1+a^2} & \dfrac{-2a}{1+a^2} \\ \dfrac{2a}{1+a^2} & \dfrac{1-a^2}{1+a^2} \end{bmatrix}$$

Therefore, $(I-A)B = \begin{bmatrix} 1 & a \\ -a & 1 \end{bmatrix}\begin{bmatrix} \dfrac{1-a^2}{1+a^2} & \dfrac{-2a}{1+a^2} \\ \dfrac{2a}{1+a^2} & \dfrac{1-a^2}{1+a^2} \end{bmatrix}$

$$= \begin{bmatrix} \dfrac{1-a^2}{1+a^2} + \dfrac{2a^2}{1+a^2} & \dfrac{-2a}{1+a^2} + \dfrac{a(1-a^2)}{1+a^2} \\ \dfrac{-a(1-a^2)}{1+a^2} + \dfrac{2a}{1+a^2} & \dfrac{2a^2}{1+a^2} + \dfrac{1-a^2}{1+a^2} \end{bmatrix}$$

$$= \begin{bmatrix} 1 & -a \\ a & 1 \end{bmatrix} = I + A$$

19. Let $X = \begin{pmatrix} a & b \\ c & d \end{pmatrix}\begin{pmatrix} a & b \\ c & d \end{pmatrix}$

$$\Rightarrow X^2 = \begin{pmatrix} a^2+bc & ab+bd \\ ac+cd & bc+d^2 \end{pmatrix} = \begin{bmatrix} a^2+bc & b(a+d) \\ c(a+d) & bc+d^2 \end{bmatrix}$$

$a^2+bc = 1$ ...(i)

$b(a+d) = 1$ ...(ii)

$c(a+d) = 2$ ...(iii)

$bc+d^2 = 3$ ...(iv)

From Eq. (i), $\quad c = \dfrac{1-a^2}{b}$ ...(v)

From Eqs. (iii) and (iv), we get

$$\left(\frac{1-a^2}{b}\right)(a+d) = 2$$

$\Rightarrow \quad a+d-a^3-a^2d = 2b$

$a^3+a^2d-a+(2b-d)-0$ is a equation of third degree and will have three solutions.

Therefore, b, c and d will also have three values corresponding to the values of a.

Therefore, the matrix equation has more than two solutions.

20. Since, matrix $A = \begin{bmatrix} \cos\theta & i\sin\theta \\ i\sin\theta & \cos\theta \end{bmatrix}$

$\Rightarrow \quad A^2 = \begin{bmatrix} \cos 2\theta & i\sin 2\theta \\ i\sin 2\theta & \cos 2\theta \end{bmatrix}$

$\Rightarrow \quad A^3 = \begin{bmatrix} \cos 3\theta & i\sin 3\theta \\ i\sin 3\theta & \cos 3\theta \end{bmatrix}$

$\Rightarrow \quad A^4 - \begin{bmatrix} \cos 4\theta & i\sin 4\theta \\ i\sin 4\theta & \cos 4\theta \end{bmatrix}$

$\Rightarrow \quad A^5 = \begin{bmatrix} \cos 5\theta & i\sin 5\theta \\ i\sin 5\theta & \cos 5\theta \end{bmatrix} = \begin{bmatrix} a & b \\ c & d \end{bmatrix}$

$\Rightarrow \quad a = d = \cos 5\theta,\; b = c = i\sin 5\theta$

$\therefore \quad a^2 - b^2 = \cos^2 5\theta - i^2\sin^2 5\theta = 1$

21. It is given that,

$$\begin{bmatrix} 2 & 1 & 1 \\ 0 & 3 & -1 \\ 1 & -1 & 1 \end{bmatrix}\begin{bmatrix} x \\ y \\ z \end{bmatrix} = \begin{bmatrix} 1 \\ 1 \\ 0 \end{bmatrix}$$

$\Rightarrow \quad 2x + y + z = 1 \quad \ldots(i)$

$3y - z = 1 \quad \ldots(ii)$

and $\quad x - y + z = 0 \quad \ldots(iii)$

From Eqs. (i) and (ii), we get

$$2x + y + 3y - 1 = 1$$

$\Rightarrow \quad 2x + 4y = 2$

$\Rightarrow \quad x + 2y = 1 \quad \ldots(iv)$

Now, from Eqs. (ii) and (iv), we have

$$\frac{x-1}{-2} = y = \frac{z+1}{3} = k \quad \text{(let)}$$

$$\therefore \quad \begin{bmatrix} x \\ y \\ z \end{bmatrix} = \begin{bmatrix} 1 \\ 0 \\ -1 \end{bmatrix} + k\begin{bmatrix} -2 \\ 1 \\ 3 \end{bmatrix}, k \in R$$

22. We have, $E(\alpha)E(\beta)$

$$= \begin{bmatrix} \cos^2\alpha & \cos\alpha\sin\alpha \\ \cos\alpha\sin\alpha & \sin^2\alpha \end{bmatrix}\begin{bmatrix} \cos^2\beta & \cos\beta\sin\beta \\ \cos\beta\sin\beta & \sin^2\beta \end{bmatrix}$$

$$= \begin{bmatrix} \cos\alpha\cos\beta\cos(\alpha-\beta) & \cos\alpha\sin\beta\cos(\alpha-\beta) \\ \sin\alpha\cos\beta\cos(\alpha-\beta) & \sin\alpha\sin\beta\cos(\alpha-\beta) \end{bmatrix}$$

As α and β differ by an odd multiple of $\pi/2$, $\alpha - \beta = (2n+1)\pi/2$ for some integer n. Thus, $\cos[(2n+1)\pi/2] = 0$

$$\therefore \quad E(\alpha)E(\beta) = O$$

23. We have,

$$AA^T = \frac{1}{9}\begin{bmatrix} 1 & 2 & 2 \\ 2 & 1 & -2 \\ x & 2 & y \end{bmatrix}\begin{bmatrix} 1 & 2 & x \\ 2 & 1 & 2 \\ 2 & -2 & y \end{bmatrix}$$

$$= \frac{1}{9}\begin{bmatrix} 1+4+4 & 2+2-4 & x+4+2y \\ 2+2-4 & 4+1+4 & 2x+2-2y \\ x+4+2y & 2x+2-2y & x^2+4+y^2 \end{bmatrix}$$

$$= \frac{1}{9}\begin{bmatrix} 9 & 0 & x+2y+4 \\ 0 & 9 & (2x+2y+2) \\ x+2y+4 & 2x-2y+2 & x^2+y^2+4 \end{bmatrix}$$

Now

$AA^T =$

$$\begin{bmatrix} 1 & 0 & (x+2y+4)/9 \\ 0 & 1 & (2x-2y+2)/9 \\ (x+2y+4)/9 & (2x-2y+2)/9 & (x^2+y^2+4)/9 \end{bmatrix}$$

$$I^3 = \begin{bmatrix} 1 & 0 & 0 \\ 0 & 1 & 0 \\ 0 & 0 & 1 \end{bmatrix}$$

Solving we get

$x + 2y + 4 = 0 \quad \ldots(i)$

$2x - 2y + 2 = 0 \quad \ldots(ii)$

$x^2 + y^2 + 4 = 9 \quad \ldots(iii)$

From Eqs. (i) and (ii), we get

$x = -2$, $y = -1$. These value also satisfy Eq. (iii).

24. $A(x)A(y) = (1-x)^{-1}(1-y)^{-1}\begin{bmatrix} 1 & -x \\ -x & 1 \end{bmatrix}\begin{bmatrix} 1 & -y \\ -y & 1 \end{bmatrix}$

$$= (1+xy-(x+y))^{-1}\begin{bmatrix} 1+xy & -(x+y) \\ -(x+y) & 1+xy \end{bmatrix}$$

$$= \left(1 - \frac{x+y}{1+xy}\right)^{-1}\begin{bmatrix} 1 & -\dfrac{x+y}{1+xy} \\ -\dfrac{x+y}{1+xy} & 1 \end{bmatrix} = A(z)$$

25. We have, $A = \begin{bmatrix} i & -i \\ -i & i \end{bmatrix} = i\begin{bmatrix} 1 & -1 \\ -1 & 1 \end{bmatrix} = iB$

$\Rightarrow \quad A^2 = (iB)^2 = i^2B^2 = -B^2$

$$= -\begin{bmatrix} 1 & -1 \\ -1 & 1 \end{bmatrix}\begin{bmatrix} 1 & -1 \\ -1 & 1 \end{bmatrix} = \begin{bmatrix} -2 & 2 \\ 2 & -2 \end{bmatrix} = -2B$$

$\Rightarrow \quad A^4 = -(-2B)^2 = 4B^2$

$= 4(2B) = 8B$

$\Rightarrow \quad A^8 = (A^4)^2 = (8B)^2$

$= 64B^2 = 128B$

26. Let $A = \begin{bmatrix} 1 & 2 & 3 \\ 1 & 2 & 3 \\ -1 & -2 & -3 \end{bmatrix}$

$$\therefore A^2 = \begin{bmatrix} 1 & 2 & 3 \\ 1 & 2 & 3 \\ -1 & -2 & -3 \end{bmatrix}\begin{bmatrix} 1 & 2 & 3 \\ 1 & 2 & 3 \\ -1 & -2 & -3 \end{bmatrix} = \begin{bmatrix} 0 & 0 & 0 \\ 0 & 0 & 0 \\ 0 & 0 & 0 \end{bmatrix}$$

Nilpotent matrix of index 2.

27. It is given that 'α' is the root of equation $x^2 + x + 1 = 0$, so $\alpha^2 + \alpha + 1 = 0$ and $\alpha^3 = 1$.

Now, matrix $A = \dfrac{1}{\sqrt{3}}\begin{bmatrix} 1 & 1 & 1 \\ 1 & \alpha & \alpha^2 \\ 1 & \alpha^2 & \alpha^4 \end{bmatrix} = \dfrac{1}{\sqrt{3}}\begin{bmatrix} 1 & 1 & 1 \\ 1 & \alpha & \alpha^2 \\ 1 & \alpha^2 & \alpha \end{bmatrix}$

$[\because \alpha^3 = 1]$

$$\therefore A^2 = \left(\frac{1}{\sqrt{3}}\right)^2\begin{bmatrix} 1 & 1 & 1 \\ 1 & \alpha & \alpha^2 \\ 1 & \alpha^2 & \alpha \end{bmatrix}\begin{bmatrix} 1 & 1 & 1 \\ 1 & \alpha & \alpha^2 \\ 1 & \alpha^2 & \alpha \end{bmatrix}$$

$$= \frac{1}{3}\begin{bmatrix} 1+1+1 & 1+\alpha+\alpha^2 & 1+\alpha^2+\alpha \\ 1+\alpha+\alpha^2 & 1+\alpha^2+\alpha^4 & 1+\alpha^3+\alpha^3 \\ 1+\alpha^2+\alpha & 1+\alpha^3+\alpha^3 & 1+\alpha^4+\alpha^2 \end{bmatrix}$$

$$= \frac{1}{3}\begin{bmatrix} 3 & 0 & 0 \\ 0 & 0 & 3 \\ 0 & 3 & 0 \end{bmatrix} = \begin{bmatrix} 1 & 0 & 0 \\ 0 & 0 & 1 \\ 0 & 1 & 0 \end{bmatrix}$$

$[\because \alpha^3 = 1 \text{ and } \alpha^2 + \alpha + 1 = 0]$

$\therefore \quad A^4 = A^2 \cdot A^2$

$$= \begin{bmatrix} 1 & 0 & 0 \\ 0 & 0 & 1 \\ 0 & 1 & 0 \end{bmatrix}\begin{bmatrix} 1 & 0 & 0 \\ 0 & 0 & 1 \\ 0 & 1 & 0 \end{bmatrix} = \begin{bmatrix} 1 & 0 & 0 \\ 0 & 1 & 0 \\ 0 & 0 & 1 \end{bmatrix} = I$$

$\therefore \quad A^8 = I = A^{16} = A^{20} = A^{24} = A^{28}$

$\because \quad A^{31} = A^{28} \cdot A^3 = I \cdot A^3 = A^3$

$\Rightarrow A^{31} = A^3$

28. The order of matrix A is 3×4, so rank $(A) \le 3$ but it is given that rank $(A) = 3$, so

$$\begin{vmatrix} 1 & r & r^2 \\ r & r^2 & 1 \\ r^2 & 1 & r \end{vmatrix} \ne 0$$

$$\Rightarrow 1(r^3 - 1) - r(r^2 - r^2) + r^2(r - r^4) \ne 0$$

$$\Rightarrow \quad r^3 - 1 + r^3 - r^6 \ne 0$$

$$\Rightarrow \quad r^6 - 2r^3 + 1 \ne 0$$

$$\Rightarrow \quad (r^3 - 1)^2 \ne 0$$

$$\Rightarrow \quad r^3 \ne 1 \Rightarrow r \ne 1$$

Hence, $r \in R - \{1\}$.

29. Z is idempotent, then

$$Z^2 = Z$$

$$\Rightarrow \quad Z^3, Z^4, \ldots, Z^n = Z$$

$$\therefore \quad (I + Z)^n = {}^nC_0 I^n + {}^nC_1 I^{n-1} Z + {}^nC_2 I^{n-2} Z^2 + + {}^nC_n Z^n$$

$$= {}^nC_0 I + {}^nC_1 Z + {}^nC_2 Z + {}^nC_3 Z + + {}^nC_n Z$$

$$= I + ({}^nC_1 + {}^nC_2 + {}^nC_3 + + {}^nC_n) Z$$

$$= I + (2^n - 1) Z$$

30. $\because \begin{bmatrix} 0 & 1 \\ 1 & 0 \end{bmatrix} \begin{bmatrix} x \\ y \end{bmatrix} = \begin{bmatrix} y \\ x \end{bmatrix} = \begin{bmatrix} X \\ Y \end{bmatrix}$

Then, $\quad X = y$ and $Y = x$

i.e. $\quad y = x$

31. Here, $P = \begin{bmatrix} 1 & 0 & 0 \\ 4 & 1 & 0 \\ 16 & 4 & 1 \end{bmatrix}$

$$\therefore \quad P^2 = \begin{bmatrix} 1 & 0 & 0 \\ 4 & 1 & 0 \\ 16 & 4 & 1 \end{bmatrix} \begin{bmatrix} 1 & 0 & 0 \\ 4 & 1 & 0 \\ 16 & 4 & 1 \end{bmatrix} = \begin{bmatrix} 1 & 0 & 0 \\ 4+4 & 1 & 0 \\ 16+32 & 4+4 & 1 \end{bmatrix}$$

$$= \begin{bmatrix} 1 & 0 & 0 \\ 4 \times 2 & 1 & 0 \\ 16(1+2) & 4 \times 2 & 1 \end{bmatrix} \quad \ldots \text{(i)}$$

$$\text{and } P^3 = \begin{bmatrix} 1 & 0 & 0 \\ 4 \times 2 & 1 & 0 \\ 16(1+2) & 4 \times 2 & 1 \end{bmatrix} \begin{bmatrix} 1 & 0 & 0 \\ 4 & 1 & 0 \\ 16 & 4 & 1 \end{bmatrix}$$

$$= \begin{bmatrix} 1 & 0 & 0 \\ 4 \times 3 & 1 & 0 \\ 16(1+2+3) & 4 \times 3 & 1 \end{bmatrix} \quad \ldots \text{(ii)}$$

From symmetry,

$$P^{50} = \begin{bmatrix} 1 & 0 & 0 \\ 4 \times 50 & 1 & 0 \\ 16(1+2+3+\ldots+50) & 4 \times 50 & 1 \end{bmatrix}$$

$$\because \quad P^{50} - Q = I \quad \text{[given]}$$

$$\therefore \begin{bmatrix} 1 - q_{11} & -q_{12} & -q_{13} \\ 200 - q_{21} & 1 - q_{22} & -q_{23} \\ 16 \times \frac{50}{2}(51) - q_{31} & 200 - q_{32} & 1 - q_{33} \end{bmatrix} = \begin{bmatrix} 1 & 0 & 0 \\ 0 & 1 & 0 \\ 0 & 0 & 1 \end{bmatrix}$$

$$\Rightarrow \quad 200 - q_{21} = 0, \frac{16 \times 50 \times 51}{2} - q_{31} = 0,$$

$$200 - q_{32} = 0$$

$$\therefore \quad q_{21} = 200, q_{32} = 200, q_{31} = 20400$$

Thus, $\dfrac{q_{31} + q_{32}}{q_{21}} = \dfrac{20400 + 200}{200} = \dfrac{20600}{200} = 103$

32. Here, $A^2 = A$

$$\Rightarrow \begin{bmatrix} 2 & -2 & -16-4x \\ -1 & 3 & 16+4x \\ 4+x & -8-2x & -12+x^2 \end{bmatrix} = \begin{bmatrix} 2 & -2 & -4 \\ -1 & 3 & 4 \\ 1 & -2 & x \end{bmatrix}$$

On comparing, $16 + 4x = 4 \Rightarrow x = -3$

33. It is given that matrix.0

$$M = \begin{bmatrix} \sin^4\theta & -1 - \sin^2\theta \\ 1 + \cos^2\theta & \cos^4\theta \end{bmatrix} = \alpha I + \beta M^{-1}, \text{ where}$$

$\alpha = \alpha(\theta)$ and $\beta = \beta(\theta)$ are real numbers and I is the 2×2 identity matrix.

Now, $\det(M) = |M| = \sin^4\theta \cos^4\theta + 1 + \sin^2\theta + \cos^2\theta + \sin^2\theta\cos^2\theta$

$$= \sin^4\theta\cos^4\theta + \sin^2\theta\cos^2\theta + 2$$

and $\begin{bmatrix} \sin^4\theta & -1 - \sin^2\theta \\ 1 + \cos^2\theta & \cos^4\theta \end{bmatrix} = \begin{bmatrix} \alpha & 0 \\ 0 & \alpha \end{bmatrix} + \dfrac{\beta}{|M|}(\text{adj}\, M)$

$$\left[\because M^{-1} = \frac{\text{adj}\, M}{|M|}\right]$$

$$\Rightarrow \begin{bmatrix} \sin^4\theta & -1 - \sin^2\theta \\ 1 + \cos^2\theta & \cos^4\theta \end{bmatrix} = \begin{bmatrix} \alpha & 0 \\ 0 & \alpha \end{bmatrix} + \frac{\beta}{|M|} \begin{bmatrix} \cos^4\theta & 1 + \sin^2\theta \\ -1 - \cos^2\theta & \sin^4\theta \end{bmatrix}$$

$$\left\{\because \text{adj} \begin{bmatrix} a & b \\ c & d \end{bmatrix} = \begin{bmatrix} d & -b \\ -c & a \end{bmatrix}\right\}$$

$\Rightarrow \beta = -|M|$ and $\alpha = \sin^4\theta + \cos^4\theta$

$\Rightarrow \alpha = \alpha(\theta) = 1 - \dfrac{1}{2}\sin^2(2\theta)$, and

$$\beta = \beta(\theta) = -\left\{\left(\sin^2\theta\cos^2\theta + \frac{1}{2}\right)^2 + \frac{7}{4}\right\}$$

$$= -\left\{\left(\frac{\sin^2(2\theta)}{4} + \frac{1}{2}\right)^2 + \frac{7}{4}\right\}$$

Now, $\alpha^* = \alpha_{\min} = \dfrac{1}{2}$ and $\beta^* = \beta_{\min} = -\dfrac{37}{16}$

$\because$ α is minimum at $\sin^2(2\theta) = 1$ and β is minimum at $\sin^2(2\theta) = 1$

So, $\alpha^* + \beta^* = \dfrac{1}{2} - \dfrac{37}{16} = -\dfrac{29}{16}$

34. Given matrix $A = \begin{bmatrix} 0 & 2y & 1 \\ 2x & y & -1 \\ 2x & -y & 1 \end{bmatrix}, (x, y \in R, x \ne y)$

for which

$$A^T A = 3I_3$$

$$\Rightarrow \begin{bmatrix} 0 & 2x & 2x \\ 2y & y & -y \\ 1 & -1 & 1 \end{bmatrix}\begin{bmatrix} 0 & 2y & 1 \\ 2x & y & -1 \\ 2x & -y & 1 \end{bmatrix} = \begin{bmatrix} 3 & 0 & 0 \\ 0 & 3 & 0 \\ 0 & 0 & 3 \end{bmatrix}$$

$$\Rightarrow \begin{bmatrix} 8x^2 & 0 & 0 \\ 0 & 6y^2 & 0 \\ 0 & 0 & 3 \end{bmatrix} = \begin{bmatrix} 3 & 0 & 0 \\ 0 & 3 & 0 \\ 0 & 0 & 3 \end{bmatrix}$$

Here, two matrices are equal, therefore equating the corresponding elements, we get

$$8x^2 = 3 \text{ and } 6y^2 = 3 \Rightarrow x = \pm\sqrt{\frac{3}{8}} \text{ and } y = \pm\frac{1}{\sqrt{2}}$$

$\because$ There are 2 different values of x and y each.

So, 4 matrices are possible such that $A^T A = 3I_3$.

35. $\because \quad A = \frac{1}{\sqrt{2}}\begin{bmatrix} 1 & i \\ -i & a \end{bmatrix}$

$$\Rightarrow \quad \bar{A} = \frac{1}{\sqrt{2}}\begin{bmatrix} 1 & -i \\ i & a \end{bmatrix}$$

$$\therefore \quad (\bar{A})' = \frac{1}{\sqrt{2}}\begin{bmatrix} 1 & i \\ -i & a \end{bmatrix} = A^\theta \Rightarrow A A^\theta = I$$

$$\Rightarrow \frac{1}{\sqrt{2}}\begin{bmatrix} 1 & i \\ -i & a \end{bmatrix}\frac{1}{\sqrt{2}}\begin{bmatrix} 1 & i \\ -i & a \end{bmatrix} = \begin{bmatrix} 1 & 0 \\ 0 & 1 \end{bmatrix}$$

$$\Rightarrow \frac{1}{2}\begin{bmatrix} 2 & i(1+a) \\ -i(1+a) & 1+a^2 \end{bmatrix} = \begin{bmatrix} 1 & 0 \\ 0 & 1 \end{bmatrix}$$

$$\therefore \quad 1 + a = 0 \Rightarrow a = -1$$

36. $3A = \begin{bmatrix} 1 & 2 & 2 \\ 2 & 1 & -2 \\ x & 2 & y \end{bmatrix}$

$$\therefore (3A)(3A') = \begin{bmatrix} 1 & 2 & 2 \\ 2 & 1 & -2 \\ x & 2 & y \end{bmatrix}\begin{bmatrix} 1 & 2 & x \\ 2 & 1 & 2 \\ 2 & -2 & y \end{bmatrix}$$

$$= \begin{bmatrix} 9 & 0 & x+2y+4 \\ 0 & 9 & 2x-2y+2 \\ x+2y+4 & 2x-2y+2 & x^2+y^2+4 \end{bmatrix}$$

$$= 9\begin{bmatrix} 1 & 0 & 0 \\ 0 & 1 & 0 \\ 0 & 0 & 1 \end{bmatrix}$$

$\Rightarrow \quad x + 2y + 4 = 0$ and $2x - 2y + 2 = 0$

Then, we get $x = -2, y = -1$

$\therefore \quad x + y = -3$

37. $\because A^2 = A, B^2 = B$

$$A + B = (A+B)^2 = A^2 + B^2 + AB + BA$$

$\Rightarrow \quad A + B = A + B + AB + BA$ or $AB + BA = O$

then $\quad AB = BA = O$.

38. $A = \begin{bmatrix} 1 & 2 & 3 \\ 2 & 4 & 6 \\ 3 & 6 & 9 \end{bmatrix}$

$$\begin{matrix} R_{21}(-2) \\ \sim \\ R_{31}(-3) \end{matrix}\begin{bmatrix} 1 & 2 & 3 \\ 0 & 0 & 0 \\ 0 & 0 & 0 \end{bmatrix},$$

$\therefore \rho(A) = 1$

39. Given matrix $A = \begin{bmatrix} 1 & 0 \\ 0 & -1 \end{bmatrix}$ is orthogonal matrix, because $AA^T = I$.

So, $\quad A^T X^{50} A = A^T X^{49}(APA^T)A$

$$= A^T X^{49} AP(A^T A) = A^T X^{49} AP$$

$$= A^T X^{48}(APA^T)AP = A^T X^{48} AP^2 \ldots$$

$$\ldots. = A^T AP^{50} = IP^{50} = P^{50}$$

$$\because \quad P = \begin{bmatrix} 1 & 1 \\ 0 & 1 \end{bmatrix} \Rightarrow P^2 = \begin{bmatrix} 1 & 2 \\ 0 & 1 \end{bmatrix}$$

$$\Rightarrow \quad P^3 = \begin{bmatrix} 1 & 3 \\ 0 & 1 \end{bmatrix} \ldots\ldots \Rightarrow P^{50} = \begin{bmatrix} 1 & 50 \\ 0 & 1 \end{bmatrix}$$

So, $\quad A^T X^{50} A = P^{50} = \begin{bmatrix} 1 & 50 \\ 0 & 1 \end{bmatrix}$

40. $$A + 2B = \begin{bmatrix} 1 & 2 & 0 \\ 6 & -3 & 3 \\ -5 & 3 & 1 \end{bmatrix} \quad \ldots\text{(i)}$$

$$2A - B = \begin{bmatrix} 2 & -1 & 5 \\ 2 & -1 & 6 \\ 0 & 1 & 2 \end{bmatrix}$$

$$\Rightarrow \quad 4A - 2B = \begin{bmatrix} 4 & -2 & 10 \\ 4 & -2 & 12 \\ 0 & 2 & 4 \end{bmatrix} \quad \ldots\text{(ii)}$$

Eq. (i) + Eq. (ii), we get

$$5A = \begin{bmatrix} 5 & 0 & 10 \\ 10 & -5 & 15 \\ -5 & 5 & 5 \end{bmatrix}$$

$$A = \begin{bmatrix} 1 & 0 & 2 \\ 2 & -1 & 3 \\ -1 & 1 & 1 \end{bmatrix} \text{ and } 2A = \begin{bmatrix} 2 & 0 & 4 \\ 4 & -2 & 6 \\ -2 & 2 & 2 \end{bmatrix}$$

$$\therefore \quad B = \begin{bmatrix} 2 & 0 & 4 \\ 4 & -2 & 6 \\ -2 & 2 & 2 \end{bmatrix} - \begin{bmatrix} 2 & -1 & 5 \\ 2 & -1 & 6 \\ 0 & 1 & 2 \end{bmatrix}$$

$$B = \begin{bmatrix} 0 & 1 & -1 \\ 2 & -1 & 0 \\ -2 & 1 & 0 \end{bmatrix}$$

$tr(A) = 1 - 1 + 1 = 1$

$tr(B) = -1$

$tr(A) = 1$ and $tr(B) = -1$

$\therefore \quad tr(A) - tr(B) = 2$

41. We have, $A^2 = \begin{pmatrix} \cos 2\alpha & \sin 2\alpha \\ -\sin 2\alpha & \cos 2\alpha \end{pmatrix}$,

$$A^3 = \begin{pmatrix} \cos 3\alpha & \sin 3\alpha \\ -\sin 3\alpha & \cos 3\alpha \end{pmatrix}$$

and $\quad A^4 = \begin{pmatrix} \cos 4\alpha & \sin 4\alpha \\ -\sin 4\alpha & \cos 4\alpha \end{pmatrix}$

We have $\cos\alpha + \cos 2\alpha + \cos 3\alpha + \cos 4\alpha$

$= \cos\alpha + \cos 2\alpha + \cos(\pi - 2\alpha) + \cos(\pi - \alpha)$ $[\because 5\alpha = \pi]$

$= \cos\alpha + \cos 2\alpha - \cos 2\alpha - \cos\alpha = 0$

and $\sin\alpha + \sin 2\alpha + \sin 3\alpha + \sin 4\alpha$

$$= \sin\alpha + \sin 2\alpha + \sin(\pi - 2\alpha) + \sin(\pi - \alpha)$$
$$= 2[\sin\alpha + \sin 2\alpha]$$
$$= 2\left\{2\sin\left(\frac{3\alpha}{2}\right)\cos\frac{\alpha}{2}\right\} = 4\sin\left(\frac{3\pi}{10}\right)\cos\frac{\pi}{10}$$
$$= 4\sin\left(\frac{\pi}{2} - \frac{\pi}{5}\right)\cos\frac{\pi}{10} = 4\cos\frac{\pi}{5}\cos\frac{\pi}{10} = a \text{ (say)}$$

Thus, $B = \begin{pmatrix} 0 & a \\ -a & 0 \end{pmatrix}$

$\therefore B$ is skew-symmetric.

42. We have,

$$A = \begin{bmatrix} 0 & 5 \\ 0 & 0 \end{bmatrix}$$
$$A^2 = \begin{bmatrix} 0 & 5 \\ 0 & 0 \end{bmatrix}\begin{bmatrix} 0 & 5 \\ 0 & 0 \end{bmatrix} = \begin{bmatrix} 0 & 0 \\ 0 & 0 \end{bmatrix}$$

Similarly, $A^3 = A^4 = \dots A^n = \begin{bmatrix} 0 & 0 \\ 0 & 0 \end{bmatrix}$

$$\therefore \quad f(A) = A + A^2 + A^3 + \dots + A^{2018}$$
$$f(A) = A + 0 + 0 + \dots + 0 = A$$
$$f(A) + I = \begin{bmatrix} 0 & 5 \\ 0 & 0 \end{bmatrix} + \begin{bmatrix} 1 & 0 \\ 0 & 1 \end{bmatrix} = \begin{bmatrix} 1 & 5 \\ 0 & 1 \end{bmatrix}$$

43. $\because \quad B = [-b \quad -a]$

$\therefore \quad -B = [b \quad a] \neq A$

$A + B = [a - b \quad b - a]$ and $A - B = [a + b \quad b + a]$

$\Rightarrow \quad A + B \neq B + A$

and $\quad AC = [a \quad b]\begin{bmatrix} a \\ -a \end{bmatrix} = [a^2 - ab]$

$$BC = [-b \quad -a]\begin{bmatrix} a \\ -a \end{bmatrix} = [a^2 - ab]$$

i.e. $\quad AC = BC.$

and $\quad CA = \begin{bmatrix} a \\ -a \end{bmatrix}[a \quad b] = \begin{bmatrix} a^2 & ab \\ -a^2 & -ab \end{bmatrix}$

$$CB = \begin{bmatrix} a \\ -a \end{bmatrix}[-b - a] = \begin{bmatrix} -ab & -a^2 \\ ab & a^2 \end{bmatrix}$$

ie, $\quad CA \neq CB.$

44. $X^2 = X \cdot X = \begin{bmatrix} 3 & -4 \\ 1 & -1 \end{bmatrix}\begin{bmatrix} 3 & -4 \\ 1 & -1 \end{bmatrix} = \begin{bmatrix} 5 & -8 \\ 2 & -3 \end{bmatrix}$

For $n = 2$,

$\neq$ (a), (b), (c).

45. $\because \quad A^2 = 2A - I$

$$A^3 = A^2 \cdot A = 2A^2 - IA$$
$$= 2A^2 - A = 2(2A - I) - A$$
$$= 3A - 2I = 3A - (3 - 1)I$$

....

....

$$A^n = nA - (n - 1)I.$$

Objective Questions Type II [One or more than one correct answer(s)]

1. Given, $\quad AB = A, BA = B \Rightarrow B \times AB = B \times A$

or $\quad (BA)B = B$ or $B^2 = B$

Also, $\quad A \times BA = AB$

$\Rightarrow \quad (AB)A = A \Rightarrow A^2 = A$

Now, $(A^T)^2 = (A^T \times A^T) = (A \times A)^T = (A^2)^T = A^T$

Similarly, $\quad (B^T)^2 = B^T$

Hence, A^T and B^T are idempotent.

2. Applying $R_2 \to R_2 - 2R_1$ and $R_3 \to R_3 - R_1$

Then, $A = \begin{bmatrix} 1 & 4 & 5 \\ \lambda - 2 & 0 & 8\lambda - 16 \\ \lambda^2 & 8\lambda & 2\lambda + 16 \end{bmatrix}$

For $\lambda = 2$,

$$A = \begin{bmatrix} 1 & 4 & 5 \\ 0 & 0 & 0 \\ 4 & 16 & 20 \end{bmatrix} \underset{\sim}{R_3 \to R_3 - 4R_1} \begin{bmatrix} 1 & 4 & 5 \\ 0 & 0 & 0 \\ 0 & 0 & 0 \end{bmatrix}$$

$\therefore \quad \rho(A) = 1$

For $\quad \lambda = -1,$

$$A = \begin{bmatrix} 1 & 4 & 5 \\ -3 & 0 & -24 \\ 1 & -8 & 14 \end{bmatrix} \quad R_2 \to R_2 + 3R_1$$

$$\underset{R_3 \to R_3 - R_1}{\sim} \begin{bmatrix} 1 & 4 & 5 \\ 0 & 12 & -9 \\ 0 & -12 & 9 \end{bmatrix}$$

$$\underset{\sim}{R_3 \to R_3 + R_2} \begin{bmatrix} 1 & 4 & 5 \\ 0 & 12 & -9 \\ 0 & 0 & 0 \end{bmatrix}$$

$\therefore \quad \rho(A) = 2$

For $\lambda \neq -1, 2$, $\rho(A) = 3$

3. $A^{n+2} - B^{n+2} = (A + B)(A^{n+1} - B^{n+1}) - AB(A^n - B^n)$

$\Rightarrow \quad A^n - B^n = (A + B)(A^n - B^n) - AB(A^n - B^n)$

$\Rightarrow \quad I = A + B - AB \quad [\because A^n - B^n$ is not invertible$]$

$\Rightarrow \quad (I - A)(I - B) = O$

As, $A, B \neq I$, we get

$I - A$ and $I - B$ are singular matrices.

4. Given matrices,

$$P_1 = I = \begin{bmatrix} 1 & 0 & 0 \\ 0 & 1 & 0 \\ 0 & 0 & 1 \end{bmatrix}, P_2 = \begin{bmatrix} 1 & 0 & 0 \\ 0 & 0 & 1 \\ 0 & 1 & 0 \end{bmatrix}, P_3 = \begin{bmatrix} 0 & 1 & 0 \\ 1 & 0 & 0 \\ 0 & 0 & 1 \end{bmatrix}$$
$$P_4 = \begin{bmatrix} 0 & 1 & 0 \\ 0 & 0 & 1 \\ 1 & 0 & 0 \end{bmatrix}, P_5 = \begin{bmatrix} 0 & 0 & 1 \\ 1 & 0 & 0 \\ 0 & 1 & 0 \end{bmatrix}, P_6 = \begin{bmatrix} 0 & 0 & 1 \\ 0 & 1 & 0 \\ 1 & 0 & 0 \end{bmatrix}$$

and $X = \sum_{K=1}^{6} P_k \begin{bmatrix} 2 & 1 & 3 \\ 1 & 0 & 2 \\ 3 & 2 & 1 \end{bmatrix} P_K^T$

$\because \; P_1^T = P_1, P_2^T = P_2, P_3^T = P_3, P_4^T = P_5, P_5^T = P_4$ and

$P_6^T = P_6$ and Let $Q = \begin{bmatrix} 2 & 1 & 3 \\ 1 & 0 & 2 \\ 3 & 2 & 1 \end{bmatrix}$ and $\because Q^T = Q$

Now, $X = (P_1QP_1^T) + (P_2QP_2^T) + (P_3QP_3^T) + (P_4QP_4^T) + (P_5QP_5^T) + (P_6QP_6^T)$

So, $X^T = (P_1QP_1^T)^T + (P_2QP_2^T)^T + (P_3QP_3^T)^T + (P_4QP_4^T)^T + (P_5QP_5^T)^T + (P_6QP_6^T)^T$

$= P_1QP_1^T + P_2QP_2^T + P_3QP_3^T + P_4QP_4^T + P_5QP_5^T + P_6QP_6^T$

$[\because (ABC)^T = C^TB^TA^T$ and $(A^T)^T = A$ and $Q^T = Q]$

$\Rightarrow X^T = X$

$\Rightarrow X$ is a symmetric matrix.

The sum of diagonal entries of $X = Tr(X)$

$$= \sum_{i=1}^{6} T_r(P_i Q P_i^T)$$

$$= \sum_{i=1}^{6} T_r(Q P_i^T P_i) \qquad [\because T_r(ABC) = T_r(BCA)]$$

$$= \sum_{i=1}^{6} T_r(QI) \qquad [\because P_i\text{'s are orthogonal matrices}]$$

$$= \sum_{i=1}^{6} T_r(Q) = 6\,T_r(Q) = 6 \times 3 = 18$$

Now, Let $R = \begin{bmatrix} 1 \\ 1 \\ 1 \end{bmatrix}$, then

$$XR = \sum_{K=1}^{6} (P_K Q P_K^T)R = \sum_{K=1}^{6} (P_K Q P_K^T R)$$

$$= \sum_{K=1}^{6} (P_K Q R) \qquad [\because P_K^T R = R]$$

$$= \sum_{K=1}^{6} P_K \begin{bmatrix} 6 \\ 3 \\ 6 \end{bmatrix} = \sum_{K=1}^{6} P_K \begin{bmatrix} 6 \\ 3 \\ 6 \end{bmatrix} = \begin{bmatrix} 2 & 2 & 2 \\ 2 & 2 & 2 \\ 2 & 2 & 2 \end{bmatrix}\begin{bmatrix} 6 \\ 3 \\ 6 \end{bmatrix}$$

$$\Rightarrow \quad XR = \begin{bmatrix} 30 \\ 30 \\ 30 \end{bmatrix} \Rightarrow XR = 30R \Rightarrow X\begin{bmatrix} 1 \\ 1 \\ 1 \end{bmatrix} = 30\begin{bmatrix} 1 \\ 1 \\ 1 \end{bmatrix}$$

$\Rightarrow (X - 30I)R = 0 \Rightarrow |X - 30I| = 0$

So, $(X - 30I)$ is not invertible and value of $\alpha = 30$.

Hence, options (a), (b) and (d) are correct.

5.
$$A^2 = A \cdot A = \begin{bmatrix} 1 & 0 \\ 1 & 1 \end{bmatrix}\begin{bmatrix} 1 & 0 \\ 1 & 1 \end{bmatrix} = \begin{bmatrix} 1 & 0 \\ 2 & 1 \end{bmatrix}$$

$$A^3 = A^2 \cdot A = \begin{bmatrix} 1 & 0 \\ 2 & 1 \end{bmatrix}\begin{bmatrix} 1 & 0 \\ 1 & 1 \end{bmatrix}\begin{bmatrix} 1 & 0 \\ 3 & 1 \end{bmatrix}$$

....

$$\therefore \quad A^n = \begin{bmatrix} 1 & 0 \\ n & 1 \end{bmatrix}$$

$$\therefore \quad A^{-n} = \begin{bmatrix} 1 & 0 \\ -n & 1 \end{bmatrix}$$

and
$$\lim_{n\to\infty} \frac{A^{-n}}{n} = \begin{bmatrix} \lim\limits_{n\to\infty} \frac{1}{n} & 0 \\ \lim\limits_{n\to\infty} \frac{-n}{n} & \lim\limits_{n\to\infty} \frac{1}{n} \end{bmatrix} = \begin{bmatrix} 0 & 0 \\ -1 & 0 \end{bmatrix}$$

and
$$\lim_{n\to\infty} \frac{A^{-n}}{n^2} = \begin{bmatrix} \lim\limits_{n\to\infty} \frac{1}{n^2} & 0 \\ \lim\limits_{n\to\infty} -\frac{1}{n} & \lim\limits_{n\to\infty} \frac{1}{n^2} \end{bmatrix} = \begin{bmatrix} 0 & 0 \\ 0 & 0 \end{bmatrix}$$

6. Assume (i). That is

$$ABC = AC - CA \qquad \text{...(i)}$$

Now, $ABC + A + B + C = (AC - CA) + AB + BC + CA = AC + AB + BC$

Therefore,

$\therefore \; (A - I)(B - I)(C - I)$

$= ABC - (AC + AB + BC) + A + B + C - I$

$= ABC - (AC + AB + BC) + AB + BC + CA - I$

$= ABC + CA - AC - I$

$= -I \qquad \text{...(ii)}$

This implies $A - I, B - I, C - I$ are invertible matrices and the inverse of $C - I$ is $-(A - I)(B - I)$.

This gives

$$(C - I)(A - I)(B - I) = -I$$

$$CAB - (CA + AB + CB) + A + B + C = 0$$

$$CAB - (CA + AB + CB) + AB + BC + CA = 0$$

$$CAB = CB - BC$$

Therefore, Eq. (i) $\Rightarrow$ Eq. (iii).

Similarly, by permuting the letters A, B, C we can show that (i), (ii) and (iii) are equivalent statements.

7. Given matrices,

$$P_1 = I = \begin{bmatrix} 1 & 0 & 0 \\ 0 & 1 & 0 \\ 0 & 0 & 1 \end{bmatrix}, P_2 = \begin{bmatrix} 1 & 0 & 0 \\ 0 & 0 & 1 \\ 0 & 1 & 0 \end{bmatrix}, P_3 = \begin{bmatrix} 0 & 1 & 0 \\ 1 & 0 & 0 \\ 0 & 0 & 1 \end{bmatrix}$$

$$P_4 = \begin{bmatrix} 0 & 1 & 0 \\ 0 & 0 & 1 \\ 1 & 0 & 0 \end{bmatrix}, P_5 = \begin{bmatrix} 0 & 0 & 1 \\ 1 & 0 & 0 \\ 0 & 1 & 0 \end{bmatrix}, P_6 = \begin{bmatrix} 0 & 0 & 1 \\ 0 & 1 & 0 \\ 1 & 0 & 0 \end{bmatrix}$$

and $X = \sum_{K=1}^{6} P_k \begin{bmatrix} 2 & 1 & 3 \\ 1 & 0 & 2 \\ 3 & 2 & 1 \end{bmatrix} P_K^T$

$\because \quad P_1^T = P_1,\ P_2^T = P_2,\ P_3^T = P_3,\ P_4^T = P_5,\ P_5^T = P_4$ and

$P_6^T = P_6$ and Let $Q = \begin{bmatrix} 2 & 1 & 3 \\ 1 & 0 & 2 \\ 3 & 2 & 1 \end{bmatrix}$ and $\because Q^T = Q$

Now, $X = (P_1QP_1^T) + (P_2QP_2^T) + (P_3QP_3^T) + (P_4QP_4^T) + (P_5QP_5^T) + (P_6QP_6^T)$

So, $X^T = (P_1QP_1^T)^T + (P_2QP_2^T)^T + (P_3QP_3^T)^T + (P_4QP_4^T)^T + (P_5QP_5^T)^T + (P_6QP_6^T)^T$

$= P_1QP_1^T + P_2QP_2^T + P_3QP_3^T + P_4QP_4^T + P_5QP_5^T + P_6QP_6^T$

$[\because (ABC)^T = C^TB^TA^T$ and $(A^T)^T = A$ and $Q^T = Q]$

$\Rightarrow X^T = X$

$\Rightarrow X$ is a symmetric matrix.

The sum of diagonal entries of $X = Tr(X)$

$$= \sum_{i=1}^{6} T_r(P_i Q P_i^T)$$

$$= \sum_{i=1}^{6} T_r(Q P_i^T P_i) \qquad [\because T_r(ABC) = T_r(BCA)]$$

$$= \sum_{i=1}^{6} T_r(QI) \qquad [\because P_i\text{'s are orthogonal matrices}]$$

$$= \sum_{i=1}^{6} T_r(Q) = 6\,T_r(Q) = 6 \times 3 = 18$$

Now, Let $R = \begin{bmatrix}1\\1\\1\end{bmatrix}$, then

$$XR = \sum_{K=1}^{6}(P_K Q P_K^T)R = \sum_{K=1}^{6}(P_K Q P_K^T R)$$

$$= \sum_{K=1}^{6}(P_K Q R) \qquad [\because P_K^T R = R]$$

$$= \sum_{K=1}^{6} P_K \begin{bmatrix}6\\3\\6\end{bmatrix} = \sum_{K=1}^{6} P_K \begin{bmatrix}6\\3\\6\end{bmatrix} = \begin{bmatrix}2&2&2\\2&2&2\\2&2&2\end{bmatrix}\begin{bmatrix}6\\3\\6\end{bmatrix}$$

$$\Rightarrow \quad XR = \begin{bmatrix}30\\30\\30\end{bmatrix} \Rightarrow XR = 30R \Rightarrow X\begin{bmatrix}1\\1\\1\end{bmatrix} = 30\begin{bmatrix}1\\1\\1\end{bmatrix}$$

$\Rightarrow (X - 30I)R = 0 \Rightarrow |X - 30I| = 0$

So, $(X - 30I)$ is not invertible and value of $\alpha = 30$.

Hence, options (a), (b) and (d) are correct.

8. $\because A^2 = \begin{bmatrix}a^2+bc & ab+bd\\ ac+cd & bc+d^2\end{bmatrix}$

Now, $A^2 + xA + yI_2 = O$ implies

$$a^2 + bc + xa + y = 0 \qquad \ldots(i)$$

$$ab + bd + xb = 0 \qquad \ldots(ii)$$

$$ac + cd + xc = 0 \qquad \ldots(iii)$$

$$bc + d^2 + xd + y = 0$$

As, one of b, d is different from zero, assume, $b \neq 0$. From Eq. (ii) we get $x = -(a+d)$. Substituting this in Eq. (1), we obtain $y = ad - bc$.

9. It is given, that matrices

$$P = \begin{bmatrix}1&1&1\\0&2&2\\0&0&3\end{bmatrix}, Q = \begin{bmatrix}2&x&x\\0&4&0\\x&x&6\end{bmatrix}$$

$$\therefore \quad P^{-1} = \frac{\text{adj}(P)}{|P|}$$

as $|P| = 6$ and $\text{adj}\,P = \begin{bmatrix}6&0&0\\-3&3&0\\0&-2&2\end{bmatrix}^T$

$$\Rightarrow \quad P^{-1} = \frac{1}{6}\begin{bmatrix}6&-3&0\\0&3&-2\\0&0&2\end{bmatrix}$$

$$\therefore \quad |R| = |PQP^{-1}| \qquad [\because R = PQP^{-1} \text{ (given)}]$$

$$\Rightarrow \quad |R| = |P||Q||P^{-1}| = |Q| \qquad [\because |P||P^{-1}| = |I| = 1]$$

$$= \begin{vmatrix}2&x&x\\0&4&0\\x&x&6\end{vmatrix} = \begin{vmatrix}2&x&x\\0&4&0\\x&x&5\end{vmatrix} + \begin{vmatrix}2&x&0\\0&4&0\\x&x&1\end{vmatrix}$$

$$= \begin{vmatrix}2&x&x\\0&4&0\\x&x&5\end{vmatrix} + 2(4-0) - x(0-0) + 0(0-4x)$$

$$= \begin{vmatrix}2&x&x\\0&4&0\\x&x&5\end{vmatrix} + 8, \text{ for all } x \in R$$

$$\because \quad PQ = \begin{bmatrix}1&1&1\\0&2&2\\0&0&3\end{bmatrix}\begin{bmatrix}2&x&x\\0&4&0\\x&x&6\end{bmatrix} = \begin{bmatrix}2+x & 4+2x & x+6\\2x & 2x+8 & 12\\3x & 3x & 18\end{bmatrix}$$

$$\text{and } QP = \begin{bmatrix}2&x&x\\0&4&0\\x&x&6\end{bmatrix}\begin{bmatrix}1&1&1\\0&2&2\\0&0&3\end{bmatrix} = \begin{bmatrix}2 & 2+2x & 2+5x\\0 & 8 & 8\\x & 3x & 3x+18\end{bmatrix}$$

There is no common value of 'x', for which each corresponding element of matrices PQ and QP is equal.

For $x = 0, Q = \begin{bmatrix}2&0&0\\0&4&0\\0&0&6\end{bmatrix}$

then, if $R\begin{bmatrix}1\\a\\b\end{bmatrix} = 6\begin{bmatrix}1\\a\\b\end{bmatrix}$

$$\Rightarrow \quad PQP^{-1}\begin{bmatrix}1\\a\\b\end{bmatrix} = 6\begin{bmatrix}1\\a\\b\end{bmatrix} \qquad [\because R = PQP^{-1}]$$

$$\Rightarrow \frac{1}{6}\begin{bmatrix}1&1&1\\0&2&2\\0&0&3\end{bmatrix}\begin{bmatrix}2&0&0\\0&4&0\\0&0&6\end{bmatrix}\begin{bmatrix}6&-3&0\\0&3&-2\\0&0&2\end{bmatrix}\begin{bmatrix}1\\a\\b\end{bmatrix} = 6\begin{bmatrix}1\\a\\b\end{bmatrix}$$

$$\Rightarrow \quad \frac{1}{6}\begin{bmatrix}2&4&6\\0&8&12\\0&0&18\end{bmatrix}\begin{bmatrix}6&-3&0\\0&3&-2\\0&0&2\end{bmatrix}\begin{bmatrix}1\\a\\b\end{bmatrix} = 6\begin{bmatrix}1\\a\\b\end{bmatrix}$$

$$\Rightarrow \quad \begin{bmatrix}12&6&4\\0&24&8\\0&0&36\end{bmatrix}\begin{bmatrix}1\\a\\b\end{bmatrix} = 36\begin{bmatrix}1\\a\\b\end{bmatrix}$$

$$\Rightarrow \quad \begin{bmatrix}12+6a+4b\\0+24a+8b\\0+0+36b\end{bmatrix} = \begin{bmatrix}36\\36a\\36b\end{bmatrix}$$

$$\Rightarrow \quad 6a + 4b = 24 \text{ and } 12a = 8b$$

$$\Rightarrow \quad 3a + 2b = 12 \text{ and } 3a = 2b$$

$$\Rightarrow \quad a = 2 \text{ and } b = 3$$

So $a + b = 5$.

Now, $R\begin{bmatrix}\alpha\\\beta\\\gamma\end{bmatrix} = \begin{bmatrix}0\\0\\0\end{bmatrix}$ and $\alpha\hat{i} + \beta\hat{j} + \gamma\hat{k}$ is a unit vector, so

det $(R) = 0$

$\Rightarrow \det(Q) = 0 \qquad [\because R = PQP^{-1} \text{ So, } |R| = |Q|]$

$$\Rightarrow \quad \begin{vmatrix}2&x&x\\0&4&0\\x&x&6\end{vmatrix} = 0$$

$$\Rightarrow \quad 2(24-0) - x(0-0) + x(0-4x) =$$

$$\Rightarrow \quad 48 - 4x^2 - 0$$

$$\Rightarrow \quad x^2 = 12$$

$$\Rightarrow \quad x = \pm 2\sqrt{3}$$

So, for $x = 1$, there does not exist a unit vector

$\alpha\hat{i} + \beta\hat{j} + \gamma\hat{k}$, for which $R\begin{bmatrix}\alpha\\\beta\\\gamma\end{bmatrix} = \begin{bmatrix}0\\0\\0\end{bmatrix}$

10. $\because \quad (A + A')' = A' + (A')' = A' + A$

$(A - A')' = A' - (A')' = A' - A = -(A - A')$

and $\quad A = \frac{1}{2}(A + A') + \frac{1}{2}(A - A')$

In skew-symmetric matrix $\begin{bmatrix} 0 & -c & b \\ c & 0 & a \\ -b & -a & 0 \end{bmatrix}$ diagonal elements must be zero.

11. Let $D_1 = \begin{bmatrix} d_1 & 0 & 0 \\ 0 & d_2 & 0 \\ 0 & 0 & d_3 \end{bmatrix}$

$\Rightarrow D_1^n = \begin{bmatrix} d_1^n & 0 & 0 \\ 0 & d_2^n & 0 \\ 0 & 0 & d_3^n \end{bmatrix}$ = Diagonal matrix

and $D_2 = \begin{bmatrix} g_1 & 0 & 0 \\ 0 & g_2 & 0 \\ 0 & 0 & g_3 \end{bmatrix}$

$\Rightarrow \quad D_2^n = \begin{bmatrix} g_1^n & 0 & 0 \\ 0 & g_2^n & 0 \\ 0 & 0 & g_3^n \end{bmatrix}$ = Diagonal matrix

$\therefore D_1D_2 = \begin{bmatrix} d_1g_1 & 0 & 0 \\ 0 & d_2g_2 & 0 \\ 0 & 0 & d_3g_3 \end{bmatrix}$ = Diagonal matrix

and $\quad D_2D_1 = \begin{bmatrix} g_1d_1 & 0 & 0 \\ 0 & g_2d_2 & 0 \\ 0 & 0 & g_3d_3 \end{bmatrix} = D_1D_2$

Also, $D_1^2 + D_2^2 = \begin{bmatrix} d_1^2 + g_1^2 & 0 & 0 \\ 0 & d_2^2 + g_2^2 & 0 \\ 0 & 0 & d_3^2 + g_3^2 \end{bmatrix}$

= Diagonal matrix.

12. $\because AoB = \frac{1}{2}(AB + BA) = \frac{1}{2}(BA + AB) = BoA$

$AoI = \frac{1}{2}(AI + IA) = \frac{1}{2}(A + A) = A$

$AoA = \frac{1}{2}(AA + AA) = \frac{1}{2}(A^2 + A^2) = A^2$

and $Ao(B + C) = \frac{1}{2}\{A(B + C) + (B + C)A\}$

$= \frac{1}{2}\{AB + AC + BA + CA\}$

$= \frac{1}{2}(AB + BA) + \frac{1}{2}(AC + CA)$

$= AoB + AoC$

Linked Comprehension Type

PASSAGE 1

1. $\because A$ is singular.

$\therefore |A| = 0 \Rightarrow \begin{vmatrix} 1 & 3 & \lambda + 2 \\ 2 & 4 & 8 \\ 3 & 5 & 10 \end{vmatrix} = 0$

$\Rightarrow 1(40 - 40) - 3(20 - 24) + (\lambda + 2)(10 - 12) = 0$

$\Rightarrow \quad 12 - 2\lambda - 4 = 0 \Rightarrow \lambda = 4$

$\therefore \quad A + B = \begin{pmatrix} 4 & 5 & 10 \\ 5 & 6 & 13 \\ 5 & 6 & 14 \end{pmatrix}$

$\Rightarrow \quad tr(A + B) = 4 + 6 + 14 = 24$

2. $\because 2A + 3B = \begin{pmatrix} 11 & 12 & 2\lambda + 16 \\ 13 & 14 & 31 \\ 12 & 13 & 32 \end{pmatrix}$

$\because \quad |2A + 3B| = 0$

we get $\quad \lambda = \frac{17}{2} \Rightarrow 2\lambda = 17$

3. For $\lambda = 3$,

$A = \begin{pmatrix} 1 & 3 & 5 \\ 2 & 4 & 8 \\ 3 & 5 & 10 \end{pmatrix}$ and $B = \begin{pmatrix} 3 & 2 & 4 \\ 3 & 2 & 5 \\ 2 & 1 & 4 \end{pmatrix}$

$\because \quad tr(AB) + tr(BA) = 2tr(AB)$

$\because tr(AB) = C_{11} + C_{22} + C_{33} = 119$

$\therefore \quad \frac{1}{7}(tr(AB) + tr(BA)) = \frac{1}{7}(238) = 34$

4. $A - 2B = \begin{pmatrix} \lambda - 6 & -1 & -5 \\ -2 & 0 & -4 \\ 2 & -5 & -1 \end{pmatrix}$

$\because \; tr(A - 2B) = 0 \quad \therefore \lambda - 6 + 0 - 1 = 0 \Rightarrow \lambda = 7$

5. $\because \quad (A + B)^2 = (A + B)(A + B)$

$= A^2 + AB + BA + B^2 = A^2 + B^2 + AB + BA$

PASSAGE 2

1. Let $A = \begin{bmatrix} a & b \\ c & d \end{bmatrix} \Rightarrow A^2 - (a + d)A + (ad - bc)I$

$= \begin{bmatrix} a & b \\ c & d \end{bmatrix}\begin{bmatrix} a & b \\ c & d \end{bmatrix} - (a + d)\begin{bmatrix} a & b \\ c & d \end{bmatrix} + (ad - bc)\begin{bmatrix} 1 & 0 \\ 0 & 1 \end{bmatrix}$

$= \begin{bmatrix} a^2 + bc & ab + bd \\ ac + cd & bc + d^2 \end{bmatrix} - \begin{bmatrix} a^2 + ad & ab + bd \\ ac + cd & ad + d^2 \end{bmatrix}$

$+ \begin{bmatrix} ad - bc & 0 \\ 0 & ad - bc \end{bmatrix} = O$

2. If $A = O$, $tr(A) = 0$. Suppose $A \neq O$ and $A = \begin{bmatrix} a & b \\ c & d \end{bmatrix}$. Then, $|A| = 0$ and $A^2 - (a + d)A + (ad - bc)I = O$

$\Rightarrow \quad a + d = 0 = tr(A)$

3. $(I + A)^{100} = {}^{100}C_0 I^{100} + {}^{100}C_1 I^{99} A$

$+ {}^{100}C_2 I^{98} A^2 + \ldots\ldots + {}^{100}C_{100} A^{100}$

$= I + 100A + O + O + \; + O = I + 100A$

Numerical Value Type

1. $\because \begin{pmatrix}1&2\\2&3\end{pmatrix}^2 - \lambda\begin{pmatrix}1&2\\2&3\end{pmatrix} - \begin{pmatrix}1&0\\0&1\end{pmatrix} = \begin{pmatrix}0&0\\0&0\end{pmatrix}$

$\Rightarrow \begin{pmatrix}5&8\\8&13\end{pmatrix} - \begin{pmatrix}\lambda&2\lambda\\2\lambda&3\lambda\end{pmatrix} - \begin{pmatrix}1&0\\0&1\end{pmatrix} = \begin{pmatrix}0&0\\0&0\end{pmatrix}$

$\Rightarrow \begin{pmatrix}4-\lambda&8-2\lambda\\8-2\lambda&12-3\lambda\end{pmatrix} = \begin{pmatrix}0&0\\0&0\end{pmatrix}$

On comparing, we get $4-\lambda=0,\ 8-2\lambda=0,\ 8-2\lambda=0,$ $12-3\lambda=0$

$\therefore \quad \lambda = 4$

$\therefore \quad 6^\lambda = 6^4 = 1296$

2. $\because \displaystyle\sum_{r=0}^{7} A_r = \sum_{r=0}^{7}\begin{pmatrix}a_r&0\\0&b_r\end{pmatrix} = \begin{pmatrix}a&0\\0&b\end{pmatrix}$

$\therefore \quad a = \displaystyle\sum_{r=0}^{7} a_r = \sum_{r=0}^{7} r\cdot({}^7C_r)$

and $\quad b = \displaystyle\sum_{r=0}^{7} b_r = \sum_{r=0}^{7} (7-r)\cdot({}^7C_r)$

$\Rightarrow \quad a+b = 7\displaystyle\sum_{r=0}^{7} {}^7C_r = 7(2^7) = 7\times128 = 896$

$\dfrac{1}{35}(a+b) = \dfrac{1}{35}\times7\times128 = \dfrac{128}{5} = 25.6$

3. Let a 3×3 matrix $A=[a_{ij}]_{3\times3}, i, j \in \{1, 2, 3\}$. So, sum of diagonal elements of AA^T

= trace of $AA^T = tr([a_{ij}][a_{ji}])$

$= a_{11}^2 + a_{12}^2 + a_{13}^2 + a_{21}^2 + a_{22}^2 + a_{23}^2 + a_{31}^2 + a_{32}^2 + a_{33}^2 = 3$ (given)

$\because a_{ij} \in \{-1, 0, 1\}$

So, possible combinations are

0, 0, 0, 0, 0, 0, 1, 1, 1
0, 0, 0, 0, 0, 0, –1, 1, 1
0, 0, 0, 0, 0, 0, –1, –1, 1
0, 0, 0, 0, 0, 0, –1, –1, –1

$\therefore$ Number of such matrices such that $tr(AA^T) = 3$ is

$= {}^9C_6 \times 2\times2\times2$

$= \dfrac{9\times8\times7}{3\times2}\times2\times2\times2 = 84\times8 = 672$

4. $A^3 = A\cdot A^2 = A(5A-7I) \qquad [\because A^2 = 5A-7I]$

$= 5A^2 - 7A$

$= 5(5A-7I) - 7A$

$= 25A - 35I - 7A$

$= 18A - 35I$

Now, $A^4 = A\cdot A^3 = A(18A-35I)$

$= 18A^2 - 35A$

$= 18(5A-7I) - 35A$

$= 55A - 126I$

Finally, $A^5 = A\cdot A^4 = A(55A-126I)$

$= 55A^2 - 126A$

$= 55(5A-7I) - 126A$

$= 149A - 385I = aA + bI$ (given)

$\therefore \quad a = 149$ and $b = -385$

$\Rightarrow \quad 2a - 3b = 2\times149 - 3\times -385$

$= 298 + 1155 = 1453$

Single Integer Type

1. Let $N = M - \dfrac{M^2}{3} + \dfrac{M^3}{9} - \dfrac{M^4}{27} + \ldots\infty$

This forms a GP to infinite series

$N = \dfrac{M}{1-\left(-\dfrac{M}{3}\right)} = \dfrac{3M}{3+M} = \dfrac{3M}{3I+M}$

$\Rightarrow \quad N = \begin{bmatrix}3&-9\\-3&3\end{bmatrix}\begin{bmatrix}4&-3\\-1&4\end{bmatrix}^{-1}$

$= \dfrac{1}{13}\begin{bmatrix}3&-27\\-9&3\end{bmatrix} = \dfrac{3}{13}\begin{bmatrix}1&-9\\-3&1\end{bmatrix}$

$\therefore \quad \alpha = -9, \beta = -3$

$\Rightarrow \quad \begin{vmatrix}\alpha\\\beta\end{vmatrix} - \begin{vmatrix}-9\\-3\end{vmatrix} = 3$

2. Since the product matrix is 3×3 matrix and the pre-multiplier of A is a 3×2 matrix, therefore A is 2×3 matrix. Let

$A = \begin{bmatrix}l&m&n\\x&y&z\end{bmatrix}$. Then, the given equation becomes

$\begin{bmatrix}2&-1\\1&0\\-3&4\end{bmatrix}\begin{bmatrix}l&m&n\\x&y&z\end{bmatrix} = \begin{bmatrix}-1&-8&-10\\1&-2&-5\\9&22&15\end{bmatrix}$

or $\begin{bmatrix}2l-x&2m-y&2n-z\\l&m&x\\-3l+4x&-3m+4y&-3n+4z\end{bmatrix} = \begin{bmatrix}-1&-8&-10\\1&-2&-5\\9&22&15\end{bmatrix}$

or $\quad 2l-x=-1,\ 2m-y=-8,\ 2n-z=-10,$

$l=1, m=-2, n=-5$

$\Rightarrow \quad x=3, y=4, z=0, l=1, m=-2, n=-5$

$\Rightarrow \quad A = \begin{bmatrix}l&m&n\\x&y&z\end{bmatrix} = \begin{bmatrix}1&-2&-5\\3&4&0\end{bmatrix}$

$\therefore$ Sum of all elements of A

$= 1 + (-2) + (-5) + 3 + 4 + 0 = 1$

3. Given, $(A+B)^2 = A^2 + B^2 + 2AB$

or $\quad (A+B)(A+B) = A^2 + B^2 + 2AB$

or $\quad A^2 + AB + BA + B^2 = A^2 + B^2 + 2AB$

or $\quad BA = AB$

or $\begin{bmatrix} a & 1 \\ b & -1 \end{bmatrix}\begin{bmatrix} 1 & -1 \\ 2 & 1 \end{bmatrix} = \begin{bmatrix} 1 & -1 \\ 2 & 1 \end{bmatrix}\begin{bmatrix} a & 1 \\ b & -1 \end{bmatrix}$

or $\begin{bmatrix} a+2 & -a+1 \\ b-2 & -b-1 \end{bmatrix} = \begin{bmatrix} a-b & 1+1 \\ 2a+b & 2-1 \end{bmatrix}$

The corresponding elements of equal matrices are equal

$$a + 2 = a - b, -a + 1 = 2$$

$\Rightarrow$ $a = -1$

$$b - 2 = 2a + b, -b - 1 = 1$$

$\Rightarrow$ $b = -2$

$\Rightarrow$ $a = -1, b = -2$

$\therefore$ $|a + b| = 3$

4. For the given matrix A we have

$$A^T = \begin{bmatrix} 0 & x & x \\ 2y & y & -y \\ z & -z & z \end{bmatrix}$$

$$AA^T = \begin{bmatrix} 0 & 2y & z \\ x & y & -z \\ x & -y & z \end{bmatrix}\begin{bmatrix} 0 & x & x \\ 2y & y & -y \\ z & -z & z \end{bmatrix}$$

$$= \begin{bmatrix} 0+4y^2+z^2 & 0+2y^2-z^2 & 0-2y^2+z^2 \\ 0+2y^2-z^2 & x^2+y^2+z^2 & x^2-y^2-z^2 \\ 0-2y^2+z^2 & x^2-y^2-z^2 & x^2+y^2+z^2 \end{bmatrix}$$

$$= \begin{bmatrix} 4y^2+z^2 & 2y^2-z^2 & -2y^2+z^2 \\ 2y^2-z^2 & x^2+y^2+z^2 & x^2-y^2-z^2 \\ -2y^2+z^2 & x^2-y^2-z^2 & x^2+y^2+z^2 \end{bmatrix}$$

But it is given that $AA^T = I$, therefore

$$AA^T = \begin{bmatrix} 1 & 0 & 0 \\ 0 & 1 & 0 \\ 0 & 0 & 1 \end{bmatrix}$$

Solving we get [taking (2-2)th entry]

$$x^2 + y^2 + z^2 = 1$$

5. Given, $A^2 = A$

Now, $I = (I - 0.4A)(I - \alpha A)$

$$= I - I\alpha A - 0.4AI + 0.4\alpha A^2$$

$$= I - A\alpha - 0.4A + 0.4\alpha A$$

$$= I - A(0.4 + \alpha) + 0.4\alpha A$$

$\Rightarrow$ $0.4\alpha = 0.4 + \alpha \Rightarrow \alpha = -2/3 \Rightarrow |9\alpha| = 6$

Matrix-Match Type

1. Let $A = [A_{ij}]_{3\times 3}$, where $a_{ij} = (k)^{i+j}$

So, $A = \begin{bmatrix} k^2 & k^3 & k^4 \\ k^3 & k^4 & k^5 \\ k^4 & k^5 & k^6 \end{bmatrix}$

(A) If A is singular, then $|A| = 0$

$$\Rightarrow \quad k^2 \cdot k^3 \cdot k^4 \begin{vmatrix} 1 & 1 & 1 \\ k & k & k \\ k^2 & k^2 & k^2 \end{vmatrix} = 0 \Rightarrow k \in I$$

(B) If A is null matrix, then $k \in \{0\}$

(C) There is no value of k for A to be skew-symmetric matrix which is not null-matrix.

$\therefore$ $k \in \phi$

(D) If $A^2 = 3A$, then

$$\begin{bmatrix} k^2 & k^3 & k^4 \\ k^3 & k^4 & k^5 \\ k^4 & k^5 & k^6 \end{bmatrix}\begin{bmatrix} k^2 & k^3 & k^4 \\ k^3 & k^4 & k^5 \\ k^4 & k^5 & k^6 \end{bmatrix} = \begin{bmatrix} 3k^2 & 3k^3 & 3k^4 \\ 3k^3 & 3k^4 & 3k^5 \\ 3k^4 & 3k^5 & 3k^6 \end{bmatrix}$$

$$\Rightarrow \begin{bmatrix} k^4+k^6+k^8 & k^5+k^7+k^9 & k^6+k^8+k^{10} \\ k^5+k^7+k^9 & k^6+k^8+k^{10} & k^7+k^9+k^{11} \\ k^6+k^8+k^{10} & k^7+k^9+k^{11} & k^8+k^{10}+k^{12} \end{bmatrix}$$

$$= \begin{bmatrix} 3k^2 & 3k^3 & 3k^4 \\ 3k^3 & 3k^4 & 3k^5 \\ 3k^4 & 3k^5 & 3k^6 \end{bmatrix}$$

$\Rightarrow k \in \{-1, 0, 1\}$

2. (A) $\because$ $A * B = \frac{1}{2}(AB + BA) = \frac{1}{2}(BA + AB) = B * A$ (P)

$$A * A = \frac{1}{2}(A \cdot A + A \cdot A) = A^2 \text{(R)}$$

$$A * I = \frac{1}{2}(AI + IA) = \frac{1}{2}(A + A) = A \text{ (S)}$$

Now, $A * (B + C) = \frac{1}{2}\{A(B + C) + (B + C)A\}$

$$= \frac{1}{2}(AB + BA) + \frac{1}{2}(AC + CA)$$

$$= A * B + A * C \text{ (Q)}$$

(B) $A * B = \frac{1}{2}(AB' + A'B) = \frac{1}{2}(A'B + AB')$

$$= \frac{1}{2}(BA' + B'A) = B * A \text{ (P)}$$

$$A * A = \frac{1}{2}(AA' + A'A) \neq A^2$$

$$A * (B + C) = \frac{1}{2}(A(B + C)' + A'(B + C))$$

$$= \frac{1}{2}(A(B' + C') + A'(B + C))$$

$$= \frac{1}{2}(AB' + A'B) + \frac{1}{2}(AC' + A'C)$$

$$= A * B + A * C \text{ (Q)}$$

$$A * I = \frac{1}{2}(AI' + A'I) = \frac{1}{2}(A + A') \neq A \text{ or } O$$

(C) $A * B = \frac{1}{2}(AB - BA) - \frac{1}{2}(BA - AB) = -(B * A)$

$$A * (B + C) = \frac{1}{2}\{A(B + C) - (B + C)A\}$$

$$= \frac{1}{2}(AB - BA) + \frac{1}{2}(AC - CA)$$

$$= A * B + A * C \text{ (Q)}$$

$$A * A = \frac{1}{2}(A^2 - A^2) = O \neq A^2$$

$$A * I = \frac{1}{2}(AI - IA) = \frac{1}{2}(A - A) = O \text{ (T)}$$

08

Logarithms and Their Properties

Objective Questions Type I [Only one correct answer]

In each of the questions below, four choices are given of which only one is correct. You have to select the correct answer which is the most appropriate.

1. If $\log_2 3 = a$, $\log_3 5 = b$ and $\log_7 2 = c$, then the logarithm of the number 63 to base 140 is

(a) $\dfrac{1+2ac}{2c+abc+1}$ (b) $\dfrac{1-2ac}{2c-abc-1}$

(c) $\dfrac{1-2ac}{2c+abc+1}$ (d) $\dfrac{1+2ac}{2c-abc-1}$

2. If $\log_5 120 + (x-3) - 2\log_5 (1-5^{x-3}) = -\log_5 (0.2 - 5^{x-4})$, then x is

(a) 1 (b) 2

(c) 3 (d) 4

3. If $x_n > x_{n-1} > \ldots > x_2 > x_1 > 1$, then the value of $\log_{x_1} \log_{x_2} \log_{x_3} \ldots \log_{x_n} x_n^{x_{n-1}^{\cdot^{\cdot^{x_1}}}}$, is

(a) 0

(b) 1

(c) 2

(d) undefined

4. The number $\log_{20} 3$ lies in

(a) $\left(\dfrac{1}{4}, \dfrac{1}{3}\right)$ (b) $\left(\dfrac{1}{3}, \dfrac{1}{2}\right)$

(c) $\left(\dfrac{1}{2}, \dfrac{3}{4}\right)$ (d) $\left(\dfrac{3}{4}, \dfrac{4}{5}\right)$

5. The number of real values of the parameter λ for which $(\log_{16} x)^2 - \log_{16} x + \log_{16} \lambda = 0$ with real coefficients will have exactly one solution is

(a) 1 (b) 2

(c) 3 (d) 4

6. If $y = 2^{\frac{1}{\log_x 4}}$, then x is equal to

(a) $\sqrt{y}$ (b) y

(c) y^2 (d) y^3

7. If $\log_2 x + \log_2 y \geq 6$, then the least value of $x + y$ is

(a) 4 (b) 8

(c) 16 (d) 32

8. If $\log_{12} 27 = a$, then $\log_6 16 =$

(a) $4\left(\dfrac{3+a}{3-a}\right)$ (b) $4\left(\dfrac{3-a}{3+a}\right)$

(c) $2\left(\dfrac{3-a}{3+a}\right)$ (d) $2\left(\dfrac{3+a}{3-a}\right)$

9. If $x = \log_5 (1000)$ and $y = \log_7 (2058)$, then

(a) $x > y$ (b) $x < y$

(c) $x = y$ (d) None of these

10. If $f(x) = \ln\left(\dfrac{1+x}{1-x}\right)$, then

(a) $f(x_1) \cdot f(x_2) = f(x_1 + x_2)$

(b) $f(x+2) - 2f(x+1) + f(x) = 0$

(c) $f(x) + f(x+1) = f(x^2 + x)$

(d) $f(x_1) + f(x_2) = f\left(\dfrac{x_1 + x_2}{1 + x_1 x_2}\right)$

11. If $\log_{10} 2$, $\log_{10} (2^x + 1)$, $\log_{10} (2^x + 3)$ are in AP, then

(a) $x = 0$ (b) $x = 1$

(c) $x = \log_{10} 2$ (d) $x = \dfrac{1}{2}\log_2 5$

12. If $A = \log_2 \log_2 \log_4 256 + 2\log_{\sqrt{2}} 2$, then A is equal to
(a) 2 (b) 3
(c) 5 (d) 7

13. $7\log\left(\frac{16}{15}\right) + 5\log\left(\frac{25}{24}\right) + 3\log\left(\frac{81}{80}\right)$ is equal to
(a) 0 (b) 1
(c) $\log 2$ (d) $\log 3$

14. If the sum of the first 20 terms of the series $\log_{(7^{1/2})} x + \log_{(7^{1/3})} x + \log_{(7^{1/4})} x + \ldots$ is 460, then x is equal to **[JEE Main 2020, 5 September]**
(a) 7^2 (b) $7^{1/2}$
(c) e^2 (d) $7^{46/21}$

15. If $\log_{10} 3 = 0.477$, the number of digit in 3^{40} is
(a) 18 (b) 19
(c) 20 (d) 21

16. If $x = \log_3 5$, $y = \log_{17} 25$, which one of the following is correct ?
(a) $x < y$ (b) $x = y$
(c) $x > y$ (d) None of these

17. Solution set of the inequality $\log_x(2x^2 + x - 1) > \log_x(2) - 1$ is
(a) $(1/2, 1)$ (b) $(1/2, 1) \cup (1, \infty)$
(c) $(1, \infty)$ (d) $(0, 1)$

18. The number $\log_2 7$ is
(a) an integer (b) a rational number
(c) an irrational number (d) a prime number

19. Solution of the equations for x and y : $(3x)^{\log 3} = (4y)^{\log 4}$, $4^{\log x} = 3^{\log y}$ is
(a) $\frac{1}{3}, \frac{1}{4}$ (b) $\frac{1}{2}, \frac{1}{3}$
(c) $\frac{1}{4}, \frac{1}{5}$ (d) $\frac{1}{4}, \frac{1}{3}$

20. Solution of the equation $\log_{x^2+6+8} \log_{2x^2+2x+3}(x^2 - 2x) = 0$.
(a) -2 (b) -1
(c) -4 (d) 0

21. If $\log_{12} 27 = a$, then $\log_6 16$ is equal to
(a) $2\left(\frac{3-a}{3+a}\right)$ (b) $3\left(\frac{3-a}{3+a}\right)$
(c) $4\left(\frac{3-a}{3+a}\right)$ (d) $5\left(\frac{3-a}{3+a}\right)$

22. $\log_7 \log_7 \sqrt{7\sqrt{(7\sqrt{7})}}$ is equal to
(a) $3\log_2 7$
(b) $3\log_7 2$
(c) $1 - 3\log_7 2$
(d) $1 - 3\log_2 7$

23. If $\frac{\log x}{b-c} = \frac{\log y}{c-a} = \frac{\log z}{a-b}$, then $x^a y^b z^c$ is equal to
(a) xyz (b) abc
(c) 0 (d) 1

24. $\frac{1}{1+\log_a bc} + \frac{1}{1+\log_b ca} + \frac{1}{1+\log_c ab}$ is equal to
(a) 0 (b) 1
(c) 2 (d) 3

25. If $(4)^{\log_9 3} + (9)^{\log_2 4} = (10)^{\log_x 83}$, then x is equal to
(a) 2 (b) 3
(c) 10 (d) 30

26. If x, y, z are in GP and $a^x = b^y = c^z$, then
(a) $\log_b a = \log_c b$ (b) $\log_c b = \log_a c$
(c) $\log_a c = \log_b a$ (d) $\log_a b = 2\log_a c$

27. If $\frac{x(y+z-x)}{\log x} = \frac{y(z+x-y)}{\log y} = \frac{z(x+y-z)}{\log z}$, then $x^y y^x = z^y y^z$ is equal to
(a) $z^x x^z$ (b) $x^z y^x$
(c) $x^y y^z$ (d) $x^x y^y$

28. If $\log_3 2$, $\log_3(2^x - 5)$, $\log_3(2^x - 7/2)$ are in AP, then x is equal to
(a) 1 (b) 2
(c) 3 (d) 4

29. If $y = a^{\frac{1}{1-\log_a x}}$ and $z = a^{\frac{1}{1-\log_a y}}$, then x is equal to
(a) $a^{\frac{1}{1+\log_a z}}$ (b) $a^{\frac{1}{2+\log_a z}}$
(c) $a^{\frac{1}{1-\log_a z}}$ (d) $a^{\frac{1}{2-\log_a z}}$

30. If $\log_{(2x+3)}(6x^2 + 23x + 21) = 4 - \log_{(3x+7)}(3x^2 + 12x + 9)$ then the value of $-4x$ is
(a) 0 (b) 1
(c) 2 (d) $-1/4$

31. If $\log_{0.3}(x-1) < \log_{0.09}(x-1)$, then x lies in the interval
(a) $(-\infty, 1)$ (b) $(1, 2)$
(c) $(2, \infty)$ (d) None of these

32. $\log_{0.04}(x-1) \geq \log_{0.2}(x-1)$, then x belongs to the interval
(a) $(1, 2)$ (b) $(-\infty, 2]$
(c) $[2, +\infty)$ (d) None of these

33. If $x^{18} = y^{21} = z^{28}$, then $3, 3\log_y x, 3\log_z y, 7\log_x z$ are in
(a) AP (b) GP
(c) HP (d) AGP

34. If $\frac{1}{\log_3 \pi} + \frac{1}{\log_4 \pi} > x$, then x be
(a) 2 (b) 3
(c) π (d) None of these

35. If $\ln\left(\frac{a+b}{3}\right)=\left(\frac{\ln a+\ln b}{2}\right)$, then $\frac{a}{b}+\frac{b}{a}$ is equal to

(a) 1 (b) 3
(c) 5 (d) 7

36. If $\log_3\{5+4\log_3(x-1)\}=2$, then x is equal to

(a) 2 (b) 4
(c) 8 (d) $\log_2 16$

37. If $2x^{\log_4 3}+3^{\log_4 x}=27$, then x is equal to

(a) 2 (b) 4
(c) 8 (d) 16

38. The interval of x in which the inequality

$$5^{\frac{1}{4}(\log_5^2 x)} \geq 5x^{\frac{1}{5}(\log_5 x)}$$

(a) $(0, 5^{-2\sqrt{5}}]$ (b) $[5^{2\sqrt{5}}, \infty)$
(c) both (a) and (b) (d) None of these

39. The solution set of the equation
$\text{og}_x\ 2\log_{2x} 2=\log_{4x} 2$ is

(a) $\{2^{-\sqrt{2}}, 2^{\sqrt{2}}\}$ (b) $\{1/2, 2\}$
(c) $\{1/4, 2^2\}$ (d) None of these

40. The least value of the expression $2\log_{10} x-\log_x 0.01$ is

(a) 2 (b) 4
(c) 6 (d) 8

41. The solution of the equation
$\log_7\log_5(\sqrt{x+5}+\sqrt{x})=0$ is

(a) 1 (b) 3
(c) 4 (d) 5

42. The number of solutions of $\log_4(x-1)=\log_2(x-3)$ is

(a) 3 (b) 1
(c) 2 (d) 0

Objective Questions Type II [One or more than one correct answer(s)]

In each of the questions below four choices of which one or more than one are correct. You have to select the correct answer(s) accordingly.

1. It is given that $x=9$ is a solution of the equation
$\ln(x^2+15a^2)-\ln(a-2)=\ln\left(\frac{8ax}{a-2}\right)$, then

(a) $a=3$
(b) $a=\frac{3}{5}$
(c) other solution is $x=3/5$
(d) other solution is $x=15$

2. The equation $\log_{x+1}(x-0.5)+\log_{x-0.5}(x+1)$ has

(a) two real solutions (b) no prime solution
(c) one integral solution (d) no irrational solution

3. If $\frac{\log_2 x}{4}=\frac{\log_2 y}{6}=\frac{\log_2 z}{3k}$ and $x^3y^2z=1$, then k is equal to

(a) -8 (b) -4
(c) 0 (d) $\log_2\left(\frac{1}{256}\right)$

4. If $\log_2(3^{2x-2}+7)=2+\log_2(3^{x-1}+1)$ then x equals

(a) 0 (b) 1
(c) 2 (d) None of these

5. The expression

$$5^{\log_{1/5}(1/2)}+\log_{\sqrt{2}}\left(\frac{4}{\sqrt{7}+\sqrt{3}}\right)+\log_{1/2}\left(\frac{1}{10+2\sqrt{21}}\right)$$

simplifies to

(a) 6 (b) 4
(c) $\sqrt{6\sqrt{6\sqrt{6\sqrt{6\ldots\infty}}}}$ (d) $3^{\log_{1/3}\left(\frac{1}{6}\right)}$

6. If $\log_a x=\alpha$, $\log_b x=\beta$, $\log_c x=\gamma$ and $\log_d x=\delta$, $x\neq 1$ and $a, b, c, d\neq 0, >1$, then $\log_{abcd} x$ equals

(a) $\leq\frac{\alpha+\beta+\gamma+\delta}{16}$ (b) $\geq\frac{\alpha+\beta+\gamma+\delta}{16}$
(c) $\frac{1}{\alpha^{-1}+\beta^{-1}+\gamma^{-1}+\delta^{-1}}$ (d) $\frac{1}{\alpha\beta\gamma\delta}$

7. $\log_p\log_p\underbrace{\sqrt[p]{\sqrt[p]{\sqrt[p]{\ldots\sqrt[p]{p}}}}}_{n\text{ times}}$, $p>0$ and $p\neq 1$, is equal to

(a) n (b) $-n$
(c) $\frac{1}{n}$ (d) $\log_{1/p}(p^n)$

8. Sum of the roots of the equation
$x+1=2\log_2(2^x+3)-2\log_4(1980-2^{-x})$ is

(a) $\log_{11} 2$ (b) $\log_2 11$
(c) $\log_{11}(0.5)$ (d) $\log_{0.5}\left(\frac{1}{11}\right)$

9. If $a^x=b$, $b^y=c$, $c^z=a$, $x=\log_b a^{k_1}$, $y=\log_c b^{k_2}$, $z=\log_a c^{k_3}$, then $k_1\, k_2\, k_3$ is equal to

(a) 1 (b) abc
(c) (xyz) (d) 0

10. If $\frac{\ln a}{b-c}=\frac{\ln b}{c-a}=\frac{\ln c}{a-b}$, $a, b, c>0$, then

(a) $a^{b+c}\cdot b^{c+a}\cdot c^{a+b}=1$
(b) $a^{b+c}+b^{c+a}+c^{a+b}\geq 3$
(c) $a^{b+c}\cdot b^{c+a}\cdot c^{a+b}=3$
(d) $a^{b+c}+b^{c+a}+c^{a+b}\geq 3(3)^{1/3}$

11. If $\dfrac{\ln a}{y-z} = \dfrac{\ln b}{z-x} = \dfrac{\ln c}{x-y}$, then

(a) $a^x \cdot b^y \cdot c^z = 1$

(b) $a^{y^2+yz+z^2} \cdot b^{z^2+zx+x^2} \cdot c^{x^2+xy+y^2} = 1$

(c) $a^{y+z} \cdot b^{z+x} \cdot c^{x+y} = 1$

(d) $abc = 1$

12. The solution of the equation $3^{\log_a x} + 3x^{\log_a 3} = 2$ is given by

(a) $a^{\log_3 a}$

(b) $(2/a)^{\log_3 2}$

(c) $a^{-\log_3 2}$

(d) $2^{-\log_3 a}$

Linked-Comprehension Type

In these questions, a passage (paragraph) has been given followed by questions based on each of the passages. You have to answer the questions based on the passage given.

PASSAGE 1

Consider $E = 2^{\left(\sqrt{\log_a \sqrt[4]{ab} + \log_b \sqrt[4]{ab}} - \sqrt{\log_a \sqrt[4]{\frac{b}{a}} + \log_b \sqrt[4]{\frac{a}{b}}}\right)\sqrt{\log_a b}}$

On the basis of above information, answer the following questions

1. The value of E if $b \geq a \geq 1$ is

(a) 1 (b) 2
(c) $2^{\log_a b}$ (d) $2^{\log_a a}$

2. The value of E if $1 < b < a$ is

(a) 1 (b) 2
(c) $2^{\log_a b}$ (d) $2^{\log_b a}$

PASSAGE 2

An equation of the form $2m \log_a f(x) = \log_a g(x)$, $a > 0$, $a \neq 1$, $m \in N$

$\begin{cases} f(x) > 0 \\ f^{2m}(x) = g(x) \end{cases}$ is equivalent to the system.

On the basis of above information, answer the following questions

1. The number of solutions of $2\log_e 2x = \log_e(7x - 2 - 2x^2)$ is

(a) 1 (b) 2
(c) 3 (d) infinite

2. The number of solutions of $\ln 2x = 2\ln(4x - 15)$ is

(a) 0 (b) 1
(c) 2 (d) infinite

3. The number of solutions of $\log(3x^2 + x - 2) = 3\log(3x - 2)$ is

(a) 1 (b) 2
(c) 3 (d) 0

4. Solution set of the equation $\log_{(x^3+6)}(x^2 - 1) = \log_{(2x^2+5x)}(x^2 - 1)$ is

(a) $\{-2\}$ (b) $\{1\}$
(c) $\{3\}$ (d) $\{-2, 1, 3\}$

5. Solution set of the equation $\log(x - 9) + 2\log\sqrt{(2x - 1)} = 2$ is

(a) $\{\phi\}$ (b) $\{1\}$
(c) $\{2\}$ (d) $\{13\}$

Numerical Value Type

Solve the following problems and write your answer.

1. The expression $(81)^{1/\log_5 3} + (25)^{\log_5 36} + (3)^{4/\log_7 9}$ equals

2. If value of $\log_p \log_p \underbrace{\sqrt[p]{\sqrt[p]{\sqrt[p]{\sqrt[p]{\ldots\ldots\sqrt[p]{p}}}}}}_{2008 \text{ times}}$ is $-\lambda$, then the value of λ is

3. If the value of $\log_4(0.25) + \log_2(0.0625)$ is λ, then the value of -1250λ is

4. If the value of $\log_8 128 - \log_9\{\cot(\pi/3)\}$ is a, then the value of $1212a$ is

5. If the value of $\log_{0.75}\log_2 - 2\sqrt{0.125}$ is a and the value of $\log_{0.01} 1000 + \log_{0.1} 0.0001$ is b, then the value of $(125)^a + b$ must be

6. If $\lambda = \dfrac{\log_2 24}{\log_{96} 2} - \dfrac{\log_2 192}{\log_{12} 2}$ and if the last two digits of λ^{100} be xy, then the value of $32xy$ must be

7. If $\lambda = \log_{x^2}(\sqrt{x\sqrt{x\sqrt{x}}})$, then the value of 8λ must be

Single Integer Type

Solve the following problems and write your answer.

1. The value of $(0.16)^{\log_{2.5}\left(\frac{1}{3}+\frac{1}{3^2}+\frac{1}{3^3}+\dots\infty\right)}$ is equal to .... **[JEE Main 2020, 3 September]**
2. If $a = \log_{12} 18$, $b = \log_{24} 54$, then the value of $ab + 5(a - b)$ is
3. Let a and b be real numbers greater than 1 for which there exists a positive real number c, real different from 1, such that $2(\log_a c + \log_b c) = 9 \log_{ab} c$. Then, the largest possible value of $\log_a b$.
4. If $\log_a bc = x$, $\log_b ca = y$ and $\log_c ab = z$, then value of $\frac{1}{x+1} + \frac{1}{y+1} + \frac{1}{z+1}$ is
5. If (x_1, y_1) and (x_2, y_2) are solutions of the system of simultaneous equations $\log_8(xy) = 3\log_8 x . \log_8 y$ $4\log_9\left(\frac{x}{y}\right) = \frac{\log_8 x}{\log_8 y}$, then $x_1x_2 + y_1y_2$ equals to .....
6. If $a = \log_{245} 175$ and $b = \log_{1715} 875$, then the value of $\frac{1-a}{a-b}$ is ....
7. Let m be the minimum possible value of $\log_3(3^{y_1} + 3^{y_2} + 3^{y_3})$, where y_1, y_2, y_3 are real numbers for which $y_1 + y_2 + y_3 = 9$. Let M be the maximum possible value of $(\log_3 x_1 + \log_3 x_2 + \log_3 x_3)$, where x_1, x_2, x_3 are positive real numbers for which $x_1 + x_2 + x_3 = 9$. Then, the value of $\log_2(m^3) + \log_3(M^2)$ is .... **[JEE Advanced 2020]**

Matrix-Match Type

Given below are Matching Type Questions, with two columns (each having some items) each . Each item of Column I has to be matched with the items of Column II, by encircling the correct match(es).

NOTE *An item of Column I can be matched with more than one items of Column II.*

1. Match the following lists:

	Column I		Column II
(A)	The value of $\log_2\log_4 256 + \log_{\sqrt{2}} 4$ is	(P)	1
(B)	If $\log_3(5x-2) - 2\log_3\sqrt{3x+1} = 1 - \log_3 4$, then $x =$	(Q)	6
(C)	Product of roots of the equation $7^{\log_7(x^2-4x+5)} = (x-1)$ is	(R)	3
(D)	Number of integers satisfying $\log_2\sqrt{x} - 2(\log_{1/4} x)^2 + 1 > 0$ are	(S)	4
		(T)	5

2. Observe the following columns :

	Column I		Column II
(A)	If $\log_{34} 5$ lies in the interval (a, b), then	(P)	$[10a + 10b] = 8$, where [.] denotes the greatest integer function
(B)	If $\log_{300} 4$ lies in the interval (a, b), then	(Q)	$(10a + 10b) = 5$ where (.) denotes the least integer function
(C)	If $\log_{400} 3$ lies in the interval (a, b), then	(R)	$[6b - 3a] = 2$ where [.] denotes the greatest integer function
		(S)	$[10a + 10b] = 3$ where [.] denotes the greatest integer function
		(T)	$(6b - 3a) = 1$ where (.) denotes the least integer function

3. Observe the following columns :

	Column I		Column II
(A)	The solution set of $\log_{100}\lvert x+y\rvert = \frac{1}{2}$, $\log_{10} y - \log_{10}\lvert x\rvert = \log_{100} 4$ is	(P)	$\{\sqrt{2}, 2\}$
(B)	The solution set of $4\log_2^2 x + 1 = 2\log_2 y$ and $\log_2 x^2 \geq \log_2 y$	(Q)	$\{1, 1\}$
(C)	The solution set of $\log_4 x - \log_2 y = 0$ and $x^2 - 5y^2 + 4 = 0$	(R)	$\{-10, 20\}$
		(S)	$\{4, 2\}$
		(T)	$\left\{\frac{10}{3}, \frac{20}{3}\right\}$

Answers

Objective Questions Type I [Only one correct answer]

1. (a)	**2.** (a)	**3.** (b)	**4.** (b)	**5.** (a)	**6.** (c)	**7.** (c)	**8.** (b)	**9.** (a)	**10.** (d)
11. (d)	**12.** (c)	**13.** (c)	**14.** (a)	**15.** (c)	**16.** (c)	**17.** (b)	**18.** (c)	**19.** (a)	**20.** (b)
21. (c)	**22.** (c)	**23.** (d)	**24.** (b)	**25.** (c)	**26.** (a)	**27.** (a)	**28.** (c)	**29.** (c)	**30.** (b)
31. (b)	**32.** (c)	**33.** (a)	**34.** (a)	**35.** (d)	**36.** (b)	**37.** (d)	**38.** (c)	**39.** (a)	**40.** (b)
41. (c)	**42.** (b)								

Objective Questions Type II [One or more than one correct answer(s)]

1. (a, d)	**2.** (b, c,)	**3.** (a, d)	**4.** (b, d)	**5.** (a, c, d)
6. (a, c)	**7.** (b, d)	**8.** (b, d)	**9.** (a, c)	**10.** (a, b)
11. (a, b, c, d)	**12.** (c, d)			

Linked-Comprehension Type

Passage 1 **1.** (b) **2.** (c)
Passage 2 **1.** (b) **2.** (b) **3.** (d) **4.** (c) **5.** (d)
Passage 3 **1.** (c) **2.** (d) **3.** (c) **4.** (d) **5.** (a)

Numerical Value Type

1. 1970 **2.** 2008 **3.** 6250 **4.** 3131 **6.** 127.5 **7.** 3201 **8.** 3.5

Single Integer Type

1. (4) **2.** (1) **3.** (2) **4.** (1) **5.** (8) **6.** (5) **7.** (8)

Matrix-Match Type

1. A → (T); B → (P); C → (Q); D → (R)
2. A → (P, R); B → (Q, T); C → (S, T)
3. A → (R, T); B → (P); C → (Q, S)

Solutions

Objective Questions Type [Only one correct answer]

1. $\because \quad a = \dfrac{\log 3}{\log 2}, b = \dfrac{\log 5}{\log 3}$ and $c = \dfrac{\log 2}{\log 7}$

$\therefore \; abc = \dfrac{\log 5}{\log 7}, 2c = \dfrac{2\log 2}{\log 7}$ and $ac = \dfrac{\log 3}{\log 7}$ ....(i)

$$\therefore \log_{140} 63 = \frac{\log 63}{\log 140} = \frac{2\log 3 + \log 7}{\log 7 + 2\log 2 + \log 5}$$

$$= \frac{2\left(\dfrac{\log 3}{\log 7}\right) + 1}{1 + 2\left(\dfrac{\log 2}{\log 7}\right) + \dfrac{\log 5}{\log 7}}$$

$$= \frac{2ac + 1}{1 + 2c + abc} \quad \text{[from Eq. (i)]}$$

2. $\because \quad 1 - 5^{x-3} > 0$

$\Rightarrow \quad 5^{x-3} < 1$

$\Rightarrow \quad x - 3 < 0 \Rightarrow x < 3$

and $\quad 0.2 - 5^{x-4} > 0$

$\Rightarrow \quad 5^{x-4} < 0.2$

$\Rightarrow \quad 5^{x-4} < 5^{-1} \Rightarrow x < 3$

Hence, $\quad x < 3$ ...(i)

or $\log_5 (24 \times 5) + (x-3) - 2\log_5(1 - 5^{x-3})$
$= -\log_5 (0.2 - 5^{x-4})$

or $\log_5 24 + 1 + (x-3) - 2\log_5(1 - 5^{x-3})$
$= -\log_5 \{0.2(1 - 5^{x-3})\}$

or $\log_5 24 + x - 2 - 2\log_5 (1 - 5^{x-3}) = -\log_5 (0.2)$
$- \log_5 (1 - 5^{x-3})$

or $\quad \log_5 24 + x - 2 - \log_5 (1 - 5^{x-3}) = 1$

or $\quad \log_5 \left\{\dfrac{1 - 5^{x-3}}{24}\right\} = (x - 3)$

or $\quad \dfrac{1 - 5^{x-3}}{24} = 5^{x-3}$

or $\quad 1 = 25 \cdot 5^{x-3}$

or $\quad 1 = 5^{x-1}$

or $\quad x - 1 = 0 \Rightarrow x = 1$

3. $\because \quad \log_{x_1} \log_{x_2} \log_{x_3} \ldots . \log_{x_{n-1}} \log_{x_n} x_n^{x_{n-1}^{\cdot^{\cdot^{x_1}}}}$

$= \log_{x_1} \log_{x_2} \log_{x_3} \ldots \log_{x_{n-1}} (x_{n-1}^{x_{n-2}^{\cdots x_1}} \cdot \log_{x_n} x_n)$

$= \log_{x_1} \log_{x_2} \log_{x_3} \ldots \log_{x_{n-1}} x_{n-1}^{x_{n-2}^{\cdot^{\cdot^{x_1}}}}$

....

$= \log_{x_1} x_1 = 1.$

4. Let $x = \log_{20} 3 = \dfrac{\log 3}{\log 20} = \dfrac{\log 3}{\log 5 + \log 4}$

or $\quad \dfrac{1}{x} = \dfrac{\log 5}{\log 3} + \dfrac{\log 4}{\log 3} > 1 + 1$

$\therefore \quad \dfrac{1}{x} > 2$ or $x < \dfrac{1}{2}$ $(\because 5 > 3$ and $4 > 3)$ ...(i)

$\therefore \quad \dfrac{\log 5}{\log 3} > 1$ and $\dfrac{\log 4}{\log 3} > 1$

Also, $\quad x = \dfrac{3 \log 3}{3 \log 20} = \dfrac{1}{3} \cdot \dfrac{\log 27}{\log 20}$

$$> \frac{1}{3} \left(\begin{array}{l} \because 27 > 20 \Rightarrow \log 27 > \log 20 \\ \therefore \dfrac{\log 27}{\log 20} > 1 \end{array} \right)$$

or $\quad x > \dfrac{1}{3}$ ....(ii)

From Eqs. (i) and (ii), we get

$$\frac{1}{3} < x < \frac{1}{2}$$

$\therefore \quad x \in \left(\dfrac{1}{3}, \dfrac{1}{2}\right).$

5. $\because \quad \lambda > 0$

$\therefore \quad \log_{16} x = \dfrac{1 \pm \sqrt{(1 - 4 \log_{16} \lambda)}}{2}$

$\because$ The given equation will have exactly one solution, if

$1 - 4 \log_{16} \lambda = 0$ or $\log_{16} \lambda = \dfrac{1}{4} = 4^{-1}$

$\therefore \; \lambda = (16)^{4^{-1}} = (2^4)^{1/4} = 2, -2, 2i, -2i$, where $i = \sqrt{-1}$

But λ is real and positive

$\therefore \quad \lambda = 2$

Number of real values = 1.

6. $\because \quad y = 2^{\frac{1}{\log_x 4}}$

$\Rightarrow \quad \log_2 y = \dfrac{1}{\log_x 4}$ $[\because x > 0, x \neq 1]$

or $\quad \log_2 y = \log_4 x = \dfrac{1}{2} \log_2 x$

or $\quad 2 \log_2 y = \log_2 x$

or $\quad \log_2 y^2 = \log_2 x$

$\therefore \quad x = y^2.$

7. $\because \quad \log_2 x + \log_2 y \geq 6$

$\Rightarrow \quad \log_2 (xy) \geq 6$

$\therefore \quad xy \geq 2^6$

or $\quad \sqrt{xy} \geq 2^3$

$\because \quad \dfrac{x + y}{2} \geq \sqrt{xy}$ or $x + y \geq 2\sqrt{xy} \geq 16$ $(\because$ AM $\geq$ GM$)$

$\therefore \quad x + y \geq 16$

8. (b) Since, $\log_6 = 16 = 4 \log_6 2 = \dfrac{4}{\log_2 6} = \dfrac{4}{1 + \log_2 3}$ ...(i)

Now, $\quad a = \log_{12} 27 = 3 \log_{12}$

$3 = \dfrac{3}{\log_3 12} = \dfrac{3}{1 + 2 \log_3 2}$

$\therefore \quad a(1 + 2 \log_3 2) = 3$

$\Rightarrow \quad 2 \log_3 2 = \dfrac{3}{a} - 1 = \dfrac{3 - a}{a}$

$\Rightarrow \quad \dfrac{2}{\log_3 3} = \dfrac{3 - a}{a} \Rightarrow \log_2 3 = \dfrac{2a}{3 - a}$

Substituting in Eq. (i), we get that

$\log_6 16 = \dfrac{4}{1 + [2a/(3 - a)]} = 4\left(\dfrac{3 - a}{3 + a}\right)$

9. $\because x = \log_5 (1000) = \log_5 (5^3 \times 8) = 3 + \log_5 8$

and $y = \log_7 (2058) = \log_7 (7^3 \times 6) = 3 + \log_7 6$

$\Rightarrow x - y = \log_5 8 - \log_7 6 > 0$ $\left[\begin{array}{l} \because \log_5 8 > 1, \log_7 6 < 1 \\ \therefore \log_5 8 - \log_7 6 > 0 \end{array}\right]$

$\therefore \quad x > y$

10. $\because f(x_1) + f(x_2) = \ln\left(\dfrac{1 + x_1}{1 - x_1}\right) + \ln\left(\dfrac{1 + x_2}{1 - x_2}\right)$

$= \ln \left\{\dfrac{(1 + x_1)(1 + x_2)}{(1 - x_1)(1 - x_2)}\right\}$

$= \ln \left\{\dfrac{1 + (x_1 + x_2) + x_1 x_2}{1 - (x_1 + x_2) + x_1 x_2}\right\}$

$= \ln \left\{\dfrac{1 + \left(\dfrac{x_1 + x_2}{1 + x_1 x_2}\right)}{1 - \left(\dfrac{x_1 + x_2}{1 + x_1 x_2}\right)}\right\} = f\left(\dfrac{x_1 + x_2}{1 + x_1 x_2}\right).$

11. $\because \log_{10} 2, \log_{10} (2^x + 1), \log_{10} (2^x + 3)$ are in AP.

$\therefore \quad 2, 2^x + 1, 2^x + 3$ are in GP.

Then, $(2^x + 1)^2 = 2(2^x + 3)$

or $\quad 2^{2x} = 5$

or $\quad 2x = \log_2 5$

or $\quad x = \dfrac{1}{2} \log_2 5.$

12. $\quad A = \log_2 \log_2 \log_4 (4)^4 + 2 \log_{2^{1/2}} 2$

$= \log_2 \log_2 4 + 2\left(\dfrac{1}{1/2}\right)$

$= \log_2 \log_2 2^2 + 4$

$= \log_2 2 + 4 = 1 + 4 = 5$

13. $7 \log\left(\dfrac{2^4}{5 \times 3}\right) + 5 \log\left(\dfrac{5^2}{2^3 \times 3}\right) + 3 \log\left(\dfrac{3^4}{2^4 \times 5}\right)$

$= 7\{4 \log 2 - \log 5 - \log 3\} + 5\{2 \log 5 - 3 \log 2 - \log 3\}$
$+ 3\{4 \log 3 - 4 \log 2 - \log 5\}$

$= \log 2$

14. (a) Given series is

$S = \log_{(7^{1/2})} x + \log_{(7^{1/3})} x + \log_{(7^{1/4})} x + \ldots.. +$ upto 20 terms

$\Rightarrow S = 2\log_7 x + 3\log_7 x + 4\log_7 x + \ldots. +$ upto 20 terms

$\Rightarrow S = (2+3+4+\ldots.+21)\log_7 x = 460$ [given]

$\Rightarrow \left(\frac{21\times 22}{2} - 1\right)\log_7 x = 460$

$\Rightarrow (231-1)\log_7 x = 460 \Rightarrow \log_7 x = 2$

$\Rightarrow x = 7^2.$

15. Let $P = 3^{40}$

$\therefore \log_{10} P = 40\log_{10} 3 = 40 \times 0.477$
$= 19.08$

$\therefore$ Number of digits in $3^{40} = 19 + 1 = 20$

16. $\frac{1}{y} = \log_{25} 17 = \frac{1}{2}\log_5 17$

and $\frac{1}{x} = \log_5 3 = \frac{1}{2}\log_5 9$

$\therefore \frac{1}{y} > \frac{1}{x} \Rightarrow x > y$

17. (b) We have,

$\log_x(2x^2 + x - 1) > \log_x(2) - 1$ ...(i)

$\Rightarrow x > 0, x \neq 1$ and $2x^2 + x - 1 > 0$

$\Rightarrow x > 0, x \neq 1$ and $(2x-1)(x+1) > 0$

$\Rightarrow x > 1/2, x \neq 1$

We can write (1) as

$$\log_x\left(\frac{2x^2 + x - 1}{2}\right) > -1$$

For $1/2 < x < 1$, (ii) can be written as

$$\frac{2x^2 + x - 1}{2} < \frac{1}{x}$$

$\Rightarrow 2x^3 + x^2 - x < 2$

$\Rightarrow 2(x^3 - 1) + x(x-1) < 0$

$\Rightarrow (x-1)(2x^2 + 3x + 2) < 0$

$\Rightarrow x < 1$

$[\because 2x^2 + 3x + 2 > 0 \ \forall \ x > 0]$

For $x > 1$, (ii) can be written as

$$\frac{2x^2 + x - 1}{2} > \frac{1}{x}$$

$\Rightarrow (x-1)(2x^2 + 3x + 2) > 0$

This is true for each $x > 1$.

Thus, (i) holds for $1/2 < x < 1, x > 1$

18. Suppose $\log_2 7$ is a rational number, then

$\log_2 7 = \left(\frac{p}{q}\right)$

$\Rightarrow 7 = 2^{p/q} \quad \Rightarrow 7^q = 2^p$

odd = even (impossible)

$\therefore \log_2 7$ is an irrational number.

19. (a) We have $(3x)^{\log 3} = (4y)^{\log 4}, 4^{\log x} = 3^{\log y}$

$\Rightarrow (\log 3)(\log 3x) = (\log 4)(\log 4y)$ and
$(\log x)(\log 4) = (\log y)(\log 3)$

$\Rightarrow (\log 3)(\log 3 + \log x) = (\log 4)(\log 4 + \log y)$ and
$(\log x)(\log 4) = (\log y)(\log 3)$

$\Rightarrow (\log 3)(\log 3 + p) = (\log 4)(\log 4 + q)$ and
$p(\log 4) = q(\log 3)$

[where $p = \log x$ and $q = \log y$]

$\Rightarrow (\log 3)\left(\log 3 + \frac{q\log 3}{\log 4}\right) = \log 4(\log 4 + q)$ (eliminating p)

$\Rightarrow (\log 3)^2 - (\log 4)^2 = \frac{(\log 4)^2 - (\log 3)^2}{\log 4} q$

$\Rightarrow q = -\log 4 \Rightarrow \log y = \log 4^{-1} \Rightarrow y = 1/4$

Now $p(\log 4) = q(\log 3)$

$\Rightarrow p(\log 4) = -(\log 4)(\log 3)$

$\Rightarrow p = -\log 3$

$\Rightarrow \log x = \log 3^{-1}$

$\Rightarrow x = 1/3$

20. (b) We have,

$\log_{(x^2+6+8)}\log_{(2x^2+2x+3)}(x^2 - 2x) = 0$

$\log_{(2x^2+2x+3)}(x^2 - 2x) = (x^2 + 6x + 8)^0$

$\Rightarrow \log_{(2x^2+2x+3)}(x^2 - 2x) = 1$

$\Rightarrow x^2 - 2x = (2x^2 + 2x + 3)^1$

$\Rightarrow x^2 + 4x + 3 = 0$

$\Rightarrow (x+1)(x+3) = 0$

$\Rightarrow x = -1, -3$

$x = -1$ and $x = -3$ satisfy the condition $x^2 - 2x > 0$

But at $x = -3$, $x^2 + 6x + 8 = 9 + 6(-3) +$
$8 = -1 < 0$ which is not possible

Hence $x = -3$ is not the solution

$x = -1$ satisfies $x^2 + 6x + 8 > 0$ and $2x^2 + 2x + 3 > 0$

Also at $x = -1$, $x^2 + 6x + 8 \neq 1$ and $2x^2 + 2x + 3 \neq 1$

Hence $x = -1$ is the only solution.

21. $\because a = \log_{12} 27 = \log_{12}(3)^3 = 3\log_{12} 3$

$$= \frac{3}{\log_3 12} = \frac{3}{1 + \log_3 4} = \frac{3}{1 + 2\log_3 2}$$

$\therefore \log_3 2 = \frac{3-a}{2a}$

Then, $\log_6 16 = \log_6 2^4 = 4\log_6 2$

$$= \frac{4}{\log_2 6} = \frac{4}{1 + \log_2 3}$$

$$= \frac{4}{1 + \frac{2a}{3-a}} = 4\left(\frac{3-a}{3+a}\right)$$

22. $\log_7 \log_7 7^{\frac{1}{2}+\frac{1}{4}+\frac{1}{8}} = \log_7\left(\frac{1}{2} + \frac{1}{4} + \frac{1}{8}\right) = \log_7\left(\frac{7}{8}\right)$

$= 1 - \log_7 8 = 1 - 3\log_7 2$

23. $\because \quad \frac{\log x}{b-c} = \frac{\log y}{c-a} = \frac{\log z}{a-b}$

$= \frac{a \log x + b \log y + c \log z}{0}$

$\Rightarrow \log (x^a y^b z^c) = 0 \Rightarrow x^a y^b z^c = 1$

24. $\frac{1}{1+\log_a bc} + \frac{1}{1+\log_b ca} + \frac{1}{1+\log_c ab}$

$= \frac{\log a}{\log a + \log b + \log c} + \frac{\log b}{\log a + \log b + \log c} + \frac{\log c}{\log a + \log b + \log c}$

$= 1$

25. $(4)^{\log_{3^2} 3} + (9)^{\log_2 2^2} = (10)^{\log_x 83}$

$\Rightarrow \quad (4)^{1/2} + 9^2 = (10)^{\log_x 83}$

$\Rightarrow \quad (83)^1 = (83)^{\log_x 10}$

$\therefore \quad 1 = \log_x 10 \Rightarrow x = 10$

26. $a^x = b^y = c^z$

$\Rightarrow \quad x \log a = y \log b = z \log c$

$\therefore \quad \frac{y}{x} = \frac{z}{y} \Rightarrow \frac{\log a}{\log b} = \frac{\log b}{\log c}$

$\Rightarrow \quad \log_b a = \log_c b$

27. Let $\frac{x(y+z-x)}{\log x} = \frac{y(z+x-y)}{\log y} = \frac{z(x+y-z)}{\log z} = \frac{1}{\lambda}$

$\therefore \quad \log x = \lambda x (y+z-x)$, $\log y = \lambda y (z+x-y)$,

$\log z = \lambda z (x+y-z)$

$\therefore y \log x + x \log y = z \log x + x \log z = y \log z + z \log y$

$\Rightarrow \quad \log (x^y y^x) = \log (x^z z^x) = \log (y^z z^y)$

Hence, $\quad x^y y^x = x^z z^x = y^z z^y$

28. $\because 2^x > 5$

$2, 2^x - 5, 2^x - 7/2$ are in GP.

Then, $2^x = 4, 8, 2^x \neq 4 \quad [\because \; 2^x > 5]$

$\therefore \quad 2^x = 8 = 2^3$

Hence, $x = 3$

29. $\log_a y = \frac{1}{1-\log_a x}$

$\therefore \quad 1 - \log_a y = 1 - \frac{1}{1-\log_a x} = \frac{-\log_a x}{1-\log_a x}$

or $\quad \frac{1}{1-\log_a y} = \frac{1-\log_a x}{-\log_a x} \quad \ldots(i)$

But $\quad z = a^{\frac{1}{1-\log_a y}}$

$\Rightarrow \quad \log_a z = \frac{1}{1-\log_a y} = -\frac{1}{\log_a x} + 1$

$\Rightarrow \quad \frac{1}{\log_a x} = 1 - \log_a z$

$\log_a x = \frac{1}{1-\log_a z}$

$\therefore \quad x = a^{\frac{1}{1-\log_a z}}$

30. (b) Here, $2x+3>0$ and $2x+3 \neq 1$,

$\Rightarrow \quad x > -3/2$ and $x \neq -1$.

Also, $\quad 3x+7>0$ and $3x+7 \neq 1$,

$\Rightarrow x > -7/3$ and $x \neq -2$. Suppose $x > -3/2$, $x \neq -1$.

Then the given equation can be written as

$\frac{\log[(2x+3)(3x+7)]}{\log(2x+3)} = 4 - \frac{2\log(2x+3)}{\log(3x+7)}$

$\Rightarrow \quad + \frac{\log(3x+7)}{\log(2x+3)} = 4 - \frac{2\log(2x+3)}{\log(3x+7)}$

Let $\quad \frac{\log(3x+7)}{\log(2x+3)} = y$

Then, we get $\quad 1 + y = 4 - \frac{2}{y} \Rightarrow y = 3 - \frac{2}{y}$

$\Rightarrow \quad y^2 - 3y + 2 = 0$

$\Rightarrow \quad (y-1)(y-2) = 0$

$\Rightarrow \quad 1$ or 2

Case 1 Suppose that $y = 1$. Then,

$\log(3x+7) = \log(2x+3)$

$\Rightarrow \quad 3x + 7 = 2x + 3$

$\Rightarrow \quad x = -4$

This is rejected because $x > -3/2$.

Case 2 Suppose that $y = 2$. Then

$\log(3x+7) = 2\log(2x+3) = \log(2x+3)^2$

$\Rightarrow \quad 3x + 7 = 4x^2 + 12x + 9$

$\Rightarrow \quad 4x^2 + 9x + 2 = 0$

$\Rightarrow \quad (4x+1)(x+2) = 0$

$\Rightarrow \quad x = -1/4$ or -2

Here $x = -1/4$ (since $x > -3/2$). So, $-4x = 1$

31. $\because \quad x - 1 > 0 \Rightarrow x > 1$

and $\quad \log_{0.3}(x-1) > \log_{(0.3)^2}(x-1)$

$\Rightarrow \quad \log_{0.3}(x-1) > \frac{1}{2}\log_{0.3}(x-1)$

$\frac{1}{2}\log_{0.3}(x-1) > 0 \Rightarrow \log_{0.3}(x-1) > 0$

$\Rightarrow \quad x - 1 < 1, \therefore x < 2$

Then, $\quad x \in (1, 2)$

32. (c) We have, $\log_{0.04}(x-1) \geq \log_{0.2}(x-1) \quad \ldots(i)$

For log to be defined $x - 1 > 0 \Rightarrow x > 1$

From (i), $\log_{(0.2)^2}(x-1) \geq \log_{0.2}(x-1)$

$\Rightarrow \quad \frac{1}{2}\log_{0.2}(x-1) \geq \log_{0.2}(x-1)$

$\Rightarrow \quad \sqrt{x-1} \leq (x-1)$

$\Rightarrow \quad \sqrt{x-1}\,(1 - \sqrt{x-1}) \leq 0 \Rightarrow 1 - \sqrt{x-1} \leq 0$

$\Rightarrow \quad \sqrt{x-1} \geq 1 \Rightarrow x \geq 2, \therefore x \in [2, \infty)$

33. $\quad x^{18} = y^{21} = z^{28}$

$\Rightarrow \quad 18 \log x = 21 \log y = 28 \log z$

$\therefore \quad \log_y x = \frac{7}{6}, \log_z y = \frac{4}{3}, \log_x z = \frac{9}{14}$

$\therefore$ 3, $3\log_y x$, $3\log_z y$, $7\log_x z$,

$\Rightarrow \quad 3, 3\cdot\frac{7}{6}, 3\cdot\frac{4}{3}, 7\cdot\frac{9}{14}$ or $3, \frac{7}{2}, 4, \frac{9}{2}$ are in AP.

34. $\frac{1}{\log_3 \pi} + \frac{1}{\log_4 \pi} > x$

$\Rightarrow \quad \log_\pi 3 + \log_\pi 4 > x$

$\Rightarrow \quad \log_\pi 12 > x$

$\Rightarrow \quad 12 > \pi^x$

$\therefore \quad x = 2$

35. $\ln\left(\frac{a+b}{3}\right) = \frac{\ln ab}{2} = \ln \sqrt{ab}$

$\Rightarrow \quad \frac{a+b}{3} = \sqrt{ab}$

$\Rightarrow \quad a^2 + 2ab + b^2 = 9ab$

$\Rightarrow \quad \frac{a}{b} + 2 + \frac{b}{a} = 9$

$\therefore \quad \frac{a}{b} + \frac{b}{a} = 7$

36. $x - 1 > 0 \Rightarrow x > 1$ ...(i)

and $\quad 5 + 4\log_3(x-1) > 0$

$\Rightarrow \quad 4\log_3(x-1) > -5$

$\Rightarrow \quad \log_3(x-1) > -\frac{5}{4}$

$\Rightarrow \quad x - 1 > 3^{-5/4}$

$\Rightarrow \quad x > 1 + 3^{-5/4}$ ...(ii)

From Eqs. (i) and (ii), $x > 1 + 3^{-5/4}$

$\therefore \quad 5 + 4\log_3(x-1) = 9$

$\Rightarrow \quad 4\log_3(x-1) = 4$

$\Rightarrow \quad \log_3(x-1) = 1$

$\Rightarrow \quad x - 1 = 3$

$\therefore \quad x = 4$

37. $2x^{\log_4 3} + 3^{\log_4 x} = 27$

$\Rightarrow \quad 2 \cdot 3^{\log_4 x} + 3^{\log_4 x} = 27$

$\Rightarrow \quad 3^{\log_4 x} = 9 = 3^2$

$\log_4 x = 2$

$\therefore \quad x = 4^2 = 16$

38. $5^{\frac{1}{4}(\log_5^2 x)} \geq 5x^{\frac{1}{5}(\log_5 x)}$

Taking logarithm on base 5, then

$\left(\frac{1}{4}\right)(\log_5^2 x) \geq 1 + \frac{1}{5}(\log_5 x)(\log_5 x)$

$\Rightarrow \quad \frac{1}{20}\log_5^2 x \geq 1$

or $\quad (\log_5 x)^2 \geq 20$

or $\quad \log_5 x \geq 2\sqrt{5}$ and $\log_5 x \leq -2\sqrt{5}$

or $\quad x \geq 5^{2\sqrt{5}}$ and $x \leq 5^{-2\sqrt{5}}$

But $\quad x > 0$

$\therefore \quad x \in (0, 5^{-2\sqrt{5}}] \cup [5^{2\sqrt{5}}, \infty)$

39. $\because \quad \log_x 2 \log_{2x} 2 = \log_{4x} 2$

$\because \quad x > 0,\ 2x > 0$ and $4x > 0$

and $\quad x \neq 1,\ 2x \neq 1,\ 4x \neq 1$

$\Rightarrow \quad x > 0$ and $x \neq 1, \frac{1}{2}, \frac{1}{4}$

Then, $\quad \frac{1}{\log_2 x} \cdot \frac{1}{\log_2 2x} = \frac{1}{\log_2 4x}$

$\Rightarrow \quad \log_2 x \cdot \log_2 2x = \log_2 4x$

$\Rightarrow \quad \log_2 x \cdot (1 + \log_2 x) = (2 + \log_2 x)$

$\Rightarrow \quad (\log_2 x)^2 = 2$

$\Rightarrow \quad \log_2 x = \pm\sqrt{2}$

$\therefore \quad x = 2^{\pm\sqrt{2}}$

$\therefore \quad x = \{2^{-\sqrt{2}}, 2^{\sqrt{2}}\}$

40. $\because 2\log_{10} x - \log_x(0.01) = 2\log_{10} x - \log_x(10^{-2})$

$= 2(\log_{10} x + \log_x 10) \quad [\because x > 0 \text{ and } x \neq 1]$

$= 2\left(\frac{\log_e x}{\log_e 10} + \frac{\log_e 10}{\log_e x}\right) \geq 2 \cdot 2 \quad [\because \text{AM} \geq \text{GM}]$

$= 4$

$\therefore \quad 2\log_{10} x - \log_x(0.01) \geq 4$

41. It is clear that $x \geq 0$ and $x \geq -5$

ie, $\quad x \geq 0$

$\because \quad \log_7 \log_5(\sqrt{x+5} + \sqrt{x}) = 0$

$\Rightarrow \quad \log_5(\sqrt{(x+5)} + \sqrt{x}) = 1$

$\Rightarrow \quad \sqrt{(x+5)} + \sqrt{x} = 5$

or $\quad \sqrt{(x+5)} = 5 - \sqrt{x}$

or $\quad (x+5) = (5 - \sqrt{x})^2$

$\Rightarrow \quad x + 5 = 25 + x - 10\sqrt{x}$ or $10\sqrt{x} = 20$

$\Rightarrow \quad \sqrt{x} = 2$

$\therefore \quad x = 4$

42. $\log_4(x-1) = \log_2(x-3)$ ...(i)

$\because \quad x - 1 > 0$ and $x - 3 > 0$

$\Rightarrow \quad x > 3$

From Eq. (i),

$\Rightarrow \quad \log_{2^2}(x-1) = \log_2(x-3)$

$\Rightarrow \quad \frac{1}{2}\log_2(x-1) = \log_2(x-3)$

$\Rightarrow \quad \log_2(x-1) = 2\log_2(x-3)$

$\Rightarrow \quad \log_2(x-1) = \log_2(x-3)^2$

or $\quad x - 1 = (x-3)^2$

$\Rightarrow \quad x^2 - 7x + 10 = 0$

$\therefore \quad x = 2, 5$

$\therefore \quad x = 5 \quad [\because x > 3]$

Objective Questions Type II [One or more than one correct answer(s)]

1. $\because \quad a-2>0$ and $\dfrac{8ax}{a-2}>0$

Then, $\quad a>2$ and $x>0 \qquad$ ..(i)

Then, the given equation can be written as

$$\ln(x^2+15a^2)=\ln\left\{\frac{8ax}{(a-2)}\cdot(a-2)\right\}$$

or $\quad \ln(x^2+15a^2)=\ln(8ax)$

or $\quad x^2+15a^2=8ax$

or $\quad x^2-8ax+15a^2=0$

or $\quad (x-5a)(x-3a)=0$

$\therefore \quad x=3a,\ 5a$

But $x=9$, then $9=3a,\ 5a$

$\therefore \quad a=3,\ 9/5$

or $\quad a=3 \qquad [\because a>2]$

Then, $x=9,\ 15$

2. (b, c, d) We have, $\dfrac{\log_2(x-0.5)}{\log_2(x+1)}=\dfrac{\log_2(x+1)}{\log_2(x-0.5)}$

$\Rightarrow \quad [\log_2(x+1)]^2=[\log_2(x-0.5)]^2$

$\therefore \quad \log_2(x+1)=\pm\log_2(x-0.5)$

If $\log_2(x+1)=\log_2(x-0.5)$.

Then, $x+1=x-0.5$, hence no solution

If $\log_2(x+1)=\log_2(x-0.5)^{-1}$.

Then, $\quad x+1=\dfrac{1}{x-(1/2)}=\dfrac{2}{2x-1}$

$\Rightarrow \quad (x+1)(2x-1)=2$

$\Rightarrow \quad 2x^2+x-3=0$

$\Rightarrow \quad 2x^2+3x-2x-3=0$

$\Rightarrow \quad (x-1)(2x+3)=0$

$\Rightarrow \quad x=1\ (x=-3/2$ is rejected)

3. $\dfrac{\log_2 x}{4}=\dfrac{\log_2 y}{6}=\dfrac{\log_2 z}{3k}$

$$=\frac{3\log_2 x+2\log_2 y+\log_2 z}{12+12+3k}$$

$$=\frac{\log_2(x^3y^2z)}{24+3k}=\frac{0}{24+3k}$$

$\therefore\ 24+3k=0$, then $k=-8$

4. (b, c) The given expression can be written as

$$\log_2(3^{2x-2}+7)=\log_2 4(3^{x-1}+1)$$

$\Rightarrow \quad 3^{2x-2}+7=4(3^{x-1}+1)$

$\Rightarrow \quad (3^{x-1})^2-4(3^{x-1})+3=0$

$\Rightarrow \quad (3^{x-1}-1)(3^{x-1}-3)=0$

$\Rightarrow \quad 3^{x-1}=1,\ 3^{x-1}=3$

$\Rightarrow \quad x-1=0,\ x-1=1$

$\Rightarrow \quad x=1,\ 2$

5. $\because \quad 5^{\log_{1/5}(1/2)}=5^{\log_5(2)}=2$

$$\log_{\sqrt{2}}\left(\frac{4}{\sqrt{7}+\sqrt{3}}\right)=\log_{\sqrt{2}}\left(\frac{4(\sqrt{7}-\sqrt{3})}{4}\right)=2\log_2(\sqrt{7}-\sqrt{3})$$

$$=\log_2(\sqrt{7}-\sqrt{3})^2=\log_2(10-2\sqrt{21})$$

and $\log_{1/2}\left(\dfrac{1}{10+2\sqrt{21}}\right)=\log_2(10+2\sqrt{21})$

$\therefore$ Given expression can be written as

$$=2+\log_2(10-2\sqrt{21})+\log_2(10+2\sqrt{21})$$

$$=2+\log_2\{(10-2\sqrt{21})(10+2\sqrt{21})\}$$

$$=2+\log_2(16)=2+4=6$$

Also, $\sqrt{6\sqrt{6\sqrt{6\sqrt{6\ldots}}}}=6^{\frac{1}{2}+\frac{1}{4}+\frac{1}{8}+\ldots}=6^{\frac{\frac{1}{2}}{1-1/2}}=6$

and $\quad 3^{\log_{1/3}(1/6)}=3^{\log_3(6)}=6$

6. $\because \log_{abcd} x=\dfrac{1}{\log_x abcd}$

$$=\frac{1}{\log_x a+\log_x b+\log_x c+\log_x d}=\frac{1}{\alpha^{-1}+\beta^{-1}+\gamma^{-1}+\delta^{-1}}$$

Also, AM $\geq$ GM

$\Rightarrow \quad \dfrac{\alpha+\beta+\gamma+\delta}{4}\geq\dfrac{4}{\dfrac{1}{\alpha}+\dfrac{1}{\beta}+\dfrac{1}{\gamma}+\dfrac{1}{\delta}}$

or $\quad \dfrac{1}{\alpha^{-1}+\beta^{-1}+\gamma^{-1}+\delta^{-1}}\leq\dfrac{\alpha+\beta+\gamma+\delta}{16}$

7. $\because \quad \underbrace{\sqrt[p]{\sqrt[p]{\sqrt[p]{\ldots\sqrt[p]{p}}}}}_{n\text{times}}=p^{1/p^n}=p^{p^{-n}}$

$$\therefore \log_p\log_p\underbrace{\sqrt[p]{\sqrt[p]{\sqrt[p]{\ldots\sqrt[p]{p}}}}}_{n\text{times}}=\log_p\log_p p^{p^{-n}}=\log_p(p^{-n})=-n$$

Also, $\log_{1/p}(p^n)=\log_{p^{-1}}(p^n)=\dfrac{n}{-1}=-n$

8. $\because \quad x+1=2\log_2(2^x+3)-2\log_{2^2}(1980-2^{-x})$

$\Rightarrow \quad x+1=\log_2(2^x+3)^2-\log_2(1980-2^{-x})$

$\Rightarrow \quad x+1=\log_2\left\{\dfrac{(2^x+3)^2}{1980-2^{-x}}\right\}$

$\Rightarrow \quad 2^x\cdot 2=\dfrac{2^{2x}+6\cdot 2^x+9}{1980-2^{-x}}$

$\Rightarrow \quad 2\times 1980\times 2^x-2=2^{2x}+6\cdot 2^x+9$

$\Rightarrow \quad 2^{2x}-3954\cdot 2^x+11=0$

If roots are α,β, then

$$2^{\alpha}\cdot 2^{\beta}=\frac{11}{1}\Rightarrow 2^{\alpha+\beta}=11$$

or $\quad \alpha+\beta=\log_2 11$

Also, $\quad \log_{0.5}\left(\dfrac{1}{11}\right)=\log_{1/2}\left(\dfrac{1}{11}\right)=\log_2 11$

9. $\quad a^x=b\Rightarrow x=\log_a b$

$\quad b^y=c\Rightarrow y=\log_b c$

and $\quad c^z=a\Rightarrow z=\log_c a$

$\therefore \quad xyz=1$

and $\quad x=k_1\log_b a,\ y=k_2\log_c b,\ z=k_3\log_a c$

$\therefore \quad xyz=k_1k_2k_3$

or $\quad k_1k_2k_3=xyz=1$

10. Since, $a > 0, b > 0, c > 0$

$$\Rightarrow \frac{(b+c)\ln a}{(b+c)(b-c)} = \frac{(c+a)\ln b}{(c+a)(c-a)}$$

$$= \frac{(a+b)\ln c}{(a+b)(a-b)}$$

$$= \frac{(b+c)\ln a + (c+a)\ln b + (a+b)\ln c}{0}$$

[using ratio and proportion]

$$= \frac{\ln a^{b+c} + \ln b^{c+a} + \ln c^{a+b}}{0}$$

$$\therefore \quad \ln a^{b+c} + \ln b^{c+a} + \ln c^{a+b} = 0$$

$$\Rightarrow \quad \ln (a^{b+c} \cdot b^{c+a} \cdot c^{a+b}) = 0$$

$$\Rightarrow \quad a^{b+c} \cdot b^{c+a} \cdot c^{a+b} = e^0 = 1$$

Again, AM $\geq$ GM

$$\Rightarrow \quad \frac{a^{b+c} + b^{c+a} + c^{a+b}}{3} \geq (a^{b+c} \cdot b^{c+a} \cdot c^{a+b})^{1/3}$$

$$= (1)^{1/3} = 1$$

$$\therefore \quad a^{b+c} + b^{c+a} + c^{a+b} \geq 3$$

11. $\because \frac{\ln a}{y-z} = \frac{\ln b}{z-x} = \frac{\ln c}{x-y} = \frac{\ln a + \ln b + \ln c}{0} = \frac{\ln (abc)}{0}$

or $\quad \ln (abc) = 0 \Rightarrow abc = e^0 = 1$

$$\frac{\ln a}{y-z} = \frac{\ln b}{z-x} = \frac{\ln c}{x-y}$$

$$= \frac{x\ln a + y\ln b + z\ln c}{0} = \frac{\ln (a^x b^y c^z)}{0}$$

or $\quad \ln (a^x \cdot b^y \cdot c^z) = 0 \Rightarrow a^x b^y c^z = e^0 = 1$

$$\frac{\ln a}{y-z} = \frac{\ln b}{z-x} = \frac{\ln c}{x-y}$$

$$= \frac{(y+z)\ln a + (z+x)\ln b + (x+y)\ln c}{0}$$

$$= \frac{\ln \{a^{y+z} \cdot b^{z+x} \cdot c^{x+y}\}}{0}$$

or $\ln \{a^{y+z} \cdot b^{z+x} \cdot c^{x+y}\} = 0$

$$\Rightarrow \quad a^{y+z} \cdot b^{z+x} \cdot c^{x+y} = e^0 = 1$$

and $\quad \frac{\ln a}{y-z} = \frac{\ln b}{z-x} = \frac{\ln c}{x-y}$

$$= \frac{(y^2+yz+z^2)\ln a + (z^2+zx+x^2)\ln b + (x^2+xy+y^2)\ln c}{0}$$

$$= \frac{\ln \{a^{y^2+yz+z^2} \cdot b^{z^2+zx+x^2} \cdot c^{x^2+xy+y^2}\}}{0}$$

or $\quad \ln \{a^{y^2+yz+z^2} \cdot b^{z^2+zx+x^2} \cdot c^{x^2+xy+y^2}\} = 0$

or $\quad a^{y^2+yz+z^2} \cdot b^{z^2+zx+x^2} \cdot c^{x^2+xy+y^2} = 1$

12. $\Rightarrow 3^{\log_a x} + 3 \cdot x^{\log_a 3} = 2$

$$\Rightarrow 3^{\log_a x} + 3 \cdot 3^{\log_a x} = 2 \Rightarrow 4 \cdot 3^{\log_a x} = 2$$

$$\Rightarrow \quad 3^{\log_a x} = \left(\frac{1}{2}\right)$$

$$\Rightarrow \quad \log_a x = -\log_3 2$$

$$\Rightarrow \quad x = a^{-\log_3 2} = a^{\log_3 (2^{-1})}$$

$$= (2^{-1})^{\log_3 a} = 2^{-\log_3 a}$$

Linked Comprehension Type

PASSAGE 1

1. (b) We have,

$$E = 2^{\left(\sqrt{\log_a \sqrt[4]{ab} + \log_b \sqrt[4]{ab}} - \sqrt{\log_a \sqrt[4]{\frac{b}{a}} + \log_b \sqrt[4]{\frac{a}{b}}}\right)\sqrt{\log_a b}}$$

$$= 2^{\frac{1}{2}(\sqrt{\log_a ab + \log_b ab} - \sqrt{\log_a b/a + \log_b a/b})\sqrt{\log_a b}}$$

$$= 2^{\frac{1}{2}(\sqrt{2 + \log_a b + \log_b a} - \sqrt{\log_a b + \log_b a - 2})\sqrt{\log_a b}}$$

$$= 2^{\frac{1}{2}(\sqrt{(\log_a b^2) + 2\log_a b + 1} - \sqrt{(\log_a b)^2 - 2\log_a b + 1})}$$

$$= 2^{\frac{1}{2}(\sqrt{(\log_a b + 1)^2} - \sqrt{(\log_a (b-1)^2})}$$

$$= 2^{\frac{1}{2}(|\log_a b + 1| - |\log_a b - 1|^2)}$$

When,

$$b \geq a > 1$$

$$\Rightarrow \log_a b \geq \log_a a$$

$$\Rightarrow \log_a b \geq 1$$

$$\Rightarrow E = 2^{\frac{1}{2}(\log_2 b + 1 - \log_a b + 1)} = 2$$

2. (c) When

$$1 < b < a$$

$$\Rightarrow 0 < \log_a b < \log_a a$$

$$\Rightarrow 0 < \log_a b < 1$$

$$\Rightarrow \quad E = 2^{\frac{1}{2}(\log_a b + 1 - 1 + \log_a b)}$$

$$= 2^{1/2(2\log_a b)} = 2\log_a b$$

PASSAGE 2

1. This equation is equivalent to the system

$$\begin{cases} 2x > 0 \\ (2x)^2 = 7x - 2 - 2x^2 \end{cases}$$

$$\Rightarrow \begin{cases} x > 0 \\ 6x^2 - 7x + 2 = 0 \end{cases}$$

$$\Rightarrow \begin{cases} x > 0 \\ (x - 1/2)(x - 2/3) = 0 \end{cases}$$

$$\Rightarrow \begin{cases} x = \frac{1}{2} \\ x = \frac{2}{3} \end{cases}$$

$\therefore$ Number of solutions $= 2$.

2. This equation is equivalent to the system

$$\begin{cases}4x-15>0\\2x=(4x-15)^2\end{cases}\Rightarrow\begin{cases}x>\dfrac{15}{4}\\2x=16x^2-120x+225\end{cases}$$

$$\Rightarrow\begin{cases}x>\dfrac{15}{4}\\16x^2-122x+225=0\end{cases}$$

$$\therefore\qquad x=\frac{9}{2}$$

$\therefore$ Number of solutions = 1.

3. This equation is equivalent to the system

$$\begin{cases}3x^2+x-2>0\\3x^2+x-2=(3x-2)^3\end{cases}$$

$$\Rightarrow\begin{cases}(x-2/3)(x+1)>0\\(x-2/3)(9x^2-13x+3)=0\end{cases}$$

$$\Rightarrow\begin{cases}x<2/3\text{ and }x<-1\\x=2/3,\ x=\dfrac{13\pm\sqrt{61}}{18}\end{cases}$$

$\therefore$ No root.

4. This equation is equivalent to the system

$$\begin{cases}x^2-1>0\\2x^2+5x>0,\neq 1\\x^3+6=2x^2+5x\end{cases}$$

$$\Rightarrow\begin{cases}x<-1\text{ and }x>1\\x<-\dfrac{5}{2}\text{ and }x>0\\x\neq\dfrac{-5\pm\sqrt{33}}{4}\\x=-2,1,3\end{cases}$$

For final solution $x<-\dfrac{5}{2}$ and $x>1$

i.e. Only $x=3$.

5. $x-9>0,\ 2x-1>0$ and $(x-9)(2x-1)=10^2=100$

or $x>9$ and $2x^2-19x-91=0$

or $x>9$ and $2x^2-26x+7x-91=0$

or $x=13$.

Numerical Value Type

1. $(81)^{1/\log_5 3}=(81)^{\log_3 5}$

$$=3^{4\log_3 5}=3^{\log_3 5^4}=5^4=625$$

$$(25)^{\log_5 36}=5^{2\log_5 6^2}=5^{4\log_5 6}=6^4=1296$$

$$(3)^{4/\log_7 9}=3^{4\log_9 7}=3^{4\log_{3^2}7}=3^{2\log_3 7}$$

$$=3^{\log_3 7^2}=7^2=49$$

$\therefore$ Value of expression $=1970$

2. $\because$ $\underbrace{\sqrt[p]{p\sqrt[p]{p\sqrt[p]{\ldots\ldots p\sqrt{p}}}}}_{2008\text{ times}}=(p)^{\frac{1}{p^{2008}}}=p^{p^{-2008}}$

$$\therefore\qquad \log_p\log_p\underbrace{\sqrt[p]{\sqrt[p]{\sqrt[p]{\ldots\ldots\sqrt[p]{p}}}}}_{2008\text{ times}}=\log_p\log_p\left(p^{p^{-2008}}\right)$$

$$=\log_p(p^{-2008})=-2008$$

$$\therefore\qquad \lambda=2008$$

3. $\because$ $\lambda=\log_4(0.25)+\log_2(0.0625)$

$$=\log_4(4^{-1})+\log_2\left(\frac{1}{2}\right)^4=-1-4=-5$$

$\therefore$ $-1250\lambda=-1250\times-5=6250$

4. $a=\log_8 128-\log_9(\cot\pi/3)$

$$=\log_{2^3}(2)^7-\log_{3^2}\left(\frac{1}{\sqrt{3}}\right)=\log_{2^3}(2)^7+\log_{3^2}(3^{1/2})$$

$$=\frac{7}{3}\log_2 2+\frac{\frac{1}{2}}{2}\log_3 3=\frac{7}{3}+\frac{1}{4}=\frac{31}{12}$$

$\therefore$ $1212a=1212\times\dfrac{31}{12}$

$$=101\times31=3131$$

5. $\because$ $a=\log_{0.75}\log_2\sqrt{\sqrt[-2]{0.125}}=\log_{(3/4)}\log_2(0.125)^{-1/4}$

$$=\log_{(3/4)}\log_2(2)^{3/4}=\log_{3/4}(3/4)=1$$

and $\quad b=\log_{0.01}1000+\log_{0.1}0.0001$

$$=\log_{10^{-2}}(10)^3+\log_{10^{-1}}(10)^{-4}=-\frac{3}{2}+4=\frac{5}{2}$$

$$(125)^a+b=125+\frac{5}{2}=125+2.5=127.5$$

6. $\because\lambda=\log_2 24\times\log_2 96-\log_2 192\times\log_2 12$

$=[\log_2(2\times12)][\log_2(2^3\times12)]-[\log_2(2^4\times12)]\times\log_2 12$

$=(1+\log_2 12)(3+\log_2 12)-(4+\log_2 12)\log_2 12$

Let $\log_2 12=a$, then

$$\lambda=(1+a)(3+a)-(4+a)a$$

$$=3+a+3a+a^2-4a-a^2$$

$$=3$$

$$\therefore\quad \lambda^{100}=3^{100}=(3^2)^{50}=(10-1)^{50}$$

$$=(10)^{50}-{}^{50}C_1(10)^{49}+{}^{50}C_2(10)^{48}\ldots$$

$$+{}^{50}C_{48}(10)^2-{}^{50}C_{49}(10)+1$$

$$=100\mu-50\times10+1\ (\mu\text{ is integer})$$

$$=100(\mu-5)+1=1000+1$$

$\therefore$ Last two digits are 01.

Hence, value of $32xy$ is 3201.

7. $\because$ $\lambda=\log_{x^2}\left(\sqrt{x\sqrt{x\sqrt{x}}}\right)$

$$=\log_{x^2}\left(x^{\frac{1}{2}+\frac{1}{4}+\frac{1}{8}}\right)$$

$$=\log_{x^2}(x^{7/8})=\frac{7/8}{2}=\frac{7}{16}$$

$$\therefore\qquad 8\lambda=8\times\frac{7}{16}=\frac{7}{2}=3.5$$

Single Integer Type

1. Since, $(0.16)^{\log_{2.5}\left(\frac{1}{3}+\frac{1}{3^2}+\frac{1}{3^3}+\ldots \text{to } \infty\right)}$

$$=(0.16)^{\log_{2.5}\left(\frac{1/3}{1-1/3}\right)}=(0.16)^{\log_{2.5}\left(\frac{1}{2}\right)}$$

$$=\frac{16}{100}^{\log_{2.5}(2)^{-1}}=\left(\frac{100}{16}\right)^{\log_{2.5}2}$$

$$=\left(\frac{10}{4}\right)^{2\log_{2.5}2}=(2.5)^{\log_{2.5}2^2}=2^2=4$$

Hence, answer is 4.00.

2. We have,

$$a=\log_{12}18=\frac{\log_2 18}{\log_2 12}=\frac{1+2\log_2 3}{2+\log_2 3}$$

and $b=\log_{24}54=\dfrac{\log_2 54}{\log_2 24}=\dfrac{1+3\log_2 3}{3+\log_2 3}$

On Putting $x=\log_2 3$, we have

$$ab+5(a-b)=\frac{1+2x}{2+x}\cdot\frac{1+3x}{3+x}+5\left(\frac{1+2x}{2+x}-\frac{1+3x}{3+x}\right)$$

$$=\frac{6x^2+5x+1+5(-x^2+1)}{(x+2)(x+3)}$$

$$=\frac{x^2+5x+6}{(x+2)(x+3)}=1$$

3. (2) We have, $a>1, b>1$

$$2(\log_a c+\log_b c)=9\log_{ab}c$$

$$\Rightarrow \quad 2\left[\frac{\log c}{\log a}+\frac{\log c}{\log b}\right]=\frac{9\log c}{\log ab}$$

$$\Rightarrow \quad 2\left\{\log c\left[\frac{\log n+\log a}{\log a\log b}\right]\right\}=\frac{9\log c}{\log a+\log b}$$

$$\Rightarrow \quad 2(\log a+\log b)^2=9(\log a)(\log b)$$

$$\Rightarrow \quad 2(\log a)^2+2(\log b)^2+4(\log a)(\log b)=9(\log a)(\log b)$$

$$\Rightarrow \quad 2\log_b a+2\log_a b+4=9$$

$$\Rightarrow \quad 2\log_b a+2\log_a b=5$$

$$\Rightarrow \quad t+\frac{1}{t}=\frac{5}{2}, \text{ where } t=\log_a b$$

$$\Rightarrow \quad 2t^2-5t+2=0$$

$$\Rightarrow \quad (2t-1)(t-2)=0$$

$$\Rightarrow \quad t=1/2, t=2$$

$$\Rightarrow \log_a b=1/2 \text{ or } \log_a b=2$$

4. (1) Given, Since $\log_a bc=x$

$\therefore x+1=\log_a bc+1=\log_a bc+\log_a a=\log_a(abc)$

Similarly, we have $y+1=\log_a abc$, $z+1=\log_a abc$

$$\therefore \quad \frac{1}{x+1}=\frac{1}{\log_a abc}=\log_{abc}a \quad [\text{since } \log_a b\cdot\log_b a=1]$$

and, similarly $\dfrac{1}{y+1}=\log_{abc}b$, $\dfrac{1}{z+1}=\log_{abc}c$

$$\therefore \quad \frac{1}{x+1}+\frac{1}{y+1}+\frac{1}{z+1}$$

$$=\log_{abc}a+\log_{abc}b+\log_{abc}c$$

$$=\log_{abc}(abc)=1$$

5. Here, clearly $x>0$, $y>0$ and $y\neq 1$. The given equations can be written as

$$\log_8 x+\log_8 y=3\log_8 x\log_8 y$$

$$4(\log_8 x-\log_8 y)=\log_8 x/\log_8 y$$

Put $\log_8 x=m$ and $\log_8 y=n\neq 0$.

Then, the above equation are

$$\left.\begin{aligned}m+n&=3mn\\4(m-n)&=m/n\end{aligned}\right\} \quad \ldots \text{(i)}$$

On Multiplying both the equations we get

$$4(m^2-n^2)=3m^2$$

Therefore, $m^2=4n^2$ or $m=\pm 2n$

Putting $m=2n$ in Eq. (i), we get that

$3n=6n^2$ or $n=\dfrac{1}{2}$ (since $n\neq 0$) and $m=1$

Now $m=1\Rightarrow \log_8 x=1\Rightarrow x=8$

$$n=\frac{1}{2}\Rightarrow \log_8 y=\frac{1}{2}\Rightarrow y=2\sqrt{2}$$

$$\therefore \quad x_1=8,\ y_1=2\sqrt{2}$$

Again by taking $m=-2n$, we get that

$$n=6n^2 \text{ or } n=1/6 \text{ and } m=-1/3$$

$$-1/3=m=\log_8 x\Rightarrow x=8^{-1/3}=(2^3)^{-1/3}=\frac{1}{2}$$

$$1/6=n=\log_8 y$$

$$\Rightarrow \quad y=8^{1/6}=(2^3)^{1/6}=\sqrt{2}$$

For $x_2=1/2$ and $y_2=\sqrt{2}$. Therefore,

$$x_1x_2+y_1y_2=8\times\frac{1}{2}+2\sqrt{2}\times\sqrt{2}=4+4=8$$

6. (5) We have, $a=\dfrac{\log_5 175}{\log_5 245}=\dfrac{2+\log_5 7}{1+2\log_5 7}$

$$\Rightarrow \quad a+2a\log_5 7=2+\log_5 7$$

$$\Rightarrow \quad \log_5 7=\frac{a-2}{1-2a}$$

Now, $b=\dfrac{\log_5 875}{\log_5 1715}=\dfrac{3+\log_5 7}{1+3\log_5 7}$

$$\Rightarrow \quad b+3b\log_5 7=3+\log_5 7$$

$$\Rightarrow \quad \log_5 7=\frac{b-3}{1-3b}$$

From Eqs. (i) and (ii), we get

$$\frac{a-2}{1-2a}=\frac{b-3}{1-3b}\Rightarrow \frac{1-ab}{a-b}=5$$

7. For real numbers y_1, y_2, y_3, the quantities 3^{y_1}, 3^{y_2} and 3^{y_3} are positive real numbers, so according to the AM-GM inequality, we have

$$\frac{3^{y_1}+3^{y_2}+3^{y_3}}{3}\geq(3^{y_1}\cdot 3^{y_2}\cdot 3^{y_3})^{1/3}$$

$$\Rightarrow \quad 3^{y_1}+3^{y_2}+3^{y_3}\geq 3(3^{y_1}\cdot 3^{y_2}\cdot 3^{y_3})^{1/3}$$

On applying logarithm with base '3', we get

$$\log_3(3^{y_1}+3^{y_2}+3^{y_3})\geq\left[1+\frac{1}{3}(y_1+y_2+y_3)\right]$$

$$=1+3$$

$$=4 \quad [\because y_1+y_2+y_3=9]$$

$$m=4$$

Now, for positive real numbers x_1, x_2 and x_3, according to AM-GM inequality, we have

$$\frac{x_1 + x_2 + x_3}{3} \geq (x_1 x_2 x_3)^{1/3}$$

On applying logarithm with base '3', we get

$$\log_3\left(\frac{x_1 + x_2 + x_3}{3}\right) \geq \frac{1}{3}(\log_3 x_1 + \log_3 x_2 + \log_3 x_3)$$

$$\Rightarrow \quad 1 \geq \frac{1}{3}(\log_3 x_1 + \log_3 x_2 + \log_3 x_3) \quad [\because x_1 + x_2 + x_3 = 9]$$

$$\therefore \quad M = 3$$

Now, $\log_2(m^3) + \log_3(M^2) = 3\log_2(4) + 2\log_3(3)$

$$= 6 + 2 = 8$$

Matrix-Match Type

1. **(A)** We have,

$\log_2\log_2\log_4 256 + \log_{\sqrt{2}}(\sqrt{2})^4$

$= \log_2\log_2\log_4 4^4 + 4$

$= \log_2\log_2 4 + 4 = 1 + 4 = 5$ (T)

(B) We have,

$$\log_3(5x - 2) - 2\log_3\sqrt{3x+1} = 1 - \log_3 4$$

$\Rightarrow \log_3(5x - 2) - \log_3(3x + 1) + \log_3 4 = 1$

$$\Rightarrow \quad \log_3\left(\frac{(5x-2)(4)}{3x+1}\right) = 1$$

$$\Rightarrow \quad \frac{(5x-2)(4)}{3x+1} = 3 \Rightarrow x = 1 \text{ (P)}$$

(C) $7^{\log_7(x^2 - 4x + 5)} = (x - 1)$

$\Rightarrow x^2 - 4x + 5 = x - 1$

$\Rightarrow x^2 - 5x + 6 = 0$

$\Rightarrow (x - 2)(x - 3) = 0$

$\Rightarrow x = 2$ or $x = 3$

Also we must have $x^2 - 4x + 5 > 0$ and $x - 1 > 0$

$\Rightarrow x > 1$ (as $x^2 - 4x + 5 > 0$ is true for all real numbers) (Q)

(D) $x > 0, \frac{1}{2}\log_2 x - 2\left(\frac{\log_2 x}{2}\right)^2 + 1 > 0$

$\Rightarrow \log_2 x - (\log_2 x)^2 + 2 > 0$

$\Rightarrow (\log_2 x)^2 - \log_2 x - 2 < 0$

$\Rightarrow (t - 2)(t + 1) < 0$

$\Rightarrow -1 < t < 2$

$\Rightarrow \frac{1}{2} < x < 4$

Hence, the number of integers is 3, i.e., {1,2, 3}. (R)

2. **(A)** We have,

$\log_2\log_2\log_4 256 + \log_{\sqrt{2}}(\sqrt{2})^4$

$= \log_2\log_2\log_4 4^4 + 4$

$= \log_2\log_2 4 + 4 = 1 + 4 = 5$ (P, R)

(B) We have,

$$\log_3(5x - 2) - 2\log_3\sqrt{3x+1} = 1 - \log_3 4$$

$\Rightarrow \log_3(5x - 2) - \log_3(3x + 1) + \log_3 4 = 1$

$$\Rightarrow \quad \log_3\left(\frac{(5x-2)(4)}{3x+1}\right) = 1$$

$$\Rightarrow \quad \frac{(5x-2)(4)}{3x+1} = 3$$

$\Rightarrow \quad x = 1$ (Q, T)

(C) $7^{\log_7(x^2 - 4x + 5)} = (x - 1)$

$\Rightarrow x^2 - 4x + 5 = x - 1$

$\Rightarrow x^2 - 5x + 6 = 0$

$\Rightarrow (x - 2)(x - 3) = 0$

$\Rightarrow x = 2$ or $x = 3$

Also we must have $x^2 - 4x + 5 > 0$ and $x - 1 > 0$

$\Rightarrow x > 1$ (as $x^2 - 4x + 5 > 0$ is true for all real numbers) (S, T)

(D) $x > 0, \frac{1}{2}\log_2 x - 2\left(\frac{\log_2 x}{2}\right)^2 + 1 > 0$

$\Rightarrow \log_2 x - (\log_2 x)^2 + 2 > 0$

$\Rightarrow (\log_2 x)^2 - \log_2 x - 2 < 0$

$\Rightarrow (t - 2)(t + 1) < 0$

$\Rightarrow -1 < t < 2$

$\Rightarrow \frac{1}{2} < x < 4$

Hence, the number of integers is 3, i.e., {1,2, 3}. (R)

2. **(A)** $\because 5^2 < 34 < 5^3$

$\Rightarrow 2\log_{34} 5 < 1 < 3\log_{34} 5$

$\therefore \log_{34} 5 < \frac{1}{2}$ and $\log_{34} 5 > \frac{1}{3}$

$\therefore \log_{34} 5 \in \left(\frac{1}{3}, \frac{1}{2}\right)$

$\therefore a = \frac{1}{3}, b = \frac{1}{2}$

$\Rightarrow [10a + 10b] = [3.33 + 5] = [8.33] = 8$ (P)

and $[6b - 3a] = [3 - 1] = 2$ (R) ($\because P, R$)

(B) $\because 4^4 < 300 < 4^5$

$\Rightarrow 4\log_{300} 4 < 1 < 5\log_{300} 4$

$\therefore \log_{300} 4 < \frac{1}{4}$ and $\log_{300} 4 > \frac{1}{5}$

$\therefore \log_{300} 4 \in \left(\frac{1}{5}, \frac{1}{4}\right)$

$\therefore a = \frac{1}{5}, b = \frac{1}{4}$

$(10a + 10b) = (2 + 2.5) = 5$ (Q)

and $(6b - 3a) = (1.5 - 0.6) = (0.9) = 1$(T)

(C) $\because 3^5 < 400 < 3^6$

$\Rightarrow 5\log_{400} 3 < 1 < 6\log_{400} 3$

$\therefore \log_{400} 3 < \frac{1}{5}$ and $\log_{400} 3 > \frac{1}{6}$

$\therefore \log_{400} 3 \in \left(\frac{1}{6}, \frac{1}{5}\right)$

$\therefore a = \frac{1}{6}, b = \frac{1}{5}$

$$(6b-3a)=\left(\frac{6}{5}-\frac{3}{6}\right)$$
$$=(1.2-0.5)=(0.7)=1 \text{ (T)}$$
and $[10a+10b]=\left[\frac{10}{6}+\frac{10}{5}\right]$
$$=[1.67+2]=[3.67]=3 \text{ (S)}$$

3. **(A)** Since, $\log_{100}|x+y|=\frac{1}{2}$

$\Rightarrow \quad |x+y|=(100)^{1/2}=10$

$\therefore \quad |x+y|=10 \quad \ldots(i)$

and $\quad \log_{10} y-\log_{10}|x|=\log_{100} 4 \quad (\because y>0)$

$\Rightarrow \log_{10}\left(\frac{y}{|x|}\right)=\log_{10^2} 2^2=\frac{2}{2}\log_{10} 2$

$\Rightarrow \quad \log_{10}\left(\frac{y}{|x|}\right)=\log_{10} 2$

$\Rightarrow \quad \frac{y}{|x|}=2 \Rightarrow y=2|x| \quad \ldots(ii)$

From Eqs. (i) and (ii),
$$|x+2|x||=10$$

Case I : $\quad x>0,\ |3x|=10$

$\Rightarrow \quad 3|x|=10$

$\Rightarrow \quad 3x=10$

$\therefore \quad x=\frac{10}{3}$

From Eq. (ii), $y=\frac{20}{3}$

$\therefore \quad \left\{\frac{10}{3}, \frac{20}{3}\right\}$ (T)

Case II : $\quad x<0,\ |x-2x|=10$

$\Rightarrow \quad |x|=10 \Rightarrow -x=10$

$\therefore \quad x=-10$

From Eq. (ii), $y=2|-10|=20$

$\therefore \quad \{-10, 20\}$ (R, T)

(B) Let $\log_2 x=A$ and $\log_2 y=B$

$\therefore \quad 4A^2+1=2B \quad \ldots(i)$

and $\quad 2A \geq B$

or $\quad 4A \geq 2B$

$\Rightarrow \quad 4A \geq 4A^2+1 \quad$ [from Eq. (i)]

$\Rightarrow \quad (2A-1)^2 \leq 0$

$\Rightarrow \quad 2A-1=0$

or $\quad A=\frac{1}{2}$ (P)

From Eq. (i), $4\left(\frac{1}{2}\right)^2+1=2B$

$\therefore \quad B=1$

$\Rightarrow \log_2 x=\frac{1}{2}$ and $\log_2 y=1$

$\Rightarrow \quad x=2^{1/2}$ and $y=2^1$

or $\quad x=\sqrt{2},\ y=2$

$\therefore \quad \{\sqrt{2}, 2\}$

(C) $\because \quad \log_4 x=\log_2 y$

$\Rightarrow \quad \log_{2^2} x=\log_2 y$

$\Rightarrow \quad \frac{1}{2}\log_2 x=\log_2 y$

$\Rightarrow \quad \log_2 x=2\log_2 y=\log_2 y^2$

$\Rightarrow \quad x=y^2 \quad \ldots$ (i)

and $\quad x^2-5y^2+4=0$

$\Rightarrow \quad x^2-5x+4=0$

$\Rightarrow \quad (x-1)(x-4)=0$

$\therefore \quad x=1,\ y=4$

Then, $y=1,\ y=\pm 2$ (but $y>0$)

$\therefore$ Solution set are $\{1, 1\}, \{4, 2\}$ (Q, S)

09

Probability

Objective Questions Type I [Only one correct answer]

In each of the questions below, four choices are given of which only one is correct. You have to select the correct answer which is the most appropriate.

1. Three of the six vertices of a regular hexagon are chosen at random. The probability that the triangle formed by these vertices is equilateral is

(a) 1/2 (b) 1/3
(c) 1/10 (d) 1/20

2. An ellipse of eccentricity $\frac{2\sqrt{2}}{3}$ is inscribed in a circle and a point within the circle is chosen at random. The probability that this point lies outside the ellipse is

(a) $\frac{1}{3}$ (b) $\frac{2}{3}$
(c) $\frac{1}{9}$ (d) $\frac{2}{9}$

3. Two small squares on a chess board are chosen at random. Probability that they have a common side is

(a) $\frac{1}{3}$ (b) $\frac{1}{9}$
(c) $\frac{1}{18}$ (d) $\frac{5}{18}$

4. A square is inscribed in a circle. If p_1 is the probability that a randomly chosen point of the circle lies within the square and p_2 is the probability that the point lies outside the square, then

(a) $p_1 = p_2$ (b) $p_1 < p_2$
(c) $p_1 > p_2$ and $p_1^2 - p_2^2 < \frac{1}{3}$ (d) None of these

5. The probability that the length of a randomly chosen chord of a circle lies between $\frac{2}{3}$ and $\frac{5}{6}$ of its diameter is

(a) 1/4 (b) 5/12
(c) 1/16 (d) 5/16

6. The coefficients of a quadratic equation $ax^2 + bx + c = 0\ (a \neq b \neq c)$ are chosen from first three prime numbers, the probability that roots of the equation are real is

(a) 1/3 (b) 2/3 (c) 1/4 (d) 3/4

7. Let a die is loaded in such a way that prime number faces are twice as likely to occur as a non prime number faces. The probability that an odd number will be show up when the die is tossed is

(a) 1/3 (b) 2/3
(c) 4/9 (d) 5/9

8. A random variable X has the following probability distribution

X	$P(X = x)$	X	$P(X = x)$
0	λ	5	11λ
1	3λ	6	13λ
2	5λ	7	15λ
3	7λ	8	17λ
4	9λ		

then, λ is equal to

(a) $\frac{1}{81}$ (b) $\frac{2}{81}$
(c) $\frac{5}{81}$ (d) $\frac{7}{81}$

9. Three randomly chosen non-negative integers x, y and z are found to satisfy the equation $x + y + z = 10$. Then, the probability that z is even, is

[JEE Advanced 2017]

(a) $\frac{1}{2}$ (b) $\frac{36}{55}$
(c) $\frac{6}{11}$ (d) $\frac{5}{11}$

10. Two dice are rolled one after another. The probability that the number on the first is less than or equal to the number on the second is

(a) $\frac{5}{12}$ (b) $\frac{7}{12}$
(c) $\frac{5}{18}$ (d) $\frac{13}{18}$

11. Let A and B be two events such that $P(\overline{A \cup B}) = \frac{1}{6}$, $P(A \cap B) = \frac{1}{4}$ and $P(\bar{A}) = \frac{1}{4}$, where $\bar{A}$ stands for the complement of the event A. Then, the events A and B are **[JEE Main 2014]**

(a) independent but not equally likely
(b) independent and equally likely
(c) mutually exclusive and independent
(d) equally likely but not independent

12. A sum of money is rounded off to the nearest rupee, the probability that the rounded off error is atleast ten paise is

(a) $\frac{1}{100}$ (b) $\frac{9}{100}$
(c) $\frac{49}{100}$ (d) $\frac{81}{100}$

13. A die is rolled three times, the probability of getting a large number than the previous number is

(a) $\frac{1}{54}$ (b) $\frac{5}{54}$
(c) $\frac{5}{108}$ (d) $\frac{13}{108}$

14. Consider the circuit,

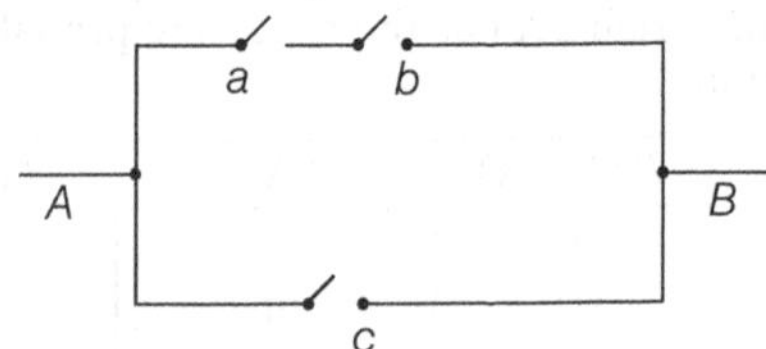

If the probability that each switch is closed is p, then find the probability of current flowing through AB

(a) $p^2 + p$ (b) $p^2 + p - 1$
(c) $p^3 + p$ (d) $p^2 + p + 1$

15. A rifle man firing at a distant target and has only 10% chance of hitting it. The minimum number of rounds he must fire in order to have 50% chance of hitting it atleast once is

(a) 6 (b) 7
(c) 8 (d) 9

16. An unbiased coin is tossed. If the outcome is a head, then a pair of unbiased dice is rolled and the sum of the numbers obtained on them is noted. If the toss of the coin results in tail, then a card from a well-shuffled pack of nine cards numbered 1, 2, 3, ..., 9 is randomly picked and the number of the card is noted. The probability that the noted number is either 7 or 8 is **[JEE Main 2019, 10 January Shift-I]**

(a) $\frac{15}{72}$ (b) $\frac{13}{36}$
(c) $\frac{19}{72}$ (d) $\frac{19}{36}$

17. Three critics review a book. Odds in favour of the book are 5 : 2, 4 : 3 and 3 : 4 respectively for the three critics. The probability that majority are in favour of the book is

(a) $\frac{35}{49}$ (b) $\frac{125}{343}$
(c) $\frac{164}{343}$ (d) $\frac{209}{343}$

18. In a game called "odd man out", n ($n > 2$) persons toss a coin to determine who will buy refreshments for the entire group. A person who gets an outcome different from that of the rest of the members of the group is called the odd man out. If the probability that there is a loser in any game is $\frac{1}{2}$, then the value of n is

(a) 4 (b) 7
(c) 8 (d) 11

19. A boy is throwing stones at a target. The probability of hitting the target at any trial is $\frac{1}{2}$. The probability of hitting the target 5th time at the 10th throw is

(a) $\frac{5}{2^{10}}$ (b) $\frac{63}{2^9}$
(c) $\frac{^{10}C_5}{2^{10}}$ (d) $\frac{^{10}C_4}{2^{10}}$

20. A dice is thrown $(2n + 1)$ times. The probability that faces with even numbers appear odd number of times is

(a) $\frac{2n+1}{2n+3}$ (b) $\frac{n+1}{2n+1}$
(c) $\frac{n}{2n+1}$ (d) None of these

21. For n independent events A_i's, $P(A_i) = 1/(1+i)$, $i = 1, 2, \ldots, n$. The probability that atleast one of the events occurs is

(a) $1/n$ (b) $1/(n+1)$
(c) $n/(n+1)$ (d) None of these

22. A fair coin is tossed a fixed number of times. If the probability of getting 7 heads is equal to probability of getting 9 heads, then the probability of getting 2 heads is **[JEE Main 2021, 26 February]**

(a) $\frac{15}{2^{12}}$ (b) $\frac{15}{2^{13}}$
(c) $\frac{15}{2^{14}}$ (d) $\frac{15}{2^8}$

23. The probability that a teacher will give an unanounced test during any class meeting is 1/5. If a student is absent twice, the probability that he will miss atleast one test, is

(a) 7/25 (b) 9/25
(c) 16/25 (d) 24/25

24. In a n-sided regular polygon, the probability that the two diagonal chosen at random will intersect inside the polygon is

(a) $\frac{2\,^nC_2}{^{(^nC_2 - n)}C_2}$ (b) $\frac{^{n(n-1)}C_2}{^{(^nC_2 - n)}C_2}$

(c) $\frac{^nC_4}{^{(^nC_2 - n)}C_2}$ (d) None of these

25. A box contains 15 transistors, 5 of which are defective. An inspector takes out one transistor at random, examines it for defects, and replaces it. After it has been replaced another inspector does the same thing, and then so does a third inspector. The probability that atleast one of the inspectors finds a defective transistor, is equal to

(a) 1/27 (b) 8/27
(c) 19/27 (d) 26/27

26. Let in a Binomial distribution, consisting of 5 independent trials, probabilities of exactly 1 and 2 successes be 0.4096 and 0.2048 respectively. Then, the probability of getting exactly 3 successes is equal to **[JEE Main 2021, 18 March]**

(a) $\frac{32}{625}$ (b) $\frac{80}{243}$

(c) $\frac{40}{243}$ (d) $\frac{128}{625}$

27. Three letters are written to three different persons and addresses on the three envelopes are also written. Without looking at the addresses, the letters are kept in these envelopes. The probability that all the letters are not placed into their right envelopes is

(a) 1/2 (b) 1/3
(c) 1/6 (d) 5/6

28. A man is known to speak truth in 75% cases. If he throws an unbiased die and tells his friend that it is a six, then the probability that it is actually a six, is

(a) 1/6 (b) 1/8
(c) 3/4 (d) 3/8

29. A bag contains 7 red and 2 white balls and another bag contains 5 red and 4 white balls. Two balls are drawn, one from each bag. The probability that both the balls are white, is

(a) 2/9 (b) 2/3
(c) 8/81 (d) 35/81

30. Three boys and two girls stand in a queue. The probability that the number of boys ahead of every girl is atleast one more that the number of girls ahead of her, is **[JEE Advanced 2014]**

(a) 1 / 2 (b) 1 / 3
(c) 2 / 3 (d) 3 / 4

31. A single letter is selected at random from the word 'PROBABILITY'. The probability that it is a vowel, is

(a) 2/11 (b) 3/11
(c) 4/11 (d) None of these

32. 10 bulbs out of a sample of 100 bulbs manufactured by a company are defective. The probability that 3 out of 4 bulbs, bought by a customer will not be defective, is

(a) $^4C_3 /^{100}C_4$ (b) $^{90}C_3 /^{96}C_4$
(c) $^{90}C_3 /^{100}C_4$ (d) $(^{90}C_3 \times {}^{10}C_1) /^{100}C_4$

33. Fifteen coupons are numbered 1, 2, 3, ...,15 respectively. Seven coupons are selected at random one at a time with replacement. The probability that the largest number appearing on a selected coupon is atmost 9, is

(a) $(1/15)^7$ (b) $(3/5)^7$
(c) $(8/15)^7$ (d) None of these

34. If 12 identical balls are to be placed in 3 different boxes, then the probability that one of the boxes contains exactly 3 balls, is **[JEE Main 2015]**

(a) $\frac{55}{3}\left(\frac{2}{3}\right)^{11}$ (b) $55\left(\frac{2}{3}\right)^{10}$

(c) $220\left(\frac{1}{3}\right)^{12}$ (d) $22\left(\frac{1}{3}\right)^{11}$

35. n letters are written to n different persons and addresses on the n envelopes are also written. If the letters are placed in the envelopes at random, the probability that atleast one letter is not placed in the right envelope, is

(a) $1-\frac{1}{n}$ (b) $1-\frac{1}{2n}$

(c) $1-\frac{1}{n^2}$ (d) $1-\frac{1}{n!}$

36. Three athletes A, B and C participate in a race. Both A and B have the same probability of winning the race and each is twice as likely to win as C. The probability that B or C wins the race is

(a) 2/3 (b) 3/5
(c) 3/4 (d) 13/25

37. A and B play a game of tennis. The situation of the game is as follows; if one scores two consecutive points after a deuce the wins, if loss of a point is followed by win of a point, it is deuce. The chance of a server to win a point is 2/3. The game is at deuce and A is serving. Probability that A will win the game is, (serves are changed after each point)

(a) 3/5 (b) 2/5
(c) 1/2 (d) 4/5

38. Out of 13 applicants for a job, there are 8 men and 5 women. It is desired to select 2 persons for the job. The probability that atleast one of the selected persons will be a woman, is

(a) 5/13 (b) 10/13
(c) 14/39 (d) 25/39

39. Two athletes A and B participate in a race along with other athletes. If the chance of A winning the race is 1/6 and that of B winning the same race is 1/8, then the chance that neither wins the race, is

(a) 1/4 (b) 7/24
(c) 17/24 (d) 35/48

40. Three players *A, B, C* in this order, cut a pack of cards, and the whole pack is reshuffled after each cut. If the winner is one who first draws a diamond, then *C*'s chance of winning is

(a) 9/28 (b) 9/37
(c) 9/64 (d) 27/64

41. 6 girls and 5 boys sit together randomly in a row, the probability that no two boys sit together, is

(a) $\frac{6!5!}{11!}$ (b) $\frac{6!6!}{11!}$
(c) $\frac{6!7!}{2!11!}$ (d) $\frac{5!7!}{2!11!}$

42. A mapping is selected at random from the set of all mappings of the set $A = (1, 2, \ldots, n)$ into itself. The probability that the mapping selected is bijective, is

(a) $\frac{1}{n!}$ (b) $\frac{1}{n^n}$
(c) $\frac{n!}{2^n}$ (d) $\frac{n!}{n^n}$

43. An unbiased die with faces marked 1, 2, 3, 4, 5 and 6 is rolled four times. Out of the four face values obtained, the probability that the minimum face value is not less than 2 and the maximum face value is not greater than 5, is

(a) 1/81 (b) 16/81
(c) 65/81 (d) 80/81

44. If two different numbers are taken from the set {0, 1, 2, 3, ..., 10}, then the probability that their sum as well as absolute difference are both multiple of 4, is **[JEE Main 2017]**

(a) $\frac{6}{55}$ (b) $\frac{12}{55}$
(c) $\frac{14}{45}$ (d) $\frac{7}{55}$

45. Fifteen coupons are numbered 1, 2, 3, ... 15. Seven coupons are selected at random one at a time with replacement. The probability that the largest number appearing on the selected coupon is 9, is

(a) $\left(\frac{9}{16}\right)^6$ (b) $\left(\frac{8}{15}\right)^7$
(c) $\left(\frac{3}{5}\right)^7$ (d) $\frac{9^7 - 8^7}{15^7}$

46. A box contains cards numbered 1 to 100. A card is drawn at random from the box. The probability of drawing a number which is a square, is

(a) 1/5 (b) 2/5
(c) 1/10 (d) None of these

47. An integer is chosen at random from the numbers 1, 2,...,25. The probability that the chosen number is divisible by 3 or 4, is

(a) 2/25 (b) 11/25
(c) 12/25 (d) 14/25

48. Two players *A* and *B* throw a die alternately for a prize of ₹ 11, which is to be won by a player who first throws a six. If *A* starts the game, their respective expectations are

(a) ₹ 6; ₹ 5 (b) ₹ 7; ₹ 4
(c) ₹ 5.50; ₹ 5.50 (d) ₹ 5.75; ₹ 5.25

49. A three-digit number is selected at random from the set of all three-digit numbers. The probability that the number selected has all the three digits same, is

(a) 1/9 (b) 1/10
(c) 1/50 (d) 1/100

50. In a college, 20% students fail in Mathematics, 25% in Physics, and 12% in both subjects. A student of this college is selected at random. The probability that this student who has failed in Mathematics would have failed in Physics too, is

(a) 1/20 (b) 3/25
(c) 12/25 (d) 3/5

51. Two integers are selected at random from the set {1, 2, ..., 11}. Given that the sum of selected numbers is even, the conditional probability that both the numbers are even is **[JEE Main 2019, 11 January]**

(a) $\frac{2}{5}$ (b) $\frac{1}{2}$
(c) $\frac{7}{10}$ (d) $\frac{3}{5}$

52. The probabilities of two events *A* and *B* are 0.3 and 0.6 respectively. The probability that both *A* and *B* occur simultaneously is 0.18. Then, the probability that neither *A* nor *B* occurs is

(a) 0.10 (b) 0.28
(c) 0.42 (d) 0.72

53. One mapping is selected at random from all the mappings of the set $A = \{1, 2, 3, \ldots, n\}$ into itself. The probability that the mapping selected is one to one is given by

(a) $\frac{1}{n^n}$ (b) $\frac{1}{n!}$
(c) $\frac{(n-1)!}{n^{n-1}}$ (d) None of these

54. Two persons each makes a single throw with a pair of dice. The probability that the throws are unequal is given by

(a) $\frac{1}{6^3}$ (b) $\frac{73}{6^3}$
(c) $\frac{51}{6^3}$ (d) None of these

55. A computer producing factory has only two plants T_1 and T_2. Plant T_1 produces 20% and plant T_2 produces 80% of the total computers produced. 7% of computers produced in the factory turn out to be defective.

It is known that P (computer turns out to be defective, given that it is produced in plant T_1) $= 10P$ (computer turns out to be defective, given that it is produced in plant T_2), where $P(E)$ denotes the probability of an event E. A computer produced in the factory is randomly selected and it does not turn out to be defective. Then, the probability that it is produced in plant T_2, is **[JEE Advanced 2016]**

(a) $\frac{36}{73}$ (b) $\frac{47}{79}$

(c) $\frac{78}{93}$ (d) $\frac{75}{83}$

56. A bag contains a white and b black balls. Two players A and B alternately draw a ball from the bag, replacing the ball each time after the draw. A begins the game. If the probability of A winning (that is drawing a white ball) is twice the probability of B winning, then the ratio $a : b$ is equal to

(a) $1 : 2$ (b) $2 : 1$

(c) $1 : 1$ (d) None of these

57. One ticket is selected at random from 100 tickets numbered 00, 01, 02, ..., 99. Suppose X and Y are the sum and product of the digit found on the ticket $P(X = 7 / Y = 0)$ is given by

(a) 2/3 (b) 2/19

(c) 1/50 (d) None of these

58. Let X be a set containing n elements. Two subsets A and B of X are chosen at random, the probability that $A \cup B = X$ is

(a) ${}^{2n}C_n/2^{2n}$

(b) $1/{}^{2n}C_n$

(c) $1 \cdot 3 \cdot 5 \cdot \ldots \cdot (2n-1)/2^n n!$

(d) $(3/4)^n$

59. A natural number x is chosen at random from the first 100 natural numbers. The probability that $x + \frac{100}{x} > 50$ is

(a) 1/10 (b) 11/50

(c) 1/20 (d) None of these

60. A person throws two fair dice. He wins ₹ 15 for throwing a doublet (same numbers on the two dice), wins ₹ 12 when the throw results in the sum of 9 and loses ₹ 6 for any other outcome on the throw. Then, the expected gain/loss (in ₹) of the person is **[JEE Main 2019, 12 April]**

(a) $\frac{1}{2}$ gain (b) $\frac{1}{4}$ loss

(c) $\frac{1}{2}$ loss (d) 2 gain

61. Eight coins are tossed at a time, the probability of getting atleast 6 heads up, is

(a) $\frac{7}{64}$ (b) $\frac{57}{64}$

(c) $\frac{37}{256}$ (d) $\frac{229}{256}$

62. 10% bulbs manufactured by a company are defective. The probability that out of a sample of 5 bulbs, none is defective, is

(a) $(1/2)^5$ (b) $(1/10)^5$

(c) $(9/10)$ (d) $(9/10)^5$

63. Four persons independently solve a certain problem correctly with probabilities $\frac{1}{2}, \frac{3}{4}, \frac{1}{4}, \frac{1}{8}$. Then, the probability that the problem is solved correctly by atleast one of them, is **[JEE Advanced 2013]**

(a) $\frac{235}{256}$ (b) $\frac{21}{256}$

(c) $\frac{3}{256}$ (d) $\frac{253}{256}$

64. The probability that atleast one of the events A and B occur is 0.6. If A and B occur simultaneously with probability 0.2, then $P(\overline{A}) + P(\overline{B})$, where $\overline{A}$ and $\overline{B}$ are complements of A and B respectively, is equal to

(a) 0.4 (b) 0.8

(c) 1.2 (d) 1.4

65. Let $A = \{1, 3, 5, 7, 9\}$ and $B = \{2, 4, 6, 8\}$. An element (a, b) of their cartesian product $A \times B$ is chosen at random. The probability that $a + b = 9$, is

(a) 1/5 (b) 2/5

(c) 3/5 (d) 4/5

66. Dialling a telephone number, a man forgot the last two digits and remembering only that they are different, dialled them at random. The probability of the number being dialled correctly is

(a) 1/2 (b) 1/45

(c) 1/72 (d) 1/90

67. Let A and B be two non-null events such that $A \subset B$. Then, which of the following statements is always correct. **[JEE Main 2019, 8 April]**

(a) $P(A / B) = P(B) - P(A)$ (b) $P(A / B) \geq P(A)$

(c) $P(A / B) \leq P(A)$ (d) $P(A / B) = 1$

68. A speaks truth in 60% cases and B speaks truth in 70% cases. The probability that they will say the same thing while describing a single event is

(a) 0.56 (b) 0.54

(c) 0.38 (d) 0.94

69. If the integers λ and μ are chosen at random between 1 to 100, then the probability that a number of the form $7^\lambda + 7^\mu$ is divisible by 5, is

(a) 1/4 (b) 1/7

(c) 1/8 (d) 1/49

70. If two events A and B are such that $P(A) > 0$ and $P(B) \neq 1$, then $P(\bar{A} / \bar{B})$ is equal to

(a) $1 - P(A/B)$ (b) $1 - P(\bar{A}/B)$
(c) $\frac{1 - P(A \cup B)}{P(\bar{B})}$ (d) $\frac{P(A)}{P(\bar{B})}$

71. A three digit number, which is multiple of 11, is chosen at random. The probability that the number so chosen is also a multiple of 9 is equal to

(a) 1/9 (b) 2/9
(c) 1/100 (d) 9/100

72. Let set 'A' has 7 elements and set B has 5 elements. If one function is selected from all possible defined functions from A to B, then the probability that it is onto is

(a) $\frac{7! \times 2}{3 \times 5^6}$ (b) $\frac{7!}{10 \times 5^6}$
(c) $\frac{7!}{5^6}$ (d) $\frac{7!}{5^7}$

73. A box contains tickets numbered 1 to 20. 3 tickets are drawn from the box with replacement. The probability that the largest number on the tickets is 7, is

(a) 7/20 (b) $1 - (7/20)^3$
(c) 2/10 (d) None of these

74. Two numbers x and y are chosen at random from the set {1, 2, 3, ... , 30}. The probability that $x^2 - y^2$ is divisible by 3 is

(a) 3/29 (b) 4/29
(c) 5/29 (d) None of these

75. A fair die is thrown until a score of less than five points is obtained. The probability of obtaining less than three points on the last throw is

(a) 3/4 (b) 4/5
(c) 5/6 (d) 1/2

76. Seven digits from the digits 1, 2, 3, 4, 5, 6, 7, 8, 9 are written in a random order. The probability that this seven digit number is divisible by 9 is

(a) 2/9 (b) 1/5
(c) 1/3 (d) 1/9

77. A wire of length l is cut into three pieces. What is the probability that the three pieces form a triangle ?

(a) 1/2 (b) 1/4
(c) 2/3 (d) None of these

78. Suppose X is a binomial variate $B(5, p)$ and $P(X = 2) = P(X = 3)$, then p is equal to

(a) 1/2 (b) 1/3
(c) 1/4 (d) 1/5

79. A bag contains four tickets numbered 00, 01, 10, 11. Four tickets are chosen at random with replacement, the probability that sum of the numbers on the tickets is 23, is

(a) 3/32 (b) 1/64
(c) 5/256 (d) 7/256

80. A natural number is selected at random from the set $X = \{x / 1 \leq x \leq 100\}$. The probability that the number satisfies the inequation $x^2 - 13x \leq 30$, is

(a) 9/50 (b) 3/20
(c) 2/11 (d) None of these

81. A bag contains 4 red and 6 black balls. A ball is drawn at random from the bag, its colour is observed an this ball along with two additional balls of the same colour are returned to the bag. If now a ball is drawn at random from the bag, then the probability that this drawn ball is red, is **[JEE Main 2018]**

(a) $\frac{3}{10}$ (b) $\frac{2}{5}$
(c) $\frac{1}{5}$ (d) $\frac{3}{4}$

82. A signal which can be green or red with probability $\frac{4}{5}$ and $\frac{1}{5}$ respectively, is received by station A and then transmitted to station B. The probability of each station receiving the signal correctly is $\frac{3}{4}$. If the signal received at station B is green, then the probability that the original signal green is

(a) $\frac{3}{5}$ (b) $\frac{6}{7}$
(c) $\frac{20}{23}$ (d) $\frac{9}{20}$

83. If $a \in [-20, 0]$, then the probability that the graph of the function $y = 16x^2 + 8(a + 5)x - 7a - 5$ is strictly above the X-axis is

(a) 1/2 (b) 1/17
(c) 17/20 (d) None of these

84. Two distinct numbers are selected at random from the first twelve natural numbers. The probability that the sum will be divisible by 3 is

(a) 1/3 (b) 23/66
(c) 1/2 (d) None of these

85. Given that $x \in [0, 1]$ and $y \in [0, 1]$. Let A be the event of (x, y) satisfying $y^2 \leq x$ and B be the event of (x, y) satisfying $x^2 \leq y$, then

(a) $P(A \cap B) = \frac{1}{3}$ (b) A, B are exclusive
(c) $P(A) = P(B)$ (d) $P(B) \leq P(A)$

86. The probability that out of 10 persons, all born in April, at least two have the same birthday is

(a) $\frac{^{30}C_{10}}{(30)^{10}}$ (b) $1 - \frac{^{30}C_{10}}{30!}$
(c) $\frac{(30)^{10} - {^{30}C_{10}}}{(30)^{10}}$ (d) None of these

87. A pair of fair dice is rolled together till a sum of either 5 or 7 is obtained, the probability that 5 comes before 7 is

(a) 0.2 (b) 0.3
(c) 0.4 (d) 0.5

88. A second order determinant is written down at random using the numbers 1, – 1 as elements. The probability that the value of the determinant is non zero is

(a) 1/2 (b) 3/8
(c) 5/8 (d) 1/3

89. A five digit number is chosen at random. The probability that all the digits are distinct and digits at odd place are odd and digits at even places are even is

(a) 1/25 (b) 25/567
(c) 1/37 (d) 1/74

90. A natural number is selected from 1 to 1000 at random, then the probability that a particular non-zero digit appears atmost once is

(a) 3/250 (b) 143/250
(c) 243/250 (d) 7/250

91. If A and B are events at the same experiments with $P(A) = 0.2$, $P(B) = 0.5$, then maximum value of $P(A' \cap B)$ is

(a) 1/4 (b) 1/2
(c) 1/8 (d) 1/16

92. $x_1, x_2, x_3, \ldots, x_{50}$ are fifty real numbers such that $x_r < x_{r+1}$ for $r = 1, 2, 3, \ldots, 49$. Five numbers out of these are picked up at random. The probability that the five numbers have x_{20} as the middle number is

(a) $\dfrac{{}^{20}C_2 \times {}^{30}C_2}{{}^{50}C_5}$ (b) $\dfrac{{}^{30}C_2 \times {}^{19}C_2}{{}^{50}C_5}$

(c) $\dfrac{{}^{19}C_2 \times {}^{30}C_3}{{}^{50}C_5}$ (d) None of these

93. A bag contains 14 balls of two colours, the number of balls of each colour being the same. 7 balls are drawn at random one by one. The ball in hand is returned to the bag before each new drawn. If the probability that atleast 3 balls of each colour are drawn is p, then

(a) $p > \dfrac{1}{2}$ (b) $p = \dfrac{1}{2}$

(c) $p < 1$ (d) $p < \dfrac{1}{2}$

94. All the spades are taken out from a pack of cards. From these cards; cards are drawn one by one without replacement till the ace of spades comes. The probability that the ace comes in the 4th draw is

(a) 1/13 (b) 12/13
(c) 4/13 (d) None of these

95. Let C_1 and C_2 be two biased coins such that the probabilities of getting head in a single toss are $\dfrac{2}{3}$ and $\dfrac{1}{3}$, respectively. Suppose α is the number of heads that appear when C_1 is tossed twice, independently, and suppose β is the number of heads that appear when C_2 is tossed twice, independently. Then the probability that the roots of the quadratic polynomial $x^2 - \alpha x + \beta$ are real and equal, is **[JEE Advanced 2020]**

(a) $\dfrac{40}{81}$ (b) $\dfrac{20}{81}$

(c) $\dfrac{1}{2}$ (d) $\dfrac{1}{4}$

96. A natural number x is chosen at random from the first one hundred natural numbers. The probability that $\dfrac{(x-20)(x-40)}{(x-30)} < 0$ is

(a) 1/50 (b) 3/50
(c) 7/25 (d) 9/50

97. A four-digit number (numbered from 0000 to 9999) is said to be lucky, if sum of its first two digits is equal to sum of its last two digits. If a four digit number is picked up at random, the probability that it is lucky number is

(a) 1.67 (b) 2.37
(c) 0.067 (d) 0.37

98. A number is chosen at random from the numbers 10 to 99. By seing the number a man will laugh if product of the digits is 12. If he chosen three numbers with replacement, then the probability that he will laugh atleast one, is

(a) $1 - \left(\dfrac{31}{45}\right)^3$ (b) $1 - \left(\dfrac{43}{45}\right)^3$

(c) $1 - \left(\dfrac{42}{43}\right)^3$ (d) $1 - \left(\dfrac{41}{45}\right)^3$

99. If X follows a binomial distribution with parameters $n = 8$ and $p = 1/2$, then $p\,(|x - 4| \le 2)$ is equal to

(a) 121/128 (b) 119/128
(c) 117/128 (d) 115/128

100. Consider $f(x) = x^3 + ax^2 + bx + c$. Parameters a, b, c are chosen, respectively, by throwing a die three times. Then, the probability that $f(x)$ is an increasing function is

(a) 5/36 (b) 8/36
(c) 4/9 (d) 1/3

101. The probabilities of different faces of a biased dice to appear are as follows

Face number	1	2	3	4	5	6
Probability	0.1	0.32	0.21	0.15	0.05	0.17

The dice is thrown and it is known that either the face number 1 or 2 will appear. Then, the probability of the face number 1 to appear is

(a) 5/21 (b) 5/13
(c) 7/23 (d) 3/10

102. S1 : 5 letters are selected from the letters of word MISSISIPPI. The probability, that 3 of them as alike and other two are also alike but their kind is different from those of 3 alike, is $\frac{23}{{}^{10}C_5}$.

S2 : A submatrix of order 2×2 is randomly selected from a matrix of order 5×4. The probability that the rows of the selected matrix are adjacent rows of the parent matrix, is $\frac{3}{5}$.

S3 : A, B and C are three shooters. Their probabilities of hitting a target are $\frac{4}{5}, \frac{3}{4}$ and $\frac{2}{3}$ respectively. If all of them make an attempt on the target and it is found that two of them have missed out then the probability that A has hit the target is equal to $\frac{2}{5}$

(a) TFT (b) FFT
(c) FTF (d) TTF

103. Sixteen players $P_1, P_2, \ldots, P_{16}$ play in a tournament. They are divided into eight pairs at random. From each pair a winner is decided on the basis of a game played between the two players of the pair. Assuming that all the players are of equal strength, the probability that exactly one of the two players P_1 and P_2 is among the eight winners is

(a) 4/15 (b) 7/15
(c) 8/15 (d) 17/30

Objective Questions Type II [One or more than one correct answer(s)]

In each of the questions below four choices of which one or more than one are correct. You have to select the correct answer(s) accordingly.

1. A fair coin is tossed n times. Let $X =$ the number of times head occurs. If $P(X = 4)$, $P(X = 5)$ and $P(X = 6)$ are in AP, then the value of n can be

(a) 7 (b) 10
(c) 12 (d) 14

2. Let X and Y be two events such that $P(X) = \frac{1}{3}$, $P(X / Y) = \frac{1}{2}$ and $P(Y / X) = \frac{2}{5}$. Then, **[JEE Advanced 2017]**

(a) $P(Y) = \frac{4}{15}$ (b) $P(X'/Y) = \frac{1}{2}$
(c) $P(X \cup Y) = \frac{2}{5}$ (d) $P(X \cap Y) = \frac{1}{5}$

3. A bag contains four tickets marked with 112, 121, 211, 222, one ticket is drawn at random from the bag. Let E_i $(i = 1, 2, 3)$ denote the event that ith digit on the ticket is 2. Then

(a) E_1 and E_2 are independent
(b) E_2 and E_3 are independent
(c) E_3 and E_1 are independent
(d) E_1, E_2, E_3 are independent

4. If A and B are independent events such that $0 < P(A) < 1$, $0 < P(B) < 1$, then

(a) A, B are mutually exclusive
(b) A and $\overline{B}$ are independent
(c) $\overline{A}, \overline{B}$ are independent
(d) $P(A/B) + P(A/B) = 1$

5. If A and B are two events such that $P(A) = 1/2$ and $P(B) = 2/3$, then

(a) $P(A \cup B) \geq 2/3$
(b) $P(A \cap B') \leq 1/3$
(c) $1/6 \leq P(A \cap B) \leq 1/2$
(d) $1/6 \leq P(A' \cap B) \leq 1/2$

6. For two events A and B, if $P(A) = P(A/B) = 1/4$ and $P(B/A) = 1/2$, then

(a) A and B are independent
(b) A and B are mutually exclusive
(c) $P(A'/B) = 3/4$
(d) $P(B'/A') = 1/2$

7. A student appears for tests I, II and III. The student is successful if he passes either in tests I and II or tests I and III. The probabilities of the student passing in tests I, II, III are p, q and $1/2$ respectively. If the probability that the student is successful is $1/2$, then

(a) $p = 1, q = 0$
(b) $p = 2/3, q = 1/2$
(c) $p = 3/5, q = 2/3$
(d) there are infinitely many values of p and q

8. Let X be a set containing n elements. If two subsets A and B of X are picked at random, the probability that A and B have the same number of elements is

(a) $\frac{{}^{2n}C_n}{2^{2n}}$ (b) $\frac{1}{{}^{2n}C_n}$
(c) $\frac{1 \cdot 3 \cdot 5 \cdot \ldots \cdot (2n-1)}{2^n \cdot n!}$ (d) $\frac{3^n}{4^n}$

9. A bag initially contains one red and two blue balls. An experiment consisting of selecting a ball at random, noting its colour and replacing it together with an additional ball of the same colour. If three such trials are made, then

(a) probability that atleast one blue ball is drawn is 0.9
(b) probability that exactly one blue ball is drawn is 0.2
(c) probability that all the drawn balls are red given that all the drawn balls are of same colour is 0.2
(d) probability that atleast one red ball is drawn is 0.6.

10. Three six faced fair dice are thrown together. The probability that the sum of the numbers appearing on the dice is $k (3 \le k \le 8)$ is

(a) $\frac{(k-1)(k-2)}{432}$ (b) $\frac{k(k-2)}{432}$

(c) ${}^{k-1}C_2 \times \frac{1}{216}$ (d) $\frac{k^2}{432}$

11. Suppose m boys and m girls take their seats randomly round a circle. The probability of their sitting is $({}^{2m-1}C_m)^{-1}$ when

(a) no two boys sit together
(b) no two girls sit together
(c) boys and girls sit alternatively
(d) all the boys sit together

12. A card is selected at random from cards numbered as 00, 01, 02,...,99. An event is said to have occured. If product of digits of the card number is 16. If card is selected 5 times with replacement each time, then the probability that the event occurs exactly three times is

(a) ${}^5C_3\left(\frac{3}{100}\right)^2\left(\frac{97}{100}\right)^3$ (b) ${}^5C_3\left(\frac{3}{100}\right)^3\left(\frac{97}{100}\right)^2$

(c) ${}^5C_3\left(\frac{0.3}{100}\right)^3\left(\frac{9.7}{100}\right)^3$ (d) $10(0.03)^3(0.97)^2$

13. The probabilities that a student in Mathematics, Physics and Chemistry are α, β and γ respectively. Of these subjects, a student has a 75% chance of passing in atleast one, a 50% chance of passing in atleast two and a 40% chance of passing in exactly two subjects. Which of the following relations are true ?

(a) $\alpha + \beta + \gamma = 19/20$ (b) $\alpha + \beta + \gamma = 27/20$

(c) $\alpha\beta\gamma = 1/10$ (d) $\alpha\beta\gamma = 1/4$

14. Let $0 < P(A) < 1$, $0 < P(B) < 1$ and $P(A \cup B) = P(A) + P(B) - P(A)P(B)$, then

(a) $P(B/A) = P(B) - P(A)$
(b) $P(\bar{A} - \bar{B}) = P(\bar{A}) - P(\bar{B})$
(c) $P(\overline{A \cup B}) = P(\bar{A})\,P(\bar{B})$
(d) $P(A/B) = P(A)$

15. If p and q are chosen randomly from the set {1, 2, 3, 4, 5, 6, 7, 8, 9, 10} with replacement, then the probability that the roots of the equation $x^2 + px + q = 0$

(a) are real is 33/50
(b) are imaginary is 19/50
(c) are real and equal is 3/50
(d) are real and distinct is 3/5

16. Let A and B are two independent events. The probability that both A and B happen is $\frac{1}{12}$ and the probability that neither A nor B happens is $\frac{1}{2}$, then

(a) $P(A) = \frac{1}{3}, P(B) = \frac{1}{4}$ (b) $P(A) = \frac{1}{2}, P(B) = \frac{1}{6}$

(c) $P(A) = \frac{1}{6}, P(B) = \frac{1}{2}$ (d) $P(A) = \frac{1}{4}, P(B) = \frac{1}{3}$

17. A random variable X follows binomial distribution with mean a and variance b. Then

(a) $a > b > 0$ (b) $\frac{a}{b} > 1$

(c) $\frac{a^2}{a-b}$ is an integer (d) $\frac{a^2}{a+b}$ is an integer

18. The letters of the word PROBABILITY are written down at random in a row. Let E_1 denotes the event that two I's are together and E_2 denotes the event that two B's are together, then

(a) $P(E_1) = P(E_2) = \frac{2}{11}$ (b) $P(E_1 \cap E_2) = \frac{2}{55}$

(c) $P(E_1 \cup E_2) = \frac{18}{55}$ (d) $P(E_1/E_2) = \frac{1}{5}$

19. Two real numbers, x and y are selected at random. Given that $0 \le x \le 1$; $0 \le y \le 1$. Let A be the event that $y^2 \le x$; B be the event that $x^2 \le y$, then

(a) $P(A \cap B) = \frac{1}{3}$
(b) A and B are exhaustive events
(c) A and B are mutually exclusive
(d) A and B are independent events

20. Three numbers are chosen at random without replacement from {1, 2, 3, ...,10}. The probability that minimum of the chosen number is 3 or their maximum is 7, cannot exceed

(a) $\frac{11}{30}$ (b) $\frac{11}{40}$

(c) $\frac{11}{50}$ (d) $\frac{11}{60}$

21. A square is inscribed in a circle. If p_1 is the probability that a randomly chosen point of the circle lies within the square and p_2 is the probability that the point lies outside the square, then

(a) $p_1 < p_2$ (b) $p_1 = p_2$

(c) $p_1 > p_2$ (d) $p_1^2 - p_2^2 < \frac{1}{3}$

22. A random variable X takes values 0, 1, 2, 3, ...with probability proportions to $(x+1)\left(\frac{1}{5}\right)^x$, then

(a) $P(X = 0) = \frac{16}{25}$ (b) $P(X \le 1) = \frac{112}{125}$

(c) $P(X \ge 1) = \frac{9}{25}$ (d) $E(X) = \frac{25}{32}$

Linked-Comprehension Type

In these questions, a passage (paragraph) has been given followed by questions based on each of the passages. You have to answer the questions based on the passage given.

PASSAGE 1

If n positive integers taken at random and multiplied together, then the chance that the last digit of the product would be

On the basis of above information, answer the following questions

1. 1, 3, 5, 7, or 9 is

(a) $\left(\frac{2}{5}\right)^n$ (b) $\left(\frac{1}{2}\right)^n$

(c) $\frac{2^n - 1}{5^n}$ (d) $\frac{5^n - 4^n}{10^n}$

2. 1, 3, 7 or 9 is

(a) $\frac{2^n - 1}{5^n}$ (b) $\frac{5^n - 4^n}{10^n}$

(c) $\left(\frac{2}{5}\right)^n$ (d) $\left(\frac{1}{2}\right)^n$

3. 5 is

(a) $\frac{5^n - 4^n}{10^n}$ (b) $\frac{2^n - 1}{5^n}$

(c) $\frac{10^n - 8^n - 5^n + 4^n}{10^n}$ (d) $\left(\frac{2}{5}\right)^n$

4. 0 is

(a) $\frac{2^n - 1}{5^n}$ (b) $\frac{5^n - 4^n}{5^n}$

(c) $\left(\frac{1}{2}\right)^n$ (d) $\frac{10^n - 8^n - 5^n + 4^n}{10^n}$

5. 2, 4, 6 or 8 is

(a) $\left(\frac{4}{10}\right)^n$ (b) $\frac{5^n - 4^n}{10^n}$ (c) $\frac{4^n - 2^n}{5^n}$ (d) $\left(\frac{1}{2}\right)^n$

PASSAGE 2

Sania Mirza is to play with Sharapova in a three set match. For a particular set, the probability of Sania winning the set is y and if she wins probability of her winning the next set becomes $\sqrt{y}$ else the probability that she wins the next one becomes y^2. There is no possibility that a set is to be abondoned. R is probability that Sania wins the first set.

On the basis of above information, answer the following questions

1. If $R = \frac{1}{2}$, then the probability that match will end in first two sets is nearly equal to

(a) 0.28 (b) 0.95

(c) 0.51 (d) 0.36

2. If $R = \frac{1}{2}$ and Sania wins the second set probability that she has won first set as well is nearly equal to

(a) 0.74 (b) 0.46

(c) 0.26 (d) 0.54

3. If Sania looses the first set, then the values of R such that her probability of winning the match is still larger than that of her loosing are given by

(a) $R \in \left(\frac{1}{2}, 1\right)$ (b) $R \in \left(\left(\frac{1}{2}\right)^{1/3}, 1\right)$

(c) $R \in \left(\left(\frac{1}{2}\right)^{3/2}, 1\right)$ (d) No values of R

PASSAGE 3

A is a set containing 10 elements. A subset P of A is chosen at random and the set A is reconstructed by replacing the elements of P. Another subset Q of A is now chosen at random. Then, the probability that if

On the basis of above information, answer the following questions

1. $P \cup Q = A$, is

(a) $\left(\frac{1}{2}\right)^{10}$ (b) $\left(\frac{2}{3}\right)^{10}$

(c) $\left(\frac{3}{4}\right)^{10}$ (d) $\left(\frac{2}{5}\right)^{10}$

2. $P \cup Q$ contains exactly 4 elements, is

(a) $\frac{5805}{2^{19}}$ (b) $\frac{8550}{2^{19}}$

(c) $\frac{5058}{2^{19}}$ (d) $\frac{8505}{2^{19}}$

3. P and Q have no common elements, is

(a) $\left(\frac{2}{3}\right)^{10}$ (b) $\left(\frac{3}{4}\right)^{10}$

(c) $\left(\frac{4}{5}\right)^{10}$ (d) $\left(\frac{5}{6}\right)^{10}$

4. $P \cap Q$ contains exactly 3 elements, is

(a) $\frac{5 \times 3^8}{2^{17}}$ (b) $\frac{15 \times 3^9}{2^{17}}$

(c) $\frac{3 \times 5^8}{2^{17}}$ (d) $\frac{7 \times 3^8}{2^{17}}$

5. $P \cap Q$ contains atleast 9 elements, is

(a) $\frac{15}{2^{20}}$ (b) $\frac{31}{2^{20}}$

(c) $\frac{30}{2^{20}}$ (d) $\frac{41}{2^{20}}$

PASSAGE 4

Football teams T_1 and T_2 have to play two games against each other. It is assumed that the outcomes of the two games are Independent. The probabilities of T_1 winning, drawing and losing a game against T_2 are $\frac{1}{2}, \frac{1}{6}$ and $\frac{1}{3}$, respectively. Each team gets 3 points for a win, 1 point for a draw and 0 point for a loss in a game. Let X and Y denote the total points scored by teams T_1 and T_2, respectively, after two games. **[JEE Advanced 2016]**

On the basis of above information, answer the following questions

1. $P(X > Y)$ is
   (a) $\frac{1}{4}$ (b) $\frac{5}{12}$
   (c) $\frac{1}{2}$ (d) $\frac{7}{12}$
2. $P(X = Y)$ is
   (a) $\frac{11}{36}$ (b) $\frac{1}{3}$
   (c) $\frac{13}{36}$ (d) $\frac{1}{2}$

PASSAGE 5

If a particular experiment be given n (a finite) independent trials.

If the probability of success in one trial (say) p.

$\therefore$ We get probability of failure, $q = (1 - p)$

The probability of r success in n trials $= {}^nC_r p^r q^{n-r}$

On the basis of above information, answer the following questions

1. The probability of a man hitting a target in one fire is $\frac{1}{4}$. Atleast n times he must fire at the target that the chances of hitting the target at least once will exceed $\frac{2}{3}$, then n is
   (a) 2 (b) 4
   (c) 6 (d) 8
2. When two dice are thrown n number of times, the probability of getting atleast one double six is greater than 99%. then the least value of n is
   (Given log 35 = 1.5441, log 36 = 1.5563)
   (a) 150 (b) 154
   (c) 160 (d) 164
3. A coin is tossed 4 times whose faces are marked with the numbers 5 and 3. The odds in favour of getting a sum less than 15 is
   (a) $\frac{5}{16}$ (b) $\frac{5}{11}$ (c) $\frac{16}{5}$ (d) $\frac{11}{5}$
4. Suppose the probability for A to win a game against B is 0.4. If A has an option of playing either a "best of 3 games" or a "best of five games" match against B. (No match in draw) the option should a choose so that the probability of his winning the match is higher
   (a) "best of 3 games" (b) "best of 5 games"
   (c) both equal (d) none of these
5. In a lot containing 100 bulbs, 10 are defective. The probability that out of a sample of 5 bulbs, none is defective is
   (a) 10^{-15} (b) $\left(\frac{1}{5}\right)^5$
   (c) $\left(\frac{9}{10}\right)^5$ (d) $\frac{9}{10}$

PASSAGE 6

A triangle is formed with the vertices of n sided regular polygon. Then, the probability that the triangle

On the basis of above information, answer the following questions

1. Have exactly one side common with the side of the polygon is
   (a) $\frac{6(n-2)(n-3)}{n(n-1)}$ (b) $\frac{6(n-3)}{(n-1)(n-2)}$
   (c) $\frac{6(n-4)}{(n-1)(n-2)}$ (d) $\frac{6(n-2)}{(n-1)(n-3)}$
2. Have exactly two sides common with the side of the polygon is
   (a) $\frac{3}{(n-1)(n-2)}$ (b) $\frac{6(n-4)}{(n-1)(n-2)}$
   (c) $\frac{6(n-3)}{(n-1)(n-2)}$ (d) $\frac{6}{(n-1)(n-2)}$
3. Does not have any side common with the sides of the polygon is
   (a) $\frac{(n-2)(n-3)}{(n-1)(n-2)}$ (b) $\frac{(n-4)(n-5)}{(n-1)(n-2)}$
   (c) $\frac{(n-1)(n-3)}{n(n-2)}$ (d) $\frac{(n-2)(n-4)}{n(n-3)}$
4. Have atleast one side common with the sides of the polygon is
   (a) $\frac{6(n-3)}{(n-1)(n-2)}$ (b) $\frac{6(n-4)}{(n-1)(n-2)}$
   (c) $\frac{6(n-2)}{n(n-1)}$ (d) $\frac{6(n-3)}{n(n-1)}$
5. Have at most one side common with the sides of the polygon is
   (a) $\frac{(n-4)(n+2)}{(n-1)(n-2)}$ (b) $\frac{(n-4)(n+1)}{(n-1)(n-2)}$
   (c) $\frac{(n-3)(n+1)}{(n-1)(n-2)}$ (d) $\frac{(n-3)(n-4)}{(n-1)(n-2)}$

Numerical Value Type

Solve the following problems and write your answer.

1. Two fair dice, each with faces numbered 1, 2, 3, 4, 5 and 6 are rolled together and the sum of the numbers on the faces is observed. This process is repeated till the sum is either a prime number or a perfect square. Suppose the sum turns out to be a perfect square before it turns out to be a prime number. If p is the probability that this perfect square is an odd number, then the value of $140\,p$ is ....
2. The probability that certain electronic component fails when first used is 0.10. If it does not fail immediately, the probability that it lasts for one year is 0.99. If the probability that a new component will last for one year is λ then, value of $10\,\lambda$ must be
3. An electric component manufactured by 'RASU electronics' is tested for its defectiveness by a sophisticated testing device. Let A denote the event "the device is defective" and B the event "the testing device reveals the component to be defective". Suppose $P(A)=\alpha$ and $P(B/A)=P(B'/A')=1-\alpha$, where $0<\alpha<1$. If the probability that the component is not defective, given that the testing device reveals it to be defective, is λ, then the value of λ must be
4. Seven digits from the numbers 1, 2, 3, 4, 5, 6, 7, 8 and 9 are written in random order. If the probability that this seven-digit number is divisible by 9 is λ, then the value of $9999\,\lambda$ must be
5. A man takes a step forward with probability 0.4 and backward with probability 0.6. If the probability that at the end of eleven steps he is one step away from the starting point is λ then the value of $5000\,\lambda$ must be
6. Cards are drawn one by one at random from a well-shuffled full pack of 52 playing cards until two aces are obtained for the first time. If N is the number of cards required to be drawn then $P(N=n)=\dfrac{(n-1)(\lambda-n)(\mu-n)}{\alpha\times\beta\times\gamma\times\delta}$ where $2\le n\le 50$, then the value of $\alpha+\beta+\gamma+\delta+2\lambda+3\mu$ must be
7. A, B, C and D cut a pack of 52 cards successively in the order given. If the person who cuts a spade first receives Rs 350, then the expectation of A is
8. A box contains coins of which are fair and rest are biased. The probability of getting a head when a fair coin is tossed is 1/2 , while it is $\frac{2}{3}$ when a biased coin is tossed. A coin is drawn from the box at random and is tossed twice. The first time it shows head and the second time it shows tail. If the probability that the coin drawn is fair is $f(N, m)$, then the value of $43\,f(20, 12)$ must be
9. Let A, B, C be three events. If the probability of occurring exactly one event out of A and B is $1-a$, out of B and C is $1-2a$, out of C and A is $1-a$ and that of occurring three events simultaneously is a^2. If the probability that atleast one out of A, B, C will occur is greater than λ, then the value of $2050\,\lambda$ must be
10. If X and Y are independent binomial variates $B\left(5, \frac{1}{2}\right)$ and $B\left(7, \frac{1}{2}\right)$. If the value of $P(X+Y=3)$ is λ, then the value of $256\,\lambda$ must be
11. An urn contains 2 white and 2 black balls. A ball is drawn at random. If it is white, it is not replaced into urn, otherwise it is replaced along with another ball of the same colour. The process is repeated. If the probability that the third ball drawn is black is $\frac{23}{\lambda}$, then the value of λ must be
12. Two integers x and y are chosen (without random) at random from the set $\{x : 0\le x\le 10,\ x \text{ is an integer}\}$. If the probability for $|x-y|\le 5$ is p, then the value of $121\,p$ must be
13. A fair coin is tossed 12 times. If the probability that two heads do not occur consecutively is p, then the value of $4096\,p$ must be
14. If four squares are chosen at random on a chess board. If the probability that they lie on a diagonal line is $\frac{\lambda}{^{64}C_4}$, then the value of λ must be
15. An artillery target may be either at point I with probability $\frac{8}{9}$ or at point II with probability $\frac{1}{9}$. We have 21 shells each of which can be fired either at point I or II. Each shell may hit the target independently of the other shell with probability $\frac{1}{2}$. If the number of shells must be fired at point I to hit the target with maximum probability is x, then the value of $12\,x$, must be

Single Integer Type

Solve the following problems and write your answer.

1. One ball is drawn from a bag containing 4 balls and is found to be white. The events that the bag contains "1 white", "2 white", "3 white" and "4 white" are equally likely. If the probability that all the balls are white is $\frac{p}{15}$, then the value of p is
2. There are $N+1$ identical boxes each containing N wall clocks. rth box contains $(r-1)$ defective and $(N-r+1)$ effective clocks for $1\le r\le N+1$. A box is selected at random and from this box a clock chosen at random and is found to be effective. The probability that it is from kth box is $\dfrac{2N-2k+\lambda}{N^2+N}$, then find λ.

3. One mapping is selected at random from all mappings of the set $S = \{1, 2, 3, \ldots, n\}$ into itself. If the probability that the mapping is one-one is 3/32, then the value of n is

4. The minimum number of times a fair coin needs to be tossed, so that the probability of getting atleast two heads is atleast 0.96, is **[JEE Main 2015]**

5. In a knockout tournament, 2^n equally skilled players; $S_1, S_2, \ldots, S_{2^n}$ are participating. In each round, players are divided in pairs at random and winner from each pair moves to the next round. If S_2 reaches the semi-final, then the probability that S_1 wins the tournament is $\frac{1}{84}$. The value of n is …… .

6. The probability of a man hitting a target is $\frac{1}{10}$. The least number of shots required, so that the probability of his hitting the target at least once is greater than $\frac{1}{4}$, is ……… . **[JEE Main 2020, 4 September]**

7. Of the three independent events E_1, E_2 and E_3, the probability that only E_1 occurs is α, only E_2 occurs is β and only E_3 occurs is γ. Let the probability p that none of events E_1, E_2 or E_3 occurs satisfy the equations $(\alpha - 2\beta)$, $p = \alpha\beta$ and $(\beta - 3\gamma)\ p = 2\beta\gamma$. All the given probabilities are assumed to lie in the interval (0, 1).

Then, $\frac{\text{probability of occurrence of } E_1}{\text{probability of occurrence of } E_3}$ is equal to

8. There are two red, two blue, two white, and certain number (greater than 0) of green socks in a drawer. If two socks are taken at random from the drawer without replacement, the probability that they are of the same color is 1/5, then the number of green socks are …… .

Matrix-Match Type

Given below are Matching Type Questions, with two columns (each having some items) each. Each item of Column I has to be matched with the items of Column II, by encircling the correct match(es).

NOTE An item of Column I can be matched with more than one items of Column II.

1. Observe the following columns :

	Column I		Column II
(A)	The probability that A, B, C solve a problem is $\frac{1}{2}, \frac{1}{3}$ and $\frac{1}{4}$. If the probability that the problem will be solved is λ and that the problem is solved by only one of them is μ, then	(P)	$\lambda + \mu = \frac{13}{24}$
(B)	The probability of hitting a target by three marks men is $\frac{1}{2}, \frac{1}{3}$ and $\frac{1}{4}$ respectively. If the probability that exactly two of them will hit the target is λ and that at least two of them hit the target is μ, then	(Q)	$\lambda + \mu = \frac{29}{24}$
(C)	A bag contains 4 white and 2 black balls. Another contains 3 white and 5 black balls. One ball is drawn from each bag. If the prabability that both are black is λ and that both are white is μ, then	(R)	$\lambda + \mu = \frac{11}{24}$
		(S)	$\lambda - \mu = \frac{7}{24}$
		(T)	$\mu - \lambda = \frac{1}{24}$

2. Observe the following columns :

	Column I		Column II
(A)	Two balls are drawn from an urn containing 2 white, 3 red and 4 black balls one by one without replacement. If the probability that both the balls are of same colour is λ and at least one ball is red is μ, then	(P)	$\lambda = \frac{8}{15}$
(B)	A bag contains 2 white and 4 black balls while another bag contains 6 white and 4 black balls. A bag is selected at random and a ball is drawn. If λ be the probability that the ball drawn is of white colour and μ be the probability that the ball drawn is black colour, then	(Q)	$\mu = \frac{7}{12}$
(C)	Bag A contains 4 red and 5 black balls and bag B contains 3 red and 7 black balls. One ball is drawn from bag A and two from bag B. If λ be the probability that out of 3 balls drawn two are black and one is red and μ be the probability that out of three balls drawn two are red and one is black, then	(R)	$\lambda = \frac{5}{18}$
		(S)	$\mu = \frac{7}{15}$
		(T)	$\mu = \frac{11}{45}$

3. A bag contains some white and some black balls, all combinations being equally likely. The total number of balls in the bag is 12. Four balls are drawn at random from the bag without replacement. Now match the entries from the following two columns

	Column I		Column II
(A)	Probability that all the four balls are black is equals to	(P)	$\frac{14}{33}$
(B)	If the bag contains 10 black and 2 white balls, then the probability that all four balls are black is equal to	(Q)	$\frac{1}{5}$
(C)	If all the four balls are black, then the probability, that the selected bag contains 10 black balls, is equal to	(R)	$\frac{70}{429}$
(D)	Probability that two balls are black and two are white	(S)	$\frac{13}{110}$
		(T)	$\frac{13}{165}$

4. Observe the following columns :

	Column I		Column II
(A)	*A*, *B* and *C* in order toss a coin. The first one to throw a head wins. If *p*, *q* and *r* are their respective chances of winning, assume that the game may continue indefinitely, then	(P)	$q - r = \frac{1}{7}$
(B)	Three persons *A*, *B* and *C* throw a dice in succession till one gets a 'six' and wins the game. If *p*, *q* and *r* are their respective chances of winning, then	(Q)	$p - r = \frac{3}{7}$
(C)	*A*, *B* and *C* play a game and chances of their winning it in an attempt are $\frac{2}{3}, \frac{1}{2}$ and $\frac{1}{4}$ respectively. *A* has the first chance, followed by *B* and then by *C*. This cycle is repeated till one of them wins the game. If *p*, *q*, *r* are their respective chances of winning the game, then	(R)	$p : q = 6 : 5$
		(S)	$p - r = \frac{5}{7}$
		(T)	$P : q = 2:1$

5. Observe the following columns :

	Column I		Column II
(A)	In a binomial distribution $\left(\frac{4}{5} + \frac{1}{5}\right)^5$ if variance *V*, standard deviation *S* , then	(P)	$S^2 = V$
(B)	In a binomial distribution $\left(\frac{3}{5} + \frac{2}{5}\right)^5$ if variance *V*, standard deviation *S*, then	(Q)	$4(S^2 - V^2) = V^2$
(C)	In a binomial distribution $\left(\frac{24}{25} + \frac{1}{25}\right)^{25}$ if variance *V*, standard deviation *S*, then	(R)	$5(V^2 - S^2) = S^2$
		(S)	$\sqrt{\left(\frac{S^2 + V^2}{S^2 - V^2}\right)} = 3$
		(T)	$\sqrt{\left(\frac{V^2 + S^2}{V^2 - S^2}\right)} = 7$

Answers

Objective Questions Type [Only one correct answer]

1. (c)	**2.** (b)	**3.** (c)	**4.** (c)	**5.** (a)	**6.** (a)	**7.** (d)	**8.** (a)	**9.** (c)	**10.** (b)
11. (a)	**12.** (d)	**13.** (b)	**14.** (a)	**15.** (b)	**16.** (c)	**17.** (d)	**18.** (a)	**19.** (b)	**20.** (d)
21. (c)	**22.** (b)	**23.** (b)	**24.** (c)	**25.** (c)	**26.** (a)	**27.** (b)	**28.** (d)	**29.** (c)	**30.** (a)
31. (b)	**32.** (d)	**33.** (b)	**34.** (a)	**35.** (d)	**36.** (b)	**37.** (c)	**38.** (d)	**39.** (c)	**40.** (b)
41. (c)	**42.** (d)	**43.** (b)	**44.** (a)	**45.** (d)	**46.** (c)	**47.** (c)	**48.** (a)	**49.** (d)	**50.** (d)
51. (a)	**52.** (b)	**53.** (c)	**54.** (d)	**55.** (c)	**56.** (c)	**57.** (b)	**58.** (d)	**59.** (d)	**60.** (c)
61. (c)	**62.** (d)	**63.** (a)	**64.** (c)	**65.** (a)	**66.** (d)	**67.** (b)	**68.** (b)	**69.** (a)	**70.** (c)
71. (a)	**72.** (a)	**73.** (d)	**74.** (d)	**75.** (d)	**76.** (d)	**77.** (b)	**78.** (a)	**79.** (a)	**80.** (b)
81. (b)	**82.** (c)	**83.** (d)	**84.** (a)	**85.** (a)	**86.** (c)	**87.** (c)	**88.** (a)	**89.** (b)	**90.** (d)
91. (b)	**92.** (c)	**93.** (a)	**94.** (a)	**95.** (b)	**96.** (c)	**97.** (c)	**98.** (b)	**99.** (b)	**100.** (c)
101. (a)	**102.** (a)	**103.** (c)							

Objective Questions Type II [One or more than one correct answer(s)]

1. (a, d)	**2.** (a, b)	**3.** (a, b, c)	**4.** (b, c, d)	**5.** (a, b, c, d)
6. (a, c, d)	**7.** (a, b, c, d)	**8.** (a, c)	**9.** (a, b, c, d)	**10.** (a, c)
11. (a, b, c)	**12.** (b, d)	**13.** (b, c)	**14.** (c, d)	**15.** (b, c, d)
16. (a, d)	**17.** (a, b, c)	**18.** (a, b, c, d)	**19.** (a, b)	**20.** (a, b)
21. (c, d)	**22.** (a, b, c, d)			

Linked-Comprehension Type

Passage 1 **1.** (b) **2.** (c) **3.** (a) **4.** (d) **5.** (c)

Passage 2 **1.** (a) **2.** (a) **3.** (b)

Passage 3 **1.** (c) **2.** (d) **3.** (b) **4.** (a) **5.** (b)

Passage 4 **1.** (b) **2.** (c)

Passage 5 **1.** (b) **2.** (d) **3.** (b) **4.** (a) **5.** (c)

Passage 6 **1.** (c) **2.** (d) **3.** (b) **4.** (a) **5.** (b)

Numerical Value Type

1. 80	**2.** 8.91	**3.** 0.5	**4.** 1111	**5.** 1850	**6.** 386	**7.** 128	**8.** 27
9. 1025	**10.** 13.75	**11.** 30	**12.** 91	**13.** 377	**14.** 364	**15.** 144	

Single Integer Type

1. (6) **2.** (2) **3.** (4) **4.** (8) **5.** (6) **6.** (3) **7.** (6) **8.** (4)

Matrix-Match Type

1. A → (Q, S); B → (P, T); C → (R, T)
2. A → (Q, R); B → (P, S); C → (S, T)
3. A → (Q); B → (P); C → (R); D → (Q);
4. A → (P, Q, T); B → (R); C → (P, S)
5. A → (P, Q, S); B → (P, R); C → (P, T)

Solutions

Objective Questions Type [Only one correct answer]

1. Let $ABCDEF$ be the regular hexagon.

Number of total triangles $= {}^6C_3 = 20$ ways

For the favourable event, the vertices should be either A, C, E or B, D, F

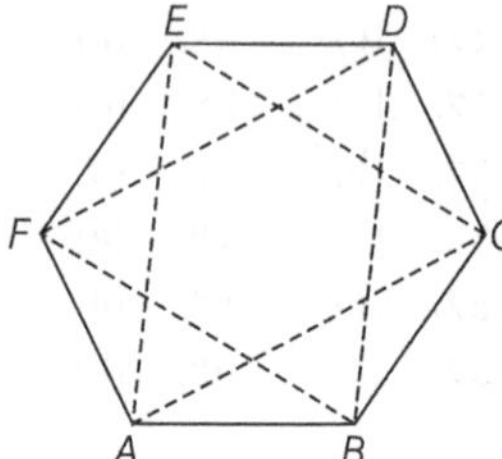

$\therefore$ The required probability $= \dfrac{\text{Favourable ways}}{\text{Total ways}} = \dfrac{2}{20} = \dfrac{1}{10}$

2. Let ellipse be $\dfrac{x^2}{a^2} + \dfrac{y^2}{b^2} = 1 \qquad (a > b)$

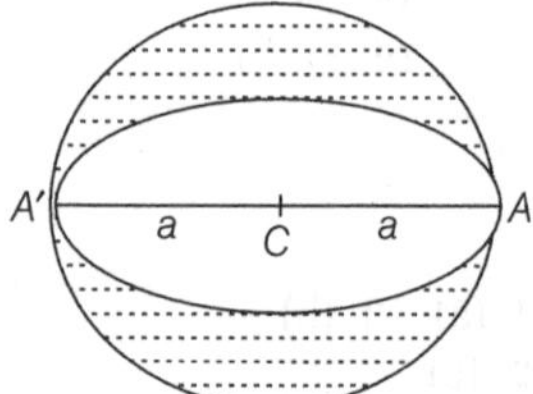

$\therefore$ Area of ellipse $= \pi ab = \pi a^2 \sqrt{(1 - e^2)}$

$$= \pi a^2 \sqrt{1 - \frac{8}{9}} = \frac{\pi a^2}{3}$$

Area of circle $= \pi a^2$

$\therefore$ The required probability

$$= \frac{\pi a^2 - \dfrac{\pi a^2}{3}}{\pi a^2} = \frac{2}{3}$$

3. Total number of ways to choose two squares

$$= {}^{64}C_2 = \frac{64 \cdot 63}{2} = 32 \cdot 63$$

For favourable ways we must chosen two consecutive small squares for any row or any columns.

$\therefore$ Number of favourable ways $= 7 \cdot 8 + 8 \cdot 7 = 2 \cdot 8 \cdot 7$

$\therefore$ Required probability $= \dfrac{2 \cdot 8 \cdot 7}{32 \cdot 63} = \dfrac{1}{18}$

4. Let a be the radius of the circle.

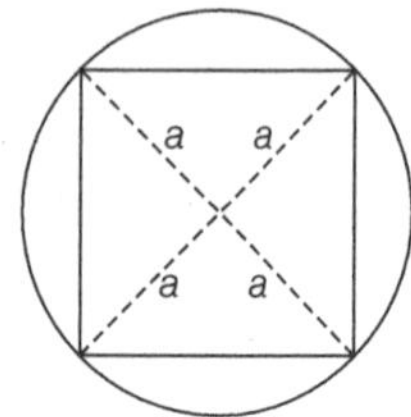

$\therefore$ The area of the inscribed square

$$= (a\sqrt{2})^2 = 2a^2$$

$\therefore \qquad p_1 = \dfrac{2a^2}{\pi a^2} = \dfrac{2}{\pi}$

and $\qquad p_2 = \dfrac{\pi a^2 - 2a^2}{\pi a^2} = \dfrac{\pi - 2}{\pi}$

$\therefore \qquad p_1 - p_2 = \dfrac{2}{\pi} - \left(\dfrac{\pi - 2}{\pi}\right) = \dfrac{4}{\pi} - 1 > 0$

$\Rightarrow \qquad p_1 > p_2$

Also, $\quad (p_1 - p_2)(p_1 + p_2) = \left(\dfrac{4 - \pi}{\pi}\right)(1)$

but $\qquad 3 < \pi < 4$

$\Rightarrow \qquad (p_1^2 - p_2^2) = \left(\dfrac{4 - \pi}{\pi}\right) < \dfrac{1}{3}$

Hence, $\quad (p_1^2 - p_2^2) < \dfrac{1}{3}$

5. Let a be the radius of the circle, l be the length of the chord and r be the distance of the mid-point of the chord from the centre of the circle.

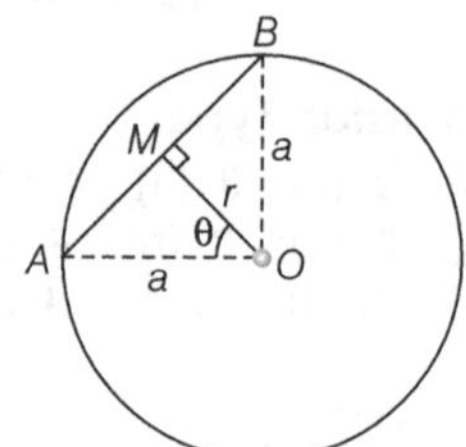

Let $\quad \angle AOM = \theta$

$\therefore \qquad \sin\theta = \dfrac{AM}{a}$ and $\cos\theta = \dfrac{r}{a}$

$\Rightarrow \qquad r = a\cos\theta,\ l = 2AM = 2a\sin\theta$

Given, $\quad (2a)\dfrac{2}{3} < AB < (2a)\dfrac{5}{6}$

$\Rightarrow \qquad \dfrac{4a}{3} < l < \dfrac{10a}{6}$

$\Rightarrow \qquad \dfrac{4a}{3} < 2a\sin\theta < \dfrac{10a}{6}$

$\Rightarrow \qquad \dfrac{2}{3} < \sin\theta < \dfrac{5}{6}$

$\Rightarrow \qquad \dfrac{4}{9} < 1 - \cos^2\theta < \dfrac{25}{36}$

or $\qquad \dfrac{4}{9} - 1 < -\cos^2\theta < \dfrac{25}{36} - 1$

$\Rightarrow \qquad \dfrac{5}{9} > \cos^2\theta > \dfrac{11}{36}$

$\Rightarrow \qquad \dfrac{\sqrt{11}}{6}a < a\cos\theta < \dfrac{\sqrt{5}}{3}a$

or $\qquad \dfrac{\sqrt{11}\,a}{6} < r < \dfrac{\sqrt{5}\,a}{3}$

∴ The given condition is satisfied, if the mid point of the chord lies within the region between the concentric circles of radius $\frac{\sqrt{11}\,a}{6}$ and $\frac{\sqrt{5}}{3}a$

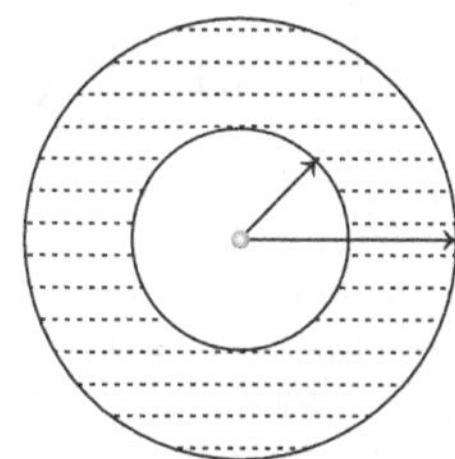

Hence, the required probability

$$= \frac{\text{the area of the circular annulus}}{\text{area of the given circle}}$$

$$= \frac{\pi\left(\frac{5}{9}a_2 - \frac{11}{36}a^2\right)}{\pi a^2} = \frac{5}{9} - \frac{11}{36}$$

$$= \frac{20-11}{36} = \frac{9}{36} = \frac{1}{4}$$

6. First three prime numbers are 2, 3 and 5

∴ Total ways of choosing $a, b, c = 3 \times 2 \times 1$

$= 6$ $[\because a \neq b \neq c]$

For roots to be real, $b^2 - 4ac \geq 0$

Value of b	Possible values of a and c
2	$4 - 4ac \geq 0 \Rightarrow ac \leq 1$ $a = c = 1$ (No values of a and c)
3	$9 - 4ac \geq 0 \Rightarrow ac \leq \frac{9}{4}$ $ac = 1, ac = 2$ $a = c = 1, a = 2, c = 1$ $a = 1, c = 2$ (no values of a and c)
5	$25 - 4ac \geq 0 \Rightarrow ac \leq \frac{25}{4}$ $\therefore ac = 1, 2, 3, 4, 5, 6$ $ac \neq 1, 2, 3, 4, 5$ for $ac = 6$ (2, 3) (3, 2)

∴ Favourable ways = 2

$$\therefore \text{Required probability} = \frac{\text{Favourable ways}}{\text{Total ways}} = \frac{2}{6} = \frac{1}{3}$$

7. Numbers on die are 1, 2, 3, 4, 5, 6

Prime numbers are 2, 3, 5 and non prime numbers are 1, 4, 6.

Now let weight assigned to non prime number is λ, then weight assigned to prime number is 2λ

$$\therefore \quad \overset{1}{\lambda} + \overset{2}{2\lambda} + \overset{3}{2\lambda} + \overset{4}{\lambda} + \overset{5}{2\lambda} + \overset{6}{\lambda} = 1$$

$$\Rightarrow \quad \lambda = \frac{1}{9}$$

∴ Probability that an odd number will be show up when the die is tossed

$= 1$ or 3 or 5

$= \lambda + 2\lambda + 2\lambda = 5\lambda = \frac{5}{9}$

8. $\sum_{x=0}^{8} P(X) = 1$

$\Rightarrow \quad P(0) + P(1) + P(2) + P(3) + P(4) + P(5) + P(6) + P(7) + P(8) = 1$

$\Rightarrow \lambda + 3\lambda + 5\lambda + 7\lambda + 9\lambda + 11\lambda + 13\lambda + 15\lambda + 17\lambda = 1$

$\Rightarrow \quad \frac{9}{2}(\lambda + 17\lambda) = 1$

$\Rightarrow \quad \frac{9}{2} \times 18\lambda = 1$

$\Rightarrow \quad 81\lambda = 1$

$\therefore \quad \lambda = \frac{1}{81}$

9. Sample space $\rightarrow {}^{12}C_2$

Number of possibilities for z is even.

$z = 0 \Rightarrow {}^{11}C_1$

$z = 2 \Rightarrow {}^{9}C_1$

$z = 4 \Rightarrow {}^{7}C_1$

$z = 6 \Rightarrow {}^{5}C_1$

$z = 8 \Rightarrow {}^{3}C_1$

$z = 10 \Rightarrow {}^{1}C_1$

Total = 36

$\therefore$ Probability $= \frac{36}{66} = \frac{6}{11}$

10. Let E_i be the event that the number i is found on the first die and E_j, the event that the number j is found on the second.

Given $i \leq j$

$\therefore$ Required probability $= \sum_{j=1}^{6} P(E_i \cap E_j), i = 1, 2, 3, \ldots, 6$

$= \sum_{j=1}^{6} P(E_i)P(E_j)$ $[\because E_i \text{ and } E_j \text{ are independent}]$

$= \sum_{j=1}^{6} \frac{1}{6} \cdot P(E_j)$

$= \frac{1}{6}\left\{\sum_{j=1}^{6} P(E_j) + \sum_{j=2}^{6} P(E_j) + \sum_{j=3}^{6} P(E_j) + \ldots + \sum_{j=1}^{6} P(E_j)\right\}$

$= \frac{1}{6}\left\{\frac{6}{6} + \frac{5}{6} + \frac{4}{6} + \frac{3}{6} + \frac{2}{6} + \frac{1}{6}\right\}$

$= \frac{6 \times 7}{2(36)} = \frac{7}{12}$

11. Given $P(\overline{A \cup B}) = \frac{1}{6}, P(A \cap B) = \frac{1}{4},$

$P(\bar{A}) = \frac{1}{4}$

$\therefore \quad P(A \cup B) = 1 - P(\overline{A \cup B}) = 1 - \frac{1}{6} = \frac{5}{6}$

and $\quad P(A) = 1 - P(\bar{A}) = 1 - \frac{1}{4} = \frac{3}{4}$

$\therefore \quad P(A \cup B) = P(A) + P(B) - P(A \cap B)$

$$\frac{5}{6} = \frac{3}{4} + P(B) - \frac{1}{4}$$

$\Rightarrow \quad P(B) = \frac{1}{3}$

$\Rightarrow$ A and B are not equally likely

$$P(A \cap B) = P(A) \cdot P(B) = \frac{1}{4}$$

So, events are independent.

12. The sample space is

$$S = \{-0.50, -0.49, -0.48, \ldots, -0.01, 0.00, 0.01, \ldots, 0.49\}$$

Let E be the event that the round off error is atleast 10 paise, then E' is the event that the round off error is atmost a paise.

$\therefore \quad E' = \{-0.09, -0.08\ldots, -0.01, 0.00, 0.01, \ldots, 0.09\}$

$\therefore \quad n(E') = 19$ and $n(S) = 100$

$\therefore \quad P(E') = \frac{n(E')}{n(S)} = \frac{19}{100}$

$\therefore$ Required probability , $P(E) = 1 - P(E')$

$$= 1 - \frac{19}{100} = \frac{81}{100}$$

13. Let S be the sample space and E be the event of getting a large number than the previous number.

$\therefore \quad n(S) = 6 \times 6 \times 6 = 216$

Now, we count the number of favourable ways.

Obviously, the second number has to be greater than 1. If the second number is i $(i > 1)$, then the number of favourable ways

$$= (i-1) \times (6-i)$$

$\therefore$ $n(E)$ = Total number of favourable ways

$$= \sum_{i=1}^{6} (i-1) \times (6-i)$$

$$= 0 + 1 \times 4 + 2 \times 3 + 3 \times 2 + 4 \times 1 + 0 = 20$$

Therefore, the required probability $= \frac{n(E)}{n(S)} = \frac{20}{216} = \frac{5}{54}$

14. Current in the upper part will flow only if both the switches a and b are closed.

$\therefore \quad$ Their probability $= p \cdot p = p^2$

Now, current in the lower part will flow only if switch c is closed. Its probability $= p$

$\therefore$ Probability of current flowing through AB

= Current flows either in upper part or flow in lower part

$= (p^2 + p)$

15. The probability of hitting in one shot $= \frac{10}{100} = \frac{1}{10}$

If he fires n shots, the probability of hitting atleast once

$$= 1 - \left(1 - \frac{1}{10}\right)^n = 1 - \left(\frac{9}{10}\right)^n = 50\% = \frac{50}{100} = \frac{1}{2}$$

$\Rightarrow \quad \left(\frac{9}{10}\right)^n = \left(\frac{1}{2}\right)$

$\therefore \quad n\{\log 9 - \log 10\} = \log 1 - \log 2$

$\Rightarrow \quad n\{2 \log 3 - 1\} = 0 - \log 2$

$\therefore \quad n = \frac{\log 2}{1 - 2\log 3} = \frac{0.3010300}{1 - 2 \times 0.4771213} = 6.6$

Hence, $\quad n = 7$

16. Clearly, $P(H)$ = Probability of getting head $= \frac{1}{2}$

and $P(T)$ = Probability of getting tail $= \frac{1}{2}$

Now, let E_1 be the event of getting a sum 7 or 8, when a pair of dice is rolled.

Then, $E_1 = \{(6, 1), (5, 2), (4, 3), (3, 4), (2, 5), (1, 6), (6, 2), (5, 3), (4, 4), (3, 5), (2, 6)\}$

$\Rightarrow P(E_1)$ = Probability of getting 7 or 8 when a pair of dice is thrown $= \frac{11}{36}$

Also, let $P(E_2)$ = Probability of getting 7 or 8 when a card is picked from cards numbered 1, 2, ..., 9 $= \frac{2}{9}$

$\therefore$ Probability that the noted number is 7 or 8

$= P((H \cap E_1)$ or $(T \cap E_2))$

$= P(H \cap E_1) + P(T \cap E_2)$

[$\because$ $(H \cap E_1)$ and $(T \cap E_2)$ are mutually exclusive]

$= P(H) . P(E_1) + P(T) \cdot P(E_2)$

[$\because$ $\{H, E_1\}$ and $\{T, E_2\}$ both are sets of independent events]

$$= \frac{1}{2} \times \frac{11}{36} + \frac{1}{2} \times \frac{2}{9} = \frac{19}{72}$$

17. Probability that the first critic favours the book,

$$P(E_1) = \frac{5}{5+2} = \frac{5}{7}$$

Probability that the second critic favours the book,

$$P(E_2) = \frac{4}{4+3} = \frac{4}{7}$$

Probability that the third critic favours the book,

$$P(E_3) = \frac{3}{3+4} = \frac{3}{7}$$

Majority will be in favour, if at least two critics favour the book

$$= P(E_1 \cap E_2 \cap \bar{E}_3) + P(E_1 \cap \bar{E}_2 \cap E_3) + P(\bar{E}_1 \cap E_2 \cap E_3) + P(E_1 \cap E_2 \cap E_3)$$

$$= P(E_1)P(E_2)P(\bar{E}_3) + P(E_1)P(\bar{E}_2)P(E_3) + P(\bar{E}_1 \cap E_2 \cap E_3) + P(E_1)P(E_2)P(E_3)$$

$$= \frac{5}{7} \times \frac{4}{7} \times \left(1 - \frac{3}{7}\right) + \frac{5}{7} \times \left(1 - \frac{4}{7}\right) \times \frac{3}{7} + \left(1 - \frac{5}{7}\right) \times \frac{4}{7} \times \frac{3}{7} + \frac{5}{7} \times \frac{4}{7} \times \frac{3}{7} = \frac{209}{343}$$

18. The total number of ways $= 2^n$

A person is odd man out if he is alone in getting a head or a tail.

The number of favourable ways

= The number of ways in which there is exactly one tail (head) and the rest are heads (tails)

$$= {}^nC_1 + {}^nC_1 = 2n$$

$$\therefore \quad P(E) = \frac{2n}{2^n} = \frac{1}{2} \text{ (given)}$$

$$\Rightarrow \quad n = 2^{n-2}$$

By trial, $n = 4$

19. The probability of hitting the target 5th time at the 10th throw = (the probability of hitting the target 4 times in the first 9 throws) × (the probability of hitting the target at the 10th throw)

$$= {}^9C_4\left(\frac{1}{2}\right)^4\left(\frac{1}{2}\right)^5 \times \frac{1}{2}$$

$$= {}^9C_4\left(\frac{1}{2}\right)^{10}$$

$$= \frac{9\cdot8\cdot7\cdot6}{(1\cdot2\cdot3\cdot4)\,2^{10}} = \frac{63}{2^9}$$

20. Even numbers are 2, 4, 6

$\therefore$ The probability that an even number appear $= \frac{3}{6} = \frac{1}{2}$

$\therefore$ The required probability

= P(that an even number occurs once or thrice or five times......or $(2n+1)$ times)

$$= {}^{2n+1}C_1\left(\frac{1}{2}\right)\left(\frac{1}{2}\right)^{2n} + {}^{2n+1}C_3\left(\frac{1}{2}\right)^3\left(\frac{1}{2}\right)^{2n-2}$$

$$+ {}^{2n+1}C_5\left(\frac{1}{2}\right)^5\left(\frac{1}{2}\right)^{2n-4} +\ldots\ldots+ {}^{2n+1}C_{2n+1}\left(\frac{1}{2}\right)^{2n+1}\left(\frac{1}{2}\right)^0$$

$$= \frac{1}{2^{2n+1}}\{{}^{2n+1}C_1 + {}^{2n+1}C_3 + {}^{2n+1}C_5 +\ldots.+ {}^{2n+1}C_{2n+1}\}$$

$$= \frac{2^{2n}}{2^{2n+1}} = \frac{1}{2}$$

21. The probability

$$= P(\bar{A}_1 \cap \bar{A}_2 \cap \bar{A}_3 \cap \ldots \cap \bar{A}_n)$$

(none of the events occur)

$$= P(\bar{A}_1)P(\bar{A}_2)P(\bar{A}_3)\ldots P(\bar{A}_n)$$

$$= \frac{1}{2}\cdot\frac{2}{3}\cdot\frac{3}{4}\cdots\frac{n}{n+1}$$

$$= \frac{1}{(n+1)}$$

$\therefore$ Probability that atleast one of the events occurs

$$= 1 - \frac{1}{n+1} = \frac{n}{n+1}$$

22. $P(x = 9) = P(x - 7)$

$${}^nC_9\left(\frac{1}{2}\right)^{n-9} \times \left(\frac{1}{2}\right)^9 = {}^nC_7\left(\frac{1}{2}\right)^{n-7} \times \left(\frac{1}{2}\right)^7$$

$${}^nC_9 \times \left(\frac{1}{2}\right)^2 = \left(\frac{1}{2}\right)^2 \times {}^nC_7$$

$$x + y = n \Rightarrow n = 16$$

$$P(x = 2) = {}^{16}C_2 \times \left(\frac{1}{2}\right)^{14} \times \left(\frac{1}{2}\right)^2$$

$$= {}^{16}C_2 \times \left(\frac{1}{2}\right)^{16} = \frac{15}{2^{13}}$$

23. Probability of student (if he miss no any test)

$$= \left(1 - \frac{1}{5}\right) \times \left(1 - \frac{1}{5}\right) = \frac{16}{25}$$

Hence, probability that he will miss atleast one test

$$= 1 - \frac{16}{25} = \frac{9}{25}$$

24. In n sides polygon, number of diagonals $= {}^nC_2 - n$

Number of ways of choosing any 2 diagonal $= {}^{({}^nC_2 - n)}C_2$

Now, out of these n sides, we shall choose 4 points so that they will form a quadrilateral and diagonal must intersect inside it.

Number of ways $= {}^nC_4$

$$\text{Probability} = \frac{{}^nC_4}{{}^{({}^nC_2 - n)}C_2}$$

25. Probability of defective transistor $= \frac{5}{15} = \frac{1}{3}$

and probability of non defective transistor $= 1 - \frac{1}{3}$

$$= \frac{2}{3}$$

Probability that the inspectors finds a non defective transistor

$$= \frac{2}{3} \times \frac{2}{3} \times \frac{2}{3}$$

$$= \frac{8}{27}$$

Hence, probability that atleast one of the inspectors finds a defective transistors

$$= 1 - \frac{8}{27} = \frac{19}{27}$$

26. $P(X = 1) = {}^5C_1 \cdot p \cdot q^4 = 0.4096$

$P(X = 2) = {}^5C_2 \cdot p^2 \cdot q^3 = 0.2048$

$$\Rightarrow \quad \frac{q}{2p} = 2$$

$$\Rightarrow \quad q = 4p \text{ and } p + q = 1$$

$$\Rightarrow \quad p = \frac{1}{5} \text{ and } q = \frac{4}{5}$$

Now, $P(X = 3) = {}^5C_3 \cdot \left(\frac{1}{5}\right)^3 \cdot \left(\frac{4}{5}\right)^2$

$$= \frac{10 \times 16}{125 \times 25} = \frac{32}{625}$$

27. Favourable ways $= 3!\left(1 - \frac{1}{1!} + \frac{1}{2!} - \frac{1}{3!}\right) = 2$

Total ways $= 3!$

$$\therefore \quad \text{Probability} = \frac{2}{3!} = \frac{1}{3}$$

28. Let E denote the event that a six occurs and A the event that the man reports that it is a six. We have $P(E) = 1/6$, $P(E') = 5/6$, $P(A/E) = \frac{3}{4}$ and $P(A/E') = 1/4$. By Baye's theorem

$$P(E/A) = \frac{P(E)\cdot P(A/E)}{P(E)\cdot P(A/E) + P(E')\cdot P(A/E')}$$

$$= \frac{\left(\frac{1}{6}\right)\left(\frac{3}{4}\right)}{\left(\frac{1}{6}\right)\left(\frac{3}{4}\right)+\left(\frac{5}{6}\right)\left(\frac{1}{4}\right)} = \frac{3}{8}$$

29. Required probability $= \frac{{}^2C_1}{{}^9C_1} \times \frac{{}^4C_1}{{}^9C_1} = \frac{8}{81}$

30. Total number of ways to arrange 3 boys and 2 girls are 5!.

According to given condition, following cases may arise.

B	G	G	B	B
G	G	B	B	B
G	B	G	B	B
G	B	B	G	B
B	G	B	G	B

So, number of favourable ways $= 5 \times 3! \times 2! = 60$

$\therefore$ Required probability $= \frac{60}{120} = \frac{1}{2}$

31. There are 11 letters in the word 'PROBABILITY' out of which 1 can be selected in ${}^{11}C_1$ ways

$\therefore$ Exhaustive number of cases $= {}^{11}C_1 = 11$

There are three vowels viz AIO.

Therefore, favourable number of cases $= {}^3C_1 = 3$

Hence, the required probability $= 3/11$

32. 100 bulbs = 10 defective + 90 non-defective

Probability that 3 out of 4 bulbs, bought by a customer will not to be defective

$$= \frac{{}^{90}C_3 \times {}^{10}C_1}{{}^{100}C_4}$$

33. Total coupons = 15

$1 \le$ selected coupon number ≤ 9

i.e. 1, 2, 3, 4, 5, 6, 7, 8, 9.

$\therefore$ Probability of one selected coupon $= \frac{9}{15} = \frac{3}{5}$

Hence, the required probability

$$= \left(\frac{3}{5}\right) \times \left(\frac{3}{5}\right) \times \ldots 7 \text{ times}$$

$$= \left(\frac{3}{5}\right)^7$$

34. We have, mentioned that boxes are different and one particular box has 3 balls.

Then, number of ways $= \frac{{}^{12}C_3 \times 2^9}{3^{12}} = \frac{55}{3}\left(\frac{2}{3}\right)^{11}$

35. Total ways $= n!$ $\quad$ ($\because$ letters different)

Favourable ways when all letters placed in the right envelope = 1

$\therefore$ Probability of all letters in right envelopes $= \frac{1}{n!}$

$\therefore$ Probability that atleast one letter is not placed in the right envelope

$$= 1 - \frac{1}{n!}$$

36. $P(A) = P(B) = 2P(C) = x$ (say)

$\therefore \quad P(A) + P(B) + P(C) = 1 \Rightarrow x = \frac{2}{5}$

$P(A) = \frac{2}{5} = P(B),\ P(C) = \frac{1}{5}$

$\therefore$ The probability that B or C wins the race

$$= P(B) + P(C) = \frac{2}{5} + \frac{1}{5} = \frac{3}{5}$$

37. Let us assume that 'A' wins after n deuces, $n \in [(0, \infty)$

Probability of deuce $= \frac{2}{3}\cdot\frac{2}{3} + \frac{1}{3}\cdot\frac{1}{3} = \frac{5}{9}$

(A wins his serve, then B wins his serve or A loses his serve, then B also loses his serve). Now, probability of 'A' winning the game

$$= \sum_{n=0}^{\infty} (5/9)^n \cdot \left(\frac{2}{3}\right)\frac{1}{3} = \frac{1}{1-(5/9)} \cdot \frac{2}{9} = \frac{1}{2}$$

38. 13 applicants = 8 men + 5 women

2 persons are selected

i.e. (1 men + 1 women) or 2 women

$\therefore$ Required probability

$$= \frac{{}^8C_1 \times {}^5C_1}{{}^{13}C_2} + \frac{{}^5C_2}{{}^{13}C_2}$$

$$= \frac{50}{{}^{13}C_2}$$

$$= \frac{25}{39}$$

39. Chance of A winning the race $= \frac{1}{6}$

and chance of B winning the race $= \frac{1}{8}$

$\therefore$ Chance of wins the race $= \frac{1}{6} + \frac{1}{8} = \frac{14}{48}$

$$= \frac{7}{24}$$

Hence, chance that neither wins the race

$$= 1 - \frac{7}{24}$$

$$= \frac{17}{24}$$

40. Probability of diamond card $= \frac{{}^{13}C_1}{{}^{52}C_1} = \frac{1}{4}$

Probability of C's winning

$= P(\bar{A})P(\bar{B})P(C) + P(\bar{A})P(\bar{B})P(\bar{C})P(\bar{A})P(\bar{B})P(C) + \ldots$

$$= \frac{P(\bar{A})P(\bar{B})P(C)}{1 - P(\bar{A})P(\bar{B})P(\bar{C})} = \frac{\frac{3}{4}\times\frac{3}{4}\times\frac{1}{4}}{1-\frac{3}{4}\times\frac{3}{4}\times\frac{3}{4}} = \frac{9}{37}$$

41. B G B G B G B G B G B G B

Places for boys = 7 (Here, 6 girls and 5 boys)

Hence, probability no two boys sit together

$$=\frac{^5P_5 \times 6!}{11!}=\frac{7!6!}{2!11!}$$

42. Since, A is a finite set, therefore every injective map from A to itself is bijective also. Total number of bijections from A to itself is n ! and total number of functions from A to itself is n^n.

$\therefore$ The required probability $=\frac{n!}{n^n}$

43. Total number = 6 (i.e. 1, 2, 3, 4, 5, 6)

Favourable number = 2, 3, 4, 5 = 4

Probability of favourable number in one draw $=\frac{4}{6}=\frac{2}{3}$

$\therefore$ Required probability $=\left(\frac{2}{3}\right)^4=\frac{16}{81}$

44. Total number of ways of selecting 2 different numbers from $\{0, 1, 2, ..., 10\} = {}^{11}C_2 = 55$

Let two numbers selected be x and y.

Then, $x + y = 4m$ ...(i)

and $x - y = 4n$ ...(ii)

$\Rightarrow$ $2x = 4(m + n)$

and $2y = 4(m - n)$

$\Rightarrow$ $x = 2(m + n)$

and $y = 2(m - n)$

$\therefore x$ and y both are even numbers.

x	y
0	4, 8
2	6, 10
4	0, 8
6	2, 10
8	0, 4
10	2, 6

$\therefore$ Required probability $=\frac{6}{55}$

45. Total ways $= 15^7$

For favourable ways, we must 7 coupons numbered from 1 to 9 so that '9' is selected atleast once. Thus, total number of favourable ways are, $9^7 - 8^7$.

$\Rightarrow$ Required probability $=\frac{9^7 - 8^7}{15^7}$

46. Square numbers are

$1^2 = 1$, $6^2 = 36$

$2^2 = 4$, $7^2 = 49$

$3^2 = 9$, $8^2 = 64$

$4^2 = 16$, $9^2 = 81$

$5^2 = 25$, $10^2 = 100$

Which are lie between 1 to 100

$\therefore$ Required probability $=\frac{10}{100}=\frac{1}{10}$

47. Numbers divisible by 3 are 3, 6, 9, 12, 15, 18, 21, 24 and numbers divisible by 4 are 4, 8, 12, 16, 20, 24 and numbers divisible by $(3 \times 4 = 12)$ are 12, 24

$\therefore$ Probability that the chosen number is divisible by 3 or 4.

$$=\frac{^8C_1}{^{25}C_1}+\frac{^6C_1}{^{25}C_1}-\frac{^2C_1}{^{25}C_1}=\frac{12}{25}$$

48. Probability through a six $=\frac{1}{6}$

$P(A)=\frac{1}{6}, P(\bar{A})=\frac{5}{6}, P(B)=\frac{1}{6}, P(\bar{B})=\frac{5}{6}$

Probability of winning A

$= P(A) + P(\bar{A})P(\bar{B})P(A) + P(\bar{A})P(\bar{B})\ P(\bar{A})P(\bar{B})P(A) + \ldots$

$=\frac{1}{6}+\frac{5}{6}\times\frac{5}{6}\times\frac{1}{6}+\frac{5}{6}\times\frac{5}{6}\times\frac{5}{6}\times\frac{5}{6}\times\frac{1}{6}+\ldots$

$$=\frac{\frac{1}{6}}{1-\frac{25}{36}}=\frac{6}{11}$$

Probability of winning $B = 1-\frac{6}{11}=\frac{5}{11}$

$\therefore$ Expectations of A and B are

$\frac{6}{11}\times 11 =$ ₹ 6 and $\frac{5}{11}\times 11 =$ ₹ 5

49. Three-digit numbers

100, 102, ...,999

Total number = 990

Three-digit number (which has all same digits)

111, 222, 333,...,999

Favourable number = 9

$\therefore$ Required probability $=\frac{9}{900}=\frac{1}{100}$

50. $P(\bar{M})=\frac{20}{100}=\frac{1}{5}, P(\bar{P})=\frac{25}{100}=\frac{1}{4}$

and $P(\bar{M}\cap\bar{P})=\frac{12}{100}=\frac{3}{25}$

$\therefore$ Required probability $=P\left(\frac{\bar{P}}{\bar{M}}\right)$

$$=\frac{P(\bar{P}\cap\bar{M})}{P(\bar{M})}=\frac{\frac{3}{25}}{\frac{1}{5}}=\frac{3}{5}$$

51. In {1, 2, 3, ..., 11} there are 5 even numbers and 6 odd numbers. The sum even is possible only when both are odd or both are even.

Let A be the event that denotes both numbers are even and B be the event that denotes sum of numbers is even. Then, $n(A) = {}^5C_2$ and $n(B) = {}^5C_2 + {}^6C_2$

Required probability

$$P(A/B)=\frac{P(A\cap B)}{P(B)}=\frac{{}^5C_2/{}^{11}C_2}{\frac{({}^6C_2+{}^5C_2)}{{}^{11}C_2}}$$

$$=\frac{{}^5C_2}{{}^6C_2+{}^5C_2}=\frac{10}{15+10}=\frac{2}{5}$$

52. $P(A) = 0.3, P(B) = 0.6$

$$P(A \cap B) = 0.18$$
$$P(\bar{A} \cap \bar{B}) = 1 - P(A \cup B)$$
$$= 1 - [P(A) + P(B) - P(A \cap B)]$$
$$= 1 - (0.3 + 0.6 - 0.18)$$
$$= 1 - (0.90 - 0.18)$$
$$= 0.10 + 0.18$$
$$= 0.28$$

53. Total number of cases $= n^n$

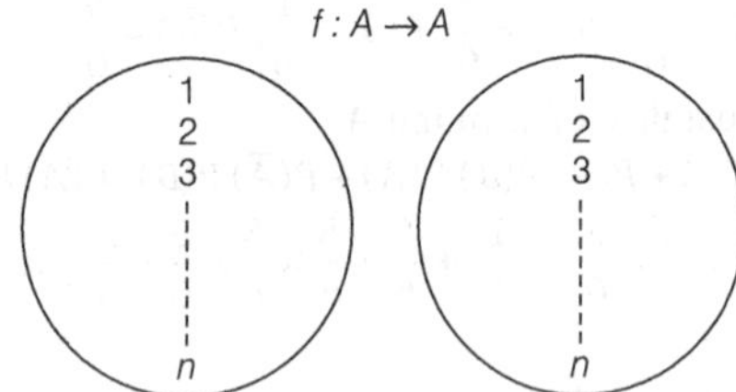

We next find the number of favourable cases. For the first element we have n choices. For the second element we have $(n - 1)$ choices and so on.

$\therefore$ The number of favourable cases

$$= n\,(n-1)\,(n-2)\ldots 2\cdot 1$$
$$= n!$$

$\therefore$ Required probability $= \dfrac{n!}{n^n}$

$$= \frac{(n-1)!}{n^{n-1}}$$

54. Let A denotes the event that the throws of the two persons are unequal. Then, A' denote the event that the throws of the two persons are equal.

The total number of cases for A' is $(36)^2$.

We now proceed to find out the number of favourable cases for A'. Suppose

$(x + x^2 + x^3 + \ldots x^6)^2 = a_2x^2 + a_3x^3 + \ldots + a_{12}x^{12}$

The number of favourable ways for

$A' = a_2^2 + a_3^2 + \ldots + a_{12}^2$

= coefficient of constant term in

$$(a_2x^2 + a_3x^3 + \ldots + a_{12}x^{12}) \times \left(\frac{a_2}{x^2} + \frac{a_3}{x^3} + \ldots + \frac{a_{12}}{x^{12}}\right)$$

= coefficient of constant term in

$$\frac{(1 - x^6)^2\,(1 - 1/x^6)^2}{(1-x)^2\,(1 - 1/x)^2}$$

= coefficient of x^{10} in $(1 - x^6)^4\,(1 - x)^{-4}$

= coefficient of x^{10} in

$(1 - 4x^6 + 6x^{12})\,(1 + {}^4C_1x + {}^5C_2x^2 + {}^6C_3x^3 + \ldots)$

$$= {}^{13}C_{10} - 4\cdot {}^7C_4 = 146$$

$\therefore$ $P(A') = \dfrac{146}{36^2} = \dfrac{73}{648}$

$\Rightarrow$ $P(A) = 1 - P(A')$

$$= \frac{575}{648}$$

55. Let $x = P$ (computer turns out to be defective, given that it is produced in plant T_2)

$\Rightarrow$ $x = P\left(\dfrac{D}{T_2}\right)$ ...(i)

where, D = Defective computer

$\therefore P$ (computer turns out to be defective given that is produced in plant T_1) $= 10x$

i.e. $P\left(\dfrac{D}{T_1}\right) = 10x$ ...(ii)

Also, $P(T_1) = \dfrac{20}{100}$ and $P(T_2) = \dfrac{80}{100}$

Given, P (defective computer) $= \dfrac{7}{100}$

i.e. $P(D) = \dfrac{7}{100}$

Using law of total probability,

$$P(D) = 9(T_1)P\left(\frac{D}{T_1}\right) + P(T_2)\cdot P\left(\frac{D}{T_2}\right)$$

$\therefore$ $\dfrac{7}{100} = \left(\dfrac{20}{100}\right)\cdot 10x + \left(\dfrac{80}{100}\right)\cdot x$

$\Rightarrow$ $7 = (280)x \Rightarrow x = \dfrac{1}{40}$ ...(iii)

$\therefore$ $P\left(\dfrac{D}{T_2}\right) = \dfrac{1}{40}$ and $P\left(\dfrac{D}{T_1}\right) = \dfrac{10}{40}$

$\Rightarrow$ $P\left(\dfrac{\bar{D}}{T_2}\right) = 1 - \dfrac{1}{40} = \dfrac{39}{40}$

and $P\left(\dfrac{\bar{D}}{T_1}\right) = 1 - \dfrac{10}{40} = \dfrac{30}{40}$ ...(iv)

Using Baye's theorem,

$$P\left(\frac{T_2}{\bar{D}}\right) = \frac{P(T_2 \cap \bar{D})}{P(T_1 \cap \bar{D}) + P(T_2 \cap \bar{D})}$$

$$= \frac{P(T_2)\cdot P\left(\frac{\bar{D}}{T_2}\right)}{P(T_1)\cdot P\left(\frac{\bar{D}}{T_1}\right) + P(T_2)\cdot P\left(\frac{\bar{D}}{T_2}\right)}$$

$$= \frac{\frac{80}{100}\cdot\frac{39}{40}}{\frac{20}{100}\cdot\frac{30}{40} + \frac{80}{100}\cdot\frac{39}{40}} = \frac{78}{93}$$

56. $P\,(W) = \dfrac{a}{a+b}$ and $P(B) = \dfrac{b}{a+b}$

Probability of A winning

$$= P(W) + P(\bar{W})\,P(\bar{B})P(W) + \ldots$$
$$= \frac{P(W)}{1 - P(\bar{W})P(\bar{B})}$$
$$= \frac{\frac{a}{a+b}}{1 - \frac{1}{a+b}\cdot\frac{a}{a+b}} = \frac{a(a+b)}{a^2 + b^2 + ab} \quad \text{(say)}$$

$$p_1 = \frac{a^2 + ab}{a^2 + b^2 + ab}$$

Probability of B winning $= 1 - p_1$

$$= 1 - \frac{a^2 + ab}{a^2 + b^2 + ab}$$

$$= \frac{b^2}{a^2 + b^2 + ab} = p_2 \text{(say)}$$

Given, $\quad p_1 = 2p_2$

$$\Rightarrow \quad \frac{a^2 + ab}{a^2 + b^2 + ab} = \frac{2b^2}{a^2 + b^2 + ab}$$

$$\Rightarrow \quad a^2 + ab - 2b^2 = 0$$

$$\Rightarrow \quad (a - b)(a + 2b) = 0$$

$$a + 2b \neq 0$$

$$\therefore \quad a - b = 0$$

$$\Rightarrow \quad a = b$$

$$\therefore \quad a : b = 1 : 1$$

57. We have,

$(X = 7) = \{07, 16, 25, 34, 43, 52, 61, 70\}$

and $(Y = 0) = \{00, 01, 02, \ldots, 10, 20, 30, \ldots, 90\}$

Thus, $(X = 7) \cap (Y = 0) = \{07, 70\}$

$$\therefore \quad P(X = 7/Y = 0) = \frac{P\{(X = 7) \cap (Y = 0)\}}{P(Y = 0)} = \frac{2}{19}$$

58. Let $X = \{x_1, x_2, \ldots, x_n\}$

For each $x_i \in X$ $(1 \leq i \leq n)$, we have the following four choices

(1) $x_i \in A$ and $x_i \in B$, (2) $x_i \in A$ and $x_i \notin B$

(3) $x_i \notin A$ and $x_i \in B$, (4) $x_i \notin A$ and $x_i \notin B$

Thus, the number of ways of choosing A and B is 4^n. Out of the four choices listed above (4) is not favourable for the occurrence of $A \cup B = X$. Therefore, the number of favourable ways is 3^n.

Hence, the probability of the required event is $(3/4)^n$.

59. $x + \dfrac{100}{x} > 50$

$$\Rightarrow \quad x^2 - 50x + 100 > 0$$

$$\Rightarrow \quad (x - 25)^2 > 525$$

$$\Rightarrow \quad x - 25 < -\sqrt{525} \text{ or } x - 25 > \sqrt{525}$$

$$\Rightarrow \quad x < 25 - \sqrt{525} \text{ or } x > 25 + \sqrt{525}$$

As, x is a positive integer and $\sqrt{525} \approx 22.91$

We must have,

$$x \leq 2 \text{ or } x \geq 48$$

Thus, the number of favourable cases is $2 + 53 = 55$. Therefore, the probability of the required event.

$$= \frac{55}{100} = \frac{11}{20}$$

60. It is given that, a person wins ₹15 for throwing a doublet (1, 1) (2, 2), (3, 3), (4, 4), (5, 5), (6, 6) and win ₹12 when the throw results in sum of 9, i.e., when (3, 6), (4, 5), (5, 4), (6, 3) occurs.

Also, losses ₹6 for throwing any other outcome, i.e., when any of the rest $36 - 6 - 4 = 26$ outcomes occurs.

Now, the expected gain/loss

$= 15 \times P$ (getting a doublet) $+ 12 \times P$ (getting sum 9) $- 6 \times P$ (getting any of rest 26 outcome)

$$= \left(15 \times \frac{6}{36}\right) + \left(12 \times \frac{4}{36}\right) - \left(6 \times \frac{26}{36}\right)$$

$$= \frac{5}{2} + \frac{4}{3} - \frac{26}{6} = \frac{15 + 8 - 26}{6}$$

$$= \frac{23 - 26}{6} = -\frac{3}{6} = -\frac{1}{2}, \text{ means loss of ₹}\frac{1}{2}$$

61. Total coins = 8

Probability of getting head $= \dfrac{1}{2}$

And probability of getting tail $= \dfrac{1}{2} = \left(\dfrac{1}{2} + \dfrac{1}{2}\right)^8$

Hence, probability of getting atleast 6 heads up

$$= {}^8C_6\left(\frac{1}{2}\right)^6\left(\frac{1}{2}\right)^2 + {}^8C_7\left(\frac{1}{2}\right)^7\left(\frac{1}{2}\right)^1 + {}^8C_8\left(\frac{1}{2}\right)^8\left(\frac{1}{2}\right)^0$$

$$= \frac{{}^8C_6 + {}^8C_7 + {}^8C_8}{2^8} = \frac{{}^8C_2 + {}^8C_1 + 1}{2^8}$$

$$= \frac{37}{256}$$

62. Probability of defective bulb $= \dfrac{10}{100} = \dfrac{1}{10}$

Probability of non-defective bulb $= 1 - \dfrac{1}{10} = \dfrac{9}{10}$

Probability that out of a sample of 5 bulbs, none is defective

$$= {}^5C_5\left(\frac{9}{10}\right)^5\left(\frac{1}{10}\right)^0 = \left(\frac{9}{10}\right)^5$$

63. Probability that the problem is solved correctly by atleast one of them = 1 − (problem is not solved by all)

$\therefore P$ (problem is solved) $= 1 - P$ (problem is not solved)

$$= 1 - P(\bar{A}) \cdot P(\bar{B}) \cdot P(\bar{C}) \cdot P(\bar{D})$$

$$= 1 - \left(\frac{1}{2} \cdot \frac{1}{4} \cdot \frac{3}{4} \cdot \frac{7}{8}\right) = 1 - \frac{21}{256} = \frac{235}{256}$$

64. $P(A \cup B) = 0.6, P(A \cap B) = 0.2$

$$\Rightarrow \quad P(A) + P(B) - P(A \cap B) = 0.6$$

$$\Rightarrow \quad P(A) + P(B) = 0.8$$

$$\Rightarrow \quad 1 - P(\bar{A}) + 1 - P(\bar{B}) = 0.8$$

$$\Rightarrow \quad P(\bar{A}) + P(\bar{B}) = 1.2$$

65. $A \times B = (1, 3, 5, 7, 9) \times (2, 4, 6, 8)$

$= (1, 2), (1, 4), (1, 6), (1, 8), (3, 2), (3, 4), (3, 6), (3, 8), (5, 2), (5, 4), (5, 6), (5, 8), (7, 2), (7, 4), (7, 6), (7, 8), (9, 2), (9, 4), (9, 6), (9, 8)$

Total ways $= 5 \times 4 = 20$

Favourable case = (1, 8), (3, 6), (5, 4), (7, 2)

$[\because a + b = 9]$

$\therefore$ Number of favourable cases = 4

$$\therefore \quad \text{Required probability} = \frac{4}{20} = \frac{1}{5}$$

66. There are 10 digits 0, 1, 2, 3, 4, 5, 6, 7, 8, 9 the last two digits can be dialled in ${}^{10}P_2 = 90$ ways out of which only one way is favourable, thus, the required probability $= \dfrac{1}{90}$

67. We know that, $P(A/B)=\dfrac{P(A\cap B)}{P(B)}$

[by the definition of conditional probability]

$$\because \quad A\subset B$$
$$\Rightarrow \quad A\cap B=A$$
$$\therefore \quad P(A/B)=\frac{P(A)}{P(B)} \quad \ldots(\text{i})$$

As we know that, $0\le P(B)\le 1$

$$\therefore \quad 1\le \frac{1}{P(B)}<\infty$$
$$\Rightarrow \quad P(A)<\frac{P(A)}{P(B)}<\infty$$
$$\Rightarrow \quad \frac{P(A)}{P(B)}\ge P(A) \quad \ldots(\text{ii})$$

Now, from Eqs. (i) and (ii), we get

$$P(A/B)\ge P(A)$$

68. $P(A)=\dfrac{60}{100}=\dfrac{3}{5}, P(\bar{A})=1-\dfrac{3}{5}=\dfrac{2}{5}$

$P(B)=\dfrac{70}{100}=\dfrac{7}{10}, P(\bar{B})=1-\dfrac{7}{10}=\dfrac{3}{10}$

Probability of single event

$$=P(A)\cdot P(B)+P(\bar{A})P(\bar{B})$$
$$=\frac{3}{5}\times\frac{7}{10}+\frac{2}{5}\times\frac{3}{10}=\frac{27}{50}=0.54$$

69. Clearly, $7^{\lambda}+7^{\mu}$ will be divisible by 5, if 7^{λ} has 3 or 7 in the unit place and 7^{μ} has 7 or 3 in the unit place or 7^{λ} has 1 or 9 in the unit place and 7^{μ} has 9 or 1 in the unit place.

$\therefore$ For any choice λ, μ the digit in the unit place of $7^{\lambda}+7^{\mu}$ is 0, 4, 6, 8.

It is divisible by 5 only when this digit is 0.

$\therefore$ The required probability $=\dfrac{1}{4}$

70. $P(\bar{A}/\bar{B})=\dfrac{P(\bar{A}\cap\bar{B})}{P(\bar{B})}=\dfrac{P(\overline{A\cup B})}{P(\bar{B})}=\dfrac{1-P(A\cup B)}{P(\bar{B})}$

71. The number of three-digit numbers, which are multiple of $11=90-9=81$. Again, the number, which are divisible of 9 also, are divisible by 99, whose number is $10-1=9$.

So, required probability $=\dfrac{9}{81}=\dfrac{1}{9}$.

72. Total number of functions from A to $B=n(S)=5^7$

Total number of onto functions from A to B is

$$P(E)=\frac{7!}{3!\,4!}\times 5!+\frac{7!}{3!\,2!}\times\frac{1}{2!\,2!}\times 5!=\frac{7!\times 20}{6}$$
$$P(E)=\frac{n(E)}{n(S)}-\frac{7!\times 2}{3\times 5^6}$$

73. Let X denotes the largest number on the 3 tickets drawn. We have $P(X\le 7)=(7/20)^3$ and $P(X\le 6)=(6/20)^3$, thus

$$P(X=7)=(7/20)^3-(6/20)^3$$

74. The total number of ways of choosing two numbers out of 1, 2, 3, …, 30 is ${}^{30}C_2=435$

Since, x^2-y^2 is divisible by 3, if either a and b are divisible by 3 or none of a and b divisible by 3.

Thus, the favourable number of cases

$$={}^{10}C_2+{}^{20}C_2=235$$

$\therefore$ Required probability $=\dfrac{235}{435}=\dfrac{47}{87}$

75. $A=$ number less than five points

$=\{1, 2, 3, 4\}$

and $B=$ numbers less than three points

$=\{1, 2\}$

$$\therefore \quad A\cap B=\{1, 2\}$$
$$\therefore \quad n(A)=4, n(B)=2 \text{ and } n(A\cap B)=2$$

Hence, $P\left(\dfrac{B}{A}\right)=\dfrac{P(A\cap B)}{P(A)}=\dfrac{n(A\cap B)}{n(A)}$

$$=\frac{2}{4}=\frac{1}{2}$$

76. 1, 2, 3, 4, 5, 6, 7, 8, 9

$\Rightarrow$ (1, 8), (2, 7), (3, 6), (4, 5), 9

Favourable ways = Taking 3 in first four system 4 last one

$$={}^4C_3\times 1=4$$

Total ways $={}^9C_7={}^9C_2=36$

$\therefore$ Required probability $=\dfrac{4}{36}=\dfrac{1}{9}$

77. Let the lengths of three parts of the rod be x, y and $l-(x+y)$. Then, $x>0$, $y>0$ and $l-(x+y)>0$

i.e., $\quad x+y<l$ or $y<l-x$

Since, in a triangle, the sum of any two sides is greater than third side, so

$$x+y>l-(x+y)$$
$$\Rightarrow \quad y>\frac{l}{2}-x$$
$$x+l-(x+y)>y \Rightarrow y<\frac{l}{2}$$
$$y+l-(x+y)>x \Rightarrow x<\frac{l}{2}$$
$$\Rightarrow \quad \frac{l}{2}-x<y<\frac{l}{2} \text{ and } 0<x<\frac{l}{2}$$

So, required probability $=\dfrac{\int_0^{l/2}\int_{l/2-x}^{l/2}dy\,dx}{\int_0^{l}\int_0^{l-x}dy\,dx}$

$$=\frac{\int_0^{l/2}\left\{\frac{l}{2}-\left(\frac{l}{2}-x\right)\right\}dx}{\int_0^{l}(l-x)\,dx}$$

$$=\frac{\int_0^{l/2}x\,dx}{\int_0^{l}(l-x)\,dx}=\frac{\frac{l^2}{8}}{\frac{l^2}{2}}=\frac{1}{4}$$

78. Since, $P(X=2)=P(X=3)$, we get

$${}^5C_2p^2q^3={}^5C_3p^3q^2$$
$$\therefore \quad q=p \quad [\because q=1-q]$$
$$1-p=p$$
$$\therefore \quad p=1/2$$

79. The total number of ways of choosing the tickets is

$$4 \times 4 \times 4 \times 4 = 256$$

The number of ways in which the sum can be 23.

$= \text{coefficients of } x^{23} \text{ in } (1 + x + x^{10} + x^{11})^4$

$= \text{coefficient of } x^{23} \text{ in } (1 + x)^4 (1 + x^{10})^4$

$= \text{coefficient of } x^{23} \text{ in } (1 + 4x + 6x^2 + 4x^3 + x^4) \times (1 + 4x^{10} + 6x^{20})$

$= 4 \times 6 = 24$

The probability of required event $= \dfrac{24}{256} = \dfrac{3}{32}$

80. Total ways $= 100$

and $x^2 - 13x \le 30$

$\Rightarrow \left(x - \dfrac{13}{2}\right)^2 \le \dfrac{289}{4}$

$\Rightarrow -\dfrac{17}{2} \le x - \dfrac{13}{2} \le \dfrac{17}{2}$

$\Rightarrow -2 = x \le 15$

$\therefore$ x is natural number

$\therefore x = (1, 2, 3, 4, \ldots, 15)$

$\therefore$ Required probability $= \dfrac{15}{100} = \dfrac{3}{20}$

81. Let E_1 = Event that first ball drawn is red

E_2 = Event that first ball drawn is black

A = Event that second ball drawn is red

$$P(E_1) = \frac{4}{10}, P\left(\frac{A}{E_1}\right) = \frac{6}{12}$$

$$\Rightarrow P(E_2) = \frac{6}{10}, P\left(\frac{A}{E_2}\right) = \frac{4}{12}$$

By law of total probability,

$$P(A) = P(E_1) \times P\left(\frac{A}{E_1}\right) + P(E_2) \times P\left(\frac{A}{E_2}\right)$$

$$= \frac{4}{10} \times \frac{6}{12} + \frac{6}{10} \times \frac{4}{12} = \frac{24 + 24}{120} = \frac{48}{120} = \frac{2}{5}$$

82. From the tree diagram, it follows that

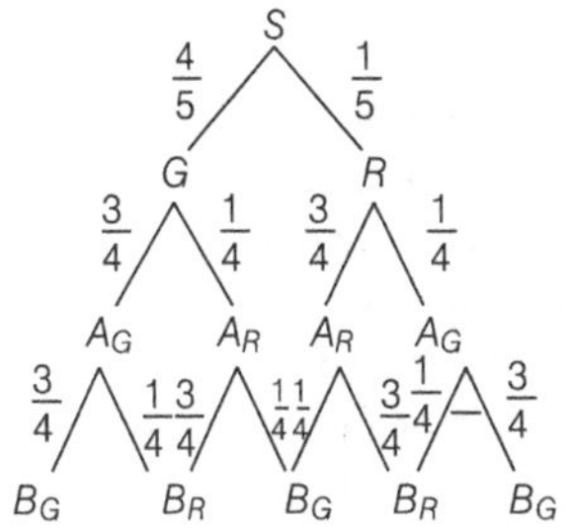

S
$\frac{4}{5}$ G, $\frac{1}{5}$ R
G: $\frac{3}{4}$ A_G, $\frac{1}{4}$ A_R; R: $\frac{3}{4}$ A_R, $\frac{1}{4}$ A_G
A_G: $\frac{3}{4}$ B_G, $\frac{1}{4}$ B_R; A_R: $\frac{3}{4}$ B_R, $\frac{1}{4}$ B_G; A_R: $\frac{1}{4}$ B_G, $\frac{3}{4}$ B_R; A_G: $\frac{1}{4}$ B_R, $\frac{3}{4}$ B_G

$$P(B_G) = \frac{46}{80}$$

$$P(B_G | G) = \frac{10}{16} - \frac{5}{8}$$

$$P(B_G \cap G) = \frac{5}{8} \times \frac{4}{5} = \frac{1}{2}$$

$$\therefore P(G | B_G) = \frac{\frac{1}{2}}{P(B_G)} = \frac{1}{2} \times \frac{80}{46} = \frac{20}{23}$$

83. Since, the graph of $y = 16x^2 + 8(a + 5)x - 7a - 5$ is strictly above X-axis, therefore $y > 0$ for all x

$\Rightarrow 16x^2 + 8(a + 5)x - 7a - 5 > 0$ for all x

$\Rightarrow$ Discriminant $< 0 \Rightarrow 64(a + 5)^2 + 64(7a + 5) < 0$

$\Rightarrow a^2 + 17a + 30 < 0 \Rightarrow -15 < a < -2$

So, required probability $= \dfrac{\int_{-15}^{-2} dx}{\int_{-20}^{0} dx} = \dfrac{13}{20}$.

84. Total ways $= {}^{12}C_{12} = 66$

E_3 = the even of the sum being 3
E_6 = the even of the sum being 6
E_9 = the even of the sum being 9
E_{12} = the even of the sum being 12
E_{15} = the even of the sum being 15
E_{18} = the even of the sum being 18
E_{21} = the even of the sum being 21
(sum maximum $= 11 + 12 = 23$)

[(Which is not divisible by 3)(Highest number divisible by 3 is 21]
[$\therefore n(E_3) = 1$, $n(E_6) = 2$, $n(E_9) = 4$,]

$n(E_{12}) = 5$, $n(E_{15}) = 5$, $n(E_{18}) = 3$, $n(E_{21}) = 3$

$\therefore$ Required probability $= \dfrac{\sum_{i=1}^{7} n(E_{3i})}{66} = \dfrac{22}{66} = \dfrac{1}{3}$

85. $P(A) = \dfrac{\int_0^1 \sqrt{x}\, dx}{\int_0^1 dx} = \dfrac{2}{3}; P(B) = \dfrac{\int_0^1 \sqrt{y}\, dy}{\int_0^1 dy} = \dfrac{2}{3}$

$$\therefore P(A \cap B) = \frac{\int_0^1 \sqrt{x}\, dx - \int_0^1 x^2\, dx}{\int_0^1 dx} = \frac{2}{3} - \frac{1}{3} = \frac{1}{3}$$

$$P(A) + P(B) = \frac{2}{3} + \frac{2}{3} = \frac{4}{3} \ne 1$$

$$P(A) \cdot P(B) = \frac{2}{3} \cdot \frac{2}{3} = \frac{4}{9} \ne P(A \cap B)$$

$$P(A \cap B) = 1, P(A) + P(B) = \frac{4}{3} \ne P(A \cup B)$$

86. There are 30 days in April but each person can have birthday in 30 days

$\therefore n(S) = 30 \times 30 \times \ldots \times 30$ (10 times) $= 30^{10}$

$n(E) = n(S)$ – The number of ways in which 10 persons can have different birthdays.

$\therefore n(E) = 30^{10} - {}^{30}C_{10}$

$\therefore P(E) = \dfrac{n(E)}{n(S)} = \dfrac{30^{10} - {}^{30}C_{10}}{30^{10}}$

87. Let A denote the event that a sum of 5 occurs, B the event that a sum of 7 occurs and C the event that neither a sum of 5 nor a sum of 7 occurs we have

$P(A) = \dfrac{4}{36} = \dfrac{1}{9}$, $P(B) = \dfrac{6}{36} = \dfrac{1}{6}$ and $P(C) = \dfrac{26}{36} = \dfrac{13}{18}$

Thus, $P(A$ occurs before $B)$

$P\{A \text{ or } (C \cap A) \text{ or } (C \cap C \cap A) \text{ or } \ldots\}$

$= P(A) + P(C)P(A) + P(C)P(C)P(A) + \ldots$

$$= \frac{P(A)}{1 - P(C)} = \frac{\frac{1}{9}}{1 - \frac{13}{18}} = \frac{2}{5} = 0.4$$

88. $n(S) = 2^4 = 16$ (because each of the four places can be filled in 2 ways). The zero determinants are

$$\begin{vmatrix} 1 & 1 \\ 1 & 1 \end{vmatrix}, \begin{vmatrix} 1 & 1 \\ -1 & -1 \end{vmatrix}, \begin{vmatrix} -1 & -1 \\ 1 & 1 \end{vmatrix}, \begin{vmatrix} -1 & 1 \\ -1 & 1 \end{vmatrix},$$

$$\begin{vmatrix} 1 & -1 \\ 1 & -1 \end{vmatrix}, \begin{vmatrix} -1 & 1 \\ -1 & 1 \end{vmatrix}, \begin{vmatrix} 1 & -1 \\ -1 & 1 \end{vmatrix}, \begin{vmatrix} -1 & -1 \\ -1 & -1 \end{vmatrix}$$

$n(E) = 8$

$\therefore \quad P(E) = \frac{n(E)}{n(S)} = \frac{8}{16} = \frac{1}{2}$

89. | | | | | |

Odd digits = 1, 3, 5, 7, 9
Even digits = 0, 2, 4, 6, 8
Since, odd digits at odd place and even digits at even place

Places of odd digits = 3
and places of even digits = 2

$\therefore$ Favourable ways $= {}^5P_3 \times {}^5P_2$
$= 1200$

Total ways $= 9 \times 9 \times 8 \times 7 \times 6$

$\therefore$ Required probability $= \frac{1200}{9 \times 9 \times 8 \times 7 \times 6}$
$= \frac{25}{567}$

90. Total ways = 1000
Favourable ways = Particular non zero digit appears atmost once
= {1, 2, 3, 4, …, 10, 20, 30,…,100, 200, 300, 400, …900, 1000}
= 28

$\therefore$ Required probability $= \frac{28}{1000} = \frac{7}{250}$

91. $P(A' \cap B) \le P(B)$
$\therefore$ Maximum value = 0.5

92. $\underbrace{x_1\ x_2 \ldots x_{19}\ x_{20}}_{2\text{ from }19}\ \underbrace{x_{21}\ x_{22} \ldots x_{50}}_{3\text{ from }30}$

Total ways $n(S) = {}^{50}C_5$

Favourable ways $n(E) = {}^{19}C_2 \times {}^{30}C_3$

$\therefore$ Required probability $= \frac{n(E)}{n(S)} = \frac{{}^{19}C_2 \times {}^{30}C_3}{{}^{50}C_5}$

93. Probability of success and failure in one draw are $\frac{7}{14} = \frac{1}{2}$ and $\frac{1}{2}$, respectively.

$\therefore$ The probability that atleast 3 balls of each colour is drawn

$= {}^7C_3\left(\frac{1}{2}\right)^3\left(\frac{1}{2}\right)^4 + {}^7C_4\left(\frac{1}{2}\right)^4\left(\frac{1}{2}\right)^3 = \frac{35}{64} > \frac{1}{2}.$

94. The probability of not drawing the ace in the first draw, in the second draw and in the third draw are (Here all spades i.e. 13 cards) $\frac{12}{13}, \frac{11}{12}, \frac{10}{11}$ respectively.

Probability of drawing ace of spades in the 4th draw $= \frac{1}{10}$ (Only one ace and remaining cards = 10)

$\therefore$ Required probability $= \frac{12}{13} \times \frac{11}{12} \times \frac{10}{11} \times \frac{1}{10} = \frac{1}{13}$

95. It is given that α is the number of heads that appear when C_1 is tossed twice, the probability distribution of random variable α is

α	0	1	2
$P(\alpha)$	$\left(\frac{1}{3}\right)^2$	$2\left(\frac{2}{3}\right)\left(\frac{1}{3}\right)$	$\left(\frac{2}{3}\right)^2$

Similarly, it is given that β is the number of heads that appear when C_2 is tossed twice, so probability distribution of random variable β is

β	0	1	2
$P(\beta)$	$\left(\frac{2}{3}\right)^2$	$\frac{4}{9}$	$\frac{1}{9}$

Now, as the roots of quadratic polynomial $x^2 - \alpha x + \beta$ are real and equal, so $D = \alpha^2 - 4\beta = 0$ and it is possible if $(\alpha, \beta) = (0, 0)$
or (2, 1)

$\therefore$ Required probability $= \left(\frac{1}{3}\right)^2\left(\frac{2}{3}\right)^2 + \left(\frac{2}{3}\right)^2\left(\frac{4}{9}\right)$
$= \frac{4}{81} + \frac{16}{81} = \frac{20}{81}$

96. $x \in N$, Total ways $= {}^{100}C_1 = 100$
$n(S) = 100$

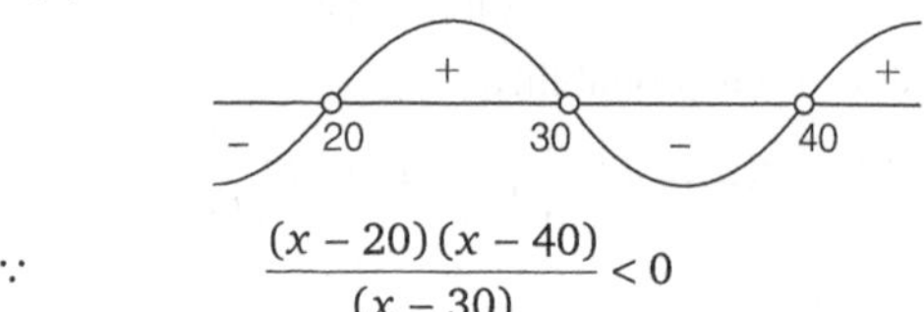

$\because \quad \frac{(x-20)(x-40)}{(x-30)} < 0$

$\therefore \quad x \in (-\infty, 20) \cup (30, 40)$
$\therefore \quad E = (1, 2, 3, \ldots 19, 31, 32, \ldots 39)$
$n(E) = 28$

$\therefore$ Required probability $= \frac{28}{100} = \frac{7}{25}$

97. The total number of ways of choosing a four digit number is $10^4 = 10000$. Let a_k denote the number of distinct non-negative integral solutions of the equation $x + y = k\ (0 \le k \le 18)$

$\therefore$ The number of favourable cases
$= a_0^2 + a_1^2 + a_2^2 + \ldots + a_{18}^2$

Suppose
$(1 + x + x^2 + \ldots + x^9)^2 = a_0 + a_1 x + a_2 x^2 + \ldots + a_{18}x^{18}$

Thus, $a_0^2 + a_1^2 + a_2^2 + \ldots + a_{18}^2$ = Coefficient of constant term in $(a_0 + a_1 x + \ldots + a_{18}x^{18}) \times \left(a_0 + \frac{a_1}{x} + \ldots + \frac{a_{18}}{x^{18}}\right)$

= Coefficient of constant term in

$(1 + x + x^2 + \ldots + x^9)^2 \times \left(1 + \frac{1}{x} + \frac{1}{x^2} + \ldots + \frac{1}{x^9}\right)^2$

$= \text{Coefficient of } x^{18} \text{ in } (1 + x + x^2 + \ldots + x^9)^4$
$= \text{Coefficient of } x^{18} \text{ in } (1 - x^{10})^4(1 - x)^{-4}$
$= \text{Coefficient of } x^{18} \text{ in } (1 - 4x^{10})(1 + {}^4C_1x + {}^5C_2x^2 + \ldots)$
$= {}^{21}C_{18} - 4.{}^{11}C_8$
$= {}^{21}C_3 - 4.{}^{11}C_3$
$= 1330 - 660 = 670$

$\therefore$ Required probability $= \dfrac{670}{10000} = 0.067$

98. Out of the numbers 10, 11, 12, 13, ..., 99 those numbers the product of whose digits is 12 are 26, 34, 43, 62 i.e. only 4.

$\therefore \quad p = P(E) = \dfrac{4}{90} = \dfrac{2}{45}$

$q = P(\bar{E}) = 1 - P(E) = 1 - \dfrac{2}{45} = \dfrac{43}{45}$

Hence, the probability that he will laugh atleast once
$= 1 - q^3$
$= 1 - \left(\dfrac{43}{45}\right)^3$

99. Here, $p = \dfrac{1}{2}$, $n = 8$

$\therefore \quad q = 1 - p = 1 - \dfrac{1}{2} = \dfrac{1}{2}$

$\therefore$ The binomial distribution is $\left(\dfrac{1}{2} + \dfrac{1}{2}\right)^8$

Also, $|x - 4| \le 2$
$\Rightarrow \quad -2 \le x - 4 \le 2$
$\Rightarrow \quad 2 \le x \le 6$

$\therefore \quad p(|x - 4| \le 2) = p(x = 2) + p(x = 3) + p(x = 4) + p(x = 5) + p(x = 6)$

$$= {}^8C_2\left(\frac{1}{2}\right)^6\left(\frac{1}{2}\right)^2 + {}^8C_3\left(\frac{1}{2}\right)^5\left(\frac{1}{2}\right)^3 + {}^8C_4\left(\frac{1}{2}\right)^4\left(\frac{1}{2}\right)^4 + {}^8C_5\left(\frac{1}{2}\right)^3\left(\frac{1}{2}\right)^5 + {}^8C_6\left(\frac{1}{2}\right)^2\left(\frac{1}{2}\right)^6$$

$$= \frac{{}^8C_2 + {}^8C_3 + {}^8C_4 + {}^8C_5 + {}^8C_6}{2^8}$$

$$= \frac{238}{256} = \frac{119}{128}$$

100. $f'(x) = 3x^2 + 2ax + b$

$y = f(x)$ is increasing
$\Rightarrow f'(x) \ge 0, \forall x$ and for $f'(x) = 0$ should not form an interval
$\Rightarrow (2a)^2 - 4 \times 3 \times b \le 0 \Rightarrow a^2 - 3b \le 0$

This is true for exactly 16 ordered pairs (a, b), $1 \le a, b \le 6$, namely (1, 1), (1, 2), (1, 3), (1, 4), (1, 5), (1, 6), (2, 2), (2, 3), (2, 4), (2, 5), (2, 6), (3, 3), (3, 4), (3, 5), (3, 6) and (4, 6). Thus, the required probability is $16/36 = 4/9$

101. The probability of the face number 1 to appear $= \dfrac{P(1)}{P(1) + P(2)} = \dfrac{0.1}{0.1 + 0.32}$

$= \dfrac{0.1}{0.42} = \dfrac{10}{42} = \dfrac{5}{21}$

102. S1. $P = \dfrac{{}^4C_3 \times {}^3C_2 + {}^4C_3 \times {}^2C_2 + {}^3C_3 \times {}^4C_2 + {}^3C_3 \times {}^2C_2}{{}^{10}C_5}$

$= \dfrac{4 \times 3 + 4 \times 1 + 1 \times 6 + 1}{{}^{10}C_5} = \dfrac{23}{{}^{10}C_5}$

S2. Two adjacent row can be selected out of 5 rows in 4 ways. Total ways of selecting 2 rows is ${}^5C_2$, hence

$P = \dfrac{4}{{}^5C_2} = \dfrac{2}{5}$

S3. Required probability

$$= \frac{P(A\bar{B}\bar{C})}{P(\bar{A}\bar{B}\bar{C}) + P(A\bar{B}\bar{C}) + P(\bar{A}\bar{B}\bar{C}) + P(\bar{A}\bar{B}C)}$$

$$= \frac{\frac{4}{5}\cdot\frac{1}{4}\cdot\frac{1}{3}}{\frac{1}{5}\cdot\frac{1}{4}\cdot\frac{1}{3} + \frac{4}{5}\cdot\frac{1}{4}\cdot\frac{1}{3} + \frac{1}{5}\cdot\frac{3}{4}\cdot\frac{1}{3} + \frac{1}{5}\cdot\frac{1}{4}\cdot\frac{2}{3}} = \frac{4}{10} = \frac{2}{5}$$

103. Let $E_1(E_2)$ denote the event that P_1 and P_2 are paired (not paired) together and let A denote the event that one of two players P_1 and P_2 is among the winners.
Since, P_1 can be paired with any of the remaining 15 players.

We have, $P(E_1) = \dfrac{1}{15}$

and $\quad P(E_2) = 1 - P(E_1) = 1 - \dfrac{1}{15} = \dfrac{14}{15}$

In case E_1 occurs, it is certain that one of P_1 and P_2 will be among the winners. In case E_2 occurs, the probability that exactly one of P_1 and P_2 is among the winners is

$P\{(P_1 \cap \bar{P}_2) \cup (\bar{P}_1 \cap P_2)\} = P(P_1 \cap \bar{P}_2) + P(\bar{P}_1 \cap P_2)$
$= P(P_1)P(\bar{P}_2) + P(\bar{P}_1)P(P_2)$

$$= \left(\frac{1}{2}\right)\left(1 - \frac{1}{2}\right) + \left(1 - \frac{1}{2}\right)\left(\frac{1}{2}\right) = \frac{1}{4} + \frac{1}{4} = \frac{1}{2}$$

i.e. $P(A/E_1) = 1$ and $P(A/E_2) = \dfrac{1}{2}$

By the total probability Rule,

$$P(A) = P(E_1)\cdot P\left(\frac{A}{E_1}\right) + P(E_2)\cdot P\left(\frac{A}{E_2}\right)$$

$$= \frac{1}{15}(1) + \frac{14}{15}\left(\frac{1}{2}\right)$$

$$= \frac{8}{15}$$

Objective Questions Type II [One or more than one correct answer(s)]

1. $\because P(X=4), P(X=5)$ and $P(X=6)$ are in AP,

$\therefore 2P(X=5) = P(X=4) + P(X=6)$

$\Rightarrow \quad 2 = \dfrac{P(X=4)}{P(X=5)} + \dfrac{P(X=6)}{P(X=5)}$

$2 = \dfrac{{}^nC_4}{{}^nC_5} + \dfrac{{}^nC_6}{{}^nC_5}$

$\Rightarrow \quad n^2 - 21n + 98 = 0 \therefore n = 7, 14$

2. $P(X) = \dfrac{1}{3}$

$P\left(\dfrac{X}{Y}\right) = \dfrac{P(X\cap Y)}{P(Y)} = \dfrac{1}{2}$

$P\left(\dfrac{Y}{X}\right) = \dfrac{P(X\cap Y)}{P(Y)} = \dfrac{2}{5}$

$P(X\cap Y) = \dfrac{2}{15}$

$P(Y) = \dfrac{4}{15}$

$P\left(\dfrac{X'}{Y}\right) = \dfrac{P(Y) - P(X\cap Y)}{P(Y)}$

$= \dfrac{\dfrac{4}{15} - \dfrac{2}{15}}{\dfrac{4}{15}} = \dfrac{1}{2}$

$P(X\cup Y) = \dfrac{1}{3} + \dfrac{4}{15} - \dfrac{2}{15} = \dfrac{7}{15} = \dfrac{7}{15}$

3. We have, $P(E_i) = \dfrac{2}{4} = \dfrac{1}{2}$ for $i = 1, 2, 3,$

Also, for $i \neq j$, $P(E_i \cap E_j) = \dfrac{1}{4} = P(E_i)P(E_j)$.

Therefore, E_i and E_j are independent for $i \neq j$,

Also, $P(E_1 \cap E_2 \cap E_3) = \dfrac{1}{4} \neq P(E_1)P(E_2)P(E_3)$

$\therefore \quad E_1, E_2, E_3$ are not independent.

4. $\therefore$ Since, A and B are independent events

$P(A\cap B) = P(A)\,P(B)$

Then, A and $\overline{B}$ are independent $\overline{A}$ and $\overline{B}$ are also independent.

Hence, $\quad P(A\cap \overline{B}) = P(A)\,P(\overline{B})$

and $\quad P(\overline{A}\cap \overline{B}) = P(\overline{A})\,P(\overline{B})$

and $P(A/B) + P(\overline{A}/B) = \dfrac{P(A\cap B)}{P(B)} + \dfrac{P(\overline{A}\cap B)}{P(B)}$

$= \dfrac{P(A)\cdot P(B)}{P(B)} + \dfrac{P(\overline{A})P(B)}{P(B)}$

$= P(A) + P(\overline{A}) = P(A) + 1 - P(A)$

$= 1$

5. $P(A) = 1/2,\ P(B) = 2/3$

Option (a) $P(A\cup B) = P(A) + P(B) - P(A\cap B)$

$= \dfrac{1}{2} + \dfrac{2}{3} - P(A\cap B)$

$\therefore \quad P(A\cap B) = \dfrac{7}{6} - P(A\cap B) \geq \dfrac{2}{3}$

$P(A\cap B) \leq \dfrac{7}{6} - \dfrac{2}{3}$

$P(A\cap B) \leq \dfrac{3}{6}$

$\Rightarrow \quad P(A\cap B) \leq \dfrac{1}{2}$ which is true.

Option (b) : $P(A\cap B') = P(A) - P(A\cap B) \leq \dfrac{1}{3}$

$\Rightarrow \quad \dfrac{1}{2} - P(A\cap B) \leq \dfrac{1}{3}$

$\therefore \quad P(A\cap B) \geq \dfrac{1}{2} - \dfrac{1}{3}$

$\therefore \quad P(A\cap B) \geq \dfrac{1}{6}$ which is true

From option (a) $\quad \dfrac{1}{6} \leq P(A\cap B) \leq \dfrac{1}{2}$ ...(i)

Option (c) is correct

From Eq. (i)

$-\dfrac{1}{6} \geq -P(A\cap B) \geq -\dfrac{1}{2}$

or $\quad P(B) - \dfrac{1}{6} \geq P(B) - P(A\cap B) \geq P(B) - \dfrac{1}{2}$

$\Rightarrow \quad \dfrac{2}{3} - \dfrac{1}{6} \geq P(A'\cap B) \geq \dfrac{2}{3} - \dfrac{1}{2}$

$\Rightarrow \therefore \quad \dfrac{1}{6} \leq P(A'\cap B) \leq \dfrac{1}{2}$

Hence, **option (d)** is correct.

6. $P(A) = P(A/B) = 1/4$ ...(i)

and $\quad P(B/A) = 1/2$ ...(ii)

From Eq. (i), $\quad P(A) = \dfrac{P(A\cap B)}{P(B)}$

$\therefore \quad P(A\cap B) = P(A)\cdot P(B)$

$\therefore \quad A$ and B are independent

Option (a) is correct.

From Eq. (ii), $\quad \dfrac{P(B\cap A)}{P(A)} = \dfrac{1}{2} \Rightarrow P(B) = \dfrac{1}{2}$

Option (b) : $P(A\cup B) = P(A) + P(B) - P(A\cap B)$

$= \dfrac{1}{2} + \dfrac{1}{2} - \dfrac{1}{4}$

$= \dfrac{3}{4} \neq P(A) + P(B)$

$\therefore \quad A$ and B are not mutually exclusive.

Option (c) : $P(A'/B) = \dfrac{P(A'\cap B)}{P(B)}$

$= \dfrac{P(B) - P(A\cap B)}{P(B)}$

$= 1 - P(A/B)$

$= 1 - 1/4 = 3/4$

Option (d) : $P(B'/A') = P(B')$

[$\because A'$ and B' are independent]

$= 1 - P(B)$

$= 1 - 1/2$

$= 1/2$

7. Let A, B, C be the events that the student is successful in tests I, II and III respectively, then P (the student is successful)

$$= P[(A \cap B \cap C') \cup (A \cap B' \cap C) \cup (A \cap B \cap C)]$$
$$= P(A \cap B \cap C') + P(A \cap B' \cap C) + P(A \cap B \cap C)$$
$$= P(A)\,P(B)\,P(C') + P(A)\,P(B')\,P(C) + P(A)\,P(B)\,P(C)$$
$$= pq\left(1-\frac{1}{2}\right) + p(1-q)\frac{1}{2} + (pq)\left(\frac{1}{2}\right) = \frac{1}{2}p(1-q)$$
$$\Rightarrow \quad \frac{1}{2} = \frac{1}{2}p(1+q)$$
$$\Rightarrow \quad p(1+q) = 1$$

8. $X = \{a_1, a_2, a_3, \ldots, a_n\}$

$\therefore$ Number of subsets $= 2^n$

The number of ways of choosing A and B is
$$= 2^n \times 2^n = 2^{2n}.$$

The number of ways of choosing A and B so that they have the same number of elements is

$${}^nC_0 \cdot {}^nC_0 + {}^nC_1 \cdot {}^nC_1 + {}^nC_2 \cdot {}^nC_2 + \ldots + {}^nC_n \cdot {}^nC_n$$
$$= {}^{2n}C_n = \frac{2^n \cdot (1\cdot 3\cdot 5 \cdot \ldots \cdot 2n-1)}{n!}$$
$$\therefore \text{Probability} = \frac{{}^{2n}C_n}{2^{2n}} = \frac{1\cdot 3\cdot 5 \ldots (2n-1)}{2^n \cdot n!}$$

9. (i) $P(E_1) = 1 - P(R\,R\,R)$
$$= 1 - \left[\frac{1}{3}\cdot\frac{2}{4}\cdot\frac{3}{5}\right] = 0.9$$

(ii) $P(E_2) = 3P(B\,R\,R) = 3\cdot\frac{2}{3}\cdot\frac{1}{4}\cdot\frac{2}{5} = 0.2$

(iii) $P(E_3) = P(R\,R\,R / R\,R\,R \cup B\,B\,B)$
$$= \frac{P(R\,R\,R)}{P(R\,R\,R) + P(B\,B\,B)}$$

but $P(B\,B\,B) = \frac{2}{3}\cdot\frac{3}{4}\cdot\frac{4}{5} = \frac{8}{20} \Rightarrow P(E_3) = \frac{0.1}{0.1 + 0.4} = 0.2$

(iv) $P(E_4) = 1 - P(B\,B\,B) = 1 - \frac{2}{5} = 0.6$

10. The total number of cases $= 6 \times 6 \times 6 = 216$

The number of favourable ways

$=$ coefficient of x^k in $(x + x^2 + \ldots + x^6)^3$

$=$ coefficient of x^{k-3} in $(1-x^6)^3(1-x)^{-3}$

$=$ coefficient of x^{k-3} in $(1-x)^{-3}$ $[0 \le k-3 \le 5]$

$$= {}^{k-3+2}C_2 = {}^{k-1}C_2 = \frac{(k-1)(k-2)}{2}$$
$$\therefore \text{Required probability} = \frac{(k-1)(k-2)}{2\times 216} = \frac{(k-1)(k-2)}{432}$$

11. The number of ways in which m boys and m girls can take their seats around a circle $= ((m+m)-1)! = (2m-1)!$

Option (a) No two boys sit together

We make the girls sit first around the table. This can be done is $(m-1)!$, After this boys can take their seats in $m!$

$\therefore$ The probability no two boys sit together
$$= \frac{m!(m-1)!}{(2m-1)!} = \frac{1}{{}^{2m-1}C_m} = ({}^{2m-1}C_m)^{-1}$$

Option (b) No two girls sit together

We make the boys sit first around the table. This can be done in $(m-1)!$, after these girls can take their seats in $(m)!$

$\therefore$ The probability no two girls sit together
$$= \frac{m!(m-1)!}{(2m-1)!} = \frac{1}{{}^{2m-1}C_m} = ({}^{2m-1}C_m)^{-1}$$

Option (c) Boys and girls sit alternatively

i.e. no two boys (girls) sit together

$\therefore$ Required probability $= ({}^{2m-1}C_m)^{-1}$

Option (d) all the boys sit together

$\because$ All the boys sit together let treat them as a single boy.

Here, $(m+1)$ objects (m girls + 1 boy)

$\therefore$ We can put $(m+1)$ objects around a circle in $m!$ ways. But boys can be tied in $m!$ boys.

$\therefore$ Required probability $= \frac{m!\,m!}{(2m-1)!} \ne ({}^{2m-1}C_m)^{-1}$

12. Out of the numbers 00, 01, 02, 03, ..., 99 those numbers the product of whose digits is 16 are 28, 44, 82 *i.e.*, only 3

$$\therefore \quad p = P(E) = \frac{3}{100}$$

and $q = P(\bar{E}) = 1 - P(E) = 1 - p = 1 - \frac{3}{100} = \frac{97}{100}$

$\therefore$ The binomial distribution are $\left(\frac{97}{100} + \frac{3}{100}\right)^5$

$\therefore$ The probability that the even occurs exactly three times
$$= {}^5C_3\left(\frac{3}{100}\right)^3\left(\frac{97}{100}\right)^2$$
$$= 10\,(0.03)^3(0.97)^2$$

13. Here, $P(M) = \alpha$, $P(P) = \beta$ and $P(C) = \gamma$

$\therefore$ The probability of passing in atleast one subject $= 0.75$ (given)

$$\Rightarrow \quad 1 - P(\bar{M}\,\bar{P}\,\bar{C}) = 0.75$$
$$\Rightarrow \quad 1 - P(\bar{M})\,P(\bar{P})P(\bar{C}) = 0.75$$
$$\Rightarrow \quad 1 - (1-\alpha)(1-\beta)(1-\gamma) = 0.75$$
$$\text{or} \quad \alpha + \beta + \gamma - \alpha\beta - \beta\gamma - \gamma\alpha + \alpha\beta\gamma = \frac{3}{4} \quad \ldots\text{(i)}$$

The probability of passing in atleast two subjects $= 0.50$ (given)

or $P(M\,P\,\bar{C}) + P(M\,\bar{P}\,C) + P(\bar{M}\,P\,C) + P(M\,P\,C) = 0.50$

$\Rightarrow P(M)\,P(P)\,P(\bar{C}) + P(M)P(\bar{P})P(C) + P(\bar{M})P(P)P(C) + P(M)P(P)P(C) = 0.50$

$$\Rightarrow \alpha\beta(1-\gamma) + \alpha(1-\beta)\gamma + (1-\alpha)\beta\gamma + \alpha\beta\gamma = \frac{1}{2}$$
$$\Rightarrow \quad 2\alpha\beta\gamma = \alpha\beta + \beta\gamma + \gamma\alpha - \frac{1}{2} \quad \ldots\text{(ii)}$$

and the probability of passing in exactly two subjects $= 0.40$ (given)

$$\Rightarrow \quad P(M\,P\,\bar{C}) + P(M\,\bar{P}\,C) + P(\bar{M}\,P\,C) = \frac{2}{5}$$

$\Rightarrow \quad P(M)P(P)P(\overline{C}) + P(M)P(\overline{P})P(C) + P(\overline{M})P(P)P(C) = \frac{2}{5}$

$\Rightarrow \quad \alpha\beta(1-\gamma) + \alpha(1-\beta)\gamma + (1-\alpha)\beta\gamma = \frac{2}{5}$

$\Rightarrow \quad \alpha\beta + \beta\gamma + \gamma\alpha - 3\alpha\beta\gamma = \frac{2}{5}$ ...(iii)

From Eqs. (ii) and (iii),

$\Rightarrow \quad 2\alpha\beta\gamma + \frac{1}{2} - 3\alpha\beta\gamma = \frac{2}{5}$

$\therefore \quad \alpha\beta\gamma = \frac{1}{2} - \frac{2}{5} = \frac{1}{10}$ (c)

From Eq. (ii), $\frac{1}{5} = \alpha\beta + \beta\gamma + \gamma\alpha - \frac{1}{2}$ ...(iv)

$\therefore \quad \alpha\beta + \beta\alpha + \gamma\alpha = \frac{7}{10}$ ...(iv)

Substituting the values of $\alpha\beta\gamma$ and $\alpha\beta + \beta\gamma + \gamma\alpha$ in Eq. (i), then

$\therefore \quad \alpha + \beta + \gamma - \frac{7}{10} + \frac{1}{10} = \frac{3}{4}$

$\Rightarrow \quad \alpha + \beta + \gamma = \frac{6}{10} + \frac{3}{4} = \frac{12+15}{20} = \frac{27}{20}$ (b)

14. $\because \quad P(A \cup B) = P(A) + P(B) - P(A)\cdot P(B)$

$\therefore \quad P(A \cap B) = P(A)\cdot P(B)$

Then, $P(A/B) = \frac{P(A \cap B)}{P(B)} = P(A)$

and $P(\overline{A \cup B}) = 1 - P(A \cup B)$

$= 1 - \{P(A) + P(B) - P(A)\cdot P(B)\}$

$= \{1 - P(A)\}\{1 - P(B)\} = P(\overline{A})P(\overline{B})$

15. Roots of $x^2 + px + q = 0$ will be real, if $p^2 \geq 4q$. The possible selections are as follows

p	*q*
1	–
2	1
3	1, 2
4	1, 2, 3, 4
5	1, 2, 3, 4, 5, 6
6	1, 2, ..., 9
7	1, 2, ..., 10
8	1, 2, ..., 10
9	1, 2, ..., 10
10	1, 2, ..., 10
Total	62

Therefore, number of favourable ways is 62 and total number of ways is $10^2 = 100$. Hence, the required probability is 62/100 = 31/50.

The probability that the roots are imaginary is $1 - 31/50 = 19/50$

Roots are equal when $(p, q) \equiv (2, 1), (4, 4), (6, 9)$. The probability that the roots are real and equal is 3/50. Hence, probability that the roots are real and distinct is 3/5.

16. Let $P(A) = x$ and $P(B) = y$

Given, $P(A \cap B) = \frac{1}{12}$

$\Rightarrow \quad P(A)P(B) = \frac{1}{12}$

or $\quad xy = \frac{1}{12}$ ...(i)

and $\quad P(\overline{A} \cap \overline{B}) = \frac{1}{2}$

$\Rightarrow \quad P(\overline{A})P(\overline{B}) = \frac{1}{2}$

$\Rightarrow \quad \{1 - P(A)\}\{1 - P(B)\} = \frac{1}{2}$

$\Rightarrow \quad (1-x)(1-y) = \frac{1}{2}$

$\Rightarrow \quad 1 - x - y + xy = \frac{1}{2}$

or $\quad x + y = \frac{1}{2} + \frac{1}{12} = \frac{7}{12}$ ...(ii)

$\therefore$ *x, y* are the roots of

$t^2 - (x+y)t + xy = 0$

$\Rightarrow \quad t^2 - \frac{7}{12}t + \frac{1}{12} = 0$

$\Rightarrow \quad 12t^2 - 7t + 1 = 0$

$\Rightarrow \quad (4t-1)(3t-1) = 0$

$\therefore \quad t = \frac{1}{3}, \frac{1}{4}$

$\therefore \quad x = \frac{1}{3}, y = \frac{1}{4}$ or $x = \frac{1}{4}, y = \frac{1}{3}$

$\Rightarrow \quad P(A) = \frac{1}{3}, P(B) = \frac{1}{4}$ [alternate (a)]

or $\quad P(A) = \frac{1}{4}, P(B) = \frac{1}{3}$ [alternate (d)]

17. Suppose $X \sim B(n, p)$ i.e. $(q+p)^n$

Here, $np = a$ and $npq = b$

$\therefore \quad q = \frac{b}{a}$, then $p = 1 - q = 1 - \frac{b}{a}$

Now, $0 < q < 1 \Rightarrow 0 < \frac{b}{a} < 1$

$\Rightarrow \quad a > b > 0$ [alternate (a)]

and $\quad \frac{a}{b} > 1$ [alternate (b)]

Also, $\frac{a^2}{a-b} = \frac{(np)^2}{np - npq} = \frac{np}{1-q} = \frac{np}{p} = n$

= Integer

18. **Option (a)** $P(E_1) = P(E_2) = \dfrac{\frac{10!}{2!}}{\frac{11!}{2!2!}} = \frac{2}{11}$

Option (b) $P(E_1 \cap E_2)$ = Probability of two I's are together and two *B*'s are together

$= \dfrac{9!}{\frac{11!}{2!2!}} = \frac{4}{110} = \frac{2}{55}$

Option (c) $P(E_1 \cup E_2) = P(E_1) + P(E_2) - P(E_1 \cap E_2)$

$$= \frac{2}{11} + \frac{2}{11} - \frac{2}{55}$$

$$= \frac{4}{11} - \frac{2}{55} = \frac{18}{55}$$

Option (d) $P(E_1/E_2) = \dfrac{P(E_1 \cap E_2)}{P(E_2)}$

$$= \frac{\frac{2}{55}}{\frac{2}{11}} = \frac{1}{5}$$

19. Area of the shaded region $= \int_0^1 (\sqrt{x} - x^2)\, dx$

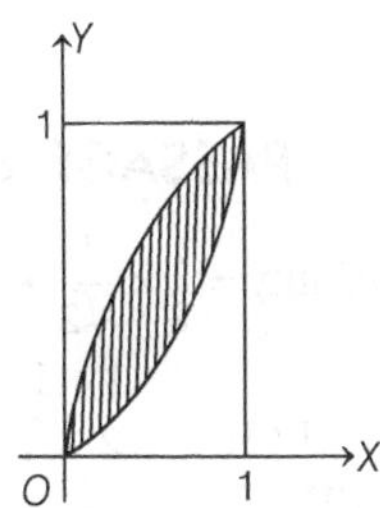

$$= \left(\frac{2}{3}x^{3/2} - \frac{x^3}{3}\right)\Bigg|_0^1 = \frac{2}{3} - \frac{1}{3} = \frac{1}{3}$$

Area of the square $= 1$

Probability $= 1/3$

$A \cup B =$ whole of the region enclosed in $0 \le x \le 1$, $0 \le y \le 1$

$\therefore A$ and B are exhaustive events.

$$P(A) = \int_0^1 \sqrt{x}\, dx = \frac{2}{3} x^{\frac{3}{2}}\Bigg|_0^1 = \frac{2}{3},\ P(B) = \frac{1}{3}$$

$$\therefore P(A)P(B) = \frac{2}{9} \ne P(A \cap B)$$

$\therefore A$ and B are not independent.

20. Let $A = \{$minimum of the chosen number is 3)

and $B = \{$maximum of the chosen number is 7$\}$

$\therefore P(A) =$ Probability of choosing 3 and two other numbers from 4 to 10.

$$= \frac{^7C_2}{^{10}C_3} = \frac{\frac{7 \cdot 6}{1 \cdot 2}}{\frac{10 \cdot 9 \cdot 8}{1 \cdot 2 \cdot 3}} = \frac{7}{40}$$

$P(B) =$ Probability of choosing 7 and two other numbers from 1 to 6

$$= \frac{^6C_2}{^{10}C_3} = \frac{\frac{6 \cdot 5}{1 \cdot 2}}{\frac{10 \cdot 9 \cdot 8}{1 \cdot 2 \cdot 3}} = \frac{1}{8}$$

Now, $P(A \cap B) =$ Probability of choosing 3 and 7 and one other number from 4 to 6.

$$= \frac{3}{^{10}C_3} = \frac{3}{\frac{10 \cdot 9 \cdot 8}{1 \cdot 2 \cdot 3}} = \frac{1}{40}$$

$$\therefore \quad P(A \cup B) = P(A) + P(B) - P(A \cap B)$$

$$= \frac{7}{40} + \frac{1}{8} - \frac{1}{40}$$

$$= \frac{11}{40} < \frac{11}{30}$$

but $\dfrac{11}{40} > \dfrac{11}{50}$ and $\dfrac{11}{40} > \dfrac{11}{60}$

21. If r be the radius of the circle and if x be the side of the square, then

$$x^2 + x^2 = (2r)^2$$

or $$x^2 = 2r^2$$

Now, $p_1 = \dfrac{\text{Area of square}}{\text{Area of circle}} = \dfrac{x^2}{\pi r^2} = \dfrac{2r^2}{\pi r^2} = \dfrac{2}{\pi}$

$$\therefore \quad p_2 = 1 - p_1 = 1 - \frac{2}{\pi} = \frac{(\pi - 2)}{\pi}$$

$$\because \quad 2 > \pi - 2$$

$$\Rightarrow \quad \frac{2}{\pi} > \frac{\pi - 2}{\pi}$$

$$\Rightarrow \quad p_1 > p_2$$

Also, $p_1^2 - p_2^2 = (p_1 + p_2)(p_1 - p_2) = 1\left(\dfrac{4 - \pi}{\pi}\right)$

$$= \frac{4}{\pi} - 1 < \frac{1}{3} \qquad \left[\because 3 < \pi < 4 \Rightarrow \frac{3}{\pi} < 1 < \frac{4}{\pi}\right]$$

22. We have, $P(X = x) \propto (x + 1)\left(\dfrac{1}{5}\right)^x$

$$= k(x + 1)\left(\frac{1}{5}\right)^x$$

Since, $\displaystyle\sum_{x=0}^{\infty} P(X = x) = 1$

$$\Rightarrow \quad k\left\{1 + 2\left(\frac{1}{5}\right) + 3\left(\frac{1}{5}\right)^2 + 4\left(\frac{1}{5}\right)^3 + \ldots \infty\right\} = 1$$

$$\Rightarrow \quad k\left\{\left(1 - \frac{1}{5}\right)^{-2}\right\} = 1$$

$$\Rightarrow \quad k\left(\frac{25}{16}\right) = 1$$

or $$k = \frac{16}{25}$$

Option (a) $P(X = 0) = k(0 + 1)\left(\dfrac{1}{5}\right)^0 = k = \dfrac{16}{25}$

Option (b) $P(X \le 1) = P(X = 0) + P(X = 1)$

$$= k + \frac{2k}{5}$$

$$= \frac{7}{5}k = \frac{7}{5} \times \frac{16}{25} = \frac{112}{125}$$

Option (c) $P(X \ge 1) = 1 - P(X = 0) = 1 - k$

$$= 1 - \frac{16}{25} = \frac{9}{25}$$

Option (d) $E(X) = \sum_{x=0}^{\infty} xP(X = x)$

$$= \sum_{x=0}^{\infty} x(x+1)\left(\frac{1}{5}\right)^x$$

$$E(X) = 1\cdot 2\left(\frac{1}{5}\right) + 2\cdot 3\left(\frac{1}{5}\right)^2 + 3\cdot 4\left(\frac{1}{5}\right)^3 + \dots \infty \quad \dots\text{(i)}$$

$$\therefore \frac{1}{5}E(X) = 1\cdot 2\left(\frac{1}{5}\right)^2 + 2\cdot 3\left(\frac{1}{5}\right)^3 + \dots \infty \quad \dots\text{(ii)}$$

Subtracting Eq. (ii) from Eq. (i), then

$$\frac{4}{5}E(X) = \frac{2}{5} + \frac{2}{5}\left\{\frac{2}{5} + \frac{3}{5^2} + \frac{4}{5^3} + \dots \infty\right\}$$

$$= \frac{2}{5} + \frac{2}{5}\left\{\left(1 + \frac{2}{5} + \frac{3}{5^2} + \frac{4}{5^3} + \dots \infty\right) - 1\right\}$$

$$= \frac{2}{5}\left(1 + \frac{2}{5} + \frac{3}{5^2} + \frac{4}{5^3} + \dots \infty\right)$$

$$= \frac{2}{5}\left(1 - \frac{1}{5}\right)^{-2} = \frac{2}{5} \times \frac{25}{16} = \frac{5}{8}$$

$$\therefore \quad E(X) = \frac{25}{32}$$

Linked Comprehension Type

PASSAGE 1

Let n positive integers be $x_1, x_2, x_3, \dots, x_n$.

Let $a = x_1 \cdot x_2 \cdot x_3 \cdot \dots \cdot x_n$.

Let S be the sample space, since the last digit in each of the numbers, $x_1, x_2, \dots, x_n$ can be any one of the digits 0, 1, 2, 3, ..., 9 (total 10)

$$\therefore \quad n(S) = (10)^n$$

Let E_1 and E_2 be the events when the last digit in a is 1, 3, 5, 7 or 9 and 1, 3, 7 or 9 respectively.

$$\therefore \quad n(E_1) = 5^n,\ n(E_2) = 4^n$$

1. $P(E_1) = \dfrac{n(E_1)}{n(S)} = \dfrac{5^n}{(10)^n} = \left(\dfrac{1}{2}\right)^n$

2. $P(E_2) = \dfrac{n(E_2)}{n(S)} = \dfrac{4^n}{(10)^n} = \left(\dfrac{2}{5}\right)^n$

3. Let E be the event that the last digit in a is 5

$$\therefore \quad n(E) = n(E_1) - n(E_2)$$
$$= 5^n - 4^n$$

$$\therefore \quad P(E) = \frac{n(E)}{n(S)} = \frac{5^n - 4^n}{10^n}$$

4. Let E_3 and E_4 be the events when the last digit in a is 1, 2, 3, 4, 6, 7, 8 or 9 and 0 respectively.

Then, $n(E_4) = n(S) - n(E_3) - n(E)$
$$= 10^n - 8^n - (5^n - 4^n)$$
$$= 10^n - 8^n - 5^n + 4^n$$

$$\therefore \quad P(E_4) = \frac{n(E_4)}{n(S)} = \frac{10^n - 8^n - 5^n + 4^n}{10^n}$$

5. Let E_5 be the event that the last digit in a is 2, 4, 6 or 8, then

$$n(E_5) = n(S) - n(E_1) - n(E_4)$$
$$= 10^n - 5^n - (10^n - 8^n - 5^n + 4^n)$$
$$= 8^n - 4^n = 2^n(4^n - 2^n)$$

$$\therefore \quad P(E_5) = \frac{n(E_5)}{n(S)} = \frac{2^n(4^n - 2^n)}{10^n}$$
$$= \frac{4^n - 2^n}{5^n}$$

PASSAGE 2

1. Required probability $= \dfrac{1}{2} \times \dfrac{1}{\sqrt{2}} + \dfrac{1}{2}\left(1 - \dfrac{1}{4}\right) = 0.728$

2. $P(\text{I}/\text{II}) = \dfrac{P(\text{I}) \cdot P(\text{II}/\text{I})}{P(\text{II})} = \dfrac{\frac{1}{2} \cdot \frac{1}{\sqrt{2}}}{\frac{1}{2} \cdot \frac{1}{\sqrt{2}} + \frac{1}{8}} = 0.74$

3. Probability Sania wins the match

$$= (1 - R)R^2R = R^3(1 - R)$$

Probability Sania looses the match

$$= (1 - R)R^2(1 - R) + (1 - R)(1 - R^2)$$

$$\therefore R^3(1 - R) > (1 - R)R^2(1 - R) + (1 - R)(1 - R^2)$$

$$\therefore \quad R^3 > R^2(1 - R) + (1 - R^2)$$
$$= 1 - R^3$$

$$\therefore \quad 2R^3 > 1$$

$$\therefore \quad R \in \left(\left(\frac{1}{2}\right)^{\frac{1}{3}}, 1\right)$$

PASSAGE 3

Let $A = \{a_1, a_2, a_3, \dots, a_{10}\}$

For each $a_i \in A (1 \le i \le 10)$ we have the following four choices

(i) $a_i \in P$ and $a_i \in Q$

(ii) $a_i \in P$ and $a_i \notin Q$

(iii) $a_i \notin P$ and $a_i \in Q$

(iv) $a_i \notin P$ and $a_i \notin Q$

Let S be the sample space and E be the event favourable.

$$\therefore \quad n(S) = 4^{10}$$

1. $\because \quad P \cup Q = A$

$$\therefore \quad n(E) = 3^{10} \quad [\because \text{(iv)} \notin A]$$

$$\Rightarrow \quad P(E) = \frac{n(E)}{n(S)} = \frac{3^{10}}{4^{10}} = \left(\frac{3}{4}\right)^{10}$$

2. $n(E)$ = Number of ways in which exactly 4 elements of A will belong to $P \cup Q$

$$= {}^{10}C_4\,(3)^4 \cdot (1)^{10-4} = {}^{10}C_4 \cdot 3^4$$

$$\therefore\; P(E) = \frac{n(E)}{n(S)} = \frac{{}^{10}C_4 \cdot 3^4}{4^{10}} = \frac{15 \times 7 \times 3^4}{2^{19}} = \frac{8505}{2^{19}}$$

3. $\because \quad P \cap Q = \phi$

$\therefore \quad n(E) = 3^{10} \qquad [\because (i) \notin P \cap Q]$

$\Rightarrow \quad P(E) = \frac{3^{10}}{4^{10}} = \left(\frac{3}{4}\right)^{10}$

4. $n(E) = {}^{10}C_3 \cdot (3)^7$ (7 elements are not common)

$$P(E) = \frac{{}^{10}C_3 \cdot 3^7}{4^{10}} = \frac{5 \times 3^8}{2^{17}}$$

5. $P \cap Q$ = 9 elements or 10 elements

$\therefore \quad n(E) = {}^{10}C_9 \cdot (3)^1 + {}^{10}C_{10}(1)^{10} = 30 + 1 = 31$

$\therefore \quad P(E) = \frac{31}{4^{10}} = \frac{31}{2^{20}}$

PASSAGE 4

1. Here, $P(X > Y) = P\,(T_1 \text{ win})\, P(T_1 \text{ win}) + P(T_1 \text{ win})\, P \text{ (draw)} + P \text{ (draw)}\, P(T_1 \text{ win})$

$$= \left(\frac{1}{2} \times \frac{1}{2}\right) + \left(\frac{1}{2} \times \frac{1}{6}\right) + \left(\frac{1}{6} \times \frac{1}{2}\right) = \frac{5}{12}$$

2. $P[X = Y] = P \text{ (draw)} \cdot P \text{ (draw)} + P(T_1 \text{win})\, P(T_2 \text{ win}) + P(T_2 \text{ win}) \cdot P(T_1 \text{ win})$

$$= (1/6 \times 1/6) + (1/2 \times 1/3) + (1/3 \times 1/2) = 13/36$$

PASSAGE 5

1. Here, $p = \frac{1}{4}, q = 1 - \frac{1}{4} = \frac{3}{4}$

$\Rightarrow$ 1 – (probability not hitting the target) $> \frac{2}{3}$

$\Rightarrow \quad 1 - {}^nC_n\left(\frac{3}{4}\right)^n > \frac{2}{3}$

or $\quad \left(\frac{1}{3}\right) > \left(\frac{3}{4}\right)^n$

$\Rightarrow$ Minimum value of $n = 4$

2. $p = \frac{1}{{}^6C_1 \times {}^6C_1} = \frac{1}{36}$

$q = 1 - p = \frac{35}{36}$

As, given

Probability of getting at least one double six $> \frac{99}{100}$

$\Rightarrow \quad 1 - \left(\frac{35}{36}\right)^n > 0.99$

$\Rightarrow \quad 0.01 > \left(\frac{35}{36}\right)^n$

$\Rightarrow \quad 10^{-2} > \left(\frac{35}{36}\right)^n$

$\Rightarrow \quad -2 > n\,(\log 35 - \log 36)$

$\Rightarrow \quad -2 > -n(0.0122)$

or $\quad n > \frac{2}{0.122} = 163.9$

$\therefore \quad n > 163.9$

$\Rightarrow \quad n = 164$

3. $p = \frac{1}{2}, q = \frac{1}{2}$

For getting total sum less than 15 in 4 trials, we will throw coins to get

(i) $3 + 3 + 3 + 3 < 15$ (ii) $5 + 3 + 3 + 3 < 15$

and so we need atleast three success

$\therefore$ Probability of this event $= {}^4C_3\left(\frac{1}{2}\right)^3\left(\frac{1}{2}\right)^1 + {}^4C_4\left(\frac{1}{2}\right)^4$

$$= \frac{4}{16} + \frac{1}{16}$$

$$P(E) = \frac{5}{16}$$

$\therefore \quad$ Odds in favour $= \frac{P(E)}{P(\bar{E})} = \frac{\frac{5}{16}}{1 - \frac{5}{16}} = \frac{5}{11}$

4. Probability of A winning, say $p = 0.4 = \frac{4}{10}$

$\therefore$ Probability of B winning $q = 1 - 0.4 = 0.6 = \frac{6}{10}$

"Best of three games" = (2 wins and 1 loss) or 3 (wins)

$$p_1 = {}^3C_2\left(\frac{4}{10}\right)^2\left(\frac{6}{10}\right) + {}^3C_3\left(\frac{4}{10}\right)^3 = \frac{18 \times 16 + 64}{1000} = \frac{44}{125}$$

"Best of five games" = (3 wins and 2 losses), (4 wins and 1 losses) or (5 wins)

$$p_2 = {}^5C_3\left(\frac{4}{10}\right)^3\left(\frac{6}{10}\right)^2 + {}^5C_4\left(\frac{4}{10}\right)^4\left(\frac{6}{10}\right) + {}^5C_5\left(\frac{4}{10}\right)^5$$

$$= \frac{10 \times 64 \times 36 + 5 \times 256 \times 6 + 1024}{10^5} = \frac{992}{3125}$$

$\therefore \quad p_1 > p_2$

5. If probability of bulb being non-defective be taken as the success,

then $p = \frac{9}{10}, q = 1 - p = \frac{1}{10}$

$\therefore$ Probability of non-defective

$$= {}^5C_5 p^5 q^0 = \left(\frac{9}{10}\right)^5$$

PASSAGE 6

1. Total number of ways $= {}^nC_3 = \frac{n(n-1)(n-2)}{1\cdot 2\cdot 3}$
$= \frac{n(n-1)(n-2)}{6}$

Favourable ways = If side AB and take one vertex out of $(n-4)$ vertices)

$= {}^{n-4}C_1 \times n$

$n(E_1) = (n-4)n$

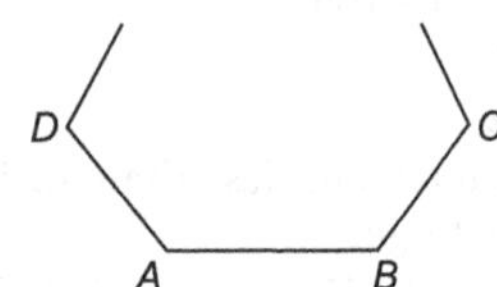

Required probability $p_1 = \frac{n(n-4)}{\frac{n(n-1)(n-2)}{6}}$
$= \frac{6(n-4)}{(n-1)(n-2)}$

2. i.e., $A_1 A_2 A_3, A_2A_3A_4, \ldots$
$A_{n-1} A_n A_1, A_n A_1 A_2$

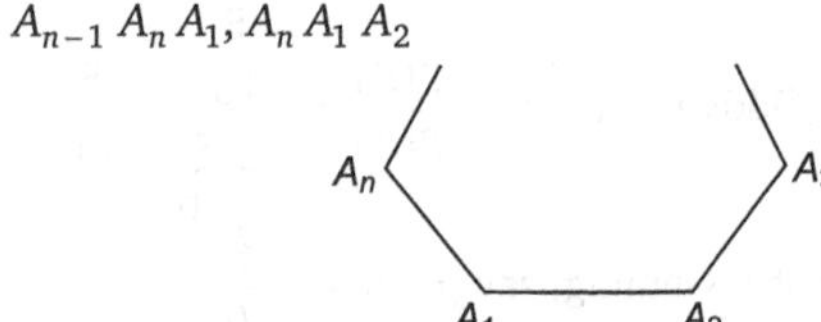

Favourable ways $= n = n(E_2)$

Required probability $p_2 = \frac{n}{{}^nC_3} = \frac{6}{(n-1)(n-2)}$

3. Favourable ways $= {}^nC_3 - n(E_1) - n(E_2)$
$= {}^nC_3 - (n-4)n - n$
$= \frac{n(n-1)(n-2)}{6} - n(n-4) - n$
$= \frac{n}{6}(n^2 - 3n + 2 - 6n + 24 - 6)$
$= \frac{n(n^2 - 9n + 20)}{6}$
$= \frac{n(n-4)(n-5)}{6}$

Required probability $= \frac{\frac{n(n-4)(n-5)}{6}}{\frac{n(n-1)(n-2)}{6}}$
$= \frac{(n-4)(n-5)}{(n-1)(n-2)}$

4. Favourable ways = One side common or two sides common
$= n(E_1) + n(E_2)$
$= (n-4)n + n$
$= n(n-3)$

$\therefore$ Required probability $= \frac{n(n-3)}{\frac{n(n-1)(n-2)}{6}} = \frac{6(n-3)}{(n-1)(n-2)}$

5. Favourable ways = At most one side common
= One side common + no side common
$= (n-4)n + \frac{n(n-4)(n-5)}{6}$
$= \frac{n(n-4)(n+1)}{6}$

$\therefore$ Required probability $= \frac{\frac{n(n-4)(n+1)}{6}}{\frac{n(n-1)(n-2)}{6}}$
$= \frac{(n-4)(n+1)}{(n-1)(n-2)}$

Numerical Value Type

1. Let an event E of sum of outputs are perfect square (i.e., 4 or 9), so
$E = \{(1, 3), (2, 2), (3, 1), (3, 6), (4, 5), (5, 4), (6, 3)\}$
and an event F of sum of outputs are prime numbers (i.e., 2, 3, 5, 7, 11) so
$F = \{(1, 1), (1, 2), (2, 1), (1, 4), (2, 3), (3, 2), (4, 1), (1, 6), (2, 5), (3, 4), (4, 3), (5, 2), (6, 1), (5, 6), (6, 5)\}$
and event T of sum of outputs are odd numbers (i.e., 3, 5, 7, 9, 11)
$T = \{(1, 2), (2, 1), (1, 4), (2, 3), (3, 2), (4, 1), (1, 6), (2, 5), (3, 4), (4, 3), (5, 2), (6, 1), (3, 6), (4, 5), (5, 4), (6, 3), (5, 6), (6, 5)\}$

Now, required probability $p = P(T/E) = \frac{P(T \cap E)}{P(E)}$

where, $P(T \cap E)$ = probability of occuring perfect square odd number before prime

$= \left(\frac{4}{36}\right) + \left(\frac{14}{36}\right)\left(\frac{4}{36}\right) + \left(\frac{14}{36}\right)^2\left(\frac{4}{36}\right) + \ldots \infty$

$= \frac{\frac{4}{36}}{1 - \frac{14}{36}} = \frac{4}{22} = \frac{2}{11}$

and $P(E)$ = probability of occuring perfect square before prime

$= \left(\frac{7}{36}\right) + \left(\frac{14}{36}\right)\left(\frac{7}{36}\right) + \left(\frac{14}{36}\right)^2\left(\frac{7}{36}\right) + \ldots \infty$

$= \frac{\frac{7}{36}}{1 - \frac{14}{36}} = \frac{7}{22}$

$\therefore\ P(T/E) = \frac{\frac{2}{11}}{\frac{7}{22}} = \frac{4}{7} = p$

Now, $140p = 80$

2. Given probability of electronic component fails when first used = 0.10

ie, $P(F) = 0.10$

$\therefore$ $P(\bar{F}) = 1 - P(F) = 0.90$

and let $P(Y)$ = probability of new component to last for one year

$\because$ $P(F) + P(\bar{F}) = 1$

Obviously the two events are mutually exclusive and exhaustive

$\therefore$ $P\left(\frac{Y}{F}\right) = 0$ and $P\left(\frac{Y}{\bar{F}}\right) = 0.99$

$$\therefore \quad P(Y) = P(F)\cdot P\left(\frac{Y}{F}\right) + P(\bar{F})\cdot P\left(\frac{Y}{\bar{F}}\right)$$
$$= 0.10 \times 0 + 0.90 \times 0.99$$
$$= 0.891 = \lambda \text{ (given)}$$
$$\therefore \quad 10\lambda = 8.91$$

3. Given that $P(A) = \alpha$, $P(B/A) = P(B'/A') = 1 - \alpha$

Thus, $P(A') = 1 - P(A)$
$= 1 - \alpha$

and $P(B/A') = 1 - P(B'/A')$
$= 1 - (1 - \alpha)$
$= \alpha$ ...(i)

$$\therefore \quad P(A'/B) = \frac{P(A' \cap B)}{P(B)}$$
$$= \frac{P(B) - P(A \cap B)}{P(B)}$$
$$= \frac{P(B) - P(A)P(B/A)}{P(B)}$$
$$\left[\because P(B/A) = \frac{P(A \cap B)}{P(A)}\right]$$
$$= \frac{P(B) - \alpha(1-\alpha)}{P(B)} \quad \text{...(ii)}$$

But $P(B) = P(A)\cdot P(B/A) + P(A')\cdot P(B/A')$
$= \alpha\cdot(1-\alpha) + (1-\alpha)\cdot\alpha$ [from Eq. (i)]
$= 2\alpha(1-\alpha)$ ...(iii)

Putting the value of $P(B)$ from Eq. (iii) in (ii), then

$$P(A'/B) = \frac{2\alpha(1-\alpha) - \alpha(1-\alpha)}{2\alpha(1-\alpha)}$$
$$= \frac{\alpha(1-\alpha)}{2\alpha(1-\alpha)}$$
$$= \frac{1}{2} = \lambda \text{ (given)}$$

4. Let $a_1, a_2, a_3, a_4, a_5, a_6, a_7$ be the seven digits and the remaining two be a_8 and a_9.

Let $a_1 + a_2 + a_3 + a_4 + a_5 + a_6 + a_7 = 9k$, $k \in I$ ...(i)

Also, $a_1 + a_2 + a_3 + a_4 + \ldots + a_9 = 1 + 2 + 3 + 4 + \ldots + 9$
$$= \frac{9 \times 10}{2}$$
$$= 45 \quad \text{...(ii)}$$

Subtracting Eq. (i) from (ii), then

$a_8 + a_9 = 45 - 9k$ ...(iii)

Since, $a_1 + a_2 + a_3 + \ldots + a_9$ and $a_1 + a_2 + \ldots + a_7$ are divisible by 9 if and only if $a_8 + a_9$ is divisible by 9.

Let S be the sample space and E be the event that the sum of the digits a_8 and a_9 is divisible by 9.

$\because$ $a_8 + a_9 = 45 - 9k$

Maximum value of $a_8 + a_9 = 17$ and minimum value of $a_8 + a_9 = 3$

$$\therefore \quad 3 \le 45 - 9k \le 17$$
$$\Rightarrow \quad -42 \le -9k \le -28$$
$$\Rightarrow \quad \frac{42}{9} \ge k \ge \frac{28}{9}$$
$$\text{or} \quad \frac{28}{9} \le k \le \frac{42}{9}$$

Hence, $k = 4$ [$\because$ k is positive integer]

$\therefore$ From Eq. (iii),

$a_8 + a_9 = 45 - 9 \times 4$

$\therefore$ $a_8 + a_9 = 9$

Now, possible pair of (a_8, a_9) can be
$\{(1, 8), (2, 7), (3, 6), (4, 5)\}$

$\therefore$ $E = \{(1, 8), (2, 7), (3, 6), (4, 5)\}$

$n(E) = 4$ and $n(S) = {}^9C_2 = 36$

$\therefore$ Required probability $P(E) = \frac{n(E)}{n(S)} = \frac{4}{36} = \frac{1}{9} = \lambda$ (given)

$$\therefore \quad 9999\lambda = 9999 \times \frac{1}{9}$$
$$= 1111$$

5. Since, the man is one step away from starting point mean that either

(i) man has taken 6 steps forward and 5 steps backward.

(ii) man has taken 5 steps forward and 6 steps backward.

Taking, movement 1 step forward as success and 1 step backward as failure.

$\therefore$ p = probability of success = 0.4

and q = probability of failure = 0.6

$\therefore$ Required probability $= P\{X = 6 \text{ or } X = 5\}$
$$= P(X = 6) + P(X = 5)$$
$$= {}^{11}C_6 p^6 q^5 + {}^{11}C_5 p^5 q^6$$
$$= {}^{11}C_5 (p^6 q^5 + p^5 q^6)$$
$$= \frac{11\cdot 10\cdot 9\cdot 8\cdot 7}{1\cdot 2\cdot 3\cdot 4\cdot 5}\{(0.4)^6(0.6)^5 + (0.4)^5(0.6)^6\}$$
$$= \frac{11\cdot 10\cdot 9\cdot 8\cdot 7}{1\cdot 2\cdot 3\cdot 4\cdot 5}(0.24)^5$$
$$= 0.37$$

Hence, the required probability = 0.37 = λ (given)

$\therefore$ $5000\lambda = 5000 \times 0.37$
$= 1850$

6. If n is the number of draws, then $2 \le n \le 50$, since nth draw must be an ace and one ace must appear in first $(n-1)$ draws.

Let S be the sample space.

Then, $n(S)$ = Total number of ways of drawing the first $(n-1)$ cards out of 52.
$$= {}^{52}C_{n-1}$$

and let E be the event of favourable cases.

$\therefore n(E)$ = number of ways of choosing 1 ace in $(n-1)$ trials × number of ways for the nth draw

$$= ({}^4C_1 \times {}^{48}C_{n-2}) \times \frac{3}{52-(n-1)}$$

{∵ number of favourable ways is 3}

$$= 4 \times \frac{48!}{(n-2)!(50-n)!} \times \frac{3}{(53-n)}$$

Hence, the required probability, $P(E) = \dfrac{n(E)}{n(S)}$

$$= \frac{\dfrac{4 \times 48! \times 3}{(n-2)! \times (50-n)! \times (54-n)}}{{}^{52}C_{n-1}}$$

$$= \frac{\dfrac{4 \times 48! \times 3}{(n-2)! \times (50-n)! \times (53-n)}}{\dfrac{52!}{(n-1)!(53-n)!}}$$

$$= \frac{12 \times 48! \times (n-1)! \times (53-n)!}{52! \times (n-2)! \times (50-n)! \times (53-n)}$$

$$= \frac{12 \times 48! \times (n-1)(n-2)! \times (53-n)(52-n)!}{52 \times 51 \times 50 \times 49 \times 48! \times (n-2)! \times (50-n)! \times (53-n)}$$

$$= \frac{12 \times (n-1) \times (52-n)!}{52 \times 51 \times 50 \times 49 \times (50-n)!}$$

$$= \frac{(n-1)(52-n)(51-n)(50-n)!}{50 \times 49 \times 17 \times 13 \times (50-n)!}$$

$$= \frac{(n-1)(52-n)(51-n)}{50 \times 49 \times 17 \times 13}$$

$\therefore \quad \lambda = 52, \mu = 51, \alpha = 50, \beta = 49, \gamma = 17, \delta = 13$

$\therefore \alpha + \beta + \gamma + \delta + 2\lambda + 3\mu = 50 + 49 + 17 + 13 + 104 + 153$

$= 386$

7. Let E be the event of any one cutting a spade in one cut and let S be the sample space then

$n(E) = {}^{13}C_1$

and $n(S) = {}^{52}C_1$

$\therefore \quad P(E) = p = \dfrac{n(E)}{n(S)} = \dfrac{{}^{13}C_1}{{}^{52}C_1} = \dfrac{13}{52} = \dfrac{1}{4}$

$\Rightarrow \quad P(E) = p = \dfrac{1}{4}$

$\therefore \quad P(\bar{E}) = q = 1 - p = \dfrac{3}{4}$

The probability of A winning (when A starts the game)

$= p + qqqp + qqqqqq\,p + \dots \infty$

$\Rightarrow \quad p + q^3p + q^6p + \dots \infty$

$= \dfrac{p}{1-q^3}$ [sum of infinite GP]

$$= \frac{\dfrac{1}{4}}{1 - \left(\dfrac{3}{4}\right)^3} = \frac{64}{175}$$

$\therefore$ Expectation of A = ₹ 350 × probability

$= ₹\, 350 \times \dfrac{64}{175} = ₹\, 128$

8. Let E_1, E_2 and A denote the following events

E_1 : coin selected is fair

E_2 : coin selected is biased

A :

The first toss results in a head and the second toss results in a tail.

$P(E_1) = \dfrac{m}{N}$, $P(E_2) = \dfrac{N-m}{N}$, $P\left(\dfrac{A}{E_1}\right) = \dfrac{1}{2} \times \dfrac{1}{2} = \dfrac{1}{4}$

and $P\left(\dfrac{A}{E_2}\right) = \dfrac{2}{3} \times \dfrac{1}{3} = \dfrac{2}{9}$

By Baye's theorem

$$P\left(\frac{E_1}{A}\right) = \frac{P(E_1) \cdot P\left(\dfrac{A}{E_1}\right)}{P(E_1) \cdot P\left(\dfrac{A}{E_1}\right) + P(E_2) \cdot P\left(\dfrac{A}{E_2}\right)}$$

$$= \frac{\dfrac{m}{N} \times \dfrac{1}{4}}{\dfrac{m}{N} \times \dfrac{1}{4} + \left(\dfrac{N-m}{N}\right) \times \dfrac{2}{9}}$$

$$= \frac{9m}{8N+m} = f(N, m) \quad \text{(given)}$$

$\therefore \quad f(20, 12) = \dfrac{9 \times 12}{8 \times 20 + 12} = \dfrac{108}{172} = \dfrac{27}{43}$

$\Rightarrow \quad 43f(20, 12) = 27$

9. Given, $P(A) + P(B) - 2P(A \cap B) = 1 - a$ ...(i)

and $P(B) + P(C) - 2P(B \cap C) = 1 - 2a$ ...(ii)

and $P(C) + P(A) - 2P(C \cap A) = 1 - a$ ...(iii)

and $P(A \cap B \cap C) = a^2$ ...(iv)

$\therefore \quad P(A \cup B \cup C) = P(A) + P(B) + P(C) - P(A \cap B) - P(B \cap C) - P(C \cap A) + P(A \cap B \cap C)$

$= \dfrac{1}{2}\{P(A) + P(B) - 2P(A \cap B) + P(B) + P(C) - 2P(B \cap C) + P(C) + P(A) - 2P(C \cap A)\} + P(A \cap B \cap C)$

$= \dfrac{1}{2}\{1 - a + 1 - 2a + 1 - a\} + a^2$

[from Eqs. (i), (ii) and (iv)]

$= \dfrac{3}{2} - 2a + a^2 = (a-1)^2 + \dfrac{1}{2} > \dfrac{1}{2}$

Here, $\lambda = \dfrac{1}{2}$

Then, $2050\,\lambda = 2050 \times \dfrac{1}{2} = 1025$

10. We have, $P(X + Y = 3) = P(X = 0, Y = 3) + P(X = 1, Y = 2) + P(X = 2, Y = 1) + P(X = 3, Y = 0)$

$P(X = 0)\,P(Y = 3) + P(X = 1)\,P(Y = 2) + P(X = 2)\,P(Y = 1) + P(X = 3)\,P(Y = 0)$

{∵ X and Y are independent}

$$= {}^5C_0\left(\frac{1}{2}\right)^5 \cdot {}^7C_3\left(\frac{1}{2}\right)^7 + {}^5C_1\left(\frac{1}{2}\right)^5 {}^7C_2\left(\frac{1}{2}\right)^7 + {}^5C_2\left(\frac{1}{2}\right)^5 {}^7C_2\left(\frac{1}{2}\right)^7 + {}^5C_3\left(\frac{1}{2}\right)^5 {}^7C_0\left(\frac{1}{2}\right)^7$$

$$= \left(\frac{1}{2}\right)^{12} [(1)(35) + (5)(21) + (10)(7) + (10)(1)]$$

$$= \frac{220}{2^{12}} = \frac{55}{1024} = \lambda \qquad \text{(given)}$$

$$\therefore \quad 256\,\lambda = 256 \times \frac{55}{1024} = \frac{55}{4} = 13.75$$

11. For the first two draw, the balls taken out may be

Let E_1 = White and White

E_2 = White and Black

E_3 = Black and White

E_4 = Black and Black

$$\therefore \quad P(E_1) = P(W)\cdot P\left(\frac{W}{W}\right) = \frac{2}{4}\cdot\frac{1}{3} = \frac{1}{6}$$

$$P(E_2) = P(W)\cdot P(B/W) = \frac{2}{4}\cdot\frac{2}{3} = \frac{1}{3}$$

$$P(E_3) = P(B)\cdot P(W/B) = \frac{2}{4}\cdot\frac{2}{5} = \frac{1}{5}$$

and $$P(E_4) = P(B)\cdot P(B/B) = \frac{2}{4}\cdot\frac{3}{5} = \frac{3}{10}$$

$$\therefore \; P(E_1) + P(E_2) + P(E_3) + P(E_4) = \frac{1}{6} + \frac{1}{3} + \frac{1}{5} + \frac{3}{10}$$

$$= \frac{10 + 20 + 12 + 18}{60}$$

$$= 1$$

Then, events E_1, E_2, E_3 and E_4 are exhaustive.

Obviously these events are mutually exclusive, then

$$P(B/E_1) = \frac{2}{2} = 1;\; P(B/E_2) = \frac{3}{4}$$

$$P(B/E_3) = \frac{3}{4} \text{ and } P(B/E_4) = \frac{4}{6} = \frac{2}{3}$$

$\therefore$ Required probability

$$P(B) = P(E_1)\cdot P\left(\frac{B}{E_1}\right) + P(E_2)\cdot P\left(\frac{B}{E_2}\right) + P(E_3)\cdot P\left(\frac{B}{E_3}\right) + P(E_4)\cdot P\left(\frac{B}{E_4}\right)$$

$$= \frac{1}{6}\times 1 + \frac{1}{3}\times\frac{3}{4} + \frac{1}{5}\times\frac{3}{4} + \frac{3}{10}\times\frac{2}{3}$$

$$= \frac{1}{6} + \frac{1}{4} + \frac{3}{20} + \frac{1}{5}$$

$$= \frac{10 + 15 + 9 + 12}{60} = \frac{46}{60}$$

$$= \frac{23}{30} = \frac{23}{\lambda} \text{ (given)}$$

$$\Rightarrow \quad \lambda = 30$$

12. There are 11 ways to choose x and 11 ways to choose y. If S be the sample space, then

$n(S)$ = Total number of choosing x and y

$$= 11 \times 11 = 121$$

The number of different values of y for a given value of x can be determined as follows

when $x = 0$, we have $|0 - y| \le 5$

$$\Rightarrow \quad |y| \le 5$$

$$\Rightarrow \quad -5 \le y \le 5$$

$$\Rightarrow \quad 0 \le y \le 5 \qquad \text{[because } y \ge 0]$$

Gives six values of y, *i.e.*, {0, 1, 2, 3, 4, 5}

When $x = 1$, we have $|1 - y| \le 5$

$$\Rightarrow \quad -5 \le 1 - y \le 5$$

$$\Rightarrow \quad 5 \ge y - 1 \ge -5$$

$$\Rightarrow \quad 6 \ge y \ge -4$$

$$\Rightarrow \quad 0 \le y \le 6 \qquad \text{[because } y \ge 0]$$

Gives seven values of y, *i.e.*, {0, 1, 2, 3, 4, 5, 6}

When $x = 2$, we have

$$|2 - y| \le 5$$

$$\Rightarrow \quad -5 \le 2 - y \le 5$$

$$\Rightarrow \quad 5 \ge -2 + y \ge -5$$

$$\Rightarrow \quad 7 \ge y \ge -3$$

$$\Rightarrow \quad 0 \le y \le 7 \qquad \text{[since } y \ge 0]$$

Gives 8 values of y, *i.e.*, {0, 1, 2, 3, 4, 5, 6, 7}

similarly we can show that when x equals 3, 4, 5, 6, 7, 8, 9, 10 there are 9, 10, 11, 10, 9, 8, 7, 6; y-values respectively.

Let E be the event of favourable cases, then

$$n(E) = 6 + 7 + 8 + 9 + 10 + 11 + 10 + 9 + 8 + 7 + 6$$

$$= 91$$

Hence, required probability,

$$P(E) = \frac{n(E)}{n(S)} = \frac{91}{121} = p \text{ (given)}$$

$$\therefore \quad 121\,p = 91$$

13. Total number of outcomes = $2 \times 2 \times 2 \ldots 12$ times

$$= 4096$$

Let a_n denote the number of outcomes in which two consecutive heads do not occur when the fair coin is tossed n times.

$$\Rightarrow \quad a_1 = 2,\; a_2 = 3$$

for $n \ge 3$, if the last outcome is T, then we can not have two consecutive heads in the first $(n-1)$ tosses. This can happen in a_{n-1} ways. If the last outcome is H, we must have T the $(n-1)^{\text{th}}$ toss and we can not have two consecutive heads in the first $(n-2)$ tosses. This can happen in a_{n-2} ways.

$$\Rightarrow \quad a_n = a_{n-1} + a_{n-2} \text{ for } n \ge 3$$

$$\Rightarrow \quad a_{10} = 144,\; a_{11} = 233,$$

$$\Rightarrow \quad a_{12} = 377.$$

Hence, the required probability is $\frac{377}{4096} = p$

$$\therefore \quad 4096\,p = 4096 \times \frac{377}{4096}$$

$$= 377$$

14. Total number of ways = ${}^{64}C_4$

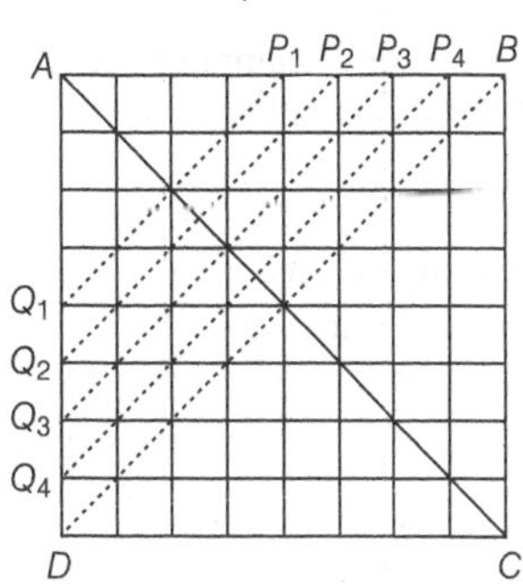

The chess board can be divided into two parts by a diagonal line BD. Now, if we begin to select four squares from the diagonal $P_1Q_1, P_2Q_2, \ldots, BD$, then we can find number squares selected

$$= 2({}^4C_4 + {}^5C_4 + {}^6C_4) + {}^7C_4$$

$$= 182$$

and similarly number of squares for the diagonals chosen parallel to $AC = 182$

$\therefore$ Total favourable ways = 364

$$\therefore \quad \text{Required probability} = \frac{364}{{}^{64}C_4}$$

$$\therefore \quad \lambda = 364$$

15. Let A denote the event that the target is hit when x shells are fired at point I. Let $E_1(E_2)$ denote the event,

we have, $P(E_1) = \frac{8}{9}, P(E_2) = \frac{1}{9}$

$$\Rightarrow \quad P(A/E_1) = 1 - \left(\frac{1}{2}\right)^x \text{ and } P(A/E_2) = 1 - \left(\frac{1}{2}\right)^{21-x}$$

Now, $$P(A) = \frac{8}{9}\left[1 - \left(\frac{1}{2}\right)^x\right] + \frac{1}{9}\left[1 - \left(\frac{1}{2}\right)^{21-x}\right]$$

$$\Rightarrow \quad \frac{dP(A)}{dx} = \frac{8}{9}\left[\left(\frac{1}{2}\right)^x \log 2\right] + \frac{1}{9}\left[-\left(\frac{1}{2}\right)^{21-x} \log 2\right]$$

Now, we must have $\frac{dP(A)}{dx} = 0$

$$\Rightarrow \quad x = 12, \text{ also } \frac{d^2P(A)}{dx^2} < 0$$

Hence, $P(A)$ is maximum when $x = 12$.

$$\therefore \quad 12x = 12 \times 12 = 144$$

Single Integer Type

1. Let W be the event that the ball drawn is white.
Then, $P(W) = \Sigma P(E_1)P(W/E_1)$

$$= \frac{1}{4}\left(\frac{1}{4} + \frac{2}{4} + \frac{3}{4} + \frac{4}{4}\right) = \frac{5}{8}$$

Now, $$P(E_4/W) = \frac{P(E_4)P(W/E_4)}{P(W)} = \frac{1/4}{5/8} = \frac{2}{5}$$

$$\therefore \quad \frac{2}{5} = \frac{p}{15} \Rightarrow p = 6$$

2. Probability of selection of any box is $= \frac{1}{N+1}$

Let E be the event that the wall clock selected is effective, then $P(E) = P(B_1)P(E/B_1)$

$$+ P(B_2)P(E/B_2) + \ldots + P(B_{N+1})P(E/B_{N+1})$$

$$= \frac{1}{(N+1)}\left[1 + \frac{N-1}{N} + \frac{N-2}{N} + \ldots + \frac{1}{N} + 0\right]$$

$$= \frac{1 + 2 + \ldots + N}{N(N+1)} = \frac{1}{2}$$

$$P(B_k/E) = \frac{P(B_k) \cdot P(E/B_k)}{P(E)}$$

$$= \frac{\frac{1}{N+1} \cdot \frac{(N-K+1)}{N}}{1/2} = \frac{2N - 2K + 2}{N^2 + N}$$

$$\therefore \quad \lambda = 2$$

3. The total number of mapping is n^n. The number of one-one mapping is $n!$. Hence, the probability is

$$\frac{n!}{n^n} = \frac{3}{32} = \frac{4!}{4^4}$$

On comparing, we get $n = 4$

4. Using Binomial distribution

$P(X \geq 2) = 1 - P(X = 0) - P(X = 1)$

$$= 1 - \left(\frac{1}{2}\right)^n - \left[{}^nC_1 \cdot \left(\frac{1}{2}\right) \cdot \left(\frac{1}{2}\right)^{n-1}\right]$$

$$= 1 - \frac{1}{2^n} - {}^nC_1 \cdot \frac{1}{2^n} = 1 - \left(\frac{1+n}{2^n}\right)$$

Given, $P(X \geq 2) \geq 0.96$

$$\therefore \quad 1 - \frac{(n+1)}{2^n} \geq \frac{24}{25} \Rightarrow \frac{n+1}{2^n} \leq \frac{1}{25}$$

$$\therefore \quad n = 8$$

5. Given, S_2 reaches the semi-finals.
Since all other players $(2^n - 1)$ are equally likely to win the finals with probability p.

$$\therefore \quad (2^n - 1)p + \frac{1}{4} = 1$$

$$\Rightarrow \quad (2^n - 1)p = \frac{3}{4}$$

$$\Rightarrow \quad p = \frac{3}{4(2^n - 1)}$$

If $p = \frac{1}{84}$, then

$$\therefore \quad \frac{1}{84} = \frac{3}{4(2^n - 1)}$$

$$\Rightarrow \quad 2^n - 1 = 63$$

$$\Rightarrow \quad 2^n = 64 \Rightarrow n = 6$$

6. Let probability of hitting a target, $p = \frac{1}{10}$

So, probability of not hitting a target $q = \frac{9}{10}$

So, probability of not hitting target in 'n' trials $= \left(\frac{9}{10}\right)^n$

$\therefore$ Probability of hitting target at least one

$$= 1 - \left(\frac{9}{10}\right)^n > \frac{1}{4} \quad \text{[given]}$$

$$\Rightarrow \quad (0.9)^n < 0.75$$

$\therefore$ Least number of shots required $= n_{\min} = 3$

7. For the events to be independent,

$$P(E_1 \cap E_2 \cap E_3) = P(E_1) \cdot P(E_2) \cdot P(E_3)$$
$$P(E_1 \cap \bar{E}_2 \cap \bar{E}_3) = P \text{ (only } E_1 \text{ occurs)}$$
$$= P(E_1) \cdot (1 - P(E_2))(1 - P(E_3))$$

Let x, y and z be probabilities of E_1, E_2 and E_3, respectively.

$\therefore \quad \alpha = x(1-y)(1-z) \quad \ldots\text{(i)}$

$\beta = (1-x).y(1-z) \quad \ldots\text{(ii)}$

$\gamma = (1-x)(1-y)z \quad \ldots\text{(iii)}$

$\Rightarrow \quad p = (1-x)(1-y)(1-z) \quad \ldots\text{(iv)}$

Given, $(\alpha - 2\beta)p = \alpha\beta$ and $(\beta - 3\gamma)p = 2\beta\gamma \quad \ldots\text{(v)}$

From above equations, $x = 2y$ and $y = 3z$

$\therefore \quad x = 6z \Rightarrow \frac{x}{z} = 6$

8. Let the number of green socks be $x > 0$.

E : Two socks drawn are of the same colour

$\Rightarrow P(E) = P(RR \text{ or } BB \text{ or } WW \text{ or } GG)$

$$= \frac{3}{{}^{6+x}C_2} = \frac{{}^xC_2}{{}^{6+x}C_2}$$
$$= \frac{6}{(x+6)(x+5)} + \frac{x(x-1)}{(x+6)(x+5)} = \frac{1}{5} \text{ (given)}$$

$\Rightarrow 5(x^2 - x + 6) = x^2 + 11x + 30$

or $\quad 4x^2 - 16x = 0$

or $\quad x = 4$

Matrix-Match Type

1. (A) $\lambda = 1 -$ Problem will not be solved

$$= 1 - P(\bar{A} \cap \bar{B} \cap \bar{C}) = 1 - P(\bar{A})P(\bar{B})P(\bar{C})$$
$$= 1 - \left(1 - \frac{1}{2}\right)\left(1 - \frac{1}{3}\right)\left(1 - \frac{1}{4}\right)$$
$$= 1 - \frac{1}{2} \times \frac{2}{3} \times \frac{3}{4} = 1 - \frac{1}{4} = \frac{3}{4}$$

and $\mu = P(A \cap \bar{B} \cap \bar{C}) + P(\bar{A} \cap B \cap \bar{C}) + P(\bar{A} \cap \bar{B} \cap C)$

$$= P(A)P(\bar{B})P(\bar{C}) + P(\bar{A})P(B)P(\bar{C}) + P(\bar{A})P(\bar{B})P(C)$$
$$= \frac{1}{2} \cdot \left(1 - \frac{1}{3}\right) \cdot \left(1 - \frac{1}{4}\right) + \left(1 - \frac{1}{2}\right) \cdot \frac{1}{3} \cdot \left(1 - \frac{1}{4}\right) + \left(1 - \frac{1}{2}\right)\left(1 - \frac{1}{3}\right) \cdot \frac{1}{4}$$
$$= \frac{6}{24} + \frac{3}{24} + \frac{2}{24} = \frac{11}{24}$$

$\therefore \lambda + \mu = \frac{3}{4} + \frac{11}{24} = \frac{29}{24}$ and $\lambda - \mu = \frac{7}{24}$ (Q, S)

(B) Here, $P(A) = \frac{1}{2}$, $P(B) = \frac{1}{3}$, $P(C) = \frac{1}{4}$

$$\therefore \lambda = P(A \cap B \cap \bar{C}) + P(\bar{A} \cap B \cap C) + P(A \cap \bar{B} \cap C)$$
$$= P(A)P(B)P(\bar{C}) + P(\bar{A})P(B)P(C) + P(A)P(\bar{B})P(C)$$
$$= \frac{1}{2} \cdot \frac{1}{3} \cdot \frac{3}{4} + \frac{1}{2} \cdot \frac{1}{3} \cdot \frac{1}{4} + \frac{1}{2} \cdot \frac{2}{3} \cdot \frac{1}{4} = \frac{6}{24}$$

and $\mu = \lambda + P(A) \cdot P(B) \cdot P(C) = \lambda + \frac{1}{24} = \frac{7}{24}$

and $\mu - \lambda = \frac{1}{24}$ and $\lambda + \mu = \frac{13}{24}$ (P, T)

(C) $\quad \lambda = \frac{2}{6} \times \frac{5}{8} = \frac{5}{24}$ and $\mu = \frac{4}{6} \times \frac{3}{8} = \frac{6}{24}$

$\therefore \quad \mu - \lambda = \frac{1}{24}$ (T)

and $\mu + \lambda = \frac{11}{24}$ (R)

2. (A) $\lambda = P(WW) + P(RR) + P(BB)$

$$= \frac{2}{9} \times \frac{1}{8} + \frac{3}{9} \times \frac{2}{8} + \frac{4}{9} \times \frac{3}{8} = \frac{20}{72} = \frac{5}{18} \text{ (R)}$$

and $\mu = 1 - P$ (neither of the two balls is red)

$$= 1 - P(WW, WB, BW, BB)$$
$$= 1 - \{P(WW) + P(WB) + P(BW) + P(BB)\}$$
$$= 1 - \left(\frac{2}{9} \times \frac{1}{8} + \frac{2}{9} \times \frac{4}{8} + \frac{4}{9} \times \frac{2}{8} + \frac{4}{9} \times \frac{3}{8}\right)$$
$$= 1 - \frac{30}{72} = \frac{42}{72} = \frac{7}{12} \text{ (Q)}$$

(B) $\lambda = \frac{1}{2}\left\{\frac{2}{6} + \frac{6}{10}\right\} = \frac{1}{6} + \frac{3}{10} = \frac{5+9}{30} = \frac{14}{30} = \frac{7}{15}$ (S)

and $\mu = \frac{1}{2}\left\{\frac{4}{6} + \frac{4}{10}\right\} = \frac{2}{6} + \frac{2}{10} = \frac{10+6}{30} = \frac{8}{15}$ (P)

(C) $\lambda = P$(One red from 1st and 2 black from 2nd) + P(One black from 1st and one red and one black from 2nd)

$$= \left(\frac{4}{9} \times \frac{{}^7C_2}{{}^{10}C_2}\right) + \left(\frac{5}{9} \times \frac{{}^3C_1 \times {}^7C_1}{{}^{10}C_2}\right)$$
$$= \left(\frac{4}{9} \times \frac{7 \cdot 6}{10 \cdot 9}\right) + \left(\frac{5}{9} \times \frac{3 \cdot 7 \cdot 2}{10 \cdot 9}\right)$$
$$= \frac{28}{135} + \frac{7}{27} = \frac{63}{135} = \frac{7}{15} \text{ (S)}$$

and $\mu = P$(one black from Ist and 2 red from 2nd) + P(One red from 1st and one red and one black from 2nd)

$$= \left(\frac{5}{9} \times \frac{{}^3C_2}{{}^{10}C_2}\right) + \left(\frac{4}{9} \times \frac{{}^3C_1 \times {}^7C_1}{{}^{10}C_2}\right)$$
$$= \left(\frac{5}{9} \times \frac{3 \cdot 2}{10 \cdot 9}\right) + \left(\frac{4}{9} \times \frac{3 \cdot 7 \cdot 2}{10 \cdot 9}\right)$$
$$= \frac{1}{27} + \frac{28}{135} = \frac{33}{135} = \frac{11}{45} \text{ (T)}$$

3. Let E_i denotes the event that the bag contains i black and $(12 - i)$ white balls ($i = 0, 1, 2, \ldots\ldots, 12$) and A denotes the event that the four balls drawn are all black. Then,

$P(E_i) = \frac{1}{13}$ ($i = 0, 1, 2, \ldots\ldots, 12$);

$P\left(\frac{A}{E_i}\right) = 0$ for $i = 0, 1, 2, 3$; $P\left(\frac{A}{E_i}\right) = \frac{{}^iC_4}{{}^{12}C_4}$ for $i \geq 4$

(A) $P(A) = \sum_{i=0}^{2} P(E_i)P\left(\frac{A}{E_i}\right) = \frac{1}{13} \times \frac{1}{{}^{12}C_4}$

$$[{}^4C_4 + {}^5C_4 + \ldots + {}^{12}C_4] = \frac{{}^{13}C_5}{13 \times {}^{12}C_4} = \frac{1}{5}$$

(B) Clearly, $P\left(\frac{A}{E_{10}}\right)=\frac{{}^{10}C_4}{{}^{12}C_4}=\frac{14}{33}$

(C) By Baye's theorem, $P\left(\frac{E_{10}}{A}\right)$

$$=\frac{P(E_{10})P\left(\frac{A}{E_{10}}\right)}{P(A)}=\frac{\frac{1}{13}\times\frac{14}{33}}{\frac{1}{5}}=\frac{70}{429}$$

(D) Let B denotes the probability of drawing 2 white and 2 black balls, then

$$P\left(\frac{B}{E_i}\right)=0 \text{ if } i=0, 1 \text{ or } 11, 12$$

$$P\left(\frac{B}{E_i}\right)=\frac{{}^iC_2\times{}^{12-i}C_2}{{}^{12}C_4} \text{ for } i=2, 3, \ldots\ldots, 10$$

$$\therefore P(B)=\sum_{i=0}^{12}P(E_i)P\left(\frac{B}{E_i}\right)=\frac{1}{13}\times\frac{1}{{}^{12}C_4}$$

$$[{}^2C_2\times{}^{10}C_2+{}^3C_2\times{}^9C_2+{}^4C_2\times{}^8C_2+\ldots\ldots+{}^{10}C_2\times{}^2C_2]$$

$$=\frac{1}{13}\times\frac{1}{{}^{12}C_4}[2\{{}^2C_2\times{}^{10}C_2+{}^3C_2\times{}^9C_2+{}^4C_2\times{}^8C_2+{}^5C_2\times{}^7C_2\}+{}^6C_2\times{}^6C_2]$$

$$=\frac{1}{13}\times\frac{1}{495}(1287)=\frac{1}{5}$$

4. (A) The chance of getting a head with a coin is $\frac{1}{2}$. A can win in 1st, 4th or 7th……throws

$\therefore p=$ The chance of A's winning

$$=\frac{1}{2}+\left(\frac{1}{2}\right)^3\times\frac{1}{2}+\left(\frac{1}{2}\right)^6\times\frac{1}{2}+\ldots=\frac{\frac{1}{2}}{1-\left(\frac{1}{2}\right)^3}=\frac{4}{7}$$

B can win in 2nd, 5th or 8th ……throws

$\therefore q=$ The chance of B's winning

$$=\left(\frac{1}{2}\right)^1\times\frac{1}{2}+\left(\frac{1}{2}\right)^4\times\frac{1}{2}+\left(\frac{1}{2}\right)^7\times\frac{1}{2}+\ldots$$

$$=\frac{\frac{1}{2}\times\frac{1}{2}}{1-\left(\frac{1}{2}\right)^3}=\frac{2}{7}$$

and $r=$ The chance of c's winning

$$=1-(p+q)=1-\left(\frac{4}{7}+\frac{2}{7}\right)=\frac{1}{7}$$

$p:q=2:1$

$p-r=\frac{3}{7},\ q-r=\frac{1}{7}$ (P, Q, T)

(B) The chance of getting a 'six' when a dice is thrown $=\frac{1}{6}$

$\therefore$ The chance of not getting a 'six', when a dice is not thrown $=1-\frac{1}{6}=\frac{5}{6}$

$\therefore p=$ The probability of A's winning

$$=\frac{1}{6}+\left(\frac{5}{6}\right)^3\times\frac{1}{6}+\left(\frac{5}{6}\right)^6\times\frac{1}{6}+\ldots\ldots=\frac{\frac{1}{6}}{1-\left(\frac{5}{6}\right)^3}=\frac{36}{91}$$

$q=$ The probability of B's winning

$$=\frac{5}{6}\times\frac{1}{6}+\left(\frac{5}{6}\right)^4\times\frac{1}{6}+\left(\frac{5}{6}\right)^7\times\frac{1}{6}+\ldots=\frac{\frac{5}{6}\times\frac{1}{6}}{1-\left(\frac{5}{6}\right)^3}=\frac{30}{91}$$

and $r=1-(p+q)=1-\left(\frac{36+30}{91}\right)=1-\frac{66}{91}=\frac{25}{91}$

$p:q=6:5$ (R)

(C) $p=$The probability of A's winning

$$=\frac{2}{3}+\left(\frac{1}{3}\times\frac{1}{2}\times\frac{3}{4}\right)\times\frac{2}{3}+\left(\frac{1}{3}\times\frac{1}{2}\times\frac{3}{4}\right)^2\times\frac{2}{3}+\ldots$$

$$=\frac{\frac{2}{3}}{1-\left(\frac{1}{3}\times\frac{1}{2}\times\frac{3}{4}\right)}=\frac{\frac{2}{3}}{\frac{7}{8}}=\frac{16}{21}$$

$q=$ The probability of B's winning

$$=\frac{1}{3}\cdot\frac{1}{2}+\left(\frac{1}{3}\cdot\frac{1}{2}\cdot\frac{3}{4}\right)\frac{1}{3}\cdot\frac{1}{2}+\left(\frac{1}{3}\cdot\frac{1}{2}\cdot\frac{3}{4}\right)^2\frac{1}{3}\cdot\frac{1}{2}+\ldots$$

$$=\frac{\frac{1}{3}\cdot\frac{1}{2}}{1-\left(\frac{1}{3}\cdot\frac{1}{2}\cdot\frac{3}{4}\right)}=\frac{\frac{1}{6}}{\frac{7}{8}}=\frac{4}{21}$$

$\therefore\quad r=1-(p+q)=1-\left(\frac{16}{21}+\frac{4}{21}\right)=\frac{1}{21}$

$\therefore\quad q-r=\frac{3}{21}=\frac{1}{7}$ (P), $p-r=\frac{15}{21}=\frac{5}{7}$ (S)

5. (A) $V=npq=5\times\frac{1}{5}\times\frac{4}{5}=\frac{4}{5}$

$S=\sqrt{npq}=\sqrt{\frac{4}{5}}=\frac{2}{\sqrt{5}}$

$V=S^2$ (P)

$\frac{V^2}{S^2}=\frac{4}{5}\Rightarrow\frac{S^2}{V^2}=\frac{5}{4}\Rightarrow 4(S^2-V^2)=V^2$ (Q)

$\Rightarrow\frac{S^2+V^2}{S^2-V^2}=\frac{9}{1}\Rightarrow\sqrt{\left(\frac{S^2+V^2}{S^2-V^2}\right)}=3$ (S)

(B) $V=npq=5\times\frac{2}{5}\times\frac{3}{5}=\frac{6}{5}$

$S=\sqrt{npq}=\sqrt{\frac{6}{5}}$

$V=S^2$ (P)

$\frac{V^2}{S^2}=\frac{6}{5}\Rightarrow 5(V^2-S^2)=S^2$ (R)

(C) $V=npq=25\times\frac{1}{25}\times\frac{24}{25}=\frac{24}{25}$

$S=\sqrt{npq}=\frac{\sqrt{24}}{5}$

$V=S^2$ (P)

$\therefore\ \frac{V^2}{S^2}=\frac{24}{25}\Rightarrow\frac{V^2+S^2}{S^2-V^2}=\frac{49}{1}\Rightarrow\sqrt{\left(\frac{V^2+S^2}{S^2-V^2}\right)}=7$ (T)

10

Sets, Relations and Functions

Objective Questions Type I [Only one correct answer]

In each of the questions below, four choices are given of which only one is correct. You have to select the correct answer which is the most appropriate.

1. If $a\,f(x) + b\,f\left(\frac{1}{x}\right) = x - 1,\ x \neq 0$ and $a \neq b$, then $f(2)$ is equal to

(a) $\frac{a}{a^2 - b^2}$ (b) $\frac{(a + 2b)}{2(a^2 - b^2)}$

(c) $\frac{(a - 2b)}{(a^2 - b^2)}$ (d) $\frac{(2a + b)}{2\,(a^2 - b^2)}$

2. If $f(x) = 2x^n + a$, if $f(2) = 26$ and $f(4) = 138$, then $f(3)$ is equal to

(a) 56 (b) 82

(c) 64 (d) 122

3. The period of the function

$f(x) = 4\sin^4\left(\frac{4x - 3\pi}{6\pi^2}\right) + 2\cos\left(\frac{4x - 3\pi}{3\pi^2}\right)$ is

(a) $\frac{3\pi^2}{4}$ (b) $\frac{3\pi^3}{4}$

(c) $\frac{4\pi^2}{3}$ (d) $\frac{4\pi^3}{3}$

4. Range of the function

$f(x) = (\tan^{-1}\sqrt{[x] + [-x]}) + \sqrt{2 - |x|} + \frac{1}{x^2}$ is

where [*] is the greatest integer function.

(a) $\left[\frac{1}{4}, \infty\right)$ (b) $\left\{\frac{1}{4}\right\} \cup [2, \infty)$

(c) $\left\{\frac{1}{4}, 2\right\}$ (d) $\left[\frac{1}{4}, 2\right]$

5. The number of roots of the equation, $(81)^{\sin^2 x} + (81)^{\cos^2 x} = 30$ in the interval $[0, \pi]$ is equal to **[JEE Main 2021, 16 March]**

(a) 3 (b) 4 (c) 8 (d) 2

6. If $f(x) = \sqrt{3|x| - x - 2}$ and $g(x) = \sin x$, then domain of definition of $(fog)\,x$ is

(a) $\left\{2n\pi + \frac{\pi}{2}\right\}_{n \in I}$

(b) $\bigcup_{n \in I}\left(2n\pi + \frac{7\pi}{6}, 2n\pi + \frac{11\pi}{6}\right)$

(c) $\left\{2n\pi + \frac{7\pi}{6}\right\}_{n \in I}$

(d) $\left\{(4m + 1)\frac{\pi}{2}, m \in I\right\} \bigcup_{n \in I}\left[2n\pi + \frac{7\pi}{6}, 2n\pi + \frac{11\pi}{6}\right]$

7. Let $f(x) = \frac{\sin 2nx}{1 + \cos^2 nx}$, $n \in N$ has $\frac{\pi}{6}$ as its fundamental period, then n is equal to

(a) 2 (b) 4

(c) 6 (d) 8

8. The inverse of $y = 5^{\log x}$ is **[JEE Main 2021, 17 March]**

(a) $x = (1/y)^{\log 5}$ (b) $x = y^{\frac{1}{\log 5}}$

(c) $x = 5^{\log y}$ (d) $x = 5^{\frac{1}{\log y}}$

9. A survey shows that 63% of the people in a city read newspaper A whereas 76% read news paper B. If x % of the people read both the newspapers, then a possible value of x can be **[JEE Main 2020, 4 September]**

(a) 55 (b) 29

(c) 65 (d) 37

10. Period of the function $f(x) = \dfrac{\sin\{\sin(nx)\}}{\tan\left(\dfrac{x}{n}\right)}$, $n \in N$, is 6π, then n is equal to
(a) 1 (b) 2
(c) 3 (d) None of these

11. If $f(x)$ is an even function and satisfies the relation $x^2 f(x) - 2f\left(\dfrac{1}{x}\right) = g(x)$, where $g(x)$ is an odd function, then the value of $f(5)$ is
(a) 0 (b) $\dfrac{37}{75}$
(c) 4 (d) $\dfrac{51}{77}$

12. The set {0, 2, 6, 12, 20} in the set-builder form is
(a) $\{x : x = n^2 - 3n + 2$, where n is a natural number and $1 \le n \le 5\}$
(b) $\{x : x = n^2 - 3n + 2$, where n is a natural number and $1 \le n \le 6)$
(c) $\{x : x = n^2 - 3n + 4$, where n is a natural number and $1 \le n \le 5\}$
(d) $\{x : x = n^2 + 5n - 6$, where n is a natural number and $1 \le n \le 5$

13. The domain of definition of the function $y = \dfrac{1}{\log_{10}(1-x)} + \sqrt{x+2}$ is
(a) $(-3, -2) \sim \{-2.5\}$ (b) $[0, 1] \sim \{0.5\}$
(c) $[-2, 1) \sim \{0\}$ (d) None of these

14. $f(x) = (\sin x^7)\, e^{x^5 \, Sgn x^9}$ is
(a) an even function (b) an odd function
(c) neither even nor odd (d) None of these

15. Let $f(x) = \sqrt{([\sin 2x] - [\cos 2x])}$ (where [.] denotes the greatest integer function), then range of $f(x)$ will be
(a) {0} (b) {1}
(c) {0, 1} (d) $\{0, 1, \sqrt{2}\}$

16. If $f : [-20, 20] \to R$, defined by $f(x) = \left[\dfrac{x^2}{a}\right] \sin x + \cos x$, (where [.] denotes the greatest integer function) is an even function, then set of values of 'a' is given by
(a) null set (b) R
(c) [0, 400] (d) $(400, \infty)$

17. Let $A \equiv \{1, 2, 3, 4\}$, $B \equiv \{a, b, c\}$, then number of functions from $A \to B$, which are not onto is
(a) 8 (b) 24 (c) 45 (d) 6

18. Let $f(x) = \sin^{-1} \sin(\tan x)$ and $g(x) = \cos^{-1} \sin \sqrt{(1 - \tan^2 x)}$ are same functions, then $x \in$
(a) $\left[0, \tan^{-1}\dfrac{\pi}{2}\right]$ (b) [0, 1]
(c) $[0, \infty]$ (d) None of these

19. If $f(x) + 2f(1-x) = x^2 + 2, \forall x \in R$, then $f(x)$ is given as
(a) $\dfrac{(x-1)^2}{3}$ (b) $\dfrac{(x-2)^2}{3}$
(c) $x^2 - 1$ (d) $x^2 - 2$

20. If $2f(x-1) - f\left(\dfrac{1-x}{x}\right) = x$, then $f(x)$ is
(a) $\dfrac{1}{3}\left\{2(1+x) + \dfrac{1}{(1+x)}\right\}$ (b) $2(x-1) - \dfrac{(1-x)}{x}$
(c) $x^2 + \dfrac{1}{x^2} + 4$ (d) $\dfrac{1}{4}\left\{(x+2) + \dfrac{1}{(x+2)}\right\}$

21. If $f : R \to R$ be a function satisfying $f(2x+3) + f(2x+7) = 2, \forall x \in R$, then period of $f(x)$ is
(a) 2 (b) 4
(c) 8 (d) 16

22. The range of the function $f(x) = \sin^{-1}\left[x^2 + \dfrac{1}{2}\right] + \cos^{-1}\left[x^2 - \dfrac{1}{2}\right]$, (where [.] denotes the greatest integer function) is
(a) $\left\{\dfrac{\pi}{2}\right\}$ (b) $\{\pi\}$
(c) $\left\{-\dfrac{1}{2}, 0\right\}$ (d) $\left(0, \dfrac{\pi}{2}\right)$

23. Let R_1 and R_2 be two relations defined as follows
$R_1 = \{(a, b) \in R^2 : a^2 + b^2 \in Q\}$
and $R_2 = \{(a, b) \in R^2 : a^2 + b^2 \notin Q\}$,
where Q is the set of all rational numbers. Then, **[JEE Main 2020, 3 September]**
(a) R_1 and R_2 are both transitive
(b) Neither R_1 nor R_2 is transitive
(c) R_1 is transitive but R_2 is not transitive
(d) R_2 is transitive but R_1 is not transitive

24. If $2 < x^2 < 3$, then the number of positive roots of $\{x^2\} = \left\{\dfrac{1}{x}\right\}$, (where $\{x\}$ denotes the fractional part of x) is
(a) 0 (b) 1
(c) 2 (d) 3

25. If $f(x)$ and $g(x)$ are periodic functions with periods 7 and 11 respectively. Then the period of $F(x) = f(x)\, g\left(\dfrac{x}{5}\right) - g(x) f\left(\dfrac{x}{3}\right)$ is
(a) 177
(b) 222
(c) 433
(d) 1155

26. Let $A = \{1, 2, 3, 4, 5, 6\}$. If f be a bijective function from A to A, then the number of such functions for which $f(\lambda) \neq \lambda$, $\lambda = 1, 2, 3, 4, 5, 6$ is

(a) 44 (b) 265
(c) 325 (d) 4585

27. Let $f : [0, 10] \to [1, 20]$ be a function defined

as $f(x) = \begin{cases} \frac{60 - 5x}{3}, & 0 \le x \le 6 \\ 10, & 6 \le x \le 7 \\ 31 - 3x, & 7 \le x \le 10 \end{cases}$, then f is

(a) bijective function
(b) One-one but not onto function
(c) onto but not one-one function
(d) neither one-one nor onto function

28. If $f(x) = -\frac{x|x|}{1 + x^2}$, then $f^{-1}(x)$ equals

(a) $\sqrt{\frac{|x|}{1-|x|}}$ (b) $(\text{sgn } x)\sqrt{\frac{|x|}{1-|x|}}$
(c) $-\sqrt{\frac{x}{1-x}}$ (d) None of these

29. Let $f : R \to R$ defined by $f(x) = \frac{e^{x^2} - e^{-x^2}}{e^{x^2} + e^{-x^2}}$, then

(a) $f(x)$ is one-one but not onto
(b) $f(x)$ is neither one-one nor onto
(c) $f(x)$ is many one but onto
(d) $f(x)$ is one-one and onto

30. Let a relation R on the set N of natural number be defined as $(x, y) \in R$. If and only If $x^2 - 4xy + 3y^2 = 0$ for all $x, y \in N$. The relation R is

(a) reflexive (b) symmetric
(c) transitive (d) an equivalence relation

31. If f is decreasing odd function, then f^{-1} is

(a) odd and decreasing (b) odd and increasing
(c) even and decreasing (d) even and increasing

32. The range of the function $f(x) = 3|\sin x| - 2|\cos x|$ is

(a) $[-2, \sqrt{13}]$ (b) $[-2, 3]$
(c) $[3, \sqrt{13}]$ (d) $[-3, 2]$

33. The real valued function $f(x) = \frac{\text{cosec}^{-1} x}{\sqrt{x - [x]}}$, where $[x]$ denotes the greatest integer less than or equal to x, is defined for all x belonging to

[JEE Main 2021, 18 March]

(a) all reals except integers
(b) all non-integers except the interval $[-1, 1]$
(c) all integers except $[0, -1]$
(d) all reals except the Interval $[-1, 1]$

34. If $g(x) = [x^2] - [x]^2$, where [.] denotes the greatest integer function and $x \in [0, 2]$, then the set of values of $g(x)$ is

(a) $\{-1, 0\}$ (b) $\{-1, 0, 1\}$
(c) $\{0\}$ (d) $\{0, 1, 2\}$

35. Which of the following functions is periodic with period π ?

(a) $f(x) = \sin 3x$ (b) $f(x) = |\cos x|$
(c) $(x) = [x + \pi]$ (d) $f(x) = x \cos x$

where $[x]$ means the greatest integer not greater than x.

36. The domain of definition of $f(x) = \sqrt{\frac{1-|x|}{2-|x|}}$ is

(a) $(-\infty, \infty) - [-2, 2]$
(b) $(-\infty, \infty) - [-1, 1]$
(c) $[-1, 1] \cup (-\infty, -2) \cup (2, \infty)$
(d) None of the above

37. Suppose $A_1, A_2, \dots A_{30}$ are thirty sets, each with five elements and $B_1, B_2, \dots B_n$ are n sets each with three elements.

Let $\bigcup_{i=1}^{30} A_i = \bigcup_{j=1}^{n} B_j = S$

If each element of S belongs to exactly ten of the A_i's and exactly nine of the B_j's then $n =$

(a) 45 (b) 35
(c) 40 (d) None of these

38. The domain of the function $y = \underbrace{\log_{10}\log_{10}\log_{10}\dots\log_{10} x}_{n \text{ times}}$ is

(a) $[10^n, \infty)$ (b) $(10^{n-1}, \infty)$
(c) $(10^{n-2}, \infty)$ (d) None of these

39. Let A, B and C be three sets and X be the set of all elements which belong to exactly two of the sets A, B and C. Then, X is equal to

(a) $(A \cap B) \cup (B \cap C) \cup (C \cap A)$
(b) $A \,\Delta\, (B \,\Delta\, C)$
(c) $(A \cup B) \cap (B \cup C) \cap (C \cup A)$
(d) $(A \cup B \cup C) - [A \,\Delta\, (B \,\Delta\, C)]$

40. If [.] denotes the greatest integer function, then the value of $\sum_{r=1}^{100} \left[\frac{1}{2} + \frac{r}{100}\right]$ is

(a) 49 (b) 50
(c) 51 (d) 52

41. In a town of 10,000 families, it was found that 40% families buy newspaper A, 20% buy newspaper B and 10% buy newspaper C. Also, 5% families buy newspapers A and B, 3% buy newspaper B and C and 4% buy newspaper A and C. If 2% families buy all the three newspaper, then number of families which buy newspaper A only is

(a) 3100 (b) 3300
(c) 2900 (d) 1400

42. If $f(x + y) = f(x) + f(y) - xy - 1$ for all x, y and $f(1) = 1$, then the number of solutions of $f(n) = n, n \in N$ is

(a) one (b) two
(c) three (d) None of these

43. The function $f(x) = \sin\left(\frac{\pi x}{n!}\right) - \cos\left(\frac{\pi x}{(n+1)!}\right)$ is
(a) non periodic
(b) periodic, with period $2(n!)$
(c) periodic, with period $(n+1)$
(d) None of the above

44. The value of b and c for which the identity $f(x+1) - f(x) = 8x + 3$ is satisfied, where $f(x) = bx^2 + cx + d$ are
(a) $b = 2, c = 11$ (b) $b = 4, c = -1$
(c) $b = -1, c = 4$ (d) $b = -1, c = 1$

45. Let A, B and C be finite sets such that $A \cap B \cap C = \phi$ and each one of the sets $A \Delta B$, $B \Delta C$ and $C \Delta A$ has 100 elements. The number of elements in $A \cup B \cup C$ is
(a) 250 (b) 200
(c) 150 (d) 300

46. Which of the following function is even function
(a) $f(x) = \left(\frac{a^x + 1}{a^x - 1}\right)$ (b) $f(x) = x\left(\frac{a^x - 1}{a^x + 1}\right)$
(c) $f(x) = \left(\frac{a^x - a^{-x}}{a^x + a^{-x}}\right)$ (d) $f(x) = \sin x$

47. If S is the set of all real x for which $1 - e^{(1/x)-1} > 0$, then S is equal to
(a) $(-\infty, 0) \cup (1, \infty)$ (b) $(-\infty, \infty)$
(c) $(-\infty, 0] \cup [1, \infty)$ (d) None of these

48. If $f(x) = \sin^2 x + \sin^2\left(x + \frac{\pi}{3}\right) + \cos x \cdot \cos\left(x + \frac{\pi}{3}\right)$ and $g(5/4) = 1$, then $(gof)\,x$ is
(a) a polynomial of the first degree in $\sin x$, $\cos x$
(b) a constant function
(c) a polynomial of the second degree in $\sin x$, $\cos x$
(d) None of the above

49. If the function $f : [1, \infty) \to [1, \infty)$ is defined by $f(x) = 2^{x(x-1)}$, then $f^{-1}(x)$ is
(a) $\left(\frac{1}{2}\right)^{x(x-1)}$ (b) $\frac{1}{2}(1 + \sqrt{(1 + 4\log_2 x)})$
(c) $\frac{1}{2}(1 - \sqrt{(1 + 4\log_2 x)})$ (d) not defined

50. Let f be a function satisfying $2f(xy) = \{f(x)\}^y + \{f(y)\}^x$ and $f(1) = k \neq 1$, then $\sum_{r=1}^{n} f(r)$ is equal to
(a) $k^n - 1$ (b) k^n
(c) $k^n + 1$ (d) None of these

51. Let R be a relation over the set $N \times N$ and it is defined by $(a, b)R(c, d) \Rightarrow a + d = b + c$. Then, R is not
(a) Reflexive (b) Symmetric
(c) Transitive (d) Anti-symmetric

52. The domain of the function $f(x) = 1/\log_{10}(1 - x) + \sqrt{(x+2)}$ is
(a) $[-3, -2]$, excluding (-2.5)
(b) $[0, 1]$, excluding 0.5
(c) $[-2, 1]$, excluding 0
(d) None of the above

53. If the functions are defined as $f(x) = \sqrt{x}$ and $g(x) = \sqrt{1 - x}$, then what is the common domain of the following functions

$$f + g,\ f - g,\ f/g,\ g/f,\ g - f$$

where $(f \pm g)(x) = f(x) \pm g(x)$, $(f/g)(x) = \frac{f(x)}{g(x)}$

[JEE Main 2021, 18 March]

(a) $0 \leq x \leq 1$ (b) $0 \leq x < 1$
(c) $(0, 1)$ (d) $[0, 1)$

54. The range of the function $f(x) = 6^x + 3^x + 6^{-x} + 3^{-x} + 2$ is
(a) $[-2, \infty)$ (b) $(-2, \infty)$
(c) $(6, \infty)$ (d) $[6, \infty)$

55. If $f : X \to Y$ defined by $f(x) = \sqrt{3}\sin x + \cos x + 4$ is one-one and onto, then Y is
(a) $[1, 4]$ (b) $[2, 5]$
(c) $[1, 5]$ (d) $[2, 6]$

56. If $f(x) = \cos^{-1}(x - x^2) + \sqrt{\left(1 - \frac{1}{|x|}\right)} + \frac{1}{[x^2 - 1]}$, then domain of $f(x)$ is (where [.] is the greatest integer)
(a) $\left(\sqrt{2}, \frac{1+\sqrt{5}}{2}\right]$ (b) $\left(-\sqrt{2}, \frac{1-\sqrt{5}}{2}\right)$
(c) $\left[\sqrt{2}, \frac{1+\sqrt{5}}{2}\right]$ (d) None of these

57. Two newspapers A and B are published in a city. It is known that 25% of the city population reads A and 20% reads B while 8% reads both A and B. Further, 30% of those who read A but not B look into advertisements and 40% of those who read B but not A also look into advertisements, while 50% of those who read both A and B look into advertisements. Then, the percentage of the population who look into advertisements is

[JEE Main 2019, 9 April]

(a) 13.5 (b) 13
(c) 12.8 (d) 13.9

58. Let $f(x)$ be a function defined on $[0, 1]$ such that

$$f(x) = \begin{cases} x, & \text{if } x \in Q \\ 1 - x, & \text{if } x \notin Q. \end{cases}$$

Then for all $x \in [0, 1]$ $(f0f)(x)$ is
(a) constant (b) $1 + x$
(c) x (d) None of these

59. $f(x) = \frac{x}{e^x - 1} + \frac{x}{2} + 2\cos^3\frac{x}{2}$ or $R - \{0\}$ is
(a) one one function (b) bijection
(c) algebraic function (d) even function

60. A function $f : R \to R$ satisfies the equation $f(x)f(y) - f(xy) = x + y\ \forall x, y \in R$ and $f(1) > 0$, then
(a) $f(x)f^{-1}(x) = x^2 - 4$
(b) $f(x)\, f^{-1}(x) = x^2 - 6$
(c) $f(x)f^{-1}(x) = x^2 - 1$
(d) None of these

61. Let $f : R - \{3\} \to R - \{1\}$ be defined by $f(x) = \dfrac{x-2}{x-3}$. Let $g : R \to R$ be given as $g(x) = 2x - 3$. Then, the sum of all the values of x for which $f^{-1}(x) + g - 1(x) = \dfrac{13}{2}$ is equal to **[JEE Main 2021, 18 March]**
(a) 7
(b) 2
(c) 5
(d) 3

62. The period of $e^{\cos^4 \pi x + x - [x] + \cos^2 \pi x}$ is ([.] denotes the greatest integer function)
(a) 2
(b) 1
(c) 0
(d) -1

63. If $f(x) = \sin^{-1}\{4 - (x-7)^3\}^{1/5}$, then its inverse is
(a) $(4 - \sin^5 x)^{1/3}$
(b) $7 - (4 - \sin^5 x)^{1/3}$
(c) $(4 - \sin^5 x)^{2/3}$
(d) $7 + (4 - \sin^5 x)^{1/3}$

64. Let $f(x) = a^x\,(a > 0)$ be written as $f(x) = g(x) + h(x)$, where $g(x)$ is an even function and $h(x)$ is an odd function. Then, the value of the $g(x+y) + g(x-y)$ is
(a) $2g(x)\,g(y)$
(b) $2g(x+y)\,g(x-y)$
(c) $2g(x)$
(d) None of these

65. Given $f(x) = \dfrac{1}{(1-x)}$, $g(x) = f\{f(x)\}$ and $h(x) = f[f\{f(x)\}]$. Then the value of $f(x) \cdot g(x) \cdot h(x)$ is
(a) 0
(b) -1
(c) 1
(d) 2

66. The inverse of the function $y = \log_a(x + \sqrt{(x^2+1)}); (a > 0, a \neq 1)$ is
(a) $\dfrac{1}{2}(a^x - a^{-x})$
(b) not defined for all x
(c) defined for only positive x
(d) None of the above

67. The domain of the function
$$f(x) = \sqrt{\sin^{-1}(\log_2 x)} + \sqrt{\cos(\sin x)} + \sin^{-1}\left(\frac{1+x^2}{2x}\right)$$
(a) $\{x : 1 \le x \le 2\}$
(b) $\{1\}$
(c) not defined for any value of x
(d) $\{-1, 1\}$

68. Let $f : R \to R$ be a function defined by $f(x) = \dfrac{x^2+2x+5}{x^2+x+1}$ is
(a) one-one and into
(b) one-one and onto
(c) many one and onto
(d) many one and into

69. Let $f(x) = \begin{cases} 1+x, & 0 \le x \le 2 \\ 3-x, & 2 < x \le 3 \end{cases}$, then $fof(x)$
(a) $= \begin{cases} 2+x, & 0 \le x \le 1 \\ 2-x, & 1 < x \le 2 \\ 4-x, & 2 < x \le 3 \end{cases}$
(b) $= \begin{cases} 2+x, & 0 \le x \le 2 \\ 4-x, & 2 < x \le 3 \end{cases}$
(c) $= \begin{cases} 2+x, & 0 \le x \le 2 \\ 2-x, & 2 < x \le 3 \end{cases}$
(d) None of these

70. Let $f(x) = (x+1)^2 - 1$, $x \ge -1$ then, the set $\{x : f(x) = f^{-1}(x)\}$ is equal to
(a) $\{0, -1\}$
(b) $\{0, 1\}$
(c) $\{-1, 1\}$
(d) $\{0\}$

71. The domain of $f(x) = \dfrac{1}{\sqrt{|\cos x| + \cos x}}$ is
(a) $[-2n\pi, 2n\pi]; \forall n \in I$
(b) $(2n\pi, \overline{2n+1}\,\pi); \forall n \in I$
(c) $\left(\dfrac{(4n+1)\pi}{2}, \dfrac{(4n+3)\pi}{2}\right); \forall n \in I$
(d) $\left(\dfrac{(4n-1)\pi}{2}, \dfrac{(4n+1)\pi}{2}\right); \forall n \in I$

72. The domain of the function $f(x) = {}^{16-x}C_{2x-1} + {}^{20-3x}P_{4x-5}$, where the symbols have their usual meanings, is the set
(a) $\{2, 3\}$
(b) $\{2, 3, 4\}$
(c) $\{1, 2, 3, 4, 5\}$
(d) None of the above

73. If $f(x) = 3\sin\sqrt{\left(\dfrac{\pi^2}{16} - x^2\right)}$, then its range is
(a) $\left[-\dfrac{3}{\sqrt{2}}, \dfrac{3}{\sqrt{2}}\right]$
(b) $\left[0, \dfrac{3}{\sqrt{2}}\right]$
(c) $\left[-\dfrac{3}{\sqrt{2}}, 0\right]$
(d) None of these

74. The domain of $f(x) = \sqrt{\{x - 4 - 2\sqrt{(x-5)}\}} - \sqrt{\{x - 4 + 2\sqrt{x-5}\}}$ is
(a) $[-5, \infty)$
(b) $(-\infty, 2]$
(c) $[5, \infty)$
(d) None of the above

75. The period of $\dfrac{|\sin x| + |\cos x|}{|\sin x - \cos x|}$ is
(a) 2π
(b) π
(c) $\pi/2$
(d) $\pi/4$

76. If [.] denotes the greatest integer function, then the domain of the real valued function $\log_{[x+1/2]}|x^2 - x - 2|$ is
(a) $\left[\dfrac{3}{2}, \infty\right)$
(b) $\left[\dfrac{3}{2}, 2\right) \cup (2, \infty)$
(c) $\left(\dfrac{1}{2}, 2\right) \cup (2, \infty)$
(d) None of these

77. Let $f(x) = \sin^2(x/2) + \cos^2(x/2)$ and $g(x) = \sec^2 x - \tan^2 x$. The two functions are equal over the set

(a) ϕ
(b) R
(c) $R - \left\{x : x = (2n+1)\frac{\pi}{2}, n \in I\right\}$
(d) None of the above

78. The domain of $f(x)$ is $(0, 1)$, therefore domain of $f(e^x) + f(\ln|x|)$ is

(a) $(-1, e)$ (b) $(1, e)$
(c) $(-e, -1)$ (d) $(-e, 1)$

79. If $f : [-4, 0] \to R$ is defined by $e^x + \sin x$, its even extension to $[-4, 4]$ is given by

(a) $-e^{-|x|} - \sin|x|$ (b) $e^{-|x|} - \sin|x|$
(c) $e^{-|x|} + \sin|x|$ (d) $-e^{-|x|} + \sin|x|$

80. If $g(x)$ be a function defined on $[-1, 1]$ if the area of the equilateral triangle with two of its vertices at $(0, 0)$ and $(x, g(x))$ is $\sqrt{3}/4$, then

(a) $g(x) = \pm\sqrt{(1-x^2)}$ (b) $g(x) = -\sqrt{(1-x^2)}$
(c) $g(x) = \sqrt{(1-x^2)}$ (d) $g(x) = \sqrt{(1+x^2)}$

81. The period of the function

$$f(x) = a^{\sin^2 x + \sin^2(x+\pi/3) + \cos x \cos(x+\pi/3)} \text{ is}$$

(where a is constant)

(a) 1 (b) $\pi/2$
(c) π (d) Cannot be determined

82. The domain of the function

$$f(x) = \sin^{-1}\left(\frac{2-|x|}{4}\right) + \cos^{-1}\left(\frac{2-|x|}{4}\right) + \tan^{-1}\left(\frac{2-|x|}{4}\right)$$

is

(a) $[0, 3]$ (b) $[-6, 6]$
(c) $[-1, 1]$ (d) $[-3, 3]$

83. Let f be a real valued function defined by

$$f(x) = \frac{e^x - e^{-|x|}}{e^x + e^{|x|}}, \text{ then range of } f \text{ is}$$

(a) R (b) $[0, 1]$
(c) $[0, 1)$ (d) $[0, 1/2)$

84. If $f(x)$ is a polynomial function of the second degree such that $f(-3) = 6$, $f(0) = 6$ and $f(2) = 11$, then the graph of the function $f(x)$ cuts the ordinate $x = 1$ at the point

(a) $(1, 8)$ (b) $(1, -2)$
(c) 1, 4 (d) None of these

85. If $f(x+y, x-y) = xy$, then the arithmetic mean of $f(x, y)$ and $f(y, x)$ is

(a) x (b) y
(c) 0 (d) None of these

86. Let $f : \{x, y, z\} \to \{1, 2, 3\}$ be a one-one mapping such that only one of the following three statements is true and remaining two are false

$f(x) \neq 2$, $f(y) = 2$, $f(z) \neq 1$, then

(a) $f(x) > f(y) > f(z)$ (b) $f(x) < f(y) < f(z)$
(c) $f(y) < f(x) < f(z)$ (d) $f(y) < f(z) < f(x)$

87. Let A, B and C be sets such that $\phi \neq A \cap B \subseteq C$. Then, which of the following statements is not true?

[JEE Main 2019, 12 April]

(a) $B \cap C \neq \phi$
(b) If $(A - B) \subseteq C$, then $A \subseteq C$
(c) $(C \cup A) \cap (C \cup B) = C$
(d) If $(A - C) \subseteq B$, then $A \subseteq B$

88. If the function $f : \mathbf{R} \to \mathbf{R}$ is defined by $f(x) = |x|(x - \sin x)$, then which of the following statements is TRUE? **[JEE Advanced 2020]**

(a) f is one-one, but NOT onto
(b) f is onto, but NOT one-one
(c) f is BOTH one-one and onto
(d) f is NEITHER one-one NOR onto

89. Let $f : R \to [0, \pi/2)$ be a function defined by $f(x) = \tan^{-1}(x^2 + x + a)$. If f is onto, then a equals

(a) 0 (b) 1
(c) 1/2 (d) 1/4

90. Let $A = R \times R$, R the real number system and $R = \{((x, y), (a, b)) \in A \times A |$ either $x < a$ or $x = a$ and $y > b\}$

Then, which one of the following is true, if $((x, y), (a, b)) \in R$ and $((a, b) \in R$ and $((a, b), (p, q)) \in R$?

(a) $((x, y), (p, q)) \in R$ (b) $((x, y), (q, p)) \in R$
(c) $((x, y), (y, q)) \in R$ (d) $((y, x), (p, q)) \in R$

91. Domain of $\sin^{-1}[\sec x]$ ([.] is greatest integer less than or equal to x) is

(a) $\{(2n+1)\pi, (2n+9)\pi\}$
$\cup \{[(2m-1)\pi, 2m\pi + \pi/3), m \in I\}$
(b) $\{2n\pi, n \in I\} \cup \{[2m\pi, (2m+1)\pi, m \in I\}$
(c) $\{(2n+1)\pi, n \in I\} \cup \{[2m\pi, 2m\pi + \pi/3], m \in I\}$
(d) None of the above

92. Let $f(x) = (x^{12} - x^9 + x^4 - x + 1)^{-1/2}$. The domain of the function is

(a) $(-\infty, -1)$ (b) $(-1, 1)$
(c) $(1, \infty)$ (d) $(-\infty, \infty)$

93. The domain of the function $f(x) = \log_e\{\text{sgn}(9 - x^2)\} + \sqrt{[x]^3 - 4[x]}$ (where [.] represents the greatest integer function is

(a) $[-2, 1) \cup [2, 3)$ (b) $[-4, 1) \cup [2, 3)$
(c) $[4, 1) \cup [2, 3)$ (d) $[2, 1) \cup [2, 3)$

94. If $f : R \to R$, $g : R \to R$ be two given functions, then $f(x) = 2 \min(f(x) - g(x), 0)$ equals

(a) $f(x) + g(x) - |g(x) - f(x)|$
(b) $f(x) + g(x) + |g(x) - f(x)|$
(c) $f(x) - g(x) + |g(x) - f(x)|$
(d) $f(x) - g(x) - |g(x) - f(x)|$

95. The domain of the function $f(x) = \ln\left(\ln \frac{x}{\{x\}}\right)$ is (where {.} denotes the fractional part function)
(a) $(0, \infty) - I$ (b) $(1, \infty) - I$
(c) $R - I$ (d) $(2, \infty) - I$

96. $\sin ax + \cos ax$ and $|\sin x| + |\cos x|$ are periodic of same fundamental period, if a equals
(a) 0 (b) 1
(c) 2 (d) 4

97. If $g(x)$ is a polynomial satisfying $g(x)\,g(y) = g(x) + g(y) + g(xy) - 2$ for all real x and y and $g(2) = 5$, then $g(3)$ is equal to
(a) 10 (b) 24
(c) 21 (d) None of these

98. The interval into which the function $y = \frac{(x-1)}{(x^2 - 3x + 3)}$ transforms the entire real line is
(a) $\left[\frac{1}{3}, 2\right]$ (b) $\left[-\frac{1}{3}, 1\right]$
(c) $\left[-\frac{1}{3}, 2\right]$ (d) None of these

99. Let the function $f(x) = x^2 + x + \sin x - \cos x + \log(1 + |x|)$ be defined over the interval [0, 1]. The odd extension of $f(x)$ in the interval [– 1, 1] is
(a) $x^2 + x + \sin x + \cos x - \log(1 + |x|)$
(b) $-x^2 + x + \sin x + \cos x - \log(1 + |x|)$
(c) $-x^2 + x + \sin x - \cos x + \log(1 + |x|)$
(d) None of the above

100. The function $f(x) = \frac{\sec^{-1} x}{\sqrt{x - [x]}}$, where [x] denotes the greatest integer less than or equal to x is defined for all x belonging to
(a) R
(b) $R - \{(-1, 1) \cup \{n, n \in I\}\}$
(c) $R^+ - (0, 1)$
(d) $R^+ - \{n, n \in N\}$

101. The period of the function $f(x) = [\sin 3x] + |\cos 6x|$ is ([.] denotes the greatest integer less than or equal to x)
(a) π (b) $2\pi/3$
(c) 2π (d) None of these

Objective Questions Type II [One or more than one correct answer(s)]

In each of the questions below four choices of which one or more than one are correct. You have to select the correct answer(s) accordingly.

1. Let $f(x) = \sin\left[\frac{\pi}{6}\sin\left(\frac{\pi}{2}\sin x\right)\right]$ for all $x \in R$ and $g(x) = \frac{\pi}{2}\sin x$ for all $x \in R$. Let $(fog)(x)$ denotes $f\{g(x)\}$ and $(gof)\ (x)$ denotes $g\{f(x)\}$. Then, which of the following is/are true? **[JEE Advanced 2015]**
(a) Range of f is $\left[-\frac{1}{2}, \frac{1}{2}\right]$
(b) Range of fog is $\left[-\frac{1}{2}, \frac{1}{2}\right]$
(c) $\lim_{x \to 0} \frac{f(x)}{g(x)} = \frac{\pi}{6}$
(d) There is an $x \in R$ such that $(gof)(x) = 1$

2. Let $f(x) = 2x - \sin x$ and $g(x) = \sqrt[3]{x}$, then
(a) range of gof is R (b) gof is one-one
(c) both f and g are one-one
(d) both f and g are onto

3. Let $f(x) = \begin{cases} 0, & \text{for } x = 0 \\ x^2 \sin\left(\frac{\pi}{x}\right), & \text{for } -1 < x < 1, (x \neq 0), \\ x|x|, & \text{for } x \geq 1 \text{ or } x \leq -1 \end{cases}$

then
(a) $f(x)$ is an odd function
(b) $f(x)$ is an even function
(c) $f(x)$ is neither odd nor even
(d) $f'(x)$ is an even function

4. Which of the following function is periodic
(a) $\text{Sgn}(e^{-x})$
(b) $\sin x + |\sin x|$
(c) $\min(\sin x, |x|)$
(d) $\left[x + \frac{1}{2}\right] + \left[x - \frac{1}{2}\right] + 2[-x]$
([x] denotes the greatest integer function)

5. Of the following functions defined from [– 1, 1] to [– 1, 1] select those which are not bijective
(a) $\sin(\sin^{-1} x)$ (b) $\frac{2}{\pi}\sin^{-1}(\sin x)$
(c) $(\text{sgn } x)/\ln(e^x)$ (d) $x^3(\text{sgn } x)$

6. If [x] denotes the greatest integer less than or equal to x, the extreme values of the function $f(x) = [1 + \sin x] + [1 + \sin 2x] + [1 + \sin 3x] + \ldots + [1 + \sin nx]$, $n \in I^+$, $x \in (0, \pi)$ are
(a) $n - 1$ (b) n
(c) $n + 1$ (d) $n + 2$

7. Domain of $f(x) = \sin^{-1}[2 - 4x^2]$ is ([.] denotes the greatest integer function)

(a) $\left[-\frac{\sqrt{3}}{2}, \frac{\sqrt{3}}{2}\right] \sim \{0\}$ (b) $\left[-\frac{\sqrt{3}}{2}, 0\right)$

(c) $\left[-\frac{\sqrt{3}}{2}, 0\right) \cup \left(0, \frac{\sqrt{3}}{2}\right]$ (d) $\left[-\frac{\sqrt{3}}{2}, 8\right]$

8. If $e^x + e^{f(x)} = e$, then for $f(x)$

(a) domain $= (-\infty, 1)$ (b) range $= (-\infty, 1)$

(c) domain $= (-\infty, 0]$ (d) range $= (-\infty, 1]$

9. Let $f(x) = \sec^{-1}[1 + \cos^2 x]$, where [.] denotes the greatest integer function, then

(a) the domain of f is R

(b) the domain of f is [1, 2]

(c) the range of f is [1, 2]

(d) the range of f is $\{\sec^{-1}1, \sec^{-1}2\}$

10. Let $f(x) = [x]^2 + [x+1] - 3$, where $[x] \le x$. Then

(a) $f(x)$ is a many-one and into function

(b) $f(x) = 0$ for infinite number of values of x

(c) $f(x) = 0$ for only two real values

(d) None of the above

11. Let $y = f(x-1) + f(x+1)$, where $f(x)$

$$= \begin{cases} 1 - |x|, & |x| \le 1 \\ 0, & |x| > 1 \end{cases}$$

The number of points of intersections with $y = \sin\alpha$ is

(a) 0 $(\pi < \alpha < 2\pi)$ (b) 4 $(0 < \alpha < \pi)$

(c) infinitely many $(\alpha = \pi)$ (d) 2$(0 < \alpha < 2\pi)$

12. If $y = f(x) = \frac{x+2}{x-1}$, then

(a) $x = f(y)$

(b) $f(1) = 3$

(c) y increases with x for $x < 1$

(d) f is rational function of x

13. If $f(x) = \cos([\pi^2]x) + \cos([-\pi^2]x)$, where $[x]$ stands for the greatest integer function, then

(a) $f\left(\frac{\pi}{2}\right) = -1$ (b) $f(\pi) = 1$

(c) $f(-\pi) = 0$ (d) $f\left(\frac{\pi}{4}\right) = 1$

14. $f(x) = \cos^2 x + \cos^2\left(\frac{\pi}{3} + x\right) - \cos x \cos\left(\frac{\pi}{3} + x\right)$ is

(a) an odd function (b) an even function

(c) a periodic function (d) $f(0) = f(1)$

15. The possible values of 'a' for which the function $f(x) = e^{x-[x]} + \cos ax$ (where [.] denotes the greatest integer function) is periodic with finite fundamental period is

(a) π (b) 2π

(c) 3π (d) 1

16. The figure illustrates the graph of the function $y = f(x)$ defined in [–3, 2]

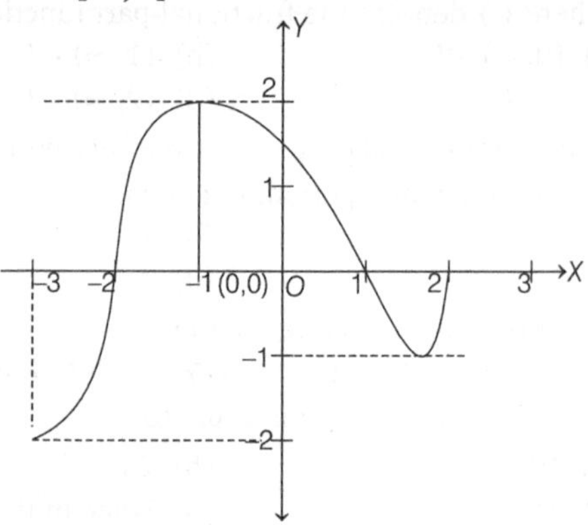

Identify the correct statements(s)?

(a) Range of $y = f(-|x|)$ is [–2, 2]

(b) Domain of $y = f(|x|)$ is [–2, 2]

(c) Domain of $y = f(|x| + 1)$ is [–1, 1]

(d) Range of $y = f(|x| + 1)$ is [–1, 0]

17. If $a \ne b \in R$ and $f : R \to R$ is a function such that

$$af(x) + bf\left(\frac{1}{x}\right) = x - 1 \ \forall\ 0 \ne x \in R,$$

Then,

(a) $f(2) = \frac{2a+b}{2(a^2 - b^2)}$ (b) $f(1) = 0$

(c) $f(-1) = -2/(a+b)$ (d) $f(-1) = 2(a-b)$

18. Let f be the greatest integer function and g be the modulus functions, then

(a) $(gof - fog)\left(-\frac{5}{3}\right) = 1$ (b) $(f + 2g)(-1) = 1$

(c) $(gof - fog)\left(\frac{5}{3}\right) = 0$ (d) $(f + 2g)(1) = 1$

19. Which of the following functions are periodic?

(a) $f(x) = \sin x + |\sin x|$

(b) $g(x) = \frac{(1 + \sin x)(1 + \sec x)}{(1 + \cos x)(1 + \operatorname{cosec} x)}$

(c) $h(x) = \max(\sin x, \cos x)$

(d) $p(x) = [x] + \left[x + \frac{1}{3}\right] + \left[x + \frac{2}{3}\right] - 3x + 10$, where [.] denotes the greatest integer function.

20. Which of the following functions are even?

(a) $f(x) = x\left(\frac{a^x + 1}{a^x - 1}\right)$

(b) $g(x) = \ln(x + \sqrt{(x^2 + a^2)})$

(c) $h(x) = \sqrt[3]{(1-x)^2} + \sqrt[3]{(1+x)^2}$

(d) $p(x) = \begin{cases} 0, & \text{if } x \text{ is rational} \\ 1, & \text{if } x \text{ is irrational} \end{cases}$

21. Which of the following functions are not identical ?

(a) $f(x) = \frac{x}{x^2}$ and $g(x) = \frac{1}{x}$

(b) $f(x) = \frac{x^2}{x}$ and $g(x) = x$

(c) $f(x) = \ln x^4$ and $g(x) = 4 \ln x$

(d) $f(x) = \ln\{(x-1)(x-2)\}$ and $g(x) = \ln(x-2) + \ln(x-3)$

22. Let $f(x) = \frac{5\sqrt{\sin x}}{1 + \sqrt[3]{\sin x}}$. If D is the domain of f, then D contains

(a) $(0, \pi)$

(b) $(-2\pi, -\pi)$

(c) $(2\pi, 3\pi)$

(d) $(4\pi, 6\pi)$

Linked-Comprehension Type

In these questions, a passage (paragraph) has been given followed by questions based on each of the passages. You have to answer the questions based on the passage given.

PASSAGE 1

Consider the functions

$$f(x) = \begin{cases} x+1, & x \le 1 \\ 2x+1, & 1 < x \le 2 \end{cases} \text{ and } g(x) = \begin{cases} x^2, & -1 \le x < 2 \\ x+2, & 2 \le x \le 3 \end{cases}$$

On the basis of above information, answer the following questions

1. The domain of the function $f(g(x))$ is

(a) $[0, \sqrt{2}]$ (b) $[-1, 2]$

(c) $[-1, \sqrt{2}]$ (d) None of these

2. The range of the function $f(g(x))$ is

(a) $[1, 5]$ (b) $[2, 3]$

(c) $[1, 2] \cup (3, 5]$ (d) None of these

3. The number of roots of the equation $f(g(x)) = 2$ is

(a) 1 (b) 2

(c) 4 (d) None of these

4. The set of all values of x, such that equation $g(x) + |g(x)| = 0$ is satisfied

(a) $[-3, -2]$ (b) $[2, 3]$

(c) $[-3, -2] \cup [2, 3]$ (d) ϕ

5. Which statement is correct for $\phi(x) = 0$

(a) one value of x is satisfied for $\phi(x) = 0$ and that x lie between 4 and 5

(b) one value of x is satisfied for $\phi(x) = 0$ and that x lie between 3 and 4

(c) two values of x is satisfied for $\phi(x) = 0$

(d) None of the above

PASSAGE 2

Let $f(x) = x^2 - 5x + 6$, $g(x) = f(|x|)$, $h(x) = |g(x)|$ and $\phi(x) = h(x) - (x)$ are four functions, where (x) is the least integral function of $x \ge x$.

On the basis of above information, answer the following questions

1. The number of solutions of the equation $g(x) = 0$ is

(a) 0 (b) 2 (c) 4 (d) 6

2. The value of λ for which the equation $g(x) - \lambda = 0$ has exactly three real and distinct roots

(a) 2 (b) 4

(c) 6 (d) None of these

3. The set of values of μ such that the equation $h(x) - \mu = 0$ has exactly eight real and distinct roots

(a) $\mu \in \left(0, \frac{1}{2}\right)$ (b) $\mu \in \left(0, \frac{1}{4}\right)$

(c) $\mu \in \left(0, \frac{1}{2}\right]$ (d) $\mu \in \left(0, \frac{1}{4}\right]$

PASSAGE 3

Let $f(x) = \min\{x - [x], -x - [-x]\}, -2 \le x \le 2$;
$g(x) = |2 - |x - 2||, -2 \le x \le 2$ and
$h(x) = \frac{|\sin x|}{\sin x}, -2 \le x \le 2$ and $x \ne 0$

(where $[x]$ denotes the greatest integer function $\le x$).

On the basis of above information, answer the following questions

1. The number of solutions of the equation $x^2 + [f(x)]^2 = 1$ is $\{-1 \le x \le 1\}$

(a) 0 (b) 2 (c) 4 (d) 6

2. The range of $f(x)$ is

(a) $\left[0, \frac{1}{2}\right]$ (b) $[0, 1]$

(c) $[0, 2]$ (d) None of these

3. The sum of all the roots of the equation $g(x) - h(x) = 0$ is $\{-2 \le x \le 2\}$

(a) positive (b) negative

(c) zero (d) None of these

4. The set of values of a such that the equation $f(x) - a = 0$ has exactly eight real and distincts roots

(a) $a \in \left(0, \frac{1}{2}\right)$ (b) $a \in \left[0, \frac{1}{2}\right)$

(c) $a \in [0,1)$ (d) $a \in (0, 1)$

5. The value of $\int_{-2}^{2} f(x)\,dx$ is
(a) 0 (b) 1
(c) 2 (d) 8

PASSAGE 4

Let f be a function satisfying
$f(x) = \dfrac{a^x}{a^x + \sqrt{a}} = g_a(x) \ (a > 0)$

On the basis of above information, answer the following questions

1. Let $f(x) = g_9(x)$, then the value of
$\left[\sum_{r=1}^{1995} f\left(\dfrac{r}{1996}\right)\right]$ is (where [.] denotes the greatest integer function)
(a) 995 (b) 996
(c) 997 (d) 998

2. Let $f(x) = g_4(x)$, then $\sum_{r=1}^{1996} f\left(\dfrac{r}{1997}\right)$ is
(a) zero (b) even
(c) odd (d) None of these

3. The value of $g_5(x) + g_5(1-x)$ is
(a) 1 (b) 5
(c) 10 (d) None of these

4. The value of $\sum_{r=1}^{2n-1} 2f\left(\dfrac{r}{2n}\right)$ is
(a) 0 (b) $2n-1$
(c) $2n$ (d) None of these

5. If the value of $\sum_{r=0}^{2n} f\left(\dfrac{r}{2n+1}\right) = \dfrac{1}{1+\sqrt{a}} + 987$, then the value of n is
(a) 493 (b) 494
(c) 987 (d) 988

PASSAGE 5

The accompanying figure shows the graph of a function $f(x)$ with domain $[-3, 4]$ and range $[-1, 2]$.

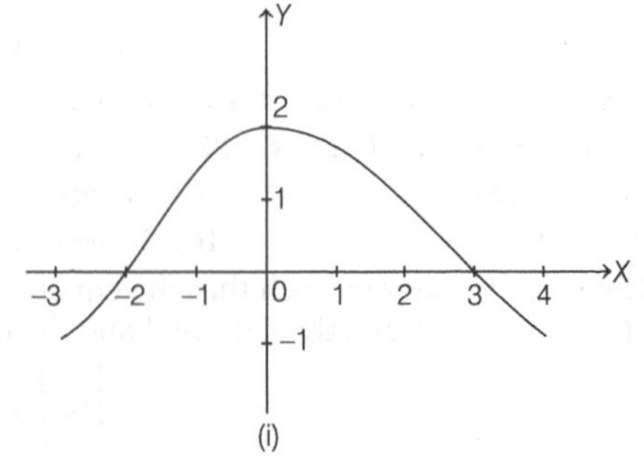

(i)

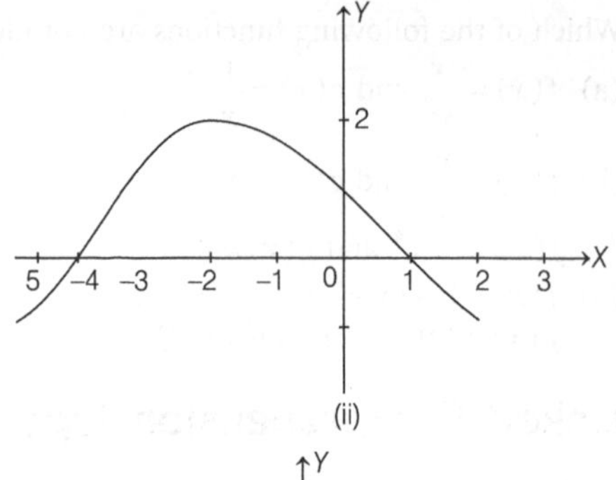

(ii)

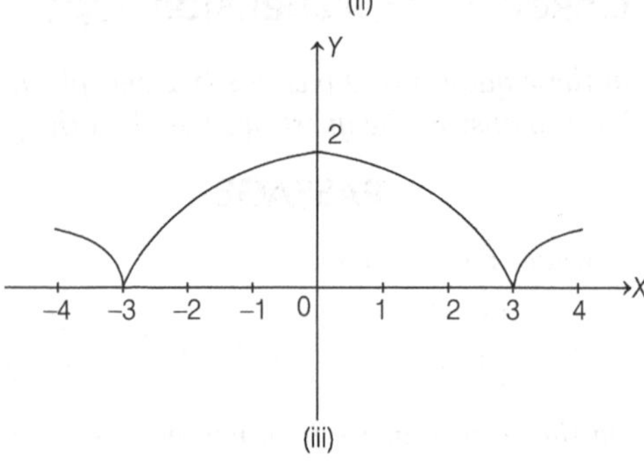

(iii)

On the basis of above information, answer the following questions

1. Figure (ii) represents the graph of the function
(a) $f(x)$
(b) $f(x-2)$
(c) $f(x+2)$
(d) $f(x-1)+1$

2. Figure (iii) represents the graph of the function
(a) $f(x)$
(b) $f(|x|)$
(c) $|f(x)|$
(d) $|f(|x|)|$

3. The domain and range respectively of
(a) $f(-x)$ are $[-4, 3]$ and $[-2, 1]$
(b) $f(x) - 1$ are $[-3, 4]$ and $[-1, 2]$
(c) $f(x) + 2$ are $[-3, 4]$ and $[-2, 4]$
(d) $-f(x+1) + 1$ are $[-4, 3]$ and $[-1, 2]$

4. $[-2, 5]$ and $[-2, 1]$ are the domain and range respectively of the function
(a) $-f(x)$
(b) $f(x-1)$
(c) $-f(x+1)+1$
(d) $-f(x+1)$

5. The number of solutions of figure (iii) and $(2x-6)^2 + 4y^2 = 49$ are
(a) 2
(b) 4
(c) 6
(d) None of the above

Numerical Value Type

Solve the following problems and write your answer.

1. A polynomial $f(x)$ satisfies the condition $f(x)\cdot f\left(\frac{1}{x}\right) = f(x)+f\left(\frac{1}{x}\right)$ and $f(10)=1001$, then the value of $f(20)$ must be

2. If maximum and minimum values of $f(x) = {}^{16-x}C_{2x-1} + {}^{20-3x}C_{4x-5}$ are λ and μ respectively, then the value of $\lambda + \mu$ must be

3. The least period of the function $\sin\left(\frac{\pi[x]}{12}\right) + \cos\left(\frac{\pi x}{4}\right) + \tan\left(\frac{\pi[x]}{3}\right)$ is λ, then the value of 20.1λ must be (where [.] denotes the greatest integer function)

4. If $f\left(2x^2 + \frac{y^2}{8}, 2x^2 - \frac{y^2}{8}\right) = xy$, then the value of $f(60, 48) + f(80, 48) + f(13, 5)$ must be

5. If $f(x+y) = f(x) + f(y) - xy - 1$ for all $x, y \in R$ and $f(1) = 1$, then the value of $-51.5 \sum_{r=1}^{5} f(r)$ must be

6. If $f(x) = a(x^n + 3)$ and $f(1) = 12$, $f(3) = 36$, then the value of $412f(2)$ must be

7. Set A has m elements and Set B has n elements. If the total number of subsets of A is 112 more than the total number of subsets of B, then the value of $m \cdot n$ is .....
[JEE Main 2020, 6 September]

8. Let f be a function from the set of positive integers to the set of real numbers *ie*, $f: N \to R$, such that

(i) $f(1) = 1$

(ii) $f(1) + 2f(2) + 3f(3) + \ldots + n\,f(n) = n(n+1)f(n)$ for $n \geq 2$, then the value of $\frac{1}{f(1004)}$ must be

9. If f be a function defined on the set of non-negative integers and taking values in the same set.
Given that,

(i) $x - f(x) = 19\left[\frac{x}{19}\right] - 90\left[\frac{f(x)}{90}\right]$ for all non-negative integers.

(ii) $1900 < f(1990) < 2000$,

then the sum of all possible values of $f(1990)$ must be

Notation : $[z]$ refers to the largest integer that is $\leq z$, *eg*, $[3.1415] = 3$.

10. The fundamental period of the function $f(x)$ satisfies the equation $f(x+1) + f(x-1) = \sqrt{3}f(x), \forall\, x \in R$ is $\frac{T}{24}$, then T' must be

11. Let $f(x) = \frac{e^x - e^{-x}}{2}$ and if $f(g(x)) = x$, then $g\left(\frac{e^{1002} - 1}{2e^{501}}\right)$ is equal to

12. Let $f(x, y)$ be a periodic function satisfying $f(x, y) = f(2x + 2y, 2y - 2x)$ for all x, y ; and $g(x) = f(2^x, 0)$, then the period of $g(x)$ must be

13. The range of the function $f(x) = \sqrt{(x-6)} + \sqrt{(12-x)}$ is an interval of length $\sqrt{\lambda} - \sqrt{\mu}$, then $\lambda^2 + \mu^2$ must be

14. The number of equivalence relations on a five element set is

Single Integer Type

Solve the following problems and write your answer.

1. Suppose that a function $f: R \to R$ satisfies $f(x+y) = f(x)f(y)$ for all $x, y \in R$ and $f(1) = 3$. If $\sum_{i=1}^{n} f(i) = 363$, then n is equal to
[JEE Main 2020, 6 September]

2. If a polynomial function $f(x)$ satisfies $f(f(f(x)) = 8x + 21$, where p and q are real numbers, then $p + q$ is equal to ....

3. The value of the limit

$$\lim_{x \to \frac{\pi}{2}} \frac{4\sqrt{2}(\sin 3x + \sin x)}{\left(2\sin 2x \sin\frac{3x}{2} + \cos\frac{5x}{2}\right) - \left(\sqrt{2} + \sqrt{2}\cos 2x + \cos\frac{3x}{2}\right)}$$

is .... **[JEE Advanced 2020]**

4. Let $f(x) = (x+1)^2 - 1, x \geq -1$, then the number of elements in the set $\{x : f(x) = f^{-1}(x)\}$ is equal to

5. If $f(x^2 - 6x + 6) + f(x^2 - 4x + 4) = 2x \;\forall\, x \in R$, then $f(-3) + f(9) - 5f(1) =$

6. Let $A = \{1, 2, 3, 4\}$, $B = \{2, 4, 6\}$. Then, the number of sets C such that $A \cap B \subseteq C \subseteq A \cup B$ is

7. Let $f: R \to R$ be a periodic function such that $f(T + x) = 1 + [1 - 3f(x) + 3(f(x))^2 - (f(x))^3]^{1/3}$ where T is a fixed positive number, if period of $f(x)$ is kT, then the value of k is ....

Matrix-Match Type

Given below are Matching Type Questions, with two columns (each having some items) each . Each item of Column I has to be matched with the items of Column II, by encircling the correct match(es).

NOTE An item of Column I can be matched with more than one items of Column II.

1. Observe the following columns

	Column I		Column II
(A)	$f(x)=\{x\}$, the fractional part of x	(P)	$f^{-1}(x)=\frac{1}{2}(4^x-4^{-x})$
(B)	$f(x)=\frac{16^x-1}{4^x}$	(Q)	f is an even function
(C)	$f(x)=\log_4(x+\sqrt{x^2+1})$	(R)	f is a periodic function
(D)	$f(x)=x\frac{3^x-1}{3^x+1}$	(S)	f is an odd function
		(T)	$f^{-1}(x)=\frac{1}{2}(3^x-3^{-x})$

2. Observe the following columns

	Column I		Column II
(A)	If $y=2\|x-2\|+3\|x+1\|$, then	(P)	$y=\begin{cases}-5x+1, & x<-1\\ x+7, & -1\le x<2\end{cases}$
(B)	If $y=2\|x-2\|-3\|x+1\|$, then	(Q)	$y=\begin{cases}5x-1, & -1\le x<2\\ x+7, & x\ge 2\end{cases}$
(C)	If $y=3\|x+1\|-2\|x-2\|$, then	(R)	$y=\begin{cases}x+7, & x<-1\\ -5x+1, & -1\le x<2\end{cases}$
		(S)	$y=\begin{cases}x+7, & -1\le x<2\\ 5x-1, & x\ge 2\end{cases}$
		(T)	$y=\begin{cases}-x-7, & x<-1\\ 5x-1, & -1\le x<2\end{cases}$

3. Observe the following columns

	Column I		Column II
(A)	The range of the function $f(x)=\frac{1}{2-\cos 3x}$ is	(P)	a subset of [0, 1]
		(Q)	a subset of [1, 2]
(B)	The range of the function $f(x)=\sqrt{(x-4)}+\sqrt{(6-x)}$ is	(R)	$\left[-\frac{3}{2},-1\right]$
(C)	If the function $f(x)$ is defined for $x\in[0,1]$, then the function $f(2x+3)$ is defined for	(S)	$\left[\frac{1}{3},1\right]$
		(T)	$[\sqrt{2},2]$

4. Observe the following columns

	Column I		Column II
(A)	The inverse function of $\sin(\tan^{-1}x)$ is	(P)	$\cot(\cos^{-1}x)$
(B)	The inverse function of $f(x)=1-2^{-x}$ is	(Q)	$\frac{\log_2 x}{\log_2 x-1}$
		(R)	$\tan(\sin^{-1}x)$
(C)	The inverse function of $f(x)=2^{x/(x-1)}$	(S)	$-\log_2(1-x)$
		(T)	$\frac{x}{\sqrt{(1-x^2)}}$

5. Observe the following columns

	Column I		Column II
(A)	Let $f(x)=\max\{1+\sin x, 1, 1-\cos x\}$, $x\in[0,2\pi]$ and $g(x)=\max\{1,\|x-1\|\}, x\in R$, then	(P)	$g(f(1))=1$
(B)	Let $f(x)=\ln\left(\frac{1+x}{1-x}\right)\ \forall\ x\in(-1,1)$	(Q)	$f(g(0))=0$
	and $g(x)=\left(\frac{3x+x^3}{1+3x^2}\right)$, then	(R)	$f(g(0))=1$
(C)	Let $f(x)=1+x^2$ and $g(x)=x-x^2$, then	(S)	$g(f(0))=1$
		(T)	$g\left(f\left(\frac{e-1}{e+1}\right)\right)=1$

Answers

Objective Questions Type I [Only one correct answer]

1. (d)	**2.** (c)	**3.** (b)	**4.** (c)	**5.** (b)	**6.** (d)	**7.** (c)	**8.** (b)	**9.** (a)	**10.** (c)
11. (a)	**12.** (b)	**13.** (c)	**14.** (b)	**15.** (c)	**16.** (d)	**17.** (c)	**18.** (a)	**19.** (b)	**20.** (a)
21. (c)	**22.** (b)	**23.** (b)	**24.** (b)	**25.** (d)	**26.** (b)	**27.** (c)	**28.** (d)	**29.** (b)	**30.** (a)
31. (a)	**32.** (d)	**33.** (b)	**34.** (d)	**35.** (b)	**36.** (c)	**37.** (a)	**38.** (d)	**39.** (d)	**40.** (c)
41. (b)	**42.** (a)	**43.** (d)	**44.** (b)	**45.** (c)	**46.** (b)	**47.** (a)	**48.** (b)	**49.** (b)	**50.** (d)
51. (d)	**52.** (d)	**53.** (c)	**54.** (d)	**55.** (d)	**56.** (c)	**57.** (d)	**58.** (c)	**59.** (d)	**60.** (c)
61. (c)	**62.** (b)	**63.** (d)	**64.** (a)	**65.** (b)	**66.** (a)	**67.** (b)	**68.** (d)	**69.** (a)	**70.** (a)
71. (d)	**72.** (a)	**73.** (b)	**74.** (c)	**75.** (b)	**76.** (b)	**77.** (c)	**78.** (c)	**79.** (b)	**80.** (a)
81. (d)	**82.** (b)	**83.** (d)	**84.** (a)	**85.** (c)	**86.** (c)	**87.** (d)	**88.** (c)	**89.** (d)	**90.** (a)
91. (c)	**92.** (d)	**93.** (a)	**94.** (d)	**95.** (b)	**96.** (d)	**97.** (a)	**98.** (b)	**99.** (b)	**100.** (b)
101. (b)									

Objective Questions Type II [One or more than one correct answer(s)]

1. (a, b, c)	**2.** (a, b, c, d)	**3.** (a, d)	**4.** (a, b, c, d)	**5.** (b, c, d)
6. (b, c)	**7.** (a, c)	**8.** (a, b)	**9.** (a, d)	**10.** (a, b)
11. (a, b, c)	**12.** (a, d)	**13.** (a, c)	**14.** (b, c, d)	**15.** (a, b, c)
16. (a, b, c, d)	**17.** (a, b, c)	**18.** (a, b, c)	**19.** (a, b, c, d)	**20.** (a, c, d)
21. (b, c, d)	**22.** (a, b, c)			

Linked-Comprehension Type

Passage 1 **1.** (c) **2.** (c) **3.** (b)
Passage 2 **1.** (c) **2.** (c) **3.** (b) **4.** (c) **5.** (b)
Passage 3 **1.** (b) **2.** (a) **3.** (a) **4.** (a) **5.** (b)
Passage 4 **1.** (c) **2.** (b) **3.** (a) **4.** (b) **5.** (c)
Passage 5 **1.** (c) **2.** (d) **3.** (d) **4.** (d) **5.** (d)

Numerical Value Type

1. 8001	**2.** 2345	**3.** 482.4	**4.** 112	**5.** 772.5	**6.** 8652
7. 28	**8.** 2008	**9.** 3898	**10.** 288	**11.** 501	**12.** 12
13. 180	**14.** 52				

Single Integer Type

1. (5) **2.** (5) **3.** (8) **4.** (2) **5.** (9) **6.** (8) **7.** (2)

Matrix-Match Type

1. A → (R); B → (S); C → (P); D → (Q)
2. A → (P, S); B → (R); C → (Q, T)
3. A → (P, S); B → (Q, T); C → (P, R)
4. A → (P, R, T); B → (S); C → (Q)
5. A → (P, S); B → (Q, T); C → (R)

Solutions

Objective Questions Type [Only one correct answer]

1. Given, $a\,f(x) + b\,f\left(\frac{1}{x}\right) = x - 1$ ...(i)

Replacing x by $\frac{1}{x}$ in Eq. (i), then

$$a\,f\left(\frac{1}{x}\right) + b\,f(x) = \frac{1}{x} - 1 \quad \text{...(ii)}$$

Now, multiplying in Eq. (i) by a and in Eq. (ii) by b, then each other subtracting, we get

$$(a^2 - b^2)\,f(x) = ax - a - \frac{b}{x} + b$$

$$\Rightarrow \quad (a^2 - b^2)\,f(2) = 2a - a - \frac{b}{2} + b = \left(a + \frac{b}{2}\right)$$

$$\therefore \quad f(2) = \frac{(2a + b)}{2(a^2 - b^2)}$$

2. $\because f(2) = 2 \cdot 2^n + a$ and $f(4) = 2 \cdot 4^n + a$

Given, $f(2) = 26$ and $f(4) = 138$

$\therefore \quad 2^{n+1} + a = 26$ and $2 \cdot 4^n + a = 138$

$\Rightarrow \quad (2 \cdot 4^n + a) - (2^{n+1} + a) = 138 - 26 = 112$

$\Rightarrow \quad 2(4^n - 2^n) = 112$

$\Rightarrow \quad 4^n - 2^n = 56$

Let $\quad 2^n = \lambda$

$\therefore \quad \lambda^2 - \lambda - 56 = 0$

$\Rightarrow \quad (\lambda - 8)(\lambda + 7) = 0$

$\lambda = 8, \lambda \neq -7$

$\Rightarrow \quad 2^n = 2^3$

$\therefore \quad n = 3$

Now, $\quad 2^{n+1} + a = 26$

$\Rightarrow \quad a = 26 - 2^4 = 10$

$\therefore \quad f(3) = 2 \cdot 3^3 + 10 = 54 + 10 = 64$

3. $$\because f(x) = \left\{2\sin^2\left(\frac{4x - 3\pi}{6\pi^2}\right)\right\}^2 + 2\cos\left(\frac{4x - 3\pi}{3\pi^2}\right)$$

$$= \left\{1 - \cos\left(\frac{4x - 3\pi}{3\pi^2}\right)\right\}^2 + 2\cos\left(\frac{4x - 3\pi}{3\pi^2}\right)$$

$$= 1 + \cos^2\left(\frac{4x - 3\pi}{3\pi^2}\right)$$

$$= \frac{1}{2}\left\{2 + 1 + \cos\left(\frac{8x - 6\pi}{3\pi^2}\right)\right\} = \frac{3}{2} + \frac{1}{2}\cos\left(\frac{8x - 6\pi}{3\pi^2}\right)$$

$$\therefore \text{Period of } f(x) = \frac{2\pi}{\left(\frac{8}{3\pi^2}\right)} = \frac{6\pi^3}{8} = \frac{3\pi^3}{4}$$

4. $[x] + [-x] \geq 0 \Rightarrow x \in I$

Also, $2 - |x| \geq 0 \Rightarrow x \in [-2, 2]$

$x \neq 0$

$\therefore \quad$ Domain of $f(x) = \{-2, -1, 1, 2\}$

$$f(x) = \sqrt{2 - |x|} + \frac{1}{x^2}$$

$\because \quad [x] + [-x] = 0 \ \forall \ x \in I$

$\therefore \quad$ Range $= \left\{\frac{1}{4}, 2\right\}$

5. $(81)^{\sin^2 x} + (81)^{\cos^2 x} = 30$

$$(81)^{\sin^2 x} + \frac{(81)^1}{(81)^{\sin^2 x}} = 30$$

$$(81)^{\sin^2 x} = t$$

$$\Rightarrow \quad t + \frac{81}{t} = 30$$

$$\Rightarrow \quad (t - 3)(t - 27) = 0 \Rightarrow t = 3, 27$$

$\therefore \quad (81)^{\sin^2 x} = 3^1$ or $(81)^{4\sin^2 x} = 3^3$

$3^{4\sin^2 x} = 3^1$ or $3^{4\sin^2 x} = 3^3$

$\sin^2 x = \frac{1}{4}$ or $\sin^2 x = \frac{3}{4}$

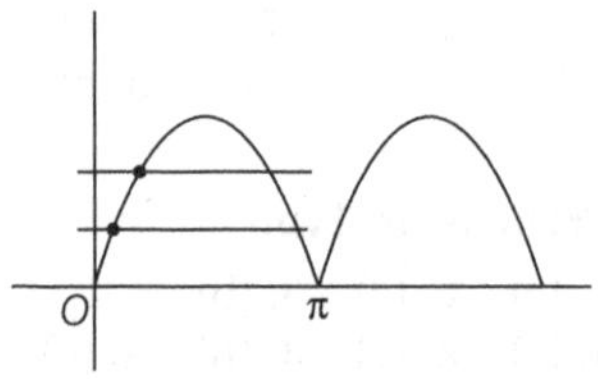

$\therefore$ Total solution = 4

6. $(fog)x = f(g(x)) = f(\sin x) = \sqrt{(3|\sin x| - \sin x - 2)}$

$(fog)x$ is defined if and only if

$$3|\sin x| - \sin x - 2 \geq 0$$

***Case* I** $0 \leq \sin x \leq 1$

Then, $\quad 3\sin x - \sin x - 2 \geq 0$

$\therefore \quad \sin x \geq 2$ (Impossible)

only $\quad \sin x = 1 \Rightarrow x = 2m\pi + \frac{\pi}{2}, m \in I$

***Case* II** $-1 \leq \sin x < 0$

Then, $-3\sin x - \sin x - 2 \geq 0$

or $\quad \sin x \leq -\frac{1}{2}$

$\Rightarrow \quad -\sin x \geq \frac{1}{2}$

or $\quad \cos(x + \pi/2) \geq \cos\frac{\pi}{3}$

$$\Rightarrow \quad 2n\pi - \frac{\pi}{3} \leq x + \frac{\pi}{2} \leq 2n\pi + \frac{\pi}{3}$$

$$2n\pi - \frac{\pi}{3} - \frac{\pi}{2} \leq x \leq 2n\pi + \frac{\pi}{3} - \frac{\pi}{2}$$

$$\Rightarrow \quad 2n\pi - \frac{5\pi}{6} \leq x \leq 2n\pi - \frac{\pi}{6}$$

or $$2n\pi + 2\pi - \frac{5\pi}{6} \leq x \leq 2n\pi + 2\pi - \frac{\pi}{6}$$

or $$2n\pi + \frac{7\pi}{6} \leq x \leq 2n\pi + \frac{11\pi}{6}$$

Hence, $x \in \{(4m + 1)\frac{\pi}{2}; m \in I\} \bigcup_{n \in I}\left[2n\pi + \frac{7\pi}{6}, 2n\pi + \frac{11\pi}{6}\right]$

7. $\because\quad f(x)=\dfrac{\sin 2n\,x}{1+\left(\dfrac{1+\cos 2nx}{2}\right)}=\dfrac{2\sin 2n\,x}{3+\cos 2n\,x}$

$\therefore\quad T_1=$ period of $\sin 2n\,x=\dfrac{2\pi}{2n}=\dfrac{\pi}{n}$

and $\quad T_2=$ period of $\cos 2nx=\dfrac{2\pi}{2n}=\dfrac{\pi}{n}$

$\therefore$ Period of $f(x)=\text{LCM}\{T_1,T_2\}$

$=\text{LCM}\left\{\dfrac{\pi}{n},\dfrac{\pi}{n}\right\}$

$=\dfrac{\pi}{n}=\dfrac{\pi}{6}$ (given)

$\therefore\quad n=6$

8. $y=5^{\log x}$

$\Rightarrow\quad y=x^{\log 5}$

$\Rightarrow\quad y^{\frac{1}{\log 5}}=x$

Replacing $x\to y$ and $y\to x$

9. Let $n(A)=$ Number of people read newspaper

$A=63\%$

$n(B)=$ Number of people read newspaper

$B=76\%$

and $n(A\cap B)=$ Number of people read both $=x\%$

$\because\quad n(A\cup B)=n(A)+n(B)-n(A\cap B)$

and $\quad 76\le n(A\cup B)\le 100$

$\therefore\quad 76\le 63+76-x\le 100$

$\Rightarrow\quad 0\le 63-x\le 24$

$\Rightarrow\quad 39\le x\le 63$

Hence, option (a) is correct.

10. $T_1=$ Period of $\sin(\sin(nx))=\dfrac{2\pi}{n}$

$T_2=$ Period of $\tan\left(\dfrac{x}{n}\right)=\dfrac{\pi}{1/n}=\pi n$

$\therefore$ Period of $f(x)=$ LCM of $\{T_1,T_2\}$

$=$ LCM of $\left(\dfrac{2\pi}{n},\pi n\right)$

$=$ LCM of $\left\{\dfrac{2n\pi}{n^2},\dfrac{2n\pi}{2}\right\}$

$=\dfrac{2n\pi}{1}=6\pi$ (given)

$\therefore\quad n=3$

11. $\because\ x^2f(x)-2f\left(\dfrac{1}{x}\right)=g(x)\qquad\ldots$(i)

$\Rightarrow$ Replacing x by $\dfrac{1}{x}$, then

$\dfrac{1}{x^2}f\left(\dfrac{1}{x}\right)-2f(x)=g\left(\dfrac{1}{x}\right)$

or $\quad 2f\left(\dfrac{1}{x}\right)-4x^2f(x)=2x^2g\left(\dfrac{1}{x}\right)\qquad\ldots$(ii)

Adding Eqs. (i) and (ii), then

$-3x^2f(x)=g(x)+2x^2g\left(\dfrac{1}{x}\right)$

or $\quad f(x)=-\left(\dfrac{g(x)+2x^2g\left(\dfrac{1}{x}\right)}{3x^2}\right)$

$\Rightarrow\quad f(-x)=-\left(\dfrac{g(-x)+2x^2g\left(-\dfrac{1}{x}\right)}{3x^2}\right)$

$f(x)=-f(x)$ ($\because g(x)$ is an odd and $f(x)$ is an even functions)

$\therefore\quad f(x)=0$

Then, $\quad f(5)=0$

12. We see that, each number in the given set follow the relation $x=n^2-3n+2$ for $1\le n\le 6$

as for

$n=1,\ x=1-3+2=0$

$n=2,\ x=4-6+2=0$

$n=3,\ x=9-9+2=2$

$n=4,\ x=16-12+2=6$

$n=5,\ x=25-15+2=12$

$n=6,\ x=36-18+2=20$

13. $\log_{10}(1-x)$ is defined for $1-x>0$ and $1/\log_{10}(1-x)$ is defined for $x\in(-\infty,1)\sim\{x:\log_{10}(1-x)=10\}$

i.e., $(-\infty,1)\sim\{0\}$. Also, $\sqrt{x+2}$ is meaningful, if $x\ge-2$

Thus, the domain of the given function is $((-\infty,1)\sim\{0\})\cap\{x:x\ge-2\}=[-2,1)\sim\{0\}$.

14. If $O=$ odd, $E=$ even

Then, $\quad f(x)=\sin x^7e^{x^5\text{sgn}\,x^9}$

$\because\quad \sin x^7\to O,\ \text{sgn}\,x^9\to O$

$x^5\to O$

$\therefore\quad e^{x^5\text{sgn}\,x^9}=E$

$\therefore\quad f(x)=O\times E=O$

15. $[\sin 2x]-[\cos 2x]\ge 0$

$\Rightarrow\quad [\sin 2x]\ge[\cos 2x]$

$\because\quad -1\le\sin 2x\le 1$

$\therefore\quad [\sin 2x]=-1,0,1$

Then, $[\cos 2x]\le-1$, $[\cos 2x]\le 0$, $[\cos 2x]\le 1$

$\downarrow\qquad\downarrow\qquad\downarrow$

$[\cos 2x]=-1$, $[\cos 2x]=-1$, $[\cos 2x]=-1,0,1$

but $[\sin 2x]=1$, $[\cos 2x]=1$ and $[\sin 2x]=1$, $[\cos 2x]=-1$ are not possible

Then, $\quad [\sin 2x]=0,[\cos 2x]=-1$

and $\quad [\sin 2x]=1,[\cos 2x]=0$

$[\sin 2x]=0,[\cos 2x]=0$

Hence, $\quad R_f=\{0,1\}$

16. Clearly, $f(x)$ will be an even number if $\left[\dfrac{x^2}{a}\right]=0$

$\forall\ x\in[-20,20]$

$\therefore\quad 0\le\dfrac{x^2}{a}<1$ or $0\le x^2\le 400$

Then, $\quad a>400$

$\therefore\quad a\in(400,\infty)$

17. Total number of functions from $A \to B = 3^4 = 81$ and number of onto mappings = coefficient of x^4 in $4!\,(e^x - 1)^3$.

$= \text{coefficient of } x^4 \text{ in } 4!\,(e^{3x} - 3e^{2x} + 3e^x - 1)$

$= 4!\left(\frac{3^4}{4!} - \frac{3\cdot 2^4}{4!} + \frac{3\cdot 1^4}{4!} - 0\right)$

$= 81 - 48 + 3 = 81 - 45 = 36$

$\therefore$ Number of functions from $A \to B$, which are not onto is $81 - 36 = 45$

18. $\because \sin^{-1}\sin(\tan x) = \cos^{-1}\cos(\tan x)\ \forall \tan x \in \left[0, \frac{\pi}{2}\right]$

$= \cos^{-1}\sin\sqrt{(1 - \tan^2 x)}$

$\therefore \quad 0 \le \tan x \le \frac{\pi}{2}$

$\Rightarrow \quad \tan^{-1} 0 \le x \le \tan^{-1}\left(\frac{\pi}{2}\right)$

$\Rightarrow \quad 0 \le x \le \tan^{-1}\left(\frac{\pi}{2}\right)$

Hence, $\quad x \in \left[0, \tan^{-1}\left(\frac{\pi}{2}\right)\right]$

19. Given, $f(x) + 2f(1 - x) = x^2 + 2$ ...(i)

Replacing x by $1 - x$ in Eq. (i), then

$f(1 - x) + 2f(x) = (1 - x)^2 + 2$ ...(ii)

Now, multiplying in Eq. (i) by 1 and in Eq. (ii) by 2, then each other subtracting, we get

$-3f(x) = x^2 + 2 - 2(1 - x)^2 - 4$

or $\quad 3f(x) = -(-x^2 + 4x - 4) = x^2 - 4x + 4 = (x - 2)^2$

$\therefore \quad f(x) = \frac{(x - 2)^2}{3}$

20. Given, $2f(x - 1) - f\left(\frac{1 - x}{x}\right) = x$ ...(i)

Replacing x by $\frac{1}{x}$, then $2f\left(\frac{1}{x} - 1\right) - f\left(\frac{1 - \frac{1}{x}}{\frac{1}{x}}\right) = \frac{1}{x}$

$\Rightarrow \quad 2f\left(\frac{1 - x}{x}\right) - f(x - 1) = \frac{1}{x}$ ...(ii)

Now, multiplying Eq. (i) by 2 and in Eq. (ii) by 1, then adding, we get

$3f(x - 1) = 2x + \frac{1}{x}$

Now replacing x by $x + 1$, then

$f(x) = \frac{1}{3}\left(2(x + 1) + \frac{1}{(x + 1)}\right)$

21. Given, $f(2x + 3) + f(2x + 7) = 2$ ...(i)

Replacing x by $x + 1$ and x by $x + 2$ in Eq. (i), then

$f(2x + 5) + f(2x + 9) = 2$ ...(ii)

and $\quad f(2x + 7) + f(2x + 11) = 2$ ...(iii)

Now, subtracting Eq. (i) from Eq. (iii), then we get

$f(2x + 11) = f(2x + 3)$

$\therefore$ Period $= 8$

22. $\because \left[x^2 - \frac{1}{2}\right] = \left[x^2 + \frac{1}{2} - 1\right] = \left[x^2 + \frac{1}{2}\right] - 1$

$\therefore \ f(x) = \sin^{-1}\left[x^2 + \frac{1}{2}\right] + \cos^{-1}\left(\left[x^2 + \frac{1}{2}\right] - 1\right)$

$\because \quad \left(x^2 + \frac{1}{2}\right) \ge \frac{1}{2}$

$\therefore \quad \left[x^2 + \frac{1}{2}\right] = 0 \text{ or } 1$

If $\quad \left[x^2 + \frac{1}{2}\right] = 0$

Then $f(x) = \sin^{-1}(0) + \cos^{-1}(0 - 1) = 0 + \pi = \pi$

and for

$\left[x^2 + \frac{1}{2}\right] = 1$

$f(x) = \sin^{-1} 1 + \cos^{-1}(1 - 1) = \sin^{-1} 1 + \cos^{-1} 0$

$= \frac{\pi}{2} + \frac{\pi}{2} = \pi$

Therefore, range of $f(x)$ is $\{\pi\}$

23. Given relation, $R_1 = \{(a, b) \in R^2 : a^2 + b^2 \in Q\}$

Let $(a, b) \in R_1 \Rightarrow a^2 + b^2 \in Q$

and $(b, c) \in R_1 \Rightarrow b^2 + c^2 \in Q$

$\therefore a^2 + 2b^2 + c^2 \in Q$, but we can not say that $a^2 + c^2 \in Q$, so $(a, c) \notin R_1$

$\therefore R_1$ is not transitive.

And another given relation

$R_2 = \{(a, b) \in R^2 : a^2 + b^2 \notin Q\}$

Let $(a, b) \in R_2 \Rightarrow a^2 + b^2 \notin Q$

and $(b, c) \in R_2 \Rightarrow b^2 + c^2 \notin Q$

$\therefore a^2 + 2b^2 + c^2 \notin Q$, but we can not say that $a^2 + c^2 \notin Q$, so $(a, c) \notin R_2$

$\therefore R_2$ is not transitive.

Hence, option (b) is correct.

24. $\because 2 < x^2 < 3 \Rightarrow 0 < x^2 - 2 < 1$

$\therefore \quad \{x^2\} = (x^2 - 2)$

Now, $\because \quad x > 0$

$\therefore \quad \sqrt{2} < x < \sqrt{3}$

or $\quad \frac{1}{\sqrt{3}} < \frac{1}{x} < \frac{1}{\sqrt{2}}$

$\therefore \quad \left\{\frac{1}{x}\right\} = \frac{1}{x}$

then $\quad x^2 - 2 = \frac{1}{x}$

or $\quad x^3 - 2x - 1 = 0$

or $\quad (x + 1)(x^2 - x - 1) = 0$

$\therefore \quad x \ne -1,$

$\therefore \quad x^2 - x - 1 = 0$

$\Rightarrow \quad x = \frac{1 \pm \sqrt{5}}{2}$

$\therefore \quad x = \frac{1 + \sqrt{5}}{2} \quad [\because x > 0]$

25. Period of $f(x)$ is 7

$\therefore$ Period of $f\left(\frac{x}{3}\right)$ is $\frac{7}{1/3} = 21$

and $\because$ Period of $g(x)$ is 11

$\therefore$ Period of $g\left(\frac{x}{5}\right)$ is $\frac{11}{1/5} = 55$

$\therefore$ $T_1 =$ Period of $f(x)g\left(\frac{x}{5}\right) = 7 \times 55 = 385$

and $T_2 =$ Period of $g(x)f\left(\frac{x}{3}\right) = 11 \times 21 = 231$

$\therefore$ Period of $F(x) = \text{LCM } \{T_1, T_2\}$

$= \text{LCM } \{385, 231\}$

$= 7 \times 11 \times 3 \times 5$

$= 1155$

26. If no, object goes to its scheduled place, then derangement

Number of such functions

$$= 6!\left(1 - \frac{1}{1!} + \frac{1}{2!} - \frac{1}{3!} + \frac{1}{4!} - \frac{1}{5!} + \frac{1}{6!}\right)$$

$$= 6!\left(\frac{1}{2} - \frac{1}{6} + \frac{1}{24} - \frac{1}{120} - \frac{1}{720}\right)$$

$$= 360 - 120 + 30 - 6 + 1$$

$$= 265$$

27. We have,

$$f(x) = \begin{cases} \frac{60-5x}{3}, & 0 \le x \le 6 \\ 10, & 6 \le x \le 7 \\ 31 - 3x, & 7 \le x \le 10 \end{cases}$$

Now, $f(x) = 10 \ \forall \ x \in [6, 7]$

So, $f(x)$ is not one-one.

Again,

$\Rightarrow \quad 0 \le x \le 6$

$\Rightarrow \quad -30 \le -5x \le 0$

$\Rightarrow \quad 60 - 30 \le 60 - 5x \le 60$

$\Rightarrow \quad \frac{30}{3} \le \frac{60-5x}{3} \le \frac{60}{3}$

$\Rightarrow \quad 10 \le \frac{60-5x}{3} \le 20$ and $7 \le x \le 10$

$\Rightarrow \quad -30 \le -3x \le -21 \Rightarrow 1 \le 31 - 3x \le 10$

$\therefore$ Range of $f(x) = [1, 20]$

It is given that co-domain of $f(x) = [1, 20]$

$\therefore$ Range of $f(x) =$ co-domain of $f(x)$

So, $f(x)$ is onto.

28. (say) $y = f(x) = \begin{cases} \frac{-x^2}{1+x^2}, & \text{if } x \ge 0 \\ \frac{x^2}{1+x^2}, & \text{if } x < 0 \end{cases}$

$$f^{-1}(y) = \begin{cases} \sqrt{\frac{-y}{1+y}}, & y < 0 \\ -\sqrt{\frac{y}{1-y}}, & y > 0 \end{cases}$$

or $\quad f^{-1}(x) = \begin{cases} \sqrt{\frac{-x}{1+x}}, & x < 0 \\ -\sqrt{\frac{x}{1-x}}, & x > 0 \end{cases}$

$\therefore \quad f^{-1}(x) = -\begin{cases} \sqrt{\left(\frac{-x}{1+x}\right)}, & x < 0 \\ \sqrt{\left(\frac{-x}{1-x}\right)}, & x > 0 \end{cases}$

$$= -(\text{sgn } x)\sqrt{\frac{|x|}{1-|x|}}$$

29. $\because \quad f(x) = \frac{e^{x^2} - e^{-x^2}}{e^{x^2} + e^{-x^2}}$

$\because \quad e^{x^2} + e^{-x^2} > 0$

and $\quad e^{x^2} - e^{-x^2} = 2\left\{x^2 + \frac{x^6}{3!} + \ldots\right\} > 0$

$\therefore \quad e^{x^2} - e^{-x^2} > 0$

Hence, $\quad f(x) > 0$

$\therefore$ $f(x)$ is into

and $\quad f'(x) = \frac{8x}{(e^{x^2} + e^{-x^2})^2}$

$f'(x) = 0$

$\Rightarrow \quad x = 0$

$\because$ $f'(x)$ has real values of x so function $f(x)$ is many one.

Hence, $f(x)$ is neither one-one nor onto.

30. We have, $R = \{(x, y); x^2 - 4xy + 3y^2 = 0, x, y \in N$

Let $\quad x \in N, x^2 - 4x \cdot x + 3x^2 = 0$

$\therefore$ $(x, x) \in R$

$\therefore R$ is reflexive

We have, $(3)^2 - 4(3)(1) + 3(1)^2 = 9 - 12 + 3 = 0$

$(3, 1) \in R$

Also, $(1)^2 - 4(1)(3) + 3(3)^2 = 1 - 12 + 27 = 16 \neq 0$

$\therefore (1, 3) \notin R$

$\therefore R$ is not symmetric $(9, 3) \in R$ because

$9^2 - 4(9)(3) + 3(3)^2 = 0$

Also, $(3, 1) \in R$ because

$(3)^2 - 4(3)(1) + 3(1)^2 = 0$

Now, $(9, 1) \in R$ if $(9)^2 - 4(9)(1) + 3(1)^2 = 0$

i.e., $\quad 48 \neq 0$

which is not so

$(9, 3), (3, 1) \in R$ and $(9, 1) \notin R$

$\therefore R$ is not transitive.

31. Since, $f(x)$ is decreasing odd function.

$\therefore \quad f(-x) = -f(x)$ and $f'(x) < 0$

Now, let $y = f^{-1}(x)$

$\therefore \quad f(y) = x$

then $\quad f'(y)\frac{dy}{dx} = 1$

$\therefore \quad \dfrac{dy}{dx} = \dfrac{1}{f'(y)} < 0 \quad [\because f'(x) < 0]$

$\therefore \quad \dfrac{dy}{dx} < 0$

$\therefore f^{-1}(x)$ is decreasing function

Let $\quad z = f^{-1}(-x) \quad \ldots(i)$

$\therefore \quad -x = f(z)$

$x = -f(z)$

$= f(-z) \quad [\because f$ is odd function$]$

$\therefore \quad -z = f^{-1}(x)$

$\Rightarrow \quad z = -f^{-1}(x)$

From Eq. (i), $f^{-1}(-x) = -f^{-1}(x)$

Hence, $f^{-1}(x)$ is odd function.

Hence, $f^{-1}(x)$ is odd and decreasing function.

32. $\because f(x)$ is continuous function and $|\sin x|$ and $|\cos x|$ are always positive.

Hence, $f(x)$ is minimum when $|\sin x| = 0$ and $|\cos x| = 1$

$\therefore$ minimum value $= 0 - 3 = -3$

And $f(x)$ is maximum when $|\sin x| = 1$ and $|\cos x| = 0$

$\therefore$ Maximum value $= 2 - 0 = 2$

$\therefore$ Required range $= [-3, 2]$

33. $f(x) = \dfrac{\text{cosec}^{-1} x}{\sqrt{\{x\}}}$

Domain $\in (-\infty, -1] \cup [1, \infty)$

$\{x\} \neq 0$

So, $x \neq$ integers.

34. $g(x) = \begin{cases} [x^2] - 0, & 0 \le x < 1 \\ [x^2] - 1, & 1 \le x < 2 \\ 0, & x = 2 \end{cases}$

$= \begin{cases} [x^2], & 0 \le x^2 < 1 \\ [x^2] - 1, & 1 \le x < 4 \\ 0, & x = 2 \end{cases}$

$= \begin{cases} [x^2], & 0 \le x^2 < 1 \\ [x^2] - 1, & 1 \le x^2 < 2 \\ [x^2] - 1, & 2 \le x^2 < 3 \\ [x^2] - 1, & 3 \le x^2 < 4 \\ 0, & x = 2 \end{cases}$

$= \begin{cases} 0, & 0 \le x^2 < 1 \\ 0, & 1 \le x^2 < 2 \\ 1, & 2 \le x^2 < 3 \\ 2, & 3 \le x^2 < 4 \\ 0, & x = 2 \end{cases}$

$\therefore \quad g(x) = \{0, 1, 2\}$

35. Option (a) Period of $\sin 3x = 2\pi / 3$

Option (b) Let $f(x) = |\cos x| = \sqrt{\left(\dfrac{1 + \cos 2x}{2}\right)}$

$\therefore$ Period of $|\cos x| = \dfrac{2\pi}{2} = \pi$

Option (c) $[x + 2\pi] \neq [x + \pi]$

Hence, period of $[x + \pi]$ is not equal to π.

Option (d) $f(x) = x \cos x$ is non-periodic.

36. $\dfrac{1 - |x|}{2 - |x|} \ge 0 \Rightarrow \dfrac{|x| - 1}{|x| - 2} \ge 0$

|x| + 1 − 2 +

$\therefore \quad |x| \le 1$ and $|x| > 2$

$\Rightarrow \quad x \in [-1, 1] \cup (-\infty, -2) \cup (2, \infty)$

37. Given A_i's are thirty sets with five elements each, so

$$\sum_{i=1}^{30} n(A_i) = 5 \times 30 = 150 \quad \ldots(i)$$

If there are m distinct elements is S and each element of S belongs to exactly 10 of the A_i's, we have

$$\sum_{i=1}^{30} n(A_i) = 10m \quad \ldots(ii)$$

$\therefore$ From Eqs. (i) and (ii), we get $10m = 150$

$\therefore \quad m = 15 \quad \ldots(iii)$

Similarly, $\sum_{j=1}^{30} n(B_j) = 3n$

and $\sum_{j=1}^{30} n(B_j) = 9m$

$\therefore \quad 3n = 9m$

$\Rightarrow \quad n = \dfrac{9m}{3} = 3m = 3 \times 15 = 45 \quad \ldots(iv)$

Hence, $n = 45$.

38. $x > 0, \log_{10} x > 0 \Rightarrow x > 1$

$\log_{10}\log_{10} x > 0$

$\Rightarrow \quad \log_{10} x > 1$

$\Rightarrow \quad x > 10$

and $\quad \log_{10} \log_{10} \log_{10} x > 0$

$\Rightarrow \quad \log_{10}\log_{10} x > 1$

$\Rightarrow \quad \log_{10} x > 10$

$\Rightarrow \quad x > 10^{10}$

Similarly, $x > 10^{10^{10^{\cdot^{\cdot(n-2)\text{times}}}}}$

39. We have,

$x \in X \Leftrightarrow x \in A \cap B$ and $x \notin C$

or $\quad x \in B \cap C$ and $x \notin A$

or $\quad x \in C \cap A$ and $x \notin B$

Therefore,

$X = [(A \cap B) - C] \cup (B \cap C) - A] \cup [(C \cap A) - B]$

$= (A \cup B \cup C) - [A \,\Delta\, (B \,\Delta\, C)]$

$[\because A \,\Delta\, (B \,\Delta\, C) = (A^c \cap B^c \cap C) \cup (A \cap B^c \cap C^c) \cup (A^c \cap B \cap C^c) \cup (A \cap B \cap C)]$

40. $\sum_{r=1}^{100} \left[\dfrac{1}{2} + \dfrac{r}{100}\right]$

$= \sum_{r=1}^{49} \left[\dfrac{1}{2} + \dfrac{r}{100}\right] + \sum_{r=50}^{100} \left[\dfrac{1}{2} + \dfrac{r}{100}\right] = 0 + 51 = 51$

41. We have, $n(A) = 40\%$ of $10000 = 4000$

$$n(B) = 20\% \text{ of } 10000 = 2000$$
$$n(C) = 10\% \text{ of } 10000 = 1{,}000$$
$$n(A \cap B) = 5\% \text{ of } 10000 = 500$$
$$n(B \cap C) = 3\% \text{ of } 10000 = 300$$
$$n(C \cap A) = 4\% \text{ of } 10000 = 400$$
$$n(A \cap B \cap C) = 2\% \text{ of } 10000 = 200$$

Number of families which buy newspaper A only

$$= n(A \cap B^c \cap C^c)$$
$$= n[A \cap (B \cup C)^c]$$
$$= n(A) - n[A \cap (B \cup C)]$$
$$= n(A) - n[(A \cap B) \cup (A \cap C)]$$
$$= n(A) - [n(A \cap B) + n(A \cap C) - n(A \cap B \cap C)]$$
$$= 4000 - [500 + 400 - 200]$$
$$= 4000 - 700 = 3300$$

42. Putting $y = 1$, $f(x+1) = f(x) + f(1) - x - 1$

$\Rightarrow \quad f(x+1) = f(x) - x \quad [\because f(1) = 1]$

$\therefore \quad f(n+1) = f(n) - n < f(n)$

$\Rightarrow \quad f(n+1) < f(n)$

So, $f(n) < f(n-1) < f(n-2) \ldots < f(3) < f(2) < f(1) = 1$

$\therefore f(n) = n$ holds only for $n = 1$

43. Period of $\sin\left(\dfrac{\pi x}{n!}\right) = \dfrac{2\pi}{\pi/n!} = 2n!$ and period of

$$\cos\left(\frac{\pi x}{(n+1)!}\right) = \frac{2\pi}{\pi/(n+1)!} = 2(n+1)!$$

$\therefore$ Period of $f(x) =$ LCM of $\{2n!, 2(n+1)!\}$

$$= 2(n+1)!$$

44. $\because f(x+1) - f(x) = 8x + 3$

$\Rightarrow \{b(x+1)^2 + c(x+1) + d\} - \{bx^2 + cx + d\} = 8x + 3$

$\Rightarrow \quad b\{(x+1)^2 - x^2\} + c = 8x + 3$

$\Rightarrow \quad b(2x+1) + c = 8x + 3$

On comparing

$$2b = 8 \text{ and } b + c = 3$$

Then, $\quad b = 4$ and $c = -1$

45. Let $n(X)$ denote the number of elements in X.

Then, $n(A \cup B \cup C) = n(A) + n(B) + n(C) - n(A \cap B)$
$- n(B \cap C) - n(C \cap A) + n(A \cap B \cap C)$
$= \Sigma n(A) - \Sigma n(A \cap B) \quad (\text{since } A \cap B \cap C = \phi)$

Now, $\quad A \Delta B = (A - B) \cup (B - A) = (A \cup B) - (A \cap B)$

Therefore,

$$n(A \Delta B) = n(A \cup B) - n(A \cap B)$$
$$= n(A) + n(B) - 2n(A \cap B)$$

and $\quad 300 = \Sigma n(A \Delta B) = \Sigma[n(A) + n(B) - 2n(A \cap B)]$
$= 2[\Sigma n(A) - \Sigma n(A \cap B)]$

Therefore,

$$n(A \cup B \cup C) = \Sigma n(A) - \Sigma n(A \cap B) = \frac{300}{2} = 150$$

46. **Option (a)** $\because f(x) = \dfrac{a^x + 1}{a^x - 1}$

$$\therefore \quad f(-x) = \frac{a^{-x} + 1}{a^{-x} - 1} = \frac{1 + a^x}{1 - a^x} = -f(x)$$

Hence, $f(x)$ is an odd function.

Option (b) $\because f(x) = x\left(\dfrac{a^x - 1}{a^x + 1}\right)$

$$\therefore \quad f(-x) = (-x)\left(\frac{a^{-x} - 1}{a^{-x} + 1}\right)$$
$$= (-x)\left(\frac{1 - a^x}{1 + a^x}\right) = x\left(\frac{a^x - 1}{a^x + 1}\right)$$
$$= f(x)$$

Hence, $f(x)$ is an even function.

Option (c) $\because f(x) = \left(\dfrac{a^x - a^{-x}}{a^x + a^{-x}}\right)$

$$\therefore \quad f(-x) = \left(\frac{a^{-x} - a^x}{a^{-x} + a^x}\right) = -\left(\frac{a^x - a^{-x}}{a^x + a^{-x}}\right) = -f(x)$$

Hence, $f(x)$ is an odd function.

Option (d) $\because f(x) = \sin x$

$$\therefore \quad f(-x) = -\sin x = -f(x)$$

Hence, $f(x)$ is an odd function.

47. $1 - e^{(1/x) - 1} > 0 \Rightarrow e^{(1/x) - 1} < 1$

$$\Rightarrow \quad \frac{1}{x} - 1 < 0,$$
$$\Rightarrow \quad \frac{x - 1}{x} > 0$$

+ 0 − 1 +

$$x \in (-\infty, 0) \cup (1, \infty)$$

48. $(gof)\,x = g\{f(x)\} = g\,[\sin^2 x + \sin^2(x + \pi/3)$
$+ \cos x \cos(x + \pi/3)]$

$$= g\left\{\sin^2 x + \left(\frac{1}{2}\sin x + \frac{\sqrt{3}}{2}\cos x\right)^2 + \cos x\left(\frac{1}{2}\cos x - \frac{\sqrt{3}}{2}\sin x\right)\right\}$$
$$= g\left\{\frac{5}{4}\sin^2 x + \frac{5}{4}\cos^2 x\right\}$$
$$= g(5/4) = 1$$

Hence, $(gof)\,x$ is a constant function.

49. $\because f : [1, \infty) \to [1, \infty)$

Let $y = f(x) = 2^{x(x-1)} \quad (\because x = f^{-1}(y))$

$$\Rightarrow \quad \log_2 y = x^2 - x$$
$$\Rightarrow \quad x^2 - x - \log_2 y = 0$$
$$\therefore \quad x = \frac{1 \pm \sqrt{(1 + 4\log_2 y)}}{2}$$
$$f^{-1}(y) = \frac{1 + \sqrt{(1 + 4\log_2 y)}}{2} \quad (\because x \geq 1)$$

Hence, $\quad f^{-1}(x) = \dfrac{1 + \sqrt{(1 + 4\log_2 x)}}{2}$

50. Putting $y = 1$, then $2f(x) = f(x) + \{f(1)\}^x$

$$\Rightarrow \quad f(x) = \{f(1)\}^x = k^x$$
$$\therefore \quad \sum_{r=1}^{n} f(r) = \sum_{r=1}^{n} k^r = k + k^2 + k^3 + \ldots + k^n = \frac{k(k^n - 1)}{(k - 1)}$$

51. We have, $(a, b) R (a, B)$ for all $(a, b) \in N \times N \Rightarrow R$ is reflexive
R is symmetric for we have $(a, b) R (c, d) \Rightarrow a + d = b + c$

$\Rightarrow \quad d + a = c + b$

$\Rightarrow \quad c + b = d + a \Rightarrow (c, d) R (a, b)$

Hence, R is symmetric.
Then, by definition of R, we have
$a + d = b + c$ and $c + f = d + e$
hence by addition, we get

$$a + d + c + f = b + c + d + e$$

or $\quad a + f = b + e$

Hence, $(a, b) R (e, f)$
Thus, $(a, b) R (c, d)$ and $(c, d) R (e, f)$

$\Rightarrow \quad (a, b) R (e, f)$

Hence, R is transitive.
Clearly R is not anti-symmetric.

52. $\log_{10}(1 - x) \neq 0,\ 1 - x > 0,\ x + 2 \geq 0$

$\therefore \quad 1 - x \neq 1,\ x < 1,\ x \geq -2$

$\Rightarrow \quad x \neq 0,\ x < 1,\ x \geq -2$

$\therefore \quad x \in [-2, 1) - \{0\}$

53. $f(x) + g(x) = \sqrt{x} + \sqrt{1 - x}$, domain $[0, 1]$
$f(x) - g(x) = \sqrt{x} - \sqrt{1 - x}$, domain $[0, 1]$
$g(x) - f(x) = \sqrt{1 - x} - \sqrt{x}$, domain $[0, 1]$

$$\frac{f(x)}{g(x)} = \frac{\sqrt{x}}{\sqrt{1 - x}}, \text{ domain } [0, 1)$$

$$\frac{g(x)}{f(x)} = \frac{\sqrt{1 - x}}{\sqrt{x}}, \text{ domain } (0, 1]$$

So, common domain is $(0, 1)$

54. $\because f(x) = 6^x + 3^x + 6^{-x} + 3^{-x} + 2$

$$= \left(6^x + \frac{1}{6^x}\right) + \left(3^x + \frac{1}{3^x}\right) + 2$$

$$\geq 2\sqrt{6^x \cdot \frac{1}{6^x}} + 2\sqrt{3^x \cdot \frac{1}{3^x}} + 2 = 6$$

$\therefore \quad f(x) \geq 6 \quad [\because \text{AM} \geq \text{GM}]$

Hence, Range $= [6, \infty)$

55. Rewrite $f(x) = 2\sin(x + \pi/6) + 4$ or

$$f(x) = 2\cos\left(x - \frac{\pi}{3}\right) + 4$$

$\therefore \quad Y = [2, 6] \quad [\because$ min and max values of $\sin\theta$ and $\cos\theta$ are -1 and $+1]$

56. Let $f_1(x) = \cos^{-1}(x - x^2)$
$f_1(x)$ is defined when $-1 \leq x - x^2 \leq 1$

$$\Rightarrow \quad \frac{1 - \sqrt{5}}{2} \leq x \leq \frac{1 + \sqrt{5}}{2} \quad \ldots\text{(i)}$$

and let $\quad f_2(x) = \sqrt{1 - \dfrac{1}{|x|}}$

$f_2(x)$ is defined when $\left(1 - \dfrac{1}{|x|}\right) \geq 0$

$\Rightarrow \quad x \leq -1$ and $x \geq 1 \quad \ldots\text{(ii)}$

and let $\quad f_3(x) = \dfrac{1}{[x^2 - 1]}$

$f_3(x)$ is defined when $x^2 < 1,\ x^2 \geq 2$

$\Rightarrow \quad x \in (-\infty, -\sqrt{2}] \cup (-1, 1) \cup [\sqrt{2}, \infty) \quad \ldots\text{(iii)}$

Then, $D_f = D_{f_1} \cap D_{f_2} \cap D_{f_3}$

$$= \left[\sqrt{2}, \frac{1 + \sqrt{5}}{2}\right]$$

57. Let the population of city is 100.
Then, $n(A) = 25$, $n(B) = 20$ and $n(A \cap B) = 8$

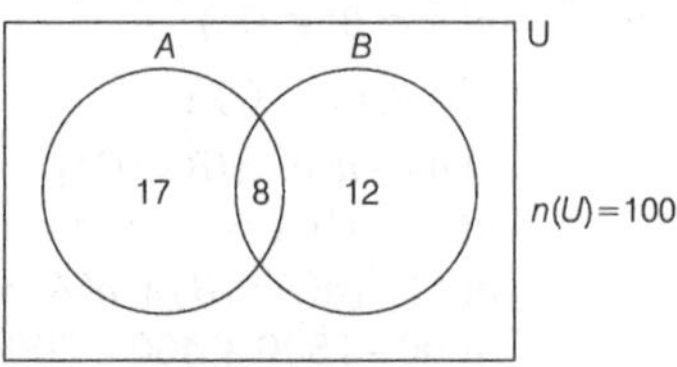

Venn diagram

So, $n(A \cap \overline{B}) = 17$ and $n(\overline{A} \cap B) = 12$
According to the question, Percentage of the population who look into advertisement is

$$= \left[\frac{30}{100} \times n(A \cap \overline{B})\right] + \left[\frac{40}{100} \times n(\overline{A} \cap B)\right] + \left[\frac{50}{100} \times n(A \cap B)\right]$$

$$= \left(\frac{30}{100} \times 17\right) + \left(\frac{40}{100} \times 12\right) + \left(\frac{50}{100} \times 8\right)$$

$$= 5.1 + 4.8 + 4 = 13.9$$

58. If $x \in Q$, then $f(x) = x$

$\therefore \quad fof(x) = f\{f(x)\} = f(x) = x$

If $x \notin Q$, then $f(x) = 1 - x$

$fof(x) = f\{f(x)\} = f(1 - x) = 1 - (1 - x) = x$
$[\because x \notin Q \Rightarrow 1 - x \notin Q]$

Thus, $fof(x) = x$ for all $x \in [0, 1]$

59. Given function,

$$f(x) = \frac{x}{e^x - 1} + \frac{x}{2} + 2\cos^3\frac{x}{2} \text{ on } R - \{0\}$$

$$\because \quad f(-x) = \frac{-x}{e^{-x} - 1} - \frac{x}{2} + 2\cos^3\left(-\frac{x}{2}\right)$$

$$= \frac{-xe^x}{1 - e^x} - \frac{x}{2} + 2\cos^3\frac{x}{2}$$

$$= \frac{xe^x}{e^x - 1} - \frac{x}{2} + 2\cos^3\frac{x}{2}$$

$$= \frac{x(e^x - 1 + 1)}{e^x - 1} - \frac{x}{2} + 2\cos^3\frac{x}{2}$$

$$= x + \frac{x}{e^x - 1} - \frac{x}{2} + 2\cos^3\frac{x}{2}$$

$$= \frac{x}{e^x - 1} + \frac{x}{2} + 2\cos^3\frac{x}{2} = f(x)$$

$\because f(-x) = f(x),\ \forall\, x \in R - \{0\}$
$\therefore f(x)$ is an even function.
Hence, option (d) is correct.

60. Taking $x = y = 1$, we get

$$f(1)f(1) - f(1) = 2$$
$$\Rightarrow \quad f^2(1) - f(1) - 2 = 0$$
$$\Rightarrow \quad f(1) - 2((f(1) + 1) = 0$$
$$\Rightarrow \quad f(1) = 2 \quad [\text{as } f(1) > 0]$$

Taking $y = 1$, we get

$$f(x).(f1) - f(x) = x + 1$$
$$\Rightarrow \quad f(x) = x + 1$$
$$\Rightarrow \quad f^{-1}(x) = x - 1$$
$$\therefore \quad f(x) \cdot f^{-1}(x) = x^2 - 1$$

61. $f(x) = y = \dfrac{x-2}{x-3}$

$$\therefore \quad x = \frac{3y-2}{y-1}$$
$$\therefore f^{-1}(x) = \frac{3x-2}{x-1}$$

and $g(x) = y = 2x - 3$

$$\therefore \quad x = \frac{y+3}{2}$$
$$\therefore g^{-1}(x) = \frac{x+3}{2}$$
$$\because f^{-1}(x) + g^{-1}(x) = \frac{13}{2}$$
$$\therefore x^2 - 5x + 6 = 0 \begin{cases} x_1 \\ x_2 \end{cases}$$

$\therefore$ Sum of roots $x_1 + x_2 = 5$

62. Let $f(x) = e^{\cos^4 \pi x + x - [x] + \cos^2 \pi x}$

Period of $x - [x] = 1$

$$\because \quad \cos^2 \pi x = \frac{1 + \cos 2\pi x}{2}$$
$$\because \text{Period of } \cos^2 \pi x = \frac{2\pi}{2\pi} = 1$$
$$\because \cos^4 \pi x = \frac{1}{8}\{3 + 4\cos 2\pi x + \cos 4\pi x\}$$
$$\because \text{Period of } \cos 2\pi x = \frac{2\pi}{2\pi} = 1$$
$$\because \text{Period of } \cos 4\pi x = \frac{2\pi}{4\pi} = \frac{1}{2}$$
$$\therefore \text{Period of } \cos^4 \pi x = \text{LCM}\left(1, \frac{1}{2}\right) = \frac{\text{LCM }(1, 1)}{\text{HCF }(1, 2)} = \frac{1}{1} = 1$$

$\therefore$ Period of $f(x) = 1$.

63. Let $y = f(x) = \sin^{-1}\{4 - (x-7)^3\}^{1/5}$

$$\Rightarrow \quad (\sin y)^5 = 4 - (x-7)^3$$
$$\Rightarrow \quad (x-7)^3 = 4 - \sin^5 y$$
$$x - 7 = (4 - \sin^5 y)^{1/3}$$
$$\therefore \quad x = 7 + (4 - \sin^5 y)^{1/3}$$
$$\Rightarrow \quad f^{-1}(y) = 7 + (4 - \sin^5 y)^{1/3}$$

Hence, $\quad f^{-1}(x) = 7 + (4 - \sin^5 x)^{1/3}$

64. Here, $g(x) = \dfrac{1}{2}(a^x + a^{-x})$

and $\quad h(x) = \dfrac{1}{2}(a^x - a^{-x})$

Now,

$$g(x+y) + g(x-y) = \frac{1}{2}(a^{x+y} + a^{-(x+y)}) + \frac{1}{2}(a^{x-y} + a^{-x+y})$$
$$= 2 \cdot \frac{1}{4}(a^x a^y + a^x a^{-y} + a^{-x} a^y + a^{-x} a^{-y})$$
$$= 2 \cdot \frac{1}{4}(a^x (a^y + a^{-y}) + a^{-x}(a^y + a^{-y}))$$
$$= 2 \cdot \left(\frac{1}{2}(a^x + a^{-x})\right)\left(\frac{1}{2}(a^y + a^{-y})\right)$$
$$= 2g(x)g(y)$$

65. $g(x) = f\{f(x)\} = f\left(\dfrac{1}{1-x}\right) = \dfrac{1}{1 - \dfrac{1}{1-x}} = -\dfrac{1-x}{x}$

and $\quad h(x) = f(f\{f(x)\}) = f\{g(x)\}$

$$= \frac{1}{1 - g(x)} = \frac{1}{1 + \dfrac{1-x}{x}} = x$$
$$\therefore \quad f(x) \cdot g(x) \cdot h(x) = \frac{1}{1-x}\left\{-\frac{1-x}{x}\right\} \cdot x = -1$$

66. $\because y = \log_a (x + \sqrt{(x^2+1)}) = f(x)$ (say)

$$\therefore \quad x = f^{-1}(y)$$

and $\quad a^y = x + \sqrt{(x^2+1)} \quad \ldots$(i)

or $\quad a^{-y} = \sqrt{(x^2+1)} - x \quad \ldots$(ii)

Subtracting Eq. (ii) from Eq. (i), then

$$x = \frac{1}{2}(a^y - a^{-y})$$
$$\Rightarrow \quad f^{-1}(y) = \frac{1}{2}(a^y - a^{-y})$$

Hence, $\quad f^{-1}(x) = \dfrac{1}{2}(a^x - a^{-x})$

67. For $f(x)$ to be defined, the following three inequalities must be satisfied.

(i) $\sin^{-1}(\log_2 x) \geq 0$

(ii) $\cos(\sin x) \geq 0$

(iii) $\left|\dfrac{1+x^2}{2x}\right| \leq 1$

For (i) $x > 0, \; -1 \leq \log_2 x \leq 1$

and $\quad \log_2 x \geq 0$

$$\therefore \quad 0 \leq \log_2 x \leq 1 \quad \text{or} \quad 1 \leq x \leq 2$$

For (ii) $\cos(\sin x) \geq 0$ is satisfied for all real x.

For (iii) $\dfrac{1+x^2}{2|x|} \leq 1 \Rightarrow \dfrac{1+|x|^2}{2|x|} - 1 \leq 0$

$$\Rightarrow \quad \frac{|x| - 1^2}{|x|} \leq 0 \quad \text{or} \quad \frac{|x| - 1^2}{|x|} = 0$$
$$\therefore \quad |x| = 1 \quad \text{or} \quad x = \pm 1$$

Hence, $f(x)$ is defined only, if $x = 1$

$\Rightarrow D_f = \{1\}$

68. $\because f(x)=\dfrac{x^2+2x+5}{x^2+x+1}=\dfrac{(x+1)^2+4}{\left(x+\frac{1}{2}\right)^2+\frac{3}{4}}>0$

$\therefore f(x)$ is into.

and $f'(x)=-\dfrac{x^2+8x+3}{(x^2+x+1)^2}$

$f'(x)=0$, has real value of x so function is many one. Hence, function is many one and into.

69. $\because \quad f(x)=\begin{cases}1+x, & 0\le x\le 2\\ 3-x, & 2<x\le 3\end{cases}$

$\therefore \quad fof(x)=f\{f(x)\}$

$$=\begin{cases}1+f(x), & 0\le f(x)\le 2\\ 3-f(x), & 2<f(x)\le 3\end{cases}$$

$$=\begin{cases}2+x, & 0\le 1+x\le 2 \text{ and } 0\le x\le 2\\ 2-x, & 2<1+x\le 3 \text{ and } 0\le x\le 2\\ 4-x, & 0\le 3-x\le 2 \text{ and } 2<x\le 3\\ x, & 2<3-x\le 3 \text{ and } 2<x\le 3\end{cases}$$

$$=\begin{cases}x+2, & 0\le x\le 1\\ 2-x, & 1<x\le 2\\ 4-x, & 2<x\le 3\\ x, & x\in\phi\end{cases}$$

$$\therefore \quad fof(x)=\begin{cases}2+x, & 0\le x\le 1\\ 2-x, & 1<x\le 2\\ 4-x, & 2<x\le 3\end{cases}$$

70. Let $y=(x+1)^2-1,\ x\ge -1$

$\Rightarrow \quad (x+1)^2=y+1$

$\Rightarrow \quad x+1=\sqrt{y+1}$ as $x\ge -1$

$\Rightarrow \quad x=-1+\sqrt{y+1},\ y\ge -1$

Thus, $\quad f^{-1}(x)=-1+\sqrt{x+1}$

So, $\quad f(x)=f^{-1}(x)$

$\Rightarrow \quad (x+1)^2-1=-1+\sqrt{x+1}$

$\Rightarrow \quad \sqrt{x+1}=0$ or $(x+1)^{3/2}=1$

$\Rightarrow \quad x=-1$ or $x=0$

71. $|\cos x|+\cos x>0$

$\Rightarrow \quad \cos x>0$

$\Rightarrow \quad 2n\pi-\dfrac{\pi}{2}<x<2n\pi+\dfrac{\pi}{2}$

$\Rightarrow \quad x\in\left(\dfrac{(4n-1)\pi}{2},\dfrac{(4n+1)\pi}{2}\right),\ n\in I$

72. $16-x\ge 2x-1$ and $20-3x\ge 4x-5$

$\Rightarrow \quad 3x\le 17$ and $7x\le 25$

$\therefore \quad x\le 17/3$ and $x\le 25/7$

$\therefore \quad x\le 25/7 \qquad [\because x\in I]$

$\therefore \quad x=2,3 \qquad [\because 2x-1\ge 0$ and $4x-5\ge 0]$

73. Let $y=f(x)=3\sin\sqrt{\left(\dfrac{\pi^2}{16}-x^2\right)}$

$\therefore \quad y\ge 0 \qquad \ldots(\text{i})$

and $\quad \dfrac{y}{3}=\sin\sqrt{\left(\dfrac{\pi^2}{16}-x^2\right)}$

$\Rightarrow \quad \left(\sin^{-1}\dfrac{y}{3}\right)^2=\dfrac{\pi^2}{16}-x^2$

$\Rightarrow \quad x^2=\dfrac{\pi^2}{16}-\left(\sin^{-1}\dfrac{y}{3}\right)^2\ge 0$

$\therefore \quad \left(\sin^{-1}\dfrac{y}{3}\right)^2\le\dfrac{\pi^2}{16}$

$\Rightarrow \quad -\dfrac{\pi}{4}\le\sin^{-1}\dfrac{y}{3}\le\dfrac{\pi}{4}$

$\Rightarrow \quad \sin\left(-\dfrac{\pi}{4}\right)\le\dfrac{y}{3}\le\sin\left(\dfrac{\pi}{4}\right)$

$\Rightarrow \quad -\dfrac{1}{\sqrt{2}}\le\dfrac{y}{3}\le\dfrac{1}{\sqrt{2}}$

$\therefore \quad -\dfrac{3}{\sqrt{2}}\le y\le\dfrac{3}{\sqrt{2}} \qquad \ldots(\text{ii})$

From Eqs. (i) and (ii),

$$0\le y\le\frac{3}{\sqrt{2}}$$

$\therefore \quad y\in\left[0,\dfrac{3}{\sqrt{2}}\right]$

74. $f(x)=|\sqrt{(x-5)}-1|-|\sqrt{(x-5)}+1|$

$\therefore \quad x\ge 5$

75. $\dfrac{|\sin x|+|\cos x|}{|\sin x-\cos x|}$

$\because$ Period of $|\sin x|+|\cos x|$ is $\pi/2$

And period of $|\sin x-\cos x|$ is π

$\therefore \quad$ Period of $\dfrac{|\sin x|+|\cos x|}{|\sin x-\cos x|}$

$$=\text{LCM}\left\{\frac{\pi}{2},\frac{\pi}{1}\right\}=\frac{\text{LCM}\{\pi,\pi\}}{\text{HCF}\{2,1\}}$$

$$=\frac{\pi}{1}=\pi$$

76. $x^2-x-2\ne 0$

$\Rightarrow \quad x\ne 2,-1$

$$\left[x+\frac{1}{2}\right]>0,\left[x+\frac{1}{2}\right]\ne 1$$

$\Rightarrow \quad x+\dfrac{1}{2}\ge 1,\ x+\dfrac{1}{2}\notin(1,2)$

$\Rightarrow \quad x\ge\dfrac{1}{2},\ x\notin\left(\dfrac{1}{2},\dfrac{3}{2}\right)$

$\therefore \quad x\in\left[\dfrac{3}{2},2\right)\cup(2,\infty)$

77. $D_f = R$,

$\therefore \quad g(x) = \dfrac{1 - \sin^2 x}{\cos^2 x}$

$\therefore \quad D_g = R - \{x : x = (2n+1)\dfrac{\pi}{2}, n \in I]$

$\therefore \quad D_f \cap D_g = D_g$

$= R - \{x : x = (2n+1)\dfrac{\pi}{2}, n \in I]$

78. Given, $D_f = (0, 1) \Rightarrow 0 < x < 1$

$\therefore \quad 0 < e^x < 1$ and $0 < \ln|x| < 1$

$\Rightarrow \quad -\infty < x < 0$ and $1 < |x| < e$

$\Rightarrow \quad x \in (-\infty, 0)$ and $x \in (-e, -1) \cup (1, e)$

$\therefore \quad x = (-e, -1)$

79. Let $f(x) = e^x + \sin x, x \in [-4, 0]$

$\therefore \quad f(-x) = e^{-x} - \sin x, x \in [0, 4]$

$\therefore$ Even extension

$\Rightarrow \quad f(x) = f(-x)$

$= e^{-x} - \sin x, x \in [0, 4]$

$= e^{-|x|} - \sin|x|$

80. Area $= \dfrac{\sqrt{3}}{4}(\text{side})^2$

$= \dfrac{\sqrt{3}}{4}[x^2 + (g(x))^2] = \dfrac{\sqrt{3}}{4}$ (given)

$\therefore \; x^2 + (g(x))^2 = 1$

$\therefore \quad g(x) = \pm\sqrt{(1 - x^2)}$

81. $\because \sin^2 x + \sin^2(x + \pi/3) + \cos x \cos(x + \pi/3)$

$= \dfrac{1 - \cos 2x}{2} + \dfrac{1 - \cos(2x + 2\pi/3)}{2} + \dfrac{\cos(2x + \pi/3) + \cos \pi/3}{2}$

$= \dfrac{1}{2}\left\{\dfrac{5}{2} - (\cos 2x + \cos(2x + 2\pi/3)) + \cos(2x + \pi/3)\right\}$

$= \dfrac{1}{2}\left\{\dfrac{5}{2} - 2\cos\left(2x + \dfrac{\pi}{3}\right)\cos\dfrac{\pi}{3} + \cos\left(2x + \dfrac{\pi}{3}\right)\right\}$

$= \dfrac{5}{4}$

$\therefore f(x) = a^{5/4}$ which is constant.

82. Let $f_1(x) = \sin^{-1}\left(\dfrac{2 - |x|}{4}\right)$

$f_1(x)$ is defined when

$-1 \le \dfrac{2 - |x|}{4} \le 1$

$\Rightarrow \quad -4 \le 2 - |x| \le 4$

$\Rightarrow \quad -6 \le -|x| \le 2$

$\Rightarrow \quad -2 \le |x| \le 6$

$\Rightarrow \quad 0 \le |x| \le 6$

$\therefore \quad D_{f_1} = [-6, 6]$

and let $f_2(x) = \cos^{-1}\left(\dfrac{2 - |x|}{4}\right)$

$f_2(x)$ is defined when

$-1 \le \dfrac{2 - |x|}{4} \le 1$

$\therefore \quad -4 \le 2 - |x| \le 4$

$\Rightarrow \quad -6 \le -|x| \le 2$

or $\quad -2 \le |x| \le 6$

$\Rightarrow \quad 0 \le |x| \le 6$

$\therefore \quad D_{f_2} = [-6, 6]$

and let $\quad f_3(x) = \tan^{-1}\left(\dfrac{2 - |x|}{4}\right)$

$D_{f_3} = R$

$\therefore \quad D_f = D_{f_1} \cap D_{f_2} \cap D_{f_3}$

$= [-6, 6]$

83. Let $y = f(x) = \dfrac{e^x - e^{-|x|}}{e^x + e^{|x|}}$

If $\quad x \ge 0, y = \dfrac{e^{2x} - 1}{2e^{2x}}$

$e^{2x} = \dfrac{1}{1 - 2y} \ge 1 \quad [\because x \ge 0]$

$\Rightarrow \quad \dfrac{1}{1 - 2y} - 1 \ge 0$

$\Rightarrow \quad \dfrac{y}{1 - 2y} \ge 0$

or $\quad \dfrac{y}{2y - 1} \le 0$

$\therefore \quad 0 \le y < \dfrac{1}{2}$

$\Rightarrow \quad y \in [0, \dfrac{1}{2})$

84. Let $f(x) = ax^2 + bx + c$

$\therefore \quad f(-3) = 9a - 3b + c = 6$

$f(0) = 0 + 0 + c = 6$

and $\quad f(2) = 4a + 2b + c = 11$

$\therefore \quad a = \dfrac{1}{2}, b = \dfrac{3}{2}, c = 6$

$f(x) = \dfrac{1}{2}x^2 + \dfrac{3}{2}x + 6,$

$\therefore \quad f(1) = \dfrac{1}{2} + \dfrac{3}{2} + 6 = 8$

$\Rightarrow \quad (1, 8)$

85. If $x + y = a$ and $x - y = b$

Then $\quad f(a, b) = \dfrac{a^2 - b^2}{4}$

$\therefore \text{AM} = \dfrac{f(x, y) + f(y, x)}{2} = \dfrac{\dfrac{x^2 - y^2}{4} + \dfrac{y^2 - x^2}{4}}{2} = 0$

86. **Case I** $f(x) \ne 2$ is true, $f(y) = 2$ and $f(z) \ne 1$ are false, then $f(x) = 1$ or 3, $f(y) = 1$ or 3 and $f(z) = 1$

$\Rightarrow f$ is not one-one

Case-II $f(x) \ne 2$ is false, $f(y) = 2$, is true, $f(z) \ne 1$ is false, then

$f(x) = 2, f(y) = 2, f(z) = 1$

$\Rightarrow$ not posible

Case-III $f(x) \neq 2$ is false, $f(y) = 2$ is false, $f(z) \neq 1$ is true, then

$$f(x) = 2, \; f(y) = 1, \text{ or } 3, \; f(z) = 2 \text{ or } 3$$

$$\Rightarrow \quad f(x) = 2, \; f(z) = 3, \; f(y) = 1$$

87. According to the question, we have the following Venn diagram.

Here, $A \cap B \subseteq C$ and $A \cap B \neq \phi$

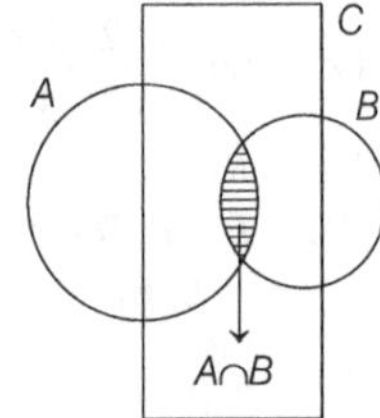

Now, from the Venn diagram, it is clear that

$$B \cap C \neq \phi, \text{ is true}$$

Also, $(C \cup A) \cap (C \cup B) = C \cup (A \cap B) = C$ is true.

If $(A - B) \subseteq C$, for this statement the Venn diagram is

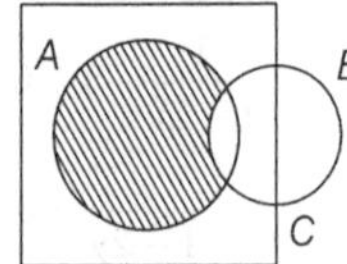

From the Venn diagram, it is clear that

if $A - B \subseteq C$, then $A \subseteq C$.

Now, if $(A - C) \subseteq B$, for this statement the Venn diagram.

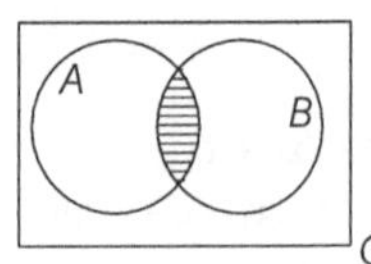

From the Venn diagram, it is clear that

$A \cap B \neq \phi$, $A \cap B \subseteq C$ and $A - C = \phi \subseteq B$ but $A \not\subseteq B$.

88. The given function $f : R \to R$ is

$$f(x) = |x| (x - \sin x) \quad \ldots \text{(i)}$$

$\because$ The function 'f' is a odd and continuous function and as $\lim\limits_{x \to \infty} f(x) = \infty$ and $\lim\limits_{x \to -\infty} f(x) = -\infty$, so range is R, therefore 'f' is a onto function.

$$\because \quad f(x) = \begin{cases} x(x - \sin x), & x \geq 0 \\ -x(x - \sin x), & x < 0 \end{cases}$$

$$\therefore \quad f'(x) = \begin{cases} 2x - \sin x - x\cos x, & x > 0 \\ -2x + \sin x + x\cos x, & x < 0 \end{cases}$$

$$\begin{cases} (x - \sin x) + x(1 - \cos x), & x > 0 \\ (-x + \sin x) - x(1 - \cos x), & x < 0 \end{cases}$$

$\because$ for $x > 0$, $x - \sin x > 0$ and $x(1 - \cos x) > 0$

$\therefore$ $f'(x) > 0 \; \forall \; x \in (0, \infty)$

$\Rightarrow f$ is strictly increasing function, $\forall \; x \in (0, \infty)$.

Similarly, for $x < 0$, $-x + \sin x > 0$ and $(-x)(1 - \cos x) > 0$, therefore, $f'(x) > 0 \; \forall \; x \in (-\infty, 0)$

$\Rightarrow f$ is strictly increasing function, $\forall \; x \in (0, \infty)$

Therefore, 'f' is a strictly increasing function for $x \in R$ and it implies that f is one-one function.

89. $\because f$ is onto

$$\therefore \quad 0 \leq x^2 + x + a < \infty$$

$$[\because \; 0 \leq \tan^{-1}(x^2 + x + a) < \pi/2]$$

$$\Rightarrow \quad 0 \leq (x + 1/2)^2 + a - 1/4 < \infty \; \forall x \in R$$

$$\Rightarrow \quad a = 1/4.$$

90. Suppose that $((x, y), (a, b)), ((a, b), (p, q)) \in R$.

Then, either $x < a$ or $x = a$ and $y > b$

and either $a < p$ or $a = p$ and $b > q$

If $x < a$ and $a < p$, then $x < p$ and hence $((x, y), (p, q)) \in R$.

Same is the case when $x < a$ and $a = p$ and also when $x = a$ and $a < p$.

If $x = a$, $y > b$, $a = p$ and $b > q$, then $x = p$ and $y > b > q$.

Therefore, $((x, y), (p, q)) \in R$.

91. $f(x) = \sin^{-1}[\sec x]$

$f(x)$ is defined when

$$-1 \leq [\sec x] \leq 1$$

$$\because \quad [\sec x] \leq 1$$

$$\Rightarrow \quad \sec x < 2$$

$$\text{But} \quad \sec x \geq 1$$

$$\therefore \quad 1 \leq \sec x < 2$$

$$\Rightarrow \quad x \in \left[2m\pi, 2m\pi + \frac{\pi}{3}\right) \quad \ldots \text{(i)}$$

$$m \in I$$

$$\text{And if} \quad [\sec x] \geq -1$$

$$\therefore \quad \sec x \geq -1$$

$$\text{But} \quad \sec x \geq 1$$

$$\text{and} \quad \sec x \leq -1$$

$$\therefore \quad \sec x = -1$$

$$x = 2n\pi + \pi$$

$$= (2n + 1)\pi, \; n \in I \quad \ldots \text{(ii)}$$

$\therefore$ From Eqs. (i) and (ii), we get

$$D_f = \{(2n + 1)\pi, n \in I\} \cup \left\{\left[2m\pi, 2m\pi + \frac{\pi}{3}\right), m \in I\right\}$$

92. $\because \; x^{12} - x^9 + x^4 - x + 1 > 0$

or $\quad x^4(x^8 + 1) - x(x^8 + 1) + 1 > 0$

or $\quad (x^8 + 1)(x^4 - x) + 1 > 0$

or $\quad (x^8 + 1)\, x\, (x^3 - 1) + 1 > 0$

For $x \leq -1$ and $x \geq 1$ the above expression is obviously positive.

If $-1 < x \leq 0$ the above hold good

And if $0 < x < 1$, $x^{12} - x(x^8 + 1) + (x^4 + 1) = 0$

Because $x^4 + 1 > x^8 + 1$

And so $\quad x^4 + 1 > x(x^8 + 1)$

93. Given,

$$f(x) = \log\{\text{sign}(9 - x^2)\} + \sqrt{[x]^3 - 4[x]}$$

We must have, $\text{sgn}(9 - x^2) > 0$

$\Rightarrow 9 - x^2 > 0$

$\Rightarrow x^2 - 9 < 0$

$\Rightarrow (x-3)(x+3) < 0$

$\Rightarrow -3 < x < 3$ ... (i)

Also, $[x]^3 - 4[x] \ge 0$

$\Rightarrow [x]([x]^2 - 4) \ge 0$

$\Rightarrow [x][([x]-2)([x]+2) \ge 0$

$\Rightarrow [x] \ge 2$ or $[x]$ lies between -2 and 0,

$\Rightarrow [x] = -2, -1$ or 0

now, $[x] \ge 2 \Rightarrow x \ge 2$

$[x] = -2 \Rightarrow -2 \le x < 1$

$[x] = -1 \Rightarrow -1 \le x < 0$

$[x] = 0 \Rightarrow 0 \le x < 1$

Hence $[x] = -2, -1, 0 \Rightarrow -2, \le x < 1$

Hence $D_f = [-2, 1) \cup [2, 3)$

94. $\because f(x) = 2\min(f(x) - g(x), 0)$

If $0 < f(x) - g(x) \Rightarrow g(x) < f(x)$

and $|f(x) - g(x)| = f(x) - g(x)$

$\therefore f(x) = 0 = (f(x) - g(x)) - (f(x) - g(x))$

$= (f(x) - g(x)) - |f(x) - g(x)|$

$= (f(x) - g(x)) - |g(x) - f(x)|$

95. $\because \dfrac{x}{\{x\}} > 0 \Rightarrow x > 0$ $[\because 0 < \{x\} < 1]$

and $\ln\left(\dfrac{x}{\{x\}}\right) > 0$

$\Rightarrow \dfrac{x}{\{x\}} > 1$

$\Rightarrow \dfrac{x - \{x\}}{\{x\}} > 0$

or $\dfrac{[x]}{\{x\}} > 0$

$\Rightarrow [x] > 0 \Rightarrow x \ge 1$

$x \in [1, \infty) - I$

or $x \in (1, \infty) - I$ $[\because 0 < \{x\} < 1]$

96. Period of $\sin ax$ is $\dfrac{2\pi}{a}$

And period of $\cos ax$ is $\dfrac{2\pi}{a}$

$\therefore$ Period of $\sin ax + \cos ax$ is $\dfrac{2\pi}{a}$

And period of $|\sin x| + |\cos x|$ is $\dfrac{\pi}{2}$

Given, $\dfrac{2\pi}{a} = \dfrac{\pi}{2}$

$\Rightarrow a = 4$

97. Putting $x = 1$, $y = 2$, then

$g(1)g(2) = g(1) + g(2) + g(2) - 2$

$\Rightarrow 5g(1) = 8 + g(1)$ $[\because g(2) = 5]$

$\therefore g(1) = 2$

Also, replacing y by $\dfrac{1}{x}$ in the given relation, then

$$g(x)\,g\left(\frac{1}{x}\right) = g(x) + g\left(\frac{1}{x}\right) + g(1) - 2$$

or $g(x)\,g\left(\dfrac{1}{x}\right) = g(x) + g\left(\dfrac{1}{x}\right)$

$\therefore g(x) = 1 \pm x^n$

$\Rightarrow g(2) = 1 \pm 2^n = 5 \Rightarrow \pm 2^n = 2^2$

Taking +ve sign

$2^n = 2^2$

$\therefore n = 2$

$\Rightarrow g(x) = 1 + x^2$

$\therefore g(3) = 1 + 3^2 = 10$

98. $y = \dfrac{x-1}{x^2 - 3x + 3}$

$\Rightarrow x^2y - 3xy + 3y = x - 1$

$\Rightarrow x^2y - x(3y+1) + 3y + 1 = 0$

$\therefore D \ge 0$

$\Rightarrow (3y+1)^2 - 4y(3y+1) \ge 0$

$\Rightarrow -3y^2 + 2y + 1 \ge 0$

$\Rightarrow 3y^2 - 2y - 1 \le 0$

$\Rightarrow y^2 - \dfrac{2y}{3} - \dfrac{1}{3} \le 0$

$\Rightarrow \left(y - \dfrac{1}{3}\right)^2 - \dfrac{1}{9} - \dfrac{1}{3} \le 0$

$\Rightarrow \left(y - \dfrac{1}{3}\right)^2 \le \dfrac{4}{9}$

$\Rightarrow -\dfrac{2}{3} \le y - \dfrac{1}{3} < \dfrac{2}{3}$

$\therefore -\dfrac{1}{3} \le y \le 1$

$\therefore y \in \left[-\dfrac{1}{3}, 1\right]$

99. $f_o(x) = \begin{cases} f(x), & x \in [0, 1] \\ -f(-x), & x \in [-1, 0] \end{cases}$

$\Rightarrow \begin{cases} x^2 + x + \sin x - \cos x + \log(1 + |x|), & x \in [0, 1] \\ -x^2 + x + \sin x + \cos x - \log(1 + |x|), & x \in [-1, 0] \end{cases}$

100. For $\sec^{-1} x$

$x \ge 1$ and $x \le -1$ or $x \in R \sim (-1, 1)$

and $x - [x] > 0$

$\Rightarrow x > [x],\ x \in R \sim I$

$D_f = R - \{(-1, 1) \cup (n, n \in I)\}$

101. Period of $[\sin 3x]$ is $\dfrac{2\pi}{3}$

and period of $|\cos 6x|$ is $\dfrac{\pi}{6}$

$\therefore$ Period of $f(x) = \text{LCM of } \left\{\dfrac{2\pi}{3}, \dfrac{\pi}{6}\right\} = \dfrac{2\pi}{3}$

Objective Questions Type II [One or more than one correct answer(s)]

1. $f(x)=\sin\left[\frac{\pi}{6}\sin\left(\frac{\pi}{2}\sin x\right)\right], x\in R$

$=\sin\left(\frac{\pi}{6}\sin\theta\right), \theta\in\left[-\frac{\pi}{2},\frac{\pi}{2}\right]$, where $\theta=\frac{\pi}{2}\sin x$

$=\sin\alpha, \alpha\in\left[-\frac{\pi}{6},\frac{\pi}{6}\right]$, where $\alpha=\frac{\pi}{6}\sin\theta$

$\therefore \quad f(x)=\left[-\frac{1}{2},\frac{1}{2}\right]$

Hence, range of $f(x)=\left[-\frac{1}{2},\frac{1}{2}\right]$

So, option (a) is correct.

(b) $f\{g(x)\}=f(t), t\in\left[-\frac{\pi}{2},\frac{\pi}{2}\right]\Rightarrow f(t)\in\left[-\frac{1}{2},\frac{1}{2}\right]$

$\therefore$ Option (b) is correct.

(c) $\lim_{x\to 0}\frac{f(x)}{g(x)}=\lim_{x\to 0}\frac{\sin\left[\frac{\pi}{6}\sin\left(\frac{\pi}{2}\sin x\right)\right]}{\frac{\pi}{2}(\sin x)}$

$=\lim_{x\to 0}\frac{\sin\left[\frac{\pi}{6}\sin\left(\frac{\pi}{2}\sin x\right)\right]}{\frac{\pi}{6}\sin\left(\frac{\pi}{2}\sin x\right)}\cdot\frac{\frac{\pi}{6}\sin\left(\frac{\pi}{2}\sin x\right)}{\left(\frac{\pi}{2}\sin x\right)}$

$=1\times\frac{\pi}{6}\times 1=\frac{\pi}{6}$

$\therefore$ Option (c) is correct.

(d) $g\{f(x)\}=1$

$\Rightarrow \quad \frac{\pi}{2}\sin\{f(x)\}=1$

$\Rightarrow \quad \sin\{f(x)\}=\frac{2}{\pi}$ ...(i)

But $\quad f(x)\in\left[-\frac{1}{2},\frac{1}{2}\right]\subset\left[-\frac{\pi}{6},\frac{\pi}{6}\right]$

$\therefore \quad \sin\{f(x)\}\in\left[-\frac{1}{2},\frac{1}{2}\right]$ ...(ii)

$\Rightarrow \sin\{f(x)\}\neq\frac{2}{\pi}$, [from Eqs. (i) and (ii)]

i.e. No solution.

$\therefore$ Option (d) is not correct.

2. $\because (gof)x=g\{f(x)\}$

$=g\{2x-\sin x\}$

$=(2x-\sin x)^{1/3}=y$ (say)

$\Rightarrow \quad y^3=2x-\sin x$

$\therefore \quad y\in R$

Range of $(gof)\,x$ is R.

Option (a) is correct.

For $\quad x_1\neq x_2\Rightarrow f(x_1)\neq f(x_2)$

and $\quad x_1\neq x_2\Rightarrow g(x_1)\neq g(x_2)$

Hence, f and g are one-one.

Option (c) is correct.

$\therefore gof$ is one-one

Option (b) is correct.

$\because$ Range of gof is R.

Hence, both f and g are onto.

Option (d) is correct.

3. $f(-x)=\begin{cases}0 & \text{for } -x=0\\ (-x)^2\sin\left(-\frac{\pi}{x}\right) & \text{for } -1<-x<1\,(-x\neq 0)\\ -x\,|-x| & \text{for } -x\geq 1 \text{ or } -x\leq -1\end{cases}$

$=\begin{cases}0 & \text{for } x=0\\ -x^2\sin\frac{\pi}{x} & \text{for } -1<x<1\,(x\neq 0)\\ -x\,|x| & \text{for } x\leq -1 \text{ or } x\geq 1\end{cases}$

$=-f(x)$

$\therefore f(x)$ is an odd function.

$\therefore \quad -f'(-x)=-f'(x)$

$\Rightarrow \quad f'(-x)=f'(x)$

$\therefore\ f'(x)$ is an even function.

Hence, **option (a)** and **(d)** are correct.

4. **Option (a)** : $Sgn\,(x)=\begin{cases}1, & x>0\\ 0, & x=0\\ -1, & x<0\end{cases}$

$\therefore \quad Sgn\,(e^{-x})=1$ $\quad [\because e^{-x}>0]$

$Sgn\,(e^{-x})$ is constant function. Hence, it is periodic.

Option (b) : $\because$ Period of $\sin x$ is 2π

and period of $|\sin x|$ is π

$\therefore$ Period of $\sin x+|\sin x|$ is LCM $\{2\pi, \pi\}$

$=2\pi$

Option (c) Let $f(x)=\min\{\sin x, |x|\}$

$=\sin x \quad [\because \sin x<|x|]$

$\sin x$ is periodic with period 2π.

Option (d)

$f(x)=\left[x+\frac{1}{2}\right]+\left[x-\frac{1}{2}\right]+2[-x]$

$=\left(x+\frac{1}{2}\right)-\left\{x+\frac{1}{2}\right\}+\left(x-\frac{1}{2}\right)$

$-\left\{x-\frac{1}{2}\right\}+2(-x-\{-x\})$

$=-\left\{x+\frac{1}{2}\right\}-\left\{x-\frac{1}{2}\right\}-2\{-x\}$

Hence, $f(x)$ is periodic.

5. **Option (a)** $\because \sin(\sin^{-1}x)=x$

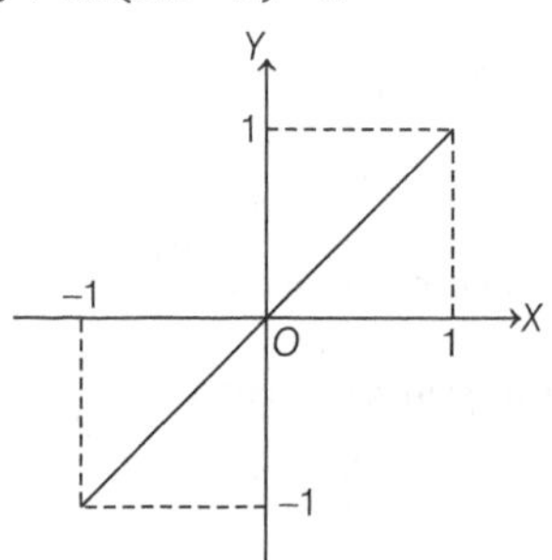

Which is bijective

Option (b) $\because \ x \in [-1, 1]$

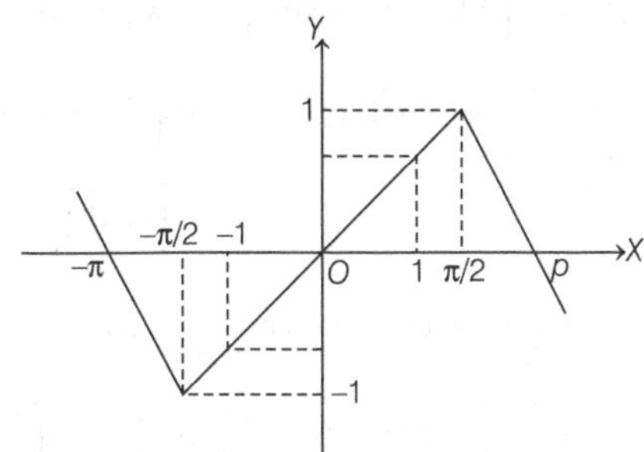

Here, range < codomain
i.e. not bijective.

Option (c)

$$\frac{\text{sgn}(x)}{\ln e^x} = \frac{\text{sgn}(x)}{x} = \begin{cases} \frac{1}{x}, & x > 0 \\ -\frac{1}{x}, & x < 0 \end{cases}$$

Which is many one into
i.e., not bijective

Option (d)

$$x^3 \,\text{Sgn}(x) = = \begin{cases} x^3, & x > 0 \\ -x^3, & x < 0 \\ 0, & x = 0 \end{cases}$$

Which is many one into
i.e., not bijective.

6. We have

$$f(x) = [1 + \sin x] + [1 + \sin 2x] + [1 + \sin 3x] + \ldots \ldots + [1 + \sin nx]$$

$$= 1 + [\sin x] + 1 + [\sin 2x] + 1 + [\sin 3x] + \ldots + 1 + [\sin nx]$$

$$= n + [\sin x] + [\sin 2x] + [\sin 3x] + \ldots + [\sin nx]$$

$\because \quad x \in (0, \pi)$

$\therefore \quad \sin x > 0$

$\therefore$ Minimum value of $f(x)$ at $x = 0$ is n

$\therefore$ Maximum value of $f(x)$ at $x = \pi / 2$ is $n + 1$ (if n is odd)

7. $\because f(x) = \sin^{-1}[2 - 4x^2]$

For the function to be defined, we must have

$$-1 \le [2 - 4x^2] \le 1$$

$\Rightarrow \quad -1 \le 2 - 4x^2 < 2$

$\Rightarrow \quad 1 \ge -2 + 4x^2 > -2$

$\Rightarrow \quad 3 \ge 4x^2 > 0$

or $\quad 0 < x^2 \le \frac{3}{4}$

$\therefore \quad x \in \left[-\frac{\sqrt{3}}{2}, \frac{\sqrt{3}}{2}\right] \sim \{0\}$

8. $e^{f(x)} = e - e^x \Rightarrow f(x) = \log_e(e - e^x)$

$$e - e^x > 0 \Rightarrow e^1 > e^x \Rightarrow x < 1$$

$$D_f = (-\infty, 1)$$

Let $\quad y = f(x) = \log_e(e - e^x) \Rightarrow e^y = e - e^x$

$\Rightarrow \quad e^x = e - e^y \Rightarrow x = \log_e(e - e^y)$

$\Rightarrow \quad e - e^y > 0 \Rightarrow e^1 > e^y$

$\therefore \quad y < 1$

$$R_f = (-\infty, 1)$$

9. $\because f(x) = \sec^{-1}[1 + \cos^2 x]$

Domain of $f(x)$ is R $\quad [\because [1 + \cos^2 x] \ge 1]$

and $\because 0 \le \cos^2 x \le 1$

$\Rightarrow \quad 1 \le (1 + \cos^2 x) \le 2$

$\therefore \quad [1 + \cos^2 x] = 1$ and 2

Then, range of $f(x)$ is $\sec^{-1} 1$ and $\sec^{-1} 2$

$\Rightarrow \{\sec^{-1} 1, \sec^{-1} 2\}$

10. $f(x) = [x]^2 + [x + 1] - 3$

$$= [x]^2 + 1 + [x] - 3$$

$$= ([x] + 2)([x] - 1)$$

$f(x) = 0$, then $[x] = -2$

$\Rightarrow \quad -2 \le x < -1$

and $\quad [x] = 1$

$\Rightarrow \quad 1 \le x < 2$

many one and into

11. We have, $f(x - 1) = \begin{cases} 0, & |x| > 1 \\ 1 - x, & 0 \le x \le 1 \\ 1 + x, & -1 \le x < 0 \end{cases}$

$$f(x + 1) = \begin{cases} 0, & x < 0 \\ x, & 0 \le x \le 1 \\ 2 - x, & 1 \le x \le 2 \\ 0, & x > 2 \end{cases}$$

So, $f(x-1)+f(x+1)=\begin{cases}0, & x<-2\\ x+2, & -2\le x\le -1\\ -x, & -1\le x\le 0\\ x, & 0\le x\le 1\\ 2-x, & 1\le x\le 2\\ 0, & x>2\end{cases}$

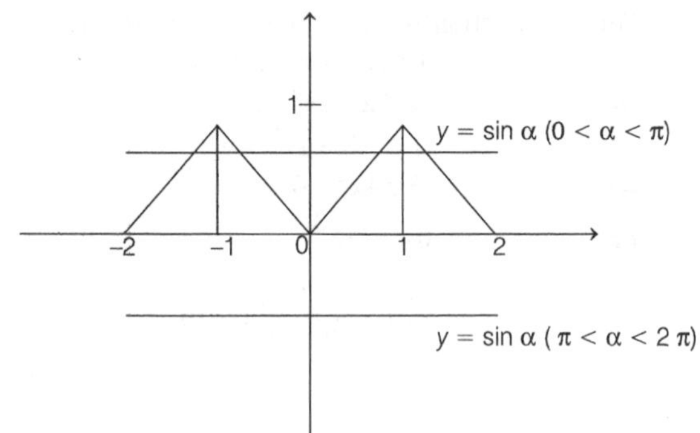

12. **Option (a)** $zy=\dfrac{x+2}{x-1}$

$\Rightarrow \quad xy-y=x+2$

$\therefore \quad x=\dfrac{y+2}{y-1}=f(y)$

Option (b) $f(1)$ is not defined

Option (c) $\because\ y=f(x)=1+\dfrac{3}{x-1}$

$\therefore \quad \dfrac{dy}{dx}=-\dfrac{3}{(x-1)^2}<0$

$\Rightarrow y$ decreases for all x except $x=1$

Option (d) $f(x)=\dfrac{x+2}{x-1}=$ rational function of x

13. $\because \quad \pi=3.14$

$\therefore \quad =0+(-1)^5=-1\pi^2=9.86,\ -\pi^2=-9.86$

$\Rightarrow \quad [\pi^2]=9,\ [-\pi^2]=-10$

$\therefore \quad f(x)=\cos 9x+\cos 10x$

Option (a) $f\left(\dfrac{\pi}{2}\right)=\cos\left(\dfrac{9\pi}{2}\right)+\cos(5\pi)$

Option (b) $f(\pi)=\cos 9\pi+\cos 10\pi=-1+1=0$

Option (c) $f(-\pi)=\cos(-9\pi)+\cos(-10\pi)$

$=\cos 9\pi+\cos 10\pi$

$=-1+1=0$

Option (d)

$$f\left(\frac{\pi}{4}\right)=\cos\left(\frac{9\pi}{4}\right)+\cos\left(\frac{5\pi}{2}\right)$$

$$=\cos\left(2\pi+\frac{\pi}{4}\right)+0$$

$$=\cos\frac{\pi}{4}=\frac{1}{\sqrt{2}}$$

14. $\because f(x)=\dfrac{1}{2}\left\{2\cos^2 x+2\cos^2\left(\dfrac{\pi}{3}+x\right)-2\cos x\cos\left(\dfrac{\pi}{3}+x\right)\right\}$

$$f(x)=\frac{1}{2}\left\{(1+\cos 2x)+\left(1+\cos\left(\frac{2\pi}{3}+2x\right)\right)-\left(\cos\left(\frac{\pi}{3}+2x\right)+\cos\frac{\pi}{3}\right)\right\}$$

$$=\frac{1}{2}\left\{\frac{3}{2}+\cos 2x+\cos\left(\frac{2\pi}{3}+2x\right)-\cos\left(\frac{\pi}{3}+2x\right)\right\}$$

$$=\frac{1}{2}\left\{\frac{3}{2}+2\cos\left(2x+\frac{\pi}{3}\right)\cos\frac{\pi}{3}-\cos\left(\frac{\pi}{3}+2x\right)\right\}$$

$$=\frac{3}{4}$$

$\therefore f(x)$ is an even function, a periodic function and $f(0)=f(1)=\dfrac{3}{4}$

15. $\therefore$ Period of $e^{x-[x]}$ is 1

and period of $\cos ax$ is $\dfrac{2\pi}{a}$

$\therefore$ Period of $f(x)=$ LCM of $\left\{1,\dfrac{2\pi}{a}\right\}$

i.e. period of $f(x)$ is possible only when a is irrational of type π

Option (a) if $a=\pi$

Then, period of $f(x)=$ LCM of $\{1, 2\}=2$

Option (b) if $a=2\pi$

Then, period of $f(x)=$ LCM of $\{1, 1\}=1$

Option (c) if $a=3\pi$

Then, period of $f(x)=$ LCM of $\left\{1,\dfrac{2}{3}\right\}$

$$=\text{LCM of }\left\{\frac{2}{2},\frac{2}{3}\right\}=\frac{2}{1}=2$$

16. Graph of $y=f(-|x|)$ is

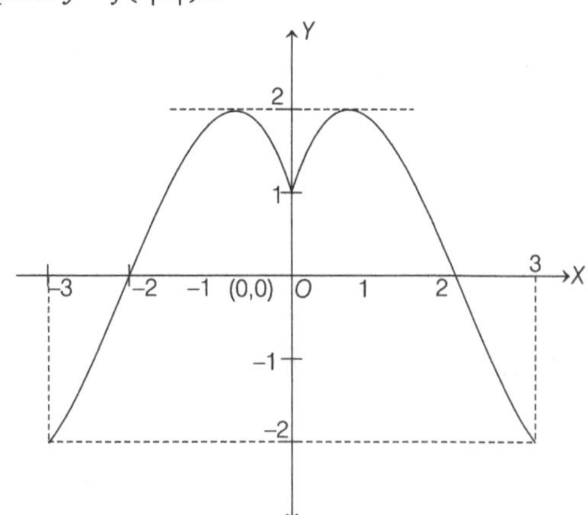

$\therefore$ range is $[-2, 2]$

(b) Domain of $y=f(|x|)$ is

$-3\le|x|\le 2$

$\Rightarrow \quad -2\le x\le 2$

(c) Domain of $y = f(|x| + 1)$ is

$$-3 \le |x| + 1 \le 2$$

$$\Rightarrow \quad -4 \le |x| \le 1$$

$$\Rightarrow \quad x \in [-1, 1]$$

(d) Graph of $y = f(|x| + 1)$ is

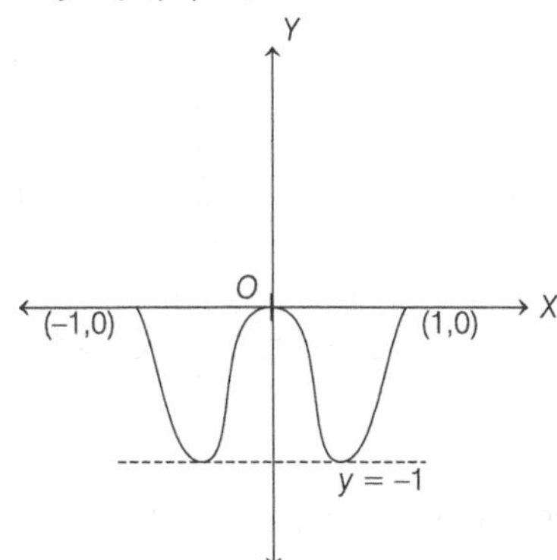

$\therefore$ Range is $[-1, 0]$

17. We are given that

$$af(x) + bf\left(\frac{1}{x}\right) = x - 1 \quad \ldots(i)$$

Replacing x with $1/x$, we get

$$bf(x) + af\left(\frac{1}{x}\right) = \frac{1}{x} - 1 \quad \ldots(ii)$$

From Eqs. (i) and (ii), we have

$$(a^2 - b^2) f(x) = a(x-1) - b\left(\frac{1}{x} - 1\right)$$

$$\therefore \quad f(2) = \frac{a + b/2}{a^2 - b^2} = \frac{2a + b}{2(a^2 - b^2)}$$

$$f(1) = 0$$

and $$f(-1) = \frac{-2a + 2b}{a^2 - b^2} = \frac{-2}{a + b}$$

18. $\because f(x) = [x]$, $[.]$ denotes the greatest integer function.

and $g(x) = |x|$

Now, $(fog)x = f(g(x)) = f(|x|) = [|x|]$

and $(gof)x = g(f(x)) = g([x]) = |[x]|$

Option (a)

$$(gof - fog)\left(-\frac{5}{3}\right) = (gof)\left(-\frac{5}{3}\right) - (fog)\left(-\frac{5}{3}\right)$$

$$= \left|\left[-\frac{5}{3}\right]\right| - \left[\left|-\frac{5}{3}\right|\right]$$

$$= |-2| - \left[\frac{5}{3}\right] = 2 - 1 = 1$$

Option (b) $(f + 2g)(-1) = f(-1) + 2g(-1)$

$$= [-1] + 2|-1|$$

$$= -1 + 2 = 1$$

Option (c) $(gof - fog)\left(\frac{5}{3}\right) = (gof)\left(\frac{5}{3}\right) - (fog)\left(\frac{5}{3}\right)$

$$= \left|\left[\frac{5}{3}\right]\right| - \left[\left|\frac{5}{3}\right|\right]$$

$$= |1| - \left[\frac{5}{3}\right] = 1 - 1 = 0$$

Option (d)

$$(f + 2g)(1) = f(1) + 2g(1) = [1] + 2|1| = 1 + 2 = 3$$

19. Option (a) $\because$ Periods of $\sin x$ and $|\sin x|$ are 2π and π respectively.

$\therefore$ Period of $f(x) =$ LCM of $\{2\pi, \pi\} = 2\pi$

Option (b) $\because$ Periods of $\sin x$, $\sec x$, $\cos x$, $\operatorname{cosec} x$ are 2π hence, period of $g(x) = 2\pi$

Option (c) Period of $h(x) =$ LCM $\{2\pi, 2\pi\} = 2\pi$

Option (d) $p(x) = [x] + \left[x + \frac{1}{3}\right] + \left[x + \frac{2}{3}\right]$

$$-\left(x + \left(x + \frac{1}{3}\right) + \left(x + \frac{2}{3}\right)\right) + 11$$

$$= 11 - \{x\} - \left\{x + \frac{1}{3}\right\} - \left\{x + \frac{2}{3}\right\}$$

$\therefore$ Period of $p(x)$ is 1.

20. Option (a) $f(-x) = (-x)\left(\frac{a^{-x} + 1}{a^{-x} - 1}\right)$

$$= (-x)\left(\frac{1 + a^x}{1 - a^x}\right)$$

$$= x\left(\frac{a^x + 1}{a^x - 1}\right) = f(x)$$

Option (b) $g'(x) = \frac{1}{\sqrt{(x^2 + a^2)}}$, is even

$\therefore g(x)$ is odd

Option (c) $h(-x) = \sqrt[3]{(1 + x)^2} + \sqrt[3]{(1 - x)^2} = h(x)$

Option (d) If x is rational, then $-x$ is also rational and if x is irrational, then $-x$ is also irrational

$$\therefore \quad p(-x) = \begin{cases} 0, & \text{if } -x \text{ is rational} \\ 1, & \text{if } -x \text{ is irrational} \end{cases}$$

$$= \begin{cases} 0, & \text{if } x \text{ is rational} \\ 1, & \text{if } x \text{ is irrational} \end{cases} = p(x)$$

21. Option (a) $D_f = R \sim \{0\}$, $D_g = R \sim \{0\}$

Also, $R_f = R_g$

then $f(x) = g(x)$

Option (b) $D_f = R \sim \{0\}$, $D_g = R$

$$D_f \ne D_g$$

Option (c) $D_f = R \sim \{0\}$, $D_g = (0, \infty)$

$$\therefore \quad D_f \ne D_g$$

Option (d) For domain of $f(x)$, $(x - 1)(x - 2) > 0$

$$D_f = (-\infty, 1) \cup (2, \infty)$$

+ – +
1 2

and $$D_g = (2, \infty) \cap (3, \infty) = (3, \infty)$$

$$\therefore \quad D_f \ne D_g$$

22. For $f(x)$ to be defined

$$\sin x \ge 0 \text{ and } \sin x \ne -1$$

$$\therefore \quad 2n\pi \le x \le (2n + 1)\pi, n \in I$$

For $n = 0, -1, 1$

$$0 \le x \le \pi, -2\pi \le x \le -\pi, 2\pi \le x \le 3\pi$$

Linked Comprehension Type

PASSAGE 1

Sol. (Q. Nos. 1-3)

$$f(x)=\begin{cases}x+1, & x\le 1\\ 2x+1, & 1<x\le 2\end{cases}$$

$$g(x)=\begin{cases}x^2, & -1\le x<2\\ x+2, & 2\le x\le 3\end{cases}$$

$$\therefore \quad f(x)=\begin{cases}g(x)+1, & g(x)\le 1\\ 2g(x)+1, & 1<g(x)\le 2\end{cases}$$

$$\text{or } f(g(x))=\begin{cases}x^2+1, & x^2\le 1, -1\le x<2\\ x+2+1, & x+2\le 1, 2\le x\le 3\\ 2x^2+1, & 1<x^2\le 2, -1\le x<2\\ 2(x+2)+1, & 1<x+2\le 2, 2\le x\le 3\end{cases}$$

$$\text{or } f(g(x))=\begin{cases}x^2+1, & -1\le x\le 1\\ 2x^2+1, & 1<x\le\sqrt{2}\end{cases}$$

1. Hence, the domain of $f(x)$ is $[-1, \sqrt{2}]$.
2. For $-1\le x\le 1$, we have $x^2\in[0, 1]$ or $x^2+1\in[1, 2]$
   For $1<x\le\sqrt{2}$, we have $x^2\in(1, 2]$ or $2x^2+1\in(3, 5]$
   Hence, the range is $[1, 2]\cup(3, 5]$.
3. For $f(g(x))=2$, $x^2+1=2$ and $2x^2+1=2$ i.e.,
   $x=\pm 1$ or $x=\pm\dfrac{1}{\sqrt{2}}$
   Therefore, $x=\pm 1$ only. Hence, two roots.

PASSAGE 2

1. $\because g(x)=0$

$$\Rightarrow \quad f(|x|)=0$$
$$\Rightarrow \quad x^2-5|x|+6=0$$
$$\Rightarrow \quad \begin{cases}x^2-5x+6=0, & x>0\\ x^2+5x+6=0, & x<0\end{cases}$$
$$\Rightarrow \quad \begin{cases}x=2, 3, & x>0\\ x=-3, -2, & x<0\end{cases}$$

$\therefore$ Number of solutions $=4$

2. $\because \quad g(x)-\lambda=0 \Rightarrow f(|x|)-\lambda=0$
   let $y=f(|x|)=\lambda$

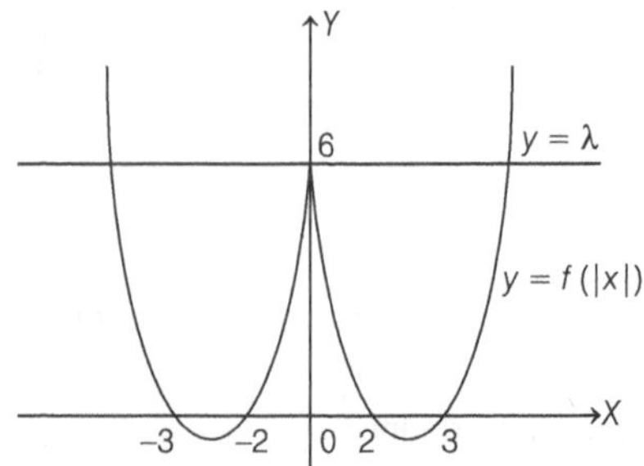

For exactly three real and distinct roots $\lambda=6$

3. $\because h(x)-\mu=0$
   $|g(x)|-\mu=0$
   let $y=|g(x)|=\mu$

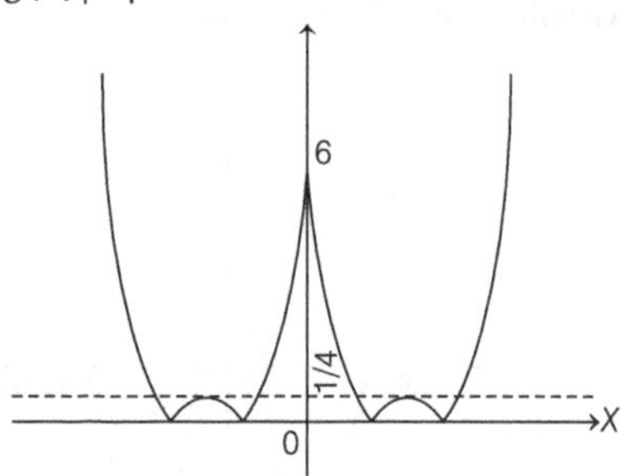

or $y=|f(|x|)|=\mu$
the vertex of $f(x)=x^2-5x+6$
is $\left(\dfrac{5}{2}, -\dfrac{1}{4}\right)$ for eight real and distinct roots
$0<\mu<\dfrac{1}{4} \Rightarrow \mu\in\left(0, \dfrac{1}{4}\right)$

4. Graph of $g(x)+|g(x)|=0$ is
   from the graph the solution set is
   $x\in[-3, -2]\cup[2, 3]$

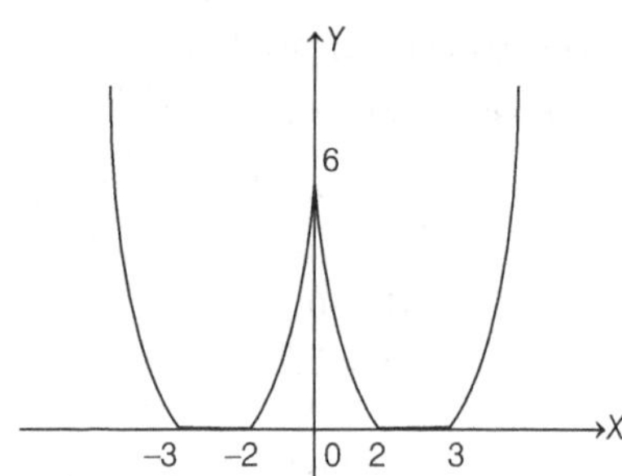

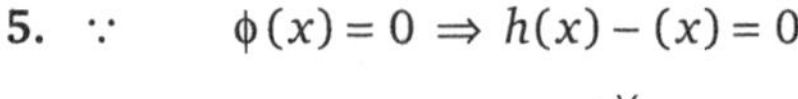
5. $\because \quad \phi(x)=0 \Rightarrow h(x)-(x)=0$

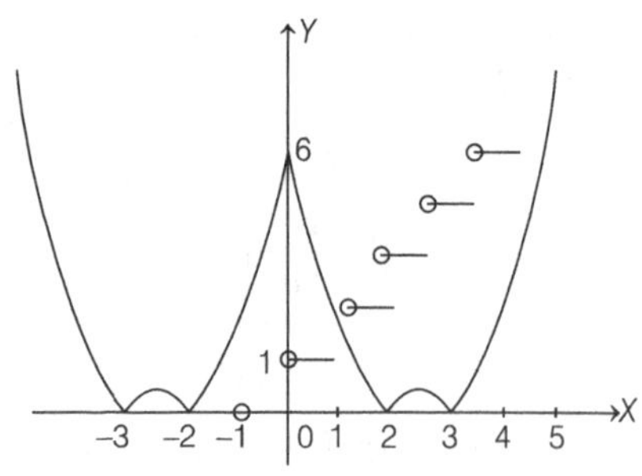

$$y=h(x)=(x)=\begin{cases}0, & -1<x\le 0\\ 1, & 0<x\le 1\\ 2, & 1<x\le 2\\ 3, & 2<x\le 3\end{cases}$$

PASSAGE 3

1. Let $y = f(x) = \min\{x - \{x\}, -x - \{-x\}\}$

$\therefore \quad x^2 + y^2 = 1$

Required number of solutions = 2

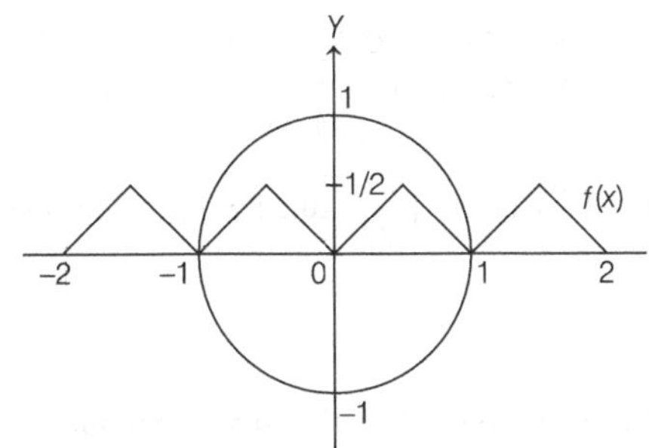

2. Range of $f(x)$ is $\left[0, \frac{1}{2}\right]$

3. $g(x) - h(x) = 0$

let $\quad y = g(x) = h(x)$

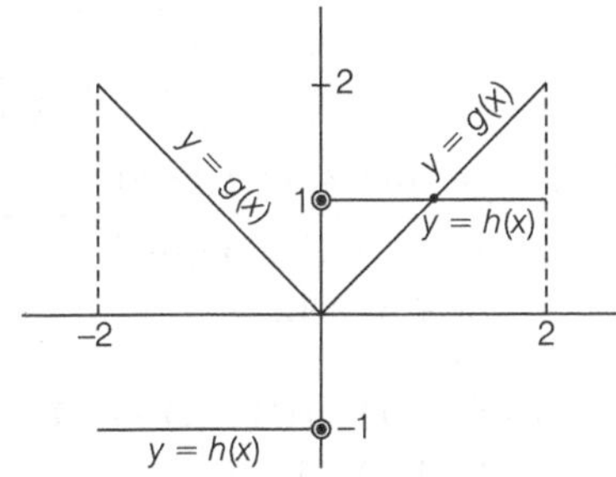

Only one +ve root

4. $a \in (0, 1/2)$

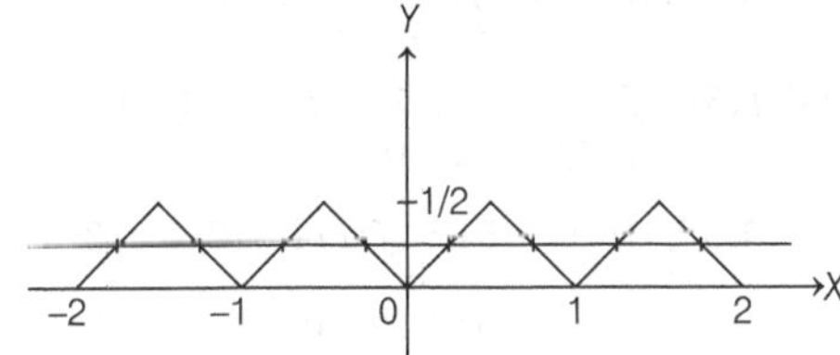

5. $\int_{-2}^{2} f(x)\,dx = 4 \times \frac{1}{2} \times 1 \times \frac{1}{2} = 1$

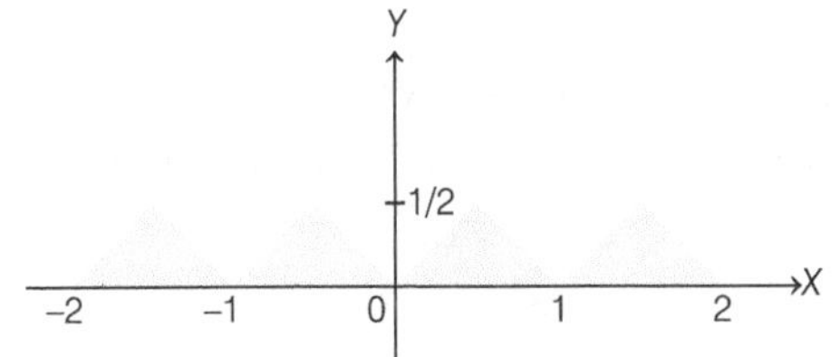

PASSAGE 4

$$\because f(x) + f(1-x) = g_a(x) + g_a(1-x)$$

$$= \frac{a^x}{a^x + \sqrt{a}} + \frac{a^{1-x}}{a^{1-x} + \sqrt{a}}$$

$$= \frac{a^x}{a^x + \sqrt{a}} + \frac{\sqrt{a}}{\sqrt{a} + a^x} = 1 \quad \ldots \text{(i)}$$

1. $$\sum_{r=1}^{1995} f\left(\frac{r}{1996}\right) = f\left(\frac{1}{1996}\right) + f\left(\frac{2}{1996}\right) + f\left(\frac{3}{1996}\right) + \ldots + f\left(\frac{1993}{1996}\right) + f\left(\frac{1994}{1996}\right) + f\left(\frac{1995}{1996}\right)$$

$$= \left\{f\left(\frac{1}{1996}\right) + f\left(\frac{1995}{1996}\right)\right\} + \left\{f\left(\frac{2}{1996}\right) + f\left(\frac{1994}{1996}\right)\right\} + \left\{f\left(\frac{3}{1996}\right) + f\left(\frac{1993}{1996}\right)\right\} + \ldots + \left\{f\left(\frac{997}{1996}\right) + f\left(\frac{999}{1996}\right)\right\} + f\left(\frac{998}{1996}\right)$$

$$= \underbrace{1 + 1 + 1 \ldots + 1 + 1}_{997 \text{ times}} + f\left(\frac{1}{2}\right)$$

$$= 997 + \frac{1}{2} \quad \text{[from Eq. (i)]}$$

$$= 997.5$$

$$\therefore \quad \left[\sum_{r=1}^{1995} f\left(\frac{r}{1996}\right)\right] = 997$$

2. $$\sum_{r=1}^{1996} f\left(\frac{r}{1997}\right) = f\left(\frac{1}{1997}\right) + f\left(\frac{2}{1997}\right) + f\left(\frac{3}{1997}\right) + \ldots + f\left(\frac{1994}{1997}\right) + f\left(\frac{1995}{1997}\right) + f\left(\frac{1996}{1997}\right)$$

$$= \left\{f\left(\frac{1}{1997}\right) + f\left(\frac{1996}{1997}\right)\right\} + \left\{f\left(\frac{2}{1997}\right) + f\left(\frac{1995}{1997}\right)\right\} + \left\{f\left(\frac{3}{1996}\right) + f\left(\frac{1994}{1996}\right)\right\} + \ldots$$

$$= 1 + 1 + 1 + \ldots 998 \text{ times} \quad \text{or} \quad = 998 \text{ (even)}$$

3. $g_5(x) + g_5(1-x) = f(x) + f(1-x) = 1$

4. $$\sum_{r=1}^{2n-1} 2f\left(\frac{r}{2n}\right)$$

$$= 2\left\{f\left(\frac{1}{2n}\right) + f\left(\frac{2}{2n}\right) + f\left(\frac{3}{2n}\right) + \ldots + f\left(\frac{2n-3}{2n}\right) + f\left(\frac{2n-2}{2n}\right) + f\left(\frac{2n-1}{2n}\right)\right\}$$

$$= 2\left\{\left\{f\left(\frac{1}{2n}\right) + f\left(\frac{2n-1}{2n}\right)\right\} + \left\{f\left(\frac{2}{2n}\right) + f\left(\frac{2n-2}{2n}\right)\right\} + \left\{f\left(\frac{3}{2n}\right) + f\left(\frac{2n-3}{2n}\right)\right\} + \ldots\right\}$$

$$= 2\left\{\underbrace{1 + 1 + 1 + \ldots + 1}_{(n-1) \text{ times}} + f\left(\frac{1}{2}\right)\right\} = 2\left((n-1) + \frac{1}{2}\right)$$

$$= 2n - 2 + 1 = (2n - 1)$$

5. $$\sum_{r=0}^{2n} f\left(\frac{r}{2n+1}\right) = f(0) + \sum_{r=1}^{2n} f\left(\frac{r}{2n+1}\right)$$

$$= \frac{1}{1 + \sqrt{a}} + n = \frac{1}{1 + \sqrt{a}} + 987 \quad \text{(given)}$$

$$\therefore \quad n = 987$$

PASSAGE 5

1. Figure (ii) is obtained by figure (i) by satisfying 2 unit in LHS.

   ∴ Figure (ii) is $f(x+2)$

2. From (i), $|f(|x|)|$

3. $\because -3 \le x+1 \le 4 \quad \Rightarrow -4 \le x \le 3$

   ∴ Domain is $[-4, 3]$

   Also $\quad -1 \le f(x+1) \le 2$

   $\Rightarrow \quad 1 \ge -f(x+1) \ge -2$

   or $\quad 2 \ge -f(x+1)+1 \ge -1$

   $\therefore \quad -f(x+1)+1 \in [-1, 2]$

4. $-2 \le x+1 \le 5 \Rightarrow -3 \le x \le 4$

   $-1 \le f(x+1) \le 2$

   $\Rightarrow \quad 1 \ge -f(x+1) \ge -2$

   $\therefore \quad -f(x+1) \in [-2, 1]$

5. $(x-3)^2 + (y)^2 = \left(\frac{7}{2}\right)^2$ cuts figure (iii) at one point.

Numerical Value Type

1. $f(x) = x^n + 1$

   $\because \quad f(10) = 10^n + 1 = 1001$ (given)

   $\therefore \quad 10^n = 1000 = 10^3$

   $\Rightarrow \quad n = 3,$

   then $\quad f(x) = x^3 + 1$

   $\therefore \quad f(20) = (20)^3 + 1 = 8001$

2. For $f(x)$ to be defined

   (i) $16 - x > 0 \Rightarrow x < 16$

   (ii) $2x - 1 \ge 0 \Rightarrow x \ge \frac{1}{2}$

   (iii) $16 - x \ge 2x - 1 \Rightarrow x \le \frac{17}{3}$

   (iv) $20 - 3x > 0 \Rightarrow x < \frac{20}{3}$

   (v) $4x - 5 \ge 0 \Rightarrow x \ge \frac{5}{4}$

   (vi) $20 - 3x \ge 4x - 5 \Rightarrow x \le \frac{25}{7}$

   Combining all we get $\frac{5}{4} \le x < \frac{25}{7}$

   $\because \quad x \in I \quad \therefore x = 2, 3$

   ∴ Domain of $f(x) = \{2, 3\}$

   ∴ Range of $\quad f(x) = \{f(2), f(3)\}$

   $= \{{}^{14}C_3 + {}^{14}C_3, {}^{13}C_5 + {}^{11}C_7\}$

   $= \{2.{}^{14}C_3, {}^{13}C_5 + {}^{11}C_4\}$

   $= \{728, 1617\}$

   $\therefore \quad \lambda = 1617, \mu = 728,$

   then $\quad \lambda + \mu = 2345$

3. The period of $\sin\left(\frac{\pi[x]}{12}\right)$ is $\frac{2\pi}{\frac{\pi}{12}} = 24$

   as $\sin\left(\frac{\pi[x+24]}{12}\right) = \sin\left(\frac{\pi([x]+24)}{12}\right)$

   $= \sin\left(2\pi + \frac{\pi[x]}{12}\right) = \sin\left(\frac{\pi[x]}{12}\right)$

   similarly the period of $\tan\left(\frac{\pi[x]}{3}\right)$ is 3 and the period of $\cos\left(\frac{\pi x}{4}\right)$ is $\frac{2\pi}{\pi/4} = 8.$

   Hence, the period of the given function

   $\lambda = \text{LCM of } 24, 8, 3 = 24$

   $\therefore \; 20.1\lambda = 20.1 \times 24 = 482.4$

4. $\therefore \; f\left(2x^2 + \frac{y^2}{8}, 2x^2 - \frac{y^2}{8}\right) = xy$

   $= \sqrt{\left(2x^2 + \frac{y^2}{8}\right)^2 - \left(2x^2 - \frac{y^2}{8}\right)^2}$

   $\therefore \quad f(60, 48) + f(80, 48) + f(13, 5)$

   $= \sqrt{(60)^2 - (48)^2} + \sqrt{(80)^2 - (48)^2} + \sqrt{(13)^2 - 5^2}$

   $= 36 + 64 + 12$

   $= 112$

5. $\because \; f(2) = f(1+1) = f(1) + (1) - 1 - 1$

   $= 1 + 1 - 1 - 1 = 0$

   $f(3) = f(2+1) = f(2) + f(1) - 2 - 1 = 0 + 1 - 2 - 1 = -2$

   $f(4) = f(2+2) = f(2) + f(2) - 4 - 1$

   $= 0 + 0 - 5 = -5$

   and $f(5) = f(1+4) = f(1) + f(4) - 4 - 1$

   $= 1 - 5 - 5 = -9$

   $\therefore \; -51.5 \sum_{r-1}^{5} f(r) = -51.5\,(f(1) + f(2) + f(3) + f(4) + f(5))$

   $= -51.5\,(1 + 0 - 2 - 5 - 9)$

   $= -51.5 \times -15$

   $= 772.5$

6. $\because \; f(1) = a\,(1^n + 3) = 4a = 12$ (given)

   $\therefore \quad a = 3$

   $\therefore \quad f(x) = 3\,(x^n + 3)$

   Also, $\quad f(3) = 3\,(3^n + 3) = 36$ (given)

   $\Rightarrow \quad 3^n + 3 = 12$

   $\Rightarrow \quad 3^n = 9 = 3^2$

   $\therefore \quad n = 2$

   then $\quad f(x) = 3\,(x^2 + 3)$

   $\therefore \quad 412 f(2) = 412 \times 3 \times (2^2 + 3)$

   $= 412 \times 21 = 8652$

7. It is given that $n(A) = m$ and $n(B) = n$ and $2^m = 2^n + 112$.

   [$\because$ number of subsets of set A and B are 2^m and 2^n respectively]

$\Rightarrow \quad 2^m - 2^n = 2^4(7)$

$\Rightarrow \quad 2^n(2^{m-n} - 1) = 2^4(2^3 - 1)$

On comparing $n = 4$ and $m - n = 3$

$\therefore \quad m = 7$

So, $\quad m \cdot n = 28$

8. Given,

$$f(1) + 2f(2) + 3f(3) + \ldots + nf(n) = n(n+1)f(n) \quad \ldots(i)$$

Replacing n by $n+1$, then

$$f(1) + 2f(2) + 3f(3) + \ldots + nf(n) + (n+1)f(n+1) = (n+1)(n+2)f(n+1) \quad \ldots(ii)$$

Subtracting Eq. (i) from Eq. (ii), then we get

$(n+1)f(n+1) = (n+1)(n+2)f(n+1) - n(n+1)f(n)$

$\Rightarrow \quad nf(n) = (n+1)f(n+1)$

From which we conclude that

$$2f(2) = 3f(3) = 4f(4) = \ldots = nf(n)$$

Substituting the values of $2f(2)$, $3f(3)$, ... in terms of $nf(n)$ in Eq. (i), we have

$f(1) + (n-1)n\,f(n) = n(n+1)f(n)$

$\Rightarrow \quad f(1) = 2nf(n)$

$$\frac{1}{f(n)} = \frac{2n}{f(1)} = \frac{2n}{1}$$

$\therefore \quad \dfrac{1}{f(1004)} = 2 \times 1004 = 2008$

9. Since, $1900 < f(1990) < 2000$

Therefore $\dfrac{1900}{90} < \dfrac{f(1990)}{90} < \dfrac{2000}{90}$

$\Rightarrow \quad 21.11 < \dfrac{f(1990)}{90} < 22.22$

$\therefore \quad \left[\dfrac{f(1990)}{90}\right] = 21, 22$

Now substituting $x = 1990$ in Eq. (i)

$$1990 - f(1990) = 19\left[\frac{1990}{19}\right] - 90\left[\frac{f(1990)}{90}\right]$$

$\Rightarrow \quad 1990 - f(1990) = 19 \times 104 - 90\ (21 \text{ or } 22)$

$\therefore \quad f(1990) = 1990 - 19 \times 104 + 90\ \ (21 \text{ or } 22)$

$= 1990 - 1976 + (1890 \text{ or } 1980)$

$= 1904 \text{ or } 1994$

$\therefore$ Possible values of $f(1990)$ are 1904 and 1994

$\therefore \quad$ Required sum $= 1904 + 1994 = 3898$

10. We have

$$f(x+1) + f(x-1) = \sqrt{3}\, f(x) \quad \ldots(i)$$

Replacing x by $x-1$ and $x+1$ in Eq. (i), then

$$f(x) + f(x-2) = \sqrt{3}f(x-1) \quad \ldots(ii)$$

and $\quad f(x+2) + f(x) = \sqrt{3}f(x+1) \quad \ldots(iii)$

Adding Eqs. (ii) and (iii), we get

$2f(x) + f(x-2) + f(x+2) = \sqrt{3}(f(x+1) + f(x-1))$

$= \sqrt{3} \cdot \sqrt{3} f(x) \quad$ (from Eq. (i))

$\Rightarrow f(x+2) + f(x-2) = f(x) \quad \ldots(iv)$

Replacing x by $x+2$ in Eq. (iv), then

$$f(x+4) + f(x) = f(x+2) \quad \ldots(v)$$

Adding Eqs. (iv) and (v), we get

$$f(x+4) + f(x-2) = 0 \quad \ldots(vi)$$

Again replacing x by $x+6$ in Eq. (vi), then

$$f(x+10) + f(x+4) = 0 \quad \ldots(vii)$$

Subtracting Eq. (vi) from Eq. (vii), we get

$$f(x+10) - f(x-2) = 0 \quad \ldots(viii)$$

Replacing x by $x+2$ in Eq. (viii), then

$$f(x+12) - f(x) = 0$$

or $\quad f(x+12) = f(x)$

Hence, period of $f(x)$ is 12

but given $\quad \dfrac{T}{24} = 12$

$\therefore \quad T = 24 \times 12 = 288$

11. Let $y = \dfrac{e^x - e^{-x}}{2}$

$\Rightarrow \quad e^{2x} - 1 = 2ye^x$

Therefore, $\quad t^2 - 2yt - 1 = 0, t = e^x$

$\Rightarrow \quad t = \dfrac{2y \pm \sqrt{4y^2 + 4}}{2}$

$= y \pm \sqrt{y^2 + 1}$

$\Rightarrow \quad x = \log(y + \sqrt{y^2 + 1})$ (since $e^x > 0$)

$\therefore \quad f^{-1}(x) = g(x) = \log(x + \sqrt{x^2 + 1})$

$$g\left(\frac{e^{1002} - 1}{2e^{501}}\right) = \log\left(\frac{e^{1002} - 1}{2e^{501}} + \frac{e^{1002} + 1}{2e^{501}}\right)$$

$= \log e^{501} = 501$

12. $\because f(x, y) = f(2x + 2y, 2y - 2x) \quad \ldots(i)$

or $f(I, II) = f(2I + 2II, 2II - 2I)$

$\therefore \quad f(2x + 2y, 2y - 2x)$

$= f(2(2x + 2y) + 2(2y - 2x), 2(2y - 2x) - 2(2x + 2y))$

$= f(8y, -8x) \quad \ldots(ii)$

From Eqs. (i) and (ii), we get

$$f(x, y) = f(8y, -8x) \quad \ldots(iii)$$

or $\quad f(I, II) = f(8II, -8I)$

$\therefore f(8y, -8x) = f(-64x, -64y) \quad \ldots(iv)$

From Eqs. (iii) and (iv), we get

$$f(x, y) = f(-64x, -64y) \quad \ldots(vi)$$

or $f(I, II) = f(-64I, -64II)$

$\therefore \quad f(-64x, -64y) = f(-64(-64x), -64(-64y))$

$= f(2^{12}x, 2^{12}y) \quad \ldots(vii)$

From Eqs. (v) and (vi) we get

$f(x, y) = f(2^{12}x, 2^{12}y)$

$\Rightarrow f(x, 0) = f(2^{12}x, 0)$

Replacing x by 2^x, then

$f(2^x, 0) = f(2^{x+12}, 0)$

$\Rightarrow \quad g(x) = g(x + 12)$

Hence, $g(x)$ is periodic with period 12.

13. The function $f(x)$ is defined, when

$x - 6 \geq 0$ and $12 - x \geq 0$

$\therefore \quad x \in [6, 12]$

$f'(x) = 0$

$\Rightarrow \frac{1}{2\sqrt{(x-6)}} - \frac{1}{2\sqrt{12-x}} = 0 \Rightarrow x = 9$

$f(6) = \sqrt{6}$

$f(9) = 2\sqrt{3} = \sqrt{12}$

$f(12) = \sqrt{6}$

$\Rightarrow$ range $= [\sqrt{6}, \sqrt{12}]$

$\Rightarrow$ Length of the interval $= \sqrt{12} - \sqrt{6}$

$\therefore \quad \lambda = 12$ and $\mu = 6$

$\Rightarrow \quad \lambda^2 + \mu^2 = 144 + 36 = 180$

14. $P_5 = \sum_{r=0}^{4} \binom{4}{r} P_r = \binom{4}{0}P_0 + \binom{4}{1}P_1 + \binom{4}{2}P_2 + \binom{4}{3}P_3 + \binom{4}{4}P_4$

Now, $P_0 = 1, P_1 = 1, P_2 = \binom{1}{0}P_0 + \binom{1}{1}P_1 = 1 + 1 = 2$

$P_3 = \binom{2}{0}P_0 + \binom{2}{1}P_1 + \binom{2}{2}P_2 = 1 + 2\cdot 1 + 1\cdot 2 = 5$

$P_4 = \binom{3}{0}P_0 + \binom{3}{1}P_1 + \binom{3}{2}P_2 + \binom{3}{3}P_3$

$= 1.1 + 3.1 + 3.2 + 1.5 = 15$

$P_5 = \binom{4}{0}P_0 + \binom{4}{1}P_1 + \binom{4}{2}P_2 + \binom{4}{3}P_3 + \binom{4}{4}P_4$

$= 1\cdot 1 + 4\cdot 1 + 6\cdot 2 + 4\cdot 5 + 1\cdot 15 = 52$

Single Integer Type

1. It is given that the function $f : R \to R$ satisfies,

$f(x + y) = f(x)\cdot f(y), \forall\, x, y \in R$

$\Rightarrow \quad f(x) = a^x$

$\because \quad f(1) = 3$

$\Rightarrow \quad a = 3$

$\therefore \quad f(x) = 3^x$

Now, it is given that $\sum_{i=1}^{n} f(i) = 363$

$\Rightarrow \quad 3 + 3^2 + 3^3 + \ldots + 3^n = 363$

$\Rightarrow \quad \frac{3(3^n - 1)}{2} = 363$

$\Rightarrow \quad 3^n = 243 = 3^5$

$\Rightarrow \quad n = 5$

2. $f(x)$ must be linear function. Therefore,

$f(x) = px + q$

$\therefore f(f(x)) = pf(x) + q = p(px + q) + q = p^2x + pq + q$

$\therefore f(f(f(x))) = p(p^2x + pq + q) + q$

$= p^3x + p^2q + pq + q$

$= 8x + 21$ [given]

$\therefore \quad p^3 = 8$ or $p = 2$

and $p^2q + pq + q = 21$ or $q = 3$

$\therefore \quad p + q = 5$

3. The limit

$$\lim_{x\to\frac{\pi}{2}} \frac{4\sqrt{2}(\sin 3x + \sin x)}{\left(2\sin 2x \sin\frac{3x}{2} + \cos\frac{5x}{2}\right) - \left(\sqrt{2} + \sqrt{2}\cos 2x + \cos\frac{3x}{2}\right)}$$

$$= \lim_{x\to\frac{\pi}{2}} \frac{4\sqrt{2}(2\sin 2x \cos x)}{2\sin 2x \sin\frac{3x}{2} + \left(\cos\frac{5x}{2} - \cos\frac{3x}{2}\right) - \sqrt{2}(1 + \cos 2x)}$$

$$= \lim_{x\to\frac{\pi}{2}} \frac{8\sqrt{2}\sin 2x \cos x}{2\sin 2x \sin\frac{3x}{2} - 2\sin 2x \sin\frac{x}{2} - \sqrt{2}(2\cos^2 x)}$$

$$= \lim_{x\to\frac{\pi}{2}} \frac{4\sqrt{2}\sin 2x \cos x}{\sin 2x\left(\sin\frac{3x}{2} - \sin\frac{x}{2}\right) - \sqrt{2}\cos^2 x}$$

$$= \lim_{x\to\frac{\pi}{2}} \frac{4\sqrt{2}\sin 2x \cos x}{2\sin 2x \cos x \sin\frac{x}{2} - \sqrt{2}\cos^2 x}$$

$$= \lim_{x\to\frac{\pi}{2}} \frac{4\sqrt{2}\sin 2x}{2\sin 2x \sin\frac{x}{2} - \sqrt{2}\cos x}$$

$$= \lim_{x\to\frac{\pi}{2}} \frac{8\sqrt{2}\sin x}{4\sin x \sin\frac{x}{2} - \sqrt{2}}$$

$$= \frac{8\sqrt{2}}{\frac{4}{\sqrt{2}} - \sqrt{2}} = \frac{16}{4-2} = 8.$$

4. Let $y = (x + 1)^2 - 1, x \geq -1$

$\Rightarrow \quad (x + 1)^2 = y + 1$

$\Rightarrow \quad x + 1 = \sqrt{y + 1}$ as $x \geq -1$

$\Rightarrow \quad x = -1 + \sqrt{y + 1}, y \geq -1$

Thus, $\quad f^{-1}(x) = -1 + \sqrt{x + 1}$

So $\quad f(x) = f^{-1}(x)$

$\Rightarrow \quad (x + 1)^2 - 1 = -1 + \sqrt{x + 1}$

$\Rightarrow \quad \sqrt{x + 1} = 0$ or $(x + 1)^{3/2} = 1$

$\Rightarrow \quad x = -1$ or $x = 0$

5. $x^2 - 6x + 6 = x^2 - 4x + 4 \Rightarrow x = 1$

Put $x = 1$ in the given relation to get $f(1) = 1$

for $\quad x^2 - 6x + 6 = 1, x = 1$ or 5

for $\quad x^2 - 4x + 4 = 1, x = 1$ or 3

Put $x = 3$ and 5 in the given relation to get

$f(-3) + f(1) = 6 \Rightarrow f(-3) = 5$

and $\quad f(1) + f(9) = 10 \Rightarrow f(9) = 9$

$\therefore \quad f(-3) + f(9) - 5f(1) = 9$

6. We have, $A \cap B = \{2, 4\}$

$\Rightarrow \quad n(A \cap B) = 2$

$A \cup B = \{1, 2, 3, 4, 6\}$

$\Rightarrow \quad n(A \cup B) = 5$

as $\quad A \cap B \subseteq C \subseteq A \cup B$

$\therefore$ Set C contains the number of elements

$= 2^{5-2} = 2^3 = 8.$

7. Given, $f(T + x) = 1 + [(1 - f(x))^3]^{1/3} = 1 + (1 - f(x))$

$\Rightarrow \quad f(T + x) + f(x) = 2 \quad \ldots \text{(i)}$

$\Rightarrow \quad f(2T + x) + f(T + x) = 2 \quad \ldots \text{(ii)}$

$(2) - (1) \Rightarrow f(2T + x) - f(x) = 0$

$\Rightarrow \quad f(2T + x) = f(x)$

Also, T is positive and least therefore period of $f(x) = 2T$

$\therefore k = 2$

Matrix-Match Type

1. **(A)** $f(x) = \{x\}$ is a period function with period 1.

(B) $f(x) = \dfrac{16^x - 1}{4^x} = 4^x - 4^{-x}$ so $f(-x) = -f(x)$.

Thus, f in this case is an odd function.

(C) If $y = \log_4 \left(x + \sqrt{x^2 + 1}\right)$ then $x + \sqrt{x^2 + 1} = 4y$

and $4^{-y} = -x + \sqrt{x^2 + 1}$, when $x = \dfrac{1}{2}(4^y - 4^{-y})$

$\Rightarrow f^{-1}(x) = \dfrac{1}{2}(4^x - 4^{-x})$

(D) $f(-x) = (-x)\dfrac{3^{-x} - 1}{3^{-x} + 1}$

$= -x\dfrac{1 - 3^x}{1 + 3^x} = x\dfrac{3^x - 1}{3x + 1} = f(x)$

$\therefore f$ is an even function.

2.

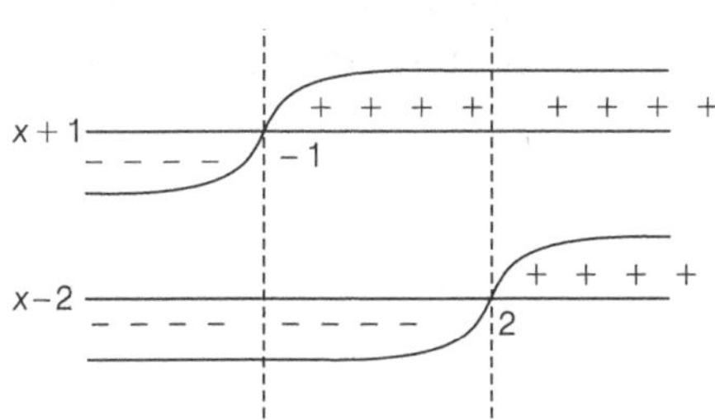

(A) $y = \begin{cases} -5x + 1, & x < -1 \\ x + 7, & -1 \le x < 2 \\ 5x - 1, & x \ge 2 \end{cases}$

(B) $y = \begin{cases} x + 7, & x < -1 \\ -5x + 1, & -1 \le x < 2 \\ -x - 7, & x \ge 2 \end{cases}$

(C) $y = \begin{cases} -x - 7, & x < -1 \\ 5x - 1, & -1 \le x < 2 \\ x + 7, & x \ge 2 \end{cases}$

3. **(A)** $\because \; -1 \le \cos 3x \le 1$

$\Rightarrow \quad 1 \ge -\cos 3x \ge -1$

$\Rightarrow \quad 2 + 1 \ge 2 - \cos 3x \ge 2 - 1$

$\Rightarrow \quad 3 \ge 2 - \cos 3x \ge 1$

or $\quad \dfrac{1}{3} \le \dfrac{1}{2 - \cos 3x} \le 1$ or $\dfrac{1}{3} \le f(x) \le 1$

$\therefore$ Range of $f(x)$ is $\left[\dfrac{1}{3}, 1\right] \subseteq [0, 1]$ (P, S)

(B) $\because \; f(x) = \sqrt{(x - 4)} + \sqrt{(6 - x)}$

$\Rightarrow f(x)$ is defined for $x - 4 \ge 0$ and $6 - x \ge 0$

i.e. $4 \le x \le 6$

Also, $f'(x) = \dfrac{1}{2\sqrt{(x - 4)}} - \dfrac{1}{2\sqrt{(6 - x)}} = 0$

or $\sqrt{(6 - x)} = \sqrt{(x - 4)}$

$\Rightarrow \quad 6 - x = x - 4$

$\therefore \quad x = 5$

$\therefore \quad f(4) = \sqrt{2}, \; f(5) = 2, \; f(6) = \sqrt{2}$

$\Rightarrow$ Range of $f(x)$ is $[\sqrt{2}, 2] \subseteq [1, 2]$ (Q, T)

(C) $\quad 0 \le 2x + 3 \le 1$

$\Rightarrow \quad -3 \le 2x \le -2$

or $\quad -\dfrac{3}{2} \le x \le -1 \;\Rightarrow\; \left[-\dfrac{3}{2}, -1\right]$ (R)

4. **(A)** Let $y = f(x) = \sin(\tan^{-1} x)$

$\therefore \quad x = f^{-1}(y) \quad \ldots \text{(i)}$

$\Rightarrow \sin^{-1} y = \tan^{-1} x$

$\Rightarrow \quad x = \tan(\sin^{-1} y) \quad \ldots \text{(ii)}$

From Eqs. (i) and (ii),

$f^{-1}(y) = \tan(\sin^{-1} y)$

$\therefore \quad f^{-1}(x) = \tan(\sin^{-1} x)$ (R)

$= \tan\left(\dfrac{\pi}{2} - \cos^{-1} x\right)$

$= \cot(\cos^{-1} x)$ (P)

$= \cot\left(\cot^{-1}\left(\dfrac{x}{\sqrt{1 - x^2}}\right)\right)$

$= \dfrac{x}{\sqrt{(1 - x^2)}}$ (T)

(B) Let $y = f(x) = 1 - 2^{-x}$

$\therefore \quad x = f^{-1}(y) \quad \ldots \text{(i)}$

and $\quad y = 1 - 2^{-x}$

$\Rightarrow \quad 2^{-x} = 1 - y$

$\Rightarrow \quad -x = \log_2 (1-y)$

$\therefore \quad x = -\log_2 (1-y) \quad \ldots(ii)$

From Eqs. (i) and (ii),

$f^{-1}(y) = -\log_2 (1-y)$

$\Rightarrow f^{-1}(x) = -\log_2 (1-x)$ (S)

(C) Let $y = f(x) = 2^{x/(x-1)}$

$\Rightarrow \quad x = f^{-1}(y) \quad \ldots(i)$

and $\quad \log_2 y = \dfrac{x}{x-1}$

$\Rightarrow x \log_2 y - \log_2 y = x$

$\therefore \quad x = \dfrac{\log_2 y}{(\log_2 y - 1)} \quad \ldots(ii)$

From Eqs. (i) and (ii),

$$f^{-1}(y) = \frac{\log_2 y}{\log_2 y - 1}$$

or $\quad f^{-1}(x) = \dfrac{\log_2 x}{\log_2 x - 1}$ (Q)

5. **(A)** $\because f(x) = \max\{1+\sin x, 1, 1-\cos x\}$

$$= \begin{cases} 1+\sin x, & 0 \le x \le \frac{3\pi}{4} \\ 1-\cos x, & \frac{3\pi}{4} \le x \le \frac{3\pi}{2} \\ 1, & \frac{3\pi}{2} \le x \le 2\pi \end{cases}$$

$$g(x) = \max\{1, |x-1|\} = \begin{cases} 1-x, & x \le 0 \\ 1, & 0 \le x \le 2 \\ x-1, & x \ge 2 \end{cases}$$

$\therefore \quad f(0) = 1 \Rightarrow g(f(0)) = g(1) = 1$

$\therefore \quad g(f(0)) = 1$ (S)

and $\quad f(1) = 1 + \sin 1 \quad \left(\because 0 < 1 < \dfrac{3\pi}{4}\right)$

$\therefore \quad g(f(1)) = g(1+\sin 1) = 1 \quad (\because 1 < 1 + \sin 1 < 2)$

$\therefore \quad g(f(1)) = 1$ (P)

(B) $\because f(g(x)) = \ln\left(\dfrac{1+g(x)}{1-g(x)}\right)$

$\therefore \quad f(g(0)) = \ln\left(\dfrac{1+g(0)}{1-g(0)}\right)$

$= \ln\left(\dfrac{1+0}{1-0}\right) = \ln 1 = 0$ (Q)

and $g\left(f\left(\dfrac{e-1}{e+1}\right)\right) = g\left(\ln\left(\dfrac{1+\frac{e-1}{e+1}}{1-\frac{e-1}{e+1}}\right)\right) = g(\ln(e)) = g(1)$

$= \dfrac{3+1}{1+3} = \dfrac{4}{4} = 1$ (T)

(C) $\quad f(g(0)) = f(0) = 1 + 0^2 = 1$ (R)

$gf(0) = g(1) = 1 - 1^2 = 0$

$g(f(1)) = g(2) = 2 - 2^2 = -2$

11

Limit

Objective Questions Type I [Only one correct answer]

In each of the questions below, four choices are given of which only one is correct. You have to select the correct answer which is the most appropriate.

1. If $f(x)=\dfrac{1}{3}\left(f(x+1)+\dfrac{5}{f(x+2)}\right)$ and $f(x)>0$ for all $x\in R$, then $\lim\limits_{x\to\infty} f(x)$ is

(a) 0 (b) $\sqrt{\dfrac{2}{5}}$

(c) $\sqrt{\dfrac{5}{2}}$ (d) ∞

2. The value of $\lim\limits_{x\to 0}\left(\left[\dfrac{100x}{\sin x}\right]+\left[\dfrac{99\sin x}{x}\right]\right)$, where [·] denotes the greatest integer function, is

(a) 197 (b) 198

(c) 199 (d) does not exist

3. $\lim\limits_{x\to 0}\left[\min(y^2-4y+11)\dfrac{\sin x}{x}\right]$ is equal to (where [.] denotes the greatest integer function)

(a) 5 (b) 6

(c) 7 (d) does not exist

4. $\lim\limits_{x\to 0}\dfrac{1-\cos x\cos 2x\cos 3x}{\sin^2 2x}$ is equal to

(a) 7/2 (b) 7/3

(c) 7/4 (d) 7/5

5. If $\lim\limits_{x\to\infty}(\sqrt{(x^2-x+1)}-ax-b)=0$, then for $k\geq 2$, $\lim\limits_{n\to\infty}\sec^{2n}(k!\pi b)$ is equal to

(a) $-a$ (b) a

(c) $-b$ (d) b

6. $\lim\limits_{y\to 0}\dfrac{\sqrt{1+\sqrt{1+y^4}}-\sqrt{2}}{y^4}$ **[JEE Main 2019]**

(a) exists and equals $\dfrac{1}{4\sqrt{2}}$

(b) does not exist

(c) exists and equals $\dfrac{1}{2\sqrt{2}}$

(d) exists and equals $\dfrac{1}{2\sqrt{2}(\sqrt{2}+1)}$

7. Let $f(x)=\dfrac{1-x(1+|1-x|)}{|1-x|}\cos\left(\dfrac{1}{1-x}\right)$ for $x\neq 1$, then

(a) $\lim\limits_{x\to 1^+} f(x)=0$

(b) $\lim\limits_{x\to 1^-} f(x)$ does not exist

(c) $\lim\limits_{x\to 1^-} f(x)=0$

(d) $\lim\limits_{x\to 1} f(x)$ does not exist

8. $\lim\limits_{n\to\infty}\dfrac{(1-2+3-4+5-6+\ldots-2n)}{\sqrt{(n^2+1)}+\sqrt{(4n^2-1)}}$ is equal to

(a) -1 (b) 1

(c) $-1/3$ (d) $1/3$

9. The value of $\lim\limits_{n\to\infty}\left(\dfrac{a-1+\sqrt[n]{b}}{a}\right)^n$; $a, b>0$ is equal to

(a) $\sqrt[a]{b}$ (b) $\sqrt[b]{a}$

(c) $\sqrt{b}$ (d) $\sqrt{a}$

10. $\lim\limits_{x\to\pi/2}\dfrac{\sin(x\cos x)}{\cos(x\sin x)}$ is equal to

(a) 1 (b) $\pi/2$

(c) $2/\pi$ (d) does not exist

11. $\lim\limits_{x\to\pi/2}\dfrac{\cot x-\cos x}{(\pi-2x)^3}$ equals **[JEE Main 2017]**

(a) $\dfrac{1}{24}$ (b) $\dfrac{1}{16}$

(c) $\dfrac{1}{8}$ (d) $\dfrac{1}{4}$

12. If $\{x\}$ and $[x]$ are the fractional part function and greatest integer functions of x respectively, then $\lim_{x \to [a]} \frac{e^{\{x\}} - \{x\} - 1}{\{x\}^2}$ is equal to

(a) 0 (b) 1/2
(c) $e - 2$ (d) does not exist

13. $k = \lim_{x \to \infty} \left(\frac{\sum_{k=1}^{1000} (x+k)^m}{x^m + 10^{1000}} \right)$ is $(m > 101)$

(a) 10 (b) 10^2 (c) 10^3 (d) 10^4

14. If $\sum_{r=1}^{k} \cos^{-1} \beta_r = \frac{k\pi}{2}$ for any $k \geq 1$ and $A = \sum_{r=1}^{k} (\beta_r)^r$, then $\lim_{x \to A} \frac{(1+x)^{1/3} - (1-2x)^{1/4}}{x + x^2}$ is equal to

(a) 0 (b) $\frac{1}{2}$
(c) $\frac{\pi}{2}$ (d) $\frac{5}{6}$

15. If α is the positive root of the equation, $p(x) = x^2 - x - 2 = 0$, then $\lim_{x \to \alpha^+} \frac{\sqrt{1 - \cos(p(x))}}{x + \alpha - 4}$ is equal to **[JEE Main 2020, 5 September]**

(a) $\frac{1}{\sqrt{2}}$ (b) $\frac{1}{2}$
(c) $\frac{3}{\sqrt{2}}$ (d) $\frac{3}{2}$

16. Let $f(x) = \lim_{n \to \infty} n\,(x^{1/n} - 1)$, $x > 0$, then $f(xy)$ is equal to

(a) $f(x) + f(y)$ (b) $f(x) \cdot f(y)$
(c) $x\,f(y) + y\,f(x)$ (d) $x\,f(y)$

17. *ABC* is an isosceles triangle inscribed in a circle of radius r. If $AB = AC$ and h is altitude from A to BC, then $\lim_{h \to 0} \frac{\Delta}{P^3}$ is equal to (where Δ is the area and P is the perimeter of the triangle *ABC*)

(a) $\frac{1}{32r}$ (b) $\frac{1}{64r}$
(c) $\frac{1}{128r}$ (d) $\frac{1}{216r}$

18. $\lim_{x \to \pi/4} (2 - \tan x)^{\frac{1}{\ln(\tan x)}}$ is equal to

(a) e (b) 1
(c) 0 (d) e^{-1}

19. $\lim_{x \to 0} \left(\frac{\ln(a+x) - \ln a}{x} \right) + k \lim_{x \to e} \frac{\ln x - 1}{x - e} = 1$ then

(a) $k = e\left(1 - \frac{1}{a}\right)$
(b) $k = e\,(1 + a)$
(c) $k = e\,(2 - a)$
(d) The equality is not possible

20. $\lim_{x \to a^-} \left(\frac{|x|^3}{a} - \left[\frac{x}{a}\right]^3 \right)$ $(a > 0)$, where $[x]$ denotes the greatest integer less than or equal to x, is

(a) $a^2 - 2$ (b) $a^2 - 1$
(c) a^2 (d) $a^2 + 1$

21. For each $t \in R$, let $[t]$ be the greatest integer less than or equal to t. Then, $\lim_{x \to 0^+} x\left(\left[\frac{1}{x}\right] + \left[\frac{2}{x}\right] + \ldots + \left[\frac{15}{x}\right]\right)$ **[JEE Main 2018]**

(a) is equal to 0 (b) is equal to 15
(c) is equal to 120 (d) does not exist (in R)

22. The value of $\lim_{x \to 0} \frac{\sqrt{1/2\,(1 - \cos 2x)}}{x}$ is equal to

(a) 1 (b) -1
(c) 0 (d) None of these

23. The value of $\lim_{x \to 0} \frac{\ln(1 + \{x\})}{\{x\}}$ is (where $\{x\}$ denotes the fractional part of x)

(a) 1 (b) 0
(c) 2 (d) does not exist

24. Let $f(x) = 1/\sqrt{(18 - x^2)}$, then value of $\lim_{x \to 3} \left(\frac{f(x) - f(3)}{x - 3} \right)$ is

(a) 0 (b) $-1/9$
(c) $-1/3$ (d) $1/9$

25. The value of $\lim_{x \to \infty} \left(\frac{x^2 \sin(1/x) - x}{1 - |x|} \right)$ is

(a) 0 (b) 1
(c) -1 (d) None of these

26. The value of the limit $\lim_{\theta \to 0} \frac{\tan(\pi \cos^2 \theta)}{\sin(2\pi \sin^2 \theta)}$ is equal to **[JEE Main 2021, 17 March]**

(a) $-\frac{1}{2}$ (b) $-\frac{1}{4}$
(c) 0 (d) $\frac{1}{4}$

27. If $x > 0$ and g is bounded function, then $\lim_{n \to \infty} \frac{f(x)\,e^{nx} + g(x)}{e^{nx} + 1}$ is

(a) $f(x)$ (b) $g(x)$
(c) 0 (d) None of these

28. The integer n for which $\lim_{x \to 0} \frac{(\cos x - 1)(\cos x - e^x)}{x^n}$ is a finite non-zero number is

(a) 1 (b) 2
(c) 3 (d) 4

29. $\lim_{x \to 0} \frac{x\sqrt{y^2 - (y - x)^2}}{\left(\sqrt{(8xy - 4x^2)} + \sqrt{(8xy)^3}\right)}$ is equal to

(a) 1/4 (b) 1/2
(c) $1/2\sqrt{2}$ (d) None of these

30. $\lim_{x \to 0} \frac{\sin(\pi \cos^2 x)}{x^2}$ is equal to

(a) $-\pi$ (b) π
(c) $\pi/2$ (d) 1

31. $\lim_{x \to \pi/2} \frac{\sin x - (\sin x)^{\sin x}}{1 - \sin x + \ln \sin x}$ is equal to

(a) 4 (b) 2
(c) 1 (d) None of these

32. $\lim_{x \to a} \frac{(a + 2x)^{1/3} - (3x)^{1/3}}{(3a + x)^{1/3} - (4x)^{1/3}}$ $(a \neq 0)$ is equal to

[JEE Main 2020, 3 September]

(a) $\left(\frac{2}{3}\right)^{4/3}$ (b) $\left(\frac{2}{9}\right)^{4/3}$
(c) $\left(\frac{2}{9}\right)\left(\frac{2}{3}\right)^{1/3}$ (d) $\left(\frac{2}{3}\right)\left(\frac{2}{9}\right)^{1/3}$

33. $\lim_{n \to \infty} \frac{n^{\alpha} \sin^2 n!}{n+1}$, $0 < \alpha < 1$, is equal to

(a) 0 (b) 1
(c) ∞ (d) None of these

34. $\lim_{x \to 0} \frac{\sin^{-1} x - \tan^{-1} x}{x^2}$ is equal to

(a) 1/2 (b) $-1/2$
(c) 0 (d) ∞

35. $\lim_{n \to \infty} \{\log_{n-1}(n) \log_n (n+1) \log_{n+1}(n+2) \ldots$

$\ldots \log_{n^k - 1}(n^x)\}$ is equal to

(a) n (b) k
(c) ∞ (d) None of these

36. For each $t \in R$, let $[t]$ be the greatest integer less than or equal to t. Then,

$$\lim_{x \to 1+} \frac{(1 - |x| + \sin|1 - x|) \sin\left(\frac{\pi}{2}[1 - x]\right)}{|1 - x|[1 - x]}$$

[JEE Main 2019, 10 January]

(a) equals 0 (b) does not exist
(c) equals -1 (d) equals 1

37. If $\lim_{x \to \infty} \left(1 + \frac{a}{x} + \frac{b}{x^2}\right)^{2x} = e^2$, then

(a) $a = 1, b = 2$ (b) $a = 2, b = 1$
(c) $a = 1, b \in R$ (d) none of these

38. $\lim_{n \to \infty} \left(\frac{1}{n} + \frac{e^{1/n}}{n} + \frac{e^{2/n}}{n} + \ldots + \frac{e^{(n-1)/n}}{n}\right)$ equals

(a) 0 (b) 1
(c) $e - 1$ (d) None of these

39. $\lim_{x \to \pi/4} \frac{2\sqrt{2} - (\cos x + \sin x)^3}{1 - \sin 2x}$ is equal to

(a) $\frac{3\sqrt{2}}{2}$ (b) $2\sqrt{2}$
(c) $\frac{4\sqrt{2}}{3}$ (d) does not exist

40. If $\lim_{x \to 0} \frac{((a - n)nx - \tan x) \sin nx}{x^2} = 0$, where n is non zero real number, then a is equal to

(a) 0 (b) $\frac{n+1}{n}$
(c) n (d) $n + \frac{1}{n}$

41. If $\lim_{x \to 0} \frac{(1 + a^3) + 8e^{1/x}}{1 + (1 - b^3) e^{1/x}} = 2$, then

(a) $a = 1,\ b = 2$ (b) $a = 1,\ b = -3^{1/3}$
(c) $a = 2,\ b = 3^{1/3}$ (d) None of these

42. If $\lim_{x \to \infty} (\sqrt{(x^4 - x^2 + 1)} - ax^2 - b) = 0$, then

(a) $a = 1, b = -2$ (b) $a = 1, b = 1$
(c) $a = 1, b = -1/2$ (d) None of these

43. If $S_n = \sum_{k=1}^{n} a_k$ and $\lim_{n \to \infty} a_n = a$, then

$\lim_{n \to \infty} \frac{S_{n+1} - S_n}{\sqrt{\sum_{k=1}^{n} k}}$ is equal to

(a) 0 (b) a
(c) $\sqrt{2}\, a$ (d) $2a$

44. The value of $\lim_{x \to 0} \frac{\cos(\sin x) - \cos x}{x^4}$ is equal to

(a) 1/5 (b) 1/6
(c) 1/4 (d) 1/2

45. Let $f(x) = \begin{cases} x^2 - 1, & \text{if } 0 < x < 2 \\ 2x + 3, & \text{if } 2 \le x < 3 \end{cases}$, the quadratic equation whose roots are $\lim_{x \to 2-0} f(x)$ and $\operatorname{Lim}_{x \to 2+0}$ is

(a) $x^2 - 6x + 9 = 0$ (b) $x^2 - 10x + 21 = 0$
(c) $x^2 - 14x + 49 = 0$ (d) None of these

46. $\lim_{x \to 1} \frac{x \sin\{x - [x]\}}{x - 1}$, where [.] denotes the greatest integer function, is

(a) 0 (b) -1
(c) not existent (d) None of these

47. The value of $\lim_{x \to 0} [x^2 + x + \sin x]$ is (where [.] denotes the greatest integer function)

(a) does not exist (b) is equal to zero
(c) -1 (d) None of these

48. $\lim_{x \to \pi/4} (2 - \tan x)^{1/\ln(\tan x)}$ equals

(a) e (b) 1
(c) 0 (d) e^{-1}

49. $\lim_{n \to \infty} \left(\frac{n!}{(mn)^n}\right)^{1/n}$ $(m \in N)$ is equal to

(a) $1/em$ (b) m/e
(c) em (d) e/m

50. If $y = 2^{-2^{1/(1-x)}}$, then $\lim_{x \to 1+} y$ is

(a) -1 (b) 1
(c) 0 (d) 1/2

51. The graph of the function $y = f(x)$ has a unique tangent at the point $(a, 0)$ through which the graph passes. Then $\lim_{x \to a} \frac{\log_e \{1 + 6f(x)\}}{3f(x)}$ is

(a) 0 (b) 1
(c) 2 (d) None of these

52. $\lim_{h \to 0} \frac{f(2h + 2 + h^2) - f(2)}{f(h - h^2 + 1) - f(1)}$, (given that $f'(2) = 6$ and $f'(1) = 4$) is equal to

(a) does not exist (b) is equal to $-3/2$
(c) is equal to 3/2 (d) is equal to 3

53. If $\lim_{x \to 0} \frac{x(1 + a\cos x) - b\sin x}{x^3} = 1$, then

(a) $a = -5/2,\ b = -1/2$ (b) $a = -3/2,\ b = -1/2$
(c) $a = -3/2,\ b = -5/2$ (d) $a = -5/2,\ b = -3/2$

54. $\lim_{x \to 0} \frac{x^n \sin^n x}{x^n - \sin^n x}$ is non zero finite, then n must be equal to

(a) 1 (b) 2
(c) 3 (d) None of these

55. The value of $\lim_{n \to \infty} \frac{[r] + [2r] + \ldots + [nr]}{n^2}$, where r is non-zero real number and $[r]$ denotes the greatest integer less than or equal to r, is equal to

[JEE Main 2021, 17 March]

(a) $\frac{r}{2}$ (b) r
(c) $2r$ (d) 0

56. If $[x]$ denotes the greatest integer less than or equal to x, then the value of $\lim_{x \to 1} (1 - x + [x - 1] + |1 - x|)$ is

(a) 0 (b) 1
(c) -1 (d) None of these

57. If $\lim_{x \to 1} \frac{x^2 - ax + b}{x - 1} = 5$, then $a + b$ is equal to

[JEE Mains 2019, 10 April]

(a) -4 (b) 1 (c) -7 (d) 5

58. If α and β be the roots of $ax^2 + bx + c = 0$, then $\lim_{x \to \alpha} (1 + ax^2 + bx + c)^{1/(x - \alpha)}$ is

(a) $a(\alpha - \beta)$ (b) $\ln|a(\alpha - \beta)|$
(c) $e^{a(\alpha - \beta)}$ (d) $e^{a|\alpha - \beta|}$

59. The value of $\lim_{x \to 0} \frac{x}{a}\left[\frac{b}{x}\right]$ $(a \neq 0)$ (where [·] denotes the greatest integer function) is

(a) a (b) b
(c) $\frac{b}{a}$ (d) $1 - \frac{b}{a}$

60. If $A_i = \frac{x - a_i}{|x - a_i|}$, $i = 1, 2, 3, \ldots, n$ and if $a_1 < a_2 < a_3 < \ldots < a_n$. Then $\lim_{x \to a_m} (A_1 A_2 \ldots A_n)$, $1 \le m \le n$

(a) is equal to $(-1)^m$ (b) is equal to $(-1)^{m+1}$
(c) is equal to $(-1)^{m-1}$ (d) does not exist

61. $\lim_{x \to 1+0} \frac{\int_1^x |t - 1|\, dt}{\sin(x - 1)}$ is equal to

(a) 0 (b) 1
(c) -1 (d) None of these

62. $\lim_{x \to \infty} \frac{\log_e [x]}{x}$, (where [.] denotes the greatest integer function) is equal to

(a) 0 (b) 1
(c) -1 (d) non-existent

63. If $[x]$ denotes the greatest integer $\le x$, then $\lim_{n \to \infty} \frac{1}{n^3}\{[1^2 x] + [2^2 x] + [3^2 x] + \ldots + [n^2 x]\}$ equals

(a) $x/2$ (b) $x/3$
(c) $x/6$ (d) 0

64. The value of the limit $\lim_{x \to 0} \{1^{1/\sin^2 x} + 2^{1/\sin^2 x} + \ldots + n^{1/\sin^2 x}\}^{\sin^2 x}$ is

(a) ∞ (b) 0
(c) $\frac{n(n+1)}{2}$ (d) n

65. $\lim_{x \to 0} \frac{x^n - \sin x^n}{x - \sin^n x}$ is non-zero finite, then n must be equal to

(a) 1 (b) 2
(c) 3 (d) None of these

66. $\lim_{x \to \infty} \frac{\log x^n - [x]}{[x]}$, $n \in N$, ($[x]$ denotes the greatest integer less than or equal to x)

(a) has value -1 (b) has value 0
(c) has value 1 (d) does not exist

67. $\lim_{x \to 0} \frac{\sin[\cos x]}{1 + [\cos x]}$ ([.] denotes the greatest integer function)

(a) equal to 1 (b) equal to 0
(c) does not exist (d) None of these

68. If α and β be the roots of $ax^2 + bx + c = 0$, then $\lim_{x \to \alpha} \frac{1 - \cos(ax^2 + bx + c)}{(x - \alpha)^2}$ is equal to

(a) 0 (b) $\frac{1}{2}(\alpha - \beta)^2$
(c) $\frac{a^2}{2}(\alpha - \beta)^2$ (d) $-\frac{a^2}{2}(\alpha - \beta)^2$

69. $\lim_{x\to 1^-} \frac{\sqrt{\pi} - \sqrt{2\sin^{-1} x}}{\sqrt{1-x}}$ is equal to **[JEE Main 2019, 12 January]**

(a) $\sqrt{\frac{\pi}{2}}$ (b) $\sqrt{\frac{2}{\pi}}$ (c) $\sqrt{\pi}$ (d) $\frac{1}{\sqrt{2\pi}}$

70. The value of $\lim_{x\to 0} \frac{(1+x)^{1/x} - e + \frac{1}{2}ex}{x^2}$ is

(a) $\frac{11}{24}e$ (b) $-\frac{11e}{24}$

(c) $\frac{e}{24}$ (d) None of these

71. $\lim_{x\to 0} \frac{\tan([-\pi^2]x^2) - \tan([-\pi^2])x^2}{\sin^2 x}$ equals (where [.] denotes the greatest integer function)

(a) 0 (b) 1

(c) tan 10 − 10 (d) ∞

72. Let $a = \min\{x^2 + 2x + 3, x \in R\}$ and $b = \lim_{\theta\to 0} \frac{1-\cos\theta}{\theta^2}$.

The value of $\sum_{r=0}^{n} a^r \cdot b^{n-r}$ is

(a) $\frac{2^{n+1}-1}{3\cdot 2^n}$ (b) $\frac{2^{n+1}+1}{3\cdot 2^n}$

(c) $\frac{4^{n+1}-1}{3\cdot 2^n}$ (d) None of these

73. If $\lim_{n\to\infty} [((2009)^{2010})^n + ((2010)^{2009})^n]^{1/n}$ is equal to a^b where $a, b \in N$, then $a - b$ is equal to

(a) 2009 (b) 1

(c) − 1 (d) 0

Objective Questions Type II [One or more than one correct answer(s)]

In each of the questions below four choices of which one or more than one are correct. You have to select the correct answer(s) accordingly.

1. Let $f(x) = \lim_{n\to\infty} \frac{x^{2n}-1}{x^{2n}+1}$, then

(a) $f(x) = 1$ for $|x| > 1$

(b) $f(x) = -1$ for $|x| < 1$

(c) $f(x)$ is not defined for any value of x

(d) $f(x) = 1$ for $|x| = 1$

2. For any positive integer n, define $f_n : (0, \infty) \to R$ as $f_n(x) = \sum_{j=1}^{n} \tan^{-1}\left(\frac{1}{1+(x+j)(x+j-1)}\right)$ for all $x \in (0, \infty)$. (Here, the inverse trigonometric function $\tan^{-1} x$ assumes values in $\left(-\frac{\pi}{2}, \frac{\pi}{2}\right)$). Then, which of the following statement(s) is (are) TRUE? **[JEE Advanced 2018]**

(a) $\sum_{j=1}^{5} \tan^2(f_j(0)) = 55$

(b) $\sum_{j=1}^{10} (1 + f'_j(0)) \sec^2(f_j(0)) = 10$

(c) For any fixed positive integer n, $\lim_{x\to\infty} \tan(f_n(x)) = \frac{1}{n}$

(d) For any fixed positive integer n, $\lim_{x\to\infty} \sec^2(f_n(x)) = 1$

3. If $\lim_{x\to 0} (\cos x + a\sin bx)^{1/x} = e^2$, then the values of a and b are

(a) $a = 1, b = 2$

(b) $a = 2, b = 1/2$

(c) $a = 2\sqrt{2}, b = \frac{1}{\sqrt{2}}$

(d) $a = 4, b = 2$

4. Let $f(x) = \lim_{n\to\infty} \left[\frac{n^n(x+n)\left(x+\frac{n}{2}\right)\dots\left(x+\frac{n}{n}\right)}{n!(x^2+n^2)\left(x^2+\frac{n^2}{4}\right)\dots\left(x^2+\frac{n^2}{n^2}\right)}\right]^{\frac{x}{n}}$, all $x = 0$. Then, **[JEE Advanced 2016]**

(a) $f\left(\frac{1}{2}\right) \ge f(1)$ (b) $f\left(\frac{1}{3}\right) \le f\left(\frac{2}{3}\right)$

(c) $f'(2) \le 0$ (d) $\frac{f'(3)}{f(3)} \ge \frac{f'(2)}{f(2)}$

5. If $a > 0, b < 0$, then $\lim_{x\to 0+} \frac{\sqrt{(1-\cos 2ax)}}{\sin bx}$ is equal to

(a) $\frac{a\sqrt{2}}{b}$ (b) $-\frac{a\sqrt{2}}{b}$

(c) $\frac{|a|\sqrt{2}}{|b|}$ (d) $-\frac{|a|\sqrt{2}}{|b|}$

6. If $\lim_{x\to 0} \left(\frac{a^x + b^x + c^x}{3}\right)^{\lambda/x}$, $(a, b, c, \lambda > 0)$ is equal to

(a) 1, if $\lambda = 1$ (b) abc, if $\lambda = 1$

(c) abc, if $\lambda = 3$ (d) $(abc)^{2/3}$, if $\lambda = 2$

7. The value of a for which

$$\lim_{x\to 0} \frac{(e^x - 1)^4}{\sin\left(\frac{x^2}{a^2}\right)\log_e\left\{1+\frac{x^2}{2}\right\}} = 8, \text{ is}$$

(a) −2 (b) −1

(c) 1 (d) 2

8. If $f(x) = \left(\frac{|x|}{2+|x|}\right)^{2x}$, then

(a) $\lim_{x \to \infty} f(x) = e^{-4}$ (b) $\lim_{x \to -\infty} f(x) = e^{4}$
(c) $\lim_{x \to \infty} f(x) = \infty$ (d) $\lim_{x \to -\infty} f(x) = 1$

9. Let $f(x) = \begin{cases} 1 + \frac{2x}{\lambda}, & 0 \le x < 1 \\ \lambda x & , 1 \le x < 2 \end{cases}$,

if $\lim_{x \to 1} f(x)$ exists, then λ is

(a) −2 (b) −1
(c) 1 (d) 2

10. If $\lim_{x \to 0}\left(2 + \frac{f(x)}{x^2}\right)^{1/x} = e^2$ and

$f(x) = a_0 + a_1 x + a_2 x^2 + a_3 x^3 + \ldots$, then

(a) $a_1 + |a_2| = 1$ (b) $|a_1| + a_2 = 1$
(c) $a_2 + |a_3| = 1$ (d) $|a_2| + a_3 = 3$

11. The possible value(s) of k for which

$$\lim_{x \to \infty} \frac{2x^3 - (\tan^{-1} x)^3}{\frac{8}{\pi} x^3 \cot^{-1}|kx| + k^2 x^6 \sin\frac{1}{x^3} - 3kx^3} = \frac{1}{2}, \text{ is}$$

(a) 0 (b) − 1
(c) 2 (d) 4

Linked-Comprehension Type

In these questions, a passage (paragraph) has been given followed by questions based on each of the passages. You have to answer the questions based on the passage given.

PASSAGE 1

Let $f(x) = \frac{\sin^{-1}(1-\{x\}) \cdot \cos^{-1}(1-\{x\})}{\sqrt{2\{x\}} \cdot (1-\{x\})}$, where $\{\cdot\}$ denotes the fractional part function.

On the basis of above information, answer the following questions

1. If $R = \lim_{x \to 0^+} f(x)$ then, the value of $\cos(100R)$ is

(a) − 1 (b) 0
(c) 1/2 (d) 1

2. If $L = \lim_{x \to 0^-} f(x)$, then the value of $\sin(99\sqrt{2}L)$ is

(a) − 1 (b) 0
(c) 1/2 (d) 1

3. The value of $[2R^2 + 4L^2]$ is [where $[\cdot]$ denotes the greatest integer function]

(a) 3 (b) 6
(c) 9 (d) 12

PASSAGE 2

Consider $P(x) = ax^2 + bx + c$, where $a, b, c \in R$ and $P(2) = 9$. Let α and β be the roots of the equation $P(x) = 0$

On the basis of above information, answer the following questions

1. If $\alpha \to \infty$ and $P^1(3) = 5$, then $\lim_{x \to \infty}\left(\frac{P(x)}{5(x-1)}\right)^x$ is equal to

(a) 1 (b) $e^{1/5}$
(c) $e^{4/5}$ (d) $e^{2/5}$

2. If α and β both tends to infinity, then $\lim_{x \to 3} \frac{\sqrt{P(x)} - 3}{\sin(x-3)}$ is equal to

(a) 0 (b) 1
(c) 9 (d) non-existent

3. If $\alpha = \beta$, then the value of

$$\lim_{\alpha \to 0}\left(\lim_{x \to \alpha} \frac{P(x)}{\left[1 - \tan^2\left(\frac{\pi}{4} - x + \alpha\right)\right](e^{x-\alpha} - 1)}\right) \text{ is equal to}$$

(a) $\frac{3}{16}$ (b) $\frac{9}{16}$ (c) $\frac{5}{4e^2}$ (d) $\frac{e^2}{4}$

PASSAGE 3

If $\lim_{x \to 0^+} f(x) =$ finite where $f(x) = \frac{\sin x + ae^x + be^{-x} + c\ln(1+x)}{x^3}$ and a, b, c are real numbers.

On the basis of above information, answer the following questions

1. The value of a is

(a) $-\frac{1}{2}$ (b) 0 (c) $\frac{1}{2}$ (d) 1

2. The value of b is

(a) $-\frac{1}{2}$ (b) $\frac{1}{2}$
(c) 0 (d) 1

3. The value of c is

(a) −1 / 2 (b) 1/2
(c) 0 (d) 2

4. The value of $\lim_{x \to 0^+} f(x)$ is

(a) − 2 (b) − 1
(c) $-\frac{1}{2}$ (d) $-\frac{1}{3}$

5. For the same values of a, b, c as obtain above, then the value of $\lim_{x \to 0^+} x^2 f(x)$ is

(a) 0 (b) 1
(c) 2 (d) 3

Numerical Value Type

Solve the following problems and write your answer.

1. The value of $\lim\limits_{x \to 1} 2008\, e\,(1 + \sin \pi x)^{\cot \pi x}$ must be

2. The value of $\lim\limits_{x \to 1} \dfrac{x + x^2 + x^3 + \ldots + x^{97} - 97}{(x - 1)}$ must be

3. The value of $\lim\limits_{x \to 1} 512\pi\,(1 - x)\,\tan\left(\dfrac{\pi x}{2}\right)$ must be

4. If $\lim\limits_{x \to 0} \dfrac{1}{x^2}(e^{\alpha x} - e^x - x) = \dfrac{3}{2}$, then the value of $1004|\alpha|$ must be

5. The value of $\lim\limits_{x \to 2} 52.9\, \dfrac{\sqrt{(x+7)} - 3\sqrt{(2x-3)}}{\sqrt[3]{(x+6)} - 2\sqrt[3]{(3x-5)}}$ must be

6. The value of $\lim\limits_{x \to 0} 27e^{-2}\left\{\tan\left(\dfrac{\pi}{4} + x\right)\right\}^{1/x}$ must be

7. Let $\sin^{-1}(1 - \{x\}) f(x) = \dfrac{\cdot \cos^{-1}(1 - \{x\})}{\sqrt{2\{x\}} \cdot (1 - \{x\})}$, then the value of $\dfrac{2008\sqrt{2}}{\pi} \lim\limits_{x \to 0^-} f(x)$ must be (where $\{x\}$ denotes the fractional part of x).

8. The value of $78.6\sqrt{e} \lim\limits_{x \to 0} \left\{\dfrac{(1 + x)^{1/x}}{e}\right\}^{\frac{1}{x}}$ must be

Single Integer Type

Solve the following problems and write your answer.

1. If $\lim\limits_{x \to 0} \left\{\dfrac{1}{x^8}\left(1 - \cos\dfrac{x^2}{2} - \cos\dfrac{x^2}{4} + \cos\dfrac{x^2}{2}\cos\dfrac{x^2}{4}\right)\right\} = 2^{-k}$, then the value of k is .... **[JEE Main 2020, 3 September]**

2. Let the sequence $< b_n >$ of real numbers satisfy the recurrence relation $b_{n+1} = \dfrac{1}{3}\left(2b_n + \dfrac{125}{b_n^2}\right)$, $b_n \neq 0$, then $\lim\limits_{n \to \infty} b_n$ is equal to

3. Let m and n be two positive integers greater than 1. If $\lim\limits_{\alpha \to 0}\left(\dfrac{e^{\cos(\alpha^n)} - e}{\alpha^m}\right) = -\left(\dfrac{e}{2}\right)$, then the value of $\dfrac{m}{n}$ is **[JEE Advanced 2015]**

4. If $\lim\limits_{x \to 0}\left(\left[\dfrac{\sin^{-1} x}{x}\right] + \left[\dfrac{2^2 \sin^{-1} 2x}{x}\right] + \left[\dfrac{3^2 \sin^{-1} 3x}{x}\right] + \ldots + \left[\dfrac{n^2 \sin^{-1} nx}{x}\right]\right) = 100$, then the value of n, is

5. The largest value of the non-negative integer a for which $\lim\limits_{x \to 1}\left\{\dfrac{-ax + \sin(x-1) + a}{x + \sin(x-1) - 1}\right\}^{\frac{1-x}{1-\sqrt{x}}} = \dfrac{1}{4}$ is **[JEE Advanced 2014]**

6. Let $\alpha, \beta \in R$ be such that $\lim\limits_{x \to 0} \dfrac{x^2 \sin(\beta x)}{\alpha x - \sin x} = 1$. Then, $6(\alpha + \beta)$ equals **[JEE Advanced 2016]**

7. For each positive integer n, let

$$y_n = \frac{1}{n}((n+1)(n+2)\ldots(n+n))^{\frac{1}{n}}.$$

For $n \in R$, let $[x]$ be the greatest integer less than or equal to x. If $\lim\limits_{n \to \infty} y_n = L$, then the value of $[L]$ is .... **[JEE Advanced 2018]**

8. If $\lim\limits_{x \to 0} \dfrac{ae^x - b\cos x + ce^{-x}}{x \sin x} = 2$, then $a + b + c$ is equal to ....

Matrix-Match Type

Given below are Matching Type Questions, with two columns (each having some items) each . Each item of Column I has to be matched with the items of Column II, by encircling the correct match(es).

NOTE *An item of Column I can be matched with more than one items of Column II.*

1. Observe the following Columns :

	Column I		Column II
A.	If $f(x) = \left(\dfrac{\lvert x\rvert}{\lvert x\rvert + 2}\right)^{-x}$, then	P.	$\lim\limits_{x \to \infty} f(x) = e^2$
B.	If $f(x) = \dfrac{(1 + x)^{1/x} - e}{x}$, then	Q.	$\lim\limits_{x \to 0} f(x) = e^2$
C.	If $f(x) = \left(\dfrac{1 + 5x^2}{1 + 3x^2}\right)^{1/x^2}$, then	R.	$\lim\limits_{x \to -\infty} f(x) = e^{-2}$
		S.	$\lim\limits_{x \to 0} f(x) = -e/2$
		T.	$\lim\limits_{x \to 0} f(x) < -1$

2. Observe the following columns :

	Column I		Column II
A.	If $\lim\limits_{x \to 0} (\cos x + a \sin bx)^{1/x} = e^2$, then	P.	$ab = 2$
B.	If $\lim\limits_{x \to 0} (1 + ax + bx^2)^{2/x} = e^3$, then	Q.	$ab = 3$
C.	If $\lim\limits_{x \to 0} \dfrac{ae^x - b}{x} = 2$, then	R.	$ab = 4$
		S.	$a + b = 3$
		T.	$a + b = 4$

3. $L = \lim_{x \to 0} \left[\frac{n_1 \sin x}{x} \right], M = \lim_{x \to 0} \left[\frac{n_2 x}{\sin x} \right] N = \lim_{x \to 0} \left[\frac{n_3 \tan x}{x} \right]$ and $P = \lim_{x \to 0} \left[\frac{n_4 x}{\tan x} \right]$

where [·] denotes the greatest integer function and $n_1, n_2, n_3, n_4 \in N$.
Observe the following columns :

	Column I		Column II
A.	For $n_1 = 3$, $n_4 = 9$, then P/L is divisible by	P.	2
B.	For $n_2 = 12, n_3 = 2$, then M/N is divisible by	Q.	3
C.	For $n_2 = 3, n_4 = 10$, then $P + M$ is divisible by	R.	4
D.	For $n_1 = 4, n_2 = 5, n_3 = 6, n_4 = 7$, then $L + N + P - M$ is divisible by	S.	5
		T.	6

Answers

Objective Questions Type [Only one correct answer]

1. (c)	**2.** (b)	**3.** (b)	**4.** (c)	**5.** (b)	**6.** (a)	**7.** (d)	**8.** (c)	**9.** (a)	**10.** (b)
11. (b)	**12.** (d)	**13.** (c)	**14.** (d)	**15.** (c)	**16.** (a)	**17.** (c)	**18.** (d)	**19.** (a)	**20.** (c)
21. (c)	**22.** (d)	**23.** (d)	**24.** (d)	**25.** (a)	**26.** (a)	**27.** (a)	**28.** (c)	**29.** (d)	**30.** (b)
31. (b)	**32.** (d)	**33.** (a)	**34.** (c)	**35.** (b)	**36.** (a)	**37.** (c)	**38.** (c)	**39.** (a)	**40.** (d)
41. (b)	**42.** (c)	**43.** (a)	**44.** (b)	**45.** (b)	**46.** (c)	**47.** (b)	**48.** (d)	**49.** (a)	**50.** (b)
51. (c)	**52.** (d)	**53.** (d)	**54.** (b)	**55.** (a)	**56.** (d)	**57.** (c)	**58.** (c)	**59.** (c)	**60.** (d)
61. (a)	**62.** (a)	**63.** (b)	**64.** (d)	**65.** (a)	**66.** (a)	**67.** (b)	**68.** (c)	**69.** (b)	**70.** (a)
71. (c)	**72.** (c)	**73.** (c)							

Objective Questions Type II [One or more than one correct answer(s)]

1. (a, b, c)	**2.** (d)	**3.** (a, c)	**4.** (b, c)	**5.** (a, d)
6. (c, d)	**7.** (a, d)	**8.** (a, b)	**9.** (b, d)	**10.** (a, c, d)
11. (a, b, d)				

Linked-Comprehension Type

Passage 1 **1.** (d) **2.** (a) **3.** (c)
Passage 2 **1.** (c) **2.** (a) **3.** (b)
Passage 3 **1.** (a) **2.** (b) **3.** (c) **4.** (d) **5.** (a)

Numerical Value Type

1. 2008 **2.** 4753 **3.** 1024 **4.** 2008 **5.** 78.2 **6.** 27 **7.** 1004 **8.** 78.6

Single Integer Type

1. (8) **2.** (5) **3.** (2) **4.** (4) **5.** (2) **6.** (7) **7.** (1) **8.** (4)

Matrix-Match Type

1. A → (P, R); B → (S, T); C → (Q)
2. A → (P, S); B → (P, Q, R, S, T); C → (R, T)
3. A → (P, R); B → (P, Q, T); C → (P, Q, R, T); D → (P, S)

Solutions

Objective Questions Type [Only one correct answer]

1. $\because \lim_{x\to\infty} f(x) = \lim_{x\to\infty} f(x+1)$

$$= \lim_{x\to\infty} f(x+2) = \lambda \text{ (say)}$$

then $\lambda = \frac{1}{3}\left(\lambda + \frac{5}{\lambda}\right)$

$\Rightarrow \quad 2\lambda = \frac{5}{\lambda}$

$\Rightarrow \quad \lambda = \sqrt{\frac{5}{2}} \qquad [\because f(x) > 0]$

$\therefore \lim_{x\to\infty} f(x) = \sqrt{\frac{5}{2}}$

2. For $x > 0$

$$\sin x < x$$

$\Rightarrow \quad \frac{\sin x}{x} < 1$

$\therefore \quad \frac{99 \sin x}{x} < 99$

$\Rightarrow \quad \left[\frac{99 \sin x}{x}\right] = 98$

and $\quad \frac{x}{\sin x} > 1$

$\therefore \quad \frac{100\, x}{\sin x} > 100$

$\Rightarrow \quad \left[\frac{100\, x}{\sin x}\right] = 100$

$\therefore \quad \lim_{x\to 0+}\left(\left[\frac{100\, x}{\sin x}\right] + \left[\frac{99 \sin x}{x}\right]\right) = 100 + 98 = 198$

For $x < 0$

$$\sin x > x$$

$\Rightarrow \quad \frac{\sin x}{x} < 1$

$\therefore \quad \frac{99 \sin x}{x} < 99$

$\Rightarrow \quad \left[\frac{99 \sin x}{x}\right] = 98$

and $\quad \frac{x}{\sin x} > 1$

$\therefore \quad \frac{100\, x}{\sin x} > 100$

$\Rightarrow \quad \left[\frac{100\, x}{\sin x}\right] - 100$

$\lim_{x\to 0-}\left(\left[\frac{100\, x}{\sin x}\right] + \left[\frac{99 \sin x}{x}\right]\right) = 100 + 98 = 198$

Hence, $\lim_{x\to 0}\left(\left[\frac{100\, x}{\sin x}\right] + \left[\frac{99 \sin x}{x}\right]\right) = 198$

3. $\because y^2 - 4y + 11 = (y-2)^2 + 7$

$\therefore \quad \min (y^2 - 4y + 11) = 7 \qquad [\text{at } y = 2]$

$\therefore \quad \lim_{x\to 0}\left[\min (y^2 - 4y + 11)\frac{\sin x}{x}\right]$

$$= \lim_{x\to 0}\left[\frac{7 \sin x}{x}\right]$$

Now, $\quad$ let $f(x) = \left[\frac{7 \sin x}{x}\right]$

For $\quad x > 0$

$$\sin x < x$$

$\therefore \quad \frac{\sin x}{x} < 1$

$\Rightarrow \quad \frac{7 \sin x}{x} < 7$

$\Rightarrow \quad \left[\frac{7 \sin x}{x}\right] = 6$

$\therefore \quad \text{RHL} = \lim_{x\to 0+} f(x) = 6$

For $\quad x < 0$

$$\sin x > x$$

$$\frac{\sin x}{x} < 1$$

$\therefore \quad \frac{7 \sin x}{x} < 7$

$\Rightarrow \quad \left[\frac{7 \sin x}{x}\right] = 6$

$\therefore \quad \text{LHL} = \lim_{x\to 0-} f(x) = 6$

Hence,

$$\lim_{x\to 0}\left[\min (y^2 - 4y + 11)\frac{\sin x}{x}\right] - 6$$

4. $\because \cos x \cos 2x \cos 3x$

$$= \frac{1}{2}\{(2\cos x \cos 2x)\cos 3x\}$$

$$= \frac{1}{2}\{(\cos 3x + \cos x)\cos 3x\}$$

$$= \frac{1}{4}\{2\cos^2 3x + 2\cos 3x \cos x\}$$

$$= \frac{1}{4}\{1 + \cos 6x + \cos 4x + \cos 2x\}$$

$\therefore \; 1 - \cos x \cos 2x \cos 3x$

$$= 1 - \frac{1}{4}\{1 + \cos 6x + \cos 4x + \cos 2x\}$$

$$= \frac{1}{4}\{4 - 1 - \cos 6x - \cos 4x - \cos 2x\}$$

$$= \frac{1}{4}\{(1-\cos 6x) + (1 - \cos 4x) + (1 - \cos 2x)\}$$

$$= \frac{1}{4}\{2\sin^2 3x + 2\sin^2 2x + 2\sin^2 x\}$$

$$= \frac{1}{2}(\sin^2 3x + \sin^2 2x + \sin^2 x)$$

$$\therefore \lim_{x\to 0} \frac{1-\cos x \cos 2x \cos 3x}{\sin^2 2x}$$

$$= \frac{1}{2} \lim_{x\to 0} \left(\frac{\sin^2 3x + \sin^2 2x + \sin^2 x}{\sin^2 2x} \right)$$

$$= \frac{1}{2} \lim_{x\to 0} \left\{ \frac{9 \cdot \frac{\sin^2 3x}{9x^2} + 4 \cdot \frac{\sin^2 2x}{4x^2} + \frac{\sin^2 x}{x^2}}{4 \cdot \frac{\sin^2 2x}{4x^2}} \right\}$$

$$= \frac{1}{2} \left\{ \frac{9 \cdot 1 + 4 \cdot 1 + 1 \cdot 1}{4} \right\}$$

$$= \frac{7}{4}$$

Alternative Best Method

$\because \; 1 - \cos x \cos 2x \cos 3x$

$= \sin^2 x + \cos^2 x - \cos x \cos 2x \cos 3x$

$= \sin^2 x + \cos x \{\cos x - \cos 2x \cos 3x\}$

$= \sin^2 x + \cos x \{\cos (3x - 2x) - \cos 2x \cos 3x\}$

$= \sin^2 x + \cos x \{\sin 3x \sin 2x\}$

$$\therefore \lim_{x\to 0} \frac{1-\cos x \cos 2x \cos 3x}{\sin^2 2x}$$

$$= \lim_{x\to 0} \frac{\sin^2 x + \cos x (\sin 3x \sin 2x)}{\sin^2 2x}$$

$$= \lim_{x\to 0} \frac{\frac{\sin^2 x}{x^2} + \cos x \cdot \frac{\sin 3x}{3x} \cdot \frac{\sin 2x}{2x} \cdot 6}{4 \cdot \frac{\sin^2 2x}{4x^2}}$$

$$= \frac{(1)^2 + 1 \cdot 1 \cdot 1 \cdot 6}{4 \cdot 1^2}$$

$$= \frac{7}{4}$$

5. Here, $a > 0$, if $a \le 0$, then limit $= \infty$

$$\therefore \lim_{x\to\infty} \frac{(\sqrt{(x^2 - x + 1)} - ax - b)(\sqrt{(x^2 - x + 1)} + ax + b)}{(\sqrt{(x^2 - x + 1)} + ax + b)} = 0$$

$$\Rightarrow \lim_{x\to\infty} \frac{(x^2 - x + 1) - (ax + b)^2}{\sqrt{(x^2 - x + 1)} + ax + b} = 0$$

$$\Rightarrow \lim_{x\to\infty} \frac{(1 - a^2)x^2 - (1 + 2ab)x + (1 - b^2)}{\sqrt{(x^2 - x + 1)} + ax + b} = 0$$

This is possible only when $1 - a^2 = 0$ and $1 + 2ab = 0$

$\therefore \quad a = \pm 1$

$\Rightarrow \quad a = 1 \quad [\because a > 0] \; \ldots \text{(i)}$

and $\quad b = -\dfrac{1}{2a} = -\dfrac{1}{2}$

Now, $k!\,\pi b = -\dfrac{\pi\, k!}{2}$

$=$ an integral multiple of $\pi \quad [\because k \ge 2]$

$\therefore \quad \sec^2 (k!\,\pi\, b) = (\pm 1)^2 = 1$

$\therefore \quad \lim_{n\to\infty} \sec^{2n} (k!\,\pi\, b) = 1^\infty = 1 = a \quad$ [from Eq. (i)]

6. Clearly, $\displaystyle \lim_{y\to 0} \frac{\sqrt{1 + \sqrt{1 + y^4}} - \sqrt{2}}{y^4}$

$$= \lim_{y\to 0} \frac{\sqrt{1 + \sqrt{1 + y^4}} - \sqrt{2}}{y^4} \times \frac{\sqrt{1 + \sqrt{1 + y^4}} + \sqrt{2}}{\sqrt{1 + \sqrt{1 + y^4}} + \sqrt{2}}$$

[rationalising the numerator]

$$= \lim_{y\to 0} \frac{(1 + \sqrt{1 + y^4}) - 2}{y^4(\sqrt{1 + \sqrt{1 + y^4}} + \sqrt{2})} \quad [\because (a + b)(a - b) = a^2 - b^2]$$

$$= \lim_{y\to 0} \frac{\sqrt{1 + y^4} - 1}{y^4(\sqrt{1 + \sqrt{1 + y^4}} + \sqrt{2})} \times \frac{\sqrt{1 + y^4} + 1}{\sqrt{1 + y^4} + 1}$$

[again, rationalising the numerator]

$$= \lim_{y\to 0} \frac{y^4}{y^4(\sqrt{1 + \sqrt{1 + y^4}} + \sqrt{2})(\sqrt{1 + y^4} + 1)}$$

$$= \frac{1}{2\sqrt{2} \times 2}$$

[by cancelling y^4 and then by direct substitution]

$$= \frac{1}{4\sqrt{2}}$$

7. $f(x) = \dfrac{1 - x(1 + |1 - x|)}{|1 - x|} \cos\left(\dfrac{1}{1 - x}\right)$

Now, $\displaystyle \lim_{x\to 1^-} f(x) = \lim_{x\to 1^-} \frac{1 - x(1 + 1 - x)}{1 - x} \cos\left(\frac{1}{1 - x}\right)$

$$= \lim_{x\to 1^+} (1 - x) \cos\left(\frac{1}{1 - x}\right) = 0$$

and $\displaystyle \lim_{x\to 1^+} f(x) = \lim_{x\to 1^+} \frac{1 - x(1 - 1 + x)}{x - 1} \cos\left(\frac{1}{1 - x}\right)$

$$= \lim_{x\to 1^+} -(x + 1) \cdot \cos\left(\frac{1}{1 - x}\right),$$

which does not exist.

8. $\because 1 - 2 + 3 - 4 + 5 - 6 + \ldots - 2n$

$= (-1) + (-1) + (-1) + \ldots n$ times

$= -n$

$$\therefore \lim_{n\to\infty} \frac{(1 - 2 + 3 - 4 + 5 - 6 + \ldots - 2n)}{\sqrt{(n^2 + 1)} + \sqrt{(4n^2 - 1)}}$$

$$= \lim_{n\to\infty} \frac{(-n)}{\sqrt{(n^2 + 1)} + \sqrt{(4n^2 - 1)}}$$

$$= \lim_{n\to\infty} \frac{-1}{\sqrt{\left(1 + \frac{1}{n^2}\right)} + \sqrt{\left(4 - \frac{1}{n^2}\right)}}$$

$$= \frac{-1}{\sqrt{1 + 0} + \sqrt{4 - 0}}$$

$$= \frac{-1}{1 + 2}$$

$$= -\frac{1}{3}$$

9. We have, $\lim_{n\to\infty}\left(\dfrac{a-1+\sqrt[n]{b}}{a}\right)^n$ [1^∞ form]

$$= \lim_{n\to\infty}\left(1+\left(\frac{\sqrt[n]{b}-1}{a}\right)\right)^n = e^{\lim_{n\to\infty} n\left(\frac{\sqrt[n]{b}-1}{a}\right)}$$

$$= e^{\frac{1}{a}\lim_{n\to\infty}\frac{b^{1/n}-1}{1/n}} = e^{\frac{1}{a}\ln b} = e^{\ln(b)^{1/a}} = b^{1/a} = \sqrt[a]{b}$$

10. $\lim_{x\to\pi/2}\dfrac{\sin(x\cos x)}{\sin\left(\dfrac{\pi}{2}-x\sin x\right)}$

$$= \lim_{x\to\frac{\pi}{2}}\frac{\sin(x\cos x)}{(x\cos x)}\cdot\frac{x\cos x}{\sin\left(\frac{\pi}{2}-x\sin x\right)}\cdot\frac{\left(\frac{\pi}{2}-x\sin x\right)}{\left(\frac{\pi}{2}-x\sin x\right)}$$

$$= 1\cdot 1\cdot\lim_{x\to\pi/2}\frac{x\cos x}{\left(\frac{\pi}{2}-x\sin x\right)}$$

Put $x = \pi/2 + h$

Then, $\lim_{h\to 0}\dfrac{\left(\dfrac{\pi}{2}+h\right)\cos\left(\dfrac{\pi}{2}+h\right)}{\dfrac{\pi}{2}-\left(\dfrac{\pi}{2}+h\right)\sin\left(\dfrac{\pi}{2}+h\right)}$

$$= \lim_{h\to 0}\frac{-\left(\frac{\pi}{2}+h\right)\sin h}{\frac{\pi}{2}(1-\cos h)-h\cos h}$$

$$= \lim_{h\to 0}\frac{-\left(\frac{\pi}{2}+h\right)\left(\frac{\sin h}{h}\right)}{\frac{\pi}{2}\frac{(1-\cos h)}{h}-\cos h}$$

[divide above and below by h]

$$= \frac{-\left(\frac{\pi}{2}+0\right)\cdot 1}{0-1} = \frac{\pi}{2}$$

11. $\lim_{x\to\pi/2}\dfrac{\cot x-\cos x}{(\pi-2x)^3}$

$$= \lim_{x\to\pi/2}\frac{1}{8}\cdot\frac{\cos x(1-\sin x)}{\sin x\left(\frac{\pi}{2}-x\right)^3}$$

$$= \lim_{h\to 0}\frac{1}{8}\cdot\frac{\cos\left(\frac{\pi}{2}-h\right)\left[1-\sin\left(\frac{\pi}{2}-h\right)\right]}{\sin\left(\frac{\pi}{2}-h\right)\left(\frac{\pi}{2}-\frac{\pi}{2}+h\right)^3}$$

$$= \frac{1}{8}\lim_{h\to 0}\frac{\sin h(1-\cos h)}{\cos h\cdot h^3}$$

$$= \frac{1}{8}\lim_{h\to 0}\frac{\sin h\left(2\sin^2\frac{h}{2}\right)}{\cos h\cdot h^3} = \frac{1}{4}\lim_{h\to 0}\frac{\sin h\,\sin^2\left(\frac{h}{2}\right)}{h^3\cos h}$$

$$= \frac{1}{4}\lim_{h\to 0}\left(\frac{\sin h}{h}\right)\left(\frac{\sin h}{\frac{h}{2}}\right)^2\cdot\frac{1}{\cos h}\cdot\frac{1}{4} = \frac{1}{4}\times\frac{1}{4} = \frac{1}{16}$$

12. $\text{LHL} = \lim_{x\to[a]^-}\dfrac{e^{\{x\}}-\{x\}-1}{\{x\}^2}$

$$= \lim_{h\to 0}\frac{e^{\{[a]-h\}}-\{[a]-h\}-1}{\{[a]-h\}^2}$$

$$= \lim_{h\to 0}\frac{e^{1-h}-(1-h)-1}{(1-h)^2} = e-2$$

$$\text{RHL} = \lim_{x\to[a]^+}\frac{e^{\{x\}}-\{x\}-1}{\{x\}^2}$$

$$= \lim_{h\to 0}\frac{e^{\{[a]+h\}}-\{[a]+h\}-1}{\{[a]+h\}^2}$$

$$= \lim_{h\to 0}\frac{e^h-h-1}{h^2} = P \text{ (say)} \quad \ldots\text{(i)}$$

Replacing h by $-h$ in P, then

$$P = \lim_{h\to 0}\frac{e^{-h}+h-1}{h^2} \quad \ldots\text{(ii)}$$

Adding Eqs. (i) and (ii), then

$$2P = \lim_{h\to 0}\frac{e^h+e^{-h}-2}{h^2}$$

$$= \lim_{h\to 0}\frac{(e^h-1)^2}{e^h\cdot h^2}$$

$$= \frac{(1)^2}{1} = 1$$

$$P = \frac{1}{2}$$

$\therefore$ $\text{RHL} = \dfrac{1}{2}$ [from Eq. (i)]

Hence, LHL $\neq$ RHL

13. Divide above and below by x^m, then

$$k = \lim_{x\to\infty}\left(\frac{\sum_{k=1}^{1000}\left(1+\frac{k}{x}\right)^m}{1+\frac{10^{1000}}{x^m}}\right)$$

$$= \frac{1+1+1+\ldots \text{ upto 1000 times}}{1+0}$$

$$= 1000 = 10^3$$

14. $\because \sum_{r=1}^{k}\cos^{-1}\beta_r = \dfrac{k\pi}{2}$

Which is possible only when

$$\cos^{-1}\beta_1 = \cos^{-1}\beta_2 = \cos^{-1}\beta_3 = \ldots = \cos^{-1}\beta_k = \frac{\pi}{2}$$

$\therefore$ $\beta_1 = \beta_2 = \beta_3 = \ldots = \beta_k = 0$

Hence, $A = \sum_{r=1}^{k}(\beta_r)^r = 0$

$$\therefore \lim_{x\to 0}\frac{(1+x)^{1/3}-(1-2x)^{1/4}}{x+x^2}$$

$$\Rightarrow \lim_{x\to 0}\frac{\left(\frac{(1+x)^{1/3}-(1)^{1/3}}{(1+x)-1}\right)-\left(\frac{(1-2x)^{1/4}-(1)^{1/4}}{(1-2x)-1}\times(-2)\right)}{(1+x)}$$

$$\Rightarrow \frac{\frac{1}{3}+2\times\frac{1}{4}}{1+0} = \frac{1}{3}+\frac{1}{2} = \frac{5}{6}$$

15. Given equation, $p(x) = x^2 - x - 2 = (x-2)(x+1)$ having a positive root α, so $\alpha = 2$.

Now, $\lim\limits_{x \to \alpha^+} \dfrac{\sqrt{1-\cos(p(x))}}{x+\alpha-4} = \lim\limits_{x \to 2^+} \dfrac{\left|\sqrt{2}\sin\left(\dfrac{p(x)}{2}\right)\right|}{x-2}$

$= \lim\limits_{x \to 2^+} \left[\sqrt{2}\, \dfrac{\sin\left(\dfrac{(x-2)(x+1)}{2}\right)}{x-2}\right]$ $[\because \sin\theta > 0, \text{ if } \theta \to 0^+]$

$= \lim\limits_{x \to 2^+} \left[\sqrt{2}\, \dfrac{\sin\left(\dfrac{(x-2)(x+1)}{2}\right)}{\dfrac{(x-2)(x+1)}{2}} \times \left(\dfrac{x+1}{2}\right)\right]$

$= \sqrt{2}\left(\dfrac{3}{2}\right) = \dfrac{3}{\sqrt{2}}$

16. $f(xy) = \lim\limits_{n \to \infty} n[(xy)^{1/n} - 1]$

$= \lim\limits_{n \to \infty} n\{y^{1/n}(x^{1/n}-1) + (y^{1/n}-1)\}$

$= \lim\limits_{n \to \infty} y^{1/n} \lim\limits_{n \to \infty} n(x^{1/n}-1) + \lim\limits_{n \to \infty} n(y^{1/n}-1)$

$= y^0 \cdot f(x) + f(y)$

$= f(x) + f(y)$

17. $\because$ $AD = h$

$\therefore$ $OD = (h-r)$

Then $BC = 2DC = 2\sqrt{r^2 - (h-r)^2}$

$\therefore$ $BC = 2\sqrt{(2hr - h^2)}$

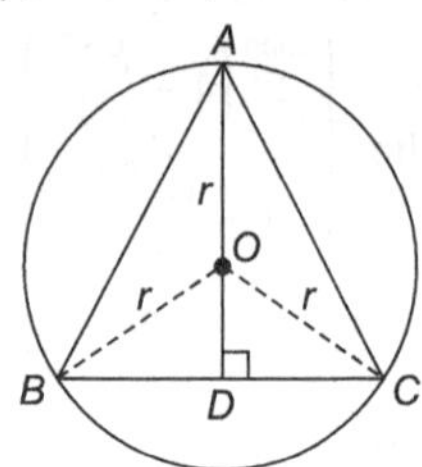

$\Delta = \dfrac{1}{2} \cdot BC \cdot h = h\sqrt{(2hr - h^2)}$

$= h^{3/2}\sqrt{(2r-h)}$

and $AB = AC = \sqrt{h^2 + (DC)^2}$

$= \sqrt{h^2 + r^2 - (h-r)^2}$

$= \sqrt{2hr}$

$\therefore$ $P = AB + BC + CA$

$= 2\sqrt{2hr} + 2\sqrt{(2hr - h^2)}$

$= 2(h)^{1/2}(\sqrt{2r} + \sqrt{(2r-h)})$

Then, $\lim\limits_{h \to 0} \dfrac{\Delta}{P^3} = \lim\limits_{h \to 0} \dfrac{h^{3/2}\sqrt{(2r-h)}}{8h^{3/2}(\sqrt{2r} + \sqrt{2r-h})^3}$

$= \dfrac{\sqrt{2r}}{8(2\sqrt{2r})^3} = \dfrac{1}{128r}$

18. $\lim\limits_{x \to \frac{\pi}{4}} (2 - \tan x)^{\frac{1}{\ln(\tan x)}}$ $[1^{1/0}$ form$]$

$= e^{\lim\limits_{x \to \frac{\pi}{4}} \left[\ln(2-\tan x)^{\frac{1}{\ln(\tan x)}}\right]}$

$= e^{\lim\limits_{x \to \frac{\pi}{4}} \frac{\ln(2-\tan x)}{\ln(\tan x)}}$

$= e^{\lim\limits_{x \to \frac{\pi}{4}} \frac{\frac{1}{2-\tan x} \times (-\sec^2 x)}{\frac{1}{\tan x} \times \sec^2 x}}$ $[e^{0/0}$ form$]$

[by L'Hospital's rule]

$= e^{\lim\limits_{x \to \frac{\pi}{4}} \frac{-\tan x}{2-\tan x}}$

$= e^{\frac{-1}{2-1}} = e^{-1}$

19. $\because$ $f'(a) = \lim\limits_{x \to a} \dfrac{f(x+a) - f(a)}{x}$

and $\lim\limits_{x \to 0} \dfrac{\ln(a+x) - \ln a}{x} + k \lim\limits_{x \to e} \dfrac{\ln x - \ln e}{x - e} = 1$

$\therefore$ $f'(a) + k f'(e) = 1$ [where $f(x) = \ln x$]

$\Rightarrow$ $\dfrac{1}{a} + \dfrac{k}{e} = 1 \Rightarrow k = e\left(1 - \dfrac{1}{a}\right)$

20. For $a - 1 < x < a$, $\left[\dfrac{x}{a}\right] = 0$

$\therefore$ $\lim\limits_{x \to a^-} \left(\dfrac{|x|^3}{a} - \left[\dfrac{x}{a}\right]^3\right) = \lim\limits_{x \to a^-} \left(\dfrac{|x|^3}{a} - 0\right)$

$= \lim\limits_{h \to 0} \left(\dfrac{|a-h|^3}{a}\right) = \dfrac{a^3}{a} = a^2$

21. $\lim\limits_{x \to 0^+} x\left(\left[\dfrac{1}{x}\right] + \left[\dfrac{2}{x}\right] + \ldots + \left[\dfrac{15}{x}\right]\right)$

We know, $[x] = x - \{x\}$

$\therefore$ $\left[\dfrac{1}{x}\right] = \dfrac{1}{x} - \left\{\dfrac{1}{x}\right\}$

Similarly, $\left[\dfrac{n}{x}\right] = \dfrac{n}{x} - \left\{\dfrac{n}{x}\right\}$

$\therefore$ Given limit

$= \lim\limits_{x \to 0^+} x\left(\dfrac{1}{x} - \left\{\dfrac{1}{x}\right\} + \dfrac{2}{x} - \left\{\dfrac{2}{x}\right\} + \ldots \dfrac{15}{x} - \left\{\dfrac{15}{x}\right\}\right)$

$= \lim\limits_{x \to 0^+} (1 + 2 + 3 + \ldots + 15) - x\left(\left\{\dfrac{1}{x}\right\} + \left\{\dfrac{2}{x}\right\} + \ldots + \left\{\dfrac{15}{x}\right\}\right)$

$= 120 - 0 = 120$ $\left[\because 0 \le \left\{\dfrac{n}{x}\right\} < 1, \text{ therefore } 0 \le x\left\{\dfrac{n}{x}\right\} < x \Rightarrow \lim\limits_{x \to 0^+} x\left\{\dfrac{n}{x}\right\} = 0\right]$

22. $\lim\limits_{x \to \infty} \dfrac{\sqrt{(1/2)(1 - \cos 2x)}}{x} = \lim\limits_{x \to 0} \dfrac{|\sin x|}{x}$

$\therefore$ LHL $= -1$ and RHL $= 1$

23. LHL $= \lim_{x\to 0^-} f(x)$

$$= \lim_{h\to 0} f(0-h)$$

$$= \lim_{h\to 0} \frac{\ln(1+\{0-h\})}{\{0-h\}}$$

$$= \lim_{h\to 0} \frac{\ln(2-h)}{(1-h)} = \ln 2$$

RHL $= \lim_{x\to 0+} f(x) = \lim_{h\to 0} f(0+h)$

$$= \lim_{h\to 0} \frac{\ln(1+\{h\})}{\{h\}} = \frac{\ln(1+h)}{h} = 1$$

LHL $\neq$ RHL

24. Since, $f(x) = \frac{1}{\sqrt{(18-x^2)}} = (18-x^2)^{-1/2}$

$\therefore \quad f'(x) = \frac{x}{(18-x^2)^{3/2}}$

$$f'(3) = \frac{1}{9}$$

$\therefore \quad \lim_{x\to 3} \frac{f(x)-f(3)}{(x-3)} = f'(3) = \frac{1}{9}$

25. $\lim_{x\to\infty} \frac{x^2 \sin(1/x) - x}{1-|x|}$

put $x = \frac{1}{t}$

$\therefore \quad \lim_{t\to 0} \frac{\frac{1}{t^2}\sin t - \frac{1}{t}}{1-\frac{1}{|t|}} = \lim_{t\to 0} \frac{\sin t - t}{t^2 - |t|} \ (t>0)$

$$= \lim_{t\to 0} \frac{\sin t - t}{t^2 - t}$$

$$= \lim_{t\to 0} \frac{\frac{\sin t}{t} - 1}{(t-1)} = \frac{1-1}{0-1} = 0$$

26. $\lim_{\theta\to 0} \frac{\tan(\pi(1-\sin^2\theta))}{\sin(2\pi\sin^2\theta)}$

$$= \lim_{\theta\to 0} \frac{-\tan(\pi\sin^2\theta)}{\sin(2\pi\sin^2\theta)}$$

$$= \lim_{\theta\to 0} -\left(\frac{\tan(\pi\sin^2\theta)}{\pi\sin^2\theta}\right)\left(\frac{2\pi\sin^2\theta}{\sin(2\pi\sin^2\theta)}\right) \times \frac{1}{2} = \frac{-1}{2}$$

27. $\because \lim_{n\to\infty} \frac{f(x)e^{nx} + g(x)}{e^{nx}+1} = \lim_{n\to\infty} \frac{f(x) + \frac{g(x)}{e^{nx}}}{1+\frac{1}{e^{nx}}}$

$$= \frac{f(x)+0}{1+0} = f(x)$$

28. $\lim_{x\to 0} \frac{(\cos x - 1)(\cos x - e^x)}{x^n}$

$$= \lim_{x\to 0} \frac{(1-\cos x)(e^x - \cos x)}{x^n}$$

$$= \lim_{x\to 0} \frac{(1-\cos x)\{(e^x-1)+(1-\cos x)\}}{x^n}$$

$$= \lim_{x\to 0} \frac{(1-\cos x)(1+\cos x)\{(e^x-1)+(1-\cos x)\}}{x^n(1+\cos x)}$$

$$= \lim_{x\to 0} \frac{\left(\frac{\sin x}{x}\right)^2 \left(\left(\frac{e^x-1}{x}\right) + \left(\frac{1-\cos x}{x}\right)\right)}{x^{n-3}(1+\cos x)}$$

For $n=3$

$$= \frac{(1)^2((1)+0)}{1\cdot(2)} = \frac{1}{2} = \text{non-zero finite}$$

29. $\lim_{x\to 0} \frac{x\sqrt{y^2-(y-x)^2}}{\left(\sqrt{(8xy-4x^2)} + \sqrt{(8xy)}\right)^3}$

$$= \lim_{x\to 0} \frac{x^{3/2}\sqrt{(2y-x)}}{x^{3/2}(2\sqrt{(2y-x)} + 2\sqrt{2y})^3}$$

$$= \frac{\sqrt{2y}}{(2\sqrt{2y}+2\sqrt{2y})^3} = \frac{\sqrt{2y}}{64\cdot 2y\sqrt{2y}} = \frac{1}{128y}$$

30. $\lim_{x\to 0} \frac{\sin(\pi\cos^2 x)}{x^2}$

$$= \lim_{x\to 0} \frac{\sin(\pi(1-\sin^2 x))}{x^2}$$

$$= \lim_{x\to 0} \frac{\sin(\pi\sin^2 x)}{(\pi\sin^2 x)} \cdot \frac{\pi\sin^2 x}{x^2}$$

$$= 1\cdot\pi\cdot(1)^2 = \pi$$

31. Let $\sin x = t$

then $\lim_{t\to 1} \frac{t-t^t}{1-t+\ln t} = \lim_{t\to 1} \frac{1-t^t(1+\ln t)}{0-1+\frac{1}{t}}$

[by L' Hospital's rule]

$$= \lim_{t\to 1} \frac{0-\{t^t(1/t)+t^t(1+\ln t)^2\}}{0-\frac{1}{t^2}} = 2$$

32. $\lim_{x\to a} \frac{(a+2x)^{1/3}-(3x)^{1/3}}{(3a+x)^{1/3}-(4x)^{1/3}}, (a\neq 0)$ $\left[\frac{0}{0} \text{ form}\right]$

Put $x = a+h$

So, $\lim_{h\to 0} \frac{(a+2a+2h)^{1/3}-(3a+3h)^{1/3}}{(3a+a+h)^{1/3}-(4a+4h)^{1/3}}$

$$= \lim_{h\to 0} \frac{(3a)^{1/3}\left[\left(1+\frac{2h}{3a}\right)^{1/3} - \left(1+\frac{3h}{3a}\right)^{1/3}\right]}{(4a)^{1/3}\left[\left(1+\frac{h}{4a}\right)^{1/3} - \left(1+\frac{4h}{4a}\right)^{1/3}\right]}$$

$$= \left(\frac{3}{4}\right)^{1/3}\left(\frac{\frac{2}{9}-\frac{3}{9}}{\frac{1}{12}-\frac{4}{12}}\right) - \left(\frac{3}{4}\right)^{1/3}\left(\frac{\frac{-1}{9}}{\frac{-3}{12}}\right)$$

$$= \left(\frac{3}{4}\right)^{1/3} \frac{4}{(3)^2} = \frac{4^{1-\frac{1}{3}}}{3^{2-\frac{1}{3}}} = \frac{4^{2/3}}{3^{5/3}} = \frac{2^{4/3}}{3^{5/3}} = \frac{2}{3}\left(\frac{2}{9}\right)^{1/3}$$

Hence, option (d) is correct.

33. $\lim_{n\to\infty} \dfrac{n^{\alpha}\sin^2 n!}{n\left(1+\dfrac{1}{n}\right)} = \lim_{n\to\infty} \dfrac{\sin^2 n!}{n^{1-\alpha}\left(1+\dfrac{1}{n}\right)} = 0$

$[\because 0 < \sin^2 n! < 1]$

34. Let $P = \lim_{x\to 0} \dfrac{\sin^{-1} x - \tan^{-1} x}{x^2}$

Replacing x by $-x$, then

$$P = \lim_{x\to 0} \frac{\sin^{-1}(-x) - \tan^{-1}(-x)}{(-x)^2} = -P$$

$\therefore \quad 2P = 0$

$\therefore \quad P = 0$

35. $\lim_{n\to\infty} \left\{\dfrac{\log_e n}{\log_e(n-1)} \times \dfrac{\log_e(n+1)}{\log_e n} \times \dfrac{\log_e(n+2)}{\log_e(n+1)} \times \ldots \times \dfrac{\log_e n^k}{\log_e(n^k-1)}\right\}$

$$\lim_{n\to\infty}\left\{\frac{\log_e n^k}{\log_e(n-1)}\right\} = k \lim_{n\to\infty} \frac{\log_e n}{\log_e(n-1)}$$

Let $n = \dfrac{1}{h}$, then $k \lim_{h\to 0} \dfrac{\log_e\left(\dfrac{1}{h}\right)}{\log_e\left(1-\dfrac{1}{h}\right)}$

$$= k \lim_{h\to 0} \frac{-\log_e(h)}{\log_e(1-h) - \log_e(h)}$$

$$= k \lim_{h\to 0} \frac{1}{\left\{1 - \dfrac{\log_e(1-h)}{\log_e(h)}\right\}}$$

$$= k \cdot \frac{1}{(1-0)} = k$$

36. Given, $\lim_{x\to 1^+} \dfrac{(1-|x| + \sin|1-x|)\sin\left(\dfrac{\pi}{2}[1-x]\right)}{|1-x|[1-x]}$

Put $x = 1+h$, then $x \to 1^+ \Rightarrow h \to 0^+$

$$\therefore \lim_{x\to 1^+} \frac{(1-|x| + \sin|1-x|)\sin\left(\dfrac{\pi}{2}[1-x]\right)}{|1-x|[1-x]}$$

$$= \lim_{h\to 0^+} \frac{(1-|h+1| + \sin|-h|)\sin\left(\dfrac{\pi}{2}[-h]\right)}{|-h|[-h]}$$

$$= \lim_{h\to 0^+} \frac{(1-(h+1) + \sin h)\sin\left(\dfrac{\pi}{2}[-h]\right)}{h[-h]}$$

$[\because |-h| = h \text{ and } |h+1| = h+1 \text{ as } h > 0]$

$$= \lim_{h\to 0^+} \frac{(-h + \sin h)\sin\left(\dfrac{\pi}{2}(-1)\right)}{h(-1)}$$

$[\because [x] = -1 \text{ for } -1 < x < 0 \text{ and } h \to 0^+ \Rightarrow -h \to 0^-]$

$$= \lim_{h\to 0^+} \frac{(-h+\sin h)}{-h} \sin\left(\frac{-\pi}{2}\right) = \lim_{h\to 0^+}\left(\frac{\sin h}{h}\right) - \lim_{h\to 0^+}\left(\frac{h}{h}\right)$$

$$= \lim_{h\to 0^+}\left(\frac{\sin h}{h}\right) - \lim_{h\to 0^+}\left(\frac{h}{h}\right) = 1 - 1 = 0 \quad \left[\because \lim_{h\to 0^+} \frac{\sin h}{h} = 1\right]$$

37. $\lim_{x\to\infty} \left(1 + \dfrac{a}{x} + \dfrac{b}{x^2}\right)^{2x} = e^2$

$$\Rightarrow \quad \lim_{x\to\infty}\left(1 + \frac{(ax+b)}{x^2}\right)^{2x} = e^2 \qquad \ldots(i)$$

$\lim_{x\to\infty} \dfrac{ax+b}{x^2}$ must be equal to zero.

$$\Rightarrow \quad \lim_{x\to\infty} \frac{ax+b}{x^2} = 0$$

$$\Rightarrow \quad \lim_{x\to\infty}\left(\frac{a}{x} + \frac{b}{x^2}\right) = 0$$

$\therefore$ a and $b \in R$

From Eq. (i), $e^{\lim_{x\to\infty} \left(1 + \frac{(ax+b)}{x^2} - 1\right) 2x} = e^2$

$$\Rightarrow \quad e^{\lim_{x\to\infty} \frac{2(ax+b)}{x}} = e^2$$

$$e^{2a} = e^2$$

$\therefore \quad a = 1$ and $b \in R$

38. $\lim_{n\to\infty}\left(\dfrac{1 + e^{1/n} + (e^{1/n})^2 + \ldots + (e^{1/n})^{n-1}}{n}\right)$

$$= \lim_{n\to\infty} \frac{1\cdot[(e^{1/n})^n - 1]}{n(e^{1/n} - 1)} = (e-1)\lim_{n\to\infty} \frac{1}{\left(\dfrac{e^{1/n}-1}{1/n}\right)}$$

$$= (e-1)\times 1 = e - 1$$

39. $\lim_{x\to\pi/4} \dfrac{2\sqrt{2}\left\{1 - \dfrac{1}{2\sqrt{2}}(\cos x + \sin x)^3\right\}}{(1-\sin 2x)}$

$$= \lim_{x\to\pi/4} \frac{2\sqrt{2}\left\{1 - \left(\dfrac{1}{\sqrt{2}}\cos x + \dfrac{1}{\sqrt{2}}\sin x\right)^3\right\}}{(1-\sin 2x)}$$

$$= \lim_{x\to\pi/4} \frac{2\sqrt{2}\left\{1 - \cos^3\left(\dfrac{\pi}{4} - x\right)\right\}}{(1-\sin 2x)}$$

Put $x = \dfrac{\pi}{4} + h$, then

$$= \lim_{h\to 0} \frac{2\sqrt{2}(1-\cos^3 h)}{(1-\cos 2h)}$$

$$= \lim_{h\to 0} \frac{2\sqrt{2}(1-\cos^3 h)}{(2\sin^2 h)}$$

$$= \lim_{h\to 0} \frac{2\sqrt{2}(1-\cos h)(1+\cos h + \cos^2 h)}{2(1-\cos h)(1+\cos h)}$$

$$= \lim_{h\to 0} \frac{\sqrt{2}(1+\cos h+\cos^2 h)}{(1+\cos h)} = \frac{3\sqrt{2}}{2}$$

40. $\lim_{x\to 0} \dfrac{((a-n)nx - \tan x)\sin nx}{x^2} = 0$

$$\Rightarrow \quad \lim_{x\to 0}\left((a-n)n - \frac{\tan x}{x}\right)\frac{\sin nx}{x} = 0$$

$$\Rightarrow \quad ((a-n)n - 1)n = 0$$

$$\Rightarrow \quad (a-n)n - 1 = 0 \qquad [\because n \neq 0]$$

$$\therefore \quad a = n + \frac{1}{n}$$

41. Divide above and below by $e^{1/x}$, then

$$\lim_{x\to 0}\frac{(1+a^3)e^{-\frac{1}{x}}+8}{e^{-\frac{1}{x}}+(1-b^3)}=2\Rightarrow\frac{0+8}{0+1-b^3}=2$$

$\Rightarrow$ $1-b^3=4$

$\therefore$ $b^3=-3\Rightarrow b=-3^{1/3}$

Then $a\in R$

42. $\lim_{x\to\infty}\left(\sqrt{(x^4-x^2+1)}-ax^2-b\right)=0$

Put $x=\frac{1}{t}$

$\therefore$ $\lim_{t\to 0}\left(\sqrt{\left(\frac{1}{t^4}-\frac{1}{t^2}+1\right)}-\frac{a}{t^2}-b\right)=0$

$\Rightarrow$ $\lim_{t\to 0}\frac{\sqrt{(1-t^2+t^4)}-a-bt^2}{t^2}=0$ ...(i)

Since, RHS is finite.

$\therefore$ Numerator must be equal to 0 at $t\to 0$

$\therefore$ $1-a=0$

$\therefore$ $a=1$

From Eq. (i),

$$\lim_{t\to 0}\frac{\sqrt{(1-t^2+t^4)}-1-bt^2}{t^2}=0$$

$$\lim_{t\to 0}(-1+t^2)\left(\frac{(1-t^2+t^4)^{1/2}-(1)^{1/2}}{(1-t^2+t^4)-1}\right)=b\Rightarrow(-1)\left(\frac{1}{2}\right)=b$$

$\therefore$ $a=1,\ b=-\frac{1}{2}$

43. $\lim_{n\to\infty}\frac{S_{n+1}-S_n}{\sqrt{\sum_{k=1}^{n}k}}=\lim_{n\to\infty}\frac{a_{n+1}}{\sqrt{\frac{n(n+1)}{2}}}=0$ $[\because a_{n+1}=a]$

44. $\lim_{x\to 0}\frac{\cos(\sin x)-\cos x}{x^4}$

$$=\lim_{x\to 0}\frac{2\sin\left(\frac{\sin x+x}{2}\right)\sin\left(\frac{x-\sin x}{2}\right)}{x^4}$$

$$=\lim_{x\to 0}\left[\frac{2\sin\left(\frac{\sin x+x}{2}\right)}{\left(\frac{\sin x+x}{2}\right)}\left(\frac{\sin x+x}{2}\right)\times\frac{\sin\left(\frac{x-\sin x}{2}\right)}{\left(\frac{x-\sin x}{2}\right)}\left(\frac{\frac{x-\sin x}{2}}{x^4}\right)\right]$$

$$=\lim_{x\to 0}\frac{\sin\left(\frac{\sin x+x}{2}\right)}{\left(\frac{\sin x+x}{2}\right)}\lim_{x\to 0}\frac{\sin\left(\frac{x-\sin x}{2}\right)}{\left(\frac{x-\sin x}{2}\right)}\times\lim_{x\to 0}\frac{(x+\sin x)(x-\sin x)}{2x^4}$$

$$=1\cdot 1\cdot\lim_{x\to 0}\frac{(x+\sin x)(x-\sin x)}{2x\cdot x^3}=\left(\frac{1+1}{2}\right)Q=Q$$

$$Q=\lim_{x\to 0}\frac{x-\sin x}{x^3}\quad\text{...(i)}$$

Replacing x by $3x$, then

$$Q=\lim_{x\to 0}\frac{3x-\sin 3x}{(3x)^3}$$

$$=\lim_{x\to 0}\left(\frac{3(x-\sin x)}{27x^3}+\frac{4\sin^3 x}{27x^3}\right)$$

$$Q=\frac{1}{9}Q+\frac{4}{27}$$

or $\frac{8}{9}Q=\frac{4}{27}\Rightarrow Q=\frac{1}{6}$

Hence, $\lim_{x\to 0}\frac{\cos(\sin x)-\cos x}{x^4}=\frac{1}{6}$

45. Let $\alpha=\lim_{x\to 2-0}f(x)$

$=\lim_{h\to 0}f(2-h)$

$=\lim_{h\to 0}(2-h)^2-1=3$

and $\beta=\lim_{x\to 2+0}f(x)$

$=\lim_{h\to 0}f(2+h)$

$=\lim_{h\to 0}2(2+h)+3=7$

$\therefore$ Equation $x^2-(3+7)x+3\cdot 7=0$

$\Rightarrow x^2-10x+21=0$

46. $\lim_{x\to 1}\frac{x\sin\{x-[x]\}}{x-1}$

RHL $=\lim_{h\to 0}\frac{(1+h)\sin\{1+h-[1+h]\}}{1+h-1}$

$=\lim_{h\to 0}\frac{(1+h)\sin(1+h-1)}{h}$

$=1$

LHL $=\lim_{h\to 0}\frac{(1-h)\sin\{1-h-[1-h]\}}{1-h-1}$

$=\lim_{h\to 0}\frac{(1-h)\sin(1-h)}{-h}$

$=\infty$

LHL $\neq$ RHL

47. $\lim_{x\to 0}[x^2+x+\sin x]$

$[0^2+0+\sin 0]=[0]=0$

$=\lim_{x\to 0}[h^2-h-\sin h]$

$=\lim_{h\to 0}\left[h^2-h-\left(h-\frac{h^3}{3!}+\frac{h^5}{5!}-\dots\right)\right]$

$=\lim_{h\to 0}\left[-2h+h^2+\frac{h^3}{3!}-\frac{h^5}{5!}+\dots\right]$

48. $\lim_{x\to\pi/4}(2-\tan x)^{1/\ln\tan x}$

$$= e^{\lim_{x\to\pi/4}(2-\tan x-1)\times\frac{1}{\ln\tan x}} = e^{\lim_{x\to\pi/4}\left(\frac{1-\tan x}{\ln\tan x}\right)}$$

$$= e^{\lim_{x\to\pi/4}\frac{-\sec^2 x}{\frac{1}{\tan x}\cdot\sec^2 x}} = e^{-\lim_{x\to\pi/4}\tan x} = e^{-1}$$

49. $\lim_{n\to\infty}\left(\frac{n!}{(mn)^n}\right)^{1/n}$

$$= \lim_{n\to\infty}\left(\frac{1\cdot2\cdot3\cdot4\cdot\ldots\cdot(n-1)\,n}{n^n}\right)^{1/n}\times\frac{1}{m}$$

$$P = \lim_{n\to\infty}\left[\left(\frac{1}{n}\right)\left(\frac{2}{n}\right)\left(\frac{3}{n}\right)\ldots\left(\frac{n-1}{n}\right)\left(\frac{n}{n}\right)\right]^{1/n}\times\frac{1}{m}$$

$$\log_e P = \lim_{n\to\infty}\frac{1}{n}\sum_{r=1}^{n}\log_e\left(\frac{r}{n}\right)+\log_e\frac{1}{m}$$

$$= \int_0^1\log_e x\,dx-\log_e m = -1-\log_e m$$

$$\log_e P = -\log_e e-\log_e m$$

$$= -\log_e(em) = \log_e\left(\frac{1}{em}\right)$$

$$\therefore\quad P = \frac{1}{em}$$

50. $\because\quad y = 2^{-2^{1/(1-x)}}$

$$\therefore\quad \lim_{x\to1+}y = \lim_{x\to1+}2^{-2^{1/(1-x)}}$$

$$= \lim_{x\to0}2^{-2^{1/(1-(1+h))}} = 2^{-2^{-\infty}} = 2^{-0} = 1$$

51. $y = f(x)$ and $f(a) = 0$

$$= \lim_{x\to a}\frac{\log_e\{1+6f(x)\}}{3f(x)}$$

$$= \lim_{x\to a}\frac{\log_e\{1+6f(x)\}}{6f(x)}\cdot2 = 1\cdot2 = 2$$

52. $$\lim_{h\to0}\left\{\frac{(f(2h+2+h^2)-f(2))}{(2h+2+h^2)-2}\cdot\frac{((h-h^2+1)-1)}{f(h-h^2+1)-f(1)}\times\frac{2h+h^2}{(h-h^2)}\right\}$$

$$= \frac{f'(2)}{f'(1)}\cdot2 = \frac{6\cdot2}{4} = 3$$

53. $$\lim_{x\to0}\frac{x\left\{1+a\left(1-\frac{x^2}{2!}+\frac{x^4}{4!}-\ldots\right)\right\}-b\left(x-\frac{x^3}{3!}+\frac{x^5}{5!}-\ldots\right)}{x^3} = 1$$

$$\lim_{x\to0}\frac{x(1+a-b)-x^3\left(\frac{a}{2!}-\frac{b}{3!}\right)+x^5\left(\frac{a}{4!}-\frac{b}{5!}\right)-\ldots}{x^3} = 1$$

Since RHS is finite, $\therefore 1+a-b = 0$

and $-\frac{a}{2}+\frac{b}{6} = 1$

$$\therefore\quad b = -3/2,\ a = -5/2$$

54. $$\lim_{x\to0}\frac{x^n\sin^n x}{x^n-\sin^n x} \Rightarrow \lim_{x\to0}\frac{x^n\left(x-\frac{x^3}{3!}+\frac{x^5}{5!}-\ldots\right)^n}{x^n-\left(x-\frac{x^3}{3!}+\frac{x^5}{5!}-\ldots\right)^n}$$

$$\Rightarrow\quad \lim_{x\to0}\frac{\left(x-\frac{x^3}{3!}+\frac{x^5}{5!}-\ldots\right)^n}{1-\left(1-\frac{x^2}{3!}+\frac{x^4}{5!}-\ldots\right)^n}$$

$$\Rightarrow\quad \lim_{x\to0}\frac{x^n\left(1-\frac{x^2}{3!}+\frac{x^4}{5!}-\ldots\right)^n}{1-\left(1-\frac{x^2}{3!}+\frac{x^4}{5!}-\ldots\right)^n}$$

For $n = 2$

$$\lim_{x\to0}\frac{x^2\left(1-\frac{x^2}{3!}+\frac{x^4}{5!}-\ldots\right)^2}{1-\left(1-\frac{x^2}{3!}+\frac{x^4}{5!}-\ldots\right)^2}$$

$$\Rightarrow\quad \lim_{x\to0}\frac{x^2\left(1-\frac{x^2}{3!}+\frac{x^4}{5!}-\ldots\right)^2}{\left(2-\frac{x^2}{3!}+\frac{x^4}{5!}-\ldots\right)\left(\frac{x^2}{3!}-\frac{x^4}{5!}+\ldots\right)}$$

$$= \frac{1\,(1-0+\ldots)^2}{(2-0+0)\left(\frac{1}{3!}-0+\ldots\right)} = 3$$

55. We know that, $r\le[r]<r+1$

and $2r\le[2r]<2r+1$

$3r\le[3r]<3r+1$

$$r+2r+3r+\ldots\le[r]+[2r]+[3r]+\ldots<(r+2r+3r+\ldots)+(1+1+1+\ldots)$$

$$\frac{n(n+1)}{2}r\le[r]+[2r]+[3r]+\ldots<\frac{n(n+1)}{2}r+n$$

$$\frac{n(n+1)}{2n^2}r\le\frac{[r]+[2r]+[3r]+\ldots}{n^2}<\frac{n(n+1)r}{2n^2}+\frac{n}{n^2}$$

Now, $\lim_{n\to\infty}\frac{n(n+1)\cdot r}{2n^2} = \frac{r}{2}$

and $\lim_{n\to\infty}\frac{\frac{n(n+1)r}{2}+n}{n^2} = \frac{r}{2}$

So, by Sandwich Theorem, we can conclude that

$$\lim_{n\to\infty}\frac{[r]+[2r]+\ldots\ldots+[nr]}{n^2} = \frac{r}{2}$$

56. RHL $= \lim_{h\to0}f(1+h)$

$$= \lim_{h\to0}(1-(1+h)+[1+h-1]+|1-(1+h)|)$$

$$= \lim_{h\to0}(-h+[h]+h) = 0$$

LHL $= \lim_{h\to0}f(1-h)$

$$= \lim_{h\to0}(1-(1-h)+[1-h-1]+|1-(1-h)|)$$

$$= \lim_{h\to0}(h+[-h]+h) = \lim_{h\to0}(2h-1) = -1$$

LHL $\ne$ RHL

57. It is given that $\lim_{x\to 1} \frac{x^2 - ax + b}{x - 1} = 5$ ...(i)

Since, limit exist and equal to 5 and denominator is zero at $x = 1$, so numerator $x^2 - ax + b$ should be zero at $x = 1$

So, $1 - a + b = 0 \Rightarrow a = 1 + b$ ...(ii)

On putting the value of 'a' from Eq. (ii) in Eq. (i), we get

$$\lim_{x\to 1} \frac{x^2 - (1+b)x + b}{x - 1} = 5$$

$$\Rightarrow \lim_{x\to 1} \frac{(x^2 - x) - b(x-1)}{x - 1} = 5$$

$$\Rightarrow \lim_{x\to 1} \frac{(x-1)(x-b)}{x-1} = 5$$

$$\Rightarrow \lim_{x\to 1}(x - b) = 5$$

$$\Rightarrow 1 - b = 5$$

$$\Rightarrow b = -4 \quad \text{...(iii)}$$

On putting value of 'b' from Eq. (iii) to Eq. (ii), we get $a = -3$

So, $a + b = -7$

58. $\lim_{x\to\alpha} (1 + a(x-\alpha)(x-\beta))^{1/(x-\alpha)}$

$$= e^{\lim_{x\to\alpha} [1 + a(x-\alpha)(x-\beta) - 1] \times \frac{1}{(x-\alpha)}} = e^{a(\alpha - \beta)}$$

59. $\lim_{x\to 0} \frac{x}{a}\left[\frac{b}{x} - \left\{\frac{b}{x}\right\}\right]$ $[\because [t] = t - \{t\}]$

$$= \frac{b}{a} - \lim_{x\to 0} \frac{x}{a}\left\{\frac{b}{x}\right\}$$

$$= \frac{b}{a} - \frac{0}{a}([0, 1)) \quad \left[\because \left\{\frac{b}{x}\right\} \in [0, 1)\right]$$

$$= \frac{b}{a} - 0 = \frac{b}{a}$$

60. We have, $A_i = \frac{x - a_i}{|x - a_i|}, i = 1, 2, 3, \dots, n$

and $a_1 < a_2 < \dots < a_{n-1} < a_n$. Let x be in the left neighbourhood of a_m. Then, $x - a_i < 0$ for $i = m, m+1, \dots, n$ and $x - a_i > 0$ for $i = 1, 2, \dots, m-1$

Therefore, $A_i = \frac{(x - a_i)}{-(x - a_i)} = -1$

For $i = m, m+1, \dots, n$

and $A_i = \frac{(x - a_i)}{(x - a_i)} = 1$

For $i = 1, 2, \dots, m-1$

Similarly, if x in the right neighbourhood of a_m, then $x - a_i < 0$ for

$i = m+1, m+2, \dots, n$

and $x - a_i > 0$ for $i = 1, 2, \dots, m$

$\therefore$ $A_i = \frac{(x - a_i)}{-(x - a_i)} = -1$ for

$i = m+1, m+2, \dots, n$

and $A_i = \frac{(x - a_i)}{(x - a_i)} = 1$

For $i = 1, 2, \dots, m$

Now, $\lim_{x\to a_{m-}} (A_1 A_2 \dots A_n) = (-1)^{n-m+1}$

and $\lim_{x\to a_{m+}} (A_1 A_2 \dots A_m) = (-1)^{n-m}$

$\because$ LHL $\neq$ RHL

Hence, $\lim_{x\to a_m} (A_1 A_2 \dots A_m)$ does not exist.

61. $\lim_{x\to 1+0} \frac{\int_1^x |t-1|\,dt}{\sin(x-1)}$

$$= \lim_{h\to 0} \frac{\int_1^{1+h} |t-1|\,dt}{\sin h} = \lim_{h\to 0} \frac{\int_1^{1+h} (t-1)\,dt}{\sin h}$$

$$= \lim_{h\to 0} \frac{\left(\frac{t^2}{2} - t\right)_1^{1+h}}{\sin h}$$

$$= \lim_{h\to 0} \frac{\left[\frac{(1+h)^2}{2} - (1+h) - \left(\frac{1}{2} - 1\right)\right]}{\sin h}$$

$$= \lim_{h\to 0} \frac{\frac{(1+h)^2 - 1}{2} - h}{\sin h}$$

$$= \lim_{h\to 0} \frac{\frac{(2+h)h}{2} - h}{\sin h} = \frac{\frac{2+0}{2} - 1}{1} = 0$$

62. Let $x = [x] + \{x\}, 0 \le \{x\} < 1$

$$\Rightarrow \lim_{[x]\to\infty} \frac{\log_e [x]}{[x] + \{x\}} = \lim_{[x]\to\infty} \frac{\frac{1}{[x]}}{1} = 0$$

63. $\because \lim_{n\to\infty} \frac{1}{n^3}\{[1^2 x] + [2^2 x] + [3^2 x] + \dots + [n^2 x]\}$

$$= \lim_{n\to\infty} \left\{\frac{\sum_{r=1}^{n} [r^2 x]}{n^3}\right\} = \lim_{n\to\infty} \left(\frac{\sum_{r=1}^{n} r^2 x - \{r^2 x\}}{n^3}\right)$$

$$= \lim_{n\to\infty} \left(\frac{x \cdot \frac{n(n+1)(2n+1)}{6}}{n^3} - \sum_{r=1}^{n} \frac{\{r^2 x\}}{n^3}\right)$$

$$= x \cdot \frac{(1)(1)(2)}{6} - 0 = \frac{x}{3}$$

64. $\lim_{x\to 0} \{1^{1/\sin^2 x} + 2^{1/\sin^2 x} + \dots + n^{1/\sin^2 x}\}^{\sin^2 x}$

Put $\frac{1}{\sin^2 x} = t \ge 1$

$\therefore$ $\lim_{t\to\infty} (1^t + 2^t + \dots + n^t)^{1/t}$

$$= \lim_{t\to\infty} (n^t)^{1/t}\left[\left(\frac{1}{n}\right)^t + \left(\frac{2}{n}\right)^t + \dots + 1\right]^{1/t}$$

$$= n \lim_{t\to\infty} \left[\left(\frac{1}{n}\right)^t + \left(\frac{2}{n}\right)^t + \dots + 1\right]^{1/t}$$

$$= n[0 + 0 + \dots + 1]^0 = n$$

65. For $n = 1$

$$\lim_{x \to 0} \frac{(x - \sin x)}{(x - \sin x)} = 1$$

66. $$\lim_{x \to \infty} \frac{\log x^n - [x]}{[x]} = \lim_{x \to \infty} \left(\frac{n \log x}{[x]} - 1 \right) = 0 - 1 = -1$$

67. $$\text{LHL} = \lim_{x \to 0^-} f(x) = \lim_{h \to 0} f(0 - h) = \lim_{h \to 0} \frac{\sin [\cos h]}{1 + [\cos h]}$$

$$= \frac{\sin (0)}{1 + 0} = 0 \quad \left[\because h > 0 \therefore \cos h < 1 \right]$$

$$\text{RHL} = \lim_{x \to 0^+} f(x) = \lim_{h \to 0} f(0 + h)$$

$$= \lim_{h \to 0} \frac{\sin [\cos h]}{1 + [\cos h]} = \frac{\sin (0)}{1 + 0} = 0 \quad \left[\because h > 0 \therefore \cos h < 1 \right]$$

$$\therefore \lim_{x \to 0} \frac{\sin [\cos x]}{1 + [\cos x]} = 0$$

68. $\because ax^2 + bx + c = a(x - \alpha)(x - \beta)$

$$\therefore \lim_{x \to \alpha} \frac{1 - \cos (ax^2 + bx + c)}{(x - \alpha)^2}$$

$$= \lim_{x \to \alpha} \frac{1 - \cos \{a(x - \alpha)(x - \beta)\}}{(x - \alpha)^2}$$

$$= \lim_{x \to \alpha} \frac{(1 - \cos \{a(x - \alpha)(x - \beta)\})(1 + \cos \{a(x - \alpha)(x - \beta)\})}{(x - \alpha)^2 (1 + \cos \{a(x - \alpha)(x - \beta)\})}$$

$$= \lim_{x \to \alpha} \frac{\sin^2 (a(x - \alpha)(x - \beta))}{a^2 (x - \alpha)^2 (x - \beta)^2} \cdot \frac{a^2 \cdot (x - \beta)^2}{(1 + \cos \{a(x - \alpha)(x - \beta)\}}$$

$$= 1^2 \cdot \frac{a^2 (\alpha - \beta)^2}{(1 + 1)} = \frac{a^2}{2} (\alpha - \beta)^2$$

69. Let $L = \lim_{x \to 1^-} \dfrac{\sqrt{\pi} - \sqrt{2 \sin^{-1} x}}{\sqrt{1 - x}}$, then

$$L = \lim_{x \to 1^-} \frac{\sqrt{\pi} - \sqrt{2 \sin^{-1} x}}{\sqrt{1 - x}} \times \frac{\sqrt{\pi} + \sqrt{2 \sin^{-1} x}}{\sqrt{\pi} + \sqrt{2 \sin^{-1} x}}$$

[on rationalization]

$$= \lim_{x \to 1^-} \frac{\pi - 2 \sin^{-1} x}{\sqrt{1 - x}} \times \frac{1}{\sqrt{\pi} + \sqrt{2 \sin^{-1} x}}$$

$$= \lim_{x \to 1^-} \frac{\pi - 2 \left(\frac{\pi}{2} - \cos^{-1} x \right)}{\sqrt{1 - x}} \times \frac{1}{\sqrt{\pi} + \sqrt{2 \sin^{-1} x}}$$

$$\left[\because \sin^{-1} x + \cos^{-1} x = \frac{\pi}{2} \right]$$

$$= \lim_{x \to 1^-} \frac{2 \cos^{-1} x}{\sqrt{1 - x}} \times \lim_{x \to 1^-} \frac{1}{\sqrt{\pi} + \sqrt{2 \sin^{-1} x}}$$

$$= \frac{1}{2\sqrt{\pi}} \lim_{x \to 1^-} \frac{2 \cos^{-1} x}{\sqrt{1 - x}} \quad \left[\because \lim_{x \to 1^-} \sin^{-1} x = \frac{\pi}{2} \right]$$

Put $x = \cos \theta$, then as $x \to 1^-$, therefore $\theta \to 0^+$

Now, $$L = \frac{1}{2\sqrt{\pi}} \lim_{\theta \to 0^+} \frac{2\theta}{\sqrt{1 - \cos \theta}}$$

$$= \frac{1}{2\sqrt{\pi}} \lim_{\theta \to 0^+} \frac{2\theta}{\sqrt{2} \sin \left(\frac{\theta}{2} \right)} \quad \left[\because 1 - \cos \theta = 2 \sin^2 \frac{\theta}{2} \right]$$

$$= \frac{1}{2\sqrt{\pi}} \cdot \sqrt{2} \lim_{\theta \to 0^+} \frac{2 \cdot \left(\frac{\theta}{2} \right)}{\sin \left(\frac{\theta}{2} \right)}$$

$$= \frac{1}{2\sqrt{\pi}} 2\sqrt{2} = \sqrt{\frac{2}{\pi}} \quad \left[\because \lim_{x \to 0^+} \frac{\theta}{\sin \theta} = 1 \right]$$

70. $$\lim_{x \to 0} \frac{(1 + x)^{1/x} - e + \frac{1}{2} ex}{x^2}$$

$$= \lim_{x \to 0} \frac{e \left(1 - \frac{x}{2} + \frac{11}{24} x^2 + \dots \right) - e + \frac{1}{2} ex}{x^2}$$

$$= \lim_{x \to 0} \frac{\frac{11}{24} e x^2 + \dots}{x^2} = \frac{11}{24} e$$

71. $\because \quad \pi = 3.14$

$\therefore \quad \pi^2 = 9.86$

$\therefore \quad [-\pi^2] = [-9.86] = -10$

Then, $$\lim_{x \to 0} \frac{\tan([-\pi^2] x^2) - \tan([-\pi^2]) x^2}{\sin^2 x}$$

$$= \lim_{x \to 0} \frac{\tan(-10x^2) - \tan(-10) x^2}{\sin^2 x}$$

$$= \lim_{x \to 0} \frac{\frac{\tan(-10x^2)}{(-10x^2)} \cdot (-10) + \tan 10 \cdot \frac{x^2}{x^2}}{\frac{\sin^2 x}{x^2}}$$

$$= \frac{(1)(-10) + \tan 10 \cdot 1}{(1)^2} = \tan 10 - 10$$

72. $$a = \min \{x^2 + 2x + 3, x \in R\}$$

$$= \min \{(x + 1)^2 + 2, x \in R\} = 2$$

and $$b = \lim_{\theta \to 0} \frac{1 - \cos \theta}{\theta^2}$$

$$= \lim_{\theta \to 0} \frac{(1 - \cos \theta)(1 + \cos \theta)}{\theta^2 (1 + \cos \theta)} = \frac{1}{2}$$

$$\therefore \sum_{r=0}^{n} a^r \cdot b^{n-r} = b^n \sum_{r=0}^{n} \left(\frac{a}{b} \right)^r$$

$$= \left(\frac{1}{2} \right)^n \sum_{r=0}^{n} (4)^r = \frac{1}{2^n} (1 + 4 + 4^2 + \dots + 4^n)$$

$$= \frac{1}{2^n} \cdot 1 \cdot \left(\frac{4^{n+1} - 1}{4 - 1} \right) = \frac{4^{n+1} - 1}{3 \cdot 2^n}$$

73. Let $x = 2009^{2010}$ and $y = 2010^{2009}$

$$\therefore \lim_{n \to \infty} (x^n + y^n)^{1/2n}$$

$$= \lim_{n \to \infty} x \left[1 + \left(\frac{y}{x} \right)^n \right]^{1/n} \quad \left[\because x > y \Rightarrow \frac{y}{x} < 1 \text{ and } \log x > \log y \right]$$

$$= \lim_{n \to \infty} x \left[1 + \left(\frac{y}{x} \right)^n \cdot \frac{1}{n} + \dots \right]$$

$$= x = (2009)^{2010} = a^b$$

$\therefore \quad a = 2009$ and $b = 2010$

$\Rightarrow \quad a - b = 2009 - 2010 = -1$

Objective Questions Type II [One or more than one correct answer(s)]

1. $f(x)=\lim_{n\to\infty}\dfrac{x^{2n}-1}{x^{2n}+1}$

Option (a) $|x|>1$

Then $f(x)=\lim_{n\to\infty}\dfrac{1-\frac{1}{x^{2n}}}{1+\frac{1}{x^{2n}}}=\dfrac{1-0}{1+0}=1$

Option (b) $|x|<1$

Then $f(x)=\lim_{n\to\infty}\dfrac{x^{2n}-1}{x^{2n}+1}=\dfrac{0-1}{0+1}=-1$

Option (c) From alternate (a) and (b), $1\neq-1$

$\therefore f(x)$ is not defined for any value of x.

Option (d) $|x|=1$

Then $f(x)=0$

2. We have,

$$f_n(x)=\sum_{j=1}^{n}\tan^{-1}\left(\frac{1}{1+(x+j)(x+j-1)}\right)\quad\text{for all } x\in(0,\infty)$$

$$\Rightarrow f_n(x)=\sum_{j=1}^{n}\tan^{-1}\left(\frac{(x+j)-(x+j-1)}{1+(x+j)(x+j-1)}\right)$$

$$\Rightarrow f_n(x)=\sum_{j=1}^{n}[\tan^{-1}(x+j)-\tan^{-1}(x+j-1)]$$

$$\Rightarrow f_n(x)=(\tan^{-1}(x+1)-\tan^{-1}x)+(\tan^{-1}(x+2)-\tan^{-1}(x+1))+(\tan^{-1}(x+3)-\tan^{-1}(x+2))+\ldots+(\tan^{-1}(x+n)-\tan^{-1}(x+n-1))$$

$$\Rightarrow f_n(x)=\tan^{-1}(x+n)-\tan^{-1}x$$

This statement is false as $x\neq0$. i.e., $x\in(0,\infty)$.

(b) This statement is also false as $0\notin(0,\infty)$

(c) $f_n(x)=\tan^{-1}(x+n)-\tan^{-1}x$

$$\lim_{x\to\infty}\tan(f_n(x))=\lim_{x\to\infty}\tan(\tan^{-1}(x+n)-\tan^{-1}x)$$

$$\Rightarrow \lim_{x\to\infty}\tan(f_n(x))=\lim_{x\to\infty}\tan\left(\tan^{-1}\frac{n}{1+nx+x^2}\right)$$

$$=\lim_{x\to\infty}\frac{n}{1+nx+x^2}=0$$

$\therefore$ (c) statement is false.

(d) $\lim_{x\to\infty}\sec^2(f_n(x))=\lim_{x\to\infty}(1+\tan^2 f_n(x))$

$$=1+\lim_{x\to\infty}\tan^2(f_n(x))=1+0=1$$

$\therefore$ (d) statement is true.

3. $\because \lim_{x\to0}(\cos x+a\sin bx)^{1/x}=e^2$ [form 1^∞]

$$\Rightarrow e^{\lim_{x\to0}(\cos x+a\sin bx-1)\frac{1}{x}}=e^2$$

$$\Rightarrow \lim_{x\to0}(a\sin bx-(1-\cos x))\cdot\frac{1}{x}=2$$

$$\Rightarrow \lim_{x\to0}\frac{a\sin bx-(1-\cos x)}{x}=2$$

$$\Rightarrow \lim_{x\to0}b\cdot\frac{a\sin bx}{bx}-\lim_{x\to0}\frac{(1-\cos x)}{x}=2$$

$$\Rightarrow (ab)(1)-0=2$$

$$\therefore ab=2$$

4. Here,

$$f(x)=\lim_{n\to\infty}\left[\frac{n^n(x+n)\left(x+\frac{n}{2}\right)\ldots\left(x+\frac{n}{n}\right)}{n!(x^2+n^2)\left(x^2+\frac{n^2}{4}\right)\ldots\left(x^2+\frac{n^2}{n^2}\right)}\right]^{\frac{x}{n}},\ x>0$$

Taking log on both sides, we get

$$\log_e\{f(x)\}=\lim_{n\to\infty}\log\left[\frac{n^n(x+n)\left(x+\frac{n}{2}\right)\ldots\left(x+\frac{n}{n}\right)}{n!(x^2+n^2)\left(x^2+\frac{n^2}{4}\right)\ldots\left(x^2+\frac{n^2}{n^2}\right)}\right]^{\frac{x}{n}}$$

$$=\lim_{n\to\infty}\frac{x}{n}\cdot\log\left[\frac{\prod_{r=1}^{n}\left(x+\frac{1}{r/n}\right)}{\prod_{r=1}^{n}\left(x^2+\frac{1}{(r/n)^2}\right)\prod_{r=1}^{n}(r/n)}\right]$$

$$=x\lim_{n\to\infty}\frac{1}{n}\sum_{r=1}^{n}\log\left[\frac{x+\frac{n}{r}}{\left(x^2+\frac{n^2}{r^2}\right)\frac{r}{n}}\right]$$

$$=x\lim_{n\to\infty}\frac{1}{n}\sum_{r=1}^{n}\log\left(\frac{\frac{r}{n}\cdot x+1}{\frac{r^2}{n^2}\cdot x^2+1}\right)$$

Converting summation into definite integration, we get

$$\log_e\{f(x)\}=x\int_0^1\log\left(\frac{xt+1}{x^2t^2+1}\right)dt$$

Put, $tx=z$

$\Rightarrow xdt=dz$

$$\therefore \log_e\{f(x)\}=x\int_0^x\log\left(\frac{1+z}{1+z^2}\right)\frac{dz}{x}$$

$$\Rightarrow \log_e\{f(x)\}=\int_0^x\log\left(\frac{1+z}{1+z^2}\right)dz$$

Using Newton-Leibnitz formula, we get

$$\frac{1}{f(x)}\cdot f'(x)=\log\left(\frac{1+x}{1+x^2}\right)\quad\ldots\text{(i)}$$

Here, at $x=1$,

$$\frac{f'(1)}{f(1)}=\log(1)=0$$

$$\therefore f'(1)=0$$

Now, sign scheme of $f'(x)$ is shown below

$$\overset{+}{\longrightarrow}\;\Big|_{x=1}\;\overset{-}{\longrightarrow}$$

$\therefore$ At $x=1$, function attains maximum.
Since, $f(x)$ increases on (0, 1).

$$\therefore \quad f(1) > f(1/2)$$

$\therefore$ Option (a) is incorrect.

$$f(1/3) < f(2/3)$$

$\therefore$ Option (b) is correct.

Also, $f'(x) < 0$, when $x > 1$

$$\Rightarrow \quad f'(2) < 0$$

$\therefore$ Option (c) is correct.

Also, $$\frac{f'(x)}{f(x)} = \log\left(\frac{1+x}{1+x^2}\right)$$

$$\therefore \quad \frac{f'(3)}{f(3)} - \frac{f'(2)}{f(2)} = \log\left(\frac{4}{10}\right) - \log\left(\frac{3}{5}\right)$$

$$= \log(2/3) < 0$$

$$\Rightarrow \quad \frac{f'(3)}{f(3)} < \frac{f'(2)}{f(2)}$$

$\therefore$ Option (d) is incorrect.

5. $$\lim_{x\to 0^+} \frac{\sqrt{(1-\cos 2ax)}}{\sin bx} = \lim_{h\to 0} \frac{\sqrt{(1-\cos 2ah)}}{\sin bh}$$

$$= \lim_{h\to 0} \frac{\sqrt{2}\,|\sin ah|}{\sin bh}$$

$$= \lim_{h\to 0} \frac{\sqrt{2}\sin ah}{\sin bh} \qquad \left[\begin{array}{l}\because a>0, h>0\\ \therefore ah>0\end{array}\right]$$

$$= \sqrt{2}\lim_{h\to 0} \frac{\frac{\sin ah}{ah}\times ah}{\frac{\sin bh}{bh}\times bh} = \frac{a}{b}\sqrt{2}$$

$$= -\frac{\sqrt{2}\,|a|}{|b|}$$

6. $$\lim_{x\to 0}\left(\frac{a^x+b^x+c^x}{3}\right)^{\lambda/x} \qquad [\text{form } 1^\infty]$$

$$= e^{\lim_{x\to 0}\left(\frac{a^x+b^x+c^x}{3}-1\right)\times\frac{\lambda}{x}}$$

$$= e^{\lambda \lim_{x\to 0}\left(\frac{a^x+b^x+c^x-3}{3x}\right)}$$

$$= e^{\lambda/3 \lim_{x\to 0}\left(\frac{(a^x-1)}{x}+\frac{(b^x-1)}{x}+\frac{(c^x-1)}{x}\right)}$$

$$= e^{\lambda/3(\ln a+\ln b+\ln c)}$$

$$= e^{\lambda/3\ln(abc)} = (abc)^{\lambda/3}$$

7. $$\lim_{x\to 0} \frac{\left(\frac{e^x-1}{x}\right)^4}{\frac{\sin\left(\frac{x^2}{a^2}\right)}{a^2\left(\frac{x^2}{a^2}\right)}\cdot\frac{\log_e\left(1+\frac{x^2}{2}\right)}{\frac{x^2}{2}\cdot 2}} = 8$$

$$= \frac{(1)^4}{\frac{1}{a^2}\cdot\frac{1}{2}} = 8$$

$$\Rightarrow \quad a^2 = 4$$

$$\therefore \quad a = \pm 2$$

8. $$\lim_{x\to\infty} f(x) = \lim_{x\to\infty}\left(\frac{|x|}{2+|x|}\right)^{2x}$$

$$= \lim_{x\to\infty}\left(\frac{2+|x|}{|x|}\right)^{-2x}$$

$$= \lim_{x\to\infty}\left(1+\frac{2}{|x|}\right)^{-2x}$$

$$= \lim_{x\to\infty}\left(1+\frac{2}{x}\right)^{-2x} \qquad \left[\begin{array}{l}\because x\to\infty\\ \therefore |x| = x\end{array}\right]$$

$$= e^{-4}$$

and $$\lim_{x\to-\infty} f(x) = \lim_{x\to-\infty}\left(1-\frac{2}{x}\right)^{-2x} \qquad \left[\begin{array}{l}\because x\to-\infty\\ \therefore |x| = -x\end{array}\right]$$

$$= e^4$$

9. LHL = RHL

$$\Rightarrow \quad \lim_{x\to 1^-} f(x) = \lim_{x\to 1^+} f(x)$$

$$\Rightarrow \quad \lim_{h\to 0} f(1-h) = \lim_{h\to 0} f(1+h)$$

$$\Rightarrow \quad \lim_{h\to 0}\left(1+\frac{2(1-h)}{\lambda}\right) = \lim_{h\to 0}\lambda(1+h)$$

$$\Rightarrow \quad 1+\frac{2}{\lambda} = \lambda$$

$$\lambda^2-\lambda-2 = 0$$

$$\Rightarrow \quad (\lambda-2)(\lambda+1) = 0$$

$$\therefore \quad \lambda = -1, 2$$

10. $\lim_{x\to 0}\left(2+\frac{f(x)}{2}\right)^{1/x}$ exists only when $\lim_{x\to 0}\left(2+\frac{f(x)}{x^2}\right) = 1$

$$\Rightarrow \quad \lim_{x\to 0}\left(2+\frac{a_0+a_1x+a_2x^2+a_3x^3+\ldots}{x^2}\right) = 1$$

$$\Rightarrow \quad \lim_{x\to 0}\left(\frac{a_0+a_1x+a_2x^2+a_3x^3+\ldots}{x^2}\right) = -1$$

$$\Rightarrow \quad \lim_{x\to 0}\left(\frac{a_0}{x^2}+\frac{a_1}{x}+a_2+a_3x+\ldots\right) = -1 \qquad \ldots\text{(i)}$$

Since, RHS is finite and equal to -1.

$\therefore$ In LHS, a_0 and a_1 must be equal to zero.

$\therefore \quad a_0 = 0,\ a_1 = 0$

Now, from Eq. (i),

$$\lim_{x \to 0} (a_2 + a_3 x + \dots) = -1 \Rightarrow a_2 = -1$$

$$\therefore \quad f(x) = -x^2 + a_3 x^3 + \dots$$

Given, $\quad \lim_{x \to 0} \left(2 + \frac{f(x)}{x^2}\right)^{1/x} = e^2$

$$\Rightarrow \quad \lim_{x \to 0} (2 - 1 + a_3 x + \dots)^{1/x} = e^2$$

$$\Rightarrow \quad \lim_{x \to 0} (1 + a_3 x + \dots)^{1/x} = e^2$$

$$\Rightarrow \quad e^{\lim_{x \to 0} (a_3 x + \dots) \times \frac{1}{x}} = e^2 \Rightarrow e^{a_3} = e^2$$

$$\therefore \quad a_3 = 2$$

Hence, $a_0 = 0,\ a_1 = 0,\ a_2 = -1,\ a_3 = 2$

11. Let $L = \lim_{x \to \infty} \dfrac{2x^3 - (\tan^{-1} x)^3}{\frac{8}{\pi} x^3 \cot^{-1} |kx| + k^2 x^6 \sin \frac{1}{x^3} - 3kx^3}$

$$\Rightarrow L = \lim_{x \to \infty} \frac{2 - \left(\dfrac{\tan^{-1} x}{x}\right)^3}{\dfrac{8}{\pi} \cot^{-1} |kx| + \dfrac{k^2 \sin \frac{1}{x^3}}{\frac{1}{x^3}} - 3k}$$

Now, if $k = 0$, then $L = \dfrac{2 - 0}{\frac{8}{\pi} \cdot \frac{\pi}{2}} = \dfrac{1}{2}$

And, if $k \neq 0$, then $L = \dfrac{2}{0 + k^2 - 3k} = \dfrac{1}{2}$

$$\Rightarrow \quad k^2 - 3k - 4 = 0$$

$$\Rightarrow \quad k = 4, -1$$

Linked Comprehension Type

PASSAGE 1

1. $f(x) = \dfrac{\sin^{-1}(1 - \{x\}) \cdot (\cos^{-1}(1 - \{x\})}{\sqrt{2\{x\}})\,(1 - \{x\})}$

$$R = \lim_{x \to 0^+} f(x) = \lim_{h \to 0} f(0 + h)$$

$$= \lim_{h \to 0} \frac{\sin^{-1}(1 - \{0 + h\} \cdot \cos^{-1}(1 - \{0 + h\})}{\sqrt{2\{0 + h\}}\,(1 - \{0 + h\}}$$

$$= \lim_{h \to 0} \frac{\sin^{-1}(1 - h) \cdot \cos^{-1}(1 - h)}{\sqrt{2h} \cdot (1 - h)}$$

$$= \lim_{h \to 0} \frac{\sin^{-1}(1 - h)}{(1 - h)} \lim_{h \to 0} \frac{\cos^{-1}(1 - h)}{\sqrt{2h}}$$

in second limit put $1 - h = \cos\theta$

$$= \lim_{h \to 0} \frac{\sin^{-1}(1 - h)}{(1 - h)} \cdot \lim_{\theta \to 0} \frac{\cos^{-1}(\cos\theta)}{\sqrt{2(1 - \cos\theta)}}$$

$$= \lim_{h \to 0} \frac{\sin^{-1}(1 - h)}{(1 - h)} \cdot \lim_{\theta \to 0} \frac{\theta}{2\sin(\theta/2)} \qquad [\because \theta > 0]$$

$$= \sin^{-1} 1 \cdot 1 = \pi/2$$

$$\Rightarrow 100R = 50\pi$$

$$\therefore \quad \cos(100R) = \cos 50\pi = (-1)^{50} = 1$$

2. $L = \lim_{x \to 0^-} f(x) = \lim_{h \to 0} f(0 - h)$

$$= \lim_{h \to 0} \frac{\sin^{-1}(1 - \{0 - h\}) \cdot \cos^{-1}(1 - \{0 - h\})}{\sqrt{(2\{0 - h\}) \cdot (1 - \{0 - h\})}}$$

$$- \lim_{h \to 0} \frac{\sin^{-1}(1 + h - 1) \cdot \cos^{-1}(1 + h - 1)}{\sqrt{2(-h + 1) \cdot (1 + h - 1)}}$$

$$= \lim_{h \to 0} \frac{\sin^{-1} h}{h} \cdot \lim_{h \to 0} \frac{\cos^{-1} h}{\sqrt{2(1 - h)}}$$

$$= 1 \cdot \frac{\pi/2}{\sqrt{2}} = \frac{\pi}{2\sqrt{2}} \quad \Rightarrow 99\sqrt{2}\,L = \frac{99\pi}{2}$$

$$\therefore \sin(99\sqrt{2}L) = \sin\left(\frac{99\,\pi}{2}\right)$$

$$= (-1)^{\frac{99 - 1}{2}} = (-1)^{49} = -1$$

3. $\because R = \dfrac{\pi}{2}$ and $L = \dfrac{\pi}{2\sqrt{2}}$

$$\therefore \quad 2R^2 + 4L^2 = \pi^2$$

$$\Rightarrow [2R^2 + 4L^2] = [\pi^2] = [9.87] = 9$$

PASSAGE 2

1. If $\alpha \to \infty$, then $a \to 0$

$$\therefore \quad P(x) = bx + c$$

Now, $P(2) = 2b + c = 9$ and $P'(3) = b = 5$

$$\Rightarrow \quad c = -1$$

$$\Rightarrow \quad P(x) = 5x - 1$$

Now, $\lim_{x \to \infty} \left(\dfrac{5x - 1}{5x - 5}\right)^x \qquad [1^\infty \text{ form}]$

$$= e^{\lim_{x \to \infty} \left(\frac{5x - 1 - 5x + 5}{5x - 5}\right) x} = e^{4/5}$$

2. When, $\alpha, \beta \to \infty$, then $a, b \to 0$

$\therefore P(x) = c$, but $P(2) = 9 \Rightarrow P(x) = 9$

Now, $\lim_{x \to 3} \dfrac{\sqrt{P(x)} - 3}{\sin(x - 3)} = \lim_{x \to 3} \dfrac{3 - 3}{\sin(x - 3)} = 0$

$$\left[\because \frac{\text{exact zero}}{\text{tending towards zero}} = 0\right]$$

3. For $\alpha = \beta$, $P(x) = a(x - \alpha)^2$

As, $P(2) = 9 \Rightarrow P(2) = a(2 - \alpha)^2 = 9$

$$\Rightarrow \quad a = \frac{9}{(2 - \alpha)^2}$$

Now, $\lim\limits_{x \to \alpha} \dfrac{\dfrac{9}{(2-\alpha)^2} \cdot (x-\alpha)^2}{\left[1 - \tan^2\left(\dfrac{\pi}{4} - (x-\alpha)\right)\right](e^{x-\alpha} - 1)}$

$= \lim\limits_{x \to \alpha} \dfrac{9}{(2-\alpha)^2} \cdot \dfrac{(x-\alpha)^2}{\left[1 + \tan\left(\dfrac{\pi}{4} - (x-\alpha)\right)\right]\left[1 - \tan\left(\dfrac{\pi}{4} - (x-\alpha)\right)\right](e^{x-\alpha} - 1)}$

Putting $x = \alpha + h$, we get

$= \lim\limits_{h \to 0} \dfrac{9}{(2-\alpha)^2} \cdot \dfrac{h^2}{\left[1 + \tan\left(\dfrac{\pi}{4} - h\right)\right]\left[1 - \tan\left(\dfrac{\pi}{4} - h\right)\right](e^h - 1)}$

$= \lim\limits_{h \to 0} \dfrac{9}{(2-\alpha)^2} \cdot \dfrac{h^2}{2 \cdot \left(1 - \dfrac{1 - \tan h}{1 + \tan h}\right)(e^h - 1)}$

$= \lim\limits_{h \to 0} \dfrac{9}{(2-\alpha)^2 \cdot 2} \cdot \dfrac{h^2(1 + \tan h)}{\dfrac{2\tan h}{h} \cdot \dfrac{(e^h - 1)}{h} \cdot h^2}$

$= \dfrac{9}{4(2-\alpha)^2}$

$= \lim\limits_{\alpha \to 0} \dfrac{9}{4(2-\alpha)^2} = \dfrac{9}{16}$

PASSAGE 3

$\because \lim\limits_{x \to 0^+} f(x) =$ finite

$\Rightarrow \lim\limits_{h \to 0} f(0 + h) =$ finite

$\Rightarrow \lim\limits_{h \to 0} \dfrac{\sin h + ae^h + be^{-h} + c\ \ln(1+h)}{h^3} =$ finite

at $h \to 0$, Numerator $\to a + b$ and Denominator $\to 0$

$\because$ RHS is finite,

$\therefore\ a + b = 0 \Rightarrow b = -a$ ...(i)

Now, $\lim\limits_{h \to 0} \dfrac{\sin h + ae^h - ae^{-h} + c\ln(1+h)}{h^3} =$ finite

$\Rightarrow \lim\limits_{h \to 0} \dfrac{\dfrac{\sin h}{h} + \dfrac{a(e^h - 1)}{h} - a\dfrac{(e^{-h} - 1)}{h} + \dfrac{c\ln(1+h)}{h}}{h^2} =$ finite

at $h \to 0$, Numerator $\to 1 + a + a + c$

and Denominator $\to 0$

$\because$ RHS is finite, $\therefore\ 1 + 2a + c = 0$

$\Rightarrow\ c = -1 - 2a$ ...(ii)

$\therefore \lim\limits_{h \to 0} \dfrac{\sin h + ae^h - ae^{-h} - (1+2a)\ln(1+h)}{h^3} =$ finite

(by L' Hospital's rule) $\left(\text{form}\left(\dfrac{0}{0}\right)\right)$

$\lim\limits_{h \to 0} \dfrac{\cos h + ae^h + ae^{-h} - \dfrac{(1+2a)}{(1+h)}}{3h^2}$ $\left[\text{form}\left(\dfrac{0}{0}\right)\right]$

$\therefore$ (by L' Hospital's rule)

$\lim\limits_{h \to 0} \dfrac{-\sin h + ae^h - ae^{-h} + \dfrac{(1+2a)}{(1+h)^2}}{6h}$

Numerator must be $= 0$

$0 + a - a + \dfrac{1 + 2a}{(1)^2} = 0$

$\therefore\ a = -\dfrac{1}{2}$

[Ans. 1. (a)]

From Eq. (i), $b = \dfrac{1}{2}$

[Ans. 2. (b)]

and from Eq. (ii), $c = -1 + 1 = 0$

[Ans. 3. (c)]

4. $\lim\limits_{x \to 0+} f(x) = \lim\limits_{h \to 0} f(h) = \lim\limits_{h \to 0} \dfrac{\sin h - \dfrac{1}{2}e^h + \dfrac{1}{2}e^{-h}}{h^3}$ $\left[\text{form}\dfrac{0}{0}\right]$

[by L'Hospital's]

$= \lim\limits_{h \to 0} \dfrac{\cos h - \dfrac{e^h}{2} - \dfrac{e^{-h}}{2}}{3h^2},$ $\left[\text{form}\dfrac{0}{0}\right]$

[by L'Hospital's]

$= \lim\limits_{h \to 0} \dfrac{-\sin h - \dfrac{e^h}{2} + \dfrac{e^{-h}}{2}}{6h}$ $\left[\text{form}\dfrac{0}{0}\right]$

$= \lim\limits_{h \to 0} \dfrac{-\cos h - \dfrac{e^h}{2} - \dfrac{e^{-h}}{2}}{6}$

$= \dfrac{-1 - \dfrac{1}{2} - \dfrac{1}{2}}{6}$

$= \dfrac{-1-1}{6} = \dfrac{-1}{3}$

5. $\lim\limits_{x \to 0^+} x^2 f(x) = \lim\limits_{h \to 0} h^2 f(h)$

$= \lim\limits_{h \to 0} \dfrac{\sin h + ae^h + be^{-h} + c\ln(1+h)}{h}$

$\left[\because a = -\dfrac{1}{2}, b = \dfrac{1}{2}, c = 0\right]$

$= \lim\limits_{h \to 0} \dfrac{\sin h - \dfrac{1}{2}e^h + \dfrac{1}{2}e^{-h}}{h}$

$= \lim\limits_{h \to 0} \dfrac{\sin h}{h} - \dfrac{1}{2}\dfrac{(e^h - 1)}{h} + \dfrac{1}{2}\dfrac{(e^{-h} - 1)}{h}$

$= 1 - \dfrac{1}{2} - \dfrac{1}{2} = 0$

Numerical Value Type

1. $\because \lim_{x\to 1}(1+\sin \pi x)^{\cot \pi x}$

$$= e^{\lim_{x\to 1}(1+\sin \pi x-1)\cot \pi x} = e^{\lim_{x\to 1}\cos \pi x} = e^{\cos \pi} = e^{-1}$$

$\therefore \lim_{x\to 1} 2008\, e(1+\sin \pi x)^{\cot \pi x}$

$$= 2008e\cdot e^{-1} = 2008$$

2. $\because \lim_{x\to 1}\dfrac{(x+x^2+x^3+\ldots+x^{97})-(1+1+1+\ldots+1)}{(x-1)}$

$$= \lim_{x\to 1}\frac{(x-1)}{(x-1)}+\frac{(x^2-1^2)}{(x-1)}+\frac{(x^3-1^3)}{x-1}+\ldots+\frac{(x^{97}-1^{97})}{(x-1)}$$

$$= 1+2+3+\ldots+97$$

$$= \frac{97}{2}(1+97) = 97\times 49 = 4753$$

3. $\because \lim_{x\to 1}(1-x)\tan\left(\dfrac{\pi x}{2}\right)$

Put $x = 1+h$

then $\lim_{h\to 0}(-h)\tan\left(\dfrac{\pi}{2}(1+h)\right)$

$$= -\lim_{h\to 0} h\left(-\cot\frac{\pi h}{2}\right) = \lim_{h\to 0}\frac{h}{\tan\left(\frac{\pi h}{2}\right)}$$

$$= \frac{2}{\pi}\cdot\lim_{h\to 0}\frac{\frac{\pi h}{2}}{\tan\left(\frac{\pi h}{2}\right)} = \frac{2}{\pi}\times 1 = \frac{2}{\pi}$$

$\therefore \lim_{x\to 1} 512\pi(1-x)\tan\left(\dfrac{\pi x}{2}\right) = 512\pi\times\dfrac{2}{\pi} = 1024$

4. Let $P = \lim_{x\to 0}\left(\dfrac{e^{\alpha x}-e^x-x}{x^2}\right)$ ...(i)

Replacing x by $-x$ in Eq. (i), then

$$P = \lim_{x\to 0}\left(\frac{e^{-\alpha x}-e^{-x}+x}{x^2}\right) \quad \ldots\text{(ii)}$$

Adding, Eqs. (i) and (ii), then

$$2P = \lim_{x\to 0}\left(\frac{e^{\alpha x}+e^{-\alpha x}-(e^x+e^{-x})}{x^2}\right)$$

$$= \lim_{x\to 0}\left(\frac{(e^{\alpha x}+e^{-\alpha x}-2)-(e^x+e^{-x}-2)}{x^2}\right)$$

$$= \lim_{x\to 0}\left\{\frac{e^{-\alpha x}(e^{2\alpha x}-2e^{\alpha x}+1)-e^{-x}(e^{2x}-2e^x+1)}{x^2}\right\}$$

$$= \lim_{x\to 0}\left\{e^{-\alpha x}\left(\frac{e^{\alpha x}-1}{x}\right)^2 - e^{-x}\left(\frac{e^x-1}{x}\right)^2\right\}$$

$$= 1\cdot\alpha^2\cdot 1^2 - 1\cdot 1 = \alpha^2-1$$

$\therefore \quad P = \dfrac{\alpha^2}{2}-\dfrac{1}{2} = \dfrac{3}{2}$ [given]

$\therefore \quad \alpha^2 = 4 \Rightarrow \alpha = \pm 2$

$\therefore \quad |\alpha| = 2$

Then, $1004|\alpha| = 2008$

5. $\because \lim_{x\to 2}\dfrac{\sqrt{(x+7)}-3\sqrt{(2x-3)}}{\sqrt[3]{(x+6)}-2\sqrt[3]{(3x-5)}}$

$$= \lim_{x\to 2}\frac{(x+7)^{1/2}-(18x-27)^{1/2}}{(x+6)^{1/3}-(24x-40)^{1/3}}$$

$$= \lim_{x\to 2}\frac{\dfrac{(x+7)^{1/2}-(18x-27)^{1/2}}{(x+7)-(18x-27)}\times(34-17x)}{\dfrac{(x+6)^{1/3}-(24x-40)^{1/3}}{(x+6)-(24x-40)}\times(46-23x)}$$

$$= \lim_{x\to 2}\frac{\frac{1}{2}(x+7)^{-1/2}}{\frac{1}{3}(x+6)^{-2/3}}\times\frac{17}{23}$$

$$= \frac{3}{2}\times\frac{9^{-1/2}}{8^{-2/3}}\times\frac{17}{23} = \frac{3}{2}\times\frac{4}{3}\times\frac{17}{23} = \frac{34}{23}$$

$\therefore \lim_{x\to 2} 52.9\dfrac{\sqrt{(x+7)}-3\sqrt{(2x-3)}}{\sqrt[3]{(x+6)}-2\sqrt[3]{(3x-5)}}$

$$= 52.9\times\frac{34}{23} = 2.3\times 34 = 78.2$$

6. $\lim_{x\to 0} 27e^{-2}\left\{\tan\left(\dfrac{\pi}{4}+x\right)\right\}^{1/x}$

$$= 27e^{-2}\cdot e^{\lim_{x\to 0}\left(\tan\left(\frac{\pi}{4}+x\right)-1\right)\frac{1}{x}}$$

[form 1^∞]

$$= 27e^{-2}\cdot e^{\lim_{x\to 0}\frac{2\tan x}{x(1-\tan x)}}$$

$$= 27e^{-2}\cdot e^2 = 27$$

7. $\because \lim_{x\to 0^-} f(x) = \lim_{h\to 0} f(0-h)$

$$= \lim_{h\to 0}\frac{\sin^{-1}(1-\{0-h\})\cdot\cos^{-1}(1-\{0-h\})}{\sqrt{2\{0-h\}}\cdot(1-\{0-h\})}$$

$$= \lim_{h\to 0}\frac{\sin^{-1}(1-(1-h))\cdot\cos^{-1}(1-(1-h))}{\sqrt{2(1-h)}\cdot(1-(1-h))}$$

$$= \lim_{h\to 0}\frac{\sin^{-1}h\cdot\cos^{-1}h}{\sqrt{2(1-h)}\cdot h} = 1\cdot\frac{\pi/2}{\sqrt{2}} = \frac{\pi}{2\sqrt{2}}$$

$\therefore \dfrac{2008\sqrt{2}}{\pi}\lim_{x\to 0^-} f(x) = \dfrac{2008\sqrt{2}}{\pi}\cdot\dfrac{\pi}{2\sqrt{2}} = 1004$

8. Let $P = \lim_{x\to 0}\left\{\dfrac{(1+x)^{1/x}}{e}\right\}^{1/x}$ [form 1^∞]

$$= e^{\lim_{x\to 0}\left(\frac{(1+x)^{1/x}}{e}-1\right)\frac{1}{x}}$$

$$= e^{\frac{1}{e}\lim_{x\to 0}\frac{(1+x)^{1/x}-e}{x}}$$

$$= e^{\frac{1}{e}(-e/2)} = e^{-1/2} = \frac{1}{\sqrt{e}}$$

$\therefore 78.6\sqrt{e}\lim_{x\to 0}\left\{\dfrac{(1+x)^{1/x}}{e}\right\}^{1/x} = 78.6\sqrt{e}\cdot\dfrac{1}{\sqrt{e}} = 78.6$

Single Integer Type

1. Given,

$$\lim_{x\to 0}\left\{\frac{1}{x^8}\left(1-\cos\frac{x^2}{2}-\cos\frac{x^2}{4}+\cos\frac{x^2}{2}\cos\frac{x^2}{4}\right)\right\}=2^{-k}$$

$$\Rightarrow \lim_{x\to 0}\frac{1}{x^8}\left(1-\cos\frac{x^2}{2}\right)\left(1-\cos\frac{x^2}{4}\right)=2^{-k}$$

$$\Rightarrow \lim_{x\to 0}\frac{2\sin^2\left(\frac{x^2}{4}\right)\cdot 2\sin^2\frac{x^2}{8}}{x^8}=2^{-k}$$

$$\Rightarrow \lim_{x\to 0} 4\left\{\left[\frac{\sin\frac{x^2}{4}}{x^2/4}\times\frac{\sin\frac{x^2}{8}}{\frac{x^2}{8}}\right]^2\frac{x^8}{(32)^2}\over x^8\right\}=2^{-k}$$

$$\Rightarrow \frac{4}{2^{10}}=2^{-k}$$

$$\Rightarrow 2^{-k}=2^{-8}$$

$$\Rightarrow k=8$$

2. $\lim_{n\to\infty} b_n=\lambda$ ...(i)

Given, $b_{n+1}=\frac{1}{3}\left(2b_n+\frac{125}{b_n^2}\right)$

$$\Rightarrow \lim_{n\to\infty} b_{n+1}=\frac{1}{3}\left\{2\lim_{n\to\infty} b_n+\frac{125}{\lim_{n\to\infty} b_n^2}\right\}$$

$$\Rightarrow \lambda=\frac{1}{3}\left(2\lambda+\frac{125}{\lambda^2}\right) \quad \text{[from Eq. (i)]}$$

$$\Rightarrow \lambda^3=125$$

$$\therefore \lambda=5$$

Hence, $\lim_{n\to\infty} b_n=5$

3. Given, $\lim_{\alpha\to 0}\left[\frac{e^{\cos(\alpha^n)}-e}{\alpha^m}\right]=-\frac{e}{2}$

$$\Rightarrow \lim_{\alpha\to 0}\frac{e\{e^{\cos(\alpha^n)-1}-1\}}{\cos(\alpha^n)-1}\cdot\frac{\cos(\alpha^n)-1}{\alpha^m}=\frac{-e}{2}$$

$$\Rightarrow \lim_{\alpha\to 0} e\left\{\frac{e^{\cos(\alpha^n)-1}-1}{\cos(\alpha^n)-1}\right\}\cdot\lim_{\alpha\to 0}\frac{-2\sin^2\frac{\alpha^n}{2}}{\alpha^m}=-e/2$$

$$\Rightarrow e\times 1\times(-2)\lim_{\alpha\to 0}\frac{\sin^2\left(\frac{\alpha^n}{2}\right)}{\frac{\alpha^{2n}}{4}}\cdot\frac{\alpha^{2n}}{4\alpha^m}=\frac{-e}{2}$$

$$\Rightarrow e\times 1\times(-2)\times 1\times\lim_{\alpha\to 0}\frac{\alpha^{2n-m}}{4}=\frac{-e}{2}$$

For this to be exists, $2n-m=0$

$$\Rightarrow \frac{m}{n}=2$$

4. As, we know that

$$\sin^{-1}x>x\Rightarrow\frac{\sin^{-1}x}{x}>1$$

We have,

$$\lim_{x\to 0}\left(\left[\frac{\sin^{-1}x}{x}\right]+\left[\frac{2^2\sin^{-1}2x}{x}\right]+\ldots\left[\frac{n^2\sin^{-1}nx}{x}\right]\right)=100$$

$$\Rightarrow \lim_{x\to 0}\frac{\sin^{-1}x}{x}+2^3\lim_{x\to 0}\frac{\sin^{-1}2x}{2x}+\ldots+n^3\lim_{x\to 0}\frac{\sin^{-1}nx}{nx}=100$$

$$\Rightarrow 1^3+2^3+3^3+\ldots n^3=100 \quad \left[\because \lim_{x\to 0}\frac{\sin^{-1}x}{x}=1\right]$$

$$\Rightarrow \left[\frac{n(n+1)}{2}\right]^2=100$$

$$=\frac{n(n+1)}{2}=10$$

$$\Rightarrow n(n+1)=20$$

$$\therefore n=4$$

5. Given, $\lim_{x\to 1}\left\{\frac{\sin(x-1)+a(1-x)}{(x-1)+\sin(x-1)}\right\}^{\frac{(1+\sqrt{x})(1-\sqrt{x})}{1-\sqrt{x}}}=\frac{1}{4}$

$$\Rightarrow \lim_{x\to 1}\left\{\frac{\frac{\sin(x-1)}{(x-1)}-a}{1+\frac{\sin(x-1)}{(x-1)}}\right\}^{1+\sqrt{x}}=\frac{1}{4}$$

$$\Rightarrow \left(\frac{1-a}{2}\right)^2=\frac{1}{4}\Rightarrow(a-1)^2=1$$

$$\Rightarrow a=2 \text{ or } 0$$

Hence, the maximum value of a is 2.

6. Here, $\lim_{x\to 0}\frac{x^2\sin(\beta x)}{\alpha x-\sin x}=1$

$$\Rightarrow \lim_{x\to 0}\frac{x^2\left(\beta x-\frac{(\beta x)^3}{3!}+\frac{(\beta x)^5}{5!}-\ldots\right)}{\alpha x-\left(x-\frac{x^3}{3!}+\frac{x^5}{5!}-\ldots\right)}=1$$

$$\Rightarrow \lim_{x\to 0}\frac{x^3\left(\beta-\frac{\beta^3x^2}{3!}+\frac{\beta^5x^4}{5!}-\ldots\right)}{(\alpha-1)x+\frac{x^3}{3!}+\frac{x^5}{5!}-\ldots}=1$$

Limit exists only, when $\alpha-1=0$

$$\Rightarrow \alpha=1 \quad \ldots\text{(i)}$$

$$\therefore \lim_{x\to 0}\frac{x^3\left(\beta-\frac{\beta^3x^2}{3!}+\frac{\beta^5x^4}{5!}-\ldots\right)}{x^3\left(\frac{1}{3!}-\frac{x^2}{5!}-\ldots\right)}=1\Rightarrow 6\beta=1 \quad \ldots\text{(ii)}$$

From Eqs. (i) and (ii), we get

$$6(\alpha+\beta)=6\alpha+6\beta=6+1=7$$

7. We have,

$$y_n = \frac{1}{n}[(n+1)(n+2)\dots(n+n)]^{1/n}$$

and $\lim_{n\to\infty} y_n = L$

$$\Rightarrow \quad L = \lim_{n\to\infty}\frac{1}{n}[(n+1)(n+2)(n+3)\dots(n+n)]^{1/n}$$

$$\Rightarrow \quad L = \lim_{n\to\infty}\left[\left(1+\frac{1}{n}\right)\left(1+\frac{2}{n}\right)\left(1+\frac{3}{n}\right)\dots\left(1+\frac{n}{n}\right)\right]^{\frac{1}{n}}$$

$$\Rightarrow \log L = \lim_{n\to\infty}\frac{1}{n}\left[\log\left(1+\frac{1}{n}\right)+\log\left(1+\frac{2}{n}\right)\dots\log\left(1+\frac{n}{n}\right)\right]$$

$$\Rightarrow \log L = \lim_{n\to\infty}\frac{1}{n}\sum_{r=1}^{n}\log\left(1+\frac{r}{n}\right)$$

$$\Rightarrow \log L = \int_0^1 \underset{\text{II}}{1}\times\underset{\text{I}}{\log(1+x)}\,dx$$

$$\Rightarrow \log L = (x\cdot\log(1+x))_0^1 - \int_0^1\left[\frac{d}{dx}(\log(1+x)\int dx\right]dx$$

[using integration by parts]

$$\Rightarrow \log L = [x\log(1+x)]_0^1 - \int_0^1\frac{x}{1+x}\,dx$$

$$\Rightarrow \log L = \log 2 - \int_0^1\left(\frac{x+1}{x+1}-\frac{1}{x+1}\right)dx$$

$$\Rightarrow \log L = \log 2 - [x]_0^1 + [\log(x+1)]_0^1$$

$$\Rightarrow \log L = \log 2 - 1 + \log 2 - 0$$

$$\Rightarrow \log L = \log 4 - \log e = \log\frac{4}{e} \Rightarrow L = \frac{4}{e} \Rightarrow [L] = \left[\frac{4}{e}\right] = 1$$

8. $\lim_{x\to 0}\dfrac{ae^x - b\cos x + ce^{-x}}{x\sin x} = 2$

$$\Rightarrow \quad \lim_{x\to 0}\frac{a\left(1+x+\frac{x^2}{2!}\dots\right)-b\left(1-\frac{x^2}{2!}+\dots\right)+c\left(1-x+\frac{x^2}{2!}\right)}{\left(\frac{x\sin x}{x}\right)x} = 2$$

$$a-b+c=0$$

$$a-c=0$$

and $\dfrac{a+b+c}{2} = 2$

$$\Rightarrow \quad a+b+c=4$$

Matrix-Match Type

1. (A) For $x\to\infty, |x| = x$

$$\therefore \quad f(x) = \left(\frac{x}{x+2}\right)^{-x} = \left(\frac{x+2}{x}\right)^x = \left(1+\frac{2}{x}\right)^x$$

$$\Rightarrow \quad \lim_{x\to\infty} f(x) = \lim_{x\to\infty}\left(1+\frac{2}{x}\right)^x = e^2 \text{ (P)}$$

and for $x\to-\infty, |x| = -x$

$$\therefore \quad f(x) = \left(\frac{-x}{-x+2}\right)^{-x} = \left(\frac{-x+2}{-x}\right)^x = \left(1-\frac{2}{x}\right)^x$$

$$\Rightarrow \quad \lim_{x\to-\infty} f(x) = \lim_{x\to-\infty}\left(1-\frac{2}{x}\right)^x = e^{-2} \text{ (R)}$$

(B) $\because f(x) = \dfrac{(1+x)^{1/x}-e}{x}$

$$= \frac{e\left\{1-\frac{x}{2}+\frac{11}{24}x^2+\dots\right\}-e}{x}$$

$$= \frac{e\left(-\frac{x}{2}+\frac{11}{24}x^2+\dots\right)}{x}$$

$$= e\left(-\frac{1}{2}+\frac{11}{24}x+\dots\right)$$

$$\Rightarrow \lim_{x\to 0} f(x) = -\frac{e}{2} < -1 \text{ (S, T)}$$

(C) $\quad f(x) = \left(\dfrac{1+5x^2}{1+3x^2}\right)^{1/x^2}$

$$\therefore \lim_{x\to 0} f(x) = \lim_{x\to 0}\left(\frac{1+5x^2}{1+3x^2}\right)^{1/x^2}$$

$$= e^{\lim_{x\to 0}\left(\frac{1+5x^2}{1+3x^2}-1\right)\frac{1}{x^2}}$$

$$= e^{\lim_{x\to 0}\left(\frac{2}{1+3x^2}\right)} = e^2 \text{ (Q)}$$

2. (A) $\because \lim_{x\to 0}(\cos x + a\sin bx)^{1/x} = e^{\lim_{x\to 0}(\cos x + a\sin bx - 1)\frac{1}{x}}$

$$= e^{\lim_{x\to 0}\left(-\frac{(1-\cos x)}{x}+\frac{a\sin bx}{x}\right)}$$

$$= e^{ab} = e^2 \qquad \text{(given)}$$

$$\Rightarrow \quad ab = 2 \text{ (P)}$$

If $a = 1, b = 2$

$$\Rightarrow \quad a+b = 3 \text{(S)}$$

(B) $\lim_{x\to 0}(1+ax+bx^2)^{2/x} = e^{\lim_{x\to 0}(1+ax+bx^2-1)\frac{2}{x}}$

$$= e^{2a} = e^3 \qquad \text{(given)}$$

$\therefore \quad a = 3/2,\ b\in R$

If $b = 4/3 \Rightarrow ab = 2$, if $b = 2$

$\Rightarrow ab = 3$, if $b = 8/3 \Rightarrow ab = 4$

If $b = 3/2 \Rightarrow a+b = 3$, if $b = 5/2 \Rightarrow a+b = 4$

(P, Q, R, S, T)

(C) $\lim_{x\to 0}\dfrac{ae^x - b}{x} = 2$

At $x\to 0$, Numerator $\Rightarrow ae^0 - b = a - b$

Since RHS is finite.

$\therefore$ Numerator must be $= 0$

$\therefore \quad a - b = 0 \Rightarrow a = b$

Then, $\lim_{x \to 0} \frac{ae^x - a}{x} = 2$

$\Rightarrow \quad a \lim_{x \to 0} \left(\frac{e^x - 1}{x} \right) = 2$

$\Rightarrow \quad a \cdot 1 = 2$

$\Rightarrow \quad a = 2 = b$

$ab = 4$ and $a + b = 4$ (R, T)

3. Since, $\frac{\sin x}{x} \to 1$ as $x \to 0$ but less than 1.

$\therefore \quad \frac{n_1 \sin x}{x} \to n_1$ as $x \to 0$ but less than n_1

Similarly,

$\frac{x}{\sin x} \to 1$ as $x \to 0$ but more than or equal to 1.

$\therefore \frac{n_2 x}{\sin x} \to n_2$ as $x \to 0$ but more than or equal to n_2

$\therefore L = \lim_{x \to 0} \left[\frac{n_1 \sin x}{x} \right] = n_1 - 1$ and $M = \lim_{x \to 0} \left[\frac{n_2 x}{\sin x} \right] = n_2$

Also, $\frac{\tan x}{x} \to 1$ as $x \to 0$ but more than 1.

$\therefore \quad \frac{n_3 \tan x}{x} \to n_3$ as $x \to 0$ but more than or equal to n_3

Similarly,

$\frac{x}{\tan x} \to 1$ as $x \to 0$ but less than 1.

$\therefore \quad \frac{n_4 x}{\tan x} \to n_4$ as $x \to 0$ but less than n_4

$\Rightarrow \quad N = \lim_{x \to 0} \left[\frac{n_3 \tan x}{x} \right] = n_3$

and $\quad P = \lim_{x \to 0} \left[\frac{n_4 x}{\tan x} \right] = n_4 - 1$

(A) For $n_1 = 3, n_4 = 9$

$\therefore \quad \frac{P}{L} = \frac{n_4 - 1}{n_1 - 1} = \frac{9-1}{3-1} = 4$ (P, R)

(B) For $n_2 = 12, n_3 = 2$

$\therefore \quad \frac{M}{N} = \frac{n_2}{n_3} = \frac{12}{2} = 6$ (P, Q, T)

(C) For $n_2 = 3, n_4 = 10$

$\therefore P + M = n_4 - 1 + n_2 = 10 - 1 + 3 = 12$ (P, Q, R, T)

(D) For $n_1 = 4, n_2 = 5, n_3 = 6, n_4 = 7$

$\therefore L + N + P - M = n_1 - 1 + n_3 + n_4 - 1 - n_2$

$= 4 - 1 + 6 + 7 - 1 - 5$

$= 10$ (P, S)

12

Continuity and Differentiability

Objective Questions Type I [Only one correct answer]

Each question in this part has four choices out of which just one is correct. Indicate your choice of correct answer for each question by writing one of the letters a, b, c, d whichever is appropriate.

1. If the function $f(x) = \left[\dfrac{(x-2)^3}{a}\right] \sin(x-2)$ $+ a\cos(x-2)$, [.] denotes the greatest integer function, is continuous and differentiable in (4, 6), then

(a) $a \in [8, 64]$ (b) $a \in (0, 8]$

(c) $a \in [64, \infty)$ (d) None of these

2. Let $f(x)$ and $g(x)$ be twice differentiable functions such that $f(x) = x^2 + g'(1)x + g''(2)$ and $g(x) = f(1)x^2 + xf'(x) + f''(x)$. Then $f(x) - g(x) =$

(a $2x + 5$ (b) $3x^2 + 6x + 1$

(c) $x^2 - 6x + 2$ (d) $x^2 - 2$

3. Let $f(x) = x^3 + x$ be function and $g(x) = \begin{cases} f(|x|), & x \geq 0 \\ f(-|x|), & x < 0 \end{cases}$, then

(a) $g(x)$ is continuous $\forall\, x \in R$

(b) $g(x)$ is continuous $\forall\, x \in R^-$ only

(c) $g(x)$ is continuous $\forall\, x \in R^+$ only

(d) $g(x)$ is discontinuous $\forall x \in R^-$

4. If a function $f(x)$ defined by

$$f(x) = \begin{cases} ae^x + be^{-x}, & -1 \leq x < 1 \\ cx^2, & 1 \leq x \leq 3 \\ ax^2 + 2cx, & 3 < x \leq 4 \end{cases}$$

be continuous for some $a, b, c \in R$ and $f'(0) + f'(2) = e$, then the value of a is **[JEE Main 2020, 2 September]**

(a) $\dfrac{e}{e^2 + 3e + 13}$ (b) $\dfrac{e}{e^2 - 3e - 13}$

(c) $\dfrac{1}{e^2 - 3e + 13}$ (d) $\dfrac{e}{e^2 - 3e + 13}$

5. If $[x]$ denotes the integral part of x and $f(x) = [n + p \sin x]$, $0 < x < \pi$, $n \in I$ and p is a prime number, then the number of points, where $f(x)$ is not differentiable is

(a) $p - 1$ (b) p

(c) $2p - 1$ (d) $2p + 1$

6. If $f(x) = (x^2 - 4)|(x^3 - 6x^2 + 11x - 6| + \dfrac{x}{1 + |x|}$, then the set of points at which the function $f(x)$ is not differentiable is

(a) $\{-2, 2, 1, 3\}$ (b) $\{-2, 0, 3\}$

(c) $\{-2, 2, 0\}$ (d) $\{1, 3\}$

7. Let $\alpha \in R$ be such that the function

$$f(x) = \begin{cases} \dfrac{\cos^{-1}(1 - \{x\}^2)\sin^{-1}(1 - \{x\})}{\{x\} - \{x\}^3}, & x \neq 0 \\ \alpha, & x = 0 \end{cases}$$

is continuous at $x = 0$, where $\{x\} = x - [x]$, $[x]$ is the greatest integer less than or equal to x.

Then, **[JEE Main 2021, 18 March]**

(a) $\alpha = \dfrac{\pi}{\sqrt{2}}$

(b) $\alpha = 0$

(c) no such α exists

(d) $\alpha = \dfrac{\pi}{4}$

8. Given, $f(x) = x^2e^{2(x-1)}, 0 \le x \le 1$

$= a\, Sgn\,(x+1)\cos(2x-2) + bx^2, 1 < x \le 2,$

$f(x)$ is differentiable at $x = 1$ provided

(a) $a = -1, b = 2$ (b) $a = 1, b = -2$

(c) $a = -3, b = 4$ (d) $a = 3, b = -4$

9. If $f(x) = \begin{cases} \dfrac{\sin\{\cos x\}}{x - (\pi/2)}, & x \ne \dfrac{\pi}{2} \\ 1, & x = \dfrac{\pi}{2} \end{cases}$, where {.} represents the fractional part function, then $f(x)$ is

(a) continuous at $x = \dfrac{\pi}{2}$

(b) $\lim\limits_{x \to \frac{\pi}{2}} f(x)$ exists, but f is not continuous at $x = \dfrac{\pi}{2}$

(c) $\lim\limits_{x \to \frac{\pi}{2}} f(x)$ does not exist

(d) $\lim\limits_{x \to \frac{\pi}{2}^-} f(x) = -1$

10. If $f(x) = \dfrac{1}{(x-1)(x-2)}$ and $g(x) = \dfrac{1}{x^2}$, then points of discontinuity of $f(g(x))$ are

(a) $\left\{-1, 0, 1, \dfrac{1}{\sqrt{2}}\right\}$ (b) $\left\{-\dfrac{1}{\sqrt{2}}, -1, 0, 1, \dfrac{1}{\sqrt{2}}\right\}$

(c) $\{0, 1\}$ (d) $\left\{0, 1, \dfrac{1}{\sqrt{2}}\right\}$

11. The value of the $f(0)$, so that the function

$$f(x) = \frac{\sqrt{(a^2 - ax + x^2)} - \sqrt{(a^2 + ax + x^2)}}{\sqrt{(a+x)} - \sqrt{(a-x)}}$$ becomes

continuous for all x, is given by

(a) $a\sqrt{a}$ (b) $\sqrt{a}$

(c) $-\sqrt{a}$ (d) $-a\sqrt{a}$

12. Let $f(x) = \tan(\pi/4 - x)/\cot 2x\ (x \ne \pi/4)$.The value which should be assigned to f at $x = \pi/4$. So that it is continuous every where, is

(a) 1/2 (b) 1

(c) 2 (d) None of these

13. If $f(x) = \dfrac{2 - (256 - 7x)^{1/8}}{(5x + 32)^{1/5} - 2}\ (x \ne 0)$; then for f to be continuous every where, $f(0)$ is equal to

(a) -1 (b) 1

(c) 2^6 (d) None of these

14. Let f be a function satisfying $f(x+y) = f(x) + f(y)$ and $f(x) = x^2 g(x)$ for all x and y, where $g(x)$ is a continuous function, then $f'(x)$ is equal to

(a) $g'(x)$ (b) $g(0)$

(c) $g(0) + g'(x)$ (d) 0

15. Let $f(x+y) = f(x)f(y)$ for all x and y. Suppose that $f(3) = 3$ and $f'(0) = 11$, then $f'(3)$ is given by

(a) 22 (b) 44

(c) 28 (d) None of these

16. If $f(x) = \begin{cases} \dfrac{1}{|x|} & ; |x| \ge 1 \\ ax^2 + b & ; |x| < 1 \end{cases}$ is differentiable at every point of the domain, then the values of a and b are respectively **[JEE Main 2021, 18 March]**

(a) $\dfrac{1}{2}, \dfrac{1}{2}$ (b) $\dfrac{1}{2}, \dfrac{-3}{2}$ (c) $\dfrac{5}{2}, \dfrac{-3}{2}$ (d) $\dfrac{-1}{2}, \dfrac{3}{2}$

17. Let $f: R \to R$ be a function such that $f\left(\dfrac{x+y}{3}\right) = \dfrac{f(x) + f(y)}{3}$, $f(0) = 3$ and $f'(0) = 3$, then

(a) $\dfrac{f(x)}{x}$ is differentiable in R

(b) $f(x)$ is continuous but not differentiable in R

(c) $f(x)$ is continuous in R

(d) $f(x)$ is bounded in R

18. If $f(x) = \dfrac{\tan \pi[(2x - 3\pi)^3]}{1 + [2x - 3\pi]^2}$ ([.] denotes the greatest integer function), then

(a) $f(x)$ is continuous in R

(b) $f(x)$ is continuous in R but not differentiable in R

(c) $f'(x)$ exists everywhere but $f''(x)$ does not exist at some $x \in R$

(d) None of the above

19. If $f(x) = \begin{cases} 1, & x \text{ is rational} \\ 2, & x \text{ is irrational} \end{cases}$, then

(a) $f(x)$ is continuous in $R \sim I$

(b) $f(x)$ is continuous in $R \sim Q$

(c) $f(x)$ is continuous in R but not differentiable in R

(d) $f(x)$ is neither continuous nor differentiable in R

20. If $f(x) = \begin{cases} x[x], & 0 \le x < 2 \\ (x-1)[x], & 2 \le x \le 4 \end{cases}$, where [·] denotes greatest integer function then

(a) neither $f'(1)$ exist nor $f'(2)$ exist

(b) $f'(1)$ exists but $f'(2)$ does not exist

(c) $f'(2)$ exists but $f'(1)$ does not exist

(d) both $f'(1)$ as well as $f'(2)$ exist

21. If the derivative of the function

$f(x) = \begin{cases} bx^2 + ax + 4; & x \ge -1 \\ ax^2 + b; & x < -1 \end{cases}$, is everywhere continuous, then

(a) $a = 2, b = 3$ (b) $a = 3, b = 2$

(c) $a = -2, b = -3$ (d) $a = -3, b = -2$

22. If $f(x) = \begin{cases} \sin x, & x \ne n\pi, n \in I \\ 2, & \text{otherwise} \end{cases}$ and

$g(x) = \begin{cases} x^2 + 1, & x \ne 0, 2 \\ 4, & x = 0 \\ 5, & x = 2 \end{cases}$, then $\lim\limits_{x \to 0} g\{f(x)\}$ is

(a) 5 (b) 6 (c) 7 (d) 1

23. The function $f(x) = |2\, Sgn\, 2x| + 2$ has
(a) jump discontinuity
(b) removal discontinuity
(c) infinite discontinuity
(d) no discontinuity at $x = 0$

24. If the function

$$f(x) = \begin{cases} (1 + |\sin x|)^{a/|\sin x|}, & -\frac{\pi}{6} < x < 0 \\ b, & x = 0 \\ e^{\tan 2x/\tan 3x}, & 0 < x < \frac{\pi}{6} \end{cases}, \text{ is continuous}$$

at $x = 0$, then
(a) $a = \log_e b, a = 2/3$
(b) $b = \log_e a, a = 2/3$
(c) $a = \log_e b, b = 2$
(d) None of the above

25. Let the functions $f: R \to R$ and $g: R \to R$ be defined as

$$f(x) = \begin{cases} x + 2, & x < 0 \\ x^2, & x \geq 0 \end{cases} \text{ and } g(x) = \begin{cases} x^3, & x < 1 \\ 3x - 2, & x \geq 1 \end{cases}$$

Then, the number of points in R where $(fog)(x)$ is NOT differentiable is equal to **[JEE Main 2021, 16 March]**
(a) 3 (b) 1
(c) 0 (d) 2

26. Let $f(x) = \begin{cases} x^n \sin\left(\frac{1}{x}\right), & x \neq 0 \\ 0, & x = 0 \end{cases}$, then $f(x)$ is continuous but not differentiable at $x = 0$ if
(a) $n \in (0, 1]$ (b) $n \in [1, \infty)$
(c) $n \in (-\infty, 0)$ (d) $n = 0$

27. Let [·] represents the greatest integer function and $f(x) = [\tan^2 x]$, then
(a) $\lim_{x \to 0} f(x)$ does not exist
(b) $f(x)$ is continuous at $x = 0$
(c) $f(x)$ is non-differentiable at $x = 0$
(d) $f'(0) = 1$

28. $$f(x) = \begin{cases} x - [x], \text{ for } 2n \leq x < 2n + 1, n \in N, \\ \text{where}[x] = \text{Integral part of } x \leq x \\ \frac{1}{2}, \text{ for } 2n + 1 \leq x < 2n + 2 \end{cases}$$

the function
(a) is discontinuous at $x = 1, 2$
(b) is periodic with period 1
(c) is periodic with period 2
(d) $\int_0^2 f(x)\, dx$ exists

29. A function, $f(x) = x\left[1 + \frac{1}{3}\sin(\log x^2)\right], x \neq 0$

$f(0) = 0$
[·] = Integral part, the function
(a) is continuous at $x = 0$
(b) is monotonic
(c) is derivable at $x = 0$
(d) cannot be defined for $x < -1$

30. If $f(x) = \begin{cases} [\cos \pi x], & x < 1 \\ |x - 2|, & 1 \leq x < 2 \end{cases}$

([·] denotes the greatest integer function), then $f(x)$ is
(a) continuous and non-differentiable at $x = -1$ and $x = 1$
(b) continuous and differentiable at $x = 0$
(c) discontinuous at $x = 1/2$
(d) continuous but not differentiable at $x = 2$

31. If $f(x) = [\sqrt{2} \sin x]$, where $[x]$ represents the greatest integer function $\leq x$, then
(a) $f(x)$ is periodic
(b) maximum value of $f(x)$ is 1 in the interval $[-2\pi, 2\pi]$
(c) $f(x)$ is discontinuous at $x = \frac{n\pi}{2} + \frac{\pi}{4}, n \in I$
(d) $f(x)$ is differentiable at $x = n\pi, n \in I$

32. The function $f(x) = \begin{cases} \frac{\pi}{4} + \tan^{-1} x, & |x| \leq 1 \\ \frac{1}{2}(|x| - 1), & |x| > 1 \end{cases}$ is

[JEE Main 2020, 4 September]
(a) both continuous and differentiable on $R - \{1\}$
(b) both continuous and differentiable on $R - \{-1\}$
(c) continuous on $R - \{-1\}$ and differentiable on $R - \{-1, 1\}$
(d) continuous on $R - \{1\}$ and differentiable on $R - \{-1, 1\}$

33. If $f(x) = \frac{x - e^x + \cos 2x}{x^2}$, $x \neq 0$ is continuous at $x = 0$,
[where [.] and {} denote the greatest integer and fractional part function, respectively]
(a) $f(0) = \frac{5}{2}$ (b) $[f(0)] = -2$
(c) $\{f(0)\} = -0.5$ (d) $[f(0)] \cdot \{f(0)\} = -1.5$

34. Let $f(x) = \begin{cases} x^p \sin\left(\frac{1}{x}\right) + x|x|, & x \neq 0 \\ 0, & x = 0 \end{cases}$, the set of values, of p for which $f''(x)$ is everywhere continuous is
(a) $[2, \infty)$ (b) $[-3, \infty)$
(c) $[5, \infty)$ (d) None of these

35. The value of p for which the function

$$f(x) = \frac{(4^x - 1)^3}{\sin(x/p) \ln\left(1 + \frac{x^2}{3}\right)}, x \neq 0$$

$$= 12(\ln 4)^3, x = 0$$

may be continuous at $x = 0$ is
(a) 1 (b) 2
(c) 3 (d) 4

36. The value of $f(0)$, so that the function $f(x) = \dfrac{1-\cos(1-\cos x)}{x^4}$ is continuous everywhere is

(a) 1/8 (b) 1/2
(c) 1/4 (d) None of these

37. The jump of the function at the point of the discontinuity of the function $f(x) = \dfrac{1-k^{1/x}}{1+k^{1/x}}\ (k>0)$ is

(a) 4 (b) 2
(c) 3 (d) None of these

38. Let $f''(x)$ be continuous at $x=0$ and $f''(0)=4$. The value of $\lim\limits_{x\to 0}\dfrac{2f(x)-3f(2x)+f(4x)}{x^2}$ is

(a) 6 (b) 10
(c) 11 (d) 12

39. Let $f(x) = [x] + \sqrt{x-[x]}$, where $[x]$ denotes the greatest integer function . Then

(a) $f(x)$ is continuous on R^+
(b) $f(x)$ is continuous on R
(c) $f(x)$ is continuous on $R \sim I$
(d) None of the above

40. Let $f(x) = \begin{cases} \dfrac{a|x^2-x-2|}{2+x-x^2}, & x<2 \\ b, & x=2 \\ \dfrac{x-[x]}{x-2}, & x>2 \end{cases}$

([.] denotes the greatest integer function)
If $f(x)$ is continuous at $x=2$, then

(a) $a=1, b=2$ (b) $a=1, b=1$
(c) $a=0, b=1$ (d) $a=2, b=1$

41. Let $f: R \to R$ be a function defined as

$$f(x) = \begin{cases} \dfrac{\sin(a+1)x+\sin 2x}{2x}, & \text{if } x<0 \\ b, & \text{if } x=0 \\ \dfrac{\sqrt{x+bx^3}-\sqrt{x}}{bx^{5/2}}, & \text{if } x>0 \end{cases}$$

If f is continuous at $x=0$, then the value of $a+b$ is equal to **[JEE Main 2021, 18 March]**

(a) $-\dfrac{5}{2}$ (b) -2
(c) -3 (d) $-\dfrac{3}{2}$

42. Let $f(x) = \begin{cases} -1, & x<0 \\ 0, & x=0 \\ 1, & x>0 \end{cases}$ and $g(x) = \sin x + \cos x$, then points of discontinuity of $f\{g(x)\}$ in $(0, 2\pi)$ is

(a) $\left\{\dfrac{\pi}{2}, \dfrac{3\pi}{4}\right\}$ (b) $\left\{\dfrac{3\pi}{4}, \dfrac{7\pi}{4}\right\}$
(c) $\left\{\dfrac{2\pi}{3}, \dfrac{5\pi}{3}\right\}$ (d) $\left\{\dfrac{5\pi}{4}, \dfrac{7\pi}{3}\right\}$

43. The points of discontinuity of the function $f(x) = \lim\limits_{n\to\infty}\dfrac{(2\sin x)^{2n}}{3^n-(2\cos x)^{2n}}$ are given by

(a) $r\pi \pm \dfrac{\pi}{12}, r \in I$ (b) $r\pi \pm \dfrac{\pi}{6}, r \in I$
(c) $r\pi \pm \dfrac{\pi}{3}, r \in I$ (d) None of these

44. Let $f(x) = [\cos x + \sin x], 0 < x < 2\pi$, where $[x]$ denotes the greatest integer less than or equal to x. The number of points of discontinuity of $f(x)$ is

(a) 6 (b) 5
(c) 4 (d) 3

45. The function $f(x) = |x^2-3x+2| + \cos|x|$ is not differentiable at x is equal to

(a) -1 (b) 0
(c) 1 (d) 2

46. Let $f(x) = \dfrac{1-\tan x}{4x-\pi}, x \neq \pi/4$ and $x \in [0, \pi/2)$
$= \lambda, x = \pi/4$
If $f(x)$ is continuous in $(0, \pi/2]$, then λ is

(a) 1 (b) 1/2
(c) $-1/2$ (d) None of these

47. If $g(x) = \begin{cases} [f(x)], & x \in \left(0, \dfrac{\pi}{2}\right) \cup \left(\dfrac{\pi}{2}, \pi\right) \\ 3, & x = \pi/2 \end{cases}$

where $[x]$ denotes the greatest integer function and $f(x) = \dfrac{2(\sin x - \sin^n x) + |\sin x - \sin^n x|}{2(\sin x - \sin^n x) - |\sin x - \sin^n x|}, n \in R$, then

(a) $g(x)$ is continuous and differentiable at $x = \pi/2$, when $0<n<1$
(b) $g(x)$ is continuous and differentiable at $x=\pi/2$, when $n>1$
(c) $g(x)$ is continuous but not differentiable at $x=\pi/2$, when $0<n<1$
(d) $g(x)$ is continuous but not differentiable, at $x=\pi/2$, when $n>1$

48. Let f be a twice differentiable function on (1, 6). If $f(2)=8$, $f'(2)=5$, $f'(x) \geq 1$ and $f''(x) \geq 4$, for all $x \in (1, 6)$, then **[JEE Main 2020, 4 September]**

(a) $f(5)+f'(5) \geq 28$
(b) $f(5) \leq 10$
(c) $f(5)+f'(5) \leq 26$
(d) $f'(5)+f''(5) \leq 20$

49. If $f(x)$ is a continuous function $\forall\, x \in R$ and the range of $f(x)$ is $(2, \sqrt{26})$ and $g(x) = \left[\dfrac{f(x)}{c}\right]$ is continuous $\forall\, x \in R$, then the least positive integral value of c is (where [.] denotes the greatest integer function)

(a) 2 (b) 3
(c) 5 (d) 6

50. If $f(x)=\begin{cases} a+\dfrac{\sin[x]}{x}, & x>0 \\ 2, & x=0 \\ b+\left[\dfrac{\sin x - x}{x^3}\right], & x<0 \end{cases}$

(where [.] denotes the greatest integer function). If $f(x)$ is continuous at $x=0$, then b is equal to
(a) $a-2$ (b) $a-1$
(c) $a+1$ (d) $a+2$

51. Let $f(x)=\begin{cases} x^2, & \text{if } x\le x_0 \\ ax+b, & \text{if } x>x_0 \end{cases}$

The values of the coefficients a and b for which the function is continuous and has a derivative at x_0, are
(a) $a=x_0, b=-x_0$ (b) $a=2x_0, b=-x_0^2$
(c) $a=x_0^2, b=-x_0$ (d) $a=x_0, b=-x_0^2$

52. Let $f(x)=\begin{cases} (1+|\cos x|)^{ab/|\cos x|}, & n\pi<x<(2n+1)\pi/2 \\ e^a\cdot e^b, & x=(2n+1)\pi/2 \\ e^{\cot 2x/\cot 8x}, & (2n+1)\pi/2<x<(n+1)\pi \end{cases}$

If $f(x)$ is continuous in $(n\pi,(n+1)\pi)$, then
(a) $a=1, b=2$ (b) $a=2, b=2$
(c) $a=2, b=3$ (d) $a=3, b=4$

Objective Questions Type II [One or more than one correct answer(s)]

Each question in this part has more than one correct answers. For each question, write the letters a, b, c, d corresponding to the correct answers.

1. Let $f(x)=\dfrac{1}{[\sin x]}$, ([.] denotes the greatest integer function) then
(a) domain of $f(x)$ is $(2n\pi+\pi, 2n\pi+2\pi)\cup\{2n\pi+\pi/2\}$, where $n\in I$
(b) $f(x)$ is continuous, when $x\in(2n\pi+\pi, 2n\pi+2\pi)$, where $n\in I$
(c) $f(x)$ is differentiable at $x=\pi/2$
(d) None of the above

2. Let $g(t)=[t(1/t)]$ for $t>0$ ([.] denotes the greatest integer function), then $g(g)$ has :
(a) discontinuities at finite number of points
(b) discontinuities at infinite number of points
(c) $g(1/2)=1$
(d) $g(3/4)=1$

3. Let $f(x)$ and $\phi(x)$ be defined by $f(x)=[x]$ and $\phi(x)=\begin{cases} 0, & x\in I \\ x^2, & x\in R-I \end{cases}$ (where [.] denotes the greatest integer function), then
(a) $\lim\limits_{x\to 1}\phi(x)$ exists, but ϕ is not continuous at $x=1$
(b) $\lim\limits_{x\to 1}f(x)$ does not exist and f is not continuous at $x=1$
(c) ϕ *of* continuous for all x
(d) *fo* ϕ is continuous for all x

4. The following functions are continuous on $(0,\pi)$:
(a) $\tan x$
(b) $\int_0^{\pi} t\sin\left(\dfrac{1}{t}\right)dt$

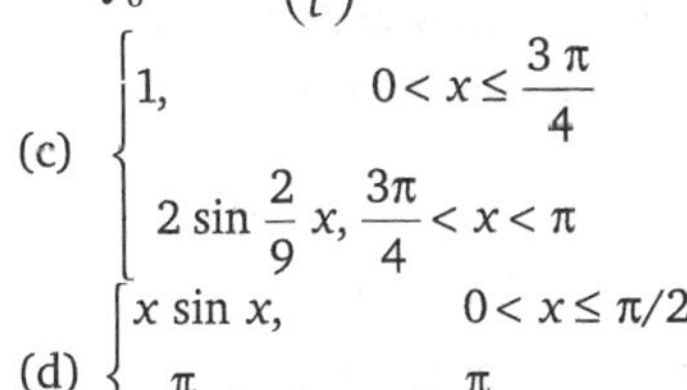

(c) $\begin{cases} 1, & 0<x\le\dfrac{3\pi}{4} \\ 2\sin\dfrac{2}{9}x, & \dfrac{3\pi}{4}<x<\pi \end{cases}$
(d) $\begin{cases} x\sin x, & 0<x\le\pi/2 \\ \dfrac{\pi}{2}\sin(\pi+x), & \dfrac{\pi}{2}<x<\pi \end{cases}$

5. Let the function $f:R\to R$ be defined by $f(x)=x^3-x^2+(x-1)\sin x$ and let $g:R\to R$ be an arbitrary functions. Let $fg:R\to R$ be the product function defined by $(fg)(x)=f(x)g(x)$. Then which of the following statements is/are TRUE?
(a) If g is continuous at $x=1$, then fg is differentiable at $x=1$
(b) If fg is differentiable at $x=1$, then g is continous at $x=1$
(c) If g is differentiable at $x=1$, then fg is differentiable at $x=1$
(d) If fg is differentiable at $x=1$, then g is differentiable at $x=1$

6. The function $f(x)=\begin{cases} 1, & |x|\ge 1 \\ \dfrac{1}{n^2}, & \dfrac{1}{n}\le|x|<\dfrac{1}{n-1}, n=2,3,\ldots \\ 0, & x=0 \end{cases}$
(a) is discontinuous at infinite points
(b) is continuous everywhere
(c) is discontinuous only at $x=\dfrac{1}{n}$, $n\in Z-\{0\}$
(d) None of the above

7. A function, which is continuous and not differentiable at the origin, is :
(a) $f(x)=x$ for $x<0$ and $f(x)=x^2$ for $x\ge 0$
(b) $g(x)=x$ for $x<0$ and $g(x)=2x$ for $x\ge 0$
(c) $h(x)=x|x|, x\in R$
(d) $k(x)=1+|x|, x\in R$

8. Let $f(x)=\lim\limits_{n\to\infty}(\sin x)^{2n}$, then f is :
(a) continuous at $x=\pi/2$
(b) discontinuous at $x=\pi/2$
(c) discontinuous at $x=-\pi/2$
(d) discontinuous at an infinite number of points

9. If $f(x) = \tan^{-1} \cot x$, then :
(a) $f(x)$ is periodic with period π
(b) $f(x)$ is discontinuous at $x = \pi/2, 3\pi/2$
(c) $f(x)$ is not differentiable at $x = \pi, 99\pi, 100\pi$
(d) $f(x) = -1$, for $2n\pi \le x \le (2n+1)\pi$

10. Let $f(x) = \dfrac{(\log(1+x) - \log 2)(3 \cdot 4^{x-1} - 3x)}{\{(7+x)^{1/3} - (1+3x)^{1/2}\} \sin \pi x}, x \neq 1$
The value of $f(1)$ so that f is continuous at $x = 1$ is
(a) an algebraic number
(b) a rational number
(c) a transcedental number
(d) $-\dfrac{9}{\pi} \log \dfrac{e}{4}$

11. $f(x) = \min\{1, \cos x, 1 - \sin x\}, -\pi \le x \le \pi$, then
(a) $f(x)$ is not differentiable at 0
(b) $f(x)$ is differentiable at $\dfrac{\pi}{2}$
(c) $f(x)$ has local maxima at 0
(d) none of the above

12. If $f(x) = \begin{cases} \dfrac{x \log \cos x}{\log(1+x^2)}, & x \neq 0 \\ 0, & x = 0 \end{cases}$, then
(a) f is continuous at $x = 0$
(b) f is continuous at $x = 0$ but not differentiable at $x = 0$
(c) f is differentiable at $x = 0$
(d) f is not continuous at $x = 0$

13. Given, $f(x) = \begin{cases} 3 - \left[\cot^{-1}\left(\dfrac{2x^3 - 3}{x^2}\right)\right] & \text{for } x > 0 \\ \{x^2\} \cos(e^{1/x}) & \text{for } x < 0 \end{cases}$,
where $\{\cdot\}$ and $[\cdot]$ denotes the fractional part and the integral part functions respectively, then which of the following statement does not hold good?
(a) $f(0^-) = 0$
(b) $f(0^+) = 3$
(c) $f(0) = 0 \Rightarrow$ continuity of f at $x = 0$
(d) irremovable discontinuity of f at $x = 0$

14. The function $f(x) = \begin{cases} |2x - 3|[x]; & x \ge 1 \\ \sin\left(\dfrac{\pi x}{2}\right); & x < 1 \end{cases}$
([.]) denotes the greatest integer function)
(a) is conerentiable at $x = 0$
(c) is continuous at $x = 0$
(b) is difftinuous but not differentiable at $x = 1$
(d) is continuous but not differentiable at $x = 3/2$

15. Let $h(x) = \min\{x, x^2\}$ for every real number x. Then
(a) h is continuous for all x
(b) h is differentiable for all x
(c) $h'(x) = 1$ for all $x > 1$
(d) h is not differentiable at two values of x

16. A function $f(x)$ is defined in the interval [1, 4] as follows :
$f(x) = \begin{cases} \log_e [x], & 1 \le x < 3 \\ |\log_e x|, & 3 \le x < 4 \end{cases}$
the graph of the function $f(x)$
(a) is broken at two points
(b) is broken at exactly one point
(c) does not have a definite tangent at two points
(d) does not have a definite tangent at more than two points

17. If $f(x) = \min(\tan x, \cot x)$, then :
(a) $f(x)$ is discontinuous at $x = 0, \dfrac{\pi}{4}, \dfrac{5\pi}{4}$
(b) $f(x)$ is continuous at $x = 0, \dfrac{\pi}{2}, \dfrac{3\pi}{2}$
(c) $\int_0^{\pi/2} f(x)\, dx = 2 \ln \sqrt{2}$
(d) $f(x)$ is periodic with period π

18. $f(x) = \dfrac{[1/2 + x] - [1/2]}{x}, -1 \le x \le 2$
has ([.] denotes the greatest integer function)
(a) discontinuity at $x = 0$
(b) discontinuity at $x = 1/2$
(c) discontinuity at $x = 1$
(d) discontinuity at $x = 3/2$

19. If $f(x) = [x] + [x + 1/3] + [x + 2/3]$, then ([.] denotes the greatest integer function)
(a) $f(x)$ is discontinuous at $x = 1, 10, 15$
(b) $f(x)$ is continuous at $x = n/3$, where n is any integer
(c) $\int_0^{2/3} f(x)\, dx = 1/3$
(d) $\lim_{x \to 2/3} f(x) = 2$

20. If $f(x) = \begin{cases} (\sin^{-1} x)^2 \cdot \cos(1/x), & x \neq 0 \\ 0, & x = 0 \end{cases}$, then
(a) $f(x)$ is continuous everywhere in $x \in [-1, 1]$
(b) $f(x)$ is continuous no where in $x \in [-1, 1]$
(c) $f(x)$ is differentiable everywhere in $x \in (-1, 1)$
(d) $f(x)$ is differentiable no where in $x \in [-1, 1]$

21. Give, a real valued function f such that
$$f(x) = \begin{cases} \dfrac{\tan^2 x}{(x^2 - [x])^2} & \text{for } x > 0 \\ 1 & \text{for } x = 0 \\ \sqrt{\{x\} \cot \{x\}} & \text{for } x < 0 \end{cases}$$

where, $[x]$ is the integral part and $\{x\}$ is the fractional part of x, then

(a) $\lim_{x \to 0} f(x) = 1$

(b) $\lim_{x \to 0^-} f(x) = \cot 1$

(c) $\cot^{-1}\left(\lim_{x \to 0^-} f(x)\right)^2 = 1$

(d) f is continuous at $x = 0$

22. The function $f(x) = \left[x^2\left[\frac{1}{x^2}\right]\right]$, $x \neq 0$ is ($[x]$ represents the greatest integer $\leq x$)

(a) continuous at $x = 1$

(b) discontinuous at $x = -1$

(c) discontinuous at infinitely many points

(d) continuous everywhere

Linked-Comprehension Type

*In this section each paragraph has some multiple choice questions have to be answered. Each question has four choices a, b, c, d out of which **Only One** is correct.*

PASSAGE 1

If $|c| \leq \frac{1}{2}$ and $f(x)$ is a differentiable function at $x = 0$ given by

$$f(x) = \begin{cases} b \sin^{-1}\left(\frac{c + x}{2}\right), & -\frac{1}{2} < x < 0 \\ \frac{1}{2}, & x = 0, \text{ then} \\ \frac{e^{ax/2} - 1}{x}, & 0 < x < \frac{1}{2} \end{cases}$$

If f is continuous at $x = 0$

On the basis of above information, answer the following questions

1. The value of a is

(a) $\frac{1}{2}$ (b) 1

(c) $\frac{3}{2}$ (d) 2

2. Which of the following is true?

(a) $16b^2 = 4 - c^2$ (b) $16b^2 = 1 - 4c^2$

(c) $64b^2 = 4 - c^2$ (d) None of these

PASSAGE 2

$$\text{Let } f(x) = \begin{cases} \frac{a(1 - x \sin x) + b \cos x + 5}{x^2} & , x < 0 \\ 3 & , x = 0 \\ \left\{1 + \left(\frac{cx + dx^3}{x^2}\right)\right\}^{1/x} & , x > 0 \end{cases}$$

If f is continuous at $x = 0$

On the basis of above information, answer the following questions

1. The value of a is

(a) -1 (b) ln 3

(c) 0 (d) -4

2. The value of b is

(a) -1 (b) ln 3

(c) 0 (d) -4

3. The value of c is

(a) 2 (b) 3

(c) 0 (d) None of these

4. The value of e^d is

(a) 0 (b) 1

(c) 2 (d) 3

5. The minimum value of $f(x)$ is

(a) 1 (b) 2

(c) 3 (d) None of these

PASSAGE 3

Let $f : R \to R$ be a differential function satisfying $f\left(\frac{x + y}{3}\right) = \frac{2 + f(x) + f(y)}{3}$ for all real x and y and $f'(2) = 2$.

On the basis of above information, answer the following questions

1. The value of $f(1)$ is

(a) 0 (b) 2

(c) 4 (d) None of these

2. The range of $f(|x|)$ is

(a) $[0, \infty)$ (b) $[1, \infty)$

(c) $[2, \infty)$ (d) None of these

3. If $g(x) = |f(|x|) - 3|$ for all $x \in R$, then for $g(x)$
(a) one non-differentiable point
(b) two non-differentiable point
(c) three non-differentiable point
(d) four non-differentiable point

4. If $x \in [-2, 3]$, then range of $f(x)$ is
(a) $[-2, 3]$ (b) $[-3, 4]$
(c) $[-2, 8]$ (d) $[-3, 10]$

5. The number of solutions of the equation $x^2 + (f(|x|)^2 = 9$ are
(a) 0 (b) 2 (c) 3 (d) 5

PASSAGE 4

Let $f(x)$ be a real valued function not identically zero such that

$$f(x + y^n) = f(x) + \{f(y)\}^n \forall\ x, y \in R$$

Where, n is odd natural number > 1 and $f'(0) \geq 0$

On the basis of above information, answer the following questions

1. The value of $f(2)$ is
(a) 0
(b) -1
(c) 2
(d) None of the above

2. The value of $f(5)$ is
(a) -5
(b) 0
(c) 5
(d) not defined

3. The value of $f'(12)$ is
(a) 1
(b) 0
(c) 12
(d) 5

4. The function $f(x)$ is
(a) even
(b) odd
(c) neither even nor odd
(d) None of the above

5. The function $f(x)$ is
(a) discontinuous at one point
(b) continuous everywhere but non-differentiable at some points
(c) discontinuous at three points
(d) continuous and differentiable everywhere

PASSAGE 5

Let $$f(x) = \begin{cases} x + a, & x < 0 \\ |x - 1|, & x \geq 0 \end{cases}$$

and $$g(x) = \begin{cases} x + 1, & x < 0 \\ (x - 1)^2 + b, & x \geq 0 \end{cases}$$

where a and b are non-negative real numbers.

On the basis of above information, answer the following questions

1. The value of a, if $(gof)x$ is continuous for all real x, is
(a) -1
(b) 0
(c) 1
(d) 2

2. The value of b, if $(gof)\,x$ is continuous for all real x, is
(a) -1 (b) 0
(c) 1 (d) 2

3. For these values of a and b, $(gof)x$ is

(a) $\begin{cases} x + 2, & x < -1 \\ (x - 2)^2, & -1 \leq x < 1 \\ x^2, & x \geq 1 \end{cases}$

(b) $\begin{cases} x^2, & x < -1 \\ (x - 2)^2, & -1 \leq x < 1 \\ x + 2, & x \geq 1 \end{cases}$

(c) $\begin{cases} x + 2, & x < -1 \\ x^2, & -1 \leq x < 1 \\ (x - 2)^2, & x \geq 1 \end{cases}$

(d) $\begin{cases} (x - 2)^2, & x < -1 \\ x^2, & -1 \leq x < 1 \\ x + 2, & x \geq 1 \end{cases}$

4. For these values of a and b, $(gof)x\ \forall x \in (-1, 1)$ is
(a) even
(b) odd
(c) neither even nor odd
(d) None of the above

5. For these values of a and b, $(gof)x$ is
(a) differentiable at $x = -1$
(b) differentiable at $x = 0$
(c) differentiable at $x = 1$
(d) non-differentiable at $x = 2$

Numerical Value Type

Solve the following problems and write your answer.

1. If $f(x)=\begin{cases}\dfrac{(1-\sin^3 x)}{3\cos^2 x}, & x<\dfrac{\pi}{2}\\ a, & x=\dfrac{\pi}{2}\\ \dfrac{b(1-\sin x)}{(\pi-2x)^2}, & x>\dfrac{\pi}{2}\end{cases}$

   is continuous at $x=\dfrac{\pi}{2}$, then the value of $\left(\dfrac{b}{a}\right)^{5/3}$ is

2. Let $f:R\to R$ satisfy the equation $f(x+y)=f(x)\cdot f(y)$ for all $x, y\in R$ and $f(x)\neq 0$ for any $x\in R$. If the function f is differentiable at $x=0$ and $f'(0)=3$, then $\lim\limits_{h\to 0}\dfrac{1}{h}(f(h)-1)$ is equal to .... **[JEE Main 2021, 18 March]**

3. Suppose a differentiable function $f(x)$ satisfies the identity $f(x+y)=f(x)+f(y)+xy^2+x^2y$, for all real x and y. If $\lim\limits_{x\to 0}\dfrac{f(x)}{x}=1$, then $f(3)$ is equal to .... **[JEE Main 2020, 4 September]**

4. If $f(x)=\begin{cases}\dfrac{36^x-9^x-4^x+1}{\sqrt{2}-\sqrt{(1+\cos x)}}, & x\neq 0\\ \lambda, & x=0\end{cases}$ is continuous at $x=0$, then $\lambda=\sqrt{\mu}\ \ln 2\cdot\ln 3$ then the value of μ must be

5. If $f(x)=\begin{cases}\dfrac{8^x-4^x-2^x+1^x}{x^2}, & x>0\\ e^x\sin x+\pi x+\lambda\ln 4, & x\le 0\end{cases}$ is continuous at $x=0$, then the value of $1000e^{\lambda}$ must be

6. If $f(x)=\begin{cases}\dfrac{(\exp\{(x+3)\ln 27\})^{\frac{1}{27}[x+1]}-9}{3^x-27}, & x<3\\ \lambda\cdot\dfrac{1-\cos(x-3)}{(x-3)\tan(x-3)}, & x>3\end{cases}$

   is continuous at $x=3$, then the value of $8100\,\lambda$ must be

7. If $f(x)+f(y)=f\left(\dfrac{x+y}{1-xy}\right)$ for all $x, y\in R$ and $xy\neq 1$ and $\lim\limits_{x\to 0}\dfrac{f(x)}{x}=2$, then the value of $\dfrac{1500}{\pi}\dfrac{f(\sqrt{3})}{f'(-2)}$ must be

Single Integer Type

Solve the following problems and write your answer.

1. Let $f(x)=x\left[\dfrac{x}{2}\right]$, for $-10<x<10$, where $[t]$ denotes the greatest integer function. Then, the number of points of discontinuity of f is equal to .... **[JEE Main 2020, 5 September]**

2. Let $f:R\to R$ and $f:R\to R$ be defined as

   $f(x)=\begin{cases}x+a, & x<0\\ |x-1|, & x\ge 0\end{cases}$ and $g(x)=\begin{cases}x+1, & x<0\\ (x-1)^2+b, & x\ge 0\end{cases}$

   where a, b are non-negative real numbers. If $(gof)\,(x)$ is continuous for all $x\in R$, then $a+b$ is equal to .... **[JEE Main 2021, 16 March]**

3. Let $f(x)=\lim\limits_{n\to\infty}\dfrac{x^{2n-1}+ax^2+bx}{x^{2n}+1}$. If $f(x)$ is continuous for all $x\in R$, then the value of $a+8b$ is ....

4. Let the functions $f:(-1,1)\to R$ and $g:(-1,1)\to(-1,1)$ be defined by $f(x)=|2x-1|+|2x+1|$ and $g(x)=x-[x]$, where $[x]$ denotes the greatest integer less than or equal to x. Let fog: $(-1,1)\to$ R be the composite function defined by $(fog)(x)=f(g(x))$. Suppose c is the number of points in the interval $(-1,1)$ at which fog is **NOT** continuous, and suppose d is the number of points in the interval $(-1,1)$ at which fog is NOT differentiable. Then the value of $c+d$ is .... **[JEE Advanced 2020]**

5. Given, $f(x)=\begin{cases}\dfrac{1-\cos ax}{x\sin x}, & x\neq 0\\ \dfrac{1}{2}, & x=0\end{cases}$

   If f is continuous at $x=0$, then the value of a^2 must be

6. Let S be the set of points where the function, $f(x)=|2-|x-3||$, $x\in R$, is not differentiable. Then, $\sum\limits_{x\in S} f(f(x))$ is equal to ..... . **[JEE Main 2020, 7 January]**

7. Let $f(x)=\dfrac{\log(1+x+x^2)+\log(1-x+x^2)}{\sec x-\cos x}$, $x\neq 0$. Then, the value of $f(0)$ so that f is continuous at $x=0$ is

8. The number of points where $f(x)=\text{Sgn}\,(x^2-3x+2)+[x-3]$, $x\in[0,4]$, is discontinuous is (where $[\cdot]$ denotes the greatest integer function)......... .

Matrix-Match Type

Given below there are certain matching type questions, where two columns (each having some items) are given. Each item of column I has to be matched with the items of column II, by encircling the correct match(es).

NOTE *An item of column I can be matched with more than one items of column II.*

1. Observe the following columns :

	Column-I		Column-II
(A)	If $f(x)=\begin{cases} -2\sin x; & -\pi\le x\le -\frac{\pi}{2} \\ a\sin x+b; & -\frac{\pi}{2}<x<\frac{\pi}{2} \\ \cos x; & \frac{\pi}{2}\le x\le \pi \end{cases}$ is continuous in $[-\pi, \pi]$, then	(P)	$\lvert a-b\rvert=0$
(B)	If $f(x)=\begin{cases} \frac{a\lvert x^2-x-2\rvert}{2+x-x^2}; & x<2 \\ b; & x=2 \\ \frac{x-[x]}{x-2}; & x>2 \end{cases}$ continuous at $x=2$, then (where $[\cdot]$ denotes the greatest integer function)	(Q)	$\lvert a-b\rvert=2$
(C)	If $f(x)=\begin{cases} \left(\frac{3}{2}\right)^{\frac{(\cot 3x)}{(\cot 2x)}}; & 0<x<\frac{\pi}{2} \\ b+3; & x=\frac{\pi}{2} \\ (1+\lvert\cos x\rvert)^{(a\lvert\tan x\rvert/b)}; & \frac{\pi}{2}<x<\pi \end{cases}$ is continuous at $x=\frac{\pi}{2}$, then	(R)	$\lvert a+b\rvert=2$
		(S)	$\lvert a+2b\rvert=3$
		(T)	$\lvert a+2b\rvert=4$

2. Observe the following columns :

	Column-I		Column-II
(A)	If $f(x)=\begin{cases} \frac{a+3\cos x}{x^2}, & x<0 \\ b\tan\left(\frac{\pi}{[x+3]}\right), & x\ge 0 \end{cases}$ is continuous at $x=0$, then (where [.] denotes the greatest integer function)	(P)	$\lvert a+b\rvert=0$
(B)	If $f(x)=\begin{cases} -2\sin x, -\pi\le x\le -\frac{\pi}{2} \\ a\sin x+b, -\frac{\pi}{2}<x<\frac{\pi}{2} \\ \cos x, \frac{\pi}{2}\le x\le \pi \end{cases}$ is continuous in $[-\pi, \pi]$, then	(Q)	$\lvert a-b\rvert=2$
(C)	If $f(x)=\begin{cases} \left(\frac{3}{2}\right)^{(\cot 3x)/(\cot 2x)}, & 0<x<\frac{\pi}{2} \\ b+3, & x=\frac{\pi}{2} \\ (1+\lvert\cos x\rvert)^{\left(\frac{a\lvert\tan x\rvert}{b}\right)}, & \frac{\pi}{2}<x<\pi \end{cases}$ is continuous at $x=\frac{\pi}{2}$, then	(R)	$\lvert a+2b\rvert=1$
		(S)	$\lvert a+2b\rvert=4$
		(T)	$[a-2b]=-2$ (where [.] denotes the greatest integer function)

3. Observe the following columns :

	Column-I		Column-II
(A)	If $f(x) = \begin{cases} x + a\sqrt{2}\sin x, & 0 \le x < \frac{\pi}{4} \\ 2x\cot x + b, & \frac{\pi}{4} \le x \le \frac{\pi}{2} \\ a\cos 2x - b\sin x, & \frac{\pi}{2} < x \le \pi \end{cases}$ is continuous in $[0, \pi]$, then	(P)	$a - b = \frac{\pi}{4}$
(B)	If $f(x) = \begin{cases} \frac{\sin(a+1)x + \sin x}{x}; & x < 0 \\ c; & x = 0 \\ \frac{(x + bx^2)^{1/2} - x^{1/2}}{bx^{3/2}}; & x > 0 \end{cases}$ is continuous at $x = 0$, then	(Q)	$a + b = \frac{\pi}{12}$
(C)	If $f(x) = \begin{cases} (1 + b\lvert\tan x\rvert)^{a/\lvert\tan x\rvert}; & \frac{-\pi}{4} < x < 0 \\ b; & x = 0 \\ e^{\cot 2x/\cot 3x}; & 0 < x < \frac{\pi}{4} \end{cases}$ is continuous at $x = 0$, then	(R)	Locus of (a, b) is $x + 2y = 0$
		(S)	Locus of (a, c) is $x - y + 2 = 0$
		(T)	Locus of (a, b) is $2xy = 3$

4. Observe the following columns :

	Column-I		Column-II
A.	The function $f(x) = \begin{cases} x^2 + 3x + a; & x \le 1 \\ bx + 2; & x > 1 \end{cases}$ is differentiable $\forall\, x \in R$, then	(P)	$a = 3$
B.	The function $f(x) = \begin{cases} \frac{1}{\lvert x\rvert}; & \lvert x\rvert \ge 1 \\ ax^2 + b; & \lvert x\rvert < 1 \end{cases}$ is differentiable everywhere, then	(Q)	$b = 5$
C.	The function $f(x) = \begin{cases} ax^2 - bx + 2; & x < 3 \\ bx^2 - 3; & x \ge 3 \end{cases}$ is differentiable everywhere then	(R)	$a = \frac{35}{9}$
		(S)	$b = \frac{3}{2}$
		(T)	$a = -\frac{1}{2}$

Answers

Objective Questions Type [Only one correct answer]

1. (c)	**2.** (d)	**3.** (a)	**4.** (d)	**5.** (c)	**6.** (d)	**7.** (c)	**8.** (a)	**9.** (c)	**10.** (b)
11. (c)	**12.** (a)	**13.** (d)	**14.** (d)	**15.** (d)	**16.** (d)	**17.** (c)	**18.** (a)	**19.** (d)	**20.** (a)
21. (a)	**22.** (d)	**23.** (b)	**24.** (a)	**25.** (b)	**26.** (a)	**27.** (b)	**28.** (b)	**29.** (a)	**30.** (c)
31. (c)	**32.** (d)	**33.** (d)	**34.** (c)	**35.** (d)	**36.** (a)	**37.** (b)	**38.** (d)	**39.** (b)	**40.** (b)
41. (d)	**42.** (b)	**43.** (b)	**44.** (c)	**45.** (c)	**46.** (c)	**47.** (b)	**48.** (a)	**49.** (d)	**50.** (c)
51. (b)	**52.** (b)								

Objective Questions Type II [One or more than one correct answer(s)]

1. (a, b)	**2.** (b, c)	**3.** (a, b, c)	**4.** (b, c)	**5.** (a, c)
6. (a, c)	**7.** (a, b, d)	**8.** (b, c, d)	**9.** (a, c)	**10.** (a, d)
11. (a, c)	**12.** (a, c)	**13.** (b, d)	**14.** (a, b, c, d)	**15.** (a, c, d)
16. (a, c)	**17.** (c, d)	**18.** (a, b, d)	**19.** (a, c)	**20.** (a, c)
21. (b, c)	**22.** (b, c)			

Linked-Comprehension Type

Passage 1 **1.** (b) **2.** (c)
Passage 2 **1.** (a) **2.** (d) **3.** (c) **4.** (d) **5.** (c)
Passage 3 **1.** (c) **2.** (c) **3.** (c) **4.** (c) **5.** (b)
Passage 4 **1.** (c) **2.** (c) **3.** (a) **4.** (b) **5.** (d)
Passage 5 **1.** (c) **2.** (b) **3.** (c) **4.** (a) **5.** (b)

Numerical Value Type

1. (32) **2.** (3/2) **3.** (10) **4.** (512) **5.** (20.4) **6.** (1800) **7.** (2500)

Single Integer Type

1. (8) **2.** (1) **3.** (8) **4.** (4) **5.** (1) **6.** (3) **7.** (1) **8.** (2)

Matrix-Match Type

1. (A) → (Q); (B) → (P, R, S); (C) → (Q, R, T)
2. A → (T); B → (P, Q, R); C → (Q, S);
3. A → (P, R, Q); B → (P, Q, R, S, T); C → (T);
4. A → (P, Q); B → (S, T); C → (R);

Solutions

Objective Questions Type [Only one correct answer]

1. Since, $[x^3]$ is not continuous and differentiable at integral point. So, $f(x)$ is continuous and differentiable in $[4, 6]$ if $\left[\frac{(x-2)^3}{a}\right] = 0 \Rightarrow a \geq 64$.

2. We have,
$$f(x) = x^2 + xg'(1) + g''(2)$$
$$\Rightarrow \quad f'(x) = 2x + g'(1) \quad \ldots(\text{i})$$
$$\Rightarrow \quad f'(x) = 2 \Rightarrow f'(x) = 0$$
Again,
$$g(x) = f(1)x^2 + xf'(x) + f'(x)$$
$$\Rightarrow \quad g'(1) = 2xf(1) + f'(x) + xf''(x) + f''(x)$$
Put $x = 1$
$$\Rightarrow \quad g'(1) = 2f(1) + f'(1) + f''(1) + 0 \quad \ldots(\text{ii})$$
Put $x = 1$ in Eq. (i), we get
$$f'(1) = 2 + g'(1) \quad \ldots(\text{iii})$$
From Eq. (ii) and Eq. (iii), we get
$$g'(1) = 2f(1) + 2 + g'(1) + 2$$
$$2f(1) + 4 = 0 \Rightarrow f(1) = -2 \quad \ldots(\text{iv})$$
Now, again
$$g''(x) = 2f(1) + f''(x) + f''(x) + xf''(x)$$
$$= 2f(1) + 4 = -4 + 4$$
$$g''(x) = 0$$
$$\therefore \quad f(x) = x^2 + xg'(1) + g''(2) = x^2 + xg'(1)$$
Put $x = 1$
$$f(1) = 1 + g'(1) \Rightarrow -2 = 1 + g'(1)$$
$$\Rightarrow \quad g'(1) = -3$$
$$\therefore \quad f(x) = x^2 + x(-3) = x^2 - 3x$$
$$\therefore \quad g(x) = f(1)x^2 + xf'(x) + f''(x)$$
$$= -2x^2 + x(2x-3) + 2$$
$$= -2x^2 + 2x^2 - 3x + 2 = 2 - 3x$$
$$\therefore f(x) - g(x) = x^2 - 3x - 2 + 3x = x^2 - 2$$

3. $g(x) = \begin{cases} x^3 + x, & x \geq 0 \\ x^3 + x, & x < 0 \end{cases}$
$$\Rightarrow \quad g(x) = x^3 + x, \forall x \in R.$$
Hence, continuous for all $x \in R$.

4. Given function
$$f(x) = \begin{cases} ae^x + be^{-x}, & -1 \leq x < 1 \\ cx^2, & 1 \leq x \leq 3 \\ ax^2 + 2cx, & 3 < x \leq 4 \end{cases}$$
Since, given function is continuous for some $a, b, c \in R$, So
$$\lim_{x \to 1^-} f(x) = \lim_{x \to 1^+} f(x) \Rightarrow ae + be^{-1} = c \quad \ldots(\text{i})$$
and $\lim_{x \to 3^-} f(x) = \lim_{x \to 3^+} f(x) \Rightarrow c(3)^2 = 9a + 6c$
$$\Rightarrow \quad 3c = 9a \Rightarrow c = 3a \quad \ldots(\text{ii})$$
Now, $f'(0) + f'(2) = a - b + 4c = e$ [given] ... (iii)

From Eqs. (i), (ii) and (iii), we get
$$a - (3a - ae)e + 12a = e \Rightarrow (13 - 3e + e^2)a = e$$
$$\Rightarrow \quad a = \frac{e}{e^2 - 3e + 13}$$

5. $[x]$ is not differentiable at integral points.
Also, $[n + p\sin x] = n + [p\sin x]$
$\therefore$ $[p\sin x]$ is not differentiable, where $p\sin x$ is an integer. But p is prime and $0 < \sin x \leq 1 [\because 0 < x < \pi]$
$\therefore$ $p\sin x$ is an integer only when $\sin x = \frac{r}{p}$, where $0 < r \leq p$ and $r \in N$
For $\quad r = p, \ \sin x = 1 \Rightarrow x = \frac{\pi}{2}$ in $(0, \pi)$
For $\quad 0 < r < p, \sin x = \frac{r}{p}$
$$\therefore \quad x = \sin^{-1}\frac{r}{p}$$
$$\text{or} \quad \pi - \sin^{-1}\frac{r}{p}$$
Number of such values of $x = p - 1 + p - 1 = 2p - 2$
$\therefore$ Total number of points, where $f(x)$ is not differentiable are $1 + 2p - 2 = 2p - 1$.

6. $\frac{x}{1 + |x|}$ is always differentiable
$\Rightarrow (x-2)(x+4)|(x-1)(x-2)(x-3)|$ is not differentiable at $x = 1, 3$. So, $f(x)$ is not differentiable at $x = 1, 3$.

7. $\lim_{x \to 0^+} f(x) = f(0) = \lim_{x \to 0^-} (x)$
$$\lim_{x \to 0^+} f(x) = \lim_{x \to 0^+} \frac{\cos^{-1}(1-x^2) \cdot \sin^{-1}(1-x)}{x(1-x)(1+x)}$$
$$= \lim_{x \to 0^+} \frac{\cos^{-1}(1-x^2)}{x \cdot 1 \cdot 1} \cdot \frac{\pi}{2}$$
Let $1 - x^2 = \cos\theta$
$$= \frac{\pi}{2} \lim_{x \to 0^+} \frac{\theta}{\sqrt{1 - \cos\theta}}$$
$$= \frac{\pi}{2} \lim_{x \to 0^+} \frac{\theta}{\sqrt{2}\sin\frac{\theta}{2}} = \frac{\pi}{\sqrt{2}}$$
Now, $\lim_{x \to 0^-} f(x) = \lim_{x \to 0^-} \frac{\cos^{-1}(1-(1+x)^2)\sin^{-1}(-x)}{(1+x)-(1+x)^3}$
$$= \lim_{x \to 0^-} \frac{\frac{\pi}{2}(-\sin^{-1}x)}{(1+x)(2+x)(-x)}$$
$$= \lim_{x \to 0^-} \frac{\frac{\pi}{2}}{1 \cdot 2} \cdot \frac{\sin^{-1}x}{x} = \frac{\pi}{4}$$
$\Rightarrow$ RHL $\neq$ LHL
Function can not be continuous
$\Rightarrow$ No value of α exist.

8. $f(1^-) = 1 = f(1)$ and $f(1^+) = a + b$

For continuity, at $x = 1$, $a + b = 1$ ...(i)

In $0 < x < 1$, $f'(x) = 2xe^{2(x-1)} + 2x^2 e^{2(x-1)}$

$\therefore \quad f'(1^-) = 4$

In $0 < x < 2$, $f'(x) = -2a \sin(2x - 2) + 2bx$

$\therefore \quad f'(1^+) = 2b$

For differentiability, at $x = 1$, $2b = 4$ ...(ii)

From Eqs. (i) and (ii), we get

$$a = -1 \text{ and } b = 2$$

9. $$\lim_{x \to \frac{\pi}{2}^-} f(x) = \lim_{h \to 0} \left(\frac{\sin(\sin h)}{-h} \right) = -1$$

$$\Rightarrow \lim_{x \to \frac{\pi}{2}^+} f(x) = \lim_{h \to 0} \frac{\sin(-\sin h)}{h} \neq -1$$

So, $\lim_{x \to \frac{\pi}{2}} f(x)$ does not exist.

10. $$f(g(x)) = \frac{1}{\left(\frac{1}{x^2} - 1\right)\left(\frac{1}{x^2} - 2\right)} = \frac{x^4}{(1 - x^2)(1 - 2x^2)}$$

$\Rightarrow f(g(x))$ is discontinuous at $x = \pm 1$, $x = \pm \frac{1}{\sqrt{2}}$ and $x = 0$

Since, $g(x)$ is discontinuous at $x = 0$.

11. $f(0) = f_+(0) = \lim_{h \to 0} f(0 + h)$

$$= \lim_{h \to 0} \frac{\sqrt{(a^2 - ah + h^2)} - \sqrt{(a^2 + ah + h^2)}}{\sqrt{(a + h)} - \sqrt{(a - h)}}$$

$$= \lim_{h \to 0} \frac{\dfrac{(a^2 - ah + h^2)^{1/2} - (a^2 + ah + h^2)^{1/2}}{(a^2 - ah + h^2) - (a^2 + ah + h^2)} \times (-2ah)}{\dfrac{(a + h)^{1/2} - (a - h)^{1/2}}{(a + h) - (a - h)} \times 2h}$$

$$= -a \lim_{h \to 0} \frac{\frac{1}{2}(a^2 + ah + h^2)^{-1/2}}{\frac{1}{2}(a + h)^{-1/2}}$$

$$= -a \cdot \frac{(a^2)^{-1/2}}{(a)^{-1/2}} = -\sqrt{a}$$

12. $f\left(\frac{\pi}{4}\right) = \text{RHL at } x = \pi/4$

$$= \lim_{x \to \pi/4^+} f(x) = \lim_{h \to 0} f\left(\frac{\pi}{4} + h\right)$$

$$= \lim_{h \to 0} \frac{\tan\left(\frac{\pi}{4} - \left(\frac{\pi}{4} + h\right)\right)}{\cot 2\left(\frac{\pi}{4} + h\right)}$$

$$= \lim_{h \to 0} \frac{-\tan h}{-\tan 2h}$$

$$= \lim_{h \to 0} \frac{h \cdot \tan h/h}{2h \tan 2h/2h} = \frac{1}{2}$$

13. $f(0) = \text{RHL of } f(x) \text{ at } x = 0$

$$= \lim_{x \to 0^+} f(x)$$

$$= \lim_{h \to 0} f(0 + h)$$

$$= \lim_{h \to 0} \frac{2 - (256 - 7h)^{1/8}}{(5h + 32)^{1/5} - 2}$$

$$= \lim_{h \to 0} \frac{(256)^{1/8} - (256 - 7h)^{1/8}}{(5h + 32)^{1/5} - (32)^{1/5}}$$

$$= \lim_{h \to 0} \frac{\dfrac{(256)^{1/8} - (256 - 7h)^{1/8}}{256 - (256 - 7h)} \times 7h}{\dfrac{(5h + 32)^{1/5} - (32)^{1/5}}{(5h + 32) - 32} \times 5h}$$

$$= \frac{7}{5} \cdot \frac{\frac{1}{8}(256)^{-7/8}}{\frac{1}{5}(32)^{-4/5}} = \frac{7}{8} \cdot 2^{-3} = \frac{7}{64}$$

14. $$f'(x) = \lim_{h \to 0} \frac{f(x + h) - f(x)}{h}$$

$$= \lim_{h \to 0} \frac{f(x) + f(h) - f(x)}{h}$$

$[\because f(x + y) = f(x) + f(y)]$

$$= \lim_{h \to 0} \frac{f(h)}{h} = \lim_{h \to 0} \frac{h^2 g(h)}{h} = 0 \quad [\because f(x) = x^2 g(x)]$$

15. $$\because f'(x) = \lim_{h \to 0} \frac{f(x + h) - f(x)}{h}$$

$$= \lim_{h \to 0} \frac{f(x) f(h) - f(x)}{h}$$

$$= f(x) \lim_{h \to 0} \frac{f(h) - 1}{h}$$

$$= f(x) f'(0)$$

$$f'(3) = f(3) f'(0)$$

$$= 3 \times 11 = 33$$

16. $$f(x) = \begin{cases} \dfrac{1}{|x|}, & |x| \geq 1 \\ ax^2 + b, & |x| < 1 \end{cases}$$

At $x = 1$ function must be continuous

So, $\quad 1 = a + b$ (i)

Differentiability at $x = 1$

$$\left(-\frac{1}{x^2}\right)_{x=1} = (2ax)_{x=1}$$

$$\Rightarrow \quad -1 = 2a \Rightarrow a = -\frac{1}{2}$$

$$\Rightarrow \quad b = 1 + \frac{1}{2} = \frac{3}{2}$$

17. Given, $f\left(\frac{x + y}{3}\right) = \frac{f(x) + f(y)}{3}$

Replacing x by $3x$ and y by zero,

then $$f(x) = \frac{f(3x) + f(0)}{3}$$

$\Rightarrow \quad f(3x) - 3f(x) = -f(0)$ ...(i)

and $$f'(x) = \lim_{h \to 0} \frac{f(x + h) - f(x)}{h}$$

$$= \lim_{h \to 0} \frac{f\left(\frac{3x + 3h}{3}\right) - f(x)}{h}$$

$$= \lim_{h \to 0} \frac{\frac{f(3x) + f(3h)}{3} - f(x)}{h}$$

$$= \lim_{h \to 0} \frac{f(3x) + f(3h) - 3f(x)}{3h}$$

$$= \lim_{h \to 0} \frac{f(3h) - f(0)}{3h} \quad \text{[from Eq. (i)]}$$

$$= f'(0) = 3$$

$\therefore \quad f(x) = 3x + c$

$\Rightarrow \quad f(0) = 0 + c = 3$

$\therefore \quad c = 3$

Then, $\quad f(x) = 3x + 3$

Hence, $f(x)$ is continuous and differentiable everywhere.

18. $\because [(2x - 3\pi)^3] =$ Integer (say n)

$\therefore \quad \tan n\pi = 0$

and $\quad 1 + [2x - 3\pi]^2 \neq 0$

Hence, $\quad f(x) = 0$

Hence, $f(x)$ is continuous and differentiable for all $x \in R$.

19. We have, $f(x) = \begin{cases} 1, & x \text{ is rational} \\ 2, & x \text{ is irrational} \end{cases}$

$\because$ Real number = Rational number + Irrational number

If x is rational, then $f(x) = 1$

and if x is irrational, then $f(x) = 2$

Hence, $f(x)$ is not continuous in R.

20. $f(x) = \begin{cases} 0, & x \in [0, 1) \\ x, & x \in (1, 2] \\ 2(x-1), & x \in (2, 3) \end{cases}$

$f(x)$ is discontinuous at $x = 1$, hence it is not differentiable at that point.

At $x = 2$

LHD $= 1$ and RHD $= 2$

$\Rightarrow f(x)$ is not differentiable at $x = 2$.

21. We have, $f(x) = \begin{cases} ax^2 + b, & x < -1 \\ bx^2 + ax + 4, & x \geq -1 \end{cases}$

$\therefore \quad f'(x) = \begin{cases} 2ax, & x < -1 \\ 2bx + a, & x \geq -1 \end{cases}$

Since, $f(x)$ is differentiable at $x = -1$, therefore it is continuous at $x = -1$ and hence,

$$\lim_{x \to -1-} f(x) = \lim_{x \to -1+} f(x)$$

$\Rightarrow \quad a + b = b - a + 4$

$\Rightarrow \quad a = 2$

and also, $\lim_{x \to -1-} f'(x) = \lim_{x \to -1+} f'(x)$

$\Rightarrow \quad -2a = -2b + a$

$\Rightarrow \quad 3a = 2b$

$\Rightarrow \quad b = 3 \quad [\because a = 2]$

Hence, $\quad a = 2, b = 3$

22. $\lim_{x \to 0} g\{f(x)\} = \lim_{x \to 0} \begin{cases} (f(x))^2 + 1, & f(x) \neq 0, 2 \\ 4, & f(x) = 0 \\ 5, & f(x) = 2 \end{cases}$

$$= \lim_{x \to 0} \begin{cases} \sin^2 x + 1, & \sin x \neq 0, 2 \text{ and } x \neq n\pi \\ 4, & \sin x = 0 \text{ and } x \neq n\pi \\ 5, & \sin x = 2 \text{ and } x \neq n\pi \\ 5, & 2 \neq 0, 2 \text{ and } x = n\pi \\ 4, & 2 = 0 \text{ and } x = n\pi \\ 5, & 2 = 2 \text{ and } x = n\pi \end{cases}$$

$$= \lim_{x \to 0} \begin{cases} 1 + \sin^2 x, & x \neq n\pi \\ 5, & x = n\pi \end{cases}$$

$= 1 + \sin^2 0$

$= 1 + 0 = 1$

23. At $x = 0$

LHL = RHL $= 4 \neq f(0) = 2$

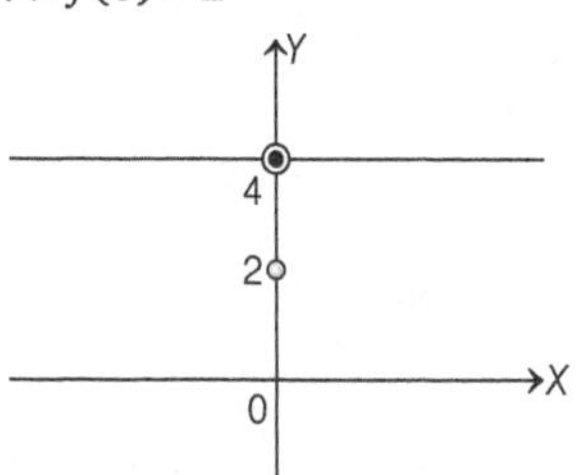

$\therefore$ Removal discontinuity

24. LHL $= \lim_{x \to 0-} f(x) = \lim_{h \to 0} f(0 - h)$

$$= \lim_{h \to 0} (1 + (\sin(-h)))^{a/|\sin(-h)|}$$

$$= \lim_{h \to 0} (1 + \sin h)^{a/\sin h}$$

$$= e^{\lim_{h \to 0} (1 + \sin h - 1)a/\sin h} = e^a$$

RHL $= \lim_{x \to 0+} f(x) = \lim_{h \to 0} f(h)$

$$= \lim_{h \to 0} e^{\tan 2h/\tan 3h}$$

$$= e^{\lim_{h \to 0} \frac{2}{3} \cdot \frac{\tan 2h}{2h} \cdot \frac{3h}{\tan 3h}} = e^{2/3}$$

$\therefore \quad$ V.F. $= f(0) = b$

$\Rightarrow \quad b = e^{2/3} = e^a$

$\therefore \quad a = 2/3, b = e^{2/3}$ or $a = \log_e b$

25. $f(g(x)) = \begin{cases} g(x) + 2, & g(x) < 0 \\ (g(x))^2, & g(x) \geq 0 \end{cases} = \begin{cases} x^3 + 2, & x < 0 \\ x^6, & x \in [0, 1) \\ (3x - 2)^2, & x \in [1, \infty) \end{cases}$

$(fog(x))' = \begin{cases} 3x, & x < 0 \\ 6x^5, & x \in (0, 1) \\ 2(3x - 2) \times 3, & x \in (1, \infty) \end{cases}$

At $x = 0$

LHL $\neq$ RHL (Discontinuous)

At $x = 1$

LHD $= 6 =$ RHD

$\Rightarrow fog(x)$ is differentiable for $x \in R - \{0\}$

26.
$$\text{RHL} = \lim_{x \to 0^+} f(x)$$
$$= \lim_{h \to 0} f(0+h)$$
$$= \lim_{h \to 0} h^n \sin\left(\frac{1}{h}\right)$$
$$= 0^n \sin(\infty)$$
$$= 0^n \{-1 \text{ to } 1\}$$
$\because$ V. F. $= f(0) = 0$

$\therefore$ $n > 0$ ..(i)

$$Rf'(0) = \lim_{h \to 0} \frac{f(0+h) - f(0)}{h}$$
$$= \lim_{h \to 0} \frac{h^n \sin\left(\frac{1}{h}\right) - 0}{h}$$
$$= \lim_{h \to 0} h^{n-1} \sin\left(\frac{1}{h}\right)$$
$$= 0^{n-1} \sin \infty$$
$$= 0^{n-1} (-1 \text{ to } 1)$$

For not differentiable

$$n - 1 \le 0$$
$\therefore$ $n \le 1$

From Eqs. (i) and (ii),

$$0 < n \le 1$$
$\therefore$ $n \in (0, 1]$

27. We have, $Lf'(0) = \lim_{h \to 0} \dfrac{f(0-h) - f(0)}{-h}$
$$= \lim_{h \to 0} \frac{[\tan^2(-h)] - (0)}{-h}$$
$$= \lim_{h \to 0} \frac{[\tan^2 h]}{h} = 0$$
$$Rf'(0) = \lim_{h \to 0} \frac{f(0+h) - f(0)}{h}$$
$$= \lim_{h \to 0} \frac{[\tan^2 h] - 0}{h} = 0$$
$$Lf'(0) = Rf'(0)$$

Hence, $f(x)$ is differentiable and hence continuous at $x = 0$ and then $f'(0) = 0$.

28.
$$\lim_{x \to (2n+1)^-} f(x) = \lim_{h \to 0} f(2n+1-h)$$
$$= \lim_{h \to 0} (2n+1-h) - [2n+1-h]$$
$$= \lim_{h \to 0} 2n+1-h-2n = 1$$
$$\lim_{x \to (2n+1)^+} f(x) = \frac{1}{2},\ f(2n+1) = \frac{1}{2}$$

$\therefore f(x)$ is discontinuous for integral values of $x = 3, 4, 5, \ldots$ do not check continuity at $x = 1$ and 2.

and $f(x) = x - [x]$ is periodic with period 1.

and $\int_0^2 f(x)\,dx$ does not exist

[$\because$ least interval of x is $2 \le x < 3$]

29. $\because f(x) = \begin{cases} x\left[1 + \frac{1}{3}\sin(\log x^2)\right], & x \ne 0 \\ 0, & x = 0 \end{cases}$

Option (a) : $\text{LHL} = \lim_{x \to 0^-} f(x)$
$$= \lim_{h \to 0} f(0-h)$$
$$= \lim_{h \to 0} \left\{-h\left[1 + \frac{1}{3}\sin(\log h^2)\right]\right\}$$
$$= 0$$
$$\text{RHL} = \lim_{x \to 0^+} f(x) = \lim_{h \to 0} f(0+h)$$
$$= \lim_{h \to 0} \left\{h\left[1 + \frac{1}{3}\sin(\log h^2)\right]\right\}$$
$$= 0$$

and V.F. $= f(0) = 0$

$\therefore$ LHL = RHL = V.F.

$\therefore f(x)$ is continuous at $x = 0$

Option (b) :

$f'(x) = \left[1 + \frac{1}{3}\sin(\log x^2)\right]$ is not defined.

Option (c) : $f'(0) = \left[1 + \frac{1}{3}\sin \infty\right]$ is not defined.

Option (d) : $f(x)$ is defined for $x < 0$, $f(x)$ is also defined for $x < -1$.

30. We have,
$$f(x) = \begin{cases} [\cos \pi x], & x < 1 \\ |x - 2|, & 1 \le x < 2 \end{cases}$$
$$= 2 - x,\ 1 \le x < 2$$
$$\begin{cases} -1, & \frac{1}{2} < x < 1 \\ 0, & 0 < x \le \frac{1}{2} \\ 1, & x = 0 \\ 0, & -\frac{1}{2} \le x < 0 \\ -1, & -\frac{3}{2} < x < -\frac{1}{2} \end{cases}$$

It is evident from the definition that $f(x)$ is discontinuous at $x = 1/2$.

31. We have, $f(x) = [\sqrt{2}\sin x]$
$$= \begin{cases} -2, & -\frac{\pi}{2} \le x < -\frac{\pi}{4} \\ -1, & -\frac{\pi}{4} \le x < 0 \\ 0, & 0 \le x < \frac{\pi}{4} \\ 1, & \frac{\pi}{4} \le x < \frac{\pi}{2} \end{cases}$$

Option (a) : $f(x)$ is non periodic.

Option (b) : maximum value of $f(x)$ is 1 in the interval $\left[-\frac{\pi}{2}, \frac{\pi}{2}\right]$.

Option (c) : $f(x)$ is discontinuous at

$$x = \frac{n\pi}{2} + \frac{\pi}{4}, n \in I$$

Option (d) :

$$Lf'(n\pi) = \lim_{h\to 0} \frac{f(n\pi - h) - f(n\pi)}{-h}$$

$$= \lim_{h\to 0} \frac{[\sqrt{2}\sin(n\pi - h)] - 0}{-h}$$

$$= \lim_{h\to 0} \frac{[\sqrt{2}(-1)^n \sin h]}{-h} \quad [\because -1 \le \sin h \le 1]$$

$$= \infty \text{ or } \frac{0}{0} \quad [\because [\sin h] = -1, 0, 1]$$

undefined.

Not differentiable at $x = n\pi, n \in I$.

32. Given function, $f(x) = \begin{cases} \frac{\pi}{4} + \tan^{-1} x, & |x| \le 1 \\ \frac{1}{2}(|x|-1), & |x| > 1 \end{cases}$

$$\Rightarrow \quad f(x) = \begin{cases} \frac{\pi}{4} + \tan^{-1} x, & x \in (-\infty, -1] \cup [1, \infty) \\ \frac{1}{2}(x-1), & x \in (-1, 0) \\ \frac{1}{2}(-x-1), & x \in (0, 1) \end{cases}$$

$\because$ RHL $= \frac{\pi}{2}$, LHL $= 0$

$\Rightarrow f$ is discontinuous at $x = 1$, so it is non-differentiable at $x = 1$

Now, at $x = -1$

$$f(-1) = \frac{\pi}{4} - \frac{\pi}{4} = 0$$

LHL $= 0$ and RHL $= 0$

$\Rightarrow f$ is continuous at $x = -1$

$\because$ LHL $= \frac{1}{2}$ and RHL $= -\frac{1}{2}$

$\therefore f$ is non-differentiable at $x = -1$

$\therefore f$ is continuous on $R - \{1\}$ and differentiable on $R - \{-1, 1\}$.

33. $\lim_{h\to 0} \frac{x - e^x + 1 - (1 - \cos 2x)}{x^2}$

$$= \lim_{h\to 0}\left[\frac{x - e^x + 1}{x^2} - \frac{1 - \cos 2x}{x^2}\right]$$

$$= \lim_{h\to 0}\left[\frac{x + 1 - \left(1 + x + \frac{x^2}{2}\right)}{x^2} - \frac{2\sin^2 x}{x^2}\right]$$

[using expansion of e^x]

$$= -\frac{1}{2} - 2 = -\frac{5}{2}$$

For continuity, $f(0) = -\frac{5}{2}$

Now, $[f(0)] = -3$

$$\{f(0)\} = f(0) - [f(0)] = -\frac{5}{2} - (-3) = \frac{1}{2}$$

Hence, $[f(0)]\{f(0)\} = -\frac{3}{5} = -1.5$

34. $\because f(x) = \begin{cases} x^p \sin\left(\frac{1}{x}\right) + x^2, & x > 0 \\ x^p \sin\left(\frac{1}{x}\right) - x^2, & x < 0 \\ 0, & x = 0 \end{cases}$

$$\Rightarrow f'(x) = \begin{cases} -x^{p-2}\cos\left(\frac{1}{x}\right) + px^{p-1}\sin\left(\frac{1}{x}\right) + 2x, & x > 0 \\ -x^{p-2}\cos\left(\frac{1}{x}\right) + px^{p-1}\sin\left(\frac{1}{x}\right) - 2x, & x < 0 \\ 0, & x = 0 \end{cases}$$

$$\therefore \quad f''(x) = \Big\{ -x^{p-4}\sin\left(\frac{1}{x}\right) - (p-2)x^{p-3}\cos\left(\frac{1}{x}\right) - px^{p-3}\cos\left(\frac{1}{x}\right) + p(p-1) x^{p-2}\sin\left(\frac{1}{x}\right) + 2, \; x > 0$$

$$\Big\{ -x^{p-4}\sin\left(\frac{1}{x}\right) - (p-2)x^{p-3}\cos\left(\frac{1}{x}\right) - px^{p-3}\cos\left(\frac{1}{x}\right) + p(p-1) x^{p-2}\sin\left(\frac{1}{x}\right) - 2, \; x < 0]$$

$$0, \; x = 0$$

RHL = LHL = V.F. $= 0$

$\because \sin\infty$ and $\cos\infty$ lies between -1 to 1.

For $p \ge 5$, RHL $= 2$

LHL $= -2$

V.F. $= 0$

For $p \in [5, \infty)$, $f''(x)$ is not continuous.

35. $f(0) =$ RHL $= \lim_{x\to 0^+} f(x) = \lim_{h\to 0} f(h)$

$$12(\ln 4)^3 = \lim_{h\to 0} \frac{(4^h - 1)^3}{\sin(h/p)\cdot \ln(1 + h^2/3)}$$

$$= \lim_{h\to 0} \frac{\left(\frac{4^h - 1}{h}\right)^3}{\frac{\sin(h/p)}{(h/p)} \cdot \frac{\ln(1 + h^2/3)}{(h^2/3)} \cdot \frac{1}{(3p)}}$$

$$= \frac{(\ln 4)^3 \cdot 3p}{1\cdot 1}$$

Hence, $\quad p = 4$

36. $f(0) = \text{RHL} = \lim_{x \to 0^+} f(x)$

$$= \lim_{h \to 0} f(h)$$

$$= \lim_{h \to 0} \frac{1 - \cos(1 - \cos h)}{h^4} \times \frac{1 + \cos(1 - \cos h)}{1 + \cos(1 - \cos h)}$$

$$= \lim_{h \to 0} \frac{\sin^2(1 - \cos h)}{h^4 \cdot (1 + \cos(1 - \cos h))} \cdot \frac{(1 - \cos h)^2}{(1 - \cos h)^2}$$

$$= \lim_{h \to 0} \left[\frac{\sin(1 - \cos h)}{(1 - \cos h)}\right]^2 \times \lim_{h \to 0} \left(\frac{1 - \cos h}{h^2}\right)^2 \times \lim_{h \to 0} \frac{1}{1 + \cos(1 - \cos h)}$$

$$= (1)^2 \times \frac{1}{4} \times \frac{1}{2} = \frac{1}{8}$$

37. Check at $x = 0$

$$\text{LHL} = \lim_{x \to 0^-} f(x) = \lim_{h \to 0} f(0 - h)$$

$$= \lim_{h \to 0} \frac{1 - k^{-1/h}}{1 + k^{-1/h}}$$

$$= \frac{1 - 0}{1 + 0} \qquad [\text{for } 0 < k < 1]$$

$$= 1$$

and $\text{RHL} = \lim_{x \to 0^+} f(x) = \lim_{h \to 0} f(0 + h)$

$$= \lim_{h \to 0} \frac{1 - k^{1/h}}{1 + k^{1/h}} = \lim_{h \to 0} \frac{k^{-1/h} - 1}{k^{-1/h} + 1}$$

$$= \frac{0 - 1}{0 + 1} = -1$$

$\therefore$ Jump = LHL − RHL = 1 − (− 1) = 2

38. $\lim_{x \to 0} \frac{2f'(x) - 6f'(2x) + 4f'(4x)}{2x}$ [by L' Hospital's rule]

$$\lim_{x \to 0} \frac{2f''(x) - 12f''(x) + 16f''(4x)}{2}$$

$$= 3f''(0) = 12$$

39. $y = [x] + \sqrt{x - [x]} = n + \sqrt{(x - n)},\ n \le x < n + 1$

The function is continuous at all numbers, check the continuity at $x = n$

$$\text{LHL} = \lim_{x \to n-} f(x) = \lim_{h \to 0} f(n - h)$$

$$= \lim_{h \to 0} [n - h] + \sqrt{n - h - [n - h]}$$

$$= (n - 1) + \sqrt{n - (n - 1)}$$

$$= n$$

$$\text{RHL} = \lim_{x \to n^+} f(x) = \lim_{h \to 0} f(n + h)$$

$$= \lim_{h \to 0} [n + h] + \sqrt{(n + h) - [n + h]}$$

$$= n + \sqrt{n - n} = n$$

and value of the function $= f(n) = n + \sqrt{n - [n]}$

$$= n + 0$$

$$= n$$

Hence, LHL = RHL = V. F.

$\therefore f(x)$ is continuous of all real numbers.

40. We have,

$$f(x) = \begin{cases} \dfrac{a|x^2 - x - 2|}{2 + x - x^2}, & x < 2 \\ b, & x = 2 \\ \dfrac{x - [x]}{x - 2}, & x > 2 \end{cases}$$

$$= \begin{cases} -\dfrac{a(x^2 - x - 2)}{2 + x - x^2}, & x < 2 \\ b, & x = 2 \\ \dfrac{x - [x]}{x - 2}, & x > 2 \end{cases}$$

$$f(x) = \begin{cases} a, & x < 2 \\ b, & x = 2 \\ \dfrac{x - [x]}{x - 2}, & x > 2 \end{cases}$$

Check at $x = 2$

LHL $= \lim_{x \to 2^-} f(x) = \lim_{h \to 0} f(2 - h)$

$$= \lim_{h \to 0} a = a$$

RHL $= \lim_{x \to 2+} f(x) = \lim_{h \to 0} f(2 + h)$

$$= \lim_{h \to 0} \frac{2 + h - [2 + h]}{2 + h - 2}$$

$$= \lim_{h \to 0} \frac{2 + h - 2}{2 + h - 2}$$

$$= 1$$

Value of the function $= f(2) = b$, so that $f(x)$ is continuous at $x = 2$

$\therefore$ LHL = RHL = V. F.

$\therefore$ $a = b = 1$

41. $f(x)$ is continuous at $x = 0$

$$\lim_{x \to 0^+} f(x) = f(0) = \lim_{x \to 0^-} f(x) \qquad \ldots\text{(i)}$$

$$f(0) = b \qquad \ldots\text{(ii)}$$

$$\lim_{x \to 0^+} f(x) = \lim_{x \to 0^+} \left(\frac{\sin(a + 1)x}{2x} + \frac{\sin 2x}{2x}\right)$$

$$= \frac{a + 1}{2} + 1 \qquad \ldots\text{(iii)}$$

$$\lim_{x \to 0^+} f(x) = \lim_{x \to 0^+} \frac{\sqrt{x + bx^3} - \sqrt{x}}{bx^{5/2}}$$

$$= \lim_{x \to 0^+} \frac{(x + bx^3 - x)}{bx^{5/2}(\sqrt{x + bx^3} + \sqrt{x})}$$

$$= \lim_{x \to 0^+} \frac{\sqrt{x}}{\sqrt{x}(\sqrt{1 + bx^2} + 1)} = \frac{1}{2} \qquad \ldots\text{(iv)}$$

Use Eqs. (ii), (iii) and (iv) in Eq. (i)

$$\frac{1}{2} = b = \frac{a + 1}{2} + 1$$

$$\Rightarrow \quad b = \frac{1}{2},\ a = -2$$

$$a + b = \frac{-3}{2}$$

42. $f\{g(x)\} = \begin{cases} 1, & 0 < x < \dfrac{3\pi}{4} \\ & \text{or } \dfrac{7\pi}{4} < x < 2\pi \\ 0, & x = \dfrac{3\pi}{4}, \dfrac{7\pi}{4} \\ -1, & \dfrac{3\pi}{4} < x < \dfrac{7\pi}{4} \end{cases}$

Clearly $f\{g(x)\}$ is not continuous at

$$x = \frac{3\pi}{4}, \frac{7\pi}{4}$$

43. The function $f(x)$ is discontinuous if $3^n - (2\cos x)^{2n} = 0$

$$\Rightarrow \quad (\cos^2 x)^n = \left(\frac{3}{4}\right)^n$$

$$\Rightarrow \quad \cos^2 x = 3/4$$

$$\therefore \quad x = r\pi \pm \pi/6, \; r \in I$$

44. $[\cos x + \sin x] = [\sqrt{2}\cos(x - \pi/4)]$

We know that $[x]$ is discontinuous at integral values of x,

Now, $\sqrt{2}\cos(x - \pi/4)$ is an integer.

At $\quad x = \pi/2, 3\pi/4, 5\pi/4, 7\pi/4.$

45. $\because f(x) = |x^2 - 3x + 2| + \cos|x|$

$$= |(x-1)||(x-2)| + \cos|x|$$

$$f(x) = \begin{cases} x^2 - 3x + 2 + \cos x, & x < 0 \\ x^2 - 3x + 2 + \cos x, & 0 \le x < 1 \\ -x^2 + 3x - 2 + \cos x, & 1 \le x < 2 \\ x^2 - 3x + 2 + \cos x, & x > 2 \end{cases}$$

$$\therefore \quad f'(x) = \begin{cases} 2x - 3 - \sin x, & x < 0 \\ 2x - 3 - \sin x, & 0 \le x < 1 \\ -2x + 3 - \sin x, & 1 \le x < 2 \\ 2x - 3 - \sin x, & x > 2 \end{cases}$$

It is clear $f(x)$ is not differentiable at $x = 1$.

$\therefore \quad f'(1^-) = -1 - \sin 1$

and $\quad f'(1^+) = 1 - \sin 1$

46. $\because \quad f(x) = \dfrac{1 - \tan x}{4x - \pi}$

$$\therefore \quad \text{RHL} = \lim_{x \to \pi/4^+} f(x) = \lim_{h \to 0} f\left(\frac{\pi}{4} + h\right)$$

$$= \lim_{h \to 0} \frac{1 - \tan\left(\frac{\pi}{4} + h\right)}{4\left(\frac{\pi}{4} + h\right) - \pi}$$

$$= \lim_{h \to 0} \frac{1 - \left(\frac{1 + \tan h}{1 - \tan h}\right)}{4h}$$

$$= \lim_{h \to 0} -\frac{2\tan h}{(1 - \tan h)4h} = -\frac{1}{2}$$

V. F. $= \lambda$

For continuous

$$\lambda = -\frac{1}{2}$$

47. For $0 < n < 1$, $\sin x < \sin^n x$

and for $n > 1$, $\sin x > \sin^n x$

Now, for $0 < n < 1$,

$$f(x) = \frac{2(\sin x - \sin^n x) - (\sin x - \sin^n x)}{2(\sin x - \sin^n x) + (\sin x - \sin^n x)} = \frac{1}{3}$$

and for $n > 1$,

$$f(x) = \frac{2(\sin x - \sin^n x) + (\sin x - \sin^n x)}{2(\sin x - \sin^n x) - (\sin x - \sin^n x)} = 3$$

For $\quad n > 1, g(x) = 3, \; x \in (0, \pi)$

$\therefore$ $g(x)$ is continuous and differentiable at

$x = \dfrac{\pi}{2}$, and for $0 < n < 1$,

$$g(x) = \begin{cases} \left[\dfrac{1}{3}\right] = 0, & x \in \left(0, \dfrac{\pi}{2}\right) \cup \left(\dfrac{\pi}{2}, \pi\right) \\ 3, & x = \dfrac{\pi}{2} \end{cases}$$

$\therefore g(x)$ is not continuous at $x = \dfrac{\pi}{2}$.

Hence, $g(x)$ is also not differentiable at $x = \dfrac{\pi}{2}$

48. $f(2) = 8, f'(2) = 5, f'(x) \ge 1, f''(x) \ge 4, \forall \; x \in (1, 6)$

$$f''(x) = \frac{f'(5) - f'(2)}{5 - 2} \ge 4$$

$$f''(5) \ge 17 \qquad \ldots\text{(i)}$$

$$f'(x) = \frac{f(5) - f(2)}{5 - 2} \ge 1$$

$$\Rightarrow \quad f(5) \ge 11 \qquad \ldots\text{(ii)}$$

$$\therefore \quad f'(5) + f(5) \ge 28$$

49. $\because 2 < f(x) < \sqrt{26}$

$$\therefore \frac{2}{c} < \frac{f(x)}{c} < \frac{\sqrt{26}}{c}$$

Since, $g(x) = \left[\dfrac{f(x)}{c}\right]$ is continuous $\forall \; x \in R$

$\Rightarrow g(x) = 0 \;$ for $c = 6$

50. $\text{RHL} = \lim_{x \to 0+} f(x) = \lim_{h \to 0} f(0 + h)$

$$= \lim_{h \to 0}\left(a + \frac{\sin[h]}{h}\right) = \lim_{h \to 0}(a + 0) = a$$

V.F. $= f(0) = 2$

$$\text{LHL} = \lim_{x \to 0-} f(x)$$

$$= \lim_{h \to 0} f(0 - h)$$

$$= \lim_{h \to 0} b + \left(\frac{-\sin h + h}{-h^3}\right)$$

$$= b + \lim_{h \to 0}\left(\frac{\sin h - h}{h^3}\right)$$

$$= b - 1 \left(\because \lim_{h \to 0} \frac{h - \sin h}{h^3} = \frac{1}{6}\right)$$

$\because f(x)$ is continuous.

$\therefore$ LHL = RHL = V.F.

$\Rightarrow \quad b - 1 = a = 2$

$\Rightarrow \quad b = a + 1$

51. For f to be continuous everywhere, we must have

$$x_0^2 = f(x_0) = \lim_{x \to x_0^+} f(x) = ax_0 + b$$

Also, f has a derivative at x_0 if $f'(x_0+) = f'(x_0-)$.

Now, $f'(x_0+) = \lim_{h \to 0^+} \frac{f(x_0+h) - f(x_0)}{h}$

$$= \lim_{h \to 0+} \frac{a(x_0+h) + b - x_0^2}{h}$$

$$= \lim_{h \to 0+} \left(\frac{ax_0 + b - x_0^2}{h} + a \right)$$

$$= \lim_{h \to 0+} a = a \qquad [\because x_0^2 = ax_0 + b]$$

and $f'(x_0-) = \lim_{h \to 0-} \frac{f(x_0+h) - f(x_0)}{h}$

$$= \lim_{h \to 0-} \frac{(x_0+h)^2 - x_0^2}{h}$$

$$= \lim_{h \to 0-} \frac{x_0^2 + h^2 + 2x_0h - x_0^2}{h}$$

$$= \lim_{h \to 0-} (h + 2x_0) = 2x_0$$

That is, $a = 2x_0$, and hence

$$b = x_0^2 - ax_0 = x_0^2 - 2x_0^2 = -x_0^2.$$

52. LHL $= \lim_{x \to (n\pi + \pi/2)-} f(x) = \lim_{h \to 0} f(n\pi + \pi/2 - h)$

$$= \lim_{h \to 0} (1 + |\cos(n\pi + \pi/2 - h)|)^{ab/|\cos[n\pi + \pi/2 - h]|}$$

$$= \lim_{h \to 0} (1 + |\sin(n\pi - h)|)^{ab/|\sin(n\pi - h)|}$$

$$= \lim_{h \to 0} (1 + |(-1)^n \sin h|)^{ab|1(-1)^n \sin h|}$$

$$= \lim_{h \to 0} (1 + \sin h)^{ab/\sin h} = e^{ab} \qquad \left[\because \lim_{h \to 0} (1+h)^{\lambda/h} = e^{\lambda}\right]$$

V.F. $= f(n\pi + \pi/2) = e^a . e^b$

RHL $= \lim_{x \to (n\pi + \pi/2)+} f(x) = \lim_{h \to 0} f\left(n\pi + \frac{\pi}{2} + h\right)$

$$= \lim_{h \to 0} e^{\cot(2n\pi + \pi + 2h)/\cot(8n\pi + 4\pi + 8h)}$$

$$= e^{\lim_{h \to 0} (\cot 2h/\cot 8h)}$$

$$= e^{\lim_{h \to 0} \tan 8h/\tan 2h}$$

$$= e^{4 \lim_{h \to 0} \tan 8h/8h \cdot \lim_{x \to 0} 2h/\tan 2h}$$

$$= e^{4 \cdot 1 \cdot 1} = e^4$$

$\because f(x)$ is continuous in $(n\pi, (n+1)\pi)$

$\therefore$ LHL = RHL = V.F.

$$e^{ab} = e^4 = e^{a+b}$$

$\Rightarrow \quad ab = 4 = a + b$

$\Rightarrow \quad a = b = 2$

Objective Questions Type II [One or more than one correct answer(s)]

1. We have, $f(x) = \frac{1}{[\sin x]}$

$f(x)$ is defined when

$$-1 < \sin x < 0 \text{ and } \sin x = 1$$

$\therefore \quad x \in ((2n+1)\pi, (2n+2)\pi) \cup \left\{2n\pi + \frac{\pi}{2}\right\}$

$n \in I$

In $\quad x \in ((2n+1)\pi, (2n+2)\pi)$

$f(x) = -1$

$\because f(x)$ is constant function.

Hence, $f(x)$ is continuous in

$$((2n+1)\pi, (2n+2)\pi)$$

$$Lf'(\pi/2) = \lim_{h \to 0} \frac{f(\pi/2 - h) - f(\pi/2)}{-h}$$

$$= \lim_{h \to 0} \frac{\frac{1}{[\sin(\pi/2 - h)]} - 1}{-h}$$

$$= \lim_{h \to 0} \frac{\frac{1}{[\cos h]} - 1}{-h} = -\infty \qquad [\because [\cos h] = 0]$$

Hence, $f(x)$ is not differentiable at $x = \frac{\pi}{2}$

2. $\because \quad g(t) = [t\,(1/t)]$

For $t > 1, g(t) = 0$

and for $0 < t < 1$

$\Rightarrow \quad 1 < \frac{1}{t} < \infty$

$$\left[\frac{1}{t}\right] = 2, 3, 4, 5, \dots.$$

$\therefore \quad g(g) = \left[g\left(\frac{1}{g}\right)\right]$

For $t > 1$, undefined

and $0 < t < 1$, undefined (Infinite numbers).

Hence, $g(g)$ is discontinuous at infinite number of points.

3. $\lim_{x \to 1+} \phi(x) = 1 = \lim_{x \to 1-} \phi(x) \neq 0$

$$\lim_{x \to 1+} f(x) = \lim_{h \to 0} f(1+h) = \lim_{h \to 0} [1+h] = 1$$

and $\lim_{x \to 1-} f(x) = \lim_{h \to 0} f(1-h)$

$$= \lim_{h \to 0} [1-h] = 0$$

$\phi \circ f(x) = \phi\{f(x)\} = \phi[x] = 0$ for all x.

So, ϕ *of* is continuous for all x.

$$f \circ \phi(x) = f\{\phi(x)\} = \begin{cases} f(0), & \text{if } x \in I \\ f(x^2), & \text{if } x \in R - I \end{cases}$$

$$= \begin{cases} 0, & x \in I \\ [x^2], & x \in R - I \end{cases}$$

Which is clearly not continuous.

4. The function $f(x) = \tan x$ is not defined at $x = \frac{\pi}{2}$, so f is not continuous on $(0, \pi)$. Since, the function $g(x) = x \sin \frac{1}{x}$ is continuous on $(0, \pi)$ and the integral function of a continuous function is continuous, therefore $F(x) = \int_0^x t \sin \frac{1}{t}\, dt$ is continuous on $(0, \pi)$.

For the function

$$f(x) = \begin{cases} 1, & 0 < x \le \frac{3\pi}{4} \\ 2 \sin \frac{2}{9} x, & \frac{3\pi}{4} < x < \pi \end{cases}$$

$$\lim_{x \to \frac{3\pi}{4}^-} f(x) = 1,$$

and $$\lim_{x \to \frac{3\pi}{4}^+} f(x) = \lim_{h \to 0} f\left(\frac{3\pi}{4} + h\right)$$

$$= \lim_{h \to 0} 2 \sin \frac{2}{9}\left(\frac{3\pi}{4} + h\right)$$

$$= 2 \sin \pi/6 = 1 \text{ and } f(3\pi/4) = 1$$

$\therefore f(x)$ is continuous at $x = 3\pi/4$ evidently $f(x)$ is continuous at all other points.

For the function

$$F(x) = \begin{cases} x \sin x, & 0 < x \le \frac{\pi}{2} \\ \frac{\pi}{2} \sin (\pi + x), & \frac{\pi}{2} < x < \pi \end{cases}$$

$$\therefore \quad F(\pi/2) = \frac{\pi}{2} \sin \frac{\pi}{2} = \frac{\pi}{2}$$

$$\lim_{x \to \frac{\pi}{2}^-} F(x) = \lim_{h \to 0} F\left(\frac{\pi}{2} - h\right)$$

$$= \lim_{h \to 0} \left(\frac{\pi}{2} - h\right) \sin \left(\frac{\pi}{2} - h\right)$$

$$= \frac{\pi}{2} \sin \frac{\pi}{2} = \frac{\pi}{2}$$

$$\lim_{x \to \frac{\pi}{2}^+} F(x) = \lim_{h \to 0} F\left(\frac{\pi}{2} + h\right)$$

$$= \lim_{h \to 0} \frac{\pi}{2} \sin \left(\frac{3\pi}{2} + h\right)$$

$$= \frac{\pi}{2} \sin \frac{3\pi}{2} = -\frac{\pi}{2}$$

$\therefore F(x)$ is not continuous at $x = \frac{\pi}{2}$

5. $f : R \to R,\ f(x) = (x^2 + \sin x)(x - 1)$

$f(1^+) = f(1^-) = f(1) = 0$

$fg(x) : f(x) \cdot g(x),\ fg : R \to R$

Let $fg(x) = h(x) = f(x) \cdot f(g),\ h : R \to R$

Option (c) $h'(x) = f'(x) g(x) + f(x) g'(x)$

$h'(1) = f'(1) g(1) + 0$ [as $f(1) = 0$, $g'(x)$ exists]

If $f(x)$ is differentiable, then $h(x)$ is also differentiable.

Option (a) If $g(x)$ is continuous at $x = 1$, then

$$g(1^+) = g(1^-) = g(1)$$

$$h'(1^+) = \lim_{r \to 0^+} \frac{h(1 + r) - h(1)}{r}$$

$$= \lim_{r \to 0^+} \frac{f(1 + r) g(1 + r) - 0}{r} = f'(1) g(1)$$

$$h'(1^-) = \lim_{r \to 0^-} \frac{f(1 - r) g(1 - r) - 0}{-r} = f'(1) g(1)$$

So, $h(x) = f(x) \cdot g(x)$ is differentiate at $x = 1$

Option (b), (d)

$$h'(1^+) = \lim_{r \to 0^+} \frac{h(1 + r) - h(1)}{-r}$$

$$= \lim_{r \to 0^+} \frac{f(1 + r) g(1 + r)}{r} = f'(1) g(1^+)$$

$$h'(1^-) = \lim_{r \to 0^+} \frac{f(1 - r) g(1 - r)}{-r} = 8'(1) g(1^-)$$

$\Rightarrow g(1^+) = g(1^-)$

So, we cannot comment on the continuity and differentiability of the function.

6. We have,

$$f(x) = \begin{cases} 1, & |x| \ge 1 \\ \frac{1}{n^2}, & \frac{1}{n} \le |x| < \frac{1}{n - 1}, \ n = 2, 3, \ldots \\ 0, & x = 0 \end{cases}$$

$$= \begin{cases} 1, & x \le -1 \text{ or } x \ge 1 \\ \frac{1}{4}, & x \in \left(-1, -\frac{1}{2}\right] \cup \left[\frac{1}{2}, 1\right) \\ \frac{1}{9}, & x \in \left(\frac{-1}{2}, \frac{-1}{3}\right] \cup \left[\frac{1}{3}, \frac{1}{2}\right) \end{cases}$$

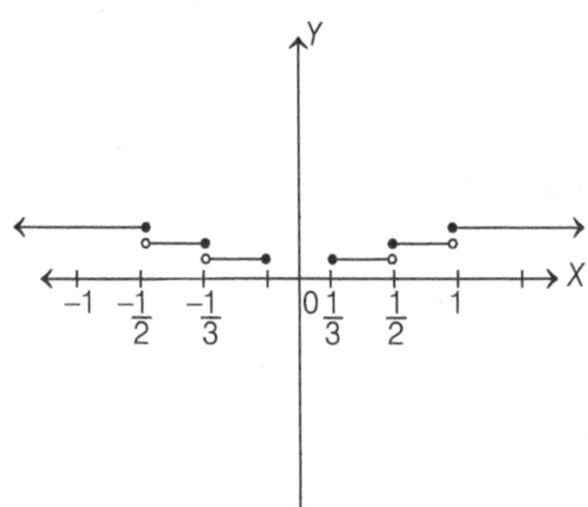

The function f is clearly continuous for $|x| > 1$.

We observe that

$$\lim_{x \to -1^-} f(x) = 1, \quad \lim_{x \to -1^+} f(x) = \frac{1}{4}$$

Also, $\lim_{x \to \frac{1}{n}^+} f(x) = \frac{1}{n^2}$, $\lim_{x \to \frac{1}{n}^-} f(x) = \frac{1}{(n + 1)^2}$

Clearly, f is discontinuous for $x = \pm \frac{1}{n}$, $n = 1, 2, 3$ ……

Also, f is continuous for $x = 0$.

Hence, (a) and (c) are the correct answers.

7. **Option (a)** : $f(x)$ is continuous but not differentiable at $x = 0$ (since at $x = 0$, corner)

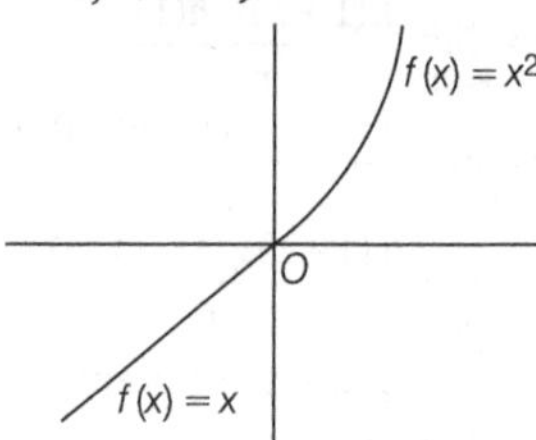

Option (b) : $\therefore g(x)$ is continuous but not differentiable at $x = 0$.

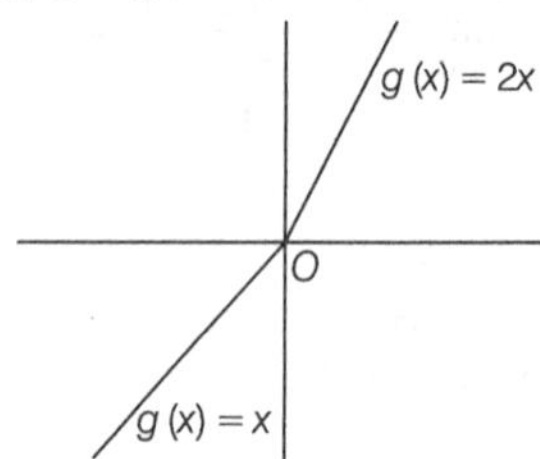

($\because$ for differentiable no break and no corner).

Option (c) : $h(x) = x|x| = \begin{cases} x^2, & x \geq 0 \\ -x^2, & x < 0 \end{cases}$

$h(x)$ is continuous and differentiable at $x = 0$.

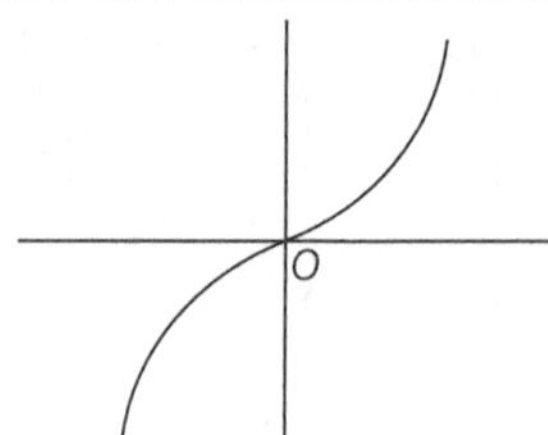

Option (d) : $k(x) = 1 + |x| = \begin{cases} 1 + x, & x \geq 0 \\ 1 - x, & x < 0 \end{cases}$

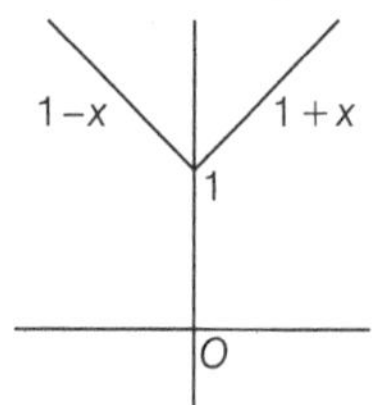

$\therefore k(x)$ is continuous but not differentiable at $x = 0$.

8. Since, $\lim\limits_{n \to \infty} x^{2m} = \begin{cases} 0, & \text{if } |x| < 1 \\ 1, & \text{if } |x| = 1 \end{cases}$

$$\therefore \quad f(x) = \lim_{n \to \infty} (\sin x)^{2n}$$

$$= \begin{cases} 0, & \text{if } |\sin x| < 1 \\ 1, & \text{if } |\sin x| = 1 \end{cases}$$

This shows that $f(x)$ is continuous $\forall\, x$, except possibility when $|\sin x| = 1$ *ie*, when

$$x = (2k + 1)\frac{\pi}{2} \, (k \in I)$$

For these points

$$\lim_{x \to (2k+1)\frac{\pi}{2}} f(x) = 0 \neq 1 = f\left[(2k+1)\frac{\pi}{2}\right]$$

$\Rightarrow f(x)$ is discontinuous for these points.

9. We have, $f(x) = \tan^{-1} \cot x$

Option (a) : Period of $f(x)$ = period of $\cot x = \pi$

Option (b) : $f(x) = \tan^{-1} \cot x$

$$= \tan^{-1} \tan\left(\frac{\pi}{2} - x\right) = \frac{\pi}{2} - x$$

$f(x)$ is continuous at $x = \pi/2, 3\pi/2, \ldots$.

Option (c) : $f'(x) = \dfrac{1}{1 + \cot^2 x}(-\operatorname{cosec}^2 x)$

$$= -\frac{\operatorname{cosec}^2 x}{1 + \cot^2 x}$$

$f'(n\pi)$ = indeterminant form

$\therefore f(x)$ is not differentiable at $x = \pi, 99\pi, 100\pi$.

Option (d) : x lies in I and II quadrant, in I quadrant $\cot x > 0$ and in II quadrant $\cot x < 0$, in I quadrant $0 \leq f(x) < \pi/2$ and in II quadrant $-\dfrac{\pi}{2} < f(x) < 0$

Hence, $f(x) \neq -1$ in $2n\pi \leq x \leq (2n+1)\pi$.

10. We have,

$$\lim_{x \to 1} \frac{(\log(1+x) - \log 2)(3 \cdot 4^{x-1} - 3x)}{\{(7+x)^{1/3} - (1+3x)^{1/2}\} \sin \pi x}$$

$$= \lim_{y \to 0} \frac{(\log(2+y) - \log 2)(3 \cdot 4^y - 3)}{\{(8+y)^{1/3} - (1+3y)^{1/2} + 3] \sin \pi / y + 1} [y = x - 1]$$

$$= -\lim_{y \to 0} \frac{\log\left(1 + \dfrac{y}{2}\right) \cdot (3(4^y - 1) - 3y)}{2\left\{\left(1 + \dfrac{y}{8}\right)^{1/3} - \left(1 + \dfrac{3}{4}y\right)^{1/2}\right\} \sin \pi y}$$

$$[\because \sin(\pi + \theta) - \sin\theta]$$

$$= -\frac{1}{\pi} \lim_{y \to 0} \frac{\log\left(1 + \dfrac{y}{2}\right)}{\dfrac{y}{2}} \cdot \left\{\frac{3(4^y - 1)}{y} - 3\right\} \times$$

$$\frac{\pi y}{\sin \pi y} \frac{y}{\left[1 + \dfrac{y}{24} - 1 - \dfrac{3}{8}y + O(y^2)\right]}$$

$$= -\frac{1}{\pi}\{3 \log 4 - 3\}(-3) = \frac{9}{\pi} \log \frac{4}{e}$$

which is a transcedental number.

11. We have,

$$f(x) = \min\{1, \cos x, 1 - \sin x\}$$

$\therefore f(x)$ can be rewritten as

$$f(x) = \begin{cases} \cos x, & -\dfrac{\pi}{2} \leq x \leq 0 \\ 1 - \sin x, & 0 < x \leq \dfrac{\pi}{2} \\ \cos x, & \dfrac{\pi}{2} < x \leq \pi \end{cases}$$

$$\Rightarrow \quad f'(x)=\begin{cases} -\sin x, & -\frac{\pi}{2}\le x\le 0 \\ -\cos x, & 0<x\le\frac{\pi}{2} \\ -\sin x, & \frac{\pi}{2}<x\le\pi \end{cases}$$

$\because \quad f'(0)=0$

Hence, $f(x)$ has local maxima at 0 and $f(x)$ is not differentiable at $x=0$. (it is clear from above).

12. $$f'(0+)=\lim_{h\to 0}\frac{f(0+h)-f(0)}{h}$$

$$=\lim_{h\to 0}\frac{f(h)}{h}$$

$$=\lim_{h\to 0}\frac{\log\cos h}{\log(1+h^2)}\times\frac{h^2}{h^2} \qquad \ldots\text{(i)}$$

$$=\lim_{h\to 0}\frac{h^2}{\log(1+h^2)}\times\lim_{h\to 0}\frac{\log\cos h}{h^2}$$

$$=\lim_{h\to 0}\frac{h^2}{\log(1+h^2)}\times\lim_{h\to 0}-\frac{\tan h}{2h}$$

$$=1\times-\frac{1}{2}$$

$$=-\frac{1}{2}$$

and $$f'(0-)=\lim_{h\to 0}\frac{f(0-h)-f(0)}{-h}$$

$$=\lim_{h\to 0}\frac{f(h)}{h}$$

$$=-\lim_{h\to 0}\frac{-h\log\cos(-h)}{\log(1+(-h)^2)h}$$

$$=\lim_{h\to 0}\frac{\log\cos h}{\log(1+h^2)}$$

$$=-\frac{1}{2} \qquad \{\text{from Eq. (i)}\}$$

$\therefore \quad f'(0+)=f'(0-)$

Hence, $f(x)$ is differentiable and continuous at $x=0$.

13. We have,

$$f(0^+)=\lim_{h\to 0}f(0+h)$$

$$=\lim_{h\to 0}3-\left[\cot^{-1}\left(\frac{2h^3-3}{h^2}\right)\right]$$

$$=3-[\cot^{-1}(-\infty)]$$

$$=3-[\pi-\cot^{-1}\infty]$$

$$=3-[\pi-0]$$

$$=3-3=0$$

$\therefore f(0^+)\ne 3$

$$f(0^-)=\lim_{h\to 0}f(0-h)=\lim_{h\to 0}\{h^2\}\cos(e^{-1/h})$$

$$=0\times\cos(e^{-\infty})$$

$$=0\times\cos(0)$$

$$=0\times 1=0$$

$\because f(0^+)=f(0^-)=f(0)=0$

$\therefore f(x)$ is continuous at $x=0$.

14. $$Lf'(0)=\lim_{h\to 0}\frac{f(0-h)-f(0)}{-h}=\lim_{h\to 0}\frac{-\sin\frac{\pi h}{2}}{-h}=\frac{\pi}{2}$$

$$Rf'(0)=\lim_{h\to 0}\frac{f(0+h)-f(0)}{h}=\lim_{h\to 0}\frac{\pi}{2}\cdot\frac{\sin\frac{\pi h}{2}}{\frac{\pi h}{2}}=\frac{\pi}{2}$$

$\therefore \quad Lf'(0)=Rf'(0)$

Hence, $f(x)$ is differentiable and continuous at $x=0$.

Hence, options (a) and (b) are correct.

Option (c) : $$Lf'(1)=\lim_{h\to 0}\frac{f(1-h)-f(1)}{-h}$$

$$=\lim_{h\to 0}\frac{\sin\left\{\frac{\pi(1-h)}{2}\right\}-1}{-h}$$

$$=\lim_{h\to 0}\frac{\cos\frac{\pi h}{2}-1}{-h}$$

$$=\lim_{h\to 0}\frac{1-\cos\frac{\pi h}{2}}{h}\times\frac{1+\cos\frac{n\pi}{2}}{1+\cos\frac{\pi h}{2}}$$

$$=\lim_{h\to 0}\frac{\sin^2\left(\frac{\pi h}{2}\right)}{h\left(1+\cos\frac{\pi h}{2}\right)}=0$$

$$Rf'(1)=\lim_{h\to 0}\frac{f(1+h)-f(1)}{h}$$

$$=\lim_{h\to 0}\frac{|-1+h|-1}{h}$$

$$=\lim_{h\to 0}\frac{1-h-1}{h}=-1$$

$$Lf'(1)\ne Rf'(1)$$

but $Lf'(1)$ and $Rf'(1)$ has finite value.

Hence, $f(x)$ is continuous but not differentiable at $x=1$.

Option (d) : $$Lf'\left(\frac{3}{2}\right)=\lim_{h\to 0}\frac{f\left(\frac{3}{2}-h\right)-f\left(\frac{3}{2}\right)}{-h}$$

$$=\lim_{h\to 0}\frac{|3-2h-3|\left[\frac{3}{2}-h\right]-0}{-h}$$

$$=-2$$

$$Rf'\left(\frac{3}{2}\right)=\lim_{h\to 0}\frac{f\left(\frac{3}{2}+h\right)-f\left(\frac{3}{2}\right)}{h}$$

$$=\lim_{h\to 0}\frac{|2h|\left[\frac{3}{2}+h\right]-0}{h}=\lim_{h\to 0}\frac{2h\cdot 1}{h}=-2$$

$\therefore f(x)$ is not differentiable.

But here $Lf'(3/2)$ and $Rf'(3/2)$ has finite value.

Hence, $f(x)$ is continuous.

15. If $x < x^2$

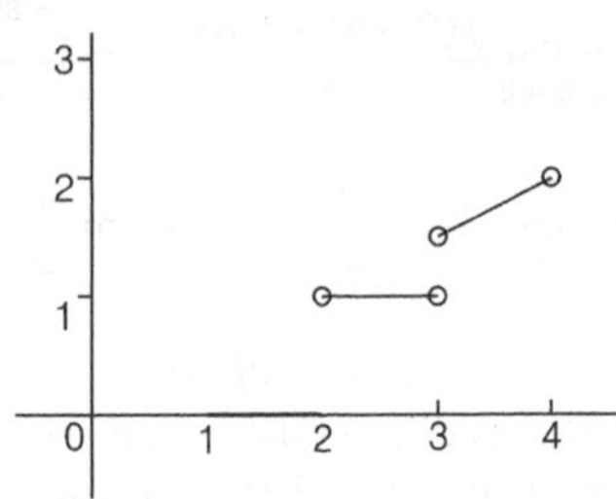

Then, $h(x) = x,\ x(x-1) > 0$

$\therefore$ $x > 1$ and $x < 0$

$\therefore$ $h'(x) = 1$

and if $x^2 < x,\ x(x-1) < 0$

$\Rightarrow$ $0 < x < 1$

Then, $h(x) = x^2$

16. $f(x) = \begin{cases} 0, & 1 \le x < 2 \\ \log_e 2, & 2 \le x < 3 \\ \log_e x, & 3 \le x < 4 \end{cases}$

$f(x)$ is continuous and differentiable everywhere except at $x = 2, 3$

$\therefore f(x)$ is not continuous at $x = 2, 3$ (i.e. the graph is broken at $x = 2, 3$)

$\Rightarrow f(x)$ is not differentiable at $x = 2, 3$ (i.e. the graph does not have a definite tangent at $x = 2, 3$).

17. $f(x)$ is discontinuous at $\frac{\pi}{2}, \pi, \frac{3\pi}{2}, 2\pi, \dots$

and $f(x)$ is periodic with period π

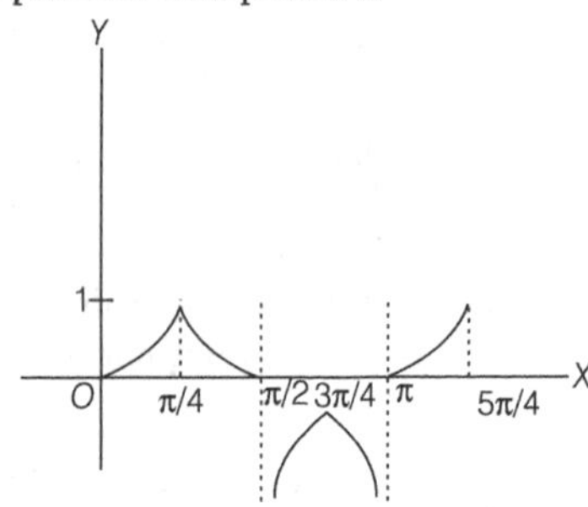

and $\int_0^{\pi/2} f(x)\,dx = \int_0^{\pi/4} f(x)\,dx + \int_{\pi/4}^{\pi/2} f(x)\,dx$

$= \int_0^{\pi/4} \tan x\,dx + \int_{\pi/4}^{\pi/2} \cot x\,dx$

$= [\ln \sec x]_0^{\pi/4} + [\ln \sin x]_{\pi/4}^{\pi/2}$

$= [\ln\sqrt{2} - 0] + \left[0 - \ln\left(\frac{1}{\sqrt{2}}\right)\right]$

$= 2\ln\sqrt{2}$

18. $f(x) = \begin{cases} -\frac{1}{x}, & -1 \le x < -\frac{1}{2} \\ 0, & -\frac{1}{2} \le x < \frac{1}{2} \\ \frac{1}{x}, & \frac{1}{2} \le x < \frac{3}{2} \\ \frac{2}{x}, & \frac{3}{2} \le x < \frac{5}{2} \end{cases}$

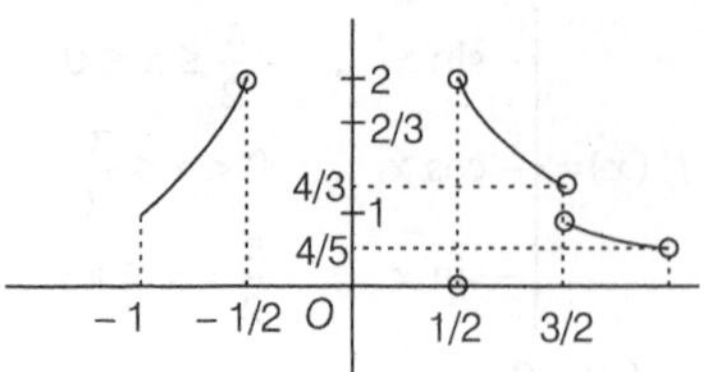

It is clear from the graph $f(x)$ is discontinuous at $x = 0, \frac{1}{2}, \frac{3}{2}$.

19. We have, $f(x) = [x] + [x + 1/3] + [x + 2/3] = [3x]$

Option (a) :

$\lim_{x \to 1-} f(x) = \lim_{h \to 0} f(1-h) = \lim_{h \to 0} [3 - 3h] = 2$

$\lim_{x \to 1+} f(x) = \lim_{h \to 0} f(1+h) = \lim_{h \to 0} [3 + 3h] = 3$

$\lim_{x \to 10-} f(x) = \lim_{h \to 0} f(10-h) = \lim_{h \to 0} [30 - 3h] = 29$

$\lim_{x \to 10^+} f(x) = \lim_{h \to 0} f(10+h) = \lim_{h \to 0} [30 + 3h] = 30$

$\lim_{x \to 15-} f(x) = \lim_{h \to 0} f(15-h) = \lim_{h \to 0} [45 - 3h] = 44$

$\lim_{x \to 15+} f(x) = \lim_{h \to 0} f(15+h) = \lim_{h \to 0} [45 + 3h] = 45$

Hence, $f(x)$ is discontinuous at $x = 1, 10, 15$.

Option (c) : $\int_0^{2/3} f(x)\,dx = \int_0^{2/3} [3x]\,dx$

$= \int_0^{1/3} [3x]\,dx + \int_{1/3}^{2/3} [3x]\,dx$

$= 0 + \int_{1/3}^{2/3} dx = \frac{1}{3}$

20. $Rf'(0) = \lim_{h \to 0} \frac{f(0+h) - f(0)}{h}$

$= -\lim_{h \to 0} \frac{(\sin^{-1} h)^2 \cdot \cos\left(\frac{1}{h}\right) - 0}{h}$

$= \lim_{h \to 0} \left(\frac{\sin^{-1} h}{h}\right)^2 . \cos\left(\frac{1}{h}\right) \cdot h$

$= (1)^2 \cdot (-1 \text{ to } 1) \cdot 0 = 0$

$Lf'(0) = \lim_{h \to 0} \frac{f(0-h) - f(0)}{-h}$

$\lim_{h \to 0} \frac{(\sin^{-1}(-h))^2 \cdot \cos\left(-\frac{1}{h}\right) - 0}{-h}$

$\lim_{h \to 0} \frac{(\sin^{-1}(h))^2 \times \cos\left(\frac{1}{h}\right)}{-h} = -\lim_{h \to 0} \left(\frac{\sin^{-1} h}{h}\right)^2 . \cos\left(\frac{1}{h}\right) \cdot h$

$= -(1)^2 \cdot (-1 + \text{to } 1) \cdot 0 = 0$

$\Rightarrow$ $Lf'(0) = Rf'(0)$

$\Rightarrow f(x)$ is differentiable everywhere.

Also, RHL $= \lim_{x \to 0} f(x)$

$= \lim_{h \to 0} f(0+h)$

$= \lim_{h \to 0} (\sin^{-1} h)^2 \cdot \cos(1/h) = 0$

Similarly, LHL $= 0$, V. $F = f(0) = 0$

21. We have,

$$\lim_{x\to 0^+} f(x)=\lim_{x\to 0^+}\frac{\tan^2\{x\}}{(x^2-[x]^2)}=\lim_{x\to 0^+}\frac{\tan^2 x}{x^2}=1 \quad ..(i)$$

[as $x\to 0^+; [x]=0 \Rightarrow \{x\}=x$]

Also, $$\lim_{x\to 0^-} f(x)=\lim_{x\to 0^-}\sqrt{\{x\}\cot\{x\}}=\sqrt{\cot 1} \quad ...(ii)$$

[As $x\to 0^-; [x]=-1 \Rightarrow \{x\}=x+1 \Rightarrow \{x\}\to 1$]

$\Rightarrow$ (a), (b) and (d) are not correct

Also, $$\cot^{-1}\left(\lim_{x\to 0^-} f(x)\right)^2=\cot^{-1}(\cot 1)=1$$

22. We have,

$$f(1)=f(-1)=1$$

Let $x>1$, then $0<\frac{1}{x^2}<1 \Rightarrow \left[\frac{1}{x^2}\right]=0 \Rightarrow f(x)=0\ \forall\ x>1$

Also, if $x<-1$, then $x^2>1 \Rightarrow 0<\frac{1}{x^2}<1$

$\Rightarrow$ $f(x)=0\ \forall\ x<-1$

Hence, $f(1)=1$, $f(-1)=1$ and $f(x)=0$ if $|x|>1$

$\therefore f(x)$ cannot be continuous at $x=1$ and $x=-1$

Again,

Let $\frac{1}{2}<x^2<1 \Rightarrow 1<\frac{1}{x^2}<2$

$\Rightarrow$ $\left[\frac{1}{x^2}\right]=1 \Rightarrow \frac{1}{2}<x^2\left[\frac{1}{x^2}\right]<1$

$\therefore$ $\left[x^2\left[\frac{1}{x^2}\right]\right]=0 \Rightarrow f(x)=0$ if $x\in\left(-1,\frac{1}{\sqrt{2}}\right)\cup\left(\frac{1}{\sqrt{2}},1\right)$

Next, let $\frac{1}{3}<x^2<\frac{1}{2}$

$\Rightarrow$ $2<\frac{1}{x^2}<3$

$\Rightarrow$ $\left[\frac{1}{x^2}\right]=2 \Rightarrow \frac{2}{3}<x^2\left[\frac{1}{x^2}\right]<1$

$\therefore$ $\left[x^2\left[\frac{1}{x^2}\right]\right]=0$

$\Rightarrow$ $f(x)=0$, if $x\in\left(-\frac{1}{\sqrt{2}},-\frac{1}{\sqrt{3}}\right)\cup\left(\frac{1}{\sqrt{3}},\frac{1}{\sqrt{2}}\right)$

At $x=\pm\frac{1}{\sqrt{2}}$, $x^2=\frac{1}{2}$, $\frac{1}{x^2}=2$

$\Rightarrow$ $f(x)=1$

Similarly at $x=\frac{1}{\sqrt{3}},\frac{1}{2},\frac{1}{\sqrt{5}},\ldots.$

$\therefore f(x)$ is discontinuous at infinite number of points given by $x\in\left\{\pm\frac{1}{\sqrt{n}}, n\in N\right\}$.

Linked Comprehension Type

PASSAGE 1

1. (a)

2. (c)

Given that,

$$f(x)=\begin{cases} b\sin^{-1}\left(\frac{c+x}{2}\right), & -\frac{1}{2}<x<0 \\ \frac{1}{2}, & x=0 \quad \text{where, } |c|\le\frac{1}{2} \\ \frac{e^{ax/2}-1}{x}, & 0<x<\frac{1}{2} \end{cases}$$

Since, $f(x)$ is differentiable at $x=0$, then $f(x)$ will also be continuous at $x=0$

$\therefore$ $$\lim_{h\to 0} f(0+h)=f(0)$$

$\Rightarrow$ $$\lim_{h\to 0}\frac{e^{ah/2}-1}{h}=\frac{1}{2}$$

$\Rightarrow$ $$\lim_{h\to 0}\frac{e^{ah/2}-1}{\frac{ah}{2}}\times\frac{a}{2}=\frac{1}{2}$$

$\Rightarrow$ $a=1$

and Also $Lf'(0)=Rf'(0)$

$$\Rightarrow \lim_{h\to 0}\frac{f(0-h)-f(0)}{-h}=\lim_{h\to 0}\frac{f(0+h)-f(0)}{h}$$

$$\Rightarrow \lim_{h\to 0}\frac{b\sin^{-1}\left(\frac{c-h}{2}\right)-\frac{1}{2}}{-h}=\lim_{h\to 0}\frac{\frac{e^{\frac{ah}{2}}-1}{h}-\frac{1}{2}}{h}$$

$$=\lim_{h\to 0}\frac{2e^{ah/2}-2-h}{2h^2}$$

For these limits to exist we must get the form 0/0 and hence using L Hospital's rule, wc get

$$\lim_{h\to 0}\frac{\left(-\frac{1}{2}\right)\frac{b}{\sqrt{1-\left(\frac{c-h}{2}\right)^2}}}{-1}$$

$$=\lim_{h\to 0}\frac{2e^{ah/2}\left(\frac{a}{2}\right)-1}{4h}$$

$$=\lim_{h\to 0}\frac{e^{h/2}-1}{8\left(\frac{h}{2}\right)} \quad \text{[Putting } a=1\text{]}$$

$\Rightarrow$ $$\frac{b}{2\sqrt{1-\frac{c^2}{4}}}=\frac{1}{8}$$

$\Rightarrow$ $$4b=\sqrt{1-\frac{c^2}{4}}$$

$\Rightarrow$ $$16b^2=\frac{4-c^2}{4} \Rightarrow 64b^2=4-c^2$$

PASSAGE 2

1. LHL $= \lim_{x \to 0-} f(x) = \lim_{h \to 0} f(0-h)$

$= \lim_{h \to 0} \frac{a(1 - h\sin h) + b\cos h + 5}{h^2}$

$\because f(x)$ is continuous

$\therefore$ At $h \to 0$, Numerator must be $= 0$

ie, $a + b + 5 = 0$ ...(i)

or $b = -(a+5)$

$\Rightarrow$ LHL $= \lim_{h \to 0} \frac{a(1 - h\sin h) - (a+5)\cos h + 5}{h^2}$

$= \lim_{h \to 0} \left(\frac{(a+5)(1-\cos h)}{h^2} - \frac{ah\sin h}{h^2} \right)$

$= (a+5)\left(\frac{1}{2}\right) - a$

$= \frac{5-a}{2} =$ V.F. $= f(0) = 3$ [$\because f(x)$ is continuous]

$\Rightarrow$ $5 - a = 6$

$\therefore$ $a = -1$

2. Now, from Eq. (i), $-1 + b + 5 = 0$

$\Rightarrow$ $b = -4$

3. Also, RHL $= \lim_{x \to 0+} f(x) = \lim_{h \to 0} f(0+h)$

$= \lim_{h \to 0} \left\{1 + \left(\frac{ch + dh^3}{h^2}\right)\right\}^{1/h}$

$= \lim_{h \to 0} \left\{1 + \left(\frac{c + dh^2}{h}\right)\right\}^{1/h}$

$\therefore$ RHL is finite

$\therefore$ $c = 0$

4. Then, RHL $= \lim_{h \to 0} \{1 + dh\}^{1/h} = e^d = 3$ (V.F.)

$\therefore$ $e^d = 3$

5. The minimum value of $f(x)$ is $f(0) = 3$.

PASSAGE 3

$\because$ $f'(x) = \lim_{h \to 0} \frac{f(x+h) - f(x)}{h}$

$= \lim_{h \to 0} \frac{f\left(\frac{3x+3h}{3}\right) - f\left(\frac{3x+0}{3}\right)}{h}$

$= \lim_{h \to 0} \frac{\frac{2 + f(3x) + f(3h)}{3} - \frac{2 + f(3x) + f(0)}{3}}{h}$

$= \lim_{h \to 0} \frac{f(3h) - f(0)}{3h - 0} = f'(0)$

$\Rightarrow$ $f'(2) = f'(0) = 2$ ($\because f'(2) = 2$)

$\Rightarrow$ $f'(x) = 2$

$\Rightarrow$ $f(x) = 2x + c$ ...(i)

Put $x = y = 0$ in $f\left(\frac{x+y}{3}\right) = \frac{2 + f(x) + f(y)}{3}$

$\Rightarrow$ $f(0) = 2$

Now, from Eq. (i), $f(0) = 0 + c = 2$

$\therefore$ $c = 2$

From Eq. (i),

$f(x) = 2x + 2$

1. $f(1) = 2 + 2 = 4$

2. $\because f(|x|) = 2|x| + 2 \geq 2$

$\therefore$ Range is $[2, \infty)$.

3. $\because f(|x|) = 2|x| + 2$

$\therefore f(|x|) - 3 = 2|x| - 1$

and $g(x) = |f(|x| - 3| = |2|x| - 1|$

Now, graphs of $y = 2x - 1$

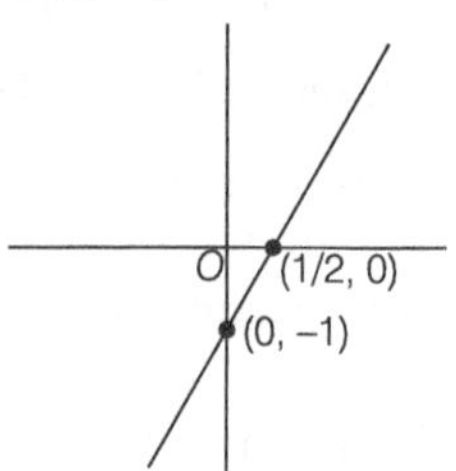

Now, graph of $y = 2|x| - 1$

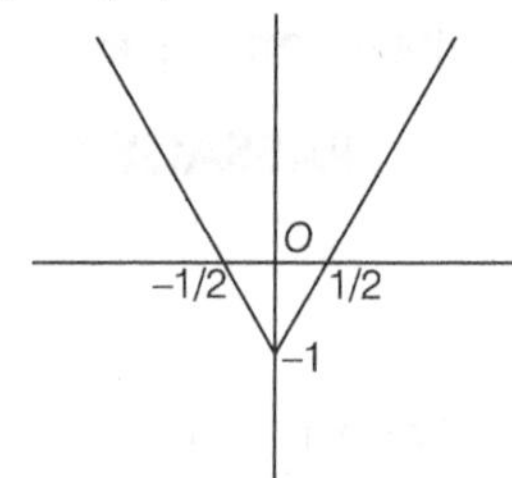

And in last graph of
$g(x) = |2|x| - 1|$

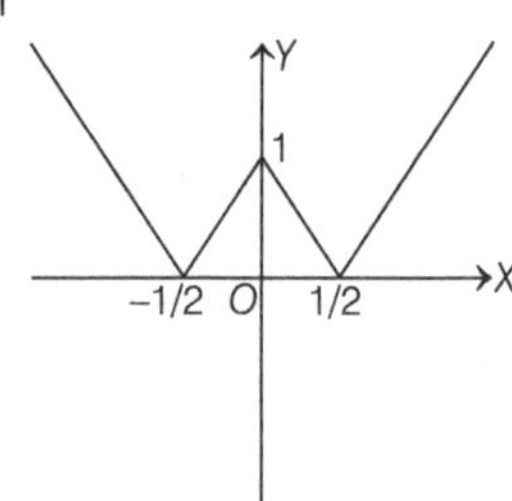

Number of non-differentiable points = 3.

4. $\because f(x) = 2x + 2$

Here, gives $x \in [-2, 3]$

or $-2 \leq x \leq 3$

$\Rightarrow$ $-4 \leq 2x \leq 6$

$\Rightarrow$ $-4 + 2 \leq 2x + 2 \leq 6 + 2$

$\therefore$ $-2 \leq f(x) \leq 8$

Hence, Range of $f(x)$ is $[-2, 8]$.

5. Let $y = f(|x|) = (2|x| + 2)$

then $x^2 + y^2 = 3^2$

$\therefore$ Number of solutions = 2

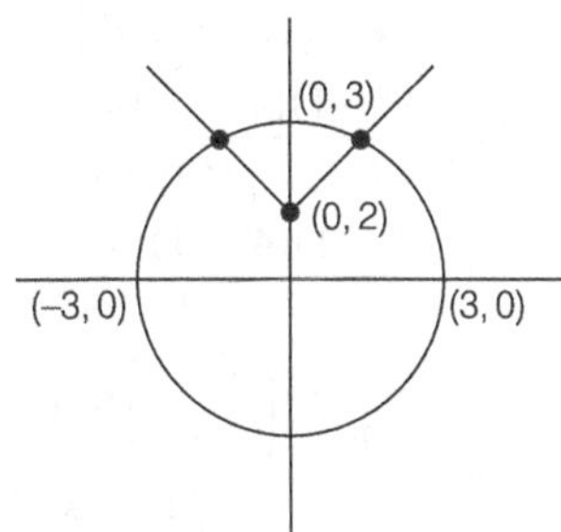

PASSAGE 4

Given that $f(x + y^n) = f(x) + (f(y))^n$

Putting $x = y = 0$

$\Rightarrow \quad f(0) = 0$

$$\therefore f'(0) = \lim_{h \to 0} \frac{f(0+h) - f(0)}{h} = \lim_{h \to 0} \frac{f(h)}{h} = \lambda \text{ (say)} \quad \ldots\text{(i)}$$

$$\text{Also, } f'(0) = \lim_{h \to 0} \frac{f(0+h) - f(0)}{h}$$

$$= \lim_{h \to 0} \frac{f(0 + (h^{1/n})^n - f(0)}{h}$$

$$= \lim_{h \to 0} \frac{f(0) + \{f(h^{1/n})\}^n - f(0)}{h}$$

$$= \lim_{h \to 0} \left\{\frac{f(h^{1/n})}{h^{1/n}}\right\}^n = \lambda^n \quad \text{[from Eq. (i)]} \ldots\text{(ii)}$$

From Eqs. (i) and (ii),

$\lambda = \lambda^n$

$\therefore \quad \lambda = -1, 0, 1$ ($\because$ n is odd and $\lambda \in R$)

$\because \quad f'(0) \geq 0$ (given)

$\Rightarrow \quad \lambda = 0, 1$

$$\text{Again, } f'(x) = \lim_{h \to 0} \frac{f(x+h) - f(x)}{h}$$

$$= \lim_{h \to 0} \frac{f(x + (h^{1/n})^n) - f(x)}{h}$$

$$= \lim_{h \to 0} \left(\frac{f(h^{1/n})}{h^{1/n}}\right)^n = \lambda^n$$

for $\lambda = 0$	for $\lambda = 1$	
$f'(x) = 0$	$f'(x) = 1$	...(iii)
$f(x) = c$	$f(x) = x + d$	
$f(0) = c = 0$	$f(0) = 0 + d = 0$	
$f(x) = 0$	$\Rightarrow \quad d = 0$	
which is impossible as $f(x)$ is not identically zero *ie*, $f(x) \neq 0$	$f(x) = x$	...(iv)

1. From Eq. (iv), $f(2) = 2$

2. From Eq. (iv), $f(5) = 5$

3. From Eq. (iii), $f'(12) = 1$

4. From Eq. (iv), $f(x) = x$

$f(-x) = -x = -f(x) \Rightarrow f(x)$ is odd.

5. From Eq. (iv), $f(x) = x$

$\Rightarrow$ Continuous and differentiable everywhere.

PASSAGE 5

1. $\because$ $(gof)x$ is continuous

$\therefore \quad g(x)$ and $f(x)$ are also continuous.

Then, $\quad f(0) = \lim_{x \to 0-} f(x) = \lim_{h \to 0} f(0 - h)$

$\Rightarrow \quad 1 = \lim_{h \to 0} (0 - h + a) = a$

$\therefore \quad a = 1$

2. and $\quad g(0) = \lim_{x \to 0-} g(x)$

$= \lim_{h \to 0} g(0 - h) = \lim_{h \to 0} (0 - h + 1)$

$\Rightarrow \quad 1 + b = 1$

$\Rightarrow \quad b = 0$

3. $\therefore \quad f(x) = \begin{cases} x + 1, & x < 0 \\ |x - 1|, & x \geq 0 \end{cases}$

and $g(x) = \begin{cases} x + 1, & x < 0 \\ (x - 1)^2, & x \geq 0 \end{cases}$

$$\Rightarrow (gof)x = g(f(x)) = \begin{cases} f(x) + 1, & f(x) < 0 \\ (f(x) - 1)^2, & f(x) \geq 0 \end{cases}$$

$$= \begin{cases} x + 2, & x + 1 < 0 \text{ and } x < 0 \\ x^2, & x + 1 \geq 0 \text{ and } x < 0 \\ |x - 1| + 1, & |x - 1| < 0 \text{ and } x \geq 0 \\ (|x - 1| - 1)^2, & |x - 1| \geq 0 \text{ and } x \geq 0 \end{cases}$$

$$= \begin{cases} x + 2, & x < -1 \\ x^2, & -1 \leq x < 0 \\ |x - 1| + 1, & x \in \phi \\ (|x - 1| - 1)^2, & x \geq 0 \end{cases}$$

$$= \begin{cases} x + 2, & x < -1 \\ x^2, & -1 \leq x < 0 \\ x^2, & 0 \leq x < 1 \\ (x - 2)^2, & x \geq 1 \end{cases}$$

$$= \begin{cases} x + 2, & x < -1 \\ x^2, & -1 \leq x < 1 \\ (x - 2)^2, & x \geq 1 \end{cases}$$

4. From above, $(gof)x = x^2 \;\; \forall \, x \in (-1, 1)$ which is even.

5. $\because (gof)x = \begin{cases} x + 2, & x < -1 \\ x^2, & -1 \leq x < 1 \\ (x - 2)^2, & x \geq 1 \end{cases}$

$$(gof)' x = \begin{cases} 1, & x < -1 \\ 2x, & -1 \leq x < 1 \\ 2(x - 2), & x \geq 1 \end{cases}$$

$(gof)x$ is non differentiable at $-1, 1$ and differentiable at $x = 0, 2$.

Numerical Value Type

1. LHL = V.F. = RHL

$$\Rightarrow \quad \lim_{x\to\frac{\pi}{2}^-} f(x) = f\left(\frac{\pi}{2}\right) = \lim_{x\to\frac{\pi}{2}^+} f(x)$$

$$\Rightarrow \quad \lim_{h\to 0} f\left(\frac{\pi}{2}-h\right) = a = \lim_{h\to 0} f\left(\frac{\pi}{2}+h\right)$$

$$\Rightarrow \quad \lim_{h\to 0} \frac{(1-\cos^3 h)}{3\sin^2 h} = a = \lim_{h\to 0} \frac{b(1-\cos h)}{4h^2}$$

$$\Rightarrow \quad \lim_{h\to 0} \frac{(1-\cos h)(1+\cos h+\cos^2 h)}{3(1-\cos h)(1+\cos h)}$$

$$a = \lim_{h\to 0} \frac{b(1-\cos h)}{4h^2}$$

$$\Rightarrow \quad \frac{3}{3(2)} = a = \frac{b}{8} \text{ or } \frac{b}{a} = 8$$

$$\therefore \quad \left(\frac{b}{a}\right)^{5/3} = (8)^{5/3} = 2^5 = 32$$

2. $f(x+y) = f(x)\cdot f(y) \forall x, y \in R$

$f(0) = f(0)\cdot f(0)$

$\Rightarrow f(0)\cdot f(0) - f(0) = 0$

$f(0)[f(0)-1] = 0$

As, $f(0) \neq 0$, $f(0) = 1$

Given, $f'(0) = 3$

$$\underset{h\to 0}{\text{Lt}} \frac{f(0+h)-f(0)}{h} = 3$$

$$\underset{h\to 0}{\text{Lt}} \frac{f(0)f(h)-f(0)}{h} = 3$$

$$\underset{h\to 0}{\text{Lt}} \frac{f(h)-1}{h} = 3$$

3. $\lim_{x\to 0} \frac{f(x)}{x} = 1$

$f(x+y) = f(x) + f(y) + x^2y + xy^2$

Put $x = y = 0$

$f(0) = f(0) + f(0) + 0 + 0$

$\Rightarrow 2f(0) = f(0)$

$\therefore f(0) = 0$

$$f'(0) = \lim_{h\to 0} \frac{f(0+h)-f(0)}{h}$$

$$= \lim_{h\to 0} \frac{f(h)-0}{h} = \lim_{h\to 0} \frac{f(h)}{h} = 1$$

$$f'(x) = \lim_{h\to 0} \frac{f(x+h)-f(x)}{h}$$

$$= \lim_{h\to 0} \frac{f(x)+f(h)+x^2h+xh^2-f(x)}{h}$$

$$= \lim_{h\to 0}\left(\frac{f(h)}{h} + \frac{x^2h}{h} + \frac{xh^2}{h}\right) = 1 + x^2 + 0 = 1 + x^2$$

$\Rightarrow f'(3) = 1 + 3^2 = 1 + 9 = 10$

4. $\because f$ is continuous at $x = 0$

$\therefore \quad f(0) = \text{RHL}$

$$\lambda = \lim_{h\to 0} f(0+h)$$

$$\Rightarrow \quad \lambda = \lim_{h\to 0}\left\{\frac{36^h - 9^h - 4^h + 1}{\sqrt{2} - \sqrt{(1+\cos h)}}\right\}$$

$$= \lim_{h\to 0} \frac{(9^h-1)(4^h-1)(\sqrt{2}+\sqrt{(1+\cos h)})}{(\sqrt{2}-\sqrt{1+\cos h})(\sqrt{2}+\sqrt{1+\cos h})}$$

$$= \lim_{h\to 0} \frac{(9^h-1)(4^h-1)(\sqrt{2}+\sqrt{(1+\cos h)})}{(1-\cos h)}$$

$$= \lim_{h\to 0} \frac{\left(\frac{9^h-1}{h}\right)\left(\frac{4^h-1}{h}\right)(\sqrt{2}+\sqrt{(1+\cos h)})}{\left(\frac{1-\cos h}{h^2}\right)}$$

$$= \frac{\ln 9 \cdot \ln 4 \cdot 2\sqrt{2}}{\left(\frac{1}{2}\right)}$$

$$= 16\sqrt{2} \ln 2 \cdot \ln 3$$

$$= \sqrt{(512)} \cdot \ln 2 \cdot \ln 3$$

Hence, $\mu = 512$

5. $\because f(x)$ is continuous at $x = 0$

$\therefore$ V.F. = RHL

$$f(0) = \lim_{h\to 0} f(0+h)$$

$$\lambda \ln 4 = \lim_{h\to 0} \frac{8^h - 4^h - 2^h + 1^h}{h^2}$$

$$= \lim_{h\to 0} \frac{(2^h-1)(4^h-1)}{h\cdot h}$$

$$= \ln 2 \cdot \ln 4$$

$\therefore \quad \lambda = \ln 2$

$\Rightarrow \quad e^\lambda = 2$

$\therefore \quad 10.f = 10.2 \times 2 = 20.4$

6. $\text{RHL} = \lim_{x\to 3+} f(x) = \lim_{h\to 0} f(3+h) = \lim_{h\to 0} \lambda\cdot\frac{(1-\cos h)}{h\tan h}$

$$= \lambda \lim_{h\to 0} \frac{\frac{1-\cos h}{h^2}}{\left(\frac{\tan h}{h}\right)}$$

$$= \lambda\frac{\left(\frac{1}{2}\right)}{1} = \frac{\lambda}{2}$$

$\text{LHL} = \lim_{x\to 3-} f(x) = \lim_{h\to 0} f(3-h)$

$$= \lim_{h\to 0} \frac{(\exp\{(6-h)\ln 27\})^{\frac{1}{27}[4-h]} - 9}{3^{3-h} - 27}$$

$$= \lim_{h\to 0} \frac{(\exp\{(6-h)\ln 27\})^{\frac{1}{9}} - 9}{3^3(3^{-h}-1)}$$

$$= \lim_{h\to 0} \frac{(27^{6-h})^{1/9} - 9}{27(3^{-h}-1)} = \lim_{h\to 0} \frac{(3^{(18-3h)})^{1/9} - 9}{27(3^{-h}-1)}$$

$$= \lim_{h\to 0} \frac{[3^2(3^{-h/3}-1)/(-h/3)]\cdot(-h/3)}{[27(3^{-h}-1)/(-h]\cdot(-h)}$$

$$= \frac{1}{9}$$

$\because$ $f(x)$ is continuous at $x=3$

$\because$ LHL = RHL

$\Rightarrow \quad \frac{1}{9}=\frac{\lambda}{2}$

$\Rightarrow \quad \lambda=\frac{2}{9}$

Then, $8100\,\lambda = 8100\times\frac{2}{9}=1800$

7. Given, $f(x)+f(y)=f\left(\frac{x+y}{1-xy}\right)$

Putting $x=0,\ y=0$, we get $f(0)=0$ ...(i)

and putting $y=-x$, we get

$f(x)+f(-x)=f(0)=0$

$\therefore \quad f(x)=-f(-x)$ ...(ii)

Now, $f'(x)=\lim_{h\to 0}\frac{f(x+h)-f(x)}{h}$

$=\lim_{h\to 0}\frac{f(x+h)+f(-x)}{h}$ [from Eq. (ii)]

$=\lim_{h\to 0}\frac{f\left(\frac{x+h-x}{1+(x+h)x}\right)}{h}$

$=\lim_{h\to 0}\frac{f\left(\frac{h}{1+x^2+xh}\right)}{\left(\frac{h}{1+x^2+xh}\right)\cdot(1+x^2+xh)}$

$f'(x)=\frac{2}{1+x^2}$ $\left[\because \lim_{x\to 0}\frac{f(x)}{x}=2\right]$

$\therefore \quad f'(-2)=\frac{2}{1+4}=\frac{2}{5}$

and $\quad f(x)=2\tan^{-1}x+c$

$f(0)=0+c=0$

$\therefore \quad c=0$

$f(x)=2\tan^{-1}x \Rightarrow f(\sqrt{3})=2\tan^{-1}\sqrt{3}=\frac{2\pi}{3}$

$\therefore \frac{1500}{\pi}\cdot\frac{f(\sqrt{3})}{f'(-2)}=\frac{1500}{\pi}\cdot\frac{\frac{2\pi}{3}}{\frac{2}{5}}=\frac{1500}{\pi}\cdot\frac{5\pi}{3}=2500$

Single Integer Type

1. $x\in(-10,10)$

$\frac{x}{2}\in(-5,5)\rightarrow 1$ integers

Check continuity at $x=0$

$\left.\begin{array}{l} f(0)=0\\ f(0^+)=0\\ f(0^-)=0\end{array}\right\}\Rightarrow f(x)$ is continuous at $x=0$

So, the function will be discontinuous when

$\frac{x}{2}=\pm 4,\pm 3,\pm 2,\pm 1$

2. $g[f(x)]=\begin{cases} f(x)+1 & f(x)<0\\ (f(x)-1)^2+b & f(x)\ge 0\end{cases}$

$g[f(x)]=\begin{cases} x+a+1 & x+a<0 \text{ and } x<0\\ |x-1|+1 & |x-1|<0 \text{ and } x\ge 0\\ (x+a-1)^2+b & x+a\ge 0 \text{ and } x<0\\ (|x-1|-1)^2+b & |x-1|\ge 0 \text{ and } x\ge 0\end{cases}$

$g[f(x)]=\begin{cases} x+a+1 & x\in(-\infty,-a) \text{ and } x\in(-\infty,0)\\ |x-1|+1 & x\in\phi\\ (x+a-1)^2+b & x\in[-a,\infty] \text{ and } x\in(-\infty,0)\\ (|x-1|-1)^2+b & x\in R \text{ and } x\in[0,\infty]\end{cases}$

$g[f(x)]=\begin{cases} x+a+1 & x\in(-\infty,-a)\\ (x+a-1)^2+b & x\in[-a,0)\\ (|x-1|-1)^2+b & x\in[0,\infty)\end{cases}$

Let $h(x)=g[f(x)]$

It is given that $h(x)$ is continuous for all $x\in R$.

At $x=-a$, LHL = RHL = $f(-a)$

$-a+a+1=(-a+a-1)^2+b=(-a+a-1)^2+b$

$1=1+b \Rightarrow b=0$

At $x=0$, LHL = RHL = $f(0)$

$(0+a-1)^2+b=(|0-1|-1)^2+b=(|0-1|-1)^2+b$

$(a-1)^2+b=b \Rightarrow a=1$

Hence, $a+b=1+0=1$

3. We have, $f(x)=\begin{cases} ax^2+bx, & \text{for } -1<x<1\\ \frac{a-b-1}{2}, & x=-1\\ \frac{a+b+1}{2}, & x=1\\ \frac{1}{x}, & \text{for } x>1 \text{ or } x<-1\end{cases}$

For continuity at $x=1$,

$a+b=\frac{a+b+1}{2} \Rightarrow a+b=1$ ...(i)

For continuity at $x=-1$, $\quad a-b=-1$

On solving Eqs. (i) and (ii), we get

$a=0$ and $b=1$ ...(ii)

Thus, $a+8b=0+8\times 1=8$

4. The given functions $f:(-1,1)\longrightarrow R$ and $g:(-1,1)\longrightarrow(-1,1)$ be defined by

$f(x)=|2x-1|+|2x+1|$ and $g(x)=x-[x]$.

As, we know the composite function $(fog)(x)$ is discontinuous at the points, where $g(x)$ is discontinuous for given domain. And, since $g(x)$ is discontinuous at $x=0$ lies in interval $(-1,1)$, so value of $c=1$.

And, since $(fog)(x)$ is not differentiable at the point where $g(x)$ is not differentiable as well as at those points also where $g(x)$ attains the values so that $f(g(x))$ is non-differentiable.

Since $g(x)$ is not continuous at $x=0\in(-1,1)$ so $fog(x)$ is not differentiable and as $f(x)=|2x-1|+|2x+1|$ is not differentiable at $x=-1/2$ and $1/2$, so $(fog)(x)$ is not differentiable for those x, for which $g(x)=-1/2$ or $1/2$.

But $g(x) \ge 0$, so $g(x)$ can be 1/2 only and for $x = -1/2$ and 1/2, $g(x) = \frac{1}{2}$.

So, $(fog)(x)$ is not differentiable at $x = -1/2, 0, 1/2$, therefore value of $d = 3$

$\therefore \quad c + d = 1 + 3 = 4$

5. $\because f$ is continuous at $x = 0$, then

$$V(0) = \text{RHL}$$

$$\Rightarrow \quad \frac{1}{2} = \lim_{x \to 0+} f(x) = \lim_{h \to 0} f(0 + h)$$

$$\Rightarrow \quad \frac{1}{2} = \lim_{h \to 0} \frac{1 - \cos ah}{h \sin h}$$

$$= \lim_{h \to 0} \frac{1 - \cos ah}{h^2 \cdot \left(\frac{\sin h}{h}\right)} = \frac{a^2}{2}$$

$$\therefore \quad a^2 = 1$$

6. Given function $f(x) = |2 - |x - 3||$

$$= \begin{vmatrix} |2 + (x - 3)|, & x < 3 \\ |2 - (x - 3)|. & x \ge 3 \end{vmatrix} = \begin{vmatrix} |x - 1|, & x < 3 \\ |5 - x|, & x \ge 3 \end{vmatrix}$$

$$= \begin{vmatrix} 1 - x, & x < 1 \\ x - 1, & 1 \le x < 3 \\ 5 - x, & 3 \le x < 5 \\ x - 5, & x \ge 5 \end{vmatrix}$$

$\because$ Function $f(x)$ is not differentiable at $x = 1, 3, 5$

$\therefore \quad S = \{1, 3, 5\}$

Now, $\sum_{x \in S} f(f(x)) = f(f(1)) + f(f(3)) + f(f(5))$

$$= f(1 - 1) + f(5 - 3) + f(5 - 5)$$
$$= f(0) + f(2) + f(0)$$
$$= (1 - 0) + (2 - 1) + (1 - 0) = 1 + 1 + 1 = 3$$

7. For f to be continuous at $x = 0$.

$$f(0) = \lim_{x \to 0} f(x)$$

$$= \lim_{x \to 0} \frac{\log(1 + x + x^2) + \log(1 - x + x^2)}{\frac{1}{\sec x} - \cos x}$$

$$= \lim_{x \to 0} \frac{\log((1 + x^2)^2 - x^2)}{1 - \cos^2 x} \cdot \cos x$$

$$= \lim_{x \to 0} \frac{\log(1 + (x^2 + x^4))}{\sin^2 x} \cdot \cos x$$

$$= \lim_{x \to 0} \frac{\log(1 + (x^2 + x^4))}{x^2 + x^4} \times \frac{x^2}{\sin^2 x} \times (1 + x^2) \cos x = 1$$

8. Given, $\text{sgn}(x^2 - 3x + 2)$ is discontinuous when $x^2 - 3x + 2 = 0$ or $x = 1, 2$.

$[x - 3] = [x] - 3$ is discontinuous at $x = 1, 2, 3, 4$.

$\therefore f(x)$ is discontinuous at $x = 3, 4$

Now, both $\text{sgn}(x^2 - 3x + 2)$ and $[x - 3]$ are discontinuous at $x = 1$ and 2.

Then, $f(x)$ may be continuous at $x = 1$ and 2.

But $f(1) = -2$ and $f(1^-) = -1 + 0 - 3 = -4$

Thus, $f(x)$ is discontinuous at $x = 1$.

Also, $f(2) = -1$ and $f(2^+) = 1 - 1 = 0$.

Hence, $f(x)$ is discontinuous at $x = 2$ also.

Matrix-Match Type

1. **(A)** Since, $f(x)$ is continuous is $[-\pi, \pi]$.

$$\therefore \quad f\left(-\frac{\pi}{2}\right) = \lim_{x \to \frac{\pi}{2}^+} f(x)$$

$$\Rightarrow \quad -2\sin\left(-\frac{\pi}{2}\right) = \lim_{h \to 0} f\left(-\frac{\pi}{2} + h\right)$$

$$\Rightarrow \quad -2 \times -1 = \lim_{h \to 0} a \sin\left(-\frac{\pi}{2} + h\right) + b$$

$$\Rightarrow \quad 2 = a \sin\left(-\frac{\pi}{2}\right) + b = 2 = -a + b \quad \text{...(i)}$$

and $f\left(\frac{\pi}{2}\right) = \lim_{x \to \frac{\pi}{2}^-} f(x)$

$$\Rightarrow \quad \cos\frac{\pi}{2} = \lim_{h \to 0} f\left(\frac{\pi}{2} - h\right)$$

$$\Rightarrow \quad 0 = \lim_{h \to 0} a \sin\left(\frac{\pi}{2} - h\right) + b$$

$$\Rightarrow \quad 0 = a \sin\frac{\pi}{2} + b$$

$$\Rightarrow \quad 0 = a + b \quad \text{...(ii)}$$

From Eqs. (i) and (ii), we get $a = -1, b = 1$

$\therefore \quad |a - b| = 2 \quad$ (Q)

(B) Since, $f(x)$ is continuous at $x = 2$

$$\therefore \quad f(2) = \lim_{x \to 2^+} f(x) = \lim_{x \to 2^-} f(x)$$

$$\Rightarrow \quad b = \lim_{h \to 0} f(2 + h) = \lim_{h \to 0} f(2 - h)$$

$$\Rightarrow \quad b = \lim_{h \to 0} \frac{2 + h - [2 + h]}{2 + h - 2}$$

$$= \lim_{h \to 0} \frac{a|(2 - h - 2)(2 - h + 1)|}{-(2 - h - 2)(2 - h + 1)}$$

$$\Rightarrow \quad b = \lim_{h \to 0} \frac{2 + h - 2}{h} = \lim_{h \to 0} \frac{a|(-h)(3 - h)|}{h(3 - h)}$$

$$\Rightarrow \quad b = 1 = \lim_{h \to 0} \frac{ah(3 - h)}{h(3 - h)}$$

$$\Rightarrow \quad b = 1 = a$$

$\therefore \quad |a - b| = 0, |a + b| = 2, |a + 2b| = 3 \quad$ (P, R, S)

(C) Since, $f(x)$ is continuous at $x = \frac{\pi}{2}$

$$\therefore \text{ LHL} = \lim_{x \to \frac{\pi}{2}^-} f(x) = \lim_{h \to 0} f\left(\frac{\pi}{2} - h\right)$$

$$= \lim_{h\to 0}\left(\frac{3}{2}\right)^{\frac{\cot\left(\frac{3\pi}{2}-3h\right)}{\cot(\pi-2h)}}$$

$$= \lim_{h\to 0}\left(\frac{3}{2}\right)^{-(\tan 3h)(\tan 2h)} = \left(\frac{3}{2}\right)^0 = 1$$

$$\text{RHL} = \lim_{x\to \frac{\pi}{2}^+} f(x) = \lim_{h\to 0} f\left(\frac{\pi}{2}+h\right)$$

$$= \lim_{h\to 0}\left(1+\left|\cos\left(\frac{\pi}{2}+h\right)\right|\right)^{\frac{a\left|\tan\left(\frac{\pi}{2}+h\right)\right|}{b}}$$

$$= \lim_{h\to 0}(1+\sin h)^{\frac{a\cot h}{b}} = e^{\lim_{h\to 0}\frac{a\cos h}{b}} = e^{a/b}$$

and $\text{V}\cdot\text{F} = f\left(\frac{\pi}{2}\right) = b+3$

$\Rightarrow \quad 1 = e^{a/b} = b+3$

$\therefore \quad b = -2$ and $e^{a/b} = 1 = e^0$

$\Rightarrow \quad \frac{a}{b} = 0$

$\Rightarrow \quad a = 0$

$\Rightarrow \quad a = 0, b = -2$

$\therefore \quad |a-b| = 2, |a+b| = 2, |a+2b| = 4$ (Q, R, T)

2. **(A)** $\text{LHL} = \lim_{x\to 0-} f(x) = \lim_{h\to 0} f(0-h)$

$$= \lim_{h\to 0}\frac{a+3\cos(0-h)}{(0-h)^2}$$

$$= \lim_{h\to 0}\frac{a+3\cos h}{h^2}$$

at $h\to 0$, Numerator $= a+3$ must be 0

$\therefore \quad a = -3$ $\quad$ [$\because f(x)$ is continuous]

$\Rightarrow \quad \text{LHL} = \lim_{h\to 0}\frac{-3(1-\cos h)}{h^2} = -\frac{3}{2}$

$\text{V.F.} = f(0) = b\tan\left(\frac{\pi}{3}\right) = b\sqrt{3}$

$\because f(x)$ is continuous, $\therefore$ LHL = V.F.

$$-\frac{3}{2} = b\sqrt{3} \Rightarrow b = -\frac{\sqrt{3}}{2}$$

Hence, $\quad a = -3, b = -\frac{\sqrt{3}}{2}$

$$[a-2b] = [-3+\sqrt{3}] = -2 \text{ (T)}$$

(B) $\because f(x)$ is continuous in $[-\pi, \pi]$

$\therefore$ at $\quad x = -\frac{\pi}{2}$

V.F. = RHL

$\Rightarrow \quad f(-\pi/2) = \lim_{x\to -\frac{\pi}{2}+} f(x)$

$$-2\sin(-\pi/2) = \lim_{h\to 0} f\left(-\frac{\pi}{2}+h\right)$$

$$-2(-1) = \lim_{h\to 0}\left(a\sin\left(-\frac{\pi}{2}+h\right)+b\right)$$

$$2 = -a+b \quad \ldots\text{(i)}$$

and $\quad$ at $x = \frac{\pi}{2}$

V.F. = LHL

$$f\left(\frac{\pi}{2}\right) = \lim_{x\to \pi/2-} f(x)$$

$$\cos\frac{\pi}{2} = \lim_{h\to 0} f\left(\frac{\pi}{2}-h\right)$$

$$0 = \lim_{h\to 0} a\sin\left(\frac{\pi}{2}-h\right)+b$$

$\Rightarrow \quad 0 = a+b \quad \ldots\text{(ii)}$

From Eqs. (i) and (ii), $b = 1, a = -1$

$a+b = 0, a-b = -2, a+2b = 1$

$|a+b| = 0, |a-b| = 2, |a+2b| = 1$ (P, Q, R)

(C) V.F. $= f(\pi/2) = b+3 \quad \ldots\text{(i)}$

$$\text{LHL} = \lim_{x\to \pi/2-} f(x) = \lim_{h\to 0} f\left(\frac{\pi}{2}-h\right)$$

$$= \lim_{h\to 0}\left(\frac{3}{2}\right)^{\cot\left(\frac{3\pi}{2}-3h\right)/\cot(\pi-2h)}$$

$$= \lim_{h\to 0}\left(\frac{3}{2}\right)^{-\frac{\tan 3h}{\cot 2h}} = \lim_{h\to 0}\left(\frac{3}{2}\right)^0 = 1 \quad \ldots\text{(ii)}$$

$$\text{RHL} = \lim_{x\to \frac{\pi}{2}+} f(x) = \lim_{h\to 0} f\left(\frac{\pi}{2}+h\right)$$

$$= \lim_{h\to 0}\left(1+\left|\cos\left(\frac{\pi}{2}+h\right)\right|\right)^{\frac{a\left|\tan\left(\frac{\pi}{2}+h\right)\right|}{b}}$$

$$= \lim_{h\to 0}(1+\sin h)^{\frac{a\cot h}{b}} = e^{\lim_{h\to 0}\frac{a\cos h}{b}} = e^{a/b}$$

$\therefore \quad$ LHL = RHL = V.F.

$\Rightarrow \quad 1 = e^{a/b} = b+3 \quad \therefore \quad b = -2, a = 0$

$a-b = 2, a+2b = -4$

$|a-b| = 2, |a+2b| = 4$ (Q, S)

3. **(A)** $\because f(x)$ is continuous in $[0, \pi]$

Then, $f\left(\frac{\pi}{4}\right) = \lim_{x\to \frac{\pi}{4}-} f(x) = \lim_{h\to 0} f\left(\frac{\pi}{4}-h\right)$

$$2\cdot\frac{\pi}{4}\cdot\cot\frac{\pi}{4}+b = \lim_{h\to 0}\left(\frac{\pi}{4}-h\right)+a\sqrt{2}\sin\left(\frac{\pi}{4}-h\right)$$

$\Rightarrow \quad \frac{\pi}{2}+b = \frac{\pi}{4}+a$

$\therefore \quad a-b = \frac{\pi}{4} \quad \ldots\text{(i)}$

and $\quad f\left(\frac{\pi}{2}\right) = \lim_{x\to \frac{\pi}{2}+} f(x) = \lim_{h\to 0} f\left(\frac{\pi}{2}+h\right)$

$$= \lim_{h\to 0} a\cos(\pi+2h) - b\sin\left(\frac{\pi}{2}+h\right)$$

$0+b = -a-b$

$\therefore \quad a = -2b \quad \ldots\text{(ii)}$

From Eqs. (i) and (ii) we get, $b = -\frac{\pi}{12}$ and $a = \frac{\pi}{6}$.

Locus of (a, b) is $x+2y = 0$ (P, Q, R).

(B) $\because f(x)$ is continuous at $x = 0$

$\therefore$ V.F. $= f(0) = c$

$$\text{LHL} = \lim_{x\to 0-} f(x) = \lim_{h\to 0} f(0-h)$$

$$= \lim_{h\to 0} \frac{\sin((a+1)(-h)) + \sin(-h)}{-h}$$

$$= \lim_{h\to 0} \frac{\sin((a+1)h)(a+1)}{(a+1)h} + \frac{\sin h}{h}$$

$$= a+1+1$$

$$= a+2$$

$$\text{RHL} = \lim_{x\to 0+} f(x) = \lim_{h\to 0} f(0+h)$$

$$= \lim_{h\to 0} \frac{(h+bh^2)^{1/2} - h^{1/2}}{bh^{3/2}}$$

$$= \lim_{h\to 0} \frac{(1+bh)^{1/2} - 1^{1/2}}{1+(bh)-1}, \ b \neq 0$$

$$= \frac{1}{2}$$

$\therefore \ c = a + 2 = \frac{1}{2}$

$\Rightarrow \quad c = \frac{1}{2}, a = -\frac{3}{2}, b \in R \sim \{0\}$

Locus of (a, c) is $x - y + 2 = 0$ (P, Q, R, S, T)

If b is $+3/4$, then $a + 2b = 0$ if b is -1, then $ab = 3/2$

if $\quad b = -\frac{3}{2} - \frac{\pi}{4}$, then $a - b = \pi/4$

if $\quad b = \frac{3}{2} + \frac{\pi}{12}$

$\Rightarrow \quad a + b = \pi/12$

(C) V.F. $= f(0) = b$

$$\text{RHL} = \lim_{x\to 0+} f(x) = \lim_{h\to 0} f(0+h)$$

$$= e^{\lim_{h\to 0}(\cot 2h/\cot 3h)} = e^{\lim_{h\to 0} \frac{\tan 3h}{\tan 2h}}$$

$$= e^{\lim_{h\to 0} \frac{\tan 3h}{3h}\cdot\frac{2h}{\tan 2h}\cdot\frac{3}{2}} = e^{3/2}$$

$$\text{LHL} = \lim_{x\to 0-} f(x) = \lim_{h\to 0} f(0-h)$$

$$= \lim_{h\to 0} (1 + b|\tan(0-h)|)^{a/|\tan(0-h)|}$$

$$= e^{ab}$$

$\therefore \quad b = e^{3/2} = e^{ab}$

$\therefore \quad b = e^{3/2}, ab = 3/2$

$\therefore \quad a = \frac{3}{2} e^{-3/2}$

Locus of (a, b) is $xy = \frac{3}{2}$ (T)

4. **(A)** $$Rf'(1) = \lim_{h\to 0} \frac{f(1+h) - f(1)}{h}$$

$$= \lim_{h\to 0} \frac{\{b(1+h) + 2\} - \{1+3+a\}}{h}$$

at $h \to 0$, Numerator $= b - a - 2$ must be zero

($\because f(x)$ is differentiable)

Then, $Rf'(1) = \lim_{h\to 0} \frac{bh}{h} = b$

$$Lf'(1) = \lim_{h\to 0} \frac{f(1-h) - f(1)}{-h}$$

$$= \lim_{h\to 0} \frac{((1-h)^2 + 3(1-h) + a) - (4+a)}{-h}$$

$$= \lim_{h\to 0} \frac{h^2 - 5h}{-h} = 5$$

$\therefore \quad Lf'(1) = Rf'(1)$

$\Rightarrow \quad b = 5$ (Q)

Also, $\quad a = b - 2 = 5 - 2 = 3$ (P)

$\therefore \quad a - b = -2$

(B) $$\because f(x) = \begin{cases} -\frac{1}{x}, & x \le -1 \\ \frac{1}{x}, & x \ge 1 \\ ax^2 + b, & -1 < x < 1 \end{cases}$$

$\because f(x)$ is differentiable, then its also continuous.

$$\therefore \quad f(1) = \lim_{x\to 1-} f(x) = \lim_{h\to 0} f(1-h)$$

$$\Rightarrow \quad 1 = \lim_{h\to 0} a(1-h)^2 + b$$

$$\Rightarrow \quad 1 = a + b$$

and $$f'(x) = \begin{cases} \frac{1}{x^2}, & x \le -1 \\ -\frac{1}{x^2}, & x \ge 1 \\ 2ax, & -1 < x < 1 \end{cases}$$

$$f'(1) = \lim_{x\to 1-} f'(x) = \lim_{h\to 0} f'(1-h)$$

$$\Rightarrow \quad -1 = \lim_{h\to 0} 2a(1-h)$$

$$\Rightarrow \quad -1 = 2a$$

$$\Rightarrow \quad a = -\frac{1}{2}$$

Then, $\quad b = \frac{3}{2} \quad \therefore \ a - b = -2$

(C) $f(x)$ is also continuous.

$$\therefore \quad f(3) = \lim_{x\to 3-} f(x)$$

$$9b - 3 = \lim_{h\to 0} f(3-h)$$

$$= \lim_{h\to 0} a(3-h)^2 - b(3-h) + 2$$

$$= 9a - 3b + 2$$

or $\quad 9a - 12b = -5 \quad$ ...(i)

Also, $$f'(x) = \begin{cases} 2ax - b, & x < 3 \\ 2bx; & x \ge 3 \end{cases}$$

$$\therefore \quad f'(3) = \lim_{x\to 3-} f(x) = \lim_{h\to 0} f(3-h)$$

$$6b = \lim_{h\to 0} 2a(3-h) - b \ = 6a - b$$

or $\quad 6a = 7b \quad$ ...(ii)

From Eqs. (i) and (ii), we get

$$a = \frac{35}{9}, b = \frac{10}{3}$$

13

Differential Coefficient

Objective Questions Type I [Only one correct answer]

In each of the questions below, four choices are given of which only one is correct. You have to select the correct answer which is the most appropriate.

1. If $y = \tan^{-1} \sqrt{\left(\frac{1 + \sin x}{1 - \sin x}\right)}, \frac{\pi}{2} < x < \pi$, then $\frac{dy}{dx}$ equals

(a) $-1/2$ (b) -1
(c) $1/2$ (d) 1

2. Let f and g be differentiable functions on R such that fog is the identity function. If for some $a, b \in R, g'(a) = 5$ and $g(a) = b$, then $f'(b)$ is equal to **[JEE Main 2020, 9 January]**

(a) $\frac{1}{5}$ (b) 5
(c) $\frac{2}{5}$ (d) 1

3. If $x = 2 \cos t - \cos 2t, y = 2 \sin t - \sin 2t$, then $\frac{d^2y}{dx^2}$ at $t = \frac{\pi}{2}$ is

(a) $-5/2$ (b) $-3/2$
(c) $3/2$ (d) $5/2$

4. If $f(x) = 2 \sin^{-1} \sqrt{(1 - x)} + \sin^{-1} (2\sqrt{x(1 - x)})$, where $x \in \left(0, \frac{1}{2}\right)$, then $f'(x)$ is

(a) $-\frac{2}{\sqrt{x(1 - x)}}$ (b) $\frac{2}{\sqrt{x(1 - x)}}$
(c) 0 (d) None of these

5. The set of values of a for which the function
$f(x) = (4a - 3)(x + \ln 5) + 2(a - 7) \cot \left(\frac{x}{2}\right) \sin^2 \left(\frac{x}{2}\right)$
does not possess critical points is

(a) $(1, \infty)$
(b) $[1, \infty)$
(c) $(-\infty, 2)$
(d) $(-\infty, -4/3) \cup (2, \infty)$

6. If $f(x) = x + \tan x$ and f is inverse of g, then $g'(x)$ is equal to

(a) $\frac{1}{1 + (g(x) - x)^2}$ (b) $\frac{1}{1 - (g(x) - x)^2}$
(c) $\frac{1}{2 + (g(x) - x)^2}$ (d) $\frac{1}{2 - (g(x) - x)^2}$

7. If $y = \sqrt{(a - x)(x - b)} - (a - b) \tan^{-1} \sqrt{\left(\frac{a - x}{x - b}\right)}$, then $\frac{dy}{dx}$ is equal to

(a) $\sqrt{(a - x)(x - b)}$ (b) $\frac{1}{\sqrt{(a - x)(x - b)}}$
(c) $\sqrt{\left(\frac{a - x}{x - b}\right)}$ (d) $\sqrt{\left(\frac{x - b}{a - x}\right)}$

8. If $\sqrt{(1 - x^{2n})} + \sqrt{(1 - y^{2n})} = a(x^n - y^n)$, then $\sqrt{\left(\frac{1 - x^{2n}}{1 - y^{2n}}\right)} \frac{dy}{dx}$ is equal to

(a) $\frac{x^{n-1}}{y^{n-1}}$ (b) $\frac{y^{n-1}}{x^{n-1}}$
(c) $\frac{x}{y}$ (d) 1

9. If $2y = \left(\cot^{-1}\left(\frac{\sqrt{3} \cos x + \sin x}{\cos x - \sqrt{3} \sin x}\right)\right)^2, x \in \left(0, \frac{\pi}{2}\right)$, then $\frac{dy}{dx}$ is equal to **[JEE Main 2019, 8 April]**

(a) $\frac{\pi}{6} - x$ (b) $x - \frac{\pi}{6}$
(c) $\frac{\pi}{3} - x$ (d) $2x - \frac{\pi}{3}$

10. If $x = \sec\theta - \cos\theta$, $y = \sec^{10}\theta - \cos^{10}\theta$ and $(x^2+4)\left(\frac{dy}{dx}\right)^2 = k(y^2+4)$, then k is equal to

(a) $\frac{1}{100}$ (b) 1
(c) 10 (d) 100

11. If $y = \ln\left(\frac{x}{a+bx}\right)^x$, then $x^3\frac{d^2y}{dx^2}$ is equal to

(a) $\left(\frac{dy}{dx}+x\right)^2$ (b) $\left(\frac{dy}{dx}-y\right)^2$
(c) $\left(x\frac{dy}{dx}+y\right)^2$ (d) $\left(x\frac{dy}{dx}-y\right)^2$

12. If $y = \left(\frac{ax+b}{cx+d}\right)$, then $2\frac{dy}{dx}\cdot\frac{d^3y}{dx^3}$ is equal to

(a) $\left(\frac{d^2y}{dx^2}\right)^2$ (b) $3\frac{d^2y}{dx^2}$
(c) $3\left(\frac{d^2y}{dx^2}\right)^2$ (d) $3\frac{d^2x}{dy^2}$

13. If $x^y = e^{x-y}$, then $\frac{dy}{dx}$ is equal to

(a) $(1+\ln x)^{-1}$ (b) $(1+\ln x)^{-2}$
(c) $\ln x\,(1+\ln x)^{-2}$ (d) None of these

14. If $\frac{d^2x}{dy^2}\left(\frac{dy}{dx}\right)^3 + \frac{d^2y}{dx^2} = k$, then k is equal to

(a) 0 (b) 1
(c) 2 (d) None of these

15. Let $y = y(x)$ be a function of x satisfying $y\sqrt{1-x^2} = k - x\sqrt{1-y^2}$ where k is a constant and $y\left(\frac{1}{2}\right) = -\frac{1}{4}$. Then $\frac{dy}{dx}$ at $x = \frac{1}{2}$, is equal to

[JEE Main 2020, 7 January]

(a) $\frac{\sqrt{5}}{2}$ (b) $-\frac{\sqrt{5}}{2}$
(c) $\frac{2}{\sqrt{5}}$ (d) $-\frac{\sqrt{5}}{4}$

16. If $y = \frac{\cos 6x + 6\cos 4x + 15\cos 2x + 10}{\cos 5x + 5\cos 3x + 10\cos x}$, then $\frac{dy}{dx} = \ldots$.

(a) $2\sin x + \cos x$ (b) $-2\sin x$
(c) $\cos 2x$ (d) $\sin 2x$

17. For $x \in \left(0, \frac{1}{4}\right)$, if the derivative of $\tan^{-1}\left(\frac{6x\sqrt{x}}{1-9x^3}\right)$ is $\sqrt{x}\cdot g(x)$, then $g(x)$ equals

[JEE Main 2017]

(a) $\frac{9}{1+9x^3}$ (b) $\frac{3x\sqrt{x}}{1-9x^3}$
(c) $\frac{3x}{1-9x^3}$ (d) $\frac{3}{1+9x^3}$

18. If $x^2 + y^2 = a^2 (a > 0)$ and $\lambda = \frac{y^n}{\sqrt{(+y'^2)^3}}$, then λ in terms of a is

(a) $-\frac{2}{a^2}$ (b) $-\frac{1}{a^2}$
(c) $-\frac{2}{a}$ (d) $-\frac{1}{a}$

19. If $y^2 = P(x)$ is a polynomial of degree 3, then $2\frac{d}{dx}\left[y^3\frac{d^2y}{dx^2}\right]$ equals

(a) $P'''(x) + P'x$ (b) $P''(x)\cdot P'''(x)$
(c) $P(x)\cdot P'''(x)$ (d) None of these

20. If $5f(x) + 3f\left(\frac{1}{x}\right) = x + 2$ and $y = x\,f(x)$, then $\left(\frac{dy}{dx}\right)_{x=1}$ is equal to

(a) 14 (b) 7/8
(c) 1 (d) None of these

21. If $2^x + 2^y = 2^{x+y}$, then the value of $\frac{dy}{dx}$ at $x = y = 1$ is

(a) 0 (b) –1 (c) 1 (d) 2

22. If $f(x) = |x-2|$ and $g(x) = fof(x)$, then for $x > 20$, $g'(x)$ is equal to

(a) 2 (b) 1
(c) 3 (d) None of these

23. Let $f(\alpha) = \cos\left[\cot^{-1}\left(\frac{\cos\alpha}{\sqrt{1-\cos 2\alpha}}\right)\right]$ where $\frac{\pi}{4} < \alpha < \frac{\pi}{2}$, then the value of $\frac{df(\alpha)}{d(\cot\alpha)}$, is

(a) 6 (b) 5
(c) 3 (d) 1

24. If $f(x) = (\log_{\cot x}\tan x)(\log_{\tan x}\cot x)^{-1} + \tan^{-1}\left(\frac{x}{\sqrt{(4-x^2)}}\right)$, then $f'(0)$ is equal to

(a) –2 (b) 2
(c) $\frac{1}{2}$ (d) 0

25. If $x^2 + y^2 = t - \frac{1}{t}$ and $x^4 + y^4 = t^2 + \frac{1}{t^2}$, then $x^3y\frac{dy}{dx}$ equals

(a) – 1 (b) 0
(c) 1 (d) None of these

26. If $e^y + xy = e$, the ordered pair $\left(\frac{dy}{dx}, \frac{d^2y}{dx^2}\right)$ at $x = 0$ is equal to

[JEE Main 2019, 12 April]

(a) $\left(\frac{1}{e}, -\frac{1}{e^2}\right)$ (b) $\left(-\frac{1}{e}, \frac{1}{e^2}\right)$
(c) $\left(\frac{1}{e}, \frac{1}{e^2}\right)$ (d) $\left(-\frac{1}{e}, -\frac{1}{e^2}\right)$

27. If $y^{1/n} = \{x + \sqrt{(1 + x^2)}\}$, then $(1 + x^2)\, y_2 + xy_1$ is equal to
(a) n^2y (b) ny^2
(c) n^2y^2 (d) None of these

28. If $f(x) = \cot^{-1}\left(\frac{x^x - x^{-x}}{2}\right)$, then $f'(1)$ is
(a) -1 (b) 1
(c) $\log 2$ (d) $-\log 2$

29. The solution set of $f'(x) > g'(x)$ where $f(x) = (1/2)\, 5^{2x+1}$ and $g(x) = 5^x + 4x \log_e 5$ is
(a) $(1, \infty)$ (b) $(0, 1)$
(c) $[0, \infty)$ (d) $(0, \infty)$

30. If $f'(x) = g(x)$ and $g'(x) = -f(x)$ for all x and $f(2) = 4 = f'(2)$, then $f^2(19) + g^2(19)$ is
(a) 16 (b) 32
(c) 64 (d) None of these

31. If $\frac{d}{dx}(f(x)) = g(x)$ and $\frac{d}{dx}(g(x)) = f(x^2)$, then $\frac{d^2}{dx^2}(f(x^3))$ can be expressed in the form $kx^a f(x^b) + px\, g(x^c)$ where $a, b, c, k, p \in N$, then the value of $(k + p + a + b + c)$ equals
(a) 26 (b) 27
(c) 28 (d) 30

32. If $y = \sin x°$ and $u = \cos x$, then $\frac{dy}{du}$ is equal to
(a) $-\operatorname{cosec} x \cdot \cos x$ (b) $\frac{\pi}{180} \operatorname{cosec} x° \cos x$
(c) $-\frac{\pi}{180} \operatorname{cosec} x \cdot \cos x°$ (d) None of these

33. The differential coefficient of $f(\log_e x)$ w.r.t. x, where $f(x) = \log_e x$ is
(a) $x/\log_e x$ (b) $\log_e x/x$
(c) $(x \log_e x)^{-1}$ (d) None of these

34. If $f(1) = 1$, $f'(1) = 3$, then the derivative of $f(f(f(x))) + (f(x))^2$ at $x = 1$ is **[JEE Main 2019, 8 April]**
(a) 12 (b) 9
(c) 15 (d) 33

35. The derivative of the function
$f(x) = \cos^{-1}\left\{\frac{1}{\sqrt{13}}(2\cos x - 3\sin x)\right\} + \sin^{-1}\left\{\frac{1}{\sqrt{13}}(2\cos x + 3\sin x)\right\}$ w.r.t. $\sqrt{(1 + x^2)}$ at $x = 3/4$ is
(a) $\frac{3}{2}$ (b) $\frac{2}{3}$
(c) $\frac{3}{10}$ (d) $\frac{10}{3}$

36. Let $f(x) = \log_e \left\{\frac{u(x)}{v(x)}\right\}$, $u'(2) = 4$, $v'(2) = 2$, $u(2) = 2$, $v(2) = 1$, then $f'(2)$ is equal to
(a) 0 (b) 1
(c) -1 (d) None of these

37. If $\sqrt{(x^2 + y^2)} = a \cdot e^{\tan^{-1}(y/x)}$ $a > 0$, then $y''(0)$ is equal to
(a) $\frac{a}{2} e^{-\pi/2}$ (b) $ae^{\pi/2}$
(c) $-\frac{2}{a} e^{-\pi/2}$ (d) not exist

38. If $f'(x) = \sin x + \sin 4x \cdot \cos x$, then $f'\left(2x^2 + \frac{\pi}{2}\right)$ at $x = \sqrt{\frac{\pi}{2}}$ is equal to
(a) -1 (b) 0
(c) $-2\sqrt{2\pi}$ (d) None of these

39. If $P(x)$ is a polynomial such that $P(x^2 + 1) = \{P(x)\}^2 + 1$ and $P(0) = 0$, then $P'(0)$ is equal to
(a) -1 (b) 0
(c) 1 (d) None of these

40. Let $f : S \to S$, where $S = (0, \infty)$ be a twice differentiable function such that $f(x + 1) = xf(x)$. If $g : S \to R$ be defined as $g(x) = \log_e f(x)$, then the value of $|g''(5) - g''(1)|$ is equal to **[JEE Main 2021, 16 March]**
(a) $\frac{205}{144}$ (b) $\frac{197}{144}$
(c) $\frac{187}{144}$ (d) 1'

41. The third derivative of a function $f(x)$ vanishes for all x. If $f(0) = 1$, $f'(1) = 2$ and $f''(1) = -1$, then $f(x)$ is equal to
(a) $(-3/2)\, x^2 + 3x + 9$ (b) $(-1/2)\, x^2 - 3x + 1$
(c) $(-1/2)\, x^2 + 3x + 1$ (d) $(-3/2)\, x^2 - 7x + 2$

42. If $y = \log_{e^x} (x - 2)^2$ for $x \neq 0, 2$, then $y'(3)$ is equal to
(a) 1/3 (b) 2/3
(c) 4/3 (d) None of these

43. If $y = \tan^{-1}\left(\frac{1}{1 + x + x^2}\right) + \tan^{-1}\left(\frac{1}{x^2 + 3x + 3}\right) + \tan^{-1}\left(\frac{1}{x^2 + 5x + 7}\right) + \ldots +$ up to n terms. then $y'(0)$ is equal to
(a) $-\frac{1}{1 + n^2}$ (b) $-\frac{n^2}{1 + n^2}$
(c) $\frac{n}{1 + n^2}$ (d) None of these

44. Let f be a function such that $f(x + y) = f(x) + f(y)$ for all x and y and $f(x) = (2x^2 + 3x)\, g(x)$ for all x where $g(x)$ is continuous and $g(0) = 3$. Then $f'(x)$ is equal to
(a) 9 (b) 3
(c) 6 (d) None of these

45. If $\sqrt{(x+y)}+\sqrt{(y-x)}=a$, then $\dfrac{d^2y}{dx^2}$ equals

(a) $2/a$ (b) $-2/a^2$
(c) $2/a^2$ (d) None of these

46. If $\sin(x+y)=\log_e(x+y)$, then $\dfrac{dy}{dx}$ is equal to

(a) 2 (b) −2
(c) 1 (d) −1

47. Let $x^{\cos y}+y^{\cos x}=5$. Then

(a) at $x=0, y=0, y'=0$ (b) at $x=0, y=1, y'=0$
(c) at $x=y=1, y'=-1$ (d) at $x=1, y=0, y'=1$

48. If $f(x)=(1+x)^n$, then the value of

$f(0)+f'(0)+\dfrac{f''(0)}{2!}+\ldots.+\dfrac{f^n(0)}{n!}$ is

(a) n (b) 2^n
(c) 2^{n-1} (d) None of these

49. If $P(x)$ be a polynomial of degree 4, with $P(2)=-1, P'(2)=0, P''(2)=2, P'''(2)=-12$ and $P^{iv}(2)=24$, then $P''(1)$ is equal to

(a) 22 (b) 24
(c) 26 (d) 28

50. If $f(x)=\sin\left\{\dfrac{\pi}{3}[x]-x^2\right\}$ for $2<x<3$ and $[x]$ denotes the greatest integer less than or equal to x, then $(f'\sqrt{\pi/3})$ is equal to

(a) $\sqrt{\pi/3}$ (b) $-\sqrt{\pi/3}$
(c) $-\sqrt{\pi}$ (d) None of these

51. If $y=\int_0^x f(t)\sin\{k(x-t)\}dt$, then $\dfrac{d^2y}{dx^2}+k^2y$ equals

(a) 0 (b) y
(c) $k\cdot f(x)$ (d) $k^2 f(x)$

52. If $\sqrt{(1-x^6)}+\sqrt{(1-y^6)}=a(x^3-y^3)$ and $\dfrac{dy}{dx}=f(x,y)\sqrt{\left(\dfrac{1-y^6}{1-x^6}\right)}$, then

(a) $f(x,y)=y/x$
(b) $f(x,y)=y^2/x^2$
(c) $f(x,y)=2y^2/x^2$
(d) $f(x,y)=x^2/y^2$

53. If $y=\tan^{-1}\left(\dfrac{\log_e(e/x^2)}{\log_e(ex^2)}\right)+\tan^{-1}\left(\dfrac{3+2\log_e x}{1-6\log_e x}\right)$, then $\dfrac{d^2y}{dx^2}$ is

(a) 2 (b) 1
(c) 0 (d) −1

54. If $x=a\cos\theta, y=b\sin\theta$, then $\dfrac{d^3y}{dx^3}$ is equal to

(a) $\left(\dfrac{-3b}{a^3}\right)\operatorname{cosec}^4\theta\cot^4\theta$ (b) $\left(\dfrac{3b}{a^3}\right)\operatorname{cosec}^4\theta\cot\theta$
(c) $\left(\dfrac{-3b}{a^3}\right)\operatorname{cosec}^4\theta\cot\theta$ (d) None of these

55. If $F(x)=\dfrac{1}{x^2}\int_4^x\{4t^2-2F'(t)\}dt$, then $F'(4)$ equals

(a) 32/9 (b) 64/3
(c) 64/9 (d) None of these

56. If $y=\sqrt{x+\sqrt{y+\sqrt{x+\sqrt{y+\ldots\infty}}}}$, then $\dfrac{dy}{dx}$ is equal to

(a) $\dfrac{1}{2y-1}$ (b) $\dfrac{y^2-x}{2y^3-2xy-1}$
(c) $(2y-1)$ (d) None of these

57. The derivative of $\cos^{-1}\left(\dfrac{x^{-1}-x}{x^{-1}+x}\right)$ at $x=-1$ is

(a) −2 (b) −1
(c) 0 (d) 1

58. Let $f(x)=\exp(x^3+x^2+x)$ for any real number x and let g be the inverse of f, then the value of $g'(e^3)$ is

(a) $\dfrac{1}{6}$ (b) 6
(c) $\dfrac{1}{6e^3}$ (d) $6e^3$

Objective Questions Type II [One or more than one correct answer(s)]

In each of the questions below four choices of which one or more than one are correct. You have to select the correct answer(s) accordingly.

1. Given $f(x)=-\dfrac{x^3}{3}+x^2\sin 1.5a-x\sin a.\sin 2a-5\sin^{-1}(a^2-8a+17)$ then

(a) $f'(x)=-x^2+2x\sin 6-\sin 4\sin 8$
(b) $f'(\sin 8)>0$
(c) $f'(x)$ is not defined at $x=\sin 8$
(d) $f'(\sin 8)<0$

2. If $f(x)=x^3+x^2f'(1)+xf''(2)+f'''(3)$ for all $x\in R$, then

(a) $f(0)+f(2)=f(1)$
(b) $f(0)+f(3)=0$
(c) $f(1)+f(3)=f(2)$
(d) None of the above

3. If $y=\log_2\{\log_2 x\}$, then $\dfrac{dy}{dx}$ is equal to

(a) $\dfrac{\log_2 e}{x\log_e x}$ (b) $\dfrac{1}{x\log_e x\log_e 2}$
(c) $\dfrac{1}{\log_e(2x)^x}$ (d) None of these

4. If $y = \dfrac{\sqrt{(1+t^2)} - \sqrt{(1-t^2)}}{\sqrt{(1+t^2)} + \sqrt{(1-t^2)}}$ and $x = \sqrt{(1-t^4)}$, then $\dfrac{dy}{dx}$ is equal to

(a) $\dfrac{-1}{t^2\{1+\sqrt{(1-t^4)}\}}$ (b) $\dfrac{\{\sqrt{(1-t^4)}-1\}}{t^6}$

(c) $\dfrac{1}{t^2\{1+\sqrt{(1-t^4)}\}}$ (d) $\dfrac{1-\sqrt{(1-t^4)}}{t^6}$

5. Let $f:(0,\infty)\to(-\infty,\infty)$ be defined as $f(x)=e^x+\ln x$ and $g=f^{-1}$, then

(a) $g''(e)=\dfrac{1-e}{(1+e)^3}$ (b) $g''(e)=\dfrac{e-1}{(1+e)^3}$

(c) $g'(e)=e+1$ (d) $g'(e)=\dfrac{1}{e+1}$

6. If $f(x)=\sin^{-1}\left(\dfrac{2x}{1+x^2}\right)$, then

(a) f is derivable for all x, with $|x|<1$
(b) f is not derivable at $x=1$
(c) f is not derivable at $x=-1$
(d) f is derivable for all x, with $|x|>1$

7. Let $f(x)=x^2+xg'(1)+g''(2)$ and $g(x)=x^2+xf'(2)+f''(3)$, then

(a) $f'(1)=4+f'(2)$ (b) $g'(2)=8+g'(1)$
(c) $g''(2)+f''(3)=4$ (d) None of these

8. If $f(x-y), f(x)\cdot f(y)$ and $f(x+y)$ are in AP, for all x, y and $f(0)\neq 0$, then

(a) $f(2)=f(-2)$ (b) $f(3)+f(-3)=0$
(c) $f'(2)+f'(-2)=0$ (d) $f'(3)=f'(-3)$

9. If $f_n(x)=e^{f_{n-1}(x)}$ for all $n\in N$ and $f_0(x)=x$, then $\dfrac{d}{dx}\{f_n(x)\}$ is equal to

(a) $f_n(x)\cdot\dfrac{d}{dx}\{f_{n-1}(x)\}$
(b) $f_n(x)\cdot f_{n-1}(x)$
(c) $f_n(x)\cdot f_{n-1}(x)\ldots f_2(x)\cdot f_1(x)$
(d) $\prod_{i=1}^{n} f_i(x)$

10. If $f(\theta)=\tan\left(\sin^{-1}\sqrt{\dfrac{2}{3+\cos 2\theta}}\right)$, then

(a) $f'\left(\dfrac{\pi}{4}\right)=\sqrt{2}$ (b) $f\left(\dfrac{\pi}{4}\right)=\sqrt{2}$

(c) $\dfrac{d(f(\theta))}{d(\cos\theta)}$ at $\theta=\dfrac{\pi}{4}$ is $-\sqrt{2}$ (d) $\dfrac{d(f(\theta))}{d(\cos\theta)}$ at $\theta=\dfrac{\pi}{4}$ is -2

11. If $2^x+2^y=2^{x+y}$, then $\dfrac{dy}{dx}$ has the value equal to

(a) $-\dfrac{2^y}{2^x}$ (b) $1-2^y$

(c) $\dfrac{1}{1-2^x}$ (d) $\dfrac{2^x(1-2^y)}{2^y(2^x-1)}$

12. If $f(x)=\sin^{-1}(\sin x)$, then

(a) $f'\left(\dfrac{3\pi}{4}\right)=1$ (b) $f'\left(\dfrac{5\pi}{4}\right)=-1$

(c) $f'\left(\dfrac{\pi}{2}\right)$ does not exist (d) $f'(\pi)$ does not exist

13. If $F(x)=f(x)g(x)$ and $f'(x)g'(x)=c$, then

(a) $F'=c\left[\dfrac{f}{f'}+\dfrac{g}{g'}\right]$ (b) $\dfrac{F''}{F}=\dfrac{f''}{f}+\dfrac{g''}{g}+\dfrac{2c}{fg}$

(c) $\dfrac{F'''}{F}=\dfrac{f'''}{f}+\dfrac{g'''}{g}$ (d) $\dfrac{F'''}{F''}=\dfrac{f'''}{f''}+\dfrac{g'''}{g''}$

14. If 1 is a twice repeated root of the equation $ax^3+bx^2+bx+d=0$ then

(a) $a=b=d$ (b) $a+b=0$
(c) $b+d=0$ (d) $a=d$

15. Differential coefficient of $\sin^{-1}x$ w.r.t. $\sin^{-1}(3x-4x^3)$ is

(a) $\dfrac{1}{3}$ if $-\dfrac{\pi}{8}<x<\dfrac{\pi}{8}$ (b) 3 if $\dfrac{-\pi}{8}<x<\dfrac{\pi}{8}$

(c) $\dfrac{1}{3}$ if $-\dfrac{\pi}{9}<x<\dfrac{\pi}{9}$ (d) 3 if $\dfrac{-\pi}{9}<x<\dfrac{\pi}{9}$

Linked-Comprehension Type

In these questions, a passage (paragraph) has been given followed by questions based on each of the passages. You have to answer the questions based on the passage given.

PASSAGE 1

If $$D^*f(x)=\lim_{h\to 0}\frac{f^2(x+h)-f^2(x)}{h}$$

where $f^2(x)=\{f(x)\}^2$

On the basis of above information, answer the following questions

1. If $u=f(x)$, $v=g(x)$, then the value of $D^*(u\cdot v)$ is

(a) $(D^*u)v+(D^*v)u$ (b) $u^2D^*v+v^2D^*u$
(c) D^*u+D^*v (d) $uvD^*(u+v)$

2. If $u=f(x)$, $v=g(x)$ then the value of $D^*\left\{\dfrac{u}{v}\right\}$ is

(a) $\dfrac{u^2D^*v-v^2D^*u}{v^4}$ (b) $\dfrac{uD^*v-vD^*u}{v^2}$

(c) $\dfrac{v^2D^*u-u^2D^*v}{v^4}$ (d) $\dfrac{vD^*u-uD^*v}{v^2}$

3. $D^*(\tan x)$ is equal to
(a) $\sec^2 x$ (b) $2\sec^2 x$
(c) $\tan x \sec^2 x$ (d) $2\tan x \sec^2 x$

4. The value of $D^* f(x)$ at the point on the curve $y = f(x)$ such that tangent to it are parallel to X-axis, then
(a) $f(x)$ (b) zero
(c) $2f(x)$ (d) $xf(x)$

5. The value of $D^* c$, where c is constant is
(a) non-zero constant (b) 2 constant
(c) does not exist (d) zero

PASSAGE 2

Consider f and g be two real-values differentiable functions on R (the set of all real numbers).

Let $f(x) = x^2 + g'(0)x + g''(3)$ and
$g(y) = f(-2)y^2 + yf''(y) + f'(y) - 2$

On the basis of above information, answer the following questions

1. $f(1)$ is equal to
(a) 0 (b) 5
(c) 13 (d) 1

2. Let $h(x) = \begin{cases} f(x), & x \geq 0 \\ g(x), & x < 0 \end{cases}$, then $h(x)$ is
(a) discontinuous at $x = 0$
(b) continuous but non-derivable at $x = 0$
(c) continuous and derivable at $x = 0$
(d) discontinuous but derivable at $x = 0$

3. If $f(\alpha)\, g(\beta) \leq 4$, where $\alpha, \beta \in [-3, 3]$ then the number of non-positive ordered pairs (α, β) is equal to
(a) 0 (b) 1
(c) 2 (d) More than two

PASSAGE 3

If $y = f(x)$ be a differentiable function of x such that whose second, third, ..., nth derivatives exist.
i.e., nth derivative of y is denoted by

$$y_n, \frac{d^n y}{dx^n}, D^n y, y^n, f^n(x)$$

$$\Rightarrow \quad \frac{d^n y}{dx^n} = \lim_{h \to 0} \frac{f^{n-1}(x+h) - f^{n-1}(x)}{h}$$

On the basis of above information, answer the following questions

1. If $y = e^{3x+7}$, then the value of $y_n(0)$ is
(a) 1 (b) 3^n
(c) $3^n \cdot e^7$ (d) $3^n \cdot e^7 \cdot 7!$

2. If $y = \dfrac{\ln x}{2 - 3x}$, then the value of $y_n(1)$ is
(a) 0 (b) $(-1)^n \cdot 3^n$
(c) $(-1)^n \cdot 3^n \cdot n!$ (d) None of these

3. If $y = \dfrac{\ln x}{x}$, then the value of $y''(e)$ is
(a) 1 (b) $-\dfrac{1}{e}$
(c) $-\dfrac{1}{e^2}$ (d) $-\dfrac{1}{e^3}$

4. If $x = \sin t$, $y = \sin kt$, then the value of $(1 - x^2)y_2 - xy_1$ is
(a) $k^2 y$ (b) $-k^2 y$
(c) ky^2 (d) $-ky^2$

5. If $n = 4p + 3$, $p \in I$ and $y = \tan^{-1} x$, then $y_n(0)$ is
(a) 0 (b) $n!$
(c) $-(n-1)!$ (d) $(n-1)!$

Numerical Value Type

Solve the following problems and write your answer.

1. If $y = \tan^{-1}\left(\dfrac{ax - b}{bx + a}\right)$, then the value of $(2008)\left.\dfrac{dy}{dx}\right|_{x=-1}$ must be

2. If $y = \dfrac{a + bx}{c + dx}$, where a, b, c, d are constants and $\lambda y_1 y_3 = \mu y_2^2$, then the value of μ^{λ^2} must be

3. If $f(x) = \sin\left[\cos^{-1}\left(\dfrac{1 - 2^{2x}}{1 + 2^{2x}}\right)\right]$ and its first derivative with respect to x is $-\dfrac{b}{a}\log_e 2$ when $x = 1$, where a and b are integers, then the minimum value of $|a^2 - b^2|$ is ....
[JEE Main 2021, 16 March]

4. If $x = t^2 + 3t - 8$, $y = 2t^2 - 2t - 5$ and if $\lambda = \dfrac{dy}{dx}$ at $(2, -1)$, then the value of 2800λ must be

5. If $f(x) = \begin{vmatrix} \sin x & \cos x \\ \tan x & \cot x \end{vmatrix}$, then the value of $-25.6\sqrt{2}\, f'\left(\dfrac{\pi}{4}\right)$ must be

6. If $y = \sin x - \cos x$ and if $f(x) = \dfrac{d^{17}y}{dx^{17}}$, then the value of $\left(\dfrac{1024}{\sqrt{2}}\right) f\left(\dfrac{\pi}{4}\right)$ must be

7. If $x = 2 + t^3$, $y = 3t^2$ and $\left(\dfrac{d^2y}{dx^2}\right) \Big/ \left(\dfrac{dy}{dx}\right)^n$ is a constant then the value of $343n$ must be

8. If $y = 2\sin^{-1}\left(\dfrac{x-2}{\sqrt{6}}\right) - \sqrt{(2 + 4x - x^2)}$, then at $x = 2$ the value of $2008\sqrt{\dfrac{3}{2}}\dfrac{dy}{dx}$ must be

Single Integer Type

1. Let $g(x)$ be differentiable for all x and let $f(x) = (ax + e^x)g(x)$, where a is some constant. If $g(0) = 5$, $g'(0) = -2$ and $f'(0) = 18$, then the value of a is

2. If $2x = (y^{1/3} + y^{-1/3})$, then find the value of $\frac{(x^2-1)}{y}\cdot\frac{d^2y}{dx^2} + \frac{x}{y}\cdot\frac{dy}{dx}$.

3. If $y = \tan^{-1}\frac{4x}{1+5x^2} + \tan^{-1}\frac{2+3x}{3-2x}$, when $0 < x < \frac{2}{3}$ and $\frac{dy}{dx} = \frac{\lambda}{1+25x^2}$, then find λ.

4. Let $f(x) = \sin^{-1}\left(\frac{2x+2}{\sqrt{4x^2+8x+13}}\right)$, then the value of $\frac{d(\tan f(x))}{d(\tan^{-1}x)}$ at $x = \frac{1}{\sqrt{2}}$.

5. Let $g(x) = f\left[\frac{x}{f(x)}\right]$ where, $f(x)$ is a differentiable positive function on $(0, \infty)$ such that, $f(1) = f'(1)$. Determine $g'(1)$.

6. Let $f(x) = \tan[\sin\{\cos^{-1}(\sin \cos^{-1} x)\} + \cos\{\sin^{-1}(\cos(\sin^{-1} x))\}]$, $x \in (0, 1)$ and $g(x) = \tan x$, $x \in \left(0, \frac{\pi}{2}\right)$, then find the value of $\frac{d(f(x))}{d(g(x))}$ at $x = \frac{\pi}{6}$.

Matrix-Match Type

Given below are Matching Type Questions, with two columns (each having some items) each . Each item of Column I has to be matched with the items of Column II, by encircling the correct match(es).

NOTE An item of Column I can be matched with more than one items of Column II.

1. Observe the following columns
Column I contains the function and Column II contains their derivatives at $x = 0$.

	Column I		Column II
A.	$f(x) = \cos^{-1}\left(\frac{2x}{1+x^2}\right)$	P.	2
B.	$g(x) = \cos^{-1}(2x^2-1)$	Q.	3
C.	$h(x) = \sin^{-1}\left(\frac{1-x^2}{1+x^2}\right)$	R.	-2
D.	$k(x) = \tan^{-1}\left[\frac{3x-x^3}{1-3x^2}\right]$	S.	non-existent

2. Observe the following columns

	Column I		Column II
A.	If $u = x^2 + y^2$ and $x = s + 3t$, $y = 2s - t$, then $\frac{d^2u}{ds^2}$ is divisible by	P.	2
B.	If $y = t^{10} + 1$ and $x = t^8 + 1$, then $\left.\frac{d^2y}{dx^2}\right\|_{t=1/2}$ is divisible by	Q.	3
C.	Let $F(x) = f(x)g(x)h(x)$ for all real x, where $f(x)$, $g(x)$ and $h(x)$ are differentiable functions. At some point x_0, $F'(x_0) = 21F(x_0)$, $f'(x_0) = 4f(x_0)$, $g'(x_0) = -7g(x_0)$ and $h'(x_0) = \lambda h(x_0)$. Then λ is divisible by	R.	4
		S.	5

3. Observe the following column.

	Column I		Column II
A.	If $y = \cos^{-1}(\cos x)$, then y' at $x = 5$ is equal to	P.	does not exist
B.	For the function $f(x) = \ln\|\tan x\|$ $f'\left(-\frac{\pi}{4}\right)$ is equal to	Q.	2
C.	The derivative of $\tan^{-1}\left(\frac{1+x}{1-x}\right)$ at $x = -1$ is	R.	$\frac{1}{2}$
D.	The derivative of $\frac{\log\|x\|}{x}$ at $x = -1$ is	S.	1
		T.	-1

Answers

Objective Questions Type [Only one correct answer]

1. (a)	**2.** (a)	**3.** (b)	**4.** (c)	**5.** (d)	**6.** (c)	**7.** (c)	**8.** (a)	**9.** (b)	**10.** (d)
11. (d)	**12.** (c)	**13.** (c)	**14.** (a)	**15.** (b)	**16.** (b)	**17.** (a)	**18.** (d)	**19.** (c)	**20.** (b)
21. (b)	**22.** (b)	**23.** (d)	**24.** (b)	**25.** (c)	**26.** (b)	**27.** (a)	**28.** (a)	**29.** (d)	**30.** (b)
31. (c)	**32.** (c)	**33.** (c)	**34.** (d)	**35.** (d)	**36.** (a)	**37.** (c)	**38.** (c)	**39.** (c)	**40.** (a)
41. (c)	**42.** (b)	**43.** (b)	**44.** (a)	**45.** (c)	**46.** (d)	**47.** (c)	**48.** (b)	**49.** (c)	**50.** (b)
51. (c)	**52.** (d)	**53.** (c)	**54.** (c)	**55.** (a)	**56.** (b)	**57.** (b)	**58.** (c)		

Objective Questions Type II [One or more than one correct answer(s)]

1. (a, d)	**2.** (a, b, c)	**3.** (a, b)	**4.** (a, b)	**5.** (a, d)
6. (a, b, c, d)	**7.** (a, b, c)	**8.** (a, c)	**9.** (a, c, d)	**10.** (a, b, d)
11. (a, b, c, d)	**12.** (b, c)	**13.** (a, b, c)	**14.** (b, c, d)	**15.** (a, c)

Linked-Comprehension Type

Passage 1 1. (b) 2. (c) 3. (d) 4. (b) 5. (d) **Passage 2** 1. (c) 2. (a) 3. (b)

Passage 3 1. (c) 2. (d) 3. (d) 4. (b) 5. (c)

Numerical Value Type Questions

1. 1004 **2.** 81 **3.** 481 **4.** 2400 **5.** 51.2 **6.** 1024 **7.** 1372 **8.** 2008

Single Integer Type Questions

1. (3) **2.** (9) **3.** (5) **4.** (1) **5.** (0) **6.** (6)

Matrix-Match Type

1. A → (R); B → (S); C → (S); D → (Q) **2.** A → (P, S); B → (P, R, S); C → (P, Q, R)

3. A → (T); B → (Q); C → (R); D → (S)

Solutions

Objective Questions Type [Only one correct answer]

1. $y = \tan^{-1}\sqrt{\left(\dfrac{1-\cos\left(\dfrac{\pi}{2}+x\right)}{1+\cos\left(\dfrac{\pi}{2}+x\right)}\right)} = \tan^{-1}\left|\tan\left(\dfrac{\pi}{4}+\dfrac{x}{2}\right)\right| \quad \ldots(\text{i})$

Now, $\dfrac{\pi}{2} < x < \pi$

$\therefore \quad \dfrac{\pi}{4} < \dfrac{x}{2} < \dfrac{\pi}{2}$

or $\quad \dfrac{\pi}{2} < \dfrac{\pi}{4} + \dfrac{x}{2} < \dfrac{3\pi}{4}$

$\therefore \quad \left|\tan\left(\dfrac{\pi}{4}+\dfrac{x}{2}\right)\right| = -\tan\left(\dfrac{\pi}{4}+\dfrac{x}{2}\right)$ ($\because$ in II quadrant)

$= \tan\left\{\pi - \left(\dfrac{\pi}{4}+\dfrac{x}{2}\right)\right\}$

From Eq. (i),

$y = \tan^{-1}\tan\left\{\pi - \left(\dfrac{\pi}{4}+\dfrac{x}{2}\right)\right\}$

$= \pi - \left(\dfrac{\pi}{4}+\dfrac{x}{2}\right) = \dfrac{3\pi}{4} - \dfrac{x}{2}$

$\left[\because \text{principal value of } \tan^{-1}\tan x \text{ in } -\dfrac{\pi}{2} \text{ to } \dfrac{\pi}{2}\right]$

$\therefore \quad \dfrac{dy}{dx} = -\dfrac{1}{2}$

2. It is given that,

$fog(x) = x$, then $gof(x) = x$

$\Rightarrow \quad g'(f(x))f'(x) = 1$

On putting $x = b$, we get

$g'(f(b))f'(b) = 1 \quad \Rightarrow \quad f'(b) = \dfrac{1}{g'(f(b))}$

$gof(x) = x$

$\Rightarrow \quad g\ f(b) = b$ (on putting $x = b$)

$\Rightarrow \quad g(f(b)) = g(a)$ ($\because g(a) = b$)

$\Rightarrow \quad f(b) = a$

$\therefore \quad f'(b) = \dfrac{1}{g'(a)} = \dfrac{1}{5}$ [given]

3. $\because \dfrac{dy}{dx}=\dfrac{\dfrac{dy}{dt}}{\dfrac{dx}{dt}}=\dfrac{2\cos t-2\cos 2t}{-2\sin t+2\sin 2t}=\dfrac{\cos t-\cos 2t}{\sin 2t-\sin t}$

$$=\frac{2\sin\left(\frac{3t}{2}\right)\sin\left(\frac{t}{2}\right)}{2\cos\left(\frac{3t}{2}\right)\sin\left(\frac{t}{2}\right)}=\tan\left(\frac{3t}{2}\right)$$

$$\therefore \quad \frac{d^2y}{dx^2}=\frac{d}{dx}\left(\frac{dy}{dx}\right)=\frac{d}{dx}\left(\tan\frac{3t}{2}\right)=\frac{d}{dt}\tan\left(\frac{3t}{2}\right)\cdot\frac{dt}{dx}$$

$$=\frac{3}{2}\cdot\sec^2\left(\frac{3t}{2}\right)\cdot\frac{1}{(-2\sin t+2\sin 2t)}$$

$$\therefore \left.\frac{d^2y}{dx^2}\right|_{t=\frac{\pi}{2}}=\frac{3}{2}\sec^2\left(\frac{3\pi}{4}\right)\cdot\frac{1}{(-2+0)}=-\frac{3}{4}(-\sqrt{2})^2=-\frac{3}{2}$$

4. Let $\sqrt{x}=\cos\theta$

$$\because \quad x\in\left(0,\frac{1}{2}\right)\Rightarrow\sqrt{x}=\cos\theta\in\left(0,\frac{1}{\sqrt{2}}\right)$$

$$\Rightarrow \quad \theta\in\left(\frac{\pi}{4},\frac{\pi}{2}\right)\Rightarrow 2\theta\in\left(\frac{\pi}{2},\pi\right)$$

Now,

$$f(x)=2\sin^{-1}\sqrt{(1-\cos^2\theta)}+\sin^{-1}(2\sqrt{\cos^2\theta\,(1-\cos^2\theta)})$$

$$=2\sin^{-1}|(\sin\theta)|+\sin^{-1}|\sin 2\theta|$$

$$=2\sin^{-1}(\sin\theta)+\sin^{-1}(\sin 2\theta)$$

[$\because$ θ and 2θ lies in 1st and 2nd quadrants respectively]

$$=2\theta+\pi-2\theta$$

$$=\pi$$

$\therefore f'(x)=0$

5. $\because f'(x)=(4a-3)(1)+(a-7)\cos x$

$$=\left(\frac{(4a-3)}{(a-7)}+\cos x\right)(a-7)$$

$$f'(x)\neq 0$$

[for non existence of critical points]

$$\frac{4a-3}{a-7}>1 \text{ or } \frac{4a-3}{a-7}<-1 \quad [\because -1\le\cos x\le 1]$$

$$\Rightarrow \quad \frac{3a+4}{a-7}>0 \text{ or } \frac{5a-10}{a-7}<0$$

$$\therefore \quad a\in\left(-\infty,-\frac{4}{3}\right)\cup(7,\infty) \text{ or } a\in(2,7)$$

Hence, $a\in(-\infty,-4/3)\cup(2,\infty)$.

[$\because$ at $a=7$, $f'(x)\neq 0$]

6. Let $y=f(x)\Rightarrow x=f^{-1}(y)$

then $\quad f(x)=x+\tan x$

$$\Rightarrow \quad y=f^{-1}(y)+\tan(f^{-1}(y))$$

$\Rightarrow y=g(y)+\tan(g(y))$ or $x=g(x)+\tan(g(x))$ ...(i)

Differentiating both sides, we get

$$1=g'(x)+\sec^2 g(x)\cdot g'(x)$$

$$\therefore \quad g'(x)=\frac{1}{1+\sec^2(g(x))}=\frac{1}{1+1+\tan^2(g(x))}$$

$$=\frac{1}{2+(x-g(x))^2} \quad \text{[from Eq. (i)]}$$

$$=\frac{1}{2+(g(x)-x)^2}$$

7. Let $x=a\cos^2\theta+b\sin^2\theta$

$\therefore \; a-x=(a-b)\sin^2\theta,\; x-b=(a-b)\cos^2\theta$

$$\therefore \quad y=(a-b)\sin\theta\cos\theta-(a-b)\theta$$

$$=\frac{(a-b)\sin 2\theta}{2}-(a-b)\theta$$

$$\Rightarrow \quad \frac{dy}{d\theta}=(a-b)\cos 2\theta-(a-b)$$

$$=-(a-b)2\sin^2\theta=-2(a-b)\sin^2\theta$$

and $\quad \dfrac{dx}{d\theta}=(b-a)\sin 2\theta$

$$\therefore \quad \frac{dy}{dx}=\frac{-2(a-b)\sin^2\theta}{(b-a)\sin 2\theta}=\tan\theta=\sqrt{\left(\frac{a-x}{x-b}\right)}$$

8. Put $x^n=\sin\theta$ and $y^n=\sin\phi$

then, $(\cos\theta+\cos\phi)=a(\sin\theta-\sin\phi)$

$$\Rightarrow \quad 2\cos\left(\frac{\theta+\phi}{2}\right)\cos\left(\frac{\theta-\phi}{2}\right)$$

$$=2a\cos\left(\frac{\theta+\phi}{2}\right)\sin\left(\frac{\theta-\phi}{2}\right)$$

$$\Rightarrow \quad \cot\left(\frac{\theta-\phi}{2}\right)=a$$

$$\Rightarrow \quad \left(\frac{\theta-\phi}{2}\right)=\cot^{-1}a\Rightarrow\theta-\phi=2\cot^{-1}a$$

or $\quad \sin^{-1}x^n-\sin^{-1}y^n=2\cot^{-1}a$

Differentiating both sides, we have

$$\frac{nx^{n-1}}{\sqrt{(1-x^{2n})}}-\frac{ny^{n-1}}{\sqrt{(1-y^{2n})}}\frac{dy}{dx}=0$$

$$\therefore \quad \sqrt{\left(\frac{1-x^{2n}}{1-y^{2n}}\right)}\frac{dy}{dx}=\frac{x^{n-1}}{y^{n-1}}$$

9. $2y=\left[\cot^{-1}\left(\dfrac{\sqrt{3}+\tan x}{1-\sqrt{3}\tan x}\right)\right]^2$

$$=\left[\cot^{-1}\left(\frac{\tan\frac{\pi}{3}+\tan x}{1-\tan\frac{\pi}{3}\tan x}\right)\right]^2$$

$$=\left[\cot^{-1}\left(\tan\left(x+\frac{\pi}{3}\right)\right)\right]^2$$

$$=\left[\cot^{-1}\left(\tan\left(\frac{\pi}{2}-\left(\frac{\pi}{6}-x\right)\right)\right)\right]^2$$

$$=\left[\cot^{-1}\left(\cot\left(\frac{\pi}{6}-x\right)\right)\right]^2 \quad \left[\because \tan\left(\frac{\pi}{2}-\theta\right)=\cot\theta\right]$$

$$=\left(\frac{\pi}{6}-x\right)^2$$

Differentiate both sides w.r.t. x, we get

$$2\frac{dy}{dx}=2\left(\frac{\pi}{6}-x\right)(-1)\Rightarrow\frac{dy}{dx}=x-\frac{\pi}{6}$$

10. $\because x^2+4=(\sec\theta-\cos\theta)^2+4=(\sec\theta+\cos\theta)^2$ ...(i)

Similarly, $y^2+4=(\sec^{10}\theta+\cos^{10}\theta)^2$ ...(ii)

Now, $\frac{dx}{d\theta}=\sec\theta\tan\theta+\sin\theta=\tan\theta(\sec\theta+\cos\theta)$

and $\frac{dy}{d\theta}=10\sec^9\theta\sec\theta\tan\theta-10\cos^9\theta(-\sin\theta)$

$$=10\tan\theta(\sec^{10}\theta+\cos^{10}\theta)$$

$$\Rightarrow\frac{dy}{dx}=\frac{\left(\frac{dy}{d\theta}\right)}{\left(\frac{dx}{d\theta}\right)}=\frac{10\tan\theta(\sec^{10}\theta+\cos^{10}\theta)}{\tan\theta(\sec\theta+\cos\theta)}$$

$$\therefore\left(\frac{dy}{dx}\right)^2=100\frac{(\sec^{10}\theta+\cos^{10}\theta)^2}{(\sec\theta+\cos\theta)^2}=\frac{100(y^2+4)}{(x^2+4)}$$

[from Eq. (i), (ii)]

or $(x^2+4)\left(\frac{dy}{dx}\right)^2=100(y^2+4)$

$\therefore$ $k=100$

11. $\because y=x\ln\left(\frac{x}{a+bx}\right)=x(\ln x-\ln(a+bx))$

or $\left(\frac{y}{x}\right)=\ln x-\ln(a+bx)$

Differentiating both sides w.r.t. x, then

$$\left(\frac{x\frac{dy}{dx}-y\cdot 1}{x^2}\right)=\frac{1}{x}-\frac{b}{a+bx}=\frac{a}{x(a+bx)}\quad\text{...(i)}$$

or $\left(x\frac{dy}{dx}-y\right)=\left(\frac{ax}{a+bx}\right)$

Again taking logarithm on both sides, then

$$\ln\left(x\frac{dy}{dx}-y\right)=\ln(ax)-\ln(a+bx)$$

Differentiating both sides w.r.t. x, then

$$\frac{x\frac{d^2y}{dx^2}+\frac{dy}{dx}-\frac{dy}{dx}}{\left(x\frac{dy}{dx}-y\right)}=\frac{1}{x}-\frac{b}{a+bx}=\frac{a}{x(a+bx)}$$

$$=\frac{\left(x\frac{dy}{dx}-y\right)}{x^2}\quad\text{[from Eq. (i)]}$$

or $x^3\frac{d^2y}{dx^2}=\left(x\frac{dy}{dx}-y\right)^2$

12. $\because$ $y=\left(\frac{ax+b}{cx+d}\right)$

or $cxy+dy=ax+b$

Differentiating both sides w.r.t. x, then

$$c\left\{x\frac{dy}{dx}+y\cdot 1\right\}+d\frac{dy}{dx}=a$$

or $x\frac{dy}{dx}+y+\left(\frac{d}{c}\right)\frac{dy}{dx}=\left(\frac{a}{c}\right)$

Again differentiating both sides w.r.t. x, then

$$x\frac{d^2y}{dx^2}+\frac{dy}{dx}+\frac{dy}{dx}+\left(\frac{d}{c}\right)\frac{d^2y}{dx^2}=0$$

or $$x+\frac{2\frac{dy}{dx}}{\left(\frac{d^2y}{dx^2}\right)}+\frac{d}{c}=0$$

Again differentiating both sides w.r.t. x, then

$$1+\frac{\left(\frac{d^2y}{dx^2}\cdot 2\frac{d^2y}{dx^2}-2\frac{dy}{dx}\cdot\frac{d^3y}{dx^3}\right)}{\left(\frac{d^2y}{dx^2}\right)^2}+0=0$$

$\therefore$ $$2\frac{dy}{dx}\cdot\frac{d^3y}{dx^3}=3\left(\frac{d^2y}{dx^2}\right)^2$$

13. Since, $x^y=e^{x-y}\Rightarrow y\ln x=x-y$

$\therefore$ $y=\frac{x}{1+\ln x}$

$\therefore$ $\frac{dy}{dx}=\frac{\ln x}{(1+\ln x)^2}$

14. $\frac{dy}{dx}=\left(\frac{dx}{dy}\right)^{-1}\Rightarrow\frac{d^2y}{dx^2}=-\left(\frac{dx}{dy}\right)^{-2}\left\{\frac{d}{dx}\left(\frac{dx}{dy}\right)\right\}$

$$\Rightarrow\frac{d^2y}{dx^2}=-\left(\frac{dx}{dy}\right)^{-2}\left\{\frac{d}{dy}\left(\frac{dx}{dy}\right)\frac{dy}{dx}\right\}$$

$$=-\left(\frac{dy}{dx}\right)^2\left\{\frac{d^2x}{dy^2}\cdot\frac{dy}{dx}\right\}=-\left(\frac{dy}{dx}\right)^3\frac{d^2x}{dy^2}$$

$$\Rightarrow\frac{d^2y}{dx^2}+\left(\frac{dy}{dx}\right)^3\frac{d^2x}{dy^2}=0$$

15. Given functional relation is $y\sqrt{1-x^2}=k-x\sqrt{1-y^2}$

On differentiating both sides w.r.t. x, we get

$$\frac{-2yx}{2\sqrt{1-x^2}}+\sqrt{1-x^2}\frac{dy}{dx}=0+\frac{2xy\frac{dy}{dx}}{2\sqrt{1-y^2}}-\sqrt{1-y^2}$$

$$\Rightarrow\frac{dy}{dx}\left\{\sqrt{1-x^2}-\frac{xy}{\sqrt{1-y^2}}\right\}=\left\{\frac{yx}{\sqrt{1-x^2}}-\sqrt{1-y^2}\right\}$$

$$\Rightarrow\frac{dy}{dx}\left\{\frac{\sqrt{1-x^2}\sqrt{1-y^2}-xy}{\sqrt{1-y^2}}\right\}=\frac{xy-\sqrt{1-x^2}\sqrt{1-y^2}}{\sqrt{1-x^2}}$$

$$\Rightarrow\frac{dy}{dx}=-\frac{\sqrt{1-y^2}}{\sqrt{1-x^2}}$$

$\because$ $y\left(\frac{1}{2}\right)=-\frac{1}{4}$ [given]

$$\therefore\left.\frac{dy}{dx}\right|_{x=\frac{1}{2}}=-\frac{\sqrt{1-\left(-\frac{1}{4}\right)^2}}{\sqrt{1-\left(\frac{1}{2}\right)^2}}=-\frac{\sqrt{\frac{16-1}{16}}}{\sqrt{\frac{4-1}{4}}}=-\frac{\sqrt{15}}{2\sqrt{3}}=-\frac{\sqrt{5}}{2}$$

16. We have, $y = \dfrac{\cos 6x + 6\cos 4x + 15\cos 2x + 10}{\cos 5x + 5\cos 3x + 10\cos x}$

Numerator $= \cos 6x + 6\cos 4x + 15\cos 2x + 10$
$= \cos 6x + (1+5)\cos 4x + (5+10)\cos 2x + 10$
$= \cos 6x + \cos 4x + 5(\cos 4x + \cos 2x) + 10(1 + \cos 2x)$
$= 2\cos 5x \cos x + 10\cos 3x \cos x + 20\cos^2 x$
$= 2\cos x[\cos 5x + 5\cos 3x + 10\cos x]$

$\therefore\ y = \dfrac{2\cos x(\cos 5x + 5\cos 3x + 10\cos x)}{\cos 5x + 5\cos 3x + 10\cos x}$

$\Rightarrow y = 2\cos x \Rightarrow \dfrac{dy}{dx} = -2\sin x$

17. Let $y = \tan^{-1}\left(\dfrac{6x\sqrt{x}}{1-9x^3}\right) = \tan^{-1}\left[\dfrac{2\cdot(3x^{3/2})}{(3x^{3/2})^2}\right]$

$= 2\tan^{-1}(3x^{3/2})$ $\quad\left[\because 2\tan^{-1} x = \tan^{-1}\dfrac{2x}{1-x^2}\right]$

$\therefore\ \dfrac{dy}{dx} = 2\cdot\dfrac{1}{1+(3x^{3/2})}\cdot 3\times\dfrac{3}{2}(x)^{1/2} = \dfrac{9}{1+9x^3}\cdot\sqrt{x}$

$\therefore\ g(x) = \dfrac{9}{1+9x^3}$

18. $x^2 + y^2 = a^2$ ...(i)

Differentiating both sides w.r.t. x, we have

$2x + 2yy' = 0$

or $x + yy' = 0$ ...(ii)

Again differentiating both sides w.r.t. x, we have

$1 + yy'' + y'^2 = 0$

$\therefore\ y'' = -\dfrac{(1+y'^2)}{y}$ ...(iii)

Now, $\lambda = \dfrac{y''}{\sqrt{(1+y'^2)^3}} = -\dfrac{(1+y'^2)}{y\sqrt{(1+y'^2)^3}}$ [From Eq. (iii)]

$= -\dfrac{1}{y\sqrt{(1+y'^2)}} = -\dfrac{1}{\sqrt{y^2+(yy')^2}}$

$= -\dfrac{1}{\sqrt{(y^2+x^2)}}$ [From Eq. (ii)]

$= -\dfrac{1}{\sqrt{a^2}} = -\dfrac{1}{|a|} = -\dfrac{1}{a}$

[From Eq. (i)} and using $a > 0$]

19. $y^2 = P(x)$

$\Rightarrow\ 2yy' = P'(x)$ ...(i)

$\Rightarrow\ (2y)y'' + y'(2y') = P''(x)$

$\Rightarrow\ 2yy'' = P''(x) - 2(y')^2$

$\Rightarrow\ 2y^3y'' = y^2P''(x) - 2(yy')^2$

$= y^2P''(x) - 2\dfrac{\{P'(x)\}^2}{4}$ [from Eq. (i)]

$\Rightarrow\ 2y^3y'' = P(x)P''(x) - \dfrac{1}{2}\{P'(x)\}^2$

$\therefore\ \dfrac{d}{dx}(2y^3y'') = P(x)P'''(x)$
$+ P''(x)P'(x) - P'(x)P''(x)]$
$= P(x)\cdot P'''(x)$

$\Rightarrow\ 2\dfrac{d}{dx}\left(y^3\dfrac{d^2y}{dx^2}\right) = P(x)P'''(x)$

20. $5f(x) + 3f\left(\dfrac{1}{x}\right) = x + 2$ ...(i)

Replacing x by $\dfrac{1}{x}$

$\therefore\ 5f\left(\dfrac{1}{x}\right) + 3f(x) = \dfrac{1}{x} + 2$ ...(ii)

From Eq. (i),

$25f(x) + 15f\left(\dfrac{1}{x}\right) = 5x + 10$ ...(iii)

and from Eq. (ii),

$9f(x) + 15f\left(\dfrac{1}{x}\right) = \dfrac{3}{x} + 6$ ...(iv)

Subtracting Eq. (iv) from Eq. (iii), we get

$\therefore\ 16f(x) = 5x - \dfrac{3}{x} + 4$

$\therefore\ xf(x) = \dfrac{5x^2 - 3 + 4x}{16} = y$

$\therefore\ \dfrac{dy}{dx} = \dfrac{10x+4}{16}$

$\left.\dfrac{dy}{dx}\right|_{x=1} = \dfrac{10+4}{16}$

$= \dfrac{7}{8}$

21. $2^x + 2^y = 2^{x+y}$

$\therefore\ 2^x\cdot\log_e 2 + 2^y\cdot\log_e 2\dfrac{dy}{dx}$

$= 2^{x+y}\cdot\log_e 2\left(1 + \dfrac{dy}{dx}\right)$

or $\quad 2^x + 2^y\dfrac{dy}{dx} = 2^{x+y}\left(1 + \dfrac{dy}{dx}\right)$

$\therefore\ \dfrac{dy}{dx} = \dfrac{2^{x+y} - 2^x}{2^y - 2^{x+y}}$

$\left.\dfrac{dy}{dx}\right|_{x=y=1} = -1$

22. We have $f(x) = |x-2| = \begin{cases} x-2, & x \ge 2 \\ -(x-2), & x < 2 \end{cases}$

$\because\ x > 20, \therefore f(x) = x - 2$

$\therefore\ g(x) = fof(x) = f\{f(x)\} = f(x-2)$
$= x - 2 - 2 = x - 4$

$\therefore\ g'(x) = 1$

23. $\because\ f(a) = \cos\left(\cot^{-1}\left(\dfrac{\cos\alpha}{\sqrt{1-\cos 2\alpha}}\right)\right)$

$= \cos\left(\cos^{-1}\left(\dfrac{\cos\alpha}{\sqrt{\cos^2\alpha + (-\cos 2\alpha)}}\right)\right)$

$\because\ \dfrac{\pi}{4} < \alpha < \dfrac{\pi}{2}$

$\therefore\ f(\alpha) = \cos\left(\cos^{-1}\left(\dfrac{\cos\alpha}{\sin\alpha}\right)\right) = \cot\alpha$

$\therefore\ \dfrac{df(\alpha)}{d(\cot\alpha)} = 1$

24. Since, $f(x)=(\log_{\cot x}\tan x)(\log_{\tan x}\cot x)^{-1} + \tan^{-1}\left(\frac{x}{\sqrt{4-x^2}}\right)]$

$$=(-1)(-1)^{-1}+\tan^{-1}\left(\frac{x}{\sqrt{4-x^2}}\right)$$

$$=1+\tan^{-1}\left(\frac{x}{\sqrt{4-x^2}}\right)$$

$$\therefore\quad f'(x)=0+\frac{1}{1+\frac{x^2}{4-x^2}}\times\frac{\sqrt{(4-x^2)}\cdot 1-x\times\frac{1}{2\sqrt{(4-x^2)}}\times -2x}{(\sqrt{4-x^2})^2}$$

$$=\frac{(4-x^2)}{4}\times\frac{4-x^2+x^2}{\sqrt{4-x^2}}=\sqrt{4-x^2}$$

$$\therefore\quad f'(0)=2$$

25. $x^2+y^2=t-\frac{1}{t},\ x^4+y^4=t^2+\frac{1}{t^2}$

$$=\left(t-\frac{1}{t}\right)^2+2$$

$$=x^4+y^4+2x^2y^2+2$$

$$\therefore\ x^2y^2=-1$$

$$\Rightarrow\quad x^2\cdot 2y\frac{dy}{dx}+y^2\cdot 2x=0$$

$$\Rightarrow\quad x^3y\frac{dy}{dx}=-x^2y^2=1$$

26. Given equation is,

$$e^y+xy=e \qquad \ldots(i)$$

On differentiating both sides w.r.t. x, we get

$$e^y\frac{dy}{dx}+x\frac{dy}{dx}+y=0 \qquad \ldots(ii)$$

$$\Rightarrow\quad \frac{dy}{dx}=-\left(\frac{y}{e^y+x}\right) \qquad \ldots(iii)$$

Again differentiating Eq. (ii) w.r.t. 'x', we get

$$e^y\frac{d^2y}{dx^2}+e^y\left(\frac{dy}{dx}\right)^2+x\frac{d^2y}{dx^2}+\frac{dy}{dx}+\frac{dy}{dx}=0 \qquad \ldots(iv)$$

Now, on putting $x=0$ in Eq. (i), we get

$$e^y=e^1\ \Rightarrow\ y=1$$

On putting $x=0,\ y=1$ in Eq. (iii), we get

$$\frac{dy}{dx}=-\frac{1}{e+0}=-\frac{1}{e}$$

Now, on putting $x=0,\ y=1$ and $\frac{dy}{dx}=-\frac{1}{e}$ in Eq. (iv), we get

$$e^1\frac{d^2y}{dx^2}+e^1\left(-\frac{1}{e}\right)^2+0\left(\frac{d^2y}{dx^2}\right)+\left(-\frac{1}{e}\right)+\left(-\frac{1}{e}\right)=0$$

$$\Rightarrow\quad \left.\frac{d^2y}{dx^2}\right|_{(0,1)}=\frac{1}{e^2}$$

So, $\left(\frac{dy}{dx},\frac{d^2y}{dx^2}\right)$ at $(0, 1)$ is $\left(-\frac{1}{e},\frac{1}{e^2}\right)$.

27. Since, $y^{1/n}=[x+\sqrt{1+x^2}]$

or $\quad y=(x+\sqrt{1+x^2})^n$

$$\therefore\quad y_1=n(x+\sqrt{1+x^2})^{n-1}\left(1+\frac{x}{\sqrt{1+x^2}}\right)$$

$$\sqrt{(1+x^2)}\,y_1=n(x+\sqrt{1+x^2})^n$$

$$\sqrt{(1+x^2)}\cdot y_1=ny$$

Squaring on both sides

$$\therefore\quad (1+x^2)y_1^2=n^2y^2$$

Differentiating both sides w.r.t. x

$$\therefore\quad (1+x^2)\,2y_1\,y_2+y_1^2\cdot 2x=2n^2yy_1$$

$$2y_1\neq 0$$

$$\therefore\quad (1+x^2)\,y_2+xy_1=n^2y$$

28. $f(x)=\cot^{-1}\left(\frac{x^x-x^{-x}}{2}\right)=\cot^{-1}\left(\frac{(x)^{2x}-1}{2x^x}\right)$

$$=\tan^{-1}\left(\frac{2x^x}{-1+(x^x)^2}\right)=-2\tan^{-1}x^x$$

$$\therefore\ f'(x)=-\frac{2}{1+(x^x)^2}\times x^x(1+\log_e x)$$

$$\therefore\ f'(1)=\frac{-2}{2}\times 1=-1$$

29. $\left(\frac{1}{2}\right)5^{2x+1}\log 5\times 2>5^x\log_e 5+4\log_e 5$

$$\Rightarrow\quad 5^{2x}\cdot 5>5^x+4$$

$$\Rightarrow\quad 5\cdot 5^{2x}-5^x-4>0$$

$$(5^x-1)(5\cdot 5^x+4)>0$$

$$\therefore\quad 5^x>1$$

$$x>0\ \Rightarrow\ x\in(0,\infty)$$

30. $f'(x)=g(x)\Rightarrow f''(x)=g'(x)=-f(x)$

$\therefore\quad 2f'(x)f''(x)=-2f(x)f'(x)$ on integrating

$\therefore\ [f'(x)]^2=-[f(x)]^2+c$

At $\quad x=2$

$$16=-16+c$$

$$\therefore\quad c=32$$

$$[f'(x)]^2+[f(x)]^2=32\Rightarrow[g(x)]^2+[f(x)]^2=32$$

$$\therefore\quad f^2(19)+g^2(19)=32$$

31. Given $\quad f'(x)=g(x)$

$$g'(x)=f(x^2)=f''(x)$$

Let $\quad y=f(x^3)$

$$\frac{dy}{dx}=f'(x^3)\cdot 3x^2$$

$$\frac{d^2y}{dx^2} = 3[3x^2x^2 . f''(x^3) + 2x . f'(x^3)]$$
$$= 9x^4 f''(x^3) + 6xf'(x^3)$$
$$= 9x^4 . f(x^6) + 6xg(x^3)$$
$$= kx^a f(x^b) + pxg(x^c)$$
$\therefore$ $(k + p + a + b + c) = 9 + 4 + 6 + 6 + 3 = 28$

32. Since, $y = \sin x^\circ$
$$= \sin\left(\frac{\pi x}{180}\right)$$
$\therefore$ $$\frac{dy}{dx} = \frac{\pi}{180}\cos\left(\frac{\pi x}{180}\right) = \frac{\pi}{180}\cos x^\circ$$
and $u = \cos x$
$\therefore$ $$\frac{du}{dx} = -\sin x$$
then $$\frac{dy}{du} = \frac{\frac{dy}{dx}}{\frac{du}{dx}} = -\frac{\frac{\pi}{180}\cos x^\circ}{\sin x}$$
$$= -\frac{\pi}{180}\cos x^\circ \operatorname{cosec} x$$

33. Let $u = f(\log_e x)$
$\therefore$ $$\frac{du}{dx} = f'(\log_e x)\cdot\frac{1}{x} \quad \ldots \text{(i)}$$
Given, $f(x) = \log_e x$
$\therefore$ $$f'(x) = \frac{1}{x}$$
$\therefore$ $$f'(\log_e x) = \frac{1}{\log_e x} \quad \ldots \text{(ii)}$$
From Eqs. (i) and (ii),
$$\frac{du}{dx} = \frac{1}{x\log_e x} = (x\log_e x)^{-1}$$

34. Let $y = f(f(f(x))) + (f(x))^2$

On differentiating both sides w.r.t. x, we get
$$\frac{dy}{dx} = f'(f(f(x)))\cdot f'(f(x))\cdot f'(x) + 2f(x)f'(x)$$
[by chain rule]

So, $$\left.\frac{dy}{dx}\right|_{\text{at } x=1} = f'(f(f(1)))\cdot f'(f(1))\cdot f'(1) + 2f(1)f'(1)$$
$\therefore$ $$\left.\frac{dy}{dx}\right|_{x=1} = f'(f(1))\cdot f'(1)\cdot(3) + 2(1)(3)$$
[$\because f(1) = 1$ and $f'(1) = 3$]
$$= f'(1)\cdot(3)\cdot(3) + 6 = (3\times 9) + 6$$
$$= 27 + 6 = 33$$

35. Given,
$$f(x) = \cos^{-1}\left\{\frac{1}{\sqrt{13}}(2\cos x - 3\sin x)\right\} + \sin^{-1}\left\{\frac{1}{\sqrt{13}}(2\cos x + 3\sin x)\right\}$$
$$= \cos^{-1}\left\{\frac{2}{\sqrt{13}}.\cos x - \frac{3}{\sqrt{13}}.\sin x\right\} + \sin^{-1}\left\{\frac{3}{\sqrt{13}}.\sin x + \frac{2}{\sqrt{13}}.\cos x\right\}$$
$$= \cos^{-1}\left[\cos\left(x + \tan^{-1}\frac{3}{2}\right)\right] + \sin^{-1}\left[\sin\left(x + \tan^{-1}\frac{2}{3}\right)\right]$$
$$= x + \tan^{-1}\left(\frac{3}{2}\right) + x + \tan^{-1}\left(\frac{2}{3}\right)$$
$\left[\because 0 < x + \tan^{-1}\frac{3}{2} < \pi \text{ and } -\frac{\pi}{2} < x + \tan^{-1}\frac{2}{3} < \frac{\pi}{2} \text{ for } x = 3/4\right]$
$$= 2x + \tan^{-1}\left(\frac{3}{2}\right) + \cot^{-1}\left(\frac{3}{2}\right) = 2x + \pi/2$$
$$\Rightarrow \frac{df(x)}{dx} = 2 \Rightarrow f'(x) = 2 \text{ or } f'(3/4) = 2$$
and let $g(x) = \sqrt{(1 + x^2)}$
$$\Rightarrow g'(x) = \frac{x}{\sqrt{1 + x^2}}$$
$\therefore$ $$g'\left(\frac{3}{4}\right) = \frac{3}{5}$$
Hence, $f'(3/4)/g'(3/4) = 2/3/5 = \frac{10}{3}$

36. $$f(x) = \log_e\left\{\frac{u(x)}{v(x)}\right\}$$
$$= \log_e u(x) - \log_e v(x)$$
$\therefore$ $$f'(x) = \frac{u'(x)}{u(x)} - \frac{v'(x)}{v(x)}$$
$$f'(2) = \frac{u'(2)}{u(2)} - \frac{v'(2)}{v(2)}$$
$$= \frac{4}{2} - \frac{2}{1}$$
$$= 2 - 2 = 0$$

37. $$\sqrt{x^2 + y^2} = ae^{\tan^{-1}(y/x)} \quad \ldots \text{(i)}$$
$$\frac{1}{2\sqrt{x^2 + y^2}}(2x + 2yy') = a\cdot e^{\tan^{-1}(y/x)} \times \frac{1}{\left(1 + \frac{y^2}{x^2}\right)} \times \frac{xy' - y}{x^2}$$
$$\Rightarrow \frac{x + yy'}{\sqrt{x^2 + y^2}} = \sqrt{x^2 + y^2} \times \frac{x^2}{(x^2 + y^2)} \times \frac{xy' - y}{x^2}$$
[from Eq. (i)]
$\therefore$ $$x + yy' = xy' - y \Rightarrow y' = \frac{x + y}{x - y}$$
$\therefore$ $$y'' = \frac{2(xy' - y)}{(x - y)^2}$$
$$y''(0) = \frac{2(0 - y(0))}{\{0 - y(0)\}^2}$$
$$= \frac{-2}{y(0)} = \frac{-2}{ae^{\pi/2}} = \frac{-2}{a}e^{-\pi/2}$$

38. $f'\left(2x^2+\frac{\pi}{2}\right)=\dfrac{df\left(2x^2+\frac{\pi}{2}\right)}{dx}$

$$=\frac{df\left(2x^2+\frac{\pi}{2}\right)}{d\left(2x^2+\frac{\pi}{2}\right)}\cdot\frac{d(2x^2+\frac{\pi}{2})}{dx}$$

$$=\left\{\sin\left(2x^2+\frac{\pi}{2}\right)+\sin(8x^2+2\pi)\cdot\cos\left(2x^2+\frac{\pi}{2}\right)\right\}4x$$

At $x=\sqrt{\frac{\pi}{2}}$

$$f'\left(2x^2+\frac{\pi}{2}\right)=4\sqrt{\frac{\pi}{2}}\{(-1)+0\}=-2\sqrt{2\pi}$$

39. Given, $P(x^2+1)=[P(x)]^2+1$

$\therefore \quad P(x)=x$

($\because P(x)$ is an identity function)

$P'(x)=1$

$\therefore \quad P'(0)=1$

40. $\ln f(x+1)=\ln(xf(x))$

$\ln f(x+1)=\ln x+\ln f(x)$

$\Rightarrow g(x+1)=\ln x+g(x)$

$\Rightarrow g(x+1)-g(x)=\ln x$

$\Rightarrow g''(x+1)-g''(x)=-\dfrac{1}{x^2}$

Put $x=1, 2, 3, 4$, we get

$$g''(2)-g''(1)=-\frac{1}{1^2} \quad \text{...(i)}$$

$$g''(3)-g''(2)=-\frac{1}{2^2} \quad \text{...(ii)}$$

$$g''(4)-g''(3)=-\frac{1}{3^2} \quad \text{...(iii)}$$

$$g''(5)-g''(4)=-\frac{1}{4^2} \quad \text{...(iv)}$$

Add all the equation, we get

$$g''(5)-g''(1)=-\frac{1}{1^2}-\frac{1}{2^2}-\frac{1}{3^2}-\frac{1}{4^2}$$

$$|g''(5)-g''(1)|=\frac{205}{144}$$

41. Given, $f'''(x)=0$

On integrating

$f''(x)=c \Rightarrow f''(1)=c \Rightarrow -1=c \quad$ (given)

$\therefore \quad f''(x)=-1$

$f'(x)=-x+d$

$\because \quad f'(1)=-1+d=2$

$\therefore \quad d=3$

$f'(x)=3-x$

or $\quad f(x)=-\dfrac{x^2}{2}+3x+e$

$\because \quad f(0)=0+0+e=1$

$\therefore \quad e=1$

Hence, $\quad f(x)=-\dfrac{x^2}{2}+3x+1$

42. $y=\dfrac{2}{x}\log_e(x-2)$

$$\therefore \quad \frac{dy}{dx}=\frac{\dfrac{2x}{(x-2)}-2\log_e(x-2)}{x^2}$$

$$\therefore \quad \left.\frac{dy}{dx}\right|_{x=3}=\frac{6}{9}=\frac{2}{3}$$

or $\quad y'(3)=\dfrac{2}{3}$

43. $y=\tan^{-1}\left\{\dfrac{1}{1+x(1+x)}\right\}+\tan^{-1}\left\{\dfrac{1}{1+(x+1)(x+2)}\right\}$

$+\tan^{-1}\left\{\dfrac{1}{1+(x+2)(x+3)}\right\}$

$+\ldots..+\tan^{-1}\left\{\dfrac{1}{1+(x+n-1)(x+n)}\right\}$

$$=\sum_{r=1}^{n}\tan^{-1}\left\{\frac{1}{1+(x+r-1)(x+r)}\right\}$$

$$=\sum_{r=1}^{n}\tan^{-1}\left\{\frac{(x+r)-(x+r-1)}{1+(x+r-1)(x+r)}\right\}$$

$$=\sum_{r=1}^{n}\{\tan^{-1}(x+r)-\tan^{-1}(x+r-1)\}$$

$$=\tan^{-1}(x+n)-\tan^{-1}x$$

$$y'=\frac{1}{1+(x+n)^2}-\frac{1}{1+x^2}$$

$$\Rightarrow \quad y'(0)=\frac{1}{1+n^2}-1=\frac{-n^2}{1+n^2}$$

44. $f'(x)=\lim\limits_{h\to0}\dfrac{f(x+h)-f(x)}{h}$

$$=\lim_{h\to0}\frac{f(x)+f(h)-f(x)}{h}=\lim_{h\to0}\frac{f(h)}{h}$$

$$=\lim_{h\to0}\frac{(2h^2+3h)g(h)}{h}=\lim_{h\to0}(2h+3)g(h)$$

$$=(0+3)g(0)=3g(0)$$

$$=3\cdot3=9$$

45. Given, $\quad \sqrt{(x+y)}+\sqrt{(y-x)}=a \quad$ ...(i)

$\therefore \quad \sqrt{x+y}-\sqrt{y-x}=\dfrac{2x}{a} \quad$ ...(ii)

Adding Eqs. (i) and (ii), then

$$2\sqrt{x+y}=a+\frac{2x}{a}$$

Squaring, $4x+4y=a^2+\dfrac{4x^2}{a^2}+4x$

$$\therefore \quad 4+4\frac{dy}{dx}=0+\frac{8x}{a^2}+4$$

$$\therefore \quad 0+4\frac{d^2y}{dx^2}=\frac{8}{a^2}$$

$$\therefore \quad \frac{d^2y}{dx^2}=\frac{2}{a^2}$$

46. $\because \sin(x+y) = \log_e(x+y)$

$$\cos(x+y)\left(1+\frac{dy}{dx}\right) = \frac{1}{(x+y)}\left(1+\frac{dy}{dx}\right)$$

$$\Rightarrow \left(1+\frac{dy}{dx}\right)\left(\cos(x+y) - \frac{1}{x+y}\right) = 0$$

$$\because \cos(x+y) - \frac{1}{(x+y)} \neq 0$$

$$\therefore 1+\frac{dy}{dx} = 0 \Rightarrow \frac{dy}{dx} = -1$$

47. $x^{\cos y} + y^{\cos x} = 5$

$$\Rightarrow e^{\cos y \log_e x} + e^{\cos x \log_e y} = 5$$

$$\therefore e^{\cos y \log_e x}\left\{\frac{\cos y}{x} - \log_e x \sin y \frac{dy}{dx}\right\}$$

$$+ e^{\cos x \log_e y}\left\{\frac{\cos x}{y}\frac{dy}{dx} - \sin x \log_e y\right\} = 0$$

Put $x = y = 1$,

then, $(\cos 1 - 0) + \left(\cos 1 \dfrac{dy}{dx} - 0\right) = 0$

$$\therefore \frac{dy}{dx} = -1 \text{ or } y' = -1$$

48. $f'(x) = n(1+x)^{n-1}$, $f''(x) = n(n-1)(1+x)^{n-2}$

$$\therefore f^n(x) = n!,\ f^n(0) = n!$$

$$\Rightarrow 1 + n + \frac{n(n-1)}{2!} + \ldots + \frac{n!}{n!}$$

$$= {}^nC_0 + {}^nC_1 + {}^nC_2 + \ldots + {}^nC_n = 2^n$$

49. Let $P(x) = ax^4 + bx^3 + cx^2 + dx + e$

$$P(2) = 16a + 8b + 4c + 2d + e = -1 \quad \ldots(\text{i})$$

$$P'(x) = 4ax^3 + 3bx^2 + 2cx + d$$

$$P'(2) = 32a + 12b + 4c + d = 0 \quad \ldots(\text{ii})$$

$$P''(x) = 12ax^2 + 6bx + 2c$$

$$P''(2) = 48a + 12b + 2c = 2 \quad \ldots(\text{iii})$$

$$\therefore P'''(x) = 24ax + 6b$$

$$\therefore P'''(2) = 48a + 6b = -12 \quad \ldots(\text{iv})$$

and $P^{iv}(x) = 24a \Rightarrow P^{iv}(2) = 24a \Rightarrow 24 = 24a$

$$\therefore a = 1,$$

From Eq. (iv), $b = -10$

From Eq. (iii), $c = 37$

$$\therefore P''(x) = 12x^2 - 60x + 74$$

$$\therefore P''(1) = 12 - 60 + 74 = -48 + 74 = 26$$

50. $\because 2 < x < 3$

$$\therefore [x] = 2$$

Then, $f(x) = \sin\left\{\dfrac{2\pi}{3} - x^2\right\}$

$$\therefore f'(x) = \cos\left\{\frac{2\pi}{3} - x^2\right\}(-2x)$$

$$\therefore f'\left(\sqrt{\frac{\pi}{3}}\right) = \cos\left(\frac{2\pi}{3} - \frac{\pi}{3}\right)\left(-2\sqrt{\frac{\pi}{3}}\right)$$

$$= \frac{1}{2}\left(-2\sqrt{\frac{\pi}{3}}\right) = -\sqrt{\frac{\pi}{3}}$$

51. $\because y = \int_0^x f(t)\sin\{k(x-t)\}\,dt$

$$= \text{In}\int_0^x f(t)\ e^{ik(x-t)}\,dt$$

$$= \ln\left\{e^{ikx}\int_0^x f(t)\,e^{-ikt}\,dt\right\} \quad \ldots(\text{i})$$

$$\therefore \frac{dy}{dx} = \ln\left\{e^{ikx}\,.\,f(x)\,e^{-ikx} + \int_0^x f(t)\,e^{-ukt}\,dt \times ike^{ikx}\right\}$$

$$= 0 + \ln\left\{ik\cdot e^{ikx}\int_0^x f(t)\,e^{-ikt}\,dt\right\}$$

Now, $\dfrac{d^2y}{dx^2} = \text{In}\left\{ik.\left(e^{ikx}.f(x).\,e^{-ikx} + \int_0^x f(t).\,e^{-ikt}dt \times ike^{ikx}\right)\right\}$

$$= k\,f(x) - k^2\,y \quad \text{[from Eq. (i)]}$$

$$\therefore \frac{d^2y}{dx^2} + k^2y = k\cdot f(x)$$

52. Put $x^3 = \sin\theta$, $y^3 = \sin\phi$, then

$(\cos\theta + \cos\phi) = a(\sin\theta - \sin\phi)$

$$\Rightarrow 2\cos\left(\frac{\theta+\phi}{2}\right)\cos\left(\frac{\theta-\phi}{2}\right) = 2a\cos\left(\frac{\theta+\phi}{2}\right)\sin\left(\frac{\theta-\phi}{2}\right)$$

$$\Rightarrow \cot\left(\frac{\theta-\phi}{2}\right) = a$$

$$\Rightarrow \frac{\theta-\phi}{2} = \cot^{-1}a \Rightarrow \sin^{-1}x^3 - \sin^{-1}y^3 = 2\cot^{-1}a$$

$$\therefore \frac{3x^2}{\sqrt{(1-x^6)}} - \frac{3y^2}{\sqrt{(1-y^6)}}\frac{dy}{dx} = 0 \Rightarrow \frac{dy}{dx} = \frac{x^2}{y^2}\sqrt{\left(\frac{1-y^6}{1-x^6}\right)}$$

$$\therefore f(x,y) = \frac{x^2}{y^2}$$

53. $y = \tan^{-1}\left(\dfrac{\log_e(e/x^2)}{\log_e(ex^2)}\right) + \tan^{-1}\left(\dfrac{3+2\log_e x}{1-6\log_e x}\right)$

$$= \tan^{-1}\left(\frac{1-2\log_e x}{1+2\log_e x}\right) + \tan^{-1}\left(\frac{3+2\log_e x}{1-3\cdot 2\log_e x}\right)$$

$$= \tan^{-1}(1) - \tan^{-1}(2\log_e x) + \tan^{-1}(3) + \tan^{-1}(2\log_e x)$$

$$= \tan^{-1}(1) + \tan^{-1}(3)$$

$$\therefore \frac{dy}{dx} = 0$$

54. $\because x = a\cos\theta \Rightarrow \dfrac{dx}{d\theta} = -a\sin\theta$

and $y = b\sin\theta \Rightarrow \dfrac{dy}{d\theta} = b\cos\theta$

$$\therefore \frac{dy}{dx} = -\frac{b}{a}\cot\theta$$

$$\Rightarrow \frac{d^2y}{dx^2} = \frac{b}{a}\text{cosec}^2\theta\,\frac{d\theta}{dx} = -\frac{b}{a^2}\text{cosec}^3\theta$$

$$\therefore \frac{d^3y}{dx^3} = -\frac{3b}{a^2}\text{cosec}^2\theta(-\text{cosec}\,\theta\cot\theta).\frac{d\theta}{dx}$$

$$= \frac{3b}{a^2}\text{cosec}^3\theta\cot\theta\left(-\frac{1}{a\sin\theta}\right)$$

$$= -\frac{3b}{a^3}\text{cosec}^4\theta\cot\theta$$

55. $\because \quad F(x)=\frac{1}{x^2}\int_4^x \{4t^2-2F'(t)\}\,dt \quad \text{...(i)}$

or $\quad x^2 F(x)=\int_4^x \{4t^2-2F'(t)\}\,dt$

Differentiating both sides w.r.t. x, then

$$x^2 F'(x)+F(x)\cdot 2x=4x^2-2F'(x)$$

Put $\quad x=4$

$$16F'(4)+8F(4)=64-2F'(4)$$

$\therefore \quad 18F'(4)+0=64 \quad [\because F(4)=0, \text{ from Eq. (i)}]$

$\therefore \quad F'(4)=\frac{32}{9}$

56. $\therefore \quad y=\sqrt{x+\sqrt{y+y}} \Rightarrow (y^2-x)=\sqrt{2y}$

or $\quad (y^2-x)^2=2y$

Differentiating both sides w.r.t. x, then

$$2(y^2-x)\left(2y\frac{dy}{dx}-1\right)=2\frac{dy}{dx}$$

$\therefore \quad \frac{dy}{dx}=\frac{(y^2-x)}{2y^3-2xy-1}$

57. Let $y=\cos^{-1}\left(\frac{x^{-1}-x}{x^{-1}+x}\right)$

$\therefore \quad \frac{dy}{dx}=\frac{-1}{\sqrt{1-\left(\frac{x^{-1}-x}{x^{-1}+x}\right)^2}}\times\frac{(x^{-1}+x)(-x^{-2}-1)-(x^{-1}-x)(-x^{-2}+1)}{(x^{-1}+x)^2}$

$$=\frac{-1\,(x^{-1}+x)(-x^{-2}-1)-(x^{-1}-x)(-x^{-2}+1)}{2|x^{-1}+x|}$$

$\therefore \quad \left.\frac{dy}{dx}\right|_{x=-1}=\frac{-\{(-1-1)(-1-1)-0\}}{2|-1-1|}$

$$=-1$$

58. Let $y=f(x)=\exp(x^3+x^2+x)$

$$=e^{x^3+x^2+x}$$

$\therefore \quad \frac{dy}{dx}=f'(x)=e^{x^3+x^2+x}.(3x^2+2x+1)$

$\Rightarrow \quad g'(y)=\frac{dx}{dy}=\frac{1}{e^{x^3+x^2+x}.(3x^2+2x+1)} \quad \text{... (i)}$

When, $\quad y=e^3$, then $e^3=e^{x^3+x^2+x}$

$\Rightarrow \quad 3=x^3+x^2+x$

$\Rightarrow x^3+x^2+x-3=0$

$\Rightarrow \quad (x-1)(x^2+2x+3)=0$

$\therefore \quad x-1=0 \quad [\because x^2+2x+3\neq 0]$

$\Rightarrow \quad x=1$

Now, put $y=e^3$ and $x=1$ in Eq. (i),we get

$$g'(e^3)=\frac{1}{e^{1+1+1}.(3+2+1)}=\frac{1}{6e^3}$$

Objective Questions Type II [One or more than one correct answer(s)]

1. $f(x)=-\frac{x^3}{3}+x^2\sin 1.5a-x\sin a\sin 2a-5\sin^{-1}(a^2-8a+17)$

$f(x)$ is defined when

$$-1\le a^2-8a+17\le 1$$

$$-1\le(a-4)^2+1\le 1\Rightarrow a=4$$

$\therefore \quad f(x)=-\frac{x^3}{3}+x^2\sin 6-x\sin 4\sin 8-\frac{5\pi}{2}$

$\Rightarrow \quad f'(x)=-x^2+2x\sin 6-\sin 4\sin 8$

Now, $f'(\sin 8)=-\sin^2 8+2\sin 8\sin 6-\sin 4\sin 8$

$$=\sin 8\,[2\sin 6-(\sin 8+\sin 4)]$$

$$=\sin 8\,[2\sin 6-2\sin 6\cos 2]$$

$$=2\sin 6\Rightarrow 6\sin 8\,(1-\cos 2)$$

Since, $\sin 6<0$, $\sin 8>0$, $1-\cos 2>0$

$\therefore \quad f'(\sin 8)<0$

2. Given, $f(x)=x^3+x^2f'(1)+xf''(2)+f'''(3)$

$\Rightarrow \quad f'(x)=3x^2+2xf'(1)+f''(2)$

$\Rightarrow \quad f'(1)=3+2f'(1)+f''(2)$

$\Rightarrow f'(1)+f''(2)=-3 \quad \text{...(i)}$

$\Rightarrow$ and $\quad f''(x)=6x+2f'(1)$

$\therefore \quad f''(2)=12+2f'(1)$

$\therefore \quad -2f'(1)+f''(2)=12 \quad \text{... (ii)}$

Solving Eqs. (i) and (ii) we get

$$f'(1)=-5 \text{ and } f''(2)=2$$

and $\quad f'''(x)=6$

$\therefore \quad f'''(3)=6 \quad \text{...(iii)}$

Substituting the values of $f'(1)$, $f''(2)$ and $f'''(3)$ from Eqs. (i), (ii) and (iii) in $f(x)$

$\therefore \quad f(x)=x^3-5x^2+2x+6$

$\Rightarrow \quad f(0)=6,\ f(1)=4,\ f(2)=-2,\ f(3)=-6$

Hence, $\quad f(0)+f(2)=f(1)$

$$f(0)+f(3)=0$$

and $\quad f(1)+f(3)=f(2)$

3. $\quad y=\log_2(\log_2 x)$

$\therefore \quad \frac{dy}{dx}=\frac{1}{\log_2 x}\times\log_2 e\times\frac{1}{x}\log_2 e$

$$=\frac{1}{x\log_2 x\cdot\log_e 2\cdot\log_e 2}$$

$$=\frac{1}{x\log_e x\log_e 2}$$

$$=\frac{\log_2 e}{x\log_e x}$$

4. $y = \dfrac{(\sqrt{1+t^2} - \sqrt{1-t^2})^2}{(\sqrt{1+t^2})^2 - (\sqrt{1-t^2})^2}$

$= \dfrac{2 - 2\sqrt{1-t^4}}{2t^2} = \dfrac{1-\sqrt{1-t^4}}{t^2}$

$\therefore \quad \dfrac{dy}{dt} = \dfrac{t^2\left\{0 - \dfrac{1}{2\sqrt{1-t^4}} \times -4t^3\right\} - \{1-\sqrt{1-t^4}\}\,2t}{t^4}$

$= \dfrac{\dfrac{2t^5}{\sqrt{1-t^4}} - 2t\,\{1-\sqrt{1-t^4}\}}{t^4} = \dfrac{2\{1-\sqrt{1-t^4}\}}{t^3\sqrt{1-t^4}}$

$\dfrac{dx}{dt} = \dfrac{-2t^3}{\sqrt{1-t^4}}$

$\therefore \quad \dfrac{dy}{dx} = \dfrac{\sqrt{1-t^4}-1}{t^6} = \dfrac{-1}{t^2\{1+\sqrt{1-t^4}\}}$

5. Given, $f : (0, \infty) \to (-\infty, \infty)$ be defined as $f(x) = e^x + \ln x$

Also, $f'(x) = e^x + \dfrac{1}{x} > 0 \; \forall \in (0, \infty)$

Also, $\lim\limits_{x \to +\infty} f(x) = \infty$ and $\lim\limits_{x \to 0^+} f(x) = -\infty$

$\Rightarrow f(x)$ is a bijective mapping

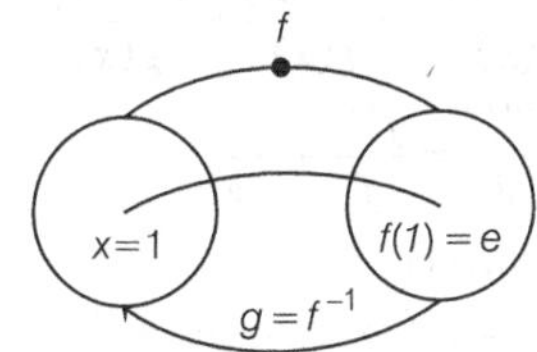

As $\quad f(1) = e \Rightarrow g(e) = 1$

Also, $\quad g'(e) = \dfrac{1}{f'(1)}$

$f'(x) = e^x + \dfrac{1}{x}$

$\Rightarrow \quad f'(1) = (e+1)$

$f''(x) = e^x - \dfrac{1}{x^2}$

$\Rightarrow \quad f''(1) = (e-1)$

So, $\quad g'(e) = \dfrac{1}{e+1}$

Also, $\quad g''(e) = \dfrac{-f''(1)}{(f''(1))^3} \quad \left[\text{As, } g''(x) = \dfrac{-f''(x)}{(f'(x))^3}\right]$

$= \dfrac{-(e-1)}{(e+1)^3} = \dfrac{1-e}{(e+1)^3}$

6. $f(x) = \sin^{-1}\left(\dfrac{2x}{1+x^2}\right)$

$f'(x) = \dfrac{1}{\sqrt{1-\left(\dfrac{2x}{1+x^2}\right)^2}} \cdot \dfrac{(1+x^2)(2-2x)2x}{(1+x^2)^2}$

$= \dfrac{1}{\sqrt{1+x^4-2x^2}} \cdot \dfrac{2(1-x^2)}{(1+x^2)} = \dfrac{2(1-x^2)}{|1-x^2|\,(1+x^2)}$

For $\quad x^2 = 1$ i.e., $x = \pm 1$

$f'(x) = \dfrac{0}{0}$ not defined

For $\quad x^2 < 1$ or $|x| < 1$

$f'(x) = \dfrac{-2}{1+x^2}$

and for $x^2 > 1$ or $|x| > 1$

$f'(x) = \dfrac{2}{1+x^2}$

7. $\quad f(x) = x^2 + xg'(1) + g''(2)$

and $\quad g(x) = x^2 + xf'(2) + f''(3)$

$f'(x) = 2x + g'(1),\; f''(x) = 2$

$g'(x) = 2x + f'(2),\; g''(x) = 2$

At $\quad x = 1,\; f'(1) = 2 + g'(1)$ and $g'(1) = 2 + f'(2)$

$\Rightarrow \quad f'(1) = 4 + f'(2)$

Putting $\quad x = 2,\; f'(2) = 4 + g'(1),\; g'(2) = 4 + f'(2)$

$\therefore \quad g'(2) = 4 + 4 + g'(1) = 8 + g'(1)$

and $\quad g''(2) + f''(3) = 2 + 2 = 4$

8. $\because \quad f(x-y),\; f(x)\cdot f(y),\; f(x+y)$ are in AP

$\therefore \quad 2f(x)\,f(y) = f(x-y) + f(x+y) \quad \ldots\text{(i)}$

Replacing x by y and y by x, then

$2f(y)\,f(x) = f(y-x) + f(y+x) \quad \ldots\text{(ii)}$

From Eqs. (i) and (ii), we get

$f(x-y) = f(y-x)$

Put $y = 0$, then $f(x) = f(-x) \Rightarrow f(2) = f(-2)$

Differentiating w.r.t. x, then

$f'(x) = -f'(-x)$

$\Rightarrow \quad f'(x) + f'(-x) = 0 \Rightarrow f'(2) + f'(-2) = 0$

9. $\because \quad f_n(x) = e^{f_{n-1}(x)} \quad \ldots\text{(i)}$

$\therefore \quad \dfrac{d}{dx}\{f_n(x)\} = e^{f_{n-1}(x)} \cdot \dfrac{d}{dx} f_{n-1}(x)$

$= f_n(x) \cdot \dfrac{d}{dx}(f_{n-1}(x))$ (Alternate (a))

$= f_n(x)\cdot f_{n-1}(x) \ldots\ldots f_2(x)\cdot f_1(x) \dfrac{d}{dx}(f_0(x))$

$= f_n(x)\cdot f_{n-1}(x) \ldots\ldots f_2(x)\cdot f_1(x)\cdot 1 = 1$

$(\because \; f_0(x) = 1)$

$= \prod\limits_{i=1}^{n} f_i(x) \quad \begin{bmatrix}\text{alternate (c)} \\ \text{alternate (d)}\end{bmatrix}$

10. $f(\theta) = \tan\left(\sin^{-1}\sqrt{\dfrac{2}{3+\cos 2\theta}}\right)$

$f(\theta) = \sqrt{\dfrac{2}{1+\cos 2\theta}} = \sec\theta$

$\Rightarrow \quad f\left(\dfrac{\pi}{4}\right) = \sqrt{2}$

$f'(\theta) = \sec\theta \tan\theta$

$f'\left(\dfrac{\pi}{4}\right) = \sqrt{2}$

(Right triangle: hypotenuse $\sqrt{3+\cos 2\theta}$, perpendicular $\sqrt{2}$, base $\sqrt{1+\cos 2\theta}$)

$\left.\dfrac{d(f(\theta))}{d(\cos\theta)}\right|_{\theta=\frac{\pi}{4}} = \left(\dfrac{-1}{\cos^2\theta}\right)_{\theta=\frac{\pi}{4}} = -2$

11. Given, $2^x + 2^y = 2^{x+y}$ ... (i)

$\Rightarrow 2^{-y} + 2^{-x} = 1$

Differentiating both sides w.r.t. 'x', we have

$$2^{-y}.\ln 2\left(-\frac{dy}{dx}\right) + 2^{-x}\ln 2(-1) = 0$$

$$\therefore \quad \frac{dy}{dx} = -\frac{2^{-x}}{2^{-y}} = -\frac{2^y}{2^x} \quad \text{...(ii)}$$

From Eq. (i), $1 + \frac{2^y}{2^x} = 2^y$ (Dividing both sides by 2^x)

$$\therefore \quad 1 - 2^y = -\frac{2^y}{2^x} \quad \text{...(iii)}$$

From Eqs. (ii) and (iii), we get

$$\frac{dy}{dx} = 1 - 2^y$$

Also from Eq. (i), $\frac{2^x}{2^y} + 1 = 2^x$

[Dividing both sides by 2^y]

$$\therefore \quad -\frac{2^y}{2^x} = \frac{1}{1-2^x} \quad \text{...(iv)}$$

From Eqs. (ii) and (iv), we get

$$\frac{dy}{dx} = \frac{1}{1-2^x}$$

Now from Eq. (ii),

$$\frac{dy}{dx} = -\frac{2^{-x}}{2^{-y}} = -\frac{(1-2^{-y})}{(1-2^{-x})} \quad \text{[from Eq. (i)]}$$

$$= -\frac{2^x(2^y-1)}{2^y(2^x-1)} = \frac{2^x(1-2^y)}{2^y(2^x-1)}$$

12. $y = \sin^{-1}(\sin x)$

$$= x, -\frac{\pi}{2} \le x \le \frac{\pi}{2}$$

$$= \pi - x, \frac{\pi}{2} < x \le \frac{3\pi}{2}$$

$$\therefore \quad \frac{dy}{dx} = 1, -\frac{\pi}{2} < x < \frac{\pi}{2},$$

$$= -1, \frac{\pi}{2} < x < \frac{3\pi}{2}$$

13. Given, $F(x) = f(x) \cdot g(x)$ ... (i)

Differentiating both sides w.r.t. x, we get

$$F'(x) = f'(x) \cdot g(x) + g'(x) \cdot f(x)$$

$$\Rightarrow \quad F'(x) = f'(x)g'(x)\left[\frac{f(x)}{f'(x)} + \frac{g(x)}{g'(x)}\right]$$

$$\Rightarrow \quad F' = c'\left[\frac{f}{f'} + \frac{g}{g'}\right]$$

$\Rightarrow$ (a) is correct

Again differentiating both sides w.r.t. x, we get

$$F''(x) = f''(x)g(x) + g''(x) \cdot f(x) + 2f'(x) \cdot g'(x)$$

$$\Rightarrow F''(x) = f''(x) \cdot g(x) + g''(x) \cdot f(x) + 2c \quad \text{...(ii)}$$

Dividing both sides by

$$F(x) = f(x) \cdot g(x) \quad [\because f'(x) \cdot g'(x) = c]$$

$$\text{Then} \quad \frac{F''(x)}{F(x)} = \frac{F''(x)}{f(x)} + \frac{g''(x)}{g(x)} + \frac{2c}{f(x)g(x)}$$

$$\text{or} \quad \frac{F''}{F} = \frac{f''}{f} + \frac{g''}{g} + \frac{2c}{fg}$$

$\Rightarrow$ (b) is correct

Again given, $f'(x)g'(x) = c$

Differentiating both sides w.r.t. x, we get

$$f'(x)g''(x) + g'(x)f''(x) = 0$$

From Eq. (ii)

$$F''(x) = f''(x) \cdot g(x) + g''(x) \cdot f(x) + 2c$$

Differentiating both sides w.r.t. x, we get

$$F'''(x) = f''(x) \cdot g'(x) + f'''(x) \cdot g(x) + g''(x) \cdot f'(x) + f(x) \cdot g'''(x) + 0$$

$$= f'''(x) \cdot g(x) + g'''(x) \cdot f(x) + 0$$

Now, dividing both sides by $F(x) = f(x)g(x)$

$$\text{Then,} \quad \frac{F'''(x)}{F(x)} = \frac{f'''(x)}{f(x)} + \frac{g'''(x)}{g(x)}$$

$$\text{or} \quad \frac{F'''}{F} = \frac{f'''}{f} + \frac{g'''}{g}$$

$\Rightarrow$ (c) is correct

14. Let $f(x) = ax^3 + bx^2 + bx + d$

Then $f(1) = 0$ and $f'(1) = 0$

$$\Rightarrow \quad a + 2b + d = 0 \quad \text{...(i)}$$

$$\text{and} \quad 3a + 2b + b = 0 \quad \text{...(ii)}$$

From Eqs. (i) and (ii), $a = d = -b$

15. $u = \sin^{-1} x$

$$v = \sin^{-1}(3x - 4x^3) = \begin{cases} -\pi - 3\sin^{-1} x, & -1 \le x \le -\frac{1}{2} \\ 3\sin^{-1} x, & -\frac{1}{2} \le x \le \frac{1}{2} \\ \pi - 3\sin^{-1} x, & \frac{1}{2} \le x \le 1 \end{cases}$$

$$\therefore \quad \frac{du}{dv} = \begin{cases} -\frac{1}{3}, & -1 \le x \le -\frac{1}{2} \\ \frac{1}{3}, & -\frac{1}{2} \le x \le \frac{1}{2} \\ -\frac{1}{3}, & \frac{1}{2} \le x \le 1 \end{cases}$$

Linked Comprehension Type

PASSAGE 1

1. $D^*(u\cdot v)=D^*(f(x)\cdot g(x))$

$$=\lim_{h\to 0}\frac{f^2(x+h)g^2(x+h)-f^2(x)g^2(x)}{h}$$

$$=\lim_{h\to 0} f^2(x+h)\left\{\frac{g^2(x+h)-g^2(x)}{h}\right\}+g^2(x)\frac{\{f^2(x+h)-f^2(x)\}}{h}$$

$$=f^2(x)D^*g(x)+g^2(x)D^*f(x)$$

$$=u^2D^*v+v^2D^*u$$

2. $$D^*\left\{\frac{u}{v}\right\}=D^*\left\{\frac{f(x)}{g(x)}\right\}=\lim_{h\to 0}\frac{\left\{\frac{f(x+h)}{g(x+h)}\right\}^2-\left\{\frac{f(x)}{g(x)}\right\}^2}{h}$$

$$=\lim_{h\to 0}\frac{f^2(x+h)\,g^2(x)-g^2(x+h)f^2(x)}{h\,g^2(x+h)g^2(x)}$$

$$=\lim_{h\to 0}\frac{g^2(x)\frac{\{f^2(x+h)-f^2(x)\}}{h}-f^2(x)\frac{\{g^2(x+h)-g^2(x)\}}{h}}{g^2(x+h)g^2(x)}$$

$$=\frac{g^2(x)D^*f(x)-f^2(x)D^*g(x)}{g^4(x)}$$

$$=\frac{v^2D^*u-u^2D^*v}{v^4}$$

3. $$D^*(\tan x)=\lim_{h\to 0}\frac{\tan^2(x+h)-\tan^2 x}{h}$$

$$=\lim_{h\to 0}\frac{\tan(x+h)-\tan x}{h}\lim_{h\to 0}(\tan(x+h)+\tan x)$$

$$=\sec^2 x\,.\,2\tan x$$

$$=2\tan x\sec^2 x$$

4. Given, $f'(x)=0$

Now, $$D^*f(x)=\lim_{h\to 0}\frac{f^2(x+h)-f^2(x)}{h}$$

$$=\lim_{h\to 0}\frac{f(x+h)-f(x)}{h}\lim_{h\to 0}(f(x+h)+f(x))$$

$$=f'(x)\cdot 2f(x)=0$$

5. $$D^*c=\lim_{h\to 0}\frac{c^2-c^2}{h}=0$$

PASSAGE 2

1. $\because f(x)=x^2+g'(0)x+g''(3)$

Let $g'(0)=a$ and $g''(3)=b$

$\therefore\quad f(x)=x^2+ax+b$

$f(-2)=4-2a+b,\ f'(x)=2x+a$

and $f''(x)=2$

$\because\quad g(y)=f(-2)y^2+yf''(y)+f'(y)-2$

$=(4-2a+b)y^2+2y+2y+(a-2)$

$=(4-2a+b)y^2+4y+(a-2)$

$\Rightarrow\quad g'(y)=2(4-2a+b)y+4$

and $g''(y)=2(4-2a+b)$

$\because\quad g''(3)=b\Rightarrow 2(4-2a+b)=b\Rightarrow 4a-b=8$

$g'(0)=a\Rightarrow 4=a$

$\therefore\quad 16-b=8\Rightarrow b=8$

$\therefore\quad f(x)=x^2+4x+8$ and $g(y)=4y^2+4y+2$

$\therefore\quad f(1)=1+4+8=13$

2. $$h(x)=\begin{cases}x^2+4x+8, & x\ge 0\\ 4x^2+4x+2, & x<0\end{cases}$$

$\because$ LHL (at $x=0$) $=2$ and RHL (at $x=0$) $=8=f(0)$

$\Rightarrow h(x)$ is discontinuous at $x=0$

3. $\because\quad f(\alpha)=\alpha^2+4\alpha+8=(\alpha+2)^2+4\ge 4$

and $\quad g(\beta)=4\beta^2+4\beta+2=(2\beta+1)^2+1\ge 1$

$\Rightarrow\quad f(\alpha)g(\beta)\ge 4$

But given $f(\alpha)\,g(\beta)\le 4$

So, $\quad f(\alpha)\,g(\beta)=4$

$\Rightarrow\quad \alpha=-2$ and $\beta=\frac{-1}{2}$

Hence, number of non-positive ordered pairs $=1$

PASSAGE 3

1. $\because\quad y=e^{3x+7}$

$\therefore\quad y_1=3e^{3x+7},\ y_2=3^2e^{3x+7}\ \ldots$

$\therefore\quad y_n(x)=3^n\cdot e^{3x+7}$

Then, $\ y_n(0)=3^n\cdot e^7$

2. $\because\quad y=(2-3x)^{-1}$

$\therefore\quad y_1=(-1)(2-3x)^{-2}(-3)$

$y_2=(-1)(-2)(2-3x)^{-3}(-3)^2$

$y_3=(-1)(-2)(-3)(2-3x)^{-4}\cdot(-3)^3$

....

$y_n=(-1)^n\cdot n!(2-3x)^{-n-1}(-3)^n$

$\therefore\quad y_n(1)=(-1)^n\cdot n!(-1)^{-n-1}(-3)^n$

$=(-1)^{n+1}\cdot 3^n\cdot n!$

3. $\because\quad y=\frac{\ln x}{x}$

$\therefore\quad y'=\frac{(1-\ln x)}{x^2}$

$$\therefore\quad y''=\frac{x^2\left(-\frac{1}{x}\right)-(1-\ln x)2x}{x^4}$$

$\therefore\ y''(e)=\frac{-e-0}{e^4}=-\frac{1}{e^3}$

4. $$\because\ \frac{dy}{dx}=\frac{\left(\frac{dy}{dt}\right)}{\left(\frac{dx}{dt}\right)}=\frac{k\cos kt}{\cos t}$$

$$y_1=\frac{k\cos kt}{\cos t}$$

On squaring both sides, we get

$$(1-\sin^2 t)y_1^2 = k^2(1-\sin^2 kt)$$

$$\Rightarrow \quad (1-x^2)y_1^2 = k^2(1-y^2)$$

Differentiating both sides w.r.t. x, then

$$(1-x^2)\,2y_1y_2 + y_1^2(-2x) = 0 - 2k^2yy_1$$

Dividing by $2y$, then

$$(1-x^2)y_2 - xy_1 = -k^2y$$

5. $\because y = \tan^{-1} x$

$$\Rightarrow \quad y_1 = \frac{1}{(1+x^2)} = \frac{1}{2i}\left(\frac{1}{x-i} - \frac{1}{x+i}\right)$$

$$= \frac{1}{2i}\{(x-i)^{-1} - (x+i)^{-1}\}$$

$\therefore \quad y_n(x) = (n-1)$th derivative of y_1

$$= \frac{1}{2i}\left\{\frac{(-1)^{n-1}(n-1)!}{(x-i)^n} - \frac{(-1)^{n-1}(n-1)!}{(x+i)^n}\right\}$$

$$\therefore \quad y_n(0) = \frac{(-1)^{n-1}\cdot(n-1)!}{2i}\left\{\frac{1}{(-i)^n} - \frac{1}{(i)^n}\right\}$$

But $\quad n = 4p+3$

$$\text{Then, } y_n(0) = \frac{(-1)^{4p+2}(4p+2)!}{2i}\left\{\frac{1}{(-i)^{4p+3}} - \frac{1}{(i)^{4p+3}}\right\}$$

$$= \frac{1}{2i}\cdot(4p+2)!\left\{\frac{1}{i} + \frac{1}{i}\right\}$$

$$= -(4p+2)! = -(n-1)!$$

Numerical Value Type

1. $\because \quad y = \tan^{-1}\left(\frac{ax-b}{bx+a}\right) = \tan^{-1}\left(\frac{x-\frac{b}{a}}{1+\frac{b}{a}\cdot x}\right)$

$$= \tan^{-1} x - \tan^{-1}\left(\frac{b}{a}\right)$$

$$\therefore \quad \frac{dy}{dx} = \frac{1}{1+x^2} - 0$$

$$\therefore \quad \left.\frac{dy}{dx}\right|_{x=-1} = \frac{1}{1+(-1)^2} = \frac{1}{1+1} = \frac{1}{2}$$

$$\Rightarrow \quad (2008)\left.\frac{dy}{dx}\right|_{x=-1} = 2008\times\frac{1}{2} = 1004$$

2. We have $(c+dx)y = a+bx$

Differentiating both sides w.r.t. x, we get

$$(c+dx)y_1 + y\cdot d = b$$

$$\frac{c}{d}y_1 + xy_1 + y = \frac{b}{a}$$

Again differentiating both sides w.r.t. x, then

$$\frac{c}{d}y_2 + xy_2 + y_1 + y_1 = 0 \Rightarrow -\frac{c}{d} = \left(\frac{2y_1 + xy_2}{y_2}\right)$$

Differentiating both sides w.r.t. x, then

$$0 = \frac{y_2(2y_2 + xy_3 + y_2) - (2y_1 + xy_2)\cdot y_3}{y_2^2}$$

$$\Rightarrow \quad 3y_2^2 = 2y_1y_3$$

$$\therefore \quad \lambda = 2, \mu = 3$$

Then, $\quad \mu^{\lambda^2} = 3^4 = 81$

3. $f(x) = \sin\left(\cos^{-1}\left(\frac{1-(2^x)^2}{1+(2^x)^2}\right)\right)$

Let $2^x = \tan\theta$

$$f(x) = \sin\left(\cos^{-1}\left(\frac{1-\tan^2\theta}{1+\tan^2\theta}\right)\right)$$

$$f(x) = \sin(\cos^{-1}\cos 2\theta)$$

$$f(x) = \sin 2\theta = \frac{2\tan\theta}{1+\tan^2\theta} \Rightarrow f(x) = \frac{2\times 2^x}{1+(2x)^2} = \frac{2^{x+1}}{1+2^{2x}}$$

$$f'(x) = \frac{(1+2^{2x})2^{x+1}\log 2 - 2^{x+1}\times 2^{2x}\log 2\times 2}{(1+2^{2x})^2}$$

$$f'(1) = \frac{5\times 4\log 2 - 4\times 4\times\log 2\times 2}{(1+4)^2}$$

$$= \frac{20\log 2 - 32\log 2}{25} = -\frac{12}{25}\log 2$$

So, $a = 25,\ b = 12$

$$|a^2 - b^2| = 25^2 - 12^2 = 625 - 144 = 481$$

4. $2 = t^2 + 3t - 8$ and $-1 = 2t^2 - 2t - 5$

$\Rightarrow \quad t^2 + 3t - 10 = 0$ and $2t^2 - 2t - 4 = 0$

or $\quad (t+5)(t-2) = 0$ and $\quad t^2 - t - 2 = 0$

or $\quad [t = -5, 2$ and $t = 2, -1 \Rightarrow t = 2$

$$\therefore \quad \lambda = \left.\frac{dy}{dx}\right|_{at\,(2,-1)} = \left.\frac{\frac{dy}{dt}}{\frac{dx}{dt}}\right|_{t=2}$$

$$= \left.\frac{4t-2}{2t+3}\right|_{t=2} = \frac{8-2}{4+3} = \frac{6}{7}$$

$$\therefore 2800\lambda = 2800\times\frac{6}{7} = 400\times 6 = 2400$$

5. $$f'(x) = \begin{vmatrix}\cos x & -\sin x\\ \tan x & \cot x\end{vmatrix} + \begin{vmatrix}\sin x & \cos x\\ \sec^2 x & -\operatorname{cosec}^2 x\end{vmatrix}$$

$$\therefore f'(\pi/4) = \begin{vmatrix}\frac{1}{\sqrt{2}} & -\frac{1}{\sqrt{2}}\\ 1 & 1\end{vmatrix} + \begin{vmatrix}\frac{1}{\sqrt{2}} & \frac{1}{\sqrt{2}}\\ 2 & -2\end{vmatrix}$$

$$= \frac{2}{\sqrt{2}} - \frac{2}{\sqrt{2}} - \frac{2}{\sqrt{2}} = -\sqrt{2}$$

$$\therefore \quad -25.6\sqrt{2}f'(\pi/4) = -25.6\sqrt{2}\times -\sqrt{2} = 51.2$$

6. $\because$ Differentiation of $\sin x$ and $\cos x$ repeat after fourth order

$$\therefore \quad \frac{d^{16}y}{dx^{16}} = \sin x - \cos x$$

$$\Rightarrow \quad f(x) = \frac{d^{17}y}{dx^{17}} = \cos x + \sin x$$

$$\Rightarrow f\left(\frac{\pi}{4}\right) = \cos\frac{\pi}{4} + \sin\frac{\pi}{4} = \frac{1}{\sqrt{2}} + \frac{1}{\sqrt{2}} = \sqrt{2}$$

$$\therefore \left(\frac{1024}{\sqrt{2}}\right)f\left(\frac{\pi}{4}\right) = \left(\frac{1024}{\sqrt{2}}\right)(\sqrt{2}) = 1024$$

7. $\because \quad \dfrac{dy}{dx} = \dfrac{\frac{dy}{dt}}{\frac{dx}{dt}} = \dfrac{6t}{3t^2} = \dfrac{2}{t}$

$\therefore \quad \dfrac{d^2y}{dx^2} = \dfrac{d}{dx}\left(\dfrac{dy}{dx}\right) = \dfrac{d}{dx}\left(\dfrac{2}{t}\right) = \dfrac{d}{dt}\left(\dfrac{2}{t}\right)\cdot\dfrac{dt}{dx}$

$$= \left(-\frac{2}{t^2}\right)\left(\frac{1}{3t^2}\right) = \frac{-2}{3t^4}$$

$$\Rightarrow \quad \frac{\left(\frac{d^2y}{dx^2}\right)}{\left(\frac{dy}{dx}\right)^n} = \frac{-\frac{2}{3t^4}}{\left(\frac{2}{t}\right)^n} = -\frac{2}{3(2^n)}t^{n-4}$$

For $n = 4$

$$\frac{\left(\frac{d^2y}{dx^2}\right)}{\left(\frac{dy}{dx}\right)^n} \text{ is constant.}$$

$\therefore \; 343n = 343 \times 4 = 1372$

8. $\dfrac{dy}{dx} = \dfrac{2}{\sqrt{1-\left(\frac{x-2}{\sqrt{6}}\right)^2}}\cdot\dfrac{1}{\sqrt{6}} - \dfrac{(4-2x)}{2\sqrt{(2+4x-x^2)}}$

$\therefore \quad \left.\dfrac{dy}{dx}\right|_{x=2} = \dfrac{2}{\sqrt{1}}\cdot\dfrac{1}{\sqrt{6}} - 0 = \sqrt{\dfrac{2}{3}}$

$\Rightarrow \quad 2008\sqrt{\dfrac{2}{3}}\left.\dfrac{dy}{dx}\right|_{x=2} = 2008$

Single Integer Type

1. $\because \quad f(x) = (ax + e^x)g(x)$

$\therefore \quad f'(x) = (ax + e^x).\,g'(x) + g(x)(a + e^x)$

$\Rightarrow \quad f'(0) = g'(0) + g(0)(a + 1)$

$\Rightarrow \quad 18 = -2 + 5(a + 1)$

[$\because g(0) = 5, g'(0 = -2$ and $f'(0) = 18$]

$\Rightarrow \quad a + 1 = 4 \quad \therefore \quad a = 3$

2. $y^{2/3} - 2xy^{1/3} + 1 = 0,\; y^{1/3} = (x \pm \sqrt{x^2-1})$

$\Rightarrow \quad \ln y = 3\ln(x \pm \sqrt{x^2-1})$

$\dfrac{y_1}{y} = \dfrac{\pm 3}{\sqrt{x^2-1}} \Rightarrow (x^2-1)y_1^2 = 9y^2$

$2xy_1^2 + (x^2-1)2y_1y_2 = 18yy_1$

$xy_1 + (x^2-1)y_2 = 9y$

[As y_1 is not equal to 0, because y is not constant]

Dividing by y, we get

$\therefore x\dfrac{y_1}{y} + (x^2-1)\dfrac{y_2}{y} = 9$

3. $y = \tan^{-1}\dfrac{4x}{1+5x^2} + \tan^{-1}\dfrac{2+3x}{3-2x}$

$= \tan^{-1}\dfrac{5x - x}{1 + 5x} + \tan^{-1}\dfrac{\frac{2}{3}+x}{1-\frac{2}{3}\cdot x}$

$= \tan^{-1}5x - \tan^{-1}x + \tan^{-1}\dfrac{2}{3} + \tan^{-1}x$

$= \tan^{-1}5x + \tan^{-1}\dfrac{2}{3}$

$\therefore \quad \dfrac{dy}{dx} = \dfrac{5}{1+25x^2}$

On comparing with $\dfrac{dy}{dx} = \dfrac{\lambda}{1+25x^2}$, we get $\lambda = 5$

4. $\because f(x) = \sin^{-1}\left(\dfrac{2x+2}{\sqrt{4x^2+8x+13}}\right)$

$= \sin^{-1}\left(\dfrac{2x+2}{\sqrt{(2x+2)^2+3^2}}\right)$

put $2x + 2 = 3\tan\theta$, then

$f(x) = \sin^{-1}\left(\dfrac{3\tan\theta}{3\sec\theta}\right) = \sin^{-1}(\sin\theta) = \theta$

$= \tan^{-1}\left(\dfrac{2x+2}{3}\right)$

$\Rightarrow \tan f(x) = \dfrac{2x+2}{3}$

$\therefore \quad \dfrac{d(\tan f(x))}{d(\tan^{-1}x)} = \dfrac{d\left(\frac{2x+2}{3}\right)}{d(\tan^{-1}x)}$

$= \dfrac{\frac{d\left(\frac{2x+2}{3}\right)}{dx}}{\frac{d(\tan^{-1}x)}{dx}} = \dfrac{\frac{2}{3}}{\frac{1}{1+x^2}}$

$= \dfrac{2}{3}(1 + x^2)$

$\Rightarrow \quad \left.\dfrac{d(\tan^{-1}f(x))}{d(\tan^{-1}x)}\right|_{x=\frac{1}{\sqrt{2}}} = \dfrac{2}{3}\left(1 + \dfrac{1}{2}\right) = 1$

5. We have, $g(x) = f\left(\dfrac{x}{f(x)}\right)$

On differentiating w.r.t. x, we get

$g'(x) = f'\left(\dfrac{x}{f(x)}\right) \times \left(\dfrac{f(x) - xf'(x)}{f^2(x)}\right)$

$\therefore \quad g'(1) = f'\left(\dfrac{1}{f(1)}\right) \times \left(\dfrac{f(1) - f'(1)}{f^2(1)}\right)$

As $\quad f(1) = f'(1)$

$\Rightarrow \quad g'(1) = 0$

6. $\cos^{-1}x = \dfrac{\pi}{2} - \sin^{1}x,$

$\sin(\cos^{-1}x) = \cos(\sin^{-1}x)$

$\sin(\cos^{-1}(\sin(\cos^{-1}x))) = \cos(\sin^{-1}(\cos(\sin^{-1}x))) = x$

$f(x) = \tan 2x$

$\dfrac{d(\tan 2x)}{d(\tan x)} = \dfrac{2\sec^2 2x}{\sec^2 x} = 6$

Matrix-Match Type

1. **(A)** $f(x) = \cos^{-1}\left(\frac{2x}{1+x^2}\right) = \frac{\pi}{2} - \sin^{-1}\left(\frac{2x}{1+x^2}\right)$

$\therefore\ f(x) = \frac{\pi}{2} - 2\tan^{-1} x, -1 \le x \le 1 \Rightarrow f'(x) = \frac{-2}{1+x^2}$

So, $f'(x)$ at $x = 0 = \frac{-2}{1+x^2} = -2$ (R)

(B) $g(x) = \cos^{-1}(2x^2 - 1)$

$\therefore g'(x) = \frac{-1}{\sqrt{1-(2x^2-1)^2}} 4(x)$

So, $g'(0)$ does not exist. (S)

(C) $h(x) = \sin^{-1}\left(\frac{1-x^2}{1+x^2}\right) = \frac{\pi}{2} - \cos^{-1}\left(\frac{1-x^2}{1+x^2}\right)$

$$= \begin{cases} \frac{\pi}{2} + 2\tan^{-1} x, & -\infty < x < 0 \\ \frac{\pi}{2} - 2\tan^{-1} x, & x \ge 0 \end{cases}$$

As $h'(0^-) = 2$ and $h'(0^+) = -2$

As LHL $\ne$ RHL

So, $h'(0)$ does not exist. (S)

(D) $k(x) = \tan^{-1}\left(\frac{3x - x^3}{1-3x^2}\right) = 3\tan^{-1} x, \frac{-1}{\sqrt{3}} < x < \frac{1}{\sqrt{3}}$

$\therefore\ k'(x) = \frac{3}{1+x^2}$, So, $k'(0) = 3$ (Q)

2. **(A)** $\because u = x^2 + y^2, x = s + 3t$ and $y = 2s - t$

$\therefore \qquad \frac{dx}{ds} = 1, \frac{dy}{ds} = 2 \qquad \ldots(\text{i})$

and $\qquad \frac{d^2x}{ds^2} = 0, \frac{d^2y}{ds^2} = 0 \qquad \ldots(\text{ii})$

Now, $u = x^2 + y^2$

$\frac{du}{ds} = 2x\frac{dx}{ds} + 2y\frac{dy}{ds}$

$\Rightarrow \frac{d^2u}{ds^2} = 2x\frac{d^2x}{ds^2} + \frac{dx}{ds}.2\frac{dx}{ds} + 2y\frac{d^2y}{ds^2} + \frac{dy}{ds}.2\frac{dy}{ds}$

$= 2x\frac{d^2x}{ds^2} + 2\left(\frac{dx}{ds}\right)^2 + 2y\frac{d^2y}{ds^2} + 2\left(\frac{dy}{ds}\right)^2$

$= 2x \times 0 + 2(1)^2 + 2y \times 0 + 2(2)^2$

[from Eqs. (i) and (ii))]

$= 10 = 2 \times 5$

which is divisible by 2 and 5. (P, S)

(B) $\because y = t^{10} + 1$ and $x = t^8 + 1$

$\therefore\ \frac{dy}{dt} = 10t^9, \frac{dx}{dt} = 8t^7 \qquad \ldots(\text{i})$

$\Rightarrow \frac{dy}{dx} = \frac{dy/dt}{dx/dt} = \frac{10t^9}{8t^7} = \frac{5}{4}t^2$

$\therefore \frac{d^2y}{dx^2} = \frac{d}{dx}\left(\frac{dy}{dx}\right) = \frac{d}{dx}\left(\frac{5}{4}t^2\right)$

$= \frac{d}{dt}\left(\frac{5}{4}t^2\right).\frac{dt}{dx} = \frac{5}{2}t \times \frac{1}{8t^7} = \frac{5}{16t^6}$

$\Rightarrow \left.\frac{d^2y}{dx^2}\right|_{t=1/2} = \frac{5}{16\left(\frac{1}{2}\right)^6} = 20 = 2 \times 2 \times 5$

$= 4 \times 5$

Which is divisible by 2, 4 and 5. (P, R S)

(C) Since, $F(x) = f(x)g(x)h(x)$

So, $F'(x) = f'(x)g(x)h(x) + f(x)g'(x)h(x) + f(x)g(x)h'(x)$

At $x = x_0$, we get

$F'(x_0) = f'(x_0)g(x_0)h(x_0) + f(x_0)g'(x_0)h(x_0) + f(x_0)g(x_0)h'(x_0)$

Using the given values of $F'(x_0), f'(x_0), g'(x_0)$ and $h'(x_0)$, we get

$21F(x_0) = 4f(x_0)g(x_0)h(x_0) + f(x_0)(-7g(x_0))h(x_0) + \lambda f(x_0)g(x_0)h(x_0)$

$= (\lambda - 3)f(x_0)g(x_0)h(x_0)$

$= (\lambda - 3)F(x_0) \quad [\because F(x_0) = f(x_0)g(x_0)h(x_0)]$

$\Rightarrow \quad 21 = \lambda - 3$

$\therefore \quad \lambda = 24$

$= 2 \times 2 \times 2 \times 3$

Which is divisible by 2, 3 and 4 (P, Q, R)

3. **(A)** $y = \cos^{-1}(\cos x)$

$y' = \frac{-1}{\sqrt{1-\cos^2 x}}.(-\sin x) = \frac{\sin x}{|\sin x|}$

$\therefore \quad y'$ at $x = 5$ is -1 (T)

(B) $y = f(x) = \ln|\tan x|$

$\therefore\ f'(x) = \frac{1}{\tan x}(\sec^2 x).\left(\frac{|\tan x|}{\tan x}\right)$

$\Rightarrow f'\left(-\frac{\pi}{4}\right) = 2$ (Q)

(C) $\frac{d}{dx}\tan^{-1}\left(\frac{1+x}{1-x}\right) = \frac{1}{1+\left(\frac{1+x}{1-x}\right)^2}.\frac{d}{dx}\left(\frac{1+x}{1-x}\right)$

$= \frac{(1-x)^2}{2(1+x^2)}.\frac{2}{(1-x)^2} = \frac{1}{1+x^2}$

At $x = -1$, we have

$\frac{d}{dx}\tan^{-1}\left(\frac{1+x}{1-x}\right) = \frac{1}{2}$ (R)

(D) $\frac{d}{dx}\frac{\ln|x|}{x} = \frac{x.\frac{1}{x} - \ln|x|}{x^2} = \frac{1 - \ln|x|}{x^2}$

$\Rightarrow \frac{d}{dx}\frac{\ln|x|}{x} = 1$ at $x = -1$ (S)

14

Applications of Derivatives

Objective Questions Type I [Only one correct answer]

In each of the questions below, four choices are given of which only one is correct. You have to select the correct answer which is the most appropriate.

1. Tangents are drawn from the origin to the curve $y = \sin x$, then their point of contact lie on the curve
(a) $x^2 + y^2 = 1$ (b) $x^2 - y^2 = 1$
(c) $\frac{1}{x^2} + \frac{1}{y^2} = 1$ (d) $\frac{1}{y^2} - \frac{1}{x^2} = 1$

2. The locus of all points on the curve $y^2 = 4a\left(x + a\sin\left(\frac{x}{a}\right)\right)$ at which the tangent is parallel to x-axis is
(a) a straight line (b) a circle
(c) a parabola (d) an ellipse

3. The angle of intersection of curves $y = [|\sin x| + |\cos x|]$ and $x^2 + y^2 = 5$, where $[.]$ denotes the greatest integer function, is
(a) $\tan^{-1} 2$ (b) $\tan^{-1}\left(\frac{1}{2}\right)$
(c) $\tan^{-1}(\sqrt{2})$ (d) $\tan^{-1}\left(\frac{1}{\sqrt{2}}\right)$

4. If the normal to the curve $y = f(x)$ at $x = 0$ be given by the equation $3x - y + 3 = 0$, then the value of $\lim_{x \to 0} x^2 \{f(x^2) - 5f(4x^2) + 4f(7x^2)\}^{-1}$ is
(a) $-1/5$ (b) $-1/4$
(c) $-1/3$ (d) $-1/2$

5. The slope of the normal at the point with abscissa $x = -2$ of the graph of the function $f(x) = |x^2 - |x||$ is
(a) $-1/6$ (b) $-1/3$
(c) $1/6$ (d) $1/3$

6. If the tangent to the curve $y = x + \sin y$ at a point (a, b) is parallel to the line joining $\left(0, \frac{3}{2}\right)$ and $\left(\frac{1}{2}, 2\right)$, then
[JEE Main 2020, 2 September]
(a) $|b - a| = 1$ (b) $|a + b| = 1$
(c) $b = a$ (d) $b = \frac{\pi}{2} + a$

7. A ladder 20 ft long has one end on the ground and the other end in contact with a vertical wall. The lower end slips along the ground. If the lower end of the ladder is 16 ft away from the wall, upper end is moving λ times as fast as the lower end, then λ is
(a) $\frac{1}{3}$ (b) $\frac{2}{3}$
(c) $\frac{4}{3}$ (d) $\frac{5}{3}$

8. A kite is moving horizontally at a height of 151.5 m. If the speed of the kite is 10 m/s, how fast is the string being let out, when the kite is 250 m from the boy who is flying the kite, the height of the boy being 1.5 m ?
(a) 4 m/s (b) 8 m/s
(c) 16 m/s (d) 32 m/s

9. Let C be the locus of the mirror image of a point on the parabola $y^2 = 4x$ with respect to the line $y = x$. Then, the equation of tangent to C at $P(2, 1)$ is
[JEE Main 2021, 16 March]
(a) $x - y = 1$ (b) $2x + y = 5$
(c) $x + 2y = 5$ (d) $x + 2y = 4$

10. Tangent of acute angle between the curves $y = |x^2 - 1|$ and $y = |x^2 - 3|$ at their points of intersection is

(a) 0 (b) $\frac{4\sqrt{2}}{3}$

(c) $\frac{4\sqrt{2}}{7}$ (d) $2\sqrt{2}$

11. The radius of a right circular cylinder increases at the rate of 0.1 cm/min, and the height decreases at the rate of 0.2 cm/min. The rate of change of the volume of the cylinder, in cm^3/min, when the radius is 2 cm and the height is 3 cm is

(a) -2π (b) $-\frac{8\pi}{5}$

(c) $-\frac{3\pi}{5}$ (d) $\frac{2\pi}{5}$

12. The approximate value of $(1.0004)^{3000}$ is

(a) 1.2 (b) 1.6
(c) 1.8 (d) 2.2

13. The chord joining the points where $x = p$ and $x = q$ on the curve $y = ax^2 + bx + c$ is parallel to the tangent at the point on the curve whose abscissa is

(a) $\frac{p+q}{2}$ (b) $\frac{p-q}{2}$

(c) $\frac{pq}{2}$ (d) None of these

14. If the tangent to the curve $xy + ax + by = 0$ at (1,1) is inclined at an angle $\tan^{-1} 2$ with x- axis, then

(a) $a = 1, b = 2$ (b) $a = 1, b = -2$
(c) $a = -1, b = 2$ (d) $a = -1, b = -2$

15. If the curves $\frac{x^2}{a^2} + \frac{y^2}{b^2} = 1$ and $\frac{x^2}{\alpha^2} - \frac{y^2}{\beta^2} = 1$ cut each other orthogonally, then

(a) $a^2 + b^2 = \alpha^2 + \beta^2$ (b) $a^2 - b^2 = \alpha^2 - \beta^2$
(c) $a^2 - b^2 = \alpha^2 + \beta^2$ (d) $a^2 + b^2 = \alpha^2 - \beta^2$

16. A spherical iron ball of 10 cm radius is coated with a layer of ice of uniform thickness that melts at a rate of 50 cm^3/min. When the thickness of ice is 5 cm, then the rate (in cm/ min.) at which of the thickness of ice decreases, is **[JEE Main 2020, 9 January]**

(a) $\frac{5}{6\pi}$ (b) $\frac{1}{54\pi}$

(c) $\frac{1}{36\pi}$ (d) $\frac{1}{18\pi}$

17. The approximate value of $\tan^{-1}(0.999)$ (upto 4 decimal places) is

(a) 0.7852 (b) 0.7102
(b) 0.8127 (d) 0.7526

18. The equation of the tangent to the curve $y = e^{-|x|}$ at the point where the curve cuts the line $x = 1$ is

(a) $x + y = e$ (b) $e(x + y) = 1$
(c) $y + ex = 1$ (d) None of these

19. If the tangent to the curve $\sqrt{x} + \sqrt{y} = \sqrt{a}$ at any point on it cuts the axes OX and OY at P and Q respectively, then $OP + OQ$ is

(a) $a/2$ (b) a
(c) $2a$ (d) $4a$

20. If the tangent at (1, 1) on $y^2 = x(2-x)^2$ meets the curve again at P, then P is

(a) (4, 4) (b) (−1, 2)
(c) (9/4, 3/8) (d) None of these

21. The tangent to the curve $x = a\sqrt{(\cos 2\theta)}\cos\theta$, $y = a\sqrt{\cos 2\theta}\sin\theta$ at the point corresponding to $\theta = \pi/6$ at

(a) parallel to the X-axis (b) parallel to the Y-axis
(c) parallel to line $y = x$ (d) None of these

22. A man of height 2 m walks directly away from a lamp of height 5 m, on a level road at 3m/s. The rate at which the length of his shadow is increasing is

(a) 1m/s (b) 2m/s
(c) 3m/s (d) 4 m/s

23. The area of the triangle formed by the positive X-axis, and the normal and tangent to the circle $x^2 + y^2 = 4$ at $(1, \sqrt{3})$ is

(a) $\sqrt{3}$ sq unit (b) $2\sqrt{3}$ sq unit
(c) $4\sqrt{3}$ sq unit (d) None of these

24. The distance between the origin and the normal to the curve $y = e^{2x} + x^2$ at $x = 0$ is

(a) $2/\sqrt{3}$ (b) $2/\sqrt{5}$
(c) $2/\sqrt{7}$ (d) None of these

25. The value of m for which the area of the triangle included between the axes and any tangent to the curve $x^m y = b^m$ is constant, is

(a) 1/2 (b) 1
(c) 3/2 (d) 2

26. The acute angles between the curves $y = |x^2 - 1|$ and $y = |x^2 - 3|$ at their points of intersection is

(a) $\pi/4$ (b) $\tan^{-1}(4\sqrt{2}/7)$
(c) $\tan^{-1}(4\sqrt{7})$ (d) None of these

27. The period T of a simple pendulum is

$$T = 2\pi\sqrt{\left(\frac{l}{g}\right)}$$

the maximum error in T due to possible errors upto 1% in l and 2.5% in g is

(a) − 1.75% (b) 2.75%
(c) 1.75% (d) 1.95%

28. A physical quantity A is related to four observable a, b, c and d as follows, $A = \frac{a^2 b^3}{c\sqrt{d}}$, the percentage errors of measurement in a, b, c and d are 1%, 3%, 2% and 2% respectively. The maximum error in the quantity A is

(a) 5% (b) 7%
(c) 12% (d) 14%

29. If the tangent at any point on the curve $x^4 + y^4 = a^4$ cuts off intercepts p and q on the coordinate axes, the value of $p^{-4/3} + q^{-4/3}$ is

(a) $a^{-4/3}$ (b) $a^{-1/2}$
(c) $a^{1/2}$ (d) None of these

30. The point of intersection of the tangents drawn to the curve $x^2y = 1 - y$ at the points where it is meet by the curve $xy = 1 - y$, is given by

(a) $(0, -1)$ (b) $(1, 1)$
(c) $(0, 1)$ (d) None of these

31. The angle of intersection of the normals at the $\left(-\frac{5}{\sqrt{2}}, \frac{3}{\sqrt{2}}\right)$ of the curves $x^2 - y^2 = 8$ and $9x^2 + 25y^2 = 225$ is

(a) 0 (b) $\frac{\pi}{2}$
(c) $\frac{\pi}{3}$ (d) $\frac{\pi}{4}$

32. The slope of the normal at the point with abscissa $x = -2$ of the graph of the function $f(x) = |x^2 - x|$ is

(a) $-1/6$ (b) $-1/3$
(c) $1/6$ (d) $1/3$

33. The subtangent, ordinate and subnormal to the parabola $y^2 = 4ax$ at a point (different from the origin) are in

(a) AP (b) GP
(c) HP (d) None of these

34. Let $P(h,k)$ be a point on the curve $y = x^2 + 7x + 2$, nearest to the line $y = 3x - 3$. Then, the equation of the normal to the curve at P is

[JEE Main 2020, 2 September]

(a) $x - 3y - 11 = 0$ (b) $x - 3y + 22 = 0$
(c) $x + 3y - 62 = 0$ (d) $x + 3y + 26 = 0$

35. The tangent to the graph of the function $y = f(x)$ at the point with abscissa $x = 1$ form an angle of $\pi/6$ and at the point $x = 2$ an angle of $\pi/3$ and at the point $x = 3$ an angle of $\pi/4$. The value of $\int_1^3 f'(x)f''(x)\,dx + \int_2^3 f''(x)\,dx$ ($f''(x)$ is suppose to be continuous) is

(a) $\frac{4\sqrt{3}-1}{3\sqrt{3}}$ (b) $\frac{3\sqrt{3}-1}{2}$
(c) $\frac{4-\sqrt{3}}{3}$ (d) None of these

36. The slope of the tangent to the curve $y = \int_x^{x^2} \cos^{-1} t^2\,dt$ at $x = \frac{1}{\sqrt[4]{2}}$ is

(a) $\left(\frac{\sqrt[4]{8}}{2} - \frac{3}{4}\right)\pi$ (b) $\left(\frac{\sqrt[4]{8}}{3} - \frac{1}{4}\right)\pi$
(c) $\left(\frac{\sqrt[5]{8}}{4} - \frac{1}{3}\right)\pi$ (d) None of these

37. The curve $y = ax^3 + bx^2 + cx + 8$ touches X-axix at $P(-2, 0)$ and cuts the Y-axis at a point Q where its gradient is 3. The values of a, b, c are respectively

(a) $-1/2, -3/4, 3$ (b) $3, -1/2, -4$
(c) $-1/2, -7/4, 2$ (d) None of these

38. For the curve $x = t^2 - 1, y = t^2 - t$, the tangent line is perpendicular to X-axis when

(a) $t = 0$ (b) $t = \infty$
(c) $t = \frac{1}{\sqrt{3}}$ (d) $t = -\frac{1}{\sqrt{3}}$

39. If the subnormal at any point on $y = a^{1-n}x^n$ is of constant length, then the value of n is

(a) -2 (b) $1/2$
(c) 1 (d) 2

40. The tangent and normal at the point $P(at^2, 2at)$ to the parabola $y^2 = 4ax$ meet the X-axis in T and G respectively, then the angle at which the tangent at P to the parabola is inclined to the tangent at P to the circle through T, P, G is

(a) $\tan^{-1} t^2$ (b) $\cot^{-1} t^2$
(c) $\tan^{-1} t$ (d) $\cot^{-1} t$

41. The value of parameter a so that the line $(3 - a)x + ay + (a^2 - 1) = 0$ is normal to the curve $xy = 1$, may lie in the interval

(a) $(-\infty, 0) \cup (3, \infty)$ (b) $(1, 3)$
(c) $(-3, 3)$ (d) None of these

Objective Questions Type II [One or more than one correct answer(s)]

In each of the questions below four choices of which one or more than one are correct. You have to select the correct answer(s) accordingly.

1. Let $f : (-1, 1) \to (-1, 1)$ be defined by $f(x) = x - x^3 + x^5 - x^7 + \ldots$ upto infinity. If the line $ax + by + c = 0$ is a tangent to the curve $y = f(x)$, then one must have

(a) $ab > 0$ (b) $ab < 0$
(c) $|a| < |b|$ (d) $|c| \le |b|$

2. The normal to the curve represented parametrically by $x = a(\cos\theta + \theta \sin\theta)$ and $y = a(\sin\theta - \theta\cos\theta)$ at any point θ, is such that it

(a) makes a constant angle with the X-axis
(b) is at a constant distance from the origin
(c) touches a fixed circle
(d) passes through the origin

3. The angle between the tangents at any point P and the line joining P to the origin, where P is a point on the curve $\ln(x^2+y^2)=c\tan^{-1}\frac{y}{x}$, c is a constant, is
(a) independent of x
(b) independent of y
(c) independent of x but dependent on y
(d) independent of y but dependent on x

4. Let the parabolas $y=x^2+ax+b$ and $y=x(c-x)$ touch each other at the point (1, 0), then
(a) $a=-3$ (b) $b=1$
(c) $c=2$ (d) $b+c=3$

5. A point on the ellipse $4x^2+9y^2=36$, where the tangent is equally inclined to the axes is
(a) $\left(\frac{9}{\sqrt{13}},\frac{4}{\sqrt{13}}\right)$
(b) $\left(-\frac{9}{\sqrt{13}},\frac{4}{\sqrt{13}}\right)$
(c) $\left(\frac{9}{\sqrt{13}},-\frac{4}{\sqrt{13}}\right)$
(d) None of the above

6. For $F(x)=\int_0^x 2|t|\,dt$, the tangent lines which are parallel to the bisector of the first coordinate angle is
(a) $y=x-1/4$ (b) $y=x+1/4$
(c) $y=x-3/2$ (d) $y=x+3/2$

7. The equations of the tangents to the curve $y=x^4$ from the point (2, 0) not on the curve, are given by
(a) $y=0$
(b) $y-1=5(x-1)$
(c) $y-\frac{4096}{81}=\frac{2048}{27}\left(x-\frac{8}{3}\right)$
(d) $y-\frac{32}{243}=\frac{80}{81}\left(x-\frac{2}{3}\right)$

8. If a line is tangent to one point and normal at another point on the curve $x=4t^2+3$, $y=8t^3-1$, then slope of such a line is
(a) -1 (b) 1 (c) $-\sqrt{2}$ (d) $\sqrt{2}$

9. If the tangent to the curve $2y^3=ax^2+x^3$ at the point (a,a) cuts off intercepts α and β on the coordinate axes, (where $\alpha^2+\beta^2=61$) then the value of a is
(a) -30 (b) 10
(c) 20 (d) 30

Linked-Comprehension Type

In these questions, a passage (paragraph) has been given followed by questions based on each of the passages. You have to answer the questions based on the passage given.

PASSAGE 1

Consider the curve $x=1-3t^2$, $y=t-3t^3$. A tangent at point $(1-3t^2, t-3t^3)$ is inclined at an angle θ to the positive X-axis another tangent at point $P(-2, 2)$ cuts the curve again at Q.

On the basis of above information, answer the following questions :

1. The value of $\tan\theta+\sec\theta$ is equal to
(a) $3t$ (b) t
(c) $t-t^2$ (d) t^2-2

2. The point Q will be
(a) $(1,-2)$ (b) $\left(-\frac{1}{3},-\frac{2}{9}\right)$
(c) $(-2, 1)$ (d) None of these

3. The angle between the tangents at P and Q will be
(a) $\frac{\pi}{4}$ (b) $\frac{\pi}{6}$
(c) $\frac{\pi}{2}$ (d) $\frac{\pi}{3}$

PASSAGE 2

The parametric equation of given curve are $x=a(2\cos t+\cos 2t)$, $y=a(2\sin t-\sin 2t)$.

On the basis of above information, answer the following questions

1. The equation of tangent at any point 't' is
(a) $x\sin\left(\frac{t}{2}\right)+y\cos\left(\frac{t}{2}\right)=a\sin\left(\frac{3t}{2}\right)$
(b) $x\cos\left(\frac{t}{2}\right)-y\sin\left(\frac{t}{2}\right)=a\sin\left(\frac{3t}{2}\right)$
(c) $x\sin\left(\frac{t}{2}\right)+y\cos\left(\frac{t}{2}\right)=3a\sin\left(\frac{3t}{2}\right)$
(d) $x\cos\left(\frac{t}{2}\right)-y\sin\left(\frac{t}{2}\right)=3a\sin\left(\frac{3t}{2}\right)$

2. The equation of normal at any point 't' is
(a) $x\cos\left(\frac{t}{2}\right)+y\sin\left(\frac{t}{2}\right)=3a\cos\left(\frac{3t}{2}\right)$
(b) $x\cos\left(\frac{t}{2}\right)-y\sin\left(\frac{t}{2}\right)=3a\cos\left(\frac{3t}{2}\right)$
(c) $x\cos\left(\frac{t}{2}\right)+y\sin\left(\frac{t}{2}\right)=3a\sin\left(\frac{3t}{2}\right)$
(d) $x\cos\left(\frac{t}{2}\right)-y\sin\left(\frac{t}{2}\right)=3a\sin\left(\frac{3t}{2}\right)$

3. The length of sub tangent at any point 't' is
(a) $|y\tan t|$ (b) $|y\cot t|$
(c) $|y\cot(t/2)|$ (d) $|y\tan(t/2)|$

4. The length of subnormal at any point 't' is

(a) $|y \tan t|$ (b) $|y \cot t|$

(c) $|y \cot(t/2)|$ (d) $|y \tan(t/2)|$

5. If length of perpendiculars from origin on tangent and normal at 't' are p and p_1, respectively, then the value of $9p^2 + p_1^2$ is equal to

(a) $9a^2$ (b) $9a^2\sin^2(t/2)$

(c) $9a^2\cos^2\left(\frac{3t}{2}\right)$ (c) a^2

PASSAGE 3

A polynomial of degree three as shown is the figure and at Q gradient is 3.

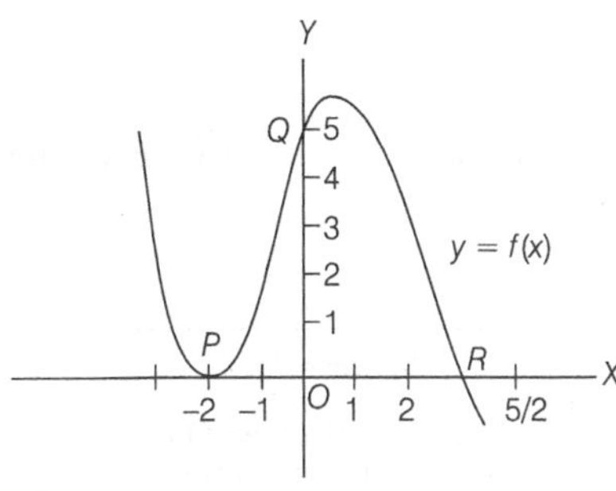

On the basis of above information, answer the following questions

1. The polynomial $f(x)$ is given by

(a) $-\frac{1}{2}x^3 - \frac{3}{4}x^2 + 3x + 5$

(b) $-\frac{1}{2}x^3 + 3x + 5$

(c) $-\frac{1}{2}x^3 - \frac{3}{4}x^2 + 5$

(d) $-\frac{1}{2}x^3 + \frac{3}{4}x^2 - 3x + 5$

2. The value of $f(2)$ is

(a) 2 (b) 4

(c) 6 (d) 8

3. The number of solutions of the equation $|f(|x|)| - 3 = 0$ is

(a) 2 (b) 3

(c) 4 (d) 5

4. The equation of normal at R is

(a) $6x - 81y - 15 = 0$ (b) $8x - 81y - 20 = 0$

(c) $2x - 81y - 5 = 0$ (d) $4x - 81y - 10 = 0$

5. The equation of tangent at Q is

(a) $4x - y + 5 = 0$ (b) $3x - 2y + 10 = 0$

(c) $3x - 2y + 10 = 0$ (d) $3x - y + 5 = 0$

Numerical Value Type

Solve the following problems and write your answer.

1. The curve $y = ax^3 + bx^2 + cx + b$ touches the x-axis at $P(-2, 0)$ and cuts the y-axis at a point Q where its gradient is 3, then the value of $-10a - 100b + 1000c$ must be

2. Tangent at a point P_1 (other than (0, 0)) on the curve $y = x^3$ meets the curve again at P_2. The tangent at P_2 meets the curve again at P_3 and so on, then we get $P_1, P_2, P_3, \ldots\ldots\ldots, P_n$ are is GP, then the ratio $\frac{\text{area }(\Delta P_1 P_2 P_3)}{\text{area }(\Delta P_2 P_3 P_4)} = \frac{\lambda}{\mu}$, then the value of $\frac{251}{2} \cdot \frac{\mu}{\lambda}$ must be

3. If the side of an equilateral triangle increases uni- formly at the rate of 3 ft/s. If λ ft^2/s be the area increasing. When the side is 10 ft. Then the value of $0.5\sqrt{3}\lambda$ must be

4. Water is dripping out from a conical funnel of semi-vertical angle $\pi/4$, at the uniform rate of 2 cm^3/s. Through a tiny hole at the vertex at the bottom when the slant height of the water is 4cm. If λ cm/s be the rate of decrease of the slant height of the water, then the value of $40\pi\sqrt{2}\lambda$ must be

5. The indicated horse power I of an engine is calculated from the formula $I = \frac{PLAN}{33000}$ where $A = \frac{\pi}{4}d^2$. Assuming that error of 10% may have been made in measuring P, L, N and d. If the greatest possible in I is λ% then λ must be

6. A balloon is in the form of right circular cylinder of radius 1.5 m and length 4 m and is surmounted by hemispherical ends. If the radius is increased by 0.01 m and the length by 0.05 m, the percentage change in the volume of the balloon is a. bcd%, then the value of $abcd$ must be

7. If the approximate value of $\log_{10}(4.04)$ is 0 $abcdef$, It is given that $\log_{10} 4 = 0.6021$ and $\log_{10} e = 0.4343$, then the value of $abcd$ must be

8. If the approximate value of $(1.999)^6$ is $ab \cdot cde\ldots$, then the value of $abcd$ must be

Single Integer Type Questions

Solve the following problems and write your answer.

1. If the tangent to the curve, $y = e^x$ at a point (c, e^c) and the normal to the parabola, $y^2 = 4x$ at the point $(1, 2)$ intersect at the same point the X-axis, then the value of c is .... **[JEE Main 2020, 3 September]**
2. A curve is defined parametrically by the equations $x = t^2$ and $y = t^3$. A variable pair of perpendicular lines through the origin O meet the curve at P and Q. If the locus of the point of intersection of the tangents at P and Q is $ay^2 = bx - 1$, then the value of $(a + b)$ is ....
3. The indicated horse power I of an engine is calculated from the formula
$$I = \frac{PLAN}{33000}$$
where $A = \frac{\pi}{4} d^2$. Assuming that errors of r percent may have been made in measuring P, L, N and d. If the greatest possible error in I is $\lambda r\%$, then the value of λ is
4. Number of different points on the curve $y = x^4$, where the tangents drawn from the point $\left(\frac{3}{4}, 0\right)$ meet the curve is
5. Let C be a curve defined by $y = e^{a + bx^2}$. The curve C passes through the point $P(1, 1)$ and the slope of the tangent at P is (-2). Then the value of $2a - 3b$ is ....

Matrix-Match Type

Given below are Matching Type Questions, with two columns (each having some items) each . Each item of Column I has to be matched with the items of Column II, by encircling the correct match(es).

NOTE *An item of Column I can be matched with more than one items of Column II.*

1. Observe the following columns

	Column I		Column II
(A)	The slope of the curve $2y^2 = ax^2 + b$ at $(1,-1)$ is -1, then	(P)	$a - b = 2$
(B)	If (a, b) be the point on the curve $9y^2 = x^3$ where normal to the curve makes equal intercepts with the axes, then	(Q)	$a - b = 7/2$
(C)	If the tangent at any point $(1, 2)$ on the curve $y = ax^2 + bx + \frac{7}{2}$ be parallel to the normal at $(-2, 2)$ on the curve $y = x^2 + 6x + 10$, then	(R)	$a - b = 4/3$
		(S)	$\alpha + \lvert b \rvert = \frac{20}{3}$
		(T)	$5a + 2b = 0$

2. Observe the following columns

	Column I		Column II
(A)	If three normals can be drawn to the curve $y^2 = x$ from the point $(c, 0)$, then c can be equal to	(P)	1
(B)	Subnormal length to $xy = c^2$ at any point varies directly as	(Q)	0
(C)	If the sides and angles of a plane triangle vary in such a way that its circum radius remains constant, then $\frac{da}{\cos A} + \frac{db}{\cos B} + \frac{dc}{\cos C} =$	(R)	$\frac{5}{4}$
		(S)	Cube of ordinate
		(T)	2

where da, db, dc are small increments in the sides a, b, c respectively

3. Observe the following columns

	Column I		Column I
A.	If the curves $y^2 = 4ax$, $a > 0$ and $xy = \sqrt{2}$ cut orthogonally, then $\frac{1}{a^2}$ is	P.	a prime number
B.	If the curves $x^2 + ay = 7, a > 0$ and $x^3 = y$ cut orthogonally at $(1, 1)$, then a is	Q.	a natural number
C.	The curve $y = 2e^{2x}$ intersects the Y-axis at an angle $\cot^{-1}\lvert(8n - 4)/3\rvert$, then $n (> 0)$ is	R.	a perfect number
		S.	a composite number

Answers

Objective Questions Type [Only one correct answer]

1. (d)	**2.** (c)	**3.** (a)	**4.** (c)	**5.** (d)	**6.** (a)	**7.** (c)	**8.** (b)	**9.** (a)	**10.** (c)
11. (d)	**12.** (d)	**13.** (a)	**14.** (b)	**15.** (a)	**16.** (d)	**17.** (a)	**18.** (d)	**19.** (b)	**20.** (c)
21. (a)	**22.** (b)	**23.** (b)	**24.** (b)	**25.** (b)	**26.** (b)	**27.** (c)	**28.** (d)	**29.** (a)	**30.** (c)
31. (b)	**32.** (d)	**33.** (b)	**34.** (d)	**35.** (d)	**36.** (b)	**37.** (d)	**38.** (a)	**39.** (b)	**40.** (c)
41. (a)									

Objective Questions Type II [One or more than one correct answer(s)]

1. (b, c) **2.** (b, c) **3.** (a, b) **4.** (a, d) **5.** (a, b, c)
6. (a, b) **7.** (a, c) **8.** (c, d) **9.** (a, d)

Linked-Comprehension Type

Passage 1 **1.** (a) **2.** (b) **3.** (c)

Passage 2 **1.** (a) **2.** (b) **3.** (c) **4.** (d) **5.** (a)

Passage 3 **1.** (a) **2.** (b) **3.** (c) **4.** (b) **5.** (d)

Numerical Value Type

1. (3080) **2.** (2008) **3.** (22.5) **4.** (20) **5.** (50)
6. (2389) **7.** (6064) **8.** (6380)

Single Integer Type

1. (4) **2.** (7) **3.** (5) **4.** (2) **5.** (5)

Matrix-Match Type

1. A → (P); B → (R, S); C → (Q, T)
2. A → (P, R,T); B → (S); C → (Q)
3. A → (Q, S); B → (Q, R, S); C → (P, Q)

Solutions

Objective Questions Type [Only one correct answer]

1. The equation of tangent at (x, y) is

$$Y - y = \cos x(X - x)$$

Its pass through (0, 0)

Then, $$0 - y = \cos x\,(0 - x)$$

or $$\cos x = \frac{y}{x} \quad \ldots \text{(i)}$$

Given, $$\sin x = y \quad \ldots \text{(ii)}$$

Squaring and adding Eqs. (i) and (ii), then

$$1 = \frac{y^2}{x^2} + y^2 \Rightarrow \frac{1}{y^2} - \frac{1}{x^2} = 1$$

2. $\because y^2 = 4a\left(x + a\sin\left(\frac{x}{a}\right)\right)$ ...(i)

$$\therefore \quad 2y\frac{dy}{dx} = 4a\left(1 + \cos\left(\frac{x}{a}\right)\right)$$

Since, tangent is parallel to X-axis

$$\therefore \quad \frac{dy}{dx} = 0$$

$$\Rightarrow \quad 1 + \cos\left(\frac{x}{a}\right) = 0$$

$$\Rightarrow \quad \cos\left(\frac{x}{a}\right) = -1 = \cos\pi$$

$$\therefore \quad \frac{x}{a} = 2n\pi \pm \pi \text{ or } x = (2n \pm 1)\pi a$$

$\therefore$ From Eq. (i), $\quad y^2 = 4ax$

3. $\because \quad 1 \le |\sin x| + |\cos x| \le \sqrt{2}$

$$\therefore \quad y = [\,|\sin x| + |\cos x|\,] = 1$$

$$\because \quad -\sqrt{2} \le \sin x + \cos x \le \sqrt{2}$$

$$|\sin x| + |\cos x| = \sqrt{\{|\sin x| + |\cos x|\}^2}$$
$$= \sqrt{(1 + |\sin 2x|)} \ge 1$$

Let P and Q be the points of intersection of given curves.

Now, solving $y = 1$ and $x^2 + y^2 = 5$

$$\therefore \quad x^2 + 1 = 5 \Rightarrow x = \pm 2$$

$\therefore \quad P \equiv (-2, 1)$ and $Q \equiv (2, 1)$

Clearly the slope of line $y = 1$ is zero

$\because \quad x^2 + y^2 = 5$

$\therefore \quad 2x + 2y\dfrac{dy}{dx} = 0$

$\Rightarrow \quad \dfrac{dy}{dx} = -\dfrac{x}{y}$

$$\left(\frac{dy}{dx}\right)_{(-2,1)} = 2 \text{ and } \left(\frac{dy}{dx}\right)_{(2,1)} = -2$$

Thus, the angle of intersection is $\tan^{-1}(2)$ and $\tan^{-1}(-2)$

4. $\because \quad y = f(x)$

$\therefore \quad \dfrac{dy}{dx} = f'(x)$

$\Rightarrow \quad \left.\dfrac{dy}{dx}\right|_{x=0} = f'(0)$

$\therefore$ Slope of normal $= -\dfrac{1}{f'(0)} = 3 \Rightarrow f'(0) = -\dfrac{1}{3}$

Now, $\displaystyle\lim_{x\to 0} \frac{x^2}{\{f(x^2) - 5f(4x^2) + 4f(7x^2)\}}$

Replacing x^2 by x, then

$$\lim_{x\to 0} \frac{x}{f(x) - 5f(4x) + 4f(7x)}$$

$$= \lim_{x\to 0} \frac{x}{(f(x) - f(0)) - 5(f(4x) - f(0)) + 4(f(7x) - f(0))}$$

$$= \lim_{x\to 0} \frac{1}{\left(\frac{f(x) - f(0)}{x - 0}\right) - 5\left(\frac{f(4x) - f(0)}{4x - 0} \times 4\right) + 4\left(\frac{f(7x) - f(0)}{7x - 0} \times 7\right)}$$

$$= \frac{1}{f'(0) - 20f'(0) + 28f'(0)} = \frac{1}{9f'(0)} = \frac{1}{9} \times -3 = \frac{-1}{3}$$

5. For $x < 0$

$$f(x) = |x^2 + x| = |x(x+1)|$$
$$= x(x+1) \quad (x < -1)$$
$$f'(x) = 2x + 1$$

$$f'(-2) = -4 + 1 = -3$$

$\therefore$ Slope of normal $= \dfrac{1}{3}$

6. Given curve is $y = x + \sin y$

$\therefore$ On differentiating both sides w.r.t. x, we get

$$\frac{dy}{dx} = 1 + \cos y \frac{dy}{dx} \Rightarrow \frac{dy}{dx} = \frac{1}{1 - \cos y} \quad \ldots \text{(i)}$$

$\because$ Tangent at point (a, b) at given curve is parallel to line joining $\left(0, \dfrac{3}{2}\right)$ and $\left(\dfrac{1}{2}, 2\right)$.

So, $\left.\dfrac{dy}{dx}\right|_{(a,\, b)} = \dfrac{2 - (3/2)}{(1/2) - 0} = 1$

$\Rightarrow \quad \dfrac{1}{1 - \cos b} = 1 \Rightarrow 1 = 1 - \cos b \Rightarrow \cos b = 0$

$\Rightarrow \quad \sin b = \pm 1$

Now, as point (a, b) on the given curve,

So, $\quad b = a + \sin b \quad \Rightarrow b - a = \sin b$

$\Rightarrow \quad |b - a| = |\sin b| \quad \Rightarrow |b - a| = 1$

7. Let OC be the wall. Let AB be the position of the ladder at any time t such that $OA = x$ and $OB = y$.

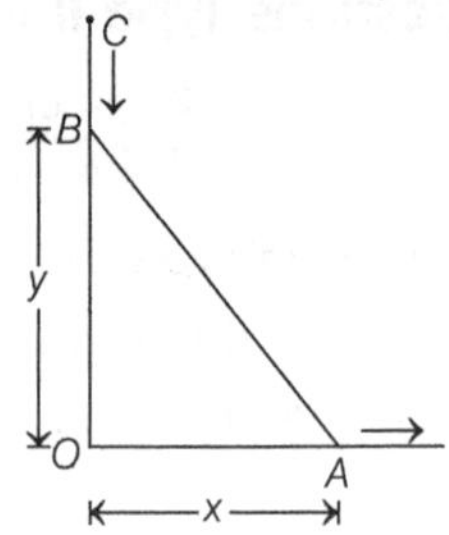

Length of the ladder $AB = 20$ ft.

In ΔAOB,

$$x^2 + y^2 = (20)^2$$

$\Rightarrow \quad 2x\dfrac{dx}{dt} + 2y\dfrac{dy}{dt} = 0$

$$\therefore \quad \frac{dy}{dt} = \frac{-x}{y} \cdot \frac{dx}{dt} = \frac{-x}{\sqrt{400 - x^2}} \cdot \frac{dx}{dt}$$

$$= -\frac{16}{\sqrt{400 - (16)^2}} \cdot \frac{dx}{dt} = -\frac{4}{3}\frac{dx}{dt}$$

–ve sign indicates, that when x increases with time, y decreases.

Hence, the upper end is moving 4/3 times as fast as the lower end.

8. Let the position of the kite at time t be at C.

$\therefore \quad BC = 151.5$ m

Let AD be the boy who is flying the kite.

Let $AB = x = DE$ and $DC = y$

$\therefore \quad CE = BC - BE = 151.5 - 1.5 = 150$ m

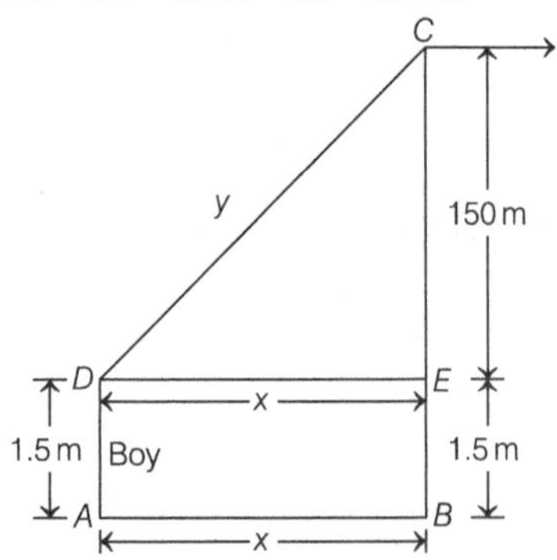

Now, in ΔCDE,

$$y^2 = x^2 + (150)^2$$

$\Rightarrow \quad 2y\dfrac{dy}{dt} = 2x\dfrac{dx}{dt}$

$$\Rightarrow \quad \frac{dy}{dt} = \frac{x}{y}\cdot(10) \qquad \left[\because \frac{dx}{dt} = 10\text{m/s}\right]$$

$$= \frac{10\sqrt{y^2-(150)^2}}{y}$$

$$= \frac{10\sqrt{(250)^2-(150)^2}}{250}$$

$$= \frac{10\times 200}{250} = 8 \text{ m/s}.$$

9. Given, $y^2 = 4x$

Mirror image on $y = x \Rightarrow C : x^2 = 4y$

$$2x = 4\cdot\frac{dy}{dx} \Rightarrow \frac{dy}{dx} = \frac{x}{2}$$

$$\left.\frac{dy}{dx}\right|_{P(2,1)} = \frac{2}{2} = 1$$

Equation of tangent at (2, 1)

$\Rightarrow y - 1 = 1(x-2) \Rightarrow x - y = 1$

10. Points of intersection of curves $y = |x^2 - 1|$ and $y = |x^2 - 3|$ are $(\pm\sqrt{2}, 1)$

$$y = |x^2-1| \text{ and } y = |x^2-3|$$

$$\Rightarrow \quad \frac{dy}{dx} = 2x \text{ and } \frac{dy}{dx} = -2x$$

$$m_1 = \left.\frac{dy}{dx}\right|_{(\sqrt{2},1)} = 2\sqrt{2}$$

and
$$m_2 = \left.\frac{dy}{dx}\right|_{(\sqrt{2},1)} = -2\sqrt{2}$$

$$\tan\theta = \left|\frac{4\sqrt{2}}{7}\right|$$

11. $V = \pi r^2 h$ ... (i)

On differentiating Eq. (i) both sides w.r.t. 't' we get

$$\frac{dV}{dt} = \pi\left(r^2\frac{dh}{dt} + 2r\frac{dr}{dt}h\right)$$

$$= \pi r\left(r\frac{dh}{dt} + 2h\frac{dr}{dt}\right)$$

Given, $\dfrac{dr}{dt} = \dfrac{1}{10}$ and $\dfrac{dh}{dt} = -\dfrac{2}{10}$

$$\therefore \quad \frac{dV}{dt} = \pi r\left(r\left(-\frac{2}{10}\right) + 2h\left(\frac{1}{10}\right)\right) = \frac{\pi r}{5}(-r+h)$$

Thus, when $r = 2$ and $h = 3$

$$\frac{dV}{dt} = \frac{\pi(2)}{5}(-2+3) = \frac{2\pi}{5}$$

12. Let $f(x) = x^{3000}$

Now, $f(x+\delta x) - f(x) = f'(x)\cdot\delta x = 3000x^{2999}\cdot\delta x$

We can write, $1.0004 = 1 + 0.0004$

Taking $x = 1$ and $\delta x = 0.0004$, we get

$f(1.0004) - f(1) = 1\cdot 2$

$\Rightarrow \quad f(1.0004) = f(1) + 1.2$

$= 1 + 1.2 = 2.2$

or $\quad (1.0004)^{3000} = 2\cdot 2$

Hence, the approximate value of $(1.0004)^{3000}$ is 2.2.

13. For $x = p$, $y = ap^2 + bp + c$

And for $\quad x = q$, $y = aq^2 + bq + c$

$$\text{Slope} = \frac{aq^2 + bq + c - ap^2 - bp - c}{q - p}$$

$$= a(q+p) + b$$

$$\frac{dy}{dx} = 2ax + b = a(q+p) + b$$

[according to the question]

$$\therefore \quad x = \frac{q+p}{2}$$

14. Given, curve $xy + ax + by = 0$ ...(i)

(1,1) lie on (i)

$\therefore \quad 1 + a + b = 0$ ...(ii)

From Eq. (i),

$$x\frac{dy}{dx} + y\cdot 1 + a + b\frac{dy}{dx} = 0$$

$$\therefore \quad \frac{dy}{dx} = -\frac{(a+y)}{(b+x)}$$

$$\therefore \quad \left(\frac{dy}{dx}\right)_{(1,1)} = -\frac{a+1}{b+1} = 2 \qquad [\because m = \tan\theta]$$

$$-\frac{(-b)}{b+1} = 2 \qquad \text{[from Eq. (ii)]}$$

$$\therefore \quad b = 2b + 2 \Rightarrow b = -2$$

From Eq. (ii), $a = 1$

Hence, $\quad a = 1, b = -2$

15. Slope of I $= \dfrac{-b^2x}{a^2y} = m_1$ (say),

And slope of II $= \dfrac{\beta^2 x}{\alpha^2 y} = m_2$ (say)

$$\therefore \quad m_1 m_2 = -1$$

$$\therefore \quad b^2\beta^2x^2 = a^2\alpha^2y^2 \qquad \text{...(i)}$$

and
$$\frac{x^2}{a^2} + \frac{y^2}{b^2} - 1 = 0$$

$$\frac{x^2}{\alpha^2} - \frac{y^2}{\beta^2} - 1 = 0$$

$$\Rightarrow \quad x^2\left(\frac{1}{a^2} - \frac{1}{\alpha^2}\right) = -y^2\left(\frac{1}{b^2} - \frac{1}{\beta^2}\right) \qquad \text{...(ii)}$$

From Eqs. (i) and (ii)

$$\frac{\frac{1}{a^2} - \frac{1}{\alpha^2}}{b^2\beta^2} = -\frac{\left(\frac{1}{b^2} - \frac{1}{\beta^2}\right)}{a^2\alpha^2}$$

$$\Rightarrow \quad \alpha^2 + \beta^2 = a^2 + b^2$$

16. It is given that, a spherical iron ball of 10 cm radius is coated with a layer of ice of uniform thickness, so let this thickness be 'x' cm, then volume of the ball is

$$V = \frac{4}{3}\pi(10+x)^3$$

On differentiating w.r.t. 't', we get

$$\frac{dV}{dt} = 4\pi(10+x)^2\frac{dx}{dt}, \qquad \text{... (i)}$$

where, t is time in min.

Given, $\frac{dV}{dt} = -50 \text{ cm}^3/\text{min}$

Now when $x = 5$ cm, then

$$-50 = 4\pi(10+5)^2 \frac{dx}{dt} \quad \text{[from Eq. (i)]}$$

$$\Rightarrow \quad \frac{dx}{dt} = -\frac{50}{4\pi(225)} = -\frac{1}{18\pi} \text{ cm/min}$$

'–' ve sign indicates the thickness of ice layer decreases with time. Hence, option (d) is correct.

17. Let $f(x) = \tan^{-1} x$

$$f'(x) = \frac{1}{1+x^2}$$

Take $a = 1$ and $h = -0.001$

$$f(a+h) = f(a) + f'(a)h$$

$$\tan^{-1}(a+h) = \tan^{-1} a + \frac{1}{1+a^2} \times h$$

$$\Rightarrow \tan^{-1}(1-0.001) = \tan^{-1}(1) + \frac{1}{1+(1)^2} \times (-0.001)$$

$$\Rightarrow \quad \tan^{-1}(0.999) = \tan^{-1}\left(\tan\frac{\pi}{4}\right) + \frac{1}{2} \times (-0.001)$$

$$\Rightarrow \quad \tan^{-1}(0.999) = \frac{\pi}{4} - \frac{0.001}{2}$$

$$= 0.7857 - 0.005 = 0.7852 \text{ radians}$$

18. At $x = 1$, $y = e^{-1}$,

$$x = 1 > 0 \quad \therefore \quad y = e^{-x}$$

$$\frac{dy}{dx} = -e^{-x} \Rightarrow \left(\frac{dy}{dx}\right)_{x=1} = -\frac{1}{e}$$

Equation, $y - e^{-1} = -\frac{1}{e}(x-1)$

$$\Rightarrow \quad ey - 1 = -x + 1 \Rightarrow x + ey = 2$$

19. $\sqrt{x} + \sqrt{y} = \sqrt{a}$ ...(i)

$$\Rightarrow \quad \frac{1}{2\sqrt{x}} + \frac{1}{2\sqrt{y}} \frac{dy}{dx} = 0$$

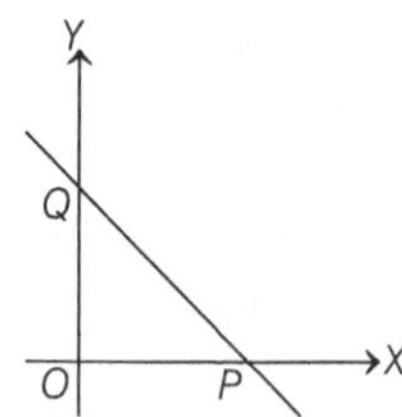

$$\therefore \quad \frac{dy}{dx} = -\frac{\sqrt{y}}{\sqrt{x}}$$

Equation of tangent at (x, y) is

$$Y - y = -\frac{\sqrt{y}}{\sqrt{x}}(X - x)$$

$$\Rightarrow \quad \frac{Y}{\sqrt{y}} - \sqrt{y} = -\frac{X}{\sqrt{x}} + \sqrt{x}$$

$$\Rightarrow \quad \frac{X}{\sqrt{x}} + \frac{Y}{\sqrt{y}} = \sqrt{x} + \sqrt{y} = \sqrt{a} \quad \text{[from Eq. (i)]}$$

$$\therefore \quad \frac{X}{\sqrt{a}\sqrt{x}} + \frac{Y}{\sqrt{a}\sqrt{y}} = 1$$

$\therefore$ $\quad OP = \sqrt{a}\sqrt{x}$

and $\quad OQ = \sqrt{a}\sqrt{y}$

then $\quad OP + OQ = \sqrt{a}\sqrt{x} + \sqrt{a}\sqrt{y}$

$$= \sqrt{a}(\sqrt{x} + \sqrt{y})$$

$$= \sqrt{a} \cdot \sqrt{a}$$

$$= a$$

20. $2y\frac{dy}{dx} = x \times -2(2-x) + (2-x)^2 \cdot 1$

$$= (2-x)(2-x-2x)$$

$$= (2-x)(2-3x)$$

$$\left(\frac{dy}{dx}\right)_{(1,1)} = \frac{(2-1)(2-3)}{2} = -\frac{1}{2}$$

Tangent at (1, 1)

$$y - 1 = -\frac{1}{2}(x-1),$$

$$\Rightarrow \quad y = 1 - \frac{x}{2} + \frac{1}{2} = \left(\frac{3}{2} - \frac{x}{2}\right)$$

$$y^2 = \frac{(3-x)^2}{4}$$

$$\Rightarrow \quad x(2-x)^2 = \frac{(3-x)^2}{4}$$

After solving $x = 9/4$, $\therefore y = 3/8$

21. $\because \quad x = a\sqrt{(\cos 2\theta)} \cos\theta$

$\therefore$

$$\frac{dx}{d\theta} = a\left\{\sqrt{(\cos 2\theta)}(-\sin\theta) + \cos\theta \times \frac{1}{2\sqrt{\cos 2\theta}} \times -2\sin 2\theta\right\}$$

$$= a\left\{\frac{\cos 2\theta(-\sin\theta) - \sin 2\theta\cos\theta}{\sqrt{\cos 2\theta}}\right\} = -\frac{a\sin 3\theta}{\sqrt{\cos 2\theta}}$$

$$\left.\frac{dx}{d\theta}\right|_{\theta=\pi/6} = -\frac{a\sin\pi/2}{\sqrt{\cos\pi/3}} = -\frac{a}{\sqrt{(1/2)}} \neq 0$$

Hence, tangent to the curve is not parallel to Y-axis.

and

$$\frac{dy}{d\theta} = a\left\{\sqrt{\cos 2\theta} \cdot \cos\theta + \sin\theta \frac{1}{2\sqrt{\cos 2\theta}} \times -2\sin 2\theta\right\}$$

$$= \frac{a}{\sqrt{\cos 2\theta}}\{\cos 3\theta\}$$

$$\left.\frac{dy}{d\theta}\right|_{\theta=\pi/6} = 0, \quad \therefore \quad \frac{dy}{dx} = 0$$

Hence, tangent to the given curve parallel to X-axis.

22. Let L be the lamp and PQ be the man and $OQ = x$ m be his shadow and let $MQ = y$ m.

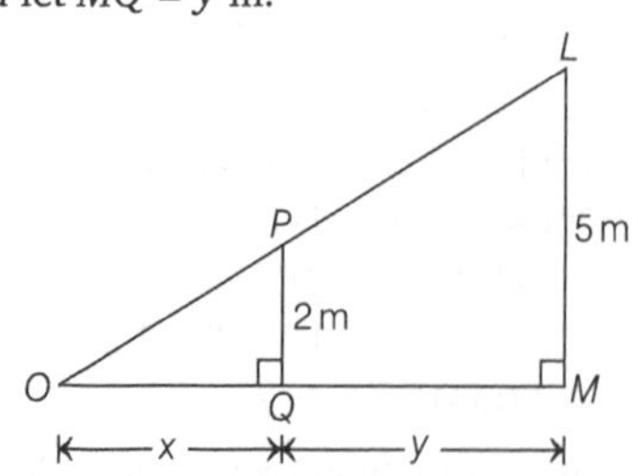

$\therefore \quad \dfrac{dy}{dt}$ = speed of the man

= 3 m/s (given)

$\because \Delta OPQ$ and ΔOLM are similar

$\therefore \quad \dfrac{OM}{OQ} = \dfrac{LM}{PQ}$

$\Rightarrow \quad \dfrac{x+y}{x} = \dfrac{5}{2}$

$\Rightarrow \quad y = \dfrac{3}{2}x$

$\therefore \quad \dfrac{dy}{dt} = \dfrac{3}{2}\dfrac{dx}{dt}$

$\Rightarrow \quad 3 = \dfrac{3}{2}\dfrac{dx}{dt} \Rightarrow \dfrac{dx}{dt} = 2$ m/s

23. $x^2 + y^2 = 4$

Equation of tangent and normal at $(1, \sqrt{3})$ to the circle are $x + \sqrt{3}y = 4$ and $y - \sqrt{3}x = 0$ respectively.

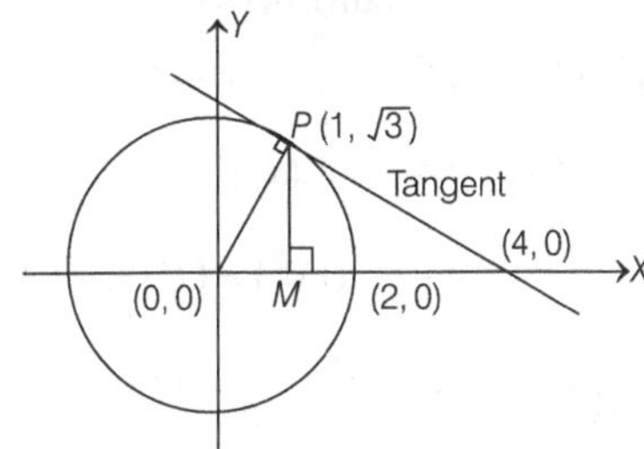

$\therefore$ Required area $= \dfrac{1}{2} \times 4 \times \sqrt{3}$

$= 2\sqrt{3}$ sq unit

24. $\dfrac{dy}{dx} = 2e^{2x} + 2x$

$\Rightarrow \quad \left(\dfrac{dy}{dx}\right)_{x=0} = 2e^0 + 0 = 2$

Slope of normal $= -\dfrac{1}{2}$, at $x = 0, y = 1$

Equation of normal, $y - 1 = -\dfrac{1}{2}(x - 0)$

$\Rightarrow x + 2y - 2 = 0$. Its distance from $(0, 0) = \dfrac{2}{\sqrt{5}}$

25. $\therefore \quad x^m y = b^m$

Taking logarithm

$m \log_e x + \log_e y = m \log_e b$

$\therefore \quad \dfrac{m}{x} + \dfrac{1}{y}\dfrac{dy}{dx} = 0$

$\therefore \quad \dfrac{dy}{dx} = -\dfrac{my}{x}$

Equation of tangent at (x, y) is

$$Y - y = -\frac{my}{x}(X - x)$$

$$xY - xy = -myX + m\,xy$$

$\Rightarrow \quad my\,X + xY = xy\,(1 + m)$

$\Rightarrow \quad \dfrac{X}{\dfrac{x\,(1+m)}{m}} + \dfrac{Y}{y\,(1+m)} = 1$

$\therefore$ Area of $\Delta OAB = \dfrac{1}{2} \cdot OA \cdot OB$

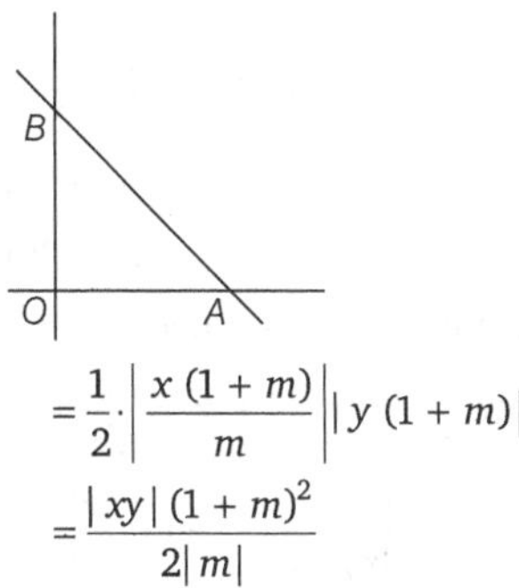

$= \dfrac{1}{2} \cdot \left|\dfrac{x\,(1+m)}{m}\right| |y\,(1+m)|$

$= \dfrac{|xy|\,(1+m)^2}{2|m|}$

For $m = 1$,

$= \dfrac{|xy|\,(4)}{2}$

$= 2|xy| \quad [\because xy = b]$

$= 2|b| =$ constant

26. The point of intersection is $x^2 = 2, y = 1$ the given equations represent four parabolas

$y = \pm(x^2 - 1)$ and $y = \pm(x^2 - 3)$

The curves intersect when $1 < x^2 < 3$

or $\quad 1 < x < \sqrt{3}$ or $-\sqrt{3} < x < -1$

$\therefore \quad y = x^2 - 1$ and $y = -(x^2 - 3)$

The point of intersection are

$(\pm\sqrt{2}, 1)$ at $(\sqrt{2}, t)$

$m_1 = 2x = 2\sqrt{2},\ m_2 = -2x = -2\sqrt{2}$

$\therefore \quad \tan\theta = \left|\dfrac{4\sqrt{2}}{1-8}\right| = \dfrac{4\sqrt{2}}{7} \Rightarrow \theta = \tan^{-1}\left(\dfrac{4\sqrt{2}}{7}\right)$

27. $\because \dfrac{\Delta T}{T} = \dfrac{1}{2}\left(\dfrac{\Delta l}{l} + \dfrac{\Delta g}{g}\right)$

or $\quad \left(\dfrac{\Delta T}{T} \times 100\right) = \dfrac{1}{2}\left(\dfrac{\Delta l}{l} \times 100\right) + \left(\dfrac{\Delta g}{g} \times 100\right)$

$= \dfrac{1}{2}(1 + 2.5) = 1.75\%$

28. $\therefore \dfrac{\Delta A}{A} = \left(2\dfrac{\Delta a}{a} + 3\dfrac{\Delta b}{b} + \dfrac{\Delta c}{c} + \dfrac{1}{2}\dfrac{\Delta d}{d}\right)$

$\therefore \quad \dfrac{\Delta A}{A} \times 100 = 2\left(\dfrac{\Delta a}{a} + 100\right) + 3\left(\dfrac{\Delta b}{b} + 100\right)$

$+ \left(\dfrac{\Delta c}{c} \times 100\right) + \dfrac{1}{2}\left(\dfrac{\Delta d}{d} + 100\right)$

$= 2 \times 1\% + 3 \times 3\% + 2\% + \dfrac{1}{2} \times 2\% = 14\%$

29. $4x^3 + 4y^3\dfrac{dy}{dx} = 0 \Rightarrow \dfrac{dy}{dx} = -\dfrac{x^3}{y^3}$

Equation of tangent,

$$Y - y = -\frac{x^3}{y^3}(X - x)$$

$\Rightarrow \quad y^3Y + x^3X = x^4 + y^4 = a^4$

$\Rightarrow \quad \dfrac{X}{a^4/x^3} + \dfrac{Y}{a^4/y^3} = 1$

Here, $p=\frac{a^4}{x^3}, q=\frac{a^4}{y^3}$

$$\Rightarrow \quad p^{-4/3}+q^{-4/3}=\frac{a^{-16/3}}{x^{-4}}+\frac{a^{-16/3}}{y^{-4}}$$
$$=a^{-16/3}(x^4+y^4)$$
$$=a^{-16/3}(a^4)=a^{-4/3}$$

30. Solving $x^2y=1-y$ and $xy=1-y$,then (0, 1) and (1, 1/2).

Now, $x^2y=1-y \Rightarrow \frac{dy}{dx}=-\frac{2xy}{x^2+1}$

$$\left.\frac{dy}{dx}\right|_{(0,1)}=0 \text{ and } \left.\frac{dy}{dx}\right|_{(1,1/2)}=-\frac{1}{2}$$

The equations of the required tangent are

$$y-1=0(x-0) \text{ and } y-1/2=-1/2(x-1)$$
$$\Rightarrow \quad y=1 \text{ and } x+2y-2=0$$

There two tangents intersect at (0,1).

31. $C_1: x^2-y^2=8$

$$\Rightarrow \quad \frac{dy}{dx}=\frac{x}{y}$$

Slope of tangent at $\left(-\frac{5}{\sqrt{2}},\frac{3}{\sqrt{2}}\right)=-\frac{5}{3}$

Slope of normal to curve C_1 at this point, $m_1=\frac{3}{5}$

$C_2: 9x^2+25y^2=225$

$$\Rightarrow \quad \frac{dy}{dx}=-\frac{9x}{25y}$$

Slope of tangent at $\left(-\frac{5}{\sqrt{2}},\frac{3}{\sqrt{2}}\right)=\frac{3}{5}$

Slope of normal to curve C_2 at this point, $m_2=-\frac{5}{3}$

$$\theta=\tan^{-1}\left|\frac{m_1-m_2}{1+m_1m_2}\right|$$
$$\Rightarrow \quad \theta=\frac{\pi}{2}$$

32. For $x<0$

$$f(x)=|x^2+x|=|x||x+1|$$

For $x<-1$

$f(x)=(-x)(-x-1)=x^2+x$

$\therefore \quad f'(x)=2x+1$

Slope of tangent $=2(-2)+1=-3$

$\therefore$ Slope of normal $=1/3$

33. $\frac{y}{dy/dx}, y, y\frac{dy}{dx}, \Rightarrow \frac{y}{4a/2y}, y, y\times\frac{4a}{2y}$

$\Rightarrow \frac{y^2}{2a}, y, 2a$ are in GP

34. As the point $P(h,k)$ is the nearest point on the curve $y=x^2+7x+2$, to the line $y=3x-3$, so the tangent to the parabola $y=x^2+7x+7$ at point $P(h,k)$ is parallel to the line $y=3x-3$

$$\therefore \quad \left.\frac{dy}{dx}\right|_P=2h+7=3\Rightarrow h=-2 \quad \text{... (i)}$$

and the point $p(h,k)$ on the curve, so

$$k=h^2+7h+2=(-2)^2+7(-2)+2$$
$$\Rightarrow \quad k=4-14+2\Rightarrow k=-8$$

$\therefore$ Point $P(-2,-8)$

Now, equation of normal to the parabola $y=x^2+7x+7$ at point $P(-2,-8)$ is

$$y+8=\frac{-1}{\left.\frac{dy}{dx}\right|_P}(x+2)$$
$$\Rightarrow \quad y+8=-\frac{1}{3}(x+2)$$
$$\Rightarrow \quad x+3y+26=0$$

35. $\because f'(1)=\tan\left(\frac{\pi}{6}\right)=\frac{1}{\sqrt{3}}$,

$$f'(2)=\tan(\pi/3)=\sqrt{3}$$

and $f'(3)=\tan(\pi/4)=1$

$$\therefore \int_1^3 f'(x)f''(x)\,dx+\int_2^3 f''(x)\,dx$$
$$=\frac{1}{2}[(f'(x))^2]_1^3+[f'(x)]_2^3$$
$$=\frac{1}{2}\{(f'(3))^2-(f'(1))^2\}+\{f'(3)-f'(2)\}$$
$$=\frac{1}{2}\left(1-\frac{1}{3}\right)+(1-\sqrt{3})=\frac{4}{3}-\sqrt{3}=\frac{4-3\sqrt{3}}{3}$$

36. $\because y=\int_x^{x^2}\cos^{-1}t^2\,dt$

$$\therefore \quad \frac{dy}{dx}=\cos^{-1}(x^4)\cdot 2x-\cos^{-1}(x^2)\cdot 1$$
$$\Rightarrow \quad \left.\frac{dy}{dx}\right|_{x=\frac{1}{\sqrt[4]{2}}}=\cos^{-1}\left(\frac{1}{2}\right)\frac{2}{\sqrt[4]{2}}-\cos^{-1}\left(\frac{1}{\sqrt{2}}\right)$$
$$=\frac{\pi}{3}\cdot 2^{3/4}-\frac{\pi}{4}=\left(\frac{\sqrt[4]{8}}{3}-\frac{1}{4}\right)\pi$$

37. We have, $\frac{dy}{dx}=3ax^2+2bx+c$

Since, the curve touches X-axis at $(-2, 0)$

$$\therefore \quad \left.\frac{dy}{dx}\right|_{(-2,0)}=0\Rightarrow 12a-4b+c=0 \quad \text{...(i)}$$

The curve cut the Y-axis at (0, 8)

$$\therefore \quad \left.\frac{dy}{dx}\right|_{(0,8)}=3\Rightarrow c=3$$

Also, the curve passes through $(-2, 0)$ so

$$0=-8a+4b-2c+8$$
$$\Rightarrow \quad -8a+4b+2=0 \quad \text{... (ii)}$$

Solving Eqs. (i) and (ii), $a=-5/4, b=-3$

38. $\because \frac{dy}{dx}=\tan 90°=\infty$

$$\therefore \quad \frac{dx}{dy}=0\Rightarrow\frac{dx}{dt}=0$$

and $\frac{dy}{dt}\neq 0$

$$2t=0$$

and $\quad 2t-1 \neq 0$

$\therefore \quad t = 0$

and $\quad t \neq \frac{1}{2}$

39. $\because y = a^{1-n} \cdot x^n$

$\therefore \quad \ln y = (1-n)\ln a + n \ln x$

$\therefore \quad \frac{1}{y}\frac{dy}{dx} = 0 + \frac{n}{x}$ or $\frac{dy}{dx} = \frac{ny}{x}$

$\therefore$ Length of sub normal $= y\frac{dy}{dx}$

$$= y \cdot \frac{ny}{x} = \frac{n}{x}y^2 = \frac{n}{x}a^{2(1-n)}x^{2n} = na^{2(1-n)}x^{2n-1}$$

Since, length of subnormal is to be constant, so x should not appear in its value i.e. $2n-1=0$

$\therefore \quad n = \frac{1}{2}$

40. Tanget and normal at $P(at^2, 2at)$ to the parabola $y^2 = 4ax$ are $t\,y = x + at^2$

and $\quad y + tx = 2at + at^3$ respectively.

$\therefore \quad T \equiv (-at^2, 0)$

and $\quad G \equiv (2a + at^2, 0)$

Equation of circle through T, P, G is

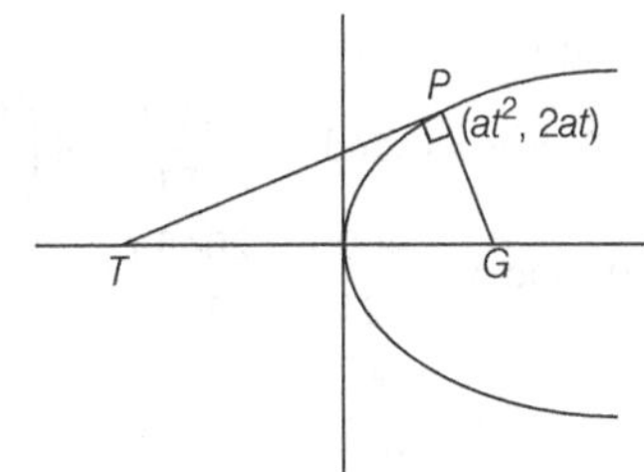

$(x + at^2)(x - 2a - at^2) + y^2 = 0 \quad$ [$\because$ TG is diameter]

$\Rightarrow \quad x^2 + y^2 - 2ax - (2a + at^2)at^2 = 0$

$\therefore$ Tangent at 'P' is

$x \times at^2 + y \times 2at - a\,(x + at^2) - (2a + at^2)at^2 = 0$

$\therefore$ Slope $= \frac{a - at^2}{2at} = \left(\frac{1-t^2}{2t}\right) = m_1$ (say)

And slope of tangent of $y^2 = 4ax$ at 'P' is $= \frac{1}{t} = m_2$ (say)

If θ be the required angle

$$\therefore \quad \tan\theta = \left|\frac{m_1 - m_2}{1 + m_1 m_2}\right| = \left|\frac{\frac{1-t^2}{2t} - \frac{1}{t}}{1 + \frac{1-t^2}{2t^2}}\right| = t$$

$\therefore \quad \theta = \tan^{-1}(t)$

41. Slope of line $(3-a)x + ay + (a^2 - 1) = 0$

is $\quad -\frac{(3-a)}{a}$ or $\left(\frac{a-3}{a}\right) \quad$ ... (i)

$\because \quad xy = 1$

$\therefore \quad x\frac{dy}{dx} + y \cdot 1 = 0$

$\frac{dy}{dx} = -\frac{y}{x} = -\frac{1}{x^2} \quad$ [$\because xy = 1$]

$\therefore$ Slope of normal $= x^2 = \left(\frac{a-3}{a}\right) \quad$ [from Eq. (i)]

$\Rightarrow \therefore \quad x^2 > 0$

$\frac{a-3}{a} > 0$

$\therefore \quad a \in (-\infty, 0) \cup (3, \infty)$

Objective Questions Type II [One or more than one correct answer(s)]

1. We have,

$f(x) = x - x^3 + x^5 - x^7 + \ldots$ upto infinity

$= \frac{x}{1 - (-x^2)} \quad$ [sum of infinite terms of a GP]

$= \frac{x}{1 + x^2}$

On differentiating w.r.t. 'x', we get

$$f'(x) = \frac{(1 + x^2)\cdot 1 - x\cdot(2x)}{(1 + x^2)^2} = \frac{(1 - x^2)}{(1 + x^2)^2}$$

Thus, slope of tangent is +ve ($\because -1 < x < 1$)

and slope of the line $(ax + by + c = 0)$ is $-\frac{a}{b}$

$\Rightarrow \quad -\frac{a}{b} > 0 \Rightarrow ab < 0$

Also, $\left|-\frac{a}{b}\right| = \left|\frac{(1 - x^2)}{(1 + x^2)^2}\right| \le 1$

$\Rightarrow \quad \frac{|a|}{|b|} \le 1$

$\Rightarrow \quad |a| \le |b|$

2. $\because x = a(\cos\theta + \theta\sin\theta)$

and $\quad y = a(\sin\theta - \theta\cos\theta)$

$\therefore \quad \frac{dx}{d\theta} = a(\theta\cos\theta), \frac{dy}{d\theta} = a(\theta\sin\theta)$

$\therefore \quad \frac{dy}{dx} = \frac{\sin\theta}{\cos\theta}$

$\therefore$ Slope of normal $= -\frac{\cos\theta}{\sin\theta}$

Equation of normal at 'θ' is

$y - a(\sin\theta - \theta\cos\theta) = -\frac{\cos\theta}{\sin\theta}(x - a(\cos\theta + \theta\sin\theta))$

$\Rightarrow y\sin\theta - a(\sin^2\theta - \theta\sin\theta\cos\theta) = -x\cos\theta +$

$a(\cos^2\theta + \theta\sin\theta\cos\theta)$

$\Rightarrow \quad x\cos\theta + y\sin\theta = a \quad$ ...(i)

Distance from origin to (i) $= a =$ constant

Hence, $x\cos\theta + y\sin\theta = a$

touches a fixed circle whose centre (0,0) and radius a, circle is

$$x^2 + y^2 = a^2$$

3. Let $P(x, y)$ be a point on the curve $\ln(x^2 + y^2) = c\tan^{-1}\frac{y}{x}$ ... (i)

On Differentiating (i) both sides w.r.t. x, we get

$$\frac{2x + 2yy'}{(x^2 + y^2)} = \frac{c(xy' - y)}{(x^2 + y^2)}$$

or $$y' = \frac{2x + cy}{cx - 2y} = m_1 \text{ (say)}$$

Slope of $OP = \frac{y}{x} = m_2$ (say) (where O is origin)

Let the angle between the tangents at P and OP be θ. Then,

$$\tan\theta = \left|\frac{m_1 - m_2}{1 + m_1m_2}\right| = \left|\frac{\frac{2x + cy}{cx - 2y} - \frac{y}{x}}{1 + \frac{2xy + cy^2}{cx^2 - 2xy}}\right| = \frac{2}{c}$$

or $$\theta = \tan^{-1}\left(\frac{2}{c}\right)$$

which is independent of x and y.

4. Slope of I curve $= 2x + a$

at (1, 0), $2 + a = m_1$ (say)

Slope of II curve $= c - 2x$

at (1, 0), $c - 2 = m_2$ (say)

For touch, $m_1 = m_2$

$$2 + a = c - 2 \qquad \text{...(i)}$$

and (1, 0) is on both the curves

$$0 = 1 + a + b \text{ and } 0 = c - 1 \qquad \text{...(ii)}$$

Solving Eqs. (i) and (ii), we get

$$a = -3,\ b = 2,\ c = 1$$

5. $8x + 18y\frac{dy}{dx} = 0$

$\therefore$ $$\frac{dy}{dx} = -\frac{4x}{9y}$$

Tangent is equally inclined to the axes.

$\Rightarrow$ $$\frac{dy}{dx} = \pm 1$$

$\therefore$ $$-\frac{4x}{9y} = \pm 1 \Rightarrow y = -\frac{4x}{9}, \frac{4x}{9}$$

$\therefore$ $$4x^2 + 9\cdot\frac{16}{81}x^2 = 36 \Rightarrow x = \pm\frac{9}{\sqrt{13}}$$

and $$y = \mp\frac{4}{\sqrt{13}}$$

6. $\because$ $$F(x) = \int_0^x 2|t|\,dt$$

$\therefore$ $$F'(x) = 2|x| = 1 \Rightarrow x = \pm\frac{1}{2}$$

$\Rightarrow$ $$F\left(\frac{1}{2}\right) = \int_0^{1/2} 2|t|\,dt = \{t^2\}_0^{1/2} = \frac{1}{4}$$

and $$F\left(-\frac{1}{2}\right) = \int_0^{-1/2} 2|t|\,dt = -2\int_0^{-1/2} t\,dt = -\frac{1}{4}$$

$\because$ Equation of tangents at $\left(\frac{1}{2}, \frac{1}{4}\right)$ and at $\left(-\frac{1}{2}, -\frac{1}{4}\right)$ are

$$y - \frac{1}{4} = 1\cdot\left(x - \frac{1}{2}\right)$$

$$y = x - \frac{1}{4}$$

and $$y + \frac{1}{4} = 1\cdot\left(x + \frac{1}{2}\right)$$

$$y = x + \frac{1}{4}$$

7. $\because$ $$y = x^4$$

$\because$ $$\frac{dy}{dx} = 4x^3$$

Equation of Tangent at (x, y) is

$$Y - y = 4x^3(X - x)$$

It pass through (2, 0), then

or $$0 - y = 4x^3(2 - x)$$

or $$-x^4 = 4x^3(2 - x) \qquad [\because y = x^4]$$

$\Rightarrow$ $$x^3(8 - 4x + x) = 0$$

$\because$ $$x = 0,\ x = \frac{8}{3}$$

$\because$ $$y = 0,\ y = \left(\frac{8}{3}\right)^4 = \frac{4096}{81}$$

$\because$ Point of tangency are (0,0)

and $\left(\frac{8}{3}, \frac{4096}{81}\right)$

Tangents are

$$y - 0 = 0 \Rightarrow y = 0 \text{ and } y - \frac{4096}{81} = 4\left(\frac{8}{3}\right)^3\left(x - \frac{8}{3}\right)$$

i.e. $y = 0$ and $y - \frac{4096}{81} = \frac{2048}{27}\left(x - \frac{8}{3}\right)$

8. $\because x = 4t^2 + 3,\ y = 8t^3 - 1$

$\therefore$ $$\frac{dy}{dt} = \frac{\frac{dy}{dt}}{\frac{dx}{dt}} = \frac{24t^2}{8t} = 3t$$

Let the tangent at $P(4t^2 + 3, 8t^3 - 1)$ be normal at

$$Q\,(4t_1^2 + 3, 8t_1^3 - 1)$$

$\therefore$ Equation of tangent at P is

$$y - (8t^3 - 1) = 3t\,(x - 4t^2 - 3) \qquad \text{...(i)}$$

Eq. (i) passes through Q, then

$$8\,(t_1^3 - t^3) = 3t\cdot 4\,(t_1^2 - t^2)$$

or $$(t_1 - t)^2\,(t + 2t_1) = 0$$

$\therefore$ $$t = -2t_1 \qquad [\because t \neq t_1] \text{ ...(ii)}$$

Given, $$\left.\frac{dy}{dx}\right|_{\text{at } t} = -\frac{1}{\left.\frac{dy}{dx}\right|_{\text{at } t_1}} \Rightarrow 3t = -\frac{1}{3t_1}$$

$\therefore$ $$t\,t_1 = -\frac{1}{9}$$

$\Rightarrow$ $$-2t_1^2 = -\frac{1}{9} \qquad \text{[from Eq. (ii)]}$$

$\therefore$ $$t_1 = \pm\frac{1}{3\sqrt{2}}$$

$\therefore$ Slope of line $= -\frac{1}{3t_1} = \mp\sqrt{2}.$

9. Since, $2y^3 = ax^2 + x^3$

$\Rightarrow \quad 6y^2 \dfrac{dy}{dx} = 2ax + 3x^2 \Rightarrow \dfrac{dy}{dx} = \left(\dfrac{2ax + 3x^2}{6y^2}\right)$

$\therefore \quad \left.\dfrac{dy}{dx}\right|_{(a,\, a)} = \dfrac{5a^2}{6a^2} = \dfrac{5}{6}$

$\Rightarrow$ Equation of tangent is,

$$y - a = \frac{5}{6}(x - a)$$

$\Rightarrow \quad 6y - 6a = 5x - 5a$

$\Rightarrow \quad -5x + 6y = a$ or $\dfrac{x}{(-a/5)} + \dfrac{y}{(a/6)} = 1$

$\therefore \quad \alpha = -\dfrac{a}{5}, \beta = \dfrac{a}{6}$

Since, $\alpha^2 + \beta^2 = 61$

$\therefore \quad \left(-\dfrac{a}{5}\right)^2 + \left(\dfrac{a}{6}\right)^2 = 61 \Rightarrow \dfrac{61a^2}{(30)^2} = 61$

$\therefore \quad a = \pm 30$

Linked Comprehension Type

PASSAGE 1

1. $\dfrac{dy}{dx} = \dfrac{1 - 9t^2}{-6t} = \tan\theta$

or $\quad 9t^2 - 6\tan\theta \cdot t - 1 = 0$

or $\quad 3t = \tan\theta \pm \sec\theta$

or $\quad \tan\theta + \sec\theta = 3t$

2. $P \equiv (-2, 2)$ or $t = -1$

$\therefore \quad \left.\dfrac{dy}{dx}\right|_{t=-1} = -\dfrac{4}{3} = m_P$

Equation of tangent is $y - 2 = -\dfrac{4}{3}(x + 2)$.

Therefore, $t - 3t^3 - 2 = -\dfrac{4}{3}(1 - 3t^2 + 2)$

$\Rightarrow \quad 9t^3 + 12t^2 - 3t - 6 = 0$

$\Rightarrow \quad 3t^3 + 4t^2 - t - 2 = 0$

$\Rightarrow \quad (3t^2 + t - 2)(t + 1) = 0$

$\Rightarrow \quad (3t - 2)(t + 1)^2 = 0$

$\Rightarrow \quad t = \dfrac{2}{3}$

$\Rightarrow \quad Q = \left(-\dfrac{1}{3}, -\dfrac{2}{9}\right)$

3. $\left.\dfrac{dy}{dx}\right|_{t=2/3} = \dfrac{3}{4} = m_Q$

$m_P m_Q = -1$

$\Rightarrow$ angle $= \dfrac{\pi}{2}$

PASSAGE 2

$\because \quad \dfrac{dx}{dt} = a(-2\sin t - 2\sin 2t)$

$= -2a \cdot 2\sin\left(\dfrac{3t}{2}\right)\cos\left(\dfrac{t}{2}\right)$

and $\quad \dfrac{dy}{dt} = a(2\cos t - 2\cos 2t)$

$= 2a \cdot 2\sin\left(\dfrac{3t}{2}\right)\sin\left(\dfrac{t}{2}\right)$

$\therefore \quad \dfrac{dy}{dx} = -\tan\left(\dfrac{t}{2}\right)$ and $\dfrac{dx}{dy} = -\cot\left(\dfrac{t}{2}\right)$

1. Equation of tangent at 't'

$y - a(2\sin t - \sin 2t)$

$= -\tan\left(\dfrac{t}{2}\right)(x - a(2\cos t + \cos 2t))$

$\Rightarrow \quad y\cos\left(\dfrac{t}{2}\right) - a\left(2\sin t\cos\dfrac{t}{2} - \sin 2t\cos\dfrac{t}{2}\right)$

$= -x\sin\dfrac{t}{2} + a\left(2\sin\dfrac{t}{2}\cos t + \sin\dfrac{t}{2}\cos 2t\right)$

$\Rightarrow \quad x\sin\left(\dfrac{t}{2}\right) + y\cos\left(\dfrac{t}{2}\right) + a\sin(2t - t/2)$

$- 2a\sin(t + t/2) = 0$

$x\sin\left(\dfrac{t}{2}\right) + y\cos\left(\dfrac{t}{2}\right) = a\sin\left(\dfrac{3t}{2}\right)$

2. Equation of normal at 't' is

$y - a(2\sin t - \sin 2t) = \cot\left(\dfrac{t}{2}\right) \quad (x - a(2\cos t + \cos 2t))$

$\Rightarrow y\sin\left(\dfrac{t}{2}\right) - a\left(2\sin t\sin\left(\dfrac{t}{2}\right) - \sin 2t\sin\left(\dfrac{t}{2}\right)\right)$

$= x\cos\left(\dfrac{t}{2}\right) - a\left(2\cos t\cos\left(\dfrac{t}{2}\right) + \cos 2t\cos\left(\dfrac{t}{2}\right)\right)$

$\Rightarrow y\sin\left(\dfrac{t}{2}\right) + 2a\cos\left(t + \dfrac{t}{2}\right) + a\cos\left(2t - \dfrac{t}{2}\right) = x\cos\left(\dfrac{t}{2}\right)$

$\Rightarrow x\cos\left(\dfrac{t}{2}\right) - y\sin\left(\dfrac{t}{2}\right) = 3a\cos\left(\dfrac{3t}{2}\right)$

3. Length of sub tangent $= \left|y\dfrac{dx}{dy}\right|$

$= |y \times (-\cot t/2)|$

$= |y\cot(t/2)|$

4. Length of sub normal $= \left|y\dfrac{dy}{dx}\right|$

$= \left|y\left(-\tan\dfrac{t}{2}\right)\right| = \left|y\tan\dfrac{t}{2}\right|$

5. $p = \dfrac{a\sin\left(\dfrac{3t}{2}\right)}{1}$ and $p_1 = \dfrac{3a\cos\left(\dfrac{3t}{2}\right)}{1}$

$\therefore \quad 9p^2 + p_1^2 = 9a^2.$

PASSAGE 3

1. Let $f(x)=a\,x^3+b\,x^2+c\,x+d$

$\because \quad f(0)=5$

$\Rightarrow\ 0+0+0+d=5$

Then, $\quad f(x)=ax^3+bx^2+c\,x+5$

$f'(x)=3ax^2+2bx+c$

$f'(-2)=12a-4b+c=0 \qquad \text{...(i)}$

and $\quad y=f(x)$ passes through $(-2, 0)$

Then, $\quad 0=-8a+4b-2c+5 \qquad \text{...(ii)}$

and $\quad f'(0)=3$

$\Rightarrow \quad c=3$

From Eq. (i),

$12a-4b+3=0 \qquad \text{...(iii)}$

From Eq. (ii),

$-8a+4b-1=0 \qquad \text{...(iv)}$

From Eqs. (iii) and (iv), we get

$$a=-\frac{1}{2}, b=-\frac{3}{4}$$

Then, $f(x)=-\frac{1}{2}x^3-\frac{3}{4}x^2+3x+5$ (1.(a))

2. (b) $\quad f(2)=-4-3+6+5=4$

3.

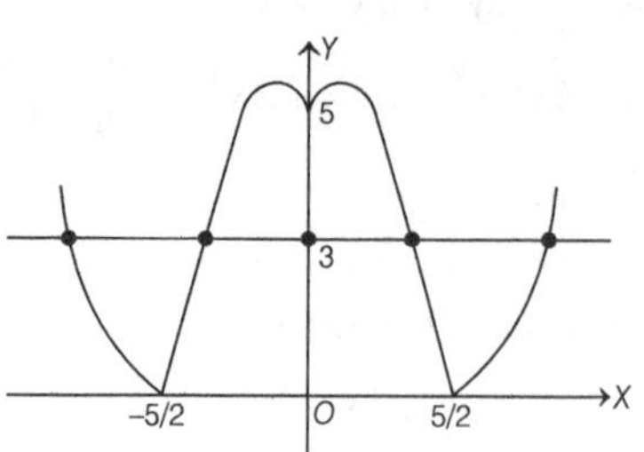

4. $f'(x)=-\frac{3}{2}x^2-\frac{3}{2}x+3$

$$f'\left(\frac{5}{2}\right)=-\frac{3}{2}\times\frac{25}{4}-\frac{3}{2}\times\frac{5}{2}+3=-\frac{75}{8}-\frac{15}{4}+3=-\frac{81}{8}$$

Slope of normal $=\frac{8}{81}$

Equation of normal

$$y-0=\frac{8}{81}\left(x-\frac{5}{2}\right) \Rightarrow 81y=8x-20$$

or $\quad 8x-81y-20=0$

5. $f'(0)=3$

$\therefore$ Equation of tangent at Q $(0, 5)$ is

$$y-5=3(x-0) \Rightarrow 3x-y+5=0$$

Numerical Value Type

1. $\because \frac{dy}{dx}=3ax^2+2bx+c \qquad \text{...(i)}$

Since, the curve $y=ax^3+bx^2+cx+5$ touches X-axis at $P(-2, 0)$, then X-axis is the tangent at $(-2, 0)$ the curve meets Y-axis in $(0, 5)$.

$\therefore \quad \left.\frac{dy}{dx}\right|_{(0,5)}=0+0+c=3$ (given) $\qquad \text{...(ii)}$

and $\left.\frac{dy}{dx}\right|_{(-2,0)}=0 \Rightarrow 12a-4b+c=0$

$\Rightarrow 12a-4b+3=0 \qquad$ [from Eq. (ii)] ...(iii)

and $(-2, 0)$ lies on the curve, then

$0=-8a+4b-2c+5$

$\Rightarrow \quad -8a+4b-1=0 \qquad [\because c=3]$

$\Rightarrow \quad 8a-4b+1=0 \qquad \text{...(iv)}$

From Eqs. (iii) and (iv) we get

$$a=-\frac{1}{2}, b=-\frac{3}{4}$$

$$\therefore -10a-100b+1000c=-10\times\frac{-1}{2}-100\times\frac{-3}{4}+1000\times 3$$

$$=5+75+3000=3080$$

2. Let $P_1(t_1, t_1^3)$ is a point on the curve $y=x^3$

$$\therefore \quad \left.\frac{dy}{dx}\right|_{(t_1,\, t_1^3)}=3t_1^2$$

Tangent at P_1 is $y-t_1^3=3t_1^2(x-t_1) \qquad \text{...(i)}$

The intersection of Eq. (i) and $y=x^3$

$\Rightarrow \quad x^3-t_1^3=3t_1^2(x-t_1) \Rightarrow (x-t_1)^2(x+2t_1)=0$

If $P_2(t_2, t_2^3)$, then $(t_2-t_1)^2(t_2+2t_1)=0$

$\therefore \quad t_2=-2t_1 (t_2 \neq t_1)$

Similarly the tangent at P_2 will meet the curve at the point $P_3(t_3, t_3^3)$ when $t_3=-2t_2=4t_1$ and so on. The abscissaes of $P_1, P_2, P_3, \ldots$ are $t_1, -2t_1, 4t_1, \ldots$ in GP.

$$\therefore \quad \frac{t_2}{t_1}=\frac{t_3}{t_2}=\frac{t_4}{t_3}=\ldots=-2 (r \text{ say})$$

$$\therefore \quad t_2=t_1r, t_3=t_2r, t_4=t_3r$$

$$\frac{\text{Area of } \Delta P_1P_2P_3}{\text{Area of } \Delta P_2P_3P_4}=\frac{\frac{1}{2}\left|\begin{vmatrix} t_1 & t_1^3 & 1 \\ t_2 & t_2^3 & 1 \\ t_3 & t_3^3 & 1 \end{vmatrix}\right|}{\frac{1}{2}\left|\begin{vmatrix} t_2 & t_2^3 & 1 \\ t_3 & t_3^3 & 1 \\ t_4 & t_4^3 & 1 \end{vmatrix}\right|}$$

$$=\frac{\left|\begin{vmatrix} t_1 & t_1^3 & 1 \\ t_2 & t_2^3 & 1 \\ t_3 & t_3^3 & 1 \end{vmatrix}\right|}{\left|\begin{vmatrix} t_1r & t_1^3r^3 & 1 \\ t_2r & t_2^3r^3 & 1 \\ t_3r & t_3^3r^3 & 1 \end{vmatrix}\right|}=\frac{1}{r^4}=\frac{1}{(-2)^4}$$

$$= \frac{1}{16} = \frac{\lambda}{\mu} \quad \text{[given]}$$

$$\therefore \quad \frac{251}{2} \cdot \frac{\mu}{\lambda} = \frac{251}{2} \times 16$$

$$= 251 \times 8 = 2008$$

3. Let a be the side of an equilateral triangle, then

$$\frac{da}{dt} = 3 \quad \text{[given]}$$

Now, area of equilateral triangle $A = \frac{\sqrt{3}}{4} a^2$

$$\therefore \quad dA = \frac{\sqrt{3}}{4} \cdot 2a \frac{da}{dt}$$

$$= \frac{\sqrt{3}}{2} \cdot 10 \cdot 3$$

$$= 15\sqrt{3} \text{ ft}^2 / \text{s}$$

$$\therefore \quad \lambda = 15\sqrt{3}$$

Then, $0.5\sqrt{3}\lambda = 0.5\sqrt{3} \times 15\sqrt{3} = 22.5$

4. Here, $r = h$

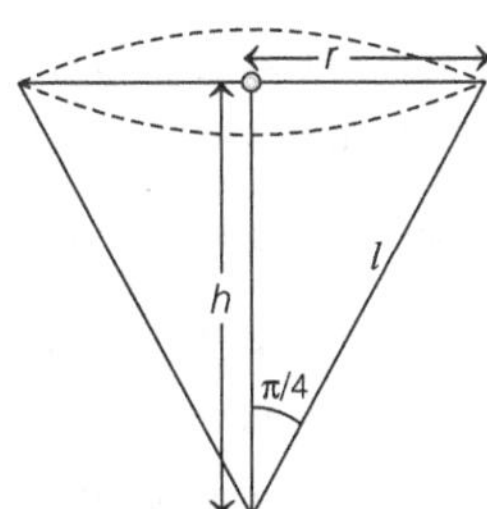

$$\therefore \quad l^2 = r^2 + h^2 = 2r^2 \quad \text{or} \quad h = \frac{l}{\sqrt{2}}, r = \frac{l}{\sqrt{2}}$$

$$V = \frac{1}{3} \pi r^2 h = \frac{1}{3} \pi \left(\frac{l}{\sqrt{2}}\right)^2 \left(\frac{l}{\sqrt{2}}\right) = \frac{\pi}{6\sqrt{2}} l^3$$

$$\therefore \quad \frac{dV}{dt} = \frac{3\pi}{6\sqrt{2}} l^2 \frac{dl}{dt}$$

$$\Rightarrow \quad 2 = \frac{3\pi}{6\sqrt{2}} (4)^2 \frac{dl}{dt}$$

$$\Rightarrow \quad \frac{dl}{dt} = \frac{\sqrt{2}}{4\pi} \text{ cm/s}$$

$\therefore$ Here, $\lambda = \frac{\sqrt{2}}{4\pi}$

Then, $40 \pi \sqrt{2} \lambda = 40 \pi \sqrt{2} \times \frac{\sqrt{2}}{4\pi} = 10 \times 2 = 20$

5. $\because \quad I = \frac{PLAN}{33000} = \frac{PL\left(\frac{\pi}{4} d^2\right) N}{33000}$

$$\therefore \quad \frac{\Delta I}{I} = \frac{\Delta P}{P} + \frac{\Delta L}{L} + \frac{2\Delta d}{d} + \frac{\Delta N}{N}$$

$$\Rightarrow \quad \left(\frac{\Delta I}{I} \times 100\right) = \left(\frac{\Delta P}{P} \times 100\right) + \left(\frac{\Delta L}{L} \times 100\right) + 2\left(\frac{\Delta d}{d} \times 100\right) + \left(\frac{\Delta N}{N} \times 100\right)$$

$$= 10\% + 10\% + 2 \times 10\% + 10\% = 50\%$$

$$\therefore \quad \lambda = 50$$

6. If r be the radius and h be the height of the cylinder.

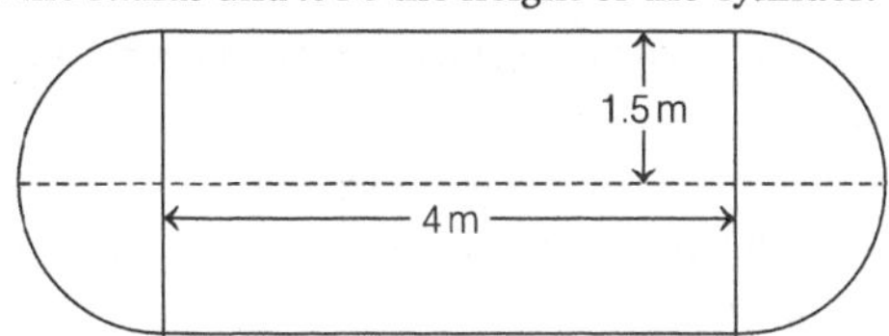

Volume $V = \pi r^2 h + \frac{2}{3} \pi r^3 + \frac{2}{3} \pi r^3$

$$= \pi r^2 h + \frac{4}{3} \pi r^3$$

$$\therefore \quad \Delta V = (\pi r^2)\Delta h + h(2\pi r \Delta r) + \frac{4}{3} \pi \cdot 3r^2 \Delta r$$

$$= \pi r^2 \Delta h + 2\pi rh \Delta r + 4\pi r^2 \Delta r$$

$$\therefore \quad \frac{\Delta V}{V} = \frac{\pi r (r\Delta h + 2h\Delta r + 4r\Delta r)}{\pi r\left(rh + \frac{4}{3} r^2\right)}$$

$$= \frac{r\Delta h + 2h\Delta r + 4r\Delta r}{rh + \frac{4}{3} r^2}$$

$$= \frac{1.5 \times 0.05 + 2 \times 4 \times 0.01 + 4 \times 1.5 \times 0.01}{1.5 \times 4 + \frac{4}{3}(1.5)^2}$$

$$= \frac{0.215}{9}$$

$$\Rightarrow \quad \frac{\Delta V}{V} \times 100 = \frac{0.215}{9} \times 100$$

$$= \frac{21.5}{9} = 2.389\%$$

$$\therefore \quad abcd = 2389$$

7. Let $f(x) = \log_{10} x \Rightarrow f'(x) = \frac{1}{x} . \log_{10} e$

Let $x = 4, \delta x = 0.04$

$\therefore \quad f(x + \delta x) = f(x) + \delta x \cdot f'(x)$

$$\log_{10}(x + \delta x) = \log_{10} x + (0.04) \times \frac{1}{x} \log_{10} e$$

$$\log_{10} 4.04 = \log_{10} 4 + (0.04) \times \frac{1}{4} \times \log_{10} e$$

$$= 0.6021 + (0.01)(0.4343)$$

$$= 0.6021 + 0.004343$$

$$= 0.606443$$

8. Let $f(x) = x^6$

Now, $f(x + \delta x) - f(x) = f'(x) \cdot \delta x = 6x^5 \cdot \delta x$

we may write $1.999 = 2 - 0.001$

Taking $x = 2$ and $\delta x = -0.001$

we have,

$f(1.999) - f(2) = 6(2)^5 \times -0.001$

$$\Rightarrow \quad f(1.999) = f(2) - 6 \times 32 \times 0.001$$

$$= 64 - 64 \times 0.003$$

$$= 64 \times 0.997$$

$$= 63.808 \text{ (approx)}$$

$$\therefore \quad abcd = 6380$$

Single Integer Type

1. The equation of tangent to the curve, $y = e^x$ at a point (c, e^c) is

$$y - e^c = e^c(x - c) \quad \text{...(i)}$$

and equation of normal to the curve, $y^2 = 4x$ at the point (1, 2) is

$$y - 2 = -1(x - 1) \quad \text{...(ii)}$$

$\because$ The lines (i) and (ii) intersect at same point on the X-axis, so put $y = 0$ in both the equation and equate, we get

$$x = 3 = c - 1 \Rightarrow c = 4$$

Hence, answer is 4.

2. Given, $x = t^2; y = t^3$

$$\frac{dx}{dt} = 2t; \frac{dy}{dt} = 3t^2$$

$$\frac{dy}{dx} = \frac{3t}{2}$$

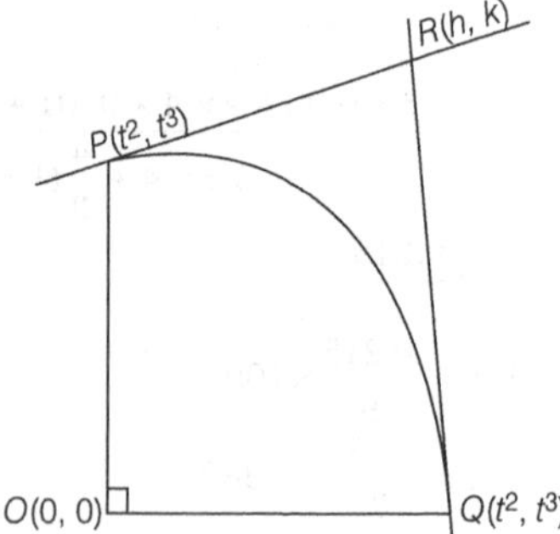

Equation of tangent at P is

$$y - t^3 = \frac{3t}{2}(x - t^2)$$

$$2k - 2t^3 = 3th - 3t^3$$

$$\therefore \quad t^3 - 3th + 2k = 0 \quad \text{...(i)}$$

Product of roots, $t_1 t_2 t_3 = -2k$

Putting $t_1 t_2 = -1, t_3 = 2k$

Now, t_3 must satisfy Eq. (i). Therefore,

$$(2k)^3 - 3(2k)h + 2k = 0$$

i.e., $4y^2 - 3x + 1 = 0$ or $4y^2 = 3x - 1$ or $a + b = 7$

3. $\because I = \dfrac{PLAN}{33000} = \dfrac{PL\left(\frac{\pi}{4}d^2\right)N}{33000}$

$$\therefore \quad \frac{\Delta I}{I} \times 100 = \left(\frac{\Delta P}{P} \times 100\right) + \left(\frac{\Delta L}{L} \times 100\right) + 2$$

$$\left(\frac{\Delta d}{d} \times 100\right) + \left(\frac{\Delta N}{N} \times 100\right) = r\% + r\% + 2r\% + r\%$$

$$= 5r\% = \lambda r\% \quad \text{(given)}$$

$$\therefore \quad \lambda = 5$$

4. Given, $y = x^4$

On differentiating w.r.t. x, we get

$$\frac{dy}{dx} = 4x^3$$

Equation of tangent at (α, α^4) is

$$y - \alpha^4 = 4\alpha^3(x - \alpha)$$

Since, it passes through the point $\left(\dfrac{3}{4}, 0\right)$

$$\therefore \quad -\alpha^4 = 3\alpha^3 - 4\alpha^4 \Rightarrow \alpha = 0, 1$$

$\therefore$ There are 2 tangents

5. Since, $y = e^{a + bx^2}$ passes through (1, 1)

$$\therefore \quad 1 = e^{a+b} \Rightarrow a + b = 0$$

Also, $\left(\dfrac{dy}{dx}\right)_{(1, 1)} = -2$

$$\therefore \quad [e^{a + bx^2} \cdot 2bx]_{(1, 1)} = -2$$

$$\Rightarrow \quad e^{a+b} \cdot 2b(1) = -2$$

$$\therefore \quad b = -1 \text{ and } a = 1$$

$$\Rightarrow \quad 2a - 3b = 5$$

Matrix-Match Type

1. **(A)** $\because \; 4y\dfrac{dy}{dx} = 2ax$

$$\Rightarrow \left.\frac{dy}{dx}\right|_{(1,-1)} = \frac{a}{-2} = -1$$

$$\Rightarrow \quad a = 2.$$

Also, $(1, -1)$ lies on $2y^2 = ax^2 + b$

$$\Rightarrow \quad 2 = a + b$$

$\therefore \quad a = 2$, then $b = 0 \Rightarrow a - b = 2$ (P)

(B) $\because$ (a, b) lies on the curve $9y^2 = x^3$

Therefore, $9b^2 = a^3$ ...(i)

$$18y\frac{dy}{dx} = 3x^2$$

and $\quad \dfrac{dy}{dx} = \dfrac{x^2}{6y} \Rightarrow \left.\dfrac{dy}{dx}\right|_{(a, b)} = \dfrac{a^2}{6b}$

Since, the normal to the curve at (a, b) make equal intercepts with the coordinate axes.

Therefore, slope of normal $= \pm 1$.

$$\Rightarrow -\frac{1}{\left.\frac{dy}{dx}\right|_{(a,b)}} = \pm 1 \quad \Rightarrow \quad \left.\frac{dy}{dx}\right|_{(a,b)} = \pm 1$$

$$\Rightarrow \quad \frac{a^2}{6b} = \pm 1 \quad \Rightarrow \quad a^2 = \pm 6b$$

$$\Rightarrow \quad a^4 = 36b^2 = 36\left(\frac{a^3}{9}\right) \quad \text{[from Eq. (i)]}$$

$\therefore \quad a = 0, 4$, then $b = 0, \pm 8/3$

But the line making equal intercepts with the coordinate axes can not pass through the origin

Hence, the required points are $(4, 8/3)$ and $(4, -8/3)$

$\therefore \quad a - b = 4/3$ (R)

and $\quad a - b = 20/3$

and $\quad a + |b| = 4 + 8/3 = 20/3$ (S)

(C) $\because$ (1, 2) lies on $y = ax^2 + bx + \frac{7}{2}$

$$\therefore \quad 2 = a + b + 7/2$$

$$\Rightarrow \quad a + b + 3/2 = 0 \quad \text{...(i)}$$

$$\left.\frac{dy}{dx}\right|_{(1,2)} = 2a(1) + b = 2a + b \quad \text{...(ii)}$$

and $$-\frac{1}{\left.\frac{dy}{dx}\right|_{(-2,2)}} = -\frac{1}{2(-2)+6} = -\frac{1}{2} \quad \text{...(iii)}$$

Given, $2a + b = -\frac{1}{2}$ ...(iv)

Solving Eqs. (i) and (iv), then we get

$$a = 1,\ b = -5/2$$

$$a - b = 7/2 \text{ (Q)}$$

and $5a + 2b = 0$ (T)

2. **(A)** $\because \frac{dy}{dx} = \frac{1}{2y}$

$$\Rightarrow \quad \left.\frac{dy}{dx}\right|_{(t^2, t)} = \frac{1}{2t}$$

$\therefore$ Slope of normal $= -2t$

$\therefore$ Equation of normal at (t^2, t) is

$$y - t = -2t(x - t^2)$$

1t passes through $(c, 0)$, then

$$0 - t = -2t(c - t^2)$$

$$\Rightarrow \quad t^2 = c - \frac{1}{2}$$

For three distinct values $t^2 > 0$

$$\Rightarrow \quad c > \frac{1}{2}$$

$$\therefore \quad c = 1, 5/4, 2 \text{ (P, R, T)}$$

(B) $\because \quad xy = c^2$

$$\therefore \quad x\frac{dy}{dx} + y \cdot 1 = 0$$

$$\Rightarrow \quad \frac{dy}{dx} = -\frac{y}{x}$$

Subnormal length $= \left|y\frac{dy}{dx}\right| = \left|y\left(\frac{-y}{x}\right)\right| = \frac{y^2}{x}$

$$\frac{y^2}{\frac{c^2}{y}} = \frac{y^3}{c} \propto y^3 \text{ (S)}$$

(C) $\because A + B + C = \pi \Rightarrow dA + dB + dC = 0$

If R is circumradius, then $\frac{a}{\sin A} = \frac{b}{\sin B} = \frac{c}{\sin C} = 2R$

$$\therefore \quad a = 2R\sin A$$

or $$da = 2R\cos A\,dA$$

$$\Rightarrow \quad \frac{da}{\cos A} = 2R\,dA$$

$$\therefore \quad \frac{da}{\cos A} + \frac{db}{\cos B} + \frac{dc}{\cos C} = 2R(dA + dB + dC) = 0 \text{ **(Q)**}$$

3. **(A)** Given curves are,

$$y^2 = 4ax \quad \text{...(i)}$$

and $$xy = \sqrt{2} \quad \text{...(ii)}$$

from Eq. (i), $2y\frac{dy}{dx} = 4a$

$$\Rightarrow \quad \frac{dy}{dx} = \frac{2a}{y}$$

[$\because$ differentiating w.r.t. 'x', we get]

From Eq. (ii), $x\frac{dy}{dx} + y \cdot 1 = 0$

$$\Rightarrow \quad \frac{dy}{dx} = -\frac{y}{x}$$

Since, curves (i) and (ii) cut orthogonally.

$$\therefore \quad \left(\frac{2a}{y}\right)\left(-\frac{y}{x}\right) = -1 \Rightarrow x = 2a \quad \text{...(iii)}$$

Also, putting the value of y from Eq. (ii) in Eq. (i), we get

$$\frac{2}{x^2} = 4ax \Rightarrow 1 = 2ax^3$$

$$\Rightarrow \quad 1 = 2a(2a)^3 \quad \text{[from (iii)]}$$

$$\therefore \quad \frac{1}{a^2} = 4 \text{ (Q, S)}$$

(B) Given curves are,

$$x^2 + ay = 7 \quad \text{...(i)}$$

and $$x^3 = y \quad \text{...(ii)}$$

from Eq. (i), $\quad 2x + a\frac{dy}{dx} = 0 \Rightarrow \frac{dy}{dx} = -\frac{2x}{a}$

[$\because$ differentiating w.r.t. 'x', we get]

From Eq. (ii), $\quad 3x^2 = \frac{dy}{dx} \Rightarrow \frac{dy}{dx} = 3x^2$

Since, curves (i) and (ii) cut orthogonally at (1, 1).

$$\therefore \quad \left(-\frac{2x}{a}\right)(3x^2) = -1$$

$$\Rightarrow \quad \left(-\frac{2}{a}\right)(3) = -1$$

$$\Rightarrow \quad a = 6 \text{ (Q, R, S)}$$

(C) Curve $y = 2e^{2x}$ intersect Y-axis at (0, 2)

$$\therefore \quad \left.\frac{dy}{dx}\right|_{(0,2)} = 4$$

$\Rightarrow$ Angle of intersection with Y-axis

$$= \frac{\pi}{2} - \tan^{-1} 4 = \cot^{-1} 4 = \cot^{-1}(|(8n-4)/3|) \text{ (given)}$$

$$\therefore \quad 4 = \left|\frac{8n-4}{3}\right|$$

$$\Rightarrow \quad 8n - 4 = \pm 12$$

$$\Rightarrow \quad 2n = 1 \pm 3$$

$$\therefore \quad n = 2, -1$$

but $\quad n > 0$

$$\therefore \quad n = 2 \text{ (P, Q)}$$

15

Monotonocity, Rolle's & Lagrange's mean Value Theorems

Objective Questions Type I [Only one correct answer]

In each of the questions below, four choices are given of which only one is correct. You have to select the correct answer which is the most appropriate.

1. Let f and g be non-increasing and non-decreasing functions respectively from $[0, \infty)$ to $[0, \infty)$ and $h(x) = f(g(x))$, $h(0) = 0$, then in $[0, \infty)$, $h(x) - h(1)$ is

(a) < 0 (b) > 0
(c) $= 0$ (d) increasing

2. If $ax^2 + \frac{b}{x} \geq c \; \forall \; x > 0$, where $a > 0, b > 0$, then

(a) $27 \, ab^2 \geq 4c^3$ (b) $27 \, ab^3 \leq 4c^3$
(c) $ab^2 \geq c^3$ (d) $ab^3 \leq c^3$

3. Let the function, $f : [-7, 0] \to R$ be continuous on $[-7, 0]$ and differentiable on $(-7, 0)$. If $f(-7) = -3$ and $f'(x) \leq 2$, for all $x \in (-7, 0)$, then for all such functions f, $f(-1) + f(0)$ lies in the interval

[JEE Main 2020, 7 January]

(a) $(-\infty, 11]$ (b) $[-3, 11]$
(c) $[-6, 20]$ (d) $(-\infty, 20]$

4. If $f(x) = x^{\alpha} \ln x$ and $f(0) = 0$, then the value of α for which Rolle's theorem can be applied in $[0, 1]$ is

(a) -2 (b) -1 (c) 0 (d) $1/2$

5. The function $f(x) = \sin^4 x + \cos^4 x$ increases, if

(a) $0 < x < \frac{\pi}{8}$ (b) $\frac{\pi}{4} < x < \frac{3\pi}{8}$
(c) $\frac{3\pi}{8} < x < \frac{5\pi}{8}$ (d) $\frac{5\pi}{8} < x < \frac{3\pi}{4}$

6. Let f be a real valued function defined on $R - \{-1, 1\}$ and given by $f(x) = 3 \log_e \left| \frac{x-1}{x+1} \right| - \frac{2}{x-1}$. Then, in which of the following intervals, function $f(x)$ is increasing?

(a) $(-\infty, -1) \cup \left(\left[\frac{1}{2}, \infty \right) - \{1\} \right)$

(b) $(-\infty, \infty) - \{-1, 1\}$

(c) $\left(-1, \frac{1}{2} \right]$

(d) $\left(-\infty, \frac{1}{2} \right] - \{-1\}$

7. The function $f(x) = \cos \left(\frac{\pi}{x} \right)$ is decreasing in the interval

(a) $(2n + 1, 2n), n \in N$ (b) $\left(\frac{1}{2n+1}, 2n \right), n \in N$

(c) $\left(\frac{1}{2n+2}, \frac{1}{2n+1} \right), n \in N$ (d) None of these

8. The set of all values of a for which the function

$$f(x) = \left(\frac{\sqrt{a+4}}{1-a} - 1 \right) x^5 - 3x + \log 5$$

decreases for all real x is

(a) $(-\infty, \infty)$

(b) $\left[-4, \frac{3 - \sqrt{21}}{2} \right] \cup (1, \infty)$

(c) $\left(-3, 5 - \frac{\sqrt{27}}{2} \right) \cup (2, \infty)$

(d) $[1, \infty)$

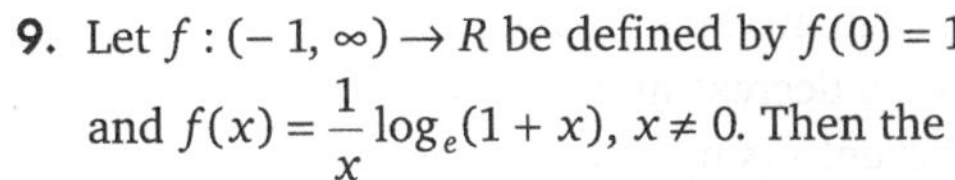

9. Let $f : (-1, \infty) \to R$ be defined by $f(0) = 1$ and $f(x) = \dfrac{1}{x}\log_e(1 + x)$, $x \neq 0$. Then the function f **[JEE Main 2020, 2 September]**

(a) decreases in $(-1, 0)$ and increases in $(0, \infty)$
(b) increases in $(-1, \infty)$
(c) increases in $(-1, 0)$ and decreases in $(0, \infty)$
(d) decreases in $(-1, \infty)$

10. If f be a continuous function on $[0, 1]$, differentiable in $(0, 1)$ such that $f(1) = 0$, then their exists some $c \in (0, 1)$ such that

(a) $cf'(c) - f(c) = 0$
(b) $f'(c) + cf(c) = 0$
(c) $f'(c) - cf(c) = 0$
(d) $cf'(c) + f(c) = 0$

11. The length of longest interval in which Rolle's theorem can be applied for the function

$$f(x) = |x^2 - a^2|, \ (a > 0) \text{ is}$$

(a) $2a$
(b) $4a^2$
(c) $a\sqrt{2}$
(d) a

12. In the function $f(x) = ax^3 + bx^2 + 11x - 6$ satisfies conditions of Rolle's theorem in $[1, 3]$ and $f'\left(2 + \dfrac{1}{\sqrt{3}}\right) = 0$, then value of a and b are respectively

(a) $1, -6$
(b) $-1, 6$
(c) $-2, 1$
(d) $-1, 1/2$

13. Let $f : [0, 2] \to R$ be a twice differentiable function such that $f''(x) > 0$, for all $x \in (0, 2)$. If $\phi(x) = f(x) + f(2 - x)$ then ϕ is **[JEE Main 2019, 8 April]**

(a) increasing on $(0, 1)$ and decreasing on $(1, 2)$
(b) decreasing on $(0, 2)$
(c) decreasing on $(0, 1)$ and increasing on $(1, 2)$
(d) increasing on $(0, 2)$

14. If $f(x) = (ab - b^2 - 2)\,x + \int_0^x (\cos^4\theta + \sin^4\theta)\,d\theta$ is decreasing function of x for all $x \in R$ and $b \in R$, b being independent of x, then

(a) $a \in (0, \sqrt{6})$
(b) $a \in (-\sqrt{6}, \sqrt{6})$
(c) $a \in (-\sqrt{6}, 0)$
(d) None of these

15. The value of a in order that $f(x) = \sqrt{3}\sin x - \cos x - 2ax + b$ decreases for all real values of x, is given by

(a) $a < 1$
(b) $a \geq 1$
(c) $a \geq \sqrt{2}$
(d) $a < \sqrt{2}$

16. If $f : R \to R$ is a twice differentiable function such that $f''(x) > 0$ for all $x \in R$, and $f\left(\dfrac{1}{2}\right) = \dfrac{1}{2}$, $f(1) = 1$, then **[JEE Advanced 2017]**

(a) $f'(1) \leq 0$
(b) $f'(1) > 1$
(c) $0 < f'(1) \leq \dfrac{1}{2}$
(d) $\dfrac{1}{2} < f'(1) \leq 1$

17. The function f defined by $f(x) = (x + 2)\,e^{-x}$ is

(a) decreasing for all x
(b) decreasing on $(-\infty, -1)$ and increasing in $(-1, \infty)$
(c) increasing for all x
(d) decreasing in $(-1, \infty)$ and increasing in $(-\infty, -1)$

18. If $a < 0$, the function $f(x) = e^{ax} + e^{-ax}$ is a monotonically decreasing function for values of x given by

(a) $x > 0$
(b) $x < 0$
(c) $x > 1$
(d) $x < 1$

19. Let $f(x)$ satisfy the requirement of Lagrange's mean value theorem in $[0, 2]$. If $f(0) = 0$ and $|f'(x)| \leq \dfrac{1}{2}$ for all $x \in [0, 2]$, then

(a) $f(x) \leq 2$
(b) $|f(x)| \leq 1$
(c) $f(x) = 2x$
(d) $f(x) = 3$ for at least one x in $[0, 2]$

20. Let f be any function continuous on $[a, b]$ and twice differentiable on (a, b). If for all $x \in (a, b)$, $f'(x) > 0$ and $f''(x) < 0$, then for any $c \in (a, b)$, $\dfrac{f(c) - f(a)}{f(b) - f(c)}$ is greater than **[JEE Main 2020, 9 January]**

(a) $\dfrac{b - c}{c - a}$
(b) 1
(c) $\dfrac{c - a}{b - c}$
(d) $\dfrac{b + a}{b - a}$

21. If the function $f(x) = \cos|x| - 2ax + b$ increases along the entire number scale, the range of values of a is given by

(a) $a \leq b$
(b) $a = b/2$
(c) $a \leq -1/2$
(d) $a \geq -3/2$

22. The function $f(x) = x\sqrt{(ax - x^2)}$, $a > 0$

(a) increases on the interval $(0, 3a/4)$
(b) decreases on the interval $(3a/4, a)$
(c) decreases on the interval $(0, 3a/4)$
(d) increases on the interval $(3a/4, a)$

23. If $f''(x) > 0, \forall x \in R$, $f'(3) = 0$ and $g(x) = f(\tan^2 x - 2\tan x + 4)$, $0 < x < \dfrac{\pi}{2}$, then $g(x)$ is increasing in

(a) $\left(0, \dfrac{\pi}{4}\right)$
(b) $\left(\dfrac{\pi}{6}, \dfrac{\pi}{3}\right)$
(c) $\left(0, \dfrac{\pi}{3}\right)$
(d) $\left(\dfrac{\pi}{4}, \dfrac{\pi}{2}\right)$

24. By LMVT, which of the following is true for $x > 1$

(a) $1 + x\ln x < x < 1 + \ln x$
(b) $1 + \ln x < x < 1 + x\ln x$
(c) $x < 1 + x\ln x < 1 + \ln x$
(d) $1 + \ln x < 1 + x\ln x < x$

25. If $f(x) = ax^3 - 9x^2 + 9x + 3$ is increasing on R, then

(a) $a < 3$
(b) $a > 3$
(c) $a \leq 3$
(d) None of these

26. Function $f(x) = |x| - |x-1|$ is monotonically increasing when

(a) $x < 0$ (b) $x > 1$
(c) $x < 1$ (d) $0 < x < 1$

27. Consider the function $f : R \to R$ defined by

$$f(x) = \begin{cases} \left(2 - \sin\left(\frac{1}{x}\right)\right)|x|, & x \neq 0 \\ 0, & x = 0 \end{cases}, \text{ then } f \text{ is}$$

(a) monotonic on $(-\infty, 0) \cup (0, \infty)$
(b) not monotonic on $(-\infty, 0)$ and $(0, \infty)$
(c) monotonic on $(0, \infty)$ only
(d) monotonic on $(-\infty, 0)$ only

28. Let $f(x) = x^3 + ax^2 + bx + 5\sin^2 x$ be an increasing function in the set of real numbers R. Then, a and b satisfy the condition

(a) $a^2 - 3b - 15 > 0$ (b) $a^2 - 3b + 15 > 0$
(c) $a^2 - 3b + 15 < 0$ (d) $a > 0$ and $b > 0$

29. If $f : R \to R$ is the function defined by

$$f(x) = \frac{e^{x^2} - e^{-x^2}}{e^{x^2} + e^{-x^2}}, \text{ then}$$

(a) $f(x)$ is an increasing function
(b) $f(x)$ is a decreasing function
(c) $f(x)$ is onto (surjective)
(d) None of the above

30. If $f'(x) = |x| - \{x\}$, where $\{x\}$ denotes the fractional part of x, then $f(x)$ is decreasing in

(a) $\left(-\frac{1}{2}, 0\right)$ (b) $\left(-\frac{1}{2}, 2\right)$
(c) $\left(-\frac{1}{2}, 2\right]$ (d) $\left(\frac{1}{2}, \infty\right)$

31. The function $f(x) = x^x$ decreases on the interval

(a) $(0, e)$ (b) $(0, 1)$
(c) $(0, 1/e)$ (d) None of these

32. Let $f(x) = \int e^x (x-1)(x-2)\, dx$, then f decreases in the interval

(a) $(-\infty, -2)$ (b) $(-2, -1)$
(c) $(1, 2)$ (d) $(2, \infty)$

33. For all $x \in (0, 1)$

(a) $e^x < 1 + x$ (b) $\log_e (1 + x) < x$
(c) $\sin x > x$ (d) $\log_e x > x$

Objective Questions Type II [One or more than one correct answer(s)]

In each of the questions below four choices of which one or more than one are correct. You have to select the correct answer(s) accordingly.

1. Let $f(x) = \sin^{-1}\left\{\frac{2g(x)}{1 + (g(x))^2}\right\}$, then

(a) $f(x)$ is decreasing if $g(x)$ is increasing and $|g(x)| > 1$
(b) $f(x)$ is increasing if $g(x)$ is increasing and $|g(x)| \leq 1$
(c) $f(x)$ is decreasing if $g(x)$ is decreasing and $|g(x)| > 1$
(d) $f(x)$ is increasing if $g(x)$ is decreasing and $|g(x)| \leq 1$

2. Let $h(x) = f(x) - \{f(x)\}^2 + \{f(x)\}^3$, for all real values of x, then

(a) $h(x)$ is increasing whenever $f(x)$ is increasing
(b) $h(x)$ is increasing whenever $f(x)$ is decreasing
(c) $h(x)$ is decreasing whenever $f(x)$ is decreasing
(d) nothing can be said in general

3. If $\phi(x) = 3f\left(\frac{x^2}{3}\right) + f(3 - x^2)\ \forall\ x \in (-3, 4)$, where $f''(x) > 0\ \forall\ x \in (-3, 4)$, then $\phi(x)$ is

(a) increasing in $\left(\frac{3}{2}, 4\right)$
(b) decreasing in $\left(-3, -\frac{3}{2}\right)$
(c) increasing in $\left(-\frac{3}{2}, 0\right)$
(d) decreasing in $\left(0, \frac{3}{2}\right)$

4. If $f : R \to R$ is a differentiable function such that $f'(x) > 2f(x)$ for all $x \in R$, and $f(0) = 1$ then **[JEE Advanced 2017]**

(a) $f(x) > e^{2x}$ in $(0, \infty)$
(b) $f'(x) < e^{2x}$ in $(0, \infty)$
(c) $f(x)$ is increasing in $(0, \infty)$
(d) $f(x)$ is decreasing in $(0\ \infty)$

5. If $\phi(x) = f(x) + f(2a - x)$ and $f''(x) > 0$, $a > 0$, $0 \leq x \leq 2a$, then

(a) $\phi(x)$ increases in $(a, 2a)$
(b) $\phi(x)$ increases in $(0, a)$
(c) $\phi(x)$ decreases in $(a, 2a)$
(d) $\phi(x)$ decreases in $(0, a)$

6. If $f(x) = 2x + \cot^{-1}x + \ln(\sqrt{1 + x^2} - x)$, then $f(x)$

(a) increases in $(0, \infty)$
(b) decreases in $[0, \infty)$
(c) neither increases nor decreases in $(0, \infty)$
(d) increases in $(-\infty, \infty)$

7. Let $f(x) = \begin{cases} (1-x)^{\alpha} \cdot x \cdot (1-\cos(2\pi x)), & 0 < x \le 1 \\ 0, & x = 0 \end{cases}$, if

Rolle's theorem is applicable to $f(x)$ for $x \in (0, 1)$, then α can be

(a) -2 (b) -1
(c) $\frac{1}{2}$ (d) 1

8. Let $f'(x) > 0$ and $g'(x) < 0$ for all $x \in R$, then

(a) $f\{g(x)\} > f\{g(x+1)\}$ (b) $f\{g(x)\} > f\{g(x-1)\}$
(c) $g\{f(x)\} > g\{f(x+1)\}$ (d) $g\{f(x)\} > g\{f(x-1)\}$

9. If $f(x) = x^3 - x^2 + 100x + 1001$, then

(a) $f(2000) > f(2001)$
(b) $f\left(\frac{1}{1999}\right) > f\left(\frac{1}{2000}\right)$
(c) $f(x+1) > f(x-1)$
(d) $f(3x-5) > f(3x)$

10. The interval to which b may belong so that the function $f(x) = \left(1 - \frac{\sqrt{(21 - 4b - b^2)}}{b+1}\right) x^3 + 5x + \sqrt{6}$ is increasing at every point of its domain is

(a) $[-7, -1)$ (b) $[-6, -2]$
(c) $[2, 2.5]$ (d) $[2, 3]$

11. For $x > 1$, $y = \log_e x$ satisfies the inequality

(a) $x - 1 > y$ (b) $x^2 - 1 > y$
(c) $y > x - 1$ (d) $\frac{x-1}{x} < y$

12. If $f(x) = \frac{x}{\sin x}$ and $g(x) = \frac{x}{\tan x}$, where $0 \le x \le 1$, then in this interval

(a) both $f(x)$ and $g(x)$ are increasing functions
(b) both $f(x)$ and $g(x)$ are decreasing functions
(c) $f(x)$ is an increasing function
(d) $g(x)$ is a decreasing function

Linked-Comprehension Type

In these questions, a passage (paragraph) has been given followed by questions based on each of the passages. You have to answer the questions based on the passage given.

PASSAGE 1

If $f(x) = |x-1| + |x-3| + |5-x|, \forall x \in R$

On the basis of above information, answer the following questions

1. $f(x)$ is increases, then $x \in$

(a) $(1, \infty)$ (b) $(3, \infty)$
(c) $(5, \infty)$ (d) $(1, 3)$

2. $f(x)$ is decreases, then $x \in$

(a) $(-\infty, 1)$ (b) $(-\infty, 3)$
(c) $(-\infty, 5)$ (d) $(3, 5)$

3. $f(x)$ is symmetrical about the line $x = \lambda$, then

(a) $\lambda = 1$ (b) $\lambda = 3$
(c) $\lambda = 5$ (d) $\lambda = 0$

4. The set of values of a such that equation $f(x) - a = 0$ has two solutions

(a) $(1, \infty)$ (b) $(3, \infty)$
(c) $(4, \infty)$ (d) $\{4\}$

5. The set of values of a such that equation $|f(x)| - a = 0$ has no solution

(a) $(4, 5)$ (b) $(4, \infty)$
(c) $(-\infty, 4)$ (d) $(5, 6)$

PASSAGE 2

Consider the cubic $f(x) = 8x^3 + 4ax^2 + 2bx + a$ where $a, b \in R$.

On the basis of above information, answer the following questions

1. For $a = 1$ if $y = f(x)$ is strictly increasing $\forall \in R$ then maximum range of values of b is

(a) $\left(-\infty, \frac{1}{3}\right]$ (b) $\left(\frac{1}{3}, \infty\right)$
(c) $\left[\frac{1}{3}, \infty\right)$ (d) $(-\infty, \infty)$

2. For $b = 1$, if $y = f(x)$ is non monotonic then the sum of all the integral values of $a \in [1, 100]$, is

(a) 4950 (b) 5049
(c) 5050 (d) 5047

3. If the sum of the base 2 logarithms of the roots of the cubic $f(x) = 0$ is 5, then the value of 'a' is

(a) -64 (b) -8
(c) -128 (d) -256

PASSAGE 3

Suppose $a, b, c, d,$ be non-zero real numbers and $ab > 0$, and $\int_0^1 (1 + e^{x^2})(ax^3 + bx^2 + cx + d)\, dx = \int_0^2 (1 + e^{x^2})(ax^3 + bx^2 + cx + d)\, dx = 0$

On the basis of above information, answer the following questions

1. Which statement is correct ?

(a) $ac > 0$ (b) $ac < 0$
(c) $ad < 0$ (d) None of these

2. Which statement is correct ?
(a) $ac > 0$ (b) $ab < 0$
(c) $ad > 0$ (d) None of these

3. The equation $ax^3 + bx^2 + cx + d = 0$ has
(a) three positive roots
(b) three negative roots
(c) two positive and one negative roots
(d) one positive and two negative roots

4. Rolle's theorem can be applied for $ax^3 + bx^2 + cx + d$ in the interval
(a) [0, 1] (b) [0, 2]
(c) [1, 2] (d) None of these

5. If $F(x) = \int f(x)\,dx = \int_0^1 f(x)\,dx = \int_0^2 f(x)\,dx = 0$,
then in which of the following can Rolle's theorem can be applied for $F(x)$, where
$f(x) = (1 + e^{x^2})(ax^3 + bx^2 + cx + d)$
(a) [0, 1] (b) [0, 2]
(c) [1, 2] (d) All of these

PASSAGE 4

Let $f(x)$ be a cubic polynomial with leading coefficient unity such that, $f(a) = b$ and $f'(a) = f''(a) = 0$. Suppose $g(x) = f(x) - f(a) + (a - x)$
$f'(x) + 3(x - a)^2$ for which conclusion of Rolle's theorem in $[a, b]$ holds at $x = 2$, where $2 \in (a, b)$.

On the basis of above information, answer the following questions

1. The value of a is equal to
(a) 2 (b) 3
(c) 4 (d) 6

2. The value of definite integral $\int_a^b f(x)\,dx$ is equal to
(a) $\frac{123}{64}$ (b) $\frac{213}{64}$
(c) $\frac{321}{64}$ (d) $\frac{481}{64}$

Numerical Value Type

Solve the following problems and write your answers.

1. Let $f(x) = \begin{cases} x^3 - x^2 + 10x - 5, & x \le 1 \\ -2x + \log_2(b^2 - 2), & x > 1 \end{cases}$
If $f(x)$ has greatest value at $x = 1$, then $[b^2 \in (2, \lambda]$.
Then, λ must be

2. Rolle's theorem holds for the function $f(x) = x^3 + bx^2 + cx, 1 \le x \le 2$ at the point $\frac{4}{3}$, then the value of $100c - 500b$ must be

3. If $u(x)$ is larger of $x + \frac{x^3}{6}$ and $\tan^{-1}x$ for $0 < x < 1$, then the value of $14.4u(1/2)$ must be

4. If $\phi(x)$ is larger of $(e^x - 1)$ and $(1 + x)\ln(1 + x)$ for $x \in (0, \infty)$,then the value of $502 \ln(\phi(4) + 1)$ must be

5. If $f(x) = \tan^{-1}(\sin x + \cos x)^3$ is an increasing function, then the value of x in $(0, 2\pi)$ is $x \in \left(\frac{a\pi}{4}, \frac{b\pi}{4}\right) \cup \left(\frac{c\pi}{4}, \frac{d\pi}{4}\right)$. Then, the value of $a + 10b + 100c + 1000d$ must be

6. If $f(x) = (x - 1)(x - 2)(x - 3)$ and $a = 0, b = 4$, then c the using LMVT is $2 \pm \lambda$. Then value of $450\sqrt{2}\lambda$ must be

7. If the function $f(x) = \sin(\ln x) - \cos(\ln x)$ strictly in creases in the interval (e^λ, e^μ), then the value of 500 $\cos(\mu - \lambda)$ must be

Single Integer Type

Solve the following problems and write your answers.

1. Let $f : R \to R$ be a function defined by $f(x) = 2x^3 - 21x^2 + 78x + 24$. Number of integers in the solutions set of x satisfying the inequality $f(f(f(x) - 2x^3)) \ge f(f(2x^3) - f(x)))$ is

2. If the function f given by $f(x) = x^3 - 3(a - 2)x^2 + 3ax + 7$, for some $a \in R$ is increasing in (0, 1] and decreasing in [1, 5), then a root of the equation, $\frac{f(x) - 14}{(x - 1)^2} = 0\ (x \ne 1)$ is

3. If all the real values of m for which the function $f(x) = \frac{x^3}{3} - (m - 3)\frac{x^2}{2} + mx - 2013$ is strictly increasing in $x \in [0, \infty)$ is $[0, k]$, then find the value of k.

4. The values of 'd' for which the function $f(x) = \sin x - a \sin 2x - \frac{1}{3}\sin 3x + 2ax$ increases throughout the number line is $[\lambda, \infty)$, then find λ.

5. If the exhaustive set of all possible values of c such that, $f(x) = e^{2x} - (c + 1)e^x + 2x + \cos 2 + \sin 1$, is monotonically increasing for all $x \in R$, is $(-\infty, \lambda]$, then find the value of λ

6. Let $f(x) = \begin{cases} xe^{ax}, & x \le 0 \\ x + ax^2 - x^3, & x > 0 \end{cases}$, where a is positive constant and the interval in which $f'(x)$ is increasing is $\left(-\frac{\lambda_1}{a}, \frac{a}{\lambda_2}\right)$, then find $(\lambda_1 + \lambda_2)$

7. If $f(x)$ is a twice differentiable function such that, $f(a)=0$, $f(b)=2$, $f(c)=-1$, $f(d)=2$, $f(e)=0$, where $a<b<c<d<e$, then the minimum number of zeroes of $g(x)=(f'(x))^2+f''(x)f(x)$ in the interval $[a, 0]$ is

Matrix-Match Type

Given below are Matching Type Questions, with two columns (each having some items) each . Each item of Column I has to be matched with the items of Column II, by encircling the correct match(es).

NOTE *An item of Column I can be matched with more than one items of Column II.*

1. Observe the following columns

	Column I		Column II
A.	The function $f(x)=\dfrac{x}{(1+x^2)}$ decreases in the interval	P.	$(-\infty,-1)$
		Q.	$(-\infty, 0)$
B.	The function $f(x)=\tan^{-1}x-x$ decreases in the interval	R.	$(0, \infty)$
		S.	$(1,\infty)$
C.	The function $f(x)=x-e^x+\tan\left(\dfrac{2\pi}{7}\right)$ increases in the interval	T.	$(-\infty,\infty)$

2. Observe the following columns

	Column I		Column II
A.	If function $f(x)=5x^3-15x^2-120x+3$ is increasing in $(-\infty, a)\cup(b,\infty)$ and decreasing in (a, b), then $(b-a)$ is	P.	a natural number
B.	If function $f(x)=8+36x+3x^2-2x^3$ is increasing in (a, b) and decreasing in $(-\infty, a)\cup(b,\infty)$, then $(b-a)$ is	Q.	a prime number
C.	If function $f(x)=\dfrac{3}{10}x^4-\dfrac{4}{5}x^3-3x^2+\dfrac{36}{5}x+11$ is increasing in $(a-b, a)\cup(b,\infty)$ and decreasing in $(-\infty, a-b)\cup(a, b)$, then $(b-a)$ is	R.	a composite number
		S.	a perfect number

3. Observe the following columns :

	Column I		Column II
A.	Let $f(x)=x^4-8x^3+22x^2-24x+21$, then	P.	$f(x)$ increases, if $x\in(2,3)$
		Q.	$f(x)$ decreases, if $x\in(0,1)\cup(2,\infty)$
B.	Let $f(x)=2\ln(x-2)-x^2+4x+1$, then	R.	$f(x)$ increases, if $x\in(1,2)\cup(3,\infty)$
C.	Let $f(x)=\dfrac{\lvert x-1\rvert}{x^2}$, then	S.	$f(x)$ decreases, if $x\in(1,2)\cup(3,\infty)$
		T.	$f(x)$ decreases, if $x\in(-\infty,1)\cup(2,3)$

4. Observe the following columns

	Column I		Column II
A.	A function f is differentiable in $[0, 5]$ and $f(0)=4$ and $f(5)=-1$. If $g(x)=\dfrac{f(x)}{x+1}$ and $c\in(0,5)$, then $g'(c)$ is equal to	P.	3
B.	Let $f(x)$ and $g(x)$ be differentiable for $0\le x\le 1$, $f(0)=2$, $g(0)=0$, $f(1)=6$. Let there exists a real number $c\in(0,1)$, such that $f'(c)=2g'(c)$, then $g(1)$ is equal to	Q.	$-5/6$
C.	The length of longest interval in which $f(x)=3\sin x-4\sin^3 x$ is increasing, is $\dfrac{\pi}{\lambda}$, then λ is	R.	15
D.	If Lagrange's mean value theorem is satisfied for $f(x)=\sqrt{25-x^2}$ and $c\in(1,5)$, then the value of c^2 is	S.	2
		T.	10

Answers

Objective Questions Type I [Only one correct answer]

1. (c)	**2.** (a)	**3.** (d)	**4.** (d)	**5.** (b)	**6.** (a)	**7.** (d)	**8.** (b)	**9.** (d)	**10.** (d)
11. (a)	**12.** (a)	**13.** (c)	**14.** (b)	**15.** (b)	**16.** (b)	**17.** (d)	**18.** (b)	**19.** (b)	**20.** (c)
21. (c)	**22.** (a)	**23.** (d)	**24.** (b)	**25.** (b)	**26.** (d)	**27.** (b)	**28.** (c)	**29.** (d)	**30.** (a)
31. (c)	**32.** (c)	**33.** (b)							

Objective Questions Type II [One or more than one correct answer(s)]

1. (a, b)	**2.** (a, c)	**3.** (a, b, c, d)	**4.** (a, c)	**5.** (a, d)
6. (a, d)	**7.** (c, d)	**8.** (a, c)	**9.** (b, c)	**10.** (a, b, c, d)
11. (a, b, d)	**12.** (c, d)			

Linked-Comprehension Type

Passage 1 **1.** (b) **2.** (b) **3.** (b) **4.** (c) **5.** (c)
Passage 2 **1.** (b) **2.** (b) **3.** (d)
Passage 3 **1.** (b) **2.** (c) **3.** (c) **4.** (b) **5.** (d)
Passage 4 **1.** (a) **2.** (c)

Numerical Value Type Questions

1. 130	**2.** 3300	**3.** 7.5	**4.** 2008	**5.** 8510
6. 600	**7.** 500			

Single Integer Type Questions

1. (5) **2.** (7) **3.** (9) **4.** (1) **5.** (3) **6.** (5) **7.** (6)

Matrix-Match Type

1. A → (P, S); B → (P, Q, R, S, T); C → (P, Q)
2. A → (P, R, S); B → (P, Q); C → (P, Q);
3. A → (R, T); B → (P, S); C → (Q)
4. A → (Q); B → (S); C → (P); D → (R)

Solutions

Objective Questions Type [Only one correct answer]

1. $\because f:[0,\infty)\to[0,\infty),\ g:[0,\infty)\to[0,\infty)$

and $\quad h(x)=f(g(x))$

$\therefore \quad h:[0,\infty)\to[0,\infty)$

$\therefore\ h'(x)=f'(g(x))g'(x)\le 0 \quad [\because f'(x)\le 0 \text{ and } g(x)\ge 0]$

$\Rightarrow h(x)$ is monotonically decreasing function

$\because \quad x\ge 0\Rightarrow h(x)\le h(0)$

$\Rightarrow \quad h(x)\le 0 \quad [\because h(0)=0]$

$\therefore \quad h(x)\le 0$ in $[0,\infty)$ $\quad \ldots$(i)

Also, $\quad h:[0,\infty)\to[0,\infty)$

$\therefore \quad h(x)\ge 0$ in $[0,\infty)$ $\quad \ldots$(ii)

From Eqs. (i) and (ii), $h(x)=0$ in $[0,\infty)$

$\therefore \quad h(x)-h(1)=0-0=0.$

2. Let $f(x)=ax^2+\dfrac{b}{x}-c$

$$\therefore \quad f'(x)=2ax-\frac{b}{x^2}$$

If $f'(x)=0\Rightarrow 2ax-\dfrac{b}{x^2}=0\Rightarrow x=\left(\dfrac{b}{2a}\right)^{1/3}>0$

But, $\quad f(x)\ge 0\ \forall\ x>0$

$$\therefore \quad f\left(\left(\frac{b}{2a}\right)^{1/3}\right)\ge 0$$

$$\Rightarrow \quad a\cdot\left(\frac{b}{2a}\right)^{2/3}+b\left(\frac{2a}{b}\right)^{1/3}\ge c$$

$$\Rightarrow \quad a\cdot\frac{b}{2a}+b\ge c\cdot\left(\frac{b}{2a}\right)^{1/3}$$

$$\Rightarrow \quad 3b\ge 2c\left(\frac{b}{2a}\right)^{1/3}$$

$$\Rightarrow \quad 27\,b^3\ge 8\,c^3\left(\frac{b}{2a}\right)$$

$$\Rightarrow \quad 27\,ab^2\ge 4c^3$$

3. If function $f:[-7,0]\to R$ be continuous on $[-7,0]$ and differentiable on $(-7,0)$, then according to LMVT, we have

$$\frac{f(-1)-f(-7)}{(-1)-(-7)}=f'(x)\le 2,\ \forall\ x\in(-7,-1)$$

$$\Rightarrow \quad \frac{f(-1)-(-3)}{6}\le 2$$

$\Rightarrow \quad f(-1)\le 9 \quad \ldots$(i)

Similarly, $\dfrac{f(0)-f(-7)}{0-(-7)}=f'(x)\le 2,\quad \forall\ x\in(-7,0)$

$$\Rightarrow \quad \frac{f(0)-(-3)}{7}\le 2$$

$\Rightarrow \quad f(0)\le 11 \quad \ldots$(ii)

From Eqs. (i) and (ii), we get

$$f(-1)+f(0)\le 20$$

$\therefore f(-1)+f(0)$ lies in the interval $(-\infty, 20]$.

4. $\because f(x)=x^{\alpha}\ln x$

In $[0,1]\Rightarrow f(0)=0=f(1)$

$\because$ The function has to be continuous in $[0,1]$

$$\Rightarrow \quad f(0)=\lim_{x\to 0+}f(x)=0\Rightarrow \lim_{h\to 0}f(0+h)=0$$

$$\Rightarrow \quad \lim_{h\to 0}h^{\alpha}\ln h=0 \quad\Rightarrow \lim_{h\to 0}\frac{\ln h}{h^{-\alpha}}=0$$

By L' Hospital's rule,

$$\lim_{h\to 0}\frac{1/h}{-\alpha\,h^{-\alpha-1}}=0$$

$$\Rightarrow \quad \lim_{h\to 0}-\frac{h^{\alpha}}{\alpha}=0\Rightarrow \alpha>0$$

$\therefore$ Option (d) is correct.

5. $\because \quad f(x)=\sin^4 x+\cos^4 x$

$\therefore \quad f'(x)=-\sin 4x$

$\because \quad f'(x)>0$

$\Rightarrow \quad -\sin 4x>0$

$\Rightarrow \quad \sin 4x<0$

$\Rightarrow \quad (2n+1)\pi<4x<(2n+2)\pi$

$$\Rightarrow \quad \frac{(2n+1)\pi}{4}<x<\frac{(n+1)\pi}{2}$$

For $\quad n=0,\ \dfrac{\pi}{4}<x<\dfrac{\pi}{2}$

Now, $\quad \dfrac{\pi}{2}=\dfrac{4\pi}{8}>\dfrac{3\pi}{8}$

$\therefore f(x)$ is increasing in $\left(\dfrac{\pi}{4},\dfrac{3\pi}{8}\right)$.

6. $f(x)=3\ln(x-1)-3\ln(x+1)-\dfrac{2}{x-1}$

$$\Rightarrow \quad f'(x)=\frac{3}{x-1}-\frac{3}{x+1}+\frac{2}{(x-1)^2}$$

$$\Rightarrow \quad f'(x)=\frac{4(2x-1)}{(x-1)^2(x+1)}$$

$\Rightarrow \quad f'(x)\ge 0$

$$\Rightarrow \quad x\in(-\infty,-1)\cup\left[\frac{1}{2},1\right)\cup(1,\infty)$$

7. $f(x)=\cos\left(\dfrac{\pi}{x}\right)$

$$\therefore \quad f'(x)=\frac{\pi}{x^2}\sin\left(\frac{\pi}{x}\right)>0$$

$$\therefore \quad \sin\left(\frac{\pi}{x}\right)>0$$

$$\Rightarrow \quad 2n\pi<\frac{\pi}{x}<(2n+1)\pi$$

$$\Rightarrow \quad \frac{1}{2n}>x>\frac{1}{2n+1}$$

$$\therefore \quad x\in\left(\frac{1}{2n+1},\frac{1}{2n}\right),\ \forall\ n\in N$$

8. $f'(x)=\left(\frac{\sqrt{a+4}}{1-a}-1\right)5x^4-3<0$

$\Rightarrow \quad \left(\frac{\sqrt{a+4}}{1-a}-1\right)x^4<\frac{3}{5}$

$\Rightarrow \quad \left(\frac{\sqrt{a+4}}{1-a}-1\right)\le 0$

$\Rightarrow \quad \frac{\sqrt{a+4}}{(1-a)}\le 1 \quad \ldots(\text{i})$

$a+4\le 1+a^2-2a \quad [\because -4\le a<1]$

$\Rightarrow \quad a^2-3a-3\ge 0$

$\therefore \quad a\in\left(-\infty,\frac{3-\sqrt{21}}{2}\right]\cup\left[\frac{3+\sqrt{21}}{2},\infty\right)$

but $-4\le a<1$

$\therefore \quad a\in\left[-4,\frac{3-\sqrt{21}}{2}\right]$

and for $a>1$, Eq. (i) is always true.

Hence, $a\in\left[-4,\frac{3-\sqrt{21}}{2}\right]\cup(1,\infty)$.

9. Given function, $f(x)=\begin{cases}\frac{1}{x}\log_e(1+x), & x\ne 0\\ 1, & x=0\end{cases}$,

for $x\in(-1,\infty)$

Now, $f'(x)=\frac{1}{x(1+x)}-\frac{\log_e(1+x)}{x^2}$, for $x\in(-1,\infty)-\{0\}$

$=\frac{x-(1+x)\log_e(1+x)}{x^2(1+x)}$

Let $g(x)=x-(1+x)\log_e(1+x)$

$\because g'(x)=1-1-\log_e(1+x)=-\log_e(1+x)$

Since, for $x\in(-1,0)$, $g'(x)>0$, So $g(x)$ is increasing function for $x\in(-1,0)$ but as

$g(x)<g(0),\ \forall x\in(-1,0)$

$\therefore \quad g(x)<0,\ \forall x\in(-1,0)$

$\therefore \quad f'(x)=\frac{g(x)}{x^2}<0$

$\Rightarrow f(x)$ is a decreasing function for all $x\in(-1,0)$.

Similarly, for $x\in(0,\infty)$, $g'(x)<0$, so $g(x)$ is decreasing function for $x\in(0,\infty)$.

So, $g(x)<g(0)\Rightarrow g(x)<0,\ \forall x\in(0,\infty)$

$\therefore \quad f'(x)=\frac{g(x)}{x^2}<0$

$\Rightarrow f(x)$ is a decreasing function for all $x\in(0,\infty)$.

$\therefore$ The given function $f(x)$ is a decreasing function in $(-1,\infty)$.

Hence, option (d) is correct.

10. Let $g(x)=xf(x)$

Then, g is continuous in [0,1] and differentiable in (0, 1)

As $\quad f(1)=0$

$\therefore \quad g(0)=0=g(1)$

Hence, Rolle's theorem is applicable for g

$\therefore \exists$ some $c\in(0,1)$ such that,

$g'(c)=0 \Rightarrow cf'(c)+f(c)=0$

11. $\because f(x)=|x^2-a^2|$

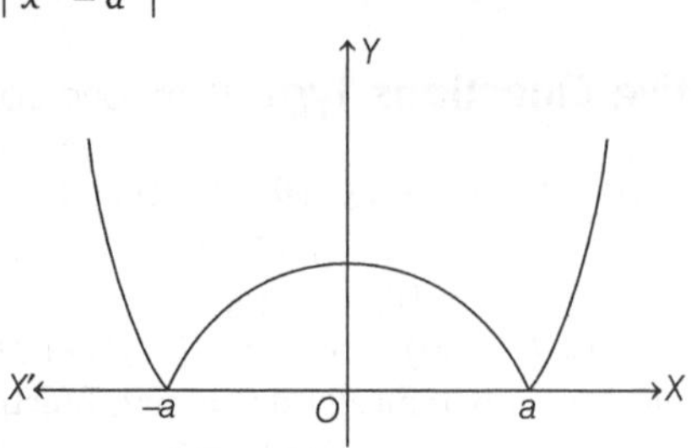

It is clear from the graph of $f(x)$.

$f(x)$ is non differentiable at $x=\pm a$

$\therefore$ Length of longest interval in which Rolle's theorem applied is $a-(-a)=2a$.

12. $\because f(1)=f(3)$

$\Rightarrow \quad a+b+11-6=27a+9b+33-6$

$\Rightarrow \quad 13a+4b=-11 \quad \ldots(\text{i})$

and $\quad f'(x)=3ax^2+2bx+11$

$\Rightarrow \quad f'\left(2+\frac{1}{\sqrt{3}}\right)=3a\left(2+\frac{1}{\sqrt{3}}\right)^2+2b\left(2+\frac{1}{\sqrt{3}}\right)+11=0$

$\Rightarrow \quad 3a\left(4+\frac{1}{3}+\frac{4}{\sqrt{3}}\right)+2b\left(2+\frac{1}{\sqrt{3}}\right)+11=0 \quad \ldots(\text{ii})$

From Eqs. (i) and (ii), we get $a=1, b=-6$.

13. Given, $\phi(x)=f(x)+f(2-x),\ \forall x\in(0,2)$

$\Rightarrow \quad \phi'(x)=f'(x)-f'(2-x)$

Also, we have $f''(x)>0\ \forall x\in(0,2)$

$\Rightarrow f'(x)$ is a strictly increasing function

Now, for $\phi(x)$ to be increasing,

$\phi'(x)\ge 0$

$\Rightarrow \quad f'(x)-f'(2-x)\ge 0 \quad$ [using Eq. (i)]

$\Rightarrow \quad f'(x)\ge f'(2-x)\Rightarrow x>2-x$

[$\because f'$ is strictly increasing function]

$\Rightarrow \quad 2x>2$

$\Rightarrow \quad x>1$

Thus, $\phi(x)$ is increasing on (1, 2).

Similarly, for $\phi(x)$ to be decreasing,

$\phi'(x)\le 0$

$\Rightarrow \quad f'(x)-f'(2-x)\le 0 \quad$ [using Eq. (i)]

$\Rightarrow \quad f'(x)\le f'(2-x)$

$\Rightarrow \quad x<2-x$

[$\because f'$ is a strictly increasing function]

$\Rightarrow \quad 2x<2$

$\Rightarrow \quad x<1$

Thus, $\phi(x)$ is decreasing on (0, 1).

14. $f'(x)=(ab-b^2-2)+\cos^4 x+\sin^4 x<0$

$=ab-b^2-2+(\sin^2 x+\cos^2 x)^2-2\sin^2 x\cos^2 x<0$

$\Rightarrow \quad ab-b^2-1<(1/2)\sin^2 2x<1/2$

$\Rightarrow \quad 2ab-2b^2-2<1$

$\Rightarrow \quad 2b^2-2ab+3>0$

$\therefore \quad (-2a)^2-4\cdot 2\cdot 3<0$

$\Rightarrow \quad a^2<6$

$\Rightarrow \quad -\sqrt{6}<a<\sqrt{6}$

15. Given, $f'(x) < 0 \ \forall \ x \in R$

$\Rightarrow \quad \sqrt{3}\cos x + \sin x - 2a < 0 \ \forall \ x \in R$

$\Rightarrow \quad \frac{\sqrt{3}}{2}\cos x + \frac{1}{2}\sin x < a \ \forall \ x \in R$

$\Rightarrow \quad \sin(x + \pi/3) < a \ \forall \ x \in R$

$\sin(x + \pi/3) < a$

$\Rightarrow \quad a \ge 1 \quad [\because \sin(x + \pi/3) \le 1]$

16. $f'(x)$ is increasing

For some x in $\left(\frac{1}{2}, 1\right)$

$f'(x) = 1$ [LMVT]

$\therefore \quad f'(1) > 1$

17. Since, $f(x) = (x + 2)e^{-x}$

$\therefore \quad f'(x) = (x + 2)(-e^{-x}) + e^{-x} \cdot 1$

$f'(x) = -e^{-x}(x + 1)$

if $f(x)$ is decreasing function, then $f'(x) < 0$

$\Rightarrow \quad -e^{-x}(x + 1) < 0 \Rightarrow x + 1 > 0$

$\therefore \quad x \in (-1, \infty)$

and if $f(x)$ is increasing function, then $f'(x) > 0$

$\Rightarrow \quad -e^{-x}(x + 1) > 0$

$\Rightarrow \quad x + 1 < 0 \Rightarrow x \in (-\infty, -1).$

18. $f(x) = e^{ax} + e^{-ax}$

$f'(x) = a(e^{ax} - e^{-ax}) < 0$

but $a < 0$

$\therefore \quad e^{ax} - e^{-ax} > 0 \Rightarrow e^{ax} > e^{-ax}$

$\Rightarrow \quad ax > -ax \Rightarrow 2ax > 0$

$\therefore \quad ax > 0$, then $x < 0 \quad [\because a < 0]$

19. Since, $f(x)$ satisfies the requirements of Lagrange's mean value theorem in [0,2].

$\therefore \quad f'(c) = \frac{f(x) - f(0)}{x - 0}$

where $0 < c < x < 2$ i.e. $0 < c < 2$

$\Rightarrow \quad f'(c) = \frac{f(x)}{x}$

$\Rightarrow \quad f(x) = xf'(c)$

$\Rightarrow \quad |f(x)| = |xf'(c)| = |x||f'(c)| \le 2, \frac{1}{2}$

$\left[\because |f'(x)| \le \frac{1}{2} \text{ and } r < 2\right]$

Hence, $|f(x)| \le 1$.

20. It is given, that a function f is continuous on $[a, b]$ and twice differentiable on (a, b), such that for all $x \in (a, b)$, $f'(x) > 0$ and $f''(x) < 0$. Now, by LMVT for $c \in (a, b)$, there is $\alpha \in (a, c)$, such that

$$f'(\alpha) = \frac{f(c) - f(a)}{c - a} \quad \dots \text{(i)}$$

and there is $\beta \in (c, b)$, such that

$$f'(\beta) = \frac{f(b) - f(c)}{b - c} \quad \dots \text{(ii)}$$

$\because f''(x) < 0 \ \forall x \in (a, b)$, then $f'(x)$ is a decreasing function, so

$f'(\beta) < f'(\alpha) \quad [\because \alpha < \beta]$

From Eqs. (i) and (ii), on putting the values of $f'(\alpha)$ and $f'(\beta)$, we get

$$\frac{f(b) - f(c)}{b - c} < \frac{f(c) - f(a)}{c - a}$$

$$\Rightarrow \quad \frac{f(c) - f(a)}{f(b) - f(c)} > \frac{c - a}{b - c}$$

[$\because f'(x) > 0$, so $f(x)$ is an increasing function $\forall x \in (a, b)$ and $a < c < b$]

Hence, option (c) is correct.

21. $f(x) = \cos|x| - 2ax + b$

$= \cos x - 2ax + b \quad [\because \cos(-x) = \cos x]$

$\therefore \quad f'(x) = -\sin x - 2a \Rightarrow f'(x) \ge 0$

$\Rightarrow -\sin x - 2a \ge 0 \ \forall \ x \in R$

$\therefore \quad \sin x + 2a \le 0 \ \forall \ x \in R$

$\Rightarrow \quad 1 + 2a \le 0$

$\therefore \quad a \le -1/2$

22. Given, $f(x) = x\sqrt{(ax - x^2)}$, $a > 0$

$$\therefore \quad f'(x) = x \cdot \frac{1}{2\sqrt{(ax - x^2)}} \cdot (a - 2x) + \sqrt{(ax - x^2)} \cdot 1$$

$$= \frac{(3ax - 4x^2)}{2\sqrt{(ax - x^2)}} = \frac{-4x(x - 3a/4)}{2\sqrt{(ax - x^2)}}$$

+ − +
0 3a/4

For $f'(x) > 0$ (increasing)

then, $x \in (0, 3a/4)$

and for $f'(x) < 0$ (decreasing)

Then, $x \in (-\infty, 0) \cup (3a/4, \infty)$.

23. $g'(x) = f'(\tan^2 x - 2\tan x + 4) \cdot (2\tan x - 2) \cdot \sec^2 x \quad \dots \text{(i)}$

$\because f''(x) > 0 \Rightarrow f'(x)$ is increasing function.

So, $f'(\tan^2 x - 2\tan x + 4)$

$= f'(\tan x - 1)^2 + 3) > f'(3) = 0$

$\forall \ x \in \left(0, \frac{\pi}{4}\right) \cup \left(\frac{\pi}{4}, \frac{\pi}{2}\right)$

Also, $(\tan x - 1) > 0$, for $x \in \left(\frac{\pi}{4}, \frac{\pi}{2}\right)$

So, $g(x)$ is increasing function in $\left(\frac{\pi}{4}, \frac{\pi}{2}\right)$.

24. Let $f(x) = 1 + x\ln x$ in $(1, x)$

$\therefore f'(x) = (1 + \ln x) > 0 \quad (\because x > 1)$

By LMVT, for $1 < c < x$

$$f'(c) = \frac{f(x) - f(1)}{x - 1}$$

$$= \frac{(1 + x\ln x) - 1}{x - 1} = \frac{x\ln x}{x - 1} > 0$$

$\Rightarrow \quad x\ln x > x - 1 \Rightarrow x\ln x + 1 > x$

Also, similarly by LMVT

Let $g(x) = 1 + \ln x$ in $(1, x)$

we get, $1 + \ln x < x$

Hence, $1 + \ln x < x < 1 + x\ln x$.

25. Since, $f(x) = ax^3 - 9x^2 + 9x + 3$

$$f'(x) = 3ax^2 - 18x + 9$$

For increasing function

$$f'(x) > 0$$

$\therefore$ $3ax^2 - 18x + 9 > 0$

$\Rightarrow$ $ax^2 - 6x + 3 > 0$

for $a > 0$

$$(-6)^2 - 4 \cdot a \cdot 3 < 0$$

$\therefore$ $a > 3$

26. $$f(x) = \begin{cases} -1, & x < 0 \\ 2x - 1, & 0 < x < 1 \\ 1, & x > 1 \end{cases}$$

$$\Rightarrow f'(x) = \begin{cases} 0, & x < 0 \\ 2, & 0 < x < 1 \\ 1, & x > 1 \end{cases}$$

$\therefore$ Monotonic increasing on $0 < x < 1$.

27. $$f(x) = \begin{cases} -x\left(2 - \sin\left(\frac{1}{x}\right)\right) & x < 0 \\ 0 & x = 0 \end{cases}$$

$$f'(x) = \begin{cases} -\left(2 - \sin\frac{1}{x}\right) - x\left(-\cos\frac{1}{x}\cdot\left(-\frac{1}{x^2}\right)\right) & x < 0 \\ \left(2 - \sin\frac{1}{x}\right) + x\left(-\cos\frac{1}{x}\cdot\left(-\frac{1}{x^2}\right)\right) & x > 0 \end{cases}$$

$$f'(x) = \begin{cases} -2 + \sin\frac{1}{x} - \frac{1}{x}\cos\frac{1}{x} & x < 0 \\ 2 - \sin\frac{1}{x} + \frac{1}{x}\cos\frac{1}{x} & x > 0 \end{cases}$$

$f'(x)$ is an oscillating function which is non-monoatomic in $(-\infty, 0) \cup (0, \infty)$.

28. $f'(x) > 0$

$\Rightarrow$ $3x^2 + 2ax + b + 5\sin 2x > 0$

$\Rightarrow$ $3x^2 + 2ax + b - 5 > 0$ $(\because \sin 2x \geq -1)$

$\therefore$ $D < 0$

$\Rightarrow$ $4a^2 - 4 \cdot 3 \cdot (b - 5) < 0$

$\Rightarrow$ $a^2 - 3b + 15 < 0$

29. $\because$ $$f(x) = \frac{e^{x^2} - e^{-x^2}}{e^{x^2} + e^{-x^2}}$$

$\therefore$ $$f'(x) = \frac{8x}{(e^{x^2} + e^{-x^2})^2} = \begin{cases} > 0, & x > 0 \\ < 0, & x < 0 \\ 0, & x = 0 \end{cases}$$

30. $\because f'(x) = |x| - \{x\}$

$\because f(x)$ is decreasing

$\therefore$ $f'(x) < 0$

$\Rightarrow$ $|x| - \{x\} < 0 \Rightarrow |x| < \{x\}$

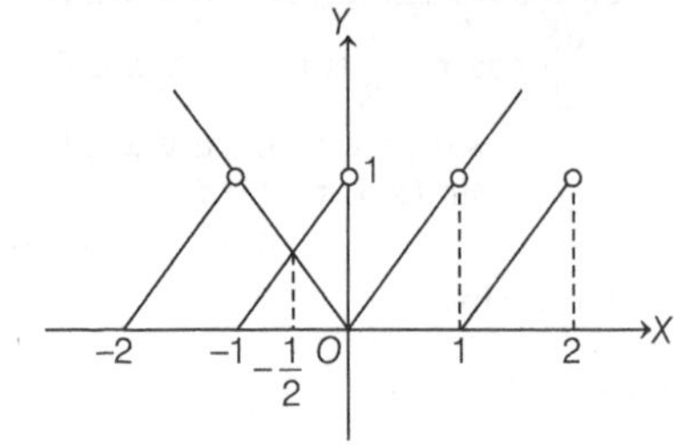

It is clear from the figure $x \in \left(-\frac{1}{2}, 0\right)$.

31. $\because$ $f(x) = x^x$

$\therefore f'(x) = x^x(1 + \ln x) < 0$ (given)

$\because$ $x^x > 0$

$\therefore$ $1 + \ln x < 0$

$\Rightarrow$ $\ln x < -1$

$\Rightarrow$ $x < e^{-1}$

$\therefore$ $x \in (0, 1/e)$ $[\because x > 0]$

32. $f'(x) = e^x(x - 1)(x - 2) < 0$

$\Rightarrow$ $(x - 1)(x - 2) < 0$

$\therefore$ $1 < x < 2$

i.e. $x \in (1, 2)$.

33. Let $f(x) = e^x - (1 + x)$

$\therefore$ $f'(x) = e^x - 1 > 0$ $[\because 0 < x < 1]$

$\therefore f(x)$ is increasing

$$f(x) > f(0)$$

$\Rightarrow$ $e^x - (1 + x) > 0$

$\Rightarrow$ $e^x > 1 + x$ $[\because x > 0]$

Now, let $g(x) = \log_e(1 + x) - x$

$\therefore$ $g'(x) = \frac{1}{1 + x} - 1 = -\frac{x}{1 + x} < 0$

$\therefore g(x)$ is decreasing

$\Rightarrow x > 0 \Rightarrow g(x) < g(0) \Rightarrow \log_e(1 + x) - x < 0$

$\Rightarrow$ $\log_e(1 + x) < x$

Again, let $h(x) = \sin x - x$

$\Rightarrow$ $h'(x) = \cos x - 1 < 0$

$h(x)$ is decreasing

$\because$ $x > 0 \Rightarrow h(x) < h(0) \Rightarrow \sin x - x < 0 \Rightarrow \sin x < x$

In last, let $t(x) = \log_e x - x$

$\therefore$ $t'(x) = \frac{1}{x} - 1 = \frac{1 - x}{x} > 0$

$\Rightarrow t(x)$ is increasing

$x < 1 \Rightarrow t(x) < t(1) \Rightarrow \log_e x - x < -1$

$\Rightarrow$ $\log_e x < x - 1$

Objective Questions Type II [One or more than one correct answer(s)]

1. $f(x) = \sin^{-1}(\sin 2\theta)$, where $g(x) = \tan\theta$

$= 2\theta, |g(x)| \le 1$

$= 2\tan^{-1}(g(x)), |g(x)| \le 1$

when, $|g(x)| > 1$

$$f(x) = \sin^{-1}\left\{\frac{2/g(x)}{1 + (1/g(x))^2}\right\}$$

$= \sin^{-1}(\sin 2\theta)$, where $\tan\theta = \dfrac{1}{g(x)}$

$= 2\theta = 2\cot^{-1}(g(x)), |g(x)| > 1$

$$\therefore \quad f'(x) = \begin{cases} \dfrac{2}{1 + (g(x))^2} \cdot g'(x), & |g(x)| \le 1 \\ \dfrac{-2}{1 + (g(x))^2} \cdot g'(x), & |g(x)| > 1 \end{cases}$$

2. $h(x) = f(x) - \{f(x)\}^2 + \{f(x)\}^3$

$\therefore \quad h'(x) = f'(x) - 2f(x)f'(x) + 3(f(x))^2 f'(x)$

$= f'(x)\{1 - 2f(x) + 3(f(x))^2\}$

$$= f'(x)\left\{\left(f(x) - \frac{1}{3}\right)^2 + \frac{2}{9}\right\}$$

if $f'(x) > 0$, then $h'(x) > 0$

and if $f'(x) < 0$, then $h'(x) < 0$.

3. $\phi'(x) = f'\left(\dfrac{x^2}{3}\right) \cdot 2x - f'(3 - x^2)\, 2x$

$$= 2x\left\{f'\left(\frac{x^2}{3}\right) - f'(3 - x^2)\right\} \forall\, x \in (-3, 4)$$

$\because f''(x) > 0 \Rightarrow f'(x)$ is increasing function.

If $\dfrac{x^2}{3} > 3 - x^2 \Rightarrow 4x^2 > 9 \Rightarrow x^2 > 9/4$

$\Rightarrow \quad x > 3/2$ and $x < -3/2$

$\Rightarrow \quad x \in \left(\dfrac{3}{2}, 4\right)$ and $x \in (-3, -3/2)$

$$\Rightarrow \quad f'\left(\frac{x^2}{3}\right) > f'(3 - x^2)$$

$$\Rightarrow \quad f'\left(\frac{x^2}{3}\right) - f'(3 - x^2) > 0$$

for $\quad x \in \left(\dfrac{3}{2}, 4\right), \phi'(x) > 0$

for $\quad x \in (-3, 3/2), \phi'(x) < 0$

and if $\dfrac{x^2}{3} < 3 - x^2 \Rightarrow 4x^2 < 9$

$\Rightarrow \quad -3/2 < x < 3/2$

$$f'\left(\frac{x^2}{3}\right) < f'(3 - x^2) \Rightarrow f'\left(\frac{x^2}{3}\right) - f'(3 - x^2) < 0$$

for $\quad x \in \left(-\dfrac{3}{2}, 0\right), \phi'(x) > 0$

for $\quad x \in \left(0, \dfrac{3}{2}\right), \phi'(x) < 0$

4. $f'(x) > 2f(x) \Rightarrow \dfrac{dy}{y} > 2dx \Rightarrow \displaystyle\int_1^{f(x)} \frac{dy}{y} > 2\int_0^x dx$

$\ln(f(x)) > 2x \Rightarrow f(x) > e^{2x}$

Also, as $f'(x) > 2f(x)$

$\Rightarrow \quad f'(x) > 2c^{2x} > 0$

5. $\phi'(x) = f'(x) - f'(2a - x)$

given, $f''(x) > 0 \Rightarrow f'(x)$ is increasing function, if

$x < 2a - x \Rightarrow x < a$

$f'(x) < f'(2a - x)$

$\Rightarrow \quad \phi'(x) < 0$

$\therefore \phi(x)$ decreases in $(0, a)$

and if $x > 2a - x \Rightarrow x > a$

$f'(x) > f'(2a - x) \Rightarrow \phi'(x) > 0$

$\therefore \phi(x)$ increases in $(a, 2a)$.

6. $f'(x) = 2 - \dfrac{1}{1 + x^2} + \dfrac{1}{(\sqrt{1 + x^2} - x)} \times \left(\dfrac{x}{\sqrt{1 + x^2}} - 1\right)$

$$= 2 - \frac{1}{1 + x^2} - \frac{1}{\sqrt{1 + x^2}}$$

$$= \frac{2 + 2x^2 - 1 - \sqrt{1 + x^2}}{(1 + x^2)}$$

$$= \frac{x^2 + \sqrt{(1 + x^2)}\,(\sqrt{1 + x^2} - 1)}{1 + x^2} > 0$$

for all x

$f(x)$ is increasing in $(-\infty, \infty)$

and in particular $(0, \infty)$.

7. Given, $f(x) = \begin{cases} (1 - x)^\alpha \cdot x \cdot (1 - \cos(2\pi x)), & 0 < x \le 1 \\ 0, & x = 0 \end{cases}$

$f(x)$ is continuous at $x = 0$ but for continuous at $x - 1$.

$f(1^-) = f(1)$

$\therefore \quad 2\lim_{h \to 0} h^\alpha \sin^2 \pi h = 0$

$\Rightarrow \quad \lim_{h \to 0} h^{\alpha + 2} = 0$

$\therefore \quad \alpha + 2 > 0 \Rightarrow \alpha > -2 \quad \ldots$(ii)

From Eqs. (i) and (ii), we get $\Rightarrow \alpha > 0$

$f(x)$ is differentiable in $(0, 1)$.

8. Since, $f'(x) > 0 \Rightarrow f(x)$ is increasing function

and $g'(x) < 0 \Rightarrow g(x)$ is decreasing function

since, $x + 1 > x$

$\therefore \quad g(x + 1) < g(x) \quad [\because g$ is decreasing]

$\therefore \quad f\{g(x + 1)\} < f\{g(x)\} \quad [\because f$ is increasing]

$\Rightarrow \quad f\{g(x)\} > f\{g(x + 1)\}$

Option (a) is correct.

Now, $x + 1 > x$

$f(x + 1) > f(x) \quad (\because f$ is increasing)

$\Rightarrow \quad g\{f(x + 1)\} < g\{f(x)\} \quad (\because g$ is decreasing)

Option (c) is correct.

Now, replacing x by $x-1$ in alternates (a) and (c), then

$$f\{g(x-1)\} > f\{g(x)\}$$

and $g\{f(x-1)\} > g\{f(x)\}$

Hence, options (b) and (d) are wrong.

9. $f(x) = x^3 - x^2 + 100x + 1001$

$\Rightarrow f(x) = 3x^2 - 2x + 100 > 0 \ \forall \ x \in R$

$\therefore f(x)$ is increasing (strictly)

$$\therefore \quad f\left(\frac{1}{1999}\right) > f\left(\frac{1}{2000}\right)$$

$$f(x+1) > f(x-1)$$

10. $f'(x) = 3\left(1 - \dfrac{\sqrt{(21-4b-b^2)}}{b+1}\right)x^2 + 5 > 0$(given)

$$\therefore \left(1 - \frac{\sqrt{(21-4b-b^2)}}{b+1}\right) \geq 0 \Rightarrow \frac{\sqrt{(21-4b-b^2)}}{(b+1)} \leq 1 \quad ...(i)$$

and $\quad 21 - 4b - b^2 \geq 0 \Rightarrow -7 \leq b \leq 3 \quad ...(ii)$

***Case* I** If $\quad b+1 < 0$ i.e. $b < -1 \quad ...(iii)$

Then, inequality Eq. (ii) is always true, then from Eqs. (ii) and (iii) we get

$$-7 \leq b < -1 \quad ...(iv)$$

***Case* II** If $\quad b+1 > 0$ i.e., $b > -1 \quad ...(v)$

From Eq. (i), $\sqrt{(21-4b-b^2)} \leq (b+1)$

$\Rightarrow \quad 21 - 4b - b^2 \leq b^2 + 2b + 1$

$\Rightarrow \quad 2b^2 + 6b - 20 \geq 0$

$\Rightarrow \quad b^2 + 3b - 10 \geq 0$

$\Rightarrow \quad (b+5)(b-2) \geq 0$

$$b \in (-\infty, -5] \cup [2, \infty)$$

From Eqs. (v) and (vi), we get

$$b \geq 2 \quad ...(vii)$$

Combining Eqs. (iv) and (vii), we get

$$b \in [-7, -1) \cup [2, \infty).$$

11. Let $f(x) = \log_e x - (x-1)$

$\Rightarrow f'(x) = \dfrac{1}{x} - 1 = \dfrac{1-x}{x} < 0$

$\because f(x)$ is decreasing function ($\because x > 1$)

$$x > 1 \Rightarrow f(x) < f(1)$$

$\Rightarrow \quad \log_e x - (x-1) < 0$

$\Rightarrow \quad (x-1) > \log_e x = y$

or $\quad (x-1) > y$

Now, let $g(x) = \log_e x - (x^2 - 1)$

$\Rightarrow \quad g'(x) = \dfrac{1}{x} - 2x = \left(\dfrac{1-2x^2}{x}\right) < 0$ (for $x > 1$)

$\therefore g(x)$ is decreasing function

$\because \quad x > 1$

$\Rightarrow \quad g(x) < g(1)$

$\Rightarrow \quad \log_e x - (x^2 - 1) < 0$

$\Rightarrow \quad y < (x^2 - 1)$

or $\quad (x^2 - 1) > y$

Again, let $h(x) = \dfrac{x-1}{x} - \log_e x$

$\therefore \quad h'(x) = 0 + \dfrac{1}{x^2} - \dfrac{1}{x} = \dfrac{1-x}{x^2} < 0 \quad$ [for $x > 1$]

$\therefore h(x)$ is decreasing function

$\because \quad x > 1 \Rightarrow h(x) < h(1)$

$\Rightarrow \quad \dfrac{x-1}{x} - \log_e x < 0$

$\Rightarrow \quad \dfrac{x-1}{x} < y$

12. Let $g(x) = \dfrac{x}{\tan x}$, if $0 \leq x \leq 1$

Now, $g(x)$ is continuous on [0, 1] and differentiable on]0, 1[.

For $0 < x < 1$,

$$g'(x) = \frac{\tan x - x \sec^2 x}{\tan^2 x}$$

Let $H(x) = \tan x - 2\sec^2 x, \ 0 \leq x \leq 1$

Now, $H(x)$ is continuous on [0, 1] and differentiable on]0, 1[.

For $0 < x < 1$,

$$H(x) = \tan x - x \sec^2 x, \ 0 \leq x \leq 1$$

$$H'(x) = \sec^2 x - \sec^2 x - 2x \sec^2 x \tan x$$

$$= -2x \sec^2 x \tan x < 0$$

$\therefore \ H(x)$ is decreasing function on [0, 1].

Thus, $H(x) < H(0)$ for $0 < x < 1$

$\Rightarrow \quad H(x) < 0 \quad$ for $0 < x < 1$

$\Rightarrow \quad g'(x) < 0 \quad$ for $0 < x < 1$

$\Rightarrow \ g(x)$ is decreasing function on [0, 1].

Therefore, $g(x) = \dfrac{x}{\tan x}$ is a decreasing function on $0 \leq x \leq 1$.

Also, $\quad g(x) < g(0) \quad$ for $0 \leq x \leq 1$

$\Rightarrow \quad g\dfrac{x}{\tan x} < 1 \quad$ for $0 \leq x \leq 1$

$\Rightarrow \quad x < \tan x$ for $0 \leq x \leq 1$

Next, $f(x) = \begin{cases} x / \sin x & \text{for } 0 \leq x \leq 1 \\ 1 & \text{for } x = 0 \end{cases}$

Now, f is continuous on [0, 1] and differentiable on]0, 1[.

For $0 < x < 1$,

$$f'(x) = \frac{\sin x - x \cos x}{\sin^2 x}$$

$$= \frac{(\tan x - x)\cos x}{\sin^2 x} \quad \text{for } 0 < x < 1$$

$\therefore \ f(x)$ increases on [0, 1].

Thus, $f(x) = \dfrac{x}{\sin x}$ increases on $0 \leq x \leq 1$.

Linked Comprehension Type

PASSAGE 1

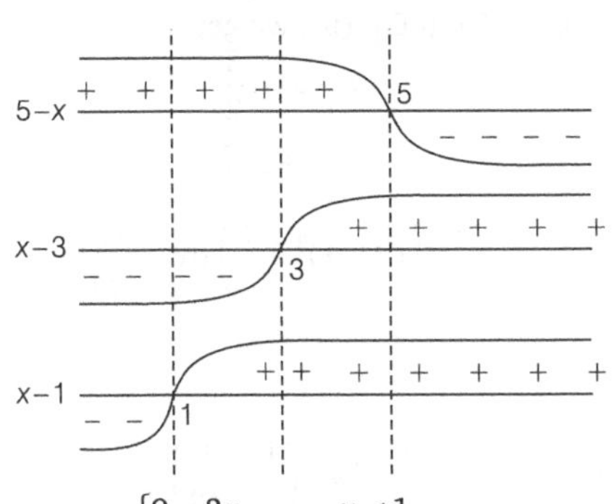

$$\therefore \quad f(x)=\begin{cases} 9-3x, & x<1 \\ 7-x, & 1\le x<3 \\ x+1, & 3\le x<5 \\ 3x-9, & x\ge 5 \end{cases}$$

$$f'(x)=\begin{cases} -3, & x<1 \\ -1, & 1\le x<3 \\ 1, & 3\le x<5 \\ 3, & x\ge 5 \end{cases}$$

1. $f'(x)>0,\ x\in(3,\infty)$.
2. $f'(x)<0,\ x\in(-\infty,3)$.
3.

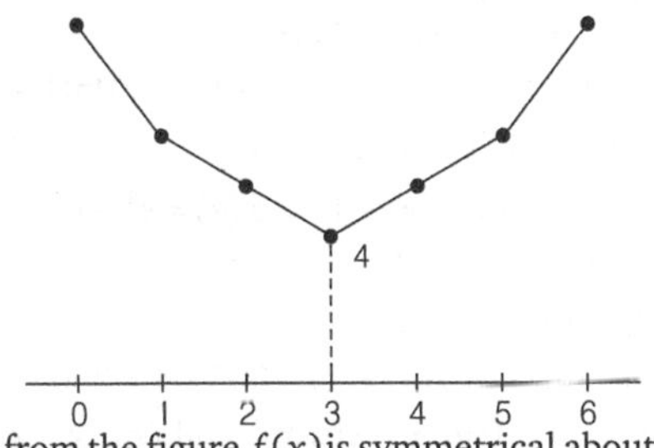

It is clear from the figure $f(x)$ is symmetrical about the line $x=3$

$\therefore \lambda = 3$

4. Any line $y=a$ cuts the graph $y=f(x)$ at two points, then $x\in(4,\infty)$.
5. $\because f(x)>0$

$\therefore |f(x)|=f(x)$

for no solution, $a<4$.

PASSAGE 2

1. $a=1$

$f(x)=8x^3+4x^2+2bx+1$

$f'(x)=24x^2+8x+2b=2(12x^2+4x+b)$

For $f(x)$ to be strictly increasing function,

$f'(x)\ge 0\ \forall\ x\in R \Rightarrow 24x^2+8x+2b>0$

$\Rightarrow b>-12x^2-4x=-12\left(x^2+2\frac{1}{6}x+\frac{1}{36}-\frac{1}{36}\right)$

$\Rightarrow b>\frac{1}{3}-12\left(x-\frac{1}{6}\right)^2$

Hence, range of b is $\left(\frac{1}{3},\infty\right)$.

2. If $b=1$

$f(x)=8x^3+4ax^2+2x+a$

$f'(x)=24x^2+8ax+2$ or $f'x=2(12x^2+4ax+1)$

For non monotonic $f'(x)=0$ must have distinct roots.

Hence, $D>0$ i.e., $16a^2-48>0$

$\Rightarrow \quad a^2>3;$

$\therefore \quad a>\sqrt{3}=1.732 \Rightarrow a\in[2, 100]$

The sum of $2+3+\ldots+100=1+2+3+\ldots+100-1$

$=100\left(\frac{100-1}{2}\right)\times 1=5049$

$\therefore$ The sum of all the integral values of $a\in[1, 100]$ is 5049.

3. If x_1, x_2 and x_3 are the roots of the given cubic equation, then $\log_2 x_1+\log_2 x_2+\log_2 x_3=5$

$\Rightarrow \quad \log_2(x_1x_2x_3)=5$

$[\because \log m+\log n=\log mn]$

$\Rightarrow \quad x_1x_2x_3=2^5=32$

$\Rightarrow \quad -\frac{a}{8}=32 \Rightarrow a=-256$

PASSAGE 3

1. $\because 1+e^{x^2}>0,\ ax^3+bx^2+cx+d=0$

For some values of x between 0 and 1 and 1 and 2.

Now, let $a^2x^3+abx^2+acx+ad=0$ has three real roots.

Now, $3a^2x^2+2abx+ac=0$ has two real roots.

$$\text{Its roots} = \frac{-2ab\pm\sqrt{4a^2b^2-12a^2(ac)}}{6a^2} \quad \ldots\text{(i)}$$

obviously both roots are real, when $ac<0$ and one of there is – ve.

2. Let $f(x)=a^2x^3+abx^2+acx+ad$

$\therefore \quad f(x)$ has one negative root

$f(0)>0 \Rightarrow ad>0$

3. $a^2x^3+abx^2+acx+ad=0$ has three real roots.

and from Eq. (i), $ax^3+bx^2+cx+d=0$

has two positive and one negative real roots.

4. $\because \int_0^1(1+e^{x^2})(ax^3+bx^2+cx+d)\,dx$

$=\int_0^2(1+e^{x^2})(ax^3+bx^2+cx+d)\,dx=0$

$\Rightarrow \quad x\in[0, 2]$.

5. $\because x\in[0, 2]$

$\therefore [0, 1]\subset[0, 2], [1, 2]\in[0, 2]$

$\therefore$ All are correct.

PASSAGE 4

We have, $f(x) = (x - a)^3 + b$ ...(i)

Since, Rolle's theorem is applicable to $g(x)$ at $x = 2$,

So, $g(a) = g(b)$ and $g'(2) = 0$

$\Rightarrow \quad 0 = f(b) - f(a) + (a - b)f'(b) + 3(b - a)^2$

and $(a - 2)f''(2) + 6(2 - a) = 0$

$\therefore \quad f'(b) = \dfrac{f(b) - f(a)}{b - a} + 3(b - a)$ ...(ii)

and $f''(2) = 6$ ...(iii) (As $a \neq 2$, think)

1. As, $f''(2) = 6$ [using Eq. (iii) in Eq. (i)]

$\Rightarrow 6(2 - a) = 6$

$\Rightarrow \quad a = 2$

2. Using Eq. (i) in Eq. (ii), we get

$$(b - a) = \frac{3}{2}$$

$$\Rightarrow \quad b = \frac{5}{2}$$

$$\therefore \quad \int_1^{5/2} \left((x - 1)^3 + \frac{5}{2}\right) dx = \frac{321}{64}$$

Numerical Value Type

1. For $x \leq 1$ $f'(x) = 3x^2 - 2x + 10$

$$= 3\left\{\left(x - \frac{1}{3}\right)^2 + \frac{29}{9}\right\} > 0$$

$\therefore f(x)$ is increasing function for $x \leq 1$ Now, for $x > 1$,

$$f'(x) = -2 < 0$$

So, $f(x)$ is decreasing function, for $x > 1$

Now, $f(x)$ will have greatest value at $x = 1$ if

$$\Rightarrow \quad \lim_{x \to 1^+} f(x) \leq f(1)$$

$$\Rightarrow \quad \lim_{h \to 0} f(1 + h) \leq 5$$

$$\Rightarrow \quad \lim_{h \to 0} -2(1 + h) + \log_2 (b^2 - 2) \leq 5$$

$$\Rightarrow \quad -2 + \log_2 (b^2 - 2) \leq 5$$

$$\Rightarrow \quad \log_2 (b^2 - 2) \leq 7$$

$$\Rightarrow \quad b^2 - 2 \leq 2^7 \Rightarrow b^2 \leq 130$$

but $\quad b^2 - 2 > 0$

$\therefore \quad 2 < b^2 \leq 130$

Hence, $\lambda = 130$

2. Now, $f(1) = f(2)$

$\Rightarrow 1 + b + c = 8 + 4b + 2c$

$\Rightarrow \quad 3b + c = -7$ ...(i)

and $f'\left(\dfrac{4}{3}\right) = 0 \Rightarrow 3\left(\dfrac{4}{3}\right)^2 + 2b\left(\dfrac{4}{3}\right) + c = 0$

$\Rightarrow \quad 16 + 8b + 3c = 0$ ...(ii)

From Eqs. (i) and (ii), we get

$$b = -5, c = 8$$

$\therefore 100c - 500b = 800 + 2500 = 3300$

3. Let $f(x) = x + \dfrac{x^3}{6} - \tan^{-1} x$

$$\therefore \quad f'(x) = 1 + \frac{x^2}{2} - \frac{1}{(1 + x^2)} = \frac{x^2(3 + x^2)}{2(1 + x^2)} > 0$$

$\Rightarrow f(x)$ is increasing.

$\because \quad x > 0$

$\therefore \quad f(x) > f(0)$

$$x + \frac{x^3}{6} - \tan^{-1} x > 0 \Rightarrow x + \frac{x^3}{6} > \tan^{-1} x$$

$$\therefore \quad u(x) = x + \frac{x^3}{6}$$

Then, $14.4u\left(\dfrac{1}{2}\right) = 14.4\left(\dfrac{1}{2} + \dfrac{1}{48}\right)$

$$= 14.4\left(\frac{25}{48}\right) = 7.5$$

4. Let $\quad f(x) = (e^x - 1) - (1 + x)\ln(1 + x)$

$\therefore \quad f'(x) = e^x - \{1 + \ln(1 + x)\} = g(x)$ (let)

$\therefore \quad g(x) = e^x - 1 - \ln(1 + x)$

$\Rightarrow \quad g'(x) = e^x - \dfrac{1}{(1 + x)} > 0$ (for $x \in (0, \infty)$)

$\Rightarrow \quad g(x)$ is increasing function

$\because \quad x > 0$

$\therefore \quad g(x) > g(0) \Rightarrow g(x) > 0$

Then, $\quad f'(x) > 0$

$\Rightarrow f(x)$ is increasing function

$f(x) > f(0) \Rightarrow e^x - 1 - (1 + x)\ln(1 + x) > 0$

$\Rightarrow \quad e^x - 1 > (1 + x)\ln(1 + x)$

$\therefore \quad \phi(x) = e^x - 1 \Rightarrow \phi(4) = e^4 - 1$

$\Rightarrow \quad \phi(4) + 1 = e^4$

$\therefore \quad \ln(\phi(4) + 1) = 4$

Then, $502 \ln(\phi(4) + 1) = 502 \times 4 = 2008$.

5. Since, $\sin^{-1} (\sin x + \cos x)^3$ and $(\sin x + \cos x)^3$ are both increasing functions.

$\therefore f(x)$ is increasing function when $\sin x + \cos x$ is an increasing function.

Let $\quad g(x) = \sin x + \cos x$

or $\quad g'(x) = \cos x - \sin x = \sqrt{2} \cos\left(x + \dfrac{\pi}{4}\right) > 0$

or $\quad \cos\left(x + \dfrac{\pi}{4}\right) > 0$

or $\quad 2n\pi - \dfrac{\pi}{2} < x + \dfrac{\pi}{4} < 2n\pi + \dfrac{\pi}{2}, n \in I$

$\Rightarrow \quad 2n\pi - \dfrac{3\pi}{4} < x < 2n\pi + \dfrac{\pi}{4}$

$$-\frac{3\pi}{4} < x < \frac{\pi}{4} \text{ and } \frac{5\pi}{4} < x < \frac{9\pi}{4}$$

but $x \in (0, 2\pi)$

$\therefore \quad x \in \left(0, \dfrac{\pi}{4}\right) \cup \left(\dfrac{5\pi}{4}, 2\pi\right)$

Then, $\quad a = 0, b = 1, c = 5, d = 8$

$\therefore a + 10b + 100c + 1000d = 0 + 10 + 500 + 8000 = 8510$

6. We have, $f(x) = (x-1)(x-2)(x-3)$

$$= x^3 - 6x^2 + 11x - 6$$

$\therefore \quad f(a) = 0 - 0 + 0 - 6 = -6$

$\therefore \quad f(b) = 4^3 - 6\cdot 4^2 + 11\cdot 4 - 6$

$$= 64 - 96 + 44 - 6 = 6$$

$\therefore \quad \dfrac{f(b)-f(a)}{b-a} = \dfrac{6-(-6)}{4-0} = 3$

Also, $f'(x) = 3x^2 - 12x + 11$

$\therefore \quad f'(c) = 3c^2 - 12c + 11$

From LMVT, $\dfrac{f(b)-f(a)}{b-a} = f'(c)$

$\Rightarrow \quad 3 = 3c^2 - 12c + 11$

$\therefore \quad 3c^2 - 12c + 8 = 0$

$\therefore \quad c = \dfrac{12 \pm \sqrt{(144-96)}}{6} = 2 \pm \dfrac{2\sqrt{2}}{3}$

As both of the values of c lie in the open interval (0, 4).

$\therefore \quad \lambda = \dfrac{2\sqrt{2}}{3}$

Then, $450\sqrt{2}\lambda = 450\sqrt{2} \times \dfrac{2\sqrt{2}}{3} = 600$

7. Since, $f(x) = \sin(\ln x) - \cos(\ln x),\ x > 0$

$$= \sqrt{2}\sin\left(\ln x - \frac{\pi}{4}\right)$$

$\therefore \quad f'(x) = \dfrac{\sqrt{2}}{x}\cos\left(\ln x - \dfrac{\pi}{4}\right)$

$$= \frac{\sqrt{2}}{x}\sin\left(\frac{\pi}{2} + \ln x - \frac{\pi}{4}\right)$$

$$= \frac{\sqrt{2}}{x}\sin\left(\ln x + \frac{\pi}{4}\right) > 0 \quad [\because x > 0]$$

or $\quad \sin\left(\ln x + \dfrac{\pi}{4}\right) > 0$

or $\quad 2n\pi < \ln x + \dfrac{\pi}{4} < (2n+1)\pi,\ n \in I$

or $\quad 2n\pi - \dfrac{\pi}{4} < \ln x < 2n\pi + \dfrac{3\pi}{4},\ n \in I$

$\Rightarrow \quad e^{2n\pi - \pi/4} < x < e^{2n\pi + 3\pi/4},\ n \in I$

$\therefore \quad x \in (e^{2n\pi - \pi/4}, e^{2n\pi + 3\pi/4}),\ n \in I$

Then, $\quad \lambda = 2n\pi - \dfrac{\pi}{4},\ \mu = 2n\pi + \dfrac{3\pi}{4}$

or $\quad \mu - \lambda = \pi$

$\because \ -500\cos(\mu - \lambda) = -500\cos\pi = -500(-1) = 500$

Single Integer Type

1. $f(x) = 2x^3 - 21x^2 + 78x + 24$

$f'(x) = 6(x^2 - 7x + 13)$

$\Rightarrow f'(x) > 0\ \forall\ x \in R$

$\Rightarrow f(x)$ is increasing function

Now, $\quad f(f(x) - 2x^3)) = f(f(2x^3 - f(x)))$

$\Rightarrow \quad f(f(x) - 2x^3) \geq f(2x^3 - f(x))$

$\Rightarrow \quad f(x) - 2x^3 \geq 2x^3 - f(x)$

$\Rightarrow \quad f(x) \geq 2x^3 \Rightarrow 7x^2 - 26x - 8 \leq 0$

$\Rightarrow \quad N \in \left[-\dfrac{2}{7}, 4\right]$

{0, 1, 2, 3, 4} are integer.

2. Given that function,

$f(x) = x^3 - 3(a-2)x^2 + 3ax + 7$, for some $a \in R$ is increasing in (0,1] and decreasing in [1, 5).

$f'(1) = 0$ [$\because$ tangent at $x = 1$ will be parallel to X-axis]

$\Rightarrow \quad (3x^2 - 6(a-2)x + 3a)_{x=1} = 0$

$\Rightarrow \quad 3 - 6(a-2) + 3a = 0$

$\Rightarrow \quad 3 - 6a + 12 + 3a = 0$

$\Rightarrow \quad 15 - 3a = 0$

$\Rightarrow \quad a = 5$

So, $\quad f(x) = x^3 - 9x^2 + 15x + 7$

$\rightarrow \quad f(x) - 14 = x^3 - 9x^2 + 15x - 7$

$\Rightarrow \quad f(x) - 14 = (x-1)(x^2 - 8x + 7)$

$$= (x-1)(x-1)(x-7)$$

$\Rightarrow \quad \dfrac{f(x)-14}{(x-1)^2} = (x-7) \quad \ldots(\text{i})$

Now, $\quad \dfrac{f(x)-14}{(x-1)^2} = 0,\ (x \neq 1) \quad$ [given]

$\Rightarrow \quad x - 7 = 0 \quad$ [from Eq. (i)]

$\Rightarrow \quad x = 7$

3. We have,

$$f(x) = \frac{x^3}{3} - (m-3)\frac{x^2}{2} + mx - 2013$$

$\therefore \quad f'(x) = (x^2 - (m-3)x + m) \geq 0,\ \forall\ x \in [0, \infty)$

***Case* I** When, $D \leq 0$

$\Rightarrow \quad m \in [1, 9] \quad \ldots(\text{i})$

***Case* II** When $D \geq 0$

$\Rightarrow \quad m \in [-\infty, 1] \cup [9, \infty]$

$$\frac{-b}{2a} \leq 0 \Rightarrow (m-3) \leq 0 \Rightarrow m \leq 3$$

and $f'(0) \geq 0 \ \Rightarrow m \geq 0$

$\therefore$ Eq. (i) $\cap$ Eq. (ii) $\cap$ Eq. (iii)

$\Rightarrow \quad m \in [0, 1] \quad \ldots(\text{ii})$

So, finally Eq. (i) $\cap$ Eq. (ii)

$\Rightarrow \quad m \in [0, 9] \equiv [0, k]$

$\therefore \quad k = 9$

4. We have, $f(x) = \sin x - a\sin 2x - \dfrac{1}{3}\sin 3x + 2ax$

$\therefore \quad f'(x) = \cos x - 2a\cos 2x - \cos 3x + 2a \geq 0\ \forall\ x \in R$

$\Rightarrow \quad \cos x - \cos 3x + 2a(1 - \cos 2x) \geq 0$

$\Rightarrow \quad 2\sin 2x + \sin x + 4a\sin^2 x \geq 0$

$\Rightarrow \quad 2\sin^2 x(\cos x + a) \geq 0$

$$a \geq -\cos x \Rightarrow a \geq 1$$

5. We have, $f(x) = e^{2x} - (c+1)e^x + 2x + \cos 2 + \sin 1$

$\therefore \quad f'(x) = 2e^{2x} - (c+1)e^x + 2$

Now, $f'(x) \geq 0 \ \forall \ x \in R$

i.e., $\quad 2\left(e^x + \frac{2}{e^x}\right) - (c+1) \geq 0 \ \forall \ x \in R$

$\Rightarrow \quad (c+1) \leq \underbrace{2\left(e^x + \frac{1}{e^x}\right)}_{\text{minimum value } = 4} \ \forall \ x \in R$

$\therefore \quad c + 1 \leq 4$

$\Rightarrow \quad c \leq 3$

$\therefore \quad c \in (-\infty, 3] \equiv (-\infty, \lambda]$

Hence, $\lambda = 3$

6. $$f(x) = \begin{cases} xe^{ax}, & x \leq 0 \\ x + ax^2 - x^3, & x > 0 \end{cases}$$

$$\Rightarrow f'(x) = \begin{cases} e^{ax} + axe^{ax}, & x \leq 0 \\ 1 + 2ax - 3x^2, & x > 0 \end{cases}$$

$$f''(x) = \begin{cases} (2a)e^{ax} + a^2xe^{ax}, & x \leq 0 \\ 2a - 6x, & x > 0 \end{cases}$$

For $\quad x < 0, \ f''(x) > 0$

$\Rightarrow \quad (2a + a^2x)e^{ax} > 0$

$\Rightarrow \quad a(2 + ax) > 0 \Rightarrow x > \frac{-2}{a}$

For $x > 0$, $2a - 6x > 0$, $x < \frac{a}{3}$

Also f is continuous at 0.

$\therefore f(x)$ is increasing in $\left(-\frac{2}{a}, \frac{a}{3}\right)$

7. $g(x) = \frac{d}{dx}(f(x).\ f'(x))$

to get the zero of $g(x)$ we take function

$$h(x) = f(x).\ f'(x)$$

between any two roots of $h(x)$ there lies at least one root of $h'(x) = 0$ i.e., $g(x) = 0$

Now $\quad h(x) = 0 \Rightarrow f(x) = 0$ or $f'(x) = 0$

As $\quad f(x) = 0$ has 4 minimum solutions.

and $\quad f'(x) = 0$ minimum 3 solutions.

$\therefore \quad h(x) = 0$ minimum 7 solutions.

$\Rightarrow \quad h'(x) = g(x) = 0$ has minimum 6 solutions.

Matrix-Match Type

1. (A) $\because \ f(x) = \frac{x}{(1+x^2)}$

$\therefore \quad f'(x) = \frac{(1-x^2)}{(1+x^2)^2}$

$\because \quad f'(x) < 0$

$\Rightarrow \quad \frac{1-x^2}{(1+x^2)^2} < 0$

$\Rightarrow \quad (1-x^2) < 0 \Rightarrow x^2 - 1 > 0$

$\therefore \quad x \in (-\infty, -1) \cup (1, \infty)$ (P, S)

(B) $\because \ f(x) = \tan^{-1} x - x$

$\therefore \quad f'(x) = \frac{1}{1+x^2} - 1 = -\frac{x^2}{1+x^2} < 0$

$f'(x) < 0 \ \forall \ x \in R$ (P, Q, R, S, T)

(C) $\because \quad f(x) = x - e^x + \tan\left(\frac{2\pi}{7}\right)$

$\therefore \quad f'(x) = 1 - e^x > 0$

or $\quad e^x < 1$

$\Rightarrow \quad e^x < e^0$

$\therefore \quad x < 0$

$x \in (-\infty, 0)$ (P, Q)

2. (A) $\because \quad f(x) = 5x^3 - 15x^2 - 120x + 3$

$\therefore \quad f'(x) = 15x^2 - 30x - 120$

$= 15(x^2 - 2x - 8) = 15(x-4)(x+2)$

For increasing function

$$f'(x) > 0$$

or $\ 15(x-4)(x+2) > 0$ or $(x-4)(x+2) > 0$

$\therefore \quad x \in (-\infty, -2) \cup (4, \infty)$

and for decreasing function $f'(x) < 0$

or $15(x-4)(x+2) < 0$ or $(x-4)(x+2) < 0$

$\therefore x \in (-2, 4)$

It is clear that $a = -2$ and $b = 4$

$\therefore b - a = 6$ (natural number, composite number and perfect number) = (P, R, S)

(B) $\because \ f(x) = 8 + 36x + 3x^2 - 2x^3$

$\therefore f'(x) = 0 + 36 + 6x - 6x^2$

$= -6(x^2 - x - 6)$

$= -6(x-3)(x+2)$

For increasing function $f'(x) > 0$

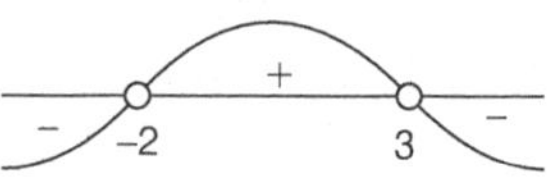

or $-6(x-3)(x+2) > 0$ or $-(x-3)(x+2) > 0$

$\therefore \quad x \in (-2, 3)$

and for decreasing function $f'(x) < 0$

or $-6(x-3)(x+2) < 0$ or $-(x-3)(x+2) < 0$

$\therefore b - a = 5$ (natural number and prime number) (P, Q)

(C) $\because \ f(x) = \frac{3}{10}x^4 - \frac{4}{5}x^3 - 3x^2 + \frac{36}{5}x + 11$

$\therefore f'(x) = \frac{12}{10}x^3 - \frac{12}{5}x^2 - 6x + \frac{36}{5}$

$= \frac{6}{5}(x^3 - 2x^2 - 5x + 6)$

$= \frac{6}{5}(x-1)(x-3)(x+2)$

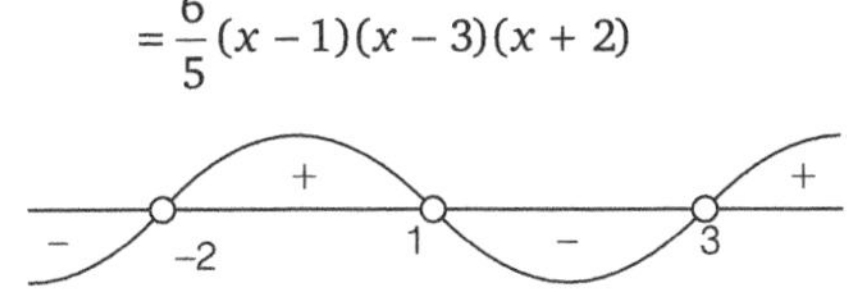

for increasing function $f'(x) > 0$

or $\frac{6}{5}(x-1)(x-3)(x+2) > 0$

or $(x-1)(x-3)(x+2) > 0$

$\therefore x \in (-2, 1) \cup (3, \infty)$

and for decreasing function

$f'(x) < 0$ or $\frac{6}{5}(x-1)(x-3)(x+2) < 0$

or $(x-1)(x-3)(x+2) < 0$

$\therefore$ $x \in (-\infty, -2) \cup (1, 3)$

it is clear that $a - b = -2$, $a = 1$, $b = 3$

$\therefore b - a = 2$ (natural number and prime number) (P, Q)

3. **(A)** $f'(x) = 4x^3 - 24x^2 + 44x - 24$

$= 4(x^3 - 6x^2 + 11x - 6)$

$= 4(x-1)(x-2)(x-3)$

for $f'(x) > 0$, $x \in (1, 2) \cup (3, \infty)$ (R)

and for $f'(x) < 0$, $x \in (-\infty, 1) \cup (2, 3)$ (T)

(B) $f'(x) = \frac{2}{(x-2)} - 2x + 4$

$= \frac{2 - 2x^2 + 4x + 4x - 8}{(x-2)}$

$= \frac{(-2x^2 + 8x - 6)}{(x-2)}$

$= \frac{-2(x^2 - 4x + 3)}{(x-2)}$

$= \frac{-2(x-1)(x-3)}{(x-2)}$

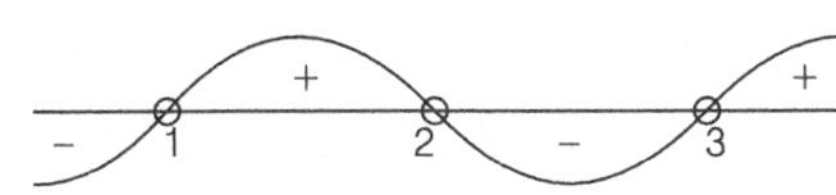

If $f'(x) > 0$

$\Rightarrow \frac{-2(x-1)(x-3)}{(x-2)} > 0$

$\Rightarrow \frac{(x-1)(x-3)}{(x-2)} < 0$

$\therefore$ $x \in (-\infty, 1) \cup (2, 3)$ but $x > 2$

$\therefore$ $x \in (2, 3)$ (P)

and if $f'(x) < 0$

$\Rightarrow \frac{-2(x-1)(x-3)}{(x-2)} < 0$

$\Rightarrow \frac{(x-1)(x-3)}{(x-2)} > 0$

$\therefore$ $x \in (1, 2) \cup (3, \infty)$ (S)

(C) $\because$ $f(x) = \begin{cases} \frac{1-x}{x^2}, & x < 1, \quad x \neq 0 \\ \frac{x-1}{x^2}, & x \geq 1, \quad x \neq 0 \end{cases}$

Case I : When $x < 1$

$f'(x) = \frac{-2}{x^3} + \frac{1}{x^2} = \frac{(x-2)}{x^3}$

for increasing, $x \in (-\infty, 0) \cup (2, \infty)$

but $x < 1$

$\therefore$ $x \in (-\infty, 0)$

for decreasing, $x \in (0, 2)$ but $x < 1$

$\therefore$ $x \in (0, 1)$

Case II : When $x \geq 1$

$f'(x) = -\frac{1}{x^2} + \frac{2}{x^3} = -\frac{(x-2)}{x^3}$

for increasing

$f'(x) > 0$

$\Rightarrow \frac{-(x-2)}{x^3} > 0$

$\Rightarrow \frac{(x-2)}{x^3} < 0 \Rightarrow x \in (0, 2)$, but $x \geq 1$, $\therefore x \in [1, 2)$

for decreasing, $f'(x) < 0$

$\Rightarrow -\frac{(x-2)}{x^3} < 0 \Rightarrow \frac{(x-2)}{x^3} > 0$

$\Rightarrow$ $x \in (-\infty, 0) \cup (2, \infty)$

but $x > 1$

$\therefore$ $x \in (2, \infty)$ (Q)

4. Using LMVT

$g'(c) = \frac{g(5) - g(0)}{5} = \frac{-1/6 - 4}{5} = -\frac{5}{6}$ (Q)

(B) Let $\phi(x) = f(x) - 2g(x)$, $x \in [0, 1]$

$\Rightarrow$ $\phi(0) = 2$, $\phi(1) = 6 - 2g(1)$

Now $\phi'(x) = f'(x) - 2g'(x)$

$\Rightarrow \phi(x)$ satisfies condition of Rolle's theorem on [0, 1]

$\Rightarrow$ $\phi(0) = \phi(1) \Rightarrow 2 = 6 - 2g(1)$

$\Rightarrow$ $g(1) = 2$ (S)

(C) $f(x) = \sin 3x$

clearly, longest length $= \frac{\pi}{6} - \left(-\frac{\pi}{6}\right) = \frac{\pi}{3}$

$\therefore$ $\lambda = 3$ (P)

(D) $f'(c) = \frac{f(5) - f(1)}{5 - 1}$

$\Rightarrow \frac{-c}{\sqrt{25 - c^2}} = \frac{-\sqrt{6}}{2}$

$\Rightarrow$ $c^2 = 15$ (R)

16

Maxima and Minima

Objective Questions Type I [Only one correct answer]

In each of the questions below, four choices are given of which only one is correct. You have to select the correct answer which is the most appropriate,

1. If the function $f(x) = 2x^3 - 9ax^2 + 12a^2x + 1$ has a local maximum at $x = x_1$ and a local minimum at $x = x_2$ such that $x_2 = x_1^2$, then a is equal to

(a) 0 (b) $\frac{1}{4}$
(c) 2 (d) Either (a) or (c)

2. The point (0, 5) is closest to the curve $x^2 = 2y$ at

(a) $(2\sqrt{2}, 0)$ (b) (2, 2)
(c) $(-2\sqrt{2}, 0)$ (d) $(2\sqrt{2}, 4)$

3. The least value of 'a' for which $\frac{4}{\sin x} + \frac{1}{1 - \sin x} = a$ has at least one solution in the interval $(0, \pi/2)$ is

(a) 9 (b) 8 (c) 4 (d) 1

4. The global maxima of $f(x) = [2\{-x^2 + x + 1\}]$ is (where $\{x\}$ denotes fractional part of x and [.] denotes greatest integer function)

(a) 2 (b) 1
(c) 0 (d) None of these

5. If $x = 1$ is a critical point of the function $f(x) = (3x^2 + ax - 2 - a)e^x$, then

[JEE Main 2020, 5 September]

(a) $x = 1$ and $x = -\frac{2}{3}$ are local minima of f

(b) $x = 1$ and $x = -\frac{2}{3}$ are local maxima of f

(c) $x = 1$ is a local maxima and $x = -\frac{2}{3}$ is a local minima of f

(d) $x = 1$ is a local minima and $x = -\frac{2}{3}$ is a local maxima of f

6. If $f(x) = \begin{cases} 1 + x^2 - 3x, & x < 0 \\ \cos x + 2x, & x \geq 0 \end{cases}$, then the global maximum and local minimum values of $f(x)$ for $x \in [-2, 2]$ are respectively

(a) $4 + \cos 2, 1$ (b) 11, 1
(c) 11, not exist (d) None of these

7. Domain of the function $f(x)$, if $3^x + 3^{f(x)} =$ minimum of $\phi(t)$, where $\phi(t) =$ minimum of $\{2t^3 - 15t^2 + 36t - 25, 2 + |\sin t|; 2 \leq t \leq 4\}$ is

(a) $(-\infty, 1)$ (b) $(-\infty, \log_3 e)$
(c) $(0, \log_3 2)$ (d) $(-\infty, \log_3 2)$

8. Let $f(x) = \begin{cases} |x^2 - 3x| + a, & 0 \leq x < \frac{3}{2} \\ -2x + 3, & x \geq \frac{3}{2} \end{cases}$

If $f(x)$ has a local maxima at $x = \frac{3}{2}$, then

(a) $a \leq 0$ (b) $a \leq -\frac{9}{4}$
(c) $a \geq \frac{9}{4}$ (d) None of these

9. The difference between the greatest and the least values of the function $f(x) = \int_0^x (at^2 + 1 + \cos t)\,dt, a > 0$ for $x \in [2, 3]$ is

(a) $\frac{19}{3}a + 1 + (\sin 3 - \sin 2)$ (b) $\frac{18}{3}a + 1 + 2\sin 3$
(c) $\frac{18}{3}a - 1 + 2\sin 3$ (d) None of these

10. The coordinate of the point on $y^2 = 8x$, which is closest from $x^2 + (y + 6)^2 = 1$ is/are

(a) (2, – 4) (b) (18, – 12)
(c) (2, 4) (d) None of these

11. The greatest value of the function $f(x) = 2\sin x + \sin 2x$ on the interval $\left[0, \frac{3\pi}{2}\right]$ is

(a) $\frac{3\sqrt{3}}{2}$ (b) 3
(c) $\frac{3}{2}$ (d) None of these

12. Let $f(x) = 5 - |x-2|$ and $g(x) = |x+1|$, $x \in R$. If $f(x)$ attains maximum value at α and $g(x)$ attains minimum value of β, then $\lim_{x \to -\alpha\beta} \frac{(x-1)(x^2-5x+6)}{x^2-6x+8}$ is equal to

[JEE Main 2019, 12 April]

(a) 1/2 (b) – 3 / 2
(c) – 1 / 2 (d) 3/2

13. A function f such that $f'(2) = f''(2) = 0$ and f has a local maximum of –17 at 2 is

(a) $(x-2)^4$ (b) $3-(x-2)^4$
(c) $-17-(x-2)^4$ (d) None of these

14. The difference between the greatest and the least value of the function

$f(x) = \int_0^x (t^2+t+1)\,dt$ on [2, 3] is

(a) 37/6 (b) 47/6
(c) 57/6 (d) 59/6

15. Let $f(x) = a - (x-3)^{8/9}$, then maxima of $f(x)$ is

(a) 3 (b) $a-3$
(c) a (d) None of these

16. The least value of $\alpha \in R$ for which $4\alpha x^2 + \frac{1}{x} \geq 1$, for all $x > 0$, is **[JEE Advanced 2016]**

(a) $\frac{1}{64}$ (b) $\frac{1}{32}$
(c) $\frac{1}{27}$ d) $\frac{1}{25}$

17. The point in the interval $[0, 2\pi]$, where $f(x) = e^x \sin x$ has maximum slope is

(a) $\frac{\pi}{4}$ (b) $\frac{\pi}{2}$
(c) π (d) None of these

18. Let

$$f(x) = \begin{cases} |x^3+x^2+3x+\sin x|\left(3+\sin\left(\frac{1}{x}\right)\right), & x \neq 0, \\ 0 & , x = 0 \end{cases}$$

then number of points (where $f(x)$ attains its minimum value) is

(a) 1 (b) 2
(c) 3 (d) infinite many

19. A function f is defined by $f(x) = |x|^m |x-1|^n \; \forall \; x \in R$. The local maximum value of the function is, $(m, n \in N)$,

(a) 1 (b) $m^n n^m$
(c) $\frac{m^m n^n}{(m+n)^{m+n}}$ (d) $\frac{(mn)^{mn}}{(m+n)^{m+n}}$

20. Let $f(x) = 1 + 2x^2 + 2^2 x^4 + \ldots + 2^{10} x^{20}$. Then $f(x)$ has

(a) more than one minimum
(b) exactly one minimum
(c) at least one maximum
(d) None of the above

21. Let $f(x) = \begin{cases} \sin^{-1}\alpha + x^2, & 0 < x < 1 \\ 2x, & x \geq 1 \end{cases}$

$f(x)$ can have a minimum at $x = 1$ is the value of α is

(a) 1 (b) –1
(c) 0 (d) None of these

22. A differentiable function $f(x)$ has a relative minimum at $x = 0$, then the function $y = f(x) + ax + b$ has a relative minimum at $x = 0$ for

(a) all a and all b (b) all b if $a = 0$
(c) all $b > 0$ (d) all $a > 0$

23. The function

$f(x) = \int_1^x \{2(t-1)(t-2)^3 + 3(t-1)^2(t-2)^2\}\,dt$

attains its maximum at x is equal to

(a) 1 (b) 2
(c) 3 (d) 4

24. Assuming that the petrol burnt in a motor boat varies as the cube of its velocity, the most economical speed, when going against a current of c km/h is

(a) $(3c/2)$ km/h (b) $(3c/4)$ km/h
(c) $(5c/2)$ km/h (d) $(c/2)$ km/h

25. N Characters of information are held on magnetic tape, in batches of x characters each, the batch processing time is $\alpha + \beta x^2$ seconds, α and β are constants. The optical value of x for fast processing is

(a) $\frac{\alpha}{\beta}$ (b) $\frac{\beta}{\alpha}$
(c) $\sqrt{\frac{\alpha}{\beta}}$ (d) $\sqrt{\frac{\beta}{\alpha}}$

26. The minimum value of the function defined by $f(x) = \text{maximum}\{x, x+1, 2-x\}$ is

(a) 0 (b) $\frac{1}{2}$
(c) 1 (d) $\frac{3}{2}$

27. On $[1, e]$, the least and greatest values of $f(x) = x^2 \ln x$ is

(a) e, 1 (b) 1, e
(c) 0, e^2 (d) None of these

28. The minimum value of $\left(1+\frac{1}{\sin^n\alpha}\right)\left(1+\frac{1}{\cos^n\alpha}\right)$ is

(a) 1 (b) 2
(c) $(1+2^{n/2})^2$ (d) None of these

29. The fuel charges for running a train are proportional to the square of the speed generated in mile/h and costs ₹ 48 per h at 16 miles/h. The most economical speed if the fixed charges i.e. salaries etc amount to ₹ 300 per h

(a) 10 mile/h (b) 20 mile/h
(c) 30 mile/h (d) 40 mile/h

30. The maximum area of the rectangle that can be inscribed in a circle of radius r is

(a) πr^2 (b) r^2
(c) $\frac{\pi r^2}{4}$ (d) $2r^2$

31. If $f(x) = x^2 + 2bx + 2c^2$ and $g(x) = -x^2 - 2cx + b^2$ are such that $\min f(x) > \max g(x)$, then the relation between b and c is

(a) no relation (b) $0 < c < b/2$
(c) $|c| < |b|\sqrt{2}$ (d) $|c| > |b|\sqrt{2}$

32. A cylindrical gas container is closed at the top and open at the bottom; if the iron plate of the top is 5/4 times as thick as the plate forming the cylindrical sides. The ratio of the radius to the height of the cylinder using minimum material for the same capacity is

(a) 2/3 (b) 1/2
(c) 4/5 (d) 1/3

33. Let $f(x) = \begin{cases} x^3 - x^2 + 10x - 5, & x \le 1 \\ -2x + \log_2(b^2 - 2), & x > 1 \end{cases}$, the set of values of b for which $f(x)$ have greatest value at $x = 1$ is given by

(a) $1 \le b \le 2$ (b) $b = \{1, 2\}$
(c) $b \in (-\infty, -1)$ (d) None of these

34. If S_1 and S_2 are respectively the sets of local minimum and local maximum points of the function, $f(x) = 9x^4 + 12x^3 - 36x^2 + 25$, $x \in R$, then **[JEE Main 2019, 8 April]**

(a) $S_1 = \{-2\}$; $S_2 = \{0, 1\}$
(b) $S_1 = \{-2, 0\}$; $S_2 = \{1\}$
(c) $S_1 = \{-2, 1\}$; $S_2 = \{0\}$
(d) $S_1 = \{-1\}$; $S_2 = \{0, 2\}$

35. If $f : R \to R$ is a twice differentiable function such that $f''(x) > 0$ for all $x \in R$, and $f\left(\frac{1}{2}\right) = \frac{1}{2}$, $f(1) = 1$, then **[JEE Advanced 2017]**

(a) $f'(1) \le 0$ (b) $f'(1) > 1$
(c) $0 < f'(1) \le \frac{1}{2}$ (d) $\frac{1}{2} < f'(1) \le 1$

36. The number of solutions of the equation $a^{f(x)} + g(x) = 0$, where $a > 0$, $g(x) \ne 0$ and has minimum value 1/2 is

(a) one (b) two
(c) infinite many (d) zero

37. Let a tangent be drawn to the ellipse $\frac{x^2}{27} + y^2 = 1$ at $(3\sqrt{3}\cos\theta, \sin\theta)$ where $\theta \in \left(0, \frac{\pi}{2}\right)$. Then, the value of θ such that the sum of intercepts on axes made by this tangent is minimum is equal to

(a) $\frac{\pi}{8}$ (b) $\frac{\pi}{4}$ (c) $\frac{\pi}{6}$ (d) $\frac{\pi}{3}$

38. Two towns A and B are 60 km apart. A school is to be built to serve 150 students in town A and 50 students in town B. If the total distance to be travelled by all 200 students is to be as small as possible, then the school should be built at

(a) town B (b) 45 km from town A
(c) town A (d) 45 km from town B

Objective Questions Type II [One or more than one correct answer(s)]

In each of the questions below four choices of which one or more than one are correct. You have to select the correct answer(s) accordingly.

1. Let $f(x) = \cos x \sin 2x$, then

(a) $\min\limits_{x \in (-\pi, \pi)} f(x) > -7/9$
(b) $\min\limits_{x \in (-\pi, \pi)} f(x) > -9/7$
(c) $\min\limits_{x \in [-\pi, \pi]} f(x) > -1/9$
(d) $\min\limits_{x \in [-\pi, \pi]} f(x) > -2/9$

2. Let $f : R \to R$ be given by $f(x) = (x-1)(x-2)(x-5)$. Define $F(x) = \int_0^x f(t)\,dt$, $x > 0$

Then, which of the following options is/are correct? **[JEE Advanced 2019]**

(a) $F(x) \ne 0$ for all $x \in (0, 5)$
(b) F has a local maximum at $x = 2$
(c) F has two local maxima and one local minimum in $(0, \infty)$
(d) F has a local minimum at $x = 1$

3. Let $f(x) = ax^3 + bx^2 + cx + 1$ have extrema at $x = \alpha, \beta$ such that $\alpha\beta < 0$ and $f(\alpha) \cdot f(\beta) < 0$, then the equation $f(x) = 0$ has

(a) three equal real roots
(b) three distinct real roots
(c) one positive root, if $f(\alpha) < 0$ and $f(\beta) > 0$
(d) one negative root, if $f(\alpha) > 0$ and $f(\beta) < 0$

4. Let $f(x) = (x^2 - 1)^n (x^2 + x + 1)$, then $f(x)$ has local extremum at $x = 1$ when

(a) $n = 2$ (b) $n = 3$
(c) $n = 4$ (d) $n = 6$

5. Let $f : R \to R$ be given by

$$f(x) = \begin{cases} x^5 + 5x^4 + 10x^3 + 10x^2 + 3x + 1, & x < 0; \\ x^2 - x + 1, & 0 \le x < 1; \\ \frac{2}{3}x^3 - 4x^2 + 7x - \frac{8}{3}, & 1 \le x < 3; \\ (x-2)\log_e(x-2) - x + \frac{10}{3}, & x \ge 3; \end{cases}$$

Then which of the following options is/are correct? **[JEE Advanced 2019]**

(a) f is increasing on $(-\infty, 0)$
(b) f' is NOT differentiable at $x = 1$
(c) f is onto
(d) f' has a local maximum at $x = 1$

6. Let $f(x) = \int_0^x \frac{\cos t}{t} dt$ $(x > 0)$; then $f(x)$ has
   (a) maxima, when $n = -2, -4, -6, \ldots$
   (b) maxima, when $n = -1, -3, -5, \ldots$
   (c) minima, when $n = 0, 2, 4, \ldots$
   (d) minima, when $n = 1, 3, 5, \ldots$

7. Let $f(x) = (x-1)^m (x-2)^n$, $x \in R$, then each critical point of $f(x)$ is either local maximum or local minimum where
   (a) $m = 2, n = 3$ (b) $m = 2, n = 4$
   (c) $m = 3, n = 4$ (d) $m = 4, n = 2$

8. Let $f(x) = -x^3 + \frac{b^3 - b^2 + b - 1}{b^2 + 3b + 2}, 0 \le x < 1$

   $= 2x - 3, 1 \le x < 3$

   If $f(x)$ has least value at $x = 1$, then
   (a) $-2 < b < -1$ (b) $-1 < b < 0$
   (c) $0 < b < 1$ (d) $1 \le b < \infty$

9. Let $f : R \to (0, \infty)$ and $g : R \to R$ be twice differentiable functions such that f'' and g'' are continuous functions on R. Suppose $f'(2) = g(2) = 0$, $f''(2) \ne 0$ and $g'(2) \ne 0$.

   If $\lim_{x \to 2} \frac{f(x)g(x)}{f'(x)g'(x)} = 1$, then **[JEE Advanced 2016]**
   (a) f has a local minimum at $x = 2$
   (b) f has a local maximum at $x = 2$
   (c) $f''(2) > f(2)$
   (d) $f(x) - f''(x) = 0$, for atleast one $x \in R$

10. The function $f(x) = 3 + 2(a+1)x + (a^2 + 1)x^2 - x^3$ has a local minimum at $x = x_1$ and local maximum at $x = x_2$ such that $x_1 < 2 < x_2$, then a belongs to the interval(s)
    (a) $\left(-\infty, -\frac{3}{2}\right)$ (b) $\left(-\frac{3}{2}, 1\right)$
    (c) $(0, \infty)$ (d) $(1, \infty)$

11. If composite function $f_1(f_2(f_3(\ldots(f_n(x))\ldots)$ is an increasing function and if r of f_i's are decreasing functions while rest are increasing, then maximum value of $r(n-r)$ is
    (a) $\frac{n^2}{4}$, when n is an even number
    (b) $\frac{n^2}{4}$, when n is an odd number
    (c) $\frac{n^2 - 1}{4}$, when n is an odd number
    (d) $\frac{n^2 - 1}{4}$, when n is an even number

Linked-Comprehension Type

In these questions, a passage (paragraph) has been given followed by questions based on each of the passages. You have to answer the questions based on the passage given.

PASSAGE 1

Let $f(x)$ be a polynomial function of degree 4, vanishes at $x = 0$. Given $f(-2) = 216$ and $f(x)$ has relative maximum/relative minimum at $x = 1, 2, 3$

On the basis of above information, answer the following questions

1. Area of triangle formed by extremum points of $f(x)$, is
   (a) $\frac{1}{8}$ (b) $\frac{1}{4}$
   (c) $\frac{1}{2}$ (d) 1

2. The value of $15\int_{-1}^{1} (f(-x) + f(x))dx$, is
   (a) 352 (b) 452
   (c) 552 (d) 652

3. Which of the following statement is correct?
   (a) $f(x)$ has two relative maximum points and one relative minimum point
   (b) $f(x)$ has exactly one inflection point.
   (c) Range of $f(x)$ contains 9 negative integers
   (d) Sum of real roots of $f(x) = 0$ is 6

PASSAGE 2

A cubic $f(x) = ax^3 + bx^2 + cx + d$ vanishes at $x = -2$ and has relative minimum/maximum at $x = -1$ and $x = \frac{1}{3}$ and if $\int_{-1}^{1} f(x)\, dx = \frac{14}{3}$.

On the basis of above information, answer the following questions

1. The value of c is
   (a) -2 (b) -1
   (c) 0 (d) 2

2. The value of d is
   (a) 5 (b) 2
   (c) 0 (d) -4

3. The function $f(x)$ is
   (a) $x^3 + x^2 + x - 2$
   (b) $x^3 - x^2 + x - 2$
   (c) $x^3 - x^2 - x + 2$
   (d) $x^3 + x^2 - x + 2$

4. $f(x)$ decreases in the interval

(a) $\left(-\frac{1}{3}, 1\right)$ (b) $\left(-\frac{1}{3}, -1\right)$

(c) $\left(-1, \frac{1}{3}\right)$ (d) $\left(1, \frac{3}{2}\right)$

5. The nature of roots of $f(x) = 3$ is

(a) one root is real and other two are distinct
(b) all roots real and distinct
(c) all roots are real; two of them are equal
(d) None of the above

Numerical Value Type

Solve the following problems and write your answer.

1. If $y = \frac{(ax+b)}{(x-4)(x-1)}$ has extreme value at $(2, -1)$, then the value of $50.5\,(a+10b)$ must be
2. If the greatest and least values of the function $f(x) = x^3 - 6x^2 + 9x + 1$ on $[0, 2]$ are λ and μ, then the value of $\lambda^4 + \mu^8$ must be
3. The three sides of a trapezium are equal each being 6 cm long. If area of trapezium when it is maximum is A, then the value of $4\sqrt{3}A$ must be
4. A wire of given length l is cut into two portions which are bent into the shapes of a circle and a square respectively. If the sum of the areas of the circle and the square will be least when (side of square) = λ (diameter of circle), then the value of 2008 λ must be
5. When travelling at x km/h, a truck burns diesel at the rate of $\frac{1}{300}\left(\frac{900}{x} + x\right)$L/km. If the diesel oil costs 40 paise/L and driver is paid Rs 1.50 per h, if the steady speed that will minimize the total cost of the trip of 500 km is λ km/h, then the value of 50 λ must be
6. If the value of maximum slope of $f(x)$ $= -x^3 + 8x^2 - 13x - 18$ is λ, then the value of 15λ must be
7. Let the function $f : (0, \pi) \to R$ be defined by
$$f(\theta) = (\sin\theta + \cos\theta)^2 + (\sin\theta - \cos\theta)^4$$
Suppose the function f has a local minimum at θ precisely when $\theta \in \{\lambda_1\pi, \dots, \lambda_r\pi\}$, where $0 < \lambda_1 < \dots \lambda_r < 1$. Then, the value of $\lambda_1 + \dots + \lambda_r$ is .... **[JEE Advanced 2020]**

Single Integer Type

Solve the following problems and write your answer.

1. The maximum value of the function $f(x) = \frac{(1+x)^{0.6}}{1+x^{0.6}}$ in the interval $[0, 1]$ is
2. A curve passes through the point $(2, 0)$ and the slope of the tangent at any point (x, y) is $x^2 - 2x$ for all values of x then $3y_{max}$ is equal to
3. Let $f(x)$ be a polynomial of degree 3 such that $f(-1) = 10$, $f(1) = -6$, $f(x)$ has a critical point at $x = -1$ and $f'(x)$ has a critical point at $x = 1$. The $f(x)$ has a local minima at $x =$ .... **[JEE Main 2020, 8 January]**
4. For $a > 0$, if the function $f(x) = 2x^3 - 9ax^2 + 12a^2x + 1$ attains its maximum value at p and minimum value at q such that $p^2 = q$, then $a =$
5. The maximum value of $\frac{x^2 - x + 1}{x^2 + x + 1}$ for all real values of x is
6. Let $f : [-1, 1] \to R$ be defined as $f(x) = ax^2 + bx + c$ for all $x \in [-1, 1]$, where $a, b, c \in R$ such that $f(-1) = 2$, $f'(-1) = 1$ and for $x \in (-1, 1)$ the maximum value of $f''(x)$ is $\frac{1}{2}$. If $f(x) \le \alpha$ $x \in [-1, 1]$, then the least value of α is equal to ....
7. For a polynomial $g(x)$ with real coefficients, let m_g denote the number of distinct real roots of $g(x)$. Suppose S is the set of polynomials with real coefficients defined by
$$S = \{(x^2 - 1)^2(a_0 + a_1x + a_2x^2 + a_3x^3) : a_0, a_1, a_2, a_3 \in R\};$$
For a polynomial f, let f' and f'' denote its first and second order derivatives, respectively. Then the minimum possible value of $(m_{f'} + m_{f''})$, where $f \in S$, is .... **[JEE Advanced 2020]**
8. A cylindrical container is to be made from certain solid material with the following constraints : It has a fixed inner volume of V mm^3, has a 2 mm thick solid wall and is open at the top. The bottom of the container is a solid circular disc of thickness 2 mm and is of radius equal to the outer radius of the container. If the volume of the material used to make the container is minimum, when the inner radius of the container is 10 mm, then the value of $\frac{V}{250\pi}$ is **[JEE Advanced 2015]**
9. The number of points of extreme of the function $F(x) = \int_0^x \frac{\sin t}{t}\,dt$ in the interval $(0, 16)$ must be

Matrix-Match Type

Given below are Matching Type Questions, with two columns (each having some items) each. Each item of Column I has to be matched with the items of Column II, by encircling the correct match(es).

NOTE *An item of Column I can be matched with more than one items of Column II.*

1. The function $f(x)=\sqrt{(ax^3+bx^2+cx+d)}$ has its non-zero local minimum and maximum values at $x=-2$ and $x=2$, respectively. If a is a root of $x^2-x-6=0$, then match the following lists

	Column I		Column II
(A)	The value/values of a	(P)	$=0$
(B)	The value/values of b	(Q)	$=24$
(C)	The value/values of c	(R)	>32
(D)	The value/values of d	(S)	-2
		(T)	-4

2. Observe the following columns :

	Column I		Column II
(A)	The maximum and minimum values of critical points of $f(x)=x^2-2\|x\|$ are λ and μ respectively, then	(P)	$3\lambda+\mu=11$
(B)	The maximum and minimum values of critical points of $f(x)=(1-x)\|x-3\|$ are λ and μ respectively, then	(Q)	$\lambda+3\mu=9$
(C)	The maximum and minimum values of critical points of $f(x)=(x^2+2x-4)\|x\|$ are λ and μ respectively, then	(R)	$3\lambda+\mu=2$
		(S)	$\lambda+3\mu=-2$
		(T)	$3\lambda+\mu=0$

3. Observe the following columns :

	Column I		Column II
(A)	If the greatest and least values of the function $f(x)=\begin{cases}2x^2+\dfrac{2}{x^2}, & x\in[-2,2]\sim\{0\}\\ 1, & x=0\end{cases}$ are G and L respectively, then	(P)	$[G+L]=1$ where $[.]$ = greatest integer function
(B)	If the greatest and least values of the function $f(x)=x^3-6x^2+9x+1$ on $[0,2]$ are G and L respectively, then	(Q)	$[G+L]=6$ where $[.]$ = greatest integer function
(C)	If the greatest and least values of the functions $f(x)=$ arc tan $x-\frac{1}{2}\ln x$ on $\left[\frac{1}{\sqrt{3}},\sqrt{3}\right]$ are G and L respectively, then	(R)	$[G+L]=9$ where $[.]$ = greatest integer function
		(S)	$(G+L)=2$ where $(.)$ = Least integer function

Answers

Objective Questions Type I [Only one correct answer]

1. (c)	**2.** (d)	**3.** (a)	**4.** (b)	**5.** (d)	**6.** (b)	**7.** (d)	**8.** (b)	**9.** (a)	**10.** (a)
11. (a)	**12.** (c)	**13.** (c)	**14.** (d)	**15.** (c)	**16.** (c)	**17.** (b)	**18.** (a)	**19.** (c)	**20.** (b)
21. (d)	**22.** (b)	**23.** (a)	**24.** (a)	**25.** (c)	**26.** (d)	**27.** (c)	**28.** (c)	**29.** (d)	**30.** (d)
31. (d)	**32.** (c)	**33.** (d)	**34.** (c)	**35.** (b)	**36.** (d)	**37.** (c)	**38.** (c)		

Objective Questions Type II [One or more than one correct answer(s)]

1. (a, b)	**2.** (a, b, d)	**3.** (b, c, d)	**4.** (a, c, d)	**5.** (b, c, d)
6. (b, d)	**7.** (b, d)	**8.** (a, d)	**9.** (a, d)	**10.** (a, d)
11. (a, c)				

Linked-Comprehension Type

Passage 1 **1.** (d) **2.** (b) **3.** (c)

Passage 2 **1.** (b) **2.** (b) **3.** (d) **4.** (c) **5.** (c)

Numerical Value Type

1. 50.5 **2.** 626 **3.** 324 **4.** 2008 **5.** 2250 **6.** 125 **7.** 0.50

Single Integer Type

1. (1)	**2.** (4)	**3.** (3)	**4.** (2)	**5.** (3)	**6.** (5)
7. (5)	**8.** (4)	**9.** (5)			

Matrix-Match Type

1. A → (S); B → (P); C → (Q); D → (R)
2. A → (R, S); B → (P, Q); C → (T)
3. A → (R, T); B → (Q); C → (P, S)

Solutions

Objective Questions Type [Only one correct answer]

1. $f'(x) = 6x^2 - 18ax + 12a^2 = 0$ (for max or min)

$$\Rightarrow \quad x = a, 2a$$
$$x_2 = x_1^2 \quad \text{[given]}$$

$\Rightarrow a$ cannot be negative

If $a > 0$, then local minimum occurs at $x = 2a$ and local maximum at $x = a$. Thus, $x_1 = a$ and $x_2 = 2a$

$$\Rightarrow \quad 2a = a^2$$
$$\Rightarrow \quad a = 2$$

[if $a = 0$, $f(x)$ is always increasing]

2. $\therefore$ Let $P\left(t, \frac{t^2}{2}\right)$ be a point on $x^2 = 2y$ and A be (0, 5)

If $\quad AP = d$

$$\Rightarrow \quad z = d^2 = t^2 + \left(\frac{t^2}{2} - 5\right)^2$$
$$\therefore \quad \frac{dz}{dt} = 2t + 2\left(\frac{t^2}{2} - 5\right) \cdot t$$
$$= t^3 - 8t$$
$$= t(t^2 - 8)$$
$$\Rightarrow \quad \frac{d^2z}{dt^2} = 3t^2 - 8$$
$$\therefore \quad \frac{dz}{dt} = 0 \Rightarrow t = 0 \text{ or } t = \pm 2\sqrt{2}$$

At $t = 0$, $\frac{d^2z}{dt^2}$ is – ive

At $t = \pm 2\sqrt{2}$, $\frac{d^2z}{dt^2}$ is + ive

Hence, the closest point is $(2\sqrt{2}, 4)$

3. $f(x) = \frac{4}{\sin x} + \frac{1}{1 - \sin x} = a \quad \ldots(i)$

$$f'(x) = -\frac{4 \cos x}{\sin^2 x} + \frac{\cos x}{(1 - \sin x)^2}$$
$$= \cos x \left(\frac{1}{(1 - \sin x)^2} - \frac{4}{\sin^2 x}\right)$$
$$\therefore \quad f'(x) = 0$$
$$\Rightarrow \quad \frac{1}{(1 - \sin x)^2} - \frac{4}{\sin^2 x} = 0 \text{ as } \cos x \neq 0 \text{ in } \left(0, \frac{\pi}{2}\right)$$

This gives $\sin x = \frac{2}{3}$.

Substituting this in Eq. (i), we get

$$a = 9$$

4. $\because 0 \leq \{-x^2 + x + 1\} < 1$

$$\Rightarrow \quad 0 \leq 2\{-x^2 + x + 1\} < 2$$
$$\therefore \quad [2\{-x^2 + x + 1\}] = 0, 1$$

$\therefore$ Global maximum of $f(x)$ is 1.

5. It is given that $x = 1$ is a critical point of the function $f(x) = (3x^2 + ax - 2 - a)e^x$

So, $f'(1) = e^x(6x + a) + e^x(3x^2 + ax - 2 - a)|_{x=1} = 0$

$$\Rightarrow \quad 6 + a + 3 + a - 2 - a = 0 \quad \Rightarrow \quad a = -7.$$
$$\therefore \quad f'(x) = e^x[3x^2 - x - 2] = 0 \Rightarrow \quad x = 1 \text{ or } \left(-\frac{2}{3}\right)$$

and $f''(x) = e^x(6x - 1 + 3x^2 - x - 2)$
$= e^x(3x^2 + 5x - 3)$

$$\because \quad f''(1) = 5e > 0$$

$\Rightarrow x = 1$ is the point of local minima.

and $f''\left(-\frac{2}{3}\right) = e^{-2/3}\left(\frac{4}{3} - \frac{10}{3} - 3\right) = -5e^{-2/3} < 0$

$\Rightarrow x = -\frac{2}{3}$ is the point of local maxima.

6. Since, the given function is continuous but not differentiable at $x = 0$ and $Lf'(0) = -3$, $Rf'(0) = 2$

$\Rightarrow x = 0$ is point of local minima and $f(0) = 1$.

Also, there is no other critical point.

Now, $\quad f(-2) = 11$, $f(2) = 4 + \cos 2$

On comparing values of function at $x = -2, 0, 2$

global maximum occurs at $x = -2 \Rightarrow f(-2) = 11$

Hence, global maximum and local minimum values are 11, 1.

7. Let $\quad g(t) = 2t^3 - 15t^2 + 36t - 25$

$$g'(t) = 6t^2 - 30t + 36$$
$$= 6(t^2 - 5t + 6)$$

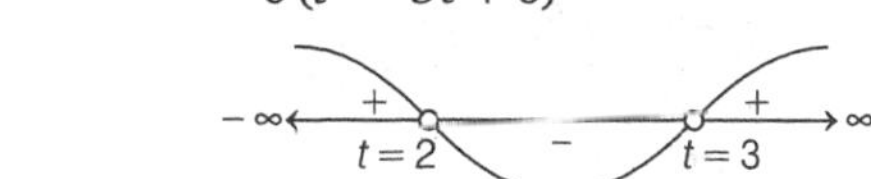

$$= 6(t - 2)(t - 3) = 0$$
$$\Rightarrow \quad t = 2, 3$$

For $\quad 2 \leq t \leq 4$

$$g(t)_{\min} = g(3) = 2 \times 27 - 15 \times 9 + 36 \times 3 - 25$$
$$= 54 - 135 + 108 - 25 = 2$$

Also, $\quad 2 + |\sin t| \geq 2$

Hence, minimum $\phi(t) = 2$

$$\therefore \quad 3^x + 3^{f(x)} = 2 \Rightarrow 3^{f(x)} = 2 - 3^x$$
$$\because \quad 3^{f(x)} > 0 \Rightarrow 2 - 3^x > 0$$
$$\Rightarrow \quad 3^x < 2$$
$$\Rightarrow \quad x < \log_3 2$$
$$\therefore \quad x \in (-\infty, \log_3 2)$$

8.
$$f\left(\frac{3}{2}\right) = 0$$
$$\Rightarrow \quad \lim_{x \to \frac{3}{2}} |x^2 - 3x| + a \leq 0$$
$$\Rightarrow \quad \frac{9}{4} + a \leq 0 \Rightarrow a \leq -\frac{9}{4}$$

9. $f'(x) = ax^2 + 1 + \cos x > 0$, for $a > 0$ and $\forall\ x \in R$

$\Rightarrow f(x)$ is an increasing function

$\Rightarrow$ max $f(x) = f(3)$ and min $f(x) = f(2)$

$\Rightarrow$ Difference $= f(3) - f(2) = \int_2^3 (at^2 + 1 + \cos t)\,dt$

$$= \frac{19}{3}a + 1 + (\sin 3 - \sin 2)$$

10. Let point on parabola $y^2 = 8x$ be $(2\alpha^2, 4\alpha)$

Centre of circle $= (0, -6)$

Distance between centre of circle and point on parabola

$$S = \sqrt{(2\alpha^2)^2 + (4\alpha + 6)^2}$$

$$D = 4\alpha^4 + (4\alpha + 6)^2$$

where $D = S^2$

$$\frac{dD}{d\alpha} = 16\alpha^3 + 2(4\alpha + 6)\cdot 4 = 0$$

$\Rightarrow \quad 2\alpha^3 + 4\alpha + 6 = 0$

$\Rightarrow \quad \alpha^3 + 2\alpha + 3 = 0$

$\Rightarrow \quad (\alpha^2 - \alpha + 3)(\alpha + 1) = 0$

$\Rightarrow \quad \alpha = -1 \qquad [\because \alpha^2 - \alpha + 3 \neq 0]$

Now, $\dfrac{d^2D}{d\alpha^2} = 48\alpha^2 + 8\cdot 4$

$$= 48\alpha^2 + 32 > 0 \text{ at } \alpha = -1$$

$\Rightarrow D$ is minimum at $\alpha = -1$. Point on parabola $(2, -4)$.

11. $\because \quad f(x) = 2\sin x + \sin 2x$

$\therefore \quad f'(x) = 2\cos x + 2\cos 2x = 2(\cos x + \cos 2x)$

$$= 4\cos\left(\frac{3x}{2}\right)\cos\left(\frac{x}{2}\right)$$

$$f'(x) = 0$$

Then, $x = \dfrac{\pi}{3}, \pi$

Now, $f(0) = 0$

$\Rightarrow \quad f\left(\dfrac{\pi}{3}\right) = \dfrac{2\sqrt{3}}{2} + \dfrac{\sqrt{3}}{2} = \dfrac{3}{2}\sqrt{3}$

$$f(\pi) = 0$$

$\Rightarrow \quad f\left(\dfrac{3\pi}{2}\right) = 2(-1) + 0 = -2$

Hence, greatest value $= \dfrac{3\sqrt{3}}{2}$

12. $f(1) = -6$

$$\lim_{x \to 1^-} \tan^{-1}\alpha - 5x^2 < -6$$

$\Rightarrow \quad \tan^{-1}\alpha - 5 < -6$

$\Rightarrow \quad \tan^{-1}\alpha < -1$

$\Rightarrow \quad \alpha < -\tan 1$

13. Since, $f'(2) = 0 = f''(2)$

Let $\quad f(x) = A + (x - 2)^n, n \geq 3$

but f has local maximum at $x = 2$

So, n may be taken as 4 and since maximum value is -17.

Therefore, $f(x) = -17 - (x - 2)^4$

14. $f(x) = \int_0^x (t^2 + t + 1)\,dt$

$$f'(x) = x^2 + x + 1 = 0$$

$\therefore \quad x = \dfrac{-1 \pm \sqrt{1 - 4}}{2}$ impossible.

$\therefore \quad f(2) = \int_0^2 (t^2 + t + 1)\,dt = \left[\dfrac{t^3}{3} + \dfrac{t^2}{2} + t\right]_0^2 = \dfrac{20}{3}$

$$f(3) = \int_0^3 (t^2 + t + 1)\,dt = \left[\frac{t^3}{3} + \frac{t^2}{2} + t\right]_0^3 = \frac{33}{2}$$

$\therefore \quad f(3) - f(2) = \dfrac{33}{2} - \dfrac{20}{3} = \dfrac{59}{6}$

15. $\because \quad f(x) = a - (x - 3)^{8/9}$

$\therefore \quad f'(x) = 0 - \dfrac{8}{9}(x - 3)^{-1/9}$

At $x = 3$, $f'(x)$ is not defined.

Hence, $x = 3$ is the point of extremum.

Hence, maximum value of $f(x) = a$ at $x = 3$.

16. Here, to find the least value of $\alpha \in R$, for which $4\alpha x^2 + \dfrac{1}{x} \geq 1$, for all $x > 0$.

i.e. to find the minimum value of α when

$y = 4\alpha x^2 + \dfrac{1}{x}$; $x > 0$ attains minimum value of α.

$\therefore \quad \dfrac{dy}{dx} = 8\alpha x - \dfrac{1}{x^2} \qquad \ldots\text{(i)}$

and $\quad \dfrac{d^2y}{dx^2} = 8\alpha + \dfrac{2}{x^3} \qquad \ldots\text{(ii)}$

When $\dfrac{dy}{dx} = 0$, then $8x^3\alpha = 1$

At $x = \left(\dfrac{1}{8\alpha}\right)^{1/3}$, $\dfrac{d^2y}{dx^2} = 8\alpha + 16\alpha = 24\alpha$, Thus, y attains minimum when $x = \left(\dfrac{1}{8\alpha}\right)^{1/3}$; $\alpha > 0$.

$\therefore y$ attains minimum when $x = \left(\dfrac{1}{8\alpha}\right)^{1/3}$.

i.e. $\quad 4\alpha\left(\dfrac{1}{8\alpha}\right)^{2/3} + (8\alpha)^{1/3} \geq 1$

$\Rightarrow \quad \alpha^{1/3} + 2\alpha^{1/3} \geq 1$

$\Rightarrow \quad 3\alpha^{1/3} \geq 1 \Rightarrow \alpha \geq \dfrac{1}{27}$

Hence, the least value of α is $\dfrac{1}{27}$.

17. (Slope) $f'(x) = e^x\cos x + \sin x\, e^x$

$$= e^x\sqrt{2}\sin(x + \pi/4)$$

$$f''(x) = \sqrt{2}\cdot e^x\{\sin(x + \pi/4) + \cos(x + \pi/4)\}$$

$$= 2e^x \cdot \sin(x + \pi/2)$$

For maximum slope, $f''(x) = 0$

$$\sin(x + \pi/2) = 0$$

$$\cos x = 0$$

$\therefore \quad x = \pi/2, 3\pi/2$

$$f'''(x) = 2e^x\cos(x + \pi/2)$$

$$f'''(\pi/2) = 2e^{\pi/2}\cdot\cos\pi = -\text{ve}.$$

Maximum slope at $x = \pi/2$

18. $f(x)=\begin{cases}|x^3+x^2+3x+\sin x|\left(3+\sin\left(\frac{1}{x}\right)\right), & x\neq 0\\ 0, & x=0\end{cases}$

let $\quad g(x)=x^3+x^2+3x+\sin x$

$\therefore \quad g'(x)=3x^2+2x+3+\cos x$

$$=3\left(x^2+\frac{2x}{3}+1\right)+\cos x$$

$$=3\left\{\left(x+\frac{1}{3}\right)^2+\frac{8}{9}\right\}+\cos x>0$$

and $\quad 2<3+\sin\left(\frac{1}{x}\right)<4.$

Hence, minimum value of $f(x)$ is 0 at $x=0$.
Hence, number of points $=1$.

19. We have

$$f(x)=\begin{cases}(-1)^{m+n}x^m(x-1)^n, & \text{if } x<0\\ (-1)^n x^m(x-1)^n, & \text{if } 0\le x<1\\ x^m(x-1)^n, & \text{if } x\ge 1\end{cases}$$

Let $g(x)=x^m(x-1)^n$.

Then, $g'(x)=mx^{m-1}(x-1)^n+nx^m(x-1)^{n-1}$

$$=x^{m-1}(x-1)^{n-1}\{mx-m+nx\}$$

Now, $g'(x)=0$, is $x=0$, 1 or $\dfrac{m}{m+n}$

$f(0)=0$, $f(1)=0$, and

$$f\left(\frac{m}{m+n}\right)=(-1)^n\frac{m^m n^n(-1)^n}{(m+n)^{m+n}}=\frac{m^m n^n}{(m+n)^{m+n}}>0$$

$\therefore$ Maximum value $=\dfrac{m^m n^n}{(m+n)^{m+n}}$

20. $f'(x)=4x+16x^3+\ldots+2^{10}\cdot 20\cdot x^{19}$

$$=4x(1+4x^2+\ldots+5\cdot 2^{10}x^{18}) \qquad \ldots\text{(i)}$$

$f''(x)=4+48x^2+\ldots.+2^{10}380\cdot x^{18}>0$

Minimum at $x=0$ [from Eq. (i)]

21. $f'(x)=\begin{cases}2x, & 0<x<1\\ 2, & x\ge 1\end{cases}$

minimum at $x=1$

$\therefore \quad f(1)=\sin^{-1}\alpha+1^2\Rightarrow 2=\sin^{-1}\alpha+1$

$\sin^{-1}\alpha=1 \quad \Rightarrow \alpha=\sin 1$

22. Given

and $\left.\begin{array}{l}f'(x)=0 \quad \text{at } x=0\\ f''(0)>0\end{array}\right\}$ ...(i)

$\because \quad y=f(x)+ax+b$

$$\frac{dy}{dx}=f'(x)+a=0+a$$

$=a$ [from Eq. (i)]

For maximum or minimum

$$a=0$$

$\dfrac{d^2y}{dx^2}=f''(x)>0$ [from Eq. (i)]

$\therefore y$ is minimum for all b if $a=0$

23. $f'(x)=2(x-1)(x-2)^3+3(x-1)^2(x-2)^2$

$$=(x-1)(x-2)^2\{2(x-2)+3(x-1)\}$$

$$=(x-1)(x-2)^2(5x-7)$$

+ + + − 1 7/5 2

Sign change of $f'(x)$ from +ve to −ve at $x=1$

$\therefore$ Maximum at $x=1$

24. Suppose p is the quantity of petrol burnt in one hour. Then $p=kv^3$. Let the distance of the journey be s km.

Duration of the journey $=\dfrac{s}{v-c}$ h

$\therefore$ Total petrol burnt for the whole journey

$$=\frac{sp}{v-c}=\frac{skv^3}{v-c}$$

Take $\quad u=\dfrac{v^3}{v-c}$ (sk = constant),

$$\frac{du}{dv}=\frac{(v-c)3v^2-v^3}{(v-c)^2}=\frac{v^2(2v-3c)}{(v-c)^2}$$

$\dfrac{du}{dv}=0\Rightarrow v=\dfrac{3c}{2},\ \dfrac{3c}{2}$ is found to give a minimum for u.

$\therefore$ Most economical speed $=\dfrac{3c}{2}$ km/h

25. x batch has characters $=N$

$\therefore$ 1 batch has characters $=N/x$

Let $\quad P=(\alpha+\beta x^2)\times\dfrac{N}{x}$

$$=N\left(\frac{\alpha}{x}+\beta x\right);$$

$\therefore \quad \dfrac{dP}{dx}=N\left(-\dfrac{\alpha}{x^2}+\beta\right)$

For max or min, $\dfrac{dP}{dx}=0$

$\therefore \quad x=\pm\sqrt{(\alpha/\beta)}$

$$\frac{d^2P}{dx^2}=N\left(\frac{2\alpha}{x^3}\right)$$

For least of P

$$x=\sqrt{\frac{\alpha}{\beta}}$$

26. $\because f(x)=\text{maximum}\{x, x+1, 2-x\}$

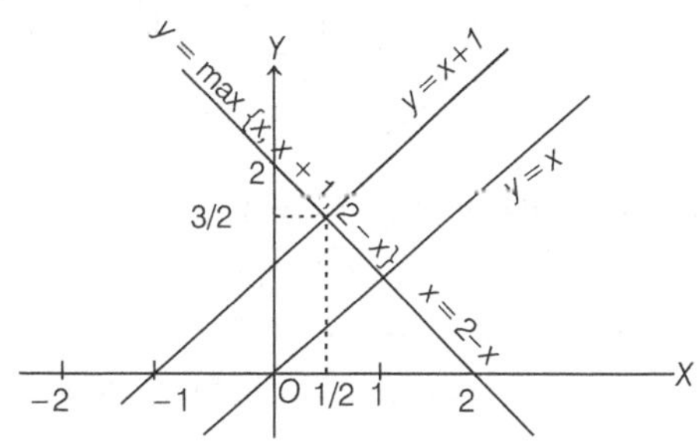

$\therefore$ Minimum value of function $=3/2$

27. $f(x) = x^2 \ln x$

$\Rightarrow \quad f'(x) = x(1 + 2\ln x)$

and $\quad f'(x) > 0$ for $\in [1, e]$

$\therefore f(x)$ is continuously increasing on $[1, e]$ with the least value zero at $x = 1$ and the greatest value e^2 at $x = e$.

28. $f(\alpha) = \left(1 + \dfrac{1}{\sin^n \alpha}\right)\left(1 + \dfrac{1}{\cos^n \alpha}\right)$

$$= 1 + \frac{1}{\sin^n \alpha} + \frac{1}{\cos^n \alpha} + \frac{1}{\sin^n \alpha \cos^n \alpha}$$

$$\therefore \quad f'(\alpha) = -\frac{n\cos\alpha}{\sin^{n+1}\alpha} + \frac{n\sin\alpha}{\cos^{n+1}\alpha} - \frac{n}{\sin^{n+1}\alpha\cos^{n+1}\alpha}\{\cos^2\alpha - \sin^2\alpha\}$$

$\therefore \quad f'(\alpha) = 0$

$\Rightarrow \quad \cos\alpha - \sin\alpha = 0,\ f''(\pi/4) = +\text{ve}$

$\therefore \quad \alpha = \pi/4$

$f(\pi/4) = (1 + 2^{n/2})^2$

29. Let the speed of the train be v and distance to be covered be s so that total time taken is s/v hours cost of fuel per hour $= kv^2$ (k is constant)

Also, $48 = k \cdot 16^2$ by given condition

$\therefore \quad k = 3/16$

$\therefore$ Cost of fuel per hour $= (3/16)\,v^2$

Other charges per hour $= 300$

Total charge $= \dfrac{3}{16}v^2 \times \dfrac{s}{v} + \dfrac{300s}{v}$

or $\quad C = s\left(\dfrac{3}{16}v + \dfrac{300}{v}\right)$

$\therefore \quad \dfrac{dC}{dv} = s\left(\dfrac{3}{16} - \dfrac{300}{v^2}\right)$

For max or min $\dfrac{dC}{dv} = 0$

$\therefore \quad v = 40$ mile/h

then $\quad \dfrac{d^2C}{dv^2} > 0$

$\therefore$ Most economical speed $v = 40$ mile/h

30.

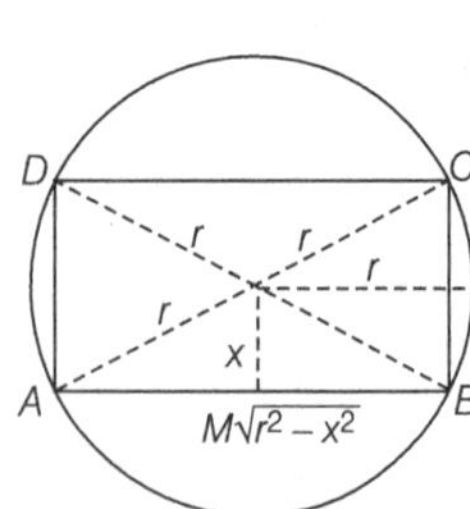

Area of rectangle, $\Delta = 2\sqrt{(r^2 - x^2)} \times 2x$

$$\Delta = 4x\sqrt{(r^2 - x^2)}$$

or $\quad \Delta^2 = 16x^2(r^2 - x^2)$

$$= (16r^2x^2 - 16x^4) = \lambda \text{ (say)}$$

$\therefore \quad \dfrac{d\lambda}{dx} = 32r^2x - 64x^3$

Now, $\dfrac{d\lambda}{dx} = 0, \quad \therefore x = \dfrac{r}{\sqrt{2}}$

$$\frac{d^2\lambda}{dx^2} = 32r^2 - 192x^2$$

$$\left.\frac{d^2\lambda}{dx^2}\right|_{x = \frac{r}{\sqrt{2}}} = -64r^2 < 0$$

Hence, area maximum when $x = \dfrac{r}{\sqrt{2}}$

$\therefore$ Maximum area $= \dfrac{4r}{\sqrt{2}}\sqrt{\left(r^2 - \dfrac{r^2}{2}\right)}$

$$= \frac{4r}{\sqrt{2}} \cdot \frac{r}{\sqrt{2}} = 2r^2$$

Alternate Method In a circle maximum area rectangle is always square, whose area $= \dfrac{(\text{diagonal})^2}{2} = \dfrac{(2r)^2}{2} = 2r^2$

31. We have,

$f(x) = x^2 + 2bx + 2c^2 = (x + b)^2 + 2c^2 - b^2$

$\Rightarrow \min f(x) = 2c^2 - b^2$

Also, $g(x) = -x^2 - 2cx + b^2 = b^2 + c^2 - (x + c)^2$

so, $\max g(x) = b^2 + c^2$

Since, $\min f(x) > \max g(x)$,

so, $2c^2 - b^2 > b^2 + c^2$

$\Rightarrow c^2 > 2b^2 \Rightarrow |c| > \sqrt{2}|b|$

32. $V = \pi r^2 h$

If k be the thickness of the sides, then that of the top will be $(5/4)\,k$.

$\therefore S = (2\pi rh)k + (\pi r^2)(5/4)k$ ('S' is volume of material used)

or $\quad S = 2\pi rk \cdot \dfrac{V}{\pi r^2} + \dfrac{5}{4}\pi r^2 k = k\left(\dfrac{2V}{r} + \dfrac{5}{4}\pi r^2\right)$

$\therefore \quad \dfrac{dS}{dr} = k\left(-\dfrac{2V}{r^2} + \dfrac{5}{2}\pi r\right),$

$\therefore \quad r^3 = 4V/5\pi$

$$\frac{d^2S}{dr^2} = k\left(\frac{4V}{r^3} + \frac{5}{2}\pi\right) = \pi k\left(5 + \frac{5}{2}\right) = +\text{ ive}$$

when $r^3 = 4V/5\pi$ or $5\pi r^3 = 4\pi r^2 h$

$\therefore \quad r/h = 4/5$

33. $f(1) \geq \lim\limits_{x \to 1+} f(x)$

$\Rightarrow \quad f(1) \geq \lim\limits_{h \to 0} f(1 + h)$

$\Rightarrow \quad 1 - 1 + 10 - 5 \geq \lim\limits_{h \to 0} -2(1 + h) + \log_2(b^2 - 2),$

$$b^2 - 2 > 0$$

$\Rightarrow \quad 5 \geq -2 + \log_2(b^2 - 2)$ and $b^2 - 2 > 0$

$$2^7 \geq b^2 - 2 \text{ and } b^2 > 2$$

or $\quad 2 < b^2 \leq 130$

$\therefore \quad b \in [-\sqrt{130}, -\sqrt{2}) \cup (\sqrt{2}, \sqrt{130}]$

34. Given function is

$$f(x) = 9x^4 + 12x^3 - 36x^2 + 25$$

For maxima or minima put $f'(x) = 0$

$\Rightarrow \quad 36x^3 + 36x^2 - 72x = 0$

$\Rightarrow \quad x^3 + x^2 - 2x = 0$

$\Rightarrow \quad x[x^2 + x - 2] = 0$

$\Rightarrow \quad x[x^2 + 2x - x - 2] = 0$

$\Rightarrow \quad x[x(x+2) - 1(x+2)] = 0$

$\Rightarrow \quad x(x-1)(x+2) = 0$

$\Rightarrow \quad x = -2, 0, 1$

By sign method, we get

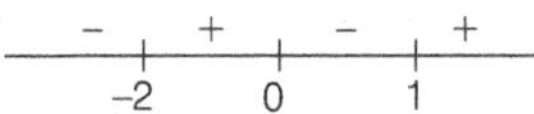

Since, $f'(x)$ changes sign from negative to positive at $x =$ '-2' and '1', so $x = -2, 1$ are points of local minima. Also, $f'(x)$ changes sign from positive to negative at $x = 0$, so $x = 0$ is point of local maxima.

$\therefore \quad S_1 = \{-2, 1\}$ and $S_2 = \{0\}$.

35. $f'(x)$ is increasing

For some x in $\left(\frac{1}{2}, 1\right)$

$f'(x) = 1$ [LMVT]

$\therefore \quad f'(1) > 1$

36. $a^{f(x)} + g(x) = 0$

$\Rightarrow \quad a^{f(x)} = -g(x)$

LHS > 0 and RHS < 0 $\left[\because g(x) \geq \frac{1}{2}\right]$

which is impossible. Hence, no solutions.

$\therefore$ Number of solutions = zero.

37. Equation of tangent be

$$\frac{x\cos\theta}{3\sqrt{3}} + \frac{y\sin\theta}{1} = 1, \ \theta \in \left(0, \frac{\pi}{2}\right)$$

intercept on X-axis

$$OA = 3\sqrt{3}\sec\theta$$

intercept on Y-axis

$$OB = \operatorname{cosec}\theta$$

Now, sum of intercept $= 3\sqrt{3}\sec\theta + \operatorname{cosec}\theta = f(\theta)$ let

$$f'(\theta) = 3\sqrt{3}\sec\theta\tan\theta - \operatorname{cosec}\theta\cot\theta$$

$$= 3\sqrt{3}\frac{\sin\theta}{\cos^2\theta} - \frac{\cos\theta}{\sin^2\theta}$$

$$= \underbrace{\frac{\cos\theta}{\sin^2\theta}}_{\oplus} \cdot 3\sqrt{3}\left[\tan^2\theta - \frac{1}{3\sqrt{3}}\right]$$

$$= 0$$

$\Rightarrow \quad \theta = \frac{\pi}{6}$

⊖↓ ⊕↑
$\theta = \frac{\pi}{6}$

$\Rightarrow$ at $\theta = \frac{\pi}{6}$, $f(\theta)$ is minimum.

38. Let the school be at a distance x from A (with 150 students), then total distance travelled by the students is

$$z = 150x + 50(60 - x) = 3000 + 100x$$

z will be least when $x = 0$

$\therefore \quad z = 3000$

i.e. school is built at A.

Objective Questions Type II [One or more than one correct answer(s)]

1. $f(x) = \cos x \sin 2x = 2\sin x - 2\sin^3 x$

$\min f(x) = \min g(t)$, where $g(t) = 2t - 2t^3$

$x \in [-\pi, \pi], t \in [-1, 1]$

$g'(t) = 2 - 6t^2 = 0$

$\Rightarrow \quad t = \pm\frac{1}{\sqrt{3}}$

$g''(t) = -12t,$

$\therefore \quad g''(1/\sqrt{3}) < 0$

and $\quad g''(-1/\sqrt{3}) > 0$

Hence, $\min g(t) = g(-1/\sqrt{3})$

$$t \in [-1, 1] = -\frac{4}{3\sqrt{3}} > -\frac{7}{9} > -\frac{9}{7}$$

2. Given, $f : R \to R$ and $f(x) = (x-1)(x-2)(x-5)$

Since, $F(x) = \int_0^x f(t)\,dt, \ x > 0$

So, $F'(x) = f(x) = (x-1)(x-2)(x-5)$

According to wavy curve method

− + − +
1 2 5

$F'(x)$ changes, it's sign from negative to positive at $x = 1$ and 5, so, $F(x)$ has minima at $x = 1$ and 5 and as $F'(x)$ changes, it's sign from positive to negative at $x = 2$, so $F(x)$ has maxima at $x = 2$.

$$\because F(2) = \int_0^2 f(t)\,dt = \int_0^2 (t^3 - 8t^2 + 17t - 10)dt$$

$$= \left[\frac{t^4}{4} - 8\frac{t^3}{3} + 17\frac{t^2}{2} - 10t\right]_0^2$$

$$= 4 - \frac{64}{3} + 34 - 20 = 38 - \frac{124}{3} = -\frac{10}{3}$$

$\because$ At the point of maxima $x = 2$, the functional value $F(2) = -\frac{10}{3}$, is negative for the interval $x \in (0, 5)$, so $F(x) \neq 0$ for any value of $x \in (0, 5)$.

Hence, options (a), (b) and (d) are correct.

3. $f'(x) = 0$ at $x = \alpha, \beta$

given $\quad \alpha\beta < 0$

then $\quad \alpha > 0, \beta < 0$ or $\alpha < 0, \beta > 0$

where $\quad \alpha < 0, \beta > 0$

then $\quad f(0) = 1 > 0$

$\therefore \quad f(\alpha) < 0, f(\beta) > 0$

$\Rightarrow f(x) = 0$ has one positive root.

and $\quad f(\alpha) > 0, f(\beta) < 0$

$\Rightarrow f(x) = 0$ has one negative root.

4. $\because f'(x) = (x^2 - 1)^n \{(x^2 - 1)(2x + 1) + 2nx(x^2 + x + 1)\}$

$= (x + 1)^{n+1}(x - 1)^{n+1}(2x + 1) + 2nx(x + 1)^n (x - 1)^n (x^2 + x + 1)$

If power even, then neither max nor min

$\therefore \quad n = 2, 4, 6 \quad [\because n + 1 = \text{odd}]$

5. Given function $f : R \longrightarrow R$ is

$$f(x) = \begin{cases} x^5 + 5x^4 + 10x^3 + 10x^2 + 3x + 1, & x < 0 \\ x^2 - x + 1, & 0 \le x < 1 \\ \frac{2}{3}x^3 - 4x^2 + 7x - \frac{8}{3}, & 1 \le x < 3 \\ (x - 2)\log_e(x - 2) - x + \frac{10}{3}, & x \ge 3 \end{cases}$$

$$\text{So, } f'(x) = \begin{cases} 5x^4 + 20x^3 + 30x^2 + 20x + 3, & x < 0 \\ 2x - 1, & 0 < x < 1 \\ 2x^2 - 8x + 7, & 1 < x < 3 \\ \log_e(x - 2), & x > 3 \end{cases}$$

At $x = 1$, $f''(1^-) = 2 > 0$ and $f''(1^+) = 4 - 8 = -4 < 0$

$\therefore f'(x)$ is not differentiable at $x = 1$ and $f'(x)$ has a local maximum at $x = 1$.

For $x \in (-\infty, 0)$

$f'(x) = 5x^4 + 20x^3 + 30x^2 + 20x + 3$

and since $f'(-1) = 5 - 20 + 30 - 20 + 3 = -2 < 0$

So, $f(x)$ is not increasing on $x \in (-\infty, 0)$.

Now, as the range of function $f(x)$ is R, so f is onto function.

Hence, options (b), (c) and (d) are correct.

6. $f'(x) = \dfrac{\cos x}{x}$

For max or min $f'(x) = 0$

$\therefore \quad \cos x = 0$

$\Rightarrow \quad x = (2n + 1)\dfrac{\pi}{2}, n \in I$

$\therefore \quad f''(x) = \dfrac{-x\sin x - \cos x}{x^2} = -\dfrac{1}{x^2}(x\sin x + \cos x)$

For $\quad n = -2, -4, -6, \ldots$

$x = -\dfrac{3\pi}{2}, -\dfrac{7\pi}{2}, -\dfrac{11\pi}{2}, \ldots$

$f''(x) > 0$ (minima)

For $\quad n = -1, -3, -5, \ldots$

$x = -\dfrac{\pi}{2}, -\dfrac{5\pi}{2}, -\dfrac{9\pi}{2}, \ldots$

$f''(x) < 0$ (maxima).

For $\quad n = 0, 2, 4, \ldots$

$x = \dfrac{\pi}{2}, \dfrac{5\pi}{2}, \dfrac{9\pi}{2}, \ldots$

$f''(x) < 0$, (maxima)

For $\quad n = 1, 3, 5, \ldots$

$x = \dfrac{3\pi}{2}, \dfrac{7\pi}{2}, \dfrac{11\pi}{2}, \ldots$

$f''(x) > 0$ (minima)

7. $f(x) = (x - 1)^m (x - 2)^n$

$\therefore \quad f'(x) = (x - 1)^{m-1}(x - 2)^{n-1}\{(m + n)x - (2m + n)\}$

For max or min $f'(x) = 0$

$\therefore \quad x = 1, 2, \dfrac{2m + n}{m + n}$

Option (a) : If $m = 2, n = 3$

$\therefore \quad f'(x) = (x - 1)(x - 2)^2(5x - 7)$

$\therefore$ Maximum at $x = 1$ and minimum at $x = \dfrac{7}{5}$, neither max. nor min. at $x = 2$.

Option (b) : For $m = 2, n = 4$

$\therefore \quad f'(x) = (x - 1)(x - 2)^3(6x - 8)$

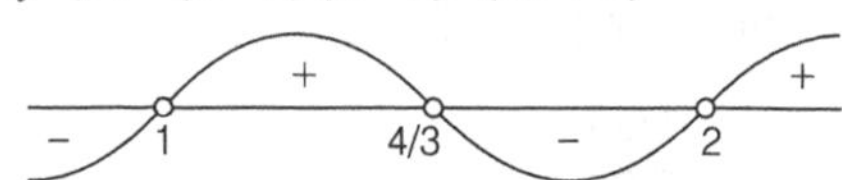

$\therefore \quad$ Minimum at $x = 1, 2$

and $\quad$ maximum at $x = \dfrac{4}{3}$

Option (c) : For $m = 3, n = 4$

$\therefore \quad f'(x) = (x - 1)^2(x - 2)^3(7x - 10)$

Neither max or min at $x = 1$

Maximum at $x = \dfrac{10}{7}$

and minimum at $x = 2$

Option (d) : For $m = 4, n = 2$

$\therefore \quad f'(x) = (x - 1)^3(x - 2)^1(6x - 10)$

Minimum at $\quad x = 1, 2$

and maximum at $x = \dfrac{5}{3}$

8. $f(x) = -x^3 + \dfrac{b^3 - b^2 + b - 1}{b^2 + 3b + 2}, \quad 0 \le x < 1$

$= 2x - 3, \quad 1 \le x < 3 \quad$ (given)

$f(x)$ has least value at $x = 1$

$\therefore$ when $x < 1$, $f(x) > f(1)$

$\therefore \quad \lim\limits_{x \to 1-0} f(x) \ge f(1)$

$$\Rightarrow \quad -1^3 + \frac{b^3 - b^2 + b - 1}{b^2 + 3b + 2} \geq -1$$

$$\Rightarrow \quad \frac{(b-1)(b^2+1)}{(b+1)(b+2)} \geq 0$$

$$\Rightarrow \quad -2 < b < -1$$

or $\quad 1 \leq b < \infty$

9. Here, $\lim_{x \to 2} \frac{f(x) \cdot g(x)}{f'(x) \cdot g'(x)} = 1$

$$\Rightarrow \lim_{x \to 2} \frac{f(x) g'(x) + f'(x) g(x)}{f''(x) g'(x) + f'(x) g''(x)} = 1$$

[using L' Hospital's rule]

$$\Rightarrow \quad \frac{f(2) g'(2) + f'(2) g(2)}{f''(2) g'(2) + f'(2) g''(2)} = 1$$

$$\Rightarrow \quad \frac{f(2) g'(2)}{f''(2) g'(2)} = 1 \quad [\because f'(2) = g(2) = 0]$$

$$\Rightarrow \quad f(2) = f''(2) \quad \ldots \text{(i)}$$

$\therefore \quad f(x) - f''(x) = 0$, for atleast one $x \in R$.

$\Rightarrow$ Option (d) is correct.

Also, $f : R \to (0, \infty) \Rightarrow f(2) > 0$

$\therefore \quad f''(2) = f(2) > 0 \quad$ [from Eq. (i)]

Since, $f'(2) = 0$ and $f''(2) > 0$

$\therefore$ $f(x)$ attains local minimum at $x = 2$.

$\Rightarrow$ Option (a) is correct.

10. 2 lies between the roots of $f'(x) = 0$

i.e. $2(a+1) + 2(a^2+1)x - 3x^2 = 0$

$$\Rightarrow \quad 2(a+1) + 2(a^2+1)(2) - 3(2)^2 > 0$$

$$\Rightarrow \quad a < -\frac{3}{2} \text{ or } a > 1$$

11. r must be an even integer because two decreasing are required to make it increasing function.

Let $\quad y = r(n-r)$

when n is odd, $r = \frac{n-1}{2}$ or $\frac{n+1}{2}$ for maximum value of y

when n is even, $r = \frac{n}{2}$ for maximum value of y

$\therefore$ Maximum $(y) = \frac{n^2-1}{4}$, when n is odd and $\frac{n^2}{4}$, when n is even.

Linked Comprehension Type

PASSAGE 1

Since, $f(x)$ be a polynomial of degree 4, $f'(x)$ is a cubic polynomial. $f(x)$ has relative maximum/relative minimum at $x = 1$, 2 and 3, so $f'(1) = f'(2) = f'(3) = 0$

Then, $f'(x) = a(x-1)(x-2)(x-3)$

$= a(x^3 - 6x^2 + 11x - 6)$, where a is constant

Integrating w.r.t. x, we get

$$f(x) = a\left(\frac{x^4}{4} - 2x^3 + \frac{11x^2}{2} - 6x\right) + b \quad \ldots \text{(i)}$$

where b is constant of integration and $f(0) = 0$

then $a(0 - 0 + 0 - 0) + b = 0$

$\therefore \quad b = 0$

from Eq. (i), $f(x) = a\left(\frac{x^4}{4} - 2x^3 + \frac{11x^2}{2} - 6x\right) \quad \ldots$(ii)

Also, $f(-2) = 216$

$\Rightarrow a(4 + 16 + 22 + 12) = 216$ or $a = 4$

from Eq. (ii), $f(x) = x^4 - 8x^3 + 22x^2 - 24x \quad \ldots$(iii)

and $\quad f'(x) = 4(x-1)(x-2)(x-3) \quad \ldots$(iv)

1. Extremum points of $f(x)$ are $(1, f(1))$, $(2, f(2))$ and $(3, f(3))$ i.e. $(1, -9)$, $(2, -8)$ and $(3, -9)$

$\therefore$ Area of triangle formed by extremum points of

$$f(x) = \frac{1}{2}\left|\begin{vmatrix} 1-3 & -9+9 \\ 2-3 & -8+9 \end{vmatrix}\right| = \frac{1}{2}\left|\begin{vmatrix} -2 & 0 \\ -1 & 1 \end{vmatrix}\right| = 1$$

2. From Eq. (iii),

$f(x) = x^4 - 8x^3 + 22x^2 - 24x$

$\therefore f(-x) = x^4 + 8x^3 + 22x^2 + 24x$

$\Rightarrow f(x) + f(-x) = 2(x^4 + 22x^2)$

Now, $15\int_{-1}^{1}(f(x) + f(-x))dx = 30\int_{-1}^{1}(x^4 + 22x^2)dx$

$$= 60\int_0^1 (x^4 + 22x^2)dx = 60\left(\frac{x^5}{5} + \frac{22x^3}{3}\right)_0^1$$

$$= 60\left(\frac{1}{5} + \frac{22}{3}\right) = 4 \times 113 = 452$$

3. $\because f'(x) = 4(x-1)(x-2)(x-3)$

Option (a)

$\Rightarrow f(x)$ has two relative minima at $x = 1, 3$ and one relative maxima at $x = 2$. option (a) is wrong

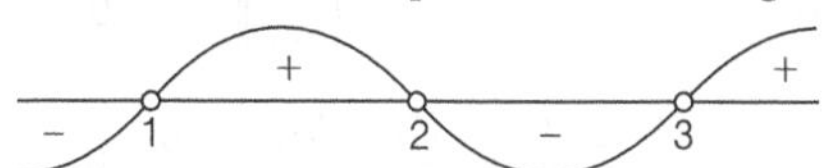

Option (b)

Now, $f''(x) = 4(3x^2 - 12x + 11) = 0 \Rightarrow x = 2 \pm \frac{1}{\sqrt{3}}$

$\Rightarrow f'''(x) = 24(x-2) = \pm \frac{24}{\sqrt{3}}$ (at $x = 2 \pm \frac{1}{\sqrt{3}}$)

i.e., two points of inflection.

(option (b) is wrong)

Option (c)

$\because f(0) = 0$, $f(1) = -9$, $f(2) = -8$ and $f(3) = -9$

$\therefore$ Range of $f(x) = [9, \infty)$

i.e., Range of $f(x)$ contains 9 negative integers.

Option (d) $\because f(x) = 0$

$\Rightarrow x^4 - 8x^3 + 22x^2 - 24x = 0$

$\therefore$ Sum of roots $= 8$

PASSAGE 2

$\because f'(-1) = f'\left(\frac{1}{3}\right) = 0$

$\therefore \quad f'(x) = d'(x+1)\left(x - \frac{1}{3}\right)$

$= d'\left(x^2 + \frac{2}{3}x - \frac{1}{3}\right)$

Integrating w.r.t. x, we get

$$f(x) = \frac{a'}{3}(x^3 + x^2 - x) + \lambda \quad \text{...(i)}$$

where λ is constant of integration.

and $f(-2) = 0$

$\Rightarrow \quad \frac{d'}{3}(-8+4+2) + \lambda = 0$

$\therefore \quad \lambda = \frac{2d'}{3}$

From Eq. (i), $f(x) = \frac{d'}{3}(x^3 + x^2 - x + 2)$

Also, $\int_{-1}^{1} f(x)\,dx = \frac{14}{3}$

$\Rightarrow \quad \frac{d'}{3}\int_{-1}^{1}(x^3 + x^2 - x + 2)\,dx = \frac{14}{3}$

$\Rightarrow \quad 0 + \frac{2d'}{3}\int_{0}^{1}(x^2 + 2)\,dx = \frac{14}{3}$

$\Rightarrow \quad \frac{2d'}{3}\left\{\frac{1}{3} + 2\right\} = \frac{14}{3}$

$\Rightarrow \quad \frac{14d'}{9} = \frac{14}{3}$

$\Rightarrow \quad d' = 3$

Then, $f(x) = x^3 + x^2 - x + 2$

Comparing with $f(x) = ax^3 + bx^2 + cx + d$

$\therefore \quad a = 1, b = 1, c = -1, d = 2$

Ans. 1. (b) **2.** (b) **3.** (d)

4. $f'(x) = 3(x+1)\left(x - \frac{1}{3}\right) < 0$

$\therefore \quad -1 < x < \frac{1}{3}$

$\Rightarrow \quad x \in \left(-1, \frac{1}{3}\right)$

5. $\because f(x) = 3$

$\Rightarrow \quad x^3 + x^2 - x + 2 = 3$

$\Rightarrow \quad x^3 + x^2 - x - 1 = 0$

$\Rightarrow \quad x^2(x+1) - (x+1) = 0$

$\Rightarrow \quad (x^2 - 1)(x+1) = 0$

$\therefore \quad x = -1, -1, 1$

Numerical Value Type

1. $\because y = \frac{ax+b}{(x-4)(x-1)} \quad \text{...(i)}$

$\therefore \quad \frac{dy}{dx} = \frac{(x-4)(x-1)a - (ax+b)(2x-5)}{(x-4)^2(x-1)^2} \quad \text{...(ii)}$

Since, $(2, -1)$ will lie on the function (i)

$\Rightarrow \quad -1 = \frac{(2a+b)}{(2-4)(2-1)}$

$\Rightarrow \quad 2a + b = 2 \quad \text{...(iii)}$

Also, $\left(\frac{dy}{dx}\right)\Big|_{x=2} = 0$

$\therefore \quad \frac{(-2)(1)a - (2a+b)(4-5)}{(-2)^2(1)^2} = 0$

$\Rightarrow \quad b = 0$

From Eq. (iii), $a = 1$

Hence, $50.5(a + 10b) = 50.5(1+0) = 50.5$

2. Given $f(x) = x^3 - 6x^2 + 9x + 1$

$\therefore \quad f'(x) = 3x^2 - 12x + 9$

$= 3(x-1)(x-3)$

For maxima or minima $f'(x) = 0$

$\therefore \quad x = 1, 3$ but $3 \notin [0, 2]$

$\therefore$ Only one critical point in $[0, 2]$ is $x = 1$

$\therefore$ Greatest value $\lambda = \max\{f(0), f(1), f(2)\}$

$= \max\{1, 5, 3\} = 5$

and least value $\mu = \min\{f(0), f(1), f(2)\}$

$= \min\{1, 5, 3\} = 1$

$\therefore \quad \lambda^4 + \mu^8 = 5^4 + 1^8 = 625 + 1 = 626$

3. Let $ABCD$ be the given trapezium.

Let $AM = BN = x$

Then $DM = CN = \sqrt{(36 - x^2)}$

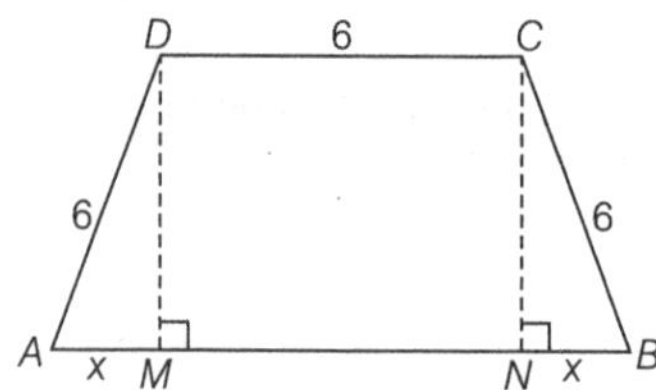

$\therefore$ Area of trapezium $ABCD$ is

$S = \frac{1}{2}(6 + x + x + 6) \times \sqrt{(36 - x^2)}$

$= (6+x)\sqrt{(36 - x^2)}$

or $\quad S^2 = (6+x)^2(36 - x^2) = (6+x)^3(6-x)$

Let $\quad y = S^2 = (6+x)^3(6-x)$

$\therefore \quad \frac{dy}{dx} = (6+x)^3 \cdot (-1) + (6-x) \cdot 3(6+x)^2$

$= (6+x)^2\{-6 - x + 18 - 3x\}$

$= (6+x)^2(-4x + 12)$

$= -4(6+x)^2(x-3)$

$= 4(-x-6)^2(3-x)$

$\Rightarrow$ y is maximum at $x = 3$

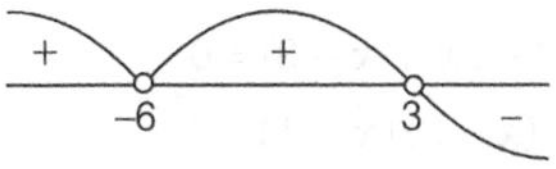

Then, S is also maximum at $x = 3$

$\therefore$ $A = (6+3)\sqrt{(36-9)} = 9\sqrt{27} = 27\sqrt{3}$

Then, $4\sqrt{3}A = 4\sqrt{3}\cdot 27\sqrt{3} = 27 \times 12$

$= 324$

4. Let r be the radius of the circle and x be the side of the square.

Then, $l = 2\pi r + 4x$ ...(i)

Let S be the sum of the areas of the circle and the square, then

$$S = \pi r^2 + x^2$$

$$= \pi\left(\frac{l-4x}{2\pi}\right)^2 + x^2 \quad \text{[from Eq. (i)]}$$

$$= \frac{1}{4\pi}(l-4x)^2 + x^2$$

$$\therefore \frac{dS}{dx} = \frac{2}{4\pi}(l-4x)(-4) + 2x$$

$$= -\frac{2}{\pi}(l-4x) + 2x$$

and $\dfrac{d^2S}{dx^2} = \dfrac{8}{\pi} + 2 > 0$

$\therefore S$ is least

For max or min $\dfrac{dS}{dx} = 0$

$\Rightarrow$ $-\dfrac{2}{\pi}(l-4x) + 2x = 0$

or $x = \dfrac{l}{4+\pi}$ ...(ii)

From Eqs. (i) and (ii),

$4x + \pi x = 2\pi r + 4x$

$\Rightarrow$ $x = 2r$

$\therefore$ $\lambda = 1$

Then, $2008\,\lambda = 2008$

5. $\because$ Speed of the truck is x km/h

$\therefore$ Time taken to cover 500 km is

$$t = \left(\frac{500}{x}\right) \text{h}$$

Total cost for a trip of 500 km (in ₹) is

= Cost of diesel + Payment to the driver

$$C \text{ (say)} = \frac{1}{300}\left(\frac{900}{x} + x\right) \times 500 \times \frac{400}{100} + \frac{500}{x} \times 1.5$$

$\Rightarrow$ $C = \dfrac{1350}{x} + \dfrac{2}{3}x$

$\therefore$ $\dfrac{dC}{dx} = -\dfrac{1350}{x^2} + \dfrac{2}{3}$; for max or min of C

$$\frac{dC}{dx} = 0$$

$\Rightarrow$ $x = 45$ km/h

Now, $\dfrac{d^2C}{dx^2} = \dfrac{2700}{x^3}$

$\therefore$ $\left.\dfrac{d^2C}{dx^2}\right|_{x=45} = \dfrac{2700}{(45)^3} > 0$

Total cost C is minimum where $x = 45$ km/h

$\therefore$ $\lambda = 45$

Then, $50\,\lambda = 50 \times 45 = 2250$

6. $\because$ Slope $= f'(x) = -3x^2 + 16x - 13 = u$ (say)

$\therefore$ $\dfrac{du}{dx} = -6x + 16$

For max or min of u, $\dfrac{du}{dx} = 0$

$\Rightarrow$ $x = \dfrac{8}{3}$

$\Rightarrow$ $\dfrac{d^2u}{dx^2} = -6 < 0$

$\therefore$ Slope is maximum

u at $x = \dfrac{8}{3}$ is $-3\left(\dfrac{8}{3}\right)^2 + 16\left(\dfrac{8}{3}\right) - 13$

$$= \frac{-64}{3} + \frac{128}{3} - 13 = \frac{64-39}{3} = \frac{25}{3}$$

$\therefore$ $\lambda = \dfrac{25}{3}$

Then, $15\lambda = 15 \times \dfrac{25}{3} = 125$

7. The given function $f: R \to R$ be defined by

$$f(\theta) = (\sin\theta + \cos\theta)^2 + (\sin\theta - \cos\theta)^4$$

$$= 1 + \sin 2\theta + (1 - \sin 2\theta)^2$$

$$= 1 + \sin 2\theta + 1 + \sin^2 2\theta - 2\sin 2\theta$$

$$= \sin^2 2\theta - \sin 2\theta + 2$$

$$= \left(\sin 2\theta - \frac{1}{2}\right)^2 + \frac{7}{4}$$

The local minimum of function 'f' occurs when

$$\sin 2\theta = \frac{1}{2}$$

$\Rightarrow$ $2\theta = \dfrac{\pi}{6}, \dfrac{5\pi}{6}, \dfrac{13\pi}{6} \ldots$

$\Rightarrow$ $\theta = \dfrac{\pi}{12}, \dfrac{5\pi}{12}, \dfrac{13\pi}{12}, \ldots$

but $\theta \in \{\lambda_1\pi, \lambda_2\pi, \ldots \lambda_r\pi\}$, where $0 < \lambda_1 < \ldots < \lambda_r < 1$.

$\therefore$ $\theta = \dfrac{\pi}{12}, \dfrac{5\pi}{12}$

So, $\lambda_1 + \ldots + \lambda_r = \dfrac{1}{12} + \dfrac{5}{12} = 0.50$

Single Integer Type

1. $f'(x) = \dfrac{0.6(1+x)^{-0.4}(1+x^{0.6}) - 0.6x^{-0.4}(1+x)^{0.6}}{(1+x^{0.6})^2}$

$$= 0.6\frac{(1+x^{0.6}) - x^{-0.4}(1+x)^1}{(1+x^{0.6})^2(1+x)^{0.4}}$$

$$= 0.6\frac{(1+x^{0.6})x^{0.4} - (1+x)}{(1+x^{0.6})^2(1+x)^{0.4}x^{0.4}}$$

$$= 0.6\frac{x^{0.4}-1}{(1+x^{0.6})^2(1+x)^{0.4}x^{0.4}} < 0 \ \forall\ x \in [0, 1]$$

Hence, $f(x)$ is a decreasing function $\forall\ x \in [0, 1]$

Thus, $f(x)_{max} = f(0) = 1$

2. $\dfrac{dy}{dx} = x^2 - 2x = \dfrac{d}{dx}\left(\dfrac{x^3}{3} - x^2 + c\right)$

$\Rightarrow y = \dfrac{x^3}{3} - x^2 + c$

Since, the curve passes through (2, 0), we get $0 = 8/3 - 4 + c$, i.e, $c = 4/3$. Hence, the equation of the curve is

$$y = \frac{x^3}{3} - x^2 + \frac{4}{3}$$

Now, from $dy/dx = 0$, we get $x = 0$ or 2. Also,

$$\frac{d^2y}{dx^2} = 2x - 2$$

$$\Rightarrow \left.\frac{d^2y}{dx^2}\right|_{x=0} = -2 \text{ and } \left.\frac{d^2y}{dx^2}\right|_{x=2} = 2$$

Hence, y has a maximum at $x = 0$ and $y_{max} = 4/3$

$\therefore$ 3y subscript = 4

3. Let a cubic polynomial

$$f(x) = ax^3 + bx^2 + cx + d$$

Since, $f(-1) = 10$

$\Rightarrow -a + b - c + d = 10$ ...(i)

Also, $f(1) = -6$

$\Rightarrow a + b + c + d = -6$ ...(ii)

Also, $f'(-1) = 0$

$\Rightarrow 3a - 2b + c = 0$ ...(iii)

Also, $f''(1) = 0$

$\Rightarrow 6a + 2b = 0$

$\Rightarrow 3a + b = 0 \Rightarrow b = -3a$ ...(iv)

From Eqs. (i) and (ii), we get

$$-2a - 2c = 16$$

$\Rightarrow a + c = -8 \Rightarrow c = -8 - a$ ...(v)

From Eqs. (iii), (iv) and (v), we get

$$3a - 2(-3a) + (-a - 8) = 0$$

$\Rightarrow 3a + 6a - a - 8 = 0$

$\Rightarrow 8a = 8 \Rightarrow a = 1$

$\therefore\ b = -3, c = -9, d = 5$

So, $f(x) = x^3 - 3x^2 - 9x + 5$

$\Rightarrow f'(x) = 3x^2 - 6x - 9 = 0$

$\Rightarrow 3(x^2 - 2x - 3) = 0$

$\Rightarrow 3(x^2 - 3x + x - 3) = 0$

$\Rightarrow 3(x-3)(x+1) = 0$

+ − +
−1 3

$\therefore f(x)$ has local minima at $x = 3$.

4. $f(x) = 2x^3 - 9ax^2 + 12a^2x + 1$

To find the extremum points are differentiate and equate it to zero.

$\Rightarrow f'(x) = 6x^2 - 18ax + 12a^2$

$\Rightarrow f'(x) = 0$

$\Rightarrow 6x^2 - 18ax + 12a^2 = 0$

$\Rightarrow (x - 2a)(x - a) = 0 \Rightarrow x = 2a, a$

Now, to find whether at the critical points we find a maxima or minima, we use the second derivative test.

$\therefore f''(x) = 12x - 18a$

$\Rightarrow f''(2a) = 24a - 18a = 6a > 0$

$\Rightarrow f''(a) = 12a - 18a = -6a < 0$

Hence, we get a minima at $x = 2a$ and a maxima at $x = c$.

$\therefore\ p = a$ and $q = 2a$

Using the given relation, we get

$$p^2 - q = 0 \Rightarrow a^2 - 2a = 0 \Rightarrow a(a-2) = 0$$

$\therefore\ a = 2$ $[\because a > 0, \text{ so } a \neq 0]$

5. Let $S = \dfrac{x^2 - x + 1}{x^2 + x + 1} = 1 - \dfrac{2x}{x^2 + x + 1}$

$\Rightarrow S'(x) = -2\dfrac{(x^2 + x + 1) - x(2x+1)}{(x^2+x+1)^2} = -2\dfrac{-x^2 + 1}{(x^2+x+1)^2}$

$S'(x) = 0 \Leftrightarrow x = \pm 1$.

Since, $S'(x) > 0$ for $x < -1$ and $S'(x) < 0$ for $-1 < x < 1$, also $S'(x) > 0$ for $x > 1$. So, S is maximum when $x = -1$. Hence, $S_{max} = 1 + 2/1 = 3$.

6. $f : [-1, 1] \to R$

$$f(x) = ax^2 + bx + c$$

$$f(-1) = a - b + c = 2$$

$$f'(-1) = -2a + b = 1$$

$$f''(x) = 2a$$

$\Rightarrow$ Max value of $f''(x) = 2a = \dfrac{1}{2}$

$\Rightarrow a = \dfrac{1}{4};\ b = \dfrac{3}{2};\ c = \dfrac{13}{4}$

$\therefore f(x) = \dfrac{x^2}{4} + \dfrac{3}{2}x + \dfrac{13}{4}$

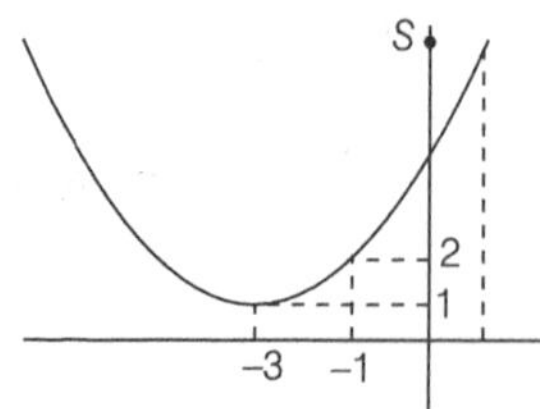

For $x \in [-1, 1] \Rightarrow 2 \le f(x) \le 5$

$\therefore$ Least value of α is 5.

7. Given set S of polynomials with real coefficients
$S = \{(x^2 - 1)^2(a_0 + a_1x + a_2x^2 + a_3x^3) : a_0, a_1, a_2, a_3 \in R\}$
and for a polynomial $f \in S$, Let
$f(x) = (x^2 - 1)^2 (a_0 + a_1x + a_2x^2 + a_3x^3)$
it have -1 and 1 as repeated roots twice, so graph of $f(x)$ touches the X-axis at $x = -1$ and $x = 1$, so $f'(x)$ having at least three roots $x = -1, 1$ and α. Where $\alpha \in (-1, 1)$ and $f''(x)$ having at least two roots in interval $(-1, 1)$
So, $m_{f'} = 3$ and $m_{f''} = 2$
$\therefore$ Minimum possible value of $(m_{f'} + m_{f''}) = 5$

8. Here, volume of cylindrical container, $V = \pi r^2 h$ ...(i)
and let volume of the material used be T.

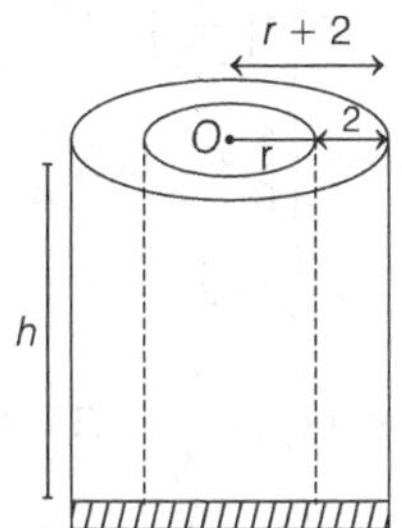

$$\therefore \quad T = \pi[(r+2)^2 - r^2]h + \pi(r+2)^2 \times 2$$

$$\Rightarrow \quad T = \pi[(r+2)^2 - r^2]\cdot\frac{V}{\pi r^2} + 2\pi(r+2)^2$$

$$\left[\because V = \pi r^2 h \Rightarrow h = \frac{V}{\pi r^2}\right]$$

$$\Rightarrow \quad T = V\left(\frac{r+2}{r}\right)^2 + 2\pi(r+2)^2 - V$$

On differentiating w.r.t. r, we get

$$\frac{dT}{dr} = 2V\cdot\left(\frac{r+2}{r}\right)\cdot\left(\frac{-2}{r^2}\right) + 4\pi(r+2)$$

At $\quad r = 10, \; \dfrac{dT}{dr} = 0$

Now, $\quad 0 = (r+2)\cdot 4\left(\pi - \dfrac{V}{r^3}\right) \Rightarrow \dfrac{V}{r^3} = \pi$

where $\quad r = 10$

$$\Rightarrow \quad \frac{V}{1000} = \pi \text{ or } \frac{V}{250\pi} = 4$$

9. $\therefore F'(x) = \dfrac{\sin x}{x}$

$\therefore \quad F'(x) = 0$

$\Rightarrow \quad \sin x = 0, x \neq 0$

Then, $\quad x = n\pi$

$\therefore$ $x = \pi, 2\pi, 3\pi, 4\pi, 5\pi$ all lies between 0 and 16.

$\therefore$ Number of points = 5

Matrix-Match Type

1. **(A)** Since, $f(x)$ is minimum at $x = -2$ and maximum at $x = 2$, let $g(x) = ax^3 + bx^2 + cx + d$.
Thus, $g(x)$ is also minimum at $x = -2$ and maximum at $x = 2$.
$\therefore \; a < 0$
Since a is a root of $x^2 - x - 6 = 0 \Rightarrow a = 3, -2, a = -2$
Then, $g(x) = -2x^3 + bx^2 + cx + d$

$$g'(x) = -6x^2 + 2bx + c = -6(x+2)(x-2)$$

[$\because g(x)$ is minimum at $x = -2$ and maximum at $x = 2$]
On comparing, we get
$b = 0$ and $c = 24$

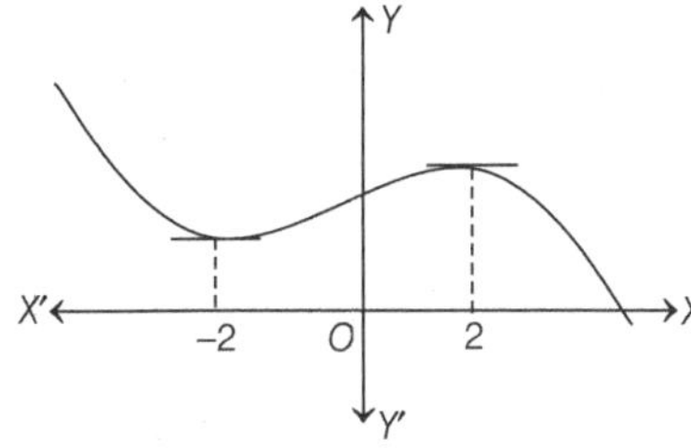

Since minimum and maximum values are positive,
$\therefore \; g(-2) > 0 \Rightarrow 16 - 48 + d > 0 \Rightarrow d > 32$
and $g(2) > 0 \Rightarrow -16 + 48 + d > 0 \Rightarrow d > -32$
It is clear that $d > 32$
Hence, $a = -2, b = 0, c = 24, d > 32$

2. **(A)** $\because f(x) = x^2 - 2|x| = \begin{cases} x^2 - 2x, & x \geq 0 \\ x^2 + 2x, & x < 0 \end{cases}$

$$\therefore \quad f'(x) = \begin{cases} 2x - 2, & x > 0 \\ 2x + 2, & x < 0 \end{cases}$$

Clearly $f'(x)$ is not defined at $x = 0$ and $f'(x) = 0$ at $x = \pm 1$
So, $\quad x = -1, 0, 1$ are critical points
$\therefore \quad \lambda = 1, \mu = -1$
$\Rightarrow \quad \lambda + 3\mu = -2, 3\lambda + \mu = 2$ (R, S)

(B) $\because \quad f(x) = (1 - x)|x - 3|$

$$= \begin{cases} (1-x)(x-3), & x \geq 3 \\ (x-1)(x-3), & x < 3 \end{cases}$$

$$\therefore \quad f'(x) = \begin{cases} 2(2-x), & x > 3 \\ 2(x-2), & x < 3 \end{cases}$$

Clearly $f'(x)$ is not defined at $x = 3$ and $f'(x) = 0$ at $x = 2$
So, $x = 2, 3$ are critical points
$\therefore \quad \lambda = 3, \mu = 2$
$\Rightarrow \lambda + 3\mu = 9, 3\lambda + \mu = 11$ (P, Q)

(C) $\because f(x) = \begin{cases} (x^2 + 2x - 4)x, & x \geq 0 \\ -(x^2 + 2x - 4)x, & x < 0 \end{cases}$

$$\therefore \quad f'(x) = \begin{cases} 3x^2 + 4x - 4, & x > 0 \\ -3x^2 - 4x + 4, & x < 0 \end{cases}$$

$$= \begin{cases} 3\left(x - \dfrac{2}{3}\right)(x+2), & x > 0 \\ 3\left(\dfrac{2}{3} - x\right)(x+2), & x < 0 \end{cases}$$

Clearly $f'(x)$ is not defined at $x = 0$ and $f'(x) = 0$ at $x = -2, \frac{2}{3}$.

So, $x = -2, 0, \frac{2}{3}$ are critical points.

$\therefore \quad \lambda = \frac{2}{3}, \mu = -2$

$\Rightarrow \quad \lambda + 3\mu = \frac{2}{3} - 6 = -\frac{16}{3}$

and $\quad 3\lambda + \mu = 0$ (T)

3. **(A)** The function $f(x)$ is differentiable except at $x = 0$

$\therefore \quad f(x)$ is not continuous at $x = 0$, also

$\because \quad f(x) = 2x^2 + \frac{2}{x^2}$

$\Rightarrow \quad f'(x) = 4x - \frac{4}{x^3}$

Critical points are $x = 1, -1$

Now, values of $f(x)$ at $x = -2, -1, 0, 1, 2$

are $\quad f(-2), f(-1), f(0), f(1), f(2)$

ie, $\quad \frac{17}{2}, 4, 1, 4, \frac{17}{2}$

$\Rightarrow \quad G = \frac{17}{2}, L = 1$

$\Rightarrow \quad [G + L] = 9, (G + L) = 10$ (R, T)

(B) Given $f(x) = x^3 - 6x^2 + 9x + 1$

$\therefore \quad f'(x) = 3x^2 - 12x + 9$

$$= 3(x-1)(x-3)$$

$\therefore \quad x = 1, 3$ but $3 \notin [0, 2]$

$G = \max\{f(0), f(1), f(2)\} = \max\{1, 5, 3\} = 5$

$L = \min\{f(0), f(1), f(2)\} = \min\{1, 5, 3\} = 1$

$\Rightarrow \quad G = 5, L = 1$

$\Rightarrow \quad [G + L] = 6, (G + L) = 6$ (Q)

(C) Given $f(x) = \text{arc tan } x - \frac{1}{2} \ln x$

$\therefore \quad f'(x) = \frac{1}{1 + x^2} - \frac{1}{2x}$

$$= -\frac{(x-1)^2}{2x(1 + x^2)}$$

$\therefore$ Critical points are $x = 0$ and $x = 1$

[$\because$ at $x = 0$, $f'(x)$ does not exist]

but $\quad 0 \notin \left[\frac{1}{\sqrt{3}}, \sqrt{3}\right]$

Then, $\quad G = \max\left\{f\left(\frac{1}{\sqrt{3}}\right), f(1), f(\sqrt{3})\right\}$

$$= \max\left\{\frac{\pi}{6} + \frac{1}{4}\ln 3, \frac{\pi}{4}, \frac{\pi}{3} - \frac{1}{4}\ln 3\right\}$$

$$= \frac{\pi}{6} + \frac{1}{4}\ln 3$$

and $\quad L = \min\left\{f\left(\frac{1}{\sqrt{3}}\right), f(1), f(\sqrt{3})\right\}$

$$= \min\left\{\frac{\pi}{6} + \frac{1}{4}\ln 3, \frac{\pi}{4}, \frac{\pi}{3} - \frac{1}{4}\ln 3\right\}$$

$$= \frac{\pi}{3} - \frac{1}{4}\ln 3$$

Hence, $\quad G = \frac{\pi}{6} + \frac{1}{4}\ln 3,$

$L = \frac{\pi}{3} - \frac{1}{4}\ln 3.$

$[G + L] = 1, (G + L) = 2$ (P, S)

17

Indefinite Integration

Objective Questions Type I [Only one correct answer]

In each of the questions below, four choices are given of which only one is correct. You have to select the correct answer which is the most appropriate.

1. $\int \frac{(x^2-1)}{x^3\sqrt{(2x^4-2x^2+1)}}\,dx$ is equal to

(a) $\frac{\sqrt{(2x^4-2x^2+1)}}{x(x^2-1)}+c$ (b) $\frac{\sqrt{(2x^4-2x^2+1)}}{x^3}+c$

(c) $\frac{\sqrt{(2x^4-2x^2+1)}}{x^2}+c$ (d) $\frac{\sqrt{(2x^4-2x^2+1)}}{2x^2}+c$

2. The integral $\int \frac{dx}{\left(\sqrt{x}+\sqrt[3]{x^2}\right)}$ represents the function

(a) $6\{\sqrt[3]{x^2}-\sqrt[3]{x}+\ln|1+\sqrt[3]{x}|\}+c$

(b) $3\sqrt[3]{x}-6\sqrt[6]{x}+6\ln|1+\sqrt[6]{x}|+c$

(c) $3\sqrt[3]{x}+6\sqrt[6]{x}+6\ln|1+\sqrt[6]{x}|+c$

(d) $6\sqrt[3]{x^2}-3\sqrt[3]{x}+6\ln|1+\sqrt[3]{x}|+c$

3. If $\int (e^{2x}+2e^x-e^{-x}-1)e^{(e^x+e^{-x})}dx = g(x)e^{(e^x+e^{-x})}+c$, where c is a constant of integration, then $g(0)$ is equal to **[JEE Main 2020, 5 September]**

(a) 2 (b) e^2

(c) e (d) 1

4. Let $x^2+1\neq n\pi,\ n\in N$, then

$\int x\sqrt{\left\{\frac{2\sin(x^2+1)-\sin\{2x^2+1\}}{2\sin(x^2+1)+\sin\{2x^2+1\}}\right\}}\,dx$ is

(a) $\ln\left|\frac{1}{2}\sec(x^2+1)\right|+c$ (b) $\ln\left|\sec\left\{\frac{1}{2}(x^2+1)\right\}\right|+c$

(c) $\frac{1}{2}\ln|\sec(x^2+1)|+c$ (d) $\ln|\sec(x^2+1)|+c$

5. $\int[\sin(101x)\cdot\sin^{99}x]\,dx$ equals

(a) $\frac{\sin(100x)(\sin x)^{100}}{100}+c$

(b) $\frac{\cos(100x)(\sin x)^{100}}{100}+c$

(c) $\frac{\cos(100x)(\cos x)^{100}}{100}+c$

(d) $\frac{\sin(100x)(\sin x)^{101}}{101}+c$

6. $\int(x-{}^{11}C_1x^2+{}^{11}C_2x^3-{}^{11}C_3x^4+\ldots-{}^{11}C_{11}x^{12})\,dx$ equals

(a) $\frac{(1-x)^{12}}{12}-\frac{(1-x)^{11}}{11}+c$

(b) $\frac{(1-x)^{13}}{13}-\frac{(1-x)^{12}}{12}+c$

(c) $\frac{(1-x)^{11}}{11}-\frac{(1-x)^{12}}{12}+c$

(d) $\frac{(1-x)^{12}}{12}-\frac{(1-x)^{13}}{13}+c$

7. If $\int \sin^{-1}\left(\sqrt{\frac{x}{1+x}}\right)dx = A(x)\tan^{-1}(\sqrt{x})+B(x)+C$, where C is a constant of integration, then the ordered pair $(A(x), B(x))$ can be **[JEE Main 2020, 3 September]**

(a) $(x-1,\ \sqrt{x})$ (b) $(x+1,\ \sqrt{x})$

(c) $(x+1,\ -\sqrt{x})$ (d) $(x-1,\ -\sqrt{x})$

8. $\int \frac{3+2\cos x}{(2+3\cos x)^2}\,dx$ is equal to

(a) $\left(\frac{\sin x}{2+3\cos x}\right)+c$ (b) $\left(\frac{2\cos x}{2+3\sin x}\right)+c$

(c) $\left(\frac{2\cos x}{2+3\cos x}\right)+c$ (d) $\left(\frac{2\sin x}{2+3\sin x}\right)+c$

9. $\int \sec^{4/9}\theta\,\mathrm{cosec}^{14/9}\theta\,d\theta$ is equal to

(a) $\frac{5}{9}(\tan\theta)^{-5/9}+c$ (b) $-\frac{9}{5}(\tan\theta)^{-5/9}+c$

(c) $\frac{9}{5}(\tan\theta)^{-9/5}+c$ (d) $-\frac{5}{9}(\tan\theta)^{-9/5}+c$

10. If $\int \frac{(\sqrt{x})^5}{(\sqrt{x})^7 + x^6} dx = \lambda \ln\left(\frac{x^a}{x^a+1}\right) + c$, then $a + \lambda$ is

(a) $=2$ (b) > 2
(c) < 2 (d) $= 1$

11. $\int |x| \ln|x| \, dx$ equals $(x \neq 0)$

(a) $\frac{x^2}{2}\ln|x| - \frac{x^2}{4} + c$

(b) $\frac{1}{2}x|x| \ln x + \frac{1}{4}x|x| + c$

(c) $-\frac{x^2}{2}\ln|x| + \frac{x^2}{4} + c$

(d) $\frac{1}{2}x|x|\ln|x| - \frac{1}{4}x|x| + c$

12. Let $\int \frac{dx}{x^{2008} + x} = \frac{1}{p}\ln\left(\frac{x^q}{1+x^r}\right) + C$, where $p, q, r \in N$ and need not be distinct, then the value of $(p+q+r)$ equals

(a) 6024 (b) 6022
(c) 6021 (d) 6020

13. Let $f(x)$ be a function such that, $f(0) = f'(0) = 0$, $f''(x) = \sec^4 x + 4$, then the function is

(a) $\ln|(\sin x)| + \frac{1}{3}\tan^3 x + x$

(b) $\frac{2}{3}\ln|(\sec x)| + \frac{1}{6}\tan^2 x + 2x^2$

(c) $\ln|\cos x| + \frac{1}{6}\cos^2 x + \frac{x^2}{5}$

(d) None of the above

14. $\int \frac{(2x^{12} + 5x^9)}{(x^5 + x^3 + 1)^3} dx$ is equal to

(a) $\frac{x^2 + 2x}{(x^5 + x^3 + 1)^2} + c$

(b) $\frac{x^{10}}{2(x^5 + x^3 + 1)^2} + c$

(c) $\ln|x^5 + x^3 + 1| + \sqrt{(2x^7 + 5x^4)} + c$

(d) None of the above

15. For $x \in \left(\frac{3\pi}{4}, \pi\right)$, $\int (\sqrt{1 + \sin 2x} + \sqrt{1 - \sin 2x})\, dx =$

(a) $-2\cos x + c$ (b) $2\sin x + c$
(c) $-2\sin x = c$ (d) $2\cos x + c$

16. For any natural number m, $\int (x^{7m} + x^{2m} + x^m)(2x^{6m} + 7x^m + 14)^{1/m} dx$ (where $x > 0$) equals

(a) $\frac{(7x^{7m} + 2x^{2m} + 14x^m)^{\frac{m+1}{m}}}{14(m+1)} + c$

(b) $\frac{(2x^{7m} + 14x^{2m} + 7x^m)^{\frac{m+1}{m}}}{14(m+1)} + c$

(c) $\frac{(2x^{7m} + 7x^{2m} + 14x^m)^{\frac{m+1}{m}}}{14(m+1)} + c$

(d) $\frac{(7x^{7m} + 2x^{2m} + x^m)^{\frac{m+1}{m}}}{14(m+1)} + c$

17. The primitive function of the function $f(x) = \frac{\sqrt{(a^2 - x^2)}}{x^4}$ is

(a) $c + \frac{\sqrt{a^2 - x^2}}{3a^2x^3}$ (b) $c - \frac{(a^2 - x^2)^{3/2}}{2a^2x^2}$

(c) $c - \frac{(a^2 - x^2)^{3/2}}{3a^2x^3}$ (d) None of these

18. The antiderivative of $f(x) = \frac{1}{3 + 5\sin x + 3\cos x}$, whose graph passes through the point (0, 0) is

(a) $\frac{1}{5}\left(\ln\left|1 - \frac{5}{3}\tan(x/2)\right|\right)$

(b) $\frac{1}{5}\left(\ln\left|1 + \frac{5}{3}\tan(x/2)\right|\right)$

(c) $\frac{1}{5}\left(\ln\left|1 + \frac{5}{3}\cot(x/2)\right|\right)$

(d) None of the above

19. $\int x^x (1 + \ln|x|)\, dx$ is equal to

(a) $x^x \ln|x| + c$ (b) $e^{x^x} + c$

(c) $x^x + c$ (d) None of these

20. $\int \frac{\sin\frac{5x}{2}}{\sin\frac{x}{2}} dx$ is equal to

(where, C is a constant of integration)

[JEE Main 2019, 8 April]

(a) $2x + \sin x + 2\sin 2x + c$
(b) $x + 2\sin x + 2\sin 2x + c$
(c) $x + 2\sin x + \sin 2x + c$
(d) $2x + \sin x + \sin 2x + c$

21. $\int \frac{1}{x^2(x^4 + 1)^{3/4}} dx$ is equal to

(a) $\left(1 + \frac{1}{x^4}\right)^{1/4} + c$ (b) $(x^4 + 1)^{1/4} + c$

(c) $\left(1 - \frac{1}{x^4}\right)^{1/4} + c$ (d) $-\left(1 + \frac{1}{x^4}\right)^{1/4} + c$

22. Let the equation of a curve passing through the point (0, 1) be given by $y = \int x^2 \cdot e^{x^3} dx$. If the equation of the curve is written in the form $x = f(y)$, then $f(y)$ is

(a) $\sqrt{\ln\left|\frac{3y - 2}{3}\right|}$ (b) $\sqrt[3]{\ln\left|\frac{2 - 3y}{3}\right|}$

(c) $\sqrt[3]{\ln\left|\frac{3y - 2}{3}\right|}$ (d) None of these

23. $\int \frac{xe^x}{(1+x)^2} dx$ is equal to

(a) $\frac{e^x}{x+1} + c$ (b) $e^x (x+1) + c$

(c) $-\frac{e^x}{(x+1)^2} + c$ (d) $\frac{e^x}{1+x^2} + c$

24. $\int \frac{y^2 + \sqrt[3]{y^4} + \sqrt[6]{y^2}}{y(1+\sqrt[3]{y^2})} dy =$

(a) $\frac{3}{4}\sqrt[3]{y^4} + 3\tan^{-1}(\sqrt[3]{y}) + c$

(b) $\frac{3}{2}y^{2/3} + 6\tan^{-1}(\sqrt[6]{y^2}) + c$

(c) $\frac{2}{3\sqrt[3]{y^2}} + 6\log(1+y^2) + c$

(d) $\frac{3}{1+y} + \tan^{-1}(\sqrt[3]{y^2}) + c$

25. $\int x^{-2/3}(1+x^{1/2})^{-5/3} dx$ is equal to

(a) $3(1+x^{-1/2})^{-1/3} + c$ (b) $3(1+x^{-1/2})^{-2/3} + c$

(c) $3(1+x^{1/2})^{-2/3} + c$ (d) None of these

26. $\int \frac{\cos 4x - 1}{\cot x - \tan x} dx$ is equal to

(a) $-\frac{1}{2}\cos 4x + c$ (b) $-\frac{1}{4}\cos 4x + c$

(c) $-\frac{1}{2}\sin 2x + c$ (d) None of these

27. The integral $\int \frac{dx}{(x+4)^{8/7}(x-3)^{6/7}}$ is equal to (where c is a constant of integration) **[JEE Main 2020, 8 January]**

(a) $-\left(\frac{x-3}{x+4}\right)^{-1/7} + c$ (b) $\frac{1}{2}\left(\frac{x-3}{x+4}\right)^{3/7} + c$

(c) $\left(\frac{x-3}{x+4}\right)^{1/7} + c$ (d) $-\frac{1}{13}\left(\frac{x-3}{x+4}\right)^{-13/7} + c$

28. The integral $\int \frac{(2x-1)\cos\sqrt{(2x-1)^2+5}}{\sqrt{4x^2-4x+6}} dx$ is equal to (where c is a constant of integration) **[JEE Main 2021, 18 March]**

(a) $\frac{1}{2}\sin\sqrt{(2x-1)^2+5} + c$

(b) $\frac{1}{2}\cos\sqrt{(2x+1)^2+5} + c$

(c) $\frac{1}{2}\cos\sqrt{(2x-1)^2+5} + c$

(d) $\frac{1}{2}\sin\sqrt{(2x+1)^2+5} + c$

29. The value of the integral $\int \frac{\cos^3 x + \cos^5 x}{\sin^2 x + \sin^4 x} dx$ is

(a) $\sin x - 6\tan^{-1}(\sin x) + c$

(b) $\sin x - 2(\sin x)^{-1} + c$

(c) $\sin x - 2(\sin x)^{-1} - 6\tan^{-1}(\sin x) + c$

(d) $\sin x - 2(\sin x)^{-1} + 5\tan^{-1}(\sin x) + c$

30. If $\int f(x)\sin x\cos x\, dx = \frac{1}{2(b^2-a^2)}\ln f(x) + c$, then $f(x)$ is equal to

(a) $\frac{1}{a^2\sin^2 x + b^2\cos^2 x}$ (b) $\frac{1}{a^2\sin^2 x - b^2\cos^2 x}$

(c) $\frac{1}{a^2\cos^2 x + b^2\sin^2 x}$ (d) None of these

31. If $\int \frac{\cos x\, dx}{\sin^3 x(1+\sin^6 x)^{2/3}} = f(x)(1+\sin^6 x)^{1/\lambda} + c$

where c is a constant of integration, then $\lambda f\left(\frac{\pi}{3}\right)$ is equal to **[JEE Main 2020, 8 January]**

(a) 2 (b) $-\frac{9}{8}$

(c) -2 (d) $\frac{9}{8}$

32. $\int \frac{(x^4-x)^{1/4}}{x^5} dx$ is equal to

(a) $\frac{4}{15}\left(1-\frac{1}{x^3}\right)^{5/4} + c$ (b) $\frac{4}{5}\left(1-\frac{1}{x^3}\right)^{5/4} + c$

(c) $\frac{4}{15}\left(1+\frac{1}{x^3}\right)^{5/4} + c$ (d) None of these

Objective Questions Type II [One or more than one correct answer(s)]

In each of the questions below four choices of which one or more than one are correct. You have to select the correct answer(s) accordingly.

1. If $\int \frac{(x^2-x+1)e^x}{(x^2+1)^{3/2}} dx = e^x f(x) + c$, then,

(a) $f(x)$ is an even function

(b) $f(x)$ is a bounded function

(c) The range of $f(x)$ is $(0, 1)$

(d) $f(x)$ has two point of extrema

2. Suppose $J = \int \frac{\sin^2 x + \sin x}{1+\sin x+\cos x} dx$ and $K = \int \frac{\cos^2 x + \cos x}{1+\sin x+\cos x} dx$. If c is an arbitrary constant of integration then which of the following is/are correct?

(a) $J = \frac{1}{2}(x - \sin x + \cos x) + c$

(b) $J = K - (\sin x + \cos x) + c$

(c) $J = x - K + c$

(d) $K = \frac{1}{2}(x - \sin x + \cos x) + c$

3. Let $\int \frac{x^{1/2}}{\sqrt{(1 - x^3)}} dx = \frac{2}{3} gof(x) + c$, then

(a) $f(x) = \sqrt{x}$ (b) $f(x) = x^{3/2}$

(c) $f(x) = x^{2/3}$ (d) $g(x) = \sin^{-1} x$

4. If $\int \text{cosec } 2x \, dx = f\{g(x)\} + c$, then

(a) range of $g(x) = (-\infty, \infty)$

(b) domain of $f(x) = (-\infty, \infty) - \{0\}$

(c) $g'(x) = \sec^2 x$

(d) $f'(x) = 1/x$ for all $x \in (0, \infty)$

5. If $\int \frac{xe^x}{\sqrt{(1 + e^x)}} dx = f(x)\sqrt{(1 + e^x)} - 2 \ln g(x) + c$, then

(a) $f(x) = x - 1$ (b) $g(x) = \frac{\sqrt{(1 + e^x)} - 1}{\sqrt{(1 + e^x)} + 1}$

(c) $g(x) = \frac{\sqrt{(1 + e^x)} + 1}{\sqrt{(1 + e^x)} - 1}$ (d) $f(x) = 2(x - 2)$

6. Let $f : R \to R$ and $g : R \to R$ be two non-constant differentiable functions. If $f'(x) = (e^{(f(x) - g(x))}) g'(x)$ for all $x \in R$ and $f(1) = g(2) = 1$, then which of the following statement(s) is (are) TRUE? **[JEE Advanced 2018]**

(a) $f(2) < 1 - \log_e 2$ (b) $f(2) > 1 - \log_e 2$

(c) $g(1) > 1 - \log_e 2$ (d) $g(1) < 1 - \log_e 2$

7. If $f(x) = \lim_{n \to \infty} e^{x \tan(1/n) \ln(1/n)}$ and $\int \frac{f(x)}{\sqrt[3]{(\sin^{11} x \cos x)}} dx = g(x) + c$, then

(a) $g\left(\frac{\pi}{4}\right) = \frac{3}{2}$

(b) $g(x)$ is continuous for all x

(c) $g\left(\frac{\pi}{4}\right) = -\frac{15}{8}$

(d) $g(x)$ is non differentiable at infinitely many points

8. If $\int (\cos^{-1} x + \cos^{-1} \sqrt{(1 - x^2)}) dx = Ax + f(x) \sin^{-1} x - 2\sqrt{(1 - x^2)} + c \, \forall \, x \in [-1, 0)$, then

(a) $f(x) = x$ (b) $f(x) = -2x$

(c) $A = \frac{\pi}{4}$ (d) $A = \frac{\pi}{2}$

9. The value of $\int_0^x \frac{(t - |t|)^2}{(1 + t^2)} dt$ is equal to

(a) $4(x - \tan^{-1} x)$, if $x < 0$ (b) 0, if $x > 0$

(c) $\ln(1 + x^2)$, if $x > 0$ (d) None of the above

10. If the primitive of $\sin(\ln x)$ is $f(x)\{\sin g(x) - \cos h(x)\} + c$ (c being the constant of integration), then

(a) $\lim_{x \to 2} f(x) = 1$

(b) $\lim_{x \to 1} \frac{g(x)}{h(x)} = 1$

(c) $g(e^3) = 3$

(d) $h(e^5) = 5$

Linked-Comprehension Type

In these questions, a passage (paragraph) has been given followed by questions based on each of the passages. You have to answer the questions based on the passage given.

PASSAGE 1

If the integrand is a rational function of x and fractional power of a linear fractional function of the form $\left(\frac{ax + b}{cx + d}\right)$

i.e. $\int f\left(x, \left(\frac{ax + b}{cx + d}\right)^{m/n}, \ldots, \left(\frac{ax + b}{cx + d}\right)^{r/s}\right) dx$

In this form substitute $\frac{ax + b}{cx + d} = t^m$ where m is the LCM of the denominators of fractional powers of $\left(\frac{ax + b}{cx + d}\right)$.

On the basis of above information, answer the following questions

1. The value of $\int \frac{dx}{(1 + x)^{1/2} - (1 + x)^{1/3}}$ is

(a) $2\lambda^{1/2} + 3\lambda^{1/3} + 6\lambda^{1/6} + 6 \ln |\lambda^{1/6} - 1| + c$

(b) $2\lambda^{1/2} - 3\lambda^{1/3} + 6\lambda^{1/6} + 6 \ln |\lambda^{1/6} - 1| + c$

(c) $2\lambda^{1/2} + 3\lambda^{1/3} - 6\lambda^{1/6} + 6 \ln |\lambda^{1/6} - 1| + c$

(d) $2\lambda^{1/2} + 3\lambda^{1/3} + 6\lambda^{1/6} - 6 \ln |\lambda^{1/6} - 1| + c$

where $\lambda = (1 + x)$

2. The value of $\int \frac{1+x^{1/2}-x^{2/3}}{1+x^{1/3}} dx$ is

(a) $\frac{3}{4}x^{4/3}+\frac{6}{7}x^{7/6}+x+\frac{6}{5}x^{5/6}+2x^{1/2} - 6x^{1/6}+6\tan^{-1}(x^{1/6})+c$

(b) $-\frac{3}{4}x^{4/3}+\frac{6}{7}x^{7/6}+x+\frac{6}{5}x^{5/6}-2x^{1/2} + 6x^{1/6}-6\tan^{-1}(x^{1/6})+c$

(c) $\frac{3}{4}x^{4/3}-\frac{6}{7}x^{7/6}+x-\frac{6}{5}x^{5/6}+2x^{1/2} - 6x^{1/6}+6\tan^{-1}(x^{1/6})+c$

(d) $-\frac{3}{4}x^{4/3}+\frac{6}{7}x^{7/6}+x-\frac{6}{5}x^{5/6}+2x^{1/2}-6x^{1/6} + 6\tan^{-1}(x^{1/6})+c$

3. The value of $\int \frac{\sqrt[4]{x}}{\sqrt{x}-1} dx$ is

(a) $-\frac{4}{3}x^{3/4}+4x^{1/4}+2\ln\left|\frac{x^{1/4}-1}{x^{1/4}+1}\right|+c$

(b) $\frac{4}{3}x^{3/4}+4x^{1/4}+2\ln\left|\frac{x^{1/4}-1}{x^{1/4}+1}\right|+c$

(c) $-\frac{4}{3}x^{3/4}-4x^{1/4}+2\ln\left|\frac{x^{1/4}-1}{x^{1/4}+1}\right|+c$

(d) $\frac{4}{3}x^{3/4}-4x^{1/4}+2\ln\left|\frac{x^{1/4}-1}{x^{1/4}+1}\right|+c$

4. The value of $\int \frac{x}{\sqrt[3]{(a+bx)}} dx$ is

(a) $-\frac{3}{5b^2}(a+bx)^{5/3}+\frac{3a}{2b^2}(a+bx)^{2/3}+c$

(b) $-\frac{3}{5b^2}(a+bx)^{5/3}-\frac{3a}{2b^2}(a+bx)^{2/3}+c$

(c) $\frac{3}{5b^2}(a+bx)^{5/3}-\frac{3a}{2b^2}(a+bx)^{2/3}+c$

(d) $\frac{3}{5b^2}(a+bx)^{5/3}+\frac{3a}{2b^2}(a+bx)^{2/3}+c$

5. The value of $\int \sqrt{\left(\frac{x+2}{2x+3}\right)}\cdot\frac{1}{x} dx$ is

(a) $\frac{1}{\sqrt{2}}\ln\lambda+\sqrt{\frac{2}{3}}\ln\mu+c$

(b) $-\frac{1}{\sqrt{2}}\ln\lambda+\sqrt{\frac{2}{3}}\ln\mu+c$

(c) $\frac{1}{\sqrt{2}}\ln\lambda-\sqrt{\frac{2}{3}}\ln\mu+c$

(d) $-\frac{1}{\sqrt{2}}\ln\lambda-\sqrt{\frac{2}{3}}\ln\mu+c$

where $\lambda=\left|\frac{\sqrt{(2x+3)}+\sqrt{2}\sqrt{(x+2)}}{\sqrt{(2x+3)}-\sqrt{2}\sqrt{(x+2)}}\right|$,

$\mu=\left|\frac{\sqrt{3}\sqrt{(x+2)}-\sqrt{2}\sqrt{(2x+3)}}{\sqrt{3}\sqrt{(x+2)}+\sqrt{2}\sqrt{(2x+3)}}\right|$

PASSAGE 2

If an integral can not be evaluated, then it is connected to another integral of lower degree but of same type. This is called reduction formula, we can derive reduction formulas for the integral of the form $\int \sin^n x dx$, $\int \cos^n x dx$, $\int \tan^n x dx$, $\int \cot^n x dx$, $\int \sec^n x dx$, $\int \text{cosec}^n x dx$ by using integration by parts. In term these reduction formulas can be used to compute integrals of sin x, cos x, tan x etc.

On the basis of above information, answer the following questions

1. If $\int \sin^5 x dx = -\frac{\sin^4 x\cos x}{5} + A\sin^2 x\cos x + B\cos x + c$ then $A+B$ is equal to

(a) $-\frac{2}{3}$ (b) $-\frac{3}{4}$

(c) $-\frac{4}{5}$ (d) $-\frac{5}{6}$

2. If $\int \sec^6 x\, dx = \frac{1}{5}\tan^5 x + B\tan^3 x + C\tan x + D$ then $B+C$ is equal to

(a) $\frac{7}{3}$ (b) $\frac{5}{3}$

(c) $\frac{11}{3}$ (d) $\frac{13}{3}$

3. If $\int \tan^6 x\, dx = \frac{\tan^5 x}{5} + A\tan^3 x + B\tan x + Cx + D$, then $A+B+C$ is equal to

(a) $-1/2$ (b) $-1/4$

(c) $-1/3$ (d) $-1/5$

4. If $\int \cos^n x dx = \frac{\cos^{n-1} x\sin x}{n} + f(n)\int \cos^{n-2} x dx$, then $f(5)$ is equal to

(a) 1/5

(b) 2/5

(c) 3/5

(d) 4/5

5. If $\int \text{cosec}^n x dx = -\frac{\text{cosec}^{n-2}\cot x}{(n-1)} + f(n)\int \text{cosec}^{n-2} x dx$ then $f(n+1)$ is equal to

(a) $\frac{n-1}{n}$

(b) $\frac{n-2}{n-1}$

(c) $\frac{n-3}{n-2}$

(d) None of the above

PASSAGE 3

Repeated application of integration by parts gives us, the reduction formula if the integrand is dependent of n, $n \in N$.

On the basis of above information, answer the following questions

1. If $I_n = \int \tan^n x\, dx$ and $I_n = -\dfrac{\tan^{n-1} x}{(n-1)} + \lambda I_{n-2}$, then λ is equal to
(a) $\dfrac{1}{(n-1)}$ (b) $\dfrac{1}{(n-2)}$
(c) $\dfrac{1}{n}$ (d) None of these

2. If $I_{m-2,\, n+2} = \int \sin^{m-2} x \cos^{n+2} x\, dx$ and $I_{m,n} = -\dfrac{\sin^{m-1} x \cos^{n+1} x}{(n+1)} + f(m, n) I_{m-2,\, n+2}$, then $f(2, 3)$ is equal to
(a) $\dfrac{1}{2}$ (b) $\dfrac{1}{3}$
(c) $\dfrac{1}{4}$ (d) $\dfrac{1}{5}$

3. If $I_{m,\, p} = \int x^m \cdot (a + bx^n)^p\, dx$ and $I_{m,\, p} = \dfrac{x^{m+1}(a+bx^n)^p}{(m+1)} - f(m, n, p, b) I_{m+n,\, p-1}$, then the value of $f(1, 2, 3, 4)$ is equal to
(a) 8 (b) 10
(c) 12 (d) None of these

4. If $I_n = \int x \sin^n x dx$ and $I_n = -\dfrac{x \sin^{n-1} x \cos x}{n} + \dfrac{\sin^n x}{n^2} + f(n) I_{n-2}$, then the value of $f(n)$ is equal to
(a) $\dfrac{n-1}{n}$ (b) $\dfrac{n-2}{n-1}$
(c) $\dfrac{n+1}{n}$ (d) $\dfrac{n+1}{n-1}$

5. If $I_n = \int e^{ax} \cdot \sin^n x dx$ and $I_n = \dfrac{e^{ax} \sin^{n-1} x(a \sin x - \cos x)}{(n + a^2)} - A \sin^n x + B\, I_{n-2}$, then $A + B$ is equal to
(a) $\dfrac{a(n^2+1)}{n(n+a^2)}$ (b) $\dfrac{a(n^2-1)}{n(n+a^2)}$
(c) $\dfrac{a(n^2+1)}{n(n^2+a^2)}$ (d) $\dfrac{a(n^2-1)}{n(n^2+a^2)}$

PASSAGE 4

Consider two differentiable function $f(x)$ and $g(x)$ satisfying
$6\int f(x)\, g(x)\, dx = x^6 + 3x^4 + 3x^2 + c$ and
$2\int \dfrac{g(x)\, dx}{f(x)} = x^2 + c$, where $f(x) > 0, \forall\, x \in R$.

1. $\int (g(x) - f(x))\, dx$ is equal to
(a) $\dfrac{x^4}{4} - \dfrac{x^2}{2} + x + c$ (b) $\dfrac{x^4}{4} + \dfrac{x^2}{2} - \dfrac{x^3}{3} - x + c$
(c) $\dfrac{x^4}{4} - \dfrac{x^3}{3} + \dfrac{x^2}{2} - x + c$ (d) $\dfrac{x^4}{4} + \dfrac{x^3}{3} + c$

2. $\lim\limits_{x \to 0} \dfrac{\log (f(x))}{g(x)}$ is equal to
(a) e (b) 2
(c) 1 (d) 0

Numerical Value Type

Solve the following problems and write your answer.

1. If $\int \dfrac{x dx}{\sqrt{(7x - 10 - x^2)^3}} = \dfrac{\lambda \cdot (7x - 20)}{\sqrt{(7x - 10 - x^2)}} + c$, then the value of 1800λ must be

2. If $\int \dfrac{dx}{(1 + \sin x)} = \tan\left(\dfrac{x}{2} + a\right) + b$, then the value of $-\dfrac{204.8a}{\pi}$ must be

3. If $\int (x^9 + x^6 + x^3)(2x^6 + 3x^3 + 6)^{1/3} dx = \dfrac{1}{a}(2x^9 + 3x^6 + 6x^3)^{4/3} + c$, then the value of $1.2a$ must be

4. If $\int (\sqrt{\tan x} + \sqrt{\cot x}) dx = a \tan^{-1}\left(\dfrac{\tan x - 1}{\sqrt{b \tan x}}\right) + c$, then the value of $a^4 + b^5$ must be

5. If $\int \dfrac{\ln x}{x^3} dx = -\dfrac{\ln x}{ax^2} - \dfrac{1}{bx^2} + c$, then the value of $a^5 + b^4$ must be

6. If $\int \left(\dfrac{\cos 8x - \cos 7x}{1 + 2\cos 5x}\right) dx = \dfrac{\sin 3x}{a} - \dfrac{\sin 2x}{b} + c$, then the value of $(a^b)^4$ must be

7. If $\int \sin 4x \cdot e^{\tan^2 x} dx = a \cos^b x \cdot e^{\tan^2 x} + c$, then the value of a^{2b} must be

Single Integer Type

Solve the following problems and write your answer.

1. If $\int \frac{(2x+3)\,dx}{x(x+1)(x+2)(x+3)+1} = c - \frac{1}{f(x)}$, where $f(x)$ is of the form ax^2+bx+c, then $(a+b+c)$ equals

2. If $\int \frac{2x^{12}+5x^9}{(1+x^3+x^5)^3}dx = \frac{x^m}{l(1+x^3+x^5)^r} + c$, then $\frac{m-l}{r} =$

3. A curve $g(x) = \int x^{27}(1+x+x^2)^6(6x^2+5x+4)\,dx$ is passing through origin, if the value of $\frac{g(1)}{g(-1)}$ is p^λ, where p is prime number and $\lambda \in N$, then the value of $|\lambda - p|$ is

4. For real numbers α, β, γ and δ, if
$$\int \frac{(x^2-1)+\tan^{-1}\left(\frac{x^2+1}{x}\right)}{(x^4+3x^2+1)\tan^{-1}\left(\frac{x^2+1}{x}\right)}dx$$
$$= \alpha \log_e\left(\tan^{-1}\left(\frac{x^2+1}{x}\right)\right) + \beta \tan^{-1}\left(\frac{\gamma(x^2-1)}{x}\right) + \delta \tan^{-1}\left(\frac{x^2+1}{x}\right) + C$$
where C is an arbitrary constant, then the value of $10(\alpha + \beta\gamma + \delta)$ is equal to ....
[JEE Main 2021, 16 March]

5. If $f(x) = \int \frac{5x^8+7x^6}{(x^2+1+2x^7)^2}dx, (x \geq 0), f(0) = 0$ and $f(1) = \frac{1}{k}$, then the value of k is

6. Let $f: R \to R$ be a differentiable function with $f(0) = 1$ and satisfying the equation $f(x+y) = f(x)f'(y) + f'(x)f(y)$ for all $x, y \in R$. Then, the value of $\log_e(f(4))$ is .... **[JEE Advanced 2018]**

Matrix-Match Type

Given below are Matching Type Questions, with two columns (each having some items) each . Each item of Column I has to be matched with the items of Column II, by encircling the correct match(es).

NOTE *An item of Column I can be matched with more than one items of Column II.*

1. Observe the following columns

	Column I		Column II
A.	$\int \frac{dx}{x(x^n+1)}; n \in N$	P.	$\frac{1}{(1-n)}(x^{-n}+1)^{1-1/n} + c$
B.	$\int \frac{dx}{x^2(x^n+1)^{(n-1)/n}}; n \in N$	Q.	$\ln\|x\| - \frac{1}{n}\ln\|1+x^n\| + c$
C.	$\int \frac{dx}{x^n(1+x^n)^{1/n}}; n \in N$	R.	$\frac{(1+x^n)^{1-1/n}\cdot x^{1-n}}{(1-n)} + C$
		S.	$-\frac{1}{n}\ln\|1+x^{-n}\| + C$
		T.	$-(1+x^{-n})^{1/n} + C$

2. Observe the following columns

	Column I		Column II
A.	$\int(\tan x)^{1/3}dx = A\ln\frac{(t^4-t^2+1)}{(t^2+1)^2}$, $+ B\tan^{-1}\left(\frac{2t^2-1}{\sqrt{3}}\right) + c$ where $t = \tan^{1/3}x$	P.	$A = \frac{1}{4}$
B.	$\int\left(\frac{\sin x + \sin^3 x}{\cos 2x}\right)dx$ $= A\cos x + B\ln\left\|\frac{\sqrt{2}\cos x + 1}{\sqrt{2}\cos x - 1}\right\| + c$	Q.	$A = \frac{1}{3}$
C.	$\int \frac{dx}{(x^2+1)(x^2+4)}$ $= A\tan^{-1}x + B\tan^{-1}\left(\frac{x}{2}\right) + c$	R.	$A = \frac{1}{2}$
		S.	$B = \frac{\sqrt{3}}{2}$
		T.	$B = \frac{3}{4\sqrt{2}}$

3. Observe the following columns

	Column I		Column II
A.	If $\int \frac{2^x}{\sqrt{1-4^x}}dx = k\sin^{-1}2^x + c$, then k is greater than	P.	0
B.	If $\int \frac{(\sqrt{x})^5}{(\sqrt{x})^7 + x^6}dx = a\ln\frac{x^k}{x^k+1} + c$, then ak is less than	Q.	1
C.	If $\int \frac{x^4+1}{x(x^2+1)^2}dx = k\ln\|x\| + \frac{m}{1+x^2} + n$, where n is the constant of integration, then mk is greater than	R.	3
D.	If $\int \frac{dx}{5+4\cos x} = k\tan^{-1}\left(m\tan\frac{x}{2}\right) + c$, then $\frac{k}{m}$ is greater than	S.	4

Answers

Objective Questions Type I [Only one correct answer]

1. (d)	2. (b)	3. (a)	4. (b)	5. (a)	6. (b)	7. (c)	8. (a)	9. (b)	10. (b)
11. (d)	12. (c)	13. (b)	14. (b)	15. (a)	16. (c)	17. (c)	18. (b)	19. (c)	20. (c)
21. (d)	22. (d)	23. (a)	24. (a)	25. (b)	26. (d)	27. (c)	28. (a)	29. (c)	30. (a)
31. (c)	32. (a)								

Objective Questions Type II [One or more than one correct answer(s)]

1. (a, b, c)	2. (b, c)	3. (b, d)	4. (a, b, c)	5. (b, d)
6. (b, c)	7. (c, d)	8. (b, d)	9. (a, b)	10. (a, b, c, d)

Linked-Comprehension Type

Passage 1 1. (a) 2. (d) 3. (b) 4. (c) 5. (a) **Passage 2** 1. (c) 2. (b) 3. (c) 4. (d) 5. (a)

Passage 3 1. (d) 2. (c) 3. (c) 4. (a) 5. (b) **Passage 4** 4. (b) 2. (d)

Numerical Value Type

1. (400) 2. (51.2) 3. (28.8) 4. (36) 5. (288) 6. (6561) 7. (256)

Single Integer Type

1. (5) 2. (4) 3. (4) 4. (6) 5. (4) 6. (2)

Matrix-Match Type

1. A → (Q, S); B → (T); C → (P, R) 2. A → (P, S); B → (R, T); C → (Q)
3. A → (P, Q, R); B → (R, S); C → (P); D → (P, Q)

Solutions

Objective Questions Type [Only one correct answer]

1. Let $I = \int \frac{(x^2 - 1)\,dx}{x^3 \cdot x^2 \sqrt{\left(2 - \frac{2}{x^2} + \frac{1}{x^4}\right)}}$

$= \int \frac{\left(\frac{1}{x^3} - \frac{1}{x^5}\right) dx}{\sqrt{\left(2 - \frac{2}{x^2} + \frac{1}{x^4}\right)}}$

Put $2 - \frac{2}{x^2} + \frac{1}{x^4} = t$

$\therefore$ $\left(\frac{4}{x^3} - \frac{4}{x^5}\right) dx = dt$

Then, $I = \frac{1}{4} \int \frac{dt}{\sqrt{t}} = \frac{1}{4} \cdot 2\sqrt{t} + c$

$= \frac{\sqrt{(2x^4 - 2x^2 + 1)}}{2x^2} + c$

2. Let $I = \int \frac{dx}{\left(\sqrt{x} + \sqrt[3]{x^2}\right)}$

Put $x = t^6$

$\therefore$ $dx = 6t^5 dt$

Then, $I = \int \frac{6t^5 dt}{(t^3 + t^4)} = 6 \int \frac{t^2 dt}{(1 + t)}$

$= 6 \int \left(t - 1 + \frac{1}{1 + t}\right) dt$

$= 6 \left\{\frac{t^2}{2} - t + \ln|1 + t|\right\} + c$

$= 3t^2 - 6t + 6 \ln|1 + t| + c$

$= 3 \cdot \sqrt[3]{x} - 6 \cdot \sqrt[6]{x} + 6 \ln|1 + \sqrt[6]{x}| + c$

3. Given, integral

$$I=\int (e^{2x}+2e^{x}-e^{-x}-1)e^{(e^{x}+e^{-x})}dx$$

Let $e^{x}=t \Rightarrow dx=\frac{dt}{t}$

So, $I=\int\left(t^{2}+2t-\frac{1}{t}-1\right)e^{\left(t+\frac{1}{t}\right)}\frac{dt}{t}$

$$=\int\left(t+2-\frac{1}{t^{2}}-\frac{1}{t}\right)e^{\left(t+\frac{1}{t}\right)}dt$$

$$=\int\left[\left(t-\frac{1}{t}\right)+\left(1-\frac{1}{t^{2}}\right)+1\right]e^{\left(t+\frac{1}{t}\right)}dt$$

$$=\int\left[(1+t)\left(1-\frac{1}{t^{2}}\right)+1\right]e^{\left(t+\frac{1}{t}\right)}dt$$

$$=\int (1+t)(1-1/t^{2})e^{\left(t+\frac{1}{t}\right)}dt+\int e^{\left(t+\frac{1}{t}\right)}dt$$

$$=(1+t)e^{\left(t+\frac{1}{t}\right)}-\int 1\cdot e^{\left(t+\frac{1}{t}\right)}dt+\int e^{\left(t+\frac{1}{t}\right)}dt$$

$$=(1+t)e^{t+\frac{1}{t}}+c=(1+e^{x})e^{(e^{x}+e^{-x})}+c$$

$$=g(x)e^{(e^{x}+e^{-x})}+c \qquad \text{[given]}$$

$\therefore\ g(x)=1+e^{x} \Rightarrow g(0)=1+1=2$

4. Let $I=\int x\sqrt{\left\{\frac{2\sin(x^{2}+1)-\sin\{2(x^{2}+1)\}}{2\sin(x^{2}+1)+\sin\{2(x^{2}+1)\}}\right\}}dx$

Put $x^{2}+1=t$

$\therefore\ x\,dx=\frac{1}{2}dt$

Then, $I=\frac{1}{2}\int\sqrt{\left(\frac{2\sin t-\sin 2t}{2\sin t+\sin 2t}\right)}dt$

$$=\frac{1}{2}\int\sqrt{\left(\frac{1-\cos t}{1+\cos t}\right)}dt=\frac{1}{2}\int\tan(t/2)\,dt$$

$$=\frac{1}{2}\cdot 2\cdot\ln|\sec(t/2)|+c$$

$$=\ln\left|\sec\left(\frac{x^{2}+1}{2}\right)\right|+c$$

$$=\ln\left|\sec\left\{\frac{1}{2}(x^{2}+1)\right\}\right|+c$$

5. $I=\int(\sin(100x+x)\cdot(\sin x)^{99}dx$

$$=\int[(\sin(100x)`\cos x+\cos 100x\cdot\sin x)](\sin x)^{99}dx$$

$$=\int \underbrace{\sin(100x)}_{\text{I}}\ \underbrace{\cos x\cdot(\sin x)^{99}dx}_{\text{II}}$$

$$+\int\cos(100x)\cdot(\sin x)^{100}dx$$

$$=\frac{\sin(100x)(\sin x)^{100}}{100}-\frac{100}{100}$$

$$\int\cos(100x)(\sin x)^{100}dx+\int\cos(100x)(\sin x)^{100}dx$$

$$=\frac{\sin(100x)(\sin x)^{100}}{100}+C$$

6. Let $I=\int x(1-{}^{11}C_{1}x+{}^{11}C_{2}x^{2}-{}^{11}C_{3}x^{3}+\ldots-{}^{11}C_{11}x^{11})\,dx$

$$=\int x(1-x)^{11}dx$$

Put $1-x=t$

$\therefore\ dx=-dt$

Then, $I=-\int(1-t)t^{11}dt=-\frac{t^{12}}{12}+\frac{t^{13}}{13}+c$

$$=\frac{(1-x)^{13}}{13}-\frac{(1-x)^{12}}{12}+c$$

7. The integral, $I=\int\sin^{-1}\sqrt{\frac{x}{1+x}}dx$

Let $x=\tan^{2}\theta \Rightarrow dx=2\tan\theta\sec^{2}\theta d\theta$,

So, $I=\int\left(\sin^{-1}\sqrt{\frac{\tan^{2}\theta}{1+\tan^{2}\theta}}\right)2\tan\theta\sec^{2}\theta d\theta$

$$=2\int\left(\sin^{-1}\left(\frac{\tan\theta}{\sec\theta}\right)\right)\tan\theta\sec^{2}\theta d\theta$$

$$=2\int(\sin^{-1}(\sin\theta))\tan\theta\sec^{2}\theta d\theta$$

$$=2\int\theta\tan\theta\sec^{2}\theta d\theta$$

Using integration by parts, we get

$$I=2\left[\theta\int\tan\theta\sec^{2}\theta d\theta-\int\left(\frac{d}{d\theta}\theta\int\tan\theta\sec^{2}\theta d\theta\right)d\theta\right]$$

$$=2\left[\theta\frac{\tan^{2}\theta}{2}-\int\frac{\tan^{2}\theta}{2}d\theta\right]$$

$$=\theta\tan^{2}\theta-\int(\sec^{2}\theta-1)d\theta$$

$$=\theta\tan^{2}\theta-\tan\theta+\theta+C$$

$$=x\tan^{-1}\sqrt{x}-\sqrt{x}+\tan^{-1}\sqrt{x}+C$$

$$=(x+1)\tan^{-1}\sqrt{x}-\sqrt{x}+C$$

$$=A(x)\tan^{-1}\sqrt{x}+B(x)+C \qquad \text{[given]}$$

So, $(A(x),B(x))=(x+1,-\sqrt{x})$

Hence, option (c) is correct.

8. Divide above and below by $\sin^{2}x$, then

$$\int\frac{(3\operatorname{cosec}^{2}x+2\operatorname{cosec}x\cot x)}{(2\operatorname{cosec}x+3\cot x)^{2}}dx$$

Put $2\operatorname{cosec}x+3\cot x=t$

$\therefore\ (-2\operatorname{cosec}x\cot x-3\operatorname{cosec}^{2}x)\,dx=dt$

$$=\int-\frac{dt}{t^{2}}=\frac{1}{t}+c=\frac{1}{2\operatorname{cosec}x+3\cot x}+c$$

$$=\frac{\sin x}{(2+3\cos x)}+c$$

9. Let $I=\int\sec^{4/9}\theta\operatorname{cosec}^{14/9}\theta\,d\theta$

$$-\int\frac{d\theta}{(\sin\theta)^{14/9}(\cos\theta)^{4/9}}$$

$$=\int\frac{d\theta}{\left(\frac{\sin\theta}{\cos\theta}\right)^{14/9}\cdot(\cos\theta)^{4/9}\cdot(\cos\theta)^{14/9}}$$

$$=\int\frac{\sec^{2}\theta\,d\theta}{(\tan\theta)^{14/9}}$$

Put $\tan\theta = t$

$\therefore \sec^2\theta \, d\theta = dt$

Then, $I = \int \frac{dt}{t^{14/9}} = \int t^{-14/9} dt = \frac{t^{-5/9}}{-5/9} + c$

$= -\frac{9}{5}(\tan\theta)^{-5/9} + c$

10. Let $I = \int \frac{(\sqrt{x})^5}{(\sqrt{x})^7 + x^6} dx = \int \frac{dx}{(\sqrt{x})^2 + (\sqrt{x})^7}$

$= \int \frac{dx}{(\sqrt{x})^7 \left(\frac{1}{(\sqrt{x})^5} + 1\right)}$

Put $\frac{1}{(\sqrt{x})^5} + 1 = t$

$\therefore \quad dt = -5/2\,(x)^{-7/2} dx = -\frac{5}{2} \cdot \frac{1}{(\sqrt{x})^7} dx$

or $\frac{dx}{(\sqrt{x})^7} = -\frac{2}{5} dt$

$\therefore \quad I = -\frac{2}{5}\int \frac{dt}{t} = -\frac{2}{5}\ln|t| + c$

$= -\frac{2}{5}\ln\left|\frac{1}{(\sqrt{x})^5} + 1\right| + c$

$= \frac{2}{5}\ln\left(\frac{(\sqrt{x})^5}{(\sqrt{x})^5 + 1}\right) + c$

$= \frac{2}{5}\ln\left(\frac{x^{5/2}}{x^{5/2} + 1}\right) + c$

On comparing, we get

$\lambda = \frac{2}{5}$ and $a = \frac{5}{2}$

$\because$ AM > GM

$\therefore \quad \frac{\lambda + a}{2} > \sqrt{\lambda a} = 1$

$\therefore \quad \lambda + a > 2$

11. **Case I** If $x > 0$, then $|x| = x$

$\therefore \quad \int |x| \ln|x| \, dx$

$= \int x \ln x \, dx = \ln x \cdot \frac{x^2}{2} - \int \frac{1}{x} \cdot \frac{x^2}{2} dx$

$= \frac{x^2}{2} \cdot \ln x - \frac{x^2}{4} + c$

$= \frac{x^2}{2}\ln|x| - \frac{x^2}{4} + c$

Case II If $x < 0$, then $|x| = -x$

$\therefore \int |x| \ln|x| \, dx = -\int x \ln(-x)\, dx$

$= -\left\{\ln(-x) \cdot \frac{x^2}{2} - \frac{x^2}{4}\right\} + c$

$= -\frac{x^2}{2}\ln|x| + \frac{x^2}{4} + c$

Combining both cases, then we get

$\frac{1}{2} x|x| \ln|x| - \frac{1}{4} x|x| + c$

12. $I = \int \frac{dx}{x(x^{2007} + 1)} = \int \frac{x^{2007} + 1 - x^{2007}}{x(x^{2007} + 1)} dx$

$= \int \left(\frac{1}{x} - \frac{x^{2006}}{1 + x^{2007}}\right) dx$

$= \ln x - \frac{1}{2007}\ln(1 + x^{2007})$

$= \frac{\ln x^{2007} - \ln(1 + x^{2007})}{2007}$

$= \frac{1}{2007}\ln\left(\frac{x^{2007}}{1 + x^{2007}}\right) + C$

$p + q + r = 6021$

13. Since, $f''(x) = \sec^4 x + 4$

$f''(x) = (1 + \tan^2 x)\sec^2 x + 4$

$\therefore \quad f'(x) = \tan x + \frac{\tan^3 x}{3} + 4x + c$

$\because \quad f'(0) = 0$

$\therefore \quad 0 = c$

Then, $f'(x) = \tan x + \frac{\tan^3 x}{3} + 4x$

$= \tan x + \frac{1}{3}\tan x(\sec^2 x - 1) + 4x$

$f'(x) = \frac{2}{3}\tan x + \frac{1}{3}\tan x \sec^2 x + 4x$

$\therefore \quad f(x) = \frac{2}{3}\ln|\sec x| + \frac{\tan^2 x}{6} + 2x^2 + d$

but $f(0) = 0$

$d = 0$

Then, $f(x) = \frac{2}{3}\ln|\sec x| + \frac{1}{6}\tan^2 x + 2x^2$

14. Let $I = \int \frac{2x^{12} + 5x^9}{(x^5 + x^3 + 1)^3} dx$

$= \int \frac{\left(\frac{2}{x^3} + \frac{5}{x^6}\right)}{\left(1 + \frac{1}{x^2} + \frac{1}{x^5}\right)^3} dx$

Put $1 + \frac{1}{x^2} + \frac{1}{x^5} = t$

$\therefore \left(-\frac{2}{x^3} - \frac{5}{x^6}\right) dx = dt$

Then, $I = -\int \frac{dt}{t^3} = \frac{1}{2t^2} + c$

$= \frac{1}{2\left(1 + \frac{1}{x^2} + \frac{1}{x^5}\right)^2} + c = \frac{x^{10}}{2(x^5 + x^3 + 1)^2} + c$

15. $\int (\sqrt{1 + \sin 2x}) + \sqrt{1 - \sin 2x})\, dx$

$\int (\sqrt{\sin^2 x + \cos^2 x + 2\sin x \cos x}$

$+ \sqrt{\sin^2 x + \cos^2 x - 2\sin x \cos x})\, dx$

$\int ((\sin x + \cos x) + (\sin x - \cos x))\, dx \qquad \left[\because x \in \left(\frac{3\pi}{4}, \pi\right)\right]$

$\int \sin x\, dx = 2\int \sin x\, dx \Rightarrow 2\int (1 - \sin x) + c = -2\cos x + c$

16. We have,

$$I = \int (x^{7m-1} + x^{2m-1} + x^{m-1})(2x^{7m} + 7x^{2m} + 14x^m)^{1/m} dx$$

Now, put $2x^{7m} + 7x^{2m} + 14x^m = t^m$

$14m(x^{7m-1} + x^{2m-1} + x^{m-1})\, dx = mt^{m-1} dt$

$$\therefore \quad I = \frac{1}{14}\int t^{m-1}(t^m)^{\frac{1}{m}}\, dt = \frac{t^{m+1}}{14(m+1)} + c$$

$$= \frac{(2x^{7m} + 7x^{2m} + 14x^m)^{\frac{m+1}{m}}}{14(m+1)} + c$$

17. Primitive function means indefinite integral.

$\therefore$ Primitive function of $f(x)$

$$I = \int \frac{\sqrt{(a^2 - x^2)}}{x^4}\, dx \quad \text{(say)}$$

$$= \int \frac{x\sqrt{\left(\frac{a^2}{x^2} - 1\right)}}{x \cdot x^3}\, dx = \int \frac{1}{x^3} \cdot \sqrt{\left(\frac{a^2}{x^2} - 1\right)}\, dx$$

Put $\quad \frac{a^2}{x^2} - 1 = t^2$

$\therefore \quad -\frac{2a^2}{x^3}\, dx = 2t\, dt$

or $\quad \frac{1}{x^3}\, dx = -\frac{1}{a^2} t\, dt$

Then, $\quad I = -\frac{1}{a^2}\int t^2 dt$

$$= -\frac{1}{3a^2} t^3 + c$$

$$= -\frac{1}{3a^2}\left(\frac{a^2}{x^2} - 1\right)^{3/2} + c$$

$$= -\frac{(a^2 - x^2)^{3/2}}{3a^2 x^3} + c$$

18.
$$y = \int \frac{dx}{(3 + 5\sin x + 3\cos x)}$$

$$= \int \frac{\sec^2 (x/2)\, dx}{10\tan (x/2) + 6}$$

$$= \frac{1}{5}\ln|5\tan x/2 + 3| + c$$

$$= \frac{1}{5}\ln\left|\frac{5}{3}\tan (x/2) + 1\right| + c$$

Passes through (0, 0)

$\therefore \quad c = 0$

Then, $\quad y = \frac{1}{5}\ln\left|1 + \frac{5}{3}\tan\left(\frac{x}{2}\right)\right|$

19. Put $\quad x^x = t$

$\Rightarrow \quad x^x(1 + \log_e|x|)\, dx = dt$

$\therefore \quad \int x^x (1 + \log_e|x|)\, dx = t + c = x^x + c$

20. Let $I = \int \frac{\sin\frac{5x}{2}}{\sin\frac{x}{2}}\, dx = \int \frac{2\sin\frac{5x}{2}\cos\frac{x}{2}}{2\sin\frac{x}{2}\cos\frac{x}{2}} dx$

[multiplying by $2\cos\frac{x}{2}$ in numerator and denominator]

$$= \int \frac{\sin 3x + \sin 2x}{\sin x} dx$$

[$\because$ $2\sin A\cos B = \sin(A + B) + \sin(A - B)$ and $\sin 2A = 2\sin A\cos A$]

$$= \int \frac{(3\sin x - 4\sin^3 x) + 2\sin x\cos x}{\sin x}\, dx$$

[$\because$ $\sin 3x = 3\sin x - 4\sin^3 x$]

$$= \int (3 - 4\sin^2 x + 2\cos x)\, dx$$

$$= \int [3 - 2(1 - \cos 2x) + 2\cos x]\, dx$$

[$\because$ $2\sin^2 x = 1 - \cos 2x$]

$$= \int [3 - 2 + 2\cos 2x + 2\cos x]\, dx$$

$$= \int [1 + 2\cos 2x + 2\cos x]\, dx$$

$$= x + 2\sin x + \sin 2x + c$$

21. $\int \frac{dx}{x^2\left\{x^4\left(1 + \frac{1}{x^4}\right)\right\}^{3/4}} = \int \frac{dx}{x^5\left(1 + \frac{1}{x^4}\right)^{3/4}}$

Put $\quad 1 + \frac{1}{x^4} = t^4,$

$\Rightarrow \quad -\frac{4}{x^5}\, dx = 4t^3 dt$

$\Rightarrow \quad \int \frac{-t^3 dt}{t^3} = -t + c$

$$= -\left(1 + \frac{1}{x^4}\right)^{1/4} + c$$

22. $\because \quad y = \int x^2 \cdot e^{x^3} dx$

$$= \frac{1}{3}\int 3x^2 \cdot e^{x^3} dx$$

$$y = \frac{1}{3} \cdot e^{x^3} + c$$

It is passing through (0, 1), then

$$1 = \frac{1}{3} + c$$

$\Rightarrow \quad c = \frac{2}{3}$

Then, $\quad y = \frac{1}{3}e^{x^3} + \frac{2}{3}$

$\Rightarrow \quad e^{x^3} = (3y - 2)$

$\therefore \quad x^3 = \ln|3y - 2|$

or $\quad x = \sqrt[3]{\ln|3y - 2|}$

23. Let $I = \int \frac{xe^x}{(1+x)^2}\,dx$

$= \int e^x \left[\frac{(1+x)-1}{(1+x)^2}\right] dx$

$= \int e^x \left\{\frac{1}{1+x} - \frac{1}{(1+x)^2}\right\} dx = \frac{e^x}{(1+x)} + c$

$[\because \int e^x \{f(x) + f'(x)\}\,dx = e^x f(x) + c]$

24. Let $I = \int \frac{y^2 + \sqrt[3]{y^4} + \sqrt[6]{y^2}}{y\,(1+\sqrt[3]{y^2})}\,dy$

$I = \int \frac{y^2 + y^{4/3} + y^{1/3}}{y\,(1+y^{2/3})}\,dy$

$= \int \frac{y^{4/3}(y^{2/3}+1) + y^{1/3}}{y(1+y^{2/3})}\,dy$

$= \int \left(y^{1/3} + \frac{y^{-2/3}}{1+y^{2/3}}\right) dy$

$= \frac{y^{4/3}}{\frac{4}{3}} + \int \frac{y^{-2/3}}{1+(y^{1/3})^2}\,dy$

$= \frac{3}{4}\sqrt[3]{y^4} + 3\int \frac{d(y^{1/3})}{1+(y^{1/3})^2}$

$= \frac{3}{4}\sqrt[3]{y^4} + 3\tan^{-1} y^{1/3} + C$

$= \frac{3}{4}\sqrt[3]{y^4} + 3\tan^{-1} \sqrt[3]{y} + C$

25. $\int x^{-2/3}\cdot\{(x^{1/2})(x^{-1/2}+1)\}^{-5/3}\,dx$

$= \int x^{-2/3}\cdot x^{-5/6}\cdot(x^{-1/2}+1)^{-5/3}\,dx$

$= \int x^{-3/2}\cdot(x^{-1/2}+1)^{-5/3}\,dx$

Put $x^{-1/2} + 1 = t$

$\therefore$ $-\frac{1}{2}x^{-3/2}\,dx = dt$

Then, $-2\int t^{-5/3}\,dt = -2\cdot\frac{t^{-2/3}}{-2/3} + c$

$= 3(x^{-1/2}+1)^{-2/3} + c$

26. $\int \frac{\cos 4x - 1}{\cot x - \tan x}\,dx = \int \frac{\frac{-2\sin^2 2x}{\cos 2x}}{\sin x \cos x}\,dx$

$= -\int \frac{\sin 2x\,(1-\cos^2 2x)}{\cos 2x}\,dx$

Put $\cos 2x = t$

$\Rightarrow$ $-\sin 2x\,dx = \frac{dt}{2}$

$= \frac{1}{2}\int \frac{(1-t^2)}{t}\,dt$

$= \frac{1}{2}\ln|t| - \frac{t^2}{4} + c$

$= \frac{1}{2}\ln \cos 2x - \frac{\cos^2 2x}{4} + c$

27. The given integral,

$I = \int \frac{dx}{(x+4)^{8/7}\,(x-3)^{6/7}}$

$= \int \frac{\frac{dx}{(x+4)^2}}{\frac{(x+4)^{8/7}(x-3)^{6/7}}{(x+4)^2}} = \int \frac{\frac{dx}{(x+4)^2}}{\left(\frac{x-3}{x+4}\right)^{6/7}}$

Now, let $\frac{x-3}{x+4} = t^7$

$\Rightarrow$ $\frac{(x+4)-(x-3)}{(x+4)^2}\,dx = 7t^6 dt$

$\Rightarrow$ $\frac{7}{(x+4)^2}\,dx = 7t^6 dt$

So, $I = \int \frac{t^6 dt}{t^6} = \int dt = t + c$

$\therefore$ $I = \left(\frac{x-3}{x+4}\right)^{1/7} + c$

Hence, option (c) is correct.

28. $\int \frac{(2x-1)\cos\sqrt{(2x-1)^2+5}}{\sqrt{(2x-1)^2+5}}\,dx$

$\Rightarrow$ $(2x-1)^2 + 5 = t^2$

$\Rightarrow$ $2(2x-1)\,2dx = 2tdt \Rightarrow 2\sqrt{t^2-5}\,dx = tdt$

So, $\int \frac{\sqrt{t^2-5}\cos t}{2\sqrt{t^2-5}}\,dt = \frac{1}{2}\sin t + c = \frac{1}{2}\sin\sqrt{(2x-1)^2+5} + c$

29. Let $I = \int \frac{\cos^3 x + \cos^5 x}{(\sin^2 x + \sin^4 x)}\,dx$

$= \int \frac{\{(1-\sin^2 x) + (1-\sin^2 x)^2\}\cos x}{(\sin^2 x + \cos^4 x)}\,dx$

Put $\sin x = t$

$\therefore$ $\cos x\,dx = dt$

Then, $I = \int \frac{\{1-t^2+(1-t^2)^2\}\,dt}{(t^2+t^4)}$

$= \int \frac{(2-3t^2+t^4)}{(t^2+t^4)}\,dt$

$= \int \frac{(t^4+t^2)+(2-4t^2)}{(t^2+t^4)}\,dt$

$= \int 1\cdot dt + \int \frac{(2-4t^2)}{t^2\,(1+t^2)}\,dt$

$= \int 1\cdot dt + \int\left(\frac{2}{t^2} - \frac{6}{1+t^2}\right) dt$

$= t - \frac{2}{t} - 6\tan^{-1} t + c$

$= \sin x - 2(\sin x)^{-1} - 6\tan^{-1}(\sin x) + c$

30. $\int f(x)\sin x\cos x\,dx = \frac{1}{2(b^2-a^2)}\ln f(x) + c$

Differentiating both sides w.r.t. x, then

$f(x)\sin x\cos x = \frac{1}{2(b^2-a^2)}\cdot\frac{1}{f(x)}\cdot f'(x)$

$\Rightarrow \quad \dfrac{f'(x)}{\{f(x)\}^2} = 2(b^2 - a^2)\sin x \cos x$

$\Rightarrow \quad -\dfrac{f'(x)}{\{f(x)\}^2} = -2b^2 \sin x \cos x + 2a^2 \sin x \cos x$

Integrating both sides w.r.t. x, then

$$\frac{1}{f(x)} = b^2\cos^2 x + a^2 \sin^2 x + c$$

for $\quad c = 0,\ f(x) = \dfrac{1}{(a^2\sin^2 x + b^2\cos^2 x)}$

31. $\displaystyle\int \frac{\cos x}{\sin^3 x(1+\sin^6 x)^{2/3}}dx$

$= f(x)(1+\sin^6 x)^{1/\alpha} + C$ [given]

Let $\sin x = t \Rightarrow \cos x\, dx = dt$

$$\int \frac{\cos x}{\sin^3 x(1+\sin^6 x)^{2/3}}dx = \int \frac{1}{t^3(1+t^6)^{2/3}}dt$$

$$= \int \frac{1}{t^3 \times t^4\left(1+\frac{1}{t^6}\right)^{\frac{2}{3}}}dt = \int \frac{1}{t^7\left(1+\frac{1}{t^6}\right)^{\frac{2}{3}}}dt$$

Put $1+\dfrac{1}{t^6} = r^3 \Rightarrow -6t^{-7}\dfrac{dt}{dr} = 3r^2$

$\Rightarrow \quad \dfrac{dt}{t^7} = -\dfrac{1}{2}r^2\,dr$

$\therefore \displaystyle\int \frac{\cos x}{\sin^3 x(1+\sin^6 x)^{2/3}}dx = -\frac{1}{2}\int r^2\frac{dr}{r^2} = -\frac{1}{2}r + c$

$$= -\frac{1}{2}\left(\frac{\sin^6 x + 1}{\sin^6 x}\right)^{1/3} + c$$

$$\left[\because r = \left(1+\frac{1}{t^6}\right)^{\frac{1}{3}} \text{ and } t = \sin x\right]$$

$$= -\frac{1}{2\sin^2 x}(1+\sin^6 x)^{1/3} + c$$

$\therefore f(x) = -\dfrac{1}{2}\text{cosec}^2 x$ and $\lambda = 3$

Now, for $x = \dfrac{\pi}{3}$, $\text{cosec}\,\dfrac{\pi}{3} = \dfrac{2}{\sqrt{3}}$

$\therefore \lambda f\left(\dfrac{\pi}{3}\right) = -\dfrac{1}{2}\times\left(\dfrac{2}{\sqrt{3}}\right)^2 \times 3 = -2$

32. Let Let $I = \displaystyle\int \frac{(x^4 - x)^{1/4}}{x^5}dx$

$$= \int \frac{x\left(1-\frac{1}{x^3}\right)^{1/4}}{x^5}dx = \int \frac{\left(1-\frac{1}{x^3}\right)^{1/4}}{x^4}dx$$

Put $\quad 1 - \dfrac{1}{x^3} = t^4 \Rightarrow \dfrac{3}{x^4}dx = 4t^3 dt$

$\therefore \quad I = \dfrac{4}{3}\displaystyle\int_0^x t \cdot t^3 dt = \frac{4}{3}\cdot\left(\frac{t^5}{5}\right) + c$

$$= \frac{4}{15}\left(1-\frac{1}{x^3}\right)^{5/4} + c$$

Objective Questions Type II [One or more than one correct answer(s)]

1. $\displaystyle\int \frac{(x^2 - x + 1)e^x\,dx}{(x^2+1)^{3/2}}$

$$= \int e^x\left(\frac{1}{\sqrt{(1+x^2)}} - \frac{x}{(1+x^2)^{3/2}}\right)dx$$

$$= \frac{e^x}{\sqrt{(1+x^2)}} + c \quad \left[\int e^x(f(x) + f'(x))\,dx = e^x f(x) + c\right]$$

$\therefore \ f(x) = \dfrac{1}{\sqrt{(1+x^2)}}$

The graph of $f(x)$ is given below

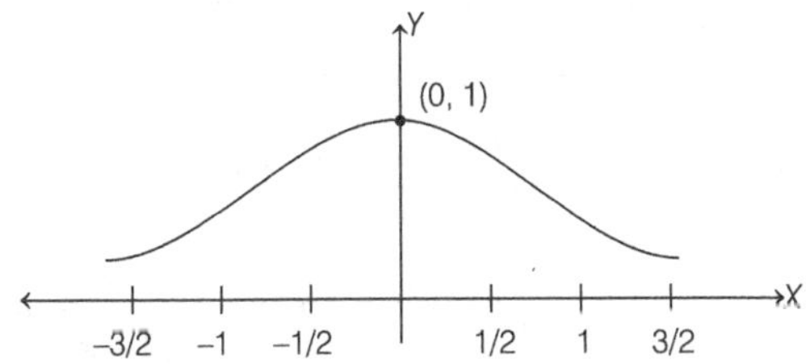

From the graph, $f(x)$ is even, bounded function and has the range (0, 1).

2. $J + K = \displaystyle\int \frac{1+\sin x + \cos x}{1 + \sin x + \cos x}dx$

$\Rightarrow \quad J + K = x + C \quad$ ...(i)

Again, $J - K = \displaystyle\int \frac{(\sin^2 x - \cos^2 x) + \sin x - \cos x}{1+\sin x + \cos x}dx$

$$= \int \frac{(\sin x - \cos x)(\sin x + \cos + 1)}{1 + \sin x + \cos x}$$

$J - K = -\cos x - \sin x + c \quad$ ...(ii)

Hence, $J = K - (\sin x + \cos x) + c$

Also, Eq. (i) + Eq. (ii)

$$2J = x - (\cos x + \sin x) + c$$

$$J = \frac{1}{2}[x - \sin x - \cos x] + c$$

and Eq. (i) – Eq. (ii)

$$2K = x + (\sin x + \cos x) + c$$

$$K = \frac{1}{2}(x + \sin x + \cos x) + c$$

From Eq. (i) $J = x - K + c$

3. Put $x^{3/2} = t, \therefore \dfrac{3}{2}x^{1/2}\,dx = dt$

$\therefore \quad \displaystyle\int \frac{x^{1/2}}{\sqrt{(1-x^3)}}dx = \frac{2}{3}\int\frac{dt}{\sqrt{(1-t^2)}} = \frac{2}{3}\sin^{-1} t + c$

$$= \frac{2}{3}\sin^{-1}(x^{3/2}) + c$$

$$= \frac{2}{3}(gof)(x) + c$$

$\Rightarrow \quad f(x) = x^{3/2},\ g(x) = \sin^{-1} x$

4. $\int \text{cosec } 2x \, dx = f\{g(x)\} + c$

$\Rightarrow \quad \text{cosec } 2x = f'\{g(x)\}g'(x)$

$\Rightarrow \quad \frac{1}{2\tan x} \times \sec^2 x = f'\{g(x)\}g'(x)$

$\Rightarrow \quad f(x) = \frac{1}{2x}$

$g'(x) = \sec^2 x$

Domain $\quad f(x) = (-\infty, \infty) - \{0\}, \; g(x) = \tan x + c$

Range $\quad g(x) = (-\infty, \infty)$

5. Let $\quad I = \int \frac{x e^x}{\sqrt{(1+e^x)}} dx = \int x \cdot \frac{e^x}{\sqrt{(1+e^x)}} dx$

$= x \cdot 2\sqrt{(1+e^x)} - \int 1 \cdot 2\sqrt{(1+e^x)}\, dx$

$= 2x\sqrt{(1+e^x)} - 2\int \sqrt{(1+e^x)}\, dx$

In second integral

Put $\quad 1 + e^x = t^2$

$\therefore \quad dx = \frac{2t\, dt}{t^2 - 1}$

Then,

$= 2x\sqrt{(1+e^x)} - 4\int \frac{t^2 - 1 + 1}{(t^2 - 1)} dt$

$= 2x\sqrt{(1+e^x)} - 4\int \left(1 + \frac{1}{t^2 - 1}\right) dt$

$= 2x\sqrt{(1+e^x)} - 4\left\{t + \frac{1}{2}\ln\left(\frac{t-1}{t+1}\right)\right\} + c$

$= 2x\sqrt{(1+e^x)} - 4\sqrt{(1+e^x)} - 2\ln\left(\frac{\sqrt{(1+e^x)} - 1}{\sqrt{(1+e^x)} + 1}\right) + c$

$= (2x - 4)\sqrt{(1+e^x)} - 2\ln\left(\frac{\sqrt{(1+e^x)} - 1}{\sqrt{(1+e^x)} + 1}\right) + c$

On comparing

$f(x) = 2x - 4, \; g(x) = \frac{\sqrt{(1+e^x)} - 1}{\sqrt{(1+e^x)} + 1}$

6. We have, $f'(x) = e^{(f(x) - g(x))} g'(x) \; \forall \; x \in R$

$\Rightarrow \quad f'(x) = \frac{e^{f(x)}}{e^{g(x)}} g'(x)$

$\Rightarrow \quad \frac{f'(x)}{e^{f(x)}} = \frac{g'(x)}{e^{g(x)}}$

$\Rightarrow \quad e^{-f(x)} f'(x) = e^{-g(x)} g'(x)$

On integrating both sides, we get

$e^{-f(x)} = e^{-g(x)} + c$

At $x = 1$

$e^{-f(1)} = e^{-g(1)} + c$

$e^{-1} = e^{-g(1)} + c \quad [\because f(1) = 1] \; \ldots(\text{i})$

At $x = 2$

$e^{-f(2)} = e^{-g(2)} + c$

$\Rightarrow \quad e^{-f(2)} = e^{-1} + c \quad [\because g(2) = 1] \; \ldots(\text{ii})$

From Eqs. (i) and (ii)

$e^{-f(2)} = 2e^{-1} - e^{-g(1)} \quad \ldots(\text{iii})$

$\Rightarrow \quad e^{-f(2)} > 2e^{-1}$

We know that, e^{-x} is decreasing

$\therefore \quad -f(2) < \log_e 2 - 1$

$f(2) > 1 - \log_e 2$

$\Rightarrow \quad e^{-g(1)} + e^{-f(2)} = 2e^{-1} \quad$ [from Eq. (iii)]

$\Rightarrow \quad e^{-g(1)} < 2e^{-1}$

$-g(1) < \log_e 2 - 1$

$\Rightarrow \quad g(1) > 1 - \log_e 2$

7. $\therefore \quad \lim_{n \to \infty} \tan(1/n) \ln(1/n)$

$= \lim_{n \to \infty} \frac{\tan(1/n)}{(1/n)} \cdot \frac{\ln(1/n)}{n}$

$= -\lim_{n \to \infty} \frac{\tan(1/n)}{(1/n)} \cdot \frac{\ln(n)}{(n)} = -1 \cdot \lim_{n \to \infty} \frac{1/n}{1} = 0$

Then, $\quad f(x) = e^0 = 1$

$\therefore \quad \int \frac{f(x)}{\sqrt[3]{(\sin^{11} x \cos x)}} dx = \int \frac{1}{\sin^{11/3} x \cos^{1/3} x} dx$

$= \int \sin^{-11/3} x \cdot \cos^{-1/3} x \, dx$

$= \int (\tan x)^{-11/3} \cos^{-4} x \, dx = \int (\tan x)^{-11/3} \cdot \sec^4 x \, dx$

$= \int (\tan x)^{-11/3} \cdot (1 + \tan^2 x) \cdot \sec^2 x \, dx$

$= \frac{(\tan x)^{\frac{-11}{3}+1}}{\left(\frac{-11}{3}+1\right)} + \frac{(\tan x)^{-2/3}}{(-2/3)} + c$

$= -\frac{3}{8}(\tan x)^{-8/3} - \frac{3}{2}(\tan x)^{-2/3} + c$

$\therefore \quad g(x) = -\frac{3}{8}(\tan x)^{-8/3} - \frac{3}{2}(\tan x)^{-2/3}$

$\therefore \quad g(\pi/4) = -\frac{3}{8} - \frac{3}{2} = -\frac{15}{8}$

and $g(x)$ is non differentiable at $\tan x = 0$

or $\quad x = n\pi, \; n \in I$

8. $\because \quad x \in [-1, 0)$ or $-1 \le x < 0$

For $\quad -1 \le x < 0$

$\cos^{-1}\sqrt{(1 - x^2)} = -\sin^{-1} x$

$\therefore \quad \int \{\cos^{-1} x + \cos^{-1}\sqrt{(1 - x^2)}\}\, dx$

$= \int (\cos^{-1} x - \sin^{-1} x)\, dx$

$= \int \left(\frac{\pi}{2} - 2\sin^{-1} x\right) dx$

$= \frac{\pi}{2} x - 2\left\{\sin^{-1} x \cdot x - \int \frac{x}{\sqrt{1 - x^2)}} dx\right\}$

$= \frac{\pi}{2} x - 2x\sin^{-1} x + 2\{-\sqrt{(1 - x^2)}\} + c$

On comparing, we get

$A = \frac{\pi}{2}, \; f(x) = -2x$

9. Let $I=\int_0^x \frac{(t-|t|)^2}{(1+t^2)}\,dt$

Case I $x>0$, then $0<t<x, |t|=t$

$\therefore \quad I=\int_0^x \frac{(t-t)^2}{1+t^2}\,dt=0$

Case II $x<0$, then $x<t<0 \Rightarrow |t|=-t$

$\therefore \quad I=\int_0^x \frac{4t^2}{1+t^2}\,dt=4\int_0^x\left(1-\frac{1}{1+t^2}\right)dt$

$=4(x-\tan^{-1}x)+c$

10. Let $I=\int \sin(\ln x)\,dx$

Put $\ln x=t$

$\therefore \quad x=e^t \Rightarrow dx=e^t\,dt$

Then, $I=\int e^t \sin t\,dt$

$=\frac{e^t}{2}(\sin t-\cos t)+c$

$=\frac{x}{2}\{\sin(\ln x)-\cos(\ln x)\}+c$

On comparing, we get

$f(x)=\frac{x}{2},\ g(x)=\ln x,\ h(x)=\ln x$

Option (a) $\lim_{x\to 2} f(x)=\lim_{x\to 2}\frac{x}{2}=\frac{2}{2}=1$

Option (b) $\lim_{x\to 1}\frac{g(x)}{h(x)}=\lim_{x\to 1}\frac{\ln x}{\ln x}=1$

Option (c) $g(e^3)=\ln e^3=3\ln e=3$

Option (d) $h(e^5)=\ln e^5=5\ln e=5$

Linked-Comprehension Type

PASSAGE 1

1. Let $I=\int \frac{dx}{(1+x)^{1/2}-(1+x)^{1/3}}$

Put $1+x=t^6$

$\Rightarrow \quad dx=6t^5dt$

Then, $I=\int \frac{6t^5dt}{(t^3-t^2)}=6\int \frac{t^3}{(t-1)}\,dt$

$=6\int \frac{(t^3-1)+1}{(t-1)}\,dt$

$=6\int\left(t^2+t+1+\frac{1}{t-1}\right)dt$

$=6\left\{\frac{t^3}{3}+\frac{t^2}{2}+t+\ln|t-1|\right\}+c$

$=2(1+x)^{1/2}+3(1+x)^{1/3}+6(1+x)^{1/6}$

$+6\ln|(1+x)^{1/6}-1|+c$

2. Let $I=\int \frac{(1+x^{1/2}-x^{2/3})}{(1+x^{1/3})}\,dx$

Put $x=t^6$

$\Rightarrow dx=6t^5dt$

Then, $I=\int \frac{(1+t^3-t^4)6t^5dt}{(1+t^2)}$

$=6\int \frac{(-t^9+t^8+t^5)}{(1+t^2)}\,dt$

$=6\left(-t^7+t^6+t^5-t^4+t^2-1+\frac{1}{t^2+1}\right)dt$

$=-\frac{6t^8}{8}+\frac{6t^7}{7}+\frac{6t^6}{6}-\frac{6t^5}{5}+\frac{6t^3}{3}-6t+6\tan^{-1}(t)+c$

$=-\frac{3}{4}x^{4/3}+\frac{6}{7}x^{7/6}+x-\frac{6}{5}x^{5/6}+2x^{1/2}$

$-6x^{1/6}+6\tan^{-1}(x^{1/6})+c$

3. Let $I=\int \frac{x^{\frac{1}{4}}}{x^{\frac{1}{2}}-1}\,dt$

Put $x=t^4 \Rightarrow dx=4t^3dt$

$\therefore \quad I=\int \frac{t\cdot 4t^3dt}{(t^2-1)}=4\int\left(\frac{t^4-1+1}{t^2-1}\right)dt$

$=4\int\left(t^2+1+\frac{1}{t^2-1}\right)dt$

$=4\left\{\frac{t^3}{3}+t+\frac{1}{2}\ln\left|\frac{t-1}{t+1}\right|\right\}+c$

$=\frac{4}{3}x^{3/4}+4x^{1/4}+2\ln\left|\frac{x^{1/4}-1}{x^{1/4}+1}\right|+c$

4. Let $I=\int \frac{x}{\sqrt[3]{(a+bx)}}\,dx$

Put $a+bx=t^3$

$\therefore \quad dx=\frac{1}{b}\cdot 3t^2dt$

Then, $I=\int \frac{(t^3-a)/b}{t}\cdot\frac{3t^2}{b}\,dt=\frac{3}{b^2}\int(t^4-at)\,dt$

$=\frac{3}{b^2}\left\{\frac{t^5}{5}-\frac{at^2}{2}\right\}+c$

$=\frac{3}{5b^2}(a+bx)^{5/3}-\frac{3a}{2b^2}(a+bx)^{2/3}+c$

5. Let $I=\int \sqrt{\left(\frac{x+2}{2x+3}\right)}\cdot\frac{1}{x}\,dx$

Put $\frac{x+2}{2x+3}=t^2 \Rightarrow x+2=2t^2x+3t^2$

$x=\left(\frac{3t^2-2}{1-2t^2}\right)$

$\therefore \quad dx=\left\{\frac{(1-2t^2)(6t)-(3t^2-2)(-4t)}{(1-2t^2)^2}\right\}dt=\frac{-2t\,dt}{(1-2t^2)^2}$

Then, $I=\int t\cdot\frac{-2t\,dt}{(1-2t^2)^2}\cdot\frac{(1-2t^2)}{(3t^2-2)}$

$$=-2\int\frac{t^2dt}{(1-2t^2)^2(3t^2-2)}$$

$$=-2\int\left(\frac{-1}{1-2t^2}-\frac{2}{3t^2-2}\right)dt$$

$$=2\int\frac{dt}{(1-2t^2)}+4\int\frac{dt}{(3t^2-2)}$$

$$=\frac{2}{2}\int\frac{dt}{\left(\frac{1}{\sqrt{2}}\right)^2-t^2}+\frac{4}{3}\int\frac{dt}{t^2-\left(\sqrt{\frac{2}{3}}\right)^2}$$

$$=\frac{1}{2\cdot\frac{1}{\sqrt{2}}}\ln\left|\frac{\frac{1}{\sqrt{2}}+t}{\frac{1}{\sqrt{2}}-t}\right|+\frac{4}{3\cdot 2}\cdot\frac{1}{\sqrt{\frac{2}{3}}}\ln\left|\frac{t-\sqrt{\frac{2}{3}}}{t+\sqrt{\frac{2}{3}}}\right|+c$$

$$=\frac{1}{\sqrt{2}}\ln\left|\frac{1+t\sqrt{2}}{1-t\sqrt{2}}\right|+\sqrt{\frac{2}{3}}\ln\left|\frac{\sqrt{3}t-\sqrt{2}}{\sqrt{3}t+\sqrt{2}}\right|+c$$

$$=\frac{1}{\sqrt{2}}\ln\left|\frac{\sqrt{2x+3}+\sqrt{2}\sqrt{x+2}}{\sqrt{2x+3}-\sqrt{2}\sqrt{x+2}}\right|$$

$$+\sqrt{\frac{2}{3}}\ln\left|\frac{\sqrt{3}\sqrt{(x+2)}-\sqrt{2}\sqrt{(2x+3)}}{\sqrt{3}\sqrt{x+2}+\sqrt{2}\sqrt{2x+3}}\right|+c$$

PASSAGE 2

1. Let $I=\int\sin^5x\,dx=\int\sin^4x\cdot\sin x\,dx$

$$=\sin^4x(-\cos x)-\int 4\sin^3x\cdot\cos x\cdot(-\cos x)dx$$

$$=-\sin^4x\cos x+4\int\sin^3x(1-\sin^2x)dx$$

$\Rightarrow$ $5I=-\sin^4x\cos x+4\int(1-\cos^2x)\sin x\,dx$

$$=-\sin^4x\cos x+4\int\cos^2x(-\sin x)dx+4\int\sin x\,dx$$

$$=-\sin^4x\cos x+\frac{4}{3}\cos^3x-4\cos x+c$$

or $I=-\frac{\sin^4x\cos x}{5}+\frac{4}{15}\cos^3x-\frac{4}{5}\cos x+\frac{c}{5}$

$$=-\frac{\sin^4x\cos x}{5}+\frac{4}{15}\cos x(1-\sin^2x)-\frac{4}{5}\cos x+c_1$$

$$=-\frac{\sin^4x\cos x}{5}-\frac{4}{15}\sin^2x\cos x-\frac{8}{15}\cos x+c$$

$\therefore$ $A=\frac{-4}{15},B=\frac{-8}{15}$

$\Rightarrow$ $A+B=\frac{-12}{15}=-\frac{4}{5}$

2. Let $I=\int\sec^6x\,dx=\int\sec^4x\cdot\sec^2x\,dx$

$$=\int(1+\tan^2x)^2\cdot\sec^2x\,dx$$

Put $\tan x=t$

$\therefore$ $\sec^2x\,dx=dt$

Then, $I=\int(1+t^2)^2dt$

$$=\int(t^4+2t^2+1)dt$$

$$=\frac{t^5}{5}+\frac{2t^3}{3}+t+D$$

$$=\frac{1}{5}\tan^5x+\frac{2}{3}\tan^3x+\tan x+D$$

Here, $B=\frac{2}{3},C=1$

$\therefore$ $B+C=\frac{5}{3}$

3. Let $I=\int\tan^6x\,dx=\int\tan^4x\cdot(\sec^2x-1)dx$

$$=\int\tan^4x\sec^2x\,dx-\int\tan^2x(\sec^2x-1)dx$$

$$=\int\tan^4x\sec^2x\,dx-\int\tan^2x\sec^2x\,dx+\int(\sec^2x-1)dx$$

$$=\frac{\tan^5x}{5}-\frac{\tan^3x}{3}+\tan x-x+D$$

Here, $A=-\frac{1}{3},B=1,C=-1$

$\therefore$ $A+B+C=-\frac{1}{3}$

4. Let $I_n=\int\cos^{n-1}x\cos x\,dx$

$$=\cos^{n-1}x\cdot\sin x-\int(n-1)\cos^{n-2}x\cdot(-\sin x)\sin x\,dx$$

$$=\cos^{n-1}x\cdot\sin x+(n-1)\int\cos^{n-2}x(1-\cos^2x)dx$$

$$=\cos^{n-1}x\cdot\sin x+(n-1)\int\cos^{n-2}x\,dx-(n-1)I_n$$

$\therefore$ $I_n=\frac{\cos^{n-1}x\sin x}{n}+\frac{(n-1)}{n}\int\cos^{n-2}x\,dx$

$\therefore$ $f(n)=\frac{n-1}{n}$

Then, $f(5)=\frac{4}{5}$

5. Let $I_n=\int\text{cosec}^{n-2}x\cdot\text{cosec}^2x\,dx$

$$=\text{cosec}^{n-2}x\cdot(-\cot x)$$

$$-\int(n-2)\text{cosec}^{n-3}x(-\text{cosec}\,x\cot x)(-\cot x)dx$$

$$=-\text{cosec}^{n-2}x\cot x-(n-2)\int\text{cosec}^{n-3}x\cdot\text{cosec}\,x(\text{cosec}^2x-1)dx]$$

$$=-\text{cosec}^{n-2}x\cdot\cot x-(n-2)I_n+(n-2)I_{n-2}$$

$\therefore$ $I_n=-\frac{\text{cosec}^{n-2}x\cot x}{(n-1)}+\frac{(n-2)}{(n-1)}I_{n-2}$

Here, $f(n)=\left(\frac{n-2}{n-1}\right)$

$\therefore$ $f(n+1)=\left(\frac{n-1}{n}\right)$

PASSAGE 3

1. $\because \quad I_n = \int \tan^{n-2}x \cdot (\sec^2 x - 1)dx$

$$= \frac{\tan^{n-1} x}{(n-1)} - I_{n-2}$$

$\therefore \quad \lambda = -1$

2. Let $\quad P = \sin^{m-1} x \cdot \cos^{n+1} x$

$$\therefore \quad \frac{dp}{dx} = \sin^{m-1} x(n+1)\cos^n x(-\sin x) + \cos^{n+1} x \cdot (m-1)\sin^{m-2} x \cdot \cos x$$

$$= -(n+1)\sin^m x \cos^n x + (m-1)\sin^{m-2} x \cos^{n+2} x$$

On integrating both sides, we get

$$(n+1)I_{m,n} = -\sin^{m-1} x \cos^{n+1} x + (m-1)I_{m-2,\, n+2}$$

or $\quad I_{m,n} = -\dfrac{\sin^{m-1} x \cos^{n+1} x}{(n+1)} + \dfrac{(m-1)}{(n+1)} I_{m-2,\, n+2}$

Here, $f(m, n) = \dfrac{(m-1)}{(n+1)}$

$\therefore \quad f(2, 3) = \dfrac{1}{4}$

3. Let $\quad P = x^{m+1}(a + bx^n)^p$

$$\therefore \quad \frac{dP}{dx} = x^{m+1} \cdot p(a + bx^n)^{p-1} \cdot (bnx^{n-1}) + (a + bx^n)^p \cdot (m+1)x^m$$

$$= bpnx^{m+n} \cdot (a + bx^n)^{p-1} + (m+1)x^m (a + bx^n)^p$$

On, integrating both sides, we get

$$P = bpnI_{m+n,\, p-1} + (m+1)I_{m,\, p}$$

or $\quad I_{m,p} = \dfrac{x^{m+1} \cdot (a + bx^n)^p}{(m+1)} - \dfrac{bnp}{(m+1)} I_{m+n,\, p-1}$

$\therefore \quad f(m, n, p, b) = \dfrac{bnp}{(m+1)}$

$\Rightarrow \quad f(1, 2, 3, 4) = \dfrac{2 \times 3 \times 4}{1+1} = 12$

4. $I_n = \int (x \sin^{n-1} x)\sin x dx$

$$= x \sin^{n-1} x(-\cos x) - \int \{x(n-1)\sin^{n-2} x \cdot \cos x + \sin^{n-1} x \cdot 1\}(-\cos x)dx$$

$$= -x \sin^{n-1} x \cos x + (n-1)\int x \sin^{n-2} x(1 - \sin^2 x)dx + \frac{\sin^n x}{n}\Bigg]$$

$n \quad I_n = -x \sin^{n-1} x \cos x + \dfrac{\sin^n x}{n} + (n-1)I_{n-2}$

or $\quad I_n = -\dfrac{x \sin^{n-1} x \cos x}{n} + \dfrac{\sin^n x}{n^2} + \dfrac{(n-1)}{n} I_{n-2}$

$\therefore f(n) = \dfrac{n-1}{n}$

5. $I_n = \int \sin^{n-1} x \cdot (e^{ax} \sin x)dx$

$$= \sin^{n-1} x \cdot \left(\frac{e^{ax}}{(1+a^2)}(a \sin x - \cos x)\right)$$

$$- \frac{1}{(1+a^2)}\int (n-1)\sin^{n-2} x \cos x[e^{ax}(a \sin x - \cos] dx$$

$$= \frac{e^{ax} \sin^{n-1} x(a \sin x - \cos x)}{(1+a^2)} - \frac{(n-1)}{(1+a^2)}\int (a \sin^{n-1} x \cos x - \sin^{n-2} x(1 - \sin^2 x))dx]$$

$$= \frac{e^{ax} \sin^{n-1} x(a \sin x - \cos x)}{(1+a^2)} - \frac{(n-1)a}{(1+a^2)} \cdot \frac{\sin^n x}{n} + \frac{(n-1)}{(1+a^2)} I_{n-2} - \frac{(n-1)}{(1+a^2)} I_n$$

or $I_n(n + a^2) = e^{ax} \sin^{n-1} x(a \sin x - \cos x) - (n-1)\dfrac{a \sin^n x}{n} + (n-1)I_{n-2}$

or $\quad I_n = \dfrac{e^{ax} \sin^{n-1} x(a \sin x - \cos x)}{(n + a^2)} - \dfrac{(n-1)a \sin^n x}{n(n+a^2)} + \dfrac{(n-1)}{(n+a^2)} I_{n-2}$

Here, $\quad A = \dfrac{(n-1)a}{n(n+a^2)},\ B = \dfrac{(n-1)}{(n+a^2)}$

$\therefore \quad A + B = \dfrac{(n-1)a + n(n-1)a}{n(n+a^2)} = \dfrac{a(n^2-1)}{n(n+a^2)}$

PASSAGE 4

We have, $6\int f(x)g(x)\, dx = x^6 + 3x^4 + 3x^2 + c$

Differentiating both sides w.r.t. x, we get

$$6f(x)g(x) = 6x^5 + 12x^3 + 6x$$

$\therefore \quad f(x)g(x) = x^5 + 2x^3 + x \quad \quad \text{...(i)}$

$$2\int \frac{g(x)dx}{f(x)} = x^2 + c$$

Differentiating both sides w.r.t. x, we get

$\dfrac{g(x)}{f(x)} = x \quad \quad \text{...(ii)}$

Multiplying Eqs. (i) and (ii), we get

$$(g(x))^2 = x^6 + 2x^4 + x^2 = (x^3 + x)^2$$

$\therefore \quad g(x) = x^3 + x$

1. Now, $\int (g(x) - f(x))\, dx = \int (x^3 + x - x^2 - 1)\, dx$

$$= \frac{x^4}{4} + \frac{x^2}{2} - \frac{x^3}{3} - x + c$$

2. $\lim\limits_{x \to 0} \dfrac{\log(f(x))}{g(x)} = \lim\limits_{x \to 0} \dfrac{\log(1 + x^2)}{x + x^3} \quad \left(\dfrac{0}{0} \text{ form}\right)$

$$= \lim_{x \to 0} \frac{\log(1 + x^2)}{x^2} \cdot \frac{x^2}{x + x^3}$$

$$= \lim_{x \to 0} \frac{x}{1 + x^2} = 0$$

Numerical Value Type

1. $\because\ 7x-10-x^2=(x-2)(5-x)$

Let $\quad x = 2\sin^2\theta + 5\cos^2\theta$

$\Rightarrow \quad x-2 = 3\cos^2\theta,\ 5-x = 3\sin^2\theta$

$$\sqrt{(7x-10-x^2)} = 3\sin\theta\cos\theta$$

and $\quad dx = -6\sin\theta\cos\theta\, d\theta$

Then, $\displaystyle\int \frac{x\,dx}{\sqrt{(7x-10-x^2)^3}} = \int \frac{(2\sin^2\theta + 5\cos^2\theta)}{27\sin^3\theta\cos^3\theta} \times -6\sin\theta\cos\theta\, d\theta$

$$= -\frac{2}{9}\int (2\sec^2\theta + 5\operatorname{cosec}^2\theta)\,d\theta$$

$$= -\frac{2}{9}\{2\tan\theta - 5\cot\theta\} + c$$

$$= \frac{2}{9}\{5\cot\theta - 2\tan\theta\} + c$$

$$= \frac{2}{9}\left\{5\sqrt{\left(\frac{x-2}{5-x}\right)} - 2\sqrt{\left(\frac{5-x}{x-2}\right)}\right\} + c$$

$$= \frac{2}{9}\left\{\frac{7x-20}{\sqrt{(7x-10-x^2)}}\right\} + c$$

$\therefore \quad \lambda = \dfrac{2}{9}$

Then, $\quad 1800\,\lambda = 1800 \times \dfrac{2}{9} = 400$

2. Let $I = \displaystyle\int \frac{dx}{1+\sin x} = \int \frac{dx}{1+\left(\dfrac{2\tan x/2}{1+\tan^2 x/2}\right)}$

$$= \int \frac{\sec^2 x/2\, dx}{\left(\tan\dfrac{x}{2}+1\right)^2}$$

Put $\quad \tan\dfrac{x}{2} + 1 = t$

$\therefore \quad \dfrac{1}{2}\sec^2\dfrac{x}{2}\,dx = dt \ \Rightarrow\ \sec^2\dfrac{x}{2}\,dx = 2dt$

$\therefore \quad I = \displaystyle\int \frac{2dt}{t^2} = \frac{-2}{t} + c = -\frac{2}{(1+\tan x/2)} + c$

$$= 1 - \frac{2}{(1+\tan x/2)} + c - 1$$

$$= \left(\frac{\tan\left(\frac{x}{2}\right)-1}{\tan\left(\frac{x}{2}\right)+1}\right) + c - 1$$

$$= \tan\left(\frac{\pi}{2} - \frac{\pi}{4}\right) + c - 1$$

Then, we get $a = -\dfrac{\pi}{4}$, $b = c-1 =$ an arbitrary constant.

$\therefore \quad \dfrac{-204.8}{\pi}a = \dfrac{-204.8}{\pi} \times \dfrac{-\pi}{4} = 51.2$

3. Let $I = \displaystyle\int (x^9 + x^6 + x^3)(2x^6 + 3x^3 + 6)^{1/3}dx$

$$= \int x^3\,(x^6 + x^3 + 1)(2x^6 + 3x^3 + 6)^{1/3}dx$$

$$= \int x^2\,(x^6 + x^3 + 1)(2x^9 + 3x^6 + 6x^3)^{1/3}dx$$

Put $\quad 2x^9 + 3x^6 + 6x^3 = t$

$\therefore \quad 18\,(x^8 + x^5 + x^2)\,dx = dt$

$$x^2\,(x^6 + x^3 + 1)\,dx = \frac{dt}{18}$$

$\therefore \quad I = \dfrac{1}{18}\displaystyle\int t^{1/3}dt = \frac{1}{18}\cdot\frac{t^{4/3}}{4/3} + c$

$$I = \frac{1}{24}\cdot(2x^9 + 3x^6 + 6x^3)^{4/3} + c$$

$\therefore \quad a = 24$

Then, $\quad 1.2a = 1.2 \times 24 = 28.8$

4. Let $I = \displaystyle\int (\sqrt{\tan x} + \sqrt{\cot x})\,dx = \int \frac{(\sin x + \cos x)}{\sqrt{(\sin x \cos x}}\,dx$

Put $\sin x - \cos x = t \Rightarrow 1 - \sin 2x = t^2$

$\therefore$ $(\cos x + \sin x)\,dx = dt$

Then, $I = \displaystyle\int \frac{dt}{\sqrt{\left(\frac{1-t^2}{2}\right)}} = \sqrt{2}\sin^{-1}t = \sqrt{2}\tan^{-1}\left(\frac{t}{\sqrt{1-t^2}}\right) + c$

$$= \sqrt{2}\tan^{-1}\left(\frac{\sin x - \cos x}{\sqrt{\sin 2x}}\right) + c$$

$$= \sqrt{2}\tan^{-1}\left(\frac{\tan x - 1}{\sqrt{2\tan x}}\right) + c$$

We get $a = \sqrt{2}$, $b = 2$

Then, $a^4 + b^5 = 4 + 32 = 36$

5. Let $I = \displaystyle\int \frac{\ln x}{x^3}\,dx = \int (\ln x)\cdot\left(\frac{1}{x^3}\right)dx$

$$= (\ln x)\cdot\left(-\frac{1}{2x^2}\right) - \int\left(\frac{1}{x}\right)\cdot\left(-\frac{1}{2x^2}\right)dx$$

$$= -\frac{\ln x}{2x^2} + \frac{1}{2}\int\frac{dx}{x^3} = -\frac{\ln x}{2x^2} - \frac{1}{4x^2} + c$$

$\therefore \quad a = 2,\ b = 4$

$\Rightarrow \quad a^5 + b^4 = 2^5 + 4^4 = 32 + 256 = 288$

6. Let $\ I = \displaystyle\int\left(\frac{\cos 8x - \cos 7x}{1 + 2\cos 5x}\right)dx$

$$= \int \frac{2\sin\left(\frac{15x}{2}\right)\sin\left(\frac{-x}{2}\right)dx}{1 + 2\left(1 - 2\sin^2\left(\frac{5x}{2}\right)\right)}$$

$$= -\int \frac{2\sin\left(\frac{15x}{2}\right)\sin\left(\frac{x}{2}\right)}{\left(3 - 4\sin^2\frac{5x}{2}\right)}\,dx$$

$$= -\int \frac{2\sin\left(\frac{15x}{2}\right)\sin\left(\frac{x}{2}\right)}{\frac{\sin\left(\frac{15x}{2}\right)}{\sin\left(\frac{5x}{2}\right)}}dx = -\int 2\sin\left(\frac{5x}{2}\right)\sin\left(\frac{x}{2}\right)dx$$

$$= -\int(\cos 2x - \cos 3x)dx = -\frac{\sin 2x}{2} + \frac{\sin 3x}{3} + c$$

$\therefore \quad a = 3$ and $b = 2$

$\therefore \quad (a^b)^4 = (3^2)^4 = 3^8 = 6561$

7. Let $I = \int \sin 4x \cdot e^{\tan^2 x}dx$

$$= \int 2\sin 2x\cos 2x \cdot e^{\tan^2 x}dx$$

$$= 4\int \sin x\cos x \cdot \left(\frac{1-\tan^2 x}{1+\tan^2 x}\right)\cdot e^{\tan^2 x}dx$$

$$= 4\int \tan x \sec^2 x \cos^6 x\,(1-\tan^2 x)\cdot e^{\tan^2 x}dx$$

Put $\tan^2 x = t$

$2\tan x\sec^2 x\,dx = dt$

Then, $I = 2\int \frac{(1-t)e^t dt}{(1+t)^3} = -2\int \frac{\{(t+1)-2\}e^t dt}{(t+1)^3}$

$$= -2\int e^t\left(\frac{1}{(1+t)^2} - \frac{2}{(1+t)^3}\right)dt = -\frac{2e^t}{(1+t)^2} + c$$

$$= -2\cos^4 x \cdot e^{\tan^2 x} + c$$

$\therefore \quad a = -2,\ b = 4$, then $a^{2b} = (-2)^8 = 256$

Single Integer Type Questions

1. $\int \frac{2x+3}{(x^2+3x)(x^2+3x+2)+1}dx$

Put $x^2 + 3x = t \Rightarrow (2x+3)\,dx = dt$

$$\int \frac{dt}{t(t+2)+1};\ \int \frac{dt}{(t+1)^2} = C - \frac{1}{t+1} = C - \frac{1}{x^2+3x+1}$$

$\Rightarrow \quad a = 1,\ b = 3,\ c = 1$

$\Rightarrow \quad a + b + c = 5$

2. Let $I = \int \frac{2x^{12}+5x^9}{(1+x^3+x^5)^3}dx = \int \frac{2x^{12}+5x^9}{\left[x^5\left(1+\frac{1}{x^2}+\frac{1}{x^5}\right)\right]^3}dx$

$$= \int \frac{2x^{12}+5x^9}{x^{15}\left(1+\frac{1}{x^2}+\frac{1}{x^5}\right)^3}dx = \int \frac{2x^{-3}+5x^{-6}}{\left(1+\frac{1}{x^2}+\frac{1}{x^5}\right)^3}dx$$

Put $1 + \frac{1}{x^2} + \frac{1}{x^5} = t$

$$\Rightarrow \left(-\frac{2}{x^3} - \frac{5}{x^6}\right)dx = dt$$

$$\therefore \quad I = -\int \frac{dt}{t^3} = \frac{1}{2t^2} + c$$

$$= \frac{1}{2\left(1+\frac{1}{x^2}+\frac{1}{x^5}\right)^2} + c$$

$$= \frac{x^{10}}{2(x^5+x^3+1)^2} + c$$

$\therefore m = 10,\ l = 2,\ r = 2$

$$\therefore \quad \frac{m-l}{r} = \frac{10-2}{2} = 4$$

3. $\because\ g(x) = \int x^{27}(1+x+x^2)^6(6x^2+5x+4)\,dx$

$$= \int (x^4+x^5+x^6)^6(6x^5+5x^4+4x^3)\,dx$$

Put $x^4 + x^5 + x^6 = t$

$\therefore \quad (4x^3+5x^4+6x^5)dx = dt$

$$\Rightarrow \quad g(x) = \int t^6 dt = \frac{t^7}{7} + c$$

$$= \frac{(x^4+x^5+x^6)^7}{7} + c$$

Given, $g(0) = 0 \Rightarrow 0 = 0 + c$

$\therefore \quad c = 0$

Then, $g(x) = \frac{(x^4+x^5+x^6)^7}{7} \Rightarrow g(1) = \frac{3^7}{7}$ and $g(-1) = \frac{1}{7}$

$\therefore \quad \frac{g(1)}{g(-1)} = 3^7 = p^\lambda$ (given)

$\therefore \quad p = 3,\ \lambda = 7$

Hence, $|\lambda - p| = 4$

4. $\int \frac{(x^2-1)dx}{(x^4+3x^2+1)\tan^{-1}\left(x+\frac{1}{x}\right)} + \int \frac{dx}{x^4+3x^2+1}$

$$= \int \frac{\left(1-\frac{1}{x^2}\right)dx}{\left(\left(x+\frac{1}{x}\right)^2+1\right)\tan^{-1}\left(x+\frac{1}{x}\right)} + \frac{1}{2}\int \frac{(x^2+1)-(x^2-1)dx}{x^4+3x^2+1}$$

Put $\tan^{-1}\left(x+\frac{1}{x}\right) = t$

$$= \int \frac{dt}{t} + \frac{1}{2}\int \frac{\left(1+\frac{1}{x^2}\right)dx}{\left(x-\frac{1}{x}\right)^2+5} - \frac{1}{2}\int \frac{\left(1-\frac{1}{x^2}\right)dx}{\left(x+\frac{1}{x}\right)^2+1}$$

Put $x - \frac{1}{x} = y,\ x + \frac{1}{x} = z$

$$= \log_e t + \frac{1}{2}\int \frac{dy}{y^2+5} - \frac{1}{2}\int \frac{dz}{z^2+1}$$

$$= \log_e \tan^{-1}\left(x+\frac{1}{x}\right) + \frac{1}{2\sqrt{5}}\tan^{-1}\left(\frac{x^2-1}{\sqrt{5}x}\right) - \frac{1}{2}\tan^{-1}\left(\frac{x^2+1}{x}\right) + c$$

$\Rightarrow \quad \alpha = 1,\ \beta = \frac{1}{2\sqrt{5}},\ \gamma = \frac{1}{\sqrt{5}},\ \delta = \frac{-1}{2}$

or $\quad \alpha = 1,\ \beta = \frac{-1}{2\sqrt{5}},\ \gamma = \frac{-1}{\sqrt{5}},\ \delta = \frac{-1}{2}$

$$\therefore\ 10(\alpha + \beta\gamma + \delta) = 10\left(1 + \frac{1}{10} - \frac{1}{2}\right) = 6$$

5. $f(x)=\int \frac{(5x^8+7x^6)dx}{x^{14}(x^{-5}+x^{-7}+2)^2}$

Let $x^{-5}+x^{-7}+2=t$

$$(-5x^{-6}-7x^{-8})dx = dt$$

$\Rightarrow$ $$f(x)=\int -\frac{dt}{t^2}=\frac{1}{t}+c$$

$\Rightarrow$ $$f(x)=\frac{x^7}{x^2+1+2x^7} \Rightarrow f(1)=\frac{1}{4}$$

6. Given, $f(x+y)=f(x)f'(y)+f'(x)f(y), \forall x, y \in R$

and $f(0)=1$

Put $x=y=0$, we get

$$f(0)=f(0)f'(0)+f'(0)f(0)$$

$\Rightarrow$ $1=2f'(0)$

$\Rightarrow$ $f'(0)=\frac{1}{2}$

Put $x=x$ and $y=0$, we get

$$f(x)=f(x)f'(0)+f'(x)f(0)$$

$\Rightarrow$ $$f(x)=\frac{1}{2}f(x)+f'(x)$$

$\Rightarrow$ $$f'(x)=\frac{1}{2}f(x) \Rightarrow \frac{f'(x)}{f(x)}=\frac{1}{2}$$

On integrating, we get

$$\log f(x)=\frac{1}{2}x+C$$

$\Rightarrow$ $f(x)=Ae^{\frac{1}{2}x}$, where $e^C=A$

If $f(0)=1$, then $A=1$

Hence, $f(x)=e^{\frac{1}{2}x}$

$\Rightarrow$ $\log_e f(x)=\frac{1}{2}x \Rightarrow \log_e f(4)=\frac{1}{2}\times 4=2$

Matrix-Match Type

1. **(A)** $\int \frac{dx}{x(x^n+1)}=\int \frac{dx}{x^{n+1}\left(1+\frac{1}{x^n}\right)}$

Put $1+\frac{1}{x^n}=t$; $\therefore \frac{-n}{x^{n+1}}dx=dt$

Then, $\frac{dx}{x(x^n+1)}=-\frac{1}{n}\int\frac{dt}{t}=-\frac{1}{n}\ln|t|$

$$=-\frac{1}{n}\ln\left|1+\frac{1}{x^n}\right|+c \text{ (S)}$$

$$=-\frac{1}{n}\{\ln|1+x^n|-n\ln|x|\}+c$$

$$=\ln|x|-\frac{1}{n}\ln|1+x^n|+c \text{ (Q)}$$

(B) $\int \frac{dx}{x^2(x^n+1)^{(n-1)/n}}=\int \frac{dx}{x^2\cdot x^{n-1}\cdot\left(1+\frac{1}{x^n}\right)^{(n-1)/n}}$

$$=\int \frac{dx}{x^{n+1}\left(1+\frac{1}{x^n}\right)^{(n-1)/n}}$$

Put $1+\frac{1}{x^n}=t \Rightarrow \frac{-n}{x^{n+1}}dx=dt$

Then, $\int\frac{dx}{x^2(x^n+1)^{(n-1)/n}}=-\frac{1}{n}\int\frac{dt}{t^{(n-1)/n}}$

$$=-\frac{1}{n}\int t^{-(n-1)/n}dt$$

$$=\frac{-1}{n}\cdot\frac{t^{\frac{-(n-1)}{n}+1}}{\frac{-(n-1)}{n}+1}+c$$

$$=-\left(1+\frac{1}{x^n}\right)^{1/n}+c$$

$$=-(1+x^{-n})^{1/n}+c \text{ (T)}$$

(C) $\int \frac{dx}{x^n(1+x^n)^{1/n}}=\int \frac{dx}{x^{n+1}\left(\frac{1}{x^n}+1\right)^{1/n}}$

Put $\frac{1}{x^n}+1=t$

$\therefore$ $-\frac{n}{x^{n+1}}dx=dt$

Then, $\int\frac{dx}{x^n(1+x^n)^{1/n}}=-\frac{1}{n}\int\frac{dt}{t^{1/n}}=-\frac{1}{n}\int t^{-1/n}dt$

$$=-\frac{1}{n}\cdot\frac{t^{-\frac{1}{n}+1}}{\left(-\frac{1}{n}+1\right)}+c$$

$$=\frac{1}{(1-n)}t^{1-1/n}+c$$

$$=\frac{1}{(1-n)}(x^{-n}+1)^{1-1/n}+c \text{ (P)}$$

$$=\frac{1}{(1-n)}\left(\frac{(1+x^n)^{1-1/n}}{x^{n-1}}\right)+c$$

$$=\frac{x^{1-n}}{(1-n)}(1+x^n)^{1-1/n}+c \text{ (R)}$$

2. **(A)** Let $I=\int(\tan x)^{1/3}dx$

Put $\tan x=t^3$

$\therefore$ $\sec^2 x\,dx=3t^2dt \Rightarrow dx=\frac{3t^2dt}{(1+t^6)}$

$\therefore$ $$I=3\int\frac{t^3dt}{(1+t^6)}=3\int\frac{t^2\cdot t\,dt}{1+(t^2)^3}$$

Put $t^2=z \Rightarrow 2tdt=dz$

Then, $$I=\frac{3}{2}\int\frac{zdz}{1+z^3}=\frac{3}{2}\int\frac{zdz}{(1+z)(1-z+z^2)}$$

$$=\frac{3}{2}\int\frac{-\frac{1}{3}}{(1+z)}dz+\frac{3}{2}\int\frac{\frac{(z+1)}{3}}{(1-z+z^2)}dz$$

$$=-\frac{1}{2}\ln(1+z)+\frac{1}{2}\int\frac{\frac{1}{2}(-1+2z)+\frac{3}{2}}{(1-z+z^2)}dz$$

$$= -\frac{1}{2}\ln(1+z) + \frac{1}{4}\ln(1-z+z^2) + \frac{3}{4}\int\frac{dz}{1-z-z^2}$$

$$= -\frac{1}{2}\ln(1+z) + \frac{1}{4}\ln(1-z+z^2) + \frac{3}{4}\int\frac{dz}{\left(z-\frac{1}{2}\right)^2+\left(\frac{\sqrt{3}}{2}\right)^2}$$

$$= \frac{1}{4}\ln\left(\frac{1-z+z^2}{(1+z)^2}\right) + \frac{3}{4}\cdot\frac{2}{\sqrt{3}}\tan^{-1}\left(\frac{2z-1}{\sqrt{3}}\right) + c$$

$$= \frac{1}{4}\ln\left(\frac{t^4-t^2+1}{(t^2+1)^2}\right) + \frac{\sqrt{3}}{2}\tan^{-1}\left(\frac{2t^2-1}{\sqrt{3}}\right) + c$$

$\therefore \quad A = \frac{1}{4}$ (P); $B = \frac{\sqrt{3}}{2}$ (S)

(B) Let $I = \int\frac{(\sin x + \sin^3 x)\,dx}{\cos 2x}$

$$= \int\frac{(1+\sin^2 x)\sin x\,dx}{(2\cos^2 x - 1)} = \int\frac{(2-\cos^2 x)\sin x}{(\sqrt{2}\cos x)^2 - 1}dx$$

Put $\quad \sqrt{2}\cos x = t$

$\therefore \quad \sin x\,dx = -\frac{dt}{\sqrt{2}}$

Then, $\quad I = \int\frac{\left(2-\frac{t^2}{2}\right)\left(-\frac{dt}{\sqrt{2}}\right)}{(t^2-1)}$

$$= \frac{1}{2\sqrt{2}}\int\frac{(t^2-4)}{(t^2-1)}dt$$

$$= \frac{1}{2\sqrt{2}}\int\left(1-\frac{3}{t^2-1}\right)dt$$

$$= \frac{1}{2\sqrt{2}}\left\{t - \frac{3}{2}\ln\left|\frac{t-1}{t+1}\right|\right\} + c$$

$$= \frac{1}{2\sqrt{2}}(\sqrt{2}\cos x) - \frac{3}{4\sqrt{2}}\ln\left|\frac{\sqrt{2}\cos x - 1}{\sqrt{2}\cos x + 1}\right| + c$$

$$= \frac{1}{2}\cos x - \frac{3}{4\sqrt{2}}\ln\left|\frac{\sqrt{2}\cos x - 1}{\sqrt{2}\cos x + 1}\right| + c$$

$\therefore \quad A = \frac{1}{2}$ (R), $B = \frac{3}{4\sqrt{2}}$ (T)

(C) Let $I = \int\frac{dx}{(x^2+1)(x^2+4)}$

$$= \frac{1}{3}\int\left(\frac{1}{x^2+1} - \frac{1}{x^2+4}\right)dx$$

$$= \frac{1}{3}\left\{\tan^{-1}x - \frac{1}{2}\tan^{-1}\left(\frac{x}{2}\right)\right\} + c$$

$$= \frac{1}{3}\tan^{-1}x - \frac{1}{6}\tan^{-1}\left(\frac{x}{2}\right) + c$$

$\therefore \quad A = \frac{1}{3}$ (Q), $B = -\frac{1}{6}$

3. **(A)** Let $I = \int\frac{2^x}{\sqrt{1-4^x}}dx$

Putting $2^x = t$, $2^x\log 2\,dx = dt$, we get

$$I = \frac{1}{\log 2}\int\frac{1}{\sqrt{1-t^2}}dt$$

$$I = \frac{1}{\log 2}\sin^{-1}\left(\frac{t}{1}\right) + c = \frac{1}{\log 2}\sin^{-1}(2^x) + c$$

$\therefore \quad k = \frac{1}{\log 2} = 3.32$

(B) $I = \int\frac{dx}{(\sqrt{x})^2 + (\sqrt{x})^7} = \int\frac{dx}{(\sqrt{x})^7\left(1+\frac{1}{(\sqrt{x})^5}\right)}$

Put $\frac{1}{(\sqrt{x})^5} = y$, $\frac{dy}{dx} = -\frac{5}{2(\sqrt{x})^7}$

$\therefore I = \int\frac{-2dy}{5(1+y)} = -\frac{2}{5}\ln|1+y| + c$

$$= \frac{2}{5}\ln\left(\frac{1}{1+\frac{1}{(\sqrt{x})^5}}\right) + c$$

$\therefore a = \frac{2}{5}$ and $k = \frac{5}{2}$

(C) $\int\frac{x^4+1}{x(x^2+1)^2}dx = K\ln|x| + \frac{m}{1+x^2} + n$

$$\frac{x^4+1}{x(x^2+1)^2} = \frac{A}{x} + \frac{Bx+C}{(x^2+1)^2} = \frac{A(x^2+1)^2 + (Bx+C)^x}{x(x^2+1)^2}$$

$\Rightarrow x^4 + 1 = A(x^4 + 2x^2 + 1) + Bx^2 + C\,x$

$\Rightarrow A = 1,\ 2A + B = 0,\ C = 0$

$B = -2A = -2$

$\int\left[\frac{1}{x} - \frac{2x}{(x^2+1)^2}\right]dx = \int\frac{1}{x} - \int\frac{dt}{t^2}$, where $t = x^2 + 1$

$$= \ln|x| + \frac{1}{t} + c = \ln|x| + \frac{1}{x^2+t} + c$$

$\therefore k = 1,\ m = 1,\ n = C$

(D) $I = \int\frac{dx}{5 + 4\cos x}$

$$= \int\frac{dx}{5\left(\sin^2\frac{x}{2} + \cos^2\frac{x}{2}\right) + 4\left(\cos^2\frac{x}{2} - \sin^2\frac{x}{2}\right)}$$

$$= \int\frac{dx}{9\cos^2\frac{x}{2} + \sin^2\frac{x}{2}} = \int\frac{\sec^2\frac{x}{2}}{9 + \tan^2\frac{x}{2}}dx$$

Let $t = \tan\frac{x}{2}$ or $2dt = \sec^2\frac{x}{2}dx$

$\therefore \quad I = \int\frac{2\,dt}{9+t^2} = \frac{2}{3}\tan^{-1}\left(\frac{t}{3}\right) + c$

$$= \frac{2}{3}\tan^{-1}\left(\frac{\tan\left(\frac{x}{2}\right)}{3}\right) + c$$

$\therefore \quad k = \frac{2}{3},\ m = \frac{1}{3}$

18

Definite Integration

Objective Questions Type I [Only one correct answer]

In each of the questions below, four choices are given of which only one is correct. You have to select the correct answer which is the most appropriate.

1. $\int_{-\pi}^{\pi} |\pi - |x|| \, dx$ is equal to

[JEE Main 2020, 3 September]

(a) $\sqrt{2}\pi^2$ (b) $2\pi^2$

(c) π^2 (d) $\frac{\pi^2}{2}$

2. If $f(x) = \int_0^x \frac{dt}{\{f(t)\}^2}$ and $\int_0^2 \frac{dt}{\{f(t)\}^2} = \sqrt[3]{6}$, $f(9)$ equals

(a) 0 (b) 1

(c) 2 (d) None of these

3. If $A = \int_0^{\pi} \frac{\cos x}{(x+2)^2} dx$, then $\int_0^{\pi/2} \frac{\sin 2x}{(x+1)} dx$ is equal to

(a) $A - \frac{1}{2} - \frac{1}{\pi + 2}$ (b) $\frac{1}{2} + \frac{1}{\pi + 2} - A$

(c) $\frac{1}{\pi + 2} - A$ (d) $1 + \frac{1}{\pi + 2} - A$

4. Consider the integral $I = \int_0^{10} \frac{[x]e^{[x]}}{e^{x-1}} dx$ where $[x]$ denotes the greatest integer less than or equal to x. Then, the value of I is equal to

[JEE Main 2021, 16 March]

(a) $9(e-1)$ (b) $45(e+1)$

(c) $45(e-1)$ (d) $9(e+1)$

5. If $I_1 = \int_0^{\pi/2} \cos(\sin x) \, dx$; $I_2 = \int_0^{\pi/2} \sin(\cos x) \, dx$ and $I_3 = \int_0^{\pi/2} \cos x \, dx$, then

(a) $I_1 > I_2 > I_3$ (b) $I_2 > I_3 > I_1$

(c) $I_3 > I_1 > I_2$ (d) $I_1 > I_3 > I_2$

6. The value of $\lim_{n \to \infty} \frac{(\Sigma n^2)(\Sigma n^3)}{(\Sigma n^6)}$ is

(a) 1 (b) 3/2

(c) 5/6 (d) 7/12

7. The value of the constant $a > 0$ such that $\int_0^a [\tan^{-1}\sqrt{x}] \, dx = \int_0^a [\cot^{-1}\sqrt{x}] \, dx$, where $[\cdot]$ denotes the greatest integer function, is

(a) $\frac{(3 + \cos 4)}{(1 - \cos 4)}$ (b) $\frac{(3 - \cos 4)}{(1 + \cos 4)}$

(c) $\frac{2(3 + \cos 4)}{(1 - \cos 4)}$ (d) $\frac{2(3 - \cos 4)}{(1 + \cos 4)}$

8. The value of the integral

$$\int_{-10}^{0} \frac{\left|\frac{2[x]}{3x - [x]}\right|}{\left(\frac{2[x]}{3x - [x]}\right)} dx,$$ where $[\cdot]$ denotes the greatest integer function, is

(a) 0 (b) -10

(c) 10 (d) None of these

9. The value of $\int_0^4 \{\sqrt{x}\} \, dx$, where $\{\cdot\}$ denotes the fractional part function, is

(a) 1/3 (b) 1

(c) 5/3 (d) 7/3

10. If $f(x)$ be a real valued function, $f(x) + f(x+4) = f(x+2) + f(x+6)$, $g(x) = \int_x^{x+8} f(t) \, dt$. Then $g'(x)$ is equal to

(a) $f(x)$ (b) $f(x+8)$

(c) 8 (d) 0

11. The value of the definite integral $\int_{a+2\pi}^{a+5\pi/2} (\sin^{-1}(\cos x) + \cos^{-1}(\sin x)) \, dx$ is equal to

(a) $\frac{\pi^2}{8}$ (b) $\frac{\pi^2}{4}$

(c) $\frac{\pi^2}{2}$ (d) π^2

12. The greatest value of $f(x) = \int_{5\pi/3}^{x} (6\cos t - 2\sin t)dt$ in the interval $\left[\frac{5\pi}{3}, \frac{7\pi}{4}\right]$, is equal to

(a) $3\sqrt{3} - 2\sqrt{2} - 1$ (b) $3\sqrt{3} + \sqrt{2} - 1$
(c) $3\sqrt{3} + 2\sqrt{2} + 1$ (d) $\sqrt{3} - 1$

13. If $f(x) = \int_0^x (\sin^4 t + \cos^4 t)\, dt$, then $f(x + \pi)$ will be equal to

(a) $f(x) + f\left(\frac{\pi}{2}\right)$
(b) $f(x) + f(\pi)$ or $f(x) + 2f\left(\frac{\pi}{2}\right)$
(c) $f(x) - f(\pi)$
(d) $f(x) - 2f\left(\frac{\pi}{2}\right)$

14. If $I_k = \int_{-2k\pi}^{2k\pi} |\sin x|\,[\sin x]\, dx\ \forall\ k \in N$, where $[\cdot]$ denotes the greatest integer function, then $\sum_{k=1}^{10} I_k$ is equal to

(a) –110 (b) –440
(c) –330 (d) –220

15. The value of

$\int_0^{\frac{3\pi}{2}} \frac{|\tan^{-1}\tan x| - |\sin^{-1}\sin x|}{|\tan^{-1}\tan x| + |\sin^{-1}\sin x|}\, dx$ is equal to

(a) $\frac{\pi}{2}$ (b) π
(c) $\frac{3\pi}{2}$ (d) None of these

16. Value of $\int_1^5 \left(\sqrt{x + 2\sqrt{(x-1)}} + \sqrt{x - 2\sqrt{(x-1)}}\right) dx$ is

(a) $\frac{8}{3}$ (b) $\frac{16}{3}$
(c) $\frac{32}{3}$ (d) $\frac{64}{3}$

17. The value of $\frac{\int_0^n [x]\, dx}{\int_0^n \{x\}\, dx}$ is, (where $[x]$ and $\{x\}$ denotes the integral part and fractional part functions of x and $n \in N$)

(a) $n + 2$ (b) $n + 1$
(c) n (d) $n - 1$

18. The value of $\int_{-1}^{15} \text{sgn}\,(\{x\})\, dx$, where $\{\cdot\}$ denotes the fractional part function, is

(a) 8 (b) 16
(c) 24 (d) 0

19. The value of $\int_0^2 \{3x\} dx$, where $\{\cdot\}$ denotes the fractional part function, is

(a) 7 (b) 5
(c) 3 (d) 1

20. The number of solutions of $I(x) = \int_0^x [t]^2 dt = 2(x - 1)\ \forall\ x > 0$, where $[\cdot]$ denotes the greatest integer function, are

(a) 2 (b) 3
(c) 4 (d) 5

21. Let $f: R \to R$ be defined as $f(x) = e^{-x} \sin x$. If $F: [0, 1] \to R$ is a differentiable function such that $F(x) = \int_0^x f(t)\, dt$, then the value of $\int_0^1 (F'(x) + f(x))\, e^x\, dx$ lies in the interval

[JEE Main 2021, 17 March]

(a) $\left[\frac{327}{360}, \frac{329}{360}\right]$ (b) $\left[\frac{330}{360}, \frac{331}{360}\right]$
(c) $\left[\frac{331}{360}, \frac{324}{360}\right]$ (d) $\left[\frac{335}{360}, \frac{336}{360}\right]$

22. The value of the integral $\int_0^{\pi} \frac{\sin\left(n + \frac{1}{2}\right)x}{\sin x/2}\, dx\ (n \in N)$ is

(a) π (b) 2π
(c) 3π (d) None of these

23. $\int_0^{\pi/3} [\sqrt{3} \tan x]\, dx$ is equal to

([.] denotes the greatest integer function)

(a) $5\pi/6$ (b) $\frac{5\pi}{6} - \tan^{-1}(2/\sqrt{3})$
(c) $\frac{\pi}{2} - \tan^{-1}\left(\frac{2}{\sqrt{3}}\right)$ (d) None of these

24. Let $f(x) = |x - 2|$ and $g(x) = f(f(x)), x \in [0, 4]$. Then $\int_0^3 (g(x) - f(x))\, dx$ is equal to

[JEE Main 2020, 4 September]

(a) $\frac{3}{2}$ (b) $\frac{1}{2}$
(c) 0 (d) 1

25. The value of $\int_{-1}^{1} [x[1 + \sin \pi x] + 1]\, dx$ is ($[\cdot]$ denotes the greatest integer function)

(a) 2 (b) 0
(c) 1 (d) None of these

26. $\int_{-\pi/4}^{n\pi - \pi/4} |\sin x + \cos x|\, dx\ (n \in N)$ is

(a) 0 (b) $2n$
(c) $2\sqrt{2}n$ (d) None of these

27. The range of the function $f(x) = \int_{-1}^{1} \frac{\sin x\, dt}{(-2t\cos x + t^2)}$ is

(a) $\left[-\frac{\pi}{2}, \frac{\pi}{2}\right]$ (b) $[0, \pi]$
(c) $\{0, \pi\}$ (d) $\left\{-\frac{\pi}{2}, \frac{\pi}{2}\right\}$

28. For any $t \in R$ and f be a continuous function,

Let $I_1 = \int_{\sin^2 t}^{1+\cos^2 t} xf(x(2-x))\,dx$

and $I_2 = \int_{\sin^2 t}^{1+\cos^2 t} f(x(2-x))dx$ then $\frac{I_1}{I_2}$ is

(a) 0 (b) 1
(c) 2 (d) 3

29. The value of the integral

$\int_0^{n\pi+t} (|\cos x| + |\sin x|)\,dx$, where $0 < t < \frac{\pi}{2}$, is

(a) n (b) $2n + \sin t + \cos t$
(c) $\cos t$ (d) $4n + \sin t - \cos t + 1$

30. If $I = \int_2^3 \frac{2x^5 + x^4 - 2x^3 + 2x^2 + 1}{(x^2+1)(x^4-1)}\,dx$, then I equals

(a) $\frac{1}{2}\log 6 + \frac{1}{10}$ (b) $\frac{1}{2}\log 6 - \frac{1}{10}$
(c) $\frac{1}{2}\log 3 - \frac{1}{10}$ (d) $\frac{1}{2}\log 2 + \frac{1}{10}$

31. $\int_{-\pi/4}^{\pi/4} \frac{e^x \cdot \sec^2 x\,dx}{e^{2x} - 1}$ is equal to

(a) 0 (b) 2
(c) e (d) None of these

32. The value of $\int_0^{2\pi} \frac{dx}{e^{\sin x} + 1}$ is

(a) π (b) 0
(c) 2π (d) $\pi/2$

33. $\int_0^{\pi/4} \sin x\, d(x - [x])$ is equal to (where [·] denotes the greatest integer function)

(a) 1/2 (b) $1 - \frac{1}{\sqrt{2}}$
(c) 1 (d) None of these

34. The value of the integral $I = \int_1^\infty \frac{(x^2-2)}{x^3\sqrt{(x^2-1)}}\,dx$ is

(a) 0 (b) 2/3
(c) 4/3 (d) None of these

35. If the integral $\int_0^{10} \frac{[\sin 2\pi x]}{e^{x-[x]}}\,dx = \alpha e^{-1} + \beta e^{-\frac{1}{2}} + \gamma$, where α, β, γ is are integers and $[x]$ denotes the greatest integer less than or equal to x, then the value of $\alpha + \beta + \gamma$ is equal to **[JEE Main 2021, 17 March]**

(a) 0 (b) 20
(c) 25 (d) 10

36. $f(x) > 0\ \forall\ x \in R$ and is bounded. If

$\lim_{n\to\infty} \sum_{r=0}^{n-1} \int_{ra}^{(r+1)a} \frac{a^r f(x)dx}{f(x) + f((2r+1)a - x)} = 7/5$ (where $a < 1$), then a is equal to

(a) $\frac{2}{7}$ (b) $\frac{1}{7}$
(c) $\frac{14}{19}$ (d) $\frac{9}{14}$

37. Let $f(x)$ be an odd function in the interval $\left[-\frac{T}{2}, \frac{T}{2}\right]$, with a period T. Then $F(x) = \int_a^x f(t)\,dt$ is

(a) periodic with period T
(b) non periodic
(c) periodic with period $2T$
(d) periodic with period aT

38. If $\int_0^{10} f(x)\,dx = 5$, then $\sum_{k=1}^{10} \int_0^1 f(k-1+x)\,dx$ is

(a) 50 (b) 10
(c) 5 (d) None of these

39. The value of the integral $\left|\int_0^{2\pi} [2\sin x]\,dx\right|$ is ([·] denotes the greatest integer function)

(a) π (b) 2π (c) 3π (d) 4π

40. Let $f(x) = \text{minimum}\left(|x|, 1-|x|, \frac{1}{4}\right), \forall\ x \in R$, then the value of $\int_{-1}^1 f(x)\,dx$ is equal to

(a) $\frac{1}{32}$ (b) $\frac{3}{8}$
(c) $\frac{3}{32}$ (d) None of these

41. If $\int_{1/2}^2 \frac{1}{x}\operatorname{cosec}^{101}\left(x - \frac{1}{x}\right)dx = k$, then the value of k is

(a) 1 (b) 1/2
(c) 0 (d) 1/101

42. If $\int_0^\infty e^{-x^2}dx = \frac{\sqrt{\pi}}{2}$ and $\int_0^\infty e^{-ax^2}dx, a > 0$ is

(a) $\frac{\sqrt{\pi}}{2}$ (b) $\frac{\sqrt{\pi}}{2a}$
(c) $2\frac{\sqrt{\pi}}{a}$ (d) $\frac{1}{2}\sqrt{\frac{\pi}{a}}$

43. If $\int_0^1 e^{x^2}(x - \alpha)\,dx = 0$, then

(a) $1 < \alpha < 2$ (b) $\alpha < 0$
(c) $0 < \alpha < 1$ (d) $\alpha = 0$

44. If $[x]$ denotes the greatest integer less than or equal to x, then $\int_0^\infty \left[\frac{2}{e^x}\right]dx$ is equal to

(a) $\log_e 2$ (b) e^2
(c) 0 (d) $\frac{2}{e}$

45. If $I_1 = \int_0^1 (1 - x^{50})^{100}dx$ and $I_2 = \int_0^1 (1 - x^{50})^{101}dx$ such that $I_2 = \alpha I_1$, then α equals to **[JEE Main 2020, 6 September]**

(a) $\frac{5049}{5050}$ (b) $\frac{5050}{5049}$
(c) $\frac{5050}{5051}$ (d) $\frac{5051}{5050}$

46. The value of $\int_{\pi/4}^{3\pi/4} \frac{x}{1+\sin x}\,dx$ is equal to

(a) $(\sqrt{2}-1)\pi$ (b) $(\sqrt{2}+1)\pi$
(c) π (d) None of these

47. Let f be a positive function. If

$$I_1 = \int_{1-k}^{k} x\, f\{x(1-x)\}\,dx,$$

$$I_2 = \int_{1-k}^{k} f\{x(1-x)\}\,dx$$

where $2k-1>0$, then $I_1 : I_2$ is equal to

(a) $2:1$ (b) $k:1$
(c) $1:2$ (d) $1:1$

48. Let $f(x) = \frac{1}{2}a_0 + \sum_{i=1}^{n} a_i \cos(ix) + \sum_{j=1}^{n} b_j \sin(jx)$, then $\int_{-\pi}^{\pi} f(x)\cos kx\,dx$ is equal to

(a) a_k (b) b_k
(c) πa_k (d) πb_k

49. The value of the definite integral $\int_0^1 \frac{x\,dx}{(x^3+16)}$ lies in the interval $[a, b]$. Then smallest such interval is

(a) $\left[0, \frac{1}{17}\right]$ (b) $[0, 1]$
(c) $\left[0, \frac{1}{27}\right]$ (d) None of these

50. Value of $\int_2^3 \frac{dx}{\sqrt{(1+x^3)}}$ is

(a) less than 1 (b) greater than 2
(c) lies between 3 and 4 (d) None of these

51. Suppose for every integer n, $\int_n^{n+1} f(x)\,dx = n^2$, the value of $\int_{-2}^{4} f(x)\,dx$ is

(a) 16 (b) 14
(c) 19 (d) None of these

52. The value of $\int_0^{\sin^2 x} \sin^{-1}\sqrt{t}\,dt + \int_0^{\cos^2 x} \cos^{-1}\sqrt{t}\,dt$ is

(a) $\pi/2$ (b) 1
(c) $\pi/4$ (d) None of these

53. If $I = \int_0^1 \cos\left(2\cot^{-1}\sqrt{\left(\frac{1-x}{1+x}\right)}\right)dx$, then

(a) $I > \frac{1}{2}$ (b) $I = -\frac{1}{2}$
(c) $0 < I < \frac{1}{2}$ (d) None of these

54. The value of $\int_0^2 [x^2-1]\,dx$, where $[x]$ denotes the greatest integer function, is given by

(a) $3-\sqrt{3}-\sqrt{2}$ (d) 2
(c) 1 (d) None of these

55. Let $\int_0^x \left(\frac{bt\cos 4t - a\sin 4t}{t^2}\right)dt = \frac{a\sin 4x}{x}$, then a and b are given by

(a) $a = 1/4, b = 1$ (b) $a = 2, b = 2$
(c) $a = -1, b = 4$ (d) $a = 2, b = 4$

56. If $f(x) = \cos x - \int_0^x (x-t)\,f(t)\,dt$, then $f''(x) + f(x)$ equals

(a) $-\cos x$ (b) 0
(c) $\int_0^x (x-t)\,f(t)\,dt$ (d) $-\int_0^{-x} (x-t)\,f(t)\,dt$

57. Let $f(x) = \max\{x+|x|, x-[x]\}$, where $[x]$ denotes the greatest integer $\le x$. Then $\int_{-2}^{2} f(x)\,dx$ is equal to

(a) 3 (b) 2
(c) 1 (d) None of these

58. The value of $\int_{-2}^{2} \frac{\sin^2 x}{[x/\pi]+[1/2]}\,dx$, where $[x]$ denotes the greatest integer $\le x$, is

(a) 1 (b) 0
(c) $4-\sin 4$ (d) None of these

59. The value of $\int_{-1}^{1} \max\{2-x, 2, 1+x\}\,dx$ is

(a) 4 (b) 9/2
(c) 2 (d) None of these

60. Let $f(x)$ be a function satisfying $f'(x) = f(x)$ with $f(0) = 1$ and g be the function satisfying $f(x) + g(x) = x^2$ the value of the integral $\int_0^1 f(x)\,g(x)\,dx$ is

(a) $\frac{1}{4}(e-7)$ (b) $\frac{1}{4}(e-2)$
(c) $\frac{1}{2}(e-3)$ (d) None of these

61. The value of $\int_0^{\pi} \left(\sum_{r=0}^{3} a_r \cos^{3-r} x \sin^r x\right)dx$ depends on

(a) a_0 and a_2 (b) a_1 and a_2
(c) a_0 and a_3 (d) a_1 and a_3

62. Let $f: R \to R$ such that, $f(x+2y) = f(x) + f(2y) + 4xy\ \forall\ x, y \in R$ and $f'(0) = 0$. If $I_1 = \int_0^1 f(x)\,dx$, $I_2 = \int_{-1}^{0} f(x)\,dx$ and $I_3 = \int_{1/2}^{2} f(x)\,dx$, then

(a) $I_1 = I_2 > I_3$ (b) $I_1 > I_2 > I_3$
(c) $I_1 = I_2 < I_3$ (d) $I_1 < I_2 < I_3$

63. If $\int_a^b \frac{x^n}{x^n + (16-x)^n}\,dx = 6$, then

(a) $a = 4, b = 12, n \in R$ (b) $a = 2, b = 14, n \in R$
(c) $a = -4, b = 20, n \in R$ (d) $a = 2, b = 8, n \in R$

64. If $f(x)$ and $g(x)$ are continuous functions, then

$$\int_{\ln \lambda}^{\ln (1/\lambda)} \frac{f\left(\frac{x^2}{4}\right)(f(x)-f(-x))}{g\left(\frac{x^2}{4}\right)(g(x)+g(-x))} dx \text{ is}$$

(a) depend on λ (b) a non-zero constant
(c) zero (d) None of these

65. Let $g(x)=\int_0^x f(t)\,dt$, where f is continuous function in $[0, 3]$ such that $\frac{1}{3} \leq f(t) \leq 1$ for all $t \in [0, 1]$ and $0 \leq f(t) \leq \frac{1}{2}$ for all $t \in (1, 3]$. The largest possible interval in which $g(3)$ lies is **[JEE Main 2021, 18 March]**

(a) $\left[-1, -\frac{1}{2}\right]$ (b) $\left[-\frac{3}{2}, -1\right]$
(c) $\left[\frac{1}{3}, 2\right]$ (d) $[1, 3]$

66. The value of $\int_0^{2\pi} \frac{x \sin^{2n} x}{\sin^{2n} x + \cos^{2n} x} dx$ is

(a) $\frac{\pi^2}{4}$ (b) $\frac{\pi^2}{2}$
(c) π^2 (d) $2\pi^2$

67. Consider the integrals

$I_1 = \int_0^1 e^{-x} \cos^2 x\, dx$, $I_2 = \int_0^1 e^{-x^2} \cos^2 x\, dx$,

$I_3 = \int_0^1 e^{-x^2} dx$ and $I_4 = \int_0^1 e^{-x^2/2} dx$

let I be the greatest integral among I_1, I_2, I_3, I_4, then

(a) $I = I_1$ (b) $I = I_2$
(c) $I = I_3$ (d) $I = I_4$

68. If $f(x) = \int_2^{x^2} \frac{(\sin^{-1} \sqrt{t})^2}{\sqrt{t}} dt$, then the value of $(1-x^2)\{f''(x)\}^2 - 2f'(x)$ at $x = \frac{1}{\sqrt{2}}$ is

(a) $2-\pi$ (b) $3+\pi$
(c) $4-\pi$ (d) None of these

69. If for $x \neq 0$, $af(x) + bf\left(\frac{1}{x}\right) = \frac{1}{x} - 5$, where $a \neq b$, then $\int_1^2 xf(x)\, dx$ is equal to

(a) $\frac{b-9a}{9(a^2-b^2)}$ (b) $\frac{b-9a}{b(a^2-b^2)}$
(c) $\frac{b-9a}{6(a^2-b^2)}$ (d) None of these

70. The value of $\int_{-\pi}^{\pi} \sum_{r=0}^{999} \cos rx \left(1 + \sum_{r=1}^{999} \sin rx\right) dx$ is,

(a) 2π (b) 999π
(c) 0 (d) π

71. Let $f(x)$ be a continuous function such that $f(a-x) + f(x) = 0$ for all $x \in [0, a]$, then $\int_0^a \frac{dx}{1+e^{f(x)}}$ is equal to

(a) a (b) $\frac{a}{2}$
(c) $f(a)$ (d) $\frac{1}{2} f(a)$

72. The equation

$$\int_{-\pi/4}^{\pi/4} \left(a|\sin x| + \frac{b \sin x}{1+\cos x} + c\right) dx = 0,$$

where a, b, c are constants, gives a relation between

(a) a, b and c (b) a and c
(c) a and b (d) b and c

73. The value of $\int_0^{16\pi/3} |\sin x|\, dx$ is

(a) 17/2 (b) 19/2
(c) 21/2 (d) None of these

74. $\int_0^{\pi/4} (\pi x - 4x^2) \ell n(1 + \tan x) dx =$

(a) $\frac{\pi^3}{192} \ln 2$ (b) $\frac{\pi^3}{191} \ln 2$
(c) $\frac{\pi^3}{192} \ln 3$ (d) $\frac{\pi^2}{192} \ln 2$

75. If $f'''(x) = k$ in $[0, a]$, then

$$\int_0^a f(x)\, dx - \left\{x f(x) - \frac{x^2}{2!} f'(x) + \frac{x^3}{3!} f''(x)\right\}_0^a \text{ is}$$

(a) $-ka^4/12$
(b) $ka^4/24$
(c) $-ka^4/24$
(d) None of the above

76. The value of $\int_0^{n^2} [\sqrt{x}]\, dx$, ([·] denotes the greatest integer function), $n \in N$, is

(a) $\left\{\frac{n(n+1)}{2}\right\}^2$ (b) $\frac{1}{6} n(n-1)(4n+1)$
(c) Σn^2 (d) $\frac{n(n+1)(n+2)}{6}$

77. $\lim_{n \to \infty} \frac{(n!)^{1/n}}{n}$ equals

(a) e (b) e^{-1}
(c) 1 (d) None of these

78. If [.] stands for the greatest integer function, the value of $\int_4^{10} \frac{[x^2]\, dx}{[x^2 - 28x + 196] + [x^2]}$ is

(a) 0 (b) 1
(c) 3 (d) None of these

79. Let $I(a)=\int_0^{\pi}\left(\frac{x}{a}+a\sin x\right)^2 dx$ where 'a' is positive real.

The value of 'a' for which $I(a)$ attains its minimum value is

(a) $\sqrt{\pi\sqrt{\frac{2}{3}}}$ (b) $\sqrt{\pi\sqrt{\frac{3}{2}}}$

(c) $\sqrt{\frac{\pi}{16}}$ (d) $\sqrt{\frac{\pi}{13}}$

80. If $na=1$ always and $n\to\infty$, then the value of $\Pi\{1+(ra)^2\}^{1/r}$ is

(a) 1 (b) $e^{\pi^2/8}$

(c) $e^{\pi^2/24}$ (d) $e^{-\pi^2/12}$

81. The value of $\int_0^{100}[\tan^{-1}x]\,dx$ is

(where [·] denotes the greatest integer function)

(a) 100 (b) $100-\tan^{-1}1$

(c) $100-\tan 1$ (d) None of these

82. If $f(a+b+1-x)=f(x)$, for all x, where a and b are fixed positive real numbers, then $\frac{1}{a+b}\int_a^b x(f(x)+f(x+1))dx$ is equal to **[JEE Main 2020, 7 January]**

(a) $\int_{a+1}^{b+1}f(x+1)dx$ (b) $\int_{a+1}^{b+1}f(x)dx$

(c) $\int_{a-1}^{b-1}f(x+1)dx$ (d) $\int_{a-1}^{b-1}f(x)dx$

83. The value of $\int_{-2}^{2}\min(x-[x],-x-[-x])\,dx$ is ([·] denotes the greatest integer function)

(a) 0 (b) 1

(c) 2 (d) None of these

84. The value of $\int_0^2[x^2-x+1]\,dx$, (where [·] denotes the greatest integer function) is given by

(a) $\frac{5-\sqrt{5}}{2}$ (b) $\frac{6-\sqrt{5}}{2}$

(c) $\frac{7-\sqrt{5}}{2}$ (d) $\frac{8-\sqrt{5}}{2}$

85. The value of $\lim_{n\to\infty}\left(\frac{1^k+2^k+\ldots+n^k}{n^{k+1}}\right)$ is

(a) $\frac{1}{k+2}$ (b) $\frac{1}{k+1}$

(c) $\frac{2}{k+3}$ (d) 0

86. The value of the integral $\int_{1/e}^{e}|\ln x|\,dx$ is

(a) $1-1/e$ (b) $2(1-1/e)$

(c) $e^{-1}-1$ (d) None of these

87. The value of $\int_{-\pi/2}^{\pi/2}\frac{x^2\cos x}{1+e^x}dx$is equal to **[JEE Advanced 2016]**

(a) $\frac{\pi^2}{4}-2$ (b) $\frac{\pi^2}{4}+2$

(c) $\pi^2-e^{-\pi/2}$ (d) $\pi^2+e^{\pi/2}$

88. $\lim_{n\to\infty}\frac{1}{n}\sum_{r=n+1}^{2n}\ln\left(1+\frac{r}{n}\right)$ equals

(a) $\ln\left(\frac{27}{4e}\right)$ (b) $\ln\left(\frac{27}{e^2}\right)$

(c) $\ln\left(\frac{4}{e}\right)$ (d) None of these

89. Given that n is odd and m is even integer. The value of $\int_0^{\pi}\cos mx\sin nx\,dx$ is

(a) $\frac{2m}{n^2-m^2}$ (b) $\frac{2n}{n^2-m^2}$

(c) $\frac{m^2+n^2}{n^2-m^2}$ (d) None of these

90. $\int_0^x\frac{2^t}{2^{[t]}}dt$, where [.] denotes the greatest integer function, and $x\in R^+$, is equal to

(a) $\frac{1}{\ln 2}([x]+2^{\{x\}}-1)$ (b) $\frac{1}{\ln 2}([x]+2^{\{x\}})$

(c) $\frac{1}{\ln 2}([x]+2^{\{x\}})$ (d) $\frac{1}{\ln 2}([x]+2^{\{x\}}+1)$

91. The value of $\int_0^2\left|\cos\left(\frac{\pi x}{2}\right)\right|dx$ is

(a) 2π (b) $\pi/2$

(c) $3/4\pi$ (d) $4/\pi$

92. If $f'(x)=f(x)+\int_0^1 f(x)\,dx$ and given $f(0)=1$, then $f(x)$ is equal to

(a) $\frac{e^x}{2-e}+\left(\frac{1+e}{1-e}\right)$ (b) $\frac{2e^x}{3-e}+\left(\frac{1-e}{3-e}\right)$

(c) $\frac{e^x}{2-e}$ (d) $\frac{2e^x}{3-e}$

93. The value of the integral $\int_0^1\frac{x^{\alpha}-1}{\ln x}dx$, is

(a) $\ln\alpha$ (b) $2\ln(\alpha+1)$

(c) $3\ln\alpha$ (d) None of these

94. The value of $\int_{\pi}^{2\pi}[2\sin x]\,dx$, (where [·] represents the greatest integer function) is

(a) $-\frac{5\pi}{3}$ (b) $-\pi$

(c) $\frac{5\pi}{3}$ (d) -2π

95. $\lim_{n\to\infty}\sum_{k=1}^{n}\frac{k^{1/a}\left\{n^{a-\frac{1}{a}}+k^{a-\frac{1}{a}}\right\}}{n^{a+1}}$ is equal to

(a) 1 (b) 2

(c) 3 (d) None of these

96. Let $f(x) = \min(\{x\}, \{-x\}) \forall x \in R$, where $\{x\}$ denotes the fractional part of x, then $\int_{-100}^{100} f(x)\,dx$ is equal to

(a) 50 (b) 100
(c) 200 (d) None of these

97. The value of $\int_{-\pi/2}^{199\pi/2} \sqrt{(1+\cos 2x)}\,dx$ is

(a) $50\sqrt{2}$ (b) $100\sqrt{2}$ (c) $150\sqrt{2}$ (d) $200\sqrt{2}$

98. The value of $\int_0^{\pi/4} (\tan^n x + \tan^{n-2} x)\,d(x-[x])$, (where [·] denotes the greatest integer function) is

(a) $\frac{1}{n-1}$ (b) $\frac{1}{n+1}$
(c) $\frac{2}{n-1}$ (d) None of these

99. If $\int_{-1}^{4} f(x)\,dx = 4$ and $\int_2^4 (3 - f(x))\,dx = 7$, the value of $\int_2^{-1} f(x)\,dx$ is

(a) 2 (b) −3
(c) −5 (d) None of these

100. If $I_1 = \int_0^x e^{zx} e^{-z^2} dz$ and $I_2 = \int_0^x e^{-z^2/4} dz$, then

(a) $I_1 = e^x I_2$ (b) $I_1 = e^{x^2} I_2$
(c) $I_1 = e^{x^2/2} I_2$ (d) None of these

101. Let $f(x) = \int_0^x g(t)dt$, where g is a non-zero even function. If $f(x+5) = g(x)$, then $\int_0^x f(t)dt$ equals

[JEE Main 2019, 8 April]

(a) $5\int_{x+5}^{5} g(t)dt$ (b) $\int_5^{x+5} g(t)dt$
(c) $2\int_5^{x+5} g(t)dt$ (d) $\int_{x+5}^{5} g(t)dt$

Objective Questions Type II [One or more than one correct answer(s)]

In each of the questions below four choices of which one or more than one are correct. You have to select the correct answer(s) accordingly.

1. If x satisfies the equation $x^2\left(\int_0^1 \frac{dt}{t^2 + 2t\cos\alpha + 1}\right)$

$$-x\left(\int_{-3}^{3} \frac{t^3 \sin 2t}{t^2+1}\,dt\right) - 2 = 0$$

$(0 < \alpha < \pi)$, then the value of x is

(a) $2\sqrt{\left(\frac{\sin\alpha}{\alpha}\right)}$ (b) $-2\sqrt{\left(\frac{\sin\alpha}{\alpha}\right)}$
(c) $4\sqrt{\left(\frac{\sin\alpha}{\alpha}\right)}$ (d) None of these

2. The value of the integral $\int_a^b \frac{|x|}{x}\,dx$, $a < b$ is

(a) $b - a$ (b) $a - b$
(c) $b + a$ (d) None of these

3. If $I_1 = \int_0^1 2^{x^2} dx$, $I_2 = \int_0^1 2^{x^3} dx$, $I_3 = \int_1^2 2^{x^2} dx$ and $I_4 = \int_1^2 2^{x^3} dx$, then

(a) $I_1 > I_2$ (b) $I_2 > I_1$
(c) $I_3 > I_4$ (d) $I_4 > I_3$

4. The absolute value of $\int_{10}^{19} \frac{\sin x}{1+x^8}\,dx$ is

(a) less than 10^{-7} (b) more than 10^{-7}
(c) less than 10^{-6} (d) None of these

5. If $I = \int_3^4 \frac{dx}{\sqrt[3]{(\ln x)}}$, then

(a) $I < 1$ (b) $I > 1$
(c) $I > 0.92$ (d) $I < 0.92$

6. If $I_1 = \int_x^1 \frac{dt}{1+t^2}$ and $I_2 = \int_1^{1/x} \frac{dt}{1+t^2}$ for $x > 0$, then

(a) $I_1 = I_2$ (b) $I_1 > I_2$
(c) $I_2 > I_1$ (d) $I_2 = \cot^{-1} x - \pi/4$

7. The value of $\int_a^{a+\pi/2} (\sin^4 x + \cos^4 x)\,dx$ is

(a) independent of a (b) $a\left(\frac{\pi}{2}\right)^2$
(c) $3\pi/8$ (d) $\frac{3}{8}\pi a^2$

8. The value α in the interval $[-\pi, 0]$ satisfying $\sin\alpha + \int_\alpha^{2\alpha} \cos 2x\,dx = 0$ is

(a) $-\pi/2$ (b) $-\pi$
(c) $-\pi/3$ (d) 0

9. If $I = \int_0^1 \sqrt{(1+x^3)}\,dx$, then

(a) $I < 1$ (b) $I \neq \sqrt{5}/2$
(c) $I < \sqrt{7}/2$ (d) None of these

10. The maximum and minimum values of the integral $\int_0^{\pi/2} \frac{dx}{(1+\sin^2 x)}$ are

(a) $\pi/4$ (b) π
(c) $\pi/2$ (d) $3\pi/4$

11. Let $f(a) > 0$ and let $f(x)$ be a non decreasing continuous function in $[a, b]$, Then $\frac{1}{b-a}\int_a^b f(x)\,dx$ has the

(a) maximum value $f(b)$ (b) minimum value $f(a)$

(c) maximum value $bf(b)$ (d) minimum value $\frac{f(a)}{b-a}$

12. Let $f : R \to (0, 1)$ be a continuous function. Then, which of the following function(s) has (have) the value zero at some point in the interval (0, 1) ? **[JEE Advanced 2017]**

(a) $e^x - \int_0^x f(t)\sin t\,dt$ (b) $f(x) + \int_0^{\frac{\pi}{2}} f(t)\sin t\,dt$

(c) $x - \int_0^{\frac{\pi}{2}-x} f(t)\cos t\,dt$ (d) $x^9 - f(x)$

13. The values of α, which satisfy $\int_{\pi/2}^{\alpha} \sin x\,dx = \sin 2\alpha$ ($\alpha \in [0, 2\pi]$) are equal to

(a) $\pi/2$ (b) $3\pi/2$

(c) $7\pi/6$ (d) $11\pi/6$

14. If $I_n = \int_0^{\pi/4} \tan^n x\,dx$, then

(a) $I_n + I_{n-2} = \frac{1}{n-1}$

(b) $I_{n+1} + I_{n-1} = \frac{1}{n}$

(c) $\frac{1}{n+1} < 2I_n < \frac{1}{n-1}$, where $n > 1$ is a natural number

(d) $(I_{n+2} + I_n)(n+1) = 1$

15. The integral $\int_0^{\pi} x f(\sin x)\,dx$ is equal to

(a) $\frac{\pi}{2}\int_0^{\pi} f(\sin x)dx$ (b) $\frac{\pi}{4}\int_0^{\pi} f(\sin x)\,dx$

(c) $\pi\int_0^{\pi/2} f(\sin x)\,dx$ (d) $\pi\int_0^{\pi/2} f(\cos x)\,dx$

16. If $I_n = \int_0^{\pi/4} \tan^n x\,dx$, then

(a) $I_5 + I_3 = \frac{1}{4}$ (b) $I_8 + I_6 = \frac{1}{7}$

(c) $I_6 - I_{10} = \frac{2}{63}$ (d) $I_6 + 2I_8 + I_{10} = \frac{16}{63}$

17. Let $\int_{\alpha}^{\beta} \frac{f(\alpha+\beta-x)}{f(x)+f(\alpha+\beta-x)}\,dx = 4$, then

(a) $\alpha = -1, \beta = 7$ (b) $\alpha = 5, \beta = 13$

(c) $\alpha = -2, \beta = 6$ (d) $\alpha = -10, \beta = -2$

18. If $\int_a^b f(t)\,g(h(t))\,dt = \int_a^b f(h(t)g(t)\,dt$, where f, g, h are non negative continuous functions on $[a, b]$, then $h(t)$ is equal to

(a) t (b) $-t$

(c) $a - b - t$ (d) $a + b - t$

19. If $G(x,t) = \begin{cases} x(t-1), \text{ where } x \le t \\ t(x-1), \text{ where } t < x \end{cases}$

and if t is continuous function of x in [0, 1]

Let $g(x) = \int_0^1 f(t)\,G(x,t)\,dt$, then

(a) $g(0) = 1$ (b) $g(0) = 0$

(c) $g(1) = 1$ (d) $g''(x) = f(x)$

20. Let $f(x) = 7\tan^8 x + 7\tan^6 x - 3\tan^4 x - 3\tan^2 x$ for all $x \in \left(-\frac{\pi}{2}, \frac{\pi}{2}\right)$. Then, the correct expression(s) is/are **[JEE Advanced 2015]**

(a) $\int_0^{\pi/4} x f(x)\,dx = \frac{1}{12}$ (b) $\int_0^{\pi/4} f(x)\,dx = 0$

(c) $\int_0^{\pi/4} x f(x)\,dx = \frac{1}{6}$ (d) $\int_0^{\pi/4} f(x)\,dx = 1$

21. $\int_{-1/2}^{1/2} \sqrt{\left\{\left(\frac{x+1}{x-1}\right)^2 + \left(\frac{x-1}{x+1}\right)^2 - 2\right\}}\,dx$ is

(a) $4\ln\left(\frac{4}{3}\right)$ (b) $4\ln\left(\frac{3}{4}\right)$

(c) $-\ln\left(\frac{81}{256}\right)$ (d) $\ln\left(\frac{256}{81}\right)$

22. Let e be the eccentricity of a hyperbola $\frac{x^2}{a^2} - \frac{y^2}{b^2} = 1$ and $f(e)$ be the eccentricity of conjugate hyperbola $-\frac{x^2}{a^2} + \frac{y^2}{b^2} = 1$, then $\int_1^3 \underbrace{fff \ldots f(e)}_{n \text{ times}}\,de$ is equal to

(a) 2, if n is even (b) 4, if n is even

(c) $2\sqrt{2}$, if n is odd (d) $4\sqrt{2}$, if n is odd

23. Let $f(x) = \int_0^x |\sin t|\,dt$, then

(a) f is continuous everywhere

(b) $f'(x)$ is differentiable at $x = \pi$

(c) $f(\pi) = 2$

(d) $f(x) \ge 0$ for all x

24. The integral $\int_{\tan^{-1}\lambda}^{\cot^{-1}\lambda} \frac{\tan x}{\tan x + \cot x}\,dx, \forall \lambda \in R$ can not take the value

(a) $-\frac{\pi}{4}$ (b) $\frac{\pi}{4}$

(c) $-\frac{\pi}{2}$ (d) $\frac{3\pi}{4}$

25. The value of $\int_{-\pi}^{\pi} \sin mx \cos nx\,dx$ ($m, n \in N$) is equal to

(a) 0 (b) $\int_{-\pi/8}^{\pi/8} x^8 \sin^9 x\,dx$

(c) $\int_{-\pi/2}^{\pi/2} \sin x\, f(\cos x)\,dx$ (d) $\int_{-1/2}^{1/2} e^{\cos x}\,dx$

26. The value of $\int_{1/e}^{\tan x} \frac{t\,dt}{1+t^2} + \int_{1/e}^{\cot x} \frac{dt}{t(1+t^2)}$ is

(a) $\frac{1}{2+\tan^2 x}$ (b) 1

(c) $\frac{\pi}{4}$ (d) $\frac{2}{\pi}\int_{-1}^{1} \frac{dt}{1+t^2}$

27. If $\frac{2x}{\pi} < \sin x < x$ for $0 < x < \frac{\pi}{2}$, then the value of the integral $\int_0^{\pi/2} \frac{\sin x}{x} dx$ is

(a) >1 (b) <1

(c) $>\frac{\pi}{2}$ (d) $<\frac{\pi}{2}$

28. The option(s) with the values of a and L that satisfy the equation $\frac{\int_0^{4\pi} e^t(\sin^6 at + \cos^4 at)dt}{\int_0^{\pi} e^t(\sin^6 at + \cos^4 at)dt} = L$, is/are

[JEE Advanced 2015]

(a) $a = 2, L = \frac{e^{4\pi}-1}{e^{\pi}-1}$ (b) $a = 2, L = \frac{e^{4\pi}+1}{e^{\pi}+1}$

(c) $a = 4, L = \frac{e^{4\pi}-1}{e^{\pi}-1}$ (d) $a = 4, L = \frac{e^{4\pi}+1}{e^{\pi}+1}$

29. If $I = \int_0^{1/2} \frac{dx}{\sqrt{(1-x^{2n})}}$ for $n \geq 1$, then the value of I is

(a) less than 1 (b) more than $\frac{1}{2}$

(c) less than $\frac{1}{4}$ (d) more than 1

30. If $I = \int_0^1 \frac{dx}{1+x^{\pi/2}}$, then

(a) $I > \ln 2$ (b) $I < \ln 2$

(c) $I < \frac{\pi}{4}$ (d) $I > \frac{\pi}{4}$

31. Let $I_n = \int_0^{\pi/2} \cos^n x\,dx, n \in N$, then

(a) $I_{n-2} > I_n$ (b) $n(I_{n-2} - I_n) = I_{n-2}$

(c) $I_n : I_{n-1} = n : (n-1)$ (d) None of these

32. For $a \in R, |a| > 1$, let

$$\lim_{n\to\infty}\left(\frac{1+\sqrt[3]{2}+\ldots+\sqrt[3]{n}}{n^{7/3}\left(\frac{1}{(an+1)^2}+\frac{1}{(an+2)^2}+\ldots+\frac{1}{(an+n)^2}\right)}\right) = 54$$

Then the possible value(s) of a is/are

[JEE Advanced 2020]

(a) -6 (b) 7 (c) 8 (d) -9

33. If $a > \sqrt{(b^2+c^2)}$ for $a, b \in R$ and $c \in R^-$, then the value of $\int_0^{2\pi} \frac{d\theta}{(a+b\cos\theta + c\sin\theta)}$ cannot be

(a) $\frac{1}{\sqrt{(a^2-b^2-c^2)}} \tan^{-1}\left(\frac{c}{\sqrt{(a^2-b^2-c^2)}}\right)$

(b) $\frac{1}{\sqrt{(a^2-b^2-c^2)}} \cot^{-1}\left(\frac{c}{\sqrt{(a^2-b^2-c^2)}}\right)$

(c) $\frac{2}{\sqrt{(a^2-b^2-c^2)}} \cot^{-1}\left(\frac{c}{\sqrt{(a^2-b^2-c^2)}}\right)$

(d) $\frac{2}{\sqrt{(a^2-b^2-c^2)}} \tan^{-1}\left(\frac{c}{\sqrt{(a^2-b^2-c^2)}}\right)$

34. Which of the following is/are true

(a) $\int_0^1 2^{x^2} dx > \int_0^1 2^{x^5} dx$

(b) $\int_1^2 2^{x^3} dx > \int_1^2 2^{x^5} dx$

(c) $\int_1^2 \ln x\,dx > \int_1^2 (\ln x)^2 dx$

(d) $\int_3^4 \ln x\,dx > \int_3^4 (\ln x)^2 dx$

Linked-Comprehension Type

In these questions, a passage (paragraph) has been given followed by questions based on each of the passages. You have to answer the questions based on the passage given.

PASSAGE 1

Let $f(x)$ be a continuous function defined on the closed interval $[a, b]$, then

$$\lim_{n\to\infty} \sum_{r=0}^{n-1} \frac{1}{n} f\left(\frac{r}{n}\right) = \int_0^1 f(x)\,dx$$

On the basis of above information, answer the following questions

1. The value of $\lim_{n\to\infty} \frac{1}{n}\left\{\frac{1}{n+1}+\frac{2}{n+2}+\ldots+\frac{3n}{4n}\right\}$ is

(a) $5 - 2\ln 2$ (b) $4 - 2\ln 2$

(c) $3 - 2\ln 2$ (d) $2 - 2\ln 2$

2. The value of $\lim_{n\to\infty}\left\{\frac{1}{n}+\frac{n^2}{(n+1)^3}+\frac{n^2}{(n+2)^3}+\ldots\ldots+\frac{1}{8n}\right\}$ is

(a) 5/4 (b) 3/4

(c) 5/8 (d) 3/8

3. The value of $\lim_{n\to\infty}\left\{\prod_{r=1}^{n}\left(1+\frac{r}{n}\right)\right\}^{1/n}$ is

(a) $3/e$

(b) $4/e$

(c) $1/e$

(d) $2/e$

4. The value of

$$\lim_{n\to\infty}\left\{\tan\left(\frac{\pi}{2n}\right)\tan\left(\frac{2\pi}{2n}\right)\tan\left(\frac{3\pi}{2n}\right)\ldots\tan\left(\frac{n\pi}{2n}\right)\right\}^{1/n} \text{ is}$$

(a) 1 (b) 2
(c) 3 (d) not defined

5. The value of

$$\lim_{n\to\infty}\left\{\frac{1}{n+m}+\frac{1}{n+2m}+\frac{1}{n+3m}+\ldots+\frac{1}{n+nm}\right\} \text{ is}$$

(a) $\frac{\ln(m)}{m}$ (b) $\frac{\ln(1+m)}{(1+m)}$
(c) $\frac{\ln(1+m)}{m}$ (d) None of these

PASSAGE 2

Let $F:R\to R$ be a thrice differentiable function. Suppose that $F(1)=0, F(3)=-4$ and $F'(x)<0$ for all $x\in(1,3)$. Let $f(x)=xF(x)$ for all $x\in R$.

[JEE Advanced 2015]

On the basis of above information, answer the following questions

1. The correct statement(s) is/are
(a) $f'(1)<0$
(b) $f(2)<0$
(c) $f'(x)\neq 0$ for any $x\in(1,3)$
(d) $f'(x)=0$ for some $x\in(1,3)$

2. If $\int_1^3 x^2 F'(x)\,dx=-12$ and $\int_1^3 x^3 F''(x)\,dx=40$, then the correct expression(s) is/are
(a) $9f'(3)+f'(1)-32=0$ (b) $\int_1^3 f(x)\,dx=12$
(c) $9f'(3)-f'(1)+32=0$ (d) $\int_1^3 f(x)\,dx=-12$

PASSAGE 3

If function $f(x)$ is continuous in the interval (a,b) and having same definition between a and b, then we can find $\int_a^b f(x)\,dx$ and if $f(x)$ is discontinuous and not having same definition between a and b, then we must break the interval such that $f(x)$ becomes continuous and having same defintion in the breaking intervals.

Now, if $f(x)$ is discontinuous at $x=c$ $(a<c<b)$, then $\int_a^b f(x)\,dx=\int_a^c f(x)\,dx+\int_c^b f(x)\,dx$ and also if $f(x)$ is discontinuous at $x=a$ in $(0,2a)$, then we can write $\int_0^{2a} f(x)\,dx=\int_0^a\{f(a-x)+f(a+x)\}\,dx$

On the basis of above information, answer the following questions

1. $\int_{-1}^1 [3x-1]\,dx$ (where [.] denotes greatest integer function) is equal to
(a) -5 (b) -3
(c) -1 (d) None of these

2. $\int_{\pi/2}^{3\pi/2}[2\sin x]\,dx$ (where [.] denotes the greatest integer function) is equal to
(a) $-\frac{\pi}{2}$ (b) $-\pi$
(c) 0 (d) $\frac{\pi}{2}$

3. $\int_0^{10}\left[\frac{x^2+2}{x^2+1}\right]dx$ (where [.] denotes the greatest integer function) is equal to
(a) 0 (b) 2
(c) 5 (d) None of these

4. $\int_0^1 \sin([x]+[2x])\,dx$ (where [.] denotes the greatest integer function) is equal to
(a) sin 1 (b) $\sin\left(\frac{3}{2}\right)$
(c) $\frac{\sin 1}{2}$ (d) $\frac{\sin 2}{3}$

5. $\int_{-1}^1 [|x|]\,d\left(\frac{1}{1+e^{-1/x}}\right)$ (where [.] denotes the greatest integer function) is equal to
(a) -3 (b) -2
(c) -1 (d) None of these

PASSAGE 4

$$\int_a^b f(x)\,d\alpha(x)+\int_a^b \alpha(x)\,d f(x) =\alpha(b)f(b)-\alpha(a)f(a)$$

On the basis of above information, answer the following questions

1. $\int_0^3 (x^2+1)d[x]$ (where [.] denotes the greatest integer function) is equal to
(a) 3 (b) $\frac{9}{2}$
(c) 17 (d) $\frac{27}{2}$

2. $\int_{-2}^3 [|x|]\,d|x|$ (where [.] denotes the greatest integer function) is equal to
(a) 0 (b) 1
(c) 2 (d) None of these

3. $\int_{1/4}^{3/4}[x]\,d[2x]$ (where [.] denotes the greatest integer function) is equal to
(a) 0 (b) 1
(c) 2 (d) None of these

4. $\int_0^3 e^x\,d\{x+[x]\}$ (where [.] denotes the greatest integer function) is equal to
(a) $e-1$ (b) e^2-1
(c) $\sqrt{e}-1$ (d) None of these

5. $\int_0^6 (x^2+[x])\,d|3-x|$ (where [.] denotes the greatest integer function) is equal to

(a) 36 (b) 72

(c) 63 (d) 126

PASSAGE 5

If $f(x)$ be an increasing function defined on $[a, b]$.

Then $\max\{f(t)|a \le t \le x, a \le x \le b\} = f(x)$

and $\min\{f(t)|a \le t \le x, a \le x \le b\} = f(a)$

and If $f(x)$ be a decreasing function defined on $[a, b]$.

Then $\max\{f(t\,|\,a \le t \le x, a \le x \le b\} = f(a)$

and $\min\{f(t)|a \le t \le x, a \le x \le b\} = f(x)$

On the basis of above information, answer the following questions

1. $\int_0^3 \min\{1, |x|, |x-2|\}\,dx$ is equal to

(a) 1 (b) 3/2

(c) 2 (d) 5/2

2. $\int_{-1}^1 \max\{x, x^3\}\,dx$ is equal to

(a) 1/2 (b) 3/2

(c) 1/4 (d) 3/4

3. $\int_{-2}^2 \min\{|x|, [x]\}dx$ (where [.] denotes the greatest integer function) is equal to

(a) – 2 (b) – 1

(c) 0 (d) 1

4. $\int_0^\pi \max\{\sin x, \cos x\}\,dx$ is equal to

(a) $1-\frac{1}{\sqrt{2}}$ (b) $1+\frac{1}{\sqrt{2}}$

(c) $\sqrt{2}-1$ (d) $\sqrt{2}+1$

5. $\int_0^\pi \min\{2\sin x, 1-\cos x, 1\}\,dx$ is equal to

(a) $\frac{\pi}{6}+2-\sqrt{3}$ (b) $\frac{5\pi}{6}+1-\sqrt{3}$

(c) $\frac{\pi}{3}+2-\sqrt{3}$ (d) $\frac{\pi}{6}+1+\sqrt{3}$

Numerical Value Type

Solve the following problems and write your answer.

1. If $\int_0^{x^2(1+x)} f(t)\,dt = x$, then the value of $9905\,f(2)$ must be

2. If $\frac{\pi}{2} < \alpha < \frac{2\pi}{3}$ and $I = \int_\delta^{\sin 2\alpha} \frac{dx}{\sqrt{(4\cos^2\alpha - x^2)}}$ then the value of $\frac{2008\,(I+\alpha)}{\pi}$ must be

3. If $I = \int_0^2 x\,[2x]\,dx$, where [.] denotes the greatest integer function, then the value of $12.8\,I$ must be

4. If $I = \int_{-1}^1 \frac{e^{|x|}}{1+a^x}\,dx$, then the value of $\frac{9997}{(e-1)} I$ must be

5. If $I = \int_{-\pi/2}^{2\pi} \sin^{-1}(\sin x)\,dx$, then the value of $-\frac{7992}{\pi^2} I$ must be

6. Let $f: R \to R$ be a continuous function such that $f(x) + f(x+1) = 2$, for all $x \in R$. If $I_1 = \int_0^8 f(x)dx$ and $I_2 = \int_{-1}^3 f(x)dx$, then the value of $I_1 + 2I_2$ is equal to

[JEE Main 2021, 16 March]

7. Let $f(x)$ and $g(x)$ be two functions satisfying $f(x^2) + g(4-x) = 4x^3$ and $g(4-x) + g(x) = 0$, then the value of $\int_{-4}^4 f(x)^2 dx$ is

[JEE Main 2021, 18 March]

8. If $\int_0^{a+2008\pi} |\cos x|\,dx = \lambda + \sin a$, where $a \in \left(-\frac{\pi}{2}, \frac{\pi}{2}\right)$, then the value of (λ) must be (where (.) denotes the least integer function)

9. If $I = \int_{-3/2}^3 f(x)\,dx$, where $f(x)$ is given by $f(x) = \max_{-3/2 \le t \le x} \{|t-1| - |t| + |t+1|\}$, then the value of $64\,I$ must be

10. If $f(x) = \int_a^x \frac{1}{f(x)}\,dx$ and $\int_a^1 \frac{1}{f(x)}\,dx = \sqrt{2}$, then the value of $99\,f(200)$ must be

11. If $\int_1^3 (\{x\})^{[x]}\,dx = \lambda$, where {.} and [.] denotes the fractional part and greatest integer functions respectively, then the value of $1296\,\lambda$ must be

12. Let $f: R^+ \to R$ be a differentiable function with $f(1) = 3$ and satisfying $\int_1^{xy} f(t)\,dt = y\int_1^x f(t) + x\int_1^y f(t)\,dt \;\forall\, x, y \in R^+$, then the value of $f(e^{37})$ must be

13. If $\int_0^{1.8} [x^2]\,dx = I$, then the value of $505(I + \sqrt{2} + \sqrt{3})$ must be (where [.] denotes the greatest integer function)

14. If $\int_0^{\pi/2} f(\sin 2x)\sin x\,dx = \frac{\lambda\sqrt{2}}{208}\int_0^{\pi/4} f(\cos x)\cos x\,dx$, then the value of λ must be

15. The value of the integral $\int_0^{\pi/2} \frac{3\sqrt{\cos\theta}}{(\sqrt{\cos\theta}+\sqrt{\sin\theta})^5}\,d\theta$ equals ....

[JEE Advanced 2019]

16. If $\int_{-\sqrt{2}}^{\sqrt{2}} \frac{(2x^7 + 3x^6 - 10x^5 - 7x^3 - 12x^2 + x + 1)}{(x^2+2)}\,dx = \frac{\sqrt{2}}{20}(5\pi - \lambda)$, then the value of λ must be

Single Integer Type

Solve the following problems and write your answer.

1. Let $[t]$ denote the greatest integer less than or equal to t. Then the value of $\int_1^2 |2x - [3x]|\, dx$ is ....

[JEE Main 2020, 2 September]

2. Let $f : (0, 2) \to R$ be defined as $f(x) = \log_2\left(1 + \tan\left(\frac{\pi x}{4}\right)\right)$.

Then, $\lim_{n\to\infty} \frac{2}{n}\left(f\left(\frac{1}{n}\right) + f\left(\frac{2}{n}\right) + \ldots + f(1)\right)$ is equal to

[JEE Main 2021, 16 March]

3. If $[\cdot]$ represents the greatest integer function, then the value of $\left|\int_0^{\sqrt{\frac{\pi}{2}}} [[x^2] - \cos x]\, dx\right|$ is

[JEE Main 2021, 17 March]

4. If $I = \frac{2}{\pi}\int_{-\pi/4}^{\pi/4} \frac{dx}{(1 + e^{\sin x})(2 - \cos 2x)}$ then $27I^2$ equals ....

[JEE Advanced 2019]

5. The value of the integral $\int_0^{1/2} \frac{1 + \sqrt{3}}{((x + 1)^2 (1 - x)^6)^{1/4}}\, dx$ is ....

[JEE Advanced 2018]

6. Let $f : R \to R$ be a function defined by $f(x) = \begin{cases} [x], & x \le 2 \\ 0, & x > 2 \end{cases}$, where $[x]$ denotes the greatest integer less than or equal to x.

If $I = \int_{-1}^{2} \frac{xf(x^2)}{2 + f(x + 1)}\, dx$, then the value of $(4I - 1)$ is

[JEE Advanced 2015]

7. If $\alpha = \int_0^1 (e^{9x + 3\tan^{-1} x})\left(\frac{12 + 9x^2}{1 + x^2}\right) dx$, where $\tan^{-1} x$ takes only principal values, then the value of $\left(\log_e|1 + \alpha| - \frac{3\pi}{4}\right)$ is

[JEE Advanced 2015]

8. Let $f : R \to R$ be a differentiable function such that its derivative f' is continuous and $f(\pi) = -6$.

If $F : [0, \pi] \to R$ is defined by $F(x) = \int_0^x f(t)dt$, and if

$$\int_0^{\pi} (f'(x) + F(x))\cos x\, dx = 2$$

then the value of $f(0)$ is ....

[JEE Advanced 2020]

Matrix-Match Type

Given below are Matching Type Questions, with two columns (each having some items) each . Each item of Column I has to be matched with the items of Column II, by encircling the correct match(es).

NOTE *An item of Column I can be matched with more than one items of Column II.*

1. Observe the following columns :

	Column I		Column II
(A)	If $\int_1^4 \|x - 3\|\, dx = 2a + b$, then	(P)	$a = 3, b = 3/2$
(B)	If $\int_{-1}^{1} \frac{\|x\|}{x} dx = a - 2b$, then	(Q)	$a = 1, b = \frac{1}{2}$
(C)	If $\int_0^{2\pi} \|\sin x\|\, dx = 5a + b$, then	(R)	$a = 1, b = -1$
		(S)	$a = 2, b = -3/2$
		(T)	$a = \frac{1}{2}, b = \frac{3}{2}$

2. If m is the least value (global minimum) and M is the greatest value (global maximum) of the function $f(x)$ on the interval $[a, b]$. (estimation of an integral).

Then,

$$m(b - a) \le \int_a^b f(x)dx \le M(b - a).$$

Match the entries from the following two columns :

	Column I		Column II
(A)	If $\mu < \int_0^2 \left(\frac{5 - x}{9 - x^2}\right) dx < \lambda$, then	(P)	$[6\lambda + 8\mu] = 3$, where [.] denotes the greatest integer function.
		(Q)	$[\lambda - 2] = 8$, where [.] denotes the greatest integer function.
(B)	If $\mu < \int_{\pi/4}^{\pi/3} \frac{\sin x}{x} dx < \lambda$, then	(R)	$[\lambda + \mu] = 2$, where [.] denotes the greatest integer function.
(C)	If $\mu < \int_1^3 \sqrt{(3 + x^3)}\, dx < \lambda$, then	(S)	$[\lambda - \mu] = 6$, where [.] denotes the greatest integer function.
		(T)	$[5\lambda - \mu] = 5$, where [.] denotes the greatest integer function.

3. If at every point x of an interval $[a, b]$ the inequalities $g(x) \le f(x) \le h(x)$ are fulfilled, then

$$\int_a^b g(x)\, dx \le \int_a^b f(x)\, dx \le \int_a^b h(x)\, dx,\ a < b$$

Match the entries from the following two columns :

	Column I		Column II
(A)	If $\mu < \int_0^1 \frac{x^7\,dx}{\sqrt[3]{(1+x^8)}} < \lambda$, then	(P)	$[\lambda+\mu]=2$, where [.] denotes the greatest integer function.
		(Q)	$[\lambda+\mu]=4$, where [.] denotes the greatest integer function.
(B)	If $\mu < \int_0^1 \frac{dx}{\sqrt{(1+x^6)}} < \lambda$, then	(R)	$[\lambda-\mu]=0$, where [.] denotes the greatest integer function.
(C)	If $\mu < \int_0^1 \frac{dx}{\sqrt{(4-x^2-x^3)}} < \lambda$, then	(S)	$[\lambda-\mu]=3$, where [.] denotes the greatest integer function.
		(T)	$[\lambda+\mu]=0$, where [.] denotes the greatest integer function.

4. Observe the following columns :

	Column I		Column II
(A)	The value of $\int_\alpha^{\pi/2-\alpha} \frac{d\theta}{1+\cot^n\theta}$ Where, $0<\alpha<\frac{\pi}{2}, n>0$ is	(P)	$\frac{\pi}{2}$
		(Q)	$\frac{\pi}{4}-\alpha$
		(R)	$2\pi^2 - 2\pi\alpha$
(B)	The value of $\int_{-\pi}^{\pi} \frac{\sin^2 x}{1+\alpha^x}dx, \alpha>0$ is	(S)	dependent of α
		(T)	independent of n
(C)	The value of $\int_\alpha^{2\pi-\alpha} \frac{x\sin^{2n} x}{\sin^{2n} x+\cos^{2n} x}dx$ is		

Answers

Objective Questions Type I [Only one correct answer]

1. (c) 2. (d) 3. (b) 4. (c) 5. (d) 6. (d) 7. (c) 8. (d) 9. (d) 10. (d)
11. (b) 12. (a) 13. (b) 14. (d) 15. (d) 16. (c) 17. (d) 18. (b) 19. (d) 20. (a)
21. (b) 22. (a) 23. (c) 24. (d) 25. (a) 26. (c) 27. (d) 28. (b) 29. (d) 30. (b)
31. (a) 32. (a) 33. (b) 34. (a) 35. (a) 36. (c) 37. (a) 38. (c) 39. (a) 40. (b)
41. (c) 42. (d) 43. (c) 44. (a) 45. (c) 46. (a) 47. (c) 48. (c) 49. (a) 50. (a)
51. (c) 52. (c) 53. (b) 54. (a) 55. (a) 56. (a) 57. (d) 58. (b) 59. (b) 60. (d)
61. (d) 62. (c) 63. (b) 64. (c) 65. (c) 66. (c) 67. (d) 68. (d) 69. (d) 70. (c)
71. (b) 72. (b) 73. (c) 74. (a) 75. (c) 76. (b) 77. (b) 78. (c) 79. (a) 80. (c)
81. (c) 82. (c) 83. (b) 84. (a) 85. (b) 86. (b) 87. (a) 88. (a) 89. (b) 90. (a)
91. (d) 92. (b) 93. (d) 94. (a) 95. (a) 96. (a) 97. (d) 98. (a) 99. (c) 100. (d)
101. (d)

Objective Questions Type II [One or more than one correct answer(s)]

1. (a, b) 2. (a, b, c) 3. (a, d) 4. (a, c) 5. (a, c)
6. (a, d) 7. (a, c) 8. (b, c, d) 9. (a, c) 10. (a, c)
11. (a, b) 12. (c, d) 13. (a, b, c, d) 14. (a, b, c, d) 15. (a, c, d)
16. (a, b, c, d) 17. (a, b, c, d) 18. (a, d) 19. (b, d) 20. (a, b)
21. (a, c, d) 22. (b, c) 23. (a, c, d) 24. (a, c, d) 25. (a, b, c)
26. (b, d) 27. (a, d) 28. (a, c) 29. (a, b) 30. (a, c)
31. (a, b) 32. (c, d) 33. (a, b, c, d) 34. (c)

Linked-Comprehension Type

Passage 1 1. (c) 2. (d) 3. (b) 4. (a) 5. (c) **Passage 2** 1. (a, b, c) 2. (c, d)
Passage 3 1. (b) 2. (a) 3. (d) 4. (c) 5. (d) **Passage 4** 1. (c) 2. (c) 3. (a) 4. (d) 5. (c)
Passage 5 1. (b) 2. (c) 3. (a) 4. (d) 5. (b)

Numerical Value Type Questions

1. (1981) 2. (2008) 3. (54.4) 4. (9997) 5. (999) 6. (16)
7. (512) 8. (4016) 9. (408) 10. (1980) 11. (1080) 12. (114)
13. (2222) 14. (208) 15. (0.50) 16. (64)

Single Integer Type Questions

1. (1) 2. (1) 3. (1) 4. (4) 5. (2) 6. (0) 7. (9) 8. (4)

Matrix-Match Type

1. A → (Q, S, T); B → (P, Q); C → (R, T)
2. A → (R, T); B → (P); C → (Q, S)
3. A → (R, T); B → (P, R); C → (Q, S)
4. A → (Q, S, T); B → (P, T); C → (R, S, T)

Solutions

Objective Questions Type [Only one correct answer]

1. Given integral $\int_{-\pi}^{\pi} |\pi - |x|| \, dx$

$$= 2\int_0^{\pi} |\pi - |x|| \, dx = 2\int_0^{\pi} |\pi - x| \, dx$$

$$= 2\int_0^{\pi} (\pi - x) \, dx \qquad [\because x \in (0, \pi)]$$

$$= 2\left[\pi x - \frac{x^2}{2}\right]_0^{\pi} = 2\left[\pi^2 - \frac{\pi^2}{2}\right] = 2\left(\frac{\pi^2}{2}\right) = \pi^2$$

2. $\because \quad f(x) = \int_0^x \frac{dt}{\{f(t)\}^2}$

$\therefore \quad f'(x) = \frac{1}{(f(x))^2}$

$\Rightarrow (f(x))^2 \cdot f'(x) = 1$

or $\quad \frac{(f(x))^3}{3} = x + c$

Put $x = 2$, then

$$\frac{(f(2))^3}{3} = 2 + c \qquad [\because f(2) = \sqrt[3]{6}]$$

$\Rightarrow \quad \frac{6}{3} = 2 + c$

$\therefore \quad c = 0$

$\Rightarrow \quad f(x) = (3x)^{1/3}$

$\therefore \quad f(9) = (27)^{1/3} = 3$

3. $\because A = \int_0^{\pi} \frac{\cos x}{(x+2)^2} dx = \left\{\cos x \cdot \left(-\frac{1}{x+2}\right)\right\}_0^{\pi}$

$$- \int_0^{\pi} (-\sin x)\left(-\frac{1}{x+2}\right) dx$$

$$= \frac{1}{\pi + 2} + \frac{1}{2} - \int_0^{\pi} \frac{\sin x}{x+2} dx$$

$$A = \frac{1}{\pi+2} + \frac{1}{2} - 2\int_0^{\pi/2} \frac{\sin 2x}{2x+2} dx \qquad \text{[by property]}$$

$$\therefore \int_0^{\pi/2} \frac{\sin 2x}{x+1} dx = \frac{1}{\pi+2} + \frac{1}{2} - A$$

4. $I = \int_0^{10} [x] \cdot e^{[x]-x+1}$

$$I = \int_0^1 0 \, dx + \int_1^2 1 \cdot e^{2-x} + \int_2^3 2 \cdot e^{3-x} + \dots + \int_9^{10} 9 \cdot e^{10-x} dx$$

$$\Rightarrow I = \Sigma_{n=0}^{9} \int_n^{n+1} n \cdot e^{n+1-x} dx$$

$$= -\Sigma_{n=0}^{9} n (e^{n+1-x})_n^{n+1}$$

$$= (e-1)\Sigma_{n=0}^{9} n$$

$$= -\Sigma_{n=0}^{9} n \cdot (e^0 - e^1)$$

$$= 45(e-1)$$

5. $\because x > 0$

$\therefore \quad \sin x < x \Rightarrow \cos(\sin x) > \cos x \qquad \dots$(i)

Also $\because \quad 0 < x < \frac{\pi}{2}$

$\therefore \quad 1 > \cos x > 0$

Now, $\quad \sin x < x$ for $x \in \left(0, \frac{\pi}{2}\right)$

$\therefore \quad \sin(\cos x) < \cos x \qquad \dots$(ii)

$\therefore$ From Eqs. (i) and (ii), we get

$$\cos(\sin x) > \cos x > \sin(\cos x)$$

or $\quad \int_0^{\pi/2} \cos(\sin x) \, dx > \int_0^{\pi/2} \cos x \, dx$

$$> \int_0^{\pi/2} \sin(\cos x) \, dx$$

$\Rightarrow \quad I_1 > I_3 > I_2$

6. $$\lim_{n \to \infty} \frac{(\Sigma n^2)(\Sigma n^3)}{(\Sigma n^6)} = \lim_{n \to \infty} \frac{\sum_{r=1}^{n} \left(\frac{r}{n}\right)^2 \sum_{r=1}^{n} \left(\frac{r}{n}\right)^3}{n \sum_{r=1}^{n} \left(\frac{r}{n}\right)^6}$$

$$= \frac{\lim_{n \to \infty} \frac{1}{n} \sum_{r=1}^{n} \left(\frac{r}{n}\right)^2 \cdot \lim_{n \to \infty} \frac{1}{n} \sum_{r=1}^{n} \left(\frac{r}{n}\right)^3}{\lim_{n \to \infty} \frac{1}{n} \sum_{r=1}^{n} \left(\frac{r}{n}\right)^6}$$

$$= \frac{\int_0^1 x^2 \, dx \cdot \int_0^1 x^3 \, dx}{\int_0^1 x^6 \, dx}$$

$$= \frac{\frac{1}{3} \cdot \frac{1}{4}}{\frac{1}{7}} = \frac{7}{12}$$

7. The given integrals can be equal only if $a > \tan^2 1$

$$\therefore \int_0^a [\tan^{-1}\sqrt{x}]\,dx = \int_0^{\tan^2 1} [\tan^{-1}\sqrt{x}]\,dx + \int_{\tan^2 1}^{a} [\tan^{-1}\sqrt{x}]\,dx$$

$$= 0 + 1\cdot\int_{\tan^2 1}^{a} dx = a - \tan^2 1$$

$$\int_0^a [\cot^{-1}\sqrt{x}]\,dx = \int_0^{\cot^2 1} [\cot^{-1}\sqrt{x}]\,dx + 0$$

$$= 1\cdot(\cot^2 1 - 0)$$

$$\therefore \quad a - \tan^2 1 = \cot^2 1$$

$$\Rightarrow \quad a = \tan^2 1 + \cot^2 1 = \frac{\sin^2 1}{\cos^2 1} + \frac{\cos^2 1}{\sin^2 1}$$

$$= \frac{1-\cos 2}{1+\cos 2} + \frac{1+\cos 2}{1-\cos 2}$$

$$= \frac{(1+\cos 2)^2 + (1-\cos 2)^2}{\sin^2 2}$$

$$= \frac{2(1+\cos^2 2)}{\sin^2 2}$$

$$= \frac{2(2+2\cos^2 2)}{2\sin^2 2} = \frac{2(3+\cos 4)}{(1-\cos 4)}$$

8. Let $f(x) = \dfrac{2[x]}{3x - [x]}$

It is clear that $f(x)$ is not defined if $x = 0$ and if $3x = [x]$

So, in $(-10, 0)$, f is not defined at $x = -\dfrac{1}{3}$

Case I : $x \in \left(-10, -\dfrac{1}{3}\right)$

$[x] < 0$ and $3x - [x] < 0$

so $$\frac{[x]}{3x-[x]} > 0$$

$$\therefore \quad \left|\frac{2[x]}{3x-[x]}\right| = \frac{2[x]}{3x-[x]}$$

$$\Rightarrow \quad \int_{-10}^{-1/3} 1\,dx = -\frac{1}{3} + 10 = \frac{29}{3}$$

Case II : $x \in \left(-\dfrac{1}{3}, 0\right)$

$[x] < 0$ and $3x - [x] > 0$

So, $$\frac{[x]}{3x-[x]} < 0$$

$$\therefore \quad \left|\frac{2[x]}{3x-[x]}\right| = -\left(\frac{2[x]}{3x-[x]}\right)$$

$$\Rightarrow \quad \int_{-\frac{1}{3}}^{0} (-1)\,dx = -\left(0+\frac{1}{3}\right) = -\frac{1}{3}$$

Hence,

$$\int_{-10}^{0} \frac{|f(x)|}{f(x)}\,dx = \int_{-10}^{-1/3} \frac{|f(x)|}{f(x)}\,dx + \int_{-1/3}^{0} \frac{|f(x)|}{f(x)}\,dx$$

$$= \frac{29}{3} - \frac{1}{3} = \frac{28}{3}$$

9. $$\int_0^4 \{\sqrt{x}\}\,dx = \int_0^4 (\sqrt{x} - [\sqrt{x}])\,dx$$

$$= \int_0^4 \sqrt{x}\,dx - \int_0^4 [\sqrt{x}]\,dx$$

$$= \frac{2}{3}\{x^{3/2}\}_0^4 - \left(\int_0^{1^2} [\sqrt{x}]\,dx + \int_{1^2}^{2^2} [\sqrt{x}]dx\right)$$

$$= \frac{2}{3}\cdot 8 - [0 + 1\cdot(2^2 - 1^2)] = \frac{16}{3} - 3 = \frac{7}{3}$$

10. $\because f(x) + f(x+4) = f(x+2) + f(x+6)$ ....(i)

Replacing x by $x+2$ in Eq. (i), then

$f(x+2) + f(x+6) = f(x+4) + f(x+8)$, ...(ii)

From Eqs. (i) and (ii), we get

$$f(x+8) = f(x)$$

$\Rightarrow \quad f(x+8) - f(x) = 0$ ...(iii)

Now, $$g(x) = \int_x^{x+8} f(t)\,dt$$

$\therefore \quad g'(x) = f(x+8) - f(x)$

$= 0$ [from Eq. (iii)]

11. $\because \sin^{-1}(\cos x) + \cos^{-1}(\sin x)$ is periodic with period 2π, then

$$\int_{a+2\pi}^{a+5\pi/2} (\sin^{-1}(\cos x) + \cos^{-1}(\sin x))\,dx$$

$$= \int_0^{\pi/2} (\sin^{-1}(\cos x) + \cos^{-1}(\sin x))\,dx$$

$$= \int_0^{\pi/2} \left(\sin^{-1}\left\{\cos\left(\frac{\pi}{2} - x\right)\right\} + \cos^{-1}\left\{\sin\left(\frac{\pi}{2} - x\right)\right\}\right) dx$$

$$= \int_0^{\pi/2} (\sin^{-1}\sin x + \cos^{-1}\cos x)\,dx$$

$$= 2\int_0^{\pi/2} x\,dx = 2\left\{\frac{x^2}{2}\right\}_0^{\frac{\pi}{2}} = \frac{\pi^2}{4}$$

12. $f(x) = [6\sin t + 2\cos t]_{5\pi/3}^{x}$

$$= (6\sin x + 2\cos x) - \left(6\times\left(\frac{-\sqrt{3}}{2}\right) + 2\times\frac{1}{2}\right)$$

$$= 6\sin x + 2\cos x + 3\sqrt{3} - 1$$

$\Rightarrow f'(x) = 6\cos x - 2\sin x = 0$

$\Rightarrow \quad 3\cos x = \sin x$

$\Rightarrow \quad \tan x = 3$

$f''(x) = -6\sin x - 2\cos x < 0$

$f(x)$ gives maximum value at $x = \dfrac{7\pi}{4}$

$$f\left(\frac{7\pi}{4}\right) = 6\sin\frac{7\pi}{4} + 2\cos\frac{7\pi}{4} + 3\sqrt{3} - 1$$

$$= 6\left(\frac{-1}{\sqrt{2}}\right) + 2\times\frac{1}{\sqrt{2}} + 3\sqrt{3} - 1$$

$$= -3\sqrt{2} + \sqrt{2} + 3\sqrt{3} - 1$$

$$= 3\sqrt{3} - 2\sqrt{2} - 1$$

13. $\because \sin^4 t + \cos^4 t$ is periodic with period $\dfrac{\pi}{2}$

Now, $$f(x+\pi) = \int_0^{\pi+x} (\sin^4 t + \cos^4 t)\,dt$$

$$= \int_0^{x} (\sin^4 t + \cos^4 t)\,dt + \int_x^{x+\pi} (\sin^4 t + \cos^4 t)\,dt$$

$$= f(x) + \int_0^{\pi} (\cos^4 t + \sin^4 t)\,dt = f(x) + f(\pi)$$

or $$= f(x) + 2\int_0^{\pi/2} (\cos^4 t + \sin^4 t)\,dt = f(x) + 2f\left(\frac{\pi}{2}\right)$$

14. $\because I_k = \int_{-2k\pi}^{0} |\sin x|[\sin x]\,dx + \int_0^{2k\pi} |\sin x|[\sin x]\,dx$

$$= -\int_{2k\pi}^{0} |\sin x|[-\sin x]\,dx + \int_0^{2k\pi} |\sin x|[\sin x]\,dx$$

$$= \int_0^{2k\pi} |\sin x|([\sin x] + [-\sin x])\,dx$$

$$= 2k\int_0^{\pi} |\sin x|([\sin x] + [-\sin x])\,dx$$

$$= 2k\int_0^{\pi} \sin x\,(0-1)\,dx$$

$$= -2k(-\cos x)_0^{\pi} \quad \left[\begin{array}{l}\because \sin x > 0,\ x \in (0,\pi) \\ \therefore -\sin x < 0,\ x \in (0,\pi)\end{array}\right]$$

$$= -4k$$

$$\therefore \sum_{k=1}^{10} I_k = -4\sum_{k=1}^{10} k = -4\cdot\frac{10\cdot 11}{2} = -220$$

15. $\because$ Integrand is discontinuous at $\frac{\pi}{2}$, then

$$\int_0^{\pi/2} 0\cdot dx + \int_{\pi/2}^{3\pi/2} 0\cdot dx = 0$$

$\because \quad 0 < x < \frac{\pi}{2}, |\tan^{-1}\tan x| = |\sin^{-1}\sin x|$

and $\frac{\pi}{2} < x < \frac{3\pi}{2}, |\tan^{-1}\tan x| = |\sin^{-1}\sin x|$

16. $\because \sqrt{x + 2\sqrt{(x-1)}} = \sqrt{(\sqrt{(x-1)})^2 + 1^2 + 2\sqrt{(x-1)}}$

$$= \sqrt{(x-1)} + 1$$

and $\sqrt{x - 2\sqrt{(x-1)}} = \sqrt{(\sqrt{(x-1)})^2 + 1^2 - 2\sqrt{(x-1)}}$

$$= |\sqrt{(x-1)} - 1|$$

Then $\int_1^5 (\sqrt{x + 2\sqrt{(x-1)}} + \sqrt{x - 2\sqrt{(x-1)}})\,dx$

$$= \int_1^5 (\sqrt{(x-1)} + 1)\,dx + \int_1^5 |\sqrt{(x-1)} - 1|\,dx$$

$$= \int_1^5 (\sqrt{(x-1)} + 1)\,dx + \int_1^2 (1 - \sqrt{(x-1)})\,dx + \int_2^5 (\sqrt{(x-1)} - 1)\,dx$$

$$= \int_0^4 (\sqrt{x} + 1)\,dx + \int_0^1 (1 - \sqrt{x})\,dx + \int_1^4 (\sqrt{x} - 1)\,dx$$

$$= \left[\frac{2}{3}(x^{3/2}) + x\right]_0^4 + \left[x - \frac{2}{3}x^{3/2}\right]_0^1 + \left[\frac{2}{3}x^{3/2} - x\right]_1^4$$

$$= \left(\frac{16}{3} + 4\right) + \left(1 - \frac{2}{3}\right) + \left(\frac{16}{3} - 4\right) + \left(\frac{2}{3} - 1\right) = \frac{32}{3}$$

17. $\because \int_0^n [x]\,dx = \int_0^n (x - \{x\})\,dx = \int_0^n x\,dx - \int_0^n \{x\}\,dx$

$$= \int_0^n x\,dx - n\int_0^1 \{x\}\,dx$$

$$= \int_0^n x\,dx - n\int_0^1 x\,dx = \left(\frac{n^2}{2} - \frac{n}{2}\right)$$

$$\therefore \frac{\int_0^n [x]\,dx}{\int_0^n \{x\}\,dx} = \frac{\frac{n^2}{2} - \frac{n}{2}}{\frac{n}{2}} = n - 1$$

18. $\int_{-1}^{15} \text{sgn}(\{x\})\,dx = \int_0^{16} \text{sgn}(\{x-1\})\,dx$ (by property)

$$= \int_0^{16} \text{sgn}(\{x\})dx = 16\int_0^1 \text{sgn}(\{x\})dx = 16\int_0^1 \text{sgn}(x)\,dx$$

$$= 16\int_0^1 1\cdot dx = 16$$

19. $\int_0^2 \{3x\}dx = \int_0^2 (3x - [3x])\,dx = 3\int_0^2 x\,dx - \int_0^2 [3x]\,dx$

$$= 3\cdot\left[\frac{x^2}{2}\right]_0^2 - \frac{1}{3}\int_0^6 [x]\,dx \quad \text{[by property]}$$

$$= 6 - \frac{1}{3}\sum_{r=0}^{5}\int_r^{r+1} [x]\,dx = 6 - \frac{1}{3}\sum_{r=0}^{5} r$$

$$= 6 - \frac{1}{3}(0 + 1 + 2 + \ldots\ldots + 5)$$

$$= 6 - \frac{1}{3}\cdot 15 = 1$$

20. $\because x = [x] + \{x\} = n + f$ (say)

where $[x] = n$, $\{x\} = f (0 \le f < 1)$

Now, $\int_0^x [t]^2 dt = 2(x-1)$

$$\Rightarrow \int_0^{n+f} [t]^2 dt = 2(n + f - 1)$$

$$\Rightarrow \sum_{r=0}^{n-1}\int_r^{r+1} [t]^2 dt + \int_n^{n+f} [t]^2 dt = 2(n + f - 1)$$

$$\Rightarrow \sum_{r=0}^{n-1} r^2 + n^2 f = 2(n + f - 1)$$

$$\Rightarrow \frac{(n-1)n(2n-1)}{6} - 2(n-1) = (2 - n^2)f$$

For $n = 1$, $0 - 0 = f \Rightarrow f = 0$

Then, $x = n + f = 1 + 0 = 1$

For $n = 2, 1 - 2 = -2f \Rightarrow f = \frac{1}{2}$

Then, $x = 2 + \frac{1}{2} = \frac{5}{2}$

For $n > 2$, $f < 0$ impossible

Hence, $x = 1, \frac{5}{2}$

21. $f(x) = e^{-x}\sin x$

Now, $F(x) = \int_0^x f(t)\,dt \Rightarrow F'(x) = f(x)$

$$I = \int_0^1 (F'(x) + f(x))e^x dx = \int_0^1 (f(x) + f(x))\cdot e^x dx$$

$$= 2\int_0^1 f(x)\cdot e^x dx = 2\int_0^1 e^{-x}\sin x\cdot e^x dx$$

$$= 2\int_0^1 \sin x\,dx = 2(1 - \cos 1)$$

$$I = 2\left\{1 - \left[1 - \left(\frac{1^2}{2!}\right) + \left(\frac{1^4}{4!}\right) - \left(\frac{1^6}{6!}\right) + \ldots\right]\right\}$$

$$\Rightarrow 2\left\{1 - \left[1 - \left(\frac{1}{2}\right) + \left(\frac{1}{2^4}\right)\right]\right\} < 2(1 - \cos 1)$$

$$< 2\left\{1 - \left[1 - \left(\frac{1}{2}\right) + \left(\frac{1}{2^4}\right) - \left(\frac{1}{720}\right)\right]\right\}$$

$$\Rightarrow \frac{330}{360} < 2(1-\cos 1) < \frac{331}{360}$$

$$\Rightarrow \frac{330}{360} < I < \frac{331}{360}$$

$$\Rightarrow \quad I \in \left[\frac{330}{360}, \frac{331}{360}\right]$$

22. Let $I_n = \int_0^{\pi} \frac{\sin\left(n+\frac{1}{2}\right)x}{\sin x/2}\,dx$

$$\therefore I_n - I_{n-1} = \int_0^{\pi} \frac{\left\{\sin\left(n+\frac{1}{2}\right)x - \sin\left(n-\frac{1}{2}\right)x\right\}}{\sin\frac{x}{2}}\,dx$$

$$= \int_0^{\pi} \frac{2\cos nx \cdot \sin(x/2)}{\sin(x/2)}\,dx$$

$$= \int_0^{\pi} 2\cos nx\,dx$$

$$= 2\left\{\frac{\sin nx}{n}\right\}_0^{\pi}$$

$$= 0 - 0$$

$$= 0$$

$$\Rightarrow \quad I_n = I_{n-1}$$

Replacing n by $n-1$, $n-2$, ..., then we get

$I_n = I_{n-1} = I_{n-2} = \ldots = I_1$

$$\Rightarrow \; I_n = I_1 = \int_0^{\pi} \frac{\sin(3x/2)}{\sin(x/2)}\,dx = \int_0^{\pi}\left(\frac{\sin 2x + \sin x}{\sin x}\right)dx$$

$$= \int_0^{\pi} (2\cos x + 1)\,dx$$

$$= \{2\sin x + x\}_0^{\pi} = \pi$$

Hence, $\quad I_n = \pi$

23. $\int_0^{\pi/3} [\sqrt{3}\tan x]\,dx = \int_0^{\pi/6} [\sqrt{3}\tan x]\,dx$

$$+ \int_{\pi/6}^{\tan^{-1}(2/\sqrt{3})} [\sqrt{3}\tan x]\,dx$$

$$+ \int_{\tan^{-1}(2/\sqrt{3})}^{\pi/3} [\sqrt{3}\tan x]\,dx$$

$$= 0 + 1\cdot\left(\tan^{-1}\left(\frac{2}{\sqrt{3}}\right) - \frac{\pi}{6}\right) + 2\left(\frac{\pi}{3} - \tan^{-1}\left(\frac{2}{\sqrt{3}}\right)\right)$$

$$= \frac{\pi}{2} - \tan^{-1}\left(\frac{2}{\sqrt{3}}\right)$$

24. Given function, $f(x) = |x-2|$

$$= \begin{bmatrix} x-2, & x \ge 2 \\ 2-x, & x < 2 \end{bmatrix}, \; x \in [0, 4]$$

and $g(x) = f(f(x)) = |f(x) - 2| = \begin{bmatrix} f(x)-2, & f(x) \ge 2 \\ 2-f(x), & f(x) < 2 \end{bmatrix}$

$$= \begin{bmatrix} 2-(2-x), & 0 \le x < 2 \\ 2-(x-2), & 2 \le x \le 4 \end{bmatrix}$$

$$= \begin{bmatrix} x, & 0 \le x < 2 \\ 4-x, & 2 \le x \le 4 \end{bmatrix}$$

$$\therefore \int_0^3 (g(x) - f(x))\,dx$$

$$= \int_0^2 [x-(2-x)]\,dx + \int_2^3 [(4-x)-(x-2)]\,dx$$

$$= [x^2 - 2x]_0^2 + [6x - x^2]_2^3$$

$$= 18 - 9 - 12 + 4 = 1$$

25. Let $I = \int_{-1}^{1} [x[1+\sin \pi x] + 1]\,dx$

$$= \int_{-1}^{0} [x[1+\sin \pi x] + 1]\,dx + \int_0^1 [x[1+\sin \pi x] + 1]\,dx$$

Now, $\quad -1 < x < 0 \Rightarrow [1 + \sin \pi x] = 0$

and $\quad 0 < x < 1 \Rightarrow [1 + \sin \pi x] = 1$

$$\therefore \quad I = \int_{-1}^{0} 1\cdot dx + \int_0^1 [x+1]\,dx$$

$$= \int_{-1}^{0} 1\cdot dx + \int_0^1 [x]\,dx + \int_0^1 1\cdot dx$$

$$= [0-(-1)] + 0 + (1-0) = 2$$

26. Let $I = \int_{-\pi/4}^{n\pi - \pi/4} |\sin x + \cos x|\,dx$

$$= \sqrt{2}\int_{-\pi/4}^{n\pi - \pi/4} |\sin(x + \pi/4)|\,dx$$

Put $\quad x + \pi/4 = t$

$\therefore \quad dx = dt$

Then $\quad I = \sqrt{2}\int_0^{n\pi} |\sin t|\,dt \quad$ [by property]

$$= n\sqrt{2}\int_0^{\pi} |\sin t|\,dt$$

$$= n\sqrt{2}\int_0^{\pi} \sin t\,dt$$

$$= -n\sqrt{2}[\cos t]_0^{\pi}$$

$$= -n\sqrt{2}\{-1-1\} = 2\sqrt{2}n$$

27. Given,

$$f(x) = \int_{-1}^{1} \frac{\sin x\,dt}{\sin^2 x + (t - \cos x)^2}$$

$$= \frac{\sin x}{\sin x}\tan^{-1}\left(\frac{t-\cos x}{\sin x}\right)\Bigg|_{-1}^{1}$$

$$= \tan^{-1}\left(\frac{1-\cos x}{\sin x}\right) - \tan^{-1}\left(\frac{-1-\cos x}{\sin x}\right)$$

$$= \tan^{-1}(\tan x/2) + \tan^{-1}(\cot x/2)$$

Now, we know that

$$\tan^{-1} x + \tan^{-1}\frac{1}{x} = \begin{cases} \frac{\pi}{2}, & x > 0 \\ -\frac{\pi}{2}, & x < 0 \end{cases}$$

Or, $\tan^{-1}\left(\tan\frac{x}{2}\right) + \tan^{-1}\left(\frac{1}{\tan\frac{x}{2}}\right) = \begin{cases} \frac{\pi}{2}, \tan\frac{x}{2} > 0 \\ -\frac{\pi}{2}, \tan\frac{x}{2} < 0 \end{cases}$

Hence, range of $f(x)$ is $\left\{-\frac{\pi}{2}, \frac{\pi}{2}\right\}$.

28. Using property $\int_a^b f(x)\,dx = \int_a^b f(a+b-x)\,dx$

Then, $I_1 = \int_{\sin^2 t}^{1+\cos^2 t} (2-x) f\{(2-x)(2-(2-x))\}\,dx$

$= \int_{\sin^2 t}^{1+\cos^2 t} (2-x) f((2-x)x)\,dx = 2I_2 - I_1$

$\Rightarrow \quad 2I_1 = 2I_2$

$\therefore \quad \frac{I_1}{I_2} = 1$

29. Let $I = \int_0^{n\pi + t} (|\cos x| + |\sin x|)\,dx$

$= \int_0^{n\pi} (|\cos x| + |\sin x|)\,dx + \int_{n\pi}^{n\pi + t} (|\cos x| + |\sin x|)\,dx$

$\because$ Period of $|\cos x| + |\sin x|$ is $\pi/2$, then

$I = \int_0^{2n(\pi/2)} (|\cos x| + |\sin x|)\,dx + \int_{2n(\pi/2)}^{2n(\pi/2)+t} (|\cos x| + |\sin x|)\,dx$

$= 2n\int_0^{\pi/2} |\cos x| + |\sin x|\,dx + \int_0^t (|\cos x| + |\sin x|)\,dx$ [by property]

$= 2n\int_0^{\pi/2} (\cos x + \sin x)\,dx + \int_0^t (\cos x + \sin x)\,dx$

$= 2n\{\sin x - \cos x\}_0^{\pi/2} + \{\sin x - \cos x\}_0^t$

$= 2n((1-0)-(0-1)) + ((\sin t - \cos t)-(0-1))$

$= 4n + \sin t - \cos t + 1$

30. We have,

$2x^5 + x^4 - 2x^3 + 2x^2 + 1 = 2x^3(x^2-1) + (x^2+1)^2$

$\therefore I = \int_2^3 \frac{2x^3(x^2-1) + (x^2+1)^2}{(x^2+1)^2(x^2-1)}\,dx$

$= \int_2^3 \frac{2x^3 dx}{(x^2+1)^2} + \int_2^3 \frac{dx}{x^2-1}$

$= I_1 + \frac{1}{2}\log\left|\frac{x-1}{x+1}\right|\Bigg]_2^3$

$= I_1 + \frac{1}{2}\left(\log\frac{2}{1} - \log\frac{4}{3}\right)$

where, $I_1 = \int_2^3 \frac{x^2}{(x^2+1)^2}(2x)\,dx$

Put $x^2 + 1 = t \Rightarrow 2x\,dx = dt$

$\therefore \quad I_1 = \int_5^{10} \frac{t-1}{t^2}\,dt = \left(\log|t| + \frac{1}{t}\right)\Bigg]_5^{10}$

$= \log 2 - \frac{1}{10}$

Thus, $\quad I = \frac{1}{2}\log 6 - \frac{1}{10}$

31. Let $I = \int_{-\pi/4}^{\pi/4} \frac{e^x \cdot \sec^2 x\,dx}{e^{2x}-1}$

If $\quad f(x) = \frac{e^x \sec^2 x}{e^{2x}-1}$

$\therefore \quad f(-x) = \frac{e^{-x}\sec^2(-x)}{e^{-2x}-1}$

$= \frac{e^x \sec^2 x}{1-e^{2x}}$

$= -\frac{e^x \sec^2 x}{e^{2x}-1}$

$= -f(x)$

$\therefore \quad I = 0 \quad [\because f(x) \text{ is odd function}]$

32. Let $I = \int_0^{2\pi} \frac{dx}{e^{\sin x}+1}$ ... (i)

$= \int_0^{2\pi} \frac{dx}{e^{\sin(2\pi - x)}+1}$ [by property]

$= \int_0^{2\pi} \frac{dx}{e^{-\sin x}+1}$

$= \int_0^{2\pi} \frac{e^{\sin x}}{e^{\sin x}+1}\,dx$ ... (ii)

Adding Eqs. (i) and (ii), we get

$2I = \int_0^{2\pi} 1\cdot dx = 2\pi$

$\therefore \quad I = \pi$

33. $\because \ 0 \le x < \pi/4$

$\therefore [x] = 0$

Then, $\int_0^{\pi/4} \sin x\, d(x - [x]) = \int_0^{\pi/4} \sin x\,dx$

$= -\{\cos x\}_0^{\pi/4}$

$= -\left(\frac{1}{\sqrt{2}} - 1\right)$

$= 1 - \frac{1}{\sqrt{2}}$

34. Since, $I = \int_1^\infty \frac{(x^2-2)}{x^3\sqrt{(x^2-1)}}\,dx$

Put $\quad x = \frac{1}{t}$

$\therefore \quad dx = -\frac{1}{t^2}\,dt$

$\therefore \quad I = \int_1^0 \frac{\left(\frac{1}{t^2}-2\right) \times -\frac{1}{t^2}\,dt}{\frac{1}{t^3}\sqrt{\left(\frac{1}{t^2}-1\right)}} = \int_0^1 \frac{(1-2t^2)}{\sqrt{(1-t^2)}}\,dt$

Put $\quad t = \sin\theta$

$\therefore \quad dt = \cos\theta\,d\theta$

$\therefore \quad I = \int_0^{\pi/2} \cos 2\theta\,d\theta$

$= \left\{\frac{1}{2}\sin 2\theta\right\}_0^{\pi/2}$

$= \frac{1}{2}(\sin\pi - \sin 0)$

$= 0$

35. Let $I = \int_0^{10} \frac{[\sin 2\pi x]}{e^{x-[x]}}\,dx$

$\Rightarrow \quad I = 10\int_0^1 [\sin 2\pi x]/e^{x-[x]}\,dx$

$= 10\int_0^1 [\sin 2\pi x]/e^x\,dx$

$= 10\left[\int_0^{1/2} \frac{[\sin 2\pi x]}{e^x}dx + \int_{1/2}^1 \frac{[\sin 2\pi x]}{e^x}dx\right]$

$= 10\left[\int_0^{1/2} 0\cdot dx + \int_{1/2}^1 \frac{-1}{e^x}dx\right] = -10\left[\frac{e^{-x}}{-1}\right]_{1/2}^1$

$I = 10[e^{-1} - e^{-1/2}]$

Comparing with the given relation,

$\alpha = 10,\ \beta = -10,\ r = 0$

$\alpha + \beta + r = 0$

36. $\lim_{n\to\infty}\left[\int_0^a \frac{f(x)}{f(x)+f(a-x)}dx + a\int_a^{2a} \frac{f(x)}{f(x)+f(3a-x)}dx + \ldots\right.$

$\left. + a^{n-1}\int_{(n-1)a}^{na} \frac{f(x)}{f(x)+f(12_{n-1})a-x}dx\right]$

$I_1 = \int_0^a \frac{f(x)}{f(x)+f(a-x)}dx = \int_b^a \frac{f(a-x)}{f(x)+f(a-x)}dx$

$\therefore\ I_1 + I_1 = \int_0^a 1\cdot dx$

$\Rightarrow\quad 2I_1 = a$

$\Rightarrow\quad I_1 = a/2$

$I_2 = \int_a^{2a} \frac{f(x)}{f(x)+f(3a-x)}dx = \int_a^{2a} \frac{f(3a-x)}{f(x)+f(3a-x)}dx$

$\therefore I_2 + I_2 = \int_a^{2a} 1\cdot dx$

$\Rightarrow\quad 2I_2 = [x]_a^{2a} = 2a - a = a$

$\Rightarrow\quad I_2 = a/2$

Similarly, $I_3 = I_4 = \ldots = I_n = a/2$

$\lim_{n\to\infty}\left[\frac{a}{2} + a\left(\frac{a}{2}\right) + a^2\left(\frac{a}{2}\right) + \ldots + a^{n-1}\left(\frac{a}{2}\right)\right] = \frac{7}{5}$

$\Rightarrow \lim_{n\to\infty}\frac{a}{2}[1 + a + a^2 + \ldots + a^{n-1} =] = \frac{7}{5}$

$\Rightarrow\quad \frac{a}{2}\lim_{n\to\infty}[1 + a + a^2 + \ldots + a^{n-1}] = \frac{7}{5}$

$\Rightarrow \frac{a}{2}\lim_{n\to\infty} 1\frac{(1-a^n)}{1-a} = \frac{7}{5}$

$\Rightarrow\quad \frac{a}{2}\times\frac{1}{1-a} = \frac{7}{5}\qquad [\because \lim_{n\to\infty} a^n = 0 \text{ when } |a| < 1]$

$\Rightarrow\quad 5a = 14 - 14a$

$\Rightarrow\quad 19a = 14 \Rightarrow a = \frac{14}{19}$

37. $F(x+T) = \int_a^{x+T} f(t)\,dt = \int_a^x f(t)dt + \int_x^{x+T} f(t)\,dt$

$= F(x) + \int_x^{x+T} f(t)\,dt$

$= F(x) + \int_x^0 f(t)\,dt + \int_0^T f(t)\,dt + \int_T^{x+T} f(t)\,dt$

Putting $t - T = y$ in the last integral, we have

$F(x+T) = F(x) + \int_x^0 f(t)\,dt + \int_0^T f(t)\,dt + \int_0^x f(y+T)\,dy$

$= F(x) - \int_0^x f(x)\,dx + \int_0^T f(t)\,dt + \int_0^x f(x+T)dx$

$= F(x) + \int_0^T f(t)\,dt \qquad [\because f(x+T) = f(x)]$

Hence, $\int_0^T f(t)dt$ is independent of x.

Again, $\int_x^{x+T} f(t)\,dt$

Putting $x = -T/2$

$\therefore\quad \int_x^{x+T} f(t)\,dt = \int_{-T/2}^{T/2} f(t)\,dt = 0$

$[\because f(x)$ is an odd function$]$

Therefore, $F(x+T) = F(x)$

Hence, $F(x)$ is periodic with period T.

38. $\sum_{K=1}^{10}\int_0^1 f(k-1+x)\,dx$

$= \int_0^1 f(x)\,dx + \int_0^1 f(1+x)\,dx + \int_0^1 f(2+x)\,dx$

$+ \int_0^1 f(3+x)\,dx + \ldots + \int_0^1 f(9+x)\,dx$

$= \int_0^1 f(x)\,dx + \int_1^2 f(x)\,dx + \int_2^3 f(x)\,dx$

$+ \ldots + \int_9^{10} f(x)\,dx$

$= \int_0^{10} f(x)\,dx = 5$ (given)

39. Let $I = \left|\int_0^{2\pi}[2\sin x]\,dx\right|$

$= \left|\int_0^{\pi/6} 0\,dx + \int_{\pi/6}^{5\pi/6} 1\,dx + \int_{5\pi/6}^{\pi} 0\,dx\right.$

$\left. + \int_{\pi}^{7\pi/6}(-1)\,dx + \int_{7\pi/6}^{11\pi/6}(-2)\,dx + \int_{11\pi/6}^{2\pi}(-1)\,dx\right|$

$= \pi$

40. $\int_{-1}^1 f(x)\,dx =$ Area of shaded region

$= 2\times\frac{1}{2}\times\left(1+\frac{1}{2}\right)\times\frac{1}{4} = \frac{3}{8}$

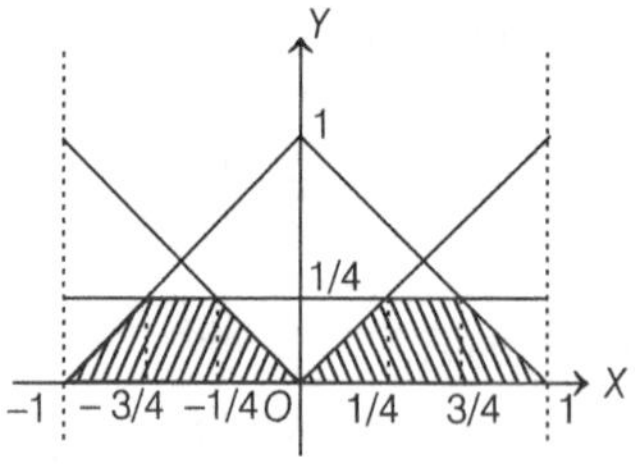

41. $\because \int_{1/2}^2 \frac{1}{x}\operatorname{cosec}^{101}\left(x - \frac{1}{x}\right)dx = k$

Put $\quad x = \frac{1}{t},\ \therefore\ dx = -\frac{1}{t^2}dt$

$\int_2^{1/2} t\cdot\operatorname{cosec}^{101}\left(\frac{1}{t} - t\right)\cdot\left(-\frac{1}{t^2}\right)dt = k$

$\int_{1/2}^2 \frac{1}{t}\operatorname{cosec}^{101}\left(\frac{1}{t} - t\right)dt = k$

$-k = k$

$\therefore\quad 2k = 0 \Rightarrow k = 0$

42. Let $I = \int_0^\infty e^{-ax^2} dx$

Put $\sqrt{a}\, x = t$

$\therefore \quad dx = \frac{dt}{\sqrt{a}}$

Then $I = \frac{1}{\sqrt{a}} \int_0^\infty e^{-t^2} dt$

$= \frac{1}{\sqrt{a}} \int_0^\infty e^{-x^2} dx$ [by property]

$= \frac{1}{\sqrt{a}} \cdot \frac{\sqrt{\pi}}{2}$ [given]

$= \frac{1}{2}\sqrt{\left(\frac{\pi}{a}\right)}$

43. We have $\int_0^1 e^{x^2}(x-\alpha)\, dx = 0$

$\Rightarrow \quad \int_0^1 xe^{x^2} dx = \int_0^1 \alpha e^{x^2} dx$

Put $x^2 = t$ in first integral

$\therefore \quad \frac{1}{2}\int_0^1 e^t\, dt = \alpha \int_0^1 e^{x^2} dx$

$\frac{1}{2}(e-1) = \alpha \int_0^1 e^{x^2} dx$ ...(i)

But $(1-0)e^0 \le \int_0^1 e^{x^2} dx \le e^1(1-0)$

$\Rightarrow \quad 1 \le \int_0^1 e^{x^2} dx \le e$ ...(ii)

$\frac{1}{e} \le \frac{1}{\int_0^1 e^{x^2} dx} \le 1$

$\Rightarrow \quad \frac{(e-1)}{2e} \le \frac{\frac{(e-1)}{2}}{\int_0^1 e^{x^2} dx} \le \left(\frac{e-1}{2}\right)$

$\Rightarrow \quad \left(\frac{e-1}{2e}\right) \le \alpha \le \left(\frac{e-1}{2}\right)$ [from Eq. (i)]

$\therefore \quad 0.312 \le \alpha \le 0.632$

44. $\left[\frac{2}{e^x}\right] = \begin{cases} 1, \text{ if } x \le \log_e 2 \\ 0, \text{ if } x > \log_e 2 \end{cases}$

$\therefore \quad I = \int_0^{\log_e 2} \left[\frac{2}{e^x}\right] dx + \int_{\log 2}^{\infty} \left[\frac{2}{e^x}\right] dx$

$= \int_0^{\log_e 2} 1 \cdot dx + 0 = \log_e 2$

45. It is given that, $I_1 = \int_0^1 (1-x^{50})^{100} dx$

and $\quad I_2 = \int_0^1 (1-x^{50})^{101} dx$

$= \int_0^1 (1-x^{50})(1-x^{50})^{100} dx$

$= \int_0^1 (1-x^{50})^{100} dx - \int_0^1 x \cdot x^{49} (1-x^{50})^{100} dx$

Now, let $I = \int_0^1 x \cdot x^{49}(1-x^{50})^{100} dx$

$= [x\int x^{49}(1-x^{50})^{100} dx - \int 1 \int x^{49}(1-x^{50})^{100} dx]_0^1$

$= \left[t^{1/50} \int \frac{(1-t)^{100}}{50} dt - \int \left(\int \frac{(1-t)^{100}}{50} dt\right) dx\right]_0^1$

[where, $x^{50} = t$]

$I = 0 - \frac{I_2}{50 \times 101}$

$\therefore \quad I_2 = I_1 - \frac{I_2}{5050} \Rightarrow \frac{5051}{5050} I_2 = I_1$

$\because \quad I_2 = \alpha I_1$

$\therefore \quad \alpha = \frac{5050}{5051}$

46. Let $I = \int_{\pi/4}^{3\pi/4} \frac{x}{1+\sin x} dx$ ...(i)

$= \int_{\pi/4}^{3\pi/4} \frac{\left(\frac{3\pi}{4} + \frac{\pi}{4} - x\right)}{1 + \sin\left(\frac{3\pi}{4} + \frac{\pi}{4} - x\right)} dx$ [by property]

$= \int_{\pi/4}^{3\pi/4} \frac{(\pi - x)\, dx}{1 + \sin x}$ ...(ii)

Adding Eqs. (i) and (ii), then

$2I = \pi \int_{\pi/4}^{3\pi/4} \frac{dx}{1+\sin x}$

$= \pi \int_{\pi/4}^{3\pi/4} \frac{(1-\sin x)}{\cos^2 x} dx$

$= \pi \int_{\pi/4}^{3\pi/4} (\sec^2 x - \sec x \tan x)\, dx$

$= \pi \{\tan x - \sec x\}_{\pi/4}^{3\pi/4}$

$= \pi \{(-1+\sqrt{2}) - (1-\sqrt{2})\}$

$2I = \pi(2\sqrt{2} - 2)$

$\therefore \quad I = \pi(\sqrt{2} - 1)$

47. Since, $I_1 = \int_{1-k}^{k} x f\{x(1-x)\}\, dx$ ...(i)

$I_1 = \int_{1-k}^{k} (1-x) f\{(1-x)x\}\, dx$ ...(ii)

[by property]

Adding Eqs. (i) and (ii), then

$2I_1 = \int_{1-k}^{k} f\{x(1-x)\}\, dx = I_2$ [given]

$\therefore \quad I_1 : I_2 = 1 : 2$

48. $f(x) = \frac{1}{2}a_0 + \sum_{i=1}^{n} a_i \cos(ix) + \sum_{j=1}^{n} b_j \sin(jx)$

let $\quad F(x) = f(x)\cos kx$

$= \frac{1}{2}a_0 \cos kx + \sum_{i=1}^{n} a_i \cos ix \cos kx$

$+ \sum_{j=1}^{n} b_j \sin(jx) \cos kx$

$\because \sum_{j=1}^{n} b_j \sin jx \cos kx$ is odd function

$\because \int_{-\pi}^{\pi} \sum_{j=1}^{n} bj \sin jx \cos kx\, dx = 0$

and $\int_{-\pi}^{\pi} \frac{1}{2} a_0 \cos kx\, dx = 2 \cdot \frac{1}{2} a_0 \int_0^\pi \cos kx\, dx$

$= a_0 \left[\frac{\sin kx}{k}\right]_0^\pi = a_0 \cdot 0 = 0$

$$\therefore \int_{-\pi}^{\pi} f(x) \cos kx\, dx = \int_{-\pi}^{\pi} \sum_{i=1}^{n} a_i \cos ix \cos kx\, dx$$

$$= \int_{-\pi}^{\pi} (a_1 \cos x \cos kx + a_2 \cos 2x \cos kx + \dots + a_k \cos^2 kx + \dots + a_n \cos nx \cos kx)\, dx$$

$$= \frac{1}{2}\int_{-\pi}^{\pi} (a_1 (\cos(k+1)x + \cos(k-1)x) + a_2(\cos(k+2)x + \cos(k-2)x) + \dots + a_k(1 + \cos 2kx) + \dots + a_n(\cos(n+k)x + \cos(n-k)x))\, dx$$

$$= \frac{1}{2}\left\{a_1\left(\frac{\sin(k+1)x}{k+1} + \frac{\sin(k-1)x}{k-1}\right) + a_2\left(\frac{\sin(k+2)x}{k+2} + \frac{\sin(k-2)x}{k-2}\right) + \dots + a_k\left(x + \frac{\sin 2kx}{2k}\right) + \dots + a_n\left(\frac{\sin(n+k)x}{n+k} + \frac{\sin(n-k)x}{n-k}\right)\right\}_{-\pi}^{\pi}$$

$$= \frac{1}{2}\{0 + 0 + \dots + a_k(2\pi + 0) + \dots + a_n(0+0)\} = \pi a_k$$

49. We have,

$$\because \quad (1-0)\left(\frac{0}{0^3+16}\right) \le \int_0^1 \frac{x\, dx}{(x^3+16)} \le (1-0)\left(\frac{1}{1^3+16}\right) \text{[by property]}$$

$$0 \le \int_0^1 \frac{x\, dx}{x^3+16} \le \frac{1}{17}$$

$$\therefore \quad \int_0^1 \frac{x\, dx}{x^3+16} \in \left[0, \frac{1}{17}\right]$$

50. Let $I = \int_2^3 \frac{dx}{\sqrt{(1+x^3)}}$

Since, $1 + x^3 > (1+x)^2$ in $x \in (2, 3)$

$$\Rightarrow \quad \sqrt{(1+x^3)} > (1+x)$$

$$\Rightarrow \quad \int_2^3 \frac{1}{\sqrt{(1+x^3)}}\, dx < \int_2^3 \frac{1}{(1+x)}\, dx$$

$$\Rightarrow \quad I < [\log_e(1+x)]_2^3 \Rightarrow I < \log_e(4/3)$$

$$\Rightarrow \quad I < 2.303 \log_{10}(4/3) \Rightarrow 1 < 0.287 < 1$$

$$\therefore \quad I < 1$$

51. Given, $\int_n^{n+1} f(x)\, dx = n^2$

$$\therefore \int_{-2}^{4} f(x)\, dx = \int_{-2}^{-1} f(x)\, dx + \int_{-1}^{0} f(x)\, dx + \int_0^1 f(x)\, dx + \int_1^2 f(x)\, dx + \int_2^3 f(x)\, dx + \int_3^4 f(x)\, dx$$

$$= (-2)^2 + (-1)^2 + (0)^2 + (1)^2 + (2)^2 + (3)^2$$

$$= 4 + 1 + 0 + 1 + 4 + 9 = 19$$

52. Let $I = \int_0^{\sin^2 x} \sin^{-1}\sqrt{t}\, dt + \int_0^{\cos^2 x} \cos^{-1}\sqrt{t}\, dt$ ...(i)

$$\therefore \quad \frac{dI}{dx} = \left(\sin^{-1}\sqrt{(\sin^2 x)}\right)(\sin 2x) + \left(\cos^{-1}\sqrt{(\cos^2 x)}\right)(-\sin 2x)$$

$$= x \sin 2x - x \sin 2x = 0$$

$\therefore$ $I = c$ (constant)

Put $x = \pi/4$ in Eq. (i), then

$$c = \int_0^{1/2} (\sin^{-1}\sqrt{t} + \cos^{-1}\sqrt{t})\, dt = \frac{\pi}{4}$$

Hence, $I = \frac{\pi}{4}$

53. Since, $I = \int_0^1 \cos\left(2\cot^{-1}\sqrt{\left(\frac{1-x}{1+x}\right)}\right) dx$

Put $x = \cos\theta$

$\therefore$ $dx = -\sin\theta\, d\theta$

$$\therefore \quad I = \int_{\pi/2}^{0} \cos\left(2\cot^{-1}\sqrt{\left(\frac{1-\cos\theta}{1+\cos\theta}\right)}\right)(-\sin\theta)\, d\theta$$

$$= \int_0^{\pi/2} \cos(2\cot^{-1}\tan\theta/2) \sin\theta\, d\theta$$

$$= \int_0^{\pi/2} \cos\left(2\left(\frac{\pi}{2} - \tan^{-1}\tan\theta/2\right)\right) \sin\theta\, d\theta$$

$$= \int_0^{\pi/2} \cos\left(2\left(\frac{\pi}{2} - \frac{\theta}{2}\right)\right) \sin\theta\, d\theta$$

$$= \int_0^{\pi/2} \cos(\pi - \theta) \sin\theta\, d\theta$$

$$= \int_0^{\pi/2} \cos\theta(-\sin\theta)\, d\theta = \left[\frac{\cos^2\theta}{2}\right]_0^{\pi/2}$$

$$\therefore \quad I = -\frac{1}{2}$$

54. Let $I = \int_0^2 [x^2 - 1]\, dx$

At $x = 2$ the value of $x^2 - 1$ is 3

and at $x = 0$ the value of $x^2 - 1$ is -1

Integers in between -1 to 3 are 0, 1, 2

$\therefore$ $x^2 - 1 = 0$, $x^2 - 1 = 1$ and $x^2 - 1 = 2$

Or, $x = 1$, $x = \sqrt{2}$, $x = \sqrt{3}$ $[\because x > 0]$

$$\therefore \quad I = \int_0^1 [x^2-1]\, dx + \int_1^{\sqrt{2}} [x^2-1]\, dx + \int_{\sqrt{2}}^{\sqrt{3}} [x^2-1]\, dx + \int_{\sqrt{3}}^{2} [x^2-1]\, dx$$

$$= (-1)\int_0^1 dx + 0 + 1\int_{\sqrt{2}}^{\sqrt{3}} dx + 2\int_{\sqrt{3}}^{2} dx$$

$$= -1 + 0 + \sqrt{3} - \sqrt{2} + 2(2 - \sqrt{3})$$

$$= 3 - \sqrt{3} - \sqrt{2}$$

55. Since, $\int_0^x \left(\frac{bt\cos 4t - a\sin 4t}{t^2}\right) dt = \frac{a \sin 4x}{x}$

Differentiating both sides w.r.t. x

$$\therefore \quad \frac{bx\cos 4x - a\sin 4x}{x^2} = \frac{a\{4x\cos 4x - \sin 4x\}}{x^2}$$

On comparing

$$b = 4a$$

$$\therefore \quad a = 1/4 \text{ and } b = 1$$

56. $f(x) = \cos x - \int_0^x (x-t) f(t)\, dt$

$f(x) = \cos x - x\int_0^x f(t)\, dt + \int_0^x t f(t)\, dt$

$\therefore f'(x) = -\sin x - \left\{x f(x) + \int_0^x f(t)\, dt\right\} + x f(x)$

$= -\sin x - \int_0^x f(t)\, dt$

$\therefore\ f''(x) = -\cos x - f(x)$

$\Rightarrow\ f''(x) + f(x) = -\cos x$

57. $\because f(x) = \max\{x + |x|, x - [x]\}$

$\therefore\ \int_{-2}^{2} f(x)\, dx = \int_{-2}^{0} f(x)\, dx + \int_0^2 f(x)\, dx$

$= \int_{-2}^{0} (x - [x])\, dx + \int_0^2 (x + |x|)\, dx$

$= \int_{-2}^{0} x\, dx - \int_{-2}^{0} [x]\, dx + \int_0^2 2x\, dx$

$= -2 - \left\{\int_{-2}^{-1} [x]\, dx + \int_{-1}^{0} [x]\, dx\right\} + 4$

$= -2 - \{(-2)(1) + (-1)(1)\} + 4$

$= -2 + 2 + 1 + 4 = 5$

58. Let $f(x) = \dfrac{\sin^2 x}{\left[\dfrac{x}{\pi}\right] + \dfrac{1}{2}}$

$\therefore\ f(-x) = \dfrac{\sin^2 x}{\left[-\dfrac{x}{\pi}\right] + \dfrac{1}{2}}$

Case I : If $x = n\pi$, then $f(-x) = 0$

Case II : If $x \neq n\pi$,

then $f(-x) = \dfrac{\sin^2 x}{-1 - \left[\dfrac{x}{\pi}\right] + \dfrac{1}{2}} = \dfrac{\sin^2 x}{-\dfrac{1}{2} - \left[\dfrac{x}{\pi}\right]} = -f(x)$

$\because f(x)$ is an odd function.

$\therefore \int_{-2}^{2} \dfrac{\sin^2 x}{\left[\dfrac{x}{\pi}\right] + \dfrac{1}{2}}\, dx = 0$

59. Let $I = \int_{-1}^{1} \max\{2 - x, 2, 1 + x\}\, dx$

= Area bounded by dark line X-axis

$= \dfrac{1}{2} \times (2 + 3) \times 1 + 1 \times 2 = \dfrac{9}{2}$

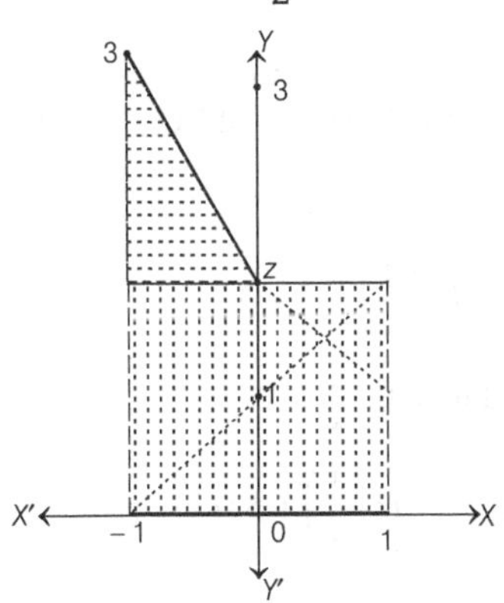

= Area bounded by dark line with X-axis

$= \dfrac{1}{2} \times (2 + 3) \times 1 + 1 \times 2 = \dfrac{9}{2}$

60. Since, $f'(x) = f(x) \Rightarrow \dfrac{f'(x)}{f(x)} = 1$

On integrating, we have

$\log_e f(x) = x + c$ ...(1)

$\because\ f(0) = 1$

$\therefore\ \log_e 1 = 0 + c$

$\therefore\ c = 0$

From Eq. (i), $f(x) = e^x$

$\because\ f(x) + g(x) = x^2$

$\therefore\ g(x) = (x^2 - e^x)$

$\therefore \int_0^1 f(x) g(x)\, dx = \int_0^1 e^x (x^2 - e^x)\, dx$

$= \int_0^1 x^2 e^x dx - \int_0^1 e^{2x} dx$

$= [x^2(e^x) - 2x(e^x) + 2e^x]_0^1 - \dfrac{1}{2}[e^{2x}]_0^1$

$= (e - 2e + 2e) - (0 - 0 - 2) - \dfrac{1}{2}(e^2 - 1)$

$= e + 2 - \dfrac{1}{2}e^2 + \dfrac{1}{2} = \dfrac{5}{2} + e - \dfrac{1}{2}e^2$

61. $\int_0^\pi \left(\sum_{r=0}^{3} a_r \cos^{3-r} x \sin^r x\right) dx$

$= \int_0^\pi (a_0 \cos^3 x + a_1 \cos^2 x \sin x$

$+ a_2 \cos x \sin^2 x + a_3 \sin^3 x)\, dx$

$= a_0 \int_0^\pi \cos^3 x\, dx + a_1 \int_0^\pi \cos^2 x \sin x\, dx$

$+ a_2 \int_0^\pi \cos x \sin^2 x\, dx + a_3 \int_0^\pi \sin^3 x\, dx$

$= 0 + 2a_1 \int_0^{\pi/2} \cos^2 x \sin x\, dx + 0 + 2a_3 \int_0^{\pi/2} \sin^3 x\, dx$ [by property]

$= (2a_1)\dfrac{(1)(1)}{3 \cdot 1}(1) + (2a_3) \cdot \dfrac{2}{3}$ [by Walli's formula]

$= \dfrac{2}{3}a_1 + \dfrac{4}{3}a_3$

62. $\because\ f'(x) = \lim_{h \to 0} \dfrac{f(x+h) - f(x)}{h}$

$= \lim_{h \to 0} \dfrac{f\left(x + 2\left(\dfrac{h}{2}\right)\right) - f(x + 2.0)}{h}$

$= \lim_{h \to 0} \dfrac{f(x) + f(h) + 2xh - f(x) - f(0) - 0}{h}$

$= f'(0) + 2x = 0 + 2x$

$\therefore\ f(x) = x^2 + c \Rightarrow f(0) = 0 + c = 0$

$\therefore\ c = 0$

Then, $f(x) = x^2$

$\therefore\ I_1 = \dfrac{1}{3}, I_2 = \dfrac{1}{3}$ and $I_3 = \dfrac{21}{8}$

Hence, $I_1 = I_2 < I_3$

63. $\int_a^b \frac{x^n}{x^n+(16-x)^n}dx=6$ ...(i)

Let $a+b=16$, then by property.

$$\int_a^b \frac{(16-x)^n}{(16-x)^n+x^n}dx=6 \quad ...(ii)$$

Adding Eqs. (i) and (ii), then

$$\int_a^b 1\cdot dx=12 \text{ or, } b-a=12$$

Solving $a+b=16$ and $b-a=12$, then we get $a=2$, $b=14$ and $n\in R$

64. Let $I=\int_{\ln\lambda}^{\ln(1/\lambda)}F(x)\,dx=\int_{\ln\lambda}^{-\ln\lambda}F(x)\,dx$

where, $F(x)=\dfrac{f\left(\frac{x^2}{4}\right)(f(x)-f(-x))}{g\left(\frac{x^2}{4}\right)(g(x)+g(-x))}$

$\because \quad F(-x)=-F(x)$ [an odd function]

$\therefore \quad I=0$

65. $\frac{1}{3}\le f(t)\le 1\,\forall t\in[0,1]$

$0\le f(t)\le\frac{1}{2}\,\forall t\in(1,3]$

Now, $g(3)=\int_0^3 f(t)\,dt=\int_0^1 f(t)dt+\int_1^3 f(t)\,dt$

$\because \int_0^1 \frac{1}{3}dt\le\int_0^1 f(t)dt\le\int_0^1 1\cdot dt$ ...(i)

and $\int_1^3 0dt\le\int_1^3 f(1)dt\le\int_1^3\frac{1}{2}dt$ ...(ii)

Adding, we get

$$\frac{1}{3}+0\le g(3)\le 1+\frac{1}{2}(3-1)$$

$$\frac{1}{3}\le g(3)\le 2$$

66. Let $I=\int_0^{2\pi}\frac{x\sin^{2n}x}{\sin^{2n}x+\cos^{2n}x}dx$ ...(i)

$$=\int_0^{2\pi}\frac{(2\pi-x)\sin^{2n}(2\pi-x)}{\sin^{2n}(2\pi-x)+\cos^{2n}(2\pi-x)}dx \text{ [by property]}$$

$$I=\int_0^{2\pi}\frac{(2\pi-x)\sin^{2n}x}{\sin^{2n}x+\cos^{2n}x}dx \quad ...(ii)$$

Adding Eqs. (i) and (ii), then

$$2I=2\pi\int_0^{2\pi}\frac{\sin^{2n}x}{\sin^{2n}x+\cos^{2n}x}dx$$

$$\therefore \quad I=\pi\int_0^{2\pi}\frac{\sin^{2n}x}{\sin^{2n}x+\cos^{2n}x}dx$$

$$=\pi\cdot 2\int_0^{\pi}\frac{\sin^{2n}x}{\sin^{2n}x+\cos^{2n}x}dx \quad \text{[by property]}$$

$$I=\pi\cdot 2\cdot 2\int_0^{\pi/2}\frac{\sin^{2n}x}{\sin^{2n}x+\cos^{2n}x}dx \quad \text{[by property]}$$

$$\therefore \quad I=4\pi\int_0^{\pi/2}\frac{\sin^{2n}x}{\sin^{2n}x+\cos^{2n}x}dx \quad ...(iii)$$

$$=4\pi\int_0^{\pi/2}\frac{\sin^{2n}\left(\frac{\pi}{2}-x\right)}{\sin^{2n}\left(\frac{\pi}{2}-x\right)+\cos^{2n}\left(\frac{\pi}{2}-x\right)}dx$$

[by property]

$$=4\pi\int_0^{\pi/2}\frac{\cos^{2n}x}{\cos^{2n}x+\sin^{2n}x}dx \quad ...(iv)$$

Adding Eqs. (iii) and (iv), then

$$2I=4\pi\int_0^{\pi/2}1\cdot dx$$

$$=4\pi\cdot\frac{\pi}{2}$$

$$\therefore \quad I=\pi^2$$

67. For $0<x<1$, $x^2<x$

$\Rightarrow \quad -x^2>-x$

$\Rightarrow \quad e^{-x^2}>e^{-x}$

$\Rightarrow \quad e^{-x^2}\cos^2 x>e^{-x}\cos^2 x$

$\Rightarrow \quad \int_0^1 e^{-x^2}\cos^2 x\,dx>\int_0^1 e^{-x}\cos^2 x\,dx$

$\therefore \quad I_2>I_1$ ...(i)

Also, $\cos^2 x<1\Rightarrow e^{-x^2}\cos^2 x<e^{-x^2}$

$\Rightarrow \quad \int_0^1 e^{-x^2}\cos^2 x\,dx<\int_0^1 e^{-x^2}dx$

$\therefore \quad I_3>I_2$ ...(ii)

and for $0<x<1$, $\int_0^1 e^{-x^2/2}dx>\int_0^1 e^{-x^2}dx$

$\therefore \quad I_4>I_3$ ...(iii)

From Eqs. (i), (ii), (iii), we get

$$I_1<I_2<I_3<I_4$$

$$\therefore \quad I=I_4$$

68. $\because f(x)=\int_2^{x^2}\frac{(\sin^{-1}\sqrt{t})^2}{\sqrt{t}}dt$

$\therefore \quad f'(x)=\frac{(\sin^{-1}x)^2}{x}\cdot(2x-0)$

$f'(x)=2(\sin^{-1}x)^2$ ...(i)

$f''(x)=\frac{4\sin^{-1}x}{\sqrt{(1-x^2)}}$

$\therefore \quad [f''(x)]^2=\frac{16(\sin^{-1}x)^2}{(1-x^2)}$ ...(ii)

From Eqs. (i) and (ii), we have

$$(1-x^2)[f''(x)]^2-2f'(x)=12(\sin^{-1}x)^2$$

$$=12\left(\sin^{-1}\frac{1}{\sqrt{2}}\right)^2=12\left(\frac{\pi}{4}\right)^2=\frac{3}{4}\pi^2$$

69. $af(x)+bf\left(\frac{1}{x}\right)=\frac{1}{x}-5$ ...(i)

Replacing x by $\frac{1}{x}$

Then, $af\left(\frac{1}{x}\right)+bf(x)=x-5$ ...(ii)

Multiplying Eq. (i) by a and (ii) by b, then

$$a^2 f(x) + ab\, f\left(\frac{1}{x}\right) = \frac{a}{x} - 5a \qquad \text{...(iii)}$$

and $$b^2 f(x) + ab\, f\left(\frac{1}{x}\right) = bx - 5b \qquad \text{...(iv)}$$

Subtracting Eq. (iv) from (iii), we get

$$\therefore\ (a^2 - b^2) f(x) = \frac{a}{x} - bx - 5a + 5b$$

$$\therefore\quad x f(x) = \frac{a - bx^2 - 5ax + 5bx}{(a^2 - b^2)}$$

$$\int_1^2 x f(x)\,dx = \frac{\left\{ax - \frac{bx^3}{3} - \frac{5ax^2}{2} + \frac{5bx^2}{2}\right\}_1^2}{(a^2 - b^2)}$$

$$= \frac{\left(2a - \frac{8b}{3} - 10a + 10b\right) - \left(a - \frac{b}{3} - \frac{5a}{2} + \frac{5b}{2}\right)}{(a^2 - b^2)}$$

$$= \frac{12a - 16b - 60a + 60b - 6a + 2b + 15a - 15b}{6(a^2 - b^2)}$$

$$= \frac{-39a + 31b}{6(a^2 - b^2)}$$

70. $$\int_{-\pi}^{\pi} \sum_{r=0}^{999} \cos rx \left(1 + \sum_{r=1}^{999} \sin rx\right) dx$$

$$= \int_{-\pi}^{\pi} \sum_{r=0}^{999} \cos rx + \int_{-\pi}^{\pi} \left(\sum_{r=0}^{999} \cos rx \sum_{r=0}^{999} \sin rx\right) dx$$

Now, $$\int_{-\pi}^{\pi} \cos rx = \left[\frac{\sin rx}{r}\right]_{-\pi}^{\pi} = 2\left[\frac{\sin rx}{r}\right]_0^{\pi} = 2\frac{\sin rx}{r}$$

$$\therefore \sum_{r=0}^{999} \int_{-\pi}^{\pi} \cos rx = 2\sum_{r=0}^{999} \frac{\sin rx}{r} = 0$$

$$\Rightarrow \int_{-\pi}^{\pi} \sum_{r=0}^{999} \cos rx \left(1 + \sum_{r=1}^{999} \sin rx\right) dx$$

$$= 0 + \int_{-\pi}^{\pi} \left(\sum_{r=0}^{999} \cos rx \cdot \sum_{r=1}^{999} \sin rx\right) dx$$

$$= \sum_{r=0}^{999} \int_{-\pi}^{\pi} \cos rx(\sin x + \sin 2x + \ldots + \sin 999x)\,dx$$

$$= \sum_{r=0}^{999} \int_{-\pi}^{\pi} 2\cos rx \sin x + 2\cos rx \sin 2x + \ldots + 2\cos rx \sin 999x)\,dx$$

$$= \frac{1}{2}\sum_{r=0}^{999} [\sin(r+1)x - \sin(r-1)x + \ldots \sin(999+r)x - \sin(999-r)x]_{-\pi}^{\pi}$$

$$= \frac{1}{2}[0 - 0 - 0 - 0] = 0$$

71. Let $$I = \int_0^a \frac{dx}{1 + e^{f(x)}} \qquad \text{... (i)}$$

$$= \int_0^a \frac{dx}{1 + e^{f(a-x)}} \qquad \text{[by property]}$$

$$= \int_0^a \frac{dx}{1 + e^{-f(1x)}} \quad \{\text{given } f(x) + f(a-x) = 0\}$$

$$= \int_0^a \frac{e^{f(x)}}{1 + e^{f(x)}}\,dx \qquad \text{... (ii)}$$

Adding Eqs. (i) and (ii), then

$$2I = \int_0^a dx = a$$

$$\therefore\quad I = a/2$$

72. $$\int_{-\pi/4}^{\pi/4} a|\sin x|\,dx + b\int_{-\pi/4}^{\pi/4} \frac{\sin x}{1 + \cos x}\,dx + \int_{-\pi/4}^{\pi/4} c\,dx = 0$$

$$\Rightarrow\quad 2a\int_0^{\pi/4} \sin x\,dx + 0 + c\left(\frac{\pi}{2}\right) = 0$$

$$\Rightarrow\quad 2a\left(1 - \frac{1}{\sqrt{2}}\right) + 0 + \frac{\pi c}{2} = 0 \qquad \text{[by property]}$$

73. Let $$I = \int_0^{16\pi/3} |\sin x|\,dx$$

$$= \int_0^{5\pi + \frac{\pi}{3}} |\sin x|\,dx$$

$$= \int_0^{5\pi} |\sin x|\,dx + \int_{5\pi}^{5\pi + \frac{\pi}{3}} |\sin x|\,dx$$

$$= 5\int_0^{\pi} |\sin x|\,dx + \int_0^{\pi/3} |\sin x|\,dx \qquad \text{[by property]}$$

$$= 5\int_0^{\pi} \sin x\,dx + \int_0^{\pi/3} \sin x\,dx$$

$$= 5[-\cos x]_0^{\pi} + [-\cos x]_0^{\pi/3}$$

$$= -5(-1-1) - \left(\frac{1}{2} - 1\right)$$

$$= 10 + 1 - \frac{1}{2} = \frac{21}{2}$$

74. $$\int_0^a f(x)\,dx = \int_0^a f(a-x)\,dx$$

$$I = \int_0^{\pi/4} \left[\left(\frac{\pi}{4} - x\right) - 4\left(\frac{\pi}{4} - x\right)^2\right] \ln\left(1 + \tan\left(\frac{\pi}{4} - x\right)\right) dx$$

$$= \int_0^{\pi/4} \left[\frac{\pi^2}{4} - \pi x - 4\left(\frac{\pi^2}{16} + x^2 - \frac{\pi x}{2}\right)\right] \ln\left(1 + \frac{1 - \tan x}{1 + \tan x}\right) dx$$

$$= \int_0^{\pi/4} \left[\frac{\pi^2}{4} - \pi x - \frac{\pi^2}{4} - 4x^2 + 2\pi x\right] \ln\left(\frac{2}{1 + \tan x}\right) dx$$

$$= \int_0^{\pi/4} (\pi x - 4x^2)[\ln 2 - \ln(1 + \tan x)]\,dx$$

$$= \int_0^{\pi/4} (\pi x - 4x^2)\ln 2 - \int_0^{\pi/4} (\pi x - 4x^2)\ln(1 + \tan x)\,dx$$

$$\Rightarrow 2I = \ln 2 \int_0^{\pi/4} (\pi x - 4x^2)\,dx$$

$$I = \frac{\ln 2}{2}\left[\pi\frac{x^2}{2} - 4\frac{x^3}{3}\right]_0^{\pi/4}$$

$$= \frac{\ln 2}{2}\left[\frac{\pi}{2} \times \frac{\pi^2}{16} - \frac{4}{3} \times \frac{\pi^3}{64} - 0\right]$$

$$= \frac{\ln 2}{2}\left(\frac{\pi^3}{32} - \frac{\pi^3}{48}\right)$$

$$= \frac{\ln 2}{2}\pi^3\left(\frac{3-2}{96}\right)$$

$$= \frac{\ln 2}{192}\pi^3$$

75. Let $I = \int_0^a f(x)\,dx = \int_0^a f(x)\cdot 1\,dx$

Integration by parts, choose unity as second function, then

$$I = \{f(x)\cdot x\}_0^a - \int_0^a f'(x)\cdot x\,dx$$

$$= \{x\,f(x)\}_0^a - \left\{\left(f'(x)\cdot\frac{x^2}{2}\right)_0^a - \int_0^a f''(x)\cdot\frac{x^2}{2}\,dx\right\}$$

[again by parts]

$$= \left\{x\,f(x) - \frac{x^2}{2!}f'(x)\right\}_0^a + \int_0^a f''(x)\cdot\frac{x^2}{2}\,dx$$

$$= \left\{x\,f(x) - \frac{x^2}{2!}f'(x)\right\}_0^a + \left\{f''(x)\cdot\frac{x^3}{3!}\right\}_0^a - \int_0^a f'''(x)\cdot\frac{x^3}{3!}\,dx$$

[again by parts]

$$= \left\{x\,f(x) - \frac{x^2}{2!}f'(x) + \frac{x^3}{3!}f''(x)\right\}_0^a - k\cdot\left\{\frac{x^4}{4!}\right\}_0^a$$

$$\therefore\ \int_0^a f(x)\,dx - \left\{x\,f(x) - \frac{x^2}{2!}f'(x) + \frac{x^3}{3!}f''(x)\right\}_0^a = -\frac{ka^4}{24}$$

76. $\int_0^{n^2}[\sqrt{x}]\,dx = \int_0^{1^2}[\sqrt{x}]\,dx + \int_{1^2}^{2^2}[\sqrt{x}]dx$

$$+ \int_{2^2}^{3^2}[\sqrt{x}]\,dx + \ldots + \int_{(n-1)^2}^{n^2}[\sqrt{x}]\,dx$$

$$= 0 + 1\cdot(2^2 - 1^2) + 2\cdot(3^2 - 2^2) + \ldots$$

$$+ \ldots + (n-1)[n^2 - (n-1)^2]$$

$$-1^2 - 2^2 - 3^2 - \ldots - (n-1)^2 + (n-1)n^2$$

$$= -(1^2 + 2^2 + 3^2 + \ldots + (n-1)^2) + (n-1)n^2$$

$$= -\frac{(n-1)n(2n-1)}{6} + (n-1)n^2$$

$$= \frac{1}{6}n(n-1)(4n+1)$$

77. Let $P = \lim_{n\to\infty}\frac{(n!)^{1/n}}{n}$

$$= \lim_{n\to\infty}\left(\frac{n!}{n^n}\right)^{1/n}$$

$$= \lim_{n\to\infty}\left(\frac{1\cdot 2\cdot 3\cdot 4\cdot\ldots\cdot n}{n\cdot n\cdot n\cdot\ldots\cdot n}\right)^{1/n}$$

$$= \lim_{n\to\infty}\left(\left(\frac{1}{n}\right)\left(\frac{2}{n}\right)\left(\frac{3}{n}\right)\cdots\left(\frac{n}{n}\right)\right)^{1/n}$$

$$\therefore \ln P = \lim_{n\to\infty}\frac{1}{n}\left[\ln\left(\frac{1}{n}\right) + \ln\left(\frac{2}{n}\right) + \ln\left(\frac{3}{n}\right) + \ldots + \ln\left(\frac{n}{n}\right)\right]$$

$$= \lim_{n\to\infty}\frac{1}{n}\sum_{r=1}^{n}\ln\left(\frac{r}{n}\right)$$

$$= \int_0^1 \ln x\,dx$$

$$= [\ln x\cdot x - x]_0^1$$

$$= 0 - 1 = -1$$

$$\therefore\ P = e^{-1}$$

78. Let $I = \int_4^{10}\frac{[x^2]\,dx}{[(x-14)^2] + [x^2]}$ ...(i)

$$I = \int_4^{10}\frac{[14-x]^2}{[(x)^2] + [(x-14)^2]}dx \quad \text{[by property] ...(ii)}$$

Adding Eqs. (i) and (ii), we get

$$2I = \int_4^{10} 1\cdot dx = 10 - 4 = 6$$

$$\therefore\ I = 3$$

79. $I(a) = \int_0^{\pi}\left(\frac{x^2}{a^2} + a^2\sin^2 x + 2x\sin x\right)dx \quad \left(\int_0^{\pi} x\sin x\,dx = \pi\right)$

$$\therefore I(a) = \frac{\pi^3}{3a^2} + \frac{\pi a^2}{2} + 2\pi = \pi\left[\frac{\pi^2}{3a^2} + \frac{a^2}{2}\right] + 2\pi$$

$$= \pi\left[\left(\frac{\pi}{\sqrt{3}a} - \frac{a}{\sqrt{2}}\right)^2 + \frac{2\pi}{\sqrt{6}}\right] + 2\pi$$

$I(a)$ is minimum when, $\frac{\pi}{\sqrt{3}a} = \frac{a}{\sqrt{2}}$

$$\Rightarrow\quad a^2 = \pi\sqrt{\frac{2}{3}} \Rightarrow a = \sqrt{\pi\sqrt{\frac{2}{3}}}$$

Also, $I(a)|_{\min} = 2\pi + \pi^2\sqrt{\frac{2}{3}}$

80. Let $P = \Pi\,\{1 + (ra)^2\}^{1/r} = \Pi\left\{1 + \left(\frac{r}{n}\right)^2\right\}^{1/r}$ $[\because na = 1]$

$$\therefore\quad \ln P = \lim_{n\to\infty}\sum_{r=1}^{n}\frac{1}{r}\ln\left\{1 + \left(\frac{r}{n}\right)^2\right\}$$

$$= \int_0^1 \frac{\ln(1+x^2)}{x}dx$$

Put $\quad x^2 = t \Rightarrow dx = \frac{dt}{2x}$

$$= \frac{1}{2}\int_0^1 \frac{\ln(1+t)}{t}dt$$

$$= \frac{1}{2}\int_0^1\left(1 - \frac{t}{2} + \frac{t^2}{3} - \frac{t^3}{4} + \ldots\right)dt$$

$$= \frac{1}{2}\left\{1 - \frac{1}{2^2} + \frac{1}{3^2} - \frac{1}{4^2} + \ldots\right]$$

$$= \frac{1}{2}\cdot\frac{\pi^2}{12} = \frac{\pi^2}{24}$$

$$\therefore\quad P = e^{\pi^2/24}$$

81. $\int_0^{100}[\tan^{-1}x]\,dx = \int_0^{\tan 1}[\tan^{-1}x]\,dx + \int_{\tan 1}^{100}[\tan^{-1}x]\,dx$

$$= \int_0^{\tan 1} 0\,dx + \int_{\tan 1}^{100} 1\cdot dx$$

$$= 0 + 100 - \tan 1 = 100 - \tan 1$$

82. Let $I = \frac{1}{a+b}\int_a^b x(f(x) + f(x+1))dx$ ...(i)

On applying property,

$\int_a^b f(x)dx = \int_a^b f(a+b-x)dx$, we get

$$I = \frac{1}{a+b}\int_a^b (a+b-x)[f(a+b-x) + f(a+b-x+1)]dx$$

$$= \frac{1}{a+b}\int_a^b (a+b-x)[f(x+1)+f(x)]dx \quad \ldots\text{(ii)}$$

$$\begin{bmatrix} \because f(a+b+1-x) = f(x) \\ \therefore f(a+b-x) = f(x+1) \end{bmatrix}$$

On adding integrals Eqs. (i) and (ii), we get

$$2I = \frac{a+b}{a+b}\int_a^b (f(x)+f(x+1))dx$$

$$\Rightarrow \quad I = \frac{1}{2}\left[\int_a^b f(x)dx + \int_a^b f(x+1)dx\right]$$

$$= \frac{1}{2}\left[\int_a^b f(x)dx + \int_a^b (a+b-x+1)dx\right]$$

$$= \frac{1}{2}\left[\int_a^b f(x)dx + \int_a^b f(x)dx\right] \Rightarrow I = \int_a^b f(x)dx$$

Put $x = t+1$, then upper limit $t = b-1$ and lower limit $t = a-1$ and $dx = dt$, so

$$I = \int_{a-1}^{b-1} f(t+1)dt = \int_{a-1}^{b-1} f(x+1)dx$$

83. The graph of $f(x) = \min(x-[x], -x-[-x])$ is shown as in figure

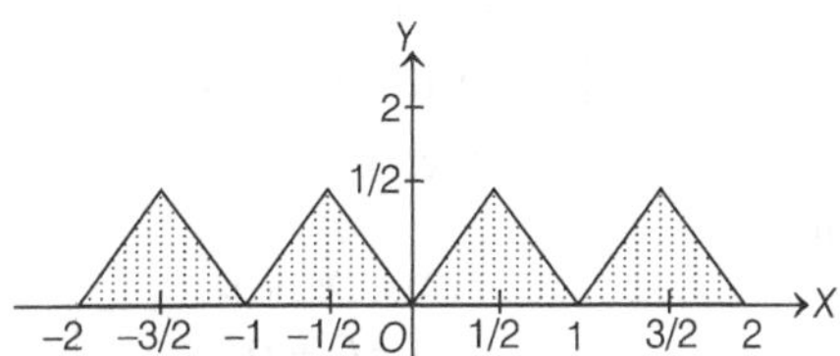

Therefore, $\int_{-2}^{2} \min(x-[x], -x-[-x])\,dx$

$$= \text{Area of shaded region} = 4\times\left(\frac{1}{2}\times 1\times\frac{1}{2}\right) = 1$$

84. Let $I = \int_0^2 [x^2-x+1]\,dx$

Let $f(x) = x^2-x+1 \Rightarrow f'(x) = 2x-1$

for $x > 1/2$, $f'(x) > 0$ and $x < 1/2$, $f'(x) < 0$

Values of $f(x)$ at $x = 1/2$ and 2 are 3/4 and 3 integers between them are 1, 2, then $x^2-x+1 = 1, 2$

we get $x = 1$, $x = \dfrac{1+\sqrt{5}}{2}$

and values of $f(x)$ at $x = 0$ and $1/2$ are 1 and 3/4

No integer between them.

$$\therefore I = \int_0^{1/2} [x^2-x+1]\,dx + \int_{1/2}^{1} [x^2-x+1]\,dx$$

$$+ \int_1^{\frac{1+\sqrt{5}}{2}} [x^2-x+1]\,dx + \int_{\frac{1+\sqrt{5}}{2}}^{2} [x^2-x+1]\,dx$$

$$= 0+0+1\int_1^{\frac{1+\sqrt{5}}{2}} 1\cdot dx + 2\int_{\frac{1+\sqrt{5}}{2}}^{1} 1\,dx$$

$$= \left(\frac{1+\sqrt{5}}{2}-1\right) + 2\left(2-\frac{1+\sqrt{5}}{2}\right) = 3-\left(\frac{1+\sqrt{5}}{2}\right)$$

$$= \left(\frac{5-\sqrt{5}}{2}\right)$$

Alternative Method (Graphical Method)

It is clear from the figure

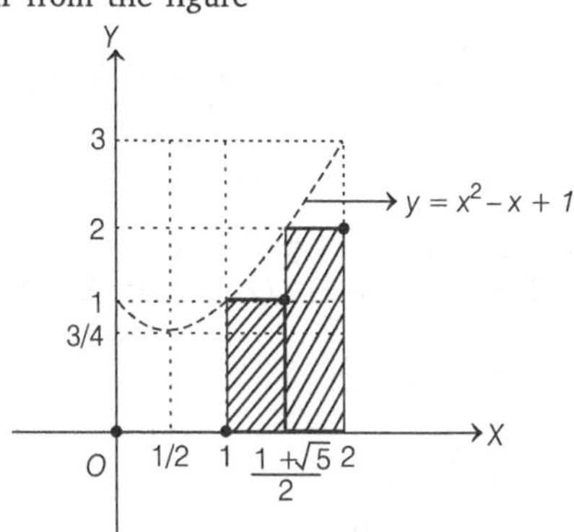

$$\int_0^2 [x^2-x+1]\,dx = \text{Area of bounded region}$$

$$= 0 + \left(\frac{1+\sqrt{5}}{2}-1\right)\times 1 + \left(2-\frac{1+\sqrt{5}}{2}\right)\times 2$$

$$= 3-\left(\frac{1+\sqrt{5}}{2}\right) = \left(\frac{5-\sqrt{5}}{2}\right)$$

85.

$$P = \lim_{n\to\infty}\left\{\frac{1^k+2^k+3^k+\ldots+n^k}{n^{k+1}}\right\}$$

$$= \lim_{n\to\infty}\frac{1}{n}\left(\left(\frac{1}{n}\right)^k + \left(\frac{2}{n}\right)^k + \left(\frac{3}{n}\right)^k + \ldots + \left(\frac{n}{n}\right)^k\right)$$

$$= \lim_{n\to\infty}\frac{1}{n}\sum_{r=1}^{n}\left(\frac{r}{n}\right)^k$$

$$= \int_0^1 x^k\,dx$$

$$= \left[\frac{x^{k+1}}{k+1}\right]_0^1 = \frac{1}{k+1}$$

86. $\int_{1/e}^{e} |\ln x|\,dx = \int_{1/e}^{1} |\ln x|\,dx + \int_1^e |\ln x|\,dx$

$$= \int_{1/e}^{1} (-\ln x)\,dx + \int_1^e (\ln x)\,dx$$

$$= -\{x(\ln x-1)\}_{1/e}^{1} + \{x(\ln x-1)\}_1^e$$

$$= -\left(-1+\frac{2}{e}\right)+1$$

$$= 2-\frac{2}{e}$$

$$= 2(1-1/e)$$

87. Let $I = \int_{-\pi/2}^{\pi/2} \dfrac{x^2\cos x}{1+e^x}\,dx \quad \ldots\text{(i)}$

$$\left[\because \int_a^b f(x)\,dx = \int_a^b f(a+b-x)\,dx\right]$$

$$\Rightarrow \quad I = \int_{-\pi/2}^{\pi/2} \frac{x^2\cos(-x)}{1+e^{-x}}\,dx \quad \ldots\text{(ii)}$$

On adding Eqs. (i) and (ii), we get

$$2I = \int_{-\pi/2}^{\pi/2} x^2\cos x\left[\frac{1}{1+e^x}+\frac{1}{1+e^{-x}}\right]dx$$

$$= \int_{-\pi/2}^{\pi/2} x^2 \cos x \cdot (1)\, dx$$

$$\left[\because \int_{-a}^{a} f(x)\,dx = 2\int_0^a f(x)\,dx, \text{ when } f(-x) = f(x)\right]$$

$$\Rightarrow \quad 2I = 2\int_0^{\pi/2} x^2 \cos x\, dx$$

Using integration by parts, we get

$$2I = 2[x^2(\sin x) - (2x)(-\cos x) + (2)(-\sin x)]_0^{\pi/2}$$

$$\Rightarrow \quad 2I = 2\left[\frac{\pi^2}{4} - 2\right]$$

$$\therefore \quad I = \frac{\pi^2}{4} - 2$$

88. Let $P = \lim_{n\to\infty} \frac{1}{n} \sum_{r=n+1}^{2n} \log_e\left(1 + \frac{r}{n}\right)$

$$= \int_1^2 \log_e(1 + x)\, dx$$

Put $1 + x = t$

$\therefore \quad dx = dt$

Then, $P = \int_2^3 \log_e t\, dt$

$$= [\log_e t \cdot t - t]_2^3$$

$$= (3\log_e 3 - 3) - (2\log_e 2 - 2)$$

$$= \log_e\left(\frac{27}{4}\right) - 1$$

$$= \log_e\left(\frac{27}{4}\right) - \log_e e$$

$$= \log_e\left(\frac{27}{4e}\right)$$

89. $\int_0^\pi \cos mx \sin nx\, dx = \frac{1}{2}\int_0^\pi 2\sin nx \cos mx\, dx$

$$= \frac{1}{2}\int_0^\pi \{\sin(n+m)x + \sin(n-m)x\}\, dx$$

$$= \frac{1}{2}\left\{-\frac{\cos(n+m)x}{n+m} - \frac{\cos(n-m)x}{n-m}\right\}_0^\pi$$

$$= \frac{1}{2}\left\{\begin{array}{l} -\frac{(-1)^{n+m}}{n+m} - \frac{(-1)^{n-m}}{n-m} \\ + \frac{1}{n+m} + \frac{1}{n-m} \end{array}\right\}$$

$$= \frac{1}{2}\left\{\frac{1}{n+m} + \frac{1}{n-m} + \frac{1}{n+m} + \frac{1}{n-m}\right\}$$

$$= \frac{2n}{n^2 - m^2}$$

90. Let $n \le x < n + 1$, where $n \in I$

$$I = \int_0^x \frac{2^t}{2^{[t]}}\, dt = \int_0^n 2^{\{t\}}\, dt + \int_n^x 2^{\{t\}}\, dt$$

$$= n\int_0^1 2^{\{t\}}\, dt + \int_n^x 2^{\{t\}}\, dt$$

$$= n\int_0^1 2^t dt + \int_n^x 2^{t-n} dt$$

$$= n\frac{2^t}{\ln 2}\Big|_0^1 + \frac{1}{2^n}\frac{2^t}{\ln 2}\Big|_n^x$$

$$= \frac{n}{\ln 2}(2-1) + \frac{1}{2^n \ln 2}(2^x - 2^n)$$

$$= \frac{n}{\ln 2} + \frac{1}{\ln 2}(2^{x-n} - 1) = \frac{[x] + 2^{\{x\}} - 1}{\ln 2}$$

91. $\int_0^2 \left|\cos\frac{\pi x}{2}\right| dx = \int_0^1 \left|\cos\frac{\pi x}{2}\right| dx + \int_1^2 \left|\cos\frac{\pi x}{2}\right| dx$

$$= \int_0^1 \cos\left(\frac{\pi x}{2}\right) dx - \int_1^2 \cos\left(\frac{\pi x}{2}\right) dx$$

$$= \frac{2}{\pi}\left[\sin\frac{\pi x}{2}\right]_0^1 - \frac{2}{\pi}\left[\sin\frac{\pi x}{2}\right]_1^2$$

$$= \frac{2}{\pi}(1 - 0) - \frac{2}{\pi}(0 - 1) = \frac{4}{\pi}$$

92. $\because \quad f'(x) = f(x) + \int_0^1 f(x)$ ....(i)

$\therefore \quad f''(x) = f'(x) + 0$

$\Rightarrow \quad \frac{f''(x)}{f'(x)} = 1$

$\Rightarrow \quad \ln f'(x) = x + \ln C$

$\Rightarrow \quad f'(x) = Ce^x$

$$\int f'(x)\, dx = \int Ce^x\, dx$$

$\therefore \quad f(x) = Ce^x + D$

at $\quad x = 0$

$$f(0) = Ce^0 + D$$

$$1 = C + D$$

From Eq. (i),

$$Ce^x = Ce^x + (1 - C) + \int_0^1 (Ce^x + 1 - C)\, dx$$

$\therefore \quad C = \frac{2}{3 - e}$

$\therefore \quad D = \frac{1 - e}{3 - e}$

Hence, $f(x) = Ce^x + D$

$$= \frac{2e^x}{(3-e)} + \left(\frac{1-e}{3-e}\right)$$

93. Let $\quad I(\alpha) = \int_0^1 \frac{x^\alpha - 1}{\ln x}\, dx$ ...(i)

$\therefore \quad I'(\alpha) = \int_0^1 \frac{x^\alpha \cdot \ln x}{\ln x}\, dx = \int_0^1 x^\alpha dx$

$$= \left[\frac{x^{\alpha+1}}{\alpha+1}\right]_0^1 = \frac{1}{(\alpha+1)}$$

Now, $I(\alpha) = \int \frac{d\alpha}{(\alpha + 1)}$

$$= \ln|\alpha + 1| + c$$

Put $\alpha = 0$, then

$I(0) = \ln 1 + c = 0$ [from Eq. (i)]

$\Rightarrow \quad 0 + c = 0$

$\therefore \quad c = 0$

Hence, $\quad I(\alpha) = \ln|\alpha + 1|$

94. $\because \quad -1 \le \sin x \le 1$

$\therefore \quad -2 \le 2\sin x \le 2$ and period 2π

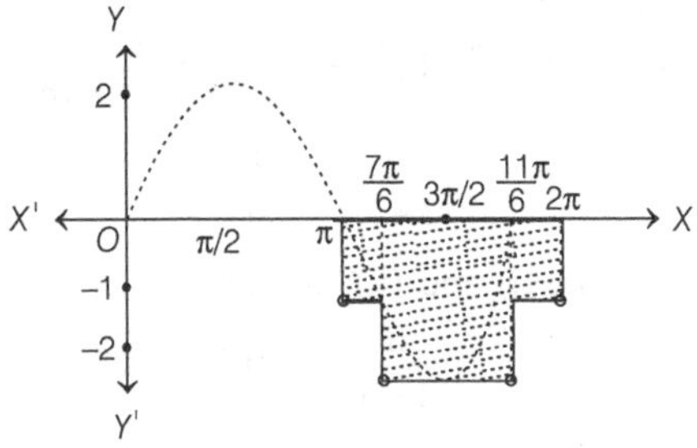

$$\therefore \int_{\pi}^{2\pi} [2\sin x]\, dx = -\left\{ 1 \times \left(\frac{7\pi}{6} - \pi\right) + 2 \times \left(\frac{11\pi}{6} - \frac{7\pi}{6}\right) + 1 \times \left(2\pi - \frac{11\pi}{6}\right) \right\}$$

$$= -\frac{10\pi}{6} = -\frac{5\pi}{3}$$

95. $$\lim_{n\to\infty} \sum_{k=1}^{n} \frac{k^{\frac{1}{a}}\left\{n^{a-\frac{1}{a}} + k^{a-\frac{1}{a}}\right\}}{n^{a+1}}$$

$$= \lim_{n\to\infty} \sum_{k=1}^{n} \frac{1}{n} \cdot \left\{ \left(\frac{k}{n}\right)^{\frac{1}{a}} + \left(\frac{k}{n}\right)^{a} \right\}$$

$$= \int_0^1 (x^{1/a} + x^a)\, dx$$

$$= \left[\frac{x^{(1/a)+1}}{\frac{1}{a}+1} + \frac{x^{a+1}}{a+1} \right]_0^1 = \frac{a}{a+1} + \frac{1}{a+1} = 1$$

96. $$\int_{-100}^{100} f(x)\, dx = \int_{-100}^{100} \min(\{x\}, \{-x\})\, dx$$

$$= 2\int_0^{100} \min(\{x\}, \{-x\})\, dx \quad \text{[by property]}$$

$$= 200 \int_0^1 \min(\{x\}, \{-x\})\, dx \quad \text{[by property]}$$

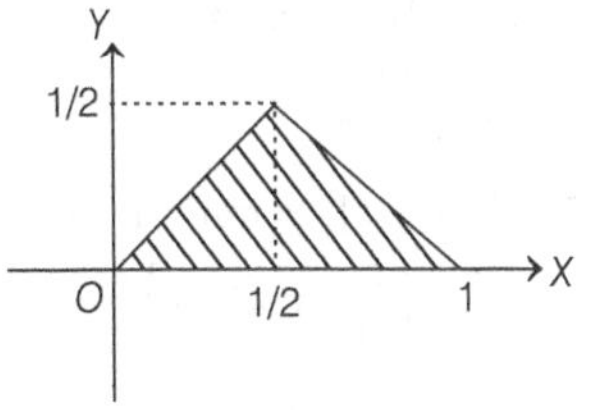

$$= 200 \times \frac{1}{2} \times 1 \times \frac{1}{2} = 50$$

97. $$\int_{-\pi/2}^{199\pi/2} \sqrt{(1+\cos 2x)}\, dx = \sqrt{2}\int_{-\pi/2}^{199\pi/2} |\cos x|\, dx$$

$$= \sqrt{2}\int_0^{100\pi} \left|\cos\left(x - \frac{\pi}{2}\right)\right| dx$$

$$= \sqrt{2}\int_0^{100\pi} |\sin x|\, dx = 100\sqrt{2}\int_0^{\pi} |\sin x|\, dx \quad \text{[by property]}$$

$$= 100\sqrt{2}\int_0^{\pi} \sin x\, dx$$

$$= 100\sqrt{2}(2) = 200\sqrt{2}$$

98. $$\int_0^{\pi/4} (\tan^n x + \tan^{n-2} x)\, dx = \int_0^{\pi/4} \tan^{n-2} x (\tan^2 x + 1)\, dx$$

$$\left[\begin{pmatrix} \because 0 < x < \frac{\pi}{4} \\ \therefore [x] = 0 \end{pmatrix}\right]$$

$$= \int_0^{\pi/4} \tan^{n-2} x \cdot \sec^2 x\, dx = \left\{\frac{\tan^{n-1} x}{n-1}\right\}_0^{\frac{\pi}{4}} = \frac{1}{(n-1)}$$

99. $\because \int_{-1}^{4} f(x)\, dx = 4$ and $\int_2^4 ((3 - f(x))\, dx = 7$

or $\quad \int_4^{-1} f(x)\, dx = -4$ and $6 - \int_2^4 f(x)\, dx = 7$

$\therefore \quad \int_4^{-1} f(x)\, dx = -4, \int_2^4 f(x)\, dx = -1$

Now, $\int_2^{-1} f(x)\, dx = \int_2^4 f(x)\, dx + \int_4^{-1} f(x)\, dx$

$$= -1 - 4 = -5$$

100. $$\because I_1 = \int_0^x e^{zx} \cdot e^{-z^2} dz = \int_0^x e^{-(z^2 - zx)} dz$$

$$= \int_0^x e^{-\left\{\left(z - \frac{x}{2}\right)^2 - \frac{x^2}{4}\right\}} dz = e^{x^2/4} \int_0^x e^{-\left(z - \frac{x}{2}\right)^2} dz$$

$$= e^{x^2/4} \int_{-x/2}^{x/2} e^{-z^2} dz = \frac{1}{2} e^{x^2/4} \int_{-x}^{x} e^{-z^2/4} dz$$

$$= e^{x^2/4} \int_0^x e^{-z^2/4} dz = e^{x^2/4} I_2$$

Hence, $\quad I_1 = e^{x^2/4} I_2$

101. Given, $f(x) = \int_0^x g(t)\, dt$

On replacing x by $(-x)$, we get

$$f(-x) = \int_0^{-x} g(t) dt$$

Now, put $t = -u$,

$$f(-x) = -\int_0^x g(-u) du = -\int_0^x g(u) du = -f(x)$$

[$\because g$ is an even function]

f is an odd function.

Now, it is given that $f(x+5) = g(x)$

$\therefore \quad f(5-x) = g(-x) = g(x) = f(x+5)$

[$\because g$ is an even function]

$\Rightarrow f(5-x) = f(x+5)$ ...(i)

Let $I = \int_0^x f(t)\, dt$

Put $t = u + 5 \Rightarrow t - 5 = u \Rightarrow dt = du$

$$\therefore \quad I = \int_{-5}^{x-5} f(u+5) du - \int_{-5}^{x-5} g(u) du = \int_{-5}^{x-5} f'(u)\, du$$

$$= f(x-5) - f(-5)$$

$$= -f(5-x) + f(5)$$

$$= f(5) - f(5+x)$$

$$= \int_{5+x}^{5} f'(t) dt = \int_{5+x}^{5} g(t) dt$$

Objective Questions Type II [One or more than one correct answer(s)]

1. Since, $\int_0^1 \frac{dt}{(t^2 + 2t\cos\alpha + 1)}$

$$= \int_0^1 \frac{dt}{(t+\cos\alpha)^2 + \sin^2\alpha}$$

$$= \frac{1}{\sin\alpha}\left\{\tan^{-1}\left(\frac{t+\cos\alpha}{\sin\alpha}\right)\right\}_0^1$$

$$= \frac{1}{\sin\alpha}\left\{\tan^{-1}\left(\frac{1+\cos\alpha}{\sin\alpha}\right) - \tan^{-1}(\cot\alpha)\right\}$$

$$= \frac{1}{\sin\alpha}\{\tan^{-1}(\cot\alpha/2) - \tan^{-1}(\cot\alpha)\}$$

$$= \frac{1}{\sin\alpha}\left[\left\{\frac{\pi}{2} - \cot^{-1}(\cot\alpha/2)\right\} - \left\{\frac{\pi}{2} - \cot^{-1}(\cot\alpha)\right\}\right]$$

$$= \frac{1}{\sin\alpha}\left\{\frac{\pi}{2} - \frac{\alpha}{2} - \frac{\pi}{2} + \alpha\right\}$$

$$= \frac{\alpha}{2\sin\alpha}$$

and $\int_{-3}^{3} \frac{t^2\sin 2t}{t^2+1}\,dt = 0$ (it is odd function)

We have,

$$x^2\left(\int_0^1 \frac{dt}{t^2 + 2t\cos\alpha + 1}\right) - x\left(\int_{-3}^{3}\frac{t^2\sin 2t}{t^2+1}dt\right) - 2 = 0$$

$$\Rightarrow \quad \frac{\alpha x^2}{2\sin\alpha} - 0 - 2 = 0$$

$$\therefore \quad x = \pm 2\sqrt{\left(\frac{\sin\alpha}{\alpha}\right)}$$

2. Case I $0 < a < b$

$\therefore$ $|x| = x$

then $\int_a^b \frac{|x|}{x}\,dx = b - a$

Case II $a < 0 < b$

then $\int_a^b \frac{|x|}{x}\,dx = \int_a^0 \frac{|x|}{x}\,dx + \int_0^b \frac{|x|}{x}\,dx$

$$= \int_a^0 (-1)\,dx + \int_0^b (1)\,dx = a + b$$

Case III $a < b < 0$

$$\int_a^b \frac{|x|}{x}\,dx = \int_a^b (-1)\,dx = -(b-a) = a - b$$

3. For $0 < x < 1$

We know that $x^2 > x^3$ and for $1 < x < 2$

we have $x^3 > x^2$

Thus, $2^{x^2} > 2^{x^3}$ for $x \in (0, 1)$

and $2^{x^3} > 2^{x^2}$ for $x \in (1, 2)$

So, $I_1 > I_2$ and $I_3 < I_4$

4. Since, $|\sin x| \le 1$ for $x \ge 10$, then

$$\left|\frac{\sin x}{1+x^8}\right| \le \frac{1}{|1+x^8|} \quad \text{...(i)}$$

but $10 \le x \le 19$

$$\therefore \quad 1 + x^8 > x^8 \ge 10^8 \Rightarrow \frac{1}{1+x^8} < \frac{1}{x^8} \le \frac{1}{10^8}$$

or $$\frac{1}{|1+x^8|} \le \frac{1}{10^8} \quad \text{...(ii)}$$

From Eqs. (i) and (ii), we get

$$\left|\frac{\sin x}{1+x^8}\right| \le 10^{-8}$$

$$\therefore \quad \left|\int_{10}^{19}\frac{\sin x}{1+x^8}\,dx\right| \le \int_{10}^{19} 10^{-8}\,dx$$

$$= (19-10)\times 10^{-8}$$

$$= 9\times 10^{-8}$$

$$= (10-1)\times 10^{-8}$$

$$= 10^{-7} - 10^{-8} < 10^{-7}$$

Hence, $\left|\int_{10}^{19}\frac{\sin x}{1+x^8}\,dx\right| < 10^{-7}$

5. If $x > e$, then $1 < \ln x < x/e$

So, $$1 > \frac{1}{\sqrt[3]{\ln x}} > \sqrt[3]{\frac{e}{x}}$$

Hence, $$\int_3^4 1\cdot dx > \int_3^4 \frac{1}{\sqrt[3]{(\ln x)}}\,dx > \sqrt[3]{e}\int_3^4 \frac{1}{\sqrt[3]{x}}\,dx$$

$$\Rightarrow \quad 1 > I > \sqrt[3]{e}\,\frac{3}{2}(16^{1/3} - 9^{1/3})$$

using logarithmic table it is easy to verify that $\sqrt[3]{e}\,\frac{3}{2}(16^{1/3} - 9^{1/3})$ is approximately equal to 0.92

6. $\because$ $$I_2 = \int_1^{1/x}\frac{dt}{1+t^2}$$

$$= [\tan^{-1} t]_1^{1/x}$$

$$= \tan^{-1}\left(\frac{1}{x}\right) - \tan^{-1}(1)$$

$$I_2 = \cot^{-1}(x) - \pi/4$$

and $$I_1 = \int_x^1 \frac{dt}{1+t^2}$$

$$= [\tan^{-1} t]_x^1 = \tan^{-1} 1 - \tan^{-1} x$$

$$= \frac{\pi}{4} - \left(\frac{\pi}{2} - \cot^{-1} x\right)$$

$$= \cot^{-1} x - \pi/4 = I_2$$

Hence, $I_1 = I_2$ and $I_2 = \cot^{-1} x - \pi/4$

7. Let $f(x) = \sin^4 x + \cos^4 x$, it is easy to show that $f(x)$ is periodic with period $\pi/2$

$$\therefore \quad \int_a^{a+\pi/2}(\sin^4 x + \cos^4 x)\,dx$$

$$= \int_0^{\pi/2}(\sin^4 x + \cos^4 x)\,dx \quad \text{[by property]}$$

$$= \frac{3\cdot 1}{4\cdot 2}\cdot\frac{\pi}{2} + \frac{3\cdot 1}{4\cdot 2}\cdot\frac{\pi}{2} \quad \text{[by Walli's formula]}$$

$$= \frac{3\pi}{8}$$

8. Since, $\sin\alpha + \int_\alpha^{2\alpha}\cos 2x\,dx = 0 \Rightarrow \sin\alpha + \left[\frac{\sin 2x}{2}\right]_\alpha^{2a} = 0$

$\Rightarrow \quad \sin\alpha + \frac{1}{2}(\sin 4\alpha - \sin 2\alpha) = 0$

$\Rightarrow \quad \sin\alpha + \frac{1}{2}\cdot 2\cos 3\alpha \sin\alpha = 0$

$\Rightarrow \quad \sin\alpha(1 + \cos 3\alpha) = 0,$

if $\sin\alpha = 0,$

then $\alpha = 0, -\pi$ $\quad [\because \alpha \in [-\pi, 0]]$

and if $1 + \cos 3\alpha = 0$

$\Rightarrow \quad \cos 3\alpha = -1 = \cos(-\pi)$

$\Rightarrow \quad 3\alpha = -\pi;\ \Rightarrow \alpha = -\pi/3$

9. Since, $I = \int_0^1 \sqrt{(1 + x^3)}\,dx$

$\because \quad 0 < x < 1$

$\therefore \quad x^3 < x$

or $\sqrt{(1 + x^3)} < \sqrt{(x + 1)}$

$\int_0^1 \sqrt{(1 + x^3)}\,dx < \int_0^1 \sqrt{(1 + x)}\,dx$

$\Rightarrow \quad I < \frac{2}{3}[(1 + x)^{3/2}]_0^1$

$\Rightarrow \quad I < \frac{2}{3}(2\sqrt{2} - 1)$

Hence, $I < 1$ and $I < \sqrt{7}/2$

10. Let $I = \int_0^{\pi/2} \frac{dx}{1 + \sin^2 x}$

Since, $0 < x < \pi/2$

$\therefore \quad 0 < \sin x < 1$

or $1 < \sin^2 x + 1 < 2$

or $\frac{1}{2} < \frac{1}{1 + \sin^2 x} < 1$

or $\int_0^{\pi/2} \frac{1}{2}\,dx < \int_0^{\pi/2} \frac{1}{(1 + \sin^2 x)}\,dx < \int_0^{\pi/2} 1\cdot dx$

$\Rightarrow \quad \frac{\pi}{4} < I < \frac{\pi}{2}$

Hence, maximum and minimum values of I are $\frac{\pi}{2}$ and $\frac{\pi}{4}$ respectively.

11. Here, $f'(x) \geq 0$ in $[a, b]$, So, $f(x)$ is monotonically increasing

Hence, $f(a) \leq f(x) \leq f(b)$

$\therefore \quad \int_a^b f(a)\,dx \leq \int_a^b f(x)\,dx \leq \int_a^b f(b)\,dx$

$\Rightarrow \quad f(a)\cdot(b - a) \leq \int_a^b f(x)\,dx \leq f(b)(b - a)$

$\therefore \quad f(a) \leq \frac{1}{(b - a)}\int_a^b f(x)\,dx \leq f(b)$

12. $\because e^x \in (1, e)$ in $(0, 1)$ and $\int_0^x f(t)\sin t\,dt \in (0, 1)$ in $(0, 1)$

$\therefore e^x - \int_0^x f(t)\sin t\,dt$ cannot be zero.

So, option (a) is incorrect.

(b) $f(x) + \int_0^{\pi/2} f(t)\sin t\,dt$ always positive

$\therefore$ Option (b) is incorrect.

(c) Let $h(x) = x - \int_0^{\frac{\pi}{2} - x} f(t)\cos t\,dt,$

$h(0) = -\int_0^{\frac{\pi}{2}} f(t)\cos t\,dt < 0$

$h(1) = 1 - \int_0^{\frac{\pi}{2} - 1} f(t)\cos t\,dt > 0$

$\therefore$ Option (c) is correct.

(d) Let $g(x) = x^9 - f(x)$

$g(0) = -f(0) < 0$

$g(1) = 1 - f(1) > 0$

$\therefore$ Option (d) is correct.

13. $\because \int_{\pi/2}^{\alpha} \sin x\,dx = \sin 2\alpha \Rightarrow -[\cos x]_{\pi/2}^{\alpha} = \sin 2\alpha$

$\Rightarrow \quad -(\cos\alpha - 0) = \sin 2\alpha \Rightarrow \cos\alpha(2\sin\alpha + 1) = 0$

$\therefore \quad \cos\alpha = 0$ and $\sin\alpha = -1/2$

$\therefore \quad \alpha = \frac{\pi}{2}, \frac{3\pi}{2}$ and $\alpha = \pi + \frac{\pi}{6}, 2\pi - \frac{\pi}{6}$

$\therefore \quad \alpha = \frac{\pi}{2}, \frac{3\pi}{2}, \frac{7\pi}{6}, \frac{11\pi}{6}$

14. $\because \quad I_n = \int_0^{\pi/4} \tan^n x\,dx$

$= \int_0^{\pi/4} \tan^{n-2} x(\sec^2 x - 1)\,dx = \frac{1}{(n - 1)} - I_{n-2}$

or $\quad I_n + I_{n-2} = \frac{1}{(n - 1)}$ ...(i)

Then option (a) is correct.

Now, replacing n by $n + 1$ in Eq. (i), then

$I_{n+1} + I_{n-1} = \frac{1}{n}$ ...(ii)

ie, option (b) is correct.

Replacing n by $n + 1$ in Eq. (ii), then

$I_{n+2} + I_n = \frac{1}{n + 1}$ ...(iii)

or $\quad (n + 1)(I_{n+2} + I_n) = 1$

i.e. option (d) is correct.

Also, $\because \quad 0 < x < \pi/4$

$\therefore \quad 0 < \tan x < 1$

$\Rightarrow \quad \tan^{n+2} x < \tan^n x < \tan^{n-2} x$

or $\tan^n x + \tan^{n+2} x < 2\tan^n x < \tan^n x + \tan^{n-2} x$

or $\int_0^{\pi/4} (\tan^n x + \tan^{n+2} x)\,dx < 2\int_0^{\pi/4} \tan^n x\,dx$

$< \int_0^{\pi/4} (\tan^n x + \tan^{n-2} x)\,dx$

$\Rightarrow \quad I_n + I_{n+2} < 2I_n < I_n + I_{n-2}$

$\Rightarrow \quad \frac{1}{n + 1} < 2I_n < \frac{1}{n - 1}$ [from Eqs. (i) and (iii)]

i.e. option (c) is correct.

15. Let $I = \int_0^\pi x f(\sin x)\, dx$ ...(i)

$= \int_0^\pi (\pi - x) f(\sin x)\, dx$ ...(ii) [by property]

Adding Eqs. (i) and (ii), we get

$$2I = \pi \int_0^\pi f(\sin x)\, dx$$

$\therefore$ $I = \frac{\pi}{2}\int_0^\pi f(\sin x)\, dx$ [option (a)]

$= \frac{\pi}{2}\cdot 2\int_0^{\pi/2} f(\sin x)\, dx$

$= \pi \int_0^{\pi/2} f(\sin x)\, dx$ [option (c)] [by property]

$= \pi \int_0^{\pi/2} f(\cos x)\, dx$ [option (d)] [by property]

16. $\because$ $I_n = \int_0^{\pi/4} \tan^{n-2}x\,(\sec^2 x - 1)\, dx$

$= \int_0^{\pi/4} \tan^{n-2}x \cdot \sec^2 x\, dx - I_{n-2}$

or $I_n + I_{n-2} = \frac{1}{(n-1)}$

For $n = 5, I_5 + I_3 = \frac{1}{4}$ ...(i)

For $n = 8,\ I_8 + I_6 = \frac{1}{7}$ ...(ii)

For $n = 10, I_{10} + I_8 = \frac{1}{9}$ ...(iii)

Subtracting Eq. (iii) from (ii), then

$$I_6 - I_{10} = \frac{2}{63}$$ ...(iv)

Adding Eqs. (ii) and (iii), we get

$$I_6 + 2I_8 + I_{10} = \frac{16}{63}$$

17. Let $I = \int_\alpha^\beta \frac{f(\alpha+\beta-x)}{f(x)+f(\alpha+\beta-x)}\, dx$ ...(i)

$= \int_\alpha^\beta \frac{f(x)}{f(\alpha+\beta-x)+f(x)}\, dx$ [by property]...(ii)

Adding Eqs. (i) and (ii), then

$$2I = \int_\alpha^\beta 1\cdot dx = \beta - \alpha$$

$\therefore$ $I = \left(\frac{\beta-\alpha}{2}\right) = 4$ [given]

$\therefore$ $\beta - \alpha = 8$

18. If $h(t) = t$, then two given integrals are equal. Also,

$$\int_a^b f(t)\, g(h(t))\, dt = \int_a^b f(a+b-t)\, g(h(a+b-t))\, dt$$

Now, If $h(t) = a + b - t$

Then, $h(a+b-t) = a + b - (a+b-t) = t$

$\therefore$ $h(t) = t, (a+b-t)$

19. For $x = 0$

$$G(0, t) = 0, t \ge 0$$

$\therefore$ $g(0) = \int_0^1 f(t)\, G(0,t)\, dt = \int_0^1 0\cdot dt = 0$

and for $x = 1$

$$G(1, t) = 0, t < 1$$

$\therefore$ $g(1) = \int_0^1 f(t) G(1,t)\, dt = 0$

Also, $g(x) = \int_0^x f(t) G(x,t)\, dt + \int_x^1 f(t) G(x,t)\, dt$

$= \int_0^x f(t)\cdot t\,(x-1)\, dt + \int_x^1 f(t)\cdot x\,(t-1)\, dt$

$= (x-1)\int_0^x t f(t)\, dt + x\int_x^1 (t-1) f(t)\, dt$

$g'(x) = (x-1)\cdot x f(x) + \int_0^x t f(t)\, dt \cdot 1$

$+ x\,\{0 - (x-1) f(x)\} + \int_x^1 (t-1) f(t)\, dt \cdot 1$

$= \int_0^x t\, f(t)\, dt + \int_x^1 (t-1) f(t)\, dt$

$\therefore$ $g''(x) = x f(x) + \{0 - (x-1) f(x)\}$

$= f(x)$

20. Here, $f(x) = 7\tan^8 x + 7\tan^6 x - 3\tan^4 x - 3\tan^2 x$

for all $x \in \left(\frac{-\pi}{2}, \frac{\pi}{2}\right)$

$\therefore$ $f(x) = 7\tan^6 x \sec^2 x - 3\tan^2 x \sec^2 x$

$= (7\tan^6 x - 3\tan^2 x)\sec^2 x$

Now, $\int_0^{\pi/4} x\, f(x) dx = \int_0^{\pi/4} \underset{\text{I}}{x}\, \underset{\text{II}}{(7\tan^6 x - 3\tan^2 x)\sec^2 x}\, dx$

$= [x(\tan^7 x - \tan^3 x)]_0^{\pi/4}$

$- \int_0^{\pi/4} 1\,(\tan^7 x - \tan^3 x) dx$

$= 0 - \int_0^{\pi/4} \tan^3 x(\tan^4 x - 1) dx$

$= -\int_0^{\pi/4} \tan^3 x\,(\tan^2 x - 1)\sec^2 x\, dx$

Put $\tan x = t \Rightarrow \sec^2 x\, dx = dt$

$\therefore$ $\int_0^{\pi/4} x\, f(x) dx = -\int_0^1 t^3(t^2-1) dt$

$= \int_0^1 (t^3 - t^5) dt = \left[\frac{t^4}{4} - \frac{t^5}{5}\right]_0^1 = \frac{1}{4} - \frac{1}{6} = \frac{1}{12}$

Also, $\int_0^{\pi/4} f(x) dx = \int_0^{\pi/4} (7\tan^6 x - 3\tan^2 x)\sec^2 x\, dx$

$= \int_0^1 (7t^6 - 3t^2) dt = [t^7 - t^3]_0^1 = 0$

21. Let $I = \int_{-1/2}^{1/2} \sqrt{\left(\frac{x+1}{x-1} - \frac{x-1}{x+1}\right)^2}\, dx$

$= \int_{-1/2}^{1/2} \left|\frac{4x}{x^2-1}\right| dx = -2\int_0^{1/2} \frac{4x}{(x^2-1)}\, dx$

$= -4\cdot\{\ln|x^2-1|\}_0^{1/2} = -4\ln\left(\frac{3}{4}\right)$

$= 4\ln\left(\frac{4}{3}\right) = \ln\left(\frac{256}{81}\right) = -\ln\left(\frac{81}{256}\right)$

22. $\because$ $b^2 = a^2(e^2-1)$ and $a^2 = b^2((f(e))^2 - 1)$

Then, we get $\frac{1}{e^2} + \frac{1}{(f(e))^2} = 1$

or $f(e) = \frac{e}{\sqrt{e^2-1}}$

$\therefore$ $f(f(e)) = e$

and $f(f(f(e)) = f(e) = e/\sqrt{e^2-1}$

$$\therefore \quad \underbrace{f\,f\,f\,f \dots f(e)}_{n\text{times}} = \begin{cases} e, \text{ if } n \text{ is even} \\ \dfrac{e}{\sqrt{e^2 - 1}}, \text{ is } n \text{ is odd} \end{cases}$$

If n is even :

$$\int_1^3 \underbrace{ffff \dots f(e)}_{n\text{ times}}\, de = \int_1^3 e\, de = \left\{\frac{e^2}{2}\right\}_1^3 = \frac{9-1}{2} = 4$$

If n is odd :

$$\int_1^3 \underbrace{ffff \dots f(e)}_{n\text{times}}\, d(e) = \int_1^3 \frac{e}{\sqrt{(e^2-1)}}\, de$$

$$= \{\sqrt{(e^2-1)}\}_1^3 = \sqrt{8} - 0 = 2\sqrt{2}$$

23. $f'(x) = |\sin x|$

$f'(\pi) = 0$ (finite values)

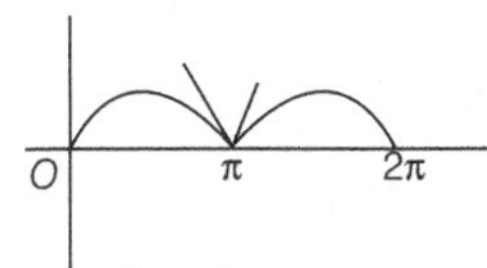

$\therefore f(x)$ is continuous everywhere and $f'(x)$ is not differentiable at $x = n\pi \;\forall\; n \in I$

Also, $f(\pi) = \int_0^{\pi} \sin t\, dt = -[\cos t]_0^{\pi} = -(-1-1) = 2$

24. Let $I = \int_{\tan^{-1}\lambda}^{\cot^{-1}\lambda} \dfrac{\tan x}{\tan x + \cot x}\, dx$ ...(i)

$$= \int_{\tan^{-1}\lambda}^{\cot^{-1}\lambda} = \frac{\cot x}{\cot x + \tan x}\, dx \quad \text{[by property]}$$

$[\because \tan^{-1}\lambda + \cot^{-1}\lambda = \pi/2]$

Adding Eqs. (i) and (ii), we get

$$2I = \int_{\tan^{-1}\lambda}^{\cot^{-1}\lambda} 1dx = \cot^{-1}\lambda - \tan^{-1}\lambda$$

$$= \left(\frac{\pi}{2} - \tan^{-1}\lambda\right) - \tan^{-1}\lambda$$

or $\quad I = \left(\dfrac{\pi}{4} - \tan^{-1}\lambda\right)$

Now $\quad -\dfrac{\pi}{2} < \tan^{-1}\lambda < \dfrac{\pi}{2}$

$\Rightarrow \quad -\dfrac{\pi}{2} < -\tan^{-1}\lambda < \dfrac{\pi}{2}$

$\therefore \quad \dfrac{\pi}{4} - \dfrac{\pi}{2} < \dfrac{\pi}{4} - \tan^{-1}\lambda < \dfrac{\pi}{4} + \dfrac{\pi}{2}$

Hence, $\quad -\dfrac{\pi}{4} < I < \dfrac{3\pi}{4}.$

25. Let $g(x) = \sin mx \cos nx$

$\therefore \quad g(-x) = -g(x)$

$g(x)$ is odd function

$$\therefore \quad \int_{-\pi}^{\pi} \sin mx \cos nx\, dx = 0$$

Also, $x^8 \sin^9 x$ and $\sin x\, f(\cos x)$ are odd functions

$$\therefore \quad \int_{-\pi/8}^{\pi/8} x^8 \sin^9 x\, dx = 0$$

and $\quad \displaystyle\int_{-\pi/2}^{\pi/2} \sin x\, f(\cos x)\, dx = 0.$

26. Let $I = \displaystyle\int_{1/e}^{\cot x} \frac{dt}{t(1+t^2)}$

put $\quad t = \dfrac{1}{z}$

$\therefore \quad dt = -\dfrac{1}{z^2}\, dz$

$$\therefore \quad I = \int_e^{\tan x} \frac{-\dfrac{1}{z^2}\, dz}{\dfrac{1}{z}\left(1 + \dfrac{1}{z^2}\right)} = \int_{\tan x}^{e} \frac{z\, dz}{(z^2+1)}$$

$$= \int_{\tan x}^{e} \frac{t\, dt}{(1+t^2)} \quad \text{(By property)}$$

$$\therefore \quad \int_{1/e}^{\tan x} \frac{t\, dt}{(1+t^2)} + \int_{1/e}^{\cot x} \frac{dt}{t(1+t^2)}$$

$$= \int_{1/e}^{\tan x} \frac{t}{(1+t^2)}\, dt + \int_{\tan x}^{e} \frac{t}{(1+t^2)}\, dt$$

$$= \int_{1/e}^{e} \frac{t\, dt}{(1+t^2)} = \frac{1}{2}[\ln(1+t^2)]_{1/e}^{e}$$

$$= \frac{1}{2}\left\{\ln(1+e^2) - \ln\left(1 + \frac{1}{e^2}\right)\right\}$$

$$= \frac{1}{2}(\ln e^2) = 1$$

Also, $\quad \dfrac{2}{\pi}\displaystyle\int_{-1}^{1} \frac{dt}{1+t^2} = \frac{4}{\pi}\int_0^1 \frac{dt}{1+t^2} = \frac{4}{\pi}\cdot \tan^{-1} 1$

$$= \frac{4}{\pi}\cdot\frac{\pi}{4} = 1$$

27. $\because \quad 0 < x < \dfrac{\pi}{2}$

$\therefore \quad \dfrac{2}{\pi} < \dfrac{\sin x}{x} < 1$

$\Rightarrow \quad \displaystyle\int_0^{\pi/2} \frac{2}{\pi}\, dx < \int_0^{\pi/2} \frac{\sin x}{x}\, dx < \int_0^{\pi/2} 1 \cdot dx$

$\Rightarrow \quad 1 < \displaystyle\int_0^{\pi/2} \frac{\sin x}{x}\, dx < \frac{\pi}{2}$

28. Let $I_1 = \displaystyle\int_0^{4\pi} e^t(\sin^6 at + \cos^4 at) dt$

$$= \int_0^{\pi} e^t(\sin^6 at + \cos^4 at)\, dt + \int_{\pi}^{2\pi} e^t(\sin^6 at + \cos^4 at)\, dt + \int_{2\pi}^{3\pi} e^t(\sin^6 at + \cos^4 at)\, dt + \int_{3\pi}^{4\pi} e^t(\sin^6 at + \cos^4 at)\, dt$$

$\therefore \quad I_1 = I_2 + I_3 + I_4 + I_5$ ...(i)

Now, $\quad I_3 = \displaystyle\int_{\pi}^{2\pi} e^t(\sin^6 at + \cos^4 at)\, dt$

Put $\quad t = \pi + x \Rightarrow dt = dx$

$\therefore \quad I_3 = \displaystyle\int_0^{\pi} e^{\pi + x}\cdot(\sin^6 at + \cos^4 at)\, dt = e^{\pi}\cdot I_2$ ...(ii)

Now, $\quad I_4 = \displaystyle\int_{2\pi}^{3\pi} e^t(\sin^6 at + \cos^4 at)\, dt$

Put $\quad t = 2\pi + x \;\Rightarrow\; dt = dx$

$\therefore \quad I_4 = \displaystyle\int_0^{\pi} e^{x+2\pi}(\sin^6 at + \cos^4 at)\, dt = e^{2\pi}\cdot I_2$ ...(iii)

and $\quad I_5 = \displaystyle\int_{3\pi}^{4\pi} e^t(\sin^6 at + \cos^4 at)\, dt$

Put $t = 3\pi + x$

$\therefore \quad I_5 = \int_0^{\pi} e^{3\pi + x} (\sin^6 at + \cos^4 at) dt = e^{3\pi} \cdot I_2 \quad \ldots(\text{iv})$

From Eqs. (i), (ii), (iii) and (iv), we get

$I_1 = I_2 + e^{\pi} \cdot I_2 + e^{2\pi} \cdot I_2 + e^{3\pi} \cdot I_2 = (1 + e^{\pi} + e^{2\pi} + e^{3\pi}) I_2$

$$\therefore \quad L = \frac{\int_0^{4\pi} e^t (\sin^6 at + \cos^4 at) dt}{\int_0^{\pi} e^t (\sin^6 at + \cos^4 at) dt}$$

$$= (1 + e^{\pi} + e^{2\pi} + e^{3\pi}) = \frac{1 \cdot (e^{4\pi} - 1)}{e^{\pi} - 1} \text{ for } a \in R$$

29. $\because \quad 0 < x < \frac{1}{2}$

$\therefore \quad x^{2n} < x^2$

$\Rightarrow \quad -x^{2n} > -x^2$

$\Rightarrow \quad (1 - x^{2n}) > (1 - x^2)$

$\Rightarrow \quad 1 > \sqrt{(1 - x^{2n})} > \sqrt{(1 - x^2)}$

$\Rightarrow \quad 1 < \frac{1}{\sqrt{(1 - x^{2n})}} < \frac{1}{\sqrt{(1 - x^2)}}$

or $\quad \int_0^{1/2} 1 \cdot dx < \int_0^{1/2} \frac{dx}{\sqrt{(1 - x^{2n})}} < \int_0^{1/2} \frac{dx}{\sqrt{(1 - x^2)}}$

$\Rightarrow \quad \frac{1}{2} < I < \sin^{-1}\left(\frac{1}{2}\right)$

or $\quad \frac{1}{2} < I < \frac{\pi}{6}$

or $\quad 0.5 < I < 0.52.$

30. $\because \frac{\pi}{2} > 1$ and $\frac{\pi}{2} < 2$

$\therefore \quad x^{\pi/2} < x^1 \ (\because 0 < x < 1)$ and $x^{\pi/2} > x^2 \ (\because 0 < x < 1)$

$\Rightarrow \quad 1 + x^{\pi/2} < 1 + x$ and $1 + x^{\pi/2} > 1 + x^2$

or $\quad \frac{1}{(1 + x^{\pi/2})} > \frac{1}{(1 + x)}$ and $\frac{1}{(1 + x^{\pi/2})} < \frac{1}{(1 + x^2)}$

or $\quad \int_0^1 \frac{dx}{1 + x^{\pi/2}} > \int_0^1 \frac{dx}{1 + x}$

and $\quad \int_0^1 \frac{dx}{1 + x^{\pi/2}} < \int_0^1 \frac{dx}{1 + x^2}$

$\therefore \quad I > \ln 2$ and $I < \pi/4.$

31. $\because I_n = \int_0^{\pi/2} \cos^{n-1} x \cdot \cos x \, dx$

$= [\cos^{n-1} x \cdot \sin x]_0^{\pi/2} - \int_0^{\pi/2} (n-1) \cos^{n-2} x (-\sin x)(\sin x) \, dx$

$= 0 + (n-1) \int_0^{\pi/2} \cos^{n-2} x (1 - \cos^2 x) \, dx$

$= (n-1)(I_{n-2} - I_n)$

$\therefore \quad n(I_{n-2} - I_n) = I_{n-2}$

Also, $\because \quad 0 < x < \pi/2$

$\therefore \quad 1 > \cos x > 0$

$\therefore \quad \cos^n x < \cos^{n-2} x$

$\Rightarrow \quad \int_0^{\pi/2} \cos^n x \, dx < \int_0^{\pi/2} \cos^{n-2} x \, dx$

or $\quad I_n < I_{n-2} \Rightarrow I_{n-2} > I_n$

32. Since,

$$\lim_{n \to \infty} \left[\frac{1 + \sqrt[3]{2} + \sqrt[3]{3} + \ldots + \sqrt[3]{n}}{n^{7/3} \left(\frac{1}{(an+1)^2} + \frac{1}{(an+2)^2} + \ldots + \frac{1}{(an+n)^2} \right)} \right],$$

$a \in R, |a| > 1$

$$= \lim_{n \to \infty} \frac{\sum_{r=1}^{n} (r^{1/3})}{n^{7/3} \sum_{r=1}^{n} \frac{1}{(an+r)^2}} = \lim_{n \to \infty} \frac{\sum_{r=1}^{n} \left(\frac{r}{n}\right)^{1/3} \frac{1}{n}}{\sum_{r=1}^{n} \frac{1}{\left(a + \frac{r}{n}\right)^2} \frac{1}{n}}$$

$$= \frac{\int_0^1 x^{1/3} dx}{\int_0^1 \frac{dx}{(a+x)^2}} = 54, \quad \text{(given)}$$

$$\Rightarrow \quad \frac{\frac{3}{4}[x^{4/3}]_0^1}{\left[-\frac{1}{x+a}\right]_0^1} = 54 \Rightarrow \frac{3/4}{-\frac{1}{a+1} + \frac{1}{a}} = 54$$

$\Rightarrow \quad \frac{3}{4 \times 54} = \frac{1}{a(a+1)}$

$\Rightarrow \quad a^2 + a = 72$

$\Rightarrow \quad a^2 + 9a - 8a - 72 = 0$

$\Rightarrow \quad a(a+9) - 8(a+9) = 0$

$\Rightarrow \quad (a-8)(a+9) = 0$

$\Rightarrow \quad a = 8$ or -9

33. Let $I = \int_0^{2\pi} \frac{d\theta}{(a + b\cos\theta + c\sin\theta)}$

$= \int_{-\pi}^{\pi} \frac{d\theta}{a + b\cos(\pi + \theta) + c\sin(\pi + \theta)}$ (by property)

$= \int_{-\pi}^{\pi} \frac{d\theta}{(a - b\cos\theta - c\sin\theta)}$

$= \int_{-\pi}^{\pi} \frac{\sec^2 \theta/2}{[a(1 + \tan^2 \theta/2) - b(1 - \tan^2 \theta/2) - 2c\tan\theta/2]} d\theta$

Put $\tan \theta/2 = t$

$\therefore \quad \sec^2 \theta/2 \, d\theta = 2 \, dt$

Then, $I = 2 \int_{-\infty}^{\infty} \frac{dt}{(a+b)t^2 - 2ct + (a-b)}$

$$= \frac{2}{(a+b)} \int_{-\infty}^{\infty} \frac{dt}{\left(t - \frac{c}{a+b}\right)^2 + \left(\frac{\sqrt{a^2 - b^2 - c^2}}{a+b}\right)^2}$$

$$= \frac{2}{\sqrt{(a^2 - b^2 - c^2)}} [\tan^{-1} \infty - \tan^{-1}(-\infty)]$$

$$= \frac{2}{\sqrt{(a^2 - b^2 - c^2)}} \left(\frac{\pi}{2} - \left(-\frac{\pi}{2}\right)\right)$$

$$= \frac{2\pi}{\sqrt{(a^2 - b^2 - c^2)}}.$$

34. In option (a), $\because 0 < x < 1$

$\Rightarrow \quad x^5 < x^2$

$\Rightarrow \quad 2^{x^5} < 2^{x^2}$

$\Rightarrow \quad \int_0^1 2^{x^5}\,dx < \int_0^1 2^{x^2}\,dx$

In option (b), $\because 1 < x < 2$

$\therefore \quad x^5 > x^3$

$\Rightarrow \quad 2^{x^5} > 2^{x^3}$

$\Rightarrow \quad \int_1^2 2^{x^5}\,dx > \int_1^2 2^{x^3}\,dx$

In option (c),

$\because \quad 1 < x < 2$

$\Rightarrow \quad 0 < \ln x < 0.693$

$\therefore \ln x > (\ln x)^2 \Rightarrow \int_1^2 \ln x\,dx > \int_1^2 (\ln x)^2\,dx$

In option (d),

$\because \quad 3 < x < 4 \Rightarrow 1.09 < \ln x < 1.386$

$\therefore \quad \ln x < (\ln x)^2$

$\therefore \quad \int_3^4 \ln x\,dx < \int_3^4 (\ln x)^2\,dx.$

Linked Comprehension Type

PASSAGE 1

1. $\because \lim_{n\to\infty} \frac{1}{n}\left\{\frac{1}{n+1}+\frac{2}{n+2}+\ldots+\frac{3n}{4n}\right\}$

$$= \lim_{n\to\infty} \frac{1}{n}\left\{\frac{1}{n+1}+\frac{2}{n+2}+\ldots+\frac{3n}{n+3n}\right\}$$

$$= \lim_{n\to\infty} \sum_{r=1}^{3n} \frac{1}{n}\left(\frac{r}{n+r}\right) = \lim_{n\to\infty} \sum_{r=1}^{3n} \frac{1}{n}\cdot\left(\frac{\frac{r}{n}}{1+\frac{r}{n}}\right)$$

$$= \int_0^3 \frac{x}{1+x}\,dx = \int_0^3 \left(1-\frac{1}{1+x}\right)dx$$

$$= [x - \ln(1+x)]_0^3 = 3 - \ln 4 = 3 - 2\ln 2$$

2. $\because \lim_{n\to\infty}\left\{\frac{1}{n}+\frac{n^2}{(n+1)^3}+\frac{n^2}{(n+2)^3}+\ldots+\frac{1}{8n}\right\}$

$$= \lim_{n\to\infty}\left\{\frac{n^2}{(n+0)^3}+\frac{n^2}{(n+1)^3}+\frac{n^2}{(n+2)^3}+\ldots+\frac{n^2}{(n+n)^3}\right\}$$

$$= \lim_{n\to\infty}\sum_{r=0}^{n} \frac{n^2}{(n+r)^3}$$

$$= \lim_{n\to\infty}\sum_{r=0}^{n} \frac{1}{n\left(1+\frac{r}{n}\right)^3} = \int_0^1 \frac{dx}{(1+x)^3}$$

$$= -\left\{\frac{1}{2(1+x)^2}\right\}_0^1 = -\left\{\frac{1}{8}-\frac{1}{2}\right\} = \frac{1}{2}-\frac{1}{8} = \frac{3}{8}$$

3. Let $P = \lim_{n\to\infty}\left\{\prod_{r=1}^{n}\left(1+\frac{r}{n}\right)\right\}^{1/n}$

$$\therefore \ln P = \lim_{n\to\infty}\sum_{r=1}^{n}\frac{1}{n}\ln\left(1+\frac{r}{n}\right)$$

$$= \int_0^1 \ln(1+x)\,dx$$

$$= [x\ln(1+x)]_0^1 - \int_0^1 \frac{x}{1+x}\,dx$$

$$= \ln 2 - \int_0^1 \left(1-\frac{1}{1+x}\right)dx$$

$$= \ln 2 - \{x - \ln(1+x)\}_0^1$$

$$= \ln 2 - \{1 - \ln 2\}$$

$$= 2\ln 2 - 1$$

$$= \ln 4 - \ln e$$

$$= \ln(4/e)$$

$\therefore \quad P = 4/e$

4. Let $P = \lim_{n\to\infty}\left\{\prod_{r=1}^{n}\tan\left(\frac{r\pi}{2n}\right)\right\}^{1/n}$

$$\therefore \ln P = \lim_{n\to\infty}\sum_{r=1}^{n}\frac{1}{n}\ln\tan\left(\frac{r\pi}{2n}\right) = \int_0^1 \ln\tan\left(\frac{\pi x}{2}\right)dx$$

$$\Rightarrow \ln P = \frac{2}{\pi}\int_0^{\pi/2}\ln\tan x\,dx \quad \ldots\text{(i)}$$

$$\text{and } \ln P = \frac{2}{\pi}\int_0^{\pi/2}\ln\tan\left(\frac{\pi}{2}-x\right)dx \quad \text{[by property]}$$

$$= \frac{2}{\pi}\int_0^{\pi/2}\ln\cot x\,dx \quad \ldots\text{(ii)}$$

Adding Eqs. (i) and (ii), then

$$2\ln P = \frac{2}{\pi}\int_0^{\pi/2}(\ln\tan x + \ln\cot x)\,dx$$

$$= \frac{2}{\pi}\int_0^{\pi/2}\ln 1\,dx = 0$$

$\Rightarrow \quad 2\ln P = 0$

or $\quad \ln P = 0$

$\therefore \quad P = e^0 = 1$

5. Let

$$P = \lim_{n\to\infty}\left\{\frac{1}{n+m}+\frac{1}{n+2m}+\frac{1}{n+3m}+\ldots+\frac{1}{n+nm}\right\}$$

$$= \lim_{n\to\infty}\sum_{r=1}^{n}\frac{1}{n+mr}$$

$$= \lim_{n\to\infty}\sum_{r=1}^{n}\frac{1}{n}\cdot\frac{1}{\left(1+m\cdot\frac{r}{n}\right)}$$

$$= \int_0^1 \frac{dx}{(1+mx)}$$

$$= \frac{1}{m}\{\ln(1+mx)\}_0^1 = \frac{1}{m}\ln(1+m)$$

PASSAGE 2

1. According to the given data,

$$F'(x) < 0, \ \forall x \in (1, 3)$$

We have, $f(x) = xF(x)$

$\Rightarrow \quad f'(x) = F(x) + xF'(x) \quad \ldots(i)$

$\Rightarrow \quad f'(1) = F(1) + F'(1) < 0$

[given $F(1) = 0$ and $F'(x) < 0$]

Also, $f(2) = 2F(2) < 0$ [using $F(x) < 0, \forall x \in (1, 3)$]

Now, $f'(x) = F(x) + xF'(x) < 0$

[using $F(x) < 0, \ \forall x \in (1, 3)$]

$\Rightarrow \quad f'(x) < 0$

2. Given, $\int_1^3 x^2 F'(x)\,dx = -12$

$\Rightarrow [x^2F(x)]_1^3 - \int_1^3 2x \cdot F(x)\,dx = -12$

$\Rightarrow 9F(3) - F(1) - 2\int_1^3 f(x)\,dx = -12$

[$\because xF(x) = f(x)$, given]

$\Rightarrow \quad -36 - 0 - 2\int_1^3 f(x)\,dx = -12$

$\therefore \quad \int_1^3 f(x)\,dx = -12$ and $\int_1^3 x^3 F''(x)dx = 40$

$\Rightarrow \quad [x^3F'(x)]_1^3 - \int_1^3 3x^2F'(x)\,dx = 40$

$\Rightarrow \quad [x^2(xF'(x))]_1^3 - 3 \times (-12) = 40$

$\Rightarrow \quad \{x^2 \cdot [f'(x) - F(x)]\}_1^3 = 4$

$\Rightarrow \quad 9[f'(3) - F(3)] - [f'(1) - F(1)] = 4$

$\Rightarrow \quad 9[f'(3) + 4] - [f'(1) - 0] = 4$

$\Rightarrow \quad 9f'(3) - f'(1) = -32$

PASSAGE 3

1. $\int_{-1}^1 [3x - 1]\,dx = \int_{-1}^1 ([3x] - 1)dx$

$= \int_{-1}^1 [3x]\,dx - \int_{-1}^1 1 \cdot dx \qquad \left[\begin{array}{l}\because -1 < x < 1 \\ \therefore -3 < 3x < 3\end{array}\right]$

$= \int_{-1}^{-2/3} [3x]\,dx + \int_{-2/3}^{-1/3} [3x]\,dx + \int_{-1/3}^{0} [3x]\,dx$

$+ \int_0^{1/3} [3x]\,dx + \int_{1/3}^{2/3} [3x]\,dx + \int_{2/3}^{1} [3x]\,dx - 2$

$= (-3)\int_{-1}^{-2/3} 1 \cdot dx + (-2)\int_{-2/3}^{-1/3} 1 \cdot dx + (-1)\int_{-1/3}^{0} 1 \cdot dx$

$+ 0 + 1\int_{1/3}^{2/3} 1 \cdot dx + 2\int_{2/3}^{1} 1 \cdot dx - 2$

$= (-3)\left(\frac{1}{3}\right) + (-2)\left(\frac{1}{3}\right) + (-1)\left(\frac{1}{3}\right) + 0 + \left(\frac{1}{3}\right) + 2 \cdot \left(\frac{1}{3}\right) - 2$

$= -1 - \frac{2}{3} - \frac{1}{3} + \frac{1}{3} + \frac{2}{3} - 2$

$= -3$

2. $\int_{\pi/2}^{3\pi/2} [2\sin x]\,dx = \int_{\pi/2}^{5\pi/6} [2\sin x]\,dx + \int_{5\pi/6}^{\pi} [2\sin x]\,dx$

$+ \int_{\pi}^{7\pi/6} [2\sin x]\,dx + \int_{7\pi/6}^{3\pi/2} [2\sin x]\,dx$

$= \int_{\pi/2}^{5\pi/6} 1 \cdot dx + 0 - \int_{\pi}^{7\pi/6} 1 \cdot dx - 2\int_{7\pi/6}^{3\pi/2} 1 \cdot dx$

$= \frac{\pi}{3} - \frac{\pi}{6} - \frac{2\pi}{3} = -\frac{\pi}{2}$

3. $\because \frac{x^2 + 2}{x^2 + 1} = 1 + \frac{1}{x^2 + 1} = 1 +$ Proper fraction

$\therefore \quad \left[\frac{x^2 + 2}{x^2 + 1}\right] = 1$

Then, $\int_0^{10} \left[\frac{x^2 + 2}{x^2 + 1}\right] dx = \int_0^{10} 1 \cdot dx = 10$

4. $\int_0^1 \sin([x] + [2x])\,dx = \int_0^{1/2} \sin([x] + [2x])\,dx$

$+ \int_{1/2}^1 \sin([x] + [2x]\,dx)$

$= 0 + \int_{1/2}^1 \sin(0 + 1)\,dx = \frac{\sin 1}{2}$

5. $\int_{-1}^1 [|x|]\,d\left(\frac{1}{1 + e^{-1/x}}\right) = \int_{-1}^0 [-x]\,d\left(\frac{1}{1 + e^{-1/x}}\right)$

$+ \int_0^1 [x]\,d\left(\frac{1}{1 + e^{-1/x}}\right)$

$= 0 + 0 = 0$

PASSAGE 4

1. Here, $f(x) = x^2 + 1$, $\alpha(x) = [x]$

$\therefore \int_0^3 (x^2 + 1)\,d[x] + \int_0^3 [x]\,2x\,dx = [3]\,(9 + 1) - 0$

$\Rightarrow \quad \int_0^3 (x^2 + 1)\,d[x] = 30 - 2\int_0^3 x[x]\,dx$

$= 30 - 2\left\{\int_0^1 0 \cdot dx + \int_1^2 x dx + \int_2^3 2x dx\right\}$

$= 30 - 2\left\{0 + \frac{3}{2} + 5\right\} = 30 - 3 - 10 = 17$

2. $\int_{-2}^3 [|x|]\,d|x| = \int_{-2}^0 [-x]\,d(-x) + \int_0^3 [x]\,dx$

$= -\int_{-2}^0 [-x]\,dx + \int_0^3 [x]\,dx$

$= \int_2^0 [x]\,dx + \int_0^3 [x]\,dx \qquad$ [by property]

$= -\int_0^2 [x]\,dx + \int_0^2 [x]\,dx + \int_2^3 [x]\,dx$

$= \int_2^3 [x]\,dx = \int_2^3 2 \cdot dx = 2\,(3 - 2) = 2$

3. Here, $f(x) = [x]$ and $\alpha(x) = [2x]$

$\int_{1/4}^{3/4} [x]\,d[2x] + \int_{1/4}^{3/4} [2x]\,d[x] = \left[\frac{3}{2}\right]\left[\frac{3}{4}\right] - \left[\frac{1}{2}\right]\left[\frac{1}{4}\right]$

$= 0 - 0 = 0$

$\Rightarrow \int_{1/4}^{3/4} [x]\,d[2x] = -\int_{1/4}^{3/4} [2x]\,d[x]$

$= -\int_{1/4}^{1/2} [2x]\,d[x] - \int_{1/2}^{3/4} [2x]\,d[x]$

$= 0 - \int_{1/2}^{3/4} d[x] = -\{[x]\}_{1/2}^{3/4}$

$= -(0 - 0) = 0$

4. $\int_0^3 e^x \, d\{x+[x]\}\,dx = \int_0^3 e^x dx = [e^x]_0^3$

$= e^3 - 1$

5. $\int_0^6 (x^2+[x])\,d\,|3-x| = \int_0^3 (x^2+[x])\,d\,(3-x)$

$+ \int_3^6 (x^2+[x])\,d\,(x-3)$

$= -\int_0^3 (x^2+[x])\,dx + \int_3^6 (x^2+[x])\,dx$

$= -\int_0^3 x^2\,dx + \int_3^6 x^2 dx - \int_0^3 [x]\,dx + \int_3^6 [x]\,dx$

$= -\left\{\frac{x^3}{3}\right\}_0^3 + \left\{\frac{x^3}{3}\right\}_3^6 - \int_0^3 [x]\,dx + \int_0^3 [x+3]\,dx$

$= -9 + (72-9) - \int_0^3 [x]\,dx$

$+ \int_0^3 [x]\,dx + 3\int_0^3 1dx$

$= 72 - 18 + 9$

$= 72 - 9 = 63$

PASSAGE 5

1. $\int_0^3 \min\{1, |x|, |x-2|\}\,dx$

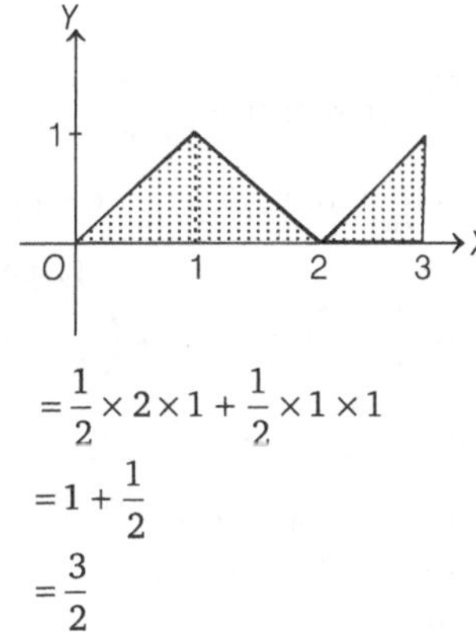

$= \frac{1}{2} \times 2 \times 1 + \frac{1}{2} \times 1 \times 1$

$= 1 + \frac{1}{2}$

$= \frac{3}{2}$

2. $\int_{-1}^1 \max\{x, x^3\}\,dx = \int_{-1}^0 x^3 dx + \int_0^1 x\,dx$

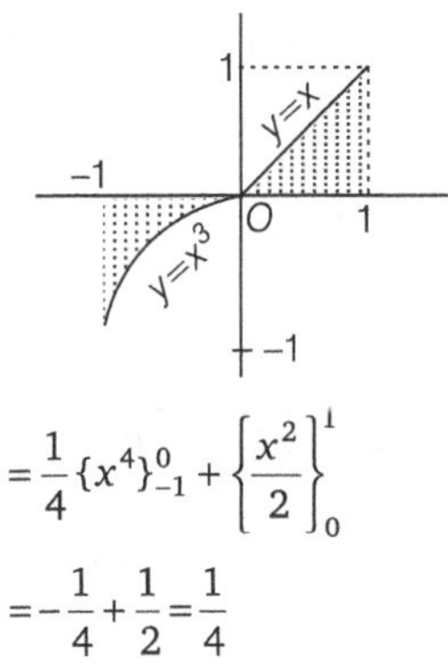

$= \frac{1}{4}\{x^4\}_{-1}^0 + \left\{\frac{x^2}{2}\right\}_0^1$

$= -\frac{1}{4} + \frac{1}{2} = \frac{1}{4}$

3. $\int_{-2}^2 \min\{|x|, [x]\}dx$

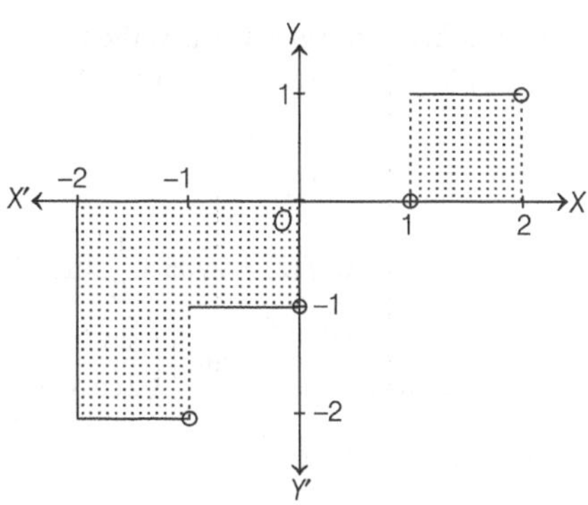

$= \int_{-2}^{-1} [x]\,dx + \int_{-1}^0 [x]\,dx + \int_0^1 [x]\,dx + \int_1^2 [x]\,dx$

$= -(2\times1) - (1\times1) + 0 + 1\times1 = -2$

4. $\int_0^\pi \max\{\sin x, \cos x\}\,dx = \int_0^{\pi/4} \cos x\,dx + \int_{\pi/4}^\pi \sin x\,dx$

$\{\sin x\}_0^{\pi/4} - \{\cos x\}_{\pi/4}^\pi$

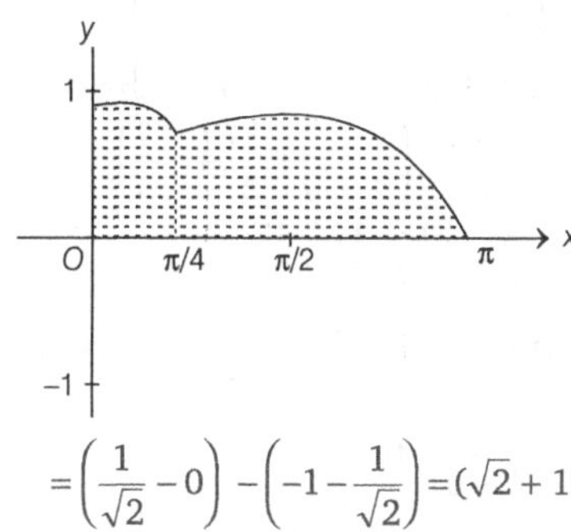

$= \left(\frac{1}{\sqrt{2}} - 0\right) - \left(-1 - \frac{1}{\sqrt{2}}\right) = (\sqrt{2}+1)$

5. $\because \min\{2\sin x, 1 \quad \cos x, 1\} = \begin{cases} 1-\cos x & , \quad 0 \le x \le \frac{\pi}{2} \\ 1 & , \quad \frac{\pi}{2} \le x \le \frac{5\pi}{6} \\ 2\sin x & , \quad \frac{5\pi}{6} \le x \le \pi \end{cases}$

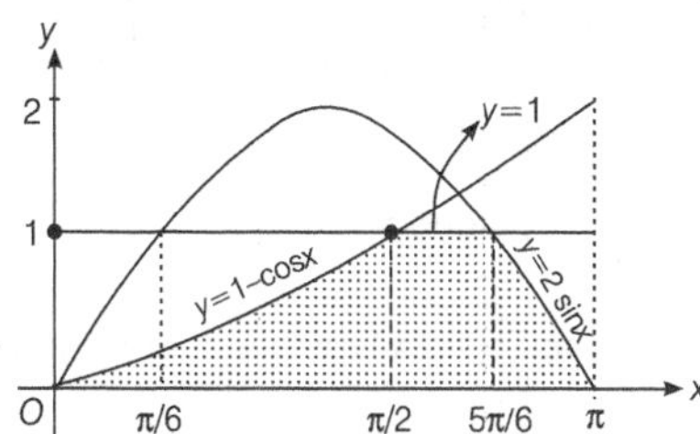

$\Rightarrow \int_0^\pi \min\{2\sin x, 1-\cos x, 1\}\,dx$

$= \int_0^{\pi/2} (1-\cos x)\,dx + \int_{\pi/2}^{5\pi/6} 1dx + \int_{5\pi/6}^{\pi} 2\sin x\,dx$

$= \{x - \sin x\}_0^{\pi/2} + \left(\frac{5\pi}{6} - \frac{\pi}{2}\right) - 2(\cos x)_{5\pi/6}^\pi$

$= \left(\frac{\pi}{2} - 1\right) + \frac{\pi}{3} - 2\left(-1 + \frac{\sqrt{3}}{2}\right) = \frac{5\pi}{6} + 1 - \sqrt{3}$

Numerical Value Type

1. Differentiating both sides w.r.t. x, then

$$f(x^2(1+x))\times(2x+3x^2)=1$$

At $x=1$

$$f(2)=\frac{1}{5}$$

$$\therefore \quad 9905f(2)=9905\times\frac{1}{5}=1981$$

2. $I=\left\{\sin^{-1}\left(\frac{x}{2\cos\alpha}\right)\right\}_0^{\sin 2\alpha}=\sin^{-1}\left(\frac{\sin 2\alpha}{2\cos\alpha}\right)$

$=\sin^{-1}(\sin\alpha)=\pi-\alpha \qquad \left[\because \pi<\alpha<\frac{2\pi}{3}\right]$

$\therefore \quad 2008\frac{(I+\alpha)}{\pi}=2008$

3. $I=\int_0^2 x[2x]\,dx$

$=\int_0^{1/2}0\cdot dx+\int_{1/2}^1 x\,dx+\int_1^{3/2}2x\,dx+\int_{3/2}^2 3x\,dx$

$=0+\left[\frac{x^2}{2}\right]_{1/2}^1+[x^2]_1^{3/2}+\left[\frac{3x^2}{2}\right]_{3/2}^2$

$=\frac{1}{2}\left(1-\frac{1}{4}\right)+\left(\frac{9}{4}-1\right)+\frac{3}{2}\left(4-\frac{9}{4}\right)$

$=\frac{3}{8}+\frac{5}{4}+\frac{21}{8}=\frac{34}{8}=\frac{17}{4}$

$\therefore \quad 12.8=12.8\times\frac{17}{4}=3.2\times17=54.4$

4. $I=\int_{-1}^1\frac{e^{|x|}}{1+a^x}dx \qquad \ldots\text{(i)}$

$I=\int_{-1}^1\frac{e^{|0-x|}}{1+a^{0-x}}dx\left[\because \int_a^b f(x)\,dx=\int_a^b f(a+b-x)\,dx\right]$

$=\int_{-1}^1\frac{a^x e^{|x|}}{1+a^x}dx \qquad \ldots\text{(ii)}$

Adding Eqs. (i) and (ii), then

$$2I=\int_{-1}^1\frac{e^{|x|}(1+a^x)}{(1+a^x)}dx=\int_{-1}^1 e^{|x|}\,dx=2\int_0^1 e^x\,dx$$

$\therefore \quad I=(e-1)$

Then, $\frac{9997}{(e-1)}I=9997$

5. $I=\int_{-\pi/2}^0\sin^{-1}(\sin x)\,dx+\int_0^\pi\sin^{-1}(\sin x)\,dx+\int_\pi^{2\pi}\sin^{-1}(\sin x)\,dx$

= Area of shaded region

$=-\left(\frac{1}{2}\times\frac{\pi}{2}\times\frac{\pi}{2}\right)+\left(\frac{1}{2}\times\pi\times\frac{\pi}{2}\right)-\left(\frac{1}{2}\times\pi\times\frac{\pi}{2}\right)=-\frac{\pi^2}{8}$

$\therefore \quad -\frac{7992}{\pi^2}I=999$

6. $f(x)+f(x+1)=2$

$x\to x+1$

$\Rightarrow f(x+1)+f(x+2)=2 \Rightarrow 2-f(x)+f(x+2)=2$

$\Rightarrow \qquad f(x+2)=f(x) \Rightarrow f(x)=f(x+2)$

$f(x)$ is periodic with $T=2$

$I_1=\int_0^{2\times4}f(x)\,dx=4\int_0^2 f(x)\,dx$

$I_2=\int_{-1}^3 f(x)\,dx=\int_0^4 f(x+1)\,dx=\int_0^4[2-f(x)]\,dx$

$I_2=8-2\int_0^2 f(x)\,dx \Rightarrow I_2=8-\frac{I_1}{2} \Rightarrow 2I_2=16-I_1$

$I_1+2I_2=16$

7. $I=2\int_0^4 f(x)^2dx$ {Even function}

$=2\int_0^4(4x^3-g(4-x))dx$

$=2\left(\left.\frac{4x^4}{4}\right|_0^4-\int_0^4 g(4-x)dx\right)=2(256-0)=512$

8. Let $I=\int_0^{a+2008\pi}|\cos x|\,dx$

$=\int_0^{2008\pi}|\cos x|\,dx+\int_{2008\pi}^{a+2008\pi}|\cos x|\,dx$

$=2008\int_0^\pi|\cos x|\,dx+\int_0^a|\cos x|\,dx$

($\because$ period of $|\cos x|$ is π)

$=2008\left\{\int_0^{\pi/2}|\cos x|\,dx+\int_{\pi/2}^\pi\cos x\,dx\right\}+\int_0^a\cos x\,dx$

$=2008\left\{\int_0^{\pi/2}\cos x\,dx-\int_{\pi/2}^\pi\cos x\,dx\right\}+\sin a$

$=2008[\sin x]_0^{\pi/2}-[\sin x]_{\pi/2}^\pi\}+\sin a$

$=2008\{(1-0)-(0-1)\}+\sin a=4016+\sin a$

$\therefore \quad \lambda=4016$

Hence, $(\lambda)=(4016)=4016$

9. Let $g(t)=|t-1|-|t|+|t+1|=\begin{cases}-t, & t<-1\\ t+2, & -1\le t<0\\ 2-t, & 0\le t<1\\ t, & t\ge1\end{cases}$

Hence, $f(x)=\begin{cases}3/2, & -3/2\le x<-1/2\\ 2+x, & -\frac{1}{2}<x\le0\\ 2, & 0<x\le2\end{cases}$

$\therefore I=\int_{-3/2}^2 f(x)\,dx=\int_{-3/2}^{-1/2}\frac{3}{2}dx+\int_{-1/2}^0(2+x)\,dx+\int_0^2 2\,dx$

$=\frac{3}{2}\left(-\frac{1}{2}+\frac{3}{2}\right)+0-\left(-1+\frac{1}{8}\right)+2(2-0)$

$=\frac{3}{2}+\frac{7}{8}+4=\frac{51}{8}$

$\therefore \quad 64I=64\times\frac{51}{8}=8\times51=408$

when $x = 0, t = 1, x = \frac{1}{2}, t = \frac{1}{3}$

$\therefore \quad I = \int_1^{1/3} \frac{(1+\sqrt{3})\, dt}{-2(t)^{6/4}} \Rightarrow I = \frac{-(1+\sqrt{3})}{2}\left[\frac{-2}{\sqrt{t}}\right]_1^{1/3}$

$\Rightarrow \quad I = (1+\sqrt{3})(\sqrt{3}-1)$

$\Rightarrow \quad I = 3 - 1 = 2$

6. Here, $f(x) = \begin{cases} [x], & x \le 2 \\ 0, & x > 2 \end{cases}$

$\therefore \quad I = \int_{-1}^{2} \frac{x\, f(x^2)}{2 + f(x+1)} dx$

$= \int_{-1}^{0} \frac{x\, f(x^2)}{2+f(x+1)} dx + \int_0^1 \frac{x\, f(x^2)}{2+f(x+1)} dx + \int_1^{\sqrt{2}} \frac{x\, f(x^2)}{2+f(x+1)} dx + \int_{\sqrt{2}}^{\sqrt{3}} \frac{x\, f(x^2)}{2+f(x+1)} dx + \int_{\sqrt{3}}^{2} \frac{x\, f(x^2)}{2+f(x+1)} dx$

$= \int_{-1}^{0} 0\, dx + \int_0^1 0\, dx + \int_1^{\sqrt{2}} \frac{x \cdot 1}{2+0} dx + \int_{\sqrt{2}}^{\sqrt{3}} 0\, dx + \int_{\sqrt{3}}^{2} 0\, dx$

$\left[\begin{array}{l} \because -1 < x < 0 \Rightarrow 0 < x^2 < 1 \Rightarrow [x^2] = 0, \\ 0 < x < 1 \Rightarrow 0 < x^2 < 1 \Rightarrow [x^2] = 0, \\ 1 < x < \sqrt{2} \Rightarrow \begin{cases} 1 < x^2 < 2 & \Rightarrow [x^2] = 1 \\ 2 < x+1 < 1+\sqrt{2} & \Rightarrow f(x+1) = 0, \end{cases} \\ \sqrt{2} < x < \sqrt{3} \Rightarrow 2 < x^2 < 3 \Rightarrow f(x^2) = 0, \\ \text{and } \sqrt{3} < x < 2 \Rightarrow 3 < x^2 < 4 \Rightarrow f(x^2) = 0 \end{array}\right]$

$\Rightarrow \quad I = \int_1^{\sqrt{2}} \frac{x}{2} dx = \left[\frac{x^2}{4}\right]_1^{\sqrt{2}} = \frac{1}{4}(2-1) = \frac{1}{4}$

$\therefore \quad 4I = 1 \Rightarrow 4I - 1 = 0$

7. Here, $\alpha = \int_0^1 e^{(9x + 3\tan^{-1} x)} \left(\frac{12 + 9x^2}{1+x^2}\right) dx$

Put $\quad 9x + 3\tan^{-1} x = t \Rightarrow \left(9 + \frac{3}{1+x^2}\right) dx = dt$

$\therefore \ \alpha = \int_0^{9 + 3\pi/4} e^t\, dt = [e^t]_0^{9+3\pi/4} = e^{9+3\pi/4} - 1$

$\Rightarrow \ \log_e |1+\alpha| = 9 + \frac{3\pi}{4} \Rightarrow \log_e |\alpha + 1| - \frac{3\pi}{4} = 9$

8. It is given that, for functions $f : R \to R$ and $F : [0, \pi] \to R$,

$F(x) = \int_0^x f(t) dt$, where $f(\pi) = -6$

$\Rightarrow \quad F'(\pi) = f(\pi) = -6 \quad \ldots \text{(i)}$

Now, $\int_0^{\pi} (f'(x) + F(x)) \cos x\, dx$

$= \int_0^{\pi} f'(x) \cos x dx + \int_0^{\pi} F(x) \cos x dx$

$= [\cos x\, f(x)]_0^{\pi} + \int_0^{\pi} f(x) \sin x\, dx + \int_0^{\pi} F(x) \cos x\, dx$

[by integration by parts]

$= (-1)(-6) - f(0) + [(\sin x) F(x)]_0^{\pi} - \int_0^{\pi} F(x) \cos x\, dx + \int_0^{\pi} F(x) \cos x\, dx$

$= 6 - f(0) = 2$ [given]

$\Rightarrow \quad f(0) = 4$

Matrix-Match Type

1. **(A)** $\int_1^4 |x-3|\, dx = \int_1^3 (3-x)\, dx + \int_3^4 (x-3)\, dx$

$= \left[3x - \frac{x^2}{2}\right]_1^3 + \left[\frac{x^2}{2} - 3x\right]_3^4$

$= \left\{\left(9 - \frac{9}{2}\right) - \left(3 - \frac{1}{2}\right)\right\} + \left\{(8-12) - \left(\frac{9}{2} - 9\right)\right\}$

$= 5/2$

$\therefore \ 2a + b = 5/2$

$\therefore \ a = 1, b = \frac{1}{2}; a = 2, b = -3/2; a = \frac{1}{2}, b = \frac{3}{2}$ (Q, S, T)

(B) $\int_{-1}^{1} \frac{|x|}{x} dx = \int_{-1}^{0} (-1) \cdot dx + \int_0^1 1 dx$

$= -(0+1) + 1(1-0) = 0$

$\therefore \quad a - 2b = 0$ or $a = 2b$

$\therefore \ a = 3, b = \frac{3}{2}; a = 1, b = \frac{1}{2}$ (P, Q)

(C) $\int_0^{2\pi} |\sin x|\, dx = \int_0^{\pi} \sin x\, dx - \int_{\pi}^{2\pi} \sin x\, dx$

$= -[\cos x]_0^{\pi} + [\cos x]_{\pi}^{2\pi}$

$= -(-1-1) + (1 - (-1)) = 4$

$\therefore \quad 5a + b = 4$

$\therefore \ a = 1, b = -1; a = 1/2, b = 3/2$ (R, T)

2. **(A)** Let $f(x) = \left(\frac{5-x}{9-x^2}\right)$,

$\therefore \ f'(x) = -\frac{(x-9)(x-1)}{(9-x^2)^2}$

$\Rightarrow f'(x) = 0$ or not defined (for critical points)

$\therefore \quad x = 1$

Then, $f(10) = \frac{5}{9}, f(1) = \frac{1}{2}, f(2) = \frac{3}{5}$

$\therefore \quad M = \frac{3}{5}$ and $m = \frac{1}{2}$

$\lambda = (b-a) M = (2-0)\frac{3}{5} = \frac{6}{5}$

and $\quad \mu = (b-a) m = (2-0)\frac{1}{2} = 1$

$\lambda + \mu = 2 \cdot 2$

$\Rightarrow \ [\lambda + \mu] = 2$ (R)

$5\lambda - \mu = 5$

$\Rightarrow \ [5\lambda - \mu] = 5$ (T)

(B) Let $f(x)=\dfrac{\sin x}{x}$

$\therefore \quad f'(x)=\dfrac{x\cos x-\sin x}{x^2}$

$=\dfrac{(x-\tan x)\cos x}{x^2}<0\ \forall\ x\in\left[\dfrac{\pi}{4},\dfrac{\pi}{3}\right]$

$\therefore$ $f(x)$ is decreases on the interval $\left[\dfrac{\pi}{4},\dfrac{\pi}{3}\right]$

Then, $\quad m=f\left(\dfrac{\pi}{3}\right)=\dfrac{\sin(\pi/3)}{(\pi/3)}=\dfrac{\sqrt{3}/2}{\pi/3}=\dfrac{3\sqrt{3}}{2\pi}$

and $\quad M=f\left(\dfrac{\pi}{4}\right)=\dfrac{\sin(\pi/4)}{\pi/4}=\dfrac{4}{\pi\sqrt{2}}=\dfrac{2\sqrt{2}}{\pi}$

$\therefore \quad \lambda=(b-a)M=\left(\dfrac{\pi}{3}-\dfrac{\pi}{4}\right)\times\dfrac{2\sqrt{2}}{\pi}=\dfrac{\sqrt{2}}{6}$

and $\quad \mu=(b-a)m=\left(\dfrac{\pi}{3}-\dfrac{\pi}{4}\right)\times\dfrac{3\sqrt{3}}{2\pi}=\dfrac{\sqrt{3}}{8}$

$\therefore \quad [6\lambda+8\mu]=[\sqrt{2}+\sqrt{3}]=3$ (P)

(C) Let $f(x)=\sqrt{(3+x^3)}$

$\therefore \quad f'(x)=\dfrac{3x^2}{2\sqrt{(3+x^3)}}>0\ \forall\ x\in[1,3]$

$\therefore$ $f(x)$ is increasing on the interval $[1, 3]$

$\Rightarrow \quad m=f(1)=\sqrt{4}=2$

and $\quad M=f(3)=\sqrt{30}$

$\therefore \quad \lambda=(b-a)M=(3-1)\sqrt{30}=2\sqrt{30}$

and $\quad \mu=(b-a)m=(3-1)2=4$

$[\lambda-2]=8$ and $[\lambda-\mu]=6$ (Q, S)

3. **(A)** Since, $0<\dfrac{x^7}{\sqrt[3]{(1+x^8)}}<x^7\ \forall\ 0<x<1$

Then, $\displaystyle\int_0^1 0\,dx<\int_0^1\frac{x^7}{\sqrt[3]{(1+x^8)}}dx<\int_0^1 x^7dx$

Hence, $\displaystyle 0<\int_0^1\frac{x^7\,dx}{\sqrt[3]{(1+x^8)}}<\frac{1}{8}$

$\therefore \quad \lambda=\dfrac{1}{8},\mu=0$

$[\lambda+\mu]=0,[\lambda-\mu]=0$ (R, T)

(B) Since, $\sqrt{(1-x^2)}<\sqrt{(1+x^6)}<\sqrt{(1+x^2)}\ \forall\ x\in(0,1)$

$\Rightarrow \quad \dfrac{1}{\sqrt{(1-x^2)}}>\dfrac{1}{\sqrt{(1+x^6)}}>\dfrac{1}{\sqrt{(1+x^2)}}\ \forall\ x\in(0,1)$

$\Rightarrow \quad \displaystyle\int_0^1\frac{dx}{\sqrt{(1+x^2)}}<\int_0^1\frac{dx}{\sqrt{(1+x^6)}}<\int_0^1\frac{dx}{\sqrt{(1-x^2)}}$

$\Rightarrow \displaystyle \ln\{x+\sqrt{(1+x^2)}\}_0^1<\int_0^1\frac{dx}{\sqrt{(1+x^6)}}<\{\sin^{-1}x\}_0^1$

$\Rightarrow \displaystyle \ln 2<\int_0^1\frac{dx}{\sqrt{(1+x^6)}}<\frac{\pi}{2}$

$\therefore \quad \lambda=\dfrac{\pi}{2}\approx 1.57,\mu=\ln 2\approx 0.693$

$[\lambda+\mu]=2,[\lambda-\mu]=0$ (P, R)

(C) Since, $4-x^2>4-x^2-x^3>4-2x^2\ \forall\ x\in(0,1)$

$\Rightarrow \sqrt{(4-x^2)}>\sqrt{(4-x^2-x^3)}>\sqrt{(4-2x^2)}\ \forall\ x\in(0,1)$

$\Rightarrow \dfrac{1}{\sqrt{(4-x^2)}}<\dfrac{1}{\sqrt{(4-x^2-x^3)}}<\dfrac{1}{\sqrt{(4-2x^2)}}\ \forall\ x\in(0,1)$

$\therefore \displaystyle \int_0^1\frac{dx}{\sqrt{(4-x^2)}}<\int_0^1\frac{dx}{\sqrt{(4-x^2-x^3)}}<\int_0^1\frac{dx}{\sqrt{(4-2x^2)}}$

$\Rightarrow \displaystyle \left[\sin^{-1}\left(\frac{x}{2}\right)\right]_0^1<\int_0^1\frac{dx}{\sqrt{(4-x^2-x^3)}}<\frac{1}{\sqrt{2}}\left[\sin^{-1}\frac{x}{\sqrt{2}}\right]_0^1$

$\Rightarrow \displaystyle \frac{\pi}{6}<\int_0^1\frac{dx}{\sqrt{(4-x^2-x^3)}}<\frac{\pi}{4\sqrt{2}}$

$\therefore\ \lambda=\dfrac{\pi}{4\sqrt{2}}\approx 4.43$ and $\mu=\dfrac{\pi}{6}\approx 0.52$

$[\lambda+\mu]=4,[\lambda-\mu]=3$ (Q, S)

4. (A) $\displaystyle I=\int_\alpha^{\pi/2-\alpha}\frac{\sin^n x}{\sin^n x+\cos^n x}dx \quad \ldots(\text{i})$

$\displaystyle I=\int_\alpha^{\pi/2-\alpha}\frac{\sin^n\left(\frac{\pi}{2}-\alpha\right)}{\sin^n\left(\frac{\pi}{2}-\alpha\right)+\cos^n\left(\frac{\pi}{2}-\alpha\right)}dx$

$\displaystyle =\int_\alpha^{\pi/2-\alpha}\frac{\cos^n\alpha}{\cos^n\alpha+\sin^n\alpha}dx \quad \ldots(\text{ii})$

Adding Eqs. (i) and (ii), then

$\displaystyle 2I=\int_\alpha^{\pi/2-\alpha}1dx=\frac{\pi}{2}-2\alpha$

$\therefore \quad I=\dfrac{\pi}{4}-\alpha$ (Q, S, T)

(B) $\displaystyle I=\int_{-\pi}^{\pi}\frac{\sin^2 x}{1+\alpha^x}dx \quad \ldots(\text{i})$

$\displaystyle I=\int_{-\pi}^{\pi}\frac{\sin^2(0-x)}{1+\alpha^{0-x}}dx$

$\displaystyle =\int_{-\pi}^{\pi}\frac{\alpha^x\sin^2 x}{1+\alpha^x}dx \quad \ldots(\text{ii})$

Adding Eqs. (i) and (ii), then

$\displaystyle 2I=\int_{-\pi}^{\pi}\sin^2 x\,dx=2\int_0^{\pi}\sin^2 x\,dx$

$\Rightarrow \displaystyle I=\int_0^{\pi}\sin^2 x\,dx=2\int_0^{\pi/2}\sin^2 x dx$

$=2\cdot\dfrac{1}{2}\cdot\dfrac{\pi}{2}=\dfrac{\pi}{2}$ (P, T)

(C) $\displaystyle I=\int_\alpha^{2\pi-\alpha}\frac{x\sin^{2n}x}{\sin^{2n}x+\cos^{2n}x}dx \quad \ldots(\text{i})$

$\displaystyle =\int_\alpha^{2\pi-\alpha}\frac{(2\pi-x)\sin^{2n}(2\pi-x)}{\sin^{2n}(2\pi-x)+\cos^{2n}(2\pi-x)}dx$

$\displaystyle =\int_\alpha^{2\pi-\alpha}\frac{(2\pi-x)\sin^{2n}x}{\sin^{2n}x+\cos^{2n}x}dx \quad \ldots(\text{ii})$

Adding Eqs. (i) and (ii), then

$\displaystyle 2I=2\pi\int_\alpha^{2\pi-\alpha}1\cdot dx=2\pi(2\pi-\alpha-\alpha)$

$\therefore \quad I=2\pi^2-2\pi\alpha$ (R, S, T)

19

Areas

Objective Questions Type I [Only one correct answer]

In each of the questions below, four choices are given of which only one is correct. You have to select the correct answer which is the most appropriate.

1. The area bounded by the curve $f(x) = x + \sin x$ and its inverse between the ordinates $x = 0$ to $x = 2\pi$ is

(a) 4 sq units (b) 8 sq units
(c) 4π sq units (d) 8π sq units

2. The area (in sq units) of the region $A = \{(x, y) : (x-1)[x] \le y \le 2\sqrt{x}, 0 \le x \le 2\}$, where $[t]$ denotes the greatest integer function, is **[JEE Main 2020, 5 September]**

(a) $\frac{8}{3}\sqrt{2} - \frac{1}{2}$ (b) $\frac{4}{3}\sqrt{2} + 1$
(c) $\frac{8}{3}\sqrt{2} - 1$ (d) $\frac{4}{3}\sqrt{2} - \frac{1}{2}$

3. Area of the region bounded by $[x]^2 = [y]^2$, if $x \in [1, 5]$, where [.] denotes the greatest integer function, is

(a) 4 sq units (b) 8 sq units
(c) 5 sq units (d) 10 sq units

4. If A_n is the area bounded by $y = (1 - x^2)^n$ and coordinate axes, $n \in N$, then

(a) $A_n = A_{n-1}$ (b) $A_n < A_{n-1}$
(c) $A_n > A_{n-1}$ (d) $A_n = 2A_{n-1}$

5. The area bounded by $\min(|x|, |y|) = 2$ and $\max(|x|, |y|) = 4$ is

(a) 8 sq units (b) 16 sq units
(c) 24 sq units (d) 32 sq units

6. Let $f(x) = 2\sqrt{x}$ and $g(x) = 2\sqrt{(1-x)}$ be two functions and let $f_1(x) = \max\{f(t), 0 \le t \le x, 0 \le x \le 1\}$ and $g_1(x) = \min\{g(t), 0 \le t \le x, 0 \le x \le 1\}$. Then the area bounded by $f_1(x) \le 0, g_1(x) \le 0$ and X-axis is

(a) $\frac{1}{3\sqrt{2}}$ sq unit (b) $\frac{2}{3\sqrt{2}}$ sq unit
(c) $\frac{1}{\sqrt{2}}$ sq unit (d) $\frac{4}{3\sqrt{2}}$ sq unit

7. Area bounded by the curve $y = \sqrt{(\sin[x] + [\sin x])}$, where [.] denotes the greatest integer function, lines $x = 1$ and $x = \frac{\pi}{2}$ and the X-axis is

(a) $\left(\frac{\pi}{2} - 1\right)$ sq unit (b) $\sqrt{\sin 1}\left(\frac{\pi}{2} - 1\right)$ sq unit
(c) $\sqrt{\cos 1}\left(\frac{\pi}{2} - 1\right)$ sq unit (d) $\sqrt{\frac{\pi}{2}}\left(\frac{\pi}{2} - 1\right)$ sq unit

8. If the area of the region bounded by $y = \cos x$, $y = \sin x$, $x = \frac{\pi}{4}$ and $x = \pi$ is bisected by the line $x = a$, then $\sin\left(a + \frac{\pi}{4}\right) =$

(a) $\frac{\sqrt{2}}{2 + \sqrt{2}}$ (b) $\frac{\sqrt{3} + 1}{2}$
(c) $\frac{\sqrt{2} - 1}{2\sqrt{2}}$ (d) $\frac{\sqrt{3} + 1}{2\sqrt{2}}$

9. The area bounded by the curve $4y^2 = x^2(4 - x)(x - 2)$ is equal to **[JEE Main 2021, 18 March]**

(a) $\frac{\pi}{8}$ (b) $\frac{3\pi}{8}$
(c) $\frac{3\pi}{2}$ (d) $\frac{\pi}{16}$

10. Area of the region bounded by the curves $y|y| \pm x|x| = 1$ and $y = |x|$ is

(a) $\frac{\pi}{8}$ sq units (b) $\frac{\pi}{4}$ sq units
(c) $\frac{\pi}{2}$ sq units (d) π sq units

11. The area of the region bounded by the curve $y = x^2$ and $y = \sec^{-1}[-\sin^2 x]$, where [·] denotes the greatest integer function, is

(a) $\frac{3\pi^{3/2}}{4}$ sq units (b) $\frac{3\pi}{4}$ sq units
(c) $\frac{4\pi^{3/2}}{3}$ sq units (d) $\frac{4\pi}{3}$ sq units

12. The area bounded by the x-axis, the curve $y = f(x)$ and the lines $x = 1$ and $x = b$ is equal to $(\sqrt{(b^2+1)} - \sqrt{2})$ for all $b > 1$, then $f(x)$ is0

(a) $\sqrt{(x-1)}$ (b) $\sqrt{(x+1)}$

(c) $\sqrt{(x^2+1)}$ (d) $\dfrac{x}{\sqrt{(1+x^2)}}$

13. The area of the region bounded by $1 - y^2 = |x|$ and $|x| + |y| = 1$ is

(a) 1/3 sq units (b) 2/3 sq units

(c) 4/3 sq units (d) 1 sq unit

14. The area (in sq units) of the region $A = \{(x, y): |x| + |y| \le 1, 2y^2 \ge |x|\}$ is **[JEE Main 2020, 6 September]**

(a) $\dfrac{1}{3}$ (b) $\dfrac{7}{6}$

(c) $\dfrac{1}{6}$ (d) $\dfrac{5}{6}$

15. The area of the figure bounded by the curves $y = |x - 1|$ and $y = 3 - |x|$ is

(a) 2 sq units (b) 3 sq units

(c) 4 sq units (d) 1 sq unit

16. Area enclosed by the curve $y = x^2 + 1$ and a normal drawn to it with gradient -1, is equal to

(a) $\dfrac{8}{12}$ (b) $\dfrac{16}{12}$

(c) $\dfrac{19}{12}$ (d) $\dfrac{43}{12}$

17. Let $f(x) = \min\{x+1, \sqrt{(1-x)}\}$, then area bounded by $f(x)$ and X-axis is

(a) $\dfrac{1}{6}$ sq unit (b) $\dfrac{5}{6}$ sq unit

(c) $\dfrac{7}{6}$ sq units (d) $\dfrac{11}{6}$ sq units

18. The area bounded by the graph $y = |[x - 3]|$, the X-axis and the lines $x = -2$ and $x = 3$ is ([.] denotes the greatest integer function)

(a) 7 sq units (b) 15 sq units

(c) 21 sq units (d) 28 sq units

19. The value of c for which the area of the figure bounded by the curve $y = 8x^2 - x^5$, the straight lines $x = 1$ and $x = c$ and the X-axis is equal to $\dfrac{16}{3}$ is

(a) 2 (b) $\sqrt{8 - \sqrt{17}}$

(c) 3 (d) -1

20. Area bounded by the curves $y = \left[\dfrac{x^2}{64} + 2\right], y = x - 1$ and $x = 0$ above X-axis is ([.] denotes the greatest integer function)

(a) 2 sq units (b) 3 sq units

(c) 4 sq units (d) None of these

21. Let $g(x) = \cos x^2$, $f(x) = \sqrt{x}$ and α, β $(\alpha < \beta)$ be the roots of the quadratic equation $18x^2 - 9\pi x + \pi^2 = 0$. Then, the area (in sq units) bounded by the curve $y = (gof)(x)$ and the lines $x = \alpha$, $x = \beta$ and $y = 0$, is **[JEE Main 2018]**

(a) $\dfrac{1}{2}(\sqrt{3} - 1)$ (b) $\dfrac{1}{2}(\sqrt{3} + 1)$

(c) $\dfrac{1}{2}(\sqrt{3} - \sqrt{2})$ (d) $\dfrac{1}{2}(\sqrt{2} - 1)$

22. The area (in sq units) bounded by the parabola $y = x^2 - 1$, the tangent at the point (2, 3) to it and the Y-axis is **[JEE Main 2019, 9 January]**

(a) $\dfrac{8}{3}$ (b) $\dfrac{56}{3}$

(c) $\dfrac{32}{3}$ (d) $\dfrac{14}{3}$

23. The area bounded by $y = x e^{|x|}$ and lines $|x| = 1, y = 0$ is

(a) 4 sq units (b) 6 sq units

(c) 1 sq unit (d) 2 sq units

24. The area of the figure bounded by $f(x) = \sin x, g(x) = \cos x$ in the first quadrant is

(a) $2(\sqrt{2} - 1)$ sq units (b) $(\sqrt{3} + 1)$ sq units

(c) $2(\sqrt{3} - 1)$ sq units (d) None of these

25. The area of the figure bounded by two branches of the curve $(y - x)^2 = x^3$ and the straight line $x = 1$ is

(a) $\dfrac{1}{3}$ sq units (b) $\dfrac{4}{5}$ sq units

(c) $\dfrac{5}{4}$ sq units (d) 3 sq units

26. If $C_1: y^2 = 2x - x^2, C_2: y = x^2$ and $C_3: x = y^2$, A_{ij} = the area bounded by the curves C_i and C_j between $x = 0$ and $x = 1, i \ne j, i, j = 1, 2, 3$ $(A_{ij} = A_{ji}$ for $i \ne j)$, then $A_{12} + A_{23} + A_{31}$ is equal to

(a) $\left(\dfrac{\pi}{2} - 1\right)$ sq unit (b) $\left(\dfrac{3\pi}{4} + \dfrac{2}{3}\right)$ sq units

(c) $\left(\dfrac{\pi}{2} - \dfrac{2}{3}\right)$ sq unit (d) $\left(\dfrac{3\pi}{4} - \dfrac{4}{3}\right)$ sq units

27. Let the functions $f: \mathbf{R} \to \mathbf{R}$ and $g: \mathbf{R} \to \mathbf{R}$ be defined by $f(x) = e^{x-1} - e^{-|x-1|}$ and $g(x) = \dfrac{1}{2}(e^{x-1} + e^{1-x})$. Then the area of the region in the first quadrant bounded by the curves $y = f(x)$, $y = g(x)$ and $x = 0$ is **[JEE Advanced 2020]**

(a) $(2 - \sqrt{3}) + \dfrac{1}{2}(e - e^{-1})$ (b) $(2 + \sqrt{3}) + \dfrac{1}{2}(e - e^{-1})$

(c) $(2 - \sqrt{3}) + \dfrac{1}{2}(e + e^{-1})$ (d) $(2 + \sqrt{3}) + \dfrac{1}{2}(e + e^{-1})$

28. The area bounded by $y = x^2$, $y = [x + 1]$, $0 \le x < 1$ and the Y-axis is ([.] denotes the greatest integer function)

(a) 1/3 sq unit (b) 2/3 sq unit

(c) 1 sq unit (d) 7/3 sq units

29. Let $f(x)$ be a continuous function such that the area bounded by the curve $y = f(x)$, the X-axis and the two ordinates $x = 0$ and $x = a$ is $\left(\frac{a^2}{2} + \frac{a}{2}\sin a + \frac{\pi}{2}\cos a\right)$ sq unit, then $f\left(\frac{\pi}{2}\right)$ is

(a) $\frac{1}{2}$ (b) $\frac{\pi^2}{8} + \frac{\pi}{4}$

(c) $\frac{\pi+1}{2}$ (d) None of these

30. The area bounded by the curve $y = x^4 - 2x^3 + x^2 + 3$, the axis of abscissas and two ordinates corresponding to the points of minimum of the function $y(x)$ is

(a) $\frac{10}{3}$ sq units (b) $\frac{27}{10}$ sq units

(c) $\frac{21}{10}$ sq units (d) None of these

31. The area bounded by the curves $y = \ln x$, $y = \ln|x|$, $y = |\ln x|$ and $y = |\ln|x||$ is

(a) 5 sq units (b) 2 sq units

(c) 4 sq units (d) None of these

32. The area bounded by the curve $y = x^2 + 2x + 1$ and tangent at (1, 4) and Y-axis is

(a) $\frac{2}{3}$ sq unit (b) $\frac{1}{3}$ sq unit

(c) 2 sq units (d) None of these

33. The triangle formed by the tangent to the curve $f(x) = x^2 + bx - b$ at the point (1, 1) and the coordinate axes, lies in the first quadrant. If its area is 2 sq unit, then the value of 'b' is

(a) –3 (b) –2

(c) –1 (d) 0

34. The area bounded by $y = 2 - |2 - x|$, $y = \frac{3}{|x|}$ is

(a) $\left(\frac{5 - 4\ln 2}{3}\right)$ sq units (b) $\left(\frac{2 - \ln 3}{2}\right)$ sq unit

(c) $\left(\frac{4 - 3\ln 3}{2}\right)$ sq units (d) None of these

35. If $f(x) = \max\left\{\sin x, \cos x, \frac{1}{2}\right\}$, then the area of the region bounded by the curves $y = f(x)$, X-axis, Y-axis and $x = 2\pi$ is

(a) $\left(\frac{5\pi}{12} + 3\right)$ sq units (b) $\left(\frac{5\pi}{12} + \sqrt{2}\right)$ sq units

(c) $\left(\frac{5\pi}{12} + \sqrt{3}\right)$ sq units (d) $\left(\frac{5\pi}{12} + \sqrt{2} + \sqrt{3}\right)$ sq units

36. The area of the region $\{(x, y) : xy \le 8, 1 \le y \le x^2\}$ is **[JEE Advanced 2019]**

(a) $8\log_e 2 - \frac{14}{3}$ (b) $8\log_e 2 - \frac{7}{3}$

(c) $16\log_e 2 - \frac{14}{3}$ (d) $16\log_e 2 - 6$

37. For $a > 0$, let the curves $C_1 : y^2 = ax$ and $C_2 : x^2 = ay$ intersect at origin O and a point P. Let the line $x = b\,(0 < b < a)$ intersect the chord OP and the X-axis at points Q and R, respectively. If the line $x = b$ bisects the area bounded by the curves, C_1 and C_2, and the area of $\Delta OQR = \frac{1}{2}$, then 'a' satisfies the equation **[JEE Main 2020, 8 January]**

(a) $x^6 + 6x^3 - 4 = 0$

(b) $x^6 - 12x^3 + 4 = 0$

(c) $x^6 - 6x^3 + 4 = 0$

(d) $x^6 - 12x^3 - 4 = 0$

38. Let $f(x) = \begin{cases} x^2, & x < 0 \\ x, & x \ge 0 \end{cases}$; area bounded by the curve $y = f(x)$, $y = 0$ and $x = \pm 3a$ is $\frac{9a}{2}$ sq units. Then, a is equal to

(a) $\frac{1}{2}$ (b) $-\frac{1}{2}$

(c) 0 (d) None of these

39. The area bounded by the curves $|x| + |y| \ge 1$ and $x^2 + y^2 \le 1$ is

(a) 2 sq units (b) π sq units

(c) $(\pi - 2)$ sq unit (d) $(\pi + 2)$ sq units

40. A polynomial P is positive for $x > 0$ and the area of the region bounded by $P(x)$, the X- axis and the vertical lines $x = 0$ and $x = \lambda$ is $\frac{\lambda^2(\lambda + 3)}{3}$ sq unit. Then polynomial $P(x)$ is

(a) $x^2 + 2x$ (b) $x^2 + 2x + 1$

(c) $x^2 + x + 1$ (d) $x^3 + 2x^2 + 2$

Objective Questions Type II [One or more than one correct answer(s)]

In each of the questions below four choices of which one or more than one are correct. You have to select the correct answer(s) accordingly.

1. The area enclosed between the curves, $x^2 = y$ and $y^2 = x$ is equal to

(a) $\frac{1}{3}$ sq unit

(b) $2\int_0^1 (x - x^2)\,dx$

(c) area of the region $\{(x, y) : x^2 \le y \le |x|\}$

(d) None of the above

2. Consider curve $y = \frac{1}{x^2}$; $y = \frac{1}{4(x-1)}$, let α be the value of $a(a > 2)$ for which area bounded by curves between $x = 2$ and $x = a$ is $\frac{1}{a}$ and β be the of $b \in (1, 2)$, for which the area bounded by curves between $x = b$ and $x = 2$ is $1 - \frac{1}{b}$, then

(a) $\alpha = e^2 + 1$ (b) $\alpha = e^2 - 2$

(c) $\beta = 1 + e^{-1}$ (d) $\beta = 1 + e^{-2}$

3. Area of the region bounded by the curve $y = \tan x$ and lines $y = 0$ and $x = 1$ is equal to

(a) $\int_0^1 \tan(1-x)\, dx$ (b) $\int_0^{\tan 1} \tan^{-1} y\, dy$

(c) $\int_0^1 \tan^{-1} x\, dx$ (d) $\tan 1 - \int_0^{\tan 1} \tan^{-1} x\, dx$

4. Area bounded by the curve $y = \ln x$, $y = 0$ and $x = 3$ is

(a) $(\ln 9 - 2)$ sq unit (b) $(\ln 27 - 2)$ sq unit

(c) $\ln\left(\frac{27}{e^2}\right)$ sq unit (d) (greater than 3) sq unit

5. The value (s) of 'a' for which the area of the triangle included between the axes and any tangent to the curve $x^a y = \lambda^a$ is constant, is/are

(a) $-\frac{1}{2}$ (b) -1

(c) $\frac{1}{2}$ (d) 1

6. If the line $x = \alpha$ divides the area of region $R = \{(x, y) \in R^2 : x^3 \le y \le x, 0 \le x \le 1\}$ into two equal parts, then **[JEE Advanced 2017]**

(a) $2\alpha^4 - 4\alpha^2 + 1 = 0$ (b) $\alpha^4 + 4\alpha^2 - 1 = 0$

(c) $\frac{1}{2} < \alpha < 1$ (d) $0 < \alpha \le \frac{1}{2}$

Linked-Comprehension Type

In these questions, a passage (paragraph) has been given followed by questions based on each of the passages. You have to answer the questions based on the passage given.

PASSAGE 1

Let $f(x) = x^2 - 3x + 2$ be a function, $\forall\, x \in R$.

On the basis of above information, answer the following questions

1. The area bounded by $f(x)$, the X-axis and Y-axis is

(a) $\frac{1}{3}$ sq unit (b) $\frac{2}{3}$ sq unit

(c) $\frac{3}{5}$ sq unit (d) $\frac{5}{6}$ sq unit

2. The area bounded by the curve $f(|x|)$ and X-axis is

(a) $\frac{2}{5}$ sq unit (b) $\frac{5}{2}$ sq units

(c) $\frac{3}{5}$ sq unit (d) $\frac{5}{3}$ sq units

3. The area bounded by the curve $|f(|x|)|$ between $1 \le |x| \le 2$ and X-axis is

(a) $\frac{1}{5}$ sq unit (b) $\frac{1}{4}$ sq unit

(c) $\frac{1}{3}$ sq unit (d) $\frac{1}{2}$ sq unit

4. If $|f(|x|)| = a$ has 6 real roots, then

(a) $a \in \left(0, \frac{1}{4}\right)$ (b) $a \in \left(\frac{1}{4}, 2\right)$

(c) $a \in \left(-\frac{1}{4}, 0\right)$ (d) None of these

5. The number of solutions of $|y| = |f(|x|)|$ and $x^2 + y^2 = 2$ is

(a) 4 (b) 6 (c) 8 (d) 5

PASSAGE 2

Consider a square with vertices at A (1, 1), B (–1, 1), C (–1, –1) and D (1, –1). Let S be the region consisting of all points inside the square which are nearer to the origin than to any edge.

On the basis of above information, answer the following questions

1. If we consider (x, y) inside the square and edge BC, then which of the following is true

(a) all points lie inside the parabola $y^2 = -2\left(x + \frac{1}{2}\right)$

(b) all points lie outside the parabola $y^2 = -2\left(x + \frac{1}{2}\right)$

(c) all points lie inside the parabola $y^2 = 2\left(x + \frac{1}{2}\right)$

(d) all points lie outside the parabola $y^2 = 2\left(x + \frac{1}{2}\right)$

2. The required area consists of common area bounded by

(a) two parabolas with vertices $\left(\pm\frac{1}{2}, 0\right)$

(b) two parabolas with vertices $\left(0, \pm\frac{1}{2}\right)$

(c) four parabolas with vertices $\left(\pm\frac{1}{2}, 0\right), \left(0, \pm\frac{1}{2}\right)$

(d) four parabolas with vertices $\left(\pm\frac{1}{2}, \pm\frac{1}{2}\right)$

3. The point of intersection of all parabolas are

(a) $\left(\pm\left(1-\frac{1}{\sqrt{2}}\right), \pm\left(1-\frac{1}{\sqrt{2}}\right)\right)$

(b) $(\pm(\sqrt{2}-1), \pm(\sqrt{2}-1))$

(c) $\left(\pm\left(1-\frac{1}{\sqrt{2}}\right), \pm(\sqrt{2}-1)\right)$

(d) $\left(\pm(\sqrt{2}-1), \pm\left(1-\frac{1}{\sqrt{2}}\right)\right)$

4. The area bounded by the parabolas with vertices $\left(\pm\frac{1}{2}, 0\right)$ is

(a) 1/3 sq unit
(b) 2/3 sq unit
(c) 3/4 sq unit
(d) 4/3 sq units

5. Required area of the region so formed is

(a) $4\left\{(\sqrt{2}-1)^2 + \int_{\sqrt{2}-1}^{1/2} ydx\right\}$

(b) $4\left\{(\sqrt{2}-1)^2 + 2\int_{\sqrt{2}-1}^{1/2} ydx\right\}$

(c) $4\left\{(\sqrt{2}-1)^2 - \int_{1-\frac{1}{\sqrt{2}}}^{1/2} ydx\right\}$

(d) $4\left\{(\sqrt{2}-1)^2 - 2\int_{1-\frac{1}{\sqrt{2}}}^{1/2} ydx\right\}$

PASSAGE 3

Let there are two functions defined by
$f(x) = \min(|x|, |x-1|, |x+1|)$ and
$g(x) = \min\{e^x, e^{-x}\}$. Now, the root of the equation $e^x - x = 0$ is $a, \forall a \in R$.

On the basis of above information, answer the following questions

1. The area bounded by $f(x)$ in $[-1, 1]$ and X-axis is

(a) 1/5 sq unit (b) 1/4 sq unit
(c) 1/3 sq unit (d) 1/2 sq unit

2. The root of the equation $e^x + x = 0$ is

(a) α (b) $-\alpha$ (c) $\frac{\alpha}{2}$ (d) $-\frac{\alpha}{2}$

3. Which statement is correct?

(a) $\alpha \in (-\infty, 0)$ (b) $\alpha \in (0, \infty)$
(c) $\alpha \in (2, \infty)$ (d) $\alpha \in (0, 1)$

4. The area bounded by $f(x), g(x)$ and $x = 0$ in first quadrant

(a) $e^{-\alpha} - 1$ (b) $2 - e^{-\alpha}$
(c) $1 + e^{-\alpha}$ (d) None of these

5. The sum of roots of equation $f(x) - g(x) = 0$ is

(a) 0 (b) 2α
(c) -2α (d) 4α

PASSAGE 4

Let A_r be the area of the region bounded between the curve $y^2 = (e^{-kr})x$, $k > 0$, $r \in N$ and the line $y = mx$, $m \neq 0$, k and m are some constants.

On the basis of above information, answer the following questions

1. $A_1, A_2, A_3, \ldots\ldots$ are in GP with common ratio

(a) e^{-k} (b) e^{-2k} (c) e^{-3k} (d) e^{-4k}

2. If $\lim_{n\to\infty} \sum_{r=1}^{n} A_r = \frac{4}{3(e^{2k}-1)}$, then the value of m is

(a) $\frac{1}{2}$ (b) 2
(c) $\frac{3}{2}$ (d) 3

Numerical Value Type

Solve the following problems and write your answers.

1. If area enclosed by the curve $|x| + |y| \leq k$ is $f(k)$ sq unit, then the value of $7f(10)$ must be

2. If area enclosed by the parabola $y^2 = 64x$ and its latusrectum is λ sq unit, then the value of 3λ must be

3. The area enclosed by the curve $xy^2 = 4(2-x)$ and Y-axis is λ sq unit, then the value of $5050\cos(1824\lambda)$ must be

4. If area between the curves $y = xe^x$ and $y = xe^{-x}$ and the line $x = 1$ is λ sq unit, then the value of $128e\lambda$ must be

5. If area enclosed between the curves $y = \ln(x+e)$ and $x = \ln\left(\frac{1}{y}\right)$ and the axis of X is λ sq unit, then the value of $\frac{\lambda^2}{5}$ must be

6. If the area bounded by the functions $f(x) - (\cos^{-1}(\cos x))^2$ and $g(x) = \cos^{-1}(\cos|x|)$ and $x = 2\pi$ is equal to $\frac{a(\pi^3+1) - b\pi^2}{c}$, where $a, b, c \in N$, then the value of $\left(\frac{b+c}{a}\right)$ is

Single Integer Type

Solve the following problems and write your answers.

1. Consider the following regions in the plane :
$R_1 = \{(x, y): 0 \le x \le 1 \text{ and } 0 \le y \le 1\}$
$R_2 = \{(x, y): x^2 + y^2 \le 4/3\}$
The area of the region $R_1 \cap R_2$ can be expressed as $\frac{a\sqrt{3} + b\pi}{9}$, where a and b are integers. Then, the value of $(a + b)$ equals

2. Find the area enclosed by the solution set of $[x] \cdot [y] = 2$. Where $[\cdot]$ represent greatest integral function of x.

3. Let $y = f(x)$ be a real-valued differentiable function on R (the set of all real numbers) such that $f(1) = 1$. If $f(x)$ satisfies $xf'(x) = x^2 + f(x) - 2$, then find the area enclosed by $f(x)$ with X-axis between coordinates $x = 0$ and $x = 3$.

4. A farmer F_1 has a land in the shape of a triangle with vetrices at $P(0, 0)$, $Q(1, 1)$ and $R(2, 0)$. From this land, a neighbouring farmer F_2 takes away the region which lies between the side PQ and a curve of the form $y = x^n$ $(n > 1)$. If the area of the region taken away by the farmer F_2 is exactly 30% of the area of ΔPQR, then the value of n is ... . **[JEE Advanced 2018]**

5. If the area bounded by the curves $f(x) = [\cos^{-1}|\cos x|]^2$, $g(x) = [\cos^{-1}|\cos x|]$ and $|x| = \frac{\pi}{2}$ is $a\pi^3 + b\pi^2 + c$, then find the minimum value of $(|a| + |b| + |c|)$.

6. Let the curve $y = y(x)$ be the solution of the differential equation, $\frac{dy}{dx} = 2(x + 1)$. If the numerical value of area bounded by the curve $y = y(x)$ and X-axis is $\frac{4\sqrt{8}}{3}$, then the value of $y(1)$ is equal to ....

Matrix-Match Type

Given below are Matching Type Questions, with two columns (each having some items) each . Each item of Column I has to be matched with the items of Column II, by encircling the correct match(es).

NOTE *An item of Column I can be matched with more than one items of Column II.*

1. Observe the following columns

	Column I		Column I
(A)	Let $y_1 = \|\sin\|\|x\| - \pi\|\|$ and $y_2 = \|\sin x\|$ be two curves $\forall x \in R$, then	(P)	y_1 is periodic with period π
		(Q)	y_2 is periodic with period π
(B)	Let $y_1 = \pi \sin \pi x$ and $y_2 = \|\pi \sin \pi x\|$ be two curves $\forall x \in R$, then	(R)	Area of region bounded by two curves with x-axis is same for all intervals $[a, b]$
(C)	Let $y_1 = \sin^2 x$ and $y_2 = \sin x$ be two curves $\forall x \in R$, then	(S)	Length of both the curves is same for all real interval $[a, b]$
		(T)	The periods of y_1 and y_2 are not same

2. Observe the following columns :

	Column I		Column II
(A)	The parabolas $y^2 = 4x$ and $x^2 = 4y$ divide the square region bounded the lines $x = 4, y = 4$ and the coordinate axes. If S_1, S_2, S_3 are respectively the areas of these parts numbered from top to bottom, then	(P)	S_1, S_2, S_3 are in AP
		(Q)	S_1, S_2, S_3 are in GP
(B)	The areas S_1, S_2, S_3 bounded by the X-axis and half waves of the curves $y = e^{-\alpha x} \sin \beta x$, $x \ge 0$ from left to right, then	(R)	$S_1^3 + S_3^3 > 2S_2^3$
(C)	Let $f(x) = \max\{x^2, (1-x)^2, 2x(1-x)\}$, where $0 \le x \le 1$ the areas of the region bounded by the curve $y = f(x)$ between $x \in \left[0, \frac{1}{3}\right], \left[\frac{1}{3}, \frac{2}{3}\right], \left[\frac{2}{3}, 1\right]$ and X-axis represented by S_1, S_2, S_3, then	(S)	$S_1 = S_3$
		(T)	$S_1 + S_2 + S_3 = \frac{17}{27}$

3. Observe the following columns

	Column I		Column II
(A)	Area enclosed by $y = \|x\|, \|x\| = 1$ and $y = 0$ is	(P)	2
(B)	Area enclosed by the curve $y = \sin x$, $x = 0$, $x = \pi$ and $y = 0$ is	(Q)	4
(C)	If the area of the region bounded by $x^2 \le y$ and $y \le x + 2$ is $\frac{k}{4}$, then $k =$	(R)	27
(D)	Area of the quadrilateral formed by tangents at the ends of latus rectum of ellipse $\frac{x^2}{9} + \frac{y^2}{5} = 1$ is	(S)	18
		(T)	1

Answers

Objective Questions Type I [Only one correct answer]

1. (b)	**2.** (a)	**3.** (b)	**4.** (b)	**5.** (b)	**6.** (d)	**7.** (b)	**8.** (c)	**9.** (c)	**10.** (b)
11. (c)	**12.** (d)	**13.** (b)	**14.** (d)	**15.** (c)	**16.** (b)	**17.** (c)	**18.** (b)	**19.** (d)	**20.** (c)
21. (a)	**22.** (a)	**23.** (d)	**24.** (a)	**25.** (b)	**26.** (d)	**27.** (a)	**28.** (b)	**29.** (a)	**30.** (d)
31. (c)	**32.** (b)	**33.** (a)	**34.** (c)	**35.** (d)	**36.** (c)	**37.** (b)	**38.** (a)	**39.** (c)	**40.** (a)

Objective Questions Type II [One or more than one correct answer(s)]

1. (a, b, c) **2.** (a, d) **3.** (a, d) **4.** (b, c) **5.** (b, d)
6. (a, c)

Linked-Comprehension Type

Passage 1 **1.** (d) **2.** (d) **3.** (c) **4.** (d) **5.** (c) **Passage 2** **1.** (a) **2.** (c) **3.** (b) **4.** (d) **5.** (b)
Passage 3 **1.** (d) **2.** (b) **3.** (d) **4.** (d) **5.** (a) **Passage 4** **1.** (b) **2.** (a)

Numerical Value Type Questions

1. 1400 **2.** 2048 **3.** 5050 **4.** 256 **5.** 0.8
6. 2.25

Single Integer Type Questions

1. (4) **2.** (4) **3.** (6) **4.** (4) **5.** (1) **6.** (2)

Matrix-Match Type

1. A → (P, Q, R, S); B → (T); C → (P, T)
2. A → (P, Q, S); B → (Q, R); C → (S, T)
3. A → (T); B → (P); C → (S); D → (R)

Solutions

Objective Questions Type [Only one correct answer]

1. Graph of $f^{-1}(x)$ is the mirror image of $f(x)$ about the line $y = x$.

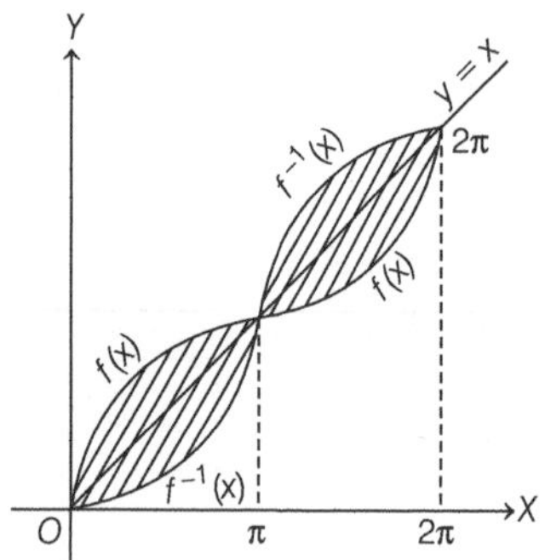

Required area $= 4A$ ...(i)

where, $$A = \int_0^{\pi} f(x)\,dx - \frac{1}{2}\cdot\pi\cdot\pi$$

$$= \int_0^{\pi} (x + \sin x)\,dx - \frac{\pi^2}{2}$$

$$= \left\{\frac{x^2}{2} - \cos x\right\}_0^{\pi} - \frac{\pi^2}{2}$$

$$= \left(\frac{\pi^2}{2} - (-1)\right) - (0 - 1) - \frac{\pi^2}{2}$$

$$= 2 \text{ sq units}$$

Required area $= 4 \times 2 = 8$ sq units

2. As we know that,

$$y = (x-1)[x] = \begin{cases} 0, & 0 \le x < 1 \\ x - 1, & 1 \le x < 2 \\ 2(x-1), & x = 2 \end{cases}$$

Now, on drawing the graph of given region with the help of equation of curves

$$y = (x-1)[x] \text{ and } y = 2\sqrt{x}$$

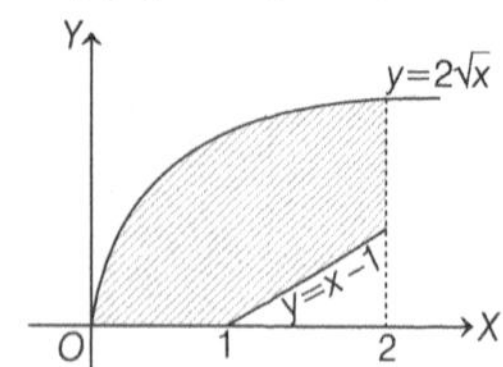

$\therefore$ Area of given region

$$= \int_0^1 2\sqrt{x}\,dx + \int_1^2 (2\sqrt{x} - x + 1)dx$$

$$= \left[\frac{4}{3}x^{3/2}\right]_0^1 + \left[\frac{4}{3}x^{3/2} - \frac{x^2}{2} + x\right]_1^2$$

$$= \frac{4}{3} + \left[\frac{8\sqrt{2}}{3} - 2 + 2 - \frac{4}{3} + \frac{1}{2} - 1\right]$$

$$= \frac{8\sqrt{2}}{3} - \frac{1}{2} \text{ sq units}$$

3. $$[x]^2 = [y]^2$$

$$\therefore \quad [y] = \pm[x]$$

$$\begin{aligned} [y] &= \pm 1, \quad 1 \le x < 2 \\ &= \pm 2, \quad 2 \le x < 3 \\ &= \pm 3, \quad 3 \le x < 4 \\ &= \pm 4, \quad 4 \le x < 5 \\ &= \pm 5, \quad x = 5 \end{aligned}$$

Now, when $x \in [1, 2)$, then $y \in [-1, 0) \cup [1, 2)$

when $x \in [2, 3)$, then $y \in [-2, -1) \cup [2, 3)$

when $x \in [3, 4)$, then $y \in [-3, -2) \cup [3, 4)$

when $x \in [4, 5)$, then $y \in [-4, -3) \cup [4, 5)$

when $x = 5$, then $y \in [-5, -4) \cup [5, 6)$

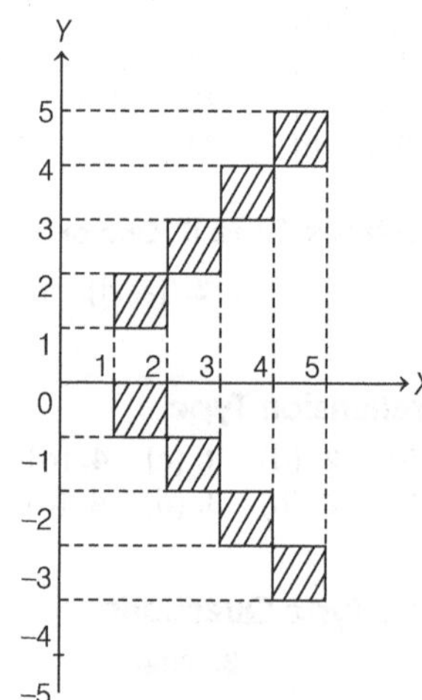

Hence, required area $= 2(4) = 8$ sq units

4. Cut off X-axis, $y = 0$

$$\Rightarrow \quad 0 = (1 - x^2)^n$$

$$\therefore \quad x = \pm 1$$

Then, $$A_n = \int_{-1}^{1} (1 - x^2)^n\,dx$$

$$= 2\int_0^1 (1 - x^2)^n\,dx \qquad \text{[by property]}$$

$$= 2[\{(1 - x^2)^n \cdot x\}_0^1 - \int_0^1 n(1 - x^2)^{n-1}(-2x)\cdot x\,dx]$$

$$= 2\{0 - 2n\int_0^1 (1 - x^2)^{n-1}(1 - x^2 - 1)\,dx\}$$

$$= 2\{-n\,A_n + n\,A_{n-1}\}$$

$$\Rightarrow \quad A_n = \left(\frac{2n}{2n+1}\right)A_{n-1}$$

$$\because \quad 2n + 1 > 2n$$

$$\therefore \quad \left(\frac{2n}{2n+1}\right) < 1 \text{ or } A_n < A_{n-1}$$

5.

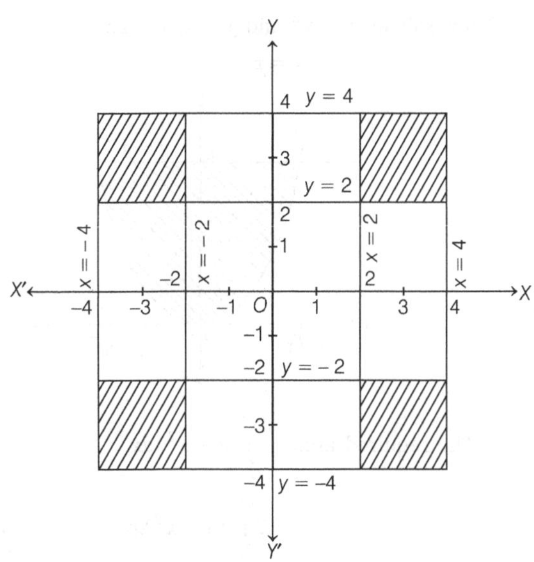

Required area $= 4 \times (2 \times 2)$

$= 16$ sq units

6. $\because \quad f(x) = 2\sqrt{x}$

$\therefore \quad f'(x) = \dfrac{1}{\sqrt{x}} > 0$

$\Rightarrow f(x)$ is an increasing function

$\therefore \quad f_1(x) = f(x) = 2\sqrt{x}$

or $\quad f_1(x) = 2\sqrt{x},\ x \in [0, 1] \qquad \ldots$(i)

and $\quad g(x) = 2\sqrt{(1-x)}$

$\therefore \quad g'(x) = -\dfrac{1}{\sqrt{(1-x)}} < 0$

$\Rightarrow g(x)$ is decreasing function

$\therefore \quad g_1(x) = g(x) = 2\sqrt{(1-x)}$

or $\quad g_1(x) - 2\sqrt{(1-x)},\ x \in [0, 1] \qquad \ldots$(ii)

From Eqs. (i) and (ii),

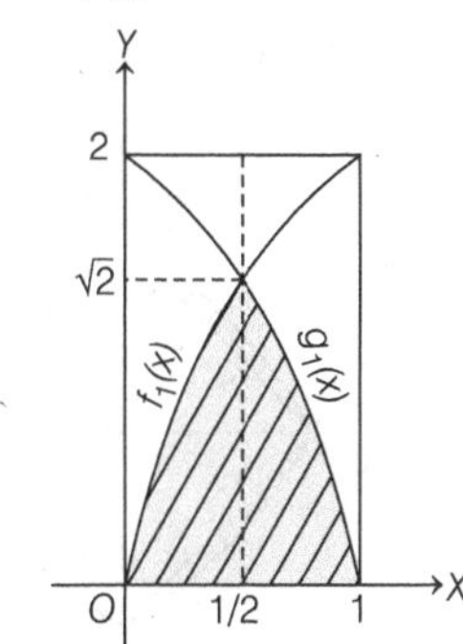

$2\sqrt{x} = 2\sqrt{(1-x)}$

or $\quad x = 1 - x$

or $\quad x = \dfrac{1}{2}$

Then, $\quad f_1\left(\dfrac{1}{2}\right) = g_1\left(\dfrac{1}{2}\right) = \sqrt{2}$

$\therefore$ Required area $= \displaystyle\int_0^{1/2} f_1(x)\,dx + \int_{1/2}^{1} g_1(x)\,dx$

$= \displaystyle\int_0^{1/2} 2\sqrt{x}\,dx + \int_{1/2}^{1} 2\sqrt{(1-x)}\,dx$

$= \dfrac{4}{3}\{x^{3/2}\}_0^{\frac{1}{2}} + \dfrac{4}{3}\{-(1-x)^{3/2}\}_{\frac{1}{2}}^{1}$

$= \dfrac{4}{3}\left(\dfrac{1}{2\sqrt{2}}\right) + \dfrac{4}{3}\left(\dfrac{1}{2\sqrt{2}}\right) = \dfrac{4}{3\sqrt{2}}$ sq unit

7. $\because \quad 1 \le x < \dfrac{\pi}{2}$

$\therefore \quad [x] = 1$ and $\sin 1 \le \sin x < 1$

and $\quad [\sin x] = 0$

Then, $\quad y = \sqrt{\sin 1 + 0} = \sqrt{\sin 1}$

$\therefore$ Required area $= \displaystyle\int_1^{\pi/2} y\,dx = \int_1^{\pi/2} \sqrt{\sin 1}\,dx$

$= \sqrt{\sin 1}\left(\dfrac{\pi}{2} - 1\right)$ sq unit

8.

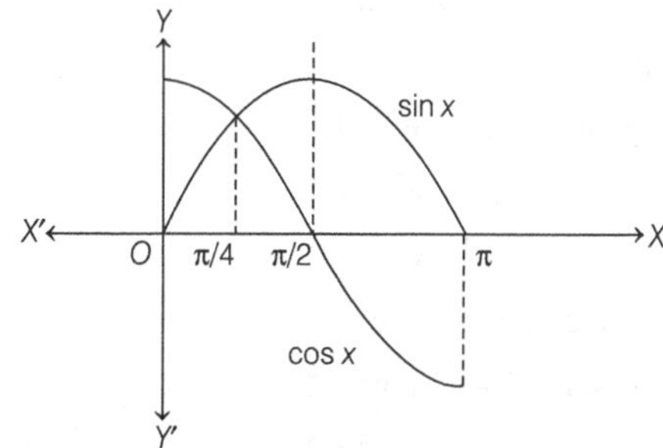

According to the question,

$$\int_{\pi/4}^{a} (\sin x - \cos x)\,dx = \int_{a}^{\pi} (\sin x - \cos x)\,dx$$

$\Rightarrow \quad [-\cos x - \sin x]_{\pi/4}^{a} = [-\cos x - \sin x]_{a}^{\pi}$

$\Rightarrow \quad -\cos a - \sin a + \dfrac{1}{\sqrt{2}} + \dfrac{1}{\sqrt{2}} - 1 + \cos a + \sin a$

$\Rightarrow \quad \sqrt{2} - 1 = 2(\cos a + \sin a)$

$\Rightarrow \quad \dfrac{\sqrt{2}-1}{2} = \sin a + \cos a$

$\Rightarrow \quad \dfrac{\sqrt{2}-1}{2\sqrt{2}} = \dfrac{1}{\sqrt{2}}\sin a + \dfrac{1}{\sqrt{2}}\cos a$

$\Rightarrow \quad \dfrac{\sqrt{2}-1}{2\sqrt{2}} = \sin a \cdot \cos\dfrac{\pi}{4} + \cos a \sin\dfrac{\pi}{4}$

$\Rightarrow \quad \dfrac{\sqrt{2}-1}{2\sqrt{2}} = \sin\left(a + \dfrac{\pi}{4}\right)$

9. $4y^2 = x^2(4-x)(x-2)$

$\Rightarrow \quad |y| = \dfrac{|x|}{2}\sqrt{(4-x)(x-2)}$

$\Rightarrow \quad y_1 = \dfrac{x}{2}\sqrt{(4-x)(x-2)}$

and $\quad y_2 = \dfrac{-x}{2}\sqrt{(4-x)(x-2)}$

$D : x \in [2, 4]$

Required area $= \int_2^4 (y_1 - y_2)\,dx$

$$= \int_2^4 x\sqrt{(4-x)(x-2)}\,dx \quad \ldots\text{(i)}$$

Applying $\int_a^b f(x)\,dx = \int_a^b f(a+b-x)\,dx$

$$\text{Area} = \int_2^4 (6-x)\sqrt{(4-x)(x-2)}\,dx \quad \ldots\text{(ii)}$$

From Eqs. (i) and (ii), we get

$$2A = 6\int_2^4 \sqrt{(4-x)(x-2)}\,dx$$

$$\Rightarrow \quad A = 3\int_2^4 \sqrt{1-(x-3)^2}\,dx$$

$$\Rightarrow \quad A = 3\cdot\frac{\pi}{2} = \frac{3\pi}{2}$$

10. In I quadrant $y^2 \pm x^2 = 1$

and in II quadrant $y^2 \pm x^2 = 1$

i.e. $x^2 + y^2 = 1$

and $y^2 - x^2 = 1$

and in III quadrant $-y^2 \pm x^2 = 1$

and in IV quadrant $-y^2 \pm x^2 = 1$

i.e. $x^2 - y^2 = 1 \quad [\because -x^2 - y^2 \neq 1]$

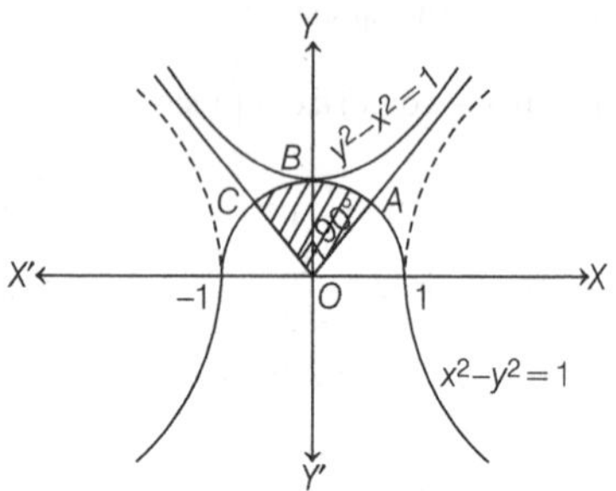

$\therefore$ Required area = area of $OABCO$

$$= \frac{1}{4}\text{ (area of circle)}$$

$$= \frac{1}{4}\pi(1)^2$$

$$= \frac{\pi}{4}\text{ sq unit}$$

11. $\because 0 \le \sin^2 x \le 1$

$\Rightarrow \quad -1 \le -\sin^2 x \le 0$

$\therefore \quad [-\sin^2 x] = -1$ or 0

But $\sec^{-1}(0)$ is not defined

$\therefore \quad y = \sec^{-1}(-1) = \pi$

Now, solving $y = x^2$ and $y = \pi$, we get

$$x = \pm\sqrt{\pi}$$

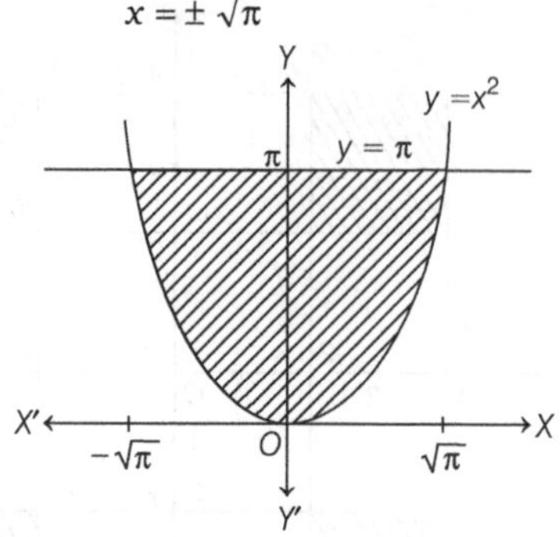

$\therefore$ The required area $= \int_{-\sqrt{\pi}}^{\sqrt{\pi}} (\pi - x^2)\,dx$

$$= 2\int_0^{\sqrt{\pi}} (\pi - x^2)dx \quad \text{(by prop.)}$$

$$= 2\left[\pi x - \frac{x^3}{3}\right]_0^{\sqrt{\pi}} = 2\left[\pi\sqrt{\pi} - \frac{\pi^{3/2}}{3}\right]$$

$$= \frac{4}{3}\pi^{3/2}\text{ sq units}$$

12. $\int_1^b f(x)\,dx = \sqrt{(b^2+1)} - \sqrt{2}$

Differentiating both sides w.r.t. b, then

$$f(b) = \frac{b}{\sqrt{(b^2+1)}}$$

Hence, $\quad f(x) = \dfrac{x}{\sqrt{(x^2+1)}}$

13. Since, $|x| + |y| = 1$

$$\Rightarrow \begin{cases} x+y=1, & x>0, y>0 \\ x-y=1, & x>0, y<0 \\ -x+y=1, & x<0, y>0 \\ -x-y=1, & x<0, y<0 \end{cases}$$

and $\quad 1 - y^2 = |x|$

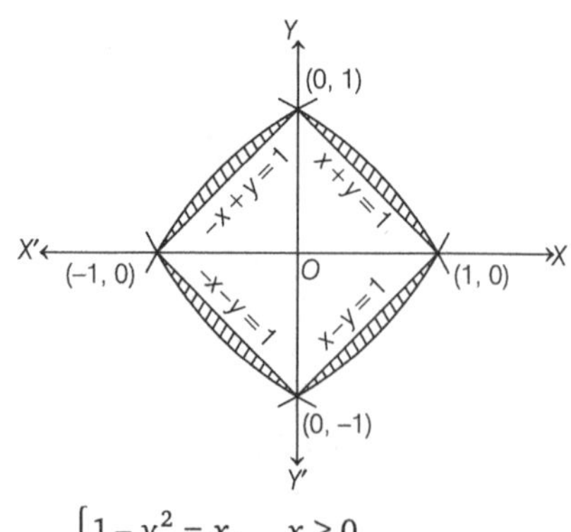

$$\Rightarrow \begin{cases} 1-y^2 = x, & x \ge 0 \\ 1-y^2 = -x, & x < 0 \end{cases}$$

$\therefore$ Required area $= \left|2\int_0^1 \sqrt{(1-x)}\,dx\right| + \left|2\int_{-1}^0 \sqrt{(x+1)}\,dx\right| - 4\left(\frac{1}{2}\cdot 1\cdot 1\right)$

$= \frac{2}{3}$ sq unit

14. The area of the given region

$$A = \{(x, y): |x| + |y| \le 1,\ 2y^2 \ge |x|\}$$

by the figure due to symmetry

$$A = 4\int_0^{1/2}\left(1 - x - \sqrt{\frac{x}{2}}\right)dx$$

$[\because 2y^2 + y - 1 = 0 \Rightarrow (2y-1)(y+1) = 0 \Rightarrow y = \frac{1}{2}]$

$$= 4\left[x - \frac{x^2}{2} - \frac{2x^{3/2}}{3\sqrt{2}}\right]_0^{1/2} = 4\left[\frac{1}{2} - \frac{1}{8} - \frac{2}{12}\right]$$

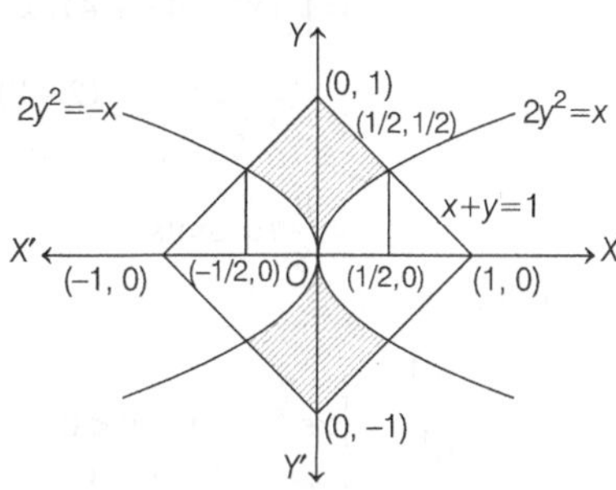

$$= 4\left[\frac{12 - 3 - 4}{24}\right] = \frac{5}{6} \text{ sq units}$$

15. Since, $y = |x - 1| = \begin{cases} x - 1, & x \ge 1 \\ 1 - x, & x < 1 \end{cases}$ ...(i)

and $y = 3 - |x| = \begin{cases} 3 - x, & x \ge 0 \\ 3 + x, & x < 0 \end{cases}$ ...(ii)

Solving Eqs. (i) and (ii), we get

$x = 2$ and $x = -1$

$\therefore$ Required area $= \left|\int_{-1}^2 (3 - |x| - |x - 1|)\,dx\right|$

$= \left|\int_{-1}^0 (2x + 2)\,dx + \int_0^1 2\,dx + \int_1^2 (4 - 2x)\,dx\right|$

$= 4$ sq units

16. $y = x^2 + 1$

$\frac{dy}{dx} = 2x_1$ ...(i)

$m_N = -\frac{1}{2x_1} = -1$

$\Rightarrow \quad x_1 = \frac{1}{2}$

$y_1 = 1 + \frac{1}{4} = \frac{5}{4}$

Equation of normal,

$$y - \frac{5}{4} = -1\left(x - \frac{1}{2}\right)$$

$$y + x = \frac{1}{2} + \frac{5}{4} = \frac{7}{4}$$

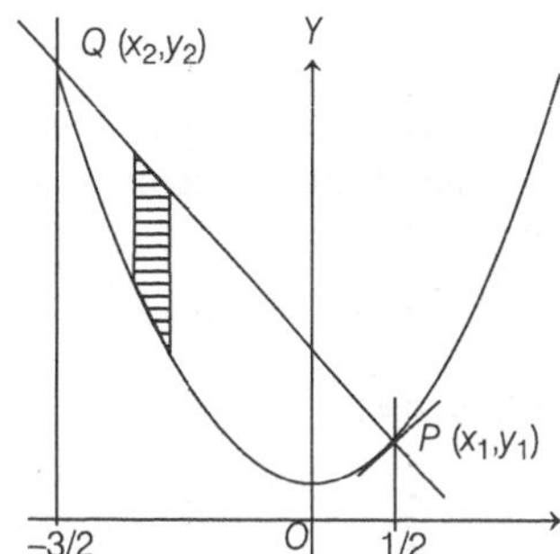

Put $y = \frac{7}{4} - x$ in Eq. (i)

$$\frac{7}{4} - x = x^2 + 1 \Rightarrow x^2 + x - \frac{3}{4} = 0$$

$\Rightarrow \quad x_1 x_2 = -\frac{3}{4};\ x_2 \cdot \frac{1}{2} = -\frac{3}{4}$

$\Rightarrow \quad x_2 = -\frac{3}{2}$

$$A = \int_{-3/2}^{1/2}\left[\left(\frac{7}{4} - x\right) - (x^2 + 1)\right]dx$$

$$= \int_{-3/2}^{1/2}\left[\left(\frac{3}{4} - x - x^2\right)dx = \left[\frac{3x}{4} - \frac{x^2}{2} - \frac{x^3}{3}\right]_{-3/2}^{1/2}\right.$$

$$= \left(\frac{3}{4}\cdot\frac{1}{2} - \frac{1}{8} - \frac{1}{24}\right) - \left(-\frac{9}{8} - \frac{9}{8} + \frac{27}{24}\right)$$

$$= \left(\frac{3}{8} - \frac{1}{8} - \frac{1}{24}\right) - \left(-\frac{18}{8} + \frac{27}{24}\right)$$

$$= \frac{2}{8} - \frac{1}{24} + \frac{18}{8} - \frac{27}{24} = \frac{20}{8} - \frac{28}{24} = \frac{32}{24} = \frac{16}{12}$$

17. $f(x) = \min\{x + 1, \sqrt{(1-x)}$

$= \begin{cases} x + 1, & -1 \le x < 0 \\ \sqrt{1 - x}, & 0 < x \le 1 \end{cases}$

$\therefore$ Required area $= \left|\int_{-1}^0 (x + 1)\,dx\right| + \left|\int_0^1 \sqrt{(1-x)}\,dx\right|$

$= \frac{7}{6}$ sq units

18. $\therefore$ Required area $= \int_{-2}^3 |[x - 3]|\,dx$

$= \int_{-2}^{-1} |[x-3]|\,dx + \int_{-1}^0 |[x-3]|\,dx + \int_0^1 |[x-3]|\,dx + \int_1^2 |[x-3]|\,dx + \int_2^3 |[x-3]|\,dx$

$= \int_{-2}^{-1} 5\cdot dx + \int_{-1}^0 4\cdot dx + \int_0^{-1} 3\cdot dx + \int_1^2 2\cdot dx + \int_2^3 1\cdot dx$

$= 5(1) + 4(1) + 3(1) + 2(1) + 1(1) = 15$ sq units

19. For $c < 1$, $\int_c^1 (8x^2 - x^5)\,dx = \frac{16}{3}$

$\Rightarrow \quad \frac{8}{3} - \frac{1}{6} - \frac{8c^3}{3} + \frac{c^6}{6} = \frac{16}{3}$

$\Rightarrow \quad c^3\left[-\frac{8}{3} + \frac{c^3}{6}\right] = \frac{16}{3} - \frac{8}{3} + \frac{1}{6} = \frac{17}{6}$

$\Rightarrow c = -1$ satisfy the above equation.

For $c \ge 1$, none of the values of c satisfy the required condition that

$$\int_1^c (8x^2 - x^5)\,dx = \frac{16}{3}$$

20. $-8 < x < 8 \Rightarrow y = 2$

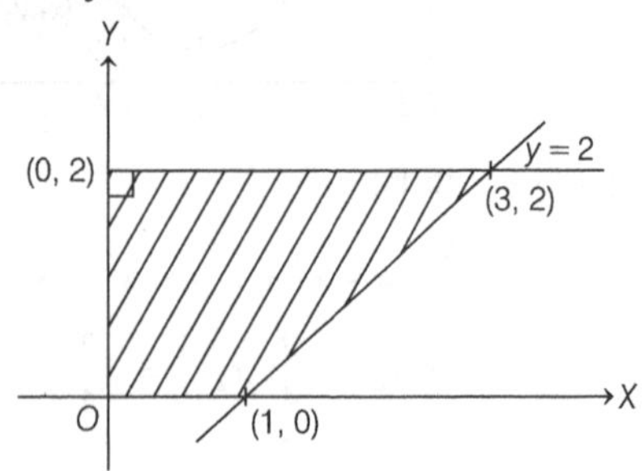

$\therefore$ Required area $= \frac{1}{2}(1 + 3) \times 2 = 4$ sq units

21. We have,

$\Rightarrow \quad 18x^2 - 9\pi x + \pi^2 = 0$

$\Rightarrow \quad 18x^2 - 6\pi x - 3\pi x + \pi^2 = 0$

$(6x - \pi)(3x - \pi) = 0$

$\Rightarrow \quad x = \frac{\pi}{6}, \frac{\pi}{3}$

Now, $\alpha < \beta$

$\therefore \quad \alpha = \frac{\pi}{6}, \beta = \frac{\pi}{3}$

Given, $\quad g(x) = \cos x^2$ and $f(x) = \sqrt{x}$

$y = gof(x)$

$\therefore \quad y = g(f(x)) = \cos x$

Area of region bounded by $x = \alpha, x = \beta$, $y = 0$ and curve $y = g(f(x))$ is

$$A = \int_{\pi/6}^{\pi/3} \cos x\,dx \Rightarrow A = [\sin x]_{\pi/6}^{\pi/3}$$

$$A = \sin\frac{\pi}{3} - \sin\frac{\pi}{6} = \frac{\sqrt{3}}{2} - \frac{1}{2}$$

$$A = \left(\frac{\sqrt{3} - 1}{2}\right)$$

22. Given, equation of parabola is $y = x^2 - 1$, which can be rewritten as $x^2 = y + 1$ or $x^2 = (y - (-1))$.

$\Rightarrow$ Vertex of parabola is $(0, -1)$ and it is open upward.

Equation of tangent at $(2, 3)$ is given by $T = 0$

$\Rightarrow \frac{y + y_1}{2} = x\,x_1 - 1$, where, $x_1 = 2$ and $y_1 = 3$.

$\Rightarrow \quad \frac{y + 3}{2} = 2x - 1$

$\Rightarrow \quad y = 4x - 5$

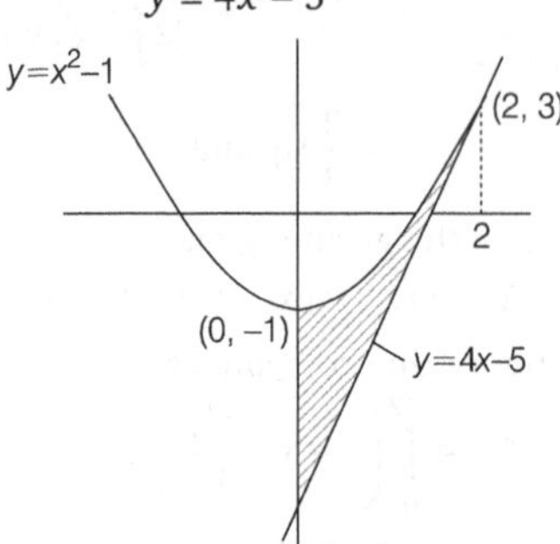

Now, required area = area of shaded region

$= \int_0^2 (y(\text{parabola}) - y(\text{tangent}))\,dx$

$= \int_0^2 [(x^2 - 1) - (4x - 5)]\,dx$

$= \int_0^2 (x^2 - 4x + 4)\,dx = \int_0^2 (x - 2)^2\,dx$

$= \left|\frac{(x - 2)^3}{3}\right|_0^2 = \frac{(2 - 2)^3}{3} - \frac{(0 - 2)^3}{3}$

$= \frac{8}{3}$ sq units

23. Since, $|x| = 1$

$\therefore \quad x = \pm 1$

$$y = xe^{|x|} = \begin{cases} xe^{-x}, & -1 < x < 0 \\ xe^{x}, & 0 \le x < 1 \end{cases}$$

$\therefore$ Required area $= \left|\int_{-1}^0 xe^{-x}dx\right| + \left|\int_0^1 xe^{x}dx\right|$

$= |\{-xe^{-x} - e^{-x}\}_{-1}^0| + |\{xe^x - e^x\}_0^1|$

$= 2$ sq unit

24. We know that, $\sin x \ge \cos x$ for $x \in [\pi/4, \pi/2]$ and $\sin x \le \cos x$ for $x \in \left[0, \frac{\pi}{4}\right]$, therefore the required area

$= \int_0^{\pi/4} (\cos x - \sin x)\,dx + \int_{\pi/4}^{\pi/2} (\sin x - \cos x)\,dx$

$= (\sin x + \cos x)_0^{\pi/4} + (-\cos x - \sin x)_{\pi/4}^{\pi/2}$

$= (\sqrt{2} - 1) + (\sqrt{2} - 1)$

$= 2(\sqrt{2} - 1)$ sq unit

25. Given curve is $(y - x)^2 = x^3$

$\Rightarrow \quad y - x = \pm x\sqrt{x}$ or $y = x \pm x\sqrt{x}$

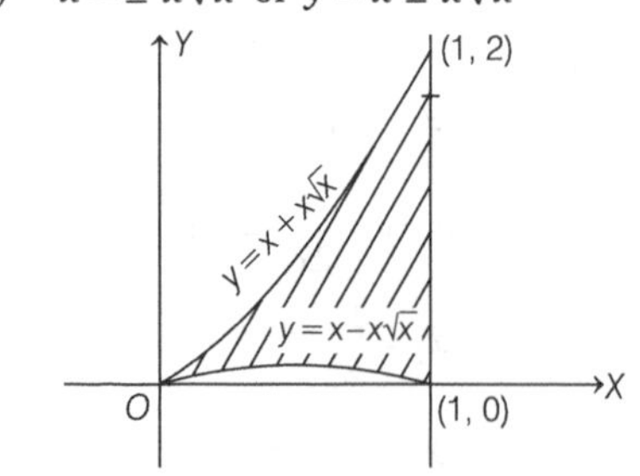

$\therefore$ Required area $= \left|\int_0^1 [(x + x\sqrt{x}) - (x - x\sqrt{x})]\, dx\right|$

$$= \left|2\int_0^1 x^{3/2} dx\right|$$

$$= \left|2 \cdot \frac{2}{5}\right| = \frac{4}{5} \text{ sq unit}$$

26. $\because$ Area of circle $= \pi(1)^2 = \pi$

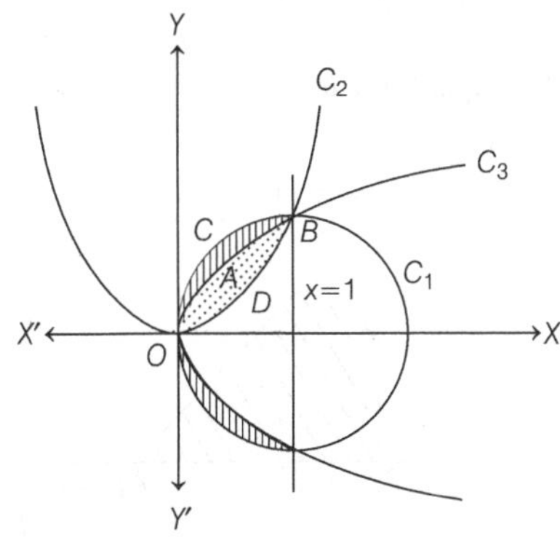

Now, A_{12} = the area bounded by the curves C_1 and C_2 between $x = 0$ and $x = 1$

$$= \frac{\pi}{4} - \int_0^1 x^2\, dx$$

$$= \left(\frac{\pi}{4} - \frac{1}{3}\right) \text{ sq unit (Area of } ODBCO \text{ region)}$$

A_{23} = the area bounded by the curves C_2 and C_3 between $x = 0$ and $x = 1$

$$= \int_0^1 (\sqrt{x} - x^2)dx = \left[\frac{2}{3}x^{3/2} - \frac{x^3}{3}\right]_0^1$$

$$= \frac{2}{3} - \frac{1}{3} = \frac{1}{3} \text{ sq units (Ara of } ODBAO \text{ region)}$$

and A_{31} = the area bounded by the curves C_1 and C_3 between $x = 0$ and $x = 1$

$$= \frac{1}{2}(\pi) - 2\int_0^1 \sqrt{x}\, dx = \frac{\pi}{2} - 2\left(\frac{x^{3/2}}{3/2}\right)_0^1$$

$$= \left(\frac{\pi}{2} - \frac{4}{3}\right) \text{ sq units} \quad \text{(Area of strip regions)}$$

$$\therefore\ A_{12} + A_{23} + A_{31} = \left(\frac{3\pi}{4} - \frac{4}{3}\right) \text{ sq units}$$

27. The given functions $f : R \to R$ and $g : R \to R$ be defined by

$$f(x) = e^{x-1} - e^{-|x-1|} = \begin{cases} e^{x-1} - e^{1-x}, & x \ge 1 \\ 0, & x < 1 \end{cases}$$

and $\quad g(x) = \frac{1}{2}(e^{x-1} + e^{1-x})$

For point of intersection of curves $f(x)$ and $g(x)$ put $f(x) = g(x)$

for $x \ge 1$, $e^{x-1} - e^{1-x} = \frac{1}{2}(e^{x-1} + e^{1-x})$

$\Rightarrow \quad e^{x-1} = 3e^{1-x}$

$\Rightarrow \quad e^{2x} = 3e^2 \Rightarrow x = \frac{1}{2}\log_e^3 + 1$

So, required area is

$$\int_0^{1/2\log_e^3 + 1} (g(x) - f(x))dx$$

$$= \int_0^{1/2\log_e^3+1} g(x)dx - \int_1^{1/2\log_e^3+1} f(x)dx$$

$$= \frac{1}{2}\int_0^{1/2\log_e^3+1} (e^{x-1} + e^{1-x})dx$$

$$- \int_1^{1/2\log_e^3+1} (e^{x-1} - e^{1-x})\, dx]$$

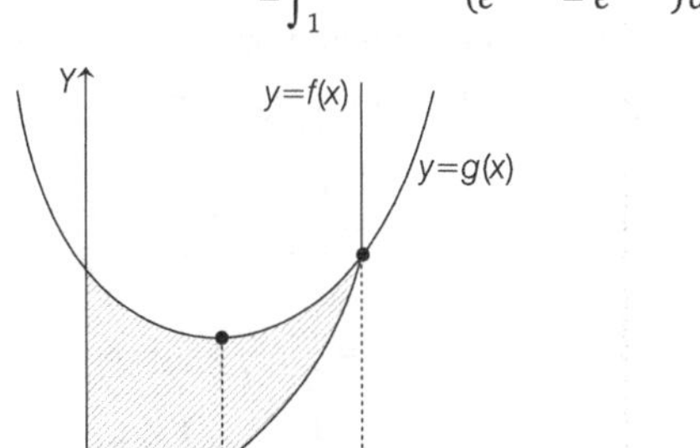

$$= \frac{1}{2}[e^{x-1} - e^{1-x}]_0^{\frac{1}{2}\log_e^3+1} - [e^{x-1} + e^{1-x}]_1^{\frac{1}{2}\log_e^3+1}$$

$$= \frac{1}{2}\left[\sqrt{3} - \frac{1}{\sqrt{3}} - e^{-1} + e\right] - \left[\sqrt{3} + \frac{1}{\sqrt{3}} - 1 - 1\right]$$

$$= \frac{1}{\sqrt{3}} + \frac{1}{2}(e - e^{-1}) - \frac{4}{\sqrt{3}} + 2$$

$$= (2 - \sqrt{3}) + \frac{1}{2}(e - e^{-1})$$

28. $\because 0 \le x < 1$

$\therefore y = [x + 1] = 1$

Hence, required area $= \int_0^1 \sqrt{y}\, dy$

$$= \frac{2}{3}[y^{3/2}]_0^1 = \frac{2}{3} \text{ sq unit}$$

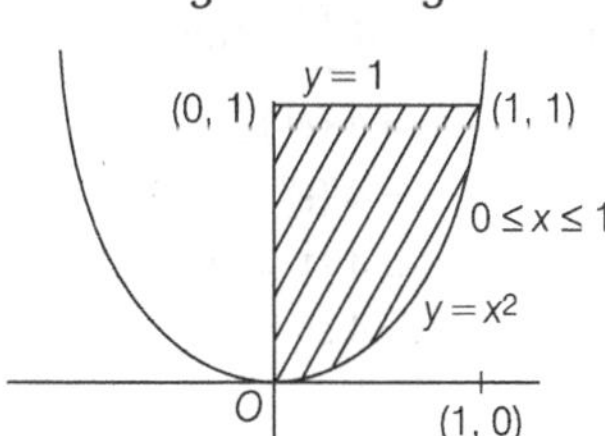

29. Since, $\int_0^a f(x)\, dx = \frac{a^2}{2} + \frac{a}{2}\sin a + \frac{\pi}{2}\cos a$

Differentiating both sides w.r.t. a, we get

$$f(a) = a + \frac{a}{2}\cos a + \frac{1}{2}\sin a - \frac{\pi}{2}\sin a$$

$$\therefore \quad f\left(\frac{\pi}{2}\right) = \frac{\pi}{2} + 0 + \frac{1}{2}\cdot 1 - \frac{\pi}{2}\cdot 1 = \frac{1}{2}$$

30. Given curve is

$$y = x^4 - 2x^3 + x^2 + 3$$

$$\therefore \quad \frac{dy}{dx} = 4x^3 - 6x^2 + 2x$$

$$\frac{d^2y}{dx^2} = 12x^2 - 12x + 2$$

For maxima and minima $\frac{dy}{dx} = 0$

$\therefore \quad 4x^3 - 6x^2 + 2x = 0$

Then, $\quad x = 0, \frac{1}{2}, 1$

$$\therefore \left.\frac{d^2y}{dx^2}\right|_{x=0} = 2, \left.\frac{d^2y}{dx^2}\right|_{x=\frac{1}{2}} = -1 \text{ and } \left.\frac{d^2y}{dx^2}\right|_{x=1} = 2$$

$\therefore$ Points of minimum are $x = 0$ and $x = 1$.

$\therefore$ Required area $= \int_0^1 (x^4 - 2x^3 + x^2 + 3)\, dx$

$$= \left[\frac{x^5}{5} - \frac{2x^4}{4} + \frac{x^3}{3} + 3x\right]_0^1$$

$$= \frac{1}{5} - \frac{2}{4} + \frac{1}{3} + 3$$

$$= \frac{1}{5} - \frac{1}{2} + \frac{1}{3} + 3$$

$$= \frac{91}{30} \text{ sq units}$$

31.

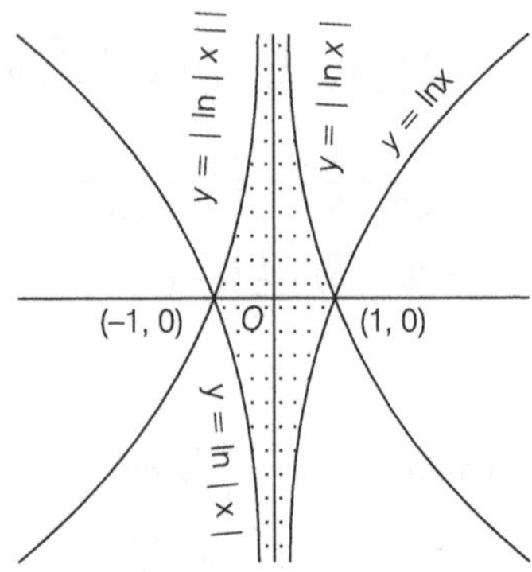

$\therefore$ Required area $= 4\int_{-\infty}^{0} e^y dy$

$$= 4\{e^y\}_{-\infty}^{0}$$

$$= 4\{e^0 - e^{-\infty}\}$$

$$= 4\{1 - 0\}$$

$$= 4 \text{ sq units}$$

32. Since, $y = x^2 + 2x + 1$

$$\therefore \quad \frac{dy}{dx} = 2x + 2$$

$$\therefore \quad \left.\frac{dy}{dx}\right|_{x=1} = 2 \times 1 + 2 = 4$$

Equation of tangent at (1, 4) is

$$y - 4 = 4(x - 1) \text{ or } 4x - y = 0$$

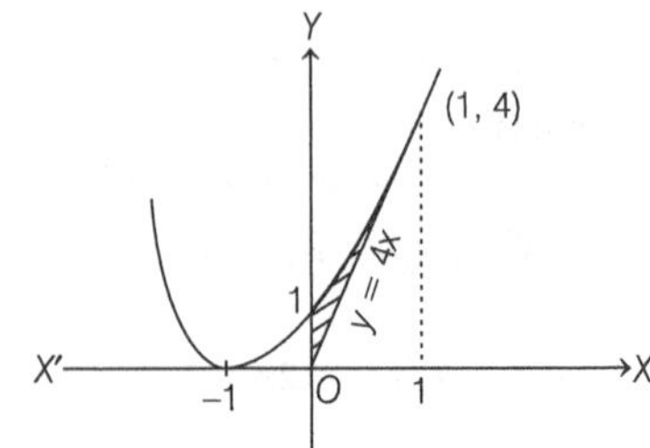

$\therefore$ Required area $= \int_0^1 y\, dx - \frac{1}{2} \times 1 \times 4$

$$= \int_0^1 (x^2 + 2x + 1)\, dx - 2$$

$$= \left[\frac{x^3}{3} + x^2 + x\right]_0^1 - 2 = \frac{1}{3} \text{ sq unit}$$

33. Equation of tangent at (1, 1) is

$$y - 1 = (2 + b)(x - 1)$$

For first quadrant $b < 0$

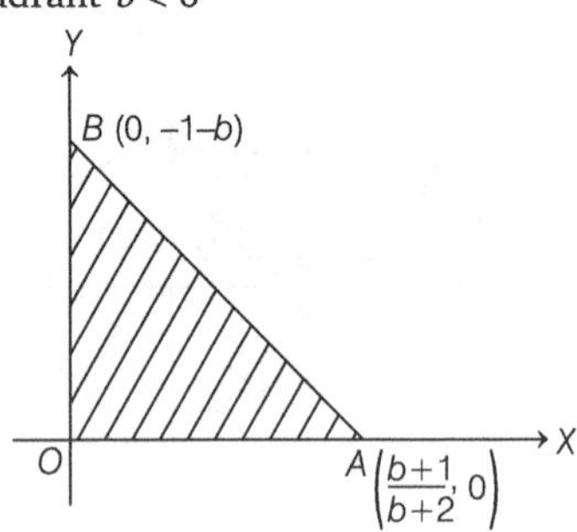

$\therefore$ Area = 2 (given)

$$\frac{1}{2} \times \left(\frac{b+1}{b+2}\right) \times (-1 - b) = 2$$

$$\Rightarrow \quad (b + 1)^2 + 4b + 8 = 0$$

$$\Rightarrow \quad b^2 + 6b + 9 = 0 \Rightarrow (b + 3)^2 = 0$$

$$\therefore \quad b = -3$$

34. $y = 2 - |2 - x|,\ y = \frac{3}{|x|}$

$$y = \begin{cases} x, & x \le 2 \\ 4 - x, & x \ge 2 \end{cases};\ y = \begin{cases} \frac{3}{x}, & x > 0 \\ -\frac{3}{x}, & x < 0 \end{cases}$$

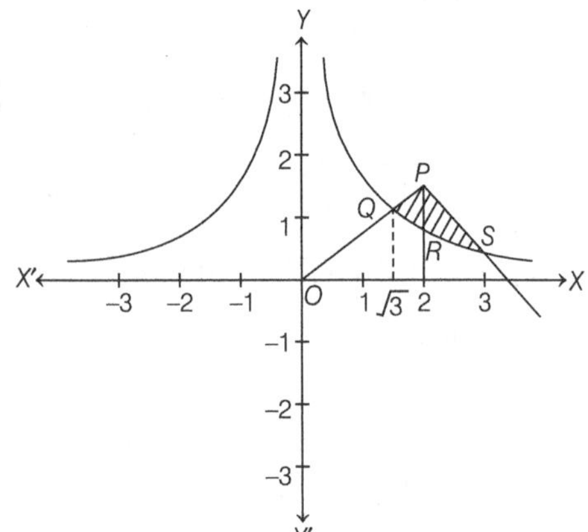

Hence, required area

PQRSP = area *PQRP* + area *PRSP*

$$= \left|\int_{\sqrt{3}}^{2}\left(x - \frac{3}{x}\right) dx\right| + \left|\int_2^3 \left((4 - x) - \frac{3}{x}\right) dx\right|$$

$$= \frac{4 - 3\ln 3}{2} \text{ sq unit}$$

35. $\because f(x) = \max\left\{\sin x, \cos x, \frac{1}{2}\right\}$

Interval value of $f(x)$

For $0 \le x < \frac{\pi}{4}$, $\cos x$

For $\frac{\pi}{4} \le x < \frac{5\pi}{6}$, $\sin x$

For $\frac{5\pi}{6} \le x < \frac{5\pi}{3}$, $\frac{1}{2}$

For $\frac{5\pi}{3} \le x < 2\pi$, $\cos x$

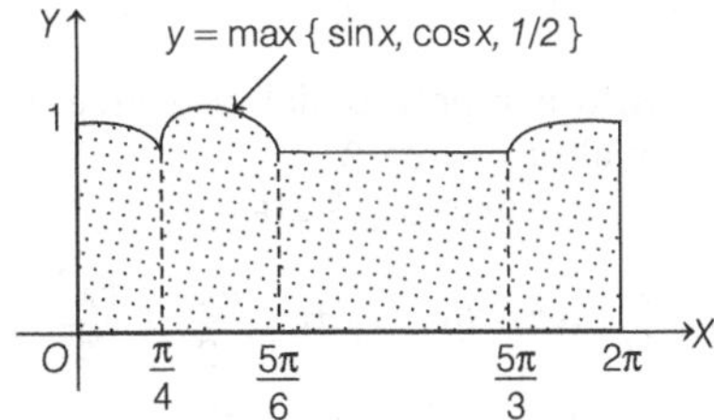

Hence, required area

$$= \int_0^{\pi/4} \cos x\, dx + \int_{\pi/4}^{5\pi/6} \sin x\, dx + \int_{5\pi/6}^{5\pi/3} (1/2)\, dx + \int_{5\pi/3}^{2\pi} \cos x\, dx$$

$$= [\sin x]_0^{\pi/4} - [\cos x]_{\pi/4}^{5\pi/6} + \frac{1}{2}[x]_{5\pi/6}^{5\pi/3} + [\sin x]_{5\pi/3}^{2\pi}$$

$$= \left(\frac{1}{\sqrt{2}} - 0\right) - \left(-\frac{\sqrt{3}}{2} - \frac{1}{\sqrt{2}}\right) + \frac{1}{2}\left(\frac{5\pi}{3} - \frac{5\pi}{6}\right) + \left(0 + \frac{\sqrt{3}}{2}\right)$$

$$= \left(\frac{5\pi}{12} + \sqrt{2} + \sqrt{3}\right) \text{ sq units}$$

36. The given region $\{(x, y): xy \le 8, 1 \le y \le x^2\}$.

From the figure, region A and B satisfy the given region, but only A is bounded region, so area of bounded region

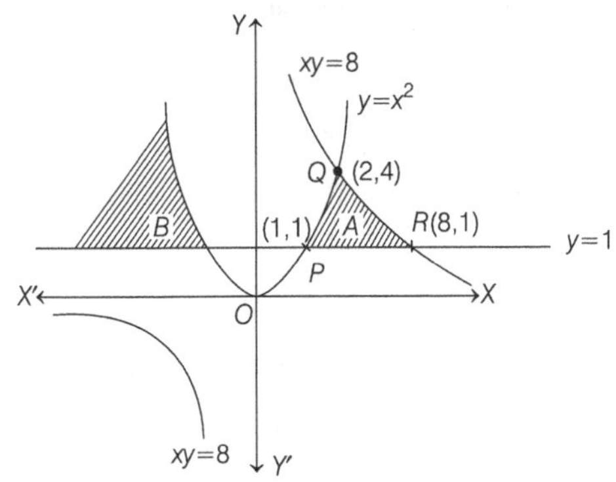

$$A = \int_1^2 (x^2 - 1)\, dx + \int_2^8 \left(\frac{8}{x} - 1\right) dx$$

[$\because$ Points $P(1, 1)$, $Q(2, 4)$ and $R(8, 1)$]

$$= \left[\frac{x^3}{3} - x\right]_1^2 + [8 \log|x| - x]_2^8$$

$$= \left(\frac{8}{3} - 2 - \frac{1}{3} + 1\right) + 8 \log 8 - 8 - 8 \log 2 + 2$$

$$= -\frac{14}{3} + 16 \log 2 = 16 \log 2 - \frac{14}{3}$$

37. Given curves $C_1 : y^2 = ax$, and $C_2 : x^2 = ay$, $(a > 0)$, intersect each other at origin 'O' and a point $P(a, a)$.

$\because O(0, 0)$, Q and $P(a, a)$ are collinear. So, $Q(b, b)$.

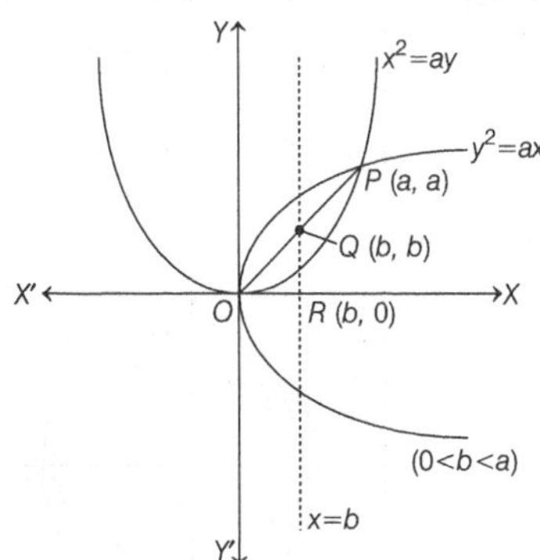

Now, as it is given area of $\Delta OQR = \frac{1}{2}$

$$\Rightarrow \quad \frac{1}{2}b^2 = \frac{1}{2} \Rightarrow b = 1 \qquad (\because b > 0)$$

$\because$ The line $x = b$ bisects the area bounded by the curves, C_1 and C_2, so

$$\int_0^1 \left(\sqrt{a}\sqrt{x} - \frac{x^2}{a}\right) dx = \int_1^a \left(\sqrt{a}\sqrt{x} - \frac{x^2}{a}\right) dx$$

$$\Rightarrow \left[\sqrt{a}\frac{x^{3/2}}{3/2} - \frac{1}{a}\frac{x^3}{3}\right]_0^1 = \left[\sqrt{a}\frac{x^{3/2}}{3/2} - \frac{1}{a}\frac{x^3}{3}\right]_1^a$$

$$\Rightarrow \frac{2\sqrt{a}}{3} - \frac{1}{3a} = \frac{2a^2}{3} - \frac{a^2}{3} - \frac{2\sqrt{a}}{3} + \frac{1}{3a}$$

$$\Rightarrow \frac{4\sqrt{a}}{3} = \frac{a^2}{3} + \frac{2}{3a} \Rightarrow 4a\sqrt{a} = a^3 + 2$$

$$\Rightarrow 16a^3 = a^6 + 4 + 4a^3 \Rightarrow a^6 - 12a^3 + 4 = 0$$

$\therefore a$ satisfies the equation $x^6 - 12x^3 + 4 = 0$

38. If $a > 0$, then

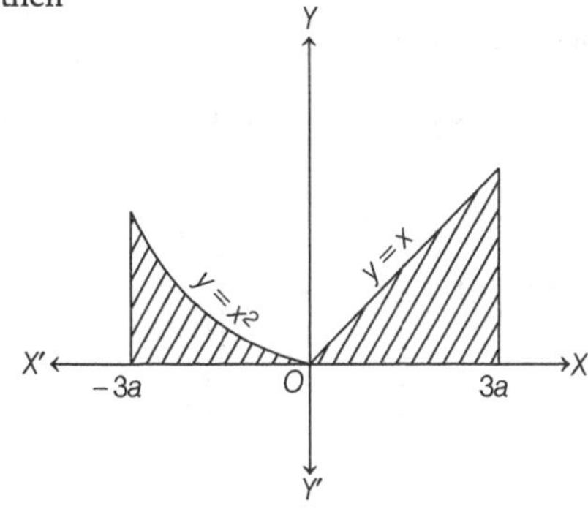

$\therefore$ Required area

$$= \int_{-3a}^0 x^2 dx + \int_0^{3a} x\, dx = \int_0^{3a} x^2 dx + \int_0^{3a} x\, dx$$

$$= \int_0^{3a} (x^2 + x)\, dx$$

$$= \frac{x^3}{3} + \frac{x^2}{2}\Bigg|_0^{3a}$$

$$= \frac{27a^3}{3} + \frac{9a^2}{2} = \frac{9a}{2} \quad \text{(given)}$$

$\Rightarrow \quad a^2 + \frac{a}{2} = \frac{1}{2} \quad (\because a \neq 0)$

$\Rightarrow \quad 2a^2 + a - 1 = 0$

We get, $\quad a = \frac{1}{2}, a = -1$

$\therefore \quad a = \frac{1}{2}, a \neq -1$

39. $\therefore$ Required area

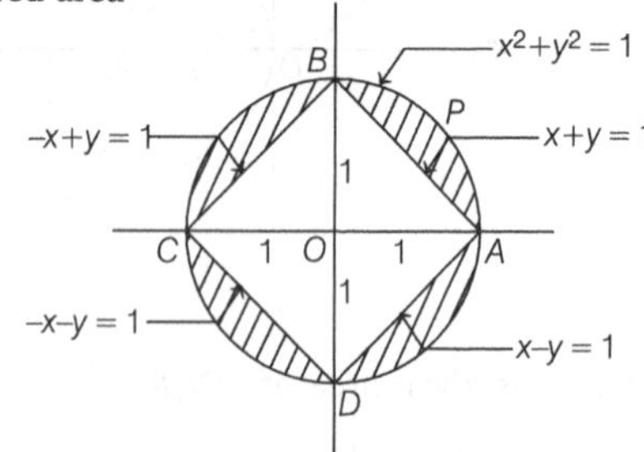

= area of circle – area of square

$= \pi(1)^2 - (\sqrt{2})^2$

$= (\pi - 2)$ sq unit

40. $\int_0^{\lambda} P(x)\, dx = \frac{\lambda^2(\lambda + 3)}{3}$ (given)

Differentiating both sides w.r.t. λ, then

$$P(\lambda) = \frac{3\lambda^2 + 6\lambda}{3} = \lambda^2 + 2\lambda$$

$\therefore \quad P(x) = x^2 + 2x$

Alternative method

$\because \int_0^{\lambda} P(x)\, dx = \frac{\lambda^2(\lambda + 3)}{3}$

Since, RHS is a polynomial of degree 3. So, $P(x)$ is a polynomial of degree 2.

Let $\quad P(x) = ax^2 + bx + c$

Then, $\int_0^{\lambda} (ax^2 + bx + c)\, dx = \frac{a\lambda^3}{3} + \frac{b\lambda^2}{2} + c\lambda$

but given, $\frac{a\lambda^3}{3} + \frac{b\lambda^2}{2} + c\lambda = \frac{\lambda^3}{3} + \lambda^2$

On comparing, we get $a = 1, b = 2, c = 0$

$\therefore \quad P(x) = x^2 + 2x$

Objective Questions Type II [One or more than one correct answer(s)]

1. $\because$ Curves $x^2 = y$ and $y^2 = x$ intersect at (0, 0) and (1, 1).

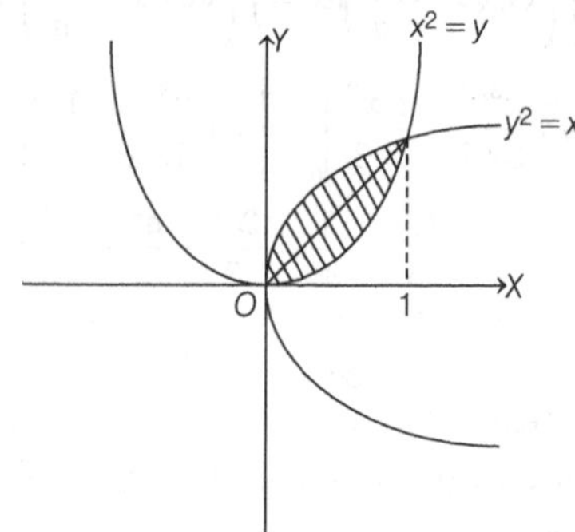

$\therefore$ Required area $= \int_0^1 (\sqrt{x} - x^2)\, dx = \frac{2}{3} - \frac{1}{3} = \frac{1}{3}$ sq unit

Also, both curves $x^2 = y$ and $y^2 = x$ are symmetrical about $y = x$.

$\therefore$ Required area $= 2\int_0^1 (x - x^2)\, dx$

Option (c) : $x^2 \leq y \leq |x|$

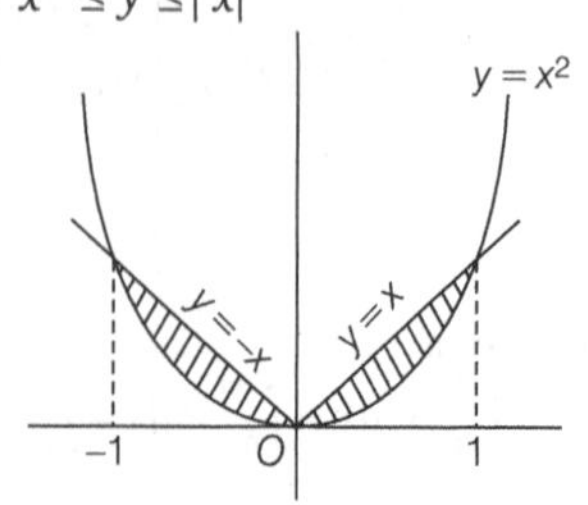

$y = x^2, y = |x|$ point of intersection is (0, 0) and (1, 1).

$\therefore$ Required area $= 2\int_0^1 (x - x^2)\, dx$

$$= 2\left(\frac{1}{2} - \frac{1}{3}\right) = 1 - \frac{2}{3} = \frac{1}{3} \text{ sq unit}$$

2. Solving given curves $y = \frac{1}{x^2}$; $y = \frac{1}{4(x-1)}$, we get

$$x^2 = 4(x-1) \Rightarrow (x-2)^2 = 0$$

So, curves touch other at $x = 2$.

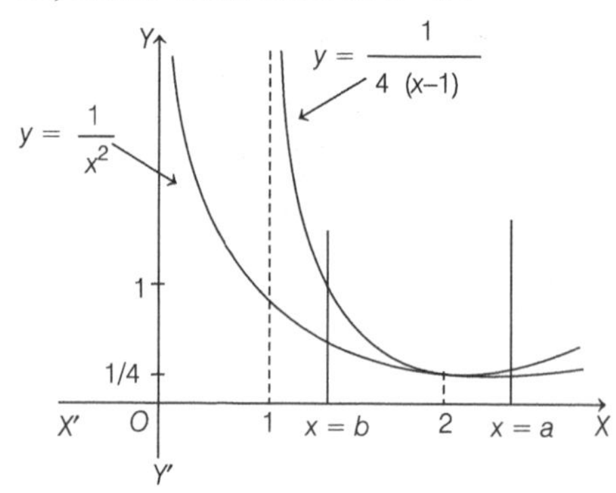

$$\therefore \quad A = \int_2^a \left(\frac{1}{4(x-1)} - \frac{1}{x^2}\right) dx = \frac{1}{a}$$

$$\Rightarrow \quad \left[\frac{1}{4}\log(x-1) + \frac{1}{x}\right]_2^a = \frac{1}{a}$$

$$\Rightarrow \quad \frac{1}{4}\log(a-1) + \frac{1}{a} - \frac{1}{2} = \frac{1}{a}$$

$\Rightarrow \qquad \log(a-1) = 2 \Rightarrow a = e^2 + 1$

Also , $\qquad 1 - \dfrac{1}{b} = \int_b^2 \left(\dfrac{1}{4(x-1)} - \dfrac{1}{x^2} \right) dx$

$\Rightarrow \qquad b = 1 + e^{-2}$

3. Required area $= \int_0^1 \tan x \, dx = \int_0^1 \tan(1-x)\, dx$

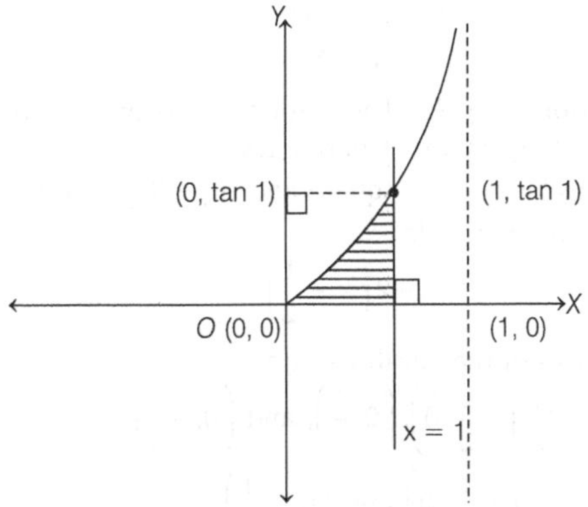

Also, required area $= \tan 1 - \int_0^{\tan 1} (\tan^{-1} y)\, dy$

$\therefore \qquad \text{Area} = \tan 1 - \int_0^{\tan 1} (\tan^{-1} x)\, dx$

4.

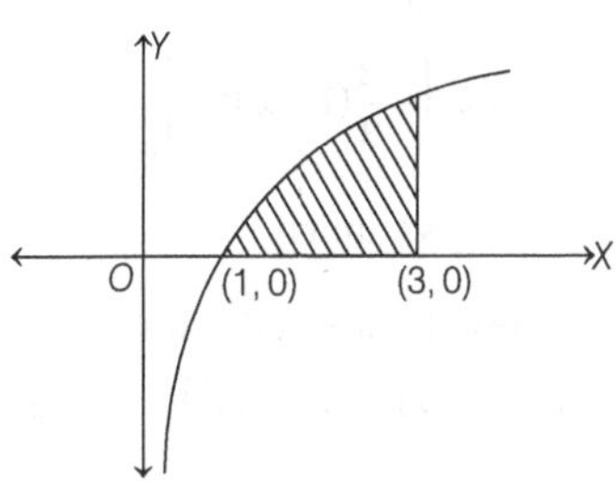

Required area $= \int_1^3 \ln x \, dx$

$= [x \ln x - x]_1^3$ (integration by parts)

$= (3 \ln 3 - 2)$

$= \ln 27 - 2$ sq unit (b is correct)

$= \ln 27 - \ln e^2$

$= \ln (27/e^2)$ sq unit (c is correct)

Also, $\ln\left(\dfrac{27}{e^2}\right) > 3 \Rightarrow \ln\left(\dfrac{27}{e^2}\right) > 3 \ln e$

$\Rightarrow \ln 27 > \ln e^5$, which is false.

5. Given curve, $x^a y = \lambda^a$ ...(i)

$(\lambda, 1)$ is a point on the given curve.

Now, differentiating Eq. (i) w.r.t. x, we get

$$a x^{a-1} y + x^a \frac{dy}{dx} = 0$$

$$\Rightarrow \quad \frac{dy}{dx} = \frac{-a x^{a-1} y}{x^a} = -\frac{ay}{x}$$

At $(\lambda, 1)\ \dfrac{dy}{dx} = -\dfrac{a}{\lambda}$

Equation of tangent at $(\lambda, 1)$

$$y - 1 = -\frac{a}{\lambda}(x - \lambda),$$

Now, $x = 0$

$$\Rightarrow \quad y = 1 + a \Rightarrow y = 0$$

$$\Rightarrow \quad x = \frac{\lambda}{a} + \lambda = \frac{\lambda(1+a)}{a}$$

Area, $A = \dfrac{1}{2} \times (1+a) \dfrac{(1+a)\lambda}{a}$

Now, $\dfrac{dA}{da} = \dfrac{1}{2}\lambda \left[\dfrac{a \cdot 2(1+a) - (1+a)^2}{a^2} \right] = 0$

$$\Rightarrow \quad (2a - 1 - a)(1 + a) = 0$$

$$\Rightarrow \quad (a-1)(a+1) = 0$$

$$\Rightarrow \quad a = 1, a = -1$$

6. $\int_0^1 (x - x^3) dx = 2\int_0^{\alpha} (x - x^3) dx$

$$\frac{1}{4} = 2\left(\frac{\alpha^2}{2} - \frac{\alpha^4}{4} \right)$$

$$2\alpha^4 - 4\alpha^2 + 1 = 0$$

$$\Rightarrow \quad \alpha^2 = \frac{4 - \sqrt{16-8}}{4} \qquad [\because \alpha \in (0, 1)]$$

$$\alpha^2 = 1 - \frac{1}{\sqrt{2}}$$

Linked Comprehension Type

PASSAGE 1

1. $\because f(x) = x^2 - 3x + 2$

Required area $= \int_0^1 f(x)\, dx = \int_0^1 (x^2 - 3x + 2)\, dx$

$= \dfrac{1}{3} - \dfrac{3}{2} + 2 = \dfrac{5}{6}$ sq unit

2. The graph of $f(|x|)$ is

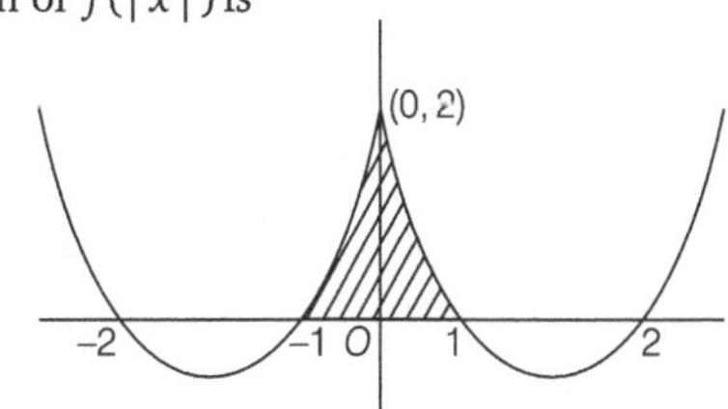

$\therefore$ Required area $= 2\int_0^1 f(x)\,dx$ (from Q. 1)

$$= 2\times\frac{5}{6} = \frac{5}{3} \text{ sq unit}$$

3. The graph of $|f(|x|)|$ is

$\because$ $1 \le |x| \le 2$

$\therefore$ $x \in [-2, -1] \cup [1, 2]$

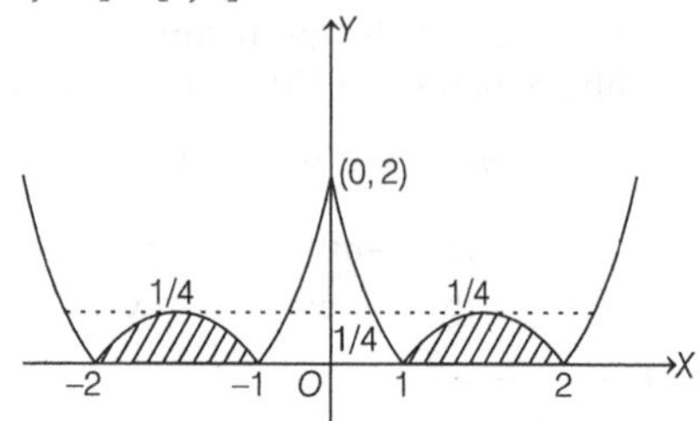

$\therefore$ Required area $= 2\left|\int_1^2 f(x)\,dx\right|$

$$= 2\left|\int_1^2 (x^2 - 3x + 2)\,dx\right|$$

$$= 2\left|\left\{\frac{x^3}{3} - \frac{3x^2}{2} + 2x\right\}_1^2\right|$$

$$= 2\left|\left\{\left(\frac{8}{3} - \frac{12}{2} + 4\right) - \left(\frac{1}{3} - \frac{3}{2} + 2\right)\right\}\right|$$

$$= 2\left|\frac{2}{3} - \frac{5}{6}\right| = \frac{1}{3} \text{ sq unit}$$

4. For six real roots of $|f(|x|)| = a$

$$a = \frac{1}{4}$$

5. Graph of $|y| = |f(|x|)|$ is

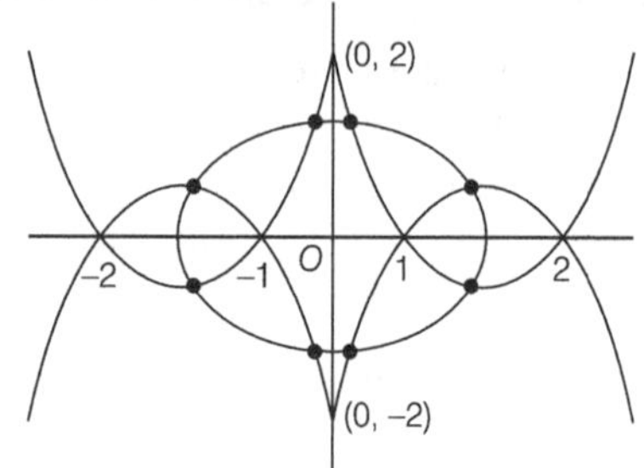

$\because$ Both graphs intersect at 8 points.
Hence, number of solutions = 8.

PASSAGE 2

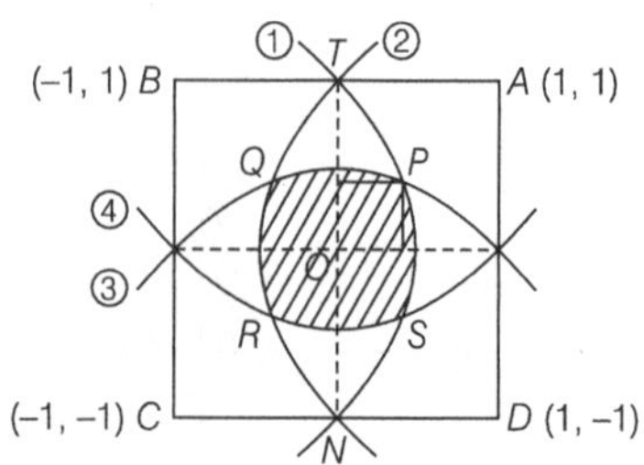

Let (x, y) be a point in the required region, then

$$\sqrt{(x^2 + y^2)} < |1 - x|; |1 + x|; |1 - y|; |1 + y|.$$

i.e. $x^2 + y^2 < (1 - x)^2$; $(1 + x)^2$; $(1 - y)^2$; $(1 + y)^2$

i.e. the required region lies inside the four curves

$$y^2 = 1 - 2x \quad \ldots\text{(i)}$$

$$y^2 = 1 + 2x \quad \ldots\text{(ii)}$$

$$x^2 = 1 - 2y \quad \ldots\text{(iii)}$$

$$x^2 = 1 + 2y \quad \ldots\text{(iv)}$$

i.e. interior of $PQRS$. These four curves are parabolas and symmetry of these parabolas.

1. Point (x, y) inside the square and edge BC, then all points lie inside the parabola

$$y^2 = -2\left(x + \frac{1}{2}\right)$$

2. The vertices of the parabolas are

$$\left(\frac{1}{2}, 0\right), \left(-\frac{1}{2}, 0\right), \left(0, \frac{1}{2}\right), \text{ and } \left(0, -\frac{1}{2}\right)$$

i.e. $\left(\pm\frac{1}{2}, 0\right)$ and $\left(0, \pm\frac{1}{2}\right)$

3. Points of intersections of (1), (2), (3) and (4) are $(\pm(\sqrt{2} - 1), \pm(\sqrt{2} - 1))$.

4. Required area of the region $NSPTQRN$

$$= 4\left|\int_0^{1/2} \sqrt{(1 - 2x)}\,dx\right| = 2\left|\int_0^1 \sqrt{(1 - x)}\,dx\right|$$

$$= 2\left|\left\{-\frac{2}{3}(1 - x)^{3/2}\right\}_0^1\right|$$

$$= \frac{4}{3} \text{ sq units}$$

5. $\because$ Coordinates of P are $(\sqrt{2} - 1, \sqrt{2} - 1)$.

Then, required area $= 4\{(\sqrt{2} - 1)(\sqrt{2} - 1) + 2\int_{\sqrt{2}-1}^{1/2} y\,dx\}$

$$= 4\{(\sqrt{2} - 1)^2 + 2\int_{\sqrt{2}-1}^{1/2} y\,dx\}$$

PASSAGE 3

$\because$ Graph of $f(x) = \min(|x|, |x - 1|, |x + 1|)$

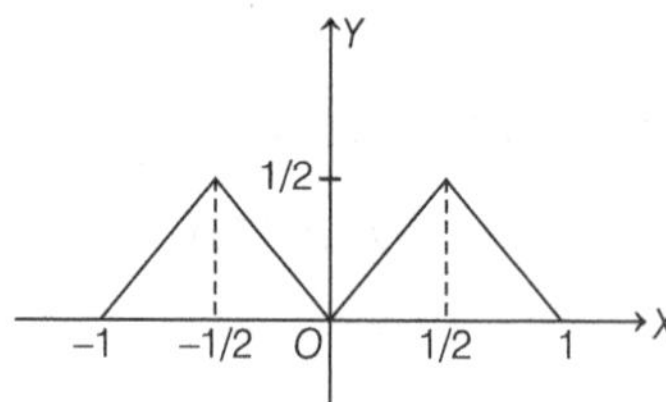

and graph of $g(x) = \min(e^x, e^{-x})$

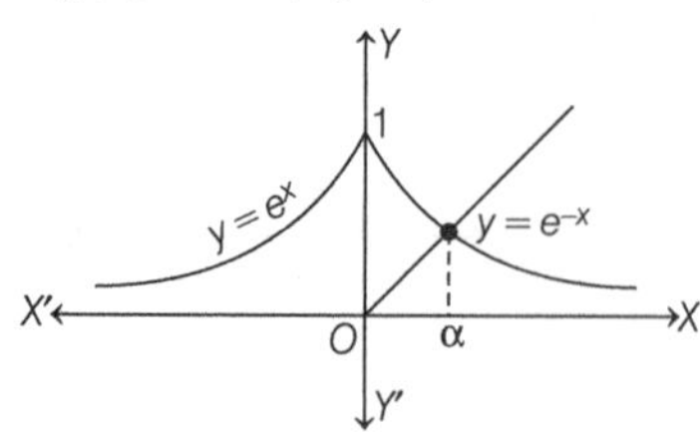

1. Required area $= 2 \times \frac{1}{2} \times 1 \times \frac{1}{2} = \frac{1}{2}$ sq unit

2. $x = -\alpha$

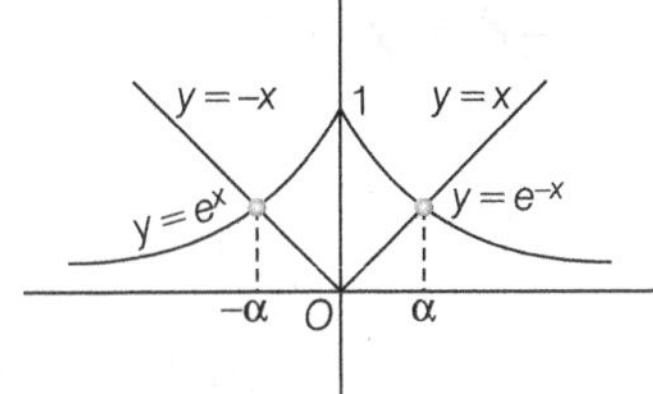

Clear from graph.

3. At $x = 0,\ e^{-x} > x$
and at $x = 1,\ e^{-x} < x$
$\therefore \quad \alpha \in (0, 1).$

4. The required area

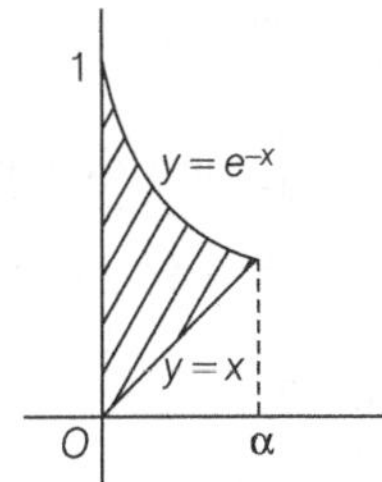

$$= \int_0^{\alpha} (e^{-x} - x)\, dx = \left[-e^{-x} - \frac{x^2}{2}\right]_0^{\alpha}$$

$$= \left(-e^{-\alpha} - \frac{\alpha^2}{2}\right) - (-e^{-1} - 0) = \left(e^{-1} - e^{-\alpha} - \frac{\alpha^2}{2}\right)$$

5. By symmetry,
Sum of roots = 0

PASSAGE 4

Solving the equations

$$y^2 = (e^{-kr})x \qquad \ldots(i)$$

and

$$y = mx \qquad \ldots(ii)$$

we get, $x = 0,\ x = \dfrac{e^{-kr}}{m^2}$

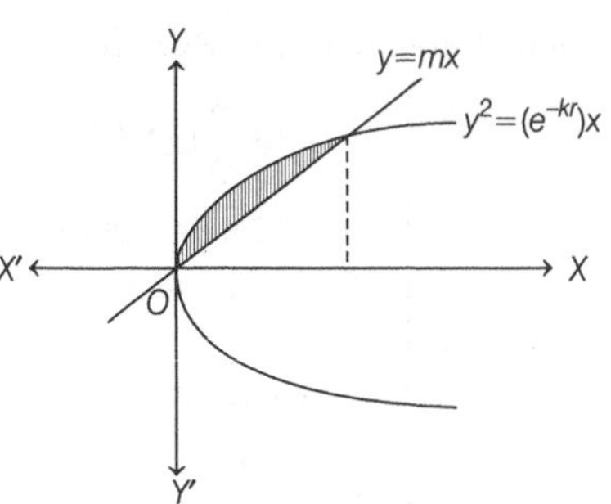

So, the area, $A_r = \int_0^{e^{-kr}/m^2} (e^{-kr/2} \cdot \sqrt{x} - mx)\, dx$

$$= \left[\frac{2}{3} e^{-kr/2}\, x^{3/2} - \frac{mx^2}{2}\right]_0^{\frac{e^{-kr}}{m^2}}$$

$$= \frac{2}{3} \cdot e^{-kr/2} \cdot \frac{e^{-3kr/2}}{m^3} - \frac{m \cdot e^{-2kr}}{2m^4} = \frac{e^{-2kr}}{6m^3}$$

Now, $\dfrac{A_{r+1}}{A_r} = \dfrac{e^{e^{-2k(r+1)}}}{e^{-2kr}} = e^{-2k}$

for $\quad r = 1, 2, 3, 4 \ldots\ldots$

$$\frac{A_2}{A_1} = \frac{A_3}{A_2} = \frac{A_4}{A_3} = \ldots\ldots = e^{-2k}$$

1. Here $A_1, A_2, A_3 \ldots$ are in GP.
$\therefore$ Common ratio $= e^{-2k}$

2. $\lim_{n \to \infty} \sum_{r=1}^{n} A_r = \lim_{n \to \infty} \sum_{r=1}^{n} \frac{e^{-2kr}}{6m^3} = \frac{1}{6m^3} \lim_{n \to \infty} \sum_{r=1}^{n} e^{-2kr}$

$$= \frac{1}{6m^3}(e^{-2k} + e^{-4k} + e^{-6k} + \ldots \text{ to } \infty)$$

$$= \frac{e^{-2k}}{6m^3(1 - e^{-2k})}$$

$$= \frac{1}{6m^3(e^{2k} - 1)} = \frac{4}{3(e^{2k} - 1)} \qquad \text{(given)}$$

$$\therefore \quad m^3 = \frac{3}{6 \times 4} = \frac{1}{8} \Rightarrow m = \frac{1}{2}$$

Numerical Value Type

1. For $x > 0, y > 0 \Rightarrow x + y = k$
For $x < 0, y > 0 \Rightarrow -x + y = k$
For $x > 0, y < 0 \Rightarrow x - y = k$
For $x < 0, y < 0 \Rightarrow -x - y = k$

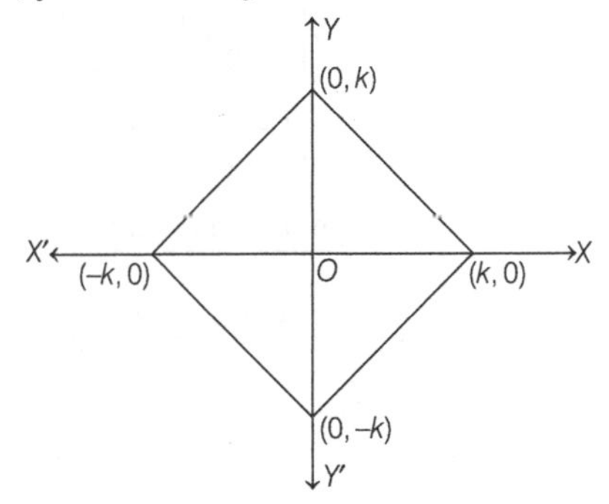

$\therefore$ Required area $= 4 \times \frac{1}{2} \times k \times k$
$= 2k^2$
$\Rightarrow \quad f(k) = 2k^2$
$\Rightarrow \quad 7f(10) = 7\{2(10)^2\} = 7 \times 200 = 1400$

2. Latusrectum is $x = 16$

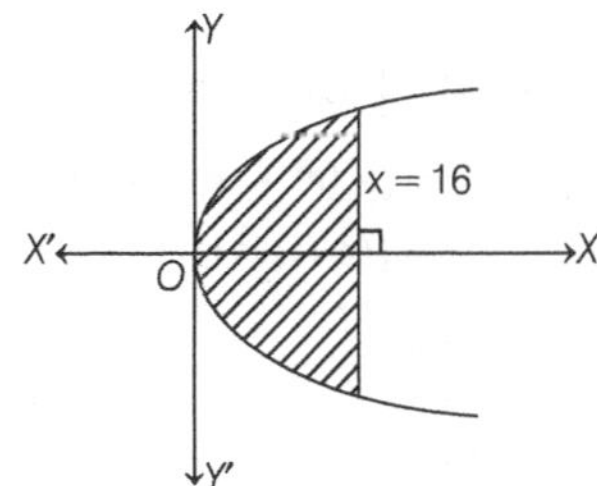

$\therefore$ Required area $(\lambda) = 2\int_0^{16} 8\sqrt{x}\, dx$

$$= 16\cdot\frac{2}{3}\{x^{3/2}\}_0^{16} = \frac{32}{3}(64-0)$$

$\therefore \quad 3\lambda = 32\times 64 = 2048$

3. The curve $xy^2 = 4(2-x)$ is symmetrical about X-axis and at $x \to 0$, $y \to \infty$

i.e. Y-axis is asymptote and cut off x-axis at $x = 2$

$\because \quad y^2 = \dfrac{4(2-x)}{x} \geq 0$

$\Rightarrow \quad \dfrac{2-x}{x} \geq 0 \Rightarrow \dfrac{(x-2)}{x} \leq 0$

$\therefore \quad x \in (0, 2]$

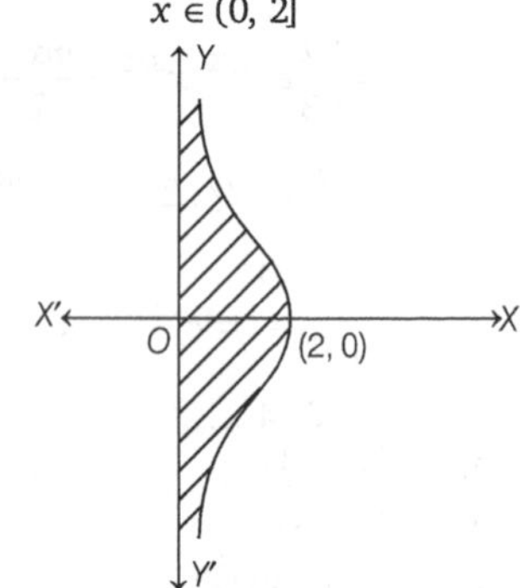

Then, $\lambda = 2\int_0^2 2\sqrt{\left(\dfrac{2-x}{x}\right)}\, dx$

Put $\quad x = 2\sin^2\theta$

$\therefore \quad dx = 4\sin\theta\cos\theta\, d\theta$

$\Rightarrow \quad \lambda = 4\int_0^{\pi/2}\sqrt{\left(\dfrac{2-2\sin^2\theta}{2\sin^2\theta}\right)} \times 4\sin\theta\cos\theta\, d\theta$

$$= 16\int_0^{\pi/2}\cos^2\theta\, d\theta = 16\cdot\frac{1}{2}\cdot\frac{\pi}{2} = 4\pi$$

Then, $\cos(1824\lambda) = \cos(1824\times 4\pi) = 1$

$\therefore \quad 5050\cos(1824\lambda) = 5050\times 1 = 5050$

4.

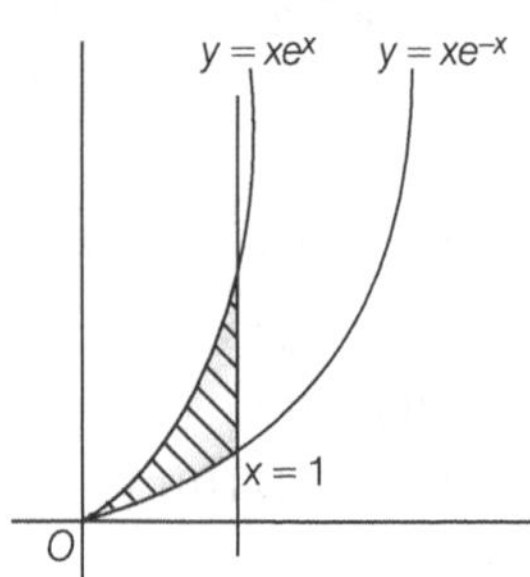

Required area $\lambda = \int_0^1 (xe^x - xe^{-x})\, dx$

$$= \{x(e^x + e^{-x}) - (e^x - e^{-x})\}_0^1$$

$$= \{(e + e^{-1}) - (e - e^{-1})\} - \{0 - (1-1)\}$$

$$= \frac{2}{e}$$

$\therefore \quad 128e\lambda = 128e \times \dfrac{2}{e} = 256$

5. The given curves are $y = \ln(x+e)$ and $x = \ln\left(\dfrac{1}{y}\right)$

$\therefore \quad \dfrac{1}{y} = e^x \Rightarrow y = e^{-x}$

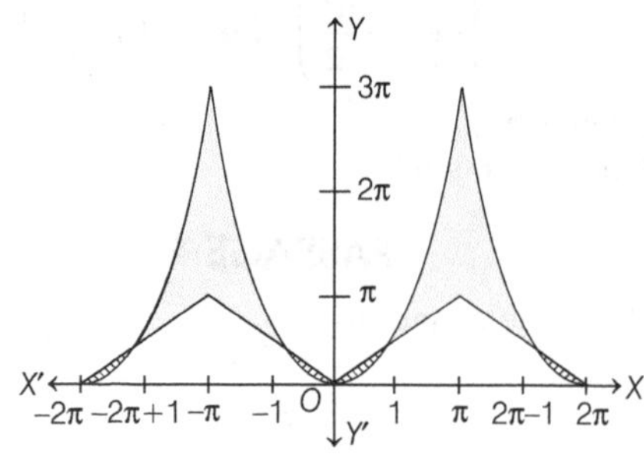

$\therefore$ Required area $= \int_{1-e}^{0} \ln(x+e)dx + \int_0^{\infty} e^{-x}dx$

$$= \int_1^e \ln x\, dx + \int_0^{\infty} e^{-x}\, dx \quad \text{(by property)}$$

$$= \{x\ln x - x\}_1^e - \{e^{-x}\}_0^{\infty}$$

$$= \{(e-e) - (0-1)\} - \{0-1\} = 2 \text{ sq unit}$$

$\therefore \quad \lambda = 2$

Then, $\quad \dfrac{\lambda^2}{5} = \dfrac{4}{5} = 0.8$

6. Given function are $f(x) = (\cos^{-1}(\cos x))^2$ and $g(x) = \cos^{-1}(\cos|x|)$

Y
3π
2π
π
X'
X
−2π −2π+1 −π −1 O 1 π 2π−1 2π
Y'

The required area $= 4\left[\int_0^1 (x - x^2)\, dx + \int_1^{\pi} (x^2 - x)\, dx\right]$

$$= 4\left[\left[\frac{x^2}{2} - \frac{x^3}{3}\right]_0^1 + \left[\frac{x^3}{3} - \frac{x^2}{2}\right]_1^{\pi}\right]$$

$$= 4\left[\frac{1}{2} - \frac{1}{3} + \frac{\pi^3}{3} - \frac{\pi^2}{2} - \frac{1}{3} + \frac{1}{2}\right]$$

$$= 4\left[1 - \frac{2}{3} + \frac{\pi^3}{3} - \frac{\pi^2}{2}\right]$$

$$= 4\left[\frac{(\pi^3+1)}{3} - \frac{\pi^2}{2}\right]$$

$$= \frac{4(\pi^3+1) - 6\pi^2}{3}$$

$$= \frac{a(\pi^3+1) - b\pi^2}{c} \quad \text{(given)}$$

$\therefore \quad a = 4,\ b = 6,\ c = 3$

$\therefore \quad \dfrac{b+c}{a} = \dfrac{9}{4} = 2.25$

Single Integer Type

1. $A=\dfrac{1}{\sqrt{3}}\times 1+A_1$

$$A_1=\int_{\frac{1}{\sqrt{3}}}^{1}\left(\sqrt{\left(\frac{4}{3}\right)-x^2}\right)dx$$

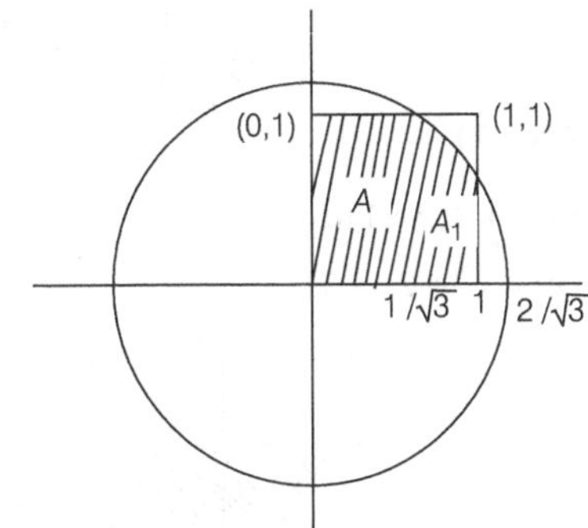

Put $x=\dfrac{2}{\sqrt{3}}\sin\theta \Rightarrow dx=\dfrac{2}{\sqrt{3}}\cos\theta\, d\theta$

$$=\int_{\frac{\pi}{6}}^{\frac{\pi}{3}}\frac{2}{\sqrt{3}}\cos\theta\cdot\frac{2}{\sqrt{3}}\cos\theta\, d\theta$$

$$=\frac{2}{3}\int_{\frac{\pi}{6}}^{\frac{\pi}{3}}(\cos 2\theta+1)d\theta=\frac{2}{3}\left[\frac{1}{2}\sin 2\theta+\theta\right]_{\frac{\pi}{6}}^{\frac{\pi}{3}}$$

$$=\frac{2}{3}\left[\frac{1}{2}\cdot\frac{\sqrt{3}}{2}+\frac{\pi}{3}-\frac{1}{2}\cdot\frac{\sqrt{3}}{2}-\frac{\pi}{6}\right]=\frac{2}{3}\cdot\frac{\pi}{6}=\frac{\pi}{9}$$

$$A=\frac{1}{\sqrt{3}}+\frac{\pi}{9}=\frac{3\sqrt{3}+\pi}{9}$$

$a=3,\ b=1$

$a+b=4$

2. $[x]\cdot[y]=2$

Here four cases arise

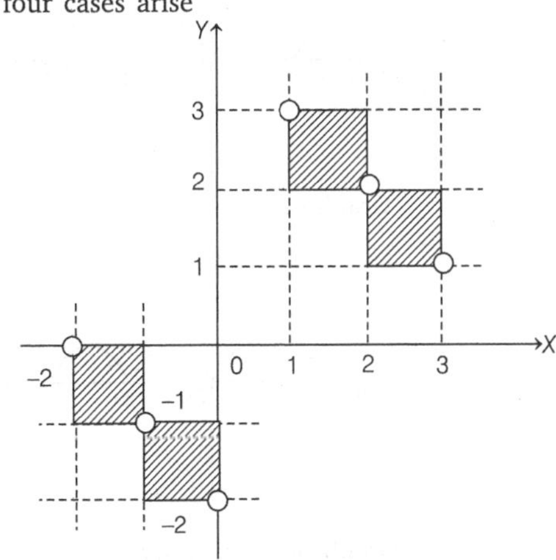

(1) $[x]=2\ \&\ [y]=1 \Rightarrow 2\le x<3\ \&\ 1\le y<2$

(2) $[x]=1\ \&\ [y]=2 \Rightarrow 1\le x<2\ \&\ 2\le y<3$

(3) $[x]=-2\ \&\ [y]=-1 \Rightarrow -2\le x<-1\ \&\ -1\le y<0$

(4) $[x]=-1\ \&\ [y]=-2 \Rightarrow -1\le x<0\ \&\ -2\le y<-1$

Area enclosed by solution set = 4

3. We have, $y=f(x)$, so $\dfrac{dy}{dx}=f'(x)$

Now, $\dfrac{dy}{dx}+\left(\dfrac{-1}{x}\right)y=\left(x-\dfrac{2}{x}\right)$

(linear differential equations)

$\therefore$ $\quad$ If $=e^{-\int\frac{1}{x}dx}=e^{-\ln x}=\dfrac{1}{x}$

Now, general solution is

$$y\left(\frac{1}{x}\right)=\int\left(x-\frac{2}{x}\right)\frac{1}{x}dx+c$$

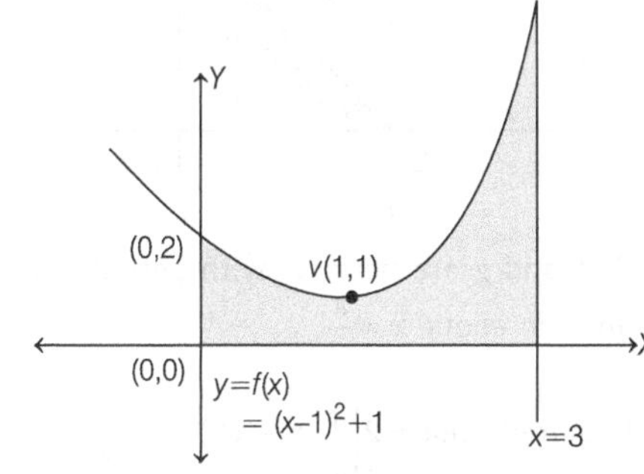

$\therefore$ $\quad \dfrac{y}{x}=x+\dfrac{2}{x}+c$

As, $\quad y(1)=1\Rightarrow c=-2$

$\therefore$ $\quad \dfrac{y}{x}=x+\dfrac{2}{x}-2$

$\Rightarrow$ $\quad f(x)=x^2-2x+2=(x-1)^2+1$

$\therefore$ Required area $=\displaystyle\int_0^3((x-1)^2+1)\,dx$

$$=\left(\frac{1}{3}(x-1)^3\right)_0^3+(x)_0^3$$

$$=\left(\frac{8}{3}+\frac{1}{3}\right)+3=3+3=6$$

4. According to the question,

$$\int_0^1(x-x^n)dx=\frac{3}{10}\left(\frac{1}{2}\times 2\times 1\right)$$

$$\Rightarrow \left|\frac{x^2}{2}-\frac{x^{n+1}}{n+1}\right|_0^1=\frac{3}{10}\ \Rightarrow\ \frac{1}{2}-\frac{1}{n+1}=\frac{3}{10}$$

$$\Rightarrow \frac{1}{n+1}=\frac{1}{2}-\frac{3}{10}=\frac{1}{5}\ \Rightarrow\ n=4$$

5. We have, $f(x)=\begin{cases} x^2; & x\in\left[0,\dfrac{\pi}{2}\right] \\ (\pi-x)^2; & x\in\left(\dfrac{\pi}{2},\pi\right)\end{cases}$;

$$g(x)=\begin{cases} x; & x\in\left[0,\dfrac{\pi}{2}\right] \\ (\pi-x); & x\in\left(\dfrac{\pi}{2},\pi\right)\end{cases};$$

Both f and g are periodic with period $=\pi$ and symmetric about $x=\dfrac{\pi}{2}$

$$\therefore \text{ Required area } = 2\left(\int_0^1 (x-x^2)dx+\int_1^{\pi/2}(x^2-x)dx\right)$$

$$=2\left(\left(\frac{1}{2}-\frac{1}{3}\right)+\left(\frac{\pi^3}{24}-\frac{1}{3}\right)+\left(\frac{\pi^2}{8}-\frac{1}{2}\right)\right)$$

$$=2\left(\frac{\pi^3}{24}-\frac{\pi^2}{8}+\frac{1}{3}\right)$$

$$=\frac{\pi^3}{12}-\frac{\pi^2}{4}+\frac{2}{3}\equiv a\pi^3+b\pi^2+c \quad \text{(given)}$$

So, $a=\dfrac{1}{12}, b=\dfrac{-1}{4}, c=\dfrac{2}{3}$

Hence, $(|a|+|b|+|c|)=\left|\dfrac{1}{12}\right|+\left|\dfrac{-1}{4}\right|+\left|\dfrac{2}{3}\right|$

$$=\frac{1}{12}+\frac{1}{4}+\frac{2}{3}=\frac{1+3+8}{12}=\frac{12}{12}=1$$

6. $\dfrac{dy}{dx}=2(x+1)$

$\Rightarrow \int dy=\int 2(x+1)\,dx$

$\Rightarrow y(x)=x^2+2x+C$

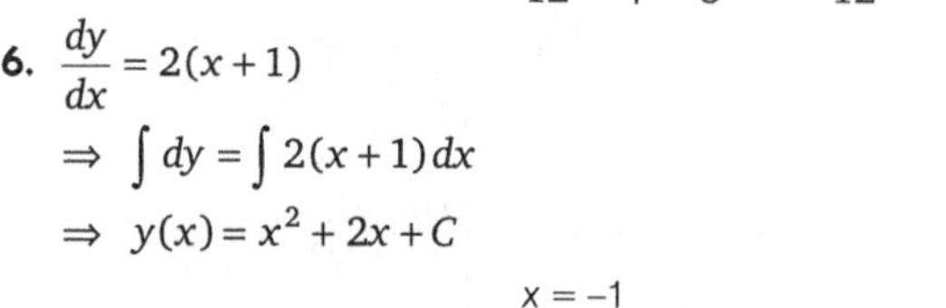

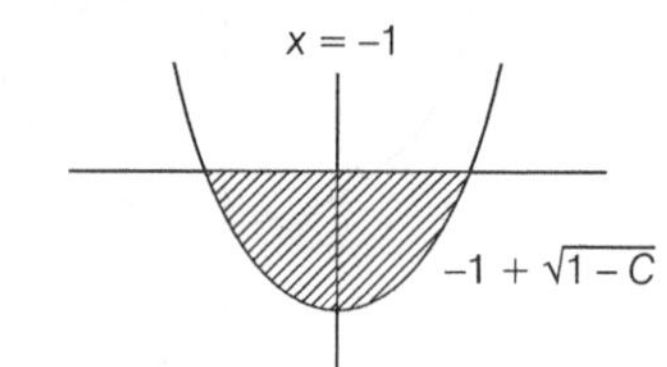

Area $=\dfrac{4\sqrt{8}}{3}$ $-1+\sqrt{1-C}$

$\Rightarrow \quad 2\displaystyle\int_{-1}^{-1+\sqrt{1-C}}(-(x+1)^2-C+1)\,dx=\frac{4\sqrt{8}}{3}$

$\Rightarrow 2\left[-\dfrac{(x+1)^3}{3}-Cx+x\right]_{-1}^{-1+\sqrt{1-C}}=\dfrac{4\sqrt{8}}{3}$

$\Rightarrow -(\sqrt{1-C})^3+3C-3C\sqrt{1-C}$

$-3+3\sqrt{1-C}-3C+3=2\sqrt{8}$

$\Rightarrow \quad C=-1$

$\Rightarrow f(x)=x^2+2x-1,\ f(1)=2$

Matrix-Match Type

1. (A)

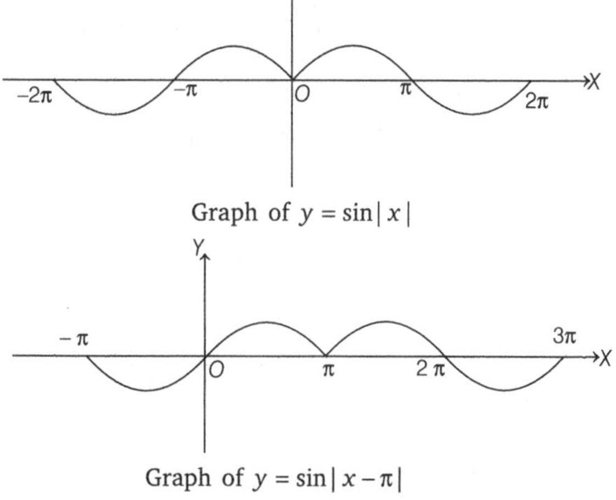

Graph of $y=\sin|x|$

Graph of $y=\sin|x-\pi|$

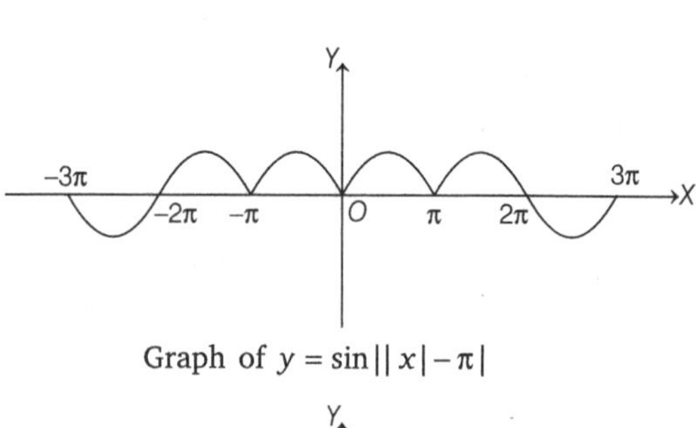

Graph of $y=\sin||x|-\pi|$

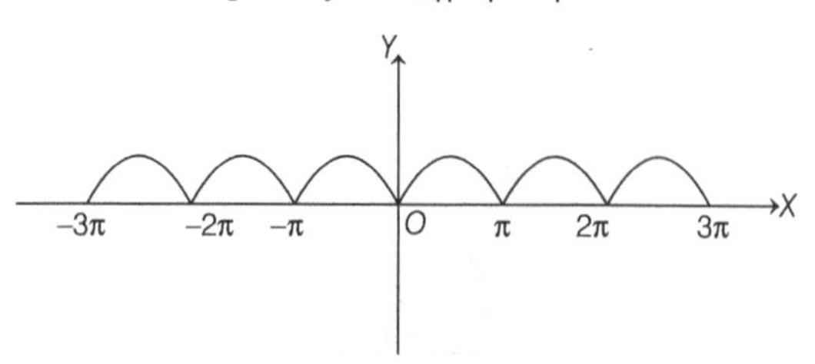

Graph of $y_1=|\sin||x|-\pi||$

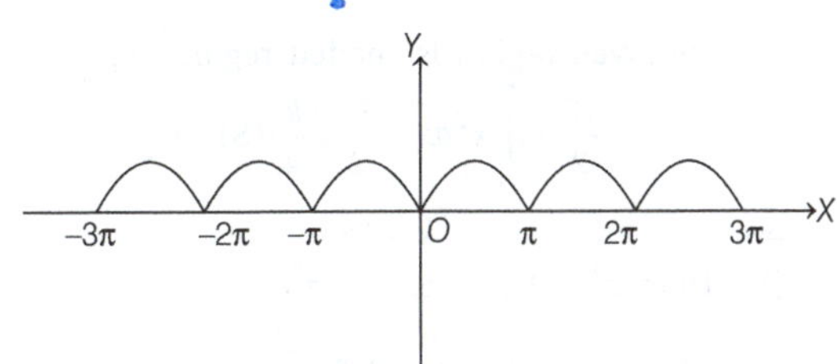

and graph of $y_2 = |\sin x|$

It is clear from the graphs, y_1 and y_2 are periodic with period π and area same.

and length of curves are also same. (P, Q, R, S)

(B) Periods of y_1 and y_2 are 2 and 1 respectively. Hence, graphs are not same. (T)

(C) Periods of y_1 and y_2 are π and 2π respectively. (P, T)

2. **(A)** $S_1 = S_3$

$$S_2 = \int_0^4 \left(2\sqrt{x} - \frac{x^2}{4}\right) dx$$

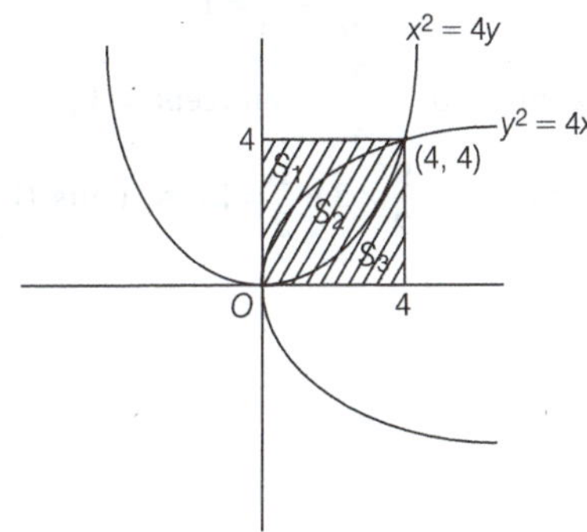

$$= \left\{\frac{4}{3}x^{3/2} - \frac{x^3}{12}\right\}_0^4$$

$$= \frac{32}{3} - \frac{16}{3} = \frac{16}{3} \text{ sq unit}$$

$$S_1 = S_3 = \frac{1}{2}\left(16 - \frac{16}{3}\right)$$

$$= 8 - \frac{8}{3} = \frac{16}{3}$$

$\therefore \quad S_1 = S_3 = \dfrac{16}{3} = S_2$

$\because \quad S_1 = S_2 = S_3$

$\therefore \quad S_1, S_2, S_3$ are in AP and GP. (P, Q, S)

(B) $\because S_n = \left|\int_{n\pi/\beta}^{(n+1)\pi/\beta} e^{-\alpha x} \sin \beta x \, dx\right|$

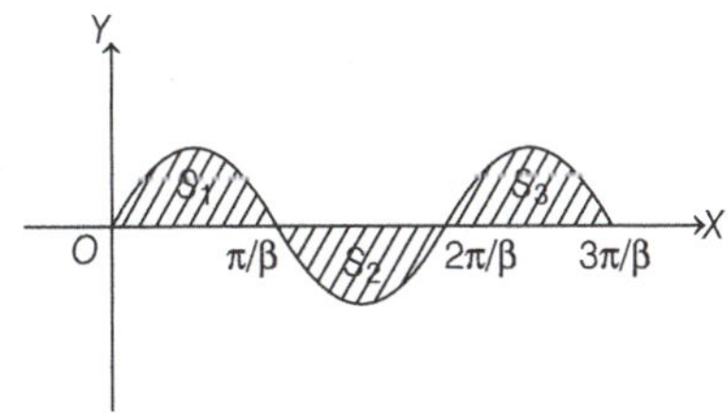

$$= \frac{1}{\beta}\left|\int_{n\pi}^{(n+1)\pi} e^{-\frac{\alpha}{\beta}x} \sin x \, dx\right|$$

$$= \frac{1}{\beta}\left|\int_0^{\pi} e^{-\frac{\alpha}{\beta}(x+n\pi)} \sin(n\pi + x) dx\right|$$

$$= \frac{1}{\beta}\left|\int_0^{\pi} e^{-\frac{\alpha(x+n\pi)}{\beta}} (-1)^n \sin x \, dx\right|$$

$$S_n = e^{\frac{-n\pi\alpha}{\beta}} I \text{ (say)}$$

$\therefore \quad \dfrac{S_2}{S_1} = \dfrac{S_3}{S_2} = e^{-\pi\alpha/\beta}$

$\therefore S_1, S_2, S_3$ are in GP.

$\Rightarrow \quad$ AM > GM

$\Rightarrow \quad \dfrac{S_1 + S_3}{2} > S_2 \Rightarrow S_1 + S_3 > 2S_2$

S_1^n, S_2^n, S_3^n are also in GP.

$\therefore \quad \dfrac{S_1^n + S_3^n}{2} > S_2^n \Rightarrow S_1^n + S_3^n > 2S_2^n$ (Q, R)

(C) $\quad S_1 = \left|\int_0^{1/3} (1-x)^2 \, dx\right|$

$$= \left|\left[-\frac{1}{3}(1-x)^3\right]_0^{1/3}\right|$$

$$= \left|-\frac{1}{3}\left\{\frac{8}{27} - 1\right\}\right| = \frac{19}{81}$$

$$S_2 = \left|\int_{1/3}^{2/3} 2x(1-x) \, dx\right|$$

$$= \left|x^2 - \frac{2x^3}{3}\right|_{1/3}^{2/3}$$

$$= \left(\frac{4}{9} - \frac{16}{81}\right) - \left(\frac{1}{9} - \frac{2}{81}\right) = \left|\frac{3}{9} - \frac{14}{81}\right| = \frac{13}{81}$$

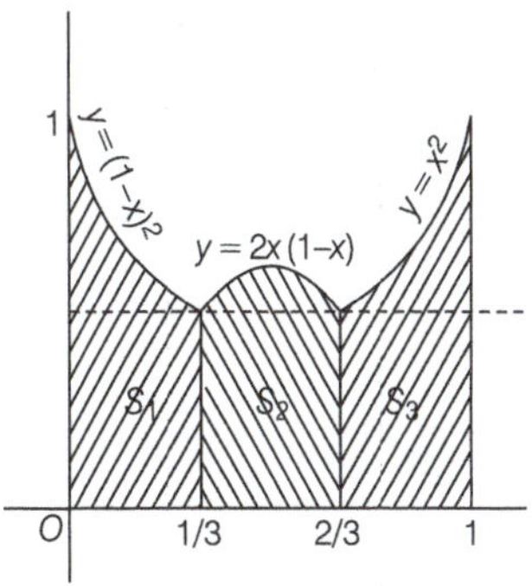

and $\quad S_3 = \left|\int_{2/3}^{1} x^2 dx\right| = \dfrac{1}{3}\left(1 - \dfrac{8}{27}\right) = \dfrac{19}{81}$

Hence, $S_1 = S_3 = \dfrac{19}{81}$ and $S_2 = \dfrac{13}{81}$

$\therefore S_1 + S_2 + S_3 = \dfrac{51}{81} = \dfrac{17}{27}$ (S,T)

3. **(A)** The area = 1 unit (T)

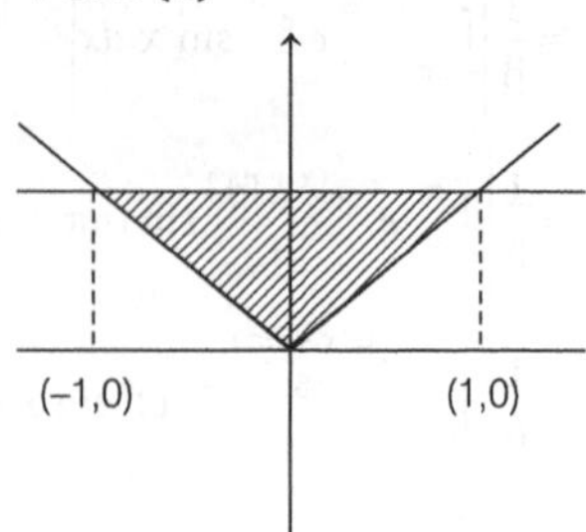

(B) Area enclosed $= \int_0^{\pi} \sin x \, dx = 2$ (P)

(C) The line $y = x + 2$ intersects $y = x^2$ at $x = -1$ and $x = 2$

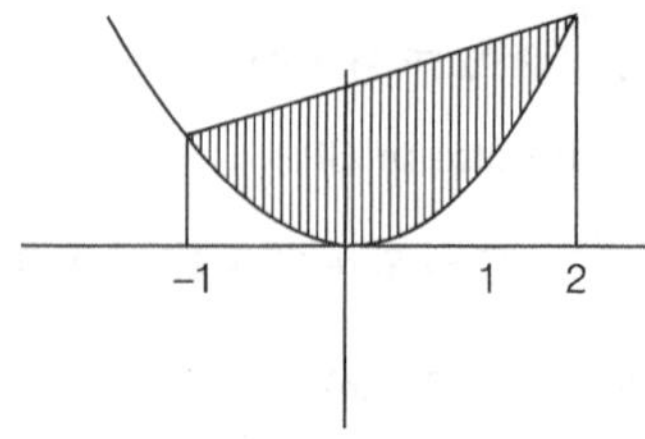

The given region is shaded region are

$$\frac{15}{2} - \int_{-1}^{2} x^2 dx = \frac{9}{2} = \frac{k}{4} \text{ (S)}$$

$$\Rightarrow \qquad k = 18$$

(D) Here $a^2 = 9$, $b^2 = 5$, $b^2 = a^2(1 - e^2)$

$$\Rightarrow \qquad e^2 = \frac{4}{9} \Rightarrow e = \frac{2}{3}$$

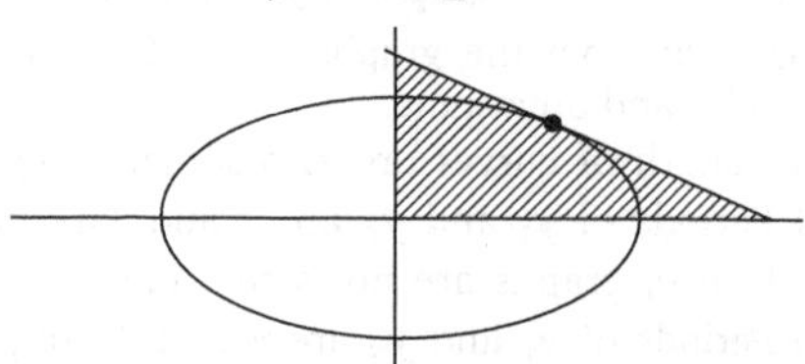

Equation of tangent at $\left(2, \frac{5}{3}\right)$ is

$$\frac{2x}{9} + \frac{y}{3} = 1$$

x-intercept $= \frac{9}{2}$, y-intercept $= 3$;

Area $= 4 \times \frac{9}{2} \times 3 \times \frac{1}{2} = 27$ sq units (R)

20

Differential Equations

Objective Questions Type I [Only one correct answer]

In each of the questions below, four choices are given of which only one is correct. You have to select the correct answer which is the most appropriate.

1. The order of the differential equation whose general solution is given by
$y = (c_1 + c_2)\cos(x + c_3) - c_4 \cdot e^{x+c_5}$, where
c_1, c_2, c_3, c_4, and c_5 are arbitrary constants, is

(a) 2 (b) 3
(c) 4 (d) 5

2. If $y = c_1 e^{2x} + c_2 e^x + c_3 e^{-x}$ satisfies the differential equation $\frac{d^3y}{dx^3} + a\frac{d^2y}{dx^2} + b\frac{dy}{dx} + cy = 0$, then $\frac{a^3 + b^3 + c^3}{abc}$ is equal to

(a) $-\frac{1}{4}$ (b) $-\frac{1}{2}$
(c) 0 (d) $\frac{1}{2}$

3. Solution of the differential equation
$x = 1 + xy\frac{dy}{dx} + \frac{(xy)^2}{2!}\left(\frac{dy}{dx}\right)^2 + \frac{(xy)^3}{3!}\left(\frac{dy}{dx}\right)^3 + \ldots$ is

(a) $y = \log_e x + C$
(b) $y = (\log_e x)^2 + C$
(c) $y = \pm\sqrt{(\log_e x)^2 + 2C}$
(d) $xy = x^y + k$

4. Let $y = y(x)$ be the solution of the differential equation, $\frac{2 + \sin x}{y + 1} \cdot \frac{dy}{dx} = -\cos x, y > 0, y(0) = 1$. If $y(\pi) = a$ and $\frac{dy}{dx}$ at $x = \pi$ is b, then the ordered pair (a, b) is equal to **[JEE Main 2020, 2 September]**

(a) (1, 1) (b) $\left(2, \frac{3}{2}\right)$
(c) (1, – 1) (d) (2, 1)

5. Solution of the differential equation
$\frac{dy}{dx} = \frac{y(x - y\ln y)}{x(x\ln x - y)}$ is

(a) $\frac{x\ln x + y\ln y}{xy} = c$ (b) $\frac{x\ln x - y\ln y}{xy} = c$
(c) $\frac{\ln x}{x} + \frac{\ln y}{y} = c$ (d) $\frac{\ln x}{x} - \frac{\ln y}{y} = c$

(where c is arbitrary constant)

6. Let $f(x)$ be continuously differentiable on the interval $(0, \infty)$ such that $f(1) = 1$, and $\lim_{t \to x} \frac{t^2 f(x) - x^2 f(1)}{t - x} = 1$ for each $x > 0$. then $f(x)$ is

(a) $\frac{1}{3x} + \frac{2x^2}{3}$
(b) $-\frac{1}{3x} + \frac{4x^2}{3}$
(c) $-\frac{1}{x} + \frac{2}{x^2}$
(d) $\frac{1}{x}$

7. The differential equation satisfied by the system of parabolas $y^2 = 4a(x + a)$ is **[JEE Main 2021, 18 March]**

(a) $y\left(\frac{dy}{dx}\right)^2 - 2x\left(\frac{dy}{dx}\right) - y = 0$
(b) $y\left(\frac{dy}{dx}\right)^2 - 2x\left(\frac{dy}{dx}\right) + y = 0$
(c) $y\left(\frac{dy}{dx}\right)^2 + 2x\left(\frac{dy}{dx}\right) - y = 0$
(d) $y\left(\frac{dy}{dx}\right) + 2x\left(\frac{dy}{dx}\right) - y = 0$

8. Solution of $\left(\frac{x+y-1}{x+y-2}\right)\frac{dy}{dx}=\left(\frac{x+y+1}{x+y+2}\right)$, given that $y=1$ when $x=1$ is

(a) $\ln\left|\frac{(x-y)^2-2}{2}\right|=2(x+y)$

(b) $\ln\left|\frac{(x+y)^2-2}{2}\right|=2(x-y)$

(c) $\ln\left|\frac{(x-y)^2+2}{2}\right|=2(x+y)$

(d) $\ln\left|\frac{(x+y)^2+2}{2}\right|=2(x-y)$

9. If the solution of $y^5x+y-x\frac{dy}{dx}=0$ is

(a) $\frac{y^5}{5}+\frac{y^4}{4x^4}=c$ (b) $\frac{x^4}{4}+\frac{x^5}{5y^5}=c$

(c) $\frac{y^4}{4}+\frac{y^5}{5x^5}=c$ (d) $\frac{x^5}{5}+\frac{x^4}{4y^4}=c$

(where c is arbitrary constant)

10. The solution of the equation $\frac{dy}{dx}+x(x+y)=x^3(x+y)^3-1$ is

(a) $\frac{1}{(x+y)^2}=x^2+1+ce^{x}$

(b) $\frac{1}{(x+y)}=x^2+1+ce^{x}$

(c) $\frac{1}{(x+y)^2}=x^2+1+ce^{x^2}$

(d) $\frac{1}{(x+y)}=x^2+1+ce^{x^2}$

(where c is arbitrary constant)

11. Solution of the differential equation $\left\{\frac{1}{x}-\frac{y^2}{(x-y)^2}\right\}dx+\left\{\frac{x^2}{(x-y)^2}-\frac{1}{y}\right\}dy=0$ is

(a) $\ln\left|\frac{x}{y}\right|+\frac{xy}{(x-y)}=c$

(b) $\ln|xy|+\frac{xy}{(x-y)}=c$

(c) $\frac{xy}{(x-y)}=ce^{x/y}$

(d) $\frac{xy}{(x-y)}=ce^{xy}$

(where c is arbitrary constant)

12. The real value of n for which the substitution $y=u^n$ will transform the differential equation $2x^4y\frac{dy}{dx}+y^4=4x^6$ into a homogeneous equation is

(a) 1/2 (b) 1
(c) 3/2 (d) 2

13. The equation of the curve in which the portion of Y-axis cut off between the origin and the tangent varies as the cube of the abscissa of the point of contact is

(a) $y=\frac{kx^3}{3}+cx$ (b) $y=-\frac{kx^2}{2}+c$

(c) $y=-\frac{kx^3}{2}+cx$ (d) $y=\frac{kx^3}{3}+\frac{cx^2}{2}$

(k is constant of proportionality)
(where c is arbitrary constant)

14. If $y=y(x)$ is the solution of the differential equation $\frac{dy}{dx}+(\tan x)y=\sin x,\ 0\le x\le\frac{\pi}{3}$, with $y(0)=0$, then $y\left(\frac{\pi}{4}\right)$ equal to **[JEE Main 2021, 16 March]**

(a) $\frac{1}{4}\log_e 2$ (b) $\left(\frac{1}{2\sqrt{2}}\right)\log_e 2$

(c) $\log_e 2$ (d) $\frac{1}{2}\log_e 2$

15. Let $y=y(x)$ be the solution of the differential equation $\frac{dy}{dx}=(y+1)((y+1)e^{x^2/2}-x),\ 0<x<2.1$, with $y(2)=0$. Then, the value of $\frac{dy}{dx}$ at $x=1$ is equal to **[JEE Main 2021, 18 March]**

(a) $\frac{-e^{3/2}}{(e^2+1)^2}$ (b) $-\frac{2e^2}{(1+e^2)^2}$

(c) $\frac{e^{5/2}}{(1+e^2)^2}$ (d) $\frac{5e^{1/2}}{(e^2+1)^2}$

16. If $y=f(x)$ passing through (1, 2) satisfies the differential equation $y(1+xy)\,dx-x\,dy=0$, then

(a) $f(x)=\frac{2x}{2-x^2}$ (b) $f(x)=\frac{x+1}{x^2+1}$

(c) $f(x)=\frac{x-1}{4-x^2}$ (d) $f(x)=\frac{4x}{1-2x^2}$

17. The solution of the differential equation $y'''-8y''=0$, where $y(0)=\frac{1}{8}$, $y'(0)=0$, $y''(0)=1$, is

(a) $y=\frac{1}{8}\left(\frac{e^{8x}}{8}+x-\frac{7}{9}\right)$ (b) $y=\frac{1}{8}\left(\frac{e^{8x}}{8}+x+\frac{7}{8}\right)$

(c) $y=\frac{1}{8}\left(\frac{e^{8x}}{8}-x+\frac{7}{8}\right)$ (d) None of these

18. If $y=y(x)$ satisfies the differential equation $8\sqrt{x}(\sqrt{9+\sqrt{x}})\,dy=\left(\sqrt{4+\sqrt{9+\sqrt{x}}}\right)^{-1}dx,\ x>0$ and $y(0)=\sqrt{7}$, then $y(256)=$ **[JEE Advanced 2017]**

(a) 16 (b) 3
(c) 9 (d) 80

19. Let a and b be respectively the degree and order of the differential equation of the family of circles touching the lines $y^2 - x^2 = 0$ and lying in the first and second quadrant, then

(a) $a = 1, b = 2$ (b) $a = 1, b = 1$
(c) $a = 2, b = 1$ (d) $a = 2, b = 2$

20. The general solution of the differential equation $\sqrt{1 + x^2 + y^2 + x^2y^2} + xy\dfrac{dy}{dx} = 0$ is
(where C is a constant of integration)
[JEE Main 2020, 6 September]

(a) $\sqrt{1+y^2} + \sqrt{1+x^2} = \dfrac{1}{2}\log_e\left(\dfrac{\sqrt{1+x^2}+1}{\sqrt{1+x^2}-1}\right) + C$

(b) $\sqrt{1+y^2} - \sqrt{1+x^2} = \dfrac{1}{2}\log_e\left(\dfrac{\sqrt{1+x^2}+1}{\sqrt{1+x^2}-1}\right) + C$

(c) $\sqrt{1+y^2} + \sqrt{1+x^2} = \dfrac{1}{2}\log_e\left(\dfrac{\sqrt{1+x^2}-1}{\sqrt{1+x^2}+1}\right) + C$

(d) $\sqrt{1+y^2} - \sqrt{1+x^2} = \dfrac{1}{2}\log_e\left(\dfrac{\sqrt{1+x^2}-1}{\sqrt{1+x^2}+1}\right) + C$

21. A differential equation associated with the primitive $y = a + be^{5x} + ce^{-7x}$ is

(a) $y_3 + 2y_2 - y_1 = 0$ (b) $y_3 + 2y_2 - 35y_1 = 0$
(c) $4y_3 + 5y_2 - 20y_1 = 0$ (d) none of these

22. Solution of the differential equation, $\dfrac{\sqrt{x}\,dx + \sqrt{y}\,dx}{\sqrt{x}\,dx - \sqrt{y}\,dx} = \sqrt{\left(\dfrac{y^3}{x^3}\right)}$ is $\ln(f(x, y)) + \tan^{-1}$ $(g(x, y)) = c$, where c is an arbitrary constant, then

(a) $f(x, y) = (x^{3/2} + y^{3/2}),\ g(x, y) = \left(\dfrac{y}{x}\right)$

(b) $f(x, y) = (x^{3/2} + y^{3/2}),\ g(x, y) = \left(\dfrac{x}{y}\right)$

(c) $f(x, y) = \sqrt{(x^3 + y^3)},\ g(x, y) = \sqrt{\left(\dfrac{y^3}{x^3}\right)}$

(d) $f(x, y) = \sqrt{(x^3 + y^3)},\ g(x, y) = \sqrt{\left(\dfrac{x^3}{y^3}\right)}$

23. Let $y = y(x)$ be the solution of the differential equation $\cos x\,(3\sin x + \cos x + 3)\,dy$ $= (1 + y\sin x(3\sin x + \cos x + 3))\,dx,\ 0 \le x \le \dfrac{\pi}{2}$, $y(0) - 0$. Then, $y\left(\dfrac{\pi}{3}\right)$ is equal to
[JEE Main 2021, 17 March]

(a) $2\log_e\left(\dfrac{2\sqrt{3}+9}{6}\right)$ (b) $2\log_e\left(\dfrac{2\sqrt{3}+10}{11}\right)$

(c) $2\log_e\left(\dfrac{\sqrt{3}+7}{2}\right)$ (d) $2\log_e\left(\dfrac{3\sqrt{3}-8}{4}\right)$

24. The form of the differential equation of all central conics is

(a) $x = y\dfrac{dy}{dx}$

(b) $x + y\dfrac{dy}{dx} = 0$

(c) $x\left(\dfrac{dy}{dx}\right)^2 + xy\dfrac{d^2y}{dx^2} = y\dfrac{dy}{dx}$

(d) None of the above

25. Let a solution $y = y(x)$ of the differential equation, $dy + xy\,dx = x\,dx$ satisfy $y(0) = 2$, then the area enclosed the curve $y = y(x)$, $x = 0$ and $y = 1$ in the first quadrant is

(a) $\int_0^\infty e^{\frac{-x^2}{2}}\,dx$ (b) $2\int_0^\infty e^{\frac{-x^2}{2}}\,dx$

(c) $\int_0^\infty\left(1 + e^{\frac{-x^2}{2}}\right)dx$ (d) $2\int_0^\infty\left(1 + e^{\frac{-x^2}{2}}\right)dx$

26. The particular solution of $\ln\left(\dfrac{dy}{dx}\right) = 3x + 4y,\ y(0) = 0$ is

(a) $3e^{3x} + 4e^{-4y} = 7$ (b) $3e^{4y} - 4e^{-3x} = 7$
(c) $4e^{3x}x + 3e^{-4y} = 7$ (d) none of these

27. If the slope of tangent to the curve is maximum at $x = 1$ and curve has a minimum value 1 at $x = 0$, then the curve which also satisfies the equation $y''' = 4x - 3$ is

(a) $y + 2x + \dfrac{dy}{dx} = 0$ (b) $y = 1 + \dfrac{x^2}{2} - \dfrac{x^3}{2} + \dfrac{x^4}{6}$
(c) $y = 1 + x + x^2 + x^3$ (d) None of these

28. The solution of differential equation $yy' = x\left(\dfrac{y^2}{x^2} + \dfrac{f(y^2/x^2)}{f'(y^2 + x^2)}\right)$ is

(a) $f(y^2/x^2) = cx^2$
(b) $x^2 f(y^2/x^2) = c^2y^2$
(c) $x^2 f(y^2/x^2) = c$
(d) $f(y^2/x^2) = cy/x$

29. Through any point (x, y) of a curve which passes through the origin, lines are drawn parallel to the coordinate axes. The curve, given that it divides the rectangle formed by the two lines and the axes into two areas, one of which is twice the other, represents a family of

(a) circle (b) parabola
(c) ellipse (d) hyperbola

30. If $y = y(x)$ is the solution of the differential equation, $e^y\left(\dfrac{dy}{dx} - 1\right) = e^x$ such that $y(0) = 0$, then $y(1)$ is equal to
[JEE Main 2020, 6 September]

(a) $2 + \log_e 2$ (b) $2e$
(c) $1 + \log_e 2$ (d) $\log_e 2$

31. Let f be a non-negative function defined on the interval [0, 1]. If $\int_0^x \sqrt{1-(f'(t))^2}\, dt = \int_0^x f(t)\, dt$, $0 \le x \le 1$ and $f(0) = 0$, then

(a) $f(1/2) < 1/2$ and $f(1/3) > 1/3$
(b) $f(1/2) > 1/2$ and $f(1/3) > 1/3$
(c) $f(1/2) < 1/2$ and $f(1/3) < 1/3$
(d) $f(1/2) > 1/2$ and $f(1/3) < 1/3$

32. The slope of the tangent at (x, y) to a curve passing through $(1, \pi/4)$ is given by $\frac{y}{x} - \cos^2\left(\frac{y}{x}\right)$, then the equation of the curve is

(a) $y = \tan^{-1}\left(\ln\left(\frac{e}{x}\right)\right)$ (b) $y = x\tan^{-1}\left(\ln\left(\frac{x}{e}\right)\right)$
(c) $y = x\tan^{-1}\left(\ln\left(\frac{e}{x}\right)\right)$ (d) None of these

33. If the curve $y = y(x)$ is the solution of the differential equation
$2(x^2 + x^{5/4})dy - y(x + x^{1/4})dx = 2x^{9/4}dx$, $x > 0$ which passes through the point $\left(1, 1 - \frac{4}{3}\log_e 2\right)$, then the value of $y(16)$ is equal to **[JEE Main 2021, 17 March]**

(a) $4\left(\frac{31}{3} + \frac{8}{3}\log_e 3\right)$ (b) $\left(\frac{31}{3} + \frac{8}{3}\log_e 3\right)$
(c) $4\left(\frac{31}{3} - \frac{8}{3}\log_e 3\right)$ (d) $\left(\frac{31}{3} - \frac{8}{3}\log_e 3\right)$

34. If $\phi(x) = \int \{\phi(x)\}^{-2} dx$ and $\phi(1) = 0$, then $\phi(x)$ is equal to

(a) $\{2(x-1)\}^{1/4}$ (b) $\{5(x-2)\}^{1/5}$
(c) $\{3(x-1)\}^{1/3}$ (d) None of these

35. Let $y = y(x)$ be the solution of the differential equation, $xy' - y = x^2(x\cos x + \sin x)$, $x > 0$. If $y(\pi) = \pi$, then $y''\left(\frac{\pi}{2}\right) + y\left(\frac{\pi}{2}\right)$ is equal to **[JEE Main 2020, 4 September]**

(a) $2 + \frac{\pi}{2} + \frac{\pi^2}{4}$ (b) $1 + \frac{\pi}{2}$
(c) $1 + \frac{\pi}{2} + \frac{\pi^2}{4}$ (d) $2 + \frac{\pi}{2}$

36. Solution of the differential equation $\sin y \frac{dy}{dx} = \cos y(1 - x\cos y)$ is

(a) $\sec y = x - 1 - ce^x$ (b) $\sec y = x + 1 + ce^x$
(c) $\sec y = x + e^x + c$ (d) None of these
(where c is arbitrary constant)

37. If $y = y(x)$ is the solution of the differential equation $\frac{dy}{dx} = (\tan x - y)\sec^2 x$, $x \in \left(-\frac{\pi}{2}, \frac{\pi}{2}\right)$, such that $y(0) = 0$, then $y\left(-\frac{\pi}{4}\right)$ is equal to **[JEE Main 2019. 6 September]**

(a) $\frac{1}{e} - 2$ (b) $\frac{1}{2} - e$
(c) $2 + \frac{1}{e}$ (d) $e - 2$

Objective Questions Type II [One or more than one correct answer(s)]

In each of the questions below four choices of which one or more than one are correct. You have to select the correct answer(s) accordingly.

1. Solution of the differential equation $x\cos x\left(\frac{dy}{dx}\right) + y(x\sin x + \cos x) = 1$ is

(a) $xy = \sin x + c\cos x$
(b) $xy\sec x = \tan x + c$
(c) $xy + \sin x + c\cos x = 0$
(d) none of the above
(where c is arbitrary constant)

2. Solution of the differential equation $(1 + y^2)\,dx = (\tan^{-1} y - x)\,dy$ is

(a) $xe^{\tan^{-1} y} = (1 - \tan^{-1} y)\,e^{\tan^{-1} y} + c$
(b) $xe^{\tan^{-1} y} = (\tan^{-1} y - 1)\,e^{\tan^{-1} y} + c$
(c) $x = e^{\tan^{-1} y} - 1 + ce^{-\tan^{-1} y}$
(d) None of the above
(where c is arbitrary constant)

3. Let $\frac{dy}{dx} + y = f(x)$ where y is a continuous function of x with $y(0) = 1$ and $f(x) = \begin{cases} e^{-x} & \text{if } 0 \le x \le 2 \\ e^{-2} & \text{if } x > 2 \end{cases}$. Which of the following hold(s) good?

(a) $y(1) = 2e^{-1}$ (b) $y'(1) = -e^{-1}$
(c) $y(3) = -2e^{-3}$ (d) $y'(3) = -2e^{-3}$

4. Solution of the equation $\frac{dy}{dx} + \frac{1}{x}\tan y = \frac{1}{x^2}\tan y \sin y$ is

(a) $2x = \sin y\,(1 + 2cx^2)$
(b) $2x = \sin y\,(1 + cx^2)$
(c) $2x + \sin y\,(1 + cx^2) = 0$
(d) None of the above
(where c is arbitrary constant)

5. Solution of the differential equation $\frac{dy}{dx} = \frac{x+y+7}{2x+2y+3}$ is
(a) $6(x+y)+11\ln(3x+3y+10) = 9x+c$
(b) $6(x+y)-11\ln(3x+3y+10) = 9x+c$
(c) $6(x+y)-11\ln(x+y+10/3) = 9x+c$
(d) None of the above
(where c is arbitrary constant)

6. Let b be a nonzero real number. Suppose $f:\mathbf{R}\to\mathbf{R}$ is a differentiable function such that $f(0)=1$. If the derivative f' of f satisfies the equation
$$f'(x) = \frac{f(x)}{b^2+x^2}$$
for all $x \in R$, then which of the following statements is/are TRUE? **[JEE Advanced 2020]**
(a) If $b>0$, then f is an increasing function
(b) If $b<0$, then f is a decreasing function
(c) $f(x)f(-x)=1$ for all $x\in R$
(d) $f(x)-f(-x)=0$ for all $x \in R$

7. If $f(x), g(x)$ is twice differential functions on $[0, 2]$ satisfying $f''(x)=g''(x)$, $f'(1)=2g'(1)=4$ and $f(2)=3g(2)=9$, then
(a) $f(4)-g(4)=10$
(b) $|f(x)-g(x)|<2 \Rightarrow -2<x<0$
(c) $f(2)=g(2)\Rightarrow x=-1$
(d) $f(x)-g(x)=2x$ has real root

8. The curve for which the area of the triangle formed by the X-axis, the tangent line and radius vector of the point of tangency is equal to a^2 is
(a) $x = cy + \frac{a^2}{y}$ (b) $y = x - cx^2$
(c) $y = cx + \frac{a^2}{x}$ (d) $x = cy - \frac{a^2}{y}$
(where c is arbitrary constant)

9. A normal is drawn at a point $P(x, y)$ of a curve. It meets the X-axis at Q. If PQ is of constant length k. Such a curve passing through $(0, k)$ is
(a) a circle with centre $(0, 0)$
(b) $x^2+y^2=k^2$
(c) $(1-k)x^2+y^2=k^2$
(d) $x^2+(1+k^2)y^2=k^2$

10. The solution of $y'(x^2y^3+xy)=1$ is
(a) $\frac{1}{x} = 2-y^2+ce^{-y^2/2}$
(b) $e^{y^2/2}\left(\frac{1-2x}{x}+y^2\right)=c$
(c) the solution of an equation which is reducible to linear equation
(d) $2/x = 1-y^2+e^{-y/2}$
(where c is arbitrary constant)

11. The tangent at any point P on $y=f(x)$ meets X-axis and Y-axis at A and B respectively. If $PA : PB = 2:1$, then the equation of the curve, is
(a) $|x|y=c$ (b) $x^2|y|=c$
(c) $|x|y^2=c$ (d) $x=cy^2$
(where c is arbitrary constant)

12. The solution of $\left(\frac{xdx+ydy}{xdy-ydx}\right) = \sqrt{\left(\frac{a^2-x^2-y^2}{x^2+y^2}\right)}$ is
(a) $\sqrt{(x^2+y^2)} = a\sin\{(\tan^{-1}y/x)+\text{constant}\}$
(b) $\sqrt{(x^2+y^2)} = a\cos\{(\tan^{-1}y/x)+\text{constant}\}$
(c) $\sqrt{(x^2+y^2)} = a\{\tan(\sin^{-1}y/x)+\text{constant}\}$
(d) $y = x\tan\{\text{constant}+\sin^{-1}\frac{1}{a}\sqrt{(x^2+y^2)}\}$

13. The solution of $\left(\frac{dy}{dx}\right)^2 + 2y\cot x\frac{dy}{dx} = y^2$ is
(a) $y + \frac{c}{1+\cos x} = 0$ (b) $y = \frac{c}{1-\cos x}$
(c) $x = 2\sin^{-1}\sqrt{c/2y}$ (d) None of these

14. Let $f:[0,\infty)\to R$ be a continuous function such that $f(x) = 1-2x+\int_0^x e^{x-t}f(t)\,dt$ for all $x\in[0,\infty)$. Then, which of the following statement(s) is (are) TRUE? **[JEE Advanced 2018]**
(a) The curve $y=f(x)$ passes through the point $(1, 2)$
(b) The curve $y=f(x)$ passes through the point $(2, -1)$
(c) The area of the region $\{(x,y)\in[0,1]\times R : f(x)\le y\le\sqrt{1-x^2}\}$ is $\frac{\pi-2}{4}$
(d) The area of the region $\{(x,y)\in[0,1]\times R : f(x)\le y\le\sqrt{1-x^2}\}$ is $\frac{\pi-1}{4}$

15. The solution of $\left(\frac{dy}{dx}\right)^2 + 2y\cot x\frac{dy}{dx} = y^2$ is
(a) $x = 2\sin^{-1}\sqrt{\left(\frac{c}{2y}\right)}$ (b) $x = 2\cos^{-1}\sqrt{\left(\frac{c}{2y}\right)}$
(c) $y = \frac{c}{1-\cos x}$ (d) $y - \frac{c}{1+\cos x} = 0$
(where c is arbitrary constant)

16. Let Γ denote a curve $y=y(x)$ which is in the first quadrant and let the point $(1, 0)$ lie on it. Let the tangent to Γ at a point P intersect the Y-axis at Y_P. If PY_P has length 1 for each point P on Γ, then which of the following options is/are correct? **[JEE Advanced 2019]**
(a) $xy' + \sqrt{1-x^2} = 0$ (b) $xy' - \sqrt{1-x^2} = 0$
(c) $y = \log_e\left(\frac{1+\sqrt{1-x^2}}{x}\right) - \sqrt{1-x^2}$
(d) $y = -\log_e\left(\frac{1+\sqrt{1-x^2}}{x}\right) + \sqrt{1-x^2}$

Linked-Comprehension Type

In these questions, a passage (paragraph) has been given followed by questions based on each of the passages. You have to answer the questions based on the passage given.

PASSAGE 1

Let $y = f(x)$ satisfies the equation

$$f(x) = (e^{-x} + e^{x})\cos x - 2x - \int_0^x (x-t)f'(t)dt.$$

1. y satisfies the differential equation
(a) $\frac{dy}{dx} + y = e^x(\cos x - \sin x) - e^{-x}(\cos x + \sin x)$
(b) $\frac{dy}{dx} - y = e^x(\cos x - \sin x) + e^{-x}(\cos x + \sin x)$
(c) $\frac{dy}{dx} + y = e^x(\cos x + \sin x) - e^{-x}(\cos x - \sin x)$
(d) $\frac{dy}{dx} - y = e^x(\cos x - \sin x) + e^{-x}(\cos x - \sin x)$

2. The value of $f'(0) + f''(0)$ equals
(a) -1 b) 2
(c) 1 (d) 0

3. $f(x)$ as a function of x equals
(a) $e^{-x}(\cos x - \sin x) + \frac{e^x}{5}(3\cos x + \sin x) + \frac{2}{5}e^{-x}$
(b) $e^{-x}(\cos x + \sin x) + \frac{e^x}{5}(3\cos x - \sin x) - \frac{2}{5}e^{-x}$
(c) $e^{-x}(\cos x - \sin x) + \frac{e^x}{5}(3\cos x - \sin x) + \frac{2}{5}e^{-x}$
(d) $e^{-x}(\cos x + \sin x) + \frac{e^x}{5}(3\cos x - \sin x) - \frac{2}{5}e^{-x}$

PASSAGE 2

The **Order** of the differential equation is the order of the highest derivative appearing in the equation and the **Degree** of a differential equation which can be written as polynomial in the derivatives in the degree of the derivative of the highest order occuring in it, after it has been expressed in a form free from radicals and fractions and if differential equation can not be written as a polynomial in the derivatives, then degree deos not defined but order defined.

On the basis of above information, answer the following questions

1. The degree of the differential equation
$$x = 1 + \left(xy\frac{dy}{dx}\right) + \frac{x^2y^2}{2!}\left(\frac{dy}{dx}\right)^2 + \frac{(xy)^3}{3!}\left(\frac{dy}{dx}\right)^3 + \ldots \text{ is}$$
(a) 0 (b) 1
(c) 2 (d) not defined

2. The degree of the differential equation
$$\left(\frac{d^3y}{dx^3}\right)^{2/3} + 3\frac{d^2y}{dx^2} + \frac{dy}{dx} + 5 = 0 \text{ is}$$
(a) 1 (b) 2
(c) 3 (d) None of these

3. The order and degree of the differential equation whose solution is $y = cx + c^3 - 5c^{3/2} + 3$ where c is arbitrary constants, are
(a) 1 and 3 (b) 1 and 4
(c) 1 and 5 (d) 1 and 6

4. The order and degree of the differential equation whose solution is
$c(y + c)^2 = x^3$, where c is arbitrary constants, are
(a) 1 and 2 (b) 1 and 3
(c) 1 and 4 (d) 1 and 5

5. The order and degree of the differential equation $\frac{d^2y}{dx^2} = \cos\left(\frac{dy}{dx}\right) + xy$ are respectively
(a) 2, 1 (b) 2, 0
(c) 2, infinite (d) 2, not defined

PASSAGE 3

Newton's law of cooling states that the rate at which a substance cools in moving air is proportional to the difference between the temperatures of the substance and that of the air. If the temperature of the air is 290K.

We can write it as $\frac{dT}{dt} = -k(T - 290)$, $k > 0$ constants.

where T is temperature of substance.

On the basis of above information, answer the following questions

1. The substance cools from 370 K to 330 K in 10 min, then
(a) $T = 290 + 160e^{-kt}$ (b) $T = 290 + 80e^{-kt}$
(c) $T = 290 + 40e^{-kt}$ (d) $T = 290 + 20e^{-kt}$

2. The value of k must be
(a) $\ln 2$ (b) $\frac{\ln 2}{40}$
(c) $\frac{\ln 2}{20}$ (d) $\frac{\ln 2}{10}$

3. The time taken when temperature be 295 K is
(a) 10 min (b) 20 min
(c) 40 min (d) 80 min

4. If $T = f(t)$, then the number of solutions of $|T| = |f(|t|)|$ and $T^2 + t^2 = 1600$ is

(a) 2 (b) 4
(c) 6 (d) 8

5. The value of $f\left(\frac{-2}{k}\right)$ must be

(a) $290 + 80e^2$ (b) $290 + 40e^2$
(c) $290 + 20e^2$ (d) $290 + 10e^2$

Numerical Value Type

Solve the following problems and write your answer.

1. A particle moves in a straight line with a velocity given by $\frac{dx}{dt} = x + 1$ (x is the distance travelled). If the time taken by a particle to traverse a distance of 99 m is λ, then the value of $20\lambda \log_{10} e$ must be

2. The solution of the differential equation $\frac{d^3y}{dx^3} - 8\frac{d^2y}{dx^2} = 0$ satisfying $y(0) = \frac{1}{8}$, $y_1(0) = 0$ and $y_2(0) = 1$ is $y_1 = \frac{e^{8x} - 8x + 7}{\lambda}$, then, the numerical value of λ must be

3. The differential equation whose solution represents the family $y = ae^{3x} + be^{5x}$ is $\frac{d^2y}{dx^2} - 8\frac{dy}{dx} + \lambda y = 0$, then the value of λ must be

4. If solution of the differential equation $\frac{dy}{dx} + \frac{\cos x\,(3\cos y - 7\sin x - 3)}{\sin y\,(3\sin x - 7\cos y + 7)} = 0$ is $(\sin x + \cos y - 1)^{\lambda} \cdot (\sin x - \cos y + 1)^{\mu} = c$, where c is arbitrary constant, then the value of λ/μ is equal to

5. Let $\frac{dy}{dx} = \frac{y\phi(x) - y^2}{\phi(x)}$, where $\phi(x)$ is a specified function satisfying $\phi(1) = 1$, $\phi(4) = 1296$. If $y(1) = 1$, then $y(4)$ is equal to

6. Let $f: R \to R$ be a differentiable function with $f(0) = 0$. If $y = f(x)$ satisfies the differential equation $\frac{dy}{dx} = (2 + 5y)(5y - 2)$, then the value of $\lim_{x \to -\infty} f(x)$ is .... **[JEE Advanced 2018]**

Single Integer Type

Solve the following problems and write your answer.

1. If the solution of the differential equation $\frac{dy}{dx} = \frac{1}{x\cos y + \sin 2y}$ is $x = ce^{\sin y} - k(1 + \sin y)$, then $k =$

2. If for $x \geq 0$, $y = y(x)$ is the solution of the differential equation, $(x+1)dy = ((x+1)^2 + y - 3)dx$, $y(2) = 0$, then $y(3)$ is equal to...... . **[JEE Main 2020, 9 January]**

3. If solution of the differential equation $\frac{dy}{dx} + y + \int_0^1 y\,dx$ and $y(0) = 1$ is $y(x)$, then the value of $[y(1) - 7]$ (where [·] denotes the greatest integer function) is

4. Let f be a continuous function satisfying the equation $\int_0^x f(t)\,dt + \int_0^x t\,f(x-t)\,dt = e^{-x} - 1$, then find the value of $e^{10} f(10)$.

5. Let $y = y(x)$ be the solution of the differential equation $xdy - ydx = \sqrt{(x^2 - y^2)}\,dx$, $x \geq 1$, with $y(1) = 0$. If the area bounded by the line $x = 1$, $x = e^{x}$, $y = 0$ and $y = y(x)$ is $\alpha e^{2\pi} + \beta$, then the value of $10(\alpha + \beta)$ is the equal to ....

6. Let $y = f(x)$ defined in [0, 2] satisfies the differential equation $y^3 y'' + 1 = 0$ where $f(x) \geq 0\ \forall\ x \in D_f$ and $f'(1) = 0$, $f(1) = 1$ then find the maximum value of $f(x)$. [**Note** : D_f denotes the domain of the function and y'' denotes the 2nd derivative of y w.r.t. x.]

7. If $x \log x \frac{dy}{dx} + y = \log x^2$ and $y(e) = 0$, then $2y(e^2)$ is equal to ....

8. If the curve $y = f(x)$ passing through the point (1, 2) and satisfies the differential equation $xdy + (y + x^3y^2)dx = 0$, then $x^3 y$ is equal to

9. If the solution of the differential equation $\sec^2 y \frac{dy}{dx} + 2x\tan y = x^3$ is $2\tan y = \lambda(x^2 - 1) + ce^{-x^2}$, c is arbitrary constant, then the numerical value of λ must be

Matrix-Match Type

Given below are Matching Type Questions, with two columns (each having some items) each . Each item of Column I has to be matched with the items of Column II, by encircling the correct match(es).

NOTE *An item of Column I can be matched with more than one items of Column II.*

1. Observe the following columns :

	Column I		Column II
(A)	The differential equation formed by differentiating and eliminating the constants from	(P)	$O + 2D = 5$
	$y = a\sin^2 x + b\cos^2 x + c\sin 2x + d\cos 2x$, where a, b, c, d are arbitrary constants.	(Q)	$2O + 3D = 5$
	If order and degree of the differential equation represented by O and D, then	(R)	$O = D$
(B)	The order and degree of the differential equation, whose general solution is given by	(S)	$O^D + D^O = 4$
	$y = (c_1 + c_2)\sin(x + c_3) - c_4 e^{x + c_5 + c_6}$, where $c_1, c_2, c_3, c_4, c_5, c_6$ are arbitrary constants, are O and D, then	(T)	$2^O + 3^D = 11$
(C)	The order and degree of the differential equation satisfying $\sqrt{(1 + x^2)} + \sqrt{(1 + y^2)} = A(x\sqrt{(1 + y^2)} + y\sqrt{(1 + x^2)})$ are O and D, then		

2. Observe the following columns :

	Column I		Column II
(A)	Solutions of the differential equation $\left(\frac{dy}{dx}\right)^2 - \frac{dy}{dx}(e^x + e^{-x}) + 1 = 0$ are given by	(P)	$x + y + 2 = ce^y$, (where c is arbitrary constant)
		(Q)	$\ln(x + y + 2) = c + y$, (where c is arbitrary constant)
(B)	Solutions of the differential equation $(x + y + 1)dy = dx$ are given by	(R)	$y + e^{-x} = c$, (where c is arbitrary constant)
(C)	Solutions of the differential equation $(x + y + 2)\,dx + (2x + 2y - 1)\,dy = 0$ are given by	(S)	$2(x + y + 2) + 5\ln(x + y - 3) = x + c$, (where c is arbitrary constant)
		(T)	$y - e^x = c$, (where c is arbitrary constant)

3. Match the following lists

	List I		List II
A.	If the function $y = e^{4x} + 2e^{-x}$ is a solution of the differential equation $\frac{\frac{d^3y}{dx^3} - 13\frac{dy}{dx}}{y} = K$, then the value of $K/3$ is	P.	3
B.	Let a curve $y = f(x)$, $f(x) \geq 0$ $\forall\, x \in R$ have property that for every point P on the curve, length of subnormal is equal to abscissa of P.If $f(1) = 3$, then $\frac{f(4)}{\sqrt{6}} =$	Q.	4
C.	If real value of m for which the substitution, $y = u^m$ will transform the differential equation, $2x^4 y\frac{dy}{dx} + y^4 = 4x^6$ into a homogeneous equation, then the value of $2m$ is	R.	2
D.	If the solution of differential equation $x^2\frac{d^2y}{dx^2} + 2x\frac{dy}{dx} = 12y$ is $y = Ax^m + Bx^{-n}$, then possible value of $\|m + n\|$ is	S.	0
		T.	8

Answers

Objective Questions Type I [Only one correct answer]

1. (b)	**2.** (a)	**3.** (c)	**4.** (a)	**5.** (a)	**6.** (a)	**7.** (c)	**8.** (b)	**9.** (d)	**10.** (c)
11. (a)	**12.** (c)	**13.** (c)	**14.** (b)	**15.** (a)	**16.** (a)	**17.** (c)	**18.** (b)	**19.** (c)	**20.** (a)
21. (b)	**22.** (c)	**23.** (b)	**24.** (c)	**25.** (a)	**26.** (c)	**27.** (b)	**28.** (a)	**29.** (b)	**30.** (c)
31. (c)	**32.** (c)	**33.** (c)	**34.** (c)	**35.** (d)	**36.** (b)	**37.** (d)			

Objective Questions Type II [One or more than one correct answer(s)]

1. (a, b)	**2.** (b, c)	**3.** (a, b, d)	**4.** (a, b)	**5.** (b, c)
6. (a, c)	**7.** (a, b, c)	**8.** (a, d)	**9.** (a, b)	**10.** (a, b, c)
11. (c, d)	**12.** (a, d)	**13.** (b, c)	**14.** (b, c)	**15.** (a, b, c, d)
16. (a, c)				

Linked-Comprehension Type

Passage 1 **1.** (a) **2.** (d) **3.** (c)
Passage 2 **1.** (b) **2.** (b) **3.** (d) **4.** (b) **5.** (d)
Passage 3 **1.** (b) **2.** (d) **3.** (c) **4.** (a) **5.** (a)

Numerical Grid-Based Problems

1. (40) **2.** (64) **3.** (15) **4.** (2.5) **5.** (324)
6. (0.4)

Single Integer type Questions

1. (2) **2.** (3) **3.** (6) **4.** (9) **5.** (4) **6.** (1)
7. (3) **8.** (2) **9.** (1)

Matrix-Match Type

1. A → (P, S, T); B → (R, S, T); C → (Q, R)
2. A → (R, T); B → (Q, P); C → (S)
3. A → Q, B → R C → P, D → S

Solutions

Objective Questions Type [Only one correct answer]

1. $\because \quad y = (c_1 + c_2)\cos(x + c_3) - (c_4 e^{c_5})e^x$

$= A\cos(x + c_3) - B\,e^x \text{ (let } c_1 + c_2 = A \text{ and } c_4 e^{c_5} = B)$

There are three independent arbitrary constants. Thus differential equation will be of order 3.

2. $\because \quad y = c_1 e^{2x} + c_2 e^x + c_3 e^{-x}$

$\Rightarrow \quad e^x y = c_1 e^{3x} + c_2 e^{2x} + c_3$

Differentiating both sides w.r.t. x, then

$$e^x \frac{dy}{dx} + y \cdot e^x = 3c_1 e^{3x} + 2c_2 e^{2x} + 0$$

$$\Rightarrow \quad e^{-x}\left(\frac{dy}{dx} + y\right) = 3c_1 e^x + 2c_2$$

Again differentiating both sides w.r.t x, then

$$e^{-x}\left(\frac{d^2y}{dx^2} + \frac{dy}{dx}\right) + \left(\frac{dy}{dx} + y\right)(-e^{-x}) = 3c_1 e^x$$

or $$e^{-2x}\left(\frac{d^2y}{dx^2} - y\right) = 3c_1$$

Finally, differentiating both sides w.r.t. x, then

$$e^{-2x}\left(\frac{d^3y}{dx^3} - \frac{dy}{dx}\right) + \left(\frac{d^2y}{dx^2} - y\right)(-2e^{-2x}) = 0$$

or $$\frac{d^3y}{dx^3} - 2\frac{d^2y}{dx^2} - \frac{dy}{dx} + 2y = 0$$

On comparing with $\frac{d^3y}{dx^3} + a\frac{d^2y}{dx^2} + b\frac{dy}{dx} + cy = 0$

we get, $a = -2$, $b = -1$, $c = 2$

$$\therefore \quad = \frac{a^3 + b^3 + c^3}{abc}$$

$$= \frac{-8 - 1 + 8}{4}$$

$$= -\frac{1}{4}$$

3. We have, $x = e^{xy\frac{dy}{dx}}$

$$\Rightarrow \quad \log x = xy\frac{dy}{dx} \Rightarrow ydy\, \frac{\log x}{x}\, dx$$

on integration, we get

$$\Rightarrow \quad \frac{y^2}{2} = \frac{(\log_e x)^2}{2} + C$$

$$\Rightarrow \quad y^2 = (\log_e x)^2 + 2C$$

$$\Rightarrow \quad y = \pm\sqrt{(\log_e x)^2 + 2C}$$

4. Given differential equation is

$$\frac{2 + \sin x}{y + 1}\frac{dy}{dx} = -\cos x$$

$$\Rightarrow \quad \int \frac{dy}{y + 1} = -\int \frac{\cos x}{2 + \sin x}\, dx$$

$$\Rightarrow \quad \log_e|y + 1| = -\log_e|2 + \sin x| + \log_e C$$

$$\because \quad y(0) = 1 \Rightarrow \log_e C = \log_e^4$$

So, $(y + 1)(2 + \sin x) = 4$

$$\Rightarrow \quad y = \frac{4}{2 + \sin x} - 1$$

$$\because \quad y(\pi) = a \Rightarrow a = \frac{4}{2} - 1 = 2 - 1 = 1$$

and $$\frac{dy}{dx} = -\frac{4\cos x}{(2 + \sin x)^2}$$

Since, $\left.\frac{dy}{dx}\right|_{x = \pi} = b \Rightarrow \frac{4}{4} = b \Rightarrow b = 1$

$\therefore$ Ordered pair $(a, b) = (1, 1)$

5. $$\frac{dy}{dx} = \frac{y\,(x - y\ln y)}{x\,(x\ln x - y)}$$

$$\Rightarrow \quad x^2 \ln x\, dy - xy\,dy = xy\,dx - y^2 \ln y\, dx$$

On dividing by x^2y^2, then

$$\Rightarrow \quad \frac{\ln x}{y^2}dy - \frac{1}{xy}dy = \frac{1}{xy}dx - \frac{\ln y}{x^2}dx$$

$$\Rightarrow \left(\ln x\left(-\frac{1}{y^2}dy\right) + \frac{1}{xy}dx\right) + \left(\ln y\left(-\frac{1}{x^2}dx\right) + \frac{1}{xy}dy\right) = 0$$

$$\Rightarrow d\left(\frac{\ln x}{y}\right) + d\left(\frac{\ln y}{x}\right) = 0$$

On integrating, we get

$$\frac{\ln x}{y} + \frac{\ln y}{x} = c \text{ or } \frac{x\ln x + y\ln y}{xy} = c$$

6. Given, $1 = \lim_{t \to x} \frac{t^2 f(x) - x^2 f(t)}{t - x}\left(\frac{0}{0}\text{ form}\right)$

$$= \lim_{t \to x} \frac{2t\, f(x) - x^2 f'(t)}{1}$$

$$\Rightarrow \quad 2x\, f(x) - x^2 f'(x) = 1$$

$$\Rightarrow x^2\frac{dy}{dx} - 2xy = -1,\ y = f(x)$$

$$\Rightarrow \quad \frac{dy}{dx} - \frac{2}{x}y = -\frac{1}{x^2}$$

which is linear equation whose IF $= e^{-2\log x} = \frac{1}{x^2}$

Multiplying both sides of the last equation by IF, we get

$$\frac{d}{dx}\left(\frac{y}{x^2}\right) = -\frac{1}{x^4}$$

$$\Rightarrow \quad \frac{y}{x^2} = \frac{1}{3x^3} + \text{K. Since, } y = f(1) = 1$$

So $\quad \text{K} = \frac{2}{3}$ Thus $y = \frac{1}{3x} + \frac{2}{3}x^2$

7. Given, $y^2 = 4ax + 4a^2$

differentiate with respect to x

$$\Rightarrow \quad 2y\frac{dy}{dx} = 4a$$

$$\Rightarrow \quad a = \left(\frac{y}{2}\frac{dy}{dx}\right)$$

So, required differential equation is

$$y^2 = \left(4\times\frac{y}{2}\frac{dy}{dx}\right)x + 4\left(\frac{y}{2}\frac{dy}{dx}\right)^2$$

$$\Rightarrow \quad y^2\left(\frac{dy}{dx}\right)^2 + 2xy\left(\frac{dy}{dx}\right) - y^2 = 0$$

$$\Rightarrow \quad y\left(\frac{dy}{dx}\right)^2 + 2x\left(\frac{dy}{dx}\right) - y = 0$$

8. $\because \left(\frac{x+y-1}{x+y-2}\right)\frac{dy}{dx} = \left(\frac{x+y+1}{x+y+2}\right)$

Put $\quad x + y = v$

$\therefore \quad 1 + \frac{dy}{dx} = \frac{dv}{dx}$

or $\left(\frac{v-1}{v-2}\right)\left(\frac{dv}{dx} - 1\right) = \left(\frac{v+1}{v+2}\right)$

$$\Rightarrow \quad \frac{dv}{dx} - 1 = \frac{(v+1)(v-2)}{(v-1)(v+2)} = \frac{v^2-v-2}{v^2+v-2}$$

or $\quad \frac{dv}{dx} = \frac{2v^2-4}{(v^2+v-2)}$

$$\Rightarrow \quad \frac{(v^2+v-2)}{(v^2-2)}\,dv = 2\,dx$$

$$\Rightarrow \quad \left(1 + \frac{v}{v^2-2}\right)dv = 2\,dx$$

On integrating, we get

$$v + \frac{1}{2}\ln|v^2-2| = 2x + c$$

or $\quad (y-x) + \frac{1}{2}\ln\left|(x+y)^2 - 2\right| = c$

Given, that $y = 1$, when $x = 1$

$\therefore \quad 0 + \frac{1}{2}\ln 2 = c$

or $\quad (y-x) + \frac{1}{2}\left|\frac{(\ln x+y)^2 - 2}{2}\right| = 0$

or $\quad \ln\left|\frac{(x+y)^2-2}{2}\right| = 2(x-y)$

9. The given equation can be written as

$$y^5 x\,dx + y\,dx - x\,dy = 0$$

or $\quad x^4 dx + \frac{x^3}{y^3}\left(\frac{y\,dx - x\,dy}{y^2}\right) = 0$

or $\quad x^4 dx + \left(\frac{x}{y}\right)^3 d\left(\frac{x}{y}\right) = 0$

On integrating, we get

$$\frac{x^5}{5} + \frac{\left(\frac{x}{y}\right)^4}{4} = c \text{ or } \frac{x^5}{5} + \frac{x^4}{4y^4} = c$$

10. $\Rightarrow \quad \left(\frac{dy}{dx} + 1\right) + x(x+y) = x^3(x+y)^3$

or $\quad \frac{(dx+dy)}{dx} + x(x+y) = x^3(x+y)^3$

Put, $\quad x + y = v$

or $\quad dx + dy = dv$

Then, $\quad \frac{dv}{dx} + vx = v^3x^3$

or $\quad v^{-3}\frac{dv}{dx} + v^{-2}x = x^3 \quad$ ...(i)

Now, put $v^{-2} = t$

$\therefore \quad -2v^{-3}\frac{dv}{dx} = \frac{dt}{dx}$

Then, from Eq. (i)

$$\frac{dt}{dx} - 2t\,x = -2x^3$$

IF $= e^{\int -2x dx} = e^{-x^2}$

$\therefore$ Solution is

$$t\,(e^{-x^2}) = \int(-2x^3)(e^{-x^2})\,dx$$

$$t\,(e^{-x^2}) = x^2e^{-x^2} + e^{-x^2} + c$$

$$\Rightarrow \quad t = x^2 + 1 + ce^{x^2}$$

$$\Rightarrow \quad \frac{1}{(x+y)^2} = x^2 + 1 + ce^{x^2}$$

11. The given equation can be written as

$$\left(\frac{dx}{x} - \frac{dy}{y}\right) + \frac{(x^2dy - y^2dx)}{(x-y)^2} = 0$$

$$\Rightarrow \quad \left(\frac{dx}{x} - \frac{dy}{y}\right) + \frac{\left(\frac{dy}{y^2} - \frac{dx}{x^2}\right)}{\left(\frac{1}{y} - \frac{1}{x}\right)^2} = 0$$

$$\Rightarrow \quad \left(\frac{dx}{x} - \frac{dy}{y}\right) + \frac{\frac{dy}{y^2} - \frac{dx}{x^2}}{\left(\frac{1}{x} - \frac{1}{y}\right)^2} = 0$$

Integrating, we get

$$\ln|x| - \ln|y| - \frac{1}{\left(\frac{1}{x} - \frac{1}{y}\right)} = c$$

$$\Rightarrow \quad \ln\left|\frac{x}{y}\right| - \frac{xy}{(y-x)} = c$$

or $\quad \ln\left|\frac{x}{y}\right| + \frac{xy}{(x-y)} = c$

12. $\because y = u^n$

$\therefore \quad \frac{dy}{dx} = nu^{n-1}\frac{du}{dx}$

Substituting the values of y and $\frac{dy}{dx}$ in the given equation, then

$$2x^4 \cdot u^n \cdot nu^{n-1}\frac{du}{dx} + u^{4n} = 4x^6$$

or $\quad \frac{du}{dx} = \frac{4x^6 - u^{4n}}{2nx^4u^{2n-1}}$

Since, it is homogeneous, then the degree of $4x^6 - u^{4n}$ and $2nx^4u^{2n-1}$ must be same

$\therefore \quad 4n = 6$ and $4 + 2n - 1 = 6$

Then, we get $n = \frac{3}{2}$

13. Equation of tangent at (x, y) is

$$Y - y = \frac{dy}{dx}(X - x)$$

for Y-axis $X = 0$,

Then, $$Y = y - x\frac{dy}{dx}$$

Given, $$\left(y - x\frac{dy}{dx}\right) \propto x^3$$

$\Rightarrow$ $$y - x\frac{dy}{dx} = kx^3$$

or $$\frac{dy}{dx} - \frac{y}{x} = -kx^2$$

$$\text{IF} = e^{\int -1/x\,dx} = e^{-\ln x} = e^{\ln(1/x)} = \frac{1}{x}$$

Then, solution is

$$y \cdot \left(\frac{1}{x}\right) = \int \frac{-kx^2}{x}\,dx$$

$\Rightarrow$ $$\frac{y}{x} = -\frac{kx^2}{2} + c$$

or $$y = -\frac{kx^3}{2} + cx$$

14. We have,

$$\frac{dy}{dx} + (\tan x)y = \sin x;\ 0 \le x \le \frac{\pi}{3}$$

$$\text{IF} = e^{\int \tan x\,dx} = e^{\ln \sec x} = \sec x$$

$\Rightarrow$ $$y \sec x = \int \tan x\,dx$$

$\Rightarrow$ $$y \sec x = \ln|\sec x| + C$$

$\Rightarrow$ $$x = 0,\ y = 0$$

$\therefore$ $$c = 0$$

Now, $$y \sec x = \ln|\sec x|$$

$\Rightarrow$ $$y = \cos x \cdot \ln|\sec x|$$

$\Rightarrow$ $$y|_{x=\frac{\pi}{4}} = \left(\frac{1}{\sqrt{2}}\right) \cdot \ln\sqrt{2}$$

$\Rightarrow$ $$y|_{x=\frac{\pi}{4}} = \frac{1}{2\sqrt{2}}\log_e 2$$

15. Let $y + 1 = Y$

$\therefore$ $$\frac{dY}{dx} = Y^2 e^{\frac{x^2}{2}} - xY$$

Put $-\frac{1}{Y} = k$

$\Rightarrow$ $$\frac{dk}{dx} + k(-x) = e^{\frac{x^2}{2}}$$

$$\text{IF} = e^{\left(-\frac{x^2}{2}\right)}$$

$\therefore$ $$k = (x + c)e^{x^2/2}$$

Put $k = -\frac{1}{y+1}$

$$\therefore y + 1 = -\frac{1}{(x+c)e^{x^2/2}} \quad \ldots\text{(i)}$$

when $x = 2,\ y = 0$, then $c = -2 - \frac{1}{e^2}$

On differentiating Eq. (i) and then put $x = 1$, we get

$$\left(\frac{dy}{dx}\right)_{x=1} = -\frac{e^{3/2}}{(1+e^2)^2}$$

16. Since, $y = f(x)$

and given differential equation

$$y(1 + xy)\,dx - x\,dy = 0$$

$\Rightarrow$ $$(y\,dx - x\,dy) + xy^2 dx = 0$$

$\Rightarrow$ $$\frac{y\,dx - x\,dy}{y^2} + x\,dx = 0$$

$\Rightarrow$ $$d\left(\frac{x}{y}\right) + x\,dx = 0$$

On integrating, then

$$\frac{x}{y} + \frac{x^2}{2} = c$$

which passes through (1, 2)

$\therefore$ $$c = 1$$

Then, curve is

$$\frac{x}{y} + \frac{x^2}{2} = 1$$

$\Rightarrow$ $$y = \frac{2x}{2 - x^2}$$

$\therefore$ $$f(x) = \frac{2x}{2 - x^2}$$

17. $\frac{y'''}{y''} = 8$ or $\log y'' = 8x + c$

When, $x = 0$, $y'' = 1$ or $\log 1 = 0$. Thus, $c = 0$

$\therefore$ $$y'' = e^{8x}$$

Integrating again, we get

$$y' = \frac{e^{8x}}{8} + \lambda$$

When, $x = 0$, $y'(0) = 0$

$\therefore$ $$\lambda = -1/8$$

$\therefore$ $$y' = \frac{e^{8x}}{8} - \frac{1}{8}$$

Integrate again. Then,

$$y = \frac{e^{8x}}{64} - \frac{x}{8} + k$$

Also, when $x = 0$, $y = \frac{1}{8}$. Thus, $k = \frac{7}{64}$

$\therefore$ $$y = \frac{1}{8}\left(\frac{e^{8x}}{8} - x + \frac{7}{8}\right)$$

18. $$\frac{dy}{dx} = \frac{1}{8\sqrt{x}\sqrt{9+\sqrt{x}}\sqrt{4+\sqrt{9+\sqrt{x}}}}$$

$\Rightarrow$ $$y = \sqrt{4 + \sqrt{9 + \sqrt{x}}} + c$$

Now, $y(0) = \sqrt{7} + c$

$\Rightarrow$ $$c = 0$$

$$y(256) = \sqrt{4 + \sqrt{9 + 16}}$$

$$= \sqrt{4 + 5} = 3$$

19. $\because$ Equation of circle

$$(x-0)^2+(y-k)^2=\left(\frac{k}{\sqrt{2}}\right)^2$$

$$\Rightarrow \quad x^2+y^2-2ky+\frac{k^2}{2}=0 \quad \ldots(i)$$

$$\therefore \quad 2x+2yy'-2ky'=0 \Rightarrow k=\frac{x+yy'}{y'}$$

From Eq. (i),

$$x^2+y^2-2y\left(\frac{x+yy'}{y'}\right)+\frac{(x+yy')^2}{2(y')^2}=0$$

$\therefore$ Order $=1$, degree $=2$

i.e., $a=2,\ b=1$

20. The given differential equation

$$\sqrt{1+x^2+y^2+x^2y^2}+xy\frac{dy}{dx}=0$$

$$\Rightarrow \quad \sqrt{(1+x^2)(1+y^2)}+xy\frac{dy}{dx}=0$$

$$\Rightarrow \quad \int\frac{\sqrt{1+x^2}}{x}dx+\int\frac{y}{\sqrt{1+y^2}}dy=0 \quad \ldots(i)$$

Now, let $I_1=\int\frac{\sqrt{1+x^2}}{x}dx$

Put, $x=\tan\theta \Rightarrow dx=\sec^2\theta d\theta$

$$\therefore \quad I_1=\int\frac{\sec\theta}{\tan\theta}\sec^2\theta d\theta=\int\frac{d\theta}{\sin\theta\cos^2\theta}$$

$$=\int\frac{\sin\theta}{\cos^2\theta(1-\cos^2\theta)}d\theta$$

Put, $\cos\theta=t \Rightarrow -\sin\theta d\theta=dt$

$$\therefore \quad I_1=\int\frac{-dt}{t^2(1-t^2)}=\int\left(\frac{-1}{1-t^2}-\frac{1}{t^2}\right)dt$$

$$=-\frac{1}{2}\log_e\left|\left(\frac{1+t}{1-t}\right)\right|+\frac{1}{t}+C_1$$

$$=\frac{-1}{2}\log_e\left|\frac{1+\cos\theta}{1\quad\cos\theta}\right|+\frac{1}{\cos\theta}+C_1$$

$$=\frac{-1}{2}\log_e\left|\frac{1+\frac{1}{\sqrt{1+x^2}}}{1-\frac{1}{\sqrt{1+x^2}}}\right|+\sqrt{1+x^2}+C_1$$

$$\Rightarrow \quad I_2=\frac{-1}{2}\log_e\left|\frac{\sqrt{1+x^2}+1}{\sqrt{1+x^2}-1}\right|+\sqrt{1+x^2}+C_1 \quad \ldots(ii)$$

$$\text{And } I_2=\frac{1}{2}\int\frac{2y}{\sqrt{1+y^2}}dy=\frac{1}{2}\frac{\sqrt{1+y^2}}{1/2}+C_2$$

$$=\sqrt{1+y^2}+C_2 \quad \ldots(iii)$$

From Eqs. (i), (ii) and (iii), we get

$$-\frac{1}{2}\log_e\left(\frac{\sqrt{1+x^2}+1}{\sqrt{1+x^2}-1}\right)+\sqrt{1+x^2}+\sqrt{1+y^2}-C=0$$

$$\Rightarrow \sqrt{1+x^2}+\sqrt{1+y^2}=\frac{1}{2}\log_e\left(\frac{\sqrt{1+x^2}+1}{\sqrt{1+x^2}-1}\right)+C$$

[where, $-C=C_1+C_2$]

21. $$y=a+be^{5x}+ce^{-7x} \quad \ldots(i)$$

$$\therefore \quad y_1=0+5be^{5x}-7ce^{-7x}$$

Dividing by $y\,e^{5x}$, then we get

$$e^{-5x}y_1=5b-7\,ce^{-12x}$$

Again differentiating both sides w.r.t. x, then

$$e^{-5x}\cdot y_2+y_1(-5)e^{-5x}=0+84\,ce^{-12x}$$

Dividing by e^{-12x}, then we get

$$e^{7x}(y_2-5y_1)=84c$$

Differentiating both sides w.r.t. x, then

$$e^{7x}\cdot(y_3-5y_2)+(y_2-5y_1)\cdot 7e^{7x}=0$$

$$\Rightarrow \quad y_3+2y_2-35y_1=0$$

22. We have,

$$\frac{\sqrt{x}\,dx+\sqrt{y}\,dx}{\sqrt{x}\,dx-\sqrt{y}\,dx}=\sqrt{\left(\frac{y^3}{x^3}\right)} \Rightarrow \frac{d(x^{3/2})+d(y^{3/2})}{d(x^{3/2})-d(y^{3/2})}=\frac{y^{3/2}}{x^{3/2}}$$

Putting $x^{3/2}=u$ and $y^{3/2}=v$, then we get,

$$\frac{du+dv}{du-dv}=\frac{v}{u}$$

$$\Rightarrow \quad u\,du+u\,dv=v\,du-v\,dv$$

$$\Rightarrow \quad u\,du+v\,dv=v\,du-u\,dv$$

$$\Rightarrow \quad \frac{u\,du+v\,dv}{u^2+v^2}=\frac{v\,du-u\,dv}{u^2+v^2}$$

$$\Rightarrow \quad \frac{d(u^2+v^2)}{2(u^2+v^2)}=-d\left(\tan^{-1}\left(\frac{v}{u}\right)\right)$$

On integrating, we get

$$\frac{1}{2}\ln(u^2+v^2)=-\tan^{-1}\left(\frac{v}{u}\right)+c$$

$$\Rightarrow \quad \ln\sqrt{(u^2+v^2)}+\tan^{-1}\left(\frac{v}{u}\right)=c$$

$$\Rightarrow \ln\sqrt{(x^3+y^3)}+\tan^{-1}\left(\frac{y}{x}\right)^{3/2}=c$$

$$\therefore \quad f(x,y)=\sqrt{(x^3+y^3)}$$

$$\text{and} \quad g(x,y)=\left(\frac{y}{x}\right)^{3/2}=\sqrt{\left(\frac{y^3}{x^3}\right)}$$

23. We have, $\cos x(3\sin x + \cos x + 3)dy$

$$= (1 + y\sin x(3\sin x + \cos x + 3))dx$$

$$\frac{dy}{dx} - (\tan x)y = \frac{1}{(3\sin x + \cos x + 3)\cos x}$$

$$\text{IF} = e^{\int -\tan x\, dx} = e^{\ln|\cos x|} = |\cos x|$$

$$= \cos x \ \forall\ x \in \left[0, \frac{\pi}{2}\right)$$

Solution of D.E.

$$y(\cos x) = \int (\cos x)\frac{1}{\cos x(3\sin x + \cos x + 3)}dx + C$$

$$y(\cos x) = \int \frac{dx}{3\sin x + \cos x + 3}dx + C$$

$$y(\cos x) = \int \frac{\left(\sec^2\frac{x}{2}\right)}{2\tan^2\frac{x}{2} + 6\tan\frac{x}{2} + 4}dx + C$$

$$= \int \frac{(\sec^2\frac{x}{2})}{2\left(\tan^2\frac{x}{2} + 3\tan\frac{x}{2} + 2\right)}dx + C$$

Now, let I_1

Put $\tan\frac{x}{2} = 1 \Rightarrow \frac{1}{2}\sec^2\frac{x}{2}dx = dt$

$$I_1 = \int \frac{dt}{t^3 + 3t + 2} = \int \frac{dt}{(t+2)(t+1)} = \int\left(\frac{1}{t+1} - \frac{1}{t+2}\right)dt$$

$$= \ln\left|\left(\frac{t+1}{t+2}\right)\right| = \ln\left|\left(\frac{\tan\frac{x}{2}+1}{\tan\frac{x}{2}+2}\right)\right|$$

So solution of D.F.

$$y(\cos x) = \ln\left|\frac{1+\tan\frac{x}{2}}{2+\tan\frac{x}{2}}\right| + C$$

$$\Rightarrow y(\cos x) = \ln\left(\frac{1+\tan\frac{x}{2}}{2+\tan\frac{x}{2}}\right) + C \text{ for } 0 \le x < \frac{\pi}{2}$$

Now, it is given $y(0) = 0$

$$\Rightarrow \quad 0 = \ln\left(\frac{1}{2}\right) + C \Rightarrow C = \ln 2$$

$$\Rightarrow \quad y(\cos x) = \ln\left(\frac{1+\tan\frac{x}{2}}{2+\tan\frac{x}{2}}\right) + \ln 2$$

For $x = \frac{\pi}{3}$

$$y\left(\frac{1}{2}\right) = \ln\left(\frac{1+\frac{1}{\sqrt{3}}}{2+\frac{1}{\sqrt{3}}}\right) + \ln 2$$

$$y = 2\ln\left(\frac{2\sqrt{3}+10}{11}\right)$$

24. Let central conic is $\frac{x^2}{a^2} + \frac{y^2}{b^2} = 1$

Differentiating both sides w.r.t. x, we get

$$\frac{2x}{a^2} + \frac{2yy_1}{b^2} = 0$$

$$\Rightarrow \quad \frac{b^2}{a^2} + \frac{yy_1}{x} = 0$$

Again differentiating both sides w.r.t. x, we get

$$0 + \frac{x\frac{d}{dx}(yy_1) - yy_1 \cdot 1}{x^2} = 0$$

$$\Rightarrow \quad x[yy_2 + y_1 \cdot y_1] - yy_1 = 0$$

$$\Rightarrow \quad x(y_1^2) + xyy_2 = yy_1$$

$$\Rightarrow \quad x\left(\frac{dy}{dx}\right)^2 + xy\frac{d^2y}{dx^2} = y\frac{dy}{dx}$$

25. Given, $dy + xy\,dx = x\,dx$

$$\frac{dy}{dx} + xy = x$$

$\Rightarrow$ Integrating factor $= e^{\frac{x^2}{2}}$

$$\therefore \quad ye^{\frac{x^2}{2}} = \int xe^{\frac{x^2}{2}}dx = e^{\frac{x^2}{2}} + C$$

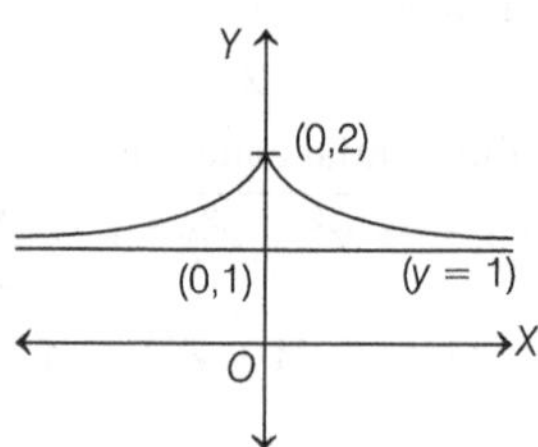

Hence, $\quad y = 1 + Ce^{\frac{-x^2}{2}}$

$$y(0) = 2 \Rightarrow C = 1$$

Hence, $\quad y(x) = 1 + e^{\frac{-x^2}{2}}$

$$A = \int_0^\infty\left[\left(e^{\frac{-x^2}{2}} + 1\right) - 1\right]dx = \int_0^\infty e^{\frac{-x^2}{2}}dx$$

26. We have, $\ln\left(\frac{dy}{dx}\right) = 3x + 4y \Rightarrow \frac{dy}{dx} = e^{3x+4y}$

$$\Rightarrow \quad e^{-4y}dy = e^{3x}dx$$

Integrating both sides, we get

$$-\frac{1}{4}e^{-4y} = \frac{e^{3x}}{3} + c \qquad \text{...(i)}$$

Given, at $x = 0$, $y = 0$

Then, $-\frac{1}{4} = \frac{1}{3} + c$

$$\therefore \quad c = -\frac{1}{4} - \frac{1}{3} \Rightarrow c = -7/12$$

From Eq. (i), $\quad -\frac{1}{4}e^{-4y} = \frac{e^{3x}}{3} - \frac{7}{12}$

$$\Rightarrow \quad 4e^{3x} + 3e^{-4y} = 7$$

27. We have, $\frac{d^3y}{dx^3} = 4x - 3$

Integrating both sides w.r.t. x, we get

$$\frac{d^2y}{dx^2} = 2x^2 - 3x + c$$

Given, slope of tangent to the curve is maximum at $x = 1$, then $\left(\frac{d^2y}{dx^2}\right)_{x=1} = 0$

$\therefore \quad 0 = 2 - 3 + c \Rightarrow c = 1$

$\Rightarrow \quad \frac{d^2y}{dx^2} = 2x^2 - 3x + 1$

Integrating both sides

$$\frac{dy}{dx} = \frac{2x^3}{3} - \frac{3x^2}{2} + x + D$$

$\because$ Curve has a minimum value 1 at $x = 0$

$\therefore \quad \left(\frac{dy}{dx}\right)_{x=0} = 0$

$\Rightarrow \quad 0 = 0 + D$

$D = 0$

Then, $\quad \frac{dy}{dx} = \frac{2x^3}{3} - \frac{3x^2}{2} + x$

Again integrating both sides, we get

$$y = \frac{x^4}{6} - \frac{x^3}{2} + \frac{x^2}{2} + E$$

Given, minimum value of curve is 1 at $x = 0$, then, $1 = 0 - 0 + 0 + E$

$\therefore \quad E = 1$

Hence, curve is $y = \frac{x^4}{6} - \frac{x^3}{2} + \frac{x^2}{2} + 1$

28. The given equation can be written as

$$\frac{y}{x}\frac{dy}{dx} = \left\{\frac{y^2}{x^2} + \frac{f(y^2/x^2)}{f'(y^2/x^2)}\right\}$$

The above equation is a homogeneous equation.
Putting $y = vx$, we get

$$v\left[v + x\frac{dv}{dx}\right] = v^2 + \frac{f(v^2)}{f'(v^2)}$$

$\Rightarrow \quad vx\frac{dv}{dx} = \frac{f(v^2)}{f'(v^2)}$ (variable separable)

$\Rightarrow \quad \frac{2vf'(v^2)}{f(v^2)}dv = 2\frac{dx}{x}$

Now, integrating both sides, we get

$\log f(v^2) = \log x^2 + \log c$ ($\log c$ = constant)

$\Rightarrow \quad \log f(v^2) = \log cx^2$

$\Rightarrow \quad f(v^2) = cx^2 \Rightarrow f(y^2/x^2) = cx^2$

29. Let $P(x, y)$ be the point on the curve and let PM and PN be the lines parallel to the coordinate axes.

$\therefore$ Area of $POM = \int_0^x y\,dx$

$\therefore$ Area of PON = area of $OMPNO$ – area of POM

$$= xy - \int_0^x y\,dx$$

and given that,

2 area of POM = area of PON

$$2\int_0^x y\,dx = xy - \int_0^x y\,dx$$

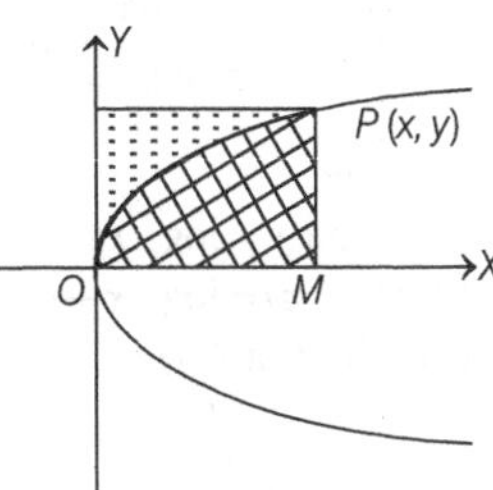

$\Rightarrow \quad 3\int_0^x y\,dx = xy$

Differentiating both sides w.r.t. x, we get

$$3y = x\frac{dy}{dx} + y \cdot 1$$

$\Rightarrow \quad 2y = x\frac{dy}{dx}$

$\Rightarrow \quad \frac{2\,dx}{x} = \frac{dy}{y}$

$\Rightarrow$ Integrating, we get $y = cx^2, c \neq 0$ parabola.

30. The given differential equation

$e^y\left(\frac{dy}{dx} - 1\right) = e^x \quad \Rightarrow \quad \frac{dy}{dx} - 1 = e^{x-y}$

Let $\quad x - y = t \quad \Rightarrow \quad 1 - \frac{dy}{dx} = \frac{dt}{dx}$

$\therefore \quad -\frac{dt}{dx} = e^t \quad \Rightarrow \quad -\int\frac{dt}{e^t} = dx$

$\Rightarrow \quad e^{-t} = x + c \Rightarrow \quad e^{-(x-y)} = x + c$

$\Rightarrow \quad e^{y-x} = x + c$

$\because \quad y(0) = 0$, so $e^{0-0} = 0 + c \Rightarrow c = 1$

$\because \quad e^{y-x} = x + 1$

$\Rightarrow \quad y = x + \log_e(x+1)$

So, $\quad y(1) = 1 + \log_e 2$

31. Differentiating w.r.t. x the given equation, we have

$$\sqrt{1 - (f'x))^2} = f(x)$$

If $y = f(x)$, then we have,

$$1 - \left(\frac{dy}{dx}\right)^2 = y^2 \frac{dy}{dx} = \pm\sqrt{1-y^2}$$

$\Rightarrow \quad \frac{dy}{\sqrt{1-y^2}} - \pm\,dx$

$\Rightarrow \quad \sin^{-1} y = C \pm x \Rightarrow y = \sin(C \pm x)$

Since, $\quad f(0) - 0$ so $0 = \sin C \Rightarrow C = 0$

Thus $f(x) = \pm \sin x$

As $\quad f(x) \geq 0$ for $x \in [0, 1]$

so we have $f(x) = \sin x$.

Since, $\sin x < x \;\forall > 0$, we get

$f(x) < x$ for $0 < x \leq 1$.

Thus, $f(1/2) < 1/2$ and $f(1/3) < 1/3$.

32. We have, $\frac{dy}{dx}=\frac{y}{x}-\cos^2\left(\frac{y}{x}\right)$

Putting $y = vx$

$\therefore \quad \frac{dy}{dx}=v+x\frac{dv}{dx} \Rightarrow v+x\frac{dv}{dx}=v-\cos^2 v$

$\Rightarrow \quad -\sec^2 v\,dv=\frac{dx}{x}$

Integrating, we get

$$-\tan v=\log_e x+c$$
$$-\tan(y/x)=\log_e x+c$$

It passes through $(1, \pi/4)$

$$-1=0+c$$

$\therefore \quad c=-1$

$\Rightarrow \quad -\tan(y/x)=\log_e x-1$

$\Rightarrow \quad \tan(y/x)=1-\log_e x=\log_e(e/x)$

$\Rightarrow \quad y=x\tan^{-1}(\ln(e/x))$

33. $\frac{dy}{dx}-\frac{y}{2x}=\frac{x^{9/4}}{x^{5/4}(x^{3/4}+1)}$

$$\text{IF}=e^{-\int\frac{dx}{2x}}=e^{\frac{1}{2}\ln x}=\frac{1}{x^{1/2}}$$

$\Rightarrow yx^{-1/2}=\int\frac{x^{9/4}\cdot x^{-1/2}}{x^{5/4}(x^{3/4}+1)}dx$

$\int\frac{x^{1/2}}{(x^{3/4}+1)}dx$

$x=t^4 \Rightarrow dx=4t^3dt$

$\int\frac{t^2 4t^3dt}{t^3+1}$

$4\int\frac{t^2(t^3+1-1)}{t^3+1}dt$

$\Rightarrow 4\int t^2dt-4\int\frac{t^2}{t^3+1}dt$

$\frac{4t^3}{3}-\frac{4}{3}\ln(t^3+1)+C$

$\Rightarrow yx^{-1/2}=\frac{4x^{3/4}}{3}-\frac{4}{3}\ln(x^{3/4}+1)+C$

$1-\frac{4}{3}\log_e 2=\frac{4}{3}-\frac{4}{3}\log_e 2+C$

$\Rightarrow \quad C=-\frac{1}{3}$

$$y=\frac{4}{3}x^{5/4}-\frac{4}{3}\sqrt{x}\ln(x^{3/4}+1)-\frac{\sqrt{x}}{3}$$

$$y(16)=\frac{4}{3}\times 32-\frac{4}{3}\times 4\ln 9-\frac{4}{3}$$
$$=\frac{124}{3}-\frac{32}{3}\ln 3$$
$$=4\left(\frac{31}{3}-\frac{8}{3}\ln 3\right)$$

34. $\phi(x)=\int[\phi(x)]^{-2}dx$

$$\phi'(x)=\frac{1}{[\phi(x)]^2}$$

$\Rightarrow \quad [\phi(x)]^2\phi'(x)=1$

$\Rightarrow \quad 3[\phi(x)]^2\phi'(x)=3$

On integrating, $[\phi(x)]^3=3x+c$

$$x=1,\ \phi(1)=0$$
$$0=3+c$$

$\therefore \quad c=-3$

Then, $\quad \phi(x)=\{3(x-1)\}^{1/3}$

35. Given differential equation,

$$xy'-y=x^2(x\cos x+\sin x),\ x>0$$

$\Rightarrow \quad \frac{dy}{dx}-\frac{1}{x}y=x(x\cos x+\sin x),$

it is in the form of linear differential equation,

of the form $\frac{dy}{dx}+py=Q$

So, $\quad \text{IF}=e^{-\int\frac{dx}{x}}=\frac{1}{x}$

$\therefore$ Solution of given differential equation

$$y(\text{IF})=\int Q(\text{IF})dx+C, \text{ where}$$
$$Q=x(x\cos x+\sin x)$$

$\Rightarrow \quad \frac{y}{x}=\int(x\cos x+\sin x)dx+C=x\sin x+C$

$$[\because \int(xf'(x)+f(x))dx=xf(x)+C]$$

$\because \quad y(\pi)=\pi \Rightarrow C=1$

$\therefore \quad y=x^2\sin x+x$

$\therefore \quad y'=2x\sin x+x^2\cos x+1$

$\Rightarrow \quad y''=2x\cos x+2\sin x+2x\cos x-x^2\sin x$

So, $y''\left(\frac{\pi}{2}\right)+y\left(\frac{\pi}{2}\right)=2-\frac{\pi^2}{4}+\frac{\pi^2}{4}+\frac{\pi}{2}=2+\frac{\pi}{2}$

Hence, option (d) is correct.

36. $\because \quad \sin y\frac{dy}{dx}=\cos y(1-x\cos y)$

$\Rightarrow \quad \tan y\frac{dy}{dx}=1-x\cos y$

$\Rightarrow \quad \sec y\tan y\frac{dy}{dx}-\sec y=-x$ ...(i)

Put, $\sec y=v$

$\therefore \quad \sec y\tan y\frac{dy}{dx}=\frac{dv}{dx}$

Then, from Eq. (i), $\frac{dv}{dx}-v=-x$

$\therefore \quad \text{IF}=e^{\int-1\cdot dx}=e^{-x}$

Then, $\quad v\cdot(e^{-x})=\int(-x)e^{-x}dx$

$\Rightarrow \quad v\cdot e^{-x}=(-x)(-e^{-x})+e^{-x}+c$

$\Rightarrow \quad v=x+1+ce^x \Rightarrow \sec y=x+1+ce^x$

37. Given differential equation

$$\frac{dy}{dx}=(\tan x-y)\sec^2 x$$

$\Rightarrow \quad \frac{dy}{dx}+(\sec^2 x)y=\sec^2 x\tan x,$

which is linear differential equation of the form $\frac{dy}{dx}+Py=Q,$

where $P=\sec^2 x$ and $Q=\sec^2 x\tan x$

IF $\quad =e^{\int\sec^2 x\,dx}=e^{\tan x}$

So, solution of given differential equation is

$$y \times \text{IF} = \int (Q \times \text{IF})dx + C$$

$$y(e^{\tan x}) = \int e^{\tan x} \cdot \sec^2 x \tan x \, dx + C$$

Let $\tan x = t \Rightarrow \sec^2 x \, dx = dt$

$$ye^{\tan x} = \int e^t \cdot t \, dt + C = te^t - \int e^t dt + C$$

[using integration by parts method]

$$= e^t(t-1) + C$$

$$\Rightarrow y \cdot e^{\tan x} = e^{\tan x}(\tan x - 1) + C \qquad [\because t = \tan x]$$

$\because \; y(0) = 0$

$\Rightarrow \; 0 = 1(0-1) + C \Rightarrow C = 1$

$\therefore \; y \cdot e^{\tan x} = e^{\tan x}(\tan x - 1) + 1$

Now, at $x = -\dfrac{\pi}{4}$

$$ye^{-1} = e^{-1}(-1-1) + 1$$

$$\Rightarrow ye^{-1} = -2e^{-1} + 1 \Rightarrow y = e - 2$$

Objective Questions Type II [One or more than one correct answer(s)]

1. We have,

$$x \cos x\left(\frac{dy}{dx}\right) + y(x \sin x + \cos x) = 1$$

$$\frac{dy}{dx} + y\left(\tan x + \frac{1}{x}\right) = \left(\frac{1}{x \cos x}\right)$$

Which is linear differential equation.

$$\therefore \quad \text{IF} = e^{\int\left(\tan x + \frac{1}{x}\right)dx} = e^{\ln \sec x + \ln x}$$

$$= e^{\ln(x \sec x)}$$

$$= (x \sec x)$$

Hence, required solution

$$y(\text{IF}) = \int (\text{IF}) \cdot \frac{1}{x \cos x} dx + c$$

$$\Rightarrow y(x \sec x) = \int x \sec x \cdot \frac{1}{x \cos x} dx + c$$

$$\Rightarrow xy \sec x = \int \sec^2 x \, dx + c$$

$$\Rightarrow xy \sec x = \tan x + c$$

(Option (b) is correct)

or $xy = \sin x + c \cos x$ (Multiplying by $\cos x$)

(Option (a) is correct)

2. We have,

$$(1 + y^2)\,dx = (\tan^{-1} y - x)\,dy$$

$$\Rightarrow \frac{dx}{dy} = \frac{\tan^{-1} y}{1+y^2} - \frac{x}{1+y^2}$$

$$\frac{dx}{dy} + \frac{1}{1+y^2} \cdot x = \frac{\tan^{-1} y}{1+y^2}$$

Which is linear differential equation.

$$\therefore \quad \text{IF} = e^{\int \frac{1}{1+y^2} dy} = e^{\tan^{-1} y}$$

Hence, required solution is

$$x(\text{IF}) = \int \frac{\tan^{-1} y}{1+y^2} (\text{IF})\, dy + c$$

$$\Rightarrow xe^{\tan^{-1} y} = \int e^{\tan^{-1} y} \cdot \frac{\tan^{-1} y}{(1+y^2)} dy + c \qquad \ldots\text{(i)}$$

Put, $\tan^{-1} y = t$

$$\therefore \quad \frac{1}{1+y^2} dy = dt$$

Then, Eq. (i) reduces to

$$x \cdot e^{\tan^{-1} y} = \int t \cdot e^t dt + c$$

$$= te^t - e^t + c$$

$$\Rightarrow xe^{\tan^{-1} y} = \tan^{-1} y \, e^{\tan^{-1} y} - e^{\tan^{-1} y} + c$$

$$\text{or} \quad xe^{\tan^{-1} y} = e^{\tan^{-1} y}\left(e^{\tan^{-1} y} - 1\right) + c$$

(Optioon (b) is correct)

Dividing both sides by $e^{\tan^{-1} y}$

Then, $x = e^{\tan^{-1} y} - 1 + c\,e^{-\tan^{-1} y}$

(Option (c) is correct)

3. $\dfrac{dy}{dx} + y = f(x)$

$$\text{IF} = e^x$$

$$ye^x = \int e^x f(x)\,dx + C$$

Now, if $0 \le x \le 2$ then $ye^x = \int e^x e^{-x} dx + C$

$$\Rightarrow \quad ye^x = x + C$$

$$x = 0,\; y(0) = 1,\; C = 1$$

$$\therefore \quad ye^x = x + 1$$

$$y = \frac{x+1}{e^x};\; y(1) = \frac{2}{e} = 2e^{-1} \;\therefore \text{(a) is correct.}$$

$$y' = \frac{e^x - (x+1)e^x}{e^{2x}}$$

$$y'(1) = \frac{e - 2e}{e^2} = \frac{-e}{e^2} = -\frac{1}{e} = -e^{-1} \;\therefore \text{(b) is correct}$$

if $x > 2$

$$ye^x = \int e^{x-2} dx$$

$$ye^x = e^{x-2} + C$$

$$y = e^{-2} + Ce^{-x}$$

as y is continuous

$$\therefore \quad \lim_{x \to 2} \frac{x+1}{e^x} = \lim_{x \to 2}(e^{-2} + Ce^{-x})$$

$$3e^{-2} = e^{-2} + Ce^{-2} \Rightarrow C = 2$$

$\therefore$ for $x > 2$

$$y = e^{-2} + 2e^{-x};\text{ hence}$$

$$y(3) = 2e^{-3} + e^{-2} = e^{-2}(2e^{-1} + 1)$$

$$y = -2e^{-x}$$

$$y'(3) = -2e^{-3} \;\therefore \text{(d) is correct.}$$

4. $\because \frac{dy}{dx}+\frac{1}{x}\tan y=\frac{1}{x^2}\tan y\sin y$

$\Rightarrow \quad \cot y\operatorname{cosec} y\frac{dy}{dx}+\operatorname{cosec} y\cdot\frac{1}{x}=\frac{1}{x^2} \quad \ldots(i)$

Put, $\operatorname{cosec} y = v$

$\therefore \quad \operatorname{cosec} y\cot y\frac{dy}{dx}=-\frac{dv}{dx}$

Then, from Eq. (i)

$$-\frac{dv}{dx}+\frac{v}{x}=\frac{1}{x^2} \Rightarrow \frac{dv}{dx}-\frac{v}{x}=-\frac{1}{x^2}$$

$\therefore \quad \text{IF}=e^{\int -1/x\,dx}=e^{-\ln x}=e^{\ln(1/x)}=\frac{1}{x}$

$\therefore$ Solution is $v\cdot\left(\frac{1}{x}\right)=\int\left(-\frac{1}{x^2}\right)\left(\frac{1}{x}\right)dx+c=\frac{1}{2x^2}+c$

$\Rightarrow \quad \frac{1}{x\sin y}=\frac{1}{2x^2}(1+2cx^2)$

$\Rightarrow \quad 2x=\sin y\,(1+2cx^2)$

or $2x=\sin y\,(1+cx^2)$ ($\because c$ is arbitrary constant)

5. Put, $x+y=v$

$\therefore \quad \frac{dy}{dx}=\frac{dv}{dx}-1$

Then, the given equation can be written as

$$\frac{dv}{dx}-1=\frac{v+7}{2v+3} \text{ or } \frac{dv}{dx}=\frac{3v+10}{2v+3}$$

or $\quad \frac{(2v+3)}{(3v+10)}dv=dx$ or $\frac{9\,(2v+3)}{(3v+10)}dv=9dx$

or $\frac{(6\,(3v+10)-33)}{(3v+10)}dv=9dx \Rightarrow \left\{6-\frac{33}{3v+10}\right\}dv=9dx$

On integrating, we get

$6v-11\ln 3v+10=9x+c$

$\Rightarrow \quad 6(x+y)-11\ln(3(x+y)+10)=9x+c$

$\Rightarrow \quad 6(x+y)-11\{\ln(x+y+10/3)+\ln 3\}=9x+c$

$\Rightarrow \quad 6(x+y)-11\ln(x+y+10/3)=9x+c$

6. Given differential equation

$$f'(x)=\frac{f(x)}{b^2+x^2} \Rightarrow \int\frac{f'(x)}{f(x)}dx=\int\frac{dx}{b^2+x^2}$$

$\Rightarrow \log_e|f(x)|=\frac{1}{b}\tan^{-1}\left(\frac{x}{b}\right)+c$

$\because \quad f(0)=1$, so $c=0$

$\therefore \quad |f(x)|=e^{\frac{1}{b}\tan^{-1}\left(\frac{x}{b}\right)}$

$\Rightarrow \quad f(x)=e^{\frac{1}{b}\tan^{-1}\left(\frac{x}{b}\right)}$ or $-e^{\frac{1}{b}\tan^{-1}\left(\frac{x}{b}\right)}$

$\because \quad f(0)=1$

$\therefore \quad f(x)=e^{\frac{1}{b}\tan^{-1}\left(\frac{x}{b}\right)}$

and $\quad f'(x)=\frac{1}{b^2}\frac{e^{\frac{1}{b}\tan^{-1}\left(\frac{x}{b}\right)}}{1+\left(\frac{x}{b}\right)^2}=\frac{e^{\frac{1}{b}\tan^{-1}\left(\frac{x}{b}\right)}}{b^2+x^2}$

$\therefore f'(x)>0\ \forall x\in R$ and $b\in R_0$.

Therefore, $f(x)$ is an increasing function $\forall\ b\in R_0$.

and $\quad f(x)\,f(-x)=e^{\frac{1}{b}\tan^{-1}\left(\frac{x}{b}\right)}\cdot e^{-\frac{1}{b}\tan^{-1}\left(\frac{x}{b}\right)}=1$

7. We have, $f''(x)=g''(x)$. On integration, we get

$$f'(x)=g'(x)+C \quad \ldots(i)$$

Putting $x=1$, we get

$f'(1)=g'(1)+C$ or $4=2+C$ or $C=2$

$\therefore \quad f'(x)=g'(x)+2$

Integrating w.r.t. x, we get $f(x)=g(x)+2x+c_1 \quad \ldots(ii)$

Putting $x=2$, we get

$f(2)=g(2)+4+c_1$ or $9=3+4+c_1$ or $c_1=2$

$\therefore \quad f(x)=g(x)+2x+2$

Putting $x=4$, we get $f(4)-g(4)=10$

$|f(x)-g(x)|<2$ or $|2x+2|<2$

$\Rightarrow \quad |x+1|<1 \Rightarrow -2<x<0$

Also, $\quad f(2)=g(2)$ or $x=-1$

$f(x)-g(x)=2x$ has no solution.

8. Equation of tangent at (x, y), $Y-y=\frac{dy}{dx}(X-x)$

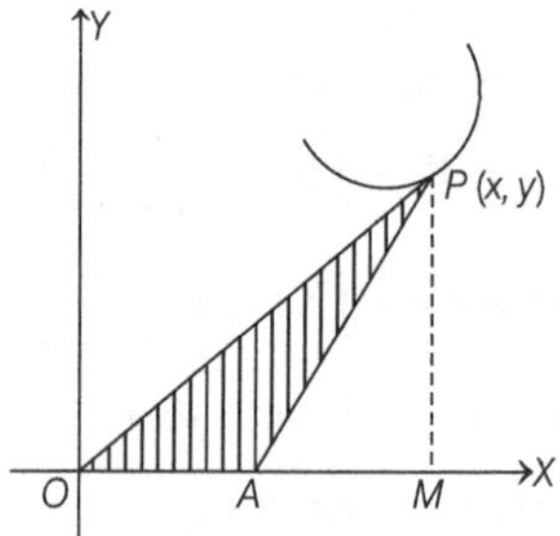

$\therefore$ Coordinate of A is $\left(x-y\frac{dx}{dy}, 0\right)$

Radius vector $OP=\sqrt{(x^2+y^2)}$

Area of $\Delta OAP=\Delta OPM-\Delta APM$

$$=\frac{1}{2}xy-\frac{1}{2}\left(y\frac{dx}{dy}\right)y=\pm a^2 \text{ (given)}$$

$\Rightarrow \quad xy-y^2\frac{dx}{dy}=\pm 2a^2$

$\Rightarrow \quad y^2\frac{dx}{dy}-xy\pm 2a^2=0$

$\Rightarrow \quad \frac{dx}{dy}-\frac{x}{y}=\pm\frac{2a^2}{y^2}$

This is a linear equation and $\text{IF}=e^{-\ln y}=\frac{1}{y}$

$\therefore$ Solution is $x\left(\frac{1}{y}\right)=\mp 2a^2\int\frac{1}{y^3}dy+c$

$\Rightarrow \quad \frac{x}{y}=\pm\frac{a^2}{y^2}+c \Rightarrow x=cy\pm\frac{a^2}{y}$

c is arbitrary constant.

9. The equation of normal at $P(x, y)$ is

$$Y - y = -\frac{dx}{dy}(X - x)$$

So, the coordinates of Q are $\left(x + y\frac{dy}{dx}, 0\right)$

Thus, $PQ^2 = (X - x)^2 + (0 - y)^2 = \left(y\frac{dy}{dx}\right)^2 + y^2$

$$\Rightarrow \quad k^2 = \left(y\frac{dy}{dx}\right)^2 + y^2$$

$$\Rightarrow \quad \left(y\frac{dy}{dx}\right)^2 = (k^2 - y^2)$$

$$\Rightarrow \quad y\frac{dy}{dx} = \pm\sqrt{(k^2 - y^2)}$$

To find the equation of the curve, we rewrite it as

$$\frac{y\,dy}{\sqrt{(k^2 - y^2)}} = \pm\,dx$$

Integrating, we get

$$\int \frac{y\,dy}{\sqrt{(k^2 - y^2)}} = \pm\int dx$$

$$\Rightarrow \quad -\sqrt{(k^2 - y^2)} = \pm\,x + c \quad \ldots\text{(i)}$$

As the curve passes through $(0, k)$, we get

$$-\sqrt{k^2 - k^2} = \pm\,(0) + c$$

$$\Rightarrow \quad c = 0$$

Therefore, Eq. (i) can be written as

$$-\sqrt{k^2 - y^2} = \pm\,x$$

$$\Rightarrow \quad k^2 - y^2 = x^2 \text{ or } x^2 + y^2 = k^2$$

10. Rewriting the given equation, we have

$$\frac{dx}{dy} = x^2y^3 + xy$$

$\Rightarrow x^{-2}\frac{dx}{dy} - x^{-1}y = y^3$, which is reducible to linear form.

Putting $x^{-1} = u$, we have $\frac{du}{dy} + yu = -y^3$

The IF of this equation is $e^{y^2/2}$. So, the solution is

$ue^{y^2/2} = -\int y^3e^{y^2/2}dy + c = -(y^2e^{y^2/2} - 2e^{y^2/2}) + c$

$$\Rightarrow \quad \frac{1}{x} = (2 - y^2) + c\,e^{-y^2/2}$$

$$\Rightarrow \quad (1 - 2x)/x = -y^2 + c\,e^{-y^2/2}$$

$$\Rightarrow \quad e^{y^2/2}\left(\frac{1 - 2x}{x} + y^2\right) = c$$

11. The equation of tangent at (x, y) is $Y - y = \frac{dy}{dx}(X - x)$

Points A and B are respectively,

$$\left(x - y\frac{dx}{dy}, 0\right) \text{ and } \left(0, y - x\frac{dy}{dx}\right)$$

Now, $\quad \frac{PA}{PB} = \frac{2}{1} \Rightarrow (PA)^2 = 4\,(PB)^2$

$$\Rightarrow \quad \left(y\frac{dx}{dy}\right)^2 + y^2 = 4\left[x^2 + x^2\left(\frac{dy}{dx}\right)^2\right]$$

$$\therefore \quad 4x^2\left(\frac{dy}{dx}\right)^4 + (4x^2 - y^2)\left(\frac{dy}{dx}\right)^2 - y^2 = 0$$

$$\therefore \quad \left(\frac{dy}{dx}\right)^2 = \frac{(y^2 - 4x^2) \pm \sqrt{(4x^2 - y^2)^2 + 16x^2y^2}}{8x^2}$$

$$= \frac{(y^2 - 4x^2) \pm (4x^2 + y^2)}{8x^2}$$

$$\therefore \quad \left(\frac{dy}{dx}\right)^2 = \frac{y^2}{4x^2} \text{ or } \left(\frac{dy}{dx}\right)^2 = -1$$

$$\therefore \quad \frac{dy}{dx} = \pm\frac{y}{2x} \Rightarrow 2\ln|y| = \pm\ln|x| + c$$

$$\Rightarrow \quad \ln\frac{y^2}{|x|} = \ln c$$

or $$\ln y^2|x| = \ln c$$

$$\Rightarrow \quad y^2 = c_1|x| \text{ or } y^2|x| = c_2$$

12. Taking $x = r\cos\theta$ and $y = r\sin\theta$, so that $x^2 + y^2 = r^2$ and $\frac{y}{x} = \tan\theta$, we have $x\,dx + y\,dy = r\,dr$ and $x\,dy - y\,dx = x^2\sec^2\theta\,d\theta = r^2d\theta$.

The given equation can be transformed into

$$\frac{rdr}{r^2d\theta} = \sqrt{\left(\frac{a^2 - r^2}{r^2}\right)} \Rightarrow \frac{dr}{d\theta} = \sqrt{(a^2 - r^2)} \Rightarrow \frac{dr}{\sqrt{(a^2 - r^2)}} = d\theta$$

Integrating both sides, then we get

$$\sin^{-1}\left(\frac{r}{a}\right) = \theta + c$$

$$\Rightarrow \quad \sin^{-1}\left(\frac{\sqrt{(x^2 + y^2)}}{a}\right) = \tan^{-1}\left(\frac{y}{x}\right) + c \quad \ldots\text{(i)}$$

$$\Rightarrow \quad \sqrt{(x^2 + y^2)} = a\sin\{\tan^{-1}(y/x) + c\}$$

$$\Rightarrow \quad \sqrt{(x^2 + y^2)} = a\sin\{\tan^{-1}(y/x) + \text{constant}\}$$

Also from Eq. (i),

$$\tan^{-1}\left(\frac{y}{x}\right) = \left\{\sin^{-1}\left(\frac{\sqrt{(x^2 + y^2)}}{a}\right) - c\right\}$$

$$\Rightarrow \quad y = x\tan\left\{\sin^{-1}\left(\frac{\sqrt{(x^2 + y^2)}}{a}\right) - c\right\}$$

$$= x\tan\left\{\sin^{-1}\left(\frac{\sqrt{(x^2 + y^2)}}{a}\right) + \text{constant}\right\}$$

13. Solving for $\frac{dy}{dx}$, we get

$$\frac{dy}{dx} = \frac{-2y\cot x \pm \sqrt{4y^2\cot^2 x + 4y^2}}{2}$$

$$= y(-\cot x \pm \operatorname{cosec} x)$$

Thus, we have,

$$\frac{dy}{y} = (-\cot x + \operatorname{cosec} x)\, dx$$

$\Rightarrow \log y = -\log \sin x + \log \tan x/2 + \log c$

$\Rightarrow \quad y = \frac{c \tan x/2}{\sin x} = \frac{c}{2\cos^2 x/2} = \frac{c}{1+\cos x}$

Solving $\frac{dy}{y} = -(\cot x + \operatorname{cosec} x)\,dx$, we get

$$y = \frac{c}{1-\cos x}$$

$\Rightarrow \quad x = 2\sin^{-1}\sqrt{c/2y}$

14. We have,

$$f(x) = 1 - 2x + \int_0^x e^{x-t} f(t)\, dt$$

On multiplying e^{-x} both sides, we get

$$e^{-x} f(x) = e^{-x} - 2xe^{-x} + \int_0^x e^{-t} f(t)\, dt$$

On differentiating both sides w.r.t. x, we get

$$e^{-x} f'(x) - e^{-x} f(x) = -e^{-x} - 2e^{-x} + 2xe^{-x} + e^{-x} f(x)$$

$\Rightarrow \quad f'(x) - 2f(x) = 2x - 3 \quad$ [dividing both sides by e^{-x}]

Let $\quad f(x) = y \Rightarrow f'(x) = \frac{dy}{dx}$

$\therefore \quad \frac{dy}{dx} - 2y = 2x - 3$

which is linear differential equation of the form $\frac{dy}{dx} + Py = Q$. Here, $P = -2$ and $Q = 2x - 3$.

Now, $\quad \text{IF} = e^{\int P\,dx} = e^{\int -2\,dx} = e^{-2x}$

$\therefore$ Solution of the given differential equation is

$$y \cdot e^{-2x} = \int \underset{\text{I}}{(2x-3)}\, \underset{\text{II}}{e^{-2x}}\, dx + C$$

$$y \cdot e^{-2x} = \frac{-(2x-3)\cdot e^{-2x}}{2} + 2\int \frac{e^{-2x}}{2}\, dx + C$$

[by using integration by parts]

$\Rightarrow \quad y \cdot e^{-2x} = \frac{-(2x-3)\, e^{-2x}}{2} - \frac{e^{-2x}}{2} + C$

$\Rightarrow \quad y = (1-x) + Ce^{2x}$

On putting $x = 0$ and $y = 1$, we get

$$1 = 1 + C$$

$\Rightarrow \quad C = 0$

$\therefore \quad y = 1 - x$

$y = 1 - x$ passes through $(2, -1)$

Now, area of region bounded by curve $y = \sqrt{1-x^2}$ and $y = 1 - x$ is shows as

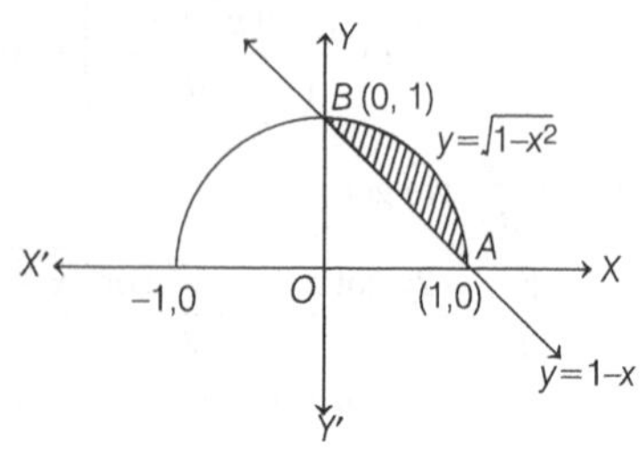

$\therefore$ Area of shaded region

= Area of 1st quadrant of a circle − Area of ΔOAB

$$= \frac{\pi}{4}(1)^2 - \frac{1}{2} \times 1 \times 1 = \frac{\pi}{4} - \frac{1}{2} = \frac{\pi - 2}{4}$$

Hence, options (b) and (c) are correct.

15. $\frac{dy}{dx} = \frac{-2y\cot x \pm \sqrt{(4y^2\cot^2 x + 4y^2)}}{2}$

$= y(-\cot x \pm \operatorname{cosec} x)$

$\therefore \quad \frac{dy}{dx} = (-\cot x + \operatorname{cosec} x)$

$\Rightarrow \quad \ln y = -\ln \sin x + \ln \tan \frac{x}{2} + \ln c$

$\Rightarrow \quad y = \frac{c \tan \frac{x}{2}}{\sin x} = \frac{c}{2\cos^2 \frac{x}{2}} \quad$ ...(i)

$$= \frac{c}{1 + \cos x}$$

Solving, $\frac{dy}{y} = -(\cot x + \operatorname{cosec} x)\,dx$, we get

$$y = \frac{c}{1-\cos x} \Rightarrow x = 2\sin^{-1}\sqrt{\frac{c}{2y}}$$

Also from Eq. (i),

$$x = 2\cos^{-1}\sqrt{\left(\frac{c}{2y}\right)}$$

16. Let a point $P(h, k)$ on the curve $y = y(x)$, so equation of tangent to the curve at point P is

$$y - k = \left(\frac{dy}{dx}\right)_{h,k}(x - h) \quad \text{...(i)}$$

Now, the tangent (i) intersect the Y-axis at Y_p, so coordinates Y_p is $\left(0, k - h\frac{dy}{dx}\right)$, where $\frac{dy}{dx} = \left(\frac{dy}{dx}\right)_{(h,k)}$

So, $PY_p = 1$ (given)

$\Rightarrow \quad \sqrt{h^2 + h^2\left(\frac{dy}{dx}\right)^2} = 1$

$\Rightarrow \quad \frac{dy}{dx} = \pm\frac{\sqrt{1-x^2}}{x} \quad$ [on replacing h by x]

$\Rightarrow \quad dy = \pm\frac{\sqrt{1-x^2}}{x}\,dx$

On putting $x = \sin\theta$, $dx = \cos\theta\, d\theta$, we get

$$dy = \pm\frac{\sqrt{1-\sin^2\theta}}{\sin\theta}\cos\theta\, d\theta = \pm\frac{\cos^2\theta}{\sin\theta}\,d\theta$$

$$= \pm(\operatorname{cosec}\theta - \sin\theta)\,d\theta$$

$\Rightarrow \quad y = \pm[\ln(\operatorname{cosec}\theta - \cot\theta) + \cos\theta] + C$

$\Rightarrow \quad y = \pm\left[\ln\left(\frac{1-\cos\theta}{\sin\theta}\right) + \cos\theta\right] + C$

$\Rightarrow \quad y = \pm\left[\ln\left(\frac{1-\sqrt{1-\sin^2\theta}}{\sin\theta}\right) + \sqrt{1-\sin^2\theta}\right] + C$

$\Rightarrow \quad y = \pm\left[\ln\left(\frac{1-\sqrt{1-x^2}}{x}\right) + \sqrt{1-x^2}\right] + C \quad [\because x = \sin\theta]$

$$= \pm\left[-\ln\frac{1+\sqrt{1-x^2}}{x} + \sqrt{1-x^2}\right] + C \quad \text{[on rationalization]}$$

$\because$ The curve is in the first quadrant so y must be positive, so

$$y = \ln\left(\frac{1+\sqrt{1-x^2}}{x}\right) - \sqrt{1-x^2} + C$$

As curve passes through (1, 0), so

$0 = 0 - 0 + c \Rightarrow c = 0$, so required curve is

$$y = \ln\left(\frac{1+\sqrt{1-x^2}}{x}\right) - \sqrt{1-x^2}$$

and required differential equation is

$$\frac{dy}{dx} = -\frac{\sqrt{1-x^2}}{x} \Rightarrow xy' + \sqrt{1-x^2} = 0$$

Hence, options (a) and (c) are correct.

Linked Comprehension Type

PASSAGE 1

1. Here, we have, $f(0) = 2$

$$f(x) = (e^x + e^{-x})\cos x - 2x - \left[x\int_0^x f'(t)dt - \int_0^x \underbrace{t}_{\text{I}}\,\underbrace{f'(t)}_{\text{II}}dt\right]$$

$$f(x) = (e^x + e^{-x})\cos x - 2x - \left[x[f(x) - f(0)] - \left\{t \cdot f(t)\Big|_0^x - \int_0^x f(t)\,dt\right\}\right]$$

$$f(x) = (e^x + e^{-x})\cos x - 2x - x\,f(x) + 2x + \left[x\,f(x) - \int_0^x f(t)\,dt\right]$$

$$f(x) = (e^x + e^{-x})\cos x - \int_0^x f(t)\,dt \qquad \text{...(i)}$$

Differentiating Eq. (i)

$$f'(x) + f(x) = \cos x\,(e^x - e^{-x}) - (e^x + e^{-x})\sin x \qquad \text{... (ii)}$$

Hence, $\dfrac{dy}{dx} + y = e^x(\cos x - \sin x) - e^{-x}(\cos x + \sin x)$

2. $f'(0) + f(0) = 0 - 2.0 = 0$

3. Integrating factor of differential Eq. (i) is e^x

$\therefore y \cdot e^x = \int e^{2x}(\cos x - \sin x)dx - \int(\cos x + \sin x)dx$

$y \cdot e^x = \int e^{2x}(\cos x - \sin x)\,dx - (\sin x - \cos x) + C$

Let $I = \int e^{2x}(\cos x - \sin x)dx = e^{2x}(A\cos x + B\sin x)$

Solving $A = 3/5$ and $B = -1/5$ and $C = 2/5$

$$\therefore y = e^x\left(\frac{3}{5}\cos x - \frac{1}{5}\sin x\right) - (\sin x - \cos x)e^{-x} + \frac{2}{5}e^{-x}$$

PASSAGE 2

1. $x = e^{xy\frac{dy}{dx}} \Rightarrow \ln x = xy\dfrac{dy}{dx}$

Hence, degree $= 1$

2. $$\left(\frac{d^3y}{dx^3}\right)^{2/3} = -\left(3\frac{d^2y}{dx^2} + \frac{dy}{dx} + 5\right)$$

$$\Rightarrow \left(\frac{d^3y}{dx^3}\right)^2 = -\left(3\frac{d^2y}{dx^2} + \frac{dy}{dx} + 5\right)^3$$

Hence, degree $= 2$

3. $\because \dfrac{dy}{dx} = c$

then, $y = x\left(\dfrac{dy}{dx}\right) + \left(\dfrac{dy}{dx}\right)^3 - 5\left(\dfrac{dy}{dx}\right)^{3/2} + 3$

$$\Rightarrow \quad 5\left(\frac{dy}{dx}\right)^{3/2} = \left\{x\frac{dy}{dx} + \left(\frac{dy}{dx}\right)^3 + 3 - y\right\}$$

$$\therefore \quad 25\left(\frac{dy}{dx}\right)^3 = \left\{x\frac{dy}{dx} + \left(\frac{dy}{dx}\right)^3 + 3 - y\right\}^2$$

$\therefore$ Order $= 1$, degree $= 6$

4. Given, $c(y + c)^2 = x^3$ ...(i)

Differentiating both sides Eq. (i) w.r.t. x, then

$$2c(y + c)y' = 3x^2 \qquad \text{...(ii)}$$

Dividing Eqs. (i) and (ii), then

$$\frac{(y + c)}{2y'} = \frac{x}{3} \Rightarrow c = \left(\frac{2xy'}{3} - y\right) \qquad \text{...(iii)}$$

Eliminating c from Eq. (i) with the help of Eq. (iii), then

$$\left(\frac{2xy'}{3} - y\right)\left(\frac{2xy'}{3}\right)^2 = x^3 \text{ or } \left(\frac{2x}{3}\left(\frac{dy}{dx}\right) - y\right)\left(\frac{dy}{dx}\right)^2 = \frac{9x}{4}$$

$\therefore$ Order $= 1$, degree $= 3$

5. Since, the given differential equation can not be written as a polynomial in all the differentials.

$\therefore$ Degree of the equation is not defined but order $= 2$.

PASSAGE 3

1. $\because \dfrac{dT}{dt} = -k(T - 290)$

$$\Rightarrow \quad \frac{dT}{(T - 290)} = -k\,dt$$

$$\Rightarrow \quad \ln(T - 290) = -kt + c \qquad \text{...(i)}$$

Initially, $T = 370K$ and $t = 0$, then

$\ln(80) = c$

From Eq. (i), $\ln(T - 290) = -kt + \ln 80$

$$\ln\left(\frac{T - 290}{80}\right) = -kt$$

$$\Rightarrow \quad \frac{T - 290}{80} = e^{-kt}$$

$$\Rightarrow \quad T = 290 + 80e^{-kt}$$

2. For $t = 10$ min, $T = 330$

Then, $330 - 290 = 80e^{-10k}$

$\Rightarrow \frac{1}{2} = e^{-10k}$

$\Rightarrow e^{10k} = 2$

$\Rightarrow 10k = \ln 2$

or $k = \frac{\ln 2}{10}$

3. At $T = 295$ K

$295 - 290 = 80e^{-kt}$

$\Rightarrow 16 = e^{kt}$

$\Rightarrow 4\ln 2 = kt = \frac{\ln 2}{10} \times t \quad \left(\because k = \frac{\ln 2}{10}\right)$

$\therefore t = 40$ min

4.

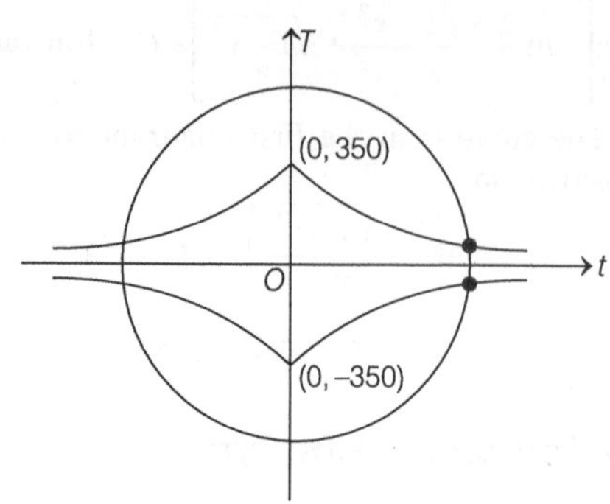

$\because$ Time can never be negative.

Number of solutions = 2 .

5. $\because T = f(t) = 290 + 80e^{-kt}$

$\therefore f\left(-\frac{2}{k}\right) = 290 + 80e^2$

Numerical Value Type

1. $\frac{dx}{dt} = x + 1 \Rightarrow \frac{dx}{x+1} = dt \Rightarrow \ln(x+1) = t + c$

Putting $t = 0$, $x = 0$, we get $c = 0 \Rightarrow t = \ln(x+1)$

for $x = 99, t = \ln 100 = 2\log_e 10$

$\therefore 20\lambda \log_{10} e = 20 \times 2\log_e 10 \times \log_{10} e = 40$

2. We have, $\frac{\frac{d^3y}{dx^3}}{\frac{d^2y}{dx^2}} = 8$

$\Rightarrow \ln \frac{d^2y}{dx^2} = 8x + c \Rightarrow \ln y_2 = 8x + c$

Putting, $x = 0$ we have, $c = \log y_2(0) = \ln 1 = 0$

$\therefore \ln y_2 = 8x \Rightarrow y_2 = e^{8x}$

i.e. $y_1 = \frac{e^{8x}}{8} + D$

Again putting $x = 0$, then $y_1(0) = \frac{1}{8} + D$

$\Rightarrow 0 = \frac{1}{8} + D$

$\therefore D = -\frac{1}{8}$

$\Rightarrow y_1 = \frac{e^{8x}}{8} - \frac{1}{8}$

$\Rightarrow y = \frac{e^{8x}}{64} - \frac{x}{8} + E$

Putting $x = 0$, we have

$y(0) = \frac{1}{64} - 0 + E = \frac{1}{8}$

$\therefore E = \frac{1}{8} - \frac{1}{64} = \frac{7}{64}$

Hence, $y = \frac{e^{8x}}{64} - \frac{x}{8} + \frac{7}{64} = \frac{(e^{8x} - 8x + 7)}{64}$

Hence, $\lambda = 64$

3. $\because y = ae^{3x} + be^{5x} \Rightarrow ae^{3x} + be^{5x} - y = 0 \quad \ldots(\text{i})$

$\therefore \frac{dy}{dx} = 3ae^{3x} + 5be^{5x}$

or $3ae^{3x} + 5be^{5x} - \frac{dy}{dx} = 0 \quad \ldots(\text{ii})$

Again differentiating both sides w.r.t. x, then

$9ae^{3x} + 25be^{5x} - \frac{d^2y}{dx^2} = 0 \quad \ldots(\text{iii})$

From Eqs. (i), (ii) and (iii), we get

$$\begin{vmatrix} 1 & 1 & y \\ 3 & 5 & \frac{dy}{dx} \\ 9 & 25 & \frac{d^2y}{dx^2} \end{vmatrix} = 0$$

Expanding w.r.t. R_1, then

$\left(5\frac{d^2y}{dx^2} - 25\frac{dy}{dx}\right) - 1\left(3\frac{d^2y}{dx^2} - 9\frac{dy}{dx}\right) + y(75 - 45) = 0$

$\Rightarrow 2\frac{d^2y}{dx^2} - 16\frac{dy}{dx} + 30y = 0$

$\therefore \frac{d^2y}{dx^2} - 8\frac{dy}{dx} + 15y = 0$

$\therefore \lambda = 15$

4. Since, $\sin y(3\sin x - 7\cos y + 7)\,dy + \cos x(3\cos y - 7\sin x - 3)\,dx = 0$

$\Rightarrow \sin y\,((5-2)\sin x - (5+2)\cos y + (5+2))\,dy + \cos x\,((5-2)\cos y - (5+2)\sin x - (5-2))\,dx = 0$

$\Rightarrow 5\sin y\,(\sin x - \cos y + 1)\,dy + 5\cos x \times (\cos y - \sin x - 1)\,dx - 2\sin y(\sin x + \cos y - 1)\,dy - 2\cos x\,(\cos y + \sin x - 1)\,dx = 0$

$\Rightarrow 5(\sin x - \cos y + 1)(\sin y\,dy - \cos x\,dx) - 2(\sin x + \cos y - 1)(\sin y\,dy + \cos x\,dx) = 0$

$$\Rightarrow \frac{5(\cos x\,dx - \sin y\,dy)}{(\sin x + \cos y - 1)} + \frac{2(\cos x\,dx + \sin y\,dy)}{(\sin x - \cos y + 1)} = 0$$

On integrating, we have

$5\ln(\sin x + \cos y - 1) + 2\ln(\sin x - \cos y + 1) = \ln c$

$\Rightarrow (\sin x + \cos y - 1)^5 (\sin x - \cos y + 1)^2 = c$

$\therefore \quad \lambda = 5, \mu = 2$

then $\quad \lambda\mu = \frac{5}{2} = 2.5$

5. The given equation can be written as,

$$\frac{dy}{dx} - y\frac{\phi'(x)}{\phi(x)} = \frac{-y^2}{\phi(x)}$$

$$\Rightarrow -\frac{1}{y^2}\frac{dy}{dx} + \frac{1}{y}\frac{\phi'(x)}{\phi(x)} = \frac{1}{\phi(x)} \quad \ldots \text{(i)}$$

Put $\frac{1}{y} = u$ so that $-\frac{1}{y^2}\frac{dy}{dx} = \frac{du}{dx}$. Now Eq. (i) becomes

$$\frac{du}{dx} + u\frac{\phi'(x)}{\phi(x)} = \frac{1}{\phi(x)}$$

This is a linear equation in u so its integrating factor is

$$e^{\int \frac{\phi'(x)}{\phi(x)}dx} = e^{\log\phi(x)} = \phi(x)$$

Multiplying both sides by the integrating factor, we have

$$\frac{d}{dx}[u\,\phi(x)] = 1 \Rightarrow u\,\phi(x) = x + \text{constant} \Rightarrow y = \frac{\phi(x)}{x + C}$$

$$\Rightarrow y(1) = \frac{1}{1+C} \Rightarrow C = 0.$$

Thus, $y = \frac{\phi(x)}{x}$

Hence $y(4) = \frac{1296}{4} = 324$

6. We have, $\frac{dy}{dx} = (2 + 5y)(5y - 2)$

$$\Rightarrow \frac{dy}{25y^2 - 4} = dx \Rightarrow \frac{1}{25}\left(\frac{dy}{y^2 - \frac{4}{25}}\right) = dx$$

On integrating both sides, we get

$$\frac{1}{25}\int \frac{dy}{y^2 - \left(\frac{2}{5}\right)^2} = \int dx$$

$$\Rightarrow \frac{1}{25} \times \frac{1}{2 \times \frac{2}{5}} \log\left|\frac{y - 2/5}{y + 2/5}\right| = x + C$$

$$\Rightarrow \log\left|\frac{5y - 2}{5y + 2}\right| = 20(x + C)$$

$$\Rightarrow \left|\frac{5y - 2}{5y + 2}\right| = Ae^{20x} \qquad [\because e^{20C} = A]$$

when $x = 0 \Rightarrow y = 0$, then $A = 1$

$$\therefore \left|\frac{5y - 2}{5y + 2}\right| = e^{20x}$$

$$\lim_{x \to -\infty}\left|\frac{5f(x) - 2}{5f(x) + 2}\right| = \lim_{x \to -\infty} e^{20x}$$

$$\Rightarrow \lim_{x \to -\infty}\left|\frac{5f(x) - 2}{5f(x) + 2}\right| = 0$$

$$\Rightarrow \lim_{x \to -\infty} 5f(x) - 2 = 0$$

$$\Rightarrow \lim_{x \to -\infty} f(x) = \frac{2}{5} = 0.4$$

Single Integer Type

1. Given, $\frac{dy}{dx} = \frac{1}{x\cos y + 2\sin y\cos y}$

$$\therefore \frac{dx}{dy} = x\cos y + 2\sin y\cos y$$

$$\Rightarrow \frac{dx}{dy} + (-\cos y)x = 2\sin y\cos y$$

$\therefore$ IF $= e^{-\int \cos y\,dy} = e^{-\sin y}$

The solution is

$$x.\,e^{-\sin y} = 2\int e^{-\sin y}.\sin y\cos y\,dy$$

$$\Rightarrow x.\,e^{\sin y} = -2\sin y\,e^{-\sin y} - 2\int (e^{-\sin y})\cos y\,dy$$

$$\Rightarrow x.\,e^{-\sin y} = -2\sin y\,e^{-\sin y} + 2$$

$$\int e^{-\sin y}\cos y\,dy = -2\sin y\,e^{-\sin y} - 2e^{-\sin y} + c$$

$$\Rightarrow x = -2\sin y - 2 + c\,e^{\sin y} = ce^{\sin y} - 2(1 + \sin y)$$

$\therefore \quad k = 2$

2. Given differential equation for $x \geq 0$ is

$$(x + 1)\,dy = ((x + 1)^2 + y - 3)\,dx$$

$$\Rightarrow \frac{dy}{dx} - \frac{1}{x+1}y = (x + 1) - \frac{3}{x+1}$$

$\because$ The above differential equation is in the form of linear, so

$$\text{IF} = e^{-\int \frac{dx}{1+x}} = \frac{1}{1+x}$$

$\therefore$ Solution of the differential equation is

$$\frac{y}{1+x} = \int\left[(x+1) - \frac{3}{x+1}\right]\frac{dx}{1+x}$$

$$\Rightarrow \frac{y}{1+x} = \int\left(1 - \frac{3}{(x+1)^2}\right)dx$$

$$\Rightarrow \frac{y}{1+x} = x + \frac{3}{x+1} + C \quad \ldots\text{(i)}$$

$\because \quad y(2) = 0$

$\therefore \quad 0 = 2 + 1 + C \Rightarrow C = -3$

So, for $y(3)$ on putting $x = 3$ and $C = -3$ in Eq. (i), we get

$$\frac{y}{4} = 3 + \frac{3}{4} - 3$$

$$\Rightarrow y = 3$$

3. The given differential equation can be written as

$$y'(x) = y(x) + \int_0^1 y(x)\,dx \quad \ldots \text{(i)}$$

Differentiating both sides w.r.t. x, then

$$y''(x) = y'(x)$$

$$\Rightarrow \quad \frac{y''(x)}{y'(x)} = 1$$

$$\Rightarrow \quad \int \frac{y''(x)}{y'(x)}\,dx = \int dx$$

$$\Rightarrow \quad \ln y'(x) = x + \ln c$$

$$\Rightarrow \quad \ln\left(\frac{y'(x)}{c}\right) = x$$

$$\Rightarrow \quad y'(x) = c\,e^x$$

and $\quad y(x) = c\,e^x + d \quad \ldots \text{(ii)}$

$\Rightarrow \quad y(0) = c + d = 1 \quad (\because y(0) = 1)$

$\therefore \quad y(x) = c\,e^x + 1 - c \quad \ldots \text{(iii)}$

From Eqs. (i), (ii) and (iii), we get

$$c\,e^x = c\,e^x + 1 - c + \int_0^1 (c\,e^x + 1 - c)dx$$

$$\Rightarrow \quad c - 1 = [ce^x + (1-c)x]_0^1$$

$$\Rightarrow \quad c - 1 = ce + 1 - c - c$$

$$\therefore \quad c = \frac{2}{3-e} \quad \ldots \text{(iv)}$$

From Eq. (iii),

$$y(1) = ce + 1 - c = \frac{e+1}{3-e} \quad \text{[From Eq. (iv)]}$$

$$= \frac{2.718+1}{3-2.718} = \frac{3.718}{0.282} = 13.18$$

$$[y(1) - 7] = [6.18] = 6$$

4. Given,

$$\int_0^x f(t)\,dt + \int_0^x (x-t)\,f(t)\,dt = e^{-x} - 1$$

$$x\int_0^x f(t)\,dt + \int_0^x f(t)\,(1-t)\,dt = e^{-x} - 1$$

Differentiate w.r.t. x

$$x.\,f(x) + \int_0^x f(t)\,dt + f(x)\,(1-x) = -e^{-x}$$

Again differentiate w.r.t. x

$$xf'(x) + f(x) + f(x) + f'(x) - xf'(x) - f(x) = e^{-x}$$

$$f'(x) + f(x) = e^{-x}$$

$$\text{IF} = e^x \Rightarrow y.\,e^x = \int dx$$

$$f(x)e^x = x + C;\ f(0) = -1$$

$$\therefore \quad f(x)e^x = x - 1$$

$$x = 10$$

$$\Rightarrow \quad e^{10}f(10) = 9$$

5. We have, $xdy - ydx = \sqrt{x^2 - y^2}\,dx$

$$\Rightarrow \frac{xdy - ydx}{x^2} = \frac{1}{2}\sqrt{1 - \frac{y^2}{x^2}}\,dx$$

$$\Rightarrow \int \frac{d\left(\frac{y}{x}\right)}{\sqrt{1-\left(\frac{y}{x}\right)^2}} = \int \frac{dx}{x}$$

$$\Rightarrow \quad \sin^{-1}\left(\frac{y}{x}\right) = \ln|x| + c$$

at $x = 1$, $y = 0 \Rightarrow c = 0$

$$y = x\sin(\ln x)dx$$

$$A = \int_1^{e^x} x\sin(\ln x)\,dx$$

$$x = e^t,\ dx = e^t dt \Rightarrow \int_0^{\pi} e^{2t}\sin(t)\,dt = A$$

$$\alpha e^{2\pi} + \beta = \left(\frac{e^{2t}}{5}(2\sin t - \cos t)\right)_0^{\pi} = \frac{1+e^{2\pi}}{5}$$

$$\alpha = \frac{1}{5},\ \beta = \frac{1}{5} \text{ so } 10(\alpha + \beta) = 4$$

6. We have, $y^3 y'' = -1$

$$y'y'' = \frac{-y'}{y^3}$$

Integrating on both sides

$$\frac{(y')^2}{2} = \frac{1}{2y^2} + C$$

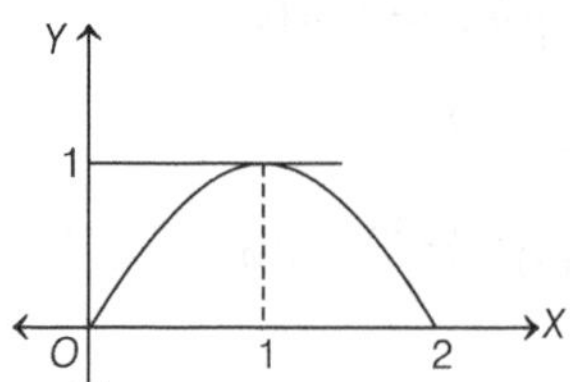

Putting $x = 1$, $y = 1$ and $y'(1) = 0$

$$0 = \frac{1}{2} + C \Rightarrow C = \frac{-1}{2}$$

$$\frac{(y')^2}{2} = \frac{1}{2y^2} - \frac{1}{2}$$

$$(y')^2 = \frac{1}{y^2} - 1 \Rightarrow (y')^2 = \frac{1-y^2}{y^2}$$

$$\Rightarrow \quad y' = \pm\frac{\sqrt{1-y^2}}{y} \Rightarrow \frac{y}{\sqrt{1-y^2}}\,y' = \pm 1$$

$$\Rightarrow \quad \int \frac{y}{1-y^2}\,y'\,dx = \int 1\,dx \text{ [Taking as + ve sign]}$$

Putting, $\quad y = t \Rightarrow y'\,dx = dt$

$$\int \frac{t}{\sqrt{1-t^2}}\,dt = x + \lambda \Rightarrow 1 - t^2 = z^2$$

$$\Rightarrow \quad -2tdt = 2zdz$$

$$-\int \frac{zdz}{z} = x + \lambda \Rightarrow -z = x + \lambda$$

$$\Rightarrow \quad -\sqrt{1-t^2} = x + \lambda \Rightarrow -\sqrt{1-y^2} = x + \lambda$$

Putting, $x = 1$, $y(1) = 1$

$\therefore \quad \lambda = -1 \Rightarrow -\sqrt{1-y^2} = x - 1$

$\Rightarrow \quad 1 - y^2 = x^2 - 2x + 1$

$y^2 = 2x - x^2$

$y = \sqrt{2x - x^2} = f(x) \quad [\because y \geq 0 \; \forall \; x \in D_f]$

Taking as negative sign we will obtain the same function.

$f'(x) = \dfrac{2-2x}{2\sqrt{2x-x^2}} = 0 \Rightarrow x = 1$ which is point of maxima.

$\therefore$ Maximum value of $f(x) = f(1) = \sqrt{2-1} = 1$

7. We have, $\quad x \log x \dfrac{dy}{dx} + y = \log x^2$

$\Rightarrow \quad \dfrac{dy}{dx} + \dfrac{y}{x \log x} = \dfrac{2 \log x}{x \log x}$

$\Rightarrow \quad \dfrac{dy}{yx} + \dfrac{y}{x \log x} = \dfrac{2}{x}$

$\text{IF} = e^{\int \frac{1}{x \log x} dx} = e^{\log(\log x)}$

$\text{IF} = \log x$

Solution of different equations is

$y \log x = \int \dfrac{2 \log x}{x} dx + c$

$y \log x = (\log x)^2 + c$

put $\quad x = e, y = 0$ we get

$0 = 1 + c \Rightarrow c = -1$

$\therefore \quad y \log x = (\log x)^2 - 1$

put $\quad x = e^2$

$y \log x = (\log x)^2 - 1$

$2y = 4 - 1 \Rightarrow y = \dfrac{3}{2}$

$\therefore \quad y(e^2) = \dfrac{3}{2} \Rightarrow 2y(e^2) = 3$

8. $xdy + (y + x^3y^2)dx = 0$

$\Rightarrow \quad xdy + ydx = -x^3y^2dx$

$\Rightarrow \quad \dfrac{xdy + ydx}{x^2y^2} = -x\,dx \Rightarrow \dfrac{d(xy)}{(xy)^2} = -x\,dx$

Integrating, we get

$\Rightarrow \quad -\dfrac{1}{xy} = -\dfrac{x^2}{2} + c$

Using (1, 2) in Eq. (i), we get

$-\dfrac{1}{2} = -\dfrac{1}{2} + c \Rightarrow c = 0$

$\therefore \quad -\dfrac{1}{xy} = -\dfrac{x^2}{2} \Rightarrow y = \dfrac{2}{x^3}$

or $\quad x^3y = 2$

is the required curve.

9. $\because \sec^2 y \dfrac{dy}{dx} + 2x \tan y = x^3 \quad \ldots$(i)

let $\quad \tan y = v$

$\therefore \quad \sec^2 y \dfrac{dy}{dx} = \dfrac{dv}{dx}$

Then from Eq. (i),

$\dfrac{dv}{dx} + 2vx = x^3$

$\therefore \; \text{IF} = e^{\int 2x\,dx} = e^{x^2}$

$\therefore$ Solution is

$v \cdot (\text{IF}) = \int x^3 \cdot (\text{IF}) dx + c$

$\Rightarrow \quad \tan y \cdot e^{x^2} = \int x^3 \cdot e^{x^2} dx + c$

Put $\quad x^2 = t \Rightarrow x\,dx = \dfrac{dt}{2}$

$\therefore \quad \tan y \cdot e^{x^2} = \dfrac{1}{2} \int te^t dt + c = \dfrac{1}{2}(te^t - e^t) + c$

$\Rightarrow \quad \tan y = \dfrac{1}{2}(t-1) + ce^{-x^2}$

$\Rightarrow \quad \tan y = \dfrac{1}{2}(x^2-1) + ce^{-x^2}$

$\Rightarrow \quad 2 \tan y = (x^2 - 1) + 2ce^{-x^2}$

$= (x^2 - 1) + ce^{-x^2}$ (Replacing $2c$ by c)

Hence, $\lambda = 1$

Matrix-Match Type

1. (A) $\because$

$y = a\left(\dfrac{1 - \cos 2x}{2}\right) + b\left(\dfrac{1 + \cos 2x}{2}\right) + c \sin 2x + d \cos 2x$

$= A + B \sin 2x + C \cos 2x$

$\therefore \quad \dfrac{dy}{dx} = 2B \cos 2x - 2C \sin 2x \quad \ldots$(i)

$\Rightarrow \quad \dfrac{d^2y}{dx^2} = -4B \sin 2x - 4C \cos 2x$

$\Rightarrow \quad \dfrac{d^3y}{dx^3} = -8B \cos 2x + 8C \sin 2x = -4\dfrac{dy}{dx}$

[From Eq. (i)]

$\Rightarrow \quad \dfrac{d^3y}{dx^3} + 4\dfrac{dy}{dx} = 0$

$\therefore \quad O = 3, D = 1$

$O + 2D = 5, O^D + D^O = 4, 2^O + 3^D$

$= 8 + 3 = 11$ (P, S, T)

(B) $\because \; y = (c_1 + c_2)\sin(x + c_3) - c_4 e^{c_5 + c_6} \cdot e^x$

or $\quad y = A \sin(x + B) + Ce^x \quad \ldots$(i)

$\therefore \quad \dfrac{dy}{dx} = A \cos(x + B) + Ce^x \quad \ldots$(ii)

Subtracting Eq. (i) from Eq. (ii), then

$\dfrac{dy}{dx} - y = A \cos(x + B) - A \sin(x + B) \quad \ldots$(iii)

$$\frac{d^2y}{dx^2} - \frac{dy}{dx} = -A\sin(x+B) - A\cos(x+B)$$

$$\frac{d^3y}{dx^3} - \frac{d^2y}{dx^2} = -A\cos(x+B) + A\sin(x+B)$$

$$= -\left(\frac{dy}{dx} - y\right) \quad \text{[from Eq. (iii)]}$$

$$\frac{d^3y}{dx^3} - \frac{d^2y}{dx^2} + \frac{dy}{dx} - y = 0$$

$$O = 3,\ D = 1$$

$$O + 2D = 5,\ O^D + D^O = 4,\ 2^O + 3^D = 11 \text{ (R, S, T)}$$

(C) Put $x = \tan\theta$, $y = \tan\phi$

Then, $(\sec\theta + \sec\phi) = A(\tan\theta\sec\phi + \tan\phi\sec\theta)$

$$\Rightarrow \left(\frac{\cos\theta + \cos\phi}{\cos\theta\cos\phi}\right) = A\left(\frac{\sin\theta + \sin\phi}{\cos\theta\cos\phi}\right)$$

$$\Rightarrow 2\cos\left(\frac{\theta+\phi}{2}\right)\cos\left(\frac{\theta-\phi}{2}\right) = A\cdot 2\sin\left(\frac{\theta+\phi}{2}\right)\cos\left(\frac{\theta-\phi}{2}\right)$$

$$\Rightarrow \cot\left(\frac{\theta+\phi}{2}\right) = A$$

$$\Rightarrow \frac{\theta+\phi}{2} = \cot^{-1}A$$

$$\Rightarrow \theta + \phi = 2\cot^{-1}A$$

$$\Rightarrow \tan^{-1}x + \tan^{-1}y = 2\cot^{-1}A$$

$$\Rightarrow \frac{1}{(1+x^2)} + \frac{1}{(1+y^2)}\frac{dy}{dx} = 0$$

$$\therefore \quad O = 1,\ D = 1$$

Then, $O = D$ and $2O + 3D = 5$ (Q, R)

2. **(A)** The given equation can be written as $\left(\frac{dy}{dx} - e^{-x}\right)\left(\frac{dy}{dx} - e^x\right) = 0$

$$\Rightarrow \frac{dy}{dx} - e^{-x} = 0 \quad \text{or} \quad \frac{dy}{dx} - e^x = 0$$

$$\Rightarrow dy - e^{-x}\,dx = 0 \quad \text{or} \quad dy - e^x\,dx = 0$$

$$\Rightarrow y + e^{-x} = c \quad \text{or} \quad y - e^x = c \text{ (R, T)}$$

(B) $\because (x + y + 1)\,dy = dx$

Put $x + y + 1 = v$

Then, $dx + dy = dv$

and the given equation reduces to

$$v(dv - dx) = dx$$

$$\Rightarrow dx = \frac{v\,dv}{v+1} = \left(1 - \frac{1}{v+1}\right)dv$$

$$\Rightarrow x + c = v - \ln(v+1)$$

$$\Rightarrow \ln(v+1) = v - x - c$$

$\Rightarrow \ln(x + y + 2) = y + 1 - c = y + c$ (Q)

Also, $x + y + 2 = e^{y+c} = e^y \cdot e^c = ce^y$ (P)

(C) $\frac{dy}{dx} + \frac{(x+y+2)}{(2x+2y-1)} = 0$

Put, $x + y = v$

$$\therefore \quad 1 + \frac{dy}{dx} = \frac{dv}{dx}, \text{ then}$$

$$\frac{dv}{dx} - 1 + \frac{v+2}{2v-1} = 0$$

$$\Rightarrow \frac{dv}{dx} - \frac{(v-3)}{(2v-1)} = 0$$

$$\text{or} \quad \frac{(2v-1)}{(v-3)}dv = dx$$

$$\Rightarrow \left(2 + \frac{5}{v-3}\right)dv = dx$$

$$\Rightarrow 2v + 5\ln(v-3) = x + c$$

$$\Rightarrow 2(x+y) + 5\ln(x+y-3) = x + c$$

$$\Rightarrow 2(x+y+2) + 5\ln(x+y-3) = x + c_1$$

$$= x + c \text{ (S)}$$

3. **(A)** We have $y = e^{4x} + 2e^{-x}$;

$$y_1 = 4e^{4x} - 2e^{-x};\ y_2 = 16e^{4x} + 2e^{-x};$$

$$y_3 = 64e^{4x} - 2e^{-x}$$

Now, $y_3 - 13y_1 = (64e^{4x} - 2e^{-x}) - 13(4e^{4x} - 2e^{-x})$

$$= 12e^{4x} + 24e^{-x}$$

$$y_3 - 13y = 12(e^{4x} + 2e^{-x}) = 12y$$

$$\therefore \quad K = 12 \text{ and } K/3 = 4 \text{ (Q)}$$

(B) We have, $y\frac{dy}{dx} = x \Rightarrow y\,dy = x\,dx$

Integrating, we get

$$y^2 = x^2 + c$$

$$\Rightarrow f(1) = 3 \Rightarrow c = 8$$

$$\Rightarrow y^2 = x^2 + 8,\ f(x) = \sqrt{x^2 + 8}$$

$$\Rightarrow f(4) = \sqrt{16+8} = 2\sqrt{6} \text{ (R)}$$

(C) We have $y = u^m \Rightarrow \frac{dy}{dx} = m\,u^{m-1}\frac{du}{dx}$

Substituting the value of y and $\frac{dy}{dx}$ in

$2x^4y\frac{dy}{dx} + y^4 = 4x^6$, we have,

$$2x^4u^m m\,u^{m-1}\frac{du}{dx} + u^{4m} = 4x^6 \Rightarrow \frac{du}{dx} = \frac{4x^6 - u^{4m}}{2m\,x^4\,u^{2m-1}}$$

For homogeneous equation, $4m = 6$ or $m = \frac{3}{2}$

and $2m - 1 = 2$ or $m = \frac{3}{2}$ (P)

(D) We have, $y = Ax^m + Bx^{-n}$

$$\therefore \quad \frac{dy}{dx} = Amx^{m-1} - nBx^{-n-1}$$

$$\therefore \quad \frac{d^2y}{dx^2} = Am(m-1)x^{m-2} + n(n+1)Bx^{-n-2} \text{ (s)}$$

Putting, these values in $x^2\frac{d^2y}{dx^2} + 2x\frac{dy}{dx} = 12y$, we get

$m(m+1)Ax^m + n(n-1)Bx^{-n} = 12(Ax^m + Bx^{-n})$

i.e., $m(m+1) = 12$ and $n(n-1) = 12$

i.e., $m = 3, -4$ and $n = 4, -3$

$\therefore \quad |m + n| = 0, 1$ (S)

21

The Straight Lines

Objective Questions Type I [Only one correct answer]

In each of the questions below, four choices are given of which only one is correct. You have to select the correct answer which is the most appropriate.

1. A triangle has a vertex at (1, 2) and the mid-points of the two sides through it are (–1, 1) and (2, 3). Then, the centroid of this triangle is **[JEE Main 2019, 12 April]**

(a) $\left(1, \frac{7}{3}\right)$ (b) $\left(\frac{1}{3}, 2\right)$

(c) $\left(\frac{1}{3}, 1\right)$ (d) $\left(\frac{1}{3}, \frac{5}{3}\right)$

2. The vertices of a triangle are $A(x_1, x_1 \tan\alpha)$, $B(x_2, x_2 \tan\beta)$ and $C(x_3, x_3 \tan\gamma)$. If the circumcentre of ΔABC coincides with the origin and $H(a, b)$ be its orthocentre, then $\frac{a}{b}$ is equal to

(a) $\frac{\cos\alpha + \cos\beta + \cos\gamma}{\cos\alpha \cos\beta \cos\gamma}$ (b) $\frac{\sin\alpha + \sin\beta + \sin\gamma}{\sin\alpha \sin\beta \sin\gamma}$

(c) $\frac{\tan\alpha + \tan\beta + \tan\gamma}{\tan\alpha \tan\beta \tan\gamma}$ (d) $\frac{\cos\alpha + \cos\beta + \cos\gamma}{\sin\alpha + \sin\beta + \sin\gamma}$

3. If in ΔABC, $A \equiv (1, 10)$, circumcentre $\equiv \left(-\frac{1}{3}, \frac{2}{3}\right)$ and orthocentre $\equiv \left(\frac{11}{3}, \frac{4}{3}\right)$ then, the coordinate of mid-point of side opposite to A is

(a) $\left(1, -\frac{11}{3}\right)$ (b) (1, 5)

(c) (1, – 3) (d) (1, 6)

4. A straight line L with negative slope passes through the point (8, 2) and cuts the positive coordinate axes at points P and Q. As L varies, the absolute minimum value of $OP + OQ$ is (O is origin)

(a) 10 (b) 18

(c) 16 (d) 12

5. Drawn from origin are two mutually perpendicular lines forming an isosceles triangle together with the straight line $2x + y = a$, then the area of this triangle is

(a) $\frac{a^2}{2}$ sq unit (b) $\frac{a^2}{3}$ sq unit

(c) $\frac{a^2}{5}$ sq unit (d) None of these

6. The number of integral values of m so that the abscissa of point of intersection of lines $3x + 4y = 9$ and $y = mx + 1$ is also an integer, is **[JEE Main 2021, 18 March]**

(a) 1 (b) 2

(c) 3 (d) 0

7. The line $3x - 4y + 7 = 0$ is rotated through an angle $\frac{\pi}{4}$ in the clockwise direction about the point (–1, 1). The equation of the line in its new position is

(a) $7y + x - 6 = 0$ (b) $7y - x - 6 = 0$

(c) $7y + x + 6 = 0$ (d) $7y - x + 6 = 0$

8. If the orthocentre of the triangle formed by the lines $2x + 3y - 1 = 0$, $x + 2y - 1 = 0$ and $ax + by - 1 = 0$ is at origin, then the point (a, b) is

(a) (8, – 8) (b) (–8, 8)

(c) (8, 8) (d) None of these

9. The distance between the circumcentre and orthocentre of the triangle whose vertices are (0, 0), (6, 8) and (– 4, 3) is

(a) $\frac{125}{8}$ unit (b) $\frac{\sqrt{5}}{2}$ unit

(c) $\frac{5\sqrt{5}}{2}$ unit (d) $5\sqrt{5}$ unit

10. For all real values of a and b lines $(2a + b)x + (a + 3b)y + (b - 3a) = 0$ and $mx + 2y + 6 = 0$ are concurrent, then m is equal to

(a) –2 (b) –3

(c) –4 (d) –5

11. A system of lines is given as $y = m_i x + c_i$, where m_i can take any value out of 0, 1,–1 and when m_i is positive, then c_i can be 1 or –1 when m_i equal 0,c_i can be 0 or 1 and when m_i equals to – 1, c_i can take 0 or 2. Then the area enclosed by all these straight lines is

(a) $\frac{3}{\sqrt{2}}(\sqrt{2}-1)$ sq unit (b) $\frac{3}{\sqrt{2}}$ sq unit

(c) $\frac{3}{2}$ sq unit (d) None of these

12. If all the 3 vertices of an isosceles right angle triangle be integral points and length of base is also an integer, then which of the point is never a rational point (A point $P(x, y)$ is integer point if both x and y are integers and point is rational, if both x and y are rational)

(a) Centroid
(b) Incentre
(c) Circumcentre
(d) Orthocentre

13. A man starts from the point $P(-3, 4)$ and reaches point $Q(0, 1)$ touching X axis at R such that $PR + RQ$ is minimum, then the point R is

(a) $\left(\frac{3}{5}, 0\right)$ (b) $\left(-\frac{3}{5}, 0\right)$

(c) $\left(-\frac{2}{5}, 0\right)$ (d) $(-2, 0)$

14. A beam of light is sent along the line $x - y = 1$, which after refracting from the X-axis enters the opposite side by turning through 30° towards the normal at the point of incidence on the X-axis. Then, the equation of the refracted ray is

(a) $(2-\sqrt{3})x - y = 2+\sqrt{3}$
(b) $(2+\sqrt{3})x - y = 2+\sqrt{3}$
(c) $(2-\sqrt{3})x + y = (2+\sqrt{3})$
(d) None of the above

15. Slope of a line passing through $P(2, 3)$ and intersecting the line, $x + y = 7$ at a distance of 4 units from P, is **[JEE Mains 2019, 9 April]**

(a) $\frac{1-\sqrt{5}}{1+\sqrt{5}}$ (b) $\frac{\sqrt{7}-1}{\sqrt{7}+1}$

(c) $\frac{1-\sqrt{7}}{1+\sqrt{7}}$ (d) $\frac{\sqrt{5}-1}{\sqrt{5}+1}$

16. Given a family of lines $a(2x + y + 4) + b(x - 2y - 3) = 0$, the number of lines belonging to the family at a distance $\sqrt{10}$ from $P(2, -3)$ is

(a) 0 (b) 1
(c) 2 (d) 4

17. The equation of one of the straight lines which passes through the point (1, 3) and makes an angles $\tan^{-1}(\sqrt{2})$ with the straight line, $y + 1 = 3\sqrt{2}x$ is **[JEE Main 2021, 18 March]**

(a) $4\sqrt{2}x + 5y - (15 + 4\sqrt{2}) = 0$
(b) $5\sqrt{2}x + 4y - (15 + 4\sqrt{2}) = 0$
(c) $4\sqrt{2}x + 5y - 4\sqrt{2} = 0$
(d) $4\sqrt{2}x - 5y - (15 + 4\sqrt{2}) = 0$

18. If $5a + 4b + 20c = t$, then the value of t for which the line $ax + by + c - 1 = 0$ always passes through a fixed point is

(a) 0 (b) 20
(c) 30 (d) None of these

19. Consider the family of lines $(x + y - 1) + \lambda(2x + 3y - 5) = 0$ and $(3x + 2y - 4) + \mu(x + 2y - 6) = 0$, equation of a straight line that belongs to both the families is

(a) $x - 2y - 8 = 0$ (b) $x - 2y + 8 = 0$
(c) $2x + y - 8 = 0$ (d) $2x - y - 8 = 0$

20. If the distance of any point (x, y) from the origin is defined as $d(x, y) = \max\{|x|, |y|\}$, $d(x, y) = a$, non zero constant, then the locus is

(a) a circle (b) a straight line
(c) a square (d) a triangle

21. The point (4, 1) undergoes the following three transformations successively

(I) Reflection about the line $y = x$
(II) Translation through a distance 2 units along the positive direction of X-axis
(III) Rotation through an angle $\pi/4$ about the origin in the anti-clockwise direction.

The final position of the point is given by the coordinates

(a) $\left(\frac{1}{\sqrt{2}}, \frac{7}{\sqrt{2}}\right)$ (b) $(-2, 7\sqrt{2})$

(c) $\left(-\frac{1}{\sqrt{2}}, \frac{7}{\sqrt{2}}\right)$ (d) $(\sqrt{2}, 7\sqrt{2})$

22. Two lines are drawn through a point A at an angle 30°. A point B is taken on one of them at a distance 1 unit from A, a perpendicular is drawn from the point B to the other line, another perpendicular is drawn from its foot to AB and so on. Then, the sum of the lengths of all these perpendiculars is

(a) $2-\sqrt{3}$ (b) $2+\sqrt{3}$

(c) $\frac{1}{4(2+\sqrt{3})}$ (d) None of these

23. Line L has intercepts a and b on the coordinate axes, when the axes are rotated through a given angle; keeping the origin fixed, the same line has intercepts p and q, then

(a) $a^2 + b^2 = p^2 + q^2$ (b) $\frac{1}{a^2} + \frac{1}{b^2} = \frac{1}{p^2} + \frac{1}{q^2}$

(c) $a^2 + p^2 = b^2 + q^2$ (d) $\frac{1}{a^2} + \frac{1}{p^2} = \frac{1}{b^2} + \frac{1}{q^2}$

24. The point A (2, 1) is translated parallel to the line $x - y = 3$ by a distance 4 units. If the new position A' is in third quadrant, then the coordinates of A' are
(a) $(2 + 2\sqrt{2}, 1 + 2\sqrt{2})$
(b) $(-2 + \sqrt{2}, -1 - 2\sqrt{2})$
(c) $(2 - 2\sqrt{2}, 1 - 2\sqrt{2})$
(d) None of the above

25. A ray of light coming from the point (1, 2) is reflected at a point A on the X-axis and then passes through the point (5, 3). The coordinates of the point A are
(a) $\left(\frac{13}{5}, 0\right)$ (b) $\left(\frac{5}{13}, 0\right)$
(c) $(-7, 0)$ (d) none of these

26. The orthocentre of the triangle formed by the lines $x + y = 1$, $2x + 3y = 6$ and $4x - y + 4 = 0$ lies in
(a) I quadrant (b) II quadrant
(c) III quadrant (d) IV quadrant

27. The reflection of the point $(4, -13)$ in the line $5x + y + 6 = 0$ is
(a) $(-1, -14)$ (b) $(3, 4)$
(c) $(1, 2)$ (d) $(-4, 13)$

28. Let k be an integer such that the triangle with vertices $(k, -3k)$, $(5, k)$ and $(-k, 2)$ has area 28 sq units. Then, the orthocentre of this triangle is at the point **[JEE Main 2017]**
(a) $\left(2, -\frac{1}{2}\right)$ (b) $\left(1, \frac{3}{4}\right)$
(c) $\left(1, -\frac{3}{4}\right)$ (d) $\left(2, \frac{1}{2}\right)$

29. A line passing through P (4, 2) meets the X and Y-axis at A and B respectively. If O is the origin, then locus of the centre of the circumcircle of ΔOAB is
(a) $x^{-1} + y^{-1} = 2$ (b) $2x^{-1} + y^{-1} = 1$
(c) $x^{-1} + 2y^{-1} = 1$ (d) $2x^{-1} + 2y^{-1} = 1$

30. Let n be the number of points having rational coordinates equidistant from the point $(0, \sqrt{3})$, then
(a) $n \le 1$ (b) $n = 1$
(c) $n \le 2$ (d) $n > 2$

31. The coordinates of the middle points of the sides of a triangle are (4, 2), (3, 3) and (2, 2), then the coordinates of its centroid are
(a) $(3, 7/3)$ (b) $(3, 3)$
(c) $(4, 3)$ (d) None of these

32. Let L denote the line in the xy-plane with x and y intercepts as 3 and 1 respectively. Then, the image of the point $(-1, -4)$ in this line is **[JEE Main 2020, 6 September]**
(a) $\left(\frac{11}{5}, \frac{28}{5}\right)$ (b) $\left(\frac{29}{5}, \frac{8}{5}\right)$
(c) $\left(\frac{8}{5}, \frac{29}{5}\right)$ (d) $\left(\frac{29}{5}, \frac{11}{5}\right)$

33. If a straight line passes through (x_1, y_1) and its segment between the axes is bisected at this point, then its equation is given by
(a) $\frac{x}{x_1} + \frac{y}{y_1} = 2$
(b) $2(xy_1 + yx_1) = x_1y_1$
(c) $xy_1 + yx_1 = x_1y_1$
(d) None of the above

34. If p_1, p_2, p_3 be the length perpendiculars from the points $(m^2, 2m)$, $(mm', m + m')$ and $(m'^2, 2m')$ respectively on the line
$x\cos\alpha + y\sin\alpha + \frac{\sin^2\alpha}{\cos\alpha} = 0$, then p_1, p_2, p_3 are in
(a) AP (b) GP
(c) HP (d) None of these

35. The acute angle θ through which the coordinate axes should be rotated for the point A (2, 4) to attain the new abscissa 4 is given by
(a) $\tan\theta = 3/4$ (b) $\tan\theta = 5/6$
(c) $\tan\theta = 7/8$ (d) None of these

36. If $\frac{2}{1!9!} + \frac{2}{3!7!} + \frac{1}{5!5!} = \frac{2^m}{n!}$, then orthocentre of the triangle having sides $x - y + 1 = 0$, $x + y + 3 = 0$ and $2x + 5y - 2 = 0$ is
(a) $(2m - 2n, m - n)$ (b) $(2m - 2n, n - m)$
(c) $(2m - n, m + n)$ (d) $(2m - n, m - n)$

37. The equation of a straight line passing through (1, 2) and having intercept of length 3 between the straight lines, $3x + 4y = 24$ and $3x + 4y = 12$ is
(a) $7x - 24y + 41 = 0$ (b) $7x + 24y = 55$
(c) $24x + 7y - 38 = 0$ (d) $24x - 7y = 10$

38. If one of the diagonal of a square is along the line $x = 2y$ and one of its vertices is (3, 0), then its sides through this vertex are given by, the equations
(a) $y - 3x + 9 = 0$, $3y + x - 3 = 0$
(b) $y + 3x + 9 = 0$, $3y + x - 3 = 0$
(c) $y - 3x + 9 = 0$, $3y - x + 3 = 0$
(d) $y - 3x + 3 = 0$, $3y + x + 9 = 0$

39. The equation $y = \sin x \sin(x + 2) - \sin^2(x + 1)$ represents a straight line lying in **[JEE Main 2019, 12 April]**
(a) II and III quadrants only
(b) I, II and IV quadrants
(c) I, III and IV quadrants
(d) III and IV quadrants only

40. P (m, n) (where m, n are natural numbers) is any point in the interior of the quadrilateral formed by the pair of lines $xy = 0$ and the two lines $2x + y - 2 = 0$ and $4x + 5y = 20$. The possible number of positions of the P is
(a) six (b) five
(c) four (d) eleven

41. If the family of straight lines $ax + by + c = 0$, where $2a + 3b = 4c$ is concurrent at the point $P(l, m)$, then the foot of the perpendicular drawn from P to the line $x + y + 1 = 0$ is

(a) $\left(\frac{-3}{8}, \frac{-5}{8}\right)$ (b) $\left(\frac{-2}{5}, \frac{-3}{5}\right)$

(c) $(3, -4)$ (d) $(-5, 4)$

42. $ABCD$ is a square whose vertices A, B, C and D are $(0, 0)$, $(2, 0)$, $(2, 2)$ and $(0, 2)$ respectively. This square is rotated in the $X - Y$ plane with an angle of 30° in anticlockwise direction about an axis passing through the vertex A the equation of the diagonal BD of this rotated square is....... If E is the centre of the square, the equation of the cirecumcircle of the ΔABE is

(a) $\sqrt{3}x + (1 - \sqrt{3})\, y = \sqrt{3}, x^2 + y^2 = 4$

(b) $(1 + \sqrt{3})\, x - (1 - \sqrt{2})\, y = 2, x^2 + y^2 = 9$

(c) $(2 - \sqrt{3})\, x + y = 2(\sqrt{3} - 1), x^2 + y^2 - x\sqrt{3} - y = 0$

(d) None of the above

43. If G be the centroid and I be the incentre of the triangle with vertices $A(-36, 7)$, $B(20, 7)$ and $C(0, -8)$ and $GI = \frac{25}{3}\sqrt{205}\,\lambda$, then $\lambda =$

(a) 25 (b) $\frac{1}{25}$ (c) $\frac{1}{5}$ (d) 5

44. The line $x + y = a$ meets the axis of x and y at A and B respectively a triangle AMN is inscribed in the triangle OAB, O being the origin, with right angle at N, M and N lie respectively on OB and AB. If the area of the triangle AMN is 3/8 of the area of the ΔOAB, then AN/BN is equal to

(a) 1 (b) 2 (c) 3 (d) 4

45. A straight line through a fixed point (2, 3) intersects the coordinate axes at distinct points P and Q. If O is the origin and the rectangle $OPRQ$ is completed, then the locus of R is **[JEE Main 2018]**

(a) $3x + 2y = 6$ (b) $2x + 3y = xy$

(c) $3x + 2y = xy$ (d) $3x + 2y = 6xy$

46. One diagonal of a square is along the line $8x - 15y = 0$ and one of its vertex is (1, 2). Then, the equation of the sides of the square passing through this vertex are

(a) $7x - 8y + 9 = 0$, $8x + 7y - 22 = 0$

(b) $9x - 8y + 7 = 0$, $8x + 9y - 26 = 0$

(c) $23x - 7y - 9 = 0, 7x + 23y - 53 = 0$

(d) None of the above

47. If the line $3x + 4y - 24 = 0$ intersects the X-axis at the point A and the Y-axis at the point B, then the incentre of the ΔOAB, where O is the origin, is **[JEE Main 2019, 10 January]**

(a) (4, 3) (b) (3, 4)

(c) (4, 4) (d) (2, 2)

48. The equations of the three sides of a triangle are $x = 2$, $y + 1 = 0$ and $x + 2y = 4$. The coordinates of the circumcentre of the triangle are

(a) (4, 0) (b) (2, – 1)

(c) (0, 4) (d) None of these

49. If $P\,(1 + \alpha/\sqrt{2}, 2 + \alpha/\sqrt{2})$ be any point on a line, then the range of values of t for which the point P lies between the parallel lines $x + 2y = 1$ and $2x + 4y = 15$ is

(a) $-\frac{4\sqrt{2}}{3} < \alpha < \frac{5\sqrt{2}}{6}$ (b) $0 < \alpha < \frac{5\sqrt{2}}{6}$

(c) $-\frac{4\sqrt{2}}{3} < \alpha < 0$ (d) None of these

50. If the point (a, a) fall between the lines $|x + y| = 2$, then

(a) $|a| = 2$ (b) $|a| = 1$

(c) $|a| < 1$ (d) $|a| < \frac{1}{2}$

51. If a ray travelling the line $x = 1$ gets reflected the line $x + y = 1$, then the equation of the line along which the reflected ray travels is

(a) $y = 0$ (b) $x - y = 1$

(c) $x = 0$ (d) None of these

52. Let a, b, c and d be non-zero numbers. If the point of intersection of the lines $4ax + 2ay + c = 0$ and $5bx + 2by + d = 0$ lies in the IV quadrant and is equidistant from the two axes, then **[JEE Main 2014]**

(a) $2bc - 3ad = 0$ (b) $2bc + 3ad = 0$

(c) $2ad - 3bc = 0$ (d) $3bc + 2ad = 0$

53. The line joining two points $A(2, 0)$ and $B(3, 1)$ is rotated about A in anticlockwise direction through an angle 15°. If B goes to C, then C is

(a) $\left(\frac{1}{2}(4 + \sqrt{2}), \sqrt{6}\right)$ (b) $\left(\frac{1}{2}(6 + \sqrt{2}), \frac{1}{2}\sqrt{6}\right)$

(c) $\left(\frac{1}{2}(2 + \sqrt{2}), \frac{1}{2}\sqrt{6}\right)$ (d) $\left(\frac{1}{2}(4 + \sqrt{2}), \frac{1}{2}\sqrt{6}\right)$

54. Consider the straight line $ax + by = c$, where $a, b, c \in R^+$ this line meets the coordinate axes at A and B respectively. If the area of the $\Delta\, OAB$, O being origin, does not depend upon a, b and c, then

(a) a, b, c are in AP

(b) a, b, c are in GP

(c) a, b, c are in HP

(d) None of the above

55. If A and B are two points having coordinates (3,4) and (5,–2) respectively and P is a point such that $PA = PB$ and area of $\Delta PAB = 10$ square unit, then the coordinates of P are

(a) (7,4) or (13,2) (b) (7,2) or (1,0)

(c) (2,7) or (4,13) (d) None of these

56. If the lines $x + ay + a = 0$, $bx + y + b = 0$ and $cx + cy + 1 = 0$ (a, b, c being distinct $\neq 1$) are concurrent, then the value of $\frac{a}{a-1} + \frac{b}{b-1} + \frac{c}{c-1}$ is

(a) -1 (b) 0
(c) 1 (d) None of these

57. The medians AD and BE of the triangle with vertices $A(0, b)$, $B(0,0)$ and $C(a,0)$ are mutually perpendicular if

(a) $b = \sqrt{2}\,a$ (b) $a = \sqrt{2}\,b$
(c) $b = -\sqrt{2}\,a$ (d) $a = 5\sqrt{2}\,b$

58. If the point $P(a^2, a)$ lies in the region corresponding to the acute angle between the lines $2y = x$ and $4y = x$, then

(a) $a \in (2,4)$ (b) $a \in (2,6)$
(c) $a \in (4,6)$ (d) $a \in (4,8)$

59. A ray of light coming along the line $3x + 4y - 5 = 0$ gets reflected from the line $ax + by - 1 = 0$ and goes along the line $5x - 12y - 10 = 0$, then

(a) $a = \frac{64}{115}, b = \frac{112}{15}$ (b) $a = -\frac{64}{115}, b = \frac{8}{115}$
(c) $a = \frac{64}{115}, b = -\frac{8}{115}$ (d) $a = -\frac{64}{115}, b = -\frac{8}{115}$

60. If line $2x + 7y - 1 = 0$ intersect the lines $L_1 \equiv 3x + 4y + 1 = 0$ and $L_2 \equiv 6x + 8y - 3 = 0$ in A and B respectively, then equation of a line parallel to L_1 and L_2 and passes through a point P such that $AP : PB = 2:1$ (internally) is (P is on the line $2x + 7y - 1 = 0$)

(a) $9x + 12y + 3 = 0$ (b) $9x + 12y - 3 = 0$
(c) $9x + 12y - 2 = 0$ (d) None of these

61. In a ΔABC, side AB has the equation $2x + 3y = 29$ and the side AC has the equation $x + 2y = 16$. If the mid-point of BC is (5,6), then the equation of BC is

(a) $2x + y = 16$ (b) $x + y = 11$
(c) $2x - y = 4$ (d) None of these

62. If the two lines $x + (a-1)y = 1$ and $2x + a^2y = 1$, ($a \in R - \{0, 1\}$) are perpendicular, then the distance of their point of intersection from the origin is **[JEE Main 2019, 9 April]**

(a) $\frac{2}{5}$ (b) $\frac{\sqrt{2}}{5}$ (c) $\frac{2}{\sqrt{5}}$ (d) $\sqrt{\frac{2}{5}}$

63. If the point $(\cos\theta, \sin\theta)$ does not fall in that angle between the lines $y = |x - 1|$ in which the origin lies, then θ belongs to

(a) $\left(\frac{\pi}{2}, \frac{3\pi}{2}\right)$ (b) $\left(-\frac{\pi}{2}, \frac{\pi}{2}\right)$
(c) $(0, \pi)$ (d) None of these

64. ABC is an equilateral triangle such that the vertices B and C lie on two parallel lines at a distance 6. If A lies between the parallel lines at a distance 4 from one of them, then the length of a side of the equilateral triangle is

(a) 8 (b) $\sqrt{\frac{88}{3}}$
(c) $\frac{4\sqrt{7}}{\sqrt{3}}$ (d) None of these

65. The four sides of a quadrilateral are given by the equation $xy(x-2)(y-3) = 0$. The equation of the line parallel to $x - 4y = 0$ that divides the quadrilateral in two equal areas is

(a) $x - 4y - 5 = 0$ (b) $x - 4y + 5 = 0$
(c) $x - 4y - 1 = 0$ (d) $x - 4y + 1 = 0$

66. $O(0, 0)$, $A(-3, -1)$ and $B(-1, -3)$ are the vertices of a ΔOAB. P is a point on the perpendicular AD drawn from A on OB such that, $\frac{AP}{PD} = \frac{3}{4}$. Then, the equation of the line L parallel to OB and passing through P, is

(a) $3x - y + 3 = 0$ (b) $21x - 7y + 32 = 0$
(c) $15x - 5y + 32 = 0$ (d) $3x - y + 35 = 0$

67. The line $x + y = 1$ meets X-axis at A and Y-axis at B, P is the mid-point of AB, P_1 is the foot of the perpendicular from P to OA; M_1 is that of P_1 from OP; P_2 is that of M_1 from OA; M_2 is that of P_2 from OP; P_3 is that of M_2 from OA and so on. If P_n denotes the nth foot of the perpendicular on OA from M_{n-1}, then OP_n is equal to

(a) $1/2n$ (b) $1/2^n$
(c) $2^n - 1$ (d) $2^n + 3$

68. On the portion of the straight line $x + y = 2$ which is intercepted between the axes, a square is constructed away from the origin, with this portion as one of its side. If p denotes the perpendicular distance of a side of this square from the origin, then the maximum value of p is

(a) $\sqrt{2}$ (b) $2\sqrt{2}$
(c) $3\sqrt{2}$ (d) $4\sqrt{2}$

69. If each of the points $(x_1, 4)$, $(-2, y_1)$ lies on the line joining the points $(2, -1)$, $(5, -3)$, then the point $P(x_1, y_1)$ lies on the line

(a) $6(x + y) - 25 = 0$ (b) $2x + 6y + 1 = 0$
(c) $2x + 3y - 6 = 0$ (d) $6(x + y) - 23 = 0$

70. Two vertices of a triangle are $(3, -2)$ and $(-2, 3)$ and its orthocentre is $(-6, 1)$. Then its third vertex is

(a) $(1, 6)$ (b) $(-1, 6)$
(c) $(1, -6)$ (d) None of these

71. The point $P(1, 1)$ is translated parallel to $2x = y$ in the first quadrant through a unit distance. The coordinates of the new position of P are

(a) $\left(1 \pm \frac{2}{\sqrt{5}}, 1 \pm \frac{1}{\sqrt{5}}\right)$ (b) $\left(1 \pm \frac{1}{\sqrt{5}}, 1 \pm \frac{2}{\sqrt{5}}\right)$
(c) $\left(\frac{1}{\sqrt{5}}, \frac{2}{\sqrt{5}}\right)$ (d) $\left(\frac{2}{\sqrt{5}}, \frac{1}{\sqrt{5}}\right)$

72. If $3a + 2b + 6c = 0$, the family of straight lines $ax + by + c = 0$ passes through a fixed point whose coordinates are given by
(a) (1/2, 1/3) (b) (2, 3)
(c) (3, 2) (d) (1/3, 1/2)

73. The base BC of a triangle ABC is bisected at the point (p, q) and the equations to the sides AB and AC are $px + qy = 1$ and $qx + py = 1$. The equation of the median through A is
(a) $(p - 2q)x + (q - 2p)y + 1 = 0$
(b) $(p + q)(x + y) - 2 = 0$
(c) $(2pq - 1)(px + qy - 1) = (p^2 + q^2 - 1)(qx + py - 1)$
(d) None of the above

74. A family of lines is given by the equation $(3x + 4y + 6) + \lambda(x + y + 2) = 0$. The line situated at the greatest distance from the point (2, 3) belonging to this family has the equation
(a) $15x + 8y + 30 = 0$
(b) $4x + 3y + 8 = 0$
(c) $5x + 3y + 6 = 0$
(d) $5x + 3y + 10 = 0$

75. The coordinates of the point P on the line $2x + 3y + 1 = 0$, such that $|PA - PB|$ is maximum, where A is (2, 0) and B is (0, 2) is
(a) (5, – 3) (b) (7, – 5)
(c) (9, – 7) (d) (11, – 9)

Objective Questions Type II [One or more than one correct answer(s)]

In each of the questions below four choices of which one or more than one are correct. You have to select the correct answer(s) accordingly.

1. Two straight lines $u = 0$ and $v = 0$ passes through the origin and angle between them is $\tan^{-1}\left(\frac{7}{9}\right)$. If the ratio of the slope of $v = 0$ and $u = 0$ is $\frac{9}{2}$, then their equations are
(a) $y = 3x$ and $3y = 2x$
(b) $2y = 3x$ and $3y = x$
(c) $y + 3x = 0$ and $3y + 2x = 0$
(d) $2y + 3x = 0$ and $3y + x = 0$

2. For all value of θ, the lines represented by the equation
$$(2\cos\theta + 3\sin\theta)x + (3\cos\theta - 5\sin\theta)y - (5\cos\theta - 2\sin\theta) = 0$$
(a) pass through a fixed point
(b) pass through the point (1, 1)
(c) pass through a fixed point whose reflection in the line $x + y = \sqrt{2}$ is $(\sqrt{2} - 1, \sqrt{2} - 1)$
(d) pass through the origin

3. The equations $(b - c)x + (c - a)y + a - b = 0$
$(b^3 - c^3)x + (c^3 - a^3)y + a^3 - b^3 = 0$
will represent the same line if
(a) $b = c$ (b) $c = a$
(c) $a = b$ (d) $a + b + c = 0$

4. The area of a triangle is 5 sq unit. Two of its vertices are (2, 1) and (3, –2). The third vertex lies on $y = x + 3$. The coordinates of the third vertex can be
(a) (–3/2, 3/2) (b) (3/4, –3/2)
(c) (7/2, 13/2) (d) (–1/4, 11/4)

5. If the point $P(x, y)$ be equidistant from the points $A(a + b, a - b)$ and $B(a - b, a + b)$, then
(a) $ax = by$ (b) $bx = ay$
(c) $x^2 - y^2 = 2(ax + by)$ (d) P can be (a, b)

6. The straight line $3x + 4y - 12 = 0$ meets the coordinate axes at A and B. An equilateral triangle ABC is constructed. The possible coordinates of vertex C are
(a) $\left(2\left(1 + \frac{3\sqrt{3}}{4}\right), \frac{3}{2}\left(1 + \frac{4}{\sqrt{3}}\right)\right)$
(b) $\left(2\left(1 - \frac{3\sqrt{3}}{4}\right), \frac{3}{2}\left(1 - \frac{4}{\sqrt{3}}\right)\right)$
(c) $\left(2(1 + \sqrt{3}), \frac{3}{2}(1 + \sqrt{3})\right)$
(d) $\left(-2(1 + \sqrt{3}), \frac{3}{2}(1 - \sqrt{3})\right)$

7. Equation of a straight line passing through the point of intersection of $x - y + 1 = 0$ and $3x + y - 5 = 0$ are perpendicular to one of them is
(a) $x + y + 3 = 0$ (b) $x + y - 3 = 0$
(c) $x - 3y - 5 = 0$ (d) $x - 3y + 5 = 0$

8. If one vertex of an equilateral triangle of side a, lies at the origin and the other lies on the line $x - \sqrt{3}y = 0$, the coordinates of the third vertex are
(a) $(0, a)$ (b) $(\sqrt{3}a/2, -a/2)$
(c) $(0, -a)$ (d) $(-\sqrt{3}a/2, a/2)$

9. If the lines $ax + by + c = 0$, $bx + cy + a = 0$ and $cx + ay + b = 0$ are concurrent $(a + b + c \neq 0)$ then
(a) $a^3 + b^3 + c^3 - 3abc = 0$
(b) $a \neq b$
(c) $a = b = c$
(d) $a^2 + b^2 + c^2 - bc - ca - ab = 0$

10. If the coordinates of the vertices of a triangle are rational numbers, then which of the following points of the triangle will always have rational coordinates
(a) centroid (b) incentre
(c) circumcentre (d) orthocentre

11. Which of the following statement(s) is/are correct?
(a) The lines $x + (\log_a b)y + (\log_a c) = 0$, $(\log_b a)\,x + y + (\log_b c) = 0$ and $(\log_c a)x + (\log_c b)y + 1 = 0$ are concurrent
(b) Equation of a straight line passing through the origin and making with positive X-axis an angle twice the size of the angle made by the line $y = 0.2\,x$ with the positive X-axis, is $y = 0.\,4x$
(c) Area of the triangle formed by the lines $y^2 - 9xy + 18x^2 = 0$ and $y = 9$ is $\frac{27}{4}$ sq units
(d) Lines through the origin and perpendicular to the lines $xy - 3y^2 + y - 2x + 10 = 0$ is $3y^2 + xy = 0$

12. If $bx + cy = a$, where a, b, c are the same sign, be a line such that the area enclosed by the line and the axes of reference is $\frac{1}{8}$ unit^2, then
(a) b, a, c are in GP (b) $b, 2a, c$ are in GP
(c) $b, \frac{a}{2}, c$ are in AP (d) $b, -2a, c$ are in GP

13. Two sides of a rhombus $OABC$ (lying entirely in the I and III quadrant) of area equal to 2 sq units are $y = \frac{x}{\sqrt{3}}$, $y = \sqrt{3}x$. Then, the possible coordinates of B is/are (O being at origin)
(a) $(1 + \sqrt{3}, 1 + \sqrt{3})$ (b) $(-1 - \sqrt{3}, -1 - \sqrt{3})$
(c) $(1 + \sqrt{3}, 3 + \sqrt{3})$ (d) $(\sqrt{3} - 1, \sqrt{3} - 1)$

14. Two roads are represented by the equations $y - x = 6$ and $x + y = 8$. An inspection bunglow has to be so constructed that it is at a distance of 100 from each of the roads. Possible location of the bunglow is given by
(a) $(100\sqrt{2} + 1, 7)$ (b) $(1 - 100\sqrt{2}, 7)$
(c) $(1, 7 + 100\sqrt{2})$ (d) $(1, 7 - 100\sqrt{2})$

15. Angles made with the X-axis by two lines drawn through the point (1, 2) cutting the line $x + y = 4$ at a distance $\sqrt{6}/3$ from the point (1, 2) are
(a) $\frac{\pi}{12}$ and $\frac{5\pi}{12}$ (b) $-\frac{7\pi}{12}$ and $-\frac{11\pi}{12}$
(c) $\frac{\pi}{8}$ and $\frac{3\pi}{8}$ (d) None of these

16. If (a, b) be an end of a diagonal of a square and the other diagonal has the equation $x - y = a$, then another vertex of the square can be
(a) $(a - b, a)$ (b) $(a, 0)$
(c) $(0, -a)$ (d) $(a + b, b)$

17. The equation of the bisectors of the angles between the two intersecting lines $\frac{x-3}{\cos\theta} = \frac{y+5}{\sin\theta}$ and $\frac{x-3}{\cos\phi} = \frac{y+5}{\sin\phi}$ are $\frac{x-3}{\cos\alpha} = \frac{y+5}{\sin\alpha}$ and $\frac{x-3}{\beta} = \frac{y+5}{\gamma}$, then
(a) $\alpha = \frac{\theta + \phi}{2}$ (b) $\beta = -\sin\alpha$
(c) $\gamma = \cos\alpha$ (d) $\beta = \sin\alpha$

18. The family of lines $(2\cos\theta + 3\sin\theta)\,x + (3\cos\theta - 5\sin\theta)\,y - (5\cos\theta - 2\sin\theta) = 0$ is concurrent at (x_1, y_1) and the family of lines $(a + 2b)\,x + (a - 3b)\,y + b - a = 0$, for different values of a and b is concurrent at (x_2, y_2), then
(a) $x_2 + y_2 = x_1$ (b) $x_2 + y_2 = y_1$
(c) $6x_2 + y_2 = 2x_1 + y_1$ (d) None of these

19. A line which makes an acute angle θ with the positive direction of X-axis is drawn through the point P (3, 4) to meet the line $x = 6$ at R and $y = 8$ at S, then
(a) $PR = 3\sec\theta$
(b) $PS = 4\operatorname{cosec}\theta$
(c) $PR + PS = \frac{2\,(3\sin\theta + 4\cos\theta)}{\sin 2\theta}$
(d) $\frac{9}{(PR)^2} + \frac{16}{(PS)^2} = 1$

20. A (1, 3) and C (7, 5) are two opposite vertices of a square. The equation of a side through A is
(a) $x + 2y - 7 = 0$ (b) $x - 2y + 5 = 0$
(c) $2x + y - 5 = 0$ (d) $2x - y + 1 = 0$

21. Equation(s) of the straight line(s), inclined at 30° to the axis of x such that the length of its (each of their) line segment(s) between the coordinate axes is 10 units, is (are)
(a) $x - \sqrt{3}y - 5\sqrt{3} = 0$ (b) $x - \frac{y}{\sqrt{3}} - \frac{5}{\sqrt{3}} = 0$
(c) $x - \sqrt{3}y + 5\sqrt{3} = 0$ (d) $x - \frac{y}{\sqrt{3}} + \frac{5}{\sqrt{3}} = 0$

22. Let L be the line $2x + y = 2$. If the axes are rotated by 45°, then the intercept made by the line L on the length of new axes are respectively
(a) $\sqrt{2}$ and 1 (b) 1 and $\sqrt{2}$
(c) $2\sqrt{2}$ and $2\sqrt{2}/3$ (d) $2\sqrt{2}/3$ and $2\sqrt{2}$

23. If m_1, m_2 are the roots of the equation $x^2 - ax - a - 1 = 0$, then the area of the triangle formed by the three straight lines $y = m_1 x$, $y = m_2 x$ and $y = a\ (a \neq -1)$ is
(a) $\frac{a^2\,(a+2)}{2\,(a+1)}$, if $a > -1$
(b) $-\frac{a^2\,(a+2)}{2\,(a+1)}$, if $a < -1$
(c) $-\frac{a^2\,(a+2)}{2\,(a+1)}$, if $-2 < a < -1$
(d) $\frac{a^2(a+2)}{2\,(a+1)}$, if $a < -2$

24. A (1, 2) and B (7, 10) are two points. If P (x, y) is a point such that the angle APB is 60° and the area of the Δ APB is maximum, then which of given is (are) true ?
(a) P lies on any line perpendicular to AB
(b) P lies on the right bisector of AB
(c) P lies on the straight line $3x + 4y = 36$
(d) P lies on the circle passing through the points (1, 2) and (7, 10) and having a radius of 10 units

25. The coordinates of the points A and B are respectively (–3, 2) and (2, 3). P and Q are points on the line joining A and B such that $AP = PQ = QB$. A square $PQRS$ is constructed on PQ as one side, the coordinates of R can be
(a) $\left(-\frac{4}{3}, \frac{7}{3}\right)$ (b) $\left(0, \frac{13}{3}\right)$
(c) $\left(\frac{1}{3}, \frac{8}{3}\right)$ (d) $\left(\frac{2}{3}, 1\right)$

Linked-Comprehension Type

In these questions, a passage (paragraph) has been given followed by questions based on each of the passages. You have to answer the questions based on the passage given.

PASSAGE 1

Let P (x, y) be the cartesian coordinates with respect to axes OX and OY, then (r, θ) be its polar coordinates with respect to pole O and initial line OX *ie*, $OP = r$ *(radius vector) and* $\angle XOP = \theta$ *(vectorial angle)*

Now let p be the length of perpendicular from O upon straight line (through A, B) i.e., $OM = p$ and $\angle XOM = \alpha$

In ΔOMP, $\angle POM = (\theta - \alpha)$

We have,

$OM = OP \cos(\theta - \alpha)$

or $p = r \cos(\theta - \alpha)$ which is the required equation to the given line.

On the basis of above information, answer the following questions

1. Cartesian form of the curve $\sqrt{r} = \sqrt{a} \sin(\theta / 2), \forall\, a > 0$ is
(a) $4(x^2 - y^2)(x^2 + y^2 + ax) = a^2y^2$
(b) $4(x^2 - y^2)(x^2 + y^2 - ax) = a^2y^2$
(c) $4(x^2 + y^2)(x^2 + y^2 + ax) = a^2y^2$
(d) $4(x^2 + y^2)(x^2 + y^2 - ax) = a^2y^2$

2. Polar form of the curve $x^3 + 3x^2y - 3xy^2 - y^3 = 5kxy$ is
(a) $2r(\cos 3\theta + \sin 3\theta) = 5k \sin\theta \cos\theta$
(b) $r(\cos 3\theta + \sin 3\theta) = 5k \sin\theta \cos\theta$
(c) $2r(\cos 3\theta - \sin 3\theta) = 5k \sin\theta \cos\theta$
(d) $r(\cos 3\theta - \sin 3\theta) = 5k \sin\theta \cos\theta$

3. Locus of the point $P(r, \theta)$, if $\frac{2}{r} = -\frac{1}{2} + 4\cos\theta$ is a/an
(a) circle
(b) parabola
(c) ellipse
(d) hyperbola

4. If $\frac{l}{r} = f(\theta)$, where $f(\theta) = a\cos(\theta + \alpha) + b\cos(\theta + \beta)$, represents a straight line and that any line perpendicular to it is
(a) $\frac{l}{r} = J(\theta + \pi/2)$
(b) $\frac{l}{r} = f(\theta + \pi/2)$
(c) $\frac{\lambda}{r} = J(\theta + \pi / 2)$, where λ is a parameter
(d) $\frac{\lambda}{r} = f(\theta + \pi / 2)$, where λ is a parameter

5. If $(1, \pi / 6), (2, \pi / 3)$ and $(3, \pi / 2)$ be the angular points of a triangle. Then, the area of the triangle is
(a) $\left(\frac{13 - 3\sqrt{3}}{4}\right)$ sq unit (b) $\left(\frac{11 - 3\sqrt{3}}{4}\right)$ sq unit
(c) $\left(\frac{10 - 3\sqrt{3}}{4}\right)$ sq unit (d) $\left(\frac{8 - 3\sqrt{3}}{4}\right)$ sq unit

PASSAGE 2

The locus of a moving point is the path traced out by that point under one or more given conditions. Technically, a *locus* represents the '*set of all points*' which lies on it.

A relation $f(x, y) = 0$ between x and y which is satisfied by each point on the locus and such that each point satisfying the equation is on the locus is called the *equation of the locus.*

On the basis of above information, answer the following questions

1. Locus of centroid of the triangle whose vertices are $(a\cos t, a\sin t)$, $(b\sin t, -b\cos t)$ and (1, 0), where t is a parameter, is
(a) $(3x - 1)^2 + (3y)^2 = a^2 - b^2$
(b) $(3x - 1)^2 + (3y)^2 = a^2 + b^2$
(c) $(3x + 1)^2 + (3y)^2 = a^2 + b^2$
(d) $(3x + 1)^2 + (3y)^2 = a^2 - b^2$

2. Let $A\ (2, -3)$ and $B\ (-2, 1)$ be vertices of a $\Delta\ ABC$. If the centroid of this triangle moves on the line $2x + 3y = 1$, then the locus of the vertex C is the line
(a) $2x + 3y = 9$ (b) $2x - 3y = 7$
(c) $3x + 2y = 5$ (d) $3x - 2y = 3$

3. If the equation of the locus of a point equidistant from the points (a_1, b_1) and (a_2, b_2) is $(a_1 - a_2)\ x + (b_1 - b_2)y + c = 0$, then the value of '$c$' is
(a) $\frac{1}{2}(a_2^2 + b_2^2 - a_1^2 - b_1^2)$
(b) $a_1^2 - a_2^2 + b_1^2 - b_2^2$
(c) $\frac{1}{2}(a_1^2 + a_2^2 + b_1^2 + b_2^2)$
(d) $\sqrt{(a_1^2 + b_1^2 - a_2^2 - b_2^2)}$

4. A straight line L through the origin meets the lines $x + y = 1$ and $x + y = 3$ at P and Q respectively. Through P and Q two straight lines L_1 and L_2 are drawn, parallel to $2x - y = 5$ and $3x + y = 5$ respectively. Lines L_1 and L_2 intersected at R. Then the locus of R is the line
(a) $x + 3y + 5 = 0$
(b) $x - 3y - 5 = 0$
(c) $x - 3y + 5 = 0$
(d) $x + 3y - 5 = 0$

5. The area of the triangle formed by the intersection of a line parallel to X-axis and passing through $P\ (h, k)$ with the lines $y = x$ and $x + y = 2$ is $4h^2$. Then the locus of the point P is
(a) $y = 3x - 2$ or $y = -3x - 2$
(b) $y = 2x - 1$ or $y = -2x - 1$
(c) $y = 3x + 2$ or $y = -3x + 2$
(d) $y = 2x + 1$ or $y = -2x + 1$

PASSAGE 3

Let $A\ (x_1, y_1), B\ (x_2, y_2)$ and $C\ (x_3, y_3)$ be three points such that $AC = AB$ and $\angle CAB = \theta$.

Let $z_1 = x_1 + iy_1, z_2 = x_2 + iy_2$ and $z_3 = x_3 + iy_3$, where $i = \sqrt{-1}$.

Therefore, $\mathbf{AB} = z_2 - z_1$, $\mathbf{AC} = z_3 - z_1$

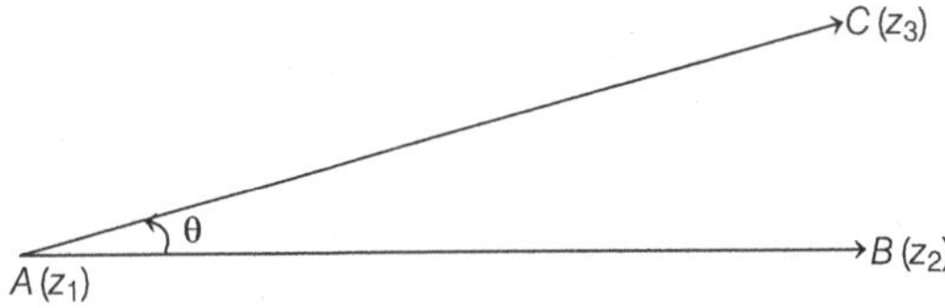

Then, **AC** will be obtained by rotating **AB** through an angle θ in anticlockwise sense and therefore $\mathbf{AC} = \mathbf{AB}\ e^{i\theta}$

or $(z_3 - z_1) = (z_2 - z_1)e^{i\theta}$

or $\left(\frac{z_3 - z_1}{z_2 - z_1}\right) = e^{i\theta}$

On the basis of above information, answer the following questions

1. The line joining the points $A\ (1, 0)$ and $B\ (1 + \sqrt{3}, 1)$ is rotated about A by an angle θ. If B goes to $C\ (2, \sqrt{3})$, then the angle θ is
(a) $15°$ (b) $30°$
(c) $45°$ (d) $90°$

2. The extremities of the diagonal of a square are $(1, 1), (-2, -1)$, then the equation of the other diagonal is
(a) $8x + 3y + 4 = 0$
(b) $8x - 3y + 4 = 0$
(c) $6x + 4y + 3 = 0$
(d) $6x - 4y + 3 = 0$

3. A point $(2, 2)$ undergoes reflection in the x-axis and then the coordinate are rotated through an angle of $\pi / 4$ in anticlockwise direction. The final position of the point in the new coordinate system is
(a) $(0, 2\sqrt{2})$ (b) $(0, -2\sqrt{2})$
(c) $(2\sqrt{2}, 0)$ (d) $(-2\sqrt{2}, 0)$

4. The line joining the points $A\ (3, 0)$ and $B\ (5, 2)$ is rotated about A in the anticlockwise direction through an angle of $15°$. If B goes to C in the new position now the line joining A and C is rotated about A in the anticlockwise direction through an angle of $45°$ of C goes to D in the new position, then the coordinates of D are
(a) $(4 - \sqrt{3}, \sqrt{3} - 1)$
(b) $(4 + \sqrt{3}, \sqrt{3} - 1)$
(c) $(4 - \sqrt{3}, \sqrt{3} + 1)$
(d) $(4 + \sqrt{3}, \sqrt{3} + 1)$

PASSAGE 4

A variable line L is drawn through $O(0, 0)$ to meet the lines L_1 and L_2 given by $y - x - 10 = 0$ and $y - x - 20 = 0$ at points A and B, respectively.

On the basis of above information, answer the following questions

1. A point P is taken on L such that, $\frac{2}{OP} = \frac{1}{OA} + \frac{1}{OB}$. Then the locus of P is
(a) $3x + 3y = 40$
(b) $3x + 3y + 40 = 0$
(c) $3x - 3y = 40$
(d) $3y - 3x = 40$

2. Locus of P, if $OP^2 = OA \times OB$, is
(a) $(y - x)^2 = 100$ (b) $(y + x)^2 = 50$
(c) $(y - x)^2 = 200$ (d) None of these

3. Locus of P, if $\left(\frac{1}{OP^2}\right) = \left(\frac{1}{OB^2}\right) + \left(\frac{1}{OA^2}\right)$, is
(a) $(y - x)^2 = 80$ (b) $(y - x)^2 = 100$
(c) $(y - x)^2 = 64$ (d) None of these

Numerical Value Type

Solve the following problems and write your answer.

1. If (α, β) be the coordinates of incentre of the triangle whose vertices are $(4, -2)$, $(-2, 4)$ and $(5, 5)$, then the value of $2\alpha + 10\beta$ must be
2. If G be the centroid and I be the incentre of the triangle with vertices $A(-36, 7)$, $B(20, 7)$ and $C(0, -8)$ and $GI = \frac{25}{3}\sqrt{(205)}\,\lambda$, then the value of $\frac{1}{\lambda}$ must be
3. If a triangle has its orthocentre at $(1, 1)$ and circumcentre at $\left(\frac{3}{2}, \frac{3}{4}\right)$ and if centroid and nine point centre are (α, β) and (γ, δ) respectively, then the value of $6\alpha + 12\beta + 4\gamma + 8\delta$ must be
4. The locus of the moving point P such that $2PA = 3PB$, where A is $(0, 0)$ and B is $(4, -3)$ is $5x^2 + 5y^2 - 72x + \lambda y + \mu = 0$, then the value of $\lambda + \mu$ must be
5. If m_1 and m_2 are the roots of the equation $x^2 + (\sqrt{3} + 2)x + (\sqrt{3} - 1) = 0$ and if area of the triangle formed by the lines $y = m_1 x$, $y = m_2 x$, and $y = c$ is $(a + b)c^2$, then the value of $2008\,(a^2 + b^2)$ must be
6. If $(\lambda, 2)$ is an interior point of ΔABC formed by $x + y = 4$, $3x - 7y = 8$ and $4x - y = 31$ then $\lambda \in (a, b)$, the value of $6a + 8b$ must be
7. Consider the two lines $L_1 \equiv x - y = 0$ and $L_2 \equiv x + y = 0$ and a moving point (x, y). Let $d(P, L_i)$, $i = 1, 2$ represents the distance of the point 'P' from L_i. If point 'P' moves in certain region 'R' in such a way $\sum_{i=1}^{2} d(P, L_i) \in [2, 4]$. If the area of region 'R' is λ sq unit, then the value of λ must be
8. A rectangle $PQRS$ has its side PQ parallel to the line $y = mx$ and vertices P, Q and S on the lines $y = a$, $x = b$ and $x = -b$ respectively. If the locus of the vertex R is $24x - 5y + 36 = 0$ (for $a = 2, b = 1$), then the value of $1 + m + m^2$ must be
9. $P_1, P_2, P_3 \ldots\ldots\ldots\ldots, P_n$ are points on the line $y = x$ lying in the positive quadrant such that $\frac{OP_n}{OP_{n-1}} = n$, O being the origin. If $OP_1 = 1$ then the coordinates of P_8 are $(3a\sqrt{2}, 3a\sqrt{2})$, then the value of a must be
10. In a triangle ABC, E is a point on AC dividing it in the ratio 1:2 and F is a point on AB dividing it in the ratio 1:2. The equation of the line DE, which is perpendicular to AC, is $x + 2y + 12 = 0$ and the equation of the line DF, which is perpendicular to AB is $x - y + 3 = 0$. If λ sq unit be the area of the triangle DBC, where A is $(2, 3)$, then λ must be
11. A $(3, 0)$ and B $(6, 0)$ are two fixed points and U (α, β) is a variable point on the plane. AU and BU meet the Y-axis at C and D respectively, and AD meets OU at V, then CV passes through $(\lambda, 0)$ for any position of U in the plane, then the value of $1004\,\lambda$ must be
12. Let $(3, 4)$ be a fixed point. A straight line passing through this point cuts the positive direction of the coordinate axes at the points P and Q. If λ sq unit be the minimum area of the triangle OPQ, O being the origin. Then the value of $\lambda/5$ must be
13. If from point P $(4, 4)$ perpendiculars to the straight lines $3x + 4y + 5 = 0$ and $y = mx + 7$ meet at Q and R and area of triangle PQR is maximum. Then the value of m must be
14. If α, β are natural numbers such that $100\alpha - 199\beta = (100)(100) + (99)(101) + (98)(102) + \ldots + (1)(199)$, then the slope of the line passing through (α, β) and origin is **[JEE Main 2021, 18 March]**

Single Integer Type

Solve the following problems and write your answer.

1. Let ABC be a triangle. Let A be the point $(1, 2)$, $y = x$ is the perpendicular bisector of AB and $x - 2y + 1 = 0$ is the angle bisector of $\angle C$. If equation of BC is given by $ax + by - 5 = 0$, then the value of $a + b$ is
2. Let $A(1, 0)$, $B(6, 2)$ and $C\left(\frac{3}{2}, 6\right)$ be the vertices of a ΔABC. If P is a point inside the ΔABC such that, the ΔAPC, ΔAPB and ΔBPC have equal areas, then the length of the line segment PQ where Q is the point $\left(-\frac{7}{6}, -\frac{1}{3}\right)$ is **[JEE Main 2020, 7 January]**
3. In a ΔABC the bisec of angles B and C lie along the lines $x = y$ and $y = 0$. If A is $(1, 2)$ then $\sqrt{10}\, d(A, BC)$ where $d(A, BC)$ represents distance of point A from side BC.
4. ΔABC with $AB = 13$, $BC = 5$, and $AC = 12$ slides on the coordinates axes with A and B on the positive X-axis and positive Y-axis respectively. The locus of vertex C is a line $12x - ky = 0$. Then, the value of k is ....
5. Th equations, of the two adjacent sides of a rhombus formed in I quadrant are represented by $7x^2 - 8xy + y^2 = 0$, then slope of its longer diagonal is
6. The line $y = \frac{3x}{4}$ meets the lines $x - y + 1 = 0$ and $2x - y - 5 = 0$ at points A and B, respectively. If P on the line $y = \frac{3x}{4}$ satisfies the condition $PA \cdot PB = 25$, then the number of possible coordinates of P is ....
7. The vertices B and C of a ΔABC lie on the lines $3y = 4x$ and $y = 0$ respectively and the side BC passes through the point $\left(\frac{2}{3}, \frac{2}{3}\right)$. If $ABOC$ is a rhombus, O being the origin. If co-ordinates of vertex A is (α, β), then find the value of $\frac{5}{2}(\alpha + \beta)$.

8. The line $x + y = a$ meets the axis of x and y at A and B respectively. A ΔAMN is inscribed in the ΔOAB, O being the origin, with right angle at N. M and N lie on OB and AB respectively. If the area of the ΔAMN is $\frac{3}{8}$ of the area of ΔOAB, then $\frac{AN}{BN}$ =

Matrix-Match Type

Given below are Matching Type Questions, with two columns (each having some items) each . Each item of Column I has to be matched with the items of Column II, by encircling the correct match(es).

NOTE *An item of Column I can be matched with more than one items of Column II.*

1. Observe the following columns

	Column I		Column II
(A)	The family of lines $x(2a + 3b) + y(3a + 2b) = 2(b - a)$ passes through the point for all values of a and b, then the point is	(P)	$(3, -2)$
		(Q)	$(3, 2)$
(B)	If $9a^2 + 4b^2 - c^2 + 12ab = 0$, then the family of st. lines $ax + by + c = 0$ is concurrent at	(R)	$(-4, -3)$
(C)	If $12a^2 - 6b^2 - c^2 + ab + 5bc + ac = 0$, then the family of lines $ax + by + c = 0$ is concurrent at	(S)	$(2, -2)$
		(T)	$(-3, -2)$

2. Observe the following columns

	Column I		Column II
(A)	For the lines $4x + 3y - 6 = 0$ and $5x + 12y + 9 = 0$, acute angle bisector and obtuse angle bisectors represented by A and O respectively, then	(P)	$A: 7x - 9y - 3 = 0$
(B)	For the lines $4x - 3y - 6 = 0$ and $5x - 12y + 9 = 0$, acute angle bisector and obtuse angle bisectors represented by A and O respectively, then	(Q)	$A: 7x - 9y + 3 = 0$
(C)	For the straight lines $4x - 3y + 6 = 0$ and $5x - 12y - 9 = 0$, acute angle bisector and obtuse angle bisectors represented by A and O respectively, then	(R)	$A: 7x + 9y - 3 = 0$
		(S)	$O: 9x + 7y - 41 = 0$
		(T)	$O: 9x - 7y - 41 = 0$

3. Observe the following columns

	Column I		Column II
(A)	The area bounded by the curve $\max\{\|x\|, \|y\|\} = 1$ is	(P)	0
(B)	If the point (a, a) lies between the lines $\|x + y\| = 6$, then $[\|a\|]$ is (where [.] denotes the greatest integer function)		
		(Q)	1
(C)	Number of integral values of b for which the origin and the point (1, 1) lie on the same side of the st. line $a^2x + aby + 1 = 0$ for all $a \in R \sim \{0\}$ is	(R)	2
		(S)	3
		(T)	4

4. Observe the following columns

	Column I		Column II
(A)	If (α, α^2) lies inside the triangle formed by the lines $2x + 3y - 1 = 0$, $x + 2y - 3 = 0$, $5x - 6y - 1 = 0$ then $[\alpha]$ is (where [.] denotes the greatest integer function)	(P)	−2
(B)	If $(0, \alpha)$ lies inside the triangle formed by the lines $y - 3x + 2 = 0$, $3y - 2x - 5 = 0$ and $4y + x - 14 = 0$ then $[\alpha]$ is (where [.] denotes the greatest in the function)	(Q)	−1
(C)	The point $(\alpha^2, \alpha + 1)$ lies in the angle between the lines $3x - y + 1 = 0$ and $x + 2y - 5 = 0$ containing the origin then $[\alpha]$ is (where [.] denotes the greatest integer function)	(R)	0
		(S)	1
		(T)	2

5. If $y = m_i x + \frac{1}{m_i}$ $(i = 1, 2, 3)$ represent three straight line whose slopes are the roots of the equation $2m^3 - 3m^2 - 3m + 2 = 0$, then

Observe the following columns

	Column I		Column II
(A)	Algebraic sum of the intercepts made by the lines on X-axis.	(P)	$\frac{(4\sqrt{2} + 9\sqrt{5})}{4}$
(B)	Algebraic sum of the intercepts made by the lines on Y-axis	(Q)	$\frac{3}{2}$
(C)	Sum of the distances of the lines from the origin.	(R)	$-\frac{21}{4}$
(D)	Sum of the lengths of the lines intercepted between the coordinate axes.	(S)	$\frac{(5\sqrt{2} + 9\sqrt{5})}{10}$

Answers

Objective Questions Type I [Only one correct answer]

1. (b)	**2.** (d)	**3.** (a)	**4.** (b)	**5.** (c)	**6.** (b)	**7.** (a)	**8.** (b)	**9.** (c)	**10.** (a)
11. (c)	**12.** (b)	**13.** (b)	**14.** (d)	**15.** (c)	**16.** (b)	**17.** (a)	**18.** (b)	**19.** (b)	**20.** (b)
21. (c)	**22.** (b)	**23.** (b)	**24.** (c)	**25.** (a)	**26.** (a)	**27.** (a)	**28.** (d)	**29.** (b)	**30.** (c)
31. (a)	**32.** (a)	**33.** (a)	**34.** (b)	**35.** (a)	**36.** (a)	**37.** (a)	**38.** (a)	**39.** (d)	**40.** (a)
41. (a)	**42.** (c)	**43.** (b)	**44.** (c)	**45.** (c)	**46.** (c)	**47.** (d)	**48.** (a)	**49.** (a)	**50.** (c)
51. (a)	**52.** (c)	**53.** (d)	**54.** (b)	**55.** (b)	**56.** (c)	**57.** (b)	**58.** (a)	**59.** (c)	**60.** (c)
61. (b)	**62.** (d)	**63.** (b)	**64.** (c)	**65.** (b)	**66.** (b)	**67.** (b)	**68.** (c)	**69.** (b)	**70.** (b)
71. (b)	**72.** (a)	**73.** (c)	**74.** (b)	**75.** (b)					

Objective Questions Type II [One or more than one correct answer(s)]

1. (a, b, c, d)	**2.** (a, b, c)	**3.** (a, b, c, d)	**4.** (a, c)	**5.** (b, d)
6. (a, b)	**7.** (b, d)	**8.** (a, b, c, d)	**9.** (a, c, d)	**10.** (a, c, d)
11. (a, c)	**12.** (b, d)	**13.** (a, b)	**14.** (a, b, c, d)	**15.** (a, b)
16. (b, d)	**17.** (a, b, c)	**18.** (a, b, c)	**19.** (a, b, c, d)	**20.** (a, d)
21. (a, c)	**22.** (c, d)	**23.** (a, c, d)	**24.** (b, c)	**25.** (b, d)

Linked-Comprehension Type

Passage 1 **1.** (c) **2.** (b) **3.** (d) **4.** (c) **5.** (d)
Passage 2 **1.** (b) **2.** (a) **3.** (a) **4.** (c) **5.** (d)
Passage 3 **1.** (b) **2.** (c) **3.** (b) **4.** (c)
Passage 4 **1.** (d) **2.** (c) **3.** (a)

Numerical Grid-Based Problems

1. 30 **2.** 25 **3.** 30 **4.** 279 **5.** 5522 **6.** 110 **7.** 24 **8.** 31 **9.** 6720 **10.** 27
11. 2008 **12.** 4.8 **13.** 1.33 **14.** 550

Single Integer Type Question

1. (2) **2.** (5) **3.** (4) **4.** (0) **5.** (2) **6.** (3) **7.** (6) **8.** (3)

Matrix-Match Type

1. A → (S); B → (Q, T); C → (P, R)
2. A → (R, T); B → (P, S); C → (Q)
3. A → (T); B → (P, Q, R); C → (S)
4. A → (P, R); B → (S, T); C → (P, Q, R)
5. A → (R); B → (Q); C → (S); D → (P)

Solutions

Objective Questions Type [Only one correct answer]

1. Let a ΔABC is such that vertices

$$A(1, 2), B(x_1 y_1) \text{ and } C(x_2, y_2)$$

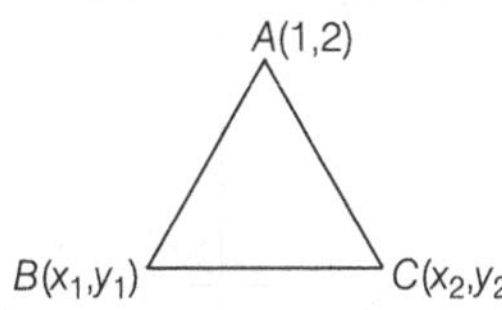

It is given that mid-point of side AB is $(-1, 1)$.

So, $\dfrac{x_1+1}{2} = -1$ and $\dfrac{y_1+2}{2} = 1$

$\Rightarrow$ $x_1 = -3$ and $y_1 = 0$

So, point B is $(-3, 0)$

Also, it is given that mid-point of side AC is $(2, 3)$, so $\dfrac{x_2+1}{2} = 2$ and $\dfrac{y_2+2}{2} = 3$

$\Rightarrow$ $x_2 = 3$ and $y_2 = 4$

So, point C is $(3, 4)$.

Now, centroid of ΔABC is

$$G\left(\frac{1+(-3)+3}{3}, \frac{2+0+4}{3}\right) = G\left(\frac{1}{3}, 2\right)$$

2. Let R be the radius of the circumcircle and O be the origin, then, $AO = \sqrt{(x_1^2 + x_1^2 \tan^2 \alpha)}$

$\Rightarrow$ $R = x_1 \sec \alpha$

$\Rightarrow$ $x_1 = R \cos \alpha$

Similarly, $x_2 = R \cos \beta$ and $x_3 = R \cos \gamma$

So, the coordinates of vertices are $A(R\cos\alpha, R\sin\alpha)$, $B(R\cos\beta, R\sin\beta)$, $C(R\cos\gamma, R\sin\gamma)$.

Hence, the coordinates of centroid G are

$$\left(\frac{\Sigma R\cos\alpha}{3}, \frac{\Sigma R\sin\alpha}{3}\right).$$

Since, the orthocentre $H(a, b)$, circumcentre $O(0, 0)$ and the centroid G are collinear therefore, slope of OH = slope of OG

$$= \frac{b}{a} = \frac{R(\sin\alpha + \sin\beta + \sin\gamma)}{R(\cos\alpha + \cos\beta + \cos\gamma)}.$$

$\therefore$ $$\frac{a}{b} = \frac{\cos\alpha + \cos\beta + \cos\gamma}{\sin\alpha + \sin\beta + \sin\gamma}$$

3. Circumcentre $O \equiv \left(\dfrac{-1}{3}, \dfrac{2}{3}\right)$ and Orthocentre $H \equiv \left(\dfrac{11}{3}, \dfrac{4}{3}\right)$

$\therefore$ coordinates of G are $\left(1, \dfrac{8}{9}\right)$

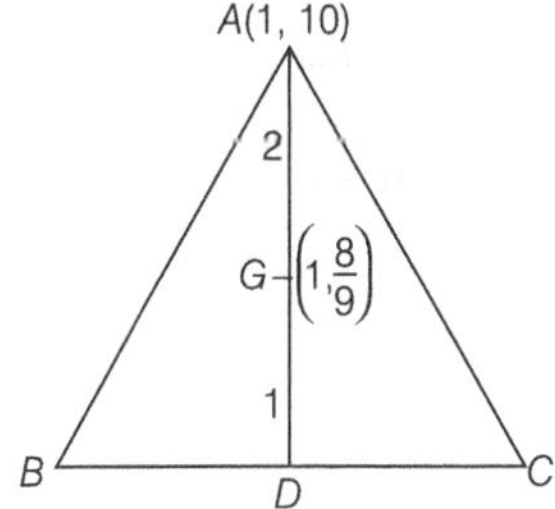

$$A(1, 10), G\left(1, \frac{8}{9}\right)$$

$$AG : GD = 2 : 1$$

$\therefore$ $$D = \left(1, -\frac{11}{3}\right)$$

4. The equation of the line L be $y - 2 = m(x - 8)$, $m < 0$ coordinates of P and Q are $P\left(8 - \dfrac{2}{m}, 0\right)$ and $Q(0, 2 - 8m)$.

So, $OP + OQ = 8 - \dfrac{2}{m} + 2 - 8m = 10 + \dfrac{2}{(-m)} + 8(-m) \geq$

$$10 + 2\sqrt{\frac{2}{(-m)} \times 8(-m)} \geq 18$$

So, absolute minimum value of $OP + OQ = 18$

5. Let the two perpendiculars through the origin intersect $2x + y = a$ at A and B so that the ΔOAB is isosceles.

OM = length of perpendicular from O to AB,

$OM = \dfrac{a}{\sqrt{5}}$.

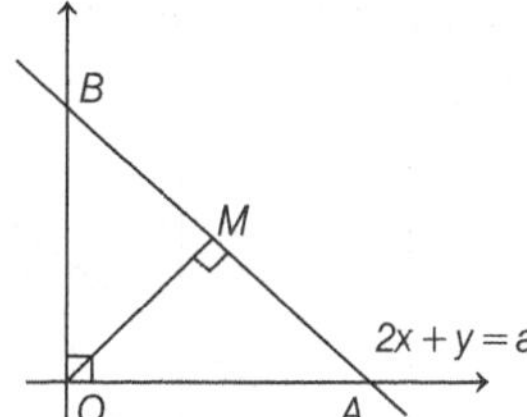

Also, $AM = MB = OM$

$\Rightarrow$ $AB = \dfrac{2a}{\sqrt{5}}$

Area of $\Delta OAB = \dfrac{1}{2} \cdot AB \cdot OM = \dfrac{1}{2} \cdot \dfrac{2a}{\sqrt{5}} \cdot \dfrac{a}{\sqrt{5}} = \dfrac{a^2}{5}$ sq unit

6. We have, $3x + 4y = 9$

$y = mx + 1$

$\Rightarrow$ $3x + 4mx + 4 = 9$

$\Rightarrow$ $(3 + 4m)x = 5$

$\Rightarrow$ x will be an integer when

$3 + 4m = 5, -5, 1, -1$

$\Rightarrow$ $m = \dfrac{1}{2}, -2, -\dfrac{1}{2}, -1$

So, number of integral values of m is 2.

7. As $(-1, 1)$ is a point on $3x - 4y + 7 = 0$, the rotation is possible.

Slope of the given line $= \dfrac{3}{4}$.

Slope of the line in its new position $= \dfrac{\dfrac{3}{4} - 1}{1 + \dfrac{3}{4}} = -\dfrac{1}{7}$

The required equation is $y - 1 = -\dfrac{1}{7}(x + 1)$

or $7y + x - 6 = 0$

8. $O(0, 0)$ is the orthocentre.

B is found to be $(-1, 1)$, $OB \perp ax + by - 1 = 0$.

$$\therefore \quad \left(-\frac{a}{b}\right)(-1) = -1 \Rightarrow a = -b \quad \ldots\text{(i)}$$

Equation of line through A is

$$(ax + by - 1) + \lambda(2x + 3y - 1) = 0.$$

It is $\perp$ to BC.

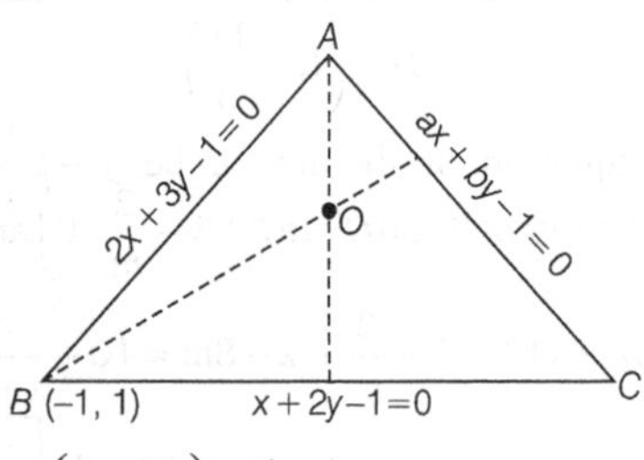

$$\therefore \quad -\left(\frac{a + 2\lambda}{b + 3\lambda}\right) \times \left(-\frac{1}{2}\right) = -1$$

$$\Rightarrow \quad a + 2\lambda = -2b - 6\lambda \quad \ldots\text{(ii)}$$

Also OA passes through origin.

$$\therefore \quad -1 - \lambda = 0 \Rightarrow \lambda = -1$$

$$\therefore \quad a - 2 = -2b + 6$$

$$\Rightarrow \quad a + 2b = 8$$

Solving Eqs. (i) and (ii), we get, $b = 8$, $a = -8$.

9. Slope of $OB = \frac{8}{6}$

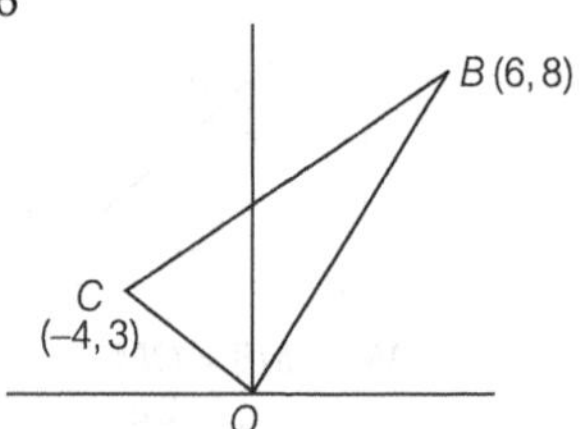

Slope of $OC = -\frac{3}{4}$

$$\therefore \quad \angle BOC = \frac{\pi}{2}$$

ΔOBC is right angled at O.

Circumcentre = mid-point of hypotenuse $= \left(1, \frac{11}{2}\right)$

Orthocentre = vertex O (0, 0).

$$\text{Distance} = \sqrt{\left(1 + \frac{121}{4}\right)} = \frac{5\sqrt{5}}{2} \text{ unit}$$

10. Given equations, $(2a + b)x + (a + 3b)y + (b - 3a) = 0$ and $mx + 2y + 6 = 0$ are concurrent for all real values of a and b, so they must represent the same line for same values of a and b.

Then, $$\frac{2a + b}{m} = \frac{(a + 3b)}{2} = \frac{b - 3a}{6}$$

Taking last two ratios,

$$\frac{a + 3b}{2} = \frac{-3a + b}{6} \Rightarrow b = -\frac{3}{4}a$$

Taking first two ratios,

$$m = \frac{2(2a + b)}{a + 3b} = \frac{2(2a - (3/4)a)}{a + 3(-3/4)a} = \frac{10}{-5} = -2$$

11. Lines are $y = 1$, $y = 0$

$$y = -x, \; y = -x + 2$$

$$y = x + 1, \; y = x - 1$$

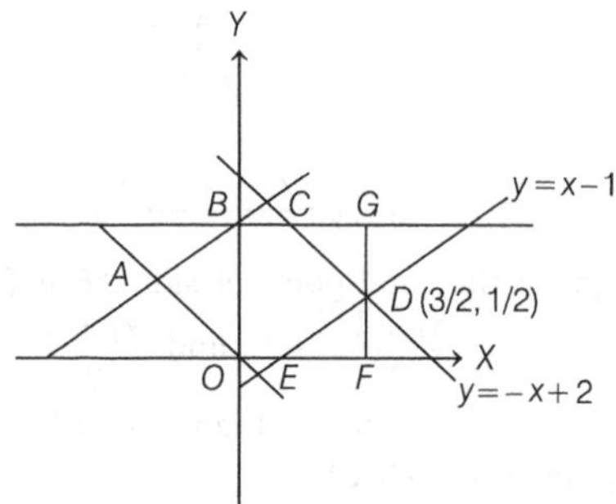

Area of $OABCDE$ = area of $OBGF$

$$= \frac{3}{2} \times 1 = \frac{3}{2} \text{ sq unit}$$

12. Since a is an integer, $\sqrt{2}\,a$ is irrational.

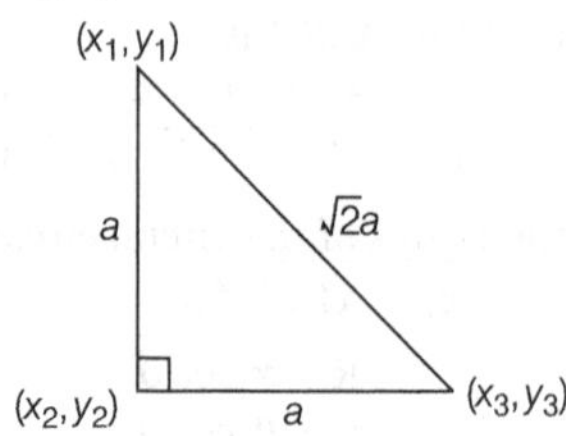

$\therefore$ Incentre will be

$$\left(\frac{ax_1 + ax_2 + \sqrt{2}\,ax_3}{2a + a\sqrt{2}}, \frac{ay_1 + ay_2 + \sqrt{2}\,ay_3}{2a + a\sqrt{2}}\right)$$

Which is clearly an irrational point.

13. Let $R = (\alpha, 0)$

For $PR + PQ$ to be minimum it should be the path of light and thus we have

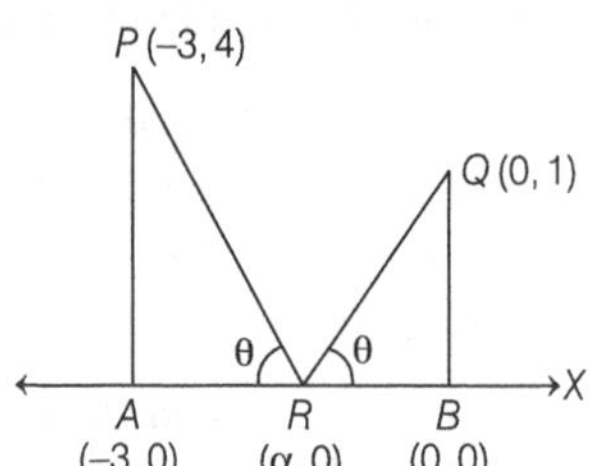

$$\Delta APR \sim \Delta BQR$$

$$\Rightarrow \quad \frac{AR}{RB} = \frac{PA}{QB}$$

$$\Rightarrow \quad \frac{\alpha + 3}{0 - \alpha} = \frac{4}{1}$$

$$\Rightarrow \quad \alpha = -\frac{3}{5}$$

14. From figure refracted ray makes an angle of 75° with positive direction of X-axis and passes through the point (1, 0)

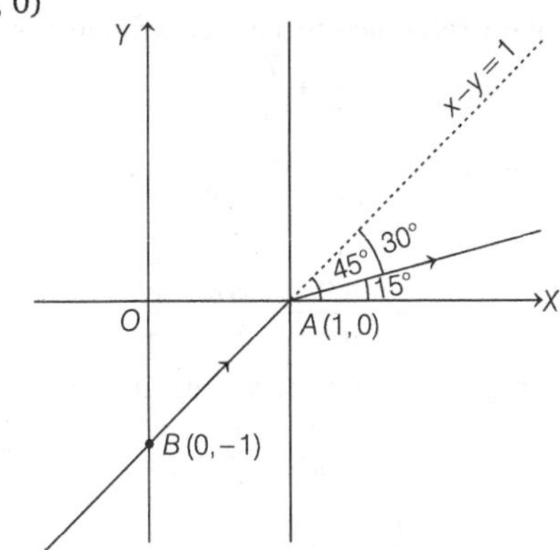

$\therefore$ Its equation is $(y-0) = \tan(45° - 30°)(x-1)$

or $\quad y = (2-\sqrt{3})(x-1)$

15. Let the slope of line is m, which is passing through $P(2, 3)$.

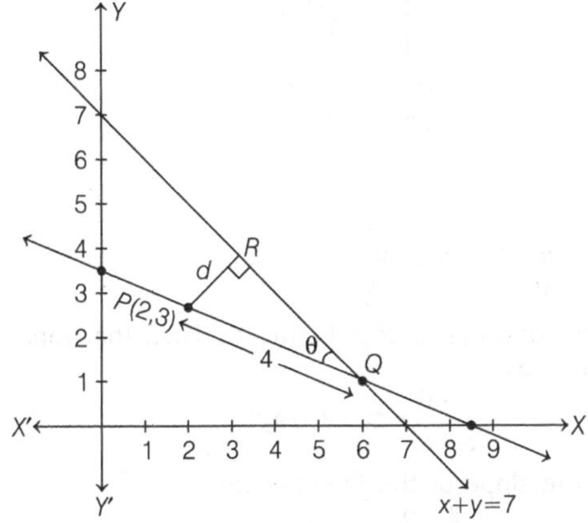

Since, the distance of a point (x_1, y_1) from the line $ax + by + c = 0$ is $d = \left|\dfrac{ax_1 + by_1 + c}{\sqrt{a^2+b^2}}\right|$.

$\therefore$ The distance of a point $P(2, 3)$ from the line $x + y - 7 = 0$, is $d = \dfrac{|2+3-7|}{\sqrt{1+1}} = \dfrac{2}{\sqrt{2}} = \sqrt{2}$

Now, in ΔPRQ,

$$QR = \sqrt{16-d^2} = \sqrt{16-2} = \sqrt{14}$$

$$\therefore \tan\theta = \frac{d}{QR} = \frac{\sqrt{2}}{\sqrt{14}} = \frac{1}{\sqrt{7}} = \left|\frac{m+1}{1-m}\right| \quad \left[\because \tan\theta = \left|\frac{m_2 - m_1}{1 + m_1 m_2}\right|\right]$$

$$\Rightarrow \quad \frac{m+1}{1-m} = \pm\frac{1}{\sqrt{7}} \Rightarrow \frac{m+1}{1-m} = \frac{1}{\sqrt{7}} \text{ or } \frac{m+1}{1-m} = -\frac{1}{\sqrt{7}}$$

$$\Rightarrow \quad m = \frac{1-\sqrt{7}}{1+\sqrt{7}} \text{ or } m = \frac{-1-\sqrt{7}}{\sqrt{7}-1}$$

16. $P = \left|\dfrac{a(4-3+4) + b(2+6-3)}{\sqrt{(2a+b)^2 + (a-2b)^2}}\right| = \sqrt{10}$

$$\Rightarrow \quad 25(a+b)^2 = 10(5a^2 + 5b^2)$$

$$\Rightarrow \quad 25(a-b)^2 = 0 \Rightarrow a = b$$

only line is $3x - y + 1 = 0$

17.
$$y = mx + c$$
$$3 = m + c$$
$$\sqrt{2} = \left|\frac{m - 3\sqrt{2}}{1 + 3\sqrt{2}m}\right|$$

$$\Rightarrow \quad 6m + \sqrt{2} = m - 3\sqrt{2}$$

$$\Rightarrow \quad 5m = -4\sqrt{2} \Rightarrow m = \frac{-4\sqrt{2}}{5}$$

$$\Rightarrow \quad 6m - \sqrt{2} = m - 3\sqrt{2}$$

$$\Rightarrow \quad 7m - 2\sqrt{2} \Rightarrow m = \frac{2\sqrt{2}}{7}$$

According to option take

$$m = \frac{-4\sqrt{2}}{5}$$

So, $\quad y = \dfrac{-4\sqrt{2}x}{5} + 3 + \dfrac{4\sqrt{2}}{5}$

18. Equation of line $\dfrac{ax}{c-1} + \dfrac{by}{c-1} + 1 = 0$ has two independent parameters. It can pass through a fixed point if it contains only one independent parameter. Now there must be one relation between $\dfrac{a}{c-1}$ and $\dfrac{b}{c-1}$ independent of a, b and c so that $\dfrac{a}{c-1}$ can be expressed in terms of $\dfrac{b}{c-1}$ and straight line contains only one independent parameter. Now that given relation can be expressed as $\dfrac{5a}{c-1} + \dfrac{4b}{c-1} = \dfrac{t-20c}{c-1}$.

RHS is independent of c if $t = 20$.

19. The family of lines

$$(x + y - 1) + \lambda(2x + 3y - 5) = 0$$

passes through a point such that

$$x + y - 1 = 0$$
$$2x + 3y - 5 = 0$$

i.e., (−2, 3) and family of lines

$$(3x + 2y - 4) + \mu(x + 2y - 6) = 0$$

Passes through a point such that

$$3x + 2y - 4 = 0$$

and $\quad x + 2y - 6 = 0$ i.e. $(-1, 7/2)$

$\therefore$ Equation of the straight line that belongs to both the families passes through (−2, 3) and (−1, 7/2) is

$$y - 3 = \frac{\frac{7}{2} - 3}{-1 + 2}(x + 2)$$

$$\Rightarrow \quad y - 3 = \frac{x+2}{2}$$

$$\Rightarrow \quad x - 2y + 8 = 0$$

20. $d(x, y) = \max\{|x|, |y|\}$ ...(i)

but $\quad d(x, y) = a$ ...(ii)

From Eqs. (i) and (ii)

$$a = \max\{|x|, |y|\}$$

if $\quad |x| > |y|$ then $a = |x|$

$\therefore \quad x = \pm a$

and if $|y| > |x|$, then $a = |y|$

$\therefore \quad y = \pm a$

Therefore locus represents a straight line.

21. Image of the point P about the line $y = x$ is $Q\,(1, 4)$

$\because$ Transformation through a 2 units along the positive direction of X-axis, then new point

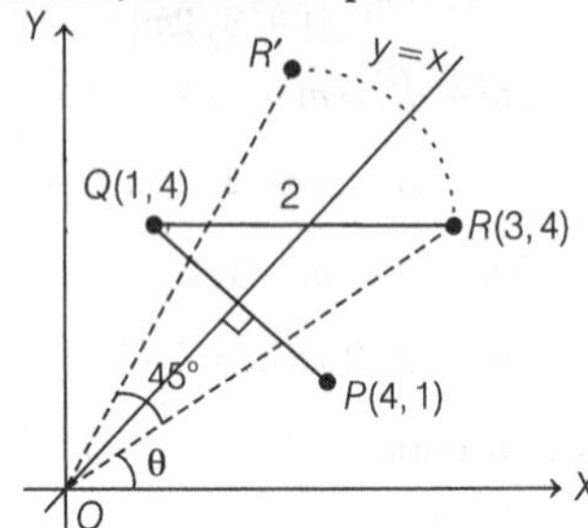

$R\,(1 + 2, 4)$ i.e., $R\,(3, 4)$

$\because \quad OR = OR' = 5$

and $\quad \tan\theta = 4/3$

$\therefore \; \sin\theta = \dfrac{4}{5}$ and $\cos\theta = \dfrac{3}{5}$

Then, final position of the point is

$[(OR' \cos(\pi/4 + \theta), OR' \sin(\pi/4 + \theta)]$

$$\Rightarrow \left[5\left(\left\{\frac{1}{\sqrt{2}}\cos\theta - \frac{1}{\sqrt{2}}\sin\theta\right\}\right), 5\left(\frac{1}{\sqrt{2}}\cos\theta + \frac{1}{\sqrt{2}}\sin\theta\right)\right]$$

$$\Rightarrow \left(-\frac{1}{\sqrt{2}}, \frac{7}{\sqrt{2}}\right)$$

22.

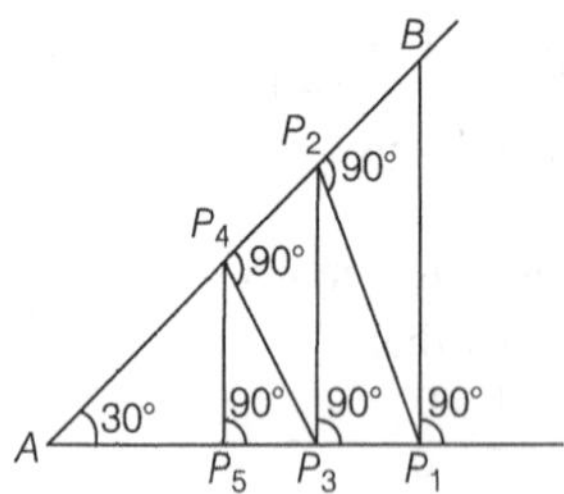

As, $\quad AB = 1, BP_1 = \sin 30° = \dfrac{1}{2}$

and $\quad AP_1 = \dfrac{\sqrt{3}}{2}$

$\quad P_1P_2 = AP_1 \sin 30° = \dfrac{\sqrt{3}}{4}$

and $\quad AP_2 = AP_1 \cos 30° = \dfrac{3}{4}$

So, $\quad P_2P_3 = AP_2 \sin 30° = \dfrac{3}{8}$ and etc.

$\therefore$ sum of perpendiculars $= BP_1 + P_1P_2 + P_2P_3 + P_3P_4 \ldots$

$$= \frac{1}{2} + \frac{\sqrt{3}}{4} + \frac{3}{8} + \ldots = \frac{\frac{1}{2}}{1 - \frac{\sqrt{3}}{2}} = \frac{1}{2 - \sqrt{3}} = 2 + \sqrt{3}.$$

23. Equation of L is $\dfrac{x}{a} + \dfrac{y}{b} = 1$ and let the axis be rotated through an angle θ and let (X, Y) be the new coordinates of any point $P\,(x, y)$ in the plane, then

$x = X\cos\theta - Y\sin\theta$, $y = X\sin\theta + Y\cos\theta$ the equation of the line with reference to original coordinates is

$$\frac{x}{a} + \frac{y}{b} = 1$$

i.e. $\dfrac{X\cos\theta - Y\sin\theta}{a} + \dfrac{X\sin\theta + Y\cos\theta}{b} = 1 \quad \ldots\text{(i)}$

and with reference to new coordinates is

$$\frac{X}{p} + \frac{Y}{q} = 1 \quad \ldots\text{(ii)}$$

Comparing Eqs. (i) and (ii), we get

$$\frac{\cos\theta}{a} + \frac{\sin\theta}{b} = \frac{1}{p} \quad \ldots\text{(iii)}$$

and $\quad -\dfrac{\sin\theta}{a} + \dfrac{\cos\theta}{b} = \dfrac{1}{q} \quad \ldots\text{(iv)}$

Squaring and adding Eqs. (iii) and (iv), we get

$$\frac{1}{a^2} + \frac{1}{b^2} = \frac{1}{p^2} + \frac{1}{q^2}$$

24. $\dfrac{x - 2}{\cos 45°} = \dfrac{y - 1}{\sin 45°} = -4$

$\Rightarrow \quad x - 2 = -2\sqrt{2},\; y - 1 = -2\sqrt{2}$

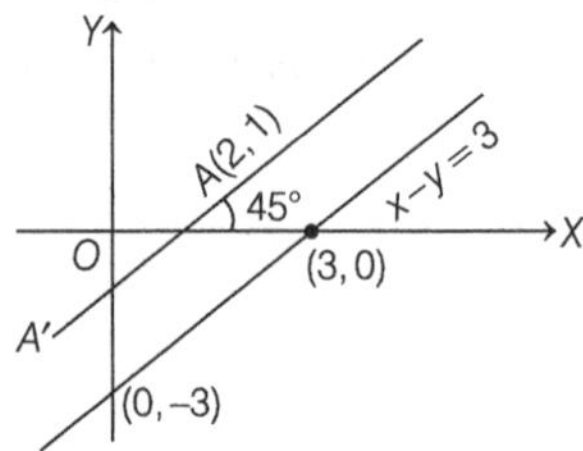

$\therefore \quad x = 2 - 2\sqrt{2},\; y = 1 - 2\sqrt{2}$

$\Rightarrow \quad A' = (2 - 2\sqrt{2}, 1 - 2\sqrt{2})$

25. Let the coordinates of A be $(a, 0)$. Then the slope of the reflected ray is

$$\frac{3 - 0}{5 - a} = \tan\theta \text{ (say)} \quad \ldots\text{(i)}$$

Then the slope of the incident ray

$$= \frac{2 - 0}{1 - a} = \tan(\pi - \theta) \quad \ldots\text{(ii)}$$

From Eqs. (i) and (ii),

$$\tan\theta + \tan(\pi - \theta) = 0$$

$$\Rightarrow \quad \frac{3}{5 - a} + \frac{2}{1 - a} = 0$$

$$\Rightarrow \quad 3 - 3a + 10 - 2a = 0;\; a = \frac{13}{5}$$

Thus, the coordinate of A is $\left(\dfrac{13}{5}, 0\right)$

26. Coordinates of A and B are

$(-3, 4)$ and $\left(-\dfrac{3}{5}, \dfrac{8}{5}\right)$

If orthocentre $P\,(h, k)$

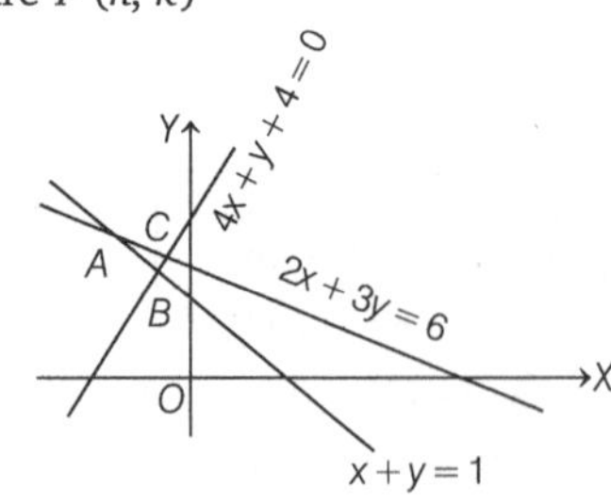

Then, slope of $PA \times$ slope of $BC = -1$

$$\frac{k-4}{h+3} \times 4 = -1$$

$$\Rightarrow \quad 4k - 16 = -h - 3$$

$$\Rightarrow \quad h + 4k = 13 \quad \ldots(i)$$

and slope of $PB \times$ slope of $AC = -1$

$$\Rightarrow \quad \frac{k - \frac{8}{5}}{h + \frac{3}{5}} \times -\frac{2}{3} = -1$$

$$\Rightarrow \quad \frac{5k-8}{5h+3} \times \frac{2}{3} = 1$$

$$\Rightarrow \quad 10k - 16 = 15h + 9$$

$$15h - 10k + 25 = 0$$

$$3h - 2k + 5 = 0 \quad \ldots(ii)$$

Solving Eqs. (i) and (ii), we get

$$h = \frac{3}{7}, k = \frac{22}{7}$$

Hence, orthocentre lies in I quadrant.

27. Let $Q(x_1, y_1)$ be the reflection of $P(4, -13)$ in the line $5x + y + 6 = 0$.

$$\frac{x_1 - 4}{5} = \frac{y_1 + 13}{1} = \frac{-2(20 - 13 + 6)}{25 + 1}$$

$$= -1$$

$$\Rightarrow \quad x_1 - 4 = -5 \text{ and } y_1 + 13 = -1$$

$$\therefore \quad x_1 = -1, y_1 = -14$$

Hence, reflection point is $(-1, -14)$.

28. Given, vertices of triangle are $(k, -3k)$, $(5, k)$ and $(-k, 2)$.

$$\therefore \quad \frac{1}{2}\begin{vmatrix} k & -3k & 1 \\ 5 & k & 1 \\ -k & 2 & 1 \end{vmatrix} = \pm 28$$

$$\Rightarrow \quad \begin{vmatrix} k & -3k & 1 \\ 5 & k & 1 \\ -k & 2 & 1 \end{vmatrix} = \pm 56$$

$$\Rightarrow \quad k(k-2) + 3k(5+k) + 1(10 + k^2) = \pm 56$$

$$\Rightarrow \quad k^2 - 2k + 15k + 3k^2 + 10 + k^2 = \pm 56$$

$$\Rightarrow \quad 5k^2 + 13k + 10 = \pm 56$$

$$\Rightarrow \quad 5k^2 + 13k - 66 = 0$$

$$\text{or} \quad 5k^2 + 13k - 46 = 0$$

$$\Rightarrow \quad k = 2 \quad [\because k \in I]$$

Thus, the coordinates of vertices of triangle are $A(2, -6)$, $B(5, 2)$ and $C(-2, 2)$.

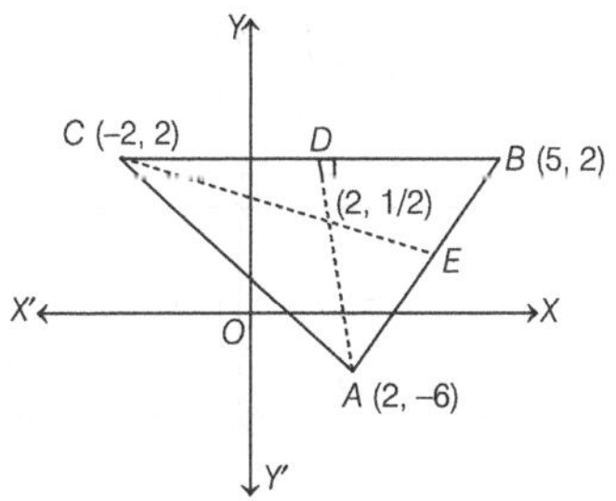

Now, equation of altitude from vertex A is

$$y - (-6) = \frac{-1}{\left(\frac{2-2}{-2-5}\right)}(x - 2) \Rightarrow x = 2 \quad \ldots(i)$$

Equation of altitude from vertex C is

$$y - 2 = \frac{-1}{\left[\frac{2-(-6)}{5-2}\right]}[x - (-2)]$$

$$\Rightarrow \quad 3x + 8y - 10 = 0 \quad \ldots(ii)$$

On solving Eqs. (i) and (ii), we get $x = 2$ and $y = \frac{1}{2}$.

$\therefore$ Orthocentre $= \left(2, \frac{1}{2}\right)$

29. Let coordinates of A and B are $(a, 0)$ and $(0, b)$ equation of AB

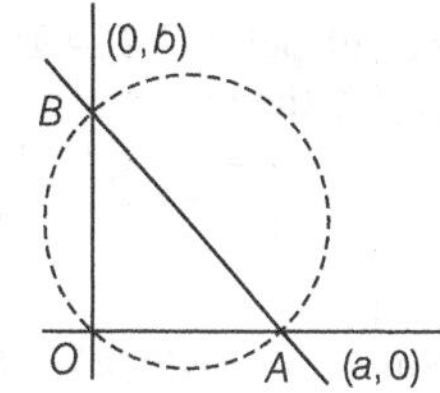

is $$\frac{x}{a} + \frac{y}{b} = 1$$

$$\Rightarrow \quad \frac{4}{a} + \frac{2}{b} = 1 \quad \ldots(i)$$

Centre of circle is $\left(\frac{a}{2}, \frac{b}{2}\right)$

Eq. (i) can be rewritten as

$$\frac{2}{(a/2)} + \frac{1}{(b/2)} = 1$$

$\therefore$ locus of circumcentre is $\frac{2}{x} + \frac{1}{y} = 1$

$$\Rightarrow \quad 2x^{-1} + y^{-1} = 1$$

30. Let $A(x_1, y_1)$ and $B(x_2, y_2)$ be two points at equal distance λ from $(0, \sqrt{3})$

Then, $\quad (x_1 - 0)^2 + (y_1 - \sqrt{3})^2 = \lambda^2 \quad \ldots(i)$

and $\quad (x_2 - 0)^2 + (y_2 - \sqrt{3})^2 = \lambda^2 \quad \ldots(ii)$

Subtracting Eq. (ii) from Eq. (i), then

$$x_1^2 - x_2^2 + (y_1 - \sqrt{3})^2 - (y_2 - \sqrt{3})^2 = 0$$

$$\Rightarrow \quad x_1^2 - x_2^2 + (y_1^2 - y_2^2) - 2\sqrt{3}(y_1 - y_2) = 0$$

On comparing rational and irrational parts on both sides

$$y_1 - y_2 = 0 \Rightarrow y_1 = y_2 = \alpha \text{ (say)}$$

and $\quad x_1^2 - x_2^2 + (y_1^2 - y_2^2) = 0$

or $\quad x_1^2 = x_2^2 = \beta^2 \quad$ (say)

$$\therefore \quad x_1 = x_2 = \pm \beta$$

Hence, points are $(\pm \beta, \alpha)$

If $\beta = 0$, then $(0, \alpha)$

$$\therefore \quad n \le 2$$

31. Let D, E, F are the mid points of BC, CA and AB

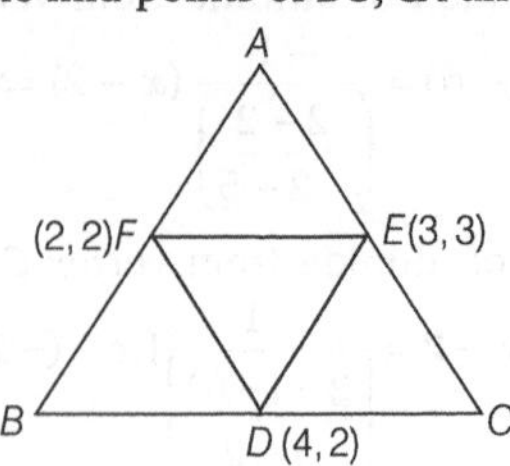

$\therefore$ Centroid of $\Delta\, ABC$ = Centroid of ΔDEF

$$= \left(\frac{1}{3}(2+4+3), \frac{1}{3}(2+2+3)\right)$$

$$= (3, 7/3)$$

32. It is given that the line 'L' made x and y intercepts 3 and 1 respectively, so equation of line L is

$$\frac{x}{3} + \frac{y}{1} = 1 \quad \Rightarrow \quad x + 3y = 3 \qquad \text{...(i)}$$

Now, let image of point $(-1, -4)$ with respect to line Eq. (i) is (x_1, y_1), then

$$\frac{x_1+1}{1} = \frac{y_1+4}{3} = -2\left(\frac{-1-12-3}{1+9}\right) = \frac{16}{5}$$

$$\left[\begin{array}{l}\text{Since, image } (h, k) \text{ or } (x_1, y_1) \text{ w.r.t. the line} \\ \text{mirror } ax + by + c = 0 \text{ is given by} \\ \dfrac{h - x_1}{a} = \dfrac{k - y_1}{b} = \dfrac{-2(ax_1 + by_1 + c)}{a^2 + b^2}\end{array}\right]$$

$$\Rightarrow \quad x_1 = \frac{16}{5} - 1 = \frac{11}{5} \text{ and } y_1 = \left(\frac{16}{5} \times 3\right) - 4 = \frac{28}{5}$$

$$\therefore \quad (x_1, y_1) = \left(\frac{11}{5}, \frac{28}{5}\right)$$

33. $\because$ C is the mid point of AB

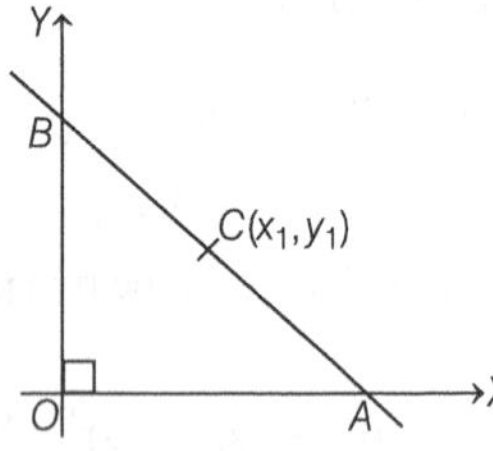

$\therefore$ Coordinates of A and B are $(2x_1, 0)$ and $(0, 2y_1)$ respectively.

$\therefore$ Equation of AB is

$$\frac{x}{2x_1} + \frac{y}{2y_1} = 1 \quad \Rightarrow \quad \frac{x}{x_1} + \frac{y}{y_1} = 2$$

34. $$p_1 = \left|m^2 \cos\alpha + 2m \sin\alpha + \frac{\sin^2\alpha}{\cos\alpha}\right| = \frac{(m\cos\alpha + \sin\alpha)^2}{|\cos\alpha|}$$

$$p_2 = \left|mm' \cos\alpha + (m + m')\sin\alpha + \frac{\sin^2\alpha}{\cos\alpha}\right|$$

$$= \frac{|(m\cos\alpha + \sin\alpha)||m'\cos\alpha + \sin\alpha|}{|\cos\alpha|}$$

and $p_3 = \left|m'^2 \cos\alpha + 2m' \sin\alpha + \dfrac{\sin^2\alpha}{\cos\alpha}\right|$

$$= \frac{(m'\cos\alpha + \sin\alpha)^2}{|\cos\alpha|}$$

$\because \quad p_2^2 = p_1 p_3$

Hence, p_1, p_2, p_3 are in GP

35. Given, $OB = 4$, $OA = \sqrt{(4+16)} = \sqrt{20}$

$$AB = \sqrt{(OA)^2 - (OB)^2} = 2$$

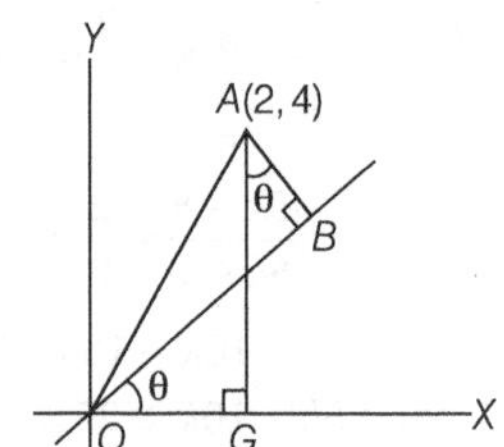

$$\therefore \quad \theta = \tan^{-1}\frac{4}{2} - \tan^{-1}\left(\frac{AB}{OB}\right) = \tan^{-1} 2 - \tan^{-1}\left(\frac{2}{4}\right)$$

$$= \tan^{-1}\left(\frac{2 - 1/2}{1+1}\right) = \tan^{-1}\left(\frac{3}{4}\right)$$

$\therefore \tan\theta = 3/4$

36. $$\because \quad \frac{2}{1!9!} + \frac{2}{3!7!} + \frac{1}{5!5!} = \frac{2^m}{n!}$$

$$\frac{1}{10!}\left\{\frac{2\times 10!}{1!9!} + \frac{2\times 10!}{3!7!} + \frac{10!}{5!5!}\right\} = \frac{2^m}{n!}$$

$$\Rightarrow \quad \frac{1}{10!}\{2\,{}^{10}C_1 + 2{}^{10}C_3 + {}^{10}C_5\} = \frac{2^m}{n!}$$

$$\frac{1}{10!}\{{}^{10}C_1 + {}^{10}C_3 + {}^{10}C_5 + {}^{10}C_7 + {}^{10}C_9\} = \frac{2^m}{n!}$$

$$\Rightarrow \quad \frac{1}{10!}(2)^{10-1} = \frac{2^m}{n!}$$

$$\therefore \quad m = 9 \text{ and } n = 10$$

Hence, $x - y + 1 = 0$ and $x + y + 3 = 0$ are perpendicular to each other, then orthocentre is the point of intersection which is $(-2, -1)$

$$\therefore \quad -2 = 2m - 2n$$

and $$-1 = m - n$$

$\therefore$ Point is $(2m - 2n, m - n)$.

37. Distance between parallel lines $= \dfrac{12}{5}$.

$$\therefore \quad \tan\theta = \frac{4}{3}$$

Let slope of the required line be m . Then,

$$\frac{m - \left(-\dfrac{3}{4}\right)}{1 + m \times \left(-\dfrac{3}{4}\right)} = \pm\frac{4}{3}$$

C
3x + 4y = 12
3
12/5
B
θ
9/5
D
3x + 4y = 24

$\Rightarrow \quad 3(4m+3) = \pm\, 4(4-3m)$

$\Rightarrow \quad m = \dfrac{7}{24}$ or ∞.

$\therefore$ Lines are $y - 2 = \dfrac{7}{24}(x-1)$ and a line parallel to Y-axis through (1, 2) i.e., $x = 1$.

$\therefore$ Required lines are $7x - 24y + 41 = 0$ and $x = 1$

38. Equation of diagonal AC is

$$y - 0 = -2(x-3)$$

$\Rightarrow \quad 2x + y = 6$

Solving $\quad 2x + y = 6$ and $x = 2y$

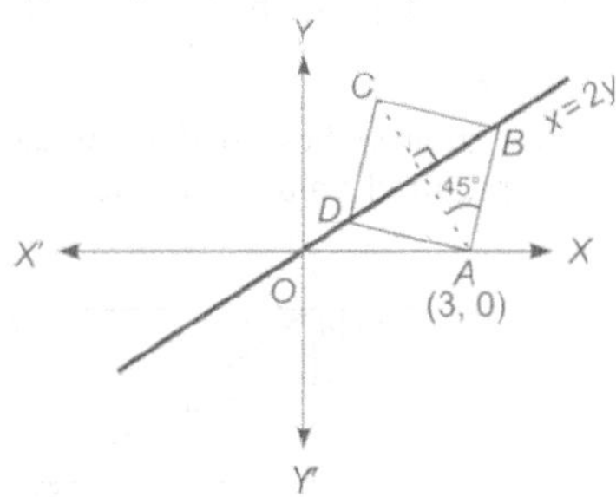

Then, we get $\quad y = \dfrac{6}{5}$ and $x = \dfrac{12}{5}$

$\therefore$ Centre of square is $\left(\dfrac{12}{5}, \dfrac{6}{5}\right)$

Let slope of side AB or AD is m then,

$$\left|\frac{m-(-2)}{1+m(-2)}\right| = 1$$

$\therefore \quad (m+2) = \pm\,(1-2m)$

or $\quad m = -\dfrac{1}{3}$ and $m = 3$

Hence, slope AB and AD are 3 and $-\dfrac{1}{3}$ respectively

$\therefore$ Equations of sides AB and AD are

$y - 0 = 3(x-3)$ and $y - 0 = -\dfrac{1}{3}(x-3)$

or $y - 3x + 9 = 0$ and $3y + x - 3 = 0$ respectively.

39. Given equation is, $y = \sin x \sin(x+2) - \sin^2(x+1)$

$= \dfrac{1}{2}[\cos 2 - \cos(2x+2)] - \dfrac{1}{2}[1 - \cos(2x+2)]$

$[\because 2\sin A \sin B = \cos(A-B) - \cos(A+B)$ and $\cos 2\theta = 1 - 2\sin^2\theta \Rightarrow 2\sin^2\theta = 1 - \cos 2\theta]$

$= \dfrac{1}{2}\cos 2 - \dfrac{1}{2}\cos(2x+2) - \dfrac{1}{2} + \dfrac{1}{2}\cos(2x+2)$

$= \dfrac{1}{2}(\cos(2) - 1) = -\dfrac{1}{2}(2\sin^2(1))$

$= -\sin^2(1) < 0 \Rightarrow y < 0$

and as we know that $y < 0$, is in III and IV quadrants only.

40. The line $2x + y = 2 \Rightarrow \dfrac{x}{1} + \dfrac{y}{2} = 1$ cuts X-axis at (1, 0) and Y-axis at (0, 2)

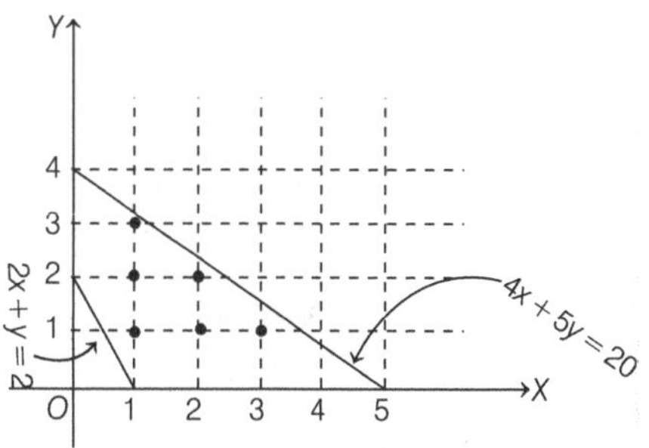

and the line $4x + 5y = 20 \Rightarrow \dfrac{x}{5} + \dfrac{y}{4} = 1$ cuts X-axis at (5, 0) and Y-axis at (0, 4). It is clear from the figure the positions of P are six which are (1, 1), (2, 1), (3, 1), (1, 2), (2, 2) and (1, 3).

41. We have, $ax + by + c = 0$ and $2a + 3b = 4c$

Putting $c = \dfrac{2a+3b}{4}$ in $ax + by + c = 0$, we get

$$ax + by + \frac{2a+3b}{4} = 0$$

$\Rightarrow \quad 4ax + 4by + 2a + 3b = 0$

$\Rightarrow \quad a(4x+2) + b(4y+3) = 0$

$\Rightarrow \quad (4x+2) + \dfrac{b}{a}(4y+3) = 0$

This is of the form $L_1 + \lambda L_2 = 0$

So, set of lines are

$$4x + 2 = 0 \text{ and } 4y + 3 = 0$$

These two lines intersect at $\left(\dfrac{-1}{2}, \dfrac{-3}{4}\right)$

$\therefore \quad P(l, m) = \left(\dfrac{-1}{2}, \dfrac{-3}{4}\right)$

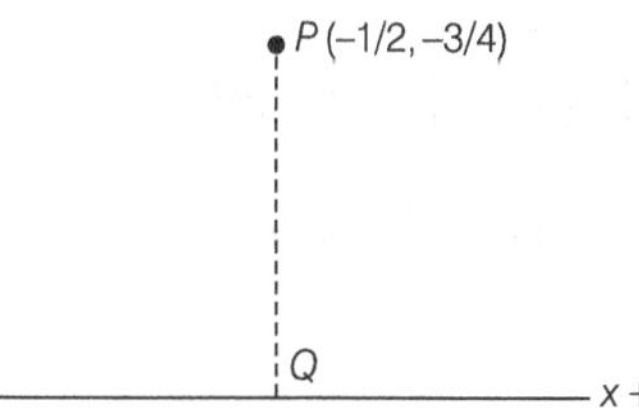

Equation of line passing through $\left(\dfrac{-1}{2}, \dfrac{-3}{4}\right)$ and perpendicular to $x + y + 1 = 0$ is,

$$y + \frac{3}{4} = 1\left(x + \frac{1}{2}\right)$$

$\Rightarrow \quad 4y = 4x - 1 \Rightarrow 4x - 4y - 1 = 0$

On solving $x + y + 1 = 0$ and $4x - 4y - 1 = 0$, we get

$$Q\left(-\frac{3}{8}, -\frac{5}{8}\right).$$

42. Side of the square = 2 unit
Coordinates of B, C and D are $(\sqrt{3}, 1)$, $(\sqrt{3}-1, \sqrt{3}+1)$ and $(-1, \sqrt{3})$ respectively.

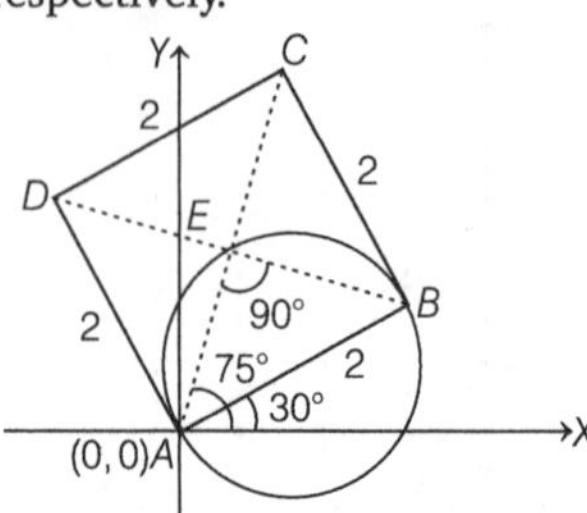

Slope of $BD = \dfrac{\sqrt{3}-1}{-1-\sqrt{3}} = \dfrac{(\sqrt{3}-1)(\sqrt{3}-1)}{-2} = \sqrt{3}-2$

∴ Equation of BD is

$$y - 1 = (\sqrt{3}-2)(x-\sqrt{3})$$

$$\Rightarrow \quad (2-\sqrt{3})x + y = 2(\sqrt{3}-1)$$

and equation of the circumcircle of the triangle ABE (Apply diametric form as AB is diameter)

$$(x-0)(x-\sqrt{3}) + (y-0)(y-1) = 0$$

$$\Rightarrow \quad x^2 + y^2 - x\sqrt{3} - y = 0$$

43. (b) $G = \left(\dfrac{-36+20+0}{3}, \dfrac{7+7-8}{3}\right) = \left(\dfrac{-16}{3}, 2\right)$

$$a = BC = \sqrt{(0-20)^2 + (-8-7)^2} = 25$$

$$b = CA = \sqrt{36^2 + (-8-7)^2} = 39$$

$$c = AB = \sqrt{(-36-20)^2 + (7-7)^2} = 56$$

Let $I = (\alpha, \beta)$, then

$$\alpha = \frac{-36\times 25 + 39\times 20 + 56\times 0}{25+39+56} = -1$$

$$\beta = \frac{25\times 7 + 39\times 7 - 8\times 56}{25+39+56} = 0$$

$$I = (-1, 0)$$

$$GI = \sqrt{\left(\frac{-16}{3}+1\right)^2 + (2-0)^2}$$

$$= \frac{1}{3}\sqrt{205} = \frac{25}{3}\sqrt{205}\,\lambda \Rightarrow \lambda = \frac{1}{25}$$

44. Let $\dfrac{AN}{BN} = \lambda$

Then, coordinate of N is $\left(\dfrac{a}{1+\lambda}, \dfrac{a\lambda}{1+\lambda}\right)$

∵ Slope of $AB = -1$

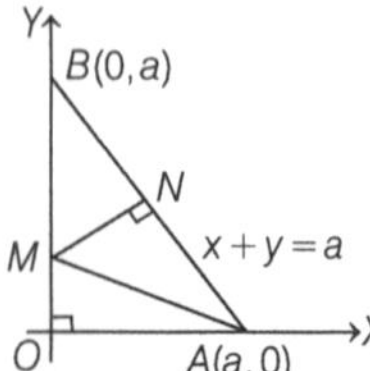

∴ Slope of $MN = 1$

∴ Equation of MN is

$$y - \frac{a\lambda}{1+\lambda} = x - \frac{a}{1+\lambda}$$

$$\Rightarrow \quad x - y = a\left(\frac{1-\lambda}{1+\lambda}\right)$$

So, the coordinates of M are $\left(0, a\left(\dfrac{\lambda-1}{\lambda+1}\right)\right)$.

Therefore, area of $\Delta AMN = \dfrac{3}{8}$ area of ΔOAB

$$\Rightarrow \quad \frac{1}{2}\cdot AN\cdot MN = \frac{3}{8}\cdot\frac{1}{2}a\cdot a$$

$$\Rightarrow \quad \frac{1}{2}\cdot\left|\frac{a\lambda\sqrt{2}}{1+\lambda}\cdot\frac{a\sqrt{2}}{1+\lambda}\right| = \frac{3}{8}\cdot\frac{1}{2}a\cdot a \Rightarrow \frac{a^2\lambda}{(1+\lambda)^2} = \frac{3}{8}\cdot\frac{1}{2}a^2$$

$$\therefore \quad \lambda = 3 \text{ or } \lambda = 1/3$$

For $\lambda = 1/3$, then M lies outside the segment OB and hence the required value of $\lambda = 3$.

45.

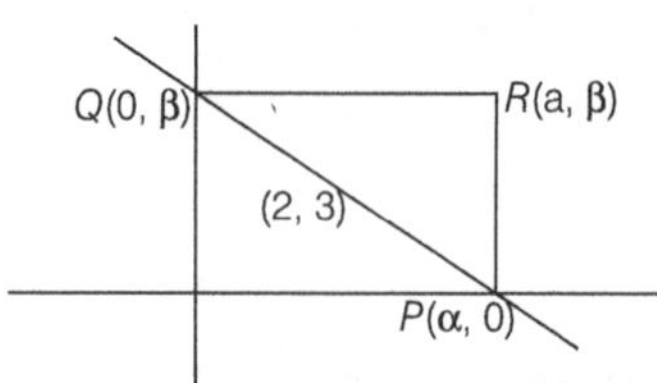

Equation of line PQ is $\dfrac{x}{\alpha} + \dfrac{y}{\beta} = 1$

Since, this line is passes through fixed point (2, 3).

$$\therefore \quad \frac{2}{\alpha} + \frac{3}{\beta} = 1$$

∴ Locus of R is $2\beta + 3\alpha = \alpha\beta$

i.e., $\quad 2y + 3x = xy \Rightarrow 3x + 2y = xy$

46. The slope of BD is $\dfrac{8}{15}$ and the angle made by BD with DC and BC is 45°. So, let the slope of DC be m. Then,

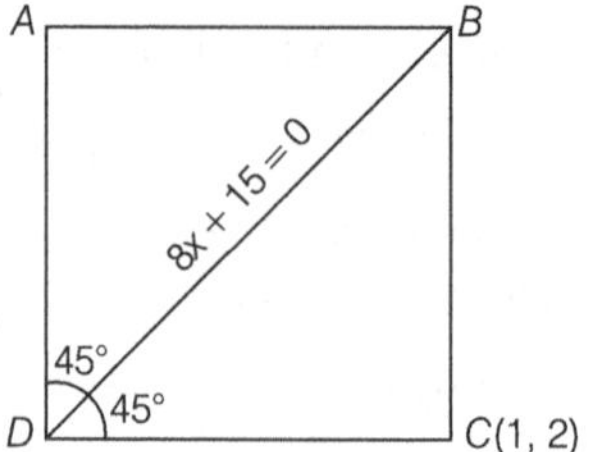

$$\tan 45° = \pm\frac{m - \frac{8}{15}}{1 + \frac{8}{15}m}$$

$$\Rightarrow \quad (15+8m) = \pm(15m-8)$$

$$\Rightarrow \quad m = \frac{23}{7} \text{ and } -\frac{7}{23}$$

Hence, the equations of DC and BC are

$$y - 2 = \frac{23}{7}(x-1)$$

$$\Rightarrow \quad 23x - 7y - 9 = 0$$

and

$$y - 2 = -\frac{7}{23}(x-1)$$

$$\Rightarrow \quad 7x + 23y - 53 = 0$$

47. Given equation of line is, $3x + 4y - 24 = 0$

For intersection with X-axis put $y = 0$

$\Rightarrow \quad 3x - 24 = 0 \Rightarrow x = 8$

For intersection with Y-axis, put $x = 0$

$\Rightarrow \quad 4y - 24 = 0 \Rightarrow y = 6$

$\therefore \quad A(8, 0)$ and $B(0, 6)$

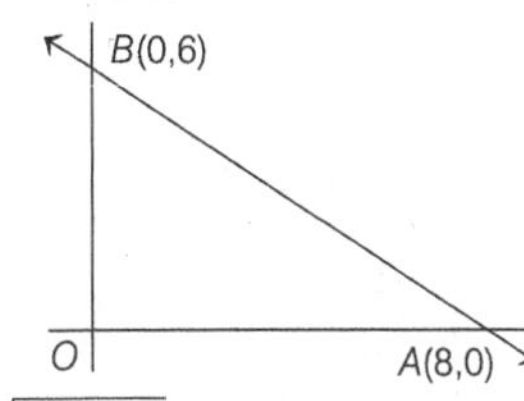

Let $AB = c = \sqrt{8^2 + 6^2} = 10$

$OB = a = 6$ and $OA = b = 8$

Also, let incentre is $(h\ k)$, then

$$h = \frac{ax_1 + bx_2 + cx_3}{a + b + c} \quad [\text{here, } x_1 = 8, x_2 = 0, x_3 = 0]$$

$$= \frac{6 \times 8 + 8 \times 0 + 10 \times 0}{6 + 8 + 10} = \frac{48}{24} = 2$$

and $$k = \frac{ay_1 + by_2 + cy_3}{a + b + c} \quad [\text{here, } y_1 = 0, y_2 = 6, y_3 = 0]$$

$$= \frac{6 \times 0 + 8 \times 6 + 10 \times 0}{6 + 8 + 10} = \frac{48}{24} = 2$$

$\therefore$ incentre is $(2, 2)$

48. $\because \angle ABC = 90°$

$\therefore$ AC is diameter of the circle.

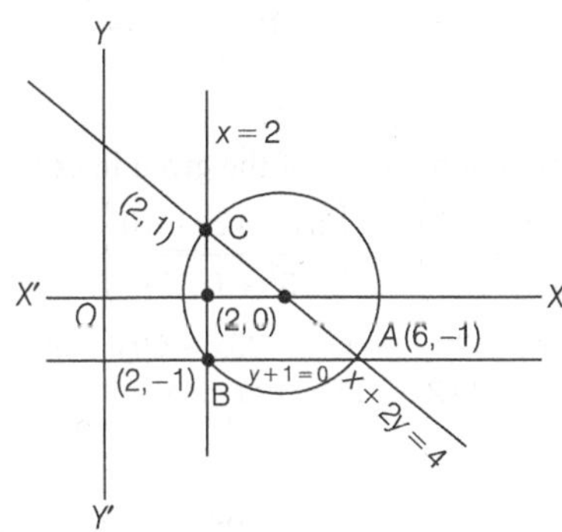

$\therefore$ Mid point of AC = coordinates of circumcentre

$$= \left(\frac{6+2}{2}, \frac{1-1}{2}\right) \text{ or } (4, 0)$$

49. $\because P\left(1 + \frac{\alpha}{\sqrt{2}}, 2 + \frac{\alpha}{\sqrt{2}}\right)$ lies between the parallel lines $x + 2y = 1$ and $2x + 4y = 15$, then

$$\frac{\left(1 + \frac{\alpha}{\sqrt{2}}\right) + 2\left(2 + \frac{\alpha}{\sqrt{2}}\right) - 1}{2\left(1 + \frac{\alpha}{\sqrt{2}}\right) + 4\left(2 + \frac{\alpha}{\sqrt{2}}\right) - 15} < 0$$

$$\Rightarrow \quad \frac{4 + \frac{3\alpha}{\sqrt{2}}}{-5 + \frac{6\alpha}{\sqrt{2}}} < 0 \Rightarrow \frac{\left(\alpha + \frac{4\sqrt{2}}{3}\right)}{\left(\alpha - \frac{5\sqrt{2}}{6}\right)} < 0$$

$$\therefore \quad -\frac{4\sqrt{2}}{3} < \alpha < \frac{5\sqrt{2}}{6}$$

50. $\because (a, a)$ fall between the lines $|x + y| = 2$, then

$$\frac{a + a - 2}{a + a + 2} < 0$$

$$\Rightarrow \quad \frac{a - 1}{a + 1} < 0 \text{ or } -1 < a < 1$$

$$\therefore \quad |a| < 1$$

51. Reflected ray is X-axis.

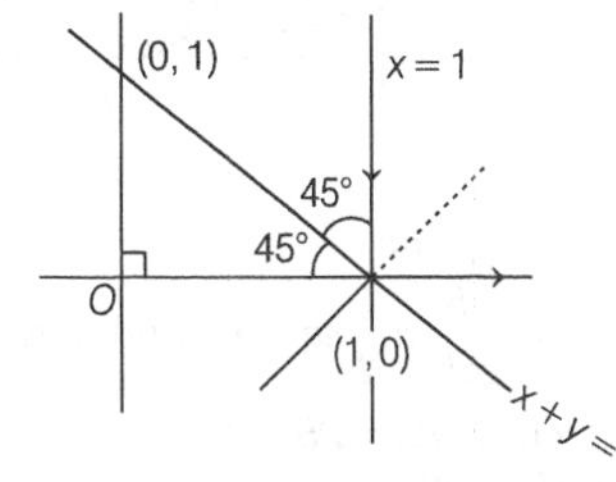

$\therefore$ equation $y = 0$

52. Let coordinate of the intersection point in IV quadrant be $(\alpha, -\alpha)$.

Since, $(\alpha, -\alpha)$ lie on both lines

$$4ax + 2ay + c = 0$$

and $$5bx + 2by + d = 0$$

$$\therefore \quad 4a\alpha - 2a\alpha + c = 0$$

$$\Rightarrow \quad \alpha = \frac{-c}{2a} \quad \ldots(\text{i})$$

Also, $$5b\alpha - 2b\alpha + d = 0$$

$$\Rightarrow \quad \alpha = \frac{-d}{3b} \quad \ldots(\text{ii})$$

From Eqs. (i) and (ii), $\frac{-c}{2a} = \frac{-d}{3b}$

$$\therefore \quad 3bc = 2ad$$

$$\Rightarrow \quad 3bc - 2ad = 0$$

53. $\angle BAX = \tan^{-1}\left(\frac{1-0}{3-2}\right) = 45°$

$$\Rightarrow \quad \angle CAX = 60°$$

$AC = AB = \sqrt{(1 + 1)} = \sqrt{2}$

$\therefore \quad x = OA + AM = 2 + \sqrt{2} \cos 60°$

$$= 2 = \frac{\sqrt{2}}{2} = \frac{(4 + \sqrt{2})}{2}$$

and $y = CM = \sqrt{2} \sin 60° = \sqrt{2} \cdot \frac{\sqrt{3}}{2} = \frac{\sqrt{6}}{2}$

$\therefore$ Coordinates of C are $\left[\frac{1}{2}(4 + \sqrt{2}), \frac{1}{2}\sqrt{6}\right]$

54. $\therefore$ Area of $\Delta OAB = \frac{1}{2} \cdot OA \cdot OB$

$$= \frac{1}{2} \cdot \frac{c}{a} \cdot \frac{c}{b} = \frac{c^2}{2ab}$$

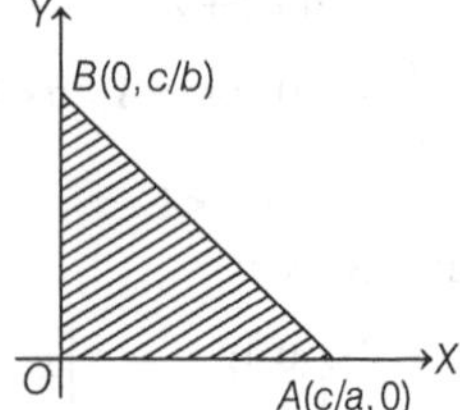

For independent a, b, c

$$c^2 = ab$$

Then, area $= \frac{ab}{2ab} = \frac{1}{2}$ sq unit and a, b, c are in GP.

55. Let $P \equiv (x, y)$, $\because PA = PB$

$\Rightarrow (PA)^2 = (PB)^2$

$\Rightarrow (x-3)^2 + (y-4)^2 = (x-5)^2 + (y+2)^2$

$\Rightarrow (x-3)^2 - (x-5)^2 + (y-4)^2 - (y+2)^2 = 0$

$\Rightarrow (2x-8)(2) + (2y-2)(-6) = 0$

$\Rightarrow x - 4 - 3y + 3 = 0$

$\Rightarrow x - 3y - 1 = 0$ ...(i)

and area of $\Delta PAB = 10$

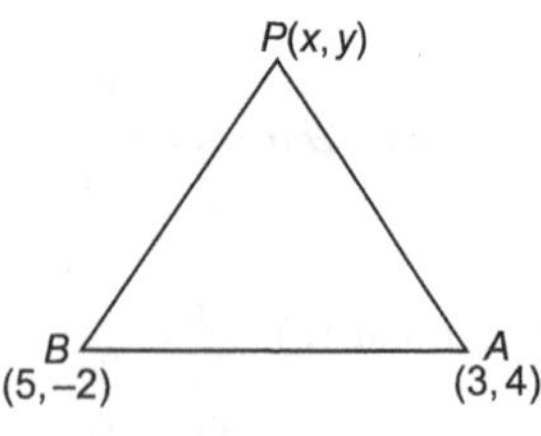

$$\Rightarrow \frac{1}{2}\left|\begin{vmatrix} 5 & -2 \\ 3 & 4 \\ x & y \\ 5 & -2 \end{vmatrix}\right| = 10$$

$\Rightarrow (20 + 3y - 2x) - (-6 + 4x + 5y) = \pm 20$

$\Rightarrow -6x - 2y + 26 = \pm 20$

$\Rightarrow 3x + y - 13 \pm 10 = 0$

or $3x + y - 3 = 0$ and $3x + y - 23 = 0$ ...(ii)

Solving Eqs. (i) and (ii), we get

$$P(x, y) \equiv (7, 2) \text{ or } (1, 0)$$

56. $\because \begin{vmatrix} 1 & a & a \\ b & 1 & b \\ c & c & 1 \end{vmatrix} = 0$

Applying $C_2 \to C_2 - C_1$ and $C_3 \to C_3 - C_1$

Then, $\begin{vmatrix} 1 & a-1 & a-1 \\ b & 1-b & 0 \\ c & 0 & 1-c \end{vmatrix} = 0$

Let $a-1 = A$, $b-1 = B$ and $c-1 = C$, then

$$\begin{vmatrix} 1 & A & A \\ 1+B & -B & 0 \\ 1+C & 0 & -C \end{vmatrix} = 0$$

$\Rightarrow 1(BC - 0) - A(-C - BC) + A(B + BC) = 0$

or $\frac{1}{A} + \frac{1}{B} + \frac{1}{C} = -2$ ...(i)

$$\therefore \frac{a}{a-1} + \frac{b}{b-1} + \frac{c}{c-1} = \frac{1+A}{A} + \frac{1+B}{B} + \frac{1+C}{C}$$

$$= \frac{1}{A} + \frac{1}{B} + \frac{1}{C} + 3$$

$= -2 + 3$ [from Eq. (i)]

$= 1$

57. $\therefore$ Slope of $AD \times$ slope of $BE = -1$

$$\Rightarrow \frac{0-b}{\frac{a}{2}-0} \times \frac{\frac{b}{2}-0}{\frac{a}{2}-0} = -1$$

$$\Rightarrow \frac{-\frac{b^2}{2}}{\frac{a^2}{4}} = -1 \Rightarrow \frac{b^2}{2} = \frac{a^2}{4}$$

$$\Rightarrow a^2 = 2b^2 \quad \therefore a = \pm b\sqrt{2}$$

58. We have, $\frac{a^2 - 2a}{a^2 - 4a} < 0 \Rightarrow \frac{a-2}{a-4} < 0$

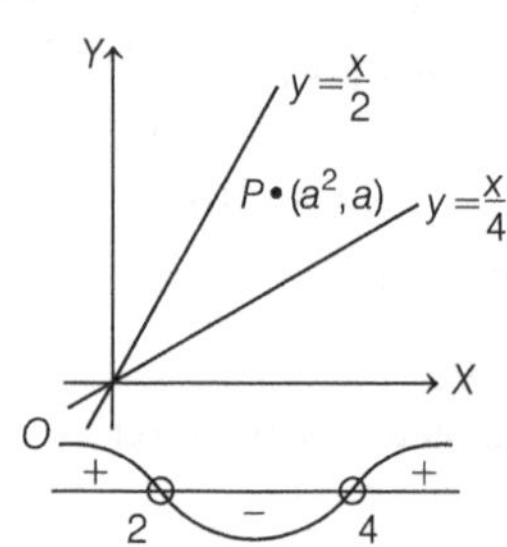

$$\therefore a \in (2, 4)$$

59. Equation of bisectors of the given lines are

$$\left(\frac{3x + 4y - 5}{\sqrt{3^2 + 4^2}}\right) = \pm\left(\frac{5x - 12y - 10}{\sqrt{5^2 + (-12)^2}}\right)$$

$\therefore (39x + 52y - 65) = \pm(25x - 60y - 50)$

$\Rightarrow 14x + 112y - 15 = 0$ or $64x - 8y - 115 = 0$

or $\frac{14}{15}x + \frac{112}{15}y - 1 = 0$

or $\frac{64}{115}x - \frac{8}{115}y - 1 = 0$

$$\therefore a = \frac{14}{15}, b = \frac{112}{15} \text{ or } a = \frac{64}{115}, b = -\frac{8}{115}$$

60. $AP : PB = 2 : 1$

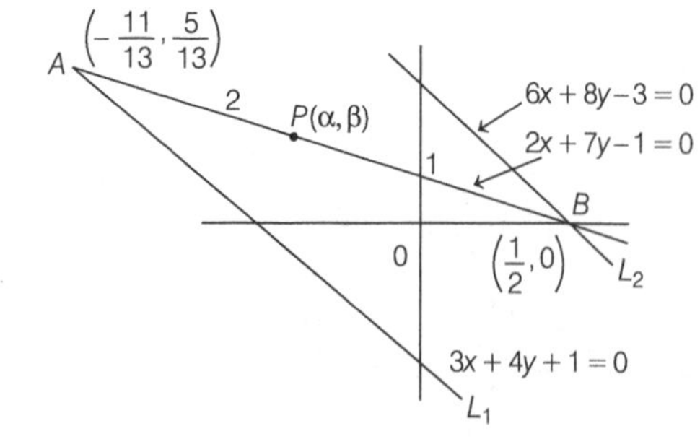

$$\alpha = \frac{2\times\frac{1}{2}+1\times\left(-\frac{11}{13}\right)}{2+1} = -\frac{2}{39}$$

$$\beta = \frac{2\times 0+1\times\frac{5}{13}}{2+1} = \frac{5}{39}$$

$\therefore$ Required line $y - \frac{5}{39} = -\frac{3}{4}\left(x - \frac{2}{39}\right)$

or $9x + 12y - 2 = 0$

61. Since, mid-point of BC is D (5, 6)

Let coordinate C (h, k) which is lie on

$$x + 2y = 16$$

$\therefore$ $h + 2k = 16$ ...(i)

$\because$ D is the mid point of B and C

$\therefore$ Coordinate of B is $(10 - h, 12 - k)$

Since B lies on $2x + 3y = 29$

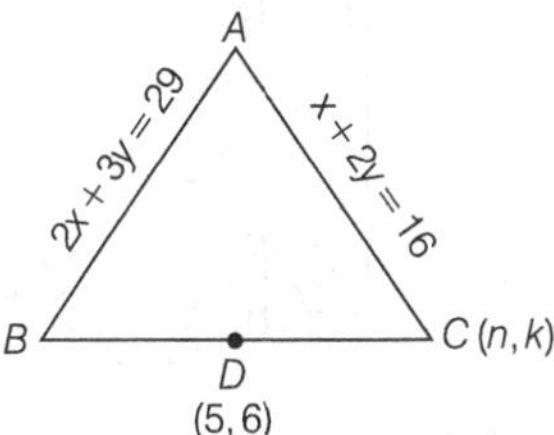

$\Rightarrow$ $2(10 - h) + 3(12 - k) = 29$

$\Rightarrow$ $20 - 2h + 36 - 3k = 29$

$\Rightarrow$ $2h + 3k = 27$ ...(ii)

Solving Eqs. (i) and (ii), we get

$$h = 6, k = 5$$

Coordinates of C are (6, 5)

$\therefore$ Equation of BC is $y - 6 = \frac{5-6}{6-5}(x - 5)$

$\Rightarrow$ $y - 6 = -x + 5$

or $x + y = 11$

62. Given, lines $x + (a-1)y = 1$

and $2x + a^2y = 1$, where $a \in R - \{0, 1\}$

are perpendicular to each other

$\therefore$ $\left(-\frac{1}{a-1}\right)\times\left(-\frac{2}{a^2}\right) = -1$

[$\because$ if lines are perpendicular, then product of their slopes is -1]

$\Rightarrow$ $a^2(a-1) = -2 \Rightarrow a^3 - a^2 + 2 = 0$

$\Rightarrow$ $(a+1)(a^2 - 2a + 2) = 0 \Rightarrow a = -1$

$\therefore$ Equation of lines are

$x - 2y = 1$ ...(i)

and $2x + y = 1$ ...(ii)

On solving Eq. (i) and Eq. (ii), we get

$$x = \frac{3}{5} \text{ and } y = -\frac{1}{5}$$

$\therefore$ Point of intersection of the lines (i) and (ii) is $P\left(\frac{3}{5}, -\frac{1}{5}\right)$.

Now, required distance of the point $P\left(\frac{3}{5}, -\frac{1}{5}\right)$ from origin $= \sqrt{\frac{9}{25} + \frac{1}{25}} = \sqrt{\frac{10}{25}} = \sqrt{\frac{2}{5}}$

63. The point $(\cos\theta, \sin\theta)$ lies on a circle

$$x^2 + y^2 = 1$$

$\because$ $y = |x - 1|$

$\therefore$ $y = x - 1$

and $y = 1 - x$

From the figure, θ can vary from

$$-\frac{\pi}{2} \text{ to } \frac{\pi}{2}$$

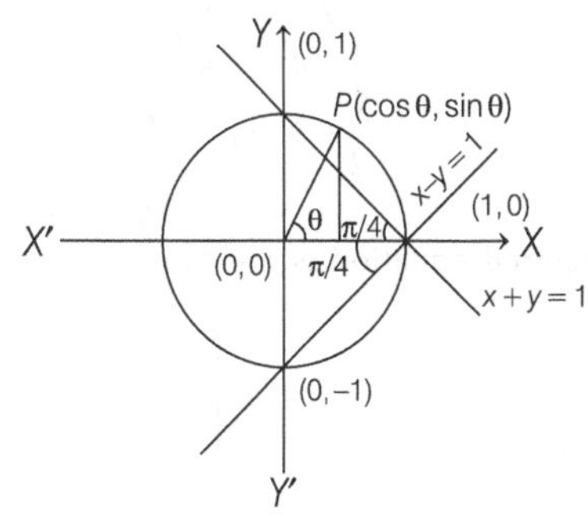

64. Let, A (0, 0),

Coordinate of B are $(2\cot\theta, 2)$ and coordinate of C are $\{4\cot(60° - \theta), -4\}$

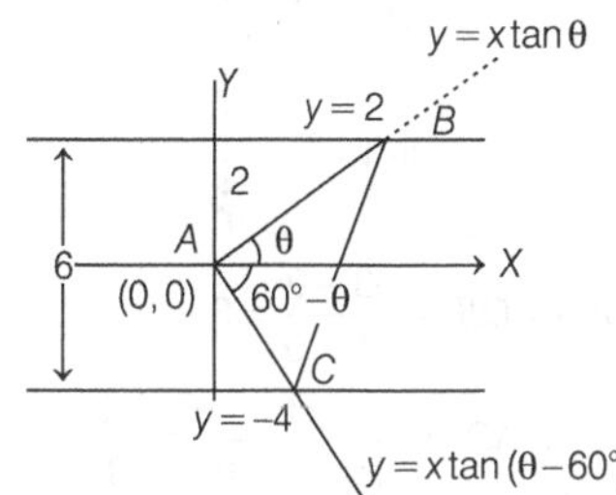

$\therefore$ $AB = AC$

$\Rightarrow$ $(AB)^2 = (AC)^2$

$$4\cot^2\theta + 4 = 16\cot^2(60° - \theta) + 16$$

$$4\operatorname{cosec}^2\theta = 16\operatorname{cosec}^2(60° - \theta)$$

$\Rightarrow$ $\sin(60° - \theta) = 2\sin\theta$

$\Rightarrow$ $\left(\frac{\sqrt{3}}{2}\cos\theta - \frac{1}{2}\sin\theta\right) = 2\sin\theta$

$\therefore$ $\tan\theta = \frac{\sqrt{3}}{5}$

$\therefore$ The required length $= AB = \sqrt{(4\cot^2\theta + 4)}$

$$= 2\sqrt{\frac{28}{3}} = \frac{4\sqrt{7}}{\sqrt{3}}$$

65. $\therefore$ Area of quadrilateral $OABC = 2 \times 3 = 6$ sq unit.

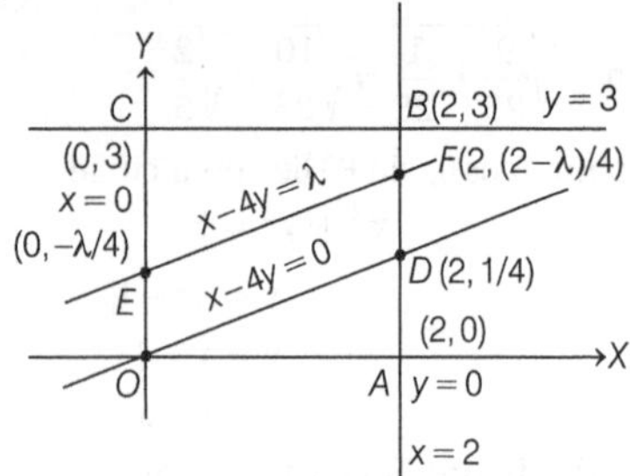

$\therefore$ Area of $OAFEO = \frac{1}{2}$ area of quadrilateral $OABC$

$$\Rightarrow \quad \frac{1}{2} \times \left(\frac{2-\lambda}{4} - \frac{\lambda}{4}\right) \times 2 = 3$$

$$\Rightarrow \quad \frac{1}{2} - \frac{\lambda}{2} = 3 \quad \text{or} \quad -\frac{\lambda}{2} = \frac{5}{2}$$

$$\therefore \quad \lambda = -5$$

$\therefore$ Required equation is $x - 4y + 5 = 0$

66.

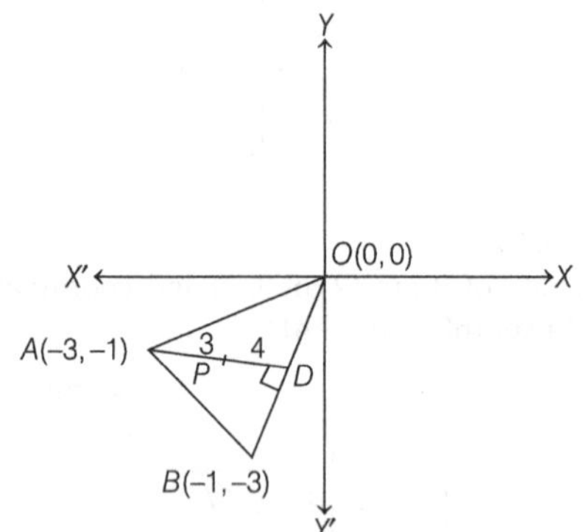

$$\text{Slope of } OB = \frac{-3-0}{-1-0} = 3$$

$$\therefore \text{ Slope of } AD = \frac{-1}{\text{Slope of } OB} = \frac{-1}{3}$$

Now, equation of OB is given by

$$y - 0 = 3((x - 0)$$

$$\Rightarrow \quad y = 3x \quad \text{...(i)}$$

And equation of AD is given by,

$$y + 1 = -\frac{1}{3}(x + 3)$$

$$3y + 3 = -x - 3$$

$$x + 3y = -6 \quad \text{...(ii)}$$

On solving Eq. (i) and Eq. (ii), we get

$$D\left(\frac{-3}{5}, \frac{-9}{5}\right)$$

Now, P divides AD in the ratio 3 : 4.

$$\underset{A(-3,-1)}{\bullet}\overset{3}{\text{———}}\underset{P(x,y)}{+}\overset{4}{\text{———}}\underset{D(-3/5,-9/5)}{\bullet}$$

$$\therefore \quad P(x, y) = \left(\frac{\frac{-9}{5} - 12}{7}, \frac{\frac{-27}{5} - 4}{7}\right) = \left(\frac{-69}{35}, \frac{-47}{35}\right)$$

$\therefore$ Equation of line passing through $P\left(\frac{-69}{36}, \frac{-47}{35}\right)$ and parallel to OB is

$$y + \frac{47}{35} = 3\left(x + \frac{69}{35}\right)$$

$$\Rightarrow \quad 35y + 47 = 3(35x + 69)$$

$$\Rightarrow \quad 35y + 47 = 105x + 207$$

$$\Rightarrow \quad 105x - 35y + 160 = 0$$

$$\Rightarrow \quad 21x - 7y + 32 = 0$$

67. $\because$ Equation AB is $x + y = 1$, then coordinates of A and B are (1, 0) and (0, 1) respectively.

$\therefore$ Coordinate of P is $\left(\frac{1}{2}, \frac{1}{2}\right)$

$\because$ PP_1 is perpendicular to OA

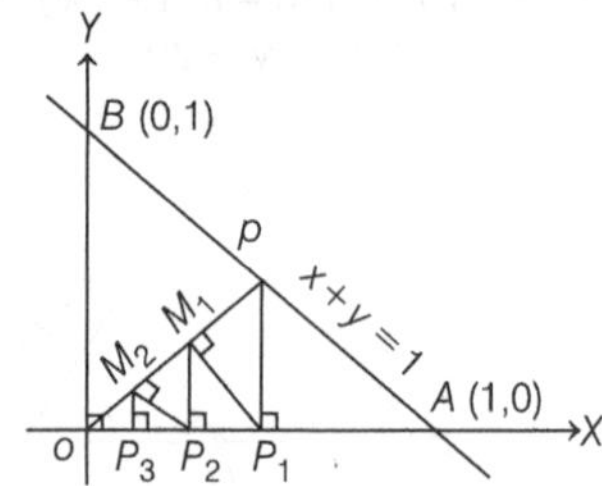

Equation of OP is $y = x$

Then, $OP_1 = PP_1 = 1/2$

We have, $(OM_{n-1})^2 = (OP_n)^2 + (P_n M_{n-1})^2$

$$= 2(OP_n)^2 \qquad [\because y = x]$$

$$= 2\alpha_n^2 \text{ (say)}$$

Also, $(OP_{n-1})^2 = (OM_{n-1})^2 + (P_{n-1} M_{n-1})^2$

$$\alpha_{n-1}^2 = 2\alpha_n^2 + \frac{1}{2}\alpha_{n-1}^2$$

$$\Rightarrow \quad \frac{1}{2}\alpha_{n-1}^2 = 2\alpha_n^2 \Rightarrow \alpha_n = \frac{1}{2}\alpha_{n-1}$$

$$\therefore \quad OP_n = \alpha_n = \frac{1}{2}\alpha_{n-1}$$

$$= \frac{1}{2^2}\alpha_{n-2} = \frac{1}{2^3}\alpha_{n-3}$$

....

....

....

$$= \frac{1}{2^{n-1}}\alpha_1 = \frac{1}{2^{n-1}}\left(\frac{1}{2}\right) = \frac{1}{2^n}$$

68. $p = ON = OM + MN$

= (Perpendicular distance from 0 to AB) + AD

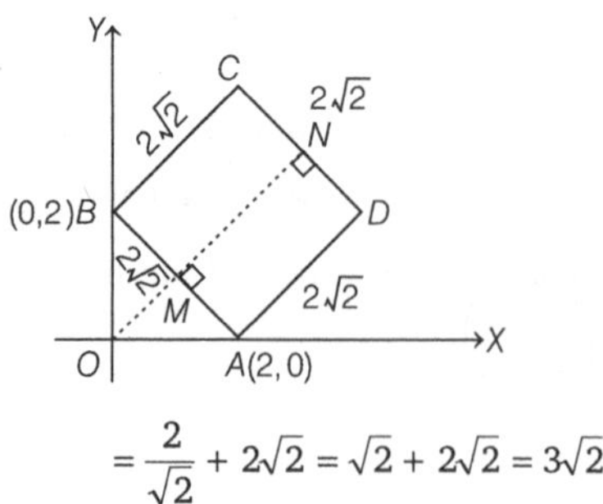

$$= \frac{2}{\sqrt{2}} + 2\sqrt{2} = \sqrt{2} + 2\sqrt{2} = 3\sqrt{2}$$

69. Equation of line passing through (2, −1) and (5, −3) is

$$y+1=\left(\frac{-3+1}{5-2}\right)(x-2)$$

$$3y+3=-2x+4$$

$$\Rightarrow \quad 2x+3y=1 \quad \text{...(i)}$$

$(x_1, 4)$ and $(-2, y_1)$ lie on Eq. (i)

Then, $2x_1+12=1 \Rightarrow x_1=-\dfrac{11}{2}$

and $-4+3y_1=1 \Rightarrow y_1=\dfrac{5}{3}$

Hence, Point $P\left(-\dfrac{11}{2}, \dfrac{5}{3}\right)$ lies an $2x+6y+1=0$.

70. ∵ Slope of OA × slope of $BC = -1$

$$\frac{\beta-1}{\alpha+6}\times\frac{3+2}{-2-3}=-1$$

$$\Rightarrow \quad \beta-1=\alpha+6$$

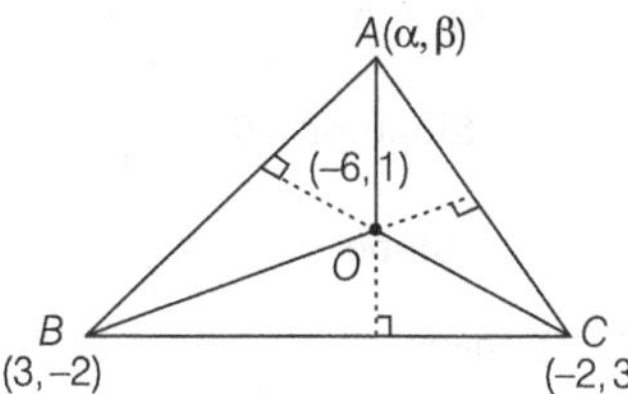

$$\therefore \quad \alpha-\beta+7=0 \quad \text{...(i)}$$

and slope of OB × slope of $AC = -1$

$$\Rightarrow \quad \frac{-2-1}{3+6}\times\frac{\beta-3}{\alpha+2}=-1$$

$$\frac{\beta-3}{\alpha+2}=3$$

$$\beta-3=3\alpha+6$$

$$\Rightarrow \quad 3\alpha-\beta+9=0 \quad \text{...(ii)}$$

Solving Eqs. (i) and (ii), we get

$$\alpha=-1, \beta=6$$

∴ Third vertex is (−1, 6).

71. $\tan\theta=2$

$$\therefore \quad \sin\theta=\frac{2}{\sqrt5}, \cos\theta=\frac{1}{\sqrt5}$$

Let new co-ordinate is (x_1, y_1)

$$\therefore \quad \frac{x_1-1}{\cos\theta}=\frac{y_1-1}{\sin\theta}=\pm 1$$

$\therefore x_1=1\pm\cos\theta, y_1=1\pm\sin\theta$

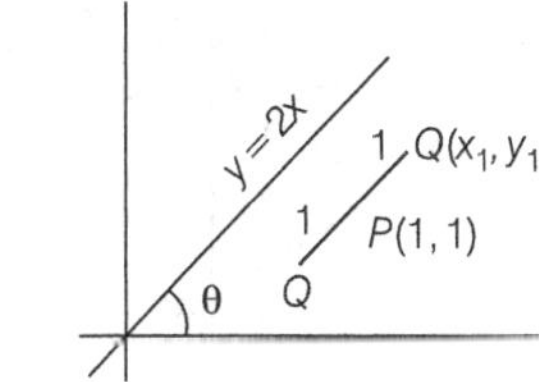

$$x_1=1\pm\frac{1}{\sqrt5}, y_1=1\pm\frac{2}{\sqrt5}$$

Hence, new position of P are $\left(1\pm\dfrac{1}{\sqrt5}, 1\pm\dfrac{2}{\sqrt5}\right)$.

72. ∵ $3a+2b+6c=0$

$$\therefore \quad c=-\frac{(3a+2b)}{6}$$

$$\therefore \quad ax+by+c=0$$

$$\Rightarrow ax+by-\frac{(3a+2b)}{6}=0$$

$$\Rightarrow \quad 6ax+6by-3a-2b=0$$

$$\Rightarrow \quad 3a(2x-1)+2b(3y-1)=0$$

$$\Rightarrow \quad (2x-1)+\frac{2b}{3a}(3y-1)=0$$

$$P+\lambda Q=0, \therefore P=0, Q=0$$

Then, $2x-1=0, 3y-1=0; x=1/2, y=1/3$

Hence, fixed points is $\left(\dfrac{1}{2}, \dfrac{1}{3}\right)$.

73. Equation of line through AB and AC is

$(px+qy-1)+\lambda(qx+py-1)=0$ ...(i)

∵ It is passes through (p, q), then

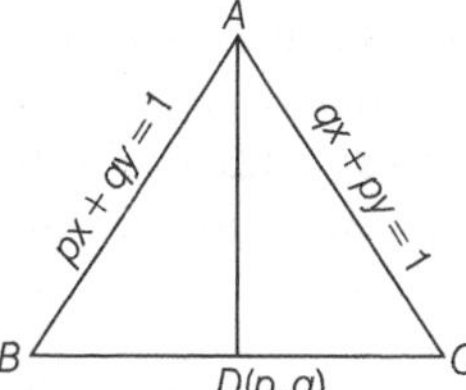

$$(p^2+q^2-1)+\lambda(pq+pq-1)=0$$

$$\therefore \quad \lambda=-\frac{(p^2+q^2-1)}{(2pq-1)}$$

From Eq. (i),

$$(px+qy-1)-\frac{(p^2+q^2-1)}{(2pq-1)}(qx+py-1)=0$$

or $(2pq-1)(px+qy-1)=(p^2+q^2-1)(qx+py-1)$

74. The system of lines (given) pass through (−2, 0). So, the required line is the one through (−2, 0) and perpendicular to join of (−2, 0) and (2, 3).

∴ Its equation is

$$y-0=-\frac{4}{3}(x+2)$$

$$\Rightarrow \quad 4x+3y+8=0$$

75. $|PA-PB|\le|AB|$

Maximum value of $|PA-PB|$ is $|AB|$, which is possible only when P, A, B are collinear

if $P(x, y)$ then equation AB is

$$\frac{x}{2}+\frac{y}{2}=1$$

$$\Rightarrow \quad x+y=2 \quad \text{...(i)}$$

Now solving Eq. (i)

and $\quad 2x+3y+1=0$ ...(ii)

There we get,

$$x=7, y=-5$$

$$\therefore \quad P\equiv(7,-5)$$

Objective Questions Type II [One or more than one correct answer(s)]

1. Let the slope of $u = 0$ be m then the slope of $v = 0$ is $\frac{9m}{2}$

$$\therefore \quad \frac{7}{9} = \left| \frac{m - \frac{9m}{2}}{1 + m \cdot \frac{9m}{2}} \right| = \left| \frac{-7m}{2 + 9m^2} \right|$$

i.e. $9m^2 - 9m + 2 = 0$

$$m = \frac{9 \pm \sqrt{81 - 72}}{18} = \frac{9 \pm 3}{18} = \frac{2}{3}, \frac{1}{3}$$

or $9m^2 + 9m + 2 = 0$

or $m = \frac{-9 \pm 3}{18} = \frac{-2}{3}, \frac{-1}{3}$

Equations of the lines are
(i) $3y = x$ and $2y = 3x$
(ii) $3y = 2x$ and $y = 3x$
(iii) $x + 3y = 0$ and $3x + 2y = 0$
(iv) $2x + 3y = 0$ and $3x + y = 0$

2. The given equation can be written as,
$(2x + 3y - 5)\cos\theta + (3x - 5y + 2)\sin\theta = 0$
or $(2x + 3y - 5) + \tan\theta(3x - 5y + 2) = 0$
This passes through the point of intersection of the lines $2x + 3y - 5 = 0$ and $3x - 5y + 2 = 0$ for all values of θ. The coordinates of the point P of intersection are $(1, 1)$. Let $Q(h, k)$ be the reflection of $P(1, 1)$ in the line.

$$x + y = \sqrt{2} \quad \ldots\text{(i)}$$

Then PQ is perpendicular to Eq. (i) and the mid-point of PQ lies on Eq. (i).

$$\therefore \quad \frac{k-1}{h-1} = 1 \Rightarrow k = h$$

and $$\frac{h+1}{2} + \frac{k+1}{2} = \sqrt{2} \Rightarrow h = k = \sqrt{2} - 1$$

3. The two lines will be identical if their exists some real number k such that

$$b^3 - c^3 = k(b - c),\ c^3 - a^3 = k(c - a)$$

and $a^3 - b^3 = k(a - b)$

$\Rightarrow$ $b - c = 0$ or $b^2 + c^2 + bc = k$

$c - a = 0$ or $c^2 + a^2 + ca = k$

and $a - b = 0$ or $a^2 + b^2 + ab = k$

$\Rightarrow$ $a = b$ or $b = c$ or $c = a$

or $b^2 + c^2 + bc = c^2 + a^2 + ca \Rightarrow b = c$ or $c = a$

or $a = b$ or $a + b + c = 0$

4. As the third vertex lies on the line $y = x + 3$, its coordinates are of the form $(x, x + 3)$. The area of the triangle with vertices $(2, 1)$, $(3, -2)$ and $(x, x + 3)$ is given by

$$\frac{1}{2}\left| \begin{vmatrix} x & x+3 & 1 \\ 2 & 1 & 1 \\ 3 & -2 & 1 \end{vmatrix} \right| = |2x - 2| = 5 \quad \text{(given)}$$

$\therefore\ 2x - 2 = \pm 5 \Rightarrow x = -3/2, 7/2$

Thus, the coordinates of the third vertex are $(7/2, 13/2)$ or $(-3/2, 3/2)$.

5. It is clear $A(a + b, a - b)$ is the image of $B(a - b, a + b)$ w.r.t. $y = x$
$\therefore P$ is the mid-point of AB i.e., (a, a)
If P be (a, b), then $bx = ay$ $[\because a = b]$

6. $AB = 5$, $D \equiv \left(2, \frac{3}{2}\right)$

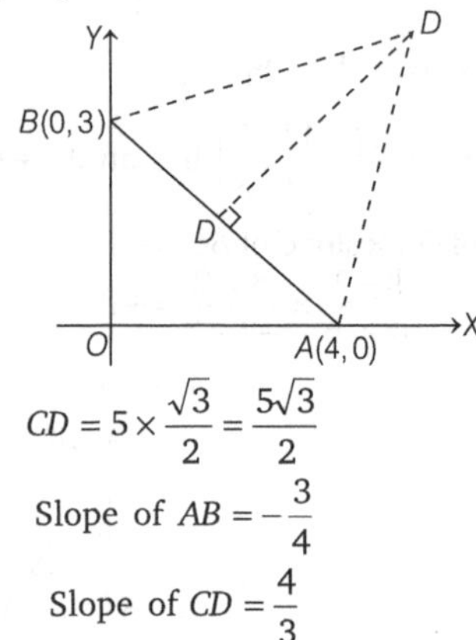

$$CD = 5 \times \frac{\sqrt{3}}{2} = \frac{5\sqrt{3}}{2}$$

$$\text{Slope of } AB = -\frac{3}{4}$$

$$\text{Slope of } CD = \frac{4}{3}$$

If $C \equiv (h, k)$, then

$$\frac{h-2}{\frac{3}{5}} = \frac{k - \frac{3}{2}}{\frac{4}{5}} = \pm \frac{5\sqrt{3}}{2}$$

or $h = 2\left(1 - \frac{3\sqrt{3}}{4}\right),\ k = \frac{3}{2}\left(1 - \frac{4}{\sqrt{3}}\right)$

or $h = 2\left(1 + \frac{3\sqrt{3}}{4}\right),\ k = \frac{3}{2}\left(1 + \frac{4}{\sqrt{3}}\right)$

7. Equation of any line through the point of intersection of the given lines is $(3x + y - 5) + \lambda(x - y + 1) = 0$ since this line is perpendicular to one of the given lines $\frac{3 + \lambda}{\lambda - 1} = -1$ or $\frac{1}{3}$
$\Rightarrow \lambda = -1$ or -5, therefore the required straight line is $x + y - 3 = 0$ or $x - 3y + 5 = 0$

8. If B lies on y-axis, then coordinates of B are $(0, a)$ or $(0, -a)$

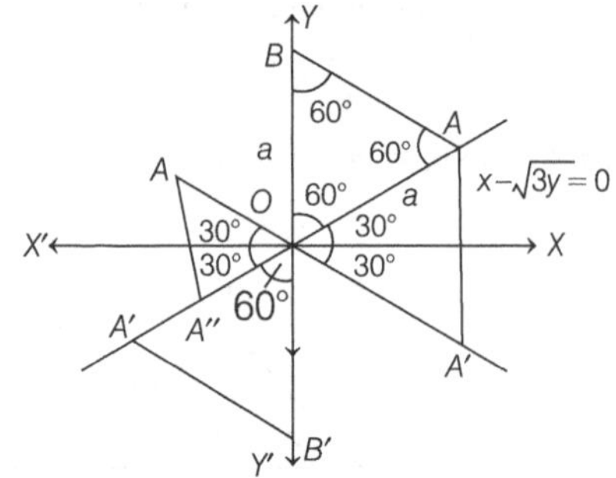

If third vertex in IV quadrant or in II quadrant, then its coordinates are $(a\cos 30°, -a\sin 30°)$
and $(-a\cos 30°, a\sin 30°)$
i.e., $\left(\frac{a\sqrt{3}}{2}, -\frac{a}{2}\right)$ and $\left(-\frac{a\sqrt{3}}{2}, \frac{a}{2}\right)$

9. Since $ax + by + c = 0$, $bx + cy + a = 0$ and $cx + ay + b = 0$ are concurrent

$$\therefore \quad \begin{vmatrix} a & b & c \\ b & c & a \\ c & a & b \end{vmatrix} = 0$$

$$\Rightarrow \quad 3abc - a^3 - b^3 - c^3 = 0$$

$$\Rightarrow \quad -(a + b + c)(a^2 + b^2 + c^2 - ab - bc - ca) = 0$$

as $\quad a + b + c \neq 0$

$$\therefore \quad a^2 + b^2 + c^2 - ab - bc - ca = 0$$

$$\frac{1}{2}\{(a - b)^2 + (b - c)^2 + (c - a)^2\} = 0$$

As a, b, c are real numbers

$$\therefore \quad b - c = 0,\ c - a = 0,\ a - b = 0$$

$$\Rightarrow \quad a = b = c.$$

10. Let vertices of a triangle ABC are

$A \equiv (x_1, y_1), B \equiv (x_2, y_2)$ and $C \equiv (x_3, y_3)$

Where, $x_1, x_2, x_3, y_1, y_2, y_3 \in Q$

$\therefore a = BC = \sqrt{(x_2 - x_3)^2 + (y_2 - y_3)^2} =$ irrational

Similarly, $b = CA$ and $c = AB$

Then, Incentre $= \left(\dfrac{ax_1 + bx_2 + cx_3}{a+b+c}, \dfrac{ay_1 + by_2 + cy_3}{a+b+c}\right)$

$\therefore$ Incentre has Irrational coordinate but centroid, circumcentre and orthocentre have rational coordinates.

11. (a) $\begin{vmatrix} 1 & \log_a b & \log_a c \\ \log_b a & 1 & \log_b c \\ \log_c a & \log_c b & 1 \end{vmatrix}$

$$= \frac{1}{ln\, a \cdot ln\, b}\begin{vmatrix} ln\, a & ln\, b & ln\, c \\ ln\, a & ln\, b & ln\, c \\ \log_c a & \log_c b & 1 \end{vmatrix} = 0$$

(b) Let $\tan\theta = 0.2$

$$\Rightarrow \quad \tan 2\theta = \frac{2 \times \cdot 2}{1 - (\cdot 2)^2} = \frac{5}{12}$$

Hence the line is $y = \dfrac{5}{12}x$.

(c) Sides of triangle are $y = 3x$, $y = 6x$ and $y = 9$ so vertices of triangle are (0, 0), (3, 9) and $\left(\dfrac{3}{2}, 9\right)$ Area

of triangle $= \dfrac{1}{2}\begin{vmatrix} 0 & 0 & 1 \\ 3 & 9 & 1 \\ \frac{3}{2} & 9 & 1 \end{vmatrix} = \dfrac{27}{4}$ sq units

(d) Equation of the line pair through the origin and perpendicular to the lines pair

$xy - 3y^2 + y - 2x + 10 = 0$ is $3x^2 + xy = 0$.

12. $bx + cy = a \Rightarrow \dfrac{x}{(a/b)} + \dfrac{y}{(a/c)} = 1$

Area of $\Delta OAB = \dfrac{1}{8}$ (given)

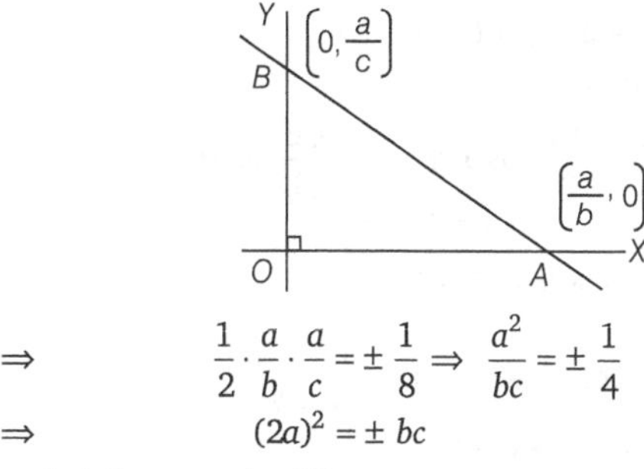

$$\Rightarrow \quad \frac{1}{2}\cdot\frac{a}{b}\cdot\frac{a}{c} = \pm\frac{1}{8} \Rightarrow \frac{a^2}{bc} = \pm\frac{1}{4}$$

$$\Rightarrow \quad (2a)^2 = \pm\, bc$$

$\therefore$ $b, \pm 2a, c$ are in GP.

13.

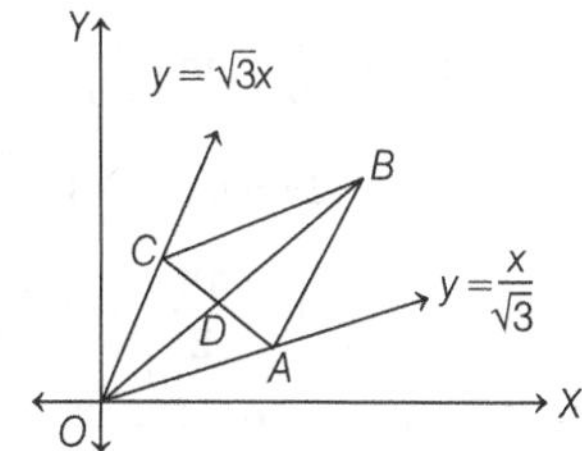

Here, $\angle COA = 30°$

Area of rhombus $= 2 \times \dfrac{1}{2} \times OA \times OC \sin 30°$

or $\quad 2 = \dfrac{1}{2}x^2$

or $\quad OA = OC = 2$

Also, $\quad \angle OAB = 150°$

$$\therefore \quad \cos 150° = \frac{OA^2 + AB^2 - OB^2}{2OA \times AB}$$

$$OB^2 = 8 + 4\sqrt{3} \text{ or } OB = \sqrt{2}(\sqrt{3} + 1)$$

Hence, the coordinates of B are $(\pm\sqrt{2}(\sqrt{3} + 1))$ $\cos 45° \pm \sqrt{2}(\sqrt{3} + 1)\sin 45°)$.

14. Let position of bunglow is $P(x_1, y_1)$, then $PM = 100$ and $PN = 100$

$$\therefore \quad \frac{x_1 + y_1 - 8}{\sqrt{2}} = \pm 100$$

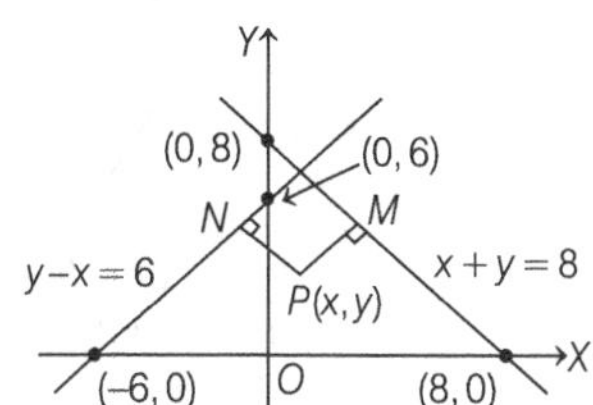

and $\quad \dfrac{x_1 - y_1 + 6}{\sqrt{2}} = \pm 100$

After solving, we get

$$x_1 = 1 \pm 100\sqrt{2}, 1$$

and $\quad y_1 = 7, 7 \pm 100\sqrt{2}$

Hence, $(1 + 100\sqrt{2}, 7)$, $(1 - 100\sqrt{2}, 7)$, $(1, 7 + 100\sqrt{2})$, $(1, 7 - 100\sqrt{2})$

15. $\dfrac{x-1}{\cos\theta}=\dfrac{y-2}{\sin\theta}=\pm\dfrac{\sqrt{6}}{3}$

$\therefore \quad x=1\pm\dfrac{\sqrt{6}}{3}\cos\theta,\ y=2\pm\dfrac{\sqrt{6}}{3}\sin\theta$

$\because \quad x+y=4$

$\therefore \ 3\pm\dfrac{\sqrt{6}}{3}(\cos\theta+\sin\theta)=4$

$\Rightarrow \quad \pm(\cos\theta+\sin\theta)=\dfrac{3}{\sqrt{6}}=\sqrt{\dfrac{3}{2}}$

Squaring on both sides,

$$1+\sin 2\theta=\frac{3}{2}$$

$$\Rightarrow \quad \sin 2\theta=\frac{1}{2}=\sin\left(\frac{\pi}{6}\right)$$

$$\therefore \quad 2\theta=n\pi+(-1)^n\frac{\pi}{6},\ n\in I$$

$$\therefore \quad \theta=\frac{n\pi}{2}+(-1)^n\frac{\pi}{12},\ n\in I$$

For $n=0,\ \theta=\dfrac{\pi}{12}$, for $n=1,\ \theta=\dfrac{\pi}{2}-\dfrac{\pi}{12}=\dfrac{5\pi}{12}$

For $n=-1,\ \theta=-\dfrac{\pi}{2}-\dfrac{\pi}{12}=-\dfrac{7\pi}{12}$;

for $n=-2,\ \theta=-\pi+\dfrac{\pi}{12}=-\dfrac{11\pi}{12}$

16. Equation of the other diagonal is $x+y=\lambda$ which pass through (a, b), then

$$a+b=\lambda$$

$\therefore$ Equation of other diagonal

$$x+y=a+b$$

i.e., then centre of the square is the point of intersection of $x-y=a$ and $x+y=a+b$ is $\left(a+\dfrac{b}{2},\dfrac{b}{2}\right)$, then vertex

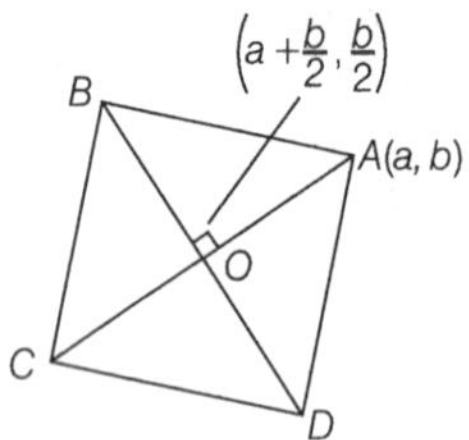

$$C\equiv(2a+b-a,\ b-b)$$

$\therefore \quad C\equiv(a+b,\ 0)$

If $\quad B\equiv z$

Then, $\dfrac{z-\left(a+\dfrac{b}{2}+\dfrac{ib}{2}\right)}{(a+ib)-\left(a+\dfrac{b}{2}+\dfrac{ib}{2}\right)}=\dfrac{BO}{AO}e^{i\pi/2}=i\ (\because BO=AO)$

$\Rightarrow \quad z-\left(a+\dfrac{b}{2}+\dfrac{ib}{2}\right)=i\left(-\dfrac{b}{2}+\dfrac{ib}{2}\right)$

$$=-\frac{ib}{2}-\frac{b}{2}$$

$\therefore \quad z=a$

$\therefore \quad B\equiv(a,\ 0)$

then, $\quad D\equiv(a+b,\ b)$

Hence, other vertices are $(a+b, 0)$, $(a, 0)$ and $(a+b, b)$.

17.

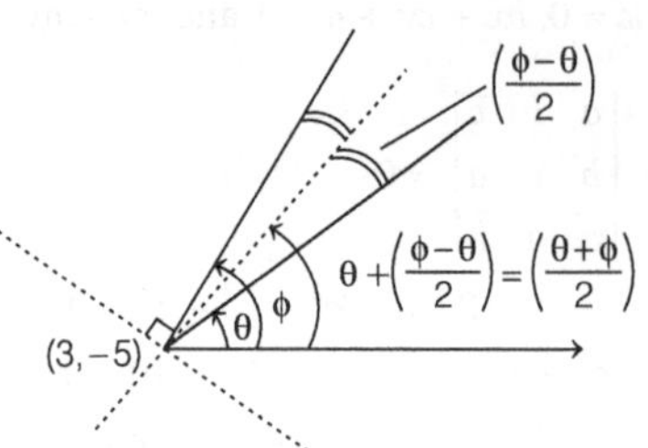

$\therefore$ One bisector makes an angle $\left(\dfrac{\theta+\phi}{2}\right)$ with X-axis, then other bisector makes an angle $90°+\left(\dfrac{\theta+\phi}{2}\right)$ with X-axis.

$\therefore$ Equations of bisectors are

$$\frac{x-3}{\cos\left(\dfrac{\theta+\phi}{2}\right)}=\frac{y+5}{\sin\left(\dfrac{\theta+\phi}{2}\right)} \quad \text{...(i)}$$

and $\dfrac{x-3}{\cos\left(\dfrac{\pi}{2}+\dfrac{\theta+\phi}{2}\right)}=\dfrac{y+5}{\sin\left(\dfrac{\pi}{2}+\dfrac{\theta+\phi}{2}\right)}$

$$\Rightarrow \quad \frac{x-3}{-\sin\left(\dfrac{\theta+\phi}{2}\right)}=\frac{y+5}{\cos\left(\dfrac{\theta+\phi}{2}\right)} \quad \text{...(ii)}$$

But given bisectors are $\dfrac{x-3}{\cos\alpha}=\dfrac{y+5}{\sin\alpha}$

$\therefore \quad \alpha=\dfrac{\theta+\phi}{2}$ [from Eq. (i)] ...(iii)

and $\quad \dfrac{x-3}{\beta}=\dfrac{y+5}{\gamma}$

$\therefore \quad \beta=-\sin\left(\dfrac{\theta+\phi}{2}\right)=-\sin\alpha$ [from Eq. (ii)]

and $\gamma=\cos\left(\dfrac{\theta+\phi}{2}\right)=\cos\alpha$

18. Given family of lines are

$\cos\theta(2x+3y-5)+\sin\theta(3x-5y+2)=0$ ...(i)

and $a(x+y-1)+b(2x-3y+1)=0$ ...(ii)

$$\left.\begin{aligned}2x+3y-5=0\\3x-5y+2=0\end{aligned}\right\}\Rightarrow x=1,\ y=1$$

$$\left.\begin{aligned}x+y-1=0\\2x-3y+1=0\end{aligned}\right\}\Rightarrow x=\frac{2}{5},\ y=\frac{3}{5}$$

$\therefore \quad x_1=1,\ y_1=1,\ x_2=\dfrac{2}{5},\ y_2=\dfrac{3}{5}$

19. Equation of any line through $P(3, 4)$ making an angle θ with the positive direction of X-axis is $\dfrac{x-3}{\cos\theta}=\dfrac{y-4}{\sin\theta}=r$ ...(i)

where r is the distance of any point on the line from P. Therefore, coordinates of any point on the line (i) are $(3+r\cos\theta,\ 4+r\sin\theta)$ ...(ii)

If Eq. (ii) represents R, then

$$3+r\cos\theta=6 \quad \Rightarrow \ r=\frac{3}{\cos\theta}=PR$$

If Eq. (ii) represents S, the $4+r\sin\theta=8$

$\Rightarrow r=4/\sin\theta=PS$

Hence, $PR = 3\sec\theta, PS = 4\text{cosec}\,\theta$

$$PR + PS = \frac{3\sin\theta + 4\cos\theta}{\sin\theta\cos\theta} = \frac{2(3\sin\theta + 4\cos\theta)}{\sin 2\theta}$$

and $(3/PR)^2 + (4/PS)^2 = \cos^2\theta + \sin^2\theta = 1$

$$\Rightarrow \frac{9}{(PR)^2} + \frac{16}{(PS)^2} = 1$$

20. Slope of $AC = \frac{5-3}{7-1} = \frac{1}{3} = \tan\theta$ (say)

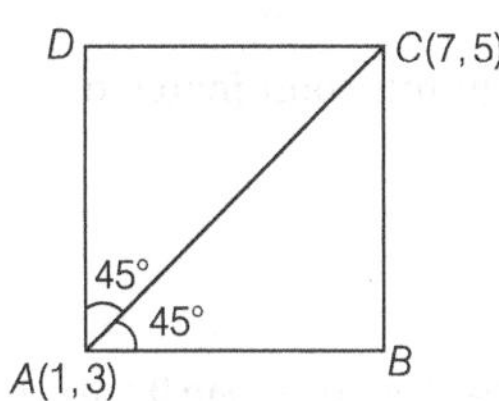

Slope of $AB = \tan(\theta - 45°)$

$$= \frac{\frac{1}{3} - 1}{1 + \frac{1}{3}\cdot 1} = -\frac{1}{2}$$

Equation of AB is $y - 3 = -\frac{1}{2}(x - 1)$

or $x + 2y - 7 = 0$

Slope of $AD = \tan(\theta + 45°) = \frac{\frac{1}{3} + 1}{1 - \frac{1}{3}\cdot 1} = 2$

Equation of AD is $y - 3 = 2(x - 1)$ or $2x - y + 1 = 0$.

21. $A \equiv (-5\sqrt{3}, 0), B \equiv (0, 5), C \equiv (5\sqrt{3}, 0), D \equiv (0, -5)$

Equation of AB is $\frac{x}{-5\sqrt{3}} + \frac{y}{5} = 1$

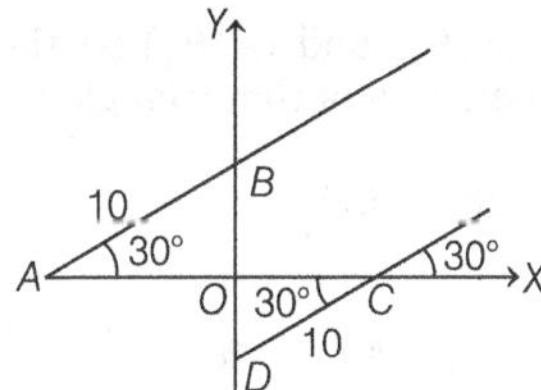

or $x - \sqrt{3}y = -5\sqrt{3}$

Equation of CD is $\frac{x}{5\sqrt{3}} - \frac{y}{5} = 1$

or $x - \sqrt{3}y = 5\sqrt{3}$

Which is, clearly an irrational point.

22. Equation of given line is $2x + y = 2$ ...(i)

When axes are rotated by an angle of 45° in anticlockwise direction then equation of the line with respect to new axes will be

$2(x\cos 45° - y\sin 45°) + (x\sin 45° + y\cos 45°) = 2$

or $3x - y = 2\sqrt{2}$ ...(ii)

length of intercepts made by this line on new axes are

$$\frac{2\sqrt{2}}{3} \text{ and } 2\sqrt{2}.$$

When the axes are rotated by an angle of 45° in cloc kwise direction, then equation of line (i) with respect to new axes will be

$2(x\cos 45° + y\sin 45°) + (-x\sin 45° + y\cos 45°) = 2$

or $x + 3y = 2\sqrt{2}$ ...(iii)

Length of intercept of line (iii) on new axes are $2\sqrt{2}$ and $\frac{2\sqrt{2}}{3}$.

23. Vertices of the given triangle are (0, 0), $\left(\frac{a}{m_1}, a\right)$ and $\left(\frac{a}{m_2}, a\right)$ so that the area of the triangle is equal to $\frac{a^2}{2}\left|\frac{(m_2 - m_1)}{m_1 m_2}\right|$.

Since m_1, m_2 are the roots of $x^2 - ax - a - 1 = 0$, so

$$m_1 + m_2 = a, m_1 m_2 = -(a+1)$$

$\Rightarrow (m_1 - m_2)^2 = a^2 + 4(a+1) = (a+2)^2$

$\Rightarrow |m_1 - m_2| = |(a+2)|$

So, the required area is $\Delta = \frac{a^2}{2}\left|\frac{a+2}{-(a+1)}\right| = \frac{a^2}{2}\left|\frac{a+2}{a+1}\right|$

Since, the area of Δ is a positive quantity.

$$\Delta = \frac{a^2(a+2)}{2(a+1)}, \text{ if } a > -1 \text{ or } a < -2$$

and $\Delta = -\frac{a^2(a+2)}{2(a+1)}$, if $-2 < a < -1$

24. Since base AB of the triangle is fixed, therefore for maximum area of ΔPAB, height of the triangle should be maximum. This is possible only when P lies on the perpendicular bisector of AB.

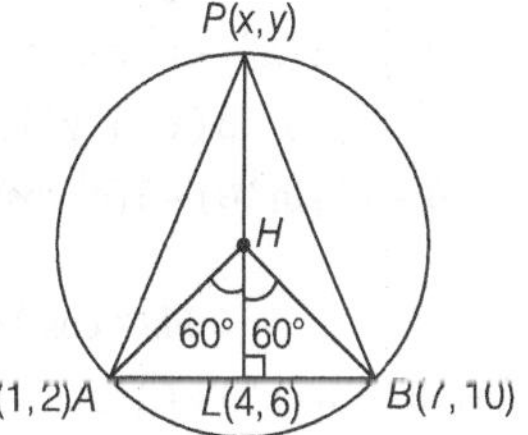

Equation of perpendicular bisector of AB is

$$y - 6 = -\frac{1-7}{2-10}(x - 4)$$

or $y - 6 = -\frac{3}{4}(x - 4)$

or $4y - 24 = -3x + 12$

or $3x + 4y - 36 = 0$

Also $AB = 10 \Rightarrow \therefore AL = 5$

$\therefore AH = AL\text{ cosec }60° = \frac{10}{\sqrt{3}}$

25. P, Q divides AB in the ratio of 1 : 2 and 2 : 1 respectively and hence the coordinates of P and Q are

$$\left[\frac{1\times 2 + 2(-3)}{3}, \frac{1\times 3 + 2\times 2}{3}\right] \text{ and}$$

$$\left[\frac{2\times 2 + 1(-3)}{3}, \frac{2\times 3 + 1\times 2}{3}\right]$$

$\Rightarrow P\left(-\frac{4}{3}, \frac{7}{3}\right)$ and $Q\left(\frac{1}{3}, \frac{8}{3}\right)$, then $PQ = \frac{\sqrt{26}}{3}$

Let the coordinates of R be (x, y).
Then, $QR = PQ$

$$\Rightarrow \left(x - \frac{1}{3}\right)^2 + \left(y - \frac{8}{3}\right)^2 = \frac{26}{9} \quad \ldots \text{(i)}$$

and QR is perpendicular to PQ.

$$\Rightarrow \frac{y - \frac{8}{3}}{x - \frac{1}{3}} \times \frac{\frac{8}{3} - \frac{7}{3}}{\frac{1}{3} + \frac{4}{3}} = -1$$

$$\Rightarrow \left(y - \frac{8}{3}\right) = -5\left(x - \frac{1}{3}\right) \quad \ldots \text{(ii)}$$

From Eqs. (i) and (ii), we get

$$26\left(x - \frac{1}{3}\right)^2 = \frac{26}{9}, \; x - \frac{1}{3} = \pm \frac{1}{3}$$

$$\Rightarrow x = 0 \text{ or } x = \frac{2}{3}$$

From Eq. (ii), when $x = 0$, $y = \frac{13}{3}$ and when

$$x = \frac{2}{3}, y = 1$$

So, the required coordinates of R are $\left(0, \frac{13}{3}\right)$ or $\left(\frac{2}{3}, 1\right)$.

Linked Comprehension Type

PASSAGE 1

1. $\because \sqrt{r} = \sqrt{a} \sin(\theta/2)$

Squaring both sides, then

$$r = a\sin^2(\theta/2)$$

$$\Rightarrow 2r = a(2\sin^2\theta/2) = a(1 - \cos\theta)$$

or $$2r^2 = a(r - r\cos\theta)$$

$$\Rightarrow \{2r^2 + a(r\cos\theta)\}^2 = a^2r^2$$

$$\Rightarrow [2(x^2 + y^2) + ax]^2 = a^2(x^2 + y^2)$$

$$4(x^2 + y^2)^2 + a^2x^2 + 4ax(x^2 + y^2) = a^2x^2 + a^2y^2$$

or $$4(x^2 + y^2)(x^2 + y^2 + ax) = a^2y^2$$

2. $x^3 + 3x^2y - 3xy^2 - y^3 = 5k\,xy$

$$\Rightarrow 4(x^3 - y^3) - 3x^3 + 3x^2y - 3xy^2 + 3y^3 = 5k\,xy$$

$$\Rightarrow 4(x^3 - y^3) - 3(x^3 - x^2y + xy^2 - y^3) = 5k\,xy$$

$$\Rightarrow 4(x^3 - y^3) - 3(x^2 + y^2)(x - y) = 5k\,xy$$

$$\Rightarrow 4(r^3\cos^3\theta - r^3\sin^3\theta) - 3(r^2\cos^2\theta + r^2\sin^2\theta)(r\cos\theta - r\sin\theta)$$

$$= 5kr^2\cos\theta\sin\theta$$

$$\Rightarrow r\{(4\cos^3\theta - 3\cos\theta) + (3\sin\theta - 4\sin^3\theta)\}$$

$$= 5k\sin\theta\cos\theta$$

$$\Rightarrow r(\cos 3\theta + \sin 3\theta) = 5k\sin\theta\cos\theta$$

3. $\because \frac{2}{r} = -\frac{1}{2} + 4\cos\theta$

$$\Rightarrow 2 = -\frac{r}{2} + 4r\cos\theta \Rightarrow 4 + r = 8r\cos\theta$$

$$\Rightarrow r = (8r\cos\theta - 4)$$

$$\Rightarrow \sqrt{(x^2 + y^2)} = (8x - 4)$$

Squaring both sides, then

$$x^2 + y^2 = 64x^2 - 64x + 16$$

$$\Rightarrow 63x^2 - y^2 - 64x + 16 = 0$$

Which is a hyperbola

4. $\because \frac{l}{r} = f(\theta) = a\cos(\theta + \alpha) + b\cos(\theta + \beta)$

$$\Rightarrow l = ar(\cos\theta\cos\alpha - \sin\theta\sin\alpha) + br(\cos\theta\cos\beta - \sin\theta\sin\beta)$$

$$\Rightarrow l = a(r\cos\theta\cos\alpha - r\sin\theta\sin\alpha) + b(r\cos\theta\cos\beta - r\sin\theta\sin\beta)$$

$$\Rightarrow l = a(x\cos\alpha - y\sin\alpha) + b(x\cos\beta - y\sin\beta)$$

$$= x(a\cos\alpha + b\cos\beta) - y(a\sin\alpha + b\sin\beta)$$

$$l = Ax + By \quad \ldots \text{(i)}$$

where A, B are constants.

The equation of any line perpendicular to the line (i) is $Bx - Ay = \lambda$, where λ is a parameter.

$$\Rightarrow Br\cos\theta - Ar\sin\theta = \lambda$$

or $$\frac{\lambda}{r} = B\cos\theta - A\sin\theta$$

or $$\frac{\lambda}{r} = A(-\sin\theta) + B(\cos\theta)$$

or $$\frac{\lambda}{r} = A\cos\left(\frac{\pi}{2} + \theta\right) + B\sin\left(\frac{\pi}{2} + \theta\right)$$

$$= J\left(\theta + \frac{\pi}{2}\right)$$

5. If (r_1, θ_1), (r_2, θ_2) and (r_3, θ_3) be the angular points of a triangle, then area of that triangle is given by

$$\frac{1}{2}\left|\sum r_1r_2\sin(\theta_2 - \theta_1)\right|$$

Here, $r_1 = 1, \theta_1 = \pi/6$; $r_2 = 2, \theta_2 = \frac{\pi}{3}$ and $r_3 = 3, \theta_3 = \pi/2$

$$\therefore \theta_2 - \theta_1 = \pi/6,$$

$$\theta_3 - \theta_2 = \pi/6$$

and $$\theta_1 - \theta_3 = -\frac{\pi}{3}$$

$$\Rightarrow \sin(\theta_2 - \theta_1) = \frac{1}{2}, \sin(\theta_3 - \theta_2) = \frac{1}{2}$$

and $$\sin(\theta_1 - \theta_3) = -\frac{\sqrt{3}}{2}$$

$$\text{Required area} = \frac{1}{2}\left|(1 \times 3)\frac{1}{2} + (2)(3) \times \frac{1}{2} + (3)(1) \times \frac{-\sqrt{3}}{2}\right|$$

$$= \left(\frac{8 - 3\sqrt{3}}{4}\right) \text{ sq unit}$$

PASSAGE 2

1. Coordinates of a triangle are $(a\cos t, a\sin t)$, $(b\sin t, -b\cos t)$ and $(1, 0)$

$\therefore$ Centroid is $x = \dfrac{a\cos t + b\sin t + 1}{3}$

$$\Rightarrow \quad 3x - 1 = a\cos t + b\sin t \qquad \ldots(i)$$

and $$y = \frac{a\sin t - b\cos t + 0}{3}$$

$$\Rightarrow \quad 3y = a\sin t - b\cos t \qquad \ldots(ii)$$

Squaring and adding Eqs. (i) and (ii), we get

$$(3x-1)^2 + (3y)^2 = (a\cos t + b\sin t)^2 + (a\sin t - b\cos t)^2$$

or $$(3x-1)^2 + (3y)^2 = a^2 + b^2$$

2. Let (x, y) be coordinate of vertex C and (x_1, y_1) be coordinates of centroid of the triangle

$$\therefore \quad x_1 = \frac{x+2-2}{3} \text{ and } y_1 = \frac{y-3+1}{3}$$

$$\Rightarrow \quad x_1 = \frac{x}{3} \text{ and } y_1 = \frac{y-2}{3}$$

$\because$ The centroid (x_1, y_1) lies on the line $2x + 3y = 1$.

So, x_1 and y_1 satisfied the equation of line.

$$\therefore \quad 2x_1 + 3y_1 = 1$$

$$\Rightarrow \quad 2\left(\frac{x}{3}\right) + 3\left(\frac{y-2}{3}\right) = 1$$

$\Rightarrow$ $2x + 3y = 9$, which is the required locus of the vertex C.

3. Mid point of (a_1, b_1) and (a_2, b_2) is $\left(\dfrac{a_1 + a_2}{2}, \dfrac{b_1 + b_2}{2}\right)$

Which lies on $(a_1 - a_2)x + (b_1 - b_2)y + c = 0$

$$\Rightarrow (a_1 - a_2)\frac{(a_1 + a_2)}{2} + (b_1 - b_2)\frac{(b_1 + b_2)}{2} + c = 0$$

$$\therefore \quad c = \frac{1}{2}(a_2^2 - a_1^2 + b_2^2 - b_1^2)$$

$$= \frac{1}{2}(a_2^2 + b_2^2 - a_1^2 - b_1^2)$$

4. The line $y = mx$ meets the given lines in $P\left(\dfrac{1}{m+1}, \dfrac{m}{m+1}\right)$

and $Q\left(\dfrac{3}{m+1}, \dfrac{3m}{m+1}\right)$

$\therefore$ Equation of L_1 is $y - \dfrac{m}{m+1} = 2\left(x - \dfrac{1}{m+1}\right)$

$$\Rightarrow \quad y - 2x - 1 = -\frac{3}{m+1} \qquad \ldots(i)$$

and equation of L_2 is

$$y - \frac{3m}{m+1} = -3\left(x - \frac{3}{m+1}\right)$$

$$\Rightarrow \quad y + 3x - 3 = \frac{6}{m+1} \qquad \ldots(ii)$$

Dividing Eq. (ii) by (i),

$$\frac{y + 3x - 3}{y - 2x - 1} = -2$$

$$\Rightarrow \quad y + 3x - 3 = -2y + 4x + 2$$

or $$x - 3y + 5 = 0$$

5. A line passing through $P(h, k)$ and parallel to x-axis is

$$y = k \qquad (i)$$

The other two given lines are

$$y = x \qquad \ldots(ii)$$

and $$x + y = 2 \qquad \ldots(iii)$$

Solving Eqs. (i), (ii); (ii), (iii) and (iii), (i), we get $A(k, k)$, $B(1, 1)$ and $C(2-k, k)$

$\therefore$ Area of $\Delta ABC = 4h^2$

$$\Rightarrow \quad \frac{1}{2}\left|\begin{vmatrix} k & k & 1 \\ 1 & 1 & 1 \\ 2-k & k & 1 \end{vmatrix}\right| = 4h^2$$

Applying $C_2 \to C_2 - C_1$, then

$$\frac{1}{2}\left|\begin{vmatrix} k & 0 & 1 \\ 1 & 0 & 1 \\ 2-k & 2k-2 & 1 \end{vmatrix}\right| = 4h^2$$

$$\Rightarrow \quad (k-1)^2 = 4h^2 \Rightarrow k - 1 = \pm 2h$$

$$\therefore \quad k = \pm 2h + 1$$

$\therefore$ locus of (h, k) is $y = 2x + 1$

or $$y = -2x + 1$$

PASSAGE 3

1. Here, $z_1 = 1 + i\cdot 0$, $z_2 = (1+\sqrt{3}) + i$

and $$z_3 = 2 + i\sqrt{3}$$

then $$\frac{(2 + i\sqrt{3}) - (1 + i\cdot 0)}{[(1+\sqrt{3}) + i] - (1 + i\cdot 0)} = e^{i\theta}$$

$$\Rightarrow \quad \frac{(1 + i\sqrt{3})}{(\sqrt{3} + i)} = e^{i\theta}$$

$$\Rightarrow \quad \frac{(1 + i\sqrt{3})(\sqrt{3} - i)}{4} = e^{i\theta}$$

$$\Rightarrow \quad \frac{2\sqrt{3} + 2i}{4} = e^{i\theta}$$

$$\Rightarrow \quad \frac{\sqrt{3}}{2} + \frac{i}{2} = e^{i\theta}$$

$$\Rightarrow \quad e^{i\pi/6} = e^{i\theta}$$

$$\therefore \quad \theta = \frac{\pi}{6} = 30^\circ$$

2. $\because A \equiv (1, 1)$

$$\therefore \quad z_A = 1 + i$$

and $$C \equiv (-2, -1)$$

$$\therefore \quad z_c = -2 - i$$

The centre of the square $E \equiv \left(-\dfrac{1}{2}, 0\right)$

$$\therefore \quad z_E = -\frac{1}{2}$$

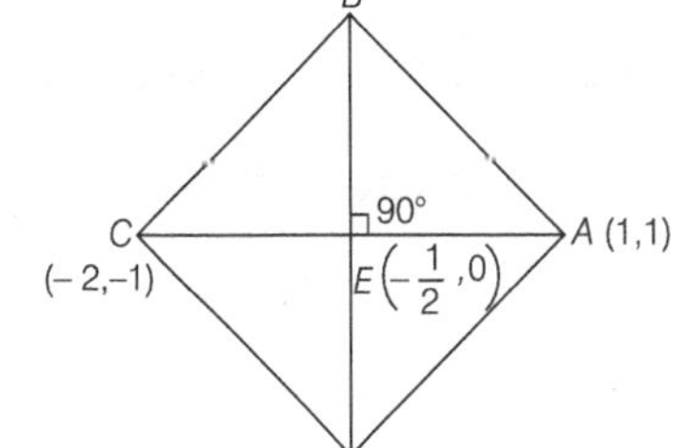

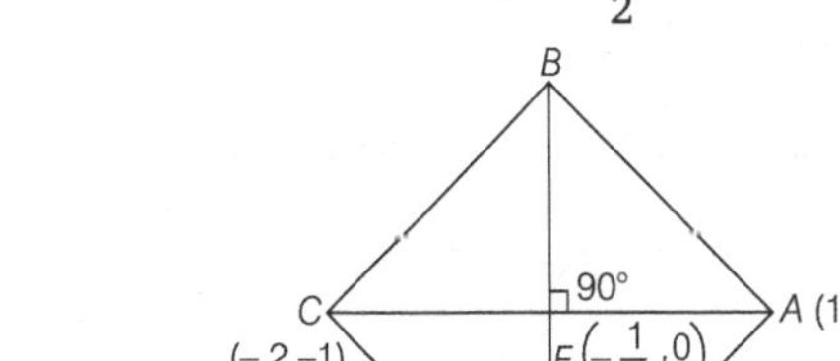

Now in ΔAEB $(EA = EB)$

$$\frac{z_B - z_E}{z_A - z_E} = e^{i\pi/2} = i$$

$$\Rightarrow \quad z_B + \frac{1}{2} = i\left(1 + i + \frac{1}{2}\right) = \frac{3}{2}i - 1$$

or
$$z_B = -\frac{3}{2} + \frac{3}{2}i$$

$$\therefore \quad B \equiv \left(-\frac{3}{2}, \frac{3}{2}\right)$$

$\Rightarrow$ Equation of other diagonal is

$$y - 0 = \frac{\frac{3}{2} - 0}{-\frac{3}{2} + \frac{1}{2}}\left(x + \frac{1}{2}\right)$$

$$\Rightarrow \quad y = -\frac{3}{2}x - \frac{3}{4} \Rightarrow 4y = -6x - 3$$

$$6x + 4y + 3 = 0$$

3. Image of the point (2, 2) in the X-axis is $(2, -2)$. If (x, y) be the coordinates of any point and (X, Y) be its new coordinates, then $X = x\cos\theta + y\sin\theta$ and $Y = -x\sin\theta + y\cos\theta$ where θ is the angle through which the axes have been rotated.

Here, $\theta = \pi/4$, $x = 2$, $y = -2$

$$\therefore \quad X = 0, \ Y = -\frac{2}{\sqrt{2}} - \frac{2}{\sqrt{2}} = -2\sqrt{2}$$

$$\therefore \quad (X, Y) \equiv (0, -2\sqrt{2})$$

4. $\because AB = AC = AD$

$$\angle BAD = 60° = \pi/3$$

$$\therefore \quad \frac{z_D - z_A}{z_B - z_A} = e^{i\pi/3}$$

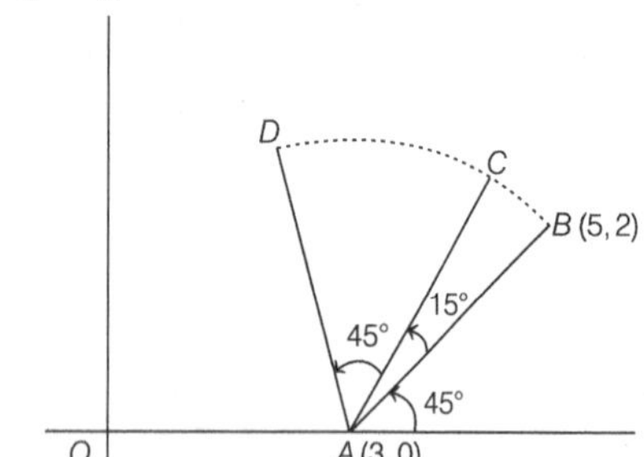

$$\Rightarrow \frac{z_D - 3}{5 + 2i - 3} = \cos\pi/3 + i\sin\pi/3 = \frac{1}{2} + \frac{i\sqrt{3}}{2}$$

$$\Rightarrow \quad z_D - 3 = (2 + 2i)\left(\frac{1}{2} + \frac{i\sqrt{3}}{2}\right)$$

$$= (1 + i)(1 + i\sqrt{3})$$

$$= 1 + i\sqrt{3} + i - \sqrt{3}$$

$$\therefore \quad z_D = 3 + 1 + i\sqrt{3} + i - \sqrt{3}$$

$$= (4 - \sqrt{3}) + i(1 + \sqrt{3})$$

$$\Rightarrow \quad D \equiv (4 - \sqrt{3}, \sqrt{3} + 1)$$

PASSAGE 4

1. Let the parametric equation of the line drawn be,

$$\frac{x}{\cos\theta} = \frac{y}{\sin\theta} = r$$

or $\quad x = r\cos\theta, \ y = r\sin\theta$

Putting it in L_1, we get

$$r\sin\theta = r\cos\theta + 10$$

or
$$\frac{1}{OA} = \frac{\sin\theta - \cos\theta}{10}$$

Similarly, putting the general point of drawn line in the equation of L_2, we get

$$\frac{1}{OB} = \frac{\sin\theta - \cos\theta}{20}$$

Let $\quad P \equiv (h, k)$ and $OP = r$.

Then, $\quad r\cos\theta = h, \ r\sin\theta = k$.

We have,
$$\frac{2}{r} = \frac{\sin\theta - \cos\theta}{10} + \frac{\sin\theta - \cos\theta}{20}$$

or $\quad 40 = 3r\sin\theta - 3r\cos\theta$

or $\quad 3y - 3x = 40$

2. (c) $r^2 = \dfrac{10 \times 20}{(\sin\theta - \cos\theta)^2}$

or $(r\cdot\sin\theta - r\cdot\cos\theta)^2 = 200$

Hence, the locus is $(y - x)^2 = 200$.

3. (a) $\dfrac{1}{r^2} = \dfrac{(\sin\theta - \cos\theta)^2}{100} + \dfrac{(\sin\theta - \cos\theta)^2}{400}$

or $400 = 5(r\sin\theta - r\cos\theta)^2$

Hence, the locus is $400 = 5(x - y)^2$,

i.e. $\quad (x - y)^2 = 80$

Numerical Value Type

1. Let $A(4, -2)$, $B(-2, 4)$ and $C(5, 5)$ be the vertices of the given triangle. Then

$$a = BC = \sqrt{(-2-5)^2 + (4-5)^2} = \sqrt{50} = 5\sqrt{2}$$

$$b = CA = \sqrt{(5-4)^2 + (5+2)^2} = \sqrt{50} = 5\sqrt{2}$$

and $\quad c = AB = \sqrt{(4+2)^2 + (-2-4)^2} = \sqrt{72} = 6\sqrt{2}$

Let (x, y) be the coordinates of incentre of ΔABC.

Then,
$$x = \frac{ax_1 + bx_2 + cx_3}{a + b + c}$$

$$= \frac{5\sqrt{2} \times 4 + 5\sqrt{2} \times (-2) + 6\sqrt{2} \times 5}{5\sqrt{2} + 5\sqrt{2} + 6\sqrt{2}}$$

$$= \frac{20\sqrt{2} - 10\sqrt{2} + 30\sqrt{2}}{16\sqrt{2}} = \frac{40}{16} = \frac{5}{2}$$

and $$y = \frac{ay_1 + by_2 + cy_3}{a + b + c}$$

$$= \frac{5\sqrt{2} \times (-2) + 5\sqrt{2} \times 4 + 6\sqrt{2} \times 5}{5\sqrt{2} + 5\sqrt{2} + 6\sqrt{2}} = \frac{40}{16} = \frac{5}{2}$$

$\therefore$ The coordinates of the incentre are $\left(\frac{5}{2}, \frac{5}{2}\right)$.

$\therefore$ $$\alpha = \frac{5}{2} \text{ and } \beta = \frac{5}{2}$$

$\Rightarrow$ $$2\alpha + 10\beta = 2 \times \frac{5}{2} + 10 \times \frac{5}{2}$$

$$= 5 + 25$$

$$= 30$$

2. Coordinates of centroid are

$$G \equiv \left(\frac{-36 + 20 + 0}{3}, \frac{7 + 7 - 8}{3}\right)$$

or $$G \equiv \left(-\frac{16}{3}, 2\right)$$

and $a = |BC| = \sqrt{(20 - 0)^2 + (7 + 8)^2} = \sqrt{625} = 25$

$b = |CA| = \sqrt{(0 + 36)^2 + (-8 - 7)^2} = \sqrt{1521} = 39$

and $c = |AB| = \sqrt{(-36 - 20)^2 + (7 - 7)^2} = \sqrt{(56)^2} = 56$

Therefore, the coordinates of incentre are

$$I \equiv \left(\frac{25 \times -(36) + 39 \times 20 + 56 \times 0}{25 + 39 + 56}, \frac{25 \times 7 + 39 \times 7 + 56 \times (-8)}{25 + 39 + 56}\right)$$

i.e., $$I \equiv (-1, 0)$$

$\therefore$ $$GI = \sqrt{\left(\frac{-16}{3} + 1\right)^2 + (2 - 0)^2} = \frac{\sqrt{205}}{3}$$

but given $GI = \frac{25}{3}\sqrt{(205)}\,\lambda$

$\therefore$ $$\frac{1}{3}\sqrt{(205)} = \frac{25}{3}\sqrt{(205)}\,\lambda$$

$\therefore$ $$\lambda = \frac{1}{25}$$

$\Rightarrow$ $$\frac{1}{\lambda} = 25$$

3. Since, centroid divides the orthocentre and circumcentre in the ratio 2 : 1 (internally) and if centroid $G(x, y)$, then

$$O(1, 1) \overset{2}{\text{———}} G(x, y) \overset{1}{\text{———}} C\left(\frac{3}{2}, \frac{3}{4}\right)$$

$$x = \frac{2 \times \frac{3}{2} + 1 \times 1}{2 + 1} = \frac{4}{3}$$

and $$y = \frac{2 \times \frac{3}{4} + 1 \times 1}{2 + 1} = \frac{5}{6}$$

$\therefore$ Centroid is $\left(\frac{4}{3}, \frac{5}{6}\right)$ and nine point centre is the mid-point of orthocentre and circumcentre.

$\therefore$ Nine point centre is $\left(\frac{1 + \frac{3}{2}}{2}, \frac{1 + \frac{3}{4}}{2}\right)$, *ie*, $\left(\frac{5}{4}, \frac{7}{8}\right)$

$\Rightarrow$ $$\alpha = \frac{4}{3}, \beta = \frac{5}{6}$$

and $$\gamma = \frac{5}{4}, \delta = \frac{7}{8}$$

$\therefore$ $$6\alpha + 12\beta + 4\gamma + 8\delta$$

$$= 6 \times \frac{4}{3} + 12 \times \frac{5}{6} + 4 \times \frac{5}{4} + 8 \times \frac{7}{8}$$

$$= 8 + 10 + 5 + 7 = 30$$

4. Let $P(x_1, y_1)$ be the moving point whose locus is required. By hypothesis.

$$2PA = 3PB$$

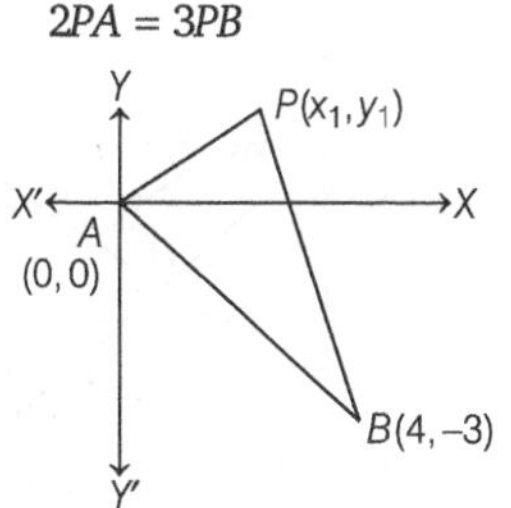

or $$4(PA)^2 = 9(PB)^2$$

$\Rightarrow$ $$4\{x_1^2 + y_1^2\} = 9\{(x_1 - 4)^2 + (y_1 + 3)^2\}$$

$\Rightarrow$ $$4(x_1^2 + y_1^2) = 9(x_1^2 + y_1^2 - 8x_1 + 6y_1 + 25)$$

or $$5x_1^2 + 5y_1^2 - 72x_1 + 54y_1 + 225 = 0$$

Changing (x_1, y_1) to (x, y), then

$$5x^2 + 5y^2 - 72x + 54y + 225 = 0$$

which is required locus of P.

On comparing we get, $\lambda = 54$ and $\mu = 225$

$\therefore$ $$\lambda + \mu = 54 + 225 = 279$$

5. Since m_1 and m_2 are the roots of the equation

$$x^2 + (\sqrt{3} + 2)x + (\sqrt{3} - 1) = 0$$

then, $m_1 + m_2 = -(\sqrt{3} + 2)$, $m_1 m_2 = (\sqrt{3} - 1)$

$\therefore$ $$m_1 - m_2 = \sqrt{(m_1 + m_2)^2 - 4m_1 m_2}$$

$$= \sqrt{(3 + 4 + 4\sqrt{3} - 4\sqrt{3} + 4)} = \sqrt{11}$$

and coordinates of the vertices of the given triangle are $(0, 0)$, $(c/m_1, c)$ and $(c/m_2, c)$.

Hence, the required area of triangle

$$= \frac{1}{2}\left|\frac{c}{m_1} \times c - \frac{c}{m_2} \times c\right|$$

$$= \frac{1}{2}c^2\left|\left(\frac{1}{m_1} - \frac{1}{m_2}\right)\right| = \frac{1}{2}c^2\frac{|m_2 - m_1|}{|m_1 m_2|}$$

$$= \frac{1}{2}c^2\frac{\sqrt{11}}{(\sqrt{3} - 1)}$$

$$= \frac{1}{2}c^2\frac{\sqrt{11}(\sqrt{3} + 1)}{(\sqrt{3} - 1)(\sqrt{3} + 1)}$$

$$= \left(\frac{\sqrt{33} + \sqrt{11}}{4}\right)c^2$$

On comparing, $a = \frac{\sqrt{33}}{4}, b = \frac{\sqrt{11}}{4}$

or $a = \frac{\sqrt{11}}{4}, b = \frac{\sqrt{33}}{4}$

$\therefore \quad a^2 + b^2 = \frac{33}{16} + \frac{11}{16} = \frac{44}{16} = \frac{11}{4}$

$\Rightarrow \quad 2008\,(a^2 + b^2) = 2008 \times \frac{11}{4}$

$= 502 \times 11 = 5522$

6. Let $P \equiv (\lambda, 2)$

First draw the exact diagram of ΔABC the point $P(\lambda, 2)$ move on the line $y = 2$ for all λ.

Now D and E are the intersection of

$$3x - 7y = 8, \ y = 2$$

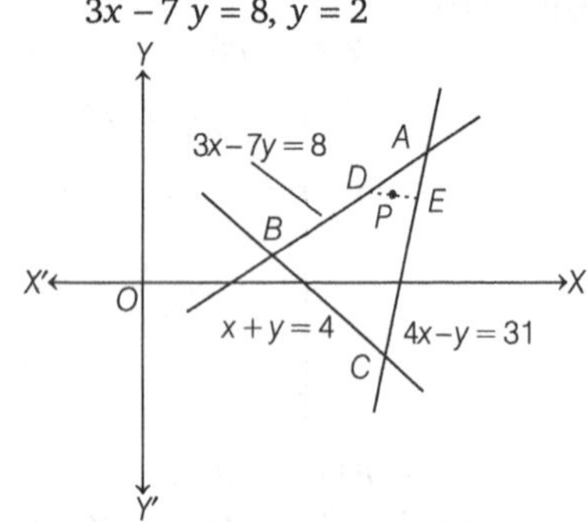

and $\quad 4x - y = 31, \ y = 2$ respectively.

$\therefore \quad D \equiv \left(\frac{22}{3}, 2\right)$ and $E \equiv \left(\frac{33}{4}, 2\right)$

Thus the points on the line $y = 2$ whose x-coordinates lies between $\frac{22}{3}$ and $\frac{33}{4}$ lie within the ΔABC.

Hence, $\quad \frac{22}{3} < \lambda < \frac{33}{4}$

ie, $\quad \lambda \in \left(\frac{22}{3}, \frac{33}{4}\right)$

$\Rightarrow \quad a = \frac{22}{3}$ and $b = \frac{33}{4}$

$\therefore \quad 6a + 8b = 6 \times \frac{22}{3} + 8 \times \frac{33}{4}$

$= 44 + 66 = 110$

7. $\because$ Given $\sum_{i=1}^{2} d(P, L_i) \in [2, 4]$

$\Rightarrow \quad 2 \le \sum_{i=1}^{2} d(P, L_i) \le 4$

$\Rightarrow \quad 2 \le d(P, L_1) + d(P, L_2) \le 4$

$\Rightarrow \quad 2 \le \frac{|x - y|}{\sqrt{2}} + \frac{|x + y|}{\sqrt{2}} \le 4$

$\Rightarrow \quad 2\sqrt{2} \le |x - y| + |x + y| \le 4\sqrt{2}$

Case I If $x - y > 0, x + y > 0$

then $\quad 2\sqrt{2} \le (x - y) + (x + y) \le 4\sqrt{2}$

$\Rightarrow \quad 2\sqrt{2} \le 2x \le 4\sqrt{2}$

or $\quad \sqrt{2} \le x \le 2\sqrt{2}$

Case II If $x - y > 0, x + y < 0$

then $\quad 2\sqrt{2} \le (x - y) - (x + y) \le 4\sqrt{2}$

$\Rightarrow \quad 2\sqrt{2} \le -2y \le 4\sqrt{2}$

or $\quad -\sqrt{2} \ge y \ge -2\sqrt{2}$

or $\quad -2\sqrt{2} \le y \le -\sqrt{2}$

Case III If $x - y < 0, x + y > 0$,

then $\quad 2\sqrt{2} \le -(x - y) + (x + y) \le 4\sqrt{2}$

$\Rightarrow \quad 2\sqrt{2} \le 2y \le 4\sqrt{2}$

or $\quad \sqrt{2} \le y \le 2\sqrt{2}$

Case IV : If $x - y < 0, x + y < 0$,

then $\quad 2\sqrt{2} \le -(x - y) - (x + y) \le 4\sqrt{2}$

$\Rightarrow \quad 2\sqrt{2} \le -2x \le 4\sqrt{2}$

or $\quad -\sqrt{2} \ge x \ge -2\sqrt{2}$

or $\quad -2\sqrt{2} \le x \le -\sqrt{2}$

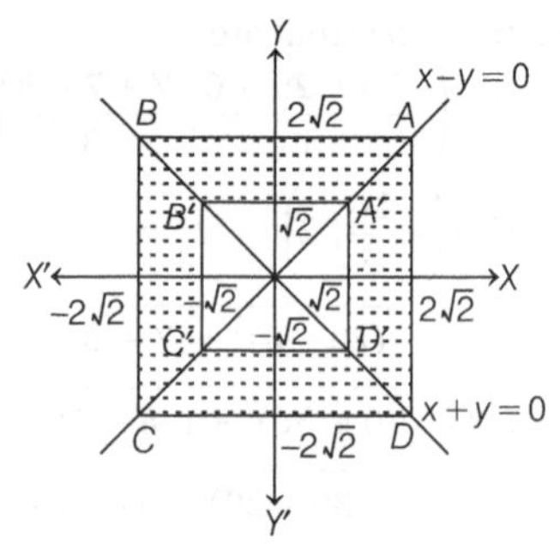

Combining all cases, we get

$$x \in [-2\sqrt{2}, -\sqrt{2}] \cup [\sqrt{2}, 2\sqrt{2}]$$

and $\quad y \in [-2\sqrt{2}, -\sqrt{2}] \cup [\sqrt{2}, 2\sqrt{2}]$

Hence, area of the required region

$= (4\sqrt{2})^2 - (2\sqrt{2})^2$

$= 32 - 8 = 24$ sq unit.

$\therefore \quad \lambda = 24$

8. $\because$ PQ is parallel to $y = mx$

$\therefore$ Equation of PQ is $y = mx + \lambda$ ...(i)

$\because$ Diagonals bisect to each other

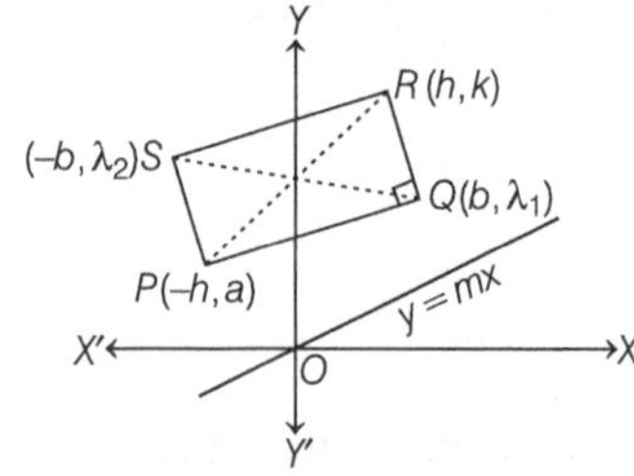

$\therefore$ x-coordinate of P is $-h$.

Suppose y-coordinate of Q and S are λ_1 and λ_2 respectively.

$\because$ Equation (i) pass through $P(-h, a)$

then $\quad a = -mh + \lambda$

$\therefore \quad \lambda = a + mh$

$\Rightarrow \quad y = mx + a + mh$

Q also lie on it, then $\lambda_1 = mb + a + mh$

$\therefore \quad Q \equiv (b, mb + a + mh)$

Also, slope of $PQ \times$ slope of $QR = -1$

$$m \times \frac{[(k - (mb + a + mh)]}{(h - b)} = -1$$

$\Rightarrow \quad mk - m^2b - am - m^2h = -h + b$

$\Rightarrow \quad (m^2 - 1)h - mk + b(m^2 + 1) + am = 0$

$\therefore$ Locus of R is $(m^2-1)x-my+b(m^2+1)+am=0$

Put $\quad a=2, b=1$, then

$$(m^2-1)x-my+m^2+1+2m=0$$

On comparing, we get

$$m^2-1=24, m=5, m^2+1+2m=36$$

$\therefore \quad m=5$

then $\quad 1+m+m^2=1+5+25=31$

9. $\because \; OP_n=nOP_{n-1}$

$$OP_8=8\cdot OP_7=8\cdot 7\, OP_6=8\cdot 7\cdot 6\, OP_5$$
$$=8\cdot 7\cdot 6\cdot 5\cdot OP_4=8\cdot 7\cdot 6\cdot 5\cdot 4\cdot OP_3$$
$$=8\cdot 7\cdot 6\cdot 5\cdot 4\cdot 3\, OP_2=8\cdot 7\cdot 6\cdot 5\cdot 4\cdot 3\cdot 2\cdot OP_1$$
$$=8\cdot 7\cdot 6\cdot 5\cdot 4\cdot 3\cdot 2\cdot 1 \quad \text{...(i)}$$

Coordinates of P_8 $(OP_8\cos 45°, OP_8 \sin 45°)$

or $\quad P_8\equiv\left(\frac{OP_8}{\sqrt{2}}, \frac{OP_8}{\sqrt{2}}\right)$

But given $\quad \frac{OP_8}{\sqrt{2}}=3a\sqrt{2}$ or $OP_8=6a \quad$ ...(ii)

From Eqs. (i) and (ii), we get

$$6a=8\cdot 7\cdot 6\cdot 5\cdot 4\cdot 3\cdot 2\cdot 1$$

$\therefore \quad a=6720$

10. Point D is $(-6,-3)$. Equation of the line AC is $y-3=2(x-2)$

$\Rightarrow \quad 2x-y-1=0 \quad$ ...(i)

and equation of the line AB is $y-3=-1(x-2)$

$\Rightarrow \quad x+y-5=0 \quad$ ...(ii)

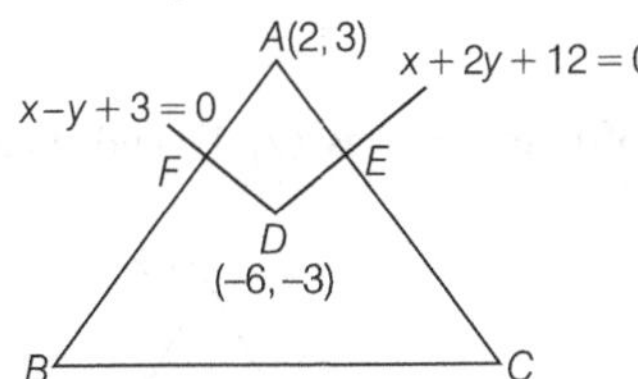

E is the point of intersection of (i) and $x+2y+12=0$.

Therefore, E is $(-2,-5)$.

Similarly, the coordinates of F is $(1, 4)$

Since, E and F divides AC and AB in the ratio 1 : 2, we get the points

$$B(-1, 6) \text{ and } C(-10,-21)$$

$$\therefore \text{Area of } \Delta ABC=\frac{1}{2}\left|\begin{vmatrix}-6 & -3 & 1\\ -1 & 6 & 1\\ -10 & -21 & 1\end{vmatrix}\right|=27 \text{ sq unit.}$$

$\therefore \quad \lambda=27$

11. The equation of BU is $y-\beta=\frac{0-\beta}{6-\alpha}(x-\alpha)$

So, that the coordinates of D are $\left(0, \frac{6\beta}{6-\alpha}\right)$

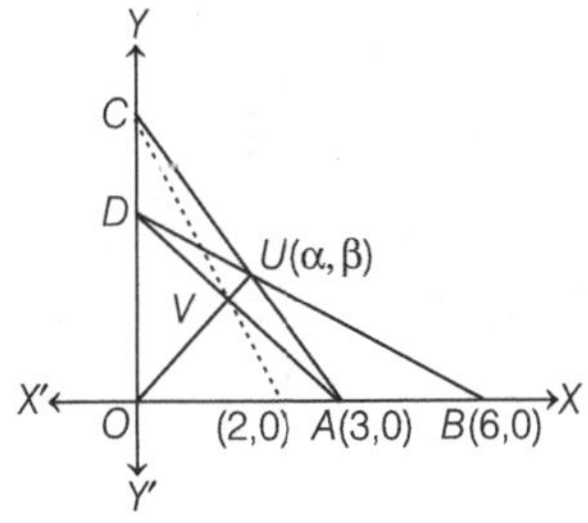

Similarly the coordinates of C are $\left(0, \frac{3\beta}{3-\alpha}\right)$

Now, the equation of AD is $\frac{x}{3}+\frac{6-\alpha}{6\beta}y=1 \quad$ ...(i)

and the equation of OU is $\beta x=\alpha y \quad$ ...(ii)

Solving Eqs. (i) and (ii), we get

$$x=\frac{6\alpha}{6+\alpha}, y=\frac{6\beta}{6+\alpha}$$

Hence, coordinates of V are $\left(\frac{6\alpha}{6+\alpha}, \frac{6\beta}{6+\alpha}\right)$

Then the equation of CV is

$$y-\frac{3\beta}{3-\alpha}=\frac{\frac{6\beta}{6+\alpha}-\frac{3\beta}{3-\alpha}}{\frac{6\alpha}{6+\alpha}-0}(x-0)$$

$$\Rightarrow \quad y-\frac{3\beta}{3-\alpha}=\frac{-9\alpha\beta}{6\alpha(3-\alpha)}x$$

$$\Rightarrow \quad y=\frac{3\beta}{(3-\alpha)}\left(1-\frac{x}{2}\right)$$

which pass through the point $(2, 0)$ for all values of (α, β).

On comparing we get, $\lambda=2$

$\therefore \quad 1004\lambda=1004\times 2=2008$

12. Equation of any line passing through the fixed point $(3, 4)$ and having slope m can be taken as

$$y-4=m(x-3) \quad \text{...(i)}$$

$$\Rightarrow \quad \frac{y}{m}-\frac{4}{m}=x-3$$

or $\quad x-\frac{y}{m}=3-\frac{4}{m}$

$$\Rightarrow \quad \frac{x}{\left(3-\frac{4}{m}\right)}-\frac{y}{m\left(3-\frac{4}{m}\right)}=1$$

$$\Rightarrow \quad \frac{x}{\left(3-\frac{4}{m}\right)}+\frac{y}{(4-3m)}=1$$

$\therefore \quad OQ=3-\frac{4}{m}$ and $OP=4-3m$

Area of $\Delta OPQ=\frac{1}{2}\cdot OQ\cdot OP$

$$=\frac{1}{2}\left(3-\frac{4}{m}\right)(4-3m)$$

$$=\frac{1}{2}\left\{24-9m-\frac{16}{m}\right\}$$

$$\Rightarrow \quad A(m)=12-\frac{9m}{2}-\frac{8}{m} \quad \text{...(ii)}$$

To minimize,

$$\frac{dA}{dm} = 0 \Rightarrow 0 - \frac{9}{2} + \frac{8}{m^2} = 0$$

$$\therefore \quad m = \pm\frac{4}{3}$$

$$\frac{d^2A}{dm^2} = -\frac{16}{m^3}$$

$$\left.\frac{d^2A}{dm^2}\right|_{m=-\frac{4}{3}} = \frac{27}{4} > 0$$

Hence, for $m = -\frac{4}{3}$, $A(m)$ is minimum.

Put $m = -\frac{4}{3}$ in Eq. (ii), we get minimum area

$\Rightarrow$ Minimum area of $\Delta OPQ = 12 + 6 + 6 = 24$

$$\therefore \quad \lambda = 24$$

then $$\frac{\lambda}{5} = 4.8$$

13. Since, PQ is of fixed length.

Area of $\Delta PQR = \frac{1}{2}|PQ||RP| \sin\theta$.

This will be maximum, if $\sin\theta = 1$ and RP is maximum.

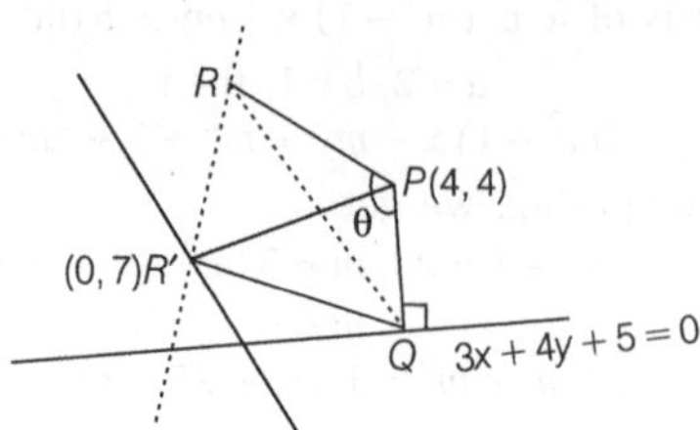

Since, line $y = mx + 7$ rotates about $(0, 7)$, if PR' is perpendicular to the line than PR' is maximum value of PR.

$$\therefore \quad m = -\left(\frac{4-0}{4-7}\right) = \frac{4}{3} = 1.33$$

14. We have,

$$S = (100)(100) + (99)(101) + (98)(102) \ldots\ldots \ldots (2)(198) + (1)(199)$$

$$S = \sum_{x=0}^{99} (100 - x)(100 + x) = \Sigma 100^2 - x^2$$

$$= 100^3 - \frac{99 \times 100 \times 199}{6}$$

Hence, $\alpha = 3$ and $\beta = 1650$

$$\therefore \quad \text{Slope} = \frac{1650}{3} = 550$$

Single Integer Type

1. The point B is $(2, 1)$ (B is image of A about line $y = x$) image of $A(1, 2)$ in the line $x - 2y + 1 = 0$ is given by $\frac{x-1}{1} = \frac{y-2}{-2} = \frac{4}{5}$ coordinates of the point are $\left(\frac{9}{5}, \frac{2}{5}\right)$.

Since, this point lies on BC equation of BC is $3x - y - 5 = 0$.

$$\therefore \quad a + b = 2$$

2. Given, vertices of ΔABC are $A(1, 0)$, $B(6, 2)$ and $C\left(\frac{3}{2}, 6\right)$ and the centroid of triangle G divides the triangles such that, area of ΔAGB, ΔBGC and ΔCGA are equal, so point P is the centroid of triangle G.

$$\therefore \text{ Coordinate of } P\left(\frac{1 + 6 + \frac{3}{2}}{3}, \frac{0 + 2 + 6}{3}\right) = P\left(\frac{17}{6}, \frac{8}{3}\right)$$

and given point $Q\left(-\frac{7}{6}, -\frac{1}{3}\right)$.

So, length of the line segment PQ

$$= \sqrt{\left(\frac{17}{6} + \frac{7}{6}\right)^2 + \left(\frac{8}{3} + \frac{1}{3}\right)^2} = \sqrt{\left(\frac{24}{6}\right)^2 + \left(\frac{9}{3}\right)^2}$$

$$= \sqrt{4^2 + 3^2}$$

$$= \sqrt{16 + 9} = \sqrt{25} = 5$$

3. Image of point $A(1, 2)$ in bisector of angle B and C lie of on the line BC.

Image of A in $x = y$ is $(2, 1)$ and image of A in $y = 0$ is $(1, -2)$.

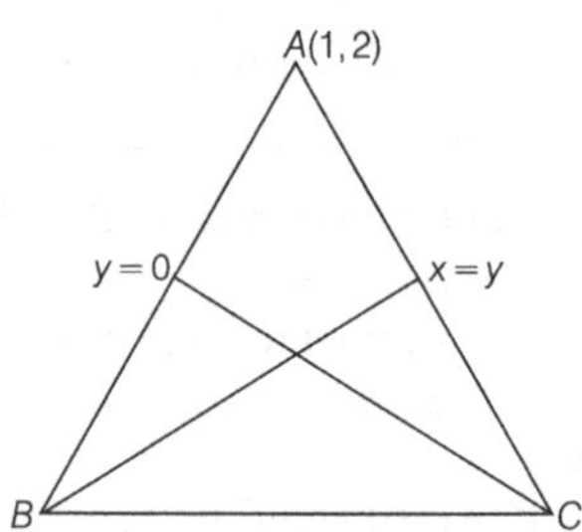

So, equation of line BC is $y = 3x - 5$

So, $$d(A, BC) = \frac{4}{\sqrt{10}}$$

So, $$\sqrt{10}\, d(A, BC) = 4$$

4.

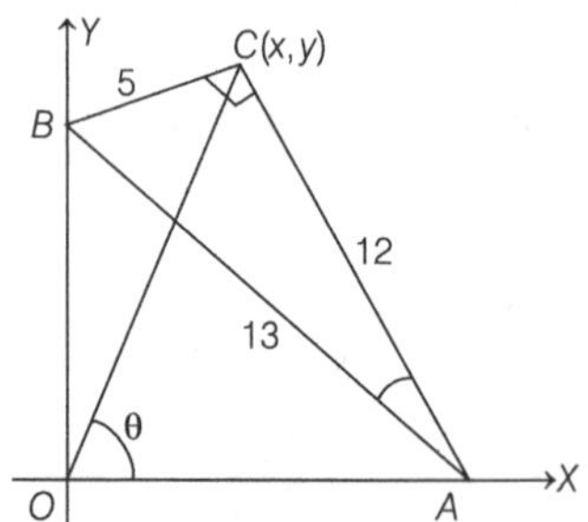

Since, $\angle BCA = 90°$, point A, O, B, C are concyclic.

Let $\angle AOC = \theta$

$$\angle BOC = \angle BAC$$

$$\tan\left(\frac{\pi}{2} - \theta\right) = \frac{5}{12}$$

$$\frac{x}{y} - \frac{5}{12} = \text{ or } 12x - 5y = 0$$

5. $\because \theta_1 - \theta = \theta - \theta_2$

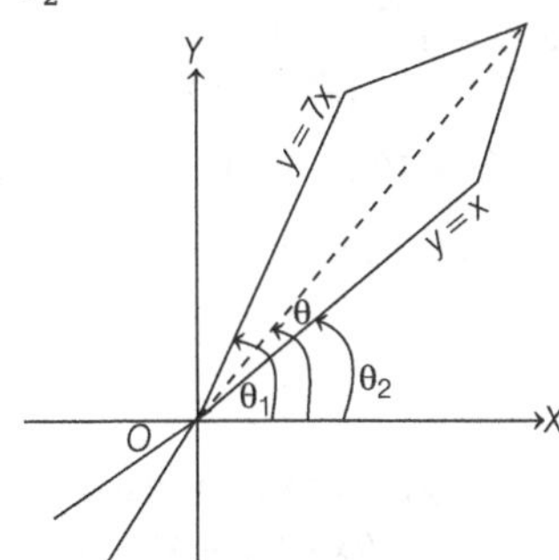

$\Rightarrow$ $2\theta = \theta_1 + \theta_2$

$\therefore$ $\tan 2\theta = \dfrac{\tan\theta_1 + \tan\theta_2}{1 - \tan\theta_1 \tan\theta_2}$

$\Rightarrow$ $\tan 2\theta = \dfrac{7+1}{1-7} \Rightarrow \tan 2\theta = \dfrac{-4}{3}$

$\Rightarrow$ $\dfrac{2\tan\theta}{1-\tan^2\theta} = -\dfrac{4}{3}$

$\Rightarrow$ $\tan\theta = 2 \text{ or } -\dfrac{1}{2}$

Hence, slope of longer diagonal is 2.

6. Point P which lies on the line $y = \dfrac{3x}{4}$ can be chosen as $P\left(h, \dfrac{3h}{4}\right)$

If θ is the angle that the line $y = \dfrac{3x}{4}$ makes with the positive direction of the X-axis, then

$$\tan\theta = \frac{3}{4} \text{ or } \cos\theta = \frac{4}{5} \text{ and } \sin\theta = \frac{3}{5}$$

Now, the coordinates of points A and B which lie on the line $y = \dfrac{3x}{4}$ can be chosen as

$$A \equiv \left(h + \frac{4r_1}{5}, \frac{3h}{4} + \frac{3r_1}{5}\right)$$

and $$B \equiv \left(h + \frac{4r_2}{5}, \frac{3h}{4} + \frac{3r_2}{5}\right)$$

Since, A lies on the line $x - y + 1 = 0$, we have

$$\left(h + \frac{4r_1}{5}\right) - \left(\frac{3h}{4} + \frac{3r_1}{5}\right) + 1 = 0$$

or $$r_1 = \frac{-5}{4}(h+4)$$

Since, B lies on the line $2x - y - 5 = 0$

We have,

$$2\left(h + \frac{4r_2}{5}\right) - \left(\frac{3h}{4} + \frac{3r_2}{5}\right) - 5 = 0$$

or $$r_2 = \frac{-5}{4}(h-4)$$

According to the given condition, we have

$$PA \cdot PB = 25 \text{ i.e. } |r_1| \cdot |r_2| = 25$$

i.e. $$\frac{25}{16}(h^2 - 16) = \pm 25$$

i.e. $$h^2 = 16 \pm 16 = 32, 0$$

i.e. $$h = \pm 4\sqrt{2}, 0$$

Hence, the required points are (0, 0), $(4\sqrt{2}, 3\sqrt{2})$ and $(-4\sqrt{2}, -3\sqrt{2})$.

7. Let $OC = a$

$\therefore$ $OC = CA = AB = BO = a$

Let $\left(x_1 \dfrac{4x_1}{3}\right)$

$\therefore$ $A\left(a + x_1, \dfrac{4x_1}{3}\right)$

$\because$ $x_1^2 + \dfrac{16x_1^2}{9} = a^2$

[$\because$ ODB is a right angle triangle]

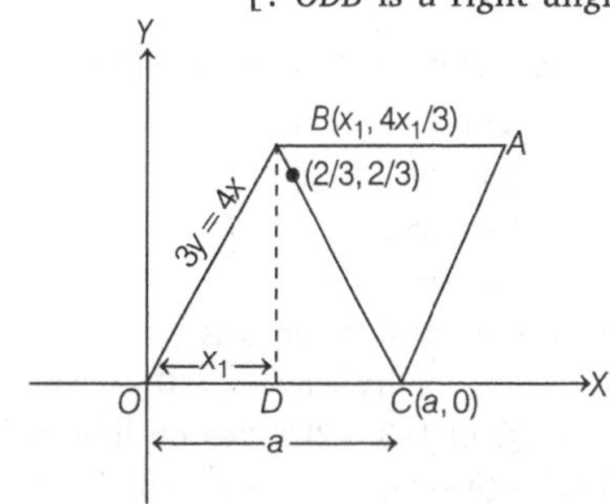

$\therefore$ $a = \dfrac{5x_1}{3}$

$\because$ equation of BC is

$$y - 0 = \frac{\frac{4x_1}{3} - 0}{x_1 - a}(x - a)$$

$\because$ $a = \dfrac{5x_1}{3}$

$\therefore$ $y = -2x + \dfrac{10x_1}{3}$;

$\therefore$ BC passes through $\left(\dfrac{2}{3}, \dfrac{2}{3}\right)$

$\therefore$ $x_1 = \dfrac{3}{5}$; $\therefore a = 1$

$\therefore$ $A\left(1 + \dfrac{3}{5}, \dfrac{4}{3} \times \dfrac{3}{5}\right)$;

$\therefore$ $A\left(\dfrac{8}{5}, \dfrac{4}{5}\right) \therefore \dfrac{5}{2}(\alpha + \beta) = 6.$

8. Let $\frac{AN}{BN} = \lambda$

Then, $N = \left(\frac{a}{\lambda+1}, \frac{a\lambda}{\lambda+1}\right)$

Equation of MN is $x - y = \frac{a(1-\lambda)}{\lambda+1}$

$\therefore \quad M = \left(0, \frac{a(\lambda-1)}{\lambda+1}\right)$

Since, M is internal point of segment OB, $\lambda > 1$

$\therefore \quad AN = \frac{a\sqrt{2}\,\lambda}{\lambda+1}, \; MN = \frac{a\sqrt{2}}{\lambda+1}$

Then, $\frac{1}{2}\cdot\frac{a\sqrt{2}\,\lambda}{\lambda+1} \times \frac{a\sqrt{2}}{\lambda+1} = \frac{3}{8} \times \frac{1}{2}a^2$

$\therefore \quad 16\lambda^2 = 3(\lambda+1)^2$

$\Rightarrow \quad 3\lambda^2 - 10\lambda + 3 = 0 \Rightarrow \lambda = 3$ or $\frac{1}{3}$

Since, $\lambda > 1$, so $\lambda = 3$

Matrix-Match Type

1. **(A)** $\because \quad x(2a+3b) + y(3a+2b) = 2(b-a)$

$\Rightarrow \quad a(2x+3y+2) + b(3x+2y-2) = 0$

Which is equation of a line passing through the point of intersection of the lines $2x+3y+2=0$ and $3x+2y-2=0$. The point of intersection of these lines is $(2, -2)$ (S)

(B) Given $9a^2 + 4b^2 - c^2 + 12ab = 0$

or $\quad (3a+2b)^2 - c^2 = 0$ or $c = \pm(3a+2b)$

$\Rightarrow \quad 3a+2b+c=0$

or $\quad 3a+2b-c=0$

$\Rightarrow \quad 3a+2b+c=0$

or $\quad -3a-2b+c=0$

and family of straight lines is

$$ax+by+c=0$$

$\Rightarrow (3, 2)$ (Q) or $(-3, -2)$ T lies on line $ax+by+c=0$

(C) $12a^2 + (b+c)a - (6b^2 + c^2 - 5bc) = 0$

$\therefore \quad a = \frac{-(b+c) \pm \sqrt{(b+c)^2 + 48(6b^2+c^2-5bc)}}{24}$

$= \frac{-(b+c) \pm (17b - 7c)}{24}$

$= \frac{16b - 8c}{24}$ or $\frac{-18b+6c}{24}$

$= \frac{2b-c}{3}$ or $\frac{-3b+c}{4}$

or $\quad 3a - 2b + c = 0$

or $\quad 4a + 3b - c = 0$

$\Rightarrow (3, -2)$ (P) or $(-4, -3)$ (R) lies on line $ax+by+c=0$

2. **(A)** $\because \; 4x+3y-6=0$

$\Rightarrow \quad -4x-3y+6=0$

and $\quad 5x+12y+9=0$

$\because (-4)(5) + (-3)(12) = -56 < 0$

$\therefore$ Bisectors are $\left(\frac{-4x-3y+6}{5}\right) = \pm\left(\frac{5x+12y+9}{13}\right)$

$\Rightarrow \quad (-52x - 39y + 78) = \pm(25x + 60y + 45)$

$O: \; (-52x - 39y + 78) = -(25x + 60y + 45)$

or $\quad 27x - 21y - 123 = 0$

or $\quad 9x - 7y - 41 = 0$

$\therefore \quad O: 9x - 7y - 41 = 0$ (T)

and $\; A: (-52x - 39y + 78) = (25x + 60y + 45)$

or $\quad 77x + 99y - 33 = 0$

or $\quad 7x + 9y - 3 = 0$

$\therefore \quad A: 7x + 9y - 3 = 0$ (R)

(B) $\because \quad 4x - 3y - 6 = 0$

$\Rightarrow \quad -4x + 3y + 6 = 0$

and $\quad 5x - 12y + 9 = 0$

$\because \quad (-4)(5) + 3(-12) = -56 < 0$

$\therefore$ Bisectors are $\left(\frac{-4x+3y+6}{5}\right) = \pm\left(\frac{5x-12y+9}{13}\right)$

$\Rightarrow \quad (-52x + 39y + 78) = \pm(25x - 60y + 45)$

$O: (-52x + 39y + 78) = -(25x - 60y + 45)$

or $\quad 27x + 21y - 123 = 0$

or $\quad 9x + 7y - 41 = 0$

$\therefore \quad O: 9x + 7y - 41 = 0$ (S)

and $\; A: (-52x + 39y + 78) = (25x - 60y + 45)$

or $\quad 77x - 99y - 33 = 0$

or $\quad 7x - 9y - 3 = 0$

$\therefore \quad A: 7x - 9y - 3 = 0$ (P)

(C) $\because \quad 4x - 3y + 6 = 0$

and $\quad 5x - 12y - 9 = 0$

or $\quad -5x + 12y + 9 = 0$

$\because \quad (4)(-5) + (-3)(12) = -56 < 0$

$\therefore$ Bisectors are $\left(\frac{4x-3y+6}{5}\right) = \pm\left(\frac{-5x+12y+9}{13}\right)$

$\Rightarrow \quad (52x - 39y + 78) = \pm(-25x + 60y + 45)$

$O: (52x - 39y + 78) = -(-25x + 60y + 45)$

or $\quad 27x + 21y + 123 = 0$

or $\quad 9x + 7y + 41 = 0$

$\therefore \quad O: 9x + 7y + 41 = 0$

and $\; A: (52x - 39y + 78) = (-25x + 60y + 45)$

or $\quad 77x - 99y + 33 = 0$

or $\quad 7x - 9y + 3 = 0$

$\therefore \quad A: 7x - 9y + 3 = 0$ (Q)

3. (A) $\because \quad \max\{|x|, |y|\} = 1$

If $\quad |x| = 1$ and if $|y| = 1$,

Then, $\quad x = \pm 1$ and $y = \pm 1$

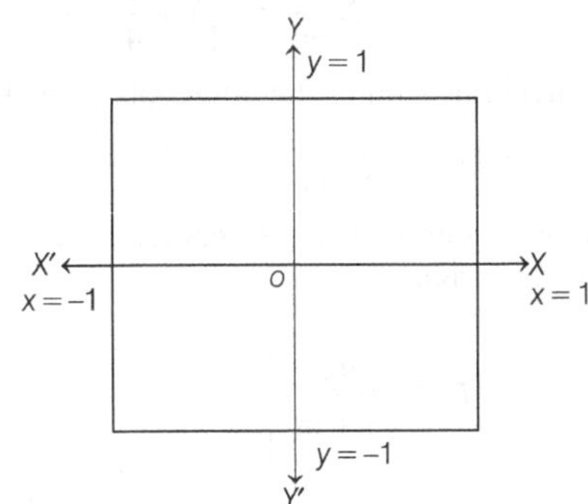

$\therefore$ Required area $= 2 \times 2$

$= 4$ sq unit (T)

(B) The line $y = x$ cuts the lines $|x + y| = 6$

i.e. $\quad x + y = \pm 6$

at $\quad x = \pm 3, \; y = \pm 3$

or $\quad (-3, -3)$ and $(3, 3)$

then $\quad -3 < a < 3$

$\therefore \quad 0 \le |a| < 3$

$\therefore \quad [|a|] = 0, 1, 2$ (P, Q, R)

(C) Since, (0, 0) and (1, 1) lie on the same side.

So, $\quad a^2 + ab + 1 > 0$

$\because$ Coefficient of a^2 is > 0

$\therefore \quad D < 0$

$b^2 - 4 < 0$

or $\quad -2 < b < 2$

$\Rightarrow \quad b = -1, 0, 1$

$\therefore$ Number of values of b is 3. (S)

4. Let $P(\alpha, \alpha^2)$ first draw the exact diagram of ΔABC.

The point $P(\alpha, \alpha^2)$ move on the curve $y = x^2$ for all α.

Now intersection of $y = x^2$ and $2x + 3y - 1 = 0$

or $\quad 2x + 3x^2 - 1 = 0$

$\therefore \quad x = -1, \; x = \frac{1}{3}$

Let intersection points $D \equiv (-1, 1)$ and $E \equiv \left(\frac{1}{3}, \frac{1}{9}\right)$

intersection of $y = x^2$ and $x + 2y - 3 = 0$

or $\quad x + 2x^2 - 3 = 0$

$\therefore \quad x = 1, \; x = -3/2$

Let intersection points $F \equiv (1, 1)$

and $G \equiv \left(-\frac{3}{2}, \frac{9}{4}\right)$ and intersection of $y = x^2$

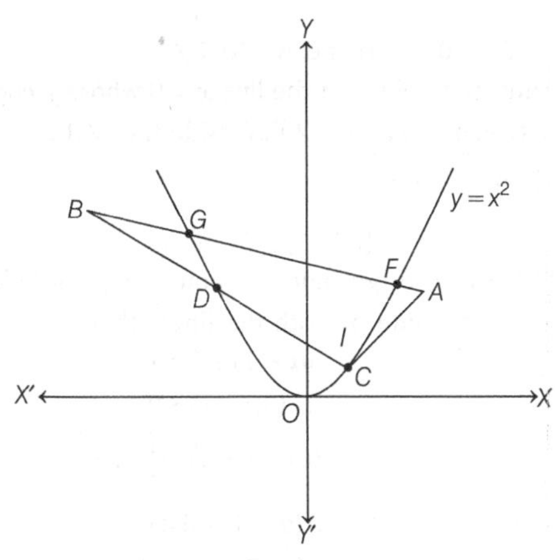

and $\quad 5x - 6y - 1 = 0$

or $\quad 5x - 6x^2 - 1 = 0$

$\therefore \quad x = \frac{1}{3}, \; x = \frac{1}{2}$

Let intersection points

$$H \equiv \left(\frac{1}{3}, \frac{1}{9}\right) \text{ and } I \equiv \left(\frac{1}{2}, \frac{1}{4}\right)$$

Thus, the points on the curve $y = x^2$ whose x-coordinates lies between $-3/2$ and -1 and $\frac{1}{2}$ and 1 lies within the triangle ABC.

Hence, $-\frac{3}{2} < \alpha < -1$ and $\frac{1}{2} < \alpha < 1$

i.e. $\quad \alpha \in \left(-\frac{3}{2}, -1\right) \cup \left(\frac{1}{2}, 1\right)$

$\therefore \quad [\alpha] = -2, 0$ (P, R)

(B) Let $P \equiv (0, \alpha)$, first draw the exact diagram of the triangle. The point $P(0, \alpha)$ move on the line $x = 0$ (Y-axis) for all α.

Now, D and E are the intersection of $3y - 2x - 5 = 0$, $x = 0$ and $4y + x - 14 = 0$, $x = 0$ respectively.

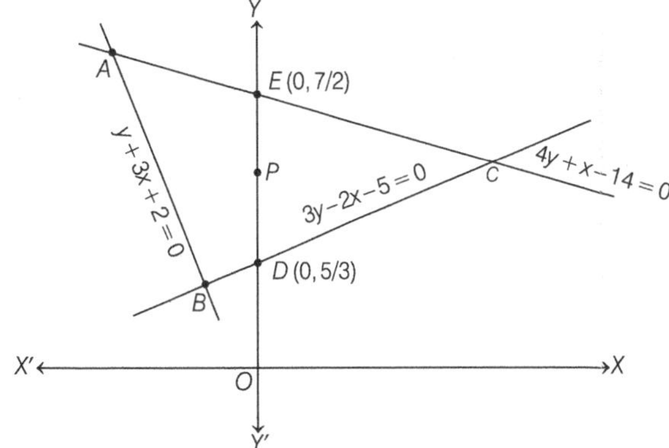

$\therefore\ D \equiv (0, 5/3)$ and $E \equiv (0, 7/2)$.

Thus, the points on the line $x = 0$ whose y-coordinates lies between 5/3 and 7/2 lies inside the ΔABC.

Then, $\frac{5}{3} < \alpha < \frac{7}{2}$

$\therefore \quad [\alpha] = 1, 2, 3$ (S, T)

(C) Since, origin and the point $(\alpha^2, \alpha + 1)$ lie on the same side of both the lines, then

$$3\alpha^2 - (\alpha + 1) + 1 > 0$$

$\Rightarrow \quad \alpha(3\alpha - 1) > 0$

or $\quad \alpha \in (-\infty, 0) \cup \left(\frac{1}{3}, \infty\right) \quad$ ...(i)

and $\quad \alpha^2 + 2(\alpha + 1) - 5 < 0$

$\Rightarrow \quad \alpha^2 + 2\alpha - 3 < 0$

$(\alpha - 1)(\alpha + 3) < 0$

$\Rightarrow \quad \alpha \in (-3, 1) \quad$ ...(ii)

From Eqs. (i) and (ii), we get

$$\alpha \in \left((-\infty, 0) \cup \left(\frac{1}{3}, \infty\right)\right) \cap (-3, 1)$$

$\therefore \quad \alpha \in (-3, 0) \cup \left(\frac{1}{3}, 1\right)$

$\therefore \quad [\alpha] = -3, -2, -1, 0$ (P, Q, R)

5. Solving the equation

$2m^3 - 3m^2 - 3m + 2 = 0$ we get

$2(m^3 + 1) - 3m(m + 1) = 0$

$\Rightarrow \quad (m + 1)(2m^2 - 5m + 2) = 0$

$\Rightarrow \quad (m + 1)(2m - 1)(m - 2) = 0$

$\Rightarrow \quad m = -1, \frac{1}{2}$ or 2

Equation of the given lines can be written as

$$m_i^2 x - m_i y = -1$$

(A) Algebraic sum of the intercepts made by the lines on X-axis

$$= -\Sigma \frac{1}{m_i^3} = -\left[1 + \frac{1}{4} + 4\right] = \frac{-21}{4} \text{ (R)}$$

(B) algebraic sum of the intercepts made by the line on

$$Y\text{-axis} = \sum \frac{1}{m_1} = -1 + 1 + 1/2 = 3/2 \text{ (Q)}$$

(C) Let p_i denote the perpendicular distance of the line from the origin

Then, $\quad p_i = \left|\frac{\frac{-1}{m_1}}{\sqrt{1 + m_i^2}}\right|$

$$\Rightarrow \quad \Sigma p_i = \frac{1}{\sqrt{1+1}} + \frac{2}{\sqrt{1 + \left(\frac{1}{4}\right)}} + \frac{\frac{1}{2}}{\sqrt{1+4}}$$

$$= \frac{1}{\sqrt{2}} + \frac{4}{\sqrt{5}} + \frac{1}{2\sqrt{5}}$$

$$= \frac{1}{10}(5\sqrt{2} + 9\sqrt{5}) \text{ (S)}$$

(D) l_i = length of the line intercepted between the coordinates axes

$$= \sqrt{\left(\frac{1}{m_i^2}\right)^2 + \left(\frac{1}{m_i}\right)^2}$$

$$\Rightarrow \quad \Sigma l_i = \sqrt{1+1} + \sqrt{16+4} + \sqrt{\frac{1}{16} + \frac{1}{4}}$$

$$= \sqrt{2} + 2\sqrt{5} + \frac{\sqrt{5}}{4}$$

$$= \frac{(4\sqrt{2} + 9\sqrt{5})}{4} \text{ (P)}$$

22

Pairs of Straight Lines

Objective Questions Type I [Only one correct answer]

In each of the questions below, four choices are given of which only one is correct. You have to select the correct answer which is the most appropriate.

1. If the angle between the lines represented by $6x^2 + 5xy - 4y^2 + 7x + 13y - 3 = 0$ is $\tan^{-1}(m)$ and $a^2 + b^2 - ab - a - b + 1 \le 0$, then $5a + 6b$ is equal to

(a) $\dfrac{1}{m}$ (b) m

(c) $\dfrac{1}{2m}$ (d) $2m$

2. The equation of the pair of lines passing through the origin and having slope $m \in I$ for which equation $(x-3)(x+m)+1=0$ has integral roots is

(a) $y^2 - 6xy + 5x^2 = 0$ (b) $y^2 + 6xy - 5x^2 = 0$

(c) $y^2 + 6xy + 5x^2 = 0$ (d) $y^2 - 6xy - 5x^2 = 0$

3. The line $x + y = 1$ meets the lines represented by the equation $y^3 - 6xy^2 + 11x^2y - 6x^3 = 0$ at the points P, Q, R. If O is the origin, then $(OP)^2 + (OQ)^2 + (OR)^2$ is equal to

(a) $\dfrac{85}{72}$ (b) $\dfrac{121}{72}$ (c) $\dfrac{211}{72}$ (d) $\dfrac{217}{72}$

4. Let P be the pair of lines represented by $2x^2 - 5xy + 2y^2 + 6x - 3y = 0$ and consider the following independent statements

(i) α is the x coordinate of the point of intersection of the pair of lines P.

(ii) β is the slope of one of the lines of P passing through origin.

(iii) γ is the constant term in the equation of the pair of angular bisectors of P.

Then,

(a) $\beta < \gamma < \alpha$ (b) $\alpha < \beta = \gamma$

(c) $\alpha = \beta < \gamma$ (d) $\gamma < \alpha < \beta$

5. If the pairs of lines $x^2 + 2xy + ay^2 = 0$ and $ax^2 + 2xy + y^2 = 0$ have exactly one line in common, then the joint equation of the other two lines is given by

(a) $3x^2 + 8xy - 3y^2 = 0$

(b) $3x^2 + 10xy + 3y^2 = 0$

(c) $y^2 + 2xy - 3x^2 = 0$

(d) $x^2 + 2xy - 3y^2 = 0$

6. If the two pairs of lines $x^2 - 2mxy - y^2 = 0$ and $x^2 - 2nxy - y^2 = 0$ are such that one of them represents the bisectors of the angles between the other, then

(a) $mn + 1 = 0$ (b) $mn - 1 = 0$

(c) $1/m + 1/n = 0$ (d) $1/m - 1/n = 0$

7. If the lines joining the origin to the points of intersection of $y = mx + 1$ with $x^2 + y^2 = 1$ are perpendicular, then

(a) $m = 1$ only (b) $m = \pm 1$

(c) $m = 0$ (d) None of these

8. If the pair of straight line given by $Ax^2 + 2Hxy + By^2 = 0$, where $(H^2 > AB)$, forms an equilateral triangle with the line $ax + by + c = 0$, then $(A + 3B)(3A + B) =$

(a) $4H^2$ (b) $2H^2$

(c) $-2H^2$ (d) $-4H^2$

9. The pair of lines joining the origin to the points of intersection of the curves

$ax^2 + 2hxy + by^2 + 2gx = 0$ and

$a'x^2 + 2h'xy + b'y^2 + 2g'x = 0$

will be at right angles to one another, if

(a) $g(a' + b') = g'(a + b)$ (b) $g(a + b) = g'(a' + b')$

(c) $gg' = (a + b)(a' + b')$ (d) None of these

10. The equation $x^2 + 2\sqrt{2}xy + 2y^2 + 4x + 4\sqrt{2}y + 1 = 0$ represents a pair of lines. The distance between them

(a) 4 (b) $4/\sqrt{3}$

(c) 2 (d) $2\sqrt{3}$

11. Let PQR be a right-angled isosceles triangle, right angled at $P(2, 1)$ at. If the equation of the line QR is $2x + y = 3$, then the equation representing the pair of lines PQ and PR is

(a) $3x^2 - 3y^2 + 8xy + 20x + 10y + 25 = 0$
(b) $3x^2 - 3y^2 + 8xy - 20x - 10y + 25 = 0$
(c) $3x^2 - 3y^2 + 8xy + 10x + 15y + 20 = 0$
(d) $3x^2 - 3y^2 - 8xy - 15y - 20 = 0$

12. The difference of the tangents of the angles which the lines $x^2(\sec^2\theta - \sin^2\theta) - 2xy \tan\theta + y^2\sin^2\theta = 0$ makes with the X-axis is

(a) $2\tan\theta$ (b) 2
(c) $2\cot\theta$ (d) $\sin 2\theta$

13. If $\alpha, \beta > 0$ and $\alpha < \beta$ and $ax^2 + 4\gamma xy + \beta y^2 + 4p(x + y + 1) = 0$ represent a pair of straight lines, then

(a) $\alpha \le p \le \beta$ (b) $p \le \alpha$
(c) $p \ge \alpha$ (d) $p \le \alpha$ or $p \ge \beta$

14. The image of the pair of lines represented by $ax^2 + 2hxy + by^2 = 0$ by the line mirror $y = 0$ is

(a) $ax^2 - 2hxy - by^2 = 0$ (b) $bx^2 - 2hxy + ay^2 = 0$
(c) $bx^2 + 2hxy + ay^2 = 0$ (d) $ax^2 - 2hxy + by^2 = 0$

15. If the equation $ax^2 + 2hxy + by^2 + 2gx + 2fy + c = 0$ represents a pair of parallel lines, then the distance between them is

(a) $2\sqrt{\left(\dfrac{g^2 - ac}{h^2 + a^2}\right)}$ (b) $2\sqrt{\left(\dfrac{g^2 + ac}{h^2 + a^2}\right)}$
(c) $3\sqrt{\left(\dfrac{g^2 + ac}{a(a + b)}\right)}$ (d) $3\sqrt{\left(\dfrac{g^2 + ac}{a(a + b)}\right)}$

16. If the lines represented by $x^2 - 2pxy - y^2 = 0$ are rotated about the origin through an angle θ, one clockwise direction and other in anticlockwise direction, then the equation of the bisectors of the angle between the lines in the new position is

(a) $px^2 + 2xy - py^2 = 0$ (b) $px^2 + 2xy + py^2 = 0$
(c) $x^2 - 2pxy + y^2 = 0$ (d) None of these

17. A pair of lines $S = 0$ together with the lines given by the equation $8x^2 - 14xy + 3y^2 + 10x + 10y - 25 = 0$ form a parallelogram. If its diagonals intersect at the point (3, 2), then the equation $S = 0$, is

(a) $6x^2 - 9xy + y^2 - 25x + 30y + 25 = 0$
(b) $8x^2 - 14xy + 3y^2 - 25x + 30y + 50 = 0$
(c) $8x^2 - 14xy + 3y^2 - 50x + 50y + 75 = 0$
(d) $6x^2 + 14xy - 3y^2 - 30x + 40y - 75 = 0$

18. The pair of straight lines joining the origin to the common points of $x^2 + y^2 = 4$ and $y = 3x + c$ are perpendicular, if c^2 is equal to

(a) 20 (b) 13
(c) 1/5 (d) 5

19. Two of the straight lines given by $3x^3 + 3x^2y - 3xy^2 + dy^3 = 0$ are at right angles, if

(a) $d = -1/3$ (b) $d = 1/3$
(c) $d = -3$ (d) $d = 3$

20. If the angle between the two lines represented by $2x^2 + 5xy + 3y^2 + 6x + 7y + 4 = 0$ is $\tan^{-1}(m)$, then m is equal to

(a) 1/5 (b) -1
(c) $-2/3$ (d) None of these

21. If the pair of straight lines $ax^2 + 2hxy + by^2 = 0$ is rotated about the origin through 90°, then their equations in the new position are given by

(a) $ax^2 - 2hxy + by^2 = 0$ (b) $ax^2 - 2hxy - by^2 = 0$
(c) $bx^2 - 2hxy + ay^2 = 0$ (d) $bx^2 + 2hxy + ay^2 = 0$

22. Product of the perpendiculars from (α, β) to the lines $ax^2 + 2hxy + by^2 = 0$ is

(a) $\dfrac{|a\alpha^2 - 2h\alpha\beta + b\beta^2|}{\sqrt{\{4h^2 + (a + b)^2\}}}$ (b) $\dfrac{|a\alpha^2 - 2h\alpha\beta + b\beta^2|}{\sqrt{\{4h^2 - (a - b)^2\}}}$
(c) $\dfrac{|a\alpha^2 - 2h\alpha\beta + b\beta^2|}{\sqrt{\{4h^2 - (a + b)^2\}}}$ (d) None of these

23. Equation of pair of straight lines drawn through (1, 1) and perpendicular to the pair of lines $3x^2 - 7xy - 2y^2 = 0$ is

(a) $2x^2 + 7xy - 11x + 6 = 0$
(b) $2(x - 1)^2 + 7(x - 1)(y - 1) - 3y^2 = 0$
(c) $2(x - 1)^2 + 7(x - 1)(y - 1) + 3(y - 1)^2 = 0$
(d) None of the above

24. A pair of straight lines is passing through the point (1,1). One of the lines makes an angle θ with the positive direction of X-axis and the other makes the same angle with the positive direction of Y-axis. If the equation of the pair of straight lines is $x^2 - (a + 2)xy + y^2 + a(x + y - 1) = 0$, $a \ne -2$, then the value of θ is

(a) $\dfrac{1}{2}\sin^{-1}\left(\dfrac{2}{a + 2}\right)$ (b) $\dfrac{1}{2}\sin\left(\dfrac{2}{a + 2}\right)$
(c) $\dfrac{1}{2}\tan^{-1}\left(\dfrac{2}{a + 2}\right)$ (d) $\dfrac{1}{2}\tan\left(\dfrac{2}{a + 2}\right)$

25. Mixed term xy is to be removed from the general equation of second degree $ax^2 + 2hxy + by^2 + 2gx + 2fy + c = 0$, one should rotate the axes through an angle θ, then $\tan 2\theta$ equal to

(a) $\dfrac{a - b}{2h}$ (b) $\dfrac{2h}{a + b}$
(c) $\dfrac{a + b}{2h}$ (d) $\dfrac{2h}{a - b}$

26. The equation $\sqrt{\{(x - 2)^2 + y^2\}} + \sqrt{\{(x + 2)^2 + y^2\}} = 4$ represents

(a) circle (b) pair of lines
(c) a parabola
(d) line segment $y = 0$, $-2 < x < 2$

Objective Questions Type II [One or more than one correct answer(s)]

In each of the questions below four choices of which one or more than one are correct. You have to select the correct answer(s) accordingly.

1. If $x^2 + \alpha y^2 + 2\beta y = a^2$ represents a pair of perpendicular straight lines, then
(a) $\alpha = 1, \beta = a$ (b) $\alpha = 1, \beta = -a$
(c) $\alpha = -1, \beta = -a$ (d) $\alpha = -1, \beta = a$

2. Type of quadrilateral formed by the two pairs of lines $6x^2 - 5xy - 6y^2 = 0$
and $6x^2 - 5xy - 6y^2 + x + 5y - 1 = 0$ is
(a) square (b) rhombus
(c) parallelogram (d) rectangle

3. If the line $y = mx$ is one of the bisector of the lines $x^2 + 4xy - y^2 = 0$, then the value of m is equal to
(a) $\frac{-1+\sqrt{5}}{2}$ (b) $\frac{1+\sqrt{5}}{2}$
(c) $\frac{-1-\sqrt{5}}{2}$ (d) $\frac{1-\sqrt{5}}{2}$

4. Two pairs of straight lines have the equations $y^2 + xy - 12x^2 = 0$ and $ax^2 + 2hxy + by^2 = 0$. One line will be common among them, if
(a) $a = -3(2h + 3b)$ (b) $a = 8(h - 2b)$
(c) $a = 2(b + h)$ (d) $a = -3(b + h)$

5. The combined equation of three sides of a triangle is $(x^2 - y^2)(2x + 3y - 6) = 0$. If $(-2, a)$ is an interior point and $(b, 1)$ is an exterior point of the triangle, then
(a) $2 < a < \frac{10}{3}$ (b) $-2 < a < \frac{10}{3}$
(c) $-1 < b < \frac{9}{2}$ (d) $-1 < b < 1$

6. If the two lines represented by $x^2(\tan^2\theta + \cos^2\theta) - 2xy\tan\theta + y^2\sin^2\theta = 0$ make angles α, β with the X-axis, then
(a) $\tan\alpha + \tan\beta = 4\operatorname{cosec} 2\theta$
(b) $\tan\alpha\tan\beta = \sec^2\theta + \tan^2\theta$
(c) $\tan\alpha - \tan\beta = 2$
(d) $\frac{\tan\alpha}{\tan\beta} = \frac{2+\sin 2\theta}{2-\sin 2\theta}$

7. If $x^2 + 2hxy + y^2 = 0\ (h > 1)$ represents the equation of the straight lines through the origin which make an angle α with the straight line $y + x = 0$, then
(a) $\sec 2\alpha = h$ (b) $\cos\alpha = \sqrt{\frac{(1+h)}{(2h)}}$
(c) $2\sin\alpha = \sqrt{\frac{(1+h)}{h}}$ (d) $\cot\alpha = \sqrt{\frac{(h+1)}{(h-1)}}$

8. If the angle between the lines $x^2 - xy + ay^2 = 0$ is 45°, then value(s) of a is/are
(a) -6 (b) 0
(c) 6 (d) 12

9. If the lines represented by $2x^2 - 5xy + 2y^2 = 0$ be the two sides of a parallelogram and the line $5x + 2y = 1$ be one of its diagonal. Final the equation of the other diagonal, and area of the parallelgram.
(a) The equation of the other diagonal is $11x - 10y = 0$
(b) Diagonal of parallelogram bisects at $\left(\frac{5}{36}, \frac{11}{72}\right)$
(c) Area of the parallelogram is $\frac{1}{21}$ sq units.
(d) Area of the parallelogram is $\frac{1}{36}$ sq units

10. If the pair of lines $ax^2 + 2hxy + by^2 + 2gx + 2fy + c = 0$ intersect on Y-axis, then
(a) $f^2 = bc$ (b) $abc = 2fgh$
(c) $bg^2 \neq ch^2$ (d) $2fgh = bg^2 + ch^2$

Linked-Comprehension Type

In these questions, a passage (paragraph) has been given followed by questions based on each of the passages. You have to answer the questions based on the passage given.

PASSAGE 1

If the lines represented by $2x^2 - 5xy + 2y^2 = 0$ be the two sides of a parallelogram and the line $5x + 2y = 1$ be one of its diagonal.

On the basis of above information, answer the following questions

1. The equation of the other diagonal is
(a) $10x - 11y = 0$ (b) $11x - 10y = 0$
(c) $3x - 2y = 0$ (d) $2x - 3y = 0$

2. The centroid of the parallelogram is
(a) $\left(\frac{5}{72}, \frac{11}{36}\right)$ (b) $\left(\frac{11}{72}, \frac{5}{36}\right)$
(c) $\left(\frac{5}{36}, \frac{11}{72}\right)$ (d) $\left(\frac{11}{36}, \frac{5}{72}\right)$

3. The area of the parallelogram is
(a) $\frac{1}{36}$ sq unit (b) $\frac{1}{18}$ sq unit
(c) $\frac{1}{9}$ sq unit (d) None of these

4. The ratio of the longer side to smaller side is
(a) 6 : 5 (b) 7 : 6
(c) 5 : 4 (d) 4 : 3

5. The ratio of the longer diagonal to smaller diagonal is
(a) 6 : 5 (b) 9 : 2
(c) 13 : 5 (d) None of these

PASSAGE 2

Let $f_1(x, y) \equiv ax^2 + 2hxy + by^2 = 0$ and let $f_{i+1}(x, y) = 0$ denotes the equation of the bisectors of $f_i(x, y) = 0$ for all $i = 1, 2, 3, \ldots$

On the basis of above information, answer the following questions

1. Equation $f_2(x, y) = 0$ is
(a) $hx^2 - (a - b)xy + hy^2 = 0$
(b) $hx^2 - (a - b)xy - hy^2 = 0$
(c) $hx^2 + (a - b)xy + hy^2 = 0$
(d) $hx^2 + (a - b)xy - hy^2 = 0$

2. Equation $f_3(x, y) = 0$ is
(a) $(a - b)x^2 - 4hxy + (a - b)y^2 = 0$
(b) $(a - b)x^2 - 4hxy - (a - b)y^2 = 0$
(c) $(a - b)x^2 + 4hxy - (b - a)y^2 = 0$
(d) $(a - b)x^2 + 4hxy - (a - b)y^2 = 0$

3. If $f_{i+1}(x, y) = 0$ represents the equation of a pair of perpendicular lines, then $f_2(x, y) = 0$ is
(a) $bx^2 - 2hxy + ay^2 = 0$ (b) $a\,x^2 + 2hxy + by^2 = 0$
(c) $ax^2 - 2hxy + by^2 = 0$ (d) None of these

4. If $f_{i+1}(x, y) = 0$ represents the equation of pair of perpendicular lines, then $f_3(x, y) = 0$ is
(a) $bx^2 - 2hxy - ay^2 = 0$ (b) $ax^2 + 2hxy + by^2 = 0$
(c) $ax^2 - 2hxy + by^2 = 0$ (d) $bx^2 - 2hxy + ay^2 = 0$

5. If $f_{i+1}(x, y) = 0$ represents the equation of pair of perpendicular lines, then $f_{n+2}(x, y) = 0, \forall n \geq 2$ is same as
(a) $f_{n+2}(x, y) = 0$ (b) $f_{n+1}(x, y) = 0$
(c) $f_n(x, y) = 0$ (d) None of these

Numerical Value Type

Solve the following problems and write your answer.

1. If the area of the triangle formed by the lines $y^2 - 9xy + 18x^2 = 0$ and $y = 9$ is λ sq unit, then the value of 4λ must be

2. If the angle between the pair of straight lines
$y^2 \sin^2\theta - xy \sin^2\theta + x^2(\cos^2\theta - 1) = 0$ is ϕ, then the value of $\sin^3\phi + 32\sin\phi + 9$ must be

3. Equation $ax^3 - 9x^2y - xy^2 + 4y^3 = 0$ represents three straight lines. If the two of the lines are perpendicular, then $\left|\dfrac{a^2 - a}{8}\right|$ must be

4. If the angle between the two lines represented by $2x^2 + 5xy + 3y^2 + 7y + 4 = 0$ is $\tan^{-1}(m)$, then the value of m must be

5. The lines joining the origin to the common points of $(x - 3)^2 + (y - 4)^2 = c^2$ and $4x + 3y = 24$ are at right angles, then the value of c^2 must be

Single Integer Type

Solve the following problems and write your answer.

1. If one of the pair of lines $4x^2 + 6xy + ky^2 = 0$ is perpendicular to one of the lines represented by $3x^2 - 5xy + 2y^2 = 0$, then twice the absolute difference of such possible values of k is

2. If the slope of one of the lines represented by $ax^2 + 2hxy + by^2 = 0$ is the square of the other then $\dfrac{a + b}{h} + \dfrac{8h^2}{ab}$ is equal to

3. If the angle between the lines joining the origin to the points of intersection of the straight line $y = 3x + 2$ with the curve $x^2 + 2xy + 3y^2 + 4x + 8y - 11 = 0$ is $\tan^{-1}\left(\dfrac{2\sqrt{2}}{k}\right)$. Then, value of k is ……… .

4. If $lx^2 + 3xy - 2y^2 - 5x + 5y + k = 0$ represents a pair of perpendicular lines, then value of $|a + l|$ is equal to …… .

5. One of the bisectors of the angle between the lines $a(x - 1) + 2h(x - 1)(y - 2) + b(y - 2)^2 = 0$ is $x + 2y - 5 = 0$. Then, the value of $4\left(\dfrac{b - a}{h}\right)$ is ……… .

Matrix-Match Type

Given below are Matching Type Questions, with two columns (each having some items) each. Each item of Column I has to be matched with the items of Column II, by encircling the correct match(es).

NOTE *An item of Column I can be matched with more than one items of Column II.*

1. Observe the following columns

	Column I		Column II
(A)	If the slope of one of the lines represented by $ax^2 - 6xy + y^2 = 0$ is square of the other, then a is	(P)	-4
(B)	If two of the lines given by the equation $ax^3 - 9x^2y - xy^2 + 4y^3 = 0$ are perpendicular, then a is	(Q)	-6
(C)	If two of the lines represented by $x^4 + x^3y + cx^2y^2 - xy^3 + y^4 = 0$ bisect the angle between the other two, then the value of c is	(R)	8
		(S)	5
		(T)	-27

2. Observe the following columns

	Column I		Column II
(A)	If the pair of lines represented by the equation $12x^2 - 10xy + 2y^2 + 11x - 5y + 2 = 0$ be $3x - y + \lambda = 0$ and $4x - 2y + \mu = 0$, then	(P)	$\lambda + \mu = 8$
(B)	If the pair of lines represented by the equation $6x^2 + 17xy + 12y^2 + 22x + 31y + 20 = 0$ be $2x + 3y + \lambda = 0$ and $3x + 4y + \mu = 0$, then	(Q)	$\lambda + \mu = 9$
(C)	If the pair of lines represented by the equation $8x^2 + 8xy + 2y^2 + 26x + 13y + 15 = 0$ be $2x + y + \lambda = 0$ and $4x + 2y + \mu = 0$, then	(R)	$\lambda + \mu = 3$
		(S)	$\lambda^2 + \mu^2 = 41$
		(T)	$\lambda^2 + \mu^2 = 34$

Answers

Objective Questions Type I [Only one correct answer]

1. (d) **2.** (c) **3.** (b) **4.** (d) **5.** (b) **6.** (a) **7.** (b) **8.** (a) **9.** (a) **10.** (c)
11. (b) **12.** (b) **13.** (d) **14.** (d) **15.** (a) **16.** (a) **17.** (c) **18.** (a) **19.** (d) **20.** (a)
21. (c) **22.** (d) **23.** (d) **24.** (a) **25.** (d) **26.** (d)

Objective Questions Type II [One or more than one correct answer(s)]

1. (c, d) **2.** (a, d) **3.** (a, c) **4.** (a, b) **5.** (a, b, d)
6. (a, c, d) **7.** (a, b, d) **8.** (a, b) **9.** (a, b, d) **10.** (a, d)

Linked-Comprehension Type

Passage 1 **1.** (b) **2.** (c) **3.** (a) **4.** (d) **5.** (d)

Passage 2 **1.** (b) **2.** (d) **3.** (a) **4.** (b) **5.** (c)

Numerical Grid-Based Problems

1. 27 **2.** 42 **3.** 2.5 **4.** 0.2 **5.** 25

Single Integer Type Questions

1. (2) **2.** (6) **3.** (3) **4.** (1) **5.** (6)

Matrix-Match Type

1. A → (R, T); B → (P, S); C → (Q)

2. A → (R); B → (Q, S); C → (P, T)

Solutions

Objective Questions Type [Only one correct answer]

1. If θ be the angle between the lines, then

$$\theta = \tan^{-1}\left\{\frac{2\sqrt{\frac{25}{4}+24}}{6-4}\right\} = \tan^{-1}\left\{\frac{11}{2}\right\}$$

$\therefore \quad m = \frac{11}{2}$

Also, $\quad a^2 + b^2 - ab - a - b + 1 \le 0$

$\Rightarrow \quad \frac{1}{2}\{2a^2 + 2b^2 - 2ab - 2a - 2b + 2\} \le 0$

$\Rightarrow \quad \frac{1}{2}\{(a-b)^2 + (a-1)^2 + (b-1)^2\} \le 0$

$\Rightarrow \quad (a-b)^2 + (a-1)^2 + (b-1)^2 \le 0$

Which is possible only when $a - b = 0$, $a - 1 = 0$, $b - 1 = 0$

$\therefore \quad a = b = 1$

$\Rightarrow \quad 5a + 6b = 5 + 6 = 11 = 2m$

2. $\because (x-3)(x+m) = -1$ has integral roots

$x - 3 = 1$ and $x + m = -1$

$\Rightarrow \quad x = 4$, then $4 + m = -1$

$\therefore \quad m = -5$

and $x - 3 = -1$ and $x + m = 1$

$\Rightarrow \quad x = 2$, then $m = -1$

$\therefore \quad m = -5, -1$

Hence, joint equation of lines through the origin having slopes -5 and -1 is

$(y + 5x)(y + x) = 0$

or $\quad y^2 + 6xy + 5x^2 = 0$

3. $\because y^3 - 6xy^2 + 11x^2y - 6x^3 = 0$

$\Rightarrow \quad (y - x)(y - 2x)(y - 3x) = 0$

$\therefore \quad$ Lines $y = x, y = 2x, y = 3x \quad \quad \ldots$(i)

Solving Eq. (i) and $x + y = 1$, then we get

$P \equiv \left(\frac{1}{2}, \frac{1}{2}\right), Q = \left(\frac{1}{3}, \frac{2}{3}\right)$

and $\quad R \equiv \left(\frac{1}{4}, \frac{3}{4}\right)$

$\therefore \quad (OP)^2 + (OQ)^2 + (OR)^2$

$= \frac{1}{4} + \frac{1}{4} + \frac{1}{9} + \frac{4}{9} + \frac{1}{16} + \frac{9}{16}$

$= \frac{1}{2} + \frac{5}{9} + \frac{5}{8}$

$= \frac{121}{72}$

4. Given pairs of lines

$2x^2 - 5xy + 2y^2 + 6x - 3y = 0$

$\Rightarrow \quad (2y - x - 3)(y - 2x) = 0$

$\therefore$ Pairs of line are

$2y - x - 3 = 0$ and $y - 2x = 0$

Solving these equation we get,

$x = 1, y = 2$

$\therefore \quad \alpha = 1$

Equation of line passing through origin is

$y - 2x = 0$

$\because$ Slope of line $m = 2$

$\therefore \quad \beta = 2$

Equation of angular bisector of line

$2y - x - 3 = 0$ and $y - 2x = 0$ is

$$\frac{2y - x - 3}{\sqrt{5}} = \pm \frac{y - 2x}{\sqrt{5}}$$

$2y - x - 3 = y - 2x$ or $2y - x - 3 = -y + 2x$

$\because \quad x + y - 3 = 0$ or $3x - 3y + 3 = 0$

$x + y - 3 = 0$ or $x - y + 1 = 0$

$\gamma = -3$

$\therefore \gamma < \alpha < \beta$ i.e. $-3 < 1 < 2$

5. Let $y = mx$ be a line common to the given pairs of lines.

$am^2 + 2m + 1 = 0$

and $\quad m^2 + 2m + a = 0$

$\therefore \quad \frac{m^2}{2(1-a)} = \frac{m}{a^2 - 1} = \frac{1}{2(1-a)}$

or $\quad m^2 = 1$ and $m = -\frac{a+1}{2}$

or $\quad (a+1)^2 = 4$

or $\quad a = 1$ or -3

But for $a = 1$, the two pairs have both the lines common $a = -3$ and the slope m of the line common to both the is 1. Now,

$\therefore \ x^2 + 2xy + ay^2 = x^2 + 2xy - 3y^2$

$= (x - y)(x + 3y)$

and $ax^2 + 2xy + y^2 = -3x^2 + 2xy + y^2 = -(x - y)(3x + y)$

So, the equation of the required pair of lines

$(x + 3y)(3x + y) = 0$

or $\quad 3x^2 + 10xy + 3y^2 = 0$

6. Since, $x^2 - 2mxy - y^2 = 0 \quad \quad \ldots$(i)

and $\quad x^2 - 2nxy - y^2 = 0 \quad \quad \ldots$(ii)

Pair of bisectors of Eq. (i) is

$$\frac{x^2 - y^2}{1 - (-1)} = \frac{xy}{-m}$$

or $\quad x^2 + \frac{2xy}{m} - y^2 = 0 \quad \quad \ldots$(iii)

Given Eqs. (ii) and (iii) are identical

$\therefore \quad -n = \frac{1}{m}$

or $\quad mn + 1 = 0$

7. Since, $y = mx + 1$

or $\quad y - mx = 1 \quad \quad \ldots$(i)

and $\quad x^2 + y^2 = 1 \quad \quad \ldots$(ii)

For combined equation of OA and OB making homogeneous Eq. (ii) with help of Eq. (i), then

$$x^2 + y^2 = (y - mx)^2$$

or $$(m^2 - 1)x^2 + 0 \cdot y^2 - 2mxy = 0$$

$\because$ $$OA \perp OB$$

Then, coefficient of x^2 + coefficient of $y^2 = 0$

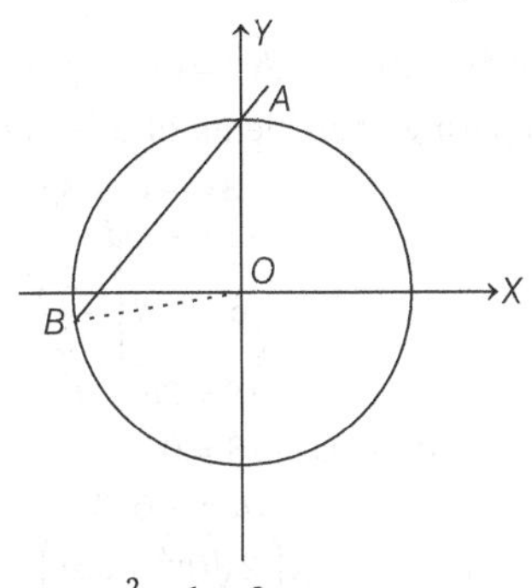

$\therefore$ $$m^2 - 1 = 0$$

or $$m = \pm 1$$

8. According to the given information in the question the angle between lines represented by the equation $Ax^2 + 2Hxy + By^2 = 0$ must be $\frac{\pi}{3}$.

So, $$\tan\frac{\pi}{3} = \frac{2\sqrt{H^2 - AB}}{|A + B|}$$

$\Rightarrow$ $3(A + B)^2 = 4(H^2 - AB)$

$\Rightarrow$ $$3A^2 + 10AB + 3B^2 = 4H^2$$

$\Rightarrow$ $$3A^2 + 9AB + AB + 3B^2 = 4H^2$$

$\Rightarrow$ $$3A(A + 3B) + B(A + 3B) = 4H^2$$

$\Rightarrow$ $$(A + 3B)(3A + B) = 4H^2$$

9. $ax^2 + 2hxy + by^2 + 2gx = 0$ ...(i)

$a'x^2 + 2h'xy + b'y^2 + 2g'x = 0$ ...(ii)

Multiplying Eq. (i) by g' and Eq. (ii) by g and then subtracting

$(ag' - a'g)x^2 + 2xy(hg' - h'g) + y^2(bg' - b'g) = 0$

For perpendicular, coefficient of x^2 + coefficient of $y^2 = 0$

$\therefore$ $$ag' - a'g + bg' - b'g = 0$$

$\Rightarrow$ $$(a + b)g' = (a' + b')g$$

10. Since, $x^2 + 2\sqrt{2}\,xy + 2y^2 + 4x + 4\sqrt{2}y + 1 = 0$

or $(x + y\sqrt{2})^2 + 4x + 4\sqrt{2}y + 1 = 0$

Let $$x + y\sqrt{2} = t$$

$\therefore$ $$t^2 + 4t + 1 = 0$$

$\therefore$ $$t = \frac{-4 \pm \sqrt{16 - 4}}{2}$$

$$t = -2 \pm \sqrt{3}$$

or $$x + y\sqrt{2} = -2 \pm \sqrt{3}$$

Lines are $$x + y\sqrt{2} + 2 + \sqrt{3} = 0$$

and $$x + y\sqrt{2} + 2 - \sqrt{3} = 0$$

$\therefore$ Distance between these lines $= \frac{|2\sqrt{3}|}{\sqrt{1 + 2}} = 2$

11.

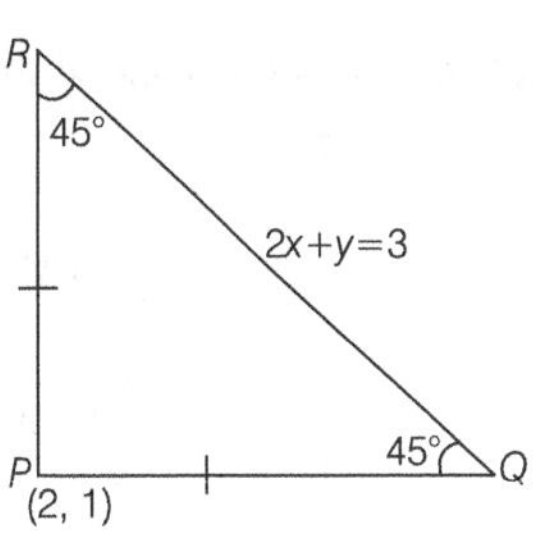

Let m be the slope of PQ. Then,

$$\tan 45° = \left|\frac{m - (-2)}{1 + m(-2)}\right| = \left|\frac{m + 2}{1 - 2m}\right|$$

$\therefore$ $$m + 2 = 1 - 2m$$

or $$-1 + 2m = m + 2$$

$\therefore$ $$m = -\frac{1}{3} \text{ or } m = 3$$

Hence, the equation of PQ is

$$y - 1 = -\frac{1}{3}(x - 2) \text{ or } x + 3y - 5 = 0$$

and the equation of PR is

$$y - 1 = 3(x - 2) \text{ or } 3x - y - 5 = 0$$

Hence, the combined equation of PQ and PR is

$$(x + 3y - 5)(3x - y - 5) = 0$$

or $3x^2 - 3y^2 + 8xy - 20x - 10y + 25 = 0$

12. Here, $x^2(\sec^2\theta - \sin^2\theta) - 2xy\tan\theta + y^2\sin^2\theta = 0$

Comparing with $ax^2 + 2hxy + by^2 = 0$

$\therefore$ $$a = \sec^2\theta - \sin^2\theta,\ b = \sin^2\theta$$

$$h = -\tan\theta$$

If slopes of lines are m_1 and m_2, then

$$m_1 + m_2 = \frac{2\tan\theta}{\sin^2\theta} = \frac{2}{\sin\theta\cos\theta}$$

and $$m_1 m_2 = \frac{\sec^2\theta - \sin^2\theta}{\sin^2\theta} = \frac{1}{\sin^2\theta\cos^2\theta} - 1$$

$\therefore$ $$m_1 - m_2 = \sqrt{(m_1 + m_2)^2 - 4m_1m_2}$$

$$= \sqrt{\left(\frac{4}{\sin^2\theta\cos^2\theta} - \frac{4}{\sin^2\theta\cos^2\theta} + 4\right)} = 2$$

13. We have, $\alpha x^2 + 4\gamma xy + \beta y^2 + 4p(x + y + 1) = 0$

$\therefore$ $$\Delta = 0$$

$$\alpha \cdot \beta \cdot 4p + 2 \cdot 2p \cdot 2p \cdot 2\gamma - \alpha(2p)^2 - \beta(2p)^2 - 4p(2\gamma)^2 = 0$$

$\Rightarrow$ $4p\alpha\beta + 16p^2\gamma - 4\alpha p^2 - 4\beta p^2 - 16p\gamma^2 = 0$

$\Rightarrow$ $$16p\gamma^2 - 16p^2\gamma + 4p(\alpha p + \beta p - \alpha\beta) = 0$$

$\because$ $$\gamma \text{ is real}$$

$\therefore$ $$B^2 - 4AC \geq 0$$

$\Rightarrow$ $(-16p^2)^2 - 4 \cdot 16p \cdot 4p(\alpha p + \beta p - \alpha\beta) \geq 0$

$$p^2 - (\alpha + \beta)p + \alpha\beta \geq 0 \rightarrow (p - \alpha)(p - \beta) \geq 0$$

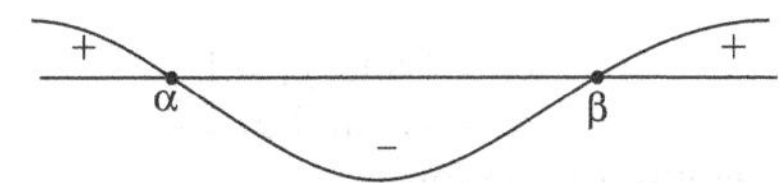

Hence, $$p \leq \alpha \text{ and } p \geq \beta$$

14. Replacing y by $-y$, then image of the pair of lines

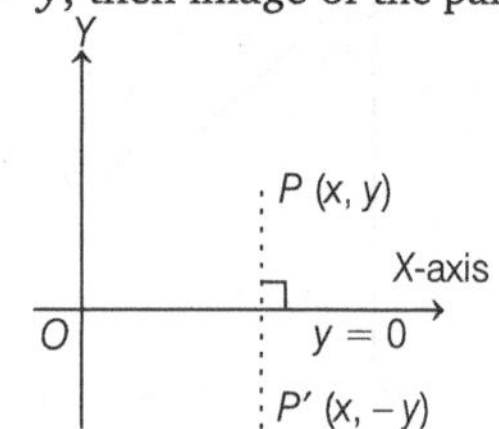

$ax^2 + 2hxy + by^2 = 0$ is

$$ax^2 + 2hx(-y) + by^2 = 0$$

or $$ax^2 - 2hxy + by^2 = 0$$

15. For parallel lines $h^2 = ab$

$\therefore$ $$ax^2 + 2hxy + by^2 + 2gx + 2fy + c$$
$$= (\sqrt{a}\,x + \sqrt{b}\,y)^2 + 2gx + 2fy + c$$

Let lines are $\sqrt{a}\,x + \sqrt{b}\,y + \alpha = 0$

and $\sqrt{a}\,x + \sqrt{b}\,y + \beta = 0$

$$(ax^2 + 2hxy + by^2 + 2gx + 2fy + c)$$
$$= (\sqrt{a}x + \sqrt{b}y + \alpha)(\sqrt{a}x + \sqrt{b}y + \beta)$$

On comparing, we get $\alpha + \beta = \dfrac{2g}{\sqrt{a}}$, $\alpha\beta = c$

$\therefore$ Distance between lines $= \left|\dfrac{\alpha - \beta}{\sqrt{(a+b)}}\right|$

$$= \left|\frac{\sqrt{(\alpha+\beta)^2 - 4\alpha\beta}}{\sqrt{a+b}}\right|$$

$$= \left|\frac{\sqrt{\dfrac{4g^2}{a} - 4c}}{\sqrt{a+b}}\right| = \frac{2\sqrt{(g^2 - ac)}}{\sqrt{a}\,\sqrt{(a+b)}}$$

$$= \frac{2\sqrt{(g^2 - ac)}}{\sqrt{(a^2 + h^2)}} = 2\sqrt{\left(\frac{g^2 - ac}{h^2 + a^2}\right)} \quad [\because ab = h^2]$$

16. The bisectors of the angles between the lines in new position are same as the bisectors of the angles between their old positions. Therefore, the required equation is

$$\frac{x^2 - y^2}{1 - (-1)} = \frac{xy}{-p}$$

$\Rightarrow$ $$px^2 + 2xy - py^2 = 0$$

17. Equation of given pair of straight lines is

$8x^2 - 14xy + 3y^2 + 10x + 10y - 25 = 0$

$\Rightarrow$ $$(4x - y - 5)(2x - 3y + 5) = 0$$

Now point of intersection of lines

$$4x - y - 5 = 0$$
$$2x - 3y + 5 = 0 \text{ is } (2, 3)$$

So equation of $S = 0$ is

$$(4x - y + c_1)(2x - 3y + c_2) = 0$$

and $S = 0$ passes through a point $P(x_1, y_1)$ and $(2, 3)$ is $(3, 2)$.

So, $x_1 = 4$ and $y_1 = 1$

$\therefore$ $c_1 = -15$ and $c_2 = -5$

So, required equation is

$$(4x - y - 15)(2x - 3y - 5) = 0$$

$\Rightarrow 8x^2 - 14xy + 3y^2 - 50x + 50y + 75 = 0$

18. Homogeneous equation is $x^2 + y^2 = 4\left(\dfrac{y - 3x}{c}\right)^2$

$\Rightarrow$ $$c^2x^2 + c^2y^2 = 4(y^2 + 9x^2 - 6xy)$$

These are perpendicular.

$\therefore$ $$c^2 - 36 + c^2 - 4 = 0$$

$\therefore$ $$c^2 = 20$$

19. $3x^3 + 3x^2y - 3xy^2 + dy^3 = (x^2 + pxy - y^2)(3x - dy)$

On comparing coefficients of x^2y, xy^2

$$3 = -d + 3p$$

and $$-3 = -pd - 3$$

or $$-6 = -d\left\{\frac{3+d}{3}\right\}$$

$\Rightarrow$ $$18 = 3d + d^2$$

$$d^2 + 3d - 18 = 0$$

$\therefore$ $$d = -6, 3$$

20. Here, $\tan^{-1} m = \tan^{-1}\left(\dfrac{2\sqrt{(h^2 - ab)}}{|a+b|}\right)$

$$= \tan^{-1}\left(\frac{2\sqrt{\left(\dfrac{25}{4} - 6\right)}}{|2+3|}\right)$$

$$= \tan^{-1}\left(\frac{1}{5}\right)$$

Hence, $m = \dfrac{1}{5}$

21. $ax^2 + 2hxy + by^2 = 0$ ...(i)

If lines represented by Eq. (i) are $y = m_1x$ and $y = m_2x$

Then, $m_1 + m_2 = -\dfrac{2h}{b}$ and $m_1m_2 = \dfrac{a}{b}$

Hence, pair of lines in new position is

$$(x + m_1y)(x + m_2y) = 0$$

$\Rightarrow$ $$x^2 + (m_1 + m_2)xy + m_1m_2y^2 = 0$$

$\Rightarrow$ $$x^2 - \frac{2h}{b}xy + \frac{a}{b}y^2 = 0$$

or $$bx^2 - 2hxy + ay^2 = 0$$

22. $ax^2 + 2hxy + by^2 = a(y - m_1x)(y - m_2x)$

$\therefore m_1 + m_2 = -\dfrac{2h}{b}$, $m_1m_2 = \dfrac{a}{b}$

Product of perpendiculars from (α, β)

$$= \frac{|\beta - m_1\alpha|}{\sqrt{(1 + m_1^2)}} \times \frac{|\beta - m_2\alpha|}{\sqrt{(1 + m_2^2)}}$$

$$= \frac{|\beta^2 - (m_1 + m_2)\alpha\beta + m_1m_2\alpha^2|}{\sqrt{\{1 + (m_1 + m_2)^2 - 2m_1m_2 + m_1^2m_2^2\}}}$$

$$= \frac{\left|\beta^2 + \dfrac{2h}{b}\alpha\beta + \dfrac{a}{b}\alpha^2\right|}{\sqrt{\left\{1 + \dfrac{4h^2}{b^2} - \dfrac{2a}{b} + \dfrac{a^2}{b^2}\right\}}}$$

$$= \frac{|a\alpha^2 + 2h\alpha\beta + b\beta^2|}{\sqrt{\{(a-b)^2 + 4h^2\}}}$$

23. $3x^2 - 7xy - 2y^2 = 0$

$(x - 2y)(3x + y) = 0$

Pair of perpendicular lines through (1, 1) is

$$(2x + y)(x - 3y) + \lambda = 0$$

or $\quad (2 + 1)(1 - 3) + \lambda = 0$

$\therefore \quad \lambda = 6$

$\therefore \quad (2x + y)(x - 3y) + 6 = 0$

$$2x^2 - 5xy - 3y^2 + 6 = 0$$

24. According to question, equation of lines

$$(y - 1) - \tan\theta(x - 1) = 0$$

and $\quad (y - 1) - \cot\theta(x - 1) = 0$

So, combined equation is

$$(y-1)^2 - (x-1)(y-1)[\tan\theta + \cot\theta] + (x-1)^2 = 0$$

$$\Rightarrow x^2 - \frac{2}{\sin 2\theta}xy + y^2 + \left(\frac{2}{\sin 2\theta} - 2\right)(x + y - 1) = 0$$

On comparing with the equation

$$x^2 - (a + 2)xy + y^2 + a(x + y - 1) = 0$$

$$\Rightarrow \quad a + 2 = \frac{2}{\sin 2\theta}$$

$$\Rightarrow \quad \sin 2\theta = \frac{2}{a + 2}$$

$$\Rightarrow \quad \theta = \frac{1}{2}\sin^{-1}\left(\frac{2}{a + 2}\right)$$

25. Replacing x by $x\cos\theta - y\sin\theta$ and y by $x\sin\theta + y\cos\theta$ in the given second degree equation and then put coefficient of $xy = 0$

we get, $\quad \tan 2\theta = \dfrac{2h}{a - b}$

26. Let $P(x, y)$ and $A(2, 0)$, $B(-2, 0)$

If $PA + PB = AB$,

then A, P, B are collinear

$\therefore$ Line segment

$y = 0, \ -2 < x < 2$

B (– 2, 0) P (x, y) A (2, 0)

Objective Questions Type II [One or more than one correct answer(s)]

1. Coefficient of x^2 + coefficient of $y^2 = 0 \Rightarrow 1 + \alpha = 0$

$\therefore \quad \alpha = -1$

Also, $\quad \Delta = 0$

$$1 \cdot \alpha \cdot (-a^2) + 0 - 1 \cdot \beta^2 - 0 - 0 = 0$$

$\therefore \quad \beta^2 = -\alpha a^2 = a^2$

$\therefore \quad \beta = \pm a$

Hence, $\quad \alpha = -1, \beta = \pm a$

2. Since, lines represented by

$$6x^2 - 5xy - 6y^2 = 0$$

and $6x^2 - 5xy - 6y^2 + x + 5y - 1 = 0$ are parallel.

Since, quadratic terms in both equations are same.

Hence, opposite sides are parallel.

But lines in both pairs are perpendiculars to each other. Hence, it is a square or rectangle.

3. Equation of bisectors

$$\Rightarrow \quad \frac{x^2 - y^2}{1 - (-1)} = \frac{xy}{2}$$

Put $\quad y = mx$

Then, $\quad x^2 - m^2x^2 = x\,mx \Rightarrow m^2 + m - 1 = 0$

$$\therefore \quad m = \frac{-1 \pm \sqrt{5}}{2}$$

4. $\quad y^2 + xy - 12x^2 = 0 \quad \ldots(i)$

$$(y + 4x)(y - 3x) = 0$$

$$\therefore \quad \frac{y}{x} = -4, 3$$

$$ax^2 + 2hxy + by^2 = 0$$

$$\Rightarrow \quad a + 2h\left(\frac{y}{x}\right) + b\left(\frac{y}{x}\right)^2 = 0 \quad \ldots(ii)$$

The two pairs will have a line common, if – 4 or 3 is a root of Eq. (ii).

Then, $a - 8h + 16b = 0$ or $a + 6h + 9b = 0$

5. The point $(-2, a)$ lies on the line $x = -2$ and $(b, 1)$ lies on the line $y = 1$, $x = -2$ cut the line

$y = -x$, $y = x$ and $2x + 3y - 6 = 0$, then $y = 2$, $y = -2$ and $y = 10/3$

$$\therefore \quad 2 < a < \frac{10}{3} \text{ and } -2 < a < \frac{10}{3}$$

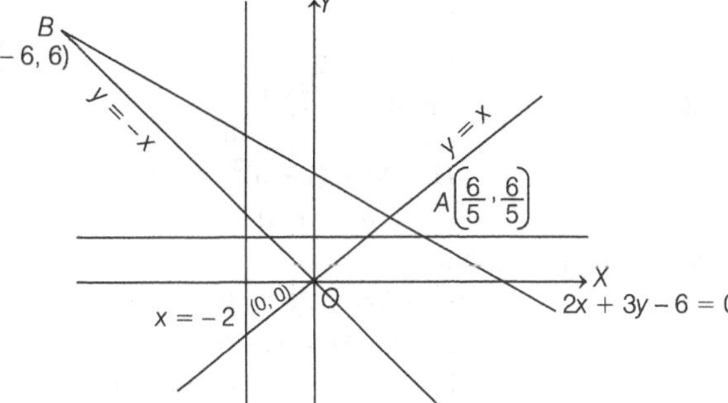

Similarly, $y = 1$ cut $y = -x$ and $y = x$, then $x = -1$ and $x = 1$

$\therefore \quad -1 < b < 1$

6. Here, $m_1 = \tan\alpha$, $m_2 = \tan\beta$

(a) $\tan\alpha + \tan\beta = m_1 + m_2 = -\dfrac{2h}{b} = \dfrac{2\tan\theta}{\sin^2\theta}$

$$= \frac{2}{\sin\theta\cos\theta} = 4\operatorname{cosec} 2\theta$$

(b) $\tan\alpha\tan\beta = m_1m_2 = \dfrac{\tan^2\theta + \cos^2\theta}{\sin^2\theta} = \sec^2\theta + \cot^2\theta$

(c) $\tan\alpha - \tan\beta = m_1 - m_2 = \dfrac{2\sqrt{(h^2 - ab)}}{b}$

$$= \frac{2\sqrt{\tan^2\theta - \sin^2\theta(\tan^2\theta + \cos^2\theta)}}{\sin^2\theta}$$

$$= \frac{2\sin\theta\sqrt{(\sec^2\theta - \tan^2\theta - \cos^2\theta)}}{\sin^2\theta}$$

$$= \frac{2\sin\theta\sqrt{(1-\cos^2\theta)}}{\sin^2\theta} = \frac{2\sin^2\theta}{\sin^2\theta} = 2$$

(d) $\because \frac{m_1 + m_2}{m_1 - m_2} = \frac{4\ \text{cosec}\ 2\theta}{2} = \frac{2}{\sin 2\theta}$

or $\frac{(m_1+m_2)+(m_1-m_2)}{(m_1+m_2)-(m_1-m_2)} = \frac{2+\sin 2\theta}{2-\sin 2\theta}$

$\Rightarrow \frac{m_1}{m_2} = \frac{2+\sin 2\theta}{2-\sin 2\theta}$

or $\frac{\tan\alpha}{\tan\beta} = \frac{2+\sin 2\theta}{2-\sin 2\theta}$

7. Let the equation of the lines given by $x^2 + 2hxy + y^2 = 0$ be $y = m_1 x$ and $y = m_2 x$. Since, these makes an angle α with $y + x = 0$ whose slope is -1, we have

$$\frac{m_1+1}{1-m_1} = \tan\alpha = \frac{-1-m_2}{1-m_2}$$

or $m_1 + m_2 = \frac{(\tan\alpha - 1)^2 + (\tan\alpha + 1)^2}{\tan^2\alpha - 1}$

$$= \frac{-2\sec^2\alpha \times \cos^2\alpha}{\cos 2\alpha}$$

$\therefore$ $-2\sec 2\alpha = -2h \Rightarrow \sec 2\alpha = h \Rightarrow \cos 2\alpha = \frac{1}{h}$

or $2\cos^2\alpha - 1 = \frac{1}{h}$

$\Rightarrow$ $\cos\alpha = \sqrt{\frac{1+h}{2h}}$ and $\cot\alpha = \sqrt{\frac{h+1}{h-1}}$

8. $\tan 45° = \frac{2\sqrt{\left(\frac{1}{4} - a\right)}}{|1+a|} = 1$

$\Rightarrow$ $(1-4a) = (1+a)^2 \Rightarrow a^2 + 6a = 0$

$\therefore$ $a = -6, 0$

9. Given pair of lines

$$2x^2 - 5xy + 2y^2 = 0$$

i.e., $(x-2y)(2x-y) = 0$

$\therefore$ $x - 2y = 0$ and $2x - y = 0$

and the diagonal $5x + 2y = 1$ does not pass through origin hence it is AC.

Solving, $x - 2y = 0$ and $5x + 2y = 1$, we get

$$A = \left(\frac{1}{6}, \frac{1}{12}\right)$$

and solving $2x - y = 0$

and $5x + 2y = 1$, we get $C \equiv \left(\frac{1}{9}, \frac{2}{9}\right)$

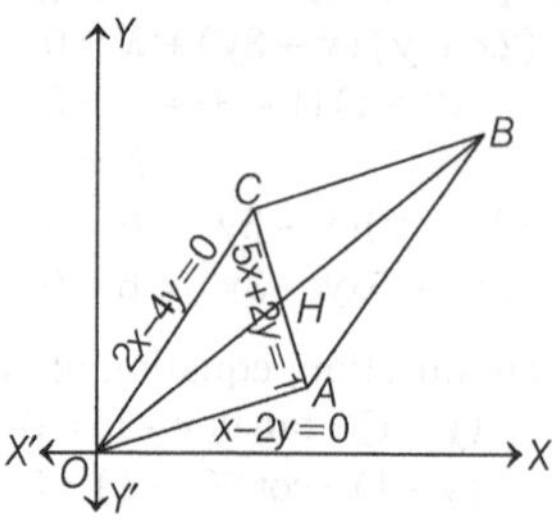

Since, diagonals of parallelogram bisect each other, if bisect at then

$$H \equiv \left\{\frac{1}{2}\left(\frac{1}{6} + \frac{1}{9}\right), \frac{1}{2}\left(\frac{1}{12} + \frac{2}{9}\right)\right\} \text{ i.e. } H = \left(\frac{5}{36}, \frac{11}{72}\right)$$

Hence, equation of other diagonal which passes through O and H is

$$y - 0 = \frac{\frac{11}{72} - 0}{\frac{5}{36} - 0}(x - 0) \Rightarrow y = \frac{11}{10}x$$

or $11x - 10y = 0$

Now, area of parallelogram $= 2 \times$ Area of ΔOAC

$$= 2 \times \frac{1}{2}\left|\frac{1}{6} \times \frac{2}{9} - \frac{1}{9} \times \frac{1}{12}\right| = \frac{1}{36} \text{ sq units}$$

10. For Y-axis, $x = 0$

$by^2 + 2fy + c = 0$

$\because$ Point of intersection of lines is

$$\left(\sqrt{\left(\frac{f^2 - bc}{h^2 - ab}\right)}, \sqrt{\left(\frac{g^2 - ac}{h^2 - ab}\right)}\right)$$

or $\left(\frac{bg - hf}{h^2 - ab}, \frac{af - gh}{h^2 - ab}\right) \Rightarrow \sqrt{\left(\frac{f^2 - bc}{h^2 - ab}\right)} = 0$

or $\left(\frac{bg - hf}{h^2 - ab}\right) = 0$

$\therefore$ $f^2 = bc$ and $bg = hf$

and $abc + 2fgh - af^2 - bg^2 - ch^2 = 0$

$\Rightarrow$ $a(bc - f^2) + 2fgh - bg^2 - ch^2 = 0$

or $0 + 2fgh - bg^2 - ch^2 = 0$

$\Rightarrow$ $bg^2 + ch^2 = 2fgh$

Linked Comprehension Type

PASSAGE 1

Given pair of lines $2x^2 - 5xy + 2y^2 = 0$

$\Rightarrow$ $(x - 2y)(2x - y) = 0$

$\therefore$ $OA : x - 2y = 0$

and $OC : 2x - y = 0$

and the diagonal $5x + 2y = 1$ does not pass through origin hence it is AC.

Solving $x - 2y = 0$ and $5x + 2y = 1$, we get

$$A \equiv \left(\frac{1}{6}, \frac{1}{12}\right)$$

and solving $2x - y = 0$ and $5x + 2y = 1$, we get

$$C \equiv \left(\frac{1}{9}, \frac{2}{9}\right)$$

Also, mid point of OB = mid point of AC

$$\left(\frac{1}{2}(\alpha + 0), \frac{1}{2}(\beta + 0)\right) = \left(\frac{1}{2}\left(\frac{1}{6} + \frac{1}{9}\right), \frac{1}{2}\left(\frac{1}{12} + \frac{2}{9}\right)\right)$$

$$\Rightarrow \left(\frac{\alpha}{2}, \frac{\beta}{2}\right) = \left(\frac{5}{36}, \frac{11}{72}\right) = \text{coordinate of } H$$

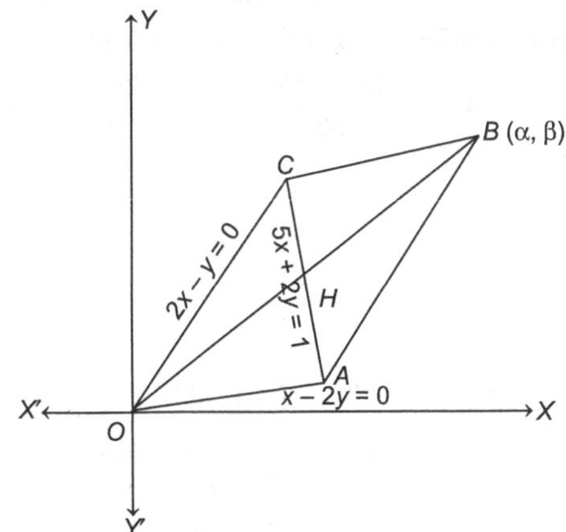

$\therefore \quad H \equiv \left(\frac{5}{36}, \frac{11}{72}\right)$ and $B = (\alpha, \beta) = \left(\frac{5}{18}, \frac{11}{36}\right)$

1. Equation of other diagonal = Equation of OB is

$$y - 0 = \frac{\frac{11}{36}}{\frac{5}{18}}(x - 0) \Rightarrow y = \frac{11x}{10}$$

$\Rightarrow \quad 11x - 10y = 0$

2. Centroid of parallelogram

$= $ point of intersection of diagonals

$= H = \left(\frac{5}{36}, \frac{11}{72}\right)$

3. Area of parallelogram $= 2 \times$ area of AOC

$$= 2 \times \frac{1}{2}\left|\frac{1}{6} \times \frac{2}{9} - \frac{1}{9} \times \frac{1}{12}\right| = \frac{1}{36} \text{ sq unit}$$

4. $$OA = \sqrt{\left(\frac{1}{36} + \frac{1}{144}\right)} = \frac{\sqrt{5}}{12}$$

and $$OC = \sqrt{\frac{1}{81} + \frac{4}{81}} = \frac{\sqrt{5}}{9}$$

$\therefore \quad \frac{OC}{OA} = \frac{\sqrt{5}/9}{\sqrt{5}/12} = \frac{4}{3}$

$\Rightarrow \quad OC : OA = 4 : 3$

5. $$AC = \sqrt{\left\{\left(\frac{1}{6} - \frac{1}{9}\right)^2 + \left(\frac{1}{12} - \frac{2}{9}\right)^2\right\}} = \frac{\sqrt{29}}{36}$$

and $$OB = \sqrt{\left(\frac{5}{18} - 0\right)^2 + \left(\frac{11}{36} - 0\right)^2} = \frac{\sqrt{221}}{36}$$

$\therefore \quad OB : OA = \sqrt{221} : \sqrt{29}$

PASSAGE 2

1. $\because f_2(x, y) = 0$ is the pair of bisectors of $f_1(x, y) = 0$

i.e. $$\frac{x^2 - y^2}{a - b} = \frac{xy}{h}$$

$\Rightarrow \quad hx^2 - (a - b)xy - hy^2 = 0$

2. Now, $f_3(x, y) = 0$ is the pair of bisectors of $f_2(x, y) = 0$

i.e. $$\frac{x^2 - y^2}{h - (-h)} = \frac{xy}{-\frac{(a-b)}{2}}$$

$$\Rightarrow \quad (x^2 - y^2) = -\frac{4hxy}{(a - b)}$$

or $(a - b)x^2 + 4hxy - (a - b)y^2 = 0$

3. $f_2(x, y) = 0$ is the pair of perpendicular lines of $f_1(x, y) = 0$

i.e. Interchange the coefficients of x^2 and y^2 and sign change of xy, then

$f_2(x, y) = 0$ is $bx^2 - 2hxy + ay^2 = 0$

4. $f_3(x, y) = 0$ is the pair of perpendicular lines of $f_2(x, y) = 0$

$\therefore \quad f_3(x, y) = 0$

is $\quad ax^2 + 2hxy + by^2 = 0$

5. $\because f_3(x, y) = f_1(x, y)$

$\Rightarrow \quad f_5(x, y) = f_3(x, y) = f_1(x, y) = 0$

$\therefore \quad f_{n+2}(x, y) = f_n(x, y)$

i.e. $\quad f_n(x, y) = 0$

Numerical Value Type

1. $\because y^2 - 9xy + 18x^2 = 0$

$\Rightarrow \quad (y - 6x)(y - 3x) = 0$

$\therefore \quad y = 6x, y = 3x$

Vertices of the triangle formed by the lines $y = 6x$, $y = 3x$ and $y = 9$ are $A(0, 0)$, $B\left(\frac{3}{2}, 9\right)$ and $C(3, 9)$

$$\therefore \quad \text{Required area } \lambda = \frac{1}{2}\left|\begin{vmatrix} 0 & 0 & 1 \\ 3/2 & 9 & 1 \\ 3 & 9 & 1 \end{vmatrix}\right|$$

$$= \frac{1}{2}\left|-\frac{27}{2}\right| = \frac{27}{4}$$

$\therefore \quad 4\lambda = 27$

2. $\because$ Coefficient of x^2 + coefficient of y^2

$= \sin^2\theta + \cos^2\theta - 1 = 1 - 1 = 0$

$\therefore \quad \phi = \pi/2 \Rightarrow \sin\phi = 1$

Then, $\sin^3\phi + 32\sin\phi + 9 = (1)^3 + 32(1) + 9 = 42$

3. $\because ax^3 - 9x^2y - xy^2 + 4y^3 = (x^2 + pxy - y^2)(ax - 4y)$

on comparing the similar terms, then

$-9 = ap - 4 \quad \text{...(i)}$

and $\quad -1 = 4p - a \quad \text{...(ii)}$

From Eqs. (i) and (ii)

$$\frac{-5}{a} = \frac{a - 1}{4} \Rightarrow -20 = a^2 - a$$

$\Rightarrow \quad a^2 - a + 20 = 0$ or $a - a^2 = 20$

$$\therefore \quad \left|\frac{a^2 - a}{8}\right| = 2.5$$

4. Angle between lines $\theta = \tan^{-1}\left\{\dfrac{2\sqrt{\left(\frac{25}{4}-6\right)}}{2+3}\right\} = \tan^{-1}\left\{\dfrac{1}{5}\right\}$

$\therefore \qquad m = \dfrac{1}{5} = 0.2$

5. Making $(x-3)^2 + (y-4)^2 = c^2$ homogeneous with the help of $4x + 3y = 24$, then

$$x^2 + y^2 - (6x + 8y)\left(\frac{4x+3y}{24}\right) + (25 - c^2)\left(\frac{4x+3y}{24}\right)^2 = 0$$

Now, coefficient of x^2 + coefficient of $y^2 = 0$

$$\Rightarrow \quad 1 - 1 + \frac{(25-c^2)}{36} + 1 - 1 + \frac{(25-c^2)}{64} = 0$$

or $\qquad 25 - c^2 = 0$

$\therefore \qquad c^2 = 25$

Single Integer Type

1. We have,

$$4x^2 + 6xy + ky^2 = 0 \qquad \ldots\text{(i)}$$

and $\qquad 3x^2 - 5xy + 2y^2 = 0 \qquad \ldots\text{(ii)}$

Let $y = mx$ be one of the lines represented by Eq. (i). Then, $y = mx$ satisfies Eq. (i)

$\therefore \quad 4x^2 + 6x \cdot mx + km^2x^2 = 0$

$\Rightarrow \qquad km^2 + 6m + 4 = 0 \qquad \ldots\text{(iii)}$

The equation of a line passing through the origin and perpendicular to $y = mx$, is

$$y - 0 = -\frac{1}{m}(x - 0)$$

$\Rightarrow \qquad my + x = 0$

Since, one of the lines given by Eq. (i) is perpendicular to one of the lines given by $3x^2 - 5xy + 2y^2 = 0$.

Therefore one of the lines given by (ii) is $my + x = 0$

Hence, $my + x = 0$ i.e., $x = -my$ satisfies Eq. (ii).

$\therefore \qquad 3(-my)^2 - 5y(-my) + 2y^2 = 0$

$\Rightarrow \qquad 3m^2y^2 + 5my^2 + 2y^2 = 0$

$\Rightarrow \qquad 3m^2 + 5m + 2 = 0 \qquad \ldots\text{(iv)}$

Using Eqs. (iii) and (iv), we get

$$(4 \times 3 - k \times 2)^2 + 4\left(3 \times 3 + \left(\frac{-5}{2}\right) \times k\right)\left(3 \times 2 + \left(\frac{-5}{2}\right) \times 4\right) = 0$$

$$\Rightarrow \quad (12 - 2k)^2 + 4\left(9 - \frac{5k}{2}\right)(6 - 10) = 0$$

$$\Rightarrow \quad (12 - 2k)^2 - 16\left(9 - \frac{5k}{2}\right) = 0$$

$\Rightarrow \qquad 144 + 4k^2 - 48k - 144 + 40k = 0$

$\Rightarrow \qquad 4k^2 - 8k = 0$

$\Rightarrow \qquad 4k^2 = 8k$

$\Rightarrow \qquad k = 2$

2. Let m and m^2 be the slopes of the lines represented by $ax^2 + 2hxy + by^2 = 0$. Then,

$$m + m^2 = -\frac{2h}{b}$$

and $\qquad mm^2 = \dfrac{a}{b}$ or $m^3 = \dfrac{a}{b}$

Now, $(m + m^2)^3 = \left(-\dfrac{2h}{b}\right)^3$

$\therefore \qquad m^3 + m^6 + 3mm^2(m + m^2) = -\dfrac{8h^3}{b^3}$

$\therefore \qquad \dfrac{a}{b^2}(a + b) + \dfrac{8h^3}{b^3} = \dfrac{6ah}{b^2}$

$\therefore \qquad \dfrac{a+b}{h} + \dfrac{8h^2}{ab} = 6$

These are the set of parallel lines and the distance between parallel lines are equal. So, the figure is a rhombus.

3. Equation of curve is

$$x^2 + 2xy + 3y^2 + 4x + 8y - 11 = 0$$

and $\qquad y = 3x + 2 \qquad \ldots\text{(i)}$

$\Rightarrow \qquad \dfrac{y - 3x}{2} = 1 \qquad \ldots\text{(ii)}$

Making (i) homogeneous with the help of (ii), then

$$x^2 + 2xy + 3y^2 + 4x\left(\frac{y-3x}{2}\right) + 8y\left(\frac{y-3x}{2}\right) - 11\left(\frac{y-3x}{2}\right)^2 = 0$$

$$x^2 + 2xy + 3y^2 + 2xy - 6x^2 + 4y^2 - 12xy - \frac{11}{4}(y - 3x)^2 = 0$$

$$\Rightarrow -5x^2 - 8xy + 7y^2 - \frac{11}{4}(y^2 - 6xy + 9x^2) = 0$$

$$\Rightarrow -20x^2 - 32xy + 28y^2 - 11y^2 + 66xy - 99x^2 = 0$$

$\Rightarrow \qquad 119x^2 - 34xy - 17y^2 = 0$

$\qquad 7x^2 - 2xy - y^2 = 0 \qquad \ldots\text{(iii)}$

This is the equation of lines joining the origin to the points of intersection of (i) and (ii) Comparing (iii) with $ax^2 + 2hxy + by^2 = 0$

$$a = 7, h = -1, b = -1$$

If θ is the acute angle between pair of lines of (iii), then

$$\tan\theta = \frac{2\sqrt{h^2 - ab}}{|a + b|} = \frac{2\sqrt{(1+7)}}{|7 - 1|}$$

$$= \frac{2\sqrt{8}}{6} = \frac{4\sqrt{2}}{6} = \frac{2\sqrt{2}}{3}$$

$$\theta = \tan^{-1}\left(\frac{2\sqrt{2}}{3}\right)$$

$\therefore \qquad k = 3$

4. Given equation of pair of line is

$$lx^2 + 3xy - 2y^2 - 5x + 5y + k = 0 \qquad \ldots(i)$$

$\therefore a = l \mid 2h = 3 \mid b = -2 \mid 2g = -5 \mid 2f = 5 \mid c = k$

If Eq. (i) represents, pair of perpendicular lines

$$a + b = 0$$
$$l + (-2) = 0$$
$$l = 2 \Rightarrow a = 2$$
$$\Delta = 0$$
$$abc + 2fhg - af^2 - bg^2 - ch^2 = 0$$
$$(2)(-2)(k) + 2\left(\frac{5}{2}\right)\left(\frac{-5}{2}\right)\left(\frac{3}{2}\right) - 1\left(\frac{5}{2}\right)^2 + 2\left(\frac{-5}{2}\right)^2 - k\left(\frac{3}{2}\right)^2 = 0$$
$$-4k - \frac{75}{4} - \frac{25}{2} + \frac{25}{2} - k\left(\frac{9}{4}\right) = 0$$
$$-4k - \frac{9k}{4} = \frac{75}{4}$$
$$\frac{-25k}{4} = \frac{75}{4}k = -3$$
$$\therefore \qquad |a + l| = 1$$

5. Shifting the origin to (1, 2), the equation of given pair of lines becomes

$$ax^2 + 2hxy + by^2 = 0$$

whose one angle bisector is $x + 2y = 0$ (given) and other is $2x - y = 0$.

$\therefore$ Combined equation of angle bisector is

$$(x + 2y)(2x - y) = 0$$

On comparing with $\frac{x^2 - y^2}{a - b} = \frac{xy}{h}$, we get

$$a - b = -3 \text{ and } h = 2$$
$$\therefore \qquad 4\left(\frac{b - a}{h}\right) = 6$$

Matrix-Match Type

1. (A) Let slopes are m, m^2

Then, $m + m^2 = 6$ and $m \cdot m^2 = a$

$$\Rightarrow \qquad m^2 + m - 6 = 0$$
$$\Rightarrow \qquad (m + 3)(m - 2) = 0$$
$$\therefore \qquad m = -3, 2$$
$$\Rightarrow \qquad a = m^2$$
$$= -27, 8 \text{(R, T)}$$

(B) $ax^3 - 9x^2y - xy^2 + 4y^3 = (x^2 + pxy - y^2)(ax - 4y)$

On comparing, we get

$$-9 = -4 + ap$$
$$\Rightarrow \qquad p = \frac{-5}{a}$$
$$\text{and} \qquad -1 = -4p - a$$
$$\Rightarrow \qquad p = \frac{1 - a}{4}$$
$$\Rightarrow \qquad \frac{-5}{a} = \frac{1 - a}{4}$$
$$\Rightarrow \qquad -20 = a - a^2$$
$$\Rightarrow \qquad a^2 - a - 20 = 0$$
$$\Rightarrow \qquad (a - 5)(a + 4) = 0$$
$$\therefore \qquad a = 5, -4 \text{ (P, S)}$$

(C) By hypothesis

$$x^4 + x^3y + cx^2y^2 - xy^3 + y^4$$
$$= (ax^2 + 2hxy + by^2)(hx^2 - 2axy - hy^2)$$
$$= ah(x^4 + y^4) + 2(h^2 - a^2)(x^3y - xy^3) - 6ahx^2y^2$$

Comparing the coefficients, then we get

$ah = 1,\ 2(h^2 - a^2) = 1,\ c = -6ah$

$$\therefore \qquad c = -6 \times 1 = -6 \text{ (Q)}$$

2. (A) $\because 12x^2 - 10xy + 2y^2 = 2(3x - y)(2x - y)$

$$= (3x - y)(4x - 2y)$$
$$\therefore \qquad 12x^2 - 10xy + 2y^2 + 11x - 5y + 2$$
$$= (3x - y + \lambda)(4x - 2y + \mu)$$

On comparing,

$$4\lambda + 3\mu = 11 \text{ and } -2\lambda - \mu = -5$$
$$\text{we get} \qquad \lambda = 2, \mu = 1$$
$$\therefore \qquad \lambda + \mu = 3,\ \lambda^2 + \mu^2 = 5 \text{ (R)}$$

(B) $6x^2 + 17xy + 12y^2 + 22x + 31y + 20$

$$= (2x + 3y + \lambda)(3x + 4y + \mu)$$

On comparing,

$$3\lambda + 2\mu = 22,\ 4\lambda + 3\mu = 31$$
$$\text{We get} \qquad \lambda = 4, \mu = 5$$
$$\therefore \qquad \lambda + \mu = 9,\ \lambda^2 + \mu^2 = 41 \text{ (Q, S)}$$

(C) $8x^2 + 8xy + 2y^2 + 26x + 13y + 15$

$$= (2x + y + \lambda)(4x + 2y + \mu)$$

On comparing,

$$2\lambda + \mu = 13,\ \lambda\mu = 15$$
$$\therefore \qquad 2\lambda - \mu = \sqrt{(2\lambda + \mu)^2 - 4 \cdot 2\lambda \cdot \mu}$$
$$= \sqrt{169 - 120} = 7$$
$$\text{We get} \qquad \lambda = 5, \mu = 3$$
$$\Rightarrow \qquad \lambda + \mu = 8$$
$$\lambda^2 + \mu^2 = 34 \text{ (P, T)}$$

23

Circle

Objective Questions Type I [Only one correct answer]

In each of the questions below, four choices are given of which only one is correct. You have to select the correct answer which is the most appropriate.

1. If the length of the chord of the circle, $x^2 + y^2 = r^2 (r > 0)$ along the line, $y - 2x = 3$ is r, then r^2 is equal to **[JEE Main 2020, 5 September]**

(a) $\frac{9}{5}$ (b) 12

(c) $\frac{24}{5}$ (d) $\frac{12}{5}$

2. Let PQ and RS be tangents at the extremities of the diameter PR of a circle of radius r. If PS and RQ intersect at a point X on the circumference of the circle, then $2r$ equals

(a) $\sqrt{PQ \cdot RS}$ (b) $\frac{PQ + RS}{2}$

(c) $\frac{2PQ \cdot RS}{PQ + RS}$ (d) $\sqrt{\frac{PQ^2 + RS^2}{2}}$

3. Let $0 < \alpha < \frac{\pi}{2}$ be a fixed angle. If $P = (\cos\theta, \sin\theta)$ and $Q = (\cos(\alpha - \theta), \sin(\alpha - \theta))$, then Q is obtained from P by

(a) clockwise rotation around origin through an angle α

(b) anti-clockwise rotation around origin through an angle α

(c) reflection in the line through origin with slope $\tan\alpha$

(d) reflection in the line through origin with slope $\tan\alpha/2$

4. If r_1 and r_2 are the radii of smallest and largest circles which passes through (5, 6) and touches the circle $(x - 2)^2 + y^2 = 4$, then $r_1 r_2$ is

(a) $\frac{4}{41}$ (b) $\frac{41}{4}$ (c) $\frac{5}{41}$ (d) $\frac{41}{6}$

5. If the tangent at the point P on the circle $x^2 + y^2 + 6x + 6y = 2$ meets the straight line $5x - 2y + 6 = 0$ at a point Q on the Y-axis, then the length of PQ is

(a) 4 (b) $2\sqrt{5}$ (c) 5 (d) $3\sqrt{5}$

6. Let AB be a chord of the circle $x^2 + y^2 = r^2$ subtanding a right angle at the centre, then the locus of the centroid of the ΔPAB as P moves on the circle is

(a) a parabola (b) a circle

(c) an ellipse (d) a pair of straight line

7. If the angle between the pair of tangents drawn to the circle $x^2 + y^2 - 2x + 4y + 3 = 0$ from $(6, -5)$ is θ, then $\tan\theta =$

(a) $\frac{5}{8}$ (b) $\frac{15}{8}$

(c) $\frac{8}{15}$ (d) $\frac{19}{8}$

8. If $\lambda x^2 + \mu y^2 + (\lambda + \mu - 4)xy - \lambda x - \mu y - 20 = 0$ represents a circle, the radius of the circle is

(a) $\sqrt{21}/2$ (b) $\sqrt{42}/2$

(c) $2\sqrt{21}$ (d) $\sqrt{22}$

9. A point P lies inside the circles $x^2 + y^2 - 4 = 0$ and $x^2 + y^2 - 8x + 7 = 0$. The point P starts moving under the conditions that its path encloses greatest possible area and it is at a fixed distance from any arbitrarily chosen fixed point in its region. The locus of P is

(a) $4x^2 + 4y^2 - 12x + 1 = 0$

(b) $4x^2 + 4y^2 + 12x - 1 = 0$

(c) $x^2 + y^2 - 3x - 2 = 0$

(d) $x^2 + y^2 - 3x + 2 = 0$

10. If (α, β) is a point on the circle whose centre is on the X-axis and which touches the line $x + y = 0$ at $(2, -2)$, then the greatest value of α is

(a) $4 - \sqrt{2}$ (b) 6

(c) $4 + 2\sqrt{2}$ (d) $4 + \sqrt{2}$

11. The set of values of 'c' so that the equations $y = |x| + c$ and $x^2 + y^2 - 8|x| - 9 = 0$ have no solution is

(a) $(-\infty, -3) \cup (3, \infty)$ (b) $(-3, 3)$

(c) $(-\infty, -5\sqrt{2}) \cup (5\sqrt{2}, \infty)$ (d) $(5\sqrt{2} - 4, \infty)$

12. If the circles $x^2 + y^2 - 16x - 20y + 164 = r^2$ and $(x-4)^2 + (y-7)^2 = 36$ intersect at two distinct points. then, **[JEE Main 2019, 9 January]**
(a) $0 < r < 1$ (b) $r > 11$
(c) $1 < r < 11$ (d) $r = 11$

13. $(-6, 0)$, $(0, 6)$ and $(-7, 7)$ are the vertices of ΔABC. The in circle of the triangle has the equation
(a) $x^2 + y^2 - 9x - 9y + 36 = 0$
(b) $x^2 + y^2 + 9x - 9y + 36 = 0$
(c) $x^2 + y^2 + 9x + 9y - 36 = 0$
(d) $x^2 + y^2 + 18x - 18y + 36 = 0$

14. Two rods of lengths a and b slide along the X-axis and Y-axis respectively in such a manner that their ends are concyclic. The locus of the centre of the circle passing through the end points is
(a) $4(x^2 + y^2) = a^2 + b^2$ (b) $x^2 + y^2 = a^2 + b^2$
(c) $4(x^2 - y^2) = a^2 - b^2$ (d) $x^2 - y^2 = a^2 - b^2$

15. The range of values of 'a' such that the angle θ between the pair of tangents drawn from $(a, 0)$ to the circle $x^2 + y^2 = 1$ satisfies $\frac{\pi}{2} < \theta < \pi$, is
(a) $(1, 2)$
(b) $(1, \sqrt{2})$
(c) $(-\sqrt{2}, -1)$
(d) $(-\sqrt{2}, -1) \cup (1, \sqrt{2})$

16. In ΔABC equation of side BC is $x - y = 0$ circumcentre and orthocentre of the triangle are $(2, 3)$ and $(5, 8)$ respectively. Equation of circumcircle of the triangle is
(a) $x^2 + y^2 - 4x + 6y - 27 = 0$
(b) $x^2 + y^2 - 4x - 6y - 27 = 0$
(c) $x^2 + y^2 + 4x + 6y - 27 = 0$
(d) $x^2 + y^2 + 4x - 6y - 27 = 0$

17. Let the tangents drawn from the origin to the circle, $x^2 + y^2 - 8x - 4y + 16 = 0$ touch it at the points A and B. The $(AB)^2$ is equal to **[JEE Main 2020, 7 January]**
(a) $\frac{56}{5}$ (b) $\frac{52}{5}$
(c) $\frac{64}{5}$ (d) $\frac{32}{5}$

18. An isosceles right angled triangle is inscribed in the circle $x^2 + y^2 = r^2$. If the coordinates of an end of the hypotenuse are (a, b), the coordinates of the vertex are
(a) $(-a, -b)$ (b) $(b, -a)$
(c) (b, a) (d) $(-b, -a)$

19. If the circle $x^2 + y^2 + 4x + 22y + c = 0$ bisects the circumference of the circle $x^2 + y^2 - 2x + 8y - d = 0$, then $c + d$ is equal to
(a) 60 (b) 50
(c) 40 (d) 56

20. If one of the diameters of the circle, given by the equation, $x^2 + y^2 - 4x + 6y - 12 = 0$, is a chord of a circle S. whose centre is at $(-3, 2)$, then the radius of S is **[JEE Main 2016]**
(a) $5\sqrt{2}$ (b) $5\sqrt{3}$
(c) 5 (d) 10

21. The locus of a point such that the tangents drawn from it to the circle $x^2 + y^2 - 6x - 8y = 0$ are perpendicular to each other is
(a) $x^2 + y^2 - 6x - 8y - 25 = 0$
(b) $x^2 + y^2 + 6x - 8y - 5 = 0$
(c) $x^2 + y^2 - 6x + 8y - 5 = 0$
(d) $x^2 + y^2 - 6x - 8y + 25 = 0$

22. Equation of chord of the circle $x^2 + y^2 - 3x - 4y - 4 = 0$, which passes through the origin such that origin divides it in the ratio 4 : 1, is
(a) $x = 0$ (b) $24x + 7y = 0$
(c) $7x + 24y = 0$ (d) $7x - 24y = 0$

23. Let $S_1 : x^2 + y^2 = 9$ and $S_2 : (x-2)^2 + y^2 = 1$. Then, the locus of center of a variable circle S which touches S_1 internally and S_2 externally always passes through the points **[JEE Main 2021, 18 March]**
(a) $(0, \pm\sqrt{3})$ (b) $\left(\frac{1}{2}, \pm\frac{\sqrt{5}}{2}\right)$
(c) $\left(2, \pm\frac{3}{2}\right)$ (d) $(1, \pm 2)$

24. If a line segment $AM = a$ moves in the plane XOY remaining parallel to OX so that the left end point A slides along the circle $x^2 + y^2 = a^2$, the locus of M is
(a) $x^2 + y^2 = 4a^2$
(b) $x^2 + y^2 = 2ax$
(c) $x^2 + y^2 = 2ay$
(d) $x^2 + y^2 - 2ax - 2ay = 0$

25. If $(2, 5)$ is an interior point of the circle $x^2 + y^2 - 8x - 12y + p = 0$ and the circle neither cuts nor touches any one of the axes of coordinates, then
(a) $p \in (36, 47)$ (b) $p \in (16, 47)$
(c) $p \in (16, 36)$ (d) None of these

26. If the lines $a_1x + b_1y + c_1 = 0$ and $a_2x + b_2y + c_2 = 0$ cut the coordinate axes in concyclic points, then
(a) $a_1b_1 = a_2b_2$ (b) $\frac{a_1}{a_2} = \frac{b_1}{b_2}$
(c) $a_1 + a_2 = b_1 + b_2$ (d) $a_1a_2 = b_1b_2$

27. AB is a diameter of a circle and C is any point on the circumference of the circle. Then
(a) the area of ΔABC is maximum when it is isosceles
(b) the area of ΔABC is minimum when it is isosceles
(c) the perimeter of ΔABC is maximum when it is isosceles
(d) None of the above

28. The four points of intersection of the lines $(2x - y + 1)(x - 2y + 3) = 0$ with the axes lie on a circle whose centre is at the point
(a) $(-7/4, 5/4)$ (b) $(3/4, 5/4)$
(c) $(9/4, 5/4)$ (d) $(0, 5/4)$

29. Origin is a limiting point of a coaxial system of which $x^2 + y^2 - 6x - 8y + 1 = 0$ is a member. The other limiting point is
(a) $(-2, -4)$ (b) $\left(\frac{3}{25}, \frac{4}{25}\right)$
(c) $\left(-\frac{3}{25}, -\frac{4}{25}\right)$ (d) $\left(\frac{4}{25}, \frac{3}{25}\right)$

30. The centres of a set of circles, each of radius 3, lie on the circle $x^2 + y^2 = 25$. The locus of any point in the set is
(a) $4 \le x^2 + y^2 \le 64$ (b) $x^2 + y^2 \le 25$
(c) $x^2 + y^2 \ge 25$ (d) $3 \le x^2 + y^2 \le 9$

31. If $A(-1, 3)$ and $B(5, 3)$ are points on a circle C and the chord AB subtends an angle $\frac{\pi}{4}$ at a point P on C, then the equation of such a circle C is
(a) $x^2 + y^2 - 4x + 12y + 22 = 0$
(b) $x^2 + y^2 - 4x - 12y + 22 = 0$
(c) $x^2 + y^2 - 4x - 12y - 22 = 0$
(d) $3x^2 + 3y^2 - 12x - 36y - 66 = 0$

32. The abscissaes of two points A and B are the roots of the equation $x^2 + 2ax - b^2 = 0$ and their ordinate are the roots of the equation $x^2 + 2px - q^2 = 0$. The radius of the circle with AB as diameter is
(a) $\sqrt{(a^2 + b^2 + p^2 + q^2)}$ (b) $\sqrt{(a^2 + p^2)}$
(c) $\sqrt{(b^2 + q^2)}$ (d) None of these

33. If the two circles $x^2 + y^2 + 2gx + 2fy = 0$ and $x^2 + y^2 + 2g_1x + 2f_1y = 0$ touch each other, then
(a) $f_1 g = fg_1$ (b) $ff_1 = gg_1$
(c) $f^2 + g^2 = f_1^2 + g_1^2$ (d) None of these

34. A line $x = mx + 1$ intersects the circle $(x - 3)^2 + (y + 2)^2 = 25$ at the points P and Q. If the mid-point of the line segment PQ has x-coordinate $-\frac{3}{5}$, then which one of the following options is correct? **[JEE Advanced 2019]**
(a) $6 \le m < 8$ (b) $-3 \le m < -1$
(c) $4 \le m < 6$ (d) $2 \le m < 4$

35. A triangle is formed by the lines whose combined equation is given by $(x + y - 4)(xy - 2x - y + 2) = 0$. The equation of its circumcircle is
(a) $x^2 + y^2 - 5x - 3y + 8 = 0$
(b) $x^2 + y^2 - 3x - 5y + 8 = 0$
(c) $x^2 + y^2 + 2x + 2y - 3 = 0$
(d) None of the above

36. If the radius of the circumcircle of triangle TPQ, where PQ is the chord of contact corresponding to the point T with respect to the circle $x^2 + y^2 - 2x + 4y - 11 = 0$, is 12 units, then the minimum distance of T from the director circle of the given circle is
(a) 6 (b) 12
(c) $6\sqrt{2}$ (d) $12 - 4\sqrt{2}$

37. Two tangents are drawn from a point P to the circle $x^2 + y^2 - 2x - 4y + 4 = 0$, such that the angle between these tangents is $\tan^{-1}\left(\frac{12}{5}\right)$, where $\tan^{-1}\left(\frac{12}{5}\right) \in (0, \pi)$. If the centre of the circle is denoted by C and these tangents touch the circle at points A and B, then the ratio of the areas of ΔPAB and ΔCAB is **[JEE Main 2021, 17 March]**
(a) 11 : 4 (b) 9 : 4
(c) 3 : 1 (d) 2 : 1

38. The circle passing through the intersection of the circles, $x^2 + y^2 - 6x = 0$ and $x^2 + y^2 - 4y = 0$, having its centre on the line, $2x - 3y + 12 = 0$, also passes through the point **[JEE Main 2020, 4 September]**
(a) $(-1, 3)$ (b) $(-3, 1)$
(c) $(1, -3)$ (d) $(-3, 6)$

39. The coordinates of the middle point of the chord cut off by $2x - 5y + 18 = 0$ by the circle $x^2 + y^2 - 6x + 2y - 54 = 0$ are
(a) $(1, 4)$ (b) $(2, 4)$
(c) $(4, 1)$ (d) $(1, 1)$

40. Let $\phi(x, y) = 0$ be the equation of a circle. If $\phi(0, \lambda) = 0$ has equal roots $\lambda = 2, 2$ and $\phi(\lambda, 0) = 0$ has roots $\lambda = \frac{4}{5}$, 5, then the centre of the circle is
(a) $(2, 29/10)$ (b) $(29/10, 2)$
(c) $(-2, 29/10)$ (d) None of these

41. Two distinct chords drawn from the point (p, q) on the circle $x^2 + y^2 = px + qy$, where $pq \ne 0$, are bisected by the X-axis. Then
(a) $|p| = |q|$ (b) $p^2 = 8q^2$
(c) $p^2 < 8q^2$ (d) $p^2 > 8q^2$

42. P is a point (a, b) in the first quadrant. If the two circles which pass through P and touch both the coordinate axes cut at right angles, then
(a) $a^2 - 6ab + b^2 = 0$ (b) $a^2 + 2ab - b^2 = 0$
(c) $a^2 - 4ab + b^2 = 0$ (d) $a^2 - 8ab + b^2 = 0$

43. The locus of the point of intersection of the tangents to the circle $x = r\cos\theta$, $y = r\sin\theta$ at points whose parametric angles differ by $\pi/3$ is
(a) $x^2 + y^2 = 4(2 - \sqrt{3})r^2$
(b) $3(x^2 + y^2) = 1$
(c) $x^2 + y^2 = (2 - \sqrt{3})r^2$
(d) $3(x^2 + y^2) = 4r^2$

44. If one circle of a coaxial system is $x^2 + y^2 + 2gx + 2fy + c = 0$ and one limiting point is (a, b), then equation of the radical axis will be
(a) $(g + a)x + (f + b)y + c - a^2 - b^2 = 0$
(b) $2(g + a)x + 2(f + b)y + c - a^2 - b^2 = 0$
(c) $2gx + 2fy + c - a^2 - b^2 = 0$
(d) None of the above

45. The mid-point of the chord of the circle $x^2 + y^2 - 6x + 4y - 12 = 0$ drawn parallel to the tangent at $(-1, 1)$ and at a distance of 1 unit from the tangent is
(a) $\left(\frac{2}{4}, \frac{1}{4}\right)$ (b) $\left(\frac{1}{4}, \frac{3}{4}\right)$
(c) $\left(\frac{-1}{5}, \frac{2}{5}\right)$ (d) $\left(\frac{3}{5}, \frac{2}{5}\right)$

46. The straight line $x + 2y = 1$ meets the coordinate axes at A and B. A circle is drawn through A, B and the origin. Then, the sum of perpendicular distances from A and B on the tangent to the circle at the origin is **[JEE Main 2019, 11 January]**
(a) $2\sqrt{5}$ (b) $\frac{\sqrt{5}}{4}$
(c) $4\sqrt{5}$ (d) $\frac{\sqrt{5}}{2}$

47. If $(1 + \alpha x)^n = 1 + 8x + 24x^2 + \dots$ and a line through $P(\alpha, n)$ cuts the circle $x^2 + y^2 = 4$ in A and B, then $PA \cdot PB$ is equal to
(a) 4 (b) 8
(c) 16 (d) 32

48. One of the diameter of the circle circumscribing the rectangle $ABCD$ is $4y = x + 7$. If A and B are the points $(-3, 4)$ and $(5, 4)$ respectively, then the area of the rectangle is
(a) 16 sq unit (b) 24 sq unit
(c) 32 sq unit (d) None of these

49. To which of the following circles, the line $y - x + 3 = 0$ is normal at the point $\left(3 + \frac{3}{\sqrt{2}}, \frac{3}{\sqrt{2}}\right)$?
(a) $\left(x - 3 - \frac{3}{\sqrt{2}}\right)^2 + \left(y - \frac{3}{\sqrt{2}}\right)^2 = 9$
(b) $\left(x - \frac{3}{\sqrt{2}}\right)^2 + \left(y - \frac{3}{\sqrt{2}}\right)^2 = 9$
(c) $x^2 + (y - 3)^2 = 9$
(d) $(x - 3)^2 + y^2 = 9$

50. A circle of radius 5 unit touches both the axes and lies in the first quadrant. If the circle makes one complete roll on X-axis along the positive direction of X-axis, then its equation in the new position is
(a) $x^2 + y^2 + 20\pi x - 10y + 100\pi^2 = 0$
(b) $x^2 + y^2 + 20\pi x + 10y + 100\pi^2 = 0$
(c) $x^2 + y^2 - 20\pi x - 10y + 100\pi^2 = 0$
(d) None of the above

51. If two circles $(x - 1)^2 + (y - 3)^2 = r^2$ and $x^2 + y^2 - 8x + 2y + 8 = 0$ intersect in two distinct points, then
(a) $2 < r < 8$ (b) $r < 2$
(c) $r = 2$ (d) $r > 2$

52. A variable circle always touches the line $y = x$ and passes through the point $(0, 0)$. The common chords of above circle and $x^2 + y^2 + 6x + 8y - 7 = 0$ will pass through a fixed point whose coordinates are
(a) $\left(-\frac{1}{2}, \frac{3}{2}\right)$ (b) $\left(-\frac{1}{2}, -\frac{1}{2}\right)$
(c) $\left(\frac{1}{2}, \frac{1}{2}\right)$ (d) None of these

53. The locus of the centres of the circles which cut the circles $x^2 + y^2 + 4x - 6y + 9 = 0$ and $x^2 + y^2 - 5x + 4y + 2 = 0$ orthogonally is
(a) $3x + 4y - 5 = 0$ (b) $9x - 10y + 7 = 0$
(c) $9x + 10y - 7 = 0$ (d) $9x - 10y + 11 = 0$

54. If from any point on the circle $x^2 + y^2 + 2gx + 2fy + c = 0$ tangents are drawn to the circle $x^2 + y^2 + 2gx + 2fy + c \sin^2\alpha + (g^2 + f^2)\cos^2\alpha = 0$, then the angle between the tangents is
(a) 2α (b) α
(c) $\alpha/2$ (d) $\alpha/4$

55. The equations of the circles which touch both the axes and the line $x = a$ are
(a) $x^2 + y^2 \pm ax \pm ay + \frac{a^2}{4} = 0$
(b) $x^2 + y^2 + ax \pm ay + \frac{a^2}{4} = 0$
(c) $x^2 + y^2 - ax \pm ay + \frac{a^2}{4} = 0$
(d) None of the above

56. A, B, C and D are the points of intersection with the coordinate axes of the lines $ax + by = ab$ and $bx + ay = ab$, then
(a) A, B, C, D are concyclic
(b) A, B, C, D form a parallelogram
(c) A, B, C, D form a rhombus
(d) None of the above

57. The common chord of $x^2 + y^2 - 4x - 4y = 0$ and $x^2 + y^2 = 16$ subtends at the origin an angle is equal to
(a) $\pi/6$ (b) $\pi/4$
(c) $\pi/3$ (d) $\pi/2$

58. The number of common tangents that can be drawn to the circles $x^2 + y^2 - 4x - 6y - 3 = 0$ and $x^2 + y^2 + 2x + 2y + 1 = 0$ is
(a) 1 (b) 2
(c) 3 (d) 4

59. If the distances from the origin of the centres of three circles $x^2 + y^2 + 2\lambda_i x - c^2 = 0\ (i = 1, 2, 3)$ are in GP, then the lengths of the tangents drawn to them from any point on the circle $x^2 + y^2 = c^2$ are in
(a) AP (b) GP
(c) HP (d) None of these

60. If $4l^2 - 5m^2 + 6l + 1 = 0$ and the line $lx + my + 1 = 0$ touches a fixed circle, then
(a) the centre of the circle is at the point (4, 0)
(b) the radius of the circle is equal to $\sqrt{5}$
(c) the circle passes through origin
(d) none of the above

61. A variable chord is drawn through the origin to the circle $x^2 + y^2 - 2ax = 0$. The locus of the centre of the circle drawn on this chord as diameter is
(a) $x^2 + y^2 + ax = 0$ (b) $x^2 + y^2 + ay = 0$
(c) $x^2 + y^2 - ax = 0$ (d) $x^2 + y^2 - ay = 0$

62. If a circle passes through the point (a, b) and cuts the circle $x^2 + y^2 = \lambda^2$ orthogonally, equation of the locus of its centre is
(a) $2ax + 2by = a^2 + b^2 + \lambda^2$
(b) $ax + by = a^2 + b^2 + \lambda^2$
(c) $x^2 + y^2 + 2ax + 2by + \lambda^2 = 0$
(d) $x^2 + y^2 - 2ax - 2by + a^2 + b^2 - \lambda^2 = 0$

63. Let the lengths of intercepts on X-axis and Y-axis made by the circle $x^2 + y^2 + ax + 2ay + c = 0\ (a < 0)$ be $2\sqrt{2}$ and $2\sqrt{5}$, respectively. Then, the shortest distance from origin to a tangent to this circle which is perpendicular to the line $x + 2y = 0$, is equal to
(a) $\sqrt{11}$ (b) $\sqrt{7}$
(c) $\sqrt{6}$ (d) $\sqrt{10}$

64. The circle passing through the distinct points $(1, t)$, $(t, 1)$ and (t, t) for all values of t, passes through the point
(a) (1, 1) (b) (−1, − 1)
(c) (1, − 1) (d) (−1, 1)

65. Equation of a circle through the origin and belonging to the coaxial system, of which the limiting points are (1, 2) , (4, 3) is
(a) $x^2 + y^2 - 2x + 4y = 0$
(b) $x^2 + y^2 - 8x - 6y = 0$
(c) $2x^2 + 2y^2 - x - 7y = 0$
(d) $x^2 + y^2 - 6x - 10y = 0$

66. Equation of the normal to the circle $x^2 + y^2 - 4x + 4y - 17 = 0$ which passes through (1, 1) is
(a) $3x + 2y - 5 = 0$ (b) $3x + y - 4 = 0$
(c) $3x + 2y - 2 = 0$ (d) $3x - y - 8 = 0$

67. α, β and γ are parametric angles of three points P, Q and R respectively, on the circle $x^2 + y^2 = 1$ and A is the point(−1, 0). If the lengths of the chords AP, AQ and AR are in GP, then $\cos\alpha/2$, $\cos\beta/2$ and $\cos\gamma/2$ are in
(a) AP (b) GP
(c) HP (d) None of these

68. The area bounded by the circles $x^2 + y^2 = r^2, r = 1, 2$ and the rays given by $2x^2 - 3xy - 2y^2 = 0$, $y > 0$ is
(a) $\frac{\pi}{4}$ sq unit (b) $\frac{\pi}{2}$ sq unit
(c) $\frac{3\pi}{4}$ sq unit (d) π sq unit

69. The equation of the circle touching the lines $|y| = x$ at a distance $\sqrt{2}$ unit from the origin is
(a) $x^2 + y^2 - 4x + 2 = 0$ (b) $x^2 + y^2 + 4x - 2 = 0$
(c) $x^2 + y^2 + 4x + 2 = 0$ (d) None of these

70. If the line $ax + by = 2$ is a normal to the circle $x^2 + y^2 - 4x - 4y = 0$ and a tangent to the circle $x^2 + y^2 = 1$, then
(a) $a = \frac{1}{2}, b = \frac{1}{2}$
(b) $a = \frac{1 + \sqrt{7}}{2}, b = \frac{1 - \sqrt{7}}{2}$
(c) $a = \frac{1}{4}, b = \frac{3}{4}$
(d) $a = 1, b = \sqrt{3}$

71. The equation of the circle passing through (2, 0) and (0, 4) and having the minimum radius is
(a) $x^2 + y^2 = 20$
(b) $x^2 + y^2 - 2x - 4y = 0$
(c) $(x^2 + y^2 - 4) + \lambda(x^2 + y^2 - 16) = 0$
(d) None of the above

72. The shortest distance from the point (2, − 7) to the circle $x^2 + y^2 - 14x - 10y - 151 = 0$ is
(a) 1 (b) 2
(c) 3 (d) 4

73. The circle $x^2 + y^2 = 4$ cuts the line joining the points $A(1, 0)$ and $B(3, 4)$ in two points P and Q. Let $\frac{BP}{PA} = \alpha$ and $\frac{BQ}{QA} = \beta$, then α and β are roots of the quadratic equation
(a) $x^2 + 2x + 7 = 0$ (b) $3x^2 + 2x - 21 = 0$
(c) $2x^2 + 3x - 27 = 0$ (d) None of these

74. The equation of the image of the circle $(x - 3)^2 + (y - 2)^2 = 1$ by the mirror $x + y = 19$ is
(a) $(x - 14)^2 + (y - 13)^2 = 1$
(b) $(x - 15)^2 + (y - 14)^2 = 1$
(c) $(x - 16)^2 + (y - 15)^2 = 1$
(d) $(x - 17)^2 + (y - 16)^2 = 1$

75. If P and Q are two points on the circle $x^2 + y^2 - 4x - 4y - 1 = 0$ which are farthest and nearest respectively from the point (6, 5), then

(a) $P \equiv \left(-\frac{22}{5}, 3\right)$ (b) $Q \equiv \left(\frac{22}{5}, \frac{19}{5}\right)$

(c) $P \equiv \left(\frac{14}{3}, -\frac{11}{5}\right)$ (d) $Q \equiv \left(-\frac{14}{3}, -4\right)$

76. If α, β are the roots of $ax^2 + bx + c = 0$ and α', β' those of $a'x^2 + b'x + c' = 0$, the equation of the circle having $A(\alpha, \alpha')$ and $B(\beta, \beta')$ as diameter is

(a) $cc'(x^2 + y^2) + ac'x + a'cy + a'b + ab' = 0$

(b) $cc'(x^2 + y^2) + a'cx + ac'y + a'b + ab' = 0$

(c) $bb'(x^2 + y^2) + a'bx + ab'y + a'c + ac' = 0$

(d) $aa'(x^2 + y^2) + a'bx + ab'y + a'c + ac' = 0$

77. A circle C_1 of radius b touches the circle $x^2 + y^2 = a^2$ externally and has it centre on the positive X-axis; another circle C_2 of radius c touches the circle C_1 externally and has its centre on the positive X-axis. Given $a < b < c$, then the three circles have a common tangent if a, b, c are in

(a) A P (b) G P
(c) H P (d) None of these

78. Equation of the circle cutting orthogonally the three circles $x^2 + y^2 - 2x + 3y - 7 = 0$, $x^2 + y^2 + 5x - 5y + 9 = 0$ and $x^2 + y^2 + 7x - 9y + 29 = 0$ is

(a) $x^2 + y^2 - 16x - 18y - 4 = 0$

(b) $x^2 + y^2 - 7x + 11y + 6 = 0$

(c) $x^2 + y^2 + 2x - 8y + 9 = 0$

(d) None of the above

79. A circle touching the X-axis at (3, 0) and making a intercept of length 8 on the Y-axis passes through the point **[JEE Main 2019, 12 April]**

(a) (3, 10) (b) (3, 5)
(c) (2, 3) (d) (1, 5)

80. A line meets the coordinate axes in A and B. A circle is circumscribed about the triangle OAB. If m and n are the distances of the tangent to the circle at the origin from the points A and B respectively, the diameter of the circle is

(a) $m(m + n)$ (b) $(m + n)$

(c) $n(m + n)$ (d) $\frac{1}{2}(m + n)$

81. If a line, $y = mx + c$ is a tangent to the circle, $(x - 3)^2 + y^2 = 1$ and it is perpendicular to a line L_1, where L_1 is the tangent to the circle, $x^2 + y^2 = 1$ at the point $\left(\frac{1}{\sqrt{2}}, \frac{1}{\sqrt{2}}\right)$, then **[JEE Main 2020, 8 January]**

(a) $c^2 + 7c + 6 = 0$ (b) $c^2 - 6c + 7 = 0$

(c) $c^2 - 7c + 6 = 0$ (d) $c^2 + 6c + 7 = 0$

82. If (2, 1) is a limiting point of a coaxial system of circles containing $x^2 + y^2 - 6x - 4y - 3 = 0$, then the other limiting point is

(a) (2, 4) (b) (−5, − 6)
(c) (3, 5) (d) (− 2, 4)

83. Lenght of the tangent drawn from any point of the circle $x^2 + y^2 + 2gx + 2fy + c = 0$ to the circle $x^2 + y^2 + 2gx + 2fy + d = 0, (d > c)$ is

(a) $\sqrt{(c - d)}$ (b) $\sqrt{(d - c)}$

(c) $\sqrt{(g - f)}$ (d) $\sqrt{(f - g)}$

84. A region in the xy-plane is bounded by the curve $y = \sqrt{(25 - x^2)}$ and the line $y = 0$. If the point $(a, a + 1)$ lies in the interior of the region, then

(a) $a \in (-4, 3)$ (b) $a \in (-\infty, -1) \cup (3, \infty)$
(c) $a \in (-1, 3)$ (d) None of these

85. The equation of the circle having its centre on the line $x + 2y - 3 = 0$ and passing through the points of intersection of the circles, $x^2 + y^2 - 2x - 4y + 1 = 0$ and $x^2 + y^2 - 4x - 2y + 1 = 0$ is

(a) $x^2 + y^2 - 6x + 1 = 0$

(b) $x^2 + y^2 - 3x + 4 = 0$

(c) $x^2 + y^2 - 2x - 2y + 1 = 0$

(d) $x^2 + y^2 + 2x - 4y + 4 = 0$

86. The locus of centre of a circle which touches externally the circle $x^2 + y^2 - 6x - 6y + 14 = 0$ and also touch the Y-axis is given by the equation

(a) $x^2 - 6x - 10y + 14 = 0$

(b) $x^2 - 10x - 6y + 14 = 0$

(c) $y^2 - 6x - 10y + 14 = 0$

(d) $y^2 - 10x - 6y + 14 = 0$

87. A square is inscribed in the circle $x^2 + y^2 - 6x + 8y - 103 = 0$ with its sides parallel to the coordinate axes. Then, the distance of the vertex of this square which is nearest to the origin is **[JEE Main 2019, 11 January]**

(a) 6 (b) 13
(c) $\sqrt{41}$ (d) $\sqrt{137}$

88. A circle of the coaxial system with limiting points (0, 0) and (1, 0) is

(a) $x^2 + y^2 - 2x = 0$ (b) $x^2 + y^2 - 6x + 3 = 0$

(c) $x^2 + y^2 = 1$ (d) $x^2 + y^2 - 2x + 1 = 0$

89. Let C be the circle with centre at (1, 1) and radius 1. If T is the circle centred at $(0, y)$ passing through origin and touching the circle C externally, then the radius of T is equal to **[JEE Main 2014]**

(a) $\frac{\sqrt{3}}{\sqrt{2}}$ (b) $\frac{\sqrt{3}}{2}$

(c) $\frac{1}{2}$ (d) $\frac{1}{4}$

90. The locus of the mid points of the chords of the circle $x^2 + y^2 + 4x - 6y - 12 = 0$ which subtend an angle of $\pi/3$ radians at its circumference is
(a) $(x + 2)^2 + (y - 3)^2 = 6.25$
(b) $(x - 2)^2 + (y + 3)^2 = 6.25$
(c) $(x + 2)^2 + (y - 3)^2 = 18.75$
(d) $(x + 2)^2 + (y + 3)^2 = 18.75$

91. The point $([P + 1], [P])$, (where [.] denotes the greatest integer function) lying inside the region bounded by the circle $x^2 + y^2 - 2x - 15 = 0$ and $x^2 + y^2 - 2x - 7 = 0$, then
(a) $P \in [-1, 0) \cup [0, 1) \cup [1, 2)$
(b) $P \in [-1, 2) - \{0, 1\}$
(c) $P \in (-1, 2)$
(d) None of the above

Objective Questions Type II [One or more than one correct answer(s)]

In each of the questions below four choices of which one or more than one are correct. You have to select the correct answer(s) accordingly.

1. A circle S passes through the point (0, 1) and is orthogonal to the circles $(x - 1)^2 + y^2 = 16$ and $x^2 + y^2 = 1$. Then, **[JEE Advanced 2014]**
(a) radius of S is 8 (b) radius of S is 7
(c) centre of S is $(-7, 1)$ (d) centre of S is $(-8, 1)$

2. If P is a point on the circle $x^2 + y^2 = 9$, Q is a point on the line $7x + y + 3 = 0$, and the line $x - y + 1 = 0$, is the perpendicular bisector of PQ, then the coordinates of P are
(a) (3, 0) (b) $\left(\frac{72}{25}, -\frac{21}{25}\right)$ (c) (0, 3) (d) $\left(-\frac{72}{25}, \frac{21}{25}\right)$

3. If a circle passes through the point $\left(3, \sqrt{\frac{7}{2}}\right)$ and touches $x + y = 1$ and $x - y = 1$, then the centre of the circle is
(a) (4, 0) (b) (4, 2) (c) (6, 0) (d) (7, 9)

4. The tangents drawn from the origin to the circle $x^2 + y^2 - 2px - 2qy + q^2 = 0$ are perpendicular, if
(a) $p = q$ (b) $p^2 = q^2$
(c) $q = -p$ (d) $p^2 + q^2 = 1$

5. An equation of a circle touching the axes of coordinates and the line $x\cos\alpha + y\sin\alpha = 2$ can be
(a) $x^2 + y^2 - 2gx - 2gy + g^2 = 0$
where $g = 2/(\cos\alpha + \sin\alpha + 1)$
(b) $x^2 + y^2 - 2gx - 2gy + g^2 = 0$
where $g = 2/(\cos\alpha + \sin\alpha - 1)$
(c) $x^2 + y^2 - 2gx + 2gy + g^2 = 0$
where $g = 2/(\cos\alpha - \sin\alpha + 1)$
(d) $x^2 + y^2 - 2gx + 2gy + g^2 = 0$
where $g = 2/(\cos\alpha - \sin\alpha - 1)$

6. Consider the circle $x^2 + y^2 - 10x - 6y + 30 = 0$. Let O be the centre of the circle and tangent at $A(7, 3)$ and $B(5, 1)$ meta at C. Let $S = 0$ represents family of circles passing through A and B, then -
(a) area of quadrilateral $OACB = 4$
(b) the radical axis for the family of circles $S = 0$ is $x + y = 10$
(c) the smallest possible circle of the family $S = 0$ is $x^2 + y^2 - 12x - 4y + 38 = 0$
(d) the coordinates of point C are (7, 1)

7. The two circles $x^2 + y^2 + ax = 0$ and $x^2 + y^2 = c^2$ touch each other, if
(a) $a + c = 0$ (b) $a - c = 0$
(c) $a^2 = c^2$ (d) None of these

8. The equation of a common tangent to the circles $x^2 + y^2 + 14x - 4y + 28 = 0$ and $x^2 + y^2 - 14x + 4y - 28 = 0$ is
(a) $x - 7 = 0$ (b) $y - 7 = 0$
(c) $28x + 45y + 371 = 0$ (d) $7x - 2y + 14 = 0$

9. If A and B are two points on the circle $x^2 + y^2 - 4x + 6y - 3 = 0$ which are farthest and nearest respectively from the point (7, 2), then
(a) $A \equiv (2 - 2\sqrt{2}, -3 - 2\sqrt{2})$
(b) $A \equiv (2 + 2\sqrt{2}, -3 + 2\sqrt{2})$
(c) $B \equiv (2 + 2\sqrt{2}, -3 + 2\sqrt{2})$
(d) $B \equiv (2 - 2\sqrt{2}, -3 - 2\sqrt{2})$

10. The equations of four circles are $(x \pm a)^2 + (y \pm a)^2 = a^2$. The radius of a circle touching all the four circles is
(a) $(\sqrt{2} - 1)a$ (b) $2\sqrt{2}\,a$
(c) $(\sqrt{2} + 1)a$ (d) $(2 + \sqrt{2})a$

11. Let T be the line passing through the points $P(-2, 7)$ and $Q(2, -5)$. Let F_1 be the set of all pairs of circles (S_1, S_2) such that T is tangent to S_1 at P and tangent to S_2 at Q, and also such that S_1 and S_2 touch each other a point, say M. Let E_1 be the set representing the locus straight line segments joining a pair of distinct points of E_1 and passing through the point $R(1, 1)$ be F_2. Let the set of all straight line segments joining a pair of distinct points of E_1 and passing through the point $R(1, 1)$ be Let E_2 be the set of the mid-points of the line segments in the set F_2. Then, which of the following statements(s is (are) TRUE? **[JEE Advanced 2018]**
(a) The point $(-2, 7)$ lines in E_1
(b) The point $\left(\frac{4}{5}, \frac{7}{5}\right)$ does NOT lie in E_2
(c) The point $\left(\frac{1}{2}, 1\right)$ lies in E_2
(d) The point $\left(0, \frac{3}{2}\right)$ does NOT lie in E_1

12. The equation of circle passing through (3, – 6) and touching both the axes is
(a) $x^2 + y^2 - 6x + 6y + 9 = 0$
(b) $x^2 + y^2 + 6x - 6y + 9 = 0$
(c) $x^2 + y^2 + 30x - 30y + 225 = 0$
(d) $x^2 + y^2 - 30x + 30y + 225 = 0$

13. The equation of the tangents drawn from the origin to the circle $x^2 + y^2 - 2rx - 2hy + h^2 = 0$, are
(a) $x = 0$ (b) $y = 0$
(c) $(h^2 - r^2)x - 2rhy = 0$ (d) $(h^2 - r^2)x + 2rhy = 0$

14. Equation of a circle with centre (4, 3) touching the circle $x^2 + y^2 = 1$ is
(a) $x^2 + y^2 - 8x - 6y - 9 = 0$
(b) $x^2 + y^2 - 8x - 6y + 11 = 0$
(c) $x^2 + y^2 - 8x - 6y - 11 = 0$
(d) $x^2 + y^2 - 8x - 6y + 9 = 0$

15. The equation of a tangent to the circle $x^2 + y^2 = 25$ passing through (–2, 11) is
(a) $4x + 3y = 25$ (b) $3x + 4y = 38$
(c) $24x - 7y + 125 = 0$ (d) $7x + 24y = 230$

16. Tangents are drawn to the circle $x^2 + y^2 = 32$ from a point A lying on the X-axis. The tangents cut the Y-axis at points B and C, then the coordinate (s) of A such that the area of ΔABC is minimum may be
(a) $(8\sqrt{2}, 0)$ (b) $(8, 0)$
(c) $(-8, 0)$ (d) $(-8\sqrt{2}, 0)$

17. If the circle $x^2 + y^2 + 2gx + 2fy + c = 0$ cuts each of the circles $x^2 + y^2 - 4 = 0$, $x^2 + y^2 - 6x - 8y + 10 = 0$ and $x^2 + y^2 + 2x - 4y - 2 = 0$ at the extremities of a diameter, then
(a) $c = -4$ (b) $g + f = c - 1$
(c) $g^2 + f^2 - c = 17$ (d) $gf = 6$

18. From the point $A(0, 3)$ on the circle $x^2 + 4x + (y - 3)^2 = 0$, a chord AB is drawn and extended to a point M, such that $AM = 2AB$. An equation of the locus of M is
(a) $x^2 + 6x + (y - 2)^2 = 0$
(b) $x^2 + 8x + (y - 3)^2 = 0$
(c) $x^2 + y^2 + 8x - 6y + 9 = 0$
(d) $x^2 + y^2 + 6x - 4y + 4 = 0$

19. If a chord of the circle $x^2 + y^2 - 4x - 2y - c = 0$ is trisected at the points (1/3, 1/3) and (8/3, 8/3), then
(a) $c = 10$ (b) $c = 20$
(c) $c = 15$ (d) $c^2 - 40c + 400 = 0$

20. Consider the circles
$C_1 \equiv x^2 + y^2 - 2x - 4y - 4 = 0$ and
$C_2 \equiv x^2 + y^2 + 2x + 4y + 4 = 0$
and the line $L \equiv x + 2y + 2 = 0$, then
(a) L is the radical axis of C_1 and C_2
(b) L is the common tangent of C_1 and C_2
(c) L is the common chord of C_1 and C_2
(d) L is perpendicular to the line joining centres of C_1 and C_2

21. The circle $C_1 : x^2 + y^2 = 3$ with centre at O intersects the parabola $x^2 = 2y$ at the point P touches other two circles C_2 and C_3 at R_2 and R_3, respectively. Suppose C_2 and C_3 have equal radii $2\sqrt{3}$ and centres Q_2 and Q_3, respectively. If Q_2 and Q_3 lie on the Y-axis, then **[JEE Advanced 2016]**
(a) $Q_2Q_3 = 12$
(b) $R_2R_3 = 4\sqrt{6}$
(c) area of the ΔOR_2R_3 is $6\sqrt{2}$
(d) area of the ΔPQ_2Q_3 is $4\sqrt{2}$

22. If point $P(x, y)$ is called a lattice point, if $x, y \in I$. Then the total number of lattice points in the interior of the circle $x^2 + y^2 = a^2$, $a \neq 0$ cannot be
(a) 202 (b) 203
(c) 204 (d) 205

23. A square is inscribed in the circle $x^2 + y^2 - 10x - 6y + 30 = 0$. One side of the square is parallel to $y = x + 3$, then one vertex of the square is
(a) (3, 3)
(b) (7,3)
(c) $(6, 3 - \sqrt{3})$
(d) $(6, 3 + \sqrt{3})$

24. The circles $x^2 + y^2 - 4x - 81 = 0$, $x^2 + y^2 + 24x - 81 = 0$ intersect each other at points A and B. A line through point A meet one circle at P and a parallel line through B meet the other circle at Q. Then the locus of the mid-point of PQ is
(a) $(x + 5)^2 + (y + 0)^2 = 25$
(b) $(x - 5)^2 + (y - 0)^2 = 25$
(c) $x^2 + y^2 + 10x = 0$
(d) $x^2 + y^2 - 10x = 0$

25. Point M moved on the circle $(x - 4)^2 + (y - 8)^2 = 20$. Then it broke away from it and moving along a tangent to the circle, cuts the X-axis at the point (–2, 0). The coordinates of a point on the circle at which the moving point broke away is
(a) $\left(-\frac{3}{5}, \frac{46}{5}\right)$ (b) $\left(-\frac{2}{5}, \frac{44}{5}\right)$
(c) (6, 4) (d) (3, 5)

26. If the area of the quadrilateral formed by the tangents from the origin to the circle $x^2 + y^2 + 6x - 10y + c = 0$ and the pair of radii at the points of contact of these tangents to the circle is 8 sq unit, then the value of c is
(a) 2 (b) 4
(c) 16 (d) 32

27. The equation of a circle in which the chord joining the points (1, 2) and (2, –1) subtends an angle of $\pi/4$ at any point on the circumference is

(a) $x^2 + y^2 - 5 = 0$

(b) $x^2 + y^2 - 6x - 2y + 5 = 0$

(c) $x^2 + y^2 + 6x + 2y - 15 = 0$

(d) $x^2 + y^2 - 2x - 4y + 4 = 0$

28. Let RS be the diameter of the circle $x^3 - y^2 = 1$, where S is the point (1, 0). Let P be a variable point (other than R and S) on the circle and tangents to the circle at S and P meet at the point Q. The normal to the circle at P intersects a line drawn through Q parallel to RS at point E. Then, the locus of E passes through the point(s) **[JEE Advanced 2016]**

(a) $\left(\frac{1}{3}, \frac{1}{\sqrt{3}}\right)$ (b) $\left(\frac{1}{4}, \frac{1}{2}\right)$

(c) $\left(\frac{1}{3}, -\frac{1}{\sqrt{3}}\right)$ (d) $\left(\frac{1}{4}, -\frac{1}{2}\right)$

29. An isosceles ΔABC is inscribed in a circle $x^2 + y^2 = a^2$ with the vertex A at $(a, 0)$ and the base angles B and C each equal to 75°, then coordinates of an end point of the base are

(a) $\left(\frac{\sqrt{3}a}{2}, \frac{a}{2}\right)$ (b) $\left(-\frac{\sqrt{3}a}{2}, \frac{a}{2}\right)$

(c) $\left(-\frac{\sqrt{3}a}{2}, -\frac{a}{2}\right)$ (d) $\left(\frac{\sqrt{3}a}{2}, -\frac{a}{2}\right)$

30. If the line $3x - 4y - \lambda = 0$ touches the circle $x^2 + y^2 - 4x - 8y - 5 = 0$ at (a, b), then $\lambda + a + b$ is equal to

(a) 20 (b) 22

(c) –30 (d) –28

Linked-Comprehension Type

In these questions, a passage (paragraph) has been given followed by questions based on each of the passages. You have to answer the questions based on the passage given.

PASSAGE 1

If $7l^2 - 9m^2 + 8l + 1 = 0$ and we have to find equation of circle having $lx + my + 1 = 0$ is a tangent and we can adjust given condition as $16l^2 + 8l + 1 = 9(l^2 + m^2)$

or $(4l + 1)^2 = 9(l^2 + m^2) \Rightarrow \frac{|4l + 1|}{\sqrt{(l^2 + m^2)}} = 3$

Centre of circle = (4, 0) and radius = 3 when any two non parallel lines touching a circle, then centre of circle lies on angle bisector of lines.

On the basis of above information, answer the following questions

1. If $16m^2 - 8l - 1 = 0$, then equation of the circle having $lx + my + 1 = 0$ is a tangent is

(a) $x^2 + y^2 + 8x = 0$ (b) $x^2 + y^2 - 8x = 0$

(c) $x^2 + y^2 + 8y = 0$ (d) $x^2 + y^2 - 8y = 0$

2. If $16l^2 + 9m^2 = 24lm + 6l + 8m + 1$ and if S be the equation of the circle having $lx + my + 1 = 0$ is a tangent when the equation of director circle of S is

(a) $x^2 + y^2 + 6x + 8y = 25$

(b) $x^2 + y^2 - 6x + 8y = 25$

(c) $x^2 + y^2 - 6x - 8y = 25$

(d) $x^2 + y^2 + 6x - 8y = 25$

3. If $4l^2 - 5m^2 + 6l + 1 = 0$, then the centre and radius of the circle which have $lx + my + 1 = 0$ is a tangent is

(a) $(0, 4); \sqrt{5}$ (b) $(4, 0); \sqrt{5}$

(c) $(0, 3); \sqrt{5}$ (d) $(3, 0); \sqrt{5}$

4. If $x + 2y = 3$ and $2x + y = 3$ touches a circle whose centre lies on a line $3x + 4y = 5$, then possible centres of circle are

(a) $\left(\frac{5}{3}, 0\right), (3, -1)$ (b) $\left(\frac{5}{7}, \frac{5}{7}\right), \left(4, -\frac{3}{4}\right)$

(c) $\left(\frac{5}{3}, 0\right), \left(4, -\frac{3}{4}\right)$ (d) $\left(\frac{5}{7}, \frac{5}{7}\right), (3, -1)$

5. Centre of circles touching both axes and passing through (–2, 3) is

(a) $(5 + 2\sqrt{3}, 5 + 2\sqrt{3}), (5 - 2\sqrt{3}, 5 - 2\sqrt{3})$

(b) $(-(5 + 2\sqrt{3}), 5 + 2\sqrt{3}), (-5 + 2\sqrt{3}, 5 - 2\sqrt{3})$

(c) $(5 + 2\sqrt{3}, -(5 + 2\sqrt{3})), (5 - 2\sqrt{3}, -5 + 2\sqrt{3})$

(d) imaginary centre

PASSAGE 2

Let O be the fixed point, called origin or pole and OX be a fixed straight line, called the initial line. Let C be the centre and a the radius of the circle.

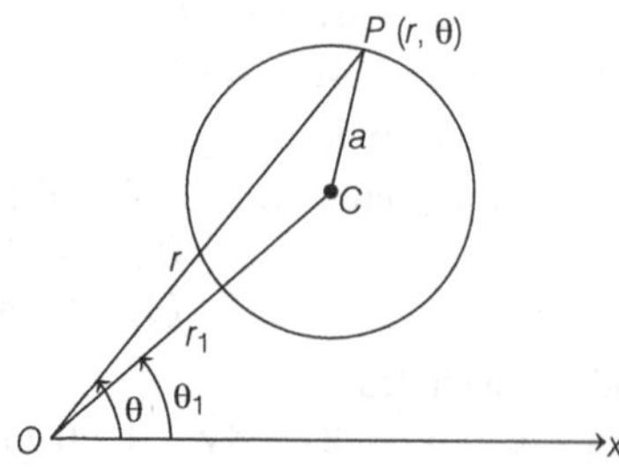

Suppose P is a point such that $OP = r$ and anticlockwise angle $\angle XOP = \theta$, then we define (r, θ) as a polar coordinates of the point P and let $OC = r_1$ and anticlockwise $\angle XOC = \theta_1$, then we define (r_1, θ_1) as a polar coordinates of the centre C.

In $\Delta\, POC$,

$$(CP)^2 = (OP)^2 + (OC)^2 - 2\, OP \cdot OC \cos(\theta - \theta_1) \qquad (\because \angle POC = \theta - \theta_1)$$

$$\Rightarrow \quad a^2 = r^2 + r_1^2 - 2rr_1 \cos(\theta - \theta_1)$$

or $\quad r^2 - 2rr_1 \cos(\theta - \theta_1) + r_1^2 - a^2 = 0 \qquad \ldots(\text{i})$

This is the general equation of circle in polar form.

Now, if the pole lies on the circle, then $r_1 = a$, then equation of circle becomes

$$r = 2a \cos(\theta - \theta_1) \qquad \ldots(\text{ii})$$

Also if the centre of the circle lies on the pole, then $r_1 = 0$, then equation of circle becomes $r^2 = a^2$ and, if $\theta_1 = 0$, in Eq. (ii), we see that if the initial line is a diameter and the pole is an extremity of the diameter, the equation of the circle of radius a is $r = 2a \cos\theta$

On the basis of above information, answer the following questions

1. A line is drawn from O and intersecting the circle $x^2 + y^2 - 6x - 8y + 16 = 0$ at P_1 and Q_1 and making angle α with positive X-axis, then $OP_1 \cdot OQ_1$ is equal to

(a) 0 (b) 9
(c) 16 (d) 25

2. The equation of circle described on the straight line joining the points $(4, \pi/6)$ and $(3, \pi/3)$ as diameter is

(a) $r^2 + r\{4\cos(\theta - \pi/6) + 3\cos(\theta - \pi/3)\} + 6\sqrt{3} = 0$
(b) $r^2 - r\{4\cos(\theta - \pi/6) + 3\cos(\theta - \pi/3)\} + 6\sqrt{3} = 0$
(c) $r^2 + r\{4\cos(\theta - \pi/6) - 3\cos(\theta - \pi/3)\} + 6\sqrt{3} = 0$
(d) $r^2 - r\{4\cos(\theta - \pi/6) + 3\cos(\theta - \pi/3)\} - 6\sqrt{3} = 0$

3. The line $\frac{1}{r} = A\cos\theta + B\sin\theta$, touches the circle $r = 2a\cos\theta$, then

(a) $a^2B^2 + 2aA = 1$ (b) $a^2A^2 + 2aB = 1$
(c) $a^2A^2 + 2aA = 1$ (d) $a^2B^2 + 2aB = 1$

4. If a circle passes through the point (r', θ') and touches the initial line at a distance c from the pole, then its polar equation is $(r^2 - 2cr\cos\theta + c^2) = \lambda(r'^2 - 2cr'\cos\theta' + c^2)$, then λ is

(a) $\frac{r\sin\theta'}{r'\sin\theta}$ (b) $\frac{r'\sin\theta'}{r\sin\theta}$
(c) $\frac{r'\sin\theta}{r\sin\theta'}$ (d) $\frac{r\sin\theta}{r'\sin\theta'}$

5. If equation of the chord joining the points on the circle $r = 2a\cos\theta$ whose vectorial angles are α and β is $r\cos(\theta - \alpha - \beta) = 2\lambda\cos\alpha$, then λ is

(a) $a\cos\beta$ (b) $a\cos\alpha$
(c) $a\cos(\alpha + \beta)$ (d) $a\cos(\alpha - \beta)$

PASSAGE 3

P is a variable point on the line $L = 0$. Tangents are drawn to the circle $x^2 + y^2 = 4$ from P to touch it at Q and R. The parallelogram $PQSR$ is completed.

On the basis of above information, answer the following questions

1. If $L \equiv 2x + y = 6$, then the locus of circumcentre of ΔPQR is

(a) $2x - y = 4$ (b) $2x + y = 3$
(c) $x - 2y = 4$ (d) $x + 2y = 3$

2. If $L \equiv y = 4$, then the locus of the circumcentre of $\Delta\, PQR$ is

(a) $y - 2 = 0$ (b) $x - 2 = 0$
(c) $y + 2 = 0$ (d) $x + 2 = 0$

3. If $P \equiv (2, 3)$, then the centre of circumcircle of triangle QRS is

(a) $\left(\frac{2}{13}, \frac{7}{26}\right)$ (b) $\left(\frac{2}{13}, \frac{3}{26}\right)$
(c) $\left(\frac{3}{13}, \frac{9}{26}\right)$ (d) $\left(\frac{3}{13}, \frac{2}{13}\right)$

4. If $P \equiv (6, 8)$, then the area of ΔQRS is

(a) $\frac{(6)^{3/2}}{25}$ sq unit (b) $\frac{(24)^{3/2}}{25}$ sq unit
(c) $\frac{48\sqrt{6}}{25}$ sq unit (d) $\frac{96\sqrt{6}}{25}$ sq unit

5. If $P \equiv (3, 4)$, then coordinate of S is

(a) $\left(-\frac{46}{25}, \frac{-63}{25}\right)$ (b) $\left(-\frac{51}{25}, -\frac{68}{25}\right)$
(c) $\left(-\frac{46}{25}, -\frac{68}{25}\right)$ (d) $\left(-\frac{68}{25}, -\frac{51}{25}\right)$

PASSAGE 4

A circle C of radius 1 is inscribed in an equailateral triangle PQR. The points of contact of C with the sides PQ, QR, RP and D, E, F respectively. The line PQ is given by the equation $\sqrt{3}x + y - 6 = 0$ and point D is $\left(\frac{3\sqrt{3}}{2}, \frac{3}{2}\right)$. Further it is given that the origin and the centre of C are on the same side of PQ.

On the basis of above information, answer the following questions

1. The equation of circle C is

(a) $(x - 2\sqrt{3})^2 + (y - 1)^2 = 1$

(b) $(x - 2\sqrt{3})^2 + \left(y + \frac{1}{2}\right)^2 = 1$

(c) $(x - \sqrt{3})^2 + (y + 1)^2 = 1$

(d) $(x - \sqrt{3})^2 + (y - 1)^2 = 1$

2. Points E and F are given by

(a) $(\sqrt{3}, 0), \left(\frac{\sqrt{3}}{2}, \frac{3}{2}\right)$ (b) $(\sqrt{3}, 0), \left(\frac{\sqrt{3}}{2}, \frac{1}{2}\right)$

(c) $\left(\frac{\sqrt{3}}{2}, \frac{1}{2}\right), \left(\frac{\sqrt{3}}{2}, \frac{3}{2}\right)$ (d) $\left(\frac{\sqrt{3}}{2}, \frac{1}{2}\right), \left(\frac{3}{2}, \frac{\sqrt{3}}{2}\right)$

3. Equations of the sides QR, RP are

(a) $y = \left(\frac{2}{\sqrt{3}}\right)x + 1, y = \left(-\frac{2}{\sqrt{3}}\right)x - 1$

(b) $y = \left(\frac{1}{\sqrt{3}}\right)x, y = 0$

(c) $y = \left(\frac{\sqrt{3}}{2}\right)x + 1, y = \left(-\frac{\sqrt{3}}{2}\right)x - 1$

(d) $y = \sqrt{3}\,x, y = 0$

PASSAGE 5

Let S be the circle in the XY-plane defined by the equation $x^2 + y^2 = 4$. *(There are two questions based on Paragraph "X", the question given below is one of them)* **[JEE Advanced 2018]**

On the basis of above information, answer the following questions

1. Let E_1E_2 and F_1F_2 be the chords of S passing through the point P_0 (1, 1) and parallel to the X-axis and the Y-axis, respectively. Let G_1G_2 be the chord of S passing through P_0 and having slope -1. Let the tangents to S at E_1 and E_2 meet at E_3, then tangents to S at F_1 and F_2 meet at F_3, and the tangents to S at G_1 and G_2 meet at G_3. Then, the points E_3, F_3 and G_3 lie on the curve

(a) $x + y = 4$

(b) $(x-4)^2 + (y-4)^2 = 16$

(c) $(x - 4)(y - 4) = 4$

(d) $xy = 4$

2. Let P be a point on the circle S with both coordinates being positive. Let the tangent to S at P intersect the coordinate axes at the points M and N. Then, the mid-point of the line segment MN must lie on the curve

(a) $(x + y)^2 = 3xy$

(b) $x^{2/3} + y^{2/3} = 2^{4/3}$

(c) $x^2 + y^2 = 2xy$

(d) $x^2 + y^2 = x^2y^2$

Numerical Value Type

Solve the following problems and write your answer.

1. If the lengths of external and internal common tangents to two circles $x^2 + y^2 + 14x - 4y + 28 = 0$ and $x^2 + y^2 - 14x + 4y - 28 = 0$ are λ and μ the value of $\lambda + \mu$ must be

2. If the radius of a circle which passes through the point (2, 0) and whose centre is the limit of the point of intersection of the lines $3x + 5y = 1$ and $(2 + c)x + 5c^2y = 1$ as $c \to 1$ is $\frac{\sqrt{\lambda}}{25}$, then the value of λ must be

3. Let $2x^2 + y^2 - 3xy = 0$ be the equation of a pair of tangents drawn from the origin O to a circle of radius 3 with centre in the first quadrant. If A is one of the points of contact. If $OA = \lambda + \sqrt{\mu}$, then the value of $\lambda^2 + \mu$ must be

4. The circle $x^2 + y^2 = 1$ cuts the X-axis at P and Q, another circle with centre at Q and variable radius intersects the first circle at R above the X-axis and the line segment PQ at S. If the maximum area of the ΔQSR is λ sq unit, then $27\lambda^2$ must be

5. If the equation of the circle whose radius is 5 and which touches the circle $x^2 + y^2 - 2x - 4y - 20 = 0$ at the point (5, 5) is $(x - h)^2 + (y - k)^2 = 25$, then the value of $h^2 + k^2$ must be

6. The length of the intercept on the straight line $4x - 3y - 10 = 0$ by the circle $x^2 + y^2 - 2x + 4y - 20 = 0$ is

7. The area of the triangle formed by tangents from the point (4, 3) to the circle $x^2 + y^2 = 9$ and the line segment joining their points of contact is λ square unit, then the value of 25λ must be

8. The angle between the circles $S : x^2 + y^2 - 4x + 6y + 11 = 0$ and $S' : x^2 + y^2 - 2x + 8y + 13 = 0$ is $\theta°$, then θ must be

9. If the minimum radius of the circle which contains the three circles

$$x^2 - y^2 - 4y - 5 = 0$$
$$x^2 + y^2 + 12x + 4y + 31 = 0$$

and $x^2 + y^2 + 6x + 12y + 36 = 0$ is $\left(3 + \frac{5}{36}\sqrt{\lambda}\right)$, then the value of λ must be

10. A circle of unit radius touches positive X-axis and Y-axis at A and B respectively. A variable line passing through origin intersects the circle in two points D and E. If the slope of the line is m and the area of ΔDEB is maximum, then the value of $\frac{3333}{m^2}$ must be

Single Integer Type

Solve the following problems and write your answer.

1. Let PQ be a diameter of the circle $x^2 + y^2 = 9$. If α and β are the lengths of the perpendiculars from P and Q on the straight line, $x + y = 2$ respectively, then the maximum value of $\alpha\beta$ is .... **[JEE Main 2020, 4 September]**
2. If a circle C passing through the point (4, 0) touches the circle $x^2 + y^2 + 4x - 6y = 12$ externally at the point (1, – 1), then the radius of C is
3. The number of integral values of k for which the line, $3x + 4y - k$ intersects the circle, $x^2 + y^2 - 2x - 4y + 4 = 0$ at two distinct points is **[JEE Main 2020, 2 September]**
4. Let $A_0\, A_1\, A_2\, A_3\, A_4\, A_5$ be a regular hexagon inscribed in a unit circle with centre at the origin. Then the product of the lengths of the line segments $A_0\,A_1$, A_0A_2 and A_0A_4 is
5. The straight line $2x - 3y = 1$ divide the circular region $x^2 + y^2 \le 6$ into two parts. If $S = \left\{\left(2, \frac{3}{4}\right), \left(\frac{5}{2}, \frac{3}{4}\right), \left(\frac{1}{4}, -\frac{1}{4}\right), \left(\frac{1}{8}, \frac{1}{4}\right)\right\}$, then the number of point(s) in S lying inside the smaller part is **[JEE Advanced 2016]**
6. The straight line $2x - 3y = 1$ divide the circular region $x^2 + y^2 \le 6$ into two parts. If $S = \left\{\left(2, \frac{3}{4}\right), \left(\frac{5}{2}, \frac{3}{4}\right), \left(\frac{1}{4}, -\frac{1}{4}\right), \left(\frac{1}{8}, \frac{1}{4}\right)\right\}$, then the number of point(s) in S lying inside the smaller part is ....
7. Let $ABCD$ be a square of side of unit length. Let a circle C_1 centered at A with unit radius is drawn. Another circle C_2 which touches C_1 and the lines AD and AB are tangent to it, is also drawn. Let a tangent line from the point C to the circle C_2 meet the side AB at E. If the length of EB is $\alpha + \sqrt{3}\beta$, where α, β are integers, then $\alpha + \beta$ is equal to ....

Matrix-Match Type

Given below are Matching Type Questions, with two columns (each having some items) each . Each item of Column I has to be matched with the items of Column II, by encircling the correct match(es).

NOTE An item of Column I can be matched with more than one items of Column II.

1. Observe the following columns :

	Column I		Column II
(A)	If the shortest and largest distance from the point (10, 7) to the circle $x^2 + y^2 - 4x - 2y - 20 = 0$ are L and M respectively, then	(P)	$M + L = 10$
(B)	If the shortest and largest distance from the point (3, – 6) to the circle $x^2 + y^2 - 16x - 12y - 125 = 0$ are L and M respectively, then	(Q)	$M + L = 20$
(C)	If the shortest and largest distance from the point (6, – 6) to the circle $x^2 + y^2 - 4x + 6y - 12 = 0$ are L and M respectively, then	(R)	$M + L = 30$
		(S)	$M - L = 10$
		(T)	$M - L = 26$

2. Observe the following columns :

	Column I		Column II
(A)	If the straight line $y = mx\ \forall\ m \in I$ touches or lies outside the circle $x^2 + y^2 - 20y + 90 = 0$, then the value of $\lvert m \rvert$ can be	(P)	0
(B)	If the straight line $3x - 4y - 5k = 0, \forall k \in I$ touches or lies inside the circle, $x^2 + y^2 - 4x - 8y - 5 = 0$, then the value of $\lvert k + 2 \rvert$ can be	(Q)	1
(C)	Two circles $x^2 + y^2 + px + py - 7 = 0$ and $x^2 + y^2 - 10x + 2py + 1 = 0$ will orthogonally, then the value of p is	(R)	2
		(S)	3
		(T)	4

3. Observe the following columns :

	Column I		Column II
(A)	If point of intersection and number of tangents of two circles $x^2 + y^2 - 2x - 6y + 9 = 0$ and $x^2 + y^2 + 6x - 2y + 1 = 0$ are λ and μ respectively, then	(P)	$\mu - \lambda = 0$
(B)	If point of intersection and number of tangents of two circles $x^2 + y^2 - 6x = 0$ and $x^2 + y^2 + 2x = 0$ are λ and μ respectively, then	(Q)	$\mu - \lambda = 2$
(C)	If point of intersection and number of tangents of two circles $x^2 + y^2 - 2x - 4y + 1 = 0$ and $x^2 + y^2 - 4x - 2y + 1 = 0$ are λ and μ respectively, then	(R)	$\mu - \lambda = 4$
		(S)	$\mu + \lambda = 4$
		(T)	$\lambda^{\mu} + \mu^{\lambda} = 4$

4. Observe the following columns :

	Column I		Column II
(A)	The length of the common chord of two circles of radii 3 and 4 units which intersect orthogonally is $\frac{k}{5}$, then k equals to	(P)	1
(B)	The circumference of the circle $x^2 + y^2 + 4x + 12y + p = 0$ is bisected by the circle $x^2 + y^2 - 2x + 8y - q = 0$, then $p + q$ is equal to -	(Q)	24
(C)	Number of distinct chords of the circle $2x(x - \sqrt{2}) + y(2y - 1) = 0$ passing through the point $\left(\sqrt{2}, \frac{1}{2}\right)$ and are bisected by X-axis, is	(R)	32
(D)	One of the diameters of the circle circumscribing the rectangle $ABCD$ is $4y = x + 7$. If A and B are the point $(-3, 4)$ and $(5, 4)$ respectively, then the area of the rectangle is equal to	(S)	2
		(T)	36

Answers

Objective Questions Type I [Only one correct answer]

1. (d)	2. (a)	3. (d)	4. (b)	5. (c)	6. (b)	7. (c)	8. (b)	9. (d)	10. (c)
11. (d)	12. (c)	13. (b)	14. (c)	15. (d)	16. (b)	17. (c)	18. (b)	19. (b)	20. (b)
21. (a)	22. (b)	23. (c)	24. (b)	25. (a)	26. (d)	27. (a)	28. (a)	29. (b)	30. (a)
31. (b)	32. (a)	33. (a)	34. (d)	35. (b)	36. (d)	37. (b)	38. (d)	39. (a)	40. (b)
41. (d)	42. (c)	43. (d)	44. (b)	45. (c)	46. (d)	47. (c)	48. (c)	49. (d)	50. (d)
51. (a)	52. (c)	53. (b)	54. (a)	55. (c)	56. (a)	57. (d)	58. (c)	59. (b)	60. (b)
61. (c)	62. (a)	63. (c)	64. (a)	65. (c)	66. (b)	67. (b)	68. (c)	69. (a)	70. (b)
71. (b)	72. (b)	73. (b)	74. (d)	75. (b)	76. (d)	77. (b)	78. (a)	79. (a)	80. (b)
81. (d)	82. (b)	83. (b)	84. (c)	85. (a)	86. (d)	87. (c)	88. (d)	89. (d)	90. (a)
91. (d)									

Objective Questions Type II [One or more than one correct answer(s)]

1. (b, c)	2. (a, d)	3. (a, c)	4. (a, b, c)	5. (a, b, c, d)
6. (a, c, d)	7. (a, b, c)	8. (b, c)	9. (b, d)	10. (a, c)
11. (a, d)	12. (a, d)	13. (a, c)	14. (c, d)	15. (a, c)
16. (b, c)	17. (a, b, c, d)	18. (b, c)	19. (b, d)	20. (a, c, d)
21. (a, b, c)	22. (a, b, c)	23. (a, b)	24. (a, c)	25. (b, c)
26. (a, d)	27. (a, b)	28. (b, d)	29. (b, c)	30. (a, d)

Linked-Comprehension Type

Passage 1	**1.** (b)	**2.** (c)	**3.** (d)	**4.** (d)	**5.** (b)
Passage 2	**1.** (c)	**2.** (b)	**3.** (a)	**4.** (d)	**5.** (a)
Passage 3	**1.** (b)	**2.** (a)	**3.** (c)	**4.** (d)	**5.** (b)
Passage 4	**1.** (d)	**2.** (a)	**3.** (d)		
Passage 5	**1.** (a)	**2.** (d)			

Numerical Value Type

1. 18 **2.** 1601 **3.** 171 **4.** 64 **5.** 145 **6.** 10 **7.** 192 **8.** 135 **9.** 949 **10.** 9999

Single Integer Type

1. 7 **2.** (5) **3.** (9) **4.** (3) **5.** (2) **6.** (2) **7.** (1)

Matrix-Match Type

1. A → (Q, S); B → (R, T); C → (P, S)

2. A → (P,Q, R, S); B → (P, Q, R, S, T); C → (R, S)

3. A → (R, S); B → (Q, S, T); C → (P, S)

4. A → (Q); B → (T); C → (P) D → (R)

Solutions

Objective Questions Type [Only one correct answer]

1. Equation of given circle

$$x^2 + y^2 = r^2 \quad \ldots(i)$$

and the chord $\quad y = 2x + 3 \quad \ldots(ii)$

From the figure

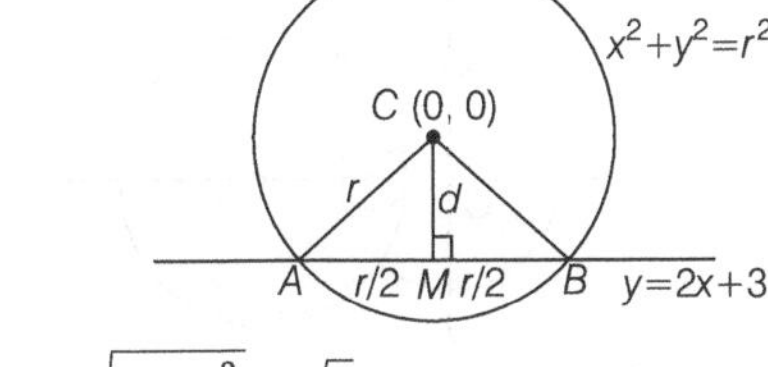

$$d = \sqrt{r^2 - \frac{r^2}{4}} = \frac{\sqrt{3}}{2} r \text{ and } d = \frac{3}{\sqrt{1+4}}$$

$$\therefore \quad \frac{3}{\sqrt{5}} = \frac{\sqrt{3}}{2} r$$

$$\Rightarrow \quad r^2 = \frac{12}{5}.$$

2. From figure

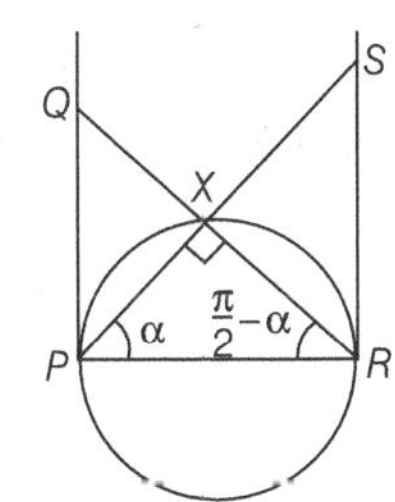

$$\tan \alpha = \frac{RS}{PR} = \frac{RS}{2r} \quad \ldots(i)$$

$$\tan\left(\frac{\pi}{2} - \alpha\right) = \frac{PQ}{PR} = \frac{PQ}{2r} \quad \ldots(ii)$$

Multiplying Eqs. (i) and (ii),

$$\tan \alpha \cot \alpha = \frac{RS.\,PQ}{(2r)^2}$$

$$\therefore \quad 2r = \sqrt{(RS)(PQ)}$$

3. See the diagram. Since there is no condition on θ, Q can be placed either at Q_1 or Q_2 for a particular position of P. So option (a) and (b) cannot be definitely true.

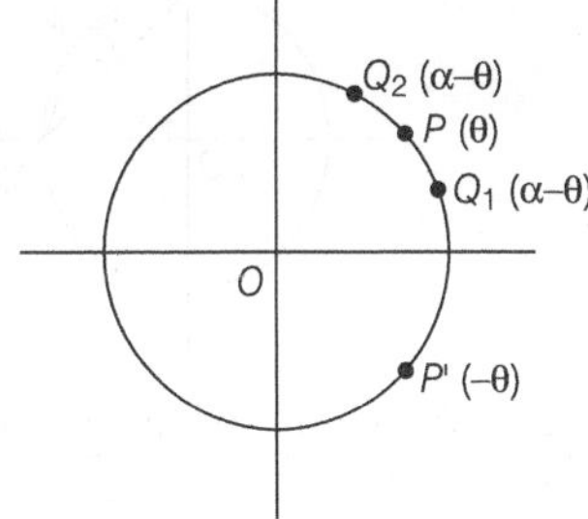

Consider a line through origin $y = mx$. If Q and P are reflection of each other with line mirror $y = mx$

$$(\text{Slope of } PQ) \times m = -1$$

or
$$m\left(\frac{\sin\theta - \sin(\alpha - \theta)}{\cos\theta - \cos(\alpha - \theta)}\right) = -1$$

$$m\left(\frac{2\cos\left(\frac{\alpha}{2}\right) \cdot \sin\left(\frac{2\theta - \alpha}{2}\right)}{2\sin\left(\frac{\alpha}{2}\right) . \sin\left(\frac{\alpha - 2\theta}{2}\right)}\right) = -1$$

or
$$m\left(-\cot\frac{\alpha}{2}\right) = -1$$

or
$$m = \tan(\alpha/2)$$

4. $(x-2)^2 + y^2 = 4$ centre is (2, 0) and radius = 2 distance between (2, 0) and (5, 6) is $\sqrt{9+36} = 3\sqrt{5}$.

$$\therefore \quad r_1 r_2 = \frac{3\sqrt{5}-2}{2} \cdot \frac{3\sqrt{5}+2}{2} = \frac{41}{4}$$

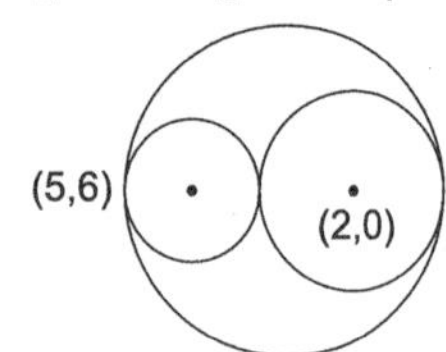

5. $\because Q$ lies on Y-axis,

Put $x = 0$ in $5x - 2y + 6 = 0$

$$\therefore \quad y = 3$$

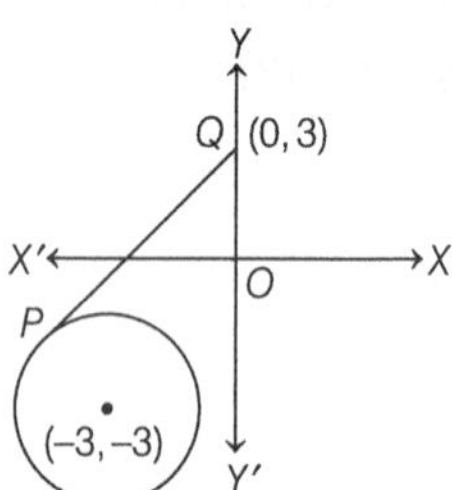

$$\Rightarrow \quad Q(0, 3)$$

$$\therefore \quad PQ = \sqrt{0^2 + 3^2 + 0 + 6 \times 3 - 2} = \sqrt{25} = 5$$

6. Let $P \equiv (r\cos\theta, r\sin\theta)$

Centroid of ΔPAB is

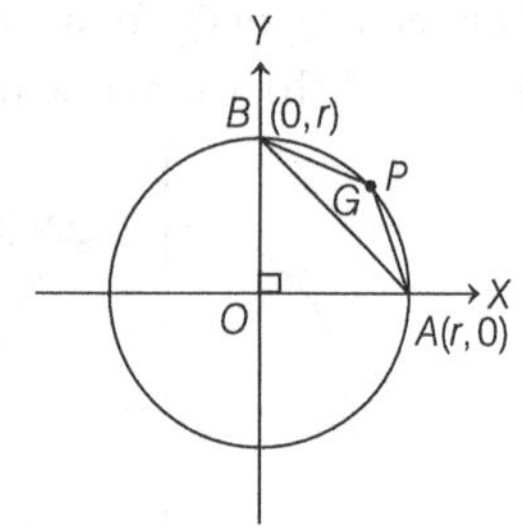

$$G\left(\frac{r+0+r\cos\theta}{3}, \frac{0+r+r\sin\theta}{3}\right)$$

Let $\quad x = \dfrac{r + r\cos\theta}{3} \Rightarrow 3x - r = r\cos\theta \quad \ldots\text{(i)}$

and $\quad y = \dfrac{r + r\sin\theta}{3} \Rightarrow 3y - r = r\sin\theta \quad \ldots\text{(ii)}$

Squaring and adding Eqs. (i) and (ii), we get

$$(3x-r)^2 + (3y-r)^2 = r^2$$

7. Equation of circle is $x^2 + y^2 - 2x + 4y + 3 = 0$

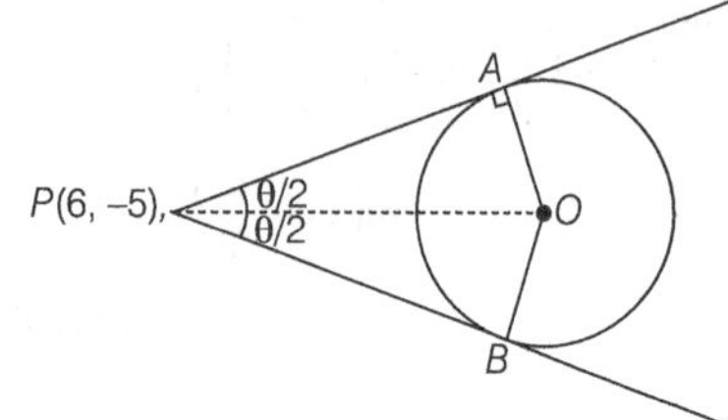

$\therefore$ Length of $PA = \sqrt{S_1}$

$$= \sqrt{(6)^2 + (-5)^2 - 2(6) + 4(-5) + 3}$$

$$= \sqrt{36 + 25 - 12 - 20 + 3} = \sqrt{32} = 4\sqrt{2}$$

and radius, $OA = \sqrt{(1)^2 + (-2)^2 - 3} = \sqrt{1 + 4 - 3} = \sqrt{2}$

$\therefore$ In ΔAOP,

$$\tan\frac{\theta}{2} = \frac{OA}{AP} = \frac{\sqrt{2}}{4\sqrt{2}} = \frac{1}{4}$$

$$\therefore \quad \tan\theta = \frac{2\tan\frac{\theta}{2}}{1-\tan^2\frac{\theta}{2}} = \frac{2\times\frac{1}{4}}{1-\left(\frac{1}{4}\right)^2} = \frac{\frac{1}{2}}{\frac{15}{16}} = \frac{8}{15}$$

8. Since the given equation represents a circle

$\therefore$ The coefficient of x^2 = coefficient of $y^2 \Rightarrow \lambda = \mu$

and the coefficient of $xy = 0$

$\Rightarrow \quad \lambda + \mu - 4 = 0 \Rightarrow \lambda = \mu = 2$

Thus the equation of the circle becomes

$$x^2 + y^2 - x - y - 10 = 0$$

whose radius is $= \sqrt{\left\{\left(\frac{1}{2}\right)^2 + \left(\frac{1}{2}\right)^2 + 10\right\}}$

$$= \sqrt{21/2} = \frac{\sqrt{42}}{2}$$

9. The circles are $x^2 + y^2 = 2^2$ and $(x-4)^2 + (y-0)^2 = 3^2$

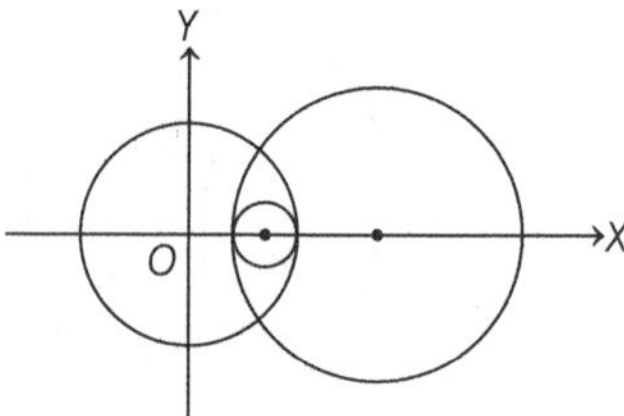

For the point P to enclose greatest area, the arbitrarily chosen point should be $\left(\frac{3}{2}, 0\right)$ and P should move in a circle of radius $\frac{1}{2}$.

The locus of P is $\left(x - \frac{3}{2}\right)^2 + (y-0)^2 = \left(\frac{1}{2}\right)^2$

$$\Rightarrow \quad x^2 + y^2 - 3x + 2 = 0.$$

10. If $(a, 0)$ is the centre C and P is $(2, -2)$, then

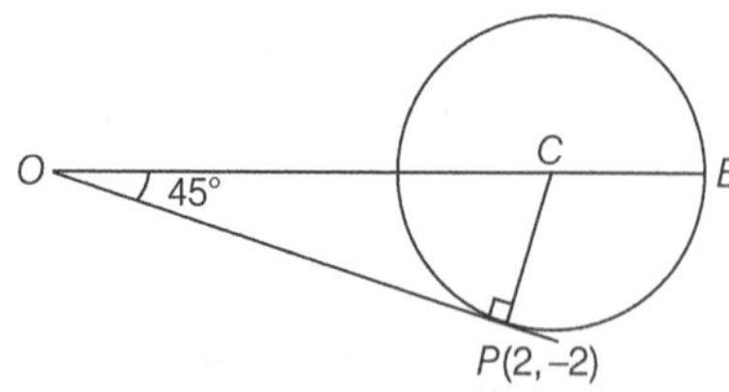

$\angle COP = 45°$, since the equation of OP is $x + y = 0$.

$\therefore OP = 2\sqrt{2} = CP$. Hence $OC = 4$

The point on the circle with the greatest x coordinate is B.

$$\alpha = x_B = OB = OC + CB = 4 + 2\sqrt{2}$$

11. Since $y = |x| + c$ and $x^2 + y^2 - 8|x| - 9 = 0$ both are symmetrical about Y-axis for $x > 0$, $y = x + c$.

Equation of tangent to circle $x^2 + y^2 - 8x - 9 = 0$

Parallel to $y = x + c$ is $y = (x - 4) + 5\sqrt{1 + 1}$

$\Rightarrow \quad y = x + (5\sqrt{2} - 4)$

for no solution $c > 5\sqrt{2} - 4$, $\therefore$ $c \in (5\sqrt{2} - 4, \infty)$

12. Circle I is $x^2 + y^2 - 16x - 20y + 164 = r^2$

$$\Rightarrow \quad (x - 8)^2 + (y - 10)^2 = r^2$$

$\Rightarrow$ $C_1(8, 10)$ is the centre of Ist circle and $r_1 = r$ is its radius

Circle II is $(x - 4)^2 + (y - 7)^2 = 36$

$\Rightarrow$ $C_2(4, 7)$ is the centre of 2nd circle and $r_2 = 6$ is its radius.

Two circles intersect if $|r_1 - r_2| < C_1C_2 < r_1 + r_2$

$$\Rightarrow \quad |r - 6| < \sqrt{(8 - 4)^2 + (10 - 7)^2} < r + 6$$

$$\Rightarrow \quad |r - 6| < \sqrt{16 + 9} < r + 6 \Rightarrow |r - 6| < 5 < r + 6$$

Now as, $5 < r + 6$ always, we have to solve only

$$|r - 6| < 5 \Rightarrow -5 < r - 6 < 5$$

$$\Rightarrow \quad 6 - 5 < r < 5 + 6 \Rightarrow 1 < r < 11$$

13. The triangle is isosceles and therefore the median through C is the bisector of $\angle C$. The equation of the angle bisector can be taken as $y = -x$ and $l = (-a, a)$, where a is positive.

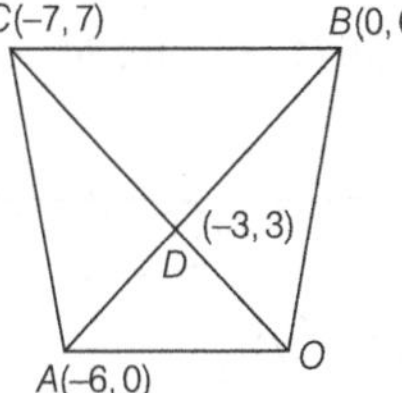

Equation of AC is $y - 0 = -7(x + 6)$ or $7x + y + 42 = 0$ and equation of AB is $x - y + 6 = 0$.

The length of the perpendicular from l to AB and AC are equal

$\therefore \left|\dfrac{-7a + a + 42}{\sqrt{50}}\right| = \left|\dfrac{-a - a + 6}{\sqrt{2}}\right|$ giving the positive value

$$a = \frac{9}{2}$$

$\therefore$ Centre is $\left(-\dfrac{9}{2}, \dfrac{9}{2}\right)$ and radius $= \dfrac{3}{\sqrt{2}}$

$\therefore$ The equation of the circle is $\left(x + \dfrac{9}{2}\right)^2 + \left(y - \dfrac{9}{2}\right)^2 = \dfrac{9}{2}$

$$\therefore \quad x^2 + y^2 + 9x - 9y + 36 = 0$$

14. Let $C(h, k)$ be the centre of the circle passing through the end points of the rod AB and PQ of lengths a and b respectively, CL and CM be perpendiculars from C on AB and PQ respectively. Then $CA = CP$ (radii of the same circle)

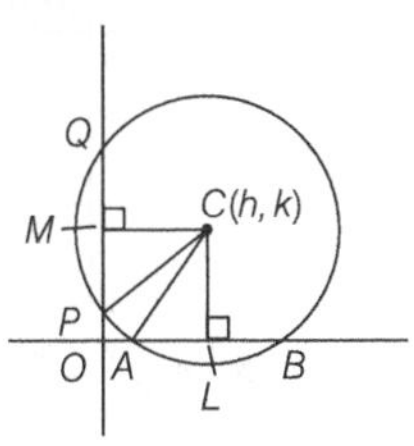

$$\Rightarrow \quad k^2 + \frac{a^2}{4} = h^2 + \frac{b^2}{4}$$

$[\because AL = a/2 \text{ and } MP = b/2]$

$$\Rightarrow \quad 4(h^2 - k^2) = a^2 - b^2$$

Hence, locus of (h, k) is $4(x^2 - y^2) = a^2 - b^2$.

15. Equation of pair of tangents by $SS_1 = T^2$ is $(a^2 - 1)y^2 - x^2 + 2ax - a^2 = 0$

If θ be the angle between the tangents, then

$$\tan\theta = \frac{2\sqrt{(h^2 - ab)}}{a + b} = \frac{2\sqrt{-(a^2 - 1)(-1)}}{a^2 - 2} = \frac{2\sqrt{a^2 - 1}}{a^2 - 2}$$

$\because$ θ lies in II quadrant, than $\tan\theta < 0$

$$\therefore \quad \frac{2\sqrt{a^2 - 1}}{a^2 - 2} < 0$$

$$\Rightarrow \quad a^2 - 1 > 0 \text{ and } a^2 - 2 < 0$$

$$\Rightarrow \quad 1 < a^2 < 2 \Rightarrow a \in (-\sqrt{2}, -1) \cup (1, \sqrt{2})$$

16. $\because$ Reflection of $(5, 8)$ in BC will lie on circumcircle.

$\therefore$ $(8, 5)$ will lie on circumcircle equation of circumcircle is $(x - 2)^2 + (y - 3)^2 = (8 - 2)^2 + (3 - 5)^2$

$\Rightarrow x^2 + y^2 - 4x - 6y - 27 = 0$

17. The equation of chord of contact AB to circle $x^2 + y^2 - 8x - 4y + 16 = 0$ w.r.t. point origin $(0, 0)$ is $T = 0$

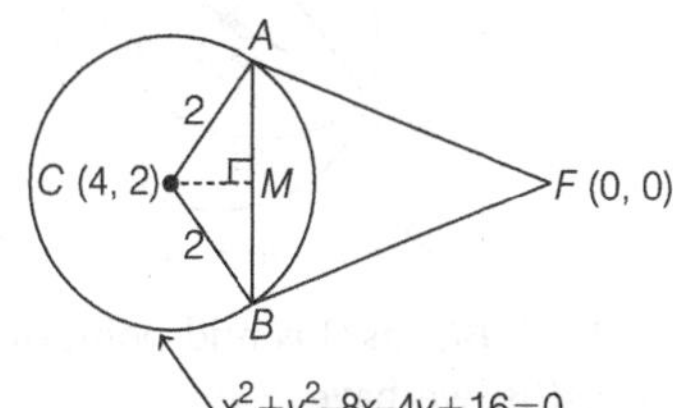

$$\Rightarrow \quad 0x + 0y - 4(x + 0) - 2(y + 0) + 16 = 0$$

$$\Rightarrow \quad 2x + y - 8 = 0 \quad \ldots\text{(i)}$$

In right angled triangle AMC,

$$CM = \frac{|2(4) + 2 - 8|}{\sqrt{2^2 + 1^2}} = \frac{2}{\sqrt{5}}$$

$$\because \quad AM^2 = CA^2 - CM^2 = 2^2 - \left(\frac{2}{\sqrt{5}}\right)^2 = 4 - \frac{4}{5} = \frac{16}{5}$$

$$\Rightarrow \quad AM = \frac{4}{\sqrt{5}}$$

$$\because \quad AB^2 = (2AM)^2 = \left(\frac{2 \times 4}{\sqrt{5}}\right)^2 = \frac{64}{5}$$

18. Since the hypotenuse of a right angled triangle inscribed in a circle is a diameter of the circle. If the coordinates of the end C of the hypotenuse BC are (a, b), the coordinates of B are $(-a, -b)$. Equation of BC is $\dfrac{y}{x} = \dfrac{b}{a}$. If A is the vertex of the

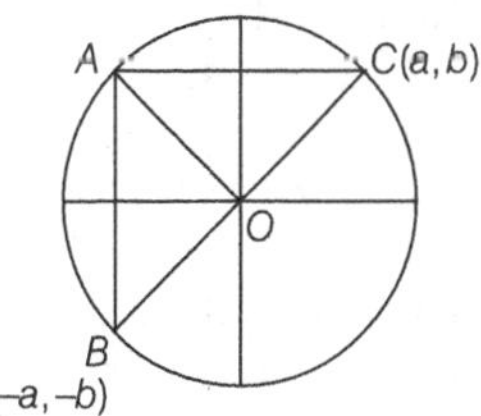

isosceles triangle, then OA is perpendicular to BC and the equation of AO is $\frac{y}{x} = -\frac{a}{b}$ which meets the circle $x^2 + y^2 = r^2$ at points for which $\left(\frac{a^2}{b^2} + 1\right)x^2 = r^2 = a^2 + b^2$

$[\because (a, b)$ lies on $x^2 + y^2 = r^2]$

$\Rightarrow x^2 = b^2 \Rightarrow x = \pm b \Rightarrow y = \mp a$

$\therefore$ Coordinates of A are $(-b, a)$ or $(b, -a)$.

19. The common chord must be a diameter of the second circle. The equation to the common chord is $S_1 - S_2 = 0$

i.e. $6x + 14y + (c + d) = 0$. The centre of the second circle $(1, -4)$ lies on it. $6(1) + 14(-4) + c + d = 0$

$\therefore c + d = 50$

20. Given equation of circle is $x^2 + y^2 - 4x + 6y - 12 = 0$, whose centre is $(2, -3)$ and radius

$$= \sqrt{2^2 + (-3)^2 + 12}$$
$$= \sqrt{4 + 9 + 12} = 5$$

Now, according to given information, we have the following figure.

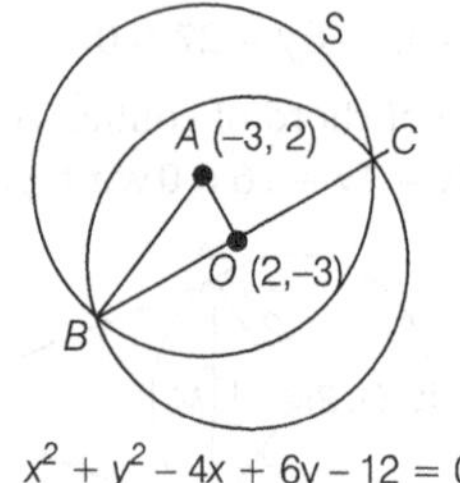

$x^2 + y^2 - 4x + 6y - 12 = 0$

Clearly, $AO \perp BC$, as O is mid-point of the chord.

Now, in ΔAOB, we have

$$OA = \sqrt{(-3-2)^2 + (2+3)^2}$$
$$= \sqrt{25 + 25} = \sqrt{50} = 5\sqrt{2}$$

and $\quad OB = 5$

$$\therefore \quad AB = \sqrt{OA^2 + OB^2} = \sqrt{50 + 25} = \sqrt{75} = 5\sqrt{3}$$

21. Given circle is $(x - 3)^2 + (y - 4)^2 = 25$

Since, locus of point of intersection of two perpendicular tangents is director circle, then its equation is

$$(x - 3)^2 + (y - 4)^2 = 50$$
$$\Rightarrow \quad x^2 + y^2 - 6x - 8y - 25 = 0$$

22. $y = mx$ be chord the points of intersection are given by $x^2(1 + m^2) - x(3 + 4m) - 4 = 0$

$$\therefore \quad x_1 + x_2 = \frac{3 + 4m}{1 + m^2} \text{ and } x_1 x_2 = \frac{-4}{1 + m^2}$$

Since, $(0, 0)$ divides the chord joining (x_1, y_1) and (x_2, y_2) in the ratio $1 : 4$

$$\therefore \quad x_2 = -4x_1$$

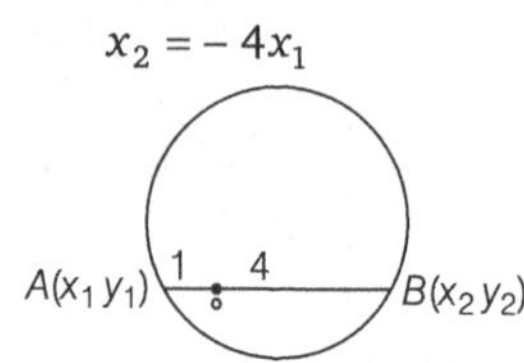

then $-3x_1 = \frac{3 + 4m}{1 + m^2}$ and $-4x_1^2 = \frac{-4}{1 + m^2}$

$$9 + 9m^2 = 9 + 16m^2 + 24m$$

i.e. $\quad m = 0, -\frac{24}{7}$

$\therefore$ Then lines are $y = 0$ and $7y + 24x = 0$

23. We have,

$S_1 : x^2 + y^2 = 9$ centre $(0, 0)$ and radius $r_1 = 3$

$S_2 : (x - 2)^2 + y^2 = 1$ centre $(2, 0)$ and radius $r_2 = 1$

$\because c_1c_2 = r_1 - r_2$

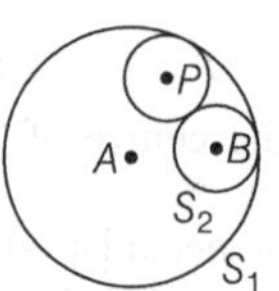

$\therefore$ Given circle are touching internally

Let a variable circle with centre P and radius r.

$\Rightarrow \quad PA = r_1 - r$ and $PB = r_2 + r$

$\Rightarrow \quad PA + PB = r_1 + r_2$

$\Rightarrow \quad PA + PB = 4\ (> AB)$

$\Rightarrow$ Locus of P is an ellipse with foci at $A(0, 0)$ and $B(2, 0)$ and length of major axis is $2a = 4$.

$$e = \frac{1}{2}$$

$\Rightarrow$ centre is at $(1, 0)$ and $b^2 = a^2(1 - e^2) = 3$ if x-ellipse

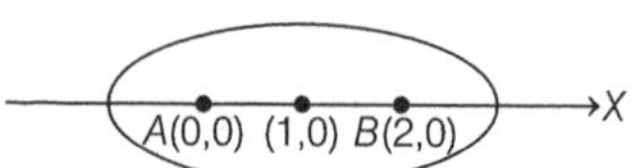

$$\Rightarrow E : \frac{(x-1)^2}{4} + \frac{y^2}{3} = 1$$

which is satisfied by $\left(2, \pm\frac{3}{2}\right)$.

24. Let $\angle AOL = \theta$

$\therefore \quad A \equiv (a\cos\theta, a\sin\theta)$

$\therefore \quad M \equiv (a + a\cos\theta, a\sin\theta)$

$\quad x = a + a\cos\theta$

$\Rightarrow \quad (x - a) = a\cos\theta \quad \ldots(i)$

and $\quad y = a\sin\theta \quad \ldots(ii)$

From Eqs. (i) and (ii),

$$(x - a)^2 + y^2 = a^2$$
$$\Rightarrow \quad x^2 + y^2 - 2ax = 0$$

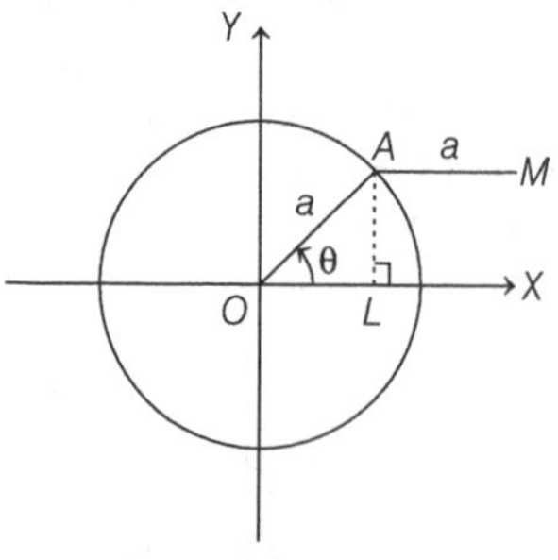

$$\Rightarrow \quad x^2 + y^2 = 2ax$$

25. We have, $x^2 + y^2 - 8x - 12y + p = 0$
then centre and radius of the circle are (4, 6) and $\sqrt{(52 - p)}$ respectively.
$\because$ Circle neither cuts nor touches any one of the axes of coordinates, then
x-coordinate of centre > radius *ie*, $4 > \sqrt{52 - p}$

$\Rightarrow \quad p > 36 \quad \ldots(i)$

and y-coordinate of centre > radius

$6 > \sqrt{52 - p} \Rightarrow p > 16 \quad \ldots(ii)$

$\because D$ is interior point of the circle, then CD < radius

$\sqrt{5} < \sqrt{52 - p} \Rightarrow p < 47 \quad \ldots(iii)$

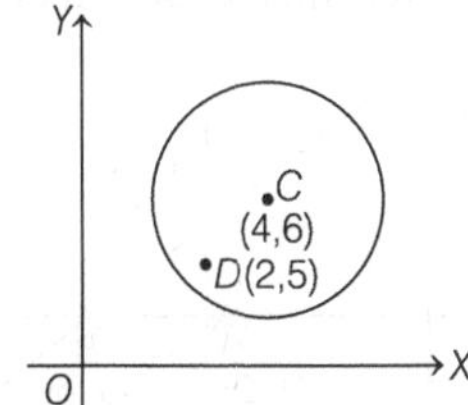

From Eqs. (i), (ii) and (iii) we obtain

$$36 < p < 47$$

$$\therefore \quad p \in (36, 47).$$

26. Let the given lines be $L_1 \equiv a_1x + b_1y + c_1 = 0$ and $L_2 \equiv a_2x + b_2y + c_2 = 0$. Suppose L_1 meets the coordinate axes at A and B and L_2 meets at C and D. Then coordinates of A, B, C, D are

$$A\left(-\frac{c_1}{a_1}, 0\right), B\left(0, -\frac{c_1}{b_1}\right), C\left(-\frac{c_2}{a_2}, 0\right), \text{ and } D\left(0, -\frac{c_2}{b_2}\right),$$

Since A, B, C, D are concyclic, therefore $OA \cdot OC = OD \cdot OB$

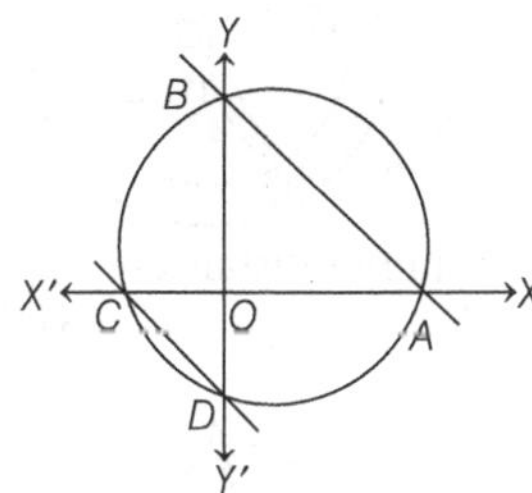

$$\Rightarrow \quad \left(-\frac{c_1}{a_1}\right)\left(-\frac{c_2}{a_2}\right) = \left(-\frac{c_2}{b_2}\right)\left(-\frac{c_1}{b_1}\right) \Rightarrow a_1a_2 = b_1b_2$$

27. $\because$ Area of triangle $= \frac{1}{2}$ (Base) (Height)

Area of triangle is maximum if height of the triangle is maximum. We know that height is maximum of the isosceles triangle.

Hence area of $\Delta\, ABC$ is maximum when it is isosceles.

28. Equation of conic is $(2x - y + 1)(x - 2y + 3) + \lambda xy = 0$
for circle coefficient of $xy = 0$

ie, $\quad -5 + \lambda = 0, \therefore \lambda = 5$

$\therefore$ Circle is $2x^2 + 2y^2 + 7x - 5y + 3 = 0$

$$\Rightarrow \quad x^2 + y^2 + \frac{7}{2}x - \frac{5}{2}y + \frac{3}{2} = 0$$

$\therefore$ Centre is $\left(-\frac{7}{4}, \frac{5}{4}\right)$

29. We know that, limiting points of the co-axial system of circles $x^2 + y^2 + 2gx + 2fy + c = 0$ (other than origin) are given by $\left(-\frac{gc}{g^2 + f^2}, -\frac{fc}{g^2 + f^2}\right)$

Here given circle $x^2 + y^2 - 6x - 8y + 1 = 0$, then $g = -3, f = -4, c = 1$.

Hence other limiting point is $\left(\frac{3}{25}, \frac{4}{25}\right)$

30. Let (h, k) be any point in the set, then equation of circle is

$$(x - h)^2 + (y - k)^2 = 9$$

But $\quad (h, k)$ lie on $x^2 + y^2 = 25$,

then $\quad h^2 + k^2 = 25$

$\therefore\ 2 \le$ Distance between the centres of two circles ≤ 8

$\Rightarrow\ 2 \le \sqrt{(h^2 + k^2)} \le 8$

$4 \le h^2 + k^2 \le 64$

$\therefore$ Locus of (h, k) is $4 \le x^2 + y^2 \le 64$

31.

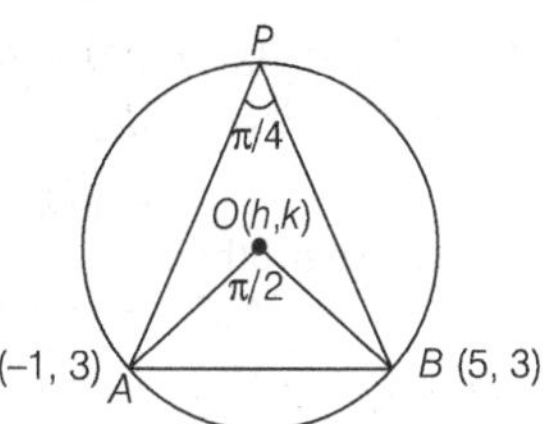

We have, $\angle APB = \frac{\pi}{4}$

$\Rightarrow \quad \angle AOB = 2\angle APB = 2 \times \frac{\pi}{4} = \frac{\pi}{2}$

$\Rightarrow \quad \angle OAB = \angle OBA = \frac{\pi}{4}$

$\Rightarrow \quad OA = OB$

$\Rightarrow \quad OA^2 = OB^2$

$\Rightarrow (h + 1)^2 + (k - 3)^2 = (h - 5)^2 + (k - 3)^2$

$\Rightarrow \quad (h + 1)^2 = (h - 5)^2$

$\Rightarrow \quad h^2 + 2h + 1 = h^2 - 10h + 25$

$\Rightarrow \quad 12h = 24 \Rightarrow h = 2$

Since, $\quad \angle AOB = \frac{\pi}{2}$

$\therefore \quad OA \perp OB$

$\Rightarrow$ Slope of $OA \times$ slope of $OB = -1$

$\Rightarrow \quad \frac{k - 3}{h + 1} \times \frac{k - 3}{h - 5} = -1$

$\Rightarrow (k - 3)^2 = -(h + 1)(h - 5)$

$\Rightarrow (k - 3)^2 = -(2 + 1)(2 - 5)$

$\Rightarrow (k - 3)^2 = 9$

$\Rightarrow \quad k - 3 = \pm 3 \Rightarrow k = 3 \pm 3$

$\Rightarrow \quad k = 0, 6$

$\therefore$ Centre of circle is (2, 0) or (2, 6)

When centre is (2, 0)

$\therefore$ Radius of circle $= OA$

$$= \sqrt{(2 + 1)^2 + (0 - 3)^2} = \sqrt{9 + 9} = \sqrt{18}$$

When centre is (2, 6)

$\therefore$ Radius of circle $= \sqrt{(2+1)^2 + (6-3)^2} = \sqrt{9+9} = 3\sqrt{2}$

$\therefore$ Equation of circle is

$$(x-2)^2 + (y-0)^2 = 18$$
$$\Rightarrow \quad x^2 + y^2 - 4x - 22 = 0$$
or
$$(x-2)^2 + (y-6)^2 = 18$$
$$\Rightarrow \quad x^2 + y^2 - 4x - 12y + 22 = 0$$

32. Let coordinates of A and B are (α, β) and (γ, δ) respectively.

$$\alpha + \gamma = -2a, \alpha\gamma = -b^2$$

and $\beta + \delta = -2p, \beta\delta = -q^2$

Equation of circle with AB as diameter.

$$(x-\alpha)(x-\gamma) + (y-\beta)(y-\delta) = 0$$
$$\Rightarrow \quad x^2 + y^2 - x(\alpha+\gamma) - y(\beta+\delta) + \alpha\gamma + \beta\delta = 0$$
$$\Rightarrow \quad x^2 + y^2 + 2ax + 2py - b^2 - q^2 = 0$$
$$\therefore \quad \text{Radius} = \sqrt{(a^2 + p^2 + b^2 + q^2)} = \sqrt{(a^2 + b^2 + p^2 + q^2)}$$

33. Centre and radius of the circle $x^2 + y^2 + 2gx + 2fy = 0$ are $C_1(-g, -f)$ and $r_1 = \sqrt{(g^2 + f^2)}$ and centre and radius of the circle $x^2 + y^2 + 2g_1x + 2f_1y = 0$ are $C_2(-g_1, -f_1)$

and $\quad r_2 = \sqrt{(g_1^2 + f_1^2)}$

$\because$ Circles touch each other.

$$\therefore \quad |C_1C_2| = |r_1 \pm r_2|$$
$$\Rightarrow \sqrt{(g-g_1)^2 + (f-f_1)^2} = |\sqrt{(g^2+f^2)} \pm \sqrt{(g_1^2 + f_1^2)}|$$

Squaring both sides then

$$g^2 + f^2 + g_1^2 + f_1^2 - 2gg_1 - 2ff_1 = g^2 + f^2 + g_1^2 + f_1^2 \pm 2\sqrt{(g^2+f^2)}\sqrt{(g_1^2+f_1^2)}$$
$$\Rightarrow \quad (gg_1 + ff_1)^2 = (g^2 + f^2)(g_1^2 + f_1^2)$$
$$\Rightarrow \quad g^2g_1^2 + f^2f_1^2 + 2gg_1ff_1 = g^2g_1^2 + g^2f_1^2 + f^2g_1^2 + f^2f_1^2$$
$$\Rightarrow \quad g^2f_1^2 + f^2g_1^2 - 2gg_1ff_1 = 0$$
$$\Rightarrow \quad (gf_1 - fg_1)^2 = 0 \Rightarrow gf_1 = fg_1$$

34. It is given that points P and Q are intersecting points of circle

$$(x-3)^2 + (y+2)^2 = 25 \quad \ldots\text{(i)}$$

Line $\quad y = mx + 1 \quad \ldots\text{(ii)}$

And, the mid-point of PQ is A having x-coordinate $\frac{-3}{5}$

so y-coordinate is $1 - \frac{3}{5}m$. So, $A\left(-\frac{3}{5}, 1 - \frac{3}{5}m\right)$

From the figure,

$\because AC \perp PQ$

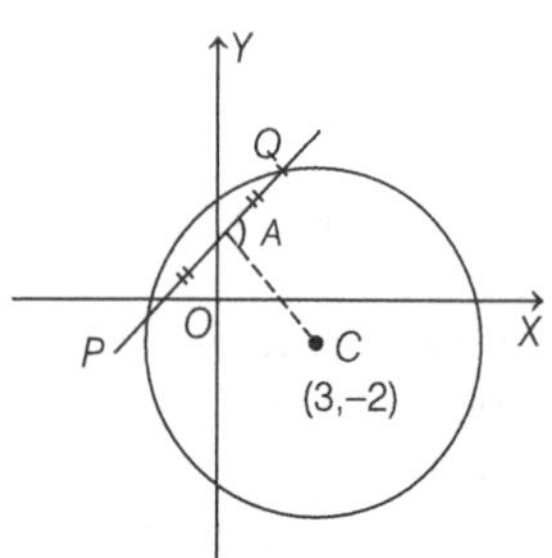

$\Rightarrow$ (slope of AC) $\times$ (slope of PQ) $= -1$

$$\Rightarrow \quad \left(\frac{-2-1+\frac{3}{5}m}{3+\frac{3}{5}}\right) \times m = -1$$
$$\Rightarrow \quad \frac{(3/5)m - 3}{18/5} m = -1 \Rightarrow \left(\frac{3m-15}{18}\right)m = -1$$
$$\Rightarrow \quad 3m^2 - 15m + 18 = 0 \Rightarrow m^2 - 5m + 6 = 0$$
$$\Rightarrow \quad m = 2 \text{ or } 3$$

35. $(x + y - 4)(xy - 2x - y + 2) = 0$

$(x + y - 4)(x-1)(y-2) = 0 \quad \because AB \perp AC$

$\therefore BC$ is diameter, then equation of circumcircle is

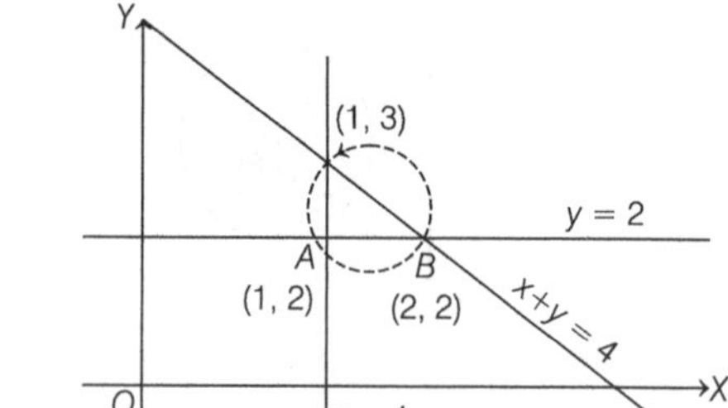

$$(x-2)(x-1) + (y-2)(y-3) = 0$$
$$\Rightarrow \quad x^2 + y^2 - 3x - 5y + 8 = 0$$

36. Given circle $(x-1)^2 + (y+2)^2 = 16$

Its director circle is $(x-1)^2 + (y+2)^2 = 32$

$$\therefore \quad OS = 4\sqrt{2}$$

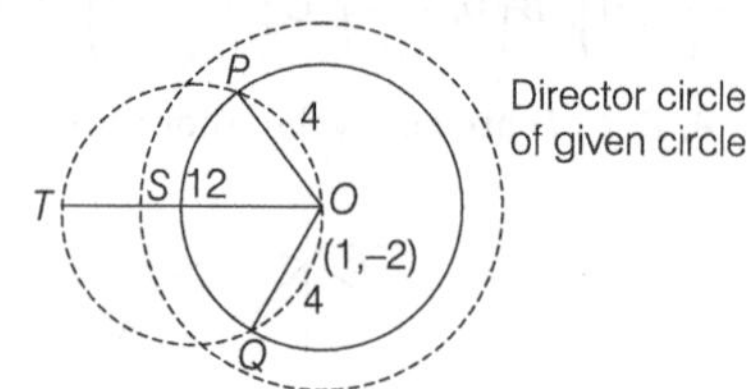

Therefore, required distance,

$$TS = OT - SO = 12 - 4\sqrt{2}$$

37. We have,

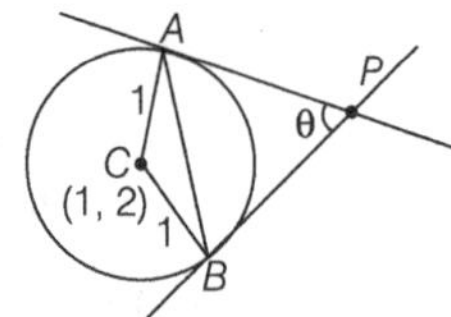

$$\tan\theta = \frac{12}{5} \text{ and } PA = \cot\frac{\theta}{2}$$

$$\therefore \text{Area of } \Delta PAB = \frac{1}{2}(PA)^2 \sin\theta = \frac{1}{2}\cot^2\frac{\theta}{2}\sin\theta$$
$$= \frac{1}{2}\left(\frac{1+\cos\theta}{1-\cos\theta}\right)\sin\theta$$
$$= \frac{1}{2}\left(\frac{1+\frac{5}{13}}{1-\frac{5}{13}}\right)\left(\frac{12}{13}\right) = \frac{1}{2} \times \frac{18}{18} \times \frac{2}{13} = \frac{27}{26}$$

$$\text{Area of } \Delta CAB = \frac{1}{2}\sin\theta = \frac{1}{2}\left(\frac{12}{13}\right) = \frac{6}{13}, \therefore \frac{\text{Area of } \Delta PAB}{\text{Area of } \Delta CAB} = \frac{9}{4}$$

38. Equation of circle passing through the intersection of the circles $x^2 + y^2 - 6x = 0$ and $x^2 + y^2 - 4y = 0$, is

$(x^2 + y^2 - 6x) + \lambda(x^2 + y^2 - 4y) = 0$, $(\lambda \neq -1)$

$\Rightarrow (1+\lambda)x^2 + (1+\lambda)y^2 - 6x - 4\lambda y = 0$, having centre $C\left(\frac{3}{1+\lambda}, \frac{2\lambda}{1+\lambda}\right)$ and the centre 'C' lies on the lies on the line $2x - 3y + 12 = 0$, so

$$\frac{6}{1+\lambda} - \frac{6\lambda}{1+\lambda} + 12 = 0 \Rightarrow 1 - \lambda + 2\lambda + 2 = 0 \Rightarrow \lambda = -3.$$

So, equation of required circle is

$$2x^2 + 2y^2 + 6x - 12y = 0 \Rightarrow x^2 + y^2 + 3x - 6y = 0 \quad \ldots(i)$$

from the given options circle (i) passes through the point (–3,6).

39. Let coordinates of mid-point AB is $D(h, k)$, D lie on

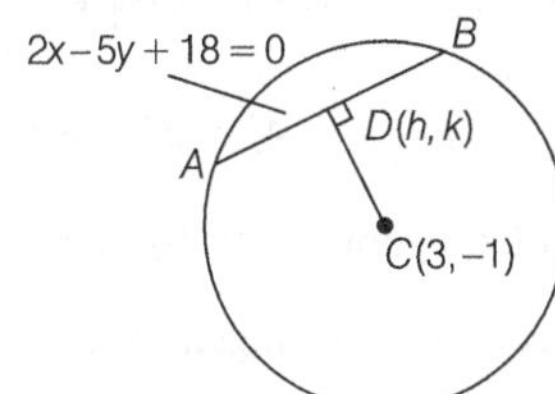

$$2x - 5y + 18 = 0$$

$\therefore \quad 2h - 5k + 18 = 0 \quad \ldots(i)$

Equation CD, (with slope – 5/2)

$$y + 1 = -\frac{5}{2}(x - 3) \Rightarrow 2y + 2 = -5x + 15$$

$$5x + 2y - 13 = 0 \quad \ldots(ii)$$

D lies on $\therefore \quad 5h + 2k - 13 = 0 \quad \ldots(iii)$

On solving Eqs. (i) and (iii), we get

$$h = 1, k = 4$$

$\therefore$ Coordinate of mid point of chord is (1, 4)

40. Let $\phi(x, y) \equiv x^2 + y^2 + 2gx + 2fy + c = 0$

$\therefore \phi(0, \lambda) = 0 + \lambda^2 + 0 + 2f\lambda + c = 0$ have equal roots,

Then $2 + 2 = -\frac{2f}{1}$ and $2 \cdot 2 = \frac{c}{1}$

$\therefore \quad f = -2$ and $c = 4$

and $\phi(\lambda, 0) \equiv \lambda^2 + 0 + 2g\lambda + 0 + c = 0$

$\therefore \quad \lambda^2 + 2g\lambda + c = 0$

Here $c = 4 \quad \therefore \lambda^2 + 2g\lambda + 4 = 0$ have roots 4/5, 5

$\therefore \quad \frac{4}{5} + 5 = -2g \Rightarrow g = -\frac{29}{10}$

$\therefore \quad$ Centre $\equiv (-g, -f) = \left(\frac{29}{10}, 2\right)$

41.

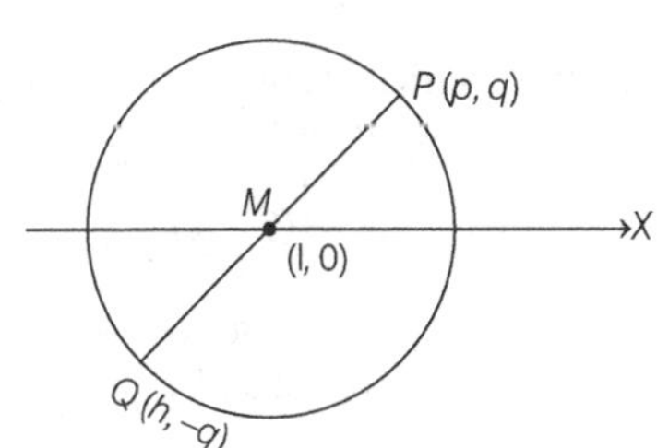

Suppose chord bisect at $M(\lambda, 0)$, then other end point of chord is $(h, -q)$

where $\lambda = \frac{p+h}{2}$

which lie on $x^2 + y^2 = px + qy$

or $h^2 + q^2 = ph - q^2$

$\Rightarrow h^2 - ph + 2q^2 = 0$

for two chords, $B^2 - 4AC > 0$

or $\quad p^2 - 4 \cdot 1 \cdot 2q^2 > 0$

or $\quad p^2 > 8q^2$

42. (c) Equation of the two circles be $(x - r)^2 + (y - r)^2 = r^2$ i.e. $x^2 + y^2 - 2rx - 2ry + r^2 = 0$ where $r = r_1$ and r_2. Condition of orthogonality gives

$$2r_1r_2 + 2r_1r_2 = r_1^2 + r_2^2$$

$$\Rightarrow \quad 4r_1r_2 = r_1^2 + r_2^2$$

Circle passes through (a, b)

$\Rightarrow \quad a^2 + b^2 - 2ra - 2rb + r^2 = 0$

i.e., $r^2 - 2r(a + b) + a^2 + b^2 = 0$

$$r_1 + r_2 = 2(a + b) \text{ and } r_1r_2 = a^2 + b^2$$

$\therefore \quad 4(a^2 + b^2) = 4(a + b)^2 - 2(a^2 + b^2)$

i.e. $\quad a^2 - 4ab + b^2 = 0$

43. Circle is $x^2 + y^2 = r^2\cos^2\theta + r^2\sin^2\theta$

$$x^2 + y^2 = r^2$$

Equation of tangent at θ is

$$x\cos\theta + y\sin\theta = r \quad \ldots(i)$$

and at $(\theta + \pi/3)$ is

$$x\cos(\theta + \pi/3) + y\sin(\theta + \pi/3) = r$$

$$\Rightarrow x\left(\frac{1}{2}\cos\theta - \frac{\sqrt{3}}{2}\sin\theta\right) + y\left(\frac{1}{2}\sin\theta + \frac{\sqrt{3}}{2}\cos\theta\right) = r$$

$$\Rightarrow \quad x\cos\theta + y\sin\theta - x\sqrt{3}\sin\theta + y\sqrt{3}\cos\theta = 2r$$

$$\Rightarrow \quad r - \sqrt{3}(x\sin\theta - y\cos\theta = 2r$$

or $\quad x\sin\theta - y\cos\theta = -\frac{r}{\sqrt{3}} \quad \ldots(ii)$

Squaring and adding Eqs. (i) and (ii), then we get

$$x^2 + y^2 = \frac{4r^2}{3}$$

$$\Rightarrow \quad 3(x^2 + y^2) = 4r^2$$

44. Given circle is

$$x^2 + y^2 + 2gx + 2fy + c = 0 \quad \ldots(i)$$

end the circle with limiting point (a, b) is

$$(x - a)^2 + (y - b)^2 = 0$$

or $\quad x^2 + y^2 - 2ax - 2by + a^2 + b^2 = 0 \quad \ldots(ii)$

$\therefore$ Equation of radical axis is

$$(x^2 + y^2 + 2gx + 2fy + c) - (x^2 + y^2 - 2ax - 2by + a^2 + b^2) = 0$$

$$\Rightarrow \quad 2(g + a)x + 2(f + b)y + c - a^2 - b^2 = 0$$

45. (c) Equation of circle is

$$x^2 + y^2 - 6x + 4y - 12 = 0$$

$\therefore \quad 2x + 2yy' - 6 + 4y' = 0$

$\Rightarrow \quad y' = \dfrac{6 - 2x}{2y + 4}$

$\therefore$ slope of tangent at $(-1, 1) = \left.\dfrac{dv}{dx}\right|_{(-1, 1)}$

$$= \frac{6 - 2(-1)}{2(1) + 4} = \frac{8}{6} = \frac{4}{3}$$

$\therefore$ Equation of tangent is

$$y - 1 = \frac{4}{3}(x + 1)$$

$\Rightarrow \quad 3y - 3 = 4x + 4$

$\Rightarrow \quad 4x - 3y + 7 = 0$

Let the equation of chord is

$$4x - 3y + k = 0$$

$\therefore \quad \dfrac{k - 7}{\sqrt{16 + 9}} = \pm 1$

$\Rightarrow \quad k - 7 = \pm 5$

$\Rightarrow \quad k = 7 \pm 5 \Rightarrow k = 12, 2$

$\therefore$ Equation of chord can be

$$4x - 3y + 12 = 0$$

or $\quad 4x - 3y + 2 = 0$

Now, Distance of both above lines from centre of circle are

$$d_1 = \left|\frac{12 + 6 + 12}{5}\right| = 6$$

and $\quad d_2 = \left|\dfrac{12 + 6 + 2}{5}\right| = 4$

and radius of circle,

$$r = \sqrt{9 + 4 + 12} = 5$$

$\therefore \quad d_2 < r$

$\therefore$ Equation of chord is $4x - 3y + 2 = 0$

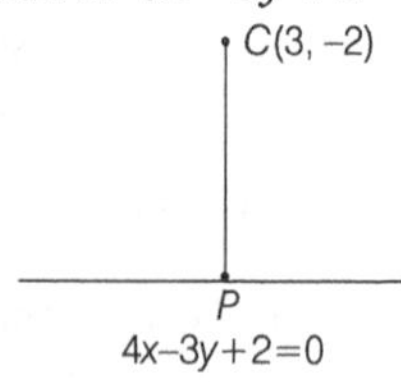

Now, slope of line $4x - 3y + 2 = 0$ is $\dfrac{4}{3}$

$\therefore$ Slope of line $CP = \dfrac{-3}{4}$

$\therefore$ Equation of CP is

$$y + 2 = \frac{-3}{4}(x - 3)$$

$\Rightarrow \quad 4y + 8 = -3x + 9$

$\Rightarrow \quad 3x + 4y = 1$

On solving $4x - 3y + 2 = 0$ and $3x + 4y = 1$, we get $P\left(-\dfrac{1}{5}, \dfrac{2}{5}\right)$.

$\therefore$ Mid-point of chord is $\left(-\dfrac{1}{5}, \dfrac{2}{5}\right)$

46. According to given information, we have the following figure.

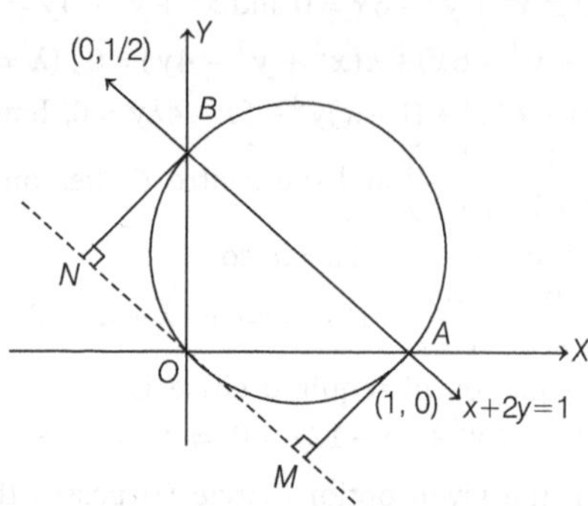

From figure, equation of circle (diameter form) is

$$(x - 1)(x - 0) + (y - 0)\left(y - \frac{1}{2}\right) = 0$$

$\Rightarrow \quad x^2 + y^2 - x - \dfrac{y}{2} = 0$

Equation of tangent at (0, 0) is $x + \dfrac{y}{2} = 0$

[$\because$ equation of tangent at (x_1, y_1) is given by $T = 0$.
Here, $T = 0 \Rightarrow xx_1 + yy_1 - \dfrac{1}{2}(x + x_1) - \dfrac{1}{4}(y + y_1) = 0$]

$\Rightarrow \quad 2x + y = 0$

Now, $\quad AM = \dfrac{|2 \cdot 1 + 1 \cdot 0|}{\sqrt{5}} = \dfrac{2}{\sqrt{5}}$

[$\because$ distance of a point $P(x_1, y_1)$ from a line $ax + by + c = 0$ is $\dfrac{|ax_1 + by_1 + c|}{\sqrt{a^2 + b^2}}$]

and $\quad BN = \dfrac{\left|2 \cdot 0 + 1\left(\dfrac{1}{2}\right)\right|}{\sqrt{5}} = \dfrac{1}{2\sqrt{5}}$

$\therefore \quad AM + BN = \dfrac{2}{\sqrt{5}} + \dfrac{1}{2\sqrt{5}} = \dfrac{4 + 1}{2\sqrt{5}} = \dfrac{\sqrt{5}}{2}$

47. Given, $(1 + \alpha x)^n = 1 + 8x + 24x^2 + \dots$

$\Rightarrow \quad 1 + n(\alpha x) + \dfrac{n(n - 1)}{1 \cdot 2}(\alpha x)^2 + \dots$

$$= 1 + 8x + 24x^2 + \dots$$

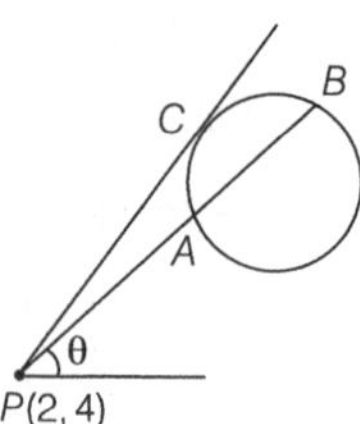

Equating the coefficients of x and x^2, we get

$$n\alpha = 8, \frac{n\alpha(n\alpha - \alpha)}{1 \cdot 2} = 24$$

or $\quad \dfrac{8(8 - \alpha)}{2} = 24$

$\Rightarrow \quad 8 - \alpha = 6$

$\therefore \quad \alpha = 2$ and $n = 4$

Equation of line is $\frac{x-2}{\cos\theta} = \frac{y-4}{\sin\theta} = r$, then point $(2 + r\cos\theta, 4 + r\sin\theta)$ lies on the circle $x^2 + y^2 = 4$,

then $(2 + r\cos\theta)^2 + (4 + r\sin\theta)^2 = 4$

or $r^2 + 4r(\cos\theta + 2\sin\theta) + 16 = 0$

$\therefore \quad PA \cdot PB = r_1 r_2 = \frac{16}{1} = 16$

Alternate Method

$$PA \cdot PB = (PC)^2$$
$$= 2^2 + 4^2 - 4 = 16$$

48. Let MN be the diameter of the circle whose equation is

$$4y = x + 7 \quad \ldots(i)$$

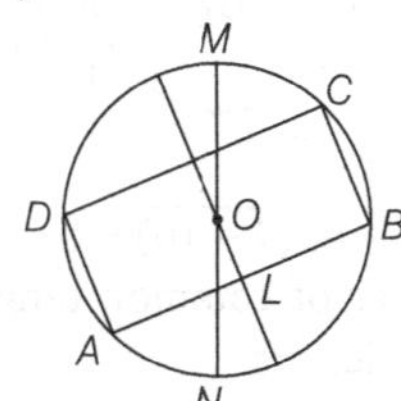

and coordinates of A and B are $(-3, 4)$ and $(5, 4)$ respectively.

Equation of $\perp$ bisector of AB is $[L \equiv (1, 4)]$

$$y - 4 = -\frac{1}{0}(x - 1) \quad [\because \text{slope of } AB = 0]$$

$\therefore \quad x = 1 \quad \ldots(ii)$

Solving Eqs. (i) and (ii) we get the coordinates of the centre of the circle as (1, 2)

$\therefore \quad OL = \sqrt{(1-1)^2 + (4-2)^2} = 2$

$\therefore \quad BC = 2OL = 4$ unit

$AB = 8$ unit

$\therefore$ Area of rectangle $ABCD = 4 \times 8$

$= 32$ sq units

49. The point $\left(3 + \frac{3}{\sqrt{2}}, \frac{3}{\sqrt{2}}\right)$ does not satisfy the given circles in (a) and (c).

$\therefore$(*a*) and (*c*) can not be the correct choices. The centre of the circle given in (*b*) is $\left(\frac{3}{\sqrt{2}}, \frac{3}{\sqrt{2}}\right)$, which does not lie on $x - y = 3$.

Hence (b) cannot be correct choice.

The centre of the circle given in (d) is (3, 0), which lie on $x - y = 3$,

then radius $= \sqrt{\left(3 + \frac{3}{\sqrt{2}} - 3\right)^2 + \left(\frac{3}{\sqrt{2}} - 0\right)^2}$

$= \sqrt{\frac{9}{2} + \frac{9}{2}} = \sqrt{9} = 3$

Hence, the correct answer is (d).

50. The x-coordinate of the centre of the new position of the circle = 5 + circumference of the first circle

$= 5 + 10\pi$

and y coordinate of the centre of the new position of the circle = 5 and radius is also 5.

$\therefore$ Equation of circle is

i.e. $(x - 5 - 10\pi)^2 + (y - 5)^2 = 25$

51. Centres and radii of the given circles are $C_1(1, 3)$, $r_1 = r$ and $C_2(4, -1)$, $r_2 = 3$ respectively since circles intersect in two distinct points, then

$$|r_1 - r_2| < C_1C_2 < r_1 + r_2$$

$\Rightarrow \quad |r - 3| < 5 < r + 3 \quad \ldots(i)$

From last two relations, $r > 2$

From first two relations

$$|r - 3| < 5$$

$\Rightarrow \quad -5 < r - 3 < 5$

$\Rightarrow \quad -2 < r < 8 \quad \ldots(ii)$

From Eqs. (i) and (ii), we get $2 < r < 8$

52. Equation of line perpendicular to $x = y$ and passing through (0, 0) is $y + x = 0$

which passes through the centre of the circle

if x-coordinate of centre $= h$,

then y-coordinate of centre $= -h$,

then radius of the circle

$= \perp$ distance from $(h, -h)$

on $x - y = 0$

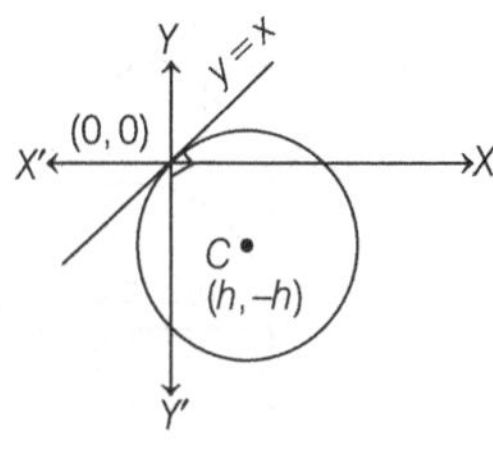

$$= \frac{2h}{\sqrt{2}} = h\sqrt{2}$$

$\therefore$ Equation of circle is

$$(x - h)^2 + (y + h)^2 = 2h^2$$

$\Rightarrow \quad x^2 + y^2 - 2hx + 2hy = 0 \quad \ldots(i)$

and other circle

$$x^2 + y^2 + 6x + 8y - 7 = 0 \quad \ldots(ii)$$

$\therefore$ Equation of common chord of Eqs. (i) and (ii) is

$$2x(h + 3) + 2y(4 - h) - 7 = 0$$

$\Rightarrow \quad (6x + 8y - 7) + 2h(x - y) = 0$

$\therefore \quad 6x + 8y - 7 = 0 \quad [\text{from } P + \lambda Q = 0]$

and $\quad x = y$,

$\Rightarrow \quad x = y = 1/2$,

Then coordinates of the fixed point is $\left(\frac{1}{2}, \frac{1}{2}\right)$

53. Let circle $x^2 + y^2 + 2gx + 2fy + c = 0$

According to the question

$$2g \times 2 + 2f \times -3 = c + 9$$

$\Rightarrow \quad 4g - 6f = c + 9 \quad \ldots(i)$

and $\quad 2g \times -\frac{5}{2} + 2f \times 2 = c + 2$

$\Rightarrow \quad -5g + 4f = c + 2 \quad \ldots(ii)$

Subtracting Eq. (ii) from Eq. (i), we get

$$9g - 10f = 7$$

$\Rightarrow$ $-9(-g) + 10(-f) = 7$

$\therefore$ Locus of centre is

$$-9x + 10y = 7$$

or $9x - 10y + 7 = 0$

54. If $\angle CPQ = \theta$

$\therefore$ $\sin\theta = \dfrac{CQ}{CP} = \sin\alpha$

$\therefore$ $\theta = \alpha$

$\angle QPR = 2\theta = 2\alpha$

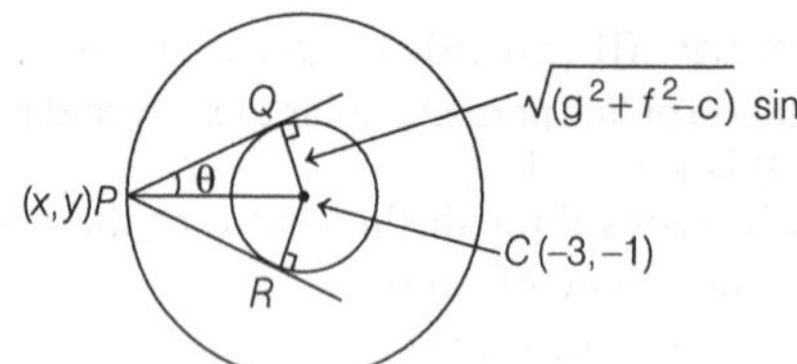

55. $\left(x - \dfrac{a}{2}\right)^2 + \left(y \pm \dfrac{a}{2}\right)^2 = \left(\dfrac{a}{2}\right)^2$

$$x^2 + y^2 - ax \pm ay + \frac{a^2}{4} = 0$$

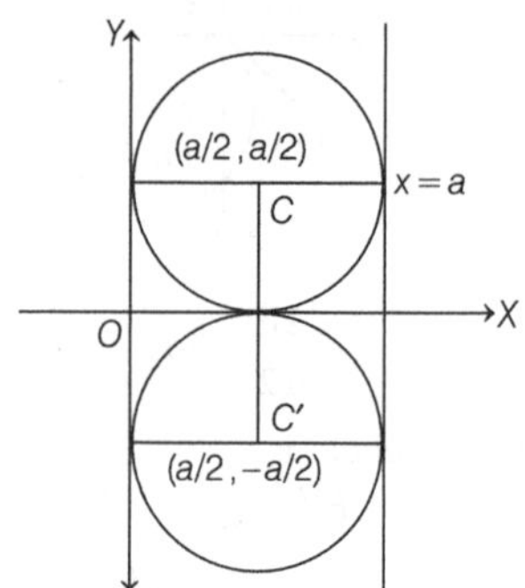

56. $\because OC \cdot OA = ab = OB \cdot OD$

$\therefore A, B, C, D$ are concyclic.

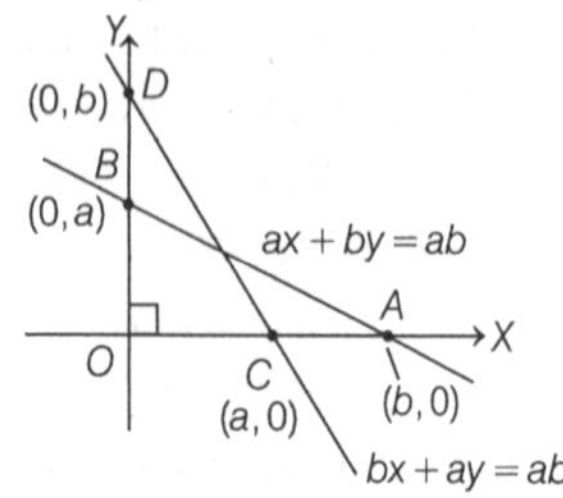

57. Equation of common chord is

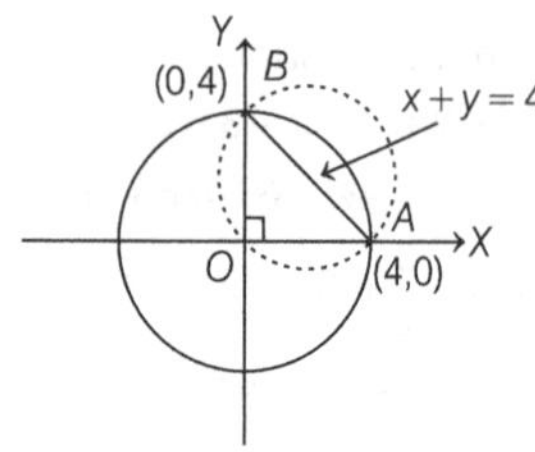

$$(x^2 + y^2 - 16) - (x^2 + y^2 - 4x - 4y) = 0$$

$\Rightarrow$ $4x + 4y - 16 = 0$

$\therefore$ $x + y = 4$

$\therefore$ $\angle AOB = \pi/2$

58. Let $S_1 \equiv x^2 + y^2 - 4x - 6y - 3 = 0$

and $S_2 \equiv x^2 + y^2 + 2x + 2y + 1 = 0$

Centre and radius of S_1 are $C_1(2, 3)$ and $r_1 = 4$ and centre and radius of S_2 are $C_2(-1, -1)$ and $r_2 = 1$

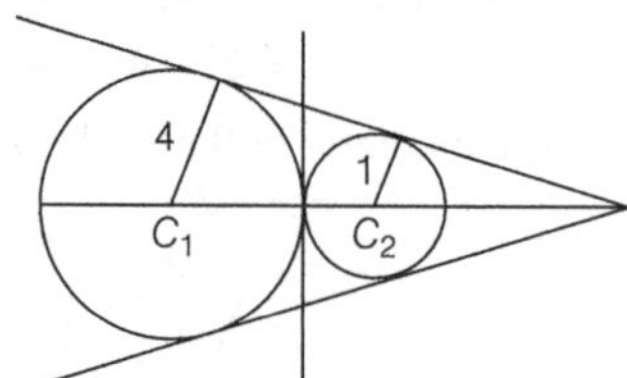

$\because$ $C_1C_2 = \sqrt{(9 + 16)} = 5 = r_1 + r_2$

Hence, number of common tangents $= 3$

59. Centres of circles are

$$C_1(-\lambda_1, 0); C_2(-\lambda_2, 0) \text{ and } C_3(-\lambda_3, 0)$$

If $O(0, 0)$,

then $OC_1 = \lambda_1, OC_2 = \lambda_2, OC_3 = \lambda_3$

$\because OC_1, OC_2$ and OC_3 are in GP

$\therefore$ $\lambda_2^2 = \lambda_1\lambda_3$ ...(i)

Let any point on $x^2 + y^2 = c^2$ is $P(c\cos\alpha, c\sin\alpha)$

$\therefore$ Length of tangents are $PT_1 = \sqrt{2\lambda_1(c\cos\alpha)}$

$$PT_2 = \sqrt{2\lambda_2(c\cos\alpha)}$$

$$PT_3 = \sqrt{2\lambda_3(c\cos\alpha)}$$

$$(PT_2)^2 = 2\lambda_2(c\cos\alpha)$$

$$= 2\sqrt{\lambda_1}\sqrt{\lambda_3}\sqrt{c\cos\alpha}\sqrt{c\cos\alpha} \quad \text{[from Eq. (i)]}$$

$$\sqrt{2\lambda_1(c\cos\alpha)}\sqrt{2\lambda_3(c\cos\alpha)} = PT_1 \cdot PT_3$$

Hence PT_1, PT_2 and PT_3 are in GP.

60. Let the circle be $(x - h)^2 + (y - k)^2 = r^2$,

then $r = \dfrac{|lh + mk + 1|}{\sqrt{(l^2 + m^2)}}$

or $l^2(h^2 - r^2) + m^2(k^2 - r^2) + 2lmhk + 2lh + 2mk + 1 = 0$

also $4l^2 - 5m^2 + 6l + 1 = 0$

Comparing the coefficients of similar terms

$$h^2 - r^2 = 4, k^2 - r^2 = -5, hk = 0,$$

$$2h = 6, 2k = 0$$

$\therefore$ $k = 0, h = 3$ and $9 - r^2 = 4$

$\therefore$ $r = \sqrt{5}$

Hence, centre $(3, 0)$ and radius $\sqrt{5}$.

61. Let OA chord is $y = mx$ ...(i)

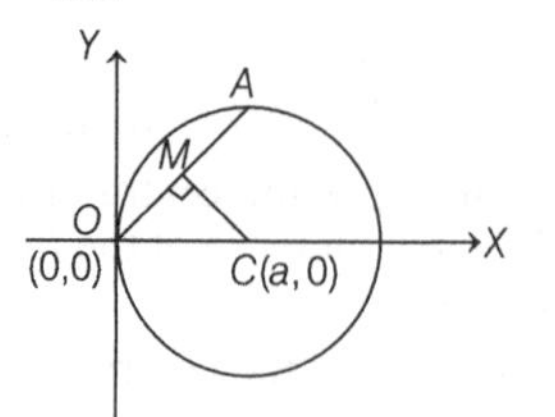

Equation of CM, $x + my = \lambda$
passing through $(a, 0)$

$$\therefore \quad x + my = a \quad \text{...(ii)}$$

From Eqs. (i) and (ii), $x + y \times \frac{y}{x} = a$

$$\Rightarrow \quad x^2 + y^2 - ax = 0$$

62. Let equation of circle is

$$x^2 + y^2 + 2gx + 2fy + c = 0 \quad \text{...(i)}$$

its pass through (a, b) then

$$a^2 + b^2 + 2ga + 2fb + c = 0 \quad \text{...(ii)}$$

and given circle (i) and $x^2 + y^2 = \lambda^2$ cuts orthogonally then

$$2g \times 0 + 2f \times 0 = c - \lambda^2$$

$$\Rightarrow \quad c = \lambda^2$$

substitute the value of c in (ii), then

$$a^2 + b^2 + 2ga + 2fb + \lambda^2 = 0$$

or

$$2a(-g) + 2b(-f) = a^2 + b^2 + \lambda^2$$

$\therefore$ Locus of centre of circle is :

$$2ax + 2by = a^2 + b^2 + \lambda^2$$

63. We have,

$$x^2 + y^2 + ax + 2ay + c = 0$$

$$2\sqrt{g^2 - c} = 2\sqrt{\frac{a^2}{4} - c} = 2\sqrt{2}$$

$$\Rightarrow \quad \frac{a^2}{4} - c = 2 \quad \text{...(i)}$$

$$2\sqrt{f^2 - c} = 2\sqrt{a^2 - c} = 2\sqrt{5}$$

$$\Rightarrow \quad a^2 - c = 5 \quad \text{...(ii)}$$

From Eqs. (i) and (ii), we get

$$\frac{3a^2}{4} = 3$$

$$\Rightarrow \quad a = -2 \quad (a < 0)$$

$$\therefore \quad c = -1$$

Circle $\Rightarrow x^2 + y^2 - 2x - 4y - 1 = 0$

$$\Rightarrow \quad (x - 1)^2 + (y - 2)^2 = 6$$

Given, $x + 2y = 0$

$$\Rightarrow \quad m = -\frac{1}{2}$$

Slope of tangent $= 2$
Equation of tangent

$\Rightarrow (y - 2) = 2(x - 1) \pm \sqrt{6}\sqrt{1 + 4}$

$\Rightarrow 2x - y \pm \sqrt{30} = 0$

Perpendicular distance from $(0, 0) = \left|\frac{\pm\sqrt{30}}{\sqrt{4 + 1}}\right| = \sqrt{6}$

64. Let circle $x^2 + y^2 + 2gx + 2fy + c = 0$ ...(A)
its passes through $(1, t)$, $(t, 1)$ and (t, t), then $1 + t^2 + 2g + 2ft + c = 0$ ...(i)

$$t^2 + 1 + 2gt + 2f + c = 0 \quad \text{...(ii)}$$

$$2t^2 + 2gt + 2ft + c = 0 \quad \text{...(iii)}$$

Subtracting Eq. (i) from Eq. (ii) and Eq. (ii) from Eq. (iii) then $2g(t - 1) + 2f(1 - t) = 0$

or $\quad g - f = 0$

and $\quad t^2 - 1 + 2f(t - 1) = 0$

$$\therefore \quad f = -\frac{(t + 1)}{2} = g$$

From Eq. (iii), $2t^2 - t(t + 1) - t(t + 1) + c = 0$

$$\therefore \quad c = 2t$$

From (A),

$$x^2 + y^2 - (t + 1)x - (t + 1)y + 2t = 0$$

$$\Rightarrow \quad (x^2 + y^2 - x - y) - t(x + y - 2) = 0$$

$$\therefore \quad P + \lambda Q = 0$$

$$\therefore \quad P = 0 \text{ and } Q = 0,$$

then $x^2 + y^2 - x - y = 0$ and $x + y - 2 = 0$

after solving we get $x = 1$ and $y = 1$

65. Point circles of the co-axial system are

$$(x - 1)^2 + (y - 2)^2 = 0$$

and $\quad (x - 4)^2 + (y - 3)^2 = 0$

or $\quad x^2 + y^2 - 2x - 4y + 5 = 0$

and $\quad x^2 + y^2 - 8x - 6y + 25 = 0$

$\therefore$ Equation of co-axial system is
$(x^2 + y^2 - 2x - 4y + 5)$

$$+ \lambda(x^2 + y^2 - 8x - 6y + 25) = 0 \quad \text{...(i)}$$

$\because$ its pass through $(0, 0)$,
then $5 + 25\lambda = 0$

$$\therefore \quad \lambda = -\frac{1}{5}$$

Then from Eq. (i)

$$5x^2 + 5y^2 - 10x - 20y + 25 - x^2$$
$$- y^2 + 8x + 6y - 25 = 0$$

$$\Rightarrow \quad 2x^2 + 2y^2 - x - 7y = 0$$

66. Normal of circle always pass through the centre of the circle *i.e.*, $(2, -2)$ but its also pass through $(1, 1)$

$\therefore$ Equation of normal

$$y + 2 = \frac{1 + 2}{1 - 2}(x - 2)$$

or $\quad -y - 2 = 3x - 6$

or $\quad 3x + y - 4 = 0$

67. Coordinates of P, Q, R are $(\cos\alpha, \sin\alpha,)$, $(\cos\beta, \sin\beta)$ and $(\cos\gamma, \sin\gamma)$ respectively.
and $A \equiv (-1, 0)$

$$\therefore \quad AP = \sqrt{(1 + \cos\alpha)^2 + \sin^2\alpha} = 2\cos\alpha/2$$

$$AQ = \sqrt{(1 + \cos\beta)^2 + \sin^2\beta} = 2\cos\beta/2$$

$$AR = \sqrt{(1 + \cos\gamma)^2 + \sin^2\gamma} = 2\cos\gamma/2$$

$\because$ AP, AQ, AR are in GP then

$\cos\alpha/2, \cos\beta/2, \cos\gamma/2$ are also in GP.

68. We have, $2x^2 - 3xy - 2y^2 = 0$

i.e. $(x - 2y)(2x + y) = 0$, which represents a pair of perpendicular lines passing through the origin

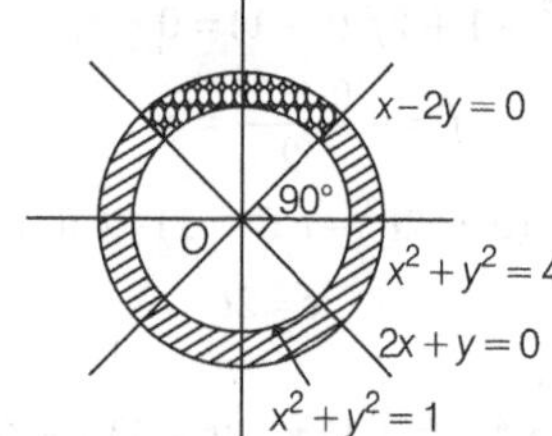

$\therefore$ Required area $= \frac{\pi (2)^2}{4} - \frac{\pi (1)^2}{4}$

$= \frac{3\pi}{4}$ sq unit.

69. $\because x = |y| > 0$

$\therefore$ Circle lies in I and IV quadrant

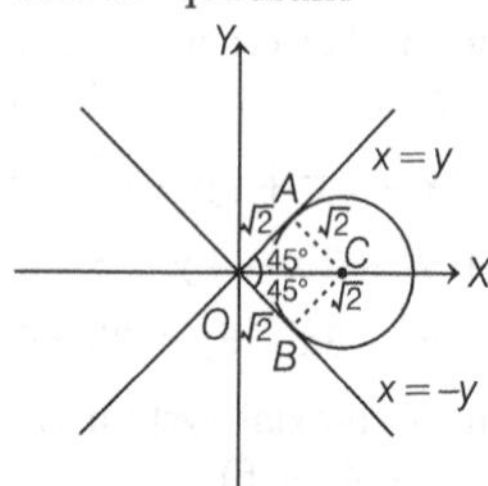

Centre of circle lies on x-axis.

$$OC = \sqrt{(OA)^2 + (AC)^2} = \sqrt{(2+2)} = 2$$

$\therefore$ Centre is (2, 0) and radius $= \sqrt{2}$

$\therefore$ Equation of circle is

$$(x - 2)^2 + (y - 0)^2 = 2$$

$$\Rightarrow \quad x^2 + y^2 - 4x + 2 = 0$$

70. Given, normal is $ax + by = 2$. Therefore,

$$2a + 2b = 2 \text{ or } a + b = 1$$

$ax + by = 2$ touches $x^2 + y^2 = 1$.

So, $$1 = \left| \frac{-2}{\sqrt{a^2 + b^2}} \right|$$

$\therefore$ $a^2 + b^2 = 4$ or $a^2 + (1 - a)^2 = 4$

or $2a^2 - 2a - 3 = 0$

$$\therefore \quad a = \frac{2 \pm \sqrt{4 + 24}}{4} = \frac{1 \pm \sqrt{7}}{2}$$

$$\therefore \quad b = 1 - a = 1 - \frac{1 \pm \sqrt{7}}{2} = \frac{1 \mp \sqrt{7}}{2}$$

71. Let Equation of circle be

$$x^2 + y^2 + 2gx + 2fy + c = 0$$

It pass through (2, 0) and (0, 4), then

$$4 + 0 + 4g + 0 + c = 0 \Rightarrow g = -\frac{(c + 4)}{4}$$

and $$0 + 16 + 0 + 8f + c = 0 \Rightarrow f = -\frac{(c + 16)}{8}$$

$\because$ radius $r = \sqrt{(g^2 + f^2 - c)}$

$$= \sqrt{\left\{\frac{(c + 4)^2}{16} + \frac{(c + 16)^2}{64} - c\right\}}$$

$$= \sqrt{\left\{\frac{4\{c^2 + 8c + 16\} + \{c^2 + 32c + 256\} - 64c}{64}\right\}}$$

$$= \sqrt{\left(\frac{5c^2 + 320}{64}\right)}$$

For minimum radius $c = 0$

$\therefore$ $g = -1, f = -2$

Required circle is $x^2 + y^2 - 2x - 4y = 0$

72. Centre $C \equiv (7, 5)$ and radius $r = \sqrt{(49 + 25 + 151)} = 15$

If $P(2, -7)$

The shortest distance $= |CP - r|$

$= |\sqrt{25 + 144} - 15|$

$= |13 - 15| = 2$

73. $\because \frac{BP}{PA} = \alpha$

$BP : PA = \alpha : 1$

$\therefore$ Coordinates of P is $\left(\frac{3 + \alpha}{1 + \alpha}, \frac{4}{\alpha + 1}\right)$

P lie on $x^2 + y^2 = 4$

$\Rightarrow$ $(\alpha + 3)^2 + 16 = 4(\alpha + 1)^2$

$\Rightarrow$ $3\alpha^2 + 2\alpha - 21 = 0$ ...(i)

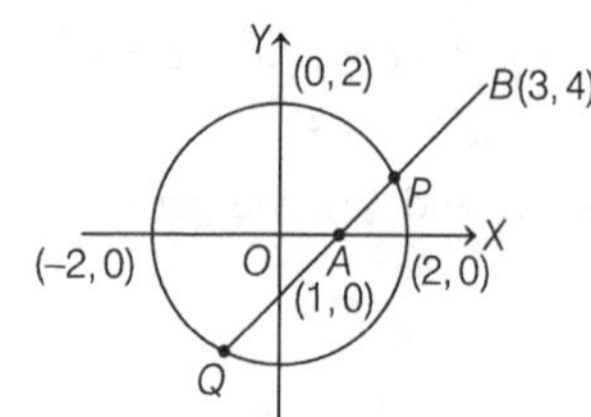

and $$\frac{BQ}{QA} = \frac{\beta}{1}$$

$$BQ : QA = \beta : 1$$

or $$\frac{BQ}{QA} - 1 = \beta - 1$$

$$\frac{AB}{QA} = \frac{(\beta - 1)}{1}$$

$$AB : QA = (\beta - 1) : 1$$

$\therefore$ Coordinates of Q is $\left(\frac{\beta - 3}{\beta - 1}, -\frac{4}{\beta - 1}\right)$

Q lie on $x^2 + y^2 = 4$

$$\therefore \quad (\beta - 3)^2 + 16 = 4(\beta - 1)^2$$

$$\Rightarrow \quad 3\beta^2 - 21\beta - 21 = 0$$

Hence α is a roots of $3x^2 + 2x - 21 = 0$

and β is a root of $3x^2 + 2x - 21 = 0$

74. The image of the circle has same radius but centre different. If centre is (α, β), then

$$\frac{\alpha-3}{1}=\frac{\beta-2}{1}=\frac{-2(3+2-19)}{1^2+1^2}$$

$\Rightarrow$ $\alpha-3=\beta-2=14$

$\therefore$ $\alpha=17, \beta=16$

$\therefore$ Required circle is $(x-17)^2+(y-16)^2=1$

75. We have $x^2+y^2-4x-4y-1=0$

Centre (2, 2) radius $=\sqrt{(4+4+1)}=3$

$$AQ = CA - CQ$$
$$= 5 - 3 = 2$$

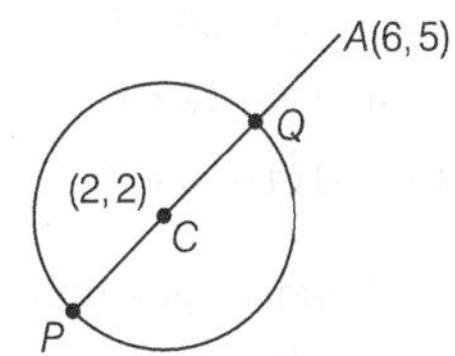

$\therefore$ $CQ : QA = 3 : 2,$

then $Q\left(\frac{3\cdot 6+2\cdot 2}{3+2}, \frac{3\cdot 5+2\cdot 2}{3+2}\right)$

i.e. $Q\left(\frac{22}{5}, \frac{19}{5}\right)$

If $P(h, k)$,

then $h+\frac{22}{5}=4$ and $k+\frac{19}{5}=4$,

then $h=-\frac{2}{5}$ and $k=-\frac{1}{5}$

$\therefore$ $P\left(-\frac{2}{5}, -\frac{1}{5}\right)$

76. $\because \alpha+\beta=-\frac{b}{a}, \alpha\beta=\frac{c}{a}$

and $\alpha'+\beta'=-\frac{b}{a'}, \alpha'\beta'=\frac{c'}{a'}$

Equation of circle having $A(\alpha, \alpha')$ and $B(\beta, \beta')$ as diameter is

$(x-\alpha)(x-\beta)+(y-\alpha')(y-\beta')=0$

$\Rightarrow$ $x^2+y^2-(\alpha+\beta)x-(\alpha'+\beta')y+\alpha\beta+\alpha'\beta'=0$

$\Rightarrow$ $x^2+y^2+\frac{bx}{a}+\frac{b'}{a'}y+\frac{c}{a}+\frac{c'}{a'}=0$

$\Rightarrow$ $aa'(x^2+y^2)+a'bx+ab'y+a'c+ac'=0$

77. The centre of C_1 is $(a+b, 0)$ and the centre of C_2 is $(a+2b+c, 0)$

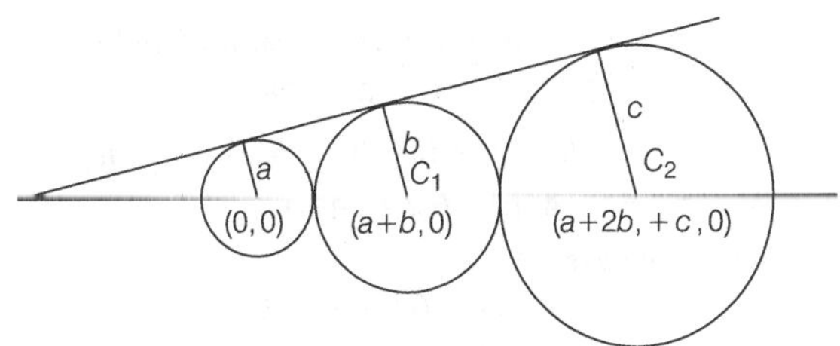

Let $y=mx+k$ be a tangent common to the three circles.

Since, it touches $x^2+y^2=a^2, C_1$ and C_2

$$\frac{k}{\sqrt{1+m^2}}=\pm a, \quad \frac{m(a+b)+k}{\sqrt{1+m^2}}=\pm b$$

and $$\frac{m(a+2b+c)+k}{\sqrt{1+m^2}}=\pm c$$

As the centre of the three circles lie on the same side of the line $y=mx+k$, taking the same sign, say positive, in the three relations we get,

$$\frac{k}{\sqrt{1+m^2}}=a=b-m(a+b)=c-m(a+2b+c)$$

$\Rightarrow$ $\frac{a+b}{b-a}=\frac{a+2b+c}{c-a}$ (eliminating m)

$\Rightarrow$ $(a+b)(c-a)=(b-a)(a+2b+c)$

$\Rightarrow$ $ac-a^2+bc-ba=ba-a^2+2b^2-2ab+bc-ac$

$\Rightarrow$ $2ac=2b^2$

$\Rightarrow$ $ac=b^2$

$\Rightarrow$ a, b, c are in G P

78. Let equation of circle is

$x^2+y^2+2gx+2fy+c=0$

According to question

$2g\times -1+2f\times\frac{3}{2}=c-7$

$\Rightarrow$ $-2g+3f=c-7$ ...(i)

$2g\times\frac{5}{2}+2f\times -\frac{5}{2}=c+9$

$\Rightarrow$ $5g-5f=c+9$ ...(ii)

and $2g\times\frac{7}{2}+2f\times -\frac{9}{2}=c+29$

$\Rightarrow$ $7g-9f=c+29$ ...(iii)

Solving Eqs. (i), (ii), and (iii), we get

$g=-8, f=-9, c=-4$

$\therefore$ Equation of circle is

$x^2+y^2-16x-18y-4=0$

79. It is given that the circle touches the X-axis at (3, 0) and making an intercept of 8 on the Y-axis.

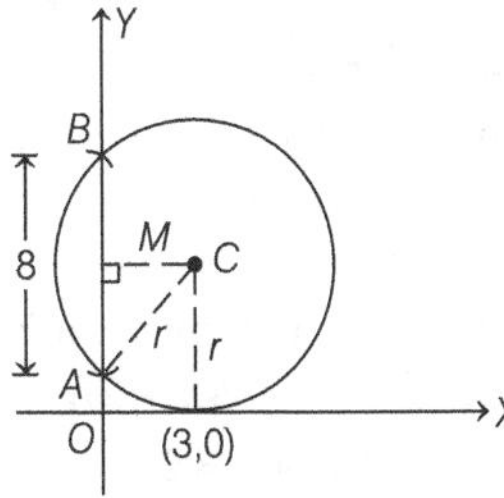

Let the radius of the cirlce is 'r', then the coordinates of centre of circle are $(3, r)$.

From the figure, we have

$CM=3$, $CA=$ radius $=r$

and $AM=BM=\frac{AB}{2}=4$

Then, $r^2=CM^2+AM^2=9+16=25$

$\Rightarrow$ $r=\pm 5$

Now, the equation of circle having centre $(3, \pm 5)$ and radius $= 5$ is $(x-3)^2 + (y \pm 5)^2 = 25$

Now, from the option (3, 10) satisfy the equation of circle $(x-3)^2 + (y-5)^2 = 25$

80. Let line is $\frac{x}{a} + \frac{y}{b} = 1$

$\therefore$ $A(a, 0)$ and $B(0, b)$

Equation of circle since AB is diameters is

$$(x-a)(x-0) + (y-0)(y-b) = 0$$

$$\Rightarrow \quad x^2 + y^2 - ax - by = 0$$

Tangent at (0, 0) is $ax + by = 0$

$$\therefore \quad AM = m = \frac{a^2}{\sqrt{a^2+b^2}}$$

and $\quad BN = n = \frac{b^2}{\sqrt{a^2+b^2}}$

$$\Rightarrow \quad m + n = \sqrt{(a^2+b^2)} = \text{diameter of circle}$$

81. Since, equation of tangent to the circle $x^2 + y^2 = 1$ at point $\left(\frac{1}{\sqrt{2}}, \frac{1}{\sqrt{2}}\right)$ is $T = 0$.

$$\Rightarrow \quad \frac{x}{\sqrt{2}} + \frac{y}{\sqrt{2}} = 1$$

$$\Rightarrow \quad x + y = \sqrt{2} \quad \text{...(i)}$$

$\because$ Line $L_1 : x + y = \sqrt{2}$ is perpendicular to line $y = mx + c$, so $m = 1$, since line $y = mx + c$ is tangent to the circle $(x-3)^2 + y^2 = 1$, where $m = 1$, then $y = x + c$ is equivalent to $y = 1(x-3) \pm 1\sqrt{1+1^2}$.

So, $\quad c = -3 \pm \sqrt{2} \Rightarrow c + 3 = \pm\sqrt{2}$

$$\Rightarrow \quad (c+3)^2 = 2 \Rightarrow c^2 + 6c + 9 = 2 \Rightarrow c^2 + 6c + 7 = 0$$

82. Point circle is $(x-2)^2 + (y-1)^2 = 0$

$$\Rightarrow \quad x^2 + y^2 - 4x - 2y + 5 = 0$$

Family of circle $(x^2 + y^2 - 6x - 4y - 3)$

$$+ \lambda(x^2 + y^2 - 4x - 2y + 5) = 0$$

$$\therefore \quad x^2 + y^2 - \frac{2x(3+2\lambda)}{(1+\lambda)} - \frac{2y(2+\lambda)}{(1+\lambda)} + \frac{(5\lambda-3)}{(1+\lambda)} = 0$$

Centre $\left(\frac{3+2\lambda}{1+\lambda}, \frac{2+\lambda}{1+\lambda}\right)$ ...(i)

$\therefore$ Radius $= 0$ then we get

$$\lambda = -8/7$$

From Eq. (i), the other limiting point is (–5,–6)

83. Let $P(x_1, y_1)$ on $x^2 + y^2 + 2gx + 2fy + c = 0$

$$\therefore x_1^2 + y_1^2 + 2gx_1 + 2fy_1 + c = 0 \quad \text{...(i)}$$

$$\therefore \quad PT = \sqrt{x_1^2 + y_1^2 + 2gx_1 + 2fy_1 + d}$$

$$= \sqrt{(d-c)} \quad \text{[from Eq. (i)]}$$

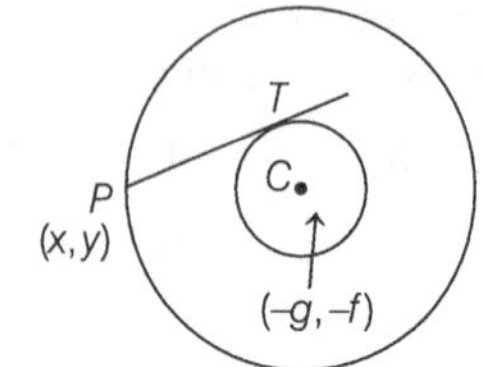

84. For interior point $OP < 5$

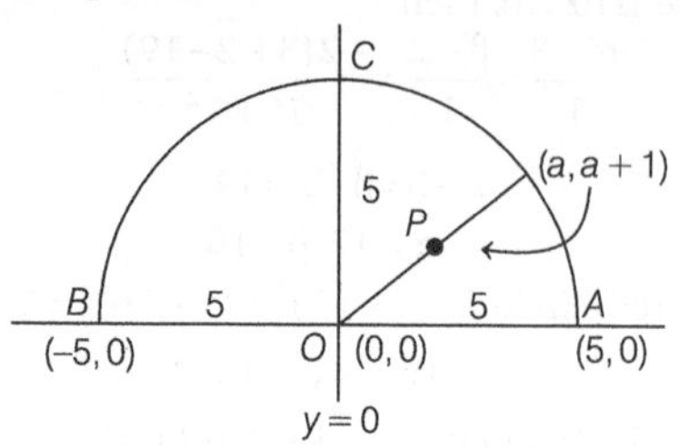

$$\Rightarrow \quad \sqrt{a^2 + (a+1)^2} < 5$$

$$\Rightarrow \quad 2a^2 + 2a - 24 < 0$$

$$\therefore \quad a^2 + a - 12 < 0$$

$$(a+4)(a-3) < 0$$

$\Rightarrow -4 < a < 3$ and I and II quadrant

$$a + 1 > 0$$

$$\therefore \quad a > -1$$

Hence, $-1 < a < 3$

85. Equation of circle through points of intersection of the circles $x^2 + y^2 - 2x - 4y + 1 = 0$ and $x^2 + y^2 - 4x - 2y + 1 = 0$ is $(x^2 + y^2 - 2x - 4y + 1) + \lambda(x^2 + y^2 - 4x - 2y + 1) = 0$

$$\Rightarrow (1+\lambda)x^2 + (1+\lambda)y^2 - 2(1+2\lambda)x - 2(2+\lambda)y + (1+\lambda) = 0$$

$$\Rightarrow x^2 + y^2 - \frac{2(1+2\lambda)}{(1+\lambda)}x - \frac{2(2+\lambda)}{(1+\lambda)}y + 1 = 0 \quad \text{...(i)}$$

$(\lambda \neq -1)$

$$\therefore \quad \text{Centre} \equiv \left(\frac{1+2\lambda}{1+\lambda}, \frac{2+\lambda}{1+\lambda}\right)$$

$\because$ Centre lie on $\quad x + 2y - 3 = 0$

$$\Rightarrow \quad 5 + 4\lambda - 3(1+\lambda) = 0$$

$$\therefore \quad \lambda = -2$$

From Eqs. (i),

$$x^2 + y^2 - 6x + 1 = 0$$

86. Let centre of the circle be (h, k)

$\because$ Circle touch Y-axis

$\therefore$ Radius $= |h|$

Equation of Circle is $(x-h)^2 + (y-k)^2 = h^2$ ...(i)

Given circle (i) touches externally the circle

$$x^2 + y^2 - 6x - 6y + 14 = 0$$

$\therefore$ Distance between centres = sum of radii

$$\Rightarrow \quad \sqrt{(h-3)^2 + (k-3)^2} = |h| + 2$$

$$\Rightarrow \quad (h-3)^2 + (k-3)^2 = h^2 + 4 + 4|h|$$

$$\Rightarrow \quad k^2 - 6h - 4|h| - 6k + 14 = 0$$

Here h always +ve

$$\therefore \quad k^2 - 10h - 6k + 14 = 0$$

$\therefore$ Locus of centre of circle is

$$y^2 - 10x - 6y + 14 = 0$$

87. Given equation of circle is

$x^2 + y^2 - 6x + 8y - 103 = 0$, which can be written as

$(x-3)^2 + (y+4)^2 = 128 = (8\sqrt{2})^2$

$\therefore$ Centre $= (3, -4)$ and radius $= 8\sqrt{2}$

Now, according to given information, we have the following figure.

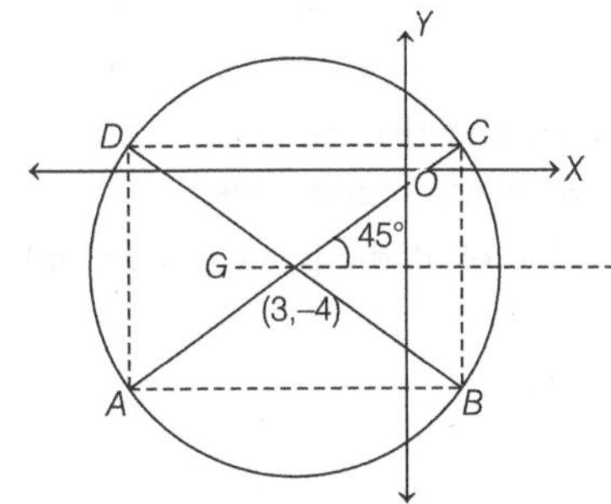

For the coordinates of A and C.

Consider, $\dfrac{x-3}{\frac{1}{\sqrt{2}}} = \dfrac{y+4}{\frac{1}{\sqrt{2}}} = \pm 8\sqrt{2}$

[using distance (parametric) form of line, $\dfrac{x-x_1}{\cos\theta} = \dfrac{y-y_1}{\sin\theta} = r$]

$\Rightarrow \quad x = 3 \pm 8,\ y = -4 \pm 8$

$\therefore \quad A(-5, -12)$ and $C(11, 4)$

Similarly, for the coordinates of B and D, consider

$\dfrac{x-3}{-\frac{1}{\sqrt{2}}} = \dfrac{y+4}{\frac{1}{\sqrt{2}}} = \pm 8\sqrt{2}$ [in this case, $\theta = 135°$]

$\Rightarrow \quad x = 3 \mp 8,\ y = -4 \pm 8$

$\therefore B\ (11, -12)$ and $D\ (-5, 4)$

Now, $OA = \sqrt{25+144} = \sqrt{169} = 13;$

$OB = \sqrt{121+144} = \sqrt{265}$

$OC = \sqrt{121+16} = \sqrt{137}$

and $OD = \sqrt{25+16} = \sqrt{41}$

88. Equation of co-axial system with limiting points are (0, 0) and (1, 0) is

$$\{(x-1)^2 + y^2\} + \lambda\{x^2 + y^2\} = 0$$

$$\Rightarrow (1+\lambda)x^2 + (1+\lambda)y^2 - 2x + 1 = 0$$

$$\Rightarrow \quad x^2 + y^2 - \frac{2x}{(1+\lambda)} + \frac{1}{(1+\lambda)} = 0 \ (\lambda \neq -1) \quad \text{...(i)}$$

Now, radius $= 0$

$$\frac{1}{(1+\lambda)^2} - \frac{1}{(1+\lambda)} = 0$$

$$\Rightarrow \quad 1 - (1+\lambda) = 0$$

$$\therefore \quad \lambda = 0$$

From Eq. (i), required circle is

$$x^2 + y^2 - 2x + 1 = 0$$

89. Let the coordinate of the centre of T be $(0, k)$.

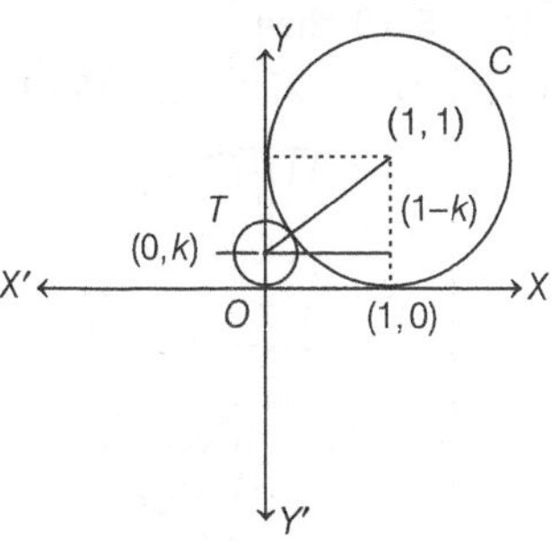

Distance between their centre

$$k+1 = \sqrt{1+(k-1)^2}$$

$$\Rightarrow \quad k+1 = \sqrt{1+k^2+1-2k}$$

$$\Rightarrow \quad k+1 = \sqrt{k^2+2-2k}$$

$$\Rightarrow \quad k^2+1+2k = k^2+2-2k$$

$$\Rightarrow \quad k = \frac{1}{4}$$

So, the radius of circle T is k i.e. $\frac{1}{4}$.

90. $\because \angle ACB = \pi/3$ (given)

$\therefore \angle AOB = \dfrac{2\pi}{3}$

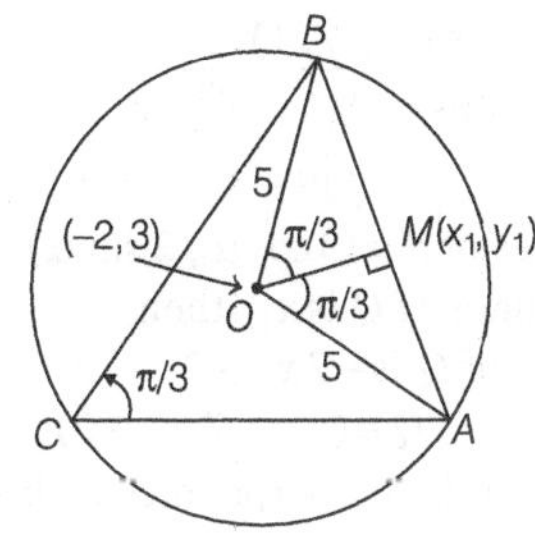

$\Rightarrow \angle AOM = \angle BOM = \dfrac{\pi}{3}$

Let mid-point of chord AB is (x_1, y_1)

$\therefore$ In ΔAOM,

$$\cos\frac{\pi}{3} = \frac{OM}{OA} = \frac{\sqrt{(x_1+2)^2 + (y_1-3)^2}}{5}$$

or $$\left(\frac{5}{2}\right)^2 = (x_1+2)^2 + (y_1-3)^2$$

$$\Rightarrow \quad \frac{25}{4} = x_1^2 + y_1^2 + 4x_1 - 6y_1 + 13$$

$$4x_1^2 + 4y_1^2 + 16x_1 - 24y_1 + 27 - 0$$

$\therefore$ Locus of Mid point is

$$x^2 + y^2 + 4x - 6y + \frac{27}{4} = 0$$

$$\Rightarrow \quad (x+2)^2 + (y-3)^2 = 6.25$$

91. $\because$ The point $([P+1],[P])$ lies inside the circle $x^2+y^2-2x-15=0$, then

$$[P+1]^2+[P]^2-2[P+1]-15<0$$

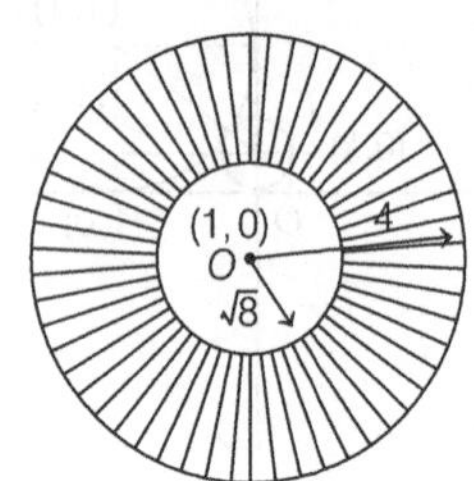

$$\Rightarrow \quad ([P]+1)^2+[P]^2-2([P]+1)-15<0$$

$$\Rightarrow \quad 2[P]^2-16<0 \Rightarrow [P]^2<8 \quad \text{...(i)}$$

$\because$ Circles are Concentric

$\therefore$ Point $([P+1],[P])$ out side the circle

$$x^2+y^2-2x-7=0$$

$$\therefore \quad ([P+1]^2+[P]^2-2([P+1])-7>0$$

$$\Rightarrow ([P]+1]^2+[P]^2-2([P]+1)-7>0$$

$$\Rightarrow \quad 2[P]^2-8>0$$

$$\therefore \quad [P]^2>4 \quad \text{...(ii)}$$

From Eqs. (i) and (ii), we get

$4<[P]^2<8$ which is impossible

$\therefore$ For no value of 'P' the point will be within the region.

Objective Questions Type II [One or more than one correct answer(s)]

1. Let circle be $x^2+y^2+2gx+2fy+c=0$

It passes through (0, 1).

$$\therefore \quad 1+2f+c=0 \quad \text{...(i)}$$

Orthogonal with $x^2+y^2-2x-15=0$

$$2g(-1)=c-15$$

$$\Rightarrow \quad c=15-2g \quad \text{...(ii)}$$

Orthogonal with $x^2+y^2-1=0$

$$c=1 \quad \text{...(iii)}$$

$$\Rightarrow \quad g=7 \text{ and } f=-1$$

Centre is $(-g,-f)=(-7,1)$

$$\therefore \quad \text{Radius} = \sqrt{g^2+f^2-c}$$

$$=\sqrt{49+1-1}=7$$

2. Let coordinates of $P \equiv (3\cos\alpha, 3\sin\alpha)$

Let x-coordinate of Q is x_1, then

y-coordinate of Q is $-7x_1-3$

$$\therefore \quad Q \equiv (x_1, -7x_1-3)$$

$\because$ $x-y+1=0$ is the perpendicular bisector of PQ, then mid point of PQ lie on $x-y+1=0$

$$\Rightarrow \quad \frac{3\cos\alpha+x_1}{2}-\frac{3\sin\alpha-7x_1-3}{2}+1=0$$

$$\Rightarrow \quad 8x_1+3\cos\alpha-3\sin\alpha+5=0$$

$$\Rightarrow \quad 24x_1+9\cos\alpha-9\sin\alpha+15=0 \quad \text{...(i)}$$

and Slope of

$$(x-y+1=0)\times \text{slope of } PQ=-1$$

$$\Rightarrow \quad 1\times\frac{3\sin\alpha+7x_1+3}{3\cos\alpha-x_1}=-1$$

$$\Rightarrow \quad 3\sin\alpha+7x_1+3=-3\cos\alpha+x_1$$

$$\Rightarrow \quad 6x_1+3\sin\alpha+3\cos\alpha+3=0$$

$$\Rightarrow \quad 24x_1+12\sin\alpha+12\cos\alpha+12=0 \quad \text{...(ii)}$$

On subtracting Eq. (i) and (ii), we obtain

$$-3\cos\alpha-21\sin\alpha+3=0$$

$$\Rightarrow \quad (1-\cos\alpha)=7\sin\alpha$$

$$\Rightarrow \quad (1-\cos\alpha)^2=49(1-\cos^2\alpha)$$

$$\Rightarrow \quad (1-\cos\alpha)^2=49(1+\cos\alpha)(1-\cos\alpha)$$

$$\Rightarrow \quad (1-\cos\alpha)(1-\cos\alpha-49-49\cos\alpha)=0$$

$$\therefore \quad \cos\alpha=1 \ \& \ \cos\alpha=-\frac{24}{25}$$

$\therefore$ Coordinates of P are $(3, 0)$ and $\left(-\frac{72}{25},\frac{21}{25}\right)$

3. Circle possible in I region and centre of circle on X-axis.

If centre is $(h, 0)$, then

$$\frac{|h-0-1|}{\sqrt{2}}=\sqrt{(h-3)^3+\frac{7}{2}}$$

$$\therefore \quad h=6, 4$$

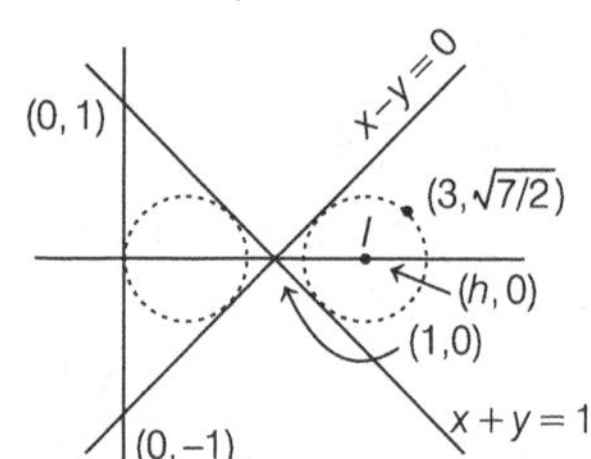

Then centres are (6, 0) and (4, 0)

4. $x^2+y^2-2px-2qy+q^2=0$

$$\text{or} \quad (x-p)^2+(y-q)^2=p^2 \quad \text{...(i)}$$

since tangents are perpendicular, then locus of point of intersection of tangents is director circle.

Director circle of (i) is

$$(x-p)^2+(y-q)^2=p^2+p^2 \quad \text{...(ii)}$$

since point of intersection of tangent is (0, 0), then from Eq. (ii)

$$p^2+q^2=2p^2$$

$$\Rightarrow \quad p^2=q^2$$

$$\text{or} \quad p=\pm q$$

5. $x\cos\alpha+y\sin\alpha=2$ or $\dfrac{x}{2\sec\alpha}+\dfrac{y}{2\operatorname{cosec}\alpha}=1$

$$2\sec\alpha=+\text{ve in IV}$$

$$\text{and} \quad 2\operatorname{cosec}\alpha=-\text{ve in IV}$$

Value of g in I quadrant

$$g = \left(\frac{g\cos\alpha + g\sin\alpha - 2}{\sqrt{1}}\right)$$

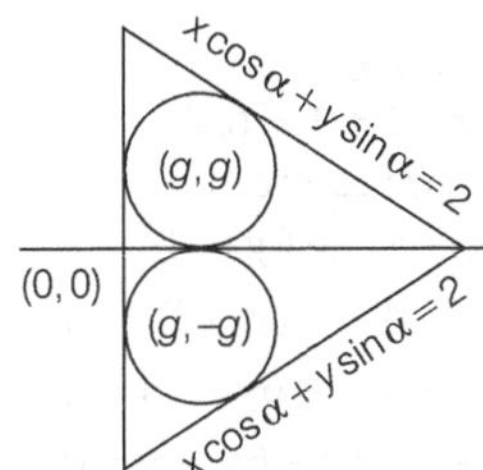

$\therefore$ $$g = \frac{2}{\cos\alpha + \sin\alpha \pm 1}$$

and value g of in IV quadrant

$$g = \left|\frac{g\cos\alpha - g\sin\alpha - 2}{1}\right|$$

$\Rightarrow$ $$\pm g = g(\cos\alpha - \sin\alpha) - 2$$

$$g = \frac{2}{\cos\alpha - \sin\alpha \pm 1}$$

Equation of circles in I quadrant

$(x-g)^2 + (y-g)^2 = g^2$

And in IV quadrant is $(x-g)^2 + (y-g)^2 = g^2$

6. Coordinates of O are (5, 3) and radius = 2

Equation of tangent at $A(7, 3)$ is

$$7x + 3y - 5(x+7) - 3(y+3) + 30 = 0$$

i.e. $2x - 14 = 0$ i.e. $x = 7$

Equation of tangent at B (5, 1) is

$5x + y - 5(x+5) - 3(y+1) + 30 = 0$

i.e. $-2y + 2 = 0$ i.e. $y = 1$

Coordinate of C are (7, 1)

$\because$ Area of $OACB = 4$

Equation of AB is $x - y = 4$ (radical axis)

Equation of the smallest circle is

$$(x-7)(x-5) + (y-3)(y-1) = 0$$

i.e., $$x^2 + y^2 - 12x - 4y + 38 = 0$$

7. Centre and radii of given circles are

$C_1\left(-\frac{a}{2}, 0\right), r_1 = a/2$ and C_2 (0, 0), $r_2 = c$

For touch $C_1C_2 = r_1 \pm r_2$

$$\sqrt{\frac{a^2}{4}} = \frac{a}{2} \pm c$$

or $$\frac{a^2}{4} = \frac{a^2}{4} + c^2 \pm ac$$

$\Rightarrow$ $c^2 \pm ac = 0$

or $c \pm a = 0$ $[\because c \neq 0]$

$\therefore$ $a + c = 0, a - c = 0$ and $a^2 = c^2$

8. Centres and radii of the given circles are

$C_1(-7, 2), r_1 = 5$ and $C_2(7, -2), r_2 = 9$

$\because$ $C_1C_2 = \sqrt{(14)^2 + (4)^2} = \sqrt{212} > r_1 + r_2$

For common tangent length of perpendicular from centre on tangent = radius

Of centre $C_1(-7, 2)$ and $r_1 = 5$, then (b) and (c) are correct.

9. Slope of $PC = \dfrac{-3-2}{2-7} = 1$

If $\tan\theta = 1 \because \theta = 45°$

Equation of PA is $\dfrac{x-7}{1/\sqrt{2}} = \dfrac{y-2}{1/\sqrt{2}} = r$

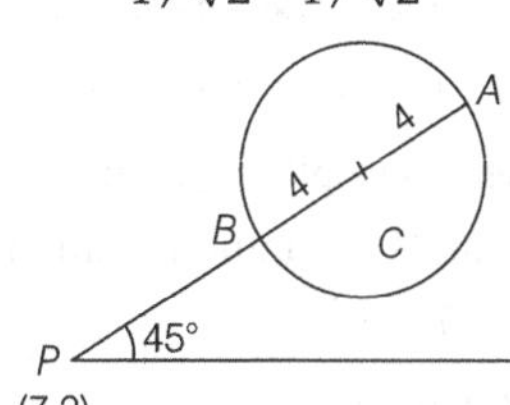

$\therefore$ $\left(7 + \frac{r}{\sqrt{2}}, 2 + \frac{r}{\sqrt{2}}\right)$ lie on circle,

Then, $$\left(7 + \frac{r}{\sqrt{2}}\right)^2 + \left(2 + \frac{r}{\sqrt{2}}\right)^2 - 4\left(7 + \frac{r}{\sqrt{2}}\right) + 6\left(2 + \frac{r}{\sqrt{2}}\right) - 3 = 0$$

$\Rightarrow r^2 + 10\sqrt{2}r + 34 = 0$

$\therefore$ $$r = -5\sqrt{2} \pm 4$$

$\therefore$ Points $\left(7 + \dfrac{-5\sqrt{2} \pm 4}{\sqrt{2}}, 2 + \dfrac{-5\sqrt{2} \pm 4}{\sqrt{2}}\right)$

$\Rightarrow$ $$(2 \pm 2\sqrt{2}, -3 \pm 2\sqrt{2})$$

Taking + ve sign for A, –ve sign for B.

10. Radius of inner circle $= OR - a = \sqrt{a^2 + a^2} - a = a(\sqrt{2} - 1)$

Radius of outer circle $= OR + RQ$

$= a\sqrt{2} + a = a(\sqrt{2} + 1)$

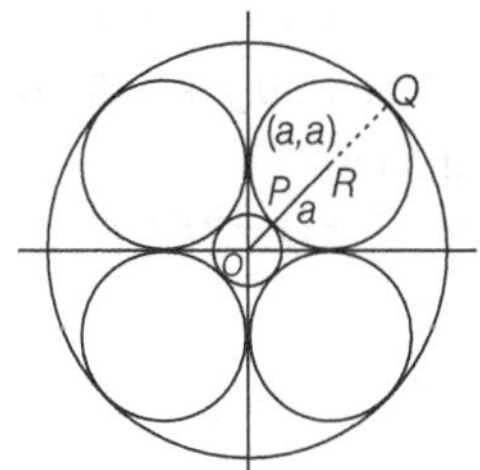

11. It is given that T is tangents to S_1 at P and S_2 at Q and S_1 and S_2 touch externally at M.

S_1 M S_2

T

P(–2, 7) N Q(2, –5)

$\therefore MN = NP = NQ$

$\therefore$ Locus of M is a circle having PQ as its diameter of circle.

$\therefore$ Equation of circle

$$(x-2)(x+2) + (y+5)(y-7) = 0$$

$\Rightarrow$ $$x^2 + y^2 - 2y - 39 = 0$$

Hence, $E_1 : x^2 + y^2 - 2y - 39 = 0, x \neq \pm 2$

Locus of mid-point of chord (h, k) of the circle E_1 is

$$xh + yk - (h+k) - 39 = h^2 + k^2 - 2k - 39$$

$\Rightarrow$ $$xh + yk\ \ y - k = h^2 + k^2 - 2k$$

Since, chord is passing through (1, 1).

$\therefore$ Locus of mid-point of chord (h, k) is

$$h+k-1-k=h^2+k^2-2k$$

$$\Rightarrow \quad h^2+k^2-2k-h+1=0$$

Locus is $E_2 : x^2+y^2-x-2y+1=0$

Now, after checking option (a) and (d) are correct.

12. Circle possible in IV quadrant. Equation of circle is

$(x-c)^2+(y+c)^2=c^2$ ...(i)

its passes through (3,–6)

$$\Rightarrow \quad (3-c)^2+(c-6)^2=c^2$$

$$\Rightarrow \quad c^2-18c+45=0$$

$$(c-15)(c-3)=0$$

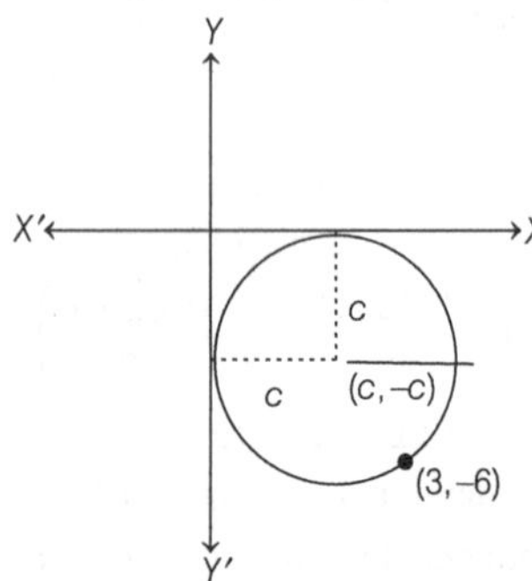

$$\therefore \quad c=3, 15$$

From Eq. (i), equation of circles are

$$x^2+y^2-6x+6y+9=0$$

and $x^2+y^2-30x+30y+225=0$

13. The given equation is

$$(x-r)^2+(y-h)^2=r^2$$

Tangents are $x=0$

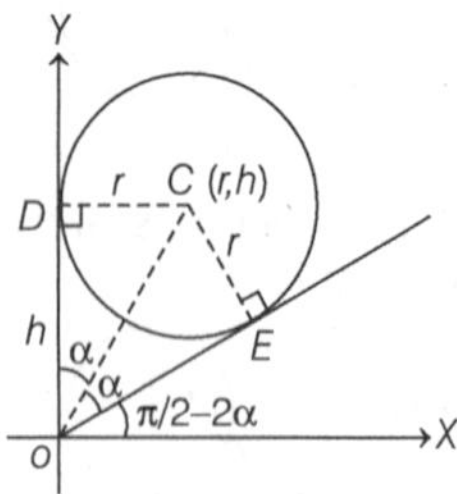

and $\quad y=x\tan\left(\dfrac{\pi}{2}-2\alpha\right)$

$$=x\cot 2\alpha$$

$$=\frac{x(1-\tan^2\alpha)}{2\tan\alpha}$$

$$y=\frac{x\left(1-\dfrac{r^2}{h^2}\right)}{2\left(\dfrac{r}{h}\right)} \quad \left[\because \text{in } \Delta ODC, \tan\alpha=\frac{r}{h}\right]$$

or $\quad (h^2-r^2)x-2rhy=0$

14. Let radius of required circle is r

$\therefore$ For touch

Distance between centres = radius of I circle $\pm$ radius of II circle

$$5=r\pm 1$$

$$\because \quad r=5\pm 1=6, 4$$

$\therefore$ Equation of circles are

$$(x-4)^2+(y-3)^2=6^2$$

and $\quad (x-4)^2+(y-3)^2=4^2$

or $\quad x^2+y^2-8x-6y-11=0$

and $\quad x^2+y^2-8x-6y+9=0$

15. The equation of tangent in terms of slope of $x^2+y^2=25$ is

$$y=mx\pm 5\sqrt{(1+m^2)} \quad \text{...(i)}$$

Given Eq. (i), pass through (– 2, 11), then

$$11=-2m\pm 5\sqrt{(1+m^2)}$$

squaring both sides, then we get

$$21m^2-44m-96=0$$

$$\Rightarrow \quad (7m-24)(3m+4)=0$$

$\therefore \quad m=-4/3,\ 24/7$

There from Eq. (i) we get required tangents are

$$24x-7y\pm 125=0 \text{ and } 4x+3y=\pm 25$$

Hence tangents are

$$24x-7y+125=0 \text{ and } 4x+3y=25$$

16. $OA=4\sqrt{2}\sec\alpha$

$BC=2OB=8\sqrt{2}\operatorname{cosec}\alpha$

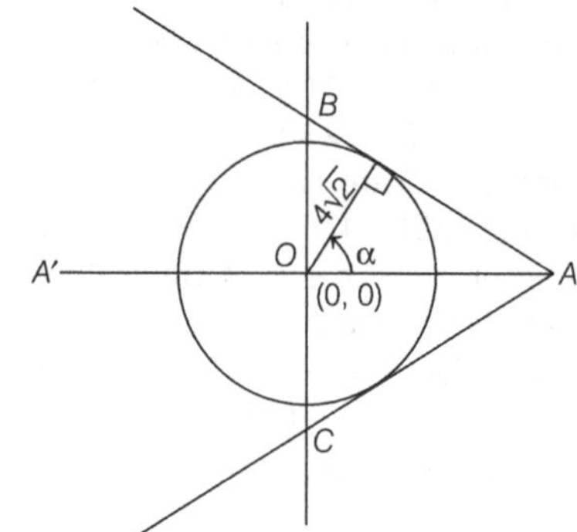

Area of $\Delta ABC=\dfrac{1}{2}\times OA\times BC$

$$=\frac{1}{2}\times 4\sqrt{2}\sec\alpha\times 8\sqrt{2}\operatorname{cosec}\alpha$$

$$=\frac{64}{\sin 2\alpha}$$

Which is minimum when $\sin 2\alpha$ is maximum

i.e. when $\alpha=\dfrac{\pi}{4}$

$\therefore \quad OA=4\sqrt{2}\times\sqrt{2}=8$

$\therefore \quad A\equiv(8, 0)$ and symmetrically on the other side

$A'\equiv(-8, 0)$

17. Let $S\equiv x^2+y^2+2gx+2fy+c=0$

$$S_1\equiv x^2+y^2-4=0$$

$$S_2\equiv x^2+y^2-6x-8y+10=0$$

$$S_3\equiv x^2+y^2+2x-4y-2=0$$

$\therefore$ Common chords are

$S - S_1 \equiv 2gx + 2fy + c + 4 = 0$ ...(i)

$S - S_2 \equiv (2g + 6)x + (2f + 8)y + c - 10 = 0$ ...(ii)

$S - S_3 \equiv (2g - 2)x + (2f + 4)y + c + 2 = 0$ ...(iii)

For cutting the extremities of diameter, chords (i), (ii) and (iii) pass through the centres of S_1, S_2 and S_3 respectively, then

$\therefore$ $c + 4 = 0$, $(2g + 6)3 + (2f + 8)4 + c - 10 = 0$

and $(2g - 2)(-1) + (2f + 4)(2) + c + 2 = 0$

after solving $c = -4, g = -2, f = -3$

18. Let $M(\alpha, \beta)$

$\because$ $AM = 2AB$

$\Rightarrow$ $AB + BM = 2AB$

$\therefore$ $AB = BM$

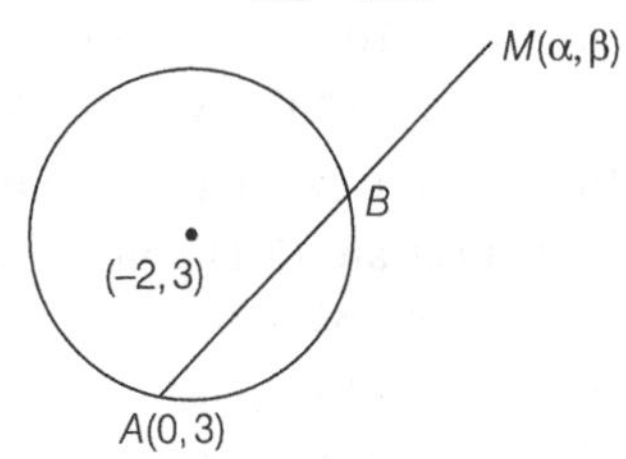

$\therefore$ B is the mid point of AM.

$\therefore$ Coordinates of B are $\left(\frac{\alpha}{2}, \frac{3+\beta}{2}\right)$

which lie on $x^2 + 4x + (y - 3)^2 = 0$

$$\frac{\alpha^2}{4} + 4 \times \frac{\alpha}{2} + \left(\frac{3+\beta}{2} - 3\right)^2 = 0$$

$$\alpha^2 + 8\alpha + (\beta - 3)^2 = 0$$

$\therefore$ Requires locus is

$$x^2 + 8x + (y - 3)^2 = 0$$

or $x^2 + y^2 + 8x - 6y + 9 = 0$

19. It is clear that N is the mid-point of M and P

$\therefore$ $\alpha = \frac{16}{3} - \frac{1}{3} = 5$ and $\beta = \frac{16}{3} - \frac{1}{3} = 5$

$(\alpha, \beta) \equiv (5, 5)$ lie on the circle

$\therefore$ $5^2 + 5^2 - 4 \times 5 - 2 \times 5 - c = 0$

$$c = 20$$

or $c^2 - 40c + 400 = 0$

20. $C_1 \equiv x^2 + y^2 - 2x - 4y - 4 = 0$ ...(i)

and $C_2 \equiv x^2 + y^2 + 2x + 4y + 4 = 0$ ...(ii)

$\therefore$ Radical axis is $C_1 - C_2 = 0$

$\Rightarrow$ $-4x - 8y - 8 = 0$

or $x + 2y + 2 = 0$ which is $L = 0$

(a) option is correct.

Centre and radius of $C_1 = 0$ are $(1, 2)$ and 3.

$\because$ Length of $\perp$ from $(1, 2)$ on $L = 0$

is $\frac{|1 + 4 + 2|}{\sqrt{1 + 4}} = \frac{7}{\sqrt{5}} \neq$ radius

$\therefore$ (b) option is wrong.

L is also the common chord of C_1 and C_2.

$\therefore$ (c) option is correct.

$\because$ Centres of $C_1 = 0$ and $C_2 = 0$ are $(1, 2)$ and $(-1, -2)$

$\therefore$ Slope of Line joining centres of circles $C_1 = 0$ & $C_2 = 0$ is $\frac{-2-2}{-1-1} = \frac{4}{2} = 2 = m_1$ (say)

and slope of $L = 0$ is $-\frac{1}{2} = m_2$ (say)

$\therefore$ $m_1 m_2 = -1$

Hence L is perpendicular to the line joining centres of C_1 and C_2.

$\therefore$ (d) option is correct.

21. Given $C_1: x^2 + y^2 = 3$ intersects in parabola $x^2 = 2y$.

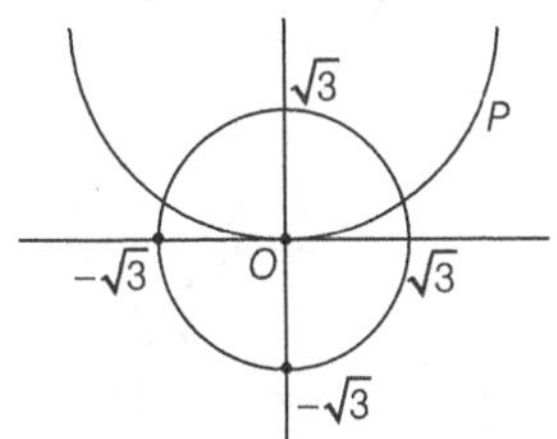

On solving $x^2 + y^2 = 3$ and $x^2 = 2y$, we get

$$y^2 + 2y = 3$$

$\Rightarrow$ $y^2 + 2y - 3 = 0$

$\Rightarrow$ $(y + 3)(y - 1) = 0$

$\therefore$ $y = 1, -3$

[neglecting $y = -3$, a– $\sqrt{3} \le y \le \sqrt{3}$]

$\therefore$ $y = 1 \Rightarrow x = \pm\sqrt{2}$

$\Rightarrow$ $P(\sqrt{2}, 1) \in$ I quadrant

Equation of tangent at $P(\sqrt{2}, 1)$ to $C_1 : x^2 + y^2 = 3$ is

$\sqrt{2}x + 1 \cdot y = 3$ ...(i)

Now, let the centres of C_2 and C_3 be Q_2 and Q_3, and tangent at P touches C_2 and C_3 at R_2 and R_3 shown as below

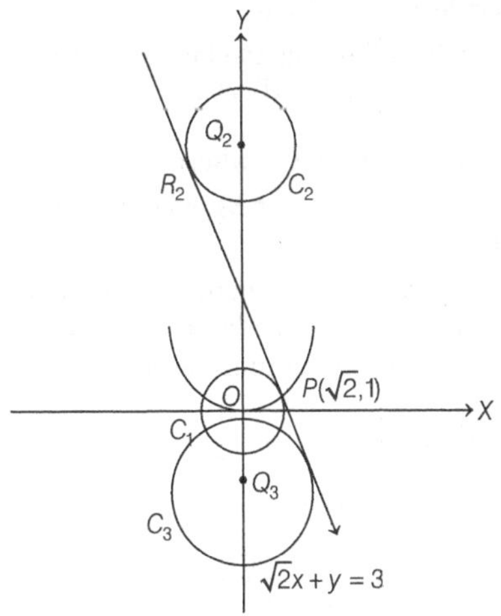

Let Q_2 be $(0, k)$ and radius is $2\sqrt{3}$.

$\therefore$ $\frac{|0 + k - 3|}{\sqrt{2+1}} = 2\sqrt{3} \Rightarrow |k - 3| = 6$

$\Rightarrow$ $k = 9, -3$

$\therefore$ $Q_2(0, 9)$ and $Q_3(0, -3)$

Hence, $Q_2Q_3 = 12$

$\therefore$ Option (a) is correct.

Also, R_2R_3 is common internal tangent to C_2 and C_3, and $r_2 = r_3 = 2\sqrt{3}$

$\therefore \quad R_2R_3 = \sqrt{d^2 - (r_1 + r_2)^2}$

$= \sqrt{12^2 - (4\sqrt{3})^2}$

$= \sqrt{144 - 48} = \sqrt{96} = 4\sqrt{6}$

$\therefore$ Option (b) is correct.

$\because$ Length of perpendicular from $O(0, 0)$ to R_2R_3 is equal to radius of $C_1 = \sqrt{3}$.

$\therefore$ Area of $\Delta OR_2R_3 = \frac{1}{2} \times R_2R_3 \times \sqrt{3}$

$= \frac{1}{2} \times 4\sqrt{6} \times \sqrt{3} = 6\sqrt{2}$

$\therefore$ Option (c) is correct.

Also, area of $\Delta PQ_2Q_3 = \frac{1}{2} Q_2Q_3 \times \sqrt{2} = \frac{\sqrt{2}}{2} \times 12 = 6\sqrt{2}$

$\therefore$ Option (d) is incorrect.

22. Given circle is $x^2 + y^2 = a^2$ ...(i)

clearly (0, 0) will belong the interior of circle (i) Also other points interior to circle (i) will have the coordinates of the form

$(\pm \lambda, 0), (0, \pm \lambda)$, where $\lambda^2 < a^2$

and $(\pm \lambda, \pm \mu)$ and $(\pm \mu, \pm \lambda)$, where $\lambda^2 + \mu^2 < a^2$ and $\lambda, \mu \in I$

$\therefore$Number of lattice points in the interior of the circle will be of the form $1 + 4r + 8t$, where $r, t = 0, 1, 2, ...$

$\therefore$Number of such points must be of the form $4n + 1$, where $n = 0, 1, 2, ...$

23. Let slope of OA is m,

Then $\left|\frac{m-1}{1+m}\right| = \tan 45^\circ$

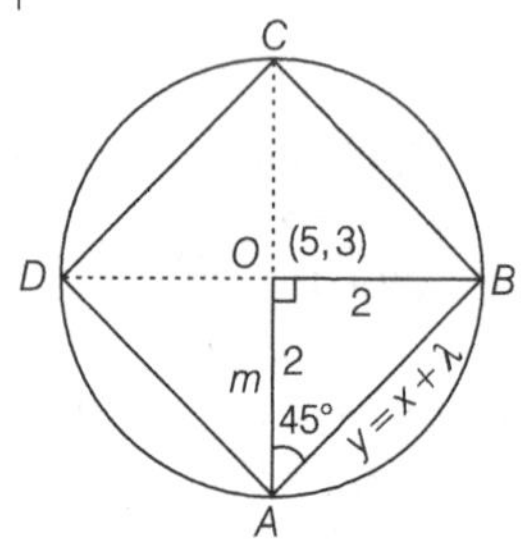

$\Rightarrow \quad \frac{m-1}{1+m} = \pm 1$ or $\frac{m-1}{m+1} = -1$

$\Rightarrow \quad m - 1 = -m - 1$

$\therefore \quad m = 0$

$\therefore$ Equation of OA is $y = 3$

Solving $y = 3$ and $x^2 + y^2 - 10x - 6y + 30 = 0$

$\Rightarrow \quad x^2 + 9 - 10x - 18 + 30 = 0$

$\Rightarrow \quad x^2 - 10x + 21 = 0$

$(x - 7)(x - 3) = 0$

or $\quad x = 3, 7$

$\therefore$Two vertices are (3, 3) and (7, 3) and other diagonal is $\perp$ to $y = 3$ and through centre (5, 3) is $x = 5$

Now solving $x = 5$

and $\quad x^2 + y^2 - 10x - 6y + 30 = 0$

$\Rightarrow \quad 25 + y^2 - 50 - 6y + 30 = 0$

$\Rightarrow \quad y^2 - 6y + 5 = 0 \Rightarrow (y - 1)(y - 5) = 0$

$\therefore$ Other two vertices are (5,1) and (5, 5)

24. Solving $x^2 + y^2 - 4x - 81 = 0$ ...(i)

and $x^2 + y^2 + 24x - 81 = 0$ ...(ii)

we get $\quad x = 0$ and $y = \pm 9$

$\therefore$ $A(0, -9)$ and $B(0, 9)$

Equation of line through $A(0, -9)$ is

$y + 9 = mx \Rightarrow y = mx - 9$ ...(iii)

Solving Eqs. (i) and (iii), then

$x^2 + (mx - 9)^2 - 4x - 81 = 0$

$x^2(1 + m^2) - (18m + 4)x = 0$

$\therefore \quad P\left(\frac{18m + 4}{1 + m^2}, \frac{9m^2 + 4m - 9}{1 + m^2},\right)$

and equation of line through $B(0, 9)$ and parallal to Eq, (iii) is

$y = mx + \lambda \Rightarrow 9 = 0 + \lambda \Rightarrow y = mx + 9$ ...(iv)

Solving Eqs. (ii) and (iv), then

$\Rightarrow \quad x^2(1 + m^2) + (18m + 24) \cdot x = 0$

$\therefore \quad Q\left(-\frac{(18m + 24)}{1 + m^2}, \frac{-9m^2 - 24m + 9}{1 + m^2},\right)$

Let mid point of PQ is (h, k), then

$2h = -\frac{20}{1 + m^2} \Rightarrow h = -\frac{10}{(1 + m^2)}$ ...(v)

and $2k = -\frac{20m}{1 + m^2} \Rightarrow k = -\frac{10m}{1 + m^2}$ ...(vi)

From Eqs. (v) and (vi), $\frac{k}{h} = m$

Substituting the value of m in Eq. (v)

Then $h\left(1 + \frac{k^2}{h^2}\right) = -10$

$\Rightarrow \quad h^2 + k^2 = -10h$

$\Rightarrow h^2 + k^2 + 10h = 0$

$\therefore$ Locus of mid point of PQ is

$x^2 + y^2 + 10x = 0$

or $\quad (x + 5)^2 + (y + 0)^2 = 25$

25. $x^2 + y^2 - 8x - 16y + 60 = 0$ ...(i)

Equation of chord of contact from (–2, 0) is

$-2x - 4(x - 2) - 8y + 60 = 0$

$3x + 4y - 34 = 0$ ...(ii)

From Eqs. (i) and (ii),

$$x^2 + \left(\frac{34 + 3x}{4}\right)^2 - 8x - 16\left(\frac{34 - 3x}{4}\right) + 60 = 0$$

$$\Rightarrow \quad 16x^2 + 1156 - 204x + 9x^2 - 128x - 2176 + 192x + 960 = 0$$

$$\Rightarrow \quad 5x^2 - 28x - 12 = 0$$

$$\Rightarrow \quad (x - 6)(5x + 2) = 0$$

$$x = 6, -\frac{2}{5}$$

$\therefore$ Points are (6, 4), $\left(-\frac{2}{5}, \frac{44}{5}\right)$.

26. Let OA, OB be the tangents from the origin to the given circle with centre $C(-3, 5)$ and radius $\sqrt{(9 + 25 - c)}$

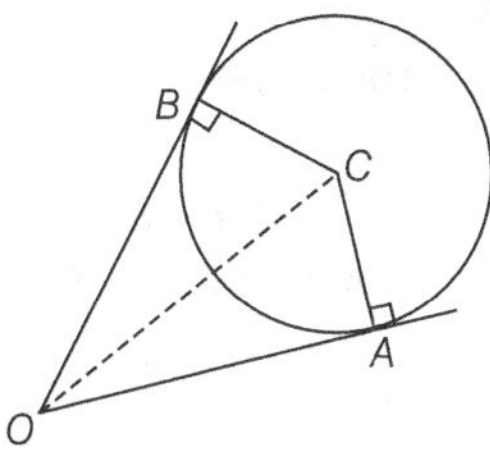

The area of the quadrilateral $OACB$

$= 2 \times$ area of ΔOAC

$= 2 \times \frac{1}{2} \times OA \times AC$

Now OA = length of tangent from the origin to the given circle = $\sqrt{c}$ and AC = radius of the circle = $\sqrt{34 - c}$

So $\sqrt{c}\sqrt{34 - c} = 8$ (given)

$\Rightarrow \quad c(34 - c) = 64$

$\Rightarrow \quad c^2 - 34c + 64 = 0$

$\Rightarrow \quad c = 2$ or $c = 32$

27. Let $P(x, y)$ be any point of the circle on the chord joining $A(1, 2)$ and $B(2, -1)$

The slope of $AP = \frac{y - 2}{x - 1}$

and slope of $BP = \frac{y + 1}{x - 2}$

Since they include an angle of $\frac{\pi}{4}$ so,

$$\frac{\frac{y + 1}{x - 2} - \frac{y - 2}{x - 1}}{1 + \frac{y - 2}{x - 1} \times \frac{y + 1}{x - 2}} = \pm \tan\frac{\pi}{4} = \pm 1$$

$\Rightarrow \quad (y + 1)(x - 1) - (x - 2)(y - 2)$
$= \pm[(x - 1)(x - 2) + (y - 2)(y + 1)]$

$\Rightarrow 3x + y - 5 = \pm[x^2 + y^2 - 3x - y]$

$\Rightarrow \quad x^2 + y^2 - 5 = 0$

or $\quad x^2 + y^2 - 6x - 2y + 5 = 0$

28. Given, RS is the diameter of $x^2 + y^2 = 1$.

Here, equation of the tangent at $P(\cos\theta, \sin\theta)$ is $x\cos\theta + y\sin\theta = 1$.

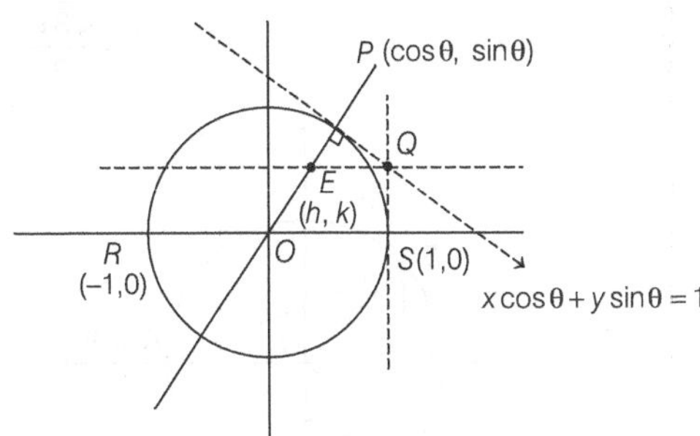

Intersecting with $x = 1$,

$$y = \frac{1 - \cos\theta}{\sin\theta}$$

$$\therefore \quad Q\left(1, \frac{1 - \cos\theta}{\sin\theta}\right)$$

$\therefore$ Equation of the line through Q parallel to RS is

$$y = \frac{1 - \cos\theta}{\sin\theta} = \frac{2\sin^2\frac{\theta}{2}}{2\sin\frac{\theta}{2}\cos\frac{\theta}{2}} = \tan\frac{\theta}{2} \quad \text{...(i)}$$

Normal at $P : y = \frac{\sin\theta}{\cos\theta} \cdot x$

$\Rightarrow \quad y = x\tan\theta$ ...(ii)

Let their point of intersection be (h, k).

Then, $\quad k = \tan\frac{\theta}{2}$ and $k = h\tan\theta$

$$\therefore \quad k = h\left(\frac{2\tan\frac{\theta}{2}}{1 - \tan^2\frac{\theta}{2}}\right) \Rightarrow k = \frac{2h \cdot k}{1 - k^2}$$

$\Rightarrow \quad k(1 - k^2) = 2hk$

$\therefore$ Locus for point $E : 2x = (1 - y^2)$ ...(iii)

When $x = \frac{1}{3}$, then

$$1 - y^2 = \frac{2}{3} \Rightarrow y^2 = 1 - \frac{2}{3} \Rightarrow y = \pm\frac{1}{\sqrt{3}}$$

$\therefore \left(\frac{1}{3}, \pm\frac{1}{\sqrt{3}}\right)$ satisfy $2x = 1 - y^2$.

When $x = \frac{1}{4}$, then

$$1 - y^2 = \frac{2}{4} \Rightarrow y^2 = 1 - \frac{1}{2} \Rightarrow y = \pm\frac{1}{\sqrt{2}}$$

$\therefore \left(\frac{1}{4}, \pm\frac{1}{2}\right)$ does not satisfy $1 - y^2 = 2x$.

29. Since $\quad \angle B = \angle C = 75°$

$\Rightarrow \quad \angle BAC = 30° \Rightarrow \angle BOC = 60°$

$\Rightarrow \Delta OBC$ is equilateral

$\Rightarrow \; BC = OB = OC = a$

If M is the middle point of BC, then $OM = \sqrt{\left(a^2 - \frac{a^2}{4}\right)}$

$$= \frac{\sqrt{3}\,a}{2}$$

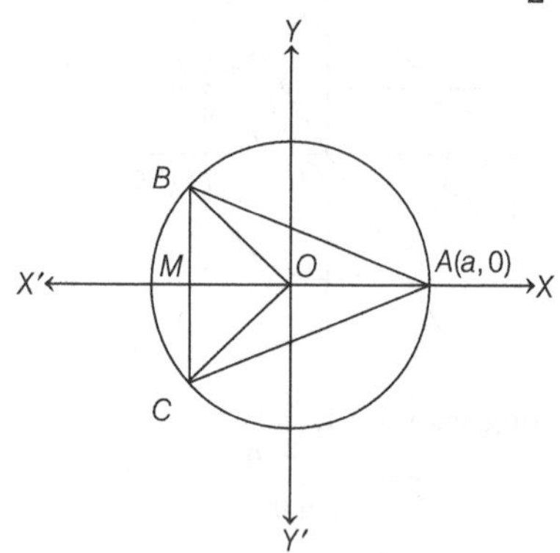

As BC is $\perp$ to MO, the x-axis.

Equation of BC is $\quad x = -\frac{\sqrt{3}\,a}{2}$

Which meets the circle $x^2 + y^2 = a^2$ at the points $\left(-\frac{\sqrt{3}a}{2}, \pm\frac{a}{2}\right)$ which are the end points of the base BC.

30. Since the given line touches the given circle, the length of the perpendicular from the centre (2, 4) of the circle from the line $3x - 4y - \lambda = 0$ is equal to the radius $\sqrt{(4 + 16 + 5)} = 5$ of the circle.

$$\Rightarrow \quad \frac{(3\times 2 - 4\times 4 - \lambda)}{\sqrt{(9+16)}} = \pm 5$$

$$\Rightarrow \quad \lambda = 15 \text{ or } -35$$

Now, equation of the tangent at (a, b) to the given circle is

$$xa + yb - 2(x + a) - 4(y + b) - 5 = 0$$

$$\Rightarrow \quad (a - 2)x + (b - 4)y - (2a + 4b + 5) = 0$$

If it represents the given line $3x - 4y - \lambda = 0$, then

$$\frac{a-2}{3} = \frac{b-4}{-4} = \frac{2a+4b+5}{\lambda} = l \text{ (say)}$$

Then $\quad a = 3l + 2,\ b = 4 - 4l$ and $2a + 4b + 5 = \lambda l \quad$ ...(i)

$\Rightarrow \quad 2(3l + 2) + 4(4 - 4l) + 5 = 15l$ (if $\lambda = 15$)

$\Rightarrow \quad l = 1 \Rightarrow a = 5, b = 0$ and $\lambda + a + b = 20$

Again, if $\lambda = -35$ [from Eq. (i)]

$$25 - 10l = -35l$$

$$\Rightarrow \quad l = -1$$

$$\Rightarrow \quad a = -1, b = 8$$

and $\quad \lambda + a + b = -35 - 1 + 8 = -28$

Linked Comprehension Type

PASSAGE 1

1. $\because 16m^2 = 8l + 1$

$\Rightarrow \quad 16(l^2 + m^2) = 16l^2 + 8l + 1 = (4l + 1)^2$

or $\quad 4\sqrt{(l^2 + m^2)} = |4l + 1|$

or $\quad \frac{|4l+1|}{\sqrt{(l^2+m^2)}} = 4$

$\therefore \quad$ Centre $\equiv (4, 0)$ and radius $\equiv 4$

Equation of circle is $(x - 4)^2 + (y - 0)^2 = 4^2$

$$\Rightarrow \quad x^2 + y^2 - 8x = 0$$

2. $\because 9l^2 + 16m^2 + 1 + 24lm + 6l + 8m = 25(l^2 + m^2)$

$$\Rightarrow \quad (3l + 4m + 1)^2 = 25(l^2 + m^2)$$

$$\Rightarrow \quad \frac{|3l + 4m + 1|}{\sqrt{(l^2+m^2)}} = 5$$

$\therefore \quad$ Centre $\equiv (3, 4)$ and radius $\equiv 5$

$\Rightarrow$ Equation of circle is $(x - 3)^2 + (y - 4)^2 = 5^2 = 25$

$\therefore$ Equation of director circle is $(x - 3)^2 + (y - 4)^2 = 50$

or $\quad x^2 + y^2 - 6x - 8y = 25$

3. $\because \quad 4l^2 - 5m^2 + 6l + 1 = 0$

$\Rightarrow \quad 4l^2 + 6l + 1 = 5m^2$

$\Rightarrow \quad 9l^2 + 6l + 1 = 5(l^2 + m^2)$

$\Rightarrow \quad (3l + 1)^2 = 5(l^2 + m^2)$

$\Rightarrow \quad \frac{|3l+1|}{\sqrt{(l^2+m^2)}} = \sqrt{5}$

$\therefore \quad$ Centre $\equiv (3, 0)$ and radius $\equiv \sqrt{5}$

4. $\because x + 2y - 3 = 0$ and $2x + y - 3 = 0$ are non parallel

Bisector of lines

$$\frac{(x + 2y - 3)}{\sqrt{5}} = \pm\frac{(2x + y - 3)}{\sqrt{5}}$$

$\Rightarrow \quad (x + 2y - 3) = (2x + y - 3)$

$\Rightarrow \quad x - y = 0 \quad$ ...(i)

and $\quad (x + 2y - 3) = -(2x + y - 3)$

$\Rightarrow \quad x + y = 2 \quad$ ...(ii)

$\because$ Centre of circle lies on angle bisector of lines.

Also centre lies on a line $3x + 4y = 5 \quad$ ...(iii)

Solving Eqs. (i) and (iii), we get $\left(\frac{5}{7}, \frac{5}{7}\right)$

and solving Eqs. (ii) and (iii), we get

$$(3, -1)$$

5. It is clear that circle lie in II quadrant

$\therefore$ Centre is $(-h, h)$,

then radius $= \sqrt{(-2 + h)^2 + (3 - h)^2} = |h|$

$$\Rightarrow \quad 2h^2 - 10h + 4 + 9 = h^2$$

$$\Rightarrow \quad h^2 - 10h + 13 = 0$$

$$\therefore \quad h = \frac{10 \pm \sqrt{100 - 52}}{2}$$

$$= (5 \pm 2\sqrt{3})$$

Centres are $(-(5 \pm 2\sqrt{3}), 5 \pm 2\sqrt{3})$

$i.e, \quad (-(5 + 2\sqrt{3}), 5 + 2\sqrt{3})$

and $\quad (-5 + 2\sqrt{3}, 5 - 2\sqrt{3})$

PASSAGE 2

1. $\because$ Centre $C \equiv (3, 4)$ radius $a = 3$
Let OC makes an angle β with OX and $OC = \gamma$,
From Eq. (i),
$$r^2 - 2r\gamma\cos(\alpha - \beta) + \gamma^2 - a^2 = 0$$
$\therefore\ OP_1 \cdot OQ_1 = r_1 r_2 = \gamma^2 - a^2 = 25 - 9 = 16$

2. $\therefore\ (AP)^2 = 4^2 + r^2 - 8r\cos(\theta - \pi/6)$
$$(BP)^2 = 3^2 + r^2 - 6r\cos\left(\frac{\pi}{3} - \theta\right)$$

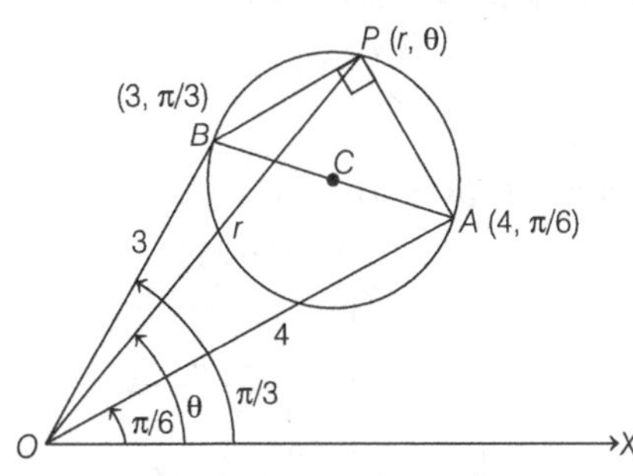

and $(AB)^2 = 3^2 + 4^2 - 24\cos\left(\frac{\pi}{3} - \frac{\pi}{6}\right) = 25 - 12\sqrt{3}$

$\because\ \angle APB = 90°$

$\therefore\ (AP)^2 + (BP)^2 = (AB)^2$

$\Rightarrow 2r^2 + 25 - 2r\{4\cos(\theta - \pi/6) + 3\cos(\pi/3 - \theta)\}$
$$= 25 - 12\sqrt{3}$$
or $r^2 - r\{4\cos(\theta - \pi/6) + 3\cos(\theta - \pi/3)\} + 6\sqrt{3} = 0$

3. $\because \frac{1}{r} = A\cos\theta + B\sin\theta, r = 2a\cos\theta$
$$\therefore\ \frac{1}{2a\cos\theta} = A\cos\theta + B\sin\theta$$
Dividing by $\cos\theta$, then
$$\frac{1}{2a\cos^2\theta} = A + B\tan\theta \Rightarrow \frac{1}{2a}(1 + \tan^2\theta) = A + B\tan\theta$$
$\Rightarrow \tan^2\theta - 2aB\tan\theta + (1 - 2aA) = 0$
Above being a quadratic in $\tan\theta$ gives us two values of θ, but if the line is to touch the circle, then the two roots of the above equation should be equal, the condition for which is
$$(-2aB)^2 - 4\cdot 1\cdot(1 - 2aA) = 0 \Rightarrow a^2B^2 + 2aA = 1$$

4. If (R, α) be the coordinates of the centre of the circle and a be its radius, then its equation is

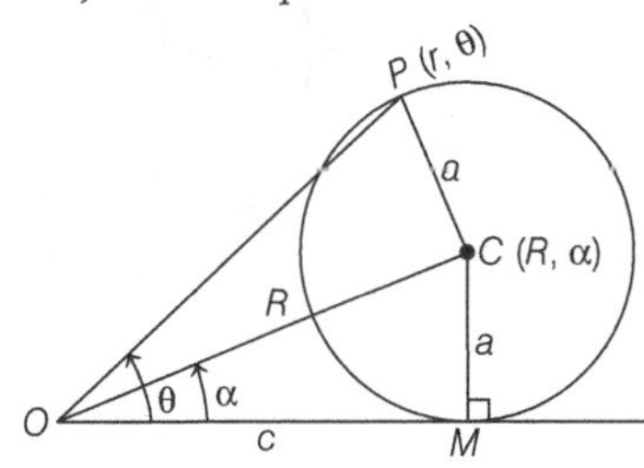

$$r^2 - 2rR\cos(\theta - \alpha) + (R^2 - a^2) = 0 \quad \ldots\text{(i)}$$
Now, In ΔOCM,
$C = R\cos\alpha$ and $a = R\sin\alpha$ and $R^2 - a^2 = c^2$ ...(ii)
Then from Eq. (i),
$$r^2 - 2Rr(\cos\theta\cos\alpha + \sin\theta\sin\alpha) + R^2 - a^2 = 0$$
$$\Rightarrow \quad r^2 - 2r(c\cos\theta + a\sin\theta) + c^2 = 0$$
[from Eq. (ii)]
$$\text{or} \quad r^2 - 2cr\cos\theta + c^2 = 2ar\sin\theta \quad \ldots\text{(iii)}$$
Since it passes through (r', θ'), we have
$$r'^2 - 2cr'\cos\theta' + c^2 = 2ar'\sin\theta' \quad \ldots\text{(iv)}$$
Dividing Eq. (iii) by Eq. (iv), we get
$$\frac{r^2 - 2cr\cos\theta + c^2}{r'^2 - 2cr'\cos\theta' + c^2} = \frac{r\sin\theta}{r'\sin\theta'}$$
$$\text{or} \quad (r^2 - 2cr\cos\theta + c^2) = \frac{r\sin\theta}{r'\sin\theta'}(r'^2 - 2cr'\cos\theta' + c^2)$$
$$\therefore \quad \lambda = \frac{r\sin\theta}{r'\sin\theta'}$$

5. The equation of any straight line is
$$x\cos\gamma + y\sin\gamma = p$$
Puting $x = r\cos\theta$ and $y = r\sin\theta$ it becomes
$$p = r\cos(\theta - \gamma) \quad \ldots\text{(i)}$$
When p and γ are unknown.
The radius vector corresponding to vectorial angle α is $2a\cos\alpha$ and that for β is $2a\cos\beta$, then
From Eq. (i),
$$p = 2a\cos\alpha\cos(\alpha - \gamma) \quad \ldots\text{(ii)}$$
$$\text{and} \quad p = 2a\cos\beta\cos(\beta - \gamma) \quad \ldots\text{(iii)}$$
From Eqs. (ii) and (iii), we get
$$2\cos\alpha\cos(\alpha - \gamma) = 2\cos\beta\cos(\beta - \gamma)$$
$$\Rightarrow \quad \cos(2\alpha - \gamma) + \cos\gamma = \cos(2\beta - \gamma) + \cos\gamma$$
$$\Rightarrow \quad \cos(2\alpha - \gamma) = \cos(2\beta - \gamma)$$
$$\therefore \quad (2\alpha - \gamma) = \pm(2\beta - \gamma)$$
(Neglecting + ve sign as $\alpha \neq \beta$)
$$\Rightarrow \quad 2\alpha - \gamma = -2\beta + \gamma$$
$$\text{or} \quad \gamma = \alpha + \beta$$
Substituting the value of γ in Eqs. (i) and (ii), then we get
$\therefore\ r\cos(\theta - \alpha - \beta) = 2a\cos\alpha\cos\beta$
$$\therefore \quad \frac{r\cos(\theta - \alpha - \beta)}{2\cos\alpha} = a\cos\beta = \lambda$$
$$\therefore \quad \lambda = a\cos\beta$$

PASSAGE 3

$\because$ Length of tangents from a point to circle are equal.
$$PQ = PR$$
Then Parallelogram PQRS is rhombus.
$\therefore$ Mid point of QR = mid point of PS
and $QR \perp PS$
$\therefore$ S is the mirror image of P w.r.t. QR

1. $\because\ L \equiv 2x + y = 6$
Let $P \equiv (\lambda, 6 - 2\lambda)$

$\because \angle PQO = \angle PRO = \pi / 2$

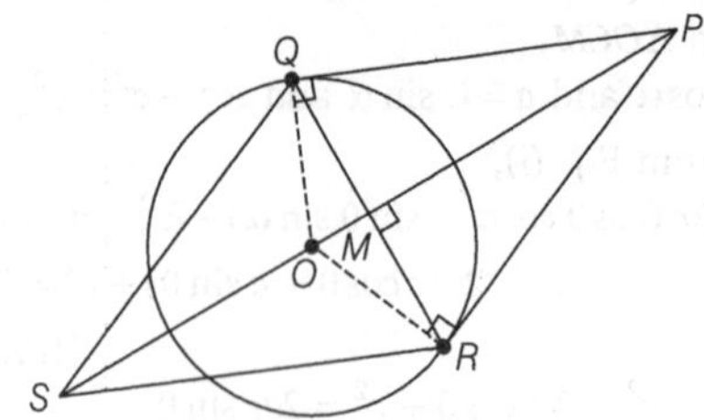

$\therefore$ OP is diameter of circumcircle PQR, then

centre is $\left(\frac{\lambda}{2}, 3-\lambda\right)$

$\therefore \quad x = \frac{\lambda}{2} \Rightarrow \lambda = 2x$

and $\quad y = 3 - \lambda,$

then $\quad 2x + y = 3$

2. $\because L \equiv y = 4$

Let $P \equiv (h, 4)$

$\therefore$ Circumcentre of ΔPQR is $\left(\frac{h}{2}, 2\right)$,

then $\quad y = 2 \Rightarrow y - 2 = 0$

3. $\because P \equiv (2, 3)$

$\therefore$ Equation of QR is $2x + 3y = 4$

Let $\quad S \equiv (\alpha, \beta)$

$$\Rightarrow \frac{\alpha - 2}{2} = \frac{\beta - 3}{3} = \frac{-2(4 + 9 - 4)}{(4 + 9)} = -\frac{18}{13}$$

$$\therefore \quad \alpha = \frac{10}{-13}, \beta = -\frac{15}{13}$$

Now, equation of circumcircle of ΔQRS is

$$(x^2 + y^2 - 4) + \lambda\,(2x + 3y - 4) = 0$$

$\because$ Its pass through $S \equiv \left(-\frac{10}{13}, -\frac{15}{13}\right)$, then

$$\left(\frac{100}{169} + \frac{225}{169} - 4\right) + \lambda\left(-\frac{20}{13} - \frac{45}{13} - 4\right) = 0$$

$$\left(-\frac{351}{169}\right) + \lambda\,(-9) = 0$$

$$\therefore \quad \lambda = -\frac{39}{169} = -\frac{3}{13}$$

$\therefore$ Circumcentre is $\left(-\lambda, -\frac{3\lambda}{2}\right) \equiv \left(\frac{3}{13}, \frac{9}{26}\right)$

4. $\because P \equiv (6, 8)$

$\therefore$ Equation of QR is

$$6x + 8y = 4$$

$$\Rightarrow \quad 3x + 4y - 2 = 0$$

$$\therefore \quad PM = \frac{|3 \times 6 + 4 \times 8 - 2|}{\sqrt{(3^2 + 4^2)}} = \frac{48}{5}$$

and $\quad PQ = \sqrt{(6^2 + 8^2 - 4)} = \sqrt{96},$

then $QM = \sqrt{(PQ)^2 - (PM)^2} = \sqrt{96 - \frac{(48)^2}{25}}$

$$= \sqrt{48\left(2 - \frac{48}{25}\right)} = \sqrt{\frac{96}{25}}$$

$\therefore$ Area of $\Delta PQR = \frac{1}{2} \cdot PM \cdot QR$

$$= \frac{1}{2} \cdot \frac{48}{5} \cdot \sqrt{\frac{96}{25}} = \frac{48\sqrt{24}}{25} = \frac{96\sqrt{6}}{25}$$

$\because$ $PQRS$ is a rhombus

$\therefore$ Area of ΔQRS = Area of ΔPQR

$$= \frac{96\sqrt{6}}{25} \text{ sq unit}$$

5. $\because P \equiv (3, 4)$

$\therefore$ Equation of QR is $3x + 4y = 4$ ...(i)

Let $\quad S \equiv (x_1, y_1)$

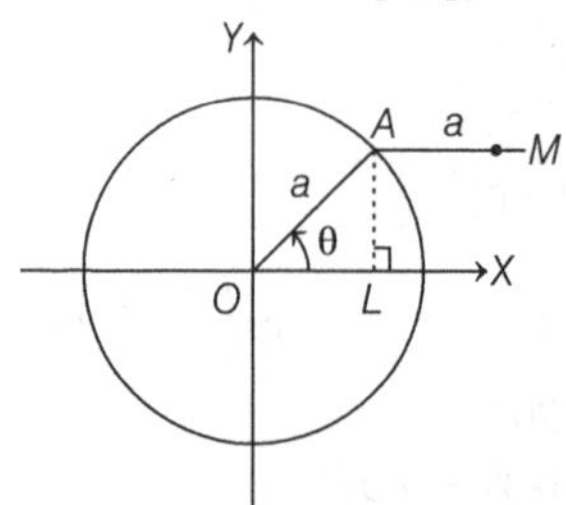

$\because$ S is the mirror image of P w.r.t. Eq. (i), then

$$\frac{x_1 - 3}{3} = \frac{y_1 - 4}{4} = \frac{-2(3 \times 3 + 4 \times 4 - 4)}{3^2 + 4^2} = -\frac{42}{25}$$

$$\therefore \quad x_1 = -\frac{51}{25} \text{ and } y_1 = -\frac{68}{25}$$

$$\therefore \quad S \equiv \left(-\frac{51}{25}, -\frac{68}{25}\right)$$

PASSAGE 4

1. Centre of C lies on the line through D perpendicular to PQ. Thus centre of C lies on

$$y - \frac{3}{2} = \left(\frac{1}{\sqrt{3}}\right)\left(x - \frac{3\sqrt{3}}{2}\right)$$

$$\Rightarrow \quad x = \sqrt{3}\, y$$

Let the centre of the circle C be $(\sqrt{3}\, y_1, y_1)$ then

$$\left(\frac{3\sqrt{3}}{2} - \sqrt{3}\, y_1\right)^2 + \left(\frac{3}{2} - y_1\right)^2 = 1$$

$$\Rightarrow \quad 4\left(\frac{3}{2} - y_1\right)^2 = 1$$

$$\Rightarrow \quad \frac{3}{2} - y_1 = \pm\frac{1}{2}$$

$$\Rightarrow \quad y_1 = 1, 2$$

Thus, centre of C can be $(\sqrt{3}, 1)$, or $(2\sqrt{3}, 2)$. Since, centre of the circle and origin lie on the same side of $\sqrt{3}x + y - 6 = 0$ and $\sqrt{3}(0) + 0 - 6 < 0$ and $\sqrt{3}(\sqrt{3}) + 1 - 6 < 0$ we get centre of the circle C to be $I(\sqrt{3}, 1)$ and hence its equation, is $(x - \sqrt{3})^2 + (y - 1)^2 = 1$

2. Next, if m is the slope of QR

$$\pm \tan\left(\frac{\pi}{3}\right) = \frac{m - (-\sqrt{3})}{1 + m(-\sqrt{3})}$$

$\Rightarrow \quad \pm \sqrt{3}(1 - m\sqrt{3}) = m + \sqrt{3}$

$\Rightarrow \quad m = 0$ or $\sqrt{3}$

Let the slope of QR be 0 and the coordinate of E be (x, y), then since, IE is perpendicular to QR $\dfrac{x - \sqrt{3}}{y - 1} = 0$.

$\Rightarrow x = \sqrt{3}$ and as $IE = 1$, $(x - \sqrt{3})^2 + (y - 1)^2 = 1$

$\Rightarrow y = 0$ or 2 so the coordinates of E are $(\sqrt{3}, 0)$ as $y = 2$ is not given in the choices.

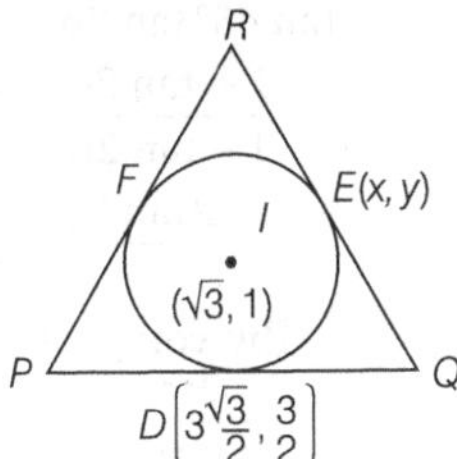

Similarly, let the slope of PR be $\sqrt{3}$. If the coordinates of F are (p, q)

$$\frac{q - 1}{p - \sqrt{3}} \times \sqrt{3} = -1$$

$\Rightarrow \quad p - \sqrt{3} = -\sqrt{3}(q - 1)$

Also, $\quad (p - \sqrt{3})^2 + (q - 1)^2 = 1$

$\Rightarrow \quad 4(q - 1)^2 = 1$

$\Rightarrow \quad q - 1 = \pm \dfrac{1}{2}$

$\Rightarrow \quad q = \dfrac{3}{2}$ or $\dfrac{1}{2}$

If $\quad q = \dfrac{3}{2}, p = \sqrt{3} - \dfrac{\sqrt{3}}{2} = \dfrac{\sqrt{3}}{2}$

If $\quad q = \dfrac{1}{2}, p = \sqrt{3} + \dfrac{\sqrt{3}}{2} = \dfrac{3\sqrt{3}}{2}$

So, the coordinates of F are $\left(\dfrac{\sqrt{3}}{2}, \dfrac{3}{2}\right)$.

3. Now equation of QR is

$$y - 0 = 0(x - \sqrt{3})$$

$\Rightarrow \quad y = 0$

Equation of RP is

$$y - \frac{3}{2} = \sqrt{3}\left(x - \frac{\sqrt{3}}{2}\right)$$

$\Rightarrow \quad y = \sqrt{3}\,x$

PASSAGE 5

1.

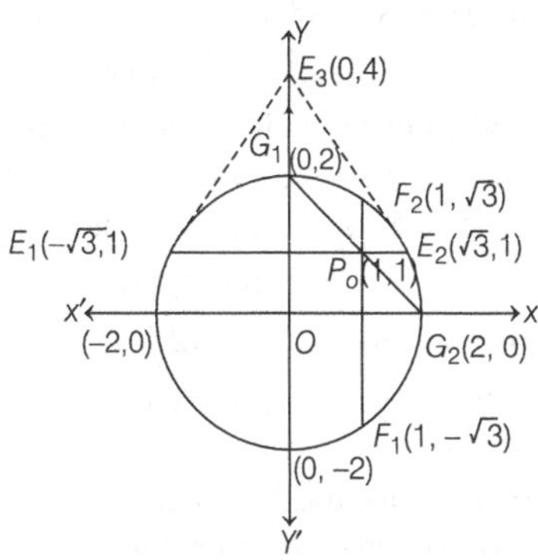

Equation of tangent at $E_1(-\sqrt{3}, 1)$ is $-\sqrt{3}x + y = 4$ and at $E_2(\sqrt{3}, 1)$ is $\sqrt{3}x + y = 4$

Intersection point of tangent at E_1 and E_2 is $(0, 4)$

$\therefore$ Coordinates of E_3 is $(0, 4)$

Similarly, equation of tangent at $F_1(1, -\sqrt{3})$ and $F_2(1, \sqrt{3})$ are $x - \sqrt{3}y = 4$ and $x + \sqrt{3}y = 4$, respectively and intersection point is $(4, 0)$, i.e., $F_3(4, 0)$ and equation of tangent at $G_1(0, 2)$ and $G_2(2, 0)$ are $2y = 4$ and $2x = 4$, respectively and intersection point is $(2, 2)$ i.e., $G_3(2, 2)$.

Point $E_3(0, 4)$, $F_3(4, 0)$ and $G_3(2, 2)$ satisfies the line $x + y = 4$.

2. We have, $\quad x^2 + y^2 = 4$

Let $P(2\cos\theta, 2\sin\theta)$ be a point on a circle.

$\therefore$ Tangent at P is $\quad 2\cos\theta\, x + 2\sin\theta\, y = 4$

$\Rightarrow \quad x\cos\theta + y\sin\theta = 2$

$\therefore$ The coordinates at $M\left(\dfrac{2}{\cos\theta}, 0\right)$ and $N\left(0, \dfrac{2}{\sin\theta}\right)$

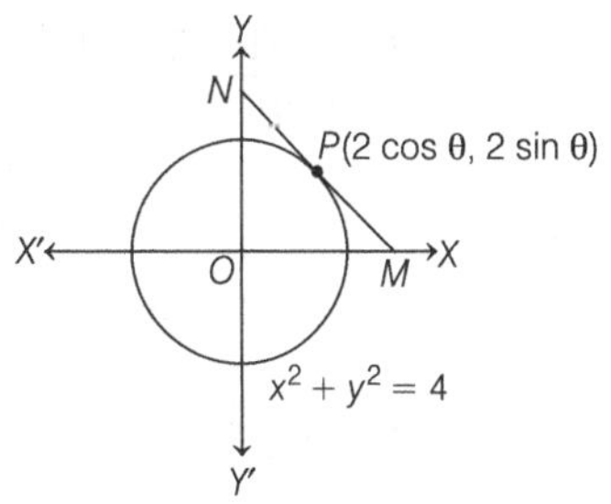

Let (h, k) is mid-point of MN

$\therefore \quad h = \dfrac{1}{\cos\theta}$ and $k = \dfrac{1}{\sin\theta}$

$\Rightarrow \quad \cos\theta = \dfrac{1}{h}$ and $\sin\theta = \dfrac{1}{k}$

$\Rightarrow \quad \cos^2\theta + \sin^2\theta = \dfrac{1}{h^2} + \dfrac{1}{k^2}$

$\Rightarrow \quad 1 = \dfrac{h^2 + k^2}{h^2 \cdot k^2} \Rightarrow h^2 + k^2 = h^2 k^2$

$\therefore$ Mid-point of MN lie on the curve

$$x^2 + y^2 = x^2 y^2$$

Numerical Value Type

1. The given circles $S_1 \equiv x^2 + y^2 + 14x - 4y + 28 = 0$

and $S_2 \equiv x^2 + y^2 - 14x + 4y - 28 = 0$

Centres and radii of circles S_1 and S_2 are

$C_1(-7, 2),\ r_1 = \sqrt{49 + 4 - 28} = 5$

and $C_2(7, -2),\ r_2 = \sqrt{49 + 4 + 28}$

$= 9$ respectively.

Here $d = C_1C_2 = \sqrt{(-7-7)^2 + (2+2)^2}$

$= \sqrt{212} > r_1 + r_2 = 5 + 9 = 14$

$\therefore \quad d > r_1 + r_2$

Hence circles don't touch or cut.

$\therefore$ Length of external common tangent

$$L_{ex} = \sqrt{d^2 - (r_2 - r_1)^2}$$
$$= \sqrt{212 - (9-5)^2} = \sqrt{212 - 16}$$
$$= \sqrt{196} = 14 = \lambda$$

and length of internal common tangent

$$L_{in} = \sqrt{d^2 - (r_1 + r_2)^2}$$
$$= \sqrt{212 - (5+9)^2} = \sqrt{212 - 196}$$
$$= \sqrt{16} = 4 = \mu$$

$\therefore \quad \lambda + \mu = 18$

2. Solving the equations

$$(2+c)x + 5c^2y = 1$$

and $$3x + 5y = 1,$$

$$(2+c)x + 5c^2\left(\frac{1-3x}{5}\right) = 1$$

then $(2+c)\,x + c^2(1-3x) = 1$

$$\therefore \quad x = \frac{1-c^2}{2+c-3c^2}$$

or $$x = \frac{(1+c)(1-c)}{(3c+2)(1-c)} = \frac{1+c}{3c+2}$$

$$\therefore \quad x = \lim_{c \to 1} \frac{1+c}{3c+2}$$

$$x = \frac{2}{5}$$

$$\therefore \quad y = \frac{1-3x}{5} = \frac{1-\frac{6}{5}}{5} = -\frac{1}{25}$$

Therefore the centre of the required circle is $\left(\frac{2}{5}, -\frac{1}{25}\right)$

but circle passes through (2, 0)

$\therefore$ Radius of the required circle

$$= \sqrt{\left(\frac{2}{5}-2\right)^2 + \left(-\frac{1}{25}-0\right)^2}$$
$$= \sqrt{\frac{64}{25} + \frac{1}{625}} = \sqrt{\frac{1601}{625}} = \frac{\sqrt{1601}}{25} = \frac{\sqrt{\lambda}}{25}$$

$\therefore \quad \lambda = 1601$

3. $2x^2 + y^2 - 3xy = 0$

$\Rightarrow (2x - y)(x - y) = 0$

$\Rightarrow \quad y = 2x,\ y = x$

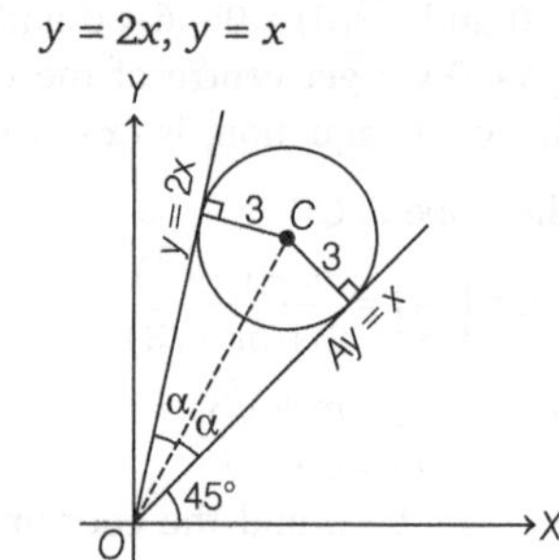

are the equations of straight lines passing through origin.

Now let the angle between tangents is 2α,

then $\tan(45° + 2\alpha) = 2$

$$\Rightarrow \quad \frac{\tan 45° + \tan 2\alpha}{1 - \tan 45° \tan 2\alpha} = 2$$

$$\Rightarrow \quad \frac{1 + \tan 2\alpha}{1 - \tan 2\alpha} = \frac{2}{1}$$

$$\Rightarrow \quad \frac{2\tan 2\alpha}{2} = \frac{1}{3}$$

[by componendo and dividendo rule]

$$\Rightarrow \quad \frac{2\tan\alpha}{1-\tan^2\alpha} = \frac{1}{3}$$

$$\Rightarrow \quad \tan^2\alpha + 6\tan\alpha - 1 = 0$$

$$\therefore \quad \tan\alpha = \frac{-6 \pm \sqrt{(36+4)}}{2} = -3 \pm \sqrt{10}$$

$$= -3 + \sqrt{10} \qquad \left[\because 0 < \alpha < \frac{\pi}{4}\right]$$

Now, in ΔOAC, $\tan\alpha = \dfrac{3}{OA} = (\sqrt{10} - 3)$

$$\therefore \ OA = \frac{3(\sqrt{10}+3)}{(\sqrt{10}-3)(\sqrt{10}+3)} = 3(3+\sqrt{10})$$

$$= 9 + \sqrt{90} = \lambda + \sqrt{\mu}$$

$\therefore\ \lambda = 9$ and $\mu = 90$, then $\lambda^2 + \mu = 81 + 90 = 171$

4. The given circle is $x^2 + y^2 = 1$ ...(i)

with centre at O (0, 0) and radius 1. It cuts X-axis at the points when $y = 0$, then $x = \pm 1$ i.e. at $P(-1, 0)$ and $Q(1, 0)$.

Equation of circle with centre at $Q(1, 0)$ and radius r is

$$(x-1)^2 + (y-0)^2 = r^2 \qquad (0 < r < 2) \ldots \text{(ii)}$$

Solving Eqs. (i) and (ii), we get

$$x = \frac{2-r^2}{2} \text{ and } y = \pm\frac{r\sqrt{(4-r^2)}}{2}$$

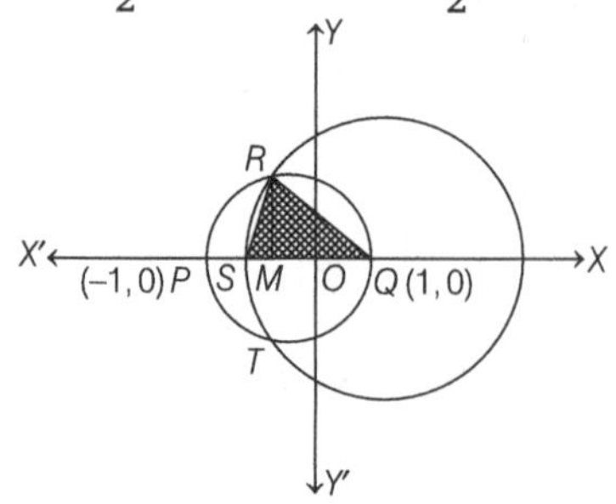

but R above the X-axis.

$$\therefore \quad R \equiv \left(\frac{2-r^2}{2}, \frac{r\sqrt{(4-r^2)}}{2} \right)$$

So, $SQ = r$ and $MR = \dfrac{r\sqrt{(4-r^2)}}{2}$

$\therefore$ Area of $\Delta QSR = \dfrac{1}{2} \cdot QS \cdot MR$

$$\therefore \quad \Delta = \frac{1}{2} \cdot r \cdot \frac{r\sqrt{(4-r^2)}}{2}$$

$$\Delta \text{ (say)} = \frac{r^2 \sqrt{(4-r^2)}}{4}$$

$$\Delta^2 = \frac{r^4}{16}(4-r^2) = A \text{ (say)}$$

$$\therefore \quad A = \frac{1}{16}(4r^4 - r^6)$$

$$\therefore \quad \frac{dA}{dr} = \frac{1}{16}(16r^3 - 6r^5)$$

and $$\frac{d^2A}{dr^2} = \frac{1}{16}(48r^2 - 30r^4)$$

For maximum and minimum area, $\dfrac{dA}{dr} = 0$

we get $$r^2 = \frac{8}{3}$$

$$\therefore \quad \left(\frac{d^2A}{dr^2} \right)_{r=\sqrt{\frac{8}{3}}} = \frac{1}{16}\left(48 \times \frac{8}{3} - 30 \times \frac{64}{9} \right) < 0$$

$\therefore$ A is maximum. Hence Δ is also maximum.

$\therefore$ Maximum value of $\Delta = \dfrac{1}{2} \times \dfrac{8}{3} \times \dfrac{1}{2}\sqrt{\left(4 - \dfrac{8}{3}\right)} = \dfrac{2}{3} \times \dfrac{2}{\sqrt{3}}$

$$= \frac{4\sqrt{3}}{9} \text{ sq unit}$$

$$\Rightarrow \therefore \quad \lambda = \frac{4\sqrt{3}}{9} = \frac{4}{3\sqrt{3}}$$

$$27\lambda^2 = 64$$

5. The given circle is $x^2 + y^2 - 2x - 4y - 20 = 0$ with centre $(1, 2)$ and radius $= \sqrt{1 + 4 + 20} = 5$
And required circle has radius 5 hence circles touch each other externally.

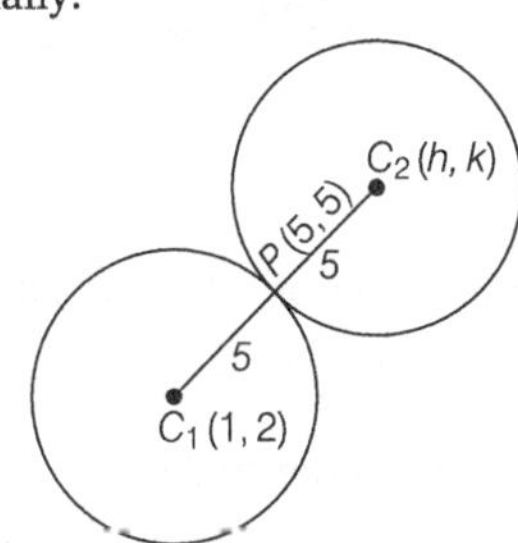

Since point of contact is $P\,(5, 5)$.
$\therefore$ P is the mid point of C_1 and C_2, let coordinates of centre C_2 is (h, k), then

$$\frac{1+h}{2} = 5$$

$$\Rightarrow \quad h = 9$$

and $$\frac{2+k}{2} = 5 \Rightarrow k = 8$$

$$\therefore \quad h^2 + k^2 = 9^2 + 8^2$$

$$= 81 + 64 = 145$$

6. Centre and radius of the circle $x^2 + y^2 - 2x + 4y - 20 = 0$ are $(1, -2)$ and $\sqrt{(1 + 4 + 20)} = 5$ respectively.

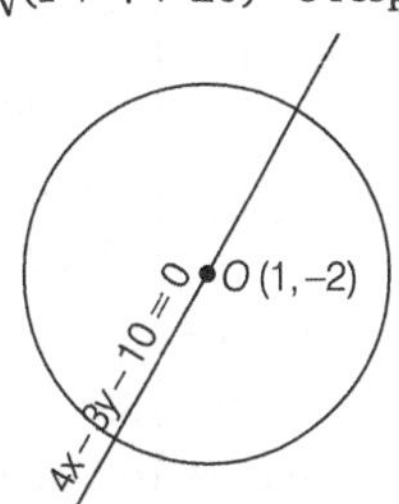

Let OM be the perpendicular from O on the line
$4x - 3y - 10 = 0$,

then $OM = \dfrac{|4 \times 1 - 3 \times (-2) - 10|}{\sqrt{4^2 + (-3)^2}} = 0$

Hence line $4x - 3y - 10 = 0$ passes through the centre of the circle.
Hence intercepted length = diameter of the circle
$= 2 \times 5 = 10$

7. Since $PQ = PR = \sqrt{(4^2 + 3^2 - 9)} = 4$ unit

$$\therefore \quad \angle CPQ = \angle CPR = \alpha \text{ (let)}$$

$$\therefore \quad PC = \sqrt{(4-0)^2 + (3-0)^2} = 5 \text{ unit}$$

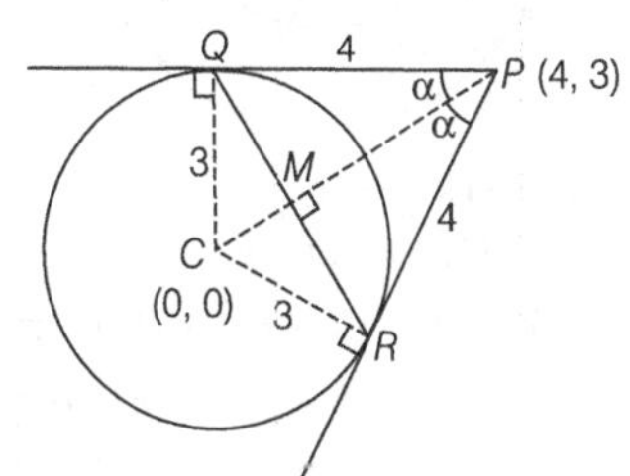

$\therefore$ In ΔPQC, $\tan\alpha = \dfrac{3}{4}$,

$$\therefore \quad \sin\alpha = \frac{3}{5} \text{ and } \cos\alpha = \frac{4}{5}$$

In ΔPMQ, $\cos\alpha = \dfrac{PM}{4} = \dfrac{4}{5}$

$$\therefore \quad PM = \frac{16}{5} \text{ and } \sin\alpha = \frac{QM}{4} = \frac{3}{5}$$

$$\therefore \quad QM = \frac{12}{5}$$

$\therefore$ Area of $\Delta PQR = \dfrac{1}{2} \cdot QR \cdot PM$

$$= \frac{1}{2}(2QM) \cdot PM = (QM)(PM)$$

$$= \left(\frac{12}{5}\right)\left(\frac{16}{5}\right) = \frac{192}{25} \text{ sq unit}$$

$$\therefore \quad \lambda = \frac{192}{25}, \text{ then } 25\lambda = 192$$

8. Centre and radii of circles S and S' are
$C_1\ (2, -3),\ r_1 = \sqrt{2},\ C_2(1, -4),\ r_2 = 2.$
Distance between centres, $d = |C_1C_2|$

$$= \sqrt{(2-1)^2 + (-3+4)^2} = \sqrt{2}$$

If angle between the circles is θ, then

$$\cos(180° - \theta) = \left(\frac{r_1^2 + r_2^2 - d^2}{2r_1r_2}\right)$$

$$\cos(180° - \theta) = \left(\frac{2+4-2}{2\cdot\sqrt{2}\cdot 2}\right) = \frac{1}{\sqrt{2}}$$

$\therefore\quad 180° - \theta = 45°$

$\therefore\quad \theta = 135°$

9. The coordinates of the centres and radii of three given circles are as given below

$$C_1 \equiv (0, 2);\ r_1 = 3$$
$$C_2 \equiv (-6, -2);\ r_2 = 3$$
and $$C_3 \equiv (-3, -6);\ r_3 = 3$$

Let $C \equiv (h, k)$ be the centre of the circle passing through the centres $C_1\ (0, 2), C_2\ (-6, -2)$ and $C_3\ (-3, -6)$.

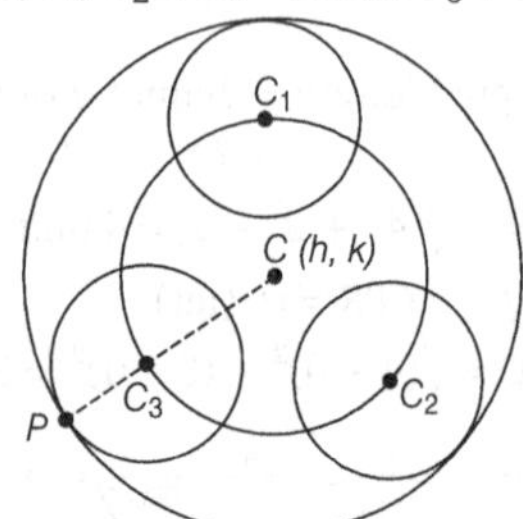

Then, $\quad CC_1 = CC_2 = CC_3$

$\Rightarrow\quad (CC_1)^2 = (CC_2)^2 = (CC_3)^2$

$\Rightarrow (h-0)^2 + (k-2)^2 = (h+6)^2 + (k+2)^2$

$$= (h+3)^2 + (k+6)^2$$

$\Rightarrow\quad -4k + 4 = 12h + 4k + 40 = 6h + 12k + 45$

$\Rightarrow\quad 12h + 8k + 36 = 0$

or $\quad 3h + 2k + 9 = 0 \quad \ldots(\text{i})$

and $\quad 6h - 8k - 5 = 0 \quad \ldots(\text{ii})$

Solving Eqs. (i) and (ii), we get $h = -\dfrac{31}{18},\ k = -\dfrac{23}{12}$

Now, $\quad CP = CC_3 + C_3P = CC_3 + 3$

$$= \sqrt{\left(-3 + \frac{31}{18}\right)^2 + \left(-6 + \frac{23}{12}\right)^2} + 3$$

$$= \left(\frac{5}{36}\sqrt{(949)} + 3\right) = \text{radius} = 3 + \frac{5}{36}\sqrt{\lambda}$$

$\therefore\quad \lambda = 949$

10. The equation of the circle is $(x-1)^2 + (y-1)^2 = 1$

$\Rightarrow\quad x^2 + y^2 - 2x - 2y + 1 = 0 \quad \ldots(\text{i})$

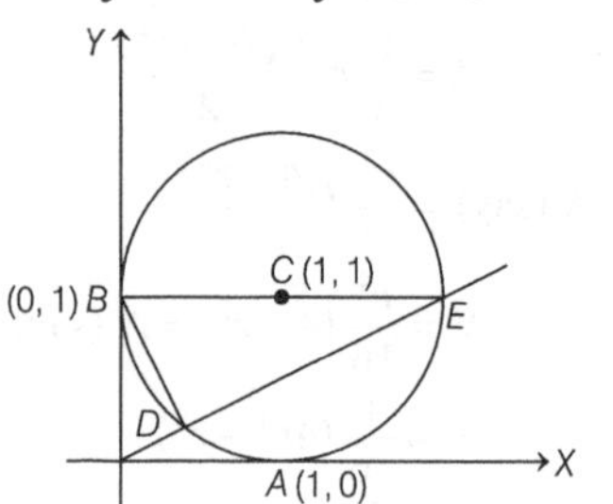

Let the equation of the variable straight line be

$$y = mx \quad \ldots(\text{ii})$$

Solving Eqs. (i) and (ii), we get

$$(1+m^2)x^2 - 2x(1+m) + 1 = 0$$

$\therefore$ Length $DE = \sqrt{\left(\dfrac{8m}{1+m^2}\right)}$

Area of ΔDEB, $A = \dfrac{1}{2} DE \times$ distance of B from DE

$\therefore\quad A^2 = \dfrac{1}{4}\cdot\left(\dfrac{8m}{1+m^2}\right) \times \dfrac{1}{(1-m^2)} = \dfrac{2m}{(1+m^2)^2}$

$\Rightarrow\quad A = \dfrac{\sqrt{2m}}{(1+m^2)}$

$$\frac{dA}{dM} = \frac{1-3m^2}{\sqrt{2m}\,(1+m^2)^2} = 0 = m = \pm\frac{1}{\sqrt{3}}$$

$$\frac{d^2A}{dm^2} < \text{if } m = \frac{1}{\sqrt{3}}$$

$\therefore$ Area is maximun for $m = \dfrac{1}{\sqrt{3}}$

$\therefore\quad \dfrac{3333}{m^2} = 3333 \times 3 = 9999$

Single Integer Type

1. Equation of given circle is $x^2 + y^2 = 9$
So, let end point of diameter PQ are $P(3\cos\theta, 3\sin\theta)$ and $Q\ (-3\cos\theta, -3\sin\theta)$.
Now, the lengths of the perpendiculars from P and Q on the straight line, $x + y = 2$ are respectively.

$$\alpha = \frac{|3\cos\theta + 3\sin\theta - 2|}{\sqrt{2}} \text{ and } \beta = \frac{|-3\cos\theta - 3\sin\theta - 2|}{\sqrt{2}}$$

$$\alpha\beta = \frac{|(3\cos\theta + 3\sin\theta)^2 - 2^2|}{2}$$

$$= \frac{|9 - 4 + 9\sin 2\theta|}{2} = \frac{|5 + 9\sin 2\theta|}{2}$$

Since, maximum value of $\alpha\beta$, if $\sin 2\theta = 1$

$\therefore\quad \alpha\,\beta|_{\max} = \dfrac{5+9}{2} = 7$

2. Equation of tangent to the circle
$x^2 + y^2 + 4x - 6y - 12 = 0$ at $(1, -1)$ is given by
$xx_1 + yy_1 + 2(x + x_1) - 3(y + y_1) - 12 = 0$,
where $x_1 = 1$ and $y_1 = -1$

$\Rightarrow\quad x - y + 2(x+1) - 3(y-1) - 12 = 0$

$\Rightarrow\quad 3x - 4y - 7 = 0$

This will also a tangent to the required circle.

Now, equation of family of circles touching the line $3x-4y-7=0$ at point $(1,-1)$ is given by
$(x-1)^2+(y+1)^2+\lambda(3x-4y-7)=0$

So, the equation of required circle will be
$(x-1)^2+(y+1)^2+\lambda(3x-4y-7)=0$,
for some $\lambda \in R$ ...(i)

$\because$ The required circle passes through (4, 0)

$\therefore (4-1)^2+(0+1)^2+\lambda(3\times 4-4\times 0-7)=0$

$\Rightarrow 9+1+\lambda(5)=0 \Rightarrow \lambda=-2$

Substituting $\lambda=-2$ in Eq. (i), we get

$(x-1)^2+(y+1)^2-2(3x-4y-7)=0$

$\Rightarrow \quad x^2+y^2-8x+10y+16=0$

On comparing it with
$x^2+y^2+2gx+2fy+c=0$, we get
$g=-4, f=5, c=16$

$\therefore$ Radius $=\sqrt{g^2+f^2-c}=\sqrt{16+25-16}=5$

3. Since, the line $3x+4y=k$ intersects the circle

$$x^2+y^2-2x-4y+4=0$$

$\Rightarrow (x-1)^2+(y-2)^2=1$ at two distinct points, so perpendicular distance from centre (1, 2) to the line $3x+4y-k=0$ must be less than the radius 1.

$$\therefore \quad \left|\frac{3+8-k}{\sqrt{9+16}}\right|<1$$

$\Rightarrow \quad |11-k|<5 \Rightarrow (k-11)\in(-5,5) \Rightarrow k\in(6,16)$

$\Rightarrow \quad k=7, 8, 9, 10, 11, 12, 13, 14, 15$

$\therefore$ number of integral values of k is 9.

4.

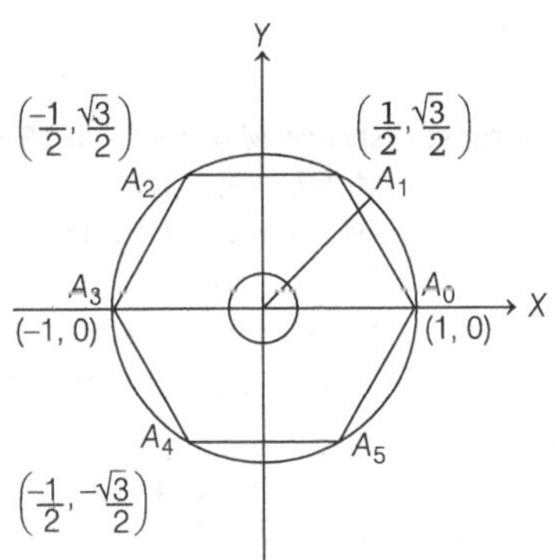

Let O be the centre of the circle of unit radius and the coordinates of A_0 be (1, 0).

Since each side of the regular hexagon makes an angle of 60° at the centre O.

Coordinate of A_1 are

$$(\cos 60°, \sin 60°)=\left(\frac{1}{2},\frac{\sqrt{3}}{2}\right)$$

A_2 are $(\cos 120°, \sin 120°)=\left(-\frac{1}{2},\frac{\sqrt{3}}{2}\right)$

A_3 are $(-1, 0)$

A_4 are $=\left(-\frac{1}{2},-\frac{\sqrt{3}}{2}\right)$ and A_5 are $\left(\frac{1}{2},-\frac{\sqrt{3}}{2}\right)$

$$\text{Now, } A_0A_1=\sqrt{\left(1-\frac{1}{2}\right)^2+\left(\frac{\sqrt{3}}{2}\right)^2}=\sqrt{\frac{1}{4}+\frac{3}{4}}=1$$

$$A_0A_2=\sqrt{\left(1+\frac{1}{2}\right)^2+\left(\frac{\sqrt{3}}{2}\right)^2}=\sqrt{\frac{9}{4}+\frac{3}{4}}=\sqrt{3}=A_0A_4$$

So that $(A_0A_1)(A_0A_2)(A_0A_4)=3$

5. $x^2+y^2\le 6$ and $2x-3y=1$ is shown as

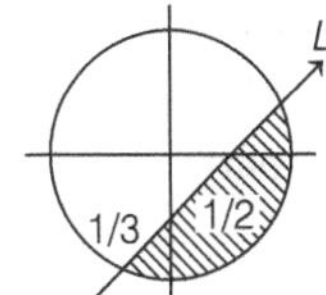

For the point to lie in the shade part, origin and the point lie on opposite side of straight line L.

$\therefore$ For any point in shaded part $L>0$ and for any point inside the circle $S<0$.

Now, for $\left(2,\frac{3}{4}\right)$ $L: 2x-3y-1$

$$L: 4-\frac{9}{4}-1=\frac{3}{4}>0$$

and $S: x^2+y^2-6, S: 4+\frac{9}{16}-6<0$

$\Rightarrow \left(2,\frac{3}{4}\right)$ lies in shaded part.

For $\left(\frac{5}{2},\frac{3}{4}\right)$, $L: 5-9-1<0$ [neglect]

For $\left(\frac{1}{4},-\frac{1}{4}\right)$, $L: \frac{1}{2}+\frac{3}{4}-1>0\}$

$\therefore \left(\frac{1}{4},-\frac{1}{4}\right)$ lies in the shaded part.

For $\left(\frac{1}{8},\frac{1}{4}\right)$, $L: \frac{1}{4}-\frac{3}{4}-1<0$ [neglect]

$\Rightarrow$ Only 2 points lie in the shaded part.

6. $x^2+y^2\le 6$ and $2x-3y=1$ is shown as

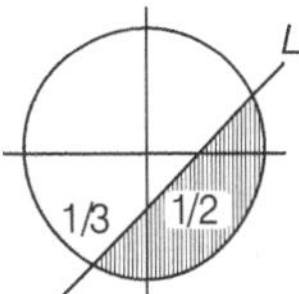

For the point to lie in the shade part, origin and the point lie on opposite side of straight line L.

$\therefore$ For any point in shaded part $L>0$ and for point inside the circle $S<0$.

Now, for $\left(2,\frac{3}{4}\right)$ $L: 2x-3y-1$

$$L: 4-\frac{9}{4}-1=\frac{3}{4}>0$$

and $S : x^2 + y^2 - 6$,

$$S : 4 + \frac{9}{16} - 6 < 0$$

$\Rightarrow \left(2, \frac{3}{4}\right)$ lies in shaded part.

For $\left(\frac{5}{2}, \frac{3}{4}\right)$, $L : 5 - 9 - 1 < 0$ [neglect]

For $\left(\frac{1}{4}, -\frac{1}{4}\right)$, $L : \frac{1}{2} + \frac{3}{4} - 1 > 0\}$

$\therefore \left(\frac{1}{4}, \frac{-1}{4}\right)$ lies in the shaded part.

For $\left(\frac{1}{8}, \frac{1}{4}\right)$, $L : \frac{1}{4} - \frac{3}{4} - 1 < 0$

$\Rightarrow$ Only 2 points lie in the shaded part.

7.

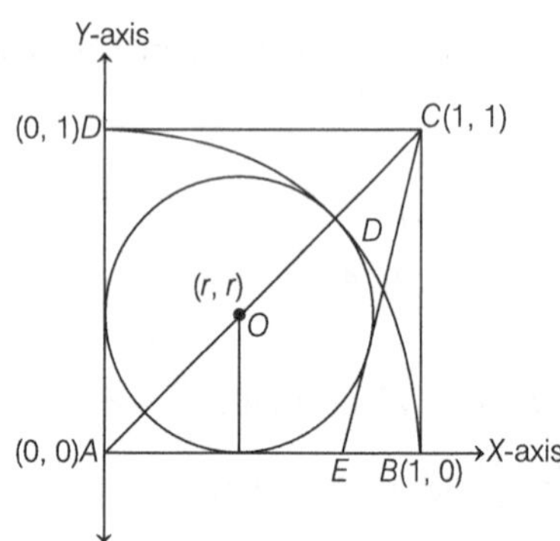

Here, $AO + OD = 1$ or $(\sqrt{2} + 1)r = 1$

$\Rightarrow \quad r = \sqrt{2} - 1$

equation of circle $(x - r)^2 + (y - r)^2 = r^2$

Equation of CE

$$y - 1 = m(x - 1) \Rightarrow mx - y + 1 - M = 0$$

It is tangent to circle

$$\therefore \quad \left|\frac{mr - r + 1 - m}{\sqrt{m^2 + 1}}\right| = r$$

$$\Rightarrow \quad \left|\frac{(m - 1)r + 1 - m}{\sqrt{m^2 + 1}}\right| = r$$

$$\Rightarrow \quad \frac{(m - 1)^2 (r - 1)^2}{m^2 + 1} = r^2$$

Put $r = \sqrt{2} - 1$

On solving $m = 2 - \sqrt{3}, 2 + \sqrt{3}$

Taking greater slope of CE as $2 + \sqrt{3}$

$$\Rightarrow \quad y - 1 = (2 + \sqrt{3})(x - 1)$$

Put $y = 0$, we get

$$-1 = (2 + \sqrt{3})(x - 1)$$

$$\Rightarrow \quad \frac{-1}{2 + \sqrt{3}} \times \left(\frac{2 - \sqrt{3}}{2 - \sqrt{3}}\right) = x - 1$$

$$\Rightarrow \quad x - 1 = \sqrt{3} - 1$$

$$\Rightarrow \quad EB = 1 - x = 1 - (\sqrt{3} - 1)$$

$$\Rightarrow \quad EB = 2 - \sqrt{3} = \alpha + \sqrt{3}\beta$$

$$\therefore \quad \alpha = 2, \beta = -1$$

$$\therefore \quad \alpha + \beta = 1$$

Matrix-Match Type

1. (A) $\because P \equiv (10, 7)$

and $S = x^2 + y^2 - 4x - 2y - 20$

$\therefore \quad S_1 = 100 + 49 - 40 - 14 - 20 > 0$

$\therefore P$ outside the circle

radius $r = \sqrt{4 + 1 + 20} = 5$

centre $C \equiv (2, 1)$

$\therefore$ Shortest distance $L = CP - r$

$$= \sqrt{(10 - 2)^2 + (7 - 1)^2} - 5$$

$$= \sqrt{(64 + 36)} - 5$$

$$= 10 - 5 = 5$$

And largest distance $M = CP + r$

$$= 10 + 5 = 15$$

$$M + L = 20, M - L = 10 \quad (Q, S)$$

(B) $\because \quad P \equiv (3, -6)$

and $S \equiv x^2 + y^2 - 16x - 12y - 125$

$\therefore \quad S_1 = 9 + 36 - 48 + 72 - 125 < 0$

$\therefore$ P inside the circle

radius $r = \sqrt{64 + 36 + 125} = 15$

centre $C \equiv (8, 6)$

$\therefore$ Shortest distance $L = r - CP$

$$= 15 - \sqrt{(8 - 3)^2 + (6 + 6)^2}$$

$$= 15 - 13 = 2$$

And largest distance $M = r + CP = 15 + 13 = 28$

$$M + L = 30, \text{ (R)}$$

$$M - L = 26 \quad \text{(R , T)}$$

(C) $\quad P \equiv (6, -6)$

and $\quad S \equiv x^2 + y^2 - 4x + 6y - 12$

$\therefore \quad S_1 = (6)^2 + (-6)^2 - 24 - 36 - 12 = 0$

P on the circle

$\therefore$ radius $\quad r = \sqrt{(4 + 9 + 12)} = 5$

$\therefore$ Shortest distance $L = 0$

and largest distance $M = 2r = 10$

$M + L = 10, M - L = 10$ (P, S)

2. (A) Distance from centre $(0, 10)$ to the line $(y - mx = 0)$

$$= \frac{10}{\sqrt{(1 + m^2)}} \geq \text{ radius}$$

$$= \sqrt{10}$$

or $\quad \frac{10}{\sqrt{(1 + m^2)}} \geq \sqrt{10}$

$$\Rightarrow \quad \sqrt{10} \geq \sqrt{1 + m^2} \Rightarrow m^2 \leq 9$$

$$\therefore \quad -3 \leq m \leq 3$$

Then $\quad 0 \leq |m| \leq 3$

$\therefore \quad |m| = 0, 1, 2, 3$ (P, Q, R, S)

(B) Distance from the centre (2, 4) to the line $(3x - 4y - 5k = 0)$

$$= \frac{|6 - 16 - 5k|}{5} \le \text{radius} = 5$$

$\Rightarrow \quad |10 + 5k| \le 25$

$\Rightarrow \quad 0 \le |2 + k| \le 5$

$\therefore \quad |2 + k| = 0, 1, 2, 3, 4, 5$ (P, Q, R, S, T)

(C) The given circles will cut orthogonally, if

$$2\left(\frac{1}{2}\right)(-5) + 2\left(\frac{p}{2}\right)(p) = -7 + 1$$

$\Rightarrow \quad -5p + p^2 = -6$

$\Rightarrow \quad p^2 - 5p + 6 = 0$

$\Rightarrow \quad (p - 2)(p - 3) = 0$

$\therefore \quad p = 2, 3$ (R, S)

3. (A) $C_1 \equiv (1, 3),\ r_1 = \sqrt{(1 + 9 - 9)} = 1$

and $C_2 \equiv (-3, 1),\ r_2 = \sqrt{(9 + 1 - 1)} = 3$

Now, $|C_1C_2| = \sqrt{(16 + 4)}$

$= \sqrt{20} = 2\sqrt{5} > r_1 + r_2$

Hence circles neither cuts nor touch each other

$\therefore$ Point of intersection = 0

$\Rightarrow \quad \lambda = 0$

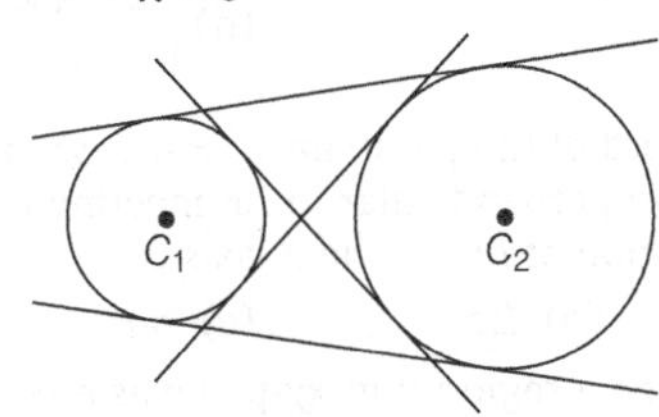

and $\mu = 4$ (i.e. 2 direct common tangents and 2 transverse common tangents)

$\therefore \quad \lambda + \mu = 4$ (S)

$\mu - \lambda = 4$ (R)

$\lambda^\mu + \mu^\lambda = 1$

(B) $C_1 \equiv (3, 0),\ r_1 = 3$

and $C_2 \equiv (-1, 0),\ r_2 = 1$

Now, $|C_1C_2| = 4 = r_1 + r_2$

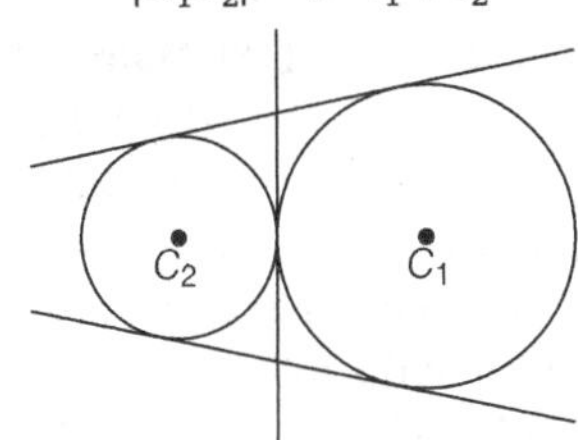

Hence circles touch each other externally

$\therefore$ Point of intersection is one

$\therefore \quad \lambda = 1$

and $\quad \mu = 3$

$\therefore \quad \lambda + \mu = 4$ (S)

$\mu - \lambda = 2$, (Q)

$\lambda^\mu + \mu^\lambda = 1^3 + 3^1 = 4$ (T)

(C) $C_1 \equiv (1, 2),\ r_1 = \sqrt{(1 + 4 - 1)} = 2$

and $C_2 \equiv (2, 1),\ r_2 = \sqrt{(4 + 1 - 1)} = 2$

$\therefore |C_1 C_2| = \sqrt{2}$

$\because |r_1 - r_2| < |c_1c_2| < r_1 + r_2$

Hence circles cuts at two points.

$\therefore$ Point of intersection is two

$\therefore \lambda = 2,\ \mu = 2;\ \lambda + \mu = 4$, (S) $\mu - \lambda = 0$ (P)

$\lambda^\mu + \mu^\lambda = 2^2 + 2^2 = 8$

4. (A) $A \to q,\ B \to t,\ C \to p,\ D \to r$

Let length of common chord be $2a$, then

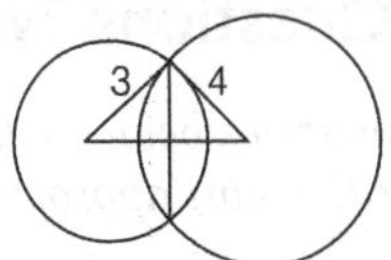

$$\sqrt{9 - a^2} + \sqrt{16 - a^2} = 5$$

i.e., $\quad \sqrt{16 - a^2} = 5 - \sqrt{9 - a^2}$

$16 - a^2 = 25 + 9 - a - 10\sqrt{9 - a^2}$

i.e. $\quad 10\sqrt{9 - a^2} = 18 \Rightarrow 100(9 - a^2) = 324$

i.e. $\quad 100a^2 = 576$

$$a = \sqrt{\frac{576}{100}} = \frac{24}{10}$$

$\therefore \quad 2a = \frac{24}{5} = \frac{k}{5} \Rightarrow k = 24$

(B) Equation of common chord is $6x + 4y + p + q = 0$ common chord pass through centre $(-2, -6)$ of circle $x^2 + y^2 + 4x + 12y + p = 0,\ p + q = 36$.

(C) Equation of the circle is

$$2x^2 + 2y^2 - 2\sqrt{2}x - y = 0$$

Let $(\alpha, 0)$ be mid-point of a chord.

Then, equation of the chord is

$$2\alpha x - \sqrt{2}(x + \alpha) - \frac{1}{2}(y + 0) = 2\alpha^2 - 2\sqrt{2}\alpha$$

Since it passes through the point $\left(\sqrt{2}, \frac{1}{2}\right)$

$\therefore\ 2\sqrt{2}\alpha - \sqrt{2}(\sqrt{2} + \alpha) - \frac{1}{4} = 2\alpha^2 - 2\sqrt{2}\alpha$

i.e. $\quad 8\alpha^2 - 12\sqrt{2}\alpha + 9 = 0$

i.e. $\quad (2\sqrt{2}\alpha - 3)^2 = 0$

i.e. $\quad \alpha = \frac{3}{2\sqrt{2}}, \frac{3}{2\sqrt{2}}$

$\therefore$ Number of chords is 1

(D) Mid-point of $AB = (1, 4)$

$\therefore$ Equation perpendicular bisector of AB is $x = 1$

A diameter is $4y = x + 7$

$\therefore$ Centre of the circle is (1, 2)

$\therefore$ Sides of the rectangle are 8 and

$\therefore$ Area = 32

24

Parabola

Objective Questions Type I [Only one correct answer]

In each of the questions below, four choices are given of which only one is correct. You have to select the correct answer which is the most appropriate.

1. The mirror image of the directrix of the parabola $y^2 = 4(x + 1)$ in the line mirror $x + 2y = 3$ is

(a) $x = -2$ (b) $4y - 3x = 16$
(c) $x - 3y = 0$ (d) $x + y = 0$

2. The shortest distance between the parabolas $y^2 = 4x$ and $y^2 = 2x - 6$ is

(a) 2 (b) $\sqrt{5}$
(c) 3 (d) None of these

3. The locus of a point which divides the line segment joining the point $(0, -1)$ and a point on the parabola, $x^2 = 4y$, internally in the ratio $1 : 2$, is

[JEE Main 2020, 8 January]

(a) $4x^2 - 3y = 2$ (b) $x^2 - 3y = 2$
(c) $9x^2 - 12y = 8$ (d) $9x^2 - 3y = 2$

4. The mirror image of the parabola $y^2 = 4x$ in the tangent to the parabola at the point (1, 2) is

(a) $(x-1)^2 = 4(y+1)$ (b) $(x+1)^2 = 4(y+1)$
(c) $(x+1)^2 = 4(y-1)$ (d) $(x-1)^2 = 4(y-1)$

5. A parabola is drawn with focus at (3, 4) and vertex at the focus of the parabola $y^2 - 12x - 4y + 4 = 0$. The equation of the parabola is

(a) $x^2 - 6x - 8y + 25 = 0$ (b) $y^2 - 8x - 6y + 25 = 0$
(c) $x^2 - 6x + 8y - 25 = 0$ (d) $x^2 + 6x - 8y - 25 = 0$

6. If the 4th term in the expansion of $\left(px + \frac{1}{x}\right)^n, n \in N$ is $\frac{5}{2}$ and three normals to the parabola $y^2 = x$ are drawn through a point $(q, 0)$, then

(a) $q = p$ (b) $q > p$
(c) $q < p$ (d) $pq = 1$

7. A circle is described whose centre is the vertex and whose diameter is 3-quarters of the latus rectum of the parabola $y^2 = 4ax$. If PQ is the common chord of the circle and the parabola and L_1L_2 is the latus rectum, then the area of the trapezium $PL_1 L_2Q$ is

(a) $3\sqrt{2}a^2$ (a) $2\sqrt{2}a^2$
(c) $4a^2$ (d) $\left(\frac{2+\sqrt{2}}{2}\right)a^2$

8. AB is a chord of the parabola $y^2 = 4ax$ with vertex at A, BC is drawn perpendicular to AB meeting the axis at C. The projection of BC on the X-axis is

(a) a (b) $2a$ (c) $4a$ (d) $8a$

9. Let us define a region R in xy-plane as a set of points (x, y) satisfying $[x^2] = [y]$ (where $[x]$ denotes greatest integer $\le x$), then the region R defines

(a) a parabola whose axis is horizontal
(b) a parabola whose axis is vertical
(c) integer point on the parabola $y = x^2$
(d) None of the above

10. The area (in sq units) of an equilateral triangle inscribed in the parabola $y^2 = 8x$, with one of its vertices on the vertex of this parabola, is **[JEE Main 2020, 2 September]**

(a) $64\sqrt{3}$ (b) $256\sqrt{3}$
(c) $192\sqrt{3}$ (d) $128\sqrt{3}$

11. If a chord, which is not a tangent, of the parabola $y^2 = 16x$ has the equation $2x + y = p$, and mid-point (h, k), then which of the following is(are) possible value(s) of p, h and k? **[JEE Advanced 2017]**

(a) $p = -1, h = 1, k = -3$ (b) $p = 2, h = 3, k = -4$
(c) $p = -2, h = 2, k = -4$ (d) $p = 5, h = 4, k = -3$

12. The radius of the circle whose centre is $(-4, 0)$ and which cuts the parabola $y^2 = 8x$ at A and B such that the common chord AB subtends a right angle at the vertex of the parabola is equal to

(a) $4\sqrt{13}$ (b) $3\sqrt{5}$
(c) $3\sqrt{2}$ (d) $2\sqrt{5}$

13. The locus of the points of trisection of the double ordinates of the parabola $y^2 = 4ax$ is

(a) $y^2 = ax$ (b) $9y^2 = 4ax$
(c) $9y^2 = ax$ (d) $y^2 = 9ax$

14. If the tangents at P and Q on a parabola meet in T, then SP, ST and SQ are in

(a) AP (b) GP
(c) HP (d) None of these

15. If the normals at two points P and Q of a parabola $y^2 = 4ax$ intersect at a third point R on the curve, then the product of ordinates of P and Q is

(a) $4a^2$ (b) $2a^2$
(c) $-4a^2$ (d) $8a^2$

16. The point $(-2m, m+1)$ is an interior point of the smaller region bounded by the circle $x^2 + y^2 = 4$ and the parabola $y^2 = 4x$. Then m belongs to the interval

(a) $-5 - 2\sqrt{6} < m < 1$ (b) $0 < m < 4$
(c) $-1 < m < \frac{3}{5}$ (d) $-1 < m < -5 + 2\sqrt{6}$

17. AB, AC are tangents to a parabola $y^2 = 4ax$, p_1, p_2, p_3 are the lengths of the perpendiculars from A, B, C on any tangent to the curve, then p_2, p_1, p_3 are in

(a) AP (b) GP
(c) HP (d) None of these

18. If the line $y - \sqrt{3}x + 3 = 0$ cuts the parabola $y^2 = x + 2$ at A and B, then $PA \cdot PB$ is equal to [where $P \equiv (\sqrt{3}, 0)$]

(a) $\frac{4(\sqrt{3}+2)}{3}$ (b) $\frac{4(2-\sqrt{3})}{3}$
(c) $\frac{4\sqrt{3}}{2}$ (d) $\frac{2(\sqrt{3}+2)}{3}$

19. The normals at three points P, Q, R of the parabola $y^2 = 4ax$ meet in (h, k). The centroid of triangle PQR lies on

(a) $x = 0$ (b) $y = 0$
(c) $x = -a$ (d) $y = a$

20. If tangents at A and B on the parabola $y^2 = 4ax$ intersect at point C, then ordinates of A, C and B are

(a) always in AP (b) always in GP
(c) always in HP (d) None of these

21. The condition that the parabolas $y^2 = 4c(x-d)$ and $y^2 = 4ax$ have a common normal other than X-axis $(a > 0, c > 0)$ is

(a) $2a < 2c + d$ (b) $2c < 2a + d$
(c) $2d < 2a + c$ (d) $2d < 2c + a$

22. Let L_1 be a tangent to the parabola $y^2 = 4(x+1)$ and L_2 be a tangent to the parabola $y^2 = 8(x+2)$ such that L_1 and L_2 intersect at right angles. Then, L_1 and L_2 meet on the straight line **[JEE Main 2020, 6 September]**

(a) $x + 3 = 0$ (b) $2x + 1 = 0$
(c) $x + 2 = 0$ (d) $x + 2y = 0$

23. If the normals at three points, P, Q, R of the parabola $y^2 = 4ax$ meet in a point O and S be its focus, then $|SP| \cdot |SQ| \cdot |SR|$ is equal to

(a) a^2 (b) $a(SO)^3$
(c) $a(SO)^2$ (d) None of these

24. The set of points on the axis of the parabola $y^2 - 4x - 2y + 5 = 0$ from which all the three normals to the parabola are real is

(a) $(k, 0); k > 1$ (b) $(k, 1); k > 3$
(c) $(k, 2); k > 6$ (d) $(k, 3); k > 8$

25. The orthocentre of a triangle formed by any three tangents to a parabola lies on

(a) focus (b) directrix
(c) vertex (d) focal chord

26. The vertex of a parabola is the point (a, b) and latusrectum is of length l. If the axis of the parabola is along the positive direction of Y-axis, then its equation is

(a) $(x-a)^2 = \frac{l}{2}(y - 2b)$ (b) $(x-a)^2 = \frac{l}{2}(y - b)$
(c) $(x-a)^2 = l(y-b)$ (d) None of these

27. If the three normals drawn to the parabola, $y^2 = 2x$ pass through the point $(a, 0)$ $a \neq 0$, then 'a' must be greater than **[JEE Main 2021, 16 March]**

(a) $\frac{1}{2}$ (b) $-\frac{1}{2}$
(c) -1 (d) 1

28. The condition that the straight line $lx + my + n = 0$ touches the parabola $x^2 = 4ay$ is

(a) $bn = am^2$ (b) $al^2 - mn = 0$
(c) $ln = am^2$ (d) $am = ln^2$

29. Consider a parabola $y^2 = 4ax$, the length of focal chord is l and the length of the perpendicular from vertex to the chord is p, then

(a) $l \cdot p$ is constant (b) $l \cdot p^2$ is constant
(c) $l^2 \cdot p$ is constant (d) None of these

30. The slope of a chord of the parabola $y^2 = 4ax$, which is normal at one end and which subtends a right angle at the origin, is

(a) $1/\sqrt{2}$ (b) $\sqrt{2}$
(c) 2 (d) None of these

31. Let α be the angle which a tangent to the parabola $y^2 = 4ax$ makes with its axis, the distance between the tangent and a parallel normal will be

(a) $a \sin^2\alpha \cos^2\alpha$ (b) $a \operatorname{cosec}\alpha \sec^2\alpha$
(c) $a \tan^2\alpha$ (d) $a \cos^2\alpha$

32. Let P be a point on the parabola, $y^2 = 12x$ and N be the foot of the perpendicular drawn from P on the axis of the parabola. A line is now drawn through the mid-point M and PN, parallel to its axis which meets the parabola at Q. If the y-intercept of the line NQ is $\frac{4}{3}$, then **[JEE Main 2020, 3 September]**

(a) $PN = 4$ (b) $MQ = \frac{1}{3}$
(c) $MQ = \frac{1}{4}$ (d) $PN = 3$

33. The equation of the line touching both the parabolas $y^2 = 4x$ and $x^2 = -32y$ is

(a) $x + 2y + 4 = 0$ (b) $2x + y - 4 = 0$
(c) $x - 2y - 4 = 0$ (d) $x - 2y + 4 = 0$

34. The normal at the point $(at^2, 2at)$ on the parabola $y^2 = 4ax$ cuts the curve again at the point whose parameter is

(a) $-\frac{1}{t}$ (b) $-\left(t + \frac{2}{t}\right)$
(c) $-2t + \frac{1}{t}$ (d) $t + \frac{2}{t}$

35. If X is the foot of the directrix on axis of the parabola. PP' is a double ordinate of the curve and PX meets the curve again in Q. Then, $P'Q$ passes through fixed point which is

(a) vertex
(b) focus
(c) mid-point of vertex and focus
(d) None of the above

36. The Harmonic mean of the segments of a focal chord of the parabola $y^2 = 4ax$ is

(a) $4a$ (b) $2a$
(c) a (d) a^2

37. A double ordinate of the parabola $y^2 = 8px$ is of length $16p$. The angle subtended by it at the vertex of the parabola is

(a) $\pi/4$ (b) $\pi/2$
(c) π (d) $\pi/3$

38. If one end of a focal chord of the parabola, $y^2 = 16x$ is at $(1, 4)$, then the length of this focal chord is **[JEE Main 2019, 9 April]**

(a) 22 (b) 25
(c) 24 (d) 20

39. Chords of the parabola $(y + 2)^2 = \frac{4}{3}(x - 1)$ which subtend right angle at the vertex pass through.

(a) $(7/3, -2)$ (b) $(1/3, 0)$
(c) $(4/3, 0)$ (d) $(0, 4/3)$

40. The length of the latus-rectum of the parabola $169\{(x - 1)^2 + (y - 3)^2\} = (5x - 12y + 17)^2$ is

(a) 14/13 (b) 12/13
(c) 28/13 (d) None of these

41. The points on the axis of the parabola $3y^2 + 4y - 6x + 8 = 0$ from when 3 distinct normals can be drawn is given by

(a) $\left(a, \frac{4}{3}\right); a > \frac{19}{9}$ (b) $\left(a, -\frac{2}{3}\right); a > \frac{19}{9}$
(c) $\left(a, \frac{1}{3}\right), a > \frac{7}{9}$ (d) None of these

42. Let the line $lx + my = 1$ cut the parabola $y^2 = 4ax$ in the points A and B. Normals at A and B meet at point C. Normal from C other than these two meet the parabola at D, then the coordinate of D are

(a) $(a, 2a)$ (b) $\left(\frac{4am}{l^2}, \frac{4a}{l}\right)$
(c) $\left(\frac{2am^2}{l^2}, \frac{2a}{l}\right)$ (d) $\left(\frac{4am^2}{l^2}, \frac{4am}{l}\right)$

43. The triangle formed by the tangent to the parabola $y = x^2$ at the point whose abscissa is $x_0 (x_0 \in [1, 2])$, the Y-axis and the straight line $y = x_0^2$ has the greatest area if x_0 is equal to

(a) 0 (b) 1
(c) 2 (d) 3

44. If the normal at P 't' on $y^2 = 4ax$ meets the curve again at Q, the point on the curve, the normal at which also passes through Q has coordinates

(a) $\left(\frac{2a}{t^2}, \frac{2a}{t}\right)$ (b) $\left(\frac{4a}{t^2}, \frac{2a}{t}\right)$
(c) $\left(\frac{4a}{t^2}, \frac{4a}{t}\right)$ (d) $\left(\frac{4a}{t^2}, \frac{8a}{t}\right)$

45. Two parabolas C and D intersect at two different points, where C is $y = x^2 - 3$ and D is $y = kx^2$. The intersection at which the x value is positive is designated point A, and $x = a$ at this intersection the tangent line l at A to the curve D intersects curve C at point B, other than A. If x-value of point B is 1, then a is equal to

(a) 1 (b) 2
(c) 3 (d) 4

46. The locus of point of intersection of tangents to the parabolas $y^2 = 4(x + 1)$ and $y^2 = 8(x + 2)$ which are perpendicular to each other is

(a) $x + 7 = 0$ (b) $x - y = 4$
(c) $x + 3 = 0$ (d) $y - x = 12$

47. If the common tangent of the parabolas, $y^2 = 4x$ and $x^2 = 4y$ also touches the circle, $x^2 + y^2 = c^2$, then c is equal to **[JEE Main 2020, 5 September]**

(a) $\frac{1}{\sqrt{2}}$ (b) $\frac{1}{4}$
(c) $\frac{1}{2}$ (d) $\frac{1}{2\sqrt{2}}$

48. If the normals from any point to the parabola $x^2 = 4y$ cuts the line $y = 2$ in points whose abscissas are in AP, then the slopes of the tangents at the three conormal points are in

(a) AP (b) GP
(c) HP (d) None of these

49. If the parabolas $y^2 = 4b(x - c)$ and $y^2 = 8ax$ have a common normal, then which one of the following is a valid choice for the ordered triad (a, b, c)? **[JEE Main 2019, 10 January]**

(a) $\left(\frac{1}{2}, 2, 0\right)$ (b) $(1, 1, 0)$
(c) $(1, 1, 3)$ (d) $\left(\frac{1}{2}, 2, 3\right)$

50. A focal chord of $y^2 = 4ax$ meets in P and Q. If S is the focus, then $\frac{1}{SP} + \frac{1}{SQ}$ is equal to

(a) $\frac{1}{a}$ (b) $\frac{2}{a}$
(c) $\frac{4}{a}$ (d) None of these

51. If b, k are intercepts of a focal chord of the parabola $y^2 = 4ax$, then k is equal to

(a) $\frac{ab}{b-a}$ (b) $\frac{b}{b-a}$
(c) $\frac{a}{b-a}$ (d) $\frac{ab}{a-b}$

52. A line L passing through the focus of the parabola $y^2 = 4(x-1)$ intersects the parabola in two distinct points. If 'm' be the slope of the line L, then

(a) $m \in (-1, 1)$ (b) $m \in (-\infty, -1) \cup (1, \infty)$
(c) $m \in R$ (d) None of these

53. The length of the latus-rectum of the parabola $x = ay^2 + by + c$ is

(a) $a/4$ (b) $a/3$
(c) $1/a$ (d) $1/4a$

54. The coordinates of the point on the parabola $y^2 = 8x$, which is at minimum distance from the circle $x^2 + (y+6)^2 = 1$, are

(a) $(2, -4)$ (b) $(18, -12)$
(c) $(2, 4)$ (d) None of these

55. The equation of the parabola whose vertex and focus lie on the axis of X, at distances a and a_1 from the origin respectively is

(a) $y^2 = 4(a_1 - a)x$ (b) $y^2 = 4(a_1 - a)(x - a)$
(c) $y^2 = 4(a_1 - a)(x - a_1)$ (d) None of these

56. If $(\alpha^2, \alpha - 2)$ be a point interior to the region of the parabola $y^2 = 2x$ bounded by the chord joining the points $(2, 2)$ and $(8, -4)$, then α belongs to the interval

(a) $(-2 + 2\sqrt{2}, 2)$ (b) $(-2 + 2\sqrt{2}, \infty)$
(c) $(-2 - 2\sqrt{2}, \infty)$ (d) None of these

57. If a circle and a parabola intersect in 4 points, then the algebraic sum of the ordinates is

(a) proportional to arithmetic mean of the radius and latus-rectum
(b) zero
(c) equal to the ratio of arithmetic mean and latus-rectum
(d) None of the above

58. If the normal to the parabola $y^2 = 4ax$ at the point $(at^2, 2at)$ cuts the parabola again at $(aT^2, 2aT)$, then

(a) $-2 \le T \le 2$ (b) $T \in (-\infty, -8) \cup (8, \infty)$
(c) $T^2 < 8$ (d) $T^2 \ge 8$

59. P is the point 't' on the parabola $y^2 = 4ax$ and PQ is a focal chord, PT is the tangent at P and QN is the normal at Q. If the angle betwen PT and QN, be α and the distance between PT and QN be d, then

(a) $0 < \alpha < 90°$ (b) $\alpha = 0°$
(c) $d = 0$ (d) $d = a\left(\sqrt{1+t^2} + \frac{1}{\sqrt{1+t^2}}\right)$

60. For parabola $x^2 + y^2 + 2xy - 6x - 2y + 3 = 0$, the focus is

(a) $(1, -1)$ (b) $(-1, 1)$
(c) $(3, 1)$ (d) None of these

61. The length of latus-rectum of the parabola whose parametric equation is $x = t^2 + t + 1$, $y = t^2 - t + 1$, where $t \in R$ is equal to

(a) 2 (b) 4
(c) 8 (d) None of these

Objective Questions Type II [One or more than one correct answer(s)]

In each of the questions below four choices of which one or more than one are correct. You have to select the correct answer(s) accordingly.

1. Consider a circle with its centre lying on the focus of the parabola $y^2 = 2px$ such that it touches the directrix of the parabola, then a point of intersection of the circle and the parabola is

(a) $\left(\frac{p}{2}, p\right)$ (b) $\left(\frac{p}{2}, -p\right)$
(c) $\left(-\frac{p}{2}, p\right)$ (d) $\left(-\frac{p}{2}, -p\right)$

2. The equation of a locus is $y^2 + 2ax + 2by + c = 0$, then

(a) it is an ellipse (b) it is a parabola
(c) its latus-rectum $= a$ (d) its latus-rectum $= 2a$

3. The equation of a tangent to the parabola $y^2 = 8x$ which makes an angle 45° with the line $y = 3x + 5$ is

(a) $2x + y + 1 = 0$
(b) $y = 2x + 1$
(c) $x - 2y + 8 = 0$
(d) $x + 2y - 8 = 0$

4. The normal $y = mx - 2am - am^3$ to the parabola $y^2 = 4ax$ subtends a right angle at the vertex, if

(a) $m = 1$
(b) $m = \sqrt{2}$
(c) $m = -\sqrt{2}$
(d) $m = 1/\sqrt{2}$

5. The locus of the mid-point of the focal distance of a variable point moving on the parabola $y^2 = 4ax$ is a parabola whose
(a) latus rectum is half the latus rectum of the original parabola.
(b) vertex is $(a / 2, 0)$.
(c) directrix is Y-axis.
(d) focus has coordinates $(a, 0)$.

6. Let PQ be a chord of the parabola $y^2 = 4x$. A circle drawn with PQ as a diameter passes through the vertex V of the parabola. If area of $\Delta PVQ = 20$ sq units, then coordinates of P are
(a) $(-16, -8)$ (b) $(-16, 8)$
(c) $(16, -8)$ (d) $(16, 8)$

7. P is a point which moves in the x-y plane such that the point P is nearer to the centre of a square than any of the sides. The four vertices of the square are $(\pm a, \pm a)$. The region in which P will move is bounded by parts of parabolas of which one has the equation
(a) $y^2 = a^2 + 2ax$ (b) $x^2 = a^2 + 2ay$
(c) $y^2 + 2ax = a^2$ (d) None of these

8. The figure shows the graph of the parabola $y = ax^2 + bx + c$, then

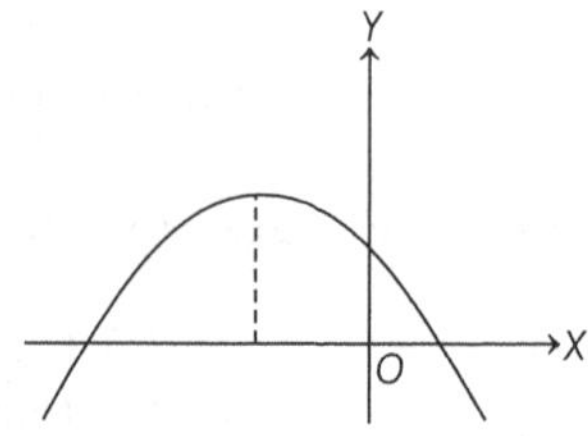

(a) $a > 0$ (b) $b < 0$
(c) $c > 0$ (d) $b^2 - 4ac > 0$

9. If $f(x + y) = f(x) \cdot f(y)$ for all x, y and $f(1) = 2$ and $a_r = f(r)$, for $r \in N$, then the coordinates of a point on the parabola $y^2 = 8x$ whose focal distance is 4, may be
(a) (a_1, a_2) (b) $(a_1, -a_2)$
(c) (a_1, a_1) (d) None of these

10. Let P be the point on the parabola $y^2 = 4x$, which is at the shortest distance from the centre S of the circle $x^2 + y^2 - 4x - 16y + 64 = 0$. Let Q be the point on the circle dividing the line segment SP internally. Then, **[JEE Advanced 2016]**
(a) $SP = 2\sqrt{5}$
(b) $SQ : QP = (\sqrt{5} + 1) : 2$
(c) the x-intercept of the normal to the parabola at P is 6
(d) the slope of the tangent to the circle at Q is $\frac{1}{2}$

11. The normals to the parabola $y^2 = 4ax$ from the point $(5a, 2a)$ are
(a) $y = x - 3a$ (b) $y = -2x + 12a$
(c) $y = -3x + 33a$ (d) $y = x + 3a$

12. If equation of directrix of the parabola $x^2 + 4y - 6x + k = 0$ is $y + 1 = 0$, then
(a) $k = 17$ (b) $k = -17$
(c) focus is $(3, -3)$ (d) vertex is $(3, -3)$

13. The tangent PT and the normal PN to the parabola $y^2 = 4ax$ at a point P on it meet its axis at T and N respectively. The locus of the centroid of the ΔPTN is a parabola whose
(a) vertex is $(2a / 3, 0)$
(b) directrix is $x = 0$
(c) latus rectum is $2a / 3$
(d) focus is $(a, 0)$

14. The equation of the tangent to the parabola $y^2 = 9x$ which goes through the point $(4, 10)$ is
(a) $x + 4y + 1 = 0$ (b) $9x + 4y + 4 = 0$
(c) $x - 4y + 36 = 0$ (d) $9x - 4y + 4 = 0$

15. Which of the following is (are) true about the parabola $y^2 = 4ax\ (a > 0)$?
(a) If t_1, t_2 are end points of a focal chord, then $t_1 t_2 = -1$
(b) Tangent at the end of a focal chord cuts at right angle at directrix
(c) Distance of any point on the parabola from directrix is equal to the sum of 'a' abscissa of the point
(d) End points of latusrectum are $(a, 2a)$, $(-a, 2a)$

16. The following line(s) is(are) tangent to the curve $y = x^2 - x$
(a) $x - y = 0$ (b) $x + y = 0$
(c) $x - y = 1$ (d) $x + y = 1$

Linked-Comprehension Type

In these questions, a passage (paragraph) has been given followed by questions based on each of the passages. You have to answer the questions based on the passage given.

PASSAGE 1

If a source of light is placed at the fixed point of a parabola and if the parabola is a reflecting surface, then the ray will bounce back in a line parallel to the axis of the parabola.

On the basis of above information, answer the following questions

1. A ray of light is coming along the line $y = 2$ from the positive direction of X-axis and strikes a concave mirror whose intersection with the $x - y$ plane is a parabola $y^2 = 8x$, then the equation of the reflected ray is
(a) $2x + 5y = 4$ (b) $3x + 2y = 6$
(c) $4x + 3y = 8$ (d) $5x + 4y = 10$

2. A ray of light moving parallel to the X-axis gets reflected from a parabolic mirror whose equation is $y^2 + 10y - 4x + 17 = 0$. After reflection, the ray must pass through the point
(a) $(-2, -5)$ (b) $(-1, -5)$
(c) $(-3, -5)$ (d) $(-4, -5)$

3. A ray of light is coming along the line $x = 2$ from the positive direction of Y-axis and strikes a concave mirror whose intersection with the x-y plane is a parabola $x^2 = 4y$, then the equation of the reflected ray after second reflection is
(a) $2x + y = 1$
(b) $3x - 2y + 2 = 0$
(c) $y - 1 = 0$
(d) None of the above

4. Two rays of light coming along the lines $y = 1$ and $y = -2$ from the positive direction of X-axis and strikes a concave mirror whose intersection with the x-y plane is a parabola $y^2 = x$ at A and B respectively. The reflected rays pass through a fixed point C, then the area of the ΔABC is
(a) $\frac{21}{8}$ sq units (b) $\frac{19}{2}$ sq units
(c) $\frac{17}{2}$ sq units (d) $\frac{15}{2}$ sq units

5. A ray of light is coming along the line $x = \lambda (\lambda < 0)$ from the positive direction of Y-axis and strikes a concave mirror $x^2 = 4(y - 1)$, the ray is reflected from it is $x + y = 2$, if this reflected ray again strikes, then the equation of new reflected ray is
(a) $y + 1 = 2\sqrt{2}$ (b) $x + 1 = 2\sqrt{2}$
(c) $y + 2 = 2\sqrt{2}$ (d) $x + 2 = 2\sqrt{2}$

PASSAGE 2

If l and m are variable real numbers such that $5l^2 + 6m^2 - 4lm + 3l = 0$, then the variable line $lx + my = 1$ always touches a fixed parabola, whose axes is parallel to the X-axis.

On the basis of above information, answer the following questions

1. The vertex of the parabola is
(a) $(-5/3, 4/3)$ (b) $(-7/4, 3/4)$
(c) $(5/6, -7/6)$ (d) $(1/2, -3/4)$

2. The focus of the parabola is
(a) $(1/6, -7/6)$ (b) $(1/3, 4/3)$
(c) $(3/2, -3/2)$ (d) $(-3/4, 3/4)$

3. The directrix of the parabola is
(a) $6x + 7 = 0$ (b) $4x + 11 = 0$
(c) $3x + 11 = 0$ (d) None of these

PASSAGE 3

Three normals AA_1, BB_1 and CC_1 are drawn from a point $P(h, k)$ to the parabola $y^2 = 4ax$, at A, B and C points. The following conditions are satisfied by the three normals :
(I) Any two of three normals are coincide
(II) $S(a, 0)$ be the focus of the parabola
(III) Three normals be real, then $h > 2a$
(IV) Slopes of the normals are m_1, m_2 and m_3. If $m_1 m_2 = \lambda$, then the locus of P is a parabola
(V) P lies on the line $y = \mu$, then the sides of the triangle ABC touch the parabola $S' = 0$

On the basis of above information, answer the following questions

1. Locus of point P is, if (I) is satisfied
(a) $ay^2 = 4(x - 2a)^3$ (b) $27ay^2 = 4(x - 2a)^3$
(c) $ax^2 = 4(y - 2a)^3$ (d) $27ax^2 = 4(y - 2a)^3$

2. $SA \cdot SB \cdot SC$ is equal to
(a) $a(SP)^2$ (b) $2a(SP)^2$
(c) $(aSP)^2$ (d) $4a(SP)^2$

3. If $a = 1$ is a parabola and (IV) is satisfied, then λ is
(a) 4 (b) 3 (c) 2 (d) 1

4. If (V) is satisfied, then the equation of $S' = 0$ is
(a) $y^2 = 4\mu x$ (b) $x^2 = 4\mu y$
(c) $y^2 = 2\mu x$ (d) $x^2 = 2\mu y$

5. If $a = -1$ and $k = 2$ and (III) is satisfied, then
(a) $h > 2$ (b) $h > -2$
(c) $h < -2$ (d) $-2 < h < 2$

Numerical Value Type

Solve the following problems and write your answers.

1. The equation to the parabola whose axis parallel to the Y-axis and which passes through the points (0, 4), (1,9) and (4, 5). If latus rectum of parabola is λ, then the value of 361λ must be

2. The normal to the parabola $y^2 = 8x$ at the point (2, 4) meets it again at (18, – 12). If length of normal chord is λ, then the value of λ^2 must be

3. The locus of a point that divides a chord of slope 2 of the parabola $y^2 = 4x$ internally in the ratio 1 : 2 is a parabola. If the vertex of parabola is (λ, μ), then the value of 729 $(\lambda + \mu)^2$ must be

4. From a point A common tangents are drawn to the circle $x^2 + y^2 = \frac{a^2}{2}$ and the parabola $y^2 = 4ax$. If the area of the quadrilateral formed by the common tangents, the chords of contact of the point A, w.r.t. the circle and the parabola is λ square unit, then the value of $\frac{256}{a^2}\lambda$ must be

5. Three normals with slopes m_1, m_2 and m_3 are drawn from a point P not on the axis of parabola $y^2 = 4x$. If $m_1 m_2 = \alpha$, results in the locus of P being a part of parabola, then the value of $(36)^\alpha$ must

6. A parabola of latusrectum $4a$, touches a fixed equal parabola, the axes of two curves being parallel; the locus of the vertex of the moving curve is a parallel of latus rectum λ, then the value of $\frac{59}{a^2}\lambda^2$ must be

7. A variable chord PQ of the parabola $y = 4x^2$ subtends a right angle at the vertex. If the locus of the point of intersection of the normals at P and Q is $y = \lambda x^2 + \mu$, then the value of $\frac{\mu}{\lambda}$ must be

8. A water jet from a fountain reaches its maximum height of 4 m at a distance 0.5m from the vertical passing through the point O of water outlet. If λ m be the height of the jet above the horizontal OX at a distance of 0.75 m from the point O, then the value of λ^6 must be

9. If the distribution of weight is uniform, then the rope of the suspended bridge takes the form of parabola. The height of the supporting towers is 20 m, the distance between these towers is 150 m and the height of the lowest point of the rope from the road is 3 m and the equation of the parabolic shape of the rope considering the floor of the bridge as X-axis and the axis of the parabola as Y-axis. If λ m be the height of that tower which supports the rope and is at a distance of 30 m from the centre of the road, then the value of λ must be

Single Integer Type

Solve the following problems and write your answers.

1. If P is a point on the parabola $y^2 = 3(2x - 3)$ and M is the foot of perpendicular drawn from P on the directrix of the parabola, then length of each side of an equilateral triangle SMP, where S is focus of the parabola, is

2. The number of points with integral coordinates that lie in the interior of the region common to the circle $x^2 + y^2 = 16$ and the parabola $y^2 = 4x$ is

3. Angle between the tangents drawn from (1, 4) the parabola $y^2 = 4x$ is π / m, where m is equal to

4. The equation of the line touching both the parabolas $y^2 = 4x$ and $x^2 = -32y$ is $ax + by + c = 0$.
   Then, the value of $a + b + c$ is

5. If the line $x - 1 = 0$ is the directrix of the parabola $y^2 - kx + 8 = 0$, then one of the values of k is

6. If the normal chord at a point 't' on the parabola $y^2 = 4ax$ subtends a right angle at the vertex, then a value of t^2 is

7. The locus of the mid-points of the portion of the normal to the parabola $y^2 = 16x$ intercepted between the curve and the axis is another parabola whose latus rectum is ______ .

8. If $mx - y + c = 0$ is a normal at a point P on the parabola $y^2 = 16x$ and the focal distance of P is 40 units, then $\frac{2}{61}|c|$.

Matrix-Match Type

Given below are Matching Type Questions, with two columns (each having some items) each . Each item of Column I has to be matched with the items of Column II, by encircling the correct match(es).

NOTE *An item of Column I can be matched with more than one items of Column II. All the items of Column II have to be matched.*

1. The equation of conics represented by the general equation of second degree $ax^2 + 2hxy + by^2 + 2gx + 2fy + c = 0$ and the discriminant of above equation is represented by Δ, where $\Delta = abc + 2fgh - af^2 - bg^2 - ch^2$

$$\text{or } \Delta = \begin{vmatrix} a & h & g \\ h & b & f \\ g & f & c \end{vmatrix}$$

Now match the entries from the following two columns :

	Column I		Column II
(A)	The conic is represented by the equation $\sqrt{\left(\frac{x}{a}\right)} + \sqrt{\left(\frac{y}{b}\right)} = 1$ is $(a \neq 0, b \neq 0)$	(P)	degenerate
(B)	The conic is represented by the equation $3x^2 + 10xy + 3y^2 - 15x - 21y + 18 = 0$ is	(Q)	non-degenerate

(C)	The conic is represented by the equation $8x^2 - 4xy + 5y^2 - 16x - 14y + 17 = 0$ is	(R)	a parabola
		(S)	an ellipse
		(T)	a pair of intersecting straight lines

2. Observe the following columns :

	Column I		Column II
(A)	The x-coordinate of points on the axis of the parabola $y^2 - 4x - 2y + 5 = 0$ from which all the three normals to the parabola are real is	(P)	4
(B)	The x-coordinate of points on the axis of the parabola $4y^2 - 32x + 4y + 65 = 0$ from which all the three normals to the parabola are real is	(Q)	5
(C)	The x-coordinate of points on the axis of the parabola $4y^2 - 16x - 4y + 41 = 0$ from which all the three normals to the parabola are real is	(R)	6
		(S)	7
		(T)	8

3. Observe the following columns

	Column I		Column II
(A)	If a parabola represented by $25(x^2 + y^2) = (4x + 3y - 12)^2$ then	(P)	Equation of directrix is $4x - 3y + 12 = 0$
(B)	If a parabola represented by $25(x^2 + y^2) = (4x - 3y + 12)^2$ then	(Q)	Equation of axis of parabola is $4x + 3y = 0$
(C)	If a parabola represented by $25(x^2 + y^2) = (3x - 4y + 12)^2$ then	(R)	Equation of directrix is $4x + 3y = 12$
		(S)	Equation of axis of parabola is $3x + 4y = 0$
		(T)	Equation of directrix is $3x - 4y + 12 = 0$

Answers

Objective Questions Type I [Only one correct answer]

1. (b)	**2.** (b)	**3.** (c)	**4.** (c)	**5.** (a)	**6.** (b)	**7.** (d)	**8.** (c)	**9.** (d)	**10.** (c)
11. (b)	**12.** (a)	**13.** (b)	**14.** (b)	**15.** (d)	**16.** (d)	**17.** (b)	**18.** (a)	**19.** (b)	**20.** (a)
21. (a)	**22.** (d)	**23.** (c)	**24.** (b)	**25.** (b)	**26.** (c)	**27.** (d)	**28.** (b)	**29.** (b)	**30.** (b)
31. (b)	**32.** (c)	**33.** (d)	**34.** (b)	**35.** (b)	**36.** (b)	**37.** (b)	**38.** (b)	**39.** (a)	**40.** (c)
41. (b)	**42.** (d)	**43.** (c)	**44.** (c)	**45.** (c)	**46.** (c)	**47.** (a)	**48.** (b)	**49.** (c)	**50.** (a)
51. (a)	**52.** (d)	**53.** (c)	**54.** (a)	**55.** (b)	**56.** (a)	**57.** (b)	**58.** (d)	**59.** (b)	**60.** (d)
61. (a)									

Objective Questions Type II [One or more than one correct answer(s)]

1. (a, b)	**2.** (b, d)	**3.** (a, c)	**4.** (b, c)	**5.** (a, b, c, d)
6. (c, d)	**7.** (a, b, c)	**8.** (b, c, d)	**9.** (a, b)	**10.** (a, c, d)
11. (a, b)	**12.** (a, c)	**13.** (a, d)	**14.** (c, d)	**15.** (a, b, c)
16. (b, c)				

Linked-Comprehension Type

Passage 1 **1.** (c) **2.** (b) **3.** (d) **4.** (a) **5.** (d)

Passage 2 **1.** (a) **2.** (b) **3.** (c)

Passage 3 **1.** (b) **2.** (a) **3.** (c) **4.** (d) **5.** (c)

Numerical Value Type Questions

1. 228	**2.** 512	**3.** 900	**4.** 960	**5.** 1296
6. 3776	**7.** 0.125	**8.** 729	**9.** 5.72	

Single Inter Type Question

1. (6) **2.** (8) **3.** (3) **4.** (3) **5.** (4) **6.** (2) **7.** (4) **8.** (4)

Matrix-Match Type

1. A → (Q, R); B → (P, T); C → (Q, S)

2. A → (P, Q,R, S, T); B → (S, T); C → (Q, R, S, T)

3. A → (R); B → (P, S); C → (Q, T)

Solutions

Objective Questions Type [Only one correct answer]

1. Directrix of $y^2 = 4(x+1)$ is $x = -2$. Any point on $x = -2$ is $(-2, k)$.
Now mirror image (x, y) of $(-2, k)$ in the line $x + 2y = 3$ is given by
$$\frac{x+2}{1} = \frac{y-k}{2} = -2\left(\frac{-2+2k-3}{5}\right)$$
$$\Rightarrow \quad x = \frac{10-4k}{5} - 2$$
$$\Rightarrow \quad x = -4k/5 \quad \ldots\text{(i)}$$
and $$y = \frac{20-8k}{5} + k$$
$$\Rightarrow \quad y = \frac{20-3k}{5} \quad \ldots\text{(ii)}$$
From Eqs. (i) and (ii)
$$y = 4 + \frac{3}{5}\left(\frac{5x}{4}\right)$$
$$y = 4 + \frac{3x}{4}$$
$\Rightarrow 4y = 16 + 3x$ is the equation of the mirror image of the directrix.

2. Shortest distance between two curves occured along the common normal.
Normal to $y^2 = 4x$ at $(m^2, 2m)$ is $y + mx - 2m - m^3 = 0$
Normal to $y^2 = 2(x-3)$ at $\left(\frac{m^2}{2} + 3, m\right)$ is
$$y + m(x-3) - m - \frac{m^3}{2} = 0$$
Both are same if $-2m - m^3 = -4m - \frac{1}{2}m^3$
$$\Rightarrow \quad m = 0, \pm 2$$
So, points will be (4, 4) and (5, 2) or (4, – 4) and (5, – 2)
Hence, shortest distance will be $\sqrt{(1+4)} = \sqrt{5}$

3. Let a point on parabola $x^2 = 4y$ is $A(x_1, y_1)$,
so $$x_1^2 = 4y_1 \quad \ldots\text{(i)}$$
and an another point $P(h, k)$ which divides the line segment joining the point $B(0, -1)$ and $A(x_1, y_1)$ internally in the ratio 1 : 2,
so $$h = \frac{(2\times 0) + (1 \times x_1)}{2+1}$$
and $$k = \frac{2\times(-1) + (1\times y_1)}{2+1}$$
$$\Rightarrow \quad x_1 = 3h \text{ and } y_1 = 3k + 2 \quad \ldots\text{(ii)}$$
Now, on the elimination of (x_1, y_1) with the help of Eq. (i) and relation (ii), we get
$$9h^2 = 4(3k+2)$$
On taking locus of point $P(h, k)$, we have
$$9x^2 = 12y + 8$$
$$\Rightarrow \quad 9x^2 - 12y = 8$$

4. Any point on the given parabola is $(t^2, 2t)$. The equation of the tangent at (1, 2) is $x - y + 1 = 0$.
The image (h, k) of the point $(t^2, 2t)$ in $x - y + 1 = 0$ is given by $\frac{h-t^2}{1} = \frac{k-2t}{-1} = \frac{-2(t^2 - 2t + 1)}{1+1}$
$$\therefore \quad h = t^2 - t^2 + 2t - 1 = 2t - 1$$
and $$k = 2t + t^2 - 2t + 1 = t^2 + 1$$
Eliminating t from $h = 2t - 1$ and $k = t^2 + 1$, we get
$$(h+1)^2 = 4(k-1)$$
The required equation of reflection is
$(x+1)^2 = 4(y-1)$.

5. $y^2 - 12x - 4y + 4 = 0$
$$\Rightarrow \quad (y-2)^2 = 12x$$
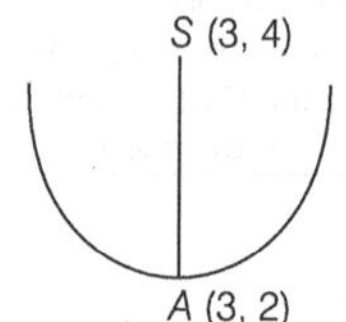

Its vertex is (0, 2) and $a = 3$,
its focus = (3, 2)
Hence, for the required parabola ; focus is (3, 4), vertex = (3, 2) and $a = 2$
Hence, the equation of the parabola is
$$(x-3)^2 = 4(2)(y-2)$$
or $\quad x^2 - 6x - 8y + 25 = 0$

6. Given, $\frac{5}{2} = {}^nC_3\,(px)^{n-3}\left(\frac{1}{x}\right)^3$
$$= {}^nC_3 \cdot p^{n-3}x^{n-6} \quad \ldots\text{(i)}$$
Since, LHS of Eq. (i) is independent of x
$$\therefore \quad n - 6 = 0$$
$$\Rightarrow \quad n = 6$$
From Eq. (i),
$$\frac{5}{2} = {}^6C_3p^3 = 20p^3$$
$$\Rightarrow \quad p^3 = \left(\frac{1}{2}\right)^3 \Rightarrow p = \frac{1}{2}$$
Given, parabola is $y^2 = x$ $\quad \ldots\text{(ii)}$
Here, $$4a = 1$$
$$\Rightarrow \quad a = \frac{1}{4}$$
Since, three normals are drawn from point $(q, 0)$,
$$\therefore \quad q > 2a \text{ or } q > \frac{1}{2} \text{ or } q > p \quad \left[\because p = \frac{1}{2}\right]$$

7. Centre (0, 0), radius $= \frac{1}{2} \cdot \frac{3}{4} \cdot 4a = \frac{3a}{2}$

Equation of the circle is $4(x^2 + y^2) = 9a^2$ ... (i)

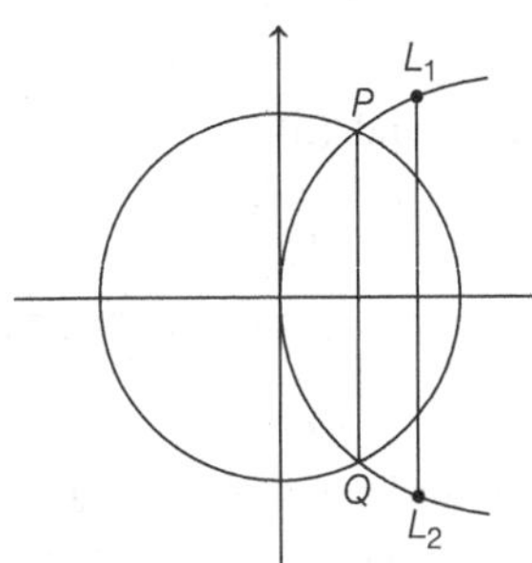

Equation of the parabola is $y^2 = 4ax$ ...(ii)

On solving Eqs. (i) and (ii), we get

$$x^2 + 4ax - \frac{9a^2}{4} = 0$$

$$x = \frac{-4a \pm \sqrt{16a^2 + 9a^2}}{2} = \frac{-4a \pm 5a}{2}$$

$\therefore$ Area $x = a/2$

For $x = a/2$. $y^2 = 4ax = 4a \cdot a/2 = 2a^2$

$\Rightarrow$ $y = \pm \sqrt{2}a$

$\therefore$ The double ordinate $= 2\sqrt{2}a$

$\therefore$ Area of the trapezium $PL_1 L_2Q = \frac{1}{2}\left(a - \frac{a}{2}\right)$

$$(4a + 2\sqrt{2}a) = \left(\frac{2+\sqrt{2}}{2}\right)a^2$$

8. Let B be $(at^2, 2at)$

Slope of $AB = \frac{2}{t}$

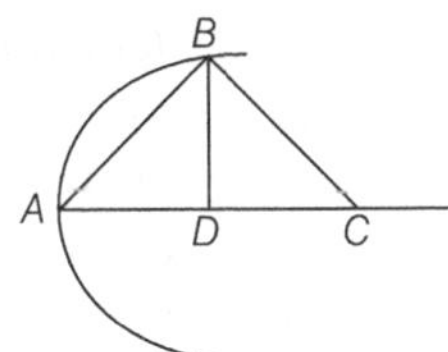

Equation of BC is $y - 2at = -\frac{t}{2}(x - at^2)$

This meets $y = 0$ at C whose x-coordinate $= 4a + at^2$

$$D = (at^2, 0)$$

$\therefore$ $DC = 4a + at^2 - at^2 = 4a$

9. $\because$ $[x^2] = [y]$

If $0 \le y < 1$, then $[y] = 0$

$\therefore$ $[x^2] = 0$

$0 \le x^2 < 1$

$\Rightarrow$ $x \in (-1, 1)$

for $1 \le y < 2$, then $[y] = 1$

$\therefore$ $[x^2] = 1$

$\Rightarrow$ $1 \le x^2 < 2$

$\Rightarrow$ $x \in (-\sqrt{2}, -1] \cup [1, \sqrt{2}]$

for $2 \le y < 3$, then $[y] = 2$

then $[x^2] = 2 \Rightarrow 2 \le x^2 < 3$

$\therefore$ $x \in (-\sqrt{3}, -\sqrt{2}] \cup [\sqrt{2}, \sqrt{3})$

The graph of the region will not only contain of the parabola $y = x^2$ but $[x^2] = [y]$ contain points within the rectangles of side 1, 2, ; 1, $\sqrt{2} - 1$; 1, $\sqrt{3} - \sqrt{2}$ etc.

Hence, a, b, c are incorrects.

10. Equation of given parabola is $y^2 = 8x$ and the equilateral triangle inscribed in the given parabola having one of its vertices on the vertex of the parabola, so according to symmetry other two vertices of the equilateral triangle, we can take as, $A\,(2t^2, 4t)$ and $B\,(2t^2, 4t)$

$\therefore$ Area of $\Delta\, OAB$ is $\Delta = \frac{1}{2}(2t^2)(8t) = \frac{\sqrt{3}}{4}(8t)^2$

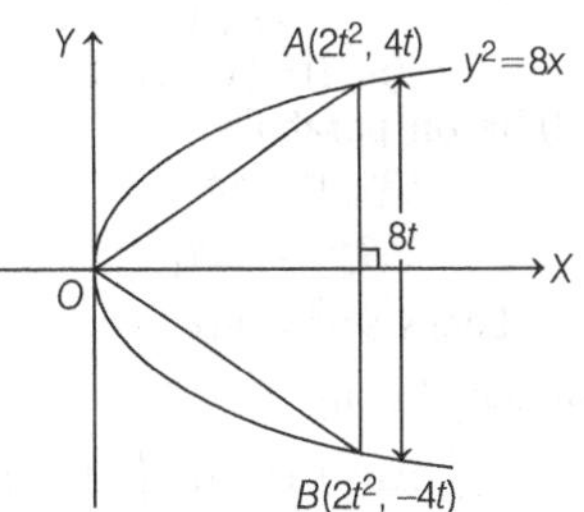

$\Rightarrow$ $t = 2\sqrt{3}$

$\therefore$ Area of equilateral $\Delta\, OAB = \frac{\sqrt{3}}{4}(8 \times 2\sqrt{3})^2$

$= 192\sqrt{3}$ sq unit

11. Equation of chord with mid-point (h, k).

$$T = S_1 \Rightarrow yk - 8x - 8h = k^2 - 16h$$

$$2x - \frac{yk}{4} = 2h - \frac{k^2}{4}$$

$\because$ $2x + y = p$

$\therefore$ $k = -4$ and $p = 2h - 4$

where $h = 3 \Rightarrow p = 2 \times 3 - 4 = 2$

12. As shown in the figure, common chord AB subtends right angle at vertex.

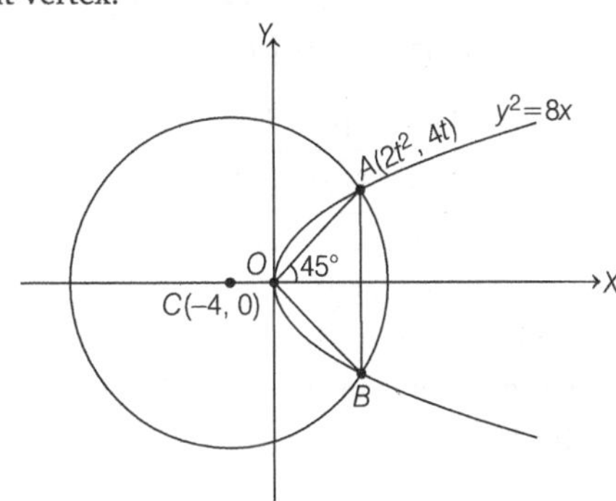

Slope of $OA = \frac{4t - 0}{2t^2 - 0} = \tan 45°$

$\Rightarrow$ $\frac{2}{t} = 1 \Rightarrow t = 2$

$\therefore$ $A \equiv (8, 8)$

So, radius of circle $= CA = \sqrt{(8+4)^2 + (8-0)^2}$

$= \sqrt{144 + 64} = 4\sqrt{13}$

13. Let B and C be the trisected points. B divides LL' in $1:2$.

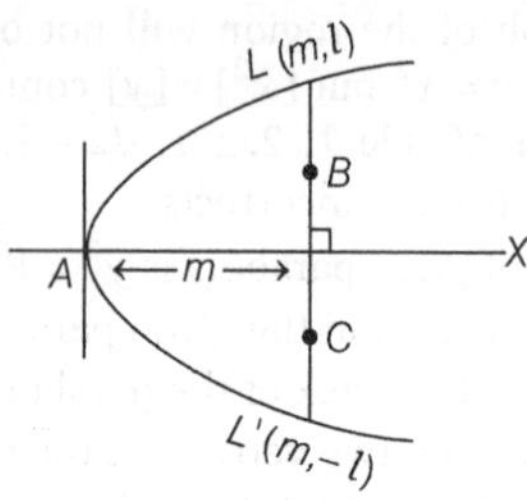

Then, coordinate of $B\left(\frac{1\cdot m+2\cdot m}{1+2}, \frac{-l+2l}{3}\right)$ or $B\,(m, l/3)$

Let $x_1 = m, y_1 = l/3$

$\therefore$ $(m, l) = (x_1, 3y_1)$

but (m, l) lie on parabola,

$$(3y_1)^2 = 4ax_1$$

$\Rightarrow$ $9y_1^2 = 4ax_1$

$\therefore$ Locus $9y^2 = 4ax$

14. Let parabola $y^2 = 4ax$

Let $P \equiv (at_1^2, 2at_1), Q \equiv (at_2^2, 2at_2)$

then, $T \equiv (at_1t_2, a\,(t_1+t_2))$

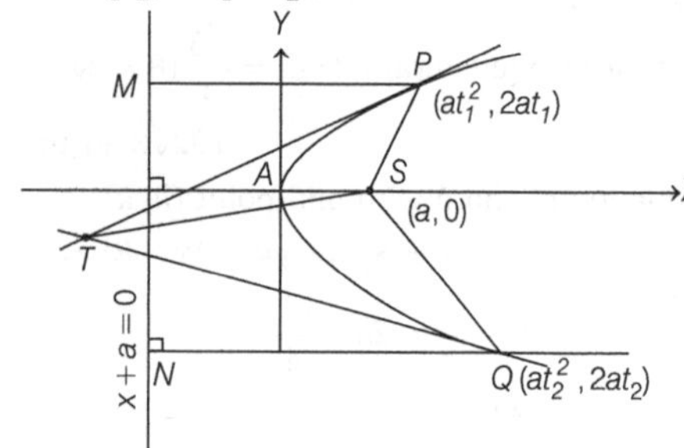

Then, $SP = PM = a + at_1^2$

$SQ = QN = a + at_2^2$

and $ST = \sqrt{(a - at_1t_2)^2 + (0 - a(t_1+t_2))^2}$

$= a\sqrt{(1-t_1t_2)^2 + (t_1+t_2)^2}$

$= a\sqrt{(1 + t_1^2t_2^2 + t_1^2 + t_2^2)}$

$= a\sqrt{(1+t_1^2)(1+t_2^2)}$

$= \sqrt{a\,(1+t_1^2)\cdot a\,(1+t_2^2)}$

$= \sqrt{(SP)\,(SQ)}$

$\therefore$ SP, ST, SQ are in GP.

15. Let $P \equiv (at_1^2, 2at_1)$ and $Q = (at_2^2, 2at_2)$

and let $R \equiv (aT^2, 2aT)$

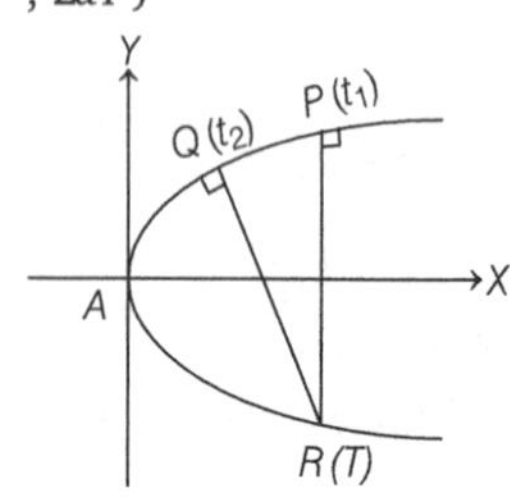

$\therefore$ $T = -t_1 - \frac{2}{t_1}$ ...(i)

and $T = -t_2 - \frac{2}{t_2}$ ...(ii)

From Eqs. (i) and (ii),

$$-t_1 - \frac{2}{t_1} = -t_2 - \frac{2}{t_2}$$

$\Rightarrow$ $(t_1 - t_2) = \frac{2}{t_2} - \frac{2}{t_1} = \frac{2\,(t_1-t_2)}{t_1t_2}$

or $t_1t_2 = 2$

$\therefore$ Product of ordinates of P and Q is

$$(2at_1)(2at_2) = 4a^2t_1t_2 = 8a^2$$

16. It is clear that point $(-2m, m+1)$ lie inside the circle and parabola, then

$$(-2m)^2 + (m+1)^2 - 4 < 0$$

and $(m+1)^2 - 4\,(-2m) < 0$

$\therefore$ $5m^2 + 2m - 3 < 0$

and $m^2 + 10m + 1 < 0$

$(m+1)(5m-3) < 0$ and $(m+5)^2 - 24 < 0$

$\Rightarrow$ $-1 < m < \frac{3}{5}$

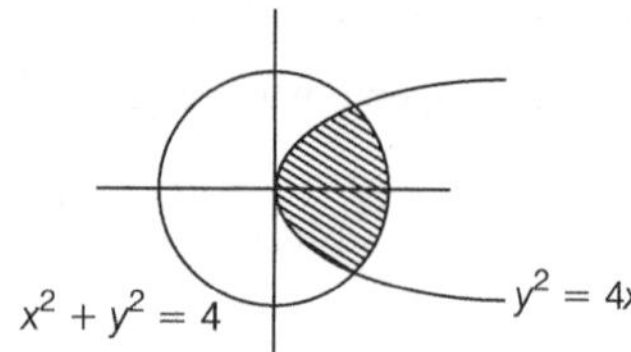

and $-5 - 2\sqrt{6} < m < -5 + 2\sqrt{6}$

Hence, $-1 < m < -5 + 2\sqrt{6}$

17. Let $B \equiv (at_1^2, 2at_1), C \equiv (at_2^2, 2at_2)$

then, $A \equiv (at_1t_2, a\,(t_1+t_2))$

Let any tangent to $y^2 = 4ax$ is

$$ty = x + at^2$$

or $x - ty + at^2 = 0$ ...(i)

$\therefore$ p_1 = length of perpendicular from A on Eq. (i)

$$= \frac{|at_1t_2 - ta\,(t_1+t_2) + at^2|}{\sqrt{(1+t^2)}}$$

$$= \frac{|a\,(t-t_1)(t-t_2)|}{\sqrt{(1+t^2)}}$$

p_2 = length of perpendicular from B on Eq. (i)

$$= \frac{|at_1^2 - t\,(2at_1) + at^2|}{\sqrt{(1+t^2)}} = \frac{|a\,(t-t_1)^2|}{\sqrt{(1+t^2)}}$$

and p_3 = length of perpendicular from C on Eq. (i)

$$= \frac{|at_2^2 - (2at_2)t + at^2|}{\sqrt{(1+t^2)}} = \frac{|a\,(t-t_2)^2|}{\sqrt{(1+t^2)}}$$

It is clear that $p_2p_3 = p_1^2$

Hence, p_2, p_1, p_3, are in GP.

18. $\because P \equiv (\sqrt{3}, 0)$

$$\frac{x-\sqrt{3}}{\cos 60°} = \frac{y-0}{\sin 60°} = r$$

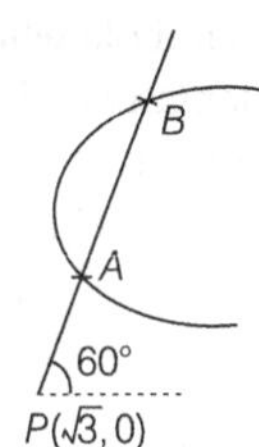

or $$x = \sqrt{3} + \frac{r}{2}, y = \frac{r\sqrt{3}}{2}$$

or $\left(\sqrt{3} + \frac{r}{2}, \frac{r\sqrt{3}}{2}\right)$ lie on $y^2 = x + 2$, then

$$\frac{3r^2}{4} = \sqrt{3} + \frac{r}{2} + 2 \Rightarrow \frac{3r^2}{4} - \frac{r}{2} - (2+\sqrt{3}) = 0$$

$$\therefore PA \cdot PB = r_1 r_2 = \left| -\frac{(2+\sqrt{3})}{\frac{3}{4}} \right| = \frac{4}{3}(2+\sqrt{3})$$

19. Equation of normal in terms of slope $y = mx - 2am - am^3$

$\because$ normals meet in (h, k), then

$$am^3 - (h-2a)m + k = 0$$

Let $P \equiv (am_1^2, -2am_1)$,

$Q \equiv (am_2^2, -2am_2)$

and $R \equiv (am_3^2, -2am_3)$

$\therefore$ $m_1 + m_2 + m_3 = 0$

or $$\frac{-2am_1 - 2am_2 - 2am_3}{3} = 0$$

$\because$ y-coordinate of centroid of triangle $PQR = 0$

$\therefore$ Centroid lies on X-axis $(y = 0)$

20. Let $A \equiv (at_1^2, 2at_1), B \equiv (at_2^2, 2at_2)$

Then, $C \equiv (at_1t_2, a(t_1 + t_2))$

$\because$ Ordinates of A, B, C are $2at_1$, $2at_2$ and $a(t_1 + t_2)$

Also, $$\frac{\text{ordinate of } A + \text{ordinate of } B}{2} = \text{ordinate of } C$$

Hence, ordinates of A, C and B are in AP.

21. Normals of parabolas $y^2 = 4ax$ and $y^2 = 4c(x-d)$ in terms of slope are

$$y = mx - 2am - am^3 \quad \ldots(i)$$

and $$y = m(x-d) - 2cm - cm^3 \quad \ldots(ii)$$

Subtracting Eqs. (ii) from (i), then

$$md - 2am - am^3 + 2cm + cm^3 = 0$$

$$m \neq 0$$

$\therefore$ $$d - 2a - am^2 + 2c + cm^2 = 0$$

$\Rightarrow$ $$(a-c)m^2 = d - 2a + 2c$$

$\Rightarrow$ $$m^2 = \frac{d-2a+2c}{(a-c)}$$

$\Rightarrow$ $$\frac{d}{a-c} - 2 > 0$$

$\Rightarrow$ $$d > 2a - 2c$$

$\Rightarrow$ $$2a < 2c + d$$

22. Since equation of tangent to the parabola $y^2 = 4(x+1)$ having slope 'm' is

$$y = m(x+1) + \frac{1}{m}$$

$\Rightarrow$ $$y = mx + \left(m + \frac{1}{m}\right) \quad \ldots(i)$$

Similarly, equation of tangent to the parabola $y^2 = 8(x+2)$ having slope m' is

$$y = m'(x+2) + \frac{2}{m'}$$

$\Rightarrow$ $$y = m'x + \left(2m' + \frac{2}{m'}\right) \quad \ldots(ii)$$

$\because$ It is given that tangents (i) and (ii) are perpendicular so $mm' = -1$

$\therefore$ Eq. (ii) becomes as

$$y = -\frac{1}{m}x - 2m - \frac{2}{m} \quad \ldots(iii)$$

Let tangents (i) and (iii) intersects each other at point $P(h, k)$, so

$$(h+1)m^2 - km + 1 = 0$$

and $$2m^2 + km + (h+2) = 0$$

$\therefore$ $$\frac{h+1}{2} = \frac{-k}{k} = \frac{1}{h+2} \Rightarrow h + 3 = 0$$

On taking locus of point $P(h, k)$, we get

$$x + 3 = 0$$

23. Equation of normal in terms of slope is $y = mx - 2am - am^3$

Let $O \equiv (h, k)$

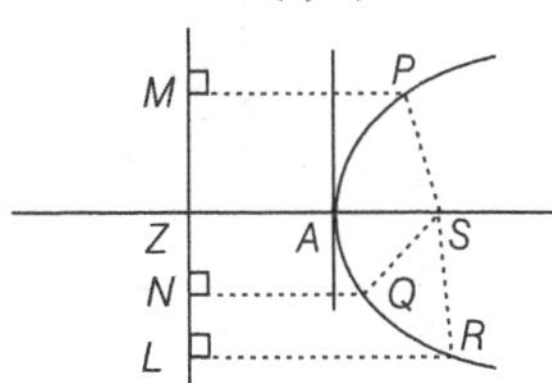

then, $am^3 - (h-2a)m + k = 0$

or $$m_1 + m_2 + m_3 = 0$$

$$m_1m_2 + m_2m_3 + m_3m_1 = -\frac{(h-2a)}{a}$$

and $$m_1m_2m_3 = -\frac{k}{a}$$

Let $P \equiv (am_1^2, -2am_1)$,

$Q \equiv (am_2^2 - 2am_2)$

and $R \equiv (am_3^2, -2am_3)$

and $S \equiv (a, 0)$

$\therefore \; |SP|\cdot|SQ|\cdot|SR| = |PM|\cdot|QN|\cdot|RL|$

$= |a + am_1^2||a + am_2^2||a + am_3^2|$

$= a^3|(1 + m_1^2)(1 + m_2^2)(1 + m_3^2)|$

$= a^3|1 + \Sigma m_1^2 + \Sigma m_1^2 m_2^2 + m_1^2 m_2^2 m_3^2|$

$= a^3|1 + (\Sigma m_1)^2 - 2\Sigma m_1 m_2) + (\Sigma m_1 m_2)^2 - 2m_1 m_2 m_3 \Sigma m_1 + (m_1 m_2 m_3)^2|$

$= a^3\left|1 + (0)^2 + \frac{2(h - 2a)}{a} + \frac{(h - 2a)^2}{a^2} - 0 + \frac{k^2}{a^2}\right|$

$= a|k^2 + 2a(h - 2a) + a^2 + (h - 2a)^2|$

$= a|k^2 + (h - 2a + a)^2|$

$= a|k^2 + (h - a)^2|$

$= a|(k - 0)^2 + (h - a)^2| = a(SO)^2$

24. $y^2 - 4x - 2y + 5 = 0 \Rightarrow (y - 1)^2 - 4x + 4 = 0$

$\Rightarrow \quad (y - 1)^2 = 4(x - 1)$

Let $y - 1 = Y$ and $x - 1 = X$,

Then, $Y^2 = 4X$ comparing with $Y^2 = 4ax$

$\therefore \quad a = 1$

$\because$ All three normals to the parabola are real and meet on the axis of parabola, then

$X > 2a$ and $Y = 0$

i.e. $\quad x - 1 > 2$ and $y - 1 = 0$

$\therefore \quad x > 3$ and $y = 1$ or $(k, 1)$; $k > 3$

25. Coordinates of A, B and C are

$(at_1t_3, a(t_1 + t_3))$, $(at_1t_2, a(t_1 + t_2))$

and $(at_2t_3, a(t_2 + t_3))$ respectively.

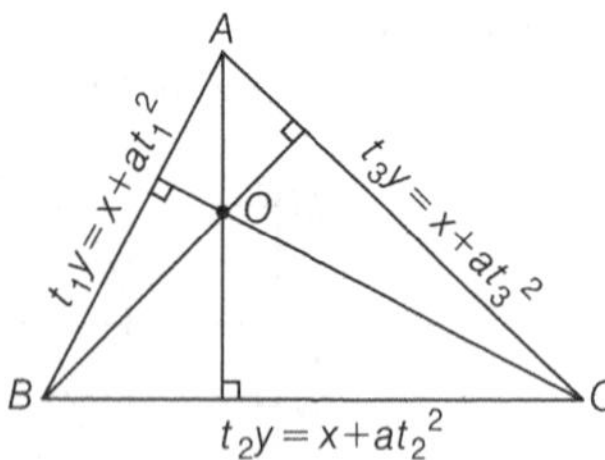

Let orthocentre is $O(h, k)$

$\because \quad AO \perp BC$

$\therefore \quad \left\{\frac{k - a(t_1 + t_3)}{h - at_1t_3}\right\} \times \frac{1}{t_2} = -1$

$\Rightarrow \quad k - a(t_1 + t_3) = -ht_2 + at_1t_2t_3 \quad \ldots(i)$

Also, $BO \perp CA$

Similarly, $k - a(t_1 + t_2) = -ht_3 + at_1t_2t_3 \quad \ldots(ii)$

Subtracting Eq. (ii) from Eq. (i), then we get

$h + a = 0$

$\therefore$ Locus of orthocentre is $x + a = 0$

Which is directrix of parabola.

26. If vertex of parabola is (0, 0), length of latus-rectum is l and axis along the positive direction of Y-axis, then equation of parabola is $x^2 = ly$.

Now, equation of parabola whose vertex is (a, b) and latus-rectum is of length l and if axis of the parabola is along the positive direction of Y-axis is

$$(x - a)^2 = l(y - b)$$

27. The equation of normal to the parabola $y^2 = 4ax$ is $y = mx - 2am - am^3$.

Here, $a = \frac{1}{2}$

Equation of the normal is $y = mx - m - \frac{m^3}{2}$

Normal passes through $(a, 0)$

$$0 = am - m - \frac{m^3}{2}$$

Either $\underset{\text{(not posiible)}}{m = 0}$ or $a - 1 - \frac{m^2}{2} = 0$

$m^2 = 2(a - 1)$

As, $m^2 > 0$,

$2(a - 1) > 0 \Rightarrow a - 1 > 0 \Rightarrow a > 1$

28. $\because lx + my + n = 0$

$\Rightarrow \quad y = -\frac{(lx + n)}{m} \quad \ldots(i)$

and parabola $\quad x^2 = 4ay \quad \ldots(ii)$

From Eqs. (i) and (ii),

$$x^2 = 4a\left\{-\frac{(lx + n)}{m}\right\}$$

$\Rightarrow \quad mx^2 + 4alx + 4an = 0$

for touching

$B^2 = 4AC$

$\Rightarrow \quad 16a^2l^2 = 4m(4an)$

$\Rightarrow \quad al^2 = mn \Rightarrow al^2 - mn = 0$

29. Let $P(at^2, 2at)$ and $Q(a/t^2, -2a/t)$ be a focal chord of the parabola (as $t_1t_2 = -1$)

Length of $PQ = l = \sqrt{(at^2 - a/t^2)^2 + (2at + 2a/t)^2}$

$= a\sqrt{(t^2 - 1/t^2)^2 + 4(t + 1/t)^2}$

$= a(t + 1/t)\sqrt{(t - 1/t)^2 + 4}$

$= a(t + 1/t)^2$

Length of the perpendicular from the vertex (0, 0) on the line PQ whose equation is

$y\left(t - \frac{1}{t}\right) = 2(x - a)$ is given by

$$p = \frac{2a}{\sqrt{(t - 1/t)^2 + 2^2}} = \frac{2a}{(t + 1/t)}$$

so that $\quad l \cdot p^2 = \frac{4a^2}{(t + 1/t)^2} \times a(t + 1/t)^2$

$= 4a^3$ which is constant.

30. Normal in terms of slope $y = mx - 2am - am^3$

$$\Rightarrow \quad \frac{mx - y}{2am + am^3} = 1 \quad \ldots(i)$$

and $$y^2 = 4ax \quad \ldots(ii)$$

making homogeneous Eq. (ii) with the help of Eq. (i)

$$\therefore \quad 4ax\left(\frac{mx - y}{2am + am^3}\right) = y^2$$

$$\Rightarrow \quad (2am + am^3)\,y^2 = 4amx^2 - 4axy$$

For right angle, Coefficient of x^2 + coefficient of $y^2 = 0$

$$\Rightarrow \quad -4am + 2am + am^3 = 0 \Rightarrow m = \pm\sqrt{2}$$

31. Tangent of parabola $y^2 = 4ax$ is

$$y = mx + \frac{a}{m} \quad \ldots(i)$$

where $m = \tan\alpha$

and equation of normal parallel to Eq. (i) is

$$y = mx - 2am - am^3 \quad \ldots(ii)$$

Distance between Eqs. (i) and (ii) is $\dfrac{\left|\dfrac{a}{m} + 2am + am^3\right|}{\sqrt{(1+m^2)}}$

$$= \frac{a(1+m^2)^2}{m\sqrt{(1+m^2)}} = \frac{a(1+m^2)^{3/2}}{m}$$

$$= \frac{a(1+\tan^2\alpha)^{3/2}}{\tan\alpha}$$

$$= \frac{a\sec^3\alpha}{\tan\alpha} = a\sec^2\alpha \cdot \operatorname{cosec}\alpha$$

32. Equation of given parabola is $y^2 = 12x$ ...(i)

Now, let a point $P(3t^2, 6t)$ on the parabola, since point N is the foot of perpendicular drawn from P on the axis of the parabola.

(i.e. X-axis) so coordinate of N is $(3t^2, 0)$ and M is mid-point of PN, so coordinate of M is $(3t^2, 3t)$. Now, equation of the parallel to the axis of parabola through point M is

$$y = 3t \quad \ldots(ii)$$

So, point of intersection of parabola (i) and line (ii) is $Q\left(\dfrac{3t^2}{4}, 3t\right)$, so equation of line NQ is

$$y - 0 = \frac{3t}{\dfrac{3t^2}{4} - 3t^2}(x - 3t^2) \quad \ldots(iii)$$

For y-intercept made by line (iii) put $x = 0$, then

$$y = \frac{1}{(1/4) - 1}(-3t) = \frac{12t}{3} = \frac{4}{3} \quad \text{[given]}$$

$$\Rightarrow t = \frac{1}{3}$$

$$\therefore \quad P\left(\frac{1}{3}, 2\right), N\left(\frac{1}{3}, 0\right), M\left(\frac{1}{3}, 1\right), Q\left(\frac{1}{12}, 1\right)$$

$$\therefore \quad PN = 2, \text{ and } MQ = \frac{1}{3} - \frac{1}{12} = \frac{1}{4}$$

33. Equation of tangent in terms of slope of parabola $y^2 = 4x$ is

$$y = mx + \frac{1}{m} \quad \ldots(i)$$

$\because$ Eq. (i), is also tangent of $x^2 = -32y$

then $$x^2 = -32\left(mx + \frac{1}{m}\right)$$

$$\Rightarrow \quad x^2 + 32mx + \frac{32}{m} = 0$$

$$\therefore \quad B^2 = 4AC \text{ (Condition of tangency)}$$

$$\Rightarrow \quad (32m)^2 = 4 \cdot 1 \cdot \frac{32}{m}$$

$$\Rightarrow \quad m^3 = \frac{1}{8} \text{ or } m = \frac{1}{2}$$

From Eq. (i), $y = \dfrac{x}{2} + 2$

$$\Rightarrow \quad x - 2y + 4 = 0$$

34. Normal at $(at^2, 2at)$ on the parabola $y^2 = 4ax$ is

$$y + tx = 2at + at^3 \quad \ldots(i)$$

Suppose normal Eq. (i) cuts the curve again at $(at_1^2, 2at_1)$, then

$$2at_1 + att_1^2 = 2at + at^3$$

$$\Rightarrow \quad 2a(t - t_1) + at(t^2 - t_1^2) = 0$$

or $$2 + t(t + t_1) = 0 \quad [\because a(t - t_1) \neq 0]$$

$$\therefore \quad t_1 = -t - \frac{2}{t} = -\left(t + \frac{2}{t}\right)$$

35. Consider parabola $y^2 = 4ax$

So, $X \equiv (-a, 0)$

Let point P be $(at^2, 2at)$.

$$\therefore \quad P' \equiv (at^2, -2at)$$

Equation of line PX is $y = \dfrac{2at - 0}{at^2 + a}(x + a)$

or $$(1 + t^2)y = 2t(x + a) \quad \ldots(i)$$

Solving this line with given parabola, we get

$$\frac{4t^2(x+a)^2}{(1+t^2)^2} = 4ax$$

$$\Rightarrow \quad t^2(x+a)^2 = ax(1+t^2)^2$$

$$\Rightarrow \quad t^2(x^2 + a^2 + 2ax) = a(t^4 + 1 + 2t^2)x$$

$$\Rightarrow \quad t^2x^2 + t^2a^2 = xat^4 + ax$$

$$\Rightarrow t^2x^2 - (a + at^4)x + a^2t^2 = 0$$

$$\Rightarrow xt^2(x - at^2) - a(x - at^2) = 0$$

$$\Rightarrow \quad (x - at^2)(xt^2 - a) = 0 \Rightarrow x = at^2, a/t^2$$

Putting $x = a/t^2$ in Eq. (i), we get

$$(1 + t^2)y = 2t\left(\frac{a}{t^2} + a\right)$$

$$\therefore \quad y = 2a/t$$

Thus, point Q is $(a/t^2, 2a/t)$.

Clearly, points Q and P' are extremities of focal chord. So, $P'Q$ passes through the focus.

36. Harmonic mean of focal segments $= \dfrac{2}{\dfrac{1}{SP} + \dfrac{1}{SQ}}$

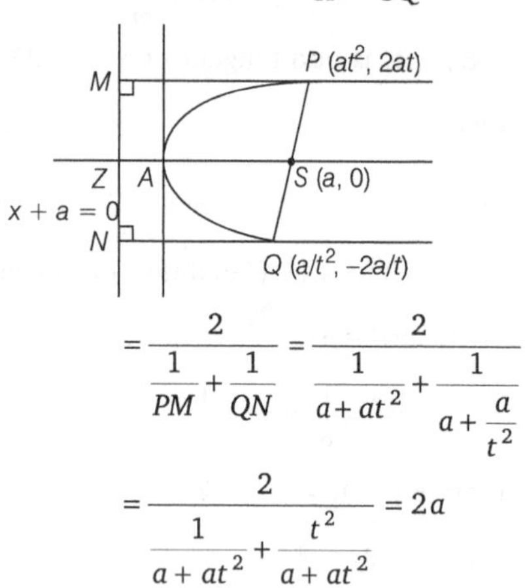

$$= \frac{2}{\dfrac{1}{PM} + \dfrac{1}{QN}} = \frac{2}{\dfrac{1}{a + at^2} + \dfrac{1}{a + \dfrac{a}{t^2}}}$$

$$= \frac{2}{\dfrac{1}{a + at^2} + \dfrac{t^2}{a + at^2}} = 2a$$

37. $\because \quad BC = 16p$

$\therefore \quad BM = CM = 8p$

On solving $y = 8p$ for B and $y^2 = 8px$, we get

$$x = 8p$$

and solving $y = -8p$ for C

and $y^2 = 8px$, we get

$$x = 8p$$

$\therefore \quad B \equiv (8p, 8p)$ and $C \equiv (8p, -8p)$

$\Rightarrow \quad AM = 8p$

$$\tan\theta = \frac{8p}{8p} = 1 \Rightarrow \theta = 45°$$

$\Rightarrow \angle BAC = 2\theta = 90° = \pi / 2$

38. Equation of given parabola is $y^2 = 16x$, its focus is (4, 0).

Since, slope of line passing through (x_1, y_1) and (x_2, y_2) is given by $m = \tan\theta = \dfrac{y_2 - y_1}{x_2 - x_1}$.

$\therefore$ Slope of focal chord having one end point is (1, 4) is

$$m = \tan\alpha = \frac{4-0}{1-4} = -\frac{4}{3}$$

[where, 'α' is the inclination of focal chord with X-axis]

Since, the length of focal chord $= 4a\operatorname{cosec}^2\alpha$

$\therefore$ The required length of the focal chord

$$= 16[1 + \cot^2\alpha] \quad [\because a = 4 \text{ and } \operatorname{cosec}^2\alpha = 1 + \cot^2\alpha]$$

$$= 16\left[1 + \frac{9}{16}\right] = 25 \text{ units} \qquad \left[\because \cot\alpha = \frac{1}{\tan\alpha} = -\frac{3}{4}\right]$$

39. Given equation of the parabola is

$$(y + 2)^2 = \frac{4}{3}(x - 1)$$

or $Y^2 = 4aX$ where $Y = y + 2$, $a = \dfrac{1}{3}$, $X = x - 1$

Equation of the chord of this parabola Joining $P(at_1^2, 2at_1)$ and $Q(at_2^2, 2at_2)$ is $Y(t_1 + t_2) = 2X + 2at_1t_2$

PQ subtends a right angle at the vertex

$$(X, Y) = (0, 0), \text{ so } \frac{2at_1}{at_1^2} \times \frac{2at_2}{at_2^2} = -1 \Rightarrow t_1t_2 = -4$$

So, the equation of the chord PQ is

$$Y(t_1 + t_2) = 2(X - 4a)$$

which passes through $X = 4a$, $Y = 0$

$\Rightarrow x - 1 = 4 \times 1/3, \ y + 2 = 0$

$\Rightarrow x = 1 + 4/3 = 7/3, \ y = -2$

40. $169[(x-1)^2 + (y-3)^2] = 13^2\left(\dfrac{5x - 12y + 17}{13}\right)^2$

$$\Rightarrow \quad (x-1)^2 + (y-3)^2 = \left(\frac{5x - 12y + 17}{13}\right)^2$$

$$SP^2 = PM^2 \Rightarrow SP = PM$$

$\therefore$Focus (1, 3) directrix $5x - 12y + 17 = 0$

The distance of the focus from the directrix

$$= \left|\frac{5 - 36 + 17}{\sqrt{25 + 144}}\right| = \frac{14}{13}$$

$\therefore$ Latusrectum $= 2 \times \dfrac{14}{13} = \dfrac{28}{13}$

41. We have $3y^2 + 4y - 6x + 8 = 0$

$$\Rightarrow \quad y^2 + \frac{4}{3}y - 2x + \frac{8}{3} = 0$$

$$\Rightarrow \quad \left(y + \frac{2}{3}\right)^2 - \frac{4}{9} - 2x + \frac{8}{3} = 0$$

$$\Rightarrow \quad \left(y + \frac{2}{3}\right)^2 = 2x - \frac{20}{9}$$

$$\Rightarrow \quad \left(y + \frac{2}{3}\right)^2 = 2\left(x - \frac{10}{9}\right)$$

Let $\quad y + \dfrac{2}{3} = Y$

and $\quad x - \dfrac{10}{9} = X \Rightarrow Y^2 = 2X$

Comparing with $Y^2 = 4aX$

$\therefore \quad a = \dfrac{1}{2}$

$\because$ All three normals to the parabola are real and meet on the axis of parabola, then

$$X > 2a \text{ and } Y = 0$$

i.e. $\quad x - \dfrac{10}{9} > 1$ and $y + \dfrac{2}{3} = 0$

$\therefore \quad x > \dfrac{19}{9}$ and $y = -\dfrac{2}{3}$

or $\quad \left(a, -\dfrac{2}{3}\right); \ a > \dfrac{19}{9}$

42. Let $A \equiv (am_1^2, -2am_1)$ and $B \equiv (am_2^2, -2am_2)$

Now, A and B lie on $lx + my = 1$

$$\Rightarrow \quad l(am_1^2) + m(-2am_1) = 1 \quad \ldots\text{(i)}$$

and $$l(am_2^2) + m(-2am_2) = 1 \quad \ldots\text{(ii)}$$

Subtracting Eq. (ii) from Eq. (i), then

$$la(m_1^2 - m_2^2) - 2am(m_1 - m_2) = 0$$

$$\Rightarrow \quad a(m_1 - m_2) \neq 0$$

$$\therefore \quad l(m_1 + m_2) - 2m = 0$$

$$\Rightarrow \quad m_1 + m_2 = \frac{2m}{l} \quad \ldots\text{(iii)}$$

Let $D \equiv (am_3^2, -2am_3)$ and $C \equiv (h, k)$

$\therefore$ Equation of normal in terms of slope

$$y = Mx - 2aM - aM^3,$$

Then, $$aM^3 - (h - 2a)M + k = 0$$

$$\therefore \; m_1 + m_2 + m_3 = 0 \Rightarrow \frac{2m}{l} + m_3 = 0$$

$$\therefore \quad m_3 = -\frac{2m}{l}$$

$$\therefore \quad D \equiv \left(a\left(\frac{-2m}{l}\right)^2, -2a\left(\frac{-2m}{l}\right)\right)$$

$$\Rightarrow \quad D \equiv \left(\frac{4am^2}{l^2}, \frac{4am}{l}\right)$$

43. Let $P(x_0, x_0^2)$ be any point on the parabola $y = x^2$

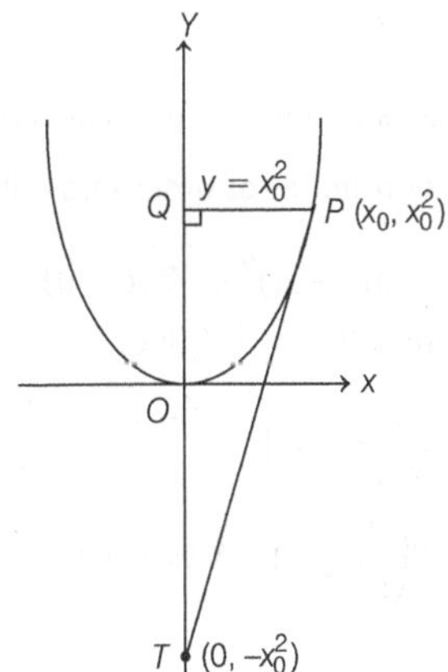

Equation of the tangent at $P(x_0, x_0^2)$ is

$$xx_0 = \frac{1}{2}(y + x_0^2)$$

$$\Rightarrow \quad 2xx_0 - y - x_0^2 = 0$$

Tangent meets the Y-axis at $T(0, -x_0^2)$.

Hence, the area of the $\Delta PTQ = \frac{1}{2} \times PQ \times QT$

$$= \frac{1}{2} \times x_0 \times 2x_0^2 = x_0^3$$

which increases in the interval [1, 2] and hence is greatest when $x_0 = 2$.

44. $\because P \equiv (at^2, 2at)$ and let $R = (at_1^2, 2at_1)$

$\because$ Normals at P and R meet parabola $y^2 = 4ax$ at Q, then

$$tt_1 = 2 \text{ or } t_1 = \frac{2}{t}$$

$$\therefore \quad R \equiv \left(a\left(\frac{2}{t}\right)^2, 2a\left(\frac{2}{t}\right)\right)$$

$$\therefore \quad R \equiv \left(\frac{4a}{t^2}, \frac{4a}{t}\right)$$

45. $C: y = x^2 - 3$ and $D: y = kx^2$

Solving C and D, then

$$kx^2 = x^2 - 3$$

$$\Rightarrow \quad x^2 = \frac{3}{(1-k)}, \text{ then } y = \frac{3k}{(1-k)}$$

$$\therefore \quad A \equiv \left(\sqrt{\left(\frac{3}{1-k}\right)}, \frac{3k}{(1-k)}\right)$$

(given x-value of A is positive)

and $$a = \sqrt{\left(\frac{3}{1-k}\right)} \quad \ldots\text{(i)}$$

Then, $$A \equiv (a, ka^2) \equiv (a, a^2 - 3)$$

tangent 'l' at A to the curve D is

$$\frac{y + a^2 - 3}{2} = kx(a)$$

$$\Rightarrow \quad y + a^2 - 3 = 2ax\left(1 - \frac{3}{a^2}\right) \quad \ldots\text{(ii)}$$

$$\therefore \quad B \equiv (1, -2) \; (a \neq 1)$$

From Eq. (ii), $-2 + a^2 - 3 = 2a\left(1 - \frac{3}{a^2}\right) = 2\frac{(a^2 - 3)}{a}$

$$\Rightarrow \quad a^3 - 5a = 2a^2 - 6$$

$$\Rightarrow \quad a^3 - 2a^2 - 5a + 6 = 0$$

$$\Rightarrow \quad (a-1)(a+2)(a-3) = 0$$

$$\therefore \quad a = 3 \quad [\because a \neq 1, a \neq -2]$$

46. $y = m(x + 1) + (1/m)$

or $$y = mx + \left(m + \frac{1}{m}\right) \quad \ldots\text{(i)}$$

is a tangent to the first parabola

and $$y = m'(x + 2) + \frac{2}{m'} = m'x + 2\left(m' + \frac{1}{m'}\right) \quad \ldots\text{(ii)}$$

is a tangent to the second parabola given

$$m \cdot m' = -1$$

or $$m' = -\frac{1}{m}$$

Then, from Eq. (ii), $y = -\frac{x}{m} + 2\left(-\frac{1}{m} - m\right)$

$$\Rightarrow \quad y = -\frac{x}{m} - 2\left(m + \frac{1}{m}\right) \quad \ldots\text{(iii)}$$

Subtracting Eq. (iii) from Eq. (i), then

$$x\left(m + \frac{1}{m}\right) + 3\left(m + \frac{1}{m}\right) = 0$$

or $$x + 3 = 0$$

47. Equation of tangent to the parabola $y^2 = 4x$ having slope 'm' is

$$y = mx + \frac{1}{m} \quad \ldots(i)$$

$\because$ Line (i) is also tangent to the parabola $x^2 = 4y$

So, the discriminant of $x^2 - 4mx - \frac{4}{m} = 0$ must be zero,

so $16m^2 + \frac{16}{m} = 0 \Rightarrow m = -1$

Therefore, the equation of common tangent to given parabolas is

$$x + y + 1 = 0 \quad \ldots(ii)$$

It is given that line (ii) is tangent to the circle $x^2 + y^2 = c^2$

$$\therefore \quad c = \frac{1}{\sqrt{2}}$$

48. Let any point on the parabola $x^2 = 4y$ is $(2t, t^2)$

$\therefore$ tangent at $(2t, t^2)$ is

$$x(2t) = 2(y + t^2) \text{ or } tx = y + t^2$$

$\therefore$ Slope of tangent is t ...(i)

$\therefore$ Slope of normal is $-\frac{1}{t}$

Equation of normal at $(2t, t^2)$ is

$$y - t^2 = -\frac{1}{t}(x - 2t)$$

$$\Rightarrow \quad ty - t^3 = -x + 2t$$

or $$t^3 - (y - 2)t - x = 0 \quad \ldots(ii)$$

$\therefore \quad t_1 + t_2 + t_3 = 0, t_1t_2t_3 = x$

Solving $y = 2$ and Eq. (ii), then we get

$$x = t^3$$

Given t_1^3, t_2^3, t_3^3 are in AP

$$\therefore \quad 2t_2^3 = t_1^3 + t_3^3 = (t_1 + t_3)^3 - 3t_1t_3(t_1 + t_3)$$
$$= (-t_2)^3 - 3t_1t_3(-t_2)$$
$$\Rightarrow \quad 3t_2^3 = 3t_1t_2t_3 \Rightarrow t_2^2 = t_1t_3$$

$\therefore t_1, t_2, t_3$ are in GP.

i.e. slopes of the tangents at the three conormal points are in GP.

49. Normal to parabola $y^2 = 4ax$ is given by

$$y = mx - 2am - am^3$$

$\therefore$ Normal to parabola

$y^2 = 4b(x - c)$ is

$$y = m(x - c) - 2bm - bm^3$$

[replacing a by b and x by $x - c$]

$$= mx - (2b + c)m - bm^3 \quad \ldots(i)$$

and normal to parabola $y^2 = 8ax$ is

$$y = mx - 4am - 2am^3 \quad \ldots(ii)$$

[replacing a by $2a$]

For common normal, we should have

$mx - 4am - 2am^3 = mx - (2b + c)m - bm^3$

[using Eqs. (i) and (ii)]

$$4am + 2am^3 = (2b + c)m + bm^3$$
$$\Rightarrow \quad (2a - b)m^3 + (4a - 2b - c)m = 0$$
$$\Rightarrow \quad m((2a - b)m^2 + (4a - 2b - c)) = 0 \Rightarrow m = 0$$

or $$m^2 = \frac{2b + c - 4a}{2a - b} = \frac{c}{2a - b} - 2$$

As, $m^2 > 0$, therefore $\frac{c}{2a - b} > 2$

Note that if $m = 0$, then all options satisfy ($\because$ $y = 0$ is a common normal) and if common normal is other than the axis, then only option (c) satisfies

[$\because$ for option (c), $2a - b = \frac{3}{2-1} = 3 > 2$]

50. $\therefore SP = PM = a + at^2, SQ = QN = a + \frac{a}{t^2}$

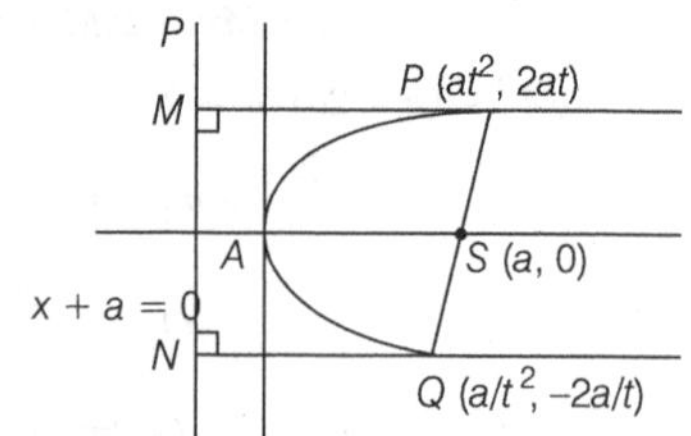

$$\frac{1}{SP} = \frac{1}{a(1 + t^2)}$$

$$\therefore \quad \frac{1}{SQ} = \frac{t^2}{a(1 + t^2)}$$

$$\therefore \quad \frac{1}{SP} + \frac{1}{SQ} = \frac{1}{a}$$

51. If coordinates of one end of focal chord are $P(at^2, 2at)$, then the coordinates of other end will be $Q\left(\frac{a}{t^2}, \frac{-2a}{t}\right)$.

$$\therefore \quad SP = \sqrt{(at^2 - a)^2 + (2at - 0)^2}$$
$$= a(t^2 + 1) = b \text{ (given)} \quad \ldots(i)$$

and $$SQ = \sqrt{\left(\frac{a}{t^2} - a\right)^2 + \left(\frac{-2a}{t} - 0\right)^2}$$
$$= a\left(\frac{1}{t^2} + 1\right) = k \text{ (given)} \quad \ldots(ii)$$

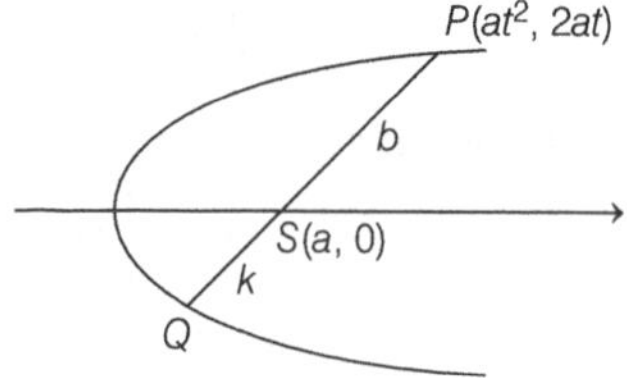

Dividing Eq. (i) by Eq. (ii), we get $t^2 = \frac{b}{k}$

Putting in (i), $\frac{b}{k} + 1 = \frac{b}{a}$

$$\Rightarrow \quad k = \frac{ab}{b - a}$$

52. Equation of L is

$$y - 0 = m(x - 2)$$

$\therefore \quad \frac{y}{m} + 2 = x \quad$ [$\because$ focus of $y^2 = 4(x-1)$ is $(2, 0)$]

$\therefore \quad y^2 = 4(x-1)$

$\Rightarrow \quad y^2 = 4 \cdot \left(\frac{y}{m} + 1\right)$

$\Rightarrow y^2 - \frac{4y}{m} - 4 = 0$

For real and distinct $B^2 - 4AC > 0$

$$\left(-\frac{4}{m}\right)^2 - 4 \cdot 1 \cdot (-4) > 0$$

$\Rightarrow \quad \frac{16}{m^2} + 16 > 0$

$\Rightarrow \quad \frac{1 + m^2}{m^2} > 0$

$\therefore \quad m \in R \sim \{0\}$

53. $\because x = ay^2 + by + c$

$\Rightarrow \quad ay^2 + by = x - c$

$\Rightarrow \quad a\left\{\left(y + \frac{b}{2}\right)^2 - \frac{b^2}{4}\right\} = x - c$

$\Rightarrow \quad a\left(y + \frac{b}{2}\right)^2 = \left(x + \frac{ab^2}{4} - c\right)$

or $\quad \left(y + \frac{b}{2}\right)^2 = \frac{1}{a}\left(x + \frac{ab^2}{4} - c\right)$

$\therefore$ Length of latusrectum $= \frac{1}{a}$

54. Let $P \equiv (2t^2, 4t)$ be any point on the parabola $y^2 = 8x$.

$\therefore$ Normal at 'P' on $y^2 = 8x$

$$y + tx = 4t + 2t^3$$

It must pass through centre of the circle $(0, -6)$ in order that it gives minimum distance between the two curves

$\therefore \quad -6 + 0 = 4t + 2t^3$

$\Rightarrow \quad t^3 + 2t + 3 = 0$

$\therefore \quad t = -1, t = \frac{1 \pm i\sqrt{11}}{2}$ (impossible)

$\therefore$ Required point is $(2, -4)$.

55. Vertex is $(a, 0)$ and focus is $(a_1, 0)$

$\therefore$ Distance between vertex and focus is $(a_1 - a)$

$\therefore$ Length of latus-rectum $= 4(a_1 - a)$

Equation of parabola with vertex $(0, 0)$, latus rectum $4A$ and axis parallel to X-axis is

$$y^2 = 4Ax$$

Now, equation of parabola with vertex $(a, 0)$ and latus-rectum $4(a_1 - a)$ and axis along X-axis is

$$y^2 = 4(a_1 - a)(x - a)$$

56. Equation of chord joining $(2, 2)$ and $(8, -4)$ is

$$y - 2 = \frac{-4-2}{8-2}(x - 2) \Rightarrow y - 2 = -x + 2$$

$\therefore \quad x + y = 4 \quad \ldots$(i)

Let $R(\alpha^2, \alpha - 2)$.

$\because R$ is interior to the parabola, O and R are on the same side of PQ

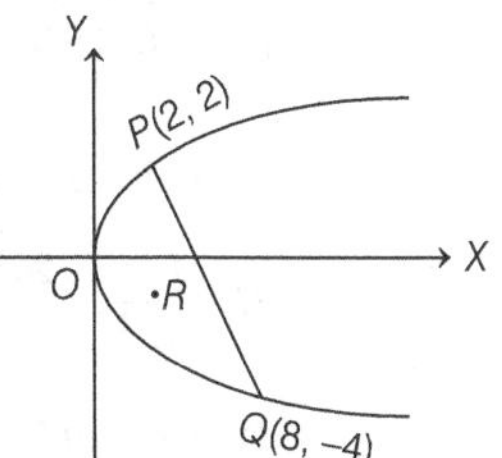

$\therefore \quad (\alpha - 2)^2 - 2\alpha^2 < 0$

$\Rightarrow \quad -\alpha^2 - 4\alpha + 4 < 0$

$\Rightarrow \quad \alpha^2 + 4\alpha - 4 > 0$

$\Rightarrow \quad (\alpha + 2)^2 > 8$

or $\quad \alpha + 2 < -2\sqrt{2}$

and $\quad \alpha + 2 > 2\sqrt{2}$

$\therefore \quad \alpha \in (-\infty, -2 - 2\sqrt{2}) \cup (2\sqrt{2} - 2, \infty) \quad \ldots$(ii)

Also from Eq. (i),

$$\frac{\alpha - 2 + \alpha^2 - 4}{0 + 0 - 4} > 0$$

$\Rightarrow \quad \alpha^2 + \alpha - 6 < 0$

$\Rightarrow \quad (\alpha + 3)(\alpha - 2) < 0$

$\therefore \quad -3 < \alpha < 2 \quad \ldots$(iii)

From Eqs. (ii) and (iii), we get

$\quad \alpha \in (2\sqrt{2} - 2, 2)$

or $\quad \alpha \in (-2 + 2\sqrt{2}, 2)$

57. Let parabola is $y^2 = 4ax \quad \ldots$(i)

and circle is $x^2 + y^2 + 2gx + 2fy + c = 0 \quad \ldots$(ii)

From Eqs. (i) and (ii),

$$\left(\frac{y^2}{4a}\right)^2 + y^2 + 2g\left(\frac{y^2}{4a}\right) + 2fy + c = 0$$

$\Rightarrow \quad y^4 + 16a^2y^2 + 8agy^2 + 32a^2fy + 16a^2c = 0$

$\Rightarrow \quad y^4 + (16a^2 + 8ag)y^2 + 32a^2fy + 16a^2c = 0$

$\therefore \quad y_1 + y_2 + y_3 + y_4 = 0$

58. $\because$ Normal at $(at^2, 2at)$ cuts the parabola again at $(aT^2, 2aT)$, then,

$$T = -t - \frac{2}{t}$$

or $\quad tT = -t^2 - 2$

$\Rightarrow \quad t^2 + tT + 2 = 0$

$\because t$ is real

$\therefore \quad T^2 - 4 \cdot 1 \cdot 2 \geq 0$ or $T^2 \geq 8$

59. Let $Q = (at_1^2, 2at_1)$

$\because$ PQ is focal chord

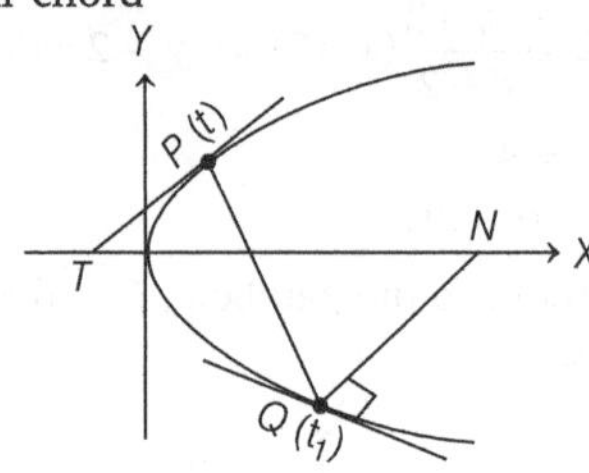

$\therefore \quad tt_1 = -1 \text{ or } t_1 = -\frac{1}{t}$

$\therefore \quad Q \equiv \left(\frac{a}{t^2}, \frac{-2a}{t}\right)$

Equation of PT is

$$ty = x + at^2$$

$\therefore \quad \text{Slope} = \frac{1}{t} = m_1 \text{ (say)}$

Equation of normal at Q is

$$y + \frac{2a}{t} = \frac{1}{t}\left(x - \frac{a}{t^2}\right)$$

$\therefore \quad \text{Slope} = \frac{1}{t} = m_2 \text{ (say)}$

$\because \quad m_1 = m_2$

$\therefore$ PT and QN are parallel

$\Rightarrow \quad \alpha = 0°$

$\therefore$ Distance between them

$$d = \frac{at^2 + 2a + \frac{a}{t^2}}{\sqrt{(1+t^2)}} = a\frac{(t^2+1)^2}{t^2\sqrt{(1+t^2)}}$$

$$= \frac{a(t^2+1)^{3/2}}{t^2}$$

60. $\because x^2 + y^2 + 2xy - 6x - 2y + 3 = 0$

$\Rightarrow \quad (x+y)^2 = (6x + 2y - 3)$

or $(x + y + \lambda)^2 = \lambda^2 + 2\lambda(x+y) + 6x + 2y - 3$ ...(i)

Where λ is such that

(slope of $x + y + \lambda = 0$) $\times$ (slope of $\lambda^2 + 2\lambda(x+y) + 6x + 2y - 3 = 0$) $= -1$

$\Rightarrow \quad (-1) \times -\frac{(2\lambda+6)}{(2\lambda+2)} = -1$

$\Rightarrow \quad 2\lambda + 6 = -2\lambda - 2$

$\therefore \quad 4\lambda = -8$

$\Rightarrow \quad \lambda = -2$

From Eq. (i),

$$(x + y - 2)^2 = 4 - 4(x+y) + 6x + 2y - 3$$
$$= (2x - 2y + 1)$$

or $\quad 2\left(\frac{x+y-2}{\sqrt{2}}\right)^2 = \sqrt{8}\left(\frac{2x-2y+1}{\sqrt{4+4}}\right)$

$\Rightarrow \quad \left(\frac{x+y-2}{\sqrt{2}}\right)^2 = \sqrt{2}\left(\frac{2x-2y+1}{\sqrt{8}}\right)$

Let $Y = \frac{x+y-2}{\sqrt{2}}$ and $X = \frac{2x-2y+1}{\sqrt{8}}$

and $\quad 4a = \sqrt{2}$

Then, focus $X = a$ and $Y = 0$

$$\frac{2x-2y+1}{\sqrt{8}} = \frac{\sqrt{2}}{4} \text{ and } \frac{x+y-2}{\sqrt{2}} = 0$$

$\Rightarrow \quad x = y \text{ and } x + y = 2$

Then, $\quad x = 1, y = 1$

$\therefore$ Focus is (1, 1).

61. Now, $t^2 + 1 = x - t$ ...(i)

and $\quad t^2 + 1 = y + t$ ...(ii)

$\Rightarrow \quad x - t = y + t$

$\therefore \quad t = \left(\frac{x-y}{2}\right)$

From Eq. (i),

$$\left(\frac{x-y}{2}\right)^2 + 1 = x - \left(\frac{x-y}{2}\right)$$
$$= \left(\frac{x+y}{2}\right)$$

$\Rightarrow \quad (x-y)^2 + 4 = 2(x+y)$

$\Rightarrow \quad (x-y)^2 = 2(x + y - 2)$

$\therefore$ Length of latusrectum = 2

Objective Questions Type II [One or more than one correct answer(s)]

1. Focus of the parabola $y^2 = 2px$ is $\left(\frac{p}{2}, 0\right)$

$\therefore$ Centre of circle is $\left(\frac{p}{2}, 0\right)$

Radius of the circle = Distance between focus and directrix

= semi-latusrectum

$= p$

$\therefore$ Equation of circle is

$$\left(x - \frac{p}{2}\right)^2 + (y-0)^2 = p^2 \quad \text{...(i)}$$

Solving Eq. (i) and $y^2 = 2px$, then

$$\left(x - \frac{p}{2}\right)^2 + 2px = p^2$$

$\Rightarrow \quad x^2 + px = \frac{3p^2}{4}$

$\Rightarrow \quad 4x^2 + 4px - 3p^2 = 0$

$\therefore \quad x = p/2 \text{ and } x = -3p/2$

$\because \quad y^2 = 2px = 2p\left(\frac{p}{2}\right)$

and $y^2 = 2p\left(\frac{-3p}{2}\right) = -3p^2$, y is imaginary (Impossible)

$\therefore \quad y^2 = p^2 \Rightarrow y = \pm p$

$\therefore$ Point of intersection are $\left(\frac{p}{2}, p\right)$ and $\left(\frac{p}{2}, -p\right)$

2. $y^2 + 2ax + 2by + c = 0$

$$(y+b)^2 = -2ax + b^2 - c$$

$$= -2a\left(x - \frac{b^2}{2a} + \frac{c}{2a}\right)$$

It is a parabola.

$\therefore$ Latusrectum $= 2a$

3. Equation of tangent in terms of slope of $y^2 = 8x$ is

$$y = mx + \frac{2}{m} \quad \ldots(\text{i})$$

$\because$ Angle between Eq. (i) and $y = 3x + 5$ is 45°, then

$\Rightarrow \quad \left|\frac{m-3}{1+3m}\right| = \tan 45° = 1$

$\Rightarrow \quad \pm(m-3) = 1 + 3m$

Taking '+' sign, then $m - 3 = 1 + 3m$

$\therefore \quad m = -2$

and taking '–' sign, then

$$-m + 3 = 1 + 3m$$

$\therefore \quad m = \frac{1}{2}$

Now, from Eq. (i) Equation of tangents are

$$y = -2x - 1 \text{ and } y = \frac{x}{2} + 4$$

or $\quad 2x + y + 1 = 0$ and $x - 2y + 8 = 0$

4. Making $y^2 = 4ax$ homogeneous with the help of $y = mx - 2am - am^3$ or $\frac{mx - y}{2am + am^3} = 1.$

Then, $\quad y^2 = 4ax\left(\frac{mx - y}{2am + am^3}\right)$

$\Rightarrow \quad y^2 = \frac{4x(mx - y)}{2m + m^3}$

$\Rightarrow \quad (2m + m^3)y^2 - 4mx^2 + 4xy = 0 \quad \ldots(\text{i})$

$\because$ Angle between the lines represented by Eq. (i) is $\pi/2$

$\therefore$ Coefficient of x^2 + coefficient of $y^2 = 0$

$\Rightarrow \quad 2m + m^3 - 4m = 0$

$\Rightarrow \quad m^3 - 2m = 0$

$$m^2 = 2, m \neq 0$$

$\therefore \quad m = \pm\sqrt{2}$

5. Any point on the parabola is $P(at^2, 2at)$.

Therefore, the mid-point of $S(a, 0)$ and $P(at^2, 2at)$ is

$$R\left(\frac{a + at^2}{2}, at\right) \equiv (h, k)$$

$\therefore \quad h = \frac{a + at^2}{2}, k = at$

Eliminate t, i.e. $2x = a\left(1 + \frac{y^2}{a^2}\right) = a + \frac{y^2}{a}$

i.e. $\quad 2ax = a^2 + y^2$

i.e. $\quad y^2 = 2a\left(x - \frac{a}{2}\right)$

It is a parabola with vertex at $(a/2, 0)$ and latus rectum $2a$.

The directrix is $x - \frac{a}{2} = -\frac{a}{2}$

i.e. $\quad x = 0$

Thus focus is $x - \frac{a}{2} = \frac{a}{2}$

i.e. $\quad x = a$

i.e. $\quad (a, 0)$

6. Slope of $VP = \frac{2t - 0}{t^2 - 0} = \frac{2}{t}$

$\therefore$ Slope of $VQ = -\frac{t}{2}$

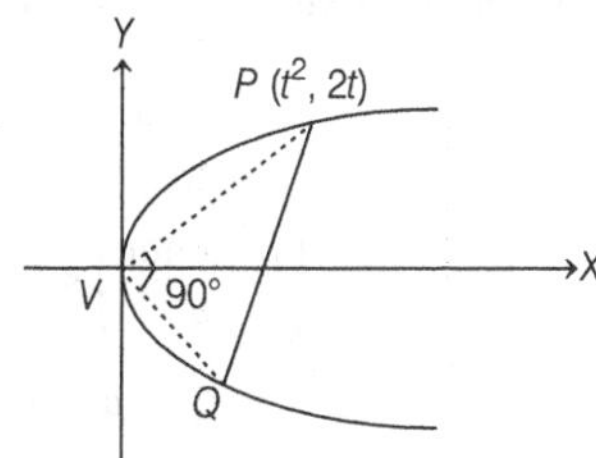

Equation of VQ is $y = -\frac{t}{2}x$

Solving it with $y^2 = 4x$

i.e. $\quad \frac{t^2x^2}{4} - 4x = 0$

$\therefore \quad x = 0, x = \frac{16}{t^2}$

Coordinates of Q is $\left(\frac{16}{t^2}, -\frac{8}{t}\right)$

$\because$ Area of $\Delta PVQ = \frac{1}{2}(PV)(VQ) = 20$

or $\quad (PV)^2(VQ)^2 = 1600$

$$(t^4 + 4t^2)\left(\frac{256}{t^4} + \frac{64}{t^2}\right) = 1600$$

$\Rightarrow \quad t^2(t^2 + 4)\cdot\frac{64}{t^4}(t^2 + 4) = 1600$

$\Rightarrow \quad (t^2 + 4)^2 = \frac{1600}{64}t^2$

or $\quad t^2 + 4 = \pm\frac{40t}{8}$

or $\quad t^2 + 4 = \pm 5t$

we get $\quad t = \pm 4, \pm 1$

Hence, coordinates of P are

$$(16, \pm 8) \text{ and } (1, \pm 2)$$

7. If $P \equiv (x, y)$, then

$$\sqrt{x^2+y^2} < |a-x|, \sqrt{x^2+y^2} < |a+x|,$$
$$\sqrt{x^2+y^2} < |a-y| \text{ and } \sqrt{x^2+y^2} < |a+y|$$

On squaring, then we get

$\therefore$ The region bounded by the curves

$$x^2+y^2=(a-x)^2,\ x^2+y^2=(a+x)^2$$
$$x^2+y^2=(a-y)^2,\ x^2+y^2=(a+y)^2$$

8. Parabola open downward

$\therefore \quad a < 0$

Also x-coordinates of vertex $-\dfrac{b}{2a} < 0$

$\Rightarrow \quad \dfrac{b}{a} > 0$

$\because \quad a < 0$

$\therefore \quad b < 0$

and y-coordinate of vertex, $-\dfrac{D}{4a} > 0$

$\Rightarrow \quad -\dfrac{(b^2-4ac)}{4a} > 0$

$\Rightarrow \quad \left(\dfrac{b^2-4ac}{a}\right) < 0$

$\because \quad a < 0$

$\therefore \quad b^2 - 4ac > 0$

and cut off Y-axis $x = 0$

$\therefore \quad y = c > 0$

$\therefore \quad c > 0$

9. Given parabola is $y^2 = 8x$ ...(i)

Here, $a = 2$.

Let $P(2t^2, 4t)$ be a point on parabola

Eq. (i), and S be the focus

Given, $SP = 4$

$\therefore \quad a(1+t^2) = 4$

$2(1+t^2) = 4$

$\Rightarrow \quad t = \pm 1$

$\therefore \quad P \equiv (2, 4)$ or $(2, -4)$

Given $f(x+y) = f(x) f(y)$ for all x and y ...(ii)

Given, $f(1) = 2$

From Eq. (ii), $f(2) = f(1+1) = f(1) \cdot f(1) = 2^2 = 4$

Similarly $f(n) = 2^n$

$\because \quad a_r = f(r)$

$\therefore \quad a_1 = f(1) = 2,$

$a_2 = f(2) = 4$

Hence, $P \equiv (a_1, a_2)$ or $(a_1, -a_2)$

10. Tangent to $y^2 = 4x$ at $(t^2, 2t)$ is

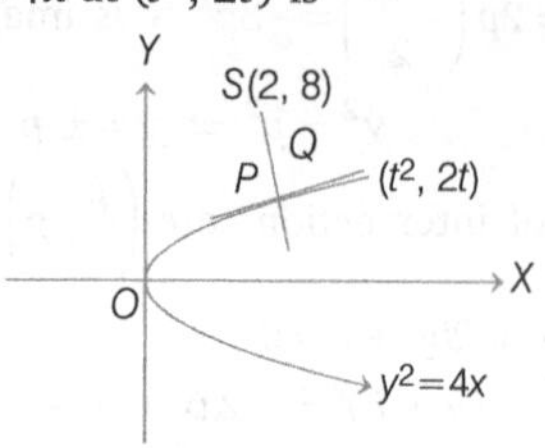

$$y(2t) = 2(x + t^2)$$

$\Rightarrow \quad yt = x + t^2$ ...(i)

Equation of normal at $P(t^2, 2t)$ is

$$y + tx = 2t + t^3$$

Since, normal at P passes through centre of circle $S(2, 8)$.

$\therefore \quad 8 + 2t = 2t + t^3$

$\Rightarrow \quad t = 2$, i.e. $P(4, 4)$

[since, shortest distance between two curves lie along their common normal and the common normal will pass through the centre of circle]

$\therefore \ SP = \sqrt{(4-2)^2 + (4-8)^2} = 2\sqrt{5}$

Also, $SQ = 2$

$\therefore \quad PQ = SP - SQ = 2\sqrt{5} - 2$

Thus, $\dfrac{SQ}{QP} = \dfrac{1}{\sqrt{5}-1} = \dfrac{\sqrt{5}+1}{4}$

Now, x-intercept of normal is $x = 2 + 2^2 = 6$

Slope of tangent $= \dfrac{1}{t} = \dfrac{1}{2}$

11. Any normal to parabola $y^2 = 4ax$ may be taken as

$$y - 2at = -t(x - at^2)$$

or $\quad y + tx - 2at - at^3 = 0$ ...(i)

If Eq. (i) passes through $(5a, 2a)$, then

$$2a + 5at - 2at - at^3 = 0$$

or $\quad t^3 - 3t - 2 = 0$

or $\quad (t+1)(t^2 - t - 2) = 0$

$\Rightarrow \quad t = -1, 2, -1$

$\therefore$ From Eq. (i) normals are

$$y - x + 3a = 0 \text{ and } y + 2x - 12a = 0$$

12. $(x-3)^2 = -4y - k + 9 = -4\left(y + \dfrac{k-9}{4}\right)$

Equation of directrix is

$$y + \frac{k-9}{4} = 1 \text{ or } y = 1 - \left(\frac{k-9}{4}\right) = \frac{13-k}{4}$$

According to question, $\dfrac{13-k}{4} = -1 \Rightarrow k = 17$

vertex is $\left(3, \dfrac{-k+9}{4}\right) = (3, -2)$

and focus is $(3, -3)$ $[\because |a| = 1]$

13. Equations of the tangents and normal at a point $P(at^2, 2at)$ to the parabola $y^2 = 4ax$ are

$$ty = x + at^2 \text{ and } y = -tx + 2at + at^3$$

These meet the axis $y = 0$ at $T(-at^2, 0)$ and $N(2a + at^2, 0)$

Let centroid of ΔPTN be (h, k),

then $h = \frac{1}{3}(at^2 - at^2 + 2a + at^2)$

and $k = \frac{1}{3}(0 + 0 + 2at) \Rightarrow h = \frac{1}{3}(2a + at^2), k = \frac{2at}{3}$

Locus of (h, k) the centroid is $y^2 = \frac{4}{3}a\left(x - \frac{2a}{3}\right)$

Vertex of this parabola is $(2a/3, 0)$.

Length of the latus rectum is $4a/3$.

Focus is $\left(\frac{2a}{3} + \frac{a}{3}, 0\right) = (a, 0)$

and directrix is $x = \frac{2a}{3} - \frac{a}{3} = \frac{a}{3}$

14. Equation of any tangent to $y^2 = 9x$ is of the form

$yt = x + \frac{9}{4}t^2$

$\because$ It passed through (4, 10)

$\therefore \quad 10t = 4 + \frac{9}{4}t^2 \Rightarrow 9t^2 - 40t + 16 = 0$

$\therefore \quad t = 4, \frac{4}{9}$

$\therefore$ Equation of tangents can be

$$x - 4y + 36 = 0 \text{ or } 9x - 4y + 4 = 0$$

15. By properties of parabola options (a, b, c) are always correct.

16. Given curve is $y = x^2 - x$ ...(i)

when $x - y = 0$ or $y = x$, from Eq. (i)

$x^2 - 2x = 0$, its $D \neq 0$

when $x + y = 0$ or $y = -x$, from Eq. (i),

$x^2 = 0$, its $D = 0$

when $x - y = 1$ or $y = x - 1$, from Eq. (i),

$x^2 - 2x + 1 = 0$, its $D = 0$

when $x + y = 1$ or $y = 1 - x$, from Eq. (i),

$x^2 - 1 = 0$, its $D \neq 0$

Hence, only $x + y = 0$ and $x - y = 1$ are the tangents to given parabola.

Linked Comprehension Type

PASSAGE 1

1. Point of intersection of $y = 2$ and $y^2 = 8x$ is $P\left(\frac{1}{2}, 2\right)$ and focus of the parabola is $S(2, 0)$

$\therefore$ Equation of the reflected ray is $y - 0 = \frac{2 - 0}{\frac{1}{2} - 2}(x - 2)$

$\Rightarrow \quad y = -\frac{4}{3}(x - 2) \Rightarrow 4x + 3y = 8$

2. $\because y^2 + 10y - 4x + 17 = 0$

$\Rightarrow (y + 5)^2 - 25 - 4x + 17 = 0$

$\Rightarrow (y + 5)^2 = 4x + 8 \Rightarrow (y + 5)^2 = 4(x + 2)$

Let $y + 5 = Y, x + 2 = X$, then $Y^2 = 4X$

Focus is $X = 1, Y = 0$ i.e. $(-1, -5)$

After reflection, the ray must pass through focus $(-1, -5)$.

3. Solving $x = 2$ and $x^2 = 4y$ i.e. (2, 1)

Focus is (0, 1).

Equation of first reflection is $y = 1$ which is perpendicular to incident ray.

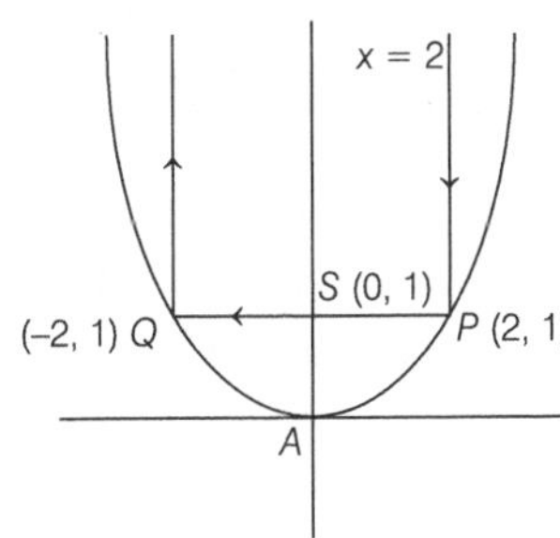

Hence, 2nd reflected ray is parallel to incident ray and pass through $(-2, 1)$ is

$$x = -2$$

or $\quad x + 2 = 0$

4. Solving $y = 1$ and $y^2 = x$

then $\quad A \equiv (1, 1)$

and solving $y = -2$ and $y^2 = x$

then $\quad B \equiv (4, -2)$

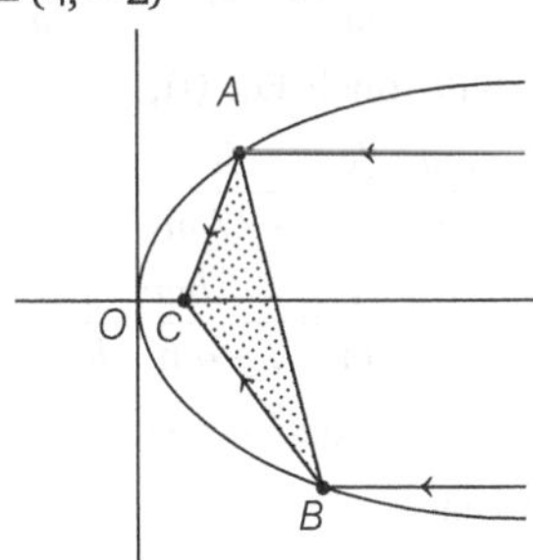

$\because$ After reflection both reflected rays pass through focus of the parabola $y^2 = x$

i.e. $\quad C \equiv \left(\frac{1}{4}, 0\right)$

$\therefore$ Required area $= \frac{1}{2}\left|\begin{vmatrix} 1 & 1 & 1 \\ 4 & -2 & 1 \\ \frac{1}{4} & 0 & 1 \end{vmatrix}\right|$

$= \frac{21}{8}$ sq units

5. Focus of the parabola

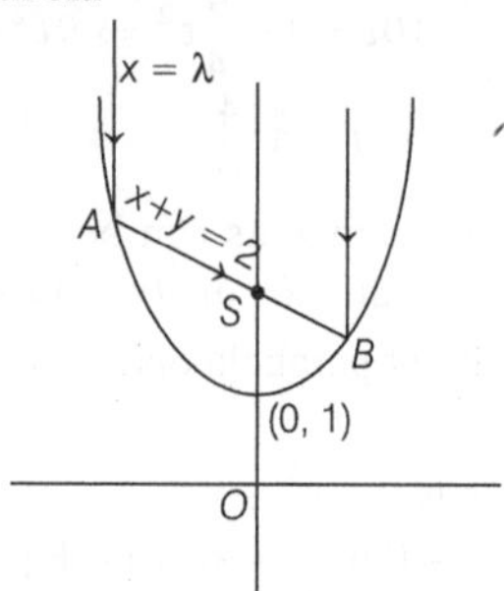

$$x^2 = 4(y-1) \text{ is } S \equiv (0, 2)$$

$\because$ First reflected ray pass through S.

Now, solving $x + y = 2$ and $x^2 = 4(y-1)$

$$\Rightarrow \quad x^2 = 4(2 - x - 1)$$

$$\Rightarrow \quad x^2 + 4x - 4 = 0$$

$$\therefore \quad x = \frac{-4 \pm \sqrt{(16+16)}}{2}$$

$$= -2 \pm 2\sqrt{2}$$

Then, equation of new reflected ray parallel to $x = \lambda$ is

$$x = -2 + 2\sqrt{2}$$

or $$x + 2 = 2\sqrt{2}$$

PASSAGE 2

1. Any parabola whose axes is parallel to the X-axis will be of the form

$$(y-a)^2 = 4b(x-c) \quad \dots(i)$$

Now, $lx + my = 1$ can be rewritten as

$$y - a = -\frac{l}{m}(x-c) + \frac{1 - am - lc}{m} \quad \dots(ii)$$

$\because$ Eq. (ii) will touch Eq. (i), if

$$\frac{1 - am - lc}{m} = \frac{b}{-l/m}$$

or $$-\frac{l}{m} = \frac{bm}{1 - am - lc}$$

or $$cl^2 - bm^2 + alm - l = 0 \quad \dots(iii)$$

But given that

$$5l^2 + 6m^2 - 4lm + 3l = 0 \quad \dots(iv)$$

Comparing Eqs. (iii) and (iv), we get

$$\frac{c}{5} = \frac{-b}{6} = \frac{a}{-4} = \frac{-1}{3}$$

or $$c = \frac{-5}{3},\ b = 2 \text{ and } a = \frac{4}{3}$$

So, the parabola is $\left(y - \frac{4}{3}\right)^2 = 8\left(x + \frac{5}{3}\right)$

whose focus is $(1/3, 4/3)$ and directrix is $3x + 11 = 0$

PASSAGE 3

The equation of the normal to $y^2 = 4ax$ is

$$y = mx - 2am - am^3 \quad \dots(i)$$

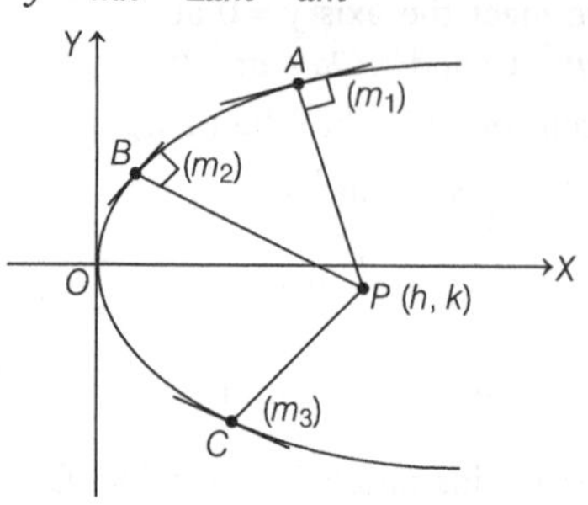

$\because$ It passes through (h, k) then $am^3 + m(2a - h) + k = 0$ ...(ii)

$\because$ Roots of Eq. (ii) be m_1, m_2, m_3

then $$m_1 + m_2 + m_3 = 0 \quad \dots(iii)$$

$$m_1m_2 + m_2m_3 + m_3m_1 = \frac{(2a-h)}{a} \quad \dots(iv)$$

and $$m_1m_2m_3 = -\frac{k}{a} \quad \dots(v)$$

1. Here, two of the three normals are given to be coincident *i.e.*, $m_1 = m_2$

Putting $m_1 = m_2$ in Eq. (iii) and (v), we get

$$2m_1 + m_3 = 0 \quad \dots(vi)$$

and $$m_1^2 m_3 = -\frac{k}{a} \quad \dots(vii)$$

From Eqs. (vi) and (vii), we get

$$m_1^3 = \frac{k}{2a}$$

since m_1 is a root of Eq. (i),

$$am_1^3 + m_1(2a-h) + k = 0$$

$$\Rightarrow \quad \left(\frac{k}{2} + k\right)^3 = -m_1^3(2a-h)^3$$

$$\Rightarrow \quad 27\frac{k^3}{8} = -\frac{k}{2a}(2a-h)^3$$

$$\Rightarrow \quad 27ak^2 = 4(h-2a)^3$$

$\therefore$ Locus of P is $27ay^2 = 4(x-2a)^3$.

2. $SA \cdot SB \cdot SC = (a + am_1^2)(a + am_2^2)(a + am_3^2)$

$$= a^3\{1 + \Sigma m_1^2 + \Sigma m_1^2 m_2^2 + m_1^2 m_2^2 m_3^2\}$$

$$= a^3\{1 + (\Sigma m_1)^2 - 2\Sigma m_1 m_2 + (\Sigma m_1 m_2)^2 - 2m_1m_2m_3\Sigma m_1 + (m_1m_2m_3)^2\}$$

$$= a^3\left\{1 + 0 - \frac{2(2a-h)}{a} + \frac{(2a-h)^2}{a^2} - 0 + \frac{k^2}{a^2}\right\}$$

$$= a\{a^2 - 2a(2a-h) + (2a-h)^2 + k^2\}$$

$$= a\{(a - (2a-h))^2 + k^2\}$$

$$= a\{(h-a)^2 + k^2\} = a(SP)^2$$

3. $\because m_1 m_2 = \lambda$

From Eq. (iii), $\lambda m_3 = -\dfrac{k}{a}$ or $m_3 = -\dfrac{k}{a\lambda}$

Here, $a = 1$

$$\therefore \quad m_3 = -\frac{k}{\lambda}$$

Put $a = 1$ in Eq. (ii), then

$$m^3 + m(2 - h) + k = 0 \quad \ldots\text{(viii)}$$

$\therefore m_3$ is a root of Eq. (viii), then

$$m_3^3 + m_3(2 - h) + k = 0$$

$$\Rightarrow \quad \left(-\frac{k}{\lambda}\right)^3 + \left(-\frac{k}{\lambda}\right)(2 - h) + k = 0$$

$$\Rightarrow \quad -\frac{k^3}{\lambda^3} - \frac{2k}{\lambda} + \frac{hk}{\lambda} + k = 0$$

$$\text{or} \quad \frac{-k^2}{\lambda^3} - \frac{2}{\lambda} + \frac{h}{\lambda} + 1 = 0$$

$\therefore$ Locus $P(h, k)$ is

$$\frac{y^2}{\lambda^3} = \frac{x}{\lambda} - \frac{2}{\lambda} + 1$$

For $\lambda = 2$, $\quad y^2 = 4x$

4. The equation of the line AB is

$$y - (-2am_1) = \frac{(-2am_2) - (-2am_1)}{am_2^2 - am_1^2}(x - am_1^2)$$

$$\Rightarrow \quad y + 2am_1 = -\frac{2}{m_1 + m_2}(x - am_1^2)$$

$$\Rightarrow y(m_1 + m_2) + 2am_1(m_1 + m_2) = -2x + 2am_1^2$$

$$\Rightarrow \quad y(m_1 + m_2) + 2am_1 m_2 = -2x$$

$$\Rightarrow \quad y(-m_3) + 2a\left(-\frac{\mu}{am_3}\right) = -2x$$

[from Eqs. (iii) and (v) replacing k by μ]

$$\Rightarrow \quad ym_3^2 - 2m_3 x + 2\mu = 0$$

which is a quadratic in m_3.

Since, AB will touch it then

$B^2 - 4AC = 0 \Rightarrow (-2x)^2 - 4y \cdot 2\mu = 0$ or $x^2 = 2\mu y$

5. Put $a = -1$ and $k = 2$ in Eq. (ii), then

$$-m^3 + m(-2 - h) + 2 = 0 \Rightarrow m^3 + (h + 2)m - 2 = 0$$

Let $f(m) = m^3 + (h + 2)m - 2$

$\therefore \quad f'(m) = 3m^2 + (h + 2)$

If $f(m) = 0$ has three distinct real roots, then necessary conditions is that $f'(m) = 0$ should have two distinct real roots i.e. $h < -2$.

Numerical Value Type

1. The equation of parabola parallel to Y-axis is

$$y = Ax^2 + Bx + C \quad \ldots\text{(i)}$$

the point (0, 4), (1, 9) and (4, 5) lie on Eq. (i), then

$$4 = 0 + 0 + C \Rightarrow C = 4 \quad \ldots\text{(ii)}$$

$$9 = A + B + C$$

$$\text{or} \quad 9 = A + B + 4 \quad [\because C = 4]$$

$$\therefore \quad A + B = 5 \quad \ldots\text{(iii)}$$

$$\text{and} \quad 5 = 16A + 4B + C \quad [\because C = 4]$$

$$\text{or} \quad 5 = 16A + 4B + 4$$

$$\therefore \ 16A + 4B = 1 \ \text{ or } \ 4A + B = \frac{1}{4} \quad \ldots\text{(iv)}$$

Solving Eqs. (iii) and (iv), we get

$$A = -\frac{19}{12}, B = \frac{79}{12} \quad \ldots\text{(v)}$$

Substituting the values of A, B and C from Eqs. (ii) and (v) in Eq. (i), then equation of parabola is

$$y = -\frac{19}{12}x^2 + \frac{79}{12}x + 4$$

Hence, length of latusrectum $= \dfrac{1}{\left|-\dfrac{19}{12}\right|} = \dfrac{12}{19} = \lambda$ (given)

$$\therefore \quad 361\lambda = 12 \times 19 = 228$$

2. Comparing the given parabola (*i.e.*, $y^2 = 8x$) with $y^2 = 4ax$

$$\therefore \quad 4a = 8$$

$$\therefore \quad a = 2$$

Since, normal at (x_1, y_1) to the parabola $y^2 = 4ax$ is

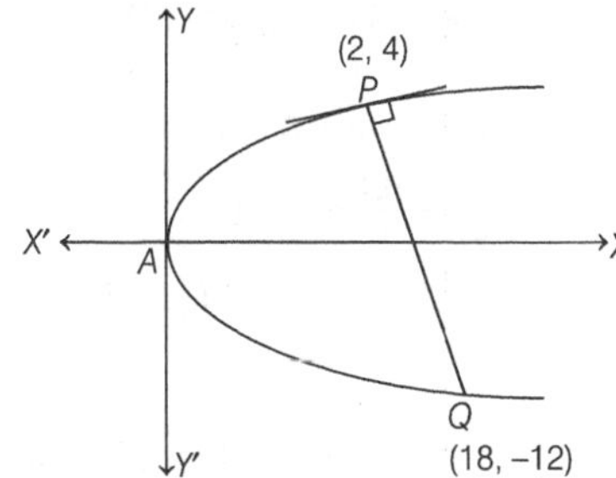

$$y - y_1 = -\frac{y_1}{2a}(x - x_1)$$

Here, $\quad x_1 = 2$ and $y_1 = 4$

$\therefore$ Equation of normal is $y - 4 = -\dfrac{4}{4}(x - 2)$

$$\Rightarrow y - 4 = -x + 2 \Rightarrow x + y - 6 = 0 \quad \ldots\text{(i)}$$

Solving Eq. (i) and $y^2 = 8x$

$$\text{then} \quad y^2 = 8(6 - y)$$

$$\Rightarrow \quad y^2 + 8y - 48 = 0$$

$$\Rightarrow \quad (y + 12)(y - 4) = 0$$

$$\therefore \quad y = -12 \text{ and } y = 4$$

$$\text{then} \quad x = 18 \quad \text{and } x = 2$$

Hence, point of intersection of normal and parabola are (18, –12) and (2, 4).

Therefore, normal meets the parabola at (18, –12) and length of normal chord is distance between their points

$$= PQ = \sqrt{(18-2)^2 + (-12-4)^2} = 16\sqrt{2}$$

$$= \lambda \text{ (given)},$$

$\therefore \quad \lambda^2 = 512$

3. Let $P(t_1^2, 2t_1)$ and $Q(t_2^2, 2t_2)$ be extremities of the chord with slope 2.

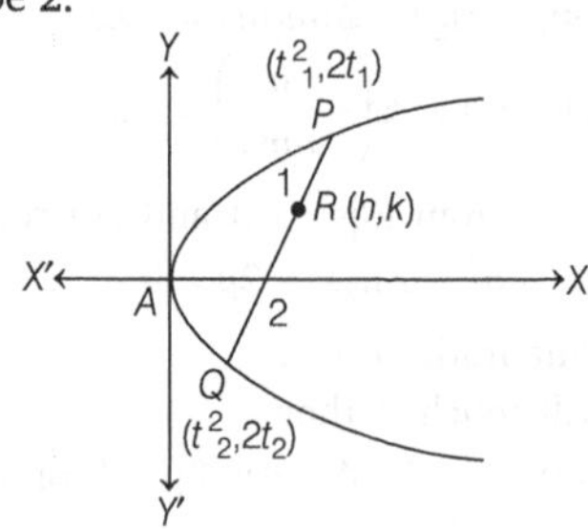

$$\therefore \quad \frac{2t_1 - 2t_2}{t_1^2 - t_2^2} = 2 \Rightarrow t_1 + t_2 = 1 \quad \text{...(i)}$$

Let $R(h, k)$ be coordinates of the point which divides PQ in the ratio 1 : 2, then

$$h = \frac{2t_1^2 + t_2^2}{3} \text{ and } k = \frac{4t_1 + 2t_2}{3}$$

$$\Rightarrow \quad 3h = 2t_1^2 + (1 - t_1)^2 \text{ and } 3k = 4t_1 + 2(1 - t_1)$$

[from Eq. (i)]

or $\quad 3h = 3t_1^2 - 2t_1 + 1$ and $3k = 2t_1 + 2$

Eliminating t_1

$$\therefore \quad 3h = 3\left(\frac{3k-2}{2}\right)^2 - 2\left(\frac{3k-2}{2}\right) + 1$$

or $\quad 9k^2 - 16k - 4h + 8 = 0$

$$\Rightarrow \quad k^2 - \frac{16k}{9} - \frac{4h}{9} + \frac{8}{9} = 0$$

$$\Rightarrow \quad \left(k - \frac{8}{9}\right)^2 = \frac{4}{9}\left(h - \frac{2}{9}\right)$$

$\therefore$ Locus of $R(h, k)$ is $\left(y - \frac{8}{9}\right)^2 = \frac{4}{9}\left(x - \frac{2}{9}\right)$

which is a parabola with vertex $\left(\frac{8}{9}, \frac{2}{9}\right)$

$$\therefore \quad \lambda = \frac{8}{9}, \mu = \frac{2}{9}$$

$$\therefore \quad \lambda + \mu = \frac{10}{9}$$

then $\quad 729(\lambda + \mu)^2 = 729 \times \frac{100}{81} = 900$

4. Here, centre of the circle is the vertex of the parabola and both circle and parabola are symmetrical about axis of parabola. In this case the point of intersection of common tangents must lie on the directrix and axis of the parabola. i.e. $A(-a, 0)$

Chord of contact of circle w.r.t. $A(-a, 0)$ is

$$x(-a) + y \cdot 0 = \frac{a^2}{2}$$

$$\therefore \quad x = -\frac{a}{2}$$

$\therefore$ Coordinates of R is $\left(-\frac{a}{2}, \frac{a}{2}\right)$ and chord of contact of parabola w.r.t. $A(-a, 0)$ is

$$y \cdot 0 = 2a(x - a) \text{ i.e. } x = a$$

$\therefore$ Coordinates of P is $(a, 2a)$

$\therefore$ Area of quadrilateral $PQRS'$

$$= 2\{\text{Area of } \Delta PAS - \text{Area of } \Delta RAN\}$$

$$= 2\left\{\frac{1}{2} \cdot 2a \cdot 2a - \frac{1}{2} \cdot \frac{a}{2} \cdot \frac{a}{2}\right\}$$

$$= 4a^2 - \frac{a^2}{4} = \frac{15}{4}a^2 \text{ sq unit}$$

$$\therefore \quad \lambda = \frac{15a^2}{4}$$

then $\quad \frac{256}{a^2}\lambda = 960$

5. Any normal of the parabola $y^2 = 4x$ with slope m is

$$y = mx - 2m - m^3$$

It is pass through P, then

$$k = mh - 2m - m^3$$

$$m^3 + (2 - h)m + k = 0 \quad \text{...(i)}$$

Thus, $\quad m_1 m_2 m_3 = -k$

$$\alpha m_3 = -k \quad [\because m_1 m_2 = \alpha]$$

$$m_3 = -\frac{k}{\alpha}$$

$\because m_3$ is a root of Eq. (i), then

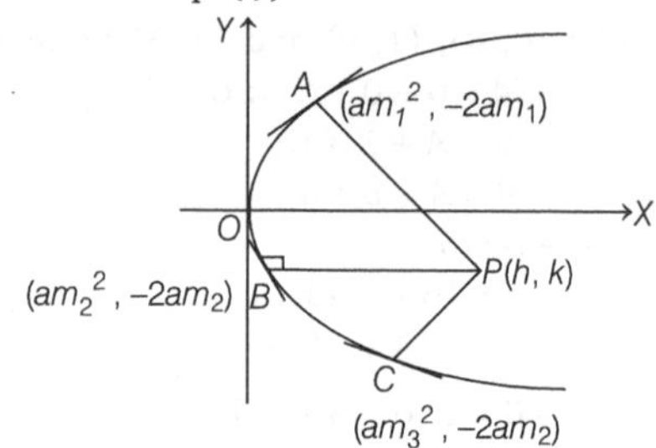

$$\frac{-k^3}{\alpha^3} + (2 - h)\left(-\frac{k}{\alpha}\right) + k = 0$$

$$k^3 + (2 - h)k\alpha^2 - k\alpha^3 = 0$$

$\therefore$ Locus of $P(h, k)$ is $y^3 + (2 - x)y\alpha^2 - y\alpha^3 = 0$

$$\Rightarrow \quad y^2 + (2 - x)\alpha^2 - \alpha^3 = 0 \quad (\because y \neq 0)$$

(P does not lie on the axis of the parabola)

$$\Rightarrow \quad y^2 = \alpha^2 x - 2\alpha^2 + \alpha^3$$

If it is a part of the parabola $y^2 = 4x$

then $\quad \alpha^2 = 4$ and $-2\alpha^2 + \alpha^3 = 0$

$$\Rightarrow \quad \alpha^2(\alpha - 2) = 0 \Rightarrow \alpha - 2 = 0, \alpha \neq 0$$

$$\therefore \quad \alpha = 2$$

$$\therefore \quad (36)^\alpha = (36)^2 = 1296$$

6. Let the given parabola is $y^2 = 4ax$ ...(i)

If the vertex of moving parabola is (α, β), then equation of moving parabola is

$$(y-\beta)^2 = -4a(x-\alpha) \quad ...(ii)$$

Substituting the value of x i.e. $x = \frac{y^2}{4a}$ in Eq. (ii)

then $$(y-\beta)^2 = -4a\left(\frac{y^2}{4a} - \alpha\right)$$

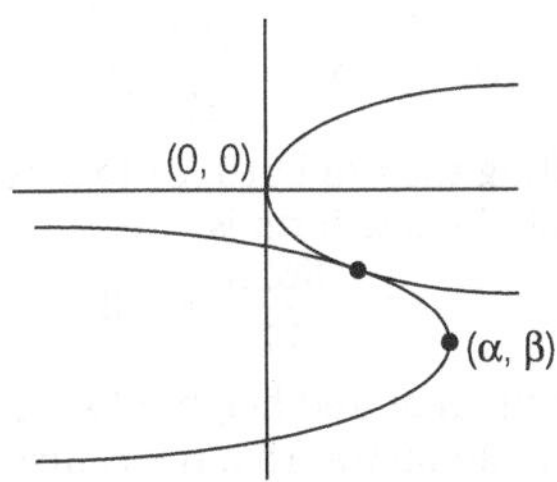

$\Rightarrow$ $y^2 - 2\beta y + \beta^2 = -y^2 + 4a\alpha$

$$2y^2 - 2\beta y + \beta^2 - 4a\alpha = 0 \quad ...(iii)$$

Since, two parabolas Eq. (i) and (ii) touch each other

Hence, roots of Eq. (iii) are equal i.e. Discriminant = 0

or $$B^2 - 4AC = 0$$

$\therefore$ $$(-2\beta)^2 = 4 \cdot 2 \cdot (\beta^2 - 4a\alpha)$$

$\Rightarrow$ $$4\beta^2 = 32\alpha \text{ or } \beta^2 = 8a\alpha$$

the required locus is $y^2 = 8ax$

$\therefore$ $$\lambda = 8a$$

then $$\frac{59\lambda^2}{a^2} = 59 \times 64 = 3776$$

7. Parametric point on the parabola

$$y = 4x^2 \text{ is } (t, 4t^2)$$

Let $P(t_1, 4t_1^2)$ and $Q(t_2, 4t_2^2)$

If A is vertex of the parabola, then

Slope of AP $\times$ Slope of $AQ = -1$ $(\because AP \perp AQ)$

$\therefore$ $$4t_1 \times 4t_2 = -1$$

$\therefore$ $$t_1 t_2 = -\frac{1}{16}$$

Equations of normals at P and Q are

$$x + 4t_1 y = t_1 + 16t_1^3 \quad ...(i)$$

and $$x + 4t_2 y = t_2 + 16t_2^3 \quad ...(ii)$$

Let the normals Eq (i) and (ii) intersect at (h, k), then

Solving Eqs. (i) and (ii), then we get

$$k = \frac{1 + 16(t_1^2 + t_2^2 + t_1 t_2)}{4}$$

$\Rightarrow$ $$k = \frac{1 + 16(t_1^2 + t_2^2) + 16t_1 t_2}{4}$$

$$= \frac{1 + 16\{(t_1 + t_2)^2 - 2t_1 t_2\} - 1}{4}$$

$[\because 16t_1 t_2 = -1]$

$$= 4\left\{(t_1 + t_2)^2 + \frac{1}{8}\right\}$$

$$k = 4(t_1 + t_2)^2 + \frac{1}{2} \quad ...(iii)$$

again $$h = -16t_1 t_2 (t_1 + t_2)$$

$$h = t_1 + t_2 \quad ...(iv)$$

Eliminating t_1 and t_2 from Eq. (iii) and (iv) we get

$$k = 4h^2 + \frac{1}{2}$$

$\Rightarrow$ $$2k = 8h^2 + 1$$

Hence, locus of (h, k) is

$$2y = 8x^2 + 1$$

$\Rightarrow$ $$y = 4x^2 + \frac{1}{2}$$

$\therefore$ $$\lambda = 4, \mu = \frac{1}{2}$$

then $\frac{\mu}{\lambda} = 0.125$

8. The path of the water jet is a parabola. Let the equation of the water jet being a parabola is

$$y = ax^2 + bx + c \quad ...(i)$$

the path is symmetrical to the line AB, the maximum height.

So, it strikes the X-axis at E such that

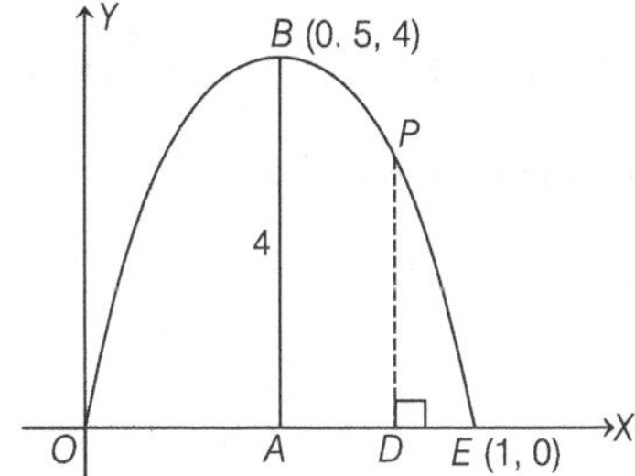

$$AE = OA = 0.5 \text{ m}$$

i.e. $$OE = 2OA = 2(0.5) = 1 \text{ m}$$

Coordinates of B and E are (0.5, 4) and (1, 0) respectively.

Since, O, B, E lie on Eq. (i)

$\therefore$ $$0 = c, \; 4 = \frac{1}{4}a + \frac{1}{2}b + c, \; 0 = a + b + c$$

Solving these we get

$$a = -16, b = 16, c = 0$$

From Eq. (i), the equation of the parabola is

$$y = -16x^2 + 16x \quad ...(ii)$$

Let P be a point on the parabola (ii), such that P is at a distance 0.75 m from y-axis and let P is at a distance h from x-axis

$\therefore$ Coordinates of P is $(0.75, h)$

$\because$ P lies on Eq. (ii) so we have

$$h = -16(0.75)^2 + 16(0.75)$$

$$= -16\left(\frac{9}{16}\right) + 16\left(\frac{3}{4}\right) = -9 + 12 = 3 \text{ m}$$

$\therefore \quad \lambda = 3$

Then, $\lambda^6 = 3^6 = 729$

9. Here, MZ is the road let V is lowest point of the rope. Given $VZ = 3\,\text{m}$.

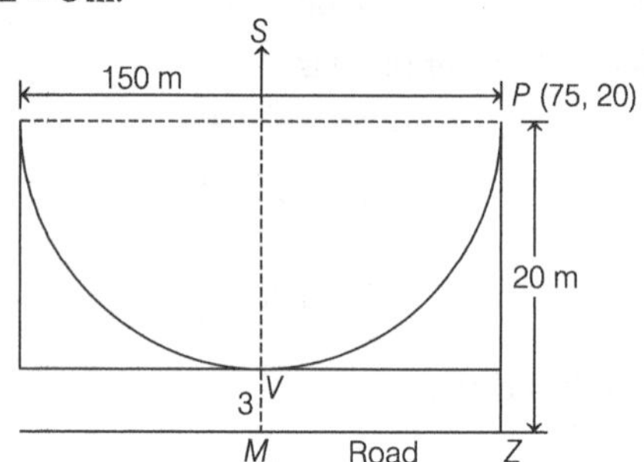

Let MZ is X-axis and let MVS is Y-axis.

Taking x and Y-axis as shown in the given figure, the equation of parabola is of the form

$$x^2 = 4ay$$

If V is origin

Coordinates of P is $\left\{\frac{1}{2}(150), 20 - 3\right\}$

i.e. $(75, 17)$ P lies on Eq. (i), then

$$(75)^2 = 4a(17) \text{ or } 4a = \frac{5625}{17}$$

From Eq. (i), $\quad x^2 = \frac{5625}{17}y$

Now shifting the origin $(0, 0)$ to $M(0, -3)$, then equation of parabola in new form is

$$x^2 = \frac{5625}{17}(y - 3) \quad \ldots\text{(ii)}$$

Again let the required height of supporting tower at a distance of 30 metre from the centre of road be h metres, then the coordinates of the top of this tower are $(30, h)$ referred to given axes,

From Eq. (ii), $(30)^2 = \frac{5625}{17}(h - 3)$

or $\quad h = \left(\frac{900 \times 17}{5625}\right) + 3 = 5.72 = \lambda \text{ m} \quad$ (given)

$\therefore \quad \lambda = 5.72$

Single Integer Type

1. $y^2 = 6\left(-x - \frac{3}{2}\right)$

Equation of directrix $x - \frac{3}{2} = -\frac{3}{2}$ i.e. $x = 0$

Let coordinate P be $\left(\frac{3}{2} + \frac{3}{2}t^2, 3t\right)$

$\therefore$ Coordinate of M are $(0, 3t)$

$$\therefore \quad MS = \sqrt{9 + 9t^2}$$

$$MP = \frac{3}{2} + \frac{3}{2}t^2$$

$$\therefore \quad 9 + 9t^2 = \left(\frac{3}{2} + \frac{3}{2}t^2\right)^2 = \frac{9}{4}(1 + t^2)^2$$

$$\therefore \quad 4 = 1 + t^2$$

$\therefore$ Length of side = 6

2. (λ, μ) is the point interior to both the curves if $\lambda^2 + \mu^2 - 16 < 0$ and $\mu^2 - 4\lambda < 0$

Now, $\mu^2 - 4\lambda < 0 \Rightarrow \lambda > \left(\frac{\mu}{2}\right)^2$

Hence, if $\mu = 0$, $\lambda = 1, 2, 3, \ldots$,
if $\mu = 1$, $\lambda = 1, 2, 3, \ldots$,
if $\mu = 2$, $\lambda = 2, 3, \ldots$,
if $\mu = 3$, $\lambda = 3, 4, \ldots$

Also $\lambda^2 + \mu^2 - 16 < 0 \Rightarrow \lambda^2 < 16 - \mu^2$

Hence, if $\mu = 0$

$\Rightarrow \quad \lambda = 1, 2, 3$; if $\mu = 1$

$\Rightarrow \quad \lambda = 1, 2, 3$; if $\mu = 2$

$\Rightarrow \quad \lambda = 2, 3$; if $\mu = 3$

$\Rightarrow \lambda$ has no integral value.

$\therefore$ (1, 0), (2, 0), (3, 0), (1, 1), (2, 1), (3, 1), (2, 2), (3, 2) are the 8 possible points.

3. Equations of the pair of tangents is $(y^2 - 4x)(16 - 4)$

$$= [4y - 2(x + 1)]^2$$

$\Rightarrow 12(y^2 - 4x) = 4(2y - x - 1)^2$

$\Rightarrow 3(y^2 - 4x) = 4y^2 + x^2 + 1 - 4xy + 2x - 4y$

$\Rightarrow x^2 + y^2 - 4xy + 14x - 4y + 1 = 0$

If θ is the angle between these lines

Then $\tan\theta = \frac{2\sqrt{(2)^2 - (1)(1)}}{(1)(1)} = \sqrt{3}$

$\Rightarrow \quad \theta = \pi/3 = \pi/m \Rightarrow m = 3$

4. The equation of tangent to parabola $y^2 = 4x$ is

$$y = mx + \frac{1}{m} \quad \ldots\text{(i)}$$

Since Eq. (i) is also the tangent of $x^2 = -32y$, we have

$$x^2 = -32\left(mx + \frac{1}{m}\right)$$

or $$x^2 + 32mx + \frac{32}{m} = 0$$

The above equation must have equal roots.
Hence, its discriminant must be zero. Therefore,

$$(32m)^2 = 4 \times 1 \times \frac{32}{m} \text{ i.e. } m^3 = \frac{1}{8} \text{ or } m = \frac{1}{2}$$

From Eq. (i),

$$y = \frac{x}{2} + 2 \text{ or } x - 2y + 4 = 0$$

$\therefore\ a = 1,\ b = -2,\ c = 4; \qquad \therefore\ a + b + c = 3$

5. The given parabola is $y^2 = kx - 8 = k\left(x - \frac{8}{k}\right)$.

Shifting the origin to $(8/k, 0)$, the parabola becomes $Y^2 = kX$ where $X = x - 8/k$ and $Y = y$.

Directrix of this parabola is $X = \frac{-k}{4}$ or $x - \frac{8}{k} = \frac{-k}{4}$

This will coincide with $x = 1$ if $\frac{8}{k} - \frac{k}{4} = 1 \Rightarrow 32 - k^2 = 4k$

$$\Rightarrow \quad k^2 + 4k - 32 = 0 \text{ or } (k+8)(k-4) = 0$$

$$\Rightarrow \quad k = -8, 4$$

Thus, $\quad k = 4$

6. Equation of the normal at 't' to the parabola.

$y^2 = 4ax$ is

$$y = -tx + 2at + at^3 \qquad \ldots\text{(i)}$$

The joint equation of the lines joining the vertex (origin) to the points of intersection of the parabola and the line (i) is

$$y^2 = 4ax\left[\frac{y + tx}{2at + at^3}\right]$$

$$\Rightarrow \quad (2t + t^3)y^2 = 4x(y + tx)$$

$$\Rightarrow \quad 4tx^2 - (2t + t^3)y^2 + 4xy = 0$$

Since these lines are at right angles coefficient of x^2 + coefficient of $y^2 = 0$

$$\Rightarrow \quad 4t - 2t - t^3 = 0 \Rightarrow t^2 = 2$$

For $t = 0$, the normal line is $y = 0$, i.e. axis of the parabola which passes through the vertex (0, 0).

7.

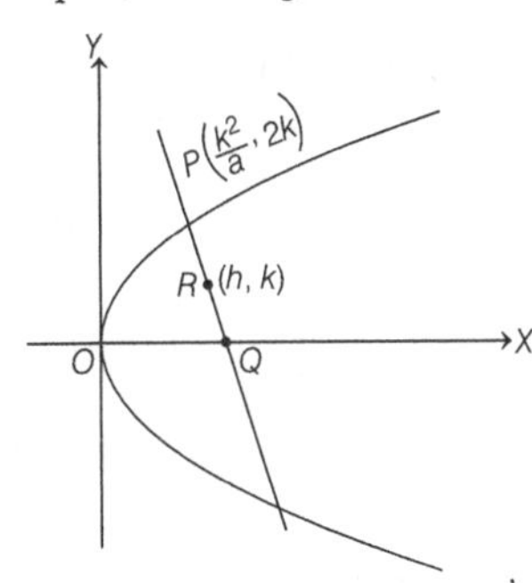

Consider the parabola $y^2 = 4ax$.

We have to find the locus of $R(h, k)$.

Since Q has ordinate 0, the ordinate of P is $2k$.

Also, P is on the curve. Then, the abscissa of P is k^2/a.

Now, PQ is normal to the curve.

Slope of tangent to the curve at any point is $\frac{dy}{dx} = \frac{2a}{y}$.

Hence, the slope of normal at point P is $-k/a$.

Also, the slope of normal joining P and $R(h, k)$ is

$$\frac{2k - k}{(k^2/a) - h} = -\frac{k}{a} \text{ or } y^2 = a(x - a)$$

For $y^2 = 16x$, $a = 4$

Hence, the locus is $y^2 = 4(x - 4)$

8. Given, equation of normal $mx - y + c = 0$

Parabola $y^2 = 16x$

comparing with general equation of parabola i.e.,

$$y^2 = 4ax$$

$$\Rightarrow \quad a = 4$$

Let the coordinates of point P on parabola be $(4t^2, 8t)$

$\because$ Slope of tanglent at point $P = \frac{1}{t}$ $\qquad \ldots$(i)

Also, slope of tangent $= -\frac{1}{\text{Slope of normal}} = -\frac{1}{m}$ $\qquad \ldots$(ii)

From Eqs. (i) and (ii), $\frac{1}{t} = -\frac{1}{m}$ or $t = -m$

$\therefore$ Coordinate of point is $P(4m^2, -8m)$.

Focurs of parabola is $(a, 0)$ or $(4, 0)$.

Given, focal distance = 40

Using distance formula,

$$\sqrt{(4m^2 - 4)^2 + (-8m - 0)^2} = 40$$

or $$\sqrt{(m^2 - 1)^2 + (-2m)^2} = 10$$

or $\sqrt{(m^2 - 1)^2 + 4m^2} = 10$ or $\sqrt{(m^2 + 1)^2} = 10$

or $m^2 + 1 = 10$ or $m^2 = 9$ or $m = \pm 3$

Therefore, coordinate of point P

$(4 \times (\pm 3)^2, -8(\pm 3)) \equiv (36, \pm 24)$

$\because$ Point P lies on line $mx - y + c = 0$

Then it must satisfy equation of the

$$(\pm 3)(36) - (\pm 48) + c = 0$$

$$\pm(3 \times 36 + 48) + c = 0$$

or $$c = \pm 132 \text{ or } |c| = 132$$

$$\therefore \quad \frac{2}{61}|c| = 4$$

Matrix-Match Type

1. (A) $\because \sqrt{\left(\frac{x}{a}\right)}+\sqrt{\left(\frac{y}{b}\right)}=1$

On squaring both sides, then

$$\frac{x}{a}+\frac{y}{b}+2\sqrt{\left(\frac{xy}{ab}\right)}=1$$

or $$\left(\frac{x}{a}+\frac{y}{b}-1\right)=-2\sqrt{\left(\frac{xy}{ab}\right)}$$

Again squaring both sides then

$$\frac{x^2}{a^2}+\frac{y^2}{b^2}+1+\frac{2xy}{ab}-\frac{2x}{a}-\frac{2y}{b}=\frac{4xy}{ab}$$

$$\Rightarrow \quad \frac{x^2}{a^2}-\frac{2xy}{ab}+\frac{y^2}{b^2}-\frac{2x}{a}-\frac{2y}{b}+1=0$$

Now, comparing with

$$Ax^2+2Hxy+By^2+2Gx+2Fy+C=0,$$

then $A=\frac{1}{a^2}, H=-\frac{1}{ab}, B=\frac{1}{b^2}, G=-\frac{1}{a}, F=-\frac{1}{b}, C=1$

$$\therefore \quad \Delta=ABC+2FGH-AF^2-BG^2-CH^2$$

$$=\frac{1}{a^2b^2}-\frac{2}{a^2b^2}-\frac{1}{a^1b^2}-\frac{1}{a^2b^2}-\frac{1}{a^1b^2}$$

$$=-\frac{4}{a^1b^2}\neq 0$$

and $H^2=AB$

$\because \quad \Delta\neq 0, H^2=AB$,

$\therefore$ Conic is non-degenerate and a parabola i.e. (Q, R)

(B) Now, comparing

$$3x^2+10xy+3y^2-15x-21y+18=0$$

with $ax^2+2hxy+by^2+2gx+2fy+c=0$

$\therefore \quad a=3, h=5, b=3, g=-\frac{15}{2}, f=-\frac{21}{2}, c=18$

$$\therefore \quad \Delta=abc+2fgh-af^2-bg^2-ch^2$$

$$=162+\frac{1575}{2}-\frac{1323}{4}-\frac{675}{4}-450$$

$$=162+288-450$$

$$=450-450$$

$$=0$$

and $\quad h^2\neq ab$

$\because \Delta=0,\ h^2\neq ab$, then conic is degenerate and a pair of intersecting straight line *i.e.*, (P, T)

(C) Now comparing

$$8x^2-4xy+5y^2-16x-14y+17=0$$

with, $ax^2+2hxy+by^2+2gx+2fy+c=0$

$\therefore \quad a=3, h=-2, b=5, g=-8, f=-7, c=17$

$$\therefore \quad \Delta=abc+2fgh-af^2-bg^2-ch^2$$

$$=255-224-147-320-68$$

$$=255-759$$

$$=-504\neq 0$$

and $\quad h^2<ab$

$\therefore \Delta\neq 0$ and $h^2<ab$ then conic is non-degenerate and an ellipse. i.e. (Q, S).

2. If three normals drawn to any parabola $y^2=4ax$ from a given point (h, k) be real, then $h>2a$.

(A) $\because \quad y^2-4x-2y+5=0$

$$\Rightarrow \quad (y-1)^2=4(x-1)$$

Let $\ y-1=Y$ and $x-1=X$

$\therefore \quad Y^2=4X$

On comparing with $Y^2=4aX$

$\therefore \quad a=1$

According to the question, $X>2a$

$\Rightarrow \quad x-1>2$ or $x>3$

$\therefore \ x=4, 5, 6, 7, 8$ (P, Q, R, S, T)

(B) $\because 4y^2-32x+4y+65=0$

$$\Rightarrow \quad 4(y^2+y)=32x-65$$

$$\Rightarrow \quad 4\left(\left(y+\frac{1}{2}\right)^2-\frac{1}{4}\right)=32x-65$$

$$\Rightarrow \quad 4\left(y+\frac{1}{2}\right)^2=32x-64$$

or $$\left(y+\frac{1}{2}\right)^2=8(x-2)$$

Let $\quad y+\frac{1}{2}=Y$ and $x-2=X$

$\therefore \quad Y^2=8X$

On comparing with $y^2=4ax$

$\therefore \quad a=2$

According to the question,

$$X>2a$$

$$\Rightarrow \quad x-2>4$$

$$\therefore \quad x>6$$

$\therefore \quad x=7, 8$ (S, T)

(C) $\because 4y^2-16x-4y+41=0$

$$\Rightarrow \quad 4(y^2-y)=16x-41$$

$$\Rightarrow \quad 4\left\{\left(y-\frac{1}{2}\right)^2-\frac{1}{4}\right\}=16x-41$$

$$\Rightarrow \quad 4\left(y-\frac{1}{2}\right)^2=16x-40$$

or $$\left(y-\frac{1}{2}\right)^2=4\left(x-\frac{5}{2}\right)$$

Let $\quad y-\frac{1}{2}=Y$ and $x-\frac{5}{2}=X$

$\therefore \quad Y^2=4X$

On comparing with $Y^2=4aX$

$\therefore \quad a=1$

According to the question,

$$X>2a$$

$\Rightarrow \quad x-\frac{5}{2}>2$ or $x>\frac{9}{2}$

$\therefore \quad x=5, 6, 7, 8 \qquad$ (Q, R, S, T)

3. **(A)** $\because 25(x^2+y^2)=(4x+3y-12)^2$

or $25x^2+25y^2=16x^2+9y^2+24xy-96x-72y+144$

$\Rightarrow 9x^2+16y^2-24xy=-96x-72y+144$

$\Rightarrow (3x-4y)^2=-24(4x+3y-6)$

$\Rightarrow 25\left(\frac{3x-4y}{5}\right)^2=-24\left(\frac{4x+3y-6}{5}\right)\times 5$

$\Rightarrow \left(\frac{3x-4y}{5}\right)^2=-\frac{24}{5}\left(\frac{4x+3y-6}{5}\right)$

Let $\frac{3x-4y}{5}=Y$ and $\frac{4x+3y-6}{5}=X$

$\therefore Y^2=-\frac{24}{5}X$

On comparing with $Y^2=-4\rho X$

$\therefore 4\rho=\frac{24}{5}\Rightarrow \rho=\frac{6}{5}$

Equation of directrix is

$X-\rho=0$

i.e. $\frac{4x+3y-6}{5}-\frac{6}{5}=0$

or $4x+3y=12$ (R)

Axis of the parabola is

$Y=0$

i.e., $\frac{3x-4y}{5}=0$

or $3x-4y=0$

(B) $\because 25(x^2+y^2)=(4x-3y+12)^2$

or $25x^2+25y^2=16x^2+9y^2-24xy+96x-72y+144$

$\Rightarrow 9x^2+16y^2+24xy=96x-72y+144$

$\Rightarrow (3x+4y)^2=24(4x-3y+6)$

or $25\times\left(\frac{3x+4y}{5}\right)^2=24\left(\frac{4x-3y+6}{5}\right)\times 5$

$\Rightarrow \left(\frac{3x+4y}{5}\right)^2=\frac{24}{5}\left(\frac{4x-3y+6}{5}\right)$

Let $\frac{3x+4y}{5}=Y$ and $\frac{4x-3y+6}{5}=X$

$\therefore Y^2=\frac{24}{5}X$

On comparing with $Y^2=4\rho X$

$\therefore 4\rho=\frac{24}{5}$

$\Rightarrow \rho=\frac{6}{5}$

Equation of directrix is

$X+\rho=0$

i.e. $\frac{4x-3y+6}{5}+\frac{6}{5}=0$

$\Rightarrow 4x-3y+12=0$ (P)

and axis of the parabola is

$Y=0$

i.e. $\frac{3x+4y}{5}=0$

or $3x+4y=0$ (S)

(C) $\because 25(x^2+y^2)=(3x-4y+12)^2$

or $25x^2+25y^2=9x^2+16y^2-24xy+72x-96y+144$

$\Rightarrow 16x^2+9y^2+24xy=72x-96y+144$

$\Rightarrow (4x+3y)^2=24(3x-4y+6)$

or $25\times\left(\frac{4x+3y}{5}\right)^2=24\left(\frac{3x-4y+6}{5}\right)\times 5$

$\Rightarrow \left(\frac{4x+3y}{5}\right)^2=\frac{24}{5}\left(\frac{3x-4y+6}{5}\right)$

Let $\frac{4x+3y}{5}=Y$ and $\frac{3x-4y+6}{5}=X$

$\therefore Y^2=\frac{24}{5}X$

On comparing with $Y^2=4\rho X$

$\therefore 4\rho=\frac{24}{5}\Rightarrow \rho=\frac{6}{5}$

$\therefore$ Equation of directrix is

$X+\rho=0$

i.e. $\frac{3x-4y+6}{5}+\frac{6}{5}=0$

or $3x-4y+12=0$ (T)

and axis of the parabola is

$Y=0$

i.e. $4x+3y=0$ (Q)

25

Ellipse

Objective Questions Type I [Only one correct answer]

In each of the questions below, four choices are given of which only one is correct. You have to select the correct answer which is the most appropriate.

1. The length of the minor axis (along Y-axis) of an ellipse in the standard form is $\frac{4}{\sqrt{3}}$. If this ellipse touches the line, $x + 6y = 8$; then its eccentricity is

(a) $\sqrt{\frac{5}{6}}$ (b) $\frac{1}{2}\sqrt{\frac{11}{3}}$

(c) $\frac{1}{3}\sqrt{\frac{11}{3}}$ (d) $\frac{1}{2}\sqrt{\frac{5}{3}}$

[JEE Main 2020, 9 January]

2. If $\frac{x^2}{f(4a)} + \frac{y^2}{f(a^2 - 5)}$ represents an ellipse with major axis as Y-axis and f is a decreasing function, then

(a) $a \in (-\infty, 1)$ (b) $a \in (5, \infty)$

(c) $a \in (1, 4)$ (d) $a \in (-1, 5)$

3. Three points A, B, C are taken on the ellipse $\frac{x^2}{a^2} + \frac{y^2}{b^2} = 1$ with accentric angles $\theta, \theta + \alpha$ and $\theta + 2\alpha$, then

(a) the area of $\Delta\, ABC$ is independent of θ

(b) the area of $\Delta\, ABC$ is independent of α

(c) the maximum value of area is $\frac{\sqrt{3}}{4}ab$

(d) the maximum value of area is $\frac{3\sqrt{3}}{4}ab$

4. The point, at shortest distance from the line $x + y = 7$ and lying on an ellipse $x^2 + 2y^2 = 6$, has coordinates

(a) $(\sqrt{2}, \sqrt{2})$ (b) $(0, \sqrt{3})$

(c) $(2, 1)$ (d) $\left(\sqrt{5}, \frac{1}{\sqrt{2}}\right)$

5. If the normal at the end of latusrectum of the ellipse $\frac{x^2}{a^2} + \frac{y^2}{b^2} = 1$ passes through $(0, -b)$, then $e^4 + e^2$ (where e is eccentricity) equals

(a) 1 (b) $\sqrt{2}$

(c) $\frac{\sqrt{5} - 1}{2}$ (d) $\frac{\sqrt{5} + 1}{2}$

6. If the line $x + 2y + 4 = 0$ cutting the ellipse $\frac{x^2}{a^2} + \frac{y^2}{b^2} = 1$ in points whose eccentric angles are 30° and 60° subtends a right angle at the origin then its equation is

(a) $\frac{x^2}{8} + \frac{y^2}{4} = 1$ (b) $\frac{x^2}{16} + \frac{y^2}{4} = 1$

(c) $\frac{x^2}{4} + \frac{y^2}{16} = 1$ (d) None of these

7. If the point of intersections of the ellipse $\frac{x^2}{16} + \frac{y^2}{b^2} = 1$ and the circle $x^2 + y^2 = 4b$, $b > 4$ lie on the curve $y^2 = 3x^2$, then b is equal to **[JEE Main 2021, 16 March]**

(a) 12 (b) 5

(c) 6 (d) 10

8. If $\cos\alpha = \frac{2}{3}$, then the range of values of ϕ for which the point ϕ on the ellipse $x^2 + 4y^2 = 4$ falls inside the circle $x^2 + y^2 + 4x + 3 = 0$ is

(a) $(-\alpha, \alpha)$ (b) $(0, \alpha)$

(c) (α, π) (d) $(\pi - \alpha, \pi + \alpha)$

9. The minimum distance of the centre of the ellipse $\frac{x^2}{16}+\frac{y^2}{9}=1$ from the chord of contact of mutually perpendicular tangents of the ellipse is
(a) $\frac{144}{5}$ (b) $\frac{16}{5}$
(c) $\frac{9}{5}$ (d) None of these

10. An arc of a bridge is semi-elliptical with major axis horizontal. If the length of the base is 9 m and the highest part of the bridge is 3 m from the horizontal; the best approximation of the height of the arch 2 m from the centre of the base is
(a) $\frac{11}{4}$ m (b) $\frac{8}{3}$ m
(c) $\frac{7}{2}$ m (d) 2n

11. From a point on the axis of x common tangents are drawn to the parabola $y^2=4x$ and the ellipse $\frac{x^2}{a^2}+\frac{y^2}{b^2}=1$ $(a>b>0)$. If these tangents form an equilateral triangle with their chord of contact w.r.t. parabola, then set of exhaustive values of a is
(a) $(0, 3)$ (b) $\left(0, \frac{3}{2}\right)$
(c) $\left(\frac{3}{2}, 3\right)$ (d) $\left(1, \frac{3}{2}\right)$

12. Let P be a variable point on the ellipse $\frac{x^2}{25}+\frac{y^2}{16}=1$ with foci at S and S'. If A be the area of $\Delta PSS'$, then the maximum value of A is
(a) 24 sq unit (b) 12 sq unit
(c) 36 sq unit (d) None of these

13. The area of a triangle inscribed in an ellipse bears a constant ratio to the area of the triangle formed by joining points on the auxiliary circle corresponding to the vertices of the first triangle. This ratio is
(a) b/a (b) $2a/b$
(c) a^2/b^2 (d) b^2/a^2

14. Let the length of the latus rectum of an ellipse with its major axis along X-axis and centre at the origin, be 8. If the distance between the foci of this ellipse is equal to the length of its minor axis, then which one of the following points lies on it?
[JEE Main 2019, 11 January]
(a) $(4\sqrt{2}, 2\sqrt{3})$ (b) $(4\sqrt{3}, 2\sqrt{2})$
(c) $(4\sqrt{2}, 2\sqrt{2})$ (d) $(4\sqrt{3}, 2\sqrt{3})$

15. Find the minimum length of a tangent to an ellipse $\frac{x^2}{a^2}+\frac{y^2}{b^2}=1$ intercept between the coordinate axes.
(a) a^2+b^2 (b) a^2-b^2
(c) $(a+b)^2$ (d) $2ab$

16. The set of values of a for which $(13x-1)^2+(13y-2)^2=a(5x+12y-1)^2$ represents an ellipse, if
(a) $1<a<2$ (b) $0<a<1$
(c) $2<a<3$ (d) None of these

17. The length of the common chord of the ellipse $\frac{(x-1)^2}{9}+\frac{(y-2)^2}{4}=1$ and the circle $(x-1)^2+(y-2)^2=1$ is
(a) zero (b) one
(c) three (d) eight

18. If CF is the perpendicular from the centre C of the ellipse $\frac{x^2}{a^2}+\frac{y^2}{b^2}=1$ on the tangent at any point P and G is the point when the normal at P meets the major axis, then $CF \cdot PG$ is equal to
(a) a^2 (b) ab
(c) b^2 (d) b^3

19. Tangents are drawn from the points on the line $x-y-5=0$ to $x^2+4y^2=4$, then all the chords of contact pass through a fixed point, whose coordinate are
(a) $\left(\frac{1}{5}, -\frac{2}{5}\right)$ (b) $\left(\frac{4}{5}, -\frac{1}{5}\right)$
(c) $\left(\frac{2}{5}, -\frac{1}{5}\right)$ (d) None of these

20. An ellipse has OB as semi-minor axis. F and F' are its focii and the angle FBF' is a right angle. Then eccentricity of the ellipse is
(a) $1/\sqrt{3}$ (b) $1/2$
(c) $1/\sqrt{2}$ (d) None of these

21. The set of positive value of m for which a line with slope m is a common tangent to ellipse $\frac{x^2}{a^2}+\frac{y^2}{b^2}=1$ and parabola $y^2=4ax$ is given by
(a) $(2, 0)$ (b) $(3, 5)$
(c) $(0, 1)$ (d) None of these

22. Let a tangent be drawn to the ellipse $\frac{x^2}{27}+y^2=1$ at $(3\sqrt{3}\cos\theta, \sin\theta)$ where $\theta \in \left(0, \frac{\pi}{2}\right)$. Then, the value of θ such that the sum of intercepts on axes made by this tangent is minimum is equal to
[JEE Main 2020, 9 January]
(a) $\frac{\pi}{8}$ (b) $\frac{\pi}{4}$
(c) $\frac{\pi}{6}$ (d) $\frac{\pi}{3}$

23. The minimum length of the intercept of any tangent on the ellipse $\frac{x^2}{a^2}+\frac{y^2}{b^2}=1$ between the coordinate axes is
(a) $2a$ (b) $2b$
(c) $a-b$ (d) $a+b$

24. If CP and CD are semi-conjugate diameters of the ellipse $\frac{x^2}{a^2}+\frac{y^2}{b^2}=1$, then CP^2+CD^2 is equals
(a) $a+b$ (b) a^2+b^2
(c) a^2-b^2 (d) $\sqrt{a^2+b^2}$

25. A man running round a race course notes that the sum of the distances of two flag posts from him is always 10m and the distance between the flag posts is 8 m. The area of the path he encloses in square metre is
(a) 15π (b) 12π (c) 18π (d) 8π

26. If the normal at the point $P(\phi)$ to the ellipse $\frac{x^2}{14}+\frac{y^2}{5}=1$ intersects it again at the point $Q(2\phi)$, then $\cos\phi$ is equal to
(a) $\frac{2}{3}$ (b) $-\frac{2}{3}$
(c) $\frac{3}{2}$ (d) $-\frac{3}{2}$

27. If α and β are eccentric angles of the ends of a focal chord of the ellipse $\frac{x^2}{a^2}+\frac{y^2}{b^2}=1$, then $\tan\frac{\alpha}{2}\tan\frac{\beta}{2}$ is equal to
(a) $\frac{1-e}{1+e}$ (b) $\frac{e-1}{e+1}$
(c) $\frac{e+1}{e-1}$ (d) $\frac{e-1}{e+3}$

28. If the normal to the ellipse $3x^2+4y^2=12$ at a point P on it is parallel to the line, $2x+y=4$ and the tangent to the ellipse at P passes through $Q(4, 4)$ then PQ is equal to **[JEE Main 2019, 12 April]**
(a) $\frac{5\sqrt{5}}{2}$ (b) $\frac{\sqrt{61}}{2}$
(c) $\frac{\sqrt{221}}{2}$ (d) $\frac{\sqrt{157}}{2}$

29. The distances from the foci of $P(a, b)$ on the ellipse $\frac{x^2}{9}+\frac{y^2}{25}=1$ are
(a) $4\pm\frac{5}{4}b$ (b) $5\pm\frac{4}{5}a$
(c) $5\pm\frac{4}{5}b$ (d) None of these

30. If $3x+4y=12\sqrt{2}$ is a tangent to the ellipse $\frac{x^2}{a^2}+\frac{y^2}{9}=1$ for some $a\in R$, then the distance between the foci of the ellipse is **[JEE Main 2020, 7 January]**
(a) $2\sqrt{7}$ (b) 4
(c) $2\sqrt{2}$ (d) $2\sqrt{5}$

31. The area of the parallelogram formed by the tangents at the ends of conjugate diameters of an ellipse is
(a) constant and is equal to the product of the axes
(b) cannot be constant
(c) constant and is equal to the two lines of the product of the axes
(d) None of the above

32. If C be the centre of the ellipse $9x^2+16y^2=144$ and S is one focus, the ratio of CS to major axis is
(a) $\sqrt{7}:16$ (b) $\sqrt{7}:4$
(c) $\sqrt{5}:\sqrt{7}$ (d) None of these

33. Let S and S' be the foci of an ellipse and B be any one of the extremities of its minor axis. If $\Delta S'BS$ is a right angled triangle with right angle at B and area $(\Delta S'BS)=8$ sq units, then the length of a latusrectum of the ellipse is **[JEE Main 2019, 12 January]**
(a) $2\sqrt{2}$ (b) $4\sqrt{2}$
(c) 2 (d) 4

34. The radius of the circle passing through the foci of the ellipse $\frac{x^2}{16}+\frac{y^2}{9}=1$, and having its centre $(0, 3)$ is
(a) 4 (b) 3
(c) $\sqrt{12}$ (d) 7/2

35. If the line $x-2y=12$ is tangent to the ellipse $\frac{x^2}{a^2}+\frac{y^2}{b^2}=1$ at the point $\left(3, \frac{-9}{2}\right)$, then the length of the latusrectum of the ellipse is **[JEE Main 2019, 10 April]**
(a) $8\sqrt{3}$ (b) 9
(c) 5 (d) $12\sqrt{2}$

36. Coordinates of the vertices B and C of a ΔABC are (2, 0) and (8, 0) respectively. The vertex A is varying in such a way that $4\tan\frac{B}{2}\cdot\tan\frac{C}{2}=1$. Then locus of A is
(a) $\frac{(x-5)^2}{25}+\frac{y^2}{16}=1$ (b) $\frac{(x-5)^2}{16}+\frac{y^2}{25}=1$
(c) $\frac{(x-5)^2}{25}+\frac{y^2}{9}=1$ (d) $\frac{(x-5)^2}{9}+\frac{y^2}{25}=1$

37. The tangent and normal drawn to the ellipse $x^2+4y^2=4$ at the point $P(\theta)$ meets the X-axis at the points A and B. If then $\cos^2\theta=$
(a) 4/9 (b) 8/9
(c) 2/9 (d) None of these

38. The locus of midpoints of a focal chord of the ellipse $\frac{x^2}{a^2}+\frac{y^2}{b^2}=1$ is
(a) $\frac{x^2}{a^2}+\frac{y^2}{b^2}=\frac{ex}{a}$ (b) $\frac{x^2}{a^2}-\frac{y^2}{b^2}=\frac{ex}{a}$
(c) $x^2+y^2=a^2+b^2$ (d) None of these

39. The locus of the point of intersection of tangents to the ellipse $\frac{x^2}{a^2}+\frac{y^2}{b^2}=1$, which meet at right angles, is
(a) a circle
(b) a parabola
(c) an ellipse
(d) a hyperbola

40. If the tangents on the ellipse $4x^2 + y^2 = 8$ at the points (1, 2) and (a, b) are perpendicular to each other, then a^2 is equal to **[JEE Main 2019, 8 April]**

(a) $\frac{128}{17}$ (b) $\frac{64}{17}$

(c) $\frac{4}{17}$ (d) $\frac{2}{17}$

41. Equation to the ellipse whose centre is (–2, 3) and whose semi-axes are 3 and 2 and major axis is parallel to the X-axis, is given by

(a) $4x^2 + 9y^2 + 16x - 54y - 61 = 0$

(b) $4x^2 + 9y^2 - 16x + 54y + 61 = 0$

(c) $4x^2 + 9y^2 + 16x - 54y + 61 = 0$

(d) None of the above

42. The eccentricity of an ellipse whose centre is at the origin is 1/2. If one of its directrices is $x = -4$, then the equation of the normal to it at $\left(1, \frac{3}{2}\right)$ is **[JEE Main 2017]**

(a) $2y - x = 2$ (b) $4x - 2y = 1$

(c) $4x + 2y = 7$ (d) $x + 2y = 4$

43. The maximum distance of perpendicular from the centre of the ellipse $4x^2 + 9y^2 = 36$ on the chord of contact of mutually perpendicular tangents of the ellipse is

(a) $\frac{6}{\sqrt{13}}$ (b) $\frac{5}{\sqrt{13}}$

(c) $\frac{9}{\sqrt{13}}$ (d) None of these

44. If the tangent to the parabola $y^2 = x$ at a point (α, β), $(\beta > 0)$ is also a tangent to the ellipse, $x^2 + 2y^2 = 1$, then α is equal to **[JEE Main 2019, 9 April]**

(a) $\sqrt{2} + 1$ (b) $\sqrt{2} - 1$

(c) $2\sqrt{2} + 1$ (d) $2\sqrt{2} - 1$

45. The sum of the square of perpendiculars on any tangent to the ellipse $\frac{x^2}{a^2} + \frac{y^2}{b^2} = 1$ from two points on the minor axis, each at a distantce ae (where e is eccentricity of the ellipse) from the centre, is

(a) $2a^2$ (b) $2b^2$

(c) $a^2 + b^2$ (d) $a^2 - b^2$

46. If $P(\theta)$ and $Q\left(\frac{\pi}{2} + \theta\right)$ are two points on the ellipse $\frac{x^2}{a^2} + \frac{y^2}{b^2} = 1$, locus of the mid point of PQ is

(a) $\frac{x^2}{a^2} + \frac{y^2}{b^2} = \frac{1}{2}$ (b) $\frac{x^2}{a^2} + \frac{y^2}{b^2} = 4$

(c) $\frac{x^2}{a^2} + \frac{y^2}{b^2} = 2$ (d) None of these

47. Tangent at a point P on the ellipse $x^2 + 4y^2 = 4$ meets the X-axis at B and AP is ordinate of P. If Q is a point on AP produced such that $AQ = AB$, then locus of Q is

(a) $x^2 + xy - 4 = 0$ (b) $x^2 - xy + 4 = 0$

(c) $x^2 + xy - 1 = 0$ (d) $x^2 - xy + 1 = 0$

48. If α, β are eccentric angles of the extremities of a focal chord of an ellipse, then eccentricity of the ellipse is

(a) $\frac{\cos\alpha + \cos\beta}{\cos(\alpha + \beta)}$ (b) $\frac{\sin\alpha - \sin\beta}{\sin(\alpha - \beta)}$

(c) $\sec\alpha + \sec\beta$ (d) $\frac{\sin\alpha + \sin\beta}{\sin(\alpha + \beta)}$

49. If e_1 and e_2 are the eccentricities of the ellipse, $\frac{x^2}{18} + \frac{y^2}{4} = 1$ and the hyperbola, $\frac{x^2}{9} - \frac{y^2}{4} = 1$ respectively and (e_1, e_2) is a point on the ellipse, $15x^2 + 3y^2 = k$, then k is equal to **[JEE Main 2020, 9 January]**

(a) 14 (b) 15 (c) 17 (d) 16

50. The area (in sq units) of the quadrilateral formed by the tangents at the end points of the latera recta to the ellipse $\frac{x^2}{9} + \frac{y^2}{5} = 1$ is **[JEE Main 2015]**

(a) $\frac{27}{4}$ (b) 18 (c) $\frac{27}{2}$ (d) 27

51. The parametric representation of a point on the ellipse whose foci are (–1, 0) and (7, 0) and eccentricity is 1/2, is

(a) $(3 + 8\cos\theta, 4\sqrt{3}\sin\theta)$ (b) $(8\cos\theta, 4\sqrt{3}\sin\theta)$

(c) $(3 + 4\sqrt{3}\cos\theta, 8\sin\theta)$ (d) None of these

52. Let E be the ellipse $\frac{x^2}{9} + \frac{y^2}{4} = 1$ and C be the circle $x^2 + y^2 = 9$. Let P and Q be the points (1, 2) and (2, 1) respectively. Then

(a) Q lies inside C but outside E

(b) Q lies outside both C and E

(c) P lies inside both C and E

(d) P lies inside C but outside E

53. The tangent at a point $P(a\cos\phi, b\sin\phi)$ of an ellipse $\frac{x^2}{a^2} + \frac{y^2}{b^2} = 1$ meets its auxiliary circle in two points, the chord joining which subtends a right angle at the centre, then the eccentricity of the ellipse is

(a) $(1 + \sin^2\phi)^{-1}$ (b) $(1 + \sin^2\phi)^{-1/2}$

(c) $(1 + \sin^2\phi)^{-3/2}$ (d) $(1 + \sin^2\phi)^{-2}$

54. Locus of foot of perpendicular drawn from centre to any tangent to the ellipse $x^2 + 2y^2 = 2$ is

(a) $(x^2 + y^2)^2 = (2x^2 - y^2)$

(b) $(x^2 + y^2)^2 = (2x^2 + y^2)$

(c) $(x^2 - y^2)^2 = (2x^2 - y^2)$

(d) $(x^2 - y^2) = (2x^2 + y^2)$

55. The eccentricity of the ellipse which meets the straight line $\frac{x}{7}+\frac{y}{2}=1$ on the axis of X and the straight line $\frac{x}{3}-\frac{y}{5}=1$ on the axis of Y and whose axes lie along the axes of coordinates, is

(a) $\frac{3\sqrt{2}}{7}$

(b) $\frac{2\sqrt{3}}{7}$

(c) $\frac{\sqrt{3}}{7}$

(d) None of the above

56. Let a, b and λ be positive real numbers. Suppose P is an end point of the latus rectum of the parabola $y^2 = 4\lambda x$, and suppose the ellipse $\frac{x^2}{a^2}+\frac{y^2}{b^2}=1$ passes through the point P. If the tangents to the parabola and the ellipse at the point P are perpendicular to each other, then the eccentricity of the ellipse is **[JEE Advanced 2020]**

(a) $\frac{1}{\sqrt{2}}$ (b) $\frac{1}{2}$

(c) $\frac{1}{3}$ (d) $\frac{2}{5}$

Objective Questions Type II [One or more than one correct answer(s)]

In each of the questions below four choices of which one or more than one are correct. You have to select the correct answer(s) accordingly.

1. The locus of extremities of the latus-rectum of the family of ellipse $b^2x^2 + y^2 = a^2b^2$ is

(a) $x^2 - ay = a^2$ (b) $x^2 - ay = b^2$

(c) $x^2 + ay = a^2$ (d) $x^2 + ay = b^2$

2. $P(x_1, y_1)$ and $Q(x_2, y_2)$, $y_1 < 0$, $y_2 < 0$ be the end points of the latus rectum of the ellipse $x^2 + 4y^2 = 4$. The equation of the parabolas with latus rectum PQ are

(a) $x^2 + 2\sqrt{3}y = 3+\sqrt{3}$ (b) $x^2 - 2\sqrt{3}y = 3+\sqrt{3}$

(c) $x^2 + 2\sqrt{3}y = 3-\sqrt{3}$ (d) $x^2 - 2\sqrt{3}y = 3-\sqrt{3}$

3. If latus rectum of the ellipse $x^2\tan^2\alpha + y^2\sec^2\alpha = 1$ is 1/2, then α $(0<\alpha<\pi)$ is equal to

(a) $\pi/12$ (b) $\pi/6$

(c) $5\pi/12$ (d) None of these

4. In the ellipse $25x^2 + 9y^2 - 150x - 90y + 225 = 0$

(a) foci are at (3, 1), (3, 9) (b) $e = 4/5$

(c) centre is (5, 3) (d) major axis is 6

5. Consider two straight lines, each of which is tangent to both the circle $x^2 + y^2 = (1/2)$ and the parabola $y^2 = 4x$. Let these lines intersect at the point Q. Consider the ellipse whose centre is at the origin $O(0, 0)$ and whose semi-major axis is OQ. If the length of the minor axis of this ellipse is $\sqrt{2}$, then which of the following statement(s) is (are) TRUE? **[JEE Advanced 2018]**

(a) For the ellipse, the eccentricity is $1/\sqrt{2}$ and the length of the latus rectum is 1

(b) For the ellipse, the eccentricity is 1/2 and the length of the latus rectum is 1/2

(c) The area of the region bounded by the ellipse between the lines $x=\frac{1}{\sqrt{2}}$ and $x=1$ is $\frac{1}{4\sqrt{2}}(\pi-2)$

(d) The area of the region bounded by the ellipse between the lines $x=\frac{1}{\sqrt{2}}$ and $x=1$ is $\frac{1}{16}(\pi-2)$

6. The coordinates (2, 3) and (1, 5) are the foci of an ellipse which passes through the origin. Then, the equation of the

(a) tangent at the origin is $(3\sqrt{2}-5)x + (1-2\sqrt{2})y = 0$

(b) tangent at the origin is $(3\sqrt{2}+5)x + (1+2\sqrt{2})y = 0$

(c) normal at the origin is $(3\sqrt{2}+5)x - (2\sqrt{2}+1)y = 0$

(d) normal at the origin is $x(3\sqrt{2}-5) - y(1-2\sqrt{2}) = 0$

7. The points, where the normals to the ellipse $x^2 + 3y^2 = 37$ be parallel to the line $6x - 5y + 7 = 0$ is/are

(a) (5, 2)

(b) (2, 5)

(c) (1, 3)

(d) (– 5, – 2)

8. Define the collections $\{E_1, E_2, E_3, \dots\}$ of ellipses and $\{R_1, R_2, R_3, \dots\}$ of rectangles as follows :

$E_1 : \frac{x^2}{9}+\frac{y^2}{4}=1$

R_1 : rectangle of largest area, with sides parallel to the axes, inscribed in E_1;

E_n : ellipse $\frac{x^2}{a_n^2}+\frac{y^2}{b_n^2}=1$ of the largest area inscribed in R_{n-1}, $n>1$;

R_n : rectangle of largest area, with sides parallel to the axes, inscribed in E_n, $n > 1$.

Then, which of the following options is/are correct? **[JEE Advanced 2019]**

(a) The eccentricities of E_{18} and E_{19} are NOT equal.

(b) The distance of a focus from the centre in E_9 is $\frac{\sqrt{5}}{32}$.

(c) $\sum_{n=1}^{N}$(area of R_n) < 24, for each positive integer N.

(d) The length of latusrectum of E_9 is $\frac{1}{6}$

9. If (5, 12) and (24, 7) are the focii of a conic passing through the origin, then the eccentricity of conic is

(a) $\frac{\sqrt{386}}{38}$ (b) $\frac{\sqrt{386}}{12}$

(c) $\frac{\sqrt{386}}{13}$ (d) $\frac{\sqrt{386}}{25}$

10. Which of the following is/are true about the ellipse $x^2 + 4y^2 - 2x - 16y + 13 = 0$?

(a) The latus rectum of the ellipse is 1.

(b) The distance between the foci of the ellipse is $4\sqrt{3}$.

(c) The sum of the focal distance of a point$P(x, y)$ on the ellipse is 4.

(d) Line $y = 3$ meets the tangents drawn at the vertices of the ellipse at points P and Q. Then PQ subtends a right angle at any of its foci.

11. The eccentric angle of a point on the ellipse $x^2 + 3y^2 = 6$ at point on the ellipse $x^2 + 3y^2 = 6$ at a distance 2 unit from origin is

(a) $\frac{\pi}{4}$ (b) $\frac{3\pi}{4}$

(c) $\frac{5\pi}{4}$ (d) $\frac{7\pi}{4}$

12. If the tangent at the point $\left(4\cos\theta, \frac{16}{\sqrt{11}}\sin\theta\right)$ to the ellipse $16x^2 + 11y^2 = 256$ is also a tangent to the circle $x^2 + y^2 - 2x = 15$, then θ equals

(a) $\frac{\pi}{3}$ (b) $\frac{2\pi}{3}$

(c) $-\frac{\pi}{3}$ (d) $\frac{5\pi}{3}$

13. A tangent to the ellipse $4x^2 + 9y^2 = 36$ is cut by the tangent at the extremities of the major axis at T and T'. The circle on TT' as diameter passes through the point

(a) $(-\sqrt{5}, 0)$ (b) $(\sqrt{5}, 0)$

(c) $(0, 0)$ (d) $(3, 2)$

14. If the tangent to the ellipse $x^2 + 4y^2 = 16$ at the point $P(\theta)$ is a normal to the circle $x^2 + y^2 - 8x - 4y = 0$, then θ equals

(a) $\pi/2$ (b) $\pi/4$

(c) 0 (d) $-\pi/4$

15. Let E_1 and E_2 be two ellipses whose centres are at the origin. The major axes of E_1 and E_2 lie along the X-axis and Y-axis, respectively. Let S be the circle $x^2 + (y - 1)^2 = 2$. The straight line $x + y = 3$ touches the curves S, E_1 and E_2 at P, Q and R, respectively. Suppose that $PQ = PR = \frac{2\sqrt{2}}{3}$. If e_1 and e_2 are the eccentricities of E_1 and E_2 respectively, then the correct expression(s) is/are **[JEE Advanced 2015]**

(a) $e_1^2 + e_2^2 = \frac{43}{40}$ (b) $e_1 e_2 = \frac{\sqrt{7}}{2\sqrt{10}}$

(c) $|e_1^2 - e_2^2| = \frac{5}{8}$ (d) $e_1 e_2 = \frac{\sqrt{3}}{4}$

Linked-Comprehension Type

In these questions, a passage (paragraph) has been given followed by questions based on each of the passages. You have to answer the questions based on the passage given.

PASSAGE 1

A sequence of ellipses $E_1, E_2, \ldots, E_n$ is constructed as follows : Ellipse E_n is drawn so as to touch ellipse E_{n-1} as the extremites of the major axis of E_{n-1} and to have its foci at the extremities of the minor axis of E_{n-1}.

On the basis of above information, answer the following questions

1. If E_n is independent of n, then the eccentricity of ellipse E_{n-2} is

(a) $\left(\frac{3-\sqrt{5}}{2}\right)$ (b) $\left(\frac{\sqrt{5}-1}{2}\right)$

(c) $\frac{2-\sqrt{3}}{2}$ (d) $\frac{\sqrt{3}-1}{2}$

2. If eccentricity of ellipse E_n is e_n then the locus of (e_n^2, e_{n-1}^2) is

(a) a parabola (b) an ellipse

(c) a hyperbola (d) a rectangular hyperbola

3. If eccentricity of ellipse E_1 is $\sqrt{3}/2$ then the eccentricity of ellipse E_4 is

(a) $1/\sqrt{5}$ (b) 2/3

(c) $3/\sqrt{23}$ (d) None of these

4. If equation of ellipse E_1 is $\frac{x^2}{9} + \frac{y^2}{16} = 1$, then the equation of ellipse E_3 is

(a) $\frac{x^2}{9} + \frac{y^2}{16} = 1$ (b) $\frac{x^2}{25} + \frac{y^2}{49} = 1$

(c) $\frac{x^2}{25} + \frac{y^2}{41} = 1$ (d) $\frac{x^2}{16} + \frac{y^2}{25} = 1$

5. If eccentricity E_n is independent of n, then the locus of mid point of chords of slope -1 of E_n is (If axis of E_n along Y-axis).

(a) $(\sqrt{5} - 1)x = 2y$

(b) $(\sqrt{5} + 1)x = 2y$

(c) $(3 - \sqrt{5})x = 2y$

(d) $(3 + \sqrt{5})\,x = 2y$

PASSAGE 2

Consider the standard equation of an ellipse whose focus and corresponding foot of directrix are $(\sqrt{7}, 0)$ and $\left(\frac{16}{\sqrt{7}}, 0\right)$ and a circle with equation $x^2 + y^2 = r^2$. If in the first quadrant, the common tangent to a circle of this family and the above ellipse meets the coordinate axes at A and B.

On the basis of above information, answer the following questions

1. The equation of the ellipse is

(a) $16x^2 + 9y^2 = 144$

(b) $9x^2 + 16y^2 = 144$

(c) $16x^2 + y^2 = 144$

(d) $x^2 + 9y^2 = 144$

2. Let P be a variable point on the ellipse with foci at S and S'. If Δ be the area of triangle PSS', then the maximum value of Δ is

(a) $\sqrt{7}$ sq unit

(b) $2\sqrt{7}$ sq unit

(c) $3\sqrt{7}$ sq unit

(d) $4\sqrt{7}$ sq unit

3. If mid point of A and B is (x_1, y_1) and slope of common tangent be m, then

(a) $2mx_1 + y_1 = 0$

(b) $2my_1 + x_1 = 0$

(c) $my_1 + x_1 = 0$

(d) $mx_1 + y_1 = 0$

4. The locus of mid point of A and B is

(a) $y = x\sqrt{\left(\frac{r^2 - 9}{16 - r^2}\right)}$

(b) $y = x\sqrt{\left(\frac{16 + r^2}{9 - r^2}\right)}$

(c) $y = x\sqrt{\left(\frac{16 - r^2}{r^2 + 9}\right)}$

(d) $y = x\sqrt{\left(\frac{r^2 - 9}{r^2 - 16}\right)}$

5. The domain of value of r $(r > 0)$ for which the above question is true is

(a) $r \in (1, 2)$

(b) $r \in (3, 4)$

(c) $r \in (4, 5)$

(d) $r \in (5, 6)$

PASSAGE 3

$C : x^2 + y^2 = 9, E : \frac{x^2}{9} + \frac{y^2}{4} = 1, L : y = 2x.$

On the basis of above information, answer the following questions

1. P is a point on the circle C, the perpendicular PQ to the major axis of the ellipse E meets the ellipse at M, then $\frac{MQ}{PQ}$ is equal to

(a) 1/3 (b) 2/3

(c) 1/2 (d) None of these

2. If L represents the line joining the point P on C to its centre O, then equation of the tangent at M to the ellipse E is

(a) $x + 3y = 3\sqrt{5}$ (b) $4x + 3y = \sqrt{5}$

(c) $3x + y + 3\sqrt{5} = 0$ (d) $4x + 3y + \sqrt{5} = 0$

3. Equation of the diameter of the ellipse E conjugate to the diameter represented by L is

(a) $9x + 2y = 0$ (b) $2x + 9y = 0$

(c) $4x + 9y = 0$ (d) $4x - 9y = 0$

4. If R is the point of intersection of the line L with the line $x = 1$, then

(a) R lies inside both C and E

(b) R lies outside both C and E

(c) R lies on both C and E

(d) R lies inside C but outside E

PASSAGE 4

Let F_1 $(x_1, 0)$ and F_2 $(x_2, 0)$, for $x_1 < 0$ and $x_2 > 0$, be the foci of the ellipse $\frac{x^2}{9} + \frac{y^2}{8} = 1$. Suppose a parabola having vertex at the origin and focus at F_2 intersects the ellipse at point M in the first quadrant and at point N in the fourth quadrant. **[JEE Advanced 2016]**

On the basis of above information, answer the following questions

1. The orthocentre of ΔF_1MN is

(a) $\left(-\frac{9}{10}, 0\right)$ (b) $\left(\frac{2}{3}, 0\right)$

(c) $\left(\frac{9}{10}, 0\right)$ (d) $\left(\frac{2}{3}, \sqrt{6}\right)$

2. If the tangents to the ellipse at M and N meet at R and the normal to the parabola at M meets the X-axis at Q, then the ratio of area of ΔMQR to area of the quadrilateral MF_1NF_2 is

(a) 3 : 4 (b) 4 : 5

(c) 5 : 8 (d) 2 : 3

Numerical Value Type

Solve the following problems and write your answer.

1. The equation of the ellipse referred to its centre whose minor axis is equal to the distance between the foci and whose latusrectum is 10 is $mx^2 + ny^2 = 100$, when $m, n \in N$, then the value of $10^m + 20^n$ must be
2. If the normal at an end of a latusrectum of an ellipse $\frac{x^2}{a^2} + \frac{y^2}{b^2} = 1$ passes through one extremity of the minor axis. If e be the eccentricity of the ellipse then the value of $625(2e^2 + 1)^2$ must be
3. A ray emanating from the point $(-3, 0)$ is incident on the ellipse $16x^2 + 25y^2 = 400$ at the point P with ordinate 4. If the equation of the reflected ray after first reflection is $4x + 3y = \lambda$, then the value of 2^λ must be
4. If e be the eccentricity of the ellipse $4(x - 2y + 1)^2 + 9(2x + y + 2)^2 = 25$, then the value of e^2 must be
5. The orbit of the earth is an ellipse with eccentricity 1/60 with the sun at one focus the major axis being approximately 186×10^6 miles in length. If the shortest and longest distances of th earth from the sun are $\lambda \times 10^5$ miles and $\mu \times 10^5$ miles then the value of $(\lambda + \mu)$ must be
6. A variable point P on an ellipse of eccentricity 3/5 is joined to its *foci* S, S', then the locus of the incentre of the $\Delta PSS'$ is an ellipse. If λ be the eccentricity of the ellipse, then the value of $2\lambda^2$ must be
7. Tangents are drawn to the ellipse $\frac{x^2}{9} + \frac{y^2}{5} = 1$ at ends of latusrectum. If the area of an quadrilateral be λ sq unit, then the value of λ must be
8. Let Δ_1 be the area of ΔPQR inscribed in an ellipse and Δ_2 be the area of the $\Delta P'Q'R'$ whose vertices are the points lying on the auxiliary circle corresponding to the points P, Q, R, respectively. If the eccentricity of the ellipse is $\frac{4\sqrt{3}}{7}$ then the ratio $343\frac{\Delta_2}{\Delta_1}$ must be
9. A parabola is drawn to pass through A and B the ends of a diameter of a given circle of radius 3 unit and to have as directrix a tangent to a concentric circle of radius 5 unit, the axes being AB and a perpendicular diameter, then the locus of the focus of the parabola is an ellipse, then the (radius)^2 of director circle of the ellipse is

Single Integer Type

Solve the following problems and write your answer.

1. The tangent and normal at any point $P(\theta)$ of an ellipse $\frac{x^2}{a^2} + \frac{y^2}{b^2}$ having eccentricity 'e' cut its major axis in points Q and R. If $QR = a$, then $e^2\cos^2\theta + \cos\theta =$
2. Suppose that the foci of the ellipse $\frac{x^2}{9} + \frac{y^2}{5} = 1$ are $(f_1, 0)$ and $(f_2, 0)$, where $f_1 > 0$ and $f_2 < 0$. Let P_1 and P_2 be two parabolas with a common vertex at $(0, 0)$ with foci at $(f_1, 0)$ and $(2f_2, 0)$, respectively. Let T_1 be a tangent to P_1 which passes through $(2f_2, 0)$ and T_2 be a tangent to P_2 which passes through $(f_1, 0)$. If m_1 is the slope of T_1 and m_2 is the slope of T_2, then the value of $\left(\frac{1}{m_1^2} + m_2^2\right)$ is **[JEE Advanced 2015]**
3. For an ellipse $\frac{x^2}{9} + \frac{y^2}{4} = 1$ with vertices A and A', tangent drawn at the point P in the first quadrant meets the Y-axis in Q and the chord $A'P$ meets the Y-axis in M. If 'O' is the origin, then $OQ^2 - MQ^2$ equals to
4. Tangents drawn from the point $P(2, 3)$ to the circle $x^2 + y^2 - 8x + 6y + 1 = 0$ touch circle at points A and B. The circumcircle of ΔPAB cuts the director circle of ellipse $\frac{(x+5)^2}{9} + \frac{(y-3)^2}{b^2} = 1$ orthogonally. Then the value of $[b]$ is ......, where $[\cdot]$ is greatest integer function.
5. Number of distinct normal lines that can be drawn to ellipse $\frac{x^2}{169} + \frac{y^2}{25} = 1$ from the point $P(0, 6)$ is
6. A stick of length 10 m slides on coordinate axes, then locus of a point dividing this stick reckoning from X-axis in the ratio 6 : 4 is a curve whose eccentricity is e, then $9e^2$ is equal to
7. Let S and S' be the foci of an ellipse and B be any one of the extremities of its minor axis. If $\Delta S'BS$ is a right angled triangle with right angle at B and area $(\Delta S'BS) = 8$ sq units, then the length of a latus rectum of the ellipse is

Matrix-Match Type

Given below are Matching Type Questions, with two columns (each having some items) each . Each item of Column I has to be matched with the items of Column II, by encircling the correct match(es).

NOTE An item of Column I can be matched with more than one items of Column II.

1. Observe the following columns :

	Column I		Column II
(A)	A ray emanating from the point $(0, -\sqrt{5})$ is incident on the ellipse $9x^2 + 4y^2 = 36$ at the point P with abscissa 2. The equations of reflected rays after first reflection and second reflection are represented by F and S respectively, then	(P)	$F: 2x + y - 2\sqrt{5} = 0$
		(Q)	$S: x + 4\sqrt{5}y - \sqrt{5} = 0$
(B)	A ray emanating from the point $(\sqrt{5}, 0)$ is incident on the ellipse $4x^2 + 9y^2 = 36$ at the point P with ordinate -2. The equations of reflected rays after first reflection and second reflection are represented by F and S respectively, then	(R)	$F: x\sqrt{5} + 2y - 2\sqrt{5} = 0$
(C)	A ray emanating from the point $(-\sqrt{5}, 0)$ is incident on the ellipse $4x^2 + 5y^2 = 100$ at the point P with ordinate $2\sqrt{5}$. The equations of reflected rays after first reflection and second reflection are represented by F and S respectively, then	(S)	$S: 4\sqrt{5}\,x + y + \sqrt{5} = 0$
		(T)	$F: 2x + y\sqrt{5} + 2\sqrt{5} = 0$

2. Observe the following columns :

	Column I		Column II
(A)	A tangent to the ellipse $\frac{x^2}{27} + \frac{y^2}{48} = 1$ having slope $-\frac{4}{3}$ cuts the X and Y-axes at the points A and B respectively. If O is the origin then area of ΔOAB is equal to	(P)	36
(B)	Product of the perpendiculars drawn from the point $(\pm 3, 0)$ to the line $y = mx - \sqrt{25m^2 + 16}$ is	(Q)	72
(C)	An ellipse passing through the origin has its foci (3, 4) and (6, 8), then length of its minor axis is	(R)	10
(D)	If PQ is focal chord of ellipse $\frac{x^2}{25} + \frac{y^2}{16} = 1$ which passes through $S \equiv (3, 0)$ and $PS = 2$ then length of chord PQ is	(S)	16
		(T)	$10\sqrt{2}$

3. Observe the following columns :

	Column I		Column II
(A)	If $E: 2x^2 + y^2 = 2$ and director circle of E is C_1, director circle of C_1 is C_2, director circle of C_2 is C_3 and so on. If $r_1, r_2, r_3, \ldots$ are the radii of $C_1, C_2, C_3 \ldots$ respectively. Then	(P)	$r_1^2, r_2^2, r_3^2 \ldots$ are in GP with common ratio 2.
		(Q)	GM of r_1^2, r_2^2, r_3^2 is 6.
(B)	If $E: 3x^2 + 2y^2 = 6$ and director circle of E is C_1, director circle of C_1 is C_2, director circle of C_2 is C_3 and so on. If $r_1, r_2, r_3, \ldots$ are the radii of $C_1, C_2, C_3, \ldots$ respectively. Then	(R)	$r_1^2 + r_2^2 + r_3^2 + \ldots + r_n^2 = r_1^2 (2^n - 1)$
(C)	If $E: 3x^2 + 4y^2 = 12$ and director circle of E is C_1, director circle of C_1 is C_2, director circle of C_2 is C_3 and so on. If $r_1, r_2, r_3, \ldots$ are the radii of $C_1, C_2, C_3, \ldots$ respectively. Then	(S)	GM of r_1^2, r_2^2, r_3^2 is 10
		(T)	GM of r_1^2, r_2^2, r_3^2 is 14

Answers

Objective Questions Type [Only one correct answer]

1. (b)	**2.** (d)	**3.** (d)	**4.** (c)	**5.** (a)	**6.** (b)	**7.** (a)	**8.** (d)	**9.** (b)	**10.** (b)
11. (c)	**12.** (b)	**13.** (a)	**14.** (b)	**15.** (c)	**16.** (b)	**17.** (a)	**18.** (c)	**19.** (b)	**20.** (c)
21. (c)	**22.** (c)	**23.** (d)	**24.** (b)	**25.** (a)	**26.** (b)	**27.** (b)	**28.** (a)	**29.** (c)	**30.** (a)
31. (a)	**32.** (d)	**33.** (d)	**34.** (a)	**35.** (b)	**36.** (a)	**37.** (a)	**38.** (a)	**39.** (a)	**40.** (d)
41. (c)	**42.** (b)	**43.** (c)	**44.** (a)	**45.** (a)	**46.** (a)	**47.** (a)	**48.** (d)	**49.** (d)	**50.** (d)
51. (a)	**52.** (d)	**53.** (b)	**54.** (b)	**55.** (d)	**56.** (a)				

Objective Questions Type II [One or more than one correct answer(s)]

1. (a, c)	**2.** (b, c)	**3.** (a, c)	**4.** (a, b)	**5.** (a, c)
6. (a, c)	**7.** (a, d)	**8.** (c, d)	**9.** (a, b)	**10.** (a, c, d)
11. (a, b, c, d)	**12.** (a, c, d)	**13.** (a, b)	**14.** (a, c)	**15.** (a, b)

Linked-Comprehension Type

Passage 1 **1.** (b) **2.** (d) **3.** (d) **4.** (c) **5.** (b)
Passage 2 **1.** (b) **2.** (c) **3.** (d) **4.** (a) **5.** (b)
Passage 3 **1.** (b) **2.** (a) **3.** (b) **4.** (d)
Passage 4 **1.** (a) **2.** (c)

Numerical Value Type

1. (410)	**2.** (3125)	**3.** (4096)	**4.** (0.56)	**5.** (1860)
6. (1.5)	**7.** (27)	**8.** (2401)	**9.** (41)	

Single Integer Type

1. (1) **2.** (4) **3.** (4) **4.** (7) **5.** (3) **6.** (5) **7.** (4)

Matrix-Match Type

1. A → (R, S,); B → (Q, T); C → (P)
2. A → (P); B → (S); C → (T); D → (R)
3. A → (P, Q, R,); B → (P, R, S); C → (P, R, T)

Solutions

Objective Questions Type [Only one correct answer]

1. According to given information,

Let the equation of ellipse $\frac{x^2}{a^2}+\frac{3y^2}{4}=1$ ...(i)

$\because$ Ellipse (i) touches the line $x+6y=8$, then

$$\left(\frac{8}{6}\right)^2=a^2\left(\frac{1}{6}\right)^2+\frac{4}{3}$$

$$\Rightarrow \quad \frac{16}{9}=\frac{a^2}{36}+\frac{4}{3} \Rightarrow \frac{a^2}{36}=\frac{4}{9}$$

$$\Rightarrow \quad a^2=16$$

Now, eccentricity $e=\sqrt{1-\frac{b^2}{a^2}}=\sqrt{1-\frac{4}{3\times16}}=\frac{1}{2}\sqrt{\frac{11}{3}}$

Hence, option (b) is correct.

2. Since Y-axis is major axis

$$\Rightarrow \quad f(4a)<f(a^2-5)$$

$$\Rightarrow \quad 4a>a^2-5 \quad [\because f \text{ is decreasing}]$$

$$\Rightarrow a^2-4a-5<0 \Rightarrow a\in(-1,5)$$

3. $A\equiv(a\cos\theta, b\sin\theta)$

$B\equiv(a\cos(\theta+\alpha), b\sin(\theta+\alpha))$

$C\equiv(a\cos(\theta+2\alpha), b\sin(\theta+2\alpha))$

$\Delta=$ Area of ΔABC

$$=\frac{1}{2}\begin{vmatrix}1 & a\cos\theta & b\sin\theta\\ 1 & a\cos(a+\alpha) & b\sin(\theta+\alpha)\\ 1 & a\cos(\theta+2\alpha) & b\sin(\theta+2\alpha)\end{vmatrix}$$

$$=2ab\sin^2\left(\frac{\alpha}{2}\right)\sin\alpha$$

$$\Delta(\alpha)=ab\sin\alpha(1-\cos\alpha)$$

$$=\frac{ab}{2}(2\sin\alpha-\sin2\alpha)$$

$$\Delta'(\alpha)=0$$

$$\Rightarrow \quad \cos\alpha=1$$

or $\cos\alpha=-\frac{1}{2}$

$\cos\alpha=1$ gives $\Delta=0$

$\cos\alpha=-\frac{1}{2}$ gives maximum value of $\Delta=\frac{3\sqrt{3}}{4}ab$

4. The tangent at the point of shortest distance from the line $x+y=7$ parallel to the given line.

Any point on the given ellipse is $(\sqrt{6}\cos\theta, \sqrt{3}\sin\theta)$.

Equation of the tangent is

$\frac{x\cos\theta}{\sqrt{6}}+\frac{y\sin\theta}{\sqrt{3}}=1$. It is parallel to $x+y=7$

$$\Rightarrow \quad \frac{\cos\theta}{\sqrt{6}}=\frac{\sin\theta}{\sqrt{3}}$$

$$\Rightarrow \quad \frac{\cos\theta}{\sqrt{2}}=\frac{\sin\theta}{1}=\frac{1}{\sqrt{3}}$$

The required point is (2, 1).

5. Normal at the extremity of latusrectum in the first quadrant $(ae, b^2/a)$ is

$$\frac{x-ae}{ae/a^2}=\frac{y-b^2/a}{b^2/ab^2}$$

As it passes through $(0,-b)$

$$\frac{-ae}{ae/a^2}=\frac{-b-b^2/a}{1/a}$$

$$\Rightarrow \quad -a^2=-ab-b^2$$

$$\Rightarrow \quad a^2-b^2=ab$$

$$\Rightarrow \quad a^2e^2=ab \text{ or } e^2=b/a$$

$$\therefore \quad e^4=\frac{b^2}{a^2}=1-e^2 \Rightarrow e^4+e^2=1$$

6. Slope of line

$$=\frac{b}{a}\left(\frac{\sin60°-\sin30°}{\cos60°-\cos30°}\right)=-\frac{b}{a}=-\frac{1}{2}$$

Homogenizing the ellipse with $x+2y+4=0$

$$\frac{x^2}{4b^2}+\frac{y^2}{b^2}=\left(\frac{x+2y}{-4}\right)^2$$

$$\Rightarrow \quad \frac{1}{b^2}\left(\frac{x^2}{4}+y^2\right)=\frac{x^2+4y^2+4xy}{16}$$

Coefficient of x^2+ coefficient of $y^2=0$

$$\frac{1}{4b^2}-\frac{1}{16}+\frac{1}{b^2}-\frac{4}{16}=0 \Rightarrow b^2=4, b=2$$

$$\Rightarrow \quad a=4$$

Ellipse is $\frac{x^2}{16}+\frac{y^2}{4}=1$

7. We have, $y^2=3x^2$ and $x^2+y^2=4b$

Solve both, we get

$$\Rightarrow \quad x^2=b$$

$$\Rightarrow \quad \frac{x^2}{16}+\frac{3x^2}{b^2}=1$$

$$\Rightarrow \quad \frac{b}{16}+\frac{3}{b}=1$$

$$\Rightarrow \quad b^2-16b+48=0$$

$$\Rightarrow \quad (b-12)(b-4)=0$$

$$\Rightarrow \quad b=12, b>4$$

8. The point ϕ on the ellipse $\frac{x^2}{4}+y^2=1$ will be $(2\cos\phi, \sin\phi)$.

It lies inside the circle $x^2+y^2+4x+3=0$ if

$$4\cos^2\phi+\sin^2\phi+8\cos\phi+3<0$$

$$\Rightarrow \quad 3\cos^2\phi+8\cos\phi+4<0$$

$$\Rightarrow \quad (\cos\phi+2)(3\cos\phi+2)<0$$

or $\quad -2<\cos\phi<-\frac{2}{3}$

or $\quad -1\le\cos\phi<\cos(\pi-\alpha) \quad [\because \cos\phi \not< -1]$

$\therefore \quad \pi-\alpha<\phi<\pi+\alpha$ is most appropriate.

9. Mutually perpendicular tangents have been drawn by taking any point on the director circle of given ellipse. Any point on the director circle can be taken as $(5\cos\theta, 5\sin\theta)$.

Equation of corresponding chord of contact is

$$\frac{5x}{16}\cos\theta + \frac{5y}{9}\sin\theta - 1 = 0$$

It's distance from the origin is equal to

$$\frac{1}{\sqrt{\left(\frac{25\cos^2\theta}{256}+\frac{25\sin^2\theta}{81}\right)}} = \frac{144}{5}$$

$$\frac{1}{\sqrt{(256-175\cos^2\theta)}} \le \frac{144}{5\cdot 9} = \frac{16}{5}$$

10. Let the equation of the semi elliptical arch be $\frac{x^2}{a^2}+\frac{y^2}{b^2}=1$

$(y > 0)$

Length of the major axis $= 2a = 9$

$\Rightarrow \quad a = \frac{9}{2}$

Length of the semi minor axis $b = 3$

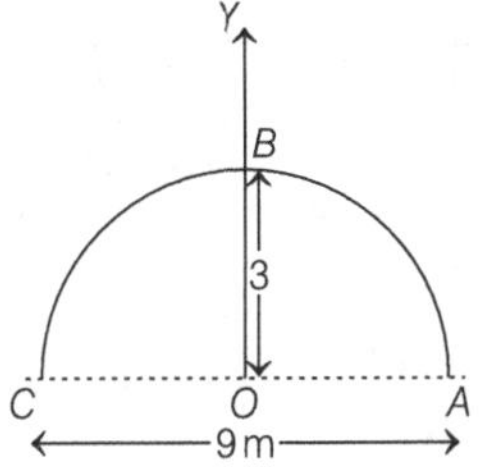

So, the equation of the arc becomes $\frac{4x^2}{81}+\frac{y^2}{9}=1$

If $\quad x = 2$, than $y^2 = \frac{65}{9}$

$\Rightarrow \quad y = \frac{\sqrt{65}}{3} = \frac{8}{3}$ (app.)

11. Let $y = mx + \frac{1}{m}$ be a tangent to the parabola, then for the required equilateral triangle.

$m = \pm\frac{1}{\sqrt{3}}$ and it is tangent to ellipse also, if

$$\left(\frac{1}{m}\right)^2 = a^2m^2 + b^2$$

$\Rightarrow \quad 3 = \frac{a^2}{3} + b^2$

$\Rightarrow \quad 3 = \frac{a^2}{3} + a^2 - a^2e^2$

$\Rightarrow \quad a^2e^2 = \frac{4a^2}{3} - 3$ or $e^2 = \frac{4}{3} - \frac{3}{a^2}$

Now $\quad 0 < e^2 < 1$

$\Rightarrow \quad 0 < \frac{4}{3} - \frac{3}{a^2} < 1$

$\Rightarrow \quad \frac{3}{2} < a < 3$

12. $\frac{x^2}{25}+\frac{y^2}{16}=1$

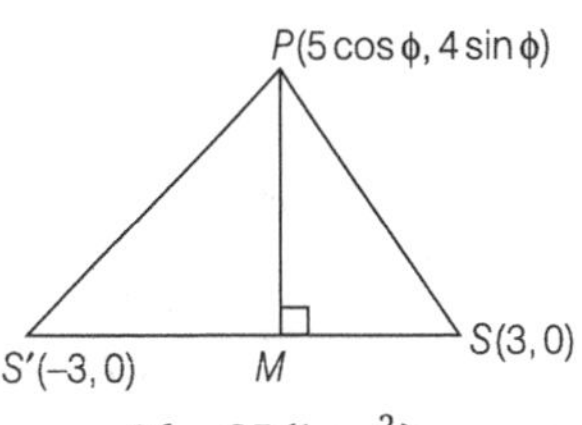

$\therefore \quad 16 = 25(1-e^2)$

$\Rightarrow \quad e^2 = \frac{9}{25} \Rightarrow e = \frac{3}{5}$

$\therefore \quad S(3, 0)$ and $S'(-3, 0)$

$$A = \frac{1}{2}\times SS'\times PM = \frac{1}{2}\times 6\times 4\sin\phi = 12\sin\phi$$

Max. Area $= 12$ sq unit

13. Let $P(a\cos\alpha, b\sin\alpha)$, $Q(a\cos\beta, b\sin\beta)$ and $R(a\cos\gamma, b\sin\gamma)$ be the vertices of the triangle inscribed in the ellipse $\frac{x^2}{a^2}+\frac{y^2}{b^2}=1$. The points on the auxiliary circle corresponding to these points are $A(a\cos\alpha, a\sin\alpha)$, $B(a\cos\beta, a\sin\beta)$ and $C(a\cos\gamma, a\sin\gamma)$

$\therefore \Delta =$ Area of

$$\Delta PQR = \frac{1}{2}\left|\begin{vmatrix} a\cos\alpha & b\sin\alpha & 1\\ a\cos\beta & b\sin\beta & 1\\ a\cos\gamma & b\sin\gamma & 1\end{vmatrix}\right|$$

$$= \frac{1}{2}ab\left|\begin{vmatrix} \cos\alpha & \sin\alpha & 1\\ \cos\beta & \sin\beta & 1\\ \cos\gamma & \sin\gamma & 1\end{vmatrix}\right|$$

and $\Delta_1 =$ Area of

$$\Delta ABC = \frac{1}{2}\left|\begin{vmatrix} a\cos\alpha & a\sin\alpha & 1\\ a\cos\beta & a\sin\beta & 1\\ a\cos\gamma & a\sin\gamma & 1\end{vmatrix}\right|$$

$$= \frac{1}{2}a^2\left|\begin{vmatrix} \cos\alpha & \sin\alpha & 1\\ \cos\beta & \sin\beta & 1\\ \cos\gamma & \sin\gamma & 1\end{vmatrix}\right|$$

$\therefore \quad \frac{\Delta}{\Delta_1} = \frac{b}{a}$

14. Let the equation of ellipse be

$$\frac{x^2}{a^2}+\frac{y^2}{b^2}=1$$

Then, according the problem, we have

$$\frac{2b^2}{a} = 8 \text{ and } 2ae = 2b$$

[length of latus rectum $= \frac{2b^2}{a}$ and length of minor axis $= 2b$]

$\Rightarrow \quad b\left(\frac{b}{a}\right) = 4$ and $\frac{b}{a} = e \Rightarrow b(e) = 4$

$\Rightarrow \quad b = 4\cdot\frac{1}{e}$ ...(i)

Also, we know that $b^2 = a^2(1-e^2)$

$\Rightarrow \quad \dfrac{b^2}{a^2} = 1 - e^2$

$\Rightarrow \quad e^2 = 1 - e^2 \qquad \left[\because \dfrac{b}{a} = e\right]$

$\Rightarrow \quad 2e^2 = 1 \Rightarrow e = \dfrac{1}{\sqrt{2}} \qquad \ldots\text{(ii)}$

From Eqs. (i) and (ii), we get

$$b = 4\sqrt{2}$$

Now, $\quad a^2 = \dfrac{b^2}{1-e^2} = \dfrac{32}{1-\dfrac{1}{2}} = 64$

$\therefore$ Equation of ellipse be $\dfrac{x^2}{64} + \dfrac{y^2}{32} = 1$

Now, check all the options.

Only $(4\sqrt{3}, 2\sqrt{2})$, satisfy the above equation.

15. Equation of tangent is $x \cdot \dfrac{\cos\theta}{a} + y \cdot \dfrac{\sin\theta}{b} = 1$

$\therefore$ The intercepts are $a \sec\theta$ and $b \operatorname{cosec}\theta$

$\therefore$ Length of intercept $= \sqrt{a^2 \sec^2\theta + b^2 \operatorname{cosec}^2\theta}$

$\therefore \quad a^2 \sec^2\theta + b^2 \operatorname{cosec}^2\theta$

$= a^2 + b^2 + a^2 \tan^2\theta + b^2 \cot^2\theta \geq (a+b)^2$

16. We have,

$(13x-1)^2 + (13y-2)^2 = a(5x+12y-1)^2 \qquad \ldots\text{(i)}$

$$\left(x - \frac{1}{13}\right)^2 + \left(y - \frac{2}{13}\right)^2 = \frac{a(5x+12y-1)^2}{\left(\sqrt{(5^2+12^2)}\right)^2}$$

or $\sqrt{\left(x - \dfrac{1}{13}\right)^2 + \left(y - \dfrac{2}{13}\right)^2} = \sqrt{a}\dfrac{|5x+12y-1|}{\sqrt{(5^2+12^2)}}$

$\therefore \quad SP = ePM$

or $\quad 0 < e < 1$

$\Rightarrow \quad 0 < \sqrt{a} < 1$

$\therefore \quad 0 < a < 1$

17. Centre of the ellipse is (1, 2) and length of major axis and minor axis are 6 and 4 respectively and centre and radius of the circle are (1, 2) and 1 respectively.

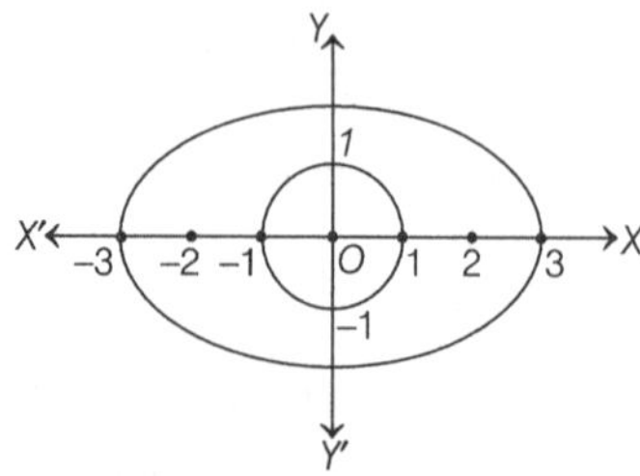

Hence, ellipse and circle do not touch or cut.

$\therefore$ Common chord impossible.

$\therefore$ Hence, length of comman chord = 0

18. Ellipse is $\dfrac{x^2}{a^2} + \dfrac{y^2}{b^2} = 1$

Tangent at 'P' is

$$\frac{x}{a}\cos\phi + \frac{y}{b}\sin\phi = 1$$

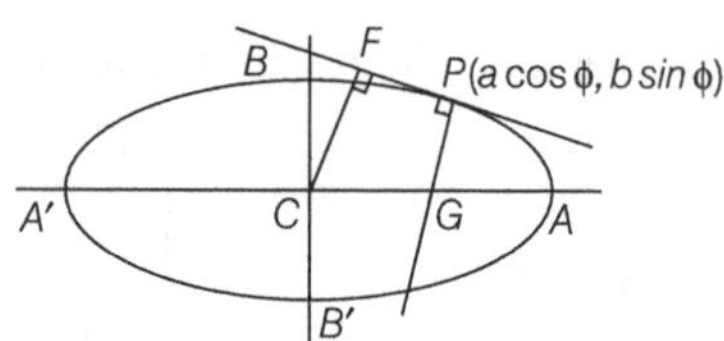

$\therefore \quad CF = \dfrac{1}{\sqrt{\left(\dfrac{\cos^2\phi}{a^2} + \dfrac{\sin^2\phi}{b^2}\right)}}$

$= \dfrac{ab}{\sqrt{(a^2\sin^2\phi + b^2\cos^2\phi)}}$

Equation of normal at P

$ax \sec\phi - by \operatorname{cosec}\phi = a^2 - b^2$

then $\quad G = \left(\dfrac{(a^2-b^2)}{2}\cos\phi, 0\right)$

$$PG = \sqrt{\left\{\left(a\cos\phi - \frac{(a^2-b^2)}{a}\cos\phi\right)^2 + (b\sin\phi - 0)^2\right\}}$$

$$= \sqrt{\left(\frac{b^4}{a^2}\cos^2\phi + b^2\sin^2\phi\right)}$$

$$= \frac{b}{a}\sqrt{(a^2\sin^2\phi + b^2\cos^2\phi)}$$

$\Rightarrow \quad CF \cdot PG = b^2$

19. Let any point on $x - y - 5 = 0$ is $(\lambda, \lambda - 5)$

$\therefore$ Chord of contact of $x^2 + 4y^2 = 4$ with respect to $(\lambda, \lambda - 5)$ is

$$x\lambda + 4y \cdot (\lambda - 5) = 4$$

$\Rightarrow \quad (4 + 20y) - \lambda(x + 4y) = 0$

For fixed point $4 + 20y = 0$ and $x + 4y = 0$

$\therefore \quad y = -\dfrac{1}{5}$ then $x = \dfrac{4}{5}$

$\therefore$ Fixed point is $\left(\dfrac{4}{5}, -\dfrac{1}{5}\right)$

20. Let the ellipse be $\dfrac{x^2}{a^2} + \dfrac{y^2}{b^2} = 1$. Then coordinates of F, B and F' are $(ae, 0)$, $(0, b)$ and $(-ae, 0)$ respectively.

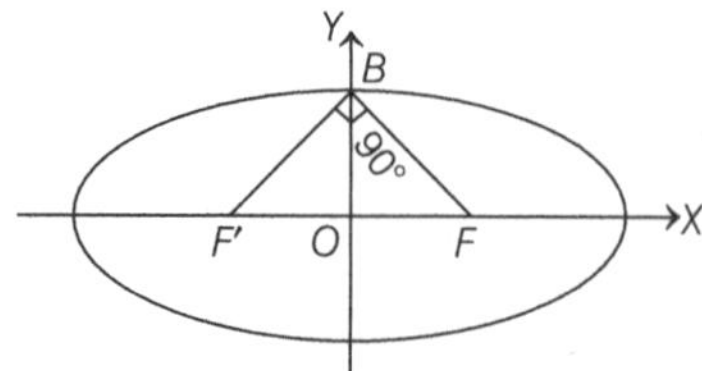

Then, slope of $BF = \dfrac{b-0}{0-ae} = -\dfrac{b}{ae}$

and slope of $BF' = \frac{0-b}{-ae-0} = \frac{b}{ae}$

$\because \quad \angle FBF' = 90°$

$\Rightarrow$ Slope of $BF \times$ slope of $BF' = -1$

$\Rightarrow \quad -\frac{b^2}{a^2e^2} = -1$

$\Rightarrow \quad b^2 = a^2e^2$

$\Rightarrow \quad a^2(1-e^2) = a^2e^2$

$\therefore \quad e = \frac{1}{\sqrt{2}}$

21. $\because$ Equation of tangent of $y^2 = 4ax$ in terms of slope (m) is

$y = mx + \frac{a}{m}$

Which is also tangent of $\frac{x^2}{a^2} + \frac{y^2}{b^2} = 1$

then $\quad \left(\frac{a}{m}\right)^2 = a^2m^2 + b^2$

$\Rightarrow \quad a^2\left(\frac{1}{m^2} - m^2\right) = b^2 \Rightarrow \left(\frac{1}{m^2} - m^2\right) = \frac{b^2}{a^2}$

$\Rightarrow \frac{(1+m^2)(1-m^2)}{m^2} = \frac{b^2}{a^2}$

$\Rightarrow \quad \left(\frac{1-m^2}{m^2}\right) = \frac{b^2}{a^2(1+m^2)} > 0$

$\Rightarrow \quad \frac{1-m^2}{m^2} > 0$

$\Rightarrow \quad \frac{m^2-1}{m^2} < 0$

$\Rightarrow \quad 0 < m^2 < 1$

$\therefore \quad m \in (-1, 0) \cup (0, 1)$

For positive values of m set is $m \in (0, 1)$

22. Equation of tangent be

$\frac{x\cos\theta}{3\sqrt{3}} + \frac{y \cdot \sin\theta}{1} = 1, \quad \left[\theta \in \left(0, \frac{\pi}{2}\right)\right]$

Intercept on X-axis

$OA = 3\sqrt{3}\sec\theta$

intercept on Y-axis

$OB = \operatorname{cosec}\theta$

Now, sum of intercept

$= 3\sqrt{3}\sec\theta + \operatorname{cosec}\theta = f(\theta)$ let

$f'(\theta) = 3\sqrt{3}\sec\theta\tan\theta - \operatorname{cosec}\theta\cot\theta$

$= 3\sqrt{3}\frac{\sin\theta}{\cos^2\theta} - \frac{\cos\theta}{\sin^2\theta}$

$= \underbrace{\frac{\cos\theta}{\sin^2\theta}}_{\oplus} \cdot 3\sqrt{3}\left[\tan^3\theta - \frac{1}{3\sqrt{3}}\right] = 0 \Rightarrow \theta = \frac{\pi}{6}$

$\ominus\downarrow \quad \theta = \frac{\pi}{6} \quad \oplus\uparrow$

$\Rightarrow$ at $\theta = \frac{\pi}{6}$, $f(\theta)$ is minimum

23. Equation of tangent at 'ϕ' of $\frac{x^2}{a^2} + \frac{y^2}{b^2} = 1$ is

$\frac{x}{a}\cos\phi + \frac{y}{b}\sin\phi = 1$

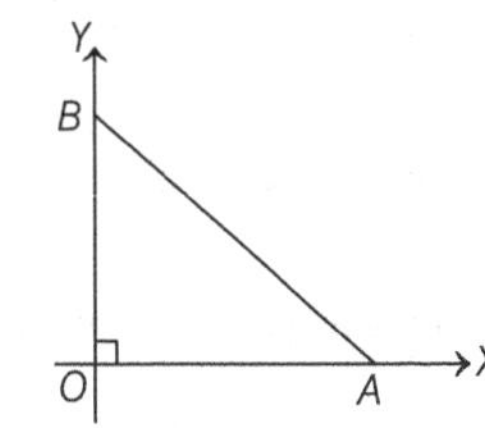

$\Rightarrow \quad \frac{x}{a\sec\phi} + \frac{y}{b\operatorname{cosec}\phi} = 1$

$\therefore \quad A \equiv (a\sec\phi, 0)$

and $\quad B \equiv (0, b\operatorname{cosec}\phi)$

$\therefore$ Length of intercept $= \sqrt{(a^2\sec^2\phi + b^2\operatorname{cosec}^2\phi)}$

Let $z = a^2\sec^2\phi + b^2\operatorname{cosce}^2\phi$

$\therefore \frac{dz}{d\phi} = 2a^2\sec^2\phi\tan\phi - 2b^2\operatorname{cosec}^2\phi\cot\phi$

For maxima or minima, $\frac{dz}{d\phi} = 0$

$\therefore \quad \frac{a^2\sin\phi}{\cos^3\phi} = \frac{b^2\cos\phi}{\sin^3\phi}$

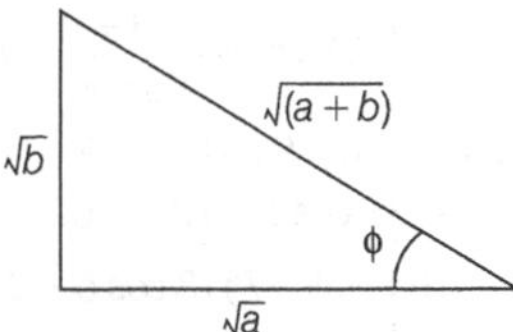

$\Rightarrow \quad \tan^2\phi = \frac{b}{a}$

$\therefore \quad \tan\phi = \sqrt{\frac{b}{a}}$

Hence, minimum length of intercept

$= \sqrt{a^2\frac{(a+b)}{a} + b^2\frac{(a+b)}{b}}$

$= \sqrt{(a+b)^2}$

$= (a+b)$

24. $\because \quad P \equiv (a\cos\phi, b\sin\phi)$

and $D \equiv (-a\sin\phi, b\cos\phi)$

and $C \equiv (0, 0)$

$\therefore \quad CP^2 + CD^2$

$= a^2\cos^2\phi + b^2\sin^2\phi + a^2\sin^2\phi + b^2\cos^2\phi$

$= a^2(\sin^2\phi + \cos^2\phi) + b^2(\sin^2\phi + \cos^2\phi)$

$= a^2(1) + b^2(1)$

$= a^2 + b^2$

25. $\therefore\ SP + S'P = 10 \Rightarrow 2a = 10$

$\therefore\quad a = 5$

$2ae = 8 \Rightarrow ae = 4$

$\because\quad e = 4/5$

$$b^2 = a^2(1-e^2) = 25\left(1-\frac{16}{25}\right) = 9$$

$\therefore$ Area $\pi ab = \pi \cdot 5 \cdot 3 \cdot \Rightarrow = 15\pi$

26. $\because\ \dfrac{x^2}{14}+\dfrac{y^2}{5}=1 \Rightarrow \dfrac{x^2}{(\sqrt{14})^2}+\dfrac{y^2}{(\sqrt{5})^2}=1$

$\therefore\quad P \equiv (\sqrt{14}\cos\phi\ \sqrt{5}\sin\phi)$

and $Q \equiv (\sqrt{14}\cos 2\phi, \sqrt{5}\sin 2\phi)$

$\therefore$ Equation of normal at 'P' is

$$\sqrt{14}\,x\sec\phi - \sqrt{5}\,y\,\text{cosec}\,\phi = 14-5$$

$$\Rightarrow\quad \frac{\sqrt{14}\,x}{\cos\phi} - \frac{\sqrt{5}\,y}{\sin\phi} = 9$$

Also it pass through 'Q' then

$$\frac{14\cos 2\phi}{\cos\phi} - \frac{5\sin 2\phi}{\sin\phi} = 9$$

$$\Rightarrow\quad \frac{14(2\cos^2\phi-1)}{\cos\phi} - 10\cos\phi = 9$$

$$\Rightarrow\quad 18\cos\phi - \frac{14}{\cos\phi} = 9$$

$$\Rightarrow\quad 18\cos^2\phi - 9\cos\phi - 14 = 0$$

$$\Rightarrow\quad 18\cos^2\phi - 21\cos\phi + 12\cos\phi - 14 = 0$$

$$\Rightarrow\quad (6\cos\phi - 7)(3\cos\phi + 2) = 0$$

$$\therefore\quad \cos\phi = -\frac{2}{3}\quad \left[\because \cos\phi \neq \frac{7}{6}\right]$$

27. Equation of PQ is

$$\frac{x}{a}\cos\left(\frac{\alpha+\beta}{2}\right) + \frac{y}{b}\sin\left(\frac{\alpha+\beta}{2}\right) = \cos\left(\frac{\alpha-\beta}{2}\right)$$

Passing through $(ae, 0)$

$$\therefore e\cos\left(\frac{\alpha+\beta}{2}\right) = \cos\left(\frac{\alpha-\beta}{2}\right) \Rightarrow \frac{\cos\left(\frac{\alpha-\beta}{2}\right)}{\cos\left(\frac{\alpha+\beta}{2}\right)} = \frac{e}{1}$$

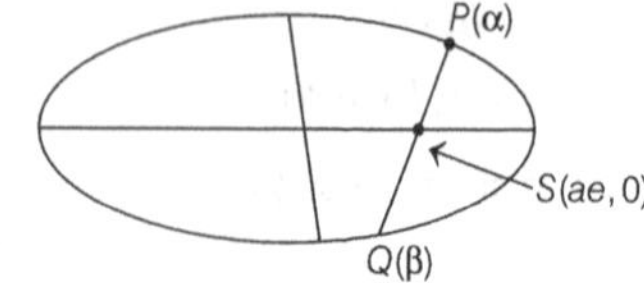

$$\frac{\cos\left(\frac{\alpha-\beta}{2}\right) - \cos\left(\frac{\alpha+\beta}{2}\right)}{\cos\left(\frac{\alpha-\beta}{2}\right) + \cos\left(\frac{\alpha+\beta}{2}\right)} = \frac{e-1}{e+1}$$

$$\Rightarrow\quad \tan(\alpha/2)\tan(\beta/2) = \left(\frac{e-1}{e+1}\right)$$

28. Equation of given ellipse is $3x^2 + 4y^2 = 12$

$$\Rightarrow\quad \frac{x^2}{4} + \frac{y^2}{3} = 1 \qquad \text{... (i)}$$

Now, let point $P(2\cos\theta, \sqrt{3}\sin\theta)$, so equation of tangent to ellipse (i) at point P is

$$\frac{x\cos\theta}{2} + \frac{y\sin\theta}{\sqrt{3}} = 1 \qquad \text{... (ii)}$$

Since, tangent (ii) passes through point $Q(4, 4)$

$$\therefore\ 2\cos\theta + \frac{4}{\sqrt{3}}\sin\theta = 1 \qquad \text{... (iii)}$$

and equation of normal to ellipse (i) at point P is

$$\frac{4x}{2\cos\theta} - \frac{3y}{\sqrt{3}\sin\theta} = 4-3$$

$$\Rightarrow\quad 2x\sin\theta - \sqrt{3}\cos\theta y = \sin\theta\cos\theta \qquad \text{... (iv)}$$

Since, normal (iv) is parallel to line, $2x + y = 4$

$\therefore$ Slope of normal (iv) = slope of line, $2x + y = 4$

$$\Rightarrow\quad \frac{2}{\sqrt{3}}\tan\theta = -2 \Rightarrow \tan\theta = -\sqrt{3} \Rightarrow \theta = 120^\circ$$

$$\Rightarrow (\sin\theta, \cos\theta) = \left(\frac{\sqrt{3}}{2}, -\frac{1}{2}\right)$$

Hence, point $P\left(-1, \dfrac{3}{2}\right)$.

Now, $PQ = \sqrt{(4+1)^2 + \left(4-\dfrac{3}{2}\right)^2}$

[given cordinates of $Q \equiv (4, 4)$]

$$= \sqrt{25 + \frac{25}{4}} = \frac{5\sqrt{5}}{2}$$

29. $\because\ \dfrac{x^2}{9} + \dfrac{y^2}{25} = 1$

If e is eccentricity, then

$$9 = 25(1-e^2)$$

$$\Rightarrow\quad e = \frac{4}{5}$$

$\therefore$ foci $(S) \equiv (0, \pm 4)$

then $SP = 5 \pm e(b)$

$$= 5 \pm \frac{4b}{5}$$

30. Since line $3x + 4y = 12\sqrt{2}$ is a tangent to the ellipse $\dfrac{x^2}{a^2} + \dfrac{y^2}{9} = 1$ for some $a \in R$, then the equation of tangent to ellipse having slope 'm' is

$$y = mx \pm \sqrt{a^2m^2 + 9} \qquad \text{...(i)}$$

$\because$ Slope of line (i), m = slope of line $3x + 4y = 12\sqrt{2}$

$$\Rightarrow\quad m = -\frac{3}{4}$$

On putting the value of m in Eq. (i), we get

$$y = -\frac{3}{4}x \pm \sqrt{\frac{9}{16}a^2 + 9}$$

$$\Rightarrow\quad 3x + 4y = \pm 3\sqrt{a^2 + 16} \qquad \text{...(ii)}$$

Since tangent $3x + 4y = 12\sqrt{2}$ represented by Eq. (ii) for some $a \in R$, therefore

$$12\sqrt{2} = \pm 3\sqrt{a^2 + 16}$$

$\Rightarrow \quad 4\sqrt{2} = \pm \sqrt{a^2 + 16}$

$\Rightarrow \quad 32 = a^2 + 16$ [on squaring both sides]

$\Rightarrow \quad a^2 = 16 \Rightarrow a = \pm 4$

$\therefore$ Eccentricity $e = \sqrt{1 - \frac{b^2}{a^2}} = \sqrt{1 - \frac{9}{16}} = \sqrt{\frac{7}{16}} = \frac{\sqrt{7}}{4}$

$\therefore$ Distance between the foci $= |2ae| = 2 \times 4 \times \frac{\sqrt{7}}{4} = 2\sqrt{7}$

31. Area of parallelogram

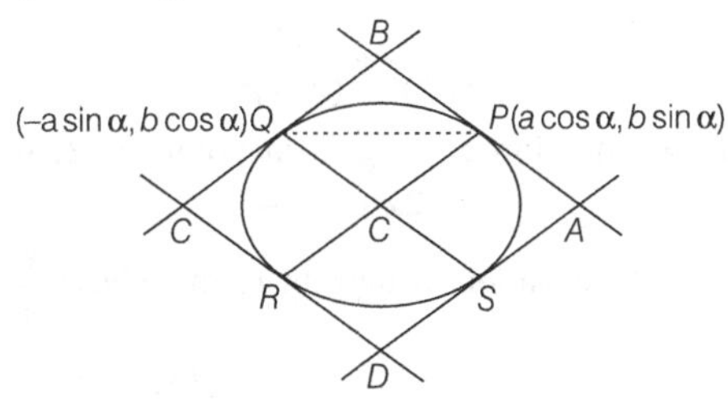

$= 4$ (Area of parallelogram $CPBQ$

$= 4$ (2 area of ΔCPQ)

$$= 8 \cdot \frac{1}{2} \begin{vmatrix} 0 & 0 & 1 \\ a\cos\alpha & b\sin\alpha & 1 \\ -a\sin\alpha & b\cos\alpha & 1 \end{vmatrix}$$

$= 4(ab\cos^2\alpha + ab\sin^2\alpha) = 4ab = (2a)(2b)$

$=$ Product of the axes of the ellipse

32. $\because 9x^2 + 16y^2 = 144$

$\Rightarrow \quad \frac{x^2}{16} + \frac{y^2}{9} = 1$ [Major axis > Minor axis]

$\therefore \quad 9 = 16(1 - e^2)$

$\Rightarrow \quad 16e^2 = 7$

or $\quad e = \frac{\sqrt{7}}{4}$

$\therefore \quad CS = 4e = \sqrt{7}$

$\Rightarrow \quad \frac{CS}{\text{major axis}} = \frac{\sqrt{7}}{2 \times 4} = \frac{\sqrt{7}}{8}$

i.e. $\quad \sqrt{7} : 8$

33. Let the ellipse be $\frac{x^2}{a^2} + \frac{y^2}{b^2} = 1$.

Then, according to given information, we have the following figure.

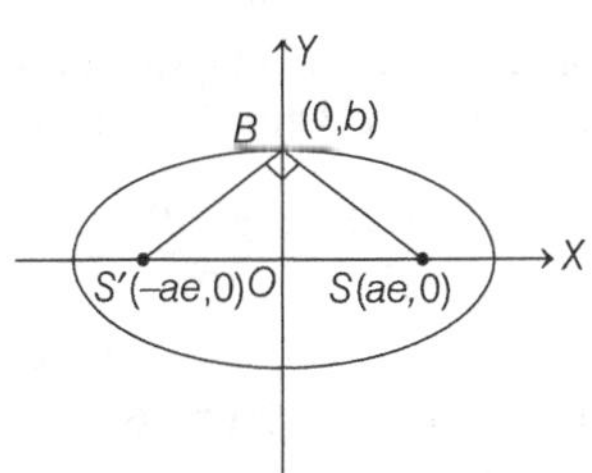

Clearly, slope of line $SB = \frac{b}{-ae}$ and slope of line $S'B = \frac{b}{ae}$

$\because$ Lines SB and $S'B$ are perpendicular, so

$$\left(\frac{b}{-ae}\right) \cdot \left(\frac{b}{ae}\right) = -1$$

[$\because$ product of slopes of two perpendicular lines is (-1)]

$\Rightarrow \quad b^2 = a^2e^2 \quad \ldots$(i)

Also, it is given that area of $\Delta S'BS = 8$

$\therefore \quad \frac{1}{2}a^2 = 8$

[$\because S'B = SB = a$ because $S'B + SB = 2a$ and $S'B = SB$]

$\Rightarrow \quad a^2 = 16 \Rightarrow a = 4 \quad \ldots$(ii)

$\because \quad e^2 = 1 - \frac{b^2}{a^2} = 1 - e^2$ [from Eq. (i)]

$\Rightarrow \quad 2e^2 = 1 \Rightarrow e^2 = \frac{1}{2} \quad \ldots$(iii)

From Eqs. (i) and (iii), we get

$b^2 = a^2\left(\frac{1}{2}\right) = 16\left(\frac{1}{2}\right)$ [using Eq. (ii)]

$\Rightarrow \quad b^2 = 8$

Now, length of latus rectum $= \frac{2b^2}{a} = \frac{2 \times 8}{4} = 4$ units

34. Here, Major axis > Minor axis

If e be the eccentricity of the ellipse, then $9 = 16(1 - e^2)$

$\therefore e = \frac{\sqrt{7}}{4}$

Then foci $(\pm\sqrt{7}, 0)$ or $S(\sqrt{7}, 0)$ and $S'(-\sqrt{7}, 0)$

$\because$ Centre of circle $= C(0, 3)$

$\therefore$ Radius of circle $= CS = CS'$

$= \sqrt{7 + 9} = \sqrt{16} = 4$

35. Equation of given ellipse is

$$\frac{x^2}{a^2} + \frac{y^2}{b^2} = 1 \quad \ldots\text{(i)}$$

Now, equation of tangent at the point $\left(3, -\frac{9}{2}\right)$ on the ellipse (i) is

$\Rightarrow \quad \frac{3x}{a^2} - \frac{9y}{2b^2} = 1 \quad \ldots$(ii)

[$\because$ the equation of the tangent to the ellipse $\frac{x^2}{a^2} + \frac{y^2}{b^2} = 1$ at the point (x_1, y_1) is $\frac{xx_1}{a^2} + \frac{yy_1}{b^2} = 1$]

$\because$ Tangent (ii) represent the line $x - 2y = 12$, so

$$\frac{1}{\frac{3}{a^2}} = \frac{2}{\frac{9}{2b^2}} = \frac{12}{1}$$

$\Rightarrow \quad a^2 = 36$ and $b^2 = 27$

Now, Length of latusrectum $= \frac{2b^2}{a} = \frac{2 \times 27}{6} = 9$ units

36. $4\tan\frac{B}{2}\cdot\tan\frac{C}{2}=4\sqrt{\frac{(s-a)(s-c)}{s(s-b)}}\sqrt{\frac{(s-a)(s-b)}{s(s-c)}}=1$

$\Rightarrow 4\frac{(s-a)}{s}=1 \Rightarrow s=\frac{4a}{3}=4\times\frac{6}{3}=8$

But $2s=a+b+c=16$

$$b+c=10$$

Hence, locus is an ellipse having center $\equiv(5,0)$

$$2ae=6 \text{ and } 2a=10$$

$$b^2=a^2-a^2e^2=25-9=15$$

Equation of ellipse,

$$\frac{(x-5)^2}{25}+\frac{y^2}{16}=1$$

37. Any point $P(\theta)$ on the ellipse $\frac{x^2}{4}+\frac{y^2}{1}=1$ is

$P(2\cos\theta, \sin\theta)$

Equation of tangent at $P(\theta)$ is

$$2x\cos\theta+4y\sin\theta=4$$

$\therefore \quad A=(2\sec\theta, 0)$

Equation of normal at $P(\theta)$ is

$$\frac{2x}{\cos\theta}-\frac{y}{\sin\theta}=4-1=3$$

$\therefore \quad B=\left(\frac{3}{2}\cos\theta, 0\right)$

$$AB=2\Rightarrow\left|2\sec\theta-\frac{3}{2}\cos\theta\right|=2$$

$\Rightarrow \quad \left(2-\frac{3}{2}\cos^2\theta\right)^2=4\cos^2\theta$

$\Rightarrow \quad 9\cos^4\theta-40\cos^2\theta+16=0$

$\Rightarrow \quad (9\cos^2\theta-4)(\cos^2\theta-4)=0$

$\Rightarrow \quad \cos^2\theta=4/9 \text{ as } \cos^2\theta\neq 4$

38. Let mid point of focal chord is (x_1, y_1),

then equation of a chord whose mid point (x_1, y_1) is $T=S_1$

$$\frac{xx_1}{a^2}+\frac{yy_1}{b^2}-1=\frac{x_1^2}{a^2}+\frac{y_1^2}{b^2}-1$$

$\Rightarrow \quad \frac{xx_1}{a^2}+\frac{yy_1}{b^2}=\frac{x_1^2}{a^2}+\frac{y_1^2}{b^2}$

Since, it is a focal chord, then its passes through focus $(\pm ae, 0)$, then

$$\pm\frac{ex_1}{a}+0=\frac{x_1^2}{a^2}+\frac{y_1^2}{b^2}$$

$\therefore$ Locus of mid point of focal chord is

$$\frac{x^2}{a^2}+\frac{y^2}{b^2}=\pm\frac{ex}{a}$$

39. $\frac{x^2}{a^2}+\frac{y^2}{b^2}=1$

Equation of tangent in terms of slope

$$y=mx\pm\sqrt{(a^2m^2+b^2)}$$

$\Rightarrow \quad (y-mx)^2=a^2m^2+b^2$

$y^2+m^2x^2-2mxy=a^2m^2+b^2$

$\Rightarrow m^2(x^2-a^2)-2mxy+y^2-b^2=0$

which is perpendicular, $\therefore m_1m_2=-1$

$\Rightarrow \quad \frac{y^2-b^2}{x^2-a^2}=-1,$

$\Rightarrow \quad x^2+y^2=a^2+b^2$

40. Equation of given ellipse is

$$4x^2+y^2=8 \quad \text{...(i)}$$

$\Rightarrow \quad \frac{x^2}{2}+\frac{y^2}{8}=1$

$\Rightarrow \quad \frac{x^2}{(\sqrt{2})^2}+\frac{y^2}{(2\sqrt{2})^2}=1$

Now, equation of tangent at point (1, 2) is

$$2x+y=4 \quad \text{...(ii)}$$

[$\because$ equation of tangent to the ellipse $\frac{x^2}{a^2}+\frac{y^2}{b^2}=1$ at (x_1, y_1) is $\frac{xx_1}{a^2}+\frac{yy_1}{b^2}=1$]

and equation of another tangent at point (a, b) is

$$4ax+by=8 \quad \text{...(iii)}$$

Since, lines (ii) and (iii) are perpendicular to each other.

$\therefore \quad \left(-\frac{2}{1}\right)\times\left(-\frac{4a}{b}\right)=-1$

[if lines $a_1x+b_1y+c_1=0$ and $a_2x+b_2y+c_2=0$ are perpendicular, then $\left(-\frac{a_1}{b_1}\right)\left(-\frac{a_2}{b_2}\right)=-1$]

$\Rightarrow \quad b=-8a \quad \text{...(iv)}$

Also, the point (a, b) lies on the ellipse (i), so

$$4a^2+b^2=8$$

$\Rightarrow \quad 4a^2+64a^2=8 \quad$ [from Eq.(iv)]

$\Rightarrow \quad 68a^2=8$

$\Rightarrow \quad a^2=\frac{8}{68}$

$\Rightarrow \quad a^2=\frac{2}{17}$

41. Equation of ellipse with centre at (0, 0) and axis parallel to X-axis and semi axes are a and b is

$$\frac{x^2}{a^2}+\frac{y^2}{b^2}=1$$

Hence, equation of the ellipse whose centre is (–2, 3) and whose semi axes are 3 and 2 and major axis is parallel to the X- axis is

$$\frac{(x+2)^2}{3^2}+\frac{(y-3)^2}{2^2}=1$$

[Shift the origin at (–2, 3)]

$\Rightarrow \quad 4(x^2+4x+4)+9(y^2-6y+9)=36$

$\Rightarrow \quad 4x^2+9y^2+16x-54y+61=0$

42. We have, $e = \frac{1}{2}$ and $\frac{a}{e} = 4$

$\therefore \quad a = 2$

Now, $b^2 = a^2(1 - e^2) = (2)^2\left[1 - \left(\frac{1}{2}\right)^2\right]$

$= 4\left(1 - \frac{1}{4}\right) = 3$

$\Rightarrow \quad b = \sqrt{3}$

$\therefore$ Equation of the ellipse s $\frac{x^2}{(2)^2} + \frac{y^2}{(\sqrt{3})^2} = 1$

$\Rightarrow \quad \frac{x^2}{4} + \frac{y^2}{3} = 1$

Now, the equation of normal at $\left(1, \frac{3}{2}\right)$ is

$$\frac{a^2x}{x_1} - \frac{b^2y}{y_1} = a^2 - b^2$$

$$\Rightarrow \quad \frac{4x}{1} - \frac{3y}{(3/2)} = 4 - 3$$

$$\Rightarrow \quad 4x - 2y = 1$$

43. Perpendicular tangents can be drawn from any point on the director circle.

Director circle of the ellipse

$x^2/9 + y^2/4 = 1$ is given by

$x^2 + y^2 = 9 + 4 = 13$

Any point on director circle is $(\sqrt{13}\cos\theta, \sqrt{13}\sin\theta)$

Equation of chord of contact of this point is

$4\sqrt{13}\cos\theta x + 9\sqrt{13}\sin\theta - 36 = 0.$

p = perpendicular distance of this chord from centre (0, 0)

$$= \frac{36}{\sqrt{13}\sqrt{16\cos^2\theta + 81\sin^2\theta}}$$

$$= \frac{36}{\sqrt{13}\sqrt{16 + 65\sin^2\theta}}$$

$\therefore$ Maximum $p = \frac{36}{\sqrt{13} \times 4} = \frac{9}{\sqrt{13}}$. [for $\sin\theta = 0$]

44. Since the point (α, β) is on the parabola $y^2 = x$, so

$$\alpha = \beta^2 \quad \text{...(i)}$$

Now, equation of tangent at point (α, β) to the parabola $y^2 = x$, is $T = 0$

$$\Rightarrow \quad y\beta = \frac{1}{2}(x + \alpha)$$

[$\because$ equation of the tangent to the parabola $y^2 = 4ax$ at a point (x_1, y_1) is given by$yy_1 = 2a(x + x_1)$]

$\Rightarrow \quad 2y\beta = x + \beta^2$ [from Eq. (i)]

$$\Rightarrow \quad y = \frac{x}{2\beta} + \frac{\beta}{2} \quad \text{...(ii)}$$

Since, line (ii) is also a tangent of the ellipse

$x^2 + 2y^2 = 1$

$$\therefore \quad \left(\frac{\beta}{2}\right)^2 = (1)^2\left(\frac{1}{2\beta}\right)^2 + \frac{1}{2}$$

[$\because$ condition of tangency of line $y = mx + c$ to ellipse $\frac{x^2}{a^2} + \frac{y^2}{b^2} = 1$ is $c^2 = a^2m^2 + b^2$,

here, $m = \frac{1}{2\beta}$, $a = 1$, $b = \frac{1}{\sqrt{2}}$ and $c = \frac{\beta}{2}$]

$\Rightarrow \quad \frac{\beta^2}{4} = \frac{1}{4\beta^2} + \frac{1}{\sqrt{2}}$

$\Rightarrow \quad \beta^4 = 1 + 2\beta^2$

$\Rightarrow \quad \beta^4 - 2\beta^2 - 1 = 0$

$\Rightarrow \quad \beta^2 = \frac{2 \pm \sqrt{4 + 4}}{2} = \frac{2 \pm 2\sqrt{2}}{2} = 1 \pm \sqrt{2}$

$\Rightarrow \quad \beta^2 = 1 + \sqrt{2}$ [$\because \beta^2 > 0$]

$\because \quad \alpha = \beta^2 = 1 + \sqrt{2}$

45. $\because$ Two points on the minor axis, each at a distance ae from the centre are $P(0, ae)$ and $Q(0, -ae)$

Equation of tanget at 'ϕ' of $\frac{x^2}{a^2} + \frac{y^2}{b^2} = 1$ is

$$\frac{x}{a}\cos\phi + \frac{y}{b}\sin\phi = 1 \quad \text{...(i)}$$

Now, Length of perpendicular from P on Eq. (i)

$$= \frac{\left|\frac{ae}{b}\sin\phi - 1\right|}{\sqrt{\left(\frac{\cos^2\phi}{a^2} + \frac{\sin^2\phi}{b^2}\right)}} = p_1 \text{ (say)}$$

and length of perpendicular from Q on Eq. (i)

$$= \frac{\left|\frac{ae\sin\phi}{b} + 1\right|}{\sqrt{\left(\frac{\cos^2\phi}{a^2} + \frac{\sin^2\phi}{b^2}\right)}} = p_2 \text{ (say)}$$

$$\therefore \quad p_1^2 + p_2^2 = \frac{\left(\frac{ae\sin\phi}{b} - 1\right)^2 + \left(\frac{ae\sin\phi}{b} + 1\right)^2}{\left(\frac{\cos^2\phi}{a^2} + \frac{\sin^2\phi}{b^2}\right)}$$

$$= \frac{2\left(\frac{a^2e^2\sin^2\phi}{b^2} + 1\right)}{\left(\frac{\cos^2\phi}{a^2} + \frac{\sin^2\phi}{b^2}\right)}$$

$$= \frac{2\left\{\left(\frac{a^2 - b^2}{b^2}\right)\sin^2\phi + 1\right\}}{\left(\frac{\cos^2\phi}{a^2} + \frac{\sin^2\phi}{b^2}\right)}$$

$$= \frac{2\left(\frac{a^2\sin^2\phi + b^2\cos^2\phi}{b^2}\right)}{\frac{(a^2\sin^2\phi + b^2\cos^2\phi)}{a^2b^2}} = 2a^2$$

46. $\because P \equiv (a\cos\theta, b\sin\theta)$

and $Q \equiv \left(a\cos\left(\frac{\pi}{2}+\theta\right), b\sin\left(\frac{\pi}{2}+\theta\right)\right)$

$$Q \equiv (-a\sin\theta, b\cos\theta)$$

Let mid point of PQ is (x, y), then

$$2x = a(\cos\theta - \sin\theta) \quad \text{...(i)}$$

and $2y = b(\sin\theta + \cos\theta)$ ...(ii)

From Eq. (i), $\frac{2x}{a} = \cos\theta - \sin\theta$ ...(iii)

and from Eq. (ii), $\frac{2y}{b} = \sin\theta + \cos\theta$ ...(iv)

Squaring and adding Eqs. (iii) and (iv), we get

$$\frac{4x^2}{a^2} + \frac{4y^2}{b^2} = 2 \Rightarrow \frac{x^2}{a^2} + \frac{y^2}{b^2} = \frac{1}{2}$$

Which is required locus of mid point of PQ

47. Ellipse is $x^2 + 4y^2 = 4$, $P = (2\cos\theta, \sin\theta)$

Equation of tangent at P is $2\cos\theta x + 4y\sin\theta = 4$.

It meets X-axis at B.

$\therefore B = (2\sec\theta, 0)$. $A = (2\cos\theta, 0)$

$$AQ = AB = 2|\sec\theta - \cos\theta| = \frac{2\sin^2\theta}{\cos\theta}$$

Let $Q = (h, k)$.

Then, $h = 2\cos\theta$, $k = \frac{2\sin^2\theta}{\cos\theta} = \frac{2(1-\cos^2\theta)}{\cos\theta}$

$$\Rightarrow \quad hk = 4(1 - h^2/4)$$

$\therefore$ Locus of Q is $xy = 4 - x^2$

48. Let $P(a\cos\alpha, b\sin\alpha)$ and $Q(a\cos\beta, b\sin\beta)$ are the ends of focal chord of the ellipse $\frac{x^2}{a^2} + \frac{y^2}{b^2} = 1$

then equation of PQ is

$$\frac{x}{a}\cos\left(\frac{\alpha+\beta}{2}\right) + \frac{y}{b}\sin\left(\frac{\alpha+\beta}{2}\right) = \cos\left(\frac{\alpha-\beta}{2}\right)$$

its pass through $(\pm ae, 0)$

$$\pm e\cos\left(\frac{\alpha+\beta}{2}\right) = \cos\left(\frac{\alpha-\beta}{2}\right)$$

Taking +ve sign, then

$$e = \frac{\cos\left(\frac{\alpha-\beta}{2}\right)}{\cos\left(\frac{\alpha+\beta}{2}\right)}$$

$$= \frac{2\sin\left(\frac{\alpha+\beta}{2}\right)\cos\left(\frac{\alpha-\beta}{2}\right)}{2\sin\left(\frac{\alpha+\beta}{2}\right)\cos\left(\frac{\alpha+\beta}{2}\right)}$$

$$= \frac{\sin\alpha + \sin\beta}{\sin(\alpha+\beta)}$$

49. It is given that e_1 is the eccentricity of the ellipse, $\frac{x^2}{18} + \frac{y^2}{4} = 1$, so

$$e_1 = \sqrt{1 - \frac{4}{18}} = \sqrt{\frac{14}{18}} = \sqrt{\frac{7}{9}}$$

and e_2 is the eccentricity of the hyperbola,

$$\frac{x^2}{9} - \frac{y^2}{4} = 1, \text{ so } e_2 = \sqrt{1 + \frac{4}{9}} = \sqrt{\frac{13}{9}}$$

Now, as $(e_1, e_2) = \left(\sqrt{\frac{7}{9}}, \sqrt{\frac{13}{9}}\right)$ is a point on the ellipse, $15x^2 + 3y^2 = k$, then

$$15\left(\frac{7}{9}\right) + 3\left(\frac{13}{9}\right) = k \Rightarrow \frac{105 + 39}{9} = k$$

$$\Rightarrow \quad \frac{144}{9} = k \Rightarrow k = 16$$

Hence, option (d) is correct.

50. Given equation of ellipse is

$$\frac{x^2}{9} + \frac{y^2}{5} = 1 \quad \text{...(i)}$$

$\therefore$ $a^2 = 9, b^2 = 5$

$\Rightarrow$ $a = 3, b = \sqrt{5}$

Now, $e = \sqrt{1 - \frac{b^2}{a^2}} = \sqrt{1 - \frac{5}{9}} = \frac{2}{3}$

foci $= (\pm ae, 0) = (\pm 2, 0)$

and $\frac{b^2}{a} = \frac{5}{3}$

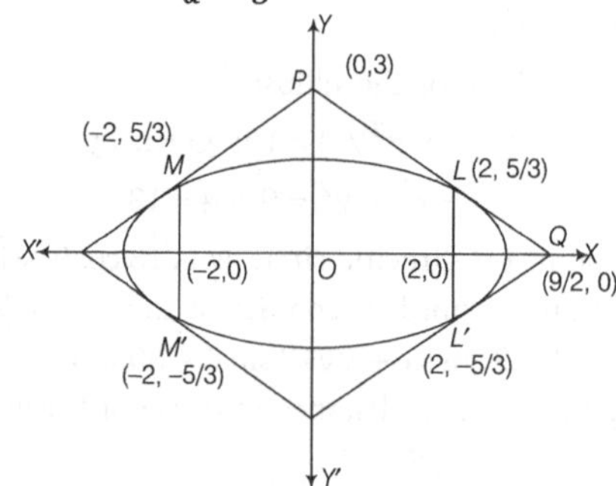

$\therefore$ Extremities of one of latusrectum arc

$$\left(2, \frac{5}{3}\right) \text{ and } \left(2, \frac{-5}{3}\right).$$

$\therefore$ Equation of tangent at $\left(2, \frac{5}{3}\right)$ is,

$$\frac{x(2)}{9} + \frac{y(5/3)}{5} = 1$$

or $2x + 3y = 9$ ...(ii)

Eq.(ii) intersects X and Y-axes at $\left(\frac{9}{2}, 0\right)$ and $(0, 3)$, respectively.

$\therefore$ Area of quadrilateral $= 4 \times$ Area of ΔPOQ

$$= 4 \times \left(\frac{1}{2} \times \frac{9}{2} \times 3\right) = 27 \text{ sq units}$$

51. Foci are $(-1, 0)$ and $(7, 0)$.

Coordinate of the centre is $(3, 0)$

Distance between focii $= \sqrt{(-1-7)^2 + 0} = 8$

$$2ae = 8$$

$$\therefore \quad ae = 4$$

Here, $e = \frac{1}{2}$

$\therefore$ $a = 8$

$\therefore$ $b^2 = a^2(1 - e^2) = 64\left(1 - \frac{1}{4}\right)$

$= 64 - 16 = 48$

Equation of ellipse with centre (3, 0) is $\frac{(x-3)^2}{64} + \frac{y^2}{48} = 1$

$\therefore$ Parametric coordinates

$(3 + 8\cos\theta, 4\sqrt{3}\sin\theta)$

52. $\because$ $E: \frac{x^2}{9} + \frac{y^2}{4} - 1 = 0$

and $C: x^2 + y^2 - 9 = 0$

and $P \equiv (1, 2)$ and $Q \equiv (2, 1)$

at P (1, 2) LHS of E is $\frac{1}{9} + \frac{4}{4} - 1 > 0$ [P lies outside of E]

at Q (2, 1) LHS of E is $\frac{4}{9} + \frac{1}{4} - 1 < 0$ [Q lies inside of E]

at P (1, 2) LHS of C is $(1)^2 + (2)^2 - 9 < 0$

[P lies inside of C]

and at Q (2, 1) LHS of C is $(2)^2 + (1)^2 - 9 < 0$

[Q lies inside of C]

53. Given ellipse $\frac{x^2}{a^2} + \frac{y^2}{b^2} = 1$ ...(i)

Auxiliary circle of Eq. (i) is $x^2 + y^2 = a^2$ ...(ii)

Equation of tangent at a point P $(a\cos\phi, b\sin\phi)$ on Eq. (i)

$$\frac{x}{a}\cos\phi + \frac{y}{b}\sin\phi = 1 \quad ...(iii)$$

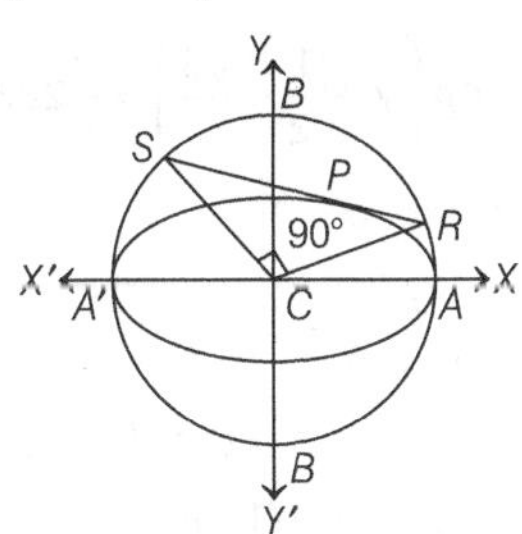

Making homogeneous Eq. (ii) with the help of Eq. (iii), then

$$x^2 + y^2 = a^2\left(\frac{x}{a}\cos\phi + \frac{y}{b}\sin\phi\right)^2$$

$$\Rightarrow x^2\sin^2\phi + \left(1 - \frac{a^2}{b^2}\sin^2\phi\right)y^2 - \frac{2xy}{ab}a^2\sin\phi\cos\phi = 0$$

$\because$ $\angle RCS = 90°$

$\therefore$ Coefficient of x^2 + coefficient of $y^2 = 0$

$\therefore$ $\sin^2\phi + 1 - \frac{a^2}{b^2}\sin^2\phi = 0$

$$1 = \left(\frac{a^2}{b^2} - 1\right)\sin^2\phi$$

$\Rightarrow$ $1 = \left(\frac{a^2 - b^2}{b^2}\right)\sin^2\phi$

$\Rightarrow$ $1 = \frac{a^2e^2}{a^2(1 - e^2)}\sin^2\phi$

$\Rightarrow$ $1 - e^2 = e^2\sin^2\phi$

$\Rightarrow$ $1 = e^2(1 + \sin^2\phi)$

$\Rightarrow$ $e^2 = \frac{1}{(1 + \sin^2\phi)} = (1 + \sin^2\phi)^{-1}$

$\therefore$ $e = (1 + \sin^2\phi)^{-1/2}$

54. (b) $\frac{x^2}{2} + \frac{y^2}{1} = 1$ equation of tangent

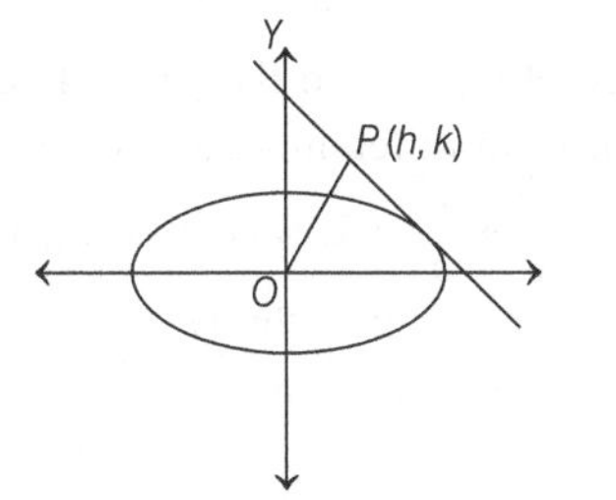

$$y = mx + \sqrt{2m^2 + 1} \quad ...(i)$$

slope of $OP = \frac{k}{h}$

slope of normal $= \frac{-h}{k}$ ...(ii)

Its passes through $P(h, k)$

$\therefore$ $k = \left(\frac{-h}{k}\right)(h) + \sqrt{2\left(\frac{-h}{k}\right)^2 + 1}$

$\Rightarrow$ $k^2 = -h^2 + \sqrt{2h^2 + k^2}$

$\Rightarrow$ $(x^2 + y^2)^2 = (2x^2 + y^2)$

55. Let ellipse be $\frac{x^2}{a^2} + \frac{y^2}{b^2} = 1$

$\because$ Ellipse meets the straight line $\frac{x}{7} + \frac{y}{2} = 1$ on the axis of x.

Then, $y = 0, \frac{x}{7} = 1 \Rightarrow x = 7$

ie, $a = 7$

and meets the straight line $\frac{x}{3} - \frac{y}{5} = 1$ on the axis of y, then

$x = 0, 0 - \frac{y}{5} = 1$

$\Rightarrow$ $y = -5$

i.e. $-b = -5$

$\Rightarrow$ $b = 5$

$\therefore$ $b^2 = a^2(1 - e^2)$

$\Rightarrow$ $25 = 49(1 - e^2)$

$\Rightarrow$ $49e^2 = 24$

$\therefore$ $e^2 = \frac{24}{49}$

$\therefore$ $e = \frac{2\sqrt{6}}{7}$

56. Equation of given parabola is

$$y^2 = 4\lambda x \qquad \dots \text{(i)}$$

So the end point of the latus rectum of the parabola (i), $P(\lambda, 2\lambda)$ and the given ellipse $\frac{x^2}{a^2}+\frac{y^2}{b^2}=1$, passes through point $P(\lambda, 2\lambda)$.

On differentiating the equation of parabola, w.r.t. 'x', we get

$$\frac{dy}{dx}=\frac{2\lambda}{y}$$

$\therefore$ Slope of tangent to the parabola at point P is $m_1 = 1$

Similarly, on differentiating the equation of given ellipse, $\frac{x^2}{a^2}+\frac{y^2}{b^2}=1$, w.r.t. x, we get

$$\frac{2x}{a^2}+\frac{2y}{b^2}\frac{dy}{dx}=0$$

$$\Rightarrow \quad \frac{dy}{dx}=-\frac{b^2x}{a^2y}$$

$\therefore$ Slope of tangent to the ellipse at point P is $m_2 = -\frac{b^2}{2a^2}$

$\because$ It is given that the tangents are perpendicular to each other. So,

$$m_1m_2=-1 \Rightarrow (1)\left(-\frac{b^2}{2a^2}\right)=-1$$

$$\Rightarrow \quad \frac{b^2}{a^2}=2 \Rightarrow b=\sqrt{2}a$$

$\therefore$ Eccentricity of ellipse $\frac{x^2}{a^2}+\frac{y^2}{b^2}=1$ will be

$$e=\sqrt{1-\frac{a^2}{b^2}}=\sqrt{1-\frac{1}{2}}=\frac{1}{\sqrt{2}} \qquad [\because b > a]$$

Objective Questions Type II [One or more than one correct answer(s)]

1. $\because b^2x^2+y^2=a^2b^2$

$$\Rightarrow \quad \frac{x^2}{a^2}+\frac{y^2}{a^2b^2}=1$$

Let $\quad a > ab,$

Then, $\quad (ab)^2 = a^2(1-e^2)$

$$\Rightarrow \quad b^2 = 1-e^2 \qquad \dots\text{(i)}$$

$\therefore$ Extremities of the latusrectum $(\pm ae, \pm ab^2)$

Let $x_1 = \pm ae$ and $y_1 = \pm ab^2 = \pm a(1-e^2)$

or $y_1 = \pm(a-ae^2)$

$$\Rightarrow ay_1 = \pm(a^2-a^2e^2)=\pm(a^2-x_1^2) \qquad [\because x_1 = \pm ae]$$

or $\quad x_1^2 \pm ay_1 = a^2$

$\therefore$ Required locus is $x^2 \pm ay = a^2$

2. Eccentricity e of the ellipse is given by

$$b^2 = a^2(1-e^2)$$

$$\Rightarrow \quad 1 = 4(1-e^2)$$

$$\Rightarrow \quad e = \sqrt{3}/2$$

Foci of the ellipse are $(\sqrt{3}, 0)$ and $(-\sqrt{3}, 0)$.

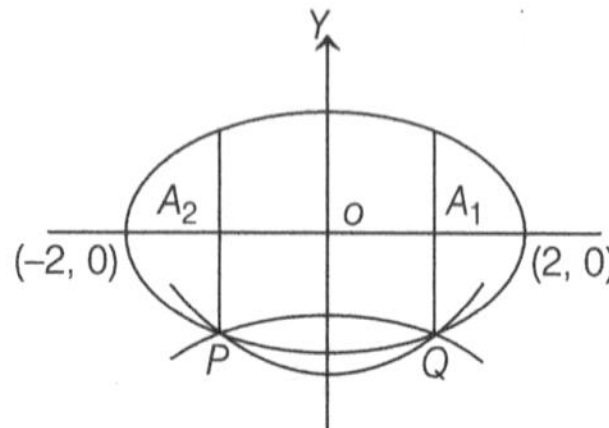

Length of a latus rectum of the ellipse is $2\frac{b^2}{a}=1$

Thus, $\quad P(x_1, y_1) = P(-\sqrt{3}, -1/2)$

and $\quad Q(x_2, y_2) = Q(\sqrt{3}, -1/2)$

Length of the latus rectum PQ of the parabola is

$$|x_2 - x_1| = 2\sqrt{3} = 4p \text{ (say)}$$

As focus of a parabola is the mid point of the latus rectum, focus of the desired parabola is $(0, -1/2)$ and hence its vertices are $(0, -1/2 \pm p)$

i.e. $\left(0, -\frac{1}{2}-\frac{\sqrt{3}}{2}\right)$ and $\left(0, -\frac{1}{2}+\frac{\sqrt{3}}{2}\right)$

Thus there are two parabolas having PQ as the latus rectum whose equations are

$$x^2 = 4p\left(y+\frac{1}{2}+\frac{\sqrt{3}}{2}\right)=2\sqrt{3}y+\sqrt{3}+3$$

and $$x^2 = -4p\left(y+\frac{1}{2}-\frac{\sqrt{3}}{2}\right)=-2\sqrt{3}y-\sqrt{3}+3$$

3. $x^2\tan^2\alpha + y^2\sec^2\alpha = 1$

$$\Rightarrow \quad \frac{x^2}{\cot^2\alpha}+\frac{y^2}{\cos^2\alpha}=1$$

$$\because \quad \cos^2\alpha = \cot^2\alpha(1-e^2)$$

$$\Rightarrow \quad \sin^2\alpha = (1-e^2)$$

$$\therefore \quad e^2 = \cos^2\alpha \ (\alpha \neq 90^\circ)$$

$$e = \cos\alpha$$

$\because$ Latusrectum $= 1/2 = 2b^2/a$

$$\Rightarrow \quad a = 4b^2,$$

$$\Rightarrow \quad \cot\alpha = 4\cos^2\alpha$$

$$\Rightarrow \quad \frac{1}{\sin\alpha}=4\cos\alpha \Rightarrow \sin 2\alpha = \frac{1}{2}$$

$$2\alpha = n\pi + (-1)^n\frac{\pi}{6} \Rightarrow \alpha = \frac{n\pi}{2}+(-1)^n\frac{\pi}{12}$$

for $n = 0$

$$\alpha = \frac{\pi}{12}$$

and For $n = 1$

$$\alpha = \frac{\pi}{2}-\frac{\pi}{12}=\frac{5\pi}{12}$$

4. $\because \quad 25x^2 + 9y^2 - 150x - 90y + 225 = 0$

$\Rightarrow 25(x^2 - 6x) + 9(y^2 - 10y) + 225 = 0$

$\Rightarrow \quad 25\{(x-3)^2 - 9\} + 9\{(y-5)^2 - 25\} + 225 = 0$

$\Rightarrow \quad 25(x-3)^2 + 9(y-5)^2 = 225$

$\Rightarrow \quad \dfrac{(x-3)^2}{3^2} + \dfrac{(y-5)^2}{5^2} = 1 \quad \ldots\text{(i)}$

Let $\quad x - 3 = X, \; y - 5 = Y$

Then, Eq. (i) becomes $\dfrac{X^2}{3^2} + \dfrac{Y^2}{5^2} = 1 \quad \ldots\text{(ii)}$

Now, comparing Eq. (ii) with $\dfrac{X^2}{a^2} + \dfrac{Y^2}{b^2} = 1$

$\therefore \quad a = 3, \; b = 5$

$\Rightarrow \quad a^2 = b^2(1 - e^2)$

$\Rightarrow \quad 9 = 25(1 - e^2)$

$\Rightarrow \quad e = \dfrac{4}{5}$

Centre : $X = 0, \; Y = 0$

$x - 3 = 0, \; y - 5 = 0$

$\therefore$ Centre $\equiv (3, 5)$

Foci : $X = 0, \; Y = \pm be$

$\Rightarrow x - 3 = 0, \; y - 5 = \pm 4$

$\Rightarrow \quad y = 5 \pm 4$

$\therefore$ Foci (3, 1) and (3, 9)

Major axis

$2b = 10$

5. We have,

Equations of circle, $x^2 + y^2 = \dfrac{1}{2}$

and Equations of parabola $y^2 = 4x$

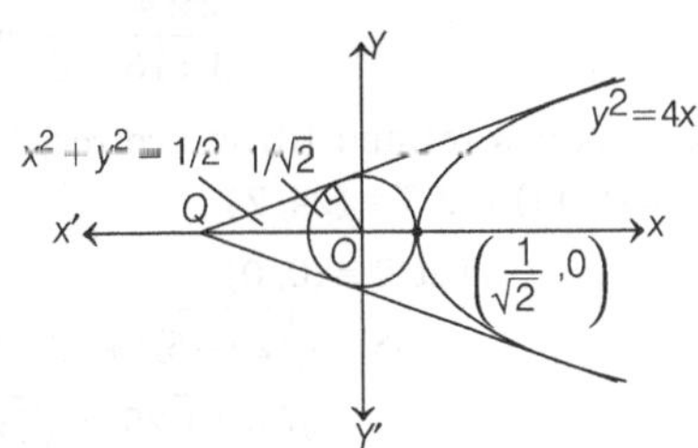

Let the equation of common tangent of parabola and circle is $y = mx + \dfrac{1}{m}$

Since, radius of circle $= \dfrac{1}{\sqrt{2}}$

$$\therefore \quad \frac{1}{\sqrt{2}} = \left|\frac{0 + 0 + \frac{1}{m}}{\sqrt{1 + m^2}}\right|$$

$\Rightarrow m^4 + m^2 - 2 = 0 \Rightarrow m = \pm 1$

$\therefore$ Equation of common tangents are

$y = x + 1$ and $y = -x - 1$

Intesection point of common tangent at $Q(-1, 0)$.

$\therefore$ Equation of ellipse, $\dfrac{x^2}{1} + \dfrac{y^2}{1/2} = 1$

where, $a^2 = 1, \; b^2 = 1/2$

Now, eccentricity

$(e) = \sqrt{1 - \dfrac{b^2}{a^2}} = \sqrt{1 - \dfrac{1}{2}} = \dfrac{1}{\sqrt{2}}$

and length of latusrectum

$$= \frac{2b^2}{a} = \frac{2\left(\frac{1}{2}\right)}{1} = 1$$

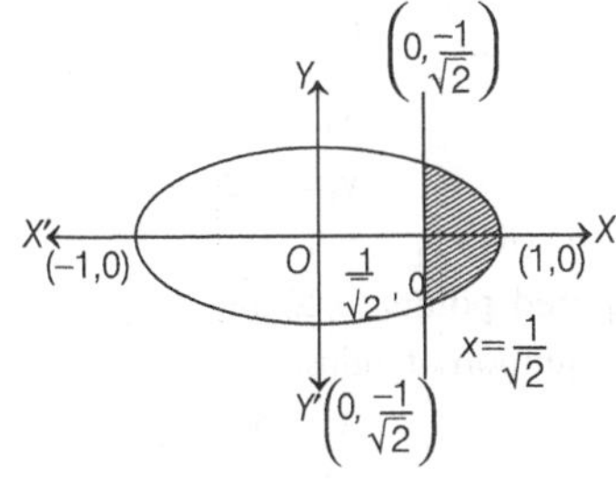

$\therefore$ Area of shaded region

$$= 2\int_{1/\sqrt{2}}^{1} \frac{1}{\sqrt{2}}\sqrt{1 - x^2}\, dx$$

$$= \sqrt{2}\left[\frac{x}{2}\sqrt{1 - x^2} + \frac{1}{2}\sin^{-1} x\right]_{1/\sqrt{2}}^{1}$$

$$= \sqrt{2}\left[\left(0 + \frac{\pi}{4}\right) - \left(\frac{1}{4} + \frac{\pi}{8}\right)\right]$$

$$= \sqrt{2}\left(\frac{\pi}{8} - \frac{1}{4}\right) = \frac{\pi - 2}{4\sqrt{2}}$$

6. Tangent and normal are bisectors of $\angle SPS'$.

Now, the equation of SP is $y = 3x/2$ and that of $S'P$ is $y = 5x$.

Then equations of angle bisectors are

$$\frac{3x - 2y}{\sqrt{13}} = \pm \frac{5x - y}{\sqrt{26}}$$

or $\quad 3x - 2y = \pm \dfrac{5x - y}{\sqrt{2}}$

Therefore, the lines are

$(3\sqrt{2} - 5)x + (1 - 2\sqrt{2})y = 0$

and $\quad (3\sqrt{2} + 5)x - (2\sqrt{2} + 1)y = 0$

Now, (2, 3) and (1, 5) lie on the same side of $(3\sqrt{2} - 5)x + (1 - 2\sqrt{2})y = 0$, which is the equation of tangent.

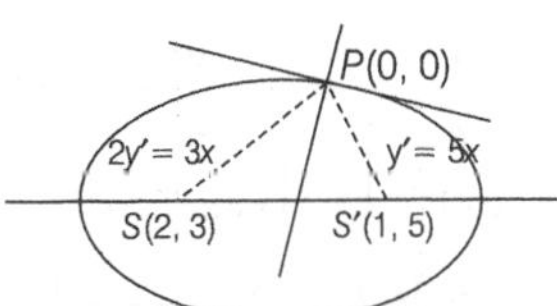

Points (2, 3) and (1, 5) lie on the different sides of $(3\sqrt{2} + 5)x - (2\sqrt{2} +)y = 0$, which is the equation of normal.

7. Let (x_1, y_1) be a point then $x_1^2 + 3y_1^2 = 37$ ...(i)

Equation of tangent at (x_1, y_1) is

$$xx_1 + 3yy_1 = 37$$

Slope of tangent $= -\dfrac{x_1}{3y_1}$

Then, slope of normal $= \dfrac{3y_1}{x_1} = \dfrac{6}{5}$ (given)

$\therefore \quad x_1 = \dfrac{5y_1}{2}$ ...(ii)

From Eqs. (i) and (ii)

$$\frac{25y_1^2}{4} + 3y_1^2 = 37$$

$\Rightarrow \quad y_1^2 = 4$

$\therefore \quad y_1 = \pm 2$

From Eq. (ii), $x_1 = \pm 5$

$\therefore$ Required points (5, 2) and (–5, – 2)

8. Given equation of ellipse

$$E_1 : \frac{x^2}{9} + \frac{y^2}{4} = 1 \quad ...(i)$$

Now, let a vertex of rectangle of largest area with sides parallel to the axes, inscribed in E_1 be $(3\cos\theta, 2\sin\theta)$.

So, area of rectangle $R_1 = 2(3\cos\theta) \times 2(2\sin\theta)$

$= 12\sin(2\theta)$

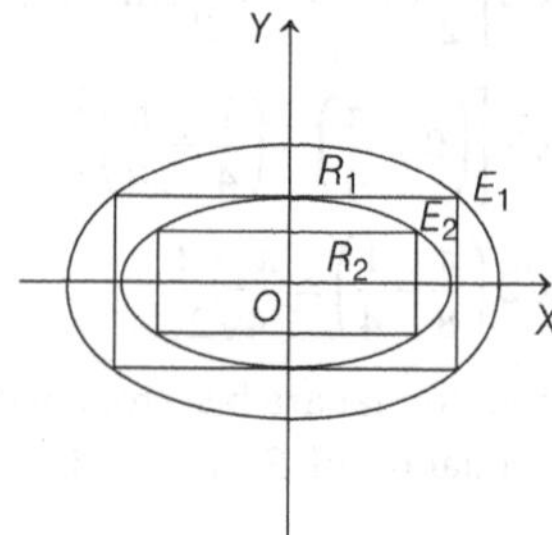

The area of R_1 will be maximum, if $\theta = \dfrac{\pi}{4}$ and maximum area is 12 square units and length of sides of rectangle R_1 are $2a\cos\theta = \sqrt{2}\,a = 3\sqrt{2}$ = length of major axis of ellipse E_2 and $2b\sin\theta = \sqrt{2}\,b = 2\sqrt{2}$ = length of minor axis of ellipse E_2.

So, $E_2 : \dfrac{x^2}{\left(\dfrac{a}{\sqrt{2}}\right)^2} + \dfrac{y^2}{\left(\dfrac{b}{\sqrt{2}}\right)^2} = 1$ and maximum area of rectangle $R_2 = 2\left(\dfrac{a}{\sqrt{2}}\right)\left(\dfrac{b}{\sqrt{2}}\right)$ and so on.

So, $E_n = \dfrac{x^2}{\left(\dfrac{a}{(\sqrt{2})^{n-1}}\right)^2} + \dfrac{y^2}{\left(\dfrac{b}{(\sqrt{2})^{n-1}}\right)^2} = 1$, and maximum area of rectangle $R_n = 2\left(\dfrac{a}{(\sqrt{2})^{n-1}}\right)\left(\dfrac{b}{(\sqrt{2})^{n-1}}\right)$

Now option (a),

Since, eccentricity of ellipse $E_n = e_n' = \sqrt{1 - \dfrac{(b_n)^2}{(a_n)^2}}$

$$= \sqrt{1 - \frac{\left(\dfrac{b}{(\sqrt{2})^{n-1}}\right)^2}{\left(\dfrac{a}{(\sqrt{2})^{n-1}}\right)^2}} = \sqrt{1 - \frac{b^2}{a^2}} = \sqrt{1 - \frac{4}{9}} = \frac{\sqrt{5}}{3}$$

is independent of 'n', so eccentricity of E_{18} and E_{19} are equal.

Option (b),

Distance between focus and centre of $E_9 = e \cdot a_9$

$$= \frac{a}{(\sqrt{2})^8}(e) = \frac{3}{2^4} \times \frac{\sqrt{5}}{3} = \frac{\sqrt{5}}{16} \text{ unit}$$

Option (c),

$\because \sum_{n=1}^{N}$ (area of R_n) < (area of R_1) + (area of R_2) + ... ∞

$$< 2ab + 2\frac{ab}{2} + 2\frac{ab}{2^2} + ...$$

$$< 2ab\left(1 + \frac{1}{2} + \frac{1}{2^2} + ...\right)$$

$$< 12\left(\frac{1}{1 - 1/2}\right)$$

$\Rightarrow \sum_{n=1}^{N}$ (area of R_n) < 24, for each positive integer N.

Option (d),

Length of latusrectum $E_9 = \dfrac{2b_9^2}{a_9} = \dfrac{2b^2}{a(\sqrt{2})^8}$

$$= \frac{2 \times 4}{3 \times 16} = \frac{1}{6} \text{ units}$$

Hence, options (c) and (d) are correct.

9. Let $S \equiv (5, 12)$ and $S' \equiv (24, 7)$

and Let $P \equiv (0, 0)$

Now, $SS' = \sqrt{(24-5)^2 + (7-12)^2}$

$= \sqrt{(361 + 25)} = \sqrt{(386)}$

For ellipse $SP + S'P = 2a$

$\Rightarrow \quad 13 + 25 = 2a$

$\therefore \quad 2a = 38$

$\because \quad SS' = 2ae$

$\Rightarrow \quad e = \dfrac{\sqrt{386}}{38}$

and for hyperbola $S'P - SP = 2a$

$\Rightarrow \quad 25 - 13 = 2a$

$\therefore \quad 2a = 12$

$SS' = 2ae$

$\Rightarrow \quad e = \dfrac{SS'}{2a} = \dfrac{\sqrt{(386)}}{12}$

10. $x^4 + 4y^2 - 2x - 16y + 13 = 0$

or $(x^2 - 2x + 1) + 4(y^2 - 4y + 4) = 4$

or $$\frac{(x-1)^2}{4} + \frac{(y-2)^2}{1} = 1$$

$\therefore$ Length of latusrectum $= \frac{2 \times 1}{2} = 1$

Also, $$e = \sqrt{1 - \frac{1}{4}} = \frac{\sqrt{3}}{2}$$

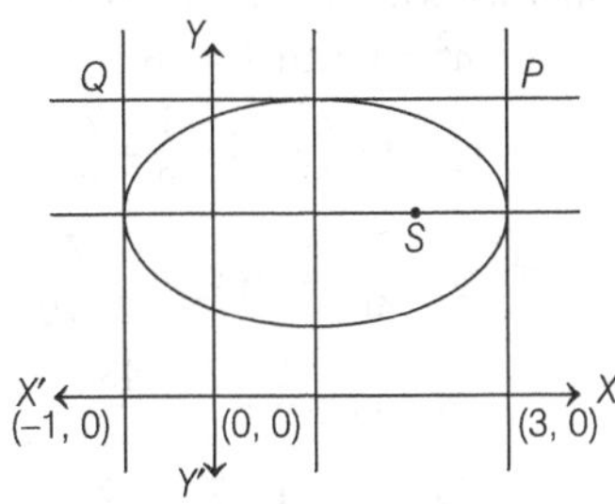

or $$2ae = 2 \times 2 \times \frac{\sqrt{3}}{2} = 2\sqrt{3}$$

Sum of focal distance $= 2a = 4$

Tangents at the vertices are

$$x - 1 = \pm 2$$

or $$x = 3, -1$$

Therefore, the line $y = 3$ intersects these at points $P(3, 3)$ and $Q(-1, 3)$.

One of the foci is $S(\sqrt{3} + 1, 2)$.

Slope of $PS = \frac{1}{2 - \sqrt{3}}$

Slope of $QS = \frac{1}{-2 - \sqrt{3}}$

$\therefore$ Product of slopes $= \frac{1}{2-\sqrt{3}} \times \frac{1}{-2-\sqrt{3}} = -1$

11. Given ellipse is

$$\frac{x^2}{6} + \frac{y^2}{2} = 1 \quad \ldots \text{(i)}$$

a point on the ellipse (i) is $(\sqrt{6}\cos\phi, \sqrt{2}\sin\phi)$

Since, its distance from the origin is 2

$\therefore$ $6\cos^2\phi + 2\sin^2\phi = (2)^2$

$\Rightarrow$ $4\cos^2\phi + 2 = 4$

$\Rightarrow$ $4\cos^2\phi = 2$

$\Rightarrow$ $\cos^2\phi = 1/2$

$\Rightarrow$ $2\cos^2\phi = 1$

$\cos 2\phi = 0$

$\Rightarrow$ $2\phi = (2n+1)\frac{\pi}{2};\ n \in I$

or $\phi = (2n+1)\pi/4;\ n \in I$

12. Equation of tangent to $16x^2 + 11y^2 = 256$ at $\left(4\cos\theta, \frac{16}{\sqrt{11}}\sin\theta\right)$ is

$$4\cos\theta\, x + \sqrt{11}\sin\theta\, y = 16$$

The perpendicular from centre (1, 0) is equal to radius

$$\sqrt{(1+15)} = 4$$

or $$\frac{|4\cos\theta - 16|}{\sqrt{(16\cos^2\theta + 11\sin^2\theta)}} = 4$$

On simplification, we get

$$\cos\theta = \frac{1}{2} \text{ or } -\frac{5}{2} \text{ (not possible)}$$

or $$\theta = \pm\frac{\pi}{3}, \frac{5\pi}{3}$$

13. Any point on the ellipse is $P(3\cos\theta, 2\sin\theta)$,

Equation of the tangent at P is

$$\frac{x}{3}\cos\theta + \frac{y}{2}\sin\theta = 1$$

Which meets the tangents $x = 3$ and $x = -3$ at the extremities of the major axis is

$$T\left(3, \frac{2(1-\cos\theta)}{\sin\theta}\right) \text{ and } T'\left(-3, \frac{2(1+\cos\theta)}{\sin\theta}\right)$$

Equation of the Circle on TT' as diameter is

$$(x-3)(x+3) + \left(y - \frac{2(1-\cos\theta)}{\sin\theta}\right)\left(y - \frac{2(1+\cos\theta)}{\sin\theta}\right) = 0$$

$\Rightarrow x^2 + y^2 - \frac{4}{\sin\theta}y - 5 = 0$, which passes through $(\pm\sqrt{5}, 0)$

14. $P(\theta)$ on the ellipse $x^2 + 4y^2 = 16$ is $(4\cos\theta, 2\sin\theta)$

Equation of tangent at P is

$$4\cos\theta\, x + 8\sin\theta\, y = 16$$

$\Rightarrow$ $x\cos\theta + 2y\sin\theta = 4$

It passes through centre (4, 2) of the given circle

$\therefore$ $4\cos\theta + 4\sin\theta = 4$

$\Rightarrow$ $\cos\theta + \sin\theta = 1,\ \theta = 0, \pi/2$

15. Here, $E_1: \frac{x^2}{a^2} + \frac{y^2}{b^2} = 1,\ (a > b)$

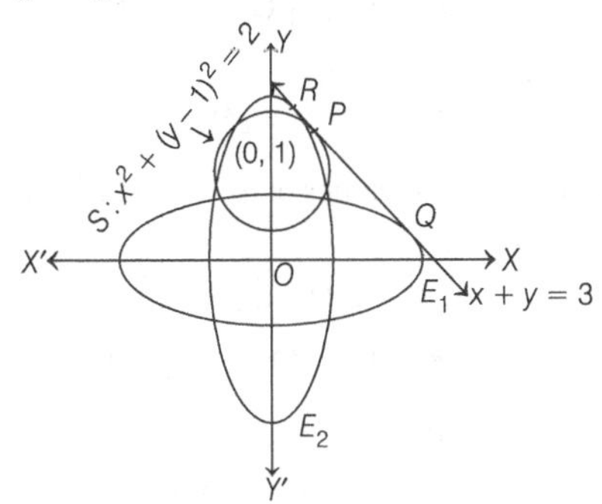

$$E_2: \frac{x^2}{c^2} + \frac{y^2}{d^2} = 1,\ (c < d)$$

and $$S: x^2 + (y-1)^2 = 2$$

as tangent to E_1, E_2 and S is $x + y = 3$.

Let the point of contact of tangent be (x_1, y_1) to S.

$\therefore$ $x \cdot x_1 + y \cdot y_1 - (y + y_1) + 1 = 2$

or $x x_1 + y y_1 - y = (1 + y_1)$, same as $x + y = 3$.

$$\Rightarrow \quad \frac{x_1}{1} = \frac{y_1 - 1}{1} = \frac{1 + y_1}{3}$$

i.e. $\quad x_1 = 1$ and $y_1 = 2$

$\therefore \quad P = (1, 2)$

Since, $PR = PQ = \dfrac{2\sqrt{2}}{3}$. Thus, by parametric form,

$$\frac{x-1}{-1/\sqrt{2}} = \frac{y-2}{1/\sqrt{2}} = \pm\frac{2\sqrt{2}}{3}$$

$$\Rightarrow \quad \left(x = \frac{5}{3}, y = \frac{4}{3}\right) \text{ and } \left(x = \frac{1}{3}, y = \frac{8}{3}\right)$$

$$\therefore \quad Q = \left(\frac{5}{3}, \frac{4}{3}\right)$$

and $\quad R = \left(\dfrac{1}{3}, \dfrac{8}{3}\right)$

Now, equation of tangent at Q on ellipse E_1 is

$$\frac{x}{a^2}\left(\frac{5}{3}\right) + \frac{y}{b^2}\left(\frac{4}{3}\right) = 1$$

On comparing with $x + y = 3$, we get

$$a^2 = 5 \text{ and } b^2 = 4$$

$$\therefore \quad e_1^2 = 1 - \frac{b^2}{a^2} = 1 - \frac{4}{5} = \frac{1}{5} \quad \ldots\text{(i)}$$

Also, equation of tangent at R on ellipse E_2 is

$$\frac{x}{a^2}\left(\frac{1}{3}\right) + \frac{y}{b^2}\left(\frac{8}{3}\right) = 1$$

On comparing with $x + y = 3$, we get

$$a^2 = 1 \text{ and } b^2 = 8$$

$$\therefore \quad e_2^2 = 1 - \frac{a^2}{b^2} = 1 - \frac{1}{8} = \frac{7}{8} \quad \ldots\text{(ii)}$$

Now, $\quad e_1^2 \cdot e_2^2 = \dfrac{7}{40} \Rightarrow e_1 e_2 = \dfrac{\sqrt{7}}{2\sqrt{10}}$

and $\quad e_1^2 + e_2^2 = \dfrac{1}{5} + \dfrac{7}{8} = \dfrac{43}{40}$

Also, $\quad \left|e_1^2 - e_2^2\right| = \left|\dfrac{1}{5} - \dfrac{7}{8}\right| = \dfrac{27}{40}$

Linked Comprehension Type

PASSAGE 1

1. If $E_n : \dfrac{x^2}{a_n^2} + \dfrac{y^2}{b_n^2} = 1$ and eccentricity of E_n is e_n

If $\quad a_n > b_n$

Then $\quad b_n^2 = a_n^2 (1 - e_n^2) \quad \ldots\text{(i)}$

According to question

$$b_n = b_{n-1} \quad \ldots\text{(ii)}$$

and $\quad a_{n-1} = a_n e_n \quad \ldots\text{(iii)}$

For ellipse E_{n-1},

$$a_{n-1}^2 = b_{n-1}^2 (1 - e^2{}_{n-1}) \quad \ldots\text{(iv)}$$

From Eqs. (i) and (ii),

$$b_{n-1}^2 = a_n^2 (1 - e_n^2) \quad \ldots\text{(v)}$$

Substituting the values of a_{n-1} and b_{n-1}^2 from Eq. (iii) and (v) in (iv), then

$$a_n^2 e_n^2 = a_n^2 (1 - e_n^2)(1 - e_{n-1}^2)$$

$$\Rightarrow \quad e_n^2 = (1 - e_n^2)\ (1 - e_{n-1}^2) \quad \ldots\text{(vi)}$$

$\because$ E_n is independent of n,

$\therefore \quad e_n = e_{n-1} = e$ (say)

$\Rightarrow \quad e^2 = (1 - e^2)^2 \Rightarrow e^4 - 3e^2 + 1 = 0$

$$\therefore \quad e^2 = \frac{3 \pm \sqrt{5}}{2} = \frac{6 \pm 2\sqrt{5}}{4} = \left(\frac{\sqrt{5} \pm 1}{2}\right)^2$$

$$\therefore \quad e = \frac{\sqrt{5} - 1}{2} \quad [\because 0 < e < 1]$$

2. From Eq. (vi), $e_n^2 = (1 - e_n^2)(1 - e_{n-1}^2)$

Locus of (e_n^2, e_{n-1}^2) is

$$x = (1 - x)(1 - y)$$

$$\Rightarrow \quad xy - 2x - y + 1 = 0$$

Here, $a = 0$, $b = 0$, $c = 1$,

$$f = -\frac{1}{2}, g = -1, h = \frac{1}{2}$$

$$\therefore \quad \Delta = 0 + 2 \times \frac{-1}{2} \times -1 \times \frac{1}{2} - 0 - 0 - 1 \times \frac{1}{4} = \frac{1}{4} \neq 0$$

$h^2 > ab$ and $a + b = 0$

$\Rightarrow$ Rectangular hyperbola

3. From Eq. (vi)

$$e_n = \sqrt{\left(\frac{1 - e_{n-1}^2}{2 - e_{n-1}^2}\right)} \quad \ldots\text{(vii)}$$

$$\therefore \quad e_2 = \sqrt{\frac{1 - e_1^2}{2 - e_1^2}} = \sqrt{\left(\frac{1 - \frac{3}{4}}{2 - \frac{3}{4}}\right)} = \frac{1}{\sqrt{5}} \quad \left[\because e_1 = \frac{\sqrt{3}}{2}\right]$$

$$e_3 = \sqrt{\left(\frac{1 - e_2^2}{2 - e_2^2}\right)} = \sqrt{\frac{1 - \frac{1}{5}}{2 - \frac{1}{5}}} = \frac{2}{3}$$

and $$e_4 = \sqrt{\left(\frac{1 - e_3^2}{2 - e_3^2}\right)} = \sqrt{\left(\frac{1 - \frac{4}{9}}{2 - \frac{4}{9}}\right)} = \sqrt{\frac{5}{14}}$$

4. $E_1 \equiv \frac{x^2}{9} + \frac{y^2}{16} = 1$

$\therefore \quad a_1 = 3, b_1 = 4$ and $9 = 16(1 - e_1^2)$

$\therefore \quad e_1 = \frac{\sqrt{7}}{4}$

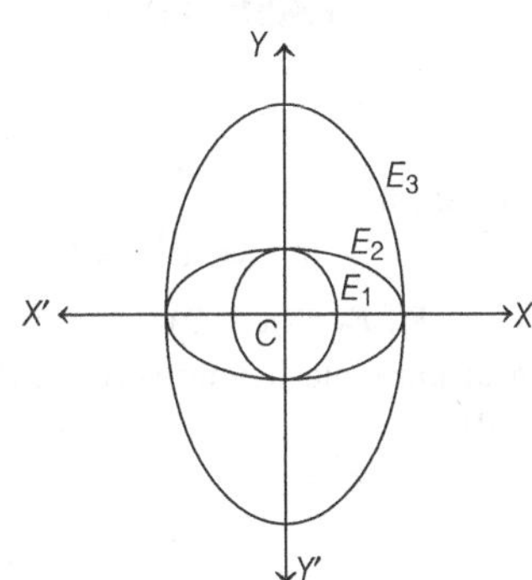

From Eq. (vii), $e_2 = \sqrt{\frac{1 - \frac{7}{16}}{2 - \frac{7}{16}}} = \frac{3}{5}$

and then $e_3 = \sqrt{\frac{1 - \frac{9}{25}}{2 - \frac{9}{25}}} = \frac{4}{\sqrt{41}}$

Also, $a_1 = a_2 e_2$ and $b_1 = b_2$

$\therefore \quad b_2 = 4 = b_3 e_3 \Rightarrow b_3 = \sqrt{41}$

and $\quad a_3^2 = b_3^2(1 - e_3^2) = 41\left(1 - \frac{16}{41}\right) = 25$

$\therefore$ Ellipse E_3 is $\frac{x^2}{25} + \frac{y^2}{41} = 1$

5. Equation of chord whose mid point (x_1, y_1) is

$$T = S_1 \Rightarrow \frac{xx_1}{a_n^2} + \frac{yy_1}{b_n^2} - 1 = \frac{x_1^2}{a_n^2} + \frac{y_1^2}{b_n^2} - 1$$

Slope $= -\frac{b_n^2 x_1}{a_n^2 y_1} = -1$ (given) ...(viii)

$\because$ Eccentricity of E_n is independent of n

$\therefore$ Axis of E_n along X-axis or along y-axis in each case

$$e = \frac{\sqrt{5} - 1}{2}$$

$$\therefore \quad e^2 = \frac{3 - \sqrt{5}}{2}$$

From Eq. (viii) $b_n^2 x_1 = a_n^2 y_1 = b_n^2(1 - e_n^2) y_1$

or $\quad x_1 = \left(1 - \frac{3 - \sqrt{5}}{2}\right) y_1$

$\Rightarrow \quad 2x_1 = (\sqrt{5} - 1) y_1$

or $2x_1(\sqrt{5} + 1) = 4y_1$

or $\quad (\sqrt{5} + 1)x_1 = 2y_1$

$\therefore$ Required locus is $(\sqrt{5} + 1)x = 2y$

PASSAGE 2

1. $\because \quad ae = \sqrt{7}, \ \frac{a}{e} = \frac{16}{\sqrt{7}}$

$\therefore \quad a^2 = 16 \Rightarrow a = 4$

Then $\quad e = \frac{\sqrt{7}}{4}$

and $\quad b^2 = a^2(1 - e^2)$

$$= 16\left(1 - \frac{7}{16}\right) = 16 - 7 = 9$$

$\therefore$ Ellipse is $\frac{x^2}{16} + \frac{y^2}{9} = 1$

$\Rightarrow \quad 9x^2 + 16y^2 = 144$

2. Let $P(4\cos\phi, 3\sin\phi)$ be any point on the ellipse $\frac{x^2}{16} + \frac{y^2}{9} = 1$ and foci are

$S \equiv (\sqrt{7}, 0)$ and $S' \equiv (-\sqrt{7}, 0)$

$$\therefore \quad \Delta = \frac{1}{2} \times S'S \times PM$$

$$= \frac{1}{2}(2\sqrt{7}) \times 3\sin\phi$$

$$= 3\sqrt{7}\sin\phi \text{ sq unit}$$

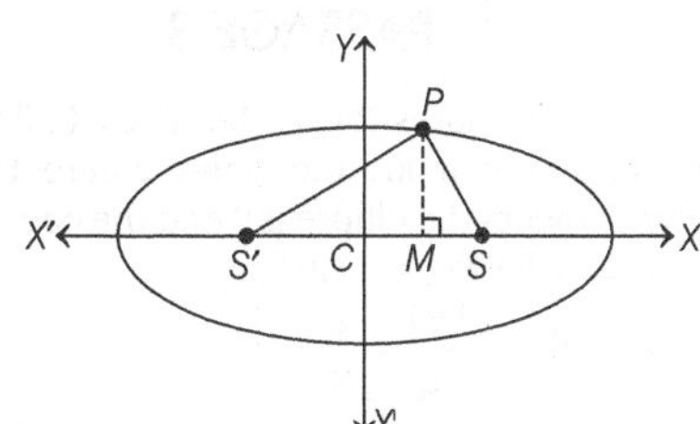

$\therefore$ Maximum value of Δ is $3\sqrt{7}$ sq unit.

3. $\because$ Mid point of A and B is (x_1, y_1)

$\therefore \quad A \equiv (2x_1, 0)$ and $B \equiv (0, 2y_1)$

$\because$ Slope of AB is m (given)

$\therefore \quad \frac{2y_1 - 0}{0 - 2x_1} = m$

$\Rightarrow \quad y_1 + mx_1 = 0$

$\Rightarrow \quad mx_1 + y_1 = 0$

4. Equation of tangent of Ellipse $\frac{x^2}{16} + \frac{y^2}{9} = 1$ in terms of slope is

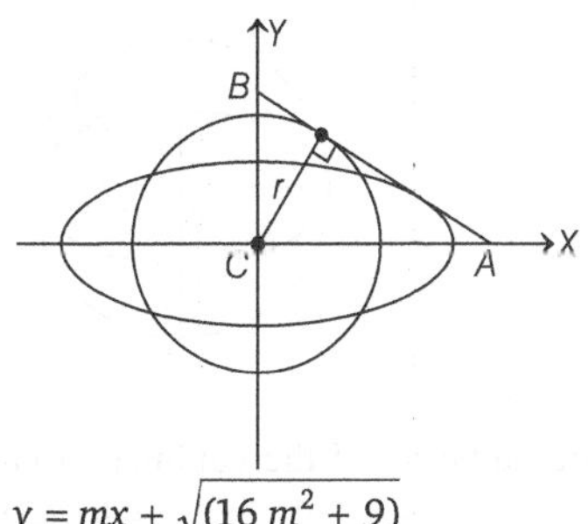

$$y = mx + \sqrt{(16m^2 + 9)} \quad \text{...(i)}$$

Here, $m < 0$ [for first quadrant]
Which is also tangent of circle, then

$$\frac{\sqrt{(16m^2 + 9)}}{\sqrt{(m^2 + 1)}} = r \Rightarrow 16m^2 + 9 = m^2r^2 + r^2$$

or $$m = \pm\sqrt{\left(\frac{r^2 - 9}{16 - r^2}\right)}$$

$\Rightarrow$ $$m = -\sqrt{\left(\frac{r^2 - 9}{16 - r^2}\right)} \quad [\because m < 0]$$

From above relation

$$y_1 = -mx_1$$

or $$y_1 = x_1\sqrt{\left(\frac{r^2 - 9}{16 - r^2}\right)}$$

$\therefore$ Locus of mid point of AB is $y = x\sqrt{\left(\frac{r^2 - 9}{16 - r^2}\right)}$

5. In above question

$$\frac{r^2 - 9}{16 - r^2} > 0 \Rightarrow 9 < r^2 < 16$$

$\therefore$ $$r \in (3, 4) \quad [\because r > 0]$$

PASSAGE 3

1. Let the coordinates of P be $(3\cos\theta, 3\sin\theta)$ then the eccentric angle of M, the point where the ordinate PQ through P meets the ellipse is θ and the coordinates of M are $(3\cos\theta, 2\sin\theta)$, $PQ = 3\sin\theta$.

$MQ = 2\sin\theta$ so $\frac{MQ}{PQ} = \frac{2}{3}$.

2. Line $L : y = 2x$ meets the circle $C : x^2 + y^2 = 9$ at points for which $x^2 + 4x^2 = 9$

$\Rightarrow$ $$x = \pm 3\sqrt{5}$$

Coordinate of P are $(\pm 3\sqrt{5}, \pm 6\sqrt{5},)$

$\Rightarrow$ coordinates of M are $(\pm 3\sqrt{5}, \pm 4\sqrt{5},)$

Equation of the tangent at M to the ellipse E is

$$\frac{x(\pm 3)}{9\sqrt{5}} + \frac{y(\pm 4)}{4\sqrt{5}} = 1$$

$$x + 3y = \pm 3\sqrt{5}$$

3. Let $y = mx$ be the diameter conjugate to the diameter $L : y = 2x$ of the ellipse E

then $2m = -4/9$ $(mm' = -b^2/a^2)$

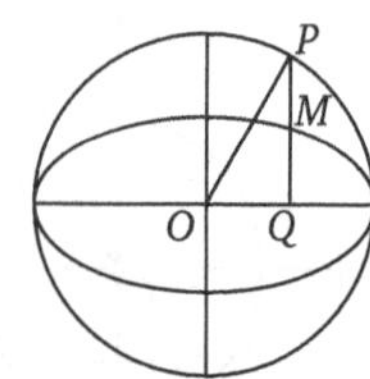

$\Rightarrow m = -2/9$

and the equation of the conjugate diameter is

$$y = (-2/9)x \text{ or } 2x + 9y = 0$$

4. Coordinate of R are $(1, 2)$

$C(1, 2) = 1 + 2^2 - 9 < 0 \Rightarrow R$ lies inside C

$E(1, 2) = \frac{1}{9} + 1 - 1 > 0 \Rightarrow R$ lies outside E.

PASSAGE 4

1. Here, $\frac{x^2}{9} + \frac{y^2}{8} = 1$ ...(i)

has foci $(\pm ae, 0)$

where, $a^2e^2 = a^2 - b^2 \Rightarrow a^2e^2 = 9 - 8 \Rightarrow ae = \pm 1$

i.e. $F_1, F_2 = (\pm 1, 0)$

Equation of parabola having vertex $O(0, 0)$ and $F_2(1, 0)$ (as, $x_2 > 0$)

$$y^2 = 4x \quad \text{...(ii)}$$

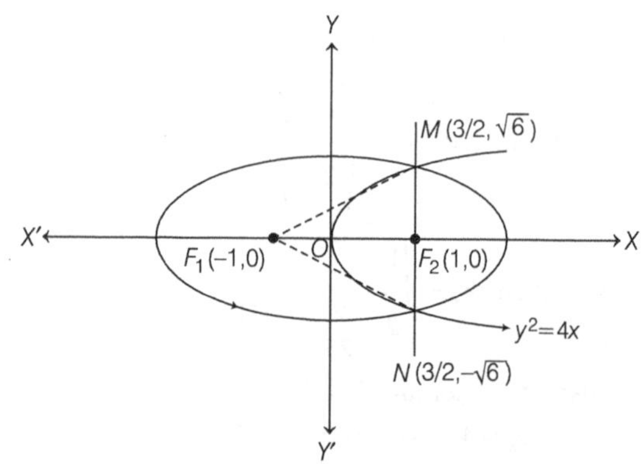

On solving $\frac{x^2}{9} + \frac{y^2}{8} = 1$ and $y^2 = 4x$, we get

$$x = 3/2 \text{ and } y = \pm\sqrt{6}$$

Equation of altitude through M on NF_1 is

$$\frac{y - \sqrt{6}}{x - 3/2} = \frac{5}{2\sqrt{6}}$$

$\Rightarrow$ $$(y - \sqrt{6}) = \frac{5}{2\sqrt{6}}(x - 3/2) \quad \text{...(iii)}$$

and equation of altitude through F_1 is $y = 0$ ...(iv)

On solving Eqs. (iii) and (iv), we get $\left(-\frac{9}{10}, 0\right)$ as orthocentre.

2. Equation of tangent at $M(3/2, \sqrt{6})$ to $\frac{x^2}{9} + \frac{y^2}{8} = 1$ is

$$\frac{3}{2} \cdot \frac{x}{9} + \sqrt{6} \cdot \frac{y}{8} = 1 \quad \text{...(i)}$$

which intersect X-axis at $(6, 0)$.

Also, equation of tangent at $N(3/2, -\sqrt{6})$ is

$$\frac{3}{2} \cdot \frac{x}{9} - \sqrt{6}\,\frac{y}{8} = 1 \quad \text{...(ii)}$$

Eqs. (i) and (ii) intersect on X-axis at

$$R(6, 0) \quad \text{...(iii)}$$

Also, normal at $M(3/2, \sqrt{6})$ is

$$y - \sqrt{6} = \frac{-\sqrt{6}}{2}\left(x - \frac{3}{2}\right)$$

On solving with $y = 0$, we get $Q(7/2, 0)$ ...(iv)

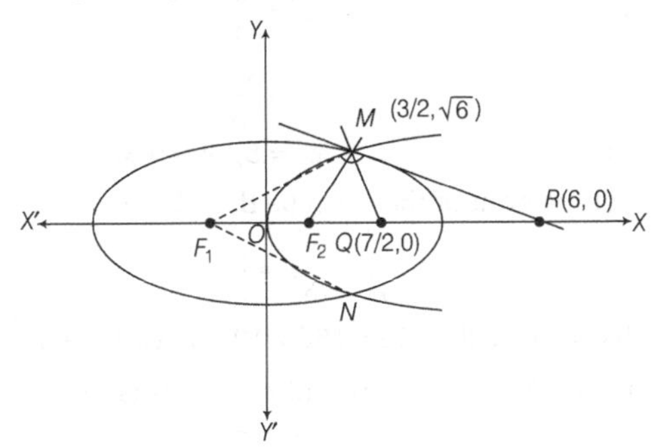

$\therefore$ Area of $\Delta MQR = \frac{1}{2}\left(6 - \frac{7}{2}\right)\sqrt{6}$

$$= \frac{5\sqrt{6}}{4} \text{ sq units}$$

and area of quadrilateral $MF_1NF_2 = 2 \times \frac{1}{2}\{1 - (-1)\}\sqrt{6}$

$$= 2\sqrt{6} \text{ sq units}$$

$$\therefore \quad \frac{\text{Area of } \Delta MQR}{\text{Area of quadrilateral } MF_1NF_2} = \frac{5}{8}$$

Numerical Value Type

1. Let the equation of the ellipse is

$$\frac{x^2}{a^2} + \frac{y^2}{b^2} = 1 \quad [\text{let } a > b]$$

Then, the foci are $S(ae, 0)$ and $S'(-ae, 0)$ length of minor axis $BB' = 2b$ and length of latusrectum $= \frac{2b^2}{a}$

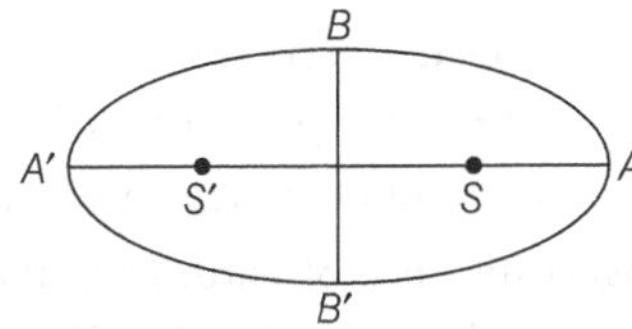

$\therefore$ According to the question

$$BB' = SS'$$

$$\Rightarrow \quad 2b = 2ae$$

$$\Rightarrow \quad b = ae \quad \text{...(i)}$$

and $$\frac{2b^2}{a} = 10 \quad \text{...(ii)}$$

$$\Rightarrow \quad b^2 = 5a$$

Also, we have $b^2 = a^2(1 - e^2)$

Putting the value of b from Eq. (i) in Eq. (iii), we have

$$a^2e^2 = a^2(1 - e^2)$$

$$\Rightarrow \quad e^2 = 1 - e^2$$

$$\Rightarrow \quad 2e^2 = 1$$

$$\therefore \quad e = \frac{1}{\sqrt{2}}$$

From Eq. (i), we have

$$b = \frac{a}{\sqrt{2}}$$

$$\therefore \quad b^2 = \frac{a^2}{2}$$

$$\Rightarrow \quad 5a = \frac{a^2}{2} \quad [\text{from Eq. (ii)}]$$

$$\Rightarrow \quad a = 10$$

From Eq. (ii),

$$b^2 = 5 \times 10 = 50$$

Putting the values of a and b in $\frac{x^2}{a^2} + \frac{y^2}{b^2} = 1$, the equation of required ellipse is

$$\frac{x^2}{100} + \frac{y^2}{50} = 1$$

$$x^2 + 2y^2 = 100$$

$$\therefore \quad m = 1, n = 2$$

Then, $10^m + 20^n = 10 + 400 = 410$

2. The coordinates of an end of the latusrectum are $(ae, b^2/a)$. The equation of normal at $P(ae, b^2/a)$ is

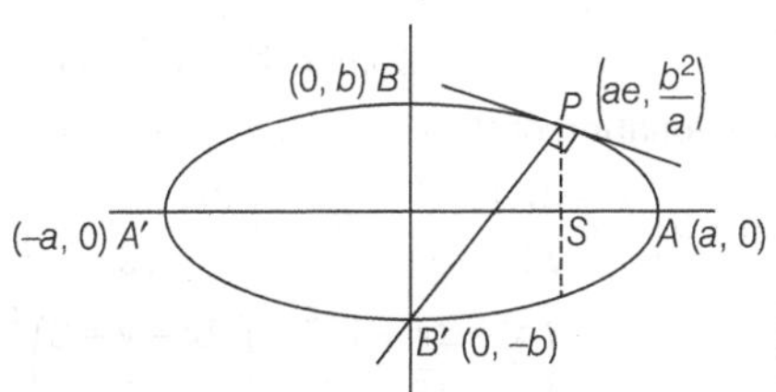

$$\frac{a^2x}{ae} - \frac{b^2(y)}{b^2/a} = a^2 - b^2$$

or $$\frac{ax}{e} - ay = a^2 - b^2$$

It passes through one extremity of the minor axis whose coordinates are $(0, -b)$

$$\therefore \quad 0 + ab = a^2 - b^2$$

or $$(a^2b^2) = (a^2 - b^2)^2$$

or $$a^2a^2(1 - e^2) = (a^2e^2)^2$$

or $$1 - e^2 = e^4$$

or $$e^4 + e^2 - 1 = 0$$

or $$(e^2)^2 + e^2 - 1 = 0$$

$$\therefore \quad e^2 - \frac{-1 \pm \sqrt{1+4}}{2}$$

$$\Rightarrow \quad e^2 = \frac{\sqrt{5} - 1}{2} \quad [\text{taking + ve sign}]$$

$$2e^2 + 1 = \sqrt{5}$$

$$\therefore \quad 625(2e^2 + 1)^2 = 625 \times 5 = 3125$$

3. For point P, y coordinate = 4

$\because$ Given ellipse is

$$16x^2 + 25y^2 = 400$$
$$16x^2 + 25(4)^2 = 400$$

Coordinate of P is (0, 4)

$$e^2 = 1 - \frac{16}{25} = \frac{9}{25}$$

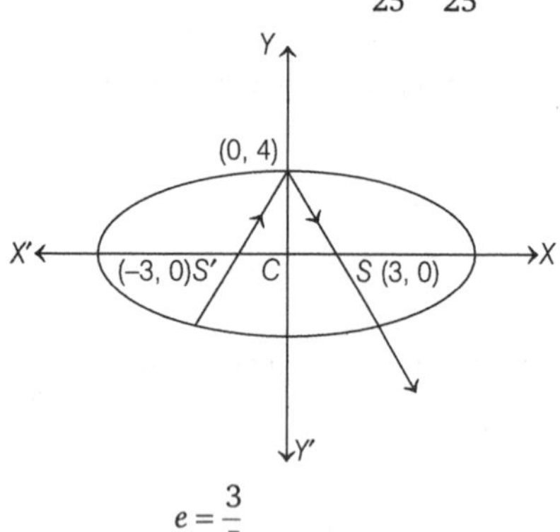

$$e = \frac{3}{5}$$

foci $(\pm\, ae, 0)$ *ie*, $(\pm\, 3, 0)$

Equation of reflected ray (*ie, PS*) is

$$\frac{x}{3} + \frac{y}{4} = 1 \text{ or } 4x + 3y = 12$$

$\therefore \qquad \lambda = 12$, then $2^{\lambda} = 2^{12} = 4096$

4. The equation of the ellipse can be written as

$$4 \times 5\left(\frac{x - 2y + 1}{\sqrt{5}}\right)^2 + 9 \times 5\left(\frac{2x + y + 2}{\sqrt{5}}\right)^2 = 25$$

or $$\frac{\left(\frac{x - 2y + 1}{\sqrt{5}}\right)^2}{(5/4)} + \frac{\left(\frac{2x + y + 2}{\sqrt{5}}\right)^2}{(5/9)} = 1$$

or $$\frac{X^2}{a^2} + \frac{Y^2}{b^2} = 1$$

Here $a > b$

$\therefore$ Equation of major axis is $Y = 0$

i.e. $\qquad 2x + y + 2 = 0$

and Equation of minor axis is $X = 0$

i.e. $\qquad x - 2y + 1 = 0$

Centre $\qquad X = 0, Y = 0$

$\Rightarrow x - 2y + 1 = 0,\ 2x + y + 2 = 0$

we get $\qquad x = -1, y = 0$

$\therefore$ Centre is $(-1, 0)$

Length of latusrectum $= \dfrac{2b^2}{a} = \dfrac{2 \times 5/9}{5/4} = \dfrac{8}{9}$

Eccentricity $\qquad b^2 = a^2(1 - e^2)$

$$\Rightarrow \qquad \frac{5}{9} = \frac{5}{4}(1 - e^2)$$

$$\Rightarrow \qquad \frac{4}{9} = 1 - e^2$$

$$\Rightarrow \qquad e^2 = \frac{5}{9}$$

$$\therefore \qquad e = \frac{\sqrt{5}}{3}$$

$$\Rightarrow \qquad e^2 = 5/9 = 0.56$$

5. Let the orbit of the earth be the ellipse

$$\frac{x^2}{a^2} + \frac{y^2}{b^2} = 1 \qquad \ldots\text{(i).}$$

Its major axis $= 2a = 186 \times 10^6$ miles (given)

ie, $a = 93 \times 10^6$ miles and $e = 1/60$ (given)

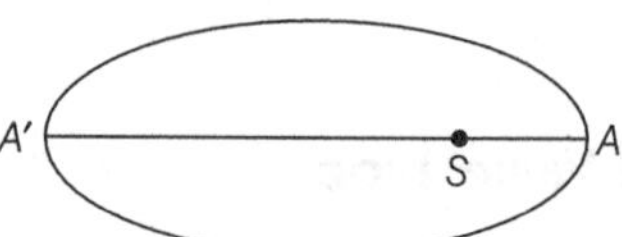

Let the sun be at the focus $S(ae, 0)$. Then the earth will be at shortest and longest distance from the sun when the earth is at the extremities of the major axis which are respectively nearest and farthest from this focus S.

$\therefore$ Shortest distance of the earth from the sun

$= SA$, where S is $(ae, 0)$ and A is $(a, 0)$

$= a - ae = a(1 - e)$

$= (93 \times 10^6)(1 + (1/60))$

$= 91450000$ miles $= 914.5 \times 10^5$ miles

and longest distance of earth from the sun

$= SA'$, where S is $(ae, 0)$ and A' is $(-a, 0)$

$= a + ae$

$= a(1 + e)$

$= (93 \times 10^6)(1 + (1/60))$

$= 94550000$ miles $= 945.5 \times 10^5$ miles

$\therefore \quad \lambda + \mu = 914.5 + 945.5 = 1860$

6. Let the given ellipse be

$$\frac{x^2}{a^2} + \frac{y^2}{b^2} = 1$$

Let the coordinates of P are $(a\cos\phi, b\sin\phi)$

By hypothesis,

$b^2 = a^2(1 - e^2)$ and $S(ae, 0)$, $S'(-ae, 0)$

$\therefore \quad SP$ = focal distance of the point P

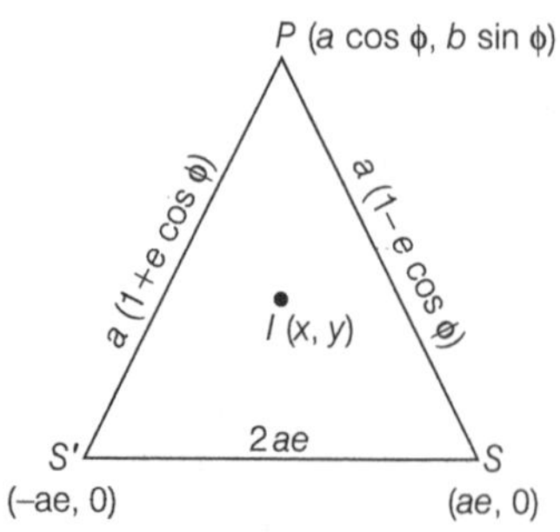

$= a - ae\cos\phi$

and $\qquad S'P = a + ae\cos\phi$

Also, $SS' = 2ae$

If (x, y) be the incentre of the $\Delta PSS'$, then

$\therefore$

$$x = \frac{(2ae) a\cos\phi + a(1 - e\cos\phi)(-ae) + 1(1 + e\cos\phi) ae}{2ae + a(1 - e\cos\phi) + a(1 + e\cos\phi)}$$

$$x = ae\cos\phi \quad \ldots(i)$$

and $$y = \frac{2ae(b\sin\phi) + a(1 + e\cos\phi)\cdot 0 + a(1 - e\cos\phi)\cdot 0}{2ae + a(1 - e\cos\phi) + a(1 + e\cos\phi)},$$

$$\Rightarrow \quad y = \frac{eb\sin\phi}{(e+1)} \quad \ldots(ii)$$

Eliminating ϕ from Eqs. (i) and (ii), we get

$$\frac{x^2}{a^2e^2} + \frac{y^2}{\left[\frac{be}{e+1}\right]^2} = 1$$

Which represents an ellipse.

$\because \lambda$ be its eccentricity.

$$\therefore \quad \frac{b^2e^2}{(e+1)^2} = a^2e^2(1 - \lambda^2)$$

$$\Rightarrow \quad \lambda^2 = 1 - \frac{b^2}{a^2(e+1)^2}$$

$$\Rightarrow \quad \lambda^2 = 1 - \frac{1-e^2}{(e+1)^2}$$

$$\Rightarrow \quad \lambda^2 = 1 - \frac{1-e}{1+e} = \frac{2e}{1+e}$$

$$= \frac{2 \times 3/5}{1 + 3/5} = \frac{3}{4}$$

$$\therefore \quad 2\lambda^2 = 1.5$$

7. $\frac{x^2}{9} + \frac{y^2}{5} = 1 \Rightarrow e^2 = 1 - \frac{5}{9} = \frac{4}{5} \Rightarrow e = 2/3$

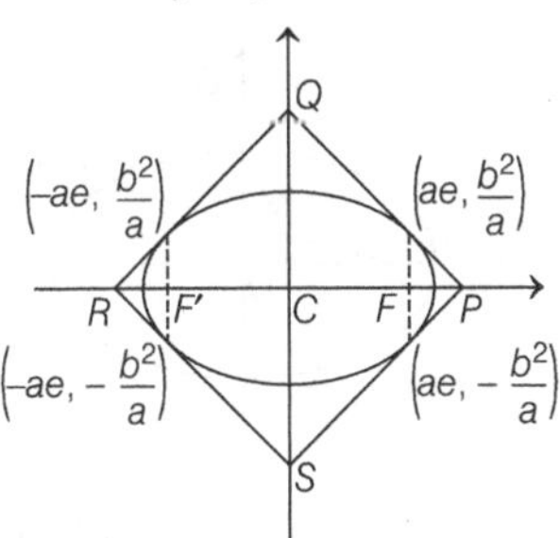

One end of latusrectum is $\left(2, \frac{5}{3}\right)$

Equation of tangent at $\left(2, \frac{5}{3}\right)$ is $\frac{2x}{9} + \frac{y}{3} = 1$

F and F' be foci

Area of $\Delta CPQ = \frac{1}{2} \times \frac{9}{2} \times 3 = \frac{27}{4}$

$\therefore$ Area of an quadrilateral $PQRS = 4 \times \frac{27}{4} = 27$ sq unit

$\therefore \lambda = 27$

8. Let $P(a\cos\theta_1, b\sin\theta_1)$, $Q(a\cos\theta_2, b\sin\theta_2)$ and $R(a\cos\theta_3, b\sin\theta_3)$ be the vertices of the triangle inscribed in the ellipse $\frac{x^2}{a^2} + \frac{y^2}{b^2} = 1$. The points on the auxiliary circle corresponding to these points are $P'(a\cos\theta_1, a\sin\theta_1)$, $Q'(a\cos\theta_2, a\sin\theta_2)$ and $R'(a\cos\theta_3, a\sin\theta_3)$

$$\therefore \quad \Delta_1 = \text{Area of } \Delta PQR$$

$$= \frac{1}{2}\begin{vmatrix} a\cos\theta_1 & b\sin\theta_1 & 1 \\ a\cos\theta_2 & b\sin\theta_2 & 1 \\ a\cos\theta_3 & b\sin\theta_3 & 1 \end{vmatrix} = \frac{1}{2}ab\begin{vmatrix} \cos\theta_1 & \sin\theta_1 & 1 \\ \cos\theta_2 & \sin\theta_2 & 1 \\ \cos\theta_3 & \sin\theta_3 & 1 \end{vmatrix}$$

and $\Delta_2 = \text{Area of } \Delta P'Q'R'$

$$= \frac{1}{2}\begin{vmatrix} a\cos\theta_1 & a\sin\theta_1 & 1 \\ a\cos\theta_2 & a\sin\theta_2 & 1 \\ a\cos\theta_3 & a\sin\theta_3 & 1 \end{vmatrix}$$

$$= \frac{1}{2}a^2\begin{vmatrix} \cos\theta_1 & \sin\theta_1 & 1 \\ \cos\theta_2 & \sin\theta_2 & 1 \\ \cos\theta_3 & \sin\theta_3 & 1 \end{vmatrix}$$

Clearly, $$\frac{\Delta_1}{\Delta_2} = \frac{b}{a} = \sqrt{(1 - e^2)} = \frac{1}{7}$$

$$\Rightarrow \quad \frac{\Delta_2}{\Delta_1} = 7$$

Then, $$343\frac{\Delta_2}{\Delta_1} = 2401$$

9. Let OB and OY be taken as x and y axes respectively. Clearly,

$$A \equiv (3, 0) \text{ and } B \equiv (3, 0)$$

Equation of circles having centres at O and radii 3 unit and 5 unit are

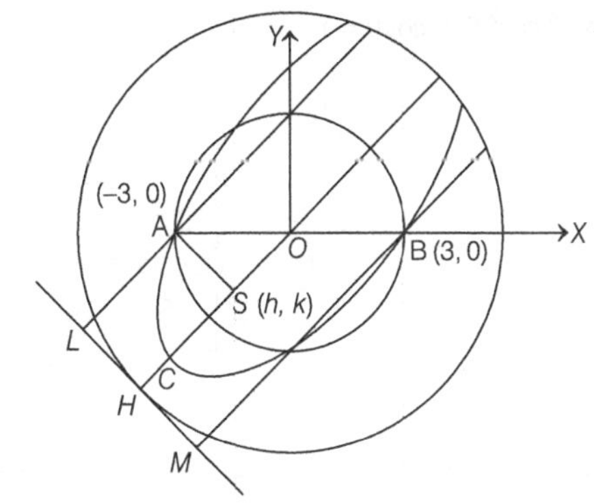

$$x^2 + y^2 = 9 \quad \ldots(i)$$

and $$x^2 + y^2 = 25 \quad \ldots(ii)$$

Let (h, k) be the focus of the variable parabola passing through A and B. Since LM is tangent to circle (ii) therefore its equation may be taken as

$$y = mx + 5\sqrt{(1 + m^2)} \quad \ldots(iii)$$

Since, A lies on parabola

$$\therefore \quad AS = AL$$

or $$AS^2 = AL^2$$

$$\therefore \quad (h+3)^2 + k^2 = \frac{(-3m + 5\sqrt{1+m^2})^2}{1+m^2} \quad \ldots\text{(iv)}$$

Similarly, since (3, 0) lies on the parabola

$$\therefore \quad (h-3)^2 + k^2 = \frac{(3m + 5\sqrt{1+m^2})^2}{1+m^2} \quad \ldots\text{(v)}$$

In order to find the locus of $S(h, k)$ we will have to eliminate m from Eq. (iv) and (v)

Subtracting Eqs. (iv) and (v)

we get $4h\,(3) = -\dfrac{4m\,(3)\,(5)\sqrt{1+m^2}}{1+m^2}$

$$\Rightarrow (1+m)^2 \cdot h^2 = 25m^2 \quad \ldots\text{(vi)}$$

Adding Eqs. (iv) and (v), then

$$2(h^2 + k^2 + 3^2) = \frac{2[m^2 3^2 + 5^2(1+m^2)]}{1+m^2}$$

or $\quad h^2 + k^2 + 9 = 9\dfrac{m^2}{1+m^2} + 25$

or $\quad h^2 + k^2 + 9 = 9 \cdot \dfrac{h^2}{25} + 25 \quad$ [from Eq. (vi)]

or $\quad h^2\left(1 - \dfrac{9}{25}\right) + k^2 = 25 - 9$

or $h^2\left(\dfrac{25-9}{25}\right) + k^2 = 16$

or $\quad \dfrac{h^2}{25} + \dfrac{k^2}{16} = 1$

$\therefore$ Locus of $S(h, k)$ is $\dfrac{x^2}{25} + \dfrac{y^2}{16} = 1$

$\therefore$ Director circle is

$$x^2 + y^2 = 25 + 16 = 41$$

$\therefore \quad (\text{radius})^2 = 41$

Single Integer Type

1. Tangent to the ellipse at $P(\theta)$ is

$$\frac{x\cos\theta}{a} + \frac{y\sin\theta}{b} = 1$$

It meets major axis (X-axis) at $Q = (a\sec\theta, 0)$

Equation of normal at $P(\theta)$ is

$ax\sec\theta - by\operatorname{cosec}\theta = a^2 - b^2$

It meets major axis at $R = \left(\dfrac{1}{a}(a^2 - b^2)\cos\theta, 0\right)$.

$$QR = \frac{a}{\cos\theta} - \frac{a^2 - b^2}{a}\cos\theta = a$$

$$\Rightarrow e^2\cos^2\theta + \cos\theta = 1$$

2.

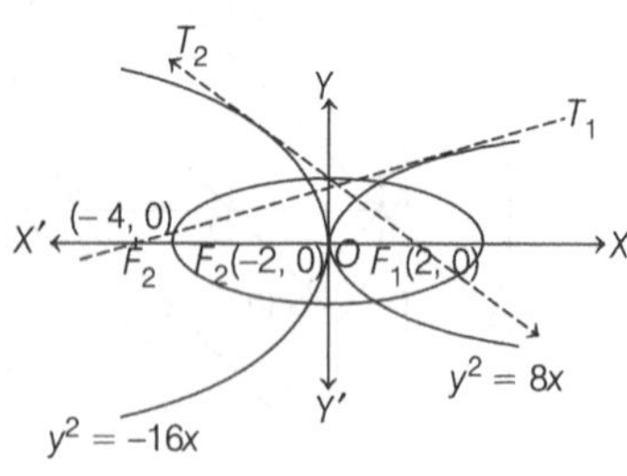

Tangent to P_1 passes through $(2f_2, 0)$, i.e. $(-4, 0)$.

$$\therefore \quad T_1 : y = m_1 x + \frac{2}{m_1}$$

$$\Rightarrow \quad 0 = -4m_1 + \frac{2}{m_1}$$

$$\Rightarrow \quad m_1^2 = \frac{1}{2} \quad \ldots\text{(i)}$$

Also, tangent to P_2 passes through $(f_1, 0)$, i.e. (2, 0).

$$\Rightarrow \quad T_2 : y = m_2 x + \frac{(-4)}{m_2}$$

$$\Rightarrow \quad 0 = 2m_2 - \frac{4}{m_2}$$

$$\Rightarrow \quad m_2^2 = 2 \quad \ldots\text{(ii)}$$

$$\therefore \quad \frac{1}{m_1^2} + m_2^2 = 2 + 2 = 4$$

3. Here $a = 3$; $b = 2$

and $T : \dfrac{x\cos\theta}{3} + \dfrac{y\sin\theta}{2} = 1$

$x = 0$; $y = 2\operatorname{cosec}\theta$

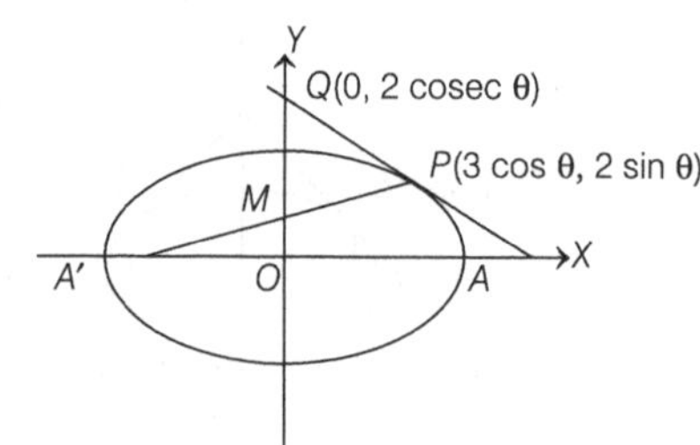

Equation of chord $A'P$, $y = \dfrac{2\sin\theta}{3(\cos\theta + 1)}(x+3)$

Put $x = 0$, $y = \dfrac{2\sin\theta}{1+\cos\theta} = OM$

Now, $OQ^2 - MQ^2 = OQ^2 - [OQ - OM]^2$

$$= 2(OQ)(OM) - OM^2 = OM\{2(OQ) - (OM)\}$$

$$= \frac{2\sin\theta}{1+\cos\theta}\left[\frac{4}{\sin\theta} - \frac{2\sin\theta}{1+\cos\theta}\right] = 4$$

4. The centre of the given circle is

$$O(4, -3)$$

The circumcircle of ΔPAB will circumscribe the quadrilateral $PBOA$ also. Hence, one of the diameters must be OP.

So, the equation of circumcircle of ΔPAB will be

$$(x-2)(x-4)+(y-3)(y+3)=0$$

or $$x^2+y^2-6x-1=0 \quad \text{... (i)}$$

Director circle of the given ellipse will be

$$(x+5)^2+(y-3)^2=9+b^2$$

or $$x^2+y^2+10x-6y+25-b^2=0 \quad \text{... (ii)}$$

So, from Eqs. (i) and (ii), by applying the condition of orthogonality, we get

$$2[-3(5)+0(-3)]=-1+25-b^2$$

or $$-30=24-b^2$$

Hence, $$b^2=54$$

$\Rightarrow$ $$b=\sqrt{54} \Rightarrow [b]=7$$

5. $\dfrac{x^2}{169}+\dfrac{y^2}{25}=1$

Equation of normal at the point $(13\cos\theta, 5\sin\theta)$

$$(y-5\sin\theta)=\frac{13\sin\theta}{5\cos\theta}(x-13\cos\theta)$$

it passes through (0, 6)

then $$\cos\theta\,(15+72\sin\theta)=0$$

or $$\cos\theta=0,\ \sin\theta+-\frac{5}{24}$$

$$\theta=\frac{\pi}{2},\ 2\pi-\sin^{-1}\frac{5}{24},\ \pi+\sin^{-1}\frac{5}{24}$$

Equation has three roots hence three normal can be drawn.

6. $x=10\cos\theta-6\cos\theta=4\cos\theta$

$$y=6\sin\theta$$

$\therefore$ Locus is $\left(\dfrac{x}{4}\right)^2+\left(\dfrac{y}{6}\right)^2=1$

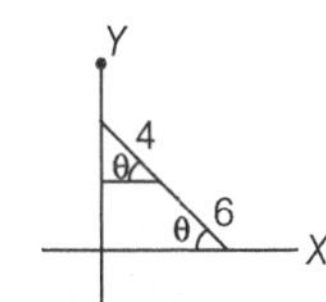

$$\frac{x^2}{16}+\frac{y^2}{36}=1$$

$\therefore$ $$e=\sqrt{\frac{36-16}{36}}=\sqrt{\frac{20}{36}}=\frac{\sqrt{5}}{3}$$

$\Rightarrow$ $$e^2=\frac{5}{9}$$

$\Rightarrow$ $$9e^2=5$$

7. Let the ellipse be $\dfrac{x^2}{a^2}+\dfrac{y^2}{b^2}=1$

Then, according to given information, we have the following figure.

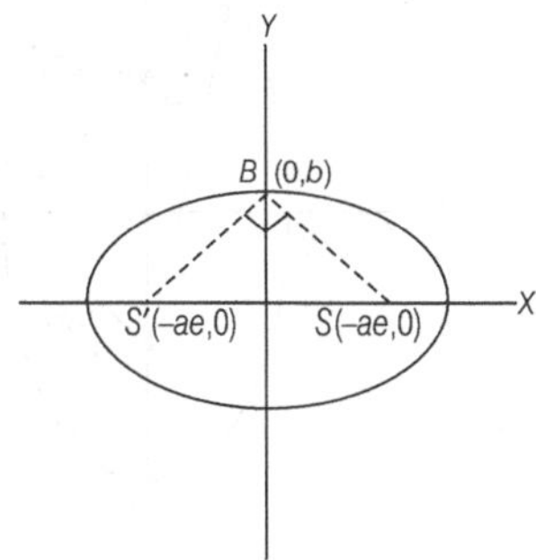

Clearly, slope of line $SB=\dfrac{b}{-ae}$ and slope of line $S'B=\dfrac{b}{ae}$

$\because$ Lines SB and $S'B$ are perpendicular so

$$\left(\frac{b}{-ae}\right)\cdot\left(\frac{b}{ae}\right)=-1$$

[$\because$ product of slopes of two perpendicular lines is (-1)]

$\Rightarrow$ $$b^2=a^2e^2 \quad \text{... (i)}$$

Also, it is given that area of $\Delta S'BS=8$

$\therefore$ $$\frac{1}{2}a^2=8$$

[$\because S'B=SB=a$ because $S'B+SB=2a$ and $S'B=SB$]

$\Rightarrow$ $$a^2=16$$

$\Rightarrow$ $$a=4 \quad \text{... (ii)}$$

$\because$ $$e^2=1-\frac{b^2}{a^2}=1-e^2 \quad \text{[from Eq. (i)]}$$

$\Rightarrow$ $$2e^2=1$$

$\Rightarrow$ $$e^2=\frac{1}{2} \quad \text{... (iii)}$$

From Eqs. (i) and (iii), we get

$$b^2=a^2\left(\frac{1}{2}\right)=16\left(\frac{1}{2}\right) \quad \text{[using Eq. (ii)]}$$

$\Rightarrow$ $$b^2=8$$

Now, length of latus rectum

$$=\frac{2b^2}{a}=\frac{2\times 8}{4}=4 \text{ units}$$

Matrix-Match Type

1. **(A)** $\because 9x^2 + 4y^2 = 36$

or $\dfrac{x^2}{4} + \dfrac{y^2}{9} = 1$

ie, ellipse along Y-axis. If e be the eccentricity, then

$$4 = 9(1 - e^2)$$

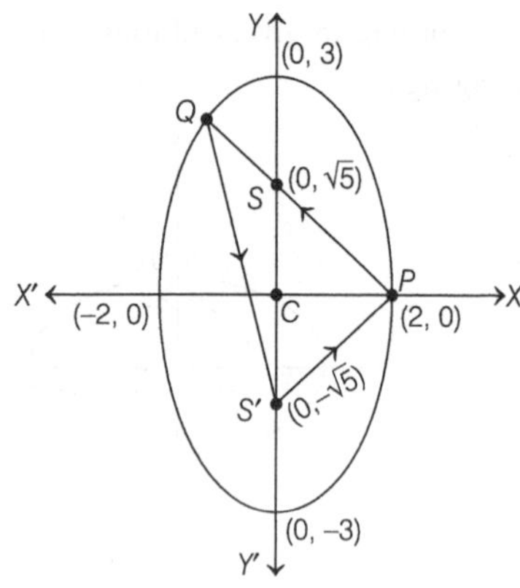

$\Rightarrow$ $e = \dfrac{\sqrt{5}}{3}$

Then, foci $(0, \pm\sqrt{5})$

and $P \equiv (2, 0)$

$\therefore$ Equation of first reflected ray is

$$\frac{x}{2} + \frac{y}{\sqrt{5}} = 1$$

or $x\sqrt{5} + 2y - 2\sqrt{5} = 0$

$\therefore$ $F : x\sqrt{5} + 2y - 2\sqrt{5} = 0$ (R)

Solving F and $9x^2 + 4y^2 = 36$

we get $Q = \left(-\dfrac{4}{7}, \dfrac{9\sqrt{5}}{7}\right)$

$\therefore$ Equation of second reflected ray is

$$y + \sqrt{5} = \frac{\dfrac{9\sqrt{5}}{7} + \sqrt{5}}{-\dfrac{4}{7} - 0}(x - 0) \text{ (Pass through Q, S')}$$

$$S : 4\sqrt{5}x + y + \sqrt{5} = 0 \text{ (S)}$$

(B) $\because 4x^2 + 9y^2 = 36$

or $\dfrac{x^2}{9} + \dfrac{y^2}{4} = 1$

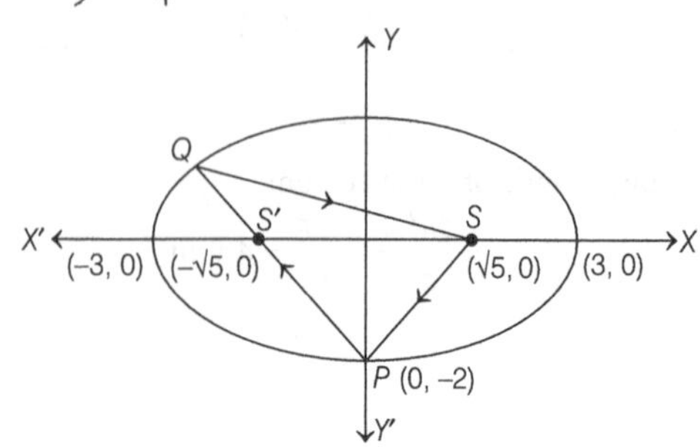

i.e. ellipse along x-axis.

If e be the eccentricity then

$$4 = 9(1 - e^2)$$

or $e = \dfrac{\sqrt{5}}{3}$

Then, foci $(\pm\sqrt{5}, 0)$

and $P \equiv (0, -2)$

$\therefore$ Equation of first reflected ray is

$$\frac{x}{-\sqrt{5}} + \frac{y}{-2} = 1$$

or $F : 2x + y\sqrt{5} + 2\sqrt{5} = 0$ (T)

Now solving F and $4x^2 + 9y^2 = 36$, we get

$$Q \equiv \left(-\frac{9\sqrt{5}}{7}, \frac{4}{7}\right)$$

$\therefore$ Equation of the second reflected ray is

$$y - 0 = \frac{\dfrac{4}{7} - 0}{\dfrac{-9\sqrt{5}}{7} - \sqrt{5}}(x - \sqrt{5}) \text{ (Pass through Q, S)}$$

$\Rightarrow$ $y = -\dfrac{1}{4\sqrt{5}}(x - \sqrt{5})$

or $S : x + 4\sqrt{5}y - \sqrt{5} = 0$ (Q)

(C) $\because 4x^2 + 5y^2 = 100$

or $\dfrac{x^2}{25} + \dfrac{y^2}{20} = 1$

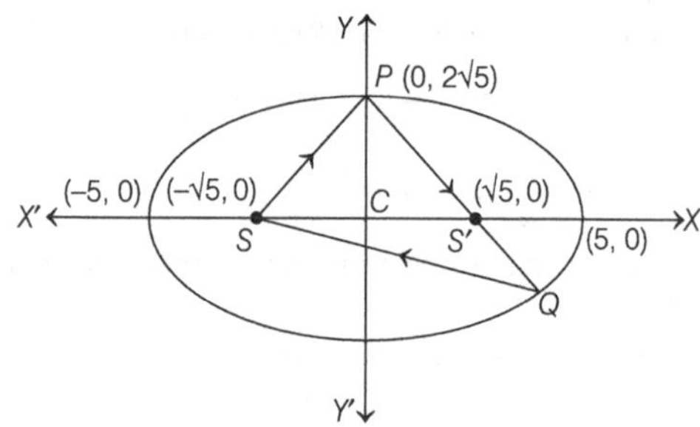

i.e. ellipse along X-axis. If e be the eccentricity then

$$20 = 25(1 - e^2)$$

$\Rightarrow$ $\dfrac{4}{5} = 1 - e^2$

$\therefore$ $e = \dfrac{1}{\sqrt{5}}$

Then, foci $(\pm\sqrt{5}, 0)$ and $P \equiv (0, 2\sqrt{5})$

$\therefore$ Equation of first reflected ray is

$$\frac{x}{\sqrt{5}} + \frac{y}{2\sqrt{5}} = 1$$

or $F : 2x + y - 2\sqrt{5} = 0$ (P)

Now, solving F and $4x^2 + 5y^2 = 100$

we get $Q \equiv \left(\frac{5\sqrt{5}}{3}, -\frac{4\sqrt{5}}{3}\right)$

$\therefore$ Equation of the second reflected ray is

$$y - 0 = \frac{\frac{-4\sqrt{5}}{3} - 0}{\frac{5\sqrt{5}}{3} + \sqrt{5}}(x + \sqrt{5}) \text{ (Pass through Q, S)}$$

$\Rightarrow$ $y = -\frac{1}{2}(x + \sqrt{5})$

or $x + 2y + \sqrt{5} = 0$

$\therefore$ $S : x + 2y + \sqrt{5} = 0$

(D) $a = 5\, b = 4$

$e = \frac{3}{5}$ $\therefore ae = 3$

$\therefore SA = 2$ also $SP = 2$

$\therefore P$ coincides with A and Q coincides with A'

$\therefore PQ = 2a = 10$

2. **(A)** Let $(\sqrt{27}\cos\theta, \sqrt{48}\sin\theta)$ be a point on the ellipse.

Thus equation of tangent on their points in

$$\frac{\sqrt{27}\cos\theta x}{27} + \frac{\sqrt{48}\sin\theta y}{48} = 1$$

$\therefore$ slope $-\frac{\cos\theta}{\sqrt{27}} \times \frac{\sqrt{48}}{\sin\theta}$

$$= -\frac{4}{3}\cot\theta = -\frac{4}{3} \text{ i.e., } \cot\theta = 1 \text{ i.e. } \theta = \frac{\pi}{4}$$

$\therefore$ Equation of the tangents is $\frac{x}{\sqrt{54}} + \frac{y}{\sqrt{96}} = 1$

Area of triangle is $\frac{x}{\sqrt{54}} + \frac{y}{\sqrt{96}} = 36$

(B) Product of perpendicular

$$= \left|\frac{3m - \sqrt{25m^2 + 16}}{\sqrt{1 + m^2}}\right| \left|\frac{-3m - \sqrt{25m^2 + 16}}{\sqrt{1 + m^2}}\right|$$

$$= \left|\frac{25m^2 + 16 - 9m^2}{1 + m^2}\right| = 16$$

(C) $2a = 5 + 10 = 15$

$\because$ Sum of focal distance of any point on ellipse is equal to length of major axis $a = \frac{15}{2}$

$$2ae = \sqrt{(6-3)^2 + (8-4)^2} = 5$$

$$e = \frac{1}{3}$$

$\therefore$ $b^2 = \frac{225}{4}\left(1 - \frac{1}{9}\right) = 50$

$\Rightarrow$ $b = 5\sqrt{2}$

$\Rightarrow$ $2b = 10\sqrt{2}$

(D) $a = 5$, $b = 4$

$$e = \frac{3}{5}$$

$\therefore$ $ae = 3$

$\therefore$ $SA = 2$ also $SP = 2$

$\therefore$ P coincides with A and Q coincides with A.

$\therefore$ $PQ = 2a = 10$

3. **(A)** $\because E : \frac{x^2}{1} + \frac{y^2}{2} = 1$

$\therefore$ $C_1 : x^2 + y^2 = 3,$

$C_2 : x^2 + y^2 = 6,$

$C_3 : x^2 + y^2 = 12, \ldots$

$\Rightarrow$ $r_1^2 = 3, r_2^2 = 6, r_3^2 = 12, \ldots$

$r_1^2, r_2^2, r_3^2, \ldots$ are in GP with common ratio 2. (P)

$$r_1^2 + r_2^2 + r_3^2 + \ldots + r_n^2 = 3 + 6 + 12 + \ldots$$

$$= \frac{3(2^n - 1)}{2 - 1} = 3(2^n - 1)$$

$$= r_1^2(2^n - 1) \text{ (R)}$$

GM of r_1^2, r_2^2, r_3^2 is $(r_1^2 r_2^2 r_3^2)^{1/3}$

i.e. $(3 \cdot 6 \cdot 12)^{1/3} = 6$ *(Q)*

(B) $\because$ $E : \frac{x^2}{2} + \frac{y^2}{3} = 1$

$\therefore$ $C_1 : x^2 + y^2 = 5,$

$C_2 : x^2 + y^2 = 10,$

$C_3 : x^2 + y^2 = 20, \ldots$

$\Rightarrow r_1^2 = 5, r_2^2 = 10, r_3^2 = 20, \ldots$

$r_1^2, r_2^2, r_3^2, \ldots$ are in GP with common ratio 2. (P)

$\therefore$ $r_1^2 + r_2^2 + r_3^2 + \ldots r_n^2 = 5 + 10 + 20 + \ldots$

$$= \frac{5(2^n - 1)}{(2 - 1)} = r_1^2(2^n - 1) \text{ (R)}$$

GM of r_1^2, r_2^2, r_3^2 is $(r_1^2 r_2^2 r_3^2)^{1/3}$

i.e. $= (5 \cdot 10 \cdot 20)^{1/3} = 10$ (S)

(C) $\because$ $E : \frac{x^2}{4} + \frac{y^2}{3} = 1$

$\therefore$ $C_1 : x^2 + y^2 = 7, C_2 : x^2 + y^2 = 14,$

$C_3 : x^2 + y^2 = 28, \ldots$

$\Rightarrow$ $r_1^2 = 7, r_2^2 = 14, r_3^2 = 28, \ldots$

$r_1^2, r_2^2, r_3^2, \ldots$ are in GP with common ratio 2. (P)

$\therefore$ $r_1^2 + r_2^2 + r_3^2 + \ldots + r_n^2 = 7 + 14 + 28 + \ldots$

$$-7\frac{(2^n - 1)}{(2 - 1)} - r_1^2(2^n - 1)\text{(R)}$$

GM of r_1^2, r_2^2, r_3^2 is $(r_1^2 r_2^2 r_3^2)^{1/3}$

i.e. $(7 \cdot 14 \cdot 28)^{1/3} = 14$ (T)

26

Hyperbola

Objective Questions Type I [Only one correct answer]

In each of the questions below, four choices are given of which only one is correct. You have to select the correct answer which is the most appropriate.

1. If the eccentricity of the hyperbola $x^2 - y^2 \sec^2 \alpha = 5$ is $\sqrt{3}$ times the eccentricity of the ellipse $x^2 \sec^2 \alpha + y^2 = 25$, then a value of α is

(a) $\frac{\pi}{6}$ (b) $\frac{\pi}{4}$

(c) $\frac{\pi}{3}$ (d) $\frac{\pi}{2}$

2. If a rectangular hyperbola $(x-1)(y-2) = 4$ cuts a circle $x^2 + y^2 + 2gx + 2fy + c = 0$ at points (3, 4), (5, 3), (2, 6) and (–1, 0), then the value of $(g + f)$ is equal to

(a) –8 (b) –9

(c) 8 (d) 9

3. If $f(x) = ax^3 + bx^2 + cx + d$, ($a, b, c, d$ are rational nos.) and roots of $f(x) = 0$ are eccentricities of a parabola and a rectangular hyperbola then $a + b + c + d$ equals

(a) –1 (b) 0

(c) 1 (d) data inadequate

4. From a point on the line $y = x + c$, c (parameter), tangents are drawn to the hyperbola $\frac{x^2}{2} - \frac{y^2}{1} = 1$ such that chords of contact pass through a fixed point (x_1, y_1). Then $\frac{x_1}{y_1}$ is equal to

(a) 2 (b) 3

(c) 4 (d) None of these

5. A hyperbola having the transverse axis of length $\sqrt{2}$ has the same foci as that of the ellipse $3x^2 + 4y^2 = 12$, then this hyperbola does not pass through which of the following points? **[JEE Main 2020, 3 Septembar]**

(a) $\left(\frac{1}{\sqrt{2}}, 0\right)$ (b) $\left(-\sqrt{\frac{3}{2}}, 1\right)$

(c) $\left(1, -\frac{1}{\sqrt{2}}\right)$ (d) $\left(\sqrt{\frac{3}{2}}, \frac{1}{\sqrt{2}}\right)$

6. A hyperbola having the transverse axis of length $2 \sin \theta$ is confocal with the ellipse $3x^2 + 4y^2 = 12$. Then its equation is

(a) $x^2 \operatorname{cosec}^2\theta - y^2 \sec^2 \theta = 1$

(b) $x^2 \sec^2 \theta - y^2 \operatorname{cosec}^2\theta = 1$

(c) $x^2 \sin^2 \theta - y^2 \cos^2 \theta = 1$

(d) $x^2 \cos^2 \theta - y^2 \sin^2 \theta = 1$

7. PQ and RS are two perpendicular chords of the rectangular hyperbola $xy = c^2$. If O is the centre of the hyperbola, then the product of the slopes of OP, OQ, OR and OS is equal to

(a) –1 (b) 1 (c) 2 (d) 4

8. A line parallel to the straight line $2x - y = 0$ is tangent to the hyperbola $\frac{x^2}{4} - \frac{y^2}{2} = 1$ at the point (x_1, y_1). Then, $x_1^2 + 5y_1^2$ is equal to **[JEE Main 2020, 2 Septembar]**

(a) 10 (b) 5 (c) 6 (d) 8

9. From the point $P(1, \sqrt{3})$ on the circle $x^2 + y^2 = 4$ a tangent is drawn to the hyperbola $\frac{x^2}{4} - \frac{y^2}{1} = 1$ which meets its transverse axis at Q. From Q a line is drawn parallel to conjugate axis, which cuts the hyperbola at R above the X-axis, then PR equals

(a) 3 (b) $\sqrt{10}$

(c) $\frac{3\sqrt{3}+4}{2}$ (d) None of these

10. Consider a hyperbola $H : x^2 - 2y^2 = 4$. Let the tangent at a point $P(4, \sqrt{6})$ meet the X-axis at Q and latusrectum at $R(x_1, y_1)$, $x_1 > 0$. If F is a focus and H which is nearer to the point P, then the area of ΔQFR is equal to **[JEE Main 2021, 18 March]**

(a) $4\sqrt{6}$ (b) $\sqrt{6} - 1$ (c) $\frac{7}{\sqrt{6}} - 2$ (d) $4\sqrt{6} - 1$

11. If chords of the hyperbola $x^2 - y^2 = a^2$ touch the parabola $y^2 = 4ax$, then the locus of the middle points of these chords is the curve

(a) $y^2 (x + a) = x^3$ (b) $y^2 (x - a) = x^3$
(c) $y^2 (x + 2a) = 3x^3$ (d) $y^2 (x - 2a) = 2x^3$

12. If the sum of the slopes of the normals from a point P on hyperbola $xy = c^2$ is constant k $(k > 0)$, then the locus of P is

(a) $y^2 = k^2 c$ (b) $x^2 = kc^2$
(c) $y^2 = ck^2$ (d) $x^2 = ck^2$

13. The area (in sq units) of the quadrilateral formed by the four common tangents drawn to the two hyperbolas $\frac{x^2}{a^2} - \frac{y^2}{b^2} = 1$ and $\frac{y^2}{a^2} - \frac{x^2}{b^2} = 1 (a > b)$ is

(a) $a^2 - b^2$ (b) $2(a^2 - b^2)$
(c) $\frac{a^2 - b^2}{\sqrt{2}}$ (d) $\frac{a^2 - b^2}{2}$

14. If PQ is a double ordinate of the hyperbola $\frac{x^2}{a^2} - \frac{y^2}{b^2} = 1$ such that OPQ is an equilateral triangle, O being the centre of the hyperbola. Then the eccentricity e of the hyperbola satisfies

(a) $1 < e < \frac{2}{\sqrt{3}}$ (b) $e = \frac{2}{\sqrt{3}}$
(c) $e = \frac{\sqrt{3}}{2}$ (d) $e > \frac{2}{\sqrt{3}}$

15. The equations of the asymptotes of the hyperbola $2x^2 + 5xy + 2y^2 - 11x - 7y - 4 = 0$ are

(a) $2x^2 + 5xy + 2y^2 - 11x - 7y - 5 = 0$
(b) $2x^2 + 4xy + 2y^2 - 7x - 11y + 5 = 0$
(c) $2x^2 + 5xy + 2y^2 - 11x - 7y + 5 = 0$
(d) None of the above

16. The normal at P to a hyperbola of eccentricity e, intersects its transverse and conjugate axes at L and M respectively. If locus of the mid point of LM is hyperbola, then eccentricity of the hyperbola is

(a) $\left(\frac{e+1}{e-1}\right)$ (b) $\frac{e}{\sqrt{(e^2 - 1)}}$
(c) e (d) None of these

17. L_1 is a tangent drawn to the curve $x^2 - 4y^2 = 16$ at $A\left(5, \frac{3}{2}\right)$. L_2 is another tangent parallel to L_1 which meets the curve at B. L_3 and L_4 are normals to the curve at A and B. Lines L_1, L_2, L_3, L_4 forms a rectangle, then which of the following is correct.

(a) Equation of tangent at B is $6y = 5x - 16$
(b) Equation of normal at b $12x + 10y + 71 = 0$
(c) Radius of largest circle inscribed in the rectangle is $\frac{32}{\sqrt{61}}$
(d) Radius of the circle circumscribing the rectangle is $\frac{\sqrt{109}}{2}$

18. Two tangents are drawn from a point on hyperbola $x^2 - y^2 = 5$ to the ellipse $\frac{x^2}{9} + \frac{y^2}{4} = 1$. If they make angles α and β with X-axis,then

(a) $\alpha - \beta = \pm \frac{\pi}{2}$ (b) $\alpha + \beta = \frac{\pi}{2}$
(c) $\alpha + \beta = \pi$ (d) $\alpha + \beta = 0$

19. If a variable straight line $x \cos \alpha + y \sin \alpha = p$, which is a chord of the hyperbola $\frac{x^2}{a^2} - \frac{y^2}{b^2} = 1 (b > a)$, subtend a right angle at the centre of the hyperbola, then it always touches a fixed circle whose radius is

(a) $\frac{ab}{\sqrt{(b - 2a)}}$ (b) $\frac{a}{\sqrt{(a - b)}}$
(c) $\frac{ab}{\sqrt{(b^2 - a^2)}}$ (d) $\frac{ab}{b\sqrt{(b + a)}}$

20. An ellipse has eccentricity 1/2 and one focus at the point $P(1/2, 1)$. Its one directrix is the common tangent nearer to the point P, to the circle $x^2 + y^2 = 1$ and the hyperbola $x^2 - y^2 = 1$. The equation of the ellipse in standard form is

(a) $9x^2 + 12y^2 = 108$
(b) $9(x - 1/3)^2 + 12(y - 1)^2 = 1$
(c) $9(x - 1/3)^2 + 4(y - 1)^2 = 36$
(d) None of the above

21. The equation of the line passing through the centre of a rectangular hyperbola is $x - y - 1 = 0$. If one of its asymptote is $3x - 4y - 6 = 0$, the equation of the other asymptote is

(a) $4x - 3y + 8 = 0$ (b) $4x + 3y + 17 = 0$
(c) $3x - 2y + 15 = 0$ (d) None of these

22. The condition that a straight line with slope m will be normal to parabola $y^2 = 4ax$ as well as a tangent to rectangular hyperbola $x^2 - y^2 = a^2$ is

(a) $m^6 - 4m^2 + 2m - 1 = 0$
(b) $m^4 + 3m^3 + 2m + 1 = 0$
(c) $m^6 - 2m = 0$
(d) $m^6 + 4m^4 + 3m^2 + 1 = 0$

23. The equation of a common tangent to the curves, $y^2 = 16x$ and $xy = -4$, is **[JEE Main 2019, 12 April]**

(a) $x - y + 4 = 0$ (b) $x + y + 4 = 0$
(c) $x - 2y + 16 = 0$ (d) $2x - y + 2 = 0$

24. If $H(x, y) = 0$ represent the equation of a hyperbola and $A(x, y) = 0$, $C(x, y) = 0$ the equations of its asymptotes and the conjugate hyperbola respectively, then for any point (α, β) in the plane; $H(\alpha, \beta)$, $A(\alpha, \beta)$ and $C(\alpha, \beta)$ are in

(a) AP (b) GP
(c) HP (d) None of these

25. The eccentricity of the conic $4(2y-x-3)^2 - 9(2x+y-1)^2 = 80$ is

(a) 2 (b) 1/2
(c) $\sqrt{13}/3$ (d) 2.5

26. Let $P(3, 3)$ be a point on the hyperbola, $\frac{x^2}{a^2} - \frac{y^2}{b^2} = 1$ If the normal to it at P intersects the X-axis at $(9, 0)$ and e is its eccentricity, then the ordered pair (a^2, e^2) is equal to **[JEE Main 2020, 4 September]**

(a) $\left(\frac{9}{2}, 3\right)$ (b) $\left(\frac{3}{2}, 2\right)$
(c) $\left(\frac{9}{2}, 2\right)$ (d) $(9, 3)$

27. The line $x \cos\alpha + y \sin\alpha = p$ touches the hyperbola $\frac{x^2}{a^2} - \frac{y^2}{b^2} = 1$ if

(a) $a^2 \cos^2\alpha - b^2 \sin^2\alpha = p^2$
(b) $a^2 \cos^2\alpha - b^2 \sin^2\alpha = p$
(c) $a^2 \cos^2\alpha + b^2 \sin^2\alpha = p^2$
(d) $a^2 \cos^2\alpha + b^2 \sin^2\alpha = p$

28. The diameter of $16x^2 - 9y^2 = 144$, which is conjugate to $x = 2y$, is

(a) $y = \frac{16}{9}x$ (b) $y = \frac{32}{9}x$
(c) $x = \frac{16}{9}y$ (d) $x = \frac{32}{9}y$

29. $P(a\sec\theta, b\tan\theta)$ and $Q(a\sec\phi, b\tan\phi)$ are the ends of a focal chord of $\frac{x^2}{a^2} - \frac{y^2}{b^2} = 1$, then $\tan\theta/2 \tan\phi/2$ equals to

(a) $\frac{e-1}{e+1}$ (b) $\frac{1-e}{1+e}$
(c) $\frac{1+e}{1-e}$ (d) $\frac{e+1}{e-1}$

30. The equation of the hyperbola whose foci are $(6, 5)$, $(-4, 5)$ and eccentricity 5/4 is

(a) $\frac{(x-1)^2}{16} - \frac{(y-5)^2}{9} = 1$
(b) $\frac{x^2}{16} - \frac{y^2}{9} = 1$
(c) $\frac{(x-1)^2}{16} + \frac{(y-5)^2}{9} = 1$
(d) None of the above

31. The locus of the middle points of chords of hyperbola $3x^2 - 2y^2 + 4x - 6y = 0$ parallel to $y = 2x$ is

(a) $3x - 4y = 4$ (b) $3y - 4x + 4 = 0$
(c) $4x - 4y = 3$ (d) $3x - 4y = 2$

32. Area of the triangle formed by the lines $x - y = 0$, $x + y = 0$ and any tangent to the hyperbola $x^2 - y^2 = a^2$ is

(a) $|a|$ (b) $\frac{1}{2}|a|$
(c) a^2 (d) $\frac{1}{2}a^2$

33. If a hyperbola passes through the point $P(10, 16)$ and it has vertices at $(\pm 6, 0)$, then the equation of the normal to it at P is **[JEE Main 2020, 8 January]**

(a) $3x + 4y = 94$ (b) $x + 2y = 42$
(c) $2x + 5y = 100$ (d) $x + 3y = 58$

34. Product of the lengths of the perpendiculars drawn from foci on any tangent to the hyperbola $x^2/a^2 - y^2/b^2 = 1$ is

(a) $\frac{1}{2}b^2$ (b) b^2
(c) a^2 (d) $\frac{1}{2}a^2$

35. A ray emanating from the point $(5, 0)$ is incident on the hyperbola $9x^2 - 16y^2 = 144$ at the point P with abscissa 8, then the equation of the reflected ray after first reflection is (P lies in first quadrant)

(a) $\sqrt{3}x - y + 7 = 0$
(b) $3\sqrt{3}x - 13y + 15\sqrt{3} = 0$
(c) $3\sqrt{3}x + 13y - 15\sqrt{3} = 0$
(d) $\sqrt{3}x + y - 14 = 0$

36. If a triangle is inscribed in a rectangular hyperbola, its orthocentre lies

(a) inside the curve (b) outside the curve
(c) on the curve (d) None of these

37. The locus of the mid-points of the chord of the circle, $x^2 + y^2 = 25$, which is tangent to the hyperbola, $\frac{x^2}{9} - \frac{y^2}{16} = 1$ is **[JEE Main 2021, 16 March]**

(a) $(x^2+y^2)^2 - 16x^2 + 9y^2 = 0$
(b) $(x^2+y^2)^2 - 9x^2 + 144y^2 = 0$
(c) $(x^2+y^2)^2 - 9x^2 - 16y^2 = 0$
(d) $(x^2+y^2)^2 - 9x^2 + 16y^2 = 0$

38. The normal to the rectangular hyperbola $xy = c^2$ at the point 't_1' meets the curve again at the point 't_2'. Then the value of $t_1^3 t_2$ is

(a) 1 (b) -1
(c) c (d) $-c$

39. If $x = 9$ is the chord of contact of the hyperbola $x^2 - y^2 = 9$, then the equation of the corresponding pair of tangents is

(a) $9x^2 - 8y^2 + 18x - 9 = 0$
(b) $9x^2 - 8y^2 - 18x + 9 = 0$
(c) $9x^2 - 8y^2 - 18x - 9 = 0$
(d) $9x^2 - 8y^2 + 18x + 9 = 0$

40. Let $P(a\sec\theta, b\tan\theta)$ and $Q(a\sec\phi, b\tan\phi)$ where $\theta+\phi=\pi/2$, be two points on the hyperbola $\frac{x^2}{a^2}-\frac{y^2}{b^2}=1$. If (h, k) is the point of intersection of normals at P and Q, then k is equal to
(a) $\frac{a^2+b^2}{a}$ (b) $-\left[\frac{a^2+b^2}{a}\right]$
(c) $\frac{a^2+b^2}{b}$ (d) $-\left[\frac{a^2+b^2}{b}\right]$

41. From any point on the hyperbola $\frac{x^2}{a^2}-\frac{y^2}{b^2}=1$, tangents are drawn to the hyperbola $\frac{x^2}{a^2}-\frac{y^2}{b^2}=2$. The area cut-off by the chord of contact on the asymptotes is equal to
(a) $a/2$ (b) ab
(c) $2ab$ (d) $4ab$

42. The points of intersection of the curves whose parametric equations are $x=t^2+1$, $y=2t$ and $x=2s$, $y=\frac{2}{s}$ is given by
(a) $(1,-3)$ (b) $(2, 2)$
(c) $(-2, 4)$ (d) $(1, 2)$

43. The equations to the common tangents to the two hyperbolas $\frac{x^2}{a^2}-\frac{y^2}{b^2}=1$ and $\frac{y^2}{a^2}-\frac{x^2}{b^2}=1$ are
(a) $y=\pm x\pm\sqrt{(b^2-a^2)}$ (b) $y=\pm x\pm\sqrt{(a^2-b^2)}$
(c) $y=\pm x\pm(a^2-b^2)$ (d) $y=\pm x\pm\sqrt{(a^2+b^2)}$

44. The asymptotes of the hyperbola $xy=hx+ky$ are
(a) $x=k, y=h$ (b) $x=h, y=k$
(c) $x=h, y=h$ (d) $x=k, y=k$

45. The equation of a hyperbola, conjugate to the hyperbola $x^2+3xy+2y^2+2x+3y=0$ is
(a) $x^2+3xy+2y^2+2x+3y+1=0$
(b) $x^2+3xy+2y^2+2x+3y+2=0$
(c) $x^2+3xy+2y^2+2x+3y+3=0$
(d) $x^2+3xy+2y^2+2x+3y+4=0$

46. If the tangent and normal to a rectangular hyperbola cut off intercepts x_1 and x_2 on one axis and y_1 and y_2 on the other axis, then
(a) $x_1y_1+x_2y_2=0$ (b) $x_1y_2+x_2y_1=0$
(c) $x_1x_2+y_1y_2=0$ (d) None of these

47. A normal to the hyperbola $\frac{x^2}{a^2}-\frac{y^2}{b^2}=1$ meets the transverse and conjugate axes in M and N and the lines MP and NP are drawn at right angles to the axes. The locus of P is
(a) the parabola $y^2=4a(x+b)$
(b) the circle $x^2+y^2=ab$
(c) the ellipse $b^2x^2+a^2y^2=a^2+b^2$
(d) the hyperbola $a^2x^2-b^2y^2=(a^2+b^2)^2$

Objective Questions Type II [One or more than one correct answer(s)]

In each of the questions below four choices of which one or more than one are correct. You have to select the correct answer(s) accordingly.

1. The equation $16x^2-3y^2-32x-12y-44=0$ represents a hyperbola with
(a) length of the transverse axis $=2\sqrt{3}$
(b) length of the conjugate axis $=8$
(c) cente at $(1,-2)$
(d) eccentricity $=\sqrt{19}$

2. Let a and b be positive real numbers such that $a>1$ and $b<a$. Let P be a point in the first quadrant that lies on the hyperbola $\frac{x^2}{a^2}-\frac{y^2}{b^2}=1$. Suppose the tangent to the hyperbola at P passes through the point $(1, 0)$, and suppose the normal to the hyperbola at P cuts off equal intercepts on the coordinate axes. Let Δ denote the area of the triangle formed by the tangent at P, the normal at P and the X-axis. If e denotes the eccentricity of the hyperbola, then which of the following statements is/are TRUE? **[JEE Advanced 2020]**
(a) $1<e<\sqrt{2}$ (b) $\sqrt{2}<e<2$
(c) $\Delta=a^4$ (d) $\Delta=b^4$

3. Equation of a tangent passing through $(2, 8)$ to the hyperbola $5x^2-y^2=5$ is
(a) $3x-y+2=0$ (b) $3x+y+14=0$
(c) $23x-3y-22=0$ (d) $3x-23y+178=0$

4. If the line $ax+by+c=0$ is a normal to the hyperbola $xy=1$, then
(a) $a>0, b>0$ (b) $a>0, b<0$
(c) $a<0, b>0$ (d) $a<0, b<0$

5. If m_1 and m_2 are the slopes of the tangents to the hyperbola $x^2/25-y^2/16=1$ which pass through the point $(6, 2)$, then
(a) $m_1+m_2=24/11$ (b) $m_1m_2=20/11$
(c) $m_1+m_2=48/11$ (d) $m_1m_2=11/20$

6. The locus of the point of intersection of two perpendicular tangents to the hyperbola $x^2/a^2-y^2/b^2=1$ is
(a) director circle (b) $x^2+y^2=a^2$
(c) $x^2+y^2=a^2-b^2$ (d) $x^2+y^2=a^2+b^2$

7. An ellipse intersects the hyperbola $2x^2 - 2y^2 = 1$ orthogonally. The eccentricity of the ellipse is reciprocal of that of the hyperbola. If the axes of the ellipse are along the coordinates axes, then
(a) Equation of ellipse is $x^2 + 2y^2 = 2$
(b) The foci of ellipse are $(\pm 1, 0)$
(c) Equation of ellipse is $x^2 + 2y^2 = 4$
(d) The foci of ellipse are $(\pm \sqrt{2}, 0)$

8. If $P(x_1, y_1)$, $Q(x_2, y_2)$, $R(x_3, y_3)$ and $S(x_4, y_4)$ are 4 concyclic points on the rectangular hyperbola $xy = c^2$, the coordinates of the orthocentre of the ΔPQR are
(a) $(x_4, -y_4)$ (b) (x_4, y_4)
(c) $(-x_4, -y_4)$ (d) $(-x_4, y_4)$

9. The line $y = x + 5$ touches
(a) the parabola $y^2 = 20x$
(b) the ellipse $9x^2 + 16y^2 = 144$
(c) the hyperbola $\frac{x^2}{29} - \frac{y^2}{4} = 1$
(d) the circle $x^2 + y^2 = 25$

10. A straight line touches the rectangular hyperbola $9x^2 - 9y^2 = 8$ and the parabola $y^2 = 32x$. The equation of the line is
(a) $9x + 3y - 8 = 0$ (b) $9x - 3y + 8 = 0$
(c) $9x + 3y + 8 = 0$ (d) $9x - 3y - 8 = 0$

11. If the normal at P to the rectangular hyperbola $x^2 - y^2 = 4$ meets the axes in G and g and C is the centre of the hyperbola, then
(a) $PG = PC$ (b) $Pg = PC$
(c) $PG = Pg$ (d) $Gg = 2PC$

12. If the tangent at the point $(a \sec \alpha, b \tan \alpha)$ to the hyperbola $\frac{x^2}{a^2} - \frac{y^2}{b^2} = 1$ meets the transverse axis at T, then the distance of T from a focus of the hyperbola is
(a) $a(e - \cos \alpha)$ (b) $b(e + \cos \alpha)$
(c) $a(e + \cos \alpha)$ (d) $\sqrt{(a^2e^2 + b^2 \cot^2 \alpha)}$

13. The coordinates of a point common to a directrix and an asymptote of the hyperbola $x^2/25 - y^2/16 = 1$ are
(a) $(25/\sqrt{41}, 20/\sqrt{41})$ (b) $(-25/\sqrt{41}, -20/\sqrt{41})$
(c) $(25/3, 20/3)$ (d) $(-25/3, -20/3)$

14. If the circle $x^2 + y^2 = 1$ cuts the rectangular hyperbola $xy = 1$ in four points (x_i, y_i), $i = 1, 2, 3, 4$, then
(a) $x_1x_2x_3x_4 = 1$
(b) $y_1y_2y_3y_4 = 1$
(c) $x_1 + x_2 + x_3 + x_4 = 0$
(d) $y_1 + y_2 + y_3 + y_4 = 0$

Linked-Comprehension Type Questions

In these questions, a passage (paragraph) has been given followed by questions based on each of the passages. You have to answer the questions based on the passage given.

PASSAGE 1

The difference between the second degree curve and pair of asymptotes is constant.

If second degree curve represented by a hyperbola $S = 0$, then the equation of its asymptotes is $S + \lambda = 0$ where λ is constant, which will be a pair of straight lines, then we get λ. Then equation of asymptotes is $A \equiv S + \lambda = 0$ and if equation of conjugate hyperbola of S represented by S_1, then A is the arithmetic mean of S and S_1.

On the basis of above information, answer the following questions

1. Pair of asymptotes of the hyperbola $xy - 3y - 2x = 0$ is
(a) $xy - 3y - 2x + 2 = 0$ (b) $xy - 3y - 2x + 4 = 0$
(c) $xy - 3y - 2x + 6 = 0$ (d) $xy - 3y - 2x + 12 = 0$

2. The asymptotes of a hyperbola having centre at the point (1, 2) are parallel to the lines $2x + 3y = 0$ and $3x + 2y = 0$. If the hyperbola passes through the point (5, 3), then its equation is
(a) $(2x + 3y - 3)(3x + 2y - 5) = 256$
(b) $(2x + 3y - 7)(3x + 2y - 8) = 156$
(c) $(2x + 3y - 5)(3x + 2y - 3) = 252$
(d) $(2x + 3y - 8)(3x + 2y - 7) = 154$

3. If angle between the asymptotes of hyperbola $\frac{x^2}{a^2} - \frac{y^2}{b^2} = 1$ is $\pi / 3$ then the eccentricity of conjugate hyperbola is
(a) $\sqrt{2}$
(b) 2
(c) $2/\sqrt{3}$
(d) $4/\sqrt{3}$

4. A hyperbola passing through origin has $3x - 4y - 1 = 0$ and $4x - 3y - 6 = 0$ as its asymptotes. Then the equation of its transverse and conjugate axes are
(a) $x - y - 5 = 0$ and $x + y + 1 = 0$
(b) $x - y = 0$ and $x + y + 5 = 0$
(c) $x + y - 5 = 0$ and $x - y - 1 = 0$
(d) $x + y - 1 = 0$ and $x - y - 5 = 0$

5. The tangent at any point of a hyperbola $16x^2 - 25y^2 = 400$ cuts off a triangle from the asymptotes and that the portion of it intercepted between the asymptotes, then the area of this triangle is
(a) 10 sq units
(b) 20 sq units
(c) 30 sq units
(d) 40 sq units

PASSAGE 2

If we rotate the axes of the rectangular hyperbola $x^2 - y^2 = a^2$ through an angle $\pi / 4$ in the clockwise direction, then the equation $x^2 - y^2 = a^2$ reduces to $xy = \frac{a^2}{2} = \left(\frac{a}{\sqrt{2}}\right)^2 = c^2$ (say). Since $x = ct, y = \frac{c}{t}$ satisfies $xy = c^2$.

$\therefore \quad (x, y) = \left(ct, \frac{c}{t}\right) (t \neq 0)$ is called a '*t*' point on the rectangular hyperbola.

On the basis of above information, answer the following questions

1. If t_1 and t_2 are the roots of the equation $x^2 - 4x + 2 = 0$, then the point of intersection of tangents at 't_1' and 't_2' on $xy = c^2$ is
   (a) $\left(\frac{c}{2}, 2c\right)$ (b) $\left(2c, \frac{c}{2}\right)$
   (c) $\left(\frac{c}{2}, c\right)$ (d) $\left(c, \frac{c}{2}\right)$
2. Let α, γ be the roots of the equation $t_1x^2 - 4x + 1 = 0$ and β, δ be the roots of $t_2x^2 - 6x + 1 = 0$ and $\alpha, \beta, \gamma, \delta$ are in HP then the point of intersection of normals at 't_1' and 't_2' on $xy = c^2$ is
   (a) $\left(\frac{327c}{264}, \frac{921c}{264}\right)$ (b) $\left(\frac{237c}{264}, \frac{291c}{264}\right)$
   (c) $\left(\frac{723c}{264}, \frac{129c}{264}\right)$ (d) None of these
3. If e_1 and e_2 are the eccentricities of the hyperbolas $xy = 9$ and $x^2 - y^2 = 25$, then (e_1, e_2) lie on a circle C_1 with centre origin then the (radius)^2 of the director circle of C_1 is
   (a) 2 (b) 4
   (c) 8 (d) 16
4. If the normal at the point 't_1' to the rectangular hyperbola $xy = c^2$ meets it again at the point 't_2,' then the value of t_1t_2 is
   (a) $-t_1^{-1}$ (b) $-t_1^{-2}$
   (c) $-t_1^{-3}$ (d) $-t_1^{-4}$
5. A triangle has its vertices on a rectagular hyperbola $xy = c^2$, then the orthocentre of this triangle lies on
   (a) $xy = c^2$ (b) $xy = 2c^2$
   (c) $xy = \frac{c^2}{2}$ (d) $xy = \frac{c^2}{4}$

PASSAGE 3

Consider the hyperbola $\frac{x^2}{9} - \frac{y^2}{a^2} = 1$ and the circle $x^2 + (y - 3)^2 = 9$.

Also, the given hyperbola and the ellipse $\frac{x^2}{41} + \frac{y^2}{16} = 1$ are orthogonal to each other.

1. Combined equation of pair of common tangents between the hyperbola and the circle is given by
   (a) $x^2 - y^2 = 0$ (b) $x^2 - 9 = 0$
   (c) $9y^2 - 16x^2 = 0$ (d) No common tangent
2. The number of points on the hyperbola and the circle from which tangents drawn to the circle and the hyperbola, respectively, are perpendicular to each other is
   (a) 0 (b) 2
   (c) 4 (d) 6
3. A variable line cuts the circle at points *A* and *B* and it cuts the hyperbola at points *C* and *D*. The locus of midpoint of *AB* such that tangents at points *C* and *D* always intersect each other at the directrix of the hyperbola, is
   (a) $x^2 + y^2 \pm 5x - 3y = 0$ (b) $x^2 + y^2 + 5x \pm 3y = 0$
   (c) $x^2 - y^2 \pm 5x - 3y = 0$ (d) $x^2 - y^2 + 5x \pm 3y = 0$

Numerical Value Type

Solve the following problems and write your answers.

1. *P* and *Q* are two variable points on the rectangular hyperbola $xy = c^2$, such that tangent at *Q* passes through the foot of the ordinate of *P*. If the locus of the intersection of tangents at *P* and *Q* is a hyperbola $xy = \lambda c^2$, then the value of 729λ must be
2. A triangle is inscribed in $xy = c^2$ and two of its sides are parallel to $y = m_1x$ and $y = m_2x$. If m_1, m_2 are two values of $x^2 - 6x + 1 = 0$ and if third side envelopes the hyperbola $xy = c^2\lambda$, then the value of $16\lambda^2$ must be
3. Let the normals are drawn from (α, β) to the hyperbola $xy = 1$, and $(x_i, y_i), i = 1, 2, 3, 4$ be the feet of the co-normal points. If the algebraic sum of the perpendicular distances drawn from (x_i, y_i), $i = 1, 2, 3, 4$ onto a variable line vanishes, then the variable line passes through the point $\left(\frac{\alpha}{\lambda}, \frac{\beta}{\lambda}\right)$ then the value of 256λ must be
4. If a circle cuts a rectangular hyperbola $xy = c^2$ in A, B, C and *D* and the parameters of these four points be t_1, t_2, t_3 and t_4 respectively, then the value of $16t_1t_2t_3t_4$ must be
5. If a hyperbola be rectangular and its equation be $xy = c^2$ then the locus of the middle points of chords of constant length $2d$ is $(x^2 + y^2)(xy - c^2) = d^{\lambda}xy$, then the value of 1000λ must be

6. The equation of the hyperbola whose asymptotes are $x + 2y + 3 = 0$ and $3x + 4y + 5 = 0$ and which passes through the point $(1, -1)$ is $3x^2 + 10xy + 8y^2 + 14x + 22y + \lambda = 0$, then the equation of the conjugate hyperbola is $3x^2 + 10xy + 8y^2 + \mu x + 22y + \nu = 0$, then the value of $\frac{\mu}{\nu}$, must be

7. Tangent are drawn from any point on the hyperbola $x^2/9 - y^2/4 = 1$ to the circle $x^2 + y^2 = 9$. If the locus of the mid-point of the chord of contact is $a(x^2 + y^2)^2 = bx^2 - cy^2$, then the value of $\frac{c}{a+b}$ is equal to

Single Integer Type

Solve the following problems and write your answers.

1. The value of b^2 in order that the foci of the hyperbola $\frac{x^2}{144} - \frac{y^2}{8125} = \frac{1}{25}$ and the ellipse $\frac{x^2}{16} + \frac{y^2}{b^2} = 1$ coincide is

2. Radii of director circles of curves $\frac{x^2}{a^2} + \frac{y^2}{b^2} = 1$ and $\frac{x^2}{a^2} - \frac{y^2}{b^2} = 1$ are $2r$ and r, respectively. If e_1 and e_2 are the eccentricities of ellipse and hyperbola, respectively, then $4e_2^2 - e_1^2 =$ ____ .

3. Chords of the circle $x^2 + y^2 = 4$, touch the hyperbola $\frac{x^2}{4} - \frac{y^2}{16} = 1$. The locus of their middle points is the curve $(x^2 + y^2)^2 = \lambda x^2 - 16y^2$, then find λ.

4. If e_1, e_2 are the eccentricities of the hyperbola $2x^2 - 2y^2 = 1$ and the ellipse $x^2 + 2y^2 = 2$ respectively. Then

5. Tangents are drawn from the point (α, β) to the hyperbola $3x^2 - 2y^2 = 6$ and are inclined at angle θ and ϕ to the x-axis. If $\tan\theta \cdot \tan\phi = 2$, then the value of $2\alpha^2 - \beta^2$ is ____ .

6. If $2x - ky + 3 = 0$, $3x - y + 1 = 0$ are conjugate lines with respect to $5x^2 - 6y^2 = 15$, then $k =$

7. The slopes of common tangents to the hyperbolas $\frac{x^2}{9} - \frac{y^2}{16} = 1$ and $\frac{y^2}{9} - \frac{x^2}{16} = 1$ then value of $|m|$ is

8. For the hyperbola $x^2 - y^2 - 4x + 2y + c = 0$, the focus is $S(2 + 2\sqrt{2}, k)$ and the directrix that is adjacent to S is $x = 2 + \sqrt{2}$, then value of $c^2$is

Matrix-Match Type Questions

Given below are Matching Type Questions, with two columns (each having some items) each . Each item of Column I has to be matched with the items of Column II, by encircling the correct match(es).

NOTE *An item of Column I can be matched with more than one items of Column II.*

1. Observe the following columns

	Column I		Column II
(A)	Position of points (3, 4) and (5, 2) with respect to hyperbola $x^2 - 4y^2 - 2x + 24y - 37 = 0$, then	(P)	(3,4) lie inside the hyperbola
(B)	Position of points (3, 4) and (5, 2) with respect to hyperbola $xy + 2x + 3y - 12 = 0$, then	(Q)	(5, 2) lie outside the hyperbola
(C)	Position of points (3, 4) and (5, 2) with respect to hyperbola $xy = 12$, then	(R)	(3, 4) lie outside the hyperbola
		(S)	(5, 2) lie inside the hyperbola
		(T)	(3, 4) lie on the hyperbola

2. Observe the following columns

	Column I		Column II
(A)	Director circles of $x^2 - 2y^2 = 2$ and $x^2 + 2y^2 = 2$ are	(P)	$x^2 + y^2 = 1$
(B)	Director circles of $3x^2 + 2y^2 = 6$ and $3x^2 - 2y^2 = 6$ are	(Q)	$x^2 + y^2 = 2$
(C)	Director circles of $5x^2 - 9y^2 = 45$ and $x^2 + y^2 = 1$ are	(R)	$x^2 + y^2 = 3$
		(S)	$x^2 + y^2 = 4$
		(T)	$x^2 + y^2 = 5$

3. Observe the following columns

	Column I		Column II
(A)	The foci of the hyperbola $12x^2 - 4y^2 - 24x + 32y - 127 = 0$ are	(P)	(10, 5)
(B)	The foci of the hyperbola $8x^2 - y^2 - 64x + 10y + 71 = 0$ are	(Q)	(2, – 2)
(C)	The foci of the hyperbola $9x^2 - 16y^2 - 36x + 96y + 36 = 0$ are	(R)	(–4, 4)
		(S)	(–2, 5)
		(T)	(6, 4)

Answers

Objective Questions Type I [Only one correct answer]

1. (b)	**2.** (a)	**3.** (b)	**4.** (a)	**5.** (d)	**6.** (a)	**7.** (b)	**8.** (c)	**9.** (a)	**10.** (c)
11. (b)	**12.** (b)	**13.** (b)	**14.** (d)	**15.** (c)	**16.** (b)	**17.** (d)	**18.** (b)	**19.** (c)	**20.** (b)
21. (b)	**22.** (d)	**23.** (a)	**24.** (a)	**25.** (c)	**26.** (a)	**27.** (a)	**28.** (b)	**29.** (b)	**30.** (a)
31. (a)	**32.** (a)	**33.** (c)	**34.** (b)	**35.** (b)	**36.** (c)	**37.** (d)	**38.** (b)	**39.** (b)	**40.** (d)
41. (d)	**42.** (b)	**43.** (b)	**44.** (a)	**45.** (b)	**46.** (c)	**47.** (d)			

Objective Questions Type II [One or more than one correct answer(s)]

1. (a, b, c)	**2.** (a, d)	**3.** (a, c)	**4.** (b, c)	**5.** (a, b)
6. (a, c)	**7.** (a, b)	**8.** (b, c)	**9.** (a, b, c)	**10.** (a, b, c, d)
11. (a, b, c, d)	**12.** (a, c)	**13.** (a, b)	**14.** (c, d)	

Linked-Comprehension Type

Passage 1 **1.** (c) **2.** (d) **3.** (b) **4.** (c) **5.** (b)

Passage 2 **1.** (d) **2.** (d) **3.** (c) **4.** (b) **5.** (a)

Passage 3 **1.** (b) **2.** (c) **3.** (a)

Numerical Value Type Questions

1. 648	**2.** 1296	**3.** 1024	**4.** 16	**5.** 2000
6. 0.61	**7.** 2.025			

Single Integer Type Questions

1. (7)	**2.** (6)	**3.** (4)
4. (1)	**5.** (7)	**6.** (6)
7. (1)	**8.** (1)	

Matrix-Match Type

1. A → (R, S); B → (P, S); C → (Q, T)

2. A → (P, R); B → T; C → (Q, S)

3. A → (R, T); B → (P, S); C → (Q)

Solutions

Objective Questions Type [Only one correct answer]

1. The given condition gives

$$\frac{5+5\cos^2\alpha}{5} = 3\cdot\frac{25-25\cos^2\alpha}{25}$$

leading then, we get

$$\cos^2\alpha = \frac{1}{2} \text{ or } \alpha = \frac{\pi}{4}, \frac{3\pi}{4}$$

2. We know that if a circle cuts a rectangular hyperbola, then arithmetic mean of points of intersections is the mid-point of centre of hyperbola and circle.

So, $\frac{3+5+2+(-1)}{4} = \frac{-g+1}{2}$

and $\frac{4+3+6+0}{4} = \frac{-f+2}{2}$

$\therefore \quad g+f=-8$

3. We know that eccentricity of a parabola and rectangular hyperbola are 1 and $\sqrt{2}$ respectively.

Also irrational roots occur in conjugate pair, thus roots of $f(x)=0$ are 1, $\sqrt{2}$ and $-\sqrt{2}$

$\therefore \quad f(x) = (x-1)(x-\sqrt{2})(x+\sqrt{2})$
$= x^3 - x^2 - 2x + 2$

$\therefore a+b+c+d = 1-1-2+2 = 0$

4. Let the point be $(\alpha, \beta) \Rightarrow \beta = \alpha + c$

Chord of contact of hyperbola $T = 0$.

$\therefore \quad \frac{x\alpha}{2} - \frac{y\beta}{1} = 1$

$\Rightarrow \quad \frac{x\alpha}{2} - y(\alpha + c) = 1$

$\Rightarrow \quad \left(\frac{x}{2} - y\right)\alpha - (yc+1) = 0$

Since, this passes through point (x_1, y_1)

$\therefore \quad x_1 = 2y_1$ and $y_1c + 1 = 0$

$\therefore \quad y_1 = \frac{x_1}{2}$. Hence, $\frac{x_1}{y_1} = 2$.

5. Equation of given ellipse is

$$3x^2 + 4y^2 = 12 \Rightarrow \frac{x^2}{4} + \frac{y^2}{3} = 1 \quad \text{... (i)}$$

$\therefore$ Eccentricity of ellipse (i) is $e_1 = \sqrt{1-\frac{3}{4}} = \frac{1}{2}$

$\therefore$ Coordinate of foci is $(\pm 1, 0)$.

Now, it is given that length of the transverse axis of hyperbola is $\sqrt{2}$, so $\frac{\sqrt{2}}{2}e_2 = 1$, where e_2 is the eccentricity of the hyperbola.

So, $\quad e_2 = \sqrt{2}$

$\therefore$ Given hyperbola is rectangular hyperbola and it's equation is

$$\frac{x^2}{\left(\frac{1}{\sqrt{2}}\right)^2} - \frac{y^2}{\left(\frac{1}{\sqrt{2}}\right)^2} = 1 \Rightarrow x^2 - y^2 = \frac{1}{2} \quad \text{... (ii)}$$

So, from the given options point $\left(\sqrt{\frac{3}{2}}, \frac{1}{\sqrt{2}}\right)$ does not pass through the hyperbola (ii).

6. The length of transverse axis is,

$$2\sin\theta = 2a \Rightarrow a = \sin\theta$$

Also, for the ellipse

$$3x^2 + 4y^2 = 12 \text{ or } \frac{x^2}{4} + \frac{y^2}{3} = 1$$

$$a^2 = 4, b^2 = 3$$

$\therefore \quad e = \sqrt{1-\frac{b^2}{a^2}} = \sqrt{1-\frac{3}{4}} = \frac{1}{2}$

Hence, the focus of ellipse is $(2 \times 1/2, 0) \equiv (1, 0)$.

As the hyperbola is confocal with the ellipse, the focus of the hyperbola is (1, 0).

$\therefore \quad ae' = 1$

or $\quad \sin\theta \times e' = 1 \Rightarrow e' = \operatorname{cosec}\theta$

$\therefore \quad b^2 = a^2(e'^2 - 1) = \sin^2\theta(\operatorname{cosec}^2\theta - 1) = \cos^2\theta$

Therefore, the equation of hyperbola is

$$\frac{x^2}{\sin^2\theta} - \frac{y^2}{\cos^2\theta} = 1 \Rightarrow x^2\operatorname{cosec}^2\theta - y^2\sec^2\theta = 1$$

7. Let P, Q, R, S be having parameters t_1, t_2, t_3, t_4

PQ is perpendicular to RS

$$\Rightarrow \left(\frac{c/t_2 - c/t_1}{ct_2 - ct_1}\right) \times \left(\frac{c/t_3 - c/t_4}{ct_3 - ct_4}\right) = -1 \Rightarrow t_1t_2t_3t_4 = -1$$

Slope of $OP = \frac{1}{t_1^2}$

$\therefore$ Product of slopes of

$OP, OQ, OR, OS = \frac{1}{t_1^2t_2^2t_3^2t_4^2} = \frac{1}{1} = 1$

8. Equation of given hyperbola is $\frac{x^2}{4} - \frac{y^2}{2} = 1$ and the point (x_1, y_1) at the hyperbola, so

$$\frac{x_1^2}{4} - \frac{y_1^2}{2} = 1 \quad \text{... (i)}$$

Now, slope of the tangent to the given hyperbola at point (x_1, y_1), is $\left.\frac{dy}{dx}\right|_{(x_1, y_1)} = \frac{x_1}{2y_1}$ and the tangent is parallel to the line $2x - y = 0$

So, $\quad \frac{x_1}{2y_1} = 2 \Rightarrow x_1 = 4y_1 \quad \text{... (ii)}$

From Eqs. (i) and (ii), we get

$$\frac{16y_1^2}{4} - \frac{y_1^2}{2} = 1 \Rightarrow \frac{14}{4}y_1^2 = 1 \Rightarrow y_1^2 = \frac{2}{7}$$

and $\quad x_1^2 = 16y_1^2 \Rightarrow x_1^2 = \frac{32}{7}$,

So, $\quad x_1^2 + 5y_1^2 = \frac{32}{7} + \frac{10}{7} = \frac{42}{7} = 6$

9. Let $P \equiv (1, \sqrt{3}) \equiv (2\cos \pi/3, 2\sin \pi/3)$

Recall the definition of parametric point on hyperbola, we have

$$R = \left(2\sec\frac{\pi}{3}, 1\cdot\tan\frac{\pi}{3}\right) = (4, \sqrt{3})$$

$\therefore \quad PR = 4 - 1 = 3$

10. $\frac{x^2}{4} - \frac{y^2}{2} = 1$

$$e = \sqrt{1 + \frac{b^2}{a^2}} = \sqrt{\frac{3}{2}}$$

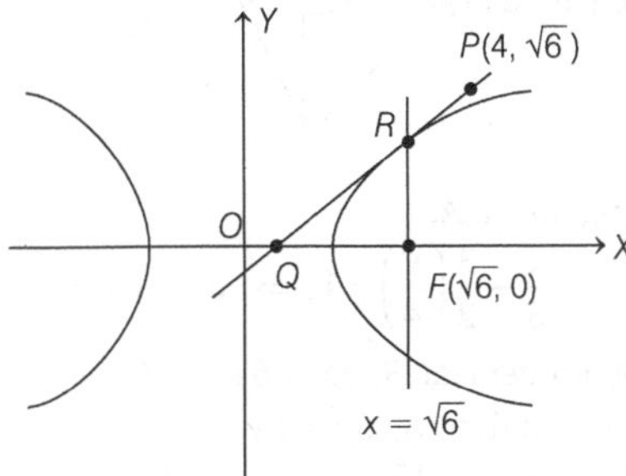

$\therefore$ Focus $F(ae, 0) \Rightarrow F(\sqrt{6}, 0)$

Equation of tangent at P to the hyperbola is $2x - y\sqrt{6} = 2$

Tangent meet X-axis at $Q(1, 0)$

and latusrectum $x = \sqrt{6}$ at $R\left(\sqrt{6}, \frac{2}{\sqrt{6}}(\sqrt{6}-1)\right)$

$\therefore$ Area of $\Delta QFR = \frac{1}{2}(\sqrt{6}-1)\cdot\frac{2}{\sqrt{6}}(\sqrt{6}-1) = \frac{7}{\sqrt{6}} - 2$

11. Let mid point of the chord is (h, k)

$\therefore$ Equation of chord of $x^2 - y^2 = a^2$ is

$$T = S_1$$
$$hx - ky = h^2 - k^2$$

or $$y = \frac{h}{k}x - \frac{(h^2 - k^2)}{k}$$

which is tangent of $y^2 = 4ax$

$\therefore$ $$-\frac{(h^2 - k^2)}{k} = \frac{a}{\frac{h}{k}}$$

or $$-(h^2 - k^2) = \frac{ak^2}{h}$$

or $$-h^3 + hk^2 = ak^2$$

or $$k^2(h - a) = h^3$$

Hence, locus of mid point is

$$y^2(x - a) = x^3$$

12. Equation of normal at $\left(ct, \frac{c}{t}\right)$ is

$$xt^3 - yt - ct^4 + c = 0 \quad \ldots(i)$$

Slope is $t^2 = k$.

Let $x = ct$ and $y = \frac{c}{t}$,

then $\frac{x}{y} = t^2 = k$ (from Eq. (i))

$\therefore \quad x = ky$

$\Rightarrow \quad x^2 = kxy \quad [\because xy = c^2]$

$\Rightarrow \quad x^2 = kc^2$

13. Equation of tangent having slope m to the hyperbola,

$$\frac{x^2}{a^2} - \frac{y^2}{b^2} = 1 \text{ is } y = mx \pm \sqrt{a^2m^2 - b^2} \quad \ldots(i)$$

$\because$ Tangent Eq. (i) is also tangent to hyperbola,

$$\frac{y^2}{a^2} - \frac{x^2}{b^2} = 1 \quad \ldots(ii)$$

From Eqs. (i) and (ii), we get

$$\frac{(mx + \sqrt{a^2m^2 - b^2})^2}{a^2} - \frac{x^2}{b^2} = 1$$

$\Rightarrow \quad b^2(m^2x^2 + a^2m^2 - b^2 + 2mx\sqrt{a^2m^2 - b^2}) - a^2x^2 - a^2b^2 = 0$

$\Rightarrow \quad (b^2m^2 - a^2)x^2 + 2mb^2(\sqrt{a^2m^2 - b^2}\,x) + (a^2b^2m^2 - b^4 - a^2b^2) = 0$

$\because$ Line Eq. (ii) is tangent to hyperbola Eq. (ii), so

$$D = 0$$

$\Rightarrow \quad 4m^2b^4(a^2m^2 - b^2) - 4(b^2m^2 - a^2)(a^2b^2m^2 - b^4 - a^2b^2) = 0$

$\Rightarrow \quad a^2b^4m^4 - b^6m^2 = a^2b^4m^4 - b^6m^2 - a^2b^4m^2 - a^4b^2m^2 + a^2b^4 + a^4b^2$

$\Rightarrow \quad a^2b^4m^2 + a^4b^2m^2 = a^2b^4 + a^4b^2$

$\Rightarrow \quad m^2(a^2b^4 + a^4b^2) = a^2b^4 + a^4b^2$

$\Rightarrow \quad m^2 = 1 \Rightarrow m = \pm 1$

So, the equation of common tangents are

$$y = \pm x \pm \sqrt{a^2 - b^2}$$

Now, required area (in Sq units) $= 4 \times \frac{1}{2}(\sqrt{a^2 - b^2})^2$

$= 2(a^2 - b^2)$

14. $\because PQ$ is the double ordinate. Let $MP = MQ = l$ given that

ΔOPQ is an equilateral, then $OP = OQ = PQ$

$\Rightarrow \quad (OP)^2 = (OQ)^2 = (PQ)^2$

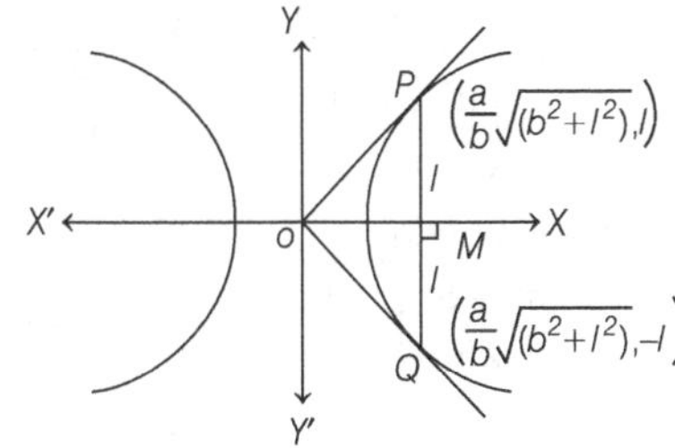

$\Rightarrow \quad \frac{a^2}{b^2}(b^2 + l^2) + l^2 = \frac{a^2}{b^2}(b^2 + l^2) + l^2 = 4l^2$

$\Rightarrow \quad \frac{a^2}{b^2}(b^2 + l^2) = 3l^2 \Rightarrow a^2 = l^2\left(3 - \frac{a^2}{b^2}\right)$

$\Rightarrow \quad l^2 = \frac{a^2b^2}{(3b^2 - a^2)} > 0$

$\therefore \quad 3b^2 - a^2 > 0 \Rightarrow 3b^2 > a^2$

$\Rightarrow \quad 3a^2(e^2 - 1) > a^2 \Rightarrow e^2 > 4/3$

$\therefore \quad e > 2/\sqrt{3}$

15. The pair of asymptotes and second degree curve differ by a constant

$\therefore$ Pair of asymptotes

$$2x^2 + 5xy + 2y^2 - 11x - 7y + \lambda = 0 \quad \ldots(i)$$

Hence, Eq. (i) represents a pair of straight lines.

$\therefore \quad \Delta = 0$

Then, $2 \times 2 \times \lambda + 2 \times -\frac{7}{2} \times -\frac{11}{2} \times \frac{5}{2}$

$$-2 \times \left(-\frac{7}{2}\right)^2 - 2 \times \left(-\frac{11}{2}\right)^2 - \lambda\left(\frac{5}{2}\right)^2 = 0$$

$\therefore \quad \lambda = 5$

From Eq. (i),

$\therefore$ Pair of asymptotes is

$2x^2 + 5xy + 2y^2 - 11x - 7y + 5 = 0$

16. Equation of normal at $P(a \sec\phi, b\tan\phi)$ is

$$ax\cos\phi + by\cot\phi = a^2 + b^2$$

Hence, coordinates of L and M are

$$L \equiv \left(\frac{a^2 + b^2}{a} \cdot \sec\phi, 0\right)$$

and $\quad M \equiv \left(0, \frac{a^2+b^2}{b}\tan\phi\right)$ respectively.

Let mid point of ML is $Q(h, k)$

then $\quad h = \frac{(a^2+b^2)}{2a}\sec\phi$

$\therefore \quad \sec\phi = \frac{2ah}{(a^2+b^2)} \quad \ldots(i)$

and $\quad k = \frac{(a^2+b^2)}{2b}\tan\phi$

$\therefore \quad \tan\phi = \frac{2bk}{(a^2+b^2)} \quad \ldots(ii)$

From Eqs. (i) and (ii)

$$\sec^2\phi - \tan^2\phi = \frac{4a^2h^2}{(a^2+b^2)^2} - \frac{4b^2k^2}{(a^2+b^2)^2}$$

Hence, locus $\dfrac{x^2}{\left(\dfrac{a^2+b^2}{2a}\right)^2} - \dfrac{y^2}{\left(\dfrac{a^2+b^2}{2b}\right)^2} = 1$

Let eccentricity is e_1

$\therefore \quad \left(\frac{a^2+b^2}{2b}\right)^2 = \left(\frac{a^2+b^2}{2a}\right)^2 (e_1^2 - 1)$

$\Rightarrow \quad a^2 = b^2(e_1^2 - 1)$

$\Rightarrow \quad a^2 = a^2(e^2-1)(e_1^2-1)$

$$[\because b^2 = a^2(e^2-1)]$$

$\Rightarrow \quad e^2e_1^2 - e^2 - e_1^2 + 1 = 1$

$\Rightarrow \quad e_1^2(e^2 - 1) = e^2$

$\therefore \quad e_1 = \frac{e}{\sqrt{(e^2-1)}}$

17. We have, given curve is $\frac{x^2}{16} - \frac{y^2}{4} = 1$

Point $\left(5, \frac{3}{2}\right)$ lies on the curve.

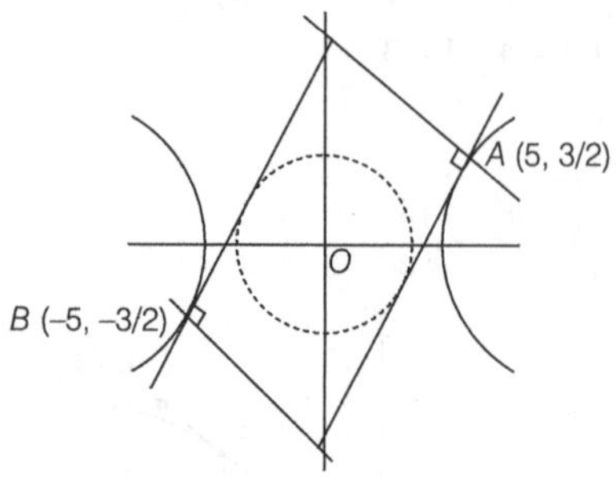

$\therefore$ Tangent at A,

$$\frac{5x}{16} - \frac{3}{2}\left(\frac{y}{4}\right) = 1 \Rightarrow 5x - 6y = 16$$

Now, tangent at B, $5x - 6y = -6$

and Normal at A, $6x + 5y = \lambda$

$$A\left(5, \frac{3}{2}\right) 30 + \frac{15}{2} = 1 \Rightarrow \lambda = \frac{75}{2} \; 6x + 5y = \frac{75}{2}$$

Similarly, normal at B is $6x + 5y = \frac{-75}{2}$

distance between tangents is $2r = \frac{32}{\sqrt{25+36}}$

$\therefore \quad r = \frac{16}{\sqrt{61}}$

Radius of circumcircle of rectangle is

$$R = \sqrt{25 + \frac{9}{4}} = \frac{\sqrt{109}}{2}$$

18. Given, equation of tangent to ellipse $\frac{x^2}{9} + \frac{y^2}{4} = 1$, having slope m is $y = mx + \sqrt{9m^2+4}$

Since, it passes through point $(\sqrt{5}\sec\theta, \sqrt{5}\tan\theta)$ on the hyperbola, we have

$$\sqrt{5}\tan\theta = m\sqrt{5}\sec\theta + \sqrt{9m^2+4}$$

$\therefore \quad (\sqrt{5}\tan\theta - m\sqrt{5}\sec\theta)^2 = 9m^2 + 4$

$\Rightarrow (9 - 5\sec^2\theta)m^2 + 10\sec\theta\tan\theta m + 4 - 5\tan^2\theta = 0$

This equation has roots m_1 and m_2, which are slopes of tangents.

$\therefore \quad m_1m_2 = \frac{4 - 5\tan^2\theta}{9 - 5\sec^2\theta} = \frac{4 - 5(\sec^2\theta - 1)}{9 - 5\sec^2\theta} = 1$

$\therefore (\tan\alpha)(\tan\beta) = 1 \Rightarrow \alpha + \beta = \frac{\pi}{2}$

19. Since, $x\cos\alpha + y\sin\alpha = p$ subtends a right angle at centre i.e. $(0, 0)$. Making homogeneous equation of hyperbola $\frac{x^2}{a^2} - \frac{y^2}{b^2} = 1$ with the help of $x\cos\alpha + y\sin\alpha = p$ and putting coefficeint of x^2 + coefficient of $y^2 = 0$.

we get $\frac{1}{a^2} - \frac{1}{b^2} = \frac{1}{p^2} \Rightarrow p = \frac{ab}{\sqrt{(b^2 - a^2)}}$

p is also the length of perpendicular drawn from (0, 0) to the line $x \cos\alpha + y \sin\alpha = p$

$\Rightarrow$ then radius of the circle $= p = \dfrac{ab}{\sqrt{(b^2 - a^2)}}$

20. It is clear from the figure the common tangent to the circle $x^2 + y^2 = 1$ and hyperbola $x^2 - y^2 = 1$ is $x = 1$ (which is nearer to P (1/2, 1) and given one focus at P (1/2, 1), so the equation of the directrix is $x = 1$.

Hence, the equation of the ellipse is

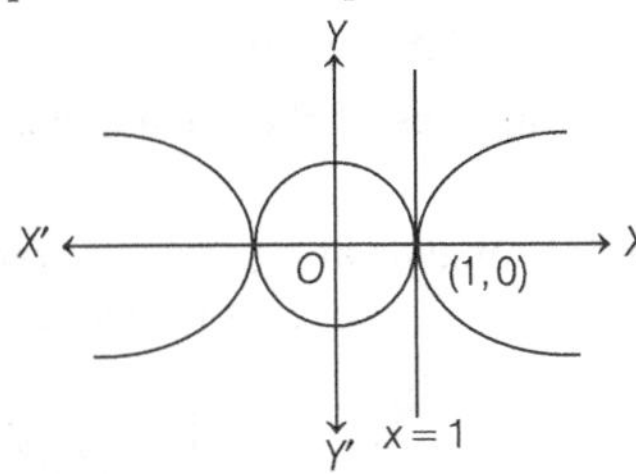

$$\sqrt{(x - 1/2)^2 + (y - 1)^2} = \frac{1}{2}|x - 1|$$

$$\Rightarrow \quad 9(x - 1/3)^2 + 12(y - 1)^2 = 1$$

21. Since, asymptotes of rectangular hyperbola are perpendicular to each other

$\because$ Given asymptote is

$$3x - 4y - 6 = 0$$

$\therefore$ Other asymptotes is $4x + 3y + \lambda = 0$ ...(i)

Given centre of hyperbola lies on $x - y - 1 = 0$ since, asymptotes pass through the centre of hyperbola

$\therefore$ Centre is the point of intersection of $x - y - 1 = 0$

and $\quad 3x - 4y - 6 = 0$

$\therefore$ Centre is $(-2, -3)$, also $(-2, -3)$ lies on Eq. (i)

then $\quad -8 - 9 + \lambda = 0$

$\therefore \quad \lambda = 17$

Hence, other asymptote is

$$4x + 3y + 17 = 0 \quad \text{[from Eq. (i)]}$$

22. Equation of normal of $y^2 = 4ax$ is

$$y = mx - 2am - am^3$$

which is tangent of $x^2 - y^2 = a^2$

$\therefore \quad (-2am - am^3)^2 = a^2m^2 - a^2$

or $\quad 4m^2 + m^6 + 4m^4 = m^2 - 1$

or $\quad m^6 + 4m^4 + 3m^2 + 1 = 0$

23. Given equation of curves are $y^2 = 16x$ (parabola) ...(i)

and $\quad xy = -4$ (rectangular hyperbola) ...(ii)

Clearly, equation of tangent having slope 'm' to parabola (i) is $y = mx + \dfrac{4}{m}$ ...(iii)

Now, eliminating y from Eqs. (ii) and (iii), we get

$$x\left(mx + \frac{4}{m}\right) = -4 \Rightarrow mx^2 + \frac{4}{m}x + 4 = 0,$$

which will give the points of intersection of tangent and rectangular hyperbola.

Since, line $y = mx + \dfrac{4}{m}$ is also a tangent to the rectangular hyperbola.

$\therefore$ Discriminant of quadratic equation $mx^2 + \dfrac{4}{m}x + 4 = 0$, should be zero.

[$\because$ there will be only one point of intersection]

$$\Rightarrow \quad D = \left(\frac{4}{m}\right)^2 - 4(m)(4) = 0$$

$$\Rightarrow \quad m^3 = 1 \Rightarrow m = 1$$

So, equation of required tangent is $y = x + 4$.

24. Let the equation of the hyperbola be

$$H(x, y) \equiv \frac{x^2}{a^2} - \frac{y^2}{b^2} - 1 = 0,$$

then $A(x, y) \equiv \dfrac{x^2}{a^2} - \dfrac{y^2}{b^2} = 0$ is the equation of asymptotes

and $C(x, y) \equiv \dfrac{x^2}{a^2} - \dfrac{y^2}{b^2} + 1 = 0$ is the equation of its conjugate hyperbola.

Since, $H(\alpha, \beta) + C(\alpha, \beta) = 2A(\alpha, \beta)$

25. $\because$ $2y - x - 3 = 0$ and $2x + y - 1 = 0$ are perpendicular to each other, then equation of conic can be written as

$$4 \times 5\left[\frac{2y - x - 3}{\sqrt{(2)^2 + (-1)^2}}\right]^2 - 9 \times 5\left[\frac{2x + y - 1}{\sqrt{(2)^2 + (1)^2}}\right]^2 = 80$$

$$\Rightarrow \quad 4\left[\frac{2y - x - 3}{\sqrt{5}}\right]^2 - 9\left[\frac{2x + y - 1}{\sqrt{5}}\right]^2 = 16$$

Putting $X = \dfrac{2y - x - 3}{\sqrt{5}}$ and $Y = \dfrac{2x + y - 1}{\sqrt{5}}$, then new equation of conic is

$$4X^2 - 9Y^2 = 16, \Rightarrow \frac{X^2}{4} - \frac{Y^2}{(16/9)} = 1$$

Here, $\quad a^2 = 4, b^2 = 16/9,$

$\therefore \quad b^2 = a^2(e^2 - 1) \Rightarrow \dfrac{16}{9} = 4(e^2 - 1),$

$\Rightarrow \quad e^2 = 13/9 \Rightarrow e = \sqrt{13}/3$

26. Equation of given hyperbola is,

$$\frac{x^2}{a^2} - \frac{y^2}{b^2} = 1 \quad \text{...(i)}$$

Since, $P(3, 3)$, be a point on Eq. (i).

So, $\quad \dfrac{9}{a^2} - \dfrac{9}{b^2} = 1$ ...(ii)

Now, equation of normal at point P to the hyperbola is

$$\frac{a^2x}{3} + \frac{b^2y}{3} = a^2 + b^2 \quad \text{...(iii)}$$

The normal Eq. (iii) intersect the X-axis at (9, 0), so

$$3a^2 = a^2 + b^2 \Rightarrow b^2 = 2a^2 \quad \text{...(iv)}$$

$\because$ The eccentricity (e) of hyperbola is,

$$e = \sqrt{1 + \frac{b^2}{a^2}} = \sqrt{1 + 2} = \sqrt{3} \Rightarrow e^2 = 3$$

From Eqs. (ii) and (iv), we get

$$\frac{9}{a^2} - \frac{9}{2a^2} = 1 \Rightarrow 18 - 9 = 2a^2 \Rightarrow a^2 = \frac{9}{2}$$

$\therefore$ The ordered pair $(a^2, e^2) = \left(\dfrac{9}{2}, 3\right)$

27. Given line is $x\cos\alpha + y\sin\alpha = p$

$\Rightarrow$ $y\sin\alpha = -x\cos\alpha + p$

$\Rightarrow$ $y = -\dfrac{\cos\alpha}{\sin\alpha}x + \dfrac{p}{\sin\alpha}$

Comparing with $y = mx + c$

$\therefore$ $m = -\dfrac{\cos\alpha}{\sin\alpha}$ and $c = \dfrac{p}{\sin\alpha}$

Condition of tangency

$$c^2 = a^2m^2 - b^2$$

$\Rightarrow$ $\dfrac{p^2}{\sin^2\alpha} = \dfrac{a^2\cos^2\alpha}{\sin^2\alpha} - b^2$

$\Rightarrow$ $p^2 = a^2\cos^2\alpha - b^2\sin^2\alpha$

28. Given hyperbola is $16x^2 - 9y^2 = 144$ or $\dfrac{x^2}{9} - \dfrac{y^2}{16} = 1$

Comparing with $\dfrac{x^2}{a^2} - \dfrac{y^2}{b^2} = 1$

$\Rightarrow$ $a = 3$ and $b = 4$

$y = m_1x$ and $y = m_2x$ are the conjugate diameters

then $m_1m_2 = \dfrac{b^2}{a^2}$ $\left[\because x = 2y, \therefore m_1 = \dfrac{1}{2}\right]$

$\Rightarrow$ $\dfrac{1}{2}\times m_2 = \dfrac{16}{9}$ or $m_2 = \dfrac{32}{9}$

$\therefore$ Diameter is $y = \dfrac{32}{9}x$

29. Equation PQ is $\dfrac{x}{a}\cos\left(\dfrac{\theta-\phi}{2}\right) - \dfrac{y}{b}\sin\left(\dfrac{\theta+\phi}{2}\right)$

$$= \cos\left(\dfrac{\theta+\phi}{2}\right)$$

Its passes through $(ae, 0)$

$\therefore$ $e\cos\left(\dfrac{\theta-\phi}{2}\right) - 0 = \cos\left(\dfrac{\theta+\phi}{2}\right)$

$\Rightarrow$ $\dfrac{\cos\left(\dfrac{\theta-\phi}{2}\right)}{\cos\left(\dfrac{\theta+\phi}{2}\right)} = \dfrac{1}{e}$

$$\dfrac{\cos\left(\dfrac{\theta-\phi}{2}\right) - \cos\left(\dfrac{\theta+\phi}{2}\right)}{\cos\left(\dfrac{\theta-\phi}{2}\right) + \cos\left(\dfrac{\theta+\phi}{2}\right)} = \dfrac{1-e}{1+e}$$

(by componendo and dividendo method)

$\Rightarrow$ $\tan\theta/2\tan\phi/2 = \dfrac{1-e}{1+e}$

30. Centre of hyperbola is $\left(\dfrac{6-4}{2}, \dfrac{5+5}{2}\right)$ i.e. $(1, 5)$

Distance between foci $= 2ae$

$\Rightarrow$ $10 = 2ae \Rightarrow 5 = a\times\dfrac{5}{4}$

$\therefore$ $a = 4$

$$b^2 = a^2(e^2-1) = 16\left(\dfrac{25}{16}-1\right) = 25 - 16 = 9$$

$\therefore$ Equation of hyperbola is

$$\dfrac{(x-1)^2}{16} - \dfrac{(y-5)^2}{9} = 1$$

31. Let the middle point of the chord is (h, k)

$\therefore$ $T = S_1$

$3xh - 2yk + 2(x+h) - 3(y+k) = 3h^2 - 2k^2 + 4h - 6k$

slope of this chord $= \dfrac{3h+2}{2k+3} = 2$ (given)

or $3h + 2 = 4k + 6$

$\Rightarrow$ $3h - 4k = 4$

Hence, locus of middle point is

$$3x - 4y = 4$$

32. $\because$ $x - y = 0$ and $x + y = 0$

are the asymptotes of the rectangular hyperbola $x^2 - y^2 = a^2$

Equation of tangent at $P(a\sec\phi, a\tan\phi)$ of $x^2 - y^2 = a^2$ is

$$ax\sec\phi - ay\tan\phi = a^2$$

or $x\sec\phi - y\tan\phi = a$ ...(i)

Solving $y = x$ and $y = -x$ with Eq. (i), then we get

and $\begin{Bmatrix} A(a(\sec\phi+\tan\phi), a(\sec\phi+\tan\phi)) \\ \text{and } B(a(\sec\phi-\tan\phi), a(\tan\phi-\sec\phi)) \end{Bmatrix}$

$\therefore$ Area of ΔCAB

$$= \dfrac{1}{2}|a(\tan^2\phi - \sec^2\phi) - a(\sec^2\phi - \tan^2\phi)|$$

$$= \dfrac{1}{2}|-a-a| = |-a| = |a|$$

33. Since, the vertices of hyperbola on X-axis at $(\pm 6, 0)$, so equation of hyperbola we can assume as $\dfrac{x^2}{a^2} - \dfrac{y^2}{b^2} = 1$ and $|a| = 6$.

and hyperbola passes through point $P(10, 16)$ so

$$\dfrac{(10)^2}{6^2} - \dfrac{(16)^2}{b^2} = 1$$

$\Rightarrow$ $\dfrac{(16)^2}{b^2} = \dfrac{100-36}{36} = \dfrac{64}{36}$

$\Rightarrow$ $b^2 = \dfrac{36\times 256}{64} = 144$

So, equation of hyperbola is $\dfrac{x^2}{36} - \dfrac{y^2}{144} = 0$, and the equation of normal to hyperbola at point P is

$$\dfrac{36x}{10} + \dfrac{144y}{16} = 36 + 144$$

$\Rightarrow$ $\dfrac{18}{5}x + 9y = 180$

$\Rightarrow$ $2x + 5y = 100$

34. Equation of tangent of hyperbola $\frac{x^2}{a^2} - \frac{y^2}{b^2} = 1$ at 'ϕ' is

$$\frac{x}{a}\sec\phi - \frac{y}{b}\tan\phi = 1$$

since foci $(\pm ae, 0)$

$$p_1 \times p_2 = \frac{(e\sec\phi - 1)(e\sec\phi + 1)}{\left(\sqrt{\left(\frac{\sec^2\phi}{a^2} + \frac{\tan^2\phi}{b^2}\right)}\right)^2}$$

$$= \frac{(e^2\sec^2\phi - 1)a^2b^2}{(a^2\tan^2\phi + b^2\sec^2\phi)}$$

$$= \frac{(a^2e^2\sec^2\phi - a^2)b^2}{(a^2\tan^2\phi + b^2\sec^2\phi)}$$

$$= \frac{[(a^2+b^2)\sec^2\phi - a^2]b^2}{(a^2\tan^2\phi + b^2\sec^2\phi)}$$

$$= \frac{(a^2\tan^2\phi + b^2\sec^2\phi)b^2}{(a^2\tan^2\phi + b^2\sec^2\phi)} = b^2$$

35. Let $P \equiv (8, y_1)$

$\therefore \quad 9(8)^2 - 16y_1^2 = 144$

$\Rightarrow \quad 9 \times 8 - 2y_1^2 = 18$

$y_1 = \pm 3\sqrt{3}$

$\therefore \quad P \equiv (8, 3\sqrt{3}) \quad [\because P$ lies in first quadrant]

Also, $\frac{x^2}{16} - \frac{y^2}{9} = 1$

$\Rightarrow \quad \frac{x^2}{4^2} - \frac{y^2}{3^2} = 1$

$\therefore \quad 3^2 = 4^2(e^2 - 1)$

$\Rightarrow \quad e = \frac{5}{4}$

then foci $\equiv (\pm 5, 0)$

$\therefore$ Equation of the reflected ray after first reflection passes through P, S' is

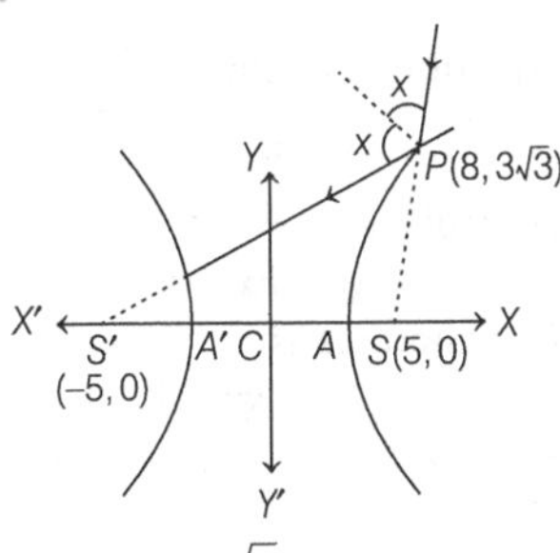

$$y - 0 = \frac{3\sqrt{3} - 0}{8 + 5}(x + 5)$$

$\Rightarrow 3\sqrt{3}x - 13y + 15\sqrt{3} = 0$

36. Let rectangular hyperbola

$$xy = c^2 \quad \ldots(\text{i})$$

Let three points on Eq. (i) are

$$A\left(ct_1, \frac{c}{t_1}\right), B\left(ct_2, \frac{c}{t_2}\right), C\left(ct_3, \frac{c}{t_3}\right)$$

Let orthocentre is $P(h, k)$ then slope of $AP \times$ slope of $BC = -1$

$$\Rightarrow \quad \frac{k - \frac{c}{t_1}}{h - ct_1} \times \frac{\frac{c}{t_3} - \frac{c}{t_2}}{ct_3 - ct_2} = -1$$

$$\Rightarrow \quad \frac{k - \frac{c}{t_1}}{h - ct_1} \times -\frac{1}{t_2t_3} = -1$$

$$\Rightarrow \quad k - \frac{c}{t_1} = ht_2t_3 - ct_1t_2t_3 \quad \ldots(\text{ii})$$

Similarly, $BP \perp AC$

then $\quad k - \frac{c}{t_2} = ht_3t_1 - ct_1t_2t_3 \quad \ldots(\text{iii})$

Subtracting Eq. (iii) from Eq. (ii), then we get

$$h = -\frac{c}{t_1t_2t_3}$$

Substituting the value h in Eq. (ii) then

$$k = -ct_1t_2t_3$$

$\therefore$ Orthocentre is $\left(-\frac{c}{t_1t_2t_3}, -ct_1t_2t_3\right)$

which lies on $xy = c^2$

37. $x^2 + y^2 = 25$

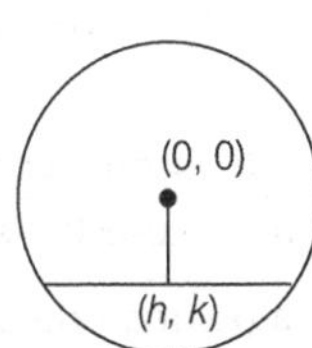

Equation of chord, $y - k = -\frac{h}{k}(x - h)$

$\Rightarrow \quad ky - k^2 = -hx + h^2$

$\Rightarrow \quad hx + ky = h^2 + k^2$

$\Rightarrow \quad y = -\frac{hx}{k} + \frac{h^2 + k^2}{k}$

Tangent to $\frac{x^2}{9} - \frac{y^2}{16} = 1$

$\Rightarrow \quad c^2 = a^2m^2 - b^2$

$\Rightarrow \quad \left(\frac{h^2 + k^2}{k}\right)^2 = 9\left(-\frac{h}{k}\right)^2 - 16$

$\Rightarrow \quad (x^2 + y^2)^2 = 9x^2 - 16y^2$

38. $\because$ Normal at t_1 meets the curve again at t_2.

So, the normal passes through $\left(ct_1, \frac{c}{t_1}\right)$ and $\left(ct_2, \frac{c}{t_2}\right)$

$$\text{Slope of normal} = \frac{\frac{c}{t_2} - \frac{c}{t_1}}{ct_2 - ct_1} = -\frac{1}{t_1t_2}$$

Also equation of normal at 't_1' is

$$xt_1^3 - yt_1 - ct_1^4 + c = 0$$

$\therefore$ Slope of normal $= t_1^2 = -\frac{1}{t_1t_2}$

$\therefore \quad t_1^3t_2 = -1$

39. Since, chord of contact $x = 9$ which is parallel to Y-axis. Hence, points of intersection of tangents must lie on X-axis. Let $(h, 0)$ is the point of tangent, then equation of chord of contact w.r.t. $(h, 0)$ is

$$h \cdot x - y \cdot 0 = 9$$

$\therefore$ $hx = 9$ but given $x = 9$

$\therefore$ $h = 1$

Coordinates of D is $(1, 0)$

Hence, combined equation of DA and DB is

$$SS_1 = T^2$$

$$(x^2 - y^2 - 9)(1 - 0 - 9) = (1 \cdot x + 0 \cdot y - 9)^2$$

then we get

$$9x^2 - 8y^2 - 18x + 9 = 0$$

40. We have, equations of the normal at P is

$$ax + by \operatorname{cosec}\theta = (a^2 + b^2)\sec\theta$$

and the equation of the normal at $Q(a\sec\phi, b\sec\phi)$ is,

$$ax + by \operatorname{cosec}\phi = (a^2 + b^2)\sec\phi \quad \dots \text{(ii)}$$

Subtracting Eq. (ii) from Eq. (i), we get

$$y = \frac{a^2 + b^2}{b} \cdot \frac{\sec\theta - \sec\phi}{\operatorname{cosec}\theta - \operatorname{cosec}\phi}$$

So, that $k = y = \dfrac{a^2 + b^2}{b} \cdot \dfrac{\sec\theta - \sec(\pi/2 - \theta)}{\operatorname{cosec}\theta - \operatorname{cosec}(\pi/2 - \theta)}$

$$[\because \theta + \phi = \pi/2]$$

$$= \frac{a^2 + b^2}{b} \cdot \frac{\sec\theta - \operatorname{cosec}\theta}{\operatorname{cosec}\theta - \sec\theta} = -\left[\frac{a^2 + b^2}{b}\right]$$

41. Let $P(x_1, y_1)$ be a point on the hyperbola,

$$\frac{x^2}{a^2} - \frac{y^2}{b^2} = 1$$

The chord of contact of tangents from P to the hyperbola $\dfrac{x^2}{a^2} - \dfrac{y^2}{b^2} = 2$, is given by

$$\frac{xx_1}{a^2} - \frac{yy_1}{b^2} = 2$$

The equations of the asymptotes are,

$$\frac{x}{a} - \frac{y}{b} = 0 \text{ and } \frac{x}{a} + \frac{y}{b} = 0$$

The points of intersection of Eq. (i) with the two asymptotes are given by,

$$x_1 = \frac{2a}{(x_1/a) - (y_1/b)},\ y_1 = \frac{2b}{(x_1/a) - (y_1/b)}$$

$$x_2 = \frac{2a}{(x_1/a) + (y_1/b)},\ y_2 = \frac{-2b}{(x_1/a) + (y_1/b)}$$

Area of the said triangle $= \dfrac{1}{2}|x_1y_2 - x_2y_1|$

$$= \frac{1}{2}\left|-\frac{4ab \times 2}{(x_1^2/a^2) - (y_1^2/b^2)}\right| = 4ab$$

42. $\because$ $x = t^2 + 1,\ y = 2t$

$\Rightarrow x = \left(\dfrac{y}{2}\right)^2 + 1$

$\Rightarrow x = \dfrac{y^2}{4} + 1 \Rightarrow y^2 = 4(x - 1) \quad \dots\text{(i)}$

Also, $x = 2s,\ y = \dfrac{2}{s}$

$\therefore$ $xy = 4 \quad \dots\text{(ii)}$

From Eqs. (i) and (ii),

$$\left(\frac{4}{x}\right)^2 = 4(x - 1)$$

$\Rightarrow$ $\dfrac{16}{x^2} = 4(x - 1)$

$\Rightarrow$ $4 = x^3 - x^2$

$\Rightarrow$ $x^3 - x^2 - 4 = 0$

$\Rightarrow (x - 2)(x^2 + x + 2) = 0$

$\therefore$ $x = 2,\ x^2 + x + 2 \neq 0$

From Eq. (ii),

$$y = \frac{4}{2} = 2$$

Point of intersection is $(2, 2)$.

43. Let $y = mx + c$ be a common tangent of hyperbolas

$$\frac{x^2}{a^2} - \frac{y^2}{b^2} = 1 \quad \dots\text{(i)}$$

and $$\frac{y^2}{a^2} - \frac{x^2}{b^2} = 1 \quad \dots\text{(ii)}$$

Condition of tangency for Eq. (i) is

$$c^2 = a^2m^2 - b^2 \quad \dots\text{(iii)}$$

and condition of tangency for Eq. (ii) is

$$c^2 = a^2 - b^2m^2 \quad \dots\text{(iv)}$$

From Eqs. (iii) and (iv),

$$a^2m^2 - b^2 = a^2 - b^2m^2$$

$\Rightarrow$ $a^2(m^2 - 1) + b^2(m^2 - 1) = 0$

$\Rightarrow$ $(a^2 + b^2)(m^2 - 1) = 0$

$\because$ $a^2 + b^2 \neq 0$

$\therefore$ $m^2 - 1 = 0$

$\Rightarrow$ $m = \pm 1$

From Eq. (iii), $c^2 = a^2 - b^2$

$\therefore$ $c = \pm\sqrt{(a^2 - b^2)}$

Hence, equations of common tangents are

$$y = \pm x \pm \sqrt{(a^2 - b^2)}$$

44. Given hyperbola is

$$xy = hx + ky$$

$\therefore$ Pair of asymptotes is

$$xy - hx - ky + \lambda = 0 \quad \dots\text{(i)}$$

Its represents a pair of tangents i.e. pair of straight lines, then $\Delta = 0$

$$abc + 2fgh - af^2 - bg^2 - ch^2 = 0$$

$$0 + 2 \times -\frac{k}{2} \times -\frac{h}{2} \times \frac{1}{2} - 0 - 0 - \lambda \times \frac{1}{4} = 0$$

$$\frac{hk}{4} - \frac{\lambda}{4} = 0$$

$\therefore$ $\lambda = hk$

From Eq. (i), $xy - hx - ky + hk = 0$

$\Rightarrow$ $x(y - h) - k(y - h) = 0$

$\Rightarrow$ $(x - k)(y - h) = 0$

$\therefore$ Asymptotes are $x = k,\ y = h$

45. $\because H : x^2 + 3xy + 2y^2 + 2x + 3y = 0$

Let pair of asymptotes is

$$x^2 + 3xy + 2y^2 + 2x + 3y + \lambda = 0$$

$\therefore$ $$\Delta = 0$$

$$1\times 2\times \lambda + 2\times \frac{3}{2}\times 1\times \frac{3}{2} - 1\times \frac{9}{4} - 2\times 1 - \lambda\times \frac{9}{4} = 0$$

$\Rightarrow$ $$-\frac{\lambda}{4} + \frac{9}{2} - \frac{9}{4} - 2 = 0 \Rightarrow \frac{\lambda}{4} = \frac{9}{4} - 2 = \frac{1}{4}$$

$\therefore$ $$\lambda = 1$$

$A : x^2 + 3xy + 2y^2 + 2x + 3y + 1 = 0$

$\because$ $H + C = 2A$

$\therefore$ $C = 2A - H = x^2 + 3xy + 2y^2 + 2x + 3y + 2$

$\therefore$ Conjugate hyperbola is

$$x^2 + 3xy + 2y^2 + 2x + 3y + 2 = 0$$

46. Let rectangular hyperbola $xy = c^2$

Equation of tangent at 't' is

$$\frac{x}{t} + yt = 2c$$

$\Rightarrow$ $$\frac{x}{2ct} + \frac{y}{\left(\frac{2c}{t}\right)} = 1 \quad \ldots(\text{i})$$

and equation of normal at 't' is

$$xt^3 - yt - ct^4 + c = 0$$

$\Rightarrow$ $$xt^3 - ty = ct^4 - c$$

$\Rightarrow$ $$\frac{x}{\left(\frac{ct^4 - c}{t^3}\right)} + \frac{y}{\left(\frac{-ct^4 + c}{t}\right)} = 1 \quad \ldots(\text{ii})$$

From Eqs. (i) and (ii) it is clear that

$$x_1 = 2ct,\ x_2 = \left(\frac{ct^4 - c}{t^3}\right)$$

and $$y_1 = \frac{2c}{t},\ y_2 = \left(\frac{-ct^4 + c}{t}\right)$$

$\therefore$ $$x_1x_2 + y_1y_2 = \frac{2ct(ct^4 - c)}{t^3} + \frac{2c}{t}\frac{(-ct^4 + c)}{t}$$

$$= \frac{2c^2(t^4 - 1)}{t^2} - \frac{2c^2(t^4 - 1)}{t^2} = 0$$

47.

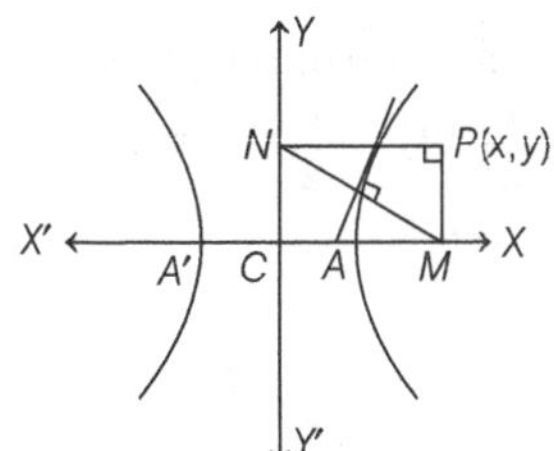

Equation of normal at 'ϕ' of $\frac{x^2}{a^2} - \frac{y^2}{b^2} = 1$ is

$ax\cos\phi + by\cot\phi = a^2 + b^2$,

then $$M \equiv \left(\frac{a^2 + b^2}{a\cos\phi}, 0\right)$$

and $$N \equiv \left(0, \frac{a^2 + b^2}{b\cot\phi}\right)$$

$P \equiv (x, y)$ then

$$x = \frac{a^2 + b^2}{a\cos\phi}$$

$$ax = (a^2 + b^2)\sec\phi \quad \ldots(\text{i})$$

$$y = \frac{(a^2 + b^2)}{b\cot\phi}$$

$\Rightarrow$ $$by = (a^2 + b^2)\tan\phi \quad \ldots(\text{ii})$$

From Eqs. (i) and (ii),

$$(ax)^2 - (by)^2 = (a^2 + b^2)(\sec^2\phi - \tan^2\phi)$$

$\Rightarrow$ $$a^2x^2 - b^2y^2 = (a^2 + b^2)^2$$

Objective Questions Type II [One or more than one correct answer(s)]

1. $16x^2 - 3y^2 - 32x - 12y - 44 = 0$

$\Rightarrow$ $$16(x^2 - 2x) - 3(y^2 + 4y) - 44 = 0$$

$\Rightarrow 16\{(x - 1)^2 - 1\} - 3\{(y + 2)^2 - 4\} - 44 = 0$

$\Rightarrow$ $$16(x - 1)^2 - 3(y + 2)^2 = 48$$

$\Rightarrow$ $$\frac{(x - 1)^2}{(\sqrt{3})^2} - \frac{(y + 2)^2}{4^2} = 1$$

Alternate. (a) : Length of transverse axis

$$= 2\times\sqrt{3} = 2\sqrt{3}$$

Alternate. (b) : Length of conjugate axis

$$= 2\times 4 = 8$$

Alternate. (c) : Centre $x - 1 = 0$

and $y + 2 = 0$, i.e. $(1, -2)$

Alternate. (d) : $4^2 = (\sqrt{3})^2(e^2 - 1)$

or $$e^2 - 1 = \frac{16}{3}$$

$\therefore$ $$e^2 = \frac{19}{3} \text{ then } e = \sqrt{\frac{19}{3}}$$

2. Equation of given hyperbola is,

$$\frac{x^2}{a^2} - \frac{y^2}{b^2} = 1,\ a > b \text{ and } a > 1$$

Let point $P(a\sec\theta, b\tan\theta)$ on the hyperbola in first quadrant i.e. $Q \in \left(0, \frac{\pi}{2}\right)$.

Now, equation of tangent to the hyperbola at point P is

$$\frac{x\sec\theta}{a} - \frac{y\tan\theta}{b} = 1 \quad \ldots(\text{i})$$

$\because$ The tangent Eq. (i) passes through point $A(1, 0)$

So, $$a = \sec\theta > 1 \quad \ldots(\text{ii})$$

and equation of normal to the hyperbola at point P having slope is $\frac{b\sec\theta}{a\tan\theta}$, as normal cuts off equal intercepts on the coordinate axes, so slope must be -1.

Therefore, $\frac{b\sec\theta}{a\tan\theta} = -1 \Rightarrow b = -\tan\theta$ {from Eq. (ii)}

$\because$ The eccentricity of hyperbola,

$$e = \sqrt{1 + \frac{b^2}{a^2}} = \sqrt{1 + \sin^2\theta}$$

$\therefore \quad 1 < e < \sqrt{2}, \quad \text{as } \theta \in \left(0, \frac{\pi}{2}\right)$

And the area of required triangle is, which is isosceles is

$$\Delta = \frac{1}{2}(AP)^2 = \frac{1}{2}[(1 - \sec^2\theta)^2 + \tan^4\theta]$$

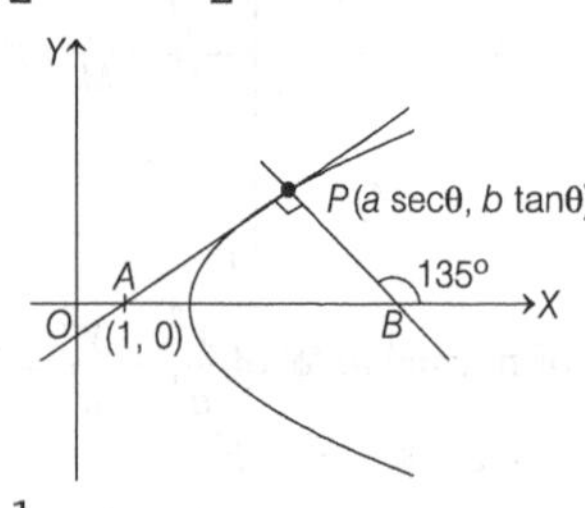

$$= \frac{1}{2}(2\tan^4\theta) = \tan^4\theta = b^4$$

3. Given hyperbola is $5x^2 - y^2 = 5$

$\Rightarrow \quad \frac{x^2}{1} - \frac{y^2}{5} = 1$

$\therefore$ Equation of tangent in terms of slope is

$$y = mx \pm \sqrt{(m^2 - 5)} \quad \ldots(i)$$

Its passes through (2, 8)

then $\quad 8 = 2m \pm \sqrt{m^2 - 5}$

or $\quad (8 - 2m)^2 = m^2 - 5$

$\Rightarrow \quad 3m^2 - 32m + 69 = 0$

$\therefore \quad m = 3, \frac{23}{3}$

From Eq. (i), Equation of tangents are

$$y = 3x \pm 2$$

and $\quad y = \frac{23x}{3} \pm \sqrt{\frac{484}{9}}$ or $3x - y \pm 2 = 0$

and $\quad 23x - 3y \pm 22 = 0$

4. $xy = 1$

Equation of normal at 't' *ie*, $\left(t, \frac{1}{t}\right)$

is $\quad xt^3 - yt - t^4 + 1 = 0$

Slope is $\quad t^2 = -\frac{a}{b}$,

$$-\frac{a}{b} > 0 \Rightarrow \frac{a}{b} < 0$$

$\therefore \quad a > 0, b < 0$

or $\quad a < 0, b > 0$

5. $\frac{x^2}{25} - \frac{y^2}{16} = 1$

Equation of tangent in terms of slope

$$y = mx \pm \sqrt{(25m^2 - 16)}$$

or $\quad (y - mx)^2 = 25m^2 - 16$

it is passing through (6, 2), then

$$(2 - 6m)^2 = 25m^2 - 16$$

$\Rightarrow \quad 4 + 36m^2 - 24m = 25m^2 - 16$

$\Rightarrow \quad 11m^2 - 24m + 20 = 0$

$\therefore \quad m_1 + m_2 = 24/11, \; m_1 m_2 = 20/11$

6. Equation of any tangent in terms of slope m is

$$y = mx + \sqrt{(a^2m^2 - b^2)}$$

its passes through (h, k), then

$$(k - mh)^2 = a^2m^2 - b^2$$

$\Rightarrow \quad m^2(h^2 - a^2) - 2mhk + k^2 + b^2 = 0$

which is quadratic in m,

Let slopes of tangents are m_1 and m_2, then

$$m_1 m_2 = -1$$

$\Rightarrow \quad \frac{k^2 + b^2}{h^2 - a^2} = -1$

$\Rightarrow \quad h^2 + k^2 = a^2 - b^2$

Hence, locus is $x^2 + y^2 = a^2 - b^2$

Which is director circle of $\frac{x^2}{a^2} - \frac{y^2}{b^2} = 1$

7. We have,

Eccentricity of the hyperbola is $\sqrt{2}$ as it is a rectangular hyperbola so eccentricity e of the ellipse is $1/\sqrt{2}$.

Let the equation of the ellipse be,

$$\frac{x^2}{a^2} + \frac{y^2}{b^2} = 1 \text{ where } b^2 = a^2(1 - e^2)$$

$$= a^2/2 \Rightarrow a^2 = 2b^2$$

So, equation of the ellipse is

$$x^2 + 2y^2 = a^2$$

Let (x_1, y_1) be a point of intersection of the ellipse and the hyperbola, then

$$2x_1^2 - 2y_1^2 = 1 \text{ and } x_1^2 + 2y_1^2 = a^2 \quad \ldots(i)$$

Equations of the tangents at (x_1, y_1) tothe two conics are

$$2xx_1 - 2yy_1 = 1 \text{ and } xx_1 + 2yy_1 = a^2$$

Since, the two conics intersect orthogonally

$$\left(\frac{x_1}{y_1}\right)\left(-\frac{x_1}{2y_1}\right) = -1$$

$\Rightarrow \quad x_1^2 = 2y_1^2$

and from Eq. (i), we get

$$x_1^2 = 1, a^2 = 2$$

Hence, the equation of the ellipse is

$x^2 + 2y^2 = 2$ and its focus is

$$(\pm ae, 0) = \left(\pm\sqrt{2} \times \frac{1}{\sqrt{2}}, 0\right) = (\pm 1, 0)$$

8. Let P, Q, R, S lie on the circle

$$x^2 + y^2 + 2gx + 2fy + c = 0 \quad \ldots(i)$$

and also lies on

$$xy = c^2 \quad \ldots(ii)$$

Solving Eqs. (i) and (ii), then

$$x^2 + \left(\frac{c^2}{x}\right)^2 + 2gx + \frac{2fc^2}{x} + c = 0$$

$\Rightarrow \quad x^4 + 2gx^3 + cx^2 + 2fc^2x + c^4 = 0$

$\therefore \quad x_1x_2x_3x_4 = c^4 \quad \ldots(iii)$

and $$P \equiv (x_1, y_1) \equiv \left(x_1, \frac{c^2}{x_1}\right)$$

$$Q \equiv \left(x_2, \frac{c^2}{x_2}\right)$$

and $$R \equiv \left(x_3, \frac{c^2}{x_3}\right)$$

Let orthocentre $O \equiv (h, k)$

Then, slope of $QR \times$ slope of $OP = -1$

$$\left(\frac{\frac{c^2}{x_3} - \frac{c^2}{x_2}}{x_3 - x_2}\right) \times \left(\frac{k - \frac{c^2}{x_1}}{h - x_1}\right) = -1$$

$$\Rightarrow \quad -\frac{c^2}{x_2x_3} \times \left(\frac{k - \frac{c^2}{x_1}}{h - x_1}\right) = -1$$

$$\Rightarrow \quad k - \frac{c^2}{x_1} = \frac{h\, x_2x_3}{c^2} - \frac{x_1x_2x_3}{c^2} \qquad \ldots\text{(iv)}$$

Also, slope of $PQ \times$ slope of $OR = -1$

$$k - \frac{c^2}{x_3} = \frac{h\, x_1x_2}{c^2} - \frac{x_1x_2x_3}{c^2} \qquad \ldots\text{(v)}$$

From Eqs. (iii) and (iv),

$$\therefore \quad h = -\frac{c^4}{x_1x_2x_3} \text{ and } k = -\frac{x_1x_2x_3}{c^2}$$

From Eq. (iii),

$$h = -x_4 \text{ and } k = -\frac{c^2}{x_4}$$

$\therefore$ Orthocentre lies on $xy = c^2$

i.e. (x_4, y_4) and $(-x_4, -y_4)$

9. $y = x + 5$

Comparing with $y = mx + c$

$\therefore \quad m = 1, \; c = 5$

Alternate. (a) : Condition of tangency

$$c = \frac{a}{m}$$

$$5 = \frac{5}{1}$$

which is true.

Alternate. (b) : $9x^2 + 16y^2 = 144$

$$\Rightarrow \quad \frac{x^2}{16} + \frac{y^2}{9} = 1$$

$\because$ Condition of tangency

$$c^2 = a^2m^2 + b^2 \Rightarrow 25 = 16 \times 1 + 9 = 25$$

Which is true.

Alternate. (c) : $\frac{x^2}{29} - \frac{y^2}{4} = 1$

$\because$ Condition of tangency

$$c^2 = a^2m^2 - b^2$$

$$25 = 29 \times 1 - 4 = 25$$

Which is true.

Alternate. (d) : Now length of perpendicular from centre (0, 0) to the line $y = x + 5$ is $\frac{|5|}{\sqrt{2}}$

i.e. $\frac{5}{\sqrt{2}} \neq$ radius (5).

10. Equation of tangent in terms of slope of $y^2 = 32x$ is

$$y = mx + \frac{8}{m} \qquad \ldots\text{(i)}$$

Which is also tangent of $9x^2 - 9y^2 = 8$

i.e. $$x^2 - y^2 = \frac{8}{9}$$

then $$\left(\frac{8}{m}\right)^2 = \frac{8}{9}m^2 - \frac{8}{9}$$

$$\Rightarrow \quad \frac{8}{m^2} = \frac{m^2}{9} - \frac{1}{9} \Rightarrow 72 = m^4 - m^2$$

$$\Rightarrow \quad m^4 - m^2 - 72 = 0$$

$$\Rightarrow \quad (m^2 - 9)(m^2 + 8) = 0$$

$$\therefore \quad m^2 = 9, \; m^2 + 8 \neq 0$$

$$\therefore \quad m = \pm 3$$

From Eq. (i), $y = \pm 3x \pm \frac{8}{3} \Rightarrow 3y = \pm 9x \pm 8$

or $\pm 9x - 3y \pm 8 = 0$

i.e. $9x - 3y + 8 = 0, \; 9x - 3y - 8 = 0$

$-9x - 3y + 8 = 0, \; -9x - 3y - 8 = 0$

or $9x - 3y + 8 = 0, \; 9x - 3y - 8 = 0$

$9x + 3y - 8 = 0$

and $9x + 3y + 8 = 0$

11. Let $P(x_1, y_1)$ be any point on the hyperbola $x^2 - y^2 = 4$, then equation of the normal at P is

$$y - y_1 = -\frac{y_1}{x_1}(x - x_1)$$

$$\Rightarrow \quad x_1y + y_1x = 2x_1y_1$$

Then, coordinates of G are $(2x_1, 0)$ and of g are $(0, 2y_1)$.

So, that $PG = \sqrt{(2x_1 - x_1)^2 + y_1^2} = \sqrt{x_1^2 + y_1^2} = PC$

$$Pg = \sqrt{x_1^2 + (2y_1 - y_1)^2} = \sqrt{x_1^2 + y_1^2} = PC$$

and $$Gg = \sqrt{(2x_1)^2 + (2y_1)^2} = 2\sqrt{x_1^2 + y_1^2} = 2PC$$

12. Equation of the tangent at $(a \sec \alpha, b \tan \alpha)$ to the hyperbola $x^2/a^2 - y^2/b^2 = 1$ is $\frac{x}{a}\sec\alpha - \frac{y}{b}\tan\alpha = 1$ which meets the transverse axis $y = 0$ at the point $T(a\cos\alpha, 0)$ whose distance from the focus $(ae, 0)$ is $ae - a\cos\alpha$ and from the focus $(-ae, 0)$ is $(ae + a\cos\alpha)$.

Note that $ae > a\cos\alpha$.

Since, $e > 1$.

13. Equation of the directrices of the given hyperbola

$$\frac{x^2}{25} - \frac{y^2}{16} = 1 \text{ are } x = \pm\frac{5 \times 5}{\sqrt{25 + 16}} = \pm\frac{25}{\sqrt{41}} \qquad \ldots\text{(i)}$$

Equations of the asymptotes of the given hyperbola are

$$\frac{x^2}{25} - \frac{y^2}{16} = 0 \qquad \ldots\text{(ii)}$$

The points of intersection of Eqs. (i) and (iii) are

$$(\pm 25/\sqrt{41}, \pm 20/\sqrt{41})$$

14. Let $(x_i, y_i) = (t_i, 1/t_i); i = 1, 2, 3, 4.$

Any point on the rectangular hyperbola $xy = 1$ is $(t, 1/t)$ which lies on the circle $x^2 + y^2 = 1$, if

$$t^2 + \frac{1}{t^2} = 1 \Rightarrow t^4 - t^2 + 1 = 0$$

The roots of the equation are t_1, t_2, t_3, t_4, where $t_1 + t_2 + t_3 + t_4 = 0 \Rightarrow x_1 + x_2 + x_3 + x_4 = 0$

$$\Sigma t_1 t_2 = -1 \Rightarrow \Sigma t_1 t_2 t_3 = 0$$

$$t_1 t_2 t_3 t_4 = 1 \Rightarrow x_1 x_2 x_3 x_4 = y_1 y_2 y_3 y_4 = 0$$

and $\quad y_1 + y_2 + y_3 + y_4 = \frac{1}{t_1} + \frac{1}{t_2} + \frac{1}{t_3} + \frac{1}{t_4}$

$$= \frac{\Sigma t_1 t_2 t_3}{t_1 t_2 t_3 t_4} = 0$$

Linked Comprehension Type

PASSAGE 1

1. Pair of asymptotes is given by

$$xy - 3y - 2x + \lambda = 0 \qquad \dots(\text{i})$$

where λ is any constant such that it represents two straight lines.

$$\therefore \quad abc + 2fgh - af^2 - bg^2 - ch^2 = 0$$

$$\Rightarrow \quad 0 + 2 \times -\frac{3}{2} \times -1 \times \frac{1}{2} - 0 - 0 - \lambda \times \left(\frac{1}{2}\right)^2 = 0$$

$$\therefore \quad \lambda = 6$$

From Eq. (i), asymptotes of given hyperbola are given by

$$xy - 3y - 2x + 6 = 0$$

2. Let the asymptotes be $2x + 3y + \lambda = 0$ and $3x + 2y + \mu = 0$.

Since, asymptotes passes through (1, 2), then

$$\lambda = -8 \text{ and } \mu = -7$$

Let the equation of hyperbola be

$$(2x + 3y - 8)(3x + 2y - 7) + \gamma = 0) \qquad \dots(\text{i})$$

$\because$ It passes through (5, 3), then

$$(10 + 9 - 8)(15 + 6 - 7) + \gamma = 0$$

$$\Rightarrow \quad 11 \times 14 + \gamma = 0$$

$$\therefore \quad \gamma = -154$$

Putting the value of γ in Eq. (i), then

$$(2x + 3y - 8)(3x + 2y - 7) = 154$$

3. $\because \; 2\tan^{-1}\left(\frac{b}{a}\right) = \frac{\pi}{3}$

$$\Rightarrow \tan^{-1}\left(\frac{b}{a}\right) = \frac{\pi}{6}$$

$$\Rightarrow \quad \frac{b}{a} = \frac{1}{\sqrt{3}} \text{ or } a = b\sqrt{3}$$

Let e be an eccentricity of conjugate hyperbola, then

$$a^2 = b^2(e^2 - 1)$$

$$\Rightarrow \quad 3b^2 = b^2(e^2 - 1) \Rightarrow e^2 = 4$$

$$\therefore \quad e = 2$$

4. The transverse axis is the bisector of the angle between asymptotes containing the origin and the conjugate axis is the other bisector. The bisectors of the angle between asymptotes are

$$\frac{(3x - 4y - 1)}{5} = \pm \frac{(4x - 3y - 6)}{5}$$

$$\Rightarrow \quad (3x - 4y - 1) = \pm (4x - 3y - 6)$$

$$\Rightarrow \quad x + y - 5 = 0$$

and $\quad x - y - 1 = 0$

Hence, transverse axis and conjugate axis are $x + y - 5 = 0$ and $x - y - 1 = 0$.

5. $\because \quad 16x^2 - 25y^2 = 400$

$$\Rightarrow \quad \frac{x^2}{5^2} - \frac{y^2}{4^2} = 1 \qquad \dots(\text{i})$$

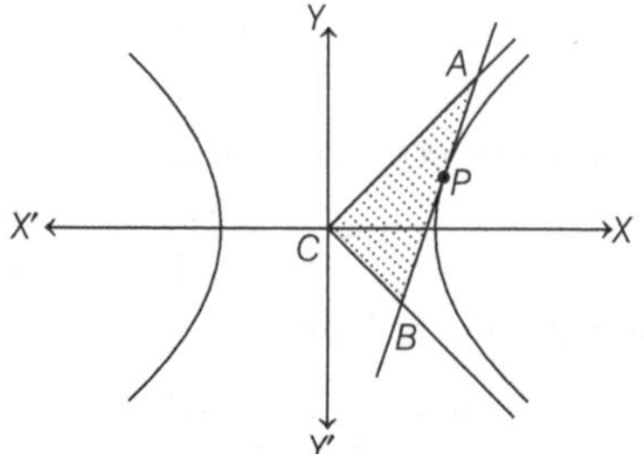

Let P $(5 \sec \phi, 4 \tan \phi)$ be any point on the hyperbola $\dots(\text{i})$

Equation of tangent at P is

$$\frac{x}{5}\sec\phi - \frac{y}{4}\tan\phi = 1 \qquad \dots(\text{ii})$$

And asymptotes of Eq. (i) are

$$y = \pm \frac{4}{5}x \qquad \dots(\text{iii})$$

Solving Eqs. (ii) and (iii), then

$$\frac{x}{5}\sec\phi \mp \frac{x}{5}\tan\phi = 1$$

or $x = \frac{5}{(\sec\phi \mp \tan\phi)} = \frac{5(\sec\phi + \tan\phi)(\sec\phi - \tan\phi)}{(\sec\phi \mp \tan\phi)}$,

then we get

$$A \equiv [5(\sec\phi + \tan\phi), 4(\sec\phi + \tan\phi)]$$

and $\; B \equiv [(5(\sec\phi - \tan\phi), -4(\sec\phi - \tan\phi)]$

$\therefore$ Area of

$$\Delta ABC = \frac{1}{2}\left|\begin{vmatrix} 5(\sec\phi + \tan\phi) & 4(\sec\phi + \tan\phi) & 1 \\ 5(\sec\phi - \tan\phi) & -4(\sec\phi - \tan\phi) & 1 \\ 0 & 0 & 1 \end{vmatrix}\right|$$

$$= \frac{1}{2}|-20 - 20| = 20 \text{ sq units}$$

PASSAGE 2

1. $\because t_1 + t_2 = 4, t_1 t_2 = 2$

tangent at t_1 is

$$\frac{x}{t_1} + y t_1 = 2c \qquad \ldots(\text{i})$$

and tangent at 't_2' is $\frac{x}{t_2} + y t_2 = 2c \qquad \ldots(\text{ii})$

Solving Eqs. (i) and (ii), we get

$$\left(\frac{2ct_1t_2}{t_1+t_2}, \frac{2c}{t_1+t_2}\right) \text{ i.e. } \left(\frac{4c}{4}, \frac{2c}{4}\right) \text{ or } \left(c, \frac{c}{2}\right)$$

2. $\because \alpha + \gamma = \frac{4}{t_1}, \alpha\gamma = \frac{1}{t_1} \qquad \ldots(\text{i})$

and $\beta + \delta = \frac{6}{t_2}, \beta\delta = \frac{1}{t_2} \qquad \ldots(\text{ii})$

Also $\alpha, \beta, \gamma, \delta$ are in HP

$$\therefore \quad \beta = \frac{2\alpha\gamma}{\alpha+\gamma} = \frac{2 \times \frac{1}{t_1}}{\frac{4}{t_1}} = \frac{1}{2} \qquad [\text{From Eq. (i)}]$$

$$\text{and} \quad \gamma = \frac{2\beta\delta}{\beta+\delta} = \frac{2 \times \frac{1}{t_2}}{\frac{6}{t_2}} = \frac{1}{3} \qquad [\text{from Eq. (ii)}]$$

Now, γ lie on $t_1 x^2 - 4x + 1 = 0$

$$\Rightarrow \quad t_1\left(\frac{1}{3}\right)^2 - 4\left(\frac{1}{3}\right) + 1 = 0$$

$$\Rightarrow \quad t_1 = 3$$

and β lie on $t_2 x^2 - 6x + 1 = 0$

$$\Rightarrow \quad t_2\left(\frac{1}{2}\right)^2 - 6\left(\frac{1}{2}\right) + 1 = 0$$

$$\Rightarrow \quad t_2 = 8$$

$\because$ Normal at 't_1' is

$$xt_1^3 - yt_1 - ct_1^4 + c = 0 \qquad \ldots(\text{iii})$$

and normal at 't_2' is

$$xt_2^3 - yt_2 - ct_2^4 + c = 0 \qquad \ldots(\text{iv})$$

Solving Eqs. (iii) and (iv), we get point of intersection is

$$\left(\frac{c\{t_1t_2(t_1^2 + t_1t_2 + t_2^2) - 1\}}{t_1t_2(t_1+t_2)}, \frac{c\{t_1^3t_2^3 + (t_1^2 + t_1t_2 + t_2^2)\}}{t_1t_2(t_1+t_2)}\right)$$

$$\text{or} \quad \left(\frac{2327c}{264}, \frac{13921c}{264}\right) \qquad [\because t_1 = 3, t_2 = 8]$$

3. Both $xy = 9$ and $x^2 - y^2 = 25$ are rectangular hyperbolas

$$\therefore \quad e_1 = e_2 = \sqrt{2}$$

$$\Rightarrow \quad e_1^2 + e_2^2 = 2 + 2 = 4$$

$$\therefore \quad (e_1, e_2) \text{ lie on } x^2 + y^2 = 4 \qquad \ldots(\text{i})$$

Director circle of Eq. (i) is

$$x^2 + y^2 = 8$$

$$\therefore \quad (\text{radius})^2 = 8$$

4. $\because$ Normal at 't_1' to the hyperbola $xy = c^2$ is

$xt_1^3 - yt_1 - ct_1^4 + c = 0$

but this passes through 't_2', then

$$ct_2t_1^3 - \frac{c}{t_2}t_1 - ct_1^4 + c = 0$$

$$\Rightarrow \quad t_2^2t_1^3 - t_1 - t_1^4t_2 + t_2 = 0$$

$$\Rightarrow \quad t_2t_1^3(t_2 - t_1) + (t_2 - t_1) = 0$$

$$\Rightarrow \quad (t_2t_1^3 + 1)(t_2 - t_1) = 0$$

$$\Rightarrow \quad t_2t_1^3 + 1 = 0 \qquad [\because t_2 \neq t_1]$$

$$\therefore \quad t_1^3t_2 = -1$$

$$\text{or} \quad t_1t_2 = -t_1^{-2}$$

5.

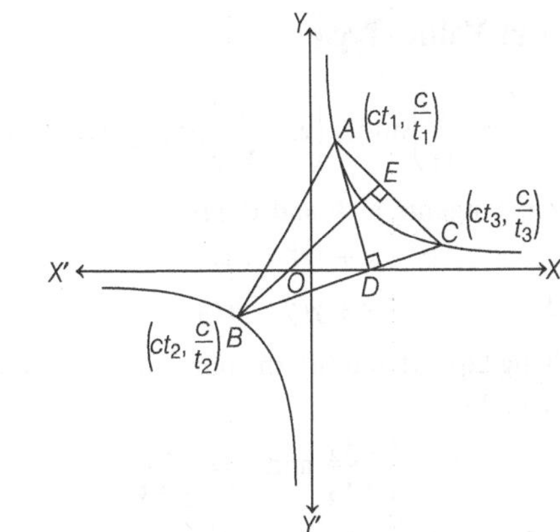

Let "t_1", "t_2" and "t_3" are vertices of the triangle ABC, described on the rectangular hyperbola $xy = c^2$.

$\therefore$ Coordinates of A, B and C are $\left(ct_1, \frac{c}{t_1}\right), \left(ct_2, \frac{c}{t_2}\right)$ and $\left(ct_3, \frac{c}{t_3}\right)$ respectively.

Now, slope of BC is $\dfrac{\frac{c}{t_3} - \frac{c}{t_2}}{ct_3 - ct_2} = -\frac{1}{t_2t_3}$

$\therefore$ Slope of AD is t_2t_3

Equation of altitude AD is

$$y - \frac{c}{t_1} = t_2t_3(x - ct_1)$$

$$\text{or} \quad t_1y - c = xt_1t_2t_3 - ct_1^2t_2t_3 \qquad \ldots(\text{i})$$

Similarly equation of altitude BE is

$$t_2y - c = xt_1t_2t_3 - ct_1t_2^2t_3 \qquad \ldots(\text{ii})$$

Solving Eqs. (i), and (ii), we get the orthocentre $\left(-\frac{c}{t_1t_2t_3}, -ct_1t_2t_3\right)$ which lies on $xy = c^2$.

PASSAGE 3

Ellipse and hyperbola are orthogonal and therefore, they are confocal.

$$\text{So,} \quad a_he_h = a_ee_e \Rightarrow a^2 + 9 = 41 - 16$$

$$\Rightarrow \quad 9 + a^2 = 25 \Rightarrow a^2 = 16$$

1. Thus, equation of hyperbola is $\frac{x^2}{9} - \frac{y^2}{16} = 1$

So, common tangents to the circle and hyperbola are

$$x = \pm 3$$

$$\Rightarrow \quad x^2 - 9 = 0$$

2. Director circle of hyperbola does not exist as $a < b$.

Director circle of circle is,

$$x^2 + (y-3)^2 = 18$$

$$\Rightarrow \quad x^2 + y^2 - 6y - 9 = 0$$

This meets the hyperbola $16x^2 - 9y^2 = 144$ at four points from where tangents drawn to the circle $x^2 + (y-3)^2 = 9$ are perpendicular to each other.

3. Let mid-point of AB be (h, k).

So, equation of line AB is

$$hx + ky - 3(y + k) = h^2 + k^2 - 6k$$

Since, tangents at C and D intersect at the directrix, CD is the focal chord of hyperbola.

So, AB passes through focus of the hyperbola and that is $(\pm 5, 0)$.

Therefore, required locus is $x^2 + y^2 \pm 5x - 3y = 0$.

Numerical Value Type

1. Let $P\left(ct_1, \frac{c}{t_1}\right)$ and $Q\left(ct_2, \frac{c}{t_2}\right)$ be any two points on $xy = c^2$.

Then, tangents at P and Q are

$$x + yt_1^2 = 2ct_1 \quad \ldots(i)$$

and

$$x + yt_2^2 = 2ct_2 \quad \ldots(ii)$$

Solving Eqs. (i) and (ii), the point of intersection, say (h, k), is given by

$$h = \frac{2ct_1t_2}{t_1 + t_2} \text{ and } k = \frac{2c}{t_1 + t_2} \quad \ldots(iii)$$

The foot of the ordinate of P is $(ct_1, 0)$ and it lies on Eq. (ii), then

$$ct_1 + 0 = 2ct_2$$

$$\therefore \quad t_1 = 2t_2$$

Then, from Eqs. (iii) and (iv),

$$h = \frac{2c \cdot 2t_2 \cdot t_2}{2t_2 + t_2}$$

and

$$k = \frac{2c}{2t_2 + t_2}$$

$$\therefore \quad h = \frac{4c}{3} t_2$$

and

$$k = \frac{2c}{3t_2}$$

$$\therefore \quad h \cdot k = \frac{4c}{3} t_2 \times \frac{2c}{3t_2}$$

$$\therefore \quad hk = \frac{8}{9} c^2$$

$\therefore$ Locus of (h, k) is $xy = \frac{8}{9} c^2$,

$$\therefore \quad \lambda = \frac{8}{9}$$

then $\quad 729\lambda = 648$

2. Let a triangle PQR be inscribed in $xy = c^2$.

Let the coordinates of the vertices of the triangle be

$$P\left(ct_1, \frac{c}{t_1}\right), Q\left(ct_2, \frac{c}{t_2}\right) \text{ and } R\left(ct_3, \frac{c}{t_3}\right)$$

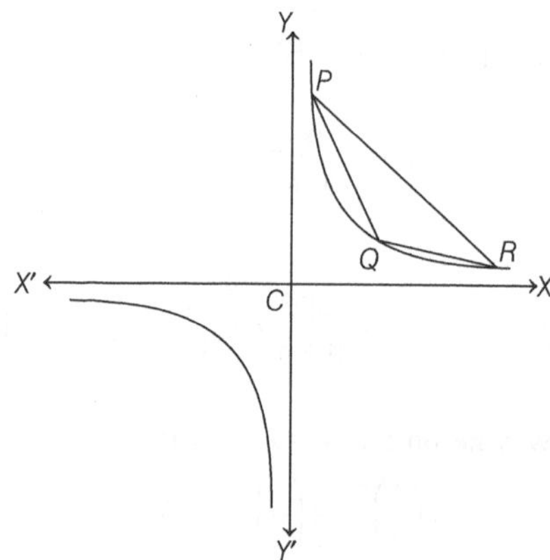

Now, the equation of chord joining P and Q is

$$x + yt_1t_2 = c(t_1 + t_2) \quad \ldots(i)$$

and the equation of chord joining Q and R is

$$x + yt_2t_3 = c(t_2 + t_3) \quad \ldots(ii)$$

Let Eq. (i) be parallel to $y = m_1x$ and Eq. (ii) be parallel to $y = m_2x$.

$$\therefore \quad m_1 = -\frac{1}{t_1t_2}$$

and

$$m_2 = -\frac{1}{t_2t_3}$$

$$\therefore \quad \frac{m_1}{m_2} = \frac{t_3}{t_1} \text{ i.e. } t_3 = \left(\frac{m_1}{m_2}\right) t_1 \quad \ldots(iii)$$

Again the equation to the third side RP is

$$x + yt_3t_1 = c(t_3 + t_1)$$

$$x + y\left(\frac{m_1}{m_2}\right) t_1^2 = c\left(\frac{m_1}{m_2} t_1 + t_1\right)$$

or $\quad ym_1t_1^2 - ct_1(m_1 + m_2) + xm_2 = 0 \quad \ldots(iv)$

t_1 being a parameter. Since, t_1 is real the envelope of Eq. (iv) is given by the discriminant of Eq. (iv) $= 0$

i.e. $\quad c^2(m_1 + m_2)^2 - 4ym_1 \cdot xm_2 = 0$

or $\quad 4m_1m_2xy = c^2(m_1 + m_2)^2 \quad \ldots(v)$

$\because \quad m_1$ and m_2 are roots of $x^2 - 6x + 1 = 0$

$\therefore \quad m_1 + m_2 = 6, m_1 m_2 = 1$

then from Eq. (v),

$$4xy = c^2 (6)^2$$

then $\quad xy = 9c^2$

$\therefore \quad \lambda = 9$

$\Rightarrow \quad 16\lambda^2 = 16 (9)^2 = 16 \times 81 = 1296$

3. Equation of the normal to the hyperbola $xy = 1$ at t is

$$xt^3 - yt - t^4 + 1 = 0 \quad \dots(\text{i})$$

$\because$ It passes through (α, β)

$\therefore \quad t^4 - \alpha t^3 + \beta t - 1 = 0$

If foot of the conormal points are

$$\left(t_1, \frac{1}{t_1}\right), \left(t_2, \frac{1}{t_2}\right), \left(t_3, \frac{1}{t_3}\right) \text{ and } \left(t_4, \frac{1}{t_4}\right)$$

$\therefore \quad t_1 + t_2 + t_3 + t_4 = \alpha$

$\Rightarrow \quad x_1 + x_2 + x_3 + x_4 = \alpha$

and $\quad \frac{1}{t_1} + \frac{1}{t_2} + \frac{1}{t_3} + \frac{1}{t_4} = \beta$

$\Rightarrow \quad y_1 + y_2 + y_3 + y_4 = \beta$

Let the variable line be

$$px + qy + r = 0$$

Since, algebraic sum of perpendicular distances from $(x_1, y_1), (x_2, y_2), (x_3, y_3)$ and (x_4, y_4) to Eq. (ii) is zero.

$$\Rightarrow p\left(\frac{x_1 + x_2 + x_3 + x_4}{4}\right) + q\left(\frac{y_1 + y_2 + y_3 + y_4}{4}\right) + r = 0$$

$$\Rightarrow \quad p \cdot \frac{\alpha}{4} + q \cdot \frac{\beta}{4} + r = 0$$

$\therefore$ Eq. (ii) passes through $\left(\frac{\alpha}{4}, \frac{\beta}{4}\right)$.

$\Rightarrow \quad \lambda = 4$

$\therefore \quad 256\lambda = 256 \times 4 = 1024$

4. Let the equation of circle

$$x^2 + y^2 + 2gx + 2fy + k = 0 \quad \dots(\text{i})$$

and the equation of the rectangular hyperbola is

$$xy = c^2 \quad \dots(\text{ii})$$

Put $x = ct$ and $y = \frac{c}{t}$ in Eq. (i)

then $c^2t^2 + \frac{c^2}{t^2} + 2gct + \frac{2fc}{t} + k = 0$

$\Rightarrow c^2t^4 + 2gct^3 + kt^2 + 2fct + c^2 = 0$

this equation being fourth degree in t. Let the roots be t_1, t_2, t_3, t_4 then

$$t_1t_2t_3t_4 = 1$$

$\therefore \quad 16t_1t_2t_3t_4 = 16 \times 1 = 16$

5. Let chord be PQ coordinates of P and Q are $\left(ct_1, \frac{c}{t_1}\right)$ and $\left(ct_2, \frac{c}{t_2}\right)$ respectively.

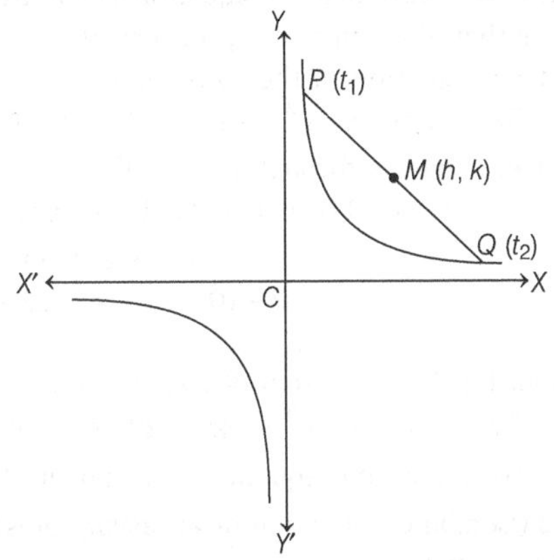

Now, let mid point of PQ is $M(h, k)$

$\therefore \quad h = \frac{1}{2}(ct_1 + ct_2)$

$\therefore \quad t_1 + t_2 = \frac{2h}{c}$

and $\quad k = \frac{1}{2}\left\{\frac{c}{t_1} + \frac{c}{t_2}\right\}$

or $\quad \frac{2k}{c} = \frac{t_1 + t_2}{t_1t_2}$

or $\quad \frac{2k}{c} = \frac{\frac{2h}{c}}{t_1t_2} \quad \left[\because t_1 + t_2 = \frac{2h}{c}\right]$

or $\quad t_1t_2 = \frac{h}{k}$

$\therefore \quad (t_1 - t_2)^2 = (t_1 + t_2)^2 - 4t_1t_2 = \frac{4h^2}{c^2} - \frac{4h}{k} = \frac{4h}{c^2k}(hk - c^2) \quad \dots(\text{i})$

Since, $|PQ| = 2d$

$\therefore \quad \sqrt{(ct_1 - ct_2)^2 + \left(\frac{c}{t_1} - \frac{c}{t_2}\right)^2} = 2d$

or $\quad c^2(t_1 - t_2)^2\left\{1 + \frac{1}{t_1^2t_2^2}\right\} = (2d)^2$

or $\quad C^2 \times \frac{4h}{c^2x}(hk - c^2)\left\{1 + \frac{k^2}{h^2}\right\} = 4d^2 \quad$ [from Eq. (i)]

or $\quad (h^2 + k^2)(hk - c^2) = d^2hk$

Hence, locus of the middle point of PQ is

$$(x^2 + y^2)(xy - c^2) = d^2xy$$

$\therefore \quad \lambda = 2$

$$1000\lambda = 1000 \times 2 = 2000$$

6. Combined equation of asymptotes is

$$(x + 2y + 3)(3x + 4y + 5) = 0$$

or $3x^2 + 10xy + 8y^2 + 14x + 22y + 15 = 0$ ...(i)

Also, we know that the equation of the hyperbola differs from that of asymptotes by a constant.

Let the equation of the hyperbola be

$$3x^2 + 10xy + 8y^2 + 14x + 22y + \lambda = 0 \quad \text{...(ii)}$$

Since, it passes through (1, –1), then

$$3(1)^2 + 10(1)(-1) + 8(-1)^2 + 14(1) + 22(-1) + \lambda = 0$$

$$\Rightarrow \quad 3 - 10 + 8 + 14 - 22 + \lambda = 0$$

$$\therefore \quad \lambda = 7$$

From Eq. (ii), equation of hyperbola is

$$3x^2 + 10xy + 8y^2 + 14x + 22y + 7 = 0 \quad \text{...(iii)}$$

But we know that equation of conjugate hyperbola = 2 (Combined equation of an asymptotes) – (equation of hyperbola)

$$\Rightarrow \quad 6x^2 + 20xy + 16y^2 + 28x + 44y + 30 - 3x^2 - 10xy - 8y^2 - 14x - 22y - 7 = 0$$

or $\quad 3x^2 + 10xy + 8y^2 + 14x + 22y + 23 = 0$

$$\therefore \quad \mu = 14 \text{ and } \nu = 23$$

$$\therefore \quad \frac{\mu}{\nu} = \frac{14}{23} = 0.61$$

7. Equation of the chord of contact of a point $P(3\sec\theta, 2\tan\theta)$ on the hyperbola with respect to the circle is

$(3\sec\theta)x + (2\tan\theta)y = 9$ ...(i)

Let $M(h, k)$ be the mid-point of Eq. (i), then equation of Eq. (i) in terms of the mid-point is

$hx + ky = h^2 + k^2$

Since, Eqs. (i) and (ii) represent the same line.

$$\sec\theta = \frac{3h}{h^2 + k^2}, \tan\theta = \frac{9k}{2(h^2 + k^2)}$$

$$\Rightarrow \text{locus of } (h, k) \text{ is } \frac{9x^2}{(x^2 + y^2)^2} - \frac{81y^2}{4(x^2 + y^2)^2} = 1$$

or $\quad 4(x^2 + y^2)^2 = 36x^2 - 81y^2$

which is same as $a(x^2 + y^2)^2 = bx^2 - cy^2$

$$\Rightarrow \quad a = 4, b = 36, c = 81$$

$$\therefore \quad \frac{c}{a + b} = \frac{81}{40} = 2.025$$

Single Integer Type

1. We have, Equation of hyperbola is

$$\frac{x^2}{144} - \frac{y^2}{81} = \frac{1}{25}$$

$$\Rightarrow \quad \frac{x^2}{\left(\frac{144}{25}\right)} - \frac{y^2}{\left(\frac{81}{25}\right)} = 1$$

$$\Rightarrow \quad \frac{x^2}{\left(\frac{12}{5}\right)^2} - \frac{y^2}{\left(\frac{9}{5}\right)^2} = 1$$

$\therefore$ Foci $= (\pm ae, 0)$

$$= \left(\pm \frac{12}{5}\sqrt{1 + \left(\frac{9/5}{12/5}\right)^2}, 0\right)$$

$$= \left(\pm \frac{12}{5} \times 3, 0\right) = (\pm 3, 0)$$

Equation of ellipse is,

$$\frac{x^2}{16} + \frac{y^2}{b^2} = 1$$

$$\Rightarrow \quad \frac{x^2}{4^2} + \frac{y^2}{b^2} = 1$$

$\therefore$ Foci $= (\pm ae, 0)$

$$= \left(\pm 4\sqrt{1 - \frac{b^2}{4^2}}, 0\right) = \left(\pm \frac{4\sqrt{16 - b^2}}{4}, 0\right)$$

$$= (\pm \sqrt{16 - b^2}, 0)$$

Since, both have same foci

$$\therefore \quad \sqrt{16 - b^2} = 3$$

$$\Rightarrow \quad 16 - b^2 = 9 \Rightarrow b^2 = 7$$

2. Equations of director circles of ellipse and hyperbola are, respectively,

$$x^2 + y^2 = a^2 + b^2$$

and $\quad x^2 + y^2 = a^2 - b^2$

According to the question, we have

$$a^2 + b^2 = 4r^2 \quad \text{... (i)}$$

$$a^2 - b^2 = r^2 \quad \text{... (ii)}$$

Solving Eqs. (i) and (ii), we get

$$a^2 = \frac{5r^2}{2},$$

$$b^2 = \frac{3r^2}{2}$$

$$e_1^2 = 1 - \frac{b^2}{a^2}$$

$$= 1 - \frac{3r^2}{2} \times \frac{2}{5r^2} = 1 - \frac{3}{5} = \frac{2}{5}$$

$$e_2^2 = 1 + \frac{b^2}{a^2}$$

$$= 1 + \frac{3}{5} = \frac{8}{5}$$

So, $\quad 4e_2^2 - e_1^2 = 4 \times \frac{8}{5} - \frac{2}{5} = 6$

3. Let (h, k) be the mid-point of the chord of the circle,

$$x^2 + y^2 = 4$$

so, by mid-point form, equation is

$$T = S_1$$

$$hx + ky = h^2 + k^2$$

or $$y = -\frac{h}{k}x + \frac{h^2 + k^2}{k}$$

$\Rightarrow$ $y = mx + c$

it will touch the hyperbola, if

$$c^2 = a^2m^2 - b^2$$

$$\Rightarrow \left(\frac{h^2 + k^2}{k}\right)^2 = 4\left(\frac{-h}{k}\right)^2 - 16$$

$$\Rightarrow (x^2 + y^2)^2 = 4x^2 - 16y^2$$

$$\therefore \lambda = 4$$

4. Given, Equation of the hyperbola is

$$\frac{x^2}{1/2} - \frac{y^2}{1/2} = 1$$

So, $e_1^2 = \frac{a^2 + b^2}{a^2}$ where, $a^2 = b^2 = \frac{1}{2}$

$\Rightarrow e_1 = \sqrt{2}$

(**Note** It is a rectangular hyperbola)

Equation of the ellipse is,

$$\frac{x^2}{2} + \frac{y^2}{1} = 1$$

So, $e_2^2 = \frac{a^2 - b^2}{a^2} = \frac{2-1}{2} = \frac{1}{2}$

$\Rightarrow$ $e_2 = 1/\sqrt{2}$ hence, $e_1e_2 = 1$

5. The given hyperbola is,

$$3x^2 - 2y^2 = 6$$

$$\Rightarrow \frac{x^2}{2} - \frac{y^2}{3} = 1$$

Now, Equation of tangent is

$$y = mx \pm \sqrt{a^2m^2 - b^2}$$

or $(y - mx)^2 = a^2m^2 - b^2$

Tangents from the point (α, β) will be,

$$(\beta - m\alpha)^2 = 2m^2 - 3 \quad (\text{Since, } a^2 = 2 \text{ and } b^2 = 3)$$

$$\Rightarrow m^2\alpha^2 + \beta^2 - 2m\alpha\beta - 2m^2 + 3 = 0$$

$$m^2(\alpha^2 - 2) - 2\alpha\beta m + \beta^2 + 3 = 0$$

$$m_1 \cdot m_2 = \frac{\beta^2 + 3}{\alpha^2 - 3} = 2$$

$$= \tan\theta \tan\phi$$

$$\therefore \beta^2 + 3 = 2(\alpha^2 - 2)$$

$$\Rightarrow 2\alpha^2 - \beta^2 = 7$$

6. We know that,

If two lines $l_1x + m_1y + n_1 = 0$ and $l_2x + m_2y + n_2 = 0$ are conjugate lines with respect to the hyperbola,

$$\frac{x^2}{a^2} - \frac{y^2}{b^2} = 1, \text{ then } a^2l_1l_2 - b^2m_1m_2 = n_1n_2$$

For hyperbola,

$$5x^2 - 6y^2 = 15 \Rightarrow \frac{x^2}{3} - \frac{y^2}{5/2} = 1$$

Here, $a^2 = 3, b^2 = \frac{5}{2}, l_1 = 2, l_2 = 3,$

$$m_1 = -k, m_2 = -1$$

and $$n_1 = 3, n_2 = 1$$

$$\therefore 3 \times 2 \times 3 - \frac{5}{2}(-k)(-1) = 3 \times 1$$

$$\Rightarrow 18 - \frac{5}{2}k = 3$$

$$\Rightarrow 15 = \frac{5}{2}k \Rightarrow k = 6$$

7. Given, two hyperbolas are

$$\frac{x^2}{9} - \frac{y^2}{16} = 1 \quad \dots \text{(i)}$$

and $$\frac{y^2}{9} - \frac{x^2}{16} = 1 \quad \dots \text{(ii)}$$

Equation of tangent to Eq. (i), having slope m is

$$y = mx \pm \sqrt{9m^2 - 16}$$

Eliminating y using Eqs. (ii) and (iii), we get

$$16(mx \pm \sqrt{9m^2 - 16})^2 - 9x^2 = 144$$

$$\Rightarrow (16m^2 - 9)x^2 \pm 32m\sqrt{9m^2 - 16x}$$
$$+ (144m^2 - 400) = 0 \quad \dots \text{(iv)}$$

For it to be a tangent, we must have $D = 0$

$$\therefore (32m\sqrt{9m^2 - 16})^2 = 4(16m^2 - 9)(144m^2 - 400)$$

$$\Rightarrow m^2 = 1$$

$$\Rightarrow m = \pm 1$$

8. Given hyperbola,

$$x^2 - y^2 - 4x + 2y + c = 0$$

$$(x - 2)^2 - (y - 1)^2 = 3 - c$$

$$\frac{(x - 2)^2}{3 - c} - \frac{(y - 1)^2}{3 - c} = 1$$

Focus of hyperbola is,

$$x - 2 = \pm \sqrt{(3 - c) + (3 - c)}$$

$$x - 2 = \pm \sqrt{6 - 2c} \quad [\because ae = \sqrt{(3 - c) + (3 - c)}]$$

$$x = 2 \pm \sqrt{6 - 2c}$$

Here, $2 + \sqrt{6 - 2c} = 2 + 2\sqrt{2}$

$$\therefore \sqrt{6 - 2c} = 2\sqrt{2}$$

$$\Rightarrow 6 - 2c = 8$$

$$\Rightarrow c = -1$$

$$\Rightarrow c^2 = 1$$

Matrix-Match Type

1. **(A)** Let $f(x, y) = x^2 - 4y^2 - 2x + 24y - 37$

$\therefore \quad f(3, 4) = 9 - 64 - 6 + 96 - 37 = -2$

$\therefore \quad f(3, 4) < 0$

$\Rightarrow$ Point (3, 4) lie outside the hyperbola (R)

and $\quad f(5, 2) = 25 - 16 - 10 + 48 - 37 = 10$

$\therefore \quad f(5, 2) > 0$

$\Rightarrow$ Point (5, 2) lie inside the hyperbola (S)

(B) Let $f(x, y) = xy + 2x + 3y - 12$

$\therefore \quad f(3, 4) = 12 + 6 + 12 - 12 = 18$

$\therefore \quad f(3, 4) > 0$

$\Rightarrow$ Point (3, 4) lie inside the hyperbola (P)

and $f(5, 2) = 10 + 10 + 6 - 12 = 14 > 0$

$\Rightarrow$ Point (5, 2) lie inside the hyperbola (S)

(C) Let $f(x, y) = xy - 12$

$\therefore \quad f(3, 4) = 12 - 12 = 0$

$\Rightarrow$ Point (3, 4) lie on the hyperbola (T)

and $\quad f(5, 2) = 5 \times 2 - 12 = -2$

$\therefore \quad f(5, 2) < 0$

$\Rightarrow$ Point (5, 2) lie outside the hyperbola (Q)

2. **(A)** $\because x^2 - 2y^2 = 2$

$$\Rightarrow \quad \frac{x^2}{2} - \frac{y^2}{1} = 1$$

$\therefore$ Director circle is $x^2 + y^2 = 2 - 1 = 1$

i.e., $\quad x^2 + y^2 = 1$ (P)

and $\quad x^2 + 2y^2 = 2$

$$\Rightarrow \quad \frac{x^2}{2} + \frac{y^2}{1} = 1$$

Director circle is $x^2 + y^2 = 2 + 1 = 3$

i.e., $\quad x^2 + y^2 = 3$ (R)

(B) $\because \quad 3x^2 + 2y^2 = 6$

$$\Rightarrow \quad \frac{x^2}{2} + \frac{y^2}{3} = 1$$

$\therefore$ Director circle is $x^2 + y^2 = 2 + 3 = 5$

i.e., $\quad x^2 + y^2 = 5$ (T)

and $\quad 3x^2 - 2y^2 = 6$

$$\Rightarrow \quad \frac{x^2}{2} - \frac{y^2}{3} = 1$$

$\therefore$ Director circle is $x^2 + y^2 = 2 - 3 = -1$ (not defined)

(C) $\because \quad 5x^2 - 9y^2 = 45$

$$\Rightarrow \quad \frac{x^2}{9} - \frac{y^2}{5} = 1$$

$\therefore$ Director circle is $x^2 + y^2 = 9 - 5 = 4$

i.e. $\quad x^2 + y^2 = 4$ (S)

and director circle of $x^2 + y^2 = 1$

is $\quad x^2 + y^2 = 2$ (Q)

3. **(A)** $\quad 12x^2 - 4y^2 - 24x + 32y - 127 = 0$

$\Rightarrow \quad 12(x^2 - 2x) - 4(y^2 - 8y) - 127 = 0$

$\Rightarrow \quad 12\{(x-1)^2 - 1\} - 4\{(y-4)^2 - 16\} = 127$

$\Rightarrow \quad 12(x-1)^2 - 4(y-4)^2 = 75$

$$\Rightarrow \quad \frac{12(x-1)^2}{75} - \frac{4(y-4)^2}{75} = 1$$

$$\Rightarrow \quad \frac{75}{4} = \frac{75}{12}(e^2 - 1)$$

$\Rightarrow \quad 3 = e^2 - 1$

$\Rightarrow \quad e^2 = 4$

$\therefore \quad e = 2$

For foci $x - 1 = \pm\left(\frac{5}{2} \times 2\right)$ and $y - 4 = 0$

$\Rightarrow \quad x = 1 \pm 5$ and $y = 4$

Foci are (–4, 4) and (6, 4) (R, T)

(B) $\quad 8x^2 - y^2 - 64x + 10y + 71 = 0$

$\Rightarrow \quad 8(x^2 - 8x) - (y^2 - 10y) + 71 = 0$

$\Rightarrow \quad 8\{(x-4)^2 - 16\} - \{(y-5)^2 - 25\} + 71 = 0$

$\Rightarrow \quad 8(x-4)^2 - (y-5)^2 = 32$

$$\Rightarrow \quad \frac{(x-4)^2}{4} - \frac{(y-5)^2}{32} = 1$$

$\Rightarrow \quad 32 = 4(e^2 - 1)$

$\Rightarrow \quad 8 = e^2 - 1$

$\therefore \quad e = 3$

For foci $\quad x - 4 = \pm(2 \times 3)$

and $\quad y - 5 = 0$

$x = 4 \pm 6$ and $y = 5$

Foci are (10, 5) and (–2, 5) (P, S)

(C) $\quad 9x^2 - 16y^2 - 36x + 96y + 36 = 0$

$\Rightarrow \quad 9(x^2 - 4x) - 16(y^2 - 6y) + 36 = 0$

$\Rightarrow \quad 9\{(x-2)^2 - 4\} - 16\{(y-3)^2 - 9\} + 36 = 0$

$\Rightarrow \quad 9(x-2)^2 - 16(y-3)^2 = -144$

$$\Rightarrow \quad -\frac{(x-2)^2}{16} + \frac{(y-3)^2}{9} = 1$$

$\Rightarrow \quad 16 = 9(e^2 - 1)$

$\Rightarrow \quad 25 = 9e^2$

$$\therefore \quad e = \frac{5}{3}$$

For foci $\quad x - 2 = 0$

and $\quad y - 3 = \pm\left(3 \times \frac{5}{3}\right)$

$\Rightarrow \quad x = 2$ and $y = 3 \pm 5$

$\therefore$ Foci are (2, – 2) and (2, 8) (Q)

27

Trigonometrical Ratios and Identities

Objective Questions Type I [Only one correct answer]

In each of the questions below, four choices are given of which only are is correct. You have to select the correct answer which is the most appropriate.

1. If $k_1 = \tan 27\theta - \tan\theta$ and

$k_2 = \dfrac{\sin\theta}{\cos 3\theta} + \dfrac{\sin 3\theta}{\cos 9\theta} + \dfrac{\sin 9\theta}{\cos 27\theta}$, then

(a) $k_1 = 2k_2$ (b) $k_1 = k_2 + 4$

(c) $k_1 = k_2$ (d) None of these

2. If $15\sin^4\alpha + 10\cos^4\alpha = 6$, for some $\alpha \in R$, then the value of $27\sec^6\alpha + 8\operatorname{cosec}^6\alpha$ is equal to

(a) 350 (b) 500

(c) 400 (d) 250

3. If $0 < x < \dfrac{\pi}{2}$ and $\sin^n x + \cos^n x \ge 1$, then

(a) $n \in [2, \infty)$ (b) $n \in (-\infty, 2]$

(c) $n \in [-1, 1]$ (d) None of these

4. If $\dfrac{\sec^8\theta}{a} + \dfrac{\tan^8\theta}{b} = \dfrac{1}{a+b}$, then for every real value of $\sin^2\theta$

(a) $ab \le 0$ (b) $ab \ge 0$

(c) $a + b = 0$ (d) None of these

5. In an acute angled ΔABC the least value of $\sec A + \sec B + \sec C$ is

(a) 6 (b) 8

(c) 3 (d) None of these

6. If $P_n = \cos^n\theta + \sin^n\theta$, then $P_n - P_{n-2} = kP_{n-4}$, where

(a) $k = 1$ (b) $k = -\sin^2\theta\cos^2\theta$

(c) $k = \sin^2\theta$ (d) $k = \cos^2\theta$

7. If $L = \sin^2\left(\dfrac{\pi}{16}\right) - \sin^2\left(\dfrac{\pi}{8}\right)$ and $M = \cos^2\left(\dfrac{\pi}{16}\right) - \sin^2\left(\dfrac{\pi}{8}\right)$, then

[JEE Main 2020, 5 September]

(a) $L = -\dfrac{1}{2\sqrt{2}} + \dfrac{1}{2}\cos\dfrac{\pi}{8}$ (b) $L = \dfrac{1}{4\sqrt{2}} - \dfrac{1}{4}\cos\dfrac{\pi}{8}$

(c) $M = \dfrac{1}{4\sqrt{2}} + \dfrac{1}{4}\cos\dfrac{\pi}{8}$ (d) $M = \dfrac{1}{2\sqrt{2}} + \dfrac{1}{2}\cos\dfrac{\pi}{8}$

8. If $U_n = \sin n\theta \sec^n\theta$, $V_n = \cos n\theta \sec^n\theta \ne 1$, then $\dfrac{V_n - V_{n-1}}{U_{n-1}} + \dfrac{1}{n}\dfrac{U_n}{V_n}$ is equal to

(a) 0 (b) $\tan\theta$

(c) $-\tan\theta + \dfrac{\tan n\theta}{n}$ (d) $\tan\theta + \dfrac{\tan n\theta}{n}$

9. If $\sin\left(x + \dfrac{4\pi}{9}\right) = a$; $\dfrac{\pi}{9} < x < \dfrac{\pi}{3}$, then $\cos\left(x + \dfrac{7\pi}{9}\right)$ equals

(a) $\dfrac{\sqrt{(1-a^2)} - a\sqrt{3}}{2}$ (b) $\dfrac{1 - a^2 + a\sqrt{3}}{2}$

(c) $\dfrac{a\sqrt{3} - \sqrt{(1-a^2)}}{2}$ (d) $\dfrac{-\sqrt{(1-a^2)} - a\sqrt{3}}{2}$

10. If $a = \sin\dfrac{\pi}{18}\sin\dfrac{5\pi}{18}\sin\dfrac{7\pi}{18}$, and x is the solution of the equation $y = 2[x] + 2$ and $y = 3[x-2]$, where $[x]$ denotes the integral part of x, then a is equal to

(a) $[x]$ (b) $\dfrac{1}{[x]}$

(c) $2[x]$ (d) $[x]^2$

11. If $A = \cos(\cos x) + \sin(\cos x)$ the least and greatest value of A are

(a) 0 and 2 (b) -1 and 1

(c) $-\sqrt{2}$ and $\sqrt{2}$ (d) 0 and $\sqrt{2}$

12. If in a ΔABC, $\tan A + \tan B + \tan C > 0$, then
(a) Δ is always obtuse angled triangle
(b) Δ is always equilateral triangle
(c) Δ is always acute angled triangle
(d) nothing can be said about the type of triangle

13. Let n be a fixed positive integer such that
$\sin \frac{\pi}{2n} + \cos \frac{\pi}{2n} = \frac{\sqrt{n}}{2}$, then
(a) $n = 4$ (b) $n = 5$
(c) $n = 6$ (d) None of these

14. In a quadrilateral if
$\sin\left(\frac{A+B}{2}\right)\cos\left(\frac{A-B}{2}\right) + \sin\left(\frac{C+D}{2}\right)\cos\left(\frac{C-D}{2}\right) = 2$,
then $\Sigma \cos\frac{A}{2}\cos\frac{B}{2}$ is equal to
(a) 0 (b) 6
(c) 3 (d) 2

15. If the mapping $f(x) = ax + b$, $a < 0$ maps $[-1, 1]$ onto $[0, 2]$, then for all values of θ, $A = \cos^2\theta + \sin^4\theta$ is such that
(a) $f\left(\frac{1}{4}\right) \le A \le f(0)$ (b) $f(0) \le A \le f(-2)$
(c) $f\left(\frac{1}{3}\right) \le A \le f(0)$ (d) $f(-1) < A \le f(-2)$

16. If $\tan\alpha$, $\tan\beta$, $\tan\gamma$ are the roots of the equation $x^3 - px^2 - r = 0$, then the value of $(1 + \tan^2\alpha)(1 + \tan^2\beta)(1 + \tan^2\gamma)$ is equal to
(a) $(p - r)^2$ (b) $1 + (p - r)^2$
(c) $1 - (p - r)^2$ (d) None of these

17. If $\alpha, \beta, \gamma, \delta$ are four solutions of the equation
$\tan\left(\theta + \frac{\pi}{4}\right) = 3\tan 3\theta$, then $\tan\alpha\tan\beta\tan\gamma\tan\delta$ equals
(a) 3 (b) 1/3
(c) $-1/3$ (d) None of these

18. If $\cos A = \frac{-60}{61}$ and $\tan B = -\frac{7}{24}$ and neither A nor B is the second quadrant, then the angle $A + \frac{B}{2}$ lies in the quadrant
(a) I (b) II
(c) III (d) IV

19. If $x \sin a + y \sin 2a + z \sin 3a = \sin 4a$
$x \sin b + y \sin 2b + z \sin 3b = \sin 4b$,
$x \sin c + y \sin 2c + z \sin 3c = \sin 4c$.
Then the roots of the equation
$t^3 - \left(\frac{z}{2}\right)t^2 - \left(\frac{y+2}{4}\right)t + \left(\frac{z-x}{8}\right) = 0$, $a, b, c \ne n\pi$, are
(a) $\sin a, \sin b, \sin c$
(b) $\cos a, \cos b, \cos c$
(c) $\sin 2a, \sin 2b, \sin 2c$
(d) $\cos 2a, \cos 2b, \cos 2c$

20. If $x = r\sin\theta\cos\phi$, $y = r\sin\theta\sin\phi$ and $z = r\cos\theta$, then the value of $x^2 + y^2 + z^2$ is independent of
(a) θ, ϕ (b) r, θ
(c) r, ϕ (d) r^2

21. If $0° < \theta < 180°$, then $\sqrt{2+\sqrt{2+\sqrt{2+\ldots+\sqrt{2(1+\cos\theta)}}}}$, then being n number of 2's, is equal to
(a) $2\cos\left(\frac{\theta}{2^n}\right)$ (b) $2\cos\left(\frac{\theta}{2^{n-1}}\right)$
(c) $2\cos\left(\frac{\theta}{2^{n+1}}\right)$ (d) None of these

22. If $\tan\alpha/2$ and $\tan\beta/2$ are the roots of the equation $8x^2 - 26x + 15 = 0$, then $\cos(\alpha+\beta)$ is equal to
(a) $-\frac{627}{725}$ (b) $\frac{627}{725}$
(c) -1 (d) None of these

23. If $a\sec\alpha - c\tan\alpha = d$ and $b\sec\alpha + d\tan\alpha = c$, then
(a) $a^2 + c^2 = b^2 + d^2$ (b) $a^2 + d^2 = b^2 + c^2$
(c) $a^2 + b^2 = c^2 + d^2$ (d) $ab = cd$

24. Let n be an odd integer. If $\sin n\theta = \sum_{r=0}^{n} b_r \sin^r\theta$ for all real θ, then
(a) $b_0 = 1, b_1 = 3$ (b) $b_0 = 0, b_1 = n$
(c) $b_0 = -1, b_1 = n$ (d) $b_0 = 0, b_1 = n^2 - 3n - 3$

25. If θ is an acute angle and $\tan\theta = \frac{1}{\sqrt{7}}$, then the value of $\frac{\operatorname{cosec}^2\theta - \sec^2\theta}{\operatorname{cosec}^2\theta + \sec^2\theta}$ is
(a) 3/4 (b) 1/2
(c) 2 (d) 5/4

26. $\tan 7\frac{1}{2}^\circ$ is equal to
(a) $\frac{2\sqrt{2} - (1+\sqrt{3})}{\sqrt{3}-1}$ (b) $\frac{1+\sqrt{3}}{1-\sqrt{3}}$
(c) $\frac{1}{\sqrt{3}} + \sqrt{3}$ (d) $2\sqrt{2} + \sqrt{3}$

27. The maximum value of $\sin(x + \pi/6) + \cos(x + \pi/6)$ in the interval $(0, \pi/2)$ is attained at
(a) $\pi/12$ (b) $\pi/6$
(c) $\pi/3$ (d) $\pi/2$

28. If x_1 and x_2 are two distinct roots of the equation $a\cos x + b\sin x = c$, then $\tan\frac{x_1 + x_2}{2}$ is equal to
(a) $\frac{a}{b}$ (b) $\frac{b}{a}$
(c) $\frac{c}{a}$ (d) $\frac{a}{c}$

29. If $\sin\alpha = \sin\beta$ and $\cos\alpha = \cos\beta$, then
(a) $\sin\left(\frac{\alpha+\beta}{2}\right) = 0$ (b) $\cos\left(\frac{\alpha+\beta}{2}\right) = 0$
(c) $\sin\left(\frac{\alpha-\beta}{2}\right) = 0$ (d) $\cos\left(\frac{\alpha-\beta}{2}\right) = 0$

30. $\left(1+\cos\frac{\pi}{8}\right)\left(1+\cos\frac{3\pi}{8}\right)\left(1+\cos\frac{5\pi}{8}\right)\left(1+\cos\frac{7\pi}{8}\right)$ is equal to
(a) 1/2 (b) $\cos \pi/8z$
(c) 1/8 (d) $\frac{1+\sqrt{2}}{2\sqrt{2}}$

31. If $\cos(\theta-\alpha)=a$ and $\sin(\theta-\beta)=b(0<\theta-\alpha, \theta-\beta<\pi/2)$, then $\cos^2(\alpha-\beta)+2ab\sin(\alpha-\beta)$ is equal to
(a) $4a^2b^2$ (b) a^2-b^2
(c) a^2+b^2 (d) $-a^2b^2$

32. If A lies in the third quadrant and $3\tan A-4=0$, then $5\sin 2A+3\sin A+4\cos A$ is equal to
(a) 0 (b) $-\frac{24}{5}$
(c) $\frac{24}{5}$ (d) $\frac{48}{5}$

33. If $x=\sum_{n=0}^{\infty}(-1)^n\tan^{2n}\theta$ and $y=\sum_{n=0}^{\infty}\cos^{2n}\theta$, for $0<\theta<\frac{\pi}{4}$, then **[JEE Main 2020, 9 January]**
(a) $y(1+x)=1$ (b) $y(1-x)=1$
(c) $x(1+y)=1$ (d) $x(1-y)=1$

34. If $\sin x+\sin^2 x=1$, then $\cos^8 x+2\cos^6 x+\cos^4 x$ is equal to
(a) -1 (b) 0
(c) 1 (d) 2

35. If $x=y\cos\frac{2\pi}{3}=z\cos\frac{4\pi}{3}$, then $xy+yz+zx$ is equal to
(a) -1 (b) 0
(c) 1 (d) 2

36. If $a\sin^2 x+b\cos^2 x=c$, $b\sin^2 y+a\cos^2 y=d$ and $a\tan x=b\tan y$, then $\frac{a^2}{b^2}$ is equal to
(a) $\frac{(b-c)(d-b)}{(a-d)(d-b)}$ (b) $\frac{(a-d)(c-a)}{(b-c)(d-b)}$
(c) $\frac{(d-a)(c-a)}{(b-c)(d-b)}$ (d) $\frac{(b-c)(b-d)}{(a-c)(a-d)}$

37. If $0<\alpha<\pi/6$ and $\sin\alpha+\cos\alpha=\sqrt{7}/2$, then $\tan\alpha/2$ is equal to
(a) $\frac{\sqrt{7}-2}{3}$ (b) $\frac{\sqrt{7}+2}{3}$
(c) $\frac{\sqrt{7}}{3}$ (d) None of these

38. The value of $\cos\frac{2\pi}{7}+\cos\frac{4\pi}{7}+\cos\frac{6\pi}{7}$ is equal to
(a) 1 (b) -1
(c) 1/2 (d) $-1/2$

39. If $\pi<\alpha<\frac{3\pi}{2}$, then the expression $\sqrt{(4\sin^4\alpha+\sin^2 2\alpha)}+4\cos^2\left(\frac{\pi}{4}-\frac{\alpha}{2}\right)$ is equal to
(a) $2+4\sin\alpha$ (b) $2-4\sin\alpha$
(c) 2 (d) None of these

40. The value of $\cos\frac{\pi}{15}\cos\frac{2\pi}{15}\cos\frac{3\pi}{15}\cos\frac{4\pi}{15}\cos\frac{5\pi}{15}\cos\frac{6\pi}{15}\cos\frac{7\pi}{15}$ is equal to
(a) $1/2^6$ (b) $1/2^7$
(c) $1/2^8$ (d) None of these

41. If $\alpha, \beta, \gamma, \delta$ are the smallest positive angles in ascending order of magnitude which have their sines equal to the positive quantity k, then the value of $4\sin\frac{1}{2}\alpha+3\sin\frac{1}{2}\beta+2\sin\frac{1}{2}\gamma+\sin\frac{1}{2}\delta$ is equal to
(a) $2\sqrt{(1-k)}$ (b) $\frac{1}{2}\sqrt{(1+k)}$
(c) $2\sqrt{(1+k)}$ (d) None of these

42. If $\frac{x}{\cos\alpha}=\frac{y}{\cos\left(\alpha-\frac{2\pi}{3}\right)}=\frac{z}{\cos\left(\alpha+\frac{2\pi}{3}\right)}$, then $x+y+z$ is equal to
(a) 1 (b) 0
(c) -1 (d) None of these

43. If $A+B+C=\frac{3\pi}{2}$, then $\cos 2A+\cos 2B+\cos 2C$ is equal to
(a) $1-4\cos A\cos B\cos C$ (b) $4\sin A\sin B\sin C$
(c) $1+2\cos A\cos B\cos C$ (d) $1-4\sin A\sin B\sin C$

44. If $\sum_{i=1}^{n}\cos\theta_i=n$, then $\sum_{i=1}^{n}\sin\theta_i$ is equal to
(a) $n-1$ (b) 0
(c) n (d) $n+1$

45. If $\cos\alpha+\cos\beta=\sin\alpha+\sin\beta$, then $\cos 2\alpha+\cos 2\beta$ is equal to
(a) $-2\sin(\alpha+\beta)$ (b) $-2\cos(\alpha+\beta)$
(c) $2\sin(\alpha+\beta)$ (d) $2\cos(\alpha+\beta)$

46. If $x_i>0$ for $1\le i\le n$ and $x_1+x_2+\ldots+x_n=\pi$ then the greatest value of the sum $\sin x_1+\sin x_2+\sin x_3+\ldots+\sin x_n$ is equal to
(a) n (b) π
(c) $n\sin\left(\frac{\pi}{n}\right)$ (d) 0

47. If $\cos(\alpha+\beta)=\frac{3}{5}$, $\sin(\alpha-\beta)=\frac{5}{13}$ and $0<\alpha,\beta<\frac{\pi}{4}$, then $\tan(2\alpha)$ is equal to **[JEE Main 2019, 8 April]**
(a) $\frac{63}{52}$ (b) $\frac{63}{16}$
(c) $\frac{21}{16}$ (d) $\frac{33}{52}$

48. If $\sin\alpha=-3/5$ and lies in the third quadrant, then the value of $\cos\alpha/2$ is
(a) 1/5 (b) $-1/\sqrt{10}$
(c) $-1/5$ (d) $1/\sqrt{10}$

49. Let α and β be such that $\pi < \alpha - \beta < 3\pi$. If $\sin\alpha + \sin\beta = -\frac{21}{65}$ and $\cos\alpha + \cos\beta = -\frac{27}{65}$, then the value of $\cos\frac{\alpha-\beta}{2}$ is

(a) $-\frac{3}{\sqrt{130}}$ (b) $\frac{3}{\sqrt{130}}$
(c) $\frac{6}{65}$ (d) $-\frac{6}{65}$

50. If $\sin^3 x \sin 3x = \sum_{m=0}^{6} c_m \cos^m x$, where $c_0, c_1, c_2, \ldots, c_6$ are constants, then

(a) $c_0 + c_2 + c_4 + c_6 = 0$ (b) $c_1 + c_3 + c_5 = 6$
(c) $2c_2 + 3c_6 = 0$ (d) $c_4 + 2c_6 = 0$

51. Let $f_k(x) = \frac{1}{k}(\sin^k x + \cos^k x)$ for $k = 1, 2, 3 \ldots$. Then, for all $x \in R$, the value of $f_4(x) - f_6(x)$ is equal to

[JEE Main 2019, 11 January]

(a) $\frac{1}{12}$ (b) $\frac{5}{12}$
(c) $\frac{-1}{12}$ (d) $\frac{1}{4}$

52. For what and only what values of α lying between 0 and π is the inequality $\sin\alpha\cos^3\alpha > \sin^3\alpha\cos\alpha$ valid?

(a) $\alpha \in (0, \pi/4)$ (b) $\alpha \in (0, \pi/2)$
(c) $\alpha \in \left(\frac{\pi}{4}, \frac{\pi}{2}\right)$ (d) None of these

53. If $x = \sum_{n=0}^{\infty} \cos^{2n}\theta$, $y = \sum_{n=0}^{\infty} \sin^{2n}\theta$, $z = \sum_{n=0}^{\infty} \cos^{2n}\theta \sin^{2n}\theta$ and $0 < \theta < \frac{\pi}{2}$, then

(a) $xz + yz = xy + z$ (b) $xyz = yz + x$
(c) $xy + z = xy + zx$ (d) $x + y + z = xyz + z$

54. If α, β, γ do not differ by a multiple of π and if $\frac{\cos(\alpha+\theta)}{\sin(\beta+\gamma)} = \frac{\cos(\beta+\theta)}{\sin(\gamma+\alpha)} = \frac{\cos(\gamma+\theta)}{\sin(\alpha+\beta)} = k$. Then, k equals

(a) ± 2 (b) $\pm 1/2$
(c) 0 (d) ± 1

55. If the expression $\frac{A\cos(\theta+\alpha) + B\sin(\theta+\beta)}{A'\sin(\theta+\alpha) + B'\cos(\theta+\beta)}$ retain the same value for all 'θ', then

(a) $(AA' - BB')\sin(\alpha-\beta) = (A'B - AB')$
(b) $AA' + BB' = (A'B + AB')\sin(\alpha-\beta)$
(c) $AA' - BB' = (A'B - AB')\sin(\alpha-\beta)$
(d) none of the above

56. If $0 < x < \pi/2$, then

(a) $\cos x > 1 - \frac{2}{\pi}x$ (b) $\cos x < 1 - \frac{2}{\pi}x$
(c) $\cos x > \frac{2}{\pi}x$ (d) $\cos x < \frac{2}{\pi}x$

57. The minimum and maximum values of $ab\sin x + b\sqrt{(1-a^2)}\cos x + c$ $(|a| < 1, b > 0)$ respectively are

(a) $\{b - c, b + c\}$
(b) $\{b + c, b - c\}$
(c) $\{c - b, b + c\}$
(d) None of the above

58. If $\cos x = \tan y$, $\cos y = \tan z$ and $\cos z = \tan x$, then $\sin x$ equals

(a) $\sin y$ (b) $\sin z$
(c) $2\sin 18°$ (d) $\sin(y + z)$

59. The value of the expression $\cos\frac{\pi}{7}\cos\frac{2\pi}{7}\cos\frac{10\pi}{7} - \sin\frac{\pi}{14}\sin\frac{3\pi}{14}\sin\frac{5\pi}{14}$ is equal to

(a) 0 (b) $-\frac{1}{4}$
(c) $\frac{1}{4}$ (d) $-\frac{1}{8}$

60. The value of $\sum_{r=1}^{18} \cos^2(5r)^\circ$, where x° denotes the x degrees, is equal to

(a) 0 (b) 7/2
(c) 17/2 (d) 25/2

61. If $4n\alpha = \pi$, then the numerical value of $\tan\alpha\tan 2\alpha\tan 3\alpha \ldots \tan(2n-1)\alpha$ is equal to

(a) -1 (b) 0
(c) 1 (d) 2

62. If $\tan\alpha$ is an integral solution of the equation $4x^2 - 16x + 15 < 0$ and $\cos\beta$ is the slope of the bisector of the angle in the first quadrant between the X and Y axes, then the value of $\sin(\alpha+\beta) : \sin(\alpha-\beta)$ is equal to

(a) -1 (b) 0
(c) 1 (d) 2

63. If $2\cos\theta + \sin\theta = 1$, then $7\cos\theta + 6\sin\theta$ equals

(a) 1 or 2 (b) 2 or 3
(c) 2 or 4 (d) 2 or 6

64. The value of $\sin 10° \sin 30° \sin 50° \sin 70°$ is

[JEE Main 2019, 9 April]

(a) $\frac{1}{36}$ (b) $\frac{1}{32}$
(c) $\frac{1}{16}$ (d) $\frac{1}{18}$

65. The ratio of the greatest value of $2 - \cos x + \sin^2 x$ to its least value is

(a) 1/4 (b) 9/4
(c) 13/4 (d) None of these

66. If $\cos x + \sin x = a\left(-\frac{\pi}{2} < x < -\frac{\pi}{4}\right)$, then $\cos 2x$ is equal to

(a) a^2 (b) $a\sqrt{(2-a)}$
(c) $a\sqrt{(2+a)}$ (d) $a\sqrt{(2-a^2)}$

67. Expression $2^{\sin\theta} + 2^{-\cos\theta}$ is minimum when $\theta = \ldots$ and its minimum value is

(a) $2n\pi + \frac{\pi}{4}, n \in I, 2^{1+1/\sqrt{2}}$

(b) $2n\pi + \frac{7\pi}{4}, n \in I, 2^{1-1/\sqrt{2}}$

(c) $n\pi \pm \pi/4, n \in I, 2^{1-1/\sqrt{2}}$

(d) None of the above

68. If in a triangle ABC, $\cos 3A + \cos 3B + \cos 3C = 1$, then one angle must be exactly equal to

(a) $\frac{\pi}{3}$ (b) $\frac{2\pi}{3}$

(c) π (d) $\frac{4\pi}{3}$

69. In a triangle ABC, angle A is greater than B. If the measures of angle A and B satisfy the equation $3\sin x - 4\sin^3 x - k = 0, 0 < k < 1$, then the measure of angle C is

(a) $-\frac{\pi}{2}$ (b) $\frac{\pi}{3}$

(c) $\frac{2\pi}{3}$ (d) $\frac{4\pi}{3}$

70. If $\tan x \tan y = a$ and $x + y = \frac{\pi}{6}$, then $\tan x$ and $\tan y$ satisfy the equation

(a) $x^2 - \sqrt{3}(1-a)x + a = 0$

(b) $\sqrt{3}x^2 - (1-a)x + a\sqrt{3} = 0$

(c) $x^2 + \sqrt{3}(1+a)x - a = 0$

(d) $\sqrt{3}x^2 + (1+a)x - a\sqrt{3} = 0$

71. If $\frac{\sin^3\theta - \cos^3\theta}{\sin\theta - \cos\theta} - \frac{\cos\theta}{\sqrt{(1+\cot^2\theta)}} - 2\tan\theta\cot\theta = -1, \theta \in [0, 2\pi]$, then

(a) $\theta \subset (0, \pi/2) - \{\pi/4\}$

(b) $\theta \in \left(\frac{\pi}{2}, \pi\right) - \{3\pi/4\}$

(c) $\theta \in \left(\pi, \frac{3\pi}{2}\right) - \{5\pi/4\}$

(d) $\theta \in (0, \pi) \sim \{\pi/4, \pi/2\}$

72. If $\sin\alpha = \frac{336}{625}$ and $450° < \alpha < 540°$, then $\sin(\alpha/4)$ is equal to

(a) $\frac{1}{5\sqrt{2}}$ (b) $\frac{7}{25}$

(c) $\frac{4}{5}$ (d) $\frac{3}{5}$

73. If $\sin(\theta + \alpha) = a$ and $\sin(\theta + \beta) = b$, then $\cos 2(\alpha - \beta) - 4ab\cos(\alpha - \beta)$ is equal to

(a) $1 - a^2 - b^2$

(b) $1 - 2a^2 - 2b^2$

(c) $2 + a^2 - b^2$

(d) $2 - a^2 - b^2$

74. The expression $3\left\{\sin^4\left(\frac{3\pi}{2} - \alpha\right) + \sin^4(3\pi + \alpha)\right\} - 2\left\{\sin^6\left(\frac{\pi}{2} + \alpha\right) + \sin^6(5\pi - \alpha)\right\}$ is equal to

(a) 0 (b) -1

(c) 1 (d) 3

75. If $\sin^4\alpha\cos^2\alpha = \sum_{k=0}^{3} C_k \cos 2k\alpha$, then $C_1 + C_3 = \ldots, C_0 + C_2 = \ldots$ and $C_1 + C_2 + C_3 + C_0 = \ldots$, are

(a) 0, 1, 2 (b) 0, 0, 0,

(c) 0, 2, 3 (d) 0, -1, 2

76. Which of the following statements are possible, a, b, m and n being non-zero real numbers ?

(a) $4\sin^2\theta = 5$

(b) $(a^2 + b^2)\cos\theta = 2ab$

(c) $(m^2 + n^2)\operatorname{cosec}\theta = m^2 - n^2$

(d) $\sin\theta = 2.375$

77. The set of values of $\lambda \in R$ such that $\tan^2\theta + \sec\theta = \lambda$ holds for some θ is

(a) $(-\infty, 1]$ (b) $(-\infty, -1]$

(c) ϕ (d) $[1, \infty)$

78. The value of $\cos^2 10° - \cos 10° \cos 50° + \cos^2 50°$ is **[JEE Main 2019, 9 April]**

(a) $\frac{3}{2}(1 + \cos 20°)$ (b) $\frac{3}{4} + \cos 20°$

(c) $3/2$ (d) $3/4$

79. The value of $e^{\log_{10}\tan 1° + \log_{10}\tan 2° + \log_{10}\tan 3° + \ldots + \log_{10}\tan 89°}$ is equal to

(a) 0 (b) e

(c) $1/e$ (d) None of these

80. For any $\theta \in \left(\frac{\pi}{4}, \frac{\pi}{2}\right)$, the expression $3(\sin\theta - \cos\theta)^4 + 6(\sin\theta + \cos\theta)^2 + 4\sin^6\theta$ equals **[JEE Main 2019, 9 January]**

(a) $13 - 4\cos^4\theta + 2\sin^2\theta\cos^2\theta$

(b) $13 - 4\cos^2\theta + 6\cos^4\theta$

(c) $13 - 4\cos^2\theta + 6\sin^2\theta\cos^2\theta$

(d) $13 - 4\cos^6\theta$

81. For any real θ, the maximum value of $\cos^2(\cos\theta) + \sin^2(\sin\theta)$ is

(a) 1 (b) $1 + \sin^2 1$

(c) $1 + \cos^2 1$ (d) does not exist

82. If $\cos^4\theta\sec^2\alpha, \frac{1}{2}$ and $\sin^4\theta\operatorname{cosec}^2\alpha$ are in AP, then $\cos^8\theta\sec^6\alpha, \frac{1}{2}$ and $\sin^8\theta\operatorname{cosec}^6\alpha$ are in

(a) AP (b) GP

(c) HP (d) None of these

83. Given that $(1+\sqrt{(1+x)})\tan x = 1+\sqrt{(1-x)}$, then $\sin 4x$ is equal to

(a) $4x$ (b) $2x$
(c) x (d) None of these

84. If $\tan\theta = n\tan\phi$, then maximum value of $\tan^2(\theta-\phi)$ is equal to

(a) $\dfrac{(n+1)^2}{4n}$ (b) $\dfrac{(n-1)^2}{4n}$
(c) $\dfrac{(2n+1)^2}{4n}$ (d) $\dfrac{(2n-1)^2}{4n}$

85. If $|\tan A| < 1$, and $|A|$ is acute, then $\dfrac{\sqrt{(1+\sin 2A)}+\sqrt{(1-\sin 2A)}}{\sqrt{(1+\sin 2A)}-\sqrt{(1-\sin 2A)}}$ is equal to

(a) $\tan A$ (b) $-\tan A$
(c) $\cot A$ (d) $-\cot A$

86. The maximum value of the expression $|\sqrt{(\sin^2 x + 2a^2)} - \sqrt{(2a^2 - 1 - \cos^2 x)}|$, where a and x are real numbers is

(a) $\sqrt{3}$ (b) $\sqrt{2}$
(c) 1 (d) $\sqrt{5}$

87. If $a_{n+1} = \sqrt{\dfrac{1}{2}(1+a_n)}$, then $\cos\left(\dfrac{\sqrt{(1-a_0^2)}}{a_1 a_2 a_3 \dots \text{to } \infty}\right)$ is equal to

(a) 1 (b) -1
(c) a_0 (d) $\dfrac{1}{a_0}$

Objective Questions Type II [One or more than one correct answer(s)]

In each of the questions below four choices of which are or more than one are correct. You have to select the correct answer(s) accordingly.

1. If $\dfrac{\tan x}{1} = \dfrac{\tan y}{2} = \dfrac{\tan z}{3} (\neq 0)$ and $x+y+z=\pi$, then

(a) maximum value of $\tan x + \tan y + \tan z$ is 6
(b) minimum value of $\tan x + \tan y + \tan z$ is -6
(c) $\tan x = \pm 1$, $\tan y = \pm 2$, $\tan z = \pm 3$
(d) $\tan x + \tan y + \tan z = 0\ \forall x, y, z \in R$

2. If $3\sin\beta = \sin(2\alpha+\beta)$, then

(a) $[\cot\alpha + \cot(\alpha+\beta)][\cot\beta - 3\cot(2\alpha+\beta)] = 6$
(b) $\sin\beta = \cos(\alpha+\beta)\sin\alpha$
(c) $2\sin\beta = \sin(\alpha+\beta)\cos\alpha$
(d) $\tan(\alpha+\beta) = 2\tan\alpha$

3. If $x = \sqrt{a^2\cos^2\alpha + b^2\sin^2\alpha} + \sqrt{a^2\sin^2\alpha + b^2\cos^2\alpha}$ then, $x^2 = a^2 + b^2 + 2\sqrt{P(a^2+b^2) - P^2}$ where P is equal to

(a) $a^2\cos^2\alpha + b^2\sin^2\alpha$
(b) $a^2\sin^2\alpha + b^2\cos^2\alpha$
(c) $(1/2)[a^2+b^2+(a^2-b^2)\cos 2\alpha]$
(d) $(1/2)[a^2+b^2-(a^2-b^2)\cos 2\alpha]$

4. If $\cos\alpha = \dfrac{3}{5}$ and $\cos\beta = \dfrac{5}{13}$, then

(a) $\cos(\alpha+\beta) = \dfrac{33}{65}$ (b) $\sin(\alpha+\beta) = \dfrac{56}{65}$
(c) $\sin^2\left(\dfrac{\alpha-\beta}{2}\right) = \dfrac{1}{65}$ (d) $\cos(\alpha-\beta) = \dfrac{63}{65}$

5. The equation $\sin^6 x + \cos^6 x = a^2$ has real solution, if

(a) $a \in [-1, 1]$ (b) $a \in \left[-1, -\dfrac{1}{2}\right]$
(c) $a \in \left[-\dfrac{1}{2}, \dfrac{1}{2}\right]$ (d) $a \in \left[\dfrac{1}{2}, 1\right]$

6. If $\tan\alpha$ and $\tan\beta$ are the roots of the equation $x^2 + px + q = 0\ (p \neq 0)$, then

(a) $\sin^2(\alpha+\beta) + p\sin(\alpha+\beta)\cos(\alpha+\beta) + q\cos^2(\alpha+\beta) = q$
(b) $\tan(\alpha+\beta) = p/(q-1)$
(c) $\cos(\alpha+\beta) = 1-q$
(d) $\sin(\alpha+\beta) = -p$

7. If $\sin\theta + \sin\phi = a$ and $\cos\theta + \cos\phi = b$, then

(a) $\cos\left(\dfrac{\theta-\phi}{2}\right) = \pm\dfrac{1}{2}\sqrt{(a^2+b^2)}$
(b) $\cos\left(\dfrac{\theta-\phi}{2}\right) = \pm\dfrac{1}{2}\sqrt{(a^2-b^2)}$
(c) $\tan\left(\dfrac{\theta-\phi}{2}\right) = \pm\sqrt{\left(\dfrac{4-a^2-b^2}{a^2+b^2}\right)}$
(d) $\cos(\theta-\phi) = \dfrac{a^2+b^2-2}{2}$

8. Let $f_n(\theta) = \tan\dfrac{\theta}{2}(1+\sec\theta)(1+\sec 2\theta)(1+\sec 4\theta)\dots(1+\sec 2^n\theta)$, then

(a) $f_2\left(\dfrac{\pi}{16}\right) = 1$ (b) $f_3\left(\dfrac{\pi}{32}\right) = 1$
(c) $f_4\left(\dfrac{\pi}{64}\right) = 1$ (d) $f_5\left(\dfrac{\pi}{128}\right) = 1$

9. For $0 < \phi \leq \pi/2$, if $x = \sum_{n=0}^{\infty}\cos^{2n}\phi$, $y = \sum_{n=0}^{\infty}\sin^{2n}\phi$, $z = \sum_{n=0}^{\infty}\cos^{2n}\phi\sin^{2n}\phi$, then

(a) $xyz = xz + y$ (b) $xyz = xy + z$
(c) $xyz = x + y + z$ (d) $xyz = yz + x$

10. If $\tan x = \dfrac{2b}{a-c}$, $(a \neq c)$

$y = a\cos^2 x + 2b\sin x\cos x + c\sin^2 x$ and

$z = a\sin^2 x - 2b\sin x\cos x + c\cos^2 x$, then

(a) $y = z$ (b) $y + z = a + c$

(c) $y - z = a - c$ (d) $y - z = (a-c)^2 + 4b^2$

11. $\left(\dfrac{\cos A + \cos B}{\sin A - \sin B}\right)^n + \left(\dfrac{\sin A + \sin B}{\cos A - \cos B}\right)^n$ (n, even or odd) is equal to

(a) $2\tan^n\left(\dfrac{A-B}{2}\right)$ (b) $2\cot^n\left(\dfrac{A-B}{2}\right)$

(c) 0 (d) None of these

12. $\dfrac{3 + \cos 76^\circ \cot 16^\circ}{\cot 76^\circ + \cot 16^\circ}$ is equal to

(a) $\tan 16^\circ$ (b) $\cot 76^\circ$

(c) $\tan 46^\circ$ (d) $\cot 44^\circ$

13. If $\cos x - \sin\alpha \cot\beta \sin x = \cos\alpha$, then the value of $\tan(x/2)$ is

(a) $-\tan(\alpha/2)\cot(\beta/2)$ (b) $\tan(\alpha/2)\tan(\beta/2)$

(c) $-\cot(\alpha/2)\tan(\beta/2)$ (d) $\cot(\alpha/2)\cot(\beta/2)$

14. If in ΔABC, $\tan A + \tan B + \tan C = 6$ and $\tan A \tan B = 2$, then $\sin^2 A : \sin^2 B : \sin^2 C$ is

(a) $8:9:5$ (b) $8:5:9$

(c) $5:9:8$ (d) $5:8:5$

15. If $\cot\theta + \tan\theta = x$ and $\sec\theta - \cos\theta = y$, then

(a) $\sin\theta\cos\theta = \dfrac{1}{x}$ (b) $\sin\theta\tan\theta = y$

(c) $(x^2y)^{2/3} - (xy^2)^{2/3} = 1$ (d) $(x^2y)^{1/3} + (xy^2)^{1/3} = 1$

16. If $\dfrac{x}{y} = \dfrac{\cos A}{\cos B}$, where $A \neq B$, then

(a) $\tan\left(\dfrac{A+B}{2}\right) = \dfrac{x\tan A + y\tan B}{x+y}$

(b) $\tan\left(\dfrac{A-B}{2}\right) = \dfrac{x\tan A - y\tan B}{x+y}$

(c) $\dfrac{\sin(A+B)}{\sin(A-B)} = \dfrac{y\sin A + x\sin B}{y\sin A - x\sin B}$

(d) $x\cos A + y\cos B = 0$

17. If $\tan\theta = \dfrac{\sin\alpha - \cos\alpha}{\sin\alpha + \cos\alpha}$, then

(a) $\sin\alpha - \cos\alpha = \pm\sqrt{2}\sin\theta$

(b) $\sin\alpha + \cos\alpha = \pm\sqrt{2}\cos\theta$

(c) $\cos 2\theta = \sin 2\alpha$

(d) $\sin 2\theta + \cos 2\alpha = 0$

18. Let $0 \le \theta \le \pi/2$ and $x = X\cos\theta + Y\sin\theta$, $y = X\sin\theta - Y\cos\theta$ such that $x^2 + 4xy + y^2 = aX^2 + bY^2$, where a, b are constants. Then

(a) $a = -1, b = 3$ (b) $\theta = \pi/4$

(c) $a = 3, b = -1$ (d) $\theta = \pi/3$

19. If $\dfrac{\tan 3A}{\tan A} = k$, $(k \neq 1)$, then

(a) $\dfrac{\cos A}{\cos 3A} = \dfrac{k^2-1}{2k}$ (b) $\dfrac{\sin 3A}{\sin A} = \dfrac{2k}{k-1}$

(c) $k < \dfrac{1}{3}$ (d) $k > 3$

20. If in ΔABC, $\angle A = 90^\circ$ and c, $\sin B$, $\cos B$ are rational numbers, then

(a) a is rational (b) a is irrational

(c) b is rational (d) b is irrational

21. If $\cos x + \sec x = -2$, then for a positive integer n, $\cos^n x + \sec^n x$ is

(a) always 2 (b) always– 2

(c) – 2, if n is odd (d) 2, if n is even

22. If $A + B = \dfrac{\pi}{3}$ and $\cos A + \cos B = 1$, then which of the following is/are true ?

(a) $\cos(A-B) = \dfrac{1}{3}$ (b) $|\cos A - \cos B| = \sqrt{\dfrac{2}{3}}$

(c) $\cos(A-B) = -\dfrac{1}{3}$ (d) $|\cos A - \cos B| = \dfrac{1}{2\sqrt{3}}$

23. If $x = \sec\phi - \tan\phi$ and $y = \operatorname{cosec}\phi + \cot\phi$, then

(a) $x = \dfrac{y+1}{y-1}$ (b) $x = \dfrac{y-1}{y+1}$

(c) $y = \dfrac{1+x}{1-x}$ (d) $xy + x - y + 1 = 0$

24. In a ΔABC

(a) $\sin A \sin B \sin C \le \dfrac{3\sqrt{3}}{8}$

(b) $\sin^2 A + \sin^2 B + \sin^2 C \le \dfrac{9}{4}$

(c) $\sin A \sin B \sin C$ is always positive

(d) $\sin^2 A + \sin^2 B \le 1 + \cos C$

25. If $x = a\cos^3\theta\sin^2\theta$, $y = a\sin^3\theta\cos^2\theta$ and $\dfrac{(x^2+y^2)^p}{(xy)^q}$ $(p, q \in N)$ is independent of θ, then

(a) $p = 4$ (b) $p = 5$

(c) $q = 4$ (d) $q = 5$

26. If $(x-a)\cos\theta + y\sin\theta = (x-a)\cos\phi + y\sin\phi = a$ and $\tan(\theta/2) - \tan(\phi/2) = 2b$, then

(a) $y^2 = 2ax - (1-b^2)x^2$ (b) $\tan\dfrac{\theta}{2} = \dfrac{1}{x}(y + bx)$

(c) $y^2 = 2bx - (1-a^2)x^2$ (d) $\tan\dfrac{\phi}{2} = \dfrac{1}{x}(y - bx)$

27. If $(a-b)\sin(\theta+\phi) = (a+b)\sin(\theta-\phi)$ and $a\tan\dfrac{\theta}{2} - b\tan\dfrac{\phi}{2} = c$, then

(a) $b\tan\phi = a\tan\theta$ (b) $a\tan\phi = b\tan\theta$

(c) $\sin\phi = \dfrac{2bc}{a^2 - b^2 - c^2}$ (d) $\sin\theta = \dfrac{2ac}{a^2 - b^2 + c^2}$

28. If $\cos^4\theta + \alpha$, $\sin^4\theta + \alpha$ are the roots of the equation $x^2 + b(2x+1) = 0$ and $\cos^2\theta + \beta$, $\sin^2\theta + \beta$ are the roots of the equation $x^2 + 4x + 2 = 0$, then b is equal to

(a) 1 (b) -1 (c) 2 (d) -2

29. If $\left(\dfrac{\sin\theta}{\sin\phi}\right)^2 = \dfrac{\tan\theta}{\tan\phi} = 3$, then

(a) $\tan\phi = 1/\sqrt{3}$ (b) $\tan\phi = -1/\sqrt{3}$
(c) $\tan\theta = \sqrt{3}$ (d) $\tan\theta = -\sqrt{3}$

30. If $\sec\theta + \tan\theta = 1$, then one root of equation $a(b-c)x^2 + b(c-a)x + c(a-b) = 0$ is

(a) $\tan\theta$ (b) $\sec\theta$
(c) $\cos\theta$ (d) $\sin\theta$

Linked-Comprehension Type

In these questions, a passage (paragraph) has been given followed by questions based on each of the passages. You have to answer the questions based on the passage given.

PASSAGE 1

Increasing product with angles are in GP

$\cos\alpha \cos 2\alpha \cos 2^2\alpha \ldots \cos 2^{n-1}\alpha$

$$= \begin{cases} \dfrac{\sin 2^n\alpha}{2^n \sin\alpha}, & \text{if } \alpha \neq n\pi \\ \dfrac{1}{2^n}, & \text{if } \alpha = \dfrac{\pi}{2^n+1} \\ -\dfrac{1}{2^n}, & \text{if } \alpha = \dfrac{\pi}{2^n-1} \end{cases}$$

Where, $n \in I$ (Integer)

On the basis of above information, answer the following questions

1. The value of $\cos\dfrac{2\pi}{7}\cos\dfrac{4\pi}{7}\cos\dfrac{6\pi}{7}$ is

(a) $-1/2$ (b) $1/2$
(c) $1/4$ (d) $1/8$

2. If $\alpha = \dfrac{\pi}{13}$, then the value of $\prod_{r=1}^{6} \cos r\alpha$ is

(a) $\dfrac{1}{64}$ (b) $-\dfrac{1}{64}$
(c) $\dfrac{1}{32}$ (d) $-\dfrac{1}{8}$

3. The value of $\sin\left(\dfrac{\pi}{14}\right)\sin\left(\dfrac{3\pi}{14}\right)\sin\left(\dfrac{5\pi}{14}\right)\sin\left(\dfrac{7\pi}{14}\right)$ $\sin\left(\dfrac{9\pi}{14}\right)\sin\left(\dfrac{11\pi}{14}\right)\sin\left(\dfrac{13\pi}{14}\right)$ is

(a) 1 (b) $\dfrac{1}{8}$
(c) $\dfrac{1}{32}$ (d) $\dfrac{1}{64}$

4. The value of $\sin\left(\dfrac{\pi}{18}\right)\sin\left(\dfrac{5\pi}{18}\right)\sin\left(\dfrac{7\pi}{18}\right)$ is

(a) $\dfrac{1}{16}$ (b) $\dfrac{1}{8}$ (c) $-\dfrac{1}{8}$ (d) -1

5. The value of $64\sqrt{3}\sin\dfrac{\pi}{48}\cos\dfrac{\pi}{48}\cos\dfrac{\pi}{24}\cos\dfrac{\pi}{12}\cos\dfrac{\pi}{6}$ is

(a) 8 (b) 6 (c) 4 (d) -1

PASSAGE 2

If $P_n = \sin^n\theta + \cos^n\theta$ where $n \in W$ (whole number) and $\theta \in R$ (real number)

On the basis of above information, answer the following questions

1. If $P_1 = m$, then the value of $4(1 - P_6)$ is

(a) $3(m-1)^2$
(b) $3(m^2-1)^2$
(c) $3(m+1)^2$
(d) $3(m^2+1)^2$

2. The value of $2P_6 - 3P_4 + 10$ is

(a) 0 (b) 6
(c) 9 (d) 15

3. The value of $6P_{10} - 15P_8 + 10P_6 + 7$ is

(a) 8 (b) 6
(c) 4 (d) 2

4. If $P_{n-2} - P_n = \sin^2\theta\cos^2\theta\, P_\lambda$, then the value of λ is

(a) $n-1$ (b) $n-2$
(c) $n-3$ (d) $n-4$

5. The value of $\dfrac{P_7 - P_5}{P_5 - P_3}$ is

(a) $\dfrac{P_7}{P_5}$ (b) $\dfrac{P_5}{P_3}$
(c) $\dfrac{P_3}{P_1}$ (d) $\dfrac{P_3}{P_5}$

PASSAGE 3

If $\cos\dfrac{\pi}{7}$, $\cos\dfrac{3\pi}{7}$, $\cos\dfrac{5\pi}{7}$ are the roots of the equation $8x^3 - 4x^2 - 4x + 1 = 0$

On the basis of above information, answer the following questions

1. The value of $\sec\dfrac{\pi}{7} + \sec\dfrac{3\pi}{7} + \sec\dfrac{5\pi}{7}$ is

(a) 2 (b) 4
(c) 8 (d) None of these

2. The value of $\sin\frac{\pi}{14}\sin\frac{3\pi}{14}\sin\frac{5\pi}{14}$ is

(a) $\frac{1}{4}$ (b) $\frac{1}{8}$

(c) $\frac{\sqrt{7}}{4}$ (d) $\frac{\sqrt{7}}{8}$

3. The value of $\cos\frac{\pi}{14}\cos\frac{3\pi}{14}\cos\frac{5\pi}{14}$ is

(a) $\frac{1}{4}$ (b) $\frac{1}{8}$

(c) $\frac{\sqrt{7}}{4}$ (d) $\frac{\sqrt{7}}{8}$

4. The equation whose roots are $\tan^2\frac{\pi}{7}$, $\tan^2\frac{3\pi}{7}$ and $\tan^2\frac{5\pi}{7}$ is

(a) $x^3-35x^2+7x-21=0$

(b) $x^3-35x^2+21x-7=0$

(c) $x^3-21x^2+35x-7=0$

(d) $x^3-21x^2+7x-35=0$

5. The value of $\sum_{r=1}^{3}\tan^2\left(\frac{2r-1}{7}\right)\sum_{r=1}^{3}\cot^2\left(\frac{2r-1}{7}\right)$ is

(a) 15 (b) 105

(c) 21 (d) 147

PASSAGE 4

The method of eliminating 'θ' from two given equations involving trigonometrical functions of 'θ'. By using given equations involving 'θ' and trigonometrical identities, we shall obtain an equation not involving 'θ'.

On the basis of above information, answer the following questions

1. If $x\sin^3\theta+y\cos^3\theta=\sin\theta\cos\theta$ and $x\sin\theta-y\cos\theta=0$ then (x, y) lie one

(a) a circle (b) a parabola

(c) an ellipse (d) a hyperbola

2. If $\frac{x}{a\cos\theta}=\frac{y}{b\sin\theta}$ and $\frac{ax}{\cos\theta}-\frac{by}{\sin\theta}=a^2-b^2$, then (x, y) lie on

(a) a circle (b) a parabola

(c) an ellipse (d) a hyperbola

3. If $\tan\theta+\sin\theta=m$ and $\tan\theta-\sin\theta=n$, then $(m^2-n^2)^2$ is

(a) $4\sqrt{mn}$ (b) $4mn$

(c) $16\sqrt{mn}$ (d) $16\,mn$

4. If $\sin\theta+\cos\theta=a$ and $\sin^3\theta+\cos^3\theta=b$, then we get $\lambda a^3+\mu b+\nu c=0$ when λ, μ, ν are independent of θ, then the value of $\lambda^3+\mu^3+\nu^3$ is

(a) -6 (b) -18

(c) -36 (d) -98

5. After eliminating 'θ' from equations $\frac{x\cos\theta}{a}+\frac{y\sin\theta}{b}=1$ and $x\sin\theta-y\cos\theta=\sqrt{(a^2\sin^2\theta+b^2\cos^2\theta)}$, we get

(a) $x^2+y^2=a^2+b^2$

(b) $\frac{x^2}{a^2}+\frac{y^2}{b^2}=1$

(c) $\frac{x^2}{a(a+b)}+\frac{y^2}{b(a+b)}=1$

(d) $x^2+y^2=(a+b)^2$

Numerical Value Type

Solve the following problems and write your answer.

1. If $4\cos 36°+\cot\left(7\frac{1}{2}°\right)=\sqrt{n_1}+\sqrt{n_2}+\sqrt{n_3}+\sqrt{n_4}+\sqrt{n_5}+\sqrt{n_6}$, then the value of $\sum_{i=1}^{6}n_i^2$ must be

2. If $4\sin 27°=\sqrt{\alpha}+\sqrt{\beta}$, then the value of $(\alpha+\beta-\alpha\beta+2)^4$ must be

3. If $(25)^2+a^2+50\,a\cos\theta=(31)^2+b^2+62\,b\cos\theta=1$ and $775+ab+(31a+25b)\cos\theta=0$, then the value of $\operatorname{cosec}^2\theta$ is

4. If $0<A<\frac{\pi}{2}$ and $\sin A+\cos A+\tan A+\cot A+\sec A+\operatorname{cosec} A-7$ and $\sin A$ and $\cos A$ are the roots of the equation $4x^2-3x+a=0$, then the value of $25a$ must be

5. If $\sec A\tan B+\tan A\sec B=53$, then the value of $(\sec A\sec B+\tan A\tan B)^2$ must be

6. If $\sin^2 A=x$ and $\prod_{r=1}^{4}\sin(rA)=ax^2+bx^3+cx^4+dx^5$, then the value of $10a-7b+15c-5d$ must be

7. If $\cot(\theta-\alpha)$, $3\cot\theta$, $\cot(\theta+\alpha)$ are in AP and θ is not an integral multiple of $\frac{\pi}{2}$, then the value of $\frac{4016\sin^2\theta}{3\sin^2\alpha}$ must be

8. If $a\tan\alpha+\sqrt{(a^2-1)}\tan\beta+\sqrt{(a^2+1)}\tan\gamma=2a$, where a is constant and α, β, γ are variable angles. Then the least value of $2727(\tan^2\alpha+\tan^2\beta+\tan^2\gamma)$ must be

9. If $A+B+C=\pi$ and $\frac{\sin 2A+\sin 2B+\sin 2C}{\sin A+\sin B+\sin C}=\lambda\sin\left(\frac{A}{2}\right)\sin\left(\frac{B}{2}\right)\sin\left(\frac{C}{2}\right)$, then the value of $1+2\lambda+3\lambda^2+4\lambda^3$ must be

10. If $A+B+C=\pi$, then the minimum value of $786\left\{\frac{\cos\left(\frac{A-B}{2}\right)}{\cos\left(\frac{A+B}{2}\right)}+\frac{\cos\left(\frac{B-C}{2}\right)}{\cos\left(\frac{B+C}{2}\right)}+\frac{\cos\left(\frac{C-A}{2}\right)}{\cos\left(\frac{C+A}{2}\right)}\right\}$ must be

11. If $\frac{1}{\cos 290^\circ}+\frac{1}{\sqrt{3}\sin 250^\circ}=\lambda$, then the value of λ^2 must be

12. If $\tan\left(\frac{3\pi}{11}\right)+4\sin\left(\frac{2\pi}{11}\right)=\lambda$, then the value of $1+\lambda^2+\lambda^4+\lambda^6$ must be

13. In any ΔABC, the minimum value of

$$2020\sum\frac{\sqrt{(\sin A)}}{(\sqrt{(\sin B)}+\sqrt{(\sin C)}-\sqrt{(\sin A)})}$$

must be

14. If $\tan^2\left(\frac{\pi}{16}\right)+\tan^2\left(\frac{2\pi}{16}\right)+\tan^2\left(\frac{3\pi}{16}\right)+\ldots+\tan^2\left(\frac{7\pi}{16}\right)=\lambda$ and if $x^y+y^x=\lambda$, then the value of $\frac{x+y}{8}$ must be

Single Integer Type Questions

Solve the following problems and write your answer.

1. If $\frac{1}{2}\left(\tan\left(\frac{\pi}{24}\right)+\cot\left(\frac{\pi}{24}\right)\right)=\sqrt{a^2+a}+\sqrt{a}$, then $a=$

2. If $\frac{\sqrt{2}\sin\alpha}{\sqrt{1+\cos 2\alpha}}=\frac{1}{7}$ and $\sqrt{\frac{1-\cos 2\beta}{2}}=\frac{1}{\sqrt{10}}$, $\alpha,\beta\in\left(0,\frac{\pi}{2}\right)$, then $\tan(\alpha+2\beta)$ is equal to .... **[JEE Main 2020, 8 January]**

3. $\cos\frac{\pi}{7}-\cos\frac{2\pi}{7}+\cos\frac{3\pi}{7}-\cos\frac{4\pi}{7}+\cos\frac{5\pi}{7}-\cos\frac{6\pi}{7}=$

4. $16\left(\cos\theta-\cos\frac{\pi}{8}\right)\left(\cos\theta-\cos\frac{3\pi}{8}\right)\left(\cos\theta-\cos\frac{5\pi}{8}\right)\left(\cos\theta-\cos\frac{7\pi}{8}\right)=\lambda\cos 4\theta$, then the value of λ is _____ .

5. If $A+B+C=\pi$, $\cos B+\cos C=4\sin^2(A/2)$, then $3\tan B/2\tan C/2$ is equal to —— .

6. Value of $\tan 9^\circ-\tan 27^\circ-\tan 63^\circ+\tan 81^\circ$ is

7. If $\sin x+\cos x+\tan x+\cot x+\sec x+\operatorname{cosec} x=7$ and $\sin 2x=a-b\sqrt{7}$, then value of $\left(\frac{a+b}{10}\right)$ is equal to

8. If $\frac{\sin^3\theta}{\sin(2\theta+\alpha)}=\frac{\cos^3\theta}{\cos(2\theta+\alpha)}$ and $\tan 2\theta=\lambda\tan(3\theta+\alpha)$, then the value of λ is ____ .

Matrix-Match Type

Given below are Matching Type Questions, with two columns (each having some items) each . Each item of Column I has to be matched with the items of Column II, by encircling the correct match(es).

NOTE *An item of Column I can be matched with more than one items of Column II.*

1. Observe the following columns

	Column I		Column II
(A)	In a Δ ABC, $\sin\left(\frac{A}{2}\right)+\sin\left(\frac{B}{2}\right)+\sin\left(\frac{C}{2}\right)=$	(P)	$-1+4\sin\left(\frac{\pi+A}{4}\right)\sin\left(\frac{\pi+B}{4}\right)\cos\left(\frac{\pi+C}{4}\right)$
(B)	In a Δ ABC, $\sin\left(\frac{A}{2}\right)+\sin\left(\frac{B}{2}\right)-\sin\left(\frac{C}{2}\right)=$	(Q)	$4\cos\left(\frac{\pi+A}{4}\right)\cos\left(\frac{\pi+B}{4}\right)\cos\left(\frac{\pi-C}{4}\right)$
(C)	In a ΔABC, $\cos\left(\frac{A}{2}\right)+\cos\left(\frac{B}{2}\right)-\cos\left(\frac{C}{2}\right)=$	(R)	$1+4\sin\left(\frac{\pi-A}{4}\right)\sin\left(\frac{\pi-B}{4}\right)\sin\left(\frac{\pi-C}{4}\right)$
		(S)	$-1+4\cos\left(\frac{\pi-A}{4}\right)\cos\left(\frac{\pi-B}{4}\right)\sin\left(\frac{\pi-C}{4}\right)$
		(T)	$1+4\cos\left(\frac{\pi+A}{4}\right)\cos\left(\frac{\pi+B}{4}\right)\sin\left(\frac{\pi-C}{4}\right)$

2. Observe the following columns

	Column I		Column II
(A)	If $\theta + \phi = \frac{\pi}{2}$, where θ and ϕ are positive, then $(\sin\theta + \sin\phi)\sin\left(\frac{\pi}{4}\right)$ is always less than	(P)	1
(B)	If $\sin\theta - \sin\phi = a$ and $\cos\theta + \cos\phi = b$, then $a^2 + b^2$ cannot exceed	(Q)	2
(C)	If $3\sin\theta + 5\cos\theta = 5$, $(\theta \neq 0)$ then the value of $5\sin\theta - 3\cos\theta$ is	(R)	3
		(S)	4
		(T)	5

3. Observe the following columns

	Column I		Column II
(A)	If maximum and minimum values of $\frac{7 + 6\tan\theta - \tan^2\theta}{(1 + \tan^2\theta)}$ for all real values of $\theta \sim \frac{\pi}{2}$ are λ and μ respectively, then	(P)	$\lambda + \mu = 2$
(B)	If maximum and minimum values of $5\cos\theta + 3\cos\left(\theta + \frac{\pi}{3}\right) + 3$ for all real values of θ are λ and μ respectively, then	(Q)	$\lambda - \mu = 6$
(C)	If maximum and minimum values of $1 + \sin\left(\frac{\pi}{4} + \theta\right) + 2\cos\left(\frac{\pi}{4} - \theta\right)$ for all real values of θ are λ and μ respectively, then	(R)	$\lambda + \mu = 6$
		(S)	$\lambda - \mu = 10$
		(T)	$\lambda - \mu = 14$

4. Observe the following columns

	Column I		Column II
(A)	If $x + y + z = xyz$, then	(P)	$\Sigma\left(\frac{x}{1 + x^2}\right) = 2\Pi\left(\frac{1}{\sqrt{(1 + x^2)}}\right)$
(B)	If $xy + yz + zx = 1$, then	(Q)	$\Sigma\left(\frac{x}{1 - x^2}\right) = 4\Pi\left(\frac{x}{1 - x^2}\right)$
(C)	If $x + y + z = \sqrt{3}$ and $x^2 + y^2 + z^2 = 1$, then	(R)	$\Sigma\left(\frac{1 - x^2}{1 + x^2}\right) = -1 - 4\Pi\left(\frac{x}{\sqrt{(1 + x^2)}}\right)$
		(S)	$\Sigma\left(\frac{x}{1 + x^2}\right) = 2\Pi\left(\frac{x}{\sqrt{(1 + x^2)}}\right)$
		(T)	$\Sigma\left(\frac{1}{1 + x^2}\right) = 1 - 2\Pi\frac{1}{\sqrt{(1 + x^2)}}$

5. Observe the following columns

	Column I		Column II
(A)	The maximum value of $\{\cos(2A + \theta) + \cos(2B + \theta)\}$, where A, B are constants, is	(P)	$2\sin(A + B)$
(B)	The maximum value of $\{\cos 2A + \cos 2B\}$, where $(A + B)$ is constant and $A, B \in (0, \pi/2)$, is	(Q)	$2\sec(A + B)$
(C)	The minimum value of $\{\sec 2A + \sec 2B\}$, where $(A + B)$ is constant and $A, B \in (0, \pi/4)$, is	(R)	$2\cos(A + B)$
(D)	The minimum value of $\sqrt{\{\tan\theta + \cot\theta - 2\cos 2(A + B)\}}$, where A, B are constants and $\theta \in (0, \pi/2)$, is	(S)	$2\cos(A - B)$
		(T)	$2\tan(A + B)$

Answers

Objective Questions Type [Only one correct answer]

1. (a)	**2.** (d)	**3.** (b)	**4.** (a)	**5.** (a)	**6.** (b)	**7.** (d)	**8.** (c)	**9.** (d)	**10.** (b)
11. (c)	**12.** (c)	**13.** (c)	**14.** (c)	**15.** (a)	**16.** (b)	**17.** (c)	**18.** (a)	**19.** (b)	**20.** (a)
21. (a)	**22.** (a)	**23.** (c)	**24.** (b)	**25.** (a)	**26.** (a)	**27.** (a)	**28.** (b)	**29.** (c)	**30.** (c)
31. (c)	**32.** (a)	**33.** (b)	**34.** (c)	**35.** (b)	**36.** (b)	**37.** (a)	**38.** (d)	**39.** (c)	**40.** (b)
41. (c)	**42.** (b)	**43.** (d)	**44.** (b)	**45.** (b)	**46.** (c)	**47.** (b)	**48.** (b)	**49.** (a)	**50.** (a)
51. (a)	**52.** (a)	**53.** (a)	**54.** (d)	**55.** (c)	**56.** (a)	**57.** (c)	**58.** (c)	**59.** (b)	**60.** (c)
61. (c)	**62.** (c)	**63.** (d)	**64.** (c)	**65.** (c)	**66.** (d)	**67.** (b)	**68.** (b)	**69.** (c)	**70.** (b)
71. (d)	**72.** (a)	**73.** (b)	**74.** (c)	**75.** (b)	**76.** (b)	**77.** (d)	**78.** (d)	**79.** (d)	**80.** (d)
81. (b)	**82.** (a)	**83.** (c)	**84.** (b)	**85.** (c)	**86.** (b)	**87.** (c)			

Objective Questions Type II [One or more than one correct answer(s)]

1. (a, b, c)	**2.** (a, b, c, d)	**3.** (a, b, c, d)	**4.** (b, c, d)	**5.** (b, d)
6. (a, b)	**7.** (a, c, d)	**8.** (a, b, c, d)	**9.** (b, c)	**10.** (b, c)
11. (b, c)	**12.** (c, d)	**13.** (a, b)	**14.** (b, c)	**15.** (a, b, c)
16. (a, b, c)	**17.** (a, b, c, d)	**18.** (b, c)	**19.** (a, b, c, d)	**20.** (a, c)
21. (c, d)	**22.** (b, c)	**23.** (b, c, d)	**24.** (a, b, c, d)	**25.** (b, c)
26. (a, b, d)	**27.** (b, c, d)	**28.** (b, c)	**29.** (a, b, c, d)	**30.** (b, c)

Linked-Comprehension Type

Passage 1 **1.** (d) **2.** (a) **3.** (d) **4.** (b) **5.** (b)

Passage 2 **1.** (b) **2.** (c) **3.** (a) **4.** (d) **5.** (c)

Passage 3 **1.** (b) **2.** (b) **3.** (d) **4.** (c) **5.** (b)

Passage 4 **1.** (a) **2.** (c) **3.** (d) **4.** (b) **5.** (c)

Numerical Value Types Questions

1. 91 **2.** 400 **3.** 1586 **4.** 28 **5.** 2810 **6.** 3448 **7.** 2008 **8.** 3636

9. 2257 **10.** 4716 **11.** 5.33 **12.** 1464 **13.** 6060 **14.** 4.375

Single Integer Type Questions

1. (2) **2.** (1) **3.** (1) **4.** (2) **5.** (1) **6.** (4) **7.** (3) **8.** (2)

Matrix-Match Type

1. A → (R, T); B → (P, S); C → (Q)

2. A → (P, Q, R, S, T); B → (S, T); C → (R)

3. A → (R, S); B → (R, T); C → (P, Q)

4. A → (Q, R, S, T); B → (P, Q); C → (P, Q)

5. A → (S); B → (S); C → (Q); D → (P)

Solutions

Objective Questions Type [Only one correct answer]

1. We have, $k_1 = \tan 27\theta - \tan\theta$

$$= (\tan 27\theta - \tan 9\theta) + (\tan 9\theta - \tan 3\theta) + (\tan 3\theta - \tan\theta)$$

Now, $\tan 3\theta - \tan\theta = \dfrac{\sin 2\theta}{\cos 3\theta \cos\theta} = \dfrac{2\sin\theta}{\cos 3\theta}$

Similarly, $\tan 9\theta - \tan 3\theta = \dfrac{2\sin 3\theta}{\cos 9\theta}$

and $\tan 27\theta - \tan 9\theta = \dfrac{2\sin 9\theta}{\cos 27\theta}$

$\therefore \quad k_1 = 2\left[\dfrac{\sin 9\theta}{\cos 27\theta} + \dfrac{\sin 3\theta}{\cos 9\theta} + \dfrac{\sin\theta}{\cos 3\theta}\right] = 2k_2$

2. $15\sin^4\alpha + 10\cos^4\alpha = 6$

$15\sin^4\alpha + 10\cos^4\alpha = 6(\sin^2\alpha + \cos^2\alpha)^2$

$(3\sin^2\alpha - 2\cos^2\alpha)^2 = 0$

$\tan^2\alpha = \dfrac{2}{3}, \cot^2\alpha = \dfrac{3}{2} \Rightarrow 27\sec^6\alpha + 8\,\text{cosec}^6\alpha$

$= 27(\sec^6\alpha)^3 + 8(\text{cosec}^6\alpha)^3$

$= 27(1+\tan^2\alpha)3 + 8(1+\cot^2\alpha)^3 = 250$

3. Since, $0 < x < \dfrac{\pi}{2}$

$\therefore \quad 0 < \sin x < 1$ and $0 < \cos x < 1$

when $\quad x = 2, \sin^n x + \cos^n x = 1$

when $n > 2$, both $\sin^n x$ and $\cos^n x$ will decreases and hence $\sin^n x + \cos^n x < 1$.

when $n > 2$, both $\sin^n x$ and $\cos^n x$ will increase and hence $\sin^n x + \cos^n x > 1$. Thus, $\sin^n x + \cos^n x \geq 1$ for $n \leq 2$.

4. $\dfrac{\sec^8\theta}{a} + \dfrac{\tan^8\theta}{b} = \dfrac{1}{a+b}$

or $a^2\sin^8\theta + ab + b^2 = ab(\cos^8\theta - \sin^8\theta)$

$= ab(1 - 2\sin^2\theta\cos^2\theta)\cos 2\theta$

$= ab(1 - 2\sin^2\theta)\left(1 - \dfrac{1}{2}\sin^2 2\theta\right)$

$\Rightarrow a^2\sin^8\theta - 2ab\sin^4\theta + b^2 = -2ab\sin^2\theta$

$+ ab\sin^2\theta\sin^2 2\theta - \dfrac{ab}{2}\sin^2 2\theta - 2ab\sin^4\theta$

$(a\sin^4\theta - b)^2 = 2ab\sin^2\theta(-\sin^2\theta - 1 + 2\sin^2\theta\cos^2\theta - \cos^2\theta)$

$\Rightarrow \quad 2ab\sin^2\theta(-2 + 2\sin^2\theta\cos^2\theta) \geq 0$

$\Rightarrow \quad 4ab\sin^2\theta(\sin^4\theta - \sin^2\theta + 1) \leq 0$

$\Rightarrow \quad ab \leq 0$

5. $\sec A, \sec B, \sec C$ are all positive

$\therefore \quad \dfrac{\sec A + \sec B + \sec C}{3} \geq \sqrt[3]{\sec A \sec B \sec C} \quad \ldots\text{(i)}$

Now, $\cos A \cos B \cos C = \dfrac{1}{2}\cos A[\cos(B - C) - \cos A]$

$\because \quad \cos(B - C) \leq 1$

$\therefore \cos A\cos B\cos C \leq \dfrac{1}{2}\cos A[1 - \cos A]$

$= \dfrac{1}{2}\left[\cos A - \cos^2 A - \dfrac{1}{4} + \dfrac{1}{4}\right]$

$= \dfrac{1}{8} - \dfrac{1}{2}\left(\cos A - \dfrac{1}{2}\right)^2$

$\therefore \quad \cos A\cos B\cos C \leq \dfrac{1}{8}$

$\Rightarrow \quad \sec A\sec B\sec C \geq 8$

$\Rightarrow \quad \sqrt[3]{\sec A \cdot \sec B \cdot \sec C} \geq 2$

$\Rightarrow \quad \sec A + \sec B + \sec C \geq 6 \quad$ [from Eq. (i)]

6. $P_n - P_{n-2} = \cos^n\theta + \sin^n\theta - \cos^{n-2}\theta - \sin^{n-2}\theta$

$= \cos^{n-2}\theta(\cos^2\theta - 1) + \sin^{n-2}\theta(\sin^2\theta - 1)$

$= -\sin^2\theta\cos^{n-2}\theta - \cos^2\theta\sin^{n-2}\theta$

$= -\sin^2\theta\cos^2\theta(\cos^{n-4}\theta + \sin^{n-4}\theta)$

$= -\sin^2\theta\cos^2\theta P_{n-4} = kP_{n-4}$

where $k = -\sin^2\theta\cos^2\theta$

7. Given, $L = \sin^2\dfrac{\pi}{16} - \sin^2\dfrac{\pi}{8}$

$= \sin\left(\dfrac{\pi}{16} + \dfrac{\pi}{8}\right)\sin\left(\dfrac{\pi}{16} - \dfrac{\pi}{8}\right)$

$= \sin\dfrac{3\pi}{16}\sin\left(-\dfrac{\pi}{16}\right) = \dfrac{1}{2}\left[2\sin\dfrac{3\pi}{16}\sin\left(-\dfrac{\pi}{16}\right)\right]$

$= \dfrac{1}{2}\left[\cos\dfrac{\pi}{4} - \cos\dfrac{2\pi}{16}\right] = \dfrac{1}{2\sqrt{2}} - \dfrac{1}{2}\cos\dfrac{\pi}{8}.$

and, $M = \cos^2\dfrac{\pi}{16} - \sin^2\dfrac{\pi}{8}$

$= \cos\left(\dfrac{\pi}{16} + \dfrac{\pi}{8}\right)\cos\left(\dfrac{\pi}{16} - \dfrac{\pi}{8}\right) = \cos\dfrac{3\pi}{16}\cos\left(-\dfrac{\pi}{16}\right)$

$= \dfrac{1}{2}\left[2\cos\dfrac{3\pi}{16}\cos\dfrac{\pi}{16}\right] = \dfrac{1}{2}\left[\cos\dfrac{\pi}{4} + \cos\dfrac{2\pi}{16}\right]$

$= \dfrac{1}{2\sqrt{2}} + \dfrac{1}{2}\cos\dfrac{\pi}{8}.$

8. We have, $\dfrac{U_n}{V_n} = \tan n\theta$

and $\dfrac{V_n - V_{n-1}}{U_{n-1}} = \dfrac{\cos n\theta\sec^n\theta - \cos(n-1)\theta\sec^{n-1}\theta}{\sin(n-1)\theta\sec^{n-1}\theta}$

$= \dfrac{\cos n\theta\sec\theta - \cos(n-1)\theta}{\sin(n-1)\theta}$

$= \dfrac{\cos n\theta - \cos(n-1)\theta\cos\theta}{\cos\theta\sin(n-1)\theta}$

$= \dfrac{\cos(n-1)\theta\cos\theta - \sin(n-1)\theta\sin\theta - \cos(n-1)\theta\cos\theta}{\cos\theta\sin(n-1)\theta}$

$= -\tan\theta$

So that, $\dfrac{V_n - V_{n-1}}{U_{n-1}} + \dfrac{1}{n}\dfrac{U_n}{V_n} = -\tan\theta + \dfrac{\tan n\theta}{n} \neq 0$

9. Given that $\sin(x+80°)=a$

$\therefore\quad \cos(x+140°)=\cos\{(x+80°)+60°\}$

$=\cos(x+80°)\cos 60° - \sin(x+80°)\sin 60°$

$=-\sqrt{(1-a^2)}\cdot\frac{1}{2}-\frac{a\sqrt{3}}{2}=\frac{-\sqrt{(1-a^2)}-\sqrt{3}\,a}{2}$

$(\because 20°<x<60°,\ \cos(x+80°)$ is $-$ ve)

10. $a=\sin\frac{\pi}{18}\sin\frac{5\pi}{18}\sin\frac{7\pi}{18}$

$=\sin 10°\sin 50°\sin 70°$

$=\frac{1}{2}[2\sin 70°\sin 10°]\sin 50°$

$=\frac{1}{2}[\cos 60°-\cos 80°]\sin 50°$

$=\frac{1}{4}\sin 50°-\frac{1}{4}(2\cos 80°\sin 50°)$

$=\frac{1}{4}\sin 50°-\frac{1}{4}(\sin 130°-\sin 30°)$

$=\frac{1}{4}\sin 50°-\frac{1}{4}\sin 50°+\frac{1}{4}\cdot\frac{1}{2}=\frac{1}{8}$

$y=2[x]+2$ and $y=3[x-2]$

$\Rightarrow 2[x]+2=3[x-2]$

$=3[x]+3[-2]$

$=3[x]-6$

$\Rightarrow\quad [x]=8$

$\therefore\quad a=\frac{1}{[x]}$

11. $A=\cos(\cos x)+\sin(\cos x)$

$=\sqrt{2}\left\{\cos(\cos x)\cos\frac{\pi}{4}+\sin(\cos x)\sin\frac{\pi}{4}\right\}$

$=\sqrt{2}\left\{\cos\left(\cos x-\frac{\pi}{4}\right)\right\}$

$\because\quad -1\le\cos\left(\cos x-\frac{\pi}{4}\right)\le 1$

$\therefore\quad -\sqrt{2}\le A\le\sqrt{2}$

12. $\tan A+\tan B+\tan C>0$

$\Rightarrow\ \tan A\tan B\tan C>0$

$\Rightarrow$ either all $\tan A$, $\tan B$, $\tan C$ are positive or two out of them are negative but if two are negative, then two angles become obtuse, which is not possible. Hence, Δ is acute angled triangle.

13. $\sin\frac{\pi}{2n}+\cos\frac{\pi}{2n}=\sqrt{2}\sin\left(\frac{\pi}{4}+\frac{\pi}{2n}\right)$

$\Rightarrow\quad \frac{\sqrt{n}}{2}=\sqrt{2}\sin\left(\frac{\pi}{4}+\frac{\pi}{2n}\right)$

So for $n>1$, $\frac{\sqrt{n}}{2\sqrt{2}}=\sin\left(\frac{\pi}{4}+\frac{\pi}{2n}\right)>\sin\frac{\pi}{4}=\frac{1}{\sqrt{2}}$ or $n>4$

Since, $\sin\left(\frac{\pi}{4}+\frac{\pi}{2n}\right)<1$ for all $n>2$, we get

$\frac{\sqrt{n}}{2\sqrt{2}}<1$ or $n<8$

So that, $4<n<8$. By actual verification we find that only $n=6$ satisfies the given relation.

14. Given

$2\left(\sin\left(\frac{A+B}{2}\right)\cos\left(\frac{A-B}{2}\right)+\sin\left(\frac{C+D}{2}\right)\cos\left(\frac{C-D}{2}\right)\right)=2\times 2$

$\sin A+\sin B+\sin C+\sin D=4$

$\Rightarrow\quad \sin A=\sin B=\sin C=\sin D=1$

$\Rightarrow\quad \angle A=\angle B=\angle C=\angle D=90°$

Now, $\Sigma\cos\frac{A}{2}\cos\frac{B}{2}=\Sigma\cos 45°\cdot\cos 45°$

$=6\times\left(\frac{1}{\sqrt{2}}\right)^2=3$

15. Given, $f(x)=ax+b$

$\therefore\quad f'(x)=a$

Since, $a<0$, $f(x)$ is a decreasing function

$\therefore\quad f(-1)=2$ and $f(1)=0$

$\Rightarrow\quad -a+b=2$ and $a+b=0$

$\therefore\quad a=-1$ and $b=1$

Thus, $\quad f(x)=-x+1$

Clearly, $f(0)=1$, $f\left(\frac{1}{4}\right)=\frac{3}{4}$, $f(-2)=3$,

$f\left(\frac{1}{3}\right)=\frac{2}{3}$, $f(-1)=2$

Also, $\quad A=\frac{1+\cos 2\theta}{2}+\left(\frac{1-\cos 2\theta}{2}\right)^2$

$=\frac{1}{2}+\frac{1}{2}\cos 2\theta+\frac{1}{4}-\frac{1}{2}\cos 2\theta+\frac{1}{4}\cos^2 2\theta$

$=\frac{3}{4}+\frac{1}{4}\left(\frac{1+\cos 4\theta}{2}\right)=\frac{7}{8}+\frac{1}{8}\cos 4\theta$

$\therefore\quad \frac{3}{4}\le A\le 1$

$\Rightarrow\quad f\left(\frac{1}{4}\right)\le A\le f(0)$

16. From the given equations we have $\Sigma\tan\alpha=p$, $\Sigma\tan\alpha\tan\beta=0$ and $\tan\alpha\tan\beta\tan\gamma=r$

So that $(1+\tan^2\alpha)(1+\tan^2\beta)(1+\tan^2\gamma)$

$=1+\Sigma\tan^2\alpha+\Sigma\tan^2\alpha\tan^2\beta+\tan^2\alpha\tan^2\beta\tan^2\gamma$

$=1+(\Sigma\tan\alpha)^2-2\Sigma\tan\alpha\tan\beta+(\Sigma\tan\alpha\tan\beta)^2$

$-2\tan\alpha\tan\beta\tan\gamma\Sigma\tan\alpha+\tan^2\alpha\tan^2\beta\tan^2\gamma$

$=1+p^2-2pr+r^2=1+(p-r)^2$

17. $\frac{1+\tan\theta}{1-\tan\theta}=3\left(\frac{3\tan\theta-\tan^3\theta}{1-3\tan^2\theta}\right)$

After simplification, we get

$3\tan^4\theta-6\tan^2\theta+8\tan\theta-1=0$

$\therefore$ Product of roots

$=\tan\alpha\cdot\tan\beta\cdot\tan\gamma\cdot\tan\delta=-\frac{1}{3}$

18. It is given that, $cosA = -\frac{60}{61}$ and $\tan B = \frac{-7}{24}$ and neighter A nor B is in the second quadrant, then A and B will be in third and fourth quadrant respectively. So, $\frac{B}{2}$ will be in second quadrant.

Now, $\tan B = -\frac{7}{24}$

$$\Rightarrow \frac{2\tan\frac{B}{2}}{1-\tan^2\frac{B}{2}} = \frac{-7}{24}$$

$$\Rightarrow 7\tan^2\frac{B}{2} - 48\tan\frac{B}{2} - 7 = 0$$

$$\Rightarrow \left(7\tan\frac{B}{2}+1\right)\left(\tan\frac{B}{2}-7\right) = 0$$

$$\Rightarrow \tan\frac{B}{2} = -\frac{1}{7} \quad \left[\because \frac{B}{2} \in \text{IInd quadrant}\right]$$

$$\because \tan\left(A+\frac{B}{2}\right) = \frac{\tan A + \tan\frac{B}{2}}{1-\tan A\tan\frac{B}{2}} = \frac{\frac{11}{60}-\frac{1}{7}}{1+\frac{11}{420}}$$

$[\because \cos A = -\frac{60}{61}$ and $A \in$ IIIrd Quadrant, $\because \tan A = \frac{11}{60}]$

$$= \frac{77-60}{420+11} = \frac{17}{431}$$

$\because A \in$ IIIrd quadrant and $\frac{B}{2} \in \left(\frac{3\pi}{4}, \pi\right)$

and $\tan\left(A+\frac{B}{2}\right) \in$ is positive.

$\therefore \left(A+\frac{B}{2}\right) \in$ Ist quadrant.

Hence, option (a) is correct.

19. Equation first can be written as

$$x\sin a + y\times 2\sin a\cos a + z\times\sin a(3-4\sin^2 a) = 2\times 2\sin a\cos a\cos 2a$$

$$\Rightarrow x + 2y\cos a + z(3+4\cos^2 a - 4) = 4\cos a(2\cos^2 a - 1) \text{ as } \sin a \neq 0$$

$$\Rightarrow 8\cos^3 a - 4z\cos^2 a - (2y+4)\cos a + (z-x) = 0$$

$$\Rightarrow \cos^3 a - \left(\frac{z}{2}\right)\cos^2 a - \left(\frac{y+2}{4}\right)\cos a + \left(\frac{z-x}{8}\right) = 0$$

Which shows that $\cos a$ is root of the equation

$$t^3 - \left(\frac{z}{2}\right)t^2 - \left(\frac{y+2}{4}\right)t + \left(\frac{z-x}{8}\right) = 0$$

Similarly, from second and third equation we can verify $\cos b$ and $\cos c$ are the roots of the given equation.

20. $\because x^2+y^2+z^2 = r^2\sin^2\theta\cos^2\phi + r^2\sin^2\theta\sin^2\phi + r^2\cos^2\theta$

$$= r^2\sin^2\theta(\cos^2\phi+\sin^2\phi) + r^2\cos^2\theta$$

$$= r^2\sin^2\theta + r^2\cos^2\theta$$

$$= r^2$$

21. $\sqrt{2+\sqrt{2+\sqrt{2+\ldots+\sqrt{2(1+\cos\theta)}}}}$

(n number of 2's)

$$\sqrt{2+\sqrt{2+\sqrt{2+\ldots+\sqrt{(2+2\cos\theta/2)}}}}$$

(($n-1$) number of 2's)

....

$$\sqrt{2+2\cos(\theta/2^{n-1})}$$

$$= \sqrt{2\{1+2\cos^2(\theta/2^n)-1\}} = 2\cos(\theta/2^n)$$

22. $\tan\alpha/2 + \tan\beta/2 = \frac{26}{8} = \frac{13}{4}$,

$\tan\alpha/2\tan\beta/2 = \frac{15}{8}$

$$\therefore \tan\left(\frac{\alpha+\beta}{2}\right) = \frac{\tan\alpha/2+\tan\beta/2}{1-\tan\alpha/2\tan\beta/2} = \frac{\frac{13}{4}}{1-\frac{15}{8}} = -\frac{26}{7}$$

$$\cos(\alpha+\beta) = \frac{1-\tan^2\left(\frac{\alpha+\beta}{2}\right)}{1+\tan^2\left(\frac{\alpha+\beta}{2}\right)} = \frac{1-\left(-\frac{26}{7}\right)^2}{1+\left(-\frac{26}{7}\right)^2} = \frac{49-676}{49+676} = -\frac{627}{725}$$

23. $\because a\sec\alpha = d + c\tan\alpha$ ...(i)

and $b\sec\alpha = c - d\tan\alpha$ ...(ii)

Squaring and adding Eqs. (i) and (ii)

$$(a^2+b^2)\sec^2\alpha = d^2 + c^2\tan^2\alpha + 2dc\tan\alpha + c^2 + d^2\tan^2\alpha - 2dc\tan\alpha$$

$$\Rightarrow (a^2+b^2)\sec^2\alpha = c^2(\tan^2\alpha+1) + d^2(1+\tan^2\alpha) = (c^2+d^2)\sec^2\alpha$$

Hence, $a^2+b^2 = c^2+d^2$

24. $\sin n\theta = \sum_{r=0}^{n} b_r\sin^r\theta$

$\sin n\theta = b_0 + b_1\sin\theta + b_2\sin^2\theta + b_3\sin^3\theta + \ldots + b_n\sin^n\theta$

when n is odd, then for possibility

Put $\theta = 0$, then $b_0 = 0$

Differentiating both sides w.r.t. θ and put $\theta = 0$, then we get $b_1 = n$

25. $\because \tan\theta = \frac{1}{\sqrt{7}}$

$$\therefore \frac{\text{cosec}^2\theta - \sec^2\theta}{\text{cosec}^2\theta + \sec^2\theta} = \frac{(1+\cot^2\theta)-(1+\tan^2\theta)}{(1+\cot^2\theta)+(1+\tan^2\theta)}$$

$$= \frac{(\cot^2\theta - \tan^2\theta)}{2+\tan^2\theta+\cot^2\theta}$$

$$= \frac{7-\frac{1}{7}}{2+\frac{1}{7}+7} = \frac{48}{14+1+49} = \frac{48}{64} = \frac{3}{4}$$

26. $\because \tan 7\frac{1^\circ}{2} = \frac{-1+\sqrt{1+\tan^2 15^\circ}}{\tan 15^\circ}$

$\because \tan 15^\circ = 2-\sqrt{3}$ or $\frac{\sqrt{3}-1}{\sqrt{3}+1}$

$\therefore \tan 7\frac{1^\circ}{2} = \frac{-1+\sqrt{1+\left(\frac{\sqrt{3}-1}{\sqrt{3}+1}\right)^2}}{\left(\frac{\sqrt{3}-1}{\sqrt{3}+1}\right)}$

$= \frac{-1(\sqrt{3}+1)+2\sqrt{2}}{(\sqrt{3}-1)}$

27. $\sin\left(x+\frac{\pi}{6}\right)+\cos\left(x+\frac{\pi}{6}\right) = \sqrt{2}\left\{\frac{1}{\sqrt{2}}\sin\left(x+\frac{\pi}{6}\right) + \frac{1}{\sqrt{2}}\cos\left(x+\frac{\pi}{6}\right)\right\}$

$= \sqrt{2}\sin\left(\frac{\pi}{4}+x+\frac{\pi}{6}\right)$

$= \sqrt{2}\sin\left(x+\frac{5\pi}{12}\right)$

Maximum value is $\sqrt{2}$.

When $\sin\left(x+\frac{5\pi}{12}\right) = 1 = \sin\frac{\pi}{2}$

$\therefore \quad x = \frac{\pi}{2}-\frac{5\pi}{12} = \frac{\pi}{12}$

28. $a\cos x + b\sin x = c$

$\Rightarrow \frac{a\left(1-\tan^2\frac{x}{2}\right)}{1+\tan^2\frac{x}{2}} + \frac{2b\tan\frac{x}{2}}{1+\tan^2\frac{x}{2}} = c$

$\Rightarrow (c+a)\tan^2\frac{x}{2} - 2b\tan\frac{x}{2} + c - a = 0$

$\Rightarrow \tan\frac{x_1}{2}+\tan\frac{x_2}{2} = \frac{2b}{c+a}$

and $\tan\frac{x_1}{2}\tan\frac{x_2}{2} = \frac{c-a}{c+a}$

$\Rightarrow \tan\left(\frac{x_1+x_2}{2}\right) = \frac{\frac{2b}{c+a}}{1-\frac{c-a}{c+a}} = \frac{2b}{2a} = \frac{b}{a}$

29. $\because \sin\alpha = \sin\beta$

$\Rightarrow \sin\alpha - \sin\beta = 0$

$\therefore 2\cos\left(\frac{\alpha+\beta}{2}\right)\sin\left(\frac{\alpha-\beta}{2}\right) = 0 \quad \ldots(i)$

and $\cos\alpha = \cos\beta$

$\Rightarrow \cos\beta - \cos\alpha = 0$

$\therefore 2\sin\left(\frac{\alpha+\beta}{2}\right)\sin\left(\frac{\alpha-\beta}{2}\right) = 0 \quad \ldots(ii)$

From Eqs. (i) and (ii), we get

$\sin\left(\frac{\alpha-\beta}{2}\right) = 0$

30. $(1+\cos\pi/8)(1+\cos 3\pi/8)(1+\cos 5\pi/8)(1+\cos 7\pi/8)$

$= 2\cos^2\pi/16\cdot 2\cos^2 3\pi/16\cdot 2\cos^2 5\pi/16\cdot 2\cos^2 7\pi/16$

$= 16(\cos\pi/16\cos 3\pi/16\cos 5\pi/16\cos 7\pi/16)^2$

$= (2\cos 7\pi/16\cos\pi/16)^2(2\cos 5\pi/16\cos 3\pi/16)^2$

$= (\cos\pi/2+\cos 3\pi/8)^2(\cos\pi/2+\cos\pi/8)^2$

$= \cos^2 3\pi/8\cos^2\pi/8$

$= \frac{1}{4}(\cos\pi/2+\cos\pi/4)^2 = \frac{1}{8}$

31. We have,

$\sin(\alpha-\beta) = \sin(\alpha-\theta+\theta-\beta)$

$= \sin[(\theta-\beta)-(\theta-\alpha)]$

$= \sin(\theta-\beta)\cos(\theta-\alpha) - \cos(\theta-\beta)\sin(\theta-\alpha)$

$= ba - \sqrt{1-b^2}\sqrt{1-a^2}$

and $\cos(\alpha-\beta) = \cos[(\theta-\beta)-(\theta-\alpha)]$

$= \cos(\theta-\beta)\cos(\theta-\alpha) + \sin(\theta-\beta)\sin(\theta-\alpha)$

$= a\sqrt{1-b^2} + b\sqrt{1-a^2}$

Substituting these value in the given expression, we get $\cos^2(\alpha-\beta) + 2ab\sin(\alpha-\beta)$

$= (a\sqrt{1-b^2}+b\sqrt{1-a^2})^2 + 2ab\,[ab - \sqrt{(1-a^2)}\sqrt{(1-b^2)}]$

$= a^2(1-b^2) + b^2(1-a^2) + 2ab\sqrt{(1-a^2)}\sqrt{(1-b^2)}$

$+ 2a^2b^2 - 2ab\sqrt{(1-a^2)}\sqrt{(1-b^2)} = a^2+b^2$

32. $\tan A = \frac{4}{3}, \sin A = -\frac{4}{5}, \cos A = -\frac{3}{5}$

(In III quadrant $\sin A < 0, \cos A < 0$)

$\therefore \quad 5\sin 2A + 3\sin A + 4\cos A$

$= 10\sin A\cos A + 3\sin A + 4\cos A$

$= 10\left(-\frac{4}{5}\right)\left(-\frac{3}{5}\right) + 3\left(-\frac{4}{5}\right) + 4\left(-\frac{3}{5}\right) = 0$

33. It is given that, $x = \sum_{n=0}^{\infty}(-1)^n\tan^{2n}\theta$

$= 1 - \tan^2\theta + \tan^4\theta - \tan^6\theta + \ldots + \text{upto } \infty$

$= \frac{1}{1+\tan^2\theta}$

$\left[\because \theta\in\left(0,\frac{\pi}{4}\right) \Rightarrow \tan^2\theta\in(0,1)\right]$

$x = \cos^2\theta \quad \ldots(i)$

and $y = \sum_{n=0}^{\infty}\cos^{2n}\theta$

$= 1 + \cos^2\theta + \cos^4\theta + \cos^6\theta + \ldots + \text{upto } \infty$

$= \frac{1}{1-\cos^2\theta} \quad \left[\because \theta\in\left(0,\frac{\pi}{4}\right)\right]$

$\Rightarrow \cos^2\theta \in \left(0,\frac{1}{2}\right)\}$

$= \frac{1}{\sin^2\theta} \Rightarrow \sin^2\theta = \frac{1}{y} \quad \ldots(ii)$

On adding Eqs. (i) and (ii), we get

$1 = x + \frac{1}{y} \Rightarrow y(1-x) = 1$

Hence, option (b) is correct.

34. $\because \sin x + \sin^2 x = 1$

$\Rightarrow \quad \sin x = 1 - \sin^2 x$

$\Rightarrow \quad \sin^2 x = \cos^4 x$

$\Rightarrow \quad 1 - \cos^2 x = \cos^4 x$

$\Rightarrow \quad \cos^4 x + \cos^2 x = 1$

Squaring both sides, then $\cos^8 x + \cos^4 x + 2\cos^6 x = 1$

Hence, $\cos^8 x + 2\cos^6 x + \cos^4 x = 1$

35. $x = y \cos \frac{2\pi}{3} = z \cos \frac{4\pi}{3} = k$

$\frac{1}{x} = \frac{1}{k}, \frac{1}{y} = \frac{\cos \frac{2\pi}{3}}{k}, \frac{1}{z} = \frac{\cos \frac{4\pi}{3}}{k}$

$\therefore$

$\frac{1}{x} + \frac{1}{y} + \frac{1}{z} = \frac{1}{k}\left(1 + \cos \frac{2\pi}{3} + \cos \frac{4\pi}{3}\right) = 0$

$\Rightarrow \quad xy + yz + zx = 0$

36. $a \sin^2 x + b \cos^2 x = c$

$\Rightarrow \quad a \tan^2 x + b = c \sec^2 x$

$\Rightarrow \quad a \tan^2 x + b = c(1 + \tan^2 x)$

$\therefore \quad \tan^2 x = \left(\frac{c-b}{a-c}\right)$

Similarly, $\quad \tan^2 y = \frac{d-a}{b-d}$

$\therefore \quad \frac{a^2}{b^2} = \frac{\tan^2 y}{\tan^2 x}$

$= \frac{\left(\frac{d-a}{b-d}\right)}{\left(\frac{c-b}{a-c}\right)}$

$= \frac{(d-a)(a-c)}{(b-d)(c-b)}$

$= \frac{(a-d)(c-a)}{(b-c)(d-b)}$

37. $\because \quad 0 < \alpha < \frac{\pi}{6}$

$\Rightarrow \quad 0 < \frac{\alpha}{2} < \frac{\pi}{12}$ (in first quadrant),

then $\tan \alpha/2$ is always positive

$\because \quad \sin \alpha + \cos \alpha = \frac{\sqrt{7}}{2}$

$\Rightarrow \quad \frac{2 \tan \frac{\alpha}{2}}{1 + \tan^2 \frac{\alpha}{2}} + \frac{1 - \tan^2 \frac{\alpha}{2}}{1 + \tan^2 \frac{\alpha}{2}} = \frac{\sqrt{7}}{2}$

$\Rightarrow \quad 2\left(2\tan \frac{\alpha}{2} + 1 - \tan^2 \frac{\alpha}{2}\right) = \sqrt{7}\left(1 + \tan^2 \frac{\alpha}{2}\right)$

$\Rightarrow (\sqrt{7} + 2)\tan^2 \frac{\alpha}{2} - 4 \tan \frac{\alpha}{2} + (\sqrt{7} - 2) = 0$

$\therefore \quad \tan \frac{\alpha}{2} = \frac{4 \pm \sqrt{16 - 4(\sqrt{7} + 2)(\sqrt{7} - 2)}}{2(\sqrt{7} + 2)}$

$= \frac{4 \pm \sqrt{16 - 12}}{2(\sqrt{7} + 2)}$

$= \frac{4 \pm 2}{2(\sqrt{7} + 2)} = \frac{2 \pm 1}{(\sqrt{7} + 2)}$

$= \frac{1}{\sqrt{7} + 2}$ or $\frac{3}{\sqrt{7} + 2}$

$= \frac{\sqrt{7} - 2}{3}$ or $\sqrt{7} - 2$

Hence, $\tan (\alpha/2) = \frac{\sqrt{7} - 2}{3} \quad \left[\because 0 < \left(\frac{\alpha}{2}\right) < \frac{\pi}{12}\right]$

38. $\cos\left(\frac{2\pi}{7}\right) + \cos\left(\frac{4\pi}{7}\right) + \cos\left(\frac{6\pi}{7}\right)$

$= \text{Re}\left\{e^{\frac{2\pi i}{7}} + e^{\frac{4\pi i}{7}} + e^{\frac{6\pi i}{7}}\right\}$

$= \frac{e^{\frac{2\pi i}{7}} + e^{\frac{4\pi i}{7}} + e^{\frac{6\pi i}{7}} + e^{\frac{-2\pi i}{7}} + e^{\frac{-4\pi i}{7}} + e^{\frac{-6\pi i}{7}}}{2}$

$= \frac{-1 + \left(1 + e^{\frac{2\pi i}{7}} + e^{\frac{4\pi i}{7}} + e^{\frac{6\pi i}{7}} + e^{\frac{-2\pi i}{7}} + e^{\frac{-4\pi i}{7}} + e^{\frac{-6\pi i}{7}}\right)}{2}$

$= \frac{-1 + (\text{sum of seven roots of unity})}{2}$

$= \frac{-1 + 0}{2} = -\frac{1}{2}$

39. $\alpha \in (\pi, 3\pi/2)$

$\therefore \quad \sin \alpha < 0 \;\&\; \cos \alpha < 0, \tan \alpha > 0$

$\sqrt{4 \sin^4 \alpha + \sin^2 2\alpha} + 4 \cos^2\left(\frac{\pi}{4} - \frac{\alpha}{2}\right)$

$= \sqrt{4 \sin^2 \alpha(\sin^2 \alpha + \cos^2 \alpha)} + 4 \cos^2\left(\frac{\pi}{4} - \frac{\alpha}{2}\right)$

$= -2\sin \alpha + 2\left(1 + \cos\left(\frac{\pi}{2} - \alpha\right)\right)$

$= -2 \sin \alpha + 2 + 2 \sin \alpha = 2$

40. $\cos \frac{\pi}{15} \cos \frac{2\pi}{15} \cos \frac{3\pi}{15} \cos \frac{4\pi}{15} \cos \frac{5\pi}{15} \cos \frac{6\pi}{15} \cos \frac{7\pi}{15}$

$= \cos \frac{\pi}{15} \cos \frac{2\pi}{15} \cos \frac{4\pi}{15} \cos\left(\pi - \frac{8\pi}{15}\right) \cos \frac{3\pi}{15} \cos \frac{5\pi}{15} \cos \frac{6\pi}{15}$

$= -\left(\cos \frac{\pi}{15} \cos \frac{2\pi}{15} \cos \frac{4\pi}{15} \cos \frac{8\pi}{15}\right) \cos \frac{\pi}{5} \cos \frac{\pi}{3} \cos \frac{2\pi}{5}$

$= -\left(\frac{\sqrt{5}+1}{4}\right)\left(\frac{1}{2}\right)\left(\frac{\sqrt{5}-1}{4}\right)\left(\cos\left(\frac{\pi}{15}\right) \cos\left(\frac{2\pi}{15}\right) \cos\left(\frac{4\pi}{15}\right) \cos\left(\frac{8\pi}{15}\right)\right)$

$= -\frac{1}{8} \cdot \frac{\sin\left(2^4 \cdot \frac{\pi}{15}\right)}{2^4 \cdot \sin\left(\frac{\pi}{15}\right)} = -\frac{1}{2^7} \cdot \frac{\sin\left(\pi + \frac{\pi}{15}\right)}{\sin\left(\frac{\pi}{15}\right)} = \frac{1}{2^7}$

41. Given that, $\alpha < \beta < \gamma < \delta$ and

$$\sin\alpha = \sin\beta = \sin\gamma = \sin\delta = k$$

Also $\alpha, \beta, \gamma, \delta$ are smallest positive angles, satisfying above two conditions.

$\therefore$ We can take

$$\beta = \pi - \alpha,\ \gamma = 2\pi + \alpha,\ \delta = 3\pi - \alpha.$$

The given expression

$$= 4\sin\frac{1}{2}\alpha + 3\sin\left(\frac{1}{2}\pi - \frac{1}{2}\alpha\right) + 2\sin\left(\pi + \frac{1}{2}\alpha\right) + \sin\left(\frac{3}{2}\pi - \frac{1}{2}\alpha\right)$$

$$= 4\sin\frac{1}{2}\alpha + 3\cos\frac{1}{2}\alpha - 2\sin\frac{1}{2}\alpha - \cos\frac{1}{2}\alpha$$

$$= 2\left(\sin\frac{1}{2}\alpha + \cos\frac{1}{2}\alpha\right)$$

$$= 2\sqrt{\left\{\left(\sin\frac{1}{2}\alpha + \cos\frac{1}{2}\alpha\right)^2\right\}}$$

$$= 2\sqrt{(1+\sin\alpha)} = 2\sqrt{(1+k)}$$

42. $\because \dfrac{x}{\cos\alpha} = \dfrac{y}{\cos\left(\alpha - \frac{2\pi}{3}\right)} = \dfrac{z}{\cos\left(\alpha + \frac{2\pi}{3}\right)} = k(\text{say})$

$$\therefore\ x + y + z = k\left\{\cos\alpha + \cos\left(\alpha - \frac{2\pi}{3}\right) + \cos\left(\alpha + \frac{2\pi}{3}\right)\right\}$$

$$= k\left\{\cos\alpha + 2\cos\alpha\cos\frac{2\pi}{3}\right\}$$

$$= k\left\{\cos\alpha + 2\cos\alpha \times -\frac{1}{2}\right\} = 0$$

43. $\because \cos 2A + \cos 2B + \cos 2C$

$$= 2\cos(A+B)\cos(A-B) + \cos 2C$$

$$= -2\sin C\cos(A-B) + 1 - 2\sin^2 C$$

$$= 1 - 2\sin C\{\cos(A-B) + \sin C\}$$

$$= 1 - 2\sin C\{\cos(A-B) - \cos(A+B)\}$$

$$= 1 - 2\sin C\{2\sin A\sin B\}$$

$$= 1 - 4\sin A\sin B\sin C$$

44. We have, $\sum_{i=1}^{n}\cos\theta_i = n$

$$\Rightarrow \cos\theta_1 + \cos\theta_2 + \cos\theta_3 + \ldots + \cos\theta_n = \underbrace{1 + 1 + 1 + \ldots + 1}_{n\text{ times}}$$

is valid only when

$$\cos\theta_1 = 1,\ \cos\theta_2 = 1,\ \cos\theta_3 = 1,\ \ldots,\ \cos\theta_n = 1$$

$$\therefore\quad \theta_1 = \theta_2 = \theta_3 = \ldots = \theta_n = 0$$

$$\therefore\quad \sum_{i=1}^{n}\sin\theta_i = 0$$

45. $\cos\alpha + \cos\beta = \sin\alpha + \sin\beta$

On squaring, $\cos^2\alpha + \cos^2\beta + 2\cos\alpha\cos\beta$

$$= \sin^2\alpha + \sin^2\beta + 2\sin\alpha\sin\beta$$

$$\Rightarrow\quad \cos 2\alpha + \cos 2\beta + 2\cos(\alpha+\beta) = 0$$

$$\therefore\quad \cos 2\alpha + \cos 2\beta = -2\cos(\alpha+\beta)$$

46. $\because 0 < x_i < \pi,\ i = 1, 2, 3, \ldots, n$

$\therefore\ \sin x_i > 0,\ i = 1, 2, 3, \ldots, n$

We know $\left(\dfrac{\sin x_1 + \sin x_2 + \ldots + \sin x_n}{n}\right)$

$$\le \sin\left(\frac{x_1 + x_2 + \ldots x_n}{n}\right)$$

$$= \sin\left(\frac{\pi}{n}\right)$$

Hence, $\sin x_1 + \sin x_2 + \ldots + \sin x_n \le n\sin\left(\dfrac{\pi}{n}\right)$

47. Given, $\sin(\alpha-\beta) = \dfrac{5}{13}$

and $\cos(\alpha+\beta) = \dfrac{3}{5}$, where $\alpha, \beta \in \left(0, \dfrac{\pi}{4}\right)$

Since, $0 < \alpha < \dfrac{\pi}{4}$ and $0 < \beta < \dfrac{\pi}{4}$

$$\therefore\quad 0 < \alpha + \beta < \frac{\pi}{4} + \frac{\pi}{4} = \frac{\pi}{2} \Rightarrow 0 < \alpha + \beta < \frac{\pi}{2}$$

Also, $-\dfrac{\pi}{4} < -\beta < 0$

$$\therefore\quad 0 - \frac{\pi}{4} < \alpha - \beta < \frac{\pi}{4} + 0$$

$$\Rightarrow\quad -\frac{\pi}{4} < \alpha - \beta < \frac{\pi}{4}$$

$\therefore\ \alpha + \beta \in \left(0, \dfrac{\pi}{2}\right)$ and $\alpha - \beta \in \left(-\dfrac{\pi}{4}, \dfrac{\pi}{4}\right)$

But $\sin(\alpha-\beta) > 0$, therefore $\alpha - \beta \in \left(0, \dfrac{\pi}{4}\right)$.

Now, $\sin(\alpha-\beta) = \dfrac{5}{13}$

$$\Rightarrow\quad \tan(\alpha-\beta) = \frac{5}{12} \quad \ldots\text{(i)}$$

and $\cos(\alpha+\beta) = \dfrac{3}{5}$

$$\Rightarrow\quad \tan(\alpha+\beta) = \frac{4}{3} \quad \ldots\text{(ii)}$$

Now, $\tan(2\alpha) = \tan[(\alpha+\beta) + (\alpha-\beta)]$

$$= \frac{\tan(\alpha+\beta) + \tan(\alpha-\beta)}{1 - \tan(\alpha+\beta)\tan(\alpha-\beta)}$$

$$= \frac{\frac{4}{3} + \frac{5}{12}}{1 - \frac{4}{3}\times\frac{5}{12}} \quad \text{[from Eqs. (i) and (ii)]}$$

$$= \frac{48 + 15}{36 - 20} = \frac{63}{16}$$

48. $\because\ \sin\alpha = -\dfrac{3}{5}$, then $\cos\alpha = -\dfrac{4}{5}$

$$\because\ \cos\alpha = -\frac{4}{5} \Rightarrow 2\cos^2\alpha/2 - 1 = -\frac{4}{5}$$

$$\Rightarrow\quad 2\cos^2\alpha/2 = \frac{1}{5} \text{ or } \cos^2\left(\frac{\alpha}{2}\right) = \frac{1}{10}$$

$$\Rightarrow\quad \cos(\alpha/2) = \pm\frac{1}{\sqrt{10}}$$

Hence, $\cos(\alpha/2) = -\dfrac{1}{\sqrt{10}}$ $\left[\begin{array}{l}\because \alpha \text{ lies in III quadrant} \\ \therefore \alpha/2 \text{ lies in II quadrant}\end{array}\right]$

49. We have,

$$\sin\alpha + \sin\beta = -\frac{21}{65} \quad \ldots \text{(i)}$$

$$\cos\alpha + \cos\beta = -\frac{27}{65} \quad \ldots \text{(ii)}$$

Squaring Eq. (i), we get

$$\sin^2\alpha + \sin^2\beta + 2\sin\alpha\sin\beta = \left(\frac{21}{65}\right)^2 \quad \ldots \text{(iii)}$$

Squaring Eq. (ii), we get

$$\cos^2\alpha + \cos^2\beta + 2\cos\alpha\cos\beta = \left(\frac{27}{65}\right)^2 \quad \ldots \text{(iv)}$$

Adding Eqs. (iii) and (iv), we get

$$2 + 2\cos(\alpha-\beta) = \frac{1}{(65)^2}[(27)^2 + (21)^2]$$

$$= \frac{1}{(65)^2}(729 + 441) = \frac{1}{(65)^2}(1170) = \frac{18}{65}$$

$$\Rightarrow 1 + \cos(\alpha-\beta) = \frac{9}{65}$$

$$\Rightarrow \quad 2\cos^2\frac{\alpha-\beta}{2} = \frac{9}{65}$$

$$\Rightarrow \quad \cos\frac{\alpha-\beta}{2} = -\frac{3}{\sqrt{130}}$$

$$\left[\because \pi < \alpha - \beta < 3\pi \Rightarrow \frac{\pi}{2} < \frac{\alpha-\beta}{2} < \frac{3\pi}{2} \Rightarrow \cos\left(\frac{\alpha-\beta}{2}\right) < 0\right]$$

50. $\sin^3 x \sin 3x = \sin^3 x\,(3\sin x - 4\sin^3 x)$

$$= \sin^4 x(3 - 4\sin^2 x)$$

$$= (1 - \cos^2 x)^2(3 - 4(1 - \cos^2 x))$$

$$= (1 - 2\cos^2 x + \cos^4 x)(4\cos^2 x - 1)$$

$$= 4\cos^6 x - 9\cos^4 x + 6\cos^2 x - 1$$

So that $\quad c_0 = -1,\ c_2 = 6,\ c_4 = -9,\ c_6 = 4$

and $\quad c_1 = c_3 = c_5 = 0$

$\Rightarrow \quad c_0 + c_2 + c_4 + c_6 = 0$

51. We have,

$$f_k(x) = \frac{1}{k}(\sin^k x + \cos^k x),\ k = 1, 2, 3, \ldots$$

$$\therefore \quad f_4(x) = \frac{1}{4}(\sin^4 x + \cos^4 x)$$

$$= \frac{1}{4}((\sin^2 x + \cos^2 x)^2 - 2\sin^2 x\cos^2 x)$$

$$= \frac{1}{4}\left(1 - \frac{1}{2}(\sin 2x)^2\right) = \frac{1}{4} - \frac{1}{8}\sin^2 2x$$

and $f_6(x) = \frac{1}{6}(\sin^6 x + \cos^6 x)$

$$= \frac{1}{6}\{(\sin^2 x + \cos^2 x)^3 - 3\sin^2 x\cos^2 x\,(\sin^2 x + \cos^2 x)\}$$

$$= \frac{1}{6}\left\{1 - \frac{3}{4}(2\sin x\cos x)^2\right\} = \frac{1}{6} - \frac{1}{8}\sin^2 2x$$

Now, $f_4(x) - f_6(x) = \frac{1}{4} - \frac{1}{6} = \frac{3-2}{12} = \frac{1}{12}$

52. We have, $\sin\alpha\cos^3\alpha > \sin^3\alpha\cos\alpha$

$$\Rightarrow \quad \sin\alpha\cos\alpha(\cos^2\alpha - \sin^2\alpha) > 0$$

$$\Rightarrow \quad \cos^3\alpha\sin\alpha(1 - \tan^2\alpha) > 0$$

$$[\because \sin\alpha > 0 \text{ for } 0 < \alpha < \pi]$$

$$\Rightarrow \quad \cos\alpha\,(1 - \tan^2\alpha) > 0$$

$$\Rightarrow \quad \cos\alpha > 0 \text{ and } 1 - \tan^2\alpha > 0$$

$$\text{or} \quad \cos\alpha < 0 \text{ and } 1 - \tan^2\alpha < 0$$

$$\Rightarrow \quad a \in (0, \pi/4) \text{ or } \alpha \in \left(\frac{3\pi}{4}, \pi\right)$$

53. $\because x = \sum_{n=0}^{\infty}\cos^{2n}\theta = \frac{1}{1-\cos^2\theta} = \frac{1}{\sin^2\theta} \quad \left[\because 0 < \theta < \frac{\pi}{2}\right]$

$$y = \sum_{n=0}^{\infty}\sin^{2n}\theta = \frac{1}{1-\sin^2\theta} = \frac{1}{\cos^2\theta}$$

$$z = \sum_{n=0}^{\infty}\cos^{2n}\theta\sin^{2n}\theta = \frac{1}{1-\cos^2\theta\sin^2\theta}$$

$$z = \frac{1}{1 - \frac{1}{xy}}$$

$$\Rightarrow \quad xyz - z = xy \quad \ldots \text{(i)}$$

$$\because \quad \frac{1}{x} + \frac{1}{y} = \sin^2\theta + \cos^2\theta = 1$$

$$\Rightarrow \quad x + y = xy \quad \ldots \text{(ii)}$$

From Eqs. (i) and (ii),

$$(x + y)z - z = xy$$

$$\Rightarrow \quad xz + yz = xy + z$$

54. Taking first two members, then

$$2\cos(\alpha + \theta)\sin(\gamma + \alpha) = 2\cos(\beta + \theta)\sin(\beta + \gamma)$$

$$\Rightarrow \quad \sin(2\alpha + \theta + \gamma) - \sin(\theta - \gamma) = \sin(2\beta + \theta + \gamma) - \sin(\theta - \gamma)$$

$$\Rightarrow \quad \sin(2\alpha + \theta + \gamma) = \sin(2\beta + \theta + \gamma)$$

$$\Rightarrow \quad 2\alpha + \theta + \gamma = n\pi + (-1)^n(2\beta + \theta + \gamma)$$

for $n = 0$, $\ 2\alpha + \theta + \gamma = 2\beta + \theta + \gamma$

$\therefore \quad \alpha = \beta$

Similarly taking last two members, we get

$$\beta = \gamma$$

Hence, $\quad \alpha = \beta = \gamma$

Also, take $\quad n = 1, -1$

Then, we get $\alpha + \beta + \gamma + \theta = \frac{\pi}{2}, -\frac{\pi}{2}$

$$\therefore \quad k = \frac{\cos(\alpha + \theta)}{\sin(\beta + \gamma)} = \frac{\cos\left(\frac{\pi}{2} - (\beta + \gamma)\right)}{\sin(\beta + \gamma)}$$

$$= \frac{\sin(\beta + \gamma)}{\sin(\beta + \gamma)} = 1$$

and $\quad k = \frac{\cos\left(-\frac{\pi}{2} - (\beta + \gamma)\right)}{\sin(\beta + \gamma)} = -1$

Hence, $k = \pm 1$

55. Put $\theta = 0°$ and $90°$, then

$$\frac{A\cos\alpha + B\sin\beta}{A'\sin\alpha + B'\cos\beta} = \frac{-A\sin\alpha + B\cos\beta}{A'\cos\alpha - B'\sin\beta}$$

$$\Rightarrow AA'(\cos^2\alpha + \sin^2\alpha) - BB'(\sin^2\beta + \cos^2\beta)$$

$$= (A'B - AB')(\sin\alpha\cos\beta - \cos\alpha\sin\beta)$$

or $\quad (AA' - BB') = (A'B - AB')\sin(\alpha - \beta)$

56. It is clear from figure

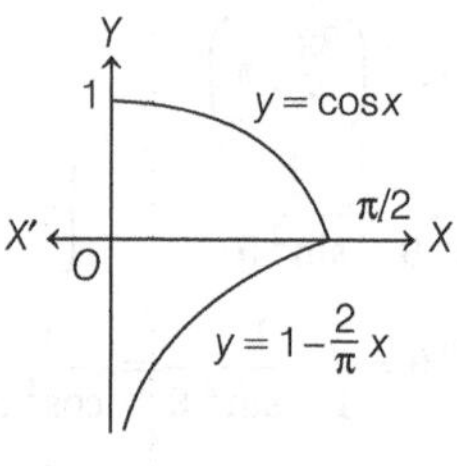

$$\cos x > 1 - \frac{2}{\pi}x$$

57. $ab\sin x + b\sqrt{(1-a^2)}\cos x$

Now, $\sqrt{(ab)^2 + (b\sqrt{(1-a^2)})^2}$

$$= \sqrt{a^2b^2 + b^2(1-a^2)} = b\sqrt{(a^2 + 1 - a^2)} = b$$

$\Rightarrow b\{(a\sin x + \sqrt{(1-a^2)}\cos x)\}$

Let $\quad a = \cos\alpha,$

$\therefore \quad \sqrt{(1-a^2)} = \sin\alpha \Rightarrow b\sin(x + a)$

$\because \quad -1 \le \sin(x+\alpha) \le 1$

$\therefore \quad c - b \le b\sin(x+\alpha) + c \le b + c$

$\therefore \quad b\sin(x+\alpha) + c \in [c-b, c+b]$

58. $\sin x = \sqrt{(1-\cos^2 x)} = \sqrt{(1-\tan^2 y)}$

$$= \sqrt{1 - (\sec^2 y - 1)} = \sqrt{(2 - \sec^2 y)}$$

$$= \sqrt{\left(2 - \frac{1}{\cos^2 y}\right)} = \sqrt{\left(2 - \frac{1}{\tan^2 z}\right)}$$

$$= \sqrt{(2 - \cot^2 z)} = \sqrt{\left(2 - \frac{\cos^2 z}{\sin^2 z}\right)}$$

$$= \sqrt{\left(2 - \frac{\cos^2 z}{1 - \cos^2 z}\right)} = \sqrt{\left(2 - \frac{\tan^2 x}{1 - \tan^2 x}\right)}$$

$$= \sqrt{\left(\frac{2 - 3\tan^2 x}{1 - \tan^2 x}\right)}$$

$$= \sqrt{\frac{2(1-\sin^2 x) - 3\sin^2 x}{(1-\sin^2 x) - \sin^2 x}}$$

$$= \sqrt{\left(\frac{2 - 5\sin^2 x}{1 - 2\sin^2 x}\right)}$$

$\Rightarrow \quad \sin^2 x(1 - 2\sin^2 x) = 2 - 5\sin^2 x$

$\Rightarrow \quad 2\sin^4 x - 6\sin^2 x + 2 = 0$

$\Rightarrow \quad \sin^4 x - 3\sin^2 x + 1 = 0$

$\therefore \quad \sin^2 x = \dfrac{3 \pm \sqrt{9-4}}{2}$

$$= \frac{3 \pm \sqrt{5}}{2} = \left(\frac{6 \pm 2\sqrt{5}}{4}\right) = \left(\frac{\sqrt{5} \pm 1}{2}\right)^2$$

$$= \left(\frac{\sqrt{5} - 1}{2}\right)^2 \qquad \left[\because \frac{\sqrt{5}+1}{2} > 1\right]$$

$$= (2\sin 18°)^2$$

$\Rightarrow \quad \sin x = 2\sin 18°$

Alternate Method

By symmetry

$$x = y = z$$

$\therefore \quad \cos x = \tan x$

$\Rightarrow \quad \cos^2 x = \sin x$

$\Rightarrow \quad 1 - \sin^2 x = \sin x$

$\Rightarrow \quad \sin^2 x + \sin x - 1 = 0$

$\therefore \quad \sin x = \dfrac{-1 \pm \sqrt{(1+4)}}{2} = \dfrac{-1 + \sqrt{5}}{2} \quad \left(\because \dfrac{-1-\sqrt{5}}{2} < -1\right)$

$$= 2\sin 18°$$

59. $\sin\left(\frac{\pi}{14}\right)\sin\left(\frac{3\pi}{14}\right)\sin\left(\frac{5\pi}{14}\right)$

$$= \cos\left(\frac{\pi}{2} - \frac{\pi}{14}\right)\cos\left(\frac{\pi}{2} - \frac{3\pi}{14}\right)\cos\left(\frac{\pi}{2} - \frac{5\pi}{14}\right)$$

$$= \cos\left(\frac{3\pi}{7}\right)\cos\left(\frac{2\pi}{7}\right)\cos\left(\frac{\pi}{7}\right)$$

$$= \cos\left(\pi - \frac{4\pi}{7}\right)\cos\left(\frac{2\pi}{7}\right)\cos\left(\frac{\pi}{7}\right)$$

$$= -\cos\left(\frac{\pi}{7}\right)\cos\left(\frac{2\pi}{7}\right)\cos\left(\frac{4\pi}{7}\right)$$

$\therefore \cos\left(\frac{\pi}{7}\right)\cos\frac{2\pi}{7}\cos\left(\frac{10\pi}{7}\right) - \sin\left(\frac{\pi}{14}\right)\sin\left(\frac{3\pi}{14}\right)\sin\left(\frac{5\pi}{14}\right)$

$$= \cos\left(\frac{\pi}{7}\right)\cos\left(\frac{2\pi}{7}\right)\cos\left(2\pi - \frac{10\pi}{7}\right) + \cos\left(\frac{\pi}{7}\right)\cos\left(\frac{2\pi}{7}\right)\cos\left(\frac{4\pi}{7}\right)$$

$$= 2\cos\left(\frac{\pi}{7}\right)\cos\left(\frac{2\pi}{7}\right)\cos\left(\frac{4\pi}{7}\right)$$

$$= \frac{2\sin\left(\frac{\pi}{7}\right)\cos\left(\frac{\pi}{7}\right)\cos\left(\frac{2\pi}{7}\right)\cos\left(\frac{4\pi}{7}\right)}{\sin\left(\frac{\pi}{7}\right)}$$

$$= \frac{2\sin\left(\frac{2\pi}{7}\right)\cos\left(\frac{2\pi}{7}\right)\cos\left(\frac{4\pi}{7}\right)}{2\sin\left(\frac{\pi}{7}\right)}$$

$$= \frac{2\sin\left(\frac{4\pi}{7}\right)\cos\left(\frac{4\pi}{7}\right)}{2\times 2\sin\left(\frac{\pi}{7}\right)}$$

$$= \frac{\sin\left(\frac{8\pi}{7}\right)}{4\sin\left(\frac{\pi}{7}\right)} = \frac{\sin\left(\pi + \frac{\pi}{7}\right)}{4\sin\left(\frac{\pi}{7}\right)} = -\frac{\sin\left(\frac{\pi}{7}\right)}{4\sin\left(\frac{\pi}{7}\right)} = -\frac{1}{4}$$

60. $\sum_{r=1}^{18} \cos^2(5r)^\circ = \cos^2 5^\circ + \cos^2 10^\circ + \cos^2 15^\circ + \ldots + \cos^2 85^\circ + \cos^2 90^\circ$

$= (\cos^2 5^\circ + \cos^2 85^\circ) + (\cos^2 10^\circ + \cos^2 80^\circ) + (\cos^2 15^\circ + \cos^2 75^\circ) + \ldots + (\cos^2 40^\circ + \cos^2 50^\circ) + \cos^2 45^\circ$

$= (\cos^2 5^\circ + \sin^2 5^\circ) + (\cos^2 10^\circ + \sin^2 10^\circ) + (\cos^2 15^\circ + \sin^2 15^\circ) + \ldots + (\cos^2 40^\circ + \sin^2 40^\circ) + \cos^2 45^\circ$

$= 1 + 1 + 1 + \ldots + 1 + \frac{1}{2} = 8 + \frac{1}{2} = \frac{17}{2}$

61. $\because 4n\alpha = \pi$

$\because \tan(2n-1)\alpha = \tan(2n\alpha - \alpha) = \tan\left(\frac{\pi}{2} - \alpha\right) = \cot\alpha$

$\tan(2n-2)\alpha = \tan(2n\alpha - 2\alpha)$

$= \tan\left(\frac{\pi}{2} - 2\alpha\right) = \cot 2\alpha$

....

$\therefore \tan\alpha \tan 2\alpha \tan 3\alpha \ldots \tan(2n-1)\alpha$

$(\tan\alpha \tan(2n-1)\alpha)(\tan 2\alpha \tan(2n-2)\alpha)\ldots \tan n\alpha$

$= (1\cdot 1\cdot 1\cdot\ldots 1)\tan n\alpha$

$= \tan n\alpha = \tan\frac{\pi}{4} = 1$

62. $4x^2 - 16x + 15 < 0$

$x^2 - 4x + \frac{15}{4} < 0 \Rightarrow (x-2)^2 - \frac{1}{4} < 0$

$\therefore \quad -\frac{1}{2} < x - 2 < \frac{1}{2} \Rightarrow \frac{3}{2} < x < \frac{5}{2}$

$\therefore \quad x = 2$

$\tan\alpha = 2, \cos\beta = \tan 45^\circ = 1$

$\frac{\sin(\alpha+\beta)}{\sin(\alpha-\beta)} = \frac{\sin\alpha\cos\beta + \cos\alpha\sin\beta}{\sin\alpha\cos\beta - \cos\alpha\sin\beta}$

$= \frac{\tan\alpha + \tan\beta}{\tan\alpha - \tan\beta} = \frac{2+0}{2-0} = 1$

63. $\because 2\cos\theta + \sin\theta = 1$

$\therefore \quad \frac{2(1-\tan^2\theta/2)}{(1+\tan^2\theta/2)} + \frac{(2\tan\theta/2)}{(1+\tan^2\theta/2)} = 1$

$\Rightarrow \quad 2 - 2\tan^2\theta/2 + 2\tan\theta/2 = 1 + \tan^2\theta/2$

$\Rightarrow \quad 3\tan^2(\theta/2) - 2\tan(\theta/2) - 1 = 0$

$\therefore \quad \tan\theta/2 = 1, \tan\theta/2 = -\frac{1}{3}$

$\Rightarrow \quad \theta = 90^\circ, \tan\theta/2 = -\frac{1}{3}$

$\therefore$ $7\cos\theta + 6\sin\theta$ at $\theta = 90^\circ$ is $0 + 6 = 6$

and $7\cos\theta + 6\sin\theta = \frac{7(1-\tan^2\theta/2)}{(1+\tan^2\theta/2)} + \frac{12\tan\theta/2}{(1+\tan^2\theta/2)}$

$\left[\because \tan\theta/2 = -\frac{1}{3}\right]$

$= \frac{7\left(1-\frac{1}{9}\right) + 12\left(-\frac{1}{3}\right)}{1+\frac{1}{9}} = 2$

64. We have, $\sin 10^\circ \sin 30^\circ \sin 50^\circ \sin 70^\circ$

$= \sin(30^\circ)[\sin(10^\circ)\sin(50^\circ)\sin(70^\circ)]$

$= \frac{1}{2}[\sin(10^\circ)\sin(60^\circ - 10^\circ)\sin(60^\circ + 10^\circ)]$

$= \frac{1}{2}\left[\frac{1}{4}\sin(3(10^\circ))\right]$

$\left[\because \sin\theta\sin(60^\circ-\theta)\sin(60^\circ+\theta) = \frac{1}{4}\sin 3\theta\right]$

$= \frac{1}{8}\sin 30^\circ = \frac{1}{8}\times\frac{1}{2} = \frac{1}{16}$

65. $2 - \cos x + 1 - \cos^2 x$

$= -\{\cos^2 x + \cos x\} + 3$

$= -\left\{\left(\cos x + \frac{1}{2}\right)^2 - \frac{1}{4}\right\} + 3 = \frac{13}{4} - \left(\cos x + \frac{1}{2}\right)^2$

Maximum value at $\cos x = -\frac{1}{2}$, *i.e.*, 13/4

Minimum value at $\cos x = 1$, *i.e.*, 1

$\therefore$ Required ratio = 13/4

66. $\because -\frac{\pi}{2} < x < -\frac{\pi}{4}$ $\quad \left[-\pi < 2x < -\frac{\pi}{2}\ ie, \text{ in III quadrant}\right]$

$\Rightarrow \quad \cos x + \sin x = a$

squaring both sides $\cos^2 x + \sin^2 x + 2\cos x\sin x = a^2$

$\Rightarrow \quad \sin 2x = (a^2 - 1)$

$\cos 2x = \sqrt{1-(a^2-1)^2} = \sqrt{a^2(2-a^2)} = a\sqrt{(2-a^2)}$

Alternate method

$\cos x + \sin x = a$

$\Rightarrow \frac{1}{\sqrt{2}}\sin x + \frac{1}{\sqrt{2}}\cos x = \frac{a}{\sqrt{2}}$

$\therefore \quad \sin\left(x + \frac{\pi}{4}\right) = \frac{a}{\sqrt{2}} \quad \ldots(i)$

$\Rightarrow \quad \cos 2x = \sin\left(\frac{\pi}{2} + 2x\right) = \sin\left(2\left(\frac{\pi}{4} + x\right)\right)$

$= 2\sin\left(\frac{\pi}{4} + x\right)\cos\left(\frac{\pi}{4} + x\right)$

$= 2\cdot\frac{a}{\sqrt{2}}\cdot\sqrt{\left(1 - \frac{a^2}{2}\right)} = a\sqrt{(2-a^2)}$

67. AM $\geq$ GM

$\Rightarrow \quad \frac{2^{\sin\theta} + 2^{-\cos\theta}}{2} \geq \sqrt{2^{\sin\theta}\cdot 2^{-\cos\theta}}$

$\Rightarrow \quad 2^{\sin\theta} + 2^{-\cos\theta} \geq 2^{1+\frac{1}{2}(\sin\theta - \cos\theta)}$

$\geq 2^{1+(\sqrt{2}/2)\sin\left(\theta - \frac{\pi}{4}\right)}$

$\geq 2^{1+(1/\sqrt{2})\sin\left(\theta - \frac{\pi}{4}\right)}$

For minimum value, $\sin\left(\theta - \frac{\pi}{4}\right) = -1$

$\theta - \frac{\pi}{4} = 2n\pi + \frac{3\pi}{2}, n \in I$

$\theta = 2n\pi + \frac{7\pi}{4}, n \in I$

and minimum value $= 2^{1-(1/\sqrt{2})}$

68. $\because \cos 3A + \cos 3B + \cos 3C = 1$

$\Rightarrow \cos 3A + \cos 3B + \cos 3C - 1 = 0$

$\Rightarrow \cos 3A + \cos 3B + \cos 3C + \cos 3\pi = 0$

$$\Rightarrow 2\cos\left(\frac{3A+3B}{2}\right)\cos\left(\frac{3A-3B}{2}\right) + 2\cos\left(\frac{3\pi+3C}{2}\right)\cos\left(\frac{3\pi-3C}{2}\right) = 0$$

$$\Rightarrow 2\cos\left(\frac{3\pi-3C}{2}\right)\left\{\cos\left(\frac{3A-3B}{2}\right) + \cos\left(\frac{3\pi+3C}{2}\right)\right\} = 0$$

$$\Rightarrow 2\cos\left(\frac{3\pi}{2} - \frac{3C}{2}\right)\cdot 2\cos\left(\frac{3\pi+3C+3A-3B}{4}\right)\cdot\cos\left(\frac{3\pi+3C-3A+3B}{4}\right) = 0$$

$$\Rightarrow 2\cos\left(\frac{3\pi}{2} - \frac{3C}{2}\right)\cdot 2\cos\left(\frac{3\pi}{2} - \frac{3B}{2}\right)\cdot\cos\left(\frac{3\pi}{2} - \frac{3A}{2}\right) = 0$$

$$\Rightarrow -4\sin\left(\frac{3A}{2}\right)\sin\left(\frac{3B}{2}\right)\sin\left(\frac{3C}{2}\right) = 0$$

$$\Rightarrow \sin\left(\frac{3A}{2}\right)\sin\left(\frac{3B}{2}\right)\sin\left(\frac{3C}{2}\right) = 0$$

$\therefore \frac{3A}{2} = \pi$ or $\frac{3B}{2} = \pi$ or $\frac{3C}{2} = \pi$

$\therefore A = \frac{2\pi}{3}$ or $B = \frac{2\pi}{3}$ or $C = \frac{2\pi}{3}$

69. $\because A > B$ [given]

and given A and B satisfy the equation

$$3\sin x - 4\sin^3 x - k = 0$$

or $\sin 3A - k = 0$ and $\sin 3B - k = 0$

$\therefore \sin 3A = \sin 3B$

$\Rightarrow 3A = \pi - 3B$

$\therefore A + B = \pi/3$, then $C = \pi - \frac{\pi}{3} = \frac{2\pi}{3}$

70. $\because \tan x \tan y = a$

and $\tan(x+y) = \tan\left(\frac{\pi}{6}\right)$

$$\Rightarrow \frac{\tan x + \tan y}{1 - \tan x \tan y} = \frac{1}{\sqrt{3}}$$

$$\Rightarrow \tan x + \tan y = \frac{1}{\sqrt{3}}(1-a)$$

Equation whose roots are $\tan x$ and $\tan y$ is

$$x^2 - \frac{(1-a)}{\sqrt{3}}x + a = 0$$

$$\Rightarrow \sqrt{3}x^2 - (1-a)x + a\sqrt{3} = 0$$

71. $\frac{\sin^3\theta - \cos^3\theta}{\sin\theta - \cos\theta} = \sin^2\theta + \cos^2\theta + \sin\theta\cos\theta, \theta \neq \frac{\pi}{4}, \frac{5\pi}{4}$

$= 1 + \sin\theta\cos\theta$

and $\frac{\cos\theta}{\sqrt{1+\cot^2\theta}} = \frac{\cos\theta}{|\operatorname{cosec}\theta|} = \sin\theta\cos\theta \ \forall\, \theta \in (0, \pi)$

and $-2\tan\theta\cot\theta = -2, \theta \neq \frac{\pi}{2}$

Hence, LHS = RHS

but $\theta \neq \frac{\pi}{4}, \frac{5\pi}{4}, \frac{\pi}{2}$

Hence, $\theta \in (0, \pi) \sim \left\{\frac{\pi}{4}, \frac{\pi}{2}\right\}$

72. $\because \sin\alpha = \frac{336}{625}$ and $450° < \alpha < 540°$

$$\therefore \sin(\alpha/4) = \sqrt{\frac{1-\cos(\alpha/2)}{2}}$$

$$= \sqrt{\frac{1}{2}\left\{1 - \sqrt{\left(\frac{1+\cos\alpha}{2}\right)}\right\}} = \sqrt{\frac{1}{2}\left\{1 - \sqrt{\frac{1+\frac{527}{625}}{2}}\right\}}$$

$$= \sqrt{\frac{1}{2}\left\{1 - \frac{24}{25}\right\}} = \sqrt{\frac{1}{50}} = \frac{1}{5\sqrt{2}}$$

73. $\sin(\theta+\alpha) = a$, $\sin(\theta+\beta) = b$

$\theta + \alpha = \sin^{-1} a$, $\quad \theta + \beta = \sin^{-1} b$

$\therefore \alpha - \beta = \sin^{-1} a - \sin^{-1} b$

$= \frac{\pi}{2} - \cos^{-1} a - \frac{\pi}{2} + \cos^{-1} b$

$= \cos^{-1} b - \cos^{-1} a$

$= \cos^{-1}\{(ab + \sqrt{1-b^2}\sqrt{1-a^2}\}$

$\cos(\alpha-\beta) = ab + \sqrt{(1-a^2-b^2+a^2b^2)}$

$\Rightarrow \cos^2(\alpha-\beta) = a^2b^2 + 1 - a^2 - b^2 + a^2b^2 + 2ab\sqrt{(1-a^2-b^2+a^2b^2)}$

$\therefore \cos 2(\alpha-\beta) - 4ab\cos(\alpha-\beta)$

$= 2\cos^2(\alpha-\beta) - 1 - 4ab\cos(\alpha-\beta)$

$= 4a^2b^2 + 2 - 2a^2 - 2b^2 + 4ab\sqrt{(1-a^2-b^2+a^2b^2)} - 1 - 4a^2b^2 - 4ab\sqrt{1-a^2-b^2}$

$= 1 - 2a^2 - 2b^2$

74. $\sin\left(\frac{3\pi}{2} - \alpha\right) = -\cos\alpha$, $\sin\left(\frac{\pi}{2} + \alpha\right) = \cos\alpha$

$\sin(3\pi + \alpha) = -\sin\alpha$, $\sin(5\pi - \alpha) = -\sin\alpha$

$$\therefore 3\left\{\sin^4\left(\frac{3\pi}{2} - \alpha\right) + \sin^4(3\pi + \alpha)\right\} - 2\left[\sin^6\left(\frac{\pi}{2} + \alpha\right) + \sin^6(5\pi - \alpha)\right]$$

$= 3\{\cos^4\alpha + \sin^4\alpha\} - 2\{\cos^6\alpha + \sin^6\alpha\}$

$= 3\{1 - 2\sin^2\alpha\cos^2\alpha\} - 2\{1 - 3\sin^2\alpha\cos^2\alpha\}$

$= 1$

75. $\because \sin^4\alpha\cos^2\alpha = \sum_{k=0}^{3} C_k \cos 2k\alpha$

$= C_0 + C_1\cos 2\alpha + C_2\cos 4\alpha + C_3\cos 6\alpha$

Put $\alpha = 0$, then $C_0 + C_1 + C_2 + C_3 = 0$ ...(i)

Put $\alpha = \frac{\pi}{2}$, then $C_0 - C_1 + C_2 - C_3 = 0$ ...(ii)

From Eqs. (i) and (ii),

$C_1 + C_3 = C_0 + C_2 = 0$

76. Alternate. (a) $\sin^2\theta = \frac{5}{4}$ (impossible)

Alternate. (b) $\because \frac{a^2+b^2}{2} \ge ab$

$\Rightarrow (a^2+b^2) \ge 2ab \Rightarrow \frac{2ab}{a^2+b^2} \le 1$

$\Rightarrow \cos\theta \le 1$, which is true

Alternate. (c) $(m^2+n^2)\operatorname{cosec}\theta = (m^2-n^2)$

$\operatorname{cosec}\theta = \frac{m^2-n^2}{m^2+n^2} = 1 - \frac{2n^2}{m^2+n^2} < 1$

which is impossible.

Alternate. (d) $\sin\theta = 2\cdot 375 > 1$

Impossible

77. $\because \tan^2\theta + \sec\theta = \lambda$

$\Rightarrow \sec^2\theta + \sec\theta - 1 - \lambda = 0$

$\therefore \sec\theta = \frac{-1 \pm \sqrt{(4\lambda+5)}}{2}$

for real $\sec\theta$, $4\lambda + 5 \ge 0$ *ie*, $\lambda \ge -\frac{5}{4}$

but we know thatm

$\sec\theta \le -1$ and $\sec\theta \ge 1$

$\therefore \frac{-1\pm\sqrt{(4\lambda+5)}}{2} \le -1$ and $\frac{-1\pm\sqrt{(4\lambda+5)}}{2} \ge 1$

$\Rightarrow -1-\sqrt{(4\lambda+5)} \le -2$ and $-1+\sqrt{(4\lambda+5)} \ge 2$

$\Rightarrow \sqrt{(4\lambda+5)} \ge 1$ and $\sqrt{(4\lambda+5)} \ge 3$

$\Rightarrow \sqrt{4\lambda+5} \ge 3 \Rightarrow 4\lambda+5 \ge 9 \Rightarrow \lambda \ge 1$

$\therefore \lambda \in [1, \infty)$

78. We have, $\cos^2 10^\circ - \cos 10^\circ\cos 50^\circ + \cos^2 50^\circ$

$= \frac{1}{2}[2\cos^2 10^\circ - 2\cos 10^\circ\cos 50^\circ + 2\cos^2 50^\circ]$

$= \frac{1}{2}[1+\cos 20^\circ - (\cos 60^\circ + \cos 40^\circ) + 1 + \cos 100^\circ]$

$[\because 2\cos^2 A = 1 + \cos 2A$ and $2\cos A\cos B = \cos(A+B) + \cos(A-B)]$

$= \frac{1}{2}\left[2 + \cos 20^\circ + \cos 100^\circ - \frac{1}{2} - \cos 40^\circ\right] \quad \left[\because \cos 60^\circ = \frac{1}{2}\right]$

$= \frac{1}{2}\left[\frac{3}{2} + (\cos 20^\circ - \cos 40^\circ) + \cos 100^\circ\right]$

$= \frac{1}{2}\left[\frac{3}{2} - 2\sin\frac{20^\circ+40^\circ}{2}\sin\frac{20^\circ-40^\circ}{2} + \cos 100^\circ\right]$

$\left[\because \cos C - \cos D = -2\sin\frac{C+D}{2}\sin\frac{C-D}{2}\right]$

$= \frac{1}{2}\left[\frac{3}{2} - 2\sin 30^\circ\sin(-10^\circ) + \cos(90^\circ + 10^\circ)\right]$

$= \frac{1}{2}\left[\frac{3}{2} + \sin 10^\circ - \sin 10^\circ\right] \quad [\because \cos(90^\circ+\theta) = -\sin\theta]$

$= \frac{1}{2} \times \frac{3}{2} = \frac{3}{4}$

79. Let $P = \log_{10}\tan 1^\circ + \log_{10}\tan 2^\circ$

$+ \log_{10}\tan 3^\circ \ldots + \log_{10}\tan 89^\circ$

$= \log_{10}(\tan 1^\circ \tan 2^\circ \tan 3^\circ \ldots \tan 45^\circ \ldots \tan 89^\circ)$

$= \log_{10} 1 = 0 \quad (\because \tan 89^\circ = \tan(90^\circ - 1^\circ) = \cot 1^\circ)$

$\therefore e^P = e^0 = 1 \quad [\therefore \tan 89^\circ \tan 1^\circ = 1]$

80. Given expression,

$= 3(\sin\theta - \cos\theta)^4 + 6(\sin\theta + \cos\theta)^2 + 4\sin^6\theta$

$= 3((\sin\theta - \cos\theta)^2)^2 + 6(\sin\theta + \cos\theta)^2 + 4(\sin^2\theta)^3$

$= 3(1 - \sin 2\theta)^2 + 6(1 + \sin 2\theta) + 4(1 - \cos^2\theta)^3$

$[\because 1 + \sin 2\theta = (\cos\theta + \sin\theta)^2$ and $1 - \sin 2\theta = (\cos\theta - \sin\theta)^2]$

$= 3(1^2 + \sin^2 2\theta - 2\sin 2\theta) + 6(1 + \sin 2\theta)$
$+ 4(1 - \cos^6\theta - 3\cos^2\theta + 3\cos^4\theta)$

$[\because (a-b)^2 = a^2 + b^2 - 2ab$ and $(a-b)^3 = a^3 - b^3 - 3a^2b + 3ab^2]$

$= 3 + 3\sin^2 2\theta - 6\sin 2\theta + 6 + 6\sin 2\theta + 4$
$- 4\cos^6\theta - 12\cos^2\theta + 12\cos^4\theta$

$= 13 + 3\sin^2 2\theta - 4\cos^6\theta - 12\cos^2\theta + 12\cos^4\theta$

$= 13 + 3(2\sin\theta\cos\theta)^2 - 4\cos^6\theta - 12\cos^2\theta(1 - \cos^2\theta)$

$= 13 + 12\sin^2\theta\cos^2\theta - 4\cos^6\theta - 12\cos^2\theta\sin^2\theta$

$= 13 - 4\cos^6\theta$

81. Let $f(\theta) = \cos^2(\cos\theta) + \sin^2(\sin\theta)$

$\because -1 \le \cos\theta \le 1$ and $-1 \le \sin\theta \le 1$

$\therefore \cos 1 \le \cos(\cos\theta) \le 1$ and $-\sin 1 \le \sin(\sin\theta) \le \sin 1$

$\therefore \cos^2 1 \le \cos^2(\cos\theta) \le 1$ and $0 \le \sin^2(\sin\theta) \le \sin^2 1$

$\therefore$ Maximum value of $f(\theta) = 1 + \sin^2 1$

82. $\because \cos^4\theta\sec^2\alpha, \frac{1}{2}$ and $\sin^4\theta\operatorname{cosec}^2\alpha$ are in AP

$1 = \cos^4\theta\sec^2\alpha + \sin^4\theta\operatorname{cosec}^2\alpha$

$\Rightarrow 1 = \frac{\cos^4\theta}{\cos^2\alpha} + \frac{\sin^4\theta}{\sin^2\alpha}$

$\Rightarrow (\sin^2\theta + \cos^2\theta)^2 = \frac{\cos^4\theta}{\cos^2\alpha} + \frac{\sin^4\theta}{\sin^2\alpha}$

$\Rightarrow \cos^4\theta\left(\frac{1}{\cos^2\alpha} - 1\right) + \sin^4\theta\left(\frac{1}{\sin^2\alpha} - 1\right)$
$- 2\sin^2\theta\cos^2\theta = 0$

$\Rightarrow \sin^4\alpha\cos^4\theta + \sin^4\theta\cos^4\alpha$
$- 2\sin^2\theta\cos^2\theta\sin^2\alpha\cos^2\alpha = 0$

$\Rightarrow (\sin^2\alpha\cos^2\theta - \cos^2\alpha\sin^2\theta)^2 = 0$

$\Rightarrow \tan^2\theta = \tan^2\alpha$

$\therefore \theta = n\pi \pm \alpha, n \in I$

Now, $\cos^8\theta \sec^6\alpha = \cos^8\alpha \sec^6\alpha = \cos^2\alpha$

and $\sin^8\theta \operatorname{cosec}^6\alpha = \sin^8\alpha \cdot \operatorname{cosec}^6\alpha = \sin^2\alpha$

Hence, $\cos^8\theta \sec^6\alpha, \frac{1}{2}, \sin^8\theta \operatorname{cosec}^6\alpha$

i.e. $\cos^2\alpha, \frac{1}{2}, \sin^2\alpha$ are in AP.

83. $\because \tan x = \dfrac{1+\sqrt{(1-x)}}{1+\sqrt{(1+x)}}$

In RHS put $x = \cos\theta$, then

$$\tan x = \frac{1+\sqrt{2}\sin(\theta/2)}{1+\sqrt{2}\cos(\theta/2)} = \frac{\frac{1}{\sqrt{2}}+\sin(\theta/2)}{\frac{1}{\sqrt{2}}+\cos(\theta/2)} = \frac{\sin\left(\frac{\pi}{4}\right)+\sin\left(\frac{\theta}{2}\right)}{\cos\left(\frac{\pi}{4}\right)+\cos\left(\frac{\theta}{2}\right)}$$

$$= \frac{2\sin\left(\frac{\pi}{8}+\frac{\theta}{4}\right)\cos\left(\frac{\pi}{8}-\frac{\theta}{4}\right)}{2\cos\left(\frac{\pi}{8}+\frac{\theta}{4}\right)\cos\left(\frac{\pi}{8}-\frac{\theta}{4}\right)} = \tan\left(\frac{\pi}{8}+\frac{\theta}{4}\right)$$

$\therefore \quad x = n\pi + \frac{\pi}{8} + \frac{\theta}{4}$

$\Rightarrow \quad 4x = 4n\pi + \frac{\pi}{2} + \theta$

$\therefore \quad \sin 4x = \cos\theta = x$

84. $\tan(\theta-\phi) = \dfrac{\tan\theta - \tan\phi}{1+\tan\theta\tan\phi} = \dfrac{(n-1)\tan\phi}{1+n\tan^2\phi} = \dfrac{(n-1)}{(\cot\phi + n\tan\phi)}$

For minimum of $\cot\phi + n\tan\phi$

$\cot\phi = n\tan\phi \Rightarrow \tan^2\phi = \frac{1}{n}$

$\therefore$ Maximum value of $\tan^2(\theta-\phi)$ is

$$\frac{(n-1)^2}{(2n\tan\phi)^2} = \frac{(n-1)^2}{4n^2\tan^2\phi} = \frac{(n-1)^2}{4n}$$

85. $\because |\tan A| < 1$

$\Rightarrow \quad -1 < \tan A < 1$

and $0 \le |A| < \frac{\pi}{2}$

$\Rightarrow \quad -\frac{\pi}{2} < A < \frac{\pi}{2}$

$\because \sqrt{(1+\sin 2A)} = \sqrt{1+\dfrac{2\tan A}{1+\tan^2 A}} = \dfrac{|1+\tan A|}{\sqrt{(1+\tan^2 A)}} = \dfrac{(1+\tan A)}{\sqrt{(1+\tan^2 A)}}$

and $\sqrt{(1-\sin 2A)} = \sqrt{1-\left(\dfrac{2\tan A}{1+\tan^2 A}\right)} = \dfrac{|1-\tan A|}{\sqrt{(1+\tan^2 A)}} = \dfrac{(1-\tan A)}{\sqrt{(1+\tan^2 A)}}$

$$\therefore \quad \frac{\sqrt{(1+\sin 2A)}+\sqrt{(1-\sin 2A)}}{\sqrt{(1+\sin 2A)}-\sqrt{(1-\sin 2A)}} = \frac{2}{2\tan A} = \cot A$$

86. For maximum value

$$2a^2 - 1 - \cos^2 x = 0$$

$\therefore \quad \cos^2 x = 2a^2 - 1$

$\Rightarrow \quad \sin^2 x = 1 - \cos^2 x = (2-2a^2)$

$\Rightarrow \quad 2a^2 + \sin^2 x = 2$

$\therefore$ Maximum value of $|\sqrt{(\sin^2 x + 2a^2)} - \sqrt{(2a^2-1-\cos^2 x)}|$

$= |\sqrt{2} - 0| = \sqrt{2}$

87. $\because \quad a_1 = \sqrt{\left(\dfrac{1+a_0}{2}\right)}$

Put, $a_0 = \cos\theta$

$\therefore \quad a_1 = \cos(\theta/2)$

$a_2 = \sqrt{\left(\dfrac{1+a_1}{2}\right)} = \cos(\theta/2^2)$

$\therefore \quad a_1 a_2 a_3 \ldots \text{to } \infty = \lim_{n\to\infty} a_1 a_2 a_3 \ldots a_n$

$$= \lim_{n\to\infty}\prod_{r=1}^{n} a_r = \lim_{n\to\infty}\prod_{r=1}^{n}\cos\left(\frac{\theta}{2^r}\right) = \lim_{n\to\infty}\prod_{r=1}^{n}\left\{\frac{2\sin\left(\frac{\theta}{2^r}\right)\cos\left(\frac{\theta}{2^r}\right)}{2\sin\left(\frac{\theta}{2^r}\right)}\right\} = \lim_{n\to\infty}\prod_{r=1}^{n}\left\{\frac{\sin\left(\frac{\theta}{2^{r-1}}\right)}{2\sin\left(\frac{\theta}{2^r}\right)}\right\}$$

$$= \lim_{n\to\infty}\left\{\frac{\sin\theta}{2\sin\left(\frac{\theta}{2}\right)}\cdot\frac{\sin\left(\frac{\theta}{2}\right)}{2\sin\left(\frac{\theta}{2^2}\right)}\cdot\frac{\sin\left(\frac{\theta}{2^2}\right)}{2\sin\left(\frac{\theta}{2^3}\right)}\cdots\frac{\sin\left(\frac{\theta}{2^{n-1}}\right)}{2\sin\left(\frac{\theta}{2^n}\right)}\right\}$$

$$= \lim_{n\to\infty}\left\{\frac{\sin\theta}{\dfrac{2^n\sin\left(\frac{\theta}{2^n}\right)}{\left(\frac{\theta}{2^n}\right)}\cdot\dfrac{\theta}{2^n}}\right\} = \frac{\sin\theta}{\theta(1)} = \frac{\sin\theta}{\theta}$$

$$\therefore \quad \cos\left(\frac{\sqrt{(1-a_0^2)}}{a_1 a_2 a_3 \ldots \infty}\right) = \cos\left(\frac{\sin\theta}{\frac{\sin\theta}{\theta}}\right) = \cos\theta = a_0$$

Objective Questions Type II [One or more than one correct answer(s)]

1. Let $\frac{\tan x}{1}=\frac{\tan y}{2}=\frac{\tan z}{3}=\lambda$ (say)

$\therefore$ $\tan x=\lambda$, $\tan y=2\lambda$, $\tan z=3\lambda$

$\because$ $x+y+z=\pi$

$\therefore$ $\tan x+\tan y+\tan z=\tan x\tan y\tan z$

$\Rightarrow$ $6\lambda=6\lambda^3$

$\therefore$ $6\lambda(\lambda^2-1)=0$

$\lambda=\pm 1(\lambda\neq 0)$ [given]

$\therefore$ $\tan x=\pm 1$, $\tan y=\pm 2$, $\tan z=\pm 3$

$\tan x+\tan y+\tan z=\pm 6$

Maximum and minimum values of $\tan x+\tan y+\tan z$ are 6 and -6 respectively.

2. $\because 3\sin\beta=\sin(2\alpha+\beta)$

$\Rightarrow 2\sin\beta=\sin(2\alpha+\beta)-\sin\beta$

$=2\cos(\alpha+\beta)\sin\alpha$

$\therefore$ $\sin\beta=\cos(\alpha+\beta)\sin\alpha$ ...(i)

$\therefore$ Alternate (b) is correct

Also, $\sin\beta=\sin((\alpha+\beta)-\alpha)$

$=\sin(\alpha+\beta)\cos\alpha-\cos(\alpha+\beta)\sin\alpha$ ...(ii)

From Eqs. (i) and (ii)

$\sin\beta=\sin(\alpha+\beta)\cos\alpha-\sin\beta$

$\therefore$ $2\sin\beta=\sin(\alpha+\beta)\cos\alpha$

[$\therefore$ Alternate. (c) is correct]

Alternate (a)

$$\text{LHS}=(\cot\alpha+\cos(\alpha+\beta))(\cot\beta-3\cot(2\alpha+\beta))$$

$$=\left(\frac{\sin(2\alpha+\beta)}{\sin\alpha\cdot\sin(\alpha+\beta)}\right)\left(\frac{\cos\beta}{\sin\beta}-\frac{3\cos(2\alpha+\beta)}{\sin(2\alpha+\beta)}\right)$$

$$=\left(\frac{3\sin\beta}{\sin\alpha\cdot\sin(\alpha+\beta)}\right)\left(\frac{\cos\beta}{\sin\beta}-\frac{3\cos(2\alpha+\beta)}{3\sin\beta}\right)$$

$(\because 3\sin\beta=\sin(2\alpha+\beta))$

$$=\left(\frac{3\sin\beta}{\sin\alpha\cdot\sin(\alpha+\beta)}\right)\left(\frac{\cos\beta-\cos(2\alpha+\beta)}{\sin\beta}\right)$$

$$=\left(\frac{3\sin\beta}{\sin\alpha\cdot\sin(\alpha+\beta)}\right)\left(\frac{2\sin(\alpha+\beta)\sin\alpha}{\sin\beta}\right)=6$$

Alternate (d)

$\because$ $\tan(\alpha+\beta)=2\tan\alpha$

$\Rightarrow$ $\frac{\sin(\alpha+\beta)}{\cos(\alpha+\beta)}=\frac{2\sin\alpha}{\cos\alpha}$

$\Rightarrow \sin(\alpha+\beta)\cos\alpha=2\cos(\alpha+\beta)\sin\alpha$

$\Rightarrow \sin(\alpha+\beta)\cos\alpha-\cos(\alpha+\beta)\sin\alpha=\cos(\alpha+\beta)\sin\alpha$

$\sin\beta=\cos(\alpha+\beta)\sin\alpha$ [Alternate. (b)]

3. Given,

$$x=\sqrt{a^2\cos^2\alpha+b^2\sin^2\alpha}+\sqrt{a^2\sin^2\alpha+b^2\cos^2\alpha}$$

$$\Rightarrow\quad x^2=a^2+b^2+2\sqrt{(a^2\cos^2\alpha+b^2\sin^2\alpha)(a^2\sin^2\alpha+b^2\cos^2\alpha)}$$

$=a^2+b^2+k$, where

$$k=2\sqrt{[(a^2+b^2-(a^2\sin^2\alpha+b^2\cos^2\alpha)]\times(a^2\sin^2\alpha+b^2\cos^2\alpha)}$$

$\therefore$ $x=a^2+b^2+2\sqrt{(a^2+b^2)P-P^2}$

where, $P=a^2\sin^2\alpha+b^2\cos^2\alpha$

$\Rightarrow$ $P=(a^2/2)(1-\cos 2\alpha)+(b^2/2)(1+\cos 2\alpha)$

$=(1/2)[a^2+b^2-(a^2-b^2)\cos 2\alpha]$

Also,

$$x^2=a^2+b^2+2\sqrt{(a^2\cos^2\alpha+b^2\sin^2\alpha)[(a^2+b^2)-(a^2\cos^2\alpha+b^2\sin^2\alpha)]}$$

$$=a^2+b^2+2\sqrt{(a^2+b^2)P-P^2}$$

where, $P=a^2\cos^2\alpha+b^2\sin^2\alpha$

$\Rightarrow$ $P=(1/2)[a^2+b^2+(a^2-b^2)\cos 2\alpha]$

4. $\because$ $\cos\alpha=\frac{3}{5}\Rightarrow\sin\alpha=\frac{4}{5}$

and $\cos\beta=\frac{5}{13}\Rightarrow\sin\beta=\frac{12}{13}$

Alternate. (a) $\cos(\alpha+\beta)=\cos\alpha\cos\beta-\sin\alpha\sin\beta$

$$=\frac{3}{5}\cdot\frac{5}{13}-\frac{4}{5}\cdot\frac{12}{13}=-\frac{33}{65}$$

Alternate. (b) $\sin(\alpha+\beta)=\sin\alpha\cos\beta+\cos\alpha\sin\beta$

$$=\frac{4}{5}\cdot\frac{5}{13}+\frac{3}{5}\cdot\frac{12}{13}=\frac{56}{65}$$

Alternate. (c) $\sin^2\left(\frac{\alpha-\beta}{2}\right)=\frac{1-\cos(\alpha-\beta)}{2}$

$$=\frac{1-\{\cos\alpha\cos\beta+\sin\alpha\sin\beta\}}{2}$$

$$=\frac{1-\left\{\frac{3}{5}\cdot\frac{5}{13}+\frac{4}{5}\cdot\frac{12}{13}\right\}}{2}=\frac{1-\frac{63}{65}}{2}=\frac{1}{65}$$

Alternate. (d)

$\cos(\alpha-\beta)=\cos\alpha\cdot\cos\beta+\sin\alpha\cdot\sin\beta$

$$=\frac{3}{5}\cdot\frac{5}{13}+\frac{4}{5}\cdot\frac{12}{13}=\frac{63}{65}$$

5. $\because$ $\sin^6 x+\cos^6 x=a^2$

$\Rightarrow (\sin^2 x)^3+(\cos^2 x)^3=a^2$

$\Rightarrow (\sin^2 x+\cos^2 x)^3-3\sin^2 x\cos^2 x(\sin^2 x+\cos^2 x)=a^2$

$\Rightarrow$ $1-3\sin^2 x\cos^2 x=a^2$

$\Rightarrow$ $1-\frac{3}{4}\sin^2 2x=a^2$

$\Rightarrow$ $\sin^2 2x=\frac{4}{3}(1-a^2)$

$\because$ $0\le\sin^2 2x\le 1$

$\therefore$ $0\le\frac{4}{3}(1-a^2)\le 1$

$\Rightarrow$ $0\le(1-a^2)\le\frac{3}{4}\Rightarrow 0\ge -1+a^2\ge-\frac{3}{4}$

$\therefore$ $\frac{1}{4}\le a^2\le 1$

$\Rightarrow$ $a\in\left[-1,-\frac{1}{2}\right]\cup\left[\frac{1}{2},1\right]$

6. $\therefore \tan\alpha + \tan\beta = -p,\ \tan\alpha\tan\beta = q\ (p \neq 0)$

$$\Rightarrow \tan(\alpha+\beta) = \frac{\tan\alpha + \tan\beta}{1 - \tan\alpha\tan\beta} = \frac{-p}{1-q}$$

$$= \frac{p}{q-1}$$ [Alternate. (b)]

Alternate. (a) :

$$\text{LHS} = \cos^2(\alpha+\beta)\{\tan^2(\alpha+\beta) + p\tan(\alpha+\beta) + q\}$$

$$= \frac{\tan^2(\alpha+\beta) + p\tan(\alpha+\beta) + q}{1 + \tan^2(\alpha+\beta)}$$

$$= \frac{\frac{p^2}{(q-1)^2} + \frac{p^2}{(q-1)} + q}{1 + \frac{p^2}{(q-1)^2}} = \frac{p^2 + p^2(q-1) + q(q-1)^2}{(q-1)^2 + p^2}$$

$$= \frac{q\{p^2 + (q-1)^2\}}{\{p^2 + (q-1)^2\}} = q$$

Alternate. (c) : $\because \tan(\alpha+\beta) = \frac{p}{q-1}$

$$\therefore \quad \cos(\alpha+\beta) = \frac{(q-1)}{\sqrt{p^2 + (q-1)^2}}$$

Alternate. (d) : $\sin(\alpha+\beta) = \frac{p}{\sqrt{p^2 + (q-1)^2}}$

7. $\sin\theta + \sin\phi = a$ and $\cos\theta + \cos\phi = b$

$$2\sin\left(\frac{\theta+\phi}{2}\right)\cos\left(\frac{\theta-\phi}{2}\right) = a \quad \text{...(i)}$$

and $$2\cos\left(\frac{\theta+\phi}{2}\right)\cos\left(\frac{\theta-\phi}{2}\right) = b \quad \text{...(ii)}$$

Squaring and additing Eqs. (i) and (ii), then

$$4\cos^2\left(\frac{\theta-\phi}{2}\right) = a^2 + b^2$$

$$\therefore \quad \cos\left(\frac{\theta-\phi}{2}\right) = \pm\frac{1}{2}\sqrt{(a^2+b^2)}$$

and $2(1 + \cos(\theta-\phi)) = a^2 + b^2$

$$\therefore \quad \cos(\theta-\phi) = \frac{a^2+b^2-2}{2}$$

$$\Rightarrow \quad \frac{1-\tan^2\left(\frac{\theta-\phi}{2}\right)}{1+\tan^2\left(\frac{\theta-\phi}{2}\right)} = \frac{a^2+b^2-2}{2}$$

By componendo and dividendo method

$$\frac{2\tan^2\left(\frac{\theta-\phi}{2}\right)}{2} = \frac{4-a^2-b^2}{a^2+b^2}$$

$$\therefore \quad \tan\left(\frac{\theta-\phi}{2}\right) = \pm\sqrt{\left(\frac{4-a^2-b^2}{a^2+b^2}\right)}$$

8. $f_n(\theta) = \tan(\theta/2)\prod_{r=0}^{n}(1 + \sec 2^r\theta)$

$$= \tan(\theta/2)\prod_{r=0}^{n}\left\{\frac{1+\cos(2^r\theta)}{\cos(2^r\theta)}\right\}$$

$$= \tan(\theta/2)\prod_{r=0}^{n}\frac{2\cos^2(2^{r-1}\theta)}{\cos(2^r\theta)}$$

$$= 2^{n+1}\cdot\tan(\theta/2)\prod_{r=0}^{n}\frac{\cos^2(2^{r-1}\theta)}{\cos(2^r\theta)}$$

$$= 2^{n+1}\cdot\tan(\theta/2)\cdot\cos^2(\theta/2)\cdot\prod_{r=0}^{n}\frac{\cos(2^r\theta)}{\cos(2^n\theta)}$$

$$= 2^n\cdot\sin\theta\cdot\frac{\sin(2^n\theta)}{2^n\cdot\sin\theta\cdot\cos(2^n\theta)}$$

$$= \tan(2^n\theta)$$

$\therefore$ **Alternate. (a) :** $f_2\left(\frac{\pi}{16}\right) = \tan\left(\frac{\pi}{4}\right) = 1$

Alternate. (b) : $f_3\left(\frac{\pi}{32}\right) = \tan\left(\frac{\pi}{4}\right) = 1$

Alternate. (c) : $f_4\left(\frac{\pi}{64}\right) = \tan\left(\frac{\pi}{4}\right) = 1$

Alternate. (d) : $f_5\left(\frac{\pi}{128}\right) = \tan\left(\frac{\pi}{4}\right) = 1$

9. $\because 0 < \phi \le \frac{\pi}{2}$

$$\therefore \quad x = \sum_{n=0}^{\infty}\cos^{2n}\phi = 1 + \cos^2\phi + \cos^4\phi + \ldots$$

$$= \frac{1}{1-\cos^2\phi} = \frac{1}{\sin^2\phi}$$

$$\therefore \quad \sin^2\phi = \frac{1}{x} \quad \text{...(i)}$$

and $$y = \sum_{n=0}^{\infty}\sin^{2n}\phi = 1 + \sin^2\phi + \sin^4\phi + \ldots$$

$$= \frac{1}{1-\sin^2\phi} = \frac{1}{\cos^2\phi}$$

$$\therefore \cos^2\phi = \frac{1}{y} \quad \text{...(ii)}$$

From Eqs. (i) and (ii),

$$\sin^2\phi + \cos^2\phi = 1$$

$$\Rightarrow \quad \frac{1}{x} + \frac{1}{y} = 1$$

$$\therefore \quad x + y = xy \quad \text{...(iii)}$$

and $$z = \sum_{n=0}^{\infty}\cos^{2n}\phi\sin^{2n}\phi = 1 + \cos^2\phi\sin^2\phi + \cos^4\phi\sin^4\phi + \ldots$$

$$= \frac{1}{1-\sin^2\phi\cos^2\phi}$$

$$= \frac{1}{1-\frac{1}{x}\cdot\frac{1}{y}}$$ [form Eqs. (i) and (ii)]

$$\Rightarrow \quad z = \frac{xy}{xy-1}$$

$\Rightarrow \quad xyz = xy + z$ [Alternate. (b)]

$\Rightarrow \quad xyz = x + y + z$ [from Eq. (iii)]

[Alternate. (c)]

10. $\because \tan x = \dfrac{2b}{a-c} (a \neq c)$

$$y = a\cos^2 x + 2b \sin x \cos x + c\sin^2 x \quad \ldots(i)$$

and $z = a\sin^2 x - 2b \sin x \cos x + \cos^2 x \quad \ldots(ii)$

Adding Eqs. (i) and (ii)

$$y + z = a + c \quad \text{[Alternate.(b)]}$$

and $y - z = a\cos 2x + 2b \sin 2x - c\cos 2x$

$$= (a-c)\cos 2x + 2b \sin 2x$$

$$= (a-c)\left\{\frac{1-\tan^2 x}{1+\tan^2 x}\right\} + \left\{\frac{4b \tan x}{1+\tan^2 x}\right\}$$

$$= (a-c)\left\{\frac{1-\dfrac{4b^2}{(a-c)^2}}{1+\dfrac{4b^2}{(a-c)^2}}\right\} + \left\{\frac{\dfrac{8b^2}{(a-c)}}{1+\dfrac{4b^2}{(a-c)^2}}\right\}$$

$$= \frac{(a-c)\{(a-c)^2 - 4b^2\}}{\{(a-c)^2 + 4b^2\}} + \frac{8b^2(a-c)}{\{(a-c)^2 + 4b^2\}}$$

$$= \frac{(a-c)\{(a-c)^2 + 4b^2\}}{\{(a-c)^2 + 4b^2\}}$$

$$= (a-c)$$

$\therefore \quad y - z = a - c \quad$ [Alternate. (c)]

11. $\left(\dfrac{\cos A + \cos B}{\sin A - \sin B}\right)^n + \left(\dfrac{\sin A + \sin B}{\cos A - \cos B}\right)^n$

$$= \cos^n\left(\frac{A-B}{2}\right) + \cot^n\left(\frac{B-A}{2}\right)$$

If n even, $2\cot^n\left(\dfrac{A-B}{2}\right)$, if n odd, 0

12. $\because \dfrac{3 + \cot 76° \cot 16°}{\cot 76° + \cot 16°}$

$$= \frac{3\sin 76° \sin 16° + \cos 76° \cos 16°}{\sin(76°+16°)}$$

$$= \frac{2\sin 76° \sin 16° + \cos(76°-16°)}{\sin(92°)}$$

$$= \frac{\cos 60° - \cos 92° + \cos 60°}{\sin 92°}$$

$$= \frac{1 - \cos 92°}{\sin 92°}$$

$= \tan 46° \quad$ [Alternate. (c)]

$= \tan(90° - 44°) = \cot 44° \quad$ [Alternate. (d)]

13. We have,

$$\cos x - \sin\alpha \cot\beta \sin x = \cos\alpha$$

$$\Rightarrow \frac{1-\tan^2\left(\dfrac{x}{2}\right)}{1+\tan^2\left(\dfrac{x}{2}\right)} - \sin\alpha\,(\delta+\beta)\frac{2\tan\left(\dfrac{x}{2}\right)}{1+\tan^2\left(\dfrac{x}{2}\right)} = \cos\alpha$$

$$\Rightarrow \tan^2\frac{x}{2}(1+\cos\alpha) + 2\sin\alpha\cot\beta\tan\frac{x}{2} - (1-\cos\alpha) = 0$$

$$\Rightarrow \tan^2\frac{x}{2} + \frac{2\sin\alpha\cot\beta}{1+\cos\alpha}\tan\frac{x}{2} - \frac{1-\cos\alpha}{1+\cos\alpha} = 0$$

$$\Rightarrow \tan^2\frac{x}{2} + 2\tan\frac{\alpha}{2}\cot\beta\tan\frac{x}{2} - \tan^2\frac{\alpha}{2} = 0$$

$$\Rightarrow \tan^2\frac{x}{2} + 2\tan\frac{\alpha}{2}\cdot\frac{1}{2}\left(\cot\frac{\beta}{2} - \tan\frac{\beta}{2}\right)\tan\frac{x}{2} - \tan^2\frac{\alpha}{2} = 0$$

$$\Rightarrow \left(\tan\frac{x}{2} + \cot\frac{\beta}{2}\tan\frac{\alpha}{2}\right)\left(\tan\frac{x}{2} - \tan\frac{\beta}{2}\tan\frac{\alpha}{2}\right) = 0$$

$$\Rightarrow \tan\left(\frac{x}{2}\right) = -\tan\left(\frac{\alpha}{2}\right)\cot\left(\frac{\beta}{2}\right)$$

$$\Rightarrow \tan\left(\frac{x}{2}\right) = \tan\left(\frac{\alpha}{2}\right)\tan\left(\frac{\beta}{2}\right)$$

14. $\tan A + \tan B + \tan C = 6 \quad \ldots(i)$

$$\Rightarrow \tan A \tan B \tan C = 6$$

$$2\tan C = 6$$

$$\therefore \quad \tan C = 3$$

$$\therefore \quad \sin^2 C = \frac{\tan^2 C}{1+\tan^2 C} = \frac{9}{1+9} = \frac{9}{10}$$

From Eq. (i), $\tan A + \tan B = 3$ and $\tan A \tan B = 2$

$$\tan A - \tan B = \pm\sqrt{\{(\tan A + \tan B)^2 - 4\tan A \tan B)\}}$$

$$= \pm 1$$

we get, $\tan A = 2, 1$ and $\tan B = 1, 2$

$$\therefore \quad \sin^2 A = \frac{4}{1+4}, \frac{1}{1+1}$$

and $\quad \sin^2 B = \dfrac{1}{1+1}, \dfrac{4}{1+4}$

$$\Rightarrow \quad \sin^2 A = \frac{8}{10}, \frac{5}{10} \text{ and } \sin^2 B = \frac{5}{10}, \frac{8}{10}$$

$\therefore \sin^2 A : \sin^2 B : \sin^2 C = 8:5:9$ or $5:8:9$

15. $\because \cot\theta + \tan\theta = x$ and $\sec\theta - \cos\theta = y$

Alternate. (a) :

$$\therefore \quad \frac{1}{x} = \frac{1}{\cot\theta + \tan\theta} = \frac{\sin\theta\cos\theta}{\cos^2\theta + \sin^2\theta}$$

$$= \frac{\sin\theta\cos\theta}{1}$$

$$\therefore \quad \sin\theta\cos\theta = \frac{1}{x}$$

Alternate. (b) :

$$\because \quad y = \sec\theta - \cos\theta = \frac{1-\cos^2\theta}{\cos\theta} = \frac{\sin^2\theta}{\cos\theta}$$

$$= \sin\theta\tan\theta$$

Hence, $\sin\theta\tan\theta = y$

Alternate. (c) :

$$\because \quad x^2y = \frac{1}{\sin^2\theta\cos^2\theta} \times \frac{\sin^2\theta}{\cos\theta} = \frac{1}{\cos^3\theta}$$

$$\Rightarrow \quad \left(\frac{1}{x^2y}\right)^{2/3} = \cos^2\theta$$

$$\Rightarrow \quad (x^2y)^{2/3} = \sec^2\theta \quad \ldots(i)$$

and $\quad xy^2 = \dfrac{1}{\sin\theta\cos\theta} \times \dfrac{\sin^4\theta}{\cos^2\theta} = \tan^3\theta$

$$\Rightarrow \quad (xy^2)^{2/3} = \tan^2\theta \quad \ldots(ii)$$

From Eqs. (i) and (ii),

$$(x^2y)^{2/3} - (xy^2)^{2/3} = 1$$

16. $\because \dfrac{x}{y} = \dfrac{\cos A}{\cos B} \Rightarrow \dfrac{x}{\cos A} = \dfrac{y}{\cos B} = \lambda$ (say)

$\therefore \quad x = \lambda \cos A, \; y = \lambda \cos B$

Alternate. (a) :

$$\text{RHS} = \frac{x \tan A + y \tan B}{x + y} = \frac{\lambda \cos A \tan A + \lambda \cos B \tan B}{\lambda \cos A + \lambda \cos B}$$

$$= \left(\frac{\sin A + \sin B}{\cos A + \cos B}\right) = \frac{2 \sin\left(\frac{A+B}{2}\right)\cos\left(\frac{A-B}{2}\right)}{2 \cos\left(\frac{A+B}{2}\right)\cos\left(\frac{A-B}{2}\right)}$$

$$= \tan\left(\frac{A+B}{2}\right) = \text{LHS}$$

Alternate. (b) :

$$\text{RHS} = \frac{x \tan A - y \tan B}{x + y} = \frac{\lambda \cos A \tan A - \lambda \cos B \tan B}{\lambda \cos A + \lambda \cos B}$$

$$= \frac{\sin A - \sin B}{\cos A + \cos B} = \frac{2 \cos\left(\frac{A+B}{2}\right)\sin\left(\frac{A-B}{2}\right)}{2 \cos\left(\frac{A+B}{2}\right)\cos\left(\frac{A-B}{2}\right)}$$

$$= \tan\left(\frac{A-B}{2}\right) = \text{LHS}$$

Alternate. (c) :

$$\text{RHS} = \frac{y \sin A + x \sin B}{y \sin A - x \sin B} = \frac{\lambda \cos B \sin A + \lambda \cos A \sin B}{\lambda \cos B \sin A - \lambda \cos A \sin B}$$

$$= \frac{\sin (A+B)}{\sin (A-B)} = \text{LHS}$$

Alternate. (d) :

$\text{LHS} = x \cos A + y \cos B = \lambda \cos^2 A + \lambda \cos^2 B \neq 0$

17. $\tan \theta = \dfrac{\sin \alpha - \cos \alpha}{\sin \alpha + \cos \alpha} = \dfrac{\tan \alpha - 1}{\tan \alpha + 1}$

$\tan \theta = \tan\left(\alpha - \dfrac{\pi}{4}\right)$

$\Rightarrow \quad \theta = n\pi + \alpha - \pi/4, \; n \in I$

or $\quad 2\theta = 2n\pi + 2\alpha - \dfrac{\pi}{2}$

$\sin 2\theta = \sin\left(2\alpha - \dfrac{\pi}{2}\right) = -\cos 2\alpha$

and $\cos 2\theta = \cos\left(2\alpha - \dfrac{\pi}{2}\right) = \sin 2\alpha$

and $\sin \alpha - \cos \alpha = \sqrt{2} \sin\left(\alpha - \dfrac{\pi}{4}\right)$

$= \sqrt{2} \sin \{\theta - n\pi\}$

$= \pm \sqrt{2} \sin \theta$

and $\sin \alpha + \cos \alpha = \sqrt{2} \sin\left(\alpha + \dfrac{\pi}{4}\right)$

$= \sqrt{2} \sin\left\{\dfrac{\pi}{2} + \theta - n\pi\right\}$

$= \sqrt{2} \cos (\theta - n\pi) = \pm \sqrt{2} \cos \theta$

18. $x^2 + 4xy + y^2 = aX^2 + bY^2$

$\Rightarrow (X \cos\theta + Y \sin\theta)^2 + 4(X \cos\theta + Y \sin\theta) \times (X \sin\theta - Y \cos\theta) + (X \sin\theta - Y \cos\theta)^2 = aX^2 + bY^2$

$\Rightarrow X^2(1 + 4 \sin\theta \cos\theta) + Y^2(1 - 4 \sin\theta \cos\theta) + 4XY(\sin^2\theta - \cos^2\theta) = aX^2 + bY^2$

On comparing

$1 + 4 \sin\theta \cos\theta = a, \; 1 - 4 \sin\theta \cos\theta = b$

and $\sin^2\theta - \cos^2\theta = 0$

$\therefore \quad \theta = \pi/4$, then $a = 3, \; b = -1$

19. Given, $\dfrac{\tan 3A}{\tan A} = k$ ...(i)

$\Rightarrow \dfrac{\tan 3A - \tan A}{\tan A} = k - 1$

$\Rightarrow \dfrac{\sin 2A}{\cos 3A \sin A} = k - 1$

$\Rightarrow \dfrac{2 \cos A}{\cos 3A} = k - 1 \Rightarrow \dfrac{\cos A}{\cos 3A} = \dfrac{k-1}{2}$

$\Rightarrow$ (a) is incorrect.

Again, $\dfrac{\tan 3A}{\tan A} = k$

$\Rightarrow \dfrac{\sin 3A}{\cos 3A} \cdot \dfrac{\cos A}{\sin A} = k$

$\Rightarrow \dfrac{\sin 3A}{\sin A} = k \cdot \dfrac{2}{k-1} = \dfrac{2k}{k-1}$

$\Rightarrow$ (b) is correct.

$\Rightarrow \dfrac{3 \sin A - 4 \sin^3 A}{\sin A} = \dfrac{2k}{k-1}$

$\Rightarrow 3 - 4 \sin^2 A = \dfrac{2k}{k-1}$ or $4 \sin^2 A = \dfrac{k-3}{k-1}$

$\Rightarrow \quad 0 < \dfrac{k-3}{k-1} < 4$ [$\sin A \neq 0$ or 1]

Now, $\dfrac{k-3}{k-1} > 0$

$\therefore \quad k \in (-\infty, 1) \cup (3, \infty)$ ...(ii)

and $\dfrac{k-3}{k-1} < 4 \Rightarrow \dfrac{3k-1}{k-1} > 0$

$\therefore \; k \in (-\infty, 1/3) \cup (1, \infty)$ ...(iii)

From Eqs. (ii) and (iii), we get

$k < \dfrac{1}{3}$ and $k > 3$

20. c, $\sin B$, $\cos B$ are rational in ΔABC,

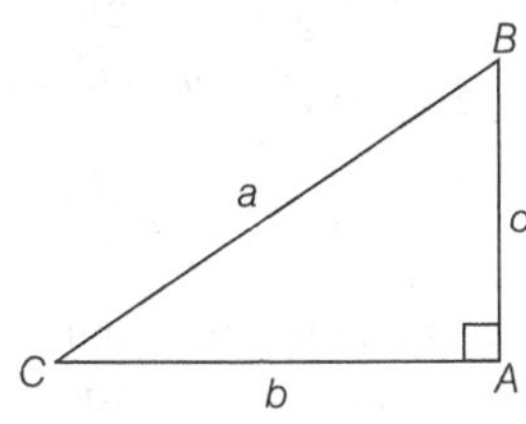

$$\cos B = \frac{c}{a} \Rightarrow a = \frac{c}{\cos B}$$

= rational [$\because$ c and $\cos B$ are rational]

and $\sin B = \frac{1}{a} \Rightarrow b = a \sin B$ = rational

[$\because a$ and $\sin B$ are rational]

and $\sin B = \frac{b}{a} \Rightarrow b = a \sin B$ = rational

[$\because a$ and $\sin B$ are rational]

21. Given, $\cos x + \sec x = -2$

$$\Rightarrow \cos^2 x + 2\cos x + 1 = 0$$

$$\Rightarrow (\cos x + 1)^2 = 0$$

$$\Rightarrow \cos x = -1$$

$$\therefore \sec x = -1$$

$$\therefore \cos^n x + \sec^n x = (-1)^n + (-1)^n = \begin{cases} 2, & \text{if } n \text{ is even} \\ -2, & \text{if } n \text{ is odd} \end{cases}$$

22. $2\cos\left(\frac{A+B}{2}\right)\cos\left(\frac{A-B}{2}\right) = 1$

$$\Rightarrow \cos\left(\frac{A-B}{2}\right) = \frac{1}{\sqrt{3}}$$

$$\Rightarrow 2\cos^2\left(\frac{A-B}{2}\right) - 1 = -\frac{1}{3}$$

$$\Rightarrow \cos(A-B) = -\frac{1}{3}$$

$$|\cos A - \cos B| = 2\sin\left(\frac{A+B}{2}\right)\sin\left(\frac{B-A}{2}\right)$$

$$= 2 \times \frac{1}{2}\sqrt{\left(1 - \frac{1}{3}\right)} = \sqrt{\frac{2}{3}}$$

23. We have, $x = \frac{1 - \sin\phi}{\cos\phi}$, $y = \frac{1 + \cos\phi}{\sin\phi}$

Multiplying, we get $xy = \frac{(1 - \sin\phi)(1 + \cos\phi)}{\cos\phi \sin\phi}$

$$\Rightarrow xy + 1 = \frac{1 - \sin\phi + \cos\phi - \sin\phi\cos\phi + \sin\phi\cos\phi}{\cos\phi\sin\phi}$$

$$= \frac{1 - \sin\phi + \cos\phi}{\cos\phi\sin\phi}$$

and $x - y = \frac{(1 - \sin\phi)\sin\phi - \cos\phi(1 + \cos\phi)}{\cos\phi\sin\phi}$

$$= \frac{\sin\phi - \sin^2\phi - \cos\phi - \cos^2\phi}{\cos\phi\sin\phi}$$

$$= \frac{\sin\phi - \cos\phi - 1}{\cos\phi\sin\phi} = -(xy+1)$$

Thus, $xy + x - y + 1 = 0$

$$\Rightarrow x = \frac{y-1}{y+1} \text{ and } y = \frac{1+x}{1-x}$$

24. $\sin^2 A + \sin^2 B + \sin^2 C$

$$= \frac{1}{2}(1 - \cos 2A) + \frac{1}{2}(1 - \cos 2B) + \sin^2 C$$

$$= 1 - \frac{1}{2}(\cos 2A + \cos 2B) + \sin^2 C$$

$$= 1 - \frac{1}{2}\{2\cos(A+B)\cdot\cos(A-B)\} + \sin^2 C$$

$$= 1 + \cos C \cdot \cos(A-B) + 1 - \cos^2 C$$

$$= 2 - \cos^2 C + \cos C \cdot \cos(A-B)$$

$$\therefore \sin^2 A + \sin^2 B + \sin^2 C \le 2 - \cos^2 C + \cos C \quad \text{...(i)}$$

$$= 2 - (\cos^2 C - \cos C)$$

$$= 2 - \left(\cos^2 C - \cos C + \frac{1}{4}\right) + \frac{1}{4} = \frac{9}{4} - \left(\cos C - \frac{1}{2}\right)^2$$

$$\therefore \sin^2 A + \sin^2 B + \sin^2 C \le \frac{9}{4}$$

Now, for positive quantities AM $\ge$ GM

$$\therefore \frac{\sin^2 A + \sin^2 B + \sin^2 C}{3} \ge \sqrt[3]{\sin^2 A \sin^2 B \sin^2 C}$$

$$\therefore \text{From Eq. (i), } \frac{9/4}{3} \ge \frac{\sin^2 A + \sin^2 B + \sin^2 C}{3}$$

$$\ge (\sin A \sin B \sin C)^{2/3}$$

$$\frac{3}{4} \ge (\sin A \sin B \sin C)^{2/3}$$

or $\sin A \sin B \sin C \le \left(\frac{3}{4}\right)^{3/2}$, *ie*, $\frac{3\sqrt{3}}{8}$

Also in ΔABC, $\sin A > 0$, $\sin B > 0$, $\sin C > 0$ and from Eq. (i)

$$\sin^2 A + \sin^2 B \le 1 + \cos C$$

25. We have, $x^2 + y^2 = a^2 \sin^4\theta \cos^4\theta(\sin^2\theta + \cos^2\theta)$

$$= a^2 \sin^4\theta\cos^4\theta \text{ and}$$

$$xy = a^2 \sin^5\theta\cos^5\theta$$

$$\therefore \frac{(x^2+y^2)^p}{(xy)^q} = \frac{a^{2p}(\sin\theta\cos\theta)^{4p}}{a^{2q}(\sin\theta\cos\theta)^{5q}}$$

which is independent of θ, if $4p = 5q$

i.e. if $p = 5$ and $q = 4$

26. Let $\tan(\theta/2) = \alpha$ and $\tan(\phi/2) = \beta$, so that $\alpha - \beta = 2b$

Also, $\cos\theta = \frac{1 - \tan^2(\theta/2)}{1 + \tan^2(\theta/2)} = \frac{1 - \alpha^2}{1 + \alpha^2}$

and $\sin\theta = \frac{2\tan(\theta/2)}{1 + \tan^2(\theta/2)} = \frac{2\alpha}{1 + \alpha^2}$

Similarly, $\cos\phi = \frac{1 - \beta^2}{1 + \beta^2}$ and $\sin\phi = \frac{2\beta}{1 + \beta^2}$

$\therefore$ We have from the given relations

$$(x - a)\frac{1 - \alpha^2}{1 + \alpha^2} + y\left(\frac{2\alpha}{1 + \alpha^2}\right) = a$$

$$\Rightarrow x\alpha^2 - 2y\alpha + 2a - x = 0$$

Similarly, $x\beta^2 - 2y\beta + 2a - x = 0$

We see that α and β are the roots of the equation $xz^2 - 2yz + 2a - x = 0$, so that $\alpha + \beta = 2y/x$ and $\alpha\beta = (2a - x)/x$. Now, from $(\alpha + \beta)^2 = (\alpha - \beta)^2 + 4\alpha\beta$, we get

$$\left(\frac{2y}{x}\right)^2 = (2b)^2 + \frac{4(2a - x)}{x}$$

$$\Rightarrow y^2 = 2ax - (1 - b^2)x^2$$

Also, from $\alpha + \beta = \frac{2y}{x}$ and $\alpha - \beta = 2b$

we get, $\alpha = \frac{y}{x} + b$ and $\beta = \frac{y}{x} - b$

$$\Rightarrow \tan\frac{\theta}{2} = \frac{1}{x}(y + bx)$$

and $\tan\frac{\phi}{2} = \frac{1}{x}(y - bx)$

27. From the first relation, we have

$a[\sin(\theta + \phi) - \sin(\theta - \phi)] = b[\sin(\theta - \phi) + \sin(\theta + \phi)]$

$$\Rightarrow 2a\sin\phi\cos\theta = 2b\sin\theta\cos\phi$$

$$\Rightarrow a\tan\phi = b\tan\theta$$

$$\Rightarrow \frac{2a\tan\frac{\phi}{2}}{1 - \tan^2\frac{\phi}{2}} = \frac{2b\tan\frac{\theta}{2}}{1 - \tan^2\frac{\theta}{2}}$$

From the second relation replacing

$$\tan\frac{\theta}{2} = \frac{1}{a}\left[b\tan\frac{\phi}{2} + c\right]$$

we have, $$\frac{a\tan\frac{\phi}{2}}{1 - \tan^2\frac{\phi}{2}} = \frac{b\,(b\tan\phi/2 + c)}{a\left[1 - \left(\frac{1}{a}(b\tan\phi/2 + c)\right)^2\right]}$$

$$\Rightarrow \tan\frac{\phi}{2}\left[a^2 - \left(b\tan\frac{\phi}{2} + c\right)^2\right] = b\left(b\tan\frac{\phi}{2} + c\right)\left(1 - \tan^2\frac{\phi}{2}\right)$$

$$\Rightarrow \tan\frac{\phi}{2}(a^2 - b^2 - c^2) = bc\left(1 + \tan^2\frac{\phi}{2}\right)$$

$$\Rightarrow \frac{2\tan\frac{\phi}{2}}{1 + \tan^2\frac{\phi}{2}} = \frac{2bc}{a^2 - b^2 - c^2}$$

$$\Rightarrow \sin\phi = \frac{2bc}{a^2 - b^2 - c^2}$$

Now, replacing ϕ by θ, b by a and c by $-c$, we get

$$\sin\theta = \frac{-2ac}{b^2 - a^2 - c^2} = \frac{2ac}{a^2 - b^2 + c^2}$$

28. We have, $\cos^4\theta - \sin^4\theta = \cos^2\theta - \sin^2\theta = \cos 2\theta$

$$\Rightarrow (-2b)^2 - 4b = (-4)^2 - 4 \times 2$$

$$\Rightarrow b^2 - b - 2 = 0$$

$$\Rightarrow b = -1, 2$$

29. $\left(\frac{\sin\theta}{\sin\phi}\right)^2 = \frac{\tan\theta}{\tan\phi}$

$$\Rightarrow \frac{\sin\theta}{\sin\phi}\cdot\frac{\sin\theta}{\sin\phi} = \frac{\sin\theta}{\sin\phi}\cdot\frac{\cos\phi}{\cos\theta}$$

$$\Rightarrow \frac{\sin\theta}{\sin\phi} = \frac{\cos\phi}{\cos\theta} \qquad \left[\because \frac{\sin\theta}{\sin\phi} = \sqrt{3} \neq 0\right]$$

$$\Rightarrow \sin 2\theta = \sin 2\phi$$

$$\Rightarrow \frac{2\tan\theta}{1 + \tan^2\theta} = \frac{2\tan\phi}{1 + \tan^2\phi}$$

$$\Rightarrow \frac{6\tan\phi}{1 + 9\tan^2\phi} = \frac{2\tan\phi}{1 + \tan^2\phi}$$

[$\because \tan\theta = 3\tan\phi$ (given)]

$$\Rightarrow \tan^2\phi = \frac{1}{3}$$

$$\Rightarrow \tan\phi = \pm\frac{1}{\sqrt{3}}$$

and $\tan\theta = \pm\sqrt{3}$

30. Given, $\sec\theta + \tan\theta = 1$

$\Rightarrow \sec\theta - \tan\theta = 1$ [$\because \sec^2\theta - \tan^2\theta = 1$]

$\therefore 2\sec\theta = 2 \Rightarrow \sec\theta = 1$

$\therefore \cos\theta = 1$

Clearly, 1 is a root of given quadratic equation.

$\therefore$ $\sec\theta$ and $\cos\theta$ are roots of given quadratic equation.

Linked Comprehension Type

PASSAGE 1

1. $\cos\frac{2\pi}{7}\cos\frac{4\pi}{7}\cos\frac{6\pi}{7}$

$$= \cos\frac{2\pi}{7}\cos\frac{4\pi}{7}\cos\left(2\pi - \frac{8\pi}{7}\right)$$

$$= \cos\frac{2\pi}{7}\cos\frac{4\pi}{7}\cos\frac{8\pi}{7}$$

$$= \frac{\sin\left(2^3\cdot\frac{2\pi}{7}\right)}{2^3\sin\left(\frac{2\pi}{7}\right)} = \frac{\sin\left(2\pi + \frac{2\pi}{7}\right)}{8\sin\left(\frac{2\pi}{7}\right)} = \frac{\sin\left(\frac{2\pi}{7}\right)}{8\sin\left(\frac{2\pi}{7}\right)} = \frac{1}{8}$$

2. $\prod_{r=1}^{6}\cos r\alpha = \cos\alpha\cos 2\alpha\cos 3\alpha\cos 4\alpha\cos 5\alpha\cos 6\alpha$ ...(i)

$\because \alpha = \frac{\pi}{13} \Rightarrow 13\alpha = \pi$

or $5\alpha = \pi - 8\alpha$

$\therefore \cos 5\alpha = -\cos 8\alpha$

and $\alpha = \pi - 12\alpha$

$\Rightarrow \cos\alpha = -\cos 12\alpha$

Then from Eq. (i),

$$\prod_{r=1}^{6}\cos r\alpha = \cos\alpha\cos 2\alpha\cos 3\alpha\cos 4\alpha\cos 5\alpha\cos 6\alpha$$

$$= (-\cos 12\alpha)\cos 2\alpha \cos 3\alpha \cos 4\alpha (-\cos 8\alpha)\cos 6\alpha$$
$$= (\cos 2\alpha \cos 4\alpha \cos 8\alpha)(\cos 3\alpha \cos 6\alpha \cos 12\alpha)$$
$$= \frac{\sin(2^3 \cdot 2\alpha)}{2^3 \sin 2\alpha} \times \frac{\sin(2^3 \cdot 3\alpha)}{2^3 \cdot \sin 3\alpha}$$
$$= \frac{\sin 16\alpha}{8 \sin 2\alpha} \times \frac{\sin 24\alpha}{8 \sin 3\alpha}$$
$$= \frac{\sin(\pi + 3\alpha)}{8 \sin 2\alpha} \times \frac{\sin(2\pi - 2\alpha)}{8 \sin 3\alpha} \qquad [\because 13\alpha = \pi]$$
$$= -\frac{\sin 3\alpha}{8 \sin 2\alpha} \times \frac{-\sin 2\alpha}{8 \sin 3\alpha} = \frac{1}{64}$$

3. $\because \sin\left(\frac{9\pi}{14}\right) = \sin\left(\pi - \frac{5\pi}{14}\right) = \sin\left(\frac{5\pi}{14}\right)$

$\sin\left(\frac{11\pi}{14}\right) = \sin\left(\pi - \frac{3\pi}{14}\right) = \sin\left(\frac{3\pi}{14}\right)$

and $\sin\left(\frac{13\pi}{14}\right) = \sin\left(\pi - \frac{\pi}{14}\right) = \sin\left(\frac{\pi}{14}\right)$

$\therefore \sin\left(\frac{\pi}{14}\right)\sin\left(\frac{3\pi}{14}\right)\sin\left(\frac{5\pi}{14}\right)\sin\left(\frac{7\pi}{14}\right)\sin\left(\frac{9\pi}{14}\right)\sin\left(\frac{11\pi}{14}\right)\sin\left(\frac{13\pi}{14}\right)$

$$= \left\{\sin\left(\frac{\pi}{14}\right)\sin\left(\frac{3\pi}{14}\right)\sin\left(\frac{5\pi}{14}\right)\right\}^2$$
$$= \left\{\cos\left(\frac{\pi}{2} - \frac{\pi}{14}\right)\cdot\cos\left(\frac{\pi}{2} - \frac{3\pi}{14}\right)\cdot\cos\left(\frac{\pi}{2} - \frac{5\pi}{14}\right)\right\}^2$$
$$= \left\{\cos\left(\frac{6\pi}{14}\right)\cos\left(\frac{4\pi}{14}\right)\cos\left(\frac{2\pi}{14}\right)\right\}^2$$
$$= \left\{\cos\left(\frac{\pi}{7}\right)\cos\left(\frac{2\pi}{7}\right)\cos\left(\frac{3\pi}{7}\right)\right\}^2$$
$$= \left\{\cos\left(\frac{\pi}{7}\right)\cos\left(\frac{2\pi}{7}\right)\cos\left(\pi - \frac{4\pi}{7}\right)\right\}^2$$
$$= \left\{-\cos\left(\frac{\pi}{7}\right)\cos\left(\frac{2\pi}{7}\right)\cos\left(\frac{4\pi}{7}\right)\right\}^2$$
$$= \left\{\cos\left(\frac{\pi}{7}\right)\cos\left(\frac{2\pi}{7}\right)\cos\left(\frac{4\pi}{7}\right)\right\}^2 \quad \left[\text{here } \alpha = \frac{\pi}{2^3 - 1}\right]$$
$$= \left(-\frac{1}{8}\right)^2$$
$$= \frac{1}{64}$$

4. $\because \sin\left(\frac{\pi}{18}\right)\sin\left(\frac{5\pi}{18}\right)\sin\left(\frac{7\pi}{18}\right)$

$$= \cos\left(\frac{\pi}{2} - \frac{\pi}{18}\right)\cos\left(\frac{\pi}{2} - \frac{5\pi}{18}\right)\cos\left(\frac{\pi}{2} - \frac{7\pi}{18}\right)$$
$$= \cos\left(\frac{8\pi}{18}\right)\cos\left(\frac{4\pi}{18}\right)\cos\left(\frac{2\pi}{18}\right)$$
$$= \cos\left(\frac{\pi}{9}\right)\cos\left(\frac{2\pi}{9}\right)\cos\left(\frac{4\pi}{9}\right) \qquad \left[\text{here } \alpha = \frac{\pi}{2^3 + 1}\right]$$
$$= \frac{1}{2^3} \quad = \frac{1}{8}$$

5. $\because 64\sqrt{3}\sin\left(\frac{\pi}{48}\right)\cos\left(\frac{\pi}{48}\right)\cos\left(\frac{\pi}{24}\right)\cos\left(\frac{\pi}{12}\right)\cos\left(\frac{\pi}{6}\right)$

$$= 32\sqrt{3}\left(2\sin\left(\frac{\pi}{48}\right)\cos\left(\frac{\pi}{48}\right)\right)\cos\left(\frac{\pi}{24}\right)\cos\left(\frac{\pi}{12}\right)\cos\left(\frac{\pi}{6}\right)$$
$$= 32\sqrt{3}\left(\sin\frac{\pi}{24}\right)\cos\left(\frac{\pi}{24}\right)\cos\left(\frac{\pi}{12}\right)\cos\left(\frac{\pi}{6}\right)$$
$$= 16\sqrt{3}\left(2\sin\left(\frac{\pi}{24}\right)\cos\left(\frac{\pi}{24}\right)\right)\cos\left(\frac{\pi}{12}\right)\cos\left(\frac{\pi}{6}\right)$$
$$= 16\sqrt{3}\sin\left(\frac{\pi}{12}\right)\cos\left(\frac{\pi}{12}\right)\cos\left(\frac{\pi}{6}\right)$$
$$= 8\sqrt{3}\left(2\sin\left(\frac{\pi}{12}\right)\cos\left(\frac{\pi}{12}\right)\right)\cos\left(\frac{\pi}{6}\right)$$
$$= 8\sqrt{3}\sin\left(\frac{\pi}{6}\right)\cdot\cos\left(\frac{\pi}{6}\right)$$
$$= 8\sqrt{3}\cdot\frac{1}{2}\cdot\frac{\sqrt{3}}{2} = 6$$

PASSAGE 2

$\because P_0 = 2, P_2 = 1$

and $P_n - P_{n-2} = (\sin^n\theta + \cos^n\theta) - (\sin^{n-2}\theta + \cos^{n-2}\theta)$

$$= -\sin^{n-2}\theta(1 - \sin^2\theta) - \cos^{n-2}\theta(1 - \cos^2\theta)$$
$$= -\sin^{n-2}\theta\cos^2\theta - \cos^{n-2}\theta\sin^2\theta$$
$$= -\sin^2\theta\cos^2\theta(\sin^{n-4}\theta + \cos^{n-4}\theta)$$
$$= -\sin^2\theta\cos^2\theta\, P_{n-4}$$

$\therefore P_n - P_{n-2} = -\sin^2\theta\cos^2\theta\, P_{n-4}$ ...(i)

for $n = 4$,

$P_4 - P_2 = -\sin^2\theta\cos^2\theta P_0$

$\Rightarrow P_4 - 1 = -2\sin^2\theta\cos^2\theta \qquad [\because P_2 = 1, P_0 = 2]$

$\therefore P_4 = 1 - 2\sin^2\theta\cos^2\theta$ ...(ii)

for $n = 6$,

$P_6 - P_4 = -\sin^2\theta\cos^2\theta P_2$

$\Rightarrow P_6 = P_4 - \sin^2\theta\cos^2\theta P_2$

$= 1 - 2\sin^2\theta\cos^2\theta - \sin^2\theta\cos^2\theta$

$\therefore P_6 = 1 - 3\sin^2\theta\cos^2\theta$ ...(iii)

1. $\because P_1 = m \Rightarrow P_1^2 = m^2$

$\sin^2\theta + \cos^2\theta + 2\sin\theta\cos\theta = m^2$

$\Rightarrow \sin\theta\cos\theta = \frac{(m^2 - 1)}{2}$

Now, from Eq. (iii),

$P_6 = 1 - 3\sin^2\theta\cos^2\theta$

$\Rightarrow (1 - P_6) = 3(\sin\theta\cos\theta)^2$

$= \frac{3(m^2 - 1)^2}{4}$

or $4(1 - P_6) = 3(m^2 - 1)^2$

2. $2P_6 - 3P_4 + 10$

$= 2(1 - 3\sin^2\theta\cos^2\theta) - 3(1 - 2\sin^2\theta\cos^2\theta) + 10$

[from Eqs. (ii) and (iii)]

$= 2 - 3 + 10 = 9$

3. Let $\sin^2\theta\cos^2\theta = k$,
then from Eq. (i),

$$P_n - P_{n-2} = -k P_{n-4}$$

From Eq. (ii), $P_4 = 1 - 2k$
and from Eq. (iii), $P_6 = 1 - 3k$
Put $n = 10$,
then $P_{10} - P_8 = -k P_6 = -k(1-3k)$

$$\therefore \quad P_{10} - P_8 = 3k^2 - k \quad \ldots\text{(iv)}$$

and put $n = 8$,
then $P_8 - P_6 = -k P_4 = -k(1-2k)$

$$\therefore \quad P_8 = P_6 + 2k^2 - k = 1 - 3k + 2k^2 - k$$

$$\Rightarrow \quad P_8 = 2k^2 - 4k + 1$$

From Eq. (iv), $P_{10} = 5k^2 - 5k + 1$

$$\therefore 6P_{10} - 15P_8 + 10P_6 + 7 = 6(5k^2 - 5k + 1) - 15(2k^2 - 4k + 1) + 10(1 - 3k) + 7 = 8$$

4. From Eq. (i),

$$P_{n-2} - P_n = \sin^2\theta\cos^2\theta\, P_{n-4}$$

$$\therefore \quad \lambda = n - 4$$

5. From Eq. (i),

$$\frac{P_n - P_{n-2}}{P_{n-4}} = -\sin^2\theta\cos^2\theta$$

Put $n = 7$, then $\dfrac{P_7 - P_5}{P_3} = -\sin^2\theta\cos^2\theta$ ...(v)

and put $n = 5$, then $\dfrac{P_5 - P_3}{P_1} = -\sin^2\theta\cos^2\theta$ ...(vi)

From Eqs. (v) and (vi), we get

$$\frac{P_7 - P_5}{P_3} = \frac{P_5 - P_3}{P_1}$$

$$\therefore \quad \frac{P_7 - P_5}{P_5 - P_3} = \frac{P_3}{P_1}$$

PASSAGE 3

Given $\cos\frac{\pi}{7}, \cos\frac{3\pi}{7}, \cos\frac{5\pi}{7}$ are the roots of the equation

$$8x^3 - 4x^2 - 4x + 1 = 0 \quad \ldots\text{(i)}$$

Replacing x by $\frac{1}{x}$ in Eq. (i), then we get

$$x^3 - 4x^2 - 4x + 8 = 0 \quad \ldots\text{(ii)}$$

1. $\sec\frac{\pi}{7}, \sec\frac{3\pi}{7}, \sec\frac{5\pi}{7}$ are the roots of Eq. (ii)

$$\therefore \quad \sec\frac{\pi}{7} + \sec\frac{3\pi}{7} + \sec\frac{5\pi}{7} = 4$$

2. From Eq. (i),
$8x^3 - 4x^2 - 4x + 1$

$$= 8\left(x - \cos\frac{\pi}{7}\right)\left(x - \cos\frac{3\pi}{7}\right)\left(x - \cos\frac{5\pi}{7}\right)$$

Put $x = 1$, then

$$1 = 8\left(1 - \cos\frac{\pi}{7}\right)\left(1 - \cos\frac{3\pi}{7}\right)\left(1 - \cos\frac{5\pi}{7}\right)$$

$$\Rightarrow \quad 1 = 8\left(2\sin^2\frac{\pi}{14}\right)\left(2\sin^2\frac{3\pi}{14}\right)\left(2\sin^2\frac{5\pi}{14}\right)$$

$$\therefore \quad \sin\left(\frac{\pi}{14}\right)\sin\left(\frac{3\pi}{14}\right)\sin\left(\frac{5\pi}{14}\right) = \frac{1}{8}$$

3. From Eq. (i),

$$8x^3 - 4x^2 - 4x + 1 = 8\left(x - \cos\frac{\pi}{7}\right)\left(x - \cos\frac{3\pi}{7}\right)\left(x - \cos\frac{5\pi}{7}\right)$$

Put $x = -1$, then

$$-8 - 4 + 4 + 1 = 8\left(-1 - \cos\frac{\pi}{7}\right)\left(-1 - \cos\frac{3\pi}{7}\right)\left(-1 - \cos\frac{5\pi}{7}\right)$$

$$\Rightarrow \quad 7 = 8\left(1 + \cos\frac{\pi}{7}\right)\left(1 + \cos\frac{3\pi}{7}\right)\left(1 + \cos\frac{5\pi}{7}\right)$$

$$= 8\left(2\cos^2\frac{\pi}{14}\right)\left(2\cos^2\frac{3\pi}{14}\right)\left(2\cos^2\frac{5\pi}{14}\right)$$

$$\therefore \cos\left(\frac{\pi}{14}\right)\cos\left(\frac{3\pi}{14}\right)\cos\left(\frac{5\pi}{14}\right) = \frac{\sqrt{7}}{8}$$

4. Let $y = \tan^2\theta = \sec^2\theta - 1 = x^2 - 1$
where, $\theta = \pi/7, 3\pi/7, 5\pi/7$
From Eq. (ii),

$$x(x^2 - 4) = 4(x^2 - 2)$$

squaring both sides

$$x^2(x^2 - 4)^2 = 16(x^2 - 2)^2$$

$$\Rightarrow \quad (y + 1)(y + 1 - 4)^2 = 16(y + 1 - 2)^2$$

$$\Rightarrow \quad (y + 1)(y^2 - 6y + 9) = 16(y^2 - 2y + 1)$$

$$\Rightarrow \quad y^3 - 21y^2 + 35y - 7 = 0$$

or we can say that

$$\tan^2\frac{\pi}{7}, \tan^2\frac{3\pi}{7}, \tan^2\frac{5\pi}{7}$$

are the roots of $x^3 - 21x^2 + 35x - 7 = 0$ ...(iii)

5. From Eq. (iii),

$$\tan^2\left(\frac{\pi}{7}\right) + \tan^2\left(\frac{3\pi}{7}\right) + \tan^2\left(\frac{5\pi}{7}\right) = 21$$

$$\Rightarrow \quad \sum_{r=1}^{3}\tan^2\left(\frac{2r-1}{7}\right) = 21$$

Also, replacing x by $\frac{1}{x}$ in Eq. (iii), then

$$\frac{1}{x^3} - \frac{21}{x^2} + \frac{35}{x} - 7 = 0$$

$$\Rightarrow \quad 7x^3 - 35x^2 + 21x - 1 = 0 \quad \ldots\text{(iv)}$$

$\therefore \cot^2\left(\frac{\pi}{7}\right), \cot^2\left(\frac{3\pi}{7}\right), \cot^2\left(\frac{5\pi}{7}\right)$ are the roots of Eq. (iv)

$$\therefore \quad \cot^2\left(\frac{\pi}{7}\right) + \cot^2\left(\frac{3\pi}{7}\right) + \cot^2\left(\frac{5\pi}{7}\right) = \frac{35}{7} = 5$$

$$\Rightarrow \quad \sum_{r=1}^{3}\cot^2\left(\frac{2r-1}{7}\right) = 5$$

Hence, $\displaystyle\sum_{r=1}^{3}\tan^2\left(\frac{2r-1}{7}\right)\sum_{r=1}^{3}\cot^2\left(\frac{2r-1}{7}\right) = 21 \times 5 = 105$

PASSAGE 4

1. $\because x\sin^3\theta + y\cos^3\theta = \sin\theta\cos\theta$ ...(i)

and $x\sin\theta - y\cos\theta = 0$ ...(ii)

From Eq. (ii), $\tan\theta = \dfrac{y}{x}$

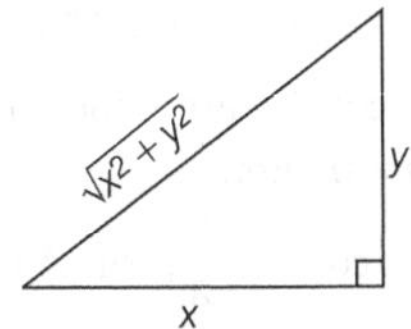

$\therefore \sin\theta = \dfrac{y}{\sqrt{(x^2+y^2)}}$ and $\cos\theta = \dfrac{x}{\sqrt{(x^2+y^2)}}$

From Eq. (i), $x \times \dfrac{y^3}{(x^2+y^2)^{3/2}} + y \times \dfrac{x^3}{(x^2+y^2)^{3/2}}$

$$= \frac{xy}{(x^2+y^2)}$$

or $\dfrac{(x^2+y^2)}{(x^2+y^2)^{3/2}} = \dfrac{1}{(x^2+y^2)}$

$\Rightarrow (x^2+y^2)^{1/2} = 1$

or $x^2+y^2 = 1$ which is a circle

2. $\because \dfrac{x}{a\cos\theta} = \dfrac{y}{b\sin\theta}$ ...(i)

and $\dfrac{ax}{\cos\theta} - \dfrac{by}{\sin\theta} = a^2 - b^2$ ...(ii)

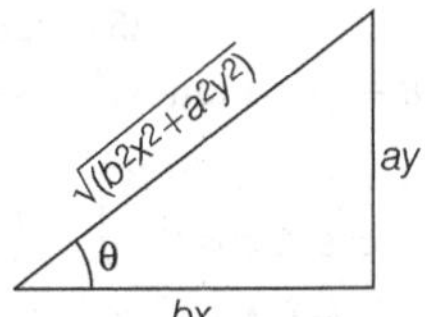

From Eq.(i), $\tan\theta = \dfrac{ay}{bx}$

From Eq. (ii),

$$\frac{ax}{\dfrac{bx}{\sqrt{(b^2x^2+a^2y^2)}}} - \frac{by}{\dfrac{ay}{\sqrt{(b^2x^2+a^2y^2)}}} = (a^2-b^2)$$

$\Rightarrow (a^2-b^2)\sqrt{(b^2x^2+a^2y^2)} = ab(a^2-b^2)$

$\Rightarrow b^2x^2 + a^2y^2 = a^2b^2$

$\therefore \dfrac{x^2}{a^2} + \dfrac{y^2}{b^2} = 1$ which is an ellipse.

3. $\because m+n = 2\tan\theta,\ m-n = 2\sin\theta$ ...(i)

and $mn = \tan^2\theta - \sin^2\theta = \sin^2\theta(\sec^2\theta - 1)$

$= \sin^2\theta\tan^2\theta$

$= \left(\dfrac{m-n}{2}\right)^2\left(\dfrac{m+n}{2}\right)^2$ [from Eq. (i)]

$\therefore (m^2-n^2)^2 = 16mn$

4. $\sin\theta + \cos\theta = a$ ...(i)

$\sin^3\theta + \cos^3\theta = b$ ...(ii)

From Eq. (i), $\sin^2\theta + \cos^2\theta + 2\sin\theta\cos\theta = a^2$

or $\sin\theta\cos\theta = \dfrac{a^2-1}{2}$ ...(iii)

From Eq. (ii),

$(\sin\theta+\cos\theta)^3 - 3\sin\theta\cos\theta(\sin\theta+\cos\theta) = b$

$\Rightarrow a^3 - \dfrac{3(a^2-1)}{2}a = b$ [from Eqs. (i) and (iii)]

$\Rightarrow 2a^3 - 3a^3 + 3a = 2b$

$\Rightarrow a^3 + 2b - 3a = 0$

On comparing, we get

$\lambda = 1, \mu = 2, \nu = -3$

$\therefore \lambda + \mu + \nu = 0$

$\therefore \lambda^3 + \mu^3 + \nu^3 = 3\lambda\mu\nu$

$= 3(1)(2)(-3) = -18$

5. $\because \dfrac{x\cos\theta}{a} + \dfrac{y\sin\theta}{b} = 1$ ...(i)

and $x\sin\theta - y\cos\theta = \sqrt{(a^2\sin^2\theta + b^2\cos^2\theta)}$ ...(ii)

Squaring Eq. (i), we get

$\dfrac{x^2}{a^2}\cos^2\theta + \dfrac{y^2}{b^2}\sin^2\theta + \dfrac{2xy}{ab}\sin\theta\cos\theta$

$= 1 = \sin^2\theta + \cos^2\theta$

or $\left(\dfrac{x^2}{a^2} - 1\right)\cos^2\theta + \left(\dfrac{y^2}{b^2} - 1\right)\sin^2\theta$

$+ \dfrac{2xy}{ab}\sin\theta\cos\theta = 0$ ...(iii)

and squaring Eq. (ii), we get

$x^2\sin^2\theta + y^2\cos^2\theta - 2xy\sin\theta\cos\theta$

$= a^2\sin^2\theta + b^2\cos^2\theta$

$(x^2-a^2)\sin^2\theta + (y^2-b^2)\cos^2\theta - 2xy\sin\theta\cos\theta = 0$

$\Rightarrow \left(\dfrac{x^2-a^2}{ab}\right)\sin^2\theta + \left(\dfrac{y^2-b^2}{ab}\right)\cos^2\theta$

$- \dfrac{2xy}{ab}\sin\theta\cos\theta = 0$ ...(iv)

Adding Eqs. (iii) and (iv), then

$\left(\dfrac{x^2-a^2}{a}\right)\left(\dfrac{\sin^2\theta}{b} + \dfrac{\cos^2\theta}{a}\right)$

$+ \left(\dfrac{y^2-b^2}{b}\right)\left(\dfrac{\sin^2\theta}{b} + \dfrac{\cos^2\theta}{a}\right) = 0$

or $\dfrac{x^2-a^2}{a} + \dfrac{y^2-b^2}{b} = 0$

or $\dfrac{x^2}{a} + \dfrac{y^2}{b} = (a+b)$

Numerical Value Type

1. $\because \quad \cot\left(7\frac{1^\circ}{2}\right) = \frac{1+\cos 15^\circ}{\sin 15^\circ}$

$$= \frac{1+\frac{\sqrt{3}+1}{2\sqrt{2}}}{\frac{\sqrt{3}-1}{2\sqrt{2}}} = \frac{2\sqrt{2}+\sqrt{3}+1}{\sqrt{3}-1}$$

$$= \frac{(2\sqrt{2}+\sqrt{3}+1)(\sqrt{3}+1)}{2}$$

$$= \frac{2\sqrt{6}+2\sqrt{2}+3+\sqrt{3}+\sqrt{3}+1}{2}$$

$$= \sqrt{6}+\sqrt{2}+2+\sqrt{3}$$

$$= \sqrt{2}+\sqrt{3}+\sqrt{4}+\sqrt{6}$$

and $4\cos 36^\circ = 4\left(\frac{\sqrt{5}+1}{4}\right) = \sqrt{5}+1 = \sqrt{5}+\sqrt{1}$

Hence, $4\cos 36^\circ + \cot\left(7\frac{1^\circ}{2}\right)$

$$= \sqrt{1}+\sqrt{2}+\sqrt{3}+\sqrt{4}+\sqrt{5}+\sqrt{6}$$

$\therefore \quad n_1 = 1, n_2 = 2, n_3 = 3, n_4 = 4,$

$n_5 = 5$ and $n_6 = 6$

$\therefore \sum_{i=1}^{6} n_i^2 = n_1^2 + n_2^2 + n_3^2 + n_4^2 + n_5^3 + n_6^2$

$$= 1^2 + 2^2 + 3^2 + 4^2 + 5^2 + 6^2 = 91$$

2. $\because (\cos 27^\circ + \sin 27^\circ)^2 = 1 + \sin 54 = 1 + \cos 36^\circ$

$\Rightarrow \quad \cos 27^\circ + \sin 27^\circ = \sqrt{(1+\cos 36^\circ)}$ $\quad [\because \text{LHS} > 0]$

Also, $\cos 27^\circ - \sin 27^\circ = \sqrt{(1-\cos 36^\circ)}$ $\quad [\because \cos 27^\circ > \sin 27^\circ]$

$\therefore \quad 2\sin 27^\circ = \sqrt{(1+\cos 36^\circ)} - \sqrt{(1-\cos 36^\circ)}$

$$= \sqrt{\left(1+\left(\frac{\sqrt{5}+1}{4}\right)\right)} - \sqrt{\left(1-\left(\frac{\sqrt{5}+1}{4}\right)\right)}$$

$\therefore 4\sin 27^\circ = \sqrt{(5+\sqrt{5})} - \sqrt{(3-\sqrt{5})}$

On comparing, we get

$$\alpha = 5+\sqrt{5}, \beta = 3-\sqrt{5}$$

$\therefore \quad \alpha+\beta = 8, \ \alpha\beta = 10 - 2\sqrt{5}$

$\alpha + \beta - \alpha\beta + 2 = 2\sqrt{5}$

$\therefore \quad (\alpha+\beta-\alpha\beta+2)^4 = 400$

3. We can write,

$(a + 25\cos\theta)^2 + (25)^2 - (25\cos\theta)^2 = 1$

$\Rightarrow \quad (a+25\cos\theta)^2 = 1 - (25\sin\theta)^2$

Similarly, $(b + 31\cos\theta)^2 = 1 - (31\sin\theta)^2$

Multiplying, we get

$[(a+25\cos\theta)(b+31\cos\theta)]^2$

$$= [1-(25\sin\theta)^2][1-(31\sin\theta)^2]$$

$\Rightarrow \quad [ab + (31a+25b)\cos\theta + 775\cos^2\theta]^2$

$$= 1 - (625+961)\sin^2\theta + (775\sin^2\theta)^2$$

$\Rightarrow \quad (-775 + 775\cos^2\theta)^2 = 1 - 1586\sin^2\theta + (775\sin^2\theta)^2$

$\Rightarrow \quad \operatorname{cosec}^2\theta = 1586$

4. $\sin A$ and $\cos A$ are the roots of the equation $4x^2 - 3x + a = 0$, then

$$\sin A + \cos A = \frac{3}{4}, \ \sin A \cos A = \frac{a}{4} \quad \ldots\text{(i)}$$

Also, $\sin A + \cos A + \tan A + \cot A + \sec A + \operatorname{cosec} A = 7$

$\Rightarrow \quad (\sin A + \cos A) + \left(\frac{\sin A}{\cos A} + \frac{\cos A}{\sin A}\right) + \left(\frac{1}{\cos A} + \frac{1}{\sin A}\right) = 7$

$\Rightarrow \quad (\sin A + \cos A) + \frac{1}{\sin A \cos A} + \frac{(\sin A + \cos A)}{\sin A \cos A} = 7$

$\Rightarrow \quad \frac{3}{4} + \frac{4}{a} + \frac{3}{a} = 7$

$\Rightarrow \quad \frac{3}{4} + \frac{7}{a} = 7$

$\Rightarrow \quad \frac{7}{a} = 7 - \frac{3}{4} = \frac{25}{4}$

$\therefore \quad 25a = 28$

5. $\because (\sec A \sec B + \tan A \tan B)^2 - (\sec A \tan B + \tan A \sec B)^2$

$$= \left(\frac{1+\sin A \sin B}{\cos A \cos B}\right)^2 - \left(\frac{\sin B + \sin A}{\cos A \cos B}\right)^2$$

$$= \frac{1 + \sin^2 A \sin^2 B - \sin^2 A - \sin^2 B}{\cos^2 A \cos^2 B}$$

$$= \frac{(1-\sin^2 A)(1-\sin^2 B)}{\cos^2 A \cos^2 B} = \frac{\cos^2 A \cos^2 B}{\cos^2 A \cos^2 B} = 1$$

$\therefore (\sec A \sec B + \tan A \tan B)^2$

$$= (\sec A \tan B + \tan A \sec B)^2 + 1$$

$$= (53)^2 + 1 = 2810$$

6. $\because \prod_{r=1}^{4} \sin(rA) = \sin A \sin 2A \sin 3A \sin 4A$

$$= \sin A \cdot 2\sin A \cos A \cdot (3\sin A - 4\sin^3 A \cdot 2\sin 2A \cos 2A)$$

$$= 2\sin^2 A \cos A \cdot \sin A(3 - 4\sin^2 A) \cdot 4\sin A \cos A \cdot (1 - 2\sin^2 A)$$

$$= 8x^2(1-x)(3-4x)(1-2x)$$

$$= 24x^2 - 104x^3 + 144x^4 - 64x^5$$

On comparing, we get $a = 24, b = -104, c = 144, d = -64$

$10a - 7b + 15c - 5d$

$$= 10 \times 24 - 7 \times -104 + 15 \times 144 - 5 \times -64$$

$$= 240 + 728 + 2160 + 320 = 3448$$

7. $\because \cot(\theta-\alpha), 3\cot\theta, \cot(\theta+\alpha)$ are in AP

$\therefore \quad 6\cot\theta = \cot(\theta-\alpha)+\cot(\theta+\alpha)$

$$\frac{6\cos\theta}{\sin\theta} = \frac{\sin 2\theta}{\sin(\theta+\alpha)\sin(\theta-\alpha)}$$

$\Rightarrow \quad 6\cos\theta\{\sin^2\theta-\sin^2\alpha\} = 2\sin^2\theta\cos\theta$

$\Rightarrow \quad 3(\sin^2\theta-\sin^2\alpha) = \sin^2\theta \qquad [\because \cos\theta \neq 0]$

$\therefore \quad 2\sin^2\theta = 3\sin^2\alpha$

or $\quad \dfrac{2\sin^2\theta}{3\sin^2\alpha} = 1$

$\therefore \quad \dfrac{4016\sin^2\theta}{3\sin^2\alpha} = 2008 \times \dfrac{2\sin^2\theta}{3\sin^2\alpha}$

$= 2008 \times 1 = 2008$

8. We have,

$$(a\tan\beta - \sqrt{(a^2-1)}\tan\alpha)^2 + (\sqrt{(a^2-1)}\tan\gamma - \sqrt{(a^2+1)}\tan\beta)^2 + (\sqrt{(a^2+1)}\tan\alpha - a\tan\gamma)^2 \geq 0$$

$\Rightarrow (a^2+a^2-1+a^2+1)(\tan^2\alpha+\tan^2\beta+\tan^2\gamma) - \{a\tan\alpha + \sqrt{(a^2-1)}\tan\beta + \sqrt{(a^2+1)}\tan\gamma\}^2 \geq 0$

[using Lagrange's identity]

$\Rightarrow \quad 3a^2(\tan^2\alpha+\tan^2\beta+\tan^2\gamma) - (2a)^2 \geq 0$

$\therefore \quad 3(\tan^2\alpha+\tan^2\beta+\tan^2\gamma) \geq 4$

Hence, $2727(\tan^2\alpha+\tan^2\beta+\tan^2\gamma) \geq 3636$

$\therefore$ Least value is 3636.

9. $\because \sin 2A + \sin 2B + \sin 2C = 4\sin A\sin B\sin C$

$= 32\sin A/2\sin B/2\sin C/2\cos A/2\cos B/2\cos C/2$

and $\sin A + \sin B + \sin C = 4\cos A/2\cos B/2\cos C/2$

[from conditional identities]

$$\therefore \quad \frac{\sin 2A+\sin 2B+\sin 2C}{\sin A+\sin B+\sin C} = 8\sin\left(\frac{A}{2}\right)\sin\left(\frac{B}{2}\right)\sin\left(\frac{C}{2}\right)$$

On comparing, we get $\lambda = 8$

Then, $1+2\lambda+3\lambda^2+4\lambda^3$

$= 1+16+3(64)+4(512)$

$= 2257$

10. $$\because \frac{\cos\left(\frac{A-B}{2}\right)}{\cos\left(\frac{A+B}{2}\right)} = \frac{2\sin\left(\frac{A+B}{2}\right)\cos\left(\frac{A-B}{2}\right)}{2\sin\left(\frac{A+B}{2}\right)\cos\left(\frac{A+B}{2}\right)}$$

$$= \frac{\sin A+\sin B}{\sin(A+B)} = \frac{\sin A+\sin B}{\sin C}$$

Similarly, $$\frac{\cos\left(\frac{B-C}{2}\right)}{\cos\left(\frac{B+C}{2}\right)} = \frac{\sin B+\sin C}{\sin A}$$

and $$\frac{\cos\left(\frac{C-A}{2}\right)}{\cos\left(\frac{C+A}{2}\right)} = \frac{\sin C+\sin A}{\sin B}$$

$$\Rightarrow \frac{\cos\left(\frac{A-B}{2}\right)}{\cos\left(\frac{A+B}{2}\right)} + \frac{\cos\left(\frac{B-C}{2}\right)}{\cos\left(\frac{B+C}{2}\right)} + \frac{\cos\left(\frac{C-A}{2}\right)}{\cos\left(\frac{C+A}{2}\right)}$$

$$= \frac{\sin A+\sin B}{\sin C} + \frac{\sin B+\sin C}{\sin A} + \frac{\sin C+\sin A}{\sin B}$$

$$= \left(\frac{\sin A}{\sin B}+\frac{\sin B}{\sin A}\right) + \left(\frac{\sin B}{\sin C}+\frac{\sin C}{\sin B}\right) + \left(\frac{\sin C}{\sin A}+\frac{\sin A}{\sin C}\right)$$

$\geq 2+2+2 \qquad [\because \text{AM} \geq \text{GM}]$

$$786\left\{\frac{\cos\left(\frac{A-B}{2}\right)}{\cos\left(\frac{A+B}{2}\right)} + \frac{\cos\left(\frac{B-C}{2}\right)}{\cos\left(\frac{B+C}{2}\right)} + \frac{\cos\left(\frac{C-A}{2}\right)}{\cos\left(\frac{C+A}{2}\right)}\right\}$$

$\geq 786 \times 6$

$= 4716$

Hence, minimum value is 4716.

11. Here, $\cos 290° = \cos(270°+20°) = \sin 20°$

and $\sin 250° = \sin(270°-20°) = -\cos 20°$

$\therefore$ The given expression $\dfrac{1}{\sin 20°} - \dfrac{1}{\sqrt{3}\cos 20°} = \lambda$

$$\Rightarrow \quad \frac{1}{\sin 20°} - \frac{\cos 60°}{\sin 60°\cos 20°} = \lambda$$

$$\Rightarrow \quad \frac{\sin 60°\cos 20° - \cos 60°\sin 20°}{\sin 20°\cos 20°\sin 60°} = \lambda$$

$$\Rightarrow \quad \frac{\sin(60°-20°)}{\frac{\sin 40°}{2}\times\frac{\sqrt{3}}{2}} = \lambda$$

$$\therefore \quad \lambda = \frac{4}{\sqrt{3}} \Rightarrow \lambda^2 = \frac{16}{3} = 5.33$$

12. Let $\lambda = \tan\left(\frac{3\pi}{11}\right) + 4\sin\left(\frac{2\pi}{11}\right)$

$$= \frac{1}{\cos\left(\frac{3\pi}{11}\right)}\left\{\sin\left(\frac{3\pi}{11}\right) + 4\sin\left(\frac{2\pi}{11}\right)\cos\left(\frac{3\pi}{11}\right)\right\}$$

$$\lambda\cos\left(\frac{3\pi}{11}\right) = \sin\left(\frac{3\pi}{11}\right) + 4\sin\left(\frac{2\pi}{11}\right)\cos\left(\frac{3\pi}{11}\right)$$

$$\lambda^2\cos^2\left(\frac{3\pi}{11}\right) = \sin^2\left(\frac{3\pi}{11}\right) + 16\sin^2\left(\frac{2\pi}{11}\right)\cos^2\left(\frac{3\pi}{11}\right) + 8\sin\left(\frac{2\pi}{11}\right)\sin\left(\frac{3\pi}{11}\right)\cos\left(\frac{3\pi}{11}\right)$$

$$\lambda^2\left(2\cos^2\left(\frac{3\pi}{11}\right)\right) = 2\sin^2\left(\frac{3\pi}{11}\right) + 32\sin^2\left(\frac{2\pi}{11}\right)\cos^2\left(\frac{3\pi}{11}\right) + 8\sin\left(\frac{2\pi}{11}\right)\sin\left(\frac{6\pi}{11}\right)$$

$$= \left(1-\cos\left(\frac{6\pi}{11}\right)\right) + 8\left(1-\cos\frac{4\pi}{11}\right)\left(1+\cos\frac{6\pi}{11}\right) + 4\left(\cos\frac{4\pi}{11} - \cos\frac{8\pi}{11}\right)$$

$$= 9 + 7\cos\left(\frac{6\pi}{11}\right) - 4\cos\left(\frac{4\pi}{11}\right) - 8\cos\left(\frac{4\pi}{11}\right)\cos\left(\frac{6\pi}{11}\right) - 4\cos\left(\frac{8\pi}{11}\right)$$

$$= 9 + 7\cos\left(\frac{6\pi}{11}\right) - 4\cos\left(\frac{4\pi}{11}\right) - 4\left(\cos\left(\frac{10\pi}{11}\right) + \cos\left(\frac{2\pi}{11}\right)\right) - 4\cos\left(\frac{8\pi}{11}\right)$$

$$= 9 + 11\cos\left(\frac{6\pi}{11}\right) - 4\left\{\cos\left(\frac{2\pi}{11}\right) + \cos\left(\frac{4\pi}{11}\right) + \cos\left(\frac{6\pi}{11}\right) + \cos\left(\frac{8\pi}{11}\right) + \cos\left(\frac{10\pi}{11}\right)\right\}$$

$$= 9 + 11\cos\left(\frac{6\pi}{11}\right) - \frac{4\cdot\cos\left(\frac{6\pi}{11}\right)\sin\left(\frac{5\pi}{11}\right)}{\sin\left(\frac{\pi}{11}\right)}$$

$$= 9 + 11\cos\left(\frac{6\pi}{11}\right) - \frac{2\left\{\sin\pi - \sin\left(\frac{\pi}{11}\right)\right\}}{\sin\left(\frac{\pi}{11}\right)}$$

$$= 9 + 11\cos\left(\frac{6\pi}{11}\right) + 2$$

$$= 11\left(1 + \cos\left(\frac{6\pi}{11}\right)\right) = 11\left(2\cos^2\left(\frac{3\pi}{11}\right)\right)$$

$\therefore\ \lambda^2 = 11$

Then, $1 + \lambda^2 + \lambda^4 + \lambda^6 = 1 + 11 + 121 + 1331 = 1464$

13. $$\frac{\sqrt{(\sin A)}}{\sqrt{(\sin B)} + \sqrt{(\sin C)} - \sqrt{(\sin A)}} = \frac{\sqrt{a}}{\sqrt{b} + \sqrt{c} - \sqrt{a}}$$

Now, $$\sqrt{b} + \sqrt{c} - \sqrt{a} = \frac{(\sqrt{b} + \sqrt{c} - \sqrt{a})(\sqrt{b} + \sqrt{c} + \sqrt{a})}{(\sqrt{b} + \sqrt{c} + \sqrt{a})}$$

$$= \frac{(\sqrt{b} + \sqrt{c})^2 - a}{(\sqrt{b} + \sqrt{c} + \sqrt{a})}$$

$$= \frac{(b + c - a) + 2\sqrt{bc}}{(\sqrt{b} + \sqrt{c} + \sqrt{a})} > 0$$

Hence, $\sqrt{b} + \sqrt{c} - \sqrt{a} > 0$

Now, let $\sqrt{b} + \sqrt{c} - \sqrt{a} = x$, $\sqrt{c} + \sqrt{a} - \sqrt{b} = y$

and $\sqrt{a} + \sqrt{b} - \sqrt{c} = z$

$$\therefore\ \frac{\sqrt{\sin A}}{\sqrt{\sin B} + \sqrt{\sin C} - \sqrt{\sin A}} = \frac{y + z}{2x}$$

$$\Rightarrow\ \sum \frac{\sqrt{(\sin A)}}{\sqrt{(\sin B)} + \sqrt{(\sin C)} - \sqrt{(\sin A)}}$$

$$= \frac{1}{2}\left\{\frac{y}{x} + \frac{z}{x}\right\} + \frac{1}{2}\left\{\frac{z}{y} + \frac{x}{y}\right\} + \frac{1}{2}\left\{\frac{x}{z} + \frac{y}{z}\right\}$$

$$= \frac{1}{2}\left(\frac{x}{y} + \frac{y}{x}\right) + \frac{1}{2}\left(\frac{y}{z} + \frac{z}{y}\right) + \frac{1}{2}\left(\frac{z}{x} + \frac{x}{z}\right)$$

$$\geq 1 + 1 + 1 \qquad [\because \text{AM} \geq \text{GM}]$$

$$2020 \sum \frac{\sqrt{(\sin A)}}{\sqrt{(\sin B)} + \sqrt{(\sin C)} - \sqrt{(\sin A)}} \geq 6060$$

$\therefore$ Minimum value is 6060.

14. $$\because\ \lambda = \left\{\tan^2\left(\frac{\pi}{16}\right) + \tan^2\left(\frac{7\pi}{16}\right)\right\} + \left\{\tan^2\left(\frac{2\pi}{16}\right) + \tan^2\left(\frac{6\pi}{16}\right)\right\} + \left\{\tan^2\left(\frac{3\pi}{16}\right) + \tan^2\left(\frac{5\pi}{16}\right)\right\} + \tan^2\left(\frac{4\pi}{16}\right)$$

$$= \left\{\tan^2\left(\frac{\pi}{16}\right) + \cot^2\left(\frac{\pi}{2} - \frac{7\pi}{16}\right)\right\} + \left\{\tan^2\left(\frac{2\pi}{16}\right) + \cot^2\left(\frac{\pi}{2} - \frac{6\pi}{16}\right)\right\} + \left\{\tan^2\left(\frac{3\pi}{16}\right) + \cot^2\left(\frac{\pi}{2} - \frac{5\pi}{16}\right)\right\} + 1$$

$$= \left\{\tan^2\left(\frac{\pi}{16}\right) + \cot^2\left(\frac{\pi}{16}\right)\right\} + \left\{\tan^2\left(\frac{2\pi}{16}\right) + \cot^2\left(\frac{2\pi}{16}\right)\right\} + \left\{\tan^2\left(\frac{3\pi}{16}\right) + \cot^2\left(\frac{3\pi}{16}\right)\right\} + 1$$

$$= \left\{\tan\left(\frac{\pi}{16}\right) + \cot\left(\frac{\pi}{16}\right)\right\}^2 + \left\{\tan\left(\frac{2\pi}{16}\right) + \cot\left(\frac{2\pi}{16}\right)\right\}^2 + \left\{\tan\left(\frac{3\pi}{11}\right) + \cot\left(\frac{3\pi}{16}\right)\right\}^2 - 2 - 2 - 2 + 1$$

$$= \frac{1}{\left\{\sin\left(\frac{\pi}{16}\right)\cos\left(\frac{\pi}{16}\right)\right\}^2} + \frac{1}{\left\{\sin\left(\frac{2\pi}{16}\right)\cos\left(\frac{2\pi}{16}\right)\right\}^2} + \frac{1}{\left\{\sin\left(\frac{3\pi}{16}\right)\cos\left(\frac{3\pi}{16}\right)\right\}^2} - 5$$

$$= \frac{4}{\sin^2\frac{\pi}{8}} + \frac{4}{\sin^2\frac{\pi}{4}} + \frac{4}{\sin^2\left(\frac{3\pi}{8}\right)} - 5$$

$$= 4\left\{\frac{1}{\sin^2\left(\frac{\pi}{8}\right)} + \frac{1}{\sin^2\left(\frac{3\pi}{8}\right)}\right\} + 4\cdot 2 - 5$$

$$= 4\left\{\frac{1}{\sin^2\left(\frac{\pi}{8}\right)} + \frac{1}{\cos^2\left(\frac{\pi}{8}\right)}\right\} + 3$$

$$= \frac{4}{\left\{\sin\left(\frac{\pi}{8}\right)\cos\left(\frac{\pi}{8}\right)\right\}^2} + 3$$

$$= \frac{16}{\left\{\sin\left(\frac{\pi}{4}\right)\right\}^2} + 3 = 32 + 3 = 35$$

$\therefore\ \lambda = 35$

Then, $x^y + y^x = 35$

$\Rightarrow\ x = 34,\ y = 1$

or $x = 1,\ y = 34$

$$\therefore\ \frac{x + y}{8} = 4.375$$

Single Integer Type

1. To get $\cot\frac{\pi}{24}$, we proceed as follows

$$\cot\frac{\pi}{24} = \frac{2\cos^2\frac{\pi}{24}}{2\sin\frac{\pi}{24}\cdot\cos\frac{\pi}{24}}$$

$$= \frac{1+\cos\frac{\pi}{12}}{\sin\frac{\pi}{12}} = \frac{1+\cos\left(\frac{\pi}{4}-\frac{\pi}{6}\right)}{1+\sin\left(\frac{\pi}{4}-\frac{\pi}{6}\right)}$$

$$= \frac{1+\frac{1}{\sqrt{2}}\times\frac{\sqrt{3}}{2}+\frac{1}{\sqrt{2}}\times\frac{1}{2}}{\frac{1}{\sqrt{2}}\times\frac{\sqrt{3}}{2}-\frac{1}{\sqrt{2}}\times\frac{1}{2}}$$

$$= 2+\sqrt{2}+\sqrt{3}+\sqrt{6}$$

and $\tan\frac{\pi}{24} = -2+\sqrt{2}-\sqrt{3}+\sqrt{6}$

So, $\frac{1}{2}\left(\tan\frac{\pi}{24}+\cot\frac{\pi}{24}\right) = \sqrt{2}+\sqrt{6} = \sqrt{2^2+2}+\sqrt{2}$

$= \sqrt{a^2+a}+\sqrt{2}$ [given]

So, $a = 2$

2. It is given that, for $\alpha, \beta \in \left(0, \frac{\pi}{2}\right)$

$$\frac{\sqrt{2}\sin\alpha}{\sqrt{1+\cos 2\alpha}} = \frac{1}{7} \Rightarrow \frac{\sqrt{2}\sin\alpha}{\sqrt{2}|\cos\alpha|} = \frac{1}{7}$$

$\Rightarrow$ $\tan\alpha = \frac{1}{7}$

$\because$ $\alpha \in \left(0, \frac{\pi}{2}\right)$ ...(i)

and $\sqrt{\frac{1-\cos 2\beta}{2}} = \frac{1}{\sqrt{10}}$

$\Rightarrow$ $|\sin\beta| = \frac{1}{\sqrt{10}}$

$\Rightarrow$ $\sin\beta = \frac{1}{\sqrt{10}}$ $\because \beta \in \left(0, \frac{\pi}{2}\right)$

$\therefore$ $\tan\beta = \frac{1}{3}$ ...(ii)

$\therefore$ $\tan 2\beta = \frac{2\tan\beta}{1-\tan^2\beta} = \frac{\frac{2}{3}}{1-\frac{1}{9}} = \frac{6}{8} = \frac{3}{4}$

Now, $\tan(\alpha+2\beta) = \frac{\tan\alpha+\tan 2\beta}{1-\tan\alpha\tan 2\beta}$

$$= \frac{\frac{1}{7}+\frac{3}{4}}{1-\left(\frac{1}{7}\times\frac{3}{4}\right)} = \frac{4+21}{28-3} = \frac{25}{25} = 1$$

Hence, answer is 1.

3. We have,

$\cos\frac{\pi}{7}-\cos\frac{2\pi}{7}+\cos\frac{3\pi}{7}-\cos\frac{4\pi}{7}+\cos\frac{5\pi}{7}-\cos\frac{6\pi}{7}$

$$= \left(\cos\frac{\pi}{7}-\cos\frac{6\pi}{7}\right)-\left(\cos\frac{2\pi}{7}-\cos\frac{5\pi}{7}\right)+\left(\cos\frac{3\pi}{7}-\cos\frac{4\pi}{7}\right)$$

$$= \left(\cos\frac{\pi}{7}+\cos\frac{\pi}{7}\right)-\left(\cos\frac{2\pi}{7}+\cos\frac{2\pi}{7}\right)+\left(\cos\frac{3\pi}{7}+\cos\frac{3\pi}{7}\right)$$

$$= 2\left(\cos\frac{\pi}{7}-\cos\frac{2\pi}{7}+\cos\frac{3\pi}{7}\right)$$

$$= 2\left(\cos\frac{\pi}{7}+\cos\frac{3\pi}{7}+\cos\frac{5\pi}{7}\right)$$

$$= \frac{2\cos\left(\frac{\pi}{7}+\frac{2\pi}{7}\right)\sin\left(\frac{3\pi}{7}\right)}{\sin\frac{\pi}{7}}$$

$$= \frac{2\cos\frac{3\pi}{7}\sin\frac{3\pi}{7}}{\sin\frac{\pi}{7}} = \frac{\sin\frac{6\pi}{7}}{\sin\frac{\pi}{7}} = \frac{\sin\frac{\pi}{7}}{\sin\frac{\pi}{7}} = 1$$

4. We have,

$$16\left(\cos\theta-\cos\frac{\pi}{8}\right)\left(\cos\theta-\cos\frac{3\pi}{8}\right)\left(\cos\theta-\cos\frac{5\pi}{8}\right)\left(\cos\theta-\cos\frac{7\pi}{8}\right)$$

$$= 16\left(\cos\theta-\cos\frac{\pi}{8}\right)\left(\cos\theta-\cos\frac{7\pi}{8}\right)\times\left(\cos\theta-\cos\frac{3\pi}{8}\right)\left(\cos\theta-\cos\frac{5\pi}{8}\right)$$

$$= 16\left(\cos\theta-\cos\frac{\pi}{8}\right)\left(\cos\theta+\cos\frac{\pi}{8}\right)\times\left(\cos\theta-\cos\frac{3\pi}{8}\right)\left(\cos\theta+\cos\frac{3\pi}{8}\right)$$

$$= 16\left(\cos^2\theta-\cos^2\frac{\pi}{8}\right)\left(\cos^2\theta-\cos^2\frac{3\pi}{8}\right)$$

$$= 16\left(\cos^2\theta-\cos^2\frac{\pi}{8}\right)\left(\cos^2\theta-\sin^2\frac{\pi}{8}\right)$$

$$= 16\left(\cos^4\theta-\cos^2\theta+\sin^2\frac{\pi}{8}\cos^2\frac{\pi}{8}\right)$$

$$= 16\left(\cos^4\theta-\cos^2\theta+\frac{1}{8}\right) = 16\left(-\cos^2\theta-\sin^2\theta-\frac{1}{8}\right)$$

$$= 16\left(\frac{-\sin^2 2\theta}{4}+\frac{1}{8}\right) = 16\left(\frac{1-2\sin^2 2\theta}{8}\right) = \frac{16\cos 4\theta}{8}$$

$\Rightarrow \lambda = 2$

5. $\cos B+\cos C = 4\sin^2(A/2)$

$$\Rightarrow 2\cos\frac{B+C}{2}\cos\frac{B-C}{2} = 4\sin\frac{A}{2}\sin\left(\frac{\pi}{2}-\frac{B+C}{2}\right)$$

$$\Rightarrow 2\sin\frac{A}{2}\cos\frac{B-C}{2} = 4\sin\frac{A}{2}\cos\frac{B+C}{2}$$

$$\Rightarrow \cos\frac{B}{2}\cos\frac{C}{2}+\sin\frac{B}{2}\sin\frac{C}{2} = 2\left(\cos\frac{B}{2}\cos\frac{C}{2}-\sin\frac{B}{2}\sin\frac{C}{2}\right)$$

$$\Rightarrow 3\sin\frac{B}{2}\sin\frac{C}{2} = \cos\frac{B}{2}\cos\frac{C}{2}$$

$$\Rightarrow 3\tan\frac{B}{2}\tan\frac{C}{2} = 1$$

6. $\tan 9^\circ - \tan 27^\circ - \tan 63^\circ + \tan 81^\circ$

$= \tan 81^\circ + \tan 9^\circ - (\tan 63^\circ + \tan 27^\circ)$

$= \cot 9^\circ + \tan 9^\circ - (\cot 27^\circ + \tan 27^\circ)$

$= \dfrac{\cos^2 9^\circ + \sin^2 9^\circ}{\sin 9^\circ \cdot \cos 9^\circ} - \dfrac{\cos^2 27^\circ + \sin^2 27^\circ}{\sin 27 \cdot \cos 27}$

$= \dfrac{1}{\sin 9^\circ \cdot \cos 9^\circ} - \dfrac{1}{\sin 27^\circ \cdot \cos 27^\circ}$

$= \dfrac{2}{\sin 18^\circ} - \dfrac{2}{\sin 54^\circ} = 2\dfrac{\sin 54 - \sin 18}{\sin 54 \sin 18}$

$= \dfrac{4\cos 36 \cdot \sin 18}{\cos 36 \cdot \sin 18} = 4$

7. Given,

$\sin x + \cos x + \tan x + \cot x + \sec x + \operatorname{cosec} x = 7$

$\Rightarrow (\sin x + \cos x) + \left(\dfrac{\sin x}{\cos x} + \dfrac{\cos x}{\sin x}\right) + \left(\dfrac{1}{\cos x} + \dfrac{1}{\sin x}\right) = 7$

$\Rightarrow (\sin x + \cos x) + \left(\dfrac{1}{\sin x \cos x}\right) + \left(\dfrac{\sin x + \cos x}{\sin x \cos x}\right) = 7$

$\Rightarrow (\sin x + \cos x)\left(1 + \dfrac{2}{\sin 2x}\right) = 7 - \dfrac{2}{\sin 2x}$

Squaring, we get

$(1 + \sin 2x)\left(1 + \dfrac{4}{\sin^2 2x} + \dfrac{4}{\sin 2x}\right)$

$= 49 + \dfrac{4}{\sin^2 2x} - \dfrac{28}{\sin 2x}$

$\Rightarrow \sin^3 2x - 44\sin^2 2x + 36\sin 2x = 0$

$\Rightarrow \sin^2 2x - 44\sin 2x + 36 = 0$ $\quad [\because \sin 2x \neq 0]$

$\Rightarrow \sin^2 2x = 22 - 8\sqrt{7}$

$\Rightarrow \quad a = 22,\ b = 8$

$\therefore \quad \dfrac{a+b}{10} = 3$

8. We have,

$$\frac{\sin^3\theta}{\sin(2\theta+\alpha)} = \frac{\cos^3\theta}{\cos(2\theta+\alpha)} = k \text{ (let)}$$

$\Rightarrow \dfrac{\sin^4\theta}{\sin\theta\sin(2\theta+\alpha)} = \dfrac{\cos^4\theta}{\cos\theta\cos(2\theta+\alpha)} = k$

$\Rightarrow \cos^4\theta - \sin^4\theta = k$

$[\cos\theta\cos(2\theta+\alpha) - \sin\theta\sin(2\theta+\alpha)]$

$\Rightarrow \cos 2\theta = k\cos(3\theta+\alpha)$ $\quad \dots$ (i)

Also, $\dfrac{\sin^3\theta\cos\theta}{\sin(2\theta+\alpha)\cos\theta} = \dfrac{\sin\theta\cos^3\theta}{\sin\theta\cos(2\theta+\alpha)} = k$

$\Rightarrow \sin^3\theta\cos\theta + \sin\theta\cos^3\theta$

$= k(\sin(2\theta+\alpha)\cos\theta + \sin\theta\cos(2\theta+\alpha))$

$\Rightarrow \sin\theta\cos\theta(\sin^2\theta + \cos^2\theta) = k\sin(3\theta+\alpha)$

$\Rightarrow \sin 2\theta = 2k\sin(3\theta+\alpha)$

From Eqs. (i) and (ii), we get

$\tan 2\theta = 2\tan(3\theta+\alpha)$

$\therefore \quad \lambda = 2$

Matrix-Match Type

1. (A) $\sin\left(\dfrac{A}{2}\right) + \sin\left(\dfrac{B}{2}\right) + \sin\left(\dfrac{C}{2}\right)$

$= 1 + \left(\sin\left(\dfrac{A}{2}\right) + \sin\left(\dfrac{B}{2}\right)\right) - \left(\sin\dfrac{\pi}{2} - \sin\dfrac{C}{2}\right)$

$= 1 + 2\sin\left(\dfrac{A+B}{4}\right)\cos\left(\dfrac{A-B}{4}\right)$

$-2\cos\left(\dfrac{\pi+C}{4}\right)\sin\left(\dfrac{\pi-C}{4}\right)$

$= 1 + 2\sin\left(\dfrac{\pi-C}{4}\right)\left\{\cos\left(\dfrac{A-B}{4}\right) - \cos\left(\dfrac{\pi+C}{4}\right)\right\}$

$[\because A+B+C=\pi]$

$= 1 + 2\sin\left(\dfrac{\pi-C}{4}\right)$

$\left\{2\sin\left(\dfrac{\pi+C+A-B}{8}\right)\sin\left(\dfrac{\pi+C+B-A}{8}\right)\right\}$

$= 1 + 4\sin\left(\dfrac{\pi-C}{4}\right)\sin\left(\dfrac{\pi-B}{4}\right)\sin\left(\dfrac{\pi-A}{4}\right)$

$= 1 + 4\sin\left(\dfrac{\pi-A}{4}\right)\sin\left(\dfrac{\pi-B}{4}\right)\sin\left(\dfrac{\pi-C}{4}\right)$ (R)

$= 1 + 4\cos\left(\dfrac{\pi}{2} - \dfrac{\pi-A}{4}\right)\cos\left(\dfrac{\pi}{2} - \dfrac{\pi-B}{4}\right)\sin\left(\dfrac{\pi-C}{4}\right)$

$= 1 + 4\cos\left(\dfrac{\pi+A}{4}\right)\cos\left(\dfrac{\pi+B}{4}\right)\sin\left(\dfrac{\pi-C}{4}\right)$ (T)

(B) $\sin\left(\dfrac{A}{2}\right) + \sin\left(\dfrac{B}{2}\right) - \sin\left(\dfrac{C}{2}\right)$

$= -1 + \left(\sin\left(\dfrac{A}{2}\right) + \sin\left(\dfrac{B}{2}\right)\right) + \left(\sin\left(\dfrac{\pi}{2}\right) - \sin\left(\dfrac{C}{2}\right)\right)$

$= -1 + 2\sin\left(\dfrac{A+B}{4}\right)\cos\left(\dfrac{A-B}{4}\right) + 2\cos\left(\dfrac{\pi+C}{4}\right)$

$\sin\left(\dfrac{\pi-C}{4}\right)$

$= -1 + 2\sin\left(\dfrac{\pi-C}{4}\right)\left\{\cos\left(\dfrac{A-B}{4}\right) + \cos\left(\dfrac{\pi+C}{4}\right)\right\}$

$[\because A+B+C=\pi]$

$= -1 + 2\sin\left(\dfrac{\pi-C}{4}\right)\left\{2\cos\left(\dfrac{\pi+C+A-B}{8}\right)\right.$

$\left.\cos\left(\dfrac{\pi+C+B-A}{8}\right)\right\}$

$= -1 + 4\sin\left(\dfrac{\pi-C}{4}\right)\cos\left(\dfrac{\pi-B}{4}\right)\cos\left(\dfrac{\pi-A}{4}\right)$

$= -1 + 4\cos\left(\dfrac{\pi-A}{4}\right)\cos\left(\dfrac{\pi-B}{4}\right)\sin\left(\dfrac{\pi-C}{4}\right)$ (S)

$= -1 + 4\sin\left(\dfrac{\pi}{2} - \dfrac{\pi-A}{4}\right)\sin\left(\dfrac{\pi}{2} - \dfrac{\pi-B}{4}\right)\cos\left(\dfrac{\pi}{2} - \dfrac{\pi-C}{4}\right)$

$= -1 + 4\sin\left(\dfrac{\pi+A}{4}\right)\sin\left(\dfrac{\pi+B}{4}\right)\cos\left(\dfrac{\pi+C}{4}\right)$ (P)

(C) $\cos\left(\frac{A}{2}\right) + \cos\left(\frac{B}{2}\right) - \cos\left(\frac{C}{2}\right)$

$$= \left(\cos\left(\frac{A}{2}\right) + \cos\left(\frac{B}{2}\right)\right) - \left(\cos\left(\frac{C}{2}\right) + \cos\left(\frac{\pi}{2}\right)\right)$$

$$= 2\cos\left(\frac{A+B}{4}\right)\cos\left(\frac{A-B}{4}\right) - 2\cos\left(\frac{\pi+C}{4}\right)\cos\left(\frac{\pi-C}{4}\right)$$

$$= 2\cos\left(\frac{\pi-C}{4}\right)\left\{\cos\left(\frac{A-B}{4}\right) - \cos\left(\frac{\pi+C}{4}\right)\right\}$$

$[\because A + B + C = \pi]$

$$= 2\cos\left(\frac{\pi-C}{4}\right)\left\{2\sin\left(\frac{\pi+C+A-B}{8}\right)\sin\left(\frac{\pi+C+B-A}{8}\right)\right\}$$

$$= 4\cos\left(\frac{\pi-C}{4}\right)\sin\left(\frac{\pi-B}{4}\right)\sin\left(\frac{\pi-A}{4}\right)$$

$$= 4\cos\left(\frac{\pi-C}{4}\right)\cos\left(\frac{\pi}{2}-\frac{\pi-B}{4}\right)\cos\left(\frac{\pi}{2}-\frac{\pi-A}{4}\right)$$

$$= 4\cos\left(\frac{\pi+A}{4}\right)\cos\left(\frac{\pi+B}{4}\right)\cos\left(\frac{\pi-C}{4}\right) \text{ (Q)}$$

2. (A) If M is mid-point of PQ, then

$$M = \left(\frac{\theta+\phi}{2}, \frac{\sin\theta+\sin\phi}{2}\right)$$

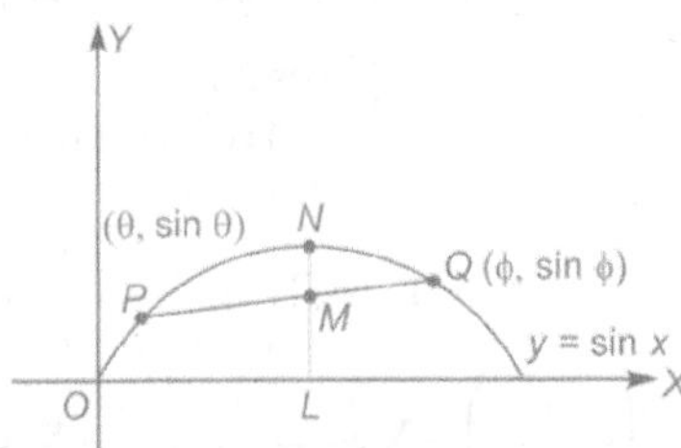

Also, $N \equiv \left(\frac{\theta+\phi}{2}, \sin\left(\frac{\theta+\phi}{2}\right)\right)$

It is clear from the figure.

$$ML \le NL$$

$$\Rightarrow \frac{\sin\theta + \sin\phi}{2} \le \sin\left(\frac{\theta+\phi}{2}\right)$$

$$\Rightarrow \sin\theta + \sin\phi \le 2\sin\left(\frac{\theta+\phi}{2}\right)$$

$$= 2\sin\left(\frac{\pi}{4}\right)$$

$$= \sqrt{2}$$

$\therefore\ \sin\theta + \sin\phi \le \sqrt{2}$

and $(\sin\theta + \sin\phi)\sin\frac{\pi}{4} \le 1$ (P, Q, R, S, T)

(B) $\because a^2 + b^2 = (\sin\theta - \sin\phi)^2 + (\cos\theta + \cos\phi)^2$

$$= 2 + 2\cos(\theta+\phi)$$

$$= 4\cos^2\left(\frac{\theta+\phi}{2}\right) \le 4 \text{ (S, T)}$$

(C) $\because 3\sin\theta + 5\cos\theta = 5$

$$\Rightarrow 3\sin\theta = 5(1-\cos\theta)$$

Squaring on both sides, then

$$9\sin^2\theta = 25(1-\cos\theta)^2$$

$$\Rightarrow 9(1-\cos\theta)(1+\cos\theta) = 25(1-\cos\theta)^2$$

$$\Rightarrow 9(1+\cos\theta) = 25(1-\cos\theta)(1-\cos\theta \ne 0)$$

$$\therefore\ 34\cos\theta = 16$$

$\cos\theta = \frac{8}{17}$, then $\sin\theta = \frac{15}{17}$

$$\therefore\ 5\sin\theta - 3\cos\theta = \frac{75}{17} - \frac{24}{17} = 3 \text{ (R)}$$

Hence, $5\sin\theta - 3\cos\theta = 3$

3. (A) Let $y = \dfrac{7 + 6\tan\theta - \tan^2\theta}{(1+\tan^2\theta)}$

$$= 7\cos^2\theta + 6\sin\theta\cos\theta - \sin^2\theta$$

$$= 7\left(\frac{1+\cos 2\theta}{2}\right) + 3\sin 2\theta - \left(\frac{1-\cos 2\theta}{2}\right)$$

$$= 3\sin 2\theta + 4\cos 2\theta + 3$$

$$-\sqrt{(3^2+4^2)} + 3 \le 3\sin 2\theta + 4\cos 2\theta + 3 \le \sqrt{(3^2+4^2)} + 3$$

$$\therefore\ -2 \le y \le 8$$

$$\Rightarrow \lambda = 8, \mu = -2$$

$$\Rightarrow \lambda + \mu = 6, \lambda - \mu = 10 \text{ (R, S)}$$

(B) Let $y = 5\cos\theta + 3\cos(\theta + \pi/3) + 3$

$$= 5\cos\theta + 3\left(\frac{1}{2}\cos\theta - \frac{\sqrt{3}}{2}\sin\theta\right) + 3$$

$$= \frac{13}{2}\cos\theta - \frac{3\sqrt{3}}{2}\sin\theta + 3$$

$$\therefore\ 3 - \sqrt{\left(\frac{13}{2}\right)^2 + \left(\frac{-3\sqrt{3}}{2}\right)^2} \le \frac{13}{2}\cos\theta - \frac{3\sqrt{3}}{2}\sin\theta + 3 \le 3 + \sqrt{\left(\frac{13}{2}\right)^2 + \left(\frac{-3\sqrt{3}}{2}\right)^2}$$

$$\Rightarrow 3 - 7 \le y \le 3 + 7$$

$$\Rightarrow -4 \le y \le 10$$

$$\therefore\ \lambda = 10, \mu = -4$$

$$\Rightarrow \lambda + \mu = 6, \lambda - \mu = 14 \text{ (R, T)}$$

(C) Let $y = 1 + \sin\left(\frac{\pi}{4} + \theta\right) + 2\cos\left(\frac{\pi}{4} - \theta\right)$

$$= 1 + \cos\left(\frac{\pi}{2} - \left(\frac{\pi}{4} + \theta\right)\right) + 2\cos\left(\frac{\pi}{4} - \theta\right)$$

$$= 1 + \cos\left(\frac{\pi}{4} - \theta\right) + 2\cos\left(\frac{\pi}{4} - \theta\right)$$

$$= 1 + 3\cos\left(\frac{\pi}{4} - \theta\right)$$

$$\therefore\ -1 \le \cos\left(\frac{\pi}{4} - \theta\right) \le 1 \Rightarrow -3 \le 3\cos\left(\frac{\pi}{4} - \theta\right) \le 3$$

$$\Rightarrow 1 - 3 \le 1 + 3\cos\left(\frac{\pi}{4} - \theta\right) \le 1 + 3$$

$$\therefore\ -2 \le y \le 4$$

$$\Rightarrow \lambda = 4, \mu = -2$$

$$\therefore\ \lambda + \mu = 2, \lambda - \mu = 6 \text{ (P, Q)}$$

4. **(A)** Let $x = \tan A$, $y = \tan B$, $z = \tan C$

$\tan A + \tan B + \tan C = \tan A \tan B \tan C$

$\Rightarrow \quad \tan A + \tan B = -\tan C\,(1 - \tan A \tan B)$

$\Rightarrow \quad \dfrac{\tan A + \tan B}{1 - \tan A \tan B} = -\tan C$

$\Rightarrow \quad \tan(A+B) = \tan(\pi - C)$

$\therefore \quad A + B = \pi - C$

$\Rightarrow \quad A + B + C = \pi$

(i) $\sum\left(\dfrac{x}{1+x^2}\right) = \sum\left(\dfrac{\tan A}{1+\tan^2 A}\right) = \dfrac{1}{2}\sum \sin 2A$

$= \dfrac{1}{2}(4\sin A \sin B \sin C)$

$= \dfrac{2xyz}{\sqrt{(1+x^2)(1+y^2)(1+z^2)}}$

$= 2\prod \dfrac{x}{\sqrt{(1+x^2)}}$ (S)

(ii) $\sum\left(\dfrac{x}{1-x^2}\right) = \sum\left(\dfrac{\tan A}{1-\tan^2 A}\right) = \dfrac{1}{2}\sum \tan 2A$

$= \dfrac{1}{2}(\tan 2A \cdot \tan 2B \cdot \tan 2C)$

$= \dfrac{1}{2} \cdot \dfrac{2x}{1-x^2} \cdot \dfrac{2y}{1-y^2} \cdot \dfrac{2z}{1-z^2}$

$= 4\prod \dfrac{x}{(1-x^2)}$ (Q)

(iii) $\sum\left(\dfrac{1-x^2}{1+x^2}\right) = \sum\left(\dfrac{1-\tan^2 A}{1+\tan^2 A}\right) = \sum \cos 2A$

$= \cos 2A + \cos 2B + \cos 2C$

$= -1 - 4\cos A \cos B \cos C$

$= -1 - 4\prod \cos A$

$= -1 - 4\prod\left(\dfrac{1}{\sqrt{(1+\tan^2 A)}}\right)$

$= -1 - 4\prod\left(\dfrac{1}{\sqrt{(1+x^2)}}\right)$ (R)

(iv) $\sum\left(\dfrac{1}{1+x^2}\right) = \sum\left(\dfrac{1}{1+\tan^2 A}\right) = \sum \cos^2 A$

$= 1 - 2\cos A \cos B \cos C$

$= 1 - 2\prod \dfrac{1}{\sec A} = 1 - 2\prod \dfrac{1}{\sqrt{(1+x^2)}}$ (T)

(B) $\because \; xy + yz + zx = 1$

Let $x = \cot A$, $y = \cot B$, $z = \cot C$

$\Rightarrow \quad \cot A \cot B + \cot B \cot C + \cot C \cot A = 1$

$\therefore \quad A + B + C = \pi$

(i) $\sum\left(\dfrac{x}{1+x^2}\right) = \sum\left(\dfrac{\cot A}{1+\cot^2 A}\right) = \dfrac{1}{2}\sum \sin 2A$

$= \dfrac{1}{2}(4\sin A \sin B \sin C)$

$= \dfrac{2}{\sqrt{(1+\cot^2 A)(1+\cot^2 B)(1+\cot^2 C)}}$

$= 2\prod\left(\dfrac{1}{\sqrt{(1+x^2)}}\right)$ (P)

(ii) $\sum\left(\dfrac{x}{1-x^2}\right) = \sum\left(\dfrac{\cot A}{1-\cot^2 A}\right)$

$= -\sum\left(\dfrac{\tan A}{1-\tan^2 A}\right)$

$= -\dfrac{1}{2}\sum \tan 2A$

$= -\dfrac{1}{2}(\tan 2A \tan 2B \tan 2C)$

$= -\dfrac{4\tan A \tan B \tan C}{(1-\tan^2 A)(1-\tan^2 B)(1-\tan^2 C)}$

$= -\dfrac{4\cot A \cot B \cot C}{(\cot^2 A - 1)(\cot^2 B - 1)(\cot^2 C - 1)}$

$= -4\prod\left(\dfrac{x}{x^2-1}\right)$

$= 4\prod\left(\dfrac{x}{1-x^2}\right)$ (Q)

(iii) $\sum\left(\dfrac{1-x^2}{1+x^2}\right) = \sum\left(\dfrac{1-\cot^2 A}{1+\cot^2 A}\right) = \sum\left(\dfrac{\tan^2 A - 1}{\tan^2 A + 1}\right)$

$= -\sum \cos 2A$

$= -(-1 - 4\cos A \cos B \cos C)$

$= 1 + 4\prod \cos A$

$= 1 + 4\prod\left(\dfrac{x}{\sqrt{(1+x^2)}}\right) \neq$ (P, Q, R, S, T)

(iv) $\sum\left(\dfrac{1}{1+x^2}\right) = \sum\left(\dfrac{1}{1+\cot^2 A}\right) = \sum \sin^2 A$

$= 2 + 2\cos A \cos B \cos C$

$= 2 + 2\prod \cos A$

$= 2 + 2\prod\left(\dfrac{x}{\sqrt{(1+x^2)}}\right) \neq$ (P, Q, R, S, T)

(C) $\because \; (x+y+z)^2 = 3$

$\Rightarrow \quad x^2 + y^2 + z^2 + 2(xy + yz + zx) = 3$

$\Rightarrow \quad 1 + 2(xy + yz + zx) = 3$

$\therefore \quad xy + yz + zx = 1$

Now, $(x-y)^2 + (y-z)^2 + (z-x)^2$

$= 2(x^2 + y^2 + z^2 - xy - yz - zx)$

$= 2(1-1)$

$= 0$

Which is possible only when

$x - y = 0,\; y - z = 0,\; z - x = 0$

$\therefore \quad x = y = z$

$\because \; x + y + z = \sqrt{3}$ and $x^2 + y^2 + z^2 = 1$

$x = y = z = \dfrac{1}{\sqrt{3}}$

(i) $\sum \frac{x}{(1+x^2)} = \frac{x}{1+x^2} + \frac{y}{1+y^2} + \frac{z}{1+z^2}$

$$= \frac{3x}{1+x^2} = \frac{3\times\frac{1}{\sqrt{3}}}{1+\frac{1}{3}} = \frac{3\sqrt{3}}{4}$$

$$2\Pi\frac{1}{\sqrt{(1+x^2)}} = 2\cdot\left(\frac{1}{\sqrt{(1+x^2)}}\right)^3$$

$$= 2\cdot\left(\frac{1}{\sqrt{1+\frac{1}{3}}}\right)^3 = 2\left(\frac{\sqrt{3}}{2}\right)^3$$

$$= \frac{2\cdot 3\sqrt{3}}{8} = \frac{3\sqrt{3}}{4}$$

$$\therefore \quad \sum\left(\frac{x}{1+x^2}\right) = 2\Pi\frac{1}{\sqrt{(1+x^2)}} \quad (P)$$

(ii) $\sum\left(\frac{x}{1-x^2}\right) = 3\left(\frac{x}{1-x^2}\right) = 3\left(\frac{\frac{1}{\sqrt{3}}}{1-\frac{1}{3}}\right) = \frac{3\sqrt{3}}{2}$

and $4\Pi\left(\frac{x}{1-x^2}\right) = 4\left(\frac{x}{1-x^2}\right)^3$

$$= 4\left(\frac{\frac{1}{\sqrt{3}}}{1-\frac{1}{3}}\right)^3 = 4\cdot\frac{\frac{1}{3\sqrt{3}}}{\frac{8}{27}}$$

$$= \frac{4\times 27}{8\times 3\sqrt{3}} = \frac{3\sqrt{3}}{2}$$

$$\sum\left(\frac{x}{1-x^2}\right) = 4\Pi\left(\frac{x}{1-x^2}\right) \quad (Q)$$

(iii) $\sum\left(\frac{1-x^2}{1+x^2}\right) = 3\left(\frac{1-x^2}{1+x^2}\right) = 3\left(\frac{1-\frac{1}{3}}{1+\frac{1}{3}}\right) = \frac{3}{2}$

and $-1-4\Pi\frac{1}{\sqrt{(1+x^2)}} = -1-4\left(\frac{1}{\sqrt{(1+x^2)}}\right)^3$

$$= -1-4\left(\frac{1}{\sqrt{1+\frac{1}{3}}}\right)^3$$

$$= -1-4\times\frac{3\sqrt{3}}{8} \neq \frac{3}{2} \neq (P, Q, R, S, T)$$

(iv) $\sum\frac{1}{(1+x^2)} = 3\left(\frac{1}{1+x^2}\right) = 3\left(\frac{1}{1+\frac{1}{3}}\right) = \frac{9}{4}$

and $1-2\Pi\frac{1}{\sqrt{(1+x^2)}} = 1-2\left(\frac{1}{\sqrt{1+x^2}}\right)^3$

$$= 1-2\left(\frac{1}{\sqrt{1+\frac{1}{3}}}\right)^3 = 1-2\left(\frac{3\sqrt{3}}{8}\right) \neq \frac{9}{4} \neq (P, Q, R, S, T)$$

5. **(A)** We have, $\{\cos(2A+\theta)+\cos(2B+\theta)\}$

$$= 2\cos(A-B)\cos(A+B+\theta)$$

Maximum value is $2\cos(A-B)$ when

$$\cos(A+B+\theta) = 1 \qquad \textbf{(S)}$$

(B) We have,

$$\{\cos 2A + \cos 2B\} = 2\cos(A+B)\cos(A-B)$$

Maximum value is $2\cos(A-B)$ when **(S)**

$$\cos(A+B) = 1$$

(C) For $y = \sec x$, $x \in (0, \pi/2)$, tangent drawn to it at any point lies completely below the graph of $y = \sec x$; thus, n **(S)**

$$\frac{\sec 2A + \sec 2B}{2} \geq 2\sec(A+B)$$

$$\Rightarrow \quad \sec 2A + \sec 2B \geq 2\sec(A+B)$$

Hence, the minimum value is $2\sec(A+B)$. **(Q)**

(D) $\sqrt{\{\tan\theta + \cot\theta - 2\cos 2(A+B)\}}$

$$= \sqrt{(\sqrt{\tan\theta} - \sqrt{\cot\theta})^2 + 2 - 2\cos 2(A+B)}$$

$$= \sqrt{(\sqrt{\tan\theta} - \sqrt{\cot\theta})^2 + 4\sin^2(A+B)}$$

Minimum value occurs when $\sqrt{\tan\theta} = \sqrt{\cot\theta}$ and minimum value is $\sqrt{4\sin^2(A+B)} = 2\sin(A+B)$ **(P)**

28

Trigonometric Equations

Objective Questions Type I [Only one correct answer]

In each of the questions below, four choices are given of which only one is correct. You have to select the correct answer which is the most appropriate.

1. Let $S = \{\theta \in [-2\pi, 2\pi] : 2\cos^2\theta + 3\sin\theta = 0\}$, then the sum of the elements of S is **[JEE Main 2019, 9 April]**

(a) 2π (b) π

(c) $\frac{5\pi}{3}$ (d) $\frac{13\pi}{6}$

2. The number of solutions of the equation $\sin x + 2\sin 2x = 3 + \sin 3x$ in the interval $[0, \pi]$ is

(a) 0 (b) 1

(c) 2 (d) 3

3. If $\sin x + \cos x = \sqrt{\left(y + \frac{1}{y}\right)}$, $x \in [0, \pi]$, then

(a) $x = \frac{\pi}{4}, y = 1$ (b) $y = 0$

(c) $y = 2$ (d) $x = \frac{3\pi}{4}$

4. The values of x between 0 and 2π which satisfy the equation $\sin x\sqrt{(8\cos^2 x)} = 1$ are in AP with common difference

(a) $\frac{\pi}{8}$ (b) $\frac{\pi}{4}$

(c) $\frac{3\pi}{8}$ (d) $\frac{5\pi}{8}$

5. $|\tan x + \sec x| = |\tan x| + |\sec x|$, $x \in [0, 2\pi]$, if and only if x belongs to the interval

(a) $[0, \pi]$ (b) $\left[0, \frac{\pi}{2}\right) \cup \left(\frac{\pi}{2}, \pi\right]$

(c) $\left[0, \frac{3\pi}{2}\right) \cup \left(\frac{3\pi}{2}, 2\pi\right]$ (d) $(\pi, 2\pi]$

6. If $[\sin x] + [\sqrt{2}\cos x] = -3$, $x \in [0, 2\pi]$ ([.] denotes the greatest integer function), then x belongs to

(a) $\left(\pi, \frac{5\pi}{4}\right)$ (b) $\left[\pi, \frac{5\pi}{4}\right]$

(c) $\left(\frac{5\pi}{4}, 2\pi\right)$ (d) $\left[\frac{5\pi}{4}, 2\pi\right]$

7. The equation $\sin(\cos x) = \cos(\sin x)$ has

(a) only one real solution

(b) infinitely many solution

(c) no real solution

(d) None of the above

8. If $\frac{\sin^3\theta - \cos^3\theta}{\sin\theta - \cos\theta} - \frac{\cos\theta}{\sqrt{(1 + \cot^2\theta)}} - 2\tan\theta\cot\theta = -1$, $\theta \in [0, 2\pi]$, then

(a) $\theta \in \left(0, \frac{\pi}{2}\right) - \left\{\frac{\pi}{4}\right\}$ (b) $\theta \in \left(\frac{\pi}{2}, \pi\right) - \left\{\frac{3\pi}{4}\right\}$

(c) $\theta \in \left(\pi, \frac{3\pi}{2}\right) - \left\{\frac{5\pi}{4}\right\}$ (d) $\theta \in (0, \pi) - \left\{\frac{\pi}{4}, \frac{\pi}{2}\right\}$

9. If $|\sin x - \cos^2 x| \geq |3 - 3\sin x + \sin^2 x| + 4|\sin x - 1|$, then $x =$

(a) $(4n + 1)\frac{\pi}{2}, n \in Z$ (b) $2n\pi + \frac{\pi}{3}, n \in Z$

(c) $n\pi + \frac{\pi}{2}, n \in Z$ (d) $2n\pi + \frac{\pi}{6}, n \in Z$

10. Let n be a positive integer such that $\sin\left(\frac{\pi}{2n}\right) + \cos\left(\frac{\pi}{2n}\right) = \frac{\sqrt{n}}{2}$, then

(a) $n = 6$ (b) $n = 2$

(c) $n = 1$ (d) $n = 3, 4, 5$

11. If the equation $\cos^4\theta + \sin^4\theta + \lambda = 0$ has real solutions for θ, then λ lies in the interval **[JEE Main 2020, 2 September]**

(a) $\left(-\frac{5}{4}, -1\right)$ (b) $\left[-1, -\frac{1}{2}\right]$

(c) $\left(-\frac{1}{2}, -\frac{1}{4}\right]$ (d) $\left[-\frac{3}{2}, -\frac{5}{4}\right]$

12. The least positive value of x satisfying the equation $8^{1+|\cos x|+\cos^2 x+|\cos^3 x|+\ldots\infty} = 4^3$ will be (where $|\cos x| < 1$)

(a) $\frac{\pi}{3}$ (b) $\frac{2\pi}{3}$

(c) $\frac{\pi}{4}$ (d) None of these

13. The number of values of θ in the interval $\left(-\frac{\pi}{2}, \frac{\pi}{2}\right)$ satisfying the equation $(\sqrt{3})^{\sec^2\theta} = \tan^4\theta + 2\tan^2\theta$ is

(a) 1 (b) 2

(c) 3 (d) None of these

14. If $\sin^4 x + \cos^4 y + 2 = 4\sin x \cos y$ and $0 \le x, y \le \frac{\pi}{2}$, then $\sin x + \cos y$ is equal to

(a) −2 (b) 0

(c) 2 (d) $\frac{3}{2}$

15. The equation $\cos^4 x - (\lambda + 2)\cos^2 x - (\lambda + 3) = 0$ possesses a solution if

(a) $\lambda > -3$

(b) $\lambda < -2$

(c) $-3 \le \lambda \le -2$

(d) λ is any positive integer

16. The arithmetic mean of the roots of the equation $4\cos^3 x - 4\cos^2 x - \cos(315\pi + x) = 1$ in the interval (0, 315) is equal to

(a) 50π (b) 51π

(c) 100π (d) 315π

17. If $0 < \theta < 2\pi$ and $2\sin^2\theta - 5\sin\theta + 2 > 0$, then the range of θ is

(a) $\left(0, \frac{\pi}{6}\right) \cup \left(\frac{5\pi}{6}, 2\pi\right)$

(b) $\left(0, \frac{5\pi}{6}\right) \cup (\pi, 2\pi)$

(c) $\left(0, \frac{\pi}{6}\right) \cup (\pi, 2\pi)$

(d) None of these

18. Set $a, b \in [-\pi, \pi]$ be such that $\cos(a - b) = 1$ and $\cos(a + b) = \frac{1}{e}$. The number of pairs of a, b satisfying the above system of equation is

(a) 0 (b) 1

(c) 2 (d) 4

19. The number of real roots of the equation $\sin^{2020} x - \cos^{2020} x + 2019 = 2020$ in the interval $\left(-\frac{3\pi}{2}, \frac{5\pi}{2}\right)$

(a) 1 (b) 3

(c) 5 (d) Infinitely many

20. The number of solutions of the equation $x^3 + x^2 + 4x + 2\sin x = 0$ in $0 \le x \le 2\pi$ is

(a) 0 (b) 1

(c) 2 (d) 4

21. Let α, β be any two positive values of x for which $2\cos x$, $|\cos x|$ and $1 - 3\cos^2 x$ are in GP. The minimum value of $|\alpha - \beta|$ is

(a) $\frac{\pi}{3}$ (b) $\frac{\pi}{4}$

(c) $\frac{\pi}{2}$ (d) None of these

22. The number of solutions of the equation $\tan x + \sec x = 2\cos x$ lying in the interval $[0, 2\pi]$ is

(a) 0 (b) 1

(c) 2 (d) 3

23. If $2\tan^2 x - 5\sec x$ is equal to 1 for exactly 7 distinct values of $x \in \left[0, \frac{n\pi}{2}\right]$, $n \in N$, then the greatest value of n is

(a) 6 (b) 12

(c) 13 (d) 15

24. The general solution of the trigonometrical equation $\sin x + \cos x = 1$ for $n = 0, \pm 1, \pm 2 \ldots$ is given by

(a) $x = 2n\pi, n \in I$

(b) $x = 2n\pi + \frac{\pi}{2}, n \in I$

(c) $x = n\pi + (-1)^n \frac{\pi}{4} - \frac{\pi}{4}, n \in I$

(d) None of the above

25. The number of solutions of the equation

$$1 + \sin^4 x = \cos^2 3x, \ x \in \left[-\frac{5\pi}{2}, \frac{5\pi}{2}\right] \text{ is}$$

[JEE Main 2019, 12 April]

(a) 3 (b) 5

(c) 7 (d) 4

26. The smallest positive root of the equation $\tan x - x = 0$, lies in

(a) $\left(0, \frac{\pi}{2}\right)$ (b) $\left(\frac{\pi}{2}, \pi\right)$

(c) $\left(\pi, \frac{3\pi}{2}\right)$ (d) $\left(\frac{3\pi}{2}, 2\pi\right)$

27. The number of solutions of the equation $\sin^5 x - \cos^5 x = \frac{1}{\cos x} - \frac{1}{\sin x}$ $(\sin x \ne \cos x)$ is

(a) 0 (b) 1

(c) infinite (d) None of these

28. The equation $(\cos p - 1)x^2 + (\cos p)x + \sin p = 0$, where x is a variable, has real roots. Then the interval of p may be any one of the followings

(a) $(0, 2\pi)$ (b) $(-\pi, 0)$

(c) $\left(-\frac{\pi}{2}, \frac{\pi}{2}\right)$ (d) $(0, \pi)$

29. The number of solutions of the equation $2(\sin^4 2x + \cos^4 2x) + 3\sin^2 x \cos^2 x = 0$ is

(a) 0 (b) 1

(c) 2 (d) 3

30. All possible values of $\theta \in [0, 2\pi]$ for which $\sin 2\theta + \tan 2\theta > 0$ lie in **[JEE Main 2021, 25 February]**

(a) $\left(0, \frac{\pi}{2}\right) \cup \left(\pi, \frac{3\pi}{2}\right)$

(b) $\left(0, \frac{\pi}{4}\right) \cup \left(\frac{\pi}{2}, \frac{3\pi}{4}\right) \cup \left(\pi, \frac{5\pi}{4}\right) \cup \left(\frac{3\pi}{2}, \frac{7\pi}{4}\right)$

(c) $\left(0, \frac{\pi}{2}\right) \cup \left(\frac{\pi}{2}, \frac{3\pi}{4}\right) \cup \left(\pi, \frac{7\pi}{6}\right)$

(d) $\left(0, \frac{\pi}{4}\right) \cup \left(\frac{\pi}{2}, \frac{3\pi}{4}\right) \cup \left(\frac{3\pi}{2}, \frac{11\pi}{6}\right)$

31. The complete solution of the equation $7\cos^2 x + \sin x \cos x - 3 = 0$ is given by

(a) $n\pi + \frac{\pi}{2} \ (n \in I)$

(b) $n\pi - \frac{\pi}{4} \ (n \in I)$

(c) $n\pi + \tan^{-1}\left(\frac{4}{3}\right) (n \in I)$

(d) $n\pi + \frac{3\pi}{4}, k\pi + \tan^{-1}\left(\frac{4}{3}\right) (k, n \in I)$

32. If $0 \le x \le \pi$ and $81^{\sin^2 x} + 81^{\cos^2 x} = 30$, then x is equal to

(a) $\frac{\pi}{6}$ (b) $\frac{\pi}{2}$

(c) π (d) $\frac{\pi}{4}$

33. If $1 + \sin\theta + \sin^2\theta + \ldots \infty = 4 + 2\sqrt{3}, 0 < \theta < \pi, \theta \ne \frac{\pi}{2}$, then

(a) $\theta = \frac{\pi}{6}$ (b) $\theta = \frac{\pi}{3}$

(c) $\theta = \frac{\pi}{3}$ or $\frac{\pi}{6}$ (d) $\theta = \frac{\pi}{3}$ or $\frac{2\pi}{3}$

34. If $\tan(\pi\cos\theta) = \cot(\pi\sin\theta)$, then the value of $\cos\left(\theta - \frac{\pi}{4}\right)$ is

(a) $\frac{1}{2}$ (b) $\frac{1}{\sqrt{2}}$

(c) $\frac{1}{2\sqrt{2}}$ (d) None of these

35. If for $x \in \left(0, \frac{\pi}{2}\right)$, $\log_{10}\sin x + \log_{10}\cos x = -1$ and $\log_{10}(\sin x + \cos x) = \frac{1}{2}(\log_{10} x - 1), x > 0$ then the value of x is equal to **[JEE Main 2021, 16 March]**

(a) 20 (b) 12

(c) 9 (d) 16

36. The most general values of x for which $\sin x + \cos x = \min_{a \in R}\{1, a^2 - 4a + 6\}$ are given by

(a) $2n\pi, n \in N$ (b) $2n\pi + \frac{\pi}{2}, n \in N$

(c) $n\pi + (-1)^n \frac{\pi}{4} - \frac{\pi}{4}, n \in N$ (d) None of these

37. If $f(x) = \sin x + \cos x$, then the most general solutions of $f(x) = \left[f\left(\frac{\pi}{10}\right)\right]$ are, (where $[x]$ is the greatest integer less than or equal to x.)

(a) $2n\pi + \frac{\pi}{2}, n \in I$ (b) $n\pi, n \in I$

(c) $2n\pi \pm \frac{2\pi}{3}, n \in I$ (d) None of these

38. The smallest positive value of x satisfying the equation $\log_2\cos x + \log_2(1 - \tan x) + \log_2(1 + \tan x) - \log_2\sin x = 1$ is

(a) $\pi/8$ (b) π

(c) $\pi/4$ (d) $\pi/6$

39. The number of roots of the equation $x + 2\tan x = \frac{\pi}{2}$ in the interval $[0, 2\pi]$ is **[JEE Main 2021, 17 March]**

(a) 1 (b) 2

(c) 3 (d) infinite

40. If sum of all the solution of the equation $8\cos x \cdot \left(\cos\left(\frac{\pi}{6} + x\right) \cdot \cos\left(\frac{\pi}{6} - x\right) - \frac{1}{2}\right) = 1$ in $[0, \pi]$ is $k\pi$, then k is equal to **[JEE Main 2018]**

(a) $\frac{2}{3}$ (b) $\frac{13}{9}$

(c) $\frac{8}{9}$ (d) $\frac{20}{9}$

41. The number of solutions of the equation $\cos(\pi\sqrt{(x-4)}\cos(\pi\sqrt{x}) = 1$ is

(a) None (b) one

(c) two (d) more than two

42. The number of solutions of the equation

$$\sin\left(\frac{\pi x}{2\sqrt{3}}\right) = x^2 - 2\sqrt{3}x + 4$$

(a) forms an empty set

(b) is only one

(c) is only two

(d) is greater than two

43. The solution of the equation $\log_{\cos x} \sin x + \log_{\sin x} \cos x = 2$ is given by

(a) $x = 2n\pi + \frac{\pi}{4}, n \in I$ (b) $x = n\pi + \frac{\pi}{2}, n \in I$

(c) $x = n\pi + \frac{\pi}{8}, n \in I$ (d) None of these

44. The general value of θ such that $\sin 2\theta = \frac{\sqrt{3}}{2}$ and $\tan \theta = \frac{1}{\sqrt{3}}$ is given by

(a) $n\pi + \frac{7\pi}{6}, n \in I$ (b) $n\pi \pm \frac{7\pi}{6}, n \in I$

(c) $2n\pi + \frac{7\pi}{6}, n \in I$ (d) None of these

45. Number of ordered pairs (a, x) satisfying the equation $\sec^2(a+2)x + a^2 - 1 = 0; -\pi < x < \pi$ is

(a) 1 (b) 2 (c) 3 (d) 4

46. Number of real roots of the equation $\sec\theta + \text{cosec}\,\theta = \sqrt{15}$ lying between 0 and 2π is

(a) 8 (b) 4 (c) 2 (d) 0

47. The solution of the equation $\sin^{10} x + \cos^{10} x = \frac{29}{16} \cos^4 2x$ is

(a) $x = \frac{n\pi}{4} + \frac{\pi}{8}, n \in I$ (b) $x = n\pi + \frac{\pi}{4}, n \in I$

(c) $x = 2n\pi + \frac{\pi}{2}, n \in I$ (d) None of these

48. Solutions of the equation $|\cos x| = 2[x]$ are (where [.] denotes the greatest integer function)

(a) nil (b) $x = \pm 1$

(c) $x = \frac{\pi}{3}$ (d) None of these

49. The sum of all the solutions in $[0, 4\pi]$ of the equation $\tan x + \cot x + 1 = \cos\left(x + \frac{\pi}{4}\right)$ is

(a) 3π (b) $\pi/2$

(c) $7\pi/2$ (d) 4π

50. The number of solutions of the equation $2^{\cos x} = |\sin x|$ in $[-2\pi, 2\pi]$ is

(a) 1 (b) 2

(c) 3 (d) 4

51. The general solution of the equation $2^{\cos 2x} + 1 = 3 \cdot 2^{-\sin^2 x}$ is

(a) $n\pi, n \in I$ (b) $n\pi + \pi, n \in I$

(c) $n\pi - \pi, n \in I$ (d) None of these

52. If $x \in (0, 1)$ the greatest root of the equation $\sin 2\pi x = \sqrt{2} \cos \pi x$ is

(a) 1/4 (b) 1/2

(c) 3/4 (d) None of these

53. Let S be the set of all $\alpha \in R$ such that the equation, $\cos 2x + \alpha \sin x = 2\alpha - 7$ has a solution. Then, S is equal to **[JEE Main 2019, 12 April]**

(a) R (b) $[1, 4]$

(c) $[3, 7]$ (d) $[2, 6]$

54. The set of all x in $(-\pi, \pi)$ satisfying $|4 \sin x - 1| < \sqrt{5}$ is given by

(a) $x \in \left(-\frac{\pi}{10}, \pi\right)$ (b) $x \in \left(-\frac{\pi}{10}, \frac{3\pi}{10}\right)$

(c) $x \in \left(-\pi, \frac{3\pi}{10}\right)$ (d) $x \in (-\pi, \pi)$

55. The solution of the inequality $\log_{1/2} \sin x > \log_{1/2} \cos x$ in $[0, 2\pi]$ is

(a) $x \in \left(0, \frac{\pi}{2}\right)$ (b) $x \in \left(0, \frac{\pi}{8}\right)$

(c) $x \in \left(0, \frac{\pi}{4}\right)$ (d) None of these

56. The set of all x in $\left(\frac{-\pi}{2}, \frac{\pi}{2}\right)$ satisfying $|4 \sin x - 1| < \sqrt{5}$ is given by

(a) $\left(-\frac{\pi}{10}, \frac{3\pi}{10}\right)$ (b) $\left(\frac{\pi}{10}, \frac{3\pi}{10}\right)$

(c) $\left(\frac{\pi}{10}, -\frac{3\pi}{10}\right)$ (d) None of these

57. The number of solutions of $\tan(5\pi \cos\alpha) = \cot(5\pi \sin\alpha)$ for α in $(0, 2\pi)$ is

(a) 7 (b) 14

(c) 21 (d) 28

58. The number of solution (s) of the equation $\sin^3 x \cos x + \sin^2 x \cos^2 x + \sin x \cos^3 x = 1$ in the interval $[0, 2\pi]$ is/are

(a) no (b) one

(c) two (d) three

59. The most general values of x for which $\sqrt{3} \sin x - \cos x = \min_{\lambda \in R} \{2, e^2, \pi, \lambda^2 - 4\lambda + 7\}$ are given by

(a) $2n\pi, n \in I$

(b) $2n\pi + \frac{2\pi}{3}, n \in I$

(c) $n\pi + (-1)^n \frac{\pi}{4} + \frac{\pi}{6}, n \in I$

(d) $n\pi + (-1)^{n+1} \frac{\pi}{4} - \frac{\pi}{3}, n \in I$

60. The solution set of the inequality $\cos^2\theta < \frac{1}{2}$ is

(a) $\left\{\theta : (8n+1)\frac{\pi}{4} < \theta < (8n+3)\frac{\pi}{4}, n \in I\right\}$

(b) $\left\{\theta : (8n-3)\frac{\pi}{4} < \theta < (8n-1)\frac{\pi}{4}, n \in I\right\}$

(c) $\left\{\theta : (4n+1)\frac{\pi}{4} < \theta < (4n+3)\frac{\pi}{4}, n \in I\right\}$

(d) None of the above

61. Complete set of values of x in $(0, \pi)$ satisfying $1 + \log_2 \sin x + \log_2 \sin 3x \geq 0$ is

(a) $\left(\frac{2\pi}{3}, \frac{3\pi}{4}\right]$ (b) $\left[\frac{\pi}{6}, \frac{\pi}{4}\right] \cup \left(\frac{2\pi}{3}, \frac{3\pi}{4}\right]$

(c) $\left[\frac{\pi}{4}, \frac{\pi}{3}\right) \cup \left(\frac{2\pi}{3}, \frac{3\pi}{4}\right]$ (d) $\left[\frac{\pi}{6}, \frac{\pi}{4}\right] \cup \left[\frac{3\pi}{4}, \frac{5\pi}{6}\right]$

62. The number of solutions of the equation $1 + \sin x \sin^2\left(\frac{x}{2}\right) = 0$ in $[-\pi, \pi]$ is

(a) 0 (b) 1

(c) 2 (d) 3

63. The number of solutions of the equation $|\cot x| = \cot x + \frac{1}{\sin x} (0 \leq x \leq 2\pi)$ is

[JEE Main 2021, 18 March]

(a) 0 (b) 1

(c) 2 (d) 3

64. Let $S = \left\{x \in (-\pi, \pi) : x \neq 0, \pm\frac{\pi}{2}\right\}$. The sum of all distinct solutions of the equation $\sqrt{3} \sec x + \operatorname{cosec} x + 2(\tan x - \cot x) = 0$ in the set S is equal to

[JEE Advanced 2016]

(a) $-\frac{7\pi}{9}$ (b) $-\frac{2\pi}{9}$ (c) 0 (d) $\frac{5\pi}{9}$

65. Number of solutions of the equations $y = \frac{1}{3}[\sin x + [\sin x + [\sin x]]]$ and $[y + [y]] = 2 \cos x$, where [.] denotes the greatest integer function is

(a) 0 (b) 1

(c) 2 (d) infinite

Objective Questions Type II [One or more than one correct answer(s)]

In each of the questions below four choices of which one or more than one are correct. You have to select the correct answer(s) accordingly.

1. The equation $81^{\sin^2 x} + 81^{\cos^2 x} = 30$ has two roots in the

(a) I quadrant (b) II quadrant

(c) III quadrant (d) IV quadrant

2. The equation $2 \sin\left(\frac{x}{2}\right) \cos^2 x - 2 \sin\left(\frac{x}{2}\right) \sin^2 x = \cos^2 x - \sin^2 x$ has a root for which

(a) $\sin 2x = 1$ (b) $\sin 2x = -1$

(c) $\cos x = \frac{1}{2}$ (d) $\cos 2x = -\frac{1}{2}$

3. $\sin x + \cos x = 1 + \sin x \cos x$, if

(a) $\sin\left(x + \frac{\pi}{4}\right) = \frac{1}{\sqrt{2}}$ (b) $\sin\left(x - \frac{\pi}{4}\right) = \frac{1}{\sqrt{2}}$

(c) $\cos\left(x + \frac{\pi}{4}\right) = \frac{1}{\sqrt{2}}$ (d) $\cos\left(x - \frac{\pi}{4}\right) = \frac{1}{\sqrt{2}}$

4. $\sin\theta + \sqrt{3}\cos\theta = 6x - x^2 - 11$, $0 \leq \theta \leq 4\pi$, $x \in R$, holds for

(a) no value of x and θ

(b) one value of x and two values of θ

(c) two values of x and two values of θ

(d) two pairs of values of (x, θ)

5. The equation $\sin x = [1 + \sin x] + [1 - \cos x]$ has (where $[x]$ is the greatest integer less than or equal to x)

(a) no solution in $\left[-\frac{\pi}{2}, \frac{\pi}{2}\right]$

(b) no solution in $\left[\frac{\pi}{2}, \pi\right]$

(c) no solution in $\left[\pi, \frac{3\pi}{2}\right]$

(d) no solution for $x \in R$

6. Let $\theta, \phi \in [0, 2\pi]$ be such that $2\cos\theta(1 - \sin\phi) = \sin^2\theta\left(\tan\frac{\theta}{2} + \cot\frac{\theta}{2}\right)\cos\phi - 1$, $\tan(2\pi - \theta) > 0$ and $-1 < \sin\theta < -\frac{\sqrt{3}}{2}$. Then ϕ cannot satisfy

(a) $0 < \phi < \frac{\pi}{2}$ (b) $\frac{\pi}{2} < \phi < \frac{4\pi}{3}$

(c) $\frac{4\pi}{3} < \phi < \frac{3\pi}{2}$ (d) $\frac{3\pi}{2} < \phi < 2\pi$

7. The solutions of the system of equations $\sin x \sin y = \frac{\sqrt{3}}{4}$, $\cos x \cos y = \frac{\sqrt{3}}{4}$ are

(a) $x_1 = \frac{\pi}{3} + \frac{\pi}{2}(2n + k)$; $n, k \in I$

(b) $y_1 = \frac{\pi}{6} + \frac{\pi}{2}(k - 2n)$; $n, k \in I$

(c) $x_2 = \frac{\pi}{6} + \frac{\pi}{2}(2n + k)$; $n, k \in I$

(d) $y_2 = \frac{\pi}{3} + \frac{\pi}{2}(k - 2n)$; $n, k \in I$

8. $2\sin^2 x + \sin^2 2x = 2, -\pi < x < \pi$, then x is equal to

(a) $\pm\frac{\pi}{2}$ (b) $\pm\frac{\pi}{4}$

(c) $\pm\frac{3\pi}{4}$ (d) None of these

9. A solution of the equation $(1 - \tan\theta)(1 + \tan\theta)\sec^2\theta + 2^{\tan^2\theta} = 0$ which θ lies in the interval $(-\pi/2, \pi/2)$ is given by

(a) $\theta = 0$ (b) $\theta = \pi/3$

(c) $\theta = -\pi/3$ (d) $\theta = \pi/6$

10. If $|\cos x|^{\sin^2 x - \frac{3}{2}\sin x + \frac{1}{2}} = 1$, then possible values of x are

(a) $n\pi$ or $n\pi + (-1)^n \frac{\pi}{6}, n \in I$

(b) $n\pi$ or $2n\pi + \frac{\pi}{2}$, or $n\pi + (-1)^n \frac{\pi}{6}, n \in I$

(c) $n\pi + (-1)^n \frac{\pi}{6}, n \in I$

(d) $n\pi, n \in I$

11. $\tan|x| = |\tan x|$, if

(a) $x \in (-\pi(2k+1)/2, -\pi k), k \in I$

(b) $x \in [\pi k, \pi(2k+1)/2], k \in I$

(c) $x \in \{-\pi k, -\pi(2k-1)/2\}, k \in I$

(d) $x \in \{\pi(2k-1)/2, \pi k\}, k \in N$

12. $\cos(\sin x) = \frac{1}{\sqrt{2}}$, then x must lie in the interval

(a) $\left(\frac{\pi}{4}, \frac{\pi}{2}\right)$ (b) $\left(-\frac{\pi}{4}, 0\right)$

(c) $\left(\pi, \frac{3\pi}{2}\right)$ (d) $\left(\frac{\pi}{2}, \pi\right)$

13. A solution of the equation
$(1 - \tan\theta)(1 + \tan\theta)\sec^2\theta + 2^{\tan^2\theta} = 0$,
where θ lies in the interval $\left(-\frac{\pi}{2}, \frac{\pi}{2}\right)$ is given by

(a) $\theta = 0$ (b) $\theta = \frac{\pi}{3}$

(c) $\theta = -\frac{\pi}{3}$ (d) $\theta = \frac{\pi}{6}$

14. The solution of the equation
$9\cos^{12} x + \cos^2 2x + 1 = 6\cos^6 x \cos 2x + 6\cos^6 x - 2\cos 2x$ is/are

(a) $x = n\pi + \frac{\pi}{2}, n \in I$

(b) $x = n\pi + \cos^{-1}\left(\sqrt[4]{\frac{2}{3}}\right), n \in I$

(c) $x = n\pi - \cos^{-1}\left(\sqrt[4]{\frac{2}{3}}\right), n \in I$

(d) None of the above

15. If $0 \le x \le 2\pi$, then $2^{\operatorname{cosec}^2 x}\sqrt{\frac{1}{2}y^2 - y + 1} \le \sqrt{2}$

(a) is satisfied by exactly one value of y

(b) is satisfied by exactly two value of x

(c) is satisfied by x for which $\cos x = 0$

(d) is satisfied by x for which $\sin x = 0$

16. A solution (x, y) of the system of equations $x - y = \frac{1}{3}$ and $\cos^2(\pi x) - \sin^2(\pi y) = \frac{1}{2}$ is given by

(a) $\left\{\frac{7}{6}, \frac{5}{6}\right\}$ (b) $\left\{\frac{2}{3}, \frac{1}{3}\right\}$

(c) $\left\{-\frac{5}{6}, -\frac{7}{6}\right\}$ (d) $\left\{\frac{13}{6}, \frac{11}{6}\right\}$

17. If $\left(\cos^2 x + \frac{1}{\cos^2 x}\right)(1 + \tan^2 2y)(3 + \sin 3z) = 4$, then

(a) x may be a multiple of π

(b) x cannot be an even multiple of π

(c) z can be a multiple of π

(d) y can be a multiple of $\frac{\pi}{2}$

18. $2\sin^2\left(\frac{\pi}{2}\cos^2 x\right) = 1 - \cos(\pi \sin 2x)$, if

(a) $x = (2n+1)\frac{\pi}{2}, n \in I$ (b) $\tan x = \frac{1}{2}, n \in I$

(c) $\tan x = -\frac{1}{2}, n \in I$ (d) $x = \frac{n\pi}{2}, n \in I$

19. $\sqrt{(\cos 2x)} + \sqrt{(1 + \sin 2x)} = 2\sqrt{(\sin x + \cos x)}$, if

(a) $\sin x + \cos x = 0$

(b) $x = 2n\pi, n \in I$

(c) $x = n\pi - \frac{\pi}{4}, n \in I$

(d) $x = 2n\pi \pm \cos^{-1}\left(-\frac{1}{5}\right), n \in I$

Linked-Comprehension Type

In these questions, a passage (paragraph) has been given followed by questions based on each of the passages. You have to answer the questions based on the passage given.

PASSAGE 1

An equation of the form
$f(\sin x \pm \cos x, \pm \sin x \cos x) = 0$
can be solved by changing variable.

Let $\sin x \pm \cos x = t$

$\Rightarrow \sin^2 x + \cos^2 x \pm 2\sin x \cos x = t^2$

$\Rightarrow \pm \sin x \cos x = \left(\frac{t^2 - 1}{2}\right)$.

Hence, reduce the given equation into $f\left(t, \frac{t^2-1}{2}\right) = 0$

On the basis of above information, answer the following questions

1. If $1 - \sin 2x = \cos x - \sin x$, then x is

(a) $2n\pi, 2n\pi - \frac{\pi}{2}, n \in I$ (b) $2n\pi, n\pi + \frac{\pi}{4}, n \in I$

(c) $2n\pi - \frac{\pi}{2}, n\pi + \frac{\pi}{4}, n \in I$ (d) None of these

2. If $\sin x + \cos x = 1 + \sin x \cos x$, then x is
(a) $2n\pi,\ 2n\pi + \frac{\pi}{2}, n \in I$ (b) $2n\pi,\ n\pi + \frac{\pi}{4}, n \in I$
(c) $2n\pi - \frac{\pi}{2},\ n\pi + \frac{\pi}{4}, n \in I$ (d) None of these

3. If $\sin^4 x + \cos^4 x = \sin x \cos x$, then x is
(a) $n\pi, n \in I$ (b) $(6n+1)\frac{\pi}{6}, n \in I$
(c) $(4n+1)\frac{\pi}{4}, n \in I$ (d) None of these

4. If $1 + \sin^3 x + \cos^3 x = \frac{3}{2}\sin 2x$, then x is
(a) $(2n+1)\pi,\ 2n\pi + \frac{\pi}{4}, n \in I$
(b) $(2n+1)\pi,\ 2n\pi - \frac{\pi}{2}, n \in I$
(c) $2n\pi + \frac{\pi}{4},\ 2n\pi + \frac{\pi}{2}, n \in I$
(d) None of the above

5. If $(\sin x + \cos x) - 2\sqrt{2}\sin x \cos x = 0$, then x is
(a) $2n\pi + \frac{\pi}{4}, n \in I$ (b) $2n\pi - \frac{\pi}{4}, n \in I$
(c) $n\pi + \frac{\pi}{4}, n \in I$ (d) $n\pi - \frac{\pi}{4}, n \in I$

PASSAGE 2

When ever the terms on the two sides of the equation are of different nature, then equations are known as Non standard form, some of them are in the form of an ordinary equation but can not be solved by standard procedures. Non standard problems require high degree of logic, they also require the use of graphs, inverse properties of functions, in equalities.

On the basis of above information, answer the following questions

1. The number of solutions of the equation $2\cos\left(\frac{x}{2}\right) = 3^x + 3^{-x}$ is
(a) 1 (b) 2
(c) 3 (d) None of these

2. The equation $2\cos^2\left(\frac{x}{2}\right)\sin^2 x = x^2 + x^{-2}, 0 < x \le \frac{\pi}{2}$ has
(a) one real solutions
(b) more than one real solutions
(c) no real solution
(d) None of the above

3. The number of real solutions of the equation $\sin(e^x) = 5^x + 5^{-x}$ is
(a) 0 (b) 1
(c) 2 (d) infinitely many

4. If $0 \le x \le 2\pi$ and $2^{\text{cosec}^2 x}\sqrt{\left(\frac{1}{2}y^2 - y + 1\right)} \le \sqrt{2}$, then number of ordered pairs of (x, y) is
(a) 1 (b) 2
(c) 3 (d) infinitely many

5. The number of solutions of the equation $\sin x = x^2 + x + 1$ is
(a) 0 (b) 1
(c) 2 (d) None of these

PASSAGE 3

Suppose equation is $f(x) - g(x) = 0$ or $f(x) = g(x) = y$ say, then draw the graphs of $y = f(x)$ and $y = g(x)$. If graphs of $y = f(x)$ and $y = g(x)$ cuts at one, two, three, ..., no points, then number of solutions are one, two, three, ... , zero respectively.

On the basis of above information, answer the following questions

1. The number of solutions of $\sin x = \frac{|x|}{10}$ is
(a) 4 (b) 6
(c) 8 (d) None of these

2. Total number of solutions of the equation $3x + 2\tan x = \frac{5\pi}{2}$ in $x \in [0, 2\pi]$ is equal to
(a) 1 (b) 2 (c) 3 (d) 4

3. Total number of solutions of $\sin\{x\} = \cos\{x\}$, where $\{\cdot\}$ denotes the fractional part, in $[0, 2\pi]$ is equal to
(a) 3 (b) 5
(c) 7 (d) None of these

4. If $1 - \sin x = \frac{\sqrt{3}}{2}\left|x - \frac{\pi}{2}\right| + a$ has no solution when $a \in R^+$, then
(a) $a \in R^+$ (b) $a > \frac{3}{2} + \frac{\pi}{\sqrt{3}}$
(c) $a \in \left(0, \frac{3}{2} + \frac{\pi}{\sqrt{3}}\right)$ (d) $a \in \left(\frac{3}{2}, \frac{3}{2} + \frac{\pi}{\sqrt{3}}\right)$

5. Total number of solutions of $\cos 2x = |\sin x|$, where $x \in \left(-\frac{\pi}{2}, \pi\right)$ is
(a) 3 (b) 4 (c) 5 (d) 6

PASSAGE 4

Consider the system of equations
$x\cos^3 y + 3x\cos y \sin^2 y = 14$
$x\sin^3 y + 3x\cos^2 y \sin y = 13$

On the basis of above information, answer the following questions

1. The value/values of x is/are
(a) $\pm 5\sqrt{5}$ (b) $\pm\sqrt{5}$
(c) $\pm 1\sqrt{5}$ (d) None of these

2. The number of values of $y \in [0, 6\pi]$ is
(a) 5 (b) 3 (c) 4 (d) 6

3. The value of $\sin^2 y + 2\cos^2 y$ is
(a) 4/5 (b) 9/5
(c) 2 (d) None of these

Numerical Value Type

Solve the following problems and write your answer.

1. The sum of the roots of equation $\cos 4x + 6 = 7 \cos 2x$ over the interval [0, 314] is $\lambda\pi$, then the numerical quantity λ must be
2. If x and y are the solutions of the equation $12 \sin x + 5 \cos x = 2y^2 - 8y + 21$, the value of $\tan\left(\frac{xy}{2}\right)$ must be
3. The number of ordered pairs (x, y) satisfying the equations $\sin x \cos y = 1$ and $x^2 + y^2 \le 9\pi^2$ must be
4. The sum of all the solutions of the equation $\cos\theta \cos\left(\frac{\pi}{3} + \theta\right) \cos\left(\frac{\pi}{3} - \theta\right) = \frac{1}{4}, \theta \in [0, 6\pi]$ is $k\pi$, then the value of k is
5. The arithmetic mean of the roots of the equation $4\cos^3 x - 4\cos^2 x - \cos(\pi + x) - 1 = 0$ in the interval (0, 315) is equal to $k\pi$, then the value of k is
6. If $\tan(\pi \cos\theta) = \cot(\pi \sin\theta)$, then $19208 \cos^2(\theta - \pi/4)$ is equal to

Single Integer Type

Solve the following problems and write your answer.

1. The number of distinct solutions of the equation, $\log_{1/2}|\sin x| = 2 - \log_{1/2}|\cos x|$ in the interval $[0, 2\pi]$, is
2. If solution of the equation $3\cos^2\theta - 2\sqrt{3}\sin\theta\cos\theta - 3\sin^2\theta = 0$ are $n\pi + \frac{\pi}{r}$ and $n\pi + \frac{\pi}{s}$, then $|r - s| =$ **[JEE Main 2020, 9 January]**
3. The number of solutions of $3\sec\theta - 5 = 4\tan\theta$ in $[0, 4\pi]$ must be
4. The number of values of x between 0 and 2π that satisfies the equation $\sin x + \sin 2x + \sin 3x = \cos x + \cos 2x + \cos 3x$ must be
5. If the equation $\sec\theta + \operatorname{cosec}\theta = c$ has two real roots between 0 and 2π, then the least integer which c^2 cannot exceed must be
6. The trigonometric equation $\operatorname{cosec} x - \operatorname{cosec} 2x = \operatorname{cosec} 4x$ has a solution $(2n+1)\frac{\pi}{7}$, (n integer), then n should not be an integer of the type $7p + \lambda$, $(0 \le \lambda \le 6)$, (p integer). The numerical quantity λ should be
7. If α be the smallest positive root of the equation $\sqrt{\sin(1-x)} = \sqrt{\cos x}$, then the approximate integral value of α must be
8. If $\sqrt{3}(\cos^2 x) = (\sqrt{3} - 1)\cos x + 1$, the number of the given equation when $x \in \left[0, \frac{\pi}{2}\right]$ is **[JEE Main 2021, 26 February]**

Matrix-Match Type

Given below are Matching Type Questions, with two columns (each having some items) each . Each item of Column I has to be matched with the items of Column II, by encircling the correct match(es).

NOTE An item of Column I can be matched with more than one items of Column II.

1. Observe the following columns

	Column I		Column II
(A)	If α, β are the solutions of $\sin x = -\frac{1}{2}$ in $[0, 2\pi]$ and α, γ are the solutions of $\cos x = -\frac{\sqrt{3}}{2}$ in $[0, 2\pi]$, then	(P)	$\alpha - \beta = \pi$
		(Q)	$\beta - \gamma = \pi$
(B)	If α, β are the solutions of $\cot x = -\sqrt{3}$ in $[0, 2\pi]$ and α, γ are the solutions of $\operatorname{cosec} x = -2$ in $[0, 2\pi]$, then	(R)	$\alpha - \gamma = \pi$
(C)	If α, β are the solutions of $\sin x = -\frac{1}{2}$ in $[0, 2\pi]$ and α, γ are the solutions of $\tan x = \frac{1}{\sqrt{3}}$ in $[0, 2\pi]$, then	(S)	$\alpha + \beta = 3\pi$
		(T)	$\beta + \gamma = 2\pi$

2. Observe the following columns

	Column I		Column II
(A)	If number of ordered pairs (x, y) satisfying $\|y\| = \cos x$ and $y = \sin^{-1}(\sin x)$, when $\|x\| \le 2\pi$ is m and when $\|x\| \le 3\pi$ is n, then	(P)	$m + n = 0$
		(Q)	$m + n = 2$
(B)	If number of solutions of $\ln\|\sin x\| = -x^2 + 2x$, when $x \in [0, \pi]$ is m and when $x \in \left[-\frac{\pi}{2}, \frac{3\pi}{2}\right]$ is n, then	(R)	$m + n = 10$
(C)	If number of solutions of $[\sin x] = \cos x$, (where [.] denotes the greatest integer function) when $x \in [0, \pi]$ is m and when $x \in [0, 3\pi]$ is n, then	(S)	$n - m = 0$
		(T)	$n - m = 2$

3. Observe the following columns

	Column I (Equation)		Column II (Solution)
(A)	If $\max_{\theta=R}\{5\sin\theta+3\sin(\theta-\alpha)\}=7$, then the set of possible values of α is	(P)	$2n\pi+3\pi/4, n\in Z$
(B)	$x\neq\frac{n\pi}{2}$ and $(\cos x)^{\sin^2 x-3\sin x+2}=1$	(Q)	$2n\pi\pm\frac{\pi}{3}, n\in Z$
(C)	$\sqrt{(\sin x)}+2^{1/4}\cos x=0$	(R)	$2n\pi+\cos^{-1}(1/3), n\in Z$
(D)	$\log_5\tan x=(\log_5 4)(\log_4(3\sin x))$	(S)	No solution

Answers

Objective Questions Type [Only one correct answer]

1. (a) **2.** (a) **3.** (a) **4.** (b) **5.** (b) **6.** (a) **7.** (c) **8.** (d) **9.** (a) **10.** (a)
11. (b) **12.** (a) **13.** (b) **14.** (c) **15.** (c) **16.** (b) **17.** (a) **18.** (d) **19.** (b) **20.** (a)
21. (d) **22.** (c) **23.** (d) **24.** (c) **25.** (b) **26.** (c) **27.** (a) **28.** (d) **29.** (a) **30.** (b)
31. (d) **32.** (a) **33.** (d) **34.** (c) **35.** (b) **36.** (c) **37.** (a) **38.** (a) **39.** (c) **40.** (b)
41. (b) **42.** (b) **43.** (a) **44.** (d) **45.** (c) **46.** (b) **47.** (a) **48.** (a) **49.** (c) **50.** (d)
51. (a) **52.** (c) **53.** (d) **54.** (b) **55.** (c) **56.** (a) **57.** (b) **58.** (a) **59.** (b) **60.** (c)
61. (d) **62.** (a) **63.** (c) **64.** (c) **65.** (a)

Objective Questions Type II [One or more than one correct answer(s)]

1. (a, b, c, d) **2.** (a, b, c, d) **3.** (a, c, d) **4.** (b, d) **5.** (a, b, c, d)
6. (a, c, d) **7.** (a, b, c, d) **8.** (b, c) **9.** (b, c) **10.** (c, d)
11. (a, b) **12.** (a, d) **13.** (b, c) **14.** (a, b, c) **15.** (a, b, c)
16. (a, c, d) **17.** (a, d) **18.** (a, b, c) **19.** (a, b, c)

Linked-Comprehension Type

Passage 1 **1.** (d) **2.** (a) **3.** (c) **4.** (b) **5.** (a)
Passage 2 **1.** (a) **2.** (c) **3.** (a) **4.** (b) **5.** (a)
Passage 3 **1.** (b) **2.** (c) **3.** (c) **4.** (b) **5.** (b)
Passage 4 **1.** (a) **2.** (d) **3.** (b)

Numerical Value Type

1. (4950) **2.** (2.4) **3.** (13) **4.** (30) **5.** (51) **6.** (2401)

Single Integer Type

1. (8) **2.** (9) **3.** (8) **4.** (2) **5.** (8)
6. (3) **7.** (6) **8.** (1)

Matrix-Match Type

1. A → (Q, S); B → (P, T); C → (R, S, T)
2. A → (R, T); B → (Q, T); C → (P, S)
3. A → Q → B → S, C → P, D → R

Solutions

Objective Questions Type [Only one correct answer]

1. We have, $\theta \in [-2\pi, 2\pi]$

and $2\cos^2\theta + 3\sin\theta = 0$

$\Rightarrow 2(1-\sin^2\theta) + 3\sin\theta = 0$

$\Rightarrow 2 - 2\sin^2\theta + 3\sin\theta = 0$

$\Rightarrow 2\sin^2\theta - 3\sin\theta - 2 = 0$

$\Rightarrow 2\sin^2\theta - 4\sin\theta + \sin\theta - 2 = 0$

$\Rightarrow 2\sin\theta(\sin\theta - 2) + 1(\sin\theta - 2) = 0$

$\Rightarrow (\sin\theta - 2)(2\sin\theta + 1) = 0$

$\Rightarrow \sin\theta = \frac{-1}{2}$ $\quad [\because (\sin\theta - 2) \neq 0]$

$\Rightarrow \theta = 2\pi - \frac{\pi}{6}, -\pi + \frac{\pi}{6}, -\frac{\pi}{6}, \pi + \frac{\pi}{6}$ $\quad [\because \theta \in [-2\pi, 2\pi]]$

Now, sum of all solutions

$$= 2\pi - \frac{\pi}{6} - \pi + \frac{\pi}{6} - \frac{\pi}{6} + \pi + \frac{\pi}{6} = 2\pi$$

2. $\sin 3x - \sin x - 2\sin 2x + 3 = 0$

$\Rightarrow 2\cos 2x \sin x - 4\sin x\cos x + 3 = 0$

$\Rightarrow \sin x(2\cos 2x - 4\cos x) + 3 = 0$

$\Rightarrow \sin x\{2(2\cos^2 x - 1) - 4\cos x\} + 3 = 0$

$\Rightarrow \sin x\{4\cos^2 x - 4\cos x - 2\} + 3 = 0$

$\Rightarrow \sin x\{(2\cos x - 1)^2 - 3\} + 3 = 0$

$\Rightarrow \sin x(2\cos x - 1)^2 + 3(1 - \sin x) = 0$

$\because 0 \le x \le \pi \Rightarrow 0 \le \sin x \le 1$

Then, $1 - \sin x \ge 0$

Also, $\sin x(2\cos x - 1)^2 \ge 0$

which is possible only if, $\sin x = 1$

then, $\cos x = 0$

$\Rightarrow \sin x(2\cos x - 1)^2 = 1 \neq 0$

i.e. no solution

$\therefore$ Number of solutions = 0.

3. $\because \frac{y + \frac{1}{y}}{2} \ge \sqrt{y \cdot \frac{1}{y}} \Rightarrow \sqrt{\left(y + \frac{1}{y}\right)} \ge \sqrt{2}$

But $|\sin x + \cos x| \le \sqrt{2}$

Which is possible only when

$$y + \frac{1}{y} = 2$$

$\therefore y = 1$ and $x = \frac{\pi}{4}$

4. We have, $\sin x\sqrt{(8\cos^2 x)} = 1$

$\Rightarrow \sin x \cdot 2\sqrt{2}|\cos x| = 1$

or $\sin x|\cos x| = \frac{1}{2\sqrt{2}}$

Case I When, $\cos x > 0$

$\Rightarrow \sin 2x = \frac{1}{\sqrt{2}} = \sin\frac{\pi}{4}$

$\Rightarrow 2x = n\pi + (-1)^n\frac{\pi}{4}$

$\Rightarrow x = \frac{n\pi}{2} + (-1)^n\frac{\pi}{8}$

$\therefore x = \frac{\pi}{8}, \frac{3\pi}{8}, \frac{9\pi}{8}, \frac{11\pi}{8}$ $\quad [\because 0 < x < 2\pi]$

But $\cos x > 0$,

$\therefore x = \frac{\pi}{8}, \frac{3\pi}{8}$

Case II When, $\cos x < 0$

$\Rightarrow \sin 2x = -\frac{1}{\sqrt{2}} = -\sin\frac{\pi}{4}$

$\therefore 2x = n\pi - (-1)^n\frac{\pi}{4} \Rightarrow x = \frac{n\pi}{2} - (-1)^n\frac{\pi}{8}$

$\therefore x = \frac{5\pi}{8}, \frac{7\pi}{8}, \frac{13\pi}{8}, \frac{15\pi}{8}$

But $\cos x < 0$

$\therefore x = \frac{5\pi}{8}, \frac{7\pi}{8}$

Hence, the values of x satisfying the given equation which lie between 0 and 2π are $\frac{\pi}{8}, \frac{3\pi}{8}, \frac{5\pi}{8}, \frac{7\pi}{8}$, these are in AP with common difference $\frac{\pi}{4}$.

5. If $|f(x) + g(x)| = |f(x)| + |g(x)|$

$\Leftrightarrow f(x) \cdot g(x) \ge 0$

$\therefore \tan x \sec x \ge 0$

$\Rightarrow \frac{\sin x}{\cos^2 x} \ge 0 \Rightarrow \sin x \ge 0$

But $\cos x \neq 0$

$\therefore x \in [0, \pi] \sim \left\{\frac{\pi}{2}\right\}$

or $x \in \left[0, \frac{\pi}{2}\right) \cup \left(\frac{\pi}{2}, \pi\right]$

6. $\because [\sin x] + [\sqrt{2}\cos x] = -3$

$\Rightarrow [\sin x] = -1$ and $[\sqrt{2}\cos x] = -2$

or $-1 \le \sin x < 0$

and $-2 \le \sqrt{2}\cos x < -1$

or $-1 \le \sin x < 0$

and $-\sqrt{2} < \cos x < -\frac{1}{\sqrt{2}}$

or $-1 \le \sin x < 0$

and $-1 \le \cos x < -\frac{1}{\sqrt{2}}$

$\therefore x \in (\pi, 2\pi)$ and $x \in \left(\frac{3\pi}{4}, \frac{5\pi}{4}\right)$

$\therefore x \in (\pi, 2\pi) \cap \left(\frac{3\pi}{4}, \frac{5\pi}{4}\right)$

$\therefore x \in \left(\pi, \frac{5\pi}{4}\right)$

7. $\therefore \quad \sin(\cos x) = \cos(\sin x)$

$\Rightarrow \cos(\sin x) - \sin(\cos x) = 0$

$\Rightarrow \quad \cos(\sin x) + \cos\left(\frac{\pi}{2} + \cos x\right) = 0$

$\Rightarrow 2\cos\left(\frac{\pi}{4} + \frac{\cos x + \sin x}{2}\right) \cdot \cos\left(\frac{\pi}{4} + \frac{\cos x - \sin x}{2}\right) = 0$

$\Rightarrow$ If $\quad \cos\left(\frac{\pi}{4} + \frac{\cos x + \sin x}{2}\right) = 0$

$$\frac{\pi}{4} + \frac{\cos x + \sin x}{2} = \left(n\pi + \frac{\pi}{2}\right)$$

$\Rightarrow \quad \sqrt{2}\left\{\sin\left(\frac{\pi}{4} + x\right)\right\} = \frac{\pi}{2}(4n+1),\ n \in I$

$\therefore \quad \sin\left(\frac{\pi}{4} + x\right) = \frac{\pi(4n+1)}{2\sqrt{2}}$

$\Rightarrow \sin\left(\frac{\pi}{4} + x\right) \notin [-1, 1]$, which does not certify the original solution,

$\therefore \quad \cos\left(\frac{\pi}{4} + \frac{\cos x + \sin x}{2}\right) \neq 0$

Then, $\quad \cos\left(\frac{\pi}{4} + \frac{\cos x - \sin x}{2}\right) = 0$

$\Rightarrow \quad \frac{\pi}{4} + \left(\frac{\cos x - \sin x}{2}\right) = (2r+1)\frac{\pi}{2}$

$\Rightarrow \quad \sqrt{2}\cos\left(\frac{\pi}{4} + x\right) = \frac{\pi}{2}(4r+1)$

$\therefore \quad \cos\left(\frac{\pi}{4} + x\right) = \frac{\pi}{2\sqrt{2}}(4r+1),\ r \in I$

$\Rightarrow \quad \cos\left(\frac{\pi}{4} + x\right) \notin [-1, 1]$

Also, $\quad \cos\left(\frac{\pi}{4} + \frac{\cos x - \sin x}{2}\right) \neq 0$

$\therefore \quad$ LHS $\neq 0$ and RHS $= 0$

Hence, no real solution.

8. Since, $\sin\theta - \cos\theta \neq 0 \Rightarrow \tan\theta \neq 1$

and also to define $\tan\theta$, $\theta \neq \frac{\pi}{2}, \frac{3\pi}{2}$

$\therefore \quad \theta \neq \frac{\pi}{4}, \frac{5\pi}{4}, \frac{\pi}{2}, \frac{3\pi}{2}$

Now, $\sin^2\theta + \cos^2\theta + \sin\theta\cos\theta$

$$-|\sin\theta|\cos\theta - 2\tan\theta\cot\theta = -1$$

$\Rightarrow \quad 1 + \cos\theta(\sin\theta - |\sin\theta|) - 2 = -1$

$\Rightarrow \quad \cos\theta(\sin\theta - |\sin\theta|) = 0$

$\therefore \quad \theta \in (0, \pi) - \left\{\frac{\pi}{4}, \frac{\pi}{2}\right\}$

9. It is given that,

$|\sin x - \cos^2 x| \geq |3 - 3\sin x + \sin^2 x| + 4|\sin x - 1|$

$\Rightarrow |\sin^2 x + \sin x - 1| \geq |\sin^2 x - 3\sin x + 3| + |4\sin x - 4|$

$\Rightarrow |(\sin^2 x - 3\sin x + 3) + (4\sin x - 4)| \geq |\sin^2 x - 3\sin x + 3| + |4\sin x - 4|$

$\because$ If $|a + b| \geq |a| + |b|$

$\Rightarrow ab \geq 0$

$\therefore (\sin^2 x - 3\sin x + 3)(4\sin x - 4) \geq 0$

$\because \sin^2 x - 3\sin x + 3 > 0\ \forall x \in R$

$\therefore \quad \sin x - 1 \geq 0 \Rightarrow \sin x = 1$

$\Rightarrow x = (4n+1)\frac{\pi}{2},\ n \in Z$

10. $\sin\left(\frac{\pi}{2n}\right) + \cos\left(\frac{\pi}{2n}\right) = \frac{\sqrt{n}}{2}$

$\Rightarrow \quad \sqrt{2}\left\{\frac{1}{\sqrt{2}}\cos\left(\frac{\pi}{2n}\right) + \frac{1}{\sqrt{2}}\sin\left(\frac{\pi}{2n}\right)\right\} = \frac{\sqrt{n}}{2}$

$\Rightarrow \quad \sqrt{2}\cos\left(\frac{\pi}{4} - \frac{\pi}{2n}\right) = \frac{\sqrt{n}}{2}$

$\Rightarrow \quad \cos\left(\frac{\pi}{4} - \frac{\pi}{2n}\right) = \frac{\sqrt{n}}{2\sqrt{2}} \quad \ldots(i)$

When, $n = 6$, LHS $= \cos\left(\frac{\pi}{6}\right) = \frac{\sqrt{3}}{2}$

and $\quad$ RHS $= \frac{\sqrt{6}}{2\sqrt{2}} = \frac{\sqrt{3}}{2}$

Eq. (i) is not satisfied for $n = 1, 2, 3, 4, 5$

11. The expression, $\cos^4\theta + \sin^4\theta$

$$= (\cos^2\theta + \sin^2\theta)^2 - 2\sin^2\theta\cos^2\theta$$

$$= 1 - \frac{1}{2}\sin^2(2\theta)$$

$\because \quad \sin^2(2\theta) \in [0, 1]$

$\Rightarrow \quad -\frac{1}{2}\sin^2(2\theta) \in \left[-\frac{1}{2}, 0\right]$

$\Rightarrow \quad 1 - \frac{1}{2}\sin^2(2\theta) \in \left[\frac{1}{2}, 1\right]$

Now, as $\cos^4\theta + \sin^4\theta + \lambda = 0$

$\Rightarrow \quad \lambda = -(\cos^4\theta + \sin^4\theta)$

$\Rightarrow \lambda \in [-1, -1/2]$ for real solution of the given equation $\cos^4\theta + \sin^4\theta + \lambda = 0$ for θ.

12. $1 + |\cos x| + \cos^2 x \ldots\ldots$

$$= \frac{1}{1 - |\cos x|}$$

$\Rightarrow \quad 8^{\frac{1}{1-|\cos x|}} = 4^3$

$\Rightarrow \quad 2^{\frac{3}{1-|\cos x|}} = 2^6$

$\Rightarrow \quad \frac{3}{1 - |\cos x|} = 6$

$\Rightarrow \quad 1 - |\cos x| = \frac{1}{2}$

$$|\cos x| = \frac{1}{2}$$

$\Rightarrow \quad \cos x = \pm\frac{1}{2}$

$\cos x = \frac{1}{2}$ will give least positive value of x

$x = \frac{\pi}{3}$

13. $(\sqrt{3})^{\sec^2\theta} = (\tan^2\theta + 1)^2 - 1 = (\sec^2\theta)^2 - 1$

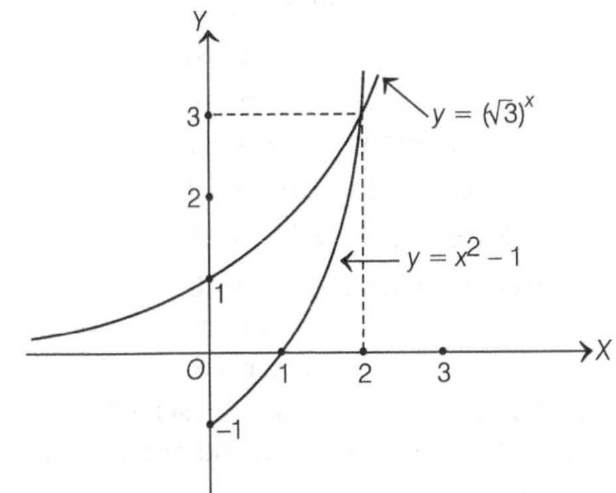

Put $\sec^2\theta = x$ $[\therefore x \geq 1]$

Then, $(\sqrt{3})^x = x^2 - 1$

Let $y = (\sqrt{3})^x = (x^2 - 1)$ $[x > 1]$

Now, Graphs of

$y = (\sqrt{3})^x$ and $y = x^2 - 1$ intersect at one point

i.e. $x = 2$, then $y = 3$

Then, $\sec^2\theta = 2$

Then, $\sec\theta = \pm\sqrt{2}$

Therefore, two values of θ in $\left(-\frac{\pi}{2}, \frac{\pi}{2}\right)$.

14. The given equation is,

$$\sin^4 x + \cos^4 y + 2 = 4\sin x \cos y$$

$$\Rightarrow (\sin^2 x - 1)^2 + (\cos^2 y - 1)^2 + 2\sin^2 x + 2\cos^2 y - 4\sin x \cos y = 0$$

$$\Rightarrow (\sin^2 x - 1)^2 + (\cos^2 y - 1)^2 + 2(\sin x - \cos y)^2 = 0$$

Which is possible only, when

$\sin^2 x - 1 = 0$, $\cos^2 y - 1 = 0$, $\sin x - \cos y = 0$

or $\sin^2 x = 1$, $\cos^2 y = 1$, $\sin x = \cos y$

As $0 \leq x, y \leq \frac{\pi}{2}$

We get $\sin x = \cos y = 1$

$\Rightarrow \sin x + \cos y = 1 + 1 = 2$

15. $\cos^4 x - (\lambda + 2)\cos^2 x - (\lambda + 3) = 0$

$$\Rightarrow (\cos^2 x)^2 - (\lambda + 2)\cos^2 x - (\lambda + 3) = 0$$

$$\therefore \cos^2 x = \frac{(\lambda + 2) \pm \sqrt{(\lambda + 2)^2 + 4(\lambda + 3)}}{2}$$

$$= \frac{(\lambda + 2) \pm (\lambda + 4)}{2} = \lambda + 3, -1$$

$\cos^2 x = \lambda + 3$ $[\cos^2 x \neq -1]$

But $0 \leq \cos^2 x \leq 1$

$\Rightarrow 0 \leq \lambda + 3 \leq 1$

$\Rightarrow -3 \leq \lambda \leq -2$

16. $\because \cos(315\pi + x) = (-1)^{315}\cos x = -\cos x$

$\therefore$ $4\cos^3 x - 4\cos^2 x - \cos(315\pi + x) = 1$

or $4\cos^3 x - 4\cos^2 x + \cos x - 1 = 0$

$\Rightarrow$ $(4\cos^2 x + 1)(\cos x - 1) = 0$

$\cos x = 1$, $4\cos^2 x + 1 \neq 0$

$\Rightarrow$ $\cos x = \cos 0$

$\therefore$ $x = 2n\pi, n \in I$

$\therefore$ $x = 2\pi, 4\pi, 6\pi, 8\pi, \ldots, 100\pi$ $[\because 0 < x < 315]$
[i.e. $100\pi < 315 < 101\pi$]

$\therefore$ Required arithmetic mean

$$= \frac{2\pi + 4\pi + 6\pi + 8\pi + \ldots + 100\pi}{50}$$

$$= \frac{2\pi(1 + 2 + 3 + 4 + \ldots + 50)}{50}$$

$$= \frac{2\pi \cdot \frac{50}{2} \cdot 51}{50} = 51\pi$$

17. $2\sin^2\theta - 5\sin\theta + 2 > 0$

$\Rightarrow$ $(2\sin\theta - 1)(\sin\theta - 2) > 0$

$\because$ $\sin\theta - 2 < 0$ (always)

$\therefore$ $2\sin\theta - 1 < 0$

$\Rightarrow$ $\sin\theta < \frac{1}{2}$

$\Rightarrow$ $-\cos\theta > -\frac{1}{2}$

$\Rightarrow$ $\cos(\pi + \theta) > \cos\frac{2\pi}{3}$

or $2n\pi - \frac{2\pi}{3} < \frac{\pi}{2} + \theta < 2n\pi + \frac{2\pi}{3}$

$\Rightarrow$ $2n\pi - \frac{7\pi}{6} < \theta < 2n\pi + \frac{\pi}{6}$

For, $n = 0$,

$$-\frac{7\pi}{6} < \theta < \frac{\pi}{6}$$

For $n = 1$,

$$\frac{5\pi}{6} < \theta < \frac{13\pi}{6}$$

$-\frac{7\pi}{6}$ —— $\frac{\pi}{6}$ $\frac{5\pi}{6}$ —— $\frac{13\pi}{6}$

0 —— 2π

$\therefore$ $\theta \in \left(0, \frac{\pi}{6}\right) \cup \left(\frac{5\pi}{6}, 2\pi\right)$

Alternative Method

$$\sin\theta < \frac{1}{2} = \sin\frac{\pi}{6} = \sin\left(\pi - \frac{\pi}{6}\right) = \sin\frac{5\pi}{6}$$

$\therefore$ $\theta \in \left(0, \frac{\pi}{6}\right)$ or $\theta \in \left(\frac{5\pi}{6}, 2\pi\right)$

$\therefore$ $\theta \in \left(0, \frac{\pi}{6}\right) \cup \left(\frac{5\pi}{6}, 2\pi\right)$

18. $\because \cos(a - b) = \cos 0$

$\therefore$ $a - b = 2n\pi, n \in I$

$a - b = -2\pi, 0, 2\pi$

$\therefore$ $a + b = 2\pi + 2a, 2a, 2a - 2\pi$

$\therefore \quad \cos(a+b) = \cos 2a, \cos 2a, \cos 2a = \frac{1}{e}$

$$y = \cos 2a = \frac{1}{e}$$

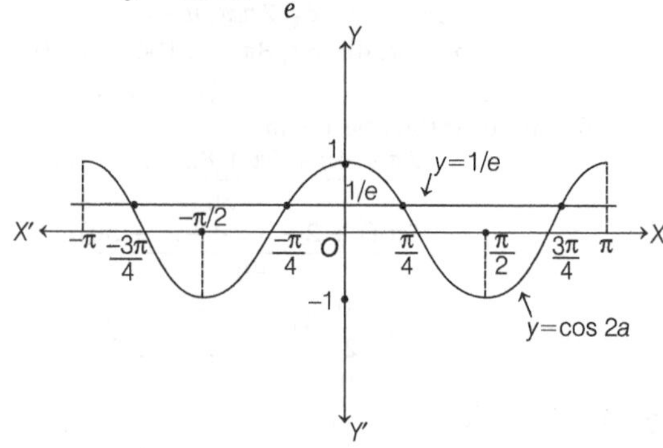

Hence, number of solutions is 4.

19. Given equation,

$\sin^{2020} x - \cos^{2020} x + 2019 = 2020$

$\Rightarrow \quad \sin^{2020} x = 1 + \cos^{2020} x$

the range of LHS is [0, 1] and

the range of RHS is [1, 2]

So, for the solution $\sin^{2020} x = 1$ and $\cos^{2020} x = 0$

and in the given internal $\left(-\frac{3\pi}{2}, \frac{5\pi}{2}\right)$, the possible values of

$$x = -\frac{\pi}{2}, \frac{\pi}{2}, \frac{3\pi}{2}$$

20. $x^3 + x^2 + 4x + 2\sin x = 0$

$x^3 + (x+2)^2 + 2\sin x = 4$

$x = 0$, satisfies the equation

Now, in $0 < x \le \pi$, $x^3 + (x+2)^2 + 2\sin x > 4$

and in $\pi < x \le 2\pi$, $x^3 + (x+2)^2 + 2\sin x > 27 + 25 - 2 = 50$

Hence, $x = 0$ is the only solution.

21. $\cos^2 x = 2\cos x \cdot (1 - 3\cos^2 x)$

$\Rightarrow 6\cos^3 x + \cos^2 x - 2\cos x = 0$

$\therefore \quad \cos x = 0, \frac{1}{2}, -\frac{2}{3}$

$\therefore \quad x = \frac{\pi}{2}, \frac{\pi}{3}, \cos^{-1}\left(-\frac{2}{3}\right)$ $\quad [\because \alpha, \beta$ are +ve$]$

If $\quad \alpha = \frac{\pi}{2}, \beta = \frac{\pi}{3}$

Then, $\quad |\alpha - \beta| = \frac{\pi}{6}$

22. Given, $\tan x + \sec x = 2\cos x$

Multiplying by $\cos x \ne 0$

$\therefore \quad \sin x + 1 = 2\cos^2 x$

$\Rightarrow \quad \sin x + 1 = 2(1 - \sin x)(1 + \sin x)$

$\Rightarrow \quad (\sin x + 1)(2\sin x - 1)$

$\therefore \quad \sin x = -1$ and $\sin x = \frac{1}{2}$

$\sin x \ne -1 \quad [\because \cos x \ne 0]$

$\therefore \quad \sin x = \frac{1}{2}$

$$x = \frac{\pi}{6}, \frac{5\pi}{6}$$

23. $2\tan^2 x - 5\sec x = 1$

$\Rightarrow \quad 2(\sec^2 x - 1) - 5\sec x = 1$

$\Rightarrow \quad 2\sec^2 x - 5\sec x - 3 = 0$

$\Rightarrow \quad (\sec x - 3)(2\sec x + 1) = 0$

$\because \quad |\sec x| \ge 1$

$\therefore \quad \sec x = 3$

Which gives two values of x in each of $[0, 2\pi]$, $(2\pi, 4\pi]$, $(4\pi, 6\pi]$ and one value in $(6\pi, 6\pi + 3\pi/2]$

$\because \quad 6\pi + \frac{3\pi}{2} = \frac{15\pi}{2}$

Hence, greatest value of n is 15.

24. Since, $\quad \sin x + \cos x = 1$

$\Rightarrow \quad \frac{1}{\sqrt{2}}\sin x + \frac{1}{\sqrt{2}}\cos x = \frac{1}{\sqrt{2}}$

$\Rightarrow \quad \sin\left(x + \frac{\pi}{4}\right) = \sin\frac{\pi}{4}$

$\Rightarrow \quad x + \frac{\pi}{4} = n\pi + (-1)^n \frac{\pi}{4}$

$\therefore \quad x = n\pi + (-1)^n \frac{\pi}{4} - \frac{\pi}{4}, n \in I$

25. Given equation is, $1 + \sin^4 x = \cos^2(3x)$

Since, range of $(1 + \sin^4 x) = [1, 2]$

and range of $\cos^2(3x) = [0, 1]$

So, the given equation holds if

$1 + \sin^4 x = 1 = \cos^2(3x)$

$\Rightarrow \quad \sin^4 x = 0$ and $\cos^2 3x = 1$

Since, $\quad x \in \left[-\frac{5\pi}{2}, \frac{5\pi}{2}\right]$

$\therefore \quad x = -2\pi, -\pi, 0, \pi, 2\pi.$

Thus, there are five different values of x is possible.

26. $\tan x - x = 0$

If $\quad x \in \left(0, \frac{\pi}{2}\right) \Rightarrow \tan x > x$

$x \in \left(\frac{\pi}{2}, \pi\right) \Rightarrow \tan x < x$

$x \in \left(\pi, \frac{3\pi}{2}\right) \Rightarrow \tan x = x$

$x \in \left(\frac{3\pi}{2}, 2\pi\right) \Rightarrow \tan x < x$

27. Given that,

$$\sin^5 x - \cos^5 x = \frac{\sin x - \cos x}{\sin x \cos x}$$

$$\sin x \cos x \left[\frac{\sin^5 x - \cos^5 x}{\sin x - \cos x}\right] = 1$$

$\sin x \cos x \{\sin^4 x + \sin^3 x \cos x + \sin^2 x \cos^2 x + \sin x \cos^3 x + \cos^4 x\} = 1$

$\sin x \cos x\,[(\sin^4 x + \cos^4 x) + \sin^2 x \cos^2 x$
$+ \sin x \cos x\,(\sin^2 x + \cos^2 x)\} = 1$

$\sin x \cos x\,\{(\sin^2 x + \cos^2 x)^2 - \sin^2 x \cos^2 x$
$+ \sin x \cos x\} = 1$

$\Rightarrow \frac{1}{2}\sin 2x\left[1 - \frac{1}{4}\sin^2 2x + \frac{1}{2}\sin 2x\right] = 1$

$\sin 2x\,(\sin^2 2x - 2\sin 2x - 4) = -8$

$\sin^3 2x - 2\sin^2 x - 4\sin 2x + 8 = 0$

$(\sin 2x - 2)^2\,(\sin 2x + 2) = 0$

$\Rightarrow \sin 2x = \pm 2$ which is impossible.

28. $\cos^2 p - 4(\cos p - 1)\sin p \geq 0$

$\Rightarrow \quad (\cos p - 2\sin p)^2 + 4\sin p\,(1 - \sin p) \geq 0 \ldots$(i)

Now, for all real values of p

$(\cos p - 2\sin p)^2 \geq 0$ and $1 - \sin p \geq 0$

inequality (i) hold only; when

$\sin p \geq 0 \;\Rightarrow\; p \in (0, \pi)$

29. Since, $2(\sin^4 2x + \cos^4 2x) + 3\sin^2 x \cos^2 x = 0$

$\Rightarrow 2\{(\sin^2 2x + \cos^2 2x)^2 - 2\sin^2 2x \cos^2 2x\}$
$+ \frac{3}{4}\sin^2 2x = 0$

$\Rightarrow \quad 2\{1 - 2\sin^2 2x\,(1 - \sin^2 2x)\} + \frac{3}{4}\sin^2 2x = 0$

$\Rightarrow \quad 2(1 - 2\sin^2 2x + 2\sin^4 2x) + \frac{3}{4}\sin^2 2x = 0$

$\Rightarrow \quad 8 - 16\sin^2 2x + 16\sin^4 2x + 3\sin^2 2x = 0$

$\Rightarrow \quad 16\sin^4 2x - 13\sin^2 2x + 8 = 0$

Here, $\quad B^2 - 4AC < 0$

Hence, no solution.

i.e. number of solutions = 0

30. $\tan 2\theta + \sin 2\theta > 0$

$\tan 2\theta\,(1 + \cos 2\theta) > 0$

$2\theta \in \left(0, \frac{\pi}{2}\right) \cup \left(\pi, \frac{3\pi}{2}\right) \cup \left(2\pi, \frac{5\pi}{2}\right) \cup \left(3\pi, \frac{7\pi}{2}\right)$

$= \theta \in \left(0, \frac{\pi}{4}\right) \cup \left(\frac{\pi}{2}, \frac{3\pi}{4}\right) \cup \left(\pi, \frac{5\pi}{4}\right) \cup \left(\frac{3\pi}{2}, \frac{7\pi}{4}\right)$

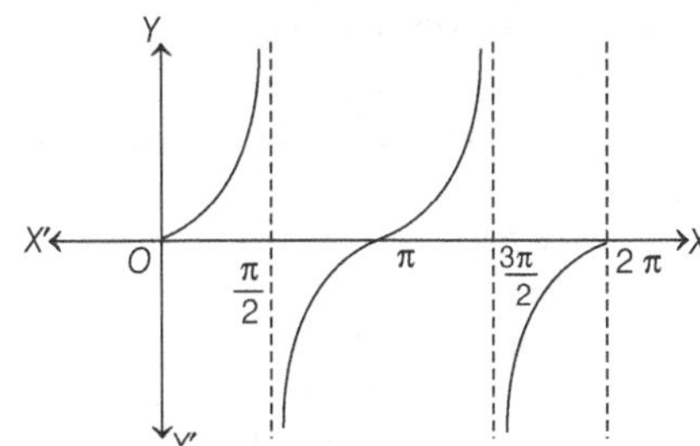

31. Since, $7\cos^2 x + \sin x \cos x - 3 = 0$

Dividing by $\cos^2 x$, then

$7 + \tan x - 3\sec^2 x = 0$

$\Rightarrow \quad 7 + \tan x - 3(1 + \tan^2 x) = 0$

$\Rightarrow \quad 3\tan^2 x - \tan x - 4 = 0$

$\Rightarrow \quad (\tan x + 1)(3\tan x - 4) = 0$

$\therefore \quad \tan x = -1$ and $\tan x = \frac{4}{3}$

Hence, $x = n\pi + \frac{3\pi}{4}$ and $x = k\pi + \tan^{-1}\left(\frac{4}{3}\right)$

where, $\quad (k, n \in I)$

32. $81^{\sin^2 x} + 81^{\cos^2 x} = 30$

$\Rightarrow \quad 81^{\sin^2 x} + 81^{1 - \sin^2 x} = 30$

$\Rightarrow \quad 81^{\sin^2 x} + \frac{81}{81^{\sin^2 x}} = 30$

Let $\quad 81^{\sin^2 x} = t$

$\therefore \quad t + \frac{81}{t} = 30$

or $\quad t^2 - 30t + 81 = 0$

$(t - 27)(t - 3) = 0$

$\therefore \quad t = 27$ and $t = 3$

or $\quad 81^{\sin^2 x} = 27$ and $81^{\sin^2 x} = 3$

or $\quad 3^{4\sin^2 x} = 3^3$ and $3^{4\sin^2 x} = 3^1$

$\therefore \quad \sin^2 x = \frac{3}{4}$ and $\sin^2 x = \frac{1}{4}$

$\therefore \quad \sin x = \pm\frac{\sqrt{3}}{2}$ and $\sin x = \pm\frac{1}{2}$

$\because \quad 0 \leq x \leq \pi$

$\therefore \quad \sin x = \frac{\sqrt{3}}{2}$ and $\sin x = \frac{1}{2}$

Then, $\quad x = \frac{\pi}{3}, \frac{\pi}{6}, \frac{2\pi}{3}, \frac{5\pi}{6}$

33. $1 + \sin\theta + \sin^2\theta + \ldots\ldots = 4 + 2\sqrt{3}$

$\frac{1}{1 - \sin\theta} = 4 + 2\sqrt{3}$

$\Rightarrow \quad 1 - \sin\theta = \frac{1}{4 + 2\sqrt{3}} = \frac{4 - 2\sqrt{3}}{4}$

$\therefore \quad \sin\theta = \frac{\sqrt{3}}{2}$

$\therefore \quad \theta = \frac{\pi}{3}$

or $\quad \frac{2\pi}{3}$

34. $\tan(\pi\cos\theta) = \cot(\pi\sin\theta)$

$\Rightarrow \quad \tan(\pi\cos\theta) = \tan\left(\frac{\pi}{2} - \pi\sin\theta\right)$

$\Rightarrow \quad \pi\cos\theta = \frac{\pi}{2} - \pi\sin\theta$

$\Rightarrow \quad \cos\theta + \sin\theta = \frac{1}{2}$

$\Rightarrow \quad \frac{1}{\sqrt{2}}\cos\theta + \frac{1}{\sqrt{2}}\sin\theta = \frac{1}{2\sqrt{2}}$

$\cos\left(\theta - \frac{\pi}{4}\right) = \frac{1}{2\sqrt{2}}$

35. We have, $\log_{10}\sin x+\log_{10}\cos x=-1$

$$\log_{10}\sin x\cos x=-1$$
$$\sin x\cos x=10^{-1}$$
$$\sin x\cos x=\frac{1}{10} \quad \ldots(i)$$

and $\log_{10}(\sin x+\cos x)=\frac{1}{2}(\log_{10}x-1)$

$$\Rightarrow (\sin x+\cos x)^2=\frac{x}{10}$$
$$\Rightarrow 1+2\sin x\cos x=\frac{x}{10} \quad \ldots(ii)$$

From Eqs. (i) and (ii), we get

$$1+\frac{2}{10}=\frac{x}{10}\Rightarrow 12=x\Rightarrow x=12$$

36. $\sin x+\cos x=\min_{a\in R}\{1, a^2-4a+6\}$

$\because$ $a^2-4a+6=(a-2)^2+2>2$

for all a

$\therefore$ $\sin x+\cos x=1$

$$\sin\left(\pi+\frac{\pi}{4}\right)=\frac{1}{\sqrt{2}}\Rightarrow x+\frac{\pi}{4}=n\pi+(-1)^n\frac{\pi}{4}$$

$$\Rightarrow x=n\pi+(-1)^n\frac{\pi}{4}-\frac{\pi}{4}$$

37. $\because$ $f(x)=\sin x+\cos x$

$$f(x)=\left[\sin\frac{\pi}{10}+\cos\frac{\pi}{10}\right]=[\sin 18°+\cos 18°]$$
$$=[\sqrt{2}\sin(45°+18°)]=[\sqrt{2}\sin 63°]$$

$\because$ $\sin 63°>\sin 45°$

$\Rightarrow$ $\sqrt{2}\sin 63°>1, -\sqrt{2}<\sqrt{2}\sin 63°<\sqrt{2}$

$\therefore$ $1<\sqrt{2}\sin 63°<\sqrt{2}$

Then, $[\sqrt{2}\sin 63°]=1$

$\therefore$ $f(x)=1$

$$\sin x+\cos x=1$$
$$\Rightarrow \cos\left(x-\frac{\pi}{4}\right)=\frac{1}{\sqrt{2}}$$

or $x-\frac{\pi}{4}=2n\pi\pm\frac{\pi}{4}$

$$x=2n\pi+\frac{\pi}{2},\ 2n\pi, n\in I$$

38. $\log_2\cos x+\log_2(1-\tan x)+\log_2(1+\tan x)-\log_2\sin x=1$

$\Rightarrow \log_2(1-\tan^2 x)-\log_2\tan x=1$

$(\cos x>0, \sin x>0, -1<\tan x<1)$

$\Rightarrow (1-\tan^2 x)\times\frac{1}{\tan x}=2\ (0<x<\pi/4)$

$\Rightarrow \tan^2 x+2\tan x-1=0$

$\Rightarrow$ $\tan x=-1\pm\sqrt{2}$

$\Rightarrow$ $\tan x=\sqrt{2}-1$ $\quad[\because 0<\tan x<1]$

$\Rightarrow x=\pi/8$ (Smallest positive angle).

39. Since, $x+2\tan x=\frac{\pi}{2}$

or $\tan x=\frac{\pi}{4}-\frac{x}{2}=y$ (say)

$\therefore$ $y=\tan x \quad \ldots(i)$

and $y=\frac{\pi}{4}-\frac{x}{2} \quad \ldots(ii)$

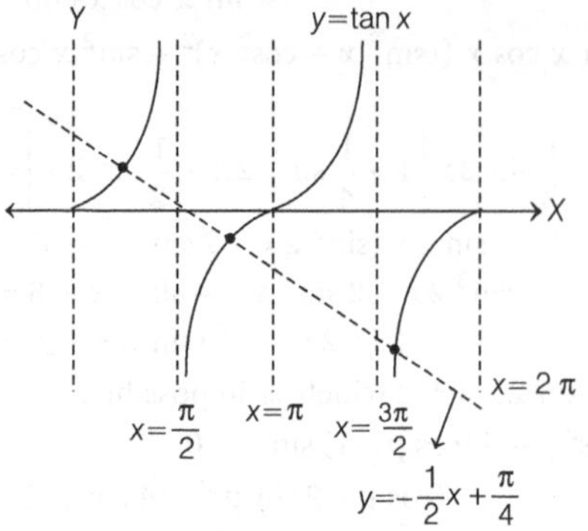

Y, y=tan x, X, x=2π, $x=\frac{\pi}{2}$, x=π, $x=\frac{3\pi}{2}$, $y=-\frac{1}{2}x+\frac{\pi}{4}$

Graphs Eqs. (i) and (ii) intersect at three points

$\therefore$ Number of solutions is 3.

40. We have, $8\cos x\left(\cos\left(\frac{\pi}{6}+x\right)\cos\left(\frac{\pi}{6}-x\right)-\frac{1}{2}\right)=1$

$$\Rightarrow 8\cos x\left(\cos^2\frac{\pi}{6}-\sin^2 x-\frac{1}{2}\right)=1$$
$$\Rightarrow 8\cos x\left(\frac{3}{4}-\sin^2 x-\frac{1}{2}\right)=1$$
$$\Rightarrow 8\cos x\left(\frac{3}{4}-\frac{1}{2}-1+\cos^2 x\right)=1$$
$$\Rightarrow 8\cos x\left(\frac{-3+4\cos^2 x}{4}\right)=1$$
$$\Rightarrow 2(4\cos^3 x-3\cos x)=1$$
$$\Rightarrow 2\cos 3x=1\Rightarrow \cos 3x=\frac{1}{2}$$
$$\Rightarrow 3x=\frac{\pi}{3},\frac{5\pi}{3},\frac{7\pi}{3} \quad [0\le 3\pi\le 3\pi]$$
$$\Rightarrow x=\frac{\pi}{9},\frac{5\pi}{9},\frac{7\pi}{9}$$

Sum $=\frac{\pi}{9}+\frac{5\pi}{9}+\frac{7\pi}{9}=\frac{13\pi}{9}\Rightarrow k\pi=\frac{13\pi}{9}$

Hence, $k=\frac{13}{9}$

41. $\cos\{\pi\sqrt{(x-4)}\}\cos(\pi\sqrt{x})=1$

It is possible only, when

$$\cos\{\pi\sqrt{(x-4)}\}=1 \text{ and } \cos(\pi\sqrt{x})=1$$

Since, $x-4\ge 0$ and $x\ge 0$

$\therefore$ $x\ge 4 \quad \ldots(i)$

From Eq. (i),

$$\pi\sqrt{(x-4)}=0$$

and $\pi\sqrt{x}=0$

$\therefore$ $x=0, 4$

But $x\ge 4$

$\therefore$ Only one solution $x=4$.

42. $\sin\left(\frac{\pi x}{2\sqrt{3}}\right)=x^2-2\sqrt{3}x+4$

$\because$ $-1\le\sin\left(\frac{\pi x}{2\sqrt{3}}\right)\le 1$

$\Rightarrow$ $-1\le x^2-2\sqrt{3}x+4\le 1$

$\therefore \quad x^2 - 2\sqrt{3}x + 4 \le 1$ and $x^2 - 2\sqrt{3}\,x + 4 \ge -1$

$\Rightarrow \quad (x-\sqrt{3})^2 \le 0$ and $\quad (x-\sqrt{3})^2 + 2 \ge 0$

Which is true. It is possible only, when

$$x - \sqrt{3} = 0$$

$$\therefore \quad x = \sqrt{3}$$

43. $\log_{\cos x} \sin x + \log_{\sin x} \cos x = 2$

$$\Rightarrow \log_{\cos x} \sin x + \frac{1}{\log_{\cos x} \sin x} = 2$$

$$\Rightarrow (\log_{\cos x} \sin x - 1)^2 = 0$$

$$\Rightarrow \log_{\cos x} \sin x = 1$$

$[\because \sin > 0, \cos x > 0, \sin x \ne 1, \cos x \ne 1]$

$$\Rightarrow \sin x = \cos x \Rightarrow \tan x = 1$$

$$\Rightarrow x = 2n\pi + \frac{\pi}{4}, n \in I$$

44. $\sin 2\theta = \dfrac{\sqrt{3}}{2} = \sin\left(\dfrac{\pi}{3}\right), \sin\left(\pi - \dfrac{\pi}{3}\right)$

$$\Rightarrow 2\theta = \frac{\pi}{3}, \frac{2\pi}{3}$$

or $\quad \theta = \dfrac{\pi}{6}, \dfrac{\pi}{3} \quad \ldots\text{(i)}$

and $\quad \tan\theta = \dfrac{1}{\sqrt{3}}$

$$\therefore \theta = \frac{\pi}{6}, \pi + \frac{\pi}{6}$$

$\theta = \dfrac{\pi}{6}, \dfrac{7\pi}{6} \quad \ldots\text{(ii)}$

From Eqs. (i) and (ii) common value of θ is $\dfrac{\pi}{6}$.

Hence, general value of θ is

$$2n\pi + \frac{\pi}{6}, \quad n \in I$$

45. Given equation, $\sec^2(a+2)x + a^2 - 1 = 0$

$\Rightarrow \tan^2(a+2)x + a^2 = 0$

$\Rightarrow \quad \tan^2(a+2)x = 0$ and $a = 0$

$$\Rightarrow \tan^2 2x = 0$$

$$\Rightarrow x = 0, \frac{\pi}{2}, -\frac{\pi}{2}$$

$\therefore$ $(0, 0)$, $(0, \pi/2)$, $(0, -\pi/2)$ are 3 ordered pairs satisfying the equation.

46. Since, $\sec\theta + \operatorname{cosec}\theta = \sqrt{15}$

$$\Rightarrow \sin\theta + \cos\theta = \sqrt{15}\sin\theta\cos\theta$$

$\Rightarrow \quad 1 + \sin 2\theta = \dfrac{15}{4}\sin^2 2\theta \quad$ [on squaring]

$$\Rightarrow 15\sin^2 2\theta - 4\sin 2\theta - 4 = 0$$

$$\Rightarrow (3\sin 2\theta - 2)(5\sin 2\theta + 2) = 0$$

$\therefore \quad \sin 2\theta = \dfrac{2}{3}$ and $\sin 2\theta = -\dfrac{2}{5}$

$\sin 2\theta = \dfrac{2}{3}$ (lies in I and II quadrants) given 2 solutions

and $\quad \sin 2\theta = -\dfrac{2}{5}$

(lies in III and IV quadrant) given 2 solutions

Hence, number of solutions = 4.

47. $\sin^{10} x + \cos^{10} x = \dfrac{29}{16}\cos^4 2x$

$$\left(\frac{1-\cos 2x}{2}\right)^5 + \left(\frac{1+\cos 2x}{2}\right)^5 = \frac{29}{16}\cos^4 2x$$

$$\Rightarrow \frac{1}{32} \times 2\{1 + 10\cos^2 2x + 5\cos^4 2x\} = \frac{29}{16}\cos^4 2x$$

$$\Rightarrow 24\cos^4 2x - 10\cos^2 2x - 1 = 0$$

$$\Rightarrow (2\cos^2 2x - 1)(12\cos^2 2x + 1) = 0$$

If $\quad \cos^2 2x = \dfrac{1}{2} \quad [\because 12\cos^2 2x + 1 \ne 0]$

$$\Rightarrow 2\cos^2 2x - 1 = 0 \Rightarrow \cos 4x = 0$$

$$4x = n\pi + \frac{\pi}{2}$$

$$\Rightarrow x = \frac{n\pi}{4} + \frac{\pi}{8}, n \in I$$

48. We have,

$$|\cos x| = 2[x] = y \text{ (say)}$$

$$\therefore \quad y = |\cos x| \text{ and } y = 2[x]$$

From graph $|\cos x|$ and $2[x]$ don't cut each other for any real value of x. Hence, number of solution is nil.

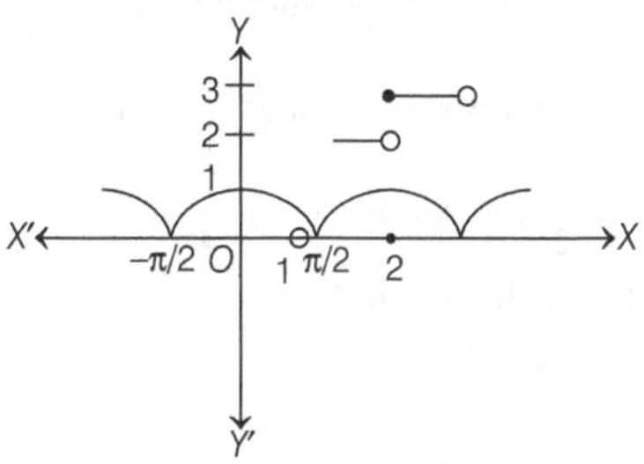

49. $\tan x + \cot x + 1 = \cos\left(x + \dfrac{\pi}{4}\right)$

or $\tan x + \cot x = \cos\left(x + \dfrac{\pi}{4}\right) - 1$

Now, $\tan x + \cot x \le -2$ and $\cos\left(x + \dfrac{\pi}{4}\right) - 1 \ge -2$

It implies that equality holds when, both are -2.

Thus, $\quad \cos\left(x + \dfrac{\pi}{4}\right) = -1$

$$\Rightarrow x + \frac{\pi}{4} = (2m+1)\pi, m \in Z$$

$$\Rightarrow x = \frac{3\pi}{4} \text{ or } \frac{11\pi}{4}$$

Therefore, the sum of the solutions is

$$\frac{3\pi}{4} + \frac{11\pi}{4} = \frac{7\pi}{2}$$

50. We have,

$$2^{\cos x} = |\sin x|$$

It is true only for $|\sin x| = 1$

$$\therefore \quad \sin x = \pm 1$$

$\Rightarrow \quad x = 2n\pi \pm \dfrac{\pi}{2}, \quad [\because x \in [-2\pi, 2\pi]]$

$$\therefore \quad x = -\frac{\pi}{2}, \frac{\pi}{2}, \frac{3\pi}{2}, -\frac{3\pi}{2}$$

i.e. number of solutions = 4.

51. Since, $2^{\cos 2x} + 1 = 3 \cdot 2^{-\sin^2 x}$

$\Rightarrow \quad 2^{1-2\sin^2 x} + 1 = 3 \cdot 2^{-\sin^2 x}$

Let $\quad 2^{-\sin^2 x} = t > 0$

$\therefore \quad 2 \cdot t^2 + 1 = 3t$

$\Rightarrow \quad 2t^2 - 3t + 1 = 0$

$\therefore \quad t = \frac{1}{2}$ and $t = 1$

$2^{-\sin^2 x} = 2^{-1}$ and $2^{-\sin^2 x} = 2^0$

Then, $\sin^2 x = 1$ and $\sin^2 x = 0$.

$\therefore$ General solutions are

$$x = n\pi + \frac{\pi}{2} \text{ and } x = n\pi$$

52. Given, $x \in (0, 1)$

and $\quad \sin 2\pi x = \sqrt{2} \cos \pi x$

$\Rightarrow \quad \cos \pi x (2 \sin \pi x - \sqrt{2}) = 0$

$\Rightarrow \quad \cos \pi x \left(\sin \pi x - \frac{1}{\sqrt{2}}\right) = 0$

$\therefore \quad \cos \pi x = 0$ and $\sin \pi x = \frac{1}{\sqrt{2}}$

$\Rightarrow \quad \pi x = n\pi + \frac{\pi}{2}$ and $\pi x = n\pi + (-1)^n \frac{\pi}{4}$

or $x = n + \frac{1}{2}$ and $x = n + (-1)^n \frac{1}{4}, n \in I$

for $n = 0$ and for $n = 0, 1$

$$x = \frac{1}{2}, x = \frac{1}{4}, \frac{3}{4}$$

Hence, greatest root $= \frac{3}{4}$.

53. The given trigonometric equation is,

$$\cos 2x + \alpha \sin x = 2\alpha - 7$$

$\Rightarrow \quad 1 - 2\sin^2 x + \alpha \sin x = 2\alpha - 7$

$[\because \cos 2x = 1 - 2\sin^2 x]$

$\Rightarrow \quad 2\sin^2 x - \alpha \sin x + 2\alpha - 8 = 0$

$\Rightarrow \quad 2(\sin^2 x - 4) - \alpha(\sin x - 2) = 0$

$\Rightarrow 2(\sin x - 2)(\sin x + 2) - \alpha(\sin x - 2) = 0$

$\Rightarrow \quad (\sin x - 2)(2\sin x + 4 - \alpha) = 0$

$\therefore \quad 2\sin x + 4 - \alpha = 0 \quad [\because \sin x - 2 \neq 0]$

$\Rightarrow \quad \sin x = \frac{\alpha - 4}{2} \quad \ldots(i)$

Now, as we know $-1 \le \sin x \le 1$

$\therefore \quad -1 \le \frac{\alpha - 4}{2} \le 1 \quad$ [from Eq. (i)]

$\Rightarrow -2 \le \alpha - 4 \le 2 \Rightarrow 2 \le \alpha \le 6 \Rightarrow \alpha \in [2, 6]$

54. We have,

$$|4 \sin x - 1| < \sqrt{5}$$

$\Rightarrow \quad -\sqrt{5} < 4 \sin x - 1 < \sqrt{5}$

$\Rightarrow \quad -\left(\frac{\sqrt{5} - 1}{4}\right) < \sin x < \frac{(\sqrt{5} + 1)}{4}$

$\Rightarrow \quad -\sin \frac{\pi}{10} < \sin x < \cos \frac{\pi}{5}$

$\Rightarrow \quad \sin\left(-\frac{\pi}{10}\right) < \sin x < \sin\left(\frac{\pi}{2} - \frac{\pi}{5}\right)$

$\Rightarrow \quad \sin\left(-\frac{\pi}{10}\right) < \sin x < \sin\left(\frac{3\pi}{10}\right)$

$x \in \left(-\frac{\pi}{10}, \frac{3\pi}{10}\right) \quad [\because x \in (-\pi, \pi)]$

55. If $\sin x > 0$

$\Rightarrow \quad x \in (0, \pi) \quad \ldots(i)$

$\cos x > 0$

$\Rightarrow \quad x \in \left(0, \frac{\pi}{2}\right) \cup \left(\frac{3\pi}{2}, 2\pi\right) \quad \ldots(ii)$

From Eqs. (i) and (ii)

$$x \in \left(0, \frac{\pi}{2}\right)$$

Now, $\log_{1/2} \sin x > \log_{1/2} \cos x$

$\Rightarrow \quad \sin x < \cos x$ in $\left(0, \frac{\pi}{2}\right)$

i.e. $\quad x \in \left(0, \frac{\pi}{4}\right)$.

56. We have, $|4 \sin x - 1| < \sqrt{5}$

$\Rightarrow \quad -\sqrt{5} < 4 \sin x - 1 < \sqrt{5}$

$\Rightarrow \quad -\left(\frac{\sqrt{5} - 1}{4}\right) < \sin x < \left(\frac{\sqrt{5} + 1}{4}\right)$

$\Rightarrow \quad -\sin \frac{\pi}{10} < \sin x < \cos \frac{\pi}{5}$

$\Rightarrow \quad \sin\left(-\frac{\pi}{10}\right) < \sin x < \sin\left(\frac{\pi}{2} - \frac{\pi}{5}\right)$

$\Rightarrow \quad \sin\left(-\frac{\pi}{10}\right) < \sin x < \sin \frac{3\pi}{10}$

$\Rightarrow \quad x \in \left(-\frac{\pi}{10}, \frac{3\pi}{10}\right)$

57. We have,

$$\tan(5\pi \cos \alpha) = \cot(5\pi \sin \alpha)$$

$\Rightarrow \quad \cot(5\pi \sin \alpha) - \tan(5\pi \cos \alpha) = 0$

$\Rightarrow \quad \frac{\cos(5\pi \sin \alpha)}{\sin(5\pi \sin \alpha)} - \frac{\sin(5\pi \cos \alpha)}{\cos(5\pi \cos \alpha)} = 0$

$\Rightarrow \quad \cos\{5\pi(\sin \alpha + \cos \alpha)\} = 0$

$\Rightarrow \quad 5\pi(\sin \alpha + \cos \alpha) = n\pi + \frac{\pi}{2}, n \in I$

$\Rightarrow \quad \sin \alpha + \cos \alpha = \frac{(2n + 1)}{10}$

$\Rightarrow \quad \frac{1}{\sqrt{2}} \sin \alpha + \frac{1}{\sqrt{2}} \cos \alpha = \frac{(2n + 1)}{10\sqrt{2}}$

$\therefore \quad \sin\left(\frac{\pi}{4} + \alpha\right) = \frac{(2n + 1)}{10\sqrt{2}} \quad \ldots(i)$

$\therefore \quad \frac{\pi}{4} + \alpha = n\pi + (-1)^n \sin^{-1}\left(\frac{2n + 1}{10\sqrt{2}}\right)$

$\Rightarrow \quad \alpha = n\pi - \frac{\pi}{4} + (-1)^n \sin^{-1}\left(\frac{2n + 1}{10\sqrt{2}}\right), n \in I \quad \ldots(ii)$

From Eq. (i)

$$-1 \le \sin\left(\frac{\pi}{4} + \alpha\right) \le 1$$

$\Rightarrow \quad -1 \le \dfrac{(2n+1)}{10\sqrt{2}} \le 1$

$\Rightarrow \quad \dfrac{-10\sqrt{2}-1}{2} \le n \le \dfrac{10\sqrt{2}-1}{2}$

$\because \quad n \in I$

$\therefore \quad n = -7, -6, -5, -4, -3, -2, -1,$
$0, 1, 2, 3, 4, 5, 6$

Now, from Eq. (ii)

Number of solutions = Number of values of $n = 14$

58. We have,

$\sin^3 x \cos x + \sin^2 x \cos^2 x + \sin x \cos^3 x = 1$

$\Rightarrow \sin x \cos x (\sin^2 x + \sin x \cos x + \cos^2 x) = 1$

$\Rightarrow \quad \dfrac{\sin 2x}{2}\left(1 + \dfrac{\sin 2x}{2}\right) = 1$

$\sin 2x\,(2 + \sin 2x) = 4$

$\Rightarrow \quad \sin^2 2x + 2\sin 2x - 4 = 0$

$\sin 2x = \dfrac{-2 \pm \sqrt{(4+16)}}{2}$

$\therefore \quad \sin 2x = -1 \pm \sqrt{5}$

Impossible $\quad [\because -1 \le \sin 2x \le 1]$

Hence, given equation has no solutions.

59. $\lambda^2 - 4\lambda + 7 = (\lambda - 2)^2 + 3 \ge 3$

$\therefore \quad \min_{\lambda \in R} \{2, e^2, \pi, \lambda^2 - 4\lambda + 7\} = 2$

$\sqrt{3}\sin x - \cos x = 2$

$\Rightarrow \quad \sin x \cdot \dfrac{\sqrt{3}}{2} - \cos x \cdot \dfrac{1}{2} = 1$

$\sin\left(x - \dfrac{\pi}{6}\right) = 1 = \sin \dfrac{\pi}{2}$

$x - \dfrac{\pi}{6} = 2n\pi + \dfrac{\pi}{2}$

$x = 2n\pi + \dfrac{\pi}{2} + \dfrac{\pi}{6} = 2n\pi + \dfrac{2\pi}{3},\ n \in I$

60. $\quad \cos^2\theta < \dfrac{1}{2}$

$\Rightarrow \quad 2\cos^2\theta - 1 < 0$

$\Rightarrow \quad \cos 2\theta < 0 \Rightarrow -\cos 2\theta > 0$

$\Rightarrow \quad \cos(\pi + 2\theta) > \cos \dfrac{\pi}{2}$

$\therefore \quad 2n\pi - \dfrac{\pi}{2} < \pi + 2\theta < 2n\pi + \dfrac{\pi}{2},\ n \in I$

$\Rightarrow \quad 2n\pi - \dfrac{3\pi}{2} < 2\theta < 2n\pi - \dfrac{\pi}{2}$

or $\quad n\pi - \dfrac{3\pi}{4} < \theta < n\pi - \dfrac{\pi}{4}$

or $\quad (4n-3)\dfrac{\pi}{4} < \theta < (4n-1)\dfrac{\pi}{4},\ n \in I$

Replacing n by $n+1$, then

$(4n+1)\dfrac{\pi}{4} < \theta < (4n+3)\dfrac{\pi}{4},\ n \in I$

61. For $1 + \log_2 \sin x + \log_2 \sin 3x \ge 0 \quad \ldots(i)$

We must have, $\sin x > 0$, i.e., $0 < x < \pi$

and $\sin 3x > 0$ i.e. $0 < 3x < \pi$ or $2\pi < 3x < 3\pi$

$\therefore \quad 0 < x < \pi, 0 < x < \dfrac{\pi}{3}$ and $\dfrac{2\pi}{3} < x < \pi$

Therefore domain of the inequation in $(0, \pi)$ is

$\left(0, \dfrac{\pi}{3}\right) \cup \left(\dfrac{2\pi}{3}, \pi\right) \quad \ldots(ii)$

Eq. (i) is equivalent to,

$\log_2(2\sin x \sin 3x) \ge 0$

$\Rightarrow \quad 2\sin x \sin 3x \ge 1$

i.e., $\quad 2\sin^2 x(3 - 4\sin^2 x) \ge 1$

i.e., $\quad 8\sin^4 x - 6\sin^2 x + 1 \le 0$

$\Rightarrow \quad (2\sin^2 x - 1)(4\sin^2 x - 1) \le 0$

$\Rightarrow \quad \dfrac{1}{4} \le \sin^2 x \le \dfrac{1}{2}$

Since, $\sin x$ is positive, we have $\dfrac{1}{2} \le \sin x \le \dfrac{1}{\sqrt{2}}$

$\therefore \quad \dfrac{\pi}{6} \le x \le \dfrac{\pi}{4}$ or $\dfrac{3\pi}{4} \le x \le \dfrac{5\pi}{6}$

Since, domain is $(0, \pi/3) \cup (2\pi/3, \pi)$

We have the solution $\left[\dfrac{\pi}{6}, \dfrac{\pi}{4}\right] \cup \left[\dfrac{3\pi}{4}, \dfrac{5\pi}{6}\right]$.

62. $1 + \sin x \sin^2 \dfrac{x}{2} = 0$

$1 + \sin x\left(\dfrac{1 - \cos x}{2}\right) = 0$

$\Rightarrow \quad 2 + \sin x - \sin x \cos x = 0$

$\Rightarrow \quad 4 + 2\sin x = \sin 2x$

LHS $\in [2, 6]$ but RHS $\in [-1, 1]$.

Hence, no solution

i.e. Number of solutions = zero

Alternative Method

Given equation is

$\sin x \sin^2\left(\dfrac{x}{2}\right) = -1$

It could be possible only, if

$\sin x = -1$ and $\sin^2\left(\dfrac{x}{2}\right) = 1 \quad [\because -1 \le \sin x \le 1]$

Which is not possible simultaneously.

63. $|\cot x| = \cot x + \dfrac{1}{\sin x}$

If x lies in I and III quadrant, then

$\cot x = \cot x + \dfrac{1}{\sin x}$

$\Rightarrow \quad \dfrac{1}{\sin x} = 0$, not possible

and, if x lies in II and IV quadrant, then

$-\cot x = \cot x + \dfrac{1}{\sin x}$

$\Rightarrow \quad 2\cos x + 1 = 0, \because \sin x \ne 0$

$\cos x = -\dfrac{1}{2} = \cos \dfrac{2\pi}{3} \Rightarrow x = 2n\pi \pm \dfrac{2\pi}{3}$

Then, $\quad x = \dfrac{2\pi}{3}, \dfrac{4\pi}{3} \quad [\because x \in [0, 2\pi]]$

Hence, number of solutions = 2.

64. Given, $\sqrt{3}\sec x + \text{cosec } x + 2(\tan x - \cot x) = 0,$

$$(-\pi < x < \pi) - \{0, \pm \pi/2\}$$

$\Rightarrow \sqrt{3}\sin x + \cos x + 2(\sin^2 x - \cos^2 x) = 0$

$\Rightarrow \quad \sqrt{3}\sin x + \cos x - 2\cos 2x = 0$

Multiplying and dividing by $\sqrt{a^2+b^2}$, *i.e.*, $\sqrt{3+1} = 2$, we get

$$2\left(\frac{\sqrt{3}}{2}\sin x + \frac{1}{2}\cos x\right) - 2\cos 2x = 0$$

$$\Rightarrow \left(\cos x \cdot \cos\frac{\pi}{3} + \sin x \cdot \sin\frac{\pi}{3}\right) - \cos 2x = 0$$

$$\Rightarrow \cos\left(x - \frac{\pi}{3}\right) = \cos 2x$$

$$\therefore \quad 2x = 2n\pi \pm \left(x - \frac{\pi}{3}\right)$$

[since, $\cos\theta = \cos\alpha \Rightarrow \theta = 2n\pi \pm \alpha$]

$$\Rightarrow \quad 2x = 2n\pi + x - \frac{\pi}{3}$$

or $\quad 2x = 2n\pi - x + \frac{\pi}{3} \Rightarrow x = 2n\pi - \frac{\pi}{3}$

or $\quad 3x = 2n\pi + \frac{\pi}{3} \Rightarrow x = 2n\pi - \frac{\pi}{3}$

or $\quad x = \frac{2n\pi}{3} + \frac{\pi}{9}$

$\therefore \quad x = \frac{-\pi}{3}$

or $\quad x = \frac{\pi}{9}, \frac{-5\pi}{9}, \frac{7\pi}{9}$

Now, sum of all distinct solutions

$$= \frac{-\pi}{3} + \frac{\pi}{9} - \frac{5\pi}{9} + \frac{7\pi}{9} = 0$$

65. $\because \; y = \frac{1}{3}[\sin x + [\sin x + [\sin x]]]$

$$= \frac{1}{3}([\sin x] + [\sin x] + [\sin x]) = [\sin x]$$

and $\quad [y + [y]] = 2\cos x$

$\Rightarrow \quad 2[y] = (2\cos x)$

$\Rightarrow \quad [y] = \cos x$

Case I when $-1 \le \sin x < 0$

$\Rightarrow \therefore \quad [\sin x] = -1$

Then, $\quad y = -1$

i.e. $\quad [-1] = \cos x$

$\Rightarrow \quad \cos x = -1$

$\Rightarrow \sin x = 0$ (impossible)

Case II when $0 \le \sin x < 1$

$\therefore \quad [\sin x] = 0$

Then, $\quad y = 0$

i.e. $\quad 0 = \cos x$

$\Rightarrow \quad \sin x = 1$ (impossible)

$\sin x = 1$

$\therefore \quad y = 1$

i.e. $\quad \cos x = 1$

$\Rightarrow \quad \sin x = 0$ (impossible)

Hence, no solution

i.e. number of solutions is zero.

Objective Questions Type II [One or more than one correct answer(s)]

1. Let $81^{\sin^2 x} = y$. Then, $81^{\cos^2 x} = 81^{1-\sin^2 x}$
$= 81 \cdot y^{-1}$, so that the given equation is equivalent to
$y + 81y^{-1} = 30$ or $y^2 - 30y + 81 = 0$

$\Rightarrow y = 3$ or $y = 27 \Rightarrow 81^{\sin^2 x} = 3$ or $81^{\sin^2 x} = 27$

$\Rightarrow \quad 3^{4\sin^2 x} = 3 \;\Rightarrow 3^{4\sin^2 x} = 3^3$

$\Rightarrow \quad \sin^2 x = \frac{1}{4} \;\Rightarrow \sin^2 x = \frac{3}{4}$

$\Rightarrow \quad \sin x = \pm\frac{1}{2} \Rightarrow \sin x = \pm\frac{\sqrt{3}}{2}$

which gives two solutions of the given equation in each quadrant.

2. $2\sin\frac{x}{2}(\cos^2 x - \sin^2 x) - (\cos^2 x - \sin^2 x) = 0$

$$\left(2\sin\frac{x}{2} - 1\right)(\cos^2 x - \sin^2 x) = 0$$

$$\Rightarrow \left(2\sin\frac{x}{2} - 1\right)\cos 2x = 0$$

$$\sin\frac{x}{2} = \frac{1}{2} \text{ and } \cos 2x = 0$$

$$x = 60° \text{ and } x = \pm 45°$$

3. Since, $\sin x + \cos x = 1 + \sin x \cdot \cos x$ ...(i)

$\sin x(1 - \cos x) - (1 - \cos x) = 0$

$\Rightarrow \quad (\sin x - 1)(1 - \cos x) = 0$

$\Rightarrow \quad \sin x = 1$ and $\cos x = 1$

Then, $\quad x = \frac{\pi}{2}$ or $x = 0$

4. Since, $\sin\theta + \sqrt{3}\cos\theta = 6x - x^2 - 11$

$$0 \le \theta \le 4\pi$$

$$\Rightarrow \frac{1}{2}\sin\theta + \frac{\sqrt{3}}{2}\cos\theta = \frac{6x - x^2 - 11}{2}$$

$$\Rightarrow \sin\left(\theta + \frac{\pi}{3}\right) = \frac{6x - x^2 - 11}{2} \quad \text{...(i)}$$

$$\because \quad -1 \le \sin\left(\theta + \frac{\pi}{3}\right) \le 1$$

$$\Rightarrow \quad -1 \le \frac{6x - x^2 - 11}{2} \le 1$$

$$\Rightarrow \quad -2 \le 6x - x^2 - 11 \le 2$$

Case I If $6x - x^2 - 11 \le 2$

$\Rightarrow \quad x^2 - 6x + 13 \ge 0$

$\Rightarrow (x - 3)^2 + 4 \ge 0$ which is always true.

Case II If $-2 \le 6x - x^2 - 11$

$\Rightarrow \quad x^2 - 6x + 9 \le 0$

$\Rightarrow \quad (x-3)^2 \le 0$

Which is possible only, when $x - 3 = 0$

$\therefore \quad x = 3$

From Eq. (i), $\sin\left(\theta + \frac{\pi}{3}\right) = -1$

$$\theta + \frac{\pi}{3} = 2n\pi - \frac{\pi}{2} \Rightarrow \theta = 2n\pi - \frac{5\pi}{6}$$

For $\quad n = 1, 2$

$$\theta = \frac{7\pi}{6}, \frac{19\pi}{6} \quad [\because 0 \le \theta \le 4\pi]$$

or $\quad (x, \theta) \equiv \left(3, \frac{7\pi}{6}\right), \left(3, \frac{19\pi}{6}\right)$

5. $\because \sin x = [1 + \sin x] + [1 - \cos x]$

$\sin x = 1 + [\sin x] + 1 + [-\cos x]$

$\Rightarrow \quad \sin x = 2 + [\sin x] + [-\cos x]$

At $\quad x = -\frac{\pi}{2}$

$\sin x = 2 - 1$

$\therefore \quad \sin x = 1$

$-1 = 1$ [impossible]

At $\quad x = \frac{\pi}{2}$

$1 = 2 + 1 - 0$

$1 = 3$ [impossible]

At $\quad x = \pi, \; 0 = 1$ [impossible]

In $\quad x \in \left(-\frac{\pi}{2}, 0\right)$,

$\sin x = 2 + (-1) + 0 = 1$ [impossible]

In $\quad x \in \left(0, \frac{\pi}{2}\right)$, $\sin x = 2 + 0 - 1 = 1$ [impossible]

In $\quad x \in \left(\frac{\pi}{2}, \pi\right)$, $\sin x = 2 + 0 + 0 = 2$ [impossible]

In $x \in (\pi, 3\pi/2)$, $\sin x = 2 - 1 + 0 = 1$ [impossible]

6. $2\cos\theta(1 - \sin\phi) = \dfrac{2\sin^2\theta}{\sin\theta}\cos\phi - 1 = 2\sin\theta\cos\phi - 1$

$\therefore \quad 2\cos\theta - 2\cos\theta\sin\phi = 2\sin\theta\cos\phi - 1$

$\therefore \quad 2\cos\theta + 1 = 2\sin(\theta + \phi)$ ...(i)

$\tan(2\pi - \theta) > 0 \Rightarrow \tan\theta < 0$

and $\quad -1 < \sin\theta < -\frac{\sqrt{3}}{2}$

$\Rightarrow \quad \theta \in \left(\frac{3\pi}{2}, \frac{5\pi}{3}\right)$

$\Rightarrow \quad 0 < \cos\theta < \frac{1}{2}$

$\Rightarrow \quad \frac{1}{2} < \sin(\theta + \phi) < 1$ (from (i))

$\Rightarrow \quad \frac{\pi}{6} + 2\pi < \theta + \phi < \frac{5\pi}{6} + 2\pi$

$\Rightarrow \quad 2\pi + \frac{\pi}{6} - \theta_{max} < \phi < 2\pi + \frac{5\pi}{6} - \theta_{min}$

$\Rightarrow \quad \frac{\pi}{2} < \phi < \frac{4\pi}{3}$

7. $\sin x \sin y = \frac{\sqrt{3}}{4}$ and $\cos x \cos y = \frac{\sqrt{3}}{4}$

Then, $\cos x \cos y + \sin x \sin y = \frac{\sqrt{3}}{2}$

$\Rightarrow \quad \cos(x - y) = \frac{\sqrt{3}}{2}$

$\Rightarrow \quad x - y = 2n\pi \pm \frac{\pi}{6}, \; n \in I$ ...(i)

and $\quad \cos x \cos y - \sin x \sin y = 0$

$\cos(x + y) = 0$

$x + y = k\pi + \frac{\pi}{2}, k \in I$ ...(ii)

From Eqs. (i) and (ii),

$$2x = 2n\pi + k\pi \pm \frac{\pi}{6} + \frac{\pi}{2}$$

$\Rightarrow \quad x = \frac{\pi}{2}(2n + k) \pm \frac{\pi}{12} + \frac{\pi}{4}$

$\therefore \quad x_1 = \frac{\pi}{2}(2n + k) + \frac{\pi}{3}$

and $\quad x_2 = \frac{\pi}{2}(2n + k) + \frac{\pi}{6}$

$\therefore \quad y_1 = \frac{\pi}{6} + \frac{\pi}{2}(k - 2n)$

and $\quad y_2 = \frac{\pi}{3} + \frac{\pi}{2}(k - 2n)$

8. $2\sin^2 x + \sin^2 2x = 2, -\pi < x < \pi$

$\Rightarrow \quad 1 - \cos 2x + 1 - \cos^2 2x = 2$

$\Rightarrow \quad \cos 2x = 0, 1$

$\Rightarrow \quad 2x = n\pi + \frac{\pi}{2}$ and $2x = 2k\pi$

$\Rightarrow \quad x = \frac{n\pi}{2} + \frac{\pi}{4}$ and $x = k\pi; n, k \in I$

for $\quad n = -2, -1, 0, 1$

$x = -\frac{3\pi}{4}, -\frac{\pi}{4}, \frac{\pi}{4}, \frac{3\pi}{4}$ and $x = 0$

$\therefore \quad x = \pm\frac{\pi}{4}$ and $\pm\frac{3\pi}{4}$ and $x = 0$

9. $(1 - \tan\theta)(1 + \tan\theta)\sec^2\theta + 2^{\tan^2\theta} = 0$

$\Rightarrow (1 - \tan^4\theta) + 2^{\tan^2\theta} = 0$

$\Rightarrow 2^t = t^2 - 1$ (where $t = \tan^2\theta$)

By inspection (or by graph) we find $y = 2^t$ and $y = t^2 - 1$ intersect in $t = 3$

$\Rightarrow \quad \tan^2\theta = 3$

$\Rightarrow \quad \tan\theta = \pm\sqrt{3} \Rightarrow \theta = \pm\pi/3$

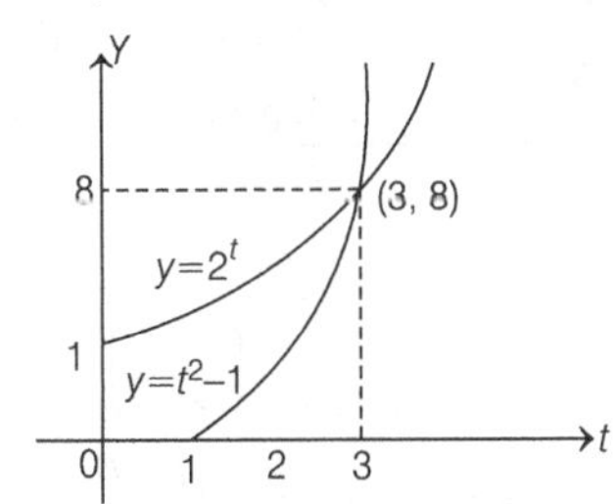

10. $\because |\cos x|^{\sin^2 x - \frac{3}{2}\sin x + \frac{1}{2}} = |\cos x|^0$

$\therefore \quad |\cos x| = 1$

or $\quad \sin^2 x - \dfrac{3}{2}\sin x + \dfrac{1}{2} = 0$

$\Rightarrow \quad \cos x = \pm 1$

or $\quad (\sin x - 1)\left(\sin x - \dfrac{1}{2}\right) = 0$

$\therefore x = n\pi$, or $x = 2n\pi + \dfrac{\pi}{2}$, $x = n\pi + (-1)^n \dfrac{\pi}{6}$, $n \in I$

11. $\tan|x| = |\tan x|$

Let $\quad y = \tan|x|$

and $\quad y = |\tan x|$

It is clear that,

$$-\frac{3\pi}{2} < x \le -\pi,\ -\frac{\pi}{2} < x < \frac{\pi}{2},\ \pi \le x < \frac{3\pi}{2}$$

$\therefore \quad x \in \left(-\dfrac{\pi(2k+1)}{2}, -\pi k\right]$

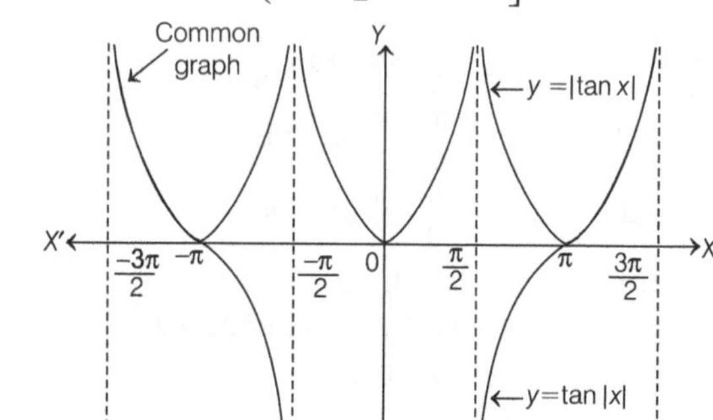

And $\quad x \in \left[\pi k, \dfrac{\pi(2k+1)}{2}\right), k \in I$

12. $\cos(\sin x) = \dfrac{1}{\sqrt{2}} = \cos\dfrac{\pi}{4}$

$\therefore \quad \sin x = 2n\pi \pm \dfrac{\pi}{4}, n \in I$

$y = \sin x = \pm\dfrac{\pi}{4}$

$y = \sin x$ and $y = \pm\dfrac{\pi}{4}$

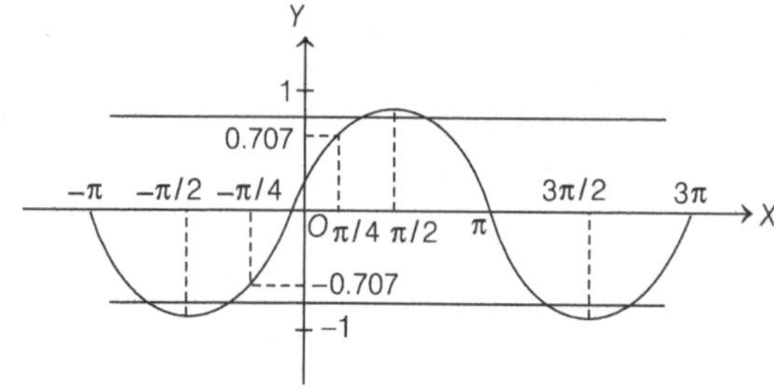

$\therefore \quad x \in \left(\dfrac{\pi}{4}, \dfrac{\pi}{2}\right) \cup \left(\dfrac{\pi}{2}, \pi\right)$

and $\quad x \in \left(-\pi, -\dfrac{\pi}{2}\right) \cup \left(-\dfrac{\pi}{2}, -\dfrac{\pi}{4}\right)$

13. $(1 - \tan\theta)(1 + \tan\theta)\sec^2\theta + 2^{\tan^2\theta} = 0$

$\Rightarrow (1 - \tan^2\theta)(1 + \tan^2\theta) + 2^{\tan^2\theta} = 0$

$\therefore \quad 2^{\tan^2\theta} = \tan^4\theta - 1$

Put $\quad \tan^2\theta = \lambda, \lambda \in [0, \infty)$

Then, $\quad y = 2^\lambda = \lambda^2 - 1$

$\therefore \quad \lambda = 3$

Then, $\quad \tan^2\theta = 3 = (\sqrt{3})^2 = \tan^2\dfrac{\pi}{3}$

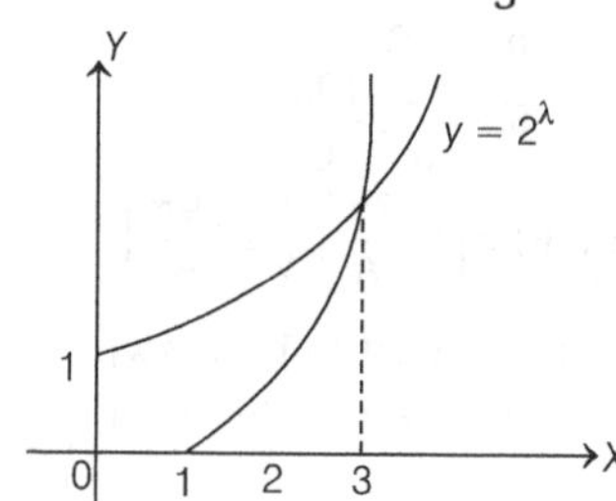

$\therefore \quad \theta = \pm\dfrac{\pi}{3}$

14. $9\cos^{12} x + \cos^2 2x + 1 - 6\cos^6 x \cos 2x$

$\qquad - 6\cos^6 x + 2\cos 2x = 0$

$\Rightarrow \quad (3\cos^6 x - 1 - \cos 2x)^2 = 0$

$\Rightarrow \quad \cos^2 x\,(3\cos^4 x - 2) = 0$

$\Rightarrow \quad \cos x = 0$

$\Rightarrow \quad x = n\pi + \dfrac{\pi}{2}, n \in I$

and $\quad \cos^4 x = \dfrac{2}{3} \Rightarrow \cos^2 x = \pm\sqrt{\dfrac{2}{3}}$

$\therefore \quad \cos x = \pm\sqrt[4]{\left(\dfrac{2}{3}\right)} \Rightarrow x = n\pi \pm \cos^{-1}\sqrt[4]{\left(\dfrac{2}{3}\right)}, n \in I$

15. The given inequality can be written as,

$$2^{\operatorname{cosec}^2 x}\sqrt{(y-1)^2 + 1} \le 2 \qquad \ldots(\text{i})$$

Since, $\operatorname{cosec}^2 x \ge 1$ for all real x, we have

$$2^{\operatorname{cosec}^2 x} \ge 2 \qquad \ldots(\text{ii})$$

Also $\quad (y-1)^2 + 1 \ge 1$

$$\Rightarrow \sqrt{(y-1)^2 + 1} \ge 1 \qquad \ldots(\text{iii})$$

From Eqs. (ii) and (iii), we get

$$2^{\operatorname{cosec}^2 x}\sqrt{(y-1)^2 + 1} \ge 2 \qquad \ldots(\text{iv})$$

Therefore, from Eqs. (i) and (iv), equality holds only when

$$2^{\operatorname{cosec}^2 x} = 2 \text{ and } \sqrt{(y-1)^2 + 1} = 1.$$

Thus, $\quad \operatorname{cosec}^2 x = 1$ and $(y-1)^2 + 1 = 1$

$\Rightarrow \quad \sin x = \pm 1$ and $y = 1$

$\Rightarrow \quad x = \dfrac{\pi}{2}, \dfrac{3\pi}{2}$ and $y = 1$

Hence, the solution of the given inequality is

$$x = \frac{\pi}{2}, \frac{3\pi}{2} \text{ and } y = 1.$$

16. $\cos^2(\pi x) - \sin^2(\pi y) = \frac{1}{2}$

$\Rightarrow \cos\{\pi(x+y)\}\cos\{\pi(x-y)\} = \frac{1}{2}$

$\Rightarrow \cos\{\pi(x+y)\}\cos\left(\frac{\pi}{3}\right) = \frac{1}{2}$ $\qquad \left[\because x - y = \frac{1}{3}\right]$

$\Rightarrow \cos\pi\{x+y)\} = 1$

$\Rightarrow \pi(x+y) = 2n\pi$

$\Rightarrow x + y = 2n, n \in I$

Now, $x + y = 2n$ and $x - y = \frac{1}{3}$

$\Rightarrow x = n + \frac{1}{6}, y = n - \frac{1}{6}$ $\qquad [n \in I]$

$\therefore (x, y) = \left\{n + \frac{1}{6}, n - \frac{1}{6}\right\}$

(a), (c), (d) correspond $n = 1, -1, 2$ respectively.

17. $\left(\cos^2 x + \frac{1}{\cos^2 x}\right)(1 + \tan^2 2y)(3 + \sin 3z) = 4$

$\therefore \cos^2 x + \frac{1}{\cos^2 x} \geq 2$

$1 + \tan^2 2y \geq 1, 2 \leq 3 + \sin 3z \leq 4$

So, the only possibility is

$\cos^2 x + \frac{1}{\cos^2 x} = 2, 1 + \tan^2 2y = 1,$

$3 + \sin 3z = 2$

$\Rightarrow \cos x = \pm 1$

$\Rightarrow x = m\pi$

$\tan 2y = 0$

$\Rightarrow y = \frac{n\pi}{2}$

$\sin 3z = -1$

$\Rightarrow z = (4p - 1)\frac{\pi}{6}; m, n, p \in I$

18. $2\sin^2\left\{\left(\frac{\pi}{2}\right)\cos^2 x\right\} = 1 - \cos(\pi \sin 2x)$

$2\sin^2\left\{\left(\frac{\pi}{2}\right)\cos^2 x\right\} = 2\sin^2\left\{\left(\frac{\pi}{2}\right)\sin 2x\right\}$

$\Rightarrow$ if $\left(\frac{\pi}{2}\right)\cos^2 x = \pm\left(\frac{\pi}{2}\right)\sin 2x$

$\Rightarrow$ either $\cos x = 0$ or $\cos x = \pm 2\sin x$

$\Rightarrow$ if $x = (2n+1)\left(\frac{\pi}{2}\right), n \in I$

$\Rightarrow \tan x = \pm\frac{1}{2}$

19. The given equation can be written as,

$\sqrt{(\cos^2 x - \sin^2 x)} + \sqrt{(\cos x + \sin x)^2}$

$= 2\sqrt{(\cos x + \sin x)}$

$\Rightarrow \sqrt{(\cos x + \sin x)}\sqrt{(\cos x - \sin x)}$

$+ \sqrt{(\cos x + \sin x)}\} = 2\sqrt{(\cos x + \sin x)}$

$\Rightarrow$ Either $\cos x + \sin x = 0$

$\Rightarrow \tan x = -1$

$\Rightarrow x = n\pi - \frac{\pi}{4}\ (n \in I)$

or $\sqrt{(\cos x - \sin x)} + \sqrt{(\cos x + \sin x)} = 2$

$\Rightarrow 2\cos x + 2\sqrt{(\cos^2 x - \sin^2 x)} = 4$

$\Rightarrow \cos^2 x - \sin^2 x = (2 - \cos x)^2$

$\Rightarrow \cos^2 x + 4\cos x - 5 = 0$

$\Rightarrow \cos x = \frac{-4 \pm \sqrt{(16 + 20)}}{2} = -5$ or 1

But $\cos x \neq -5$ or $\cos x = 1$

$\Rightarrow x = 2n\pi, n \in I$

Linked Comprehension Type

PASSAGE 1

1. Let $\cos x - \sin x = t$

$\therefore 1 - 2\sin x\cos x = t^2$

Then, the given equation can be written as

$t^2 = t \Rightarrow t(t - 1) = 0$

$\therefore t = 0, t = 1$

$\Rightarrow \cos x - \sin x = 0, \cos x - \sin x = 1$

$\therefore \tan x = 1, \frac{1}{\sqrt{2}}\cos x - \frac{1}{\sqrt{2}}\sin x = \frac{1}{\sqrt{2}}$

$\Rightarrow \tan x = 1, \cos\left(x + \frac{\pi}{4}\right) = \cos\frac{\pi}{4}$

$\therefore x = n\pi + \frac{\pi}{4}, x + \frac{\pi}{4} = 2n\pi \pm \frac{\pi}{4}$

Hence, $x = 2n\pi, 2n\pi - \frac{\pi}{2}, n\pi + \frac{\pi}{4}, n \in I$

2. Let $\sin x + \cos x = t$

$\therefore 1 + 2\sin x\cos x = t^2$

Then, the given equation can be written as

$t = 1 + \frac{t^2 - 1}{2}$

$\Rightarrow 2t = 2 + t^2 - 1$

$\Rightarrow (t - 1)^2 = 0$ or $t = 1$

$\Rightarrow \sin x + \cos x = 1$

$\Rightarrow \frac{1}{\sqrt{2}}\cos x + \frac{1}{\sqrt{2}}\sin x = \frac{1}{\sqrt{2}}$

$\Rightarrow \cos\left(x - \frac{\pi}{4}\right) = \cos\frac{\pi}{4},$

or $x - \frac{\pi}{4} = 2n\pi \pm \frac{\pi}{4}, n \in I$

$\therefore x = 2n\pi, 2n\pi + \frac{\pi}{2}$

3. $\because \sin^4 x + \cos^4 x = \sin x \cos x$

$\Rightarrow (\sin^2 x + \cos^2 x)^2 - 2\sin^2 x \cos^2 x = \sin x \cos x$

$\Rightarrow \quad 1 - 2\sin^2 x \cos^2 x = \sin x \cos x$

Let $\sin x \cos x = \lambda$, then

$1 - 2\lambda^2 = \lambda$

$\Rightarrow \quad 2\lambda^2 + \lambda - 1 = 0$

$\Rightarrow \quad (\lambda + 1)(2.\lambda - 1) = 0$

$\therefore \quad \lambda = -1, \frac{1}{2}$

$\Rightarrow \quad \sin x \cos x = -1, \frac{1}{2}$

$\Rightarrow \quad \sin 2x = -2, 1$

$\sin 2x \neq -2$

$\therefore \quad \sin 2x = 1$

or $\quad 2x = 2n\pi + \frac{\pi}{2}, n \in I$

$\therefore \quad x = (4n + 1)\frac{\pi}{4}, n \in I$

4. $\because 1 + \sin^3 x + \cos^3 x = \frac{3}{2}\sin 2x$

$\Rightarrow 1 + (\sin x + \cos x)^3 - 3\sin x \cos x (\sin x + \cos x)$

$= \frac{3}{2}\sin 2x \quad \ldots(i)$

Let $\quad \sin x + \cos x = t$

$\therefore \quad \sin x \cos x = \frac{t^2 - 1}{2}$

Then, from Eq. (i),

$1 + t^3 - \frac{3(t^2 - 1)}{2}t = \frac{3}{2}(t^2 - 1)$

$\Rightarrow \quad 2 + 2t^3 - 3t^3 + 3t = 3t^2 - 3$

$\Rightarrow \quad t^3 + 3t^2 - 3t - 5 = 0$

$\Rightarrow \quad (t + 1)(t^2 + 2t - 5) = 0$

$\Rightarrow \therefore \quad t = -1, -1 \pm \sqrt{6}$

$\Rightarrow \quad \sin x + \cos x = -1, \sin x + \cos x \neq -1 \pm \sqrt{6}$

$[\because -\sqrt{2} \le \sin x + \cos x \le \sqrt{2}]$

$\therefore \quad \sin x + \cos x = -1$

$\Rightarrow \quad \frac{1}{\sqrt{2}}\cos x + \frac{1}{\sqrt{2}}\sin x = -\frac{1}{\sqrt{2}}$

$\Rightarrow \quad \cos\left(x - \frac{\pi}{4}\right) = \cos\frac{3\pi}{4}$

$\therefore \quad x - \frac{\pi}{4} = 2n\pi \pm \frac{3\pi}{4}$

or $\quad x = 2n\pi \pm \frac{3\pi}{4} + \frac{\pi}{4}$

$\therefore \quad x = (2n + 1)\pi, (4n - 1)\pi/2, n \in I$

5. $\because (\sin x + \cos x) - 2\sqrt{2}\sin x \cos x = 0 \quad \ldots(i)$

Let $\quad \sin x + \cos x = t$

$\therefore \quad 1 + \sin 2x = t^2 \quad \ldots(ii)$

From Eq. (i), $\quad t - \sqrt{2}(t^2 - 1) = 0$

or $\quad \sqrt{2}t^2 - t - \sqrt{2} = 0$

$\Rightarrow \quad \sqrt{2}t^2 - 2t + t - \sqrt{2} = 0$

$\therefore \quad t = \sqrt{2}, -\frac{1}{\sqrt{2}}$

From Eq. (ii),

$1 + \sin 2x = 2, \frac{1}{2}$

$\therefore \quad \sin 2x = 1, -\frac{1}{2}$

or $\quad 2x = 2n\pi + \frac{\pi}{2}$ and $2x = n\pi + (-1)^n\left(\frac{-\pi}{6}\right)$

or $\quad x = n\pi + \frac{\pi}{4}, x = \frac{n\pi}{2} - \frac{(-1)^n \pi}{12}, n \in I$

Hence, $\quad x = 2n\pi + \frac{\pi}{4}, n \in I$

PASSAGE 2

1. $\because \quad AM \ge GM$

$\therefore \quad 3^x + 3^{-x} \ge 2$

$\Rightarrow \quad 2\cos\left(\frac{x}{2}\right) \ge 2$ or $\cos\left(\frac{x}{2}\right) \ge 1$

$\therefore \quad \cos\left(\frac{x}{2}\right) = 1 \quad \left[\because \cos\frac{x}{2} \text{ is never } > 1\right]$

$\Rightarrow \quad \frac{x}{2} = 2n\pi, n \in I$

$\therefore \quad x = 4n\pi$

Hence (a) corresponding to $n = 0$,

because other values of n do not satisfy the equation.

2. Let $y = 2\cos^2\left(\frac{x}{2}\right)\sin^2 x = x^2 + x^{-2}$

$\therefore \quad y = x^2 + x^{-2} \ge 2 \quad [\because AM \ge GM]$

$\Rightarrow \quad y \ge 2 \quad \ldots(i)$

and $\quad y = 2\cos^2\left(\frac{x}{2}\right)\sin^2 x = (1 + \cos x)\cdot \sin^2 x$

$= (\text{a number} < 2)\cdot(\text{a number} \le 1) < 2$

i.e. $\quad y < 2 \quad \ldots(ii)$

No value of y can be obtained satisfying Eqs. (i) and (ii) simultaneously.

$\Rightarrow$ No real solution of the equation exists.

3. $\because AM \ge GM$

$\therefore \quad 5^x + 5^{-x} \ge 2 \quad [\because \sin e^x = 5^x + 5^{-x}]$

$\Rightarrow \quad \sin e^x \ge 2$

But the value of $\sin(e^x)$ can never be > 1.

Hence, the given equation has no solution.

4. $\because \quad 2^{\operatorname{cosec}^2 x}\cdot\sqrt{\left(\frac{1}{2}y^2 - y + 1\right)} \le \sqrt{2}$

$\Rightarrow \quad 2^{\operatorname{cosec}^2 x}\sqrt{(y^2 - 2y + 2)} \le 2$

$\Rightarrow \quad 2^{\operatorname{cosec}^2 x}\cdot\sqrt{\{(y - 1)^2 + 1\}} \le 2 \quad \ldots(i)$

Since, $\operatorname{cosec}^2 x \ge 1$ for all real x.

$\therefore \quad 2^{\operatorname{cosec}^2 x} \ge 2 \quad \ldots(ii)$

Also, $\quad (y - 1)^2 + 1 \ge 1$

$\Rightarrow \quad \sqrt{(y - 1)^2 + 1} \ge 1 \quad \ldots(iii)$

From Eqs. (ii) and (iii),

$2^{\operatorname{cosec}^2 x}\cdot\sqrt{(y - 1)^2 + 1} \ge 2 \quad \ldots(iv)$

Now, Eqs. (i) and (iv), equality holds only when

$$2^{\text{cosec}^2 x} = 2 \text{ and } \sqrt{\{(y-1)^2+1\}} = 1$$

or $\quad \text{cosec}^2 x = 1 \text{ and } (y-1)^2 + 1 = 1$

$\Rightarrow \quad \sin x = \pm 1 \text{ and } y = 1$

$\Rightarrow \quad x = \pi/2, 3\pi/2 \text{ and } y = 1$

Hence, the solution of the given inequality is

$$(x, y) \equiv \left(\frac{\pi}{2}, 1\right), \left(\frac{3\pi}{2}, 1\right)$$

5. Let $y = \sin x = x^2 + x + 1$ or $y = \sin x$ and $y = x^2 + x + 1$

$\because$ Graphs of both sides of the equation do not intersect, so the equation has no roots.

$\therefore$ Number of solutions is zero.

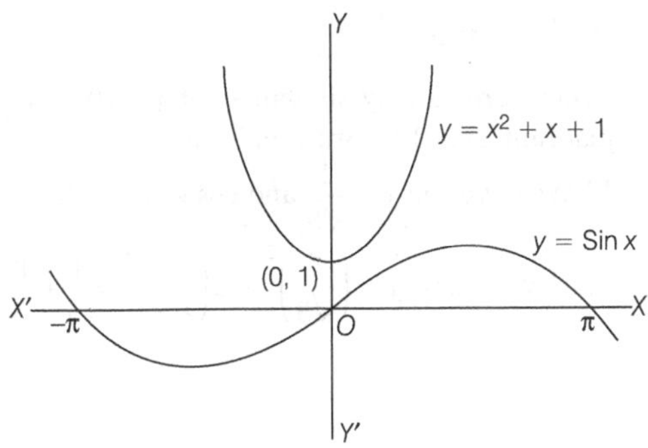

PASSAGE 3

1. Graphs of $y = \sin x$ and $y = \frac{|x|}{10}$ meet exactly six times.

Hence,
Number of solutions = 6

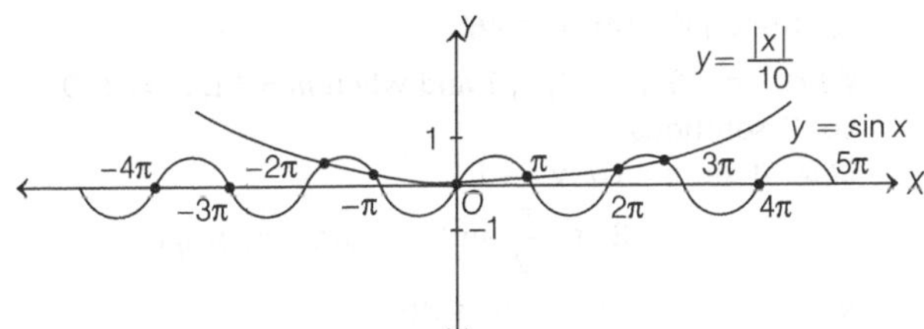

2. $3x + 2\tan x = \frac{5\pi}{2} \Rightarrow \tan x = \frac{5\pi}{4} - \frac{3x}{2}$

Graphs of $y = \frac{5\pi}{4} - \frac{3x}{2}$ and $y = \tan x$, meet exactly three times in $[0, 2\pi]$.

Thus number of solutions = 3.

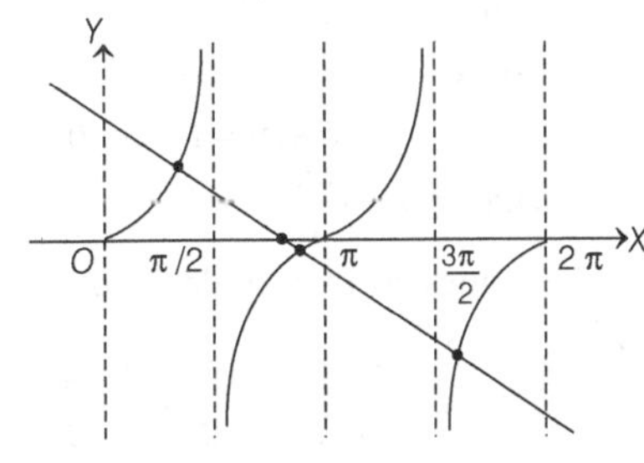

3. $\sin\{x\} = \cos\{x\}$

Graphs of $y = \sin\{x\}$ and $y = \cos\{x\}$ meet exactly 7 times in $[0, 2\pi]$

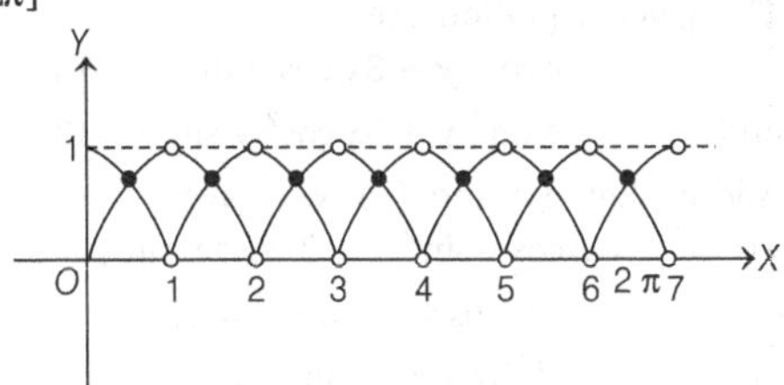

4. Slope of $y = 1 - \sin x$ is $-\cos x$. Slope of $y = \frac{\sqrt{3}}{2}\left|x - \frac{\pi}{2}\right| + a$

for $x > \frac{\pi}{2}$ is $\frac{\sqrt{3}}{2}$

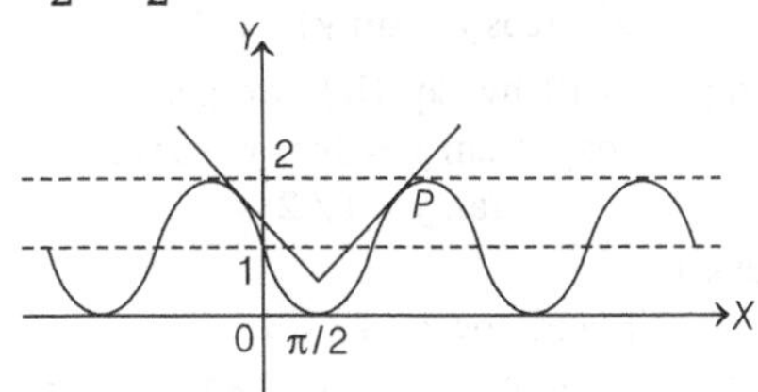

$\Rightarrow \quad \cos x = -\frac{\sqrt{3}}{2}$

$\Rightarrow \quad x = \frac{7\pi}{6}$

$\Rightarrow \quad P \equiv \left(\frac{7\pi}{6}, \frac{3}{2}\right)$

If $1y = \frac{\sqrt{3}}{2}\left|x - \frac{\pi}{2}\right| + a$, passes through '$P$' then

$$a = \frac{3}{2} - \frac{\pi}{\sqrt{3}}$$

$$\Rightarrow a > \frac{3}{2} - \frac{\pi}{\sqrt{3}}$$

5. $\sin 2x = |\sin x|$

$$\cos 2x = \sin x$$

$\Rightarrow \quad 2\sin^2 x + \sin x - 1 = 0$

$\Rightarrow \quad \sin x = -1, 1/2$

But $\sin x \neq -1 \Rightarrow \sin x = \frac{1}{2}$

$\Rightarrow \quad x = \pi/6, 5\pi/6$

Thus solution set is $x \in \left(-\frac{\pi}{6}, \frac{\pi}{6}\right) \cup \left(\frac{5\pi}{6}, \pi\right)$

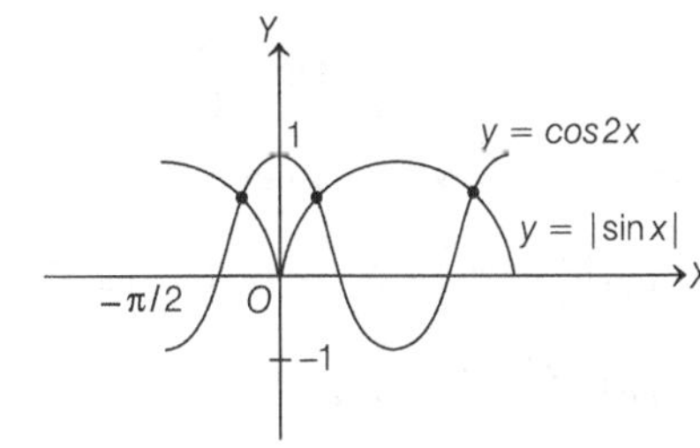

PASSAGE 4

The given equation are,

$$x\cos^3 y + 3x\cos y\sin^2 y = 14 \quad \ldots(i)$$

and $$x\sin^3 y + 3x\cos^2 y\sin y = 13 \quad \ldots(ii)$$

Adding Eqs. (i) and (ii), we have

$$x(\cos^3 y + 3\cos y\sin^2 y + 3\cos^2 y\sin y + \sin^3 y) = 27$$

or $$x(\cos y + \sin y)^3 = 27$$

or $$x^{1/3}(\cos y + \sin y) = 3 \quad \ldots(iii)$$

Subtracting Eq. (ii) from Eq. (i), we have

$$x(\cos^3 y + 3\cos y\sin^2 y - 3\cos^2 y\sin y - \sin^3 y) = 1$$

or $$x(\cos y - \sin y)^3 = 1$$

$\Rightarrow$ $$x^{1/3}(\cos y - \sin y) = 1 \quad \ldots(iv)$$

Dividing Eq. (iii) by Eq. (iv), we get

$$\cos y + \sin y = 3\cos y - 3\sin y$$

or $$\tan y = 1/2$$

1. (a) **Case I**

$\sin y = 1/\sqrt{5}$ and $\cos y = 2/\sqrt{5}$

$y = 2n\pi + \alpha$, where $0 < \alpha < \pi/2$ and $\sin\alpha = 1/\sqrt{5}$

i.e., y lies in the 1st quadrant

From Eqn. (iii), $x^{1/3}(3/\sqrt{5}) = 3$ or $x = 5\sqrt{5}$

Case II

$$\sin y = -1/\sqrt{5}$$

and $$\cos y = -2/\sqrt{5}$$

$y = 2n\pi + (\pi + \alpha)$, where $0 < \alpha < \pi/2$ and $\sin\alpha = -1/\sqrt{5}$

i.e., y lies in the 3rd quadrant

Therefore, from Eq. (iii), $x^{1/3}(-3/\sqrt{5}) = 3$

or $$x = -5\sqrt{5}$$

$\therefore$ $$x = \pm 5\sqrt{5}$$

2. (d) Now, $\tan y = \frac{1}{2}$

$\therefore$ There are exactly six values of $y \in [0, 6\pi]$, 3 in 1st quadrant and 3 in 3rd quadrant.

3. (b) We have, $\sin y = \frac{1}{\sqrt{5}}$ and $\cos y = 2/\sqrt{5}$

$$\therefore \sin^2 y + 2\cos^2 y = \left(\frac{1}{\sqrt{5}}\right)^2 + 2\left(\frac{2}{\sqrt{5}}\right)^2 = \frac{1}{5} + \frac{8}{5} = \frac{9}{5}$$

Numerical Value Type

1. On putting $\cos 2x = t$ we get

$$2t^2 - 1 + 6 = 7t$$

$\Rightarrow$ $$t = 1, \frac{5}{2}$$

$$t = \frac{5}{2} \quad \text{[impossible]}$$

$\therefore$ $$t = 1$$

$\Rightarrow$ $$\cos 2x = 1$$

$\Rightarrow$ $$2x = 2n\pi \Rightarrow x = n\pi, n \in I$$

The roots over $[0, 314]$ are $\pi, 2\pi, 3\pi, \ldots, 99\pi$

$\because$ $$100\pi > 314$$

$\Rightarrow$ Sum of roots $= \pi + 2\pi + 3\pi + \ldots + 99\pi = 4950\pi$

$\Rightarrow$ $$\lambda = 4950$$

2. We can write,

$$r\sin(x + \alpha) = 2y^2 - 8y + 21$$

where, $r = \sqrt{(144 + 25)} = 13$ and $\tan\alpha = \frac{5}{12}$

LHS ≤ 13 and RHS $= 2(y-2)^2 + 13 \geq 13$

Roots of the equation exist, if LHS = RHS = 13

$\Rightarrow$ $$y = 2 \text{ and } \sin\left\{x + \tan^{-1}\left(\frac{5}{12}\right)\right\} = 1$$

$\Rightarrow$ $$x = \frac{\pi}{2} - \tan^{-1}\frac{5}{12}$$

So, that $$xy = \pi - 2\tan^{-1}\frac{5}{12}$$

$$\tan\left(\frac{\pi}{2} - \frac{xy}{2}\right) = \frac{5}{12} \Rightarrow \cot\left(\frac{xy}{2}\right) = \frac{5}{12}$$

$\therefore$ $$\tan\left(\frac{xy}{2}\right) = \frac{12}{5} = 2.4$$

3. $\sin x\cos y = 1 \Leftrightarrow \sin x = 1$ and $\cos y = 1$

or $\sin x = -1$ and $\cos y = -1$

Case I $\sin x = \cos y = 1$

$\Rightarrow$ $$x = 2n\pi + \frac{\pi}{2} \text{ and } y = 2m\pi,\ m, n \in I$$

Now, $$x^2 + y^2 \leq 9\pi^2$$

$\Rightarrow$ $(4n+1)^2 + (4m)^2 \leq 36$

When, $m = 0$, $n = -1, 0, 1$ and when $m = \pm 1$, $n = -1, 0$

So, 7 solutions

Case II $\sin x = \cos y = -1$

$\Rightarrow$ $$x = 2p\pi - \frac{\pi}{2} \text{ and } y = 2q\pi + \pi,\ p, q \in I$$

Now, $(4p-1)^2 + (4q+2)^2 \leq 36$

when $q = -1$, $p = -1, 0, 1$

and when $q = 0$, $p = -1, 0, 1$

So, 6 solutions.

Thus, total $7 + 6 = 13$ pairs are possible

4. $2\cos\theta[\cos 120^\circ + \cos 2\theta] = 1$

$$\Rightarrow 2\cos\theta\left(-\frac{1}{2} + 2\cos^2\theta - 1\right) = 1$$

$\Rightarrow$ $$4\cos^3\theta - 3\cos\theta - 1 = 0$$

$\Rightarrow$ $$\cos 3\theta = 1 = \cos 0$$

$\Rightarrow$ $$3\theta = 2n\pi \text{ or } \theta = \frac{2n\pi}{3}, n \in Z$$

Given the values so that $2n$ does not exceed 18.

$n = 0, 1, 2, 3, \ldots, 9$

Hence, the sum $= \frac{2\pi}{3}\sum_{1}^{9} n = \frac{2\pi}{3} \times \frac{9(9+1)}{2} = 30\pi$

$\therefore$ $$k = 30$$

5. The given equation can be written as,

$$(4\cos^2 x + 1)(\cos x - 1) = 0$$

$\Rightarrow \quad \cos x = 1$

$\Rightarrow x = 2\pi, 4\pi, \ldots 100\pi$

as $100\pi < 315 < 101\pi$

So, the required arithmetic mean

$$= \frac{2(\pi + 2\pi + \ldots + 50\pi)}{50} = \frac{2\times 50\times 51}{2\times 50}\pi = 51\pi$$

$\therefore \quad k = 51$

6. $\tan(\pi\cos\theta) = \tan(\pi/2 - \pi\sin\theta)$

$\Rightarrow \quad \pi\cos\theta = n\pi + \pi/2 - \pi\sin\theta \qquad (n \in I)$

$\Rightarrow \pi(\sin\theta + \cos\theta) = (2n+1)\pi/2$

$\Rightarrow \quad \sin\theta + \cos\theta = \frac{2n+1}{2} \Rightarrow \cos(\pi/4 - \theta) = \frac{2n+1}{2\sqrt{2}}$

Since, $-1 \le \cos(\pi/4 - \theta) \le 1$

$\Rightarrow \quad -1 \le \frac{2n+1}{2\sqrt{2}} \le 1$

$\Rightarrow n = 0$ or -1 as n is an integer

$\Rightarrow \cos(\pi/4 - \theta) = \pm(1/2\sqrt{2})$

$\Rightarrow 8\cos^2(\pi/4 - \theta) = 1 \Rightarrow 19208\cos^2(\pi/4 - \theta) = 2401$

Single Integer type

1. Given equation,

$$\log_{1/2}|\sin x| = 2 - \log_{1/2}|\cos x|$$

$\Rightarrow \quad -\log_2|\sin x| = 2 + \log_2|\cos x|$

$\Rightarrow \quad \log_2|\sin x| + \log_2|\cos x| + \log_2 4 = 0$

$\Rightarrow \quad \log_2(4|\sin x||\cos x|) = 0$

$\Rightarrow \quad 4|\sin x||\cos x| = 1$

$\Rightarrow \quad \sin 2x = \pm\frac{1}{2}$

$\because \quad x \in [0, 2\pi]$

$\Rightarrow \quad 2x \in [0, \pi]$

$\therefore$ Total number of solutions are 8.

2. $3\cos^2\theta - 2\sqrt{3}\sin\theta\cos\theta - 3\sin^2\theta = 0$

$\Rightarrow (\sqrt{3}\cos\theta + \sin\theta)(\cos\theta - \sqrt{3}\sin\theta) = 0$

$\Rightarrow \quad \tan\theta = 1/\sqrt{3}$

or $\quad \tan\theta = -\sqrt{3}$

$\therefore$ Solution are $n\pi + \left(\frac{\pi}{6}\right)$ and $n\pi - \frac{\pi}{3}$

$\therefore \quad |r - s| = |-3 - 6| = 9$

3. The given equation can be written as,

$$\frac{3}{\cos\theta} - 5 = \frac{4\sin\theta}{\cos\theta}$$

$\Rightarrow 4\sin\theta + 5\cos\theta = 3, \cos\theta \ne 0$

$\Rightarrow \quad \frac{4}{\sqrt{41}}\sin\theta + \frac{5}{\sqrt{41}}\cos\theta = \frac{3}{\sqrt{41}} \qquad \ldots(i)$

Let $\quad \frac{5}{\sqrt{41}} = \cos\alpha,$

then $\quad \frac{4}{\sqrt{41}} = \sin\alpha$

Now, from Eq. (i),

$$\cos(\theta - \alpha) = \frac{3}{\sqrt{41}} = \cos\beta$$

$\therefore \quad \theta - \alpha = 2n\pi \pm \beta$, where $\cos\beta = \frac{3}{\sqrt{41}}$

or $\quad \theta = 2n\pi + \alpha + \beta$

or $\quad \theta = 2n\pi + \alpha - \beta$

Now, in $[0, 2\pi]$, $\theta = 2n\pi + (\alpha + \beta)$ gives two solutions.

So, in total in $[0, 4\pi]$

we have, $2 + 2 + 2 + 2 = 8$ solutions

4. $(\sin 3x + \sin x) + \sin 2x = (\cos 3x + \cos x) + \cos 2x$

or $2\sin 2x\cos x + \sin 2x = 2\cos 2x\cos x + \cos 2x$

$\sin 2x(2\cos x + 1) - \cos 2x(2\cos x + 1) = 0$

or $\quad (2\cos x + 1)(\sin 2x - \cos 2x) = 0$

or $\quad \cos x = -\frac{1}{2}, \tan 2x = 1$

$\therefore \quad x = \frac{2\pi}{3}, \frac{4\pi}{3}$ or $\frac{\pi}{8}, \frac{5\pi}{8}$

5. Given, $\sec\theta + \text{cosec}\,\theta = c$

$\Rightarrow \sqrt{(1 + \tan^2\theta)} + \sqrt{(1 + \cot^2\theta)} = c$

$\Rightarrow \quad \sqrt{(1+\lambda^2)} + \sqrt{\left(1 + \frac{1}{\lambda^2}\right)} = c$, where $\tan\theta = \lambda$

$\Rightarrow \quad \sqrt{(1+\lambda^2)}\left(\frac{1+\lambda}{\lambda}\right) = c$

$\Rightarrow \quad (1+\lambda^2)^2 + 2\lambda(1+\lambda^2) + \lambda^2 = (c^2+1)\lambda^2$

$\Rightarrow \quad (\lambda^2 + \lambda + 1)^2 = (c^2+1)\lambda^2$

$\Rightarrow \quad \lambda^2 + \lambda + 1 \pm \lambda\sqrt{(c^2+1)} = 0$

$\Rightarrow \quad \lambda^2 + \lambda + 1 + \lambda\sqrt{(c^2+1)} = 0 \qquad \ldots(i)$

or $\quad \lambda^2 + \lambda + 1 - \lambda\sqrt{(c^2+1)} = 0 \qquad \ldots(ii)$

Discriminant of Eq. (i) is

$$D_1 = \{1 + \sqrt{(c^2+1)}\}^2 - 4$$

Discriminant of Eq. (ii) is

$$D_2 = \{(1 - \sqrt{(c^2+1)}\}^2 - 4$$

Now, $D_1 = \{3 + \sqrt{(c^2+1)}\}\{\sqrt{(c^2+1)} - 1\} > 0$

$\therefore$ The equation always has two real roots. It will have only two roots provided

$$D_2 < 0$$

$\Rightarrow \quad 1 + c^2 + 1 - 2\sqrt{(c^2+1)} - 4 < 0$

$\Rightarrow \quad c^2 - 2 < 2\sqrt{(c^2+1)}$

It is always true, if $c^2 \le 2$. If $c^2 > 2$, then on squaring we get

$$c^4 - 4c^2 + 4 < 4c^2 + 4$$

$\Rightarrow \quad c^2(c^2 - 8) < 0$

$\Rightarrow \quad 0 < c^2 < 8 \Rightarrow c^2 < 8$

6. The given equation is not defined for

$$x = m\pi,\ x = \frac{m\pi}{2},\ x = \frac{m\pi}{4}$$

$$\Rightarrow \quad x = \frac{m\pi}{4},\ m \in I$$

Thus, the solution chain $(2n+1)\frac{\pi}{7}$ must not become $\frac{m\pi}{4}$ for integers n and m

Now, let $(2n+1)\frac{\pi}{7} = \frac{m\pi}{4} \Rightarrow 8n+4 = 7m$

Since, $\quad n = 7p+0, 7p+1, 7p+2, \ldots, 7p+6$
Hence, $\quad n = 7p+3$ only

$[\because 8(7p+3)+4 = 7p+28 = a$ multiple of 7]

Thus, $\lambda = 3$

7. The given equation possible, if $\sin(1-x) \geq 0$ and $\cos x \geq 0$ from the given equation, on squaring both the sides, we get

$$\sin(1-x) = \cos x$$

$$\Rightarrow \quad \cos x = \cos\left\{\frac{\pi}{2} - (1-x)\right\}$$

$$\Rightarrow \quad x = 2n\pi \pm \left(\frac{\pi}{2} - 1 + x\right),\ n \in I$$

Taking –ve sign

$$x = 2n\pi - \frac{\pi}{2} + 1 - x$$

$$x = n\pi - \frac{\pi}{4} + \frac{1}{2}$$

for $\quad n = 2$

$x = \frac{1}{2} + \frac{7\pi}{4}$ which is the smallest +ve root of the given equation.

So, $\alpha = \frac{1}{2} + \frac{7\pi}{4} = \frac{1}{2} + \frac{7}{4} \times \frac{22}{7} = 6$

8. We have

$$\sqrt{3}\cos^2 x = (\sqrt{3}-1)\cos x + 1$$

$$\Rightarrow \quad \sqrt{3}\cos^2 x - (\sqrt{3}-1)\cos x - 1 = 0$$

$$\sqrt{3}y^2 - (\sqrt{3}-1)y - 1 = 0 \qquad [\because \cos y = \cos x]$$

$$y = \frac{(\sqrt{3}-1) \pm \sqrt{(\sqrt{3}-1)^2 + 4\sqrt{3}}}{2\sqrt{3}}$$

$$\Rightarrow \quad y = \frac{(\sqrt{3}-1) \pm \sqrt{4+2\sqrt{3}}}{2\sqrt{3}}$$

$$\Rightarrow \quad y = \frac{(\sqrt{3}-1) \pm \sqrt{(\sqrt{3}+1)^2}}{2\sqrt{3}}$$

$$\Rightarrow \quad y = \frac{(\sqrt{3}-1) \pm (\sqrt{3}+1)}{2\sqrt{3}}$$

$$\Rightarrow \quad y = 1 \text{ or } -\frac{1}{\sqrt{3}}$$

$$\because \quad \cos x = 1 \text{ or } -\frac{1}{\sqrt{3}}$$

Here $\quad x \in \left[0, \frac{\pi}{2}\right)$

$\because \quad \cos x = 1$

$\therefore$ Number of solution $= 1$

Matrix-Match Type

1. **(A)** $\because \sin x = -\frac{1}{2} = -\sin\frac{\pi}{6}$

$$= \sin\left(\pi + \frac{\pi}{6}\right), \sin\left(2\pi - \frac{\pi}{6}\right)$$

$$\therefore \quad x = \frac{7\pi}{6}, \frac{11\pi}{6} \qquad \ldots(\text{i})$$

and $\cos x = -\frac{\sqrt{3}}{2} = -\cos\frac{\pi}{6}$

$$= \cos\left(\pi - \frac{\pi}{6}\right), \cos\left(\pi + \frac{\pi}{6}\right)$$

$$\therefore \quad x = \frac{5\pi}{6}, \frac{7\pi}{6} \qquad \ldots(\text{i})$$

From Eqs. (i) and (ii). It is clear that

$$\alpha = \frac{7\pi}{6}, \beta = \frac{11\pi}{6}, \gamma = \frac{5\pi}{6}$$

$$\Rightarrow \quad \alpha + \beta = 3\pi\ (\text{S}),\ \beta - \gamma = \pi\ (\text{Q})$$

(B) $\because \cot x = -\sqrt{3} = -\cot\frac{\pi}{6}$

$$= \cot\left(\pi - \frac{\pi}{6}\right), \cot\left(2\pi - \frac{\pi}{6}\right)$$

$$\therefore \quad x = \frac{5\pi}{6}, \frac{11\pi}{6} \qquad \ldots(\text{i})$$

and $\operatorname{cosec} x = -2 = -\operatorname{cosec}\frac{\pi}{6}$

$$= \operatorname{cosec}\left(\pi + \frac{\pi}{6}\right), \operatorname{cosec}\left(2\pi - \frac{\pi}{6}\right)$$

$$\therefore \quad x = \frac{7\pi}{6}, \frac{11\pi}{6} \qquad \ldots(\text{ii})$$

From Eqs. (i) and (ii), It is clear that

$$\alpha = \frac{11\pi}{6}, \beta = \frac{5\pi}{6}, \gamma = \frac{7\pi}{6}$$

$$\Rightarrow \quad \beta + \gamma = 2\pi\ (\text{T}),\ \alpha - \beta = \pi\ (\text{P})$$

(C) $\because \sin x = -\frac{1}{2}$

$$= -\sin\frac{\pi}{6}$$

$$= \sin\left(\pi + \frac{\pi}{6}\right), \sin\left(2\pi - \frac{\pi}{6}\right)$$

$$\therefore \quad x = \frac{7\pi}{6}, \frac{11\pi}{6} \qquad \ldots(\text{i})$$

and $\tan x = \frac{1}{\sqrt{3}} = \tan\frac{\pi}{6}$

$$= \tan\frac{\pi}{6}, \tan\left(\pi + \frac{\pi}{6}\right)$$

$\therefore \quad x = \frac{\pi}{6}, \frac{7\pi}{6}$ ...(ii)

From Eqs. (i) and (ii), it is clear that

$$\alpha = \frac{7\pi}{6}, \beta = \frac{11\pi}{6}, \gamma = \frac{\pi}{6}$$

$\Rightarrow \quad \alpha + \beta = 3\pi$, (S), $\beta + \gamma = 2\pi$ (T), $\alpha - \gamma = \pi$ (R)

2. (A)

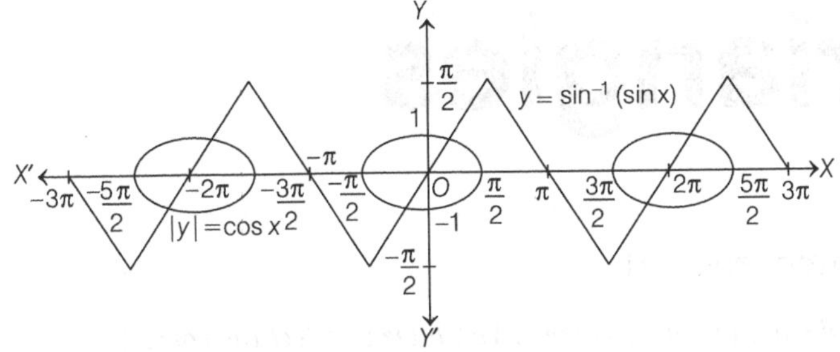

Case I If $|x| \le 2\pi \Rightarrow -2\pi \le x \le 2\pi$

Both graphs cuts at four points

$\therefore \quad m = 4$

Case II If $|x| \le 3\pi \Rightarrow -3\pi \le x \le 3\pi$

Both graphs cuts at six points

$\therefore \quad n = 6$

$\Rightarrow m + n = 10$, (R), $n - m = 2$ (T)

(B)

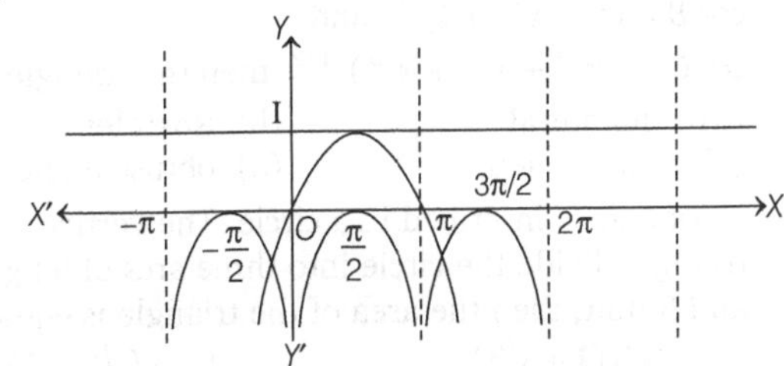

Graphs of $y = \ln|\sin x|$ and $y = -x^2 + 2x$ meet no any point in $[0, \pi]$

$\therefore \; m = 0$

and meet at two points in $\left[-\frac{\pi}{2}, \frac{3\pi}{2}\right]$

$\therefore \; n = 2$

$\Rightarrow m + n = 2$ (Q), $n - m = 2$ (T)

(C) $\because [\sin x] = \cos x$

$\Rightarrow \quad -1 \le \cos x \le 1$

$\therefore \quad [\sin x] = \cos x$

$\Rightarrow \quad \cos x = -1, 0, 1$

$\Rightarrow \quad x = \pi, \pi/2, 2\pi$

$\Rightarrow \quad [\sin x] = 0, 1, 0$

Thus, $[\sin x] = \cos x$ has no solution.

$\therefore \quad m = 0, n = 0$

$\Rightarrow \quad m + n = 0$ (P), $n - m = 0$ (S)

3. (A) $5\sin\theta + 3(\sin\theta\cos\alpha - \cos\theta\sin\alpha)$

$$= (5 + 3\cos\alpha)\sin\theta - 3\sin\alpha\cos\theta$$

$\Rightarrow \max_{\theta \in R} \{5\sin\theta + 3\sin(\theta - \alpha)\}$

$$= \sqrt{(5 + 3\cos\alpha)^2 + 9\sin^2\alpha}$$

$$= \sqrt{34 + 30\cos\alpha}$$

Therefore, the given equation is

$$34 + 30\cos\alpha = 49$$

or $\quad \cos\alpha = 1/2$

$\Rightarrow \alpha = 2n\pi \pm \pi/3, n \in Z$ (Q)

(B) $(\cos x)^{\sin^2 x - 3\sin x + 2} = 1$

$\Rightarrow \quad \sin^2 x - 3\sin x + 2 = 0$

$\Rightarrow \quad (\sin x - 2)(\sin x - 1) = 0$

$\Rightarrow \quad \sin x = 1$

But this does not satisfy the equation because $0^0 = 1$ is absurd. (S)

(C) $\sqrt{(\sin x)} + 2^{1/4}\cos x = 0$ ...(i)

$\therefore \sqrt{(\sin x)} > 0$ and so $\cos x < 0$,

Also, $\sin x > 0 \Rightarrow x$ lies in the 2nd quadrant

Eq. (i) can be rewritten as,

$$2^{1/4}\cos x = -\sqrt{(\sin x)}$$

On squaring, we get

$$\sqrt{2}\cos^2 x = \sin x$$

or $\sqrt{2}\sin^2 x + \sin x - \sqrt{2} = 0$

or $(\sqrt{2}\sin x - 1)(\sin x + \sqrt{2}) = 0$

$\sin x \ne -\sqrt{2}$, therefore $\sin x = 1/\sqrt{2}$ ...(ii)

$\Rightarrow x = 3\pi/4$ and so the general value of x is given by

$x = 2n\pi + 3\pi/4, n \in Z$ (P)

(D) $\log_5 \tan x = (\log_5 4)(\log_4(3\sin x))$

$\Rightarrow \quad \log_5 \tan x = \log_5(3\sin x)$ ...(i)

Since, $\log x$ is real when $x > 0$, we have

$$\tan x > 0, \sin x > 0$$

Therefore, x lies in the 1st quadrant. Now Eq. (i) gives

$$\tan x = 3\sin x \text{ or } \cos x = 1/3$$

$\therefore \quad x = 2n\pi + \cos^{-1}(1/3), n \in Z$ (R)

29

Solution of Triangles

Objective Questions Type I [Only one correct answer]

In each of the questions below, four choices are given of which only one is correct. You have to select the correct answer which is the most appropriate.

1. In a $\Delta\, ABC$ bisector of angle C meets the side AB at D and circumcircle at E. The maximum value of $CD \cdot DE$ is equal to

(a) $\frac{a^2}{4}$ (b) $\frac{b^2}{4}$

(c) $\frac{c^2}{4}$ (d) $\frac{(a+b)^2}{4}$

2. In a triangle ABC, $r^2 + r_1^2 + r_2^2 + r_3^2 + a^2 + b^2 + c^2$ is equal to (where r is inradius and r_1, r_2, r_3 are exradii a, b, c are the sides of ΔABC)

(a) $2R^2$ (b) $4R^2$

(c) $8R^2$ (d) $16R^2$

3. In any ΔABC, minimum value of $\sum \frac{\sin^2 A + \sin A + 1}{\sin A} =$

(a) 9 (b) 6

(c) 8 (d) Does not exist

4. The cosine of the obtuse angle formed by medians drawn from the vertices of the acute angles of an isosceles right angled triangle is

(a) $-\frac{1}{5}$ (b) $-\frac{2}{5}$

(c) $-\frac{3}{5}$ (d) $-\frac{4}{5}$

5. Sides of a triangle ABC are in AP. If $a < \min\{b, c\}$, then $\cos A$ may be equal to

(a) $\frac{3c-4b}{2b}$ (b) $\frac{3c-4b}{2c}$

(c) $\frac{4c-3b}{2b}$ (d) $\frac{4c-3b}{2c}$

6. If triangle be the area of a ΔABC and length of its two sides are 3 and 5. If c is the third side, then

(a) $\Delta \le \frac{(c^2+16c+64)}{12\sqrt{3}}$ (b) $\Delta = \frac{(c^2+16c+64)}{8\sqrt{3}}$

(c) $\Delta > \frac{(c^2+16c+64)}{4\sqrt{3}}$ (d) None of these

7. In a $\Delta\, ABC$, angles A, B, C are in AP. Then

$$\lim_{x \to C} \frac{\sqrt{(3 - 4\sin A \sin C)}}{|A - C|} \text{ is}$$

(a) 1 (b) 2

(c) 3 (d) 4

8. In a triangle ABC, if $\cot A = (x^3 + x^2 + x)^{1/2}$, $\cot B = (x + x^{-1} + 1)^{1/2}$ and $\cot C = (x^{-3} + x^{-2} + x^{-1})^{-1/2}$, then the triangle is

(a) equilateral (b) isosceles

(c) right angled (d) obtuse angled

9. A triangle is inscribed in a circle. The vertices of the triangle divide the circle into three arcs of length 3, 4 and 5 unit, then the area of the triangle is equal to

(a) $\frac{9\sqrt{3}(1+\sqrt{3})}{\pi^2}$ sq unit (b) $\frac{9\sqrt{3}(\sqrt{3}-1)}{\pi^2}$ sq unit

(c) $\frac{9\sqrt{3}(1+\sqrt{3})}{2\pi^2}$ sq unit (d) $\frac{9\sqrt{3}(\sqrt{3}-1)}{2\pi^2}$ sq unit

10. If the sine of the angles of $\Delta\, ABC$ satisfy the equation $c^3x^3 - c^2(a+b+c)x^2 + lx + m = 0$ (where a, b, c are the sides of $\Delta\, ABC$), then $\Delta\, ABC$ is

(a) always right angled for any l, m

(b) right angled only when $l = c(ab + bc + ca) = c\Sigma ab,\ m = -abc$

(c) right angled only when $l = \frac{c\Sigma ab}{4},\ m = -\frac{abc}{8}$

(d) never right angled

11. If in a $\Delta\, ABC$, $a^4 + b^4 + c^4 = 2a^2b^2 + b^2c^2 + 2c^2a^2$, then $\sin A$ is equal to

(a) $\frac{1}{\sqrt{2}}$ (b) $\frac{1}{2}$

(c) $\frac{\sqrt{3}}{2}$ (d) $\frac{\sqrt{3}+1}{2\sqrt{2}}$

12. If r and R are respectively the radii of the inscribed and circumscribed circles of a regular polygon of n sides such that $\frac{R}{r} = \sqrt{5} - 1$, then n is equal to
(a) 5 (b) 10
(c) 6 (d) 18

13. A point P is inside an equilateral triangle. If the distances of P from the sides are 4, 5, 6 units, the length of the side of the triangle is
(a) $5\sqrt{3}$ (b) $10\sqrt{3}$
(c) $15\sqrt{3}$ (d) $20\sqrt{3}$

14. Given an isosceles triangle, whose one angle is 120° and radius of its incircle = $\sqrt{3}$. Then the area of the triangle in square unit is
(a) $(7 + 12\sqrt{3})$ (b) $(12 - 7\sqrt{3})$
(c) $(12 + 7\sqrt{3})$ (d) 4π

15. Let a, b, c be the sides of a triangle. No two of them are equal and $\lambda \in R$. If the roots of the equation
$$x^2 + 2(a + b + c)x + 3\lambda (ab + bc + ca) = 0$$
are real and distinct, then
(a) $\lambda < \frac{4}{3}$ (b) $\lambda > \frac{5}{3}$
(c) $\lambda \in \left(\frac{1}{3}, \frac{5}{3}\right)$ (d) $\lambda \in \left(\frac{4}{3}, \frac{5}{3}\right)$

16. Three circular coins each of radii 1 cm are kept in an equilateral triangle so that all the three coins touch each other and also the sides of the triangle. Area of the triangle is
(a) $(4 + 2\sqrt{3})\text{ cm}^2$ (b) $\frac{1}{4}(12 + 7\sqrt{3})\text{ cm}^2$
(c) $\frac{1}{4}(48 + 7\sqrt{3})\text{ cm}^2$ (d) $(6 + 4\sqrt{3})\text{ cm}^2$

17. With the usual notation, in ΔABC, if $\angle A + \angle B = 120°$, $a = \sqrt{3} + 1$ and $b = \sqrt{3} - 1$, then the ratio $\angle A : \angle B$, is **[JEE Main 2019, 10 January]**
(a) 7 : 1 (b) 3 : 1
(c) 9 : 7 (d) 5 : 3

18. In a ΔABC, $2a^2 + 4b^2 + c^2 = 4ab + 2ac$, then the numerical value of $\cos B$ is equal to
(a) 0 (b) $\frac{3}{8}$
(c) $\frac{5}{8}$ (d) $\frac{7}{8}$

19. In a right angled ΔABC of maximum area is inscribed within a circle of radius R, then (where Δ is area and s is semi-perimeter of ΔABC)
(a) $\Delta = 2R^2$
(b) $\frac{1}{r_1} + \frac{1}{r_2} + \frac{1}{r_3} = \frac{\sqrt{2}+1}{R}$
(c) $r = (\sqrt{2} - 1)R$
(d) $s = (\sqrt{2} + 2)R$

20. If G is the centroid of a ΔABC, then $GA^2 + GB^2 + GC^2$ is equal to
(a) $(a^2 + b^2 + c^2)$ (b) $\frac{1}{3}(a^2 + b^2 + c^2)$
(c) $\frac{1}{2}(a^2 + b^2 + c^2)$ (d) $\frac{1}{3}(a + b + c)^2$

21. In an ambiguous case of solving a triangle when $a = \sqrt{5}, b = 2, \angle A = \frac{\pi}{6}$ and the two possible values of third side are c_1 and c_2, then
(a) $|c_1 - c_2| = 2\sqrt{6}$ (b) $|c_1 - c_2| = 4\sqrt{6}$
(c) $|c_1 - c_2| = 4$ (d) $|c_1 - c_2| = 6$

22. In a ΔABC, sides a, b, c are in AP and
$\frac{2}{1!9!} + \frac{2}{3!7!} + \frac{1}{5!5!} = \frac{8^a}{(2b)!}$, then the maximum value of $\tan A \tan B$ is equal to
(a) $\frac{1}{2}$ (b) $\frac{1}{3}$ (c) $\frac{1}{4}$ (d) $\frac{1}{5}$

23. If a, b, c be the sides of a triangle ABC and if roots of the equation $a(b - c)x^2 + b(c - a)x + c(a - b) = 0$ are equal, then $\sin^2\left(\frac{A}{2}\right), \sin^2\left(\frac{B}{2}\right), \sin^2\left(\frac{C}{2}\right)$ are in
(a) AP (b) GP
(c) HP (d) AGP

24. In an isosceles triangle ABC, $AB = AC$. If vertical angle A is 20°, then $a^3 + b^3$ is equal to
(a) $3a^2b$ (b) $3b^2c$
(c) $3c^2a$ (d) abc

25. Which of the following pieces of data does not uniquely determine acute angled ΔABC (R = circum radius)
(a) a, $\sin A$, $\sin B$ (b) a, b, c
(c) a, $\sin B$, R (d) a, $\sin A$, R

26. A quadrilateral $ABCD$ in which $AB = a$, $BC = b$, $CD = c$ and $DA = d$ is such that one circle can be inscribed in it and another circle circumscribed about it, then $\cos A$ is equal to
(a) $\frac{ad + bc}{ad - dc}$ (b) $\frac{ad - bc}{ad + bc}$
(c) $\frac{ac + bd}{ac - bd}$ (d) $\frac{ac - bd}{ac + bd}$

27. If A, B, C, D are the angles of quadrilateral, then $\frac{\Sigma \tan A}{\Sigma \cot A}$ is equal to
(a) $\Pi \tan A$ (b) $\Pi \cot A$
(c) $\Sigma \tan^2 A$ (d) $\Sigma \cot^2 A$

28. If in ΔABC, $A = \frac{\pi}{7}, B = \frac{2\pi}{7}, C = \frac{4\pi}{7}$ then $a^2 + b^2 + c^2$ must be
(a) R^2 (b) $3R^2$
(c) $4R^2$ (d) $7R^2$

29. In ΔABC, AD and BE are medians drawn from A and B. If $AD = \frac{7}{2}$, $\angle DAB = \frac{\pi}{8}$ and $\angle ABE = \frac{\pi}{4}$, then the area (in sq. units) of ΔABC is

(a) $\frac{7}{12}$ (b) $\frac{49}{36}$
(c) $\frac{49}{12}$ (d) $\frac{7}{36}$

30. If the area of a triangle ABC is given by $\Delta = a^2 - (b-c)^2$, then $\tan\left(\frac{A}{2}\right)$ is equal to

(a) -1 (b) 0
(c) $\frac{1}{4}$ (d) $\frac{1}{2}$

31. The perimeter of a triangle ABC is 6 times the arithmetic mean of the sines of its angles. If the side a is 1, then $\angle A$ is

(a) 30° (b) 60°
(c) 90° (d) 120°

32. $a^3 \cos(B-C) + b^3 \cos(C-A) + c^3 \cos(A-B)$ is equal to

(a) $3abc$ (b) $3(a+b+c)$
(c) $abc(a+b+c)$ (d) 0

33. If $\frac{\cos A}{a} = \frac{\cos B}{b} = \frac{\cos C}{c}$ and the side $a = 2$, then area of triangle is

(a) 1 sq units (b) 2 sq units
(c) $\frac{\sqrt{3}}{2}$ sq units (d) $\sqrt{3}$ sq units

34. If in a ΔABC, $\cos A + 2\cos B + \cos C = 2$, then a, b, c are in

(a) AP (b) GP
(c) HP (d) None of these

35. In a ΔABC, $\angle B = \frac{\pi}{3}$ and $\angle C = \frac{\pi}{4}$, let D divide BC internally in the ratio 1 : 3. Then, $\frac{\sin(\angle BAD)}{\sin(\angle CAD)}$ is equal to

(a) $\frac{1}{\sqrt{6}}$ (b) $\frac{1}{3}$ (c) $\frac{1}{\sqrt{3}}$ (d) $\sqrt{\frac{2}{3}}$

36. If D is the mid-point of side BC of a ΔABC and AD is perpendicular to AC, then

(a) $3a^2 = b^2 - 3c^2$ (b) $3b^2 = a^2 - c^2$
(c) $b^2 = a^2 - c^2$ (d) $a^2 + b^2 = 5c^2$

37. If f, g, h are the internal bisectors of a ΔABC, then $\frac{1}{f}\cos\left(\frac{A}{2}\right) + \frac{1}{g}\cos\left(\frac{B}{2}\right) + \frac{1}{h}\cos\left(\frac{C}{2}\right)$ is equal to

(a) $\frac{1}{a} + \frac{1}{b} - \frac{1}{c}$ (b) $\frac{1}{a} - \frac{1}{b} + \frac{1}{c}$
(c) $\frac{1}{a} + \frac{1}{b} + \frac{1}{c}$ (d) None of these

38. If a, b, c, d be the sides of a quadrilateral and $g(x) = f[f\{f(x)\}]$, where $f(x) = \frac{1}{1-x}$, then $\frac{d^2}{a^2 + b^2 + c^2}$ is equal to

(a) $> g(3)$ (b) $< g(3)$
(c) $> g(2)$ (d) $< g(4)$

39. If in a ΔABC,

$$\sin^3 A + \sin^3 B + \sin^3 C = 3 \sin A \cdot \sin B \cdot \sin C,$$

then the value of the determinant $\begin{vmatrix} a & b & c \\ b & c & a \\ c & a & b \end{vmatrix}$ is equal to

(a) 0 (b) $(a+b+c)^3$
(c) $(a+b+c)(ab+bc+ca)$ (d) None of these

40. In a ΔABC, if $r = r_2 + r_3 - r_1$ and $\angle A > \frac{\pi}{3}$, then the range of $\frac{s}{a}$ is equal to

(a) $\left(\frac{1}{2}, 2\right)$ (b) $\left(\frac{1}{2}, \infty\right)$
(c) $\left(\frac{1}{2}, 3\right)$ (d) $(3, \infty)$

41. If the bisector of angle A of triangle ABC makes an angle θ with BC, then $\sin\theta$ is equal to

(a) $\cos\left(\frac{B-C}{2}\right)$ (b) $\sin\left(\frac{B-C}{2}\right)$
(c) $\sin\left(B - \frac{A}{2}\right)$ (d) $\sin\left(C - \frac{A}{2}\right)$

42. With usual notations, if in a ΔABC, $\frac{b+c}{11} = \frac{c+a}{12} = \frac{a+b}{13}$, then $\cos A : \cos B : \cos C$ is equal to

(a) 7 : 19 : 25 (b) 19 : 7 : 25
(c) 12 : 14 : 20 (d) 19 : 25 : 20

43. In a triangle, the sum of lengths of two sides is x and the product of the lengths of the same two sides is y. If $x^2 - c^2 = y$, where c is the length of the third side of the triangle, then the circumradius of the triangle is **[JEE Main 2019, 11 January]**

(a) $\frac{c}{3}$ (b) $\frac{c}{\sqrt{3}}$
(c) $\frac{3}{2}y$ (d) $\frac{y}{\sqrt{3}}$

44. If in a ΔABC, $\cos A \cos B + \sin A \sin B \sin C = 1$, then the sides are proportional to

(a) $1:1:\sqrt{2}$ (b) $1:\sqrt{2}:1$
(c) $\sqrt{2}:1:1$ (d) None of these

45. In an equilateral triangle, $R : r : r_2$ is equal to

(a) 1 : 1 : 1 (b) 1 : 2 : 3
(c) 2 : 1 : 3 (d) 3 : 2 : 4

46. If in a $\Delta\, ABC$, $a^2 + b^2 + c^2 = 8R^2$, where R = circumradius, then the triangle is

(a) equilateral (b) isosceles
(c) right angled (d) None of these

47. Let $A_0A_1A_2A_3A_4A_5$ be a regular hexagon inscribed in a circle of unit radius. The product of the length of the line segments A_0A_1, A_0A_2 and A_0A_4 is

(a) $\frac{3}{4}$ (b) $3\sqrt{3}$ (c) 3 (d) $\frac{3\sqrt{3}}{2}$

48. If α, β, γ are length of the altitudes of a ΔABC, then $\frac{1}{\alpha^2} + \frac{1}{\beta^2} + \frac{1}{\gamma^2}$ is equal to

(a) $\frac{\cot A + \cot B + \cot C}{\Delta}$ (b) $\frac{\Delta}{\cot A + \cot B + \cot C}$
(c) $\Delta(\cot A + \cot B + \cot C)$ (d) None of these

49. In a $\Delta\, ABC$, the value of $\frac{a\cos A + b\cos B + c\cos C}{a + b + c}$ is equal to

(a) $\frac{R}{r}$ (b) $\frac{R}{2r}$
(c) $\frac{r}{R}$ (d) $\frac{2r}{R}$

50. In a $\Delta\, ABC$, the sides a, b, c are the roots of the equation $x^3 - 11x^2 + 38x - 40 = 0$. Then, $\frac{\cos A}{a} + \frac{\cos B}{b} + \frac{\cos C}{c}$ is equal to

(a) 1 (b) $\frac{3}{4}$
(c) $\frac{9}{16}$ (d) None of these

51. If the base angles of a triangle are $22\frac{1}{2}°$ and $112\frac{1}{2}°$, then height of the triangle is equal to

(a) half the base (b) the base
(c) twice the base (d) four times the base

52. In ΔABC, the median AD divides $\angle BAC$ such that $\angle BAD : \angle CAD = 2:1$. Then $\cos\left(\frac{A}{3}\right)$ is equal to

(a) $\frac{\sin B}{2\sin C}$ (b) $\frac{\sin C}{2\sin B}$
(c) $\frac{2\sin B}{\sin C}$ (d) None of these

53. The area of the circle and the area of a regular polygon inscribed the circle of n sides and of perimeter equal to that of the circle are in the ratio of

(a) $\tan\left(\frac{\pi}{n}\right) : \frac{\pi}{n}$ (b) $\cos\left(\frac{\pi}{n}\right) : \frac{\pi}{n}$
(c) $\sin\left(\frac{\pi}{n}\right) : \frac{\pi}{n}$ (d) $\cot\left(\frac{\pi}{n}\right) : \frac{\pi}{n}$

54. The exradii of a triangle r_1, r_2, r_3 are in HP, then the sides a, b, c are

(a) in HP (b) in AP
(c) in GP (d) None of these

55. In any ΔABC, $\sum \frac{\sin^2 A + \sin A + 1}{\sin A}$ is always greater than

(a) 9 (b) 3
(c) 27 (d) None of these

56. If twice the square of the diameter of a circle is equal to half the sum of the squares of the sides of inscribed ΔABC, then $\sin^2 A + \sin^2 B + \sin^2 C$ is equal to

(a) 1 (b) 2
(c) 4 (d) 8

57. If in a ΔABC, $2\frac{\cos A}{a} + \frac{\cos B}{b} + 2\frac{\cos C}{c} = \frac{a}{bc} + \frac{b}{ca}$, then the value of the angle A is

(a) $\frac{\pi}{3}$ (b) $\frac{\pi}{4}$
(c) $\frac{\pi}{2}$ (d) $\frac{\pi}{6}$

58. If in a triangle $\left(1 - \frac{r_1}{r_2}\right)\left(1 - \frac{r_1}{r_3}\right) = 2$, then the triangle is

(a) right angled (b) isosceles
(c) equilateral (d) None of these

59. In a ΔABC, $\sin A$ and $\sin B$ satisfy $c^2x^2 - c(a+b)x + ab = 0$, then

(a) the triangle is acute angled
(b) the triangle is obtuse angled
(c) $\sin C = \frac{\sqrt{3}}{2}$
(d) $\sin A + \cos A = \frac{a+b}{c}$

60. In any $\Delta\, ABC$, the distance of the orthocentre from the vertices A, B, C are in the ratio

(a) $\sin A : \sin B : \sin C$
(b) $\cos A : \cos B : \cos C$
(c) $\tan A : \tan B : \tan C$
(d) None of the above

61. In a $\Delta\, ABC$, I is the incentre. The ratio $IA : IB : IC$ is equal to

(a) $\operatorname{cosec}\left(\frac{A}{2}\right) : \operatorname{cosec}\left(\frac{B}{2}\right) : \operatorname{cosec}\left(\frac{C}{2}\right)$
(b) $\sin\left(\frac{A}{2}\right) : \sin\left(\frac{B}{2}\right) : \sin\left(\frac{C}{2}\right)$
(c) $\sec\left(\frac{A}{2}\right) : \sec\left(\frac{B}{2}\right) : \sec\left(\frac{C}{2}\right)$
(d) None of the above

62. Given, $\dfrac{b+c}{11}=\dfrac{c+a}{12}=\dfrac{a+b}{13}$ for a ΔABC with usual notation. If $\dfrac{\cos A}{\alpha}=\dfrac{\cos B}{\beta}=\dfrac{\cos C}{\gamma}$, then the ordered triad (α, β, γ) has a value **[JEE Main 2019, 11 January]**

(a) $(19, 7, 25)$ (b) $(3, 4, 5)$
(c) $(5, 12, 13)$ (d) $(7, 19, 25)$

63. In a ΔABC, $a = 2b$ and $|A - B| = \dfrac{\pi}{3}$. Then the $\angle C$ is

(a) $\dfrac{\pi}{4}$ (b) $\dfrac{\pi}{3}$
(c) $\dfrac{\pi}{6}$ (d) None of these

64. In a ΔABC, $\tan A \tan B \tan C = 9$. For such triangles, if $\tan^2 A + \tan^2 B + \tan^2 C = \lambda$, then

(a) $9 \cdot \sqrt[3]{3} < \lambda < 27$ (b) $\lambda \nleq 27$
(c) $\lambda < 9 \cdot \sqrt[3]{3}$ (d) $\lambda \nless 27$

65. The two adjacent sides of a cyclic quadrilateral are 2 and 5 and the angle between them is 60°. If the area of the quadrilateral is $4\sqrt{3}$ sq unit, then the remaining two sides are

(a) 2, 3 (b) 1, 2
(c) 3, 4 (d) 2, 2

66. If in a ΔABC, $a^2 \cos^2 A = b^2 + c^2$, then

(a) $A < \dfrac{\pi}{4}$ (b) $\dfrac{\pi}{4} < A < \dfrac{\pi}{2}$
(c) $A > \dfrac{\pi}{2}$ (d) $A = \dfrac{\pi}{2}$

67. In a ΔABC, the tangent of half the difference of two angles is one third the tangent of half the sum of the two angles. The ratio of the sides opposite the angles are

(a) 2 : 3 (b) 1 : 3
(c) 2 : 1 (d) 3 : 4

68. If p_1, p_2, p_3 are altitudes of a ΔABC from the vertices A, B, C and Δ the area of the triangle, then $p_1^{-1} + p_2^{-1} - p_3^{-1}$ is equal to

(a) $\dfrac{s-a}{\Delta}$ (b) $\dfrac{s-b}{\Delta}$
(c) $\dfrac{s-c}{\Delta}$ (d) $\dfrac{s}{\Delta}$

69. If the median of ΔABC through A is perpendicular to AB, then

(a) $\tan A + \tan B = 0$ (b) $2 \tan A + \tan B = 0$
(c) $\tan A + 2 \tan B = 0$ (d) None of these

70. In a ΔABC, $\cos A + \cos B + \cos C = \dfrac{3}{2}$, then the triangle is

(a) isosceles (b) right angled
(c) equilateral (d) None of these

71. If $A_1A_2A_3 \ldots A_n$ be a regular polygon of n sides and $\dfrac{1}{A_1A_2} = \dfrac{1}{A_1A_3} + \dfrac{1}{A_1A_4}$, then

(a) $n = 5$ (b) $n = 6$
(c) $n = 7$ (d) None of these

72. If p is the product of the sines of angles of a triangle and q the product of their cosines, the tangents of the angle are roots of the equation

(a) $qx^3 - px^2 + (1+q)x - p = 0$
(b) $px^3 - qx^2 + (1+p)x - q = 0$
(c) $(1+q)x^3 - px^2 + qx - q = 0$
(d) None of the above

73. In a ΔABC, $\tan\left(\dfrac{A}{2}\right) = \dfrac{5}{6}$ and $\tan\left(\dfrac{C}{2}\right) = \dfrac{2}{5}$, then

(a) a, c, b are in AP (b) a, b, c are in AP
(c) b, a, c are in AP (d) a, b, c are in GP

74. In a triangle, the line joining the circumcentre to the incentre is parallel to BC, then $\cos B + \cos C$ is equal to

(a) 3/2 (b) 1
(c) 3/4 (d) 1/2

75. If the angles of a triangle are in the ratio 1 : 2 : 3, the corresponding sides are in the ratio

(a) 2 : 3 : 1 (b) $\sqrt{3} : 2 : 1$
(c) $2 : \sqrt{3} : 1$ (d) $1 : \sqrt{3} : 2$

76. In a triangle with one angle $\dfrac{2\pi}{3}$, the lengths of the sides, form an A.P. If the length of the greatest side is 7 cm, the radius of the circumcircle of the triangle is

(a) $\dfrac{7\sqrt{3}}{3}$ cm (b) $\dfrac{5\sqrt{3}}{3}$ cm
(c) $\dfrac{2\sqrt{3}}{3}$ cm (d) $\sqrt{3}$ cm

77. In a triangle, the lengths of the two larger sides are 10 and 9. If the angles are in AP, then the length of the third side can be

(a) $5 \pm \sqrt{6}$ (b) $3\sqrt{3}$
(c) 5 (d) $\sqrt{5} \pm 6$

78. In a triangle, $a^2 + b^2 + c^2 = ca + ab\sqrt{3}$. Then the triangle is

(a) equilateral
(b) right angled and isosceles
(c) right angled with $A = 90°$, $B = 60°$, $C = 30°$
(d) None of the above

79. Three equal circles each of radius r touch one another. The radius of the circle touching all the three given circles internally is

(a) $(2 + \sqrt{3})r$ (b) $\dfrac{(2+\sqrt{3})}{\sqrt{3}} r$
(c) $\dfrac{(2-\sqrt{3})}{\sqrt{3}} r$ (d) $(2 - \sqrt{3})r$

80. In a triangle ABC; AD, BE and CF are the altitudes and R is the circum radius, then the radius of the circle DEF is
(a) $2R$ (b) R
(c) $\frac{R}{2}$ (d) None of these

81. A right angled trapezium is circumscribed about a circle. The radius of the circle. If the lengths of the bases (i.e., parallel sides) are equal to a and b is
(a) $a+b$ (b) $\frac{ab}{a+b}$
(c) $\frac{a+b}{ab}$ (d) $|a-b|$

82. If a, b, c, d are the sides of a quadrilateral, then the minimum value of $\frac{a^2+b^2+c^2}{d^2}$ is
(a) 1 (b) 1/2
(c) 1/3 (d) 1/4

83. If r_1, r_2, r_3 are the radii of the escribed circles of a ΔABC and if r is the radius of its incircle, then $r_1r_2r_3 - r(r_1r_2 + r_2r_3 + r_3r_1)$ is equal to
(a) 0 (b) 1 (c) 2 (d) 3

84. If in $\Delta\, ABC$, $\frac{\sin A}{c} + \frac{\sin C}{a} = \frac{a}{bc} + \frac{c}{ab}$, then the triangle is
(a) right angled (b) isosceles
(c) equilateral (d) None of these

85. A triangle ABC exists such that
(a) $(b+c+a)(b+c-a) = 5bc$
(b) the sides are of lengths $\sqrt{19}, \sqrt{38}, \sqrt{116}$
(c) $\left(\frac{b^2-c^2}{a^2}\right) + \left(\frac{c^2-a^2}{b^2}\right) + \left(\frac{a^2-b^2}{c^2}\right) = 0$
(d) $\cos\left(\frac{B-C}{2}\right) = (\sin B + \sin C)\cos\left(\frac{B+C}{2}\right)$

86. The ratio of the areas of two regular octagons which are respectively inscribed and circumscribed to a circle of radius r is
(a) $\cos\left(\frac{\pi}{8}\right)$ (b) $\sin^2\left(\frac{\pi}{8}\right)$
(c) $\cos^2\left(\frac{\pi}{8}\right)$ (d) $\tan^2\left(\frac{\pi}{8}\right)$

87. In a ΔABC right angled at B, the inradius is
(a) $AB + BC - AC$ (b) $AB + AC - BC$
(c) $\frac{AB + BC - AC}{2}$ (d) None of these

88. If in $\Delta\, ABC$, $\frac{a\cos A + b\cos B + c\cos C}{a\sin B + b\sin C + c\sin A} = \frac{a+b+c}{9R}$, then the triangle ABC is
(a) isosceles (b) equilateral
(c) right angled (d) None of these

89. In a ΔABC, AD is the altitude from A. Given $b > c$, $\angle C = 23°$ and $AD = \frac{abc}{b^2-c^2}$, then $\angle B$ is equal to
(a) 53° (b) 113°
(c) 87° (d) None of these

90. If a, b and c are the sides of a triangle such that $b \cdot c = \lambda^2$, then the relation in a, λ and A is
(a) $c \geq 2\lambda \sin\left(\frac{C}{2}\right)$ (b) $b \geq 2\lambda \sin\left(\frac{A}{2}\right)$
(c) $a \geq 2\lambda \sin\left(\frac{A}{2}\right)$ (d) None of these

91. If A, B, C are angles of a triangle such that the angle A is obtuse, then $\tan B \tan C <$
(a) 0 (b) 1
(c) 2 (d) 3

92. If the sines of the angles of a triangle are in the ratios 3 : 5 : 7 their cotangent are in the ratio
(a) 2 : 3 : 7 (b) 33 : 65 : –15
(c) 65 : 33 : –15 (d) None of these

93. Let $f(x+y) = f(x)\cdot f(y)$ for all x and y and $f(1) = 2$. If in a triangle ABC, $a = f(3)$, $b = f(1) + f(3)$, $c = f(2) + f(3)$, then $2A$ is equal to
(a) C (b) $2C$
(c) $3C$ (d) $4C$

94. In a $\Delta\, ABC$, $b^2 + c^2 = 1999a^2$, then $\frac{\cot B + \cot C}{\cot A}$ is equal to
(a) $\frac{1}{999}$ (b) $\frac{1}{1999}$
(c) 999 (d) 1999

95. In a $\Delta\, ABC$, $2\cos A = \frac{\sin B}{\sin C}$ and $2^{\tan^2 B}$ is a solution of equation $x^2 - 9x + 8 = 0$, then $\Delta\, ABC$ is
(a) equilateral (b) isosceles
(c) scalene (d) right angled

96. If $\frac{\sin A}{\sin C} = \frac{\sin(A-B)}{\sin(B-C)}$, then a^2, b^2, c^2 are in
(a) AP (b) GP
(c) HP (d) none of these

97. In a $\Delta\, ABC$, $A : B : C = 3 : 5 : 4$. Then, $a + b + c\sqrt{2}$ is equal to
(a) $2b$ (b) $2c$
(c) $3b$ (d) $3a$

98. If in an obtuse angled triangle the obtuse angle is $\frac{3\pi}{4}$ and the other two angles are equal to two values of θ satisfying $a\tan\theta + b\sec\theta = c$, when $|b| \leq \sqrt{(a^2+c^2)}$, then $a^2 - c^2$ is equal to
(a) ac (b) $2ac$
(c) $\frac{a}{c}$ (d) None of these

99. In a ΔABC, $A=\frac{\pi}{3}$ and $b:c=2:3$. If $\tan\theta=\frac{\sqrt{3}}{5}$, $0<\theta<\frac{\pi}{2}$, then

(a) $B=60°+\theta$ (b) $C=60°+\theta$
(c) $B=60°-\theta$ (d) $C=60°-\theta$

100. If A, A_1, A_2, A_3 are the areas of incircle and the excircles of a triangle, then $\frac{1}{\sqrt{A_1}}+\frac{1}{\sqrt{A_2}}+\frac{1}{\sqrt{A_3}}$ is equal to

(a) $\frac{2}{\sqrt{A}}$ (b) $\frac{1}{\sqrt{A}}$
(c) $\frac{1}{2\sqrt{A}}$ (d) $\frac{3}{\sqrt{A}}$

101. In any ΔABC, $\Pi\left(\frac{\sin^2 A+\sin A+1}{\sin A}\right)$ is always greater than

(a) 9 (b) 3
(c) 27 (d) None of these

102. If A is the area and $2s$ the sum of three sides of a triangle, then

(a) $A\le\frac{s^2}{3\sqrt{3}}$ (b) $a\le\frac{s^2}{2}$
(c) $A>\frac{s^2}{\sqrt{3}}$ (d) None of these

103. In a triangle if $r_1>r_2>r_3$, then

(a) $a>b>c$ (b) $a<b<c$
(c) $a>b$ and $b<c$ (d) $a<b$ and $b>c$

104. If there are only two linear functions f and g which map [1, 2] on [4, 6] and in a ΔABC, $c=f(1)+g(1)$ and a is the maximum value of r^2, where r is the distance of a variable point on the curve $x^2+y^2-xy=10$ from the origin, then $\sin A:\sin C$ is

(a) $1:2$ (b) $2:1$
(c) $1:1$ (d) None of these

105. The radius of the circle passing through the centre of incircle of ΔABC and through the end points of BC is given by

(a) $\left(\frac{a}{2}\right)\cos A$ (b) $\left(\frac{a}{2}\right)\sec\left(\frac{A}{2}\right)$
(c) $\left(\frac{a}{2}\right)\sin A$ (d) $a\sec\left(\frac{A}{2}\right)$

106. If ΔABC, if $\cot\frac{A}{2}:\cot\frac{B}{2}:\cot\frac{C}{2}=4:3:2$, then, $a:b:c=$

(a) $2:3:4$ (b) $6:5:7$
(c) $4:5:6$ (d) $5:6:7$

Objective Questions Type II [One or more than one correct answer(s)]

In each of the questions below four choices of which one or more than one are correct. You have to select the correct answer(s) accordingly.

1. CF is the internal bisector of angle C of ΔABC, then CF is equal to

(a) $\frac{2ab}{a+b}\cos\frac{C}{2}$ (b) $\frac{a+b}{2ab}\cos\frac{C}{2}$
(c) $\frac{b\sin A}{\sin\left(B+\frac{C}{2}\right)}$ (d) None of these

2. Given an isosceles triangle with equal sides of length b, base angle $\alpha<\frac{\pi}{4}$ and R, r the radii and O, I the centres of the circumcircle and incircle, respectively. Then

(a) $R=\frac{1}{2}b\operatorname{cosec}\alpha$ (b) $\Delta=2b^2\sin 2\alpha$
(c) $r=\frac{b\sin 2\alpha}{2(1+\cos\alpha)}$ (d) $OI=\left|\frac{b\cos\left(\frac{3\alpha}{2}\right)}{2\sin\alpha\cos\left(\frac{\alpha}{2}\right)}\right|$

3. In ΔABC, $A=15°$, $b=10\sqrt{2}$ cm the value of 'a' for which these will be a unique triangle meeting these requirement is

(a) $10\sqrt{2}$ cm (b) 15 cm
(c) $5(\sqrt{3}+1)$ cm (d) $5(\sqrt{3}-1)$ cm

4. In ΔABC; $a=5$, $b=4$, $A=\frac{\pi}{2}+B$ for the value of angle C

(a) cannot be evaluated (b) $\tan^{-1}\left(\frac{9}{40}\right)$
(c) $\tan^{-1}\left(\frac{1}{40}\right)$ (d) $2\tan^{-1}\left(\frac{1}{9}\right)$

5. If $\tan A$, $\tan B$ are the roots of the quadratic $abx^2-c^2x+ab=0$, where a, b, c are the sides of a triangle, then

(a) $\tan A=\frac{a}{b}$ (b) $\tan B=\frac{b}{a}$
(c) $\cos C=0$ (d) $\tan A+\tan B=\frac{c^2}{ab}$

6. In a ΔXYZ, let x, y, z be the lengths of sides opposite to the angles, X, Y, Z respectively and $2s=x+y+z$. If $\frac{s-x}{4}=\frac{s-y}{3}=\frac{s-z}{2}$ and area of incircle of the ΔXYZ is $\frac{8\pi}{3}$, then **[JEE Advanced 2016]**

(a) area of the ΔXYZ is $6\sqrt{6}$
(b) the radius of circumcircle of the ΔXYZ is $\frac{35}{6}\sqrt{6}$
(c) $\sin\frac{X}{2}\sin\frac{Y}{2}\sin\frac{Z}{2}=\frac{4}{35}$
(d) $\sin^2\left(\frac{X+Y}{2}\right)=\frac{3}{5}$ small

7. In a ΔABC, $2\cos\left(\frac{A-C}{2}\right)=\frac{a+c}{\sqrt{(a^2+c^2-ac)}}$, then
(a) $B=\frac{\pi}{3}$
(b) $B=C$
(c) A, B, C are in AP
(d) $B+C=A$

8. In a ΔABC, $\tan C<0$. Then
(a) $\tan A \tan B<1$
(b) $\tan A \tan B>1$
(c) $\tan A+\tan B+\tan C<0$
(d) $\tan A+\tan B+\tan C>0$

9. If the sines of the angles A and B of a ΔABC satisfy the equation $c^2x^2-c(a+b)x+ab=0$, then the triangle
(a) is acute angled
(b) is right angled
(c) is obtuse angled
(d) satisfies $\sin A+\cos A=\frac{(a+b)}{c}$

10. In a ΔABC $\tan A$ and $\tan B$ satisfy the inequation $\sqrt{3}x^2-4x+\sqrt{3}<0$, then
(a) $a^2+b^2+ab>c^2$
(b) $a^2+b^2-ab<c^2$
(c) $a^2+b^2>c^2$
(d) None of these

11. Let x, y and z be positive real numbers. Suppose x, y and z are the lengths of the sides of a triangle opposite to its angles X, Y and Z, respectively. If

$$\tan\frac{X}{2}+\tan\frac{Z}{2}=\frac{2y}{x+y+z},$$

then which of the following statements is/are TRUE?
[JEE Advanced 2020]
(a) $2Y=X+Z$
(b) $Y=X+Z$
(c) $\tan\frac{X}{2}=\frac{x}{y+z}$
(d) $x^2+z^2-y^2=xz$

12. If the angles of a triangle are in the ratio 2 : 3 : 7, then the sides opposite these angles are in the ratio
(a) $\sqrt{2}:2:(\sqrt{3}+1)$
(b) $2:\sqrt{2}:(\sqrt{3}+1)$
(c) $1:\sqrt{2}:\left(\frac{\sqrt{2}}{\sqrt{3}-1}\right)$
(d) $\frac{1}{\sqrt{2}}:1:\left(\frac{\sqrt{3}+1}{2}\right)$

13. Let ABC be an isosceles triangles with base BC. If 'r' is the radius of the circle inscribed in the ΔABC and ρ be the radius of the circle escribed opposite to the angle A, then the product $\rho\, r$ can be equal to
(a) $R^2\sin^2 A$
(b) $R^2\sin^2 2B$
(c) $\frac{1}{2}a^2$
(d) $\frac{a^2}{4}$

14. If $\cos(\theta-\alpha)$, $\cos\theta$, $\cos(\theta+\alpha)$ are in HP, then $\cos\theta\sec\frac{\alpha}{2}$ is equal to
(a) -1
(b) $-\sqrt{2}$
(c) $\sqrt{2}$
(d) 2

15. If $\sin\beta$ is the GM between $\sin\alpha$ and $\cos\alpha$, then $\cos 2\beta$ is equal to
(a) $2\sin^2\left(\frac{\pi}{4}-\alpha\right)$
(b) $2\cos^2\left(\frac{\pi}{4}-\alpha\right)$
(c) $2\cos^2\left(\frac{\pi}{4}+\alpha\right)$
(d) $2\sin^2\left(\frac{\pi}{4}+\alpha\right)$

16. If l is the length of median from the vertex A to the side BC of a ΔABC, then
(a) $4l^2=2b^2+2c^2-a^2$
(b) $4l^2=b^2+c^2+2bc\cos A$
(c) $4l^2=a^2+4bc\cos A$
(d) $4l^2=(2s-a)^2-4bc\sin^2\left(\frac{A}{2}\right)$

17. If in a ΔABC, $r_1=2r_2=3r_3$, then
(a) $\frac{a}{b}=\frac{4}{5}$
(b) $\frac{a}{b}=\frac{5}{4}$
(c) $\frac{a}{c}=\frac{3}{5}$
(d) $\frac{a}{c}=\frac{5}{3}$

18. In ΔABC, which of the following statements are true
(a) maximum value of $\sin 2A+\sin 2B+\sin 2C$ is same as the maximum value of $\sin A+\sin B+\sin C$
(b) $R\geq 2r$, where R is circumradius and r is inradius
(c) $R^2\geq\frac{abc}{(a+b+c)}$
(d) ΔABC is right angled if $r+2R=s$, where s is semiperimeter

19. In a ΔABC, $\frac{\sin A}{\sin C}=\frac{\sin(A-B)}{\sin(B-C)}$, then
(a) $\cot A, \cot B, \cot C$ in AP
(b) $\sin 2A, \sin 2B, \sin 2C$ in AP
(c) $\cos 2A, \cos 2B, \cos 2C$ in AP
(d) $a\sin A, b\sin B, c\sin C$ in AP

20. If in a ΔABC, $\sin C+\cos C+\sin(2B+C)-\cos(2B+C)=2\sqrt{2}$, then ΔABC is
(a) equilateral
(b) isosceles
(c) right angled
(d) obtuse angled

21. Let a, b, c be the sides of triangle whose perimeter is P and area is A, then
(a) $P^3\leq 27(b+c-a)(c+a-b)(a+b-c)$
(b) $P^2\leq 3(a^2+b^2+c^2)$
(c) $a^2+b^2+c^2\geq 4\sqrt{3}A$
(d) $P^4\leq 25<A$

22. In a triangle ABC, if the radii of ex-circles r_1, r_2 and r_3 are given by $r_1=8$, $r_2=12$ and $r_3=24$, then
(a) in radius $r=8$
(b) side $a=12$
(c) side $b=16$
(d) side $c=20$

23. In a ΔABC, if D is the mid point of BC and AD is perpendicular to AC, then

(a) $\cos A = -\frac{b}{c}$ (b) $\cos A \cos C = -\frac{2b^2}{ac}$

(c) $\cos B = \left(\frac{b^2+c^2}{bc}\right)$ (d) $a^2 - 3b^2 - c^2 = 0$

24. If in $\Delta ABC, A = 90°$ and c, $\sin B$ and $\cos B$ are rational number, then

(a) a is rational (b) a is irrational

(c) b is rational (d) b is irrational

25. In a non-right angled ΔPQR, let p, q, r denote the lengths of the sides opposite to the angles at P, Q, R respectively. The median from R meets the side PQ at S, the perpendicular from P meets the side QR at E, and RS and PE intersect at O. If $p = \sqrt{3}$, $q = 1$, and the radius of the circumcircle of the ΔPQR equals 1, then which of the following options is/are correct? **[JEE Advanced 2019]**

(a) Length of $OE = \frac{1}{6}$

(b) Length of $RS = \frac{\sqrt{7}}{2}$

(c) Area of $\Delta SOE = \frac{\sqrt{3}}{12}$

(d) Radius of incircle of $\Delta PQR = \frac{\sqrt{3}}{2}(2-\sqrt{3})$

26. If in a ΔABC, a, b, c are in AP and p_1, p_2, p_3 are the altitudes from the vertices A, B, C respectively, then

(a) p_1, p_2, p_3 are in AP (b) p_1, p_2, p_3 are in HP

(c) $p_1 + p_2 + p_3 \le \frac{3R}{\Delta}$ (d) $\frac{1}{p_1} + \frac{1}{p_2} + \frac{1}{p_3} \le \frac{3R}{\Delta}$

27. In a ΔPQR, let $\angle PQR = 30°$ and the sides PQ and QR have lengths $10\sqrt{3}$ and 10, respectively. Then, which of the following statement(s) is (are) TRUE? **[JEE Advanced 2018]**

(a) $\angle QPR = 45°$

(b) The area of the ΔPQR is $25\sqrt{3}$ and $\angle QRP = 120°$

(c) The radius of the incircle of the ΔPQR is $10\sqrt{3} - 15$

(d) The area of the circumcircle of the ΔPQR is 100π

28. If in a ΔABC, θ is the angle determined by $\cos\theta = \frac{(a-b)}{c}$, then

(a) $\frac{(a+b)\sin\theta}{2\sqrt{ab}} = \cos\frac{A-B}{2}$

(b) $\frac{(a+b)\sin\theta}{2\sqrt{ab}} = \cos\frac{A+B}{2}$

(c) $\frac{c\sin\theta}{2\sqrt{ab}} = \cos\frac{A-B}{2}$

(d) $\frac{c\sin\theta}{2\sqrt{ab}} = \cos\frac{A+B}{2}$

29. If a, b, A be given in a triangle and c_1 and c_2 two possible values of third side such that $c_1^2 + c_1c_2 + c_2^2 = a^2$, then A is equal to

(a) 30° (b) 60°

(c) 90° (d) 120°

Linked-Comprehension Type

In these questions, a passage (paragraph) has been given followed by questions based on each of the passages. You have to answer the questions based on the passage given.

PASSAGE 1

If p_1, p_2, p_3 are altitudes of a ΔABC from the vertices A, B, C respectively and triangle is the area of the triangle and s is semi perimeter of the triangle.

On the basis of above information, answer the following questions

1. If $\frac{1}{p_1} + \frac{1}{p_2} + \frac{1}{p_3} = \frac{1}{2}$ then the least value of $p_1p_2p_3$ is

(a) 8 (b) 27

(c) 125 (d) 216

2. The value of $\frac{\cos A}{p_1} + \frac{\cos B}{p_2} + \frac{\cos C}{p_3}$ is

(a) $\frac{1}{r}$ (b) $\frac{1}{R}$

(c) $\frac{a^2+b^2+c^2}{2R}$ (d) $\frac{1}{\Delta}$

3. The minimum value of $\frac{b^2p_1}{c} + \frac{c^2p_2}{a} + \frac{a^2p_3}{b}$ is

(a) Δ

(b) 2Δ

(c) 3Δ

(d) 6Δ

4. The value of $p_1^{-2} + p_2^{-2} + p_3^{-2}$ is

(a) $\frac{(\Sigma a)^2}{4\Delta^2}$ (b) $\frac{(\Pi a)^3}{8\Delta^3}$

(c) $\frac{\Sigma a^2}{4\Delta^2}$ (d) $\frac{\pi a^2}{8\Delta^2}$

5. In the triangle ABC, the altitudes are in AP, then

(a) a, b, c are in AP

(b) a, b, c are in HP

(c) a, b, c are in GP

(d) angles A, B, C are in AP

PASSAGE 2

A polygon has n sides. If all the sides and all the angles are same then this polygon is called regular polygon. Let $A_1, A_2, A_3 \ldots, A_n$ be a regular polygon of n sides. R be the radius of circumscribed circle of regular polygon and r be the radius of inscribed circle of regular polygon.

If $A_1A_2 = A_2A_3 = A_3A_4 = \ldots = A_nA_1 = a$

On the basis of above information, answer the following questions

1. The value of A_1A_j ($j = 1, 2, 3, \ldots, n$) is
   (a) $2R \sin\left\{(j-1)\dfrac{\pi}{2n}\right\}$ (b) $R \sin\left\{(j-1)\dfrac{2\pi}{n}\right\}$
   (c) $2R \sin\left\{(j-1)\dfrac{\pi}{n}\right\}$ (d) $R \sin\left\{(j-1)\dfrac{\pi}{n}\right\}$
2. The value of r is
   (a) $\dfrac{a}{2}\tan\left(\dfrac{\pi}{n}\right)$ (b) $\dfrac{a}{2}\cot\left(\dfrac{\pi}{n}\right)$
   (c) $\dfrac{a}{2}\tan\left(\dfrac{2\pi}{n}\right)$ (d) $\dfrac{a}{2}\cot\left(\dfrac{2\pi}{n}\right)$
3. The value of $r + R$ is
   (a) $\dfrac{a}{2}\cot\left(\dfrac{\pi}{n}\right)$ (b) $a\cot\left(\dfrac{\pi}{2n}\right)$
   (c) $\dfrac{a}{4}\cot\left(\dfrac{\pi}{2n}\right)$ (d) $\dfrac{a}{2}\cot\left(\dfrac{\pi}{2n}\right)$
4. The area of a regular polygon of n sides is
   (a) $\dfrac{nR^2}{2}\sin\left(\dfrac{2\pi}{n}\right)$ (b) $nR^2\tan\left(\dfrac{\pi}{n}\right)$
   (c) $\dfrac{nr^2}{2}\sin\left(\dfrac{2\pi}{n}\right)$ (d) $nr^2\tan\left(\dfrac{2\pi}{n}\right)$
5. A regular pentagon and a regular decagon have the same perimeter then their areas are in the ratio
   (a) $\sqrt{5}:1$ (b) $\sqrt{5}:2$
   (c) $1:\sqrt{5}$ (d) $2:\sqrt{5}$

PASSAGE 3

When any two sides and one of the opposite acute angle are given, under certain additional conditions two triangles are possible. The case when two triangles are possible is called the ambiguous case.

In fact when any two sides and the angle opposite to one of them are given either no triangle is possible or only one triangle is possible or two triangles are possible.

In the ambiguous case, let a, b and $\angle A$ are given and c_1, c_2 are two values of the third side c.

On the basis of above information, answer the following questions

1. Two different triangles are possible when
   (a) $b\sin A < a$ (b) $b\sin A < a$ and $b > a$
   (c) $b\sin A < a$ and $b < a$ (d) $b\sin A < a$ and $a = b$
2. The difference between two values of c is
   (a) $2\sqrt{(a^2 - b^2)}$
   (b) $\sqrt{(a^2 - b^2)}$
   (c) $2\sqrt{(a^2 - b^2\sin^2 A)}$
   (d) $\sqrt{(a^2 - b^2\sin^2 A)}$
3. The value of $c_1^2 - 2c_1c_2\cos 2A + c_2^2$ is
   (a) $4a\cos A$ (b) $4a^2\cos A$
   (c) $4a\cos^2 A$ (d) $4a^2\cos^2 A$
4. If $\angle A = 45°$ and in ambiguous case (a, b, A are given) c_1, c_2 are two values of c and if θ be the angle between the two positions of the ambiguous side c then $\cos\theta$ is
   (a) $\dfrac{c_1c_2}{c_1^2 + c_2^2}$ (b) $\dfrac{2c_1c_2}{c_1^2 + c_2^2}$
   (c) $\dfrac{\sqrt{c_1c_2}}{(c_1 + c_2)}$ (d) $\dfrac{2\sqrt{c_1c_2}}{(c_1 + c_2)}$
5. If $2b = (m+1)a$ and $\cos A = \dfrac{1}{2}\sqrt{\left(\dfrac{(m-1)(m+3)}{m}\right)}$, where $1 < m < 3$, then $\dfrac{c_1}{c_2}$ is
   (a) m or $\dfrac{1}{m}$ (b) $(m-1)$ or $\dfrac{1}{(m-1)}$
   (c) $(m+1)$ or $\dfrac{1}{(m+1)}$ (d) $(m+3)$ or $\dfrac{1}{(m+3)}$

PASSAGE 4

Let ABC be an acute angled triangle with orthocentre, H, D, E and F are the feet of perpendicular from A, B and C, respectively, on opposite sides. Also, let R be the circumradius of ΔABC. Given $AH \cdot BH \cdot CH = 3$ and $(AH)^2 + (BH)^2 + (CH)^2 = 7$.

On the basis of above information, answer the following questions

1. Value of $\dfrac{\cos A \cdot \cos B \cdot \cos C}{\cos^2 A + \cos^2 B + \cos^2 C}$ is
   (a) $\dfrac{3}{14R}$ (b) $\dfrac{3}{7R}$
   (c) $\dfrac{7}{3R}$ (d) $\dfrac{14}{3R}$
2. Value of R is
   (a) 1 (b) $\dfrac{3}{2}$
   (c) $\dfrac{5}{2}$ (d) None of these
3. Value of $HD \cdot HE \cdot HF$ is
   (a) $\dfrac{9}{64R^3}$ (b) $\dfrac{9}{8R^3}$
   (c) $\dfrac{8}{9R^3}$ (d) $\dfrac{64}{9R^3}$

PASSAGE 5

AL, *BM* and *CN* are perpendiculars from angular points of a triangle *ABC* on the opposite sides *BC*, *CA* and *AB* respectively. Δ is the areas of triangle *ABC*, *r* and *R* are the inradius and circumradius.

On the basis of above information, answer the following questions

1. If perimeters of $\Delta\, LMN$ and $\Delta\, ABC$ and λ and μ, then the value of $\frac{\lambda}{\mu}$ is
   (a) $\frac{r}{R}$ (b) $\frac{R}{r}$
   (c) $\frac{rR}{\Delta}$ (d) $\frac{\Delta}{rR}$
2. If areas of Δ' s *AMN*, *BNL* and *CLM* are Δ_1, Δ_2 and Δ_3 respectively, then the value of $\Delta_1 + \Delta_2 + \Delta_3$ is
   (a) $\Delta(2 + 2\cos A \cos B \cos C)$
   (b) $\Delta(2 + 2\sin A \sin B \sin C)$
   (c) $\Delta(1 - 2\cos A \cos B \cos C)$
   (d) $\Delta(1 - 2\sin A \sin B \sin C)$
3. If area of $\Delta\, LMN$ is Δ', then the value of $\frac{\Delta'}{\Delta}$ is
   (a) $2\sin A \sin B \sin C$ (b) $2\cos A \cos B \cos C$
   (c) $\sin A \sin B \sin C$ (d) $\cos A \cos B \cos C$
4. Radius of the circum circle of $\Delta\, LMN$ is
   (a) $2R$ (b) R
   (c) $\frac{R}{2}$ (d) $\frac{R}{4}$
5. If radius of the incircle of $\Delta\, LMN$ is r', then the value of $r' \sec A \sec B \sec C$ is
   (a) $4R$ (b) $3R$
   (c) $2R$ (d) R

PASSAGE 6

ABCD be a cyclic quadrilateral inscribed in a circle of radius *R*. Then

$$\cos B = \frac{(a^2 + b^2 - c^2 - d^2)}{2\,(ab + cd)}$$

and Area of quadrilateral $= \frac{1}{2}(ab + cd)\sin B$

$$= \sqrt{(s-a)(s-b)(s-c)(s-d)}$$

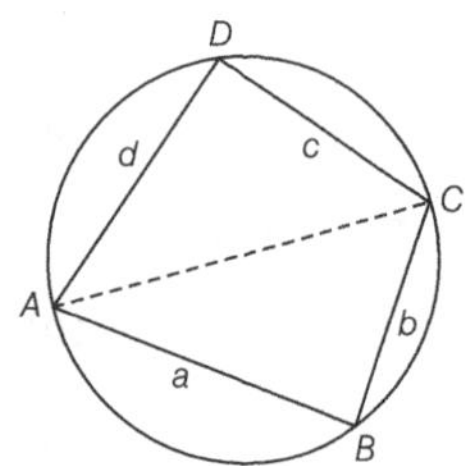

where $s = \frac{a+b+c+d}{2}$

Also, $AC^2 \cdot BD^2 = (ac + bd)^2$

i.e. $AC \cdot BD = AB \cdot CD + BC \cdot AD$

and $R = \frac{AC}{2\sin B}$

On the basis of above information, answer the following questions

1. The sides of a quadrilateral which can be inscribed in a circle are 6, 6, 8 and 8 cm. Then radius of circum circle is
   (a) $\frac{5}{2}$ cm (b) $\frac{24}{7}$ cm
   (c) $\frac{11}{7}$ cm (d) None of these
2. The sides of a quadrilateral which can be inscribed in a circle are 5, 5, 12 and 12 cm. Then the radius of incircle of the quadrilateral is
   (a) $\frac{15}{17}$ cm
   (b) $\frac{30}{17}$ cm
   (c) $\frac{60}{17}$ cm
   (d) incircle does not exist
3. A quadrilateral *ABCD* is such that one circle can be inscribed in it and another circle circumscribed about it, then $\tan^2(A/2)$ is equal to
   (a) $\frac{bc}{ad}$ (b) $\frac{ab}{cd}$
   (c) $\frac{ad - bc}{ad + bc}$ (d) $\frac{a + c}{b + d}$
4. If a quadrilateral with side lengths *a*, *b*, *c*, *d* can be inscribed in one circle and circumscribed about another circle, then its area is
   (a) $\sqrt{(ab\,cd)}$
   (b) $\sqrt{(2abcd)}$
   (c) $2\sqrt{abcd}$
   (d) None of the above
5. In previous problem the radius of the latter circle is
   (a) $\frac{\sqrt{(abcd)}}{(a+b+c+d)}$
   (b) $\frac{\sqrt{2abcd}}{(a+b+c+d)}$
   (c) $\frac{2\sqrt{(abcd)}}{(a+b+c+d)}$
   (d) $\frac{2\sqrt{(a+c)(b+d)}}{(a+b+c+d)}$

Numerical Value Type

Solve the following problems and write your answer.

1. In triangle ABC, $5050\dfrac{(r_1+r_2)(r_2+r_3)(r_3+r_1)}{R(s)^2}$ must be

2. If in a triangle, $\dfrac{s-a}{11}=\dfrac{s-b}{12}=\dfrac{s-c}{13}$. Then $\lambda\tan^2(A/2)=455$, if λ must be

3. If p_1, p_2, p_3 are the altitudes of a triangle which circumscribes a circle of diameter $\dfrac{16}{3}$ unit, then the least value of $p_1+p_2+p_3$ must be

4. The sides of triangle are three consecutive natural numbers and its largest angle is twice the smaller one. The perimeter of the triangle must be

5. In a triangle ABC; $A=\pi/3$, $b=40$, $c=30$, AD is a median through A, then $4(AD)^2$ must be

6. In a triangle ABC, right angled at A. The radius of the inscribed circle is 2 cm. Radius of the circle touching the side BC and also sides AB and AC produced is 15 cm.The length of the side BC measured in cm is

7. In a ΔABC, $a=5$, $b=4$ and $\cos(A-B)=31/32$, then 12c must be

8. If in a ΔABC, $BC=5$, $CA=4$, $AB=3$ and D, E are the points on BC such that $BD=DE=EC$, then $\tan(\angle CAE)$ must be

9. Three circles touch one another externally. The tangents at their points of contact meet at a point whose distance from the point of contact is 4. If the ratio of the product of the radii to the sum of the radii of the circles must be

10. The area of a right triangle is 6864 sq units. If the ratio of its legs is 143 : 24, then the value of r is

11. In a ΔABC, the maximum value of $\left(\dfrac{a\cos^2(A/2)+b\cos^2(B/2)+c\cos^2(C/2)}{a+b+c}\right)$ must be

12. In ΔABC, the lengths of sides AC and AB are 12 cm and 5 cm, respectively. If the area of ΔABC is 30 cm^2 and R and r are respectively the radii of circumcircle and incircle of ΔABC, then the value of $(2R+r)$ in cm is equal to .... **[JEE Main 2021, 16 March]**

13. In ΔABC, $\dfrac{r}{r_1}=\dfrac{1}{2}$, then the value of $\left(\tan\dfrac{A}{2}+\tan\dfrac{B}{2}+\tan\dfrac{C}{2}\right)$ must be

14. In a triangle ABC, the incircle touches the sides BC, CA and AB at D, E, F respectively. If radius of incircle is 4 unit and BD, CE and AF be consecutive the sum of the cubes of the length of the sides must be natural numbers, then

15. In triangle ABC, $a=5$, $b=4$, $c=3$.G is the centroid of triangle. If R_1 be the circumradius of ΔGAB then the value of $90R_1^2$ must be

16. In the given figure 'P' is any arbitrary interior point of the triangle ABC. H_a, H_b and H_c are the length of altitudes drawn from vertices A, B and C respectively. If x_a, x_b and x_c represent the distance of 'P' from sides BC, CA and AB respectively, then the minimum value of $191\left(\dfrac{H_a}{x_a}+\dfrac{H_b}{x_b}+\dfrac{H_c}{x_c}\right)$ must be

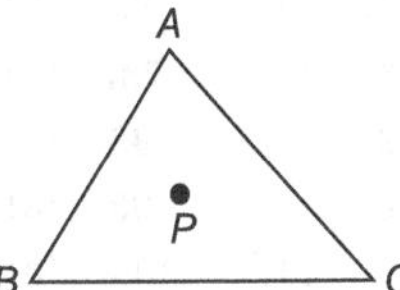

Single Integer Type

Solve the following problems and write your answer.

1. Let f, g, h be the lengths of the perpendiculars from the circumcentre of the ΔABC on the sides BC, CA and AB respectively. If $\dfrac{a}{f}+\dfrac{b}{g}+\dfrac{c}{h}=\lambda\dfrac{abc}{fgh}$, then the value of 4λ is

2. In a ΔABC, if $\tan A:\tan B:\tan C=1:2:3$ and $\sin A:\sin B:\sin C=\sqrt{5}:2\sqrt{2}:k$, then $k=$

3. If in a ΔABC, $a=5$, $b=4$ and $\cos(A-B)=\dfrac{31}{32}$, then the third side c is equal to

4. If a, b and c represents the lengths of sides of a triangle, then the possible, integral value of $\dfrac{a}{b+c}+\dfrac{b}{c+a}+\dfrac{c}{a+b}$ is...

5. The area of cyclic quadrilateral $PQRS$ is $3\sqrt{3}$ sq units. The radius of the circumcircle of ΔPQR is 2. If $PQ=2$, $QS=2\sqrt{3}$, then $QR\times RS$ is

6. If a, b and c are the sides of ΔABC for which $r_1=8$, $r_2=12$ and $r_3=24$ then value of $\left(\dfrac{b+c}{a}\right)$ is equal to

7. In ΔABC, $\angle C=2\angle A$, and $AC=2BC$, then the value of $\dfrac{a^2+b^2+c^2}{R^2}$ (where R is circumradius of triangle) is

8. In a ΔABC, $2B=A+C$ and $b^2=ac$, then $\dfrac{a(a+b+c)}{3bc}=$

Matrix-Match Type

Given below there are certain matching type questions, where two columns (each having some items) are given. Each item of column I has to be matched with the items of column II, by encircling the correct match(es).

NOTE *An item of column I can be match with more than one items of column II.*

1. Observe the following Columns

	Column I		Column II
(A)	In a triangle ABC, if $a^4 - 2(b^2 + c^2)a^2 + b^4 + b^2c^2 + c^4 = 0$, then $\angle A$ is	(P)	30°
(B)	In a triangle ABC, If $a^4 + b^4 + c^4 = a^2b^2 + 2b^2c^2 + 2c^2a^2$, then $\angle C$ is	(Q)	60°
		(R)	90°
(C)	In a triangle ABC, If $a^4 + b^4 + c^4 + 2a^2c^2 = 2a^2b^2 + 2b^2c^2$, then $\angle B$ is	(S)	120°
		(T)	150°

2. Observe the following Columns

	Column I		Column II
(A)	In a ΔABC, if $2a^2 + b^2 + c^2 = 2ac + 2ab$, then	(P)	ΔABC is equilateral triangle
(B)	In a ΔABC, if $a^2 + b^2 + c^2 = \sqrt{2}b(c + a)$, then	(Q)	ΔABC is right angled triangle
		(R)	ΔABC is scalane triangle
(C)	In a ΔABC, If $a^2 + b^2 + c^2 = bc + ca\sqrt{3}$, then	(S)	ΔABC is isosceles right angled triangle
		(T)	Angles B, C, A are in AP.

Answers

Objective Questions Type [Only one correct answer]

1. (c)	**2.** (d)	**3.** (d)	**4.** (d)	**5.** (d)	**6.** (a)	**7.** (a)	**8.** (c)	**9.** (a)	**10.** (b)
11. (b)	**12.** (a)	**13.** (b)	**14.** (c)	**15.** (a)	**16.** (d)	**17.** (a)	**18.** (d)	**19.** (b)	**20.** (b)
21. (c)	**22.** (b)	**23.** (c)	**24.** (c)	**25.** (d)	**26.** (b)	**27.** (a)	**28.** (d)	**29.** (d)	**30.** (c)
31. (a)	**32.** (a)	**33.** (d)	**34.** (a)	**35.** (a)	**36.** (b)	**37.** (c)	**38.** (b)	**39.** (a)	**40.** (a)
41. (a)	**42.** (a)	**43.** (b)	**44.** (a)	**45.** (c)	**46.** (c)	**47.** (c)	**48.** (a)	**49.** (c)	**50.** (c)
51. (a)	**52.** (b)	**53.** (a)	**54.** (b)	**55.** (a)	**56.** (c)	**57.** (c)	**58.** (a)	**59.** (d)	**60.** (b)
61. (a)	**62.** (a)	**63.** (b)	**64.** (b)	**65.** (a)	**66.** (c)	**67.** (c)	**68.** (c)	**69.** (c)	**70.** (c)
71. (c)	**72.** (a)	**73.** (b)	**74.** (b)	**75.** (d)	**76.** (a)	**77.** (a)	**78.** (c)	**79.** (b)	**80.** (c)
81. (a)	**82.** (c)	**83** (a)	**84.** (a)	**85.** (d)	**86.** (c)	**87.** (c)	**88.** (b)	**89.** (b)	**90.** (c)
91. (b)	**92.** (c)	**93** (a)	**94.** (a)	**95.** (a)	**96.** (a)	**97.** (c)	**98.** (b)	**99.** (b)	**100.** (b)
101. (c)	**102.** (a)	**103** (a)	**104.** (c)	**105.** (b)	**106.** (d)				

Objective Questions Type II [One or more than one correct answer(s)]

1. (a, c)	**2.** (a, c, d)	**3.** (a, b, d)	**4.** (b, d)	**5.** (a, b, c, d)
6. (a, c, d)	**7.** (a, c)	**8.** (a, c)	**9.** (b, d)	**10.** (a, b)
11. (b, c)	**12.** (a, c, d)	**13.** (a, b, d)	**14.** (b, c)	**15.** (a, c)
16. (a, b, c, d)	**17.** (b, d)	**18.** (a, b, c, d)	**19.** (a, c, d)	**20.** (b, c)
21. (b, c)	**22.** (b, c, d)	**23.** (a, b, d)	**24.** (a, c)	**25.** (a, b, d)
26. (b, d)	**27.** (b, c, d)	**28.** (a, d)	**29.** (b, d)	

Linked-Comprehension Type

Passage 1 1. (d) 2. (b) 3. (d) 4. (c) 5. (b) **Passage 2** 1. (a) 2. (b) 3. (d) 4. (a) 5. (d)

Passage 3 1. (b) 2. (c) 3. (d) 4. (b) 5. (a) **Passage 4** 1. (a) 2. (b) 3. (b)

Passage 5 1. (a) 2. (c) 3. (b) 4. (c) 5. (c) **Passage 6** 1. (d) 2. (c) 3. (a) 4. (a) 5. (c)

Numerical Value Type

1. 5050	**2.** 1155	**3.** 24	**4.** 15	**5.** 3700	**6.** 13	**7.** 72	**8.** 0.375
9. 16	**10.** 22	**11.** 0.75	**12.** 15	**13.** 0.5	**14.** 8316	**15.** 3250	**16.** 1719

Single Integer Type Question

1. (1) 2. (3) 3. (6) 4. (2) 5. (8) 6. (3) 7. (8) 8. (1)

Matrix-Match Type

1. A → (Q, S); B → (P, T); C → (R)
2. A → (P, T); B → (Q, S); C → (Q, R, T)

Solutions

Objective Questions Type [Only one correct answer]

1. In ΔABC,

$$\frac{AD}{BD} = \frac{AC}{BC} = \frac{b}{a} \quad \text{...(i)}$$

and $CD \cdot DE = AD \cdot BD$ ...(ii)

Now, from Eq. (i)

$$\frac{AD}{b} = \frac{BD}{a} = \frac{AD + BD}{b + a}$$

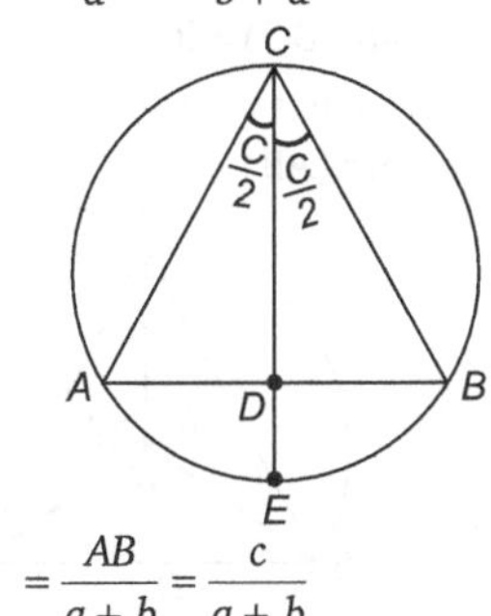

$$= \frac{AB}{a+b} = \frac{c}{a+b}$$

$\therefore \quad AD = \dfrac{bc}{a+b}, \; BD = \dfrac{ac}{a+b}$ ...(iii)

From Eqs. (ii) and (iii), then

$$CD \cdot DE = \frac{abc^2}{(a+b)^2}$$

$$\leq \frac{c^2}{4} \quad \left[\because \frac{a+b}{2} \geq \sqrt{ab} \Rightarrow \frac{ab}{(a+b)^2} \leq \frac{1}{4}\right]$$

2. $\because (r_1 + r_2 + r_3 - r)^2$

$$= r_1^2 + r_2^2 + r_3^2 + r^2 + 2(r_1r_2 + r_2r_3 + r_3r_1) - 2r(r_1 + r_2 + r_3)$$

$$\Rightarrow (4R)^2 = r^2 + r_1^2 + r_2^2 + r_3^2 + 2s^2 - 2\{(s-a)(s-b) + (s-b)(s-c) + (s-c)(s-a)\}$$

$$\Rightarrow 16R^2 = r^2 + r_1^2 + r_2^2 + r_3^2 + 2s^2 - 2\{3s^2 - 2s(a+b+c) + ab + bc + ca\}$$

$$\Rightarrow 16R^2 = r^2 + r_1^2 + r_2^2 + r_3^2 + 4s^2 - 2(ab + bc + ca)$$

$$= r^2 + r_1^2 + r_2^2 + r_3^2 + (a+b+c)^2 - 2(ab + bc + ca)$$

$$= r^2 + r_1^2 + r_2^2 + r_3^2 + a^2 + b^2 + c^2$$

Hence, $r^2 + r_1^2 + r_2^2 + r_3^2 + a^2 + b^2 + c^2 = 16R^2$

3. Let $x = \sum \dfrac{\sin^2 A + \sin A + 1}{\sin A}$

$$= \sum\left[\left(\sin A + \frac{1}{\sin A}\right) + 1\right] = \sum\left(\sin A + \frac{1}{\sin A}\right) + 3$$

In ΔABC, $0 < \sin A \leq 1 \Rightarrow \sin A + \dfrac{1}{\sin A} \geq 2$

$$\Rightarrow \quad x \geq 6 + 3 = 9$$

$\therefore$ Minimum value of x seems to be 9 but minimum $x = 9$ only when $\sin A = \sin B = \sin C = 1$

i.e. $A = B = C = \dfrac{\pi}{2}$ which is impossible in ΔABC

$\therefore$ Minimum x does not exist.

4. Let $AB = BC = \lambda$,

$\therefore \quad AC = \lambda\sqrt{2}$ and $BE = \dfrac{\lambda}{2}$

$\because$ AD and CE are medians G is centroid.

Then, $\quad GA = GC = \dfrac{2}{3} CE$

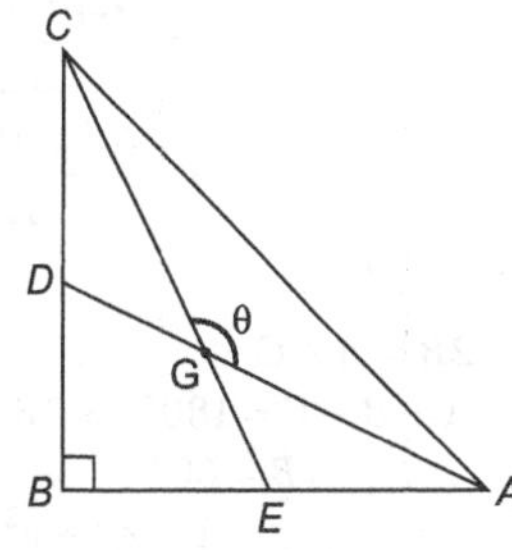

$$= \frac{2}{3}\sqrt{\{(BC)^2 + (BE)^2\}}$$

$$= \frac{2}{3}\sqrt{\lambda^2 + \left(\frac{\lambda}{2}\right)^2} = \frac{2}{3} \cdot \frac{\lambda}{2} \sqrt{5} = \frac{\lambda\sqrt{5}}{3}$$

$$\therefore \quad \cos\theta = \frac{(GA)^2 + (GC)^2 - (AC)^2}{2GA \cdot GC}$$

$$= \frac{\frac{5}{9}\lambda^2 + \frac{5}{9}\lambda^2 - 2\lambda^2}{2 \cdot \frac{\lambda\sqrt{5}}{3} \cdot \frac{\lambda\sqrt{5}}{3}} = \frac{-8\lambda^2}{10\lambda^2} = -\frac{4}{5}$$

5. $\because$ Sides are in AP and $a < \min\{b, c\}$

Case I If $\min\{b, c\} = b$

Then a, b, c are in AP.

i.e. $\quad 2b = a + c$

$$\therefore \quad \cos A = \frac{b^2 + c^2 - a^2}{2bc} = \frac{b^2 + c^2 - (2b - c)^2}{2bc}$$

$$= \frac{4bc - 3b^2}{2bc} = \frac{4c - 3b}{2c}$$

Case II If min $\{b, c\} = c$

Then a, c, b are in AP.

$$2c = a + b$$

$$\therefore \quad \cos A = \frac{b^2 + c^2 - a^2}{2bc}$$

$$= \frac{b^2 + c^2 - (2c - b)^2}{2bc}$$

$$= \frac{4bc - 3c^2}{2bc} = \frac{4b - 3c}{2b}$$

6. $\because \quad 2s = 3 + 5 + c = 8 + c$

$$\therefore \quad s = \frac{(8 + c)}{2}$$

$\because$ AM $\geq$ GM

$$\Rightarrow \frac{(s-a) + (s-b) + (s-c)}{3} \geq ((s-a)(s-b)(s-c))^{1/3}$$

$$\Rightarrow \left(\frac{s}{3}\right)^3 \geq \frac{\Delta^2}{s}$$

$$\Rightarrow \Delta^2 \leq \frac{s^4}{3^3}$$

$$\Rightarrow \Delta \leq \frac{s^2}{3\sqrt{3}}$$

$$= \frac{(8 + c)^2}{12\sqrt{3}}$$

$$= \frac{(64 + 16c + c^2)}{12\sqrt{3}}$$

$$\therefore \quad \Delta \leq \frac{(c^2 + 16c + 64)}{12\sqrt{3}}$$

7. $\because A, B, C$ are in AP.

$$\therefore \quad 2B = A + C$$

and $A + B + C = 180° \Rightarrow 3B = 180°$

$$\therefore \quad B = 60°$$

$$\therefore \quad \cos B = \frac{1}{2} = \frac{a^2 + c^2 - b^2}{2ac}$$

$$\Rightarrow \quad a^2 + c^2 = b^2 + ac$$

$$\Rightarrow \quad (a - c)^2 = b^2 - ac$$

$$\Rightarrow \quad |a - c| = \sqrt{(b^2 - ac)}$$

or $|\sin A - \sin C| = \sqrt{(\sin^2 B - \sin A \sin C)}$

$$\Rightarrow 2\cos\left(\frac{A + C}{2}\right)\left|\sin\left(\frac{A - C}{2}\right)\right| = \sqrt{\left(\frac{3}{4} - \sin A \sin C\right)}$$

$$\Rightarrow \quad 2\left|\sin\left(\frac{A - C}{2}\right)\right| = \sqrt{\left(\frac{3}{4} - \sin A \sin C\right)}$$

$$\therefore \quad \lim_{x \to C} \frac{\sqrt{(3 - 4\sin A \sin C)}}{|A - C|} = \lim_{x \to C} \frac{2\left|\sin\left(\frac{A - C}{2}\right)\right|}{|A - C|}$$

$$= \lim_{x \to C} \left|\frac{\sin\left(\frac{A - C}{2}\right)}{\left(\frac{A - C}{2}\right)}\right|$$

$$= |1| = 1$$

8. We have,

$$\cot(A + B) = \frac{\cot A \cot B - 1}{\cot B + \cot A}$$

$$= \frac{x^{1/2}(x^2 + x + 1)^{1/2} \cdot x^{-1/2}(x^2 + x + 1)^{1/2} - 1}{x^{-1/2}(x^2 + x + 1)^{1/2} + x^{1/2}(x^2 + x + 1)^{1/2}}$$

$$= \frac{x^2 + x + 1 - 1}{(x^2 + x + 1)^{1/2}(x^{1/2} + x^{-1/2})}$$

$$= \frac{x(x + 1)x^{1/2}}{(x^2 + x + 1)^{1/2}(x + 1)}$$

$$= x^{3/2}(x^2 + x + 1)^{-1/2}$$

$$= (x^{-1} + x^{-2} + x^{-3})^{1/2}$$

$$= \cot C$$

$$\Rightarrow \quad A + B = C \text{ and } A + B + C = \pi$$

$$\therefore \quad C = \frac{\pi}{2}$$

i.e. triangle is right angled.

9.

$$\text{arc}\,(BC) = 3$$
$$\text{arc}\,(CA) = 4$$
$$\text{arc}\,(AB) = 5$$

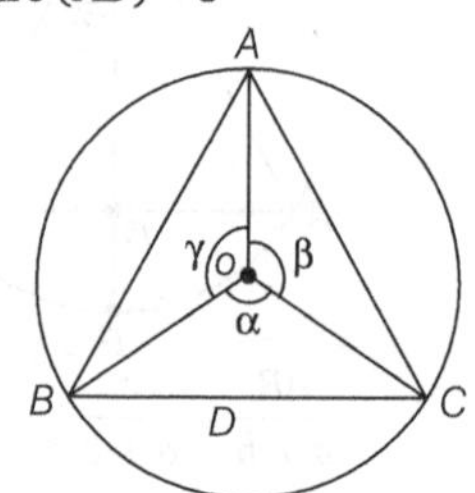

Let r be the radius of the circle, then

$$3 = r\alpha,\ 4 = r\beta,\ 5 = r\gamma \qquad \left[\because \text{angle} = \frac{\text{arc}}{\text{radius}}\right]$$

Now, $3 + 4 + 5 = r(\alpha + \beta + \gamma) = r \cdot 2\pi$

$$\therefore \quad r = \frac{6}{\pi}$$

$$\Delta ABC = \Delta OBC + \Delta OCA + \Delta OAB$$

$$= \frac{1}{2}r^2 \sin\alpha + \frac{1}{2}r^2 \sin\beta + \frac{1}{2}r^2 \sin\gamma$$

$$= \frac{1}{2}r^2\left\{\sin\left(\frac{3}{r}\right) + \sin\left(\frac{4}{r}\right) + \sin\left(\frac{5}{r}\right)\right\}$$

$$= \frac{1}{2} \times \frac{36}{\pi^2}\left[\sin\frac{\pi}{2} + \sin\frac{2\pi}{3} + \sin\frac{5\pi}{6}\right]$$

$$= \frac{18}{\pi^2}\left\{1 + \frac{\sqrt{3}}{2} + \frac{1}{2}\right\}$$

$$= \frac{9\sqrt{3}(1 + \sqrt{3})}{\pi^2} \text{ sq unit}$$

10. $\sin A, \sin B, \sin C$ are the roots of the equation

$$c^3x^3 - c^2(a + b + c)x^2 + lx + m = 0$$

$$\therefore \quad \sin A + \sin B + \sin C = \frac{c^2(a + b + c)}{c^3}$$

$$= \frac{a + b + c}{c}$$

or $\quad \frac{a}{2R}+\frac{b}{2R}+\frac{c}{2R}=\frac{a+b+c}{c}$

then $\quad c=2R,\ a+b+c\neq 0$

i.e. $\quad 2R\sin C=2R$

$\therefore \quad \sin C=1 \Rightarrow \angle C=90^\circ$

$\Rightarrow$ Triangle is a right angled triangle

and $\quad \Sigma \sin A \sin B=\frac{l}{c^3}$

and $\quad \sin A\sin B\sin C=-\frac{m}{c^3}$

$\Rightarrow \quad \Sigma\frac{ab}{4R^2}=\frac{l}{c^3}$ and $\frac{abc}{8R^3}=-\frac{m}{c^3}$

$\Rightarrow \quad l=\frac{c^3}{4R^2}\Sigma ab$ and $m=-\frac{(abc)c^3}{8R^3}$

$\Rightarrow \quad l=\frac{c^3\,\Sigma ab}{c^2}$ and $m=-\frac{(abc)c^3}{(c)^3}$

Hence, $\quad l=c\Sigma ab=c(ab+bc+ca)$

and $\quad m=-abc$

11. $\because \quad \cos A=\frac{b^2+c^2-a^2}{2bc}$

$\Rightarrow b^2+c^2-a^2=2bc\cos A$

Squaring both sides, then

$$b^4+c^4+a^4+2b^2c^2-2a^2b^2-2c^2a^2=4b^2c^2\cos^2 A$$

$\Rightarrow \quad 3b^2c^2=4b^2c^2\cos^2 A$

$[\because$ given $a^4+b^4+c^4=2a^2b^2+b^2c^2+2c^2a^2]$

$\therefore \quad \cos^2 A=\frac{3}{4}\Rightarrow \sin^2 A=\frac{1}{4}$

Hence, $\quad \sin A=\pm\frac{1}{2}$

12. Let a be length of side of regular polygon, then

$$R=\frac{a}{2}\operatorname{cosec}\left(\frac{\pi}{n}\right) \text{ and } r=\frac{a}{2}\cot\left(\frac{\pi}{n}\right)$$

$\therefore \quad \frac{R}{r}=\frac{\operatorname{cosec}\left(\frac{\pi}{n}\right)}{\cot\left(\frac{\pi}{n}\right)}=\frac{1}{\cos\left(\frac{\pi}{n}\right)}$

$=\sqrt{5}-1$ [given]

$\therefore \quad \cos\left(\frac{\pi}{n}\right)=\frac{1}{\sqrt{5}-1}=\frac{\sqrt{5}+1}{4}=\cos\left(\frac{\pi}{5}\right)$

$\therefore \quad n=5$

13. $\because \Delta ABC=\Delta PBC+\Delta PCA+\Delta PAB$

$$\frac{a^2\sqrt{3}}{4}=\frac{1}{2}\cdot a\cdot 4+\frac{1}{2}\cdot a\cdot 5+\frac{1}{2}\cdot a\cdot 6=\frac{15a}{2}$$

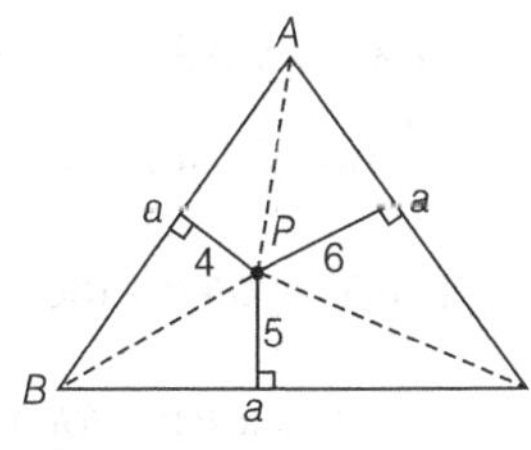

or $\quad a=\frac{30}{\sqrt{3}}=10\sqrt{3}$ unit

14. In ΔIBD, $\quad \tan 15^\circ=\frac{\sqrt{3}}{\left(\frac{a}{2}\right)}$

$$(2-\sqrt{3})=\frac{2\sqrt{3}}{a}$$

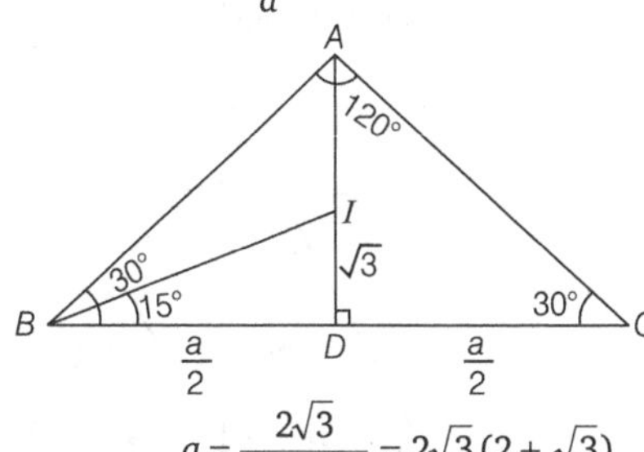

$\therefore \quad a=\frac{2\sqrt{3}}{(2-\sqrt{3})}=2\sqrt{3}(2+\sqrt{3})$

and in ΔABD,

$$\tan 30^\circ=\frac{AD}{\frac{a}{2}}\Rightarrow AD=\frac{a}{2}\cdot\frac{1}{\sqrt{3}}=\frac{a}{2\sqrt{3}}$$

$\therefore \quad \Delta ABC=\frac{1}{2}\cdot a\cdot AD$

$=\frac{1}{2}\cdot a\cdot\frac{a}{2\sqrt{3}}=\frac{a^2}{4\sqrt{3}}$

$=\frac{12(2+\sqrt{3})^2}{4\sqrt{3}}$

$=\sqrt{3}(7+4\sqrt{3})$

$=12+7\sqrt{3}$ sq unit

15. $\because$ Roots are real and distinct.

$\therefore \quad \Delta>0$

$\Rightarrow \quad 4(a+b+c)^2-12\lambda(ab+bc+ca)>0$

$\Rightarrow (a^2+b^2+c^2+2(ab+bc+ca))-3\lambda(ab+bc+ca)>0$

$\Rightarrow \quad \frac{\Sigma a^2}{\Sigma ab}>(3\lambda-2) \quad \ldots(\text{i})$

Now, in a triangle

Difference of two sides < third side

i.e. $\quad |a-b|<c,\ |b-c|<a$ and $|c-a|<b$

$\Rightarrow \quad (a-b)^2+(b-c)^2+(c-a)^2<a^2+b^2+c^2$

$\Rightarrow \quad a^2+b^2+c^2<2(ab+bc+ca)$

or $\quad \frac{\Sigma a^2}{\Sigma ab}<2 \quad \ldots(\text{ii})$

From Eqs. (i) and (ii), we get

$$3\lambda-2<\frac{\Sigma a^2}{\Sigma ab}<2$$

$\Rightarrow \quad 3\lambda-2<2\Rightarrow\lambda<\frac{4}{3}$

16. Let $\quad r=1$ cm

$a=BC=BD+DE+EC$

$=BD+C_1C_2+BD$

$=2BD+C_1C_2$

$=2r\cot 30^\circ+2r$

$=2r(\sqrt{3}+1)$

$=2(\sqrt{3}+1)$ cm

$\therefore$ Area of triangle

$$ABC = \frac{\sqrt{3}}{4}a^2 = \frac{\sqrt{3}}{4}\cdot 4\,(\sqrt{3}+1)^2 \text{ cm}^2$$
$$= \sqrt{3}\,(4+2\sqrt{3}) \text{ cm}^2$$
$$= (6+4\sqrt{3}) \text{ cm}^2$$

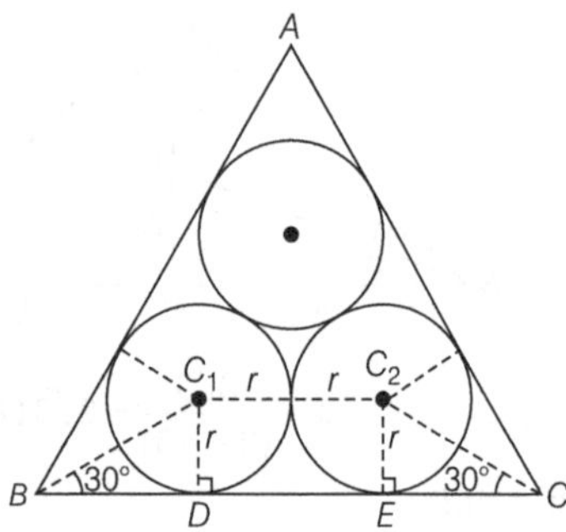

17. For a ΔABC, it is given that $a = \sqrt{3}+1$, $b = \sqrt{3}-1$ and $\angle A + \angle B = 120°$

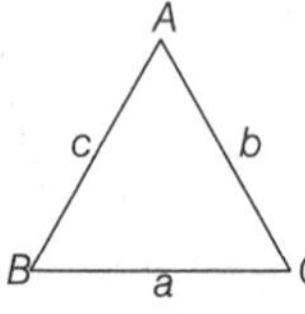

Clearly, $\angle C = 60°$ $\quad [\because \angle A + \angle B + \angle C = 180°]$

Now, by tangent law, we have

$$\tan\frac{A-B}{2} = \frac{a-b}{a+b}\cot\frac{C}{2}$$
$$= \frac{(\sqrt{3}+1)-(\sqrt{3}-1)}{(\sqrt{3}+1)+(\sqrt{3}-1)}\cot\left(\frac{60°}{2}\right)$$
$$= \frac{2}{2\sqrt{3}}\cot(30°) = \frac{1}{\sqrt{3}}\times\sqrt{3} = 1$$

$\Rightarrow$ $\tan\left(\frac{A-B}{2}\right) = 1 = \tan 45°$

$\Rightarrow$ $\frac{A-B}{2} = 45° \Rightarrow \angle A - \angle B = 90°$

On solving $\angle A - \angle B = 90°$ and $\angle A + \angle B = 120°$, we get

$$\angle A = 105° \text{ and } \angle B = 15°$$

So, $\angle A : \angle B = 7 : 1$

18. $\because$ $2a^2 + 4b^2 + c^2 = 4ab + 2ac$

$\Rightarrow$ $(a-2b)^2 + (a-c)^2 = 0$

Which is possible only when

$$a - 2b = 0 \text{ and } a - c = 0$$

or $\frac{a}{1} = \frac{b}{\frac{1}{2}} = \frac{c}{1} = \lambda$ (say)

$\therefore$ $a = \lambda,\ b = \frac{\lambda}{2},\ c = \lambda$

$\therefore$ $\cos B = \frac{a^2+c^2-b^2}{2ac} = \frac{\lambda^2+\lambda^2-\frac{\lambda^2}{4}}{2\lambda^2}$

$$= 1 - \frac{1}{8} = \frac{7}{8}$$

19. $\because$ Area of $\Delta\, ABC$ is maximum, then $\Delta\, ABC$ is isosceles with each side.

$$AB = AC = R\sqrt{2}$$

$\therefore$ $s = \left(\frac{R\sqrt{2}+R\sqrt{2}+2R}{2}\right)$

$\Rightarrow$ $s = R\,(\sqrt{2}+1)$

$\therefore$ $\Delta = \frac{1}{2}\cdot 2R\cdot R = R^2$

$\Rightarrow$ $r = \frac{\Delta}{s} = \frac{R^2}{R(\sqrt{2}+1)}$

$\Rightarrow$ $\frac{1}{r} = \frac{(\sqrt{2}+1)}{R}$

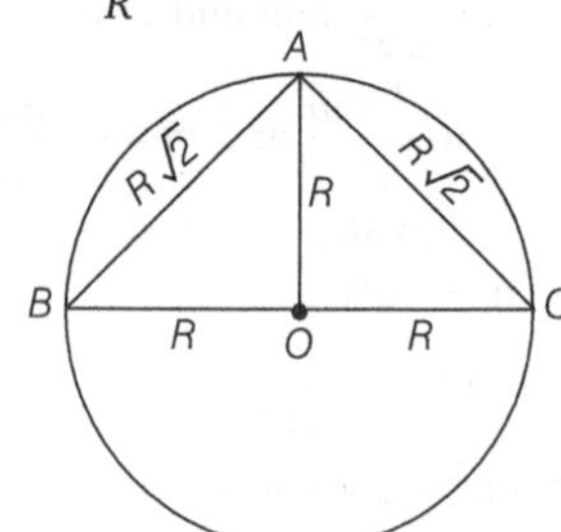

$\Rightarrow$ $\frac{1}{r_1} + \frac{1}{r_2} + \frac{1}{r_3} = \frac{(\sqrt{2}+1)}{R}$

20. By Appolloneous theorem

$$(GB)^2 + (GC)^2 = 2\{(GD)^2 + (DC)^2\}$$

$\Rightarrow$ $(GB)^2 + (GC)^2 = 2\left\{\left(\frac{1}{2}GA\right)^2 + \left(\frac{a}{2}\right)^2\right\}$

$\Rightarrow$ $(GB)^2 + (GC)^2 = \frac{(GA)^2}{2} + \frac{a^2}{2}$ ... (i)

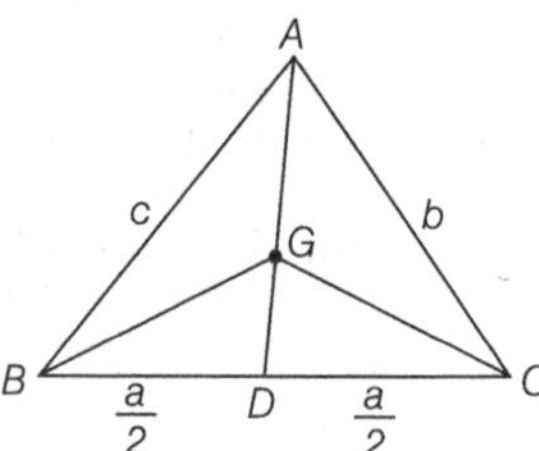

Similarly, $(GC)^2 + (GA)^2 = \frac{(GB)^2}{2} + \frac{b^2}{2}$ ...(ii)

and $(GA)^2 + (GB)^2 = \frac{(GC)^2}{2} + \frac{c^2}{2}$ ...(iii)

Adding Eqs. (i), (ii) and (iii), we get

$$2\{(GA)^2 + (GB)^2 + (GC)^2\}$$
$$= \frac{1}{2}\{(GA)^2 + (GB)^2 + (GC)^2\} + \frac{(a^2+b^2+c^2)}{2}$$

or $\frac{3}{2}\{(GA)^2 + (GB)^2 + (GC)^2\} = \frac{(a^2+b^2+c^2)}{2}$

$\Rightarrow$ $(GA)^2 + (GB)^2 + (GC)^2 = \left(\frac{a^2+b^2+c^2}{3}\right)$

21. $\because \quad \cos A = \dfrac{b^2 + c^2 - a^2}{2bc}$

or $\quad c^2 - 2bc\cos A + b^2 - a^2 = 0$

$\therefore \quad c_1 + c_2 = 2b\cos A = 2 \times 2 \times \dfrac{\sqrt{3}}{2} = 2\sqrt{3}$

and $\quad c_1 c_2 = b^2 - a^2 = 4 - 5 = -1$

$\therefore \quad |c_1 - c_2| = \sqrt{(c_1 + c_2)^2 - 4c_1c_2}$

$= \sqrt{12 + 4} = \sqrt{16} = 4$

22. $\because \quad \dfrac{2}{1!9!} + \dfrac{2}{3!7!} + \dfrac{1}{5!5!} = \dfrac{8^a}{(2b)!}$

$\Rightarrow \quad \dfrac{1}{1!9!} + \dfrac{1}{3!7!} + \dfrac{1}{5!5!} + \dfrac{1}{3!7!} + \dfrac{1}{9!1!} = \dfrac{8^a}{(2b)!}$

$\Rightarrow \quad \dfrac{1}{10!}\left(\dfrac{10!}{1!9!} + \dfrac{10!}{3!7!} + \dfrac{10!}{5!5!} + \dfrac{10!}{7!3!} + \dfrac{10!}{9!1!}\right) = \dfrac{8^a}{(2b)!}$

$\Rightarrow \quad \dfrac{1}{10!}({}^{10}C_1 + {}^{10}C_3 + {}^{10}C_5 + {}^{10}C_7 + {}^{10}C_9) = \dfrac{8^a}{(2b)!}$

$\Rightarrow \quad \dfrac{2^9}{10!} = \dfrac{8^a}{(2b)!} = \dfrac{2^{3a}}{(2b)!} \Rightarrow a = 3, b = 5$

Also, $\quad 2b = a + c \Rightarrow 10 = 3 + c \Rightarrow c = 7$

$\therefore \quad a = 3, b = 5, c = 7$

$\because \quad \dfrac{\tan A + \tan B}{2} \geq \sqrt{\tan A \tan B} \quad \ldots \text{(i)}$

Also, $\quad \cos C = \dfrac{a^2 + b^2 - c^2}{2ab} = \dfrac{9 + 25 - 49}{30} = -\dfrac{1}{2}$

$\therefore \quad C = 120^\circ$ and $A, B < 60^\circ$

$\tan A + \tan B + \tan C = \tan A \tan B \tan C$

$\Rightarrow \quad \tan A + \tan B - \sqrt{3} = -\sqrt{3}\tan A \tan B$

$\therefore \quad \tan A + \tan B = \sqrt{3}(1 - \tan A \tan B) \quad \ldots \text{(ii)}$

Also, $\quad \tan A + \tan B > 0$

$\Rightarrow \quad \sqrt{3}(1 - \tan A \tan B) > 0$

$\Rightarrow \quad \tan A \tan B < 1 \quad \ldots \text{(iii)}$

From Eqs. (i) and (ii),

$\dfrac{\sqrt{3}(1 - \tan A \tan B)}{2} \geq \sqrt{(\tan A \tan B)}$

Let $\quad \tan A \tan B = \lambda$

$\therefore \quad \sqrt{3}(1 - \lambda) \geq 2\sqrt{\lambda}$

$\Rightarrow \quad 3\lambda^2 - 10\lambda + 3 \geq 0$

$\Rightarrow \quad (3\lambda - 1)(\lambda - 3) \geq 0$

$\because \quad \lambda - 3 < 0 \quad$ [from Eq. (iii)]

$\therefore \quad 3\lambda - 1 \leq 0 \Rightarrow \lambda \leq \dfrac{1}{3}$

$\tan A \tan B \leq \dfrac{1}{3}$

23. $\because a(b - c) + b(c - a) + c(a - b) = 0$

$\therefore$ $x = 1$ is a root of the equation

$a(b - c)x^2 + b(c - a)x + c(a - b) = 0$

Then, other root $= 1 \quad$ [$\because$ roots are equal]

$\therefore \quad |X| = \dfrac{c(a - b)}{a(b - c)} \Rightarrow ab - ac = ca - bc$

$\therefore \quad b = \dfrac{2ac}{a + c}$

$\therefore$ a, b, c are in HP

Then, $\dfrac{1}{a}, \dfrac{1}{b}, \dfrac{1}{c}$ are in AP.

$\Rightarrow \dfrac{s}{a}, \dfrac{s}{b}, \dfrac{s}{c}$ are in AP.

$\Rightarrow \dfrac{s}{a} - 1, \dfrac{s}{b} - 1, \dfrac{s}{c} - 1$ are in AP.

$\Rightarrow \dfrac{(s - a)}{a}, \dfrac{(s - b)}{b}, \dfrac{(s - c)}{c}$ are in AP.

Multiplying in each by $\dfrac{abc}{(s - a)(s - b)(s - c)}$

then, $\quad \dfrac{bc}{(s - b)(s - c)}, \dfrac{ca}{(s - c)(s - a)}, \dfrac{ab}{(s - a)(s - b)}$

are in AP.

$\Rightarrow \quad \dfrac{(s - b)(s - c)}{bc}, \dfrac{(s - c)(s - a)}{ca}, \dfrac{(s - a)(s - b)}{ab}$

are in HP.

or $\quad \sin^2\left(\dfrac{A}{2}\right), \sin^2\left(\dfrac{B}{2}\right), \sin^2\left(\dfrac{C}{2}\right)$ are in HP.

24. $\because \quad \angle A = 20^\circ$

$\therefore \quad \angle B = \angle C = 80^\circ$

Then, $\quad b = c$

$\therefore \quad \dfrac{a}{\sin 20^\circ} = \dfrac{b}{\sin 80^\circ} = \dfrac{c}{\sin 80^\circ}$

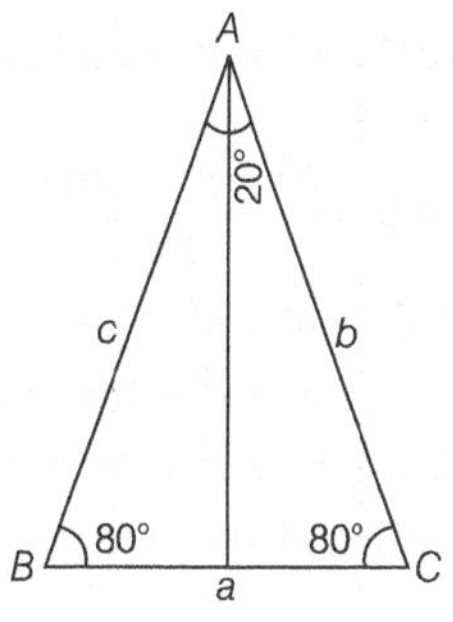

or $\quad \dfrac{a}{\sin 20^\circ} = \dfrac{b}{\cos 10^\circ}$

$\Rightarrow \quad a = 2b\sin 10^\circ \quad \ldots \text{(i)}$

$\therefore \quad a^3 + b^3 = 8b^3\sin^3 10^\circ + b^3$

$= b^3\{2(4\sin^3 10^\circ) + 1\}$

$= b^3\{2(3\sin 10^\circ - \sin 30^\circ) + 1\}$

$= b^3\{6\sin 10^\circ\} = 3b^2(2b\sin 10^\circ)$

$= 3b^2a \quad$ [from Eq. (i)]

$= 3ac^2 \quad [\because b = c]$

25. $\because$ In a ΔABC,

$\dfrac{a}{\sin A} = \dfrac{b}{\sin B} = \dfrac{c}{\sin\{\pi - (A + B)\}} = 2R$

$\Rightarrow \quad \dfrac{a}{\sin A} = \dfrac{b}{\sin B} = \dfrac{c}{\sin(A + B)} = 2R$

Alternate (a) : If we know a, $\sin A$, $\sin B$ then we can find b, c, A, B and C.

Alternate (b) : We can find A, B, C by using cosine rule .

Alternate (c) : $\because a, \sin B, R$ are given then we can find $\sin A$, b and hence

$$\sin C = \sin \{\pi - (A + B)\} = \sin C$$

Alternate (d) : $a, \sin A, R$ are given then we know only the ratio $\frac{b}{\sin B}$ or $\frac{c}{\sin (A+B)}$; we cannot determine the values of $b, c, \sin B, \sin C$ separately

$\therefore$ Triangle ABC cannot be determined in this case.

26. Since, circle is inscribed in the quadrilateral $ABCD$, then

$$a + c = b + d \quad \ldots\text{(i)}$$

and quadrilateral is cyclic then

$$A + C = \pi \text{ or } C = \pi - A \quad \ldots\text{(ii)}$$

Now in ΔABD,

$$\cos A = \frac{a^2 + d^2 - (BD)^2}{2ad}$$

or $\quad BD^2 = a^2 + d^2 - 2ad \cos A \quad \ldots\text{(iii)}$

and in ΔBCD,

$$\cos C = \frac{b^2 + c^2 - (BD)^2}{2bc}$$

or $\quad (BD)^2 = b^2 + c^2 - 2bc \cos C$

$$= b^2 + c^2 - 2bc \cos (\pi - A) \text{ [from Eq. (ii)]}$$

$$(BD)^2 = b^2 + c^2 + 2bc \cos A \quad \ldots\text{(iv)}$$

From Eqs. (iii) and (iv), we get

$$\cos A = \frac{a^2 + d^2 - b^2 - c^2}{2(bc + ad)} \quad \ldots\text{(v)}$$

Now, from Eq. (i),

$$(a - d)^2 = (b - c)^2$$

$\Rightarrow \quad a^2 + d^2 - b^2 - c^2 = 2(ad - bc) \quad \ldots\text{(vi)}$

$\therefore$ From Eqs. (v) and (vi),

$$\cos A = \left(\frac{ad - bc}{ad + bc}\right)$$

27. $\because \quad A + B + C + D = 2\pi$

or $\quad \tan (A + B + C + D) = 0$

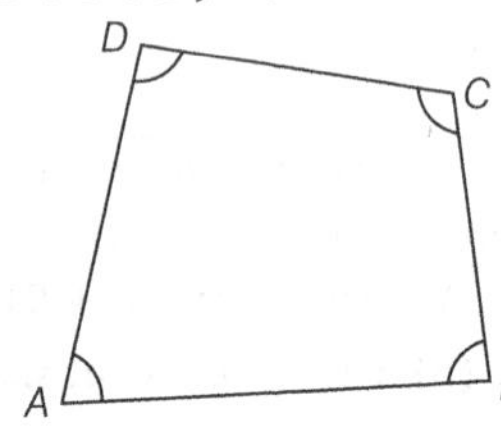

or $\quad \dfrac{\Sigma \tan A - \Sigma \tan A \tan B \tan C}{1 - \Sigma \tan A \tan B + \tan A \tan B \tan C \tan D} = 0$

or $\quad \Sigma \tan A - \Sigma \tan A \tan B \tan C = 0$

or $\quad \Sigma \tan A = \tan A \tan B \tan C \tan D \, \Sigma \cot A$

or $\quad \dfrac{\Sigma \tan A}{\Sigma \cot A} = \Pi \tan A$

28. $a^2 + b^2 + c^2 = 4R^2 (\sin^2 A + \sin^2 B + \sin^2 c)$

$$= 2R^2(3 - (\cos 2A + \cos 2B + \cos 2c))$$

Now, $\cos 2A + \cos 2B + \cos 2C$

$$= \cos \frac{2\pi}{7} + \cos \frac{4\pi}{7} + \cos \frac{8\pi}{7}$$

$$= \cos \frac{2\pi}{7} + \cos \frac{4\pi}{7} + \cos \frac{6\pi}{7}$$

$$= \frac{\sin \frac{3\pi}{7}}{\sin \frac{\pi}{7}} \cos \frac{4\pi}{7}$$

$$= \frac{-2 \sin \frac{3\pi}{7} \cos \frac{3\pi}{7}}{2 \sin \frac{\pi}{7}} = \frac{-\sin \frac{6\pi}{7}}{2 \sin \frac{\pi}{7}} = \frac{-1}{2}$$

$$\therefore a^2 + b^2 + c^2 = 2R^2 \left(3 - \left(\frac{-1}{2}\right)\right) = 7R^2$$

29. In ΔABC, AD and BE are medians of ΔABC,

$$AD = \frac{7}{2} \Rightarrow \angle DAB = \frac{\pi}{8}$$

$$\angle ABE = \frac{\pi}{4} \Rightarrow AO = \frac{2}{3} AD = \frac{2}{3} \times \frac{7}{2} = \frac{7}{3}$$

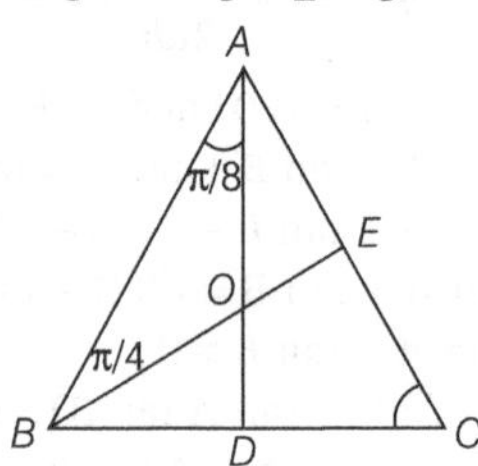

$$\angle AOB = \pi - \left(\frac{\pi}{8} + \frac{\pi}{4}\right) = \left(\pi - \frac{3\pi}{8}\right)$$

By sine rule, $\dfrac{AO}{\sin \frac{\pi}{4}} = \dfrac{BO}{\sin \frac{\pi}{8}}$

$$BO = \frac{\frac{7}{3} \times \sin \frac{\pi}{8}}{\sin \frac{\pi}{4}} = \frac{7}{3} \times \frac{\sin \frac{\pi}{8}}{2 \sin \frac{\pi}{8} \cos \frac{\pi}{8}} = \frac{7}{6 \cos \frac{\pi}{8}}$$

$$\text{Area of } \Delta AOB = \frac{1}{2} AO \times BO \times \sin \angle AOB$$

$$= \frac{1}{2} \times \frac{7}{3} \times \frac{7}{6} \times \frac{\sin\left(\pi - \frac{3\pi}{8}\right)}{\cos \frac{\pi}{8}}$$

$$= \frac{7}{36} \times \frac{\sin 3\frac{\pi}{8}}{\cos \frac{\pi}{8}} = \frac{7}{36} \quad \left[\because \sin \frac{3\pi}{8} = \cos \frac{3\pi}{8}\right]$$

$$\text{Area of } \Delta ABC = 3 \times \frac{7}{36} = \frac{7}{12}$$

30. $\because \quad \Delta = a^2 - (b-c)^2$

$$= (a+b-c)(a-b+c)$$
$$= 2(s-c)\cdot 2(s-b)$$
$$\sqrt{s(s-a)(s-b)(s-c)} = 4(s-b)(s-c)$$
$$\Rightarrow \quad \frac{1}{4} = \sqrt{\frac{(s-b)(s-c)}{s(s-a)}} = \tan\frac{A}{2}$$
$$\therefore \quad \tan\frac{A}{2} = \frac{1}{4}$$

31. $2s = 6 \times \left(\frac{\sin A + \sin B + \sin C}{3}\right)$

$$\therefore \quad \sin A + \sin B + \sin C = s$$
$$\Rightarrow \quad ak + bk + ck = s$$
$$\therefore \quad k(2s) = s$$
$$\therefore \quad k = \frac{1}{2}$$
$$\sin A = ak = 1 \times \frac{1}{2} \qquad [\text{given } a = 1]$$
$$\therefore \quad \angle A = 30°$$

32. $\because\ a^3 \cos(B-C) = a^2 \cdot a\cos(B-C)$

$$= a^2 \cdot k \sin A \cos(B-C)$$
$$= a^2 \cdot k \sin(B+C)\cos(B-C)$$
$$= a^2 \cdot \frac{k}{2}(\sin 2B + \sin 2C)$$
$$= a^2 \cdot \frac{k}{2}(2\sin B\cos B + 2\sin C\cos C)$$
$$= a^2(b\cos B + c\cos C)$$

Similarly,

$$b^3\cos(C-A) = b^2(c\cos C + a\cos A)$$

and $\quad c^3\cos(A-B) = c^2(a\cos A + b\cos B)$

$$\Rightarrow a^3\cos(B-C) + b^3\cos(C-A) + c^3\cos(A-B)$$
$$= ab(a\cos B + b\cos A) + bc(b\cos C + c\cos B) + ca(c\cos A + a\cos C)$$
$$= abc + bca + cab = 3abc$$

33. Since, $\frac{\cos A}{a} = \frac{\cos B}{b} = \frac{\cos C}{c}$

$$\Rightarrow \quad \frac{b^2+c^2-a^2}{2abc} = \frac{c^2+a^2-b^2}{2abc} = \frac{a^2+b^2-c^2}{2abc}$$
$$\Rightarrow \quad b^2+c^2-a^2 = c^2+a^2-b^2 = a^2+b^2-c^2$$

From I and II relation

$$2b^2 = 2a^2 \Rightarrow b^2 = a^2$$
$$\Rightarrow \quad b = a = 2$$

and from II and III relation

$$2c^2 = 2b^2$$
$$c^2 = b^2$$
$$c = b = 2$$

Hence, $\quad a = b = c = 2$

Hence, ΔABC is equilateral

$$\therefore \text{ Area} = \frac{\sqrt{3}}{4}(\text{side})^2$$
$$= \frac{\sqrt{3}}{4}(2)^2 = \sqrt{3} \text{ sq unit}$$

34. $\cos A + 2\cos B + \cos C = 2$

$$\Rightarrow \quad \cos A + \cos C = 2(1-\cos B)$$
$$\Rightarrow \quad 2\cos\left(\frac{A+C}{2}\right)\cos\left(\frac{A-C}{2}\right) = 2\cdot 2\sin^2\frac{B}{2}$$
$$\Rightarrow \quad 2\sin\frac{B}{2}\cos\left(\frac{A-C}{2}\right) = 4\sin^2\frac{B}{2}$$
$$\Rightarrow \quad \cos\left(\frac{A-C}{2}\right) = 2\sin\frac{B}{2}$$
$$\Rightarrow \quad \cos\left(\frac{A-C}{2}\right) = 2\cos\left(\frac{A+C}{2}\right)$$
$$\Rightarrow \quad \frac{\cos\left(\frac{A-C}{2}\right)}{\cos\left(\frac{A+C}{2}\right)} = \frac{2}{1}$$
$$\Rightarrow \quad \frac{\cos\left(\frac{A-C}{2}\right) - \cos\left(\frac{A+C}{2}\right)}{\cos\left(\frac{A-C}{2}\right) + \cos\left(\frac{A+C}{2}\right)} = \frac{1}{3}$$

(by componendo and dividendo method)

$$\Rightarrow \quad \tan\frac{A}{2}\tan\frac{C}{2} = \frac{1}{3}$$
$$\Rightarrow \quad \sqrt{\frac{(s-b)(s-c)}{s(s-a)}}\sqrt{\frac{(s-a)(s-b)}{s(s-c)}} = \frac{1}{3}$$
$$\Rightarrow \quad \frac{s-b}{s} = \frac{1}{3} \quad \Rightarrow \quad 2s = 3b$$
$$a + c = 2b$$

Hence, a, b, c are in AP.

35. $\because \quad BD : DC = 1 : 3$

$$\therefore \quad BD = \frac{a}{4}$$

and $\quad DC = \frac{3a}{4}$

In ΔABD sine rule and in ΔADC,

$$\frac{AD}{\sin\frac{\pi}{3}} = \frac{\frac{a}{4}}{\sin\alpha}, \quad \frac{\frac{3a}{4}}{\sin\beta} = \frac{AD}{\sin\frac{\pi}{4}}$$

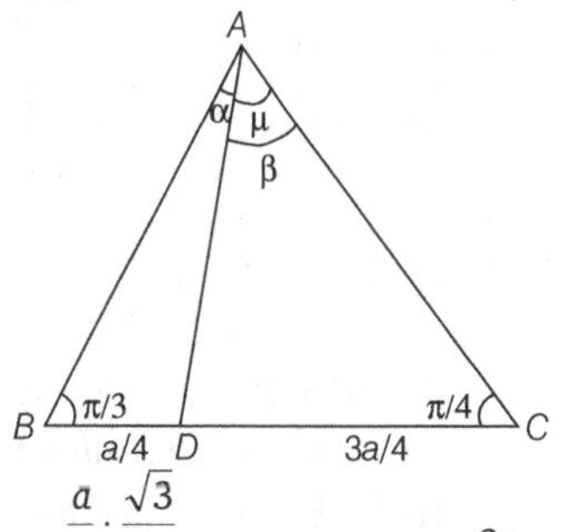

$$\therefore \quad \sin\alpha = \frac{\frac{a}{4}\cdot\frac{\sqrt{3}}{2}}{AD} \text{ and } \sin\beta = \frac{3a}{4\sqrt{2}\,AD}$$
$$\therefore \quad \frac{\sin\alpha}{\sin\beta} = \frac{a\sqrt{3}}{8AD}\cdot\frac{4\sqrt{2}\,AD}{3a} = \frac{1}{\sqrt{6}}$$

36. In ΔACD, $\cos C = \dfrac{b}{\left(\dfrac{a}{2}\right)}$

$$\frac{a^2+b^2-c^2}{2ab} = \frac{2b}{a}$$

$\therefore \quad a^2+b^2-c^2 = 4b^2$

$\Rightarrow \quad 3b^2 = a^2 - c^2$

37. Let $AF = f$, $BG = g$ and $CH = h$

Area of ΔABC = area of ΔABF + area of ΔACF

$$\frac{1}{2} bc \sin A = \frac{1}{2} cf \sin \frac{A}{2} + \frac{1}{2} bf \sin \frac{A}{2}$$

$\Rightarrow \quad f = \dfrac{2bc}{(b+c)} \cos \dfrac{A}{2}$

$\therefore \quad \dfrac{1}{f} \cos \dfrac{A}{2} = \dfrac{b+c}{2bc} = \dfrac{1}{2}\left(\dfrac{1}{b}+\dfrac{1}{c}\right)$

Similarly, $\dfrac{1}{g} \cos \dfrac{B}{2} = \dfrac{1}{2}\left(\dfrac{1}{c}+\dfrac{1}{a}\right)$

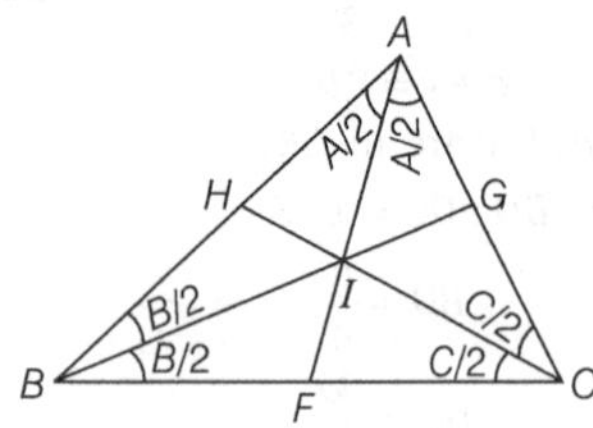

and $\dfrac{1}{h} \cos \dfrac{C}{2} = \dfrac{1}{2}\left(\dfrac{1}{a}+\dfrac{1}{b}\right)$

$\therefore \quad \dfrac{1}{f} \cos \dfrac{A}{2} + \dfrac{1}{g} \cos \dfrac{B}{2} + \dfrac{1}{h} \cos \dfrac{C}{2} = \dfrac{1}{a}+\dfrac{1}{b}+\dfrac{1}{c}$

38. $g(x) = f[f\{f(x)\}]$

$$= f\left\{f\left(\frac{1}{1-x}\right)\right\} = f\left(\frac{1}{1-\frac{1}{1-x}}\right)$$

$$= f\left(\frac{1-x}{-x}\right) = \frac{1}{1-\frac{1-x}{-x}} = x$$

$\because \quad (a-b)^2+(b-c)^2+(c-a)^2 \ge 0$

$\Rightarrow \quad 2(a^2+b^2+c^2) \ge 2(ab+bc+ca)$

$\Rightarrow \quad 3(a^2+b^2+c^2) \ge (a+b+c)^2 > d^2$

$\therefore \quad \dfrac{d^2}{a^2+b^2+c^2} < 3 = g(3)$

$\Rightarrow \quad \dfrac{d^2}{a^2+b^2+c^2} < g(3)$

39. $\sin^3 A + \sin^3 B + \sin^3 C = 3 \sin A \sin B \sin C$

From sine rule,

$$a^3k^3 + b^3k^3 + c^3k^3 = 3abc\, k^3$$

$\therefore \quad a^3+b^3+c^3 = 3abc$

$\therefore \quad a+b+c = 0$

In determinant $C_1 \to C_1 + C_2 + C_3$

Then, put $a+b+c=0$

Then, all elements of C_1 are zero.

$\therefore$ Value of determinant is zero.

40. $r = r_2 + r_3 - r_1$

$\Rightarrow \quad \dfrac{\Delta}{s} = \dfrac{\Delta}{s-b} + \dfrac{\Delta}{s-c} - \dfrac{\Delta}{s-a}$

$\Rightarrow \quad \dfrac{1}{s} + \dfrac{1}{s-a} = \dfrac{1}{s-b} + \dfrac{1}{s-c}$

$\Rightarrow \quad \dfrac{2s-a}{2s-b-c} = \dfrac{s(s-a)}{(s-b)(s-c)}$

$\Rightarrow \quad \dfrac{2s-a}{a} = \cot^2 \dfrac{A}{2}$

$\Rightarrow \quad \dfrac{s}{a} = \dfrac{1}{2}\left(\cot^2 \dfrac{A}{2} + 1\right) \Rightarrow \dfrac{s}{a} \in \left(\dfrac{1}{2}, 2\right)$

41. $\theta = B + \dfrac{A}{2} = B + \dfrac{180° - (B+C)}{2}$

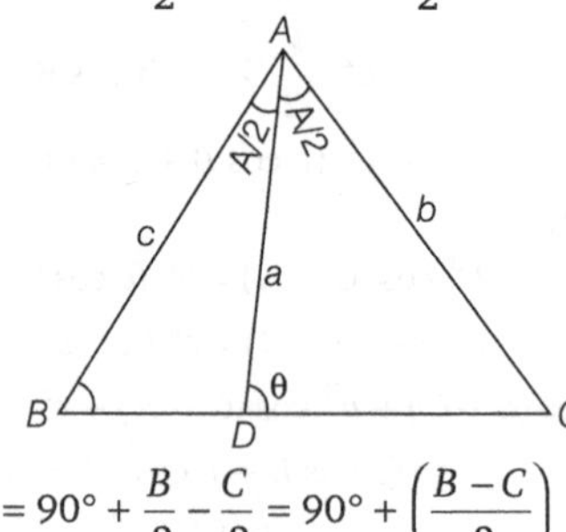

$$= 90° + \frac{B}{2} - \frac{C}{2} = 90° + \left(\frac{B-C}{2}\right)$$

$\therefore \quad \sin\theta = \cos\left(\dfrac{B-C}{2}\right)$

42. $\because \quad \dfrac{b+c}{11} = \dfrac{c+a}{12} = \dfrac{a+b}{13} = \dfrac{2(a+b+c)}{36}$

or $\quad \dfrac{b+c}{11} = \dfrac{c+a}{12} = \dfrac{a+b}{13}$

$$= \frac{a+b+c}{18} = k \text{ (say)}$$

$\therefore \quad b+c = 11k$

$c+a = 12k$

$a+b = 13k$

$a+b+c = 18k$

$\therefore \quad a = 7k,\ b = 6k,\ c = 5k$

$\Rightarrow \quad \cos A = \dfrac{(6k)^2+(5k)^2-(7k)^2}{2\cdot 6k \cdot 5k} = \dfrac{1}{5}$

$\cos B = \dfrac{(7k)^2+(5k)^2-(6k)^2}{2\cdot 7k\cdot 5k} = \dfrac{19}{35}$

$\cos C = \dfrac{(7k)^2+(6k)^2-(5k)^2}{2\cdot 7k\cdot 6k} = \dfrac{5}{7}$

$\therefore \quad \cos A : \cos B : \cos C = 7 : 19 : 25$

43. We know that, $\frac{a}{\sin A} = \frac{b}{\sin B} = \frac{c}{\sin C} = 2R$

and given that, $a + b = x$, $ab = y$ and $x^2 - c^2 = y$

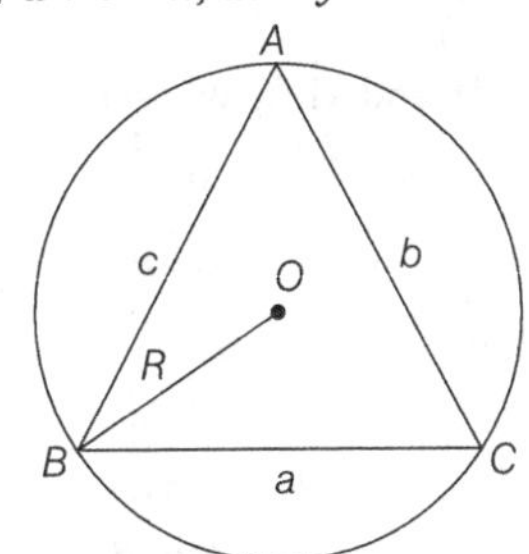

$\therefore \quad (a+b)^2 - c^2 = ab$

$\Rightarrow \quad a^2 + b^2 - c^2 = -2ab + ab$

$\Rightarrow \quad a^2 + b^2 - c^2 = -ab$

$\Rightarrow \quad \frac{a^2 + b^2 - c^2}{2ab} = \frac{-ab}{2ab} = -\frac{1}{2}$

$\therefore \quad \cos C = -\frac{1}{2} \Rightarrow C = 120°$

[using cosine rule, $\cos C = \frac{a^2 + b^2 - c^2}{2ab}$]

Now, $\frac{c}{\sin C} = 2R \Rightarrow R = \frac{1}{2} \frac{c}{\sin(120°)} = \frac{c}{2} \frac{2}{\sqrt{3}}$

$\therefore \quad R = \frac{c}{\sqrt{3}}$

44. $\cos A \cos B + \sin A \sin B \sin C = 1$

$\sin C = \frac{1 - \cos A \cos B}{\sin A \sin B} \leq 1 \quad \ldots(i)$

$\Rightarrow \quad 1 \leq \cos A \cos B + \sin A \sin B$

or $\quad \cos(A - B) \geq 1$

$\therefore \quad \cos(A - B) = 1$

$\Rightarrow \quad A = B$

From Eq. (i),

$\sin C = \frac{1 - \cos^2 A}{\sin^2 A} = 1$

$\therefore \quad \angle C = 90°$

$\therefore \quad \angle A = \angle B = 45°$

$a : b : c = \sin A : \sin B : \sin C$

$= \frac{1}{\sqrt{2}} : \frac{1}{\sqrt{2}} : 1$

or $\quad a : b : c = 1 : 1 : \sqrt{2}$

45. $\because A = B = C = 60°$, if length of side is λ.

$\therefore \quad R = \frac{\lambda}{2\sin 60°} = \frac{\lambda}{\sqrt{3}} = \frac{2\lambda}{2\sqrt{3}}$

$r = (s - a)\tan\left(\frac{60°}{2}\right)$

$= \left(\frac{3\lambda}{2} - \lambda\right)\tan 30° = \frac{\lambda}{2} \cdot \frac{1}{\sqrt{3}}$

and $\quad r_1 = s\tan\frac{A}{2} = \frac{3\lambda}{2} \cdot \tan 30°$

$= \frac{3\lambda}{2} \cdot \frac{1}{\sqrt{3}} = \frac{3\lambda}{2\sqrt{3}}$

$\therefore \quad R : r : r_1 = 2 : 1 : 3$

46. Since, $a^2 + b^2 + c^2 = 8R^2$

$\Rightarrow \quad (2R\sin A)^2 + (2R\sin B)^2 + (2R\sin C)^2 = 8R^2$

$\Rightarrow \quad \sin^2 A + \sin^2 B + \sin^2 C = 2$

$\Rightarrow \quad \cos^2 A - \sin^2 B + \cos^2 C = 0$

$\Rightarrow \quad \cos(A + B)\cos(A - B) + \cos^2 C = 0$

$\Rightarrow \quad \cos(\pi - C)\cos(A - B) + \cos^2 C = 0$

$\Rightarrow \quad -\cos C\{\cos(A - B) - \cos C\} = 0$

$\Rightarrow \quad -\cos C\{\cos(A - B) + \cos(A + B)\} = 0$

$\Rightarrow \quad -2\cos A\cos B\cos C = 0$

$\therefore \quad \cos A = 0$ or $\cos B = 0$ or $\cos C = 0$

$\therefore \quad A = \frac{\pi}{2}$ or $B = \frac{\pi}{2}$ or $C = \frac{\pi}{2}$

47. $A_0A_1 = 2 \cdot 1 \cdot \cos\frac{\pi}{3} = 1 = A_1A_2$

$\cos\frac{2\pi}{3} = \frac{(A_0A_1)^2 + (A_1A_2)^2 - (A_0A_2)^2}{2 \cdot A_0A_1 \cdot A_1A_2}$

$-\frac{1}{2} = \frac{1 + 1 - (A_0A_2)^2}{2 \cdot 1 \cdot 1}$

$\therefore \quad A_0A_2 = \sqrt{3} = A_0A_4$

$\therefore \quad A_0A_1 \times A_0A_2 \times A_0A_4 = 1 \times \sqrt{3} \times \sqrt{3} = 3$

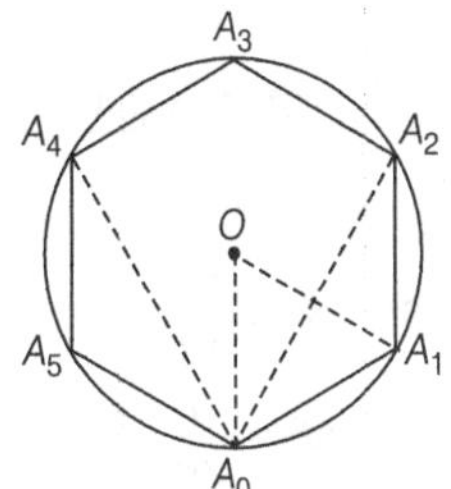

48. $\Delta = \frac{1}{2}a\alpha = \frac{1}{2}b\beta = \frac{1}{2}c\gamma$

$\Rightarrow \quad \frac{1}{\alpha^2} + \frac{1}{\beta^2} + \frac{1}{\gamma^2} = \frac{a^2 + b^2 + c^2}{4\Delta^2}$

and $\cot A + \cot B + \cot C$

$= \frac{2R}{2abc}[b^2 + c^2 - a^2 + c^2 + a^2 - b^2 + a^2 + b^2 - c^2]$

$= \frac{R(a^2 + b^2 + c^2)}{abc} = \frac{a^2 + b^2 + c^2}{4\Delta}$

Hence, $\frac{1}{\alpha^2} + \frac{1}{\beta^2} + \frac{1}{\gamma^2} = \frac{\cot A + \cot B + \cot C}{\Delta}$

49. $\frac{a\cos A + b\cos B + c\cos C}{a + b + c}$

$= \frac{R(2\sin A\cos A + 2\sin B\cos B + 2\sin C\cos C)}{2s}$

$= \frac{R(\sin 2A + \sin 2B + \sin 2C)}{2s}$

$= \frac{R \times 4\sin A\sin B\sin C}{2s}$

$= \frac{4R \times \frac{a}{2R} \times \frac{b}{2R} \times \frac{c}{2R}}{2s} = \frac{abc}{4R^2 s} = \frac{\Delta}{sR} = \frac{r}{R}$

50. Since a, b, c are roots of $x^3 - 11x^2 + 38x - 40 = 0$

$\therefore \quad a + b + c = 11,\ ab + bc + ca = 38$

and $\quad abc = 40$

$$\therefore \frac{\cos A}{a} + \frac{\cos B}{b} + \frac{\cos C}{c} = \frac{a^2 + b^2 + c^2}{2abc}$$

$$= \frac{(a + b + c)^2 - 2(ab + bc + ca)}{2abc}$$

$$= \frac{(11)^2 - 2(38)}{2(40)} = \frac{121 - 76}{80}$$

$$= \frac{45}{80} = \frac{9}{16}$$

51. Let $\quad A = 22\frac{1}{2}° \Rightarrow \therefore 4A = 90°$

$A = 90° - 3A$

$\sin A = \cos 3A \quad \ldots(i)$

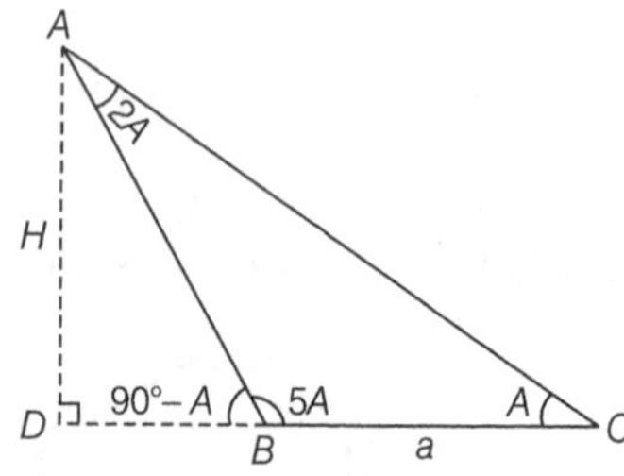

$\therefore$ In ΔABD

$$\sin(90° - A) = \frac{H}{AB}$$

or $\quad AB = \dfrac{H}{\cos A} \quad \ldots(i)$

and in ΔABC

$$\frac{a}{\sin 2A} = \frac{AB}{\sin A}$$

or $\quad \dfrac{a}{2\cos A} = \dfrac{H}{\cos A}$

$\therefore \quad H = \dfrac{a}{2}$

52. In ΔABC, the median AD divides $\angle BAC$ such $\angle BAD : \angle CAD = 2 : 1$.

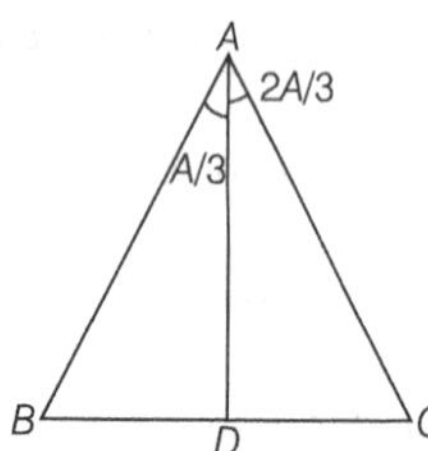

In ΔABD, $\dfrac{BD}{\sin \frac{A}{3}} = \dfrac{AD}{\sin B} \quad \ldots(i)$

In ΔACD, $\dfrac{CD}{\sin \frac{2A}{3}} = \dfrac{AD}{\sin C} \quad \ldots(ii)$

$BD = CD \quad$ [$\because$ AD is median]

$\because \quad \dfrac{BD}{\sin \frac{2A}{3}} = \dfrac{AD}{\sin C} \quad \ldots(iii)$

From Eqs. (i) and (iii), we get

$$\frac{\sin \frac{2A}{3}}{\sin C} = \frac{\sin \frac{A}{3}}{\sin B}$$

$$\Rightarrow \frac{2\sin \frac{A}{3} \cos \frac{A}{3}}{\sin C} = \frac{\sin \frac{A}{3}}{\sin B}$$

$$\Rightarrow \cos \frac{A}{3} = \frac{\sin C}{2\sin B}$$

53. Let r be the radius of the circle and A be its area

$\therefore \quad A = \pi r^2$

Let length of side of polygon is a according to the question.

$2\pi r = na$

$\Rightarrow \quad a = \dfrac{2\pi r}{n} \quad \ldots(i)$

If A_1 be the area of polygon, then

$$A_1 = \frac{1}{4} na^2 \cot\left(\frac{\pi}{n}\right)$$

$$= \frac{1}{4} n \cdot \frac{4\pi^2 r^2}{n^2} \cot\left(\frac{\pi}{n}\right)$$

$$= \frac{\pi^2 r^2}{n} \cot\left(\frac{\pi}{n}\right)$$

$$\therefore \frac{A}{A_1} = \frac{\pi r^2}{\frac{\pi^2 r^2}{n} \cot\left(\frac{\pi}{n}\right)}$$

$$\therefore A : A_1 = \tan\left(\frac{\pi}{n}\right) : \frac{\pi}{n}$$

54. $\because r_1, r_2, r_3$ are in HP.

$\Rightarrow \quad \dfrac{1}{r_1}, \dfrac{1}{r_2}, \dfrac{1}{r_3}$ are in AP.

$\therefore \quad \dfrac{s-a}{\Delta}, \dfrac{s-b}{\Delta}, \dfrac{s-c}{\Delta}$ are in AP.

Subtract in each term by $\dfrac{s}{\Delta}$

Then, $-\dfrac{a}{\Delta}, -\dfrac{b}{\Delta}, -\dfrac{c}{\Delta}$ are in AP

Multiplying in each term by $-\Delta$

Then a, b, c are in AP.

55. $$\frac{\sin^2 A + \sin A + 1}{\sin A} = \sin A + \frac{1}{\sin A} + 1$$

$$= \left(\sqrt{\sin A} - \frac{1}{\sqrt{\sin A}}\right)^2 + 3 \geq 3$$

Minimum value of $\dfrac{\sin^2 A + \sin A + 1}{\sin A} = 3$

$\therefore$ Minimum value of

$$\Sigma \frac{\sin^2 A + \sin A + 1}{\sin A} = 3 + 3 + 3 = 9$$

56. $2(2R)^2 = \frac{1}{2}(a^2+b^2+c^2)$

$\Rightarrow a^2+b^2+c^2 = 16R^2$

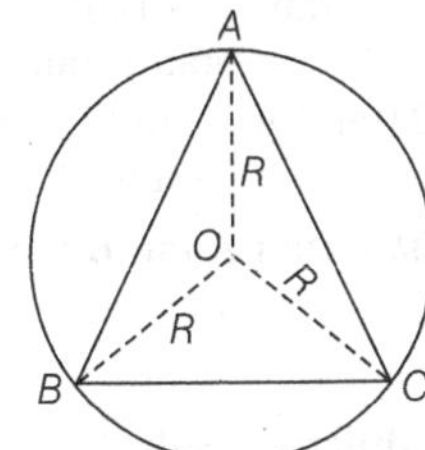

$\therefore \quad (2R\sin A)^2 + (2R\sin B)^2 + (2R\sin C)^2 = 16R^2$

$\Rightarrow \quad \sin^2 A + \sin^2 B + \sin^2 C = 4$

57. We have,

$$2\frac{\cos A}{a} + \frac{\cos B}{b} + 2\frac{\cos C}{c} = \frac{a}{bc} + \frac{b}{ca}$$

$$\Rightarrow 2\left(\frac{b^2+c^2-a^2}{2abc}\right) + \left(\frac{c^2+a^2-b^2}{2abc}\right) + 2\left(\frac{a^2+b^2-c^2}{2abc}\right) = \frac{a^2+b^2}{abc}$$

$$\Rightarrow b^2+c^2 = a^2 \Rightarrow \angle A = \frac{\pi}{2}$$

58. $\left(1-\frac{r_1}{r_2}\right)\left(1-\frac{r_1}{r_3}\right) = 2$

$$\Rightarrow \left(1-\frac{\frac{\Delta}{s-a}}{\frac{\Delta}{s-b}}\right)\left(1-\frac{\frac{\Delta}{s-a}}{\frac{\Delta}{s-c}}\right) = 2$$

$$\Rightarrow \left(1-\frac{s-b}{s-a}\right)\left(1-\frac{s-c}{s-a}\right) = 2$$

$\Rightarrow \quad (b-a)(c-a) = 2(s-a)^2$

$\Rightarrow \quad 2(bc-ab-ac+a^2) = (2s-2a)^2$

$\Rightarrow \quad 2bc-2ab-2ac+2a^2 = (b+c-a)^2$

$\Rightarrow \quad 2bc-2ab-2ac+2a^2 = b^2+c^2+a^2+2bc-2ac-2ab$

$\Rightarrow \quad b^2+c^2 = a^2$

$\Rightarrow \quad 2bc\cos A = 0$

$\therefore \quad \angle A = 90°$

59. We have, $c^2x^2 - c(a+b)x + ab = 0$

$\Rightarrow \quad c^2x^2 - cax - cbx + ab = 0$

$\Rightarrow \quad cx(cx-a) - b(cx-a) = 0$

$\Rightarrow \quad (cx-a)(cx-b) = 0$

$\Rightarrow \quad cx-a = 0$ or $cx-b=0$

$\Rightarrow \quad x = \frac{a}{c}$ and $x = \frac{b}{c}$

As, $\sin A$ and $\sin B$ are roots of the given equation.

Let $\sin A = \frac{a}{c}$ and $\sin B = \frac{b}{c}$

$\Rightarrow \quad c = \frac{a}{\sin A}$ and $c = \frac{b}{\sin B}$

$\Rightarrow \quad \frac{a}{\sin A} = \frac{b}{\sin B} = c \qquad \text{...(i)}$

Using sine law, $\frac{a}{\sin A} = \frac{b}{\sin B} = \frac{c}{\sin C} \qquad \text{...(ii)}$

So, $\frac{c}{\sin C} = c \qquad$ [from Eqs. (i) and (ii)]

$\Rightarrow \quad \sin C = \frac{c}{c} \Rightarrow \sin C = 1 \qquad \left[\because \sin\frac{\pi}{2} = 1\right]$

$\Rightarrow \quad \angle C = \frac{\pi}{2}$

Since, ΔABC is right angled at $\angle C$.

In ΔABC,

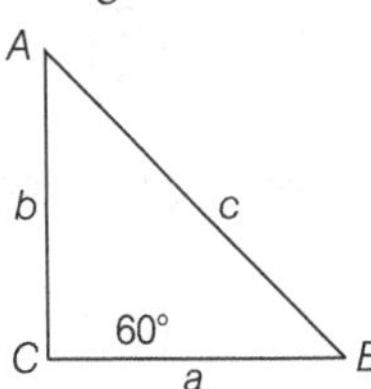

$\sin A = \frac{a}{c}$ and $\cos A = \frac{b}{c}$

$\therefore \sin A + \cos A = \frac{a}{c} + \frac{b}{c} = \frac{a+b}{c}$

60. $\because \quad OA = 2R\cos A,$

$OB = 2R\cos B$

and $\quad OC = 2R\cos C$

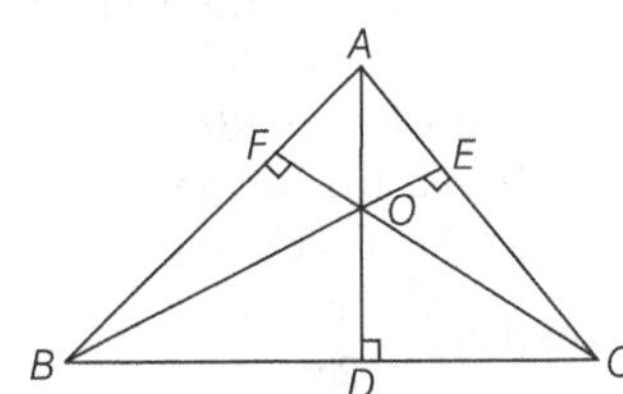

$\therefore \quad OA : OB : OC = \cos A : \cos B : \cos C$

61. $\because BD : DC = AB : AC = c : b$

$\frac{BD}{DC} = \frac{c}{b}$ or $1 + \frac{BD}{DC} = 1 + \frac{c}{b}$

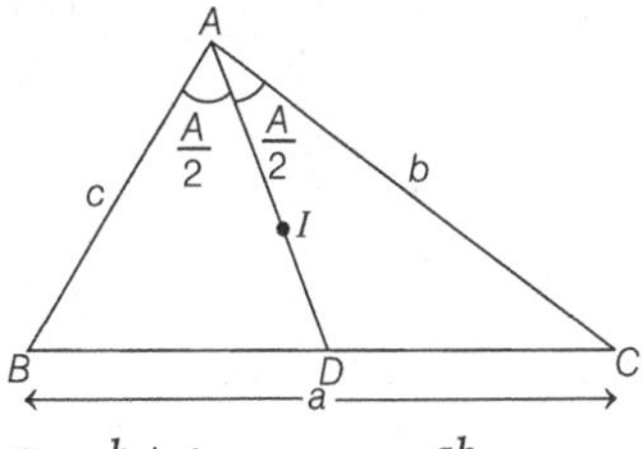

$\frac{a}{DC} = \frac{b+c}{b} \Rightarrow DC = \frac{ab}{b+c}$

and $\quad BD = \frac{ac}{b+c}$

In ΔABD, $\frac{BD}{\sin\left(\frac{A}{2}\right)} = \frac{AD}{\sin B}$

$$\Rightarrow AD = \frac{\left(\frac{ac}{b+c}\right)\sin B}{\sin\left(\frac{A}{2}\right)} = \frac{2\Delta}{(b+c)}\operatorname{cosec}\left(\frac{A}{2}\right)$$

$\because \quad \frac{AI}{AD} = \frac{(b+c)}{(a+b+c)}$

$\therefore \quad AI = \frac{(b+c)}{(a+b+c)} \cdot \frac{2\Delta}{(b+c)} \operatorname{cosec}\left(\frac{A}{2}\right)$

$= \frac{2\Delta}{2s} \operatorname{cosec}\left(\frac{A}{2}\right)$

$= \frac{\Delta}{s} \operatorname{cosec}\left(\frac{A}{2}\right)$

Similarly, $BI = \frac{\Delta}{s} \operatorname{cosec}\left(\frac{B}{2}\right)$

and $CI = \frac{\Delta}{s} \operatorname{cosec}\left(\frac{C}{2}\right)$

$\therefore IA : IB : IC = \operatorname{cosec}\left(\frac{A}{2}\right) : \operatorname{cosec}\left(\frac{B}{2}\right) : \operatorname{cosec}\left(\frac{C}{2}\right)$

62. Given, $\frac{b+c}{11} = \frac{c+a}{12} = \frac{a+b}{13} = \lambda$ (say)

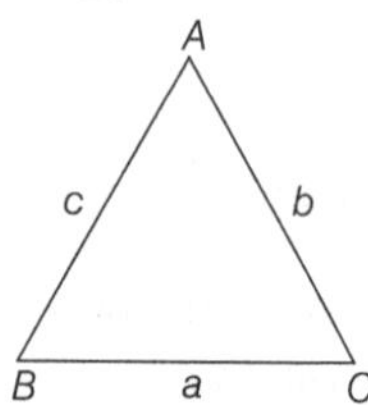

$b + c = 11\lambda$, $c + a = 12\lambda$ and $a + b = 13\lambda$ ...(i)

$\Rightarrow \quad 2(a+b+c) = 36\lambda$

$\Rightarrow \quad a + b + c = 18\lambda$ ...(ii)

From Eqs. (i) and (ii), we get

$a = 7\lambda,\ b = 6\lambda,\ c = 5\lambda$

Now, $\cos A = \frac{b^2 + c^2 - a^2}{2bc}$

$= \frac{\lambda^2[36 + 25 - 49]}{60\lambda^2} = \frac{12}{60} = \frac{1}{5}$

$\cos B = \frac{a^2 + c^2 - b^2}{2ac} = \frac{\lambda^2[49 + 25 - 36]}{70\lambda^2} = \frac{19}{35}$

and $\cos C = \frac{a^2 + b^2 - c^2}{2ab} = \frac{\lambda^2[49 + 36 - 25]}{84\lambda^2}$

$= \frac{60}{84} = \frac{5}{7}$

Thus, $\cos A = \frac{1}{5} = \frac{7}{35}$, $\cos B = \frac{19}{35}$, $\cos C = \frac{25}{35}$

$\frac{\cos A}{7} = \frac{\cos B}{19} = \frac{\cos C}{25} = \frac{1}{35}$

$\Rightarrow \quad (\alpha, \beta, \gamma) = (7, 19, 25)$

63. Here, $A > B$

$\therefore \quad \tan\left(\frac{A-B}{2}\right) = \frac{a-b}{a+b} \cot\frac{C}{2}$

$\Rightarrow \quad \tan 30^\circ = \frac{1}{3} \cot\frac{C}{2}$

$\Rightarrow \quad \cot\frac{C}{2} = \sqrt{3}$ or $\frac{C}{2} = \frac{\pi}{6}$

$\therefore \quad C = \frac{\pi}{3}$

64. $\because (\tan A - \tan B)^2 + (\tan B - \tan C)^2 + (\tan C - \tan A)^2 > 0$

$\Rightarrow \quad 2(\tan^2 A + \tan^2 B + \tan^2 C - \tan A \tan B - \tan B \tan C - \tan C \tan A) > 0$

$\Rightarrow \quad 3(\tan^2 A + \tan^2 B + \tan^2 C) - (\tan A + \tan B + \tan C)^2 > 0$

$\Rightarrow \quad 3\lambda - (\tan A \tan B \tan C)^2 > 0$

$3\lambda - 81 > 0$

$\therefore \quad \lambda > 27$

65. $\because$ Area of quadrilateral $= 4\sqrt{3}$

$\Rightarrow \quad \frac{1}{2} \cdot 2 \cdot 5 \cdot \sin 60^\circ + \frac{1}{2} \cdot a \cdot b \cdot \sin 120^\circ = 4\sqrt{3}$

$\Rightarrow \quad \frac{5\sqrt{3}}{2} + \frac{ab\sqrt{3}}{4} = 4\sqrt{3}$

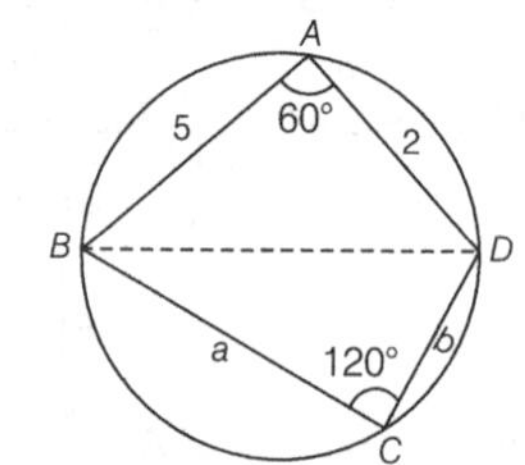

$\Rightarrow \quad \frac{5}{2} + \frac{ab}{4} = 4 \Rightarrow ab = 16 - 10 = 6$

and $\cos A = \frac{2^2 + 5^2 - a^2 - b^2}{2(10 + ab)} = \frac{1}{2}$

or $29 - a^2 - b^2 = 16$

or $a^2 + b^2 = 13$

$\therefore \quad (a+b) = \sqrt{a^2 + b^2 + 2ab} = \sqrt{(13 + 12)} = 5$

i.e. $a + b = 5$, $ab = 6$

a, b are the roots of

$t^2 - 5t + 6 = 0$

$\therefore \quad t = 2, 3$

$\Rightarrow \quad a = 2, b = 3$ or $a = 3, b = 2$

66. $a^2 \cos^2 A = b^2 + c^2$ ...(i)

$\because \quad \cos A = \frac{b^2 + c^2 - a^2}{2bc}$

$= \frac{a^2 \cos^2 A - a^2}{2bc}$ [from Eq. (i)]

$= -\frac{a^2}{2bc} \sin^2 A < 0$

$\Rightarrow \quad \cos A < 0$

$\therefore \quad A > \frac{\pi}{2}$

67. $\tan\left(\frac{A-B}{2}\right) = \frac{1}{3} \tan\left(\frac{A+B}{2}\right) = \frac{1}{3} \cot\frac{C}{2}$

$\because \quad \tan\left(\frac{A-B}{2}\right) = \left(\frac{a-b}{a+b}\right) \cot\frac{C}{2}$

$\therefore \quad \frac{a-b}{a+b} = \frac{1}{3} \Rightarrow 3a - 3b = a + b$

$\Rightarrow \quad 2a = 4b$

Then, $a : b = 2 : 1$

68. $\because \quad \Delta = \frac{1}{2} \cdot a \cdot p_1$

$\therefore \quad \frac{1}{p_1} = \frac{a}{2\Delta}$

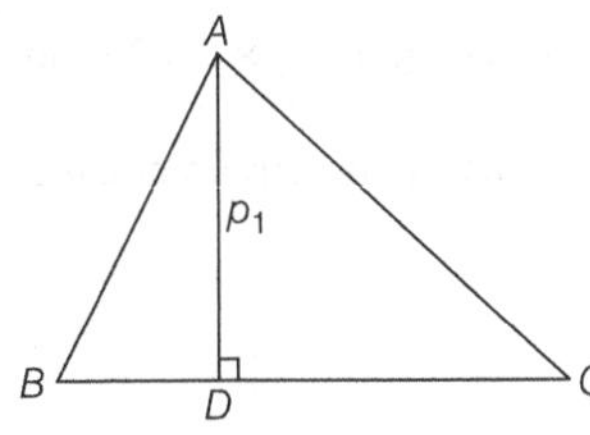

Similarly, $\frac{1}{p_2} = \frac{b}{2\Delta}, \frac{1}{p_3} = \frac{c}{2\Delta}$

$\therefore \quad p_1^{-1} + p_2^{-1} - p_3^{-1} = \frac{a+b-c}{2\Delta}$

$= \frac{2(s-c)}{2\Delta} = \left(\frac{s-c}{\Delta}\right)$

69. Applying m–n cotangent theorem,

$\left(\frac{a}{2} + \frac{a}{2}\right) \cot (90° + B)$

$= \frac{a}{2} \cot 90° - \frac{a}{2} \cot (A - 90°)$

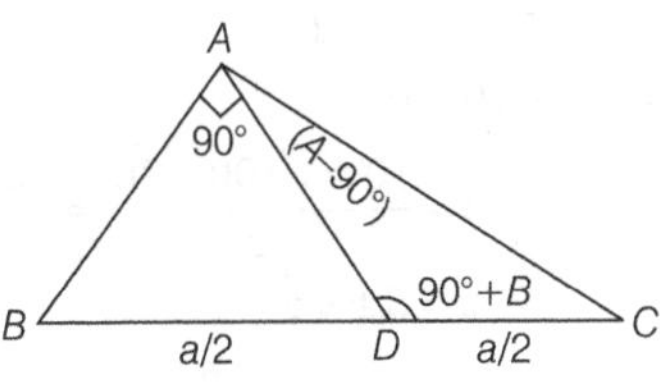

$\Rightarrow \quad -a \tan B = 0 + \frac{a}{2} \tan A$

$\therefore \quad \tan A + 2 \tan B = 0$

70. $\cos A + \cos B + \cos C = \frac{3}{2}$

$\Rightarrow \quad 1 + 4 \sin \frac{A}{2} \sin \frac{B}{2} \sin \frac{C}{2} = \frac{3}{2}$

$\Rightarrow \quad \sin \frac{A}{2} \sin \frac{B}{2} \sin \frac{C}{2} = \frac{1}{8}$

It is possible only when

$\sin \frac{A}{2} = \frac{1}{2}, \sin \frac{B}{2} = \frac{1}{2}, \sin \frac{C}{2} = \frac{1}{2}$

$\therefore \quad \angle A = \angle B = \angle C = 60°$

71. If radius of circle is r, then

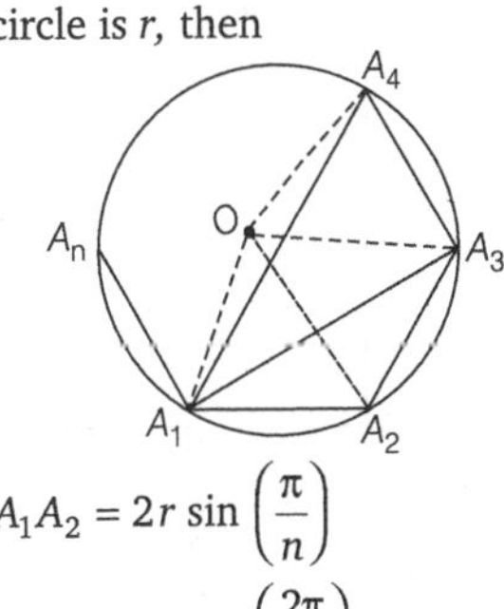

$A_1A_2 = 2r \sin \left(\frac{\pi}{n}\right)$

$A_1A_3 = 2r \sin \left(\frac{2\pi}{n}\right)$

$A_1A_4 = 2r \sin \left(\frac{3\pi}{n}\right)$

$\because \quad \frac{1}{A_1A_2} = \frac{1}{A_1A_3} + \frac{1}{A_1A_4}$

$\Rightarrow \quad \frac{1}{2r \sin \left(\frac{\pi}{n}\right)} = \frac{1}{2r \sin \left(\frac{2\pi}{n}\right)} + \frac{1}{2r \sin \left(\frac{3\pi}{n}\right)}$

$\Rightarrow \quad \sin \left(\frac{2\pi}{n}\right) \sin \left(\frac{3\pi}{n}\right) = \sin \left(\frac{3\pi}{n}\right) \sin \left(\frac{\pi}{n}\right) + \sin \left(\frac{2\pi}{n}\right) \sin \left(\frac{\pi}{n}\right)$

$\Rightarrow \quad \sin \left(\frac{2\pi}{n}\right) \left\{\sin \left(\frac{3\pi}{n}\right) - \sin \left(\frac{\pi}{n}\right)\right\} = \sin \left(\frac{3\pi}{n}\right) \sin \left(\frac{\pi}{n}\right)$

$\Rightarrow \quad \sin \left(\frac{2\pi}{n}\right) \cdot 2 \cos \left(\frac{2\pi}{n}\right) \sin \left(\frac{\pi}{n}\right) = \sin \left(\frac{3\pi}{n}\right) \sin \left(\frac{\pi}{n}\right)$

$\Rightarrow \quad \sin \left(\frac{4\pi}{n}\right) = \sin \left(\frac{3\pi}{n}\right) \left(\sin \left(\frac{\pi}{n}\right) \neq 0\right)$

$\Rightarrow \quad \frac{4\pi}{n} = r\pi + (-1)^r \frac{3\pi}{n}, r \in I$

$\Rightarrow \quad \frac{4}{n} = r + (-1)^r \frac{3}{n}$

for $\quad r = 1, n = 7$

72. Here, $\quad p = \sin A \sin B \sin C$

and $\quad q = \cos A \cos B \cos C$

$\therefore \quad \tan A \tan B \tan C = \frac{p}{q}$

and $\tan A \tan B + \tan B \tan C + \tan C \tan A$

$= \frac{\sin A \sin B \cos C + \sin B \sin C \cos A + \sin C \sin A \cos B}{\cos A \cos B \cos C}$

$= \frac{\sin B (\sin A \cos C + \cos A \sin C) + \sin C (\sin A \cos B)}{\cos A \cos B \cos C}$

$= \frac{\sin B \sin (A + C) + \sin C \sin A \cos B}{\cos A \cos B \cos C}$

$= \frac{\sin^2 B + \sin C \sin A \cos B}{\cos A \cos B \cos C}$

$= \frac{1 - \cos B (\cos B - \sin C (\sin A)}{q}$

$= \frac{1 - \cos B (-\cos (A + C) - \sin A \sin C}{q}$

$= \left(\frac{1+q}{q}\right)$

and $\tan A + \tan B + \tan C$

$= \tan A \tan B \tan C$ (in a ΔABC)

$= \frac{p}{q}$

$\therefore$ Required equation is

$x^3 - \frac{p}{q} x^2 + \left(\frac{1+q}{q}\right) x - \frac{p}{q} = 0$

$\Rightarrow \quad qx^3 - px^2 + (1+q)x - p = 0$

73. $\because \tan\frac{A}{2} = \frac{5}{6}, \tan\frac{C}{2} = \frac{2}{5}$

$\therefore \quad \cot\frac{A}{2}\cot\frac{C}{2} = 3$

$\Rightarrow \quad \sqrt{\frac{s(s-a)}{(s-b)(s-c)}} \cdot \sqrt{\frac{s(s-c)}{(s-a)(s-b)}} = 3$

$\Rightarrow \quad \frac{s}{s-b} = 3$

$\Rightarrow \quad 2b = a + c$

Hence a, b, c are in AP.

74. M is the mid point of BC

$\therefore \quad MC = \frac{a}{2}$

$OC = R$

$OM = ID = r$

$\therefore \Delta OMC \quad R^2 = r^2 + \frac{a^2}{4}$

$R^2 = r^2 + \frac{(2R\sin A)^2}{4}$

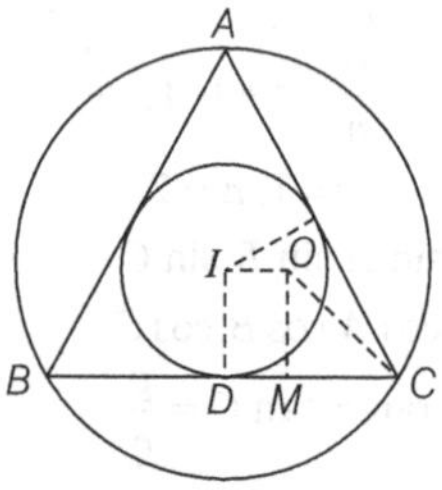

$\Rightarrow \quad R^2\cos^2 A = r^2$

$\Rightarrow \quad \cos A = \frac{r}{R} \Rightarrow r = R\cos A$

$\because \cos A + \cos B + \cos C$

$= 1 + 4\sin\frac{A}{2}\sin\frac{B}{2}\sin\frac{C}{2}$

$= 1 + \frac{r}{R} = 1 + \cos A$

$\therefore \cos B + \cos C = 1$

75. Let $A = \theta, B = 2\theta, C = 3\theta$

$\because \quad A + B + C = 180°$

$\Rightarrow \quad 6\theta = 180°$

$\therefore \quad \theta = 30°$

$\therefore A = 30°, B = 60°, C = 90°$

$a : b : c = \sin 30° : \sin 60° : \sin 90°$

$= \frac{1}{2} : \frac{\sqrt{3}}{2} : 1$

$\therefore a : b : c = 1 : \sqrt{3} : 2$

76. Let the sides of the triangle be $7, 7-d, 7-2d$

Since, the given angle is the greatest (being obtuse) angle of the triangle, it is opposite to the greatest side of the triangle and we have,

$7^2 = (7-d)^2 + (7-2d)^2 - 2(7-d)(7-2d)\cos\frac{2\pi}{3}$

$\Rightarrow \quad 7^2 = 2\times 7^2 - 42d + 5d^2 - 2(7^2 - 21d + 2d^2)(-1/2)$

$\Rightarrow \quad d^2 - 9d + 14 = 0$

$\Rightarrow \quad (d-7)(d-2) = 0$

$\Rightarrow \quad d = 2 \ (d = 7 \text{ is not possible})$

Therefore, the sides of the triangle are 7 cm, 5 cm and 3 cm.

Area of the triangle $= \frac{1}{2}\times 5\times 3\times \sin\frac{2\pi}{3} = \frac{15\sqrt{3}}{4}$ cm^2

and the radius of the circumcircle $R = \frac{7\times 5\times 3}{4(15\sqrt{3}/4)}$

$= \frac{7\times 5\times 3}{15\sqrt{3}}$

$= \frac{7\sqrt{3}}{3}$ cm

77. Let $a > b > c$

$\therefore \quad a = 10$ and $b = 9$

According to the question,

$2B = A + C$

or $\quad 2B = 180° - B$

$\therefore \quad B = 60°$

$\therefore \quad \cos B = \frac{a^2 + c^2 - b^2}{2ac}$

$\frac{1}{2} = \frac{100 + c^2 - 81}{20c}$

$\Rightarrow \quad 10c = c^2 + 19$

$\Rightarrow c^2 - 10c + 19 = 0$

$c = \frac{10 \pm \sqrt{100 - 76}}{2} = \frac{10 \pm \sqrt{24}}{2}$

$c = 5 \pm \sqrt{6}$

78. We have,

$a^2 + b^2 + c^2 = ca + ab\sqrt{3}$

$\Rightarrow \quad a^2 + b^2 + c^2 - ca - ab\sqrt{3} = 0$

$\Rightarrow \quad \left(\frac{a\sqrt{3}}{2} - b\right)^2 + \left(\frac{a}{2} - c\right)^2 = 0$

It is possible only when

$\frac{a\sqrt{3}}{2} - b = 0$ and $\frac{a}{2} - c = 0$

$\Rightarrow \quad \sqrt{3}a = 2b = 2c\sqrt{3} = k$ (let)

$\Rightarrow \quad a = \frac{k}{\sqrt{3}}, b = \frac{k}{2}, c = \frac{k}{2\sqrt{3}}$

$\because \quad b^2 + c^2 = a^2$

$\therefore \quad \angle A = 90°$

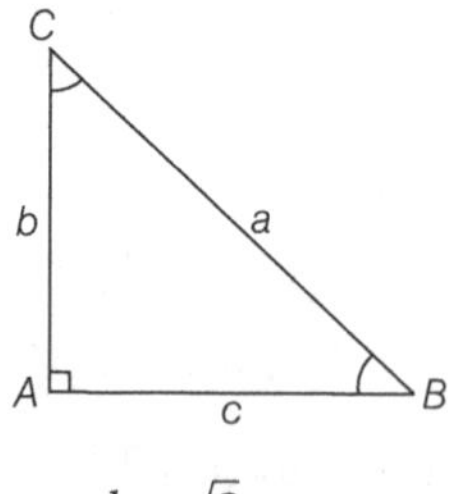

$\therefore \quad \sin B = \frac{b}{a} = \frac{\sqrt{3}}{2}$

$\angle B = 60°, \angle C = 30°$

79. ΔDEF is equilateral with side $2r$

If radius of circumcircle DEF is R_1

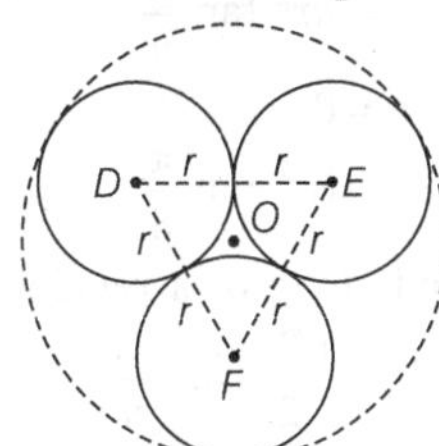

Then, area of $\Delta DEF = \frac{\sqrt{3}}{4}(2r)^2 = \sqrt{3}r^2$

$$\sqrt{3}r^2 = \frac{2r \cdot 2r \cdot 2r}{4R_1} \Rightarrow R_1 = \frac{2r}{\sqrt{3}}$$

$\therefore$ Radius of the circle touching all the three given circles $= r + R_1 = r + \frac{2r}{\sqrt{3}}$

$$= \frac{(2+\sqrt{3})r}{\sqrt{3}}$$

80. In pedal ΔDEF

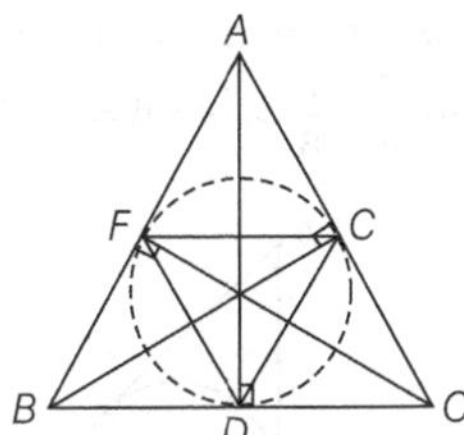

$$EF = a\cos A$$
$$DE = c\cos C$$
$$DF = b\cos B$$

If circum radius of ΔDEF is R_1

Then, $R_1 = \frac{(a\cos A)(b\cos B)(c\cos C)}{4\cdot\frac{1}{2}\cdot DF\cdot DE\sin(\angle EDF)}$

(Here, $\angle EDF = 180° - 2A$)

$$= \frac{abc\cos A\cos B\cos C}{4\cdot\frac{1}{2}\cdot b\cos B\cdot c\cos C\cdot\sin(180°-2A)}$$

$$= \frac{a\cos A}{2\sin 2A}$$

$$= \frac{2R\sin A\cos A}{4\sin A\cos A}$$

$\therefore \quad R_1 = \frac{R}{2}$

81. Let $ABCD$ is a right angled trapezium. Let radius of circle is R and let $BD = BE = x$ and $CE = CF = y$

$\because \quad AB = a$

$R + x = a \Rightarrow x = a - R \quad \ldots$(i)

and $\quad DC = b$

$R + y = b \Rightarrow y = b - R \quad \ldots$(ii)

In ΔODB,

$\tan\alpha = \frac{R}{x} \quad \ldots$(iii)

Now, in ΔCHB

$$\sin 2\alpha = \frac{CH}{BC}$$

$$\Rightarrow \frac{2\tan\alpha}{1+\tan^2\alpha} = \frac{2R}{x+y}$$

$$\Rightarrow \frac{\frac{2R}{x}}{1+\frac{R^2}{x^2}} = \frac{2R}{x+y} \Rightarrow \frac{2Rx}{x^2+R^2} = \frac{2R}{x+y}$$

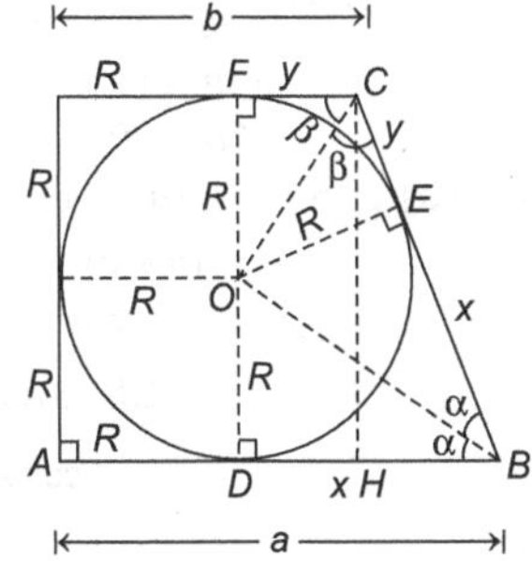

$$\Rightarrow x^2 + xy = x^2 + R^2$$
$$\Rightarrow xy = R^2$$
$$\Rightarrow (a-R)(b-R) = R^2 \quad \text{[from Eqs. (i) and (ii)]}$$
$$\Rightarrow ab - (a+b)R + R^2 = R^2$$
$$\Rightarrow ab - (a+b)R = 0$$
$$\therefore R = \frac{ab}{a+b}$$

82. $\because \quad (a-b)^2 + (b-c)^2 + (c-a)^2 \geq 0$

$$\Rightarrow 2(a^2+b^2+c^2) \geq 2ab + 2bc + 2ca$$
$$\Rightarrow 3(a^2+b^2+c^2) \geq a^2+b^2+c^2+2ab+2bc+2ca$$
$$\Rightarrow 3(a^2+b^2+c^2) > (a+b+c)^2 > d^2$$

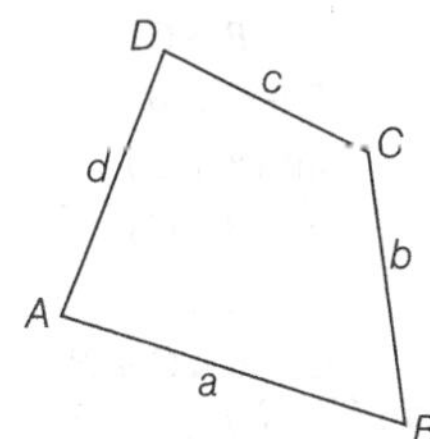

$$\Rightarrow 3(a^2+b^2+c^2) > d^2$$
$$\Rightarrow \frac{a^2+b^2+c^2}{d^2} > \frac{1}{3}$$

$\therefore$ Minimum value of $\frac{a^2+b^2+c^2}{d^2}$ is $\frac{1}{3}$.

83. $r_1r_2r_3 - r[r_1r_2 + r_2r_3 + r_3r_1]$

$$\Rightarrow \frac{\Delta}{s-a}\cdot\frac{\Delta}{s-b}\cdot\frac{\Delta}{s-c} - \frac{\Delta}{s}\left\{\frac{\Delta^2}{(s-a)(s-b)} + \frac{\Delta^2}{(s-b)(s-c)} + \frac{\Delta^2}{(s-c)(s-a)}\right]$$

$$\Rightarrow \frac{\Delta^3\cdot s}{\Delta^2} - \frac{\Delta}{s}\{s^2\}$$

$$\Rightarrow \Delta s - \Delta s = 0$$

84. $\frac{\sin A}{c} + \frac{\sin C}{a} = \frac{a}{bc} + \frac{c}{ab}$

$$\Rightarrow \quad ab \sin A + bc \sin C = a^2 + c^2 \quad \left[\because \frac{a}{\sin A} = \frac{c}{\sin C}\right]$$

$$ab \times \frac{a \sin C}{c} + bc \sin C = a^2 + c^2$$

$$\Rightarrow \quad \frac{b}{c} \sin C\, (a^2 + c^2) = (a^2 + c^2)$$

$$\therefore \quad \sin C = \frac{c}{b} = \frac{k \sin C}{k \sin B}$$

$$\therefore \quad \sin B = 1$$

$$\angle B = 90°$$

85. Alternate. (a) : $(b + c + a)(b + c - a) = 5bc$

$$\Rightarrow \quad (b + c)^2 - a^2 = 5bc$$

$$\Rightarrow \quad b^2 + c^2 + 2bc - a^2 = 5bc$$

$$\Rightarrow \quad 2bc \cos A = 3bc$$

$$\cos A = \frac{3}{2}$$

impossible.

Alternate. (b) : Let $a = \sqrt{19}, b = \sqrt{38}, c = \sqrt{116}$

$$\cos A = \frac{19 + 38 - 116}{2\sqrt{19}\sqrt{38}} = -\frac{59}{2\sqrt{722}} = -\frac{29.5}{\sqrt{722}} < -1$$

impossible.

Alternate. (c) : $\because \frac{b^2 - c^2}{a^2} = \frac{\sin^2 B - \sin^2 C}{\sin^2 A}$

$$= \frac{\sin (B + C) \sin (B - C)}{\sin^2 A}$$

$$= \frac{\sin (B - C)}{\sin A}$$

$$= \frac{\sin (B - C)}{\sin (B + C)} \neq 0 \quad [\because B \neq C]$$

$\because$ If $\quad B > C$

Then, $\quad B + C > B - C$

$$\frac{\sin (B + C)}{\sin (B - C)} > 1$$

and if $\quad B < C$

$$C - B < C + B$$

or $\quad \frac{\sin (C - B)}{\sin (C + B)} < 1$ or $\frac{\sin(B - C)}{\sin (B + C)} > -1$

Hence, $\quad \frac{b^2 - c^2}{a^2} + \frac{c^2 - a^2}{b^2} + \frac{a^2 - b^2}{c^2} \neq 0$

Alternate. (d) :

$$\cos \left(\frac{B - C}{2}\right) = (\sin B + \sin C) \cos \left(\frac{B + C}{2}\right)$$

$$\cos \left(\frac{B - C}{2}\right) = 2 \sin \left(\frac{B + C}{2}\right) \cos \left(\frac{B - C}{2}\right) \cos \left(\frac{B + C}{2}\right)$$

$$\cos \left(\frac{B - C}{2}\right) \neq 0$$

$$\therefore \quad \sin (B + C) = 1$$

$$\therefore \quad B + C = \frac{\pi}{2}$$

then, $\quad \angle A = \frac{\pi}{2}$

86. Inscribed circle of a regular polygon of n sides

$$A_1 = nr^2 \tan \frac{\pi}{n}$$

Here, $\quad n = 8$

$$\therefore \quad A_1 = 8r^2 \tan \frac{\pi}{8}$$

Circumscribed circle of a regular polygon of n sides

$$A_2 = \frac{nR^2}{2} \sin \frac{2\pi}{n}$$

for $\quad n = 8$

$$A_2 = \frac{8r^2}{2} \sin \frac{\pi}{4} \quad [\text{given } R = r]$$

$$= 8r^2 \sin \frac{\pi}{8} \cos \frac{\pi}{8}$$

$$\therefore \quad \frac{A_2}{A_1} = \cos^2 \frac{\pi}{8}$$

87. Let $AE = AF = \alpha,\ BD = BE = \beta,\ CD = CF = \gamma$

$$\because \quad AB + BC + CA = 2(\alpha + \beta + \gamma)$$

$$2s = 2(\alpha + \beta + \gamma)$$

$$\therefore \quad \alpha + \beta + \gamma = s$$

$$\alpha = s - a, \beta = s - b, \gamma = s - c$$

In ΔBID, $\tan \frac{\pi}{4} = \frac{r}{\beta} \Rightarrow r = \beta = s - b = \frac{a + b + c}{2} - b$

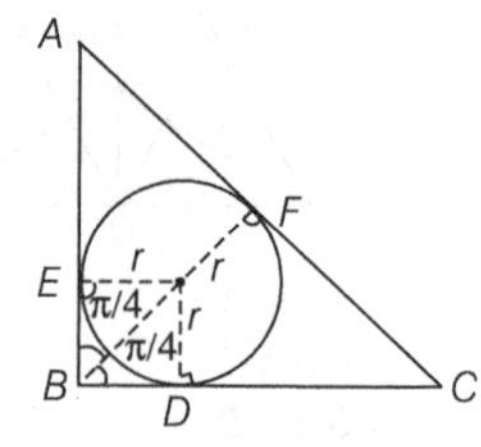

$$= \frac{a + c - b}{2} = \frac{AB + BC - CA}{2}$$

88. Since, $\frac{a \cos A + b \cos B + c \cos C}{a \sin B + b \sin C + c \sin A} = \frac{a + b + c}{9R}$

$$\Rightarrow \quad \frac{2R \sin A \cos A + 2R \sin B \cos B + 2R \sin C \cos C}{a \times \frac{b}{2R} + b \times \frac{c}{2R} + c \times \frac{a}{2R}}$$

$$= \frac{a + b + c}{9R}$$

$$\Rightarrow \quad \frac{R (\sin 2A + \sin 2B + \sin 2C)}{\frac{(ab + bc + ca)}{2R}} = \frac{a + b + c}{9R}$$

$$\Rightarrow \quad \frac{R (4 \sin A \sin B \sin C) \cdot 2R}{(ab + bc + ca)} = \frac{a + b + c}{9R}$$

$$\Rightarrow \quad \frac{2R^2 \left(4 \cdot \frac{a}{2R} \cdot \frac{b}{2R} \cdot \frac{c}{2R}\right)}{(ab + bc + ca)} = \frac{a + b + c}{9R}$$

$$9abc = (a + b + c)(ab + bc + ca)$$

$$\Rightarrow \quad a (b - c)^2 + b (c - a)^2 + c(a - b)^2 = 0$$

Which is possible only when

$$a - b = 0,\ b - c = 0,\ c - a = 0$$

$$\therefore \quad a = b = c$$

89. $\because AD = \dfrac{abc}{b^2 - c^2} = \dfrac{k \sin A \cdot b \cdot k \sin C}{k^2 (\sin^2 B - \sin^2 C)}$

$AD = \dfrac{b \sin C}{\sin (B - C)} = b \sin C$ (in ΔADC)

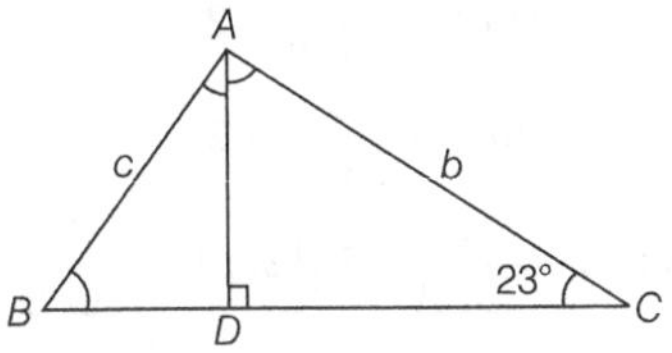

$\therefore \quad \sin (B - C) = 1$

$\therefore \quad B - C = 90°$

$\therefore \quad B = 90° + 23° = 113°$

90. $\because \quad \dfrac{b + c}{2} \geq \sqrt{bc}$

$\Rightarrow \quad \dfrac{b + c}{2} \geq \lambda \quad \ldots(i)$

and $\quad \dfrac{a}{\sin A} = \dfrac{b}{\sin B} = \dfrac{c}{\sin C} = \dfrac{b + c}{\sin B + \sin C}$

$= \dfrac{(b + c)}{2 \sin \left(\dfrac{B + C}{2}\right) \cos \left(\dfrac{B - C}{2}\right)}$

$= \dfrac{(b + c)}{2 \cos \dfrac{A}{2} \cos \left(\dfrac{B - C}{2}\right)}$

$\therefore \quad a = \left(\dfrac{b + c}{2}\right) \cdot \dfrac{\sin A}{\cos \dfrac{A}{2} \cos \left(\dfrac{B - C}{2}\right)}$

$= \left(\dfrac{b + c}{2}\right) \cdot \dfrac{2 \sin \dfrac{A}{2}}{\cos \left(\dfrac{B - C}{2}\right)}$

$\geq \left(\dfrac{b + c}{2}\right) \cdot 2 \sin \dfrac{A}{2} \quad \left[\because \sec \left(\dfrac{B - C}{2}\right) \geq 1\right]$

$\geq 2\lambda \sin \dfrac{A}{2} \quad$ [from Eq. (i)]

Hence, $\quad a \geq 2\lambda \sin \dfrac{A}{2}$

91. $\because$ A is obtuse angle.

Then, $\quad 90° < A < 180°$

$\Rightarrow \quad 90° < 180° - (B + C) < 180°$

$\Rightarrow \quad -90° > (B + C) - 180° > -180°$

$\Rightarrow \quad 90° > (B + C) > 0$

$\because \quad B + C < 90°$

$\Rightarrow \quad B < (90° - C)$

$\therefore \quad \tan B < \tan (90° - C)$

$\Rightarrow \quad \tan B < \cot C$

$\therefore \quad \tan B \tan C < \tan C \cot C$

$\Rightarrow \quad \tan B \tan C < 1$

92. Given, $\sin A : \sin B : \sin C :: 3 : 5 : 7$

$\therefore$ Sides $\quad a = 3k, b = 5k, c = 7k$

$\therefore \quad \cot A = \dfrac{\cos A}{\sin A} = \dfrac{\cos A}{\dfrac{2\Delta}{bc}} = \dfrac{bc \cos A}{2\Delta}$

$\Rightarrow \quad \cot A = \dfrac{b^2 + c^2 - a^2}{4\Delta} = \dfrac{65}{4\Delta}$

Similarly, $\quad \cot B = \dfrac{c^2 + a^2 - b^2}{4\Delta} = \dfrac{33}{4\Delta}$

and $\quad \cot C = \dfrac{a^2 + b^2 - c^2}{4\Delta} = \dfrac{-15}{4\Delta}$

$\therefore \quad \cot A : \cot B : \cot C :: 65 : 33 : -15$

93. $\therefore f(x + y) = f(x) \cdot f(y)$

for $\quad x = y = 1, \; f(2) = 2^2$

for $\quad x = 1, y = 2$

$f(3) = f(1) f(2) = 2^3$

$\therefore \quad f(n) = 2^n$

$\therefore \quad a = f(3) = 2^3 = 8,$

$b = f(1) + f(3) = 2 + 8 = 10$

and $\quad c = f(2) + f(3) = 2^2 + 2^3 = 12$

$\cos A = \dfrac{3}{4}$ and $\cos C = \dfrac{1}{8}$

$\cos 2A = 2 \cos^2 A - 1 = 2 \times \dfrac{9}{16} - 1 = \dfrac{1}{8} = \cos C$

$\therefore \quad 2A = C$

94. $\dfrac{\cot B + \cot C}{\cot A} = \dfrac{\dfrac{ac \cos B}{2\Delta} + \dfrac{ab \cos C}{2\Delta}}{\dfrac{bc \cos A}{2\Delta}}$

$= \dfrac{a (c \cos B + b \cos C)}{bc \cos A} = \dfrac{2a^2}{(b^2 + c^2 - a^2)}$

$= \dfrac{2}{\dfrac{b^2 + c^2}{a^2} - 1} = \dfrac{2}{1999 - 1} = \dfrac{2}{1998} = \dfrac{1}{999}$

95. $\because x^2 - 9x + 8 = 0$

$\therefore \quad x = 1, 8$

$\Rightarrow \quad 2^{\tan^2 B} = 1, 8$

$\Rightarrow \quad 2^{\tan^2 B} = 2^0, 2^3$

$\tan^2 B \neq 0$

$\therefore \quad \tan^2 B = 3 \Rightarrow \tan B = \sqrt{3}$

$\therefore \quad \angle B = 60°$

Also given, $2 \cos A = \dfrac{\sin B}{\sin C}$

$\Rightarrow \quad 2 \cos A \sin C = \sin B$

$\Rightarrow \quad \sin (A + C) - \sin (A - C) = \sin B$

$\Rightarrow \quad \sin (A - B) - \sin (A - C) = \sin B$

Then, $\quad \sin (A - C) = 0$

$\Rightarrow \quad A = C$

$\Rightarrow \quad A + B + C = 180°$

$A + 60° + A = 180°$

$A = 60° = C, B = 60°$

96. $\dfrac{\sin A}{\sin C} = \dfrac{\sin (A - B)}{\sin (B - C)}$

$\Rightarrow \quad \dfrac{\sin \{180° - (B + C)\}}{\sin \{180° - (A + B)\}} = \dfrac{\sin (A - B)}{\sin (B - C)}$

$\Rightarrow \quad \dfrac{\sin (B + C)}{\sin (A + B)} = \dfrac{\sin (A - B)}{\sin (B - C)}$

$\Rightarrow \quad \sin(B+C)\sin(B-C) = \sin(A+B)\sin(A-B)$

$\Rightarrow \quad \sin^2 B - \sin^2 C = \sin^2 A - \sin^2 B$

$\Rightarrow \quad \frac{b^2}{k^2} - \frac{c^2}{k^2} = \frac{a^2}{k^2} - \frac{b^2}{k^2}$

$\Rightarrow \quad 2b^2 = a^2 + c^2$

97. $A = 3k, B = 5k, C = 4k$

$\because \quad A + B + C = 180°$

$\Rightarrow \quad 12k = 180° \Rightarrow k = 15°$

$\therefore \quad A = 45°, B = 75°, C = 60°$

$a + b + c\sqrt{2} = k(\sin A + \sin B + \sqrt{2}\sin C)$

$= k\left(\frac{1}{\sqrt{2}} + \frac{\sqrt{3}+1}{2\sqrt{2}} + \sqrt{2}\frac{\sqrt{3}}{2}\right)$

$= k\left(\frac{1}{\sqrt{2}} + \frac{\sqrt{3}+1}{2\sqrt{2}} + \frac{\sqrt{3}}{\sqrt{2}}\right)$

$= k\left(\frac{2+\sqrt{3}+1+2\sqrt{3}}{2\sqrt{2}}\right) = 3k\left(\frac{\sqrt{3}+1}{2\sqrt{2}}\right)$

$= 3k \sin 75° = 3b$

98. $A + B + C = \pi \Rightarrow A + C = \frac{\pi}{4}$

$\tan(A + C) = 1$

$\Rightarrow \quad \tan A + \tan C = 1 - \tan A \tan C \quad \ldots(i)$

$\because \quad a\tan\theta + b\sec\theta = c$

$\Rightarrow \quad (a\tan\theta - c)^2 = b^2\sec^2\theta$

$\Rightarrow \quad a^2\tan^2\theta + c^2 - 2ac\tan\theta = b^2 + b^2\tan^2\theta$

$\Rightarrow \quad (a^2 - b^2)\tan^2\theta - 2ac\tan\theta + c^2 - b^2 = 0$

$\because \quad \tan A + \tan C = \frac{2ac}{a^2 - b^2}$

$\tan A \tan C = \frac{c^2 - b^2}{a^2 - b^2}$ [from Eq.(i)]

$\frac{2ac}{a^2 - b^2} = 1 - \frac{c^2 - b^2}{a^2 - b^2}$

$\Rightarrow \quad 2ac = a^2 - c^2$

99. $A = \pi/3$

$\tan\frac{C-B}{2} = \frac{c-b}{c+b}\cot\frac{A}{2}$

$\tan\left(\frac{C-B}{2}\right) = \frac{1}{5}\cot 30° = \frac{\sqrt{3}}{5} = \tan\theta$

$\therefore \quad C - B = 2\theta$

$C + B = 180° - A = 120°$

$\therefore \quad 2C = 120° + 2\theta$

$\therefore \quad C = 60° + \theta$

100. $\frac{1}{\sqrt{A_1}} + \frac{1}{\sqrt{A_2}} + \frac{1}{\sqrt{A_3}} = \frac{1}{\sqrt{\pi r_1^2}} + \frac{1}{\sqrt{\pi r_2^2}} + \frac{1}{\sqrt{\pi r_3^2}}$

$= \frac{1}{\sqrt{\pi}}\left(\frac{1}{r_1} + \frac{1}{r_2} + \frac{1}{r_3}\right) = \frac{1}{\sqrt{\pi}}\left(\frac{1}{r}\right)$

$= \frac{1}{\sqrt{\pi r^2}} = \frac{1}{\sqrt{A}}$

101. $\Pi\left(\sin A + 1 + \frac{1}{\sin A}\right)$

$\because \quad \sin A + \frac{1}{\sin A} \geq 2$ [if $A = 90°$]

Then, $\quad \sin B + \frac{1}{\sin B} > 2$

and $\quad \sin C + \frac{1}{\sin C} > 2$

$\therefore \quad \left(\sin A + \frac{1}{\sin A} + 1\right)\left(\sin B + \frac{1}{\sin B} + 1\right)\left(\sin C + \frac{1}{\sin C} + 1\right) > 27$

$\therefore \quad \Pi\left(\sin A + \frac{1}{\sin A} + 1\right) > 27$

or $\quad \Pi\left(\frac{\sin^2 A + \sin A + 1}{\sin A}\right) > 27$

102. $\because$ AM $\geq$ GM

$\therefore \quad \frac{(s-a)+(s-b)+(s-c)}{3} \geq \{(s-a)(s-b)(s-c)\}^{1/3}$

$\Rightarrow \quad \left(\frac{s}{3}\right)^3 \geq \frac{A^2}{s}$ or $A^2 \leq \frac{s^4}{27}$

$\therefore \quad A \leq \frac{s^2}{3\sqrt{3}}$

103. $\because \quad r_1 > r_2 > r_3$

$\Rightarrow \quad \frac{\Delta}{s-a} > \frac{\Delta}{s-b} > \frac{\Delta}{s-c} \Rightarrow s-a < s-b < s-c$

$\Rightarrow \quad -a < -b < -c \Rightarrow a > b > c$

104. Let linear function is $F(x) = Ax + B$

$\because \quad [1, 2] \to [4, 6]$

$\Rightarrow \quad F(1) = 4 \Rightarrow A + B = 4$

and $\quad F(2) = 6 \Rightarrow 2A + B = 6$

$\therefore \quad A = 2, B = 2$

Then, one function is $F(x) = 2x + 2 = f(x)$ (say)

$\left.\begin{array}{l} F(1) = 6 \Rightarrow A + B = 6 \\ F(2) = 4 \Rightarrow 2A + B = 4 \end{array}\right\} A = -2, B = 8$

then other function is $F(x) = -2x + 8 = g(x)$ (say)

$\therefore \quad c = f(1) + g(1) = 4 + 6 = 10$

Now, $\quad x^2 + y^2 - xy = 10$

$\Rightarrow \quad \frac{\left(x - \frac{y}{2}\right)^2}{(\sqrt{10})^2} + \frac{y^2}{\left(\frac{\sqrt{10}}{\sqrt{3}}\right)^2} = 1$

is an ellipse whose centre (0, 0).

Maximum distance from origin on any point on ellipse = Semi major axis = $\sqrt{10}$

$\therefore \quad r = \sqrt{10}$

Then, $\quad a = r^2 = 10$

$\because \quad a = c = 10$

$\therefore \quad \sin A : \sin C = 1 : 1$

105. $\therefore \quad \angle BIC = 180° - \left(\frac{B}{2} + \frac{C}{2}\right) = \left(90° + \frac{A}{2}\right)$

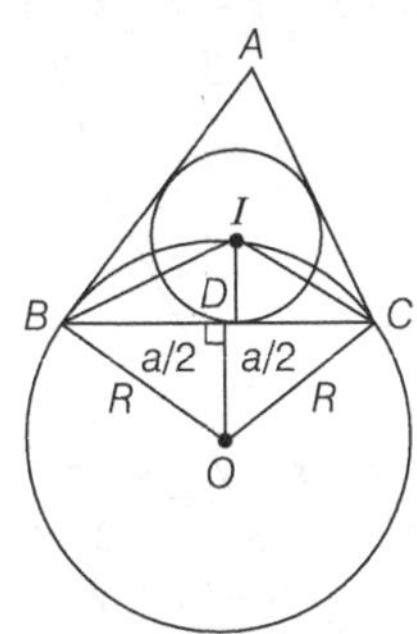

$\therefore \quad \angle BOC = 2\angle BIC = (180° + A)$

Then, $\angle BOD = \left(90° + \frac{A}{2}\right)$

$\therefore \quad \sin\left(90° + \frac{A}{2}\right) = \frac{\frac{a}{2}}{R} = \cos\left(\frac{A}{2}\right)$

$\therefore \quad R = \frac{a}{2}\sec\left(\frac{A}{2}\right)$

106. Given that,

$$\cot\frac{A}{2} : \cot\frac{B}{2} : \cot\frac{C}{2} = 4:3:2 \quad \text{...(i)}$$

$$\therefore \sqrt{\frac{s(s-a)}{(s-b)(s-c)}} : \sqrt{\frac{s(s-b)}{(s-a)(s-c)}} : \sqrt{\frac{s(s-c)}{(s-a)(s-b)}} = 4:3:2$$

On multiplying $\sqrt{(s-a)(s-b)(s-c)}$

$(s-a):(s-b):(s-c) = 4:3:2$

$\Rightarrow \frac{a+b+c}{2} - a : \frac{a+b+c}{2} - b : \frac{a+b+c}{2} - c = 4:3:2$

$\Rightarrow \frac{b+c-a}{2} : \frac{a+c-b}{2} : \frac{a+b-c}{2} = 4:3:2$

$\Rightarrow b+c-a : a+c-b : a+b-c = 4:3:2$ ...(ii)

So, from Eqs. (i) and (ii),

$b+c-a = 4k$ ...(iii)

$a+c-b = 3k$ ...(iv)

$a+b-c = 2k$ ...(v)

From Eqs. (iii) and (iv),

$2c = 7k,\ c = \frac{7k}{2}$

From Eqs. (iv) and (v),

$2a = 5k$

$a = \frac{5k}{2}$

From Eqs. (iii) and (v),

$2b = 6k \Rightarrow b = \frac{6k}{2}$

Now, $a:b:c$

$= \frac{5k}{2} : \frac{6k}{2} : \frac{7k}{2}$

$\Rightarrow \quad 5:6:7$

Objective Questions Type II [One or more than one correct answer(s)]

1. $\Delta = \Delta_1 + \Delta_2$

$\Rightarrow \frac{1}{2}ab\sin C = \frac{1}{2}b(CF)\sin\frac{C}{2} + \frac{1}{2}a(CF)\sin\frac{C}{2}$

or $\quad CF = \frac{ab\sin C}{(a+b)\sin\frac{C}{2}}$

$= \frac{2ab\cos\frac{C}{2}}{a+b}$

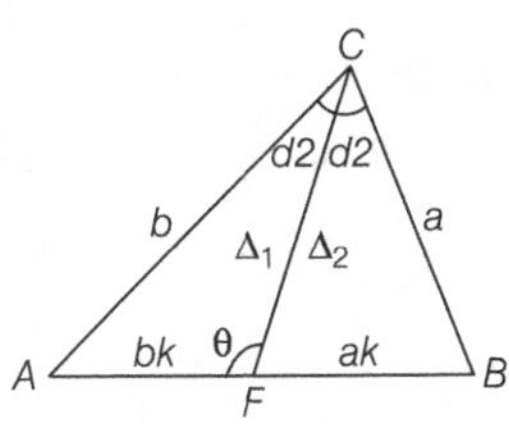

Again, in ΔCFB, by the sine rule, we have

$\frac{CF}{\sin B} = \frac{a}{\sin(\pi - \theta)} = \frac{a}{\sin\theta} = \frac{a}{\sin\left(B + \frac{C}{2}\right)} \left(\because \theta + B + \frac{C}{2} = \pi\right)$

or $CF = \frac{a\sin B}{\sin\left(B + \frac{C}{2}\right)} = \frac{b\sin A}{\sin\left(B + \frac{C}{2}\right)}$ [by sine rule]

2. $R = \frac{b}{2\sin\alpha} = \frac{b}{2}\operatorname{cosec}\alpha$

$\Delta = \frac{1}{2}b^2\sin(180° - 2\alpha) = \frac{1}{2}b^2\sin 2\alpha$

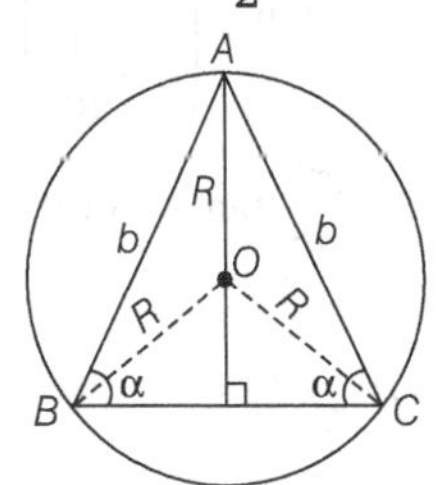

and $\quad r = 4R\sin\frac{\alpha}{2}\cdot\sin\frac{\alpha}{2}\cdot\sin(90° - \alpha)$

$= \frac{2b}{\sin\alpha}\cdot\sin^2\frac{\alpha}{2}\cdot\cos\alpha\cdot\frac{\sin\alpha}{\sin\alpha}$

$= \frac{b(1-\cos\alpha)\sin 2\alpha}{2(1-\cos^2\alpha)}$

$= \frac{b\sin 2\alpha}{2(1+\cos\alpha)}$

and $\quad OI = \sqrt{(R^2 - 2Rr)} = R\sqrt{\left(1 - \frac{2r}{R}\right)}$

$= R\sqrt{(1 - 4\cos\alpha + 4\cos^2\alpha)}$

$= R(2\cos\alpha - 1)$

$$= R\left\{2\left(2\cos^2\frac{\alpha}{2}-1\right)-1\right\}$$

$$= R\left(4\cos^2\frac{\alpha}{2}-3\right)$$

$$= \frac{R\left(4\cos^3\frac{\alpha}{2}-3\cos\frac{\alpha}{2}\right)}{\cos\frac{\alpha}{2}} = \frac{R\cos\left(\frac{3\alpha}{2}\right)}{\cos\frac{\alpha}{2}}$$

$$= \left|\frac{b\cos\left(\frac{3\alpha}{2}\right)}{2\sin\alpha\cos\left(\frac{\alpha}{2}\right)}\right|$$

3. $A = 15°$, $b = 10\sqrt{2}$ and for unique triangle $b \le a$

$\because \quad \sin B = \dfrac{b\sin A}{a} = \dfrac{10\sqrt{2}\times\dfrac{\sqrt{3}-1}{2\sqrt{2}}}{a} = \dfrac{5(\sqrt{3}-1)}{a}$

For triangle $\sin B \le 1$

$\therefore \quad \dfrac{5(\sqrt{3}-1)}{a} \le 1 \Rightarrow a \ge 5(\sqrt{3}-1)$

$\therefore \quad a = 10\sqrt{2}, 15, 5(\sqrt{3}-1)$

4. $a = 5, b = 4, A = \dfrac{\pi}{2} + B$

$\therefore \tan\left(\dfrac{A-B}{2}\right) = \dfrac{a-b}{a+b}\cot\dfrac{C}{2}$

$\Rightarrow \quad \tan\left(\dfrac{\pi}{4}\right) = \dfrac{1}{9}\cot\dfrac{C}{2}$

$\tan\dfrac{C}{2} = \dfrac{1}{9}$

$\therefore \quad C = 2\tan^{-1}\left(\dfrac{1}{9}\right)$

$= \tan^{-1}\left(\dfrac{\frac{2}{9}}{1-\frac{1}{81}}\right) = \tan^{-1}\left(\dfrac{9}{40}\right)$

5. $\tan A + \tan B = \dfrac{c^2}{ab}$, $\tan A\tan B = 1$

$\because \tan(A+B) = \dfrac{\frac{c^2}{ab}}{1-1} = \infty$

$\therefore \quad A + B = \dfrac{\pi}{2}$, $\therefore C = \dfrac{\pi}{2}$

$\therefore \quad \cos C = \dfrac{a^2+b^2-c^2}{2ab} = 0$

$\therefore \quad a^2 + b^2 = c^2$

$(2R\sin A)^2 + (2R\sin B)^2 = (2R\sin C)^2$

$\sin^2 A + \sin^2 B + \sin^2 C = 2\sin^2 C = 2 \left(\because C = \dfrac{\pi}{2}\right)$

$\tan A + \tan B = \dfrac{c^2}{ab} = \dfrac{a^2+b^2}{ab} = \dfrac{a}{b} + \dfrac{b}{a}$

But, $\quad \tan A\tan B = 1 = \dfrac{a}{b}\cdot\dfrac{b}{a}$

$\therefore \quad \tan A = \dfrac{a}{b}, \tan B = \dfrac{b}{a}$

6. Given a ΔXYZ, where $2s = x + y + z$

and $\quad \dfrac{s-x}{4} = \dfrac{s-y}{3} = \dfrac{s-z}{2}$

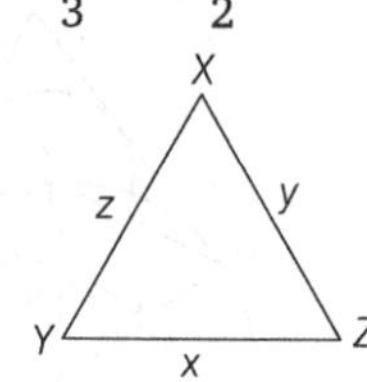

$\therefore \quad \dfrac{s-x}{4} = \dfrac{s-y}{3} = \dfrac{s-z}{2}$

$= \dfrac{3s-(x+y+z)}{4+3+2} = \dfrac{s}{9}$

or $\quad \dfrac{s-x}{4} = \dfrac{s-y}{3} = \dfrac{s-z}{2} = \dfrac{s}{9} = \lambda$ (let)

$\Rightarrow \quad s = 9\lambda,\ s = 4\lambda + x,\ s = 3\lambda + y$

and $\quad s = 2\lambda + z$

$\therefore \quad s = 9\lambda,\ x = 5\lambda,\ y = 6\lambda,\ z = 7\lambda$

Now, $\quad \Delta = \sqrt{s(s-x)(s-y)(s-z)}$ [Heron's formula]

$= \sqrt{9\lambda\cdot 4\lambda\cdot 3\lambda\cdot 2\lambda} = 6\sqrt{6}\lambda^2$ ...(i)

Also, $\quad \pi r^2 = \dfrac{8\pi}{3}$

$\Rightarrow \quad r^2 = \dfrac{8}{3}$ ...(ii)

and $\quad R = \dfrac{xyz}{4\Delta} = \dfrac{(5\lambda)(6\lambda)(7\lambda)}{4\cdot 6\sqrt{6}\lambda^2} = \dfrac{35\lambda}{4\sqrt{6}}$ ...(iii)

Now, $\quad r^2 = \dfrac{8}{3} = \dfrac{\Delta^2}{S^2} = \dfrac{216\lambda^4}{81\lambda^2}$

$\Rightarrow \quad \dfrac{8}{3} = \dfrac{8}{3}\lambda^2$ [from Eq. (ii)]

$\Rightarrow \quad \lambda = 1$

(a) $\Delta XYZ = 6\sqrt{6}\lambda^2 = 6\sqrt{6}$

$\therefore$ Option (a) is correct.

(b) Radius of circumcircle $(R) = \dfrac{35}{4\sqrt{6}}\lambda = \dfrac{35}{4\sqrt{6}}$

$\therefore$ Option (b) is incorrect.

(c) Since, $r = 4R\sin\dfrac{X}{2}\cdot\sin\dfrac{Y}{2}\cdot\sin\dfrac{Z}{2}$

$\Rightarrow \dfrac{2\sqrt{2}}{\sqrt{3}} = 4\cdot\dfrac{35}{4\sqrt{6}}\sin\dfrac{X}{2}\cdot\sin\dfrac{Y}{2}\cdot\sin\dfrac{Z}{2}$

$\Rightarrow \dfrac{4}{35} = \sin\dfrac{X}{2}\cdot\sin\dfrac{Y}{2}\cdot\sin\dfrac{Z}{2}$

$\therefore$ Option (c) is correct.

(d) $\sin^2\left(\dfrac{X+Y}{2}\right) = \cos^2\left(\dfrac{Z}{2}\right)$

as $\dfrac{X+Y}{2} = 90° - \dfrac{Z}{2}$

$= \dfrac{s(s-z)}{xy} = \dfrac{9\times 2}{5\times 6} = \dfrac{3}{5}$

$\therefore$ Option (d) is correct.

7. $2\cos\left(\frac{A-C}{2}\right) = \frac{2\sin\left(\frac{A+C}{2}\right)\cos\left(\frac{A-C}{2}\right)}{\sin\left(\frac{A+C}{2}\right)}$

$$= \frac{\sin A + \sin C}{\sin\left(\frac{A+C}{2}\right)}$$

$$\cos B = \frac{a^2 + c^2 - b^2}{2ac}, \text{ if } B = 60°$$

Then, $a^2 + c^2 - ac = b^2$

$$= \frac{(a+c)}{k \cdot \sin(60°)} = \frac{a+c}{b} = \frac{a+c}{\sqrt{a^2 + c^2 - ac}}$$

8. $\because \tan C < 0$

$\Rightarrow \quad \tan\{\pi - (A+B)\} < 0$

$\Rightarrow \quad -\tan(A+B) \leq 0$

$\Rightarrow \quad \tan(A+B) > 0$

$\Rightarrow \quad \frac{\tan A + \tan B}{1 - \tan A \tan B} > 0$

$\therefore \quad \tan A + \tan B > 0$

and $\quad 1 - \tan A \tan B > 0,$

then, $\quad \tan A \tan B < 1$

and Now, $\quad \tan A + \tan B > 0$

Since, in triangle

$\tan A + \tan B + \tan C = \tan A \tan B \tan C$

$\Rightarrow \quad \tan A + \tan B + \tan C < \tan C < 0$

$[\because \tan A \tan B < 1]$.

Hence, $\tan A + \tan B + \tan C < 0$

9. $\because \sin A$ and $\sin B$ are the roots of

$$c^2x^2 - c(a+b)x + ab = 0$$

then $\quad \sin A + \sin B = \frac{a+b}{c}$

and $\quad \sin A \sin B = \frac{ab}{c^2} \Rightarrow \frac{a}{2R} + \frac{b}{2R} = \frac{a+b}{c}$

and $\quad \frac{a}{2R} \times \frac{b}{2R} = \frac{ab}{c^2}$

$\therefore \quad c = 2R$

$\Rightarrow \quad 2R \sin C = 2R$

$\therefore \quad \sin C = 1$

$\therefore \quad \angle C = 90°$

$A + B = 90°$

$B = 90° - A$

$\because \quad \sin A + \sin B = \frac{a+b}{c}$

$\Rightarrow \quad \sin A + \sin(90° - A) = \frac{a+b}{c}$

$\Rightarrow \quad \sin A + \cos A = \frac{a+b}{c}$

10. $\sqrt{3}x^2 - 4x + \sqrt{3} < 0$

$(x - \sqrt{3})(\sqrt{3}x - 1) < 0$

$\therefore \quad \frac{1}{\sqrt{3}} < x < \sqrt{3}$

$\frac{\pi}{6} < A < \frac{\pi}{3}$ or $\frac{\pi}{6} < B < \frac{\pi}{3}$

$\frac{\pi}{3} < A + B < \frac{2\pi}{3}$

$\Rightarrow \quad \frac{\pi}{3} < \pi - C < \frac{2\pi}{3}$

$\Rightarrow \quad -\pi + \frac{\pi}{3} < -C < -\pi + \frac{2\pi}{3}$

$\Rightarrow \quad \pi - \frac{\pi}{3} > C > \pi - \frac{2\pi}{3}$

$\therefore \quad \frac{\pi}{3} < C < \frac{2\pi}{3}$

$\because \quad C > \frac{\pi}{3}$

$\cos C < \frac{1}{2} \Rightarrow \frac{a^2 + b^2 - c^2}{2ab} < \frac{1}{2}$

$\Rightarrow \quad a^2 + b^2 - ab < c^2$

and $\quad C < \frac{2\pi}{3}$

$\cos C > -\frac{1}{2} \Rightarrow \frac{a^2 + b^2 - c^2}{2ab} > -\frac{1}{2}$

$\Rightarrow \quad a^2 + b^2 + ab > c^2.$

11. For a ΔXYZ, it is given that

$$\tan\frac{X}{2} + \tan\frac{Z}{2} = \frac{2y}{x+y+z}$$

$\Rightarrow \quad \frac{\Delta}{s(s-x)} + \frac{\Delta}{s(s-z)} = \frac{y}{s}$

$\Rightarrow \quad \Delta\frac{(s-z+s-x)}{(s-x)(s-z)} = y$

$\Rightarrow \quad \Delta = (s-x)(s-z)$

$\Rightarrow \quad s(s-x)(s-y)(s-z) = (s-x)^2(s-z)^2$

$\Rightarrow \quad s^2 - sy = s^2 - (x+z)s + xz$

$\Rightarrow \quad s(x+z-y) = xz$

$\Rightarrow \quad (x+z)^2 - y^2 = 2xz$

$\Rightarrow \quad x^2 + z^2 = y^2$

$\therefore \quad y = \frac{\pi}{2}$

$\because \quad X + Y + Z = \pi \Rightarrow X + Z = \frac{\pi}{2} = Y$

$\Rightarrow \quad X + Z = Y$

$\because \quad \tan\frac{X}{2} = \sqrt{\frac{1-\cos x}{1+\cos x}} = \sqrt{\frac{1 - \frac{z}{y}}{1 + \frac{z}{y}}}$

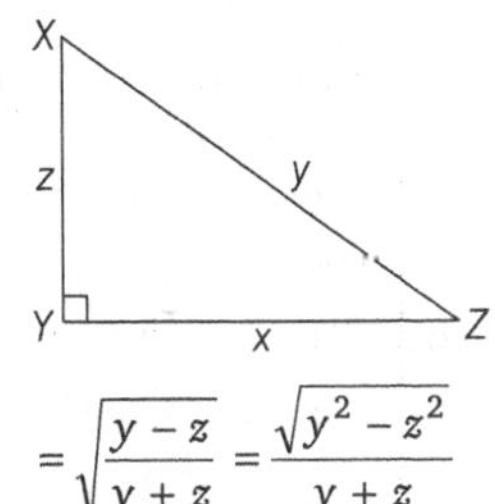

$$= \sqrt{\frac{y-z}{y+z}} = \frac{\sqrt{y^2 - z^2}}{y+z}$$

$\therefore \quad \tan\frac{X}{2} = \frac{x}{y+z} \quad [\because x^2 + z^2 = y^2]$

12. Let $A = 2\alpha$, $B = 3\alpha$ and $C = 7\alpha$

$\therefore \quad A + B + C = 180°$

$\Rightarrow \quad 2\alpha + 3\alpha + 7\alpha = 180°$

$\Rightarrow \quad 12\alpha = 180°$

$\therefore \quad \alpha = 15°$

$\therefore \quad A = 30°, B = 45°, C = 105°$

$\therefore \quad a : b : c = \sin 30° : \sin 45° : \sin 105°$

$$= \frac{1}{2} : \frac{1}{\sqrt{2}} : \frac{\sqrt{3}+1}{2\sqrt{2}}$$

or $\quad a : b : c = \sqrt{2} : 2 : \sqrt{3}+1$

or $\quad \dfrac{1}{\sqrt{2}} : 1 : \dfrac{\sqrt{3}+1}{2}$

or $\quad 1 : \sqrt{2} : \dfrac{\sqrt{2}(\sqrt{3}+1)(\sqrt{3}+1)}{2(\sqrt{3}-1)}$

$\therefore \quad 1 : \sqrt{2} : \dfrac{\sqrt{2}}{\sqrt{3}-1}$

13. We have,

$$r = \frac{\Delta}{s};\ \rho = \frac{\Delta}{s-a}$$

$$\rho r = \frac{\Delta^2}{s(s-a)} = \frac{s(s-a)(s-b)(s-c)}{s(s-a)}$$

$= (s-b)(s-c) = (s-b)^2 \qquad [\because b = c]$

$= \dfrac{(2s-2b)^2}{4} = \dfrac{(a+b+c-2b)^2}{4} \qquad [\because b = c]$

$= \dfrac{a^2}{4} = \dfrac{4R^2 \sin^2 A}{4} = R^2 \sin^2 A$

Also, if $\angle B = \theta \Rightarrow \angle A = \pi - 2\theta$

$\therefore \quad \rho r = R^2 \sin^2 (\pi - 2\theta) = R^2 \sin^2 2\theta = R^2 \sin^2 2B$

14. Since, $\cos(\theta - \alpha), \cos\theta, \cos(\theta + \alpha)$ are in HP

$\therefore \quad \dfrac{1}{\cos(\theta-\alpha)}, \dfrac{1}{\cos\theta}, \dfrac{1}{\cos(\theta+\alpha)}$ are in AP

$\Rightarrow \quad \dfrac{2}{\cos\theta} = \dfrac{1}{\cos(\theta-\alpha)} + \dfrac{1}{\cos(\theta+\alpha)}$

$= \dfrac{\cos(\theta+\alpha) + \cos(\theta-\alpha)}{\cos(\theta-\alpha)\cos(\theta+\alpha)}$

$= \dfrac{2\cos\theta\cos\alpha}{\cos(\theta-\alpha)\cos(\theta+\alpha)}$

$\Rightarrow \quad \cos^2\theta \cos\alpha = \cos^2\theta - \sin^2\alpha$

$\Rightarrow \quad \cos^2\theta (1 - \cos\alpha) = \sin^2\alpha$

$\Rightarrow \quad \cos^2\theta (1 - \cos\alpha) = (1 + \cos\alpha)(1 - \cos\alpha)$

$\Rightarrow \quad \cos^2\theta = (1 + \cos\alpha)$

$\Rightarrow \quad \cos^2\theta = 2\cos^2\dfrac{\alpha}{2}$

$\Rightarrow \quad \cos^2\theta \sec^2\dfrac{\alpha}{2} = 2$

$\therefore \quad \cos\theta \sec\dfrac{\alpha}{2} = \pm\sqrt{2}$

15. $\sin^2\beta = \sin\alpha\cos\alpha$

$\dfrac{1 - \cos 2\beta}{2} = \sin\alpha\cos\alpha$

$\cos 2\beta = 1 - 2\sin\alpha\cos\alpha$

$= 1 - \sin 2\alpha$

$= 1 + \cos\left(\dfrac{\pi}{2} + 2\alpha\right)$ or $1 - \cos\left(\dfrac{\pi}{2} - 2\alpha\right)$

$= 2\cos^2\left(\dfrac{\pi}{4} + \alpha\right)$ or $2\sin^2\left(\dfrac{\pi}{4} - \alpha\right)$

16. $\because D$ is the mid-point

$\therefore \quad AB^2 + AC^2 = 2[AD^2 + BD^2]$

$$c^2 + b^2 = 2\left\{l^2 + \left(\frac{a}{2}\right)^2\right\}$$

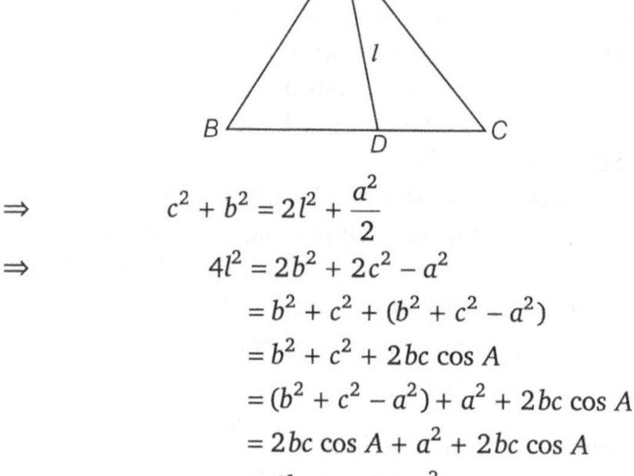

$\Rightarrow \quad c^2 + b^2 = 2l^2 + \dfrac{a^2}{2}$

$\Rightarrow \quad 4l^2 = 2b^2 + 2c^2 - a^2$

$= b^2 + c^2 + (b^2 + c^2 - a^2)$

$= b^2 + c^2 + 2bc\cos A$

$= (b^2 + c^2 - a^2) + a^2 + 2bc\cos A$

$= 2bc\cos A + a^2 + 2bc\cos A$

$= 4bc\cos A + a^2$

17. Since, $r_1 = 2r_2 = 3r_3 = k$ (say)

$\therefore \quad \dfrac{\Delta}{s-a} = k, \dfrac{2\Delta}{s-b} = k, \dfrac{3\Delta}{s-c} = k$

$\therefore \quad s - a = \dfrac{\Delta}{k}, s - b = \dfrac{2\Delta}{k}, s - c = \dfrac{3\Delta}{k}$

$\Rightarrow \quad s - a + s - b + s - c = \dfrac{\Delta}{k} + \dfrac{2\Delta}{k} + \dfrac{3\Delta}{k}$

$\therefore \quad s = \dfrac{6\Delta}{k}$

$\Rightarrow \quad a = \dfrac{5\Delta}{k}, b = \dfrac{4\Delta}{k}, c = \dfrac{3\Delta}{k}$

$\therefore \quad \dfrac{a}{b} = \dfrac{5}{4}$ and $\dfrac{a}{c} = \dfrac{5}{3}$

18. Alternate. (a) : $\because$ Maximum value of $\sin 2A + \sin 2B + \sin 2C$ and $\sin A + \sin B + \sin C$ is same that is 3.

Alternate. (b) : $\because \sin\dfrac{A}{2}\sin\dfrac{B}{2}\sin\dfrac{C}{2} \le \dfrac{1}{8}$

$\Rightarrow \quad \dfrac{r}{4R} \le \dfrac{1}{8}$

$\Rightarrow \quad R \ge 2r$

Alternate. (c) : $\because \dfrac{abc}{(a+b+c)} = \dfrac{4R\Delta}{2s}$

$$= 2R \cdot r = R(2r) \le R^2$$

$\therefore \quad R^2 \ge \dfrac{abc}{(a+b+c)} (\because R \ge 2r)$

Alternate. (d) : $\angle B = 90°$

$\therefore \quad r = (s-b)\tan\dfrac{B}{2} = s - b$

$$R = \frac{b}{2\sin B} = \frac{b}{2} \Rightarrow 2R = b$$

$\therefore \quad r + 2R = s$

19. $\because \dfrac{\sin A}{\sin C} = \dfrac{\sin(A-B)}{\sin(B-C)}$ ...(i)

$\Rightarrow \sin(B+C)\sin(B-C) = \sin(A+B)\sin(A-B)$

$\Rightarrow \sin^2 B - \sin^2 C = \sin^2 A - \sin^2 B$

$\therefore \sin^2 A, \sin^2 B, \sin^2 C$ are in AP ...(ii)

$\Rightarrow 2\sin^2 A, 2\sin^2 B, 2\sin^2 C$ are in AP

or $-2\sin^2 A, -2\sin^2 B, -2\sin^2 C$ are in AP

$\Rightarrow 1-2\sin^2 A, 1-2\sin^2 B, 1-2\sin^2 C$ are in AP

or $\cos 2A, \cos 2B, \cos 2C$ are in AP

From Eq. (ii)

$\sin A \cdot \sin A, \sin B \cdot \sin B, \sin C \cdot \sin C$ are in AP

$\Rightarrow \dfrac{a}{2R}\sin A, \dfrac{b}{2R}\sin B, \dfrac{c}{2R}\sin C$ are in AP

or $a\sin A, b\sin B, c\sin C$ are in AP

Now from Eq. (i),

$$\frac{\sin(B-C)}{\sin B \sin C} = \frac{\sin(A-B)}{\sin A \sin B}$$

$\Rightarrow \cot C - \cot B = \cot B - \cot A$

$\Rightarrow 2\cot B = \cot A + \cot C$

Hence, $\cot A, \cot B, \cot C$ are in AP.

20. $\because \sin C + \cos C + \sin(2B+C) - \cos(2B+C) = 2\sqrt{2}$

$\Rightarrow \{\sin C + \sin(2B+C)\} + \{\cos C - \cos(2B+C)\} = 2\sqrt{2}$

$\Rightarrow 2\sin(B+C)\cos B + 2\sin(B+C)\sin B = 2\sqrt{2}$

$\Rightarrow \sin A(\sin B + \cos B) = \sqrt{2}$

$\Rightarrow \sin A\left\{\sin\left(\dfrac{\pi}{4} + B\right)\right\} = 1$

Which is possible only,

$$\sin A = 1, \sin\left(\frac{\pi}{4} + B\right) = 1$$

$\Rightarrow A = 90°, \dfrac{\pi}{4} + B = 90° \Rightarrow B = 45°$

or $A = 90°, B = 45°, C = 45°$

21. In ΔABC, $b+c-a>0, c+a-b>0, a+b-c>0$, so

$$\frac{(b+c-a)+(c+a-b)+(a+b-c)}{3} \ge \{(b+c-a)(c+a-b)(a+b-c)\}^{1/3}$$

$\Rightarrow P^3 \ge 27(b+c-a)(c+a-b)(a+b-c)$

Also, $\dfrac{(a+b+c)+(b+c-a)+(c+a-b)+(a+b-c)}{4}$

$$> \{(a+b+c)(b+c-a)(c+a-b)(a+b-c)\}^{1/4}$$

$\Rightarrow \dfrac{2P}{4} > (16A)^{1/4}$

$\Rightarrow P > 4A^{1/4}$ or $P^4 = 256A$

For a given parameter, equilateral triangle has the largest area, so the area of triangle

$$A \le \frac{\sqrt{3}}{4}\left(\frac{p}{3}\right)^2$$

[for equilateral triangle, $a = b = c = \dfrac{p}{3}$]

Now, $p^2 = (a+b+c)^2$

$= a^2 + b^2 + c^2 + 2(ab+bc+ca)$

$= 3(a^2+b^2+c^2) - 2(a^2+b^2+c^2-ab-bc-ca)$

$\le 3(a^2+b^2+c^2)$

[equality holds iff $a = b = c$]

Thus, $A \le \dfrac{\sqrt{3}}{4}\dfrac{a^2+b^2+c^2}{3}$

$\Rightarrow a^2+b^2+c^2 \ge 4\sqrt{3}A$

22. $a = \dfrac{r_1(r_2+r_3)}{\sqrt{r_1r_2+r_2r_3+r_3r_1}} = \dfrac{8\times 36}{24} = 12$

$b = \dfrac{r_2(r_3+r_1)}{\sqrt{r_1r_2+r_2r_3+r_3r_1}} = \dfrac{12\times 32}{24} = 16$

$c = \dfrac{r_3(r_1+r_2)}{\sqrt{r_1r_2+r_2r_3+r_3r_1}} = \dfrac{24\times 20}{24} = 20$

Also, $r_1r_2r_3 = rs^2 \Rightarrow r = \dfrac{8\times 12\times 24}{(24)^2} = 4$

23. Draw BE perpendicular to CA produced.

Then, $BD = DC = \dfrac{a}{2}$ and $EA = AC = b$

From ΔADC, $\cos C = \dfrac{b}{\frac{a}{2}} = \dfrac{2b}{a}$

From ΔABC, $\cos(\pi - A) = \dfrac{b}{c}$

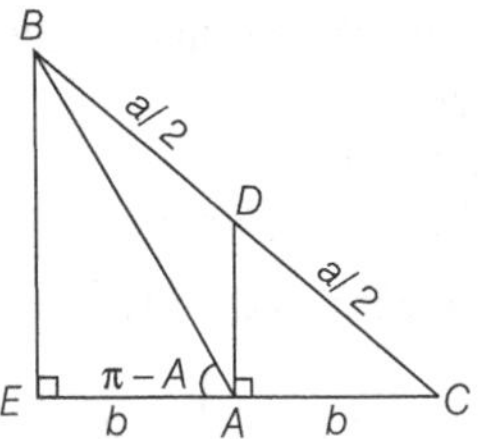

$\Rightarrow \cos A = -\dfrac{b}{c}$

So that, $\cos A \cos C = -\dfrac{2b^2}{ac}$

Also $\cos A = -\dfrac{b}{c}$

$\Rightarrow \dfrac{b^2+c^2-a^2}{2bc} = -\dfrac{b}{c} \Rightarrow a^2 - 3b^2 - c^2 = 0$

and $\cos B = \dfrac{c^2 + a^2 - b^2}{2ca}$

$$= \frac{c^2 + 3b^2 + c^2 - b^2}{2ca}$$

$$= \frac{b^2 + c^2}{ca}$$

24. Let AD be a perpendicular from A on BC.

Then, $\dfrac{BD}{c} = \cos B$

$\Rightarrow BD$ is rational, similarly AD is rational.

Now, $\sin C = \cos B = \dfrac{AD}{b}$

$\Rightarrow b$ is rational.

Since, $\cos C = \sin B = \dfrac{DC}{b}$

$\Rightarrow DC$ is rational.

Hence, $a = BD + DC$ is rational

Hence, both a and b are rational numbers.

25. Let a non-right angled ΔPQR.

Now, by sine rule

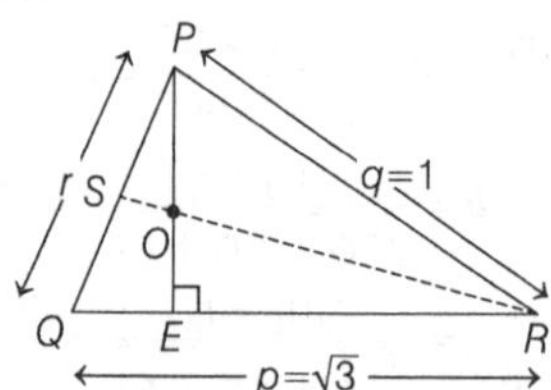

$$\frac{p}{\sin P} = \frac{q}{\sin Q} = \frac{r}{\sin R} = 2 \times \text{circumradius}$$

$$\Rightarrow \frac{\sqrt{3}}{\sin P} = \frac{1}{\sin Q} = \frac{r}{\sin R} = 2 \times 1 \quad [\text{circumradius} = 1 \text{ unit}]$$

$$\Rightarrow \sin P = \frac{\sqrt{3}}{2} \text{ and } \sin Q = \frac{1}{2}$$

$\Rightarrow P = 120^\circ$ and $Q = 30^\circ$

[$\because \Delta PQR$ is non-right angled triangle]

So, $R = 30^\circ$

$\Rightarrow r = 1$, so ΔPQR is an isosceles triangle. And, since RS and PE are the median of ΔPQR, so 'O' is centroid of the ΔPQR.

Now, Option (a),

From Apollonius theorem,

$$2(PE^2 + QE^2) = PQ^2 + PR^2 \Rightarrow 2\left(PE^2 + \frac{3}{4}\right) = 1 + 1$$

$$\Rightarrow \quad PE^2 = 1 - \frac{3}{4} \Rightarrow PE^2 = \frac{1}{4} \Rightarrow PE = \frac{1}{2} \text{ units}$$

and $\quad OE = \dfrac{1}{3} PE = \dfrac{1}{6}$ units

[$\because O$ divides PE is $2:1$]

Option (b), Again from Apollonius theorem,

$$2(PS^2 + RS^2) = PR^2 + QR^2 \Rightarrow 2\left(\frac{1}{4} + RS^2\right) = 1 + 3$$

$$\Rightarrow \quad RS^2 = 2 - \frac{1}{4} \Rightarrow RS^2 = \frac{7}{4} \Rightarrow RS = \frac{\sqrt{7}}{2} \text{ units}$$

Option (c),

$$\text{Area of } \Delta SOE = \frac{1}{2}(OE)(ST)$$

$$= \frac{1}{2} \times \frac{1}{6}[(PS)\sin 60^\circ]$$

$$= \frac{1}{12} \times \frac{1}{2} \times \frac{\sqrt{3}}{2} = \frac{\sqrt{3}}{48} \text{ sq units}$$

Option (d),

$$\because \text{Inradius of } \Delta PQR = \frac{\Delta}{s} = \frac{\frac{1}{2}pq\sin R}{\frac{1}{2}(p+q+r)} = \frac{\frac{1}{2}(\sqrt{3})(1)\frac{1}{2}}{\frac{1}{2}(\sqrt{3}+1+1)}$$

$$= \frac{\sqrt{3}}{2}(2 - \sqrt{3}) \text{ sq units}$$

Hence, options (a), (b) and (d) are correct.

26. $\Delta = \dfrac{1}{2} ap_1 = \dfrac{1}{2} bp_2 = \dfrac{1}{2} cp_3$

$$\Rightarrow \quad p_1 = \frac{2\Delta}{a}, \ p_2 = \frac{2\Delta}{b}, \ p_3 = \frac{2\Delta}{c}$$

p_1, p_2, p_3 are in HP

Now, $\quad p_1 = \dfrac{2\Delta}{2R \sin A}, \ p_2 = \dfrac{2\Delta}{2R \sin B},$

$$p_3 = \frac{2\Delta}{2R \sin C}$$

$$\Rightarrow \quad \frac{1}{p_1} = \frac{R \sin A}{\Delta}, \ \frac{1}{p_2} = \frac{R \sin B}{\Delta},$$

$$\frac{1}{p_3} = \frac{R \sin C}{\Delta}$$

Since, $\quad \sin A, \sin B, \sin C$ are in AP

$\sin A + \sin C = 2 \sin B$

$$\therefore \quad \frac{1}{p_1} + \frac{1}{p_2} + \frac{1}{p_3} = \frac{R}{\Delta}(\sin A + \sin B + \sin C)$$

$$= \frac{3R}{\Delta} \sin B \le \frac{3R}{\Delta}$$

27. We have,

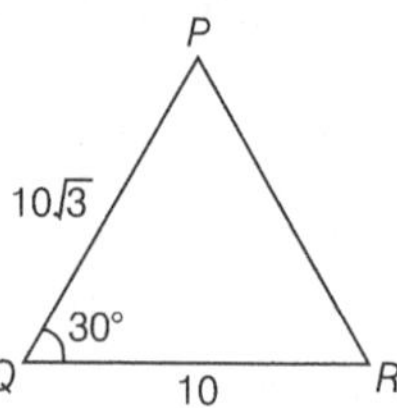

In ΔPQR,

$\angle PQR = 30^\circ$

$PQ = 10\sqrt{3}$

$QR = 10$

By cosine rule,

$$\cos 30^\circ = \frac{PQ^2 + QR^2 - PR^2}{2PQ \cdot QR}$$

$$\Rightarrow \quad \frac{\sqrt{3}}{2} = \frac{300 + 100 - PR^2}{200\sqrt{3}}$$

$$\Rightarrow \quad 300 = 300 + 100 - PR^2 \Rightarrow PR = 10$$

Since, $PR = QR = 10$

$\therefore \quad \angle QPR = 30^\circ$ and $\angle QRP = 120^\circ$

$$\text{Area of } \Delta PQR = \frac{1}{2} PQ \cdot QR \cdot \sin 30^\circ$$

$$= \frac{1}{2} \times 10\sqrt{3} \times 10 \times \frac{1}{2} = 25\sqrt{3}$$

Radius of incircle of $\Delta PQR = \dfrac{\text{Area of } \Delta PQR}{\text{Semi-perimetre of } \Delta PQR}$

i.e. $$r = \frac{\Delta}{s} = \frac{25\sqrt{3}}{\frac{10\sqrt{3} + 10 + 10}{2}} = \frac{25\sqrt{3}}{5(\sqrt{3} + 2)}$$

$\Rightarrow \quad r = 5\sqrt{3}(2 - \sqrt{3}) = 10\sqrt{3} - 15$

and radius of circumcircle $(R) = \dfrac{abc}{4\Delta}$

$$= \frac{10\sqrt{3} \times 10 \times 10}{4 \times 25\sqrt{3}} = 10$$

$\therefore$ Area of circumcircle of $\Delta PQR = \pi R^2 = 100\pi$

Hence, option (b), (c) and (d) are correct answer.

28. $\cos\theta = \dfrac{a-b}{c} = \dfrac{\sin A - \sin B}{\sin C}$ [sine rule]

$$= \frac{2\cos\left[\left(\frac{1}{2}\right)(A+B)\right]\sin\left[\left(\frac{1}{2}\right)(A-B)\right]}{2\sin\left[\left(\frac{1}{2}\right)C\right]\cos\left[\left(\frac{1}{2}\right)C\right]}$$

$$= \frac{2\sin\left[\left(\frac{1}{2}\right)C\right]\sin\left[\left(\frac{1}{2}\right)(A-B)\right]}{2\sin\left[\left(\frac{1}{2}\right)C\right]\sin\left[\left(\frac{1}{2}\right)(A+B)\right]} = \frac{\sin\left[\left(\frac{1}{2}\right)(A-B)\right]}{\sin\left[\left(\frac{1}{2}\right)(A+B)\right]}$$

$$\Rightarrow \sin\theta = \frac{\sqrt{\sin^2\left[\left(\frac{1}{2}\right)(A+B)\right] - \sin^2\left[\left(\frac{1}{2}\right)(A-B)\right]}}{\sin\left[\left(\frac{1}{2}\right)(A+B)\right]}$$

$$= \frac{\sqrt{\sin A \sin B}}{\sin\left[\left(\frac{1}{2}\right)(A+B)\right]}$$

$$\Rightarrow \frac{(a+b)\sin\theta}{2\sqrt{ab}} = \frac{\sin A + \sin B}{2\sqrt{\sin A \sin B}} \cdot \frac{\sqrt{\sin A \sin B}}{\sin\left[\left(\frac{1}{2}\right)(A+B)\right]}$$

$$= \frac{2\sin\left[\left(\frac{1}{2}\right)(A+B)\right]\cos\left[\left(\frac{1}{2}\right)(A-B)\right]}{2\sin\left[\left(\frac{1}{2}\right)(A+B)\right]}$$

$$= \cos\left[\left(\frac{1}{2}\right)(A-B)\right]$$

and $$\frac{c\sin\theta}{2\sqrt{ab}} = \frac{\sin C}{2\sqrt{\sin A \sin B}} \cdot \frac{\sqrt{\sin A \sin B}}{\sin\left[\left(\frac{1}{2}\right)(A+B)\right]}$$

$$= \frac{2\sin\left(\frac{C}{2}\right)\cos\left(\frac{C}{2}\right)}{2\sin\left[\frac{(A+B)}{2}\right]} = \cos\left(\frac{(A+B)}{2}\right)$$

29. $\cos A = \dfrac{b^2 + c^2 - a^2}{2bc}$

$\Rightarrow \quad c^2 - 2bc\cos A + b^2 - a^2 = 0$

$\therefore \quad c_1 + c_2 = 2b\cos A,\ c_1c_2 = b^2 - a^2$

Given, $\quad c_1^2 + c_1c_2 + c_2^2 = a^2$

$\Rightarrow \quad (c_1 + c_2)^2 - c_1c_2 = a^2$

$\Rightarrow \quad 4b^2\cos^2 A - (b^2 - a^2) = a^2$

$\Rightarrow \quad 4\cos^2 A = 1$

$\Rightarrow \quad 2(1 + \cos 2A) = 1$

or $\quad \cos 2A = -\dfrac{1}{2} = \cos\dfrac{2\pi}{3}$

$\therefore \quad 2A = 2n\pi \pm \dfrac{2\pi}{3}$

$\therefore \quad A = n\pi \pm \dfrac{\pi}{3},\ n \in I$

$\therefore \quad A = \dfrac{\pi}{3}$ or $\dfrac{2\pi}{3}$

i.e. $\quad A = 60^\circ$ or 120°

Linked Comprehension Type

PASSAGE 1

$\because \quad AD = p_1,$

$BE = p_2$ and $CF = p_3$

$\because \quad \Delta = \dfrac{1}{2} \cdot a \cdot p_1 = \dfrac{1}{2} \cdot b \cdot p_2 = \dfrac{1}{2} c \cdot p_3$

$\therefore \quad p_1 = \dfrac{2\Delta}{a},\ p_2 = \dfrac{2\Delta}{b},\ p_3 = \dfrac{2\Delta}{c}$

1. Given, $\dfrac{1}{p_1} + \dfrac{1}{p_2} + \dfrac{1}{p_3} = \dfrac{1}{2}$

$\because \quad \text{AM} \geq \text{GM}$

$$\therefore \quad \frac{\frac{1}{p_1} + \frac{1}{p_2} + \frac{1}{p_3}}{3} \geq \left(\frac{1}{p_1} \cdot \frac{1}{p_2} \cdot \frac{1}{p_3}\right)^{1/3} \text{ or } \left(\frac{\frac{1}{2}}{3}\right)^3 \geq \frac{1}{p_1p_2p_3}$$

$\therefore \quad p_1p_2p_3 \geq 216$

2. $\because \quad \dfrac{\cos A}{p_1} + \dfrac{\cos B}{p_2} + \dfrac{\cos C}{p_3}$

$$= \frac{a}{2\Delta}\cos A + \frac{b}{2\Delta}\cos B + \frac{c}{2\Delta}\cos C$$

$$= \frac{R(\sin 2A + \sin 2B + \sin 2C)}{2\Delta}$$

$$= \frac{R\,(4\sin A \sin B \sin C)}{2\Delta}$$
$$= \frac{2R}{\Delta}\left(\frac{a}{2r}\cdot\frac{b}{2r}\cdot\frac{c}{2r}\right)$$
$$= \frac{abc}{4R^2\Delta} = \frac{4R\Delta}{4R^2\Delta} = \frac{1}{R}$$

3. $\because$ AM $\geq$ GM

$$\therefore \frac{\frac{b^2p_1}{c}+\frac{c^2p_2}{a}+\frac{a^2p_3}{b}}{3} \geq \left(\frac{b^2p_1}{c}\cdot\frac{c^2p_2}{a}\cdot\frac{a^2p_3}{b}\right)^{1/3}$$
$$= (abcp_1p_2p_3)^{1/3}$$
$$= \left(abc\cdot\frac{8\Delta^3}{abc}\right)^{1/3} = 2\Delta$$
$$\therefore \frac{b^2p_1}{c}+\frac{c^2p_2}{a}+\frac{a^2p_3}{b} \geq 6\Delta$$

4. $p_1^{-2}+p_2^{-2}+p_3^{-2} = \left(\frac{2\Delta}{a}\right)^{-2}+\left(\frac{2\Delta}{b}\right)^{-2}+\left(\frac{2\Delta}{c}\right)^{-2}$
$$= \frac{a^2+b^2+c^2}{4\Delta^2}$$
$$= \frac{\Sigma a^2}{4\Delta^2}$$

5. $\because p_1, p_2, p_3$ are in AP

$\therefore \frac{2\Delta}{a}, \frac{2\Delta}{b}, \frac{2\Delta}{c}$ are in AP

Dividing each by 2Δ then $\frac{1}{a}, \frac{1}{b}, \frac{1}{c}$ are in AP

$\therefore a, b, c$ are in HP.

PASSAGE 2

1. $\angle A_1OA_2 = \frac{2\pi}{n}$

$$\therefore \quad A_1A_2 = \sqrt{R^2+R^2-2R\cdot R\cos\left(\frac{2\pi}{n}\right)}$$

$$= 2R\sin\left(\frac{\pi}{n}\right) \qquad \dots\text{(i)}$$

$$A_1A_3 = \sqrt{R^2+R^2-2\cdot R\cdot R\cdot\cos\left(\frac{4\pi}{n}\right)} = 2R\sin\left(\frac{2\pi}{n}\right)$$
$$A_1A_4 = 2R\sin\left(\frac{3\pi}{n}\right)$$
$$\therefore \quad A_1A_j = 2R\sin\left\{(j-1)\frac{\pi}{n}\right\},$$
$$j = 1, 2, 3, 4, \dots, n$$

2. $\because \angle A_1IA_2 = \frac{2\pi}{n}$

$\therefore \angle A_1ID = \angle A_2ID = \frac{\pi}{n}$

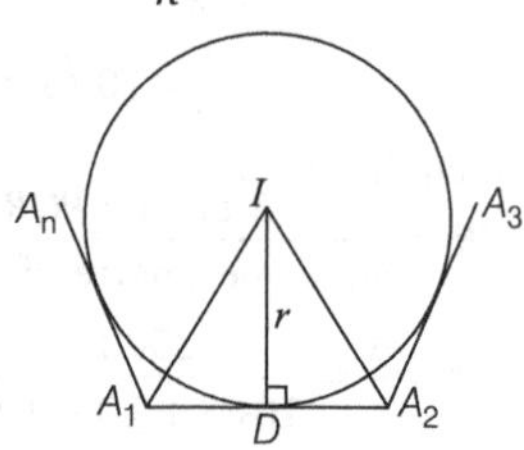

In ΔDIA_1, $\tan\left(\frac{\pi}{n}\right) = \frac{A_1D}{r} = \frac{a/2}{r}$

$$\therefore \quad r = \frac{a}{2}\cot\left(\frac{\pi}{n}\right) \qquad \dots\text{(ii)}$$

3. From Eq. (i),

$$A_1A_2 = 2R\sin\left(\frac{\pi}{n}\right) \text{ or } a = 2R\sin\left(\frac{\pi}{n}\right)$$
$$\therefore \quad R = \frac{a}{2\sin\left(\frac{\pi}{n}\right)}$$

From Eq. (ii), $r = \dfrac{a\cos\left(\frac{\pi}{n}\right)}{2\sin\left(\frac{\pi}{n}\right)}$

$$\therefore \quad r + R = \frac{a}{2\sin\left(\frac{\pi}{n}\right)}\left\{1+\cos\left(\frac{\pi}{n}\right)\right\}$$
$$= \frac{a\left\{2\cos^2\left(\frac{\pi}{2n}\right)\right\}}{4\sin\left(\frac{\pi}{2n}\right)\cos\left(\frac{\pi}{2n}\right)} = \frac{a}{2}\cot\left(\frac{\pi}{2n}\right)$$

4. Area of regular polygon of n sides inscribed in a circle

$= n \times$ Area of ΔA_1OA_2 [from Fig. (i)]

$$= n \times \frac{1}{2}R^2\sin\left(\frac{2\pi}{n}\right)$$
$$= \frac{nR^2}{2}\sin\left(\frac{2\pi}{n}\right) \qquad \dots\text{(iii)}$$

and Area of regular polygon of n sides circumscribed about a circle

$= n \times$ Area of ΔA_1IA_2 [from Fig. (ii)]

$$= n \times \frac{1}{2}A_1A_2 \times r$$
$$= n \times \frac{r}{2} \times 2r\tan\left(\frac{\pi}{n}\right) = nr^2\tan\left(\frac{\pi}{n}\right) \qquad \dots\text{(iv)}$$

5. Let the perimeter of the pentagon and decagon be $10x$. Then, each side of the pentagon is $2x$ and its area is $5x^2\cot\left(\frac{\pi}{5}\right)$ and each side of the decagon is x and its area is $\frac{5}{2}x^2\cot\left(\frac{\pi}{10}\right)$.

$$\therefore \quad \frac{\text{Area of pentagon}}{\text{Area of decagon}} = \frac{5x^2 \cot\left(\frac{\pi}{5}\right)}{\frac{5}{2}x^2 \cot\left(\frac{\pi}{10}\right)} = \frac{2\cot 36^\circ}{\cot 18^\circ}$$

$$= \frac{2\cos 36^\circ}{2\cos^2 18^\circ} = \frac{2\cos 36^\circ}{(1+\cos 36^\circ)}$$

$$= \frac{2\left(\frac{\sqrt{5}+1}{4}\right)}{\left(1+\frac{\sqrt{5}+1}{4}\right)}$$

$$= \frac{2(\sqrt{5}+1)}{(5+\sqrt{5})} = \frac{2}{\sqrt{5}}$$

PASSAGE 3

1. $\because \cos A = \dfrac{b^2+c^2-a^2}{2bc}$

or $\quad c^2 - 2bc\cos A + b^2 - a^2 = 0$

For real roots, $D > 0$

$\Rightarrow \quad 4b^2\cos^2 A - 4\cdot 1\cdot(b^2-a^2) > 0$

$\Rightarrow \quad a^2 > b^2 \sin^2 A$

$\therefore \quad b\sin A < a$

Also, $\quad c_1 + c_2 = 2b\cos A \quad$ ...(i)

and $\quad c_1c_2 = b^2 - a^2 \quad$ ...(ii)

Then, $\quad |c_1 - c_2| = \sqrt{(c_1+c_2)^2 - 4c_1c_2}$

$= \sqrt{4b^2\cos^2 A - 4(b^2-a^2)}$

$= 2\sqrt{(a^2 - b^2\sin^2 A)} \quad$ ...(iii)

Consider smaller root, say c_2.

Then, $\quad 2c_2 = (c_1+c_2) - (c_1-c_2)$

$= 2b\cos A - 2\sqrt{(a^2-b^2\sin^2 A)}$

$\Rightarrow \quad c_2 = b\cos A - \sqrt{(a^2-b^2\sin^2 A)} > 0$

$\Rightarrow \quad b\cos A > \sqrt{(a^2-b^2\sin^2 A)}$

$\Rightarrow \quad b^2\cos^2 A > a^2 - b^2\sin^2 A$

$\Rightarrow \quad b^2 > a^2$

or $\quad b > a$

Hence, two different triangles are possible if

$b\sin A < a$ and $b > a$

2. From Eq. (iii),

$|c_1 - c_2| = 2\sqrt{(a^2 - b^2\sin^2 A)}$

3. $c_1^2 - 2c_1c_2\cos 2A + c_2^2$

$= (c_1+c_2)^2 - 2c_1c_2(1+\cos 2A)$

$= (c_1+c_2)^2 - 4c_1c_2\cos^2 A$

$= (2b\cos A)^2 - 4(b^2-a^2)\cos^2 A$

[from Eqs. (i) and (ii)]

$= 4\cos^2 A\{b^2 - (b^2-a^2)\}$

$= 4a^2\cos^2 A$

4. $B_1D = B_2D = \left(\dfrac{c_1-c_2}{2}\right)$

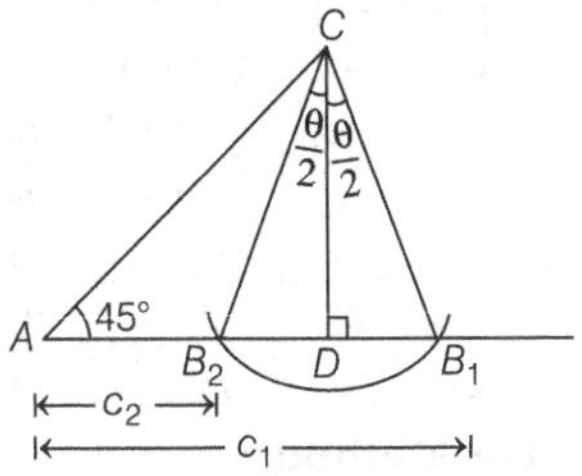

$\therefore \quad AD = AB_2 + B_2D = c_2 + \left(\dfrac{c_1-c_2}{2}\right)$

$\therefore \quad AD = \left(\dfrac{c_1+c_2}{2}\right)$

Now, in ΔACD, $\tan 45^\circ = \dfrac{CD}{AD} = 1$

$\therefore \quad CD = AD = \left(\dfrac{c_1+c_2}{2}\right)$

$\therefore$ In ΔB_2CD, $\tan(\theta/2) = \dfrac{B_2D}{CD} = \dfrac{\left(\frac{c_1-c_2}{2}\right)}{\left(\frac{c_1+c_2}{2}\right)} = \left(\dfrac{c_1-c_2}{c_1+c_2}\right)$

$\therefore \quad \cos\theta = \dfrac{1-\tan^2(\theta/2)}{1+\tan^2(\theta/2)}$

$= \dfrac{1-\left(\frac{c_1-c_2}{c_1+c_2}\right)^2}{1+\left(\frac{c_1-c_2}{c_1+c_2}\right)^2}$

$= \dfrac{2c_1c_2}{(c_1^2+c_2^2)}$

5. $\because \cos A = \dfrac{b^2+c^2-a^2}{2bc}$

$c^2 - 2bc\cos A + b^2 - a^2 = 0$

$\Rightarrow \quad c^2 - 2(m+1)\lambda\, c\cos A + (m^2+2m-3)\lambda^2 = 0$

$\left[\because \dfrac{a}{2} = \dfrac{b}{m+1} = \lambda \text{ (say)}\right]$

$\therefore \quad c_1 + c_2 = 2(m+1)\lambda\cos A$

or $\quad (c_1+c_2)^2 = 4(m+1)^2\lambda^2 \cdot \dfrac{(m-1)(m+3)}{4m}$

$= \dfrac{(m+1)^2(m-1)(m+3)}{m} \quad$ ...(iv)

and $\quad (c_1-c_2)^2 = (c_1+c_2)^2 - 4c_1c_2$

$= 4(m+1)^2\lambda^2\cos^2 A - 4(m^2+2m-3)\lambda^2$

$= 4\lambda^2\left\{(m+1)^2 \cdot \dfrac{(m-1)(m+3)}{4m} - (m+3)(m-1)\right\}$

$= 4\lambda^2\dfrac{(m-1)(m+3)}{4m}(m-1)^2$

$= \dfrac{\lambda^2(m-1)^3(m+3)}{m} \quad$ ...(v)

From Eqs. (iv) and (v),

$$\left(\frac{c_1+c_2}{c_1-c_2}\right)^2=\left(\frac{m+1}{m-1}\right)^2$$

$$\Rightarrow \quad \frac{c_1+c_2}{c_1-c_2}=\pm\frac{m+1}{m-1}=\frac{m+1}{m-1} \text{ or } \frac{m+1}{1-m}$$

At last using, componendo and dividendo rule, we get

$$\frac{2c_1}{2c_2}=\frac{2m}{2} \text{ or } \frac{2}{m}=m \text{ or } \frac{1}{m}$$

Alternative Method :

We have, $\cos A=\dfrac{b^2+c^2-a^2}{2bc}$

$$\Rightarrow \quad 2bc\cos A=b^2+c^2-a^2$$

$$\Rightarrow \quad (m+1)\,ac\cos A=\frac{(m+1)^2}{4}a^2+c^2-a^2$$

$[\because 2b=(m+1)a]$

$$\Rightarrow \quad (m+1)\,ac\cos A=\{(m+1)^2-4\}\frac{a^2}{4}+c^2$$

$$mac\cos A+ac\cos A=(m-1)(m+3)\frac{a^2}{4}+c^2$$

$$=ma^2\cos^2 A+c^2$$

$$\left[\because \frac{(m-1)(m+3)}{4}=m\cos^2 A\right]$$

$$\Rightarrow \quad ma\cos A\,(c-a\cos A)-c\,(c-a\cos A)=0$$

$$\Rightarrow \quad (c-a\cos A)(ma\cos A-c)=0$$

$\therefore \quad c=a\cos A$ and $c=ma\cos A$

If $\quad c_2=a\cos A$ and $c_1=ma\cos A$

$\therefore \quad \dfrac{c_1}{c_2}=m$

and if $\quad c_1=a\cos A$ and $c_2=ma\cos A$

Then, $\quad \dfrac{c_1}{c_2}=\dfrac{1}{m}$

$\therefore \quad \dfrac{c_1}{c_2}=m \text{ or } \dfrac{1}{m}$

PASSAGE 4

1. We have,

$AH=2R\cos A$

$BH=2R\cos B$

$CH=2R\cos C$

$HD=2R\cos B\cos C$

$HE=2R\cos A\cos C$

$HF=2R\cos A\cos B$

Now, $AH\cdot BH\cdot CH=3$ [given]

$$\Rightarrow \cos A\cdot\cos B\cdot\cos C=\frac{3}{8R^3} \quad \text{...(i)}$$

Now, $AH^2+BH^2+CH^2=7$ [given]

$$\cos^2 A+\cos^2 B+\cos^2 C=\frac{7}{4R^2}$$

Then, $$\frac{\cos A\cdot\cos B\cdot\cos C}{\cos^2 A\cdot\cos^2 B\cdot\cos^2 C}=\frac{\frac{3}{8R^3}}{\frac{7}{4R^2}}=\frac{3}{14R}$$

2. Now, we know

$$\cos^2 A+\cos^2 B+\cos^2 C=1-2\cos A\cos B\cos C$$

$$\frac{7}{4R^2}=1-2\times\frac{3}{8R^3}$$

$$\Rightarrow \quad 4R^3-7R-3=0$$

$$\Rightarrow (R+1)(2R+1)(2R-3)=0$$

$$\Rightarrow \quad R=\frac{3}{2}$$

3. Now, $HD\cdot HE\cdot HF$

$=(2R\cos B\cos C)(2R\,Cos\,A\cos C)(2R\cos A\cos B)$

$=8R^3\cos^2 A\cos^2 B\cos^2 C$

$=8R^3\times\dfrac{9}{64R^6}=\dfrac{9}{8R^3}$

PASSAGE 5

Here, ΔMNL is pedal triangle, then

$$MN=a\cos A$$

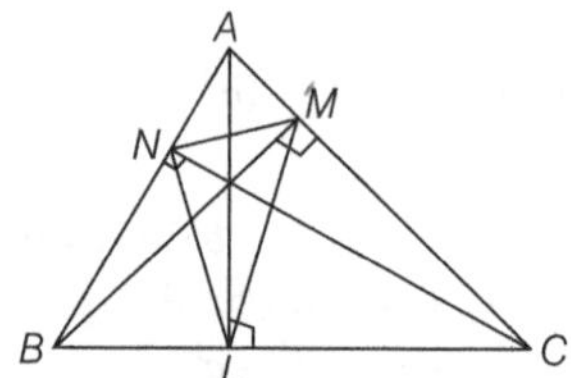

$$NL=b\cos B$$

and $\quad ML=c\cos C$

and $\quad \angle MLN=\pi-2A,\ \angle LMN=\pi-2B$

and $\quad \angle MNL=\pi-2C$

1. $\lambda=MN+NL+LM$

$=a\cos A+b\cos B+c\cos C$

$=R\,(\sin 2A+\sin 2B+\sin 2C)$

$=4R\sin A\sin B\sin C$

$=4R\left(\dfrac{a}{2R}\right)\left(\dfrac{b}{2R}\right)\left(\dfrac{c}{2R}\right)$

$=\dfrac{abc}{2R^2}=\dfrac{4R\Delta}{2R^2}=\dfrac{2\Delta}{R}$

and $\quad \mu=2s=\dfrac{2\Delta}{r}$

$\therefore \quad \dfrac{\lambda}{\mu}=\dfrac{r}{R}$

2. In ΔABM,

$$\cos A=\frac{AM}{AB}=\frac{AM}{c}$$

$\therefore \quad AM=c\cos A$ similarly $AN=b\cos A$

$\therefore \quad \Delta_1=$ Area of ΔAMN

$=\dfrac{1}{2}\cdot AM\cdot AN\cdot\sin A$

$=\dfrac{1}{2}(c\cos A)(b\cos A)(\sin A)$

$=\left(\dfrac{1}{2}bc\sin A\right)\cos^2 A=\Delta\cos^2 A$

Similarly, $\Delta_2 = \Delta \cos^2 B$ and $\Delta_3 = \Delta \cos^2 C$

$\therefore \Delta_1 + \Delta_2 + \Delta_3 = \Delta(\cos^2 A + \cos^2 B + \cos^2 C)$

$= \Delta(1 - 2\cos A \cos B \cos C)$ [from identity]

3. $\Delta' =$ Area of ΔLMN

$$= \frac{1}{2} \cdot ML \cdot NL \cdot \sin(\pi - 2A)$$

$$= \frac{1}{2} \cdot (c \cos C)(b \cos B)(\sin 2A)$$

$$= \frac{1}{2} bc \cos B \cos C \, (2 \sin A \cos A)$$

$$= \left(\frac{1}{2} bc \sin A\right)(2 \cos A \cos B \cos C)$$

$\therefore \quad \Delta' = \Delta(2 \cos A \cos B \cos C)$

$\Rightarrow \dfrac{\Delta'}{\Delta} = 2 \cos A \cos B \cos C$

4. Let R' be the circumradius of ΔLMN, then

$$R' = \frac{MN}{2 \sin(\angle MLN)} = \frac{a \cos A}{2 \sin(\pi - 2A)} = \frac{2R \sin A \cos A}{2 \sin 2A} = \frac{R \sin 2A}{2 \sin 2A} = \frac{R}{2}$$

5. $\because \; r' = 4R' \sin\left(\frac{L}{2}\right) \sin\left(\frac{M}{2}\right) \sin\left(\frac{N}{2}\right)$

Here, $L = \angle MLN = \pi - 2A$,

$M = \angle LMN = \pi - 2B$

and $N = \angle MNL = \pi - 2C$

and $R' = \dfrac{R}{2}$ (from Q. 4)

PASSAGE 6

1. Here, $AB = BC = 6$ cm, $CD = DA = 8$ cm.

$\therefore \quad \Delta ABD = \Delta BDC$ are congruent.

So, $\angle A = \angle C = \dfrac{180°}{2} = 90°$

$\therefore \quad BD = \sqrt{(AB)^2 + (AD)^2} = \sqrt{(6^2 + 8^2)}$

$= 10$ cm

$\therefore$ Radius of circumcircle $= \dfrac{BD}{2} = \dfrac{10}{2} = 5$ cm

2. Here, $AB = BC = 5$ cm, $CD = DA = 12$ cm

$\therefore \; \Delta ABD = \Delta BDC$ are congruent

So $\angle A = \angle C = \dfrac{180°}{2} = 90°$

$\therefore$ Area of quadrilateral $ABCD$,

$=$ Area of ΔABD + Area of ΔCBD

$= \dfrac{1}{2} \cdot 5 \cdot 12 + \dfrac{1}{2} \cdot 5 \cdot 12 = 60 \text{ cm}^2$

$\because \; AB + AD = BC + CD$

So, incircle can be drawn

Now, if radius of incircle is r, then

Area of quadrilateral $= r \cdot s$

$$\Rightarrow \quad 60 = r\left(\frac{5 + 5 + 12 + 12}{2}\right) = 17r$$

$$\therefore \quad r = \frac{60}{17} \text{ cm}$$

3. If a circle can be inscribed in a quadrilateral, then $a + c = b + d$

Also the quadrilateral is cyclic

$$\therefore \quad \cos A = \frac{a^2 + d^2 - b^2 - c^2}{2(ad + bc)} \quad \text{...(i)}$$

Now, $a - d = b - c$

$\Rightarrow \quad a^2 - 2ad + d^2 = b^2 - 2bc + c^2$

$\Rightarrow \; a^2 + d^2 - b^2 - c^2 = 2(ad - bc)$ ...(ii)

From Eqs. (i) and (ii)

$$\cos A = \frac{ad - bc}{ad + bc}$$

$$\Rightarrow \quad \frac{1 - \tan^2 A/2}{1 + \tan^2 A/2} = \frac{ad - bc}{ad + bc} = \frac{1 - \frac{bc}{ad}}{1 + \frac{bc}{ad}}$$

[by componendo and dividendo rule]

$$\therefore \quad \tan^2 A/2 = \frac{bc}{ad}$$

4. If the quadrilateral $ABCD$ can have a circle inscribed in it so as to touch the sides AB, BC, CD and DA in the points P, Q, R and S, we should have

$AP = AS$, $BP = BQ$, $CQ = CR$ and $DR = DS$

[$\because$ Length of tangents from outside point on a circle are equal]

$\therefore \; AP + BP + CR + DR = AS + BQ + CQ + DS$

i.e. $AB + CD = BC + DA$

i.e. $a + c = b + d$

Hence, $s = \dfrac{a + b + c + d}{2} = a + c = b + d$

$\therefore \; s - a = c, \; s - b = d, \; s - c = a, \; s - d = b$

$\therefore$ Area of cyclic quadrilateral

$$= \sqrt{(s - a)(s - b)(s - c)(s - d)} = \sqrt{(abcd)}$$

5. $\because$ Area of quadrilateral $= \sqrt{(abcd)}$

$$= \frac{1}{2} r \cdot (a + b + c + d)$$

$$\therefore \quad r = \frac{2\sqrt{(abcd)}}{(a + b + c + d)}$$

Numerical Value Type

1. $(r_1 + r_2)$

$$= 4R\left(\sin\frac{A}{2}\cdot\cos\frac{B}{2}\cdot\cos\frac{C}{2} + \sin\frac{B}{2}\cdot\cos\frac{A}{2}\cdot\cos\frac{C}{2}\right)$$

$$= 4R\left[\cos\frac{C}{2}\left(\sin\frac{A}{2}\cdot\cos\frac{B}{2} + \sin\frac{B}{2}\cdot\cos\frac{A}{2}\right)\right]$$

$$= 4R\cos^2\frac{C}{2}$$

Similarly, $(r_2 + r_3) = 4R\cos^2\frac{A}{2}$

and $(r_3 + r_1) = 4R\cos^2\frac{B}{2}$

Also, $a + b + c = 4R(\sin A + \sin B + \sin C)$

$$= 16R\cos\frac{A}{2}\cdot\cos\frac{B}{2}\cdot\cos\frac{C}{2}$$

$\Rightarrow$ $4s^2 = 256R^2\cdot\cos^2\frac{A}{2}\cdot\cos^2\frac{B}{2}\cdot\cos^2\frac{C}{2}$

$Rs^2 = 64R^3\cdot\cos^2\frac{A}{2}\cdot\cos^2\frac{B}{2}\cdot\cos^2\frac{C}{2}$

$\therefore$ $5050\frac{\Pi(r_1 + r_2)}{Rs^2} = 5050 \times 1 = 5050$

2. Given that, $\frac{s-a}{11} = \frac{s-b}{12} = \frac{s-c}{13}$

$$= \frac{s-a+s-b+s-c}{11+12+13} = \frac{3s-2s}{36} = \frac{s}{36}$$

$\therefore$ $s - a = \frac{11s}{36}, s - b = \frac{12s}{36}, s - c = \frac{13s}{36}$

$\therefore$ $\tan^2\frac{A}{2} = \frac{(s-b)(s-c)}{s(s-a)}$

$$= \frac{\frac{12s}{36}\times\frac{13s}{36}}{s\times\frac{11s}{36}} = \frac{13}{33}$$

$\because$ $\lambda\tan^2\frac{A}{2} = 455 \Rightarrow \lambda\times\frac{13}{33} = 455$

$\Rightarrow$ $\lambda = 35\times 33 = 1155$

3. $p_1 = \frac{2\Delta}{a}$, $p_2 = \frac{2\Delta}{b}$, $p_3 = \frac{2\Delta}{c}$

$\therefore$ $p_1 + p_2 + p_3 \ge 3(p_1p_2p_3)^{1/3}$

$$= 3\left\{\frac{(2\Delta)^3}{(abc)}\right\}^{1/3} = 6\Delta\left(\frac{1}{abc}\right)^{1/3}$$

$$= 6rs\left(\frac{1}{abc}\right)^{1/3}$$

$$= 6r\frac{a+b+c}{2}\left(\frac{1}{abc}\right)^{1/3}$$

$$= 9r\left(\frac{a+b+c}{3}\right)\left(\frac{1}{abc}\right)^{1/3}$$

$\ge 9r$ $\quad\left[\because \frac{\text{AM}}{\text{GM}} \ge 1\right]$

$$= 9\times\frac{8}{3} = 24$$

$\Rightarrow$ $p_1 + p_2 + p_3 \ge 24$

4. Let $AB = n$, $AC = n + 1$, $BC = n + 2$

Further,

Let $A = 2C$ (since AB is the smallest and BC is largest)

By sine rule, we have

$$\frac{\sin A}{n+2} = \frac{\sin B}{n+1} = \frac{\sin C}{n}$$

$\because$ $A = 2C$

and $B = 180° - (A + C)$

$\Rightarrow$ $B = 180° - 3C$

$\therefore$ $\frac{\sin 2C}{n+2} = \frac{\sin 3C}{n+1} = \frac{\sin C}{n}$

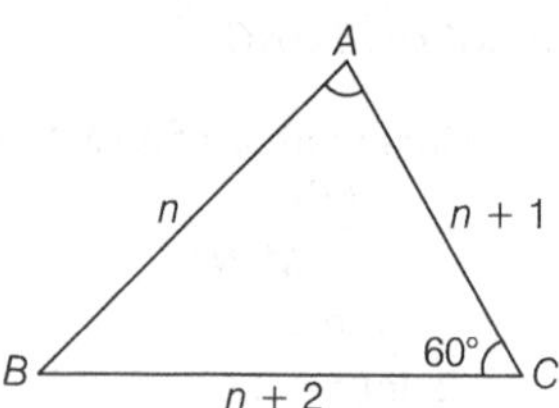

$\Rightarrow$ $\frac{2\cos C}{n+2} = \frac{3 - 4\sin^2 C}{n+1} = \frac{1}{n}$

$\therefore$ $\cos C = \frac{n+2}{2n}$ and $\sin^2 C = \frac{2n-1}{4n}$

$\therefore$ $\left(\frac{n+2}{2n}\right)^2 + \left(\frac{2n-1}{4n}\right) = 1 \Rightarrow n^2 - 3n - 4 = 0$

$\therefore$ $n = 4, n = -1$

$\Rightarrow$ $n = 4, n \ne -1$

Then, sides are 4, 5, 6.

$\therefore$ Perimeter of triangle $= 4 + 5 + 6 = 15$

5. $\cos A = \frac{b^2 + c^2 - a^2}{2bc}$

$\Rightarrow$ $\frac{1}{2} = \cos\frac{\pi}{3} = \frac{b^2 + c^2 - a^2}{2bc}$

$\Rightarrow$ $b^2 + c^2 - a^2 = bc$ $\quad$ ...(i)

In ΔABD, $AD^2 = c^2 + \frac{a^2}{4} - 2c\times\frac{a}{2}\cos B$

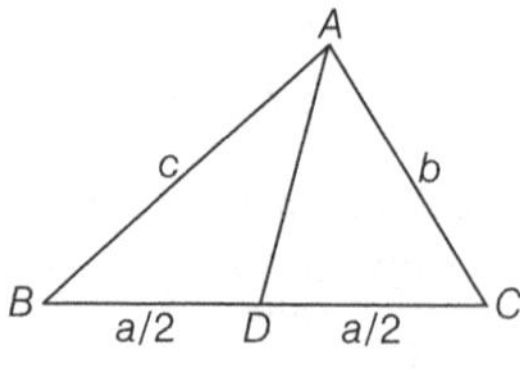

$\Rightarrow$ $4AD^2 = 4c^2 + a^2 - 4ca\left(\frac{c^2 + a^2 - b^2}{2ca}\right)$

$$= 2c^2 + 2b^2 - a^2$$

$$= 2c^2 + 2b^2 - (b^2 + c^2 - bc)$$

$= b^2 + c^2 + bc$ $\quad$ [using Eq. (i)]

$$= 40^2 + 30^2 + 40\times 30$$

$$= 1600 + 900 + 1200 = 3700$$

6. $\because\ r = \dfrac{\Delta}{s} = 2$ and $r_1 = \dfrac{\Delta}{s-a} = 15$

Thus, $\dfrac{s-a}{s} = \dfrac{r}{r_1} = \dfrac{2}{15}$ or $s = \dfrac{15}{13}a$

So, $\dfrac{\Delta}{s} = 2 \Rightarrow \Delta = \dfrac{30}{13}a$

$\Rightarrow \dfrac{1}{2}bc = \dfrac{30}{13}a$

$\Rightarrow bc = \dfrac{60}{13}a$ ...(i)

$[\because A = \pi/2]$

Also, $b + c = 2s - a = \dfrac{17}{13}a$ ...(ii)

Solving Eqs. (i) and (ii), we get

$$b = \frac{12}{13}a \text{ and } c = \frac{5}{13}a$$

Also, $\dfrac{\Delta}{s} = \dfrac{\frac{bc}{2}}{\frac{a+b+c}{2}} = \dfrac{bc}{a+b+c}$

$\Rightarrow 2 = \dfrac{2}{13}a \Rightarrow a = 13$

7. $\cos(A-B) = \dfrac{1-\tan^2\left(\frac{A-B}{2}\right)}{1+\tan^2\left(\frac{A-B}{2}\right)}$

$\Rightarrow \dfrac{1-\tan^2\left(\frac{A-B}{2}\right)}{1+\tan^2\left(\frac{A-B}{2}\right)} = \dfrac{31}{32}$

$\Rightarrow \dfrac{2\tan^2\left(\frac{A-B}{2}\right)}{2} = \dfrac{1}{63}$

[by componendo and dividendo rule]

$\Rightarrow \tan\left(\dfrac{A-B}{2}\right) = \dfrac{1}{\sqrt{63}}$

Again, $\tan\left(\dfrac{A-B}{2}\right) = \dfrac{a-b}{a+b}\cot\dfrac{C}{2}$

$\Rightarrow \dfrac{1}{\sqrt{63}} = \dfrac{1}{9}\cot\dfrac{C}{2}$

$\Rightarrow \cot\dfrac{C}{2} = \dfrac{9}{\sqrt{63}}$

$\Rightarrow \tan\dfrac{C}{2} = \dfrac{\sqrt{63}}{9} = \dfrac{3\sqrt{7}}{9} = \dfrac{\sqrt{7}}{3}$

$\Rightarrow \cos C = \dfrac{1-\tan^2 C/2}{1+\tan^2 C/2} = \dfrac{1-7/9}{1+7/9}$

$= \dfrac{2}{16} = \dfrac{1}{8}$

$\therefore c^2 = a^2 + b^2 - 2ab\cos C$

$= 25 + 16 - 2\cdot 20 \cdot \dfrac{1}{8}$

$= 41 - 5 = 36$

$\therefore c = 6$

$\Rightarrow 12c = 12 \times 6 = 72$

8. Obviously $3^2 + 4^2 = 5^2$

$\therefore \angle BAC = 90°$

In ΔCAE, let $\angle CAE = \theta$ and $CE = \dfrac{5}{3}$

We have, $\cos C = \dfrac{4^2 + (5/3)^2 - AE^2}{2\cdot 4\cdot 5/3} = \dfrac{16 + 25/9 - AE^2}{40/3}$

$\Rightarrow \dfrac{4}{5} = \dfrac{169/9 - AE^2}{40/3}$

$\Rightarrow \dfrac{169}{9} - AE^2 = \dfrac{4}{5}\times\dfrac{40}{3} = \dfrac{32}{3}$

$\therefore AE^2 = \dfrac{169}{9} - \dfrac{32}{3} = \dfrac{169-96}{9} = \dfrac{73}{9}$

Again from the ΔCAE,

$$\cos\theta = \frac{4^2 + AE^2 - (5/3)^2}{2\cdot 4\cdot AE}$$

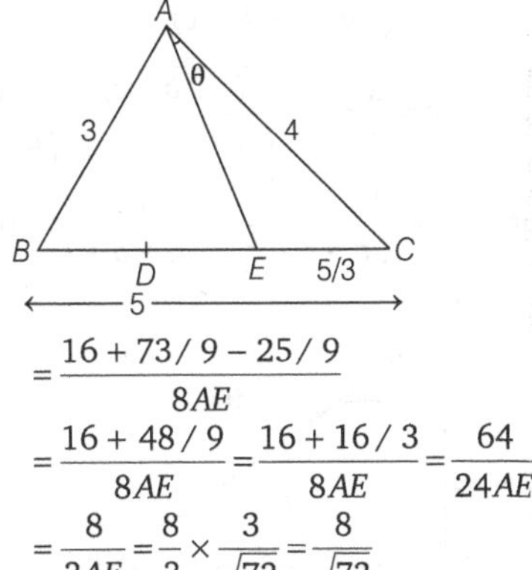

$= \dfrac{16 + 73/9 - 25/9}{8AE}$

$= \dfrac{16 + 48/9}{8AE} = \dfrac{16 + 16/3}{8AE} = \dfrac{64}{24AE}$

$= \dfrac{8}{3AE} = \dfrac{8}{3}\times\dfrac{3}{\sqrt{73}} = \dfrac{8}{\sqrt{73}}$

$\therefore \sec\theta = \dfrac{\sqrt{73}}{8}$

$\Rightarrow \tan^2\theta = \sec^2\theta - 1 = \dfrac{73}{64} - 1 = \dfrac{9}{64}$

$\therefore \tan\theta = \dfrac{3}{8} = 0.375$

9. Let the centres of the circles be C_1, C_2 and C_3 and their radii be a, b and c respectively. Let the circles touch one another at P, Q and R. Let the tangents at their points of contact meet at O.

Since, OP and OQ are two tangents from O to the circle C_2, they are equal *i.e.*, $OP = OQ$.

Similarly, $OQ = OR$

$\Rightarrow OP = OQ = OR$

Also, $OP \perp C_1C_2$, $OQ \perp C_2C_3$, $OR \perp C_1C_3$

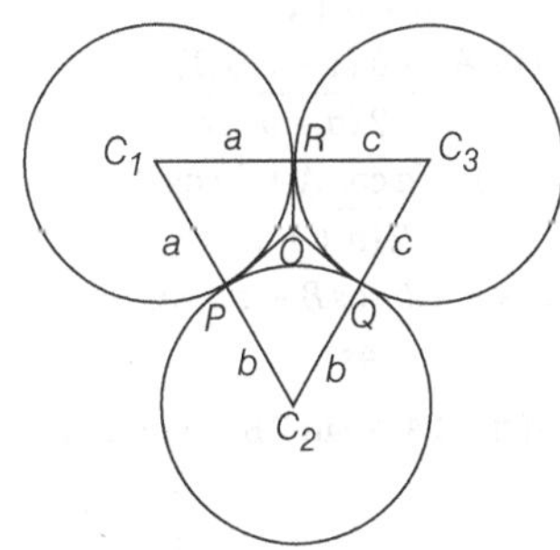

$\therefore$ OP, OQ and OR are $\perp$ distances of O from the sides of $\Delta C_1C_2C_3$.
Hence, OP or OQ or OR is inradius of $\Delta C_1C_2C_3$.
Let $OP = OQ = OR = r$
From question, $r = 4$
we have to find the value of $\dfrac{abc}{a+b+c}$

Now, $r = \dfrac{\Delta}{s}$, where s = semiperimeter of $\Delta C_1C_2C_3$
and Δ = area of $\Delta C_1C_2C_3$.

Now, $s = \dfrac{(a+b)+(b+c)+(c+a)}{2} = a+b+c$

and $\Delta = \sqrt{s\{s-(a+b)\}\{s-(b+c)\}\{s-(c+a)\}}$

$= \sqrt{(a+b+c)\cdot c\cdot a\cdot b}$ $\quad [\because s = a+b+c]$

$= \sqrt{abc\,(a+b+c)}$

$\therefore \quad r = \dfrac{\Delta}{s} = \dfrac{\sqrt{abc\,(a+b+c)}}{a+b+c} = \sqrt{\dfrac{abc}{a+b+c}}$

or $\quad 4 = \sqrt{\dfrac{abc}{a+b+c}} \Rightarrow \dfrac{abc}{a+b+c} = 16$

10.

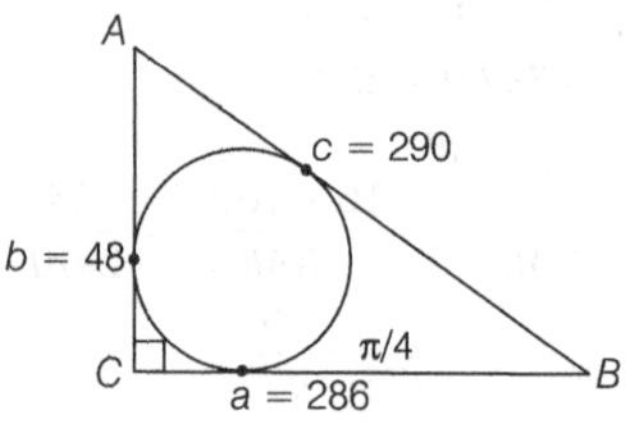

$$\frac{a}{b} = \frac{143}{24} \Rightarrow \frac{a}{143} = \frac{b}{24} = k \text{ (let)}$$

Area of triangle $(\Delta) = \dfrac{1}{2}ab = 6864$

$\Rightarrow \quad 143k \times 24k = 2 \times 6864$

$\Rightarrow \quad 12k^2 \times 143 = 6864$

$\Rightarrow \quad k^2 = 4 \Rightarrow k = 2$

$\Rightarrow \quad a = 286,\ b = 48$

Now, $\quad c^2 = 48^2 + (286)^2 = 290^2 \Rightarrow c = 290$

Hence, the radius of the circle inscribed in the triangle is given by

$$r = \frac{\Delta}{s} = 22$$

11. $\dfrac{a\cos^2 A/2 + b\cos^2 B/2 + c\cos^2 C/2}{a+b+c}$

$= \dfrac{a(1+\cos A) + b(1+\cos B) + c(1+\cos C)}{2(a+b+c)}$

$= \dfrac{(a+b+c) + (a\cos A + b\cos B + c\cos C)}{2(a+b+c)}$

$= \dfrac{1}{2} + \dfrac{a\cos A + b\cos B + c\cos C}{4s}$

$= \dfrac{1}{2} + \dfrac{R}{4s}(\sin 2A + \sin 2B + \sin 2C)$

$$\left[\because \frac{a}{\sin A} = \frac{b}{\sin B} = \frac{c}{\sin C} = 2R\right]$$

$= \dfrac{1}{2} + \dfrac{R}{4s}(4\sin A\sin B\sin C)$

$= \dfrac{1}{2} + \dfrac{R}{4s} \times \left(4\cdot\dfrac{a}{2R}\cdot\dfrac{b}{2R}\cdot\dfrac{c}{2R}\right) = \dfrac{1}{2} + \dfrac{abc}{8R^2s}$

$= \dfrac{1}{2} + \dfrac{4R\Delta}{8R^2s} = \dfrac{1}{2} + \dfrac{r}{2R} = \dfrac{1}{2}\left(1 + \dfrac{r}{R}\right)$

$\le \dfrac{1}{2}\left(1 + \dfrac{1}{2}\right)$ $\quad \left[\because \dfrac{r}{R} \le \dfrac{1}{2}\right]$

Hence, $\dfrac{a\cos^2 A/2 + b\cos^2 B/2 + c\cos^2 C/2}{a+b+c} \le \dfrac{3}{4}$

$\therefore \dfrac{(a\cos^2 A/2 + b\cos^2 B/2 + c\cos^2 C/2)}{(a+b+c)} \times \dfrac{3}{4} = 0.75$

12. Given, in ΔABC

$AC = 12$ cm $AB = 5$ cm

Area of $\Delta ABC = 30$ cm^2

Area of $\Delta ABC = \dfrac{1}{2} AB \times AC \sin A$

$\Rightarrow 30 = \dfrac{1}{2} \times 5 \times 12 \times \sin A$

$\sin A = 1$

$A = 90°$

$\because BC = \sqrt{AB^2 + AC^2} = \sqrt{5^2 + 12^2} = 13$

$\Rightarrow \quad 2R = BC = 13$

$\Rightarrow \quad r = \dfrac{\Delta}{s} = \dfrac{30}{\dfrac{(5+12+13)}{2}} = 2$

$\because \quad 2R + r = 13 + 2 = 15$

13. $\dfrac{r}{r_1} = \dfrac{4R\sin A/2\cdot\sin B/2\cdot\sin C/2}{4R\sin A/2\cdot\cos B/2\cdot\cos C/2} = \tan\dfrac{B}{2}\tan\dfrac{C}{2} = \dfrac{1}{2}$

Now,

$\tan\dfrac{A}{2}\left(\tan\dfrac{B}{2} + \tan\dfrac{C}{2}\right) = \tan\dfrac{A}{2}\cdot\tan\dfrac{B}{2} + \tan\dfrac{A}{2}\cdot\tan\dfrac{C}{2}$

$= 1 - \tan\dfrac{B}{2}\cdot\tan\dfrac{C}{2} = 1 - \dfrac{1}{2} = \dfrac{1}{2} = 0.5$

14. Let BD, CE and AF be of lengths $n-1$, n and $n+1$ respectively.

Since, the lengths of the tangent, from an external point to the circle are equal.

Then, $BF = BD = n-1$, $CD = CE = n$ and $AE = AF = n+1$

$\Rightarrow \quad BC = 2n-1$, $CA = 2n+1$ and $\quad AB = 2n \Rightarrow s = 3n$

Now, $\tan\dfrac{C}{2} = \dfrac{r}{n}$ and $\tan\dfrac{B}{2} = \dfrac{r}{n-1}$

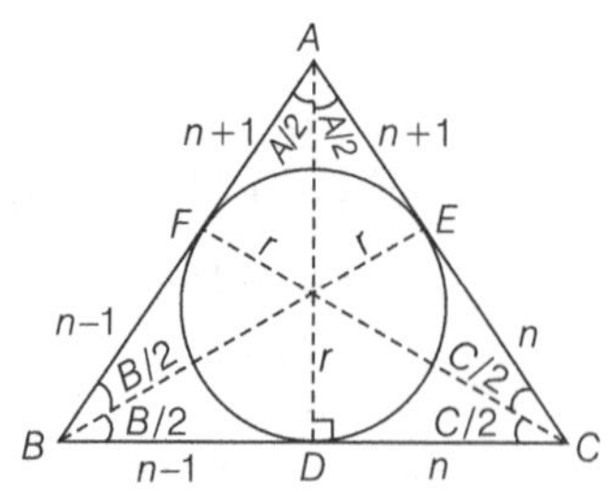

So that $\tan\left(\frac{B}{2}+\frac{C}{2}\right)=\dfrac{\frac{r}{n-1}+\frac{r}{n}}{1-\frac{r^2}{n(n-1)}}$

$\Rightarrow \quad \cot A/2=\dfrac{2rn-r}{n^2-n-r^2} \quad \left[\because \cot\frac{A}{2}=\frac{n+1}{r}\right]$

$\Rightarrow \quad n^3-3r^2n-n=0$

$\Rightarrow \quad n^3-48n-n=0 \quad [\because r=4\text{ (given)}]$

$n=0$ or $n^2=49 \Rightarrow n=7 \quad [\text{as } n\neq 0]$

So, the sides of the triangle are 13, 14 and 15 unit and the required sum is equal to

$$(13)^3+(14)^3+(15)^3=8316$$

15. $AG=\frac{2}{3}AA_1,\ BG=\frac{2}{3}BB_1$

$\Rightarrow \quad AG=\frac{2}{3}\sqrt{2b^2+2c^2-a^2}$

$BG=\frac{2}{3}\sqrt{2a^2+2c^2-b^2}$

$\Rightarrow AG=\frac{2}{3}a,\ BG=\frac{2}{3}\sqrt{b^2+4c^2}$ as $a^2=b^2+c^2$

$\Rightarrow AG=\frac{10}{3},\ BG=\frac{2}{3}\sqrt{16+36}=\frac{4}{3}\sqrt{13}$

Also, $AB=c=3$ and $\Delta_{GAB}=\frac{1}{3}\Delta_{ABC}=2$

If R_1 be the circumradius of ΔGAB, then

$$R_1=\frac{(AG)(BG)(AB)}{4\Delta_{GAB}}=\frac{10}{3}\cdot\frac{4}{3}\sqrt{13}\cdot 3\cdot\frac{1}{4\cdot 2}$$

$$=\frac{5\sqrt{13}}{3}\text{ unit}$$

$\therefore \quad 90R_1^2=90\times\frac{25\times 13}{9}=3250$

16. We have,

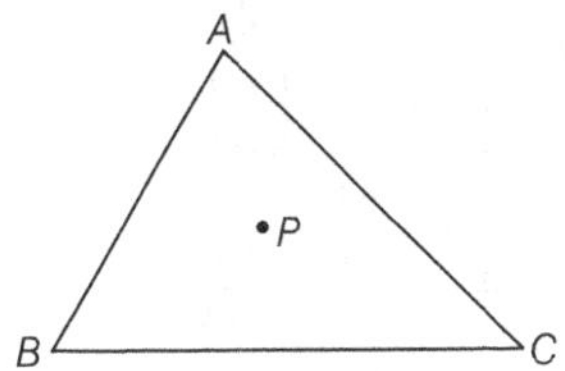

$\Delta=\Delta_{BPC}+\Delta_{CPA}+\Delta_{APB}$

$\Rightarrow \quad \Delta=\frac{1}{2}\cdot a\cdot x_a+\frac{1}{2}\cdot b\cdot x_b+\frac{1}{2}\cdot c\cdot x_c$

$\Rightarrow \quad \Delta=\frac{1}{2}(ax_a+bx_b+cx_c) \quad \ldots\text{(i)}$

Also, $\quad \Delta=\frac{1}{2}aH_a=\frac{1}{2}bH_b=\frac{1}{2}cH_c \quad \ldots\text{(ii)}$

From Eqs. (i) and (ii), we get

$$\frac{x_a}{H_a}+\frac{x_b}{H_b}+\frac{x_c}{H_c}=1$$

Now, AM ≥ HM

$$\therefore \quad \frac{\frac{x_a}{H_a}+\frac{x_b}{H_b}+\frac{x_c}{H_c}}{3}\geq\frac{3}{\left(\frac{H_a}{x_a}+\frac{H_b}{x_b}+\frac{H_c}{x_c}\right)}$$

$$\Rightarrow \quad \frac{1}{3}\geq\frac{3}{\left(\frac{H_a}{x_a}+\frac{H_b}{x_b}+\frac{H_c}{x_c}\right)}$$

or $\quad \frac{H_a}{x_a}+\frac{H_b}{x_b}+\frac{H_c}{x_c}\geq 9$

$\therefore \ 191\left(\frac{H_a}{x_a}+\frac{H_b}{x_b}+\frac{H_c}{x_c}\right)\geq 1719$

Single Integer Type

1. $R\cos A,\ g=R\cos B,\ h=R\cos C.$

$\therefore \frac{a}{f}+\frac{b}{g}+\frac{c}{h}=\frac{2R\sin A}{R\cos A}+\frac{2R\sin B}{R\cos B}+\frac{2R\sin C}{R\cos C}$

$=2(\Sigma\tan A)$

$\therefore \quad \frac{abc}{fgh}=8\,(\Pi\tan A)$

$\therefore \frac{a}{f}+\frac{b}{g}+\frac{c}{h}=\lambda\frac{abc}{fgh}$

$\Rightarrow 2\,\Sigma\tan A=\lambda\cdot 8\,(\Pi\tan A)\Rightarrow\lambda=\frac{1}{4}$

2. Given, $\tan A:\tan B:\tan C=1:2:3$

$$\tan A=\frac{\sin A}{\cos A}=\frac{\frac{a}{2R}}{\frac{b^2+c^2-a^2}{2bc}}=\left(\frac{1}{b^2+c^2-a^2}\right)\frac{2abc}{2R}$$

$$\therefore \tan A:\tan B:\tan C=\frac{1}{b^2+c^2-a^2}:\frac{1}{a^2+c^2-b^2}:\frac{1}{a^2+b^2-c^2}$$

Clearly, $\tan A:\tan B:\tan C=l:m:n$

Then, $a:b:c=\sqrt{\frac{1}{m}+\frac{1}{n}}:\sqrt{\frac{1}{n}+\frac{1}{l}}:\sqrt{\frac{1}{l}+\frac{1}{m}}$

Here, $l:m:n=1:2:3$

$$a:b:c=\sqrt{\frac{1}{2}+\frac{1}{3}}:\sqrt{\frac{1}{3}+1}:\sqrt{1+\frac{1}{2}}$$

$$a:b:c=\frac{\sqrt{5}}{\sqrt{6}}:\frac{2\sqrt{2}}{\sqrt{6}}:\frac{3}{\sqrt{6}}$$

Given, $\sin A:\sin B:\sin C=\sqrt{5}:2\sqrt{2}:k \therefore k=3$

3. Given, $\frac{31}{32}=\cos(A-B)=\dfrac{1-\tan^2\frac{A-B}{2}}{1+\tan^2\frac{A-B}{2}}$

$\Rightarrow \quad 63\tan^2\frac{A-B}{2}=1\Rightarrow\tan\frac{A-B}{2}=\frac{1}{\sqrt{63}}$

Now, $\quad \tan\frac{A-B}{2}=\frac{a-b}{a+b}\cot\frac{C}{2}$

$\Rightarrow \quad \frac{1}{\sqrt{63}}=\frac{5-4}{5+4}\cot\frac{C}{2}\Rightarrow\tan\frac{C}{2}=\frac{\sqrt{63}}{9}$

Also, $\cos C = \dfrac{1-\tan^2\left(\frac{C}{2}\right)}{1+\tan^2\left(\frac{C}{2}\right)}$

$$= \frac{1-\frac{63}{81}}{1+\frac{63}{81}} = \frac{18}{144} = \frac{1}{8}$$

$\therefore \quad c^2 = a^2 + b^2 - 2ab\cos C$

$$= 25 + 16 - 2\cdot 5\cdot 4\left(\frac{1}{8}\right) = 36$$

Hence, $c = 6$

4. $0 < a < b + c$ or $\dfrac{a}{b+c} < 1$

$\therefore \dfrac{a}{b+c} + \dfrac{b}{c+a} + \dfrac{c}{a+b} < 3$

Also, $[(b+c)+(c+a)+(a+b)]$

$$\times\left[\frac{1}{b+c} + \frac{1}{c+a} + \frac{1}{a+b}\right] \geq 9 \quad \text{[AM} \geq \text{HM]}$$

$\Rightarrow \quad (a+b+c)\left(\dfrac{1}{b+c} + \dfrac{1}{c+a} + \dfrac{1}{a+b}\right) \geq \dfrac{9}{2}$

$\Rightarrow \quad \dfrac{a}{b+c} + \dfrac{b}{c+a} + \dfrac{c}{a+b} \geq \dfrac{3}{2}$

$\Rightarrow \quad \dfrac{3}{2} \leq \dfrac{a}{b+c} + \dfrac{b}{c+a} + \dfrac{c}{a+b} < 3$

Therefore, integral values of

$$\frac{a}{b+c} + \frac{b}{c+a} + \frac{c}{a+b} = 2$$

5.

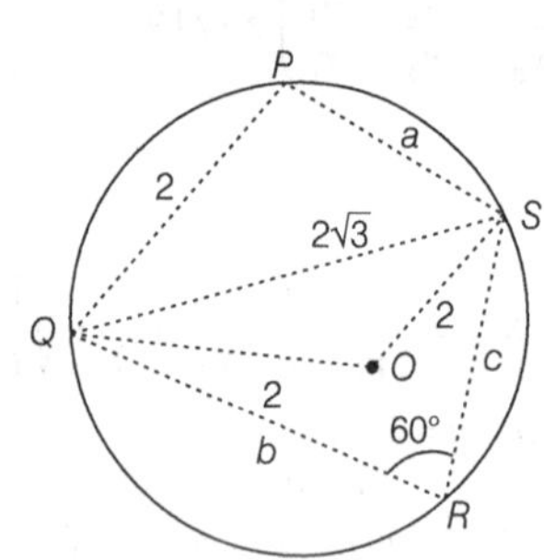

$\angle QOS = 2\angle R$

$\therefore \cos(2R) = \dfrac{4+4-12}{2\cdot 2\cdot 2} = -\dfrac{1}{2} \Rightarrow R = 60°$

$\therefore \quad \angle P = 120°$

$\cos(< P) = \dfrac{a^2+4-12}{4a} = -\dfrac{1}{2}$

$\Rightarrow a^2 + 2a - 8 = 0$

$\Rightarrow \quad a = 2$

$\therefore$ Area of $\Delta PQS = \dfrac{1}{2}\cdot 2\cdot 2\sin 120° = \sqrt{3}$

$\therefore$ Area of $\Delta QRS = \dfrac{1}{2}\cdot b\cdot c\sin 60° = 3\sqrt{3} - \sqrt{3}$

$\Rightarrow bc = 8 \Rightarrow QR\cdot RS = 8$

6. It is given that,

$$r_1 = 8 = \frac{\Delta}{s-a} \quad \text{...(i)}$$

$$r_2 = 12 = \frac{\Delta}{s-b} \quad \text{...(ii)}$$

and $$r_3 = 24 = \frac{\Delta}{s-c} \quad \text{...(iii)}$$

From Eqs. (i) and (ii), we get

$$\frac{s-b}{s-a} = \frac{2}{3} \Rightarrow 3a - 3b = 2s - 2a \Rightarrow 5a + c = 5b \quad \text{...(iv)}$$

From Eqs. (ii) and (iii), we get

$$\frac{s-c}{s-b} = \frac{1}{2} \Rightarrow 2s - 2c = s - b \Rightarrow a + 3b = 3c \quad \text{...(v)}$$

and from Eqs. (i) and (iii), we get

$$\frac{s-c}{s-a} = \frac{1}{3} \Rightarrow 3a - 3c = s - a \Rightarrow 2a + b = 2c \quad \text{...(vi)}$$

On solving Eqs. (iv), (v) and (vi), we get

$$a = 12,\ b = 16,\ c = 20$$

$\therefore \quad \dfrac{b+c}{a} = 3$

7. $A + B + C = \pi$

Given, $C = 2A \Rightarrow B = \pi - 3A$

As $0 < C < \pi \Rightarrow 0 < 2A < \pi \Rightarrow 0 < A < \dfrac{\pi}{2}$

By sine rule, $\dfrac{a}{\sin A} = \dfrac{b}{\sin B}$

$\Rightarrow \dfrac{a}{\sin A} = \dfrac{2a}{\sin(\pi - 3A)} \Rightarrow \dfrac{a}{1} = \dfrac{2a}{3 - 4\sin^2 A}$

$\Rightarrow 3 - 4\sin^2 A = 2$

$\Rightarrow \sin^2 A = \dfrac{1}{4} \Rightarrow \sin A = \dfrac{1}{2} \Rightarrow A = \dfrac{\pi}{6}$ or $\dfrac{5\pi}{6}$

But, $0 < A < \dfrac{\pi}{2}$

$\Rightarrow A = \dfrac{\pi}{6}, \angle B = \dfrac{\pi}{2}$ and $\angle C = \dfrac{\pi}{3}$

$\Rightarrow a^2 + b^2 + c^2 = 4R^2[\sin^2 A + \sin^2 B + \sin^2 C]$

$$= 4R^2\left[\sin^2\frac{\pi}{6} + \sin^2\frac{\pi}{2} + \sin^2\frac{\pi}{3}\right]$$

$$= 4R^2\left[\frac{1}{4} + 1 + \frac{3}{4}\right] = 8R^2$$

$\Rightarrow \quad \dfrac{a^2+b^2+c^2}{R^2} = 8$

8. We have, $2B = A + C \Rightarrow 3B = 180° \Rightarrow B = 60°$

$$b^2 = ac \Rightarrow \sin^2 B = \sin A \sin C = \frac{3}{4}$$

$\Rightarrow \cos(A - C) - \cos(A + C) = \dfrac{3}{2}$

$\Rightarrow \quad \cos(A - C) = \dfrac{3}{2} - \dfrac{1}{2} = 1 \quad \left[A + C = \dfrac{2\pi}{3}\right]$

$\Rightarrow \quad A - C = 0 \Rightarrow A = C$

$\therefore \quad A = C = B = 60°$

$\Rightarrow \quad a = b = c.$

Hence, $\dfrac{a(a+b+c)}{3bc} = \dfrac{3a^2}{3a^2} = 1$

Matrix-Match Type

1. (A) $\because a^4 - 2(b^2+c^2)a^2 + b^4 + b^2c^2 + c^4 = 0$

$\Rightarrow (b^2+c^2-a^2)^2 - (b^2+c^2)^2 + b^4 + b^2c^2 + c^4 = 0$

$\Rightarrow \quad (b^2+c^2-a^2)^2 = b^2c^2$

$\Rightarrow \quad \left(\dfrac{b^2+c^2-a^2}{2bc}\right)^2 = \dfrac{1}{4}$

$\Rightarrow \quad \cos^2 A = \dfrac{1}{4}$

$\therefore \quad \cos A = \pm\dfrac{1}{2}$

$\therefore \quad A = 60°$ or $120°$ (Q, S)

(B) $\because \quad \cos C = \dfrac{a^2+b^2-c^2}{2ab}$

$\Rightarrow \quad a^2+b^2-c^2 = 2ab\cos C$

Squaring both sides, then

$a^4+b^4+c^4+2a^2b^2-2a^2c^2-2b^2c^2$

$= 4a^2b^2\cos^2 C$

$\Rightarrow \quad 3a^2b^2 = 4a^2b^2\cos^2 C$

$[\because a^4+b^4+c^4 = a^2b^2+2b^2c^2+2c^2a^2]$

$\Rightarrow \quad \cos^2 C = \dfrac{3}{4}$

$\cos C = \pm\dfrac{\sqrt{3}}{2}$

$\therefore \quad C = 30°$ or $150°$ (P, T)

(C) $\because \quad a^4+b^4+c^4+2a^2c^2 = 2a^2b^2+2b^2c^2$

$\Rightarrow \quad b^4 - 2(a^2+c^2)b^2 + (a^2+c^2)^2 = 0$

$\Rightarrow \quad \{b^2-(a^2+c^2)\}^2 = 0$

$\Rightarrow \quad \therefore\ a^2+c^2 = b^2$

$\Rightarrow \quad \cos B = 0$

$\therefore \quad B = 90°$ (R)

2. (A) $\because 2a^2+b^2+c^2 = 2ac+2ab$

$\Rightarrow \quad (a^2+b^2-2ab)+(a^2+c^2-2ac)=0$

$\Rightarrow \quad (a-b)^2+(a-c)^2 = 0$

Which is possible only when

$a-b=0,\ a-c=0$

$\therefore \quad a=b=c$

$\Rightarrow \Delta ABC$ is equilateral

$\therefore \quad \angle A = \angle B = \angle C = 60°$ (P, T)

(B) $\because a^2+b^2+c^2 = b\sqrt{2}(c+a)$

$\Rightarrow \quad a^2+b^2+c^2-bc\sqrt{2}-ab\sqrt{2}=0$

$\Rightarrow \quad 2a^2+2b^2+2c^2-2bc\sqrt{2}-2ab\sqrt{2}=0$

$\Rightarrow (b^2-2bc\sqrt{2}+2c^2)+(b^2-2ab\sqrt{2}+2a^2)=0$

$\Rightarrow \quad (b-c\sqrt{2})^2+(b-a\sqrt{2})^2=0$

Which is possible only when

$b-c\sqrt{2}=0,\ b-a\sqrt{2}=0$

$\therefore \quad b=c\sqrt{2},\ b=a\sqrt{2}$

i.e. $\quad b^2+b^2=(c\sqrt{2})^2+(a\sqrt{2})^2$

$\Rightarrow \quad c^2+a^2=b^2$

$\therefore \quad \cos B = \dfrac{c^2+a^2-b^2}{2ac}=0$

$\therefore \quad \angle B = 90°$ also $c\sqrt{2}=a\sqrt{2}$

$\Rightarrow \quad c=a$

Then, $\quad \angle A = \angle C = 45°$

Hence, $\quad \angle A = \angle C = 45°$

and $\quad \angle B = 90°$ (Q, S)

(C) $\because \quad a^2+b^2+c^2 = bc+ca\sqrt{3}$

$\Rightarrow \quad a^2+b^2+c^2-bc-ca\sqrt{3}=0$

$\Rightarrow \quad \left(\dfrac{c\sqrt{3}}{2}-a\right)^2+\left(\dfrac{c}{2}-b\right)^2=0$

Which is possible only when

$\dfrac{c\sqrt{3}}{2}-a=0$ and $\dfrac{c}{2}-b=0$

$\therefore \quad a=\dfrac{c\sqrt{3}}{2}$ and $b=\dfrac{c}{2}$

Then, $\quad a^2+b^2=\dfrac{3c^2}{4}+\dfrac{c^2}{4}=c^2$

$\because \quad \cos C = \dfrac{a^2+b^2-c^2}{2ab}=0$

$\because \quad \angle C = 90°$

Also, $\quad \dfrac{a}{\sin A}=\dfrac{c}{\sin C}$

$\Rightarrow \quad \dfrac{\frac{c\sqrt{3}}{2}}{\sin A}=\dfrac{c}{1}$

$\therefore \quad \sin A = \dfrac{\sqrt{3}}{2} = \sin 60°$

$\Rightarrow \quad \angle A = 60°,\ \angle B = 30°,$

$\angle C = 90°$

$\because \quad 2C = A+B$ (Q, R, T)

30

Height and Distance

Objective Questions Type I [Only one correct answer]

In each of the questions below, four choices are given of which only one is correct. You have to select the correct answer which is the most appropriate.

1. A ladder rests against a wall at an angle α to the horizontal. Its foot is pulled away from the wall through a distance x, so that it slides a distance y down the wall making an angle β with the horizontal then, the value of $\dfrac{\cos\beta - \cos\alpha}{\sin\alpha - \sin\beta}$ is

(a) $\dfrac{x}{y}$ (b) $\dfrac{y}{x}$

(c) 1 (d) $x + y$

2. Two vertical poles $AB = 15$ m and $CD = 10$ m are standing apart on a horizontal ground with points A and C on the ground. If P is the point of intersection of BC and AD, then the height of P (in meters) above the line AC is **[JEE Main 2020, 4 September]**

(a) 20/3 (b) 6

(c) 10/3 (d) 5

3. A tower of height b subtends an angle at a point O on the ground level through the foot of the tower and at a distance a from the foot of the tower. A pole mounted on the top of the tower also subtends an equal angle at O. The height of the pole is

(a) $a\left(\dfrac{a^2 - b^2}{a^2 + b^2}\right)$ (b) $a\left(\dfrac{a^2 + b^2}{a^2 - b^2}\right)$

(c) $b\left(\dfrac{a^2 - b^2}{a^2 + b^2}\right)$ (d) $b\left(\dfrac{a^2 + b^2}{a^2 - b^2}\right)$

4. Two vertical poles of height a and b subtend the same angle 45° at a point on the line joining their feet, the square of the distance between their tops is

(a) $\left(\dfrac{1}{2}\right)(a^2 + b^2)$ (b) $a^2 + b^2$

(c) $2(a^2 + b^2)$ (d) $(a + b)^2$

5. The angle of elevation of a Cloud C from a point P, 200 m above a still lake is 30°. If the angle of depression of the image of C in the lake from the point P is 60°, then PC (in m) is equal to **[JEE Main 2020, 4 September]**

(a) 100 (b) 400

(c) $200\sqrt{3}$ (d) $400\sqrt{3}$

6. ABC is a triangular park with $AB = AC = 100$ m. A TV tower stands vertically at the middle point of BC. The angle of elevation of the top of the tower at A, B, C are 45°, 60°, 60° respectively. Find the height of the tower.

(a) 50 m (b) $50\sqrt{3}$ m

(c) $\dfrac{50}{\sqrt{3}}$ m (d) $25\sqrt{3}$ m

7. A ladder rest against a wall making an angle α with the horizontal. The foot of the ladder is pulled away from the wall through a distance x, so that it slides a distance y down the wall making an angle β with the horizontal. The correct relation is

(a) $y = x \tan\dfrac{\alpha + \beta}{2}$ (b) $x = y \tan\dfrac{\alpha + \beta}{2}$

(c) $x = y \tan(\alpha + \beta)$ (d) $y = x \tan(\alpha + \beta)$

8. A monument $ABCD$ stands on a level ground. At a point P on the ground the portions AB, AC, AD subtend angles α, β, γ respectively.
If $AB = a$, $AC = b$, $AD = c$, $AP = x$ and $\alpha + \beta + \gamma = 180°$, then x^2 is equal to

(a) $\dfrac{a}{a + b + c}$ (b) $\dfrac{b}{a + b + c}$

(c) $\dfrac{c}{a + b + c}$ (d) $\dfrac{abc}{a + b + c}$

9. A flagstaff stands in the centre of a rectangular field whose diagonal is 1200 m, and subtends angles 15° and 45° at the mid-points of the sides of the field. The height of the flagstaff is

(a) 200 m (b) $300\sqrt{2 + \sqrt{3}}$ m

(c) $300\sqrt{2 - \sqrt{3}}$ m (d) 400 m

10. The angle of elevation of the summit of a mountain from a point on the ground is 45°. After climbing up 1 km towards the summit at an inclination of 30° from the ground, the angle of elevation of the summit is found to be 60°. Then, the height (in km) of the summit from the ground is **[JEE Main 2020, 6 September]**

(a) $\frac{\sqrt{3}-1}{\sqrt{3}+1}$ (b) $\frac{\sqrt{3}+1}{\sqrt{3}-1}$ (c) $\frac{1}{\sqrt{3}-1}$ (d) $\frac{1}{\sqrt{3}+1}$

11. If each side of length a of an equilateral triangle subtends an angle of 60° at the top of a tower h m high situated at the centre of triangle, then

(a) $3a^2 = 2h^2$ (b) $2a^2 = 3h^2$
(c) $a^2 = 3h^2$ (d) $3a^2 = h^2$

12. From a man, the angle of elevation of the highest point of a tower situated west to him is 60°. On waking 240 m to north, the angle of elevation reduces to 30°. The height of the tower is

(a) $50\sqrt{3}$ m (b) $30\sqrt{6}$ m
(c) $60\sqrt{6}$ m (d) 60 m

13. A vertical pole *PO* is standing at the centre *O* of a square *ABCD*. If *AC* subtends an angle 90° at the top, *P*, of the pole then the angle subtended by a side of the square at *P* is

(a) 45° (b) 30°
(c) 60° (d) 90°

14. Two vertical poles of heights, 20 m and 80 m stand apart on a horizontal plane. The height (in m) of the point of intersection of the lines joining the top of each pole to the foot of the other, from this horizontal plane is **[JEE Main 2019, 8 April]**

(a) 15 (b) 16 (c) 12 (d) 18

15. If the angles of depression of the top and bottom of a short building from the top of tall building are 30° and 60° respectively, then, the ratio of the heights of shorts and tall buildings is

(a) 2 : 3 (b) 1 : 2
(c) 1 : 3 (d) 1 : 4

16. The longer side of a parallelogram is 10 cm and the shorter is 6 cm. If the longer diagonal makes an angle 30° with the longer side, the length of the longer diagonal is

(a) $(5\sqrt{3}+\sqrt{11})$ cm (b) $(4\sqrt{3}+\sqrt{11})$ cm
(c) $(5\sqrt{3}+\sqrt{13})$ cm (d) None of these

17. A lamp post standing at a point *A* on a circular path of radius r subtends an angle α at some point *B* on the path, and *AB* subtends an angle of 45° at any other point on the path, then height of the lampost is

(a) $\sqrt{2}\, r \cot\alpha$ (b) $\left(\frac{r}{\sqrt{2}}\right)\tan\alpha$
(c) $\sqrt{2}\, r \tan\alpha$ (d) $\left(\frac{r}{\sqrt{2}}\right)\cot\alpha$

18. A pole stands vertically inside a triangular park *ABC*. Let the angle of elevation of the top of the pole from each corner of the park be $\frac{\pi}{3}$. If the radius of the circumcircle of ΔABC is 2, then the height of the pole is equal to **[JEE Main 2021, 18 March]**

(a) $\frac{2\sqrt{3}}{3}$ (b) $2\sqrt{3}$
(c) $\sqrt{3}$ (d) $\frac{1}{\sqrt{3}}$

19. A 5 m high pole stands on a building of height 25 m. The pole and the building subtend equal angle at an antenna placed at a height of 30 m. The distance of the antenna from the top of the pole is

(a) $5\sqrt{\frac{2}{3}}$ (b) $\frac{5\sqrt{3}}{2}$
(c) $5\sqrt{\frac{3}{2}}$ (d) $5\sqrt{6}$

20. Two poles standing on a horizontal ground are of heights 5 m and 10 m, respectively. The line joining their tops makes an angle of 15° with ground. Then, the distance (in m) between the poles, is **[JEE Main 2019, 9 April]**

(a) $5(\sqrt{3}+1)$ (b) $\frac{5}{2}(2+\sqrt{3})$
(c) $10(\sqrt{3}-1)$ (d) $5(2+\sqrt{3})$

21. A flag is mounted on the semi-circular dome of radius r. The elevation of the top of the flag at any point on the ground is 30°. Moving d distance towards the dome, when the flag is just visible the angle of elevation is 45°. The relation between r and d is

(a) $r = \frac{d}{\sqrt{2}(\sqrt{3}-1)}$ (b) $r = d\frac{2\sqrt{2}}{\sqrt{3}+1}$
(c) $r = \frac{d}{\sqrt{2}(\sqrt{3}+1)}$ (d) $r = d\frac{2\sqrt{2}}{\sqrt{3}-1}$

22. *AB* is a vertical pole. The end *A* is on the ground. *C* is the middle point of *AB*. *P* is a point on the level ground. The portion *BC* subtends an angle α at *P*. If $AP = nAB$, then $\tan\alpha =$

(a) $\frac{n}{2n^2+1}$ (b) $\frac{n}{n^2-1}$
(c) $\frac{n}{n^2+1}$ (d) None of these

23. A river flows due North, and a tower stands on its left bank. From a point *A* upstream and on the same bank as the tower, the elevation of the tower is 60°, and from a point *B* just opposite *A* on the other bank the elevation is 45°. If the tower is 360 m high, the breadth of the river is

(a) $120\sqrt{6}$ m (b) $\frac{240}{\sqrt{3}}$ m
(c) $240\sqrt{3}$ m (d) $240\sqrt{6}$ m

24. ABC is a triangular park with $AB = AC = 100$ m. A vertical tower is situated at the mid-point of BC. If the angles of elevation of the top of the tower at A and B are $\cot^{-1}(3\sqrt{2})$ and $\text{cosec}^{-1}(2\sqrt{2})$ respectively, then the height of the tower (in m) is **[JEE Main 2019, 10 April]**

(a) 25 (b) 20

(c) $10\sqrt{5}$ (d) $\dfrac{100}{3\sqrt{3}}$

25. A stationary balloon is observed from three points A, B and C on the plane ground and is found that its angle of elevation from each point is α. If $\angle ABC = \beta$ and $AC = b$, the height of the balloon is

(a) $\dfrac{b}{2}\tan\alpha\,\text{cosec}\,\beta$ (b) $\dfrac{b}{2}\tan\alpha\,\sin\beta$

(c) $\dfrac{b}{2}\cot\alpha\,\text{cosec}\,\beta$ (d) $\dfrac{b}{2}\cot\alpha\,\sin\beta$

26. A car is moving at a constant speed at an angle θ east of north. Observations of the car are made from a fixed point. It is due north at some instant. 10 min earlier its bearing was α west of north, whereas 10 min afterwards its bearing is β east of north. Find the value of $\tan\theta$.

(a) $\dfrac{1}{\cot\beta - \cos\alpha}$ (b) $\dfrac{2}{\cos\beta - \cot\alpha}$

(c) $\dfrac{2}{\cot\beta - \cot\alpha}$ (d) $\dfrac{2}{\tan\beta - \tan\beta}$

27. The length of the shadow of a pole inclined at 10° to the vertical towards the sun is 2.05 m, when the elevation of the sun is 38°. The length of the pole is

(a) $\dfrac{2.05\sin 42^\circ}{\sin 38^\circ}$ (b) $\dfrac{2.05\cos 38^\circ}{\cos 42^\circ}$

(c) $\dfrac{2.05\sin 38^\circ}{\sin 42^\circ}$ (d) None of these

28. A, B, C are three points on a horizontal line through the base O of a pillar OP, such that OA, OB, OC are in AP If α, β, γ the angles of elevation of the top of the pillar at A, B, C respectively, are also in AP, then $\sin\alpha, \sin\beta, \sin\gamma$ are in

(a) AP (b) GP

(c) HP (d) None of these

29. Let a vertical tower AB have its end A on the level ground. Let C be the mid-point of AB and P be a point on the ground such that $AP = 2AB$. If $\angle BPC = \beta$, then $\tan\beta$ is equal to **[JEE Main 2017]**

(a) $\dfrac{6}{7}$ (b) $\dfrac{1}{4}$ (c) $\dfrac{2}{9}$ (d) $\dfrac{4}{9}$

30. A tower is standing in the centre of an elliptic field. If Adya observes that the angle of elevation of the top of the tower at an extremity of the major axis of the field is α, at its focus is β and an extremity of the minor axis is γ, then

(a) $\cot^2\alpha = \cot^2\beta - \cot^2\gamma$ (b) $\cot^2\beta = \cot^2\gamma - \cot^2\alpha$

(c) $\cot^2\gamma = \cot^2\alpha - \cot^2\beta$ (d) None of these

31. The angle of elevation of the top of a vertical pole when observed from each vertex of a regular hexagon is $\dfrac{\pi}{3}$. If the area of the circle circumscribing the hexagon be A m^2 then, the height of the tower is

(a) $\dfrac{2A}{\sqrt{3\pi}}$ m (b) $\dfrac{A}{\sqrt{3\pi}}$ m

(c) $2\sqrt{\dfrac{A}{3\pi}}$ m (d) $\sqrt{\dfrac{A}{3\pi}}$ m

32. An observer at O notices that the angle of elevation of the top of a vertical tower is 30°. At O the bearing of the tower is $\tan^{-1}\dfrac{1}{\sqrt{2}}$ east of north. The observer travels a distance of 300 m toward north to a point A and finds the tower to his east. The angle of elevation of the top of the tower at A is θ. Then value of θ is

(a) 30° (b) 45°

(c) 60° (d) 15°

33. An isosceles triangle of wood is placed in a vertical plane, vertex upwards and faces the sun, if $2a$ be the base of the triangle, h its height and 30° the altitude of the sun, then the tangent of the angle at the apex of the shadow is

(a) $\dfrac{2ah\sqrt{3}}{3h^2 - a^2}$ (b) $\dfrac{2ah\sqrt{3}}{3h^2 + a^2}$

(c) $\dfrac{ah\sqrt{3}}{h^2 - a^2}$ (d) None of these

34. A man is walking towards a vertical pillar in a straight path, at a uniform speed. At a certain point A on the path, he observes that the angle of elevation of the top of the pillar is 30°. After walking for 10 min from A in the same direction, at a point B, he observes that the angle of elevation of the top of the pillar is 60°. Then, the time taken (in min) by him, from B to reach the pillar, is

(a) 6 (b) 10

(c) 20 (d) 5

35. A tower and a flag staff on its top subtend equal angles at the observer's eye. If the heights of flagstaff, tower and the eye of the observer are respectively a, b and h, then the distance of the observer's eye from the base of the tower is

(a) $a\sqrt{\dfrac{a+b-2h}{a+b}}$ (b) $b\sqrt{\dfrac{a+b-2h}{a-b}}$

(c) $a\sqrt{\dfrac{a-h}{a+b}}$ (d) None of these

36. Two objects at the points P and Q subtend an angle of 30° at a point A. Lengths $AR = 20$ m and $AS = 10$ m are measured from A at right angles to AP and AQ respectively. If PQ subtends equal angles of 30°, at R and S, then length of PQ is

(a) $\sqrt{300 - 200\sqrt{3}}$ (b) $\sqrt{500 - 200\sqrt{3}}$

(c) $\sqrt{500\sqrt{3} - 200}$ (d) $\sqrt{300}$

37. The elevation of a steeple at a place due south of it is 45° and at a place B due west of A the elevation is 15°. If $AB = 2a$, the height of the steeple is

(a) $a\dfrac{(\sqrt{3}-1)}{\sqrt{2}}$ (b) $a\dfrac{(\sqrt{3}+1)}{\sqrt{2}}$

(c) $a\left[3^{\frac{1}{4}} - 3^{-\frac{1}{4}}\right]$ (d) $a\left[3^{\frac{1}{4}} + 3^{-\frac{1}{4}}\right]$

38. OAB is a triangle in the horizontal plane through the foot P of the tower at the middle point of the side OB of the triangle. If $OA = 2$ m, $OB = 6$ m, $AB = 5$ m and $\angle AOB$ is equal to the angle subtended by the tower at A then the height of the tower is

(a) $\sqrt{\dfrac{11\times 39}{25\times 3}}$ (b) $\sqrt{\dfrac{11\times 39}{25\times 2}}$

(c) $\sqrt{\dfrac{11\times 25}{39\times 2}}$ (d) None of these

39. A harbour lies in a direction 60° south-west from a fort and at a distance 30 km from it. A ship sets out from the harbour at noon and sails due east at 10 km/h. The ship will be 70 km from the fort at

(a) 7 pm
(b) 8 pm
(c) 5 pm
(d) 10 pm

Objective Questions Type II [One or more than one correct answer(s)]

In each of the questions below four choices of which one or more than one are correct. You have to select the correct answer(s) accordingly.

1. A pole 50 m high stands on a building 250 m high. To an observer at a height of 300 m, the building and the pole subtend equal angle θ. If the horizontal distance of the observer from the pole is x, then

(a) $x = 25\sqrt{6}$ m (b) $x = 25\sqrt{3}$ m

(c) $\tan\theta = \sqrt{\dfrac{2}{3}}$ (d) $\tan\theta = 2\sqrt{6}$

2. ABC is triangular park with $AB = AC = 100$ m. A clock tower is situated at the mid-point of BC. The angles of elevation of the top of the tower at A and B are $\cot^{-1} 3.2$ and $\text{cosec}^{-1} 2.6$ respectively. The height of the tower is

(a) height of tower is 50 m (b) height of tower is 25 m

(c) $\cot^2\beta = 5.76$ (d) $\cot^2\beta = 7.1$

Linked-Comprehension Type

In these questions, a passage (paragraph) has been given followed by questions based on each of the passages. You have to answer the questions based on the passage given.

PASSAGE 1

A tower OP of height b has a flagstaff PQ on its top. The tower and the flagstaff subtend equal angles at a point A, distance a from the foot of the tower on the ground.

1. The length of the flagstaff is

(a) $\dfrac{a(a^2+b^2)}{a^2-b^2}$ (b) $\dfrac{b(a^2+b^2)}{a^2-b^2}$

(c) $\dfrac{b(a^2-b^2)}{a^2+b^2}$ (d) $\dfrac{a(a^2-b^2)}{a^2+b^2}$

2. The angle subtended by the upper half of the tower at the point A is

(a) $\tan^{-1}\dfrac{2a^2+b^2}{ab}$ (b) $\cos^{-1}\dfrac{ab}{2a^2+b^2}$

(c) $\tan^{-1}\dfrac{ab}{2a^2+b^2}$ (d) $\tan^{-1}\dfrac{2ba^2}{a^2+b^2}$

PASSAGE 2

A diving stand (vertical) is situated at a point on the edge of a circular swimming pool of radius r. At another point P on the circular edge to the angle of elevation of the top of the diving stand in 30°. After moving 10 m along the circular edge, the angle of elevation in again 30°, and further going along the edge 5 m, the angle of elevation is 45°.

1. Height of the diving stand is

(a) r (b) $2r$

(c) $\left(\dfrac{2}{3}\right)r$ (d) $3r$

2. The perimeter of the swimming pool is

(a) 15 m (b) 20 m
(c) 30 m (d) 35 m

Numerical Value Type

Solve the following problems and write your answer.

1. A regular pyramid on a square base has an edge long 150 m long and the length of the side of its base is 200 m. Vertical height of the pyramid is
2. The angle of elevation of the top of a hill from a point on the horizontal plane passing through the foot of the hill is found to be 45°. After walking a distance of 80 m towards the top, up a slope inclined at an angle of 30° to the horizontal plane, the angle of elevation of the top of the hill becomes 75°. Then, the height of the hill (in m) is ………… **[JEE Main 2020, 6 September]**
3. $ABCD$ is a trapezium such that AB and CD are parallel and CB is perpendicular to them. If $\angle ADB = \theta$, $BC = 45$, $CD = 30$ and $\tan\theta = \frac{1}{5}$, then 51 AB is equal to
4. ABC is a triangular park with $AB = AC = 100$ m. A clock tower is situated at the mid-point of BC. The angles of elevation of the top of the tower at A and B are $\cot^{-1} 3.2$ and $\operatorname{cosec}^{-1} 2.6$ respectively. The height of the tower is
5. In a triangular plot ABC with $BC = 7$ m, $CA = 8$ m and $AB = 9$ m. A lamp post is situated at the middle point E of the side AC and subtends and angle $\tan^{-1} 3$ at the point B, the height of the lamp post is

Single Integer Type

Solve the following problems and write your answer.

1. Let AD and BC be two vertical poles at A and B respectively on a horizontal ground. If $AD = 8$ m, $BC = 11$ m and $AB = 10$ m; then the distance (in meters) of a point M on AB from the point A such that, $MD^2 + MC^2$ is minimum is ……. **[JEE Main 2020, 6 September]**
2. The upper three-fourths of ship's mast subtends, at a point on the deck, an angle whose tangent is 0.75. If the whole mast subtends an angle θ at the same point, then $\tan\theta$ is equal to
3. If the angles of depresession of the top and bottom of a short building from the top of tall building are 30° and 60° respectively then, the ratio of the heights of short and tall buildings is $2 : k$ then value of k^2 is equal to

Answers

Objective Questions Type [Only one correct answer]

1. (a)	**2.** (b)	**3.** (d)	**4.** (c)	**5.** (b)	**6.** (b)	**7.** (b)	**8.** (d)	**9.** (c)	**10.** (c)
11. (b)	**12.** (c)	**13.** (c)	**14.** (b)	**15.** (a)	**16.** (a)	**17.** (c)	**18.** (b)	**19.** (c)	**20.** (d)
21. (a)	**22.** (a)	**23.** (a)	**24.** (b)	**25.** (a)	**26.** (b)	**27.** (c)	**28.** (b)	**29.** (c)	**30.** (c)
31. (d)	**32.** (b)	**33.** (a)	**34.** (d)	**35.** (b)	**36.** (b)	**37.** (c)	**38.** (b)	**39.** (b)	

Objective Questions Type II [One or more than one correct answer(s)]

1. (a, c) **2.** (b, c)

Linked-Comprehension Type

Passage 1 **1.** (b) **2.** (c)
Passage 2 **1.** (a) **2.** (c)

Numerical Value Type Questions

1. 50 **2.** 80 **3.** 585 **4.** 25 **5.** 21

Single Integer Type Questions

1. (5 m) **2.** (2) **3.** (9)

Solutions

Objective Questions Type [Only one correct answer]

1. Let AB be the original position of the ladder. After sliding, it takes the position $A'B'$. $AA' = x$, $BB' = y$, $AB = A'B' = l$ (say)

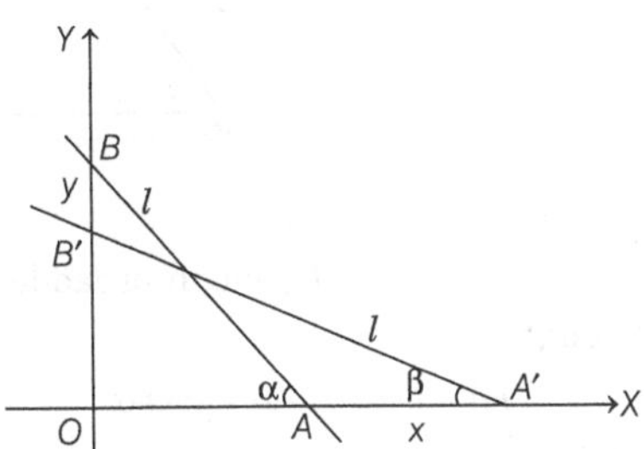

Now, $\cos\alpha = \dfrac{OA}{l}$ and $\cos\beta = \dfrac{OA'}{l}$

$\therefore \quad \cos\beta - \cos\alpha = \dfrac{OA' - OA}{l} = \dfrac{x}{l}$

Also, $\sin\alpha = \dfrac{OB}{l}$ and $\sin\beta = \dfrac{OB'}{l}$

$\Rightarrow \quad \sin\alpha - \sin\beta = \dfrac{OB - OB'}{l} = \dfrac{y}{l}$

Then, $\dfrac{\cos\beta - \cos\alpha}{\sin\alpha - \sin\beta} = \dfrac{\frac{x}{l}}{\frac{y}{l}} = \dfrac{x}{y}$

2. The diagram of given situation is

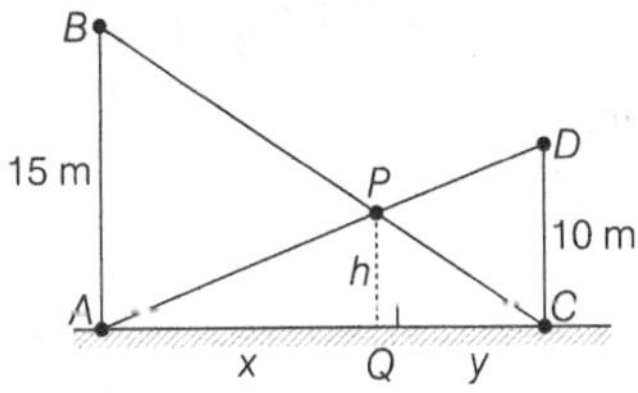

Let height of P above the line AC is h m and $AQ = x$ m and $CQ = y$ m, so

$$\frac{10}{x+y} = \frac{h}{x} \qquad \ldots\text{(i)}$$

and
$$\frac{15}{x+y} = \frac{h}{y} \qquad \ldots\text{(ii)}$$

From Eqs. (i) and (ii), we get

$$\frac{10}{15} = \frac{y}{x} \Rightarrow 2x = 3y \qquad \ldots\text{(iii)}$$

Now, from Eqs. (i) and (iii), we get

$$\frac{10}{x + \frac{2x}{3}} = \frac{h}{x} \Rightarrow \frac{10}{\frac{5}{3}} = h$$

$\Rightarrow \quad h = 6$ m

3. Let AB be the tower and PB be the pole.

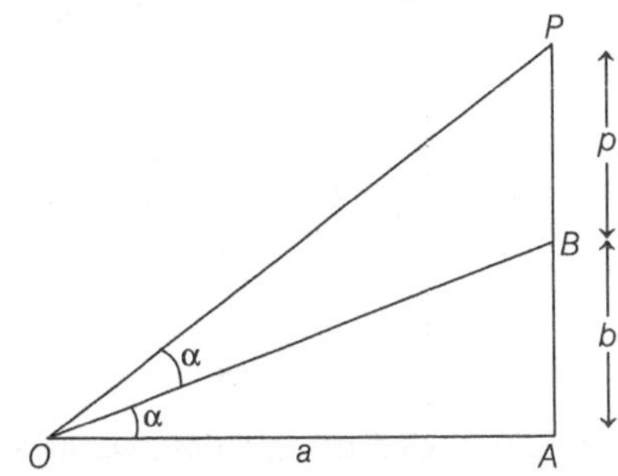

From the figure, $\tan\alpha = \dfrac{b}{a}$

and $\tan 2\alpha = \dfrac{p+b}{a}$

$\therefore \quad \dfrac{2\tan\alpha}{1-\tan^2\alpha} = \dfrac{p+b}{a}$

$\Rightarrow \quad \dfrac{2\left(\frac{b}{a}\right)}{1-\left(\frac{b}{a}\right)^2} = \dfrac{p+b}{a}$

$\Rightarrow \quad \dfrac{2ba}{a^2-b^2} = \dfrac{p+b}{a}$

$\Rightarrow \quad \dfrac{2ba^2 - a^2b + b^3}{a^2-b^2} = p$

$\Rightarrow \quad p = b\left(\dfrac{a^2+b^2}{a^2-b^2}\right)$

4. Let OP and QR be the poles of height a and b respectively subtending 45° at S, a point between O and Q.
then, $OS = a$, $QS = b$

$$PS^2 = 2a^2 \text{ and } RS^2 = 2b^2$$

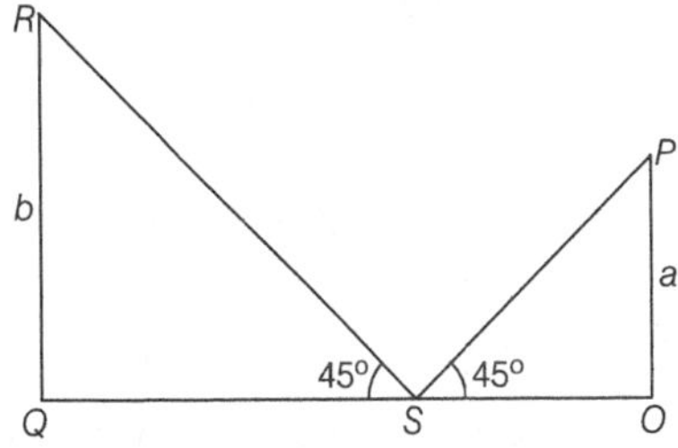

Since, $\angle PSR = 180° - (45° + 45°) = 90°$

$$PR^2 = PS^2 + RS^2$$
$$= 2(a^2 + b^2)$$

5. According to given information in question, from the diagram, where $PQ = MN = 200$m and let $CM = l$ m. So, $NC' = (l + 200)$m

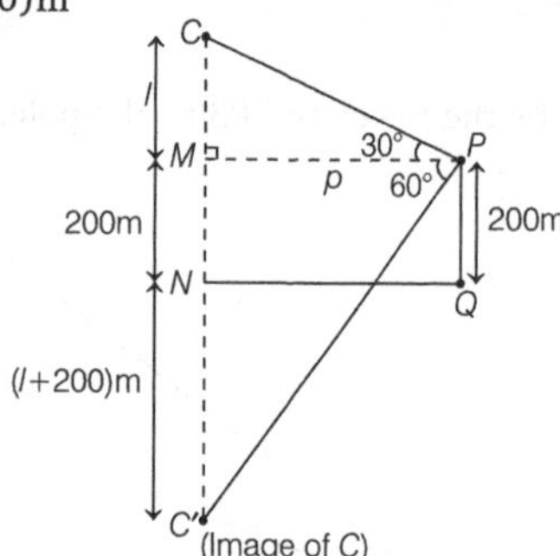

So, in right angled triangle CMP, we have

$$\tan 30° = \frac{l}{p}$$

$$\Rightarrow \quad p = \sqrt{3}l \quad \ldots(i)$$

and in right angle triangled PMC', we have

$$\tan 60° = \frac{200 + l + 200}{p}$$

$$\Rightarrow \quad \sqrt{3}p = 400 + l \quad \ldots(ii)$$

From Eqs. (i) and (ii), we get

$$2l = 400$$

$$\Rightarrow \quad l = 200 \text{ m and } p = 200\sqrt{3} \text{ m}$$

$$\therefore \quad PC = \sqrt{l^2 + p^2} = 400 \text{ m}$$

6. Let $PQ = h$ m

From ΔPQA, $QA = h \cot 45° = h$

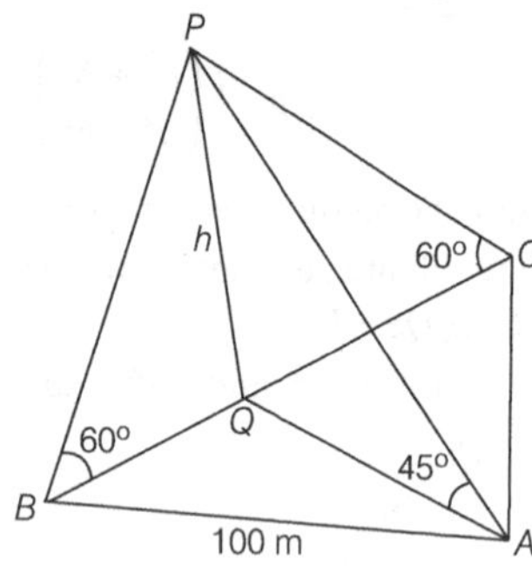

From ΔPQB ; $BQ = h \cot 60° = \dfrac{h}{\sqrt{3}}$

From ΔPQC ; $CQ = h \cot 60° = \dfrac{h}{\sqrt{3}}$

$\therefore \quad BQ = CQ \Rightarrow Q$ is the middle point of BC

Also $AB = AC = 100$ m,

So, ΔABC is isosceles.

$\therefore$ from geometry $AQ \perp BC$.

From ΔAQB, $BQ^2 + AQ^2 = AB^2$

$$\Rightarrow \quad \frac{h^2}{3} + h^2 = 100^2 \Rightarrow h = 50\sqrt{3} \text{ m}$$

7. Let AB be the wall, P be the initial position and Q be the final position of the foot of the ladder on the ground.

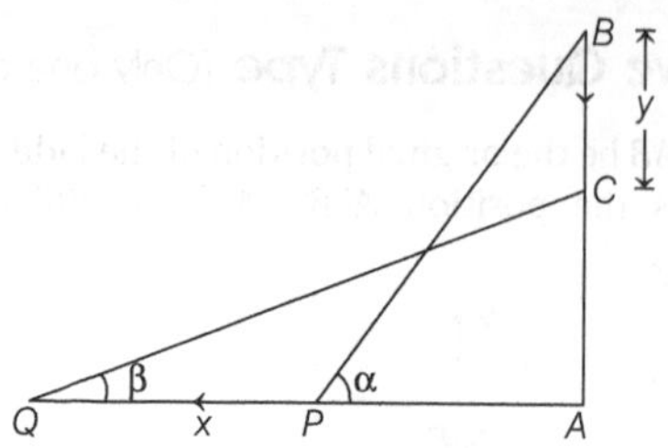

In the figure,

$$PB = QC = l \text{ (Length of ladder)}$$

In ΔBAP,

$$PA = l\cos\alpha,\ AB = l\sin\alpha$$

In ΔCAQ,

$$QA = l\cos\beta,\ AC = l\sin\beta$$

Now, $\quad y = BC = AB - AC = l(\sin\alpha - \sin\beta)$

and $\quad x = PQ = AQ - AP = l(\cos\beta - \cos\alpha)$

$$\Rightarrow \quad \frac{y}{x} = \frac{\sin\alpha - \sin\beta}{\cos\beta - \cos\alpha}$$

$$= \frac{2\sin\left(\frac{\alpha-\beta}{2}\right)\cos\left(\frac{\alpha+\beta}{2}\right)}{2\sin\left(\frac{\alpha+\beta}{2}\right)\sin\left(\frac{\alpha-\beta}{2}\right)}$$

$$\Rightarrow \quad \frac{y}{x} = \cot\left(\frac{\alpha+\beta}{2}\right)$$

$$\Rightarrow \quad x = y\tan\left(\frac{\alpha+\beta}{2}\right)$$

8. We have,

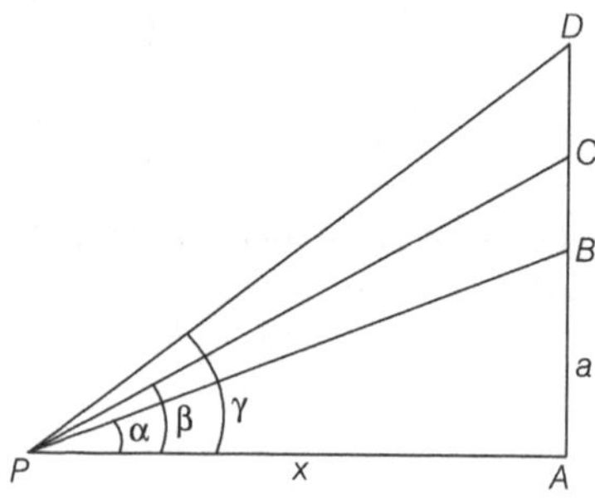

$$a = AB = AP\tan\alpha = x\tan\alpha$$

$$b = AC = x\tan\beta$$

and $\quad c = AD = x\tan\gamma$

so, that $a + b + c = x(\tan\alpha + \tan\beta + \tan\gamma)$

$$= x\tan\alpha\tan\beta\tan\gamma \quad [\because \alpha + \beta + \gamma = 180°]$$

and $abc = x^3\tan\alpha\tan\beta\tan\gamma$

$$\Rightarrow \quad x^2 = \frac{abc}{a + b + c}$$

9.

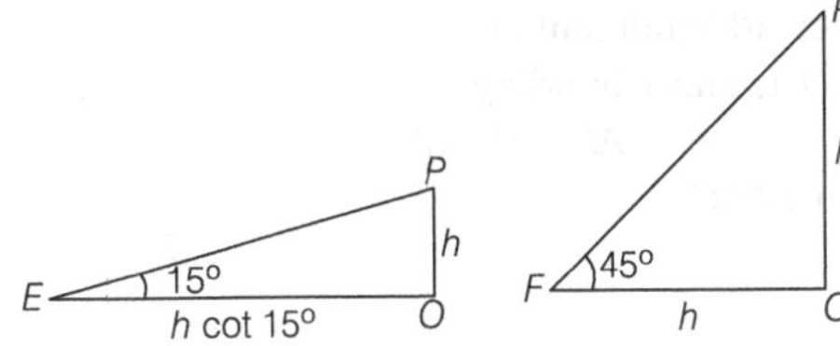

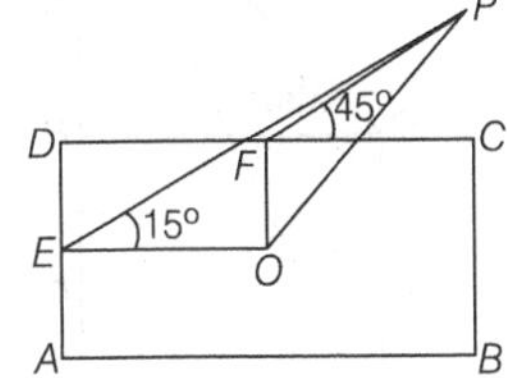

$$EF^2 = h^2[1 + \cot^2 15°]$$
$$= h^2[1 + (2+\sqrt{3})^2]$$
$$\therefore \quad EF = 2h\sqrt{2+\sqrt{3}}$$
$$1200 = AC = 2EF = 4h\sqrt{2+\sqrt{3}}$$
$$\Rightarrow \quad h = \frac{300}{\sqrt{2+\sqrt{3}}}$$
$$= 300\sqrt{2-\sqrt{3}} \text{ m}$$

10. From the diagram as $\angle PRQ = 45°$

$\therefore \quad PQ = RQ$

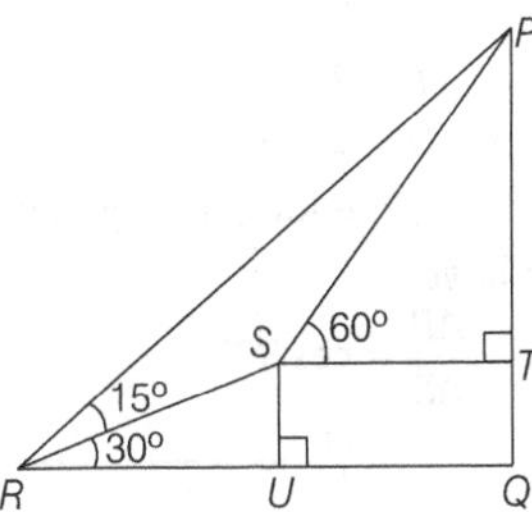

and $\quad \angle SPT = 30°$
and $\quad \angle RPQ = 45°$
$\therefore \quad \angle PRS = \angle SPR$
$\Rightarrow \quad SR = SP = 1$ km

In ΔRSU,

$$\sin 30° = \frac{SU}{RS}$$
$$\Rightarrow \quad SU = \frac{1}{2} \text{ km}$$

$\therefore$ Height of mountain $= PT + TQ$

$$= PT + SU$$
$$= \frac{\sqrt{3}}{2} + \frac{1}{2}$$
$$= \frac{\sqrt{3}+1}{2} \text{ km}$$

or $\frac{1}{\sqrt{3}-1}$ km

11. Let O be the centre of the equilateral ΔABC and OP be the tower of height h.
Then, each of the $\Delta\ PAB$, ΔPBC and ΔPCA are equilateral and thus,

$$PA = PB = AB = a.$$

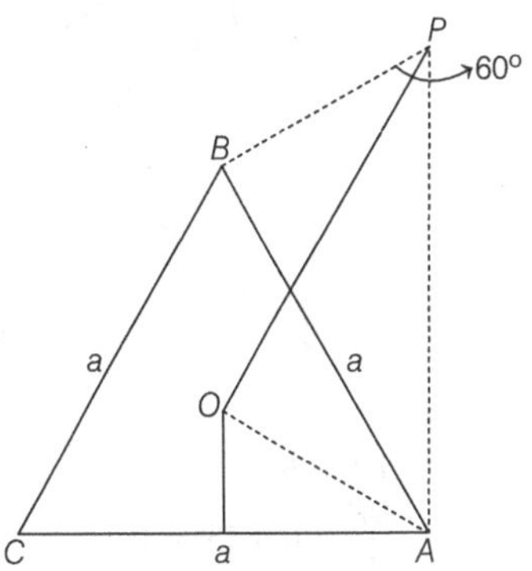

In $\Delta\ ABC$, OA is the bisector of the angle A,

So, $\quad \frac{OA}{a/2} = \sec 30°$

$$\Rightarrow \quad OA = \frac{a}{2} \cdot \frac{2}{\sqrt{3}} = \frac{a}{\sqrt{3}}$$

Now, from right angled ΔPOA.

$$PA^2 = OP^2 + OA^2$$
$$\Rightarrow \quad a^2 = h^2 + \frac{a^2}{3}$$
$$\Rightarrow \quad \frac{2a^2}{3} = h^2$$
$$\Rightarrow \quad 2a^2 = 3h^2$$

12. Let OA be the tower of height h. Also, let B and C be the initial and final positions, respectively, of the man.

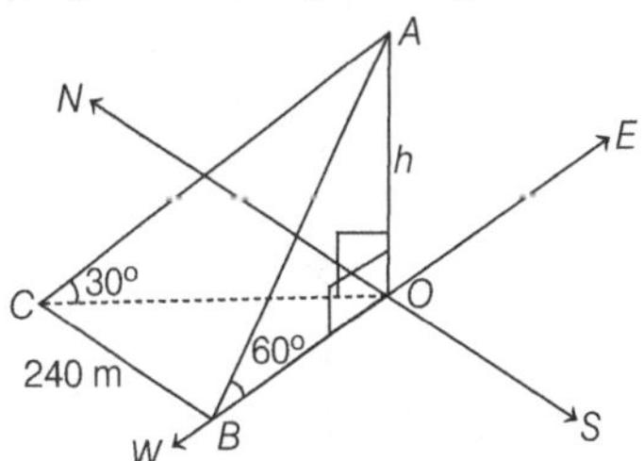

In $\Delta\ AOB$,

$$OB = h \cot 60° = \frac{h}{\sqrt{3}}$$

In $\Delta\ AOC$,

$$OC = h \cot 30° = \sqrt{3}h$$

Now in $\Delta\ OBC$,

$$OB^2 + BC^2 = OC^2$$
$$\therefore \quad \frac{h^2}{3} + (240)^2 = 3h^2$$
$$\Rightarrow \quad \frac{8h^2}{3} = (240)^2$$
$$\Rightarrow \quad h = 60\sqrt{6} \text{ m}$$

13. Clearly $\angle OPA = \angle OPC = \frac{\pi}{4}$

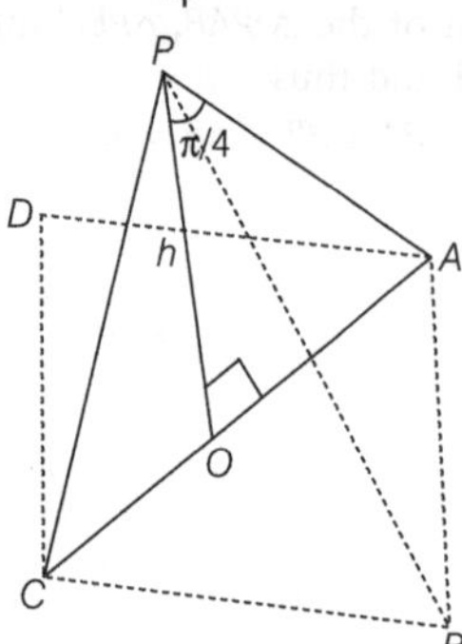

Let	$OP = h$
So,	$\frac{OP}{OA} = \tan 45° = 1$
$\therefore$	$OA = h$ and $AC = 2h$
Also	$\frac{OA}{PA} = \cos\frac{\pi}{4}$
$\Rightarrow$	$PA = \sqrt{2}h$
Also	$PA = PB$ and $AC = \sqrt{2}AB$
$\therefore$	$2h = \sqrt{2}\,AB \Rightarrow AB = \sqrt{2}h$
i.e.	$AB = PA$
$\therefore$	$PA = AB = PB$ [similarly]

So ΔBPA is equilateral triangle

Hence, required angle = 60°

14. Let a first pole AB having height 20 m and second pole PQ having height 80 m

and $\angle PBQ = \alpha$, $\angle AQB = \beta$

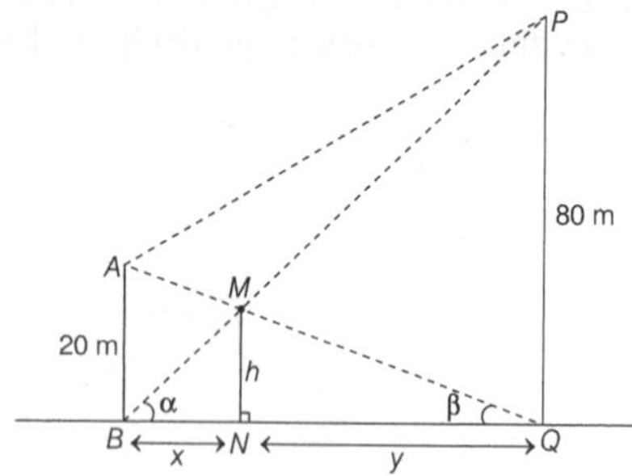

and $MN = hm$ is the height of intersection point from the horizontal plane

$\because$ $$\tan\alpha = \frac{h}{x} = \frac{80}{x+y}$$

[in ΔMNB and ΔPQB] ...(i)

and $$\tan\beta = \frac{h}{y} = \frac{20}{x+y}$$

[in ΔMNQ and ΔABQ] ...(ii)

From Eqs. (i) and (ii), we get

$$\frac{y}{x} = 4 \Rightarrow y = 4x \quad \text{...(iii)}$$

From Eqs. (i) and (iii), we get

$$\frac{h}{x} = \frac{80}{x+4x} \Rightarrow h = \frac{80}{5} = 16 \text{ m}$$

15. Let AB is tall building

CQ is short building

$$AB = H, \quad CQ = h$$

In ΔABC,

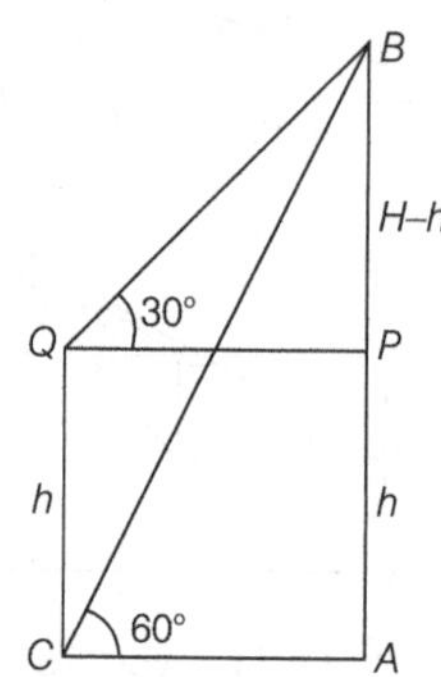

$$\tan 60° = \frac{H}{AC} \Rightarrow AC = \frac{H}{\sqrt{3}}$$

In ΔBPQ

$$\tan 30° = \frac{H-h}{PQ}$$

$\Rightarrow$ $PQ = (H-h)\sqrt{3}$

$AC = PQ$

$\therefore$ $\frac{H}{\sqrt{3}} = (H-h)\sqrt{3}$

$\Rightarrow$ $H = 3H - 3h \Rightarrow AC = PQ$

$\Rightarrow$ $\frac{h}{H} = \frac{2}{3}$

$\therefore$ $h : H = 2 : 3$

16. Draw $BP \perp AC$, the longer diagonal. From right angle ΔAPB, we have

$$\frac{AP}{AB} = \cos 30°$$

$\Rightarrow$ $AP = AB\cos 30°$

$\therefore$ $AP = 10 \times \frac{\sqrt{3}}{2} = 5\sqrt{3}$ cm

Also, $\frac{BP}{AB} = \sin 30°$

$\therefore$ $BP = AB\sin 30° = 10 \times \frac{1}{2} = 5$ cm

From right $\angle d$ ΔBPC,

we have, $BP^2 + PC^2 = BC^2$

$\Rightarrow$ $5^2 + PC^2 = 6^2$

$\Rightarrow$ $PC^2 = 6^2 - 5^2 = (6-5)(6+5) = 11$

$\therefore$ $PC = \sqrt{11}$ cm

$\therefore$ Longer diagonal

$$AC = AP + PC = (5\sqrt{3} + \sqrt{11}) \text{ cm}$$

17. Let AP be the lamp-post of height h at a point A on a circular path of radius r and centre C. Let B be the point on this path such that $\angle PBA = \alpha$.

$\Rightarrow \qquad AB = h \cot \alpha$

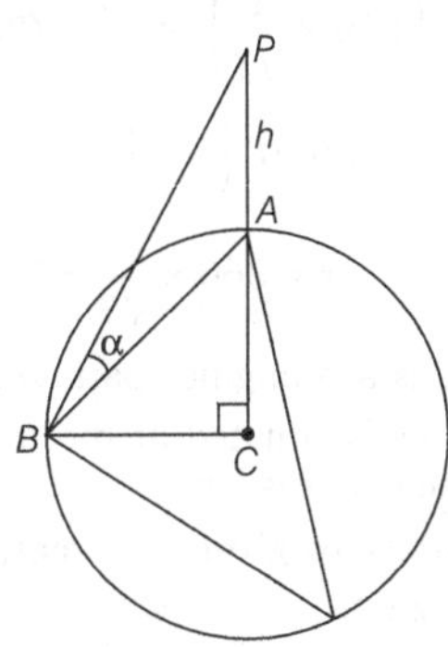

Since, AB subtends an angle 45° at another point of the path, it subtends an angle of 90° at the centre C so that $\angle BCA = 90°$

Also $\qquad CA = CB = r$

$\Rightarrow \qquad AB = \sqrt{2}\, r$

and, then $h \cot \alpha = \sqrt{2}\, r$

$\Rightarrow \qquad h = \sqrt{2}\, r \tan \alpha$

18. Let $PD = h$, $R = 2$

As angle of elevation of top of pole from A, B, C are equal.

So, D must be circumcentre of ΔABC

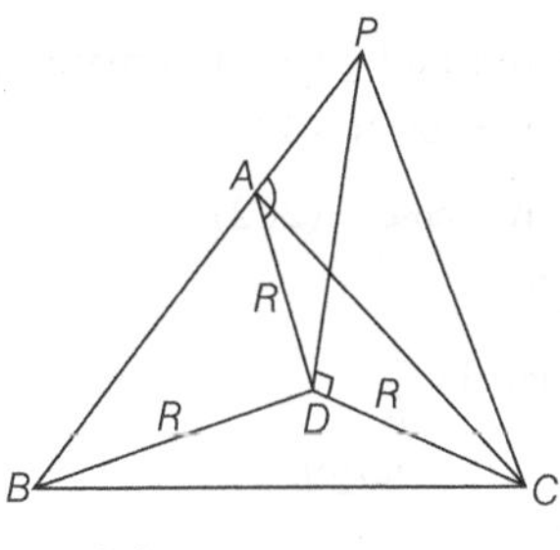

$$\tan\left(\frac{\pi}{3}\right) = \frac{PD}{R} = \frac{h}{R}$$

$$h = R \tan\left(\frac{\pi}{3}\right)$$

$$= 2\sqrt{3}$$

19. Let AB be the pole standing on the building BC and O be the position of the antenna.

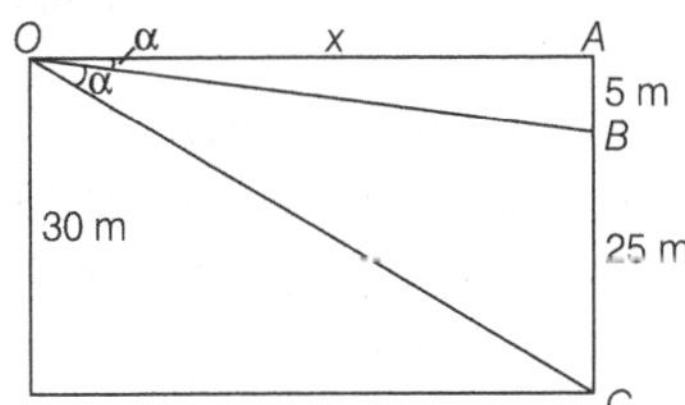

In $\Delta\, OAB$,

$$\tan \alpha = \frac{5}{x} \qquad \ldots\text{(i)}$$

In $\Delta\, OAC$,

$$\tan 2\alpha = \frac{30}{x} \qquad \ldots\text{(ii)}$$

From Eqs. (i) and (ii),

$$\tan 2\alpha = 6 \tan \alpha$$

$$\Rightarrow \quad 3 - 3\tan^2 \alpha = 1$$

$$\Rightarrow \quad \tan \alpha = \sqrt{\frac{2}{3}}$$

From Eq. (i), $x = 5 \cot \alpha = 5\sqrt{\dfrac{3}{2}}$

20. Given heights of two poles, are 5 m and 10 m.

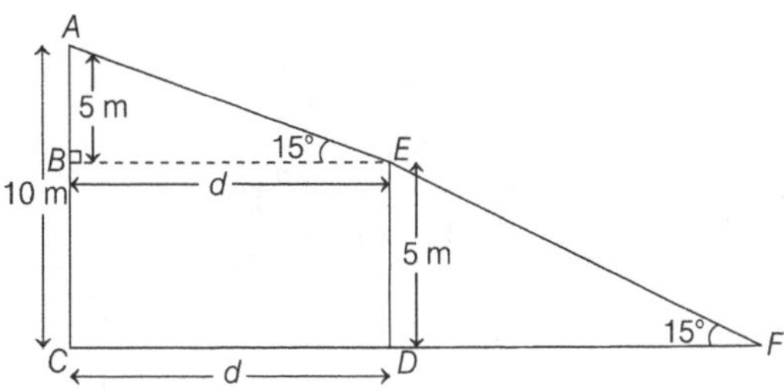

i.e., from figure $AC = 10$ m, $DE = 5$ m

$\therefore \quad AB = AC - DE = 10 - 5 = 5$ m

Let d be the distance between two poles.

Clearly, $\Delta ABE \sim \Delta ACF$ [by AA-similarity criterion]

$\therefore \quad \angle AEB = 15°$

In ΔABE, we have

$$\tan 15° = \frac{AB}{BE}$$

$$\Rightarrow \quad \frac{\sqrt{3}-1}{\sqrt{3}+1} = \frac{5}{d} \qquad \left[\because \tan 15° = \frac{\sqrt{3}-1}{\sqrt{3}+1}\right]$$

$$\Rightarrow \quad d = \frac{5(\sqrt{3}+1)}{(\sqrt{3}-1)}$$

$$\Rightarrow \quad d = 5\frac{\sqrt{3}+1}{\sqrt{3}-1} \times \frac{\sqrt{3}+1}{\sqrt{3}+1}$$

$$= \frac{5(3 + 2\sqrt{3} + 1}{3-1} = \frac{5(2\sqrt{3}+4)}{2}$$

$$= \frac{2 \times 5(\sqrt{3}+2)}{2} = 5(2 + \sqrt{3}) \text{ m}$$

21. In ΔACP, we have

$$\tan 45° = \frac{AC}{AP}$$

$$\Rightarrow \quad AP = AC = r + h$$

Also, $AM = MP = MC = r$

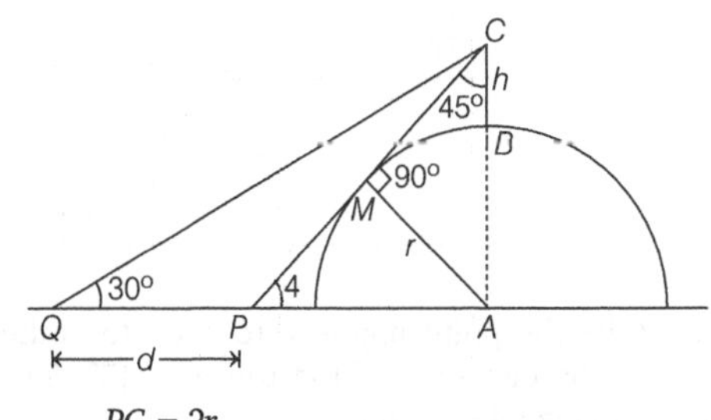

$\therefore \quad PC = 2r$

Now, from ΔACP, we have

$$PC^2 = PA^2 + AC^2$$

$$\Rightarrow \quad 4r^2 = (r+h)^2 + (r+h)^2$$

$$\Rightarrow \quad r + h = r\sqrt{2}$$

$$\Rightarrow \quad h = r(\sqrt{2}-1)$$

$$\Rightarrow \quad r + h = AQ \tan 30°$$

$$r + h = (r + h + d)\frac{1}{\sqrt{3}}$$

$$\Rightarrow \quad \sqrt{3}(r+h) = (r+h) + d$$

$$\Rightarrow \quad d = (r+h)(\sqrt{3}-1)$$

$$\Rightarrow \quad d = (\sqrt{3}-1)r\sqrt{2}$$

$$\therefore \quad r = \frac{d}{\sqrt{2}(\sqrt{3}-1)}$$

22. $\frac{h}{nh} = \tan(\alpha + \theta) = \frac{\tan\alpha + \frac{h}{2nh}}{1 - \tan\alpha \times \frac{h}{2nh}}$

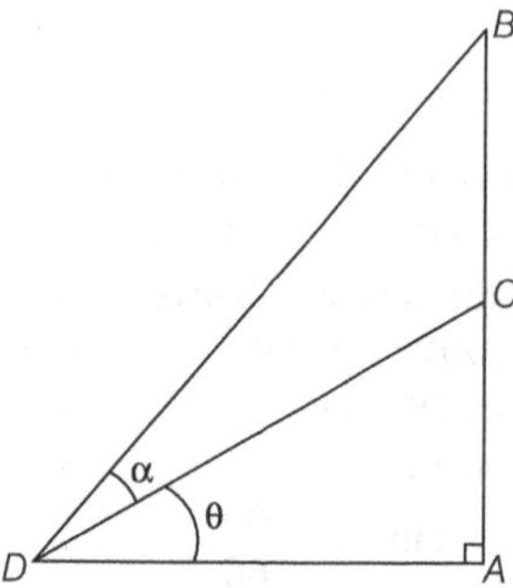

$$\Rightarrow \quad \frac{1}{n} = \frac{\tan\alpha + \frac{1}{2n}}{1 - \tan\alpha \times \frac{1}{2n}}$$

$$= \frac{2n\tan\alpha + 1}{(2n - \tan\alpha)}$$

$$\Rightarrow \quad 2n - \tan\alpha = 2n^2 \tan\alpha + n$$

$$\Rightarrow \quad (2n^2+1)\tan\alpha = n \text{ i.e. } \tan\alpha = \frac{n}{(2n^2+1)}$$

23. Let PQ be the tower of height 360 m standing on the left bank of the river subtending an angle 60° at A on the same bank as the tower. Then $PA = 360° \cot 60°$.

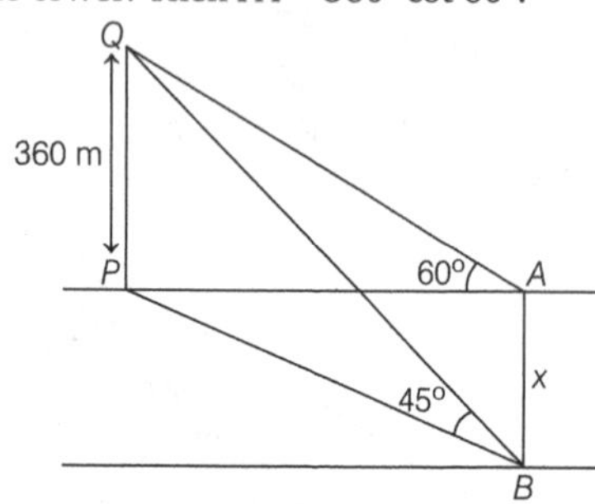

Let B be the point opposite to A on the other bank where the elevation of the tower is 45°. So that $PB = 360 \cot 45° = 360°$.

If x be the width of the river, then $AB = x$ and from the right angled ΔPAB,

$$(PA)^2 + (AB)^2 = (PB)^2$$

$$\Rightarrow \quad \left(360 \times \frac{1}{\sqrt{3}}\right)^2 + x^2 = (360)^2$$

$$\Rightarrow \quad (360)^2\left(1 - \frac{1}{3}\right) = x^2$$

$$\Rightarrow \quad x = 360 \times \frac{\sqrt{2}}{\sqrt{3}} = 120\sqrt{6} \text{ m}$$

24. Given, ABC is a triangular park with $AB = AC = 100$ m. A vertical tower is situated at the mid-point of BC. Let the height of the tower is h m.

Now, according to given information, we have the following figure.

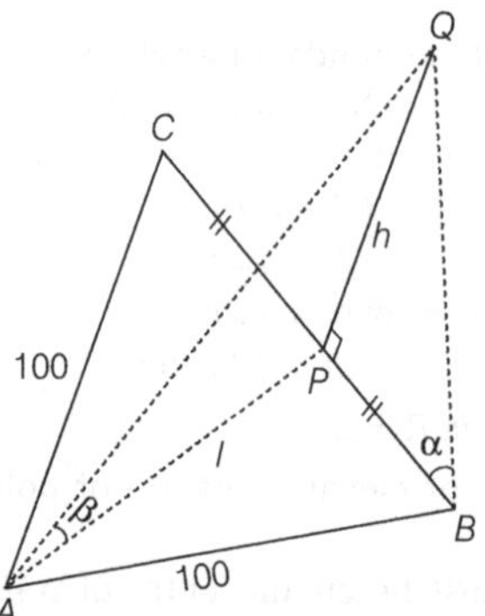

From the figure and given information, we have

$$\beta = \cot^{-1}(3\sqrt{2})$$

and $\quad \alpha = \text{cosec}^{-1}(2\sqrt{2})$

Now, In ΔQPA,

$$\cot\beta = \frac{l}{h}$$

$$\Rightarrow \quad l = (3\sqrt{2})h \quad \text{...(i)}$$

and in ΔBPQ,

$$\tan\alpha = \frac{h}{BP}$$

$$\Rightarrow \quad \cot\alpha = \frac{BP}{h}$$

$$= \frac{\sqrt{(100)^2 - l^2}}{h}$$

[$\because$ p is mid-point of isosceles ΔABC, $AP \perp BC$]

$$\Rightarrow \quad h^2\cot^2\alpha = (100)^2 - l^2$$

$$\Rightarrow \quad h^2(\text{cosec}^2\alpha - 1) = (100)^2 - (3\sqrt{2}h)^2 \quad \text{[from Eq. (i)]}$$

$$\Rightarrow \quad h^2(8-1) = (100)^2 - 18h^2$$

$$\Rightarrow \quad 25h^2 = (100)^2$$

$$h^2 = \left(\frac{100}{5}\right)^2$$

$$\Rightarrow \quad h = 20 \text{ m}$$

25. Let PQ be vertical and at its end Q there is a balloon P is on the ground

$$PA = PB = PC = h \cot \alpha$$

$\therefore P$ is circumcentre of ΔABC and its circumradius $(R) = h \cot \alpha$

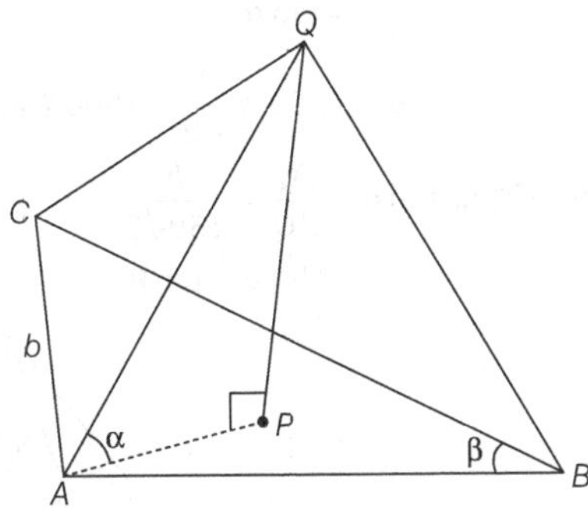

$$\Rightarrow \quad 2R = \frac{b}{\sin \beta}$$

$$\Rightarrow \quad 2h \cot \alpha = \frac{b}{\sin \beta}$$

$$\therefore \quad h = \frac{b}{2} \tan \alpha \operatorname{cosec} \beta$$

26. The car is moving on the line $AB \cdot O$ is the point of observation. At the time of observation the car is at 10 min earlier it was at A and 10 min afterwards it is at B. So $AP = PB$.

Also $\angle POA = \alpha$, $\angle POB = \beta$ and $\angle NPB = \theta$.

In the ΔOPB,

$$\frac{PB}{\sin \beta} = \frac{OP}{\sin \angle OBP}$$

$$\Rightarrow \quad \frac{PB}{\sin \beta} = \frac{OP}{\sin(\theta - \beta)} \quad \ldots(i)$$

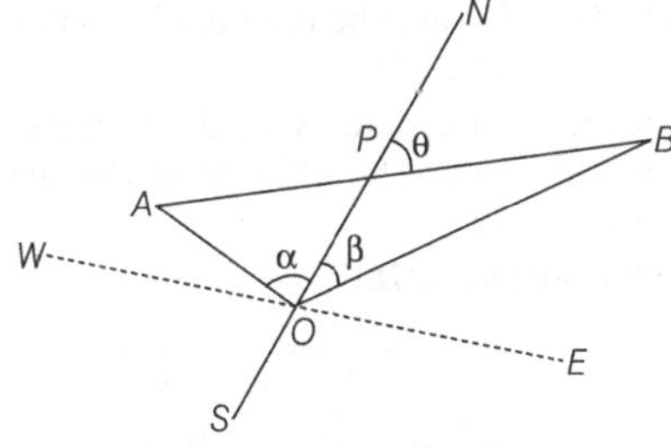

In the ΔOPA,

$$\frac{PA}{\sin \alpha} = \frac{OP}{\sin OAP}$$

$$\Rightarrow \quad \frac{PA}{\sin \alpha} = \frac{OP}{\sin(180° - (\alpha + \theta))} \quad \ldots(ii)$$

From Eqs. (i) and (ii)

$$\frac{\sin(\theta - \beta)}{\sin \beta} = \frac{\sin(\alpha + \theta)}{\sin \alpha} \quad [\because PA = PB]$$

$$\Rightarrow \quad 2\cos\theta = \sin\theta (\cot\beta - \cot\alpha)$$

$$\Rightarrow \quad \tan\theta = \frac{2}{\cot\beta - \cot\alpha}$$

27. Let PO be the tower and OS be its shadow.

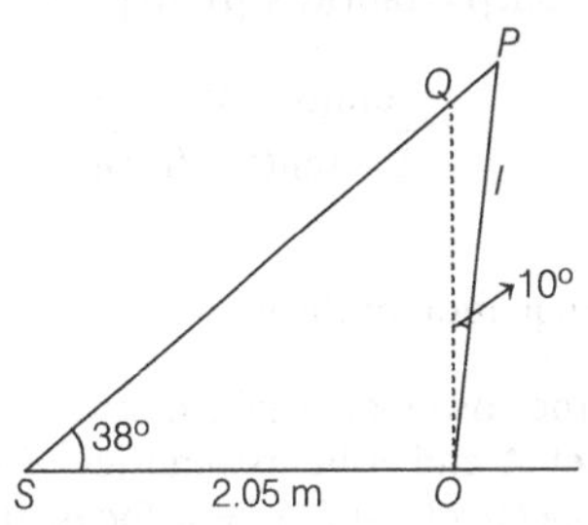

In Δ SOP, using sine rule,

$$\frac{1}{\sin 38°} = \frac{2.05}{\sin(\angle SPO)}$$

$$\angle SPO = 180° - 38° - 90° - 10° = 42°$$

So, from Eq. (i), $l = \dfrac{2.05 \sin 38°}{\sin 42°}$

28. $OA = h \cot \alpha$, $OB = h \cot \beta$, $OC = h \cot \gamma$

OA, OB, OC are in AP and α, β, γ are also in AP

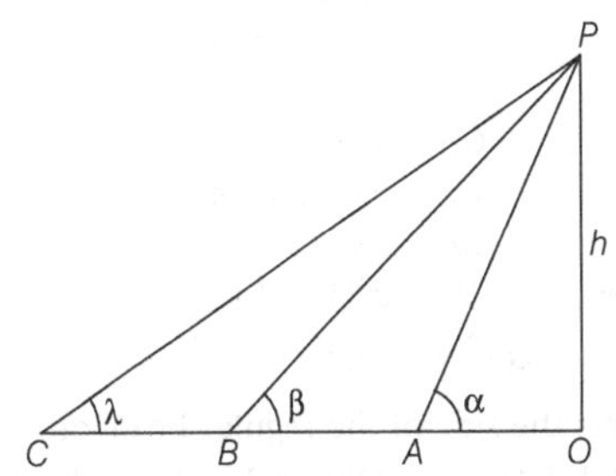

$$\therefore \quad h(\cot\alpha + \cot\gamma) = 2h \cot\beta$$

$$\Rightarrow \quad \cot\alpha + \cot\gamma = 2\cot\beta \quad \ldots(i)$$

and $\quad \alpha + \gamma = 2\beta \quad \ldots(ii)$

$\therefore \quad (\cos\alpha \sin\gamma + \sin\alpha \cos\gamma)\sin\beta = 2\cos\beta \sin\alpha \sin\gamma$

$\Rightarrow \quad \sin(\alpha + \gamma)\sin\beta = 2\cos\beta \sin\alpha \sin\gamma$

$\Rightarrow \quad \sin 2\beta \sin\beta = 2\cos\beta \sin\alpha \sin\gamma$

$\Rightarrow \quad 2\sin^2\beta \cos\beta = 2\cos\beta \sin\alpha \sin\gamma$

$\Rightarrow \quad \sin^2\beta = \sin\alpha \sin\gamma$

$\Rightarrow$ $\sin\alpha, \sin\beta, \sin\gamma$ are in GP.

29. Let $AB = h$, then $AD = 2h$ and $AC = BC = \dfrac{h}{2}$

Again, let $\angle CPA = \alpha$

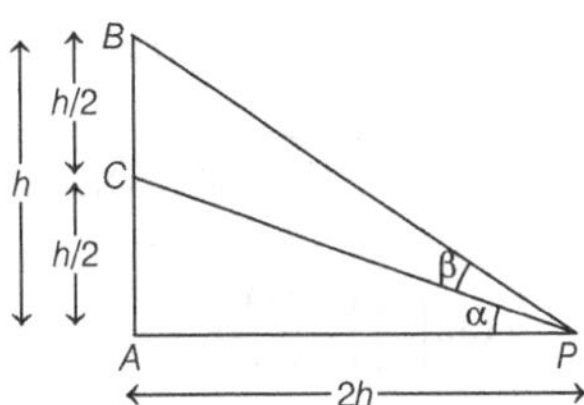

Now, in ΔABP, $\tan(\alpha + \beta) = \dfrac{AB}{AP} = \dfrac{h}{2h} = \dfrac{1}{2}$

Also, in ΔACP, $\tan\alpha = \dfrac{AC}{AP} = \dfrac{\frac{h}{2}}{2h} = \dfrac{1}{4}$

Now, $\tan\beta = \tan[(\alpha+\beta)-\alpha]$

$$= \frac{\tan(\alpha+\beta)-\tan\alpha}{1+\tan(\alpha+\beta)\tan\alpha} = \frac{\frac{1}{2}-\frac{1}{4}}{1+\frac{1}{2}\times\frac{1}{4}} = \frac{\frac{1}{4}}{\frac{9}{8}} = \frac{2}{9}$$

30. Let the equation of the elliptic field be $\frac{x^2}{a^2}+\frac{y^2}{b^2}=1.$

OP be the tower of height h at O, the centre of the field. Let A and B be extremities of major and minor axis respectively and S be a focus, then

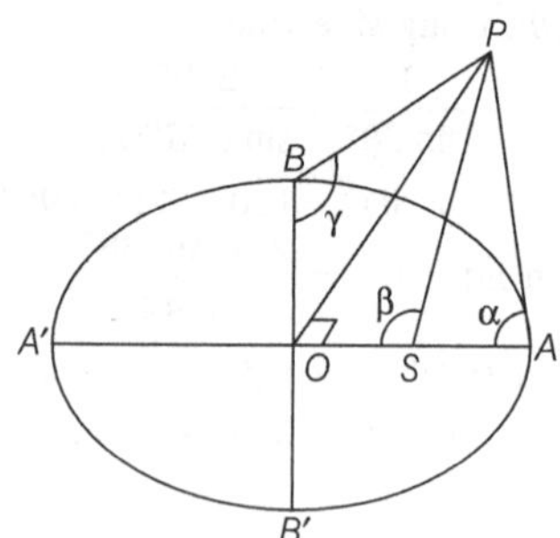

$$OA = h\cot\alpha = a$$
$$OS = h\cot\beta = ae$$
$$OB = h\cot\gamma = b$$

Since, $b^2 = a^2(1-e^2)$

$\Rightarrow \cot^2\gamma = \cot^2\alpha - \cot^2\beta$

31. Let a is the side of the regular hexagon

Now, for the equilateral ΔFOA, we have

$$OF = OA = AF = a$$

Hence, Area of Circle $= A = \pi a^2$

$\Rightarrow a = \sqrt{\frac{A}{\pi}}$

In right angled ΔAOT, right angled at O, we have

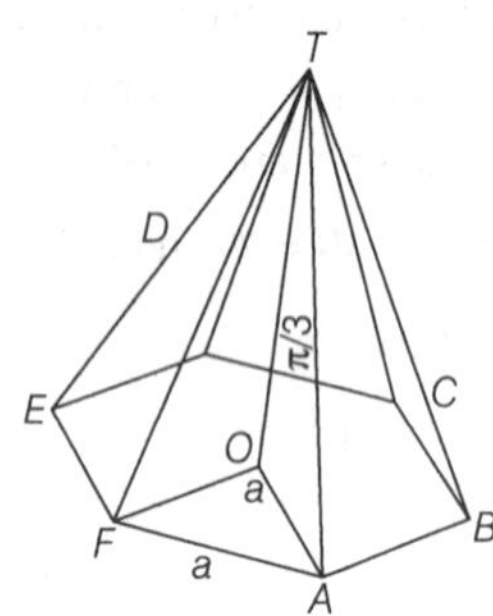

$$\tan\left(\frac{\pi}{3}\right) = \frac{OA}{OT}$$

$\Rightarrow OT = \frac{OA}{\tan\left(\frac{\pi}{3}\right)}$

$$= \frac{1}{\sqrt{3}}\sqrt{\frac{A}{\pi}}$$

$\therefore$ Height of Tower $= \sqrt{\frac{A}{3\pi}}$

32. Let the height of tower $PQ = h$ m

Then, $\angle POQ = 30°$, $\angle QOA = \alpha = \tan^{-1}\frac{1}{\sqrt{2}}$, $\angle PAQ = \theta$

From the right angled ΔQAO,

$$\frac{AQ}{OA} = \tan\alpha$$

$\therefore AQ = 300\text{m}\ \frac{1}{\sqrt{2}} = 150\sqrt{2}$ m.

From ΔPQA, $\tan\theta = \frac{PQ}{AQ} = \frac{h}{150\sqrt{2}}$

$\Rightarrow h = 150\sqrt{2}\cdot\tan\theta$ ...(i)

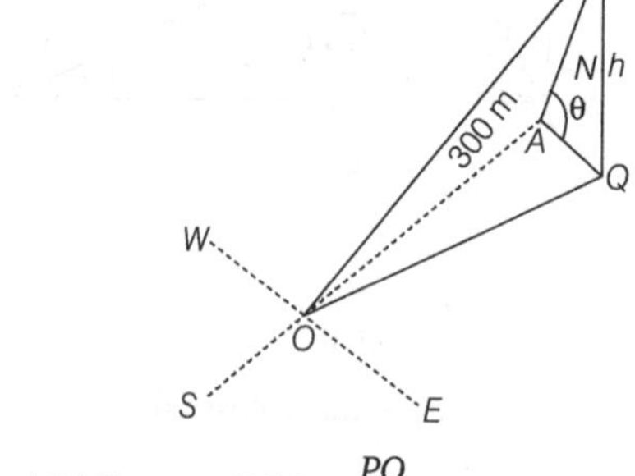

From ΔPQO, $\tan\angle POQ = \frac{PQ}{OQ}$

$\Rightarrow \tan 30° = \frac{h}{\sqrt{OA^2+AQ^2}}$

$\Rightarrow \frac{1}{\sqrt{3}} = \frac{h}{\sqrt{300^2+(150\sqrt{2})^2}} \Rightarrow h = 150\sqrt{2}$

$\Rightarrow 150\sqrt{2} = 150\sqrt{2}\cdot\tan\theta$

$\Rightarrow \tan\theta = 1 \Rightarrow \theta = 45°$

33. ABC is an isosceles triangle of wood placed in a vertical plane with the vertex A upwards. The base $BC = 2a$, and height $AD = h$. Let D be the middle point of base BC.

$\therefore BD = CD = a$

Let $A'BC$ be the shadow of the triangle ABC on the horizontal plane. The altitude of the sun is 30°.

$\therefore \angle AA'D = 30°$.

In right angled $\Delta ADA'$.

$$\tan 30° = \frac{AD}{A'D} \Rightarrow \frac{1}{\sqrt{3}} = \frac{h}{A'D} \Rightarrow A'D = \sqrt{3}h.$$

Obviously, shadow $A'BD$ is isosceles triangle.

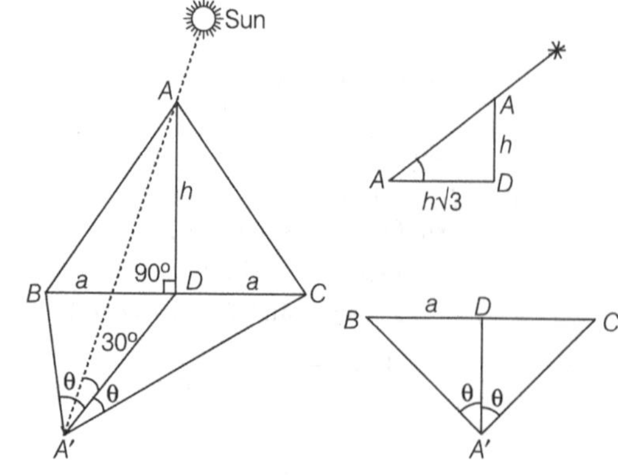

$\therefore$ Let $\angle BA'C = 2\theta$

$\therefore$ $\angle BA'D = \angle CA'D = \theta$

In right angled$\Delta A'CD$,

$$\tan\theta = \frac{CD}{A'D} = \frac{a}{(\sqrt{3}h)}$$

$$\therefore \quad \tan 2\theta = \frac{2\tan\theta}{1-\tan^2\theta} = \frac{\frac{2a}{(h\sqrt{3})}}{1-\frac{a^2}{3h^2}} = \frac{2ah\sqrt{3}}{3h^2-a^2}$$

34. According to given information, we have the following figure

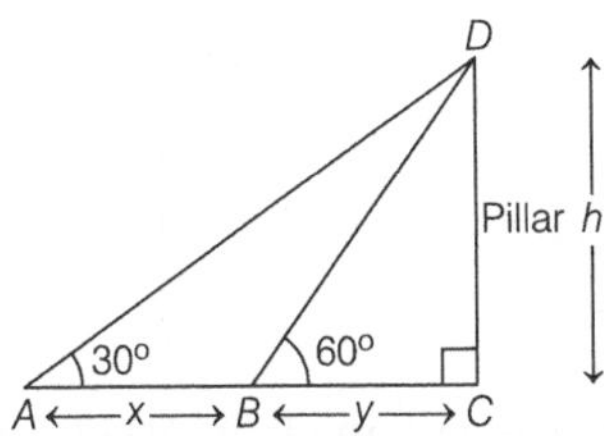

Now, from Right ΔACD and Right ΔBCD, we have

$$\tan 30° = \frac{h}{x+y}$$

and $\tan 60° = \frac{h}{y} \Rightarrow h = \frac{x+y}{\sqrt{3}}$ ...(i)

and $h = \sqrt{3}\,y$ ...(ii)

From Eqs. (i) and (ii),

$$\frac{x+y}{\sqrt{3}} = \sqrt{3}\,y$$

$\Rightarrow$ $x + y = 3y$

$\Rightarrow$ $x - 2y = 0$

$\Rightarrow$ $y = \frac{x}{2}$

$\because$ Speed is uniform.

$\therefore$ Distance y will be cover in 5 min.

$\because$ Distance x covered in 10 min.

$\therefore$ Distance $\frac{x}{2}$ will be cover in 5 min.

35.

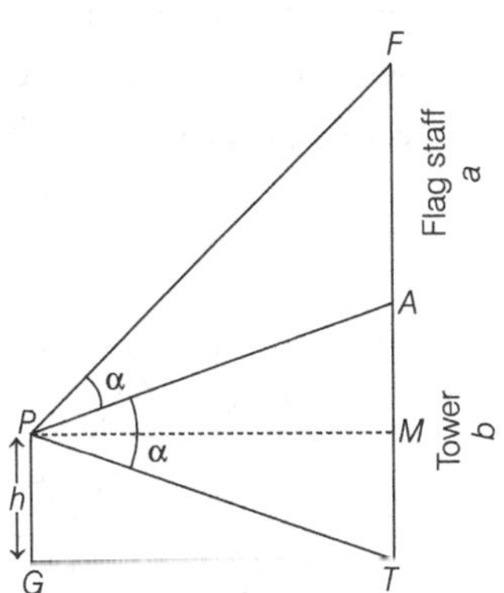

Since, the tower and flag staff subtend equal angles at the observers eye

$\therefore$ AP is the angle bisector

$\Rightarrow$ $\frac{PT}{PF} = \frac{AT}{AF} \Rightarrow PF = \frac{a}{b}PT$ ...(i)

In right angled ΔPFM, we have,

$$(PF)^2 = (PM)^2 + (FM)^2$$

$\Rightarrow$ $(PF)^2 = (GT)^2 + (FM)^2$

Now in right angled ΔPGT, we have

$$(PT)^2 = (PG)^2 + (GT)^2$$

$\Rightarrow$ $(GT)^2 = (PT)^2 - h^2$

Substituting the values of Eqs. (iii) and (i) in and Eq. (ii), we have

$$\frac{a^2}{b^2}(PT)^2 = (PT)^2 - h^2 + (a+(b-h)))^2$$

$$\Rightarrow \left(\frac{a^2}{b^2}-1\right)(PT)^2 = (a+b-h+h)(a+b-h-h)$$

$$\Rightarrow \frac{(a+b)(a-b)}{b^2}(PT)^2 = (a+b)(a+b-2h)$$

$$\therefore \quad PT = b\sqrt{\frac{a+b-2h}{a-b}}$$

36. Since, PQ subtends the same angle of 30° at each of the points A, R and S, the points A, P, Q, R and S lie on a circle.

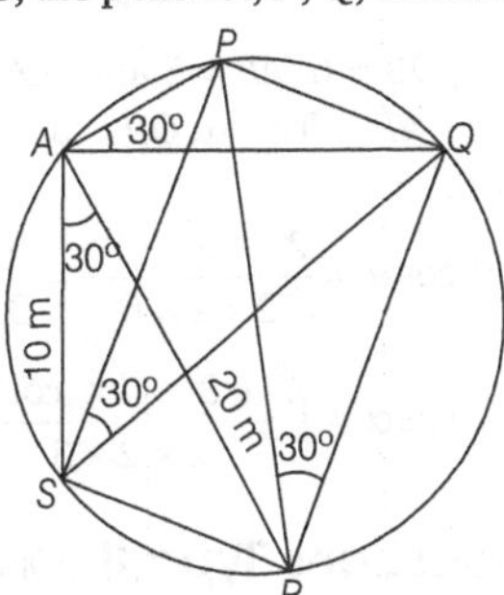

Also, $AR \perp AP$ and $AS \perp AQ$

$\Rightarrow \angle RAS = 90° - \angle RAQ = \angle PAQ = 30°$

As it is given,

$$AR = 20 \text{ m}, AS = 10\text{m}.$$

Then, PQ and RS are chords of the same circle making an angle of 30° at a point A on the circumference of the circle and hence are equal in length.

Now, from ΔSAR,

$$RS^2 = 20^2 + 10^2 - 2\times 20\times 10\cos 30°$$

$$\Rightarrow \quad PQ^2 = 500 - 200\sqrt{3} \Rightarrow PQ = \sqrt{500-200\sqrt{3}}$$

37. Let OP be the steeple of height h. A be due south of OP and $\angle OAP = 45°$ and B be due west of A and $\angle OBP = 15°$ then $OA = h\cot 45° = h$ and $OB = h\cot 15°$ and $AB = 2a$ (given) so that, from right angled ΔOAB, $OB^2 = OA^2 + AB^2$.

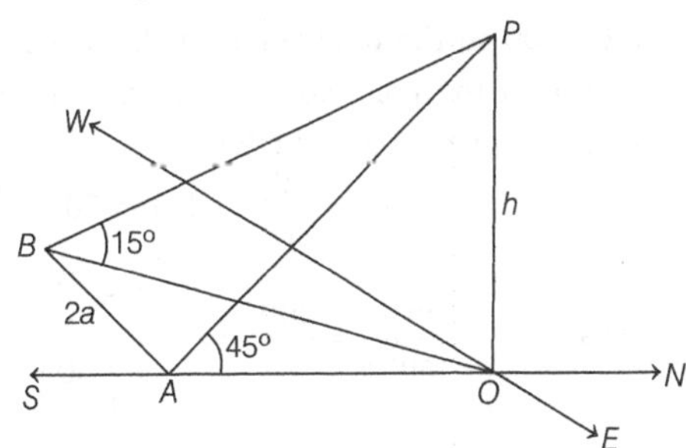

$$\Rightarrow \quad h^2 \cot^2 15 = h^2 + 4a^2$$

$$\Rightarrow h^2 \left(\left(\frac{\sqrt{3}+1}{\sqrt{3}-1} \right)^2 - 1 \right) = 4a^2$$

$$\Rightarrow \quad h^2 \times 4\sqrt{3} = 4a^2 (\sqrt{3}-1)^2$$

$$\Rightarrow \quad h = \frac{\sqrt{3}-1}{(\sqrt{3})^{1/2}} a = [3^{\frac{1}{4}} - 3^{-\frac{1}{4}}] a$$

38. Let PQ be the tower of the height h at the middle point P of the side OB of the triangle OAB.

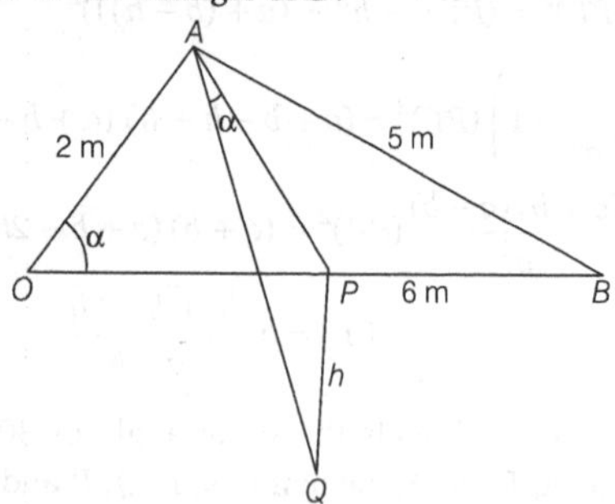

$OA = 2, OB = 6, AB = 5$ and $\angle AOB = \angle PAQ = \alpha$,

then, $\quad AP = h \cot \alpha, OP = 3$

From ΔOAB,

$$\cos \alpha = \frac{2^2 + 6^2 - 5^2}{2 \times 2 \times 6} = \frac{5}{8}$$

$$\cos \alpha = \frac{2^2 + 3^2 - h^2 \cot^2 \alpha}{2 \times 2 \times 3}$$

$$\Rightarrow \quad \frac{5}{8} = \frac{13 - h^2 \times \frac{25}{39}}{12}$$

$$\Rightarrow \quad h^2 = \frac{11 \times 39}{25 \times 2} \Rightarrow h = \sqrt{\frac{11 \times 39}{25 \times 2}}$$

39. Let A be the position of the harbour and O be the fort, $OA = 30, OB = 70$. Let $AB = x$.

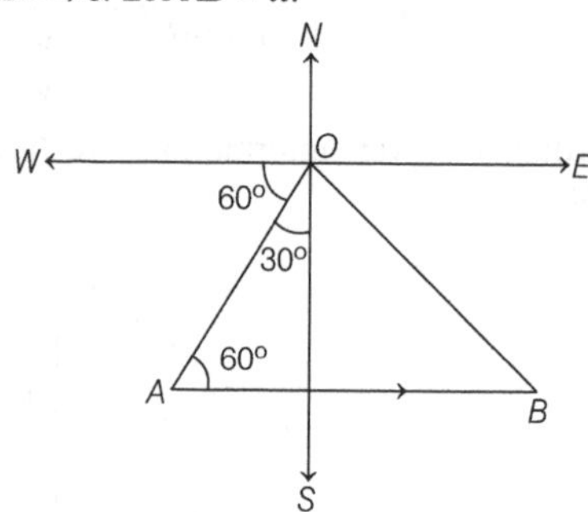

Applying cosine formula in ΔOAB, we get

$$\cos 60° = \frac{OA^2 + AB^2 - OB^2}{2OA \cdot AB}$$

$$\Rightarrow \quad \frac{1}{2} = \frac{900 + x^2 - 4900}{2 \times 30 \times x}$$

$$\Rightarrow \quad x^2 - 30x - 4000 = 0 \Rightarrow x = 80$$

$$AB = 80 \text{ km}$$

Speed of ship $= 10$ km/h

$\therefore$ Time $= 8$ pm

Objective Questions Type II [One or more than one correct answer(s)]

1. Let PQ be the pole on the building QR and O be the observer.

Then, $PQ = 50, QR = 250$

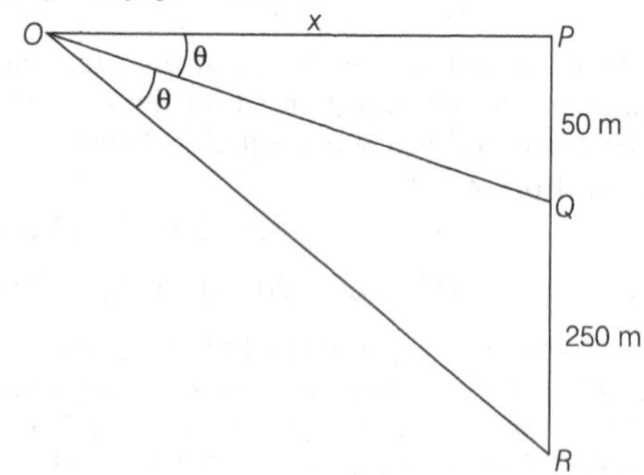

$\Rightarrow PR = 300$ so the observer is at the same height as the top P of the pole. So $OP = x$. Then, from right angled triangles OPQ and OPR.

$$\tan \theta = \frac{50}{x} \text{ and } \tan 2\theta = \frac{300}{x}$$

So, that
$$\frac{2 \tan \theta}{1 - \tan^2 \theta} = \frac{300}{x} \Rightarrow \frac{2 \times \frac{50}{x}}{1 - \left(\frac{50}{x}\right)^2} = \frac{300}{x}$$

$$\Rightarrow \quad 3\left\{1 - \left(\frac{50}{x}\right)^2\right\} = 1 \Rightarrow \left(\frac{50}{x}\right)^2 = \frac{2}{3}$$

$$\Rightarrow \quad x = \frac{50\sqrt{3}}{\sqrt{2}} = 25\sqrt{6} \text{ and } \tan \theta = \frac{50}{25\sqrt{6}} = \sqrt{\frac{2}{3}}$$

2. DP is a clock tower standing at the middle point D of BC.

$\angle PAD = \alpha = \cot^{-1} 3.2 \Rightarrow \cot \alpha = 3.2$

and $\quad \angle PBD = \beta = \operatorname{cosec}^{-1} 2.6 \Rightarrow \operatorname{cosec} \beta = 2.6$

$\therefore \quad \cot \beta = \sqrt{\operatorname{cosec}^2 \beta - 1} = \sqrt{5.76} = 2.4$

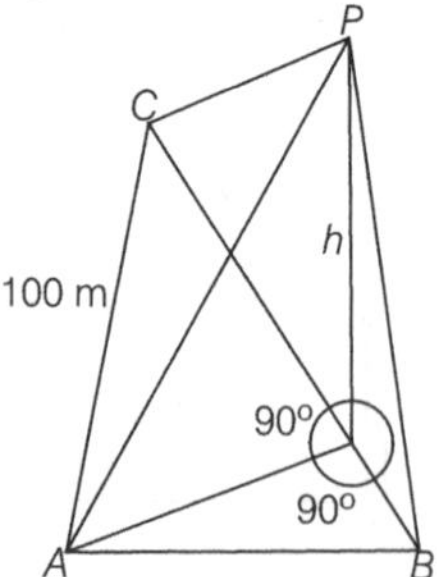

In ΔPAD, $AD = h \cot \alpha = 2.3\,h$

In ΔPBD, $BD = h \cot \beta = 2.4\,h$

In the right angled ΔABD, $AB^2 = AD^2 + BD^2$

$$\Rightarrow (100)^2 = [(3.2)^2 + (2.4)^2] h^2 = 16h^2 \Rightarrow h = 25 \text{ m}$$

Linked Comprehension Type

PASSAGE 1

1. $\tan\theta = \frac{b}{a},\ \tan 2\theta = \frac{b+h}{a}$

$$\Rightarrow \frac{2\tan\theta}{1-\tan^2\theta} = \frac{h+b}{a}$$

$$\Rightarrow h = \frac{a\times 2\left(\frac{b}{a}\right)}{1-\left(\frac{b^2}{a^2}\right)} - b$$

$$= \frac{b(a^2+b^2)}{a^2-b^2}$$

2. $\tan(\theta-\alpha) = \frac{\tan\theta - \tan\alpha}{1+\tan\theta\tan\alpha}$

$$\Rightarrow \frac{b}{2a} = \frac{\left(\frac{b}{a}\right) - \tan\alpha}{1+\left(\frac{b}{a}\right)\tan\alpha}$$

$$\Rightarrow \tan\alpha\left[\frac{b^2}{2a^2}+1\right] = \frac{b}{a} - \frac{b}{2a}$$

$$\Rightarrow \tan\alpha = \frac{ab}{b^2+2a^2}$$

PASSAGE 2

1. Let OR be the diving stand of height h and C the centre of the swimming pool of radius r.

Let P and Q be the two points on the edge of the pool where the angle of elevation of R the top of the stand is 30° and R' be the middle point of PQ,

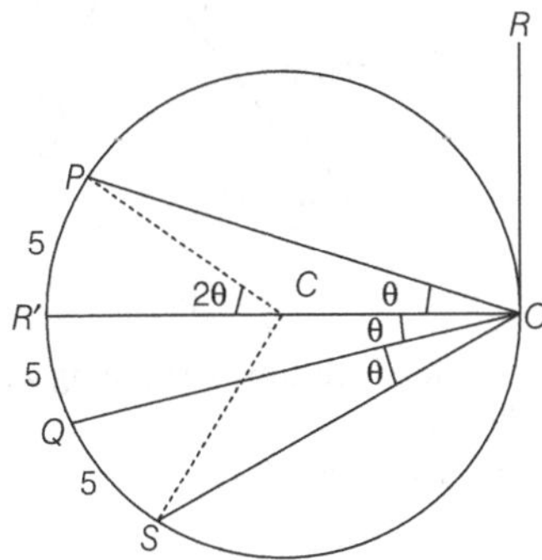

Then, are PR' = arc $R'Q = 5$ m, as arc $PQ = 10$ m.

Let S be the point where the angle of elevation of R is 45°.

Then, arc $QS = 5$ m

We have, $OP = OQ = h\cot 30° = h\sqrt{3}$

and $OS = OR\cot 45° = OR = h$

Let $\angle COP = \angle COQ = QOS = \theta$

From isosceles ΔCOP,

$$\frac{OP}{\sin(180° - 2\theta)} = \frac{CP}{\sin\theta}$$

$$\Rightarrow \frac{\sqrt{3}\,h}{2\sin\theta\cos\theta} = \frac{r}{\sin\theta}$$

$$\Rightarrow \cos\theta = \frac{\sqrt{3}\,h}{2r} \quad \ldots\text{(i)}$$

From isosceles ΔCOS,

$$\frac{OS}{\sin(180° - 4\theta)} = \frac{CS}{\sin 2\theta}$$

$$\Rightarrow \frac{h}{2\sin 2\theta\cos 2\theta} = \frac{r}{\sin 2\theta}$$

$$\Rightarrow \cos 2\theta = \frac{h}{2r}$$

$$\Rightarrow 2\cos^2\theta - 1 = \frac{h}{2r}$$

$$\Rightarrow 2\times\frac{3h^2}{4r^2} - \frac{h}{2r} - 1 = 0$$

$$\Rightarrow 3\left(\frac{h}{r}\right)^2 - \frac{h}{r} - 2 = 0$$

$$\Rightarrow (3h+2r)(h-r) = 0$$

$$\Rightarrow h = r$$

2. From Eq. (i) $\cos\theta = \frac{\sqrt{3}}{2}$

$$\Rightarrow \theta = \frac{\pi}{6}$$

$$2\theta r = PR' = 5$$

$$\Rightarrow r = \frac{15}{\pi}$$

So, the perimeter of the pool is

$$2\pi\times\frac{15}{\pi} = 30 \text{ m}$$

Numerical Value Type

1. Here, $OA = OB = OC = OD = 150$ m

$$AB = BC = CD = DA = 200 \text{ m}$$

Let $OM \perp$ base.

Then, M is the centre of the square.

$MN \perp AB$.

Clearly, $AN = NB = 100$; $MN = \frac{1}{2} AD = 100$ m.

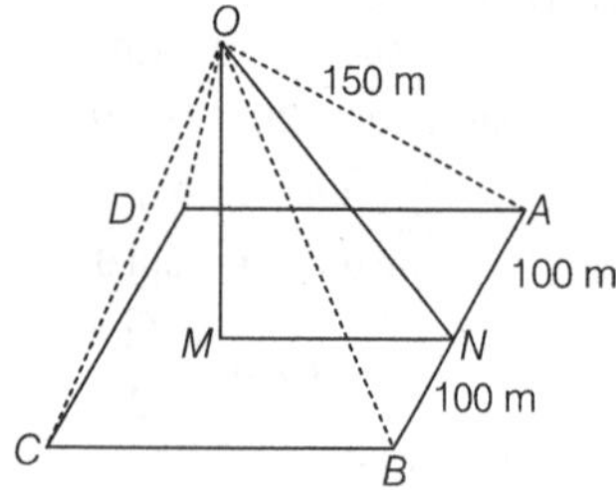

As ΔOAB is isosceles, $ON \perp NA$

In ΔOAN, $OA = 150$ m and $NA = 100$ m

$$ON = \sqrt{150^2 - 100^2} = 50\sqrt{5} \text{ m}$$

In ΔOMN,

$$OM = \sqrt{ON^2 - MN^2}$$

$$= \sqrt{12500 - 10000} = 50 \text{ m}$$

2. According to the given situation if we draw the diagram

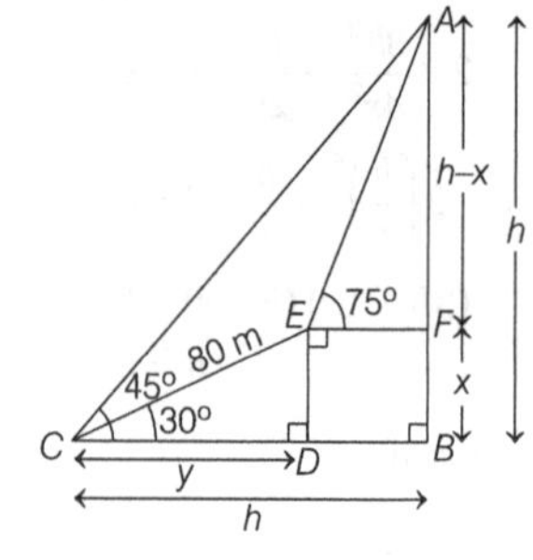

$\because \quad \angle ACB = 45°$

$\therefore \quad AB = BC = h$ [let]

Now, from the diagram

$$\sin 30° = \frac{x}{80} \text{ (in } \Delta CDE)$$

$$\Rightarrow \quad x = 40 \text{ m}$$

and $$\cos 30° = \frac{y}{80}$$

$$\Rightarrow \quad y = 40\sqrt{3} \text{ m}$$

Now, in ΔAEF,

$$\tan 75° = \frac{h - x}{h - y}$$

$$\Rightarrow \quad \frac{\sqrt{3}+1}{\sqrt{3}-1} = \frac{h-40}{h-40\sqrt{3}}$$

$$\Rightarrow \quad \sqrt{3}h + h - 120 - 40\sqrt{3} = \sqrt{3}h - h - 40\sqrt{3} + 40$$

$$\Rightarrow \quad 2h = 160$$

$$\Rightarrow \quad h = 80 \text{ m}$$

3.

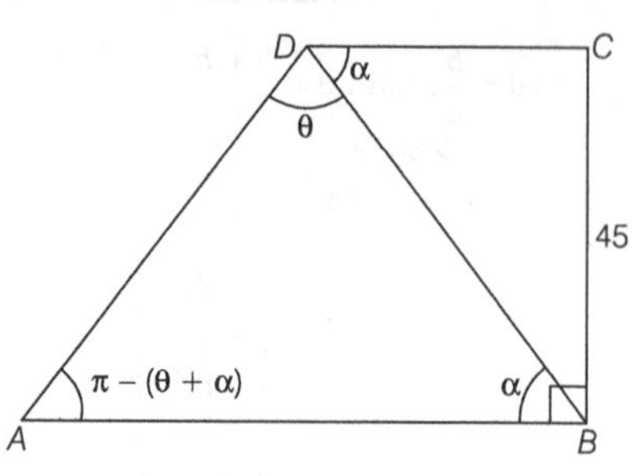

$$\frac{AB}{\sin\theta} = \frac{BD}{\sin(\pi - (\theta + \alpha))}$$

$$\Rightarrow \quad AB = \frac{BD \sin\theta}{\sin(\theta + \alpha)}$$

$$= \frac{BD \sin\theta}{\sin\theta \cos\alpha + \cos\theta \sin\alpha}$$

$$= \frac{BD \tan\theta}{\sin\alpha + \cos\alpha + \tan\theta}$$

$$= \frac{(BD)^2 \tan\theta}{45 + 30 \times \frac{1}{5}}$$

$$\left[\because \sin\alpha = \frac{45}{BD}, \cos\alpha = \frac{30}{BD}\right]$$

$$\Rightarrow \quad 51AB = [(30)^2 + (45)^2] \times \frac{1}{5} = 585$$

4.

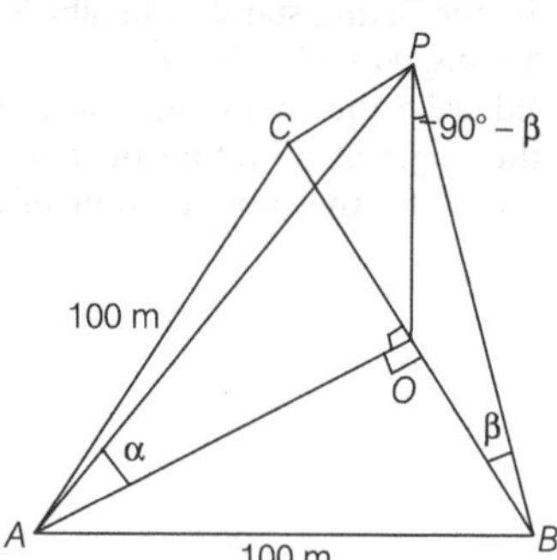

Let OP be the clock-tower standing at the mid-point O of side BC of ΔABC.

Let $\alpha = \angle PAO = \cot^{-1} 3.2$ [given]

and $\beta = \angle PBO = \text{cosec}^{-1} 2.6$

$\therefore \cot\alpha = 3.2$ and $\text{cosec}\,\beta = 2.6$

Now, $\cot\beta = \sqrt{\text{cosec}^2\beta - 1} = \sqrt{(2.6)^2 - 1} = 2.4$

In ΔPAO, we have

$$OA = h \cot\alpha = 3.2h$$

In ΔAOB, right angled at O, we have

$$AB^2 = OA^2 + OB^2$$

$$\Rightarrow \quad 100^2 = (3.2h)^2 + (2.4h)^2$$

$$\Rightarrow \quad 100^2 = 16h^2$$

$$\therefore \quad h^2 = 625$$

$$\Rightarrow \quad h = 25 \text{ m}$$

5. In ΔABC,

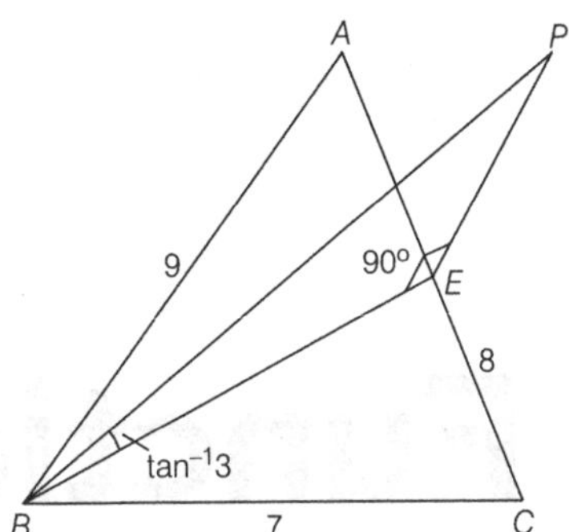

$a = BC = 7,\ b = CA = 8$ and $c = AB = 9$

so that, $\cos C = \dfrac{a^2 + b^2 - c^2}{2ab} = \dfrac{49 + 64 - 81}{2 \times 7 \times 8} = \dfrac{2}{7}$

In $\Delta\ BEC$,

$BC = 7,\ CE = 4$ and $\cos C = \dfrac{2}{7}$

so that, $BE^2 = 4^2 + 7^2 - 2 \times 4 \times 7 \times \dfrac{2}{7} = 49$

$\Rightarrow \quad BE = 7$

If h is the height of the lamp post EP at E

Then, $\dfrac{PE}{BE} = \dfrac{h}{7} = \tan(\tan^{-1} 3) = 3 \Rightarrow h = 21$ m

Single Integer Type

1. On drawing the diagram of given situation

$$MD^2 + MC^2 = 64 + x^2 + 121 + (10 - x)^2$$
$$= f(x) \quad \text{[say]}$$
$$\Rightarrow \quad f(x) = 2x^2 - 20x + 285$$

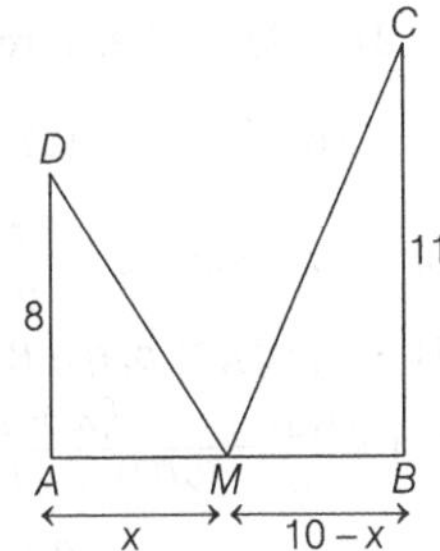

For minima, $f'(x) = 0 \Rightarrow 4x - 20 = 0 \Rightarrow x = 5$

As $f''(x) = 4 > 0$

$\therefore$ $AM = 5$ m

2. $\tan\alpha = \dfrac{h}{4x}$

$\tan\beta = \dfrac{3}{4}$

$\tan\beta = \tan(\theta - \alpha) = \dfrac{\tan\theta - \tan\alpha}{1 + \tan\theta\tan\alpha}$

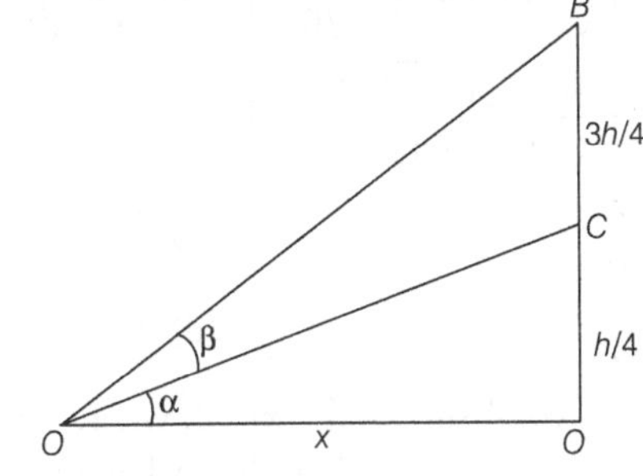

$$\Rightarrow \quad \frac{3}{4} = \frac{\frac{h}{x} - \frac{h}{4x}}{1 + \frac{h}{x} \times \frac{k}{4x}} = \frac{3hx}{(4x^2 + h^2)}$$

$$\Rightarrow \quad 4hx = 4x^2 + h^2$$

i.e. $\quad (2x - h)^2 = 0$

$$\therefore \quad \tan\theta = \frac{h}{x} = \frac{2x}{x} = 2$$

3. Let AB is tall building CQ is short building

$AB = H$

$CQ = h$

In ΔABC,

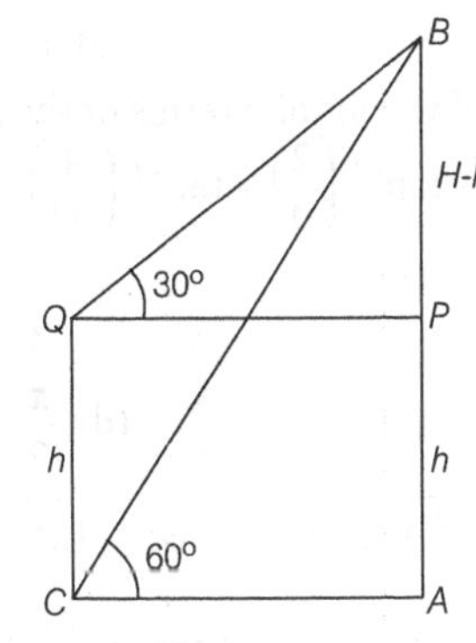

$$\tan 60° = \frac{H}{AC}$$
$$AC = \frac{H}{\sqrt{3}}$$

In ΔBPQ,

$$\tan 30° = \frac{H - h}{PQ}$$
$$\Rightarrow \quad PQ = (H - h)\sqrt{3}$$
$$AC = PQ$$
$$\therefore \quad \frac{H}{\sqrt{3}} = (H - h)\sqrt{3}$$
$$\Rightarrow \quad H = 3H - 3h \Rightarrow 2H = 3h$$
$$\Rightarrow \quad \frac{h}{H} = \frac{2}{3}$$
$$\therefore \quad h : H = 2 : 3$$
$$\therefore \quad k^2 = 9$$

31

Inverse Circular Functions

Objective Questions Type I [Only one correct answer]

In each of the questions below, four choices are given of which only are is correct. You have to select the correct answer which is the most appropriate.

1. If $x \in \left(\frac{3\pi}{2}, 2\pi\right)$, then the value of the expression $\sin^{-1}[\cos\{\cos^{-1}(\cos x)\} + \sin^{-1}(\sin x)]$, is

   (a) $-\frac{\pi}{2}$ (b) $\frac{\pi}{2}$

   (c) 0 (d) π

2. The sum of the infinite terms of the series $\tan^{-1}\left(\frac{1}{3}\right) + \tan^{-1}\left(\frac{2}{9}\right) + \tan^{-1}\left(\frac{4}{33}\right) + \dots$ is equal to

   (a) $\frac{\pi}{6}$ (b) $\frac{\pi}{4}$

   (c) $\frac{\pi}{3}$ (d) $\frac{\pi}{2}$

3. If $\sin^{-1}\sqrt{(x^2 + 2x + 1)} + \sec^{-1}\sqrt{(x^2 + 2x + 1)} = \frac{\pi}{2}$, $x \neq 0$, then the value of $2\sec^{-1}\left(\frac{x}{2}\right) + \sin^{-1}\left(\frac{x}{2}\right)$ is equal to

   (a) $-\frac{3\pi}{2}$ (b) $\frac{3\pi}{2}$

   (c) $-\frac{\pi}{2}$ (d) $\frac{\pi}{2}$

4. Range of the function $f(x) = (\tan^{-1}\sqrt{[x] + [-x]} + \sqrt{2 - |x|} + \frac{1}{x^2}$ is where [*] is the greatest integer function.

   (a) $\left[\frac{1}{4}, \infty\right)$ (b) $\left\{\frac{1}{4}\right\} \cup [2, \infty)$

   (c) $\left\{\frac{1}{4}, 2\right\}$ (d) $\left[\frac{1}{4}, 2\right]$

5. The value of 'a' for which $ax^2 + \sin^{-1}(x^2 - 2x + 2) + \cos^{-1}(x^2 - 2x + 2) = 0$ has a real solution, is

   (a) $-\frac{2}{\pi}$ (b) $\frac{2}{\pi}$ (c) $-\frac{\pi}{2}$ (d) $\frac{\pi}{2}$

6. If $\sin^{-1}x + \sin^{-1}y + \sin^{-1}z = \frac{3\pi}{2}$ and $f(1) = 2$, $f(p + q) = f(p) \cdot f(q)$, $\forall\ p, q \in R$, then $x^{f(1)} + y^{f(2)} + z^{f(3)} - \frac{(x + y + z)}{x^{f(1)} + y^{f(2)} + z^{f(3)}}$ is equal to

   (a) 0 (b) 1

   (c) 2 (d) 3

7. If the mapping $f(x) = ax + b$, $a > 0$ maps $[-1, 1]$ onto $[0, 2]$, then $\cot[\cot^{-1}7 + \cot^{-1}8 + \cot^{-1}18]$ is equal to

   (a) $f(-1)$ (b) $f(0)$

   (c) $f(1)$ (d) $f(2)$

8. The sum of the infinite series $\sin^{-1}\left(\frac{1}{\sqrt{2}}\right) + \sin^{-1}\left(\frac{\sqrt{2} - 1}{\sqrt{6}}\right) + \sin^{-1}\left(\frac{\sqrt{3} - \sqrt{2}}{\sqrt{12}}\right) + \dots + \dots + \sin^{-1}\left(\frac{\sqrt{n} - \sqrt{(n-1)}}{\sqrt{\{n(n+1)\}}}\right) + \dots$ is

   (a) $\frac{\pi}{8}$ (b) $\frac{\pi}{4}$

   (c) $\frac{\pi}{2}$ (d) π

9. If $[\sin^{-1}\cos^{-1}\sin^{-1}\tan^{-1}x] = 1$, where [.] denotes the greatest integer function, then x belongs to the interval

   (a) [tan sin cos 1, tan sin cos sin 1]

   (b) (tan sin cos 1, tan sin cos sin 1)

   (c) [– 1, 1]

   (d) [sin cos tan 1, sin cos sin tan 1]

10. Solution set of $[\sin^{-1} x] > [\cos^{-1} x]$, where [.] denotes the greatest integer function, is

(a) $\left[\frac{1}{\sqrt{2}}, 1\right]$ (b) $(\cos 1, \sin 1)$

(c) $[\sin 1, 1]$ (d) None of these

11. If $[\cot^{-1} x] + [\cos^{-1} x] = 0$, where x is a non-negative real number and [.] denotes the greatest integer function, then complete set of values of x is

(a) $(\cos 1, 1]$ (b) $(\cot 1, 1)$

(c) $(\cos 1, \cot 1)$ (d) None of these

12. The number of solutions for the equation $2\sin^{-1}\sqrt{(x^2 - x + 1)} + \cos^{-1}\sqrt{(x^2 - x)} = \frac{3\pi}{2}$ is

(a) 1 (b) 2

(c) 3 (d) infinite

13. The number of solutions of the equation $\cos^{-1}(1 - x) + m\cos^{-1} x = \frac{n\pi}{2}$, where $m > 0$, $n \le 0$, is

(a) 0 (b) 1

(c) 2 (d) infinite

14. $\sin[\cot^{-1}\{\cos(\tan^{-1} x)\}]$, is equal to

(a) $\sqrt{\left(\frac{x^2+1}{x^2+2}\right)}$ (b) $\sqrt{\left(\frac{x^2+2}{x^2+1}\right)}$

(c) $\sqrt{\left(\frac{x+2}{x^2+1}\right)}$ (d) $\sqrt{\left(\frac{x^2-1}{x^2+2}\right)}$

15. If $\cos^{-1}\left(\frac{x}{3}\right) + \cos^{-1}\left(\frac{y}{2}\right) = \frac{\theta}{2}$, then the value of $4x^2 - 12xy\cos\left(\frac{\theta}{2}\right) + 9y^2$ is equal to

(a) $18(1 + \cos\theta)$ (b) $18(1 - \cos\theta)$

(c) $36(1 + \cos\theta)$ (d) $36(1 - \cos\theta)$

16. The domain of the function $f(x) = \sin^{-1}\left(\frac{|x| + 5}{x^2 + 1}\right)$ is $(-\infty, -a] \cup [a, \infty)$. Then, a is equal to **[JEE Main 2020, 2 September]**

(a) $\frac{\sqrt{17}}{2} + 1$ (b) $\frac{\sqrt{17} - 1}{2}$

(c) $\frac{1 + \sqrt{17}}{2}$ (d) $\frac{\sqrt{17}}{2}$

17. The number of solutions of the equation $\sin^{-1}\left[x^2 + \frac{1}{3}\right] + \cos^{-1}\left[x^2 - \frac{2}{3}\right] = x^2$ for $x \in [-1, 1]$ and $[x]$ denotes the greatest integer less than or equal to x, is **[JEE Main 2021, 17 March]**

(a) 2 (b) 0

(c) 4 (d) Infinite

18. If $\sin^{-1} x + \sin^{-1} y = \frac{2\pi}{3}$, then $\cos^{-1} x + \cos^{-1} y$ is equal to

(a) $\frac{2\pi}{3}$ (b) $\frac{\pi}{3}$

(c) $\frac{\pi}{6}$ (d) π

19. The value of $\frac{\alpha^3}{2}\operatorname{cosec}^2\left(\frac{1}{2}\tan^{-1}\left(\frac{\alpha}{\beta}\right)\right) + \frac{\beta^3}{2}\sec^2\left(\frac{1}{2}\tan^{-1}\left(\frac{\beta}{\alpha}\right)\right)$ is equal to

(a) $(\alpha - \beta)(\alpha^2 + \beta^2)$ (b) $(\alpha + \beta)(\alpha^2 - \beta^2)$

(c) $(\alpha + \beta)(\alpha^2 + \beta^2)$ (d) None of these

20. If $\cos^{-1} x + \cos^{-1} y + \cos^{-1} z = 3\pi$, then $xy + yz + zx$ is equal to

(a) -3 (b) 0

(c) 3 (d) -1

21. The value of $\sin^{-1}\left[\cot\left\{\sin^{-1}\sqrt{\left(\frac{2-\sqrt{3}}{4}\right)} + \cos^{-1}\left(\frac{\sqrt{12}}{4}\right) + \sec^{-1}(\sqrt{2})\right\}\right]$ is equal to

(a) 0 (b) $\pi/4$

(c) $\pi/6$ (d) $\pi/2$

22. The number of real solutions of $\tan^{-1}\sqrt{\{x(x+1)\}} + \sin^{-1}\sqrt{\{x^2 + x + 1\}} = \frac{\pi}{2}$ is

(a) zero (b) one

(c) two (d) infinite

23. Let $S_k = \sum_{r=1}^{k} \tan^{-1}\left(\frac{6^r}{2^{2r+1} + 3^{2r-1}}\right)$, then $\lim_{k\to\infty} S_k$ equals to **[JEE Main 2021, 16 March]**

(a) $\tan^{-1}\left(\frac{3}{2}\right)$ (b) $\frac{\pi}{2}$

(c) $\cot^{-1}\left(\frac{3}{2}\right)$ (d) $\tan^{-1} 3$

24. If x_1, x_2, x_3, x_4 are roots of the equation $x^4 - x^3\sin 2\beta + x^2\cos 2\beta - x\cos\beta - \sin\beta = 0$, then $\sum_{i=1}^{4}\tan^{-1} x_i$ is equal to

(a) β (b) $\frac{\pi}{2} - \beta$

(c) $\pi - \beta$ (d) $-\beta$

25. If $\tan^{-1}\left\{\frac{\sqrt{(1+x^2)} - \sqrt{(1-x^2)}}{\sqrt{(1+x^2)} + \sqrt{(1-x^2)}}\right\} = \alpha$, then x^2 is equal to

(a) $\cos 2\alpha$ (b) $\sin 2\alpha$

(c) $\tan 2\alpha$ (d) $\cot 2\alpha$

26. $\sin^{-1}\left(\sin\left(\frac{2x^2 + 4}{1 + x^2}\right)\right) < \pi - 3$, if

(a) $-1 \le x \le 0$ (b) $0 \le x \le 1$

(c) $-1 < x < 1$ (d) $x > 1$

27. If $\frac{\sin^{-1} x}{a} = \frac{\cos^{-1} x}{b} = \frac{\tan^{-1} y}{c}$; $0 < x < 1$, then the value of $\cos\left(\frac{\pi c}{a+b}\right)$ is **[JEE Main 2021, 26 February]**

(a) $\frac{1-y^2}{2y}$ (b) $\frac{1-y^2}{1+y^2}$

(c) $1-y^2$ (d) $\frac{1-y^2}{y\sqrt{2}}$

28. If S is the sum of the first 10 terms of the series

$\tan^{-1}\left(\frac{1}{3}\right) + \tan^{-1}\left(\frac{1}{7}\right) + \tan^{-1}\left(\frac{1}{13}\right) + \tan^{-1}\left(\frac{1}{21}\right) + \dots,$

then $\tan(S)$ is equal to **[JEE Main 2020, 5 September]**

(a) $\frac{6}{5}$ (b) $\frac{5}{6}$

(c) $\frac{10}{11}$ (d) $\frac{5}{11}$

29. If $\sum_{i=1}^{2n} \sin^{-1} x_i = n\pi$, then $\sum_{i=1}^{2n} x_i$ is equal to

(a) n (b) $2n$

(c) $\frac{n(n+1)}{2}$ (d) None of these

30. The inequality $\sin^{-1} (\sin 5) > x^2 - 4x$ holds, if

(a) $x = 2 - \sqrt{(9 - 2\pi)}$

(b) $x = 2 + \sqrt{(9 - 2\pi)}$

(c) $x \in (2 - \sqrt{(9-2\pi)}, 2 + \sqrt{(9-2\pi)})$

(d) $x > 2 + \sqrt{(9 - 2\pi)}$

31. The sum of the infinite series $\cot^{-1} 2 + \cot^{-1} 8 + \cot^{-1} 18 + \cot^{-1} 32 + \dots$ is equal to

(a) π (b) $\frac{\pi}{2}$

(c) $\frac{\pi}{4}$ (d) None of these

32. $\sin \{\cot^{-1}(\tan \cos^{-1} x)\}$ is equal to

(a) x (b) $\sqrt{(1-x^2)}$

(c) $\frac{1}{x}$ (d) None of these

33. The value of $\tan^2(\sec^{-1}2) + \cot^2 (\operatorname{cosec}^{-1}3)$ is

(a) 13 (b) 15

(c) 11 (d) None of these

34. The equation $\sin^{-1} x = 2 \sin^{-1} a$ has a solution for

(a) all real values of a (b) $a < 1$

(c) $-\frac{1}{\sqrt{2}} \le a \le \frac{1}{\sqrt{2}}$ (d) $-1 < a < 1$

35. The number of real solutions of (x, y), where $|y| = \sin x$, $y = \cos^{-1} (\cos x)$, $-2\pi \le x \le 2\pi$, is

(a) 2 (b) 1

(c) 3 (d) 4

36. The number of positive integral solutions of $\tan^{-1} x + \cot^{-1} y = \tan^{-1} 3$ is

(a) one (b) two

(c) three (d) four

37. The value of $\cos^{-1}(\cos 12) - \sin^{-1}(\sin 12)$ is

(a) 0 (b) π

(c) $8\pi - 24$ (d) None of these

38. The smallest and the largest values of $\tan^{-1}\left(\frac{1-x}{1+x}\right)$, $0 \le x \le 1$ are

(a) $0, \pi$ (b) $0, \frac{\pi}{4}$

(c) $-\frac{\pi}{4}, \frac{\pi}{4}$ (d) $\frac{\pi}{4}, \frac{\pi}{2}$

39. If $-1 < x < 0$, then $\sin^{-1} x$ equals

(a) $\pi - \cos^{-1}\{\sqrt{1-x^2}\}$ (b) $\tan^{-1}\left\{\frac{x}{\sqrt{(1-x^2)}}\right\}$

(c) $-\cot^{-1}\left\{\frac{\sqrt{(1-x^2)}}{x}\right\}$ (d) $\operatorname{cosec}^{-1}x$

40. The value of $\sin^{-1} (\sin 10)$ is

(a) 10 (b) $10 - 3\pi$

(c) $3\pi - 10$ (d) None of these

41. If a, b are positive quantities and, if

$$a_1 = \frac{a+b}{2}, b_1 = \sqrt{a_1 b},$$

$a_2 = \frac{a_1 + b_1}{2}$, $b_2 = \sqrt{a_2 b_1}$ and so on, then

(a) $a_\infty = \frac{\sqrt{(b^2 - a^2)}}{\cos^{-1}\left(\frac{a}{b}\right)}$ (b) $b_\infty = \frac{\sqrt{(b^2 - a^2)}}{\cos^{-1}\left(\frac{a}{b}\right)}$

(c) $b_\infty = \frac{\sqrt{(a^2 - b^2)}}{\cos^{-1}\left(\frac{b}{a}\right)}$ (d) None of these

42. The value of $\tan^{-1}\left(\frac{c_1 x - y}{c_1 y + x}\right) + \tan^{-1}\left(\frac{c_2 - c_1}{1 + c_2 c_1}\right)$

$+ \tan^{-1}\left(\frac{c_3 - c_2}{1 + c_3 c_2}\right) + \dots + \tan^{-1}\left(\frac{1}{c_n}\right)$ is equal to

(a) $\tan^{-1}\left(\frac{y}{x}\right)$ (b) $\tan^{-1}\left(\frac{x}{y}\right)$

(c) $-\tan^{-1}\left(\frac{x}{y}\right)$ (d) None of these

43. The value of $\sin^{-1}\left\{(\sin \pi / 3) \frac{x}{\sqrt{(x^2+k^2-kx)}}\right\}$

$- \cos^{-1}\left\{\cos \frac{\pi}{6} \frac{x}{\sqrt{(x^2+k^2-kx)}}\right\}$

$\left(\text{where } \frac{k}{2} < x < 2k, k > 0\right)$ is

(a) $\tan^{-1}\left(\frac{2x^2+xk-k^2}{x^2-2xk+k^2}\right)$

(b) $\tan^{-1}\left(\frac{x^2+2xk-k^2}{x^2-2xk+k^2}\right)$

(c) $\tan^{-1}\left(\frac{x^2+2xk-2k^2}{2x^2-2xk+2k^2}\right)$

(d) None of these

44. If $\cos^{-1}\sqrt{p} + \cos^{-1}\sqrt{1-p} + \cos^{-1}\sqrt{1-q} = \frac{3\pi}{4}$, then the value of q is

(a) 1 (b) $\frac{1}{\sqrt{2}}$

(c) $\frac{1}{3}$ (d) $\frac{1}{2}$

45. Sum infinite terms of the series

$\cot^{-1}\left(1^2+\frac{3}{4}\right) + \cot^{-1}\left(2^2+\frac{3}{4}\right) + \cot^{-1}\left(3^2+\frac{3}{4}\right) + \ldots$ is

(a) $\frac{\pi}{4}$

(b) $\tan^{-1} 2$

(c) $\tan^{-1} 3$

(d) None of the above

46. If $2\tan^{-1} x + \sin^{-1}\left(\frac{2x}{1+x^2}\right)$ is independent of x, then

(a) $x \in [1, \infty)$

(b) $x \in [-1, 1]$

(c) $x \in (-\infty, -1]$

(d) None of the above

47. If $\cot^{-1}\left(\frac{n}{\pi}\right) > \left(\frac{\pi}{6}\right)$, $n \in N$, then the maximum value of n is

(a) 1

(b) 5

(c) 9

(d) None of the above

48. If $(\tan^{-1} x)^2 + (\cot^{-1} x)^2 = \frac{5\pi^2}{8}$, then x equals

(a) 0 (b) −1

(c) −2 (d) −3

49. If $0 < a, b < 1$ and $\tan^{-1} a + \tan^{-1} b = \frac{\pi}{4}$, then the value of $(a+b) - \left(\frac{a^2+b^2}{2}\right) + \left(\frac{a^3+b^3}{3}\right) - \left(\frac{a^4+b^4}{4}\right) + \ldots$ is

[JEE Main 2021, 26 February]

(a) $\log_e 2$ (b) $\log_e\left(\frac{e}{2}\right)$

(c) e (d) $e^2 - 1$

50. If $a_1, a_2, a_3, \ldots, a_n$ is an AP with common difference d, then

$\tan\left[\tan^{-1}\left(\frac{d}{1+a_1a_2}\right) + \tan^{-1}\left(\frac{d}{1+a_2a_3}\right) + \ldots + \tan^{-1}\left(\frac{d}{1+a_{n-1}a_n}\right)\right]$ is equal to

(a) $\frac{(n-1)d}{a_1+a_n}$ (b) $\frac{(n-1)d}{1+a_1a_n}$

(c) $\frac{nd}{1+a_1a_n}$ (d) $\frac{a_n-a_1}{a_n+a_1}$

51. If $\cot^{-1}(\alpha) = \cot^{-1} 2 + \cot^{-1} 8 + \cot^{-1} 18 + \cot^{-1} 32 + \ldots$ upto 100 terms, then α is **[JEE Main 2021, 17 March]**

(a) 1.01 (b) 1

(c) 1.02 (d) 1.03

52. If $\cos^{-1}\left(\frac{2}{3x}\right) + \cos^{-1}\left(\frac{3}{4x}\right) = \frac{\pi}{2}\left(x > \frac{3}{4}\right)$, then x is equal to **[JEE Main 2019, 9 January]**

(a) $\frac{\sqrt{145}}{10}$ (b) $\frac{\sqrt{146}}{12}$

(c) $\frac{\sqrt{145}}{12}$ (d) $\frac{\sqrt{145}}{11}$

53. If $-\frac{\pi}{2} \le x \le \frac{\pi}{2}$, then the two curves $y = \cos x$ and $y = \sin 3x$ intersect at

(a) $\left(\frac{\pi}{4}, \frac{1}{\sqrt{2}}\right)$ and $\left(\frac{\pi}{8}, \cos\frac{\pi}{8}\right)$

(b) $\left(-\frac{\pi}{4}, \frac{1}{\sqrt{2}}\right)$ and $\left(-\frac{\pi}{8}, \cos\frac{\pi}{8}\right)$

(c) $\left(\frac{\pi}{4}, -\frac{1}{\sqrt{2}}\right)$ and $\left(\frac{\pi}{8}, -\cos\frac{\pi}{8}\right)$

(d) $\left(-\frac{\pi}{4}, \frac{1}{\sqrt{2}}\right)$

54. The solution of the inequality $(\cot^{-1} x)^2 - 5\cot^{-1} x + 6 > 0$ is

(a) $(\cot 3, \cot 2)$

(b) $(-\infty, \cot 3) \cup (\cot 2, \infty)$

(c) $(\cot 2, \infty)$

(d) None of the above

Objective Questions Type II [One or more than one correct answer(s)]

In each of the questions below four choices of which are or more than one are correct. You have to select the correct answer(s) according.

1. The values of x satisfying $\sin^{-1} x + \sin^{-1} (1 - x) = \cos^{-1} x$ are

(a) 0 (b) $\frac{1}{2}$ (c) 1 (d) 2

2. If $\frac{1}{2} < |x| < 1$, then which of the following are real?

(a) $\sin^{-1} x$ (b) $\tan^{-1} x$ (c) $\sec^{-1} x$ (d) $\cos^{-1} x$

3. $\sin^{-1} x > \cos^{-1} x$ holds for

(a) all values of x (b) $x \in \left(0, \frac{1}{\sqrt{2}}\right)$

(c) $x \in \left(\frac{1}{\sqrt{2}}, 1\right)$ (d) $x = 0.75$

4. For any positive integer n, define $f_n : (0, \infty) \to R$ as $f_n(x) = \sum_{j=1}^{n} \tan^{-1}\left(\frac{1}{1 + (x + j)(x + j - 1)}\right)$ for all $x \in (0, \infty)$. (Here, the inverse trigonometric function $\tan^{-1} x$ assumes values in $\left(-\frac{\pi}{2}, \frac{\pi}{2}\right)$. Then, which of the following statement(s) is (are) TRUE? **[JEE Advanced 2018]**

(a) $\sum_{j=1}^{5} \tan^2(f_j(0)) = 55$

(b) $\sum_{j=1}^{10} (1 + f'_j(0)) \sec^2(f_j(0)) = 10$

(c) For any fixed positive integer n, $\lim_{x \to \infty} \tan(f_n(x)) = \frac{1}{n}$

(d) For any fixed positive integer n, $\lim_{x \to \infty} \sec^2(f_n(x)) = 1$

5. Let $f(x) = e^{\cos^{-1} \sin \left(x + \frac{\pi}{3}\right)}$, then

(a) $f\left(\frac{8\pi}{9}\right) = e^{\frac{5\pi}{18}}$ (b) $f\left(\frac{8\pi}{9}\right) = e^{\frac{13\pi}{18}}$

(c) $f\left(-\frac{7\pi}{4}\right) = e^{\frac{\pi}{12}}$ (d) $f\left(-\frac{7\pi}{4}\right) = e^{\frac{11\pi}{12}}$

6. If $z = \sec^{-1}\left(x + \frac{1}{x}\right) + \sec^{-1}\left(y + \frac{1}{y}\right)$, where $xy < 0$, then the possible values of z is (are)

(a) $\frac{8\pi}{10}$ (b) $\frac{7\pi}{10}$ (c) $\frac{9\pi}{10}$ (d) $\frac{11\pi}{10}$

7. The greatest and least values of $(\sin^{-1} x)^3 + (\cos^{-1} x)^3$ are

(a) $\frac{\pi^3}{32}$ (b) $-\frac{\pi^3}{8}$ (c) $\frac{7\pi^3}{8}$ (d) $\frac{\pi}{2}$

8. The value of $\tan^{-1} \sqrt{\frac{a(a + b + c)}{bc}} + \tan^{-1} \sqrt{\frac{b(a + b + c)}{ca}} + \tan^{-1} \sqrt{\frac{c(a + b + c)}{ab}}$ is equal to

(a) $\frac{\pi}{4}$ (b) $\frac{\pi}{2}$ (c) π (d) 0

9. The solution of the equation $\sin [2 \cos^{-1}\{\cot (2 \tan^{-1} x)\}] = 0$ are

(a) ± 1 (b) $1 \pm \sqrt{2}$ (c) $-1 \pm \sqrt{2}$ (d) None of these

10. α, β and γ are the angles given by $\alpha = 2 \tan^{-1}(\sqrt{2} - 1)$, $\beta = 3 \sin^{-1}\left(\frac{1}{\sqrt{2}}\right) + \sin^{-1}\left(-\frac{1}{2}\right)$ and $\gamma = \cos^{-1}\left(\frac{1}{3}\right)$, then

(a) $\alpha > \beta$ (b) $\beta > \gamma$ (c) $\gamma > \alpha$ (d) None of these

11. If $\alpha = 3 \sin^{-1}\left(\frac{6}{11}\right)$ and $\beta = 3 \cos^{-1}\left(\frac{4}{9}\right)$, where the inverse trigonometric functions take only the principal values, then the correct option(s) is/are **[JEE Advanced 2015]**

(a) $\cos \beta > 0$ (b) $\sin \beta < 0$ (c) $\cos (\alpha + \beta) > 0$ (d) $\cos \alpha < 0$

12. $\cos^{-1}\left(\sqrt{\frac{a - x}{a - b}}\right) = \sin^{-1}\left(\sqrt{\frac{x - b}{a - b}}\right)$ is possible, if

(a) $a > x > b$
(b) $a < x < b$
(c) $a = x = b$
(d) $a > b$ and x, takes any value

13. $\theta = \tan^{-1}(2 \tan^2 \theta) - \tan^{-1}\left\{\left(\frac{1}{3}\right) \tan \theta\right\}$, if

(a) $\tan \theta = -2$
(b) $\tan \theta = 0$
(c) $\tan \theta = 1$
(d) $\tan \theta = 2$

14. If the numerical value of $\tan \left\{\cos^{-1}\left(\frac{4}{5}\right) + \tan^{-1}\left(\frac{2}{3}\right)\right\}$ is $\frac{a}{b}$, then

(a) $a + b = 23$ (b) $a - b = 11$ (c) $3b = a + 1$ (d) $2a = 3b$

15. $2 \tan^{-1} (-3)$ is equal to

(a) $-\cos^{-1}\left(-\frac{4}{5}\right)$ (b) $-\pi + \cos^{-1}\left(\frac{4}{5}\right)$

(c) $-\frac{\pi}{2} + \tan^{-1}\left(-\frac{4}{3}\right)$ (d) $\cot^{-1}\left(\frac{4}{3}\right)$

16. $2\cot^{-1}7 + \cos^{-1}\left(\frac{3}{5}\right)$ is equal to

(a) $\cot^{-1}\left(\frac{44}{117}\right)$

(b) $\operatorname{cosec}^{-1}\left(\frac{125}{117}\right)$

(c) $\tan^{-1}\left(\frac{4}{117}\right)$

(d) $\cos^{-1}\left(\frac{44}{125}\right)$

17. If the equation $\sin^{-1}(x^2 + x + 1) + \cos^{-1}(\lambda x + 1) = \frac{\pi}{2}$ has exactly two solutions, then λ cannot have the integral value

(a) -1 (b) 0
(c) 1 (d) 2

18. The value(s) of x satisfying the equation $\sin^{-1}|\sin x| = \sqrt{\sin^{-1}|\sin x|}$ is/are given by (n is any integer)

(a) $n\pi - 1$ (b) $n\pi$
(c) $n\pi + 1$ (d) $2n\pi + 1$

19. If $\tan^{-1} y = 4\tan^{-1} x$, then y is infinite, if

(a) $x^2 = 3 + 2\sqrt{2}$
(b) $x^2 = 3 - 2\sqrt{2}$
(c) $x^4 = 6x^2 - 1$
(d) $x^4 = 6x^2 + 1$

20. If $\cos^{-1} x = \tan^{-1} x$, then

(a) $x^2 = (\sqrt{5} - 1)/2$
(b) $x^2 = (\sqrt{5} + 1)/2$
(c) $\sin(\cos^{-1} x) = (\sqrt{5} - 1)/2$
(d) $\tan(\cos^{-1} x) = (\sqrt{5} - 1)/2$

Linked-Comprehension Type

In these questions, passage (paragraph) has been given followed by questions based on each of the passages. You have to answer the questions based on the passage given.

PASSAGE 1

Let $f : A \to B$ be a function defined by $y = f(x)$ such that f is both one-one (Injective) and onto (surjective) (*ie*, bijective), then there exists a unique function $g : B \to A$ such that

$f(x) = y \Leftrightarrow g(y) = x, \forall\, x \in A$ and $y \in B$, then g is said to be inverse of f. Thus, $g = f^{-1} : B \to A = [\{f(x), x\} : \{x, f(x)\} \in f^{-1}]$.

If no branch of an inverse trigonometric function is mentioned, then it means the principal value branch of that function.

On the basis of above information, answer the following questions

1. The value of $\cos(\tan^{-1}\tan 4)$ is

(a) $\frac{1}{\sqrt{17}}$ (b) $-\frac{1}{\sqrt{17}}$
(c) $\cos 4$ (d) $-\cos 4$

2. If x takes negative permissible value, then $\sin^{-1} x$ is equal to

(a) $\cos^{-1}\sqrt{(1 - x^2)}$ (b) $\cot^{-1}\left(\frac{\sqrt{1 - x^2}}{x}\right)$

(c) $\pi - \cos^{-1}\sqrt{(1 - x^2)}$ (d) $-\pi + \cot^{-1}\left(\frac{\sqrt{1 - x^2}}{x}\right)$

3. If $\frac{3\pi}{2} \le x \le \frac{5\pi}{2}$, then $\sin^{-1}(\sin x)$ is equal to

(a) x (b) $-x$
(c) $2\pi - x$ (d) $x - 2\pi$

4. If $x > 1$, then the value of $2\tan^{-1} x + \sin^{-1}\left(\frac{2x}{1 + x^2}\right)$ is

(a) $\frac{\pi}{4}$ (b) $\frac{\pi}{2}$
(c) π (d) $\frac{3\pi}{2}$

5. If $0 \le x \le 1$, then the least and greatest values of $\tan^{-1}\left(\frac{1 + x}{1 - x}\right)$ are

(a) $-\frac{\pi}{4}, \frac{\pi}{4}$ (b) $0, \frac{\pi}{4}$
(c) $\frac{\pi}{4}, \frac{\pi}{2}$ (d) $0, \pi$

PASSAGE 2

$$\sum_{r=1}^{n} \tan^{-1}\left(\frac{x_r - x_{r-1}}{1 + x_{r-1}x_r}\right) = \sum_{r=1}^{n} (\tan^{-1} x_r - \tan^{-1} x_{r-1})$$
$$= \tan^{-1} x_n - \tan^{-1} x_0, \forall\, n \in N$$

On the basis of above information, answer the following questions

1. The sum to infinite terms of the series $\tan^{-1}\left(\frac{1}{3}\right) + \tan^{-1}\left(\frac{2}{9}\right) + \ldots + \tan^{-1}\left(\frac{2^{n-1}}{1 + 2^{2n-1}}\right) + \ldots$ is

(a) $\frac{\pi}{4}$ (b) $\frac{\pi}{2}$
(c) π
(d) None of the above

2. The value of
$\text{cosec}^{-1}\sqrt{5} + \text{cosec}^{-1}\sqrt{65} + \text{cosec}^{-1}\sqrt{(325)} + \dots$ to ∞ is

(a) π (b) $\frac{3\pi}{4}$

(c) $\frac{\pi}{2}$ (d) $\frac{\pi}{4}$

3. The sum to infinite terms of the series
$\cot^{-1}\left(2^2 + \frac{1}{2}\right) + \cot^{-1}\left(2^3 + \frac{1}{2^2}\right) + \cot^{-1}\left(2^4 + \frac{1}{2^3}\right) + \dots$ is

(a) $\frac{\pi}{4}$ (b) $\frac{\pi}{2}$

(c) $\cot^{-1} 2$ (d) $-\cot^{-1} 2$

4. The sum to infinite terms of the series
$\cot^{-1} 3 + \cot^{-1} 7 + \cot^{-1} 13 + \dots$ is

(a) $\frac{\pi}{2}$ (b) $\cot^{-1} 2$

(c) $\tan^{-1} 2$ (d) None of these

5. The sum to infinite terms of the series
$$\tan^{-1}\left(\frac{2}{1 - 1^2 + 1^4}\right) + \tan^{-1}\left(\frac{4}{1 - 2^2 + 2^4}\right) + \tan^{-1}\left(\frac{6}{1 - 3^2 + 3^4}\right) + \dots \text{ is}$$

(a) $\frac{\pi}{4}$ (b) $\frac{\pi}{2}$

(c) $\frac{3\pi}{4}$ (d) None of these

PASSAGE 3

Principal values for inverse circular functions :

$x < 0$	$x \geq 0$
$-\frac{\pi}{2} \leq \sin^{-1} x < 0$	$0 \leq \sin^{-1} x \leq \frac{\pi}{2}$
$\frac{\pi}{2} < \cos^{-1} x \leq \pi$	$0 \leq \cos^{-1} x \leq \frac{\pi}{2}$
$-\frac{\pi}{2} < \tan^{-1} x < 0$	$0 \leq \tan^{-1} x < \frac{\pi}{2}$
$\frac{\pi}{2} < \cot^{-1} x < \pi$	$0 < \cot^{-1} x \leq \frac{\pi}{2}$
$\frac{\pi}{2} < \sec^{-1} x \leq \pi$	$0 \leq \sec^{-1} x < \frac{\pi}{2}$
$-\frac{\pi}{2} \leq \text{cosec}^{-1} x < 0$	$0 < \text{cosec}^{-1} x \leq \frac{\pi}{2}$
$\frac{\pi}{2} < \sec^{-1} x \leq \pi$	$0 \leq \sec^{-1} x < \frac{\pi}{2}$
$-\frac{\pi}{2} \leq \text{cosec}^{-1} x < 0$	$0 < \text{cosec}^{-1} x \leq \frac{\pi}{2}$

Ex. $\sin^{-1}\left(\frac{\sqrt{3}}{2}\right) = \frac{\pi}{3}$ not $\frac{2\pi}{3}$, $\tan^{-1}(-\sqrt{3}) = -\frac{\pi}{3}$ not $\frac{2\pi}{3}$

On the basis of above information, answer the following questions

1. The principal value of $\sin^{-1}\left(\sin\frac{4\pi}{3}\right) + \cos^{-1}\left(\cos\frac{4\pi}{3}\right)$ is

(a) $\frac{8\pi}{3}$ (b) $\frac{4\pi}{3}$

(c) $\frac{2\pi}{3}$ (d) $\frac{\pi}{3}$

2. The principal value of $\sin^{-1}(\sin 5) - \cos^{-1}(\cos 5)$ is

(a) 0 (b) $2\pi - 10$

(c) $-\pi$ (d) $3\pi - 10$

3. The principal value of
$\tan^{-1}\left(\tan\left(-\frac{3\pi}{4}\right)\right) + \cot^{-1}\cot\left(-\frac{3\pi}{4}\right)$ is

(a) $\frac{\pi}{2}$ (b) π

(c) $\frac{-3\pi}{2}$ (d) 0

4. The value of $\sin^{-1}[\cos\{\cos^{-1}(\cos x) + \sin^{-1}(\sin x)\}]$, where $x \in \left(\frac{\pi}{2}, \pi\right)$ is

(a) $\frac{\pi}{2}$ (b) $-\pi$

(c) π (d) $-\frac{\pi}{2}$

5. The number of solutions of the equation
$\cos^{-1}\left(\frac{x^2 - 1}{x^2 + 1}\right) + \sin^{-1}\left(\frac{2x}{x^2 + 1}\right) + \tan^{-1}\left(\frac{2x}{x^2 - 1}\right) = \frac{2\pi}{3}$ is

(a) 1 (b) 2

(c) 3 (d) infinite

Numerical Value Type

Solve the following problems and write your answer.

1. If $\sin^{-1} x + \sin^{-1} y + \sin^{-1} z = \frac{3\pi}{2}$, then the value of $2500(x + y + z) - \frac{216}{(x^3 + y^3 + z^3)}$ must be

2. The value of $\frac{1}{\pi}\left\{216 \sin^{-1}\left(\sin \frac{7\pi}{6}\right) + 27 \cos^{-1}\left(\cos \frac{2\pi}{3}\right) + 28 \tan^{-1}\left(\tan \frac{5\pi}{4}\right) + 200 \cot^{-1}\left(\cot\left(\frac{-\pi}{4}\right)\right)\right\}$ must be

3. If $S = \sum_{r=1}^{50} \tan^{-1}\left(\frac{2r}{2 + r^2 + r^4}\right)$, then the value of 2550 cot S must be

4. If $\tan^{-1}(x + 1) + \tan^{-1} x + \tan^{-1}(x - 1) = \tan^{-1} 3$, then the value of (for $x < 0$) $500x^4 + 270x^2 + 997$ must be

5. If $\lambda = \cos^4[\tan^{-1}\{\sin(\cot^{-1} 5)\}]$, then the value of 3645 λ must be

6. If $\theta = \cot^{-1} 7 + \cot^{-1} 8 + \cot^{-1} 18$, then the value of $\tan 2\theta$ must be

7. If $\cos^{-1} x - \cos^{-1}\left(\frac{y}{2}\right) = \alpha$, then $400(4x^2 - 4xy \cos\alpha + y^2) + 328 \sin^2 \alpha$ is equal to $a \sin^2 \alpha$ where $a =$

8. If $\sin^{-1} x + \sin^{-1} y = \pi$ and, if $x = \lambda y$, then the value of $39^{2\lambda} + 5^{\lambda}$ must be

Single Integer Type

Solve the following problems and write your answer.

1. If $\tan^{-1}\frac{1}{5} + \frac{1}{2}\sec^{-1} x + \tan^{-1}\frac{1}{8} = \frac{\pi}{8}$, then $\frac{49}{25}x^2 =$

2. The value of $\sec^{-1}\left(\frac{1}{4}\sum_{k=0}^{10} \sec\left(\frac{7\pi}{12} + \frac{k\pi}{2}\right) \sec\left(\frac{7\pi}{12} + \frac{(k+1)\pi}{2}\right)\right)$ in the interval $\left[-\frac{\pi}{4}, \frac{3\pi}{4}\right]$ equals .... **[JEE Advanced 2019]**

3. If in the ΔABC, $\angle C = \frac{\pi}{2}$ and $\sin^{-1} x^1 = \sin^{-1}\left(\frac{ax}{c}\right) + \sin^{-1}\left(\frac{bx}{c}\right)$, where a, b, c are the sides of triangle, then total number of different values of x are

4. The number of real solutions of the equation $\sin^{-1}\left(\sum_{i=1}^{\infty} x^{i+1} - x\sum_{i=1}^{\infty}\left(\frac{x}{2}\right)^i\right) = \frac{\pi}{2} - \cos^{-1}\left(\sum_{i=1}^{\infty}\left(-\frac{x}{2}\right)^i - \sum_{i=1}^{\infty}(-x)^i\right)$ lying in the interval $\left(-\frac{1}{2}, \frac{1}{2}\right)$ is ....

(Here, the inverse trigonometric functions $\sin^{-1} x$ and $\cos^{-1} x$ assume values in $\left[-\frac{\pi}{2}, \frac{\pi}{2}\right]$ and $[0, \pi]$, respectively.) **[JEE Advanced 2018]**

5. The number of solutions of $\cos(2\sin^{-1}(\cot(\tan^{-1}(\sec(6 \operatorname{cosec}^{-1} x))))) + 1 = 0$, where $x > 0$ is ......

6. If $\theta = \tan^{-1}(2\tan^2\theta) - \tan^{-1}\left(\left(\frac{1}{3}\right)\tan\theta\right)$, then positive value of $\tan\theta$ is equal to

7. If range of the function $f(x) = \sin^{-1} x + 2\tan^{-1} x + x^2 + 4x + 1$ is $[p, q]$, then the value of $(p + q)$ is ....

Matrix-Match Type

Given below are Matching Type Questions, with two columns (each having some items) each . Each item of Column I has to be matched with the items of Column II, by encircling the correct match(es).

NOTE *An item of Column I can be matched with more than one items of Column II.*

1. Observe the following columns

	Column I		Column II
(A)	If principal values of $\sin^{-1}\left(-\frac{1}{2}\right)+\tan^{-1}(\sqrt{3})$ and $\cos^{-1}\left(-\frac{1}{2}\right)$ are λ and μ respectively, then	(P)	$\lambda+\mu=\frac{\pi}{2}$
		(Q)	$\mu-\lambda=\frac{\pi}{2}$
(B)	If Principal values of $\sin^{-1}\left(\sin\frac{7\pi}{6}\right)$ and $\cos^{-1}\left\{-\sin\left(\frac{5\pi}{6}\right)\right\}$ are λ and μ respectively, then	(R)	$\lambda+\mu=-\frac{\pi}{6}$
		(S)	$\mu-\lambda=\frac{5\pi}{6}$
(C)	If principal values of $\sin^{-1}\left(-\frac{\sqrt{3}}{2}\right)$ and $\sin^{-1}\left\{\cos\left(\sin^{-1}\frac{\sqrt{3}}{2}\right)\right\}$ are λ and μ respectively, then	(T)	$\lambda+\mu=\frac{5\pi}{6}$

2. Observe the following columns

	Column I		Column II
(A)	If $\tan^{-1}x+\tan^{-1}y+\tan^{-1}z=\pi$, then	(P)	$x=y=z$
		(Q)	$xyz\geq 3\sqrt{3}$
(B)	If $\tan^{-1}x+\tan^{-1}y+\tan^{-1}z=\frac{\pi}{2}$, then	(R)	$x+y+z=xyz$
(C)	If $\tan^{-1}x+\tan^{-1}y+\tan^{-1}z=\frac{\pi}{2}$ and $x+y+z=\sqrt{3}$, then	(S)	$xyz\leq\frac{1}{3\sqrt{3}}$
		(T)	$xy+yz+zx=1$

Answers

Objective Questions Type I [Only one correct answer]

1. (b)	2. (b)	3. (b)	4. (c)	5. (c)	6. (c)	7. (d)	8. (c)	9. (a)	10. (c)
11. (b)	12. (b)	13. (a)	14. (a)	15. (b)	16. (c)	17. (b)	18. (b)	19. (c)	20. (c)
21. (a)	22. (c)	23. (a)	24. (b)	25. (b)	26. (c)	27. (b)	28. (b)	29. (b)	30. (c)
31. (c)	32. (a)	33. (c)	34. (c)	35. (c)	36. (b)	37. (c)	38. (b)	39. (b)	40. (c)
41. (b)	42. (b)	43. (d)	44. (d)	45. (b)	46. (a)	47. (b)	48. (b)	49. (a)	50. (b)
51. (a)	52. (c)	53. (a)	54. (b)						

Objective Questions Type II [One or more than one correct answer(s)]

1. (a, b)	2. (a, b, d)	3. (c, d)	4. (d)	5. (b, c)
6. (c, d)	7. (a, c)	8. (c, d)	9. (a, b, c)	10. (b, c)
11. (b, c, d)	12. (a, b)	13. (a, b, c)	14. (a, b, c)	15. (a, b, c)
16. (a, b, d)	17. (a, c, d)	18. (a, b, c)	19. (a, b, c)	20. (a, c)

Linked-Comprehension Type

Passage 1 1. (d) 2. (d) 3. (d) 4. (c) 5. (c)

Passage 2 1. (a) 2. (d) 3. (c) 4. (d) 5. (b)

Passage 3 1. (d) 2. (c) 3. (a) 4. (d) 5. (b)

Numerical Value Type

1. (7428) 2. (139) 3. (2552) 4. (1767) 5. (3380) 6. (0.75)
7. (1928) 8. (1526)

Single Integer Type

1. (2) 2. (0) 3. (3) 4. (2) 5. (3) 6. (1) 7. (4)

Matrix-Match Type

1. A →(Q, T) B →(P, S) C → (Q, R)

2. A → (Q, R); B → (S, T); C → (P, S, T)

Solutions

Objective Questions Type [Only one correct answer]

1. $\because x \in \left(\frac{3\pi}{2}, 2\pi\right)$

Then, $\cos^{-1}(\cos x) = 2\pi - x$

and $\sin^{-1}(\sin x) = x - 2\pi$

$\therefore \cos^{-1}(\cos x) + \sin^{-1}(\sin x) = 0$

Therefore, $\sin^{-1}[\cos\{\cos^{-1}(\cos x) + \sin^{-1}(\sin x)\}]$

$$= \sin^{-1}\{\cos(0)\} = \sin^{-1}(1) = \frac{\pi}{2}$$

2. $\because$ $T_r = \tan^{-1}\left(\frac{2^{r-1}}{1 + 2^{2r-1}}\right)$

$$= \tan^{-1}\left(\frac{2^r - 2^{r-1}}{1 + 2^r \cdot 2^{r-1}}\right)$$

$$= \tan^{-1}(2^r) - \tan^{-1}(2^{r-1})$$

$\therefore S_n = \sum_{r=1}^{n} T_r = \sum_{r=1}^{n} \tan^{-1}(2^r) - \tan^{-1}(2^{r-1})$

$$= \tan^{-1}(2^n) - \tan^{-1}(2^0)$$

$$= \tan^{-1}(2^n) - \tan^{-1}(1)$$

$$= \tan^{-1}(2^n) - \frac{\pi}{4}$$

Hence, $S_\infty = \tan^{-1}(2^\infty) - \frac{\pi}{4}$

$$= \tan^{-1}(\infty) - \frac{\pi}{4}$$

$$= \frac{\pi}{2} - \frac{\pi}{4} = \frac{\pi}{4}$$

3. $\because \sin^{-1}\sqrt{(x^2 + 2x + 1)} + \sec^{-1}\sqrt{(x^2 + 2x + 1)} = \frac{\pi}{2}$

or $\sin^{-1}|x + 1| + \cos^{-1}\left(\frac{1}{|x + 1|}\right) = \frac{\pi}{2}$

or $|x + 1| = \frac{1}{|x + 1|} \ (x \neq -1)$

$\Rightarrow |x + 1|^2 = 1$

$\Rightarrow |x + 1| = 1$

$\Rightarrow x + 1 = \pm 1$

or $x = 0, -2$

But given $x \neq 0$

$\therefore x = -2$

Then, $2\sec^{-1}\left(\frac{x}{2}\right) + \sin^{-1}\left(\frac{x}{2}\right)$

$$= 2\sec^{-1}(-1) + \sin^{-1}(-1) = 2\pi - \frac{\pi}{2} = \frac{3\pi}{2}$$

4. $[x] + [-x] \geq 0 \Rightarrow x \in I$

Also, $2 - |x| \geq 0 \Rightarrow x \in [-2, 2]$

$x \neq 0$

$\therefore$ Domain of $f(x) = \{-2, -1, 1, 2\}$

$$f(x) = \sqrt{2 - |x|} + \frac{1}{x^2}$$

$\because [x] + [-x] = 0 \ \forall \ x \in I$

$\therefore$ Range $= \left\{\frac{1}{4}, 2\right\}$

5. Here, $x^2 - 2x + 2 = (x - 1)^2 + 1 \geq 1$

But $-1 \leq (x^2 - 2x + 2) \leq 1$

Which is possible only when

$$x^2 - 2x + 2 = 1$$

$\therefore x = 1$

Then, $a(1)^2 + \sin^{-1}(1) + \cos^{-1}(1) = 0$

$\Rightarrow a + \frac{\pi}{2} + 0 = 0$

$\therefore a = -\frac{\pi}{2}$

6. $\because -\frac{\pi}{2} \leq \sin^{-1} x \leq \frac{\pi}{2}, -\frac{\pi}{2} \leq \sin^{-1} y \leq \frac{\pi}{2}$

and $-\frac{\pi}{2} \leq \sin^{-1} z \leq \frac{\pi}{2}$

Given $\sin^{-1} x + \sin^{-1} y + \sin^{-1} z = \frac{3\pi}{2}$

Which is possible only when

$$\sin^{-1} x = \sin^{-1} y = \sin^{-1} z = \frac{\pi}{2}$$

or $x = y = z = 1$

Put $p = q = 1$

Then, $f(2) = f(1) f(1) = 2 \cdot 2 = 4$

and put $p = 1, q = 2$

Then, $f(3) = f(1) f(2) = 2 \cdot 2^2 = 8$

$\therefore x^{f(1)} + y^{f(2)} + z^{f(3)} - \frac{x + y + z}{x^{f(1)} + y^{f(2)} + z^{f(3)}}$

$$= 1 + 1 + 1 - \frac{3}{1 + 1 + 1} = 3 - 1 = 2$$

7. $\because f(x) = ax + b$

$\therefore f'(x) = a > 0$

$\therefore f(x)$ is an increasing function.

$\Rightarrow f(-1) = 0$ and $f(1) = 2$

or $-a + b = 0$

and $a + b = 2$

Then, $a = b = 1$

$\Rightarrow f(x) = x + 1$

Now, $\cot[\cot^{-1} 7 + \cot^{-1} 8 + \cot^{-1} 18]$

$$= \cot\left\{\tan^{-1}\left(\frac{1}{7}\right) + \tan^{-1}\left(\frac{1}{8}\right) + \tan^{-1}\left(\frac{1}{18}\right)\right\}$$

$$= \cot\left\{\tan^{-1}\left(\frac{\frac{1}{7} + \frac{1}{8}}{1 - \frac{1}{7} \cdot \frac{1}{8}}\right) + \tan^{-1}\left(\frac{1}{18}\right)\right\}$$

$$= \cot \left\{\tan^{-1}\left(\frac{15}{55}\right) + \tan^{-1}\left(\frac{1}{18}\right)\right\}$$

$$= \cot \left\{\tan^{-1}\left(\frac{3}{11}\right) + \tan^{-1}\left(\frac{1}{18}\right)\right\}$$

$$= \cot \left\{\tan^{-1}\left(\frac{\frac{3}{11} + \frac{1}{18}}{1 - \frac{3}{1 \cdot 1} \cdot \frac{1}{18}}\right)\right\}$$

$$= \cot \left\{\tan^{-1}\left(\frac{65}{195}\right)\right\} = \cot \left\{\tan^{-1}\left(\frac{1}{3}\right)\right\}$$

$$= \cot(\cot^{-1} 3) = 3 = 1 + 2 = f(2)$$

8. $\because$ $$T_r = \sin^{-1}\left(\frac{\sqrt{r} - \sqrt{(r-1)}}{\sqrt{r(r+1)}}\right)$$

$$= \tan^{-1}\left(\frac{\sqrt{r} - \sqrt{(r-1)}}{1 + \sqrt{r}\sqrt{(r-1)}}\right)$$

$\therefore$ $$S_n = \sum_{r=1}^{n} \tan^{-1}\left(\frac{\sqrt{r} - \sqrt{r-1}}{1 + \sqrt{r}\sqrt{r-1}}\right)$$

$$= \sum_{r=1}^{n} \tan^{-1}\sqrt{r} - \tan^{-1}\sqrt{(r-1)}$$

$$= \tan^{-1}\sqrt{n} - \tan^{-1}\sqrt{0}$$

$$= \tan^{-1}\sqrt{n} - 0$$

$\therefore$ $$S_\infty = \tan^{-1}\infty = \frac{\pi}{2}$$

9. We have, $1 \le \sin^{-1}\cos^{-1}\sin^{-1}\tan^{-1} x \le \frac{\pi}{2}$

$\Rightarrow$ $\sin 1 \le \cos^{-1}\sin^{-1}\tan^{-1} x \le 1$

$\Rightarrow$ $\cos \sin 1 \ge \sin^{-1} \tan^{-1} x \ge \cos 1$

$\Rightarrow$ $\sin \cos \sin 1 \ge \tan^{-1} x \ge \sin \cos 1$

$\Rightarrow$ $\tan \sin \cos \sin 1 \ge x \ge \tan \sin \cos 1$

$\therefore$ $x \in [\tan \sin \cos 1, \tan \sin \cos \sin 1]$

10. $\because$ $[\sin^{-1} x] > [\cos^{-1} x]$

$\Rightarrow$ $x > 0$

It is clear from the figure

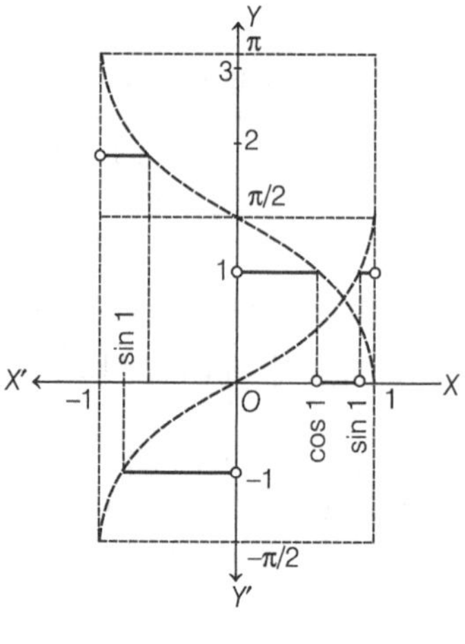

$$[\cos^{-1} x] = \begin{cases} 0, & x \in (\cos 1, 1) \\ 1, & x \in (0, \cos 1) \end{cases}$$

and $$[\sin^{-1} x] = \begin{cases} 0, & x \in [0, \sin 1) \\ 1, & x \in [\sin 1, 1] \end{cases}$$

$\because$ $[\sin^{-1} x] > [\cos^{-1} x]$

$\therefore$ $x \in [\sin 1, 1)$

$\because$ $\left[\frac{\pi}{2}\right] = 1$

or we say that $x \in [\sin 1, 1]$

11. $\because$ $0 \le \cos^{-1} x \le \pi$

and $0 < \cot^{-1} x < \pi$

$$[\cot^{-1} x] + [\cos^{-1} x] = 0$$

$\Rightarrow$ $[\cot^{-1} x] = 0$ and $[\cos^{-1} x] = 0$

$\Rightarrow$ $0 < \cot^{-1} x < 1$ and $0 \le \cos^{-1} x < 1$

$\therefore$ $x \in (\cot 1, \infty)$ and $x \in (\cos 1, 1)$

$\Rightarrow$ $x \in (\cot 1, 1)$

12. $\because x^2 - x + 1 = \left(x - \frac{1}{2}\right)^2 + \frac{3}{4} > 0$

For $\sin^{-1}\sqrt{(x^2 - x + 1)}$

$\Rightarrow$ $0 < x^2 - x + 1 \le 1$

or $x^2 - x \le 0$ ...(i)

For $\cos^{-1}\sqrt{(x^2 - x)}$

$\Rightarrow$ $x^2 - x \ge 0$ ...(ii)

From Eqs. (i) and (ii), we get

$$x^2 - x = 0$$

$\therefore$ $x = 0, 1$

13. For $\cos^{-1}(1 - x)$

$\Rightarrow$ $-1 \le 1 - x \le 1$

$\Rightarrow$ $1 \ge -1 + x \ge -1$

or $2 \ge x \ge 0$

or $x \in [0, 2]$ ...(i)

For $\cos^{-1} x$

$\Rightarrow$ $-1 \le x \le 1$ ...(ii)

From Eqs. (i) and (ii), we get

$$0 \le x \le 1$$

$\therefore$ LHS $= \cos^{-1}(1 - x) + m \cos^{-1} x$ is defined in $[0, 1]$

So, LHS ≥ 0, but RHS ≤ 0 $(\because x \le 0)$

Equality holds, if

LHS $= 0$ and RHS $= 0$

$\therefore$ $\cos^{-1}(1 - x) + m \cos^{-1} x = 0$

$\Rightarrow$ $\cos^{-1}(1 - x) = 0$ and $\cos^{-1} x = 0$

Which is impossible

i.e., no solution.

Hence, number of solution = 0.

14. $\sin [\cot^{-1}\{\cos (\tan^{-1} x)\}]$

$$= \sin \{\cot^{-1}(\cos \theta)\}$$

where $\theta = \tan^{-1} x$ or $x = \tan \theta$

$$= \sin\left\{\cot^{-1}\left(\frac{1}{\sqrt{(1+x^2)}}\right)\right\}$$

$$= \sin(\tan^{-1}\sqrt{(1+x^2)})$$

$$= \sin\phi, \text{ where } \phi = \tan^{-1}\sqrt{(1+x^2)}$$

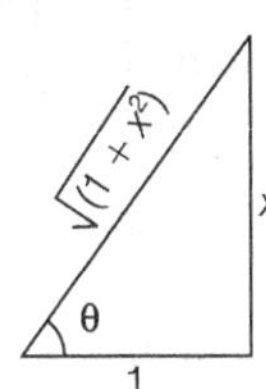

or $\quad \tan\phi = \sqrt{(1+x^2)}$

$$= \sqrt{\left(\frac{1+x^2}{2+x^2}\right)}$$

15. $\because \cos^{-1}\left(\frac{x}{3}\right) + \cos^{-1}\left(\frac{y}{2}\right) = \frac{\theta}{2}$

$$\Rightarrow \cos\left\{\cos^{-1}\left(\frac{x}{3}\right) + \cos^{-1}\left(\frac{y}{2}\right)\right\} = \cos\left(\frac{\theta}{2}\right)$$

$$\Rightarrow \cos\left\{\cos^{-1}\left(\frac{x}{3}\right)\right\}\cdot\cos\left\{\cos^{-1}\left(\frac{y}{2}\right)\right\}$$

$$-\sin\left\{\cos^{-1}\left(\frac{x}{3}\right)\right\}\cdot\sin\left\{\cos^{-1}\left(\frac{y}{2}\right)\right\} = \cos\frac{\theta}{2}$$

$$\Rightarrow \frac{x}{3}\cdot\frac{y}{2} - \sqrt{\left(1-\frac{x}{3}\right)^2}\cdot\sqrt{1-\left(\frac{y}{2}\right)^2} = \left(\cos\frac{\theta}{2}\right)$$

$$\Rightarrow \left(\frac{xy}{6} - \cos\frac{\theta}{2}\right)^2 = 1 - \frac{x^2}{9} - \frac{y^2}{4} + \frac{x^2y^2}{36}$$

$$\Rightarrow \frac{x^2y^2}{36} + \cos^2\left(\frac{\theta}{2}\right) - \frac{2xy}{6}\cos\left(\frac{\theta}{2}\right)$$

$$= 1 - \frac{x^2}{9} - \frac{y^2}{4} + \frac{x^2y^2}{36}$$

$$\Rightarrow 4x^2 - 12xy\cos\left(\frac{\theta}{2}\right) + 9y^2$$

$$= 36\left(1 - \cos^2\frac{\theta}{2}\right)$$

$$= 36\left(1 - \frac{1+\cos\theta}{2}\right) = 18(1-\cos\theta)$$

16. Given function, $f(x) = \sin^{-1}\left(\frac{|x|+5}{x^2+1}\right)$

For, domain of 'f'

$$\frac{|x|+5}{x^2+1} \le 1, \text{ as } \frac{|x|+5}{x^2+1} > 0\ \forall\ x \in R$$

$$\Rightarrow |x| + 5 \le x^2 + 1 \text{ as } x^2 + 1 \text{ is positive } \forall\ x \in R$$

$$\Rightarrow \quad x^2 - |x| - 4 \ge 0$$

$$\Rightarrow |x|^2 - |x| - 4 \ge 0 \qquad [\because \forall\ x \in R,\ x^2 = |x|^2] \quad \ldots(\text{i})$$

$\because$ For $|x|^2 - |x| - 4 = 0 \Rightarrow |x| = \frac{1 \pm \sqrt{1+16}}{2}$

$$\Rightarrow |x| = \frac{1 \pm \sqrt{17}}{2} \Rightarrow |x| = \frac{1+\sqrt{17}}{2} \text{ as } |x| \ge 0$$

$$\therefore \quad |x| \ge \frac{\sqrt{17}+1}{2}$$

$$\Rightarrow \quad x \in \left(-\infty, -\frac{\sqrt{17}+1}{2}\right] \cup \left[\frac{\sqrt{17}+1}{2}, \infty\right)$$

$$\therefore \quad a = \frac{\sqrt{17}+1}{2}$$

17. Given equation

$$\sin^{-1}\left[x^2 + \frac{1}{3}\right] + \cos^{-1}\left[x^2 - \frac{2}{3}\right] = x^2$$

Now, $\sin^{-1}\left[x^2 + \frac{1}{3}\right]$ is defined if

$$-1 \le x^2 + \frac{1}{3} < 2 \Rightarrow -\frac{4}{3} \le x^2 < \frac{5}{3}$$

$$\Rightarrow \quad 0 \le x^2 \le \frac{5}{3} \qquad \ldots(\text{i})$$

and $\cos^{-1}\left[x^2 - \frac{2}{3}\right]$ is defined if

$$-1 \le x^2 - \frac{2}{3} < 2 \Rightarrow -\frac{1}{3} \le x^2 < \frac{8}{3}$$

$$\Rightarrow \quad 0 \le x^2 < \frac{8}{3} \qquad \ldots(\text{ii})$$

So, from Eqs. (i) and (ii), we conclude the

$$x^2 \in \left[0, \frac{5}{3}\right) \text{ or } 0 \le x^2 < \frac{5}{3}$$

Case I If $0 \le x^2 < \frac{2}{3}$

$$\sin^{-1}(0) + \cos^{-1}(-1) = x^2$$

$$\Rightarrow \quad 0 + \pi = x^2 \Rightarrow x^2 = \pi$$

but $\quad \pi \notin \left[0, \frac{2}{3}\right)$

No value of x

Case II If $\frac{2}{3} \le x^2 < \frac{5}{3}$

$$\sin^{-1}(1) + \cos^{-1}(0) = x^2$$

$$\frac{\pi}{2} + \frac{\pi^2}{2}x^2 \Rightarrow x^2 = \pi$$

but $\quad \pi \notin \left[\frac{2}{3}, \frac{5}{3}\right]$

No value of x

So, number of solution of the equation is zero.

18. $\cos^{-1}x + \cos^{-1}y = \frac{\pi}{2} - \sin^{-1}x + \frac{\pi}{2} - \sin^{-1}y$

$$= \pi - (\sin^{-1}x + \sin^{-1}y)$$

$$= \pi - \frac{2\pi}{3} \qquad \text{(given)}$$

$$= \frac{\pi}{3}$$

19. We have, $\frac{\alpha^3}{2}\operatorname{cosec}^2\left(\frac{1}{2}\tan^{-1}\frac{\alpha}{\beta}\right)+\frac{\beta^3}{2}\sec^2\left(\frac{1}{2}\tan^{-1}\frac{\beta}{\alpha}\right)$

$$=\alpha^3\frac{1}{1-\cos\left(\tan^{-1}\left(\frac{\alpha}{\beta}\right)\right)}+\beta^3\frac{1}{1+\cos\left(\tan^{-1}\frac{\beta}{\alpha}\right)}$$

$$=\alpha^3\frac{1}{1-\cos\left(\cos^{-1}\left(\frac{\beta}{\sqrt{\alpha^2+\beta^2}}\right)\right)}+\beta^3\frac{1}{1+\cos\left(\cos^{-1}\frac{\alpha}{\sqrt{\alpha^2+\beta^2}}\right)}$$

$$=\alpha^3\frac{1}{1-\frac{\beta}{\sqrt{\alpha^2+\beta^2}}}+\beta^3\frac{1}{1+\frac{\alpha}{\sqrt{\alpha^2+\beta^2}}}$$

$$=\sqrt{\alpha^2+\beta^2}\left(\frac{\alpha^3}{\sqrt{\alpha^2+\beta^2}-\beta}+\frac{\beta^3}{\sqrt{\alpha^2+\beta^2}+\alpha}\right)$$

$$=\sqrt{\alpha^2+\beta^2}\left(\alpha^3\frac{(\sqrt{\alpha^2+\beta^2}+\beta)}{\alpha^2}+\beta^3\frac{(\sqrt{\alpha^2+\beta^2}-\alpha)}{\beta^2}\right)$$

$$=\sqrt{\alpha^2+\beta^2}\,[\alpha(\sqrt{\alpha^2+\beta^2}+\beta]+\beta(\sqrt{\alpha^2+\beta^2}-\alpha)]$$

$$=\sqrt{\alpha^2+\beta^2}\,(\alpha+\beta)\sqrt{\alpha^2+\beta^2}$$

$$=(\alpha+\beta)(\alpha^2+\beta^2)$$

20. $\because$ $0\le\cos^{-1}x\le\pi,\ 0\le\cos^{-1}y\le\pi,$

$0\le\cos^{-1}z\le\pi$

But given, $\cos^{-1}x+\cos^{-1}y+\cos^{-1}z=3\pi$

Which is possible only when

$\cos^{-1}x=\pi,\ \cos^{-1}y=\pi$ and $\cos^{-1}z=\pi$

$\therefore$ $x=-1,\ y=-1$ and $z=-1$

Then, $xy+yz+zx=3$

21. $\sin^{-1}\left[\cot\left\{\sin^{-1}\sqrt{\left(\frac{2-\sqrt{3}}{4}\right)}+\cos^{-1}\left(\frac{\sqrt{12}}{4}\right)+\sec^{-1}(\sqrt{2})\right\}\right]$

$$=\sin^{-1}\left[\cot\left\{\sin^{-1}\left(\frac{\sqrt{3}-1}{2\sqrt{2}}\right)+\cos^{-1}\left(\frac{\sqrt{3}}{2}\right)+\sec^{-1}(\sqrt{2})\right\}\right]$$

$$=\sin^{-1}[\cot\{15^\circ+30^\circ+45^\circ\}]$$

$$=\sin^{-1}[\cot\pi/2]$$

$$=\sin^{-1}(0)=0$$

22. Since, $\tan^{-1}\sqrt{x(x+1)}+\sin^{-1}\sqrt{(x^2+x+1)}=\frac{\pi}{2}$

$$\Rightarrow\quad\cos^{-1}\left(\frac{1}{\sqrt{(x^2+x+1)}}\right)+\sin^{-1}\sqrt{(x^2+x+1)}=\frac{\pi}{2}$$

Hence, $\frac{1}{\sqrt{(x^2+x+1)}}=\sqrt{(x^2+x+1)}$

or $x^2+x=0$

or $x=-1,\ 0$

23. We have,

$$s_k=\sum_{r=1}^{k}\tan^{-1}\frac{\left(\frac{6^r}{2^{2r+1}+3^{2r+1}}\right)}{3^r}$$

$$=\sum_{r=1}^{k}\tan^{-1}\left(\frac{\left(\frac{2}{3}\right)^r}{\left(\frac{2}{3}\right)^{2r}\cdot 2+3}\right)$$

$$=\sum_{r=1}^{k}\tan^{-1}\left[\frac{\left(\frac{2}{3}\right)^r}{3\left(\left(\frac{2}{3}\right)^{2r+1}+1\right)}\right]$$

Let $\left(\frac{2}{3}\right)^r=x$

$$\therefore\quad S_k=\sum_{r=1}^{k}\tan^{-1}\left[\frac{\frac{x}{3}}{1+\frac{2}{3}x^2}\right]$$

$$S_k=\sum_{r=1}^{k}\tan^{-1}\left[\frac{x-\frac{2x}{3}}{1+x\left(\frac{2x}{3}\right)}\right]$$

$$S_k=\sum_{r=1}^{k}\left(\tan^{-1}x-\tan^{-1}\frac{2x}{3}\right)$$

$$S_k=\sum_{r=1}^{k}\left[\tan^{-1}\left(\frac{2}{3}\right)^r-\tan^{-1}\left(\frac{2}{3}\right)^{r+1}\right]$$

$$S_k=\tan^{-1}\left(\frac{2}{3}\right)-\tan^{-1}\left(\frac{2}{3}\right)^{k+1}$$

$$S_\infty=\tan^{-1}\frac{2}{3}-\tan^{-1}0$$

$$=\tan^{-1}\frac{2}{3}$$

24. $\Sigma x_1=\sin 2\beta,\ \Sigma x_1x_2=\cos 2\beta,\ \Sigma x_1x_2x_3=\cos\beta$

and $x_1x_2x_3x_4=-\sin\beta$

$\therefore$ $\tan^{-1}x_1+\tan^{-1}x_2+\tan^{-1}x_3+\tan^{-1}x_4$

$$=\tan^{-1}\left(\frac{\Sigma x_1-\Sigma x_1x_2x_3}{1-\Sigma x_1x_2+x_1x_2x_3x_4}\right)$$

$$=\tan^{-1}\left(\frac{\sin 2\beta-\cos\beta}{1-\cos 2\beta-\sin\beta}\right)$$

$$=\tan^{-1}\left\{\frac{\cos\beta\,(2\sin\beta-1)}{\sin\beta\,(2\sin\beta-1)}\right\}=\tan^{-1}\cot\beta$$

$$=\tan^{-1}\tan\left(\frac{\pi}{2}-\beta\right)=\frac{\pi}{2}-\beta$$

25. $\tan^{-1}\left(\frac{\sqrt{(1+x^2)}-\sqrt{(1-x^2)}}{\sqrt{(1+x^2)}+\sqrt{(1-x^2)}}\right)=\alpha$

$$\Rightarrow\quad\frac{\sqrt{(1+x^2)}-\sqrt{(1-x^2)}}{\sqrt{(1+x^2)}+\sqrt{(1-x^2)}}=\frac{\tan\alpha}{1}$$

Using componendo and dividendo method

$$= \frac{2\sqrt{(1+x^2)}}{2\sqrt{(1-x^2)}} = \frac{1+\tan\alpha}{1-\tan\alpha}$$

$$= \tan\left(\frac{\pi}{4}+\alpha\right)$$

$$\Rightarrow \therefore \qquad \frac{1+x^2}{1-x^2} = \frac{\tan^2\left(\frac{\pi}{4}+\alpha\right)}{1}$$

Again, use componendo and dividendo method

$$\Rightarrow \frac{(1+x^2)-(1-x^2)}{(1+x^2)+(1-x^2)} = \frac{\tan^2\left(\frac{\pi}{4}+\alpha\right)-1}{\tan^2\left(\frac{\pi}{4}+\alpha\right)+1}$$

$$\Rightarrow \qquad x^2 = -\cos\left(\frac{\pi}{2}+2\alpha\right) = \sin 2\alpha$$

26. Let $t = \dfrac{2x^2+4}{1+x^2}$

$$\Rightarrow x^2 = \frac{4-t}{t-2}$$

Since, $x^2 \geq 0$ we get

$$2 < t \leq 4$$

so we have to solve $\sin^{-1}(\sin t) < \pi - 3$ and $2 < t \leq 4$ simultaneously.

Now, $\frac{\pi}{2} < 2 < t \leq 4 < \frac{3\pi}{2}$

So, $\sin^{-1}(\sin t) = \pi - t$

and $\pi - t < \pi - 3 \Rightarrow t > 3$

$\therefore$ we get, $3 < t \leq 4$

$$\Rightarrow \qquad 3 < \frac{2x^2+4}{x^2+1} \leq 4$$

$$\Rightarrow \qquad -1 < x < 1$$

which is the required solution

27. We have,

$$\frac{\sin^{-1}x}{a} = \frac{\cos^{-1}x}{b} = \frac{\tan^{-1}y}{c}$$

$$\Rightarrow \frac{\sin^{-1}x+\cos^{-1}x}{a+b} = \frac{\tan^{-1}y}{c}$$

$$\Rightarrow \frac{\pi/2}{a+b} = \frac{\tan^{-1}y}{c} = 2\tan^{-1}y = \frac{\pi c}{a+b}$$

$$\Rightarrow \qquad \cos\left(\frac{\pi c}{a+b}\right) = \cos(2\tan^{-1}y) = \frac{1-y^2}{1+y^2}$$

28. Given, series is

$$S = \tan^{-1}\left(\frac{1}{3}\right) + \tan^{-1}\left(\frac{1}{7}\right) + \tan^{-1}\frac{1}{13} + \tan^{-1}\left(\frac{1}{21}\right) + \ldots + \text{ up to 10 terms}$$

$$= \tan^{-1}\frac{1}{1+2} + \tan^{-1}\frac{1}{1+6} + \tan^{-1}\frac{1}{1+12} + \tan^{-1}\frac{1}{1+20} + \ldots + \text{ up to 10 terms}$$

$$= \tan^{-1}\frac{2-1}{1+(1\cdot 2)} + \tan^{-1}\frac{3-2}{1+(2\cdot 3)} + \tan^{-1}\frac{4-3}{1+(3\cdot 4)} + \tan^{-1}\frac{5-4}{1+(4\cdot 5)} + \ldots + \text{ up to 10 terms}$$

$$= [\tan^{-1}2 - \tan^{-1}1] + [\tan^{-1}3 - \tan^{-1}2] + [\tan^{-1}4 - \tan^{-1}3] + [\tan^{-1}5 - \tan^{-1}4] + \ldots\ldots + \text{ upto 10 terms}$$

$$= \tan^{-1}11 - \tan^{-1}1 = \tan^{-1}\left(\frac{11-1}{1+(11\times 1)}\right)$$

$$\Rightarrow \qquad S = \tan^{-1}\left(\frac{10}{12}\right)$$

$$\Rightarrow \qquad \tan(S) = \frac{5}{6}$$

29. $\because \qquad \sum_{i=1}^{2n} \sin^{-1} x_i = n\pi = \sum_{i=1}^{2n}\left(\frac{\pi}{2}\right)$

Which is possible only when

$$\sin^{-1} x_i = \frac{\pi}{2}\ \forall i$$

$$\therefore \qquad x_i = 1\ \forall i$$

then $\quad \sum_{i=1}^{2n} x_i = \sum_{i=1}^{2n} 1 = 2n$

30. $\sin^{-1}(\sin 5) > x^2 - 4x$

$$\Rightarrow \qquad \sin^{-1}\{\sin(5-2\pi)\} > x^2 - 4x$$

$$\Rightarrow \qquad 5 - 2\pi > x^2 - 4x$$

$$\Rightarrow \qquad 9 - 2\pi > (x-2)^2$$

$$\Rightarrow \qquad (x-2)^2 < 9 - 2\pi$$

$$\Rightarrow \qquad -\sqrt{(9-2\pi)} < x - 2 < \sqrt{(9-2\pi)}$$

$$\Rightarrow \qquad 2 - \sqrt{(9-2\pi)} < x < 2 + \sqrt{(9-2\pi)}$$

$$\therefore\ x \in (2 - \sqrt{(9-2\pi)},\ 2 + \sqrt{(9-2\pi)})$$

31. $\therefore \qquad \cot^{-1}2 + \cot^{-1}8 + \cot^{-1}18 + \cot^{-1}32 + \ldots$

$$T_n - \cot^{-1}2n^2$$

$$= \tan^{-1}\left(\frac{1}{2n^2}\right)$$

$$= \tan^{-1}\left(\frac{2}{1+4n^2-1}\right)$$

$$= \tan^{-1}\left(\frac{(2n+1)-(2n-1)}{1+(2n+1)(2n-1)}\right)$$

$$= \tan^{-1}(2n+1) - \tan^{-1}(2n-1)$$

$$\therefore \qquad S = \lim_{n\to\infty}\sum_{n=1}^{n} T_n$$

$$= \lim_{n\to\infty.}\sum_{n=1}^{n}\{\tan^{-1}(2n+1) - \tan^{-1}(2n-1)\}$$

$$= \lim_{n\to\infty}(\tan^{-1}(2n+1) - \tan^{-1}1)$$

$$= \tan^{-1}\infty - \tan^{-1}1$$

$$= \frac{\pi}{2} - \frac{\pi}{4} = \frac{\pi}{4}$$

32. Let $P = \sin\{\cot^{-1}(\tan\cos^{-1} x)\}$

Let $\cos^{-1} x = \alpha$

$\Rightarrow \quad \cos\alpha = x,\ \tan\alpha = \dfrac{\sqrt{(1-x^2)}}{x}$

$\therefore \quad P = \sin\{\cot^{-1}\tan\alpha\}$

$$= \sin\left\{\cot^{-1}\left(\frac{\sqrt{1-x^2}}{x}\right)\right\}$$

$$= \sin\sin^{-1}\left(\frac{x}{1}\right) = x$$

33. $\tan^2(\sec^{-1}2) + \cot^2(\text{cosec}^{-1}3)$

$= \sec^2(\sec^{-1}2) - 1 + \text{cosec}^2(\text{cosec}^{-1}3) - 1$

$= (\sec\sec^{-1}2)^2 - 1 + (\text{cosec}\,\text{cosec}^{-1}3)^2 - 1$

$= 2^2 - 1 + 3^2 - 1$

$= 4 - 1 + 9 - 1 = 11$

34. Since, $\sin^{-1} x = 2\sin^{-1} a$

$\because \quad -\dfrac{\pi}{2} \le \sin^{-1} x \le \dfrac{\pi}{2}$

$\Rightarrow \quad -\dfrac{\pi}{2} \le 2\sin^{-1} a \le \dfrac{\pi}{2}$

$\Rightarrow \quad -\dfrac{\pi}{4} \le \sin^{-1} a \le \dfrac{\pi}{4}$

$\Rightarrow \quad \sin\left(-\dfrac{\pi}{4}\right) \le a \le \sin\left(\dfrac{\pi}{4}\right)$

$\Rightarrow \quad -\dfrac{1}{\sqrt{2}} \le a \le \dfrac{1}{\sqrt{2}}$

35. In $[\pi, 2\pi], |y| = \sin x,\ y = \cos^{-1}\cos(2\pi - x)$

$= 2\pi - x$

One solution : $x = 2\pi,\ y = 0$

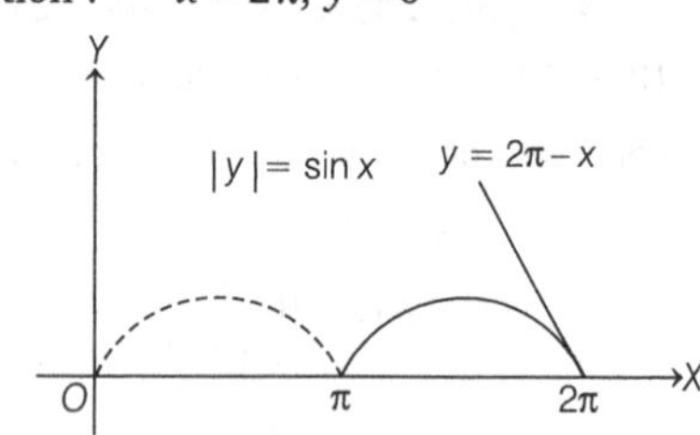

In $[-2\pi, -\pi]$, $|y| = \sin x$,

$y = \cos^{-1}[\cos(2\pi + x)] = 2\pi + x$

One solution : $x = -2\pi,\ y = 0$

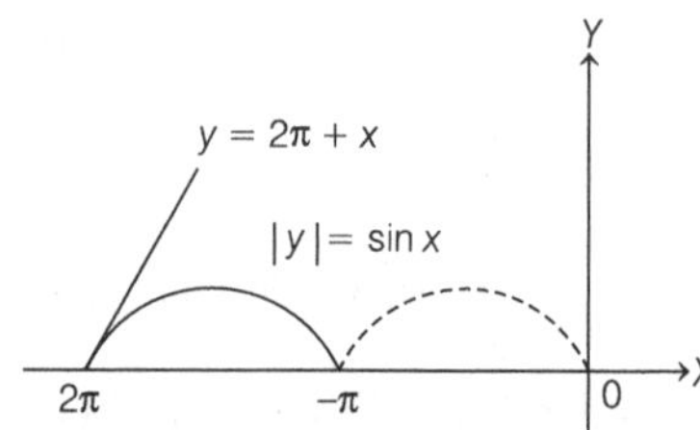

In $[0, \pi], |y| = \sin x$

$y = \cos^{-1}(\cos x) = x$

One solution : $x = 0,\ y = 0$

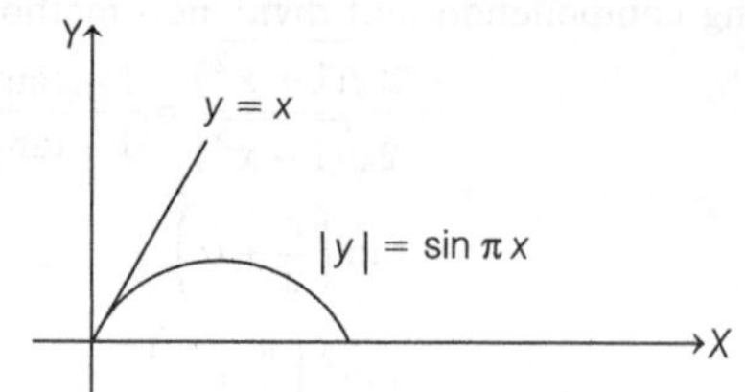

In $[-\pi, 0], |y| = \sin x$

$y = \cos^{-1}(\cos(-x)) = -x$

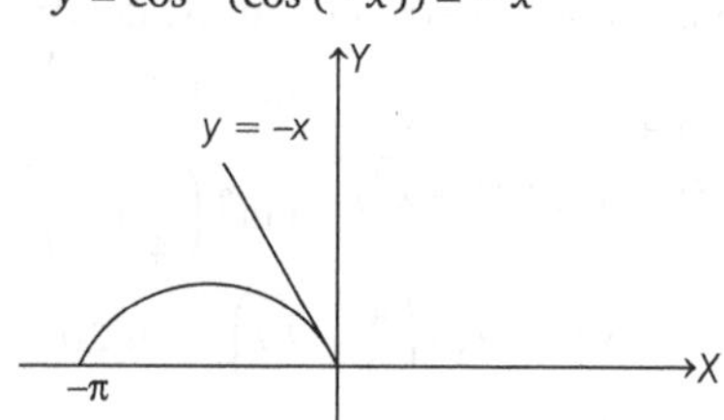

One solution : $x = 0,\ y = 0$

$\therefore$ There are 3 solutions $(0, 0), (2\pi, 0), (-2\pi, 0)$.

36. We have, $\tan^{-1} x + \cot^{-1} y = \tan^{-1} 3, (x, y) \in N$

$\Rightarrow \quad \cot^{-1} y = \tan^{-1} 3 - \tan^{-1} x$

$\Rightarrow \quad \tan^{-1}\left(\dfrac{1}{y}\right) = \tan^{-1}\left(\dfrac{3-x}{1+3x}\right) \Rightarrow y = \dfrac{1+3x}{3-x}$

Hence, $x = 1, 2$, then $y = 2, 7$

$\therefore$ Solutions (1, 2) and (2, 7).

37. Let $P = \cos^{-1}(\cos 12) - \sin^{-1}(\sin 12)$

$\because \quad \cos 12 > 0$ and $\sin 12 < 0$

$\therefore \quad P = \cos^{-1}\cos(4\pi - 12) - \sin^{-1}\sin(12 - 4\pi)$

$= (4\pi - 12) - (12 - 4\pi) = 8\pi - 24$

38. We have,

$$\tan^{-1}\left(\frac{1-x}{1+x}\right) = \tan^{-1} 1 - \tan^{-1} x = \frac{\pi}{4} - \tan^{-1} x$$

Since, $0 \le x \le 1$

Therefore, $0 \le \tan^{-1} x \le \dfrac{\pi}{4} \Rightarrow 0 \ge -\tan^{-1} x \ge -\dfrac{\pi}{4}$

$\Rightarrow \quad \dfrac{\pi}{4} \ge \dfrac{\pi}{4} - \tan^{-1} x \ge 0$

$\Rightarrow \quad \dfrac{\pi}{4} \ge \tan^{-1}\{(1-x)/(1+x)\} \ge 0$

39. $\because -1 < x < 0$, then $-\dfrac{\pi}{2} < \sin^{-1} x < 0$

Let $\sin^{-1} x = \alpha,\ \sin\alpha = x$

Then, $\tan\alpha = \dfrac{x}{\sqrt{(1-x^2)}}$

$\therefore \quad \alpha = \tan^{-1}\left(\dfrac{x}{\sqrt{(1-x^2)}}\right)$

$\therefore \quad \sin^{-1} x = \tan^{-1}\left(\dfrac{x}{\sqrt{(1-x^2)}}\right)$

40. By definition, $\alpha = \arcsin(\sin 10)$ is an angle that satisfies two conditions : $\sin\alpha = \sin 10$ and $-\frac{\pi}{2} \le \alpha \le \frac{\pi}{2}$ we find α with the help of graph. Let $y = \sin x$ plot the number 10 on the X-axis and find sin 10 geometrically.

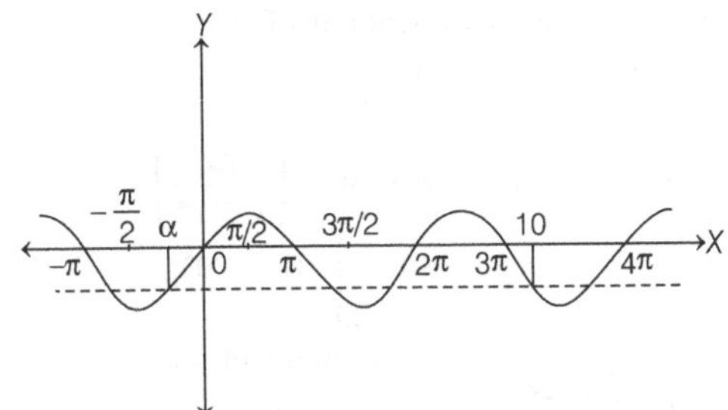

It is easy to see that the points α and 10 are symmetric about the point $\frac{3\pi}{2}$,

So, that $$10 - \frac{3\pi}{2} = \frac{3\pi}{2} - \alpha$$

$\Rightarrow$ $$\alpha = 3\pi - 10.$$

Hence, $\arcsin(\sin 10) = 3\pi - 10$.

41. Let $a = b\cos\phi$

Then, $$a_1 = \frac{a+b}{2} = \frac{b(1+\cos\phi)}{2} = b\cos^2\frac{\phi}{2}$$

$\Rightarrow$ $$b_1 = \sqrt{a_1 b} = b\cos\frac{\phi}{2}$$

Now, $$a_2 = \frac{a_1 + b_1}{2} = b\cos\frac{\phi}{2}\cos^2\frac{\phi}{4}$$

$\therefore$ $$b_2 = b\cos\frac{\phi}{2}\cos\frac{\phi}{2^2}$$

Similarly, $b_3 = b\cos\left(\frac{\phi}{2}\right)\cos\left(\frac{\phi}{2^2}\right)\cos\left(\frac{\phi}{2^3}\right)$

and so on.

Now, $$b_n = b\cos\frac{\phi}{2}\cos\left(\frac{\phi}{2^2}\right)\cos\left(\frac{\phi}{2^3}\right)\ldots\cos\left(\frac{\phi}{2^n}\right)$$

Now, $$b_\infty = \lim_{n\to\infty} b\cos\left(\frac{\phi}{2}\right)\cos\left(\frac{\phi}{2^2}\right)\cos\left(\frac{\phi}{2^3}\right)\ldots\cos\left(\frac{\phi}{2^n}\right)$$

$$= \lim_{n\to\infty}\frac{b\sin\phi}{2^n\sin\left(\frac{\phi}{2^n}\right)}$$

$$= \lim_{n\to\infty}\frac{b\sin\phi}{\phi\,\dfrac{\sin\left(\frac{\phi}{2^n}\right)}{\left(\frac{\phi}{2^n}\right)}}$$

$$= \frac{b\sin\phi}{\phi\left\{\lim_{n\to\infty}\sin\dfrac{\left(\frac{\phi}{2^n}\right)}{\left(\frac{\phi}{2^n}\right)}\right\}}$$

$$= \frac{b\sin\phi}{\phi} = b\frac{\sqrt{(1-\cos^2\phi)}}{\cos^{-1}\left(\frac{a}{b}\right)} = \frac{\sqrt{(b^2-a^2)}}{\cos^{-1}\left(\frac{a}{b}\right)}$$

42. We have,

$$\tan^{-1}\left(\frac{c_1x-y}{c_1y+x}\right) + \tan^{-1}\left(\frac{c_2-c_1}{1+c_2c_1}\right) + \tan^{-1}\left(\frac{c_3-c_2}{1+c_3c_2}\right) + \ldots + \tan^{-1}\left(\frac{1}{c_n}\right)$$

$$= \tan^{-1}\left(\frac{\frac{x}{y}-\frac{1}{c_1}}{1+\frac{x}{y}\frac{1}{c_1}}\right) + \tan^{-1}\left(\frac{\frac{1}{c_1}-\frac{1}{c_2}}{1+\frac{1}{c_1c_2}}\right) + \tan^{-1}\left(\frac{\frac{1}{c_2}-\frac{1}{c_3}}{1+\frac{1}{c_2c_3}}\right) + \ldots + \tan^{-1}\left(\frac{1}{c_n}\right)$$

$$= \tan^{-1}\left(\frac{x}{y}\right) - \tan^{-1}\left(\frac{1}{c_1}\right) + \tan^{-1}\left(\frac{1}{c_1}\right) - \tan^{-1}\left(\frac{1}{c_2}\right) + \tan^{-1}\left(\frac{1}{c_2}\right) - \tan^{-1}\left(\frac{1}{c_3}\right) + \ldots - \tan^{-1}\left(\frac{1}{c_n}\right) + \tan^{-1}\left(\frac{1}{c_n}\right) = \tan^{-1}\left(\frac{x}{y}\right)$$

43. We have,

$$\sin^{-1}\left\{\left(\sin\frac{\pi}{3}\right)\frac{x}{\sqrt{(x^2+k^2-kx)}}\right\} - \cos^{-1}\left\{(\cos\pi/6)\frac{x}{\sqrt{(x^2+k^2-kx)}}\right\}$$

$$= \sin^{-1}\left\{\frac{\sqrt{3}\,x}{2\sqrt{(x^2+k^2-kx)}}\right\} - \cos^{-1}\left\{\frac{\sqrt{3}}{2\sqrt{(x^2+k^2-kx)}}\right\}$$

$$= \frac{\pi}{2} - 2\cos^{-1}\left(\frac{\sqrt{3}\,x}{\sqrt{(4x^2-4kx+4k^2)}}\right)$$

$$= \frac{\pi}{2} - \cos^{-1}\left\{\frac{6x^2}{4x^2-4kx+4k^2} - 1\right\}$$

$$= \sin^{-1}\left[\frac{2x^2+4kx-4k^2}{4x^2-4kx+4k^2}\right] = \sin^{-1}\left[\frac{x^2+2kx-2k^2}{2x^2-2kx+2k^2}\right]$$

44. Let $\alpha = \cos^{-1}\sqrt{p}$, $\beta = \cos^{-1}\sqrt{1-p}$

and $\gamma = \cos^{-1}\sqrt{1-q}$

or $\cos\alpha = \sqrt{p}$, $\cos\beta = \sqrt{1-p}$

and $\cos\gamma = \sqrt{1-q}$

Therefore, $\sin\alpha = \sqrt{1-p}$, $\sin\beta = \sqrt{p}$, $\sin\gamma = \sqrt{q}$
The given equation may be written as

$$\alpha + \beta + \gamma = \frac{3\pi}{4}$$

or $\alpha + \beta = \frac{3\pi}{4} - \gamma$ or $\cos(\alpha+\beta) = \cos\left(\frac{3\pi}{4} - \gamma\right)$

$$\Rightarrow \cos\alpha\cos\beta - \sin\gamma\sin\beta = \cos\left\{\pi - \left(\frac{\pi}{4} + \gamma\right)\right\}$$
$$= -\cos\left(\frac{\pi}{4} + \gamma\right)$$
$$\Rightarrow \quad \sqrt{p}\sqrt{1-p} - \sqrt{1-p}\sqrt{p} = -\left(\frac{1}{\sqrt{2}}\sqrt{1-q} - \frac{1}{\sqrt{2}}\sqrt{q}\right)$$
$$\Rightarrow \quad 0 = \sqrt{1-q} - \sqrt{q} \Rightarrow 1 - q = q \Rightarrow q = \frac{1}{2}$$

45. $\because \quad T_n = \cot^{-1}\left(n^2 + \frac{3}{4}\right)$

$$= \tan^{-1}\left(\frac{1}{n^2 + \frac{3}{4}}\right) = \tan^{-1}\left(\frac{1}{1 + \left(n^2 - \frac{1}{4}\right)}\right)$$
$$= \tan^{-1}\left(\frac{1}{1 + \left(n + \frac{1}{2}\right)\left(n - \frac{1}{2}\right)}\right)$$
$$= \tan^{-1}\left(\frac{\left(n + \frac{1}{2}\right) - \left(n - \frac{1}{2}\right)}{1 + \left(n + \frac{1}{2}\right)\left(n - \frac{1}{2}\right)}\right)$$
$$= \tan^{-1}\left(n + \frac{1}{2}\right) - \tan^{-1}\left(n - \frac{1}{2}\right)$$
$$\therefore \quad S_n = \sum_{n=1}^{n} T_n$$
$$= \sum_{n=1}^{n}\left\{\tan^{-1}\left(n + \frac{1}{2}\right) - \tan^{-1}\left(n - \frac{1}{2}\right)\right\}$$
$$= \tan^{-1}\left(n + \frac{1}{2}\right) - \tan^{-1}\left(\frac{1}{2}\right)$$
$$\therefore \quad S_\infty = \tan^{-1}(\infty) - \tan^{-1}\left(\frac{1}{2}\right)$$
$$= \frac{\pi}{2} - \tan^{-1}\left(\frac{1}{2}\right)$$
$$= \cot^{-1}\left(\frac{1}{2}\right) = \tan^{-1} 2$$

46. Let $x = \tan\theta$

Then, $2\tan^{-1} x + \sin^{-1}\left(\frac{2x}{1+x^2}\right)$

$$= 2\theta + \sin^{-1}\left(\frac{2\tan\theta}{1+\tan^2\theta}\right)$$
$$= 2\theta + \sin^{-1}(\sin 2\theta)$$

If $\quad -\frac{\pi}{2} \le 2\theta \le \frac{\pi}{2}$

Then, $2\tan^{-1} x + \sin^{-1}\left(\frac{2x}{1+x^2}\right) = 2\theta + 2\theta$

$$= 4\theta = 4\tan^{-1} x$$

Which is not independent of x

and if $-\frac{\pi}{2} \le \pi - 2\theta \le \frac{\pi}{2}$

Then, $\quad 2\tan^{-1} x + \sin^{-1}\left(\frac{2x}{1+x^2}\right)$

$$= 2\theta + \sin^{-1}\sin(\pi - 2\theta)$$
$$= 2\theta + \pi - 2\theta$$
$$= \pi = \text{independent of } x$$

$\therefore \ \theta \in \left[\frac{\pi}{4}, \frac{3\pi}{4}\right]$ principal value of

$$\theta \in \left(-\frac{\pi}{2}, \frac{\pi}{2}\right)$$
$$\therefore \quad \theta \in \left[\frac{\pi}{4}, \frac{\pi}{2}\right)$$

Hence, $x \in [1, \infty)$

47. $\because \ \cot^{-1}\left(\frac{n}{\pi}\right) > \frac{\pi}{6}$

$$\Rightarrow \quad \cot\left\{\cot^{-1}\left(\frac{n}{\pi}\right)\right\} < \cot\left(\frac{\pi}{6}\right)$$
$$\Rightarrow \quad \frac{n}{\pi} < \sqrt{3}$$
$$\Rightarrow \quad n < \pi\sqrt{3} = 5.5 \text{ (approx)}$$
$$\therefore \quad n < 5.5$$

Hence, maximum value of n is 5.

48. $(\tan^{-1} x)^2 + \left(\frac{\pi}{2} - \tan^{-1} x\right)^2 = \frac{5\pi^2}{8}$

$$\Rightarrow \quad 2(\tan^{-1} x)^2 - \pi\tan^{-1} x - \frac{3\pi^2}{8} = 0$$
$$\therefore \quad \tan^{-1} x = \frac{\pi \pm \sqrt{4\pi^2}}{4}$$
$$= \frac{3\pi}{4} \text{ or } -\frac{\pi}{4}$$
$$\therefore \quad x = -1$$

49. We have,

$$\tan^{-1} a + \tan^{-1} b = \frac{\pi}{4}$$
$$\tan^{-1}\left(\frac{a+b}{1-ab}\right) = \frac{\pi}{4}$$
$$\Rightarrow \quad a + b = 1 - ab$$
$$\Rightarrow \quad a + b + ab + 1 = 1 + 1 = (1+a)(1+b) = 2$$

Now, $(a+b) - \left(\frac{a^2+b^2}{2}\right) + \left(\frac{a^3+b^3}{3}\right) - \left(\frac{a^4+b^4}{4}\right) + \ldots$

$$= \left[a - \frac{a^2}{2} + \frac{a^3}{3} \ldots\right] + \left[b - \frac{b^2}{2} + \frac{b^3}{3} + \ldots\right]$$
$$= \log_e(1+a) + \log_e(1+b)$$
$$= \log_e(1+a)(1+b) = \log_e 2$$

50. $\because\ a_2 - a_1 = a_3 - a_2 = \ldots = a_n - a_{n-1} = d$

$$\therefore \tan\left[\tan^{-1}\left(\frac{d}{1+a_1a_2}\right) + \tan^{-1}\left(\frac{d}{1+a_2\,a_3}\right) + \ldots\ldots\ldots + \tan^{-1}\left(\frac{d}{1+a_{n-1}a_n}\right)\right]$$

$$= \tan\left[\sum_{r=2}^{n} \tan^{-1}\left(\frac{d}{1+a_{r-1}\,a_r}\right)\right]$$

$$= \tan\left[\sum_{r=2}^{n} \tan^{-1}\left(\frac{a_r - a_{r-1}}{1+a_{r-1}\,a_r}\right)\right]$$

$$= \tan\left[\sum_{r=2}^{n} (\tan^{-1} a_r - \tan^{-1} a_{r-1})\right]$$

$$= \tan(\tan^{-1} a_n - \tan^{-1} a_1)$$

$$= \tan\left(\tan^{-1}\left(\frac{a_n - a_1}{1+a_1a_n}\right)\right)$$

$$= \frac{a_n - a_1}{1+a_1a_n} = \frac{a_1 + (n-1)d - a_1}{1+a_1a_n} = \frac{(n-1)d}{1+a_1a_n}$$

51. $\cot^{-1}\alpha = \cot^{-1} 2 + \cot^{-1} 8 + \cot^{-1} 18 + \ldots$

$$\cot^{-1}\alpha = \sum_{n=1}^{100} \tan^{-1}\left(\frac{2}{4n^2}\right)$$

$$= \sum_{n=1}^{100} \tan^{-1}\left[\frac{(2n+1)-(2n-1)}{1+(2n+1)(2n-1)}\right]$$

$$= \sum_{n=1}^{100} [\tan^{-1}(2n+1) - \tan^{-1}(2n-1)]$$

$$= \tan^{-1}(201) - \tan^{-1} 1$$

$$= \tan^{-1}\left(\frac{200}{202}\right)$$

$$= \cot^{-1}\left(\frac{202}{200}\right)$$

$$\Rightarrow \cot^{-1}\alpha = \cot^{-1}\left(\frac{101}{100}\right)$$

$$\therefore \quad \alpha = 1.01$$

52. We have, $\cos^{-1}\left(\frac{2}{3x}\right) + \cos^{-1}\left(\frac{3}{4x}\right) = \frac{\pi}{2}$

$$\Rightarrow \cos^{-1}\left(\frac{2}{3x}\cdot\frac{3}{4x} - \sqrt{1-\frac{4}{9x^2}}\sqrt{1-\frac{9}{16x^2}}\right) = \frac{\pi}{2}$$

$[\because \cos^{-1} x + \cos^{-1} y = \cos^{-1}(xy - \sqrt{1-x^2}\,\sqrt{1-y^2})]$

$$\Rightarrow \cos^{-1}\left(\frac{1}{2x^2} - \frac{\sqrt{9x^2-4}\,\sqrt{16x^2-9}}{12x^2}\right) = \frac{\pi}{2}$$

$$\Rightarrow \frac{6 - \sqrt{9x^2-4}\,\sqrt{16x^2-9}}{12x^2} = \cos\frac{\pi}{2} = 0$$

$$\Rightarrow \sqrt{9x^2-4}\,\sqrt{16x^2-9} = 6$$

On squaring both sides, we get

$$(9x^2-4)(16x^2-9) = 36$$

$$\Rightarrow 144x^4 - 81x^2 - 64x^2 + 36 = 36$$

$$\Rightarrow 144x^4 - 145x^2 = 0$$

$$\Rightarrow x^2(144x^2 - 145) = 0$$

$$\Rightarrow x = 0 \text{ or } x = \pm\sqrt{\frac{145}{144}} = \pm\frac{\sqrt{145}}{12}$$

But $x > \frac{3}{4}$

$$\therefore \quad x = \frac{\sqrt{145}}{12}$$

53. $y = \cos x$

$y = \sin 3x$

$$\therefore \sin 3x = \cos x = \sin\left(\frac{\pi}{2} - x\right)$$

$$\Rightarrow 3x = n\pi + (-1)^n\left(\frac{\pi}{2} - x\right)$$

Case I When n is even, then

$$3x = n\pi + \frac{\pi}{2} - x \Rightarrow x = \frac{n\pi}{4} + \frac{\pi}{8}$$

$$\therefore \quad x = \frac{\pi}{8}, \text{ at } n = 0$$

Then, $y = \cos\frac{\pi}{8} \Rightarrow \left(\frac{\pi}{8}, \cos\frac{\pi}{8}\right)$

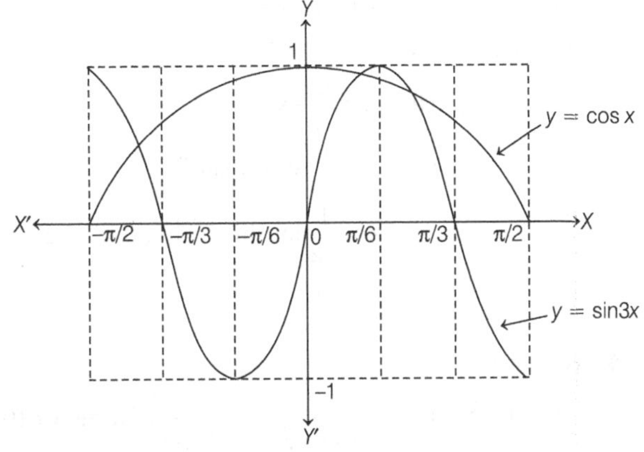

Case II When n is odd, then

$$3x = n\pi - \frac{\pi}{2} + x$$

or $$x = \frac{n\pi}{2} - \frac{\pi}{4}$$

$$= \frac{\pi}{4} \text{ (for } n = 1)$$

Then, $y = \cos\frac{\pi}{4} = \frac{1}{\sqrt{2}}$

$$\Rightarrow \left(\frac{\pi}{4}, \frac{1}{\sqrt{2}}\right)$$

Hence, point of intersection is

$$\left(\frac{\pi}{4}, \frac{1}{\sqrt{2}}\right) \text{ and } \left(\frac{\pi}{8}, \cos\frac{\pi}{8}\right)$$

54. $(\cot^{-1} x)^2 - 5\cot^{-1} x + 6 > 0$

$\Rightarrow (\cot^{-1} x - 3)(\cot^{-1} x - 2) > 0$

Then, $\cot^{-1} x < 2$ and $\cot^{-1} x > 3$

$\Rightarrow x > \cot 2$ and $x < \cot 3$

Hence, $x \in (-\infty, \cot 3) \cup (\cot 2, \infty)$

Objective Questions Type II [One or more than one correct answer(s)]

1. $\sin^{-1} x + \sin^{-1}(1-x) = \cos^{-1} x$

$\Rightarrow \quad \frac{\pi}{2} - \cos^{-1} x + \frac{\pi}{2} - \cos^{-1}(1-x) = \cos^{-1} x$

$\Rightarrow \quad 2\cos^{-1} x = \pi - \cos^{-1}(1-x)$

$\Rightarrow \quad \cos^{-1}(2x^2-1) = \cos^{-1}(x-1)$

$\Rightarrow \quad 2x^2 - 1 = x - 1$

$\Rightarrow \quad x(2x-1) = 0$

$\Rightarrow \therefore \quad x = 0, \frac{1}{2}$

2. For $\sin^{-1} x$ and $\cos^{-1} x$

$-1 \le x \le 1$

Given, $\frac{1}{2} < |x| < 1$ or $\frac{1}{2} < x < 1$

and $-1 < x < -\frac{1}{2}$ and $\tan^{-1} x, x \in R$

But for $\sec^{-1} x$, $x \ge 1$ and $x \le -1$

3. $\sin^{-1} x > \cos^{-1} x$

$\Rightarrow \quad \sin^{-1} x > \frac{\pi}{2} - \sin^{-1} x$

$\Rightarrow \quad \sin^{-1} x > \frac{\pi}{4} \Rightarrow x > \sin\frac{\pi}{4}$

$\therefore \quad x > \frac{1}{\sqrt{2}}$ but in $\sin^{-1} x$, $0 < x < 1$

$\therefore \quad x \in \left(\frac{1}{\sqrt{2}}, 1\right)$

4. We have,

$f_n(x) = \sum_{j=1}^{n} \tan^{-1}\left(\frac{1}{1+(x+j)(x+j-1)}\right)$ for all $x \in (0, \infty)$

$\Rightarrow \quad f_n(x) = \sum_{j=1}^{n} \tan^{-1}\left(\frac{(x+j)-(x+j-1)}{1+(x+j)(x+j-1)}\right)$

$\Rightarrow \quad f_n(x) = \sum_{j=1}^{n} [\tan^{-1}(x+j) - \tan^{-1}(x+j-1)]$

$\Rightarrow f_n(x) = (\tan^{-1}(x+1) - \tan^{-1} x) + (\tan^{-1}(x+2) - \tan^{-1}(x+1)) + (\tan^{-1}(x+3) - \tan^{-1}(x+2)) + \ldots + (\tan^{-1}(x+n) - \tan^{-1}(x+n-1))$

$\Rightarrow \quad f_n(x) = \tan^{-1}(x+n) - \tan^{-1} x$

This statement is false as $x \ne 0$. i.e., $x \in (0, \infty)$

(b) This statement is also false as $0 \notin (0, \infty)$

(c) $f_n(x) = \tan^{-1}(x+n) - \tan^{-1} x$

$\lim_{x\to\infty} \tan(f_n(x)) = \lim_{x\to\infty} \tan(\tan^{-1}(x+n) - \tan^{-1} x)$

$\Rightarrow \quad \lim_{x\to\infty} \tan(f_n(x)) = \lim_{x\to\infty} \tan\left(\tan^{-1}\frac{n}{1+nx+x^2}\right)$

$= \lim_{x\to\infty} \frac{n}{1+nx+x^2} = 0$

$\therefore$ (c) statement is false.

(d) $\lim_{x\to\infty} \sec^2(f_n(x)) = \lim_{x\to\infty} (1 + \tan^2 f_n(x))$

$= 1 + \lim_{x\to\infty} \tan^2(f_n(x)) = 1 + 0 = 1$

$\therefore$ (d) statement is true

5. $f\left(\frac{8\pi}{9}\right) = e^{\cos^{-1}\sin\left(\frac{8\pi}{9}+\frac{\pi}{3}\right)} = e^{\cos^{-1}\sin\left(\frac{11\pi}{9}\right)}$

$= e^{\cos^{-1}\cos\frac{13\pi}{18}} = e^{\frac{13\pi}{18}}$

and $\quad f\left(-\frac{7\pi}{4}\right) = e^{\cos^{-1}\sin\left(-\frac{7\pi}{4}+\frac{\pi}{3}\right)}$

$= e^{\cos^{-1}\sin\left(-\frac{17\pi}{12}\right)} = e^{\cos^{-1}\cos\frac{\pi}{12}} = e^{\frac{\pi}{12}}$

6. We have,

$xy < 0$

$\Rightarrow \quad x + \frac{1}{x} \ge 2, \; y + \frac{1}{y} \le -2$

or $\quad x + \frac{1}{x} \le -2, \; y + \frac{1}{y} \ge 2 \Rightarrow x + \frac{1}{x} \ge 2$

$\Rightarrow \quad \sec^{-1}\left(x + \frac{1}{x}\right) \in \left[\frac{\pi}{3}, \frac{\pi}{2}\right) \Rightarrow y + \frac{1}{y} \le -2$

$\Rightarrow \quad \sec^{-1}\left(y + \frac{1}{y}\right) \in \left(\frac{\pi}{2}, \frac{2\pi}{3}\right] \Rightarrow z \in \left(\frac{5\pi}{6}, \frac{7\pi}{6}\right)$

7. Let $y = (\sin^{-1} x)^3 + (\cos^{-1} x)^3$

$= (\sin^{-1} x + \cos^{-1} x)\{(\sin^{-1} x)^2 + (\cos^{-1} x)^2 - \sin^{-1} x \cos^{-1} x\}$

$= \frac{\pi}{2}\left\{(\sin^{-1} x)^2 + \left(\frac{\pi}{2} - \sin^{-1} x\right)^2 - \sin^{-1} x\left(\frac{\pi}{2} - \sin^{-1} x\right)\right\}$

$= \frac{\pi}{2}\left\{3(\sin^{-1} x)^2 - \frac{3\pi}{2}\sin^{-1} x + \frac{\pi^2}{4}\right\}$

Since, $\quad -1 \le \sin^{-1} x \le 1$

or $\quad -\frac{\pi}{2} \le x \le \frac{\pi}{2}$

$y_{-\frac{\pi}{2}} = \frac{7\pi^3}{8}$ ...(i)

$y_{\pi/2} = \frac{\pi^3}{8}$ ...(ii)

and $\quad 3(\sin^{-1} x)^2 - \frac{3\pi}{2}\sin^{-1} x + \frac{\pi^2}{4} = \frac{2y}{\pi}$

$\Rightarrow \quad 3(\sin^{-1} x)^2 - \frac{3\pi}{2}\sin^{-1} x + \frac{\pi^2}{4} - \frac{2y}{\pi} = 0$

For real, $B^2 - 4AC \ge 0$

$\therefore \quad \frac{9\pi^2}{4} - 4\cdot 3\left(\frac{\pi^2}{4} - \frac{2y}{\pi}\right) \ge 0$

$\Rightarrow \quad \frac{9\pi^2}{4} - 3\pi^2 + \frac{24y}{\pi} \ge 0 \Rightarrow \frac{-3\pi^2}{4} + \frac{24y}{\pi} \ge 0$

$\therefore \quad y \ge \frac{\pi^3}{32}$ ...(iii)

From Eqs. (i), (ii) and (iii),

Least value of y is $\frac{\pi^3}{32}$

and greatest value of y is $\frac{7\pi^3}{8}$.

8. Let $\frac{a+b+c}{abc}=t^2$

$$\therefore \quad \tan^{-1}\sqrt{\frac{a(a+b+c)}{bc}}+\tan^{-1}\sqrt{\frac{b(a+b+c)}{ca}}+\tan^{-1}\sqrt{\frac{c(a+b+c)}{ab}}$$

$$=\tan^{-1}at+\tan^{-1}bt+\tan^{-1}ct$$

$$=\tan^{-1}\left\{\frac{at+bt+ct-at\cdot bt\cdot ct}{1-t^2(ab+bc+ca)}\right\}$$

$$=\tan^{-1}\left\{\frac{(a+b+c)t-abct\cdot t^3}{1-t^2(ab+bc+ca)}\right\}$$

$$=\tan^{-1}\left[\frac{(a+b+c)t-(a+b+c)t}{1-t^2(ab+bc+ca)}\right]$$

$$=\tan^{-1}(0)=n\pi,\ n\in I$$

$\therefore \quad 0, \pi, 2\pi, \ldots$

9. $\because \sin[2\cos^{-1}\{\cot(2\tan^{-1}x)\}]=0$

$\Rightarrow \quad 2\cos^{-1}\{\cot(2\tan^{-1}x)\}=n\pi,\ n\in I$

$\Rightarrow \quad \cos^{-1}(\cot\{2\tan^{-1}x\})=\frac{n\pi}{2}=0, \frac{\pi}{2}, \pi$ $\quad [\because 0\le\cos^{-1}x\le\pi]$

$\Rightarrow \quad \cot(2\tan^{-1}x)=1, 0, -1$

$\Rightarrow \quad \cot\left\{\tan^{-1}\left(\frac{2x}{1-x^2}\right)\right\}=1, 0, -1$

$\Rightarrow \quad \cot\left\{\cot^{-1}\left(\frac{1-x^2}{2x}\right)\right\}=1, 0, -1$

$\Rightarrow \quad \frac{1-x^2}{2x}=1, 0, -1$

$\Rightarrow \quad 1-x^2=2x, 0, -2x$

or $x^2+2x-1=0,\ x^2-1=0,\ x^2-2x-1=0$

or $\quad x=-1\pm\sqrt{2},\ x=\pm1,\ x=1\pm\sqrt{2}$

10. $\because \alpha=2\tan^{-1}(\sqrt{2}-1)=2\cdot 22\frac{1°}{2}=45°$

$\Rightarrow \quad \cos\alpha=\frac{1}{\sqrt{2}}$

$$\beta=3\sin^{-1}\left(\frac{1}{\sqrt{2}}\right)+\sin^{-1}\left(-\frac{1}{2}\right)$$

$$=3\cdot\left(\frac{\pi}{4}\right)-\frac{\pi}{6}=105°$$

$\Rightarrow \quad \cos\beta=-\frac{(\sqrt{3}-1)}{2\sqrt{2}}$

and $\quad \gamma=\cos^{-1}\left(\frac{1}{3}\right)\Rightarrow\cos\gamma=\frac{1}{3}$

$\therefore \quad \beta>\alpha$

and $\quad \cos\beta<\cos\gamma\Rightarrow\beta>\gamma$

and $\quad \cos\alpha>\cos\gamma\Rightarrow\alpha<\gamma$

11. Here, $\alpha=3\sin^{-1}\left(\frac{6}{11}\right)$ and $\beta=3\cos^{-1}\left(\frac{4}{9}\right)$ as $\frac{6}{11}>\frac{1}{2}$

$\Rightarrow \quad \sin^{-1}\left(\frac{6}{11}\right)>\sin^{-1}\left(\frac{1}{2}\right)=\frac{\pi}{6}$

$\therefore \quad \alpha=3\sin^{-1}\left(\frac{6}{11}\right)>\frac{\pi}{2}\Rightarrow\cos\alpha<0$

Now, $\quad \beta=3\cos^{-1}\left(\frac{4}{9}\right)$

As $\quad \frac{4}{9}<\frac{1}{2}\Rightarrow\cos^{-1}\left(\frac{4}{9}\right)>\cos^{-1}\left(\frac{1}{2}\right)=\frac{\pi}{3}$

$\therefore \quad \beta=3\cos^{-1}\left(\frac{4}{9}\right)>\pi$

$\therefore \quad \cos\beta<0$ and $\sin\beta<0$

Now, $\alpha+\beta$ is slightly greater than $\frac{3\pi}{2}$.

$\therefore \quad \cos(\alpha+\beta)>0$

12. LHS and RHS of the given equation are defined, if $\frac{a-x}{a-b}>0$ and $\frac{x-b}{a-b}>0$. Either $a>x>b$ or $a<x<b$. For $a=x=b$, LHS and RHS are not defined.

13. $\theta=\tan^{-1}(2\tan^2\theta)-\tan^{-1}\left\{\left(\frac{1}{3}\right)\tan\theta\right\}$

$$\Rightarrow \quad \tan\theta=\frac{2\tan^2\theta-\left(\frac{1}{3}\right)\tan\theta}{1+\left(\frac{2}{3}\right)\tan^3\theta}$$

$$\Rightarrow \quad \tan\theta\left[\frac{2\tan\theta-\left(\frac{1}{3}\right)}{1+\left(\frac{2}{3}\right)\tan^3\theta}-1\right]=0$$

Which is true, if $\tan\theta=0$

$$\text{or} \quad \frac{2\tan\theta-\left(\frac{1}{3}\right)}{1+\left(\frac{2}{3}\right)\tan^3\theta}=1$$

$\Rightarrow \quad \tan^3\theta-3\tan\theta+2=0$

$\Rightarrow \quad (\tan\theta-1)^2(\tan\theta+2)=0$

Which holds, if $\tan\theta=1$ or $\tan\theta=-2$

14. Let $\cos^{-1}\left(\frac{4}{5}\right)=\alpha$, that is $\cos\alpha=\frac{4}{5}$, so that

$$\tan\alpha=\sqrt{\left\{\left(\frac{5}{4}\right)^2-1\right\}}=\frac{3}{4}\ [\because 0<\alpha<\pi \text{ and } \cos\alpha>0]$$

$$\text{and} \quad \tan\left(\cos^{-1}\frac{4}{5}+\tan^{-1}\frac{2}{3}\right)=\frac{\tan\alpha+\frac{2}{3}}{1-\tan\alpha\cdot\frac{2}{3}}$$

$$=\frac{\frac{3}{4}+\frac{2}{3}}{1-\frac{2}{3}\cdot\frac{3}{4}}=\frac{17}{6}=\frac{a}{b}\ \text{(given)}$$

So, $\quad a=17, b=6, a+b=23,$
$a-b=11$ and $3b=a+1$

15. Let $\tan^{-1}(-3) = \alpha \Rightarrow \tan\alpha = -3$

and $\qquad -\dfrac{\pi}{2} < \alpha < 0$

$\Rightarrow \qquad -\pi < 2\alpha < 0$

$\Rightarrow \cos(-2\alpha) = \cos 2\alpha = \dfrac{1-\tan^2\alpha}{1+\tan^2\alpha} = \dfrac{1-9}{1+9} = -\dfrac{4}{5}$

$\Rightarrow \cos(-2\alpha) = -\dfrac{4}{5}$

$\Rightarrow \quad -2\alpha = \cos^{-1}\left(-\dfrac{4}{5}\right) \Rightarrow 2\tan^{-1}(-3) = -\cos^{-1}\left(-\dfrac{4}{5}\right)$

Again, $-\pi < 2\alpha < 0 \Rightarrow 0 < 2\alpha + \pi < \pi$

So, $\cos(\pi + 2\alpha) = -\cos 2\alpha = \dfrac{4}{5}$

$\Rightarrow \quad \pi + 2\alpha = \cos^{-1}\dfrac{4}{5}$

$\Rightarrow 2\tan^{-1}(-3) = -\pi + \cos^{-1}\dfrac{4}{5}$

Finally, $-\pi < 2\alpha < 0$

$\Rightarrow \quad -\dfrac{\pi}{2} < 2\alpha + \dfrac{\pi}{2} < \dfrac{\pi}{2}$

So, $\tan\left(\dfrac{\pi}{2} + 2\alpha\right) = -\cot 2\alpha = \dfrac{-1}{\tan 2\alpha}$

$= \dfrac{-1+\tan^2\alpha}{2\tan\alpha} = \dfrac{-4}{3}$

$\therefore \quad \dfrac{\pi}{2} + 2\alpha = \tan^{-1}\left(-\dfrac{4}{3}\right)$

$\Rightarrow \quad 2\tan^{-1}(-3) = -\dfrac{\pi}{2} + \tan^{-1}\left(-\dfrac{4}{3}\right)$

16. $2\cot^{-1}7 = 2\tan^{-1}\left(\dfrac{1}{7}\right)$

$= \cos^{-1}\left(\dfrac{1-\dfrac{1}{49}}{1+\dfrac{1}{49}}\right) = \cos^{-1}\dfrac{24}{25}$

Now, $2\cot^{-1}7 + \cos^{-1}\dfrac{3}{5}$

$= \cos^{-1}\dfrac{24}{25} + \cos^{-1}\dfrac{3}{5}$

$= \cos^{-1}\left(\dfrac{24}{25}\cdot\dfrac{3}{5} - \dfrac{7}{25}\cdot\dfrac{4}{5}\right) = \cos^{-1}\dfrac{44}{125}$

Since, $\qquad \dfrac{44}{125} > 0$

$\therefore \qquad 0 < \cos^{-1}\dfrac{44}{125} < \dfrac{\pi}{2}$

Let $\cos^{-1}\dfrac{44}{125} = \theta$, then $\cos\theta = \dfrac{44}{125}$

$\therefore \quad \operatorname{cosec}\theta = \dfrac{125}{117}$ or $\theta = \operatorname{cosec}^{-1}\dfrac{125}{117}$

Also, $\qquad \cot\theta = \dfrac{44}{117}$

$\therefore \qquad \theta = \cot^{-1}\dfrac{44}{117}$

17. The given equation holds, if $x^2 + x + 1 = \lambda x + 1$ and

$-1 \le x^2 + x + 1 \le 1$

$\Rightarrow \quad x(x + 1 - \lambda) = 0$ and $-1 \le x \le 0$

$\Rightarrow \quad x = 0$ or $\lambda - 1$ and $-1 \le x \le 0$

$\therefore x = 0$ is one solution and for another different solution $-1 \le \lambda - 1 < 0$.

$\Rightarrow 0 \le \lambda < 1$, so only integral value λ can have is 0.

18. The solution of $y = \sqrt{y}$ is $y = 0$ or $y = 1$

If $\qquad \sin^{-1}|\sin x| = 1$

$\Rightarrow \qquad x = 1$ or $\pi - 1$ $\qquad$ [in the interval $(0, \pi)$]

But $y = \sin^{-1}|\sin x|$ is periodic with period π, so $x = n\pi + 1$ or $n\pi - 1$.

Again, if $\quad \sin^{-1}|\sin x| = 0$

$\Rightarrow \qquad x = n\pi.$

19. If we put $x = \tan\theta$, the given equality becomes

$\tan^{-1} y = 4\theta$

$\Rightarrow \qquad y = \tan 4\theta = \dfrac{2\tan 2\theta}{1-\tan^2 2\theta}$

$= \dfrac{2\left[\dfrac{2\tan\theta}{1-\tan^2\theta}\right]}{1-\left(\dfrac{2\tan\theta}{1-\tan^2\theta}\right)^2}$

$= \dfrac{2\times 2x(1-x^2)}{(1-x^2)^2 - 4x^2} = \dfrac{4x(1-x^2)}{1-6x^2+x^4}$

So, that y is infinite, if $x^4 - 6x^2 + 1 = 0$

$\Rightarrow \qquad x^2 = \dfrac{6 \pm \sqrt{36-4}}{2} = 3 \pm 2\sqrt{2}$

20. $\qquad \cos^{-1} x = \tan^{-1} x$

$\Rightarrow \qquad x = \cos\theta = \tan\theta$

$\Rightarrow \qquad \cos^2\theta = \sin\theta$

$\Rightarrow \sin^2\theta + \sin\theta - 1 = 0$

$\Rightarrow \qquad \sin\theta = \dfrac{-1 \pm \sqrt{1+4}}{2}$

$\Rightarrow \qquad \sin\theta = \dfrac{\sqrt{5}-1}{2}$

$\Rightarrow \qquad x^2 = \cos^2\theta$

$= \dfrac{\sqrt{5}-1}{2}$

So, option (a) is correct.

and $\quad \sin(\cos^{-1} x) = \sin\theta = \left(\dfrac{1}{2}\right)(\sqrt{5}-1)$

So, option (c) is correct.

and $\quad \tan(\cos^{-1} x) = \tan\theta \ne \left(\dfrac{1}{2}\right)(\sqrt{5}-1)$

So, option (d) is not correct.

Hence, options (a) and (c) are correct.

Linked Comprehension Type

PASSAGE 1

1. $\because$ 4 radian means 229° (approx.)
Which lies in III quadrant

For $\pi \le x < \frac{3\pi}{2}$

$$\tan^{-1}(\tan x) = x - \pi$$

$\therefore \quad \tan^{-1}(\tan 4) = 4 - \pi$

Then, $\cos\{\tan^{-1}(\tan 4)\} = \cos(4 - \pi)$
$= \cos(\pi - 4) = -\cos 4$

2. $$\sin^{-1} x = \begin{cases} \cos^{-1}\sqrt{(1-x^2)} & , \quad 0 \le x \le 1 \\ -\cos^{-1}\sqrt{(1-x^2)} & , \quad -1 \le x < 0 \end{cases}$$

$$= \begin{cases} \cot^{-1}\left(\frac{\sqrt{(1-x^2)}}{x}\right) & , \quad 0 < x < 1 \\ -\pi + \cot^{-1}\left(\frac{\sqrt{(1-x^2)}}{x}\right) & , \quad -1 \le x < 0 \end{cases}$$

3. For $\frac{3\pi}{2} \le x \le \frac{5\pi}{2}$

$\sin^{-1}(\sin x) = x - 2\pi$

4. $$\because 2\tan^{-1} x = \begin{cases} \sin^{-1}\left(\frac{2x}{1+x^2}\right) & , \quad -1 \le x \le 1 \\ \pi - \sin^{-1}\left(\frac{2x}{1+x^2}\right) & , \quad x > 1 \\ -\pi - \sin^{-1}\left(\frac{2x}{1+x^2}\right) & , \quad x < -1 \end{cases}$$

$\therefore$ For $x > 1$

$$2\tan^{-1} x = \pi - \sin^{-1}\left(\frac{2x}{1+x^2}\right)$$

$$\Rightarrow \quad 2\tan^{-1} x + \sin^{-1}\left(\frac{2x}{1+x^2}\right) = \pi$$

5. $\because \tan^{-1}\left(\frac{1+x}{1-x}\right) = \tan^{-1} 1 + \tan^{-1} x = \frac{\pi}{4} + \tan^{-1} x$

Given, $0 \le x \le 1$

$$\Rightarrow \quad \tan^{-1} 0 \le \tan^{-1} x \le \tan^{-1} 1$$

$$\Rightarrow \quad 0 \le \tan^{-1} x \le \frac{\pi}{4}$$

$$\Rightarrow \quad \frac{\pi}{4} + 0 \le \frac{\pi}{4} + \tan^{-1} x \le \frac{\pi}{4} + \frac{\pi}{4}$$

$$\Rightarrow \quad \frac{\pi}{4} \le \tan^{-1}\left(\frac{1+x}{1-x}\right) \le \frac{\pi}{2}$$

Hence, least and greatest values of $\tan^{-1}\left(\frac{1+x}{1-x}\right)$ are $\frac{\pi}{4}$ and $\frac{\pi}{2}$.

PASSAGE 2

1. $$\lim_{n\to\infty} \sum_{r=1}^{n} \tan^{-1}\left(\frac{2^{r-1}}{1+2^{2r-1}}\right)$$

$$= \lim_{n\to\infty} \sum_{r=1}^{n} \tan^{-1}\left(\frac{2^r - 2^{r-1}}{1 + 2^r \cdot 2^{r-1}}\right)$$

$$= \lim_{n\to\infty} \sum_{r=1}^{n} \{\tan^{-1}(2^r) - \tan^{-1}(2^{r-1})\}$$

$$= \lim_{n\to\infty} (\tan^{-1} 2^n - \tan^{-1} 2^\circ)$$

$$= \tan^{-1} 2^\infty - \tan^{-1} 1$$

$$= \tan^{-1} \infty - \tan^{-1} 1$$

$$= \frac{\pi}{2} - \frac{\pi}{4} = \frac{\pi}{4}$$

2. $\because \operatorname{cosec}^{-1}\sqrt{5} + \operatorname{cosec}^{-1}\sqrt{65} + \operatorname{cosec}^{-1}\sqrt{(325)} + \dots \infty$

$$= \cot^{-1} 2 + \cot^{-1} 8 + \cot^{-1} 18 + \dots \infty$$

$$= \lim_{n\to\infty} \sum_{r=1}^{n} \cot^{-1}(2r^2)$$

$$= \lim_{n\to\infty} \sum_{r=1}^{n} \tan^{-1}\left(\frac{2}{4r^2}\right)$$

$$= \lim_{n\to\infty} \sum_{r=1}^{n} \tan^{-1}\left(\frac{2}{1+4r^2-1}\right)$$

$$= \lim_{n\to\infty} \sum_{r=1}^{n} \tan^{-1}\left(\frac{(2r+1)-(2r-1)}{1+(2r+1)(2r-1)}\right)$$

$$= \lim_{n\to\infty} \sum_{r=1}^{n} (\tan^{-1}(2r+1) - \tan^{-1}(2r-1))$$

$$= \lim_{n\to\infty} \{(\tan^{-1}(2n+1) - \tan^{-1} 1)\}$$

$$= \tan^{-1} \infty - \tan^{-1} 1$$

$$= \frac{\pi}{2} - \frac{\pi}{4} = \frac{\pi}{4}$$

3. $$\lim_{n\to\infty} \sum_{r=1}^{n} \cot^{-1}\left(2^{r+1} + \frac{1}{2^r}\right)$$

$$= \lim_{n\to\infty} \sum_{r=1}^{n} \tan^{-1}\left(\frac{2^r}{1 + 2^r \cdot 2^{r+1}}\right)$$

$$= \lim_{n\to\infty} \sum_{r=1}^{n} \tan^{-1}\left(\frac{2^{r+1} - 2^r}{1 + 2^r \cdot 2^{r+1}}\right)$$

$$= \lim_{n\to\infty} \sum_{r=1}^{n} (\tan^{-1} 2^{r+1} - \tan^{-1} 2^r)$$

$$= \lim_{n\to\infty} [\tan^{-1} 2^{n+1} - \tan^{-1} 2]$$

$$= \tan^{-1} \infty - \tan^{-1} 2$$

$$= \frac{\pi}{2} - \tan^{-1} 2$$

$$= \cot^{-1} 2$$

4. $\because\ 3 = 1 + 1 + 1^2$

$7 = 1 + 2 + 2^2$

$13 = 1 + 3 + 3^2$

....

$\therefore\ \cot^{-1} 3 + \cot^{-1} 7 + \cot^{-1} 13 + \ldots\ldots$

$$= \lim_{n\to\infty} \sum_{r=1}^{n} \cot^{-1}(1 + r + r^2)$$

$$= \lim_{n\to\infty} \sum_{r=1}^{n} \tan^{-1}\left(\frac{1}{1 + r + r^2}\right)$$

$$= \lim_{n\to\infty} \sum_{r=1}^{n} \tan^{-1}\left(\frac{(r+1) - r}{1 + r(r+1)}\right)$$

$$= \lim_{n\to\infty} \sum_{r=1}^{n} [\tan^{-1}(r+1) - \tan^{-1} r]$$

$$= \lim_{n\to\infty} [\tan^{-1}(n+1) - \tan^{-1} 1]$$

$$= \tan^{-1}\infty - \frac{\pi}{4} = \frac{\pi}{2} - \frac{\pi}{4} = \frac{\pi}{4}$$

5. $\lim_{n\to\infty} \sum_{r=1}^{n} \tan^{-1}\left(\frac{2r}{1 - r^2 + r^4}\right)$

$$= \lim_{n\to\infty} \sum_{r=1}^{n} \tan^{-1}\left(\frac{(r^2 + r) - (r^2 - r)}{1 + (r^2 + r)(r^2 - r)}\right)$$

$$= \lim_{n\to\infty} \sum_{r=1}^{n} [\tan^{-1}(r^2 + r) - \tan^{-1}(r^2 - r)]$$

$$= \lim_{n\to\infty} [\tan^{-1}(n^2 + n) - \tan^{-1} 0]$$

$$= \tan^{-1}\infty - 0$$

$$= \frac{\pi}{2} - 0 = \frac{\pi}{2}$$

PASSAGE 3

1. $\because\ \cos\left(\frac{4\pi}{3}\right) = \cos\left(\pi + \frac{\pi}{3}\right) = -\cos\frac{\pi}{3} = -\frac{1}{2} < 0$

$\therefore\ \frac{\pi}{2} < \cos^{-1}\left(\cos\frac{4\pi}{3}\right) \le \pi$

$\Rightarrow\ \cos^{-1}\left(\cos\frac{4\pi}{3}\right) = \cos^{-1}\cos\left(2\pi - \frac{2\pi}{3}\right)$

$$= \cos^{-1}\cos\frac{2\pi}{3} = \frac{2\pi}{3}$$

and $\sin\left(\frac{4\pi}{3}\right) = \sin\left(\pi + \frac{\pi}{3}\right) = -\sin\frac{\pi}{3} = \frac{-\sqrt{3}}{2} < 0$

$\therefore\ -\frac{\pi}{2} \le \sin^{-1}\left(\sin\frac{4\pi}{3}\right) < 0$

$\therefore\ \sin^{-1}\left(\sin\frac{4\pi}{3}\right) = \sin^{-1}\left\{\sin\left(\pi - \frac{4\pi}{3}\right)\right\} = -\frac{\pi}{3}$

Hence, $\sin^{-1}\left(\sin\frac{4\pi}{3}\right) + \cos^{-1}\left(\cos\frac{4\pi}{3}\right) = -\frac{\pi}{3} + \frac{2\pi}{3} = \frac{\pi}{3}$

2. $\because\ 5 \approx 5 \times 57°$

$\approx 285°$

$\therefore\ \sin 5 < 0$ and $\cos 5 < 0$

$\Rightarrow\ \sin^{-1}(\sin 5) = \sin^{-1}\{\sin(\pi - 5)\} = \pi - 5$

and $\cos^{-1}(\cos 5) = \cos^{-1}\{\cos(2\pi - 5)\} = 2\pi - 5$

$\therefore\ \sin^{-1}(\sin 5) - \cos^{-1}(\cos 5) = (\pi - 5) - (2\pi - 5)$

$$= -\pi$$

3. $\because\ \tan^{-1}\tan\left(-\frac{3\pi}{4}\right) = \tan^{-1}\tan\left(-\frac{3\pi}{4} + \pi\right) = -\frac{3\pi}{4} + \pi = \frac{\pi}{4}$

and $\cot^{-1}\cot\left(-\frac{3\pi}{4}\right) = \cot^{-1}\cot\left(\pi - \frac{3\pi}{4}\right) = \pi - \frac{3\pi}{4} = \frac{\pi}{4}$

$\therefore\ \tan^{-1}\tan\left(-\frac{3\pi}{4}\right) + \cot^{-1}\left\{\cot\left(\frac{-3\pi}{4}\right)\right\} = \frac{\pi}{4} + \frac{\pi}{4} = \frac{\pi}{2}$

4. $\because\ \sin^{-1}[\cos\{\cos^{-1}(\cos x) + \sin^{-1}(\sin x)\}]$

$$= \sin^{-1}\{\cos(x + \pi - x)\} \text{ as } x \in \left(\frac{\pi}{2}, \pi\right)$$

$$= \sin^{-1}(\cos\pi) = \sin^{-1}(-1) = -\frac{\pi}{2}$$

5. **Case I** $|x| < 1$

or $x^2 < 1$

we have, $\cos^{-1}\left\{-\left(\frac{1 - x^2}{1 + x^2}\right)\right\} + \sin^{-1}\left(\frac{2x}{1 + x^2}\right) + \tan^{-1}\left\{-\left(\frac{2x}{1 - x^2}\right)\right\} = \frac{2\pi}{3}$

$\Rightarrow$

$\pi - \cos^{-1}\left(\frac{1 - x^2}{1 + x^2}\right) + \sin^{-1}\left(\frac{2x}{1 + x^2}\right) - \tan^{-1}\left(\frac{2x}{1 - x^2}\right) = \frac{2\pi}{3}$

$\Rightarrow\ \pi - 2\tan^{-1} x + 2\tan^{-1} x - 2\tan^{-1} x = \frac{2\pi}{3}$

$\Rightarrow\ 2\tan^{-1} x = \frac{\pi}{3}$

$\therefore\ \tan^{-1} x = \frac{\pi}{6}$

$\Rightarrow\ x = \tan\frac{\pi}{6} = \frac{1}{\sqrt{3}}$

Case II $|x| > 1$ or $x^2 > 1$

Then, $\cos^{-1}\left(\frac{1 - \frac{1}{x^2}}{1 + \frac{1}{x^2}}\right) + \sin^{-1}\left(\frac{2 \cdot 1/x}{1 + (1/x)^2}\right) + \tan^{-1}\left(\frac{2 \cdot 1/x}{1 - (1/x)^2}\right) = \frac{2\pi}{3}$

$\Rightarrow\ 2\tan^{-1}\left(\frac{1}{x}\right) + 2\tan^{-1}\left(\frac{1}{x}\right) + 2\tan^{-1}\left(\frac{1}{x}\right) = \frac{2\pi}{3}$

$\therefore\ \tan^{-1}\left(\frac{1}{x}\right) = \frac{\pi}{9}$

$\cot^{-1} x = 20°$

$\therefore\ x = \cot 20° > 1$

Hence, $x = \frac{1}{\sqrt{2}}$ and $\cot 20°$

$\therefore$ Number of solutions is 2.

Numerical Value Type

1. $\because |\sin^{-1} x| \le \frac{\pi}{2}$, then

$\sin^{-1} x = \sin^{-1} y = \sin^{-1} z = \frac{\pi}{2}$

$\Rightarrow \quad x = y = z = 1$

Then, $2500(x+y+z) - \frac{216}{(x^3+y^3+z^3)}$

$= 2500(1+1+1) - \frac{216}{(1^3+1^3+1^3)}$

$= 7500 - \frac{216}{3} = 7500 - 72 = 7428$

2. $\because \sin^{-1}\left(\sin\frac{7\pi}{6}\right) = \pi - \frac{7\pi}{6} = -\frac{\pi}{6}$

$\cos^{-1}\left(\cos\frac{2\pi}{3}\right) = \frac{2\pi}{3}$

$\tan^{-1}\left(\tan\frac{5\pi}{4}\right) = \frac{5\pi}{4} - \pi = \frac{\pi}{4}$

and $\cot^{-1}\left\{\cot\left(-\frac{\pi}{4}\right)\right\} = \pi - \frac{\pi}{4} = \frac{3\pi}{4}$

Hence, required value is

$\frac{1}{\pi}\left\{216 \times -\frac{\pi}{6} + 27 \times \frac{2\pi}{3} + 28 \times \frac{\pi}{4} + 200 \times \frac{3\pi}{4}\right\}$

$= -36 + 18 + 7 + 150$

$= 139$

3. $\because S = \sum_{r=1}^{50} \tan^{-1}\left(\frac{(1+r+r^2)-(1-r+r^2)}{1+(1+r+r^2)(1-r+r^2)}\right)$

$= \sum_{r=1}^{50} \tan^{-1}(1+r+r^2) - \tan^{-1}(1-r+r^2)$

$= \{at\ r = 50 \text{ value of } \tan^{-1}(1+r+r^2)\}$

$- \{\text{at } r = 1 \text{ value of } \tan^{-1}(1-r+r^2)\}$

$= \tan^{-1}(1+50+50^2) - \tan^{-1} 1$

$= \tan^{-1}\left(\frac{1+50+50^2-1}{1+1+50+50^2}\right)$

$= \tan^{-1}\left(\frac{2550}{2552}\right)$

$\therefore \quad \tan S = \frac{2550}{2552}$

$\therefore \quad 2550 \cot S = 2552$

4. $\tan^{-1}(x+1) + \tan^{-1}(x-1) = \tan^{-1} 3 - \tan^{-1} x$

$\Rightarrow \tan^{-1}\left\{\frac{(x+1)+(x-1)}{1-(x+1)(x-1)}\right\} = \tan^{-1}\left\{\frac{3-x}{1+3x}\right\}$

$\Rightarrow \quad \frac{2x}{2-x^2} = \frac{3-x}{1+3x}$

$\Rightarrow \quad 2x(1+3x) = (2-x^2)(3-x)$

$\Rightarrow \quad x^3 - 9x^2 - 4x + 6 = 0$

$\Rightarrow \quad (x+1)(x^2 - 10x + 6) = 0$

$\because \quad x = -1, 5 \pm \sqrt{19}$

$\because \quad x < 0 \therefore x = -1$

Then, $500x^4 + 270x^2 + 997$

$= 500(-1)^4 + 270(-1)^2 + 997$

$= 500 + 270 + 997 = 1767$

5. $\because \lambda = [\cos\{\tan^{-1}(\sin(\cot^{-1} 5))\}]^4$

$= \left[\cos\left\{\tan^{-1}\left(\sin\left(\sin^{-1}\frac{1}{\sqrt{26}}\right)\right)\right\}\right]^4$

$= \left[\cos\left(\tan^{-1}\left(\frac{1}{\sqrt{26}}\right)\right)\right]^4$

$= \left[\cos\left(\cos^{-1}\sqrt{\frac{26}{27}}\right)\right]^4 = \left(\sqrt{\frac{26}{27}}\right)^4 = \left(\frac{26}{27}\right)^2$

$\therefore \quad 3645\lambda = 3645 \times \frac{(26)^2}{(27)^2}$

$= 3380$

6. $\because \theta = \cot^{-1} 7 + \cot^{-1} 8 + \cot^{-1} 18$

$= \tan^{-1}\left(\frac{1}{7}\right) + \tan^{-1}\left(\frac{1}{8}\right) + \tan^{-1}\left(\frac{1}{18}\right)$

$= \tan^{-1}\left(\frac{\frac{1}{7}+\frac{1}{8}+\frac{1}{18}-\frac{1}{7}\cdot\frac{1}{8}\cdot\frac{1}{18}}{1-\frac{1}{7}\cdot\frac{1}{8}-\frac{1}{8}\cdot\frac{1}{18}-\frac{1}{18}\cdot\frac{1}{7}}\right)$

$= \tan^{-1}\left(\frac{1}{3}\right) \Rightarrow \tan\theta = \frac{1}{3}$

$\therefore \tan 2\theta = \frac{2\tan\theta}{1-\tan^2\theta} = \frac{2/3}{1-\frac{1}{9}} = \frac{3}{4} = 0.75$

7. $\cos^{-1} x = \alpha + \cos^{-1}\left(\frac{y}{2}\right)$

$\Rightarrow x = (\cos\alpha)\left(\frac{y}{2}\right) - \sin\alpha\sqrt{1-\frac{y^2}{4}}$

$\Rightarrow (2x - y\cos\alpha)^2 = (4-y)^2 \sin^2\alpha$

$\Rightarrow 4x^2 - 4xy\cos\alpha + y^2 = 4\sin^2\alpha$

and hence the given expression is equal to $1928 \sin^2 \alpha$.

$\therefore \quad a = 1928$

8. $\because \frac{\pi}{2} - \cos^{-1}\pi + \frac{\pi}{2} - \cos^{-1} y = \pi$

or $\quad \cos^{-1} x + \cos^{-1} y = 0$

$\Rightarrow \quad \cos^{-1}\{xy - \sqrt{(1-x^2)}\sqrt{(1-y)^2}\} = 0$

$\Rightarrow \quad xy - \sqrt{(1-x^2)(1-y^2)} = 1$

$\Rightarrow \quad (xy-1)^2 = (1-x^2)(1-y^2)$

$\Rightarrow \quad x^2y^2 + 1 - 2xy = 1 - x^2 - y^2 + x^2y^2$

or $\quad x^2 + y^2 - 2xy = 0$

$\Rightarrow \quad (x-y)^2 = 0$

$\therefore \quad x = y$

$\therefore \quad \lambda = 1$

Then, $39^{2\lambda} + 5^{\lambda} = 39^2 + 5 = 1521 + 5$

$= 1526$

Single Integer Type

1. We have,

$$\tan^{-1}\frac{1}{5}+\frac{1}{2}\sec^{-1}x+\tan^{-1}\frac{1}{8}=\frac{\pi}{8}$$

$$\Rightarrow \left(\tan^{-1}\frac{1}{5}+\tan^{-1}\frac{1}{8}\right)+\frac{1}{2}\sec^{-1}x=\frac{\pi}{8}$$

$$\Rightarrow \tan^{-1}\left[\frac{\frac{1}{5}+\frac{1}{8}}{1-\frac{1}{5}\times\frac{1}{8}}\right]+\frac{1}{2}\sec^{-1}x=\frac{\pi}{8}$$

$$\Rightarrow \frac{1}{2}\tan^{-1}\frac{13}{39}+\frac{1}{2}\sec^{-1}x=\frac{\pi}{8}$$

$$\Rightarrow 2\tan^{-1}\frac{13}{39}+\sec^{-1}x=\frac{\pi}{4}$$

$$\Rightarrow 2\tan^{-1}\frac{1}{3}+\sec^{-1}x=\frac{\pi}{4}$$

$$\Rightarrow \tan^{-1}\frac{\frac{2}{3}}{1-\frac{1}{9}}+\sec^{-1}x=\frac{\pi}{4}$$

$$\Rightarrow \tan^{-1}\frac{3}{4}+\tan^{-1}\sqrt{x^2-1}=\frac{\pi}{4}$$

$$\Rightarrow \tan^{-1}\left[\frac{\frac{3}{4}+\sqrt{x^2-1}}{1-\frac{3}{4}\sqrt{x^2-1}}\right]=\frac{\pi}{4}$$

$$\Rightarrow \frac{3+4\sqrt{x^2-1}}{4-3\sqrt{x^2-1}}=1$$

$$\Rightarrow 3+4\sqrt{x^2-1}=4-3\sqrt{x^2-1}$$

$$\Rightarrow 7\sqrt{x^2-1}=1$$

$$\Rightarrow \sqrt{x^2-1}=\frac{1}{7}\Rightarrow x^2-1=\frac{1}{49}$$

$$\Rightarrow x^2=\frac{50}{49}\Rightarrow\frac{49}{25}x^2=2$$

2. $\because \sum_{k=0}^{10}\sec\left(\frac{7\pi}{12}+\frac{k\pi}{2}\right)\sec\left(\frac{7\pi}{12}+\frac{(k+1)\pi}{2}\right)$

$$=\sum_{k=0}^{10}\frac{1}{\cos\left(\frac{7\pi}{12}+\frac{k\pi}{2}\right)\cos\left(\frac{7\pi}{12}+\frac{(k+1)\pi}{2}\right)}$$

$$=\sum_{k=0}^{10}\frac{\sin\left[\left(\frac{7\pi}{12}+\frac{(k+1)\pi}{2}\right)-\left(\frac{7\pi}{12}+\frac{k\pi}{2}\right)\right]}{\cos\left(\frac{7\pi}{12}+\frac{k\pi}{2}\right)\cos\left(\frac{7\pi}{12}+\frac{(k+1)\pi}{2}\right)}$$

$$\left[\because \frac{7\pi}{12}+\frac{(k+1)\pi}{2}-\left(\frac{7\pi}{12}+\frac{k\pi}{2}\right)=\frac{\pi}{2}\text{ and }\sin\frac{\pi}{2}=1\right]$$

$$=\sum_{k=0}^{10}\frac{\sin\left(\frac{7\pi}{12}+\frac{(k+1)\pi}{2}\right)\cos\left(\frac{7\pi}{12}+\frac{k\pi}{2}\right)-\sin\left(\frac{7\pi}{12}+\frac{k\pi}{2}\right)\cos\left(\frac{7\pi}{12}+\frac{(k+1)\pi}{2}\right)}{\cos\left(\frac{7\pi}{12}+\frac{k\pi}{2}\right)\cos\left(\frac{7\pi}{12}+\frac{(k+1)\pi}{2}\right)}$$

$$=\sum_{k=0}^{10}\left[\tan\left(\frac{7\pi}{12}+\frac{(k+1)\pi}{2}\right)-\tan\left(\frac{7\pi}{12}+\frac{k\pi}{2}\right)\right]$$

$$=\tan\left(\frac{7\pi}{12}+\frac{\pi}{2}\right)-\tan\left(\frac{7\pi}{12}\right)$$

$$+\tan\left(\frac{7\pi}{12}+\frac{2\pi}{2}\right)-\tan\left(\frac{7\pi}{12}+\frac{\pi}{2}\right)$$

$$\vdots$$

$$+\tan\left(\frac{7\pi}{12}+\frac{11\pi}{2}\right)-\tan\left(\frac{7\pi}{12}+\frac{10\pi}{2}\right)$$

$$=\tan\left(\frac{7\pi}{12}+\frac{11\pi}{2}\right)-\tan\frac{7\pi}{12}=\tan\frac{\pi}{12}+\cot\frac{\pi}{12}$$

$$=\frac{1}{\sin\frac{\pi}{12}\cos\frac{\pi}{12}}=\frac{2}{\sin\frac{\pi}{6}}=4$$

So, $\sec^{-1}\left(\frac{1}{4}\sum_{k=0}^{10}\sec\left(\frac{7\pi}{12}+\frac{k\pi}{2}\right)\sec\left(\frac{7\pi}{12}+\frac{(k+1)\pi}{2}\right)\right)$

$$=\sec^{-1}(1)=0$$

3. Given, $\angle C=\frac{\pi}{2}\Rightarrow a^2+b^2=c^2$

Clearly, $|x|\le 1$ and $\frac{a^2x^2}{c^2}+\frac{b^2x^2}{c^2}=x^2\le 1$

From the given equation,

$$\Rightarrow x=\frac{ax}{c}\sqrt{1-\frac{b^2x^2}{c^2}}+\frac{bx}{c}\sqrt{1-\frac{a^2x^2}{c^2}}$$

$$\Rightarrow x=0\text{ or }c^2=a\sqrt{c^2-b^2x^2}+b\sqrt{c^2-a^2x^2}$$

$$\Rightarrow x=0\text{ or }c^4=a^2(c^2-b^2x^2)+b^2(c^2-a^2x^2)+2ab\sqrt{c^2-b^2x^2}\sqrt{c^2-a^2x^2}$$

$$\Rightarrow a^2b^2x^4=c^4-(a^2+b^2)c^2x^2+a^2b^2x^4\text{ or }x=0$$

$$\Rightarrow c^4=c^4x^2\text{ or }x=0$$

$$\Rightarrow x=0\text{ or }\pm 1$$

Hence, total number of different values of x are three; $x=0,-1,1$. It is easy to see that all these values satisfy the equation.

4. We have,

$$\sin^{-1}\left(\sum_{i=1}^{\infty}x^{i+1}-x\sum_{i=1}^{\infty}\left(\frac{x}{2}\right)^i\right)=\frac{\pi}{2}-\cos^{-1}\left(\sum_{i=1}^{\infty}\left(\frac{-x}{2}\right)^i-\sum_{i=1}^{\infty}(-x)^i\right)$$

$$\Rightarrow \sin^{-1}\left[\frac{x^2}{1-x}-\frac{x\cdot\frac{x}{2}}{1-\frac{x}{2}}\right]=\frac{\pi}{2}-\cos^{-1}\left[\frac{\frac{-x}{2}}{1+\frac{x}{2}}-\frac{(-x)}{1+x}\right]$$

$$\left[\because \sum_{i=1}^{\infty}x^{i+1}=x^2+x^3+x^4+\ldots=\frac{x^2}{1-x}\right.$$
using sum of infinite terms of GP]

$$\Rightarrow \sin^{-1}\left[\frac{x^2}{1-x}-\frac{x^2}{2-x}\right]=\frac{\pi}{2}-\cos^{-1}\left[\frac{x}{1+x}-\frac{x}{2+x}\right]$$

$$\Rightarrow \sin^{-1}\left[\frac{x^2}{1-x}-\frac{x^2}{2-x}\right]=\sin^{-1}\left(\frac{x}{1+x}-\frac{x}{2+x}\right)$$

$$\left[\because \sin^{-1}x=\frac{\pi}{2}-\cos^{-1}x\right]$$

$$\Rightarrow \frac{x^2}{1-x}-\frac{x^2}{2-x}=\frac{x}{1+x}-\frac{x}{2+x}$$

$$\Rightarrow x^2\left(\frac{2-x-1+x}{(1-x)(2-x)}\right)=x\frac{(2+x-1-x)}{(1+x)(2+x)}$$

$$\Rightarrow \frac{x}{2-3x+x^2}=\frac{1}{2+3x+x^2}$$

or $x=0$

$\Rightarrow x^3+3x^2+2x=x^2-3x+2$

$\Rightarrow x^3+2x^2+5x-2=0$ or $x=0$

Let $f(x)=x^3+2x^2+5x-2$

$f'(x)=3x^2+4x+5$

$f'(x)>0,\ \forall\, x\in R$

$\therefore$ x^3+2x^2+5x-2 has only one real roots

Therefore, total number of real solution is 2.

5. We have, $\cos(2\sin^{-1}(\cot(\tan^{-1}(\sec(6\,\text{cosec}^{-1}x)))))=-1$

$$\Rightarrow \sin^{-1}(\cot(\tan^{-1}(\sec(6\,\text{cosec}^{-1}x))))=\pm\frac{\pi}{2}$$

$$\Rightarrow \cot(\tan^{-1}(\sec(6\,\text{cosec}^{-1}x)))=\pm 1$$

$$\Rightarrow \tan^{-1}(\sec(6\,\text{cosec}^{-1}x))=\frac{\pi}{4},\frac{3\pi}{4}$$

$$\Rightarrow \sec(6\,\text{cosec}^{-1}x)=\pm 1$$

$$\Rightarrow 6\,\text{cosec}^{-1}x=\pm 3\pi, \pm 2\pi, \pm\pi$$

$$\Rightarrow \text{cosec}^{-1}x=\pm\frac{\pi}{2},\pm\frac{\pi}{3},\pm\frac{\pi}{6}$$

$$\Rightarrow x=1,\frac{2}{\sqrt{3}},2 \qquad [\because x>0]$$

6. $\theta=\tan^{-1}(2\tan^2\theta)-\tan^{-1}\left(\left(\frac{1}{3}\right)\tan\theta\right)$

$$\Rightarrow \tan\theta=\frac{2\tan^2\theta-\left(\frac{1}{3}\right)\tan\theta}{1+\left(\frac{2}{3}\right)\tan^3\theta}$$

$$\Rightarrow \tan\theta\left[\frac{2\tan\theta-\left(\frac{1}{3}\right)}{1+\left(\frac{2}{3}\right)\tan^3\theta}-1\right]=0$$

which is true if $\tan\theta=0$ or $\dfrac{2\tan\theta-\left(\frac{1}{3}\right)}{1+\left(\frac{2}{3}\right)\tan^3\theta}=1$

$\Rightarrow \tan^3\theta-3\tan\theta+2=0$

$\Rightarrow (\tan\theta-1)^2(\tan\theta+2)=0$ which holds, if $\tan\theta=1$ or $\tan\theta=-2$

7. $f(x)=\sin^{-1}x+2\tan^{-1}x+(x+2)^2-3$

Domain of $f(x)$ is $[-1,1]$

Also, $f(x)$ is an increasing function in the domain. Therefore,

$$p=f_{\min}(x)=f(-1)=-\frac{\pi}{2}+2\left(\frac{-\pi}{4}\right)+1-3=-\pi-2$$

and $q=f_{\max}(x)=f(1)=\dfrac{\pi}{2}+2\left(\dfrac{\pi}{4}\right)+9-3=\pi+6$

Therefore, the range of $f(x)$ is $[-\pi-2,\pi+6]$

Hence, $(p+q)=4$

Matrix-Match Type

1. (A) $\lambda=-\dfrac{\pi}{6}+\dfrac{\pi}{3}=\dfrac{\pi}{6}$ and $\mu=\dfrac{2\pi}{3}$

$\therefore$ $\lambda+\mu=\dfrac{\pi}{6}+\dfrac{2\pi}{3}=\dfrac{5\pi}{6}$ (T)

and $\mu-\lambda=\dfrac{2\pi}{3}-\dfrac{\pi}{6}=\dfrac{\pi}{2}$ (Q)

(B) $\lambda=\pi-\dfrac{7\pi}{6}=-\dfrac{\pi}{6}$

and $\mu=\cos^{-1}\left(-\sin\dfrac{5\pi}{6}\right)$

$$=\cos^{-1}\left(-\frac{1}{2}\right)$$

$$=\cos^{-1}\left(\cos\frac{2\pi}{3}\right)=\frac{2\pi}{3}$$

$\therefore$ $\lambda+\mu=-\dfrac{\pi}{6}+\dfrac{2\pi}{3}=\dfrac{\pi}{2}$ (P)

and $\mu-\lambda=\dfrac{2\pi}{3}+\dfrac{\pi}{6}=\dfrac{5\pi}{6}$ (S)

(C) $\lambda=\sin^{-1}\left(-\dfrac{\sqrt{3}}{2}\right)=\sin^{-1}\left(\sin\left(-\dfrac{\pi}{3}\right)\right)=-\dfrac{\pi}{3}$

$$\mu=\sin^{-1}\left(\cos\left(\sin^{-1}\frac{\sqrt{3}}{2}\right)\right)=\sin^{-1}\left(\cos\frac{\pi}{3}\right)$$

$$=\sin^{-1}\left(\frac{1}{2}\right)=\frac{\pi}{6}$$

$\therefore$ $\lambda+\mu=-\dfrac{\pi}{6}$ (R) and $\mu-\lambda=\dfrac{\pi}{2}$ (Q)

2. (A) $\because \tan^{-1}x+\tan^{-1}y+\tan^{-1}z=\pi$

$$\Rightarrow \tan^{-1}\left(\frac{x+y+z-xyz}{1-xy-yz-zx}\right)=\pi$$

$$\Rightarrow \frac{x+y+z-xyz}{1-xy-yz-zx}=0$$

$\therefore$ $x+y+z-xyz=0$

or $x+y+z=xyz$ (R)

Also, $\text{AM} \geq \text{GM}$

$\Rightarrow \quad \frac{x+y+z}{3} \geq (xyz)^{1/3} \Rightarrow \frac{xyz}{3} \geq (xyz)^{1/3}$

$[\because x+y+z = xyz]$

$\Rightarrow \quad (xyz)^{2/3} \geq 3$

or $\quad xyz \geq 3\sqrt{3}$ (Q)

(B) $\because \tan^{-1} x + \tan^{-1} y + \tan^{-1} z = \frac{\pi}{2}$

$\Rightarrow \quad \tan^{-1}\left(\frac{x+y+z-xyz}{1-xy-yz-zx}\right) = \frac{\pi}{2}$

$\Rightarrow \quad \frac{x+y+z-xyz}{1-xy-yz-zx} = \infty$

or $\quad 1-xy-yz-zx = 0$

$\therefore \quad xy+yz+zx = 1$ (T)

Also, $\text{AM} \geq \text{GM}$

$\Rightarrow \quad \frac{xy+yz+zx}{3} \geq (xy \cdot yz \cdot zx)^{1/3}$

$\Rightarrow \quad \frac{1}{3} \geq (x^2y^2z^2)^{1/3} \quad [\because xy+yz+zx = 1]$

$\Rightarrow \quad \left(\frac{1}{3}\right)^{3/2} \geq xyz$ or $xyz \leq \frac{1}{3\sqrt{3}}$ (S)

(C) $\because \tan^{-1} x + \tan^{-1} y + \tan^{-1} z = \frac{\pi}{2}$

Then $xy + yz + zx = 1$ (T) (from above)

and $\quad x+y+z = \sqrt{3}$

$\because \quad (x+y+z)^2 = x^2+y^2+z^2+2(xy+yz+zx)$

$\Rightarrow \quad 3 = x^2+y^2+z^2+2$

$\therefore \quad x^2+y^2+z^2 = 1$

Hence, $(x-y)^2+(y-z)^2+(z-x)^2 = 2(x^2+y^2+z^2-xy-yz-zx)$

$= 2(1-1) = 0$

$\therefore \quad x-y=0,\ y-z=0$

and $\quad z-x=0$

$\therefore \quad x=y=z$ (P)

$\because \quad \text{AM} \geq \text{GM}$

$\Rightarrow \quad \frac{x+y+z}{3} \geq (xyz)^{1/3}$

$\Rightarrow \quad \frac{\sqrt{3}}{3} \geq (xyz)^{1/3}$

$\therefore \quad xyz \leq \frac{1}{3\sqrt{3}}$ (S)

32

Vectors

Objective Questions Type I [Only one correct answer]

In each of the questions below, four choices are given of which only one is correct. You have to select the correct answer which is the most appropriate.

1. If $|\mathbf{a}| = 2$ and $|\mathbf{b}| = 3$ and $\mathbf{a} \cdot \mathbf{b} = 0$, then $(\mathbf{a} \times (\mathbf{a} \times (\mathbf{a} \times (\mathbf{a} \times \mathbf{b}))))$ is equal to

(a) $48\,\mathbf{b}$ (b) $-48\,\mathbf{b}$
(c) $48\,\mathbf{a}$ (d) $-48\,\mathbf{a}$

2. A rigid body is spinning about a fixed point $(3, -2, -1)$ with angular velocity of 4 rad/s, the axis of rotation being in the direction of $(1, 2, -2)$, then the velocity of the particle at the point $(4, 1, 1)$ is

(a) $\frac{4}{3}(\hat{\mathbf{i}} - 4\hat{\mathbf{j}} + 10\hat{\mathbf{k}})$ (b) $\frac{4}{3}(4\hat{\mathbf{i}} - 10\hat{\mathbf{j}} + \hat{\mathbf{k}})$
(c) $\frac{4}{3}(10\hat{\mathbf{i}} - 4\hat{\mathbf{j}} + \hat{\mathbf{k}})$ (d) $\frac{4}{3}(10\hat{\mathbf{i}} + 4\hat{\mathbf{j}} + \hat{\mathbf{k}})$

3. If $\mathbf{a}, \mathbf{b}, \mathbf{c}$ are such that $[\mathbf{a}, \mathbf{b}, \mathbf{c}] = 1$, $\mathbf{c} = \lambda \mathbf{a} \times \mathbf{b}$, $\mathbf{a} \wedge \mathbf{b} < \frac{2\pi}{3}$ and $|\mathbf{a}| = \sqrt{2}, |\mathbf{b}| = \sqrt{3}, |\mathbf{c}| = \frac{1}{\sqrt{3}}$, then the angle between $\mathbf{a}$ and $\mathbf{b}$ is

(a) $\frac{\pi}{6}$ (b) $\frac{\pi}{4}$
(c) $\frac{\pi}{3}$ (d) $\frac{\pi}{2}$

4. The values of x for which the angle between the vectors $2x^2\hat{\mathbf{i}} + 4x\hat{\mathbf{j}} + \hat{\mathbf{k}}$ and $7\hat{\mathbf{i}} - 2\hat{\mathbf{j}} + x\hat{\mathbf{k}}$ are obtuse and the angle between the Z-axis and $7\hat{\mathbf{i}} - 2\hat{\mathbf{j}} + x\hat{\mathbf{k}}$ is acute and less than $\frac{\pi}{6}$ is given by

(a) $0 < x < \frac{1}{2}$
(b) $x > \frac{1}{2}$ or $x < 0$
(c) $\frac{1}{2} < x < 15$
(d) there is no such value for x

5. Let $\mathbf{a}, \mathbf{b}$ and $\mathbf{c}$ be mutually perpendicular vectors of the same magnitude. If a vector $\mathbf{x}$ satisfies the equation $\mathbf{a} \times [(\mathbf{x} - \mathbf{b}) \times \mathbf{a}] + \mathbf{b} \times [(\mathbf{x} - \mathbf{c}) \times \mathbf{b}] + \mathbf{c} \times [(\mathbf{x} - \mathbf{a}) \times \mathbf{c}] = 0$, then $\mathbf{x}$ is given by

(a) $\frac{1}{2}(\mathbf{a} + \mathbf{b} + \mathbf{c})$ (b) $\frac{1}{3}(2\,\mathbf{a} + \mathbf{b} + \mathbf{c})$
(c) $\frac{1}{3}(\mathbf{a} + \mathbf{b} + \mathbf{c})$ (d) $\frac{1}{2}(\mathbf{a} + \mathbf{b} - 2\,\mathbf{c})$

6. A unit tangent vector at $t = 2$ on the curve $x = t^2 + 2, y = 4t - 5, z = 2t^2 - 6t$ is

(a) $\frac{1}{\sqrt{3}}(\hat{\mathbf{i}} + \hat{\mathbf{j}} + \hat{\mathbf{k}})$ (b) $\frac{1}{3}(2\hat{\mathbf{i}} + 2\hat{\mathbf{j}} + \hat{\mathbf{k}})$
(c) $\frac{1}{\sqrt{6}}(2\hat{\mathbf{i}} + \hat{\mathbf{j}} + \hat{\mathbf{k}})$ (d) None of these

7. When a right handed rectangular cartesian system $OXYZ$ is rotated about the Z-axis through an angle $\frac{\pi}{4}$ in the counter-clockwise direction it is found that a vector $\mathbf{a}$ has the components $2\sqrt{2}$, $3\sqrt{2}$ and 4. The components of $\mathbf{a}$ in the $OXYZ$ coordinate system are

(a) 5, 1, 4 (b) 5, -1, $4\sqrt{2}$
(c) -1, -5, $4\sqrt{2}$ (d) None of these

8. Given three vectors $\mathbf{a} = 6\hat{\mathbf{i}} - 3\hat{\mathbf{j}}$, $\mathbf{b} = 2\hat{\mathbf{i}} - 6\hat{\mathbf{j}}$ and $\mathbf{a} = -2\hat{\mathbf{i}} + 21\hat{\mathbf{j}}$ such that $\alpha = \mathbf{a} + \mathbf{b} + \mathbf{c}$. Then the resolution of the vector α into components with respect to $\mathbf{a}$ and $\mathbf{b}$ is given by

(a) $3\,\mathbf{a} - 2\mathbf{b}$ (b) $2\,\mathbf{a} - 3\,\mathbf{b}$
(c) $3\,\mathbf{b} - 2\,\mathbf{a}$ (d) None of these

9. For any four points P, Q, R, S, $|\mathbf{PQ} \times \mathbf{RS} - \mathbf{QR} \times \mathbf{PS} + \mathbf{RP} \times \mathbf{QS}|$ is equal to 4 times the area of the Δ

(a) PQR (b) QRS
(c) PRS (d) PQS

10. If $\sum_{i=1}^{n} \mathbf{a}_i = 0$ where $|\mathbf{a}_i| = 1 \,\forall\, i$, then the value of $\sum_{1 \le i < j \le n} \Sigma\, \mathbf{a}_i \cdot \mathbf{a}_j$ is

(a) $-n/2$ (b) $-n$ (c) $n/2$ (d) n

11. If **a**, **b**, **c** are unit vectors, then $|\mathbf{a}-\mathbf{b}|^2+|\mathbf{b}-\mathbf{c}|^2+|\mathbf{c}-\mathbf{a}|^2$ does not exceed

(a) 4 (b) 9
(c) 8 (d) 6

12. Let $a, b, c \in R$ be such that $a^2+b^2+c^2=1$. If $a\cos\theta = b\cos\left(\theta+\frac{2\pi}{3}\right) = c\cos\left(\theta+\frac{4\pi}{3}\right)$, where $\theta=\frac{\pi}{9}$, then the angle between the vectors $a\hat{\mathbf{i}}+b\hat{\mathbf{j}}+c\hat{\mathbf{k}}$ and $b\hat{\mathbf{i}}+c\hat{\mathbf{j}}+a\hat{\mathbf{k}}$ is **[JEE Main 2020, 3 September Shift-II]**

(a) $\frac{\pi}{2}$ (b) 0
(c) $\frac{2\pi}{3}$ (d) $\frac{\pi}{9}$

13. If $\mathbf{a}\times\mathbf{b}=\mathbf{c}$ and $\mathbf{b}\times\mathbf{c}=\mathbf{a}$, then

(a) $|\mathbf{a}|=1, |\mathbf{b}|=|\mathbf{c}|$
(b) $|\mathbf{c}|=1, |\mathbf{a}|=1$
(c) $|\mathbf{b}|=2, |\mathbf{b}|=2|\mathbf{a}|$
(d) $|\mathbf{b}|=1, |\mathbf{c}|=|\mathbf{a}|$

14. Let **u** and **v** are unit vectors such that $\mathbf{u}\times\mathbf{v}+\mathbf{u}=\mathbf{w}$ and $\mathbf{w}\times\mathbf{u}=\mathbf{v}$, then the value of $[\mathbf{u}\ \mathbf{v}\ \mathbf{w}]$ is

(a) 1 (b) –1
(c) 0 (d) None of these

15. Let **a** and **b** be two non-zero vectors perpendicular to each other and $|\mathbf{a}|=|\mathbf{b}|$. If $|\mathbf{a}\times\mathbf{b}|=|\mathbf{a}|$, then the angle between the vectors $(\mathbf{a}+\mathbf{b}+(\mathbf{a}\times\mathbf{b}))$ and **a** is equal to **[JEE Main 2021, 18 March]**

(a) $\sin^{-1}\left(\frac{1}{\sqrt{3}}\right)$ (b) $\cos^{-1}\left(\frac{1}{\sqrt{3}}\right)$
(c) $\cos^{-1}\left(\frac{1}{\sqrt{2}}\right)$ (d) $\sin^{-1}\left(\frac{1}{\sqrt{6}}\right)$

16. If $\alpha+\beta+\gamma=a\delta$ and $\beta+\gamma+\delta=b\alpha$ and α, β, γ are non-coplanar and α is not parallel to δ, then $\alpha+\beta+\gamma+\delta$ equals

(a) $a\alpha$ (b) $b\delta$
(c) 0 (d) $(a+b)\gamma$

17. If unit vector **c** makes an angle $\frac{\pi}{3}$ with $\hat{\mathbf{i}}+\hat{\mathbf{j}}$, then minimum and maximum values of $(\hat{\mathbf{i}}\times\hat{\mathbf{j}})\cdot\mathbf{c}$ respectively are

(a) $0, \frac{\sqrt{3}}{2}$ (b) $-\frac{\sqrt{3}}{2}, \frac{\sqrt{3}}{2}$
(c) $-1, \frac{\sqrt{3}}{2}$ (d) None of these

18. Let O be the origin. Let $\mathbf{OP}=x\hat{\mathbf{i}}+y\hat{\mathbf{j}}-\hat{\mathbf{k}}$ and $\mathbf{OQ}=-\hat{\mathbf{i}}+2\hat{\mathbf{j}}+3x\hat{\mathbf{k}}$, $x, y\in R$, $x>0$ be such that $|\mathbf{PQ}|=\sqrt{20}$ and the vector **OP** is perpendicular to **OQ**. If $\mathbf{OR}=3\hat{\mathbf{i}}+z\hat{\mathbf{j}}-7\hat{\mathbf{k}}$, $z\in R$ is coplanar with **OP** and **OQ**. Then, the value of $x^2+y^2+z^2$ is **[JEE Main 2021, 17 March]**

(a) 7 (b) 9
(c) 2 (d) 1

19. A non-zero vectors **a** is such that its projections along the vectors $\frac{\hat{\mathbf{i}}+\hat{\mathbf{j}}}{\sqrt{2}}$ and $\frac{-\hat{\mathbf{i}}+\hat{\mathbf{j}}}{\sqrt{2}}$ and $\hat{\mathbf{k}}$ are equal then unit vector along **a** is

(a) $\frac{\sqrt{2}\hat{\mathbf{j}}-\hat{\mathbf{k}}}{\sqrt{3}}$ (b) $\frac{\hat{\mathbf{j}}-\sqrt{2}\hat{\mathbf{k}}}{\sqrt{3}}$
(c) $\frac{\sqrt{2}}{\sqrt{3}}\hat{\mathbf{j}}+\frac{\hat{\mathbf{k}}}{\sqrt{3}}$ (d) $\frac{\hat{\mathbf{j}}-\hat{\mathbf{k}}}{\sqrt{2}}$

20. $\mathbf{a}=\hat{\mathbf{i}}-\hat{\mathbf{j}}+\hat{\mathbf{k}}$, $\mathbf{b}=\hat{\mathbf{i}}-2\hat{\mathbf{j}}+\hat{\mathbf{k}}$, $\mathbf{c}=p\hat{\mathbf{i}}+2\hat{\mathbf{j}}+q\hat{\mathbf{k}}$ and $\mathbf{d}=p\hat{\mathbf{i}}+q\hat{\mathbf{j}}+2\hat{\mathbf{k}}$ are given vectors. If the projection of **c** on a is $5\sqrt{3}$ units and, if **a**, **b** and **c** form a parallelopiped of volume 5 cu units, then $\tan^{-1}(\mathbf{b}\cdot\mathbf{d})=$

(a) $\frac{\pi}{2}$ (b) $\frac{\pi}{3}$
(c) $\frac{\pi}{4}$ (d) $\frac{\pi}{6}$

21. If **p** and **q** are two unit vectors inclined at an angle α to each other, then $|\mathbf{p}+\mathbf{q}|<1$ if

(a) $\frac{2\pi}{3}<\alpha<\frac{4\pi}{3}$ (b) $\alpha<\frac{\pi}{3}$
(c) $\alpha>\frac{2\pi}{3}$ (d) $\alpha=\frac{\pi}{2}$

22. If **a** and **b** are unit vectors and θ is the angle between them, then $\left|\frac{\mathbf{a}-\mathbf{b}}{2}\right|$ is

(a) $\sin\frac{\theta}{2}$ (b) $\sin\theta$
(c) $2\sin\theta$ (d) $\sin 2\theta$

23. If $\mathbf{a}+\mathbf{b}+\mathbf{c}=0, |\mathbf{a}|=3, |\mathbf{b}|=5, |\mathbf{c}|=7$, then the angle between **a** and **b** is

(a) $\pi/6$ (b) $\pi/3$
(c) $2\pi/3$ (d) $5\pi/3$

24. $\mathbf{AB}=3\hat{\mathbf{i}}+\hat{\mathbf{j}}-\hat{\mathbf{k}}$ and $\mathbf{AC}=\hat{\mathbf{i}}-\hat{\mathbf{j}}+3\hat{\mathbf{k}}$. If the point P on the line segment BC is equidistant from AB and AC, then **AP** is

(a) $2\hat{\mathbf{i}}-\hat{\mathbf{k}}$ (b) $\hat{\mathbf{i}}-2\hat{\mathbf{k}}$
(c) $2\hat{\mathbf{i}}+\hat{\mathbf{k}}$ (d) None of these

25. Vectors **a** and **b** are inclined at an angle $\theta=120°$. If $|\mathbf{a}|=1, |\mathbf{b}|=2$, then $\{(\mathbf{a}+3\mathbf{b})\times(3\mathbf{a}-\mathbf{b})\}^2$ is equal to

(a) 225 (b) 275
(c) 325 (d) 300

26. Let O be the origin and let PQR be an arbitrary triangle. The point S is such that

$$\mathbf{OP}\cdot\mathbf{OQ}+\mathbf{OR}\cdot\mathbf{OS}=\mathbf{OR}\cdot\mathbf{OP}+\mathbf{OQ}\cdot\mathbf{OS}=\mathbf{OQ}\cdot\mathbf{OR}+\mathbf{OP}\cdot\mathbf{OS}$$

Then the ΔPQR has S as its **[JEE Advanced 2017]**

(a) centroid
(b) orthocentre
(c) incentre
(d) circumcentre

27. If **a**, **b**, **c** are three non-coplanar vectors, then $[\mathbf{a}\times\mathbf{b}, \mathbf{b}\times\mathbf{c}, \mathbf{c}\times\mathbf{a}]$ is equal to
(a) 0 (b) $[\mathbf{a}\ \mathbf{b}\ \mathbf{c}]$
(c) $[\mathbf{a}\ \mathbf{b}\ \mathbf{c}]^2$ (d) $2[\mathbf{a}\ \mathbf{b}\ \mathbf{c}]$

28. If **a**, **b**, **c** are non-coplanar vectors, then $\frac{\mathbf{a}\cdot(\mathbf{b}\times\mathbf{c})}{\mathbf{b}\cdot(\mathbf{c}\times\mathbf{a})}+\frac{\mathbf{b}\cdot(\mathbf{c}\times\mathbf{a})}{\mathbf{c}\cdot(\mathbf{a}\times\mathbf{b})}+\frac{\mathbf{c}\cdot(\mathbf{b}\times\mathbf{a})}{\mathbf{a}\cdot(\mathbf{b}\times\mathbf{c})}$ is equal to
(a) 0 (b) 1
(c) 2 (d) None of these

29. If **a**, **b**, **c** are three non-coplanar unit vectors, then $[\mathbf{a}\ \mathbf{b}\ \mathbf{c}]$ is
(a) ± 1 (b) 0
(c) ± 3 (d) 2

30. The value of c so that for all real x, the vectors $cx\hat{\mathbf{i}}-6\hat{\mathbf{j}}+3\hat{\mathbf{k}}$, $x\hat{\mathbf{i}}+2\hat{\mathbf{j}}+2cx\hat{\mathbf{k}}$ make an obtuse angle are
(a) $c<0$ (b) $0<c<4/3$
(c) $-4/3<c<0$ (d) $c>0$

31. If $\hat{\mathbf{a}}$, $\hat{\mathbf{b}}$ and $\hat{\mathbf{c}}$ are three unit vectors, such that $\hat{\mathbf{a}}+\hat{\mathbf{b}}+\hat{\mathbf{c}}$ is also a unit vectors θ_1, θ_2 and θ_3 are angles between the vectors $\hat{\mathbf{a}}$, $\hat{\mathbf{b}}$; $\hat{\mathbf{b}}$, $\hat{\mathbf{c}}$ and $\hat{\mathbf{c}}$, $\hat{\mathbf{a}}$ respectively, then among θ_1, θ_2 and θ_3
(a) all are acute angles
(b) all are right angles
(c) at least one is obtuse angle
(d) None of the above

32. If $\hat{\mathbf{i}}, \hat{\mathbf{j}}, \hat{\mathbf{k}}$ are unit orthonormal vectors and **a** is a vector, if $\mathbf{a}\times\mathbf{r}=\hat{\mathbf{j}}$, then $\mathbf{a}\cdot\mathbf{r}$ is
(a) -1 (b) 0
(c) 1 (d) None of these

33. Let $\mathbf{a}=\hat{\mathbf{i}}+2\hat{\mathbf{j}}-3\hat{\mathbf{k}}$ and $\mathbf{b}=2\hat{\mathbf{i}}-3\hat{\mathbf{j}}+5\hat{\mathbf{k}}$, if $\mathbf{r}\times\mathbf{a}=\mathbf{b}\times\mathbf{r}$, $\mathbf{r}\cdot(\alpha\hat{\mathbf{i}}+2\hat{\mathbf{j}}+\hat{\mathbf{k}})=3$ and $\mathbf{r}\cdot(2\hat{\mathbf{i}}+5\hat{\mathbf{j}}-\alpha\hat{\mathbf{k}})=-1$, $\alpha\in R$, then the value of $\alpha+|\mathbf{r}|^2$ is equal to
(a) 19 (b) 15
(c) 13 (d) 11

34. If $(\mathbf{a}\times\mathbf{b})^2+(\mathbf{a}\cdot\mathbf{b})^2=144$ and $|\mathbf{a}|=4$, then $|\mathbf{b}|$ is equal to
(a) 16 (b) 8
(c) 3 (d) 12

35. In a right angled ΔABC, the hypotenuse $AB=p$, then $\mathbf{AB}\cdot\mathbf{AC}+\mathbf{BC}\cdot\mathbf{BA}+\mathbf{CA}\cdot\mathbf{CB}$ equals
(a) $2p^2$ (b) p^2
(c) $\frac{p^2}{2}$ (d) None of these

36. For non-zero vectors **a**, **b**, **c** $|(\mathbf{a}\times\mathbf{b})\cdot\mathbf{c}|=|\mathbf{a}||\mathbf{b}||\mathbf{c}|$ holds if
(a) $\mathbf{a}\cdot\mathbf{b}=0, \mathbf{b}\cdot\mathbf{c}=0$
(b) $\mathbf{b}\cdot\mathbf{c}=0, \mathbf{c}\cdot\mathbf{a}=0$
(c) $\mathbf{c}\cdot\mathbf{a}=0, \mathbf{a}\cdot\mathbf{b}=0$
(d) $\mathbf{a}\cdot\mathbf{b}=\mathbf{b}\cdot\mathbf{c}=\mathbf{c}\cdot\mathbf{a}=0$

37. If **a** is a unit vector such that $\mathbf{a}\times(\hat{\mathbf{i}}+\hat{\mathbf{j}}+\hat{\mathbf{k}})=\hat{\mathbf{i}}-\hat{\mathbf{k}}$, then **a** is equal to
(a) $-\frac{1}{3}(2\hat{\mathbf{i}}-\hat{\mathbf{j}}+2\hat{\mathbf{k}})$ (b) $\hat{\mathbf{j}}$
(c) $\frac{1}{3}(\hat{\mathbf{i}}+2\hat{\mathbf{j}}+2\hat{\mathbf{k}})$ (d) $\hat{\mathbf{i}}$

38. If vectors $\mathbf{a}_1=x\hat{\mathbf{i}}-\hat{\mathbf{j}}+\hat{\mathbf{k}}$ and $\mathbf{a}_2=\hat{\mathbf{i}}+y\hat{\mathbf{j}}+z\hat{\mathbf{k}}$ are collinear, then a possible unit vector parallel to the vectors $x\hat{\mathbf{i}}+y\hat{\mathbf{j}}+z\hat{\mathbf{k}}$ is
(a) $\frac{1}{\sqrt{2}}(-\hat{\mathbf{j}}+\hat{\mathbf{k}})$ (b) $\frac{1}{\sqrt{2}}(\hat{\mathbf{i}}-\hat{\mathbf{j}})$
(c) $\frac{1}{\sqrt{3}}(\hat{\mathbf{i}}-\hat{\mathbf{j}}+\hat{\mathbf{k}})$ (d) $\frac{1}{\sqrt{3}}(\hat{\mathbf{i}}+\hat{\mathbf{j}}-\hat{\mathbf{k}})$

39. If $\mathbf{a}+2\mathbf{b}+3\mathbf{c}=0$ and $\mathbf{a}\times\mathbf{b}+\mathbf{b}\times\mathbf{c}+\mathbf{c}\times\mathbf{a}$ is equal to $\lambda(\mathbf{b}\times\mathbf{c})$, then λ is equal to
(a) 3 (b) 4
(c) 5 (d) None of these

40. If $\mathbf{OA}=\hat{\mathbf{i}}+3\hat{\mathbf{j}}-2\hat{\mathbf{k}}$ and $\mathbf{OB}=3\hat{\mathbf{i}}+\hat{\mathbf{j}}-2\hat{\mathbf{k}}$, then **OC** which bisects the $\angle AOB$ is given by
(a) $2\hat{\mathbf{i}}-2\hat{\mathbf{j}}-2\hat{\mathbf{k}}$ (b) $2\hat{\mathbf{i}}+2\hat{\mathbf{j}}+2\hat{\mathbf{k}}$
(c) $-2\hat{\mathbf{i}}+2\hat{\mathbf{j}}-2\hat{\mathbf{k}}$ (d) $2\hat{\mathbf{i}}+2\hat{\mathbf{j}}-2\hat{\mathbf{k}}$

41. If $\mathbf{a}=\hat{\mathbf{i}}+\hat{\mathbf{j}}+\hat{\mathbf{k}}$, $\mathbf{b}=4\hat{\mathbf{i}}+3\hat{\mathbf{j}}+4\hat{\mathbf{k}}$, and $\mathbf{c}=\hat{\mathbf{i}}+\alpha\hat{\mathbf{j}}+\beta\hat{\mathbf{k}}$ are linearly dependent vectors and $|\mathbf{c}|=\sqrt{3}$, then
(a) $\alpha=1, \beta=-1$ (b) $\alpha=1, \beta=\pm 1$
(c) $\alpha=-1, \beta=\pm 1$ (d) $\alpha=\pm 1, \beta=1$

42. If **a** any vector, then $|\mathbf{a}\times\hat{\mathbf{i}}|^2+|\mathbf{a}\times\hat{\mathbf{j}}|^2+|\mathbf{a}\times\hat{\mathbf{k}}|^2$ is equal to
(a) $(\mathbf{a})^2$ (b) $2(\mathbf{a})^2$
(c) $3(\mathbf{a})^2$ (d) 0

43. If the vectors $a\hat{\mathbf{i}}+\hat{\mathbf{j}}+\hat{\mathbf{k}}$, $\hat{\mathbf{i}}+b\hat{\mathbf{j}}+\hat{\mathbf{k}}$ and $\hat{\mathbf{i}}+\hat{\mathbf{j}}+c\hat{\mathbf{k}}$ $(a\neq b, c\neq 1)$ are coplanar, then the value of $\frac{1}{1-a}+\frac{1}{1-b}+\frac{1}{1-c}$ is
(a) 1 (b) -1
(c) 2 (d) None of these

44. For any vector **A**, the value of $\hat{\mathbf{i}}\times(\mathbf{A}\times\hat{\mathbf{i}})+\hat{\mathbf{j}}\times(\mathbf{A}\times\hat{\mathbf{j}})+\hat{\mathbf{k}}\times(\mathbf{A}\times\hat{\mathbf{k}})$ is equal to
(a) **0** (b) 2**A**
(c) -2**A** (d) None of these

45. If **b** is a vector whose initial point divides the join of $5\hat{\mathbf{i}}$ and $5\hat{\mathbf{j}}$ in the ratio $k:1$ and whose terminal point is the original and $|\mathbf{b}|\leq\sqrt{37}$, then k lies in the interval
(a) $\left[-6, -\frac{1}{6}\right]$ (b) $(-\infty, -6]\cup\left[-\frac{1}{6}, \infty\right)$
(c) [0, 6] (d) None of these

46. If **a**, **b** and **c** are unit coplanar vectors, then the scalar triple product $[2\mathbf{a}-\mathbf{b}\ 2\mathbf{b}-\mathbf{c}\ 2\mathbf{c}-\mathbf{a}]$ is equal to
(a) 0 (b) 1
(c) $-\sqrt{3}$ (d) $\sqrt{3}$

47. If **a**, **b**, **c** are non-coplanar unit vectors such that $\mathbf{a} \times (\mathbf{b} \times \mathbf{c}) = \frac{\mathbf{b}+\mathbf{c}}{\sqrt{2}}$, then the angle between **a** and **b** is
(a) $\pi/4$ (b) $\pi/2$
(c) π (d) $3\pi/4$

48. Let $\mathbf{a}(x) = (\sin x)\hat{\mathbf{i}} + (\cos x)\hat{\mathbf{j}}$ and $\mathbf{b}(x) = (\cos 2x)\hat{\mathbf{i}} + (\sin 2x)\hat{\mathbf{j}}$ be two variable vectors $(x \in R)$, then $\mathbf{a}(x)$ and $\mathbf{b}(x)$ are
(a) collinear for unique value of x,
(b) perpendicular for infinitely many values of x
(c) zero vectors for unique values of x
(d) None of the above

49. Let the vectors **a**, **b**, **c** and **d** be such that $(\mathbf{a} \times \mathbf{b}) \times (\mathbf{c} \times \mathbf{d}) = \mathbf{0}$. Let P_1 and P_2 be planes determined by the pairs of vectors **a**, **b** and **c**, **d** respectively. Then, the angle between P_1 and P_2
(a) 0 (b) $\pi/4$
(c) $\pi/3$ (d) $\pi/2$

50. Let **a**, **b** and **c** be three unit vectors such that $\mathbf{a}+\mathbf{b}+\mathbf{c} = 0$. If $\lambda = \mathbf{a}\cdot\mathbf{b} + \mathbf{b}\cdot\mathbf{c} + \mathbf{c}\cdot\mathbf{a}$ and $\mathbf{d} = \mathbf{a}\times\mathbf{b} + \mathbf{b}\times\mathbf{c} + \mathbf{c}\times\mathbf{a}$, then the ordered pair, $(\lambda, \mathbf{d})$ is equal to
(a) $\left(\frac{3}{2}, 3\mathbf{b}\times\mathbf{c}\right)$ (b) $\left(-\frac{3}{2}, 3\mathbf{c}\times\mathbf{b}\right)$
(c) $\left(\frac{3}{2}, 3\mathbf{a}\times\mathbf{c}\right)$ (d) $\left(-\frac{3}{2}, 3\mathbf{a}\times\mathbf{b}\right)$

51. If **a** and **b** are two vectors of magnitude 2 inclined at an angle 60°, then the angle between **a** and **a** + **b** is
(a) 30° (b) 60°
(c) 45° (d) None of these

52. $(\mathbf{r}\cdot\hat{\mathbf{i}})(\mathbf{r}\times\hat{\mathbf{i}}) + (\mathbf{r}\cdot\hat{\mathbf{j}})(\mathbf{r}\times\hat{\mathbf{j}}) + (\mathbf{r}\cdot\hat{\mathbf{k}})(\mathbf{r}\times\hat{\mathbf{k}})$ is equal to
(a) $3\mathbf{r}$ (b) $\mathbf{r}$
(c) $\mathbf{0}$ (d) None of these

53. **a**, **b**, **c** are non-coplanar vectors and **p**, **q**, **r** are defined as $\mathbf{p} = \frac{\mathbf{b}\times\mathbf{c}}{[\mathbf{b}\,\mathbf{c}\,\mathbf{a}]}, \mathbf{q} = \frac{\mathbf{c}\times\mathbf{a}}{[\mathbf{c}\,\mathbf{a}\,\mathbf{b}]}, \mathbf{r} = \frac{\mathbf{a}\times\mathbf{b}}{[\mathbf{a}\,\mathbf{b}\,\mathbf{c}]}$, then
$(\mathbf{a}+\mathbf{b})\cdot\mathbf{p} + (\mathbf{b}+\mathbf{c})\cdot\mathbf{q} + (\mathbf{c}+\mathbf{a})\cdot\mathbf{r}$ is equal to
(a) 0 (b) 1
(c) 2 (d) 3

54. Let $\mathbf{a} = 2\hat{\mathbf{i}} + \hat{\mathbf{j}} - 2\hat{\mathbf{k}}$ and $\mathbf{b} = \hat{\mathbf{i}} + \hat{\mathbf{j}}$. If **c** is a vector such that $\mathbf{a}\cdot\mathbf{c} = |\mathbf{c}|, |\mathbf{c}-\mathbf{a}| = 2\sqrt{2}$ and the angle between $\mathbf{a}\times\mathbf{b}$ and **c** is 30°, then $|(\mathbf{a}\times\mathbf{b})\times\mathbf{c}|$ is equal to
(a) $\frac{2}{3}$ (b) $\frac{3}{2}$
(c) 2 (d) 3

55. If **a** and **b** are perpendicular vectors, then $\mathbf{a}\times(\mathbf{a}\times(\mathbf{a}\times(\mathbf{a}\times\mathbf{b})))$ is equal to
(a) $\frac{1}{2}|\mathbf{a}|^4\,\mathbf{b}$ (b) $\mathbf{a}\times\mathbf{b}$
(c) $|\mathbf{a}|^4\,\mathbf{b}$ (d) $\mathbf{0}$

56. If **u**, **v**, **w** are non-coplanar vectors and p, q are real numbers, then the equality $[3\mathbf{u}\; p\mathbf{v}\; p\mathbf{w}] - [p\mathbf{v}\; \mathbf{w}\; q\mathbf{u}] - [2\mathbf{w}\; q\mathbf{v}\; q\mathbf{u}] = 0$ holds for
(a) more than two but not all values of (p, q)
(b) all values of (p, q)
(c) exactly one values of (p, q)
(d) exactly two values of (p, q)

57. If $\mathbf{d} = \lambda(\mathbf{a}\times\mathbf{b}) + \mu(\mathbf{b}\times\mathbf{c}) + \nu(\mathbf{c}\times\mathbf{a})$ and $[\mathbf{a}\,\mathbf{b}\,\mathbf{c}] = \frac{1}{8}$, then $\lambda+\mu+\nu$ is equal to
(a) $\mathbf{d}\cdot(\mathbf{a}+\mathbf{b}+\mathbf{c})$ (b) $2\mathbf{d}\cdot(\mathbf{a}+\mathbf{b}+\mathbf{c})$
(c) $4\mathbf{d}\cdot(\mathbf{a}+\mathbf{b}+\mathbf{c})$ (d) $8\mathbf{d}\cdot(\mathbf{a}+\mathbf{b}+\mathbf{c})$

58. If **a**, **b** and **c** are any three vectors, then $\mathbf{a}\times(\mathbf{b}\times\mathbf{c}) = (\mathbf{a}\times\mathbf{b})\times\mathbf{c}$, if and only if
(a) **b** and **c** are collinear
(b) **a** and **c** are collinear
(c) **a** and **b** are collinear
(d) None of the above

59. The position vectors of the points A, B and C are $\hat{\mathbf{i}}+\hat{\mathbf{j}}+\hat{\mathbf{k}}$, $\hat{\mathbf{i}}+5\hat{\mathbf{j}}-\hat{\mathbf{k}}$ and $2\hat{\mathbf{i}}+3\hat{\mathbf{j}}+5\hat{\mathbf{k}}$ respectively. The greatest angle of the ΔABC is
(a) 90° (b) 135°
(c) $\cos^{-1}\left(\frac{2}{3}\right)$ (d) $\cos^{-1}\left(\frac{5}{7}\right)$

60. Let **a** and **b** are two vectors making angles θ with each other, then unit vectors along bisector of **a** and **b** is
(a) $\pm\frac{\hat{\mathbf{a}}+\hat{\mathbf{b}}}{2}$ (b) $\pm\frac{\hat{\mathbf{a}}+\hat{\mathbf{b}}}{2\cos\theta}$
(c) $\pm\frac{\hat{\mathbf{a}}+\hat{\mathbf{b}}}{2\cos\theta/2}$ (d) $\pm\frac{\hat{\mathbf{a}}+\hat{\mathbf{b}}}{|\hat{\mathbf{a}}+\hat{\mathbf{b}}|}$

61. The volume of the tetrahedron whose vertices are with position vectors $\hat{\mathbf{i}}-6\hat{\mathbf{j}}+10\hat{\mathbf{k}}, -\hat{\mathbf{i}}-3\hat{\mathbf{j}}+7\hat{\mathbf{k}}, 5\hat{\mathbf{i}}-\hat{\mathbf{j}}+\lambda\hat{\mathbf{k}}$ and $7\hat{\mathbf{i}}-4\hat{\mathbf{j}}+7\hat{\mathbf{k}}$ is 11 cu unit if λ equals
(a) −3 (b) 3 (c) 7 (d) −1

62. If **a**, **b**, **c** be three vectors such that $\mathbf{a}\times\mathbf{b} = \mathbf{c}$ and $\mathbf{b}\times\mathbf{c} = \mathbf{a}$ then
(a) **a**, **b**, **c** are orthogonal in pairs
(b) $|\mathbf{a}| = |\mathbf{b}| = |\mathbf{c}| = 1$
(c) $|\mathbf{a}| = |\mathbf{b}| = |\mathbf{c}| \neq 1$
(d) $|\mathbf{a}| \neq |\mathbf{b}| \neq |\mathbf{c}|$

63. A vector **a** has components $2p$ and 1 with respect to a rectangular cartesian system. This system is rotated through a certain angle about the origin in the clockwise sense. If with respect to new system, **a** has components $p+1$ and 1, then
(a) $p = 0$ (b) $p = 1$ or $p = -\frac{1}{3}$
(c) $p = -1$ or $p = \frac{1}{3}$ (d) $p = 1$ or $p = -1$

64. If **a** and **b** are parallel, then the value of $(\mathbf{a}\times\mathbf{b})\times(\mathbf{c}\times\mathbf{d}) + (\mathbf{a}\times\mathbf{c})\times(\mathbf{b}\times\mathbf{d})$ is equal to
(a) $\{(\mathbf{a}\times\mathbf{c})\cdot\mathbf{b}\}\,\mathbf{d}$ (b) $\{(\mathbf{b}\times\mathbf{c})\cdot\mathbf{a}\}\,\mathbf{d}$
(c) $\{(\mathbf{a}\times\mathbf{b})\cdot\mathbf{c}\}\,\mathbf{d}$ (d) None of these

65. If $\hat{\mathbf{i}} \times (\mathbf{r} \times \hat{\mathbf{i}}) + \hat{\mathbf{j}} \times (\mathbf{r} \times \hat{\mathbf{j}}) + \hat{\mathbf{k}} \times (\mathbf{r} \times \hat{\mathbf{k}})$
$= \mathbf{a} \times \mathbf{b}$ $(\mathbf{a} \neq 0, \mathbf{b} \neq 0)$, then

(a) $\mathbf{r} = \mathbf{a} \times \mathbf{b}$ (b) $\mathbf{r} = \dfrac{\mathbf{a} \times \mathbf{b}}{2}$
(c) $\mathbf{r} = \mathbf{0}$ (d) None of these

66. Let **a**, **b**, **c** be three vectors such that $\mathbf{a} \neq 0$ and $\mathbf{a} \times \mathbf{b} = 2\mathbf{a} \times \mathbf{c}$, $|\mathbf{a}| = |\mathbf{c}| = 1$, $|\mathbf{b}| = 4$ and $|\mathbf{b} \times \mathbf{c}| = \sqrt{15}$, if $\mathbf{b} - 2\mathbf{c} = \lambda\,\mathbf{a}$. Then λ equals

(a) 1 (b) −1
(c) 2 (d) −4, 4

67. Let AD be the angle bisector of the angle A of $\Delta\, ABC$, then
$\mathbf{AD} = \alpha\, \mathbf{AB} + \beta\, \mathbf{AC}$, where

(a) $\alpha = \dfrac{|\mathbf{AB}|}{|\mathbf{AB} + \mathbf{AC}|}, \beta = \dfrac{|\mathbf{AC}|}{|\mathbf{AB} + \mathbf{AC}|}$
(b) $\alpha = \dfrac{|\mathbf{AB}| + |\mathbf{AC}|}{|\mathbf{AB}|}, \beta = \dfrac{|\mathbf{AB}| + |\mathbf{AC}|}{|\mathbf{AC}|}$
(c) $\alpha = \dfrac{|\mathbf{AC}|}{|\mathbf{AB}| + |\mathbf{AC}|}, \beta = \dfrac{|\mathbf{AB}|}{|\mathbf{AB}| + |\mathbf{AC}|}$
(d) $\alpha = \dfrac{|\mathbf{AB}|}{|\mathbf{AC}|}, \beta = \dfrac{|\mathbf{AC}|}{|\mathbf{AB}|}$

68. $p\,\hat{\mathbf{i}} + 3\,\hat{\mathbf{j}} + 4\,\hat{\mathbf{k}}$ and $\sqrt{q}\,\hat{\mathbf{i}} + 4\,\hat{\mathbf{k}}$ are two vectors, where $p, q \geq 0$ are two scalars, then the length of the vectors is equal to

(a) all values of (p, q)
(b) only finite number of values of (p, q)
(c) infinite number of values of (p, q)
(d) no value of (p, q)

69. Let **p**, **q**, **r** be three non-coplanar vectors and $[\mathbf{p\ q\ r}]\mathbf{a} = \mathbf{q} \times \mathbf{r}$, $[\mathbf{p\ q\ r}]\mathbf{b} = \mathbf{r} \times \mathbf{p}$, $[\mathbf{p\ q\ r}]\mathbf{c} = \mathbf{p} \times \mathbf{q}$. If **a**, **b**, **c** denote the coterminous edges of a parallelopiped, then its height with the base having **a** and **c** is

(a) $|\mathbf{p}|$ (b) $\dfrac{1}{|\mathbf{a}|}$ (c) $\dfrac{1}{|\mathbf{b}|}$ (d) $\dfrac{1}{|\mathbf{q}|}$

70. If $\hat{\mathbf{i}} \times (\mathbf{a} \times \hat{\mathbf{i}}) + \hat{\mathbf{j}} \times (\mathbf{a} \times \hat{\mathbf{j}}) + \hat{\mathbf{k}} \times (\mathbf{a} \times \hat{\mathbf{k}})$
$= \ldots. \{(\mathbf{a} \cdot \hat{\mathbf{i}})\hat{\mathbf{i}} + (\mathbf{a} \cdot \hat{\mathbf{j}})\,\hat{\mathbf{j}} + (\mathbf{a} \cdot \hat{\mathbf{k}})\,\hat{\mathbf{k}}\}$

(a) −1 (b) 0
(c) 2 (d) None of these

71. If **a**, **b**, **c** are position vectors of three non-collinear points A, B, C respectively, the shortest distance from A to BC is

(a) $\mathbf{a} \cdot |\mathbf{b} - \mathbf{c}|\, \mathbf{a}$ (c) $\sqrt{|\mathbf{b} - \mathbf{a}|^2 - \left\{\dfrac{\mathbf{a} \cdot \mathbf{b}}{\mathbf{a}}\right\}^2}$
(d) $|\mathbf{b} - \mathbf{a}|$ (d) None of these

72. If the non-zero vectors **a** and **b** are perpendicular to each other then the solution of the equation $\mathbf{r} \times \mathbf{a} = \mathbf{b}$ is

(a) $\mathbf{r} = x\,\mathbf{a} + \dfrac{1}{\mathbf{a} \cdot \mathbf{a}}(\mathbf{a} \times \mathbf{b})$ (b) $\mathbf{r} = x\,\mathbf{b} - \dfrac{1}{\mathbf{b} \cdot \mathbf{b}}(\mathbf{a} \times \mathbf{b})$
(c) $\mathbf{r} = x\,(\mathbf{a} \times \mathbf{b})$ (d) None of these

73. Let $\mathbf{OA} = \mathbf{a}$, $\mathbf{OB} = 10\mathbf{a} + 2\mathbf{b}$ and $\mathbf{OC} = \mathbf{b}$ where A and C are non-collinear points. Let p denote the area of the quadrilateral $OABC$, and let q denote the area of the parallelogram with OA and OC as adjacent sides. If $p = kq$, then k is equal to

(a) 2 (b) 4
(c) 6 (d) None of these

74. Let **r**, **a**, **b** and **c** be four non zero vectors such that $\mathbf{r} \cdot \mathbf{a} = 0$, $|\mathbf{r} \times \mathbf{b}| = |\mathbf{r}||\mathbf{b}|$, $|\mathbf{r} \times \mathbf{c}| = |\mathbf{r}||\mathbf{c}|$, then $[\mathbf{a\ b\ c}]$ is equal to

(a) −1 (b) 0
(c) 1 (d) 2

75. Given a cube $ABCDA_1B_1C_1D_1$ with lower base $ABCD$, upper base $A_1B_1C_1D_1$ and the lateral edges AA_1, BB_1, CC_1 and DD_1; M and M_1 are the centres of the faces $ABCD$ and $A_1B_1C_1D_1$ respectively. O is a point on the line MM_1 such that $\mathbf{OA} + \mathbf{OB} + \mathbf{OC} + \mathbf{OD} = \mathbf{OM}_1$, then $\mathbf{OM} = \lambda \mathbf{OM}_1$, if λ is equal to

(a) $\dfrac{1}{16}$ (b) $\dfrac{1}{8}$
(c) $\dfrac{1}{4}$ (d) $\dfrac{1}{2}$

76. Let **a**, **b** and **c** be three non-zero and non-coplanar vectors and **p**, **q** and **r** be three vectors given by $\mathbf{p} = \mathbf{a} + \mathbf{b} - 2\,\mathbf{c}$, $\mathbf{q} = 3\,\mathbf{a} - 2\,\mathbf{b} + \mathbf{c}$ and $\mathbf{r} = \mathbf{a} - 4\,\mathbf{b} + 2\,\mathbf{c}$
If the volume of the parallelopiped determined by **a**, **b** and **c** is V_1 and that of the parallelopiped determined by **p**, **q** and **r** is V_2 then $V_2 : V_1$ is equal to

(a) 3 : 1 (b) 7 : 1
(c) 11 : 1 (d) 15 : 1

77. A line L_1 passes through the point $3\hat{\mathbf{i}}$ and parallel to the vector $-\,\hat{\mathbf{i}} + \hat{\mathbf{j}} + \hat{\mathbf{k}}$ and another line L_2 passes through the point $\hat{\mathbf{i}} + \hat{\mathbf{j}}$ and parallel to the vector $\hat{\mathbf{i}} + \hat{\mathbf{k}}$, then point of intersection of the lines is

(a) $2\hat{\mathbf{i}} + \hat{\mathbf{j}} + \hat{\mathbf{k}}$ (b) $2\hat{\mathbf{i}} - 2\hat{\mathbf{j}} + \hat{\mathbf{k}}$
(c) $\hat{\mathbf{i}} + 2\hat{\mathbf{j}} - \hat{\mathbf{k}}$ (d) None of the above

78. The line joining the points $6\,\mathbf{a} - 4\,\mathbf{b} - 5\,\mathbf{c}$, $-\,4\,\mathbf{c}$ and the line joining the points $-\mathbf{a} - 2\,\mathbf{b} - 3\,\mathbf{c}$, $\mathbf{a} + 2\,\mathbf{b} - 5\,\mathbf{c}$ intersect at

(a) $2\,\mathbf{c}$ (b) $-4\,\mathbf{c}$
(c) $8\,\mathbf{c}$ (d) None of these

79. If **a**, **b**, **c** and **p**, **q**, **r** are reciprocal system of vectors, then $\mathbf{a} \times \mathbf{p} + \mathbf{b} \times \mathbf{q} + \mathbf{c} \times \mathbf{r}$ is equal to

(a) $[\mathbf{a}, \mathbf{b}, \mathbf{c}]$ (b) $3(\mathbf{a} + \mathbf{b} + \mathbf{c})$
(c) 0 (d) None of these

80. If I be the incentre of the $\Delta\, ABC$ and a, b, c be the lengths of the sides then the force $a\mathbf{IA} + b\,\mathbf{IB} + c\,\mathbf{IC}$ is equal to

(a) −1 (b) 2
(c) 0 (d) None of these

81. If $3\mathbf{a} - 5\mathbf{b}$ and $2\mathbf{a} + \mathbf{b}$ are perpendicular to each other and $\mathbf{a} + 4\mathbf{b}, -\mathbf{a} + \mathbf{b}$ are also mutually perpendicular then the cosine of the angle between $\mathbf{a}$ and $\mathbf{b}$ is

(a) $\frac{17}{5\sqrt{43}}$ (b) $\frac{19}{5\sqrt{43}}$

(c) $\frac{21}{5\sqrt{43}}$ (d) None of these

82. A vector $\mathbf{a} = (x, y, z)$ makes an obtuse angle with Y-axis, equal angles with $\mathbf{b} = (y, -2z, 3x)$ and $\mathbf{c} = (2z, 3x, -y)$ and $\mathbf{a}$ is perpendicular to $\mathbf{d} = (1, -1, 2)$ if $|\mathbf{a}| = 2\sqrt{3}$, then vector $\mathbf{a}$ is

(a) (1, 2, 3) (b) (2, – 2, – 2)

(c) (–1, 2, 4) (d) None of these

83. Let $\mathbf{a}$ and $\mathbf{b}$ be non collinear vectors of which $\mathbf{a}$ is a unit vector. The angles of the triangle whose two sides are represented by $\sqrt{3}\,(\mathbf{a} \times \mathbf{b})$ and $\mathbf{b} - (\mathbf{a} \cdot \mathbf{b})\,\mathbf{a}$ are

(a) $\frac{\pi}{2}, \frac{\pi}{3}, \frac{\pi}{6}$ (b) $\frac{\pi}{2}, \frac{\pi}{4}, \frac{\pi}{4}$

(c) $\frac{\pi}{3}, \frac{\pi}{3}, \frac{\pi}{3}$ (d) data insufficient

84. In a parallelogram $ABCD$, $|\mathbf{AB}| = a$, $|\mathbf{AD}| = b$ and $|\mathbf{AC}| = c$, then $\mathbf{DB} \cdot \mathbf{AB}$ has the value

(a) $\frac{3a^2 + b^2 - c^2}{2}$ (b) $\frac{a^2 + 3b^2 - c^2}{2}$

(c) $\frac{a^2 - b^2 + 3c^2}{2}$ (d) $\frac{a^2 + 3b^2 + c^2}{2}$

85. The position vectors $\mathbf{a}, \mathbf{b}, \mathbf{c}$ and $\mathbf{d}$ of four points A, B, C and D on a plane are such that $(\mathbf{a} - \mathbf{d}) \cdot (\mathbf{b} - \mathbf{c}) = (\mathbf{b} - \mathbf{d}) \cdot (\mathbf{c} - \mathbf{a}) = 0$, then the point D is

(a) centroid of ΔABC

(b) orthocentre of ΔABC

(c) circumcentre of ΔABC

(d) None of the above

86. If $\mathbf{a}, \mathbf{b}, \mathbf{c}$ are three non-zero vectors such that $\mathbf{b}$ is not perpendicular to both $\mathbf{a}$ and $\mathbf{c}$ and $(\mathbf{a} \times \mathbf{b}) \times \mathbf{c} = \mathbf{a} \times (\mathbf{b} \times \mathbf{c})$ then

(a) $\mathbf{a}$ and $\mathbf{c}$ are always non-collinear

(b) $\mathbf{a}$ and $\mathbf{c}$ are always collinear

(c) $\mathbf{a}$ and $\mathbf{c}$ are always perpendicular

(d) $\mathbf{a}, \mathbf{b}$ and $\mathbf{c}$ are always non-coplanar

87. Image of the point P with position vector $7\hat{\mathbf{i}} - \hat{\mathbf{j}} + 2\hat{\mathbf{k}}$ in the line whose vector equation is $\mathbf{r} = 9\hat{\mathbf{i}} + 5\hat{\mathbf{j}} + 5\hat{\mathbf{k}} + \lambda(\hat{\mathbf{i}} + 3\hat{\mathbf{j}} + 5\hat{\mathbf{k}})$ has the position vector

(a) $-9\hat{\mathbf{i}} + 5\hat{\mathbf{j}} + 2\hat{\mathbf{k}}$ (b) $9\hat{\mathbf{i}} + 5\hat{\mathbf{j}} - 2\hat{\mathbf{k}}$

(c) $9\hat{\mathbf{i}} - 5\hat{\mathbf{j}} - 2\hat{\mathbf{k}}$ (d) $9\hat{\mathbf{i}} + 5\hat{\mathbf{j}} + 2\hat{\mathbf{k}}$

88. Consider a tetrahedron with faces F_1, F_2, F_3, F_4. Let $\mathbf{V}_1, \mathbf{V}_2, \mathbf{V}_3, \mathbf{V}_4$ be the vectors whose magnitudes are respectively equal to areas of F_1, F_2, F_3, F_4 and whose directions are perpendicular to their faces in outward direction. Then $|\mathbf{V}_1 + \mathbf{V}_2 + \mathbf{V}_3 + \mathbf{V}_4|$, equals

(a) 1 (b) 4

(c) 0 (d) None of these

89. Let the unit vectors $\mathbf{a}$ and $\mathbf{b}$ be perpendicular to each other and the unit vectors $\mathbf{c}$ be inclined at an angle θ to both $\mathbf{a}$ and $\mathbf{b}$. If $\mathbf{c} = x\,\mathbf{a} + y\,\mathbf{b} + z(\mathbf{a} \times \mathbf{b})$, then

(a) $x = \cos\theta, y = \sin\theta, z = \cos 2\theta$

(b) $x = \sin\theta, y = \cos\theta, z = -\cos 2\theta$

(c) $x = y = \cos\theta, z^2 = \cos 2\theta$

(d) $x = y = \cos\theta, z^2 = -\cos 2\theta$

90. If $\mathbf{a}, \mathbf{b}, \mathbf{c}$ are coplanar vectors and if

$$\Delta = \begin{vmatrix} \mathbf{a} & \mathbf{b} & \mathbf{c} \\ \mathbf{a} \cdot \mathbf{a} & \mathbf{a} \cdot \mathbf{b} & \mathbf{a} \cdot \mathbf{c} \\ \mathbf{a} \cdot \mathbf{c} & \mathbf{b} \cdot \mathbf{c} & \mathbf{c} \cdot \mathbf{c} \end{vmatrix}, \text{ then}$$

(a) $\Delta = 0$

(b) $\Delta = 1$

(c) Δ = any non-zero value

(d) None of the above

91. Let $\mathbf{a} = 3\hat{\mathbf{i}} + 2\hat{\mathbf{j}} + x\hat{\mathbf{k}}$ and $\mathbf{b} = \hat{\mathbf{i}} - \hat{\mathbf{j}} + \hat{\mathbf{k}}$, for some real x. Then, $|\mathbf{a} \times \mathbf{b}| = r$ is possible if **[JEE Main 2019, 8 April]**

(a) $0 < r \le \sqrt{\frac{3}{2}}$ (b) $\sqrt{\frac{3}{2}} < r \le 3\sqrt{\frac{3}{2}}$

(c) $3\sqrt{\frac{3}{2}} < r < 5\sqrt{\frac{3}{2}}$ (d) $r \ge 5\sqrt{\frac{3}{2}}$

92. Let O be the circumcentre, G be the centroid and O' be the orthocentre of a $\Delta\, ABC$. Three vectors are taken through O and are represented by $\mathbf{a} = \mathbf{OA}$, $\mathbf{b} = \mathbf{OB}$ and $\mathbf{c} = \mathbf{OC}$, then $\mathbf{a} + \mathbf{b} + \mathbf{c}$ is

(a) $\mathbf{OG}$ (b) $2\mathbf{OG}$

(c) $\mathbf{OO'}$ (d) None of these

93. In ΔABC, $\angle A = 30°$, H is the orthocentre and D is the mid-point of BC. Segment HD is produced to T such that $HD = DT$. The length AT is equal to

(a) $2BC$ (b) $3BC$

(c) $\frac{4}{3}BC$ (d) None of these

94. The equation of the plane containing the line $\mathbf{r} = \mathbf{a} + k\,\mathbf{b}$ and perpendicular to the plane $\mathbf{r} \cdot \mathbf{n} = q$ is

(a) $(\mathbf{r} - \mathbf{b}) \cdot (\mathbf{n} \times \mathbf{a}) = 0$

(b) $(\mathbf{r} - \mathbf{a}) \cdot \{\mathbf{n} \times (\mathbf{a} \times \mathbf{b})\} = 0$

(c) $(\mathbf{r} - \mathbf{a}) \cdot (\mathbf{n} \times \mathbf{b}) = 0$

(d) $(\mathbf{r} - \mathbf{b}) \cdot \{\mathbf{n} \times (\mathbf{a} \times \mathbf{b})\} = 0$

95. The value of λ so that the points P, Q, R, S on the sides OA, OB, OC and AB of a regular tetrahedron are coplanar. When $\frac{OP}{OA} = \frac{1}{3}; \frac{OQ}{OB} = \frac{1}{2}, \frac{OR}{OC} = \frac{1}{3}$ and $\frac{OS}{AB} = \lambda$ is equal to

(a) $\lambda = \frac{1}{2}$ (b) $\lambda = -1$

(c) $\lambda = 0$ (d) for no value of λ

96. $\hat{\mathbf{x}}$ and $\hat{\mathbf{y}}$ are two mutually perpendicuar unit vectors, if the vectors $a\,\hat{\mathbf{x}} + a\,\hat{\mathbf{y}} + c(\hat{\mathbf{x}} \times \hat{\mathbf{y}}), \hat{\mathbf{x}} + (\hat{\mathbf{x}} \times \hat{\mathbf{y}})$ and $c\,\hat{\mathbf{x}} + c\,\hat{\mathbf{y}} + b(\hat{\mathbf{x}} \times \hat{\mathbf{y}})$, lie in a plane then c is

(a) AM of a and b (b) GM of a and b

(c) HM of a and b (d) equal to zero

Objective Questions Type II [One or more than one correct answer(s)]

In each of the questions below four choices of which one or more than one are correct. You have to select the correct answer(s) accordingly.

1. If a vector $\mathbf{r}$ satisfies the equation $\mathbf{r} \times (\hat{\mathbf{i}} + 2\hat{\mathbf{j}} + \hat{\mathbf{k}}) = \hat{\mathbf{i}} - \hat{\mathbf{k}}$, then $\mathbf{r}$ is equal to

(a) $\hat{\mathbf{i}} + 3\hat{\mathbf{j}} + \hat{\mathbf{k}}$

(b) $3\hat{\mathbf{i}} + 7\hat{\mathbf{j}} + 3\hat{\mathbf{k}}$

(c) $\hat{\mathbf{j}} + t\,(\hat{\mathbf{i}} + 2\hat{\mathbf{j}} + \hat{\mathbf{k}})$, where t is any scalar

(d) $\hat{\mathbf{i}} + (t + 3)\,\hat{\mathbf{j}} + \hat{\mathbf{k}}$, where t is any scalar

2. If $\mathbf{DA} = \mathbf{a}$, $\mathbf{AB} = \mathbf{b}$ and $\mathbf{CB} = k\,\mathbf{a}$, where $k > 0$ and X, Y are the mid points of DB and AC respectively such that $|\mathbf{a}| = 17$ and $|\mathbf{XY}| = 4$, then k is equal to

(a) 8/17 (b) 9/17
(c) 25/17 (d) 4/17

3. $\mathbf{a}$ and $\mathbf{c}$ are unit vectors and $|\mathbf{b}| = 4$ with $\mathbf{a} \times \mathbf{b} = 2\,\mathbf{a} \times \mathbf{c}$. The angle between $\mathbf{a}$ and $\mathbf{c}$ is $\cos^{-1}(1/4)$. Then $\mathbf{b} - 2\mathbf{c} = \lambda\mathbf{a}$, if λ is

(a) 3 (b) 1/4
(c) -4 (d) $-1/4$

4. The vector $\mathbf{c}$ directed along the bisectors of the angle between the vectors $\mathbf{a} = 7\hat{\mathbf{i}} - 4\hat{\mathbf{j}} - 4\hat{\mathbf{k}}$ and $\mathbf{b} = -2\hat{\mathbf{i}} - \hat{\mathbf{j}} + 2\hat{\mathbf{k}}$ if $|\mathbf{c}| = 3\sqrt{6}$ is given by

(a) $\hat{\mathbf{i}} - 7\hat{\mathbf{j}} + 2\hat{\mathbf{k}}$ (b) $\hat{\mathbf{i}} + 7\hat{\mathbf{j}} - 2\hat{\mathbf{k}}$
(c) $-\hat{\mathbf{i}} + 7\hat{\mathbf{j}} - 2\hat{\mathbf{k}}$ (d) $\hat{\mathbf{i}} - 7\hat{\mathbf{j}} - 2\hat{\mathbf{k}}$

5. Let $\mathbf{a}$ and $\mathbf{b}$ be two non-collinear unit vectors If $\mathbf{u} = \mathbf{a} - (\mathbf{a} \cdot \mathbf{b})\,\mathbf{b}$ and $\mathbf{v} = \mathbf{a} \times \mathbf{b}$, then $|\mathbf{v}|$ is

(a) $|\mathbf{u}|$ (b) $|\mathbf{u}| + |\mathbf{u} \cdot \mathbf{a}|$
(c) $|\mathbf{u}| + |\mathbf{u} \cdot \mathbf{b}|$ (d) $|\mathbf{u}| + \mathbf{u} \cdot (\mathbf{a} + \mathbf{b})$

6. If A, B, C are three points with position vectors $\hat{\mathbf{i}} + \hat{\mathbf{j}}, \hat{\mathbf{i}} - \hat{\mathbf{j}}$ and $p\hat{\mathbf{i}} + q\hat{\mathbf{j}} + r\hat{\mathbf{k}}$ respectively, then the points are collinear if

(a) $p = q = r = 1$ (b) $p = q = r = 0$
(c) $p = q, r = 0$ (d) $p = 1, q = 2, r = 0$

7. If $|\mathbf{a}| = 4$, $|\mathbf{b}| = 2$ and the angle between $\mathbf{a}$ and $\mathbf{b}$ is $\frac{\pi}{6}$, then $(\mathbf{a} \times \mathbf{b})^2$ is

(a) 48 (b) $(\mathbf{a})^2$
(c) 16 (d) 32

8. If $\mathbf{a}, \mathbf{b}, \mathbf{c}$ are non-coplanar non-zero vectors and $\mathbf{r}$ is any vector in space then $[\,\mathbf{b}\,\mathbf{c}\,\mathbf{r}]\mathbf{a} + [\mathbf{c}\,\mathbf{a}\,\mathbf{r}]\,\mathbf{b} + [\mathbf{a}\,\mathbf{b}\,\mathbf{r}]\,\mathbf{c}$ is equal to

(a) $3\,[\mathbf{abc}]\,\mathbf{r}$ (b) $[\mathbf{abc}]\,\mathbf{r}$
(c) $[\mathbf{bca}]\,\mathbf{r}$ (d) $[\mathbf{cab}]\,\mathbf{r}$

9. If the unit vectors $\mathbf{a}$ and $\mathbf{b}$ are inclined at an angle 2θ such that $|\mathbf{a} - \mathbf{b}| < 1$ and $0 \le \theta \le \pi$, then θ lies in the interval

(a) $[0, \pi/6]$ (b) $(5\pi/6, \pi]$
(c) $[\pi/6, \pi/2]$ (d) $[\pi/2, 5\pi/6]$

10. If $\mathbf{a} \times \mathbf{b} = 2\hat{\mathbf{i}} + 2\hat{\mathbf{j}} - \hat{\mathbf{k}}$, $\mathbf{a} + \mathbf{b} = \hat{\mathbf{i}} - 2\hat{\mathbf{j}} + 4\hat{\mathbf{k}}$

(a) $\mathbf{a} = \frac{1}{4}\hat{\mathbf{i}} + \frac{1}{4}\hat{\mathbf{j}} + \hat{\mathbf{k}}$; $\mathbf{b} = \frac{3}{4}\hat{\mathbf{i}} - \frac{13}{4}\hat{\mathbf{j}} - 5\hat{\mathbf{k}}$

(b) $\mathbf{a} = \hat{\mathbf{j}} + 2\hat{\mathbf{k}}$; $\mathbf{b} = \hat{\mathbf{i}} - 4\hat{\mathbf{j}} - 6\hat{\mathbf{k}}$

(c) $\mathbf{a} = -\frac{1}{4}\hat{\mathbf{i}} + \frac{7}{4}\hat{\mathbf{j}} + 2\hat{\mathbf{k}}$; $\mathbf{b} = \frac{5}{4}\hat{\mathbf{i}} - \frac{19}{4}\hat{\mathbf{j}} - 7\hat{\mathbf{k}}$

(d) $\mathbf{a} = \frac{1}{2}\hat{\mathbf{i}} - \frac{1}{2}\hat{\mathbf{j}}$; $\mathbf{b} = \frac{3}{4}\hat{\mathbf{i}} - \frac{13}{4}\hat{\mathbf{j}} - 4\hat{\mathbf{k}}$

11. Let a and b be positive real numbers. Suppose $\mathbf{PQ} = a\hat{\mathbf{i}} + b\hat{\mathbf{j}}$ and $\mathbf{PS} = a\hat{\mathbf{i}} - b\hat{\mathbf{j}}$ are adjacent sides of a parallelogram $PQRS$. Let $\mathbf{u}$ and $\mathbf{v}$ be the projection vectors of $\mathbf{w} = \hat{\mathbf{i}} + \hat{\mathbf{j}}$ along $\mathbf{PQ}$ and $\mathbf{PS}$, respectively. If $|\mathbf{u}| + |\mathbf{v}| = |\mathbf{w}|$ and if the area of the parallelogram $PQRS$ is 8, then which of the following statements is/are TRUE? **[JEE Advanced 2020]**

(a) $a + b = 4$

(b) $a - b = 2$

(c) The length of the diagonal PR of the parallelogram $PQRS$ is 4

(d) $\mathbf{w}$ is an angle bisector of the vectors $\mathbf{PQ}$ and $\mathbf{PS}$

12. Let $\mathbf{a} = 2\hat{\mathbf{i}} - \hat{\mathbf{j}} + \hat{\mathbf{k}}$, $\mathbf{b} = \hat{\mathbf{i}} + 2\hat{\mathbf{j}} - \hat{\mathbf{k}}$ and $\mathbf{c} = \hat{\mathbf{i}} + \hat{\mathbf{j}} - 2\hat{\mathbf{k}}$ be three vectors. A vector in the plane of $\mathbf{b}$ and $\mathbf{c}$ whose projection at $\mathbf{a}$ is of magnitude $\sqrt{(2/3)}$ is

(a) $2\hat{\mathbf{i}} + 3\hat{\mathbf{j}} - 3\hat{\mathbf{k}}$ (b) $2\hat{\mathbf{i}} + 3\hat{\mathbf{j}} + 3\hat{\mathbf{k}}$
(c) $-2\hat{\mathbf{i}} - \hat{\mathbf{j}} + 5\hat{\mathbf{k}}$ (d) $2\hat{\mathbf{i}} + \hat{\mathbf{j}} + 5\hat{\mathbf{k}}$

13. The vectors $(x, x + 1, x + 2)$, $(x + 3, x + 4, x + 5)$ and $(x + 6, x + 7, x + 8)$ are coplanar for

(a) all values of x (b) $x < 0$
(c) $x > 0$ (d) None of these

14. $\mathbf{a}$, $\mathbf{b}$ and $\mathbf{c}$ are three coplanar unit vectors such that $\mathbf{a} + \mathbf{b} + \mathbf{c} = 0$. If three vectors $\mathbf{p}$, $\mathbf{q}$ and $\mathbf{r}$ are parallel to $\mathbf{a}$, $\mathbf{b}$ and $\mathbf{c}$, respectively, and have integral but different magnitudes, then among the following options, $|\mathbf{p} + \mathbf{q} + \mathbf{r}|$ can take a value equal to

(a) 1 (b) 0
(c) $\sqrt{3}$ (d) 2

15. A parallelogram is constructed on the vectors $\mathbf{a} = 3\alpha - \beta$, $\mathbf{b} = \alpha + 3\beta$, if $|\alpha| = |\beta| = 2$ and angle between α and β is $\frac{\pi}{3}$, then the length of a diagonal of the parallelogram is

(a) $4\sqrt{5}$ (b) $4\sqrt{3}$
(c) $4\sqrt{7}$ (d) None of these

16. The vector $\mathbf{a}+\mathbf{b}$ bisects the angle between the vectors $\hat{\mathbf{a}}$ and $\hat{\mathbf{b}}$, if
(a) $|\mathbf{a}|=|\mathbf{b}|$
(b) angle between $\mathbf{a}$ and $\mathbf{b}$ is zero
(c) $|\mathbf{a}|+|\mathbf{b}|=0$
(d) None of the above

17. Let ΔPQR be a triangle, let $\mathbf{a}=\mathbf{QR}$, $\mathbf{b}=\mathbf{RP}$ and $\mathbf{c}=\mathbf{PQ}$. If $|\mathbf{a}|=12$, $|\mathbf{b}|=4\sqrt{3}$ and $\mathbf{b}\cdot\mathbf{c}=24$ then which of the following is/are true? **[JEE Advanced 2015]**
(a) $\dfrac{|\mathbf{c}|^2}{2}=-|\mathbf{a}|=12$
(b) $\dfrac{|\mathbf{c}|^2}{2}+|\mathbf{a}|=30$
(c) $|\mathbf{a}\times\mathbf{b}+\mathbf{c}\times\mathbf{a}|=48\sqrt{3}$
(d) $\mathbf{a}\cdot\mathbf{b}=-72$

18. A, B, C and D are four points such that $\mathbf{AB}=m(2\hat{\mathbf{i}}-6\hat{\mathbf{j}}+2\hat{\mathbf{k}})$, $\mathbf{BC}=m(\hat{\mathbf{i}}-2\hat{\mathbf{j}})$ and $\mathbf{CD}=n(-6\hat{\mathbf{i}}+15\hat{\mathbf{j}}-3\hat{\mathbf{k}})$. If CD intersects AB at some point E, then
(a) $m\geq\dfrac{1}{2}$ (b) $n\geq\dfrac{1}{3}$
(c) $m=n$ (d) $m<n$

19. If the vectors $\mathbf{b}=(\tan\alpha, -1, 2\sqrt{\sin\alpha/2})$ and $\mathbf{c}=\left(\tan\alpha, \tan\alpha, -\dfrac{3}{\sqrt{\sin\alpha/2}}\right)$ are orthogonal and a vector $\mathbf{a}=(1, 3, \sin 2\alpha)$ makes an obtuse angle with the Z-axis, then the value of α is
(a) $\alpha=(4n+1)\pi-\tan^{-1}2$
(b) $\alpha=(4n+2)\pi-\tan^{-1}2$
(c) $\alpha=(4n+1)\pi+\tan^{-1}2$
(d) $\alpha=(4n+2)\pi+\tan^{-1}2$

20. The resolved part of the vector $\mathbf{a}$ along the vector $\mathbf{b}$ is λ and that perpendicular to $\mathbf{b}$ is μ. Then
(a) $\lambda=\dfrac{(\mathbf{a}\cdot\mathbf{b})\,\mathbf{a}}{(\mathbf{a})^2}$ (b) $\vec{\lambda}=\dfrac{(\mathbf{a}\cdot\mathbf{b})\,\mathbf{b}}{(\mathbf{b})^2}$
(c) $\mu=\dfrac{(\mathbf{b}\cdot\mathbf{b})\,\mathbf{a}-(\mathbf{a}\cdot\mathbf{b})\,\mathbf{b}}{(\mathbf{b})^2}$ (d) $\mu=\dfrac{\mathbf{b}\times(\mathbf{a}\times\mathbf{b})}{(\mathbf{b})^2}$

21. Let the unit vectors $\mathbf{a}$ and $\mathbf{b}$ be perpendicular and the unit vector $\mathbf{c}$ be inclined at an angle θ to both $\mathbf{a}$ and $\mathbf{b}$. If $\mathbf{c}=\alpha\,\mathbf{a}+\beta\,\mathbf{b}+\gamma\,(\mathbf{a}+\mathbf{b})$, then
(a) $\alpha=\beta$ (b) $\gamma^2=1-2\alpha^2$
(c) $\gamma^2=-\cos 2\theta$ (d) $\beta^2=\dfrac{1+\cos 2\theta}{2}$

22. If $\mathbf{a}=\hat{\mathbf{i}}+\hat{\mathbf{j}}+\hat{\mathbf{k}}$ and $\mathbf{b}=\hat{\mathbf{i}}-\hat{\mathbf{j}}$, then the vectors
$(\mathbf{a}\cdot\hat{\mathbf{i}})\,\hat{\mathbf{i}}+(\mathbf{a}\cdot\hat{\mathbf{j}})\,\hat{\mathbf{j}}+(\mathbf{a}\cdot\hat{\mathbf{k}})\,\hat{\mathbf{k}}$,
$(\mathbf{b}\cdot\hat{\mathbf{i}})\,\hat{\mathbf{i}}+(\mathbf{b}\cdot\hat{\mathbf{j}})\,\hat{\mathbf{j}}+(\mathbf{b}\cdot\hat{\mathbf{k}})\,\hat{\mathbf{k}}$, and $\hat{\mathbf{i}}+\hat{\mathbf{j}}-2\hat{\mathbf{k}}$
(a) are mutually perpendicular
(b) are coplanar
(c) form a parallelopiped of volume 6 units
(d) form a parallelopiped of volume 3 units

23. If unit vectors $\hat{\mathbf{i}}$ and $\hat{\mathbf{j}}$ are at right angles to each other and $\mathbf{p}=3\hat{\mathbf{i}}+4\hat{\mathbf{j}}$, $\mathbf{q}=5\hat{\mathbf{i}}$, $4\mathbf{r}=\mathbf{p}+\mathbf{q}$,
then $2\mathbf{s}=\mathbf{p}-\mathbf{q}$
(a) $|\mathbf{r}+k\mathbf{s}|=|\mathbf{r}-k\mathbf{s}|$ for all real k
(b) $\mathbf{r}$ is perpendicular to $\mathbf{s}$
(c) $\mathbf{r}+\mathbf{s}$ is perpendicular to $\mathbf{r}-\mathbf{s}$
(d) $|\mathbf{r}|=|\mathbf{s}|=|\mathbf{p}|=|\mathbf{q}|$

24. If $\mathbf{a}$, $\mathbf{b}$ and $\mathbf{c}$ be non-coplanar unit vectors equally inclined to one another at an acute angle θ. If $\mathbf{a}\times\mathbf{b}+\mathbf{b}\times\mathbf{c}=p\mathbf{a}+q\mathbf{b}+r\mathbf{c}$
Then,
(a) $p=r$
(b) $p=\dfrac{1}{\sqrt{1+2\cos\theta}}$, $q=\dfrac{2\cos\theta}{\sqrt{1+2\cos\theta}}$
(c) $r=\dfrac{1}{\sqrt{1+2\cos\theta}}$,
(d) $p=\dfrac{2\cos\theta}{\sqrt{1+2\cos\theta}}$

25. Let $\mathbf{u}$, $\mathbf{v}$, $\mathbf{w}$ be three unit vectors such that $\mathbf{u}+\mathbf{v}+\mathbf{w}=\mathbf{a}$, $\mathbf{a}\cdot\mathbf{u}=\dfrac{3}{2}$, $\mathbf{a}\cdot\mathbf{v}=\dfrac{7}{4}$ and $|\mathbf{a}|=2$, then
(a) $\mathbf{u}\cdot\mathbf{v}=\dfrac{3}{4}$
(b) $\mathbf{v}\cdot\mathbf{w}=0$
(c) $\mathbf{u}\cdot\mathbf{w}=-\dfrac{1}{4}$
(d) None of the above

26. Non-zero vectors $\mathbf{a}$ and $\mathbf{b}$ are equally inclined to vector $\mathbf{c}$, if $\mathbf{c}$ is equal to
(a) $\dfrac{|\mathbf{b}|}{|\mathbf{a}|+|\mathbf{b}|}\,\mathbf{a}+\dfrac{|\mathbf{a}|}{|\mathbf{a}|+|\mathbf{b}|}\,\mathbf{b}$
(b) $\dfrac{|\mathbf{a}|}{|\mathbf{a}|+|\mathbf{b}|}\,\mathbf{a}+\dfrac{|\mathbf{b}|}{|\mathbf{a}|+|\mathbf{b}|}\,\mathbf{b}$
(c) $\dfrac{|\mathbf{a}|}{|\mathbf{a}|+2|\mathbf{b}|}\,\mathbf{a}+\dfrac{|\mathbf{b}|}{|\mathbf{a}|+2|\mathbf{b}|}\,\mathbf{b}$
(d) $\dfrac{|\mathbf{b}|}{2|\mathbf{a}|+|\mathbf{b}|}\,\mathbf{a}+\dfrac{|\mathbf{a}|}{2|\mathbf{a}|+|\mathbf{b}|}\,\mathbf{b}$

27. Vectors $\mathbf{A}$ and $\mathbf{B}$ satisfying the vector equation $\mathbf{A}+\mathbf{B}=\mathbf{a}$, $\mathbf{A}\times\mathbf{B}=\mathbf{b}$ and $\mathbf{A}\cdot\mathbf{a}=1$, where $\mathbf{a}$ and $\mathbf{b}$ are given vectors, are
(a) $\mathbf{A}=\dfrac{(\mathbf{a}\times\mathbf{b})-\mathbf{a}}{a^2}$
(b) $\mathbf{B}=\dfrac{(\mathbf{b}\times\mathbf{a})+\mathbf{a}(a^2-1)}{a^2}$
(c) $\mathbf{A}=\dfrac{(\mathbf{a}\times\mathbf{b})+\mathbf{a}}{a^2}$
(d) $\mathbf{B}=\dfrac{(\mathbf{b}\times\mathbf{a})-\mathbf{a}(a^2-1)}{a^2}$

28. The position vectors of the points P and Q are $5\hat{\mathbf{i}}+7\hat{\mathbf{j}}-2\hat{\mathbf{k}}$ and $-3\hat{\mathbf{i}}+3\hat{\mathbf{j}}+6\hat{\mathbf{k}}$ respectively. The vector $\mathbf{A}=3\hat{\mathbf{i}}-\hat{\mathbf{j}}+\hat{\mathbf{k}}$ passes through the point P and the vector $\mathbf{B}=-3\hat{\mathbf{i}}+2\hat{\mathbf{j}}+4\hat{\mathbf{k}}$ passes through the point Q. A third vector $2\hat{\mathbf{i}}+7\hat{\mathbf{j}}-5\hat{\mathbf{k}}$ intersects vectors **A** and **B**. The position vectors of the points of intersection are

(a) $2\hat{\mathbf{i}}+8\hat{\mathbf{j}}-3\hat{\mathbf{k}}$ (b) $\hat{\mathbf{i}}+\hat{\mathbf{j}}+2\hat{\mathbf{k}}$

(c) $2\hat{\mathbf{i}}+\hat{\mathbf{j}}+2\hat{\mathbf{k}}$ (d) $\hat{\mathbf{j}}+2\hat{\mathbf{k}}$

29. The vectors $\mathbf{AB}=3\hat{\mathbf{i}}+2\hat{\mathbf{j}}+3\hat{\mathbf{k}}$ and $\mathbf{BC}=-\hat{\mathbf{i}}-2\hat{\mathbf{k}}$ are the adjacent sides of parallelogram. The angle between its diagonals is

(a) $\frac{\pi}{4}$ (b) $\frac{\pi}{3}$

(c) $\frac{3\pi}{4}$ (d) $\frac{2\pi}{3}$

30. The vectors $a\hat{\mathbf{i}}+2a\hat{\mathbf{j}}-3a\hat{\mathbf{k}}$, $(2a+1)\hat{\mathbf{i}}+(2a+3)\hat{\mathbf{j}}+(a+1)\hat{\mathbf{k}}$ and $(3a+5)\hat{\mathbf{i}}+(a+5)\hat{\mathbf{j}}+(a+2)\hat{\mathbf{k}}$ are non-coplanar for a in

(a) $\{0\}$

(b) $(0,\infty)$

(c) $(-\infty,1)$

(d) $(1,\infty)$

31. The volume of the tetrahedron whose vertices are the points with position vectors $\hat{\mathbf{i}}-6\hat{\mathbf{j}}+10\hat{\mathbf{k}}$, $-\hat{\mathbf{i}}-3\hat{\mathbf{j}}+7\hat{\mathbf{k}}$, $5\hat{\mathbf{i}}-\hat{\mathbf{j}}+\lambda\hat{\mathbf{k}}$ and $7\hat{\mathbf{i}}-4\hat{\mathbf{j}}+7\hat{\mathbf{k}}$ is 11 cu units, then the value of λ is

(a) -1

(b) 1

(c) -7

(d) 7

Linked-Comprehension Type

In these questions, a passage (paragraph) has been given followed by questions based on each of the passages. You have to answer the questions based on the passage given.

PASSAGE 1

Let **a**, **b**, **c** be three vectors such that $|\mathbf{a}|=|\mathbf{b}|=|\mathbf{c}|=4$ and angle between **a** and **b** is $\pi/3$, angle between **b** and **c** is $\pi/3$ and angle between **c** and **a** is $\pi/3$.

On the basis of above information, answer the following questions

1. The volume of the parallelopiped whose adjacent edges are represented by the vectors **a**, **b** and **c** is

(a) $24\sqrt{2}$ (b) $24\sqrt{3}$

(c) $32\sqrt{2}$ (d) $32\sqrt{3}$

2. The height of the parallelopiped whose adjacent edges are represented by the vectors **a**, **b** and **c** is

(a) $4\sqrt{\frac{2}{3}}$ (b) $3\sqrt{\frac{2}{3}}$

(c) $4\sqrt{\frac{3}{2}}$ (d) $3\sqrt{\frac{3}{2}}$

3. The volume of the tetrahedron whose adjacent edges are represented by the vectors **a**, **b** and **c** is

(a) $\frac{4\sqrt{3}}{2}$ (b) $\frac{8\sqrt{2}}{3}$

(c) $\frac{16}{\sqrt{3}}$ (d) $\frac{16\sqrt{2}}{3}$

4. The height of the tetrahedron whose adjacent edges are represented by the vectors **a**, **b** and **c** is

(a) $2\sqrt{\frac{2}{3}}$ (b) $4\sqrt{\frac{2}{3}}$

(c) $3\sqrt{\frac{2}{3}}$ (b) $\sqrt{\frac{2}{3}}$

5. The volume of the triangular prism whose adjacent edges are represented by the vectors **a**, **b** and **c** is

(a) $12\sqrt{2}$

(b) $12\sqrt{3}$

(c) $16\sqrt{2}$

(d) $16\sqrt{3}$

PASSAGE 2

If **a**, **b** and **c** be any three non-coplanar vectors. Then the system $\mathbf{a}'$, $\mathbf{b}'$ and $\mathbf{c}'$ which satisfies

$$\mathbf{a}\cdot\mathbf{a}'=\mathbf{b}\cdot\mathbf{b}'=\mathbf{c}\cdot\mathbf{c}'=1$$
$$\mathbf{a}\cdot\mathbf{b}'=\mathbf{a}\cdot\mathbf{c}'=\mathbf{b}\cdot\mathbf{a}'$$
$$=\mathbf{b}\cdot\mathbf{c}'=\mathbf{c}\cdot\mathbf{a}'=\mathbf{c}\cdot\mathbf{b}'=0$$

is called the reciprocal system to the vectors **a**, **b** and **c**.

On the basis of above information, answer the following questions

1. The value of $[\mathbf{a}'\ \mathbf{b}'\ \mathbf{c}']^{-1}$ is

(a) $<[\mathbf{a}\ \mathbf{b}\ \mathbf{c}]$ (b) $=[\mathbf{a}\ \mathbf{b}\ \mathbf{c}]$

(c) $>[\mathbf{a}\ \mathbf{b}\ \mathbf{c}]$ (d) 0

2. The value of $(\mathbf{a}\times\mathbf{a}')+(\mathbf{b}\times\mathbf{b}')+(\mathbf{c}\times\mathbf{c}')$ is

(a) $\mathbf{a}+\mathbf{b}+\mathbf{c}$ (b) $\mathbf{a}'+\mathbf{b}'+\mathbf{c}'$

(c) $\neq 0$ (d) $=0$

3. If value of $(\mathbf{a}'\times\mathbf{b}')+(\mathbf{b}'\times\mathbf{c}')+(\mathbf{c}'\times\mathbf{a}')$ is

(a) 0 (b) 1

(c) $\frac{[\mathbf{a}+\mathbf{b}+\mathbf{c}]}{[\mathbf{a}\ \mathbf{b}\ \mathbf{c}]}$ (d) $\frac{(\mathbf{a}'+\mathbf{b}'+\mathbf{c}')}{[\mathbf{a}'\ \mathbf{b}'\ \mathbf{c}']}$

4. If the system of vectors $\mathbf{e}^1,\mathbf{e}^2,\mathbf{e}^3$ is reciprocal to the system $\mathbf{e}_1$, $\mathbf{e}_2$, $\mathbf{e}_3$, then the value of $(\mathbf{a}\cdot\mathbf{e}^1)\,\mathbf{e}_1+(\mathbf{a}\cdot\mathbf{e}^2)\,\mathbf{e}_2+(\mathbf{a}\cdot\mathbf{e}^3)\,\mathbf{e}_3$ is

(a) 0 (b) 1

(c) **a** (d) $3\mathbf{a}$

5. The value of
$[\mathbf{a}\,\mathbf{b}\,\mathbf{c}]\{(\mathbf{a}'\cdot\mathbf{a}')\,\mathbf{a}+(\mathbf{a}'\cdot\mathbf{b}')\,\mathbf{b}+(\mathbf{a}'\cdot\mathbf{c}')\,\mathbf{c}\}$ is
(a) 0 (b) $\mathbf{a}\times\mathbf{b}$
(c) $\mathbf{b}\times\mathbf{c}$ (d) $[\mathbf{a}\,\mathbf{b}\,\mathbf{c}]^{-2}$

PASSAGE 3

Let O, N, G, and O' are the circumcentre, nine point centre, centroid and orthocentre of a ΔABC respectively. AL and BM are perpendiculars from A and B on sides BC and CA respectively. Let AD be the median and OD is perpendicular to side BC. Let R be the circumradius of $\Delta\, ABC$, then $OA = OB = OC = R$.

Now, in ΔOBD, $OD = R\cos A$, in $\Delta\, ABM$,
$AO' = AM\sec(90° - C)$ $[\because \angle O'AM = 90° - C]$

$$= AM \operatorname{cosec} C = \frac{C\cos A}{\sin C} = 2R\cos A$$

$\therefore$ $AO' = 2OD$

If S be any point in the plane of $\Delta\, ABC$ and AP is the diameter of the circumcircle.

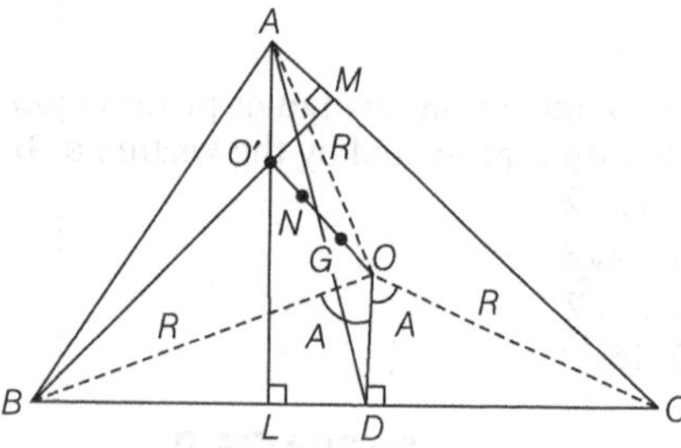

On the basis of above information, answer the following questions

1. $\mathbf{OA}+\mathbf{OB}+\mathbf{OC}$ is equal to
(a) $\mathbf{OO}'$ (b) $2\mathbf{O}'\mathbf{O}$
(c) $2\mathbf{AO}$ (d) $\mathbf{ON}$

2. $\mathbf{O}'\mathbf{A}+\mathbf{O}'\mathbf{B}+\mathbf{O}'\mathbf{C}$ is equal to
(a) $\mathbf{OO}'$ (b) $2\mathbf{O}'\mathbf{O}$
(c) $2\mathbf{AO}'$ (d) $2\mathbf{O}'\mathbf{N}$

3. $\mathbf{AO}'+\mathbf{O}'\mathbf{B}+\mathbf{O}'\mathbf{C}$ is equal to
(a) $\mathbf{OO}'$ (b) $2\mathbf{O}'\mathbf{O}$
(c) $2\mathbf{AO}$ (d) $3\mathbf{SG}$

4. $\mathbf{NA}+\mathbf{NB}+\mathbf{NC}$ is equal to
(a) $\frac{1}{2}\mathbf{OO}'$ (b) $\mathbf{OO}'$
(c) $2\mathbf{O}'\mathbf{O}$ (d) $2\mathbf{NG}$

5. $\mathbf{SA}+\mathbf{SB}+\mathbf{SC}$ is equal to
(a) $\mathbf{SG}$ (b) $2\mathbf{SG}$
(c) $3\mathbf{SG}$ (d) $\frac{1}{2}\mathbf{SG}$

PASSAGE 4

Let the lines $B'B$ and $C'C$ intersect at A. Let $\mathbf{a}$ be the P.V. of A with respect to origin O.

Now, let equations of $B'B$ and $C'C$ be respectively

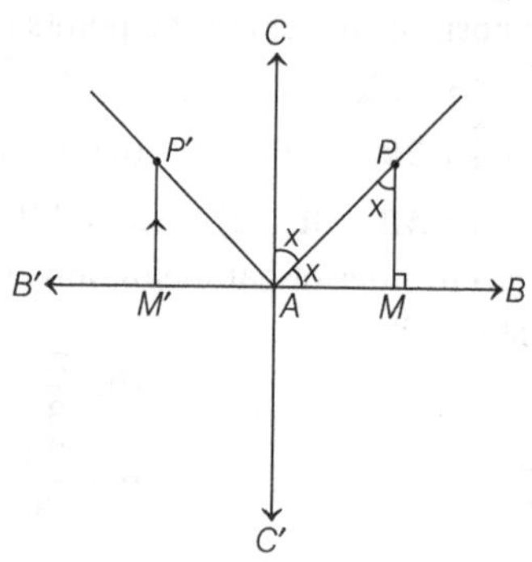

$$\mathbf{r} = \mathbf{a} + t\,\mathbf{b} \quad \ldots\text{(i)}$$

and
$$\mathbf{r} = \mathbf{a} + s\,\mathbf{c} \quad \ldots\text{(ii)}$$

where t and s are scalars.

and $\mathbf{AB}$ is parallel to $\mathbf{b}$ and $\mathbf{AC}$ is parallel to $\mathbf{c}$.

Let P be any point on the bisector of CAB.

Draw $PM \perp AB$ and let $\mathbf{OP} = \mathbf{r}$

$\therefore$ $\angle CAP = \angle PAM = \angle APM$

$\Rightarrow$ $AM = MP = \lambda$ (say)

$\therefore$ $\mathbf{AM} = AM\,\hat{\mathbf{b}} = \dfrac{\lambda\mathbf{b}}{|\mathbf{b}|}$

and $\mathbf{MP} = MP\,\hat{\mathbf{c}} = \dfrac{\lambda\,\mathbf{c}}{|\mathbf{c}|}$

$\therefore$ $\mathbf{AP} = \mathbf{AP} + \mathbf{MP} = \lambda\left(\dfrac{\mathbf{b}}{|\mathbf{b}|} + \dfrac{\mathbf{c}}{|\mathbf{c}|}\right)$

But $\mathbf{OP} = \mathbf{OA} + \mathbf{AP}$

$\therefore$ $\mathbf{r} = \mathbf{a} + \lambda\left(\dfrac{\mathbf{b}}{|\mathbf{b}|} + \dfrac{\mathbf{c}}{|\mathbf{c}|}\right)$

which is the equation of internal bisector of $\angle CAB$.

The bisector AP' of the external $\angle B'AC$ is the bisector of the angle between straight line whose directions are $-\mathbf{b}$ and $\mathbf{c}$.

Therefore, its equation is

$$\mathbf{r} = \mathbf{a} + \mu\left(\frac{\mathbf{b}}{|\mathbf{b}|} - \frac{\mathbf{c}}{|\mathbf{c}|}\right),$$

$$\mathbf{r} = \mathbf{a} + \mu\left(\frac{\mathbf{c}}{|\mathbf{c}|} - \frac{\mathbf{b}}{|\mathbf{b}|}\right)$$

where λ is some scalar.

On the basis of above information, answer the following questions

1. A vector $\mathbf{d}$, directed along the internal bisector of the angle between the vectors $\mathbf{b} = 7\hat{\mathbf{i}} - 4\hat{\mathbf{j}} - 4\hat{\mathbf{k}}$ and $\mathbf{c} = -2\hat{\mathbf{i}} - \hat{\mathbf{j}} + 2\hat{\mathbf{k}}$ and $|\mathbf{d}| = 5\sqrt{6}$ is
   (a) $\frac{5}{3}(\hat{\mathbf{i}} + 7\hat{\mathbf{j}} + 2\hat{\mathbf{k}})$ (b) $\frac{5}{3}(5\hat{\mathbf{i}} + 5\hat{\mathbf{j}} - 2\hat{\mathbf{k}})$
   (c) $\frac{5}{3}(\hat{\mathbf{i}} - 7\hat{\mathbf{j}} + 2\hat{\mathbf{k}})$ (d) $\frac{5}{3}(-5\hat{\mathbf{i}} + 5\hat{\mathbf{j}} + 2\hat{\mathbf{k}})$

2. If ABC be a triangle of sides a, b, c with position vectors of A, B, C as $\mathbf{a}$, $\mathbf{b}$, and $\mathbf{c}$ respectively, then the position vector of its incentre is
   (a) $\left(\frac{\mathbf{a}+\mathbf{b}+\mathbf{c}}{3}\right)$ (b) $\left(\frac{a\,\mathbf{a}+b\,\mathbf{b}+c\,\mathbf{c}}{a+b+c}\right)$
   (c) $\left(\frac{a\,\mathbf{a}+b\,\mathbf{b}+c\,\mathbf{c}}{3}\right)$ (d) $\left(\frac{\mathbf{a}\times\mathbf{b}+\mathbf{b}\times\mathbf{c}+\mathbf{c}\times\mathbf{a}}{a^2+b^2+c^2}\right)$

3. If the interior and exterior bisectors of the angle A of a ΔABC meet the base BC at D and E, then
   (a) $\frac{2}{BC} = \frac{1}{BD} + \frac{1}{BE}$ (b) $2BC = BD + BE$
   (c) $(BC)^2 = BD \times BE$ (d) $2BD = BC + BE$

4. Let ABC be a triangle and $\mathbf{a}$, $\mathbf{b}$, $\mathbf{c}$ be the position vectors of the points A, B, C respectively. External bisectors of $\angle B$ and $\angle C$ meet at P with the sides of the triangle as a, b, c, then the position vectors of P becomes
   (a) $-\left(\frac{b\,\mathbf{b}+c\,\mathbf{c}}{b+c}\right)$ (b) $\left(\frac{a\,\mathbf{a}-b\,\mathbf{b}-c\,\mathbf{c}}{a-b-c}\right)$
   (c) $abc\left(\frac{\mathbf{a}+\mathbf{b}+\mathbf{c}}{3}\right)$ (d) $\left(\frac{a\,\mathbf{a}+b\,\mathbf{b}+c\,\mathbf{c}}{a+b+c}\right)$

5. AD, AD' are internal and external bisectors of angle A of ΔABC, meeting BC at D and D' respectively. A' is the mid point of DD' and B', C' are similar points on CA and AB. Then A', B', C'
   (a) lie on a plane
   (b) form an equilateral triangle
   (c) form an isosceles triangle
   (d) form a right angled isosceles triangle

PASSAGE 5

Let A, B, C, D, E represent vertices of a regular pentagon $ABCDE$. Given the position vector of these vertices be $\mathbf{a}$, $\mathbf{a}+\mathbf{b}$, $\mathbf{b}$, $\lambda\mathbf{a}$ and $\lambda\mathbf{b}$, respectively.

On the basis of above information, answer the following questions

1. The ratio $\frac{AD}{BC}$ is equal to
   (a) $1 - \cos\frac{3\pi}{5} : \cos\frac{3\pi}{5}$ (b) $1 + 2\cos\frac{2\pi}{5} : \cos\frac{\pi}{5}$
   (c) $1 + 2\cos\frac{\pi}{5} : 2\cos\frac{\pi}{5}$ (d) None of these

2. AD divides EC in the ratio
   (a) $\cos\frac{2\pi}{5} : 1$ (b) $\cos\frac{3\pi}{5} : 1$
   (c) $1 : 2\cos\frac{2\pi}{5}$ (d) $1 : 2$

PASSAGE 6

Let O be the origin and $\mathbf{OX}$, $\mathbf{OY}$, $\mathbf{OZ}$ be three unit vectors in the directions of the sides $\mathbf{QR}$, $\mathbf{RP}$, $\mathbf{PQ}$ respectively, of a ΔPQR. **[JEE Advanced 2016]**

On the basis of above information, answer the following questions

1. If the ΔPQR varies, then the minimum value of $\cos(P+Q) + \cos(Q+R) + \cos(R+P)$ is
   (a) $-\frac{3}{2}$ (b) $\frac{3}{2}$
   (c) $\frac{5}{3}$ (d) $-\frac{5}{3}$

2. $|\mathbf{OX} \times \mathbf{OY}| =$
   (a) $\sin(P+Q)$ (b) $\sin(P+R)$
   (c) $\sin(Q+R)$ (d) $\sin 2R$

Numerical Value Type Questions

Solve the following problems and write your answer.

1. $ABCD$ is a parallelogram. A_1 and B_1 are the mid points of sides BC and CD respectively. If $\mathbf{AA_1} + \mathbf{AB_1} = \lambda\,\mathbf{AC}$, then the value of λ must be

2. A, B, C, D are any four points in the space. If $|\mathbf{AB}\times\mathbf{CD} + \mathbf{BC}\times\mathbf{AD} + \mathbf{CA}\times\mathbf{BD}| = \lambda$ (area of ΔABC), then the value of 125λ must be

3. The position vectors of the points A, B, C and D are $3\hat{\mathbf{i}} - 2\hat{\mathbf{j}} - \hat{\mathbf{k}}$, $2\hat{\mathbf{i}} + 3\hat{\mathbf{j}} - 4\hat{\mathbf{k}}$, $-\hat{\mathbf{i}} + \hat{\mathbf{j}} + 2\hat{\mathbf{k}}$ and $4\hat{\mathbf{i}} + 5\hat{\mathbf{j}} + \lambda\hat{\mathbf{k}}$. It is known that these points are coplanar, then the value of -17λ must be

4. Let $\mathbf{a}$ and $\mathbf{b}$ be unit vectors such that $|\mathbf{a}+\mathbf{b}| = \sqrt{3}$, then the value of $(2\mathbf{a} + 5\mathbf{b}) \cdot (3\mathbf{a} + \mathbf{b} + \mathbf{a}\times\mathbf{b})$ must be

5. 'P' is any arbitrary point on the circumcircle of the equilateral triangle of side length 26 unit, then the value of $|\mathbf{PA}|^2 + |\mathbf{PB}|^2 + |\mathbf{PC}|^2$ must be

6. E and F are the points on the sides BC and CD of a parallelogram $ABCD$. If $\mathbf{BE} = 4\mathbf{EC}$ and $\mathbf{CF} = 4\,\mathbf{FD}$. The line EF cut the diagonal AC in G then $\mathbf{AG} = \lambda\,\mathbf{AC}$, then the value of $400\,\lambda$ must be

7. Consider the cube in the first octant with sides OP, OQ and OR of length 1, along the X-axis, Y-axis and Z-axis, respectively, where $O(0, 0, 0)$ is the origin. Let $S\left(\frac{1}{2}, \frac{1}{2}, \frac{1}{2}\right)$ be the centre of the cube and T be the vertex of the cube opposite to the origin O such that S lies on the diagonal OT. If $\mathbf{p} = \mathbf{SP}$, $\mathbf{q} = \mathbf{SQ}$, $\mathbf{r} = \mathbf{SR}$ and $\mathbf{t} = \mathbf{ST}$, then the value of $|(\mathbf{p}\times\mathbf{q})\times(\mathbf{r}\times\mathbf{t})|$ is ... **[JEE Advanced 2018]**

8. $\mathbf{a} = \hat{\mathbf{i}} + \hat{\mathbf{j}} + \hat{\mathbf{k}}$, $\mathbf{b} = 4\hat{\mathbf{i}} + 3\hat{\mathbf{j}} + 4\hat{\mathbf{k}}$ and $\mathbf{c} = \hat{\mathbf{i}} + x_1\hat{\mathbf{j}} + x_2\hat{\mathbf{k}}$ are linearly dependent vectors and $|\mathbf{c}| = 2$, then the value of $729x_1^4 + 343x_1^2x_2^2 + 81x_2^2 + 15$ must be

9. $\mathbf{b}$ and $\mathbf{c}$ are non-collinear vectors. If $\mathbf{a}\times(\mathbf{b}\times\mathbf{c}) + (\mathbf{a}\cdot\mathbf{b})\,\mathbf{b} = (4 - 2x - \sin y)\,\mathbf{b} + (x^2 - 1)\,\mathbf{c}$ and $(\mathbf{c}\cdot\mathbf{c})\,\mathbf{a} = \mathbf{c}$, then the value of $x^{\sin y} + (4\sin y)^{4x} + 15$ must be

10. If $\hat{\mathbf{i}} - 3\hat{\mathbf{j}} + 5\hat{\mathbf{k}}$ bisects the angle between $\hat{\mathbf{a}}$ and $-\hat{\mathbf{i}} + 2\hat{\mathbf{j}} + 2\hat{\mathbf{k}}$ when $\hat{\mathbf{a}}$ is a unit vector, then $\hat{\mathbf{a}} = \frac{\lambda\hat{\mathbf{i}} + \mu\hat{\mathbf{j}} + \nu\hat{\mathbf{k}}}{105}$, then the value of $\lambda^2 - 2\mu + \nu$ must be

11. If G_1, G_2, G_3 are the centroids of the triangular faces OBC, OCA, OAB of a tetrahedron $OABC$. If λ be the ratio of the volume of the tetrahedron to the volume of the parallelopiped with OG_1, OG_2, OG_3 as coterminus edges. Then the value of $2008\,\lambda$ must be

12. Let $\mathbf{a} = 2\hat{\mathbf{i}} + \hat{\mathbf{j}} - \hat{\mathbf{k}}$ and $\mathbf{b} = \hat{\mathbf{i}} + 2\hat{\mathbf{j}} + \hat{\mathbf{k}}$ be two vectors. Consider a vector $\mathbf{c} = \alpha\mathbf{a} + \beta\mathbf{b}$, $\alpha, \beta \in R$. If the projection of $\mathbf{c}$ on the vector $(\mathbf{a}+\mathbf{b})$ is $3\sqrt{2}$, then the minimum value of $(\mathbf{c} - (\mathbf{a}\times\mathbf{b}))\cdot\mathbf{c}$ equals .... **[JEE Advanced 2019]**

13. Two points A and B are given in the rectangular cartesian system of coordinates OXY on the curve $y = x^2$, the points being such that $\mathbf{OA}\cdot\hat{\mathbf{i}} = 1$ and $\mathbf{OB}\cdot\hat{\mathbf{i}} = -2$, $\hat{\mathbf{i}}$ is the unit vector of the OX-axis, then the value of $|2\,\mathbf{OA} - 3\,\mathbf{OB}|^2$ must be

14. In a ΔPQR, let $\mathbf{a} = \mathbf{QR}$, $\mathbf{b} = \mathbf{RP}$ and $\mathbf{c} = \mathbf{PQ}$. If $|\mathbf{a}| = 3$, $|\mathbf{b}| = 4$ and $\frac{\mathbf{a}\cdot(\mathbf{c}-\mathbf{b})}{\mathbf{c}\cdot(\mathbf{a}-\mathbf{b})} = \frac{|\mathbf{a}|}{|\mathbf{a}|+|\mathbf{b}|}$, then the value of $|\mathbf{a}\times\mathbf{b}|^2$ is ..... **[JEE Advanced 2020]**

15. If $\mathbf{d} = \lambda(\mathbf{a}\times\mathbf{b}) + \mu(\mathbf{b}\times\mathbf{c}) + \nu(\mathbf{c}\times\mathbf{a})$, $[\mathbf{a}\ \mathbf{b}\ \mathbf{c}] = \frac{1}{8}$ and $\mathbf{d}\cdot(\mathbf{a}+\mathbf{b}+\mathbf{c}) = 8$, then $\lambda + \mu + \nu$ is equal to

16. If the distance of the point $B\,(\hat{\mathbf{i}} + 2\hat{\mathbf{j}} + 3\hat{\mathbf{k}})$ from the line which is passing through $A\,(4\hat{\mathbf{i}} + 2\hat{\mathbf{j}} + 2\hat{\mathbf{k}})$ and which is parallel to the vector $\vec{\mathbf{c}} = 2\hat{\mathbf{i}} + 3\hat{\mathbf{j}} + 6\hat{\mathbf{k}}$ is λ, then the value of $\lambda^6 + \lambda^4 + \lambda^2 + 1$ must be

Single Integer Type Questions

Solve the following problems and write your answer.

1. If A, B, C, D are four points in a space and $|\mathbf{AB}\times\mathbf{CD} + \mathbf{BC}\times\mathbf{AD} + \mathbf{CA}\times\mathbf{BD}| = \lambda$ (area of the ΔABC). Then the value of λ is

2. $\mathbf{a}$, $\mathbf{b}$ and $\mathbf{c}$ be three unit vectors such that $|\mathbf{a}-\mathbf{b}|^2 + |\mathbf{a}-\mathbf{c}|^2 = 8$. Then $|\mathbf{a} + 2\mathbf{b}|^2 + |\mathbf{a} + 2\mathbf{c}|^2$ is equal to .... **[JEE Main 2020, 2 September Shift-I]**

3. If $\mathbf{p}$, $\mathbf{q}$, $\mathbf{r}$ are mutually perpendicular vectors, where $|\mathbf{p}| = |\mathbf{q}| = |\mathbf{r}|$ and $\mathbf{p}\times\{(\mathbf{x}-\mathbf{q})\times\mathbf{p}\} + \mathbf{q}\times\{(\mathbf{x}-\mathbf{r})\times\mathbf{q}\} + \mathbf{r}\times\{(\mathbf{x}-\mathbf{p})\times\mathbf{r}\} = \mathbf{0}$ and $\mathbf{x} = \frac{1}{\lambda}(\mathbf{p}+\mathbf{q}+\mathbf{r})$, then find the value of λ.

4. Find the absolute value of parameter t for which the area of the triangle whose vertices are $A(-1, 1, 2)$; $B(1, 2, 3)$ and $C(t, 1, 1)$ is minimum.

5. If $\mathbf{a}$ and $\mathbf{b}$ are unit vectors, then the greatest value of $\sqrt{3}\,|\mathbf{a}+\mathbf{b}| + |\mathbf{a}-\mathbf{b}|$ is ...... **[JEE Main 2020, 6 September]**

6. If the point of intersection of the lines $\mathbf{r} = \hat{\mathbf{i}} - 6\hat{\mathbf{j}} + (p\sec\alpha)\hat{\mathbf{k}} + t(\hat{\mathbf{i}} + 2\hat{\mathbf{j}} + \hat{\mathbf{k}})$ and $\mathbf{r} = 4\hat{\mathbf{j}} + \hat{\mathbf{k}} + \lambda(2\hat{\mathbf{i}} + (p\tan\alpha)\hat{\mathbf{j}} + 2\hat{\mathbf{k}})$ is $8\hat{\mathbf{i}} + 8\hat{\mathbf{j}} + 9\hat{\mathbf{k}}$, (where $0 < \alpha < \frac{\pi}{2}$), then p^2 is equal to

7. $\mathbf{a}$ and $\mathbf{c}$ are unit vectors and $|\mathbf{b}| = 4$. The angle between $\mathbf{a}$ and $\mathbf{c}$ is $\cos^{-1}\left(\frac{1}{4}\right)$. If $\mathbf{b} - 2\mathbf{c} = \lambda\mathbf{a}$, then positive value of λ is equal

8. Let the vectors $\mathbf{a}$, $\mathbf{b}$, $\mathbf{c}$ be such that $|\mathbf{a}| = 2$, $|\mathbf{b}| = 4$ and $|\mathbf{c}| = 4$. If the projection of $\mathbf{b}$ on $\mathbf{a}$ is equal to the projection of $\mathbf{c}$ on $\mathbf{a}$ and $\mathbf{b}$ is perpendicular to $\mathbf{c}$, then the value of $|\mathbf{a}+\mathbf{b}-\mathbf{c}|$ is ... **[JEE Main 2020, 5 September]**

9. Let $\mathbf{a}$ and $\mathbf{b}$ be two unit vectors such that $\mathbf{a}\cdot\mathbf{b} = 0$. For some $x, y \in R$, let $\mathbf{c} = x\mathbf{a} + y\mathbf{b} + (\mathbf{a}\times\mathbf{b})$. If $|\mathbf{c}| = 2$ and the vector $\mathbf{c}$ is inclined at the same angle α to both $\mathbf{a}$ and $\mathbf{b}$, then the value of $8\cos^2\alpha$ is .... **[JEE Advanced 2018]**

Matrix-Match Type

Given below are Matching Type Questions, with two columns (each having some items) each . Each item of Column I has to be matched with the items of Column II, by encircling the correct match(es).

NOTE *An item of Column I can be matched with more than one items of Column II.*

1. Observe the following columns

	Column I		Column II
(A)	The value of $(\mathbf{a}\times\mathbf{b})\times(\mathbf{c}\times\mathbf{d})$ is	(P)	$[\mathbf{acd}]\mathbf{b}+[\mathbf{abc}]\mathbf{d}$
(B)	The value of $(\mathbf{a}\times\mathbf{c})\times(\mathbf{b}\times\mathbf{d})$ is	(Q)	$[\mathbf{a\,b\,c}]\mathbf{d}-[\mathbf{bcd}]\mathbf{a}$
(C)	The value of $(\mathbf{a}\times\mathbf{d})\times(\mathbf{b}\times\mathbf{c})$ is	(R)	$[\mathbf{acd}]\mathbf{b}-[\mathbf{bcd}]\mathbf{a}$
		(S)	$[\mathbf{abd}]\mathbf{c}-[\mathbf{acd}]\mathbf{b}$
		(T)	$[\mathbf{abd}]\mathbf{c}+[\mathbf{bcd}]\mathbf{a}$

2. Observe the following columns

	Column I		Column II
(A)	If W_1, be the workdone if two forces $\mathbf{F}_1 = 3\hat{\mathbf{i}}-2\hat{\mathbf{j}}+\hat{\mathbf{k}}$ and $\mathbf{F}_2 = \hat{\mathbf{i}}+3\hat{\mathbf{j}}-5\hat{\mathbf{k}}$ act on a particle whose P.V. is $-2\hat{\mathbf{i}}+5\hat{\mathbf{k}}$ and displace it to another point whose P.V. is $3\hat{\mathbf{i}}-7\hat{\mathbf{j}}+2\hat{\mathbf{k}}$ and W_2 be the workdone if three forces $\mathbf{P}=2\hat{\mathbf{i}}-5\hat{\mathbf{j}}+6\hat{\mathbf{k}}$, $\mathbf{Q}=-\hat{\mathbf{i}}+2\hat{\mathbf{j}}-\hat{\mathbf{k}}$ and $\mathbf{R}=2\hat{\mathbf{i}}+7\hat{\mathbf{j}}$ act on another particle whose P.V. is $4\hat{\mathbf{i}}-3\hat{\mathbf{j}}-2\hat{\mathbf{k}}$ and displace it to another point whose P.V. is $6\hat{\mathbf{i}}+\hat{\mathbf{j}}-3\hat{\mathbf{k}}$, then	(P)	$W_1 - W_2 = 3$
(B)	If W_1 be the workdone if a particle is displaced from the point whose P.V. is (–2, 5, 7) to the point whose P.V. is (3, 7, 2) under the action of constant forces (2,– 3, 1) and (1, 5, –3) and W_2 be the workdone if other particle displaced from the point whose P.V. is (4,– 3,– 2) to the point whose P.V. is (6, 1, –3) under the action of constant forces (1, –1, 1), (–1,2,–1) and (1, 0, –1), then	(Q)	$W_1 - 3W_2 = 8$
(C)	If W_1 be the workdone if forces acting on a particle having magnitudes 5, 3, 1 units and act in the directions of the vectors $6\hat{\mathbf{i}}+2\hat{\mathbf{j}}+3\hat{\mathbf{k}}$, $3\hat{\mathbf{i}}-2\hat{\mathbf{j}}+6\hat{\mathbf{k}}$ and $2\hat{\mathbf{i}}-3\hat{\mathbf{j}}-6\hat{\mathbf{k}}$ respectively. There remain constant while the particle is displaced from the point (2, –1, –3) to (5, –1, 1) and W_2 be the workdone if forces on another particle having magnitudes 1, 2, 3 units and act in the directions of the vectors $\hat{\mathbf{i}}+2\hat{\mathbf{j}}+2\hat{\mathbf{k}}$, $2\hat{\mathbf{i}}-2\hat{\mathbf{j}}+\hat{\mathbf{k}}$ and $2\hat{\mathbf{i}}-\hat{\mathbf{j}}-2\hat{\mathbf{k}}$ respectively. There remain constant while the particle is displaced from the point (–2,–1, 5) to (3,– 4,– 5), then	(R)	$2W_2 - W_1 = 9$
		(S)	$3W_2 - 2W_1 = 1$
		(T)	$W_1 - 4W_2 = 1$

3. Observe the following columns

	Column I		Column II
(A)	If V_1, V_2, V_3 are the volumes of parallelopiped, triangular prism and tetrahedron respectively. The three coterminus edges of all three figures are the vectors $\hat{\mathbf{i}}-\hat{\mathbf{j}}-6\hat{\mathbf{k}}$, $\hat{\mathbf{i}}-\hat{\mathbf{j}}+4\hat{\mathbf{k}}$ and $2\hat{\mathbf{i}}-5\hat{\mathbf{j}}+3\hat{\mathbf{k}}$, then	(P)	$2V_1 + 3V_3 = 5V_2$
(B)	If V_1, V_2, V_3 are the volumes of parallelopiped, triangular prism and tetrahedron respectively. The three cotermius edges of all three figures are the vectors $-2\hat{\mathbf{i}}+3\hat{\mathbf{j}}-3\hat{\mathbf{k}}$, $4\hat{\mathbf{i}}+5\hat{\mathbf{j}}-3\hat{\mathbf{k}}$ and $6\hat{\mathbf{i}}+2\hat{\mathbf{j}}-3\hat{\mathbf{k}}$, then	(Q)	$V_1 + V_2 + V_3 = 60$
(C)	If V_1, V_2, V_3 are the volumes of parallelopiped, triangular prism and tetrahedron respectively. The three coterminus edges of all three figures are the vectors $-3\hat{\mathbf{i}}+\hat{\mathbf{j}}+\hat{\mathbf{k}}$, $4\hat{\mathbf{i}}+2\hat{\mathbf{j}}+4\hat{\mathbf{k}}$ and $2\hat{\mathbf{i}}+2\hat{\mathbf{j}}$, then	(R)	$V_1 + 3V_3 = 3V_2$
		(S)	$V_1 + V_2 + V_3 = 50$
		(T)	$V_1 : V_2 : V_3 = 6:3:1$

4. Observe the following columns

	Column I		Column II
(A)	If $\mathbf{a}+\mathbf{b}$, $\mathbf{b}+\mathbf{c}$ and $\mathbf{c}+\mathbf{a}$ are coplanar, then	(P)	$\mathbf{a}, \mathbf{b}, \mathbf{c}$ are coplanar
(B)	If $\mathbf{a}\times\mathbf{b}$, $\mathbf{b}\times\mathbf{c}$ and $\mathbf{c}\times\mathbf{a}$ are coplanar, then	(Q)	$\mathbf{a}+\mathbf{b}+\mathbf{c}=0$
(C)	If $2\mathbf{a}+\mathbf{b}+\mathbf{c}$, $\mathbf{a}+2\mathbf{b}+\mathbf{c}$ and $\mathbf{a}+\mathbf{b}+2\mathbf{c}$ are coplanar, then	(R)	$[\mathbf{a}+\mathbf{b}\ \ \mathbf{b}+\mathbf{c}\ \ \mathbf{c}+\mathbf{a}]=0$
		(S)	$\mathbf{a}\times\mathbf{b}+\mathbf{b}\times\mathbf{c}+\mathbf{c}\times\mathbf{a}=0$
		(T)	$[\mathbf{a}\ \mathbf{b}\ \mathbf{c}]=0$

Answers

Objective Questions Type [Only one correct answer]

1. (a)	**2.** (c)	**3.** (b)	**4.** (d)	**5.** (a)	**6.** (b)	**7.** (d)	**8.** (b)	**9.** (b)	**10.** (a)
11. (b)	**12.** (a)	**13.** (d)	**14.** (a)	**15.** (b)	**16.** (c)	**17.** (b)	**18.** (b)	**19.** (c)	**20.** (c)
21. (a)	**22.** (a)	**23.** (b)	**24.** (c)	**25.** (d)	**26.** (b)	**27.** (c)	**28.** (b)	**29.** (a)	**30.** (c)
31. (c)	**32.** (d)	**33.** (b)	**34.** (c)	**35.** (b)	**36.** (d)	**37.** (a)	**38.** (c)	**39.** (d)	**40.** (d)
41. (d)	**42.** (b)	**43.** (a)	**44.** (b)	**45.** (b)	**46.** (a)	**47.** (d)	**48.** (b)	**49.** (a)	**50.** (d)
51. (a)	**52.** (c)	**53.** (d)	**54.** (b)	**55.** (c)	**56.** (c)	**57.** (d)	**58.** (b)	**59.** (a)	**60.** (c)
61. (c)	**62.** (b)	**63.** (b)	**64.** (d)	**65.** (b)	**66.** (d)	**67.** (c)	**68.** (c)	**69.** (d)	**70.** (c)
71. (d)	**72.** (a)	**73.** (c)	**74.** (b)	**75.** (c)	**76.** (d)	**77.** (a)	**78.** (b)	**79.** (c)	**80.** (c)
81. (b)	**82.** (b)	**83.** (a)	**84.** (a)	**85.** (b)	**86.** (b)	**87.** (b)	**88.** (c)	**89.** (d)	**90.** (c)
91. (d)	**92.** (c)	**93.** (a)	**94.** (c)	**95.** (b)	**96.** (b)				

Objective Questions Type II [One or more than one correct answer(s)]

1. (a, b, c)	**2.** (b, c)	**3.** (a, c)	**4.** (a, c)	**5.** (a, c)
6. (c, d)	**7.** (b, c)	**8.** (b, c, d)	**9.** (a, b)	**10.** (a, b, c)
11. (a, b, c, d)	**12.** (a, c, d)	**13.** (a, b, c)	**14.** (c, d)	**15.** (a, b, c)
16. (a, b)	**17.** (a, c, d)	**18.** (a, b)	**19.** (a, b)	**20.** (b, c, d)
21. (a, b, c, d)	**22.** (a, c)	**23.** (a, b, c)	**24.** (a, b, c)	**25.** (a, b, c)
26. (a, d)	**27.** (b, c)	**28.** (a, d)	**29.** (a, c)	**30.** (b, d)

Linked-Comprehension Type

Passage 1 **1.** (c) **2.** (a) **3.** (d) **4.** (b) **5.** (c)
Passage 2 **1.** (b) **2.** (d) **3.** (c) **4.** (c) **5.** (c)
Passage 3 **1.** (a) **2.** (b) **3.** (c) **4.** (a) **5.** (c)
Passage 4 **1.** (c) **2.** (b) **3.** (a) **4.** (b) **5.** (a)
Passage 5 **1.** (c) **2.** (c)
Passage 6 **1.** (a) **2.** (a)

Numerical Value Type Questions

1. (1.5)	**2.** (500)	**3.** (146)	**4.** (19.5)	**5.** (1352)`	**6.** (336)
7. (0.5)	**8.** (3698)	**9.** (272)	**10.** (1817)	**11.** (4518)	**12.** (18)
13. (164)	**14.** (108)	**15.** (64)	**16.** (1111)		

Single Integer Type Questions

1. (4)	**2.** (2)	**3.** (2)	**4.** (2)	**5.** (4)	**6.** (3)
7. (3)	**8.** (6)	**9.** (3)			

Matrix-Match Type

1. A → (R); B → (P, T); C → (Q, S)
2. A → (R, S); B → (Q, T); C → (P)
3. A → (P, R, S, T); B → (P, R, T); C → (P, Q, R, T)
4. A → (P, Q, R, S, T); B → (P, Q, R, S, T); C → (P, Q, R, S, T)

Solutions

Objective Questions Type [Only one correct answer]

1. $\mathbf{a} \cdot \mathbf{b} = 0$

$\therefore$ **a** and **b** are mutually perpendicular

Also, $|\mathbf{a}| = 2$ and $|\mathbf{b}| = 3$

Let $\mathbf{a} = 2\hat{\mathbf{i}}$ then $\mathbf{b} = 2\hat{\mathbf{i}}$

such that $\mathbf{a} = \hat{\mathbf{i}}, \mathbf{b} = \hat{\mathbf{j}}$

$$\therefore (\mathbf{a} \times (\mathbf{a} \times (\mathbf{a} \times (\mathbf{a} \times \mathbf{b})))) = (2\hat{\mathbf{i}} \times (2\hat{\mathbf{i}} \times (2\hat{\mathbf{i}} \times (2\hat{\mathbf{i}} \times 3\hat{\mathbf{j}}))))$$

$$= 48\,\hat{\mathbf{j}} = 48\,\mathbf{b}$$

2. Given, $\omega = 4\left(\dfrac{\hat{\mathbf{i}} + 2\hat{\mathbf{j}} - 2\hat{\mathbf{k}}}{\sqrt{1+4+4}}\right) = \dfrac{4}{3}(\hat{\mathbf{i}} + 2\hat{\mathbf{j}} - 2\hat{\mathbf{k}})$

$$\mathbf{r} = \mathbf{AP} = \mathbf{OP} - \mathbf{OA}$$
$$= (4\hat{\mathbf{i}} + \hat{\mathbf{j}} + \hat{\mathbf{k}}) - (3\hat{\mathbf{i}} - 2\hat{\mathbf{j}} - \hat{\mathbf{k}})$$
$$= \hat{\mathbf{i}} + 3\hat{\mathbf{j}} + 2\hat{\mathbf{k}}$$

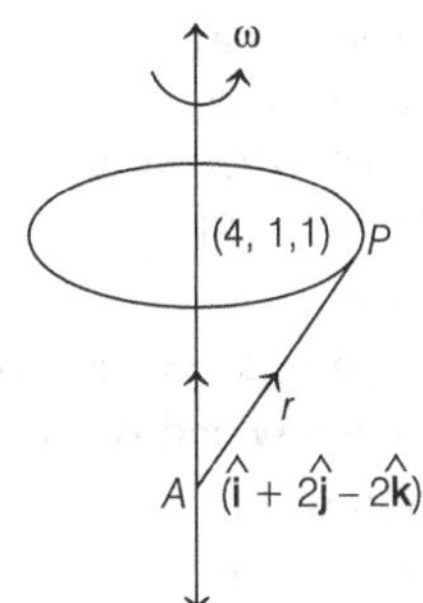

$$\therefore \quad \mathbf{v} = \omega \times \mathbf{r}$$
$$= \frac{4}{3}(\hat{\mathbf{i}} + 2\hat{\mathbf{j}} - 2\hat{\mathbf{k}}) \times (\hat{\mathbf{i}} + 3\hat{\mathbf{j}} + 2\hat{\mathbf{k}})$$
$$= \frac{4}{3}\begin{vmatrix} \hat{\mathbf{i}} & \hat{\mathbf{j}} & \hat{\mathbf{k}} \\ 1 & 2 & -2 \\ 1 & 3 & 2 \end{vmatrix}$$
$$= \frac{4}{3}(10\hat{\mathbf{i}} - 4\hat{\mathbf{j}} + \hat{\mathbf{k}})$$

3. $\mathbf{c} = \lambda(\mathbf{a} \times \mathbf{b})$

$$1 = (\mathbf{a} \times \mathbf{b}) \cdot \mathbf{c} = \frac{\mathbf{cc}}{\lambda} = \frac{|\mathbf{c}|^2}{\lambda} = \frac{1}{3\lambda}$$
$$\lambda = \frac{1}{3}$$
$$\mathbf{c}^2 = \lambda^2 (\mathbf{a} \times \mathbf{b})^2$$
$$\frac{1}{3} = \frac{1}{9}(a^2 b^2 \sin^2\theta) = \frac{1}{9} \times 2 \times 3 \sin^2\theta$$
$$\sin^2\theta = \frac{1}{2}$$
$$\therefore \quad \theta = \frac{\pi}{4}$$

4. Let $\mathbf{a} = 2x^2\hat{\mathbf{i}} + 4x\hat{\mathbf{j}} + \hat{\mathbf{k}}$ and $\mathbf{b} = 7\hat{\mathbf{i}} - 2\hat{\mathbf{j}} + x\hat{\mathbf{k}}$. The angle between **a** and **b** is obtuse

$$\Rightarrow \quad \mathbf{a} \cdot \mathbf{b} < 0$$
$$\Rightarrow \quad 14x^2 - 8x + x < 0$$
$$\Rightarrow \quad x(2x - 1) < 0$$
$$\Rightarrow \quad x \in \left(0, \frac{1}{2}\right) \quad \ldots \text{(i)}$$

Also, it is given $\mathbf{b} \cdot \hat{\mathbf{k}} = x$ and $\dfrac{\mathbf{b} \cdot \hat{\mathbf{k}}}{|\mathbf{b}|} < \cos\dfrac{\pi}{6} = \dfrac{\sqrt{3}}{2}$

$$\Rightarrow \quad 2x > \sqrt{3}\sqrt{53 + x^2}$$
$$\Rightarrow \quad x^2 > 159 \quad \ldots \text{(ii)}$$

There is no common value for Eqs. (i) and (ii)

5. Let $|\mathbf{a}| = |\mathbf{b}| = |\mathbf{c}| = \lambda$, then

$$a^2(\mathbf{x} - \mathbf{b}) - [\mathbf{a} \cdot (\mathbf{x} - \mathbf{b})]\,\mathbf{a} + b^2(\mathbf{x} - \mathbf{c}) - [\mathbf{b} \cdot (\mathbf{x} - \mathbf{c})]\mathbf{b} + c^2(\mathbf{x} - \mathbf{a}) - \mathbf{c}(\mathbf{x} - \mathbf{a})\}\,\mathbf{c} = 0$$
$$\Rightarrow \lambda^2(3\mathbf{x} - \mathbf{a} - \mathbf{b} - \mathbf{c}) - [(\mathbf{a} \cdot \mathbf{x})\mathbf{a} - (\mathbf{b} \cdot \mathbf{x})\mathbf{b} - (\mathbf{c} \cdot \mathbf{x})\,\mathbf{c}] - (\mathbf{a} \cdot \mathbf{b})\mathbf{a} + (\mathbf{b} \cdot \mathbf{c})\mathbf{b} + (\mathbf{c} \cdot \mathbf{a})\,\mathbf{c} = 0$$
$$\Rightarrow \lambda^2(3\mathbf{x} - \mathbf{a} - \mathbf{b} - \mathbf{c}) - \lambda^2\mathbf{x} = 0$$
$$\left[\because \mathbf{a} = \frac{\mathbf{a} \cdot \mathbf{x}}{|\mathbf{a}|^2}\mathbf{a} + \frac{\mathbf{b} \cdot \mathbf{x}}{|\mathbf{b}|^2}\mathbf{b} + \frac{\mathbf{c} \cdot \mathbf{x}}{|\mathbf{c}|^2}\mathbf{x}\right]$$
$$\Rightarrow \quad \mathbf{x} = \frac{1}{2}(\mathbf{a} + \mathbf{b} + \mathbf{c})$$

6. The position vector of any point at t is

$$\mathbf{r} = (2 + t^2)\hat{\mathbf{i}} + (4t - 5)\hat{\mathbf{j}} + (2t^2 - 6t)\hat{\mathbf{k}}$$
$$\frac{d\mathbf{r}}{dt} = 2t\,\hat{\mathbf{i}} + 4\hat{\mathbf{j}} + (4t - 6)\hat{\mathbf{k}}$$
$$\left.\frac{d\mathbf{r}}{dt}\right|_{t=2} = 4\hat{\mathbf{i}} + 4\hat{\mathbf{j}} + 2\hat{\mathbf{k}}$$
$$\left|\frac{d\mathbf{r}}{dt}\right|_{t=2} = \sqrt{16 + 16 + 4} = 6$$

Hence, the required unit tangent vector at $t = 2$ is $\dfrac{1}{3}(2\hat{\mathbf{i}} + 2\hat{\mathbf{j}} + \hat{\mathbf{k}})$.

7. If $\hat{\mathbf{i}}, \hat{\mathbf{j}}, \hat{\mathbf{k}}$ are unit vectors in the $OXYZ$ system and $\hat{\mathbf{i}}_1, \hat{\mathbf{j}}_1, \hat{\mathbf{k}}_1$ are the unit vectors in the system $OX'Y'Z'$ obtained after rotation, then $\hat{\mathbf{i}}_1 = \dfrac{1}{\sqrt{2}}\hat{\mathbf{i}} + \dfrac{1}{\sqrt{2}}\hat{\mathbf{j}}$, $\hat{\mathbf{j}}_1 = -\dfrac{1}{\sqrt{2}}\hat{\mathbf{i}} + \dfrac{1}{\sqrt{2}}\hat{\mathbf{j}}$ and

$\hat{\mathbf{k}}_1 = \hat{\mathbf{k}}$

$$\mathbf{a} = 2\sqrt{2}\,\hat{\mathbf{i}}_1 + 3\sqrt{2}\,\hat{\mathbf{j}}_1 + 4\hat{\mathbf{k}}_1$$
$$= 2\sqrt{2}\left(\frac{1}{\sqrt{2}}\hat{\mathbf{i}} + \frac{1}{\sqrt{2}}\hat{\mathbf{j}}\right) + 3\sqrt{2}\left(\frac{-1}{\sqrt{2}}\hat{\mathbf{i}} + \frac{1}{\sqrt{2}}\hat{\mathbf{j}}\right) + 4\hat{\mathbf{k}}$$
$$= -\hat{\mathbf{i}} + 5\hat{\mathbf{j}} + 4\hat{\mathbf{k}}$$

So, that the components in the $OXYZ$ system are $-1, 5, 4$.

8. $\alpha = \mathbf{a} + \mathbf{b} + \mathbf{c}$

$= 6\hat{\mathbf{i}} + 12\hat{\mathbf{j}}$

Let $\alpha = x\,\mathbf{a} + y\,\mathbf{b}$

$\Rightarrow \quad 6x + 2y = 6$

$-3x - 6y = 12$

$\therefore \quad x = 2, y = -3$

$\therefore \quad \alpha = 2\mathbf{a} - 3\mathbf{b}$

9. $\mathbf{PQ} \times \mathbf{RS} - \mathbf{QR} \times \mathbf{PS} + \mathbf{RP} \times \mathbf{QS} = (\mathbf{PS} + \mathbf{SQ}) \times \mathbf{RS} - \mathbf{QR} \times \mathbf{PS} + (\mathbf{RS} + \mathbf{SP}) \times \mathbf{QS}$

$= \mathbf{PS} \times (\mathbf{RS} + \mathbf{SQ} + \mathbf{QR}) - \mathbf{SQ} \times \mathbf{SR} - \mathbf{SQ} \times \mathbf{SR}$

$= \mathbf{PS} \times \mathbf{O} - 2\mathbf{SQ} \times \mathbf{SR}$

$= 4 \times$ area of ΔQRS

10. $\sum_{i=1}^{n} \mathbf{a}_i = 0$

$$\left(\sum_{i=1}^{n} \mathbf{a}_i\right) \cdot \left(\sum_{i=1}^{n} \mathbf{a}_i\right) = \sum_{i=1}^{n} |\mathbf{a}_i|^2 + 2 \sum_{1 \le i < j \le n} \sum \mathbf{a}_i \cdot \mathbf{a}_j$$

$$\sum_{1 \le i < j \le n} \sum \mathbf{a}_i \cdot \mathbf{a}_j = -\frac{n}{2}$$

11. Given expression $= 2(1 + 1 + 1) - 2\Sigma(\mathbf{a} \cdot \mathbf{b})$

$= 6 - 2\Sigma(\mathbf{a} \cdot \mathbf{b}) \quad \ldots(i)$

But $(\mathbf{a} + \mathbf{b} + \mathbf{c})^2 \ge 0$

$\therefore \quad (1 + 1 + 1) + 2\Sigma\, \mathbf{a} \cdot \mathbf{b} \ge 0$

$\therefore \quad 3 \ge -2\Sigma\, \mathbf{a} \cdot \mathbf{b} \quad \ldots(ii)$

From Eqs. (i) and (ii),

Given expression $\le 6 + 3 = 9$

12. It is given that,

$$a\cos\theta = b\cos\left(\theta + \frac{2\pi}{3}\right) = c\cos\left(\theta + \frac{4\pi}{3}\right) = k \text{ (let)}$$

$$\Rightarrow a = \frac{k}{\cos\theta},\ b = \frac{k}{\cos\left(\theta + \frac{2\pi}{3}\right)} \text{ and } c = \frac{k}{\cos\left(\theta + \frac{4\pi}{3}\right)} \quad \ldots(i)$$

Now, angle between vectors $a\hat{\mathbf{i}} + b\hat{\mathbf{j}} + c\hat{\mathbf{k}}$ and $b\hat{\mathbf{i}} + c\hat{\mathbf{j}} + a\hat{\mathbf{k}}$, where $a^2 + b^2 + c^2 = 1$ for $a, b, c \in R$ is

$\alpha = \cos^{-1}|(ab + bc + ca)|$

$\because \quad ab + bc + ca$

$$= k^2 \left[\frac{1}{\cos\theta\cos\left(\theta + \frac{2\pi}{3}\right)} + \frac{1}{\cos\left(\theta + \frac{2\pi}{3}\right)\cos\left(\theta + \frac{4\pi}{3}\right)} + \frac{1}{\cos\left(\frac{4\pi}{3} + \theta\right)\cos\theta}\right]$$

$$= k^2 \left[\frac{\cos\left(\frac{4\pi}{3} + \theta\right) + \cos\left(\frac{2\pi}{3} + \theta\right) + \cos\theta}{\cos\theta\cos\left(\frac{2\pi}{3} + \theta\right)\cos\left(\frac{4\pi}{3} + \theta\right)}\right]$$

$$= k^2 \frac{2\cos(\pi + \theta)\cos\left(\frac{\pi}{3}\right) + \cos\theta}{\cos\theta\cos\left(\frac{2\pi}{3} + \theta\right)\cos\left(\frac{4\pi}{3} + \theta\right)}$$

$$= k^2 \frac{-\cos\theta + \cos\theta}{\cos\theta\cos\left(\frac{2\pi}{3} + \theta\right)\cos\left(\frac{4\pi}{3} + \theta\right)} = 0$$

$\therefore \alpha = \cos^{-1}(0) = \dfrac{\pi}{2}$

Hence, option (a) is correct.

13. We have, $\mathbf{a} \times \mathbf{b} = \mathbf{c}$

$\Rightarrow \mathbf{c}$ is $\perp$ to $\mathbf{a}$ and $\mathbf{b}$

and $\quad \mathbf{b} \times \mathbf{c} = \mathbf{a}$

$\Rightarrow \mathbf{a}$ is $\perp$ to $\mathbf{b}$ and $\mathbf{c}$

$\Rightarrow \mathbf{a}, \mathbf{b}, \mathbf{c}$ are mutually perpendicular

Again $\quad \mathbf{a} \times \mathbf{b} = \mathbf{c}$

$|\mathbf{a} \times \mathbf{b}| = |\mathbf{c}|$

$= |\mathbf{a}||\mathbf{b}| \cdot \sin 90^\circ = |\mathbf{c}|$

$\Rightarrow \quad |\mathbf{a}||\mathbf{b}| = |\mathbf{c}| \quad \ldots(i)$

Also, $\quad \mathbf{b} \times \mathbf{c} = |\mathbf{c}|$

$|\mathbf{b}||\mathbf{c}| \sin 90^\circ = |\mathbf{a}|$

$\Rightarrow \quad |\mathbf{b}||\mathbf{c}| = |\mathbf{a}| \quad \ldots(ii)$

From Eqs. (i) and (ii), we get

$|\mathbf{b}|^2 |\mathbf{c}| = |\mathbf{c}|$

$\therefore \quad |\mathbf{b}|^2 = 1 \quad [\because |\mathbf{c}| \ne 0]$

or $\quad |\mathbf{b}| = 1 \Rightarrow |\mathbf{a}| = |\mathbf{c}|$

14. Given, $\mathbf{u} \times \mathbf{v} + \mathbf{u} = \mathbf{w}$ and $\mathbf{w} \times \mathbf{u} = \mathbf{v}$

$\Rightarrow (\mathbf{u} \times \mathbf{v} + \mathbf{u}) \times \mathbf{u} = \mathbf{v}$

$\Rightarrow \quad (\mathbf{u} \times \mathbf{v}) \times \mathbf{u} = \mathbf{v}$

$\Rightarrow \quad \mathbf{v} - (\mathbf{u} \cdot \mathbf{v}) = \mathbf{v}$

$\Rightarrow \quad (\mathbf{u} \cdot \mathbf{v})\,\mathbf{u} = 0 \Rightarrow (\mathbf{u} \cdot \mathbf{v}) = 0$

Now, $\quad [\mathbf{uvw}] = \mathbf{u} \cdot (\mathbf{v} \times \mathbf{w})$

$= \mathbf{u} \cdot (\mathbf{v} \times (\mathbf{u} \times \mathbf{v} + \mathbf{u}))$

$= \mathbf{u} \cdot (\mathbf{v} \times (\mathbf{u} \times \mathbf{v}) + \mathbf{v} \times \mathbf{u})$

$= \mathbf{u}\,(\mathbf{v}^2\,\mathbf{u} - (\mathbf{u} \cdot \mathbf{v})\,\mathbf{v} + \mathbf{v} \times \mathbf{u})$

$= \mathbf{v}^2\mathbf{u}^2 = 1$

15. We have $|\mathbf{a}| = |\mathbf{b}|, |\mathbf{a} \times \mathbf{b}| = |\mathbf{a}|$ and $\mathbf{a} \cdot \mathbf{b} = 0$

$|\mathbf{a} \times \mathbf{b}| = |\mathbf{a}||\mathbf{b}| \sin 90^\circ$

$|\mathbf{a}| = |\mathbf{a}||\mathbf{b}|$

$\Rightarrow \quad |\mathbf{b}| = 1 = |\mathbf{a}| = |\mathbf{a} \times \mathbf{b}|$

$|\mathbf{a} + \mathbf{b} + (\mathbf{a} \times \mathbf{b})|^2 = |\mathbf{a}|^2 + |\mathbf{b}|^2 + |\mathbf{a} \times \mathbf{b}|^2 + 2(\mathbf{a} \cdot \mathbf{b} + \mathbf{a} \cdot (\mathbf{a} \times \mathbf{b}) + \mathbf{b} \cdot (\mathbf{a} \times \mathbf{b}))$

$\Rightarrow |\mathbf{a} + \mathbf{b} + (\mathbf{a} \times \mathbf{b})|^2 = 1 + 1 + 1 + 0 = 3$

Now, angle bwtween $(\mathbf{a} + \mathbf{b} + (\mathbf{a} \times \mathbf{b}))$ and $\mathbf{a}$ is

$$\cos\theta = \frac{(\mathbf{a} + \mathbf{b} + (\mathbf{a} \times \mathbf{b})) \cdot \mathbf{a}}{|\mathbf{a} + \mathbf{b} + (\mathbf{a} \times \mathbf{b})||\mathbf{a}|}$$

$$\cos\theta = \frac{|\mathbf{a}|^2 + \mathbf{a} \cdot \mathbf{b} + \mathbf{a} \cdot (\mathbf{a} \times \mathbf{b})}{\sqrt{3} \times 1}$$

$$\cos\theta = \frac{1 + 0 + 0}{\sqrt{3}}$$

$$\theta = \cos^{-1}\frac{1}{\sqrt{3}}$$

16. Given, $\alpha + \beta + \gamma = a\delta$ ...(i)

$\beta + \gamma + \delta = b\alpha$ ...(ii)

From Eq. (i),

$$\alpha + \beta + \gamma + \delta = (a+1)\delta \quad \text{...(iii)}$$

From Eq. (ii),

$$\alpha + \beta + \gamma + \delta = (b+1)\alpha$$

From Eqs. (iii) and (iv),

$$(a+1)\delta = (b+1)\alpha \quad \text{...(v)}$$

Since, α is not parallel to δ

$\therefore$ From Eq. (v),

$$a+1=0 \text{ and } b+1=0$$

$\therefore$ From Eq. (iii),

$$\alpha + \beta + \gamma + \delta = 0$$

17. $(\hat{\mathbf{i}} \times \hat{\mathbf{j}}) \cdot \mathbf{c} \le |\hat{\mathbf{i}} \times \hat{\mathbf{j}}||\mathbf{c}| \cos\frac{\pi}{6}$

$$\Rightarrow \quad -\frac{\sqrt{3}}{2} \le (\hat{\mathbf{i}} \times \hat{\mathbf{j}}) \cdot \mathbf{c} \le \frac{\sqrt{3}}{2}$$

18. Given $\mathbf{OP} = x\hat{\mathbf{i}} + y\hat{\mathbf{j}} - \hat{\mathbf{k}}$, $\mathbf{OQ} = -\hat{\mathbf{i}} + 2\hat{\mathbf{j}} + 3x\hat{\mathbf{k}}$

$|\mathbf{PQ}| = \sqrt{20}$ and $\mathbf{OP}$ is perpendicular to $\mathbf{OQ}$

$\mathbf{OR} = 3\hat{\mathbf{i}} + z\hat{\mathbf{j}} - 7\hat{\mathbf{k}}$

Also, **OP**, **OQ**, **OR** are coplanar

$\mathbf{OP} \cdot \mathbf{OQ} = -x + 2y - 3x = 0 = y = 2x$

$|\mathbf{PQ}|^2 = 20 = (x+1)^2 + (y-2)^2 + (1+3x)^2$

$\Rightarrow \quad 20 = (x+1)^2 + (2x-2)^2 + (1+3x)^2$

$\Rightarrow \quad x = 1 \Rightarrow y = 2$

OP, **OQ**, **OR** are coplanar

$$\therefore \quad \begin{vmatrix} x & y & -1 \\ -1 & 2 & 3x \\ 3 & z & -7 \end{vmatrix} = 0$$

$$\begin{vmatrix} 1 & 2 & -1 \\ -1 & 2 & 3 \\ 3 & z & -7 \end{vmatrix} = 0$$

$\Rightarrow \quad (-14-3z) - 2(7-9) - 1(-z-6) = 0 \quad \Rightarrow z = -2$

$\therefore \quad x^2 + y^2 + z^2 = 1 + 4 + 4 = 9$

19. Let projection be x, then

$$\mathbf{a} = \frac{x(\hat{\mathbf{i}} + \hat{\mathbf{j}})}{\sqrt{2}} + \frac{x(-\hat{\mathbf{i}} + \hat{\mathbf{j}})}{\sqrt{2}} + x\hat{\mathbf{k}}$$

$$\therefore \quad \mathbf{a} = \frac{2x\hat{\mathbf{j}}}{\sqrt{2}} + x\hat{\mathbf{k}}$$

$$\Rightarrow \quad \hat{\mathbf{a}} = \frac{\sqrt{2}}{\sqrt{3}}\hat{\mathbf{j}} + \frac{\hat{\mathbf{k}}}{\sqrt{3}}$$

20. Given, $\mathbf{a} = \hat{\mathbf{i}} - \hat{\mathbf{j}} + \hat{\mathbf{k}}$, $\mathbf{b} = \hat{\mathbf{i}} - 2\hat{\mathbf{j}} + \hat{\mathbf{k}}$

$\mathbf{c} = p\hat{\mathbf{i}} + 2\hat{\mathbf{j}} + q\hat{\mathbf{k}}$, $\mathbf{d} = p\hat{\mathbf{i}} + q\hat{\mathbf{j}} + 2\hat{\mathbf{k}}$

and projection of $\mathbf{c}$ on $\mathbf{a} = 5\sqrt{3}$ units

$$\frac{\mathbf{a} \cdot \mathbf{c}}{|\mathbf{a}|} = 5\sqrt{3}$$

$$\Rightarrow \quad \frac{p+q-2}{\sqrt{3}} = 5\sqrt{3}$$

$$\Rightarrow \quad p+q-2 = 15$$

$$\Rightarrow \quad p+q = 17 \quad \text{...(i)}$$

Also, given that $[\mathbf{a}\ \mathbf{b}\ \mathbf{c}] = 5$

$$\Rightarrow \quad \begin{bmatrix} 1 & -1 & 1 \\ 1 & -2 & 1 \\ p & 2 & q \end{bmatrix} = 5$$

$\Rightarrow \quad 1(-2q-2) + 1(q-p) + 1(2+2p) = 5$

$\Rightarrow \quad -2q - 2 + q - p + 2 + 2p = 5$

$\Rightarrow \quad p - q = 5$ ...(ii)

By solving Eqs. (i) and (ii), we get

$$p = 11, q = 6$$

Now, $\mathbf{b} \cdot \mathbf{d} = (\hat{\mathbf{i}} - 2\hat{\mathbf{j}} + \hat{\mathbf{k}}) \cdot (p\hat{\mathbf{i}} + q\hat{\mathbf{j}} + 2\hat{\mathbf{k}})$

$= p - 2q + 2$

$= 11 - 12 + 2 = 1$

$\therefore \quad \tan^{-1}(\mathbf{b} \cdot \mathbf{d}) = \tan^{-1}(1)$

$$= \tan^{-1}\left(\tan\frac{\pi}{4}\right) = \frac{\pi}{4}$$

21. $|\mathbf{p} + \mathbf{q}| < 1$

$\Rightarrow \quad |\mathbf{p} + \mathbf{q}|^2 < 1$

$\Rightarrow \quad \mathbf{p} \cdot \mathbf{p} + \mathbf{q} \cdot \mathbf{q} + 2\mathbf{p} \cdot \mathbf{q} < 1$

$\Rightarrow \quad 2 + 2\cos\alpha < 1$

$\Rightarrow \quad 4\cos^2\left(\frac{\alpha}{2}\right) < 1$

$\Rightarrow \quad -\frac{1}{2} < \cos\left(\frac{\alpha}{2}\right) < \frac{1}{2}$

$\Rightarrow \quad \frac{\pi}{3} < \frac{\alpha}{2} < \frac{2\pi}{3}$ or $\frac{4\pi}{3} < \frac{\alpha}{2} < \frac{5\pi}{3}$

$\Rightarrow \quad \frac{2\pi}{3} < \alpha < \frac{4\pi}{3}$ or $\frac{8\pi}{3} < \alpha < \frac{10\pi}{3}$

$\Rightarrow \quad \frac{2\pi}{3} < \alpha < \frac{4\pi}{3}$ as $0 < \alpha < 2\pi$

22. $\mathbf{a} \cdot \mathbf{b} = |\mathbf{a}||\mathbf{b}|\cos\theta = 1 \cdot 1 \cos\theta = \cos\theta$

$$\therefore \quad \left|\frac{\mathbf{a} - \mathbf{b}}{2}\right|^2 = \frac{(\mathbf{a} - \mathbf{b})^2}{4}$$

$$= \frac{(\mathbf{a})^2 + (\mathbf{b})^2 - 2\mathbf{a} \cdot \mathbf{b}}{4}$$

$$= \frac{1 + 1 - 2\mathbf{a} \cdot \mathbf{b}}{4}$$

$$= \frac{1 - \cos\theta}{2} = \sin^2\left(\frac{\theta}{2}\right)$$

$$\therefore \quad \left|\frac{\mathbf{a} - \mathbf{b}}{2}\right| = \sin\left(\frac{\theta}{2}\right)$$

23. Since, $\mathbf{a} + \mathbf{b} + \mathbf{c} = 0$

$\Rightarrow \quad \mathbf{a} + \mathbf{b} = -\mathbf{c}$

On squaring

$$(\mathbf{a})^2 + (\mathbf{b})^2 + 2\mathbf{a} \cdot \mathbf{b} = (\mathbf{c})^2$$

$\Rightarrow a^2 + b^2 + 2|\mathbf{a}||\mathbf{b}|\cos(\pi - C) = c^2$

$\Rightarrow \quad 9 + 25 - 2 \cdot 3 \cdot 5\cos C = 49$

$\therefore \quad \cos C = -\frac{1}{2}$

$\therefore \quad \angle C = 120°$

Hence, angle between $\mathbf{a}$ and $\mathbf{b}$ is $\pi - \frac{2\pi}{3} = \frac{\pi}{3}$

24. A point equidistant from AB and AC is on the bisector of the $\angle BAC$.

A vector along the internal bisector of the $\angle BAC$

$$= \frac{\mathbf{AB}}{|\mathbf{AB}|} + \frac{\mathbf{AC}}{|\mathbf{AC}|}$$
$$= \frac{3\hat{\mathbf{i}} + \hat{\mathbf{j}} - \hat{\mathbf{k}}}{\sqrt{9+1+1}} + \frac{\hat{\mathbf{i}} - \hat{\mathbf{j}} + 3\hat{\mathbf{k}}}{\sqrt{1+1+9}}$$
$$= \frac{1}{\sqrt{11}}(4\hat{\mathbf{i}} + 2\hat{\mathbf{k}})$$

$\therefore \quad \mathbf{AP} = t(2\hat{\mathbf{i}} + \hat{\mathbf{k}})$

$\therefore \quad \mathbf{BP} = \mathbf{AP} - \mathbf{AB} = t(2\hat{\mathbf{i}} + \hat{\mathbf{k}}) - (3\hat{\mathbf{i}} + \hat{\mathbf{j}} - \hat{\mathbf{k}})$
$= (2t-3)\hat{\mathbf{i}} - \hat{\mathbf{j}} + (t+1)\hat{\mathbf{k}}$

Also, $\mathbf{BC} = \mathbf{AC} - \mathbf{AB} = (\hat{\mathbf{i}} - \hat{\mathbf{j}} + 3\hat{\mathbf{k}}) - (3\hat{\mathbf{i}} + \hat{\mathbf{j}} - \hat{\mathbf{k}})$
$= -2\hat{\mathbf{i}} - 2\hat{\mathbf{j}} + 4\hat{\mathbf{k}}$

But $\mathbf{BP} = s\mathbf{BC}$

$\therefore \quad (2t-3)\hat{\mathbf{i}} - \hat{\mathbf{j}} + (t+1)\hat{\mathbf{k}} = s(-2\hat{\mathbf{i}} - 2\hat{\mathbf{j}} + 4\hat{\mathbf{k}})$

$\therefore \quad 2t - 3 = -2s, \; -1 = -2s, \; t + 1 = 4s$

$\therefore \quad s = \frac{1}{2}$ and $t = 1 \; \therefore \mathbf{AP} = 2\hat{\mathbf{i}} + \hat{\mathbf{k}}.$

25. Since, $\mathbf{a} \cdot \mathbf{b} = |\mathbf{a}||\mathbf{b}| \cos 120°$
$$= 1 \cdot 2 \cdot \left(-\frac{1}{2}\right) = -1$$

$\because \{(\mathbf{a} + 3\mathbf{b}) \times (3\mathbf{a} - \mathbf{b})\}^2$
$$= \{3\mathbf{a} \times \mathbf{a} - \mathbf{a} \times \mathbf{b} + 9\mathbf{b} \times \mathbf{a} - 3\mathbf{b} \times \mathbf{b}\}^2$$
$$= [0 - \mathbf{a} \times \mathbf{b} - 9\mathbf{a} \times \mathbf{b} - 0]^2$$
$$= [-10(\mathbf{a} \times \mathbf{b})]^2$$
$$= 100(\mathbf{a} \times \mathbf{b})^2$$
$$= 100\{a^2b^2 - (\mathbf{a} \times \mathbf{b})^2\}$$
$$= 100\{4 - 1\} = 300$$

26. $\mathbf{OP} \cdot \mathbf{OQ} + \mathbf{OR} \cdot \mathbf{OS} = \mathbf{OR} \cdot \mathbf{OP} + \mathbf{OQ} \cdot \mathbf{OS}$

$\Rightarrow \quad \mathbf{OP}(\mathbf{OQ} - \mathbf{OR}) + \mathbf{OS}(\mathbf{OR} - \mathbf{OQ}) = 0$

$\Rightarrow \quad (\mathbf{OP} - \mathbf{OS})(\mathbf{OQ} - \mathbf{OR}) = 0$

$\Rightarrow \quad \mathbf{SP} \cdot \mathbf{RQ} = 0$

Similarly, $\mathbf{SR} \cdot \mathbf{PQ} = 0$ and $\mathbf{SQ} \cdot \mathbf{PR} = 0$

$\therefore S$ is orthocentre.

27. $[\mathbf{a} \times \mathbf{b}, \mathbf{b} \times \mathbf{c}, \mathbf{c} \times \mathbf{a}]$
$$= (\mathbf{a} \times \mathbf{b}) \cdot \{(\mathbf{b} \times \mathbf{c}) \times (\mathbf{c} \times \mathbf{a})\}$$
$$= (\mathbf{a} \times \mathbf{b}) \cdot \{\{(\mathbf{b} \times \mathbf{c}) \cdot \mathbf{a}\}\mathbf{c} - \{(\mathbf{b} \times \mathbf{c}) \cdot \mathbf{c}\}\mathbf{a}\}$$
$$= (\mathbf{a} \times \mathbf{b}) \cdot \{[\mathbf{b}\,\mathbf{c}\,\mathbf{a}]\mathbf{c} - 0\}$$
$$= \{(\mathbf{a} \times \mathbf{b}) \cdot \mathbf{c}\}[\mathbf{b}\,\mathbf{c}\,\mathbf{a}]$$
$$[\mathbf{a}\,\mathbf{b}\,\mathbf{c}][\mathbf{a}\,\mathbf{b}\,\mathbf{c}] = [\mathbf{a}\,\mathbf{b}\,\mathbf{c}]^2$$

28. Since, $[\mathbf{a}\,\mathbf{b}\,\mathbf{c}] \neq 0$

$$\therefore \quad \frac{\mathbf{a} \cdot (\mathbf{b} \times \mathbf{c})}{\mathbf{b} \cdot (\mathbf{c} \times \mathbf{a})} + \frac{\mathbf{b} \cdot (\mathbf{c} \times \mathbf{a})}{\mathbf{c} \cdot (\mathbf{a} \times \mathbf{b})} + \frac{\mathbf{c} \cdot (\mathbf{b} \times \mathbf{a})}{\mathbf{a} \cdot (\mathbf{b} \times \mathbf{c})}$$
$$= \frac{[\mathbf{abc}]}{[\mathbf{bca}]} + \frac{[\mathbf{bca}]}{[\mathbf{cab}]} + \frac{[\mathbf{cba}]}{[\mathbf{abc}]}$$
$$= \frac{[\mathbf{abc}]}{[\mathbf{abc}]} + \frac{[\mathbf{abc}]}{[\mathbf{abc}]} - \frac{[\mathbf{abc}]}{[\mathbf{abc}]}$$
$$= 1 + 1 - 1 = 1$$

29. If $\mathbf{a} = \hat{\mathbf{i}}, \mathbf{b} = \hat{\mathbf{j}}$ and $\mathbf{c} = \hat{\mathbf{k}}$

$\therefore \quad [\mathbf{abc}] = \hat{\mathbf{i}} \cdot (\hat{\mathbf{j}} \times \hat{\mathbf{k}}) = \hat{\mathbf{i}} \cdot \hat{\mathbf{i}} = 1$

if $\quad \mathbf{a} = \hat{\mathbf{j}}, \mathbf{b} = \hat{\mathbf{i}}, \mathbf{c} = \hat{\mathbf{k}}$

then $\quad [\mathbf{abc}] = \mathbf{a} \cdot (\mathbf{b} \times \mathbf{c})$
$= \hat{\mathbf{j}} \cdot (\hat{\mathbf{i}} \times \hat{\mathbf{k}}) = \hat{\mathbf{j}} \cdot (-\hat{\mathbf{j}}) = -1$

$\therefore \quad [\mathbf{abc}] = \pm 1$

30. For obtuse angle

$(cx\,\hat{\mathbf{i}} - 6\,\hat{\mathbf{j}} + 3\,\hat{\mathbf{k}}) \cdot (x\,\hat{\mathbf{i}} + 2\hat{\mathbf{j}} + 2cx\,\hat{\mathbf{k}}) < 0$

$\Rightarrow \quad cx^2 - 12 + 6cx < 0$

$\Rightarrow \quad cx^2 + 6cx - 12 < 0$

$\therefore \quad (6c)^2 - 4 \cdot c(-12) < 0$

$3c^2 + 4c < 0$

$\Rightarrow \quad c\left(c + \frac{4}{3}\right) < 0$

$\therefore \quad -\frac{4}{3} < c < 0$

31. Given condition

$\Rightarrow (\hat{\mathbf{a}} + \hat{\mathbf{b}} + \hat{\mathbf{c}}) \cdot (\hat{\mathbf{a}} + \hat{\mathbf{b}} + \hat{\mathbf{c}}) = 1$

$|\mathbf{a}|^2 + |\mathbf{b}|^2 + |\mathbf{c}|^2 + 2|\mathbf{a}||\mathbf{b}|\cos\theta_1$
$+ 2|\mathbf{b}||\mathbf{c}|\cos\theta_2 + 2|\mathbf{c}||\mathbf{a}|\cos\theta_3 = 1$

$\Rightarrow \cos\theta_1 + \cos\theta_2 + \cos\theta_3 = 1$

$\Rightarrow$ one of θ_1, θ_2 and θ_3 should be obtuse angle.

32. $\mathbf{a} \times \mathbf{r} = \hat{\mathbf{j}}$

$\therefore \hat{\mathbf{j}}$ is perpendicular to both $\mathbf{a}$ and $\mathbf{r}$

$\therefore \quad \mathbf{a} \cdot \hat{\mathbf{j}} = 0$ and $\mathbf{r} \cdot \hat{\mathbf{j}} = 0$

If angle between $\mathbf{a}$ and $\mathbf{r}$ is θ

then $\mathbf{a} \cdot \mathbf{r} = |\mathbf{a}||\mathbf{r}|\cos\theta = ar\cos\theta$
$$= \sqrt{a^2r^2\cos^2\theta}$$
$$= \sqrt{a^2r^2(1 - \sin^2\theta)}$$
$$= \sqrt{a^2r^2(\mathbf{a} \times \mathbf{r})^2}$$
$$= \sqrt{(a^2r^2 - 1)}$$

33. We have, $\mathbf{r} \times \mathbf{a} = \mathbf{b} \times \mathbf{r}$

$\mathbf{r} \times (\mathbf{a} + \mathbf{b}) = 0$

$\mathbf{r} = \lambda(\mathbf{a} + \mathbf{b})$

$\mathbf{r} = \lambda(\hat{\mathbf{i}} + 2\hat{\mathbf{j}} - 3\hat{\mathbf{k}} + 2\hat{\mathbf{i}} - 3\hat{\mathbf{j}} + 5\hat{\mathbf{k}})$

$\mathbf{r} = \lambda(3\hat{\mathbf{i}} - \hat{\mathbf{j}} + 2\hat{\mathbf{k}})$ ...(i)

$\mathbf{r} \cdot (\alpha\hat{\mathbf{i}} + 2\hat{\mathbf{j}} + \hat{\mathbf{k}}) = 3$ ...(ii)

Puting the value of $\mathbf{r}$ from Eq. (i) in (ii), we get

$\alpha\lambda = 1$ ...(iii)

Also, $\quad \mathbf{r} \cdot (2\hat{\mathbf{i}} + 5\hat{\mathbf{j}} - \alpha\hat{\mathbf{k}}) = -1$ ...(iv)

From Eqs. (i) and (iv) we get,

$2\lambda\alpha - \lambda = 1$ ...(v)

From Eqs. (iii) and (v) we get

$\alpha = 1, \lambda = 1$

$\Rightarrow \quad \mathbf{r} = 3\hat{\mathbf{i}} - \hat{\mathbf{j}} + 2\hat{\mathbf{k}}$

$|\mathbf{r}|^2 = 14$

$\therefore \quad \alpha + |\mathbf{r}|^2 = 1 + 14 = 15$

34. Since, $(\mathbf{a} \times \mathbf{b})^2 + (\mathbf{a} \cdot \mathbf{b})^2 = 144$

If angle between **a** and **b** is θ

then, $|\mathbf{a}|^2 |\mathbf{b}|^2 \sin^2\theta + |\mathbf{a}|^2 |\mathbf{b}|^2 \cos^2\theta = 144$

$\Rightarrow \quad |\mathbf{a}|^2 |\mathbf{b}|^2 = 144$

$\Rightarrow \quad |\mathbf{a}||\mathbf{b}| = 12$

$\Rightarrow \quad 4|\mathbf{b}| = 12$

$|\mathbf{b}| = 3$

35. We have, $\mathbf{AB} \cdot \mathbf{AC} = |\mathbf{AB}||\mathbf{AC}| \cos(90° - \theta) = AB \cdot AC \sin\theta$

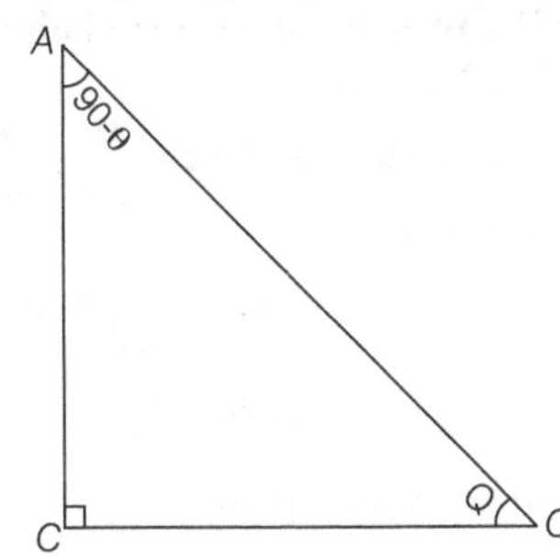

$\mathbf{BC} \cdot \mathbf{BA} = |\mathbf{BC}| \cdot |\mathbf{BA}| \cos\theta = BC \cdot BA \cos\theta$

$\mathbf{CA} \cdot \mathbf{CB} = |\mathbf{CA}| \cdot |\mathbf{CB}| \cos 90° = 0$

$\therefore \mathbf{AB} \cdot \mathbf{AC} + \mathbf{BC} \cdot \mathbf{BA} + \mathbf{CA} \cdot \mathbf{CB}$

$= AB \cdot AC \sin\theta + BC \cdot BA \cos\theta$

$= AB\left[AC \cdot \frac{AC}{AB} + BC \cdot \frac{BC}{AB}\right]$

$= AC^2 + BC^2 = AB^2 = p^2$

36. $|(\mathbf{a} \times \mathbf{b}) \cdot \mathbf{c}| = |\mathbf{a}||\mathbf{b}||\mathbf{c}|$

$|\mathbf{a} \times \mathbf{b}||\mathbf{c}| \cos\alpha = |\mathbf{a}||\mathbf{b}||\mathbf{c}|$

[α is the angle between $\mathbf{a} \times \mathbf{b}$ and **c**]

$\Rightarrow \quad |\mathbf{a} \times \mathbf{b}| \cos\alpha = |\mathbf{a}||\mathbf{b}|$

$\Rightarrow \quad |\mathbf{a}||\mathbf{b}| \sin\beta \cos\alpha = |\mathbf{a}||\mathbf{b}|$

(β is the angle between **a** and **b**)

$\therefore \quad \cos\alpha \cos\beta = 1$

It is possible when $\cos\alpha = 1, \sin\beta = 1$

$\therefore \quad \alpha = 0$ and $\beta = \pi/2$

$\therefore \quad \mathbf{a} \cdot \mathbf{b} = 0, \mathbf{b} \cdot \mathbf{c} = 0, \mathbf{c} \cdot \mathbf{a} = 0$

37. Let $\mathbf{a} = x\hat{\mathbf{i}} + y\hat{\mathbf{j}} + z\hat{\mathbf{k}}$

Given, $|\mathbf{a}| = \sqrt{(x^2 + y^2 + z^2)} = 1$

$\therefore \quad x^2 + y^2 + z^2 = 1 \qquad \ldots(i)$

but given $\mathbf{a} \times (\hat{\mathbf{i}} + \hat{\mathbf{j}} + \hat{\mathbf{k}}) = \hat{\mathbf{i}} - \hat{\mathbf{k}}$

$$\Rightarrow \begin{vmatrix} \hat{\mathbf{i}} & \hat{\mathbf{j}} & \hat{\mathbf{k}} \\ x & y & z \\ 1 & 1 & 1 \end{vmatrix} = \hat{\mathbf{i}} - \hat{\mathbf{k}}$$

$\Rightarrow \hat{\mathbf{i}}(y - z) - \hat{\mathbf{j}}(x - z) + \hat{\mathbf{k}}(x - y) = 1 \cdot \hat{\mathbf{i}} + 0 \cdot \hat{\mathbf{j}} - 1 \cdot \hat{\mathbf{k}}$

On comparing, we get

$y - z = 1, x - z = 0, x - y = -1$

$\therefore z = x, y = 1 + x$

From Eq. (i), $x^2 + (1 + x)^2 + x^2 = 1$

$\Rightarrow \quad 3x^2 + 2x = 0$

$x \neq 0$

$\therefore x = -\frac{2}{3}$, then $y = \frac{1}{3}$ and $z = -\frac{2}{3}$

$\therefore \quad \mathbf{a} = -\frac{2}{3}\hat{\mathbf{i}} + \frac{1}{3}\hat{\mathbf{j}} - \frac{2}{3}\hat{\mathbf{k}} = -\frac{1}{3}(2\hat{\mathbf{i}} - \hat{\mathbf{j}} + 2\hat{\mathbf{k}})$

38. We have $\mathbf{a}_1 = x\hat{\mathbf{i}} - \hat{\mathbf{j}} + \hat{\mathbf{k}}$ and $\mathbf{a}_2 = \hat{\mathbf{i}} + y\hat{\mathbf{j}} + z\hat{\mathbf{k}}$ are collinear

$\because \quad \frac{x}{1} = \frac{-1}{y} = \frac{1}{z} = \lambda \Rightarrow x = \lambda, y = \frac{-1}{\lambda}, z = \frac{1}{\lambda}$

Unit vector parallel to $x\hat{\mathbf{i}} + y\hat{\mathbf{j}} + z\hat{\mathbf{k}}$ is

$$\pm \frac{\left(\lambda\hat{\mathbf{i}} - \frac{1}{\lambda}\hat{\mathbf{j}} + \frac{1}{\lambda}\hat{\mathbf{k}}\right)}{\sqrt{\lambda^2 + \frac{1}{\lambda^2} + \frac{1}{\lambda^2}}}$$

$\Rightarrow \quad \pm\frac{1}{\sqrt{3}}(\hat{\mathbf{i}} - \hat{\mathbf{j}} + \hat{\mathbf{k}}) \qquad$ [from $\lambda = 1$]

39. Since, $\mathbf{a} + 2\mathbf{b} + 3\mathbf{c} = 0 \qquad \ldots(i)$

$\therefore \quad \mathbf{a} \times \mathbf{c} + 2\mathbf{b} \times \mathbf{c} + 3\mathbf{c} \times \mathbf{c} = 0$

$\Rightarrow \quad -\mathbf{c} \times \mathbf{a} + 2(\mathbf{b} \times \mathbf{c}) + 0 = 0$

$\therefore \quad \mathbf{c} \times \mathbf{a} = 2(\mathbf{b} \times \mathbf{c}) \qquad \ldots(ii)$

Again from Eq. (i),

$\mathbf{b} \times \mathbf{a} + 2\mathbf{b} \times \mathbf{b} + 3(\mathbf{b} \times \mathbf{c}) = 0$

$\therefore \quad -\mathbf{a} \times \mathbf{b} + 0 + 3(\mathbf{b} \times \mathbf{c}) = 0$

$\therefore \quad \mathbf{a} \times \mathbf{b} = 3(\mathbf{b} \times \mathbf{c}) \qquad \ldots(iii)$

Given, $\quad \mathbf{a} \times \mathbf{b} + \mathbf{b} \times \mathbf{c} + \mathbf{c} \times \mathbf{a} = \lambda(\mathbf{b} \times \mathbf{c})$

$\Rightarrow 3(\mathbf{b} \times \mathbf{c}) + (\mathbf{b} \times \mathbf{c}) + 2(\mathbf{b} \times \mathbf{c}) = \lambda(\mathbf{b} \times \mathbf{c})$

[from Eqs. (ii), (iii)]

$\Rightarrow \quad 6(\mathbf{b} \times \mathbf{c}) = \lambda(\mathbf{b} \times \mathbf{c})$

$\therefore \quad \lambda = 6$

40. $\frac{AC}{CB} = \frac{|\mathbf{OA}|}{|\mathbf{OB}|} = \frac{\sqrt{1 + 9 + 4}}{\sqrt{9 + 1 + 4}} = 1$

$\therefore C$ is the mid point of A and B.

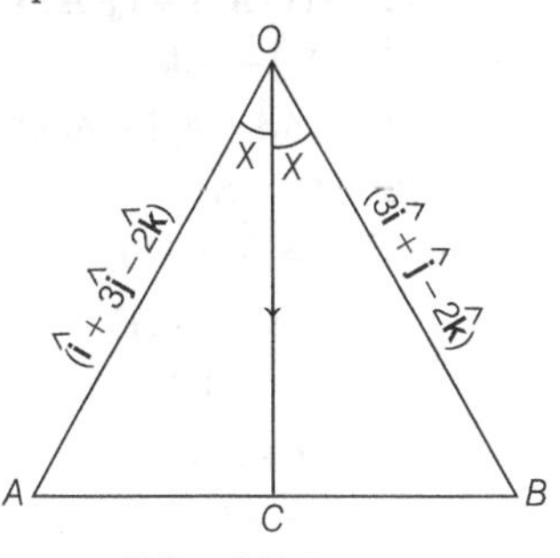

$\therefore \quad \mathbf{OC} = \frac{\mathbf{OA} + \mathbf{OB}}{2} = 2\hat{\mathbf{i}} + 2\hat{\mathbf{j}} - 2\hat{\mathbf{k}}$

41. Linear combination

$1 \cdot (\hat{\mathbf{i}} + \hat{\mathbf{j}} + \hat{\mathbf{k}}) + \lambda(4\hat{\mathbf{i}} + 3\hat{\mathbf{j}} + 4\hat{\mathbf{k}})$

$+ \mu(\hat{\mathbf{i}} + \alpha\hat{\mathbf{j}} + \beta\hat{\mathbf{k}}) = 0 \cdot \hat{\mathbf{i}} + 0 \cdot \hat{\mathbf{j}} + 0 \cdot \hat{\mathbf{k}}$

$1 + 4\lambda + \mu = 0$

and $\quad |\mathbf{c}| = \sqrt{1 + \alpha^2 + \beta} = \sqrt{3}$

$1 + 3\lambda + \alpha\mu = 0$

$\therefore \quad \alpha^2 + \beta^2 = 2$

and $\quad 1 + 4\lambda + \beta\mu = 0$

Solve any two then putting the value in remaining third equation.

42. Let $P = |\mathbf{a} \times \hat{\mathbf{i}}|^2 + |\mathbf{a} \times \hat{\mathbf{j}}|^2 + |\mathbf{a} + \hat{\mathbf{k}}|^2$

If $\mathbf{a} = x\hat{\mathbf{i}} + y\hat{\mathbf{j}} + z\hat{\mathbf{k}}$

$\therefore \mathbf{a} \times \hat{\mathbf{i}} = x(\hat{\mathbf{i}} \times \hat{\mathbf{i}}) + y(\hat{\mathbf{j}} \times \hat{\mathbf{i}}) + z(\hat{\mathbf{k}} \times \hat{\mathbf{i}}) = 0 - \hat{\mathbf{k}}y + z\hat{\mathbf{j}}$

$\therefore |\mathbf{a} \times \hat{\mathbf{i}}|^2 = y^2 + z^2$

Similarly, $|\mathbf{a} \times \hat{\mathbf{j}}|^2 = z^2 + x^2$

and $|\mathbf{a} \times \hat{\mathbf{k}}|^2 = x^2 + y^2$

then, $P = 2(x^2 + y^2 + z^2) = 2|\mathbf{a}|^2 = 2(\mathbf{a})^2$

43. $\begin{vmatrix} a & 1 & 1 \\ 1 & b & 1 \\ 1 & 1 & c \end{vmatrix} = 0$

Applying $C_2 \to C_2 - C_1$ and $C_3 \to C_3 - C_2$

then, $\begin{vmatrix} a & 1-a & 0 \\ 1 & b-1 & 1-b \\ 1 & 0 & c-1 \end{vmatrix} = 0$

$\Rightarrow a(b-1)(c-1) - (1-a)(c-1) + (1-a)(1-b) = 0$

$\Rightarrow \frac{a}{1-a} + \frac{1}{1-b} + \frac{1}{1-c} = 0$

$\Rightarrow -1 + \frac{1}{1-a} + \frac{1}{1-b} + \frac{1}{1-c} = 0$

$\therefore \frac{1}{1-a} + \frac{1}{1-b} + \frac{1}{1-c} = 1$

44. $\hat{\mathbf{i}} \times (\mathbf{A} \times \hat{\mathbf{i}}) + \hat{\mathbf{j}} \times (\mathbf{A} \times \hat{\mathbf{j}}) + \hat{\mathbf{k}} \times (\mathbf{A} \times \hat{\mathbf{k}})$

$= (\hat{\mathbf{i}} \cdot \hat{\mathbf{i}})\mathbf{A} - (\hat{\mathbf{i}} \cdot \mathbf{A})\hat{\mathbf{i}} + (\hat{\mathbf{j}} \cdot \hat{\mathbf{j}})\mathbf{A} - (\hat{\mathbf{j}} \cdot \mathbf{A})\hat{\mathbf{j}}$

$+ (\hat{\mathbf{k}} \cdot \hat{\mathbf{k}})\mathbf{A} - (\hat{\mathbf{k}} \cdot \mathbf{A})\hat{\mathbf{k}}$

$= 3\mathbf{A} - \{(\hat{\mathbf{i}} \cdot \mathbf{A})\hat{\mathbf{i}} + (\hat{\mathbf{j}} \cdot \mathbf{A})\hat{\mathbf{j}} + (\hat{\mathbf{k}} \cdot \mathbf{A})\hat{\mathbf{k}}\}$

Let $\mathbf{A} = A_1\hat{\mathbf{i}} + A_2\hat{\mathbf{j}} + A_3\hat{\mathbf{k}}$

$= 3\mathbf{A} - \{A_1\hat{\mathbf{i}} + A_2\hat{\mathbf{j}} + A_3\hat{\mathbf{k}}\}$

$= 3\mathbf{A} - \mathbf{A} = 2\mathbf{A}$

45. The point that divides $5\hat{\mathbf{i}}$ and $5\hat{\mathbf{j}}$ in the ratio of

$k : 1$ is $\frac{(5\hat{\mathbf{j}})k + (5\hat{\mathbf{i}})1}{k+1}$

$\therefore \mathbf{b} = \frac{5\hat{\mathbf{i}} + 5k\hat{\mathbf{j}}}{k+1}$

Also, $|\mathbf{b}| \le \sqrt{37}$

+ −6 − −1/6 +

$\Rightarrow \frac{1}{k+1}\sqrt{25 + 25k^2} \le \sqrt{37}$

$\Rightarrow 5\sqrt{1 + k^2} \le \sqrt{37}(k+1)$

Squaring on both sides, we get

$25(1 + k^2) \le 37(k^2 + 2k + 1)$

$\Rightarrow 6k^2 + 37k + 6 \ge 0$ or $(6k+1)(k+6) \ge 0$

$k \in (-\infty, -6] \cup \left[-\frac{1}{6}, \infty\right)$

46. Given, $[\mathbf{abc}] = 0$

and $|\mathbf{a}| = 1, |\mathbf{b}| = 1$ and $|\mathbf{c}| = 1$

$\because [2\mathbf{a} - \mathbf{b}\ 2\mathbf{b} - \mathbf{c}\ 2\mathbf{c} - \mathbf{a}]$

$(2\mathbf{a} - \mathbf{b}) \cdot \{(2\mathbf{b} - \mathbf{c}) \times (2\mathbf{c} - \mathbf{a})\}$

$= (2\mathbf{a} - \mathbf{b}) \cdot \{4\mathbf{b} \times \mathbf{c} - 2\mathbf{b} \times \mathbf{a} - 2\mathbf{c} \times \mathbf{c} + \mathbf{c} \times \mathbf{a}\}$

$(2\mathbf{a} - \mathbf{b}) \cdot \{4(\mathbf{b} \times \mathbf{c}) + 2(\mathbf{a} \times \mathbf{b}) - 0 + (\mathbf{c} \times \mathbf{a})\}$

$= 8\mathbf{a} \cdot (\mathbf{b} \times \mathbf{c}) + 4\mathbf{a} \cdot (\mathbf{a} \times \mathbf{b}) + 2\mathbf{a} \cdot (\mathbf{c} \times \mathbf{a})$

$- 4\mathbf{b} \cdot (\mathbf{b} \times \mathbf{c}) - 2\mathbf{b} \cdot (\mathbf{a} \times \mathbf{b}) - \mathbf{b}(\mathbf{c} \times \mathbf{a})$

$= 8[\mathbf{abc}] + 0 + 0 - 0 - 0 - [\mathbf{abc}]$

$= 7[\mathbf{abc}] = 0$ $[\because [\mathbf{abc}] = 0]$

47. We have, $(\mathbf{a} \cdot \mathbf{c})\mathbf{b} - (\mathbf{a} \cdot \mathbf{b})\mathbf{c} = \frac{\mathbf{b}}{\sqrt{2}} + \frac{\mathbf{c}}{\sqrt{2}}$

On comparing, we get

$\mathbf{a} \cdot \mathbf{c} = \frac{1}{\sqrt{2}}$ and $\mathbf{a} \cdot \mathbf{b} = -\frac{1}{\sqrt{2}}$

$\therefore$ Angle between $\mathbf{a}$ and $\mathbf{b}$ is $\frac{3\pi}{4}$.

48. If $(\mathbf{a})x$ and $\mathbf{b}(x)$ are $\perp$ then

$\mathbf{a} \cdot \mathbf{b} = 0$

$\Rightarrow \sin x \cos 2x + \cos x \sin 2x = 0$

$\sin(3x) = 0 = \sin 0$

$3x = 4\pi + (-1)^n \times 0 \Rightarrow x = \frac{n\pi}{3}$

$\therefore$ Perpendicular for infinitely many value of 'x'.

49. Since, $(\mathbf{a} \times \mathbf{b}) \times (\mathbf{c} \times \mathbf{d}) = 0$

Hence, $\mathbf{a} \times \mathbf{b}$ and $\mathbf{c} \times \mathbf{d}$ are parallel.

Since, planes P_1 and P_2 determined by the pair of vectors $\mathbf{a}$, $\mathbf{b}$ and $\mathbf{c}$, $\mathbf{d}$.

Hence, $\mathbf{a} \times \mathbf{b}$ is $\perp$ to $\mathbf{a}$ and $\mathbf{b}$ and $\mathbf{c} \times \mathbf{d}$ is $\perp$ to $\mathbf{c} \times \mathbf{d}$.

Hence, plane P_1 and P_2 are parallel

$\therefore$ Angle between P_1 and $P_2 = 0$

50. Given, three unit vectors $\mathbf{a}$, $\mathbf{b}$ and $\mathbf{c}$ such that

$\mathbf{a} + \mathbf{b} + \mathbf{c} = 0$ ...(i)

If we do dot product in Eq. (i) with $\mathbf{a}$, $\mathbf{b}$ and $\mathbf{c}$ then, we get relations

$\mathbf{a} \cdot \mathbf{a} + \mathbf{a} \cdot \mathbf{b} + \mathbf{a} \cdot \mathbf{c} = 0$

$\Rightarrow \mathbf{a} \cdot \mathbf{b} + \mathbf{a} \cdot \mathbf{c} = -1$, $\mathbf{b} \cdot \mathbf{a} + \mathbf{b} \cdot \mathbf{c} = -1$

and $\mathbf{c} \cdot \mathbf{a} + \mathbf{c} \cdot \mathbf{b} = -1$

$\therefore$ On adding above three obtained relations, we get

$2(\mathbf{a} \cdot \mathbf{b} + \mathbf{b} \cdot \mathbf{c} + \mathbf{c} \cdot \mathbf{a}) = -3$ $(\because \mathbf{a} \cdot \mathbf{b} = \mathbf{b} \cdot \mathbf{a})$

$\Rightarrow \lambda = \mathbf{a} \cdot \mathbf{b} + \mathbf{b} \cdot \mathbf{c} + \mathbf{c} \cdot \mathbf{a} = -\frac{3}{2}$

If we do cross product in Eq. (i) with $\mathbf{a}$, $\mathbf{b}$ and $\mathbf{c}$ then, we get relation

$0 + \mathbf{a} \times \mathbf{b} + \mathbf{a} \times \mathbf{c} = 0$

$\Rightarrow \mathbf{a} \times \mathbf{b} + \mathbf{a} \times \mathbf{c} = 0$

and $\mathbf{b} \times \mathbf{a} + \mathbf{b} \times \mathbf{c} = 0$

and $\mathbf{c} \times \mathbf{a} + \mathbf{c} \times \mathbf{b} = 0$

$\therefore \mathbf{a} \times \mathbf{b} + \mathbf{b} \times \mathbf{c} + \mathbf{c} \times \mathbf{a} = 3(\mathbf{a} \times \mathbf{b}) = 3(\mathbf{b} \times \mathbf{c})$

$= 3(\mathbf{c} \times \mathbf{a})$

$\therefore \mathbf{d} = 3\mathbf{a} \times \mathbf{b} = 3\mathbf{b} \times \mathbf{c} = 3\mathbf{c} \times \mathbf{a}$

From the given options the ordered pair,

$(\lambda, \mathbf{d}) = \left(-\frac{3}{2}, 3\mathbf{a} \times \mathbf{b}\right)$

51. Given, $|\mathbf{a}| = 2, |\mathbf{b}| = 2$ and the angle between $\mathbf{a}$ and $\mathbf{b}$ is 60°.

Now, let angle between $\mathbf{a}$ and $\mathbf{a}+\mathbf{b}$ is α

Then, $\mathbf{a}\cdot(\mathbf{a}+\mathbf{b}) = |\mathbf{b}||\mathbf{a}+\mathbf{b}|\cos\alpha$

$\Rightarrow \quad a^2 + \mathbf{a}\cdot\mathbf{b} = a\sqrt{(\mathbf{a})^2 + (\mathbf{b})^2 + 2\mathbf{a}\cdot\mathbf{b}}\cos\alpha$

$\Rightarrow \quad a^2 + ab\cos 60°$

$= a\sqrt{(a^2 + b^2 + 2ab\cos 60°)}\cos\alpha$

$\Rightarrow \quad (4 + 4\times 1/2) = 2\sqrt{(4+4+8\times 1/2)}\cos\alpha$

$\Rightarrow \quad 3 = \sqrt{12}\cos\alpha$

$\therefore \quad \cos\alpha = \dfrac{\sqrt{3}}{2}$

$\therefore \quad \alpha = \dfrac{\pi}{6} = 30°$

52. Let $\mathbf{r} = r_1\hat{\mathbf{i}} + r_2\hat{\mathbf{j}} + r_3\hat{\mathbf{k}}$

$\therefore \quad \mathbf{r}\cdot\hat{\mathbf{i}} = r_1, \mathbf{r}\cdot\hat{\mathbf{j}} = r_2, \mathbf{r}\cdot\hat{\mathbf{k}} = r_3$

and $\quad \mathbf{r}\times\hat{\mathbf{i}} = 0 + r_2(\hat{\mathbf{j}}\times\hat{\mathbf{i}}) + r_3(\hat{\mathbf{k}}\times\hat{\mathbf{i}})$

$= -r_2\hat{\mathbf{k}} + r_3\hat{\mathbf{j}}$

$\therefore \quad (\mathbf{r}\cdot\hat{\mathbf{i}})(\mathbf{r}\times\hat{\mathbf{i}}) = -r_1r_2\hat{\mathbf{k}} + r_3r_1\hat{\mathbf{j}}$

Similarly,

$(\mathbf{r}\cdot\hat{\mathbf{j}})(\mathbf{r}\times\hat{\mathbf{j}}) = -r_2r_3\hat{\mathbf{i}} + r_2r_1\hat{\mathbf{k}}$

and $(\mathbf{r}\cdot\hat{\mathbf{k}})(\mathbf{r}\times\hat{\mathbf{k}}) = -r_3r_1\hat{\mathbf{j}} + r_2r_3\hat{\mathbf{i}}$

$\therefore \quad (\mathbf{r}\cdot\hat{\mathbf{i}})(\mathbf{r}\times\hat{\mathbf{i}}) + (\mathbf{r}\cdot\hat{\mathbf{j}})(\mathbf{r}\times\hat{\mathbf{j}}) + (\mathbf{r}\cdot\hat{\mathbf{k}})(\mathbf{r}\times\hat{\mathbf{k}}) = \mathbf{0}$

53. Given, $[\mathbf{abc}] \neq 0$

$\therefore \quad (\mathbf{a}+\mathbf{b})\cdot\mathbf{p} = \mathbf{a}\cdot\mathbf{p} + \mathbf{b}\cdot\mathbf{p} = 1 + 0 = 1$

Similarly, $\quad (\mathbf{b}+\mathbf{c})\cdot\mathbf{q} = 1, (\mathbf{c}+\mathbf{a})\cdot\mathbf{r} = 1$

$\therefore \; (\mathbf{a}+\mathbf{b})\cdot\mathbf{p} + (\mathbf{b}+\mathbf{c})\cdot\mathbf{q} + (\mathbf{c}+\mathbf{a})\cdot\mathbf{r} = 3$

54. $\mathbf{a}\times\mathbf{b} = 2\hat{\mathbf{i}} - 2\hat{\mathbf{j}} + \hat{\mathbf{k}}$

$\therefore |(\mathbf{a}\times\mathbf{b})\times\mathbf{c}| = |\mathbf{c}\times\mathbf{b}||\mathbf{c}|\sin 30°$

$= \sqrt{2^2 + (-2)^2 + 1^2}\,|\mathbf{c}|\,\dfrac{1}{2}$

$= \dfrac{3}{2}|\mathbf{c}|$

Given, $|\mathbf{c}-\mathbf{a}| = 2\sqrt{2}$

$(\mathbf{c}-\mathbf{a})^2 = 8$

$\Rightarrow \quad (\mathbf{c})^2 + (\mathbf{a})^2 - 2\mathbf{c}\cdot\mathbf{a} = 8$

$\Rightarrow \quad |\mathbf{c}|^2 + 9 - 2|\mathbf{c}| = 8$

$\Rightarrow \quad (|\mathbf{c}| - 1)^2 = 0$

$\therefore \quad |\mathbf{c}| = 1$

$\therefore \quad |(\mathbf{a}\times\mathbf{b})\times\mathbf{c}| = \dfrac{3}{2}\times 1 = \dfrac{3}{2}$

55. $\mathbf{a}\times(\mathbf{a}\times(\mathbf{a}\times(\mathbf{a}\times\mathbf{b})))$

$= \mathbf{a}\times(\mathbf{a}\times((\mathbf{a}\cdot\mathbf{b})\mathbf{a} - (\mathbf{a}\cdot\mathbf{a})\mathbf{b}))$

$= \mathbf{a}\times(\mathbf{a}\times(0 - |\mathbf{a}|^2\mathbf{b})) \quad [\because \mathbf{a}\cdot\mathbf{b} = 0]$

$\Rightarrow -\mathbf{a}\times(\mathbf{a}\times\mathbf{b})|\mathbf{a}|^2$

$\Rightarrow -|\mathbf{a}|^2((\mathbf{a}\cdot\mathbf{b})\mathbf{a} - |\mathbf{a}|^2\mathbf{b})$

$\Rightarrow -|\mathbf{a}|^2(0 - |\mathbf{a}|^2\mathbf{b}) \Rightarrow |\mathbf{a}|^4\mathbf{b}$

56. $0 = [3\mathbf{u}\; p\mathbf{v}\; p\mathbf{w}] - [p\mathbf{v}\;\mathbf{w}\; q\mathbf{u}] - [2\mathbf{w}\; q\mathbf{v}\; q\mathbf{u}]$

$= (3p^2 - pq + 2q^2)[\mathbf{u}\;\mathbf{v}\;\mathbf{w}]$

As $\mathbf{u}, \mathbf{v}, \mathbf{w}$ are non-coplanar so $[\mathbf{u}\;\mathbf{v}\;\mathbf{w}] \neq 0$, so

$3p^2 - pq + 2q^2 = 0$

$\Rightarrow \quad 2\left(q - \dfrac{1}{4}p\right)^2 + \dfrac{23}{8}p^2 = 0$

$\Rightarrow \quad p = 0,\; q - \dfrac{1}{4}p = 0$

Thus, there is exactly one value of (p, q).

57. Since, $\mathbf{d} = \lambda(\mathbf{a}\times\mathbf{b})\mu(\mathbf{b}\times\mathbf{c})\nu(\mathbf{c}\times\mathbf{a})$

$\therefore \; \mathbf{d}\cdot\mathbf{a} = 0 + \mu[\mathbf{abc}] + 0 = 1/8\,\mu \qquad [\because [\mathbf{abc}] = 1/8]$

$\therefore \quad \mu = 8\,\mathbf{d}\cdot\mathbf{a}$

Similarly $\lambda = 8\,\mathbf{d}\cdot\mathbf{c}$ and $\nu = 8\,\mathbf{d}\cdot\mathbf{b}$

$\therefore \; \lambda + \mu + \nu = 8\,\mathbf{d}\cdot(\mathbf{a}+\mathbf{b}+\mathbf{c})$

58. $\mathbf{a}\times(\mathbf{b}\times\mathbf{c}) = (\mathbf{a}\times\mathbf{b})\times\mathbf{c}$

$\Rightarrow \quad (\mathbf{a}\cdot\mathbf{c})\mathbf{b} - (\mathbf{a}\cdot\mathbf{b})\mathbf{c} = (\mathbf{a}\cdot\mathbf{c})\mathbf{b} - (\mathbf{b}\cdot\mathbf{c})\mathbf{a}$

$\Rightarrow \quad (\mathbf{b}\cdot\mathbf{c})\,\mathbf{a} = (\mathbf{a}\cdot\mathbf{b})\,\mathbf{c}$

$\Rightarrow \quad \mathbf{a} = \dfrac{(\mathbf{a}\cdot\mathbf{b})}{(\mathbf{b}\cdot\mathbf{c})}\mathbf{c}$

$\therefore$ $\mathbf{a}$ and $\mathbf{c}$ are collinear.

59. Since,

$\mathbf{OA} = \hat{\mathbf{i}} + \hat{\mathbf{j}} + \hat{\mathbf{k}}$

$\mathbf{OB} = \hat{\mathbf{i}} + 5\hat{\mathbf{j}} - \hat{\mathbf{k}}$

$\mathbf{OC} = 2\hat{\mathbf{i}} + 3\hat{\mathbf{j}} + 5\hat{\mathbf{k}}$

$a = BC = |\mathbf{BC}| = |\mathbf{OC} - \mathbf{OB}|$

$= |\hat{\mathbf{i}} - 2\hat{\mathbf{j}} + 6\hat{\mathbf{k}}| = \sqrt{41}$

$b = CA = |\mathbf{CA}| = |\mathbf{OA} - \mathbf{OC}|$

$= |-\hat{\mathbf{i}} - 2\hat{\mathbf{j}} - 4\hat{\mathbf{k}}| = \sqrt{21}$

and $c = AB = |\mathbf{AB}| = |\mathbf{OB} - \mathbf{OA}|$

$= |0\hat{\mathbf{i}} + 4\hat{\mathbf{j}} - 2\hat{\mathbf{k}}|$

$= \sqrt{16 + 4} = \sqrt{20}$

$\because \quad a > b > c$

$\therefore A$ is the greatest angle.

$\cos A = \dfrac{b^2 + c^2 - a^2}{2bc} = \dfrac{21 + 20 - 41}{2\cdot\sqrt{21}\cdot\sqrt{20}} = 0$

$\therefore \quad \angle A = 90°$

60. Now, in ΔABC

$\dfrac{BD}{DC} = \dfrac{a}{b}$

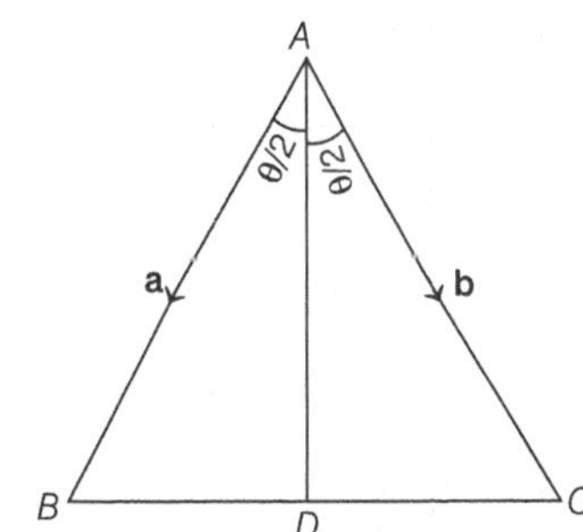

$\therefore \quad BD = ak, DC = bk$

$BC = (a+b)k$

$(BC)^2 = (AB)^2 + (AC)^2 - 2AB \cdot AC \cos\theta$

$\Rightarrow \quad (a+b)^2 k^2 = a^2 + b^2 - 2ab\cos\theta$

$$\therefore \quad k^2 = \frac{a^2 + b^2 - 2ab\cos\theta}{(a+b)^2} \quad \ldots\text{(i)}$$

In ΔADC and ΔABD

$$\cos\left(\frac{\theta}{2}\right) = \frac{b^2 + (AD)^2 - b^2k^2}{2bAD} = \frac{a^2 + (AD)^2 - a^2k^2}{2aAD}$$

$\Rightarrow \quad (AD)^2 = ab(1-k^2)$

$$\therefore \quad = ab\left\{1 - \frac{a^2 + b^2 - 2ab\cos\theta}{(a+b)^2}\right\}$$

[from Eq. (i)]

$$= \frac{4a^2b^2\cos^2\theta/2}{(a+b)^2}$$

$$\therefore \quad AD = \frac{2ab\cos\theta/2}{(a+b)}$$

$$\therefore \quad \mathbf{AD} = \pm\frac{(a\mathbf{b} + b\mathbf{a})}{(a+b)}$$

$$= \pm\frac{ab}{(a+b)}\left(\frac{\mathbf{a}}{a} + \frac{\mathbf{b}}{b}\right) = \pm\frac{ab}{(a+b)}(\hat{\mathbf{a}} + \hat{\mathbf{b}})$$

$$\therefore \quad \widehat{\mathbf{AD}} = \frac{\mathbf{AD}}{AD} = \pm\frac{(\hat{\mathbf{a}} + \hat{\mathbf{b}})}{2\cos\theta/2}$$

61. Let vertices are A, B, C, D and O is origin

$\therefore \quad \mathbf{OA} = \hat{\mathbf{i}} - 6\hat{\mathbf{j}} + 10\hat{\mathbf{k}}$

$\mathbf{OB} = -\hat{\mathbf{i}} - 3\hat{\mathbf{j}} + 7\hat{\mathbf{k}}$

$\mathbf{OC} = 5\hat{\mathbf{i}} - \hat{\mathbf{j}} + \lambda\hat{\mathbf{k}}$

$\mathbf{OD} = 7\hat{\mathbf{i}} - 4\hat{\mathbf{j}} + 7\hat{\mathbf{k}}$

$\therefore \quad \mathbf{AB} = \mathbf{OB} - \mathbf{OA} = -2\hat{\mathbf{i}} + 3\hat{\mathbf{j}} - 3\hat{\mathbf{k}}$

$\mathbf{AC} = \mathbf{OC} - \mathbf{OA} = 4\hat{\mathbf{i}} + 5\hat{\mathbf{j}} + (\lambda - 10)\hat{\mathbf{k}}$

$\mathbf{AD} = \mathbf{OD} - \mathbf{OA} = 6\hat{\mathbf{i}} + 2\hat{\mathbf{j}} - 3\hat{\mathbf{k}}$

Volume of tetrahedron

$$= \frac{1}{6}[\mathbf{AB}\ \mathbf{AC}\ \mathbf{AD}]$$

$$= \frac{1}{6}\begin{vmatrix} -2 & 3 & -3 \\ 4 & 5 & \lambda - 10 \\ 6 & 2 & -3 \end{vmatrix}$$

$$= \frac{1}{6}\{-2(-15 - 2\lambda + 20) - 3(-12 - 6\lambda + 60) - 3(8 - 30)\}$$

$$= \frac{1}{6}\{4\lambda - 10 - 144 + 18\lambda + 66\}$$

$$= \frac{1}{6}(22\lambda - 88) = 11 \quad \text{(given)}$$

$\Rightarrow \quad 2\lambda - 8 = 6$

$\therefore \quad \lambda = 7$

62. $\mathbf{a} \times \mathbf{b} = \mathbf{c}$

i.e. $\mathbf{c}$ is $\perp$ to both $\mathbf{a}$ and $\mathbf{b}$

$\mathbf{b} \times \mathbf{c} = \mathbf{a}$

i.e. $\mathbf{a}$ is $\perp$ to both $\mathbf{b}$ and $\mathbf{c}$

Hence, $\mathbf{a}$, $\mathbf{b}$, $\mathbf{c}$ are mutually perpendicular.

63. Here, $\mathbf{a} = 2p\hat{\mathbf{i}} + \hat{\mathbf{j}}$

After rotation, let the vector be $\mathbf{c}$ and let the unit vectors along the new axes be $\hat{\mathbf{i}}_1$ and $\hat{\mathbf{j}}_1$, then

$\mathbf{c} = (p+1)\hat{\mathbf{i}}_1 + \hat{\mathbf{j}}_1$.

It is clear

$|\mathbf{a}| = |\mathbf{c}|$ (since magnitude does not change)

$\Rightarrow \sqrt{4p^2 + 1} = \sqrt{(p+1)^2 + 1}$

$\Rightarrow \quad 2p = \pm(p+1) \Rightarrow p = 1$ or $p = -\frac{1}{3}$

64. Since, $\mathbf{a}$ and $\mathbf{b}$ are parallel.

$\therefore \quad \mathbf{a} \times \mathbf{b} = 0 \quad \ldots\text{(i)}$

$\therefore (\mathbf{a} \times \mathbf{b}) \times (\mathbf{c} \times \mathbf{d}) + (\mathbf{a} \times \mathbf{c}) \times (\mathbf{b} \times \mathbf{d})$

$= [\mathbf{abd}]\,\mathbf{c} - [\mathbf{abc}]\,\mathbf{d} + [\mathbf{acd}]\,\mathbf{b} - [\mathbf{acb}]\,\mathbf{d}$

$= [\mathbf{abd}]\mathbf{c} - [\mathbf{abc}]\,\mathbf{d} + [\mathbf{acd}]\,\mathbf{b} + [\mathbf{acb}]\mathbf{d}$

$= [\mathbf{abd}]\,\mathbf{c} + [\mathbf{acd}]\mathbf{b}$

$= [\mathbf{dab}]\mathbf{c} + [\mathbf{acd}]\mathbf{b}$

$= \{\mathbf{d} \cdot (\mathbf{a} \times \mathbf{b})\}\,\mathbf{c} + [\mathbf{acd}]\mathbf{b}$

$= 0 + [\mathbf{acd}]\mathbf{b}$ [from Eq. (i)]

$= [\mathbf{acd}]\mathbf{b}$

$= \{(\mathbf{a} \times \mathbf{c}) \cdot \mathbf{d}\}\mathbf{b}$

65. Since, $\hat{\mathbf{i}} \times (\mathbf{r} \times \hat{\mathbf{i}}) + \hat{\mathbf{j}} \times (\mathbf{r} \times \hat{\mathbf{j}}) + \hat{\mathbf{k}} \times (\mathbf{r} \times \hat{\mathbf{k}}) = \mathbf{a} \times \mathbf{b}$

$\Rightarrow \quad (\hat{\mathbf{i}} \cdot \hat{\mathbf{i}})\mathbf{r} - (\hat{\mathbf{i}} \cdot \mathbf{r})\hat{\mathbf{i}} + (\hat{\mathbf{j}} \cdot \hat{\mathbf{j}})\mathbf{r} - (\hat{\mathbf{j}} \cdot \mathbf{r})\hat{\mathbf{j}} + (\hat{\mathbf{k}} \cdot \hat{\mathbf{k}})\mathbf{r} - (\hat{\mathbf{k}} \cdot \mathbf{r})\hat{\mathbf{k}} = \mathbf{a} \times \mathbf{b}$

$\Rightarrow 3\mathbf{r} - \{(\hat{\mathbf{i}} \cdot \mathbf{r})\hat{\mathbf{i}} + (\hat{\mathbf{j}} \cdot \mathbf{r})\hat{\mathbf{j}} + (\hat{\mathbf{k}} \cdot \mathbf{r})\hat{\mathbf{k}}\} = \mathbf{a} \times \mathbf{b} \quad \ldots\text{(i)}$

Now, let

$\mathbf{r} = x\hat{\mathbf{i}} + y\hat{\mathbf{j}} + z\hat{\mathbf{k}}$

$\therefore \quad \hat{\mathbf{i}} \cdot \mathbf{r} = x, \hat{\mathbf{j}} \cdot \mathbf{r} = y$ and $\hat{\mathbf{k}} \cdot \mathbf{r} = z$

then, Eq. (i) reduces

$3\mathbf{r} - \{x\hat{\mathbf{i}} + y\hat{\mathbf{j}} + z\hat{\mathbf{k}}\} = \mathbf{a} \times \mathbf{b}$

$\Rightarrow \quad 3\mathbf{r} - \mathbf{r} = \mathbf{a} \times \mathbf{b}$

$$\therefore \quad \mathbf{r} = \frac{\mathbf{a} \times \mathbf{b}}{2}$$

66. If angle between $\mathbf{b}$ and $\mathbf{c}$ is α,

then $\quad |\mathbf{b} \times \mathbf{c}| = \sqrt{15}$

$\Rightarrow \quad |\mathbf{b}||\mathbf{c}|\sin\alpha = \sqrt{15}$

$$\Rightarrow \quad \sin\alpha = \frac{\sqrt{15}}{4}$$

$\therefore \quad \cos\alpha = 1/4$

$\mathbf{b} - 2\mathbf{c} = \lambda\,\mathbf{a}$

$\Rightarrow \quad (\mathbf{b} - 2\mathbf{c})^2 = \lambda^2(\mathbf{a})^2$

$\Rightarrow (\mathbf{b})^2 + 4(\mathbf{c})^2 - 4\mathbf{b} \cdot \mathbf{c} = \lambda^2(\mathbf{a})^2$

$\Rightarrow \quad 16 + 4 - 4\{|\mathbf{b}||\mathbf{c}|\cos\alpha\} = \lambda^2$

$\therefore \quad \lambda^2 = 16 \Rightarrow \lambda = \pm 4$

67. $\because \quad \dfrac{BD}{DC} = \dfrac{|\mathbf{AB}|}{|\mathbf{AC}|}$

$BD : DC = |\mathbf{AB}| : |\mathbf{AC}|$

$\therefore \quad \mathbf{AD} = \dfrac{BD\mathbf{AD} + DC\mathbf{AB}}{(BD + DC)} = \dfrac{\dfrac{BD}{DC}\mathbf{AC} + \mathbf{AB}}{\left(\dfrac{BD}{DC} + 1\right)}$

$= \dfrac{\mathbf{AB}\,|\mathbf{AC}| + \mathbf{AC}\,|\mathbf{AB}|}{|\mathbf{AB}| + |\mathbf{AC}|}$

$= \dfrac{|\mathbf{AC}|}{|\mathbf{AB}| + |\mathbf{AC}|}\mathbf{AB} + \dfrac{|\mathbf{AB}|}{|\mathbf{AB}| + |\mathbf{AC}|}\mathbf{AC}$

$= \alpha\, \mathbf{AB} + \beta\, \mathbf{AC}$ (given)

$\therefore \quad \alpha = \dfrac{|\mathbf{AC}|}{|\mathbf{AB}| + |\mathbf{AC}|}$ and $\beta = \dfrac{|\mathbf{AB}|}{|\mathbf{AB}| + |\mathbf{AC}|}$

68. $|p\hat{\mathbf{i}} + 3\hat{\mathbf{j}} + 4\hat{\mathbf{k}}| = \sqrt{p^2 + 9 + 16} = \sqrt{p^2 + 25}$

and $|\sqrt{q}\,\hat{\mathbf{i}} + 4\hat{\mathbf{k}}| = \sqrt{(q + 16)}$

$\because \quad p, q \geq 0$

Hence, infinite number of value of (p, q).

69. For three non-coplanar vector $\mathbf{p}$, $\mathbf{q}$ and $\mathbf{r}$, it is given that

$[\mathbf{p\,q\,r}]\mathbf{a} = \mathbf{q} \times \mathbf{r}$

$[\mathbf{p\,q\,r}]\mathbf{b} = \mathbf{r} \times \mathbf{p}$

and $[\mathbf{p\,q\,r}]\mathbf{c} = \mathbf{p} \times \mathbf{q}$

$\therefore$ $\mathbf{a}$, $\mathbf{b}$, $\mathbf{c}$ are reciprocal system vectors of vectors $\mathbf{p}$, $\mathbf{q}$, $\mathbf{r}$ so

$[\mathbf{a\,b\,c}]\mathbf{p} = \mathbf{b} \times \mathbf{c}$ ...(i)

$[\mathbf{a\,b\,c}]\mathbf{q} = \mathbf{c} \times \mathbf{a}$ ...(ii)

and $[\mathbf{a\,b\,c}]\mathbf{r} = \mathbf{a} \times \mathbf{b}$ ...(iii)

$\because$ $\mathbf{a}$, $\mathbf{b}$ and $\mathbf{c}$ are coterminous edges of a parallelopiped, so height with the base have $\mathbf{a}$ and $\mathbf{c}$ is

$\dfrac{[\mathbf{a\,b\,c}]}{|\mathbf{a} \times \mathbf{c}|} = \dfrac{1}{|\mathbf{q}|}$ [From Eq. (ii)]

70. If $\mathbf{a} = a_1\hat{\mathbf{i}} + a_2\hat{\mathbf{j}} + a_3\hat{\mathbf{k}}$

then, $(\mathbf{a} \cdot \hat{\mathbf{i}})\hat{\mathbf{i}} + (\mathbf{a} \cdot \hat{\mathbf{j}})\hat{\mathbf{j}} + (\mathbf{a} \cdot \hat{\mathbf{k}})\hat{\mathbf{k}}$

$= a_1\hat{\mathbf{i}} + a_2\hat{\mathbf{j}} + a_3\hat{\mathbf{k}} = \mathbf{a}$ ...(i)

and $\hat{\mathbf{i}} \times (\mathbf{a} \times \hat{\mathbf{i}}) + \hat{\mathbf{j}} \times (\mathbf{a} \times \hat{\mathbf{j}}) + \hat{\mathbf{k}} \times (\mathbf{a} \times \hat{\mathbf{k}})$

$= 3\mathbf{a} - \{(\mathbf{a} \cdot \mathbf{i})\hat{\mathbf{i}} + (\mathbf{a} \cdot \hat{\mathbf{j}})\hat{\mathbf{j}} + (\mathbf{a} \cdot \hat{\mathbf{k}})\hat{\mathbf{k}}\}$

$= 3\mathbf{a} - \mathbf{a}$

$= 2\mathbf{a}$ [from Eq. (i)]

71. Area of $\Delta\, ABC = \dfrac{1}{2} \cdot BC \cdot p$

$\Rightarrow \quad \dfrac{1}{2}|\mathbf{a} \times \mathbf{b} + \mathbf{b} \times \mathbf{c} + \mathbf{c} \times \mathbf{a}|$

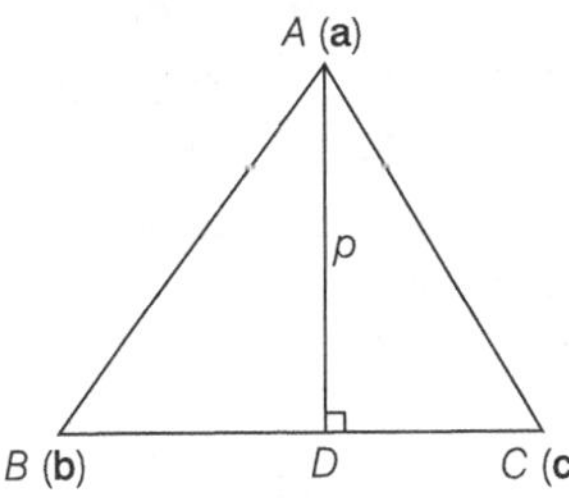

$= \dfrac{1}{2}|\mathbf{c} - \mathbf{b}| \times p$

$\therefore \quad p = \dfrac{|\mathbf{a} \times \mathbf{b} + \mathbf{b} \times \mathbf{c} + \mathbf{c} \times \mathbf{a}|}{|\mathbf{c} - \mathbf{b}|}$

72. Since, $\mathbf{a}$, $\mathbf{b}$ and $(\mathbf{a} \times \mathbf{b})$ are non coplanar

$\therefore \quad \mathbf{r} = x\,\mathbf{a} + y\,\mathbf{b} + z(\mathbf{a} \times \mathbf{b})$ ...(i)

For some scalars x, y, z

Now, $\mathbf{b} = \mathbf{r} \times \mathbf{a}$

$\therefore \quad \mathbf{b} = \{x\,\mathbf{a} + y\,\mathbf{b} + z(\mathbf{a} \times \mathbf{b})\} \times \mathbf{a}$

$= x\,(\mathbf{a} \times \mathbf{a}) + y\,(\mathbf{b} \times \mathbf{a}) + z\{(\mathbf{a} \times \mathbf{b}) \times \mathbf{a}\}$

$= 0 + y\,(\mathbf{b} \times \mathbf{a}) + z\,\{(\mathbf{a} \cdot \mathbf{a})\mathbf{b} - (\mathbf{a} \cdot \mathbf{b})\mathbf{a}\}$

$\therefore \quad \mathbf{b} = y\,(\mathbf{b} \times \mathbf{a}) + z(\mathbf{a} \cdot \mathbf{a})\mathbf{b}$ $[\because \mathbf{a} \cdot \mathbf{b} = 0]$

Comparing the coefficients, we get

$y = 0$ and $z = \dfrac{1}{(\mathbf{a} \cdot \mathbf{a})}$

Putting the values of y and z in Eq. (i), we get

$\mathbf{r} = x\,\mathbf{a} + \dfrac{1}{(\mathbf{a} \cdot \mathbf{a})}(\mathbf{a} \times \mathbf{b})$

73. p = area of the quadrilateral

$\therefore \quad p$ = Area of ΔOAB + Area of ΔOBC

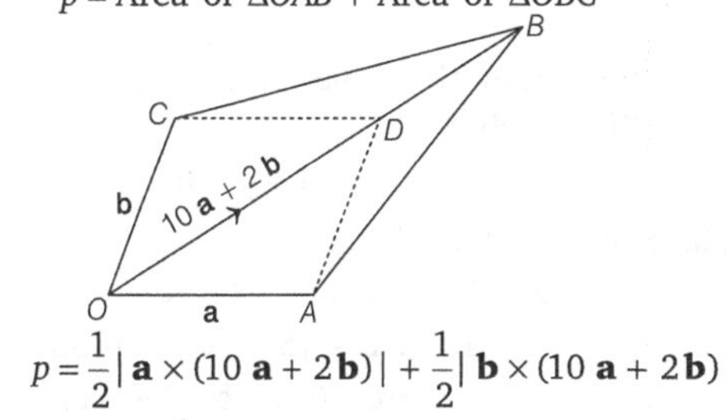

$p = \dfrac{1}{2}|\mathbf{a} \times (10\,\mathbf{a} + 2\mathbf{b})| + \dfrac{1}{2}|\mathbf{b} \times (10\,\mathbf{a} + 2\mathbf{b})|$

$\therefore \quad p = \dfrac{1}{2} \cdot 2|\mathbf{a} \times \mathbf{b}| + 5|\mathbf{a} \times \mathbf{b}|$

$\therefore \quad p = 6|\mathbf{a} \times \mathbf{b}|$ ...(i)

and q = area of the parallelogram with OA and OC as adjacent sides

$q = |\mathbf{a} \times \mathbf{b}|$ ...(ii)

From Eqs. (i) and (ii), $p = 6q$

But given $p = kq$

$\therefore \quad k = 6$

74. Given, $\mathbf{r} \cdot \mathbf{a} = 0$ and $|\mathbf{r} \times \mathbf{b}| = |\mathbf{r}||\mathbf{b}|$

let angle between $\mathbf{r}$ and $\mathbf{b}$ is θ

$\therefore \quad = |\mathbf{r} \times \mathbf{b}| = |\mathbf{r}||\mathbf{b}| \sin\theta = |\mathbf{r}||\mathbf{b}|$

$\therefore \quad \theta = \dfrac{\pi}{2}$

and let angle between $\mathbf{r} \times \mathbf{c}$ is ϕ,

Then, $|\mathbf{r} \times \mathbf{c}| = |\mathbf{r}||\mathbf{c}| \sin\phi = |\mathbf{r}||\mathbf{c}|$ are coplanar $= \dfrac{\pi}{2}$

$\therefore$ $\mathbf{r} \perp \mathbf{a}$, $\mathbf{r} \perp \mathbf{b}$ and $\mathbf{r} \perp \mathbf{c}$

Hence, $\mathbf{a}$, $\mathbf{b}$ and $\mathbf{c}$

$\therefore \quad [\mathbf{abc}] = 0$

75. $\mathbf{OM}_1 = \mathbf{OA} + \mathbf{OB} + \mathbf{OC} + \mathbf{OD}$ [given]

$= \mathbf{OM} + \mathbf{MA} + \mathbf{OM} + \mathbf{MB} + \mathbf{OM} + \mathbf{MC} + \mathbf{OM} + \mathbf{MD}$

$= 4\,\mathbf{OM} + (\mathbf{MA} + \mathbf{MC}) + (\mathbf{MB} + \mathbf{MD})$

$= 4\,\mathbf{OM}$ $[\because \mathbf{MA} = -\mathbf{MC},\ \mathbf{MB} = -\mathbf{MD}]$

$\Rightarrow \quad \mathbf{OM} = (1/4)\mathbf{OM}_1$

$\therefore \quad \lambda = \dfrac{1}{4}$

76. Given, $\mathbf{p} = \mathbf{a} + \mathbf{b} - 2\mathbf{c}$

$$\mathbf{q} = 3\mathbf{a} - 2\mathbf{b} + \mathbf{c}$$

and $\mathbf{r} = \mathbf{a} - 4\mathbf{b} + 2\mathbf{c}$

Given, $V_1 = [\mathbf{abc}]$ ...(i)

$$\therefore \quad V_2 = [\mathbf{pqr}] = \begin{vmatrix} 1 & 1 & -2 \\ 3 & -2 & 1 \\ 1 & -4 & 2 \end{vmatrix} [\mathbf{abc}]$$

$$\therefore \quad V_2 = 15[\mathbf{abc}] \quad \text{...(ii)}$$

From Eqs. (i) and (ii), $V_2 : V_1 = 15 : 1$

77. Equation of line L_1 is

$$\mathbf{r} = 3\hat{\mathbf{i}} + \lambda(-\hat{\mathbf{i}} + \hat{\mathbf{j}} + \hat{\mathbf{k}})$$

$$\mathbf{r} = (3 - \lambda)\hat{\mathbf{i}} + \lambda\hat{\mathbf{j}} + \lambda\hat{\mathbf{k}} \quad \text{...(i)}$$

and equation of line L_2 is

$$\mathbf{r} = (\hat{\mathbf{i}} + \hat{\mathbf{j}}) + \mu(\hat{\mathbf{i}} + \hat{\mathbf{k}}) = \hat{\mathbf{i}}(1 + \mu) + \hat{\mathbf{j}} + \mu\hat{\mathbf{k}} \quad \text{...(ii)}$$

For point of intersection, comparing Eqs. (i) and (ii),

$$\therefore \quad 3 - \lambda = 1 + \mu$$

$$\lambda = 1$$

and $\mu = \lambda$

$$\therefore \quad \lambda = \mu = 1$$

Substituting the values of λ and μ in Eq. (i) the point of intersection is

$$\mathbf{r} = 2\hat{\mathbf{i}} + \hat{\mathbf{j}} + \hat{\mathbf{k}}$$

78. Equation of line joining

$6\mathbf{a} - 4\mathbf{b} - 5\mathbf{c}$ and $-4\mathbf{c}$ is

$$\mathbf{r} = (6\mathbf{a} - 4\mathbf{b} - 5\mathbf{c}) + \lambda(-6\mathbf{a} + 4\mathbf{b} + \mathbf{c})$$

$$= \mathbf{a}(6 - 6\lambda) + \mathbf{b}(-4 + 4\lambda) + \mathbf{c}(-5 + \lambda) \quad \text{...(i)}$$

and equation of line joining $-\mathbf{a} - 2\mathbf{b} - 3\mathbf{c}$ and $\mathbf{a} + 2\mathbf{b} - 5\mathbf{c}$ is

$$\mathbf{r}\,(-\mathbf{a} - 2\mathbf{b} - 3\mathbf{c}) + \mu\,(2\mathbf{a} + 4\mathbf{b} - 2\mathbf{c})$$

$$= \mathbf{a}(-1 + 2\mu) + \mathbf{b}(-2 + 4\mu) + \mathbf{c}(-3 - 2\mu) \quad \text{...(ii)}$$

Comparing Eqs. (i) and (ii), then

$$6 - 6\lambda = -1 + 2\mu$$

$$-4 + 4\lambda = -2 + 4\mu$$

and $-5 + \lambda = -3 - 2\mu$

After solving, we get

$$\lambda = 1 \text{ and } \mu = \frac{1}{2}$$

Substituting the value of λ in Eq. (i) then point of intersection is

$$\mathbf{r} = -4\mathbf{c}$$

79. $\mathbf{p} = \dfrac{\mathbf{b} \times \mathbf{c}}{[\mathbf{abc}]}, \mathbf{q} = \dfrac{\mathbf{c} \times \mathbf{a}}{[\mathbf{abc}]}, \mathbf{r} = \dfrac{\mathbf{a} \times \mathbf{b}}{[\mathbf{abc}]}$

$$\therefore \quad \mathbf{a} \times \mathbf{p} = \frac{1}{[\mathbf{abc}]} \mathbf{a} \times (\mathbf{b} \times \mathbf{c})$$

$$= \frac{1}{[\mathbf{abc}]} \{(\mathbf{a} \cdot \mathbf{c})\mathbf{b} - (\mathbf{a} \cdot \mathbf{b})\mathbf{c}\}$$

Similarly, $\mathbf{b} \times \mathbf{q} = \dfrac{1}{[\mathbf{abc}]} \{(\mathbf{b} \cdot \mathbf{a})\mathbf{c} - (\mathbf{b} \cdot \mathbf{c})\mathbf{a}\}$

and $\mathbf{c} \times \mathbf{r} = \dfrac{1}{[\mathbf{abc}]} \{(\mathbf{c} \cdot \mathbf{b})\mathbf{a} - (\mathbf{c} \cdot \mathbf{a})\mathbf{b}\}$

Hence, $\mathbf{a} \times \mathbf{p} + \mathbf{b} \times \mathbf{q} + \mathbf{c} \times \mathbf{r} = 0$

80. If I be the origin, then

$$\frac{a\mathbf{IA} + b\mathbf{IB} + c\mathbf{IC}}{a + b + c}$$

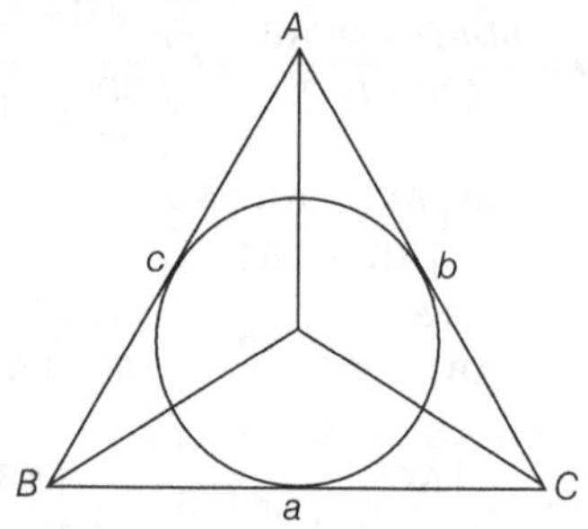

$\therefore \; a\mathbf{IA} + b\mathbf{IB} + c\mathbf{IC} = 0$

81. Given, $(3\mathbf{a} - 5\mathbf{b}) \cdot (2\mathbf{a} + \mathbf{b}) = 0$

$$\Rightarrow 6a^2 + 3\mathbf{a} \cdot \mathbf{b} - 10\mathbf{a} \cdot \mathbf{b} - 5b^2 = 0$$

$$\Rightarrow \quad 6a^2 - 5b^2 - 7\mathbf{a} \cdot \mathbf{b} = 0 \quad \text{...(i)}$$

and $(\mathbf{a} + 4\mathbf{b}) \cdot (-\mathbf{a} + \mathbf{b}) = 0$

$$\Rightarrow -a^2 + \mathbf{a} \cdot \mathbf{b} - 4\mathbf{a} \cdot \mathbf{b} + 4b^2 = 0$$

$$\therefore \quad a^2 - 4b^2 = 3\mathbf{a} \cdot \mathbf{b} = 0 \quad \text{...(ii)}$$

Eliminate $\mathbf{a} \cdot \mathbf{b}$ from Eqs. (i) and (ii), we get

$$5a = \sqrt{43}b \quad \text{...(iii)}$$

From Eq. (ii),

$$\mathbf{a} \cdot \mathbf{b} = \frac{4b^2 - a^2}{3}$$

$$= \frac{4b^2 - \dfrac{43b^2}{25}}{3}$$

$$= \frac{57b^2}{75} = \frac{19b^2}{25}$$

If angle between $\mathbf{a}$ and $\mathbf{b}$ is α

$$\therefore \quad \cos\alpha = \frac{\mathbf{a} \cdot \mathbf{b}}{|\mathbf{a}||\mathbf{b}|}$$

$$= \frac{\dfrac{19b^2}{25}}{\dfrac{\sqrt{43}b}{5} \times b} = \frac{19}{5\sqrt{43}}$$

82. $\therefore \; \mathbf{a} = (x, y, z)$

$\mathbf{a}$ makes an obtuse angle with Y-axis

$\therefore \; y < 0$ and given

$$\frac{\mathbf{a} \cdot \mathbf{b}}{|\mathbf{a}||\mathbf{b}|} = \frac{\mathbf{a} \cdot \mathbf{c}}{|\mathbf{a}||\mathbf{c}|}$$

$$\Rightarrow \quad \frac{\mathbf{a} \cdot \mathbf{b}}{|\mathbf{b}|} = \frac{\mathbf{a} \cdot \mathbf{c}}{|\mathbf{c}|}$$

$$\Rightarrow \quad \frac{xy - 2yz + 3zx}{\sqrt{(y^2 + 4z^2 + 9x^2)}} = \frac{2zx + 3xy - yz}{\sqrt{4z^2 + 9x^2 + y^2}}$$

$$\Rightarrow \quad xy - 2yz + 3zx = 2zx + 3xy - yz$$

$$\Rightarrow \quad 2xy + yz - zx = 0 \quad \text{...(i)}$$

and given $\mathbf{a} \cdot \mathbf{d} = 0$

$$\therefore \quad x - y + 2z = 0$$

$$z = \frac{y - x}{2} \quad \text{...(ii)}$$

From Eqs. (i) and (ii),

$$2xy + \frac{(y-x)^2}{2} = 0$$

$\therefore \quad (y+x)^2 = 0$

$\therefore \quad y = -x \quad \ldots(iii)$

$\therefore \quad z = -x \quad$ [from Eq. (ii)]

$\therefore \quad \mathbf{a} = (x, -x, -x)$

$$|\mathbf{a}| = \sqrt{x^2 + x^2 + x^2} = x\sqrt{3} = 2\sqrt{3}$$

$\therefore \quad x = 2, y = -2, z = -2$

$\therefore \quad \mathbf{a} = (2, -2, -2)$

83. $\mathbf{b} - (\mathbf{a}\cdot\mathbf{b})\mathbf{a} = (\mathbf{a}\cdot\mathbf{a})\mathbf{b} - (\mathbf{a}\cdot\mathbf{b})\mathbf{a} \quad [\because \mathbf{a}\cdot\mathbf{a} = 1]$

$= \mathbf{a}\times(\mathbf{b}\times\mathbf{a})$

Here, $\mathbf{a}\times\mathbf{b} \perp \mathbf{a}\times(\mathbf{b}\times\mathbf{a})$ i.e. $\angle ABC = 90°$

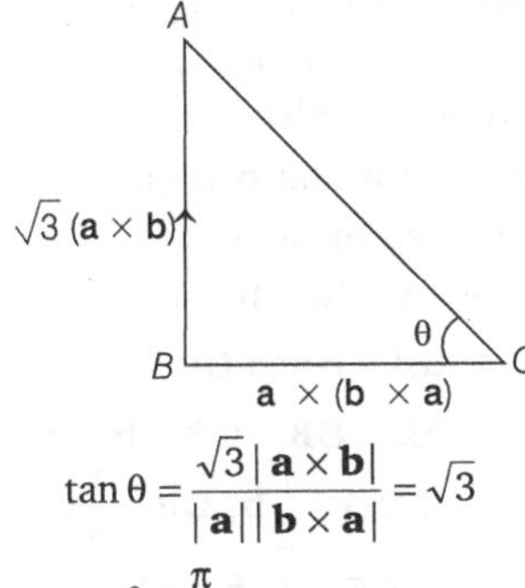

$$\tan\theta = \frac{\sqrt{3}|\mathbf{a}\times\mathbf{b}|}{|\mathbf{a}||\mathbf{b}\times\mathbf{a}|} = \sqrt{3}$$

$\therefore \quad \theta = \frac{\pi}{3}$

$\therefore \quad \angle ABC = \frac{\pi}{2}, \angle BAC = \frac{\pi}{6}$

$\therefore \quad \angle ACB = \frac{\pi}{3}$.

84. $\because$ **DB** = **DA** + **AB** or **DA** = **DA** = **DB** − **AB**

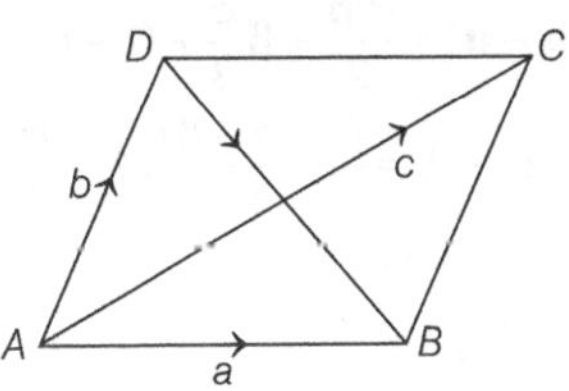

$\therefore (\mathbf{DA})^2 = (\mathbf{DB})^2 + (\mathbf{AB})^2 - 2\mathbf{DB}\cdot\mathbf{AB} \quad \ldots(i)$

In parallelogram,

$$2(a^2 + b^2) = c^2 + DB^2$$

$\therefore \quad (DB)^2 = 2a^2 + 2b^2 - c^2$

$\therefore$ From Eq. (i), $\quad b^2 = 2a^2 + 2b^2 - c^2 + a^2 - 2\mathbf{AB}\cdot\mathbf{DB}$

$$\therefore \quad \mathbf{AB}\cdot\mathbf{DB} = \frac{3a^2 + b^2 - c^2}{2}$$

85. $\therefore AD \perp BC$

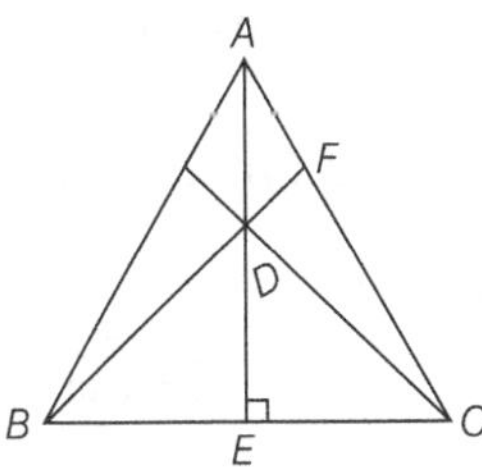

$\therefore \quad (\mathbf{a} - \mathbf{d})\cdot(\mathbf{d} - \mathbf{c}) = 0$ and $BD \perp AC$

$\therefore \quad (\mathbf{b} - \mathbf{d})\cdot(\mathbf{c} - \mathbf{a}) = 0$

$\therefore D$ is orthocentre.

86. $\mathbf{a}\times(\mathbf{b}\times\mathbf{c}) = (\mathbf{a}\times\mathbf{b})\times\mathbf{c}$

$\Rightarrow \quad (\mathbf{a}\cdot\mathbf{b})\,\mathbf{c} = (\mathbf{b}\cdot\mathbf{c})\,\mathbf{a}$

As, **b** is not perpendicular to both **a** and **c**

$\Rightarrow$ At least one of $\mathbf{a}\cdot\mathbf{b}$ and $\mathbf{b}\cdot\mathbf{c}$ is non-zero.

Hence, **a** and **c** are always collinear.

87. Let image of P w.r.t the given line be $Q(\alpha, \beta, \gamma)$. Then mid point of PQ i.e. $\left(\frac{\alpha+7}{2}, \frac{\beta-1}{2}, \frac{\gamma+2}{2}\right)$ lies on the line

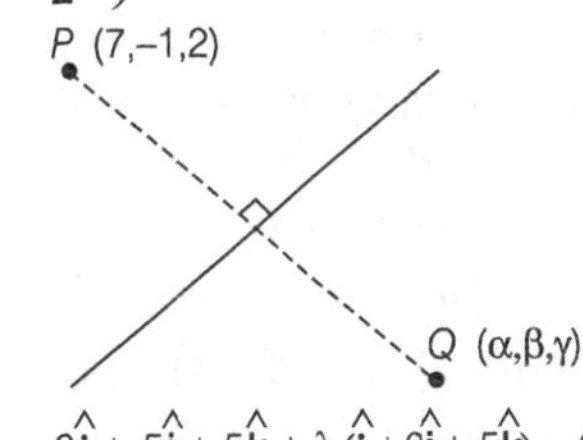

$$\therefore \quad \frac{\alpha+7}{2}\hat{\mathbf{i}} + \frac{\beta-1}{2}\hat{\mathbf{j}} + \frac{\gamma+2}{2}\hat{\mathbf{k}}$$

$$= 9\hat{\mathbf{i}} + 5\hat{\mathbf{j}} + 5\hat{\mathbf{k}} + \lambda(\hat{\mathbf{i}} + 3\hat{\mathbf{j}} + 5\hat{\mathbf{k}})$$

On comparing

$$\left.\begin{aligned} \frac{\alpha+7}{2} &= 9 + \lambda \Rightarrow \alpha = 11 + 2\lambda \\ \frac{\beta-1}{2} &= 5 + 3\lambda \Rightarrow \beta = 11 + 6\lambda \\ \text{and } \frac{\gamma+2}{2} &= 5 + 5\lambda \Rightarrow \gamma = 8 + 10\lambda \end{aligned}\right\} \quad \ldots(i)$$

Also, PQ and given line are perpendicular

i.e. $\quad (\alpha - 7)\cdot 1 + (\beta + 1)\cdot 3 + (\gamma - 2)\cdot 5 = 0$

$\Rightarrow \quad \alpha - 7 + 3\beta + 3 + 5\gamma - 10 = 0$

$\therefore \quad \alpha + 3\beta + 5\gamma = 14 \quad \ldots(ii)$

From Eqs. (i) and (ii),

$$11 + 2\lambda + 3(11 + 6\lambda) + 5(8 + 10\lambda) = 14$$

$\therefore \quad \lambda = -1$

From Eq. (i), $(\alpha, \beta, \gamma) = (9, 5, -2)$.

88. We have, $\mathbf{V}_1 = \frac{1}{2}(\mathbf{a}\times\mathbf{b}), \mathbf{V}_2 = \frac{1}{2}(\mathbf{b}\times\mathbf{c}), \mathbf{V}_3 = \frac{1}{3}(\mathbf{c}\times\mathbf{a})$

and $\quad \mathbf{V}_4 = \frac{1}{2}\{(\mathbf{c} - \mathbf{a})\times(\mathbf{b} - \mathbf{a})\}$

$\therefore \quad \mathbf{V}_1 + \mathbf{V}_2 + \mathbf{V}_3 + \mathbf{V}_4 = \mathbf{0}$

$\therefore \quad |\mathbf{V}_1 + \mathbf{V}_2 + \mathbf{V}_3 + \mathbf{V}_4| = 0$

89. Given, $\mathbf{a}\cdot\mathbf{b} = 0$ and $|\mathbf{a}| = 1, |\mathbf{b}| = 1$

and also given

$$\mathbf{a}\cdot\mathbf{c} = |\mathbf{a}||\mathbf{c}|\cos\theta - \cos\theta$$

and $\quad \mathbf{b}\cdot\mathbf{c} = |\mathbf{b}||\mathbf{c}|\cos\theta = \cos\theta$

and $\quad \mathbf{c} = x\,\mathbf{a} + y\,\mathbf{b} + z(\mathbf{a}\times\mathbf{b}) \quad \ldots(i)$

$\Rightarrow \mathbf{a}\cdot\mathbf{c} = x(\mathbf{a}\cdot\mathbf{a}) + y(\mathbf{b}\cdot\mathbf{a}) + z\,\mathbf{a}\cdot(\mathbf{a}\times\mathbf{b})$

$\Rightarrow \quad \cos\theta = x\cdot 1 + 0 + 0$

$\therefore \quad x = \cos\theta \quad \ldots(ii)$

From Eq. (i),

$$\mathbf{b}\cdot\mathbf{c} = x(\mathbf{a}\cdot\mathbf{b}) + y(\mathbf{b}\cdot\mathbf{b}) + z\,\mathbf{b}\cdot(\mathbf{a}\times\mathbf{b})$$

$\Rightarrow \quad \cos\theta = 0 + y\cdot 1 + 0$

$\therefore \quad y = \cos\theta \quad \ldots\text{(iii)}$

and $\quad \mathbf{c}\cdot\mathbf{c} = x^2 + y^2 + z^2 \Rightarrow 1 = 2\cos^2\theta + z^2$

$\therefore \quad z^2 = -\cos_2 2\theta$

90. Since, **a**, **b** and **c** are three coplanar vectors, then there must exist three scalars x, y, z such that

$$x\,\mathbf{a} + y\,\mathbf{b} + z\,\mathbf{c} = 0 \quad \ldots\text{(i)}$$

Multiplying both sides of Eq. (i) scalarly by **a** and **c** respectively, we get

$$x\,\mathbf{a}\cdot\mathbf{a} + y\,\mathbf{a}\cdot\mathbf{b} + z\,\mathbf{a}\cdot\mathbf{c} = 0 \quad \ldots\text{(ii)}$$

and $\quad x\,\mathbf{a}\cdot\mathbf{c} + y\,\mathbf{b}\cdot\mathbf{c} + z\,\mathbf{c}\cdot\mathbf{c} = 0 \quad \ldots\text{(iii)}$

Eliminating x, y, z from Eqs. (i), (ii) and (iii), we get

$$\begin{vmatrix} \mathbf{a} & \mathbf{b} & \mathbf{c} \\ \mathbf{a}\cdot\mathbf{a} & \mathbf{a}\cdot\mathbf{b} & \mathbf{a}\cdot\mathbf{c} \\ \mathbf{a}\cdot\mathbf{c} & \mathbf{b}\cdot\mathbf{c} & \mathbf{c}\cdot\mathbf{c} \end{vmatrix} = 0$$

91. Given vectors are $\mathbf{a} = 3\hat{\mathbf{i}} + 2\hat{\mathbf{j}} + x\hat{\mathbf{k}}$

and $\mathbf{b} = \hat{\mathbf{i}} - \hat{\mathbf{j}} + \hat{\mathbf{k}}$

$$\therefore \quad \mathbf{a}\times\mathbf{b} = \begin{vmatrix} \hat{\mathbf{i}} & \hat{\mathbf{j}} & \hat{\mathbf{k}} \\ 3 & 2 & x \\ 1 & -1 & 1 \end{vmatrix}$$

$$= \hat{\mathbf{i}}(2 + x) - \hat{\mathbf{j}}(3 - x) + \hat{\mathbf{k}}(-3 - 2)$$

$$= (x + 2)\hat{\mathbf{i}} + (x - 3)\hat{\mathbf{j}} - 5\hat{\mathbf{k}}$$

$$\Rightarrow |\mathbf{a}\times\mathbf{b}| = \sqrt{(x + 2)^2 + (x - 3)^2 + 25}$$

$$= \sqrt{2x^2 - 2x + 4 + 9 + 25}$$

$$= \sqrt{2\left(x^2 - x + \frac{1}{4}\right) - \frac{1}{2} + 38} = \sqrt{2\left(x - \frac{1}{2}\right)^2 + \frac{75}{2}}$$

So, $|\mathbf{a}\times\mathbf{b}| \geq \sqrt{\frac{75}{2}} \quad$ [at $x = \frac{1}{2}$, $|\mathbf{a}\times\mathbf{b}|$ is minimum]

$\Rightarrow \quad r \geq 5\sqrt{\frac{3}{2}}$

92. $\because O', G, O$ are collinear and

$$\frac{OG}{OO'} = \frac{1}{3} \Rightarrow 3\mathbf{OG} = \mathbf{OO'}$$

$$\therefore \quad \mathbf{OO'} = 3\left(\frac{\mathbf{a} + \mathbf{b} + \mathbf{c}}{3}\right)$$

$\therefore \quad \mathbf{a} + \mathbf{b} + \mathbf{c} = \mathbf{OO'}$

93. Let the origin of reference be the circumcentre of the triangle.

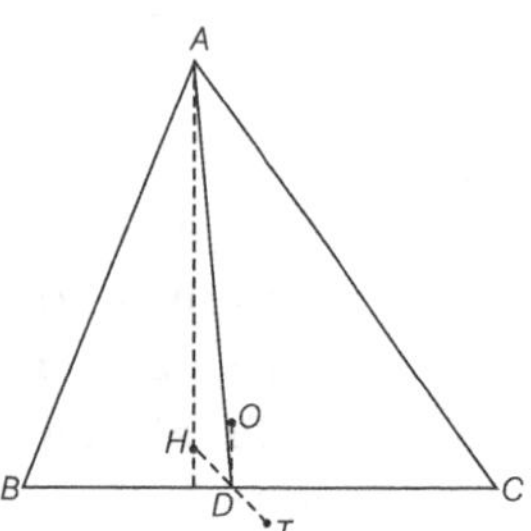

Let $\quad \mathbf{OA} = \mathbf{a}$, $\mathbf{OB} = \mathbf{b}$, $\mathbf{OC} = \mathbf{c}$ and $\mathbf{OT} = \mathbf{t}$

Then, $|\mathbf{a}| = |\mathbf{b}| = |\mathbf{c}| = R$ (circumradius)

Again, $\mathbf{OA} + \mathbf{OB} + \mathbf{OC} = \mathbf{OA} + 2\mathbf{OD} = \mathbf{OA} + \mathbf{AH} = \mathbf{OH}$

Therefore, the PV of H is $\mathbf{a} + \mathbf{b} + \mathbf{c}$. Since, D is the midpoint of HT, we have

$$\frac{\mathbf{a} + \mathbf{b} + \mathbf{c} + \mathbf{t}}{2} = \frac{\mathbf{b} + \mathbf{c}}{2} \Rightarrow \mathbf{t} = -\mathbf{a}$$

$\therefore \quad \mathbf{AT} = -2\mathbf{a}$ or $\mathbf{AT} = |-2\mathbf{a}| = 2|\mathbf{a}| = 2R.$

But $\quad BC = 2R\sin A = R$; therefore,

$$AT = 2BC$$

94. Equation of plane containing the line

$$\mathbf{r} = \mathbf{a} + k\,\mathbf{b}$$

and perpendicular to the plane

$$\mathbf{r}\cdot\mathbf{n} = q$$

$\therefore$ **b** and **n** are parallel

$\Rightarrow \perp$ vector to **n** and **b** is $(\mathbf{n}\times\mathbf{b})$

Hence, required plane is

$$(\mathbf{r} - \mathbf{a})\cdot(\mathbf{n}\times\mathbf{b}) = 0$$

95. Let $\mathbf{OA} = \mathbf{a}$, $\mathbf{OB} = \mathbf{b}$ and $\mathbf{OC} = \mathbf{c}$, then

$$\mathbf{AB} = \mathbf{OB} - \mathbf{OB} = \mathbf{b} - \mathbf{a}$$

and $\mathbf{OP} = \frac{1}{3}\mathbf{a}$, $\mathbf{OQ} = \frac{1}{2}\mathbf{b}$, $\mathbf{OR} = \frac{1}{3}\mathbf{c}$

and $\quad \mathbf{OS} = \lambda\,(\mathbf{b} - \mathbf{a})$

$\because P, Q, R$ and S are coplanar, then

$\mathbf{PS} = \alpha\,\mathbf{PQ} + \beta\,\mathbf{PR}$ (**PS** can be written as a linear combination of **PQ** and **PR**)

$$\mathbf{OS} - \mathbf{OP} = \alpha(\mathbf{OQ} - \mathbf{OQ}) + \beta(\mathbf{OR} - \mathbf{OP})$$

$$\Rightarrow \quad \mathbf{OS} = \alpha\,\mathbf{OQ} + \beta\,\mathbf{OR} + (1 - \alpha - \beta)\,\mathbf{OP}$$

$$\Rightarrow \quad \lambda\,(\mathbf{b} - \mathbf{a}) = \frac{\alpha\,\mathbf{b}}{2} + \beta\frac{\mathbf{c}}{3} + (1 - \alpha - \beta)\frac{\mathbf{a}}{3}$$

On comparing, $-\lambda = \frac{(1 - \alpha - \beta)}{3}$, $\lambda = \frac{\alpha}{2}$

and $\quad \beta = 0$

$$\therefore \quad -\lambda = \frac{1 - \alpha}{3}, \lambda = \frac{\alpha}{2}$$

$\Rightarrow \quad -3\lambda = 1 - 2\lambda$

$\therefore \quad \lambda = -1$

96. $\mathbf{x}\cdot\mathbf{y} = 0$, $|\mathbf{x}| = 1$, $|\mathbf{y}| = 1$

Any vector can be written as a linear combination of the other two, then

$$a\hat{\mathbf{x}} + a\hat{\mathbf{y}} + c(\hat{\mathbf{x}}\times\hat{\mathbf{y}}) = \lambda\,(\hat{\mathbf{x}} + (\hat{\mathbf{x}}\times\hat{\mathbf{y}})) + \mu\,(c\hat{\mathbf{x}} + c\hat{\mathbf{y}} + b(\hat{\mathbf{x}}\times\hat{\mathbf{y}}))$$

$\Rightarrow \quad a = \lambda + c\mu \quad \ldots\text{(i)}$

$\quad a = 0 + c\mu \quad \ldots\text{(ii)}$

and $\quad c = \lambda + b\mu \quad \ldots\text{(iii)}$

From Eqs. (i) and (ii), $\lambda = 0$, $\mu = \frac{a}{c}$

From Eq. (iii), $\quad c = \frac{ba}{c}$

$\Rightarrow \quad c = \sqrt{ab}$

Objective Questions Type II [One or more than one correct answer(s)]

1. $\mathbf{r}\times(\hat{\mathbf{i}}+2\hat{\mathbf{j}}+\hat{\mathbf{k}})=\hat{\mathbf{i}}-\hat{\mathbf{k}}$

Let $\mathbf{r}=x\hat{\mathbf{i}}+y\hat{\mathbf{j}}+z\hat{\mathbf{k}}$

$\therefore \quad (x\hat{\mathbf{i}}+y\hat{\mathbf{j}}+z\hat{\mathbf{k}})\times(\hat{\mathbf{i}}+2\hat{\mathbf{j}}+\hat{\mathbf{k}})=\hat{\mathbf{i}}-\hat{\mathbf{k}}$

$$\Rightarrow \begin{vmatrix} \hat{\mathbf{i}} & \hat{\mathbf{j}} & \hat{\mathbf{k}} \\ x & y & z \\ 1 & 2 & 1 \end{vmatrix}=\hat{\mathbf{i}}-\hat{\mathbf{k}}$$

$\Rightarrow \hat{\mathbf{i}}\,(y-2z)-\hat{\mathbf{j}}\,(x-z)+\hat{\mathbf{k}}\,(2x-y)=\hat{\mathbf{i}}-\hat{\mathbf{k}}$

On comparing

$$y-2z=1$$

$$x-z=0 \text{ and } 2x-y=-1$$

$\therefore \quad z=x,\ y=2x+1$

$\therefore \quad \vec{r}=x\hat{\mathbf{i}}+(2x+1)\hat{\mathbf{j}}+x\hat{\mathbf{k}} \quad \ldots\text{(i)}$

For $x=1$ and $x=3$

$\mathbf{r}=\hat{\mathbf{i}}+3\hat{\mathbf{j}}+\hat{\mathbf{k}}$ (alternate (a) is correct)

and $\mathbf{r}=3\hat{\mathbf{i}}+7\hat{\mathbf{j}}+3\hat{\mathbf{k}}$

(alternate (b) is correct)

Now, from Eq. (i)

$$\mathbf{r}=\hat{\mathbf{j}}+x\,(\hat{\mathbf{i}}+2\hat{\mathbf{j}}+\hat{\mathbf{k}})$$

or $\mathbf{r}=\hat{\mathbf{j}}+t\,(\hat{\mathbf{i}}+2\hat{\mathbf{j}}+\hat{\mathbf{k}})$

(alternate (c) is correct)

where t is scalar.

2. We have, $\mathbf{DA}=\mathbf{a}$, $\mathbf{AB}=\mathbf{b}$ and $\mathbf{CB}=k\,\mathbf{a}$. Given, X and Y are the mid points of DB and AC. Then,

$$\mathbf{OX}=\frac{\mathbf{OB}+\mathbf{OD}}{2},\ \mathbf{OY}=\frac{\mathbf{OA}+\mathbf{OC}}{2}$$

$\therefore \quad \mathbf{XY}=\mathbf{OY}-\mathbf{OX}$

$$=\frac{\mathbf{OA}+\mathbf{OC}-\mathbf{OB}-\mathbf{OD}}{2}$$

$$=\frac{\mathbf{DA}+\mathbf{BC}}{2}=\frac{\mathbf{a}-k\,\mathbf{a}}{2}$$

$$\mathbf{XY}=\frac{(1-k)\,\mathbf{a}}{2}$$

$\therefore \quad |\mathbf{XY}|=\pm\left(\frac{1-k}{2}\right)a=4 \quad$ (given)

$\Rightarrow \quad \left(\frac{1-k}{2}\right)17=4 \quad$ (taking + ve sign)

$\Rightarrow \quad 1-k=\frac{8}{17}$

$\therefore \quad k=\frac{9}{17}$

and $\quad -\left(\frac{1-k}{2}\right)a=4 \quad$ (taking – ve sign)

$\Rightarrow \quad (k-1)\,17=8$

$\Rightarrow \quad k=\frac{25}{17}$

3. Given, $|\mathbf{a}|=1, |\mathbf{c}|=1$

and $\quad |\mathbf{b}|=4$

and $\quad \mathbf{a}\times\mathbf{b}=2\,\mathbf{a}\times\mathbf{c}$

Since, angle between $\mathbf{a}$ and $\mathbf{c}$ is $\cos^{-1}(1/4)$

$\therefore \quad \mathbf{a}\cdot\mathbf{c}=|\mathbf{a}||\mathbf{c}|\frac{1}{4}$

$$=1\cdot1\cdot\frac{1}{4}$$

$\Rightarrow \quad \mathbf{a}\cdot\mathbf{c}=\frac{1}{4}$

and $\quad \mathbf{b}-2\,\mathbf{c}=\lambda\,\mathbf{a}$

or $\quad \mathbf{b}=2\,\mathbf{c}+\lambda\,\mathbf{a}$

Squaring $b^2=4c^2+\lambda^2a^2+4\lambda\,\mathbf{c}\cdot\mathbf{a}$

$\Rightarrow \quad 16=4+\lambda^2+\lambda$

$\Rightarrow \quad \lambda^2+\lambda-12=0$

$\therefore \quad (\lambda+4)(\lambda-3)=0$

$\Rightarrow \quad \lambda=-4, 3$

4. $$\mathbf{c}=\lambda\left(\frac{7\hat{\mathbf{i}}-4\hat{\mathbf{j}}-4\hat{\mathbf{k}}}{9}+\frac{-2\hat{\mathbf{i}}-\hat{\mathbf{j}}+2\hat{\mathbf{k}}}{3}\right)$$

$$\mathbf{c}=\frac{\lambda}{9}(\hat{\mathbf{i}}-7\hat{\mathbf{j}}+2\hat{\mathbf{k}}) \quad \ldots\text{(i)}$$

$$|\mathbf{c}|=\frac{\lambda^2}{81}(1+49+4)=\frac{\lambda^2}{81}\times54$$

$\Rightarrow \quad 54=\frac{\lambda^2\times54}{81}\Rightarrow\lambda=\pm9$

Putting $\lambda=\pm9$ in Eq. (i),

Then, $\quad \mathbf{c}=\pm(\hat{\mathbf{i}}-7\hat{\mathbf{j}}+2\hat{\mathbf{k}})$

5. $|\mathbf{v}|=1\cdot1\cdot\sin\theta=\sin\theta$

$\therefore \quad \mathbf{a}\cdot\mathbf{b}=\cos\theta$

$|\mathbf{u}|^2=1+\cos^2\theta\cdot1-2\cos\theta\cos\theta=\sin^2\theta$

$\therefore \quad \mathbf{u}\cdot\mathbf{a}=1-(\mathbf{a}\cdot\mathbf{b})(\mathbf{a}\cdot\mathbf{b})$

$$=1-\cos^2\theta=\sin^2\theta$$

$\therefore \quad \mathbf{u}\cdot\mathbf{b}=(\mathbf{a}\cdot\mathbf{b})-(\mathbf{a}\cdot\mathbf{b})=0$

$\therefore \quad |\mathbf{v}|=\sin\theta=|\mathbf{u}|\Rightarrow a=c$

as $\quad \mathbf{u}\cdot\mathbf{b}=0$

6. Let O be the origin.

$\therefore \quad \mathbf{OA}=\hat{\mathbf{i}}+\hat{\mathbf{j}}$

$$\mathbf{OB}=\hat{\mathbf{i}}-\hat{\mathbf{j}}$$

$$\mathbf{OC}=p\hat{\mathbf{i}}+q\hat{\mathbf{j}}+r\hat{\mathbf{k}}$$

$\therefore \quad \mathbf{AB}=\mathbf{OB}-\mathbf{OA}=-2\hat{\mathbf{j}}$

and $\quad \mathbf{AC}=\mathbf{OC}-\mathbf{OA}$

$$=(p-1)\hat{\mathbf{i}}+(q-1)\hat{\mathbf{j}}+r\hat{\mathbf{k}}$$

$\because$ A, B, C are collinear

$\therefore \quad \mathbf{AC}=\lambda\,\mathbf{AB}$

$\Rightarrow \quad (p-1)\hat{\mathbf{i}}+(q-1)\hat{\mathbf{j}}+r\hat{\mathbf{k}}=\lambda(-2\hat{\mathbf{j}})$

On comparing $p-1=0,\ q-1=-2\lambda$

and $\quad r=0$

$\therefore \quad p=1, r=0$ and $q=1-2\lambda$

For $\quad \lambda=-\frac{1}{2}, q=2$

and for $\lambda=0$, $q=1$

$\therefore \quad p=1-1, r=0$

and $\quad p=1, q=2, r=0$.

7. $(\mathbf{a}\times\mathbf{b})^2=|\mathbf{a}\times\mathbf{b}|^2$

$=(|\mathbf{a}||\mathbf{b}|\sin\pi/6)^2$

$=16\times 4\times\frac{1}{4}=16$

$=(\mathbf{a})^2$

8. Since, $(\mathbf{a}\times\mathbf{b})\times(\mathbf{c}\times\mathbf{r})$ [taking **c** is middle point]

$=(\mathbf{a}\times\mathbf{b}\cdot\mathbf{r})\,\mathbf{c}-[(\mathbf{a}\times\mathbf{b}\cdot\mathbf{c})]\,\mathbf{r}$

$=[\mathbf{a}\ \mathbf{b}\ \mathbf{r}]\,\mathbf{c}-[\mathbf{a}\ \mathbf{b}\ \mathbf{c}]\,\mathbf{r}$ ...(i)

is taking **b** is middle point

Then $=\{\mathbf{a}\cdot(\mathbf{c}\times\mathbf{r})\}\,\mathbf{b}=\{\mathbf{b}\cdot(\mathbf{c}\times\mathbf{r})\}\,\mathbf{a}$

$=[\mathbf{a}\ \mathbf{c}\ \mathbf{r}]\,\mathbf{b}-[\mathbf{b}\ \mathbf{c}\ \mathbf{r}]\,\mathbf{a}$ ...(ii)

From Eqs. (i) and (ii), we get

$[\mathbf{a}\ \mathbf{b}\ \mathbf{r}]\,\mathbf{c}-[\mathbf{a}\ \mathbf{b}\ \mathbf{c}]\,\mathbf{r}$

$=[\mathbf{a}\ \mathbf{c}\ \mathbf{r}]\,\mathbf{b}-[\mathbf{b}\ \mathbf{c}\ \mathbf{r}]\,\mathbf{a}$

$\Rightarrow [\mathbf{a}\ \mathbf{b}\ \mathbf{r}]\,\mathbf{c}-[\mathbf{a}\ \mathbf{b}\ \mathbf{c}]\,\mathbf{r}$

$=-[\mathbf{c}\ \mathbf{a}\ \mathbf{r}]\,\mathbf{b}-[\mathbf{b}\ \mathbf{c}\ \mathbf{r}]\,\mathbf{a}$

$\therefore [\mathbf{b}\ \mathbf{c}\ \mathbf{r}]\,\mathbf{a}+[\mathbf{c}\ \mathbf{a}\ \mathbf{r}]\,\mathbf{b}+[\mathbf{a}\ \mathbf{b}\ \mathbf{r}]\,\mathbf{c}=[\mathbf{a}\ \mathbf{b}\ \mathbf{c}]\,\mathbf{r}$

9. Since, **a** and **b** are unit vectors, we have

$|\mathbf{a}-\mathbf{b}|=\sqrt{(\mathbf{a}-\mathbf{b})^2}$

$\therefore \quad \sqrt{(\mathbf{a})^2+(\mathbf{b})^2-2\mathbf{a}\cdot\mathbf{b}}=\sqrt{1+1-2\cos 2\theta}$

$=2|\sin\theta|$

Therefore, $\quad |\mathbf{a}-\mathbf{b}|<1$

$\Rightarrow \quad 2|\sin\theta|<1$

$|\sin\theta|<\frac{1}{2}$

$\Rightarrow \quad \theta\in\left(0,\frac{\pi}{6}\right)$ or $\left(\frac{5\pi}{6},\pi\right)$

10. Let $\mathbf{a}=a_1\hat{\mathbf{i}}+a_2\hat{\mathbf{j}}+a_3\hat{\mathbf{k}}$ and $\mathbf{b}=b_1\hat{\mathbf{i}}+b_2\hat{\mathbf{j}}+b_3\hat{\mathbf{k}}$. Taking cross product with **a** on both sides of $\mathbf{a}+\mathbf{b}=\hat{\mathbf{i}}-3\hat{\mathbf{j}}+4\hat{\mathbf{k}}$, we get,

$$\mathbf{a}\times(\mathbf{a}+\mathbf{b})=\mathbf{a}\times\mathbf{b}=\mathbf{a}\times(\hat{\mathbf{i}}-3\hat{\mathbf{j}}-4\hat{\mathbf{k}})=\begin{vmatrix}\hat{\mathbf{i}} & \hat{\mathbf{j}} & \hat{\mathbf{k}}\\ a_1 & a_2 & a_3\\ 1 & -3 & -4\end{vmatrix}$$

$$=(-4a_2+3a_3)\hat{\mathbf{i}}+(a_3+4a_1)\hat{\mathbf{j}}+(-3a_1-a_2)\hat{\mathbf{k}}$$

Hence, $2\hat{\mathbf{i}}+2\hat{\mathbf{j}}-\hat{\mathbf{k}}=(-4a_2+3a_3)\hat{\mathbf{i}}+(a_3+4a_1)\hat{\mathbf{j}}+(-3a_1-a_2)\hat{\mathbf{k}}$

So, $\quad 2=-4a_2+3a_3,\ 2=a_3+4a_1,\ -1=-3a_1-a_2$.

Solving these equations, we have

$a_3=k,\ a_2=\frac{3k-2}{4},\ a_1=\frac{2-k}{4}$, k

being any real numbers.

Also, $\quad b_1=1-a_1=1-\frac{2-k}{4}=\frac{2+k}{4}$,

$b_2=-3-a_2=-3-\frac{3k-2}{4}$

$=-\frac{-10-3k}{4}$

$b_3=-4,\ a_3=-4-k$

Hence, $\quad \mathbf{a}=\frac{2-k}{4}\hat{\mathbf{i}}+\frac{3k-2}{4}\hat{\mathbf{j}}+k\hat{\mathbf{k}}$

$\mathbf{b}=\frac{2+k}{4}\hat{\mathbf{i}}-\frac{(10+3k)}{4}\hat{\mathbf{j}}-(4+k)\hat{\mathbf{k}}$

11. Given vectors $\mathbf{PQ}=a\hat{\mathbf{i}}+b\hat{\mathbf{j}}$ and $\mathbf{PS}=a\hat{\mathbf{i}}-b\hat{\mathbf{j}}$ are adjacent sides of a parallelogram $PQRS$, so area of parallelogram $PQRS=|\mathbf{PQ}\times\mathbf{PS}|$

$=2ab=8$ (given)

$\therefore \quad ab=4$ ... (i)

According to the question,

$|\mathbf{u}|=|$ projection vector of $\mathbf{w}=\hat{\mathbf{i}}+\hat{\mathbf{j}}$ along $\mathbf{PQ}|$

$$=\left|\frac{(\hat{\mathbf{i}}+\hat{\mathbf{j}})\cdot(a\hat{\mathbf{i}}+b\hat{\mathbf{j}})}{\sqrt{a^2+b^2}}\right|=\frac{|a+b|}{\sqrt{a^2+b^2}}$$

and, similarly, $|\mathbf{v}|=|$ projection vector of $\mathbf{w}=\hat{\mathbf{i}}+\hat{\mathbf{j}}$ along $\mathbf{PS}|$

$$=\frac{|(\hat{\mathbf{i}}+\hat{\mathbf{j}})\cdot(a\hat{\mathbf{i}}+b\hat{\mathbf{j}})|}{\sqrt{a^2+b^2}}=\frac{|a-b|}{\sqrt{a^2+b^2}}$$

$\because \quad |\mathbf{u}|+|\mathbf{v}|=|\mathbf{w}|$

$\Rightarrow \quad \frac{|a+b|}{\sqrt{a^2+b^2}}+\frac{|a-b|}{\sqrt{a^2+b^2}}=\sqrt{2}$

$\Rightarrow \quad |a+b|+|a-b|=\sqrt{2}\sqrt{a^2+b^2}$

If $a\ge b>0$, then $2a=\sqrt{2}\sqrt{a^2+b^2}$

$\Rightarrow \quad 4a^2=2a^2+2b^2 \Rightarrow a^2=b^2$

$\Rightarrow \quad a=b$ ... (ii)

From Eqs. (i) and (ii), we get

$a=2=b$

$\therefore \quad a+b=4$

and the length of diagonal PR

$=|a\hat{\mathbf{i}}+b\hat{\mathbf{j}}+a\hat{\mathbf{i}}-b\hat{\mathbf{j}}|=|2a\hat{\mathbf{i}}|$

$=2a=4$

And, the angle bisector of vector **PQ** and **PS** is along the vector

$$\pm\frac{\lambda}{\sqrt{a^2+b^2}}((a\hat{\mathbf{i}}+b\hat{\mathbf{j}})\pm(a\hat{\mathbf{i}}-b\hat{\mathbf{j}}))$$

12. Let $\mathbf{r}=\mathbf{b}+t\,\mathbf{c}$

or $\mathbf{r}=(1+t)\hat{\mathbf{i}}+(2+t)\hat{\mathbf{j}}-(1+2t)\hat{\mathbf{k}}$ ...(i)

$\because$ Projection of **r** on **a** is $\sqrt{(2/3)}$

$\therefore \quad \frac{\mathbf{r}\cdot\mathbf{a}}{|\mathbf{a}|}=\sqrt{\frac{2}{3}}$

or $\quad \frac{2(1+t)-(2+t)-(1+2t)}{\sqrt{6}}=\pm\sqrt{\frac{2}{3}}$

$\therefore \quad -t-1=\pm 2$

$\therefore \quad t=-3, 1$

Putting in Eq. (i), we get

$\mathbf{r}=-2\hat{\mathbf{i}}-\hat{\mathbf{j}}+5\hat{\mathbf{k}}$

or $\quad \mathbf{r}=2\hat{\mathbf{i}}+3\hat{\mathbf{j}}-3\hat{\mathbf{k}}$

13. Since, vectors $(x, x+1, x+2)$, $(x+3, x+4, x+5)$ and $(x+6, x+7, x+8)$ are coplanar.

$$\therefore \quad \begin{vmatrix} x & x+1 & x+2 \\ x+3 & x+4 & x+5 \\ x+6 & x+7 & x+8 \end{vmatrix} = 0$$

Applying $C_2 \to C_2 - C_1$
and $\quad C_3 \to C_3 - C_1$

$$\therefore \quad \begin{vmatrix} x & 1 & 2 \\ x+3 & 1 & 2 \\ x+6 & 1 & 2 \end{vmatrix} = 0$$

$0 = 0 \; (C_2 = C_3)$
$\therefore \quad x \in R$

14. Let **a**, **b** and **c** lie in the $x-y$ plane.

Let $\quad \mathbf{a} = \hat{\mathbf{i}},\ \mathbf{b} = -\frac{1}{2}\hat{\mathbf{i}} + \frac{\sqrt{3}}{2}\hat{\mathbf{i}}$ and $\mathbf{c} = -\frac{1}{2}\hat{\mathbf{i}} - \frac{\sqrt{3}}{2}\hat{\mathbf{j}}$.

Therefore, $|\mathbf{p}+\mathbf{q}+\mathbf{r}| = |\lambda\mathbf{a} + \mu\mathbf{b} + \nu\mathbf{c}|$

$$= \left| \lambda\hat{\mathbf{i}} + \mu\left(-\frac{1}{2}\hat{\mathbf{i}} + \frac{\sqrt{3}}{2}\hat{\mathbf{j}}\right) + \nu\left(-\frac{1}{2}\hat{\mathbf{i}} - \frac{\sqrt{3}}{2}\hat{\mathbf{j}}\right) \right|$$

$$= \left| \left(\lambda - \frac{\mu}{2} - \frac{\nu}{2}\right)\hat{\mathbf{i}} + \frac{\sqrt{3}}{2}(\mu - \nu)\hat{\mathbf{j}} \right|$$

$$= \sqrt{\left(\lambda - \frac{\mu}{2} - \frac{\nu}{2}\right)^2 + \frac{3}{4}(\mu - \nu)^2}$$

$$= \sqrt{\lambda^2 + \mu^2 + \nu^2 - \lambda\mu - \lambda\nu - \mu\nu}$$

$$= \frac{1}{\sqrt{2}}\sqrt{(\lambda - \mu)^2 + (\mu - \nu)^2 + (\nu - \lambda)^2} \geq \frac{1}{\sqrt{2}}\sqrt{1+1+4} = \sqrt{3}$$

Hence, $|\mathbf{p}+\mathbf{q}+\mathbf{r}|$ can take a value equal to $\sqrt{3}$ and 2.

15. $\therefore \quad \mathbf{AC} = \mathbf{a} + \mathbf{b}$

Then, $|\mathbf{AC}| = |\mathbf{a} + \mathbf{b}|$

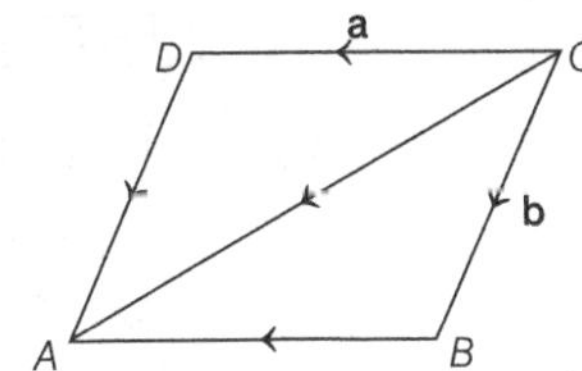

$$|\mathbf{AC}|^2 = (\mathbf{a})^2 + (\mathbf{b})^2 + 2\mathbf{a}\cdot\mathbf{b}$$
$$= \{(3\alpha - \beta)^2 + (\alpha + 3\beta)^2 + 2(3\alpha - \beta)\cdot(\alpha + 3\beta)$$
$$= 9\alpha^2 + \beta^2 - 6\alpha\cdot\beta + \alpha^2 + 9\beta^2 + 6\alpha\cdot\beta + 6\alpha^2 - 6\beta^2 + 16\alpha\cdot\beta$$
$$= 16\alpha^2 + 4\beta^2 + 16\alpha\cdot\beta$$
$$= 64 + 16 + 16|\alpha||\beta|\cos\frac{\pi}{3}$$
$$= 64 + 16 + 16 \times 2 \times 2 \times \frac{1}{2}$$
$$= 64 + 16 + 32 = 112$$

$\therefore \quad AC = 4\sqrt{7}$, similarly $BD = 4\sqrt{3}$.

16. $\therefore \hat{\mathbf{a}}\cdot(\mathbf{a}+\mathbf{b}) = \hat{\mathbf{b}}\cdot(\mathbf{a}+\mathbf{b}) \Rightarrow (\hat{\mathbf{a}} - \hat{\mathbf{b}})\cdot(\mathbf{a}+\mathbf{b}) = 0$

or $\quad \left(\frac{\mathbf{a}}{|\mathbf{a}|} - \frac{\mathbf{b}}{|\mathbf{b}|}\right)\cdot(\mathbf{a}+\mathbf{b}) = 0$

which is possible when

$$\mathbf{a} = \mathbf{b}$$

$\therefore \quad |\mathbf{a}| = |\mathbf{b}|$

If angle between **a** and **b** is θ, then

$$\mathbf{a}\cdot\mathbf{a} = |\mathbf{a}||\mathbf{a}|\cos\theta$$

$\therefore \quad \cos\theta = 1$
$\therefore \quad \theta = 0°$

17. Given, $|\mathbf{a}| = 12, |\mathbf{b}| = 4\sqrt{3}$

$$\mathbf{a} + \mathbf{b} + \mathbf{c} = \mathbf{0}$$

$\Rightarrow \quad \mathbf{a} = -(\mathbf{b} + \mathbf{c})$

We have, $\quad |\mathbf{a}|^2 = |\mathbf{b} + \mathbf{c}|^2$

$\Rightarrow \quad |\mathbf{a}|^2 = |\mathbf{b}|^2 + |\mathbf{c}|^2 + 2\mathbf{b}\cdot\mathbf{c}$

$\Rightarrow \quad 144 = 48 + |\mathbf{c}|^2 + 48$

$\Rightarrow \quad |\mathbf{c}|^2 = 48$

$\Rightarrow \quad |\mathbf{c}| = 4\sqrt{3}$

Also, $\quad |\mathbf{c}|^2 = |\mathbf{a}|^2 + |\mathbf{b}|^2 + 2\,\mathbf{a}\cdot\mathbf{b}$

$\Rightarrow \quad 48 = 144 + 48 + 2\,\mathbf{a}\cdot\mathbf{b}$

$\Rightarrow \quad \mathbf{a}\cdot\mathbf{b} = -72$

$\therefore$ Option (d) is correct.

Also, $\quad \mathbf{a}\times\mathbf{b} = \mathbf{c}\times\mathbf{a}$

$\Rightarrow \quad \mathbf{a}\times\mathbf{b} + \mathbf{c}\times\mathbf{a} = 2\mathbf{a}\times\mathbf{b}$

$\Rightarrow |\mathbf{a}\times\mathbf{b} + \mathbf{c}\times\mathbf{a}| = 2|\mathbf{a}\times\mathbf{b}| = 2\sqrt{|\mathbf{a}|^2|\mathbf{b}|^2 - (\mathbf{a}\cdot\mathbf{b})^2}$

$$= 2\sqrt{(144)(48) - (-72)^2}$$
$$= 2(12)\sqrt{48 - 36} = 48\sqrt{3}$$

$\therefore \quad$ Option (c) is correct.

Also, $\quad \frac{|\mathbf{c}|^2}{2} - |\mathbf{a}| = 24 - 12 = 12$

$\therefore \quad$ Option (a) is correct.

and $\quad \frac{|\mathbf{c}|^2}{2} + |\mathbf{a}| = 24 + 12 = 36$

$\therefore$ Option (b) is not correct.

18. Let $\mathbf{EB} = p$, $\mathbf{AB}$ and $\mathbf{CE} = q\,\mathbf{CD}$

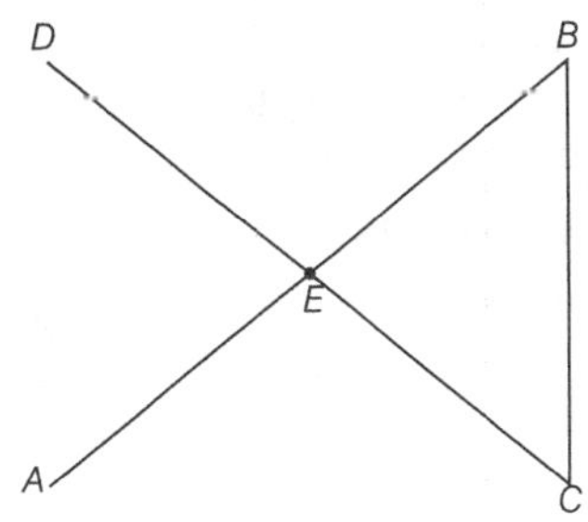

Then, $\quad 0 < p$ and $q \leq 1$

Since, $\quad \mathbf{EB} + \mathbf{BC} + \mathbf{CE} = 0$

$$pm(2\hat{\mathbf{i}} - 6\hat{\mathbf{j}} + 2\hat{\mathbf{k}}) + (\hat{\mathbf{i}} - 2\hat{\mathbf{j}}) + qn(-6\hat{\mathbf{i}} + 15\hat{\mathbf{j}} - 3\hat{\mathbf{k}}) = 0$$

or $\quad (2pm + 1 - 6qn)\hat{\mathbf{i}} + (-6pm - 2 + 15qn)\hat{\mathbf{j}} + (2pm - 6qn)\hat{\mathbf{k}} = 0$

$\Rightarrow 2pm - 6pn + 1 = \mathbf{0},\ -6pm - 2 + 15qn = \mathbf{0},$
$2pm - 6qn = \mathbf{0}$

Solving these, we get

$$p = \frac{1}{(2m)} \text{ and } q = \frac{1}{(3n)}$$

$\therefore \quad 0 < \dfrac{1}{(2m)} \leq 1$ and $0 < \dfrac{1}{(3n)} \leq 1$

$\Rightarrow \quad m \geq \dfrac{1}{2}$ and $n \geq \dfrac{1}{3}$

19. $\because \mathbf{a} = (1, 3, \sin 2\alpha)$

makes an obtuse angle with the Z-axis, therefore its z-component is negative.

$$\therefore \quad \sin 2\alpha < 0$$

$$\Rightarrow \quad -1 \leq \sin 2\alpha < 0 \quad \ldots\text{(i)}$$

But $\quad \mathbf{b} \cdot \mathbf{c} = 0 \quad$ [$\because$ orthogonal]

$$\tan^2 \alpha - \tan \alpha - 6 = 0$$

$$\therefore \quad (\tan \alpha - 3)(\tan \alpha + 2) = 0$$

$$\Rightarrow \quad \tan \alpha = 3, -2$$

Now, $\quad \tan \alpha = 3$

$$\Rightarrow \quad \sin 2\alpha = \frac{2 \tan \alpha}{1 + \tan^2 \alpha} = \frac{6}{1+9} = \frac{3}{5} > 0$$

Impossible $\because \sin 2\alpha < 0$

Now, if $\quad \tan \alpha = -2$

$$\Rightarrow \quad \sin 2\alpha = \frac{2 \tan \alpha}{1 + \tan^2 \alpha} = \frac{-4}{1+4} = -\frac{4}{5} < 0$$

which is true.

$$\therefore \quad \tan 2\alpha > 0$$

$\therefore$ 2α is in third quadrant also $\sqrt{\sin \alpha/2}$ is meaningful, if $0 < \sin \alpha/2 < 1$, then

$$\alpha = (4n+1)\pi - \tan^{-1} 2$$

and $$\alpha = (4n+2)\pi - \tan^{-1} 2$$

20. $\lambda = (OA \cos\theta)\, \hat{\mathbf{b}}$

$$= (|\mathbf{a}| \cos\theta)\, \hat{\mathbf{b}}$$

$$= \frac{(\mathbf{a} \cdot \mathbf{b})}{|\mathbf{b}|} \frac{\mathbf{b}}{|\mathbf{b}|} = \frac{(\mathbf{a} \cdot \mathbf{b})\, \mathbf{b}}{|\mathbf{b}|^2}$$

and $\mu = \mathbf{BA} = \mathbf{OA} - \mathbf{OB}$

$$= \mathbf{a} - \frac{(\mathbf{a} \cdot \mathbf{b})\, \mathbf{b}}{|\mathbf{b}|^2}$$

$$= \frac{(\mathbf{b} \cdot \mathbf{b})\, \mathbf{a} - (\mathbf{a} \cdot \mathbf{b})\, \mathbf{b}}{(\mathbf{b})^2}$$

$$= \frac{\mathbf{b} \times (\mathbf{a} \times \mathbf{b})}{(\mathbf{b})^2}$$

21. $|\mathbf{a}| = |\mathbf{b}| = |\mathbf{c}| = 1$

Given that,

Angle between **a** and **c** = angle between **b** and **c**

$$\Rightarrow \quad \frac{\mathbf{a} \cdot \mathbf{c}}{|\mathbf{a}||\mathbf{c}|} = \mathbf{a} \cdot \mathbf{c} = \cos\theta$$

$$= \frac{\mathbf{b} \cdot \mathbf{c}}{|\mathbf{b}||\mathbf{c}|} = \mathbf{b} \cdot \mathbf{c}$$

Since, $\quad \mathbf{a} \cdot \mathbf{b} = 0$

$$\because \quad \mathbf{c} = \alpha\, \mathbf{a} + \beta\, \mathbf{b} + \gamma\, (\mathbf{a} \times \mathbf{b})$$

$$\therefore \quad \mathbf{a} \cdot \mathbf{c} = \alpha\, (\mathbf{a} \cdot \mathbf{a}) + \beta\, (\mathbf{a} \cdot \mathbf{b}) + \gamma\, \mathbf{a} \cdot (\mathbf{a} \times \mathbf{b})$$

$$= \alpha \cdot 1 + 0 + 0$$

i.e. $\quad \alpha = \cos\theta$

Similarly, $\mathbf{b} \cdot \mathbf{c} = \cos\theta = \beta$

$$\therefore \quad \alpha = \beta = \cos\theta$$

Again, $\quad 1 = \mathbf{c} \cdot \mathbf{c} = 2\alpha^2 + \gamma^2 |\mathbf{a} \times \mathbf{b}|^2$

$$= 2\alpha^2 + \gamma^2 \{|\mathbf{a}|^2 |\mathbf{b}|^2 - (\mathbf{a} \cdot \mathbf{b})^2\}$$

$$= 2\alpha^2 + \gamma^2$$

$$\therefore \quad \gamma^2 = 1 - 2\alpha^2 = 1 - 2\cos^2\theta = -\cos 2\theta$$

$$\Rightarrow \quad \alpha^2 = \beta^2 = \frac{1 - \gamma^2}{2} = \frac{1 + \cos 2\theta}{2}$$

22. Let $\mathbf{A} = (\mathbf{a} \cdot \hat{\mathbf{i}})\, \hat{\mathbf{i}} + (\mathbf{a} \cdot \hat{\mathbf{j}})\, \hat{\mathbf{j}} + (\mathbf{a} \cdot \hat{\mathbf{k}})\, \hat{\mathbf{k}}$

$$= \hat{\mathbf{i}} + \hat{\mathbf{j}} + \hat{\mathbf{k}}$$

and $\mathbf{B} = (\mathbf{b} \cdot \hat{\mathbf{i}})\, \hat{\mathbf{i}} + (\mathbf{b} \cdot \hat{\mathbf{j}})\, \hat{\mathbf{j}} + (\mathbf{c} \cdot \hat{\mathbf{k}})\, \hat{\mathbf{k}} = \hat{\mathbf{i}} - \hat{\mathbf{j}} + 0$

$$\mathbf{C} = \hat{\mathbf{i}} + \hat{\mathbf{j}} - 2\hat{\mathbf{k}}$$

$$\mathbf{A} \cdot \mathbf{B} = 0,\ \mathbf{B} \cdot \mathbf{C} = 0,\ \mathbf{C} \cdot \mathbf{A} = 0$$

$$[\mathbf{A}\ \mathbf{B}\ \mathbf{C}] = \begin{vmatrix} 1 & 1 & 1 \\ 1 & -1 & 1 \\ 1 & 1 & -2 \end{vmatrix}$$

$$= 1\,(2-1) - 1\,(-2-1) + 1\,(1+1) = 6$$

$\therefore$ **A**, **B**, **C** are mutually perpendicular and form a parallelopiped of volume 6 unit.

23. $\because \quad \mathbf{p} = 3\hat{\mathbf{i}} + 4\hat{\mathbf{j}},\ \mathbf{q} = 5\hat{\mathbf{i}},\ \mathbf{r} = \dfrac{1}{4}(\mathbf{p} + \mathbf{q}) = 2\hat{\mathbf{i}} + \hat{\mathbf{j}}$

and $\mathbf{s} = \dfrac{1}{2}(\mathbf{p} - \mathbf{q}) = -\hat{\mathbf{i}} + 2\hat{\mathbf{j}}$

Alternate (a) : $|\mathbf{r} + k\,\mathbf{s}| = |2\hat{\mathbf{i}} + \hat{\mathbf{j}} - k\hat{\mathbf{i}} + 2k\,\hat{\mathbf{j}}|$

$$= |(2-k)\,\hat{\mathbf{i}} + (1+2k)\,\hat{\mathbf{j}}|$$

$$= \sqrt{(2-k)^2 + (1+2k)^2}$$

$$= \sqrt{(5k^2 + 5)}$$

and $\quad |\mathbf{r} - k\,\mathbf{s}| = |2\hat{\mathbf{i}} + \hat{\mathbf{j}} + k\hat{\mathbf{i}} - 2k\,\hat{\mathbf{j}}|$

$$= |(2+k)\,\hat{\mathbf{i}} + (1-2k)\,\hat{\mathbf{j}}|$$

$$= \sqrt{(2+k)^2 + (1-2k)^2}$$

$$= \sqrt{(5 + 5k^2)}$$

Hence, $\quad |\mathbf{r} + k\,\mathbf{s}| = |\mathbf{r} - k\,\mathbf{s}|$

Alternate (b) :

$$\because \quad \mathbf{r} \cdot \mathbf{s} = (2\hat{\mathbf{i}} + \hat{\mathbf{j}}) \cdot (-\hat{\mathbf{i}} + 2\hat{\mathbf{j}})$$

$$= -2 + 2 = 0$$

$\therefore$ **r** is perpendicular to **s**.

Alternate (c):

$$\because \quad (\mathbf{r} + \mathbf{s}) \cdot (\mathbf{r} - \mathbf{s}) = r^2 - s^2$$

$$= 5 - 5 = 0$$

$\therefore$ **r** + **s** is perpendicular to **r** − **s**

Alternate (d) :

$\therefore \quad |\mathbf{p}| = 5, |\mathbf{q}| = 5, |\mathbf{r}| = \sqrt{5}$

and $\quad |\mathbf{s}| = \sqrt{5}$

$\therefore \quad |\mathbf{p}| = |\mathbf{q}| \neq |\mathbf{r}| = |\mathbf{s}|$

24. We are given that

$$\mathbf{a}\times\mathbf{b}+\mathbf{b}\times\mathbf{c}=p\mathbf{a}+q\mathbf{b}+r\mathbf{c} \quad \text{...(i)}$$

Taking dot product with **a**, we have

$$[\mathbf{a}\ \mathbf{b}\ \mathbf{c}]=p\mathbf{a}\cdot\mathbf{a}+q\mathbf{a}\cdot\mathbf{b}+r\mathbf{a}\cdot\mathbf{c}$$
$$=p+q\cos\theta+r\cos\theta \quad \text{...(ii)}$$

Since, $\mathbf{a}\cdot\mathbf{a}=|\mathbf{a}|^2=1$

$$\mathbf{a}\cdot\mathbf{b}=|\mathbf{a}||\mathbf{b}|\cos\theta=\cos\theta$$

and $\mathbf{a}\cdot\mathbf{c}=\cos\theta$.

Similarly, taking dot product with **b** and **c** we get

$$0=p\cos\theta+q+r\cos\theta \quad \text{...(iii)}$$

and $[\mathbf{a}\ \mathbf{b}\ \mathbf{c}]=p\cos\theta+q\cos\theta+r$ ...(iv)

Therefore, from Eqs. (ii), (iv) we get

$$p=r$$

Adding Eqs. (ii), (iii) and (iv), we get

$$2[\mathbf{a}\ \mathbf{b}\ \mathbf{c}]=(2\cos\theta+1)(p+q+r)$$
$$\Rightarrow \frac{2[\mathbf{a}\ \mathbf{b}\ \mathbf{c}]}{2\cos\theta+1}=p+q+r \quad \text{...(v)}$$

Multiplying Eq. (v) by $\cos\theta$ and subtracting from (ii) we get

$$[\mathbf{a}\ \mathbf{b}\ \mathbf{c}]-\frac{2\cos\theta[\mathbf{a}\ \mathbf{b}\ \mathbf{c}]}{2\cos\theta+1}=p(1-\cos\theta)$$
$$\Rightarrow \frac{[\mathbf{a}\ \mathbf{b}\ \mathbf{c}]}{(2\cos\theta+1)(1-\cos\theta)}=p$$

Similarly, $q=\dfrac{-2[\mathbf{a}\ \mathbf{b}\ \mathbf{c}]\cos\theta}{(2\cos\theta+1)(1-\cos\theta)}$

Now, $[\mathbf{a}\ \mathbf{b}\ \mathbf{c}]^2=\begin{vmatrix}\mathbf{a}\cdot\mathbf{a} & \mathbf{a}\cdot\mathbf{b} & \mathbf{a}\cdot\mathbf{c}\\ \mathbf{b}\cdot\mathbf{a} & \mathbf{b}\cdot\mathbf{b} & \mathbf{b}\cdot\mathbf{c}\\ \mathbf{c}\cdot\mathbf{a} & \mathbf{c}\cdot\mathbf{b} & \mathbf{c}\cdot\mathbf{c}\end{vmatrix}=\begin{vmatrix}1 & \cos\theta & \cos\theta\\ \cos\theta & 1 & \cos\theta\\ \cos\theta & \cos\theta & 1\end{vmatrix}$

$$=(1+2\cos\theta)\begin{vmatrix}1 & 1 & 1\\ \cos\theta & 1 & \cos\theta\\ \cos\theta & \cos\theta & 1\end{vmatrix}$$
$$=(1+2\cos\theta)\begin{vmatrix}1 & 0 & 0\\ \cos\theta & 1-\cos\theta & 0\\ \cos\theta & 0 & 1-\cos\theta\end{vmatrix}$$
$$=(1+2\cos\theta)(1-\cos\theta)^2$$
$$\Rightarrow [\mathbf{a}\ \mathbf{b}\ \mathbf{c}]=(\sqrt{1+2\cos\theta})(1-\cos\theta)$$

(since θ is acute, $0<\cos\theta<1$)

Hence, $p=r=\dfrac{1}{\sqrt{1+2\cos\theta}},\ q=-\dfrac{2\cos\theta}{\sqrt{1+2\cos\theta}}$

25. $\because\ |\mathbf{u}|=|\mathbf{v}|=|\mathbf{w}|=1$

$\because\ \mathbf{u}+\mathbf{v}+\mathbf{w}=\mathbf{a}$

$$\Rightarrow \mathbf{a}\cdot\mathbf{u}+\mathbf{a}\cdot\mathbf{v}+\mathbf{a}\cdot\mathbf{w}=\mathbf{a}\cdot\mathbf{a}$$
$$\Rightarrow \frac{3}{2}+\frac{7}{4}+\mathbf{a}\cdot\mathbf{w}=4$$
$$\therefore \mathbf{a}\cdot\mathbf{w}=\frac{3}{4}$$
$$\Rightarrow (\mathbf{u}+\mathbf{v})^2=(\mathbf{a}-\mathbf{w})^2$$
$$\Rightarrow u^2+v^2+2\mathbf{u}\cdot\mathbf{v}=a^2+w^2-2\mathbf{a}\cdot\mathbf{w}$$
$$\Rightarrow 1+1+2\mathbf{u}\cdot\mathbf{v}=4+1-\frac{3}{2} \quad \left[\because \mathbf{a}\cdot\mathbf{w}=\frac{3}{4}\right]$$
$$\therefore \mathbf{u}\cdot\mathbf{v}=\frac{3}{4}$$

and $(\mathbf{v}+\mathbf{w})^2=(\mathbf{a}-\mathbf{u})^2$

$$\Rightarrow v^2+w^2+2\mathbf{v}\cdot\mathbf{w}=a^2+u^2-2\mathbf{a}\cdot\mathbf{u}$$
$$\Rightarrow 1+1+2\mathbf{v}\cdot\mathbf{w}=4+1-3 \quad \left[\because \mathbf{a}\cdot\mathbf{u}=\frac{3}{2}\right]$$
$$\therefore \mathbf{v}\cdot\mathbf{w}=0$$

Also, $(\mathbf{w}+\mathbf{u})^2=(\mathbf{a}-\mathbf{v})^2$

$$\therefore w^2+u^2+2\mathbf{w}\cdot\mathbf{u}=a^2+v^2-2\mathbf{a}\cdot\mathbf{v}$$
$$\Rightarrow 1+1+2\mathbf{w}\cdot\mathbf{u}=4+1-\frac{7}{2} \quad \left[\because \mathbf{a}\cdot\mathbf{v}=\frac{7}{4}\right]$$
$$\Rightarrow 2\mathbf{w}\cdot\mathbf{u}=3-\frac{7}{2}=-\frac{1}{2}$$
$$\therefore \mathbf{w}\cdot\mathbf{u}=-\frac{1}{4}$$

26. Since, **a** and **b** are equally inclined to **c**, therefore **c** must be of the form

$$t\left(\frac{\mathbf{a}}{|\mathbf{a}|}+\frac{\mathbf{b}}{|\mathbf{b}|}\right)$$

Now, $\dfrac{|\mathbf{b}|}{|\mathbf{a}|+|\mathbf{b}|}\mathbf{a}+\dfrac{|\mathbf{a}|}{|\mathbf{a}|+|\mathbf{b}|}\mathbf{b}$

$$=\frac{|\mathbf{a}||\mathbf{b}|}{|\mathbf{a}|+|\mathbf{b}|}\left(\frac{\mathbf{a}}{|\mathbf{a}|}+\frac{\mathbf{b}}{|\mathbf{b}|}\right)$$

Also, $\dfrac{|\mathbf{b}|}{2|\mathbf{a}|+|\mathbf{b}|}\mathbf{a}+\dfrac{|\mathbf{a}|}{2|\mathbf{a}|+|\mathbf{b}|}\mathbf{b}$

$$=\frac{|\mathbf{a}||\mathbf{b}|}{2|\mathbf{a}|+|\mathbf{b}|}\left(\frac{\mathbf{a}}{|\mathbf{a}|}+\frac{\mathbf{b}}{|\mathbf{b}|}\right)$$

Other two vectors cannot be written in the form $t\left(\dfrac{\mathbf{a}}{|\mathbf{a}|}+\dfrac{\mathbf{b}}{|\mathbf{b}|}\right)$.

27. We have, $\mathbf{A}+\mathbf{B}=\mathbf{a}$

or $\mathbf{A}\cdot\mathbf{a}+\mathbf{B}\cdot\mathbf{a}=\mathbf{a}\cdot\mathbf{a}$

or $1+\mathbf{B}\cdot\mathbf{a}=a^2$ [given $\mathbf{A}\cdot\mathbf{a}=1$]

or $\mathbf{B}\cdot\mathbf{a}=a^2-1$ ...(i)

Also, $\mathbf{A}\times\mathbf{B}=\mathbf{b}$

or $\mathbf{a}\times(\mathbf{A}\times\mathbf{B})=\mathbf{a}\times\mathbf{b}$

or $(\mathbf{a}\cdot\mathbf{B})\mathbf{A}-(\mathbf{a}\cdot\mathbf{A})\mathbf{B}=\mathbf{a}\times\mathbf{b}$

or $(a^2-1)\mathbf{A}-\mathbf{B}=\mathbf{a}\times\mathbf{b}$

(using (i) and $\mathbf{a}\cdot\mathbf{A}=1$) ...(ii)

and $\mathbf{A}+\mathbf{B}=a$ ...(iii)

From Eqs. (ii) and (iii), we have

$$\mathbf{A}=\frac{(\mathbf{a}\times\mathbf{b})+\mathbf{a}}{a^2}$$
$$\mathbf{B}=\mathbf{a}-\left\{\frac{(\mathbf{a}\times\mathbf{b})+\mathbf{a}}{a^2}\right\}$$
$$=\frac{(\mathbf{b}\times\mathbf{a})+\mathbf{a}(a^2-1)}{a^2}$$

Thus, $\mathbf{A}=\dfrac{(\mathbf{a}\times\mathbf{b})+\mathbf{a}}{a^2}$

and $\mathbf{B}=\dfrac{(\mathbf{b}\times\mathbf{a})+\mathbf{a}(a^2-1)}{a^2}$

28. The vector equation of the line through P and parallel to **A** is

$$\mathbf{r} = (5\hat{\mathbf{i}} + 7\hat{\mathbf{j}} - 2\hat{\mathbf{k}}) + t\,(3\hat{\mathbf{i}} - \hat{\mathbf{j}} + \hat{\mathbf{k}})$$

$$= (5+3t)\hat{\mathbf{i}} + (7-t)\hat{\mathbf{j}} - (2-t)\hat{\mathbf{k}} \quad \ldots \text{(i)}$$

The vector equation of the line through Q and parallel to **B** is

$$\mathbf{r} = (-3\hat{\mathbf{i}} + 3\hat{\mathbf{j}} + 6\hat{\mathbf{k}}) + t'\,(-3\hat{\mathbf{i}} + 2\hat{\mathbf{j}} + 4\hat{\mathbf{k}})$$

$$= -(3+3t')\hat{\mathbf{i}} + (3+2t')\hat{\mathbf{j}} + (6+4t')\hat{\mathbf{k}} \quad \ldots \text{(ii)}$$

Let the third vector $2\hat{\mathbf{i}} + 7\hat{\mathbf{j}} - 5\hat{\mathbf{k}}$ intersects line (i) and (ii) in L and M. So,

$$m\,(2\hat{\mathbf{i}} + 7\hat{\mathbf{j}} - 5\hat{\mathbf{k}}) = \mathbf{LM} = \text{PV of } M - \text{PV of } L$$

$$= (-3-3t'-5-3t)\hat{\mathbf{i}} + (3+2t'-7+t)\hat{\mathbf{j}} + (6+4t'+2-t)\hat{\mathbf{k}}$$

Thus, $-8-3t-3t' = 2m$, $-4+2t'+t = 7m$,

$8+4t'-t = -5m$

Adding Eqs. (ii) and (iii), we get

$$4+6t' = 2m$$

Multiplying the Eq. (ii) with (iii) and adding to Eq (i)

We have, $-20+3t' = 23m$

Solving these we get

$$m = -1, t = -1, t' = -1$$

Thus, the PV of L is $2\hat{\mathbf{i}} + 8\hat{\mathbf{j}} - 3\hat{\mathbf{k}}$ and the PV of M is $\hat{\mathbf{j}} + 2\hat{\mathbf{k}}$.

29. The two diagonals are given by

$\mathbf{AB} - \mathbf{BC} = 4\hat{\mathbf{i}} - 2\hat{\mathbf{j}} + 4\hat{\mathbf{k}}$ and $\mathbf{AB} + \mathbf{BC} = 2\hat{\mathbf{i}} - 2\hat{\mathbf{j}}$

These vectors have magnitudes 6 and $2\sqrt{2}$ respectively and their dot product is 12.

Therefore, the angle between them is

$$\cos^{-1}\frac{12}{6\,(2\sqrt{2})} = \cos^{-1}\frac{1}{\sqrt{2}} = \frac{\pi}{4}$$

or π minus this values i.e. $\frac{3\pi}{4}$.

30. $$\begin{vmatrix} a & 2a & -3a \\ 2a+1 & 2a+3 & a+1 \\ 3a+5 & a+5 & a+2 \end{vmatrix}$$

$$= a\begin{vmatrix} 1 & 2 & -3 \\ 2a+1 & 2a+3 & a+1 \\ 3a+5 & a+5 & a+2 \end{vmatrix}$$

$$= a\begin{vmatrix} 1 & 2 & -3 \\ 0 & -2a+1 & 7a+4 \\ 0 & -5(a+1) & 10a+17 \end{vmatrix}$$

$$= a\,(15a^2 + 31a + 37) \neq 0 \text{ for } a \neq 0$$

Hence, the three vectors are coplanar only for $a = 0$

31. Let A, B, C and D be the points with the given position vectors. Then, $\mathbf{AB} = -2\hat{\mathbf{i}} + 3\hat{\mathbf{j}} - 3\hat{\mathbf{k}}$,

$\mathbf{AC} = 4\hat{\mathbf{i}} + 5\hat{\mathbf{j}} + (\lambda - 10)\hat{\mathbf{k}}$ and $\mathbf{AD} = 6\hat{\mathbf{i}} + 2\hat{\mathbf{j}} - 3\hat{\mathbf{k}}$, so that the volume of the tetrahedron is

$$\frac{1}{6}|(\mathbf{AB} \times \mathbf{AC}) \cdot \mathbf{AD}| = \frac{1}{6}\left|\begin{vmatrix} -2 & 3 & -3 \\ 4 & 5 & \lambda-10 \\ 6 & 2 & -3 \end{vmatrix}\right|$$

$$= \frac{1}{6}|(-2)[-15 - 2(\lambda - 10)] - 3[-12 - 6(\lambda - 10)] - 3(8 - 30)|$$

$$= \frac{1}{6}|-88 + 22\lambda| = 11 \text{ (given)}$$

This gives $\lambda = 1$ or 7.

Linked Comprehension Type

PASSAGE 1

$$\because \quad [\mathbf{abc}][\mathbf{uvw}] = \begin{vmatrix} \mathbf{a}\cdot\mathbf{u} & \mathbf{b}\cdot\mathbf{u} & \mathbf{c}\cdot\mathbf{u} \\ \mathbf{a}\cdot\mathbf{v} & \mathbf{b}\cdot\mathbf{v} & \mathbf{c}\cdot\mathbf{v} \\ \mathbf{a}\cdot\mathbf{w} & \mathbf{b}\cdot\mathbf{w} & \mathbf{c}\cdot\mathbf{w} \end{vmatrix}$$

$$\therefore \quad [\mathbf{abc}]^2 = [\mathbf{abc}][\mathbf{abc}]$$

$$= \begin{vmatrix} \mathbf{a}\cdot\mathbf{a} & \mathbf{b}\cdot\mathbf{a} & \mathbf{c}\cdot\mathbf{a} \\ \mathbf{a}\cdot\mathbf{b} & \mathbf{b}\cdot\mathbf{b} & \mathbf{c}\cdot\mathbf{b} \\ \mathbf{a}\cdot\mathbf{c} & \mathbf{b}\cdot\mathbf{c} & \mathbf{c}\cdot\mathbf{c} \end{vmatrix} \quad \ldots \text{(i)}$$

Now, $\mathbf{a}\cdot\mathbf{a} = a^2 = |\mathbf{a}|^2 = 16$

$$\mathbf{a}\cdot\mathbf{b} = \mathbf{b}\cdot\mathbf{a} = |\mathbf{a}||\mathbf{b}|\cos\pi/3 = 4\cdot4\cdot\frac{1}{2} = 8$$

$$\mathbf{a}\cdot\mathbf{c} = \mathbf{c}\cdot\mathbf{a} = |\mathbf{a}||\mathbf{c}|\cos\pi/3 = 4\cdot4\cdot\frac{1}{2} = 8$$

$$\mathbf{b}\cdot\mathbf{b} = b^2 = |\mathbf{b}|^2 = 16$$

$$\mathbf{b}\cdot\mathbf{c} = \mathbf{c}\cdot\mathbf{b} = |\mathbf{b}||\mathbf{c}|\cos\pi/3 = 4\cdot4\cdot\frac{1}{2} = 8$$

$$\mathbf{c}\cdot\mathbf{c} = |\mathbf{c}|^2 = 4^2 = 16$$

From Eq. (i),

$$[\mathbf{a}\ \mathbf{b}\ \mathbf{c}]^2 = \begin{vmatrix} 16 & 8 & 8 \\ 8 & 16 & 8 \\ 8 & 8 & 16 \end{vmatrix}$$

$$= 8^3\begin{vmatrix} 2 & 1 & 1 \\ 1 & 2 & 1 \\ 1 & 1 & 2 \end{vmatrix}$$

$$= 8^3 \cdot 4 = 64 \times 32$$

$$\therefore \quad |[\mathbf{a}\ \mathbf{b}\ \mathbf{c}]| = 32\sqrt{2}$$

1. Volume of the parallelopiped $= |[\mathbf{a}\ \mathbf{b}\ \mathbf{c}]| = 32\sqrt{2}$

2. Volume of the parallelopiped = (base area) × (height)

$$32\sqrt{2} = 2 \times \left(\frac{1}{2} \times 4 \times 4 \times \sin\pi/3\right)h$$

$$= 8\sqrt{3} \times h$$

$$\therefore \quad h = 4\sqrt{\frac{2}{3}}$$

3. Volume of the tetrahedron $= \frac{1}{6}[\mathbf{a}\ \mathbf{b}\ \mathbf{c}]$

$$= \frac{1}{6} \times 32\sqrt{2} = \frac{16\sqrt{2}}{3}$$

4. Volume of the tetrahedron $= \frac{1}{3}$ (base area) $\times$ (height)

$$\frac{16\sqrt{2}}{3} = \frac{1}{3}\left(\frac{\sqrt{3}}{4}(4)^2\right) \times h$$

$$\therefore \quad h = \frac{\frac{16\sqrt{2}}{3}}{\frac{4\sqrt{3}}{3}}$$

$$= \frac{4\sqrt{2}}{\sqrt{3}} = 4\sqrt{\frac{2}{3}}$$

5. Volume of the triangular prism $= \frac{1}{2}[\mathbf{a}\ \mathbf{b}\ \mathbf{c}]$

$$= \frac{1}{2} \times (32\sqrt{2}) = 16\sqrt{2}$$

PASSAGE 2

By definition of reciprocal system

$$\mathbf{b} \cdot \mathbf{a}' = \mathbf{c} \cdot \mathbf{a}' = 0$$

i.e. $\mathbf{a}' \perp \mathbf{b}$ and $\mathbf{a}' \perp \mathbf{c}$

$$\Rightarrow \quad \mathbf{a}' \,||\, (\mathbf{b} \times \mathbf{c})$$

$$\therefore \quad \mathbf{a}' = t\,(\mathbf{b} \times \mathbf{c}) \qquad \ldots\text{(i)}$$

Also, $\quad \mathbf{a} \cdot \mathbf{a}' = 1$

$$\Rightarrow \quad 1 = \mathbf{a} \cdot t\,(\mathbf{b} \times \mathbf{c})$$

$$\therefore \quad t = \frac{1}{[\mathbf{a}\ \mathbf{b}\ \mathbf{c}]}$$

Now, from Eq. (i), $\mathbf{a}' = \frac{(\mathbf{b} \times \mathbf{c})}{[\mathbf{a}\ \mathbf{b}\ \mathbf{c}]}$

Similarly, $\quad \mathbf{b}' = \frac{(\mathbf{c} \times \mathbf{a})}{[\mathbf{a}\ \mathbf{b}\ \mathbf{c}]}$ and $\mathbf{c}' = \frac{(\mathbf{a} \times \mathbf{b})}{[\mathbf{a}\ \mathbf{b}\ \mathbf{c}]}$

1. $[\mathbf{a}'\ \mathbf{b}'\ \mathbf{c}'] = \mathbf{a}' \cdot (\mathbf{b}' \times \mathbf{c}')$

$$= \frac{(\mathbf{b} \times \mathbf{c}) \cdot \{(\mathbf{c} \times \mathbf{a}) \times (\mathbf{a} \times \mathbf{b})\}}{[\mathbf{a}\ \mathbf{b}\ \mathbf{c}]^3}$$

$$= \frac{[\mathbf{b} \times \mathbf{c}\ \ \mathbf{c} \times \mathbf{a}\ \ \mathbf{a} \times \mathbf{b}]}{[\mathbf{a}\ \mathbf{b}\ \mathbf{c}]^3}$$

$$= \frac{[\mathbf{a}\ \mathbf{b}\ \mathbf{c}]^2}{[\mathbf{a}\ \mathbf{b}\ \mathbf{c}]^3} = \frac{1}{[\mathbf{a}\ \mathbf{b}\ \mathbf{c}]}$$

$$\therefore \quad [\mathbf{a}\ \mathbf{b}\ \mathbf{c}][\mathbf{a}'\ \mathbf{b}'\ \mathbf{c}'] = 1$$

or $\quad [\mathbf{a}'\ \mathbf{b}'\ \mathbf{c}']^{-1} = [\mathbf{a}\ \mathbf{b}\ \mathbf{c}]$

2. $\because \mathbf{a} \times \mathbf{a}' = \mathbf{a} \times \frac{(\mathbf{b} \times \mathbf{c})}{[\mathbf{a}\ \mathbf{b}\ \mathbf{c}]}$

$$= \frac{(\mathbf{c} \cdot \mathbf{a})\mathbf{b} - (\mathbf{a} \cdot \mathbf{b})\mathbf{c}}{[\mathbf{a}\ \mathbf{b}\ \mathbf{c}]}$$

Similarly, $\mathbf{b} \times \mathbf{b}' = \frac{(\mathbf{a} \cdot \mathbf{b})\mathbf{c} - (\mathbf{b} \cdot \mathbf{c})\mathbf{a}}{[\mathbf{a}\ \mathbf{b}\ \mathbf{c}]}$

and $\quad \mathbf{c} \times \mathbf{c}' = \frac{(\mathbf{b} \cdot \mathbf{c})\mathbf{a} - (\mathbf{c} \cdot \mathbf{a})\mathbf{b}}{[\mathbf{a}\ \mathbf{b}\ \mathbf{c}]}$

Adding all, we get

$$\mathbf{a} \times \mathbf{a}' + \mathbf{b} \times \mathbf{b}' + \mathbf{c} \times \mathbf{c}' = 0$$

3. $\therefore \mathbf{a}' \times \mathbf{b}' = \frac{(\mathbf{b} \times \mathbf{c}) \times (\mathbf{c} \times \mathbf{a})}{[\mathbf{a}\ \mathbf{b}\ \mathbf{c}]^2}$

$$= \frac{(\mathbf{b} \times \mathbf{c} \cdot \mathbf{a})\mathbf{c} - (\mathbf{b} \times \mathbf{c} \cdot \mathbf{c})\mathbf{a}}{[\mathbf{a}\ \mathbf{b}\ \mathbf{c}]^2}$$

$$= \frac{[\mathbf{a}\ \mathbf{b}\ \mathbf{c}]\mathbf{c} - 0}{[\mathbf{a}\ \mathbf{b}\ \mathbf{c}]^2} = \frac{\mathbf{c}}{[\mathbf{a}\ \mathbf{b}\ \mathbf{c}]}$$

Similarly, $\quad \mathbf{b}' \times \mathbf{c}' = \frac{\mathbf{a}}{[\mathbf{a}\ \mathbf{b}\ \mathbf{c}]}$

and $\quad \mathbf{c}' \times \mathbf{a}' = \frac{\mathbf{b}}{[\mathbf{a}\ \mathbf{b}\ \mathbf{c}]}$

$$\therefore \ (\mathbf{a}' \times \mathbf{b}') + (\mathbf{b}' \times \mathbf{c}') + (\mathbf{c}' \times \mathbf{a}') = \frac{[\mathbf{a} + \mathbf{b} + \mathbf{c}]}{[\mathbf{a}\ \mathbf{b}\ \mathbf{c}]}$$

4. $\because \quad \mathbf{e}^1 = \frac{\mathbf{e}_2 \times \mathbf{e}_3}{[\mathbf{e}_1\ \mathbf{e}_2\mathbf{e}_3]}$

$$\therefore \quad \mathbf{a} \cdot \mathbf{e}^1 = \frac{[\mathbf{a}\ \mathbf{e}_2\mathbf{e}_3]}{[\mathbf{e}_1\ \mathbf{e}_2\mathbf{e}_3]}$$

$$\Rightarrow \quad (\mathbf{a} \cdot \mathbf{e}^1)\mathbf{e}_1 = \frac{[\mathbf{a}\mathbf{e}_2\mathbf{e}_3]}{[\mathbf{e}_1\ \mathbf{e}_2\mathbf{e}_3]}\mathbf{e}_1$$

Similarly,

$$(\mathbf{a} \cdot \mathbf{e}^2)\mathbf{e}_2 = \frac{[\mathbf{a}\mathbf{e}_2\mathbf{e}_3]}{[\mathbf{e}_1\ \mathbf{e}_2\mathbf{e}_3]}\mathbf{e}_2$$

and $\quad (\mathbf{a} \cdot \mathbf{e}^3)\mathbf{e}_3 = \frac{[\mathbf{a}\mathbf{e}_1\mathbf{e}_2]}{[\mathbf{e}_1\ \mathbf{e}_2\mathbf{e}_3]}\mathbf{e}_3$

$$\therefore \ (\mathbf{a} \cdot \mathbf{e}^1)\mathbf{e}_1 + (\mathbf{a} \cdot \mathbf{e}^2)\mathbf{e}_2 + (\mathbf{a} \cdot \mathbf{e}^3)\mathbf{e}_3$$

$$= \frac{[\mathbf{a}\mathbf{e}_2\mathbf{e}_3]\,\mathbf{e}_1 + [\mathbf{a}\mathbf{e}_3\mathbf{e}_1]\,\mathbf{e}_2 + [\mathbf{a}\mathbf{e}_1\mathbf{e}_2]\,\mathbf{e}_3}{[\mathbf{e}_1\ \mathbf{e}_2\mathbf{e}_3]}$$

$$= \frac{[\mathbf{e}_1\mathbf{e}_2\mathbf{e}_3]\,\mathbf{a}}{[\mathbf{e}_1\ \mathbf{e}_2\mathbf{e}_3]}$$

$$= \mathbf{a}$$

5. Let $\quad \mathbf{b} \times \mathbf{c} = x\,\mathbf{a} + y\,\mathbf{b} + z\,\mathbf{c} \qquad \ldots\text{(i)}$

$$\therefore \quad (\mathbf{b} \times \mathbf{c}) \cdot (\mathbf{b} \times \mathbf{c}) = x\,(\mathbf{b} \times \mathbf{c}) \cdot \mathbf{a} + y\ (\mathbf{b} \times \mathbf{c}) \cdot \mathbf{b} + z\,(\mathbf{b} \times \mathbf{c}) \cdot \mathbf{c}$$

$$[\mathbf{a}\ \mathbf{b}\ \mathbf{c}]^2\,(\mathbf{a}' \cdot \mathbf{a}') = x[\mathbf{a}\ \mathbf{b}\ \mathbf{c}] + 0 + 0$$

$$\therefore \quad x = (\mathbf{a}' \cdot \mathbf{a}')[\mathbf{a}\ \mathbf{b}\ \mathbf{c}]$$

Similarly, $\quad y = (\mathbf{a}' \cdot \mathbf{b}')[\mathbf{a}\ \mathbf{b}\ \mathbf{c}]$

and $\quad z = (\mathbf{a}' \cdot \mathbf{c}')[\mathbf{a}\ \mathbf{b}\ \mathbf{c}]$

From Eq. (i),

$$\mathbf{b} \times \mathbf{c} = [\mathbf{a}\ \mathbf{b}\ \mathbf{c}]\,[(\mathbf{a}' \cdot \mathbf{a}')\mathbf{a} + (\mathbf{a}' \cdot \mathbf{b}')\mathbf{b} + (\mathbf{a}' \cdot \mathbf{c}')\mathbf{c}]$$

PASSAGE 3

1. $\frac{\mathbf{OA} + \mathbf{OB} + \mathbf{OC}}{3} = \mathbf{OG} = \frac{1}{3}\ \mathbf{OO'}$

$$\therefore \mathbf{OA} + \mathbf{OB} + \mathbf{OC} = \mathbf{OO'}$$

2. $\frac{\mathbf{O'A} + \mathbf{O'B} + \mathbf{O'C}}{3} = \mathbf{O'G} = \frac{2}{3}\ \mathbf{O'O}$

$$\therefore \quad \mathbf{O'A} + \mathbf{O'B} + \mathbf{O'C} = 2\mathbf{O'O}$$

3. $\mathbf{O'B} + \mathbf{O'C} = 2\mathbf{O'D}$

$$= 2(\mathbf{O'A} + \mathbf{AO} + \mathbf{OD})$$

$$= 2\mathbf{O'A} + 2\mathbf{AO} + \mathbf{AO'} \qquad [\because \mathbf{AO'} = 2\mathbf{OD}]$$

$$= -2\mathbf{AO'} + 2\mathbf{AO} + \mathbf{AO'}$$

$$= -\mathbf{AO'} + 2\mathbf{AO}$$

$$\therefore \quad \mathbf{AO'} + \mathbf{O'B} + \mathbf{O'C} = 2\,\mathbf{AO}$$

4. $\dfrac{\mathbf{NA}+\mathbf{NB}+\mathbf{NC}}{3}=\mathbf{NG}$

$$\Rightarrow \quad \mathbf{NA}+\mathbf{NB}+\mathbf{NC}=3\mathbf{NG}=3(\mathbf{NO}-\mathbf{GO})=3\left(\frac{\mathbf{O'O}}{2}-\frac{\mathbf{O'O}}{3}\right)=\frac{1}{2}\mathbf{O'O}$$

5. $\dfrac{\mathbf{SA}+\mathbf{SB}+\mathbf{SC}}{3}=\mathbf{SG}$

$\therefore\ \mathbf{SA}+\mathbf{SB}+\mathbf{SC}=3\mathbf{SG}$

PASSAGE 4

1. The vector along the bisector of the given angle is

$$\mathbf{d}=\lambda\left(\frac{\mathbf{b}}{|\mathbf{b}|}+\frac{\mathbf{c}}{|\mathbf{c}|}\right)=\lambda\left(\frac{7\hat{\mathbf{i}}-4\hat{\mathbf{j}}-4\hat{\mathbf{k}}}{\sqrt{81}}+\frac{-2\hat{\mathbf{i}}-\hat{\mathbf{j}}+2\hat{\mathbf{k}}}{\sqrt{9}}\right)=\lambda\left(\frac{\hat{\mathbf{i}}-7\hat{\mathbf{j}}+2\hat{\mathbf{k}}}{9}\right) \quad \ldots(i)$$

$$\therefore \quad |\mathbf{d}|=\lambda\sqrt{\left(\frac{54}{81}\right)}$$

$$\Rightarrow \quad 5\sqrt{6}=\lambda\cdot\frac{3\sqrt{6}}{9}$$

$$\therefore \quad \lambda=15$$

Then, from Eq. (i),

$$\mathbf{d}=\frac{5}{3}(\hat{\mathbf{i}}-7\hat{\mathbf{j}}+2\hat{\mathbf{k}})$$

2. In centre is the point of intersection of the internal bisectors of angles A, B and C. Equations of the bisectors of angles A and B are

$$\mathbf{r}=\mathbf{a}+\lambda\left(\frac{\mathbf{c}-\mathbf{a}}{b}+\frac{\mathbf{b}-\mathbf{a}}{c}\right) \quad \ldots(i)$$

and $\mathbf{r}=\mathbf{b}+\mu\left(\dfrac{\mathbf{a}-\mathbf{b}}{c}+\dfrac{\mathbf{c}-\mathbf{b}}{a}\right)$ $\quad\ldots$(ii)

Now, equating the coefficients of **a**, **b** and **c**, then

$$1-\frac{\lambda}{b}-\frac{\lambda}{c}=\frac{\mu}{c} \quad \ldots(iii)$$

$$\frac{\lambda}{c}=\frac{\mu}{c}-\frac{\mu}{a} \quad \ldots(iv)$$

and $\dfrac{\lambda}{b}=\dfrac{\mu}{a}$ $\quad\ldots$(v)

From Eqs. (iii) and (v), we get

$$\therefore \quad 1-\frac{\lambda}{b}-\frac{\lambda}{c}=\frac{1}{c}\left(\frac{a\lambda}{b}\right)$$

$$\Rightarrow \quad 1=\lambda\left(\frac{1}{b}+\frac{1}{c}+\frac{a}{bc}\right)$$

$$\therefore \quad \lambda=\left(\frac{bc}{c+b+a}\right)$$

Now, from Eq. (i), we get

$$\mathbf{r}=\left(\frac{a\,\mathbf{a}+b\,\mathbf{b}+c\,\mathbf{c}}{a+b+c}\right)$$

3. Let A be the initial point and the PV of B and C be **b** and **c** respectively. Hence, let $AB=c$ and $AC=b$. The internal bisector is $\mathbf{r}=t\left(\dfrac{\mathbf{b}}{c}+\dfrac{\mathbf{c}}{b}\right)$ and the equation of the line BC is $\mathbf{r}=\mathbf{b}+\lambda(\mathbf{c}-\mathbf{b})$

and hence the position vector of D is $\left(\dfrac{b\,\mathbf{b}+c\,\mathbf{c}}{b+c}\right)$

$$\Rightarrow \quad \mathbf{BD}=\left(\frac{b\,\mathbf{b}+c\,\mathbf{c}}{b+c}\right)-\mathbf{b}=\frac{c(\mathbf{c}-\mathbf{b})}{(b+c)}$$

$$\Rightarrow \quad BD=|\mathbf{BD}|=\frac{c}{(b+c)}|\mathbf{c}-\mathbf{b}|=\frac{ca}{(b+c)}$$

Similarly, $BE=|\mathbf{BE}|=\dfrac{c}{(c-b)}|\mathbf{c}-\mathbf{b}|=\dfrac{ca}{(c-b)}$

$$\therefore \quad \frac{1}{BE}+\frac{1}{BD}=\frac{c-b}{ca}+\frac{b+c}{ca}=\frac{2c}{ca}=\frac{2}{a}=\frac{2}{BC}$$

$$\Rightarrow \quad \frac{2}{BC}=\frac{1}{BE}+\frac{1}{BD}$$

4. $\mathbf{AB}=\mathbf{b}-\mathbf{a}$, $\mathbf{BC}=\mathbf{c}-\mathbf{b}$, $\mathbf{CA}=\mathbf{a}-\mathbf{c}$ the bisectors of the angular B and C are

$$\mathbf{r}=\mathbf{b}+\lambda\left(\frac{\mathbf{b}-\mathbf{a}}{c}+\frac{\mathbf{c}-\mathbf{b}}{a}\right) \quad \ldots(i)$$

and $\mathbf{r}=\mathbf{c}+\mu\left(\dfrac{\mathbf{b}-\mathbf{c}}{a}+\dfrac{\mathbf{c}-\mathbf{a}}{b}\right)$ $\quad\ldots$(ii)

So, that for the point P

$$\mathbf{b}+\lambda\left(\frac{\mathbf{b}-\mathbf{a}}{c}+\frac{\mathbf{c}-\mathbf{b}}{a}\right)=\mathbf{c}+\mu\left(\frac{\mathbf{b}-\mathbf{c}}{a}+\frac{\mathbf{c}-\mathbf{a}}{b}\right)$$

$$\therefore \quad -\frac{\lambda}{c}=-\frac{\mu}{b} \text{ and } 1+\frac{\lambda}{c}-\frac{\lambda}{a}=\frac{\mu}{a}$$

Then we get, $\lambda=\dfrac{ac}{(b+c-a)}$

From Eq. (i),

$$\mathbf{r}=\mathbf{b}+\frac{ac}{(b+c-a)}\left(\frac{\mathbf{b}-\mathbf{a}}{c}+\frac{\mathbf{c}-\mathbf{b}}{a}\right)=\frac{\mathbf{b}(b+c-a)+a\mathbf{b}-a\mathbf{a}+c\mathbf{c}-c\mathbf{b}}{(b+c-a)}=\frac{b\,\mathbf{b}+c\,\mathbf{c}-a\,\mathbf{a}}{b+c-a}=\left(\frac{a\,\mathbf{a}-b\,\mathbf{b}-c\,\mathbf{c}}{a-b-c}\right)$$

5. Since, the three points A, B, C are coplanar. Then, all the points lying as the linear AB, BC, CD are really coplanar.

PASSAGE 5

1. (c), **2.** (c)

Given $ABCDE$ is a regular pentagon

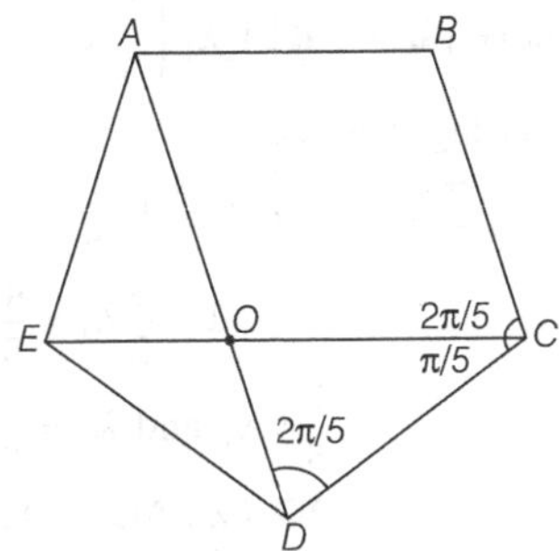

Let position vector of point A and C be $\mathbf{a}$ and $\mathbf{b}$, respectively.

AD is parallel to BC and AB is parallel to EC. Therefore,

$AOCB$ is a parallelogram

and position vector of B is $\mathbf{a}+\mathbf{b}$

The position vectors of E and D are $\lambda\mathbf{b}$ and $\lambda\mathbf{a}$ respectively.

Also $\quad OA = BC = AB = OC = 1 \quad$ (let)

Therefore, $AOCB$ is rhombus.

$$\angle ABC = \angle AOC = \frac{3\pi}{5}$$

and $$\angle OAB = \angle BCO = \pi - \frac{3\pi}{5} = \frac{2\pi}{5}$$

Further, $OA = AE = 1$ and $OC = CD = 1$

Thus, ΔEAO and ΔOCD are isosceles.

In ΔOCD, using sine rule we get

$$\frac{OC}{\sin\frac{2\pi}{5}} = \frac{OD}{\sin\frac{\pi}{5}} \Rightarrow OD = \frac{1}{2\cos\frac{\pi}{5}} = OE$$

$$\Rightarrow \quad AD = OA + OD = 1 + \frac{1}{2\cos\frac{\pi}{5}}$$

$$\Rightarrow \quad \frac{AD}{BC} = 1 + \frac{1}{2\cos\frac{\pi}{5}} = \frac{1+2\cos\frac{\pi}{5}}{2\cos\frac{\pi}{5}}$$

And $$\frac{OE}{OC} = \frac{1}{2\cos\frac{\pi}{5}}$$

PASSAGE 6

1. $\cos(P+Q)+\cos(Q+R)+\cos(R+P)$

$$= -(\cos R + \cos P + \cos Q)$$

Maximum of $\cos P + \cos Q + \cos R = \frac{3}{2}$

Minimum of $\cos(P+Q)+\cos(Q+R)+\cos(R+P)$ is $-\frac{3}{2}$

2. $|\mathbf{OX}\times\mathbf{OY}| = |\mathbf{OX}||\mathbf{OY}|\sin R$

$|\mathbf{OX}\times\mathbf{OY}| = \sin R$

$\because \quad \sin R = \sin(P+Q)$

$\therefore |\mathbf{OX}\times\mathbf{OY}| = \sin(P+Q)$

Numerical Value Type

1. Let position vectors of A, B, D be $\mathbf{0}$, $\mathbf{b}$ and $\mathbf{d}$ respectively.

$$\therefore \quad \mathbf{AC} = \mathbf{AB} + \mathbf{BC}$$
$$= \mathbf{AB} + \mathbf{AD}$$
$$= \mathbf{b} + \mathbf{d}$$
$$\mathbf{AA_1} = \frac{\mathbf{AB} + \mathbf{AC}}{2}$$
$$= \mathbf{b} + \frac{\mathbf{d}}{2}$$

and $$\mathbf{AB_1} = \frac{\mathbf{AD} + \mathbf{AC}}{2} = \mathbf{d} + \frac{\mathbf{b}}{2}$$

$$\therefore \quad \mathbf{AA_1} + \mathbf{AB_1} = \frac{3}{2}(\mathbf{b}+\mathbf{d}) = \frac{3}{2}\mathbf{AC}$$

$$\therefore \quad \lambda = \frac{3}{2} = 1.5$$

2. Let PV of A, B, C and D be $\mathbf{a}$, $\mathbf{b}$, $\mathbf{c}$ and $\mathbf{0}$

$\therefore \mathbf{AB}\times\mathbf{CD} = (\mathbf{b}-\mathbf{a})\times -\mathbf{c} = -\mathbf{b}\times\mathbf{c} + \mathbf{a}\times\mathbf{c}$

$\mathbf{BC}\times\mathbf{AD} = (\mathbf{c}-\mathbf{b})\times -\mathbf{a} = -\mathbf{c}\times\mathbf{a} + \mathbf{b}\times\mathbf{a}$

and $\mathbf{CA}\times\mathbf{BD} = (\mathbf{a}-\mathbf{c})\times -\mathbf{b} = -\mathbf{a}\times\mathbf{b} + \mathbf{c}\times\mathbf{b}$

Adding all, we get

$$\mathbf{AB}\times\mathbf{CD} + \mathbf{BC}\times\mathbf{AD} + \mathbf{CA}\times\mathbf{BD}$$
$$= -2(\mathbf{a}\times\mathbf{b} + \mathbf{b}\times\mathbf{c} + \mathbf{c}\times\mathbf{a})$$

$$\therefore \quad |\mathbf{AB}\times\mathbf{CD} + \mathbf{BC}\times\mathbf{AD} + \mathbf{CA}\times\mathbf{BD}|$$
$$= 2|\mathbf{a}\times\mathbf{b} + \mathbf{b}\times\mathbf{c} + \mathbf{c}\times\mathbf{a}|$$
$$= 2|(\mathbf{c}-\mathbf{a})\times(\mathbf{b}-\mathbf{a})|$$
$$= 2|\mathbf{AC}\times\mathbf{AB}| = 4\cdot\frac{1}{2}|\mathbf{AC}\times\mathbf{AB}|$$
$$= 4 \text{ (Area of } \Delta ABC)$$

$$\therefore \quad \lambda = 4$$

Then, $125\lambda = 500$

3. $\because \mathbf{AB} = (2\hat{\mathbf{i}} + 3\hat{\mathbf{j}} - 4\hat{\mathbf{k}}) - (3\hat{\mathbf{i}} - 2\hat{\mathbf{j}} - \hat{\mathbf{k}})$

$$= -\hat{\mathbf{i}} + 5\hat{\mathbf{j}} - 3\hat{\mathbf{k}}$$

$$\mathbf{AC} = (-\hat{\mathbf{i}} + \hat{\mathbf{j}} + 2\hat{\mathbf{k}}) - (3\hat{\mathbf{i}} - 2\hat{\mathbf{j}} - \hat{\mathbf{k}})$$
$$= -4\hat{\mathbf{i}} + 3\hat{\mathbf{j}} + 3\hat{\mathbf{k}}$$

and $\mathbf{AD} = (4\hat{\mathbf{i}} + 5\hat{\mathbf{j}} + \lambda\hat{\mathbf{k}}) - (3\hat{\mathbf{i}} - 2\hat{\mathbf{j}} - \hat{\mathbf{k}})$

$$= \hat{\mathbf{i}} + 7\hat{\mathbf{j}} + (\lambda+1)\hat{\mathbf{k}}$$

Vectors **AB**, **AC** and **AD** will be coplanar

$$\Rightarrow \begin{vmatrix} -1 & 5 & -3 \\ -4 & 3 & 3 \\ 1 & 7 & \lambda+1 \end{vmatrix} = 0 \Rightarrow 17\lambda + 146 = 0$$

$\therefore \quad -17\lambda = 146$

4. $\because \quad |\mathbf{a}+\mathbf{b}| = \sqrt{3} \Rightarrow |\mathbf{a}+\mathbf{b}|^2 = 3$

$\Rightarrow a^2 + b^2 + 2\mathbf{a}\cdot\mathbf{b} = 3$

$\therefore \quad \mathbf{a}\cdot\mathbf{b} = \frac{1}{2}$

$\therefore (2\mathbf{a}+5\mathbf{b})\cdot(3\mathbf{a}+\mathbf{b}+\mathbf{a}\times\mathbf{b})$

$$= (6a^2 + 2\mathbf{a}\cdot\mathbf{b} + 0 + 15\mathbf{a}\cdot\mathbf{b} + 5b^2 + 0)$$

$$= \left(6 + 1 + \frac{15}{2} + 5\right) = \left(\frac{39}{2}\right) = 19.5$$

5. Let PV of P, A, B and C are $\mathbf{p}$, $\mathbf{a}$, $\mathbf{b}$ and $\mathbf{c}$ respectively and O $(\mathbf{0})$ be the circumcircle of the equilateral ΔABC.

Then, $$\frac{\mathbf{a}+\mathbf{b}+\mathbf{c}}{3} = 0 \quad \text{...(i)}$$

and $$|\mathbf{p}| = |\mathbf{a}| = |\mathbf{b}| = |\mathbf{c}| = \frac{26}{\sqrt{3}} \text{ unit} \quad \text{...(ii)}$$

Now, $|\mathbf{PA}|^2 = |\mathbf{a}-\mathbf{p}|^2 = a^2 + p^2 - 2\mathbf{a}\cdot\mathbf{p}$

Similarly,

$$|\mathbf{PB}|^2 = b^2 + p^2 - 2\mathbf{b}\cdot\mathbf{p}$$

and $$|\mathbf{PC}|^2 = c^2 + p^2 - 2\mathbf{c}\cdot\mathbf{p}$$

$\therefore \Sigma|\mathbf{PA}|^2 = a^2 + b^2 + c^2 + 3p^2 - 2\mathbf{p}\cdot(\mathbf{a}+\mathbf{b}+\mathbf{c})$

$$= 6p^2 - 0 \quad \text{[from Eqs. (i) and (ii)]}$$

$$= 6\frac{(26)^2}{3} = 2(26)^2 = 1352$$

6. Let PV of A, B and D be o, $\mathbf{b}$, $\mathbf{d}$

$\therefore \quad \mathbf{AC} = \mathbf{AB} + \mathbf{BC} = \mathbf{AB} + \mathbf{AD}$

$$= \mathbf{b} + \mathbf{d}$$

$$\mathbf{AE} = \frac{4\,\mathbf{AC} + 1\cdot\mathbf{AB}}{4+1}$$

$$= \frac{4(\mathbf{b}+\mathbf{d}) + \mathbf{b}}{5}$$

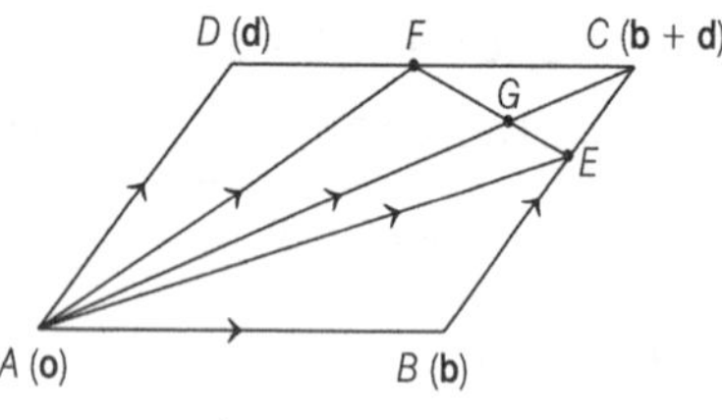

$$= \mathbf{b} + \frac{4}{5}\mathbf{d}$$

and $$\mathbf{AF} = \frac{1.\mathbf{AC} + 4.\mathbf{AD}}{4+1}$$

$$= \frac{1(\mathbf{b}+\mathbf{d}) + 4\mathbf{d}}{5} = \mathbf{d} + \frac{1}{5}\mathbf{b}$$

Equation of EF

$$\mathbf{r} = (\mathbf{b} + \frac{4}{5}\mathbf{d}) + \lambda_1\left(\mathbf{d} + \frac{1}{5}\mathbf{b} - \mathbf{b} - \frac{4}{5}\mathbf{d}\right)$$

or $$\mathbf{r} = \left(\mathbf{b} + \frac{4}{5}\mathbf{d}\right) + \lambda_1\left(\frac{1}{5}\mathbf{d} - \frac{4}{5}\mathbf{b}\right)$$

and equation of AC : $\mathbf{r} = \lambda_1(\mathbf{b}+\mathbf{d})$

For point G, $\left(\mathbf{b} + \frac{4}{5}\mathbf{d}\right) + \lambda_1\left(\frac{1}{5}\mathbf{d} - \frac{4}{5}\mathbf{b}\right) = \lambda_2(\mathbf{b}+\mathbf{d})$

On comparing, we get

$$1 - \frac{4}{5}\lambda_1 = \lambda_2 \text{ and } \frac{4}{5} + \frac{\lambda_1}{5} = \lambda_2$$

or $$1 - \frac{4}{5}\lambda_1 = \frac{4}{5} + \frac{\lambda_1}{5}$$

$\Rightarrow$ $$\frac{1}{5} = \lambda_1 \text{ and } \lambda_2 = \frac{21}{25}$$

$\therefore$ $$\mathbf{AG} = \lambda_2\,\mathbf{AC} \quad [\because \text{Given } \mathbf{AG} = \lambda\,\mathbf{AC}]$$

$\therefore$ $$\lambda_2 = \frac{21}{25}$$

Here $$\lambda = \lambda_2 = \frac{21}{25} = 0.84$$

$\Rightarrow$ $$400\lambda = 336$$

7. Here, $P(1, 0, 0)$, $Q(0, 1, 0)$, $R(0, 0, 1)$, $T = (1, 1, 1)$ and $S = \left(\frac{1}{2}, \frac{1}{2}, \frac{1}{2}\right)$.

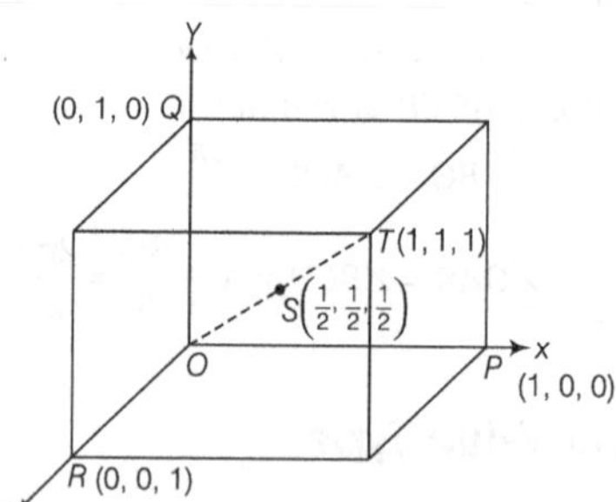

Now, $$\mathbf{p} = \mathbf{SP} = \mathbf{OP} - \mathbf{OS}$$

$$= \left(\frac{1}{2}\hat{\mathbf{i}} - \frac{1}{2}\hat{\mathbf{j}} - \frac{1}{2}\hat{\mathbf{k}}\right) = \frac{1}{2}(\hat{\mathbf{i}} - \hat{\mathbf{j}} - \hat{\mathbf{k}})$$

$$\mathbf{q} = \mathbf{SQ} = \frac{1}{2}(-\hat{\mathbf{i}} + \hat{\mathbf{j}} - \hat{\mathbf{k}})$$

$$\mathbf{r} = \mathbf{SR} = \frac{1}{2}(-\hat{\mathbf{i}} - \hat{\mathbf{j}} + \hat{\mathbf{k}})$$

and $$\mathbf{t} = \mathbf{ST} = \frac{1}{2}(\hat{\mathbf{i}} + \hat{\mathbf{j}} + \hat{\mathbf{k}})$$

$$\mathbf{p}\times\mathbf{q} = \frac{1}{4}\begin{vmatrix} \hat{\mathbf{i}} & \hat{\mathbf{j}} & \hat{\mathbf{k}} \\ 1 & -1 & -1 \\ -1 & 1 & -1 \end{vmatrix} = \frac{1}{4}(2\hat{\mathbf{i}} + 2\hat{\mathbf{j}})$$

and $$\mathbf{r}\times\mathbf{t} = \frac{1}{4}\begin{vmatrix} \hat{\mathbf{i}} & \hat{\mathbf{j}} & \hat{\mathbf{k}} \\ -1 & -1 & 1 \\ 1 & 1 & 1 \end{vmatrix} = \frac{1}{4}(-2\hat{\mathbf{i}} + 2\hat{\mathbf{j}})$$

Now, $$(\mathbf{p}\times\mathbf{q})\times(\mathbf{r}\times\mathbf{t}) = \frac{1}{16}\begin{vmatrix} \hat{\mathbf{i}} & \hat{\mathbf{j}} & \hat{\mathbf{k}} \\ 2 & 2 & 0 \\ -2 & 2 & 0 \end{vmatrix} = \frac{1}{16}(8\hat{\mathbf{k}}) = \frac{1}{2}\hat{\mathbf{k}}$$

$$\therefore |(\mathbf{p}\times\mathbf{q})\times(\mathbf{r}\times\mathbf{t})| = \left|\frac{1}{2}\hat{\mathbf{k}}\right| = \frac{1}{2} = 0.5$$

8. Clearly **a**, **b** and **c** must be coplanar

$\Rightarrow \quad \mathbf{c} = \lambda \mathbf{a} + \mu \mathbf{b}$

$\Rightarrow \quad \hat{\mathbf{i}} + x_1 \hat{\mathbf{j}} + x_2 \hat{\mathbf{k}} = \lambda (\hat{\mathbf{i}} + \hat{\mathbf{j}} + \hat{\mathbf{k}}) + \mu (4\hat{\mathbf{i}} + 3\hat{\mathbf{j}} + 4\hat{\mathbf{k}})$

$\therefore \quad 1 = \lambda + 4\mu, \ x_1 = \lambda + 3\mu$ and $x_2 = \lambda + 4\mu$

$\therefore \quad x_2 = 1$

Also, given

$|\mathbf{c}| = 2$

$\Rightarrow \quad \sqrt{1 + x_1^2 + x_2^2} = 2$

$\Rightarrow \quad 2 + x_1^2 = 4$

$\therefore \quad x_1^2 = 2$

Then, $729x_1^4 + 343x_1^2x_2^2 + 81x_2^2 + 15$

$= 729\,(2)^2 + 343\,(2)\,(1)^2 + 81\,(1)^2 + 15$

$= 2916 + 686 + 81 + 15 = 3698$

9. $\because \mathbf{a} \times (\mathbf{b} \times \mathbf{c}) + (\mathbf{a} \cdot \mathbf{b})\,\mathbf{b} = (4 - 2x - \sin y)\,\mathbf{b} + (x^2 - 1)\,\mathbf{c}$

$\Rightarrow \quad (\mathbf{a} \cdot \mathbf{c})\,\mathbf{b} - (\mathbf{a} \cdot \mathbf{b})\,\mathbf{c} + (\mathbf{a} \cdot \mathbf{b})\,\mathbf{b} = (4 - 2x - \sin y)\,\mathbf{b} + (x^2 - 1)\,\mathbf{c}$

On comparing, we get

$\mathbf{a} \cdot \mathbf{c} + \mathbf{a} \cdot \mathbf{b} = 4 - 2x - \sin y \quad \ldots \text{(i)}$

and $\quad -(\mathbf{a} \cdot \mathbf{b}) = (x^2 - 1) \quad \ldots \text{(ii)}$

Adding Eqs. (i) and (ii), then

$\mathbf{a} \cdot \mathbf{c} = x^2 - 1 + 4 - 2x - \sin y$

$\therefore \quad \mathbf{a} \cdot \mathbf{c} = (x - 1)^2 + 2 - \sin y \quad \ldots \text{(iii)}$

Also, given $\quad (\mathbf{c} \cdot \mathbf{c})\,\mathbf{a} = \mathbf{c}$

$\Rightarrow \quad (\mathbf{c} \cdot \mathbf{c})\,(\mathbf{a} \cdot \mathbf{c}) = (\mathbf{c} \cdot \mathbf{c})$

$\therefore \quad \mathbf{a} \cdot \mathbf{c} = 1 \quad [\because \mathbf{c} \cdot \mathbf{c} \neq 0]$

Now, from Eq. (iii),

$(x - 1)^2 + 1 - \sin y = 0$

or $\quad \sin y = (x - 1)^2 + 1 \geq 1$

Which is possible only when $x = 1$ and, then $\sin y = 1$

$\therefore \quad x^{\sin y} + (4 \sin y)^{4x} + 15 = (1)^1 + 4^4 + 15$

$= 1 + 256 + 15 = 272$

10. We must have,

$$\alpha\,(\hat{\mathbf{i}} - 3\hat{\mathbf{j}} + 5\hat{\mathbf{k}}) = \hat{\mathbf{a}} + \frac{2\hat{\mathbf{k}} + 2\hat{\mathbf{j}} - \hat{\mathbf{i}}}{3}$$

$\Rightarrow \quad 3\hat{\mathbf{a}} = \alpha\,(3\hat{\mathbf{i}} - 9\hat{\mathbf{j}} + 15\hat{\mathbf{k}}) - 2\hat{\mathbf{k}} - 2\hat{\mathbf{j}} + \hat{\mathbf{i}}$

$\Rightarrow \quad 3\hat{\mathbf{a}} = (3\alpha + 1)\,\hat{\mathbf{i}} - (9\alpha + 2)\,\hat{\mathbf{j}} + (15\alpha - 2)\,\hat{\mathbf{k}}$

$\therefore \quad 3|\hat{\mathbf{a}}| = |(3\alpha + 1)\,\hat{\mathbf{i}} - (9\alpha + 2)\,\hat{\mathbf{j}} + (15\alpha - 2)\,\hat{\mathbf{k}}|$

$\Rightarrow \quad 3 = \sqrt{\{(3\alpha + 1)^2 + (9\alpha + 2)^2 + (15\alpha - 2)^2}$

$\Rightarrow \quad 9 = 315\alpha^2 - 18\alpha + 9$

$\therefore \quad \alpha = 0, \dfrac{2}{35}$

For $\quad \alpha = 0, \ \hat{\mathbf{a}} = \dfrac{\hat{\mathbf{i}} - 2\hat{\mathbf{j}} - 2\hat{\mathbf{k}}}{3}$ (not acceptable)

For $\alpha = \dfrac{2}{35}, \quad \mathbf{a} = \dfrac{2}{35}(\hat{\mathbf{i}} - 3\hat{\mathbf{j}} + 5\hat{\mathbf{k}}) - \dfrac{(2\hat{\mathbf{k}} + 2\hat{\mathbf{j}} - \hat{\mathbf{i}})}{3}$

$= \dfrac{1}{105}(41\hat{\mathbf{i}} - 88\hat{\mathbf{j}} - 40\hat{\mathbf{k}})$

On comparing, we get

$\lambda = 41, \mu = -88, \nu = -40$

$\therefore \quad \lambda^2 - 2\mu + \nu = (41)^2 + 2\,(88) - 40$

$= 1681 + 176 - 40 = 1817$

11. Let PV of vertices A, B, C with reference origin O be **a**, **b**, **c** respectively. Then

$\mathbf{OG}_1 = \dfrac{\mathbf{b} + \mathbf{c}}{3}, \ \mathbf{OG}_2 = \dfrac{\mathbf{c} + \mathbf{a}}{3}, \ \mathbf{OG}_3 = \dfrac{\mathbf{a} + \mathbf{b}}{3}$

Let V_1 = volume of the tetrahedron $OABC$

$= \dfrac{1}{6}[\mathbf{abc}]$

and let V_2 = volume of the parallelopiped with OG_1, OG_2, OG_3 as coterminus edges

$= [OG_1\ OG_2\ OG_3]$

$= \left[\dfrac{\mathbf{b} + \mathbf{c}}{3}\ \dfrac{\mathbf{c} + \mathbf{a}}{3}\ \dfrac{\mathbf{a} + \mathbf{b}}{3}\right]$

$= \dfrac{1}{27}[\mathbf{b} + \mathbf{c}\ \mathbf{c} + \mathbf{a}\ \mathbf{a} + \mathbf{b}] = \dfrac{2}{27}[\mathbf{abc}]$

$\therefore \quad \lambda = \dfrac{V_1}{V_2} = \dfrac{\frac{1}{6}[\mathbf{abc}]}{\frac{2}{27}[\mathbf{abc}]} = \dfrac{9}{4}$

$\therefore \quad 2008\,\lambda = 2008 \times \dfrac{9}{4} = 502 \times 9 = 4518$

12. Given vectors $\mathbf{a} = 2\hat{\mathbf{i}} + \hat{\mathbf{j}} - \hat{\mathbf{k}}$

and $\quad \mathbf{b} = \hat{\mathbf{i}} + 2\hat{\mathbf{j}} + \hat{\mathbf{k}}$

So, $\quad \mathbf{a} + \mathbf{b} = 3\hat{\mathbf{i}} + 3\hat{\mathbf{j}} \Rightarrow |\mathbf{a} + \mathbf{b}| = 3\sqrt{2}$

Since, it is given that projection of $\mathbf{c} = \alpha\mathbf{a} + \beta\mathbf{b}$ on the vector $(\mathbf{a} + \mathbf{b})$ is $3\sqrt{2}$, then

$\dfrac{(\mathbf{a} + \mathbf{b}).\,\mathbf{c}}{|\mathbf{a} + \mathbf{b}|} = 3\sqrt{2}$

$\Rightarrow \quad (\mathbf{a} + \mathbf{b}).\,(\alpha\mathbf{a} + \beta\mathbf{b}) = 18$

$\Rightarrow \quad \alpha\,(\mathbf{a}.\,\mathbf{a}) + \beta\,(\mathbf{a}.\,\mathbf{b}) + \alpha\,(\mathbf{b}.\,\mathbf{a}) + \beta\,(\mathbf{a}.\,\mathbf{b}) = 18$

$\Rightarrow \quad 6\alpha + 3\beta + 3\alpha + 6\beta = 18$

$\Rightarrow \quad 9\alpha + 9\beta = 18 \Rightarrow (\alpha + \beta) = 2 \quad \ldots \text{(i)}$

Now, for minimum value of $(\mathbf{c} - (\mathbf{a} \times \mathbf{b})).\,\mathbf{c}$

$= (\alpha\mathbf{a} + \beta\mathbf{b} - (\mathbf{a} \times \mathbf{b})).\,(\alpha\,\mathbf{a} + \beta\mathbf{b})$

$= \alpha^2\,(\mathbf{a}.\,\mathbf{a}) + \alpha\beta(\mathbf{a}.\,\mathbf{b}) + \alpha\beta(\mathbf{a}.\,\mathbf{b}) + \beta^2(\mathbf{b}.\,\mathbf{b})$

$[\because (\mathbf{a} \times \mathbf{b}).\,\mathbf{a} = 0 = (\mathbf{a} \times \mathbf{b}).\,\mathbf{b}]$

$= 6\alpha^2 + 6\alpha\beta + 6\beta^2 = 6\,(\alpha^2 + \beta^2 + \alpha\beta)$

$= 6\,[(\alpha + \beta)^2 - \alpha\beta] = 6\,[4 - \alpha\beta] = 6\,[4 - \alpha\,(2 - \alpha)]$

$= 6\,[4 - 2\alpha + \alpha^2]$

The minimum value of $6\,(4 - 2\alpha + \alpha^2) = 6(3) = 18$

[As minimum value of $ax^2 + bx + c = -\dfrac{D}{4a}$, if $a > 0$]

13. Let $\mathbf{OA} = x_1\hat{\mathbf{i}} + y_1\hat{\mathbf{j}}$ and $\mathbf{OB} = x_2\hat{\mathbf{i}} + y_2\hat{\mathbf{j}}$

$\therefore \quad \mathbf{OA}\cdot\hat{\mathbf{i}} = x_1 = 1,\ \mathbf{OB}\cdot\hat{\mathbf{i}} = x_2 = -2$

Since, $(x_1, y_1), (x_2, y_2)$ lies on $y = x^2$

$\therefore \quad y_1 = x_1^2,\ y_2 = x_2^2$

$\therefore \quad y_1 = 1,\ y_2 = 4$

$\Rightarrow$ Coordinates of A and B are (1, 1) and (–2, 4)

$\therefore \quad 2\mathbf{OA} - 3\mathbf{OB} = 2(\hat{\mathbf{i}} + \hat{\mathbf{j}}) - 3(-2\hat{\mathbf{i}} + 4\hat{\mathbf{j}})$

$= 8\hat{\mathbf{i}} - 10\hat{\mathbf{j}}$

$|(2\mathbf{OA} - 3\mathbf{OB})| = |8\hat{\mathbf{i}} - 10\hat{\mathbf{j}}|$

$= \sqrt{64 + 100} = 2\sqrt{41}$

$\therefore |2\mathbf{OA} - 3\mathbf{OB}|^2 = 164$

14. It is given that, in a ΔPQR

$\mathbf{PQ} = \mathbf{c},\ \mathbf{QR} = \mathbf{a}$ and $\mathbf{RP} = \mathbf{b}$ so $\mathbf{a} + \mathbf{b} + \mathbf{c} = 0$

$\therefore \quad |\mathbf{a}|^2 + \mathbf{a}\cdot\mathbf{b} + \mathbf{a}\cdot\mathbf{c} = 0 \quad \ldots\text{(i)}$

and $\quad \mathbf{a}\cdot\mathbf{b} + |\mathbf{b}|^2 + \mathbf{c}\cdot\mathbf{b} = 0 \quad \ldots\text{(ii)}$

and, also given that $|\mathbf{a}| = 3,\ |\mathbf{b}| = 4$

and $\quad \dfrac{\mathbf{a}\cdot\mathbf{c} - \mathbf{a}\cdot\mathbf{b}}{\mathbf{c}\cdot\mathbf{a} - \mathbf{c}\cdot\mathbf{b}} = \dfrac{|\mathbf{a}|}{|\mathbf{a}| + |\mathbf{b}|} = \dfrac{3}{7}$

$\Rightarrow \quad 7[(\mathbf{a}\cdot\mathbf{c}) - (\mathbf{a}\cdot\mathbf{b})] = 3[(\mathbf{c}\cdot\mathbf{a}) - (\mathbf{c}\cdot\mathbf{b})]$

From Eqs. (i) and (ii), on putting the values of $\mathbf{a}\cdot\mathbf{c}$ and $\mathbf{c}\cdot\mathbf{b}$, we get

$\Rightarrow 7[-9 - (\mathbf{a}\cdot\mathbf{b}) - (\mathbf{a}\cdot\mathbf{b})] = 3[-9 - (\mathbf{a}\cdot\mathbf{b}) + 16 + (\mathbf{a}\cdot\mathbf{b})]$

$\Rightarrow \quad 7[-9 - 2(\mathbf{a}\cdot\mathbf{b})] = 3\times 7$

$\Rightarrow \quad \mathbf{a}\cdot\mathbf{b} = -6$

$\because \quad |\mathbf{a}\times\mathbf{b}|^2 = (|\mathbf{a}||\mathbf{b}|)^2 - (\mathbf{a}\cdot\mathbf{b})^2$

$= 144 - 36 = 108$

15. $\mathbf{d}\cdot\mathbf{c} = \lambda(\mathbf{a}\times\mathbf{b})\cdot\mathbf{c} + \mu(\mathbf{b}\times\mathbf{c})\cdot\mathbf{c} + \nu(\mathbf{c}\times\mathbf{a})\cdot\mathbf{c}$

$= \lambda[\mathbf{a}\ \mathbf{b}\ \mathbf{c}] + 0 + 0 = \lambda[\mathbf{a}\ \mathbf{b}\ \mathbf{c}] = \dfrac{\lambda}{8}$

Hence, $\lambda = 8(\mathbf{d}\cdot\mathbf{c})$.

Similarly, $\mu = 8(\mathbf{d}\cdot\mathbf{a})$ and $\nu = 8(\mathbf{d}\cdot\mathbf{b})$

$\therefore \quad \lambda + \mu + \nu = 8\mathbf{d}\cdot\mathbf{c} + 8\mathbf{d}\cdot\mathbf{a} + 8\mathbf{d}\cdot\mathbf{b}$

$= 8\mathbf{d}\cdot(\mathbf{a} + \mathbf{b} + \mathbf{c}) = 64$

16. $\mathbf{AB} = \mathbf{OB} - \mathbf{OA}$

$= (\hat{\mathbf{i}} + 2\hat{\mathbf{j}} + 3\hat{\mathbf{k}}) - (4\hat{\mathbf{i}} + 2\hat{\mathbf{j}} + 2\hat{\mathbf{k}})$

$= -3\hat{\mathbf{i}} + 0\hat{\mathbf{j}} + \hat{\mathbf{k}}$

$|\mathbf{AB}| = \sqrt{9 + 0 + 1} = \sqrt{10}$

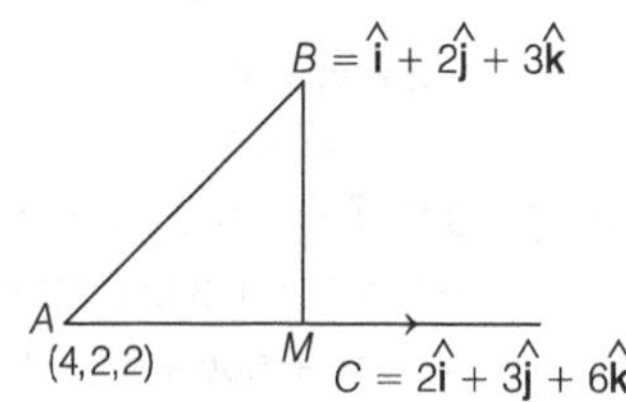

Now, AM = Projection of AB along $\mathbf{c}$

$= \dfrac{\mathbf{AB}\cdot\mathbf{c}}{|\mathbf{c}|}$

$= \dfrac{(-3\hat{\mathbf{i}} + 0\hat{\mathbf{j}} + \hat{\mathbf{k}})\cdot(2\hat{\mathbf{i}} + 3\hat{\mathbf{j}} + 6\hat{\mathbf{k}})}{\sqrt{(4 + 9 + 36)}}$

$= \dfrac{-6 + 0 + 6}{7} = 0 \quad \Rightarrow \quad BA \perp AM$

$\Rightarrow$ Perpendicular distance of B from

$AM = |\mathbf{AB}| = \sqrt{10} = \lambda$ (given)

$\therefore \quad \lambda^6 + \lambda^4 + \lambda^2 + 1 = 1000 + 100 + 10 + 1 = 1111$

Single Integer Type

1. Let D be the origin of reference and $\mathbf{DA} = \mathbf{a}$, $\mathbf{DB} = \mathbf{b}$, $\mathbf{DC} = \mathbf{c}$

$|\mathbf{AB}\times\mathbf{CD} + \mathbf{BC}\times\mathbf{AD} + \mathbf{CA}\times\mathbf{BD}|$

$= |(\mathbf{b} - \mathbf{a})\times(-\mathbf{c}) + (\mathbf{c} - \mathbf{b})\times(-\mathbf{a}) + (\mathbf{a} - \mathbf{c})\times(-\mathbf{b})|$

$= |\mathbf{b}\times\mathbf{c} - \mathbf{a}\times\mathbf{c} + \mathbf{c}\times\mathbf{a} - \mathbf{b}\times\mathbf{a} + \mathbf{a}\times\mathbf{b} - \mathbf{c}\times\mathbf{b}|$

$= 2|\mathbf{a}\times\mathbf{b} + \mathbf{b}\times\mathbf{c} + \mathbf{c}\times\mathbf{a}|$

= 2 (2 area of ΔABC) = 4 area of ΔABC.

Hence, $\lambda = 4$

2. Given, for unit vectors $\mathbf{a}$, $\mathbf{b}$ and $\mathbf{c}$

$|\mathbf{a} - \mathbf{b}|^2 + |\mathbf{a} - \mathbf{c}|^2 = 8$

$\Rightarrow \quad |\mathbf{a}|^2 + |\mathbf{b}|^2 - 2\mathbf{a}\cdot\mathbf{b} + |\mathbf{a}|^2 + |\mathbf{c}|^2 - 2\mathbf{a}\cdot\mathbf{c} = 8$

$\Rightarrow \quad -2[\mathbf{a}\cdot\mathbf{b} + \mathbf{a}\cdot\mathbf{c}] = 4$

$\Rightarrow \quad \mathbf{a}\cdot\mathbf{b} + \mathbf{a}\cdot\mathbf{c} = -2 \quad \ldots\text{(i)}$

Now, $\quad |\mathbf{a} + 2\mathbf{b}|^2 + |\mathbf{a} + 2\mathbf{c}|^2$

$= |\mathbf{a}|^2 + 4|\mathbf{b}|^2 + 4\mathbf{a}\cdot\mathbf{b} + |\mathbf{a}|^2 + 4|\mathbf{c}|^2 + 4\mathbf{a}\cdot\mathbf{c}$

$= 10 + 4[\mathbf{a}\cdot\mathbf{b} + \mathbf{a}\cdot\mathbf{c}]$

$= 10 + 4(-2)$ [from Eq. (i)]

$= 10 - 8 = 2$

3. $\Sigma[\mathbf{p}\times\{(\mathbf{x} - \mathbf{q}\}\times\mathbf{p}] = \mathbf{0}$

$\Rightarrow \Sigma[\mathbf{p}\times(\mathbf{x}\times\mathbf{p})] - \Sigma[\mathbf{p}\times(\mathbf{q}\times\mathbf{p})] = \mathbf{0}$

$\Rightarrow \Sigma\mathbf{p}^2\vec{\mathbf{x}} - \Sigma(\mathbf{p}\cdot\mathbf{x})\mathbf{p} - \Sigma\mathbf{p}^2\mathbf{q} + \Sigma(\mathbf{p}\cdot\mathbf{q})\mathbf{p} = 0$

$\Rightarrow 3\mathbf{p}^2\mathbf{x} - \mathbf{p}^2\mathbf{x} - \mathbf{p}^2(\mathbf{p} + \mathbf{q} + \mathbf{r}) = \mathbf{0}$

$\Rightarrow 2\mathbf{p}^2\mathbf{x} = \mathbf{p}^2(\mathbf{p} + \mathbf{q} + \mathbf{r})$

$\Rightarrow \mathbf{x} = \dfrac{1}{2}(\mathbf{p} + \mathbf{q} + \mathbf{r})$

4. $\mathbf{AB} = 2\hat{\mathbf{i}} + \hat{\mathbf{j}} + \hat{\mathbf{k}},\ \mathbf{AC} = (t + 1)\hat{\mathbf{i}} + 0\hat{\mathbf{j}} - \hat{\mathbf{k}}$

$$\mathbf{AB}\times\mathbf{AC} = \begin{vmatrix} \hat{\mathbf{i}} & \hat{\mathbf{j}} & \hat{\mathbf{k}} \\ 2 & 1 & 1 \\ t+1 & 0 & -1 \end{vmatrix}$$

$= -\hat{\mathbf{i}} + (t + 3)\hat{\mathbf{j}} - (t + 1)\hat{\mathbf{k}}$

$= \sqrt{1 + (t + 3)^2 + (t + 1)^2}$

$= \sqrt{2t^2 + 8t + 11}$

Area of $\Delta ABC = \dfrac{1}{2}|\mathbf{AB}\times\mathbf{AC}|$

$= \dfrac{1}{2}\sqrt{2t^2 + 8t + 11}$

Let $\quad f(t) = \Delta^2 = \dfrac{1}{4}(2t^2 + 8t + 11)$

$f'(t) = 0 \Rightarrow t = -2$

At $\quad t = -2,\ f''(t) > 0$

So, Δ is minimum at $t = -2 \Rightarrow |t| = 2$

5. Let angle between unit vectors $\mathbf{a}$ and $\mathbf{b}$ is $\theta \in [0, \pi]$.

Then, $|\mathbf{a}+\mathbf{b}|^2 = |\mathbf{a}|^2 + |\mathbf{b}|^2 + 2\mathbf{a}\cdot\mathbf{b}$

$$= 1 + 1 + 2\cos\theta = 2(1+\cos\theta) = 4\cos^2\frac{\theta}{2}$$

$$\Rightarrow \quad |\mathbf{a}+\mathbf{b}| = 2\cos\left(\frac{\theta}{2}\right) \quad \left[\because \theta\in[0,\pi] \Rightarrow \cos\left(\frac{\theta}{2}\right) \ge 0\right]$$

Similarly, $|\mathbf{a}-\mathbf{b}|^2 = |\mathbf{a}|^2 + |\mathbf{b}|^2 - 2\mathbf{a}\cdot\mathbf{b}$

$$= 1 + 1 - 2\cos\theta$$

$$= 2(1-\cos\theta) = 4\sin^2\frac{\theta}{2}$$

$$\Rightarrow \quad |\mathbf{a}-\mathbf{b}| = 2\sin\left(\frac{\theta}{2}\right)$$

So, $\sqrt{3}|\mathbf{a}+\mathbf{b}| + |\mathbf{a}-\mathbf{b}| = 2\left[\sqrt{3}\cos\left(\frac{\theta}{2}\right) + \sin\left(\frac{\theta}{2}\right)\right]$

Having greatest value $= 2\sqrt{3+1} = 4$

[$\because$ greatest value of $a\cos\theta + b\sin\theta$ is $\sqrt{a^2+b^2}$]

6. It is given that the lines

$$\mathbf{r} = \hat{\mathbf{i}} - 6\hat{\mathbf{j}} + (p\sec\alpha)\hat{\mathbf{k}} + t(\hat{\mathbf{i}} + 2\hat{\mathbf{j}} + \hat{\mathbf{k}}) \text{ and}$$

$$\mathbf{r} = 4\hat{\mathbf{j}} + \hat{\mathbf{k}} + \lambda(2\hat{\mathbf{i}} + (p\tan\alpha)\hat{\mathbf{j}} + 2\hat{\mathbf{k}})$$

Intersected at point $8\hat{\mathbf{i}} + 8\hat{\mathbf{j}} + 9\hat{\mathbf{k}}$, so

$$1 + t = 8,\ -6 + 2t = 8 \text{ and } p\sec\alpha + t = 9$$

$\therefore \quad t = 7$ and $p\sec\alpha = 2$ ...(i)

and $\quad 2\lambda = 8,\ 4 + \lambda p\tan\alpha = 8$ and $1 + 2\lambda = 9$

$\therefore \quad \lambda = 4$ and $p\tan\alpha = 1$ ...(ii)

From Eqs. (i) and (ii), we get

$$p^2\sec^2\alpha - p^2\tan^2\alpha = 4 - 1$$

$$\Rightarrow \quad p^2 = 3 \Rightarrow p = \pm\sqrt{3} \Rightarrow p^2 = 3$$

7. Given, $|\mathbf{a}| = 1, |\mathbf{c}| = 1$ and $|\mathbf{b}| = 4$

$$\Rightarrow |\mathbf{a}\cdot\mathbf{c}| = 1\cdot 1\cdot\frac{1}{4} = \frac{1}{4}$$

Now, $\quad \mathbf{b} - 2\mathbf{c} = \lambda\mathbf{a}$

$$\Rightarrow \quad \mathbf{a}\cdot\mathbf{b} - 2\mathbf{a}\cdot\mathbf{c} = \lambda a^2$$

$$\Rightarrow \quad \mathbf{a}\cdot\mathbf{b} - 2\cdot\frac{1}{4} = \lambda$$

$$\Rightarrow \quad \mathbf{a}\cdot\mathbf{b} = \lambda + \frac{1}{2}$$

Again, $\quad \mathbf{b} - 2\mathbf{c} = \lambda\mathbf{a} \Rightarrow \mathbf{b}\cdot\mathbf{b} - 2\mathbf{b}\cdot\vec{\mathbf{c}} = \lambda\mathbf{b}\cdot\mathbf{a}$

$$\Rightarrow \quad 16 - 2\mathbf{b}\cdot\mathbf{c} = \lambda\left(\lambda + \frac{1}{2}\right)$$

$$\Rightarrow \quad \mathbf{b}\cdot\mathbf{c} = 8 - \frac{\lambda^2}{2} - \frac{\lambda}{4}$$

Also, $\quad \mathbf{b} - 2\mathbf{c} = \lambda\mathbf{a} \Rightarrow \mathbf{b}\cdot\mathbf{c} - 2\mathbf{c}\cdot\mathbf{c} = \lambda\mathbf{b}\cdot\mathbf{a}$

$$\Rightarrow \quad 8 - \frac{\lambda^2}{2} - \frac{\lambda}{4} - 2(1) = \lambda\left(\frac{1}{4}\right)$$

$$\Rightarrow \quad \lambda^2 + \lambda - 12 = 0$$

$$\therefore \quad \lambda = -4, 3$$

8. It is given that projection of $\mathbf{b}$ on $\mathbf{a}$ is equal to the projection of $\mathbf{c}$ on $\mathbf{a}$, where $|\mathbf{a}| = 2, |\mathbf{b}| = 4$ and $|\mathbf{c}| = 4$.

So, $\quad \dfrac{\mathbf{a}\cdot\mathbf{b}}{2} = \dfrac{\mathbf{a}\cdot\mathbf{c}}{2} \Rightarrow \mathbf{a}\cdot\mathbf{b} = \mathbf{a}\cdot\mathbf{c}$

and $\mathbf{b}$ is perpendicular to $\mathbf{c}$, so $\mathbf{b}\cdot\mathbf{c} = 0$

Now, $|\mathbf{a}+\mathbf{b}-\mathbf{c}|^2$

$$= |\mathbf{a}|^2 + |\mathbf{b}|^2 + |\mathbf{c}|^2 + 2\mathbf{a}\cdot\mathbf{b} - 2\mathbf{b}\cdot\mathbf{c} - 2\mathbf{a}\cdot\mathbf{c}$$

$$= 4 + 16 + 16 = 36$$

$$\therefore \quad |\mathbf{a}+\mathbf{b}-\mathbf{c}| = 6$$

9. We have,

$$\mathbf{c} = x\mathbf{a} + y\mathbf{b} + \mathbf{a}\times\mathbf{b} \text{ and } \mathbf{a}\cdot\mathbf{b} = 0$$

$$|\mathbf{a}| = |\mathbf{b}| = 1 \text{ and } |\mathbf{c}| = 2$$

Also, given $\mathbf{c}$ is inclined on $\mathbf{a}$ and $\mathbf{b}$ with same angle α.

$$\therefore \quad \mathbf{a}\cdot\mathbf{c} = x|\mathbf{a}|^2 + y(\mathbf{a}\cdot\mathbf{b}) + \mathbf{a}\cdot(\mathbf{a}\times\mathbf{b})$$

$$|\mathbf{a}||\mathbf{c}|\cos\alpha = x + 0 + 0$$

$$x = 2\cos\alpha$$

Similarly,

$$|\mathbf{b}||\mathbf{c}|\cos\alpha = 0 + y + 0$$

$$\Rightarrow \quad y = 2\cos\alpha$$

$$|\mathbf{c}|^2 = x^2 + y^2 + |\mathbf{a}\times\mathbf{b}|^2$$

$$4 = 8\cos^2\alpha + |a|^2|b|^2\sin^2 90^\circ$$

$$4 = 8\cos^2\alpha + 1$$

$$\Rightarrow \quad 8\cos^2\alpha = 3$$

Matrix-Match Type

1. (A) Let $\quad \mathbf{a}\times\mathbf{b} = \mathbf{1}$

$$\therefore \quad (\mathbf{a}\times\mathbf{b})\times(\mathbf{c}\times\mathbf{d}) = \mathbf{1}\times(\mathbf{c}\times\mathbf{d}) = (\mathbf{1}\cdot\mathbf{d})\,\mathbf{c} - (\mathbf{1}\cdot\mathbf{c})\,\mathbf{d}$$

$$= ((\mathbf{a}\times\mathbf{b})\cdot\mathbf{d})\,\mathbf{c} - ((\mathbf{a}\times\mathbf{b})\cdot\mathbf{c})\mathbf{d}$$

$$= [\mathbf{a}\,\mathbf{b}\,\mathbf{d}]\,\mathbf{c} - [\mathbf{a}\,\mathbf{b}\,\mathbf{c}]\,\mathbf{d}$$

Hence, $(\mathbf{a}\times\mathbf{b})\times(\mathbf{c}\times\mathbf{d}) = [\mathbf{abd}]\,\mathbf{c} - [\mathbf{a}\,\mathbf{b}\,\mathbf{c}]\,\mathbf{d}$

Also let $\quad \mathbf{c}\times\mathbf{d} = \mathbf{m}$

Then

$$(\mathbf{a}\times\mathbf{b})\times(\mathbf{c}\times\mathbf{d}) = (\mathbf{a}\times\mathbf{b})\times\mathbf{m}$$

$$= (\mathbf{a}\cdot\mathbf{m})\,\mathbf{b} - (\mathbf{b}\cdot\mathbf{m})\,\mathbf{a}$$

$$= (\mathbf{a}\cdot(\mathbf{c}\times\mathbf{d}))\cdot\mathbf{b} - (\mathbf{b}\cdot(\mathbf{c}\times\mathbf{d}))\mathbf{a}$$

$$= [\mathbf{a}\,\mathbf{c}\,\mathbf{d}]\,\mathbf{b} - [\mathbf{b}\,\mathbf{c}\,\mathbf{d}]\,\mathbf{a}$$

Hence, $(\mathbf{a}\times\mathbf{b})\times(\mathbf{c}\times\mathbf{d}) = [\mathbf{a}\,\mathbf{c}\,\mathbf{d}]\,\mathbf{b} - [\mathbf{b}\,\mathbf{c}\,\mathbf{d}]\,\mathbf{a}$ (R)

(B) Let $\mathbf{a}\times\mathbf{c} = \mathbf{1}$

$$\therefore \quad (\mathbf{a}\times\mathbf{c})\times(\mathbf{b}\times\mathbf{d}) = \mathbf{1}\times(\mathbf{b}\times\mathbf{d})$$

$$= (\mathbf{1}\cdot\mathbf{d})\,\mathbf{b} - (\mathbf{1}\cdot\mathbf{b})\,\mathbf{d}$$

$$= ((\mathbf{a}\times\mathbf{c})\cdot\mathbf{d})\,\mathbf{b} - ((\mathbf{a}\times\mathbf{c})\cdot\mathbf{b})\mathbf{d}$$

$$= [\mathbf{a}\,\mathbf{c}\,\mathbf{d}]\,\mathbf{b} - [\mathbf{a}\,\mathbf{c}\,\mathbf{b}]\,\mathbf{d}$$

$$= [\mathbf{a}\,\mathbf{c}\,\mathbf{d}]\,\mathbf{b} + [\mathbf{a}\,\mathbf{b}\,\mathbf{c}]\,\mathbf{d}$$

Hence, $(\mathbf{a}\times\mathbf{c})\times(\mathbf{b}\times\mathbf{d}) = [\mathbf{a}\,\mathbf{c}\,\mathbf{d}]\,\mathbf{b} + [\mathbf{a}\,\mathbf{b}\,\mathbf{c}]\,\mathbf{d}$ (P)

Also, let $\mathbf{b}\times\mathbf{d} = \mathbf{m}$

$$\therefore \quad (\mathbf{a}\times\mathbf{c})\times(\mathbf{b}\times\mathbf{d}) = (\mathbf{a}\times\mathbf{c})\times\mathbf{m}$$

$$= (\mathbf{a}\cdot\mathbf{m})\,\mathbf{c} - (\mathbf{c}\cdot\mathbf{m})\,\mathbf{a}$$

$$= (\mathbf{a}\cdot(\mathbf{b}\times\mathbf{d}))\,\mathbf{c} - (\mathbf{c}\cdot(\mathbf{b}\times\mathbf{d}))\mathbf{a}$$

$$= [\mathbf{a}\,\mathbf{b}\,\mathbf{d}]\,\mathbf{c} - [\mathbf{a}\,\mathbf{d}\,\mathbf{c}]\,\mathbf{a}$$

$$= [\mathbf{a}\,\mathbf{b}\,\mathbf{d}]\,\mathbf{c} + [\mathbf{b}\,\mathbf{c}\,\mathbf{d}]\,\mathbf{a}$$

Hence, $(\mathbf{a}\times\mathbf{c})\times(\mathbf{b}\times\mathbf{d}) = [\mathbf{a}\,\mathbf{b}\,\mathbf{d}]\,\mathbf{c} + [\mathbf{b}\,\mathbf{c}\,\mathbf{d}]\,\mathbf{a}$ (T)

(C) Let $\mathbf{a}\times\mathbf{d}=\mathbf{1}$

$$\therefore\quad (\mathbf{a}\times\mathbf{d})\times(\mathbf{b}\times\mathbf{c})=\mathbf{1}\times(\mathbf{b}\times\mathbf{c})$$
$$=(\mathbf{1}\cdot\mathbf{c})\,\mathbf{b}-(\mathbf{1}\cdot\mathbf{b})\,\mathbf{c}$$
$$=((\mathbf{a}\times\mathbf{d})\cdot\mathbf{c})\,\mathbf{b}-((\mathbf{a}\times\mathbf{d})\cdot\mathbf{b})\mathbf{c}$$
$$=[\mathbf{a}\ \mathbf{d}\ \mathbf{c}]\,\mathbf{b}-[\mathbf{a}\ \mathbf{d}\ \mathbf{b}]\,\mathbf{c}$$
$$=-[\mathbf{a}\ \mathbf{c}\ \mathbf{d}]\,\mathbf{b}+[\mathbf{a}\ \mathbf{b}\ \mathbf{d}]\,\mathbf{c}$$
$$=[\mathbf{a}\ \mathbf{b}\ \mathbf{d}]\,\mathbf{c}-[\mathbf{a}\ \mathbf{c}\ \mathbf{d}]\,\mathbf{b}$$

Hence, $(\mathbf{a}\times\mathbf{d})\times(\mathbf{b}\times\mathbf{c})$
$$=[\mathbf{a}\ \mathbf{b}\ \mathbf{d}]\,\mathbf{c}-[\mathbf{a}\ \mathbf{c}\ \mathbf{d}]\,\mathbf{b}\ \text{(S)}$$

Now, also let $\mathbf{b}\times\mathbf{c}=\mathbf{m}$

$$\therefore\quad (\mathbf{a}\times\mathbf{d})\times(\mathbf{b}\times\mathbf{c})=(\mathbf{a}\times\mathbf{d})\times\mathbf{m}$$
$$=(\mathbf{a}\cdot\mathbf{m})\,\mathbf{d}-(\mathbf{d}\cdot\mathbf{m})\,\mathbf{a}$$
$$=(\mathbf{a}\cdot(\mathbf{b}\times\mathbf{c}))\,\mathbf{d}-(\mathbf{d}\cdot(\mathbf{b}\times\mathbf{c}))\,\mathbf{a}$$
$$=[\mathbf{a}\ \mathbf{b}\ \mathbf{c}]\,\mathbf{d}-[\mathbf{b}\ \mathbf{c}\ \mathbf{d}]\,\mathbf{a}$$

Hence, $(\mathbf{a}\times\mathbf{d})\times(\mathbf{b}\times\mathbf{c})=[\mathbf{a}\ \mathbf{b}\ \mathbf{c}]\,\mathbf{d}-[\mathbf{b}\ \mathbf{c}\ \mathbf{d}]\,\mathbf{a}$ (Q)

2. (A) $\mathbf{F}=\mathbf{F_1}+\mathbf{F_2}=(3\hat{\mathbf{i}}-2\hat{\mathbf{j}}+\hat{\mathbf{k}})+(\hat{\mathbf{i}}+3\hat{\mathbf{j}}-5\hat{\mathbf{k}})$

$$\therefore\quad \mathbf{F}=4\hat{\mathbf{i}}+\hat{\mathbf{j}}-4\hat{\mathbf{k}}$$

and let $\mathbf{OA}=-2\hat{\mathbf{i}}+5\hat{\mathbf{k}}$,

$$\mathbf{OB}=3\hat{\mathbf{i}}-7\hat{\mathbf{j}}+2\hat{\mathbf{k}}$$

$\therefore\ \mathbf{AB}=\mathbf{OB}-\mathbf{OB}=5\hat{\mathbf{i}}-7\hat{\mathbf{j}}-3\hat{\mathbf{k}}$

Then $W_1=\mathbf{F}\cdot\mathbf{AB}=20-7+12=25$ unit

And also $\mathbf{F'}=\mathbf{P}+\mathbf{Q}+\mathbf{R}$

$$=(2\hat{\mathbf{i}}-5\hat{\mathbf{j}}+6\hat{\mathbf{k}})+(-\hat{\mathbf{i}}+2\hat{\mathbf{j}}-\hat{\mathbf{k}})+(2\hat{\mathbf{i}}+7\hat{\mathbf{j}})$$
$$=3\hat{\mathbf{i}}+4\hat{\mathbf{j}}+5\hat{\mathbf{k}}$$

and $\mathbf{OM}=4\hat{\mathbf{i}}-3\hat{\mathbf{j}}-2\hat{\mathbf{k}}$,

$$\mathbf{ON}=6\hat{\mathbf{i}}+\hat{\mathbf{j}}-3\hat{\mathbf{k}}$$

$\therefore\ \mathbf{MN}=\mathbf{ON}-\mathbf{OM}=2\hat{\mathbf{i}}+4\hat{\mathbf{j}}-\hat{\mathbf{k}}$

Then, $W_2=\mathbf{F'}\cdot\mathbf{MN}=6+16-5=17$ unit

$\because\quad W_1=25,\ W_2=17$

$\therefore\quad 3W_2-2W_1=1$ (S) and $2W_2-W_1=9$ (R)

(B) Let $\mathbf{OA}=-2\hat{\mathbf{i}}+5\hat{\mathbf{j}}+7\hat{\mathbf{k}}$ and

$$\mathbf{OB}=3\hat{\mathbf{i}}+7\hat{\mathbf{j}}+2\hat{\mathbf{k}}$$

$\therefore\quad \mathbf{AB}=\mathbf{OB}-\mathbf{OA}=5\hat{\mathbf{i}}+2\hat{\mathbf{j}}-5\hat{\mathbf{k}}$

and let $\mathbf{F_1}=2\hat{\mathbf{i}}-3\hat{\mathbf{j}}+\hat{\mathbf{k}},\ \mathbf{F_2}=\hat{\mathbf{i}}+5\hat{\mathbf{j}}-3\hat{\mathbf{k}}$

$$\therefore\quad \mathbf{F}=\mathbf{F_1}+\mathbf{F_2}=3\hat{\mathbf{i}}+2\hat{\mathbf{j}}-2\hat{\mathbf{k}}$$

Then, $W_1=\mathbf{F}\cdot\mathbf{AB}=15+4+10=29$ unit

and also, let

$$\mathbf{OM}=4\hat{\mathbf{i}}-3\hat{\mathbf{j}}-2\hat{\mathbf{k}}\ \text{ and }\ \mathbf{ON}=6\hat{\mathbf{i}}+\hat{\mathbf{j}}-3\hat{\mathbf{k}}$$

$$\therefore\quad \mathbf{MN}=\mathbf{ON}-\mathbf{OM}=2\hat{\mathbf{i}}+4\hat{\mathbf{j}}-\hat{\mathbf{k}}$$

and also let $\mathbf{P}=\hat{\mathbf{i}}-\hat{\mathbf{j}}+\hat{\mathbf{k}}$,

$$\mathbf{Q}=-\hat{\mathbf{i}}+2\hat{\mathbf{j}}-\hat{\mathbf{k}}\ \text{ and }\ \mathbf{R}=\hat{\mathbf{i}}+0\hat{\mathbf{j}}-\hat{\mathbf{k}}$$

$$\therefore\quad \mathbf{F'}=\mathbf{P}+\mathbf{Q}+\mathbf{R}=\hat{\mathbf{i}}+\hat{\mathbf{j}}-\hat{\mathbf{k}}$$

Then, $W_2=\mathbf{F'}\cdot\mathbf{MN}=2+4+1=7$ unit

$\because\quad W_1=29,\ W_2=7$

$$W_1-3W_2=8\ \text{(Q)}$$
and $W_1-4W_2=1$ (T)

(C) Let $F_1=5$ unit $F_2=3$ unit, $F_3=1$ unit

and let $\mathbf{a}=6\hat{\mathbf{i}}+2\hat{\mathbf{j}}+3\hat{\mathbf{k}}$,

$$\mathbf{b}=3\hat{\mathbf{i}}-2\hat{\mathbf{j}}+6\hat{\mathbf{k}}\ \text{ and }\ \mathbf{c}=2\hat{\mathbf{i}}-3\hat{\mathbf{j}}-6\hat{\mathbf{k}}$$

$$\therefore\quad \hat{\mathbf{a}}=\frac{(6\hat{\mathbf{i}}+2\hat{\mathbf{j}}+3\hat{\mathbf{k}})}{7},$$
$$\hat{\mathbf{b}}=\frac{(3\hat{\mathbf{i}}-2\hat{\mathbf{j}}+6\hat{\mathbf{k}})}{7}$$
and
$$\hat{\mathbf{c}}=\frac{(2\hat{\mathbf{i}}-3\hat{\mathbf{j}}-6\hat{\mathbf{k}})}{7}$$

$$\therefore\quad \mathbf{F_1}=F_1\,\hat{\mathbf{a}}=\frac{5}{7}(6\hat{\mathbf{i}}+2\hat{\mathbf{j}}+3\hat{\mathbf{k}})$$
$$\mathbf{F_2}=F_2\,\hat{\mathbf{b}}=\frac{3}{7}(3\hat{\mathbf{i}}-2\hat{\mathbf{j}}+6\hat{\mathbf{k}})$$
and
$$\mathbf{F_3}=F_3\,\hat{\mathbf{c}}=\frac{1}{7}(2\hat{\mathbf{i}}-3\hat{\mathbf{j}}-6\hat{\mathbf{k}})$$

$$\therefore\quad \mathbf{F}=\mathbf{F_1}+\mathbf{F_2}+\mathbf{F_3}=\frac{1}{7}(41\hat{\mathbf{i}}+\hat{\mathbf{j}}+27\hat{\mathbf{k}})$$

and $\mathbf{OA}=2\hat{\mathbf{i}}-\hat{\mathbf{j}}-3\hat{\mathbf{k}},\ \mathbf{OB}=5\hat{\mathbf{i}}-\hat{\mathbf{j}}+\hat{\mathbf{k}}$

$$\therefore\quad \mathbf{AB}=\mathbf{OB}-\mathbf{OA}=3\hat{\mathbf{i}}+4\hat{\mathbf{k}}$$

Then, $W_1=\mathbf{F}\cdot\mathbf{AB}$
$$=\frac{1}{7}(123+0+108)=\frac{1}{7}(231)=33\text{ unit}$$

and also, let

$$P=1\text{ unit},\ Q=2\text{ unit},\ R=3\text{ unit}$$

and let $\mathbf{p}=\hat{\mathbf{i}}+2\hat{\mathbf{j}}+2\hat{\mathbf{k}}$,

$$\mathbf{q}=2\hat{\mathbf{i}}-2\hat{\mathbf{j}}+\hat{\mathbf{k}},\ \mathbf{r}=2\hat{\mathbf{i}}-\hat{\mathbf{j}}-2\hat{\mathbf{k}}$$

$$\therefore\quad \hat{\mathbf{p}}=\frac{1}{3}(\hat{\mathbf{i}}+2\hat{\mathbf{j}}+2\hat{\mathbf{k}}),$$
$$\hat{\mathbf{q}}=\frac{1}{3}(2\hat{\mathbf{i}}-2\hat{\mathbf{j}}+\hat{\mathbf{k}})$$
and
$$\hat{\mathbf{r}}=\frac{1}{3}(2\hat{\mathbf{i}}-\hat{\mathbf{j}}-2\hat{\mathbf{k}})$$

$$\therefore\quad \mathbf{p}=P\,\hat{\mathbf{p}}=\frac{1}{3}(\hat{\mathbf{i}}+2\hat{\mathbf{j}}+2\hat{\mathbf{k}})$$
$$\mathbf{Q}=Q\,\hat{\mathbf{q}}=\frac{2}{3}(2\hat{\mathbf{i}}-2\hat{\mathbf{j}}+\hat{\mathbf{k}})$$
and
$$\mathbf{R}=R\,\hat{\mathbf{r}}=\frac{3}{3}(2\hat{\mathbf{i}}-\hat{\mathbf{j}}-2\hat{\mathbf{k}})$$

$$\therefore\quad \mathbf{F'}=\mathbf{P}+\mathbf{Q}+\mathbf{R}=\frac{1}{3}(11\hat{\mathbf{i}}-5\hat{\mathbf{j}}-2\hat{\mathbf{k}})$$

and $\mathbf{OM}=-2\hat{\mathbf{i}}-\hat{\mathbf{j}}+5\hat{\mathbf{k}}$,

$$\mathbf{ON}=3\hat{\mathbf{i}}-4\hat{\mathbf{j}}-5\hat{\mathbf{k}}$$

$$\therefore\quad \mathbf{MN}=\mathbf{ON}-\mathbf{OM}=5\hat{\mathbf{i}}-3\hat{\mathbf{j}}-10\hat{\mathbf{k}}$$

Then, $W_2=\mathbf{F'}\cdot\mathbf{MN}=\frac{1}{3}(55+15+20)=30$ unit

$\because\quad W_1=33,\ W_2=30$

$\therefore\quad W_1-W_2=3$ (P)

3. If **a**, **b** and **c** are three coterminus edges of parallelopiped, triangular prism and tetrahedron, then

$$V_1 = [\mathbf{a}\ \mathbf{b}\ \mathbf{c}],\ V_2 = \frac{1}{2}[\mathbf{a}\ \mathbf{b}\ \mathbf{c}] \text{ and}$$

$$V_3 = \frac{1}{6}[\mathbf{a}\ \mathbf{b}\ \mathbf{c}]$$

$$\therefore\quad V_1 : V_2 : V_3 = 1 : \frac{1}{2} : \frac{1}{6} = 6 : 3 : 1$$

(A) Here, $\mathbf{a} = \hat{\mathbf{i}} - \hat{\mathbf{j}} - 6\hat{\mathbf{k}},\ \mathbf{b} = \hat{\mathbf{i}} - \hat{\mathbf{j}} + 4\hat{\mathbf{k}}$

and $\quad \mathbf{c} = 2\hat{\mathbf{i}} - 5\hat{\mathbf{j}} + 3\hat{\mathbf{k}}$

$$\therefore\quad V_1 = \begin{vmatrix} 1 & -1 & -6 \\ 1 & -1 & 4 \\ 2 & -5 & 3 \end{vmatrix}$$

$$= 1(-3+20) + 1(3-8) - 6(-5+2)$$

$$= 17 - 5 + 18 = 30$$

$$\therefore\quad V_2 = 15 \text{ and } V_3 = 5$$

$$V_1 + V_2 + V_3 = 50 \text{ (S)}$$

$$V_1 + 3V_3 = 30 + 15 = 45 = 3V_2 \text{ (R)}$$

and $\quad V_1 : V_2 : V_3 = 6 : 3 : 1$ (T)

(B) Here, $\mathbf{a} = -2\hat{\mathbf{i}} + 3\hat{\mathbf{j}} - 3\hat{\mathbf{k}},$

$$\mathbf{b} = 4\hat{\mathbf{i}} + 5\hat{\mathbf{j}} - 3\hat{\mathbf{k}}$$

and $\quad \mathbf{c} = 6\hat{\mathbf{i}} + 2\hat{\mathbf{j}} - 3\hat{\mathbf{k}}$

$$\therefore\ V_1 = \begin{vmatrix} -2 & 3 & -3 \\ 4 & 5 & -3 \\ 6 & 2 & -3 \end{vmatrix}$$

$$= -2(-15+6) - 3(-12+18) - 3(8-30)$$

$$= 18 - 18 + 66 = 66$$

$$\therefore V_2 = 33 \text{ and } V_3 = 11$$

$$\because\quad 2V_1 + 3V_3 = 132 + 33 = 165 = 5V_2 \text{ (P)}$$

$$V_1 + V_2 + V_3 = 110$$

$$V_1 + 3V_3 = 66 + 33 = 3V_2 \text{ (R)}$$

and $\quad V_1 : V_2 : V_3 = 6 : 3 : 1$ (T)

(C) Here, $\mathbf{a} = -3\hat{\mathbf{i}} + \hat{\mathbf{j}} + \hat{\mathbf{k}},\ \mathbf{b} = 4\hat{\mathbf{i}} + 2\hat{\mathbf{j}} + 4\hat{\mathbf{k}}$

and $\quad \mathbf{c} = 2\hat{\mathbf{i}} + 2\hat{\mathbf{j}}$

$$\therefore\quad V_1 = \begin{vmatrix} -3 & 1 & 1 \\ 4 & 2 & 4 \\ 2 & 2 & 0 \end{vmatrix} = -3(0-8) - 1(0-8) + 1(8-4)$$

$$= 24 + 8 + 4 = 36$$

$$\therefore\quad V_2 = 18 \text{ and } V_3 = 6$$

$$2V_1 + 3V_3 = 72 + 18 = 90 = 5V_2 \text{ (P)},$$

$$V_1 + V_2 + V_3 = 60 \text{ (Q)}$$

$$V_1 + 3V_3 = 36 + 18 = 54 = 3V_2 \text{ (R)}$$

$$V_1 : V_2 : V_3 = 6 : 3 : 1 \text{ (T)}$$

4. **(A)** $\mathbf{A} = \mathbf{b} + \mathbf{c},\ \mathbf{B} = \mathbf{c} + \mathbf{a}$ and $\mathbf{C} = \mathbf{a} + \mathbf{b}$

$$\mathbf{A} + \mathbf{B} + \mathbf{C} = 2(\mathbf{a} + \mathbf{b} + \mathbf{c})$$

$\because\quad \mathbf{a} + \mathbf{b},\ \mathbf{b} + \mathbf{c}$ and $\mathbf{c} + \mathbf{a}$ are coplanar

$$\therefore\quad [\mathbf{a}+\mathbf{b}\ \ \mathbf{b}+\mathbf{c}\ \ \mathbf{c}+\mathbf{a}] = 0 \text{ (R)}$$

$$\Rightarrow\quad (\mathbf{a}+\mathbf{b})\cdot\{(\mathbf{b}+\mathbf{c})\times(\mathbf{c}+\mathbf{a})\} = 0$$

$$\Rightarrow\quad (\mathbf{a}+\mathbf{b})\cdot\{\mathbf{b}\times\mathbf{c} + \mathbf{b}\times\mathbf{a} + \mathbf{c}\times\mathbf{c} + \mathbf{c}\times\mathbf{a}\} = 0$$

$$\Rightarrow\quad (\mathbf{a}+\mathbf{b})\cdot\{\mathbf{b}\times\mathbf{c} - \mathbf{a}\times\mathbf{b} + 0 + \mathbf{c}\times\mathbf{a}\} = 0$$

$$\Rightarrow \mathbf{a}\cdot(\mathbf{b}\times\mathbf{c}) - \mathbf{a}\cdot(\mathbf{a}\times\mathbf{b}) + \mathbf{a}\cdot(\mathbf{c}\times\mathbf{a})$$

$$+ \mathbf{b}\cdot(\mathbf{b}\times\mathbf{c}) - \mathbf{b}\cdot(\mathbf{a}\times\mathbf{b}) + \mathbf{b}\cdot(\mathbf{c}\times\mathbf{a}) = 0$$

$$\Rightarrow\quad [\mathbf{a}\ \mathbf{b}\ \mathbf{c}] - 0 + 0 + 0 - 0 + [\mathbf{b}\ \mathbf{c}\ \mathbf{a}] = 0$$

$$\Rightarrow\quad 2[\mathbf{a}\ \mathbf{b}\ \mathbf{c}] = 0$$

$$\therefore\quad [\mathbf{a}\ \mathbf{b}\ \mathbf{c}] = 0 \text{ (T)}$$

$\therefore$ **a**, **b**, **c** are coplanar (P)

$\because\quad \mathbf{a} + \mathbf{b},\ \mathbf{b} + \mathbf{c},\ \mathbf{c} + \mathbf{a}$ are coplanar

$$\therefore\quad \mathbf{a}+\mathbf{b}+\mathbf{b}+\mathbf{c}+\mathbf{c}+\mathbf{a} = 0$$

$$\Rightarrow\quad 2(\mathbf{a}+\mathbf{b}+\mathbf{c}) = 0$$

$$\therefore\quad \mathbf{a}+\mathbf{b}+\mathbf{c} = 0 \text{ (Q)}$$

Also, $\mathbf{a}\times\mathbf{b} + \mathbf{b}\times\mathbf{c} + \mathbf{c}\times\mathbf{a} = [\mathbf{a}\ \mathbf{b}\ \mathbf{c}]^2 = 0$ (S)

(B) $\because\ \mathbf{a}\times\mathbf{b},\ \mathbf{b}\times\mathbf{c}$ and $\mathbf{c}\times\mathbf{a}$ are coplanar

$$\Rightarrow\quad [\mathbf{a}\times\mathbf{b}\ \ \mathbf{b}\times\mathbf{c}\ \ \mathbf{c}\times\mathbf{a}] = 0$$

Also, $\quad \mathbf{a}\times\mathbf{b} + \mathbf{b}\times\mathbf{c} + \mathbf{c}\times\mathbf{a} = 0$ (S)

$$\Rightarrow\quad [\mathbf{a}\ \mathbf{b}\ \mathbf{c}]^2 = 0 \Rightarrow [\mathbf{a}\ \mathbf{b}\ \mathbf{c}] = 0 \text{ (T)}$$

$\Rightarrow$ **a**, **b**, **c** are coplanar (P) and $\mathbf{a}+\mathbf{b}+\mathbf{c} = 0$ (Q)

Also, $[\mathbf{a}+\mathbf{b}\ \ \mathbf{b}+\mathbf{c}\ \ \mathbf{c}+\mathbf{a}] = 2[\mathbf{a}\ \mathbf{b}\ \mathbf{c}] = 0$ (R)

(C) Let $\mathbf{A} = 2\mathbf{a} + \mathbf{b} + \mathbf{c},\ \mathbf{B} = \mathbf{a} + 2\mathbf{b} + \mathbf{c}$

and $\quad \mathbf{C} = \mathbf{a} + \mathbf{b} + 2\mathbf{c}$

$\therefore\quad \mathbf{A} + \mathbf{B} + \mathbf{C} = 0 \qquad$ [$\because$ **A**, **B**, **C** are coplanar]

$$\Rightarrow\quad 3(\mathbf{a}+\mathbf{b}+\mathbf{c}) = 0$$

$$\Rightarrow\quad \mathbf{a}+\mathbf{b}+\mathbf{c} = 0 \text{ (Q)}$$

$\therefore$ **a**, **b**, **c** are coplanar (P)

$$[\mathbf{a}+\mathbf{b}\ \ \mathbf{b}+\mathbf{c}\ \ \mathbf{c}+\mathbf{a}] = 2[\mathbf{a}\ \mathbf{b}\ \mathbf{c}] = 0 \text{ (R)}$$

$\mathbf{a}\times\mathbf{b} + \mathbf{b}\times\mathbf{c} + \mathbf{c}\times\mathbf{a} = 0$ (S) from above

Also, $\quad [\mathbf{a}\ \mathbf{b}\ \mathbf{c}] = 0$ (T)

33

Three Dimensional Geometry

Objective Questions Type I [Only one correct answer]

In each of the questions below, four choices are given of which only one is correct. You have to select the correct answer which is the most appropriate.

1. A mirror and a source of light are situated at the origin O and a point on OX respectively. A ray of light from the source strikes the mirror and is reflected. If the DRs of the normal to the plane of mirror are 1, – 1, 1, then DCs for the reflected ray are

(a) $\frac{1}{3}, \frac{2}{3}, \frac{2}{3}$ (b) $-\frac{1}{3}, \frac{2}{3}, \frac{2}{3}$

(c) $-\frac{1}{3}, -\frac{2}{3}, -\frac{2}{3}$ (d) $-\frac{1}{3}, -\frac{2}{3}, \frac{2}{3}$

2. The direction cosines of a line satisfy the relations $\lambda(l + m) = n$ and $mn + nl + lm = 0$. The value of λ, for which the two lines are perpendicular to each other, is

(a) 1 (b) 2

(c) 1/2 (d) None of these

3. Let P be any point on the plane $lx + my + nz = p$ and Q be a point on the line OP such that $OP \cdot OQ = p^2$. The locus of the point Q is

(a) $lx + my + nz - p = x^2 + y^2 + z^2$

(b) $lx + my + nz = p(x^2 + y^2 + z^2)$

(c) $p(lx + my + nz) = x^2 + y^2 + z^2$

(d) $x^2 + y^2 + z^2 = p^2$

4. The coordinates of a point on the line $\frac{x-1}{2} = \frac{y+1}{-3} = z$ at a distance $4\sqrt{14}$ from the point (1, – 1, 0) nearer the origin are

(a) (9, – 13, 4) (b) $(8\sqrt{14}, -12, -1)$

(c) $(-8\sqrt{14}, 12, 1)$ (d) (–7, 11, – 4)

5. The length of the perpendicular from the point (2, – 1, 4) on the straight line, $\frac{x+3}{10} = \frac{y-2}{-7} = \frac{z}{1}$ is

[JEE Main 2019, 8 April]

(a) greater than 3 but less than 4

(b) less than 2

(c) greater than 2 but less than 3

(d) greater than 4

6. The locus of the point, the sum of squares of whose distances, from the planes $x - z = 0$, $x - 2y + z = 0$ and $x + y + z = 0$ is 36 is

(a) $x^2 + y^2 + z^2 = 6$

(b) $x^2 + y^2 + z^2 = 36$

(c) $x^2 + y^2 + z^2 = 216$

(d) $x^{-2} + y^{-2} + z^{-2} = \frac{1}{36}$

7. The line $\frac{x-2}{3} = \frac{y+1}{2} = \frac{z-1}{-1}$ intersects the curve $xy = c^2$, $z = 0$ if c is equal to

(a) ± 1 (b) $\pm\frac{1}{3}$

(c) $\pm\sqrt{5}$ (d) None of these

8. Equation of plane which passes through the point of intersection of lines $\frac{x-1}{3} = \frac{y-2}{1} = \frac{z-3}{2}$ and $\frac{x-3}{1} = \frac{y-1}{2} = \frac{z-2}{3}$ and at greatest distance from the point (0, 0, 0) is

(a) $4x + 3y + 5z = 25$ (b) $4x + 3y + 5z = 50$

(c) $3x + 4y + 5z = 49$ (d) $x + 7y - 5z = 2$

9. The distance of the point (1, –2, 3) from the plane $x - y + z = 5$ measured paralel to line $\frac{x}{2} = \frac{y}{3} = \frac{z}{-6}$ is

[JEE Main 2020, 4 September]

(a) $\frac{1}{7}$ (b) 1

(c) 7 (d) $\frac{7}{5}$

10. The pair of lines whose direction cosines are given by the equations $3l + m + 5n = 0$, $6mn - 2nl + 5lm = 0$, are

(a) parallel (b) perpendicular

(c) inclined at $\cos^{-1}\left(\frac{1}{6}\right)$ (d) None of these

11. The distance of the point $A(-2, 3, 1)$ from the line PQ through $P(-3, 5, 2)$ which make equal angles with the axes is
(a) $\frac{2}{\sqrt{3}}$ (b) $\sqrt{\frac{14}{3}}$
(c) $\frac{16}{\sqrt{3}}$ (d) $\frac{5}{\sqrt{3}}$

12. The equation of the plane through the point $(2, 5, -3)$ perpendicular to the planes $x + 2y + 2z = 1$ and $x - 2y + 3z = 4$ is
(a) $3x - 4y + 2z - 20 = 0$ (b) $7x - y + 5z = 30$
(c) $x - 2y + z = 11$ (d) $10x - y - 4z = 27$

13. The equation of the plane through the point $(0, -4, -6)$ and $(-2, 9, 3)$ and perpendicular to the plane $x - 4y - 2z = 8$ is
(a) $3x + 3y - 2z = 0$ (b) $x - 2y + z = 2$
(c) $2x + y - z = 2$ (d) $5x - 3y + 2z = 0$

14. The equation of the plane passing through the points $(3, 2, -1)$, $(3, 4, 2)$ and $(7, 0, 6)$ is $5x + 3y - 2z = \lambda$, where λ is
(a) 23 (b) 21
(c) 19 (d) 27

15. Reflection of the line $\frac{x-1}{-1} = \frac{y-2}{3} = \frac{z-4}{1}$ in the plane $x + y + z = 7$ is
(a) $\frac{x-1}{3} = \frac{y-2}{1} = \frac{z-4}{1}$ (b) $\frac{x-1}{-3} = \frac{y-2}{-1} = \frac{z-4}{1}$
(c) $\frac{x-1}{-3} = \frac{y-2}{1} = \frac{z-4}{-1}$ (d) $\frac{x-1}{3} = \frac{y-2}{1} = \frac{z-4}{1}$

16. If the equation of plane passing through the mirror image of a point $(2, 3, 1)$ with respect to line $\frac{x+1}{2} = \frac{y-3}{1} = \frac{z+2}{-1}$ and containing the line $\frac{x-2}{3} = \frac{1-y}{2} = \frac{z+1}{1}$ is $\alpha x + \beta y + \gamma z = 24$, then $\alpha + \beta + \gamma$ is equal to **[JEE Main 2021, 17 March]**
(a) 20 (b) 19
(c) 18 (d) 21

17. The point on the line $\frac{x-2}{1} = \frac{y+3}{-2} = \frac{z+5}{-2}$ at a distance of 6 from the point $(2, -3, -5)$ is
(a) $(3, -5, -3)$ (b) $(4, -7, -9)$
(c) $(0, 2, -1)$ (d) $(-3, 5, 3)$

18. The plane passing through the point $(5, 1, 2)$ perpendicular to the line $2(x - 2) = y - 4 = z - 5$ will meet the line in the point
(a) $(1, 2, 3)$ (b) $(2, 3, 1)$
(c) $(1, 3, 2)$ (d) $(3, 2, 1)$

19. If for some $\alpha \in R$, the lines
$L_1 : \frac{x+1}{2} = \frac{y-2}{-1} = \frac{z-1}{1}$ and
$L_2 : \frac{x+2}{\alpha} = \frac{y+1}{5-\alpha} = \frac{z+1}{1}$ are coplanar, then the line L_2 passes through the point **[JEE Main 2020, 5 September]**
(a) $(10, 2, 2)$ (b) $(2, -10, -2)$
(c) $(10, -2, -2)$ (d) $(-2, 10, 2)$

20. Let $A(1, 1, 1)$, $B(2, 3, 5)$, $C(-1, 0, 2)$ be three points, then equation of a plane parallel to the plane ABC which is at a distance 2 from origin, is
(a) $2x - 3y + z + 2\sqrt{14} = 0$ (b) $2x - 3y + z - \sqrt{14} = 0$
(c) $2x - 3y + z + 2 = 0$ (d) $2x - 3y + z - 2 = 0$

21. The angle between the lines whose direction cosines satisfy the equation $l + m + n = 0$ and $l^2 = m^2 + n^2$ is
(a) $\frac{\pi}{3}$ (b) $\frac{\pi}{4}$
(c) $\frac{\pi}{6}$ (d) $\frac{\pi}{2}$

22. The acute angle between two lines whose direction cosines are given by the relation between $l + m + n = 0$ and $l^2 + m^2 - n^2 = 0$ is
(a) $\pi / 2$ (b) $\pi / 3$
(c) $\pi / 4$ (d) None of these

23. If the direction cosines of a variable line in two adjacent positions be l, m, n and $l + a, m + b, n + c$ and the small angle between the two positions be θ, then
(a) $\theta = a + b + c$ (b) $\theta^2 = a^2 + b^2 + c^2$
(c) $|\theta| = |a| + |b| + |c|$ (d) None of these

24. The vertices B and C of a ΔABC lie on the line, $\frac{x+2}{3} = \frac{y-1}{0} = \frac{z}{4}$ such that $BC = 5$ units. Then, the area (in sq units) of this triangle, given that point $A(1, -1, 2)$ is **[JEE Main 2019, 9 April]**
(a) $\sqrt{34}$ (b) $2\sqrt{34}$
(c) $5\sqrt{17}$ (d) 6

25. The direction ratios of a normal to the plane passing through $(1, 0, 0)$, $(0, 1, 0)$ and making an angle $\pi / 4$ with the plane $x + y = 3$ are
(a) $(1, \sqrt{2}, 1)$ (b) $(1, 1, \sqrt{2})$
(c) $(1, 1, 2)$ (d) $(\sqrt{2}, 1, 1)$

26. The intercepts made on the axes by the plane which bisects the line joining the points $(1, 2, 3)$ and $(-3, 4, 5)$ at right angles are
(a) $\left(-\frac{9}{2}, 9, 9\right)$ (b) $\left(\frac{9}{2}, 9, 9\right)$
(c) $\left(9, -\frac{9}{2}, 9\right)$ (d) $\left(9, \frac{9}{2}, 9\right)$

27. A lines makes angles $\alpha, \beta, \gamma, \delta$ with the four diagonals of a cube. Then $\cos^2 \alpha + \cos^2 \beta + \cos^2 \gamma + \cos^2 \delta$ is
(a) $\frac{4}{3}$ (b) $\frac{2}{3}$
(c) 3 (d) None of these

28. The equation of the plane passing through the point (1, 1, 1) and perpendicular to the planes $2x + y - 2z = 5$ and $3x - 6y - 2z = 7$ is **[JEE Advanced 2017]**
(a) $14x + 2y - 15z = 1$ (b) $-14x + 2y + 15z = 3$
(c) $14x - 2y + 15z = 27$ (d) $14x + 2y + 15z = 31$

29. A plane moves such that its distance from the origin is a constant p. If it intersects the coordinate axes at A, B, C then the locus of the centroid of the triangle ABC is
(a) $\frac{1}{x^2} + \frac{1}{y^2} + \frac{1}{z^2} = \frac{1}{p^2}$ (b) $\frac{1}{x^2} + \frac{1}{y^2} + \frac{1}{z^2} = \frac{9}{p^2}$
(c) $\frac{1}{x^2} + \frac{1}{y^2} + \frac{1}{z^2} = \frac{2}{p^2}$ (d) $\frac{1}{x^2} + \frac{1}{y^2} + \frac{1}{z^2} = \frac{4}{p^2}$

30. Let P be a plane passing through the points (2, 1, 0), (4, 1, 1) and (5, 0, 1) and R be any point (2, 1, 6). Then the image of R in the plane P is **[JEE Main 2020, 7 January]**
(a) (6, 5, 2) (b) (4, 3, 2)
(c) (6, 5, –2) (d) (3, 4, –2)

31. The line $\frac{x}{2} = -\frac{y}{3} = \frac{z}{1}$ is vertical. The direction cosines of the line of greatest slope in the plane $3x - 2y + z = 5$ are proportional to
(a) (16, 11, – 1) (b) (–11, 16, 1)
(c) (16, 11, 1) (d) (11, 16, – 1)

32. The symmetric form of the equations of the line $x + y - z = 1$, $2x - 3y + z = 2$ is
(a) $\frac{x}{2} = \frac{y}{3} = \frac{z}{5}$ (b) $\frac{x}{2} = \frac{y}{3} = \frac{z-1}{5}$
(c) $\frac{x-1}{2} = \frac{y}{3} = \frac{z}{5}$ (d) $\frac{x}{3} = \frac{y}{2} = \frac{z}{5}$

33. The equation of the plane which passes through the X-axis and perpendicular to the line $\frac{(x-1)}{\cos\theta} = \frac{(y+2)}{\sin\theta} = \frac{(z-3)}{0}$ is
(a) $x \tan\theta + y \sec\theta = 0$ (b) $x \sec\theta + y \tan\theta = 0$
(c) $x \cos\theta + y \sin\theta = 0$ (d) $x \sin\theta - y \cos\theta = 0$

34. The edge of a cube is of length of a. The shortest distance between the diagonal of a cube and an edge skew to it is
(a) $a\sqrt{2}$ (b) a
(c) $\frac{\sqrt{2}}{a}$ (d) $\frac{a}{\sqrt{2}}$

35. The equation of the plane passing through the intersection of the planes $2x - 5y + z = 3$ and $x + y + 4z = 5$ and parallel to the plane $x + 3y + 6z = 1$ is $x + 3y + 6z = k$, where k is
(a) 5 (b) 3
(c) 7 (d) 2

36. The lines which intersect the skew lines $y = mx, z = c$; $y = -mx, z = -c$ and the X-axis lie on the surface
(a) $cz = mxy$
(b) $cy = mxz$
(c) $xy = cmz$
(d) None of the above

37. The length of projection of the line segment joining the points (1, 0, – 1) and (– 1, 2, 2) on the plane $x + 3y - 5z = 6$ is equal to
(a) 2 (b) $\sqrt{\frac{271}{53}}$
(c) $\sqrt{\frac{472}{31}}$ (d) $\sqrt{\frac{474}{35}}$

38. The plane $4x + 7y + 4z + 81 = 0$ is rotated through a right angle about its line of intersection with the plane $5x + 3y + 10z = 25$. The equation of the plane in its new position is $x - 4y + 6z = k$, where k is
(a) 106 (b) –89 (c) 73 (d) 37

39. The reflection of the point $P(1, 0, 0)$ in the line $\frac{x-1}{2} = \frac{y+1}{-3} = \frac{z+10}{8}$ is
(a) (3, – 4, – 2) (b) (5, – 8, – 4)
(c) (1, – 1, – 10) (d) (2, – 3, 8)

40. If from the point $P(a, b, c)$ perpendiculars PL, PM be drawn to YOZ and ZOX planes, then the equation of the plane OLM is
(a) $\frac{x}{a} + \frac{y}{b} + \frac{z}{c} = 0$ (b) $\frac{x}{a} - \frac{y}{b} + \frac{z}{c} = 0$
(c) $\frac{x}{a} + \frac{y}{b} - \frac{z}{c} = 0$ (d) $\frac{x}{a} - \frac{y}{b} - \frac{z}{c} = 0$

41. A variable plane makes with the coordinate planes, a tetrahedron of constant volume $64k^3$. Then the locus of the centroid of tetrahedron is the surface
(a) $xyz = 6k^3$ (b) $xy + yz + zx = 6k^2$
(c) $x^2 + y^2 + z^2 = 8k^2$ (d) None of these

42. If the line $\frac{x-1}{2} = \frac{y+1}{3} = \frac{z-2}{4}$ meets the plane, $x + 2y + 3z = 15$ at a point P, then the distance of P from origin is **[JEE Main 2019, 9 April]**
(a) $7/2$ (b) $9/2$
(c) $\sqrt{5}/2$ (d) $2\sqrt{5}$

43. The perpendicular distance of the origin from the plane which makes intercepts 12, 3 and 4 on X, Y, Z axes respectively, is
(a) 13 (b) 11
(c) 17 (d) None of these

44. A plane meets the coordinate axes at A, B, C and the foot of the perpendicular from the origin O to the plane is P, $OA = a$, $OB = b$, $OC = c$. If P is the centroid of the ΔABC, then
(a) $a + b + c = 0$ (b) $|a| = |b| = |c|$
(c) $\frac{1}{a} + \frac{1}{b} + \frac{1}{c} = 0$ (d) None of these

45. $ABCD$ is a tetrahedron. A_1, B_1, C_1, D_1 are respectively the centroids of the triangles BCD, ACD, ABD and ABC; AA_1, BB_1, CC_1, DD_1 divide one another in the ratio
(a) 1 : 1 (b) 2 : 1
(c) 3 : 1 (d) 1 : 3

46. A plane makes intercepts OA, OB, OC whose measurements are a, b, c on the axes OX, OY, OZ. The area of the triangle ABC is

(a) $\frac{1}{2}(ab+bc+ca)$ (b) $\frac{1}{2}(a^2b^2+b^2c^2+c^2a^2)^{1/2}$

(c) $\frac{1}{2}abc\,(a+b+c)$ (d) $\frac{1}{2}(a+b+c)^2$

47. Let P be the image of the point (3, 1, 7) with respect to the plane $x-y+z=3$. Then, the equation of the plane passing through P and containing the straight line $\frac{x}{1}=\frac{y}{2}=\frac{z}{1}$ is **[JEE Advanced 2016]**

(a) $x+y-3z=0$ (b) $3x+z=0$
(c) $x-4y+7z=0$ (d) $2x-y=0$

48. The distance of the point $P(-2, 3, 1)$ from the line QR, through $Q(-3, 6, 2)$ which makes equal angles with the axes is

(a) 3 (b) 8
(c) $\sqrt{2}$ (d) $2\sqrt{2}$

49. If for $a>0$, the feet of perpendiculars from the points $A(a, -2a, 3)$ and $B(0, 4, 5)$ on the plane $lx+my+nz=0$ are points $C(0, -a, -1)$ and D respectively, then the length of line segment CD is equal to **[JEE Main 2021, 16 March]**

(a) $\sqrt{31}$ (b) $\sqrt{41}$
(c) $\sqrt{55}$ (d) $\sqrt{66}$

50. The coordinates of the foot of the perpendicular drawn from the origin to the line joining the points $(-9, 4, 5)$ and $(10, 0, -1)$ will be

(a) $(-3, 2, 1)$ (b) $(1, 2, 2)$
(c) $(4, 5, 3)$ (d) None of these

51. The straight lines whose direction cosines are given by $al+bm+cn=0$, $fmn+gnl+hlm=0$ are perpendicular if

(a) $\frac{f}{a}+\frac{g}{b}+\frac{h}{c}=0$

(b) $\frac{a^2}{f}+\frac{b^2}{g}+\frac{c^2}{h}=0$

(c) $a^2(g+h)+b^2(h+f)+c^2(f+g)=0$

(d) None of the above

52. If $A(3, 2, 3)$, $B(1, 4, 6)$ and $C(7, 4, 5)$ are the three vertices of a parallelogram $ABCD$, then the angle between its diagonal through D and the side DC is

(a) $\cos^{-1}\left(\frac{16}{\sqrt{357}}\right)$ (b) $\cos^{-1}\left(\frac{5}{\sqrt{126}}\right)$

(c) $\cos^{-1}\left(\frac{5}{\sqrt{21}}\right)$ (d) $\cos^{-1}\left(\frac{2}{\sqrt{357}}\right)$

53. The line $\frac{x+1}{2}=\frac{y+1}{3}=\frac{z+1}{4}$ meets the plane $x+2y+3z=14$, in the point

(a) $(3, -2, 5)$ (b) $(3, 2, -5)$
(c) $(2, 0, 4)$ (d) $(1, 2, 3)$

54. The foot of the perpendicular from $P(1, 0, 2)$ to the line $\frac{x+1}{3}=\frac{y-2}{-2}=\frac{z+1}{-1}$ is the point

(a) $(1, 2, -3)$ (b) $\left(\frac{1}{2}, 1, -\frac{3}{2}\right)$

(c) $(2, 4, -6)$ (d) $(2, 3, 6)$

55. Let P be a plane $lx+my+nz=0$ containing the line, $\frac{1-x}{1}=\frac{y+4}{2}=\frac{z+2}{3}$. If plane P divides the line segment AB joining points $A(-3, -6, 1)$ and $B(2, 4, -3)$ in ratio $k:1$ then the value of k is equal to **[JEE Main 2021, 16 March]**

(a) 1.5 (b) 3
(c) 2 (d) 4

56. The plane containing the two lines $\frac{x-3}{1}=\frac{y-2}{4}=\frac{z-1}{5}$ and $\frac{x-2}{1}=\frac{y+3}{-4}=\frac{z+1}{5}$ is $11x+my+nz=28$, where

(a) $m=-1, n=3$ (b) $m=1, n=-3$
(c) $m=-1, n=-3$ (d) $m=1, n=3$

57. The projection of the line $\frac{x+1}{-1}=\frac{y}{2}=\frac{z-1}{3}$ on the plane $x-2y+z=6$ is the line of intersection of this plane with the plane

(a) $2x+y+2=0$ (b) $3x+y-z=2$
(c) $2x-3y+8z=3$ (d) None of these

58. A variable plane passes through a fixed point $(1, -2, 3)$ and meets the coordinate axes in A, B, C. The locus of the point of intersection of the planes through A, B, C parallel to the coordinate planes is the surface

(a) $xy-\frac{1}{2}yz+\frac{1}{3}zx=6$ (b) $yz-2zx+3xy=xyz$

(c) $xy-2yz+3zx=3xyz$ (d) None of these

59. The distance of the point $(2, 1, -2)$ from the line $\frac{x-1}{2}=\frac{y+1}{1}=\frac{z-3}{-3}$ measured parallel to the plane $x+2y+z=4$ is

(a) $\sqrt{10}$ (b) $\sqrt{20}$ (c) $\sqrt{5}$ (d) $\sqrt{30}$

60. The shortest distance between the lines $\frac{x-3}{2}=\frac{y+15}{-7}=\frac{z-9}{5}$ and $\frac{x+1}{2}=\frac{y-1}{1}=\frac{z-9}{-3}$ is

(a) $2\sqrt{3}$ (b) $4\sqrt{3}$ (c) $3\sqrt{6}$ (d) $5\sqrt{6}$

61. Equation of the line of shortest distance between the lines $\frac{x}{2}=\frac{y}{-3}=\frac{z}{1}$ and $\frac{x-2}{3}=\frac{y-1}{-5}=\frac{z+2}{2}$ is

(a) $3(x-21)=3y+92=3z-32$

(b) $\frac{x-(62/3)}{1/3}=\frac{y-31}{1/3}=\frac{z+(32/3)}{1/3}$

(c) $\frac{x-21}{1/3}=\frac{y-(92/3)}{1/3}=\frac{z+(32/3)}{1/3}$

(d) $\frac{x-2}{1/3}=\frac{y+3}{1/3}=\frac{z-1}{1/3}$

62. The equation to the plane through the points $(2, -1, 0)$, $(3, -4, 5)$ parallel to a line with direction cosines proportional to 2, 3, 4 is $9x - 2y - 3z = k$, where k is

(a) 20 (b) -20
(c) 10 (d) -10

63. Through a point $P(f, g, h)$ a plane is drawn at right angles to OP, to meet the axes in A, B, C. If $OP = r$, the centroid of the ΔABC is

(a) $\left(\frac{f}{3r}, \frac{g}{3r}, \frac{h}{3r}\right)$ (b) $\left(\frac{r^2}{3f^2}, \frac{r^2}{3g^2}, \frac{r^2}{3h^2}\right)$

(c) $\left(\frac{r^2}{3f}, \frac{r^2}{3g}, \frac{r^2}{3h}\right)$ (d) None of these

64. The plane $lx + my = 0$ is rotated about its line of intersection with the xOy plane through an angle α. Then the equation of the plane is $lx + my + nz = 0$, where n is

(a) $\pm\sqrt{(l^2 + m^2)}\cos\alpha$ (b) $\pm\sqrt{(l^2 + m^2)}\sin\alpha$
(c) $\pm\sqrt{(l^2 + m^2)}\tan\alpha$ (d) None of these

65. The plane containing the line $\frac{x-3}{2} = \frac{y+2}{-1} = \frac{z-1}{3}$ and also containing its projection on the plane $2x + 3y - z = 5$, contains which one of the following points? **[JEE Main 2019, 11 January]**

(a) $(-2, 2, 2)$ (b) $(2, 2, 0)$
(c) $(2, 0, -2)$ (d) $(0, -2, 2)$

Objective Questions Type II [One or more than one correct answer(s)]

In each of the questions below four choices of which one or more than one are correct. You have to select the correct answer(s) accordingly.

1. The direction ratios of the bisector of the angle between the lines whose direction cosines are $l_1, m_1, n_1; l_2, m_2, n_2$ are

(a) $l_1 + l_2, m_1 + m_2, n_1 + n_2$
(b) $l_1 - l_2, m_1 - m_2, n_1 - n_2$
(c) $l_1m_2 - l_2m_1, m_1n_2 - m_2n_1, n_1l_2 - n_2l_1$
(d) $l_1m_2 + l_2m_1, m_1n_2 + m_2n_1, n_1l_2 + n_2l_1$

2. The points $A(4, 5, 10)$, $B(2, 3, 4)$ and $C(1, 2, -1)$ are three vertices of a parallelogram $ABCD$, then

(a) Vector equation of AB is $2\hat{\mathbf{i}} + 3\hat{\mathbf{j}} + 4\hat{\mathbf{k}} + \lambda(\hat{\mathbf{i}} + \hat{\mathbf{j}} + 3\hat{\mathbf{k}})$
(b) Cartesian equation of BC is $\frac{x-2}{1} = \frac{y-3}{1} = \frac{z-4}{5}$
(c) Coordinates of D are $(3, 4, 5)$
(d) $ABCD$ is a rectangle

3. The lines $\frac{x-2}{1} = \frac{y-3}{1} = \frac{z-4}{-k}$ and $\frac{x-1}{k} = \frac{y-4}{2} = \frac{z-5}{1}$ are coplanar if

(a) $k = 0$ (b) $k = -1$
(c) $k = -3$ (d) $k = 3$

4. The plane $x - 2y + 7z + 21 = 0$

(a) contains the line $\frac{x+1}{-3} = \frac{y-3}{2} = \frac{z+2}{1}$
(b) contains the point $(0, 7, -1)$
(c) is perpendicular to the line $\frac{x}{1} = \frac{y}{-2} = \frac{z}{7}$
(d) is parallel to the plane $x - 2y + 7z = 0$

5. Let PM be the perpendicular from the point $(1, 2, 3)$ to xy-plane. If OP makes an angle θ with the positive direction of the Z-axis and OM makes an angle ϕ with the positive direction of X-axis, where O is the origin then (θ and ϕ are acute angles)

(a) $\tan\theta = \frac{\sqrt{5}}{3}$ (b) $\sin\theta\sin\phi = \frac{2}{\sqrt{14}}$
(c) $\tan\phi = 2$ (d) $\cos\theta\cos\phi = \frac{1}{\sqrt{14}}$

6. If p_1, p_2, p_3 denote the distances of the plane $2x - 3y + 4z + 2 = 0$ from the planes $2x - 3y + 4z + 6 = 0$, $4x - 6y + 8z + 3 = 0$ and $2x - 3y + 4z - 6 = 0$ respectively. Then

(a) $p_1 + 8p_2 - p_3 = 0$ (b) $p_3 = 16p_2$
(c) $8p_2 = p_1$ (d) $p_1 + 2p_2 + 3p_3 = \sqrt{29}$

7. Let $\alpha, \beta, \gamma, \delta$ be real numbers such that $\alpha^2 + \beta^2 + \gamma^2 \neq 0$ and $\alpha + \gamma = 1$. Suppose the point $(3, 2, -1)$ is the mirror image of the point $(1, 0, -1)$ with respect to the plane $\alpha x + \beta y + \gamma z = \delta$. Then which of the following statements is/are TRUE? **[JEE Advanced 2020]**

(a) $\alpha + \beta = 2$ (b) $\delta - \gamma = 3$
(c) $\delta + \beta = 4$ (d) $\alpha + \beta + \gamma = \delta$

8. Let P_1 denote the equation of a plane to which the vector $(\hat{\mathbf{i}} + \hat{\mathbf{j}})$ is normal and which contains the line whose equation is $\mathbf{r} = \hat{\mathbf{i}} + \hat{\mathbf{j}} + \hat{\mathbf{k}} + \lambda(\hat{\mathbf{i}} - \hat{\mathbf{j}} - \hat{\mathbf{k}})$ and P_2 denote the equation of the plane containing the line L and a point with position vector $\hat{\mathbf{j}}$. Which of the following holds good?

(a) The equation of P_1 is $x + y = 2$
(b) The equation of P_2 is $\mathbf{r}\cdot(\hat{\mathbf{i}} - 2\hat{\mathbf{j}} + \hat{\mathbf{k}}) = 2$
(c) The acute angle between P_1 and P_2 is $\cot^{-1}(\sqrt{3})$.
(d) The angle between the plane P_2 and the line L is $\tan^{-1}\sqrt{3}$.

9. The value(s) of λ, for which the triangle with vertices $(6, 10, 10)$, $(1, 0, -5)$ and $(6, -10, \lambda)$ will be a right angled triangle is/are

(a) 1 (b) $\frac{70}{3}$
(c) 35 (d) 0

10. If $P(2, 3, 1)$ is a point and $L \equiv x - y - z - 2 = 0$ is a plane then
(a) origin and P lie on the same side of the plane
(b) distance of P from the plane is $\frac{4}{\sqrt{3}}$
(c) foot of perpendicular is $\left(\frac{10}{3}, \frac{5}{3}, -\frac{1}{3}\right)$
(d) image of point P by the plane $\left(\frac{10}{3}, \frac{5}{3}, -\frac{1}{3}\right)$

11. In R^3, let L be a straight line passing through the origin. Suppose that all the points on L are at a constant distance from the two planes $P_1 : x + 2y - z + 1 = 0$. and $P_2 : 2x - y + z - 1 = 0$. Let M be the locus of the feet of the perpendiculars drawn from the points on L to the plane P_1. Which of the following points lie(s) on M?
(a) $\left(0, -\frac{5}{6}, -\frac{2}{3}\right)$ (b) $\left(-\frac{1}{6}, -\frac{1}{3}, \frac{1}{6}\right)$
(c) $\left(-\frac{5}{6}, 0, \frac{1}{6}\right)$ (d) $\left(-\frac{1}{3}, 0, \frac{2}{3}\right)$

12. The lines $\frac{x-1}{3} = \frac{y-1}{-1} = \frac{z+1}{0}$ and $\frac{x-4}{2} = \frac{y+0}{0} = \frac{z+1}{3}$
(a) do not intersect (b) intersect
(c) intersect at $(4, 0, -1)$ (d) intersect at $(1, 1, -1)$

13. If $OABC$ is a tetrahedron such that $OA^2 + BC^2 = OB^2 + CA^2 = OC^2 + AB^2$, then
(a) $OA \perp BC$ (b) $OB \perp CA$
(c) $OC \perp AB$ (d) $AB \perp BC$

14. The extremities of a diameter of a sphere lie on positive Y and positive Z-axes at distances 2 and 4 from the origin, respectively, then
(a) sphere passes through the origin
(b) centre of the sphere is $(0, 1, 2)$
(c) radius of the sphere is $\sqrt{5}$
(d) equation of a diameter is $\frac{x}{0} = \frac{y-2}{1} = \frac{z-4}{-2}$

15. Consider a pyramid $OPQRS$ located in the first octant $(x \geq 0, y \geq 0, z \geq 0)$ with O as origin, and OP and OR along the X-axis and the Y-axis, respectively. The base $OPQR$ of the pyramid is a square with $OP = 3$. The point S is directly above the mid-point T of diagonal OQ such that $TS = 3$. Then, **[JEE Advanced 2016]**
(a) the acute angle between OQ and OS is $\frac{\pi}{3}$
(b) the equation of the plane containing the ΔOQS is $x - y = 0$
(c) the length of the perpendicular from P to the plane containing the ΔOQS is $\frac{3}{\sqrt{2}}$
(d) the perpendicular distance from O to the straight line containing RS is $\sqrt{\frac{15}{2}}$

16. The equation of a line $4x - 4y - z + 11 = 0 = x + 2y - z - 1$ can be put as
(a) $\frac{x}{2} = \frac{y-2}{1} = \frac{z-3}{4}$ (b) $\frac{x-4}{-2} = \frac{y-4}{2} = \frac{z-11}{2}$
(c) $\frac{x-2}{2} = \frac{y}{1} = \frac{z-3}{4}$ (d) $\frac{x-2}{2} = \frac{y-2}{1} = \frac{z}{4}$

17. Let $P_1 : 2x + y - z = 3$ and $P_2 : x + 2y + z = 2$ be two planes. Then, which of the following statement(s) is (are) TRUE? **[JEE Advanced 2018]**
(a) The line of intersection of P_1 and P_2 has direction ratios 1, 2, –1
(b) The line $\frac{3x-4}{9} = \frac{1-3y}{9} = \frac{z}{3}$ is perpendicular to the line of intersection of P_1 and P_2
(c) The acute angle between P_1 and P_2 is 60°
(d) If P_3 is the plane passing through the point $(4, 2, -2)$ and perpendicular to the line of intersection of P_1 and P_2, then the distance of the point $(2, 1, 1)$ from the plane P_3 is $\frac{2}{\sqrt{3}}$

18. The equation of a sphere which passes through $(1, 0, 0)$, $(0, 1, 0)$ and $(0, 0, 1)$ and whose centre lies on the curve $4xy = 1$ is
(a) $x^2 + y^2 + z^2 - x - y - z = 0$
(b) $x^2 + y^2 + z^2 + x + y + z - 2 = 0$
(c) $x^2 + y^2 + z^2 + x + y + z = 0$
(d) $x^2 + y^2 + z^2 - x - y - z - 2 = 0$

Linked-Comprehension Type

In these questions, a passage (paragraph) has been given followed by questions based on each of the passages. You have to answer the questions based on the passage given.

PASSAGE 1

Suppose direction cosines of two lines are given by $ul + vm + wn = 0$ and $al^2 + bm^2 + cn^2 = 0$, where u, v, w, a, b, c are arbitrary constants and l, m, n are direction cosines of the lines.

On the basis of above information, answer the following questions

1. For $u = v = w = 1$, both lines satisfies the relation

(a) $(b+c)\left(\frac{n}{l}\right)^2 + 2b\left(\frac{n}{l}\right) + (a+b) = 0$

(b) $(c+a)\left(\frac{l}{m}\right)^2 + 2c\left(\frac{l}{m}\right) + (b+c) = 0$

(c) $(a+b)\left(\frac{m}{n}\right)^2 + 2a\left(\frac{m}{n}\right) + (c+a) = 0$

(d) All of the above

2. For $u = v = w = 1$, If $\frac{n_1 n_2}{l_1 l_2} = \left(\frac{a+b}{b+c}\right)$, then

(a) $\frac{m_1 m_2}{l_1 l_2} = \frac{(b+c)}{(c+a)}$ (b) $\frac{m_1 m_2}{l_1 l_2} = \frac{(c+a)}{(b+c)}$

(c) $\frac{m_1 m_2}{l_1 l_2} = \frac{(a+b)}{(c+a)}$ (d) $\frac{m_1 m_2}{l_1 l_2} = \frac{(c+a)}{(a+b)}$

3. For $u = v = w = 1$ and if lines are perpendicular, then

(a) $a + b + c = 0$ (b) $ab + bc + ca = 0$

(c) $ab + bc + ca = 3abc$ (d) $ab + bc + ca = abc$

4. The given lines will be parallel if

(a) $\sum u^2(b+c) = 0$ (b) $\sum \frac{a^2}{u} = 0$

(c) $\sum \frac{u^2}{a} = 0$ (d) $\sum \frac{(b+c)}{u^2} = 0$

5. The given lines will be perpendicular, if

(a) $\sum u^2(a-b) = 0$ (b) $\sum u^2(b-c) = 0$

(c) $\sum u^2(a+b) = 0$ (d) $\sum u^2(b+c) = 0$

PASSAGE 2

Consider two lines $L_1 : \frac{x-2}{1} = \frac{y-1}{0} = \frac{z+1}{2}$ and $L_2 : \frac{x-3}{1} = \frac{y-1}{1} = \frac{z-0}{-1}$.

Let P be the plane which contains the line L_1 and is parallel to L_2 and intersecting coordinates axes at A, B, C respectively.

On the basis of the above information, answer the following questions

1. The shortest distance between the lines L_1 and L_2 is

(a) $\frac{1}{\sqrt{5}}$ (b) $\frac{1}{\sqrt{6}}$

(c) $\frac{1}{\sqrt{8}}$ (d) $\frac{1}{\sqrt{14}}$

2. 2. Image of origin in the plane P is

(a) $\left(\frac{2}{7}, \frac{-3}{7}, \frac{-1}{7}\right)$ (b) $\left(\frac{4}{7}, \frac{-6}{7}, \frac{-2}{7}\right)$

(c) $\left(\frac{-2}{7}, \frac{3}{7}, \frac{1}{7}\right)$ (d) $\left(\frac{-4}{7}, \frac{6}{7}, \frac{2}{7}\right)$

3. Volume of the tetrahedron $OABC$ (where 'O' is origin) is

(a) $\frac{2}{3}$ (b) $\frac{4}{9}$

(c) $\frac{2}{9}$ (d) $\frac{4}{3}$

PASSAGE 3

Let two planes $P_1 : 2x - y + z = 2$ and $P_2 : x + 2y - z = 3$ are given.

On the basis of the above information, answer the following questions

1. The equation of the plane through the intersection of P_1 and P_2 and the point (3, 2, 1) is

(a) $3x - y + 2z - 9 = 0$ (b) $x - 3y + 2z + 1 = 0$

(c) $2x - 3y + z - 1 = 0$ (d) $4x - 3y + 2z - 8 = 0$

2. Equation of the plane which passes through the point (–1, 3, 2) and is perpendicular to each of the planes P_1 and P_2 is

(a) $x + 3y - 5z + 2 = 0$ (b) $x + 3y + 5z - 18 = 0$

(c) $x - 3y - 5z + 20 = 0$ (d) $x - 3y + 5z = 0$

3. The equation of the acute angle bisector of planes P_1 and P_2 is

(a) $x - 3y + 2z + 1 = 0$ (b) $3x + y - 5 = 0$

(c) $x + 3y - 2z + 1 = 0$ (d) $3x + z + 7 = 0$

4. The equation of the bisector of angle of the planes P_1 and P_2 which not containing origin is

(a) $x - 3y + 2z + 1 = 0$ (b) $x + 3y = 5$

(c) $x + 3y + 2z + 2 = 0$ (d) $3x + y = 5$

5. The image of plane P_1 in the plane mirror P_2 is

(a) $x + 7y - 4x + 5 = 0$ (b) $3x + 4y - 5z + 9 = 0$

(c) $7x - y + 4z - 9 = 0$ (d) None of these

Numerical Value Type

Solve the following problems and write your answer.

1. If the area of the triangle whose vertices are A (1, 2, 3), B (2, – 1, 1) and C(1, 2, – 4) is λ sq unit then $20\sqrt{10}\,\lambda$ must be
2. A variable plane is at a constant distance p from the origin and meets the axes in A, B and C. If the locus of the centroid of the tetrahedron $OABC$ is $x^{-2}+y^{-2}+z^{-2}=\lambda p^{-2}$ then the value of 160 λ must be
3. If the distance of the point P (4, 3, 5) from the axis of Y is λ unit, then the value of $\frac{\lambda^2}{5}$ must be
4. If the volume of tetrahedron formed by planes whose equations are $y+z=0, z+x=0, x+y=0$ and $x+y+z=1$ is λ cubic unit then the value of λ must be
5. If the angle of intersection of the sphere $x^2+y^2+z^2-2x-4y-6z+10=0$ with the sphere, the extremities of whose diameter are (1, 2, – 3) and (5, 0, 1) is $\cos^{-1}(\lambda)$, then the value of $9999|\lambda|$ must be
6. If the limiting points of coaxial systems defined by the spheres $x^2+y^2+z^2+2x+2y+4z+2=0$ and $x^2+y^2+z^2+x+y+2z+2=0$ are $(\alpha_1, \beta_1, \gamma_1)$ and $(\alpha_2, \beta_2, \gamma_2)$, then the value of $1004|\alpha_1\alpha_2+\beta_1\beta_2+\gamma_1\gamma_2|$ must be

Single Integer Type Questions

1. If the equation of a plane P, passing through the intersection of the planes, $x+4y-z+7=0$ and $3x+y+5z=8$ is $ax+by+6z=15$ for some $a, b \in \mathbf{R}$, then the distance of the point $(3, 2, -1)$ from the plane P is ....
[JEE Main 2020, 4 September]
2. If the distance of the point (1, – 2, 3) from the plane $x+2y-3z+10=0$ measured parallel to the line, $\frac{x-1}{3}=\frac{2-y}{m}=\frac{z+3}{1}$ is $\sqrt{\frac{7}{2}}$, then the value of $|m|$ is equal to ....
[JEE Main 2021, 16 March]
3. If the line, $\frac{x-3}{2}=\frac{y+2}{-1}=\frac{z+4}{3}$ lies in the plane, $lx+my-z=9$, then l^2+m^2 is equal to
4. If the foot of the perpendicular drawn from the point (1, 0, 3) on a line passing through $(\alpha, 7, 1)$ is $\left(\frac{5}{3}, \frac{7}{3}, \frac{17}{3}\right)$, then α is equal to .... .
[JEE Main 2020, 7 January]
5. Let P be a point in the first octant, whose image Q in the plane $x+y=3$ (that is, the line segment PQ is perpendicular to the plane $x+y=3$ and the mid-point of PQ lies in the plane $x+y=3$) lies on the Z-axis. Let the distance of P from the X-axis be 5. If R is the image of P in the XY-plane, then the length of PR is ....
[JEE Advanced 2018]
6. The position vectors of the four angular points of a tetrahedron $OABC$ are (0, 0, 0), (0, 0, 2), (0, 4, 0) and (6, 0, 0), respectively. A point P inside the tetrahedron is at the same distance r from the four plane faces of the tetrahedron. Find the value of $9r$.

Matrix-Match Type

Given below are Matching Type Questions, with two columns (each having some items) each . Each item of Column I has to be matched with the items of Column II, by encircling the correct match(es).

NOTE *An item of Column I can be matched with more than one items of Column II.*

1. Observe the following columns :

	Column I		Column II
(A)	If Acute and obtuse angle bisectors of the planes $2x-y+2z+3=0$ and $3x-2y+6z+8=0$ are represented by A and O, then	(P) (Q)	$A: 32x+13y-23z-3=0$ $O: x-5y-4z-45=0$
(B)	If acute and obtuse angle bisectors of the planes $x-2y+2z-3=0$ and $2x-3y+6z+8=0$ are represented by A and O, then	(R)	$A: 23x-13y+32z+45=0$
(C)	The acute and obtuse angle bisectors of the planes $2x+y-2z+3=0$ and $6x+2y-3z-8=0$ are represented by A and O, then	(S) (T)	$O: 4x-y+5z-45=0$ $A: 13x-23y+32z+3=0$

2. Observe the following columns :

	Column I		Column II
(A)	Foot of perpendicular drawn from point (1, 2, 3) to the line $\frac{x-2}{2}=\frac{y-1}{3}=\frac{z-2}{4}$ is	(P)	$\left(\frac{107}{29},\frac{30}{29},\frac{69}{29}\right)$
(B)	Image of line point (1, 2, 3) in the line $\frac{x-2}{2}=\frac{y-1}{3}=\frac{z-2}{4}$ is	(Q)	$\left(\frac{88}{29},\frac{125}{29},\frac{69}{29}\right)$
(C)	Foot of perpendicular from the point (2, 3, 5) to the plane $2x+3y-4z+17=0$ is	(R)	$\left(\frac{107}{29},\frac{125}{29},\frac{185}{29}\right)$
(D)	Image of the point (2, 5, 1) in the plane $3x-2y+4z-5=0$ is	(S)	$\left(\frac{68}{29},\frac{44}{29},\frac{78}{29}\right)$
		(T)	$\left(\frac{38}{29},\frac{57}{29},\frac{185}{29}\right)$

3. Observe the following columns

	Column-I		Column-II
(A)	If a line makes an augh α with X and Y-axis then $\cot\alpha$ can be equal to	(P) (Q)	-3 $\frac{1}{2}$
(B)	If the straight lines $\frac{x-2}{1}=\frac{y-3}{1}=\frac{4-z}{\lambda}$ and $\frac{x-1}{\lambda}=\frac{y-4}{2}=\frac{z-5}{1}$ intersect then the volume of λ is	(R)	0
(C)	If the plane $\lambda x-\mu y+\lambda z=\phi$ contains two line $\frac{x-\lambda}{\lambda}=\frac{y-2\phi}{\mu}=\frac{z-\lambda}{\lambda}$, then the value of $\frac{\mu}{\phi}$ is	(S) (T)	1 2

Answers

Objective Questions Type [Only one correct answer]

1. (d)	**2.** (b)	**3.** (c)	**4.** (d)	**5.** (a)	**6.** (b)	**7.** (c)	**8.** (b)	**9.** (b)	**10.** (c)
11. (b)	**12.** (d)	**13.** (c)	**14.** (a)	**15.** (c)	**16.** (b)	**17.** (b)	**18.** (a)	**19.** (b)	**20.** (a)
21. (a)	**22.** (b)	**23.** (b)	**24.** (a)	**25.** (b)	**26.** (a)	**27.** (a)	**28.** (d)	**29.** (b)	**30.** (c)
31. (d)	**32.** (c)	**33.** (c)	**34.** (d)	**35.** (c)	**36.** (b)	**37.** (d)	**38.** (a)	**39.** (b)	**40.** (c)
41. (a)	**42.** (b)	**43.** (d)	**44.** (b)	**45.** (c)	**46.** (b)	**47.** (c)	**48.** (d)	**49.** (d)	**50.** (d)
51. (a)	**52.** (a)	**53.** (d)	**54.** (b)	**55.** (c)	**56.** (c)	**57.** (a)	**58.** (b)	**59.** (d)	**60.** (b)
61. (a)	**62.** (a)	**63.** (c)	**64.** (c)	**65.** (c)					

Objective Questions Type II [One or more than one correct answer(s)]

1. (a, b)	**2.** (a, b, c)	**3.** (a, c)	**4.** (a,b, c, d)	**5.** (a, b, c)
6. (a, b, c, d)	**7.** (a, c)	**8.** (a, c)	**9.** (b, d)	**10.** (a, b, c)
11. (a, b)	**12.** (b, c)	**13.** (a, b, c)	**14.** (a, b, c)	**15.** (b, c, d)
16. (a, b)	**17.** (c, d)	**18.** (a, b)		

Linked-Comprehension Type

Passage 1 **1.** (d) **2.** (b) **3.** (a) **4.** (c) **5.** (d)

Passage 2 **1.** (d) **2.** (b) **3.** (c)

Passage 3 **1.** (b) **2.** (c) **3.** (a) **4.** (d) **5.** (c)

Numerical Grid-Based Problems

1. 700 **2.** 2560 **3.** 8.2 **4.** 0.67 **5.** 6666 **6.** 2008

Single Integer Type Problem

1. (3) **2.** (2) **3.** (2) **4.** (4) **5.** (8) **6.** (6)

Matrix-Match Type

1. A → (R); B → (Q, T); C → (P, S)

2. A → (S); B → (P); C → (T); D → (Q)

3. A → (Q, R, S); B → (P, R); C → (T)

Solutions

Objective Questions Type [Only one correct answer]

1. Let the source of light be situated at $A(a, 0, 0)$, where $a \neq 0$. Let OA be the incident ray and OB the reflected ray ON is the normal to the mirror at O

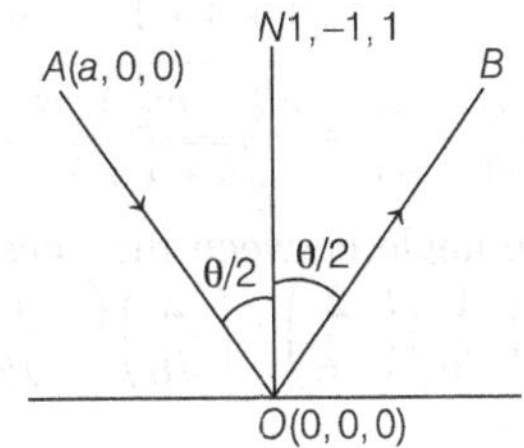

$$\angle AON = \angle NOB = \frac{\theta}{2} \quad \text{(say)}$$

DR's of OA are a, 0, 0 and so its DC's are 1,0, 0

DC's of ON are $\frac{1}{\sqrt{3}}, -\frac{1}{\sqrt{3}}, \frac{1}{\sqrt{3}}$

$$\therefore \quad \cos\frac{\theta}{2} = \frac{1}{\sqrt{3}}$$

Let l, m, n be the DC's of the reflected ray OB.

Then $\dfrac{n+0}{2\cos\frac{\theta}{2}} = \dfrac{1}{\sqrt{3}}$

$$\Rightarrow \quad l = \frac{2}{3} - 1, m = -\frac{2}{3}, n = \frac{2}{3}$$

$$\Rightarrow \quad l = -\frac{1}{3}, m = -\frac{2}{3}, n = \frac{2}{3}$$

Hence, DC's of the reflected ray are

$$-\frac{1}{3}, -\frac{2}{3}, \frac{2}{3}$$

2. Eliminating n, we get

$$\lambda(l+m)^2 + lm = 0$$

$$\Rightarrow \quad \frac{\lambda l^2}{m^2} + (2\lambda + 1)\frac{l}{m} + \lambda = 0$$

$$\Rightarrow \quad \frac{l_1 l_2}{m_1 m_2} = 1$$

where $\frac{l_1}{m_1}$ and $\frac{l_2}{m_2}$ are the roots of this equation, Further eliminating m, we get

$$\lambda l^2 - ln - n^2 = 0$$

$$\Rightarrow \quad \frac{l_1 l_2}{n_1 n_2} = -\frac{1}{\lambda}$$

Since, the lines with direction cosines (l_1, m_1, n_1) and (l_2, m_2, n_2) are perpendicular.

$$l_1 l_2 + m_1 m_2 + n_1 n_2 = 0$$

$$\Rightarrow \quad 1 + 1 - \lambda = 0$$

$$\Rightarrow \quad \lambda = 2$$

3. Let P be the point (x_1, y_1, z_1) on the given line

$$lx_1 + my_1 + nz_1 = p \quad \ldots\text{(i)}$$

Let Q be (α, β, γ) and O, P, Q are collinear, so

$$\frac{x_1}{\alpha} = \frac{y_1}{\beta} = \frac{z_1}{\gamma} = k \text{ (say)} \quad \ldots\text{(ii)}$$

Now, $OP \cdot OQ = p^2$

$$\Rightarrow \quad \sqrt{(x_1^2 + y_1^2 + z_1^2)}\,\sqrt{(\alpha^2 + \beta^2 + \gamma^2)} = p^2$$

$$\Rightarrow \quad k(\alpha^2 + \beta^2 + \gamma^2) = p^2 \quad \ldots\text{(iii)}$$

From Eqs. (i) and (ii),

$$k(l\alpha + m\beta + n\gamma) = p \quad \ldots\text{(iv)}$$

From Eqs. (iii) and (iv),

$$p\,(l\alpha + m\beta + n\gamma) = \alpha^2 + \beta^2 + \gamma^2$$

$\therefore$ Locus of $Q = (\alpha, \beta, \gamma)$ is $p(lx + my + nz) = x^2 + y^2 + z^2$

4. Any point on given line is $(2r + 1, -3r - 1, r)$, its distance from $(1, -1, 0)$.

$$\Rightarrow \quad (2r)^2 + (-3r)^2 + r^2 = (4\sqrt{14})^2$$

$$\Rightarrow \quad r = \pm 4$$

$\Rightarrow$ Coordinates are $(9, -13, 4)$ and $(-7, 11, -4)$ and nearer to the origin is $(-7, 11, -4)$.

5. Equation of given line is

$$\frac{x+3}{10} = \frac{y-2}{-7} = \frac{z}{1} = r \text{ (let)} \quad \ldots\text{(i)}$$

Coordinates of a point on line (i) is

$$A(10r - 3, -7r + 2, r)$$

Now, let the line joining the points $P(2, -1, 4)$ and $A(10r - 3, -7r + 2, r)$ is perpendicular to line (i).

Then, $\quad \mathbf{PA} \cdot (10\hat{\mathbf{i}} - 7\hat{\mathbf{j}} + \hat{\mathbf{k}}) = 0$

[$\because$ vector along line (i) is $(10\hat{\mathbf{i}} - 7\hat{\mathbf{j}} + \hat{\mathbf{k}})$]

$$\Rightarrow \quad [(10r - 5)\hat{\mathbf{i}} + (-7r + 3)\hat{\mathbf{j}} + (r - 4)\hat{\mathbf{k}}] \cdot [10\hat{\mathbf{i}} - 7\hat{\mathbf{j}} + \hat{\mathbf{k}}] = 0$$

$$\Rightarrow \quad 10(10r - 5) - 7(3 - 7r) + (r - 4) = 0$$

$$\Rightarrow \quad 100r - 50 - 21 + 49r + r - 4 = 0$$

$$\Rightarrow \quad 150r = 75 \Rightarrow r = \frac{1}{2}$$

So, the foot of perpendicular is $A\left(2, -\frac{3}{2}, \frac{1}{2}\right)$

[put $r = \frac{1}{2}$ in the coordinates of point A]

Now, perpendicular distance of point $P(2, -1, 4)$ from the line (i) is

$$PA = \sqrt{(2-2)^2 + \left(-\frac{3}{2} + 1\right)^2 + \left(\frac{1}{2} - 4\right)^2}$$

$$= \sqrt{\frac{1}{4} + \frac{49}{4}} = \sqrt{\frac{50}{4}} = \frac{5}{\sqrt{2}}$$

which lies in (3, 4).

6. Given that, $x - z = 0, x - 2y + z = 0, x + y + z = 0$,
Let the point whose locus is required be $P(\alpha, \beta, \gamma)$.
According to question

$$\frac{|\alpha-\gamma|^2}{2}+\frac{|\alpha-2\beta+\gamma|^2}{6}+\frac{|\alpha+\beta+\gamma|^2}{3}=36$$

or $3(\alpha^2+\gamma^2-2\alpha\gamma)+\alpha^2+4\beta^2+\gamma^2-4\alpha\beta-4\beta\gamma+2\alpha\gamma+2(\alpha^2+\beta^2+\gamma^2+2\alpha\beta+2\beta\gamma+2\alpha\gamma)=36\times 6$

or $6\alpha^2+6\beta^2+6\gamma^2=36\times 6$

$\Rightarrow \alpha^2+\beta^2+\gamma^2=36$

Hence, the required equation of locus is

$$x^2+y^2+z^2=36$$

7. We have, $z = 0$ for the point where the line intersects the curve

$\therefore \frac{x-2}{3}=\frac{y+1}{2}=\frac{0-1}{-1}$

$\Rightarrow \frac{x-2}{3}=1$ and $\frac{y+1}{2}=1$

$\Rightarrow x=5$ and $y=1$

Put these values in $xy=c^2$,

we get $5=c^2$

$\Rightarrow c=\pm\sqrt{5}$

8. Let a point $(3\lambda+1, \lambda+2, 2\lambda+3)$ of the first line also lies on the second line.

Then, $\frac{3\lambda+1-3}{1}=\frac{\lambda+2-1}{2}=\frac{2\lambda+3-2}{3}$

$\Rightarrow \lambda=1$

Hence, the point of intersection P of the two lines is $(4, 3, 5)$.
Equation of plane perpendicular to OP where O is $(0, 0, 0)$ and passing through P is $4x+3y+5z=50$

9. Equation of the line parallel to the given line $\frac{x}{2}=\frac{y}{3}=\frac{z}{-6}$ and passes through point $A(1, -2, 3)$ is

$$\frac{x-1}{2}=\frac{y+2}{3}=\frac{z-3}{-6}=r \text{ (let)} \quad \ldots(i)$$

$\therefore$ The coordinate of point on line (i) is

$$P(2r+1, 3r-2, 3-6r)$$

Let point P on the given plane $x-y+z=5$.

So, $2r+1-3r+2+3-6r=5$

$7r=1 \Rightarrow r=\frac{1}{7}$

$\therefore$ Required distance $AP=\sqrt{(2r)^2+(3r)^2+(-6r)^2}$

$=7r=1$

10. $3l+m+5n=0 \quad \ldots(i)$

$6mn-2nl+5ml=0 \quad \ldots(ii)$

Substituting the value of n from Eq. (i) in Eq. (ii), then

$6l^2+9lm-6m^2=0$

$\Rightarrow 6\left(\frac{l}{m}\right)^2+9\left(\frac{l}{m}\right)-6=0$

$\therefore \frac{l_1}{m_1}=\frac{1}{2}$ and $\frac{l_2}{m_2}=-2$

From Eq. (i), we get,

$\frac{l_1}{n_1}=-1$ and $\frac{l_2}{n_2}=-2$

$\therefore \frac{l_1}{1}=\frac{m_1}{2}=\frac{n_1}{-1}=\frac{\sqrt{l_1^2+m_1^2+n_1^2}}{\sqrt{1+4+1}}=\frac{1}{\sqrt{6}}$

and $\frac{l_2}{2}=\frac{m_2}{-1}=\frac{n_2}{-1}=\frac{\sqrt{l_2^2+m_2^2+n_2^2}}{\sqrt{4+1+1}}=\frac{1}{\sqrt{6}}$

If θ be the angle between the lines

$\therefore \cos\theta=\left(\frac{1}{\sqrt{6}}\right)\left(\frac{2}{\sqrt{6}}\right)+\left(\frac{2}{\sqrt{6}}\right)\left(-\frac{1}{\sqrt{6}}\right)+\left(-\frac{1}{\sqrt{6}}\right)\left(-\frac{1}{\sqrt{6}}\right)$

$=\frac{1}{6}$

$\therefore \theta=\cos^{-1}\left(\frac{1}{6}\right)$

11. Here, $\alpha=\beta=\gamma$

$\because \cos^2\alpha+\cos^2\beta+\cos^2\gamma=1$

$\therefore \cos\alpha=\frac{1}{\sqrt{3}}$

DC's of PQ are $\left(\frac{1}{\sqrt{3}}, \frac{1}{\sqrt{3}}, \frac{1}{\sqrt{3}}\right)$

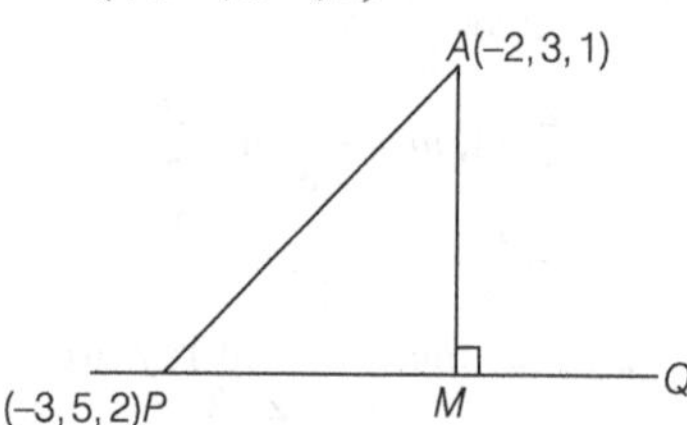

PM = Projection of AP on PQ

$=\left|(-2+3)\frac{1}{\sqrt{3}}+(3-5)\cdot\frac{1}{\sqrt{3}}+(1-2)\cdot\frac{1}{\sqrt{3}}\right|$

$=\frac{2}{\sqrt{3}}$

and $AP=\sqrt{(-2+3)^2+(3-5)^2+(1-2)^2}=\sqrt{6}$

$AM=\sqrt{(AP)^2-(PM)^2}$

$=\sqrt{6-\frac{4}{3}}=\sqrt{\frac{14}{3}}$

12. Equation of plane through $(2, 5, -3)$ is

$a(x-2)+b(y-5)+c(z+3)=0 \quad \ldots(i)$

which is perpendicular to

$x+2y+2z=1$

and $x-2y+3z=4$

then $a+2b+2c=0 \quad \ldots(ii)$

and $a-2b+3c=0 \quad \ldots(iii)$

Eliminating a, b, c from Eqs. (i), (ii) and (iii), we get

$$\begin{vmatrix} x-2 & y-5 & z+3 \\ 1 & 2 & 2 \\ 1 & -2 & 3 \end{vmatrix}=0$$

i.e. $10x-y-4z-27=0$

13. Equation of plane through $(0, -4, -6)$ is

$$a(x-0)+b(y+4)+c(z+6)=0 \quad \ldots(\text{i})$$

point $(-2, 9, 3)$ lies on Eq. (i), then

$$-2a+13b+9c=0 \quad \ldots(\text{ii})$$

also required plane is perpendicular to $x-4y-2z=8$

$$a-4b-2c=0 \quad \ldots(\text{iii})$$

From Eqs. (i), (ii), (iii) we get

$$\begin{vmatrix} x & y+4 & z+6 \\ -2 & 13 & 9 \\ 1 & -4 & -2 \end{vmatrix}=0$$

i.e. $2x+y-z-2=0$

14. Equation of plane through $(3, 2, -1)$ is

$$a(x-3)+b(y-2)+c(z+1)=0 \quad \ldots(\text{i})$$

also $(3, 4, 2)$ and $(7, 0, 6)$ lie on Eq. (i), then

$$0\cdot a+2b+3c=0 \quad \ldots(\text{ii})$$

and $4a-2b+7c=0 \quad \ldots(\text{iii})$

Eliminating a, b, c from Eqs. (i), (ii), and (iii), we get

$$\begin{vmatrix} x-3 & y-2 & z+1 \\ 0 & 2 & 3 \\ 4 & -2 & 7 \end{vmatrix}=0$$

We get, $5x+3y-2z=23$

$\therefore$ $\lambda=23$

15. Given, line passes through the point $A(1, 2, 4)$ and this point also lies in the plane. To find the reflection of the line, we need one more point of the line. Clearly $P(0, 5, 5)$ also lies in the line.

Let $Q(\alpha, \beta, \gamma)$ be the reflection of P in the plane $x+y+z=7$

Then, $\frac{\alpha}{2}=\frac{\beta+5}{2}=\frac{\gamma+5}{2}=7 \Rightarrow \alpha+\beta+\gamma=4$

Also, $PQ \perp$ to the plane, i.e., parallel to the normal of the plane.

$\therefore$ $\frac{\alpha-0}{1}=\frac{\beta-5}{1}=\frac{\gamma-5}{1}=\lambda$

$\Rightarrow$ $\alpha=\lambda, \beta=\lambda+5, \gamma=\lambda+5$

$\therefore$ $\lambda+\lambda+5+\lambda+5=4 \Rightarrow \lambda=-2$

$\therefore Q$ is $(-2, 3, 3)$

Now AQ will be the reflection of the given line, equation of AQ is

$$\frac{x-1}{-3}=\frac{y-2}{1}=\frac{z-4}{-1}$$

16. Line $\frac{x+1}{2}=\frac{y-3}{1}=\frac{z+2}{-1}$

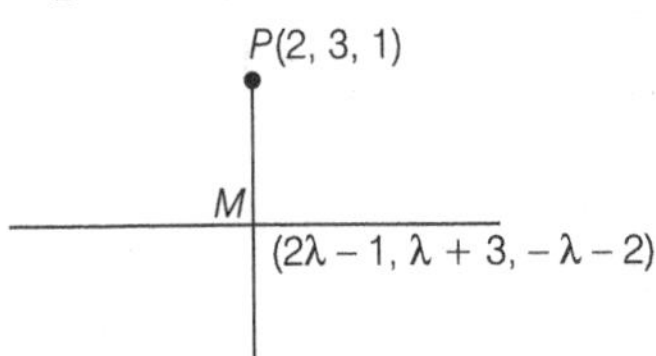

$\mathbf{PM}=(2\lambda-3, \lambda, -\lambda-3)$

$\mathbf{PM} \perp (2\hat{\mathbf{i}}+\hat{\mathbf{j}}-\hat{\mathbf{k}})$

$4\lambda-6+\lambda+\lambda+3=0 \Rightarrow \lambda=\frac{1}{2}$

$\therefore$ $M \equiv \left(0, \frac{7}{2}, \frac{-5}{2}\right)$

$\therefore$ Reflection $(-2, 4, -6)$

$$\text{Plane} \begin{vmatrix} x-2 & y-1 & z+1 \\ 3 & -2 & 1 \\ 4 & -3 & 5 \end{vmatrix}=0$$

$\Rightarrow$ $(x-2)(-10+3)-(y-1)(15-4)+(z+1)(-1)=0$

$\Rightarrow$ $-7x+14-11y+11-z-1=0$

$\Rightarrow$ $7x+11y+z=24$

$\therefore$ $\alpha=7, \beta=11, \gamma=1$

$\alpha+\beta+\gamma=19$

17. DC's of the given line is $\frac{1}{3}, -\frac{2}{3}, -\frac{2}{3}$

Hence, the equation of line can be point in the form

$$\frac{x-2}{1/3}=\frac{y+3}{-2/3}=\frac{z+5}{-2/3}=r$$

$\therefore$ Point is $\left(2+\frac{r}{3}, -3-\frac{2r}{3}, -5-\frac{2r}{3}\right)$ $\therefore$ $r=\pm 6$

Points are $(4, -7, -9)$ and $(0, 1, -1)$

18. Equation of the plane through $(5, 1, 2)$ is

$$a(x-5)+b(y-1)+c(z-2)=0 \quad \ldots(\text{i})$$

given plane (i) is perpendicular to the line

$$\frac{x-2}{1/2}=\frac{y-4}{1}=\frac{z-5}{1}$$

$\therefore$ Equation of normal of Eq. (i) and straight line (ii) are parallel

i.e. $\frac{a}{1/2}=\frac{b}{1}=\frac{c}{1}=k$ (say)

$\therefore$ $a=\frac{k}{2}, b=k, c=k$

From Eq. (i), $\frac{k}{2}(x-5)+k(y-1)+k(z-2)=0$

or $x+2y+2z=11$

Any point on Eq. (ii) is $\left(2+\frac{\lambda}{2}, 4+\lambda, 5+\lambda\right)$

Which lies on Eq. (iii), then $\lambda=-2$

$\therefore$ Required point is $(1, 2, 3)$.

19. It is given that lines $L_1 : \frac{x+1}{2}=\frac{y-2}{-1}=\frac{z-1}{1}$

and $L_2 : \frac{x+2}{\alpha}=\frac{y+1}{5-\alpha}=\frac{z+1}{1}$ are coplanar, so

$$\begin{vmatrix} -2+1 & -1-2 & -1-1 \\ \alpha & 5-\alpha & 1 \\ 2 & -1 & 1 \end{vmatrix}=0$$

$$\Rightarrow \begin{vmatrix} -1 & -3 & -2 \\ \alpha & 5-\alpha & 1 \\ 2 & -1 & 1 \end{vmatrix}=0$$

$\Rightarrow -1(5-\alpha+1)+3(\alpha-2)-2(-\alpha-10+2\alpha)=0$

$\Rightarrow$ $2\alpha+8=0 \Rightarrow \alpha=-4$

$\therefore$ Equation of line $L_2 : \frac{x+2}{-4}=\frac{y+1}{9}=\frac{z+1}{1}$

Now, from the options the point $(2, -10, -2)$ satisfy the line L_2.

20. $A(1, 1, 1)$, $B(2, 3, 5)$, $C(-1, 0, 2)$ directions ratios of AB are $< 1, 2, 4 >$ direction ratios of AC are $< 2, -1, 1 >$

$\therefore$ Direction ratios of normal to plane ABC are $< 2, -3, 1 >$

$\therefore$ Equation of the plane ABC is $2x - 3y + z = 0$

Let the equation of the required plane be

$2x - 3y + z = k$, then $\left|\frac{k}{\sqrt{4+9+1}}\right| = 2$

$$k = \pm 2\sqrt{14}$$

$\therefore$ Equation of the required plane is

$2x - 3y + z + 2\sqrt{14} = 0$

21. We know that, angle between two lines is

$$\cos\theta = \frac{a_1a_2 + b_1b_2 + c_1c_2}{\sqrt{a_1^2 + b_1^2 + c_1^2}\sqrt{a_2^2 + b_2^2 + c_2^2}}$$

Given, $\quad l + m + n = 0$

$\Rightarrow \quad l = -(m + n)$

$\Rightarrow \quad (m + n)^2 = l^2$

$\Rightarrow \quad m^2 + n^2 + 2mn = m^2 + n^2 \quad [\because l^2 = m^2 + n^2, \text{ given}]$

$\Rightarrow \quad 2mn = 0$

Case I When $m = 0$, then $l = -n$

Hence, (l, m, n) is $(1, 0, -1)$.

Case II When $n = 0$, then $l = -m$

Hence, (l, m, n) is $(1, -1, 0)$.

$$\therefore \quad \cos\theta = \frac{1 + 0 + 0}{\sqrt{2}\times\sqrt{2}} = \frac{1}{2} \Rightarrow \theta = \frac{\pi}{3}$$

22. $l + m + n = 0 \quad \ldots(i)$

$l^2 + m^2 - n^2 = 0 \quad \ldots(ii)$

From Eqs. (i) and (ii),

$$l^2 + m^2 - (-l - m)^2 = 0$$

$$lm = 0$$

$\therefore \quad l = 0$ or $m = 0$

Case I : If, $l = 0$

Then, $\quad m + n = 0$

$$\therefore \quad \frac{l}{0} = \frac{m}{-1} = \frac{n}{1} = \frac{1}{\sqrt{2}}$$

Case II : If, $m = 0$

Then, $\quad l + n = 0$

$$\therefore \quad \frac{l}{-1} = \frac{m}{0} = \frac{n}{1} = \frac{1}{\sqrt{2}}$$

If angle between lines is θ

$$\therefore \quad \cos\theta = 0 + 0 + \frac{1}{2}$$

$$\therefore \quad \theta = \frac{\pi}{3}$$

23. $l^2 + m^2 + n^2 = 1$,

$(l + a)^2 + (m + b)^2 + (n + c)^2 = 1$

$\Rightarrow al + bm + cn = -\frac{1}{2}(a^2 + b^2 + c^2)$

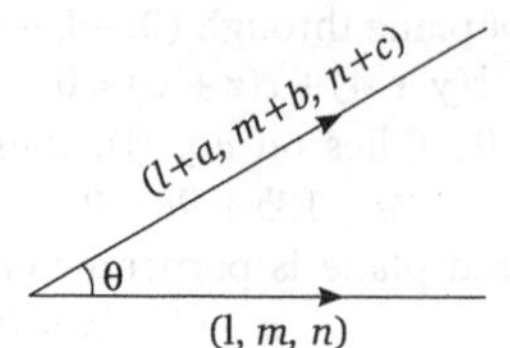

$$\cos\theta = l(a + l) + m(b + m) + n(c + n)$$

$$= 1 + al + bm + cn$$

$$\Rightarrow \quad a^2 + b^2 + c^2 = 2(1 - \cos\theta)$$

$$= 4\sin^2(\theta/2)$$

Since θ is small, $\sin(\theta/2) \approx \theta/2$

$\therefore \quad \theta^2 = a^2 + b^2 + c^2 \quad [\because \sin^2(\theta/2) \approx \theta^2/4]$

24. Given line is $\frac{x+2}{3} = \frac{y-1}{0} = \frac{z}{4}$

Vector along line is, $\mathbf{a} = 3\hat{\mathbf{i}} + 4\hat{\mathbf{k}}$

and vector joining the points $(1, -1, 2)$ to $(-2, 1, 0)$ is

$$\mathbf{b} = (1 + 2)\hat{\mathbf{i}} + (-1 - 1)\hat{\mathbf{j}} + (2 - 0)\hat{\mathbf{k}} = 3\hat{\mathbf{j}} - 2\hat{\mathbf{j}} + 2\hat{\mathbf{k}}$$

and $|\mathbf{BC}| = 5$ units

Now, area of required

$$\Delta ABC = \frac{1}{2}|\mathbf{BC}||\mathbf{b}||\sin\theta| \quad \ldots(ii)$$

[where θ is angle between vectors $\mathbf{a}$ and $\mathbf{b}$]

$$\because \quad |\mathbf{b}|\sin\theta = \frac{|\mathbf{a}\times\mathbf{b}|}{|\mathbf{a}|},$$

$$\because \quad |\mathbf{a}\times\mathbf{b}| = \begin{vmatrix} \hat{\mathbf{i}} & \hat{\mathbf{j}} & \hat{\mathbf{k}} \\ 3 & 0 & 4 \\ 3 & -2 & 2 \end{vmatrix} = 8\hat{\mathbf{i}} + 6\hat{\mathbf{j}} - 6\hat{\mathbf{k}}$$

$$\Rightarrow \quad |\mathbf{a}\times\mathbf{b}| = \sqrt{64 + 36 + 36} = \sqrt{136} = 2\sqrt{34}$$

and $\quad |\mathbf{a}| = \sqrt{9 + 16} = 5$

$$\therefore \quad |\mathbf{b}|\sin\theta = \frac{2\sqrt{34}}{5}$$

On substituting these values in Eq. (i), we get

Required area $= \frac{1}{2}\times 5\times\frac{2\sqrt{34}}{5} = \sqrt{34}$ sq units

25. Any plane through $(1, 0, 0)$, $(0, 1, 0)$ can be taken as

$$\frac{x}{1} + \frac{y}{1} + \frac{z}{c} = 1$$

DR's of normal of this plane are $1, 1, 1/c$.

This plane is inclined at an angle $\pi/4$ with the plane $x + y = 3$.

$$\therefore \quad \cos(\pi/4) = \frac{1 + 1 + 0\cdot 1/c}{\sqrt{1^2 + 1^2 + 1/c^2}\sqrt{1 + 1}}$$

$$\Rightarrow \quad \frac{4c^2}{(2c^2 + 1)^2} = \frac{1}{2}$$

$$\Rightarrow \quad c^2 = 1/2 \text{ i.e. } c = \pm\frac{1}{\sqrt{2}}$$

$\therefore$ DR's of normal of the required plane are $(1, 1, \sqrt{2})$ or $(1, 1, -\sqrt{2})$.

26. Let plane is $\frac{x}{a}+\frac{y}{b}+\frac{z}{c}=1$ ...(i)

mid point of $P(1, 2, 3)$ and $Q(-3, 4, 5)$

ie, $-1, 3, 4$ lie on Eq. (i)

$$\therefore \quad -\frac{1}{a}+\frac{3}{b}+\frac{4}{c}=1 \quad \text{...(ii)}$$

Also, PQ is parallel to normal of the plane (i)

$$\frac{1/a}{-4}=\frac{1/b}{2}=\frac{1/c}{2}$$

$$\Rightarrow \quad \frac{1}{-2a}=\frac{1}{b}=\frac{1}{c}=\lambda \text{ (say)}$$

$$\therefore \quad \frac{1}{a}=-2\lambda, \frac{1}{b}=\lambda, \frac{1}{c}=\lambda$$

$\therefore$ From Eq. (ii), $2\lambda+3\lambda+4\lambda=1$

$$\therefore \quad \lambda=\frac{1}{9}$$

$$a=-\frac{1}{2\lambda}, b=\frac{1}{\lambda}, c=\frac{1}{\lambda}$$

$$\therefore \quad a=-\frac{9}{2}, b=9, c=9$$

Intercepts are $\left(-\frac{9}{2}, 9, 9\right)$.

27. DC's of diagonals OP, BQ, CS, and AR are

$$\frac{1}{\sqrt{3}}, \frac{1}{\sqrt{3}}, \frac{1}{\sqrt{3}}; \frac{1}{\sqrt{3}}, -\frac{1}{\sqrt{3}}, \frac{1}{\sqrt{3}}; \frac{1}{\sqrt{3}}, \frac{1}{\sqrt{3}}, -\frac{1}{\sqrt{3}}; -\frac{1}{\sqrt{3}}, \frac{1}{\sqrt{3}}, \frac{1}{\sqrt{3}}$$

Let dc's of a line be l, m, n

$$\therefore \quad \cos\alpha=\frac{l+m+n}{\sqrt{3}}$$

$$\cos\beta=\frac{l-m+n}{\sqrt{3}}$$

$$\cos\gamma=\frac{l+m-n}{\sqrt{3}}$$

$$\cos\delta=\frac{-l+m+n}{\sqrt{3}}$$

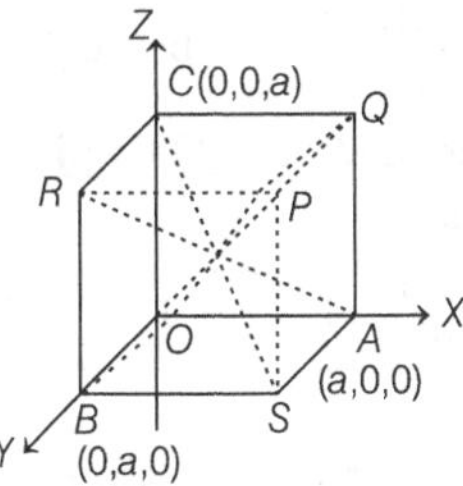

$\therefore \cos^2\alpha+\cos^2\beta+\cos^2\gamma+\cos^2\delta$

$$=\frac{(l+m+n)^2+(l-m+n)^2+(l+m-n)^2+(-l+m+n)^2}{3}$$

$$=\frac{4}{3}$$

28. Let the equation of plane be $ax+by+cz=1$.

Then, $\quad a+b+c=1$

$$2a+b-2c=0$$

$$3a-6b-2c=0$$

$$\Rightarrow \quad a=7b$$

$$c=\frac{15b}{2}$$

$$b=\frac{2}{31}, a=\frac{14}{31}, c=\frac{15}{31}$$

$$\therefore \quad 14x+2y+15z=31$$

29. Let equation of plane be $lx+my+nz=p$

$$\text{or} \quad \frac{x}{p/l}+\frac{y}{p/m}+\frac{z}{p/n}=1$$

Coordinates of A, B, C are $\left(\frac{p}{l}, 0, 0\right)$, $\left(0, \frac{p}{m}, 0\right)$ and $\left(0, 0, \frac{p}{n}\right)$ respectively.

If centroid of triangle ABC is (x_1, y_1, z_1)

Then, $\quad x_1=\frac{p}{3l}, y_1=\frac{p}{3m}, z_1=\frac{p}{3n}$

$$\Rightarrow \quad l^2+m^2+n^2=\frac{p^2}{9x_1^2}+\frac{p^2}{9y_1^2}+\frac{p^2}{9z_1^2}$$

$$\text{or} \quad \frac{1}{x_1^2}+\frac{1}{y_1^2}+\frac{1}{z_1^2}=\frac{9}{p^2}$$

$\therefore$ Required locus is $\frac{1}{x^2}+\frac{1}{y^2}+\frac{1}{z^2}=\frac{9}{p^2}$

30. Equation of plane passing through the points $(2, 1, 0)$, $(4, 1, 1)$ and $(5, 0, 1)$ is

$$\begin{vmatrix} x-2 & y-1 & z-0 \\ 4-2 & 1-1 & 1-0 \\ 5-2 & 0-1 & 1-0 \end{vmatrix}=0 \Rightarrow \begin{vmatrix} x-2 & y-1 & z \\ 2 & 0 & 1 \\ 3 & -1 & 1 \end{vmatrix}=0$$

$$\Rightarrow (x-2)(0+1)-(y-1)(2-3)+z(-2-0)=0$$

$$\Rightarrow \quad x-2+y-1-2z=0$$

$$\Rightarrow \quad x+y-2z-3=0 \quad \text{...(i)}$$

Now, let the image of point $R(2, 1, 6)$ w.r.t. plane (i) is (x_1, y_1, z_1), then

$$\frac{x_1-2}{1}=\frac{y_1-1}{1}=\frac{z_1-6}{-2}=-2\left(\frac{2+1-12-3}{1^2+1^2+(-2)^2}\right)$$

$$\Rightarrow \quad \frac{x_1-2}{1}=\frac{y_1-1}{1}=\frac{z_1-6}{-2}=\frac{2\times 12}{6}=4$$

$$\Rightarrow \quad x_1=6, y_1=5, z_1=-2$$

$\therefore$ Image of point $R(2, 1, 6)$ w.r.t., plane (i) is $(6, 5, -2)$.

31. The equation of the horizontal plane through the origin will be $2x-3y+z=0$. The direction ratios of the line of intersection of the planes $2x-3y+z=0$ and $3x-2y+z=5$ are given by

$$2a-3b+c=0$$

$$\text{and} \quad 3a-2b+c=0$$

$\therefore$ By cross multiplication

$$\frac{a}{-1}=\frac{b}{1}=\frac{c}{5}$$

Let l, m, n be the DC's of the line of greatest slope, because the line of greatest slope is perpendicular to the line of intersection of the given plane with the horizontal plane.

$$\therefore \quad -l+m+5n=0 \quad \text{...(i)}$$

$$\text{and} \quad 3l-2m+n=0 \quad \text{...(ii)}$$

From Eqs. (i) and (ii), we have

$$\frac{l}{11}=\frac{m}{16}=\frac{n}{-1}$$

$\therefore$ DC's are proportional to (11, 16, −1).

32. $y = z = 0$, then $x = 1$

Suppose direction cosines of line of intersection are l, m, n Then ,

$$l+m-n=0$$
$$2l-3m+n=0,$$

then
$$\frac{l}{2}=\frac{m}{3}=\frac{n}{5}$$

$\therefore$ Equation of line in symmetric form is $\frac{x-1}{2}=\frac{y}{3}=\frac{z}{5}$

33. Equation of plane through X-axis is

$$ax+by+cz=0 \quad \ldots(\text{i})$$

Which is perpendicular to the line

$$\frac{x-1}{\cos\theta}=\frac{y+2}{\sin\theta}=\frac{z-3}{0}$$

$$\therefore \quad \frac{a}{\cos\theta}=\frac{b}{\sin\theta}=\frac{c}{\theta}=k \text{ (say)}$$

$$\therefore \quad a=k\cos\theta,\ b=k\sin\theta,\ c=0$$

$\therefore$ From Eq. (i), we get

$x\cos\theta + y\sin\theta = 0$

34. Required distance $= KL$

$$=\sqrt{\left(a-\frac{a}{2}\right)^2+0^2+\left(0-\frac{a}{2}\right)^2}$$

$$=\sqrt{\left(\frac{a^2}{4}+\frac{a^2}{4}\right)}=\sqrt{\left(\frac{a^2}{2}\right)}=\frac{a}{\sqrt{2}}$$

Where K and L are the mid points of OH and AE respectively.

35. Equation of plane passing through the intersection of the planes $2x-5y+z=3$ and $x+y+4z=5$ is

$$(2x-5y+z-3)+\lambda(x+y+4z-5)=0$$

$$\Rightarrow (2+\lambda)x+(-5+\lambda)y+(1+4\lambda)-3-5\lambda=0 \quad \ldots(\text{i})$$

Which is parallel to the plane $x+3y+6z=1$.

Then,
$$\frac{2+\lambda}{1}=\frac{-5+\lambda}{3}=\frac{1+4\lambda}{6}$$

$$\therefore \quad \lambda=\frac{-11}{2}$$

From Eq. (i),

$$-\frac{7}{2}x-\frac{21}{2}y-21z+\frac{49}{2}=0$$

$$\therefore \quad x+3y+6z=7$$

Hence, $k=7$

36. Equation of the planes through

$y=mx,\ z=c$ and $y=-mx,\ z=-c$ are respectively.

$$(y-mx)+\lambda_1(z-c)=0 \quad \ldots(\text{i})$$

and
$$(y+mx)+\lambda_2(z+c)=0 \quad \ldots(\text{ii})$$

It meets at X-axis *i.e.*, $y=0=z$

$$\therefore \quad \lambda_2=\lambda_1$$

From Eqs. (i) and (ii), $\frac{y-mx}{z-c}=\frac{y+mx}{z+c}$

$$\therefore \quad cy=mzx$$

37. Let $A(1, 0, -1), B(-1, 2, 2)$

Direction ratios of segment AB are $< 2, -2, -3 >$

$$\therefore \quad \cos\theta=\frac{|2\times1+3(-2)-5(-3)|}{\sqrt{1+9+25}\sqrt{4+4+9}}$$

$$=\frac{11}{\sqrt{17}\sqrt{35}}=\frac{11}{\sqrt{595}}$$

Length of projection $=(AB)\sin\theta$

$$=\sqrt{(2)^2+(2)^2+(3)^2}\times\sqrt{1-\frac{121}{595}}$$

$$=\sqrt{17}\,\frac{\sqrt{474}}{\sqrt{17}\sqrt{35}}=\sqrt{\frac{474}{35}} \text{ units}$$

38. The equation of the plane through the line of intersection of the planes.

$$4x+7y+4z+81=0 \text{ and } 5x+3y+10z=25$$

if $(4x+7y+4z+81)+\lambda(5x+3y+10z-25)=0$

$\Rightarrow (4+5\lambda)x+(7+3\lambda)y+(4+10\lambda)z+81-25\lambda=0$

Which is parallel to $x-4y+6z=k$,

then
$$\frac{4+5\lambda}{1}=\frac{7+3\lambda}{-4}=\frac{4+10\lambda}{6}$$

We get, $\lambda=-1$

Then from Eq. (i),

$$-x+4y-6z+106=0$$

$$\therefore \quad x-4y+6z=106$$

Hence, $k=106$

39. Let reflection of $P(1, 0, 0)$ in the line $\frac{x-1}{2}=\frac{y+12}{-3}=\frac{z+10}{8}$ be (α, β, γ) then $\left(\frac{\alpha+1}{2}, \frac{\beta}{2}, \frac{\gamma}{2}\right)$ lies on the line and $(\alpha-1)\hat{\mathbf{i}}+\beta\hat{\mathbf{j}}+\gamma\hat{\mathbf{k}}$ is perpendicular to $2\hat{\mathbf{i}}-3\hat{\mathbf{j}}+8\hat{\mathbf{k}}$

$$\therefore \quad \frac{\frac{\alpha+1}{2}-1}{2}=\frac{\frac{\beta}{2}+1}{-3}=\frac{\frac{\gamma}{2}+10}{8}=\lambda$$

and
$$2(\alpha-1)-3(\beta)+\gamma(8)=0$$

$$\Rightarrow \quad \alpha=5,\ \beta=-8,\ \gamma=-4$$

40. Coordinates of L and M are $(0, b, c)$ and $(a, 0, c)$ respectively.

$\therefore$ Equation of plane through $(0, 0, 0)$, $(0, b, c)$ and $(a, 0, c)$ is

$$\begin{vmatrix} x-0 & y-0 & z-0 \\ 0 & b & c \\ a & 0 & c \end{vmatrix}=0$$

i.e.
$$\frac{x}{a}+\frac{y}{b}-\frac{z}{c}=0$$

41. Let the equation of the variable plane be

$$lx + my + nz = p$$

$\therefore$ Coordinates of the vertices of tetrahedron is $(0, 0,0)$; $\left(\frac{p}{l}, 0, 0\right)$; $\left(0, \frac{p}{m}, 0\right)$ and $\left(0, 0, \frac{p}{n}\right)$. If (x_1, y_1, z_1) be the centroid of tetrahedron, then

$$x_1 = \frac{p}{4l}, y_1 = \frac{p}{4m}, z_1 = \frac{p}{4n}$$

$\therefore$ $$\frac{p}{l} = 4x_1, \frac{p}{m} = 4y_1, \frac{p}{n} = 4z_1$$

Also, volume of tetrahedron $= 64k^3$

i.e. $$\frac{1}{6}\begin{vmatrix} p/l & 0 & 0 \\ 0 & p/m & 0 \\ 0 & 0 & p/n \end{vmatrix} = 64k^3$$

$\Rightarrow$ $$\frac{1}{6}\begin{vmatrix} 4x_1 & 0 & 0 \\ 0 & 4y_1 & 0 \\ 0 & 0 & 4z_1 \end{vmatrix} = 64k^3$$

$\Rightarrow$ $$\frac{64\, x_1y_1z_1}{6} = 64k^3 \Rightarrow x_1y_1z_1 = 6k^3$$

$\therefore$ Locus of centroid is $xyz = 6k^3$

42. Equation of given plane is

$$x + 2y + 3z = 15 \quad \ldots(i)$$

and line is, $$\frac{x-1}{2} = \frac{y+1}{3} = \frac{z-2}{4} = r \text{ (let)} \quad \ldots(ii)$$

So, the coordinates of any point on line (ii) is

$$P(1+2r, -1+3r, 2+4r).$$

$\because$ Point P is intersecting point of plane (i) and line (ii)

$\therefore$ $$(1+2r)+2(-1+3r)+3(2+4r) = 15$$

$$\Rightarrow 1+2r-2+6r+6+12r = 15 \Rightarrow 20r = 10 \Rightarrow r = \frac{1}{2}$$

$\therefore$ Coordinates of $P = \left(1+1, -1+\frac{3}{2}, 2+2\right) = \left(2, \frac{1}{2}, 4\right)$

Now, distance of the point P from the origin

$$= \sqrt{4+\frac{1}{4}+16} = \sqrt{20+\frac{1}{4}} = \sqrt{\frac{81}{4}} = \frac{9}{2} \text{ units}$$

43. Let equation of plane be $lx + my + nz = p$

or $$\frac{x}{(p/l)} + \frac{y}{(p/m)} + \frac{z}{(p/n)} = 1$$

According to question,

$$\frac{p}{l} = 12, \frac{p}{m} = 3, \frac{p}{n} = 4$$

or $$\frac{p}{12} = l, \frac{p}{3} = m, \frac{p}{4} = n$$

or $$\frac{p^2}{144} + \frac{p^2}{9} + \frac{p^2}{16} = l^2 + m^2 + n^2 = 1$$

$\Rightarrow$ $$p^2\left(\frac{1}{144} + \frac{1}{9} + \frac{1}{16}\right) = 1$$

$$p^2\left(\frac{1+16+9}{144}\right) = 1$$

or $$p^2 = \frac{144}{26} = \frac{72}{13}$$

$\therefore$ $$p = \frac{6\sqrt{2}}{\sqrt{13}}$$

44. Let equation of the plane is

$$\frac{x}{a} + \frac{y}{b} + \frac{z}{c} = 1$$

Coordinates of P is $\left(\frac{a}{3}, \frac{b}{3}, \frac{c}{3}\right)$

$\because$ OP is perpendicular to the plane,

$\therefore$ OP is parallel to normal to the plane

$\therefore$ $$\frac{a/3}{1/a} = \frac{b/3}{1/b} = \frac{c/3}{1/c}$$

$\Rightarrow$ $$a^2 = b^2 = c^2$$

$\Rightarrow$ $$|a| = |b| = |c|$$

45. It is clear, AA_1, BB_1, CC_1, DD_1 meet centroid of the tetrahedron $ABCD$. If centroid is G, then

$$\frac{AG}{GA_1} = \frac{BG}{GB_1} = \frac{CG}{GC_1} = \frac{DG}{GD_1} = \frac{3}{1}$$

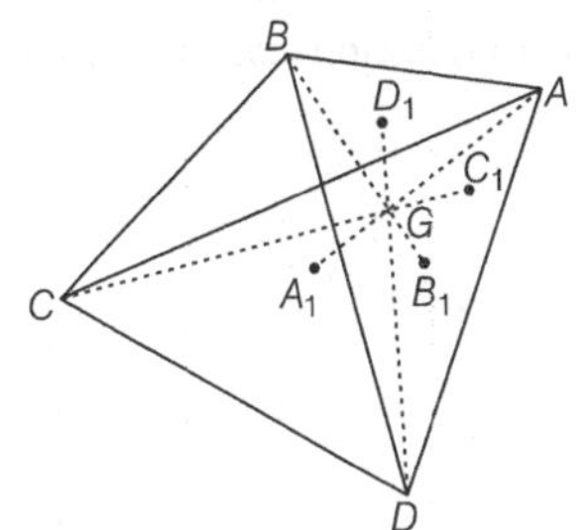

46. Required area $= \sqrt{(\Delta^2 xy + \Delta^2 yz + \Delta^2 zx)} \quad \ldots(i)$

where $$\Delta xy = \frac{1}{2}\left|\begin{vmatrix} a & 0 & 1 \\ 0 & b & 1 \\ 0 & 0 & 1 \end{vmatrix}\right| = \frac{1}{2}ab$$

$$\Delta yz = \frac{1}{2}\left|\begin{vmatrix} 0 & 0 & 1 \\ b & 0 & 1 \\ 0 & c & 1 \end{vmatrix}\right| = \frac{1}{2}bc$$

and $$\Delta zx = \frac{1}{2}\left|\begin{vmatrix} 0 & c & 1 \\ a & 0 & 1 \\ 0 & 0 & 1 \end{vmatrix}\right| - \frac{1}{2}ac$$

From Eq. (i), $\frac{1}{2}\sqrt{(a^2b^2 + b^2c^2 + c^2a^2)}$

47. Let image of $Q(3, 1, 7)$ w.r.t. $x - y + z = 3$ be $P(\alpha, \beta, \gamma)$.

$\therefore$ $$\frac{\alpha-3}{1} = \frac{\beta-1}{-1} = \frac{\gamma-7}{1} = \frac{-2(3-1+7-3)}{1^2+(-1)^2+(1)^2}$$

$\Rightarrow$ $$\alpha - 3 = 1 - \beta = \gamma - 7 = -4$$

$\therefore$ $$\alpha = -1, \beta = 5, \gamma = 3$$

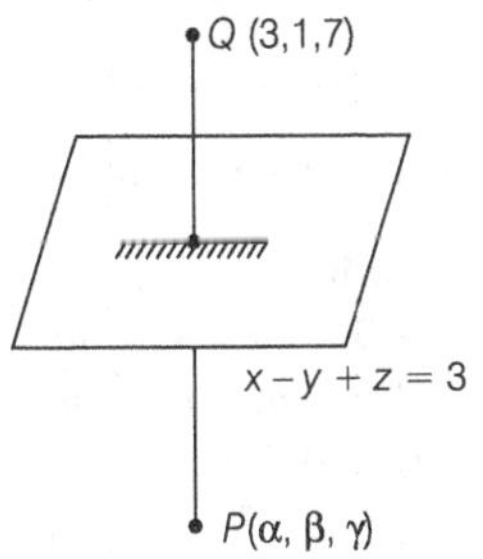

Hence, the image of $Q(3, 1, 7)$ is $P(-1, 5, 3)$.

To find equation of plane passing through $P(-1, 5, 3)$ and containing $\frac{x}{1} = \frac{y}{2} = \frac{z}{1}$

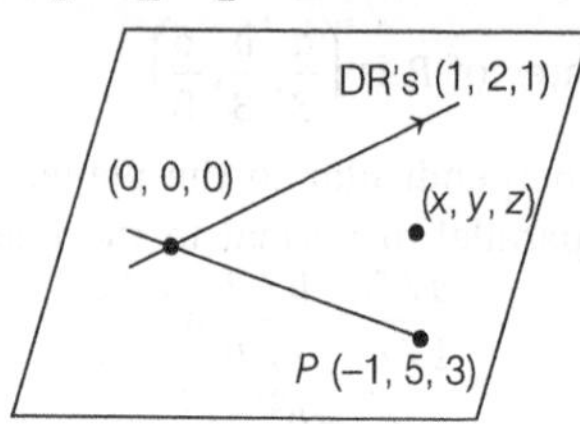

$$\Rightarrow \begin{vmatrix} x-0 & y-0 & z-0 \\ 1-0 & 2-0 & 1-0 \\ -1-0 & 5-0 & 3-0 \end{vmatrix} = 0$$

$$\Rightarrow \quad x(6-5) - y(3+1) + z(5+2) = 0$$

$$\therefore \quad x - 4y + 7z = 0$$

48. According to question $\cos^2\alpha + \cos^2\alpha + \cos^2\alpha = 1$

$$\therefore \quad \cos\alpha = \frac{1}{\sqrt{3}}$$

$\therefore$ DC's are $\frac{1}{\sqrt{3}}, \frac{1}{\sqrt{3}}, \frac{1}{\sqrt{3}}$

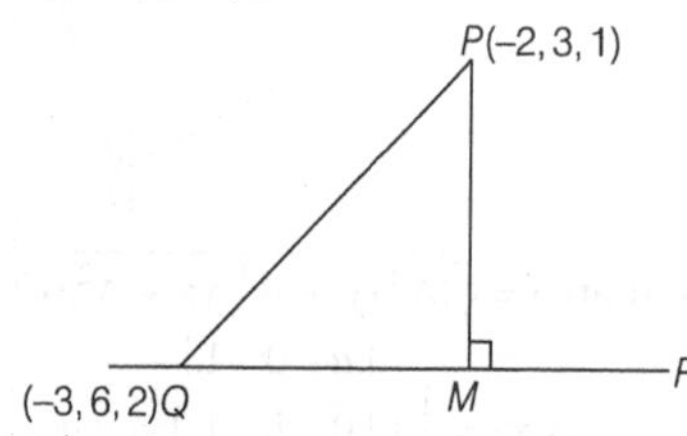

$$PQ = \sqrt{1+9+1} = \sqrt{11}$$

QM = Projection of PQ on QR

$$= \left|(-2+3)\frac{1}{\sqrt{3}} + (3-6)\frac{1}{\sqrt{3}} + (1-2)\frac{1}{\sqrt{3}}\right|$$

$$= \left|\frac{1}{\sqrt{3}} - \frac{3}{\sqrt{3}} - \frac{1}{\sqrt{3}}\right| = \sqrt{3}$$

$$\therefore \quad PM = \sqrt{(QP)^2 - (QM)^2}$$

$$= \sqrt{11-3} = 2\sqrt{2}$$

49.

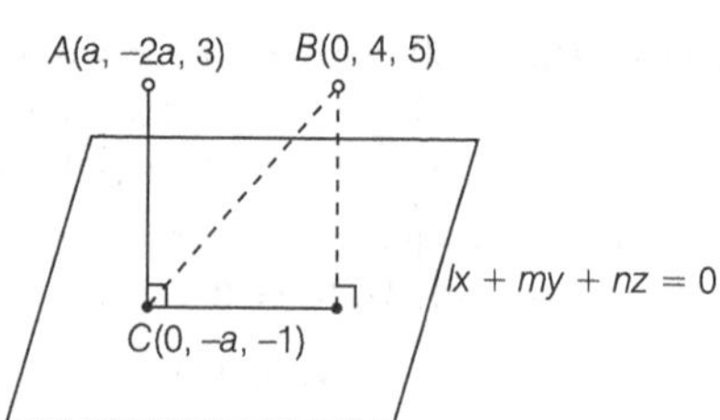

C lies on plane $\Rightarrow -ma - n = 0 \Rightarrow \frac{m}{n} = -\frac{1}{a}$ ...(i)

$\overline{CA} \parallel l\hat{\mathbf{i}} + m\hat{\mathbf{j}} + n\hat{\mathbf{k}}$

$$\frac{a-0}{l} = \frac{-a}{m} = \frac{4}{n} \Rightarrow \frac{m}{n} = -\frac{a}{4} \quad \text{...(ii)}$$

From Eqs. (i) and (ii)

$$-\frac{1}{a} = \frac{-a}{4} \Rightarrow a^2 = 4 \Rightarrow a = 2 \quad [\text{since } a > 0]$$

From Eq. (ii)

$$\frac{m}{n} = \frac{-1}{2}$$

Let $m = -t \Rightarrow n = 2t$

$$\frac{2}{l} = \frac{-2}{-t}$$

$$\Rightarrow \quad l = t$$

So plane : $t(x - y + 2z) = 0$

$$BD = \frac{6}{\sqrt{6}} = \sqrt{6}$$

$$C \equiv (0, -2, -1)$$

$$CD = \sqrt{BC^2 - BD^2}$$

$$= \sqrt{(0^2 + 6^2 + 6^2) - (\sqrt{6})^2} = \sqrt{66}$$

50. Let AD be the perpendicular and D be the foot of the perpendicular which divides BC in the ratio $\lambda : 1$, then

$$D\left(\frac{10\lambda - 9}{\lambda + 1}, \frac{4}{\lambda + 1}, \frac{-\lambda + 5}{\lambda + 1}\right) \quad \text{... (i)}$$

A(0, 0, 0)

B (–9, 4, 5) — λ — D — 1 — C (10, 0, –1)

The direction ratios of AD are $\frac{10\lambda - 9}{\lambda + 1}, \frac{4}{\lambda + 1}$ and $\frac{-\lambda + 5}{\lambda + 1}$ and direction ratios of BC are 19, – 4 and – 6.

Since $AD \perp BC$, we get

$$19\left(\frac{10\lambda - 9}{\lambda + 1}\right) - 4\left(\frac{4}{\lambda + 1}\right) - 6\left(\frac{-\lambda + 5}{\lambda + 1}\right) = 0$$

$$\text{or} \quad \lambda = \frac{31}{28}$$

Hence, on putting the value of λ in Eq. (i), we get required foot of the perpendicular, i.e., $\left(\frac{58}{59}, \frac{112}{59}, \frac{109}{59}\right)$.

51. $al + bm + cn = 0$ ...(i)

$$fmn + gnl + hlm = 0 \quad \text{...(ii)}$$

Eliminating n from Eqs. (i) and (ii),

$$(fm + gl)\left(-\frac{al + bm}{c}\right) + hlm = 0$$

$$\Rightarrow \quad -afml - bfm^2 - agl^2 - bgml + chlm = 0$$

$$\Rightarrow \quad agl^2 + bfm^2 + (af + bg - ch)lm = 0$$

$$\Rightarrow \quad ag\left(\frac{l}{m}\right)^2 + (af + bg - ch)\left(\frac{l}{m}\right) + bf = 0$$

$$\therefore \quad \frac{l_1}{m_1}\cdot\frac{l_2}{m_2} = -\frac{bf}{ag} = \frac{f/a}{g/b}$$

$$\text{or} \quad \frac{l_1 l_2}{f/a} = \frac{m_1 m_2}{g/b} \quad \text{...(iii)}$$

Similarly, $\dfrac{m_1m_2}{g/b} = \dfrac{n_1n_2}{h/c}$ ...(iv)

From Eqs. (iii) and (iv),

$$\frac{l_1l_2}{f/a} = \frac{m_1m_2}{g/b} = \frac{n_1n_2}{h/c} = \frac{0}{f/a + g/b + h/c}$$

Hence, $\dfrac{f}{a} + \dfrac{g}{b} + \dfrac{h}{c} = 0$

52. $A(3, 2, 3)$, $B(1, 4, 6)$ and $C(7, 4, 5)$ are the three vertices of parallelogram $ABCD$

Let M is mid-point of AC

$\therefore$ Point $M = (5, 3, 4)$

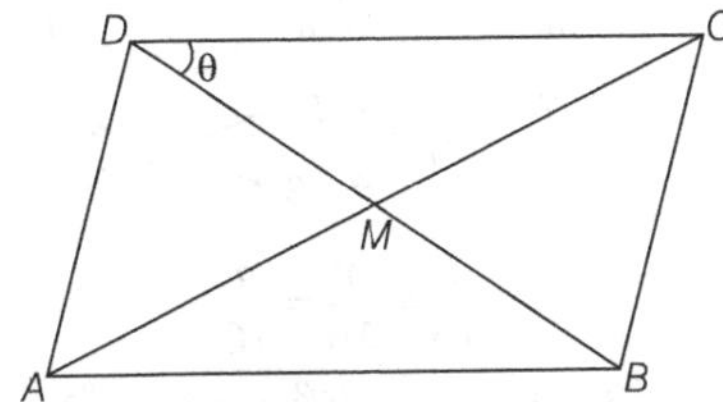

Direction ratio of DC is equal to direction ratio of AB and direction ratio of DB is equal to direction ratio of MB.

$\therefore$ Dr's of $DC = (-2, 2, 3)$

and Dr's of $DB = (-4, 1, 2)$

Angle between DC and DB is

$$\theta = \cos^{-1} \frac{8 + 2 + 6}{\sqrt{4 + 4 + 9}\sqrt{16 + 1 + 4}}$$

$$\theta = \cos^{-1} \frac{16}{\sqrt{17}\sqrt{21}}$$

$$\theta = \cos^{-1} \frac{16}{\sqrt{357}}$$

53. $\dfrac{x+1}{2} = \dfrac{y+1}{3} = \dfrac{z+1}{4} = r$ (say)

Any point of a distance r is $(2r - 1, 3r - 1, 4r - 1)$ which lies on

$$(2r - 1) + 2(3r - 1) + 3(4r - 1) = 14$$

$\Rightarrow$ $20r = 20$

$\therefore$ $r = 1$

$\therefore$ Required point is (1, 2, 3).

54. $\dfrac{x+1}{3} = \dfrac{y-2}{-2} = \dfrac{z+1}{-1} = r$

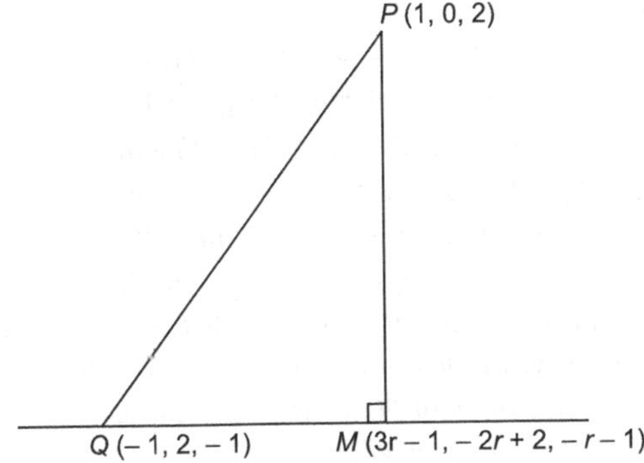

$\therefore$ $M(3r - 1, -2r + 2, -r - 1)$

DR's of PM are $3r - 2, -2r + 2, -r - 3$

DR's of line QM are $3, -2, -1$

$\therefore$ $9r - 6 + 4r - 4 + r + 3 = 0$

$\Rightarrow$ $14r - 7 = 0 \Rightarrow r = \dfrac{1}{2}$

$\therefore$ $M\left(\dfrac{1}{2}, 1, -\dfrac{3}{2}\right)$

55.

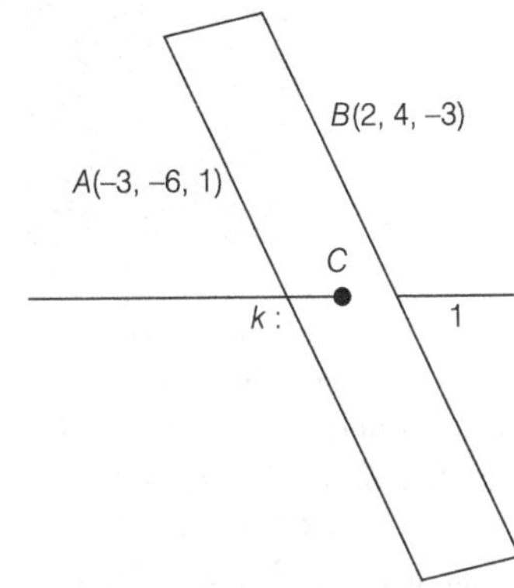

Point C is $\left(\dfrac{2k-3}{k+1}, \dfrac{4k-6}{k+1}, \dfrac{-3k+1}{k+1}\right)$

$$\frac{x-1}{-1} = \frac{y+4}{2} = \frac{z+2}{3}$$

Plane $lx + my + nz = 0$

$l(-1) + m(2) + n(3) = 0$

$\Rightarrow$ $-l + 2m + 3n = 0$ ...(i)

It is also satisfy point $(1, -4, -2)$

$l - 4m - 2n = 0$ ...(ii)

Solving Eqs. (i) and (ii),

$2m + 3n = 4m + 2n$

$\Rightarrow$ $n = 2m$

$\Rightarrow$ $l - 4m - 4m = 0$

$\Rightarrow$ $l = 8m$

$$\frac{l}{8} = \frac{m}{1} = \frac{n}{2}$$

$l : m : n = 8 : 1 : 2$

Plane is $8x + y + 2z = 0$

It will satisfy point C

$$8\left(\frac{2k-3}{k+1}\right) + \left(\frac{4k-6}{k+1}\right) + 2\left(\frac{-3k+1}{k+1}\right) = 0$$

$\Rightarrow$ $16k - 24 + 4k - 6 - 6k + 2 = 0$

$14k = 28$

$\therefore$ $k = 2$

56. Point $(3, 2, 1)$ and $(2, -3, -1)$

lies on $11x + my + nz = 28$

ie, $33 + 2m + n = 28$

$\Rightarrow$ $2m + n = -5$ ...(i)

and $22 - 3m - n = 28$

$\Rightarrow$ $-3m - n = 6$ ...(ii)

From Eqs. (i) and (ii),

$m = -1$ and $n = -3$

57. Equation of plane through (–1, 0, 1) is

$$a(x+1)+b(y-0)+c(z-1)=0 \quad \ldots(i)$$

Which is parallel to given line and perpendicular to given plane

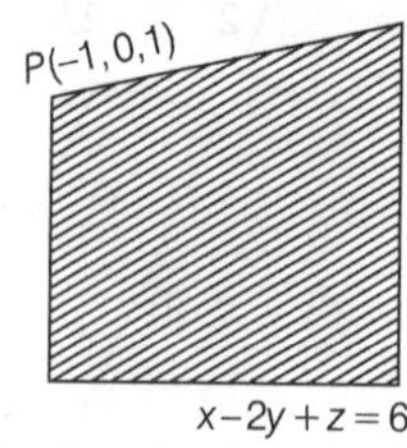

$$-a+2b+3c=0 \quad \ldots(ii)$$

and
$$a-2b+c=0 \quad \ldots(iii)$$

From Eqs. (ii) and (iii), $c=0$, $a=2b$

From Eqs. (i), $2b(x+1)+by=0$

$$\Rightarrow \quad 2x+y+2=0$$

58. Let variable plane be $\frac{x}{a}+\frac{y}{b}+\frac{z}{c}=1$

which passes through a fixed point (1, – 2, 3), then

$$\frac{1}{a}-\frac{2}{b}+\frac{3}{c}=1 \quad \ldots(i)$$

$\because$ Point of intersection of the planes through A, B, C parallel to the coordinate planes is (a, b, c)

$\therefore$ Locus of (a, b, c) from Eq. (i) is

$$\frac{1}{x}-\frac{2}{y}+\frac{3}{z}=1$$

$$\Rightarrow \quad yz-2zx+3xy=xyz$$

59. Let direction ratios of PQ are a, b, c

$$\frac{x-1}{2}=\frac{y+1}{1}=\frac{z-3}{-3}$$

Then,
$$\frac{x-2}{a}=\frac{y-1}{b}=\frac{z+2}{c}=r \quad (\text{Let } PQ=r)$$

$$Q\equiv(ar+2,\ br+1,\ cr-2)$$

Which lie on $\frac{x-1}{2}=\frac{y+1}{2}=\frac{z-3}{-3}$, then

$$\frac{ar+2-1}{2}=\frac{br+1+1}{2}=\frac{cr-2-3}{-3}$$

$$\Rightarrow \quad \frac{ar+1}{2}=\frac{br+2}{2}=\frac{cr-5}{-3}=\lambda \text{ (say)}$$

$$\therefore \ a=\frac{2\lambda-1}{r},\ b=\frac{2\lambda-2}{r},\ c=\frac{-3\lambda+5}{r}$$

Given, PQ is parallel to $x+2y+z=4$, then

$$a+2b+c=0$$

or
$$\frac{2\lambda-1}{r}+\frac{4\lambda-4}{r}+\frac{-3\lambda+5}{r}=0$$

$$\Rightarrow \quad 3\lambda=0\Rightarrow\lambda=0$$

then, $a=-\frac{1}{r},\ b=-\frac{2}{r},\ c=\frac{5}{r}$

$$\therefore \quad Q=(1,-1,3)$$

$$\therefore \quad PQ\equiv\sqrt{(2-1)^2+(1+1)^2+(3+2)^2}=\sqrt{30}$$

60. Let DC's of shortest distance line are l, m, n which is perpendicular to both the given lines

$$\therefore \quad 2l-7m+5n=0 \quad \ldots(i)$$

and
$$2l+m-3n=0 \quad \ldots(ii)$$

From Eqs. (i) and (ii),

l	m	n	l	m
2	–7	5	2	–7
2	1	–3	2	1

$$\therefore \quad \frac{l}{16}=\frac{m}{16}=\frac{n}{16}$$

$$\Rightarrow \quad \frac{l}{1}=\frac{m}{1}=\frac{n}{1}=\frac{\sqrt{(l^2+m^2+n^2)}}{\sqrt{(1^2+1^2+1^2)}}=\frac{1}{\sqrt{3}}$$

$$\Rightarrow \quad l=\frac{1}{\sqrt{3}},\ m=\frac{1}{\sqrt{3}},\ n=\frac{1}{\sqrt{3}}$$

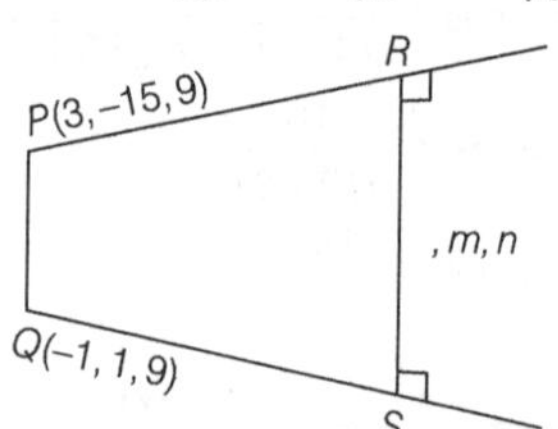

$\therefore$ Required shortest distance = Projection of PQ on RS

$$=|(3-(-1))l+(-15-1)m+(9-9)n|$$

$$=|4l-16m|=\left|\frac{-12}{\sqrt{3}}\right|$$

$$=4\sqrt{3}$$

61. Let $P(2r_1, -3r_1, r_1)$ and $Q(3r_2+2, -5r_2+1, 2r_2-2)$ be the points on the given lines so that PQ is the line of shortest distance between the given lines. Now direction ratios of PQ are

$$2r_1-3r_2-2,\ -3r_1+5r_2-1,\ r_1-2r_2+2$$

Since, it is perpendicular to the given lines

$$2(2r_1-3r_2-2)-3(-3r_1+5r_2-1)+(r_1-2r_2+2)=0$$

and $3(2r_1-3r_2-2)-5(-3r_1+5r_2-1)+2(r_1-2r_2+2)=0$

$$\Rightarrow \quad 14r_1-23r_2+1=0,\ 23r_1-38r_2+3=0$$

$$\Rightarrow \quad r_1=31/3,\ r_2=19/3$$

$\Rightarrow P$ is $(62/3, -31, 31/3)$ and Q is $(21, -92/3, 32/3)$ and direction ratios of PQ are 1/3, 1/3, 1/3.

Hence, equations of PQ as it passes through Q are

$$\frac{x-21}{1/3}=\frac{y+92/3}{1/3}=\frac{z-32/3}{1/3}$$

$$\Rightarrow \quad 3(x-21)=3y+92=3z-32$$

62. Equation of plane through $(2, -1, 0)$ is

$$a(x-2)+b(y+1)+c(z-0)=0 \quad \ldots(i)$$

It also passes through $(3, -4, 5)$, then

$$a-3b+5c=0 \quad \ldots(ii)$$

Given plane (i) is parallel to a line with direction cosines proportional to 2, 3, 4

$$\therefore \quad 2a+3b+4c=0 \quad \ldots(iii)$$

From Eqs. (ii) and (iii), we get

$$\frac{a}{9}=\frac{b}{-2}=\frac{c}{-3}=\lambda \quad \text{(say)}$$

$$\therefore \quad a=9\lambda,\ b=-2\lambda,\ c=-3\lambda$$

Now, from Eq. (i)

$$9\lambda(x-2)-2\lambda(y+1)-3\lambda(z-0)=0$$

$$\Rightarrow \quad 9x-2y-3z=20$$

$$\therefore \quad k=20$$

63. Equation of plane in normal form is

$$lx+my+nz=r \quad \ldots(i)$$

Also, DR's of OP are f, g, h

$\therefore$ DC's of OP are $\frac{f}{r}, \frac{g}{r}, \frac{h}{r}$

From Eq. (i),

$$\frac{f}{r}x+\frac{g}{r}y+\frac{h}{r}z=r$$

$$\therefore \quad \frac{x}{\left(\frac{r^2}{f}\right)}+\frac{y}{\left(\frac{r^2}{g}\right)}+\frac{z}{\left(\frac{r^2}{h}\right)}=1$$

$$\Rightarrow \therefore A \equiv \left(\frac{r^2}{f}, 0, 0\right); B \equiv \left(0, \frac{r^2}{g}, 0\right);$$

$$C \equiv \left(0, 0, \frac{r^2}{h}\right)$$

$\therefore$ Centroid of ΔABC is

$$\left(\frac{r^2}{3f}, \frac{r^2}{3g}, \frac{r^2}{3h}\right)$$

64. Given planes are

$$lx+my=0 \quad \ldots(i)$$

and

$$z=0 \quad \ldots(ii)$$

$\therefore$ Equation of any plane passing through the line of intersection section of planes (i) and (ii) is

$$lx+my+\lambda z=0 \quad \ldots(iii)$$

DC's of normal to the plane (iii) are

$$\frac{l}{\sqrt{(l^2+m^2+\lambda^2)}}, \frac{m}{\sqrt{(l^2+m^2+\lambda^2)}}, \frac{\lambda}{\sqrt{(l^2+m^2+\lambda^2)}}$$

and DC's of normal to the plane (i) are

$$\frac{l}{\sqrt{l^2+m^2)}}, \frac{m}{\sqrt{(l^2+m^2)}}, 0$$

$$\Rightarrow \quad \cos\alpha = \frac{l \cdot l + m \cdot m + 0}{\sqrt{(l^2+m^2+\lambda^2)}\sqrt{(l^2+m^2)}}$$

$$= \frac{\sqrt{(l^2+m^2)}}{\sqrt{(l^2+m^2+\lambda^2)}}$$

$$\therefore \quad \sec^2\alpha = \frac{l^2+m^2+\lambda^2}{l^2+m^2}$$

$$=1+\frac{\lambda^2}{l^2+m^2}$$

$$\Rightarrow \quad \tan^2\alpha = \frac{\lambda^2}{l^2+m^2}$$

$$\Rightarrow \quad \lambda = \pm\sqrt{(l^2+m^2)}\tan\alpha$$

On comparing, we get $n=\lambda=\pm\sqrt{(l^2+m^2)}\tan\alpha$

65. Let the direction vector of the line

$$\frac{x-3}{2}=\frac{y+2}{-1}=\frac{z-1}{3} \text{ is}$$

$$\mathbf{b}=2\hat{\mathbf{i}}-\hat{\mathbf{j}}+3\hat{\mathbf{k}}$$

Since, the required plane contains this line and its projection along the plane $2x+3y-z=5$, it will also contain the normal of the plane $2x+3y-z=5$

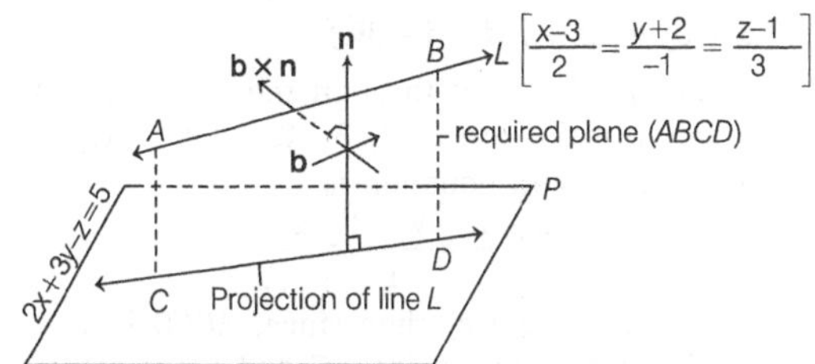

Normal vector of the plane $2x+3y-z=5$ is

$$\mathbf{n}=2\hat{\mathbf{i}}+3\hat{\mathbf{j}}-\hat{\mathbf{k}}$$

Now, the required plane contains

$$\mathbf{b}=2\hat{\mathbf{i}}-\hat{\mathbf{j}}+3\hat{\mathbf{k}}$$

and

$$\mathbf{n}=2\hat{\mathbf{i}}+3\hat{\mathbf{j}}-\hat{\mathbf{k}}$$

$\therefore$ Normal of the required plane is $\mathbf{b}\times\mathbf{n}$.

Since, the plane contains the line $\frac{x-3}{2}=\frac{y+2}{-1}=\frac{z-1}{3}$, therefore it also contains the point $\mathbf{a}=3\hat{\mathbf{i}}-2\hat{\mathbf{j}}+\hat{\mathbf{k}}$

Now, the equation of required plane is

$$(\mathbf{r}-\mathbf{a})\cdot(\mathbf{b}\times\mathbf{n})=0$$

$$\begin{vmatrix} x-3 & y+2 & z-1 \\ 2 & -1 & 3 \\ 2 & 3 & -1 \end{vmatrix}=0$$

$$\Rightarrow (x-3)[1-9]-(y+2)[-2-6]+(z-1)[6+2]=0$$

$$\Rightarrow \quad -8x+8y+8z+32=0$$

$$\Rightarrow \quad x-y-z=4$$

Note that $(2, 0, -2)$ is the only point which satisfy above equation.

Objective Questions Type II [One or more than one correct answer(s)]

1. Let $OP = OQ = r$

M is the mid point of PQ, then coordinates of

$$M\left[\left(\frac{l_1+l_2}{2}\right)r, \left(\frac{m_1+m_2}{2}\right)r \left(\frac{n_1+n_2}{2}\right)r\right]$$

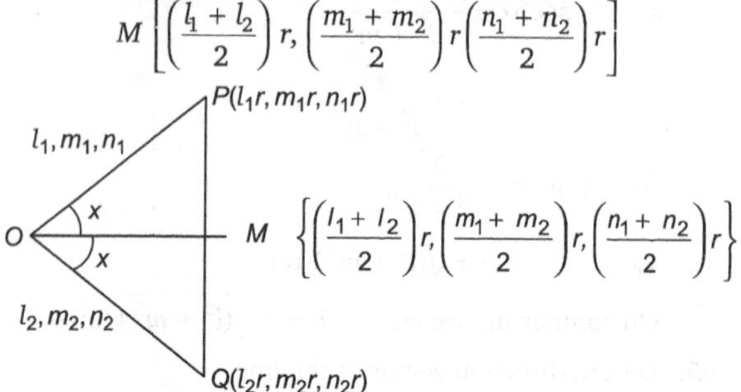

$\therefore$ DR's of the bisector are

$$l_1 + l_2,\ m_1 + m_2,\ n_1 + n_2$$

DR's of other bisector are $l_1 - l_2,\ m_1 - m_2,\ n_1 - n_2$

2. Direction ratios of AB are $4-2, 5-3, 10-4$ or $1, 1, 3$. So AB is parallel to the vector $\hat{\mathbf{i}} + \hat{\mathbf{j}} + 3\hat{\mathbf{k}}$ and passes through $B(2, 3, 4)$, the vector $2\hat{\mathbf{i}} + 3\hat{\mathbf{j}} + 4\hat{\mathbf{k}}$, its equation is $2\hat{\mathbf{i}} + 3\hat{\mathbf{j}} + 4\hat{\mathbf{k}} + \lambda(\hat{\mathbf{i}} + \hat{\mathbf{j}} + 3\hat{\mathbf{k}})$.

Similarly, BC passes through the points $B(2, 3, 4)$ and its direction ratios are $2-1, 3-2, 4+1$ or $1, 1, 5$.

So, its cartesian equation is

$$\frac{x-2}{1} = \frac{y-3}{1} = \frac{z-4}{5}$$

Next, if D is (a, b, c), then since, $ABCD$ is a parallellogram mid-point AC and BD is same, (diagonals of a parallelogram bisect each other)

$$\Rightarrow \quad \left(\frac{a+2}{2}, \frac{b+3}{2}, \frac{c+4}{2}\right) = \left(\frac{5}{2}, \frac{7}{2}, \frac{9}{2}\right)$$

$$\Rightarrow \quad (a, b, c) = (3, 4, 5)$$

AB is not perpendicular to BC because

$$1 \times 1 + 1 \times 1 + 3 \times 5 \neq 0$$

$ABCD$ is not a rectangle.

3. The given lines are coplanar, if

$$0 = \begin{vmatrix} 2-1 & 3-4 & 4-5 \\ 1 & 1 & -k \\ k & 2 & 1 \end{vmatrix}$$

$$= \begin{vmatrix} 1 & -1 & -1 \\ 1 & 1 & -k \\ k & 2 & 1 \end{vmatrix}$$

Applying $C_2 \to C_2 + C_1$ and $C_3 \to C_3 + C_1$

$$= \begin{vmatrix} 1 & 0 & 0 \\ 1 & 2 & 1-k \\ k & k+2 & 1+k \end{vmatrix}$$

or $\quad 2(1+k) - (k+2)(1-k) = 0$

or $\quad k^2 + 3k = 0$

$\Rightarrow \quad k = 0, -3$

4. (a) We know that the plane $ax + by + cz + d = 0$ contains the line $\frac{x-\alpha}{l} = \frac{y-\beta}{m} = \frac{z-\gamma}{n}$ if $a\alpha + b\beta + c\gamma + d = 0$ and $al + bm + cn = 0$.

Now, since $(-1) - 2(3) + 7(-2) + 21 = 0$

and $\quad (-3)(1) + 2(-2) + 1(7) = 0$

The line given in (a) lies on the given plane.

(b) Since, $0 - 2(7) + 7(-1) + 21 = 0$

The point $(0, 7, -1)$ lies on the plane.

(c) Direction ratios of the normal to the given plane are $(1, -2, 7)$ which are same as those of the given in (c). So, the plane is perpendicular to the lines.

(d) The direction ratios of the normal to the planes given in (d) are same as those of the given plane. So, the plane in (d) is parallel to the given plane.

5. Here, P be (x, y, z)

Then, $x = r\sin\theta\cos\phi,\ y = r\sin\theta\sin\phi,\ z = r\cos\theta$

$\Rightarrow 1 = r\sin\theta\cos\phi,\ 2 = r\sin\theta\sin\phi,\ 3 = r\cos\theta \quad \ldots$(i)

$$\Rightarrow \quad 1^2 + 2^2 + 3^2 = r^2\sin^2\theta\cos^2\phi + r^2\sin^2\theta\sin^2\phi + r^2\cos^2\theta$$

$$= r^2\sin^2\theta(\cos^2\phi + \sin^2\phi) + r^2\cos^2\theta = r^2$$

$$\Rightarrow \quad r = \pm\sqrt{14}$$

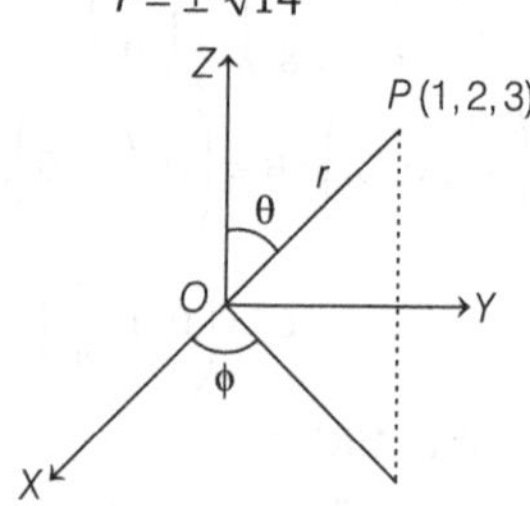

$\therefore$ From Eq. (i), we have

$$\sin\theta\cos\phi = \pm\frac{1}{\sqrt{14}}$$

$$\sin\theta\sin\phi = \frac{2}{\sqrt{14}},\ \cos\theta = \frac{3}{\sqrt{14}}$$

(neglecting $-$ve sign assuming acute angles)

$$\therefore \quad \frac{\sin\theta\sin\phi}{\sin\theta\cos\phi} = \frac{2}{1}$$

and $\quad \tan\theta = \frac{\sin\theta}{\cos\theta} = \frac{\sqrt{5}}{3}$

$\Rightarrow \tan\phi = 2$ and $\tan\theta = \frac{\sqrt{5}}{3}$

6. Since, the planes are all parallel planes.

$$p_1 = \frac{|2-6|}{\sqrt{2^2+3^2+4^2}}$$

$$= \frac{4}{\sqrt{4+9+16}} = \frac{4}{\sqrt{29}}$$

Equation of the plane $4x - 6y + 8z + 3 = 0$ can be written as $2x - 3y + 4z + \frac{3}{2} = 0$

So, $$p_2 = \frac{\left|2 - \frac{3}{2}\right|}{\sqrt{2^2 + 3^2 + 4^2}} = \frac{1}{2\sqrt{29}}$$

and $$p_3 = \frac{|2 + 6|}{\sqrt{2^2 + 3^2 + 4^2}} = \frac{8}{\sqrt{29}}$$

we find $p_3 = 16p_2 = 2p_1$

$$p_1 + 2p_2 + 3p_3 = \frac{4}{\sqrt{29}} + \frac{1}{\sqrt{29}} + \frac{24}{\sqrt{29}} = \frac{29}{\sqrt{29}} = \sqrt{29}$$

and $$p_1 + 8p_2 - p_3 = \frac{4}{\sqrt{29}} + \frac{4}{\sqrt{29}} - \frac{8}{\sqrt{29}} = 0$$

7. Since, the point $A(3, 2, -1)$ is the mirror image of the point $B(1, 0, -1)$ with respect to the plane $\alpha x + \beta y + \gamma z = \delta$, then

$$\frac{\alpha}{3-1} = \frac{\beta}{2-0} = \frac{\gamma}{(-1)-(-1)}$$

$$\Rightarrow \frac{\alpha}{1} = \frac{\beta}{1} = \frac{\gamma}{0}$$

$$\Rightarrow \gamma = 0$$

It is given that $\alpha + \gamma = 1 \Rightarrow \alpha = 1$, so $\beta = 1$.

And, the mid-point of AB, $M(2, 1, -1)$ lies on the given plane, so

$$2\alpha + \beta - \gamma = \delta$$

$$\Rightarrow 2(1) + (1) - 0 = \delta$$

$$\Rightarrow \delta = 3$$

$\therefore\ \alpha + \beta = 2,\ \delta - \gamma = 0,\ \delta + \beta = 4.$

8. Plane P_1 contains the line $\mathbf{r} = \hat{\mathbf{i}} + \hat{\mathbf{j}} + \hat{\mathbf{k}} + \lambda(\hat{\mathbf{i}} - \hat{\mathbf{j}} - \hat{\mathbf{k}})$, hence contains the point $\hat{\mathbf{i}} + \hat{\mathbf{j}} + \hat{\mathbf{k}}$ and is normal to vector $(\hat{\mathbf{i}} + \hat{\mathbf{j}})$.

Hence, equation of plane is

$$(\mathbf{r} - (\hat{\mathbf{i}} + \hat{\mathbf{j}} + \hat{\mathbf{k}}))\cdot(\hat{\mathbf{i}} + \hat{\mathbf{j}}) = 0$$

or $$x + y = 2$$

Plane P_2 contains the line

$$\mathbf{r} = \hat{\mathbf{i}} + \hat{\mathbf{j}} + \hat{\mathbf{k}} + (\hat{\mathbf{i}} - \hat{\mathbf{j}} - 3\hat{\mathbf{k}}) \text{ and point } \hat{\mathbf{j}}$$

Hence, equation of plane is

$$\begin{vmatrix} x-0 & y-1 & z-0 \\ 1-0 & 1-1 & 1-0 \\ 1 & -1 & -1 \end{vmatrix} = 0$$

or $$x + 2y - z = 2$$

If θ is the acute angle between P_1 and P_2, then

$$\cos\theta = \frac{\mathbf{n}_1\cdot\mathbf{n}_2}{|\mathbf{n}_1||\mathbf{n}_2|} = \left|\frac{(\hat{\mathbf{i}} + \hat{\mathbf{j}})\cdot(\hat{\mathbf{i}} + 2\hat{\mathbf{j}} - \hat{\mathbf{k}})}{\sqrt{2}\cdot\sqrt{6}}\right|$$

$$= \frac{3}{\sqrt{2}\cdot\sqrt{6}} = \frac{\sqrt{3}}{2}$$

$$\theta = \cos^{-1}\frac{\sqrt{3}}{2} = \frac{\pi}{6}$$

As L is contained in $P_2 \Rightarrow \theta = 0$

9. Let the given points A, B and C respectively. Then $AB^2 = 350$, $AC^2 = 500 - 20\lambda + \lambda^2$, $BC^2 = 150 + 10\lambda + \lambda^2$

Now, $$AB^2 + AC^2 = BC^2$$

$$\Rightarrow 350 + 500 - 20\lambda + \lambda^2 = 150 + 10\lambda + \lambda^2$$

$$\Rightarrow \lambda = \frac{70}{3}$$

Next, $AB^2 + BC^2 = AC^2$

$$\Rightarrow 350 + 150 + 10\lambda + \lambda^2 = 500 - 20\lambda + \lambda^2$$

$$\Rightarrow \lambda = 0$$

Further, $BC^2 + AC^2 = AB^2$

$$\Rightarrow 150 + 10\lambda + \lambda^2 + 500 - 20\lambda + \lambda^2 = 350$$

$$\Rightarrow \lambda^2 - 5\lambda + 150 = 0,$$

Which have no real solution.

$\therefore$ The triangle is right angles for $\lambda = 0, \frac{70}{3}$.

10. At $(0, 0, 0)$, $x - y - z - 2 = -2 = (-\text{ve})$

at $(2, 3, 1)$, $x - y - z - 2 = 2 - 3 - 1 - 2 = -4$ $(-\text{ve})$

Since, both have same sign $(0,0,0)$ and $(2, 3, 1)$ lie on the same side of the plane.

$$\text{Distance} = \left|\frac{2-3-1-2}{\sqrt{1^2+1^2+1^2}}\right| = \frac{4}{\sqrt{3}}$$

Equation of a line perpendicular to the plane $x - y - z - 2 = 2$ and passing through the point $(2, 3, 1)$ is

$$\frac{x-2}{1} = \frac{y-3}{-1} = \frac{z-1}{-1} = \lambda$$

A point on the line is $(\lambda + 2, 3 - \lambda, 1 - \lambda)$ and it lies on the plane $x - y - z - 2 = 0$

If $\lambda + 2 - 3 + \lambda - 1 + \lambda - 2 = 0 \Rightarrow \lambda = \frac{4}{3}$

$\therefore$ Foot of perpendicular on the plane is

$$\left(\frac{4}{3} + 2, 3 - \frac{4}{3}, 1 - \frac{4}{3}\right) = \left(\frac{10}{3}, \frac{5}{3}, -\frac{1}{3}\right)$$

11. Let $\mathbf{v}$ be the vector along L.

Clearly, line L is parallel to given planes

$P_1 : x + 2y - z + 1 = 0$ and $P_2 : 2x - y + z - 1 = 0$

$$\therefore \quad \mathbf{v} = \begin{vmatrix} \hat{\mathbf{i}} & \hat{\mathbf{j}} & \hat{\mathbf{k}} \\ 1 & 2 & -1 \\ 2 & -1 & 1 \end{vmatrix} = \hat{\mathbf{i}} - 3\hat{\mathbf{j}} - 5\hat{\mathbf{k}}$$

Since, L is passing through origin, any point on line L is $A(\lambda, -3\lambda, -5\lambda)$.

Foot of perpendicular from A to P_1, is

$$\frac{h-\lambda}{1} = \frac{k+3\lambda}{2} = \frac{l+5\lambda}{-1}$$

$$= -\frac{(\lambda - 6\lambda + 5\lambda + 1)}{1+4+1} = -\frac{1}{6}$$

$$\therefore \quad h = \lambda - \frac{1}{6},\ k = -3\lambda - \frac{1}{3},\ l = -5\lambda + \frac{1}{6}$$

So, foot or point on locus M is

$$\left(\lambda - \frac{1}{6}, -3\lambda - \frac{1}{3}, -5\lambda + \frac{1}{6}\right)$$

$\therefore$ Option (a) and (b) lie on this locus.

12. For the given lines

$$\begin{vmatrix} 4-1 & 0-1 & -1-(-1) \\ 3 & -1 & 0 \\ 2 & 0 & 3 \end{vmatrix}$$

$$= \begin{vmatrix} 3 & -1 & 0 \\ 3 & -1 & 0 \\ 2 & 0 & 3 \end{vmatrix} = 0$$

So, the given lines intersect.

Any point on the first line is $(3r_1 + 1, -r_1 + 1, -1)$ and any point on the second line is $(2r_2 + 4, 0, 3r_2 - 1)$.

Since, the lines intersect, at the point of intersection.

$$3r_1 + 1 = 2r_2 + 4, -r_1 + 1 = 0, -1 = 3r_2 - 1$$

$$\Rightarrow \quad r_1 = 1, r_2 = 0$$

Hence, the point of intersection is $(4, 0, -1)$.

13. Let $\vec{OA} = \vec{a}$, $\vec{OB} = \vec{b}$, $\vec{OC} = \vec{c}$, then

$$\vec{a} \cdot \vec{a} + (\vec{b} - \vec{c}) \cdot (\vec{b} - \vec{c})$$

$$= \vec{b} \cdot \vec{b} + (\vec{c} - \vec{a}) \cdot (\vec{c} - \vec{a})$$

$$\Rightarrow \quad -2\vec{b} \cdot \vec{c} = -2\vec{c} \cdot \vec{a}$$

$$\Rightarrow (\vec{a} - \vec{b}) \cdot \vec{c} = 0 \text{ or } \vec{BA} \cdot \vec{OC} = 0$$

Hence, $AB \perp OC$

Similarly, $BC \perp OA$ and $CA \perp OB$.

14. Extremities of a diameter of the sphere are given as (0, 2, 0) and (0, 0, 4)

$\Rightarrow$ centre is (0, 1, 2) and radius $= \sqrt{5}$

Equation of the sphere is

$$(x-0)(x-0) + (y-2)(y-0) + (z-0)(z-4) = 0$$

or $\quad x^2 + y^2 + z^2 - 2y - 4z = 0$

which passes through the origin

So, (a, b, c) are correct

Now, $\frac{x}{0} = \frac{y-2}{1} = \frac{z-4}{-2}$ represents a diameter if the centre (0, 1, 2) lies on it.

$\therefore$ There exists a value of r for which

$$(0, r+2, -2r+4) = (0, 1, 2)$$

$$\Rightarrow \quad r + 2 = 1 \text{ and } -2r + 4 = 2$$

Which is not possible.

Hence, (d) is not correct.

15. Given, square base $OP = OR = 3$

$\therefore \quad P(3, 0, 0), R = (0, 3, 0)$

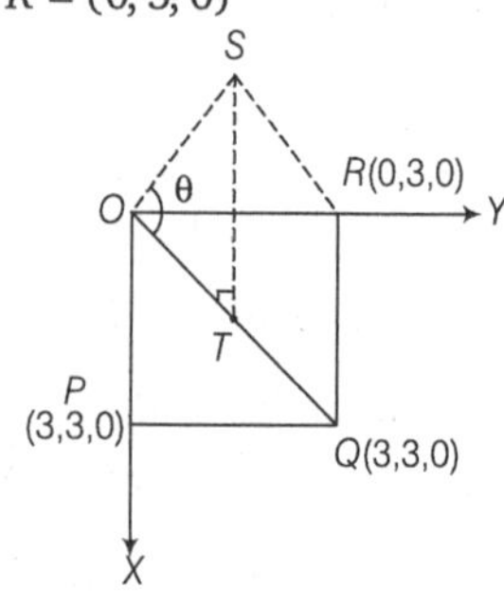

Also, mid-point of OQ is $T\left(\frac{3}{2}, \frac{3}{2}, 0\right)$.

Since, S is directly above the mid-point T of diagonal OQ and $ST = 3$.

i.e. $\quad S\left(\frac{3}{2}, \frac{3}{2}, 3\right)$

Here, DR's of OQ (3, 3, 0) and DR's of $OS\left(\frac{3}{2}, \frac{3}{2}, 3\right)$.

$$\therefore \cos\theta = \frac{\frac{9}{2} + \frac{9}{2}}{\sqrt{9+9+0}\sqrt{\frac{9}{4} + \frac{9}{4} + 9}} = \frac{9}{\sqrt{18} \cdot \sqrt{\frac{27}{2}}} = \frac{1}{\sqrt{3}}$$

$\therefore$ Option (a) is incorrect.

Now, equation of the plane containing the ΔOQS is

$$\begin{vmatrix} x & y & z \\ 3 & 3 & 0 \\ 3/2 & 3/2 & 3 \end{vmatrix} = 0 \Rightarrow \begin{vmatrix} x & y & z \\ 1 & 1 & 0 \\ 1 & 1 & 2 \end{vmatrix} = 0$$

$$\Rightarrow \quad x(2-0) - y(2-0) + z(1-1) = 0$$

$$\Rightarrow \quad 2x - 2y = 0 \text{ or } x - y = 0$$

$\therefore$ Option (b) is correct.

Now, length of the perpendicular from $P(3, 0, 0)$ to the plane containing ΔOQS is

$$\frac{|3-0|}{\sqrt{1+1}} = \frac{3}{\sqrt{2}}$$

$\therefore$ Option (c) is correct.

Here, equation of RS is

$$\frac{x-0}{3/2} = \frac{y-3}{-3/2} = \frac{z-0}{3} = \lambda$$

$$\Rightarrow \quad x = \frac{3}{2}\lambda, \; y = -\frac{3}{2}\lambda + 3, \; z = 3\lambda$$

To find the distance from $O(0, 0, 0)$ to RS.

Let M be the foot of perpendicular.

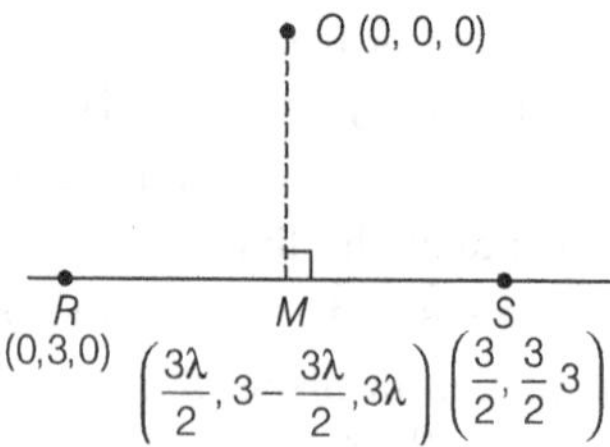

$$\because \quad \overline{OM} \perp \overline{RS} \Rightarrow \overline{OM} \cdot \overline{RS} = 0$$

$$\Rightarrow \quad \frac{9\lambda}{4} - \frac{3}{2}\left(3 - \frac{3\lambda}{2}\right) + 3(3\lambda) = 0 \Rightarrow \lambda = \frac{1}{3}$$

$$\therefore \quad M\left(\frac{1}{2}, \frac{5}{2}, 1\right) \Rightarrow OM = \sqrt{\frac{1}{4} + \frac{25}{4} + 1} = \sqrt{\frac{30}{4}} = \sqrt{\frac{15}{2}}$$

$\therefore$ Option (d) is correct.

16. The given equations are

$$4x - 4y - z + 11 = 0 \quad \ldots(i)$$

$$x + 2y - z - 1 = 0 \quad \ldots(ii)$$

The DR's of normals to the planes (i) and (ii) are $4, -4, -1$ and $1, 2, -1$ respectively.

Let DR 's of line of intersection of plane be l, m, n

As the line of intersection of the planes is perpendiuclar to the normals of the both planes, we get

$$4l - 4m - n = 0$$

and $$l + 2m - n = 0$$

By cross multiplication

$$\frac{l}{6} = \frac{m}{3} = \frac{n}{12} \text{ or } \frac{l}{2} = \frac{m}{1} = \frac{n}{4}$$

If $x = 0$, Eqs. (i) and (ii) becomes

$$-4y - z + 11 = 0$$
$$2y - z - 1 = 0$$

Solving, we get $y = 2$, $z = 3$

$\therefore$ Equation of line is $\frac{x}{2} = \frac{y-2}{1} = \frac{z-3}{4}$

also $x = 4$, $y = 4$, $z = 11$ satisfies Eqs. (i) and (ii)

Hence, (b) is also the correct option.

17. We have, $P_1 : 2x + y - z = 3$

and $P_2 : x + 2y + z = 2$

Here, $\vec{\mathbf{n}}_1 = 2\hat{\mathbf{i}} + \hat{\mathbf{j}} - \hat{\mathbf{k}}$

and $\vec{\mathbf{n}}_2 = \hat{\mathbf{i}} + 2\hat{\mathbf{j}} + \hat{\mathbf{k}}$

(a) Direction ratio of the line of intersection of P_1 and P_2 is θ $\vec{\mathbf{n}}_1 \times \vec{\mathbf{n}}_2$

i.e. $\begin{vmatrix} \hat{i} & \hat{j} & \hat{k} \\ 2 & 1 & -1 \\ 1 & 2 & 1 \end{vmatrix} = (1+2)\hat{\mathbf{i}} - (2+1)\hat{\mathbf{j}} + (4-1)\hat{\mathbf{k}}$

$$= 3(\hat{\mathbf{i}} - \hat{\mathbf{j}} + \hat{\mathbf{k}})$$

Hence, statement a is false.

(b) We have, $\frac{3x-4}{9} = \frac{1-3y}{9} = \frac{z}{3}$

$$\Rightarrow \quad \frac{x - \frac{4}{3}}{3} = \frac{\left(y - \frac{1}{3}\right)}{-3} = \frac{z}{3}$$

This line is parallel to the line of intersection of P_1 and P_2. Hence, statement (b) is false.

(c) Let acute angle between P_1 and P_2 be θ.

We know that, $\cos\theta = \frac{\vec{\mathbf{n}}_1 \cdot \vec{\mathbf{n}}_2}{|\vec{\mathbf{n}}_1||\vec{\mathbf{n}}_2|}$

$$= \frac{(2\hat{\mathbf{i}} + \hat{\mathbf{j}} - \hat{\mathbf{k}}) \cdot (\hat{\mathbf{i}} + 2\hat{\mathbf{j}} + \hat{\mathbf{k}})}{|2\hat{\mathbf{i}} + \hat{\mathbf{j}} - \hat{\mathbf{k}}||\hat{\mathbf{i}} + 2\hat{\mathbf{j}} + \hat{\mathbf{k}}|}$$

$$= \frac{2+2-1}{\sqrt{6} \times \sqrt{6}} = \frac{1}{2}$$

$$\theta = 60°$$

Hence, statement (c) is true.

(d) Equation of plane passing through the point $(4, 2, -2)$ and perpendicular to the line of intersection of P_1 and P_2 is

$$3(x-4) - 3(y-2) + 3(z+2) = 0$$

$$\Rightarrow \quad 3x - 3y + 3z - 12 + 6 + 6 = 0$$

$$\Rightarrow \quad x - y + z = 0$$

Now, distance of the point $(2, 1, 1)$ from the plane $x - y + z = 0$ is

$$D = \left|\frac{2-1+1}{\sqrt{1+1+1}}\right| = \frac{2}{\sqrt{3}}$$

Hence, statement (d) is true.

18. Let the equation of the sphere be

$x^2 + y^2 + z^2 + 2ux + 2vy + 2wz + d = 0$

Since, it passes through $(1, 0, 0)$, $(0, 1, 0)$ and $(0, 0, 1)$

we have, $u = v = w = -\frac{d+1}{2}$

since the centre $(-u, -v, -w)$ lies on $4xy = 1$

We have, $4\left(\frac{d+1}{2}\right)\left(\frac{d+1}{2}\right) = 1$

$$\Rightarrow \quad (d+1)^2 = 1$$

$$\Rightarrow \quad d + 1 = \pm 1$$

$$\Rightarrow \quad d = 0 \text{ or } d = -2$$

and the equation of the sphere is

$$x^2 + y^2 + z^2 - x - y - z = 0$$

or $$x^2 + y^2 + z^2 + x + y + z - 2 = 0$$

Linked Comprehension Type

PASSAGE 1

The DC's of the lines are given by $ul + vm + wn = 0$ and $al^2 + bm^2 + cn^2 = 0$. Eliminating n between them, we get

$$al^2 + bm^2 + c\left\{-\frac{(ul+vm)}{w}\right\}^2 = 0$$

$$\Rightarrow (aw^2 + cu^2)l^2 + (bw^2 + cv^2)m^2 + 2cuvlm = 0$$

$$\Rightarrow (aw^2 + cu^2)\left(\frac{l}{m}\right)^2 + 2cuv\left(\frac{l}{m}\right)$$

$$+ (bw^2 + cv^2) = 0 \quad \text{...(i)}$$

1. Put $u = v = w = 1$ in Eq. (i), then

$$(a+c)\left(\frac{l}{m}\right)^2 + 2c\left(\frac{l}{m}\right) + (b+c) = 0$$

Similarly, $(a+b)\left(\frac{m}{n}\right)^2 + 2a\left(\frac{m}{n}\right) + (c+a) = 0$

and $(b+c)\left(\frac{n}{l}\right)^2 + 2b\left(\frac{n}{l}\right) + (a+b) = 0 \quad$ ...(ii)

2. From Eq. (ii), $\frac{n_1}{l_1} \cdot \frac{n_2}{l_2} = \frac{(a+b)}{(b+c)}$

Similarly, $\frac{l_1 l_2}{(b+c)} = \frac{m_1 m_2}{(c+a)} = \frac{n_1 n_2}{(a+b)} \quad$ ...(iii)

$$\therefore \quad \frac{m_1 m_2}{l_1 l_2} = \frac{(c+a)}{(b+c)}$$

3. From Eq. (iii),

$$\frac{l_1 l_2}{(b+c)} = \frac{m_1 m_2}{(c+a)} = \frac{n_1 n_2}{(a+b)}$$

$$= \frac{l_1l_2 + m_1m_2 + n_1n_2}{(b+c)+(c+a)+(a+b)}$$ [by law of proportion]

$\because$ Lines are perpendicular

$\therefore \quad l_1l_2 + m_1m_2 + n_1n_2 = 0$

Then, $(b+c)+(c+a)+(a+b)$ must be zero

$\therefore \quad 2a + 2b + 2c = 0$

$\Rightarrow \quad a + b + c = 0$

4. From Eq. (i),

$$(aw^2 + cu^2)\left(\frac{l}{m}\right)^2 + 2cuv\left(\frac{l}{m}\right) + (bw^2 + cv^2) = 0$$

Since, the given lines are parallel, then DC's are equal, then the roots of Eq. (i) are equal.

$\therefore \quad B^2 = 4AC$

$\Rightarrow \quad (2cuv)^2 = 4\,(aw^2 + cu^2)\,(bw^2 + cv^2)$

$\Rightarrow \quad 4c^2u^2v^2 = 4\,(abw^4 + acv^2w^2 + bcu^2w^2 + c^2u^2v^2)$

$\Rightarrow \quad abw^2 + acv^2 + bcu^2 = 0$

or $\quad \dfrac{u^2}{a} + \dfrac{v^2}{b} + \dfrac{w^2}{c} = 0$

$\Rightarrow \quad \sum \dfrac{u^2}{a} = 0$

5. From Eq. (i),

$$\frac{l_1l_2}{m_1m_2} = \text{product of roots} = \frac{(bw^2 + cv^2)}{(aw^2 + cu^2)}$$

or $\quad \dfrac{l_1l_2}{bw^2 + cv^2} = \dfrac{m_1m_2}{aw^2 + cu^2} = \dfrac{n_1n_2}{av^2 + bu^2}$

by symmetry

$\because$ lines are perpendicular,

$\therefore \quad l_1l_2 + m_1m_2 + n_1n_2 = 0$

i.e. $(bw^2 + cv^2) + (aw^2 + cu^2) + (av^2 + bu^2) = 0$

$\Rightarrow \quad u^2(b+c) + v^2(c+a) + w^2(a+b) = 0$

$\Rightarrow \quad \sum u^2(b+c) = 0$

PASSAGE 2

1. Shortest distance between both lines

$$= \left| \frac{\begin{vmatrix} 3-2 & 1-1 & 0+1 \\ 1 & 0 & 2 \\ 1 & 1 & -1 \end{vmatrix}}{\begin{vmatrix} \hat{\mathbf{i}} & \hat{\mathbf{j}} & \hat{\mathbf{k}} \\ 1 & 0 & 2 \\ 1 & 1 & -1 \end{vmatrix}} \right|$$

$$= \left| \frac{1(0-2) - 0(-1-2) + 1(1-0)}{-2\hat{\mathbf{i}} + 3\hat{\mathbf{j}} + \hat{\mathbf{k}}} \right|$$

$$= \frac{1}{\sqrt{4+9+1}} = \frac{1}{\sqrt{14}}$$

2. Equation of plane P

$$\begin{vmatrix} x-2 & y-1 & z+1 \\ 1 & 0 & 2 \\ 1 & 1 & -1 \end{vmatrix} = 0$$

$-2(x-2) + (y-1)(3) + (z+1) = 0$

$\Rightarrow \quad -2x + 3y + z + 2 = 0$

$\Rightarrow \quad 2x - 3y - z - 2 = 0$

Now image of point $O(0, 0, 0)$ in plane P

$$\frac{x-0}{2} = \frac{y-0}{-3} = \frac{z-0}{-1}$$

$$= \frac{-2(-2)}{4+9+1}$$

$$\frac{x}{2} = \frac{y}{-3} = \frac{z}{-1} = \frac{4}{14}$$

Image point $\left(\dfrac{4}{7}, \dfrac{-6}{7}, \dfrac{-2}{7}\right)$.

3. $O\,(0, 0, 0)$, $A(1, 0, 0)$, $B\left(0, \dfrac{-2}{3}, 0\right)$, $C(0, 0, -2)$

Volume of tetrahetron $OABC = \dfrac{1}{6}\,|\,[\mathbf{OA\ OB\ OC}]\,|$

$$= \frac{1}{6}\begin{vmatrix} 1 & 0 & 0 \\ 0 & -2/3 & 0 \\ 0 & 0 & -2 \end{vmatrix}$$

$$= \frac{4}{18} = \frac{2}{9} \text{ cu unit}$$

PASSAGE 3

1. The equation of any plane through the intersection of P_1 and P_2 is

$$P_1 + \lambda P_2 = 0$$

$\Rightarrow \quad (2x - y + z - 2) + \lambda\,(x + 2y - z - 3) = 0 \quad$ ...(i)

Since, it passes through $(3, 2, 1)$, then

$$(6 - 2 + 1 - 2) + \lambda\,(3 + 4 - 1 - 3) = 0$$

$\therefore \quad \lambda = -1$

From Eq. (i),

$$x - 3y + 2z + 1 = 0$$

which is the required plane.

2. The equation of any plane through $(-1, 3, 2)$ is

$a(x+1) + b\,(y-3) + c(z-2) = 0 \quad$ ...(ii)

If this plane (ii) is perpendicular to P_1, then

$2a - b + c = 0 \quad$ ...(ii)

and if the plane (ii) is perpendicular to P_2 then

$a + 2b - c = 0 \quad$ ...(iii)

From Eqs. (ii) and (iii), we get

$$\frac{a}{-1} = \frac{b}{3} = \frac{c}{5}$$

Substituting these proportionate values of a, b, c in Eq. (ii), we get the required equation as

$$-(x+1) + 3(y-3) + 5(z-2) = 0$$

or $\quad x - 3y - 5z + 20 = 0$

3. The given planes can be written as $-2x+y-z+2=0$ and $-x-2y+z+3=0$

Here, $(-2)(-1)+(1)(-2)+(-1)(1)=-1<0$

Equation of bisectors

$$\frac{(-2x+y-z+2)}{\sqrt{(4+1+1)}}=\pm\frac{(-x-2y+z+3)}{\sqrt{1+4+1}}$$

$\therefore$ Acute angle bisector is

$$(-2x+y-z+2)=(-x-2y+z+3)$$

$\Rightarrow\quad x-3y+2z+1=0$

4. Equation of bisector which not containing the origin is

$$\frac{(-2x+y-z+2)}{\sqrt{(4+1+1)}}=-\frac{(-x-2y+z+3)}{\sqrt{(1+4+1)}}$$

$\Rightarrow\quad 3x+y-5=0$

5. The image of plane P_1 in the plane mirror P_2, then

$$2(2\cdot1+(-1)\cdot2+1\cdot(-1))(x+2y-z-3)$$
$$=(1+4+1)(2x-y+z-2)$$

$\Rightarrow\quad -(x+2y-z-3)=3(2x-y+z-2)$

$\Rightarrow\quad 7x-y+4z-9=0$

Numerical Value Type

1. The coordinates of the projections of A, B, C on the yz-plane are $(0, 2, 3)$, $(0,-1,1)$ and $(0, 2, -4)$ respectively

$\therefore\ \Delta_x$ = area of projection of ΔABC on yz-plane

$$=\frac{1}{2}\left|\begin{vmatrix}2&3&1\\-1&1&1\\2&-4&1\end{vmatrix}\right|$$

Applying $R_1\to R_1+2R_2$ and $R_3\to R_3+2R_2$

$$\text{then } \Delta_x=\frac{1}{2}\left|\begin{vmatrix}0&5&3\\-1&1&1\\0&-2&3\end{vmatrix}\right|=\frac{21}{2}\text{ sq units}$$

Similarly, the projection of A, B and C on zx and xy-planes are $(1, 0, 3)$, $(2, 0, 1)$, $(1, 0, -4)$ and $(1, 2, 0)$, $(2, -1, 0)$, $(1, 2, 0)$ respectively. Also, let Δ_y and Δ_z be the areas of the projection of the ΔABC on zx and xy-planes respectively.

Then,

$$\Delta_y=\frac{1}{2}\left|\begin{vmatrix}1&3&1\\2&1&1\\1&-4&1\end{vmatrix}\right|$$

Applying $R_2\to R_2-R_1$ and $R_3\to R_3-R_1$

$$\text{Then,}\quad \Delta_y=\frac{1}{2}\left|\begin{vmatrix}1&3&1\\-1&-2&0\\0&-7&0\end{vmatrix}\right|$$

$$=\frac{7}{2}\text{ sq units}$$

$$\text{and}\quad \Delta_z=\frac{1}{2}\left|\begin{vmatrix}1&2&1\\2&-1&1\\1&2&1\end{vmatrix}\right|$$

Applying $C_1\to C_1-C_3$

$$\text{Then, } \Delta_z=\frac{1}{2}\left|\begin{vmatrix}0&2&1\\1&-1&1\\0&2&1\end{vmatrix}\right|=0$$

$\therefore$ The required area $=\sqrt{\Delta_x^2+\Delta_y^2+\Delta_z^2}$

$$=\sqrt{\left(\frac{21}{2}\right)^2+\left(\frac{7}{2}\right)^2+(0)^2}$$

$$=\frac{7}{2}\sqrt{10}\text{ sq units}$$

$$\lambda=\frac{7}{2}\sqrt{10}\text{ sq units}$$

$\Rightarrow\quad 20\sqrt{10}\lambda=20\sqrt{10}\times\frac{7}{2}\sqrt{10}=700$

2. Let the equation of the variable plane be

$$\left(\frac{x}{a}\right)+\left(\frac{y}{b}\right)+\left(\frac{z}{c}\right)=1 \qquad \text{...(i)}$$

Given that the plane is at a distance p from $(0, 0, 0)$

$$\therefore\quad p=\frac{1}{\sqrt{\left\{\left(\frac{1}{a}\right)^2+\left(\frac{1}{b}\right)^2+\left(\frac{1}{c}\right)^2\right\}}}$$

$$\text{or}\quad \frac{1}{p^2}=\frac{1}{a^2}+\frac{1}{b^2}+\frac{1}{c^2} \qquad \text{...(ii)}$$

Also, the plane (i) meets the axes in A, B and C. So the coordinates of O, A, B and C are $(0, 0, 0)$, $(a, 0, 0)$, $(0, b, 0)$ and $(0, 0, c)$ respectively.

Let (x, y, z) be the centroid of the tetrahedron $OABC$, then

$$x=\frac{1}{4}(0+a+0+0)$$
$$=\frac{1}{4}a$$

Similarly $\quad y=\frac{1}{4}b$ and $z=\frac{1}{4}c$

or $\quad a=4x,\ b=4y,\ c=4z$

Substituting these values in Eq. (ii), we get

$$\frac{1}{p^2}=\frac{1}{16x^2}+\frac{1}{16y^2}+\frac{1}{16z^2}$$

or $x^{-2}+y^{-2}+z^{-2}=16p^{-2}$

$\therefore\quad \lambda=16$

$\Rightarrow\quad 160\lambda=160\times16=2560$

3. The equations of Y-axis are $\frac{x}{0}=\frac{y}{1}=\frac{z}{0}$,

$\therefore$ Any point N on Y-axis is $(0, r, 0)$ ...(i)

$\therefore$ The direction cosines of the line PN are $0-4,\ r-3,\ 0-5$

i.e. $\quad -4,\ r-3,\ -5$ ...(ii)

Let N be the foot of the perpendicular from P to Y-axis, then PN is $\perp$ to the Y-axis whose direction cosines are 0, 1, 0 and so from Eq. (ii), we have

$$0\cdot(-4)+1\cdot(r-3)+0\cdot(-5)=0$$

$\Rightarrow\quad r=3$

∴ From Eq. (i) the coordinates of N are (0, 3, 0)

∴ Required distance $= PN$

$$= \sqrt{(4-0)^2 + (3-3)^2 + (5-0)^2}$$

$$= \sqrt{41} \text{ units}$$

∴ $\lambda = \sqrt{41}$

⇒ $\lambda^2/5 = 8.2$

4. The planes are

$$y + z = 0 \quad \text{...(i)}$$
$$z + x = 0 \quad \text{...(ii)}$$
$$x + y = 0 \quad \text{...(iii)}$$
$$x + y + z = 1 \quad \text{...(iv)}$$

The point of intersection of the planes (i), (ii) and (iii) is obviously the origin i.e. (0, 0, 0). Solving Eqs. (ii), (iii) and (iv) we get $y = 1, z = 1$ and *ie,* these planes intersect in (−1, 1, 1). Similarly the other two vertices of the tetrahedron are (1, − 1, 1) and (1, 1, − 1).

∴ Required volume $= \dfrac{1}{6}\begin{vmatrix} x_1 & y_1 & z_1 \\ x_2 & y_2 & z_2 \\ x_3 & y_3 & z_3 \end{vmatrix}$

$$= \frac{1}{6}\begin{vmatrix} -1 & 1 & 1 \\ 1 & -1 & 1 \\ 1 & 1 & -1 \end{vmatrix}$$

Applying $C_2 \to C_2 + C_1$ and $C_3 \to C_3 + C_1$, then

$$= \frac{1}{6}\begin{vmatrix} -1 & 0 & 0 \\ 1 & 0 & 2 \\ 1 & 2 & 0 \end{vmatrix},$$

$$= -\frac{1}{6}\begin{vmatrix} 0 & 2 \\ 2 & 0 \end{vmatrix} = -\left(\frac{1}{6}\right)(-4) = \frac{2}{3}$$

∴ $\lambda = \dfrac{2}{3} = 0.67$

5. Let C_1, C_2 be the centres of the spheres and P be their point of intersection. Then the angle between the spheres is the angle between their radi C_1P and C_2P.

∴ In $\Delta C_1 PC_2$, $C_1P = r_1$, $C_2P = r_2$ and $C_1C_2 = d$

∴ If θ be the required angle, then

$$\cos\theta = \cos\angle C_1PC_2 = \frac{C_1P^2 + C_2P^2 - C_1C_2^2}{2C_1P \cdot C_2P}$$

$$= \frac{r_1^2 + r_2^2 - d^2}{2r_1r_2}$$

Now, the given spheres are

$$x^2 + y^2 + z^2 - 2x - 4y - 6z + 10 = 0 \quad \text{...(i)}$$

and $(x-1)(x-5) + (y-2)(y-0) + (z+3)(z-1) = 0$

or $x^2 + y^2 + z^2 - 6x - 2y + 2z + 2 = 0$ ...(ii)

Centre and radius of Eq. (i) are (1, 2, 3) and 2

Centre and radius of Eq. (ii) are (3, 1, − 1) and 3

∴ $r_1 = 2, r_2 = 3, d^2 = [(3-1)^2 + (1-2)^2 + (-1-3)^2] = 21$

∴ $\cos^{-1}\left[\dfrac{r_1^2 + r_2^2 - d^2}{2r_1r_2}\right] = \cos^{-1}\left[\dfrac{4+9-21}{2\cdot 2\cdot 3}\right] = \cos^{-1}\left(-\dfrac{2}{3}\right)$

∴ $\lambda = -\dfrac{2}{3}$

⇒ $9999|\lambda| = 9999 \times \dfrac{2}{3} = 6666$

6. The equation of the coaxial system of spheres is $(x^2 + y^2 + z^2 + 2x + 2y + 4z + 2) +$
$\lambda(x^2 + y^2 + z^2 + x + y + 2z + 2) = 0$

or $x^2 + y^2 + z^2 + \left(\dfrac{2+\lambda}{1+\lambda}\right)x + \left(\dfrac{2+\lambda}{1+\lambda}\right)y + \left(\dfrac{4+2\lambda}{1+\lambda}\right)$

$z + 2 = 0$

Its centre is $\left(-\dfrac{(2+\lambda)}{2(1+\lambda)}, -\dfrac{(2+\lambda)}{2(1+\lambda)}, -\dfrac{(2+\lambda)}{(1+\lambda)}\right)$ ...(i)

and Equating its radius to zero, we get

$$\frac{(2+\lambda)^2}{4(1+\lambda)^2} + \frac{(2+\lambda)^2}{4(1+\lambda)^2} + \frac{(2+\lambda)^2}{(1+\lambda)^2} - 2 = 0$$

⇒ $\left(\dfrac{2+\lambda}{1+\lambda}\right)^2\left(\dfrac{1}{4} + \dfrac{1}{4} + 1\right) = 2$

⇒ $\dfrac{4 + \lambda^2 + 4\lambda}{1 + \lambda^2 + 2\lambda} = \dfrac{4}{3}$

⇒ $\lambda^2 - 4\lambda - 8 = 0$

∴ $\lambda = \dfrac{4 \pm \sqrt{16+32)}}{2} = (2 \pm 2\sqrt{3})$

Substituting the values of λ in the coordinates of the above Eq. (i), the required limiting points are

$$\left(-\frac{1}{\sqrt{3}}, -\frac{1}{\sqrt{3}}, -\frac{2}{\sqrt{3}}\right) \text{ and } \left(\frac{1}{\sqrt{3}}, \frac{1}{\sqrt{3}}, \frac{2}{\sqrt{3}}\right)$$

∴ $1004|\alpha_1\alpha_2 + \beta_1\beta_2 + \gamma_2\gamma_2|$

$$= 1004\left|-\frac{1}{3} - \frac{1}{3} - \frac{4}{3}\right| = 2008$$

Single Integer Type

1. Since, the equation of plane passes through the intersection of two given planes

$x + 4y - z + 7 = 0$ and $3x + y + 5z = 8$

is $(x + 4y - z + 7) + \lambda(3x + y + 5z - 8) = 0$

⇒ $(3\lambda + 1)x + (4 + \lambda)y + (5\lambda - 1)z + (7 - 8\lambda) = 0$, which represents the plane $ax + by + 6z = 15$

$$\frac{a}{3\lambda + 1} = \frac{b}{4 + \lambda} = \frac{6}{5\lambda - 1} = \frac{15}{8\lambda - 7}$$

So, $48\lambda - 42 = 75\lambda - 15 \Rightarrow 27\lambda = -27 \Rightarrow \lambda = -1$

∴ Equation of required plane P is

$$-2x + 3y - 6z + 15 = 0$$

or $2x - 3y + 6z - 15 = 0$

So, required distance

$$= \frac{|(2\times 3) - (3\times 2) - (6\times 1) - 15|}{\sqrt{4 + 9 + 36}} = \frac{21}{7} = 3 \text{ units}$$

Hence, answer is 3.00.

2. $\frac{x-1}{3}=\frac{y-2}{-m}=\frac{z+3}{1}$

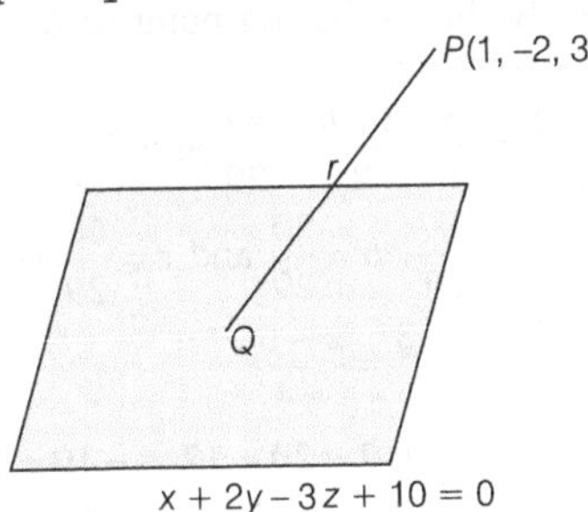

$$\text{DC of line} \equiv \left(\frac{3}{\sqrt{m^2+10}}, \frac{-m}{\sqrt{m^2+10}}, \frac{1}{\sqrt{m^2+10}}\right)$$

$$Q \equiv \left(1+\frac{3r}{\sqrt{m^2+10}}, -2+\frac{-mr}{\sqrt{m^2+10}}, 3+\frac{r}{\sqrt{m^2+10}}\right)$$

Q lies on $x+2y-3z+10=0$

$$1+\frac{3r}{\sqrt{m^2+10}}-4-\frac{2mr}{\sqrt{m^2+10}}-9-\frac{3r}{\sqrt{m^2+10}}+10=0$$

$$\Rightarrow \frac{r}{\sqrt{m^2+10}}(3-2m-3)=2$$

$$\Rightarrow \frac{r}{\sqrt{m^2+10}}(-2m)=2$$

$$\Rightarrow r^2m^2=m^2+10$$

$$\Rightarrow \frac{7}{2}m^2=m^2+10$$

$$\Rightarrow \frac{5}{2}m^2=10$$

$$\Rightarrow m^2=4$$

$$|m|=2$$

3. Since, the line $\frac{x-3}{2}=\frac{y+2}{-1}=\frac{z+4}{3}$ lies in the plane $lx+my-z=9$.

Therefore, we have

$$2l-m-3=0$$

[∵ normal will be perpendicular to the line]

$$\Rightarrow 2l-m=3 \quad \ldots\text{(i)}$$

and $3l-2m+4=9$

[∵ point $(3,-2,-4)$ lines on the plane]

$$\Rightarrow 3l-2m=5 \quad \ldots\text{(ii)}$$

On solving Eqs. (i) and (ii), we get

$$l=1 \text{ and } m=-1$$

$$\therefore l^2+m^2=2$$

4. It is given that point $Q\left(\frac{5}{3},\frac{7}{3},\frac{17}{3}\right)$ is foot of perpendicular of point $P(1, 0, 3)$ to a line passes through point $A(\alpha, 7, 1)$, so $PQ \perp AQ$

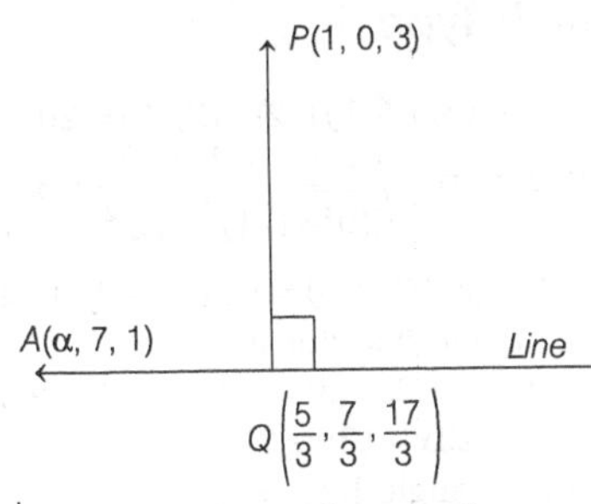

∵ Dr's of line segment PQ is $\left(\frac{2}{3},\frac{7}{3},\frac{8}{3}\right)$ and Dr's of line AQ is $\left(\alpha-\frac{5}{3},\frac{14}{3},-\frac{14}{3}\right)$

$$\therefore \frac{2}{3}\left(\alpha-\frac{5}{3}\right)+\frac{7}{3}\left(\frac{14}{3}\right)+\frac{8}{3}\left(-\frac{14}{3}\right)=0$$

$$\Rightarrow (3\alpha-5)+49-56=0$$

$$\Rightarrow 3\alpha-5-7=0$$

$$\Rightarrow \alpha=4$$

5. Let $P(\alpha,\beta,\gamma)$ and R is image of P in the xy-plane.

$$\therefore R(\alpha,\beta,-\gamma)$$

Also, Q is the image of P in the plane

$$x+y=3$$

$$\therefore \frac{x-\alpha}{1}=\frac{y-\beta}{1}=\frac{z-\gamma}{0}=\frac{-2(\alpha+\beta-3)}{2}$$

$$x=3-\beta, y=3-\alpha, z=\gamma$$

Since, Q is lies on Z-axis

$$\therefore \beta=3, \alpha=3, z=\gamma$$

$$\therefore P(3,3,\gamma)$$

Given, distance of P from X-axis be 5

$$\therefore 5=\sqrt{3^2+\gamma^2}$$

$$25-9=\gamma^2 \Rightarrow \gamma=\pm 4$$

Then, $PR=|2\gamma|=|2\times 4|=8$

6. The given points are $O(0, 0, 0)$, $A(0, 0, 2)$, $B(0, 4, 0)$ and $C(6, 0, 0)$.

Here, three faces of tetrahedron are xy, yz, zx plane.

Since, point P is equidistance from zx, xy and yz planes, its coordinates are $P(r, r, r)$

Equation of plane ABC is

$$2x+3y+6z=12 \quad \text{(from intercept form)}$$

P is also at distance r from plane ABC. Thus,

$$\frac{|2r+3r+6r-12|}{\sqrt{4+9+36}}=r$$

or $|11r-12|=7r$

or $11r-12=\pm 7r$

or $r=\frac{12}{18}, 3$

$$\therefore r=2/3 \quad [\text{as } r<2]$$

$$\Rightarrow 9r=9\times\frac{2}{3}=6$$

Matrix-Match Type

1. **(A)** $\because$ $(2)(3)+(-1)(-2)+(2)(6)=20>0$

$\therefore$ Bisectors are $\dfrac{(2x-y+2z+3)}{\sqrt{(2)^2+(-1)^2+(2)^2}}=\pm\dfrac{(3x-2y+6z+8)}{\sqrt{(3)^2+(-2)^2+(6)^2}}$

or $7(2x-y+2z+3)=\pm 3(3x-2y+6z+8)$

$\therefore$ Acute angle bisector is

$$7(2x-y+2z+3)=-3(3x-2y+6z+8)$$

$\Rightarrow$ $23x-13y+32z+45=0$

and Obtuse angle bisector is

$$7(2x-y+2z+3)=3(3x-2y+6z+8)$$

$\Rightarrow$ $5x-y-4z-3=0$

$\therefore$ $A: 23x-13y+32z+45=0$ (R)

and $O: 5x-y-4z-3=0$

(B) The given planes can be written as

$-x+2y-2z+3=0$ and $2x-3y+6z+8=0$

$\because$ $(-1)(2)+(2)(-3)+(-2)(6)=-2-6-12=-20<0$

$\therefore$ Bisectors are,

$$\frac{(-x+2y-2z+3)}{\sqrt{(-1)^2+(2)^2+(-2)^2}}=\pm\frac{(2x-3y+6z+8)}{\sqrt{(2)^2+(-3)^2+(6)^2}}$$

$\Rightarrow 7(-x+2y-2z+3)=\pm 3(2x-3y+6z+8)$

$\therefore$ Acuts angle bisector is $7(-x+2y-2z+3)$

$$=3(2x-3y+6z+8)$$

$\Rightarrow 13x-23y+32z+3=0$

and obtuse bisector is $7(-x+2y-2z+3)$

$$=-3(2x-3y+6z+8)$$

$\Rightarrow$ $x-5y-4z-45=0$

$\Rightarrow \therefore A: 13x-23y+32z+3=0$ (T)

and $O: x-5y-4z-45=0$ (Q)

(C) The given planes can be written as

$2x+y-2z+3=0$ and $-6x-2y+3z+8=0$

$\because$ $(2)(-6)+(1)(-2)+(-2)(3)=-20<0$

$\therefore$ Bisectors are $\dfrac{(2x+y-2z+3)}{\sqrt{\{(2)^2+(1)^2+(-2)^2\}}}$

$$=\pm\frac{(-6x-2y+3z+8)}{\sqrt{\{(-6)^2+(-2)^2+(3)^2\}}}$$

$\Rightarrow$ $7(2x+y-2z+3)=\pm 3(-6x-2y+3z+8)$

$\therefore$ Acute angle bisector is

$$7(2x+y-2z+3)=3(-6x-2y+3z+8)$$

$\Rightarrow 32x+13y-23z-3=0$

and obtuse bisector is $7(2x+y-2z+3)$

$$=-3(-6x-2y+3z+8)$$

$\Rightarrow 4x-y+5z-45=0$

$\therefore$ $A: 32x+13y-23z-3=0$ (P)

and $O: 4x-y+5z-45=0$ (S)

2. **(A)** Let the foot of perpendicular be $Q(2+2\lambda, 1+3\lambda, 2+4\lambda)$

$\therefore$ $2(2\lambda+1)+3(3\lambda-1)+4(4\lambda-1)=0$

$$29\lambda=5\Rightarrow\lambda=\frac{5}{29}$$

$\therefore$ Foot $=\left(\dfrac{68}{29},\dfrac{44}{29},\dfrac{78}{29}\right)$ (S)

(B) Let the image be the point (a, b, c), then from previous solution

$$\frac{1+a}{2}=\frac{68}{29},\ \frac{2+b}{2}=\frac{44}{29}\text{ and }\frac{3+c}{2}=\frac{78}{29}$$

i.e., $a=\dfrac{107}{29}$, $b=\dfrac{30}{29}$ and $c=\dfrac{68}{29}$ (P)

(C) $\dfrac{x-2}{2}=\dfrac{y-3}{3}=\dfrac{z-5}{-4}$

$$=-\frac{4+9-20+17}{4+9+16}=\frac{-10}{29}$$

$\therefore$ $a=\dfrac{38}{29}$, $b=\dfrac{57}{29}$ and $c=\dfrac{185}{29}$ (T)

(D) $\dfrac{x-2}{3}=\dfrac{y-5}{-2}=\dfrac{z-1}{2}$

$$=-2\left(\frac{6-10+4-5}{29}\right)=\frac{10}{29}$$

$$x=2+\frac{30}{29}=\frac{88}{29},\ y=5-\frac{20}{29}=\frac{125}{29},$$

$$z=1+\frac{40}{29}=\frac{69}{29}\ \text{(Q)}$$

3. **(A)** Let the line makes an angle θ with Z-axis

Then, $\cos^2\alpha+\cos^2\alpha+\cos^2\theta=1$

$\Rightarrow$ $\cos^2\theta=-\cos 2\alpha$

$\therefore$ $0\le\cos^2\theta\le 1$

$\Rightarrow$ $0\le-\cos 2\alpha\le 1$

$\Rightarrow$ $-1\le\cos 2\alpha\le 0$

$\Rightarrow$ $\cos^{-1}(0)\le 2\alpha\le\cos^{-1}(-1)$

$\Rightarrow$ $\dfrac{\pi}{2}\le 2\alpha\le\pi$ or $\dfrac{\pi}{4}\le\alpha\le\dfrac{\pi}{2}$

Then, $0\le\cot\alpha\le 1$ (Q, R, S)

(B) Let $\dfrac{x-2}{1}=\dfrac{y-3}{1}=\dfrac{4-z}{\lambda}=r_1$ (say)

then general point on Ist line is $(2+r_1, 3+r_1, 4-\lambda r_1)$

and Let $\dfrac{x-1}{\lambda}=\dfrac{y-4}{2}=\dfrac{z-5}{1}=r_2$ (say)

Then, general point on IInd line is

$$(1+\lambda r_2, 4+2r_2, 5+r_2)$$

$\because$ Lines intersect, then

$$2+r_1=1+\lambda r_2,\ 3+r_1=4+2r_2,\ 4-\lambda r_1=5+r_2$$

Eliminating r_1 and r_2, we get, $\lambda^2+3\lambda=0$

$\therefore$ $\lambda=0, -3$ (P, R)

(C) $\because$ Line is perpendicular to normal of the plane

$\therefore$ $\lambda\cdot\lambda+(-\mu)\cdot\mu+\nu\cdot\nu=0$

$\Rightarrow$ $\lambda^2-\mu^2+\nu^2=0$ ...(i)

and point $(\lambda, 2\phi, \nu)$ lies on the plane

then, $\lambda^2-2\mu\phi+\nu^2=0$ ...(ii)

From Eqs. (i) and (ii), we get

$$2\mu\phi=\mu^2$$

or $2\phi=\mu$

$\therefore$ $\dfrac{\mu}{\phi}=2$ (T)

34

Mathematical Reasoning

Objective Questions Type I [Only one correct answer]

In each of the questions below, four choices are given of which only are is correct. You have to select the correct answer which is the most appropriate.

1. Let p and q be two statements. Then, $(\sim p \vee q) \wedge (\sim p \wedge \sim q)$ is
(a) tautology
(b) contradiction
(c) neither tautology nor contradiction
(d) either tautology or contradiction

2. If each of the statements $p \to q$, $q \to r$ and $\sim r$ is true, then
(a) p is true (b) p is false
(c) q is true (d) None of these

3. The contrapositive of the statement "If I reach the station in time, then I will catch the train" is **[JEE Main 2020, 2 September]**
(a) If I do not reach the station in time, then I will catch the train
(b) If I will not catch the train, then I do not reach the station in time
(c) If I do not reach the station in time, then I will not catch the train
(d) If I will catch the train, then I reach the station in time

4. The proposition $p \Rightarrow \sim (p \wedge \sim q)$ is a
(a) Contradiction (b) Tautology
(c) either (a) or (b) (d) neither (a) nor (b)

5. If p, q and r are simple propositions with truth values T, F and T, respectively, then the truth value of $(\sim p \vee q) \wedge \sim r \to p$ is
(a) True (b) False
(c) True if r is false (d) True if q is true

6. Which of the following is a tautology? **[JEE Main 2020, 25 September]**
(a) $(\sim p) \wedge (p \vee q) \to q$ (b) $(q \to p) \vee \sim (p \to q)$
(c) $(\sim q) \vee (p \wedge q) \to q$ (d) $(p \to q) \wedge \sim (q \to p)$

7. Conclusion of the argument $p, q, - p \to r, q \to - r | \to$
(a) r (b) q
(c) $\sim r$ (d) none of these

8. The statement $p \to (q \to p)$ is equivalent to
(a) $p \to (p \to q)$ (b) $p \to (p \vee q)$
(c) $p \to (p \wedge q)$ (d) $p \to (p \leftrightarrow q)$

9. The negation of $\sim s \vee (\sim r \wedge s)$ is equivalent to :
(a) $s \wedge \sim r$ (b) $s \wedge (r \wedge \sim s)$
(c) $s \vee (r \vee \sim s)$ (d) $s \wedge r$

10. Which one of the following is a tautology? **[JEE Main 2020, 8 January]**
(a) $(P \wedge (P \to Q)) \to Q$ (b) $P \wedge (P \vee Q)$
(c) $P \vee (P \wedge Q)$ (d) $Q \to (P \wedge (P \to Q))$

11. $\sim (p \vee (\sim p \vee q))$ is equal to
(a) $\sim p \wedge (p \wedge \sim q)$ (b) $(p \wedge \sim q) \vee \sim p$
(c) $(p \vee \sim q) \vee \sim p$ (d) None of these

12. The proposition $S : (p \Rightarrow q) \Leftrightarrow (\sim p \vee q)$ is
(a) a Tautology
(b) a Contradiction
(c) either (a) or (b)
(d) neither (a) nor (b)

13. Which one of the following Boolean expressions is a tautology ? **[JEE Main 2019, 10 April]**
(a) $(p \vee q) \vee (p \vee \sim q)$
(b) $(p \wedge q) \vee (p \wedge \sim q)$
(c) $(p \vee q) \wedge (p \vee \sim q)$
(d) $(p \vee q) \wedge (\sim p \vee \sim q)$

14. If $p \to (q \vee r)$ is false, then the truth values of p, q and r are, respectively,
(a) T, T, F (b) F, F, F
(c) F, T, T (d) T, F, F

15. The statement $p \to (q \vee r)$ is not equivalent to
(a) $(p \to q) \vee (p \to r)$
(b) $p \wedge (\sim q) \to r$
(c) $p \wedge (\sim r) \to q$
(d) $p \wedge q \to (p \wedge r) \vee (q \wedge r)$

16. Which of the following Boolean expression is a tautology? **[JEE Main 2021, 16 March]**
(a) $(p \wedge q) \vee (p \vee q)$ (b) $(p \wedge q) \vee (p \rightarrow q)$
(c) $(p \wedge q) \wedge (p \rightarrow q)$ (d) $(p \wedge q) \rightarrow (p \rightarrow q)$

17. The conditional $(p \wedge q) \Rightarrow p$ is
(a) A tautology
(b) A fallacy i.e., contradiction
(c) Neither tautology nor fallacy
(d) Fallacy if p is always true

18. Given the following two statements:
$(S_1) : (q \vee p) \rightarrow (p \leftrightarrow \sim q)$ is a tautology.
$(S_2) : \sim q \wedge (\sim p \leftrightarrow q)$ is a fallacy. Then **[JEE Main 2020, 4 September]**
(a) both (S_1) and (S_2) are not correct
(b) only (S_1) is correct
(c) both (S_1) and (S_2) are correct
(d) only (S_2) is correct

19. The negation of the statement
"If I become teacher, then I will open a school."
(a) Either I will not become a teacher or I will not open school.
(b) Neither I will become a teacher nor I will open a school.
(c) I will not become a teacher or I will not open a school
(d) I will become a teacher and I will not open a school.

20. The proposition $S : \sim (p \Rightarrow q) \Rightarrow (\sim p \vee \sim q)$ is
(a) a Tautology (b) a Contradiction
(c) either (a) or (b) (d) neither (a) nor (b)

21. The statement $(p \rightarrow (q \rightarrow p)) \rightarrow (p \rightarrow (p \vee q))$ is **[JEE Main 2020, 5 September]**
(a) equivalent to $(p \wedge q) \vee (\sim q)$
(b) a contradiction
(c) equivalent to $(p \vee q) \wedge (\sim p)$
(d) a tautology

22. $(p \wedge \sim q) \wedge (\sim p \wedge q)$ is
(a) a tautology
(b) a contradiction
(c) neither a tautology nor a contradiction
(d) None of the above

23. Contrapositive of the statement 'If a function f is differentiable at a, then it is also continuous at a', is **[JEE Main 2020, 4 September]**
(a) If a function f is not continuous at a, then it is differentiable at a
(b) If a function f is continuous at a, then it is differentiable at a
(c) If a function f is continuous at a, then it is not differentiable at a
(d) If a function f is not continuous at a, then it is not differentiable at a

24. If $(p \wedge \sim r) \Rightarrow (q \vee r)$ is false, then p is
(a) True (b) False
(c) may be True or False (d) None of these

25. Let p, q, r be three statements such that the truth value of $(p \wedge q) \rightarrow (\sim q \vee r)$ is F. Then the truth values of p, q, r are respectively. **[JEE Main 2020, 3 September]**
(a) T, F, T (b) F, T, F
(c) T, T, F (d) T, T, T

26. Contrapositive of $p \rightarrow (q \rightarrow r)$ is logically equivalent to
(a) $p \rightarrow (q \rightarrow r)$ (b) $(q \rightarrow r) \rightarrow \sim p$
(c) $p \vee q \rightarrow r$ (d) $(q \rightarrow r) \rightarrow p$

27. The logical statement $[\sim (\sim p \vee q) \vee (p \wedge r)] \wedge (\sim q \wedge r)$ is equivalent to
(a) $\sim p \vee r$ (b) $(p \wedge \sim q) \vee r$
(c) $(p \wedge r) \wedge \sim q$ (d) $(\sim p \wedge \sim q) \wedge r$

28. The statement $p \rightarrow (q \rightarrow p)$ is equivalent to
(a) $p \rightarrow (q \leftrightarrow p)$ (b) $p \rightarrow (p \rightarrow q)$
(c) $p \rightarrow (p \wedge q)$ (d) $p \rightarrow (p \vee q)$

29. If the Boolean expression $(p \Rightarrow q) \Leftrightarrow (q * (\sim p))$ is a tautology, then the Boolean expression $p * (\sim q)$ is equivalent to **[JEE Main 2021, 17 March]**
(a) $q \Rightarrow p$ (b) $\sim q \Rightarrow p$
(c) $p \Rightarrow \sim q$ (d) $p \Rightarrow q$

30. The only statement among the following that is a tautology is
(a) $A \wedge (A \vee B)$ (b) $A \vee (A \wedge B)$
(c) $[A \wedge (A \rightarrow B)] \rightarrow B$ (d) $B \rightarrow [A \wedge (A \rightarrow B)]$

31. Which one of the following statements is not a tautology? **[JEE Main 2019, 8 April]**
(a) $(p \wedge q) \rightarrow (\sim p) \vee q$ (b) $(p \wedge q) \rightarrow p$
(c) $p \rightarrow (p \vee q)$ (d) $(p \vee q) \rightarrow (p \vee (\sim q))$

32. Consider the following three statements:
P : 5 is a prime number.
Q : 7 is a factor of 192.
R : LCM of 5 and 7 is 35.
Then, the truth value of which one of the following statements is true ?
(a) $(P \wedge Q) \vee (\sim R)$ (b) $P \vee (\sim Q \wedge R)$
(c) $(\sim P) \vee (Q \wedge R)$ (d) $(\sim P) \wedge (\sim Q \wedge R)$

33. The false statement in the following is
(a) $p \wedge (\sim q)$ is a contradiction
(b) $(p \Rightarrow q) \Leftrightarrow (\sim q \Rightarrow \sim p)$ is a tautology
(c) $\sim (\sim p) \Leftrightarrow p$ is a tautology
(d) $p \vee (\sim p) \Leftrightarrow$ is a tautology

34. If p, q, r are simple propositions with truth values T, F, T then the truth value of $(\sim p \vee q) \wedge \sim r \Rightarrow p$ is
(a) True (b) False
(c) True if r is false (d) True if q is true

35. For any two statements p and q, the negation of the expression $p \vee (\sim p \wedge q)$ is **[JEE Main 2019, 9 April]**
(a) $\sim p \wedge \sim q$ (b) $\sim p \vee \sim q$
(c) $p \wedge q$ (d) $p \leftrightarrow q$

36. Let p and q be two statements, then
$\sim (\sim p \wedge q) \wedge (p \vee q)$ is logically equivalent to
(a) q (b) $p \wedge q$
(c) p (d) $p \vee \sim q$

37. If p, q and r are simple propositions, then $(p \wedge q) \wedge (q \wedge r)$ is true, then
(a) p, q and r are true
(b) p, q are true and r is false
(c) p is true and q, r are false
(d) p, q and r are false

38. If the truth value of the statement $p \rightarrow (\sim q \vee r)$ is false (F), then the truth values of the statements p, q and r are respectively **[JEE Main 2019, 12 April]**
(a) T, T and F (b) T, F and F
(c) T, F and T (d) F, T and T

39. If P and Q are two statements, then which of the following compound statement is a tautology? **[JEE Main 2021, 18 March]**
(a) $((P \Rightarrow Q) \wedge \sim Q) \Rightarrow Q$
(b) $((P \Rightarrow Q) \wedge \sim Q) \Rightarrow \sim P$
(c) $((P \Rightarrow Q) \wedge \sim Q) \Rightarrow P$
(d) $((P \Rightarrow Q) \wedge \sim Q) \Rightarrow (P \wedge Q)$

40. Which of the following statements is a tautology? **[JEE Main 2020, 8 January]**
(a) $\sim (p \wedge \sim q) \rightarrow p \vee q$ (b) $\sim (p \vee \sim q) \rightarrow p \wedge q$
(c) $p \vee (\sim q) \rightarrow p \wedge q$ (d) $\sim (p \vee \sim q) \rightarrow p \vee q$

41. $\sim (p \vee q) \vee (\sim p \wedge q)$ is equivalent to
(a) q (b) p
(c) $\sim p$ (d) $\sim q$

42. If q is false and $p \wedge q \longleftrightarrow r$ is true, then which one of the following statements is a tautology? **[JEE Main 2019, 11 January]**
(a) $p \vee r$ (b) $(p \wedge r) \rightarrow (p \vee r)$
(c) $(p \vee r) \rightarrow (p \wedge r)$ (d) $p \wedge r$

43. Let p, q, r be three statements, then $(p \rightarrow (q \rightarrow r)) \leftrightarrow ((p \wedge q) \rightarrow r)$, is a
(a) Tautology
(b) Contradiction
(c) Fallacy
(d) None of the above

44. The negation of the boolean expression $\sim s \vee (\sim r \wedge s)$ is equivalent to
(a) $s \wedge r$ (b) $\sim s \wedge \sim r$
(c) $s \vee r$ (d) r

45. Consider the following argument
"If it is a good watch then it is a Titan watch. It is a Titan watch therefore it is a good watch". This argument is
(a) Valid
(b) invalid
(c) may be valid or invalid
(d) invalid if conditional connective is replaced by biconditional connective

46. The Boolean expression $((p \wedge q) \vee (p \vee \sim q)) \wedge (\sim p \wedge \sim q)$ is equivalent to **[JEE Main 2019, 12 January]**
(a) $p \wedge q$ (b) $p \vee (\sim q)$
(c) $p \wedge (\sim q)$ (d) $(\sim p) \wedge (\sim q)$

47. In the truth table for the statement $(p \rightarrow q) \leftrightarrow (\sim p \vee q)$, the last column has the truth value in the following order
(a) T T T T (b) F T F T
(c) T T F F (d) F F F F

48. The logical statement $(p \Rightarrow q) \wedge (q \Rightarrow \sim p)$ is equivalent to **[JEE Main 2020, 7 January]**
(a) $\sim p$ (b) q
(c) p (d) $-q$

49. The statement $(p \rightarrow q) \rightarrow [(\sim p \rightarrow q) \rightarrow q]$ is **[JEE Main 2017]**
(a) a tautology (b) equivalent to $\sim p \rightarrow q$
(c) equivalent to $p \rightarrow \sim q$ (d) a fallacy

50. Negation of the statement $(p \wedge r) \rightarrow [(\sim p \rightarrow q) \rightarrow q]$ is
(a) $(p \wedge r) \wedge (\sim r \wedge \sim q)$
(b) $\sim (p \wedge r) \rightarrow \sim (r \vee q)$
(c) $\sim (p \vee r) \rightarrow \sim (r \wedge q)$
(d) $(p \wedge r) \vee (r \vee q)$

Answers

Objective Questions Type I [Only one correct answer]

1. (c)	2. (b)	3. (b)	4. (d)	5. (a)	6. (a)	7. (c)	8. (b)	9. (d)	10. (a)
11. (a)	12. (a)	13. (a)	14. (d)	15. (d)	16. (d)	17. (a)	18. (a)	19. (d)	20. (a)
21. (d)	22. (b)	23. (d)	24. (a)	25. (c)	26. (a)	27. (c)	28. (d)	29. (c)	30. (c)
31. (d)	32. (b)	33. (b)	34. (a)	35. (a)	36. (c)	37. (a)	38. (a)	39. (c)	40. (d)
41. (c)	42. (b)	43. (a)	44. (a)	45. (b)	46. (d)	47. (a)	48. (a)	49. (a)	50. (a)

Answers

1.

p	q	$\sim p \vee q$	$\sim p \wedge \sim q$	$(\sim p \vee q) \wedge (\sim p \wedge \sim q)$
T	T	T	F	F
T	F	F	F	F
F	T	T	F	F
F	F	T	T	T

∵ Neither tautology nor contradiction.

2. Since, $\sim r$ is true, therefore, r is false.

Again $q \rightarrow r$ is true and r is false, therefore, q has to be false.

(∵ A true statement cannot lead to a false statement)

Further $p \rightarrow q$ is true and q is false, therefore, p is also false.

3. Let the statements

p : I reach the station is time.

q : I will catch the train.

As we know contrapositive of

$$p \Rightarrow q \equiv (\sim q) \Rightarrow (\sim p)$$

So contrapositive of given statement, "If I reach the station in time, then I will catch the train" is "If I will not catch the train, then I do not reach the station in time".

4. Let S : $p \Rightarrow \sim (p \wedge \sim q)$, hence

p	q	$\sim q$	$\sim p \vee q$	S
F	F	T	T	T
F	T	F	T	T
T	F	T	F	F
T	T	F	T	T

which is neither a tautology nor contradiction.

5. $(\sim p \vee q) \wedge \sim r \rightarrow p$

$\equiv (\text{F} \vee \text{F}) \wedge \text{F} \rightarrow \text{T}$

$\equiv \text{F} \rightarrow \text{T}$

$\equiv \text{T}$

6. By truth table

p	q	$\sim p$	$\sim q$	$p \vee q$	$p \wedge q$	$p \rightarrow q$	$q \rightarrow p$	$(\sim p) \wedge (p \vee q)$	$(\sim p) \wedge (p \vee q) \rightarrow q$
T	T	F	F	T	T	T	T	F	T
T	F	F	T	T	F	F	T	F	T
F	T	T	F	T	F	T	F	T	T
F	F	T	T	F	F	T	T	F	T

∴ The statement $(\sim p) \wedge (p \vee q) \rightarrow q$ is a tautology.

7. $p, q, -p \rightarrow r, q \rightarrow -r$ are true.

q is true, $q \rightarrow -r$ is true, therefore $-r$ is true, i.e., r is false also p is true $-p \rightarrow r$. Therefore r must be false.

Hence, $\sim r$ is most appropriate conclusion.

8. $p \rightarrow (q \rightarrow p) \equiv \sim p \vee (q \rightarrow p)$

$\equiv (\sim p) \vee (\sim q \vee p)$

$\equiv (\sim q) \vee (p \vee \sim p)$

$\equiv (\sim q) \vee \text{T} = \text{T}$

∴ $p \rightarrow (p \rightarrow q)$ is a tautology,

Also, $p \rightarrow (q \vee p) \equiv \sim p \vee (p \vee q)$

$\equiv (\sim p \vee p) \vee q \equiv \text{T} \vee q = \text{T}$

∴ $p \rightarrow (p \vee q)$ is also a tautology.

Thus, $p \rightarrow (q \rightarrow p)$ is equivalent to $p \rightarrow (p \vee q)$.

9. $\sim (\sim s \vee (\sim r \wedge s))$

$\equiv s \wedge \sim (\sim r \wedge s)$

$\equiv s \wedge (r \vee \sim s)$

$\equiv (s \wedge r) \vee (s \wedge \sim s)$

$\equiv s \wedge r$

10. From the truth table

P	Q	$P \vee Q$	$P \wedge Q$	$P \rightarrow Q$	$P \wedge (P \rightarrow Q)$	$P \wedge (P \vee Q)$	$P \vee (P \wedge Q)$
T	T	T	T	T	T	T	T
T	F	T	F	F	F	T	T
F	T	T	F	T	F	F	F
F	F	F	F	T	F	F	F

$P \wedge (P \rightarrow Q) \rightarrow Q$	$Q \rightarrow (P \wedge (P \rightarrow Q))$
T	T
T	T
T	F
T	T

∴ $P \wedge (P \rightarrow Q) \rightarrow Q$ is a tautology.

11. $\sim [p \vee (\sim p \vee q)] \equiv \sim p \wedge \sim (\sim p \vee q)$

$\equiv \sim p \wedge (\sim (\sim p) \wedge \sim q)$

$\equiv \sim p \wedge (p \wedge \sim q)$

12. Let $S : (p \Rightarrow q) \Leftrightarrow (\sim p \vee q)$, hence

p	q	$p \Rightarrow q$	$\sim p \vee q$	S
F	F	T	T	T
F	T	T	T	T
T	F	F	F	T
T	T	T	T	T

Since, for all possible combination of p and q the statement has truth value T, hence, it is a tautology.

13. Option (a) $(p \vee q) \vee (p \vee (\sim q))$

$\equiv p \vee (q \vee \sim q)$ is tautology,

[∵ $q \vee (\sim q) \equiv \text{T}$ and $p \vee \text{T} \equiv \text{T}$]

Option (b)

$(p \wedge q) \vee (p \wedge (\sim q)) \equiv p \wedge (q \vee \sim q)$

not a tautology, [∵ $q \vee \sim q \equiv \text{T}$ and $p \wedge \text{T} \equiv p$]

Option (c)

$(p \vee q) \wedge (p \vee (\sim q)) \equiv p \vee (q \wedge \sim q)$

not a tautology [∵ $q \wedge \sim q \equiv \text{F}$ and $p \vee \text{F} \equiv p$]

Option (d)

$(p \vee q) \wedge ((\sim p) \vee (\sim q)) \equiv (p \vee q) \wedge (\sim (p \wedge q))$

not a tautology.

14. $p \to q$ is false only when p is true and q is false. Therefore, $p \to (q \vee r)$ is false when p is true and $q \vee r$ is false. $q \vee r$ is false only when both q and r are false. Hence, truth values of p, q and r are, respectively, T, F and F.

15. $p \to (q \vee r) \equiv (\sim p) \vee (q \vee r)$

$\equiv (\sim p \vee q) \vee (\sim p \vee r) \equiv (p \to q) \vee (p \to r)$

Also, $p \to (q \vee r) \equiv (\sim p) \vee (q \vee r) \equiv (\sim p \vee q) \vee r$

$\equiv \sim (p \wedge (\sim q)) \vee r \equiv p \wedge (\sim q) \to r$

Interchanging the roles of q and r in the above paragraph, we find

$$p \to (q \vee r) \equiv p \wedge (\sim q) \to r \equiv p \wedge (\sim r) \to q$$

For $p = T$, $q = F$, $r = F$, $p \to (q \vee r)$ is F, but

$(p \wedge q) \to (p \vee r) \vee (q \wedge r)$ is T

$\therefore$ $p \to (q \vee r)$

and $p \wedge q \to (p \wedge r) \vee (q \wedge r)$ are not equivalent.

16.

p	q	$p \wedge q$	$p \to q$	$(p \wedge q) \to (p \to q)$
T	T	T	T	T
T	F	F	F	T
F	T	F	T	T
F	F	F	T	T

$(p \wedge q) \to (p \to q)$ is tautology.

17.

p	q	$p \wedge q$	$(p \wedge q) \Rightarrow p$
T	T	T	T
T	F	F	T
F	T	F	T
F	F	F	T

$\therefore$ $(p \wedge q) \Rightarrow p$ is a tautology.

18. Given statement $S_1 : (q \vee p) \to (p \leftrightarrow (\sim q))$ is a tautology.

For S_1:

p	q	$\sim q$	$q \vee p$	$p \leftrightarrow (\sim q)$	$(q \vee p) \to (p \leftrightarrow (\sim q))$
T	T	F	T	F	F
T	F	T	T	T	T
F	T	F	T	T	T
F	F	T	F	F	T

$\Rightarrow S_1$ is not a tautology, so S_1 is not correct.

Now, $S_2 : (\sim q) \wedge ((\sim p) \leftrightarrow q)$ is a fallacy.

For S_2 :

p	q	$\sim p$	$\sim q$	$(\sim p) \leftrightarrow q$	$(\sim q) \wedge ((\sim p) \leftrightarrow (q)$
T	T	F	F	F	F
T	F	F	T	T	T
F	T	T	F	T	F
F	F	T	T	F	F

$\Rightarrow S_2$ is not a fallacy, so S_2 is not correct.

19. Let p : I become a teacher and q : I will open a school. The given statement is $p \to q \equiv (\sim p) \vee q$. It negation is $\sim (\sim p) \vee q) \equiv (p \wedge (\sim q)$. Thus, negation of the given statement is I will become a teacher and I will not open a school.

20. Let $S \sim (p \Rightarrow q) \Rightarrow (\sim p \vee \sim q)$,

Also, $S : S_1 \Rightarrow S_2$; where, $S_1 \sim (p \Rightarrow q)$ and $S_2 : (\sim p \vee \sim q)$ for $S_1 : \sim (p \Rightarrow q)$

p	q	$p \Rightarrow q$	S_1
F	F	T	F
F	T	T	F
T	F	F	T
T	T	T	F

For $S_2 : (\sim p \vee \sim q)$

p	q	$\sim p$	$\sim q$	S_2
F	F	T	T	T
F	T	T	F	T
T	F	F	T	T
T	T	F	F	F

For $S : S_1 \Rightarrow S_2$

p	q	S_1	S_2	S
F	F	F	T	T
F	T	F	T	T
T	F	T	T	T
T	T	F	F	T

Hence, the statement S is a Tautology.

21. The truth table

p	q	$p \vee q$	$q \to p$	$p \to (q \to p)$	$p \to (p \vee q)$	$(p \to (q \to p)) \to (p \to (p \vee q))$
T	T	T	T	T	T	T
T	F	T	T	T	T	T
F	T	T	F	T	T	T
F	F	F	T	T	T	T

$\therefore$ The statement $p \to (q \to p) \to (p \to (p \vee q))$ is a tautology.

22. $(p \wedge \sim q) \wedge (\sim p \wedge q)$

$\equiv (p \wedge \sim p) \wedge (\sim q \wedge q) \equiv f \wedge f$

$\equiv f$ (by using associative laws and commutative laws)

Therefore, $(p \wedge \sim q) \wedge (\sim p \wedge q)$ is a contradiction.

23. As the contrapositive of statement $p \to q$ is $\sim q \to \sim p$

$\therefore$ Contrapositive of statement 'If a function f is differentiable at a, then it is also continuous at a' is if a function f is not continuous at a, then it is not differentiable at a.

24. Let $S : (p \wedge \sim r) \Rightarrow (q \wedge r)$

p	q	r	$\sim r$	$p \wedge \sim r$	$q \vee r$	S
F	F	F	T	F	F	T
F	F	T	F	F	T	T
F	T	F	T	F	T	T
F	T	T	F	F	T	T
T	F	F	T	T	F	F
T	F	T	F	F	T	T
T	T	F	T	T	T	T
T	T	T	F	F	T	T

So, value of p should be 'True'.

25. Given statement $(p \wedge q) \longrightarrow (\sim q \vee r)$
$= \sim (p \wedge q) \vee (\sim q \vee r) = ((\sim p) \vee (\sim q)) \vee (\sim q \vee r)$
$= (\sim p) \vee (\sim q) \vee r$, is give false, so $\sim p$, $\sim q$ and r must be false, so p, q must be true and r should be false, so truth value of p, q, r are T, T, F respectively.

26. Contrapositive of $p \rightarrow (q \rightarrow r)$ is $\sim (q \rightarrow r) \rightarrow \sim p$
$\equiv \sim [\sim (q \rightarrow r)] \vee (\sim p)$
$\equiv (q \rightarrow r) \vee (\sim p) \equiv (\sim p) \vee (q \rightarrow r) \equiv p \rightarrow (q \rightarrow r)$

27. Clearly, $[\sim (\sim p \vee q) \vee (p \wedge r)] \wedge (\sim q \wedge r)$
$\equiv [(p \wedge \sim q) \vee (p \wedge r)] \wedge (\sim q \wedge r)$
($\because \sim (\sim p \vee q) \equiv \sim (\sim p) \wedge \sim q \equiv p \wedge \sim q$ by De Morgan's law)
$\equiv [p \wedge (\sim q \vee r)] \wedge (\sim q \wedge r)]$ (distributive law)
$\equiv p \wedge [(\sim q \vee r) \wedge (\sim q \wedge r)]$ (associative law)
$\equiv p \wedge [(\sim q \wedge r) \wedge (\sim q \vee r)]$ (commutative law)
$\equiv p \wedge [\{(\sim q \wedge r) \wedge (\sim q)\} \wedge \{(\sim q \wedge r) \wedge r]$ (distributive law)
$\equiv p \wedge [(\sim q \wedge r) \vee (\sim q \wedge r)]$ (idempotent law)
$\equiv p \wedge [\sim q \wedge r]$ (idempotent law)
$\equiv p \wedge \sim q \wedge r \equiv (p \wedge r) \wedge (\sim q)$ (associative law)

28.

p	q	$q \rightarrow p$	$p \rightarrow (q \rightarrow p)$	$p \rightarrow (q \leftrightarrow p)$	$p \rightarrow (p \rightarrow q)$	$p \rightarrow (p \vee q)$
T	F	T	T	F	F	T
F	T	F	T	T	T	T
T	T	T	T	T	T	T
F	F	T	T	T	T	T

So, from the above table $P \rightarrow (q \rightarrow p) = p \rightarrow (p \vee q)$

29. $\because$ $p \rightarrow q \equiv \sim p \vee q$. So, $* \equiv \vee$
Thus, $p * (\sim q) \equiv p \vee (\sim q) \equiv q \rightarrow p$

30. We know that, $A \wedge (A \vee B)$ is F when $A = F$
$A \vee (A \wedge B)$ is F when $A = F$, $B = F$,
and $B \rightarrow [A \wedge (A \rightarrow B)]$ is F when $A = F$, $B = T$
$\therefore$ We check only (c) $[A \wedge (A \rightarrow B)] \rightarrow B$
$\equiv [A \wedge (\sim A \vee B)] \rightarrow B$
$\equiv [(A \wedge (\sim A))] \vee (A \wedge B)] \rightarrow B$
$\equiv A \wedge B \rightarrow B \equiv \sim (A \wedge B) \vee B$
$\equiv \sim [(A \wedge B) \wedge (\sim B)]$
$\equiv \sim [A \wedge (B \wedge \sim B)]$
$\equiv \sim [A \wedge F] \equiv \sim F \equiv T$
Thus, $[A \wedge (A \rightarrow B)] \rightarrow B$ is a tautology.

31. $(p \wedge q) \rightarrow (\sim p) \vee q$
$\equiv (\sim (p \wedge q)) \vee ((\sim p) \vee q)$
$\equiv ((\sim p) \vee (\sim q)) \vee ((\sim p) \vee q)$
$\equiv (\sim p) \vee (\sim q) \vee q$ $[\because (\sim p) \vee (\sim p) \equiv \sim p]$
$\equiv (\sim p) \vee T$ $[\because \sim q \vee q \equiv T]$
$\equiv T$
So, it is a tautology $[\because ((\sim q) \vee q)$ is tautology$]$

(b) $(p \wedge q) \rightarrow p \equiv (p \wedge q) \vee p$
$\equiv ((\sim p) \vee (\sim q)) \vee p$ $[\because \sim (p \wedge q) \equiv (\sim p) \vee (\sim q)]$
$\equiv (\sim p \vee p) \vee (\sim q)$ is tautology.
$[\because \sim p \vee p$ is a tautology and $(\sim q) \vee T \equiv T]$

(c) $\because$ $p \rightarrow (p \vee q) \equiv (\sim p) \vee (p \vee q)$
$[\because$ $p \rightarrow q$ is equivalent to $(\sim p \vee q)]$
$\equiv (\sim p \vee p) \vee q$ is tautology.
$[\because (\sim p \vee p)$ is tautology and $q \vee T \equiv T]$

(d) $(p \vee q) \rightarrow (p \vee (\sim q))$
$\equiv (\sim (p \vee q)) \vee (p \vee (\sim q))$
$\equiv ((\sim p) \wedge (\sim q)) \vee (p \vee (\sim q))$
$\equiv (p \vee (\sim q) \vee ((\sim p) \wedge (\sim q))$
$\equiv (p \vee (\sim q) \vee (\sim p)) \wedge (p \vee (\sim q) \vee (\sim q))$
$\equiv (T \vee (\sim q)) \wedge (p \wedge (\sim q)) \equiv T \wedge (p \wedge (\sim q))$
$\equiv p \wedge (\sim q)$, which is not a tautology.

32. Since, the statements
P : 5 is a prime number, is true statement.
Q : 7 is a factor of 192, is false statement
and R : LCM of 5 and 7 is 35, is true statement.
So, truth value of P is T, Q is F, R is T.
Now let us check all the options.

P	Q	R	$\sim P$	$\sim Q$	$\sim R$	$P \wedge Q$	$Q \wedge R$	$\sim Q \wedge R$
T	F	T	F	T	F	F	F	T

$(P \wedge Q) \vee (\sim R)$	$P \vee (\sim Q \wedge R)$	$(\sim P) \vee (Q \wedge R)$	$(\sim P) \wedge (\sim Q \wedge R)$
F	T	F	F

Clearly, the truth value of $P \vee (\sim Q \wedge R)$ is T.

33. $p \Rightarrow q$ is logically equivalent to $\sim q \Rightarrow \sim p$
$\therefore$ $(p \Rightarrow q) \Leftrightarrow (\sim q \Rightarrow \sim p)$ is a tautology but not a contradiction
$\therefore$ $(p \Rightarrow q) \Leftrightarrow (\sim q \Rightarrow \sim p)$ is a tautology

34. $\sim p \vee q$ means $F \vee F = F$, $\sim r$ means
$F (\sim p \vee q) \wedge \sim r$ means F
$\because [(\sim p \vee q) \wedge \sim r] \Rightarrow p$ means T.
$[\because$ In $p \Rightarrow q$ we have, FTT$]$

35. $\because p \vee ((\sim p) \wedge q)$
$= (p \vee (\sim p)) \wedge (p \vee q)$ [by Distributive law]
$= p \vee q$ $[\because p \vee (\sim p)$ is tautology$]$
So negation of $p \vee ((\sim p) \wedge q)$
$= \sim [p \vee (\sim p) \wedge q] = \sim (p \vee q)$
$= (\sim p) \wedge (\sim q)$ [by Demorgan's law]

36. $\sim (\sim p \wedge q) \wedge (p \vee q) \equiv [\sim (\sim p) \vee (\sim q)] \wedge (p \vee q)$
$\equiv [p \vee (\sim q) \wedge (p \vee q]$
$\equiv p \vee [(\sim q) \wedge q] \equiv p \vee F \equiv p$

37. Let $S : (p \wedge q) \wedge (q \wedge r)$

p	q	r	$p \wedge q$	$q \wedge r$	S
F	F	F	F	F	F
F	F	T	F	F	F
F	T	F	F	F	F
F	T	T	F	T	F
T	F	F	F	F	F
T	F	T	F	F	F
T	T	F	T	F	F
T	T	T	T	T	T

Here, clearly from above table if S is true, then p, q and r are true.

38. Given statement is $p \rightarrow (\sim q \vee r) = \sim p \vee (\sim q \vee r)$
Now, from the options
(a) When $p = T$, $q = T$ and $r = F$
then $\sim p \vee (\sim q \vee r) = F \vee (F \vee F) = F$

(b) When $p = T$, $q = F$ and $r = F$

then $\sim p \vee (\sim q \vee r) = F \vee (T \vee F) = T$

(c) When $p = T$, $q = F$ and $r = T$

then $\sim p \vee (\sim q \vee r) = F \vee (T \vee T) = T$

(d) When $p = F$, $q = T$ and $r = T$

then $\sim p \vee (\sim q \vee r) = T \vee (F \vee T) = T$

39. $((P \to Q) \wedge \sim Q)$

$\equiv (\sim P \vee Q) \wedge \sim Q$

$\equiv (\sim P \wedge \sim Q) \vee (Q \wedge \sim Q)$

$\equiv \sim P \wedge \sim Q$

For option (a)

$(P \wedge \sim Q) \to Q$

$\equiv \sim (\sim P \wedge \sim Q) \vee Q$

$\equiv (P \vee Q) \vee Q \neq$ tautology

For option (b)

$(\sim P \wedge \sim Q) \to \sim P$

$\equiv \sim (\sim P \wedge \sim Q) \vee \sim P$

$\equiv (P \vee Q) \vee \sim P \neq$ tautology

For option (c)

$(\sim P \wedge \sim Q) \to P$

$\equiv (P \vee Q) \vee P \neq$ tautology

For option (d)

$(\sim P \wedge \sim Q) \to (P \wedge Q)$

$\equiv (P \vee Q) \vee (P \wedge Q) \neq$ tautology

40. From the truth table

p	q	$\sim p$	$\sim q$	$p \vee (\sim q)$	$p \wedge q$	$\sim (p \vee (\sim q))$	$p \vee q$	$(\sim (p \vee (\sim q))) \to p \vee q$
T	T	F	F	T	T	F	T	T
T	F	F	T	T	F	F	T	T
F	T	T	F	F	F	T	T	T
T	F	T	T	T	F	F	F	T

So, $(\sim (p \vee (\sim q))) \to p \vee q$ is a tautology.

41. $\sim (p \vee q) \vee (\sim p \wedge q)$

$\equiv (\sim p \wedge \sim q) \vee (\sim p \wedge q)$

$\equiv \sim p \wedge (\sim q \vee q)$

$\equiv \sim p \wedge t$

$\equiv \sim p$

42. $(p \wedge q) \leftrightarrow r$ is true. This is possible under two cases

Case I When both $p \wedge q$ and r are true, which is not possible because q is false.

Case II When both $(p \wedge q)$ and r are false.

$\Rightarrow$ $p \equiv T$ or F; $q \equiv F$, $r \equiv F$

In this case,

(a) $p \vee r$ is T or F

(b) $(p \wedge r) \to (p \vee r)$ is F $\to$ (T or F), which always result in T.

(c) $(p \vee r) \to (p \wedge r)$ is (T or F) $\to$ F, which may be T or F.

(d) $p \wedge r$ is F

43. $(p \to (q \to r)) \equiv \sim p \vee (q \to r)$

$\equiv \sim p \vee (\sim q \vee r)$

$\equiv [(\sim p) \vee (\sim q)] \vee r$

$\equiv \sim (p \wedge q) \vee r$

$\equiv p \wedge q \to r$

$\therefore$ Given statement is a tautology.

44. The given boolean expression is $\sim s \vee ((\sim r) \wedge s)$

Now, the negation of given boolean expression is

$$\sim (\sim s \vee ((\sim r) \wedge s))$$

$= s \wedge (\sim ((\sim r) \wedge s))$ $\quad [\because \sim (p \wedge q) = \sim p \vee \sim q]$

$= s \wedge (r \vee (\sim s))$ $\quad [\because \sim (p \vee q) = \sim p \wedge \sim q]$

$= (s \wedge r) \vee (s \wedge (\sim s))$ $\quad [\because p \wedge (q \vee r) \equiv (p \wedge q) \vee (p \wedge r)]$

$= (s \wedge r)$ $\quad [\because p \wedge \sim p \equiv F]$

45. p: it is a good watch, q: it is a Titan watch

$S_1 : p \to q$ $\qquad S_2 : q$ $\qquad S : p$

p	q	$p \to q$	S_1	S_2	S
T	F	F	F	F	T
F	T	T	T	T	F
T	T	T	T	T	T
F	F	T	T	F	F

As, S_1 : T, S_2 : T $\Rightarrow S$: T so invalid argument.

46. Let the given Boolean expression

$((p \wedge q) \vee (p \vee \sim q)) \wedge (\sim p \wedge \sim q) \equiv r$

Now, let us construct the following truth table

p	q	$\sim p$	$\sim q$	$p \wedge q$	$p \vee \sim q$	$\sim p \wedge \sim q$	$(p \wedge q) \vee (p \vee \sim q)$	r
T	T	F	F	T	T	F	T	F
T	F	F	T	F	T	F	T	F
F	T	T	F	F	F	F	F	F
F	F	T	T	F	T	T	T	T

Clearly, $r \equiv \sim p \wedge \sim q$

47.

p	q	$p \to q$	$\sim p$	$\sim p \vee q$	$(p \to q) \leftrightarrow (\sim p \vee q)$
T	T	T	F	T	T
T	F	F	F	F	T
F	T	T	T	T	T
F	F	T	T	T	T

48. From the truth table

p	q	$p \Rightarrow q$	$\sim p$	$q \Rightarrow \sim p$	$(p \Rightarrow q) \wedge (q \Rightarrow \sim p)$
T	T	T	F	F	F
T	F	F	F	T	F
F	T	T	T	T	T
F	F	T	T	T	T

$(p \Rightarrow q) \wedge (q \Rightarrow \sim p)$ is equivalent to $\sim p$.

49. The truth table of the given expression is given below,

p	q	$x \equiv p \to q$	$\sim p$	$\sim p \to q$ $y \equiv$	$p \to q \to q$	$x \to y$
T	T	T	F	T	T	T
T	F	F	F	T	F	T
F	T	T	T	T	T	T
F	F	T	T	F	T	T

Hence, it is a tautology.

50. We know that,

$$\sim (p \to q) \cong p \wedge \sim q$$

$\therefore \quad \sim ((p \wedge r) \to (r \vee q)) \cong (p \wedge r) \wedge (\sim (r \vee q))$

$\cong (p \wedge r) \wedge (\sim r \wedge \sim q)$

35

Statistics

Objective Questions Type I [Only one correct answer]

In each of the questions below, four choices are given of which only are is correct. You have to select the correct answer which is the most appropriate.

1. The mean of n items is $\overline{X}$. If the first item is increased by 1, second by 2 and so on, then the new mean is

(a) $\overline{X}+n$ (b) $\overline{X}+\frac{n}{2}$

(c) $\overline{X}+\frac{n+1}{2}$ (d) None of these

2. Let $X=\{x\in N : 1\leq x\leq 17\}$ and $Y=\{ax+b : x\in X \text{ and } a, b\in R, a>0\}$. If mean and variance of elements of Y are 17 and 216 respectively, then $a+b$ is equal to **[JEE Main 2020, 2 September]**

(a) 9 (b) 7

(c) –7 (d) –27

3. If $\overline{X}_1$ and $\overline{X}_2$ are the means of two distributions such that $\overline{X}_1<\overline{X}_2$ and $\overline{X}$ is the mean of the combined distribution, then

(a) $\overline{X}<\overline{X}_1$ (b) $\overline{X}>\overline{X}_2$

(c) $\overline{X}=\frac{\overline{X}_1+\overline{X}_2}{2}$ (d) $\overline{X}_1<\overline{X}<\overline{X}_2$

4. For a certain frequency table which has been partly reproduced here, the arithmetic mean was found to be Rs 28.07

Income (in ₹)	15	20	25	30	35	40
Number of workers	8	12	?	16	?	10

If the total number of workers is 75, then the missing frequencies are

(a) 14, 15 (b) 15, 16

(c) 13, 16 (d) 12, 17

5. If $\sum_{i=1}^{n}(x_i-a)=n$ and $\sum_{i=1}^{n}(x_i-a)^2=na$, $(n, a>1)$, then the standard deviation of n observations $x_1, x_2, \ldots, x_n$ is **[JEE Main 2020, 6 September]**

(a) $a-1$ (b) $n\sqrt{a-1}$

(c) $\sqrt{n(a-1)}$ (d) $\sqrt{a-1}$

6. The mean of the numbers $\frac{{}^{50}C_0}{1}, \frac{{}^{50}C_2}{3}, \frac{{}^{50}C_4}{5} \ldots, \frac{{}^{50}C_{50}}{51}$ equals

(a) $\frac{2^{50}}{51}$ (b) $\frac{2^{49}}{51}$

(c) $\frac{2^{49}}{39\times 17}$ (d) None of these

7. Consider three observations a, b and c such that $b=a+c$. If the standard deviation of $a+2$, $b+2$, $c+2$ is d, then which of the following is true? **[JEE Main 2021, 16 March]**

(a) $b^2=3(a^2+c^2)+9d^2$ (b) $b^2=a^2+c^2+3d^2$

(c) $b^2=3(a^2+c^2+d^2)$ (d) $b^2=3(a^2+c^2)-9d^2$

8. If the variable takes values 0, 1, 2, 3 ..., n with frequencies proportional to ${}^nC_0, {}^nC_1, {}^nC_2, \ldots, {}^nC_n$ respectively, the variance is

(a) $\frac{n}{4}$ (b) $\frac{n}{3}$

(c) $\frac{2n}{5}$ (d) None of these

9. If a variable takes values 0, 1, 2, ..., n with frequencies $q^n, \frac{n}{1}q^{n-1}p, \frac{n(n-1)}{1\cdot 2}q^{n-2}p^2, \ldots, p^n$, where $p+q=1$, then the mean is

(a) np (b) nq

(c) $n(p+q)$ (d) None of these

10. If the mean and the standard deviation of the data 3, 5, 7, a, b are 5 and 2 respectively, then a and b are the roots of the equation **[JEE Main 2020, 5 September]**

(a) $x^2-10x+18=0$ (b) $2x^2-20x+19=0$

(c) $x^2-10x+19=0$ (d) $x^2-20x+18=0$

11. The means of two groups of observations A and B are $\bar{x}, \bar{y}$ respectively and their standard deviations are respectively 2 and 3. In order that the group A is to be more consistent than the group B $\frac{\bar{x}}{\bar{y}}>$

(a) $\frac{3}{2}$ (b) $\frac{5}{1}$

(c) $\frac{2}{3}$ (d) $\frac{6}{5}$

12. If the mean of the set of numbers $x_1, x_2, x_3 \dots x_n$ is $\bar{x}$, then the mean of the numbers $x_1 + 2i, 1 \le i \le n$ is
(a) $\bar{x} + 2n$ (b) $\bar{x} + n + 1$
(c) $\bar{x} + 2$ (d) $\bar{x} + n$

13. Coefficients of variation of two distributions are 50 and 60, and their arithmetic means are 30 and 25, respectively. Difference of their standard deviations is
(a) 0 (b) 1
(c) 1.5 (d) 2.5

14. The AM of the observations $1 \cdot 3 \cdot 5, 3 \cdot 5 \cdot 7, 5 \cdot 7 \cdot 95 \dots (2n-1)(2n+1)(2n+3)$ is
(a) $2n^3 + 6n^2 + 7n - 2$ (b) $n^3 + 8n^2 + 7n - 2$
(c) $2n^3 + 5n^2 + 6n - 1$ (d) $2n^3 + 8n^2 + 7n - 2$

15. The mean and variance of 20 observations are found to be 10 and 4, respectively. On rechecking, it was found that an observation 9 was incorrect and the correct observation was 11. Then the correct variance is **[JEE Main 2020, 8 January]**
(a) 4.01 (b) 3.99
(c) 3.98 (d) 4.02

16. If α, β are respectively the mean deviation about the mean and variance of the first five prime numbers, then the ordered pair (α, β)
(a) (2.27, 10.42) (b) (2.27, 10.24)
(c) (2.72, 10.24) (d) (2.72, 10.42)

17. The mean and SD of the marks of 200 candidates were found to be 40 and 15 respectively. Later, it was discovered that a score of 40 was wrongly read as 50. The correct mean and SD respectively are
(a) 14.98, 39.95 (b) 39.95, 14.98
(c) 39.95, 24.5 (d) 39.00, 14.00

18. Let in a series of $2n$ observations, half of them are equal to a and remaining half are equal to $-a$. Also by adding a constant b in each of these observations, the mean and standard deviation of new set become 5 and 20, respectively. Then, the value of $a^2 + b^2$ is equal to **[JEE Main 2021, 18 March]**
(a) 425 (b) 650
(c) 250 (d) 925

19. Let r be the range and $S^2 = \dfrac{1}{n-1}\sum_{i=1}^{n}(x_1 - \bar{x})^2$ be the SD of a set of observations $x_1, x_2 \dots, x_n$, then
(a) $S \le r\sqrt{\dfrac{n}{n-1}}$ (b) $S = r\sqrt{\dfrac{n}{n-1}}$
(c) $S \ge r\sqrt{\dfrac{n}{n-1}}$ (d) None of these

20. The mean of two samples of sizes 200 and 300 were found to be 25,10 respectively. Their standard deviations were 3 and 4 respectively. The variance of combined sample of size 500 is
(a) 64 (b) 65.2
(c) 67.2 (d) 64.2

21. The mean and variance of 7 observations are 8 and 16, respectively. If five observations are 2, 4, 10, 12, 14, then the absolute difference of the remaining two observations is **[JEE Main 2020, 5 September]**
(a) 1 (b) 3
(c) 4 (d) 2

22. For a data consisting of 15 observations x_i, $i = 1, 2, 3, \dots, 15$ the following results are obtained : $\sum_{i=1}^{15} x_i = 170$; $\sum_{i=1}^{15} x_1^2 = 2830$. If one of the observation namely 20 was found wrong and was replaced by its correct value 30, then the corrected variance is
(a) 80 (b) 78
(c) 76 (d) 75

23. The mean of the series $x_1, x_2, \dots, x_n$ is $\bar{x}$. If x_2 is replaced by λ, then the new mean is
(a) $\bar{x} - x_2 + \lambda$ (b) $\dfrac{\bar{x} - x_2 - \lambda}{n}$
(c) $\dfrac{(n-1)\bar{x} + \lambda}{n}$ (d) $\dfrac{n\bar{x} - x_2 + \lambda}{n}$

24. A student scores the following marks in five tests 45, 54, 41, 57, 43. His score is not known for the sixth test. If the mean score is 48 in the six tests, then the standard deviation of the marks in six tests is **[JEE Main 2019, 8 April]**
(a) $\dfrac{10}{3}$ (b) $\dfrac{10}{\sqrt{3}}$
(c) $\dfrac{100}{\sqrt{3}}$ (d) $\dfrac{100}{3}$

25. The mean and variance of 5 observations of an experiment are 4 and 5.2 respectively. If from these observations three are 1, 2 and 6, then the remaining will be
(a) 2, 9 (b) 5, 6
(c) 4, 7 (d) 3, 8

26. The mean and variance of seven observations are 8 and 16, respectively. If 5 of the observations are 2, 4, 10, 12, 14, then the product of the remaining two observations is **[JEE Main 2019, 8 April]**
(a) 45 (b) 49
(c) 48 (d) 40

27. The variance of the following frequency distribution is

Class Interval	0-4	4-8	8-12	12-16	16-20
Frequency	2	4	6	3	1

(a) $\dfrac{295}{16}$ (b) $\dfrac{304}{16}$
(c) $\dfrac{37}{4}$ (d) $\dfrac{97}{4}$

28. In any discrete series (when all the values are not same) the relationship between MD about mean and SD is
(a) MD = SD (b) MD > SD
(c) MD < SD (d) MD $\le$ SD

29. Consider the first 10 positive integers. If we mulitiply each number by –1 and then add 1 to each number, the variance of the numbers so obtained is
(a) 8.25 (b) 6.5
(c) 3.87 (d) 2.87

30. If the standard deviation of the numbers –1, 0, 1, k is $\sqrt{5}$ where $k > 0$, then k is equal to **[JEE Main 2019, 9 April]**
(a) $2\sqrt{\frac{10}{3}}$ (b) $2\sqrt{6}$
(c) $4\sqrt{\frac{5}{3}}$ (d) $\sqrt{6}$

31. The mean and standard deviation of a distribution of weights of a group of 20 boys are 40 kg and 5 kg respectively. If two boys of weights 43 kg and 37 kg are excluded from this group, then the variance of the distribution of weights of the remaining group of boys is
(a) 26.18 (b) 5.27
(c) 26.78 (d) 5.17

32. Mean of the numbers 1, 2, 3, ...,n with respective weights $1^2 + 1, 2^2 + 2, 3^2 + 3,..., n^2 + n$ is
(a) $\frac{3n(n+1)}{2(2n+1)}$ (b) $\frac{2n+1}{3}$
(c) $\frac{3n+1}{4}$ (d) $\frac{3n+1}{2}$

33. Let the observations $x_i (1 \le i \le 10)$ satisfy the equations, $\sum_{i=1}^{10}(x_i - 5) = 10$ and $\sum_{i=1}^{10}(x_i - 5)^2 = 40$. If μ and λ are the mean and the variance of the observations, $x_1 - 3, x_2 - 3, \ldots, x_{10} - 3$, then the ordered pair (μ, λ) is equal to **[JEE Main 2020, 9 January]**
(a) (6, 3) (b) (3, 6)
(c) (3, 3) (d) (6, 6)

34. Consider the numbers 1, 2, 3, 4, 5, 6, 7, 8, 9 and 10. If 1 is added to each number, the variance of the numbers so obtained is
(a) 6.5 (b) 2.87
(c) 3.87 (d) 8.25

35. A sequence of odd positive integers is written as

1
3 5 7 9
11, 13, 15, 17, 19, 21, 23, 25, 27

The mean of the nth row is
(a) $\frac{n^3(2n^2+1)}{3}$ (b) $\frac{n^3(4n^2+2)}{6}$
(c) $\frac{n(n-1)(2n-1)}{6}$ (d) $\frac{n(2n^2+1)}{3}$

36. The coefficient of variation of the following distribution is

Class Interval	0-5	5-10	10-15	15-20	20-25
Frequency	4	1	10	3	2

(a) $\frac{25\sqrt{139}}{12}$ (b) $\frac{25\sqrt{139}}{6}$
(c) $\frac{139}{6}$ (d) $\frac{25 \times 139}{12}$

37. The mean and the variance of five observations are 4 and 5.20, respectively. If three of the observations are 3, 4 and 4, then the absolute value of the difference of the other two observations, is **[JEE Main 2019, 12 January]**
(a) 1 (b) 7 (c) 5 (d) 3

38. The mean square deviation of a set of n observations $x_1, x_2, \ldots, x_n$ about a point c is defined as $\frac{1}{n}\sum_{i=1}^{n}(x_i - c)^2$.
The mean square deviation about –2 and 2 are 18 and 10 respectively, then standard deviation of this set of observations is
(a) 3 (b) 2
(c) 1 (d) None of these

39. Let $x_1, x_2, \ldots, x_n$ be n observations. Let $w_i = lxi + k$ for $i = 1, 2, \ldots, n$, where l and k are constants. The mean of x_i's is 48 and their standard deviation is 12. Also, the mean of w_i's is 55 and standard deviation of w_i's is 15. The values of l and k should be
(a) $l = 2.5, k = 5$ (b) $l = -1.25, k = 5$
(c) $l = 2.5, k = -5$ (d) $l = 1.25, k = -5$

Numerical Value Type

Solve the following problems and write your answer.

1. In an experiment with 15 observations on x, the following results were available.

$$\Sigma x^2 = 2830, \Sigma x = 170$$

One observation, 20, was found to be wrong and was replaced by the correct value 30. Then the corrected variance is

2. If the variance of the first n natural numbers is 10 and the variance of the first m even natural numbers is 16, then $m + n$ is equal to ……… . **[JEE Main 2020, 7 January]**

3. The coefficient of variation and standard deviation of an ungrouped data are 60 and 21 respectively. If 15 is added to every observation of the data, then the coefficient of variation of the new data is

4. The variance of first 50 even natural numbers is

5. If the mean and variance of eight numbers 3, 7, 9, 12, 13, 20, x and y be 10 and 25 respectively, then $x \cdot y$ is equal to …… . **[JEE Main 2020, 7 January]**

6. Consider the frequency distribution, where A is a positive integer.

x	A	$2A$	$3A$	$4A$	$5A$	$6A$
f	2	1	1	1	1	1

If the variance is 160, then the value of $A^2 + A$ is

7. Consider a set of 3 n numbers having variance 4. In this set, the mean of first $2n$ numbers is 6 and the mean of the remaining n numbers is 3. A new set is constructed by adding 1 into each of first $2n$ numbers, and subtracting 1 from each of the remaining n numbers. If the variance of the new set is k, then $9k$ is equal to .... **[JEE Main 2021, 17 March]**

8. The mean age of 25 teachers in a school is 40 yr. A teacher retires at the age of 60 yr and a new teacher is appointed in his place. If the mean age of the teachers in this school now is 39 yr, then the age (in yr) of the newly appointed teacher is .... **[JEE Main 2021, 18 March]**

Single Integer Type

Solve the following problems and write your answer.

1. If the variance of the terms in an increasing AP, $b_1, b_2, b_3, \ldots\ldots, b_{11}$ is 90, then the common difference of this AP is .... **[JEE Main 2020, 2 September]**
2. In a series of $2n$ observations, half of them equal a and remaining half equal $-a$. If the standard deviation of the observations is 2, then $|a|$ equals
3. The standard deviation and mean of five observations are 0 and 9 respectively. If one of the observations is changed such that the mean of the new set of five observations becomes 10, then their standard deviation is
4. For a group of 100 observations, the arithmetic mean and standard deviation are 8 and $\sqrt{10.5}$ respectively. The mean and standard deviation of 50 items selected from these 100 observations are 10 and 2 respectively. Then the standard deviation f the remaining 50 observations is
5. The mean square deviation of a set of n observations $x_1, x_2 \ldots x_n$ about a point c is defined as $\frac{1}{n}\sum_{i=1}^{n}(x_1 - c)^2$.

   The mean square deviation about -2 and 2 are 18 and 10 respectively, then standard deviation of this set of observations is
6. If the mean deviation about the median of the numbers $a, 2a, \ldots. 50a$ is 50 then $|a|$ equal.
7. Consider the data on x taking the values 0, 2, 4, 8, ..., 2^n with frequencies ${}^nC_0, {}^nC_1, {}^nC_2, \ldots, {}^nC_n$ respectively. If the mean of this data is $\frac{728}{2^n}$, then n is equal to **[JEE Main 2020, 6 September]**
8. Consider the statistics of two sets of observations as follows:

	Size	Mean	Variance
Observations I	10	2	2
Observations II	n	3	1

If the variance of the combined set of these two observations is $\frac{17}{9}$, then the value of n is equal to .... **[JEE Main 2021, 16 March]**

Answers

Objective Questions Type [Only one correct answer]

1. (c)	2. (c)	3. (d)	4. (a)	5. (d)	6. (c)	7. (d)	8. (a)	9. (a)	10. (c)
11. (c)	12. (b)	13. (a)	14. (d)	15. (b)	16. (c)	17. (b)	18. (a)	19. (a)	20. (c)
21. (d)	22. (b)	23. (d)	24. (b)	25. (c)	26. (c)	27. (a)	28. (c)	29. (a)	30. (b)
31. (c)	32. (c)	33. (c)	34. (d)	35. (d)	36. (b)	37. (b)	38. (a)	39. (d)	

Numerical Value Type Questions

1. 78	2. 18	3. 42	4. 833	5. 54	6. 56	7. 68	8. 35

Single Integer Type Questions

1. 3	2. 2	3. 2	4. 3	5. 3	6. 4	7. 6	8. 5

Solutions

1. Let $x_1, x_2, \ldots, x_n$ be n items. Then, $\bar{X} = \frac{1}{n}\Sigma x_i$.

Let $y_1 = x_1 + 1, y_2 = x_2 + 2, y_3 = x_3 + 3 \ldots, y_n = x_n + n$.

Then, the mean of the new series is

$$\frac{1}{n}\Sigma(x_i + i) = \frac{1}{n}\Sigma x_i + \frac{1}{n}(1 + 2 + 3 + \ldots + n)$$

$$= \bar{X} + \frac{1}{n}\frac{n(n+1)}{2}$$

$$= \bar{X} + \frac{n+1}{2}$$

2. Given, sets

$X = \{x \in N : 1 \le x \le 17\}$

and $Y = \{ax + b : x \in X \text{ and } a, b \in R, a > 0\}$

Now, $\frac{\sum_{x=1}^{17}(ax+b)}{17} = 17$

$\Rightarrow \quad a\sum_{x=1}^{17} x + 17b = 17 \times 17 \Rightarrow 9a + b = 17 \quad \ldots \text{(i)}$

and $\frac{\sum_{x=1}^{17}(ax+b)^2}{17} - (17)^2 = 216$

$\Rightarrow \quad \sum_{x=1}^{17}(ax+b)^2 = 17[216 + 289]$

$= 17 \times 505 = 8585$

$\Rightarrow \quad a^2\sum_{x=1}^{17} x^2 + 2ab\sum_{x=1}^{17} x + 17b = 8585$

$\Rightarrow a^2\left(\frac{17 \times 18 \times 35}{6}\right) + 2ab\left(\frac{17 \times 18}{2}\right) + 17b = 8585$

$\Rightarrow \quad 105a^2 + 18ab + b^2 = 505 \quad \ldots \text{(ii)}$

From Eqs. (i) and (ii), we get

$105a^2 + 18a(17 - 9a) + (17 - 9a)^2 = 505$

$\Rightarrow 105a^2 + 306a - 162a^2 + 289 + 81a^2 - 306a = 505$

$\Rightarrow \quad 24a^2 = 216 \Rightarrow a^2 = 9$

$\Rightarrow \quad a = 3$ as $a > 0$

So, $\quad b = -10$

$\therefore \quad a + b = 3 - 10 = -7$

3. Let n_1 and n_2 be the number of observations in two groups having means $\bar{X}_1$ and $\bar{X}_2$ respectively.

Then, $\quad \bar{X} = \frac{n_1\bar{X} + n_2\bar{X}_2}{n_1 + n_2}$

Now, $\quad \bar{X} - \bar{X}_1 = \frac{n_1\bar{X}_1 + n_2\bar{X}_2}{n_1 + n_2} - \bar{X}_1$

$= \frac{n_2(\bar{X}_1 - \bar{X}_1)}{n_1 + n_2} > 0 \quad [\because \bar{X}_2 > \bar{X}_1]$

$\Rightarrow \quad \bar{X} > \bar{X}_1 \quad \ldots \text{(i)}$

And, $\quad \bar{X} - \bar{X}_2 = \frac{n(\bar{X}_1 - \bar{X}_2)}{n_1 + n_2} < 0 \quad [\because \bar{X}_2 > \bar{X}_1]$

$\Rightarrow \quad \bar{X} < \bar{X}_2 \quad \ldots \text{(ii)}$

From Eqs. (i) and (ii),

$\bar{X}_1 < \bar{X} < \bar{X}_2$

4. Let the missing frequencies be f_1 and f_2 respectively

Income (x) (in ₹)	Number of workers (f)	fx
15	8	120
20	12	240
25	f_1	$25f_1$
30	16	480
35	f_2	$35f_2$
40	10	400

$\therefore \quad \Sigma f = 46 + f_1 + f_2,$

$\Sigma fx = 1240 + 25f_1 + 35f_2 \quad$ [given]

Since, $\quad \Sigma f = 75$

$\Rightarrow \quad 46 + f_1 + f_2 = 75 \quad \ldots \text{(i)}$

$\Rightarrow \quad f_1 + f_2 = 29$

Also $\quad \bar{x} = 28.07 \quad$ [given]

$\Rightarrow \quad \frac{\Sigma fx}{\Sigma f} = 28.07$

$\Rightarrow \quad \frac{1240 + 25f_1 + 35f_2}{75} = 28.07$

$\Rightarrow \quad 5f_1 + 7f_2 + 421.05 - 248 = 173.05 \quad \ldots \text{(ii)}$

Solving Eqs. (i) and (ii) we get

$f_2 = 14.025$ and $f_1 = 14.975$

Since, the frequency cannot be in fraction,

$\therefore$ We have, $f_1 = 15$ and $f_2 = 14$

5. It is given that $\sum_{i=1}^{n}(x_i - a) = n$

and $\quad \sum_{i=1}^{n}(x_i - a)^2 = na, \ (n, a > 1)$

$\therefore$ Standard deviation of 'n' observations

$$= \sqrt{\frac{\sum_{i=1}^{n}(x_i - a)^2}{n} - \frac{\sum_{i=1}^{n}(x_i - a)}{n}}$$

$$= \sqrt{\frac{na}{n} - \frac{n}{n}} = \sqrt{a - 1}$$

6. Consider $(1 + x)^{50} = {}^{50}C_0 + {}^{50}C_1x^1 + \ldots + {}^{50}C_{50}x^{50} \quad \ldots \text{(i)}$

and $\quad (1 - x)^{50} = {}^{50}C_0 - {}^{50}C_1x^1 + \ldots {}^{50}C_{50}x^{50} \quad \ldots \text{(ii)}$

Adding Eqs. (i) and (ii), we get on integrating with limits 0 to 1,

$$\left[{}^{50}C_0x + \frac{{}^{50}C_2}{3}x^3 + \frac{{}^{50}C_4}{5}x^5 + \ldots + \frac{{}^{50}C_{50}x^{51}}{51}\right]_0^1$$

$$= \frac{1}{2}\left[\frac{(1+x)^{51}}{51} - \frac{(1-x)^{51}}{51}\right]_0^1$$

$$\therefore {}^{50}C_0 + \frac{{}^{50}C_2}{3} + \frac{{}^{50}C_4}{5} + \ldots + \frac{{}^{50}C_{50}}{51} = \frac{1}{2}\frac{2^{51}}{51} = \frac{2^{50}}{51}$$

Now, number of terms from ${}^{50}C_0$ to ${}^{50}C_{50}$ are 26 (items)

$\therefore$ Required mean

$$= \frac{\frac{^{50}C_0}{1} + \frac{^{50}C_2}{3} + \frac{^{50}C_4}{5} + \ldots + \frac{^{50}C_{50}}{51}}{26}$$

$$= \frac{2^{50}}{51} \times \frac{1}{26} = \frac{2^{49}}{39 \times 17}$$

7. For a, b, c

$$\text{Mean} = \frac{a+b+c}{3} (= \bar{x})$$

$$b = a + c$$

$$\Rightarrow \quad \bar{x} = \frac{2b}{3}$$

$$\text{SD}(a+2, b+2, c+2) = \text{SD} \cdot (a, b, c) = d$$

$$\Rightarrow \quad d^2 = \frac{a^2+b^2+c^2}{3} - (\bar{x})^2$$

$$\Rightarrow \quad d^2 = \frac{a^2+b^2+c^2}{3} - \frac{4b^2}{9}$$

$$\Rightarrow \quad 9d^2 = 3(a^2+b^2+c^2) - 4b^2$$

$$\Rightarrow \quad b^2 = 3(a^2+c^2) - 9d^2$$

8. $\text{Variance} = \sigma^2 = \frac{\sum fx^2}{\sum f} - \left(\frac{\sum fx}{\sum f}\right)^2$

$$\sum f = {^nC_0} + {^nC_1} + \ldots + {^nC_n} = 2^n$$

$$\sum fx^2 = \sum_{r=1}^{n} r^2 C_r = n(n-1) \sum {^{n-2}C_{r-2}} + n \sum {^{n-1}C_{r-1}}$$

$$= n(n-1)2^{n-2} + 2^{n-1}$$

$$\therefore \quad \frac{\sum fx^2}{\sum f} = \frac{2^{n-2}[n^2 - n + 2n]}{2^n} = \frac{n(n+1)}{4}$$

$$\frac{\sum fx}{\sum f} = \frac{\sum r\,{^nC_r}}{\sum {^nC_r}} = \frac{n \cdot 2^{n-1}}{2^n} = \frac{n}{2}$$

$$\therefore \quad \sigma^2 = \frac{n^2+n}{4} - \frac{n^2}{4} = \frac{n}{4}.$$

9. The required mean is

$$\bar{x} = \frac{0 \cdot q^n + 1 \cdot \frac{n}{1} q^{n-1} p + 2 \cdot \frac{n(n-1)}{2!} q^{n-2} p^2 + \ldots + n \cdot p^n}{q^n + \frac{n}{1} q^{n-1} p + \frac{n(n-1)}{2!} q^{n-2} p^2 + \ldots + p^n}$$

$$\Rightarrow \bar{x} = \frac{0 \cdot {^nC_0} q^n p^0 + 1 \cdot {^nC_1} q^{n-1} p + \ldots + n \cdot {^nC_n} q^0 p^n}{{^nC_0} q^n p^0 + {^nC_1} q^{n-1} p^1 + \ldots + {^nC_n} q^{n-n} p^n}$$

$$\Rightarrow \bar{x} = \frac{\sum_{r=0}^{n} r \cdot {^nC_r} q^{n-r} p^r}{\sum_{r=0}^{n} {^nC_r} q^{n-r} p^r} = \frac{\sum_{r=1}^{n} r \cdot \frac{n}{r} {^{n-1}C_{1-1}} q^{n-r} p \cdot p^{r-1}}{\sum_{r=0}^{n} {^nC_r} q^{n-r} p^r}$$

$$\Rightarrow \bar{x} = \frac{np\left(\sum_{r=1}^{n} {^{n-1}C_{r-1}} p^{r-1} q^{(n-1)-(r-1)}\right)}{\sum_{r=1}^{n} {^nC_r} q^{n-r} p^r}$$

$$\Rightarrow \bar{x} = \frac{np(q+p)^{n-1}}{(q+p)^n}$$

$$\therefore \quad \bar{x} = np \qquad [\because q + p = 1]$$

10. Given data 3, 5, 7, a, b having mean = S

$$\Rightarrow \quad \frac{3+5+7+a+b}{5} = 5$$

$$\Rightarrow \quad 15 + a + b = 25$$

$$\Rightarrow \quad a + b = 10 \qquad \ldots(\text{i})$$

$$\text{and SD} = 2 \Rightarrow \sqrt{\frac{9+25+49+a^2+b^2}{5} - 25} = 2$$

$$\Rightarrow \quad a^2 + b^2 - 42 = 20$$

$$\Rightarrow \quad a^2 + b^2 = 62 \qquad \ldots(\text{ii})$$

From Eqs. (i) and (ii), we have

$$ab = \frac{(a+b)^2 - (a^2+b^2)}{2} = \frac{100-62}{2} = 19$$

$\therefore$ Equation of having roots a and b is

$$x^2 - (a+b)x + ab = 0$$

$$\Rightarrow \quad x^2 - 10x + 19 = 0$$

11. We have,

$$CV_A = \frac{\sigma_x}{x} \times 100 = \frac{2}{x} \times 100$$

$$\text{and} \quad CV_B = \frac{\sigma_y}{y} \times 100 = \frac{3}{y} \times 100$$

Now, A is more consistent than B

$$\therefore \quad CV_A < CV_B$$

$$\Rightarrow \quad \frac{2}{\bar{x}} \times 100 < \frac{3}{y} \times 100$$

$$\Rightarrow \quad \frac{2}{x} < \frac{3}{y} \Rightarrow \frac{\bar{x}}{y} > \frac{2}{3}$$

12. We know that, $\bar{x} = \frac{\sum_{i=1}^{n} x_i}{n}$, i.e., $\sum_{i=1}^{n} x_i = n\bar{x}$

$$\therefore \quad \frac{\sum_{i=1}^{n}(x_i + 2i)}{n} = \frac{\sum_{i=1}^{n} x_i + 2\sum_{i=1}^{n} i}{n}$$

$$= \frac{n\bar{x} + 2(1 + 2 + \ldots n)}{n} = \frac{n\bar{x} + 2\frac{n(n+1)}{2}}{n}$$

$$= \bar{x} + n + 1$$

13. Here, $(CV)_1 = 50$, $(CV)_2 = 60$, $\bar{x}_1 = 30$ and $\bar{x}_2 = 25$

$$\therefore \quad (CV)_1 = \frac{\sigma_1}{\bar{x}_1} \times 100$$

$$\Rightarrow \quad 50 = \frac{\sigma_1}{30} \times 100$$

$$\therefore \quad \sigma_1 = \frac{30 \times 50}{100} = 15$$

$$\text{and} \quad CV_2 = \frac{\sigma_2}{\bar{x}_2} \times 100$$

$$\Rightarrow \quad 60 = \frac{\sigma_2}{25} \times 100$$

$$\therefore \quad \sigma_2 = \frac{60 \times 25}{100} = 15$$

$$\text{Now,} \quad \sigma_1 - \sigma_2 = 15 - 15 = 0$$

14. $\text{AM} = \frac{1}{n}[1\cdot3\cdot5 + 3\cdot5\cdot7 + \ldots + (2n-1)(2n+1)(2n+3)]$

$$= \frac{1}{n}\left[\sum_{r=1}^{n}(2r-1)(2r+1)(2r+3)\right]$$

$$= \frac{1}{n}\sum_{r=1}^{n}(8r^3 + 12r^2 - 2r - 3)$$

$$= \frac{1}{n}\left[8\left(\frac{n(n+1)}{2}\right)^2 + 12\frac{n(n+1)(2n+1)}{6} - 2\frac{n(n+1)}{2} - 3n\right]$$

$$= 2n(n+1)^2 + 2(n+1)(2n+1) - (n+1) - 3$$

$$= 2n^3 + 8n^2 + 7n - 2.$$

15. Let the observations are $x_1, x_2, x_3, \ldots, x_{20}$, then mean = 10

$$\Rightarrow \quad \frac{x_1 + x_2 + x_3 + \ldots + x_{20}}{20} = 10$$

$$\Rightarrow \quad x_1 + x_2 + x_3 + \ldots + x_{20} = 200 \quad \ldots(i)$$

and, variance = 4

$$\Rightarrow \quad \frac{x_1^2 + x_2^2 + x_3^2 + \ldots + x_{20}^2}{20} - \left(\frac{x_1 + x_2 + x_3 + \ldots + x_{20}}{20}\right)^2 = 4$$

$$\Rightarrow \quad \frac{x_1^2 + x_2^2 + x_3^2 + \ldots + x_{20}^2}{20} - 100 = 4$$

$$\Rightarrow \quad x_1^2 + x_2^2 + x_3^2 + \ldots + x_{20}^2 = 2080 \quad \ldots(ii)$$

Let observation $x_1 = 9$ was incorrect and its correct value was 11.

So, mean is $\bar{x} = \frac{x_1 + x_2 + x_3 + \ldots + x_{20} - 9 + 11}{20}$

$$= \frac{200 - 9 + 11}{20} \quad \text{[from Eq. (i)]}$$

$$= \frac{202}{20}$$

and variance is

$$\sigma = \frac{x_1^2 + x_2^2 + x_3^2 + \ldots + x_{20}^2 - 9^2 + 11^2}{20} - \left(\frac{202}{20}\right)^2$$

$$= \frac{2080 - 81 + 121}{20} - \left(\frac{202}{20}\right)^2 \quad \text{[from Eq. (ii)]}$$

$$= \frac{2120}{20} - \left(\frac{202}{20}\right)^2 = \frac{2120 \times 20 - (202)^2}{(20)^2}$$

$$= \frac{42400 - 40804}{400} = \frac{1596}{400} = 3.99$$

16. Means of first five prime numbers

$$= \frac{2 + 3 + 5 + 7 + 11}{5} = \frac{28}{5} = 56$$

$\therefore$ Mean deviation about mean

$$= \frac{|2 - 5.6|\,3 - 5.6| + |5 - 5.6| + |7 - 5.6| + |11 + 5.6|}{5}$$

$$= \frac{3.6 + 2.6 + 0.6 + 1.4 + 5.4}{5} = \frac{13.6}{5} = 2.72$$

$$\text{Variance} = \frac{\Sigma x^2}{n} = \left(\frac{\Sigma\, xi}{n}\right)^2 = \frac{4 + 9 + 25 + 49 + 121}{5} - (5.6)^2$$

$$= 41.6 - 31.36 = 10.24$$

17. Corrected $\Sigma\, x = 40 \times 200 - 50 + 40 = 7990$

$$\text{corrected } \bar{x} = \frac{7990}{200} = 39.95$$

Incorrect $\Sigma\, x^2 = n(\sigma^2 + \bar{x}^2) = 200(5^2 + 40^2) = 365000$

correct $\Sigma\, x^2 = 365000 - 2500 + 1600 = 364100$

$$\text{corrected } \sigma = \sqrt{\frac{364100}{200} - (39.95)^2}$$

$$= \sqrt{(1820.5 - 1596)} = \sqrt{224.5}$$

$$= 14.98$$

18. Let observations are denoted by x_i for $1 \le i < 2n$

$$\bar{x} = \frac{\Sigma x_i}{2n} = \frac{(a + a + \ldots + a) - (a + a + \ldots + a)}{2n}$$

$$\Rightarrow \quad \bar{x} = 0$$

and $\quad \sigma_x^2 = \frac{\Sigma x_i^2}{2n} - (\bar{x})^2 = \frac{a^2 + a^2 + \ldots + a^2}{2n} - 0 = a^2$

$$\Rightarrow \quad \sigma_x = a$$

Now, adding a constant b, then $\bar{y} = \bar{x} + b = 5$

$\Rightarrow \quad b = 5$

and $\quad \sigma_y = \sigma_x$ (No change in S.D.)

$\Rightarrow \quad a = 20 \Rightarrow a^2 + b^2 = 425$

19. We have, $r = \max|x_i - x_j|$

and $\quad S^2 = \frac{1}{n-1}\sum_{i=1}^{n}(x_i - \bar{x})^2$

Now, $\quad (x_i - \bar{x})^2 = \left(x_i - \frac{x_1 + x_2 + \ldots + x_n}{n}\right)^2$

$$= \frac{1}{n^2}[(x_i - x_1) + (x_i - x_2) + \ldots + (x_i - x_i - 1)$$

$$+ (x_i - x_1 + 1) + \ldots + (x_i - x_n)] \le \frac{1}{n^2}[(n-1)r]^2,$$

$[\because\, x_i - x_j | \le r|]$

$$\Rightarrow \quad (x_i - \bar{x})^2 \le r^2$$

$$\Rightarrow \quad \sum_{i=1}^{n}(x_i \bar{x})^2 \le nr^2$$

$$\Rightarrow \quad \frac{1}{n-1}\sum_{i=1}^{n}(x_i - \bar{x})^2 \le \frac{nr^2}{(n-1)}$$

$$\Rightarrow \quad S^2 \le \frac{nr^2}{(n-1)}$$

$$\Rightarrow \quad S \le r\sqrt{\frac{n}{n-1}}$$

20. Combined mean $\bar{x} = \frac{n_1\bar{x}_1 + n_2\bar{x}_2}{n_1 + n_2}$

$$= \frac{200 \times 25 + 300 \times 10}{500} = 16$$

Let $\quad d_1 = \bar{x}_1 - \bar{x} = 25 - 16 = 9$ and

$d_2 = \bar{x}_2 - \bar{x} = 10 - 16 = -6$

Now we know that,

$$\sigma^2 = \frac{n_1(\sigma_1^2 + d_1^2) + n_2(\sigma_2^2 + d_2^2)}{n_1 + n_2}$$

$$= \frac{200(9 + 81) + 300(16 + 36)}{500}$$

$$= \frac{33600}{500} = 67.2$$

21. Let remaining two observations are x and y, so the variance

$$= \frac{2^2 + 4^2 + 10^2 + 12^2 + 14^2 + x^2 + y^2}{7} - 8^2$$

$$\Rightarrow \frac{4 + 16 + 100 + 144 + 196 + x^2 + y^2}{7} - 64 = 16 \quad \text{[given]}$$

$$\Rightarrow 460 + x^2 + y^2 - 448 = 112$$

$$\Rightarrow x^2 + y^2 = 100 \quad \ldots\text{(i)}$$

And, mean $= \dfrac{2 + 4 + 10 + 12 + 14 + x + y}{7} = 8$ [given]

$$\Rightarrow 42 + x + y = 56$$

$$\Rightarrow x + y = 14 \quad \ldots\text{(ii)}$$

From Eqs. (i) and (ii), we get remaining observations as 6 and 8.

$\therefore$ Absolute difference of remaining observations = 2

22. Given,

$$\Sigma x^2 = 2830 \text{ and } \Sigma x = 170$$

Increase in $\Sigma x = 10$

and increase in $\Sigma x^2 = (30)^2 - (20)^2$

$$= 900 - 400 = 500$$

$$\Sigma x = 170 + 10 = 180$$

and $\Sigma x^2 = 2830 + 500$

$$= 3330$$

We know that, variance $= \dfrac{\Sigma x^2}{n} - \left(\dfrac{\Sigma x}{n}\right)^2$

$$= \frac{3330}{15} - \left(\frac{180}{15}\right)^2 \quad [\text{her, } n = 15]$$

$$= 222 - 144 = 78$$

23. We have, $\bar{x} = \dfrac{x_1 + x_2 + \ldots + x_{n-1} + x_n}{n}$

$$\Rightarrow n\bar{x} = x_1 + x_2 + \ldots + x_{n-1} + x_n$$

Let $\bar{y}$ be the new mean when x_2 is replaced by λ. Then,

$$\bar{y} = \frac{x_1 + \lambda + x_3 + \ldots + x_{n-1} + x_n}{n}$$

$$\Rightarrow \bar{y} = \frac{(x_1 + x_2 + \ldots + x_n) - x_2 + \lambda}{n}$$

$$\therefore \bar{y} = \frac{n\bar{x} - x_2 + \lambda}{n}$$

24. Let the marks in sixth tests is 'x', so

Mean $= \dfrac{41 + 45 + 43 + 54 + 57 + x}{6} = 48$ (given)

$$\Rightarrow \frac{240 + x}{6} = 48$$

$$\Rightarrow 40 + \frac{x}{6} = 48$$

$$\Rightarrow \frac{x}{6} = 8$$

$$\Rightarrow x = 48$$

Now, standard deviation of these marks

$$= \sqrt{\frac{41^2 + 45^2 + 43^2 + 54^2 + 57^2 + 48^2}{6} - 48^2}$$

$$\left[\because \text{standard deviation (SD)} = \sqrt{\frac{\Sigma x_i^2}{6} - (\bar{x})^2}\right]$$

$$= \sqrt{\frac{(41^2 - 48^2) + (45^2 - 48^2) + (43^2 - 48^2) + (54^2 - 48^2) + (57^2 - 48^2)}{6}}$$

$$= \sqrt{\frac{(-7 \times 89) + (-3 \times 93) + (-5 \times 91) + (6 \times 102) + (9 \times 105)}{6}}$$

$$= \sqrt{\frac{945 + 612 - 455 - 279 - 623}{6}}$$

$$= \sqrt{\frac{1557 - 1357}{6}} = \sqrt{\frac{200}{6}} = \sqrt{\frac{100}{3}} = \frac{10}{\sqrt{3}}$$

25. As given $\bar{x} = 4$, $n = 5$ and $\sigma^2 = 5.2$

If the remaining observations are x_1, x_2 then,

$$\sigma^2 = \frac{\Sigma(x_i - \bar{x})^2}{n} = 5.2$$

$$\Rightarrow \frac{(x_1 - 4)^2 + (x_2 - 4)^2 + (1 - 4)^2 + (2 - 4)^2 + (6 - 4)^2}{5} = 5.2$$

$$\Rightarrow (x_1 - 4)^2 + (x_2 - 4)^2 = 9 \quad \ldots\text{(i)}$$

Also $\bar{x} = 4$

$$\Rightarrow \frac{x_1 + x_2 + 1 + 2 + 6}{5} = 4$$

$$\Rightarrow x_1 + x_2 = 11 \quad \ldots\text{(ii)}$$

From Eqs. (i) and (ii) $\Rightarrow x_1, x_2 = 4, 7$

26. Let the remaining two observations are a and b. According to the question,

Mean $= \dfrac{2 + 4 + 10 + 12 + 14 + a + b}{7} = 8$

$$\Rightarrow 42 + a + b = 56$$

$$\Rightarrow a + b = 14 \quad \ldots\text{(i)}$$

and variance

$$= \frac{a^2 + b^2 + 4 + 16 + 100 + 144 + 196}{7} - 8^2 = 16$$

$$\Rightarrow \frac{a^2 + b^2 + 460}{7} - 64 = 16$$

$$\Rightarrow \frac{a^2 + b^2 + 460}{7} = 80$$

$$\Rightarrow a^2 + b^2 + 460 = 560$$

$$\Rightarrow a^2 + b^2 = 100 \quad \ldots\text{(ii)}$$

We know that,

$$(a + b)^2 = (a^2 + b^2) + 2ab$$

$$\Rightarrow (14)^2 = 100 + 2ab$$

[from Eqs. (i) and (ii)]

$$\Rightarrow 196 = 100 + 2ab$$

$$\Rightarrow 2ab = 96$$

$$\Rightarrow ab = 48$$

27. Given data,

Class Interval	Mid-value x_i	f_i	$d_i = \frac{x_i - A}{h}$	f_id_i	$f_id_i^2$
0-4	2	2	−2	−4	8
4-8	6	4	−1	−4	4
8-12	10	6	0	0	0
12-16	14	3	1	3	3
16-20	18	1	2	2	4
		$\Sigma f_i = 16$		$\Sigma f_id_i = -3$	$\Sigma f_id_i^2 = 19$

Here, $A = 10, h = 4$

$$\sigma^2 = h^2\left(\frac{\Sigma f_id_i^2}{\Sigma f_i} - \left(\frac{\Sigma f_id_i}{\Sigma f_i}\right)^2\right) = 16\left(\frac{19}{16} - \frac{9}{256}\right)$$

$$\sigma^2 = 16\left(\frac{304-9}{256}\right) = \frac{295}{16}$$

28. $\text{MD} = \frac{1}{n}\sum_{i=1}^{n}|x_i - \bar{x}|$ and $\text{SD} = \sqrt{\frac{1}{n}\sum_{i=1}^{n}(x_1 - \bar{x})^2}$.

Let $\quad a_i = |x_i - \bar{x}|$

So $\quad (\text{MD})^2 = \left(\frac{1}{n}\sum_{i=1}^{n} a_i\right)^2$ and $(\text{SD})^2 = \frac{1}{n}\sum_{i=1}^{n} a_1^2$

By Cauchy Schwarz inequality

$$\left(\sum_{i=1}^{n} a_i\right)^2 < n\sum_{i=1}^{n} a_i^2$$

$$\Rightarrow \quad \left(\frac{1}{n}\Sigma a_i\right) < \frac{1}{n}\Sigma a_i^2 \Rightarrow \text{MD} < \text{SD}$$

29. The first 10 positive integers are 1, 2, 3, 4, 5, 6, 7, 8, 9 and 10.

On multiplying each number by −1, we get

−1, −2, −3, −4, −5, −6, −7, −8, −9, −10

On adding 1 to each number, we get

0, −1, −2, −3, −4, −5, −6, −7, −8, −9

$$\therefore \quad \Sigma x_i = -\frac{9 \times 10}{2} = -45$$

and $\quad \Sigma x_i^2 = 0^2 + (-1)^2 + (-2)^2 + \ldots + (-9)^2$

$$= \frac{9 \times 10 \times 19}{6} = 28.5$$

$$\therefore \quad \text{SD} = \sqrt{\frac{285}{10} - \left(\frac{-45}{10}\right)^2} = \sqrt{\frac{285}{10} - \frac{2025}{100}}$$

$$= \sqrt{\frac{2850 - 2025}{100}} = \sqrt{8.25}$$

Now, variance = $(\text{SD})^2 = (\sqrt{8.25})^2 = 8.25$

30. Given observations are −1, 0, 1 and k.

Also, standard deviation of these four observations = $\sqrt{5}$

$$\therefore \sqrt{\frac{(-1)^2 + (0)^2 + (1)^2 + k^2}{4} - \left(\frac{-1+0+1+k}{4}\right)^2} = \sqrt{5}$$

[$\because$ if $x_1, x_2, \ldots, x_n$ are n observation, then standard deviation = $\sqrt{\frac{1}{n}\sum_{i=1}^{n} x_i^2 - \left(\frac{1}{n}\sum_{i=1}^{n} x_i\right)^2}$]

$$\Rightarrow \quad \frac{2+k^2}{4} - \frac{k^2}{16} = 5 \quad \text{[squaring both sides]}$$

$$\Rightarrow \quad \frac{8+4k^2-k^2}{16} = 5 \Rightarrow \frac{8+3k^2}{16} = 5$$

$$\Rightarrow \quad 8 + 3k^2 = 80 \Rightarrow 3k^2 = 72$$

$$\Rightarrow \quad k^2 = 24 \Rightarrow k = 2\sqrt{6} \text{ or } -2\sqrt{6}$$

$$\Rightarrow \quad k = 2\sqrt{6} \quad [\because k > 0]$$

31. Here, $n = 20$; Mean = 40

$$\text{Mean} = \frac{\text{Sum of observation}}{n}$$

$$\Rightarrow \quad 40 = \frac{\text{Sum of weights}}{20}$$

Sum of weight = 20 × 40 = 800

Now, standard deviation, $\sigma = 5$

$\because$ Variance = $\sigma^2 = 5^2 = 25$

Now, two boys are excluded of wight 43 kg and 37 kg.

$$\text{New mean} = \frac{\text{Remaning weight}}{20-2} = \frac{800 - 43 - 37}{18}$$

$\bar{X}_{new}$ = new mean = 40

Now, variance $\sigma^2 = 25$

$$\Rightarrow \quad \frac{\Sigma x^2}{n} - (\bar{x})^2 = 25 \Rightarrow \frac{\Sigma x^2}{20} - 40^2 = 25$$

$$\Rightarrow \quad \frac{\Sigma x^2}{20} = 1600 + 25 \Rightarrow \Sigma x^2 = 1625 \times 20$$

$$\Rightarrow \quad \Sigma x^2 = 32500$$

When two boys are removed

$$\Sigma x_{new}^2 = 32500 - (43)^2 - (37)^2 = 29282$$

So, new varience $\sigma_{new}^2 = \frac{\Sigma x_{new}^2}{18} - (\bar{X}_{new})^2$

$$= \frac{29282}{18} - (40)^2$$

$$= 26.78$$

32. Here, for each $x_i = i$, weight $w_i = i^2 + i$

Hence, the required mean $\frac{\Sigma w_ix_i}{\Sigma w_i} = \frac{\sum_{i=1}^{n} i(i^2+i)}{\sum_{i=1}^{n}(i^2+i)}$

$$= \frac{\sum_{i=1}^{n} i^3 + \sum_{i=1}^{n} i^2}{\sum_{i=1}^{n} i^2 + \sum_{i=1}^{n} i} = \frac{\frac{n^2(n+1)^2}{4} + \frac{n(n+1)(2n+1)}{6}}{\frac{n(n+1)(2n+1)}{6} + \frac{n(n+1)}{2}}$$

$$= \frac{\frac{n(n+1)}{2}\left\{\frac{n(n+1)}{2} + \frac{2n+1}{3}\right\}}{\frac{n(n+1)}{2}\left\{\frac{2n+1}{3} + 1\right\}} = \frac{3n^2 + 7n + 2}{2(2n+4)}$$

$$= \frac{(3n+1)(n+2)}{4(n+2)} = \frac{3n+1}{4}$$

33. For the observations $x_i (1 \le i \le 10)$, we have

$$\sum_{i=1}^{10} (x_i - 5) = 10$$

$$\Rightarrow \sum_{i=1}^{10} x_i - 50 = 10$$

$$\Rightarrow \sum_{i=1}^{10} x_i = 60 \quad \ldots \text{(i)}$$

and $$\sum_{i=1}^{10} (x_i - 5)^2 = 40$$

$$\Rightarrow \sum_{i=1}^{10} x_i^2 - 10\sum_{i=1}^{10} x_i + 10(5^2) = 40$$

$$\Rightarrow \sum_{i=1}^{10} x_i^2 - 10(60) + 250 = 40, \quad \text{[from Eq. (i)]}$$

$$\Rightarrow \sum_{i=1}^{10} x_i^2 = 390 \quad \ldots \text{(ii)}$$

Now, mean $\mu = \dfrac{\sum_{i=1}^{10} (x_i - 3)}{10} = \dfrac{\sum_{i=1}^{10} x_i - 30}{10}$

$$= \frac{60 - 30}{10} = \frac{30}{10} = 3$$

and variance $\lambda = \dfrac{\sum_{i=1}^{10} (x_i - 3)^2}{10} - (\mu)^2$

$$= \frac{\sum_{i=1}^{10} x_i^2 - 6\sum_{i=1}^{10} x_i + 10(3^2)}{10} - 9$$

$$= \frac{390 - 6(60) + 90 - 90}{10} = \frac{30}{10} = 3$$

So, the ordered pair $(\mu, \lambda) = (3, 3)$

34. Given numbers are 1, 2, 3, 4, 5, 6, 7, 8, 9 and 10.
If 1 is added to each number, then observations will be 2, 3, 4, 5, 6, 7, 8, 9, 10 and 11.

$$\therefore \quad \Sigma x_i = 2 + 3 + 4 + \ldots + 11$$

$$= \frac{10}{2}[2 \times 2 + 9 \times 1] = 5[4 + 9] - 65$$

and $$\Sigma x_i^2 = 2^2 + 3^2 + 4^2 + 5^2 + \ldots + 11^2$$

$$= (1^2 + 2^2 + 3^2 + \ldots + 11^2) - (1^2)$$

$$= \frac{11 \times 12 \times 23}{6} - 1 = 505$$

$$\therefore \quad s^2 = \frac{\Sigma x_i^2}{n} - \left(\frac{\Sigma x_i}{n}\right)^2 = \frac{505}{10} - \left(\frac{65}{10}\right)^2$$

$$= 50.5 - (6.5)^2 = 50.5 - 42.25 = 8.25$$

35. The number of numbers in the n th row $= n^2$
Sequence of first terms in different row is

1, 3, 11, 29, 61 ...

$\therefore T_n$ of 1, 3, 11, 29, 61,... $= \dfrac{1}{3}(2n^3 - 3n^2 + n + 3) =$ first element of nth row.

Similarly, sequence of last terms of each row = 1, 9, 27, 59,...

$$\therefore \quad t_n = \frac{1}{3}[2n^3 + 3n^2 + n - 3]$$

= last element of the nth row.

Hence, in the nth row elements can be written as $\dfrac{1}{3}(2n^3 - 3n^2 + n + 3), \ldots \dfrac{1}{3}[2n^3 + 3n^2 + n - 3]$

(**Note** adding 2 in the preceding number to get the succeeding number)

$\therefore$ Sum of the elements of nth row (using sum of n terms of AP)

$$= \frac{N}{2}(A + L) = \frac{n^2}{2}\left[\frac{4n^3 + 2n}{3}\right] = \frac{n^3}{3}(2n^2 + 1)$$

$\therefore$ Mean of the number in the nth row

$$= \frac{n^3(2n^2 + 1)}{3 \times n^2} = \frac{n(2n^2 + 1)}{3}$$

Here, $\left\{\begin{array}{l} N = n^2, A = T_n, L = t_n \\ \therefore A + L = T_n + t_n = \dfrac{4n^3 + 2n}{3} \end{array}\right\}$

36. According to given data,

Class Interval	Frequency f_i	x_i	$x_i f_i$	$\lvert \bar{x} - x_i \rvert$
0-5	4	2.5	10	9.5
5-10	1	7.5	7.5	4.5
10-15	10	12.5	125	0.5
15-20	3	17.5	52.5	5.5
20-25	2	22.5	45	10.5
	$N = \Sigma f_i = 20$		$\Sigma x_i f_i = 240$	

$$\therefore \quad \bar{x} = \frac{\Sigma x_i f_i}{N} = \frac{240}{20} = 12$$

$$\therefore \quad \sigma^2 = \frac{\Sigma f_i(\bar{x} - x_i)^2}{N}$$

$$= \frac{4(9.5)^2 + 1(4.5)^2 + 10(0.5)^2 + 3(5.5)^2 + 21(0.5)^2}{20} = \frac{139}{4}$$

$$\therefore \text{ CV} = \frac{\sigma}{\bar{x}} \times 100 = \frac{\sqrt{139}}{2 \times 12} \times 100 = \frac{25\sqrt{139}}{6}$$

37. Given, mean $\bar{x} = 4$; variance $\sigma^2 = 5.20$
and numbers of observation $n = 5$
Let $x_1 = 3, x_2 = 4, x_3 = 4$ and x_4, x_5 be the five observations

So, $$\sum_{i=1}^{5} x_i = 5 \cdot \bar{x} = 5 \times 4 = 20$$

$$\Rightarrow \quad x_1 + x_2 + x_3 + x_4 + x_5 = 20$$

$$\Rightarrow \quad 3 + 4 + 4 + x_4 + x_5 = 20 \Rightarrow x_4 + x_5 = 9 \quad \ldots \text{(i)}$$

Now, variance $\sigma^2 = \dfrac{\sum_{i=1}^{5} x_i^2}{5} - (\bar{x})^2$

$$\Rightarrow \quad \frac{x_1^2 + x_2^2 + x_3^2 + x_4^2 + x_5^2}{5} - (4)^2 = 5.20$$

$\Rightarrow \quad \dfrac{9+16+16+x_4^2+x_5^2}{5} = 16+5.20$

$\Rightarrow \quad 41+x_4^2+x_5^2 = 5\times 21.20$

$\Rightarrow \quad x_4^2+x_5^2 = 106-41$

$\Rightarrow \quad x_4^2+x_5^2 = 65 \quad \text{...(ii)}$

$\because \quad (x_4+x_5)^2 = x_4^2+x_5^2+2x_4x_5$

$\therefore \quad 81 = 65+2x_4x_5$ [from Eqs. (i) and (ii)]

$\Rightarrow \quad 16 = 2x_4x_5$

$\Rightarrow \quad x_4x_5 = 8 \quad \text{...(iii)}$

Now, $(|x_4-x_5|)^2 = x_4^2+x_5^2-2x_4x_5$

$= 65-16 = 49$ [from Eqs. (ii) and (iii)]

$\Rightarrow \quad |x_4-x_5| = 7$

38. Since, deviation about point c is

$\dfrac{1}{n}\sum_{i=1}^{n}(x_i-c)^2$ (given)

$\therefore$ Deviation about -2 is

$\dfrac{1}{n}\sum_{i=1}^{n}\{x_i-(-2)\}^2 = 18 \quad \text{...(i)}$

Also, deviation about 2 is

$\dfrac{1}{n}\sum_{i=1}^{n}(x_i-2)^2 = 10 \quad \text{...(ii)}$

Adding and subtracting Eqs. (i) and (ii) we obtain

$\dfrac{1}{n}\sum_{i=1}^{n}(x_i^2+2^2) = 28$ or $\dfrac{2}{n}\sum_{i=1}^{n}x_i^2 = 10$

and $\quad \dfrac{8}{n}\sum_{i=1}^{n}x_i = 8$ or $\dfrac{1}{n}\sum_{i=1}^{n}x_i = 1$

Thus, $\sigma^2 \dfrac{1}{n}\left\{\sum_{i=1}^{n}x_i^2-\left(\sum_{i=1}^{n}x_i\right)^2\right\} = 10-1 = 9$

$\therefore$ SD $= \sigma = 3$

39. Given, $w_i = lx_i+k$, $\bar{x}_1 = 48$, SD of x_i's $= 12$, $w_i = 55$ and SD of w_i's $= 15$

Then, $\bar{w}_i = \sqrt{l\,\bar{x}_i}+k$ [where, $\bar{w}_i$ is mean of w_i's and $\bar{x}_i$ is mean of x_i's]

$\Rightarrow \quad 55 = 48l+k \quad \text{...(i)}$

Now, SD of w_i's $= l\times$ (SD of x_i's)

$\Rightarrow \quad 15 = l\times 12$

$\Rightarrow \quad l = \dfrac{15}{12} = 1.25$

From Eqs. (i), we get

$k = 55-1.25\times 48 = -5$

Numerical Value Type

1. $\Sigma x = 170, \Sigma x^2 = 2830$

The increase in Σx is 10, then

$\Sigma x' = 170+10 = 180$

The increase in Σx^2 is $900-400 = 500$, then

$\Sigma x^2 = 2830+500 = 3330$

Variance $= \dfrac{1}{n}\Sigma x^2-\left(\dfrac{1}{n}\Sigma x'\right)^2$

$= \dfrac{1}{15}\times 3330-\left(\dfrac{1}{15}\times 180\right)^2 = 222-144 = 78$

2. Let the variance of the first 'n' natural numbers

$= \dfrac{1^2+2^2+3^2+\ldots+n^2}{n}-\left(\dfrac{1+2+3+\ldots+n}{n}\right)^2 = 10$ [given]

$\Rightarrow \dfrac{n(n+1)(2n+1)}{6n}-\left(\dfrac{n(n+1)}{2n}\right)^2 = 10$

$\Rightarrow \dfrac{n+1}{12}[2(2n+1)-3(n+1)] = 10$

$\Rightarrow (n+1)(n-1) = 120 \Rightarrow n^2-1 = 120$

$\Rightarrow \quad n^2 = 121 \Rightarrow n = 11 \quad [\because n\in N]$

Similarly, the variance of the first 'm' even natural numbers

$= \dfrac{2^2+4^2+6^2+\ldots+\text{upto } m \text{ terms}}{m}-\left(\dfrac{2+4+6+\ldots+\text{upto } m \text{ terms}}{m}\right)^2 = 16$ [given]

$\Rightarrow 4\left(\dfrac{1^2+2^2+3^2+\ldots+m^2}{m}\right)-4\left(\dfrac{1+2+3\ldots+m}{m}\right)^2 = 16$

$\Rightarrow \left(\dfrac{m(m+1)(2m+1)}{6m}\right)-\left(\dfrac{m(m+1)}{2m}\right)^2 = 4$

$\Rightarrow \dfrac{m+1}{12}[2(2m+1)-3(m+1)] = 4$

$\Rightarrow (m+1)(m-1) = 48$

$\Rightarrow m^2-1 = 48$

$\Rightarrow m^2 = 49 \Rightarrow m = 7 \quad [\because m\in N]$

$\therefore m+n = 7+11 = 18$

3. According to the given information,

Cofficient of variance $(CV) = \dfrac{\sigma}{\mu}\times 100$

Where σ is standard deviation and μ is mean of an ungrouped data.

$\because \quad \dfrac{\sigma}{\mu}\times 100 = 60$ and $\sigma = 21$

So, $\quad \mu = \dfrac{21\times 10}{6} = 35$

After adding 15 to each observation of the data, the new mean $\mu' = 35+15 = 50$ but

$\Sigma|\mu'-x'_i| = \Sigma|\mu-x_i|$ [where $x'_i = x_i+15$]

Now, $\sigma = \sqrt{\frac{\Sigma(\mu - x_i)^2}{n}} = 21 \Rightarrow \Sigma(\mu - x_i)^2 = (21)^2 \times n$

So, new standard deviation

$$\sigma' = \sqrt{\frac{\Sigma(\mu' - x'_i)^2}{n}} \sqrt{\frac{(21)^2 \times n}{n}} = 21$$

$\therefore$ New coefficient of varience

$$= \frac{\sigma'}{\mu'} \times 100 = \frac{21}{50} \times 100 = 21 \times = 42$$

4. $\sigma^2 = \frac{\Sigma(x_i - \bar{x})^2}{N}$

Required variance $(\sigma)^2 = 2^2 \times$ Variance of (1, 2, 3, ... 50)

$$= 4 \times \frac{n^2 - 1}{12} = 4 \times \frac{2500 - 1}{12} = 833$$

5. Given eight numbers are 3, 7, 9, 12, 13, 20, x and y, then their mean

$$\bar{x} = 10 \qquad \text{[given]}$$

$$\Rightarrow \frac{3 + 7 + 9 + 12 + 13 + 20 + x + y}{8} = 10$$

$$\Rightarrow \quad 64 + x + y = 80 \Rightarrow x + y = 16 \qquad \ldots\text{(i)}$$

and, variance = 25 (given)

$$\Rightarrow \frac{\sum_{i=1}^{8} (x_i)^2}{8} - (\bar{x})^2 = 25$$

$$\Rightarrow \frac{9 + 49 + 81 + 144 + 169 + 400 + x^2 + y^2}{8} - 100 = 25$$

$\Rightarrow \; 852 + x^2 + y^2 = 1000 \Rightarrow x^2 + y^2 = 148$

$\Rightarrow \; (x + y)^2 - 2xy = 148$ [from Eq. (i)]

$\Rightarrow \; 256 - 2xy = 148$

$\Rightarrow \; 2xy = 256 - 148 = 108 \Rightarrow xy = 54$

6. $\sigma^2 = \frac{\Sigma f_i x_i^2}{n} - \left(\frac{\Sigma f_i x_i}{n}\right)^2$

$$\Rightarrow \quad 160 = \frac{92A^2}{7} - \left(\frac{22A}{7}\right)^2$$

$$\Rightarrow \quad 160 = \frac{92A^2}{7} - \left(\frac{484A^2}{49}\right)$$

$$\Rightarrow \quad 160 = (644 - 484)\frac{A^2}{49} \Rightarrow 160 = \frac{160A^2}{49}$$

$$\Rightarrow \quad A^2 = 49 \Rightarrow A = 7 \Rightarrow A^2 + A = 56$$

7. Let first $2n$ observations are $x_1, x_2, \ldots, x_{2n}$ and last n observations are $y_1, y_2, \ldots, y_n$

Now, $\frac{\Sigma x_i}{2n} = 6, \; \frac{\Sigma y_i}{n} = 3$

$\Rightarrow \quad \Sigma x_i = 12n, \; \Sigma y_i = 3n$

$\therefore \quad \frac{\Sigma x_i + \Sigma y_i}{3n} = \frac{15n}{3n} = 5$

Now, $\frac{\Sigma x_i^2 + \Sigma y_i^2}{3n} - 5^2 = 4$

$\Rightarrow \quad \Sigma x_i^2 + \Sigma y_i^2 = 29 \times 3n = 87n$

Now, mean is $\frac{\Sigma (x_i + 1) + \Sigma (y_i - 1)}{3n} = \frac{15n + 2n - n}{3n} = \frac{16}{3}$

Now, variance is $\frac{\Sigma (x_i + 1)^2 + \Sigma (y_i - 1)^2}{3n} - \left(\frac{16}{3}\right)^2$

$$= \frac{\Sigma x_i^2 + \Sigma y_i^2 + 2(\Sigma x_i - \Sigma y_i) + 3n}{3n} - \left(\frac{16}{3}\right)^2$$

$$= \frac{87n + 2 \times 9n + 3n}{3n} - \left(\frac{16}{3}\right)^2$$

$$= 29 + 6 + 1 - \left(\frac{16}{3}\right)^2 = \frac{324 - 256}{9} = \frac{68}{9} = k$$

$\Rightarrow 9k = 68$

8. $\frac{\Sigma x_i}{25} = 40$ and $\frac{\Sigma x_i - 60 + N}{25} = 39$

Let age of newly appointed teacher is N

$\Rightarrow \quad 1000 - 60 + N = 975$

$\Rightarrow \quad N = 35$ yr

Single Integer Type

1. Let the AP is $a, a + d, a + 2d, \ldots, a + 10d$.

According to the question,

$$Var = \frac{a^2 + (a + d)^2 + (a + 2d)^2 + \ldots + (a + 10d)^2}{11} - \left[\frac{a + (a + d) + \ldots + (a + 10d)}{11}\right]^2$$

$$= \frac{11a^2 + d^2 (1^2 + 2^2 + 3^2 + \ldots + 10^2) + 2ad (1 + 2 + \ldots + 10)}{11} - \left(\frac{11a + d (1 + 2 + 3 + \ldots + 10)}{11}\right)^2$$

$$= \frac{11a^2 + \frac{10 \times 11 \times 21}{6} d^2 + 2ad \frac{10 \times 11}{2}}{11} - \left(\frac{11a + \frac{10 \times 11}{2} d}{11}\right)^2$$

$$= a^2 + 35d^2 + 10ad - a^2 - 25d^2 - 10ad = 10d^2$$

$\because \quad$ Var = 90 [given]

$\therefore \quad 10d^2 = 90$

$\Rightarrow \quad d = \pm 3$

$\because$ AP is an increasing AP, so $d = 3$.

Hence, answer is 3.00.

2. In the $2n$ observations, half of them equal to a and the remaining half equal to $-a$. Then the mean of total $2n$ observations is equal to zero. Therefore,

$$\text{SD} = \sqrt{\frac{\Sigma(x-\bar{x})^2}{N}} \Rightarrow 2 = \sqrt{\frac{\Sigma x^2}{2n}}$$

$$\Rightarrow \quad 4 = \frac{\Sigma x^2}{2n} \Rightarrow 4 = \frac{2a^2}{2n}$$

$$\Rightarrow \quad a^2 = 4 \Rightarrow |a| = 2$$

3. Let the five observations are $x_1\ x_2, x_y\ x_4, x_5$.

So, $\quad \text{mean} = \dfrac{x_1 + x_2 + x_3 + x_4 + x_5}{5} = 9 \quad$ [given]

$$\Rightarrow \quad x_1 + x_2 + x_3 + c_4 + x_5 = 45 \quad \ldots\text{(i)}$$

and standard deviation $= \sum\limits_{\infty}^{5} \sqrt{\dfrac{(9-x_i)^2}{5}} = 0 \quad$ [given]

$$\Rightarrow \quad \sum_{i=1}^{5} (9 - x_1)^2 = 0 \Rightarrow x_i = 9\ \forall i$$

Now, let observation x_5 is changed with y, so mean

$= \dfrac{x_1 + x_2 + x_3 + x_4 + y}{5} = 10 \quad$ [given]

$$\Rightarrow \quad x_1 + x_2 + x_3 + c_4 + y = 50$$

$$\Rightarrow \quad 45 - x_5 + y = 50 \quad \text{[from Eq. (i)]}$$

$$\Rightarrow \quad x_5 + 5 = y \Rightarrow y = 14 \quad [\because x, = 9]$$

Now, the change standard deviation is

$$\sqrt{\frac{(10-x_1)^2 + (10-x_2)^2 + (10-x_3)^2 + (10-x_4)^2 + (10-y)^2}{5}}$$

$[\because x_1 = x_2 = x_3 = x_4 = 9]$

$$= \sqrt{\frac{1+1+1+1+16}{5}} = \sqrt{\frac{20}{5}} = \sqrt{4} = 2$$

4. Let σ_1 and $\bar{X}_1$ are the standard deviation and mean of 50 observation and σ_2 and $\bar{X}_2$ are standard deviation one mean of another 50 observation respectively.

$\bar{X}_1\ 10, \sigma_1 = 2, n_1 = 50$

$\bar{X}$ = mean of 100 observation = 8

σ = SD of 100 observation = $\sqrt{10.5}$

$$\bar{X} = \frac{50\bar{x}_1 + 60\bar{X}_2}{50+50} \Rightarrow 800 = 500 + 50\bar{X}_2$$

$$\Rightarrow \quad \bar{X} = 6$$

$$\sigma^2 = \frac{n_1(\sigma_1)^2 + n_2(\sigma_2)^2}{n_1+n_2} + \frac{n_1 n_2 (Xb_1 - \bar{X}_2)^2}{(n_1+n_2)}$$

$$10.5 = \frac{50(4) + 50(\sigma_2)^2}{(100)^2} + \frac{2500(10-6)^2}{(100)^2}$$

$$10.5 = \frac{200 + 50(\sigma_2)^2}{100} + \frac{25 \times 16}{100}$$

$$\Rightarrow \quad 1050 = 600 + 50(\sigma)^2$$

$$\Rightarrow \quad \sigma_2^2 = \frac{1050-600}{50} = 9 \Rightarrow \sigma_2 = 3$$

5. $\because \dfrac{1}{n}\Sigma(x_i + 2)^2 = 18$ and $\dfrac{1}{n}\Sigma(x_i - 2)^2 = 10$

$$\Rightarrow \Sigma(x_1+2)^2 = 18n \text{ and } \Sigma(x_1-2)^2 = 10n$$

$$\Rightarrow \Sigma(x_i+2)^2 + \Sigma(x_i-2)^2 = 28n$$

and $\Sigma(x_i+2)^2 - \Sigma(x_i-2)^2 = 8n$

$$\Rightarrow 2\Sigma x_i^2 + 8n = 28\text{n and } 8\Sigma x_i = 8n$$

$$\Rightarrow \Sigma x_i^2 = 10n \text{ and } \Sigma x_i = n \Rightarrow \frac{\Sigma x_i^2}{n} = 10 \text{ and } \frac{\Sigma x_i}{n} = 1$$

$$\therefore \quad \sigma = \sqrt{\frac{\Sigma_x^2}{n} - \left(\frac{\Sigma x_i}{n}\right)^2} = \sqrt{10-(1)^2} = 3$$

6. Median of $a, 2a\ldots, 50a$ is $\dfrac{1}{2}(25a + 26a = (25.5)a$

MD of given numbers from median

$$= \frac{1}{50}\sum_{k=1}^{50} |ka - (25.5)a| = \frac{|a|}{50}\sum_{k=1}^{50} k - 25.5|$$

$$= \frac{|a|}{50}(2(24.5 + 23.5 + \ldots + 0.5)$$

$$= \frac{|a|}{50}(1 + 3 + \ldots + 49)$$

$$= \frac{|a|}{50} \times \frac{25}{2}(1+49) = \frac{25}{2}|a| = 50$$

$$\Rightarrow \quad |a| = 4$$

7. It is given that the data on 'x' taking the values 0, 2, 4, 8, ..., 2^n with frequencies ${}^nC_0$, ${}^nC_1$, ${}^nC_2$, ..., ${}^nC_n$ respectively, so

mean $= \dfrac{\Sigma x_i f_i}{\Sigma f_i}$

$$= \frac{{}^nC_0(0) + {}^nC_1(2) + {}^nC_2(2^2) + {}^nC_3(2^3) + \ldots + {}^nC_n(2^n)}{{}^nC_0 + {}^nC_1 + {}^nC_2 + \ldots + {}^nC_n}$$

$$= \frac{{}^nC_1(2) + {}^nC_2(2^2) + {}^nC_3(2^3) + \ldots + {}^nC_n(2^n)}{2^n}$$

$$\because \ (1+x)^n = {}^nC_0 + {}^nC_1 x + {}^nC_2 x^2 + \ldots + {}^nC_n x^n$$

At $x = 2$, we get

$$3^n = 1 + {}^nC_1(2) + {}^nC_2(2^2) + \ldots + {}^nC_n(2^n)$$

$$\Rightarrow {}^nC_1(2) + {}^nC_2(2^2) + {}^nC_3(2^3) + \ldots + {}^nC_n(2^n) = 3^n - 1$$

$$\therefore \quad \text{Mean} = \frac{3^n - 1}{2^n} = \frac{728}{2^n} \quad \text{[given]}$$

$$\Rightarrow \quad 3^n = 729 = 3^6 \Rightarrow n = 6$$

8. $$\sigma^2 = \frac{n_1\sigma_1^2 + n_2\sigma_2^2}{n_1+n_2} + \frac{n_1 n_2}{(n_1+n_2)}(\bar{x}_1 - \bar{x}_2)^2$$

$$n_1 = 10, n_2 = n, \sigma_1^2 = 2, \sigma_2^2 = 1$$

$$\bar{x}_1 = 2, \bar{x}_2 = 3, \sigma^2 = \frac{17}{9}$$

$$\frac{17}{9} = \frac{10 \times 2 + n}{n+10} + \frac{10n}{(n+10)^2}(3-2)^2$$

$$\Rightarrow \frac{17}{9} = \frac{(n+20)(n+10) + 10n}{(n+10)^2}$$

$$\Rightarrow 17n^2 + 1700 + 340n = 90n + 9(n^2 + 30n + 200)$$

$$\Rightarrow \quad 8n^2 - 20n - 100 = 0$$

$$\Rightarrow \quad 2n^2 - 5n - 25 = 0$$

$$\Rightarrow \quad (2n+5)(n-5) = 0$$

$$\Rightarrow \quad n = \frac{-5}{2} \text{ (Rejected), } 5$$

Hence, $n = 5$